Cytokine Reference

Cytokine Reference

A compendium of cytokines and other mediators of host defense

Editors in Chief

Joost J. Oppenheim
Marc Feldmann

Editors

Scott K. Durum
Toshio Hirano
Jan Vilcek
Nicos A. Nicola

A Harcourt Science and Technology Company

San Diego San Francisco New York Boston
London Sydney Tokyo

This book is printed on acid-free paper.

Academic Press
A Harcourt Science and Technology Company
Harcourt Place, 32 Jamestown Road, London NW1 7BY, UK
http://www.academicpress.com

Academic Press
A Harcourt Science and Technology Company
525 B Street, Suite 1900, San Diego, California 92101-4495, USA
http://www.academicpress.com

Institutional edition: ISBN 0-12-252673-2
Individual edition: ISBN 0-12-252670-8

A catalogue record for this book is available from the British Library

Typeset by Newgen Imaging Systems (P) Ltd, Chennai, India
Printed and bound in Great Britain by MPG Books Ltd, Bodmin, Cornwall

00 01 02 03 04 05 MP 9 8 7 6 5 4 3 2 1

Contents

Preface

Scientific fields grow at a variable rate, increasing to a peak and then mature. As they do so, the rate of new publications, exciting breakthroughs and clinical applications wanes and the understanding permeates to the graduate and then undergraduate textbooks.

Cytokine research is still in the first, rapidly increasing phase, with publications increasing at a frightening rate. The relative newness of this field, its complexity augmented by erratic nomenclature and its increasing importance in medical research and now medical practice prompted us to attempt to make the field more accessible. This is important if the expanding knowledge of cytokines is to increasingly illuminate our knowledge of both physiology and pathophysiology.

The format chosen was the World Wide Web as it permits both rapid and frequent updates, and an expansion to meet the ever increasing number of cytokines and receptors. The power of the internet has made it possible to build links between chapters in the www version, and to include links to databases with DNA sequences of cytokines and receptors (Genbank, EMBL) and to obtain abstracts of references cited (Pubmed). While all researchers in the West have access to the www, this is not the case in some countries, and so Academic Press decided to publish it also in book form. There are advantages to having access to the information in both formats.

We were faced by several difficulties in organizing this on-line reference. Since some of the cytokines use multiple receptors and a minority of cytokines behave like receptors, we decided to have independent chapters and sections devoted to cytokine ligands and their receptors. Non-cytokine molecules that function as intercellular signals are too numerous to cover in an initial version of the reference. We therefore decided to limit the coverage to mediators involved in host defense including neuropeptides, relevant growth factors, selected endocrine hormones, complement components, etc. We may broaden the scope of the reference in the next millenium and include other growth factors, and mediators active in development, repair and homeostasis. Even in its present state, the reference has taken advantage of the on-line format to include much more information than is usually included in a print format.

The genesis of this project has involved discussion with many experts in the field of cytokine research. We are very pleased, and grateful that so many of our friends and colleagues have taken part in this project, helping us edit and writing the individual chapters. Without their invaluable contributions this Cytokine Reference could not exist, and with their help we hope it will mature and have a long life as a very useful information tool for laboratory scientists, research clinicians, physicians, biotechnology and pharmaceutical scientists, or anyone else with an interest in this field.

The generation of this Cytokine Reference owes much to Gillian Griffiths at Churchill Livingstone, who was an early champion of this project. With mergers and acquisitions, it has ended up as an Academic Press project, masterminded by Sarah Stafford who has poured a lot of energy and skill to bringing the project to fruition.

This is a collective enterprise, aided by the managerial skills of Cheryl Fogle. The success of the Cytokine Reference depends on fulfilling a need. We welcome suggestions, and comments. We also welcome criticisms, notification of errors, misrepresentations, omissions and so on, in order to improve the value of this work to the scientific and medical community and to more effectively provide the compendium of this expanding field we felt was sorely needed.

Joost Oppenheim and Marc Feldmann

Dedication

The Israeli Chapter of ISICR regrets to inform the untimely death of Professor Samuel Salzberg on Saturday, April 15, 2000. Born in 1940, Samuel Salzberg received his Ph.D. at the Weizmann Institute of Science, Rehovot, Israel in 1970. During 1970–1973 he did his post-doctoral studies at the St. Louis University Medical School.

In 1973 he joined the Department of Life Sciences in Bar-Ilan University, Ramat-Gan, Israel, where he stayed for the rest of his career. In 1986 Professor Salzberg became the Head of the Department of Life Sciences and in 1998 he was appointed the Dean of the Faculty of Life Sciences at Bar-Ilan University. He was a Visiting Scientist at the National Institute of Health, Bethesda, MD and at the Argonne National Laboratory, Argonne, IL.

Professor Salzberg was extensively engaged in studies on the antiretroviral effect of interferon, both in chronic and exogenously infected cells. Numerous basic findings on the subject were published. More recently he concentrated on the anti-proliferative effects of IFN, and in the last few years he studied the involvement of PKR and 2–5A synthetase in cell growth and differentiation. During his career, Professor Salzberg published 67 papers and supervised 25 students. He left a wife, three children and many friends who cherish his memory.

Contributors

Bharat B. Aggarwal
Cytokine Research Laboratory,
Department of Bioimmunotherapy,
The University of Texas M.D.,
Anderson Cancer Center,
1515 Holcombe Boulevard,
P.O. Box 143,
Houston, TX 77030, USA

Sunil K. Ahuja
Department of Medicine,
University of Texas Health Science Center
at San Antonio,
7703 Floyd Curl Drive,
San Antonio,
TX 78229-3900, USA

Gal Akiri
Department of Biology,
Technion, Israel Institute of Technology,
Technion City,
Haifa, 32000, Israel

Christine Ambrose
Molecular Genetics,
Biogen, 12 Cambridge Center,
Cambridge, MA 02142, USA

Songzhu An
University of California at San Francisco,
Departments of Medicine and Microbiology,
533 Parnassus at 4th, UB8B,
Box 0711,
San Francisco,
CA 94143-0711, USA

Francesco Annunziato
Department of Internal Medicine,
Section of Immunoallergology and
Respiratory Diseases,
University of Florence,
Florence, Italy

Avi Ashkenazi
Department of Molecular Oncology,
Genentech Inc., 1 DNA Way,
South San Francisco,
CA 94080-4918, USA

K. Frank Austen
Department of Medicine,
Harvard Medical School,
1 Jimmy Fund Way,
Rm 628,
Boston,
MA 02115, USA

Christopher J. Bagley
Protein Laboratory,
Hanson Centre for Cancer Research,
Frome Road, Adelaide,
SA, 5000, Australia

Human Immunology,
Institute of Medical and Veterinary Science,
Frome Road, Adelaide,
SA, 5000, Australia

Elizabeth Bates
Laboratory for Immunological Research,
Schering-Plough,
27 Chemin des Peupliers,
Dardilly, 69572, France

Alfons Billiau
Rega Institute,
University of Leuven,
Minderbroedersstraat 10,
B-3000 Leuven, Belgium

J. Edwin Blalock
Department of Physiology and Biophysics,
University of Alabama at Birmingham,
1918 University Boulevard MCLM896,
Birmingham,
AL 35294-0005, USA

Cédric Blanpain
Institute of Interdisciplinary Research,
Université Libre de Bruxelles,
Campus Erasme, 808 route de Lennik,
Brussels, B-1070,
Belgium

François Boulay
Department of Molecular and Structural Biology,
DBMS/BBSI UMR 314 CEA-CNRS CEA,
17 Rue des Martyrs,
Grenoble, Cedex 9, F 38054, France

William J. Boyle
Department of Cell Biology,
Amgen, Inc., One Amgen Center Drive,
Thousand Oaks,
CA 91320-1799, USA

Ernst Brandt
Research Center Borstel,
Center for Medicine and Biosciences,
D-23845 Borstel, Germany

Fionula M. Brennan
Cytokine and Cellular Immunology Division,
Kennedy Institute of Rheumatology,
Imperial College, School of Medicine,
1 Aspenlea Road, Hammersmith,
London, W6 8LH, UK

Jeffrey L. Browning
Cell Biology, Inflammation and Immunology,
Biogen, 12 Cambridge Center,
Cambridge,
MA 02142, USA

A. Gregory Bruce
Department of Pathobiology,
School of Public Health and Community Medicine,
University of Washington,
Box 357238,
Seattle, WA 98195, USA

Richard Bucala
The Picower Institute for Medical Research,
350 Community Drive,
Manhasset,
NY 11030, USA

Danielle Burger
Division of Immunology and Allergy,
Department of Internal Medicine,
University Hospital, CH-1211,
Geneva 14, Switzerland

Daniel J. J. Carr
Department of Microbiology and Immunology,
LSU Medical Center,
New Orleans,
LA 70112-1393, USA

Daniel J. Catron
Department of Immunobiology,
DNAX Research Institute of Molecular and
Cellular Biology, Inc.,
901 California Avenue,
Palo Alto,
CA 94304-1104, USA

David M. Center
Pulmonary and Critical Care Division,
Department of Medicine,
Boston University School of Medicine,
715 Albany Street, R-304,
Boston,
MA 02118, USA

Scott Chappel
Serono Corp., 100 Longwater Circle,
PO Box B,
Norwell,
MA 02061, USA

Nan Chiang
Center for Experimental Therapeutics and
Reperfusion Injury,
Brigham and Women's Hospital,
75 Francis Street,
Boston,
MA 02115, USA

Kenneth R. Chien
Department of Medicine,
Center for Molecular Genetics,
and the American Heart Association-Bugher
Foundation,
Center for Molecular Biology,
University of California,
San Diego School of Medicine,
9500 Gilman Drive,
La Jolla,
CA 92093, USA

Edward A. Clark
Primate Center,
University of Washington,
Box 357330,
Seattle,
WA 98195-0001, USA

Clary B. Clish
Center for Experimental Therapeutics and Reperfusion Injury,
Brigham and Women's Hospital and Harvard Medical School,
75 Francis Street,
Boston, MA 02132, USA

William W. Cruikshank
Pulmonary and Critical Care Division,
Department of Medicine,
Boston University School of Medicine,
715 Albany Street, R-304,
Boston,
MA 02118, USA

Peter Czabotar
School of Biomedical Sciences,
CURTIN University of Technology,
Perth, Western Australia

Jean-Michel Dayer
Division of Immunology and Allergy,
Department of Internal Medicine,
University Hospital, CH-1211,
Geneva 14, Switzerland

Rene de Waal Malefyt
Department of Molecular Biology,
DNAX Research Institute,
901 California Avenue,
Palo Alto,
CA 94304-1104, USA

Charles A. Dinarello
Department of Infectious Diseases,
University of Colorado
Health Sciences Center,
4200 East Ninth Avenue,
B168, Denver,
CO 80262, USA

Glenn Dorsam
Immunology and Allergy,
University of California,
San Francisco, 533 Parnassus Avenue,
Room UB8B, Box 0711,
San Francisco,
CA 94143-0711, USA

Steve K. Dower
Division of Molecular & Genetic Medicine,
University of Sheffield,
Royal Hallamshire Hospital,
Sheffield, S10 2JF, UK

Scott K. Durum
Laboratory of Molecular Immunoregulation,
National Cancer Institute,
FCRDC, DBS Building,
Room 31-73,
Frederick,
MD 21702-1201, USA

Julia A. Ember
Pharmingen,
10975 Torreyana Road,
San Diego,
CA 92121, USA

Clemens Esche
Biological Therapeutics Program,
University of Pittsburgh Cancer Institute,
Pittsburgh,
PA 15213, USA

Joshua Marion Farber
Laboratory of Clinical Investigation,
National Institute of Allergy and Infectious Diseases,
NIH, 10 Center Drive,
Room 11N-228, MSC 1888,
Bethesda,
MD 20892-1888, USA

Marc Feldmann
Kennedy Institute of Rheumatology,
1 Aspenlea Road,
Hammersmith,
London,
W6 8LH, UK

Cytokine and Cellular Immunology Division,
Kennedy Institute of Rheumatology,
1 Aspenlea Road,
Hammersmith,
London,
W6 8LH, UK

Napoleone Ferrara
Department of Molecular Oncology,
Genentech Inc.,
1 DNA Way,
South San Francisco,
CA 94080, USA

Hans-Dieter Flad
Research Center Borstel,
Center for Medicine and Biosciences,
D-23845 Borstel,
Germany

Kathleen C. Flanders
Laboratory of Cell Regulation
and Carcinogenesis,
National Cancer Institute,
Building 41, Room C629,
41 Library Drive MSC 5055,
Bethesda,
MD 20892-5055, USA

Reinhold Förster
Molecular Tumor Genetics and
Immunogenetics,
Max-Delbruck-Center for Molecular
Medicine,
Robert Rossle Street 10,
Berlin, 13092, Germany

François Fossiez
Laboratory for Immunological Research,
Schering-Plough,
27 Chemin des Peupliers,
Dardilly, 69572, France

Régis Fournier
School of Biomedical Sciences,
CURTIN University of Technology,
Perth, Western Australia

Giovanni Franchin
The Picower Institute for
Medical Research,
350 Community Drive,
Manhasset,
NY 11030, USA

Toshiyuki Fukada
Division of Molecular Oncology,
Department of Oncology,
Biomedical Research Center,
Osaka University Graduate
School of Medicine(C7), 2-2,
Yamada-oka, Suita,
Osaka, 565, Japan

F. Shawn Galin
Department of Physiology and
Biophysics,
University of Alabama at Birmingham,
Birmingham,
AL 35294, USA

Tomas Ganz
Medicine and Pathology,
School of Medicine,
University of California
Los Angeles,
10833 Le Conte Ave.,
Los Angeles,
CA 90095-1690, USA

Ji-Liang Gao
Laboratory of Host Defenses,
National Institute of Allergy and
Infectious Diseases,
NIH, Building 10, Room 11N111,
Bethesda,
MD 20892, USA

Elizabeth Geras-Raaka
Department of Medicine,
Division of Molecular Medicine,
Weill Medical College and
Graduate School of Medical Sciences of
Cornell University,
1300 York Avenue,
New York,
NY 10021, USA

Marvin C. Gershengorn
Department of Medicine,
Division of Molecular Medicine,
Weill Medical College and
Graduate School of Medical Sciences of
Cornell University,
1300 York Avenue,
New York,
NY 10021, USA

Edward J. Goetzl
Immunology and Allergy,
University of California, San Francisco,
533 Parnassus Avenue, Room UB8B,
Box 0711,
San Francisco,
CA 94143-0711, USA

Vincent Goffin
INSERM Unit 344,
Faculty of Medicine Necker,
156 rue de Vaugirard,
Paris, Cedex 15, 75730,
France

Gerry Graham
Cancer Research Campaign Laboratories,
The Beaton Institute for Cancer Research,
Garscube Estate Switchback Road,
Bearsdon,
Glasgow,
G61 1BD, UK

David R. Greaves
Sir William Dunn School of Pathology,
South Parks Road,
Oxford,
OX1 3RE, UK

Shawn J. Green
EntreMed Inc.,
9640 Medical Center Drive,
Rockville,
MD 20850, USA

Davida K. Grella
EntreMed Inc.,
9640 Medical Center Drive,
Rockville,
MD 20850, USA

Karsten Gronert
Center for Experimental Therapeutics and Reperfusion Injury,
Brigham and Women's Hospital,
75 Francis Street,
Boston,
MA 02115, USA

Julio Gutiérrez
Departamento de Inmunología y Oncología,
Centro Nacional de Biotecnología,
Campus Universidad Autónoma,
Cantoblanco,
Madrid,
28049, Spain

S. Jaharul Haque
Department of Cancer Biology,
Lerner Research Institute,
Cleveland Clinic Foundation,
9500 Euclid Avenue,
Cleveland,
OH 44195, USA

Department of Pulmonary and Critical Care Medicine,
Cleveland Clinic Foundation,
9500 Euclid Avenue,
Cleveland,
OH 44195, USA

Dominik Haudenschild
Department of Orthopedic Surgery,
University of California, Davis,
4635 Second Avenue,
Sacramento,
CA 95817, USA

Caroline Hebert
Department of Molecular Oncology,
Genentech,
1 DNA Way South San Francisco,
San Francisco,
CA 94080, USA

Carl-Henrik Heldin
Ludwig Institute for Cancer Research,
Box 595,
Uppsala,
S-75124, Sweden

Peter M. Henson
Program in Cell Biology,
National Jewish Medical Research Center,
1400 Jackson Street,
Denver,
CO 80206, USA

Catherine Hession
Molecular Genetics,
Biogen,
12 Cambridge Center,
Cambridge,
MA 02142, USA

Masahiko Hibi
Division of Molecular Oncology,
Department of Oncology,
Biomedical Research Center,
Osaka University Graduate School of Medicine(C7), 2-2,
Yamada-oka, Suita,
Osaka, 565, Japan

Toshio Hirano
Division of Molecular Oncology,
Biomedical Research Center,
Osaka University Graduate School of Medicine,
2-2 Yamada-oka. Suita,
Osaka, 565-0871, Japan

Richard Horuk
Department of Immunology,
Berlex Bioscience,
15049 San Pablo Avenue,
Richmond,
CA 94804, USA

O. M. Zack Howard
Frederick Cancer Research and Development Center,
Frederick,
MA 21702-1201, USA

Tony E. Hugli
Department of Immunology,
M/S IMM18 The Scripps Research Institute,
10550 North Torrey Pines Road,
La Jolla,
CA 92037-1092, USA

The Scripps Research Institute,
10550 N. Torrey Pines Road,
La Jolla,
CA 92037, USA

Hisashi Iizasa
Department of Pharmaceutics,
Kyoritsu College of Pharmacy,
Shibakoen 1-5-30, Minato-ku,
Tokyo, 105-8512, Japan

Toshio Imai
Kan Research Institute, Science Center Building,
3 Kyoto Research Park,
Kyoto, Chu-douji Kurita-Chou,
600-8815, Japan

Department of Bacteriology,
Kinki University School of Medicine,
377-2 Ohno Higashi,
Osaka,
Osaka-Sayama, 589-8511,
Japan

Shin-ichiro Kashiwamura
Hyogo College of Medicine,
Nishinomiya,
Japan

Achsah D. Keegan
Immunology Department,
Holland Laboratory of the
American Red Cross,
15601 Crabbs Branch Way,
Rockville,
MD 20855, USA

James C. Keith Jr
The Genetics Institute,
1 Burtt Road,
Andover,
MA 01810, USA

James Keith
Genetics Institute,
One Burtt Road,
Andover,
MA 01810, USA

Jonathan Roy Keller
Intramural Research and Support Program-SAIC,
Frederick Cancer Research and
Development Center-NCI,
Building 560, Room 12-03,
PO Box B, Frederick,
MD 21702-1201, USA

Paul A. Kelly
INSERM Unit 344,
Faculty of Medicine Necker,
156 rue de Vaugirard,
Paris, Cedex 15, 75730, France

Hardy Kornfeld
Pulmonary and Critical Care Division,
Department of Medicine,
Boston University School of Medicine,
715 Albany Street,
R-304, Boston, MA 02118, USA

Michael Steven Krangel
Immunology,
Duke University Medical Center,
PO Box 3010, Durham,
NC 27710, USA

Myriam S. Kunzi
Oncology Center,
Johns Hopkins University Medical School,
418 N, Bond Street,
Baltimore, MD 21231, USA

David J. Kuter
Hematology and Oncology Division,
Massachusetts General Hospital Cancer Center,
100 Blossom Street,
Boston,
MA 02114, USA

Byoung S. Kwon
The Immunomodulation Research Center,
University of Ulsan,
Ulsan, Korea

Department of Ophthalmology,
LSUMC, 2020 Gravier Street Suite B,
New Orleans, LA 70112, USA

Bing K. Lam
Department of Medicine,
Harvard Medical School,
1 Jimmy Fund Way,
Rm 628, Boston,
MA 02115, USA

Sherry LaPorte
Department of Biochemistry and Biophysics,
University of California at San Francisco,
513 Parnassus Avenue, S-960,
San Francisco,
CA 941143-0448, USA

Serge Lebecque
Laboratory for Immunological Research,
Schering-Plough, 27 Chemin des Peupliers,
Dardilly, 69572,
France

Hsinyu Lee
University of California at San Franscisco,
Departments of Medicine and Microbiology,
533 Parnassus at 4th, UB8B, Box 0711,
San Francisco,
CA 94143-0711, USA

Warren J. Leonard
Laboratory of Molecular Immunology,
NHLBI, National Institutes of Health,
Building 10, Room 7N252,
Bethesda,
MD 20892-1674, USA

Bruce D. Levy
Center for Experimental Therapeutics and Reperfusion Injury,
Brigham and Women's Hospital and Harvard Medical School,
75 Francis Street,
Boston, MA 02132, USA

Junzhi Li
Hematology Unit,
COX 640, Massachusetts General Hospital,
100 Blossom Street,
Boston,
MA 02114, USA

Diana Marie Linnekin
Basic Research Laboratory-DBS,
Frederick Cancer Research and Development Center-NCI,
Building 567, PO Box B,
Frederick, MD 21702-1201, USA

Diana Linnekin
Basic Research Laboratory,
Division of Basic Sciences,
National Cancer Institute-Frederick Cancer Research & Development Center,
Frederick, MD 21702, USA

James M. Lipton
Department of Physiology,
University of Alabama at Birmingham,
1918 University Blvd, BHSB 896,
Birmingham,
AL 35294-0005, USA

W. A. M. Loenen
Department of Medical Microbiology,
University Hospital Maastricht,
PO Box 5800,
Maastricht, 6202 AZ,
The Netherlands

Angel F. Lopez
Human Immunology,
Institute of Medical and Veterinary Science,
Frome Road,
Adelaide,
SA, 5000, Australia

Cytokine Receptor Laboratory,
Hanson Centre for Cancer Research,
Frome Road,
Adelaide,
SA, 5000, Australia

Michael T. Lotze
Biological Therapeutics Program,
University of Pittsburgh Cancer Institute,
Pittsburgh,
PA 15213, USA

Andreas Ludwig
Research Center Borstel,
Center for Medicine and Biosciences,
D-23845 Borstel, Germany

Andrew D. Luster
Massachusetts General Hospital,
Harvard Medical School,
Boston,
MA 02114, USA

Infectious Disease Unit,
Massachusetts General Hospital,
East 149 13th Street,
Charlestown,
MA 02129, USA

Gabriel Márquez
Departamento de Inmunología y Oncología,
Centro Nacional de Biotecnología,
Campus Universidad Autónoma, Cantoblanco,
Madrid, 28049,
Spain

Charles R. Mackay
The Garvan Institute of Medical Research,
384 Victoria Street,
Darlinghurst (Sydney),
New South Wales,
2010, Australia

Carlos Martinez-A.
Department of Immunology and Oncology,
Centro Nacional de Biotecnología, CSIC,
Campus de Cantoblanco,
Madrid, 28049, Spain

Manuela Martins-Green
Department of Biology,
University of California,
Riverside,
CA 92521, USA

Tadashi Matsuda
Department of Immunology,
Toyama Medical and Pharmaceutical University,
2630 Sugitani,
Toyama, 930-0194, Japan

Kouji Matsushima
Department of Molecular Preventive Medicine,
School of Medicine, University of Tokyo,
7-3-1 Hongo, Bunkyo-ku,
Tokyo, 113, Japan

David J. Matthews
Laboratory of Molecular Biology,
Hills Road,
Cambridge, CB2 2QH, UK

Carmelo Mavilia
Department of Internal Medicine,
Section of Immunoallergology and
Respiratory Diseases, University of Florence,
Viale Morgagni 85,
Florence, 50134, Italy

Grant McFadden
The John P. Robarts Research Institute and
Department of Microbiology and Immunology,
The University of Western Ontario,
1400 Western Road,
London, Ontario, N6G 2V4, Canada

Andrew N. J. McKenzie
Laboratory of Molecular Biology,
Hills Road,
Cambridge,
CB2 2QH, UK

Mario Mellado
Department of Immunology and Oncology,
Centro Nacional de Biotecnología, CSIC,
Campus de Cantoblanco,
Madrid, 28049,
Spain

Christine N. Metz
The Picower Institute for Medical Research,
350 Community Drive,
Manhasset,
NY 11030, USA

Anthony R. Mire-Sluis
Division of Immunobiology,
National Institute for
Biological Standards and Controls,
Blanche Lane, South Mimms,
Potters Bar,
Hertfordshire,
EN6 3QG, UK

Atsushi Miyajima
Institute of Molecular and Cellular Biosciences,
The University of Tokyo,
1-1-1 Yayoi,
Tokyo, Bunkyo-Ku, 113-0032,
Japan

Viatcheslav A. Mordvinov
School of Biomedical Sciences,
CURTIN University of Technology,
Perth,
Western Australia

Bernhard Moser
Theodor-Kocher Institute,
University of Bern, Freiestrasse 1,
Bern, 3012,
Switzerland

Richard Moyer
Department of Molecular Genetics and
Microbiology,
University of Florida College of Medicine,
PO Box 100266,
Gainesville,
FL 32610-0266, USA

William Murphy
Transplantation Biology Section,
SAIC Frederick, Building 567,
Room 210, PO Box B,
Frederick,
MD 21702-1201, USA

Philip M. Murphy
Molecular Signaling Section,
Laboratory of Host Defenses,
National Institute of Allergy and Infectious Diseases,
National Institutes of Health,
Bethesda,
MD 20892, USA

Shigekazu Nagata
Department of Genetics,
Osaka University Medical School,
2-2 Yamada-oka Suita,
Osaka, 565-0871, Japan

Kenji Nakanishi
Hyogo College of Medicine,
Nishinomiya, Japan

Monica Napolitano
Laboratory of Vascular Pathology,
Istituto Dermopatico dell'Immacolata-IRCCS,
Via Monti di Creta 104,
Rome, 00167, Italy

Gerard J. Nau
Infectious Diseases Unit,
Massachusetts General Hospital,
Boston,
MA 02114, USA

Gera Neufeld
Department of Biology,
Technion, Israel Institute of Technology,
Technion City,
Haifa, 32000, Israel

Rob Nibbs
Cancer Research Campaign Laboratories,
The Beaton Institute for Cancer Research,
Garscube Estate Switchback Road, Bearsdon,
Glasgow, G61 1BD, UK

Nicos A. Nicola
The Walter and Eliza Hall Institute of Medical Research and the Cooperative Research Centre for Cellular Growth Factors,
PO Royal Melbourne Hospital,
Parkville, Victoria, 3050, Australia

Hisayuki Nomiyama
Department of Biochemistry,
Kumamoto University School of Medicine,
2-2-1 Honjo,
Kumamoto, 860-0811, Japan

Luke A. J. O'Neill
Department of Biochemistry & Biotechnology,
Trinity College Dublin,
Dublin,
Ireland

Haruki Okamura
Hyogo College of Medicine,
Nishinomiya, Japan

Joost J. Oppenheim
Laboratory of Molecular Immunoregulation,
Intramural Research Support Program,
Building 560, Room 21-89A,
Frederick,
MA 21702-1201, USA

Adonia E. Papathanassiu
EntreMed Inc.,
9640 Medical Center Drive,
Rockville,
MD 20850, USA

Marc Parmentier
Institute of Interdisciplinary Research,
Université Libre de Bruxelles,
Campus Erasme, 808 route de Lennik,
Brussels, B-1070,
Belgium

Susanne Peroni
School of Biomedical Sciences,
CURTIN University of Technology,
Perth, Western Australia

Paula Pitha Rowe
Oncology Center,
Johns Hopkins University Medical School,
418 N. Bond Street,
Baltimore,
MD 21231, USA

Gianni Pizzolo
Department of Clinical and Experimental Medicine,
Section of Hematology,
University of Verona,
Verona, Italy

Christine A. Power
Serono Phamaceutical Research Institute,
14 Chemin des Aulx,
1228 Plan les Ouates,
Geneva,
Switzerland

Paul Proost
Laboratory of Molecular Immunology,
Rega Institute – University of Leuven,
Minderbroedersstraat 10,
Leuven, 3000,
Belgium

Fei-Hua Qiu
Center for Experimental Therapeutics and Reperfusion Injury,
Brigham and Women's Hospital,
75 Francis Street,
Boston,
MA 02115, USA

Marie-Josèphe Rabiet
Department of Molecular and Structural Biology,
DBMS/BBSI UMR 314 CEA-CNRS CEA,
17 Rue des Martyrs,
Grenoble, Cedex 9,
F 38054, France

E. W. Raines
Department of Pathology,
University of Washington,
Health Science Building J507 Box 357470,
Seattle,
WA 98195-7470, USA

A. Hari Reddi
Department of Orthopaedic Surgery,
Center for Tissue Regeneration and Repair,
University of California, Davis,
4635 Second Avenue, Room 2000,
Sacramento,
CA 95817, USA

Daniel G. Remick
The University of Michigan Medical School,
Catherine Road,
Ann Arbor,
MI 48109-0602, USA

Jean-Christophe Renauld
Ludwig Institute for Cancer Research and Experimental Medicine Unit,
Université Catholique de Louvain,
74 Avenue Hippocrate,
Brussels, B-1200, Belgium

Ann Richmond
Department of Cell Biology,
Vanderbilt University School of Medicine,
Department of Veterans Affairs,
Nashville, TN 37232, USA

Anita B. Roberts
Laboratory of Cell Regulation and Carcinogenesis,
National Cancer Institute, Building 41,
Room C629, 41 Library Drive MSC 5055,
Bethesda,
MD 20892-5055, USA

Jose Miguel Rodriguez-Frade
Department of Immunology and Oncology,
Centro Nacional de Biotecnología, CSIC,
Campus de Cantoblanco,
Madrid, 28049, Spain

Barrett J. Rollins
Adult Oncology,
Dana-Farber Cancer Institute,
Harvard Medical School, 44 Binney Street,
Boston,
MA 02115, USA

Paola Romagnani
Department of Physiopathology,
Endocrinology Unit,
University of Florence, Viale Morgagni 85,
Florence, 50134, Italy

Sergio Romagnani
Department of Internal Medicine,
Section of Immunoallergology and Respiratory Diseases,
University of Florence,
Viale Morgagni 85,
Florence, 50134, Italy

Timothy M. Rose
Department of Pathobiology,
School of Public Health and Community Medicine, University of Washington,
Box 357238,
Seattle,
WA 98195, USA

Iris Roth
Department of Molecular Oncology,
Genentech, 1 DNA Way South San Francisco,
San Francisco,
CA 94080, USA

Nancy Ruddle
Department of Epidemiology and Public Health Immunology,
Yale University School of Medicine,
815 LEPH,
New Haven,
CT 06520-8034, USA

Hallgeir Rui
Department of Pathology,
Uniformed Services University,
4301 Jones Bridge Road,
Bethesda,
MD 20814, USA

Ian Sabore
Leukocyte Biology Section,
Sir Alexander Fleming Building Imperial College School of Medicine,
Sir Alexander Fleming Building,
London, South Kensington,
SW7 2AZ, UK

Jeremy Saklatvala
Cytokine and Cellular Immunology Division,
Kennedy Institute of Rheumatology,
1 Aspenlea Road, Hammersmith,
London,
W6 8LH, UK

Theodora W. Salcedo
Human Genome Sciences, Inc.,
9410 Key West Avenue,
Rockville,
MA 20850, USA

Ajoy Samanta
Cytokine Research Laboratory,
Department of Bioimmunotherapy,
The University of Texas M.D.,
Anderson Cancer Center,
1515 Holcombe Boulevard,
PO Box 143,
Houston,
TX 77030, USA

Colin J. Sanderson
School of Biomedical Sciences,
CURTIN University of Technology,
Perth, Western Australia

Angela Santoni
Department of Experimental Medicine and Pathology,
University of Rome "La Sapienza",
Via Regina Elena 324,
Rome, 00161, Italy

Thomas J. Schall
ChemoCentryx Inc.,
1539 Industrial Road,
San Carlos,
CA, 94070, USA

John W. Schrader
The Biomedical Research Centre,
University of British Columbia,
2222 Health Sciences Mall,
Vancouver, British Columbia,
Canada V6T 1Z3

Robert D. Schreiber
Center for Immunology and Department of Pathology,
Washington University School of Medicine,
660 South Euclid Avenue,
Mailstop 8118, St. Louis,
MO 63108, USA

Gretchen T. F. Schwenger
School of Biomedical Sciences,
CURTIN University of Technology,
Perth, Western Australia

Charles N. Serhan
Center for Experimental Therapeutics and Reperfusion Injury,
Brigham and Women's Hospital and Harvard Medical School, 75 Francis Street,
Boston,
MA 02132, USA

Vijay Shankaran
Center for Immunology and Department of Pathology,
Washington University School of Medicine,
660 South Euclid Avenue, Mailstop 8118,
St. Louis,
MO 63108, USA

Barbara Sherry
The Picower Institute for Medical Research,
350 Community Drive,
Manhasset,
NY 11030, USA

Michael R. Shurin
Biological Therapeutics Program,
University of Pittsburgh Cancer Institute,
Pittsburgh, PA 15213, USA

B. Kim Lee Sim
EntreMed Inc., 9640 Medical Center Drive,
Rockville, MD 20850, USA

Kendall A. Smith
The Weill Medical College of Cornell University,
1300 York Avenue,
New York, NY 10021, USA

Division of Immunology,
Cornell University Medical College,
525 East 68th Street Box 41,
New York,
NY 10021, USA

Hergen Spits
Division of Immunology,
Netherlands Cancer Institute,
Plesmanlaan 121,
1066 CX Amsterdam,
The Netherlands

Jerry L. Spivak
Department of Medicine,
Johns Hopkins University School of Medicine,
Traylor 924 720 Rutland Avenue,
Baltimore,
MD 21205, USA

E. Richard Stanley
Department of Developmental and Molecular Biology,
Albert Einstein College of Medicine,
Bronx, NY 10461, USA

Harald Stein
Institute of Pathology, UK Benjamin Franklin,
Freie University of Berlin,
Berlin, Germany

Robert M. Stroud
Department of Biochemistry and Biophysics,
University of California at San Francisco,
513 Parnassus Avenue, S-960,
San Francisco,
CA 941143-0448, USA

Yutaka Tagaya
Metabolism Branch,
National Cancer Institute, NIH Building 10,
Room 4N115, 10 Center Drive, MSC 1374,
Bethesda, MD 20892-1374, USA

Andrew M. Tager
Infectious Disease Unit,
Massachusetts General Hospital,
East 149 13th Street,
Charlestown, MA 02129, USA

Marianne Tardif
Department of Molecular and Structural Biology,
DBMS/BBSI UMR 314 CEA-CNRS CEA,
17 Rue des Martyrs,
Grenoble, Cedex 9, F 38054, France

Jan Tavernier
Department of Medical Protein Research,
Flanders' Interuniversity Institute for Biotechnology,
University of Ghent,
Ghent, Belgium

Peter ten Dijke
Division of Cellular Biochemistry,
The Netherlands Cancer Institute,
Plesmanlaan 121,
Amsterdam, 1066 CX,
The Netherlands

Robin C. Thorpe
Division of Immunobiology,
National Institute for Biological Standards and Controls, Blanche Lane, South Mimms,
Potters Bar, Hertfordshire,
EN6 3QG, UK

G. J. Tigyi
University of California at San Franscisco,
Departments of Medicine and Microbiology,
533 Parnassus at 4th, UB8B, Box 0711,
San Francisco,
CA 94143-0711, USA

Jurg Tschopp
Institute of Biochemistry,
University of Lausanne, Ch. des Boveresses 155,
Epalinges, CH-1066,
Switzerland

Hiroko Tsutsui
Hyogo College of Medicine,
Nishinomiya, Japan

Zehava Vadasz
Department of Biology,
Technion, Israel Institute of Technology,
Technion City,
Haifa, 32000, Israel

Jo Van Damme
Laboratory of Molecular Immunology,
Rega Institute – University of Leuven,
Minderbroedersstraat 10,
Leuven, 3000, Belgium

Koen Vandenbroeck
Rega Institute,
University of Leuven, Minderbroedersstraat 10,
B-3000 Leuven,
Belgium

Rosa Varona
Departamento de Inmunología y Oncología,
Centro Nacional de Biotecnología,
Campus Universidad Autónoma, Cantoblanco,
Madrid, 28049, Spain

Jan Vilcek
Department of Microbiology and
Kaplan Cancer Center,
New York University School of Medicine,
550 First Avenue,
New York, NY 10016-6402, USA

Dass S. Vinay
Department of General Surgery,
University of Michigan Medical School,
1516 MSRB I, 1150 West Medical Center Drive,
Ann Arbor, MI 48109, USA

Julia K. Voice
Immunology and Allergy,
University of California,
San Francisco,
Room UB8B, Box 0711,
533 Parnassus Avenue, San Francisco,
CA 94143-0711, USA

Meenu Wadhwa
Division of Immunobiology,
National Institute for Biological Standards and Controls, Blanche Lane, South Mimms,
Potters Bar, Hertfordshire,
EN6 3QG, UK

Thomas A. Waldmann
Metabolism Branch,
National Cancer Institute, NIH Building 10,
Room 4N115, 10 Center Drive, MSC 1374,
Bethesda, MD 20892-1374, USA

David Wallach
Department of Biological Chemistry,
Weizmann Institute of Science,
Rehovot, 76100, Israel

Alfred Walz
Theodor Kocher Institute,
University of Bern, Freiestrasse 1,
3012 Bern, Switzerland

Dingzhi Wang
Department of Cell Biology,
Vanderbilt University School of Medicine,
Department of Veterans Affairs,
Nashville, TN 37232, USA

Carl F. Ware
Division of Molecular Immunology,
La Jolla Institute for Allergy and Immunology,
10355 Science Center Drive,
San Diego,
CA 92121, USA

Douglas A. Weigent
Department of Physiology and Biophysics,
University of Alabama at Birmingham,
1918 University Boulevard MCLM896,
Birmingham,
AL 35294-0005, USA

James A. Wells
Sunesis Pharmaceuticals,
3696 Haven Avenue,
Suite C,
Redwood City, CA 94063, USA

John R. White
Department of Molecular Immunology,
Smithkline Beecham, 709 Swedeland Road,
PO Box 1539,
King of Prussia,
PA 19406, USA

Theresa L. Whiteside
Pathology Department,
University of Pittsburgh School of Medicine and University of Pittsburgh Cancer Institute,
Pittsburgh,
Pennsylvania, USA

Timothy J. Williams
Leukocyte Biology Section,
Sir Alexander Fleming Building Imperial College School of Medicine,
Sir Alexander Fleming Building,
London, South Kensington,
SW7 2AZ, UK

B. R. G. Williams
Department of Cancer Biology,
Lerner Research Institute,
Cleveland Clinic Foundation,
9500 Euclid Avenue,
Cleveland, OH 44195, USA

Joanna M. Woodcock
Human Immunology,
Institute of Medical and Veterinary Science,
Frome Road,
Adelaide, SA, 5000,
Australia

Cytokine Receptor Laboratory,
Hanson Centre for Cancer Research,
Frome Road,
Adelaide,
SA, 5000, Australia

Teresa K. Woodruff
Department of Neurobiology and Physiology,
Northwestern University,
2153 N. Campus Drive,
Evanston,
IL 60208-2850, USA

Anja Wuyts
Laboratory of Molecular Immunology,
Rega Institute – University of Leuven,
Minderbroedersstraat 10,
Leuven, 3000, Belgium

Osamu Yoshie
Department of Bacteriology,
Kinki University School of Medicine,
377-2 Ohno-Higashi,
Osaka-Sayama,
Osaka, 589-8511, Japan

Tomohiro Yoshimoto
Hyogo College of Medicine,
Nishinomiya, Japan

Byung-S. Youn
Department of Microbiology and Immunology,
Indiana University School of Medicine,
635 Barnhill Drive,
Indiapolis,
IN 46202, USA

Angel Zaballos
Departamento de Inmunología y Oncología,
Centro Nacional de Biotecnología,
Campus Universidad Autónoma, Cantoblanco,
Madrid, 28049,
Spain

Albert Zlotnik
Department of Immunobiology,
DNAX Research Institute of Molecular and Cellular Biology, Inc.,
901 California Avenue,
Palo Alto,
CA 94304-1104, USA

Immunomodulating Interleukin Receptors

IL-2 Family Cytokines and their Receptors

Warren J. Leonard*

Laboratory of Molecular Immunology, NHLBI, National Institutes of Health, Building 10, Room 7N252, Bethesda, MD 20892-1674, USA

*corresponding author e-mail: wjl@helix.nih.gov

DOI: 10.1006/rwcy.2000.02003.

SUMMARY

In this chapter, the 'IL-2 family of cytokines' is defined to include IL-2, IL-4, IL-7, IL-9, IL-13, IL-15, and thymic stromal lymphopoietin (TSLP). We define the 'immediate' IL-2 family as IL-2, IL-4, IL-7, IL-9, and IL-15, as these five cytokines share the common cytokine receptor γ chain, γc, which was first discovered as a component of the IL-2 receptor. Moreover, IL-2 was the first of these five cytokines to be identified and cloned. IL-13 and TSLP can be considered as slightly more distant relatives as their receptors do not share γc. As discussed below, IL-13 is particularly closely related to IL-4 whereas TSLP is closely related to IL-7. Details of the basic properties of both human and murine versions of these cytokines, including cellular source, molecular weight, number of amino acids, major actions, chromosomal location, and genomic organization are summarized in Table 1. A number of general properties of four α-helical bundle 'type I' cytokines (which include the IL-2 family cytokines) will be discussed. Then, the bulk of this chapter will focus specifically on the IL-2 family of cytokines, their receptors, the significance of shared receptor components, and aspects of signaling.

GENOMIC ORGANIZATION OF IL-2 FAMILY CYTOKINES AND THEIR RECEPTORS

The basis for the IL-2 'family' of cytokines is determined in part by shared receptor component considerations. The sharing of receptor components suggests a progressive coevolution of different cytokines with their receptors. Thus, it is interesting that some but not all of the genes encoding these cytokines colocalize. The genes for IL-2, IL-4, IL-7, IL-9, IL-13, and IL-15 are located on chromosomes 4q26-27, 5q31.1, 8q12, 5q31.1, 5q31.1, and 4q31 respectively. Thus IL-2 and IL-15, which share both the IL-2 receptor β chain (IL-2Rβ) and the common cytokine receptor γ chain (γc), are both on chromosome 4q, whereas IL-4, IL-9, and IL-13 are all located on chromosome 5q. The genomic organization of these cytokines is not rigidly conserved. For example, whereas IL-2 has four exons, IL-4, IL-7, IL-9, and IL-15 are encoded by 4, 6, 5, and 5 exons, respectively (**Table 1**). In the four α-helical structure of IL-2 (Figure 1), helix A is encoded by exon 1, helices B and C are in exon 3, and helix D is in exon 4. The genes encoding the various receptor chains are located at widely dispersed locations, as shown in **Table 2** and parenthetically after each receptor chain for human and mouse (human/murine): IL-2Rα (10p14-p15/2), IL-2Rβ (22q11.2-q13/15), γc (*IL2RG*, Xq13/X), IL-4Rα (16p11.2-p12.1/7), IL-7Rα (5p13/?), IL-9Rα (Xq28/11), IL-13Rα1 (X/X), IL-15Rα (10p14-p15/2). Thus, the only two receptor chains that colocalize are the genes encoding IL-2Rα and IL-15Rα, although three of the receptors map to chromosome X.

IL-2 FAMILY CYTOKINES ARE SHORT-CHAIN FOUR α-HELICAL BUNDLE 'TYPE I' CYTOKINES

Each member of the IL-2 family of cytokines is believed to have the structure of a four α-helical

Table 1 Features of cytokines whose receptors share γc

Cytokine/Major source	Size[a]	Actions	Chromosomal location	Genomic organization
IL-2				
Activated T cells (TH1 cells)	Human: 153 aa/20 aa	T cell growth	4q26-q27/3	4 exons
	Mouse: 169 aa/20 aa	B cell growth		
	15.5 kDa	Ig production		
		J chain expression		
		Induces LAK and tumor-infiltrating lymphocyte activity		
		Augments NK activity		
		Stimulates macrophage/ monocytes		
		Antitumor effects		
		Role in antigen-induced cell death		
IL-4				
Activated T cells (TH2 cells)	Human: 153/24 aa	B cell proliferation	5q31.1/11	4 exons
Mast cells	Mouse: 140 aa/20 aa	IgG1, IgE production		
Basophils	18 kDa	Augments MHCII,		
CD4+ NK1.1+ natural T cells		FcϵRII, IL-4Rα and IL-2Rβ expression		
		TH2 cell differentiation		
		Antitumor effects		
		T cell growth		
IL-7				
Stromal cells	Human: 177 aa/25 aa	Thymocyte growth	8q12-q13/3	6 exons
Keratinocytes	Mouse: 154 aa/25 aa	Thymocyte/T-cell growth		
	17–25 kDa	Pre-B cell growth in mice but not humans		
		Anti-apoptotic factor		
IL-9				
Activated helper T cells	Human: 144 aa/18 aa	Stimulation of T helper clones, erythroid progenitors, B cells, mast cells, fetal thymocytes	5q31.1/13	5 exons
	Mouse: 144 aa/18 aa			
	14 kDa			
		Anti-apoptotic factor		
IL-15				
Monocytes and many cells outside the immune system[b]	Human: 162 aa/48 aa	Mast cell growth	4q31/8	8 exons
	Mouse: 162 aa/48 aa	NK cell development and activity		
	14–15 kDa	T cell proliferation		

Table 1 (*Continued*)

Cytokine/Major source	Size[a]	Actions	Chromosomal location	Genomic organization
IL-13				
T cells, NK cells, mast cells, EBV-transformed B cells		B cell proliferation IgG4, IgE production	5q31/	
TSLP				
Thymic stromal cells		Development of IgM+ B cells	?	Unknown

[a]The sizes of the entire open reading frame/length of signal peptide are shown. For IL-15, residues 1–29 have been identified as a signal peptide and residues 30–48 are a propeptide.

[b]More IL-15 mRNA is widely expressed without always having corresponding production of IL-15 protein.

Modified and expanded from Leonard (1999).

Table 2 Properties of IL-2 family receptor chains

	IL-2Rα	IL-2Rβ	γc	IL-4Rα	IL-7Rα	IL-9Rα	IL-15Rα	IL-13Rα1
Chromosomal location	10p14-15	22q11.2-q13	Xq13	16p11.2-p12.1 5p13	Xq28	10p14-p15	X	
Size (kDa)	55	70–75	64	140	75	65	60	65
Amino acids								
Signal peptide	21	26	22	25	20	39	30	21
Extracellular	219	214	232	207	220	231	175	322
TM	19	25	29	24	25	21	21	24
Cytoplasmic	13	286	86	569	195	231	41	60
Number of exons	8	19	8	?	8	10	7	?

bundle ‘type I’ cytokine (reviewed in Leonard, 1999). The structures for IL-2 and IL-4 were determined by NMR and X-ray crystallographic studies. While the structures for IL-7, IL-9, IL-13, and IL-15 are unknown, they are predicted to have similar structures to IL-2 and IL-4 based on sequence and/or molecular homology modeling. The TSLP sequence has not yet been published, but based on (1) its sharing a number of biological actions with IL-7, (2) the fact that it is produced by stromal cells, and (3) the fact that its receptor contains IL-7Rα as one of its components, it is also assumed to exhibit a similar structure.

In the structures for IL-2 and IL-4, there are four α helices that achieve an ‘up-up-down-down’ organization (Bazan, 1991; Wlodawer *et al.*, 1993; Rozwarski *et al.*, 1994; Sprang and Bazan, 1993; Davies and Wlodawer, 1995; Leonard, 1999). As is typical of all type I cytokines, the first two and the last two α helices are connected by long-overhand loops. This feature allows the first two helices (A and B) to be oriented in an ‘up’ orientation while the last two helices (C and D) are oriented in a ‘down’ orientation, as viewed from the N- to C-terminal direction. Topologically, the N- and C-termini of the cytokines are positioned on the same part of the molecule.

Four α-helical bundle cytokines can be divided into ‘short chain’ and ‘long chain’ subfamilies (Sprang and Bazan, 1993; Davies and Wlodawer, 1995). In all short-chain cytokines, including the IL-2 family of cytokines, the helices are approximately 15 amino acids in length; in contrast in the long-chain α-helical bundle cytokines (typified by molecules such as IL-6, growth hormone, erythropoietin, etc.), the helices are approximately 25 amino acids long. Another difference between short- and long-chain cytokines is that the AB loop is ‘under’ the CD loop in the short-chain cytokines, as opposed to ‘over’ the CD loop in the long-chain cytokines (**Figure 1**) (see Sprang and

Figure 1 IL-2 family cytokines are four α-helical bundle cytokines. Schematic drawing showing typical short chain and long chain four helical bundle cytokines. These both exhibit an 'up-up-down-down' topology to their four α helices, but the short-chain cytokines have the AB loop in behind the CD loop, whereas in the long-chain cytokines the situation is reversed. The figure was provided by Dr Alex Wlodawer, National Cancer Institute. Reprinted from "Fundamental Immunology", 4th edn, 1999 (ed W. E. Paul, Lippincott Raven Publishers, Philadelphia, PA), p. 742.

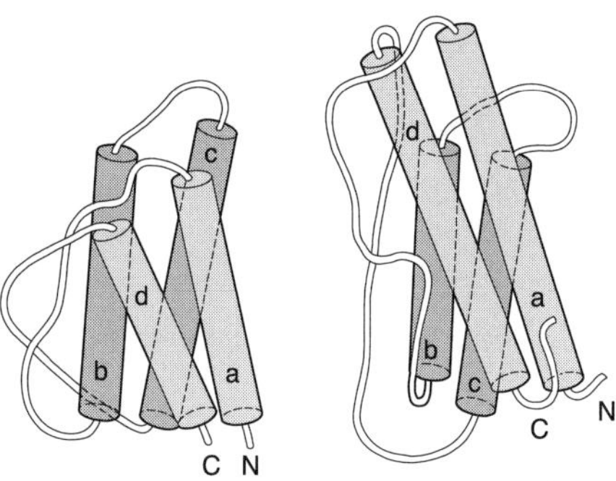

Bazan, 1993; Rozwarski *et al.*, 1994; Davies and Wlodawer, 1995; Leonard, 1999). Moreover, in the short-chain cytokines, there are β structures in the AB and CD loops. Approximately 61 residues comprise the family framework; these include most of the 31 residues that contribute to the buried inner core.

The structural similarities and differences between IL-2 and IL-4 have been carefully analyzed (Wlodawer *et al.*, 1993). It is interesting that there is considerable variation in the intrachain disulfide bonds that stabilize the structures of type I cytokines. Whereas IL-4 has three intrachain disulfide bonds, IL-2 has only one. In IL-4, the disulfide bonds connect residues 24 and 65 (and thus loop AB to BC), residues 46 and 99 (and thus helix B and loop CD), and residues 3 and 127 (and thus connects the residue preceding helix B with helix D). In contrast, IL-2 contains only a single essential disulfide bond between residues 58 and 105 which serves to connect helix B to strand CD. Thus, it is evident that within the IL-2 family, different cytokines have evolved to utilize distinct disulfide bonds to stabilize their respective structures. Nevertheless, for both IL-2 and IL-4, as well as for other short-chain cytokines, helix B is connected to the loop between helices C and D. In general, the structures formed by helices A and D are more rigorously conserved than those formed by helices B and C. The interhelical angle between helix A and D is particularly well conserved, with helix D and the connecting region being the most highly conserved region between IL-2 and IL-4 (Wlodawer *et al.*, 1993).

The structure of the growth hormone/growth hormone receptor complex revealed that helices A and D of growth hormone and residues in the connecting loops are the most important for contacts between ligand and receptor (de Vos *et al.*, 1992). Growth hormone is a long-chain four α-helical bundle cytokine, but the structure for its receptor (de Vos *et al.*, 1992) was assumed to represent a general paradigm for short-chain cytokine receptors as well as for long-chain cytokine receptors. A structure for the binding of IL-4 to IL-4Rα now exists (Hage *et al.*, 1999). While this provides valuable information, a complete structure of a functional IL-4 receptor (including either γc or IL-13Rα1) is not yet available.

The conservation of the A and D helices can help us to understand the ability of different cytokines to share various receptor components. This property helps to define the IL-2 family of cytokines. Nevertheless, there are functional properties that also help to define these cytokines as members of a subfamily of type I cytokines. The receptor interactions for these cytokines will be discussed in detail, as this helps to clarify our understanding of part of the basis for functional overlap among different members of the IL-2 family of cytokines.

IL-2 FAMILY CYTOKINES BIND TO TYPE I CYTOKINE RECEPTORS: STRUCTURAL CONSIDERATIONS

Almost all receptor components of IL-2 family cytokines are type I cytokine receptors. Type I cytokine receptors were also historically referred to as cytokine receptor superfamily members or hematopoietin receptors. The type I cytokine receptors include most interleukin and cytokine receptors and are to be distinguished from type II cytokine receptors, which include the receptors for type 1 and type 2 interferons and the IL-10 receptor components (Bazan, 1990; Leonard, 1999). It is important to recognize that type I and type II cytokines and cytokine receptors refer to structurally defined families and should not be confused with TH1 and TH2 nomenclatures. Confusingly, 'type 1' and 'type 2' nomenclature is sometimes used to refer to cytokines produced by TH1 and TH2 cells, respectively.

Table 3 Components of functional receptors for the IL-2 family of cytokines

Cytokine	IL-2Rα	IL-2Rβ	IL-4Rα	IL-7Rα	IL-9Rα	IL-13Rα1	IL-15Rα	γc	TSLPR
IL-2	+	+	–	–	–	–	–	+	–
IL-4	–	–	+	–	–	–	–	+	–
IL-4	–	–	+	–	–	+	–	–	–
IL-7	–	–	–	+	–	–	–	+	–
IL-9	–	–	–	–	+	–	–	+	–
IL-13	–	–	+	–	–	+	–	–	–
IL-15	–	+	–	–	–	–	+	+	–
TSLP	–	–	–	+	–	–	–	–	+

The IL-2 family cytokine receptor chains that are type I receptor molecules include IL-2Rβ, IL-4Rα, IL-7Rα, IL-9Rα, IL-13Rα, and γc (**Table 3**). The TSLPR protein has not yet been published but is generally presumed to be a member of this family. IL-13Rα2 is another type I cytokine receptor protein that can bind IL-13 but whose role in IL-13 signaling is unclear (Caput *et al.*, 1996); it is possible that it serves a negatively regulatory function as a 'decoy receptor'. The two chains that are different are IL-2Rα (Leonard *et al.*, 1984; Nikaido *et al.*, 1984) and IL-15Rα (Giri *et al.*, 1995), which are distinctive proteins that contain 'sushi' domains and together are the only two members of a novel family of cytokine receptor proteins. Other proteins, however, such as certain complement proteins, also have sushi domains (Giri *et al.*, 1995; He and Malek, 1998; Leonard, 1999). All of these proteins (the type I cytokine receptor proteins as well as IL-2Rα and IL-15Rα) are type I transmembrane proteins (N-terminus extracellular, C-terminus intracellular, with a single transmembrane crossing). Like all type I cytokine receptor proteins, in their extracellular domains, IL-2Rβ, IL-4Rα, IL-7Rα, IL-9Rα, IL-13Rα1, and γc have two pairs of conserved cysteine residues that form two intrachain disulfide bonds, and a Trp-Ser-X-Trp-Ser (WSXWS) motif more proximal to the membrane. Within their cytoplasmic domains, a membrane proximal region known as the Box 1/Box 2 region is conserved (**Table 4**), with a proline-rich Box 1 region being the most conserved; this region is particularly important in mediating the interaction of JAK family tyrosine kinases. Typically, each pair of conserved cysteines is in a different exon, with the WSXWS motif being located in the exon immediately upstream of the exon encoding the transmembrane domain (**Figure 2**).

Table 4 Shared features in type I cytokine receptors

Extracellular domain

- Fibronectin type III modules
- Four conserved cysteine residues that form intrachain disulfide bonds
- WSXWS motif

Cytoplasmic domain

- Box 1/Box 2 regions. The Box 1 region is proline-rich and mediates interaction with JAK kinases

As noted above, all type I cytokines including IL-2, IL-4, IL-7, IL-9, IL-13, and IL-15, have receptors that form dimeric or higher order complexes. For the IL-2 family, the receptors all contain two type I cytokine receptor chains, with IL-2 and IL-15 additionally having their unique α chains. The stoichiometry has not been rigorously studied, but only single copies of each cytokine receptor component are believed to be included in the functional receptors. Signaling is believed to be triggered by ligand-induced heterodimerization of cytoplasmic domains, as has been demonstrated for the IL-2 receptor (Nakamura *et al.*, 1994; Nelson *et al.*, 1994).

INTERACTIONS OF IL-2 FAMILY CYTOKINES WITH THEIR RECEPTORS

Each member of the IL-2 family of cytokines shares a receptor component with at least one other member of this family (Table 3). For example, the receptors for IL-2, IL-4, IL-7, IL-9, and IL-15 all share the common cytokine receptor γ chain, γc (originally denoted as the IL-2 receptor γ chain; *see* **Figure 3**) (reviewed in Leonard, 1999; see also Takeshita *et al.*,

Figure 2 Schematic of IL-2 receptor chains. Shown are the exons of IL-2Rα, IL-2Rβ, and γc, subdivided according to 5′ UTR, extracellular domain, transmembrane domain, cytoplasmic domain and 3′ UTR. Also shown are the positions of the stop codon, conserved cysteine residues and WSXWS motifs in IL-2Rβ and γc, and the sushi domains in IL-2Rα. IL-2Rβ and γc are representative of all type I cytokine receptors. IL-2Rα is related to IL-15Rα. Unlike IL-2Rα, IL-15Rα has only a single sushi domain.

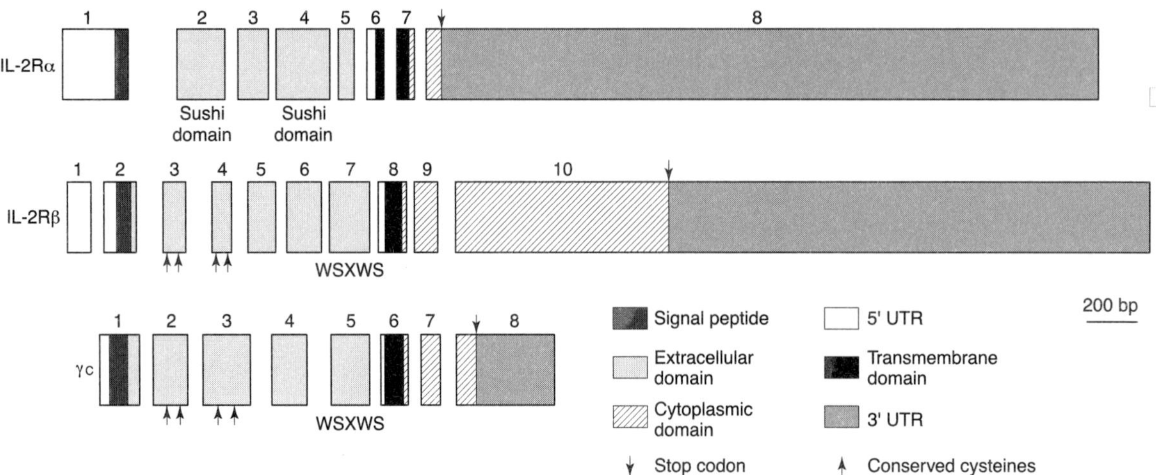

Figure 3 γc is shared by IL-2, IL-4, IL-7, IL-9, and IL-15. The cartoon shows that IL-2, IL-4, IL-7, IL-9, and IL-15 all share γc and IL-2 and IL-15 share IL-2Rβ. IL-2Rα and IL-15Rα are not shown. Whereas the distinctive chains associate with JAK1, γc associates with JAK3. Mutations in γc cause XSCID, whereas mutations in JAK3 cause an autosomal recessive form of SCID.

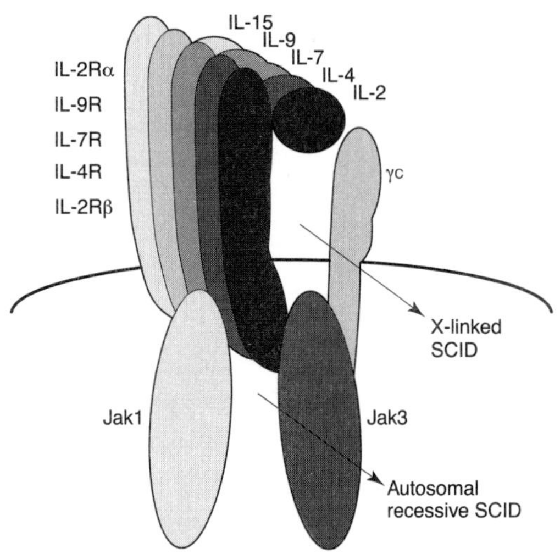

1992; Kondo *et al.*, 1993, 1994; Noguchi *et al.*, 1993a; Russell *et al.*, 1993, 1994; Kimura *et al.*, 1995; Giri *et al.*, 1994; Sugamura *et al.*, 1996); IL-2 and IL-15 additionally share IL-2Rβ (Giri *et al.*, 1994; Bamford *et al.*, 1994); the type I and type II IL-4 receptors and the IL-13 receptor share IL-4Rα (Lin *et al.*, 1995; Hilton *et al.*, 1996; Aman *et al.*, 1996), the type II IL-4 receptor and IL-13 receptor share IL-13Rα1 (Lin *et al.*, 1995; Hilton *et al.*, 1996; Aman *et al.*, 1996); and the IL-7 and TSLP receptors share IL-7Rα (Peschon *et al.*, 1994; Levin *et al.*, 1999; Pandey *et al.*, 2000).

For each of the IL-2 family cytokines, the D helix of the cytokine associates with the 'major' signaling molecule (IL-2Rβ, IL-4Rα, IL-7Rα, or IL-9Rα) whereas the A helix associates with what is generally viewed as the 'secondary' signaling molecule (γc, IL-13Rα1, and TSLPR). The sharing of receptor components presumably had its origins in the evolution of these different cytokines. For example, whereas the A helices of IL-2, IL-4, IL-7, IL-9, and IL-15 presumably share certain features that allow the association of each of these cytokines with γc, the D helices are more divergent, allowing the associations of each cytokine with its distinctive chain. Note that for IL-2 and IL-15 the situation is somewhat more complex than for the other cytokines in that IL-2 and IL-15 share both IL-2Rβ and γc but differ in their binding to distinctive α chains (IL-2Rα and IL-15Rα) that are not type I cytokine receptor proteins (Leonard *et al.*, 1984; Nikaido *et al.*, 1984; Giri *et al.*, 1995; reviewed in He and Malek, 1998; Leonard, 1999).

In contrast to the conservation of binding properties of the A helices of IL-2, IL-4, IL-7, IL-9, and IL-15, two other pairs of cytokines (IL-4/IL-13 and

IL-7/TSLP) share certain binding properties of their D helices. Specifically, given that IL-4 and IL-13 both interact with IL-4Rα via their relatively conserved D helices, they must have diverged in their A helices so that IL-4 can interact with either γc or IL-13Rα1 while IL-13 interacts only with IL-13Rα1. A similar situation presumably exists for TSLP and IL-7, both of which associate with IL-7Rα but only the latter of which associates with γc. Thus, even within the IL-2 family of cytokines, there are examples wherein the extent of functional similarity is greater for the A than D helix and vice versa.

Although the D helices of γc-dependent cytokines clearly are sufficiently conserved structurally so that they can all interact with γc, it is important to recognize that these cytokines do not all contact γc in exactly the same way. This is demonstrated by the ability of different monoclonal antibodies to γc to differentially affect signaling by various γc-dependent cytokines (Malek *et al.*, 1999). Thus, within the general framework of functionally similar D helix structures there are distinctive contacts that form for different cytokines. This suggests that in the context of their distinctive receptor chains, the affinities of these cytokines for γc may differ, resulting in differences in their relative efficiencies of recruiting γc.

For the γc family of cytokines (IL-2, IL-4, IL-7, IL-9, and IL-15), the receptor chain with the greater binding activity for each ligand is the distinctive chain (IL-2Rβ, IL-4Rα, IL-7Rα, IL-9Rα, and IL-2Rβ, respectively), which also coordinates with most downstream signaling pathways. In contrast, as an isolated component, the shared γc exhibits no detectable binding activity for the ligands. This situation is somewhat different for other sets of cytokines, such as the hematopoietic cytokines IL-3, IL-5, and GM-CSF and the IL-6 family of cytokines. IL-3, IL-5, and GM-CSF each bind primarily to distinctive α chains but the common β chain, βc, is the major signaling molecule. For IL-6, IL-11, CNTF, LIF, oncostatin M, and cardiotropin 1, the shared chain, gp130, is again the major signal transducing molecule.

LIGAND–RECEPTOR INTERACTIONS: DYNAMICS AND MORE THAN ONE CLASS OF RECEPTOR FOR CERTAIN CYTOKINES

As summarized in Table 3, each IL-2 family receptor contains more than one type I receptor protein; physiologically, none of these cytokines appears to be able to signal via homodimers. Although the receptors in at least some cases may be preformed, they are more stably formed in the presence of ligand. This was first demonstrated for the IL-2 receptor where an antibody to IL-2Rβ was found to be capable of coprecipitating γc in the presence but not absence of IL-2 (Takeshita *et al.*, 1990). This implies that each cytokine might have a primary interacting partner and that a complex comprising the cytokine/cytokine receptor chain might then recruit a second receptor chain, as was originally suggested for the growth hormone system (de Vos *et al.*, 1992). IL-2, IL-4, IL-7, IL-9, and IL-15 primarily interact with IL-2Rα/IL-2Rβ, IL-4Rα, IL-7Rα, IL-9Rα, and IL-15Rα/IL-2Rβ, respectively. Note that IL-2 and IL-15 are distinctive in that there are the accessory α chains which do not exist for the other cytokines. Subsequently, γc is recruited.

The IL-2 receptor system is particularly well studied. Three distinct classes of receptors exist, binding IL-2 with low ($K_d = 10^{-8}$ M), intermediate ($K_d = 10^{-9}$ M), and high affinity ($K_d = 10^{-11}$ M). The low-affinity receptor contains IL-2Rα, intermediate affinity receptor contains IL-2R$\beta + \gamma$c, and the high affinity receptor contains all three chains (**Table 5**). The functional forms of the receptor are the intermediate- and high-affinity forms, consistent with heterodimerization of IL-2Rβ and γc being necessary and sufficient for transducing an IL-2 signal. More than one affinity for the IL-7 receptor has also been observed on peripheral blood lymphocytes. In addition to the high-affinity receptor with a K_d of 30–65 pM, there is an extremely low-affinity receptor

Table 5 Classes of IL-2 receptors

Affinity	K_d	Where expressed	Composition	Functionality
Low	10^{-8} M	Activated cells	IL-2Rα	No
Intermediate	10^{-9} M	Resting cells	IL-2Rβ and γc	Yes
High	10^{-11} M	Activated cells	IL-2Rα, IL-2Rβ, and γc	Yes

with a K_d of 100–150 nM (Goodwin *et al.*, 1990; Noguchi *et al.*, 1993a). The biological significance and biochemical basis for this class of binding sites is unknown. Fibroblasts transfected with single chains, such as IL-7Rα and IL-9Rα, can bind ligand with low to intermediate affinities, but it is unclear whether such proteins ever physiologically exist in isolation without γc.

It has been reported that artificial homodimerization of IL-4Rα can trigger signals, and this finding has potential use in mapping functional residues for signaling within the IL-4Rα cytoplasmic domain (Kammer *et al.*, 1996; Lai *et al.*, 1996). However, such a form of the receptor has not been shown to exist (Hoffman *et al.*, 1995; Hage *et al.*, 1999). Instead, it appears that IL-4 coordinates one molecule of IL-4Rα with γc (Hoffman *et al.*, 1995) or presumably IL-13Rα1.

IL-2 FAMILY CYTOKINES EXHIBIT CYTOKINE PLEIOTROPY AND REDUNDANCY

IL-2, IL-4, IL-7, IL-9, IL-13, IL-15, and TSLP each have distinctive actions (see individual chapters on these cytokines and Table 1 for a detailed summary of their actions); however, these cytokines nevertheless collectively exhibit a number of overlapping actions as well (see He and Malek, 1998; Leonard, 1999). The terms 'cytokine pleiotropy' and 'cytokine redundancy' are used to describe respectively the ability of a single cytokine to exert multiple actions and the ability of multiple cytokines to exert similar actions (Paul, 1989, 1991). Both of these phenomena are used by the IL-2 family of cytokines. IL-2 was originally described as a T cell growth factor (Morgan *et al.*, 1976) and indeed appears to be the most important single T cell growth factor. Nevertheless, the other four cytokines that share γc (IL-4, IL-7, IL-9, and IL-15) each also exert important actions on T cells (see Table 1). While IL-4 is a B cell growth factor that is particularly vital for immunoglobulin class switching, it is also vital for the evolution of TH2 cells and can act as a T cell growth factor (Lee *et al.*, 1986; Noma *et al.*, 1986; Yokota *et al.*, 1986; Yoshimoto *et al.*, 1995; Paul, 1997). IL-7 is an essential factor for thymic development (Watson *et al.*, 1989; Murray *et al.*, 1989; Peschon *et al.*, 1994; von Freeden-Jeffrey *et al.*, 1995; Puel *et al.*, 1998) but also exerts effects on mature T cells (Chazen *et al.*, 1989; Morrissey *et al.*, 1989; Murray *et al.*, 1989; Watson *et al.*, 1989). Finally, IL-9 (van Snick *et al.*, 1989) and IL-15 (Tagaya *et al.*, 1996; Zhang *et al.*, 1998) can also stimulate T cell growth.

A number of these cytokines can also act on B cells and NK cells. For example, IL-4 is a major B cell growth factor, but IL-2 is known to augment immunoglobulin synthesis (Mingari *et al.*, 1984; Blackman *et al.*, 1986; Waldmann, 1989) and IL-7 is vital for the development of murine pre-B cells (Namen *et al.*, 1988; Peschon *et al.*, 1994; von Freeden-Jeffry *et al.*, 1995). Interestingly, in humans, this pre-B cell growth function is either redundantly served by IL-7 and another factor or alternatively might be uniquely served by another cytokine (Puel *et al.*, 1998). Although IL-2 can potently augment the cytolytic activity of NK cells (Siegel *et al.*, 1987; Lanier and Phillips, 1992), it is now clear that IL-15 is vital for the development of these cells and, like IL-2, potently boosts the activity of these cells once developed (Carson *et al.*, 1994; Cavazzana *et al.*, 1996; Leclercq *et al.*, 1996; Puzanov *et al.*, 1996; Lodolce *et al.*, 1998).

IL-4 and IL-13 are particularly closely related cytokines (Smerz-Bertling and Duschl, 1994; Zurawski *et al.*, 1994; Lin *et al.*, 1995). The main difference between them is that whereas IL-4 acts on T cells as well as non-T cells, the principal actions of IL-13 are exerted on non-T cells. *In vitro*, these cytokines exhibit extensive overlap in their functions. However, based on knockout models and other supporting data, IL-4 was found to be vital for TH2 development, whereas IL-13 plays particularly important roles for the expulsion of worms as well as for antigen-induced asthma (Urban *et al.*, 1998; Wills-Karp, 1999).

Finally, IL-7 and TSLP are related cytokines that are both made by stromal cells (Levin *et al.*, 1999). TSLP can promote the development of IgM+ B cells, although the full range of actions of TSLP is not yet clear. Thus, considerable overlap of actions exists for the IL-2 family of cytokines for T cells, B cells, and NK cells. Although not all of the cytokines act on each lineage, this family provides excellent examples of cytokine pleiotropy and redundancy. Part of these actions can be understood based on the sharing of receptor chains. The phenomenon of 'cytokine receptor pleiotropy' refers to the ability of certain proteins such as γc to act in the context of five different cytokines. Other examples wherein one receptor chain is part of more than one receptor include IL-2Rβ (part of IL-2 and IL-15 receptors), IL-4Rα (part of type I and type II IL-4 receptors as well as IL-13 receptors), IL-13Rα (part of the type I IL-4 receptor and the IL-13 receptor), and IL-7Rα (shared by IL-7 and TSLP receptors).

'Cytokine receptor redundancy' can help to explain the pleiotropic and/or redundant actions of cytokines.

An example of this phenomenon includes the IL-13Rα1 chain in the type II IL-4 receptor, which can function as an alternative to the γc chain in the type I IL-4 receptor. It has also been suggested that IL-15 can signal via two completely different types of receptors (Tagaya *et al.*, 1995). One of these includes IL-15Rα, IL-2Rβ, and γc, whereas the IL-15 receptor on mast cells has been designated IL-15RX, although this protein has not yet been cloned. One of the key issues related to the concept of cytokine receptor redundancy is whether the signals induced are truly the same and thus whether there is precise redundancy.

SIGNALING THROUGH CYTOKINE RECEPTORS

Dimerization and the JAK/STAT pathway

IL-2 family cytokines exhibit at least partial sharing of signaling pathways. In particular, each is known to activate the JAK/STAT pathway (Horvath and Darnell, 1997; Leonard and O'Shea, 1998). The JAK/STAT pathway is a very rapid cytoplasmic to nuclear signaling pathway that is used by all type I cytokines, including the IL-2 family of cytokines, and all type II cytokines (the interferons and IL-10). In this pathway, the binding of the ligand induces oligomerization of receptor components. Janus family tyrosine kinases (JAK kinases) that are associated with receptor components are activated, at least in part by transphosphorylation of critical tyrosines with the kinase activation loop (Leonard and O'Shea, 1998). In some cases, ligand augments recruitment of the JAK kinases; for example, cytokine stimulation augments the association of JAK3 with γc (Leonard and O'Shea, 1998). The JAK kinases then phosphorylate critical tyrosines on the receptors, which serve as docking sites for signal transducer and activator of transcription (STAT) proteins; in the case of IL-2 family cytokines, the receptor components that are critical for STAT protein docking are IL-2Rβ (Lin *et al.*, 1995), IL-4Rα (Hou *et al.*, 1994; Ryan *et al.*, 1996), IL-7Rα (Lin *et al.*, 1995; Pallard *et al.*, 1999), and IL-9Rα (Demoulin *et al.*, 1996), whereas γc and IL-13Rα1 have not been reported to have docking sites, nor have the non-type I cytokine receptor proteins IL-2Rα and IL-15Rα.

The STAT proteins associate with phosphorylated receptor chains via their SH2 domains. They then in turn are phosphorylated on a critical tyrosine, dissociate from the receptor and dimerize. The dimeric 'activated' STAT proteins can then translocate to the nucleus and promote the transcription of target genes. STAT proteins were the first transcription factors recognized to be phosphorylated on tyrosine. Previously, tyrosine phosphorylation was principally associated with membrane proximal events. In fact, even for STAT proteins, the tyrosine phosphorylation is a membrane proximal event, but this phosphorylation then allows dimerization, which is a prerequisite for nuclear localization and DNA binding. In addition to the docking of STATs on phosphorylated receptor components, some STATs can also interact directly with JAK kinases. This was first shown for STAT5 and JAK3 (Migone *et al.*, 1995), and this observation not only suggests that STATs may at times dock directly on JAK kinases but also makes it easy to understand that STATs are substrates for JAKs.

Of the four JAK kinases (JAK1, JAK2, JAK3, and TYK2), IL-2, IL-4, IL-7, IL-9, and IL-15 all signal via JAK1 and JAK3 (Leonard and O'Shea, 1998). This is explained by the ability of JAK1 to associate with the comparatively distinctive chain of each receptor, namely IL-2Rβ (Boussiotis *et al.*, 1994; Miyazaki *et al.*, 1994; Russell *et al.*, 1994), IL-4Rα (Yin *et al.*, 1994), IL-7Rα, and IL-9Rα (Demoulin *et al.*, 1996), whereas JAK3 associates with primarily with γc (Boussiotis *et al.*, 1994; Miyazaki *et al.*, 1994; Russell *et al.*, 1994). γc can also interact with IL-2Rβ, which raises the possibility that JAK3 might help to stabilize IL-2Rβ–γc heterodimers from within the cell (Zhu *et al.*, 1998). Interestingly, members of the γc family of cytokines are the only cytokines to use both JAK1 and JAK3. The extended IL-2 family members, IL-13 and TSLP, signal through receptors that do not contain γc and hence do not activate JAK3; this is also true of the type II IL-4 receptor, which uses IL-4Rα and IL-13Rα1.

It is noteworthy that JAK1 is the JAK kinase that is always coactivated with JAK3. It is unclear whether this is coincidence, perhaps related to how the receptors have evolved, and/or whether JAK1 might be more effective than other JAKs in cooperating with JAK3 under physiological circumstances. What is clear, however, is that because IL-2, IL-4, IL-7, IL-9, and IL-15 all activate the same JAK kinases, the JAK kinases by themselves do not determine specificity. Instead, these proteins presumably must act together with other signaling molecules to effect this function.

A principal action of JAK kinases is to mediate the activation of STAT proteins. Whereas IL-2, IL-7, IL-9, and IL-15 each activate both STAT5a and STAT5b, some of these cytokines can additionally activate STAT3 (Leonard and O'Shea, 1998). Like IL-7, TSLP can also activate STAT5 proteins (Levin *et al.*, 1999), a finding that makes sense as the TSLP receptor contains IL-7Rα (mentioned in Peschon *et al.*, 1994; see below). IL-4 and IL-13 primarily activate STAT6, providing at least a degree of relative specificity to IL-4 as compared to the other γc-dependent cytokines.

STAT proteins contain a number of conserved functional domains (Horvath and Darnell, 1997; Leonard and O'Shea, 1998). These include an N-terminal oligomerization (tetramerization) domain, a DNA-binding domain, a linker region, an SH2 domain, a phosphorylated tyrosine that is vital for STAT dimerization based on bivalent SH2-phosphotyrosine interactions, and a C-terminal transactivation domain (**Figure 4**). In at least some STATs, such as STAT1 and STAT3, this latter region is phosphorylated on serine (Horvath and Darnell, 1997).

Because different STATs differ in the fine binding specificities of their SH2 domains, different phosphorylated tyrosine residues exhibit preferences for different STATs. Thus, the microenvironments of these phosphorylated tyrosines select the STATs that are recruited. In this fashion, it is possible to understand why the same STATs are activated by both IL-2 and IL-15 (which share IL-2Rβ) or by IL-4 and IL-13 (which share IL-4Rα). In both IL-2Rβ and IL-4Rα, it is interesting that more than one tyrosine (two for IL-2Rβ (Y392 and Y510) (Lin *et al.*, 1995; Friedmann *et al.*, 1996; Ascherman *et al.*, 1997, Fujii *et al.*, 1995) and three for IL-4Rα (Y575, Y603, and Y631)) recruit STAT proteins in apparently redundant fashion. The reason for this redundancy is unclear, but it is conceivable that the recruitment of STAT proteins to more than one site could allow for a greater local concentration of activated STAT monomers that could facilitate dimerization. It is also possible that in addition to recruiting STAT proteins, phosphorylation of these tyrosines may also result in the recruitment of other signaling molecules in a redundant or non-redundant fashion.

Figure 4 Schematic of STAT proteins. Shown are the locations of the N-terminal tetramerization domain, the DNA-binding domain, the linker domain, the SH2 domain that mediates STAT docking on receptors and STAT homodimerization/heterodimerization following tyrosine phosphorylation, the conserved tyrosine phosphorylation site, and the C-terminal transactivation domain. In some STAT proteins, such as STAT1 and STAT3, serine 727, which is C-terminal to the conserved tyrosine, is phosphorylated and this phosphorylation is needed for maximal activation.

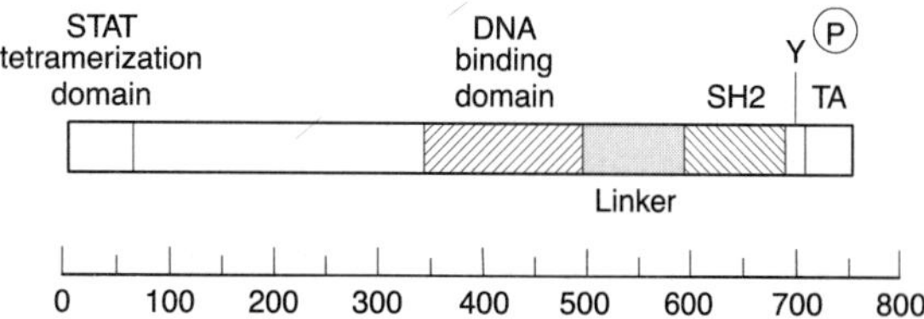

The schematic shows regions conserved in all STAT proteins. The scale is approximately correct for Stat1, Stat3, Stat4 Stat5a, and Stat5b. Stat2 and Stat6 are 50–100 amino acids longer.

Interestingly, it has been demonstrated that IL-7Rα shares a similar tyrosine motif (Y429) with IL-2Rβ, explaining why the same STATs are activated by IL-2 and IL-7 (Lin *et al.*, 1995). Y429 appears to be the only STAT docking site for IL-7Rα. Like IL-7Rα, IL-9Rα also has a single tyrosine that is needed for STAT activation (Demoulin *et al.*, 1996).

Other signaling pathways

In addition to the JAK/STAT pathway, a number of other pathways have been found to be activated by IL-2 family cytokines. These include Shc coupling to the Ras pathway, phosphatidyloinositol 3-kinase (PI-3 kinase), IRS family proteins, p70 S6 kinase, and others. However, these pathways are not well-investigated for all of the cytokines, so it cannot be concluded at present which of these are universally shared (reviewed in Leonard, 1999 and in specific chapters). The best-studied systems include IL-2, IL-4, and IL-7, whereas less is known for IL-9, IL-13, IL-15, and TSLP.

IRS-1 was originally discovered in the context of insulin signaling (Myers and White, 1996). It was later demonstrated that IL-4 not only also induced phosphorylation of an IRS-1-like protein but that IRS-1 was required for proliferation of 32D-IL-4Rα cells in response to IL-4 (Wang *et al.*, 1993). Subsequently, IRS-2 was demonstrated to also exhibit this function. While IL-2, IL-7, and IL-15 can also induce the tyrosine phosphorylation of IRS-1 and IRS-2 in T cells (Johnston *et al.*, 1995), the biological importance of these observations is less clear. For example, 32D-IL-2Rβ cells can proliferate in response to IL-2 in the absence of transfected IRS-1 or IRS-2 (Otani *et al.*, 1992). Although our understanding of IRS family proteins for cytokine signaling is obviously incomplete, these proteins contain a large number of phosphotyrosine-docking sites (Myers and White, 1996), and can serve to recruit signaling molecules, including PI-3 kinase. PI-3 kinase, which can also be recruited via YXXM motifs on receptors (Kapeller and Cantley, 1994) or by the participation of JAK1 (Migone *et al.*, 1998b) is clearly very important for IL-2 (Karnitz *et al.*, 1995; Migone *et al.*, 1998b) and IL-7 (Pallard *et al.*, 1999) signaling.

Besides JAK kinases, a number of other tyrosine kinases have been shown to be activated by IL-2 family kinases. Most well studied is the activation of Src family kinases in response to IL-2 itself. IL-2 can activate p56lck (Horak *et al.*, 1991; Hatakeyama *et al.*, 1991) in T cells and p59fyn and p53/p56lyn in B cell lines (Taniguchi, 1995). However, the biological role of Src family kinase activation by IL-2 remains unclear (Taniguchi, 1995; Leonard, 1999). Similarly, IL-2Rβ can interact with Syk, but a role for Syk in IL-2 signaling has not been defined (Minami *et al.*, 1995). In the case, of IL-4 signaling, IL-4Rα has been shown to interact with Fes (Izuhara *et al.*, 1994). Overall, relatively little is known about the significance of tyrosine kinases, other than JAK kinases, in the signaling by IL-2 family cytokines.

Another kinase pathway used by IL-2 is the Ras pathway (Cantrell *et al.*, 1993). This is less well-studied in the context of IL-4 but it appears to be more vital for IL-2 than for IL-4 signaling.

Other signaling molecules downstream of JAKs

Pyk2 is a nonreceptor protein tyrosine kinase that is also activated in response to IL-2 in a JAK-dependent manner and which physically associates with JAK3. Interestingly, a dominant negative Pyk2 construct could inhibit IL-2-induced proliferation without affecting STAT5 activation (Miyazaki *et al.*, 1998). It is not yet clear if Pyk2 is important for other IL-2 family cytokines. STAM is a putative signal transducing adapter molecule that contains an Src homology 3 (SH3) domain and immunoreceptor tyrosine-based activation motif (ITAM) (Takeshita *et al.*, 1997). STAM associates with JAK3 and JAK2 via its ITAM region and is phosphorylated by JAK3 or JAK2 following stimulation with IL-2 and GM-CSF, respectively. A dominant negative construct lacking the STAM SH3 domain inhibits DNA synthesis mediated by IL-2 and GM-CSF as well as Myc expression.

Negative regulation of signaling

To counter positive signals, negative regulation is also important. Several levels of control are possible:

- Regulation based on production of the cytokine. For example, production of IL-2 and IL-4 are induced following T cell activation in a transient, self-limited fashion.
- Regulation based on degradation/turnover rates in the receptor, following internalization into the cell.
- Regulation at the level of phosphatases. For example, Shp1 associates with IL-2Rβ and diminishes its phosphorylation (Migone *et al.*, 1998a).
- Regulation at the level of CIS/SOCS/JAB/SSI family proteins. The prototype CIS protein can inhibit IL-2, IL-3, GM-CSF, and erythropoietin signaling. SOCS1/JAB can inhibit IL-2, IL-6, and IL-4-dependent signals. While studies are incomplete, this family of proteins appears to broadly affect cytokine signaling (Yoshimura, 1998; Starr and Hilton, 1999) and it will be important to determine which CIS/SOCS/SSI family members selectively affect signaling in response to IL-2 family cytokines.

WHY SHARE RECEPTOR COMPONENTS?

As noted above, γc is shared by five related cytokines (IL-2, IL-4, IL-7, IL-9, and IL-15), whereas IL-2Rβ is shared by IL-2 and IL-15, IL-4Rα and IL-13Rα1 are shared by IL-4 and IL-13, and IL-7Rα is shared by IL-7 and TSLP. It is reasonable to consider why receptor components are shared. There are at least two possible explanations, which are not mutually exclusive (Leonard, 1999). First, shared receptor components might facilitate sharing of actions. In this model, the activation of a given component could allow the activation of a critical molecule that specifies the type of signaling. Thus, cytokines such as IL-4 and IL-13 activate STAT6 and IRS family proteins based on the existence of critical docking sites for both STAT6 and IRS proteins on IL-4Rα, which is a component of both the IL-4 and IL-13 receptors. Similarly, IL-2 and IL-15 both activate STAT5 proteins, given that docking sites for STAT5 proteins are found on IL-2Rβ, which is shared by the receptors for IL-2 and IL-15. Second, the sharing of receptor components might provide a mechanism by which different cytokines modulate each other's activities. This model is based on the fact that γc is a component that can be differentially recruited from one receptor to another based on the presence of different ligands. When two ligands are present, it is therefore possible that they may compete for γc and that the more abundant or the higher affinity/avidity ligand might not only induce its own signal via its cognate receptor, but simultaneously could inactivate the signal of the other cytokine by depriving the second cytokine of γc. Note that these two models are not mutually exclusive and both could be operative, depending on the physiological circumstances.

DEFECTS ASSOCIATED WITH MUTATIONS IN OR TARGETED HOMOLOGOUS RECOMBINATION OF IL-2 FAMILY CYTOKINES OR THEIR RECEPTORS

X-linked severe combined immunodeficiency

X-linked severe combined immunodeficiency disease (XSCID) (also known as the 'bubble boy disease') is a profound immunodeficiency in which both cellular and humoral immunity are defective (**Table 6**) (Conley, 1992; Leonard, 1996; Fischer *et al.*, 1997). In this disease, neither T cells nor NK cells develop; in contrast, B cells are normal or even increased in numbers, but they are nonfunctional, resulting in hypogammaglobulinemia. Noguchi *et al.* (1993b) discovered that mutations in γc resulted in XSCID. This finding was totally unexpected as IL-2-deficient patients and mice had normal T cell and NK cell development, making it seem that mutations in the IL-2 receptor system should not cause the XSCID phenotype. This conundrum was resolved when it was shown that γc was in fact shared by multiple cytokine receptors. Given that JAK3 was immediately downstream of γc, it was predicted that mutations in JAK3 would result in a similar clinical and immunological phenotype and that JAK3 activation was vital for T cell and NK cell development.

Added support for this hypothesis came from a family pedigree with a moderate X-linked immunodeficiency that resulted from mutations in γc that diminished but did not abrogate the association of JAK3 with γc (Russell *et al.*, 1994). This allowed a correlation between the extent of T cell and NK cell development/function and the degree of JAK3 association with γc and JAK3 activation. Accordingly, XSCID is a disease of defective cytokine signaling (Leonard, 1996). Patients with an autosomal recessive form of SCID resulting from JAK3 mutations were then identified (Macchi *et al.*, 1995; Russell *et al.*, 1995). Corresponding to the similarity between XSCID and SCID resulting from JAK3 deficiency, mice that were deficient in either γc (Cao *et al.*, 1995; DiSanto *et al.*, 1995; Ohbo *et al.*, 1996) or JAK3 (Thomis *et al.*, 1995; Nosaka *et al.*, 1995; Park *et al.*, 1995) also had very similar phenotypes. Collectively, these data suggest that JAK3 is vital for most or all γc functions and that JAK3 function may be completely restricted to γc-dependent pathways. In view of the essential γc–JAK3 connection, it was predicted that agents that prevent the γc–JAK3 association or inhibit JAK3 catalytic activity could be immunosuppressive (Russell *et al.*, 1994; Leonard, 1996).

Table 6 Immunological features of XSCID

- Greatly diminished numbers of T cells and defective mitogen responses
- No NK cells
- Normal B cell numbers, but defective B cell responses
- IgM normal or modestly decreased; other immunoglobulins are greatly diminished

As mutations in IL-2 did not affect T cell or NK cell development, it was important to determine which of the γc-dependent cytokines could explain these defects. In view of the known role of IL-7 in thymic development based on *in vitro* studies and from the phenotypes of mice lacking expression of IL-7 or IL-7Rα, it was predicted that defective IL-7 signaling could explain the T cell defect in XSCID. The critical role of the IL-7 signaling pathway in XSCID is now well-established, as patients with mutations in IL-7Rα have a form of SCID with a similar T cell defect to that found in XSCID. This form of SCID is a T cell− B cell+ NK cell+ variant in which NK cell numbers are normal or increased, whereas XSCID and SCID associated with JAK3 deficiency exhibit T cell− B cell+ NK cell− SCID.

IL-7-deficient humans have not yet been identified, but it is reasonable to hypothesize that a similar T cell− B cell+ NK cell+ phenotype might also result from mutations in the IL-7 gene. Because IL-7 is produced by the stroma, however, it is possible that T cell reconstitution might not occur after bone marrow engraftment given that the bone marrow and thymic stromal environment might not support T cell growth.

While humans with mutations in IL-15, IL-15Rα, or IL-2Rβ have not been identified, a large amount of *in vitro* data have suggested an important role for IL-15 for NK cell development (Carson *et al.*, 1994; Cavazzana *et al.*, 1996; Leclercq *et al.*, 1996; Puzanov *et al.*, 1996); moreover, mice lacking IL-15Rα (Lodolce *et al.*, 1998) or IL-2Rβ (Suzuki *et al.*, 1995) do not have NK cells.

In summary, mutations in γc or JAK3 interrupt signaling from IL-2, IL-4, IL-7, IL-9, and IL-15 (and perhaps other as yet undiscovered cytokines as well). This in turn results in defective T cell and NK cell development in humans. Based on the T cell defect in humans with SCID resulting from mutations in IL-7Rα, it appears that defective IL-7Rα-dependent signaling explains the T cell defect in XSCID.

However, until humans with mutations in IL-4, IL-9, and IL-15 are found, it remains a formal possibility that these cytokines could partially contribute to T cell development in humans.

The essential role of γc for the development of T cells and NK cells but not B cells is also supported by the X-chromosome inactivation patterns in XSCID carrier females who are heterozygous for a mutant γc allele (Conley, 1992). In these individuals, strictly nonrandom X-inactivation patterns are found in the T cell and NK cell lineages, indicating that only cells with a functional γc can mature. Interestingly, the relatively nonmature IgM+ B cells exhibit random inactivation, indicating that γc is not required for the development of these cells; however, nonrandom inactivation patterns are seen in the terminally differentiated B cell, indicating a role for γc in B cell maturation. Interestingly, patients have also been found with mutations in IL-2α (Sharfe *et al.*, 1997), indicatin the importance of this protein *in vivo*.

In murine knockout models, analogous to XSCID and JAK3-deficient SCID, as noted above, the absence of either γc or JAK3 results in very similar phenotypes to each other. However, two principal differences are seen from the situation in humans. First, whereas B cell development is normal in humans with XSCID or JAK3 deficiency, essentially no conventional B cells develop in the mouse. This is consistent with the literature indicating that IL-7 is a pre-B cell growth factor in mice. Second, whereas the thymuses in γc or JAK3-deficient mice are very small as in XSCID, there is an age-dependent accumulation of CD4+ T cells in the periphery. The significance of this latter finding is unclear, and it is possible that it is not so much a species difference as a reflection of the sterile conditions in which γc-deficient mice are maintained.

Phenotypes of mice in which IL-2 family cytokines or their receptors have been targeted by homologous recombination

Most of the IL-2 family cytokines and cytokine receptor chains have now been deleted by homologous recombination in mice, as have JAK1, JAK3, and the STATs that are activated by these cytokines. **Table 7** summarizes the major phenotypes of these

Table 7 Phenotypes of mice deficient in type I and type II cytokines, their receptors, JAKs, and STATs

Type I cytokines and their receptors		
γc family		
IL-2	Normal central and peripheral T cell development. Decreased polyclonal T cell responses *in vitro*, but good *in vivo* responses to pathogenic challenges. The mice exhibit autoimmunity with alterations in serum immunoglobulin levels. Ulcerative colitis-like inflammatory bowel disease	Schorle *et al.*, 1991; Kundig *et al.*, 1993; Sadlack *et al.*, 1993
IL-4	Defective TH2 cytokine responses, defective class switch; defective IgG1 and IgE production	Kuhn *et al.*, 1991; Kopf *et al.*, 1993
IL-2/IL-4	Features of both IL-2 and IL-4 knockout mice. No obvious abnormalities in T cell development	Sadlack *et al.*, 1994
IL-7	Defective thymic and peripheral T cell development and B-lymphopoiesis	von Freeden-Jeffrey *et al.*, 1995
IL-2Rα	Normal initial lymphoid development, but then enlargement of peripheral lymphoid organs, polyclonal T and B cell expansion. The mice have many activated T cells, with impaired activation induced cell death. Autoimmunity develops with increasing age, including hemolytic anemia and inflammatory bowel disease	Willerford *et al.*, 1995

Table 7 (*Continued*)

IL-2Rβ	Severe autoimmune hemolytic anemia and other manifestations of autoimmunity. The mice die within approximately 3 months. Deregulated T cell activation. Dysregulated B cell differentiation and altered immunoglobulin profile. Absent NK cells	Suzuki *et al.*, 1995
γc	Greatly diminished thymic development but all thymic populations are represented. Age-dependent accumulation of peripheral $CD4^+$ T cells with an 'activated-memory phenotype'. Greatly diminished numbers of conventional B cells, although B1 cells are present. No NK cells or $\gamma\delta$ cells. Absent gut-associated lymphoid tissue	DiSanto *et al.*, 1995; Cao *et al.*, 1995; Ohbo *et al.*, 1996; Nakajima *et al.*, 1997
IL-7Rα	Greatly diminished thymic and peripheral T cell and B cell development	Peschon *et al.*, 1994
IL-15Rα	Defective NK development and lymphocyte homing	Lodolce *et al.*, 1998
STATs and JAKs		
STAT3	Embryonic lethal. Embryos implant but cannot grow. The fact that this phenotype is even more severe than that seen with gp130 suggests a role for STAT3 via a gp130-independent cytokine. In a Cre-Lox system using Lck Cre, it was shown that the absence of STAT3, like STAT5a and STAT5b, is vital for IL-2-induced IL-2Rα expression	Takeda *et al.*, 1997; Akaishi *et al.*, 1998
STAT5a	Defective lobuloalveolar development in the mammary gland due to defective prolactin signaling. Defective IL-2-induced IL-2Rα expression. Defective superantigen-induced Vβ8 expansion *in vivo*. Modest decrease in NK cells	Liu *et al.*, 1997; Nakajima *et al.*, 1997
STAT5b	Defective growth analogous to Laron dwarfism, a disease of defective growth hormone signaling. Diminished NK development and marked defect in NK cytolytic activity. Diminished proliferation to IL-2 and to PMA + IL-4	Udy *et al.*, 1997; Imada *et al.*, 1998
STAT5a/STAT5b	Defective T cell proliferation. Lack of development of NK cells	Moriggl *et al.*, 1999
Stat6	Defective TH2 development, essentially the same phenotype as IL-4-deficient mice	Kaplan *et al.*, 1996; Takeda *et al.*, 1996; Shimoda *et al.*, 1996
JAK1	Perinatal lethality. Multiple defects including defects in signaling in response to IL-2, IL-4, and IL-7	Rodig *et al.*, 1998
JAK3	Very similar and possibly identical to γc-deficient mice	Thomis *et al.*, 1995; Nosaka *et al.*, 1995; Park *et al.*, 1995

mice. So far, knockout mice have not been reported for IL-9, IL-15, IL-9Rα, TSLP, IL-13, IL-13Rα1, or IL-13Rα2.

SOLUBLE RECEPTORS

Soluble forms exist for many cytokine receptors (Kurman *et al.*, 1992; Fernandez-Botran *et al.*, 1996). These include receptors for IL-1, TNFα, interferons, and also certain type I cytokine receptors. For IL-2 family cytokines, soluble receptors have been reported for IL-2, IL-4, and IL-7. The principal soluble proteins for these cytokines are IL-2Rα, IL-2Rβ, and IL-7Rα. Whereas IL-2Rα is generated by proteolytic cleavage, for IL-4Rα and IL-7Rα, the soluble forms result from alternative splicing.

There is relatively little available information on the potential physiological roles of these proteins. For soluble IL-2Rα, given that it binds IL-2 with a K_d of 10^{-8} M, it is relatively unlikely that it could effectively compete with binding to the high-affinity IL-2 receptor ($K_d = 10^{-11}$ M); nevertheless, levels of soluble IL-2Rα have been found in a number of disease states and is of potential correlative use in this regard. Similarly, no physiological role has been reported for soluble IL-4Rα or IL-7Rα.

SPECIES SPECIFICITY OF IL-2 FAMILY CYTOKINES

There are no overall rules for the species-specific actions of human and murine cytokines. IL-2 seems to work down the phylogenetic tree, with human IL-2 having the ability to stimulate both human and murine cells, whereas murine IL-2 exhibits little action on human cells (Waldmann, 1989). Other cytokines, such as IL-4 exhibits comparatively strict species specificity so that human and murine IL-4 are needed for responses in human and murine cells, respectively (Paul, 1991). While murine IL-9 is active on human cells, human IL-9 is not biologically active on murine cells (the opposite situation from that for IL-2) (Leonard, 1999).

CONCLUSIONS

The IL-2 family of cytokines is a set of proteins with vital roles related to lymphoid development, differentiation, and proliferation. The cells that produce most of these cytokines are T cells, but IL-7 instead is primarily produced by stromal cells, and IL-4, while produced by T cells, is also produced by mast cells. Collectively, these cytokines exert actions on T cells, B cells, and NK cells, as well as cells of certain other hematopoietic lineages. The actions of these cytokines are partially overlapping, and as such they exhibit both pleiotropic and redundant actions. The receptors for these cytokines are partially overlapping in their composition, helping to explain both overlapping actions as well as potentially competing actions of some of the different cytokines. This strongly suggests a coevolution of cytokines and receptors. Correspondingly, signaling pathways are also overlapping between various cytokines. Nevertheless, certain nonredundant actions are clearly indicated by the phenotype of the various knockout mice. A major challenge for the future is to achieve a clearer understanding of the basis for the specificity of each IL-2 family cytokine.

References

Akaishi, H., Takeda, K., Kaisho, T., Shineha, R., Satomi, S., Takeda, J., and Akira, S. (1998). Defective IL-2-mediated IL-2 receptor alpha chain expression in STAT3-deficient T lymphocytes. *Int. Immunol.* **10**, 1747–1751.

Aman, M. F., Tayebi, N., Obiri, N. I., Puri, R. K., Modi, W. S., and Leonard, W. J. (1996). cDNA cloning and characterization of the human interleukin-13 receptor α chain. *J. Biol. Chem.* **271**, 29265.

Ascherman, D. P., Migone, T.-S., Friedmann, M., and Leonard, W. J. (1997). Interleukin-2 (IL-2)-mediated induction of the IL-2 receptor α chain gene: critical role of two functionally redundant tyrosine residues in the IL-2 receptor β chain cytoplasmic domain and suggestion that these residues mediate more than Stat5 activation. *J. Biol. Chem.* **272**, 8704.

Bamford, R. N., Grant, A. J., Burton, J. D., Peters, C., Kurys, G., Goldman, C. K., Brennan, J., Roessler, E., and Waldmann, T. A. (1994). The interleukin (IL) 2 β chain is shared by IL-2, and a cytokine, provisionally designated IL-T, that stimulates T-cell proliferation and the induction of lymphokine-activated killer cells. *Proc. Natl Acad. Sci. USA* **91**, 4940.

Bazan, J. F. (1990). Structural design and molecular evolution of a cytokine receptor superfamily. *Proc. Natl Acad. Sci. USA* **87**, 6934.

Bazan, J. F. (1991). Neurotropic cytokines in the hematopoietic fold. *Neuron* **7**, 1.

Blackman, M. A., Tigges, M. A., Minie, M. E., and Koshland, M. E. (1986). A model system for peptide hormone action in differentiation: interleukin 2 induces a B lymphoma to transcribe the J chain gene. *Cell* **47**, 609.

Boussiotis, V. A., Barber, D. L., Nakarai, T., Freeman, G. J., Gribben, J. G., Bernstein, G. M., d'Andrea, A. D., Ritz, J., and Nadler, L. M. (1994). Prevention of T cell anergy by signaling through the γc chain of the IL-2 receptor. *Science* **266**, 1039.

Cantrell, D. A., Izquierdo, M., Reif, K., and Woodrow, M. (1993). Regulation of PtdIns-3-kinase and the guanine nucleotide binding proteins p21ras during signal transduction by the T cell antigen receptor and the interleukin-2 receptor. *Semin. Immunol.* **5**, 319–326.

Cao, X., Shores, E. W., Hu-Li, J., Anver, M. R., Kelsall, B. L., Russell, S. M., Drago, J., Noguchi, M., Grinberg, A., Bloom, E. T., Paul, W. E., Katz, S. I., Love, P. E., and Leonard, W. J. (1995). Defective lymphoid development in mice lacking expression of the common cytokine receptor γ chain. *Immunity* **2**, 223.

Caput, D., Laurent, P., Kaghad, M., Lelias, J.-M., Lefort, S., Vita, N., and Ferrara, P. (1996). Cloning and characterization of a specific interleukin (IL)-13 binding protein structurally related to the IL-5 receptor α chain. *J. Biol. Chem.* **271**, 16921.

Carson, W. E., Giri, J. G., Lindemann, M. J., Linett, M. L., Ahdieh, M., Paxton, R., Anderson, D., Eisenmann, J., Grabstein, K., and Caligiuri, M. A. (1994). Interleukin (IL) 15 is a novel cytokine that activates human natural killer cells via components of the IL-2 receptor. *J. Exp. Med.* **180**, 1395.

Cavazzana-Calvo, M., Hacein-Bey, S., de Saint Basile, G., de Coene, C., Selz, F., Le Deist, F., and Fischer, A. (1996). Role of interleukin-2 (IL-2), IL-7, and IL-15 in natural killer cell differentiation from cord blood hematopoietic progenitor cells and from γc transduced severe combined immunodeficiency X1 bone marrow cells. *Blood* **88**, 3901.

Chazen, G. D., Pereira, G. M. B., LeGros, G., Gillis, S., and Shevach, E. M. (1989). Interleukin 7 is a T-cell growth factor. *Proc. Natl Acad. Sci. USA* **86**, 5923.

Conley, M. E. (1992). Molecular approaches to analysis of X-linked immunodeficiencies. *Annu. Rev. Immunol.* **10**, 215.

Davies, D. R., and Wlodawer, A. (1995). Cytokines and their receptor complexes. *FASEB J.* **9**, 50.

Demoulin, J.-P., Uyttenhove, C., van Roost, E., de Lestre, B., Donckers, D., van Snick, J., and Renauld, J.-C. (1996). A single tyrosine of the interleukin-9 (IL-9) receptor is required for STAT activation, antiapoptoic activity, and growth regulation by IL-9. *Mol. Cell Biol.* **16**, 4710.

de Vos, A. M., Ultsch, M., and Kossiakoff, A. A. (1992). Human growth hormone and extracellular domain of its receptor: crystal structure of the complex. *Science* **255**, 306.

DiSanto, J. P., Muller, W., Guy-Grand, D., Fischer, A., and Rajewsky, K. (1995). Lymphoid development in mice with a targeted deletion of the interleukin 2 receptor γ chain. *Proc. Natl Acad. Sci. USA* **92**, 377–381.

Fernandez-Botran, R., Chilton, P. M., and Ma, Y. (1996). Soluble cytokine receptors: their roles in immunoregulation, disease, and therapy. *Adv. Immunol.* **63**, 269.

Fischer, A., Cavazzana-Calvo, M., de Saint Basile, G., DeVillartay, J. P., DiSanto, J. P., Hivroz, C., Rieux-Laucat, F., and Le Diest, F. (1997). Naturally occurring primary deficiencies of the immune system. *Annu. Rev. Immunol.* **15**, 93.

Friedmann, M. C., Migone, T.-S., Russell, S. M., and Leonard, W. J. (1996). Different interleukin 2 receptor β-chain tyrosines couple to at least two signaling pathways and synergistically mediate interleukin 2-induced proliferation. *Proc. Natl Acad. Sci. USA* **93**, 2077.

Fujii, H., Nakagawa, Y., Schindler, U., Kawahara, A., Mori, H., Gouilleux, F., Groner, B., Ihle, J. N., Minami, Y., Miyazaki, T., and Taniguchi, T. (1995). Activation of Stat5 by interleukin 2 requires a carboxyl-terminal region of the interleukin 2 receptor beta chain but is not essential for the proliferative signal transmission. *Proc. Natl Acad. Sci. USA* **92**, 5482–5486.

Giri, J. G., Ahdieh, M., Eisenman, J., Shanebeck, K., Grabstein, K., Kumaki, S., Namen, A., Park, L. S., Cosman, D., and Anderson, D. (1994). Utilization of the β and γ chains of the IL-2 receptor by the novel cytokine IL-15. *EMBO J.* **13**, 2822.

Giri, J. G., Kumaki, S., Ahdieh, M., Friend, D. J., Loomis, A., Shanebeck, K., DuBose, R., Cosman, D., Park, L. S., and Anderson, D. M. (1995). Identification and cloning of a novel IL-15 binding protein that is structurally related to the α chain of the IL-2 receptor. *EMBO J.* **14**, 3654.

Goodwin, R. G., Friend, D., Ziegler, S. F., Jerzy, R., Falk, B. A., Gimpel, S., Cosman, D., Dower, S. K., March, C. J., Namen, A. E., and Park, L. S. (1990). Cloning of the human and murine interleukin-7 receptors: demonstration of a soluble form and homology to a new receptor superfamily. *Cell* **60**, 940.

Hage, T., Sebald, W., and Reinemer, P. (1999). Crystal structure of the interleukin-4/receptor alpha chain complex reveals a mosaic binding interface. *Cell* **97**, 271–281.

Hatakeyama, M., Kono, T., Kobayash, N., Kawahara, A., Levin, S. D., Perlmutter, R. M., and Taniguchi, T. (1991). Interaction of the IL-2 receptor with the src-family kinase $p56^{lck}$: identification of novel intermolecular association. *Science* **252**, 1523.

He, Y.-W., and Malek, T. R. (1998). The structure and function of γc-dependent cytokines and receptors: regulation of T lymphocyte development and homeostasis. *Crit. Rev. Immunol.* **18**, 503–524.

Hilton, D. J., Zhang, J. G., Metcalf, D., Alexander, W. S., Nicola, N. A., and Willson, T. A. (1996). Cloning and characterization of a binding subunit of the interleukin 13 receptor that is also a component of the interleukin 4 receptor. *Proc. Natl Acad. Sci. USA* **93**, 497–501.

Horak, I. D., Gress, R. E., Lucas, P. J., Horak, E. M., Waldmann, T. A., and Bolen, J. B. (1991). T-lymphocyte interleukin 2-dependent tyrosine protein kinase signal transduction involves the activation of $p56^{lck}$. *Proc. Natl Acad. Sci. USA* **88**, 1996.

Hoffman, R. C., Castner, B. J., Gerhart, M., Gibson, M. G., Rasmussen, B. D., March, C. J., Weatherbee, J., Tsang, M., Gustchina, A., Schalk-Hihi, C., Reshetnikova, L., and Wlodawer, A. (1995). Direct evidence of a heterotrimeric complex of human interleukin-4 with its receptors. *Protein Sci.* **4**, 382.

Horvath, C. M., and Darnell, J. E. (1997). The state of the STATs: recent developments in the study of signal transduction to the nucleus. *Curr. Opin. Cell Biol.* **9**, 233.

Hou, J., Schindler, U., Henzel, W. J., Ho, T. C., Brasseur, M., and McKnight, S. L. (1994). An interleukin-4 induced transcription factor: IL-4 Stat. *Science* **265**, 1701.

Imada, K., Bloom, E. T., Nakajima, H., Horvath-Arcidiacono, J. A., Udy, G. B., Davey, H. W., and Leonard, W. J. (1998). Stat5b is essential for natural killer cell-mediated proliferation and cytolytic activity. *J. Exp. Med.* **188**, 2067–2074.

Izuhara, K., Feldman, R. A., Greer, P., and Harada, N. (1994). Interaction of the c-fes proto-oncogene product with the interleukin-4 receptor. *J. Biol. Chem.* **269**, 18623.

Johnston, J. A., Wang, L. M., Hanson, E. P., Sun, X. J., White, M. F., Oakes, S. A., Pierce, J. H., and O'Shea, J. J. (1995). Interleukins 2, 4, 7, and 15 stimulate tyrosine phosphorylation of insulin receptor substrates 1 and 2 in T cells. Potential role of JAK kinases. *J. Biol. Chem.* **270**, 28527.

Kammer, W., Lischke, A., Moriggl, R., Groner, B., Ziemiecki, A., Gurniak, C. B., Berg, L. J., and Friedrich, K. (1996). Homodimerization of interleukin-4 receptor alpha chain can induce intracellular signaling. *J. Biol. Chem.* **271**, 23634.

Kapeller, R., and Cantley, L. C. (1994). Phosphatidylinositol 3-kinase. *Bioessays* **16**, 565–576.

Kaplan, M. H., Schindler, U., Smiley, S. T., and Grusby, M. J. (1996). Stat6 is required for mediating responses to IL-4 and for the development of Th2 cells. *Immunity* **4**, 313.

Karnitz, L. M., Burns, L. A., Sutor, S. L., Blenis, J., and Abraham, R. T. (1995). Interleukin-2 triggers a novel phosphatidylinositol 3-kinase-dependent MEK activation pathway. *Mol. Cell Biol.* **15**, 3049.

Kimura, M., Ishii, N., Nakamura, M., Van Snick, J., and Sugamura, K. (1995). Sharing of the IL-2 receptor γ chain with the functional IL-9 receptor complex. *Int. Immunol.* **7**, 115.

Kondo, M., Takeshita, T., Ishii, N., Nakamura, M., Watanabe, S., Arai, K., and Sugamura, K. (1993). Sharing of the interleukin-2 (IL-2) γ chain between receptors for IL-2 and IL-4. *Science* **262**, 1874.

Kondo, M., Takeshita, T., Higuchi, M., Nakamura, M., Sudo, T., Nishikawa, S.-I., and Sugamura, K. (1994). Functional participation of the IL-2 receptor γ chain in IL-7 receptor complexes. *Science* **263**, 1453.

Kopf, M., Le Gros, G., Bachmann, M., Lamers, M. C., Bluethmann, H., and Kohler, G. (1993). Disruption of the murine IL-4 gene blocks Th2 cytokine responses. *Nature* **362**, 245.

Kuhn, R., Rajewsky, K., and Muller, W. (1991). Generation and analysis of interleukin-4 deficient mice. *Science* **254**, 707.

Kundig, T. M., Schorle, H., Bachmann, M. F., Hengartner, H., Zinkernagel, R. M., and Horak, I. (1993). Immune responses in interleukin-2-deficient mice. *Science* **262**, 1059–1061.

Kurman, C. C., Rubin, L. A., and Nelson, D. L. (1992). In "Manual of Clinical Laboratory Immunology" (ed N. R. Rose, E. C. deMacario, J. L. Fahey, H. Friedman, and G. M. Penn), Soluble products of immune activation: soluble interleukin-2 receptor (sIL-2R, Tac protein), p. 256. American Society for Microbiology. Washington DC.

Lai, S. Y., Molden, J., Liu, K. D., Puck, J. M., White, M. D., and Goldsmith, M. A. (1996). Interleukin-4-specific signal transduction events are driven by homotypic interactions of the interleukin-4 receptor a subunit. *EMBO J.* **15**, 4506.

Lanier, L. L., and Phillips, J. H. (1992). Natural killer cells. *Curr. Opin. Immunol.* **4**, 38.

Leclercq, G., Debacker, V., de Smedt, M., and Plum, J. (1996). Differential effects of interleukin-15 and interleukin-2 on the differentiation of bipotential T/natural killer progenitor cells. *J. Exp. Med.* **184**, 325.

Lee, F., Yokota, T., Otsuka, T., Meyerson, P., Villaret, D., Coffman, R., Mosmann, T., Rennick, D., Roehm, N., Smith, C., Zlotnik, A., and Arai, K. (1986). Isolation and characterization of a mouse interleukin cDNA clone that expresses B-cell stimulatory factor-1 activities and T cell and mast cell stimulating activities. *Proc. Natl Acad. Sci. USA* **83**, 2061.

Leonard, W. J. (1996). The molecular basis of X-linked combined immunodeficiency: defective cytokine receptor signaling. *Annu. Rev. Med.* **47**, 229.

Leonard, W. J. (1999). In "Fundamental Immunology, 4th edn" (ed W. E. Paul), Type I cytokines and interferons and their receptors, pp. 741–774. Lippincott Raven, Philadelphia, PA.

Leonard, W. J., and O'Shea, J. J. (1998). Jaks and STATs: Biological implications. *Annu. Rev. Immunol.* **16**, 293–322.

Leonard, W. J., Depper, J. M., Crabtree, G. R., Rudikoff, S., Pumphrey, J., Robb, R. J., Kronke, M., Svetlik, P. B., Peffer, N. J., Waldmann, T. A., and Greene, W. C. (1984). Molecular cloning and expression of cDNAs for the human interleukin-2 receptor. *Nature* **311**, 625.

Leonard, W. J., Shores, E. W., and Love, P. E. (1995). Role of the common cytokine receptor γ chain in cytokine signaling and lymphoid development. *Immunol. Rev.* **148**, 97.

Levin, S. D., Koelling, R. M., Friend, S. L., Isaksen, D. E., Ziegler, S. F., Perlmutter, R. M., and Farr, A. G. (1999). Thymic stromal lymphopoietin: a cytokine that promotes the development of IgM+ B cells in vitro and signals via a novel mechanism. *J. Immunol.* **162**, 677–683.

Lin, J.-X., Migone, T.-S., Tsang, M., Friedmann, M., Weatherbee, J. A., Zhou, L., Yamauchi, A., Bloom, E. T., Mietz, J., John, S., and Leonard, W. J. (1995). The role of shared receptor motifs and common Stat proteins in the generation of cytokine pleiotropy and redundancy by IL-2, IL-4, IL-7, IL-13, and IL-15. *Immunity* **2**, 331.

Liu, X., Robinson, G. W., Wagner, K.-U., Garrett, L., Wynshaw-Boris, A., and Hennighausen, L. (1997). Stat5a is mandatory for adult mammary gland development and lactogenesis. *Genes Dev.* **1**, 179.

Lodolce, J. P., Boone, D. L., Chai, S., Swain, R. E., Dassopoulos, T., Trettin, S., and Ma, A. (1998). IL-15 receptor maintains lymphoid homeostasis by supporting lymphocyte homing and proliferation. *Immunity* **9**, 669–676.

Macchi, P., Villa, A., Gillani, S., Sacco, M. G., Frattini, A., Porta, F., Ugazio, A. G., Johnston, J. A., Candotti, F., O'Shea, J. J., Vezzoni, P., and Notarangelo, L. D. (1995). Mutations of Jak-3 gene in patients with autosomal severe combined immune deficiency (SCID). *Nature* **377**, 65.

Malek, T. R., Porter, B. O., and He, Y. W. (1999). Multiple gamma c-dependent cytokines regulate T-cell development. *Immunol. Today* **20**, 71–76.

Migone, T.-S., Lin, J.-X., Cereseto, A., Mulloy, J. C., O'Shea, J. J., Franchini, G., and Leonard, W. J. (1995). Constitutively activated Jak-Stat pathway in T cells transformed with HTLV-I. *Science* **269**, 79.

Migone, T. S., Cacalano, N. A., Taylor, N., Yi, T., Waldmann, T. A., and Johnston, J. A. (1998a). Recruitment of SH2-containing protein tyrosine phosphatase SHP-1 to the interleukin 2 receptor; loss of SHP-1 expression in human T-lymphotropic virus type I-transformed T cells. *Proc. Natl Acad. Sci. USA* **95**, 3845–3850.

Migone, T.-S., Rodig, S., Cacalano, N. A., Berg, M., Schreiber, R. D., and Leonard, W. J. (1998b). Functional cooperation of the interleukin-2 receptor β chain and Jak1 in phosphatidylinositol 3-kinase recruitment. *Mol. Cell. Biol.* **18**, 6416–6422.

Minami, Y., Nakagawa, Y., Kawahara, A., Miyazak, I. T., Sada, K., Yamamura, H., and Taniguchi, T. (1995). Protein tyrosine kinase Syk is associated with and activated by the IL-2 receptor: possible link with the c-myc induction pathway. *Immunity* **2**, 89.

Mingari, M. C., Gerosa, F., Carra, G., Accolla, R. S., Moretta, A., Zubler, R. H., Waldmann, T. A., and Moretta, L. (1984). Human interleukin-2 promotes proliferation of activated B cells via surface receptors similar to those of activated T cells. *Nature* **312**, 641.

Miyazaki, T., Kawahara, A., Fujii, H., Nakagawa, Y., Minami, Y., Liu, Z.-J., Oishi, I., Silvennoinen, O., Witthuhn, B. A., Ihle, J. N., and Taniguchi, T. (1994). Functional activation of Jak1 and Jak3 by selective association with IL-2 receptor subunits. *Science* **266**, 1045.

Miyazaki, T., Takaoka, A., Nogueira, L., Dikic, I., Fujii, H., Tsujino, S., Mitani, Y., Maeda, M., Schlessinger, J., and Taniguchi, T. (1998). Pyk2 is a downstream mediator of the IL-2 receptor-coupled Jak signaling pathway. *Genes Dev.* **12**, 770–775.

Morgan, D. A., Ruscetti, F. W., and Gallo, R. (1976). Selective in vitro growth of T lymphocytes from normal human bone marrows. *Science* **193**, 1007.

Moriggl, R., Topham, D. J., Teglund, S., Sexl, V., McKay, C., Wang, D., Hoffmeyer, A., van Deursen, J., Sangster, M. Y., Bunting, K. D., Grosveld, G. C., and Ihle, J. N. (1999). Stat5 is required for IL-2-induced cell cycle progression of peripheral T cells. *Immunity* **10**, 249–259.

Morrissey, P. J., Goodwin, R. G., Nordan, R. P., Anderson, D., Grabstein, K. H., Cosman, D., Sims, J., Lupton, S., Acres, B., Reed, S. G., Mochizuki, D., Eisenman, J., Conlon, P. J., and Namen, A. E. (1989). Recombinant interleukin 7, pre-B cell growth factor, has costimulatory activity on purified mature T cells. *J. Exp. Med.* **169**, 707.

Murray, R., Suda, T., Wrighton, N., Lee, F., and Zlotnik, A. (1989). IL-7 is a growth and maintenance factor for mature and immature thymocyte subsets. *Int. Immunol.* **1**, 526.

Myers Jr, M. G., and White, M. F. (1996). Insulin signal transduction and the IRS proteins. *Annu. Rev. Pharmacol. Toxicol.* **36**, 615.

Nakajima, H., Shores, E. W., Noguchi, M., and Leonard, W. J. (1997). The common cytokine receptor γ chain plays an essential role in regulating lymphoid homeostasis. *J. Exp. Med.* **185**, 189–196.

Nakamura, Y., Russell, S. M., Mess, S. A., Friedmann, M., Erdos, M., Francois, C., Jacques, Y., Adelstein, S., and Leonard, W. J. (1994). Heterodimerization of the interleukin-2 receptor β and γ cytoplasmic domains is required for signaling. *Nature* **369**, 330.

Namen, A. E., Lupton, S., Hjerrild, K., Wignall, J., Mochizuki, D. Y., Schmierer, A., Mosley, B., March, C. J., Urdal, D., Gillis, S., Cosman, D., and Goodwin, R. G. (1988). Stimulation of B-cell progenitors by cloned murine interleukin-7. *Nature* **333**, 571.

Nelson, B., Lord, J. D., and Greenberg, P. D. (1994). Cytoplasmic domains of the interleukin-2 receptor β and γ chains mediate the signal for T-cell proliferation. *Nature* **369**, 333.

Nikaido, T., Shimizu, A., Ishida, N., Sabe, H., Teshigawara, K., Maeda, M., Uchiyama, T., Yodoi, J., and Honjo, T. (1984). Molecular cloning of cDNA encoding human interleukin-2 receptor. *Nature* **311**, 631.

Noguchi, M., Nakamura, Y., Russell, S. M., Ziegler, S. F., Tsang, M., Cao, X., and Leonard, W. J. (1993a). Interleukin-2 receptor γ chain: a functional component of the interleukin-7 receptor. *Science* **262**, 1877.

Noguchi, M., Yi, H., Rosenblatt, H. M., Filipovich, A. H., Adelstein, S., Modi, W. S., McBride, O. W., and Leonard, W. J. (1993b). Interleukin-2 receptor γ chain mutation results in X-linked severe combined immunodeficiency in humans. *Cell* **73**, 147.

Noma, Y., Sideras, P., Naito, T., Bergstedt-Lindquist, S., Azuma, C., Severinson, E., Tanabe, T., Kinash, T., Matsuda, F., Yaoita, Y., and Honjo, T. (1986). Cloning of cDNA encoding the murine IgG1 induction factor by a novel strategy using ST6 promoter. *Nature* **319**, 640.

Nosaka, T., van Deursen, J. M. A., Tripp, R. A., Thierfelder, W. E., Witthuhn, B. A., McMickle, A. P., Doherty, P. C., Grosveld, G. C., and Ihle, J. N. (1995). Defective lymphoid development in mice lacking Jak3. *Science* **270**, 800–802.

Ohbo, K., Suda, T., Hashiyama, M., Mantani, A., Ikebe, M., Miyakawa, K., Moriyama, M., Nakamura, M., Katsuki, M., Takahash, K., Yamamura, K.-I., and Sugamura, S. (1996). Modulation of hematopoiesis in mice with a truncated mutant of the interleukin-2 receptor γ chain. *Blood* **87**, 956.

Otani, H., Siegel, J. P., Erdos, M., Gnarra, J. R., Toledano, M. B., Sharon, M., Mostowski, H., Feinberg, M. B., Pierce, J. H., and Leonard, W. J. (1992). Interleukin (IL)-2 and IL-3 induce distinct but overlapping responses in murine IL-3 dependent 32D cells transduced with human IL-2 receptor β chain: involvement of tyrosine kinases other than $p56^{lck}$. *Proc. Natl Acad. Sci. USA* **89**, 2789.

Pallard, C., Stegmann, A. P., van Kleffens, T., Smart, F., Venkitaraman, A., and Spits, H. (1999). Distinct roles of the phosphatidylinositol 3-kinase and STAT5 pathways in IL-7-mediated development of human thymocyte precursors. *Immunity* **10**, 525–535.

Pandey, A., Ozaki, K., Baumann, H., Levin, S. D., Puel, A., Farr, A. G., Ziegler, S. F., Leonard, W. J., and Lodish, H. F. (2000). Cloning of a novel receptor subunit required for signaling by thymic stromal lymphopoietin. *Nature Immunol.* (in press).

Park, S. Y., Saijo, K., Takahash, T., Osawa, M., Arase, H., Hirayama, N., Miyake, K., Nakauchi, H., Shirasawa, T., and Saito, T. (1995). Developmental defects of lymphoid cells in Jak3 kinase-deficient mice. *Immunity* **3**, 771–782.

Paul, W. E. (1989). Pleiotropy and redundancy: T cell-derived lymphokines in the immune response. *Cell* **57**, 521–524.

Paul, W. E. (1991). Interleukin-4: A prototypic immunoregulatory lymphokine. *Blood* **77**, 1859.

Paul, W. E. (1997). Interleukin 4: signalling mechanisms and control of T cell differentiation. *Ciba Foundation Symp.* **204**, 208–216; discussion 216–219.

Peschon, J., Morrissey, P. J., Grabstein, K. H., Ramsdell, F. J., Maraskovsky, E., Gliniak, B. C., Park, L. S., Ziegler, S. F., Williams, D. E., Ware, C. B., Meyer, J. D., and Davison, B. L. (1994). aEarly lymphocyte expansion is severely impaired in interleukin 7 receptor-deficient mice. *J. Exp. Med.* **180**, 1955–1960.

Puel, A., Ziegler, S. F., Buckley, R. H., and Leonard, W. J. (1998). Defective IL7R expression in T(−)B(+)NK(+) severe combined immunodeficiency. *Nature Genet.* **20**, 394–397.

Puzanov, I. J., Bennett, M., and Kumar, V. (1996). IL-15 can substitute for the marrow microenvironment in the differentiation of natural killer cells. *J. Immunol.* **157**, 4282.

Rodig, S. J., Meraz, M. A., White, J. M., Lampe, P. A., Riley, J. K., Arthur, C. D., King, K. L, Sheehan, K. C., Yin, L., Pennica, D., Johnson Jr, E. M., and Schreiber, R. D. (1998). Disruption of the Jak1 gene demonstrates obligatory and nonredundant roles of the Jaks in cytokine-induced biologic responses. *Cell* **93**, 373–383.

Rozwarski, D. A., Gronenborn, A. M., Clore, G. M., Bazan, J. F., Bohm, A., Wlodawer, A., Hatada, M., and Karplus, P. A. (1994). Structural comparisons among the short-chain helical cytokines. *Structure* **2**, 159.

Russell, S. M., Keegan, A. D., Harada, N., Nakamura, Y., Noguchi, M., Leland, P., Friedmann, M. C., Miyajima, A., Puri, R., Paul, W. E., and Leonard, W. J. (1993). Interleukin-2 receptor γ chain: a functional component of the interleukin-4 receptor. *Science* **262**, 1880.

Russell, S. M., Johnston, J. A., Noguchi, M., Kawamura, M., Bacon, C. M., Friedmann, M., Berg, M., McVicar, D. W., Witthuhn, B. A., Silvennoinen, O., Goldman, A. S., Schmalstieg, F. C., Ihle, J. N., O'Shea, J. J., and Leonard, W. J. (1994). Interaction of IL-2Rβ and γc chains with Jak1 and Jak3: Implications for XSCID and XCID. *Science* **266**, 1042.

Russell, S. M., Tayebi, N., Nakajima, H., Riedy, M. C., Roberts, J. L., Aman, M. J., Migone, T.-S., Noguchi, M., Markert, M. L., Buckley, R. H., O'Shea, J. J., and Leonard, W. J. (1995). Mutation of Jak3 in a patient with SCID: essential role of Jak3 in lymphoid development. *Science* **270**, 797.

Ryan, J. J., McReynolds, L. J., Keegan, A., Wang, L. H., Garfein, E., Rothman, P., Nelms, K., and Paul, W. E. (1996). Growth and gene expression are predominantly controlled by distinct regions of the human IL-4 receptor. *Immunity* **4**, 123–132.

Sadlack, B., Merz, H., Schorle, H., Schimpl, A., Feller, A. C., and Horak, I. (1993). Ulcerative colitis-like disease in mice with a disrupted interleukin-2 gene. *Cell* **75**, 253–261.

Sadlack, B., Kuhn, R., Schorle, H., Rajewsky, K., Muller, W., and Horak, I. (1994). Development and proliferation of lymphocytes in mice deficient for both interleukins-2 and -4. *Eur. J. Immunol.* **24**, 281.

Schorle, H., Holtschke, T., Hunig, T., Schimpl, A., and Horak, I. (1991). Development and function of T cells in mice rendered interleukin-2 deficient by gene targeting. *Nature* **352**, 621.

Sharfe, N., Dadi, H. K., Shahar, M., and Roifman, C. M. (1997). Human immune disorder arising from mutation of the α chain of the interleukin-2 receptor. *Proc. Natl Acad. Sci. USA* **94**, 3168.

Shimoda, K., van Deursen, J., Sangster, M. Y., Sarawar, S. R., Carson, R. T., Tripp, R. A., Chu, C., Quelle, F. W., Nosaka, T., Vignali, D. A. A., Doherty, P. C., Grosveld, G., Paul, W. E., and Ihle, J. N. (1996). Lack of IL-4-induced Th2 response and IgE class switching in mice with disrupted *Stat6* gene. *Nature* **380**, 630.

Siegel, J. P., Sharon, M., Smith, P. L., and Leonard, W. J. (1987). The IL-2 receptor β chain (p70): role in mediating signals for LAK, NK, and proliferative activities. *Science* **238**, 75.

Smerz-Bertling, C., and Duschl, A. (1994). Both interleukin 4 and interleukin 13 induce tyrosine phosphorylation of the 140-kDa subunit of the interleukin 4 receptor. *J. Biol. Chem.* **270**, 966.

Sprang, S. R., and Bazan, J. F. (1993). Cytokine structural taxonomy and mechanisms of receptor engagement. *Curr. Opin. Struct. Biol.* **3**, 815.

Starr, R., and Hilton, D. J. (1999). Negative regulation of the JAK/STAT pathway. *Bioessays* **21**, 47–52.

Sugamura, K., Asao, H., Kondo, M., Tanaka, N., Ishii, N., Ohbo, K., Nakamura, M., and Takeshita, T. (1996). The interleukin-2 receptor gamma chain: its role in the multiple cytokine receptor complexes and T cell development in XSCID. *Annu. Rev. Immunol.* **14**, 179.

Suzuki, H., Kundig, T. M., Furlonger, C., Wakeham, A., Timms, E., Matsuyama, T., Schmits, R., Simard, J. J., Ohashi, P. S., Griesser, H., Mak, T. W., Taniguchi, T., Paige, C. J., and Mak, T. W. (1995). Deregulated T cell activation and autoimmunity in mice lacking interleukin-2 receptor β. *Science* **268**, 1472.

Tagaya, T., Burton, J. D., Miyamoto, Y., and Waldmann, T. A. (1995). Identification of a novel receptor signal transduction pathway for IL-15/T in mast cells. *EMBO J.* **15**, 4928.

Tagaya, T., Bamford, R. N., DeFilippis, A. P., and Waldmann, T. A. (1996). IL-15: a pleiotropic cytokine with diverse receptor signaling pathways whose expression is controlled at multiple levels. *Immunity* **4**, 329.

Takeda, K., Tanaka, T., Shi, W., Matsumoto, M., Minami, M., Kashiwamura, S., Nakanishi, K., Yoshida, N., Kishimoto, T., and Akira, S. (1996). Essential role of Stat6 in IL-4 signalling. *Nature* **380**, 627.

Takeda, K., Noguchi, K., Shi, W., Tanaka, T., Matsumoto, M., Yoshida, N., Kishimoto, T., and Akira, S. (1997). Targeted disruption of the mouse Stat3 gene leads to early embryonic lethality. *Proc. Natl Acad. Sci. USA* **94**, 3801.

Takeshita, T., Asao, H., Suzuki, J., and Sugamura, K. (1990). An associated molecule, p64, with high-affinity interleukin 2 receptor. *Int. Immunol.* **2**, 477.

Takeshita, T., Asao, H., Ohtani, K., Ishii, N., Kumaki, S., Tanaka, N., Munakata, H., Nakamura, M., and Sugamura, K. (1992). Cloning of the γ chain of the human IL-2 receptor. *Science* **257**, 379.

Takeshita, T., Arita, T., Higuchi, M., Asao, H., Endo, K., Kuroda, H., Tanaka, N., Murata, K., Ishii, N., and Sugamura, K. (1997). STAM, signal transducing adaptor molecule, is associated with Janus kinases and involved in signaling for cell growth and c-myc induction. *Immunity* **6**, 449–457.

Taniguchi, T. (1995). Cytokine signaling through nonreceptor protein tyrosine kinases. *Science* **268**, 251.

Thomis, D. C., Gurniak, C. B., Tivol, E., Sharpe, A. H., and Berg, L. J. (1995). Defects in B lymphocyte maturation and T lymphocyte activation in mice lacking Jak3. *Science* **270**, 794–797.

Udy, G. B., Towers, R. P., Snell, R. G., Wilkins, R. J., Park, S.-H., Ram, P. A., Waxman, D. J., and Davey, H. W. (1997). Requirement of STAT5b for sexual dimorphism of body growth rates and liver gene expression. *Proc. Natl Acad. Sci. USA* **94**, 7239–7244.

Urban Jr, J. F., Noben-Trauth, N., Donaldson, D. D., Madden, K. B., Morris, S. C., Collins, M., and Finkelman, F. D. (1998). IL-13, IL-4Rα, and Stat6 are required for the expulsion of the gastrointestinal nematode parasite *Nippostrongylus brasiliensis*. *Immunity* **8**, 255–264.

Van Snick, J., Goethals, A., Renauld, J.-C., Van Roost, E., Uyttenhove, C., Rubira, M. R., Moritz, R. L., and Simpson, R. J. (1989). Cloning and characterization of a cDNA for a new mouse T cell growth factor, (P40). *J. Exp. Med.* **169**, 363.

von Freeden-Jeffry, U., Vieira, P., Lucian, L. A., McNeil, T., Burdach, S. E., and Murray, R. (1995). Lymphopenia in interleukin (IL)-7 gene-deleted mice identifies IL-7 as a nonredundant cytokine. *J. Exp. Med.* **181**, 1519–1526.

Waldmann, T. A. (1989). The multi-subunit interleukin-2 receptor. *Annu. Rev. Biochem.* **58**, 875.

Wang, L.-M., Myers Jr, M. G., Sun, S.-J., Aaronson, S. A., White, M., and Pierce, J. H. (1993). IRS-1: essential for insulin-and IL-4- stimulated mitogenesis in hematopoietic cells. *Science* **261**, 1591.

Watson, J. D., Morrissey, P. J., Namen, A. E., Conlon, P. J., and Widmer, M. B. (1989). Effect of IL-7 on the growth of fetal thymocytes in culture. *J. Immunol.* **143**, 1215.

Willerford, D. M., Chen, J., Ferry, J. A., Davidson, L., Ma, A., and Alt, F. W. (1995). Interleukin-2 receptor a chain regulates the size and content of the peripheral lymphoid compartment. *Immunity* **3**, 521.

Wills-Karp, M. (1999). Immunologic basis of antigen-induced airway hyperresponsiveness. *Annu. Rev. Immunol.* **17**, 255–281.

Wlodawer, A., Pavlovsky, A., and Gustchina, A. (1993). Hematopoietic cytokines: similarities and differences in the structures, with implications for receptor binding. *Protein Sci.* **2**, 1373.

Yin, T., Tsang, M. L., and Yang, Y. C. (1994). JAK1 kinase forms complexes with interleukin-4 receptor and 4PS/insulin receptor substrate-1-like protein and is activated by interleukin-4 and interleukin-9 in T lymphocytes. *J. Biol. Chem.* **269**, 26614.

Yokota, T., Otsuka, T., Mosmann, T., Banchereau, J., DeFrance, T., Blanchard, J., de Vries, J. E., Lee, F., and Arai, K. (1986). Isolation and characterization of a human interleukin cDNA clone, homologous to mouse B-cell stimulatory factor-1, that expresses B-cell and T cell stimulating activities. *Proc. Natl Acad. Sci. USA* **83**, 5894.

Yoshimoto, T., Bendelac, A., Watson, C., Hu-Li, J., and Paul, W. E. (1995). Role of NK1.1+ T cells in a TH2 response and in immunoglobulin E production. *Science* **270**, 1845.

Yoshimura, A. (1998). The CIS/JAB family: novel negative regulators of JAK signaling pathways. *Leukemia* **12**, 1851–1857.

Zhang, X., Sun, S., Hwang, I., Tough, D. F., and Sprent, J. (1998). Potent and selective stimulation of memory-phenotype CD8+ T cells in vivo by IL-15. *Immunity* **8**, 591–599.

Zhu, M-h., Berry, J. A., Russell, S. M., and Leonard, W. J. (1998). Delineation of the regions of interleukin-2 (IL-2) receptor β chain important for association of Jak1 and Jak3. *J. Biol. Chem.* **273**, 10719–10725.

Zurawski, G., and de Vries, J. E. (1994). Interleukin 13, an interleukin 4-like cytokine that acts on monocytes and B cells, but not on T cells. *Immunol. Today* **15**, 19.

IL-2 Receptor

Kendall A. Smith*

Division of Immunology, Cornell University Medical College, 525 East 68th Street Box 41, New York, NY 10021, USA

*corresponding author tel: 212-746-4608, fax: 212-746-8167, e-mail: kasmith@mail.med.cornell.edu

DOI: 10.1006/rwcy.2000.14001.

SUMMARY

The interleukin 2 receptor (IL-2R) was the first cytokine receptor to be discovered, characterized and cloned. Therefore, the IL-2R has served as a prototypic cytokine receptor, with which all subsequent cytokine receptors have been compared and contrasted. The discovery of the IL-2R depended upon demonstrating that purified radiolabeled IL-2 binds to IL-2-responsive cells with all of the characteristics of true hormone receptors, i.e. high affinity, specificity, and saturability. As well, the IL-2 concentrations that bind to the receptor are identical to the IL-2 concentrations that promote the characteristic IL-2 response, i.e. T cell proliferation. The high-affinity IL-2R, which is expressed by antigen-activated T cells, binds IL-2 with an equilibrium dissociation constant (K_d) of 10 pM, and comprises three separate noncovalently linked type I transmembrane proteins, designated α, β, and γ. The intermediate-affinity IL-2R ($K_d = 1$ nM) comprises only the β and γ chains, and is expressed by 90% of natural killer (NK) cells. The low-affinity IL-2R ($K_d = 10$ nM) is composed of isolated α chains, and is not found on any normal cells, but has been detected on some human leukemia cells. The β and γ chains cooperate to signal the interior of the cell by activating two tyrosine-specific kinases termed JAK1 and JAK3. In turn, these kinases phosphorylate specific tyrosines on the β and γ chains of the IL-2R, which then serve as docking sites for downstream effector molecules. The receptor functions as an 'on–off' switch which is regulated by IL-2 binding. Subsequent to IL-2 binding the receptor is switched 'on'. Upon dissociation of IL-2 from the receptor the signal to the JAKS is extinguished.

There are a finite number of IL-2/IL-2R interactions that are necessary to promote movement of the T cells past the restriction point in the G_1 phase of the cell cycle, which ultimately depends upon the concentrations of second messengers generated at the cytoplasmic domains of the β and γ chains. There are three known second messenger pathways activated: the JAK/STAT, the Ras/Raf/MAPK, and the PI-3 kinase/Akt pathways. Ultimately these pathways converge on transcription factors, which stimulate the expression of specific genes that mediate the characteristic biological responses of cell proliferation, survival, differentiation and activation-induced cell death. Some of these genes are known, but many have yet to be discovered.

The main physiologic roles of the IL-2Rs are to promote the proliferative expansion of T cells and NK cells upon activation. In particular, the CD8+ T cells are very dependent upon signals from the IL-2R to be able to respond to antigenic stimulation maximally. In addition, the IL-2R is responsible for conveying survival signals to antigen-activated cells, thereby promoting the persistence of antigen-selected memory T cells. Accordingly, upon disappearance of antigen, and the consequent diminution of the IL-2 concentration available, cells that were activated undergo cytokine withdrawal cell death. There are also important negative feedback signals imparted via the IL-2R that function to promote the return of activated cells to a quiescent state, thereby promoting homeostasis of the immune system after it has successfully responded to an antigen. Because of these critical functions of the IL-2/IL-2R system, if there is a deficiency of signaling via the IL-2R, either produced pharmacologically or genetically, there are profound phenotypes generated. Deficiency of the α and β chains result in normal numbers of T cells in the thymus and in the periphery, but there are activation defects, so that immunodeficiency results early, followed by the accumulation of cells with an activated phenotype and autoantibodies later. Deficiency of the γ chain results in severe combined immunodeficiency. Given these findings, it is not surprising that agents that block the IL-2/IL-2R interaction are potent immunosuppressives.

BACKGROUND

Discovery

Like the discovery of interleukin 2 (IL-2), the discovery of the IL-2 receptor (IL-2R) proceeded in three phases: the discovery of the activity, the discovery of the molecules, and the discovery of the genes encoding the receptor molecules.

IL-2R Activity

Classic receptors described by endocrinologists and pharmacologists have two distinct activities: (1) they bind the appropriate ligand with high specificity and affinity in a saturable fashion, and (2) they signal the cell or tissue at ligand concentrations that are relevant physiologically.

At the time that IL-2 activity was first described in the mid-1970s, only a few classic receptors had been discovered. Thus, the receptors for insulin, epidermal growth factor (EGF), and nerve growth factor had been identified and characterized as cell surface molecules that were capable of binding their radiolabeled ligands. As well, the steroid hormone receptors, e.g. for estrogen and glucocorticoids, had been identified via similar methods, but were found to be intracellular receptors.

These experiments established the principles for the recognition of receptors, and also established the methods of radiolabeled ligand-binding assays necessary to demonstrate true receptor activities. However, before one could contemplate these approaches applied to IL-2, it was necessary to obtain pure, homogeneous, native IL-2 and to radiolabel the molecules so that binding assays could be performed.

Before we had purified IL-2, the IL-2 bioassay (Gillis *et al.*, 1978) was the only method we had available to detect it. Therefore, to provide initial data regarding the IL-2R, we performed adsorption experiments that were patterned after adsorption experiments that we had performed previously with erythropoietin and erythropoietin-responsive cells (Fredrickson *et al.*, 1977). Using the IL-2-dependent T cell clones (Baker *et al.*, 1979), we found that IL-2 activity was adsorbed in a time-, temperature-, and cell concentration-dependent manner. Moreover, there was target cell specificity, in that only IL-2-responsive cells, e.g. mitogen/antigen-activated T cells, had adsorptive capacity. In particular, resting T cells, and lipopolysaccharide (LPS)-activated B cells did not adsorb IL-2 (Smith, 1980).

To proceed beyond these descriptive experiments, it was necessary to purify and radiolabel IL-2. Initially, we accomplished this painstakingly using standard biochemical techniques, such as gel filtration and isoelectric focusing (Robb *et al.*, 1981). Subsequently, we developed the first IL-2-reactive monoclonal antibodies that could be used as affinity adsorbents, which enabled the purification of large amounts, i.e. milligrams, of homogeneous IL-2 (Smith *et al.*, 1983). To ensure that the molecules were not denatured during the radiolabeling procedures we first used radiolabeled amino acids, such as $[^{35}S]$Met, $[^{3}H]$Lys, and $[^{3}H]$Leu. Thus, biosynthetic labeling ensured that the molecules were not damaged in the labeling process. Later, we switched to external labeling with ^{125}I, which facilitated the binding experiments, in that it has a higher energy so that a higher specific activity could be attained.

Using techniques that we had already perfected to perform radiolabeled glucocorticoid (Smith *et al.*, 1977) and Fc receptor-binding assays (Crabtree *et al.*, 1979), the very first experiments we performed with radiolabeled IL-2 were definitive. IL-2 binds with high affinity ($K_d = 5$–$10\,pM$) in a saturable fashion to IL-2-responsive cells, whether the cytolytic T lymphocyte lines (CTLLs), or later mitogen/antigen-activated normal T cells (Robb *et al.*, 1981). Just as in our adsorptive experiments, IL-2-unresponsive cells, such as resting T cells or LPS-activated B cells had no detectable IL-2 binding. Most important from the standpoint of receptor definitions, the IL-2 concentrations that bound to the cells were identical to the concentrations that promoted T cell proliferation, i.e. 50% effective concentration (EC_{50}) = 5–10 pM. Thus, the IL-2 binding and biological response curves are coincident, and there are no 'spare receptors'.

These experiments established the validity of true hormone-like receptors as responsible for mediating cytokine effects, and the IL-2-binding assay became prototypic for the cytokine field.

IL-2R Molecules: the α Chain, the β Chain, and the γ Chain

Some receptor molecules are responsible for both receptor activities, i.e. the ligand-binding activities and the signaling activities. In the case of the IL-2R, and most of the cytokine receptors, separate molecules function to bind the ligands versus those that signal the cell. Consequently, the discovery of all of these molecules has taken a long time, more than 15 years, and has involved many investigators. Soon after we reported the discovery of IL-2R activity, Warren Leonard, Warner Greene, and Tom Waldmann contacted us. Takashi Uchiyama, working in their laboratory, had generated monoclonal antibodies (mAbs) that reacted with leukemic cells derived from patients with *adult T cell leukemia* (ATL) (Uchiyama *et al.*, 1981). We had already found that these ATL cells had high levels of IL-2Rs. Moreover, the mAbs recognized activated T cells but not resting T cells, hence the name, anti-Tac, to designate activated T cells. The very first experiments that we performed with these mAbs were clear-cut. Anti-Tac competed for IL-2 binding and IL-2-promoted T cell proliferation in a concentration-dependent fashion (Leonard *et al.*, 1982).

Anti-Tac precipitated a 55 kDa molecule from the surface of the ATL cells and from activated normal T cells. However, soon thereafter, Richard Robb (Robb *et al.*, 1984) found evidence for two distinct IL-2-binding sites on normal T cells, and we noted that not all anti-Tac+ cells could bind radiolabeled IL-2 with the same affinity. Some cell lines appeared to have two distinct binding sites, one with a high affinity similar to activated normal T cells, and another binding moiety that had a 100-fold lower affinity. Subsequently, we identified cell lines that were capable of binding IL-2, but did not react with anti-Tac. Therefore, it appeared that there might be another molecule capable of binding IL-2.

Since mAbs that reacted with this putative second chain had yet to be discovered, we performed experiments to chemically cross-link radiolabeled IL-2 to various cells. We found that a natural killer (NK) leukemic cell line (YT) that did not react with anti-Tac expressed an IL-2-binding chain that was 75 kDa, and thus distinct from the anti-Tac reactive protein that was 55 kDa (Teshigawara *et al.*, 1987). Tom Waldmann's group also identified a molecule of similar size on a leukemic cell line derived from a gibbon ape (MLA-144) (Tsudo *et al.*, 1986), and Warren Leonard's group produced evidence that normal T cells expressed a similar chain (Sharon *et al.*, 1986).

With the availability of leukemic cell lines that expressed solely the α chain, the β chain or both together, we performed kinetic and equilibrium IL-2-binding experiments (Wang and Smith, 1987). These experiments showed that the α chain bound IL-2 with a low affinity, $\sim$10 nM, and the β chain by itself also had a low affinity for IL-2 binding ($\sim$1 nM). Only when both α and β chains were expressed on the same cell, did we detect high-affinity binding (5–10 pM). In addition, kinetic binding experiments revealed that IL-2 has a very rapid association rate with the α chain, approximately 100-fold faster than IL-2 binding to the β chain. However, the dissociation rate of IL-2 from the α chain is also very fast, while its dissociation from the β chain is slow. Therefore, when α and β chains are expressed together on the cell surface a very efficient receptor is created, with a fast association rate contributed by the α chain, and a slow dissociation rate contributed by the β chain.

These experiments appeared to solidify the structure of the IL-2R as a heterodimer. However, when the cDNA encoding the β chain was isolated a few years later (Hatakeyama *et al.*, 1989) and cells were constructed to express both the α and β chains, it became obvious that these two chains formed a receptor that only had a 'pseudo high-affinity' binding site for IL-2. Thus, the $K_d = 100$ pM rather than 10 pM. Accordingly, the search was on for a putative third IL-2R chain.

Kazuo Sugamura and his group from Sendai won the race to identify the γ chain of the IL-2R (Takeshita *et al.*, 1992b). They did so by careful biochemical approaches aided by a panel of monoclonal antibodies reactive with the β chain that they had generated. They found that they were able to precipitate a $\sim$65 kDa chain together with the 55 kDa α chain and the 75 kDa β chain from the cell surface, provided that they first chemically crosslinked IL-2 to the IL-2R. Because the three chains are so similar in size, the only way that the 65 kDa γ chain could be separated and distinguished from the 55 kDa α chain and the 75 kDa β chain was by two-dimensional SDS-PAGE.

The Genes Encoding the IL-2R Chains

Once the IL-2R binding proteins were identified, it was fairly straightforward to clone the cDNAs encoding each of the proteins. However, in each instance, the key to success proved to be the identification of cell lines that expressed high levels of each of the chains, and to the generation of specific mAbs reactive with the respective chains.

In the case of the α chain, the anti-Tac mAbs were used by Warren Leonard and his coworkers Tom Waldmann, Warner Greene, and Gerald Crabtree (Leonard *et al.*, 1984), and also by Toshio Nikaido and his coworkers Takashi Uchiyama, Tasuka Honjo and Junji Yodoi (Nikaido *et al.*, 1984), to simultaneously clone the α chain cDNA using ATL cell lines that expressed high levels of the α chain. In the case of the β chain, Mitsuro Tsudo developed the first mAbs reactive with this chain, and in collaboration with Hatekeyama and Taniguchi the β chain was cloned from the YT clones that we had generated

(Hatakeyama *et al.*, 1989). The γ chain was cloned by Kazuo Sugamura and his group (Takeshita *et al.*, 1992a), after they had generated the first mAbs reactive with the γ chain.

IL-2R Signaling Molecules

While the IL-2R-binding chains were being identified, and mAbs as well as the cDNAs were under investigation during the 1980s and early 1990s, the molecules responsible for IL-2R signaling remained enigmatic. As early as 1990, it was reported that within a few minutes of IL-2 binding to cells that tyrosine-specific phosphorylation of several cytoplasmic proteins could be identified. However, as the IL-2R-binding proteins were identified one by one, it was realized that these proteins themselves were probably not responsible for these phosphorylation events. The IL-2R-binding chains simply did not contain primary sequences that could be protein tyrosine kinases (PTK).

In the early and mid-1990s, as receptor-signaling molecules were identified in other receptor systems, many investigators attempted to ascribe IL-2 signaling of PTK activity to each of the new molecules. For example, when the PTKs of the src family were the only PTKs known, these molecules were reported to be involved in IL-2 signaling. However, it was not until the discovery of the Janus kinase (JAK) family that definitive experiments revealed that both JAK1 and JAK3 were involved in the early events of IL-2R signaling (Miyazaki *et al.*, 1994). Both of these PTKs are already associated with the IL-2R prior to ligand binding. Apparently, after IL-2 binds to the combined $\alpha\beta\gamma$ heterotrimer, the receptor complex is stabilized. Then, JAK1 in association with the β chain and JAK3 in association with the γ chain are brought into close enough proximity so that they begin to phosphorylate one another, as well as both of the receptor chains. Specific tyrosine residues are phosphorylated on each of these chains, and these phosphotyrosine residues then serve as docking sites for the downstream effector molecules. The IL-2R is known to activate three signaling pathways: the JAK/STAT pathway, the Ras/Raf/MAPK pathway, and the PI-3 kinase/Akt pathway. In essence, all three of these pathways depend upon the initial PTK activities of the JAKs.

Alternative names

The only alternative names for the IL-2R relate to the names given to the α chain, as the Tac antigen, and to the γ chain as the common γ chain. This chain was subsequently found to be a part of the receptors for IL-4, IL-7, IL-9, IL-13, and IL-15. The β chain is sometimes referred to as the common β chain, because it is shared by the IL-15R.

Structure

The high-affinity IL-2R is composed of three noncovalently linked chains: α (p55), β (p75), and γ (p65). When expressed on activated T cells there is a 10–20-fold excess of α chains as compared with $\beta\gamma$ chains. Thus, there are $\sim$1000 high-affinity IL-2Rs on activated T cells and an excess of $\sim$10,000–20,000 α chains. The IL-2R has yet to be crystallized, so that the three-dimensional structure is only a conjecture at this time, based upon the structure of the human growth hormone receptor. However, it is clear that the α chain binds to different residues on the IL-2 molecule compared with the $\beta\gamma$ dimer. Consequently, the three chains cooperate to form the high-affinity IL-2R in much the way that the heavy and light chains cooperate to form the antigen-binding region of the antibody molecule.

Main activities and pathophysiological roles

Functional IL-2Rs are only expressed transiently on antigen-activated T cells and B cells (Smith, 1989). Accordingly, IL-2Rs only function during the adaptive immune response briefly after antigen activation, and when antigen is cleared via the reticuloendothelial system the IL-2R expression disappears. By comparison, NK cells express IL-2Rs constitutively. Approximately 10% of NK cells express trimeric high-affinity IL-2Rs, while $\sim$90% only express intermediate affinity IL-2Rs, which are comprised of $\beta\gamma$ dimers (Caligiuri *et al.*, 1990). It is unlikely that this intermediate-affinity IL-2R plays any role in immune responses, because IL-2 concentrations high enough to bind to this receptor are never generated *in vivo*.

IL-2Rs promote four distinct cellular changes that are fundamental to the generation of an effective immune response. First, the IL-2/IL-2R interaction is responsible for mediating cell cycle progression from early G_1 to the G_1/S phase interface, thereby accounting for the proliferative expansion of antigen-selected clones (Cantrell and Smith, 1984). Second, the IL-2/IL-2R interaction is crucial in imparting survival signals to the cell (Gillis *et al.*, 1978), such that if IL-2 is withdrawn prematurely after antigen activation, apoptosis occurs rapidly via the 'cytokine withdrawal' pathways. Third, the IL-2/IL-2R interaction activates and potentiates cellular differentiation programs within the target cells, which permits effector mechanisms to respond to the antigenic stimulation (Le Gros *et al.*, 1990; Seder *et al.*, 1994; Swain, 1994). Finally, the IL-2/IL-2R interaction primes the cell for 'activation-induced cell death' (AICD), a phenomenon that may be one mechanism operating to limit the duration and magnitude of the immune response, and that may play a role in the generation and maintenance of peripheral tolerance (Lenardo, 1991; Singer and Abbas, 1994; Zheng *et al.*, 1995).

Early in our investigation of IL-2Rs, it was evident that although both CD4+ T cells and CD8+ T cells express IL-2Rs after antigen activation, the proliferative response on the part of the CD8+ T cells is more long-lasting (Gullberg and Smith, 1986). Thus, *in vitro*, CD4+ T cells cease proliferating after ~ 7 days, while CD8+ T cells proliferate exponentially for several more days. Consequently, CD8+ T cells will predominate after 10–14 days of culture. This same phenomenon is observed *in vivo* after antigen activation. CD8+ T cells expand massively, while there is a much more modest increase in CD4+ T cells. Until recently, the contribution of the expansion of the number of antigen-reactive T cells, especially of CD8+ T cells, was not appreciated. This was due to a technical difficulty in detecting antigen-specific T cells *in vivo* after the injection of antigen. It has been appreciated for over 20 years that there is a transient expansion of T cells, but the vast majority of the expanded cells, i.e. $> 90\%$, were thought to be antigen nonreactive, so-called 'innocent bystanders'. This view originated in experiments that attempted to quantify antigen-specific cells via limiting dilution analysis (LDA), using the ^{51}Cr-release cytotoxicity assay. However, now that it is possible to enumerate antigen-specific cells directly using the MHC tetramer assay, it has become appreciated that all of the expanded cells are actually antigen-specific (Murali-Krishna *et al.*, 1998). Thus, in experimental viral infections in the mouse, antigen-specific CD8+ T cells have been found to expand as much as 100,000-fold in only 8 days. Accordingly, the calculated doubling times are extremely rapid, ~ 6 hours.

It has not been generally appreciated that the IL-2/IL-2R interaction is responsible for this massive proliferative clonal expansion. Soon after the creation of the first IL-2 knockout mouse by Ivan Horak (Schorle *et al.*, 1991), experiments were done to assess the effect of the IL-2 gene deletion on the response of these mice to experimental viral infections (Kundig *et al.*, 1993). Although there was a ~ 10-fold diminution of CTL activity as detected by the ^{51}Cr-release assay, it was concluded that IL-2 must not be absolutely obligatory for the generation of cytolytic cells. Also, the fact that CTL activity was detectable at all was interpreted as evidence that cellular proliferation was intact and relatively normal in these animals. Therefore, it was interpreted that *in vivo* there had to be other cytokines that could substitute for IL-2, even though *in vitro* only IL-2 was capable of correcting the defect in proliferation observed upon polyclonal activation of the IL-2 knockout cells.

Subsequent experiments where the proliferation of the CD8+ T cells was monitored directly after experimental infection with *lymphocytic choriomeningitis virus* (LCMV), revealed that $> 90\%$ of the proliferation detectable in IL-2 knockout mice was attenuated (Cousens *et al.*, 1995). This is extremely important for our view of the importance of the IL-2/IL-2R interaction for the generation of immune responses. It indicates that *in vivo*, as *in vitro*, IL-2 is the principal T cell growth factor, and that the role of IL-2 in the immune response is not redundant. Although other cytokines, such as IL-4, IL-7, IL-9, IL-13, and IL-15 are all capable of promoting T cell cycle progression *in vitro*, only IL-2 is produced in sufficient amounts *in vivo* in response to antigenic stimulation to mediate the rapid clonal expansion necessary to effect an adequate immune response.

Once the IL-2-responsive cells have expanded, their longevity is dependent on a continued supply of IL-2. In a self-limited infection when the antigen is cleared, the antigen/TCR-dependent triggering of IL-2 production ceases. Subsequently, the antigen-activated, IL-2-responsive cells undergo cytokine withdrawal apoptosis. In large part, the IL-2-induced survival is attributable to the induction of survival genes of the Bcl-2 family (Haldar *et al.*, 1990). However, there are most likely other such survival genes that are also induced by IL-2, so that upon withdrawal of IL-2 the cessation of their expression contributes to the initiation of the apoptotic pathways.

IL-2-induced differentiation influences the secretion of cytokines by both CD4+ and CD8+ target cells, and also the secretion of cytolytic molecules such as perforin, the granzymes, and the Fas/FasL pathway. In fact, the differentiation of T helper cells into the TH1 and TH2 pathways is dependent on IL-2. If IL-2 is excluded from the cultures, neither TH1 nor TH2 CD4+ T cells differentiate. Likewise, the differentiation of CTL is dependent upon IL-2. Accordingly, IL-2 is not only obligatory for the expansion of the number of antigen-selected cells, it is also required for their survival as well as their differentiated effector functions.

GENE

Accession numbers

IL-2R α chain: gene M10322, cDNA X01057
IL-2R β chain: gene X53093, cDNA M26062
IL-2R γ chain: gene AH002843, cDNA NM_00026

PROTEIN

Accession numbers

IL-2R α chain: P01589
IL-2R β chain: P14784
IL-2R γ chain: P31785

Description of protein

The α chain, the β chain, and the γ chain are all type I transmembrane proteins. There are only 13 amino acid residues in the intracellular domain of the α chain; the β chain intracellular domain contains 286 amino acid residues; and the γ chain intracellular domain is composed of 86 amino acid residues.

Relevant homologies and species differences

The α chain is homologous with the α chain of the IL-15R. The β chain is homologous with the receptor chains of the interleukin/hematopoietic cytokine family, especially the external domains, including receptors of IL-3, IL-4, IL-5, IL-6, IL-7, IL-9, IL-12, IL-13, IL-15, erythropoietin, G-CSF, GM-CSF, prolactin, and human growth hormone. The γ chain is homologous with the IL-2R β chain, and the other members of the interleukin/hematopoietic receptor family.

Affinity for ligand(s)

IL-2R α chain: 10 nM
IL-2R β chain: 1 nM
IL-2R γ chain: immeasurable

Cell types and tissues expressing the receptor

The α chain is expressed in recently antigen-activated T cells and B cells and $\sim$10% of NK cells, leukemia and lymphoma cells, especially from ATL patients. The β chain is expressed in antigen-activated T cells and B cells, and NK cells, as well as leukemic cells and cell lines from ATL patients and NK cell leukemia patients. The γ chain is expressed in all major lymphocyte subsets.

Regulation of receptor expression

The α chain is expressed in response to activation via the TCR-induced activation of NFκB/Rel, and by signals from the IL-2R itself via the JAK/STAT pathway. The β chain was reported initially to be expressed constitutively by T cells. However, this result was based upon crosslinkage experiments, using radiolabeled IL-2 and cell populations that contained NK cells. Subsequently, it was shown that T cells only express the β chain when activated via the TCR. By comparison, NK cells express the β chain constitutively. The γ chain is constitutively expressed by T cells, B cells, and NK cells, and does not appear to be regulated by the TCR.

Release of soluble receptors

The α chain can be found in tissue culture media of IL-2R+ cells, and in the serum of experimental animals and humans undergoing an immune response. The α chain is cleaved from the cell surface via nonspecific proteolysis. There is no alternatively spliced mRNA accounting for a secreted versus a membranous form.

SIGNAL TRANSDUCTION

Associated or intrinsic kinases

Members of the Janus family of tyrosine-specific kinases are the initiators of IL-2R signal transduction. JAK1 is associated with the β chain and JAK3 is associated with the γ chain. JAK3 expression is under the regulatory control of the TCR, so that resting T cells are unresponsive to IL-2 because they lack expression of both the α and β chains of the IL-2R, and they lack the signaling molecule associated with the γ chain.

Cytoplasmic signaling cascades

IL-2R signaling occurs once tyrosine residues on both the β chain and γ chain become phosphorylated.

Three cytoplasmic signaling cascades become activated: (1) STAT5a and STAT5b, (2) Ras/Raf/MAPK, and (3) PI-3 kinase/Akt.

DOWNSTREAM GENE ACTIVATION

Transcription factors activated

The JAKs phosphorylate tyrosine residues and thereby activate STAT5a and STAT5b. Initially it was reported that these transcription factors were not required for the IL-2-induced proliferative responses. However, this interpretation was based upon complicated experiments with IL-2R mutants that were subject to alternative interpretations. Recently, using genetic approaches of gene deletion of both STAT5a and STAT5b, more definitive data indicate that STAT5 activation is obligatory for the proliferative response (Beadling *et al.*, 1994; Moriggl *et al.*, 1999a,b).

The Ras/Raf/MAPK pathway eventually activates the AP-1 family of transcription factors, while the PI-3K pathway has been reported to eventually lead to the activation of the E2F family of transcription factors.

Genes induced

Early reports of the induction of new gene expression by activated T cells did not distinguish carefully between TCR signals and IL-2 signals. Therefore, from the literature, often it is difficult to be absolutely sure whether a gene expressed after T cell activation is due to IL-2 or due to the TCR. In addition, experimental results have often been interpreted as indicative of activation via the so-called 'second signals' generated by activation of the accessory molecule CD28. However, activation via this receptor markedly potentiates the production of IL-2 and other cytokines as well. Therefore, unless the experiments investigating TCR±CD28 triggering were done in the presence of a protein synthesis inhibitor, such as cycloheximide, it is impossible to accurately ascribe new gene expression to the triggering of the TCR, or the CD28 molecule, or IL-2, or even one of the other cytokines released early on after TCR activation.

c-Myb was the first gene found to be expressed by IL-2R signaling, distinct from the TCR and/or CD28 (Stern and Smith, 1986). Other genes tested early on, and found to be expressed after stimulation with phytohemagglutinin (PHA), and augmented by IL-2, were the tumor suppressor p53, the proto-oncogene N-Ras, and the transferrin receptor. By comparison, the proto-oncogenes c-Myc and c-Fos are expressed both after TCR triggering and after IL-2 triggering.

Proceeding beyond simply screening for the expression of known genes after IL-2 triggering proved more difficult. First, a system had to be developed to trigger the IL-2R independently of the TCR/CD28 complex. This was accomplished in my laboratory based upon the work of Doreen Cantrell (Cantrell and Smith, 1984). Using human T cells grown for 10–14 days in IL-2, then removed from IL-2 for 36–48 hours before restimulation, it was possible to promote a semi-synchronous entry of all of the cells into the cell cycle. Next, Carol Beading developed a method using thiol-derivatized uridine to label newly synthesized mRNA, which could then be affinity purified using a mercury-Sepharose column (Beadling *et al.*, 1993).

Using this technique, called SLAP (for sulfhydryl labeling and affinity purification), eight cytokine response (CR) genes were identified, only one of which was also expressed after TCR triggering. CR1 encodes a regulator of G protein signaling (RGS), which is a GTPase-activating protein (GAP) for the heterotrimeric G proteins (Beadling *et al.*, 1999). CR2 is novel and remains unknown. CR3 encodes the prostaglandin E_2 receptor, while CR4 encodes a matrix-associated region (MAR) protein. CR5 was also cloned by Miyajima and coworkers and termed CIS, for cytokine inhibitor of signaling. This gene product was found to be a member of a larger family of SH2-containing proteins that appear to function as feedback inhibitors of cytokine receptor signaling. CR6 encodes a new member of a three-gene family that appear to function as regulators of both G_1 and G_2 phases of the cell cycle (Fan *et al.*, 1999). CR7 encodes the cellular proto-oncogene c-Pim, which is a serine/threonine kinase. The v-Pim counterpart is the oncogene of Moloney leukemia virus, which causes T cell leukemias. CR8 encodes a new basic helix-loop-helix (bHLH)-containing protein that functions to regulate the length of G_1, and inhibits apoptosis. Thus, all of these newly discovered CR genes promise to give us insight as to how IL-2 promotes its effects on its target cells.

Others accomplished the separation of TCR signals from IL-2R signals by using IL-2-dependent T cell clones. In particular, Prystowsky's group was one of the first to report the isolation of IL-2-induced genes (Sabath *et al.*, 1990). However, these investigators focused on genes expressed 24 hours after activation with IL-2, after the early events of IL-2R triggering were already completed. Consequently, these investigators identified many structural genes of the cytoskeleton, as well as genes involved in the metabolic pathways in oxidation and energy production that are expressed in response to IL-2 stimulation.

More recently, Jacque Theze and coworkers (Herblot *et al.*, 1999) have focused on IL-2R-induced genes that are distinct from IL-4R-induced genes. Using a representational display subtractive approach, this group has isolated 66 IL-2-induced genes, only 16 of which correspond to already known genes. The known genes include cytoskeleton proteins, transcription factors, nuclear proteins, ribosomal proteins, and ion transporters.

James Ihle's group (Moriggl *et al.*, 1999a,b) recently reported an extremely important contribution to our knowledge of IL-2-induced gene expression involved in signaling cell cycle progression. They found that the STAT5a/b double knockout mouse is unable to mount a proliferative response to polyclonal TCR activators, and that the defect is secondary to the lack of signals that normally emanate from the IL-2R. In particular, the expression of the genes encoding cyclins D2, D3, E, A, and cdk6, were all either absent or deficient in cells from the double knockout mice. By comparison, the expression of cdk4 and cdk2 were not found to be dependent on STAT5, nor was the G_1 degradation of the cyclin-dependent kinase inhibitor p27.

All of these data support the interpretation that T cell proliferation after TCR activation is actually mediated by IL-2 and its receptor, and that the signals and genes activated by the TCR and its accessory molecule CD28 are not involved in promoting G_1 progression or S phase transition.

Promoter regions involved

Only a few of the promoter regions of the known IL-2R-induced genes have been characterized thus far. One of the first to be investigated was the IL-2R α chain promoter (Nakajima *et al.*, 1997), which was of interest because it seemed to be regulated by both the TCR and the IL-2R (Smith and Cantrell, 1985). Also, it was somewhat paradoxical that a hormone would upregulate its own receptor. It has been found that there are canonical NFκB/Rel sites immediately upstream of the transcriptional start site. These response elements are under the control of the TCR. By comparison, there are canonical STAT5 sites much further upstream that account for the observed IL-2 regulation of α chain expression.

BIOLOGICAL CONSEQUENCES OF ACTIVATING OR INHIBITING RECEPTOR AND PATHOPHYSIOLOGY

Unique biological effects of activating the receptors

One of the most important aspects of IL-2R signaling, and of signaling of cell surface receptors in general, is the way in which the signal is received by the cell, such that it is able to make the all-or-none decision to respond. Thus, binding of only one IL-2 molecule to one IL-2R molecule is insufficient to trigger a response. There are a finite number of IL-2/IL-2R interactions that must occur before the cell makes the irrevocable decision to respond (Cantrell and Smith, 1984; Smith, 1989, 1995). For the most part, this quantal nature of IL-2R signaling has been studied by monitoring the proliferative response to IL-2, but the other biological responses directed by IL-2 behave by the same rules. As well, there is a growing awareness that the TCR behaves in the same quantal fashion when triggering of the expression of cytokine genes, such as the IL-2 genes or the IFNγ genes, are monitored. Thus, there is a threshold of activation that must be surpassed before a cell can respond, and a population of cells that differ by the number of IL-2Rs will display a heterogeneous response over time when exposed to a receptor saturating IL-2 concentration. Cells with a low number of IL-2Rs will take longer to reach the critical threshold compared with cells that express high levels of IL-2Rs. Accordingly, normal cells are under the control of this ligand/receptor threshold, while abnormal cells result when this threshold is no longer operative.

The IL-2R shares with other cytokine receptors the capability of promoting cell cycle progression, survival, and differentiation. Accordingly, the most unique effect of IL-2 signaling is its ability to provide negative feedback signals that seem to be so important for the normal functioning of the immune system (Parijs *et al.*, 1999). Thus, as discussed below, the phenotype produced when either IL-2 or its receptors are deleted indicate that the IL-2/IL-2R system is very important for maintaining the integrity of the system as a whole. There are a few reports that focus on the possible molecular mechanisms that could be responsible for the negative feedback effect. However, we still do not have a very complete picture as to exactly what is responsible, nor how the apparent paradoxical effects of pro-survival and growth/differentiation are regulated in relationship with the pro-apoptotic effects of the IL-2R.

Phenotypes of receptor knockouts and receptor overexpression mice

The phenotype of the IL-2 knockout was a paradox when first uncovered (Schorle *et al.*, 1991). Thus, instead of immunodeficiency, IL-2 knockout mice develop an accumulation of T cells with an activated surface phenotype, and they develop an *autoimmune hemolytic anemia* as well as a diffuse *ulcerative colitis* (Sadlack *et al.*, 1993). The etiology of this syndrome remains obscure, but a similar phenotype results when the IL-2R genes are deleted.

Deletion of the IL-2R α chain does not influence lymphocyte development, so that at birth, normal numbers and proportions of the major lymphocyte subsets are present in both primary and secondary lymphoid organs (Willerford *et al.*, 1995). However, with aging, these mice develop progressive enlargement of peripheral lymphoid organs associated with polyclonal T and B cell expansion. Older α chain knockout mice also develop autoimmune disorders, including *hemolytic anemia* and *inflammatory bowel disease*. The α chain of the IL-2R is not involved in signaling, and only imparts its rapid association rate to the high-affinity heterotrimeric IL-2R. Therefore, the fact that the phenotype of the α chain knockout is identical to the IL-2 knockout indicates that levels of IL-2 high enough to bind to the intermediate affinity IL-2R are not produced *in vivo*.

Deletion of the β chain results in a syndrome very similar to that seen in α chain knockout mice (Suzuki *et al.*, 1995). There is normal T and B cell development at birth, but thereafter there is progressive accumulation of activated T cells and B cells and autoimmune phenomena, including diffuse *hypergammaglobulinemia*, *autoimmune hemolytic anemia*, and *inflammatory bowel disease*, leading to premature death. However, these mice suffer from immunodeficiencies before this autoimmune phenotype becomes evident. Thus, when young 3-week-old mice are infected with *vesicular stomatitis virus* (VSV), they fail to make either IgM or IgG antibodies. Also, when infected with *lymphocyte choriomeningitis virus* (LCMV), these mice fail to mount either CD4+ or CD8+ T cell responses. Thus, 14 days after infection with LCMV no CD8+ CTL activity could be detected in cells from β chain knockout mice. In addition, when activated *in vitro* with polyclonal T cell mitogens such as Con A, staphylococcal enterotoxin, and PMA/ionomycin there were no proliferative responses, underscoring that the TCR cannot generate signals that are capable of moving the cell through the cell cycle.

The phenotype of the γ chain knockout mice is distinctly different from the phenotypes of the knockouts discussed thus far (Leonard *et al.*, 1995). These mice suffer from *severe combined immunodeficiency* (SCID). The discovery that the γ chain gene is on the X chromosome led to the discovery that $\sim$50% of X-linked SCID is attributable to mutations of the γ chain gene. Subsequently, it was shown that defects of IL-7 or the IL-7R result in a similar phenotype. Therefore, the lack of the common γ chain prevents the IL-7-dependent proliferation of both B cell and T cell progenitors.

Human abnormalities

Humans with X-SCID due to γ chain mutations have a somewhat different phenotype compared with γ chain knockout mice (Leonard *et al.*, 1995). Thus, they are markedly deficient in T cells, but have normal numbers of B cells. By comparison, γ chain knockout mice accumulate T cells over time, but almost entirely lack B cells.

THERAPEUTIC UTILITY

Effects of inhibitors (antibodies) to receptors

The original monoclonal antibody found to be reactive with the α chain, anti-Tac, has now been 'humanized' and has been used in clinical trials as an immune suppressant for patients who have had a *renal allograft* (Waldmann and O'Shea, 1998). Compared with pan-reactive T cell monoclonal antibodies, the anti-Tac appears to be just as effective in preventing *graft rejection*. The obvious advantage of the IL-2R α chain mAb versus a pan T cell reactive mAb is the selectivity of the anti-Tac mAb for recently antigen-activated T cells.

References

Baker, P. E., Gillis, S., and Smith, K. A. (1979). Monoclonal cytolytic T-cell lines. *J. Exp. Med.* **149**, 273–278.

Beadling, C., Johnson, K. W., and Smith, K. A. (1993). Isolation of interleukin 2-induced immediate-early genes. *Proc. Natl Acad. Sci. USA* **90**, 2719–2723.

Beadling, C., Guschin, D., Witthuhn, B. A., Ziemiecki, A., Ihle, J. N., Kerr, I. M., and Cantrell, D. A. (1994). Activation of JAK kinases and STAT proteins by interleukin-2 and interferon alpha, but not the T cell antigen receptor, in human T lymphocytes. *EMBO J.* **13**, 5605–5615.

Beadling, C., Druey, K. M., Richter, G., Kehrl, J. H., and Smith, K. A. (1999). Regulators of G protein signaling exhibit distinct patterns of gene expression and target G protein specificity in human lymphocytes. *J. Immunol.* **162**, 2677–82.

Caligiuri, M. A., Zmuidzinas, A., Manley, T. J., Levine, H., Smith, K. A., and Ritz, J. (1990). Functional consequences of interleukin 2 receptor expression on resting human lymphocytes. Identification of a novel natural killer cell subset with high affinity receptors. *J. Exp. Med.* **171**, 1509–1526.

Cantrell, D. A., and Smith, K. A. (1984). The interleukin-2 T-cell system: a new cell growth model. *Science* **224**, 1312–1316.

Cousens, L. P., Orange, J. S., and Biron, C. A. (1995). Endogenous IL-2 contributes to T cell expansion and IFN-gamma production during lymphocytic choriomeningitis virus infection. *J. Immunol.* **155**, 5690–5699.

Crabtree, G. R., Munck, A., and Smith, K. A. (1979). Glucocorticoids inhibit expression of Fc receptors on the human granulocytic cell line HL-60. *Nature* **279**, 338–339.

Fan, W., Richter, G., Cereseto, A., Beadling, C., and Smith, K. A. (1999). Cytokine response gene 6 induces p21 and regulates both cell growth and arrest. *Oncogene* **18**, 6573–6582.

Fredrickson, T. N., Smith, K. A., Cornell, C. J., Jasmin, C., and McIntyre, O. R. (1977). The interaction of erythropoietin with fetal liver cells I. Measurement of proliferation by tritiated thymidine incorporation. *Exp. Hematol.* **5**, 254–265.

Gillis, S., Ferm, M. M., Ou, W., and Smith, K. A. (1978). T cell growth factor: parameters of production and a quantitative microassay for activity. *J. Immunol.* **120**, 2027–2032.

Gullberg, M., and Smith, K. A. (1986). Regulation of T cell autocrine growth. T4+ cells become refractory to interleukin 2. *J. Exp. Med.* **163**, 270–284.

Haldar, S., Reed, J. C., Beatty, C., and Croce, C. M. (1990). Role of bcl-2 in growth factor triggered signal transduction. *Cancer Res.* **50**, 7399–7401.

Hatakeyama, M., Tsudo, M., Minamoto, S., Kono, T., Doi, T., Miyata, T., Miyasaka, M., and Taniguchi, T. (1989). Interleukin-2 receptor beta chain gene: generation of three receptor forms by cloned human alpha and beta chain cDNAs. *Science* **244**, 551–556.

Herblot, S., Chastagner, P., Samady, L., Moreau, J. -L., Demaison, C., Froussard, P., Liu, X., Bonnet, J., and Theze, J. (1999). IL-2-dependent expression of genes involved in cytoskeleton organization, oncogene regulation and transcriptional control. *J. Immunol.* **162**, 3280–3288.

Kundig, T. M., Schorle, H., Bachmann, M. F., Hengartner, H., Zinkernagel, R. M., and Horak, I. (1993). Immune responses in interleukin-2-deficient mice. *Science* **262**, 1059–1061.

Le Gros, G., Ben-Sasson, S. Z., Seder, R., Finkelman, F. D., and Paul, W. E. (1990). Generation of interleukin 4 (IL4)-producing cells *in vivo* and *in vitro*: IL-2 and IL4 are required for *in vitro* generation of IL4-producing cells. *J. Exp. Med.* **172**, 921–929.

Lenardo, M. J. (1991). Interleukin-2 programs mouse alpha beta T lymphocytes for apoptosis. *Nature* **353**, 858–861.

Leonard, W. J., Depper, J. M., Uchiyama, T., Smith, K. A., Waldmann, T. A., and Greene, W. C. (1982). A monoclonal antibody that appears to recognize the receptor for human T-cell growth factor; partial characterization of the receptor. *Nature* **300**, 267–269.

Leonard, W. J., Depper, J. M., Crabtree, G. R., Rudikoff, S., Pumphrey, J., Robb, R. J., Kronke, M., Svetlik, P. B., Peffer, N. J., and Waldmann, T. A. (1984). Molecular cloning and expression of cDNAs for the human interleukin-2 receptor. *Nature* **311**, 626–631.

Leonard, W. J., Shores, E. W., and Love, P. E. (1995). Role of the common cytokine receptor gamma chain in signaling and lymphoid development. *Immunol. Rev.* **148**, 97–114.

Miyazaki, T., Kawahara, A., Fujii, H., Nakagawa, Y., Minami, Y., Liu, Z. J., Oishi, I., Silvennoinen, O., Witthuhn, B. A., and Ihle, J. N. (1994). Functional activation of Jak1 and Jak3 by selective association with IL-2 receptor subunits. *Science* **266**, 1045–1047.

Moriggl, R., Sexl, V., Piekorz, R., Topham, D., and Ihle, J. (1999a). Stat5 activation is uniquely associated with cytokine signaling in peripheral T cells. *Immunity* **11**, 225–230.

Moriggl, R., Topham, D. J., Teglund, S., Sexl, V., McKay, C., Wang, D., Hoffmeyer, A., van Deursen, J., Sangster, M. Y., Bunting, K. D., Grosveld, G. C., and Ihle, J. N. (1999b). Stat5 is required for IL-2-induced cell cycle progression of peripheral T cells. *Immunity* **10**, 249–259.

Murali-Krishna, K., Altman, J. D., Suresh, M., Sourdive, D. J., Zajac, A. J., Miller, J. D., Slansky, J., and Ahmed, R. (1998). Counting antigen-specific CD8 T cells: a reevaluation of bystander activation during viral infection. *Immunity* **8**, 177–187.

Nakajima, H., Liu, X. -W., Wynshaw-Boris, A., Rosenthal, L. A., Imada, K., Finbloom, L. H., Henninghausen, L., and Leonard, W. J. (1997). An indirect effect of Stat5a in IL-2-induced proliferation: a critical role for Stat5a in IL-2-mediated IL-2 receptor alpha chain induction. *Immunity* **7**, 691–701.

Nikaido, T., Shimizu, A., Ishida, N., Sabe, H., Teshigawara, K., Maeda, M., Uchiyama, T., Yodoi, J., and Honjo, T. (1984). Molecular cloning of cDNA encoding human interleukin-2 receptor. *Nature* **311**, 631–635.

Parijs, L. V., Refaeli, Y., Lord, J. D., Nelson, B. H., Abbas, A., and Baltimore, D. (1999). Uncoupling IL-2 signals that regulate T cell proliferation, survival, and Fas-mediated activation-induced cell death. *Immunity* **11**, 281–288.

Robb, R. J., Munck, A., and Smith, K. A. (1981). T cell growth factor receptors. Quantitation, specificity, and biological relevance. *J. Exp. Med.* **154**, 1455–1474.

Robb, R. J., Greene, W. C., and Rusk, C. M. (1984). Low and high affinity cellular receptors for interleukin 2. Implications for the level of Tac antigen. *J. Exp. Med.* **160**, 1126–1146.

Sabath, D. E., Podolin, P. L., Comber, P. G., and Prystowsky, M. B. (1990). cDNA cloning and characterization of interleukin 2-induced genes in a cloned T helper lymphocyte. *J. Biol. Chem.* **265**, 12671–12678.

Sadlack, B., Merz, H., Schorle, H., Schimpl, A., Feller, A. C., and Horak, I. (1993). Ulcerative colitis-like disease in mice with a disrupted interleukin-2 gene [see comments]. *Cell* **75**, 253–261.

Schorle, H., Holtschke, T., Hunig, T., Schimpl, A., and Horak, I. (1991). Development and function of T cells in mice rendered interleukin-2 deficient by gene targeting. *Nature* **352**, 621–624.

Seder, R. A., Germain, R. N., Linsley, P. S., and Paul, W. E. (1994). CD28-mediated costimulation of interleukin 2 (IL-2) production plays a critical role in T cell priming for IL-4 and interferon gamma production. *J. Exp. Med.* **179**, 299–304.

Sharon, M., Klausner, R. D., Cullen, B. R., Chizzonite, R., and Leonard, W. J. (1986). Novel interleukin-2 receptor subunit detected by cross-linking under high affinity conditions. *Science* **234**, 859–863.

Singer, G. G., and Abbas, A. K. (1994). The fas antigen is involved in peripheral but not thymic deletion of T lymphocytes in T cell receptor transgenic mice. *Immunity* **1**, 365–371.

Smith, K. A. (1980). T-cell growth factor. *Immunol. Rev.* **51**, 337–357.

Smith, K. A. (1989). The interleukin 2 receptor. *Annu. Rev. Cell Biol.* **5**, 397–425.

Smith, K. A. (1995). Cell growth signal transduction is quantal. *Ann. N.Y. Acad. Sci.* **766**, 263–271.

Smith, K. A., and Cantrell, D. A. (1985). Interleukin 2 regulates its own receptors. *Proc. Natl Acad. Sci. USA* **82**, 864–868.

Smith, K. A., Crabtree, G. R., Kennedy, S. J., and Munck, A. U. (1977). Glucocorticoid receptors and glucocorticoid sensitivity of mitogen stimulated and unstimulated human lymphocytes. *Nature* **267**, 523–526.

Smith, K. A., Favata, M. F., and Oroszlan, S. (1983). Production and characterization of monoclonal antibodies to human interleukin 2: strategy and tactics. *J. Immunol.* **131**, 1808–1815.

Stern, J. B., and Smith, K. A. (1986). Interleukin-2 induction of T-cell G1 progression and c-myb expression. *Science* **233**, 203–206.

Suzuki, H., Kundig, T. M., Furlonger, C., Wakeham, A., Timms, E., Matsuyama, T., Schmits, R., Simard, J. J., Ohashi, P. S., Griesser, H., and Mak, T. (1995). Deregulated T cell activation and autoimmunity in mice lacking interleukin-2 receptor beta. *Science* **268**, 1472–1476.

Swain, S. L. (1994). Generation and *in vivo* persistence of polarized Th1 and Th2 memory cells. *Immunity* **1**, 543–552.

Takeshita, T., Asao, H., Ohtani, K., Ishii, N., Kumaki, S., Tanaka, N., Manukata, H., Nakamura, M., Sugamura, K. (1992a). Cloning of the gamma chain of the human IL-2 receptor. *Science* **257**, 379–382.

Takeshita, T., Ohtani, K., Asao, H., Kumaki, S., Nakamura, M., and Sugamura, K. (1992b). An associated molecule, p64, with IL-2 receptor beta chain. Its possible involvement in the formation of the functional intermediate-affinity IL-2 receptor complex. *J. Immunol.* **148**, 2154–2158.

Teshigawara, K., Wang, H. M., Kato, K., and Smith, K. A. (1987). Interleukin 2 high-affinity receptor expression requires two distinct binding proteins. *J. Exp. Med.* **165**, 223–238.

Tsudo, M., Kozak, R. W., Goldman, C. K., and Waldmann, T. A. (1986). Demonstration of a non-Tac peptide that binds interleukin 2: a potential participant in a multichain interleukin 2 receptor complex. *Proc. Natl Acad. Sci. USA* **83**, 9694–9698.

Uchiyama, T., Broder, S., and Waldmann, T. A. (1981). A monoclonal antibody (anti-Tac) reactive with activated and functionally mature human T cells. I. Production of anti-Tac monoclonal antibody and distribution of Tac (+) cells. *J. Immunol.* **126**, 1393–1397.

Waldmann, T. A., and O'Shea, J. (1998). The use of antibodies against the IL-2 receptor in transplantation. *Curr. Opin. Immunol.* **10**, 507–512.

Wang, H. M., and Smith, K. A. (1987). The interleukin 2 receptor. Functional consequences of its bimolecular structure. *J. Exp. Med.* **166**, 1055–1069.

Willerford, D. M., Chen, J., Ferry, J. A., Davidson, L., Ma, A., and Alt, F. W. (1995). Interleukin 2 receptor alpha chain regulates the size and content of the peripheral lymphoid compartment. *Immunity* **4**, 521–530.

Zheng, L., Fisher, G., Miller, R. E., Peschon, J., Lynch, D. H., and Lenardo, M. J. (1995). Induction of apoptosis in mature T cells by tumour necrosis factor. *Nature* **377**, 348–351.

IL-4 Receptor

Achsah D. Keegan*

Immunology Department, Holland Laboratory of the American Red Cross, 15601 Crabbs Branch Way, Rockville, MD 20855, USA

*corresponding author tel: 301-517-0326, fax: 301-517-0344, e-mail: keegana@usa.redcross.org

DOI: 10.1006/RWCY.2000.14002.

SUMMARY

The cell surface receptor for IL-4 is composed of two polypeptide proteins that span the plasma membranes. One of these proteins chains, the IL-4Rα, binds to IL-4 with high affinity. Binding of IL-4 to the IL-4Rα on the cell surface results in its association with a second protein. In the type I IL-4 receptor complex the associating chain is the common γ chain. In the type II IL-4 receptor complex it is the IL-13Rα chain. Over the last 8 years, much work has focused on understanding the discrete signal transduction pathways activated by the IL-4 receptor and the coordination of these individual pathways in the regulation of a final biological outcome. Experiments focused on signaling pathways have delineated the mechanism by which IL-4 regulates cell growth, survival, and gene expression. Strategies to block binding of IL-4 to its receptor and to target specific signaling pathways are being tested to treat diseases associated with IL-4 such as *allergy*.

BACKGROUND

Discovery

The structure of the IL-4 receptor complex was first discovered using chemical crosslinking of [^{125}I]IL-4 to molecules on the surface of IL-4-responsive cells (Ohara and Paul, 1987). Generally, two major complexes were found crosslinked to [^{125}I]IL-4. One contained [^{125}I]IL-4 bound to a protein of ~140 kDa; the second variably contained 80 and 70 kDa molecules crosslinked to [^{125}I]IL-4, depending on the type of cell used for analysis. Subsequently, 8 years of study revealed the complex nature of the IL-4 receptor.

Alternative names

Initially, the term IL-4R was used to describe the receptor for IL-4, but in 1993 it became clear that the IL-4 receptor consists of two separate chains (Russell *et al.*, 1993): the 140 kDa IL-4-binding chain and the common gamma chain, γc (**Figure 1**). The 140 kDa chain was then termed the IL-4Rα, although other groups initially referred to it as the IL-4Rβ due to its analogy to the IL-2Rβ chain. Numerous studies suggested a relationship between the receptor for IL-4 and the receptor for IL-13 (Callard *et al.*, 1996; Murata *et al.*, 1998). One of these ligands could suppress the binding of the other to cells capable of responding to both cytokines.

Figure 1 IL-4 receptor complexes. The type I IL-4 receptor complex comprises the IL-4-binding chain termed the IL-4Rα. The binding of IL-4 to this chain allows heterodimerization with the common γ chain called the γc. The type II IL-4 receptor complex consists of the IL-4Rα chain and the IL-13Rα1. The IL-4Rα is associated with the tyrosine kinase JAK1. The γc is associated with the JAK3 while the IL-13Rα1 probably associates with either JAK2 or TYK2, but not JAK3.

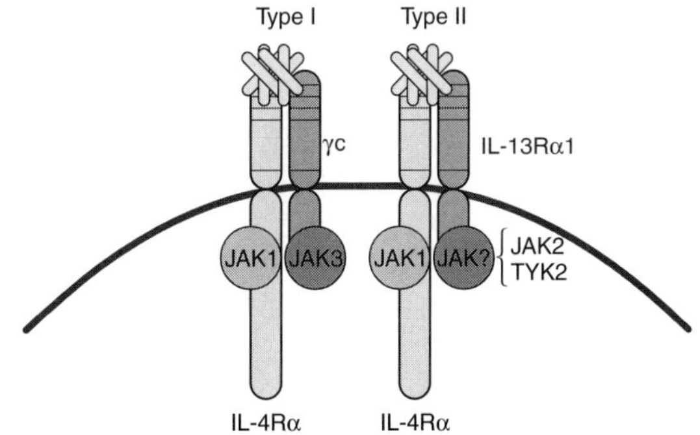

Additionally, a mutant form of human IL-4 in which Tyr124 has been replaced by Asp (Y124D), that is capable of binding to the human IL-4 receptor with high affinity but fails to signal an IL-4-specific response, is able to antagonize competitively both IL-4- and IL-13-induced proliferation. Recent analysis of the IL-13 receptor structure indicates that a functional IL-4 receptor complex can also be comprised of the IL-4Rα chain and the IL-13Rα chain (Hilton *et al.*, 1996). The complex containing the IL-4Rα and the γc is called the type I IL-4 receptor. The IL-4 receptor containing the IL-13Rα is termed type II (Callard *et al.*, 1996) (Figure 1).

Structure

As shown in Figure 1, there are two types of IL-4 receptor complexes. These are thought to function as heterodimers. The type I receptor consists of the IL-4Rα and the γc, while the type II receptor consists of the IL-4Rα and the IL-13Rα. The type I receptor is a structure specific for IL-4 (Russell *et al.*, 1993). However, the structure that makes up the type II IL-4 receptor can act as a receptor complex for IL-4 or IL-13 (Callard *et al.*, 1996). The IL-4Rα has affinity for IL-4 but not for IL-13, while the IL-13Rα has affinity for IL-13 but not for IL-4. The binding of ligand with the appropriate chain induces dimerization with the other chain. Studies using engineered chimeric receptor complexes have indicated that homodimers of the cytoplasmic region of the IL-4Rα can mediate IL-4-specific responses (for example, see Fujiwara *et al.*, 1997). It is not yet clear whether a full-length IL-4Rα ever signals as a homodimer naturally.

Main activities and pathophysiological roles

The main activities of the IL-4 receptor are very closely linked to the major activities of IL-4. As the receptor for IL-4, it plays a large role in the regulation of the differentiation of naive CD4+ T cells, driving them into a helper phenotype called TH2. In addition, it plays a major role in class switching to IgG1 and IgE. However, since the IL-4Rα chain is used as a receptor component by both IL-4 and IL-13 (Callard *et al.*, 1996), its loss results in more severe defects than the loss of IL-4 alone.

GENE

Accession numbers

The gene encoding the mouse IL-4Rα has been characterized in detail (Wrighton *et al.*, 1992) (**Figure 2**). The mouse IL-4Rα gene is located on chromosome 7 and the human IL-4Rα gene is localized on chromosome 16. In addition, cDNA encoding the human, mouse, and rat IL-4Rα have been cloned (accession numbers: human X52425, mouse M29854, rat X69903).

Sequence

The mouse IL-4Rα is encoded by 12 exons (accession numbers: M64868, M64869, M64870, M64871, M64872, M64873, M64874, M64875, M64876, M64877, M64878, M64879) (Figure 2). Exons 1–2 make up the 5′ untranslated region. Exon 3 encodes the signal peptide which is comprised of ~25 amino acids. Exons 4–7 encode the extracellular domain which contains the cysteine pairs and WSXWS motif that are the hallmark signs of the hematopoietic receptor superfamily. Exon 8 codes for the C-terminus of the soluble form of the mouse IL-4Rα. In the cell surface form, exon 8 is spliced out, leaving exons 7 and 9 adjoining. Exon 9 encodes the transmembrane region. Exon 8 contains a 114 bp sequence that encodes six novel amino acids followed by a stop codon. This sequence is inserted just prior to the end of the extracellular domain by alternative splicing of the mRNA and results in the production of a 40 kDa soluble receptor with a C-terminus that is not found in the full-length IL-4Rα. Exons 10–12 encode the cytoplasmic region. While soluble IL-4Rα capable of

Figure 2 The mouse IL-4Rα gene. The gene encoding mouse IL-4Rα contains 12 exons and 11 introns. Exons 1 and 2 code for the 5′ untranslated region (5′ UT). Exon 3 codes for the signal peptide (Sig). Exons 4–7 encode the extracellular domain. Exon 8 codes for the carboxy tail of the soluble version of the IL-4Rα (Sol) and contains a translational stop, while exon 9 contains the transmembrane domain (TM). Exons 10–12 encode the cytoplasmic domain.

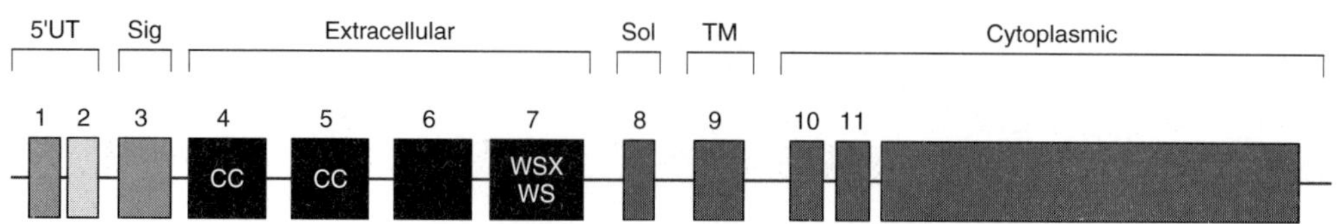

binding human IL-4 with high affinity has been found in the serum of humans, it is thought that it originates from proteolytic cleavage of the full-length receptor rather than from alternative splicing.

Interestingly, both the γc and the IL-13Rα chains are localized to the X chromosome. Both are also members of the hematopoietin receptor superfamily.

PROTEIN

Accession numbers

The amino acid sequences for the human, mouse, and rat IL-4Rα can be found in the SwissProt database. Accession numbers for the protein sequence are human A60386, mouse AAA39299, and rat S31575.

Description of protein

The IL-4Rα is a transmembrane protein of greater than 800 amino acids including its leader sequence. For the human IL-4Rα, there are a total of 825 amino acids with the first 25 comprising the leader sequence. The extracellular domain consists of $\sim$220 amino acids. There are five potential *N*-linked glycosylation sites in the extracellular region, and biochemical studies suggest that about three are in fact glycosylated (for reviews see Nelms *et al.*, 1999 and Keegan *et al.*, 1996).

The extracellular domain of the IL-4Rα (termed IL-4BP) has been co-crystallized with IL-4 (Hage *et al.*, 1999). The coordinates for this structure have been deposited in the Protein Data Bank under the code 1IAR. This analysis shows that the extracellular region is comprised of two domains arranged in an L shape. Both domains are related to the fibronectin type III domain. Both domains contain seven antiparallel β strands separated by loops and an occasional short helix. The N-terminal domain (D1) lies about 45 degrees from horizontal. It consists of amino acids 1–91 and is arranged in an H-type subclass of the Ig-type fold. D1 contains three pairs of cysteine residues engaged in disulfide bonding (Cys9–Cys19, Cys29–Cys59, Cys49–Cys61). The first two pairs are characteristic of members of the type I cytokine or hematopoietin receptor superfamily. The third pair appears novel to the IL-4BP. The domain closest to the transmembrane region (D2) is arranged in an S-type fibronectin type III domain in a vertical orientation. The D1 and D2 domains are close to perpendicular to one another. D2 does not contain any disulfide bonds, but does contain the WSXWS motif prior to the βG strand. The WSXWS sequence helps to maintain proper folding and transport to the cell surface.

Loops in D1 (L1, L2, and L3) and loops in D2 (L5 and L6) make contacts with helices A and C in IL-4. Interestingly, IL-4 bound to the IL-4BP shows small changes in its structure, especially in helices A and D. The changes are localized to regions implicated in binding to the γc, suggesting that binding of IL-4 to the IL-4Rα creates the γc-binding epitope.

The cytoplasmic domain of the human IL-4Rα contains more than 550 amino acids. There are no consensus sequences characteristic of serine/threonine or tyrosine kinases. The cytoplasmic region contains two acidic regions and is generally serine- and proline-rich. There is a Box-1 sequence close to the transmembrane region, originally described in the gp130 chain of the IL-6 receptor that is shared among many of the hematopoietin receptor family members. There are five tyrosine residues in the cytoplasmic region whose surrounding amino acid sequences are highly conserved among the rat, mouse, and human IL-4Rα (**Figure 3**). This high degree of homology between the species suggests some functional importance for these residues that has been borne out by mutagenesis studies.

Relevant homologies and species differences

Human, mouse, and rat IL-4Rα are generally similar, with virtually 90–100% identity in key sequence motifs

Figure 3 Conserved cytoplasmic tyrosine residues. The cytoplasmic domains of the rat, mouse, and human IL-4Rα contain five tyrosines whose surrounding sequences are highly conserved. These are numbered 1–5 starting with the most membrane-proximal tyrosine.

Y1	hu	488	PLViagNPAYRSFSnsl
	mu	491	PLViadNPAYRSFSdcc
	rt	491	PLVisdNPAYRSFSdfs
Y2	hu	570	APtsGYQEFVhAVeQG
	mu	570	APagGYQEFVqAVkQG
	rt	570	APtsGYQEFVqAVkQG
Y3	hu	597	PpGeaGYKAFSSLLaS
	mu	597	PsGdpGYKAFSSLLsS
	rt	597	PsGdtGYKAFSSLLsS
Y4	hu	627	GeeGYKPFQdli
	mu	627	GhgGYKPFQ---
	rt	627	GcgGYKPFQnpv
Y5	hu	705	DsLGsGIVYSaLTCH
	mu	701	DdLGfGIVYSsLTCH
	rt	701	DdLGlGIVYSsLTCH

and modules (Figure 3). Overall amino acid identity between the human and mouse is 48%, between human and rat is 46%, and between rat and mouse is 72%. The extracellular domain of the IL-4Rα shows between 20% and 30% homology to a number of other cytokine receptor family members.

Affinity for ligand(s)

IL-4 binds to its receptor with high affinity. Binding sites for [^{125}I]IL-4 with K_d values of 20–300 pM have been detected on many hematopoietic and nonhematopoietic cell types (Ohara and Paul, 1987; Lowenthal *et al.*, 1988). The number of receptors expressed on the surfaces of cells is generally quite low, ranging from 50 to 5000 sites per cell. The IL-4Rα contains the vast majority of binding affinity for IL-4. The soluble version of the IL-4Rα can bind IL-4 with high affinity also. The additional contribution of the γc to cell surface binding affinity has been estimated at 3-fold (Russell *et al.*, 1993) while the contribution of the IL-13Rα has been estimated to be between 5- and 10-fold (Hilton *et al.*, 1996).

Cell types and tissues expressing the receptor

The vast majority of cell types in the body express one or the other (or both) of the IL-4 receptor complexes. IL-4 receptors have been identified on most cells of hematopoietic origin, cells of the brain, muscle, kidney, and placenta, and in primary cell types including fibroblasts, epithelial cells, endothelial cells, and a number of different tumors, as well as many others (Ohara and Paul, 1987; Lowenthal *et al.*, 1988; Doucet *et al.*, 1998; Dubois *et al.*, 1998; Henriques *et al.*, 1998; Kotsimbos *et al.*, 1998; Mehrotra *et al.*, 1998; van der Velden *et al.*, 1998).

Regulation of receptor expression

There is not much information available on the regulation of expression of IL-4 receptor complexes. All components are expressed to some degree on resting cells. Stimulation of lymphocytes through the antigen receptor results in the increase of IL-4Rα expression (Ohara and Paul, 1988). Polyclonal activators such as LPS also induce expression. Interestingly, IL-4 itself acts to regulate the levels of expression of IL-4Rα. Preliminary studies in mice lacking expression of the active tyrosine phosphatase SHP-1 indicate that this phosphatase may also play some role in regulating the expression of IL-4Rα in lymphocytes (Huang *et al.*, 1999), although the mechanism by which this might occur is not clear. Treatment of patients with steroids has been shown to diminish IL-4Rα expression (Wright *et al.*, 1999).

Release of soluble receptors

In the murine system, soluble receptors capable of binding IL-4 with high affinity can arise by alternative splicing of the mRNA (Wrighton *et al.*, 1992). The soluble form contains novel amino acids at the C-terminus. Evidence for alternative splicing has not been provided for the human IL-4Rα. Since soluble receptors for IL-4 have been found in human sera, it is likely they arise by a proteolytic mechanism.

SIGNAL TRANSDUCTION

Several comprehensive reviews on signaling by the IL-4 receptor have been published and provide detailed discussion of the experimental data and extensive references supporting the current model of signal transduction (**Figure 4**) (see Wang *et al.*, 1995; Keegan *et al.*, 1996; Nelms *et al.*, 1999).

Associated or intrinsic kinases

Although there is no intrinsic enzyme activity encoded in the IL-4Rα, treatment of cells with IL-4 activates cytoplasmic tyrosine kinases (Ihle, 1995; Johnston *et al.*, 1996). The Janus family of tyrosine kinases (JAKs) are constitutively associated with the IL-4 receptor. JAK1 is associated with IL-4Rα while JAK3 is coupled to the γc (Figure 1). The association of the JAK1 kinase to the IL-4Rα is thought to be via the conserved Box-1 site. Mutation of this site or mutation of JAK1 itself abolishes all functional activity of the IL-4 receptor. Both JAK1 and JAK3 become tyrosine phosphorylated upon treatment of hematopoietic cells with IL-4. IL-13Rα does not associate with JAK3. However, it associates with either JAK2 or TYK2 depending on the cell type, and these kinases can be tyrosine phosphorylated in response to IL-4 in some cell types (Murata *et al.*, 1998). Heterodimerization of the receptor subunits is thought to activate the kinases and initiate the signaling cascade (Figure 4).

IL-4 is able to signal in cells expressing the IL-4Rα and either the γc or the IL-13Rα. IL-4 treatment is able to induce the tyrosine phosphorylation of a transcription factor (STAT6) in cell lines derived from *severe combined immunodeficiency* (SCID) patients lacking the γc or JAK3, albeit less efficiently than in normal cells. Therefore, in the absence of γc, or in the absence of

Figure 4 Signaling by the IL-4 receptor. Several of the IL-4-activated signal transduction pathways are shown in diagram form. The binding of IL-4 to the IL-4Rα induces heterodimerization with the γc (shown) or the IL-13Rα1. The dimerization activates the Janus kinases that initiate the phosphorylation cascade. Tyrosine residues in the cytoplasmic tail of the IL-4Rα become phosphorylated and act as docking sites for signaling molecules. The first cytoplasmic tyrosine residue (Y1, green) is in a sequence motif called the I4R-motif that interacts with protein tyrosine-binding domains (PTB). Members of the insulin receptor substrate family (IRS) and Shc dock to this site. Tyrosine residues 2–4 (orange) interact with the SH2-domain of STAT6. The fifth cytoplasmic residue (red) lies in a consensus motif termed an ITIM shown in other receptors to dock the SH2-domains of tyrosine phosphatases. After being recruited to the receptor complex, these signaling molecules are tyrosine phosphorylated. Phosphorylated STAT6 (green) dimerizes, migrates to the nucleus, and binds to promoters of genes such as CD23 and MHC class II. Phosphorylated IRS (dark blue) binds to the p85 subunit of PI-3 kinase (light blue) and to GRB2 (orange). The IRS pathway has been linked to cellular proliferation in response to IL-4. The activated PI-3 kinase regulates the p70S6 kinase and akt, both of which have been shown to participate in antiapoptosis. PI-3 kinase activity also regulates the IL-4-induced phosphorylation of the DNA-binding protein HMGI(Y). (Full colour figure may be viewed online.)

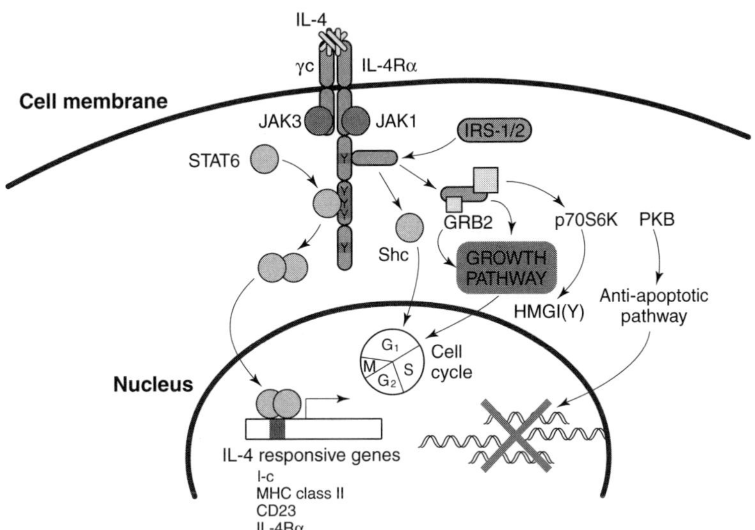

the kinase to which γc associates, IL-4 signaling can occur via the type II IL-4 receptor (Johnston *et al.*, 1996).

Another tyrosine kinase that interacts with the IL-4Rα is c-*fes*. The importance of c-*fes* in IL-4 signaling is not clear, but some evidence suggests that it may regulate growth and production of IgE in response to IL-4.

Cytoplasmic signaling cascades

One of the signaling pathways initiated by IL-4 is a latent cytoplasmic transcription factor, termed STAT6, that is a member of the signal transducers and activators of transcription (STAT) family (Ihle, 1995). STAT6 is recruited to the IL-4Rα by binding to the second, third, or fourth cytoplasmic tyrosine residues via its SH2 domain after they become phosphorylated (Figure 3). STAT6 then becomes tyrosine phosphorylated, dimerizes, migrates to the nucleus, and binds to consensus sequences found within promoters of IL-4-regulated genes. This factor plays a major role in gene regulation and in many aspects of the allergic response including TH2 differentiation, IgE production (reviewed in Nelms *et al.*, 1999) and in models of allergic lung inflammation (Kuperman *et al.*, 1998).

A second signaling pathway activated by the IL-4 receptor is the insulin receptor substrate (IRS) family (IRS-1, IRS-2, IRS-3, and IRS-4) pathway (White, 1998). These proteins are large cytoplasmic docking proteins which contain a protein tyrosine binding (PTB) domain and many sites for serine/threonine and tyrosine phosphorylation. The IRS proteins are recruited to the IL-4Rα by the first cytoplasmic tyrosine residue (Tyr1) that lies within a consensus motif also found in receptors for insulin and the insulin-like growth factor type I (IGF-1) called the I4R motif (Figure 3). The PTB domain of the IRS protein interacts directly with the I4R-motif of the IL-4Rα. The structure of this interaction has been solved (Zhou *et al.*, 1996) and can be observed by going to the molecular modeling database of the National Center for Biotechnology Information (www.ncbi.nlm.nih.gov/Structure) and searching for the identifier 5977. The Protein Data Base ID is 1IRS.

Tyrosine phosphorylated sites within the IRS proteins associate with cellular proteins that contain SH2 domains. One signaling molecule with which all IRS family members interact is the p85 subunit of phosphatidylinositol 3′-kinase (PI-3 kinase) (White, 1998). Numerous studies have shown that the interaction of IRS proteins with the p85 subunit results in the activation of the p110 catalytic subunit of the PI-3 kinase enzyme. Active PI-3 kinase catalyzes the transfer of phosphate from ATP to the D3 position of the inositol ring in membrane-bound phosphatidylinositol. PI-3 kinase activity is important for growth, survival, and regulation of gene expression in response to IL-4. Several signaling molecules whose activities are downstream of PI-3 kinase activity have been implicated in IL-4 responses, including p70S6 kinase, the Akt kinase, and the nonhistone high-mobility group DNA-binding protein HMGI(Y) (for literature review, see Nelms *et al.*, 1999) (Figure 4).

Other signaling molecules that have been shown to interact with IRS family members are the growth factor receptor-bound protein 2 (GRB2), the SH2 and SH3 domain containing adapter protein nck, the src-family kinase fyn, and the src-homology protein tyrosine phosphatase 2. The importance of these other interactions in mediating IL-4 responses is not clear. GRB2 is constitutively associated with the guanine nucleotide exchange factor called SOS whose primary function is to catalyze exchange of GDP bound to Ras with GTP. The GTP–Ras complex is active and results in the activation of the Raf kinase and MAP kinases. Although it has been clearly demonstrated that IL-4 induces the association of GRB2 with IRS, it has been more difficult to demonstrate any IL-4-induced activation of the Ras/Raf/MAP pathway.

IRS1 and IRS2 have been shown to signal similarly. However, there are some differences in the spectrum of proteins recruited to IRS1 and IRS2. The mechanism of these differences and their significance are not clear. Most cells of hematopoietic origin express IRS2, but not IRS1, whereas other cell types may express either one or both of these family members. Human thymocytes and peripheral T cells express both. It is possible that the activation of IRS1 or IRS2 by IL-4 may have subtle consequences for downstream signaling pathways. Very few studies have yet examined the function of IRS3 or IRS4 in IL-4 signaling. The role of the IRS family members in IL-4 signaling will probably be complex.

IL-4 signaling also results in the tyrosine phoshorylation of Shc. Shc contains a PTB domain and is recruited to the IL-4Rα by the I4R motif. The importance of Shc in IL-4 signaling is still unclear. In some cell types IL-4 does not induce the tyrosine phosphorylation of Shc. However, IL-4 is able to induce the tyrosine phosphorylation of Shc in mouse B cells. In addition, a cell line which is highly responsive to IL-4 shows strong Shc phosphorylation in response to IL-4. Shc acts as an adapter for GRB2/SOS and functions to activate the Ras pathway. An additional protein, FRIP-1, also binds to the I4R motif via its PTB domain. FRIP is tyrosine phosphorylated in response to IL-4, interacts with RasGAP, and therefore is thought to inactivate the Ras pathway (Nelms *et al.*, 1999). The delineation of the contribution of these molecules in mediating IL-4 responses will require further study.

The C-terminus of the IL-4Rα contains the fifth conserved cytoplasmic tyrosine residue (Figure 3), which lies in a consensus motif termed an ITIM (immunoreceptor tyrosine-based inhibitory motif). This sequence (Muta *et al.*, 1994) was identified in negative signaling receptors and has been shown to bind to the SH2 domains of the tyrosine and lipid phosphatases, SHP-1 and SHIP. While binding of SHP-1 to a phosphopeptide derived from the IL-4Rα ITIM sequence has been demonstrated, it is not yet clear whether SHP-1 or SHIP actually dock to this site of the IL-4Rα in cells.

DOWNSTREAM GENE ACTIVATION

Transcription factors activated

Two different types of transcription factors are activated by IL-4. STAT6 is directly recruited and activated by binding to phosphorylated tyrosine residues found in the cytoplasmic domain of the IL-4Rα. As discussed above, STAT6 becomes tyrosine phosphorylated itself, dimerizes, migrates to the nucleus, and binds to DNA sequences found in the promoter elements of IL-4-responsive genes (reviewed in Nelms *et al.*, 1999). IL-4 also activates the small, nonhistone chromosomal protein HMGI(Y) by inducing its serine phosphorylation in an IRS-dependent manner (Wang *et al.*, 1997). The serine phosphorylation of HMGI(Y) in response to IL-4 also depends on PI-3 kinase and p70S6 kinase.

Genes induced

Signaling by the IL-4 receptor complex results in the regulation of a number of genes. Many, but not all, are regulated by the activation of STAT6. IL-4 alone can induce/enhance the expression of CD23, MHC class II, and IL-4Rα. In the presence of some other stimulus such as LPS, TNF, or antigen receptor stimulation, IL-4 induces/enhances expression of the germline transcripts for the heavy chain of IgG1 (Gγ1) and IgE (Gε), VCAM-1, and Bcl-xL. In STAT6 knockout mice, IL-4 is not able to regulate expression of CD23, MHC class II, IL-4Rα, Gγ1 or Gε (Nelms *et al.*, 1999). However, induction of VCAM-1 by IL-4 is maintained (Kuperman *et al.*, 1998).

Promoter regions involved

The promoters of several IL-4-responsive genes have been characterized. These promoters (promoters for CD23, IL-4Rα, Gγ1, Gε) have in common the presence of a sequence motif specific for STAT6 binding termed as N4-GAS (TTCXXXXGAA). Interestingly, the promoter for IL-4 also contains a STAT6 consensus-binding site although it is not thought to play a major role in the induction of transcription of IL-4 *in vivo*. Although the IL-4-activated STAT6 is clearly important in the regulation of transcription of these genes, other transcription factors cooperate to fully regulate expression. In the case of the promoter for Gε, the factors STAT6, NFκB, cEBP, and BSAP cooperate to positively regulate transcription (Delphin and Stavnezer, 1995). Additionally, the IL-4-induced serine phosphorylation of HMGI(Y) results in the de-repression of the Gε promoter, resulting in the enhancement of transcription (Kim *et al.*, 1995).

BIOLOGICAL CONSEQUENCES OF ACTIVATING OR INHIBITING RECEPTOR AND PATHOPHYSIOLOGY

Unique biological effects of activating the receptors

Unique biological effects of activating the IL-4 receptor are similar to the unique responses to IL-4. These include the differentiation of T cells to the TH2 type and immunoglobulin class switching to IgE. In addition, activating the receptor by administration of IL-4 or IL-13 directly into the lungs elicits symptoms of *asthma* including eosinophilia, mucus production, and airway hyperresponsiveness (Grunig *et al.*, 1998). There is evidence to support the notion that signaling through the IL-4Rα chain by IL-4 or IL-13 can directly induce mucus secretion by goblet cells and eotaxin release by lung epithelial cells (Dabbagh *et al.*, 1999; Li *et al.*, 1999).

Phenotypes of receptor knockouts and receptor overexpression mice

In general terms, the IL-4Rα knockout mice are similar to the IL-4 knockout mice. They are TH2 deficient and have reduced IgE responses (Noben-Trauth *et al.*, 1997). However, the IL-4Rα knockout mice are unable to expel the gastrointestinal nematode parasite *Nippostrongylus brasiliensis*, while this parasite is expelled in IL-4 knockout mice (Barner *et al.*, 1998; Urban *et al.*, 1998). These results indicate that IL-13 signaling through the IL-4Rα may be critical for nematode expulsion. Interestingly, IL-4Rα knockout mice have greatly reduced *asthma*-like symptoms in response to allergen, even when TH2 cells are provided exogenously (Grunig *et al.*, 1998). These results suggest that IL-4Rα on lung-derived cells other than T cells is important in the development of *asthma* pathology. There are no IL-4Rα overexpressing mice.

Human abnormalities

The important role that IL-4 and IL-13 play in the regulation of T cell responses and IgE production logically suggest that the IL-4Rα may play a role in human disease. This idea is supported by the finding that the region of human chromosome 16 containing the IL-4Rα is associated with human *allergy*

(Deichmann *et al.*, 1998). There are naturally and frequently occurring polymorphisms in both the human and mouse IL-4Rα (Schulte *et al.*, 1997; Hershey *et al.*, 1997; Mitsuyasu *et al.*, 1998). Several of these polymorphisms have been associated with *atopy* in humans, although there is controversy in this field. One group has found an association between a Val50 to Ile change in the extracellular domain of the IL-4Rα in their patient population (Mitsuyasu *et al.*, 1998). An association between a Gln576 to Arg change (Figure 3) and allergy in US patients has been found (Hershey *et al.*, 1997), but other studies did not find a correlation between this polymorphism and *allergy* in Japanese and European patient populations (Mitsuyasu *et al.*, 1999; Noguchi *et al.*, 1999). Analysis of the human IL-4Rα bearing the Arg576 showed that there is no direct effect of this change on IL-4-mediated signaling in transfected cell lines (Mitsuyasu *et al.*, 1999; Wang *et al.*, 1999). The role that these polymorphisms play in IL-4 signaling and allergic disease in diverse patient populations will need to be examined closely.

THERAPEUTIC UTILITY

Effect of treatment with soluble receptor domain

Phase I/II clinical trial results have been reported by Immunex Research and Development Corporation (www.immunex.com) on the use of the human soluble IL-4Rα (NUVANCE™) for the treatment of *asthma*. Preliminary studies indicate that the administration of nebulized sIL-4Rα is well tolerated and results in the reduction in use of other medication by asthma patients. Phase II clinical trials will begin in 1999 to test the safety and efficacy of NUVANCE™ for the long-term control of asthma. Treatment of peripheral blood mononuclear cells (PBMCs) from patients with *hyper-IgE syndrome* with soluble IL-4Rα blocks the spontaneous release of IgE and IgG4 *in vitro* (Garraoud *et al.*, 1999), suggesting a new therapeutic target for sIL-4Rα.

Effects of inhibitors (antibodies) to receptors

Treatment with anti-IL-4Rα antibodies that block IL-4 binding have effects similar to treatment with soluble IL-4Rα. It completely suppresses IgE and IgG4 production by peripheral blood mononuclear cells isolated from *hyper-IgE syndrome* patients (Garraoud *et al.*, 1999). In mouse models, treatment with the anti-IL-4Rα monoclonal (M1) has many of the same effects as anti-IL-4 antibody treatment or as those observed in the IL-4 knockout mice. These include inhibition of TH2 differentiation and IgE production. In some cases, injecting mice with IL-4–M1 complexes increases the bioavailability of IL-4 since it increases the serum half-life.

References

Barner, M., Mohrs, M., Brombacher, F., and Kopf, M. (1998). Differences between IL-4R alpha-deficient and IL-4-deficient mice reveal a role for IL-13 in the regulation of Th2 responses. *Curr. Biol.* **8**, 669–672.

Callard, R. E., Matthews, D. J., and Hibbert, L. (1996). IL-4 and IL-13 receptors: are they one and the same? *Immunol. Today* **17**, 108–110.

Dabbagh, K., Takeyama, K., Lee, H., Ueki, I. F., Lausier, J. A., and Nadel, J. A. (1999). IL-4 induces mucin gene expression and goblet cell metaplasia *in vitro* and *in vivo*. *J. Immunol.* **162**, 6233–6237.

Deichmann, K. A., Henzmann, A., Forster, J., Dischinger, S., Mehl, E., Brueggenolte, E., Hildebrandt, M., Moseler, M., and Kuehr, J. (1998). Linkage and allelic association of atopy and markers flanking the IL-4 receptor gene. *Clin. Exp. Allergy* **28**, 151–153.

Delphin, S., and Stavnezer, J. (1995). Characterization of an IL-4 responsive region in the immunoglobulin heavy chain germline epsilon promoter: regulation by NF-IL-4, a C/EBP family member and NF-kappa B/p50. *J. Exp. Med.* **181**, 181–192.

Doucet, C., Brouty-Boye, D., Pottin-Clemenceau, C., Canonica, G. W., Jasmin, C., and Azzarone, B. (1998). IL-4 and IL-13 act on human lung fibroblasts. Implications in asthma. *J. Clin. Invest.* **101**, 2129–2139.

Dubois, G. R., Schweizer, R. C., Versluis, C., Bruijnzeel-Koomen, C. A., and Bruijnzeel, P. L. (1998). Human eosinophils constitutively express a functional IL-4 receptor: IL-4-induced priming of chemotactic responses and induction of PI-3 kinase activity. *Am. J. Respir. Cell. Mol. Biol.* **19**, 691–699.

Fujiwara, H., Hanissian, S. H., Tsytsykova, A., and Geha, R. S. (1997). Homodimerization of the human IL-4 receptor alpha chain induced Cepsilon germline transcripts in B cells in the absence of the IL-2 receptor gamma chain. *Proc. Natl Acad. Sci. USA* **94**, 5866–58671.

Garraoud, O., Mollis, S. N., Holland, S. M., Sneller, M. C., Malech, H. L., Gallin, J. I., and Nutman, T. B. (1999). Regulation of immunoglobulin production in hyper-IgE (Job's) syndrome. *J. Allergy Clin. Immunol.* **103**, 333–340.

Grunig, G., Warnock, M., Wakil, A. E., Venkayya, R., Brombacher, F., Rennick, D. M., Sheppard, D., Mohrs, M., Donaldson, D. D., Locksley, R. M., and Corry, D. B. (1998). Requirement for IL-13 independently of IL-4 in experimental asthma. *Science* **282**, 2261–2263.

Hage, T., Sebald, W., and Reinemer, P. (1999). Crystal structure of the IL-4/receptor α chain complex reveals a mosaic binding interface. *Cell* **97**, 271–281.

Henriques, C. U., Rice, G. E., Wong, M. H., and Bendtzen, K. (1998). Immunolocalisation of IL-4 and IL-4 receptor in placenta and fetal membranes in association with pre-term labour and pre-eclampsia. *Gynecol. Obstet. Invest.* **46**, 172–177.

Hershey, G. K. K., Freiderich, L. A., Esswein, L. A., Thomas, M. L., and Chatilia, T. A. (1997). The association of atopy with a gain-of-function mutation in the α-subunit of the IL-4 receptor. *N. Engl. J. Med.* **337**, 1720–1723.

Hilton, D. J., Zhang, J. G., Metcalf, D., Alexander, W. S., Nicola, N. A., and Willson, T. A. (1996). Cloning and characterization of a binding subunit of the IL-13 receptor that is also a component of the IL-4 receptor. *Proc. Natl Acad. Sci. USA* **93**, 497–501.

Huang, H., Niblack, E., Shultz, L., and Paul, W. E. (1999). Basal and IL-4-induced IL-4 receptor expression is controlled by PTPases. *FASEB J.* **13**, A1144.

Ihle, J. N. (1995). Cytokine receptor signaling. *Nature* **377**, 591–594.

Johnston, J. A., Bacon, C. M., Riedy, M. C., and O'Shea, J. J. (1996). Signaling by IL-2 and related cytokines: JAKs, STATs, and relationship to immunodeficiency. *J. Leukoc. Biol.* **60**, 441–452.

Keegan, A. D., Ryan, J. J., and Paul, W. E. (1996). IL-4 regulates growth and differentiation by distinct mechanisms. *The Immunologist* **4**, 194–198.

Kim, J., Reeves, R., Rothman, P., and Boothby, M. (1995). The non-histone chromosomal protein HMG-I(Y) contributes to repression of the immunoglobulin heavy chain germ-line epsilon RNA promoter. *Eur. J. Immunol.* **25**, 798–808.

Kotsimbos, T. C., Ghaffar, O., Minshall, E. M., Humbert, M., Durham, S. R., Pfister, R., Menz, G., Kay, A. B., and Hamid, Q. A. (1998). Expression of the IL-4 receptor alpha-subunit is increased in bronchial biopsy specimens from atopic and nonatopic asthmatic subjects. *J. Allergy Clin. Immunol.* **102**, 859–866.

Kuperman, D., Schofield, B., Wills-Karp, M., and Grusby, M. J. (1998). STAT6-deficient mice are protected from antigen-induced airway hyperresponsiveness and mucus production. *J. Exp. Med.* **187**, 939–945.

Li, L., Xia, Y., Nguyen, A., Lai, Y. H., Feng, L., Mosmann, T. R., and Lo, D. (1999). Effects of Th2 cytokines on chemokine expression in the lung: IL-13 potently induces eotaxin expression by airway epithelial cells. *J. Immunol.* **162**, 2477–2487.

Lowenthal, J. W., Castle, B. E., Christiansen, J., Schreurs, J., Rennick, D., Arai, N., Hoy, P., Takebe, Y., and Howard, M. (1988). Expression of high affinity receptors for murine IL-4 on hemopoietic and non-hemopoietic cells. *J. Immunol.* **140**, 456–464.

Mehrotra, R., Varricchio, F., Husain, S. R., and Puri, R. K. (1998). Head and neck cancers, but not benign lesions, express IL-4 receptors in situ. *Oncol. Rep.* **5**, 45–48.

Mitsuyasu, H., Izuhara, K., Mao, X., Gao, P., Arinobu, Y., Enomoto, T., Kawai, M., Sasaki, S., Dake, Y., Hamasaki, N., Shirakawa, T., and Hopkin, J. M. (1998). Ile50Val variant of the IL-4Rα upregulated IgE synthesis and associates with atopic asthma. *Nature Genet.* **19**, 119–121.

Mitsuyasu, H., Yanagihara, Y., Mao, Y., Gao, P., Arinobu, Y., Ihara, K., Takabauashi, A., Hara, T., Enomoto, T., and Sasaki, S. (1999). Cutting Edge: Dominant effect of Ile50Val variant of the human IL-4 receptor α chain in IgE synthesis. *J. Immunol.* **162**, 1227–1231.

Murata, T., Obiri, N. I., and Puri, R. K. (1998). Structure of and signal transduction through IL-4 and IL-13 receptors. *Int. J. Mol. Med.* **1**, 551–557.

Muta, T., Kurosaki, T., Misulovin, Z., Sanchez, M., Nussenzweig, M. C., and Ravetch, J. V. (1994). A 13-amino acid motif in the cytoplasmic domain of the Fcgamma RIIB modulates B-cell receptor signaling. *Nature* **269**, 340–341.

Nelms, K., Keegan, A. D., Zamorano, J., Ryan, J. J., and Paul, W. E. (1999). The IL-4 receptor: signaling mechanisms and biologic functions. *Annu. Rev. Immunol.* **17**, 701–738.

Noben-Trauth, N., Shultz, L. D., Brombacher, F., Urban, J. F. Jr., Gu, H., and Paul, W. E. (1997). An IL-4-independent pathway for CD4+ T cell IL-4 production is revealed in IL-4 receptor-deficient mice. *Proc. Natl Acad. Sci. USA* **94**, 10838–10843.

Noguchi, E., Shibasaki, M., Arinami, T., Takeda, K., Yokouchi, Y., Kobayashi, K., Imoto, N., Nakahara, S., Matsui, A., and Hamaguchi, H. (1999). Lack of association of atopy/asthma and the IL-4 receptor alpha gene in Japanese. *Clin. Exp. Allergy* **29**, 228–233.

Ohara, J., and Paul, W. E. (1987). Receptors for B-cell stimulatory factor-1 expressed on cells of haematopoietic lineage. *Nature* **325**, 537–540.

Ohara, J., and Paul, W. E. (1988). Up-regulation of IL-4/BSF-1 receptor expression. *Proc. Natl Acad. Sci. USA* **85**, 8221–8225.

Russell, S. M., Keegan, A. D., Harada, N., Nakamura, Y., Noguchi, M., Leland, P., Friedmann, M. C., Miyajima, A., Puri, R. K., Paul, W. E., and Leonard, W. J. (1993). The interleukin-2 receptor γ chain is a functional component of the interleukin-4 receptor. *Science* **262**, 1880–1883.

Schulte, T., Roellinghoff, M., and Gessner, A. (1997). Molecular characterization and functional analysis of murine IL-4 receptor allotypes. *J. Exp. Med.* **186**, 1419–1429.

Urban, J.F. Jr., Noben-Trauth, N., Donaldson, D. D., Madden, K. B., Morris, S. C., Collins, M., and Finkelman, F. D. (1998). IL-13, IL-4Ralpha, and STAT6 are required for the expulsion of the gastrointestinal nematode parasite *Nippostrongylus brasiliensus*. *Immunity* **8**, 255–264.

van der Velden, V. H., Naber, B. A., Wierenga-Wolf, A. F., Debets, R., Savelkoul, H. F., Overbeek, S. E., Hoogsteden, H. C., and Versnel, M. A. (1998). IL-4 receptors on human bronchial epithelial cells. An *in vivo* and *in vitro* analysis of expression and function. *Cytokine* **10**, 803–813.

Wang, D., Zamorano, J., Keegan, A. D., and Boothby, M. (1997). HMGI(Y) phosphorylation status as a nuclear target regulated through IRS1 and the I4R-motif of the IL-4 receptor. *J. Biol. Chem.* **272**, 25083–25090.

Wang, H. Y., Shelburne, C. P., Zamorano, J., Kelly, A. E., Ryan, J. J., and Keegan, A, D. (1999). Cutting Edge: Effects of an allergy-associated mutation in the human IL-4Rα (Q576R) on human IL-4-induced signal transduction. *J. Immunol.* **162**, 4385–4389.

Wang, L-M., Keegan, A. D., Frankel, M., Paul, W. E., and Pierce, J.H. (1995). Signal transduction through the IL-4 and insulin receptor families. *Stem Cells* **13**, 360–368.

White, M. F. (1998). The IRS-signaling system: a network of docking proteins that mediate insulin and cytokine action. *Recent Prog. Horm. Res.* **53**, 119–138.

Wright, E. D., Christodoulopoulos, P., Small, P., Frenkiel, S., and Hamid, Q. (1999). Th2 cytokine receptors in allergic rhinitis and in response to topical steroids. *Laryngoscope* **109**, 551–556.

Wrighton, N., Campbell, L. A., Harada, N., Miyajima, A., and Lee, F. (1992). The murine IL-4 receptor gene: genomic structure, expression, and potential for alternative splicing. *Growth Factors* **6**, 103–118.

Zhou, M. M., Huang, B., Olejniczak, E. T., Meadows, R. P., Shuker, S. B., Miyazaki, M., Trub, T., Shoelson, S. E., and Fesik, S. W. (1996). Structural basis for IL-4 receptor phosphopeptide recognition by the IRS-1 PTB domain. *Nature Struct. Biol.* **3**, 388–393.

LICENSED PRODUCTS

R & D Products

Fluorokine Cytokine Receptor Kit for human IL-4Rα

Monoclonal antibody to human and mouse IL-4Rα

Polyclonal antibody to human IL-4Rα

Soluble human IL-4Rα

ELISA kit for soluble human IL-4Rα

IL-7 Receptor

Hergen Spits*

Division of Immunology, Netherlands Cancer Institute, Plesmanlaan 121, 1066 CX, Amsterdam, The Netherlands

*corresponding author tel: 31-20-5122063, fax: 31-20-5122057, e-mail: hergen@nki.nl

DOI: 10.1006/rwcy.2000.14003.

SUMMARY

The presently accepted configuration of the functional interleukin 7 (IL-7) receptor is one of a two-chain receptor consisting of an α chain (IL-7Rα) and a γ chain (γ common, γc). The latter polypeptide is denoted γ common because it is also a component of the receptors for IL-2, IL-4, IL-9, and IL-15. The IL-7Rα chain is a member of the cytokine receptor superfamily.

The IL-7Rα chain plays an essential, nonredundant role in T and B cell development in the mouse. In addition, the IL-7R is involved in survival and proliferation of functional peripheral T cells. The IL-7Rα chain is also essential for the formation of Peyer's patches in the mouse. In human the IL-7Rα chain appears to be critical for T but not for B cell development, as two IL-7R-deficient *severe combined immunodeficiency* patients presented with B and NK cells but lacked T cells. The receptor has at least two ligands: IL-7 and TSLP-1. The latter cytokine interacts with a complex of the IL-7Rα chain and an as yet to be defined receptor distinct from γc.

BACKGROUND

Discovery

The cloning of the IL-7Rα chain was reported in 1990. cDNA clones encoding the murine (IL-7) receptor were isolated from an IL-7-dependent pre-B cell line (Park *et al.*, 1990) and that for human IL-7Rα from an SV40-transformed lung cell line (Goodwin *et al.*, 1990). IL-7R cDNAs expressed in COS-7 cells bound radiolabeled IL-7, producing curvilinear Scatchard plots containing high- and low-affinity classes. The functional receptor for IL-7 is a complex of IL-Rα and γc. The discovery of this latter receptor component is described in the chapter on the IL-2 receptor.

Main activities and pathophysiological roles

The IL-7Rα chain plays an essential, nonredundant role in T and B cell development in the mouse (Peschon *et al.*, 1994) (see also for reviews, Akashi *et al.*, 1998; DiSanto and Rodewald, 1998; Maeurer *et al.*, 1998). In addition, the IL-7R plays a role in survival and proliferation of functional peripheral T cells (Maraskovsky *et al.*, 1996). The IL-7Rα chain is also essential for the formation of Peyer's patches in the mouse (Adachi *et al.*, 1998). In humans, the IL-7Rα chain appears to be critical for T but not for B cell development as two IL-7R-deficient *severe combined immunodeficiency* patients presented with B and NK cells but lacked T cells (Puel *et al.*, 1998).

GENE

Accession numbers

Human gene: AF043123-9, AF043124-9, AF043125-9, AF043126-9, AF043127-9, AF043128-9 (these accession numbers denote the different exons)
Human cDNA: NPI_002185
Mouse cDNA: M29697

Chromosome location and linkages

Human IL-7Rα maps to chromosome 5p13 (Lynch *et al.*, 1992). The chromosomal localization of murine IL-7Rα is unknown. Human and mouse genes both contain eight exons and seven introns with a total size of 19 kb in the human and 24 kb in the mouse. The first exon contains the 5′ UTR, signal peptide, and the N-terminus of the mature protein. The remainder of the extracellular region is encoded by exons 2 to 6. In addition, exon 6 also encodes the transmembrane part of the cytoplasmic region and exons 7 and 8 encode the rest of the cytoplasmic region. The entire 3′ UTR is encoded by exon 8. Differential splicing results in mRNA encoding a secreted form of the human IL-7Rα chain. This form lacks the sequences in exon 6 that encode the transmembrane region.

Inspection of the entire 5′ region up to the start of translation of the murine IL-7Rα revealed a TATA box and a CAAT sequence. Furthermore an AP-1 and an AP-2 sequence were present. Interestingly an interferon response element (IRE)-like sequence was found. The 5′ region contains several consensus binding sites for the glucocorticoid receptor. The region −2495 to +5 contains promotor sequences that are active in a murine pre-B cell line (Pleiman *et al.*, 1991).

PROTEIN

Accession numbers

Human: NP_002176
Mouse: AAA39304

Description of protein

Murine IL-7Rα gene encodes a protein of 439 amino acids with a calculated molecular weight of 49.5 kDa. On T cells the protein has a molecular weight of 90 kDa. The associated γc chain has a molecular weight of 74 kDa. IL-7Rα is a type I membrane protein with a single transmembrane domain. The extracellular domain contains features of the cytokine receptor superfamily. The cytoplasmic domain with 195 amino acids does not contain consensus sequences of protein kinases. The crystal structure of the IL-7Rα has not been determined. Based on the model of the prolactin receptor, intrachain disulfide bridges are most probably formed between the first and second and between the third and fourth cysteine residues. The receptor contains a WS motif (WS × WS) at the C-terminus of the extracellular domain.

Relevant homologies and species differences

The IL-7Rα protein is a member of the cytokine receptor superfamily which is characterized by conserved extracellular domains fit to bind helical cytokines. IL-7Rα is similar to other receptors of the γc family including γc, IL-2Rβ, IL-4Rα, and IL-9Rα. These are characterized by two fibronectin type III-like domains containing four cysteine residues in the N-terminus domain and a WS motif at the C-terminus. Human and mouse IL-7Rα are not species-specific as both murine and human IL-7 interact with human T cells and vice versa.

Affinity for ligand(s)

Reconstitution experiments with human IL-7Rα and γc revealed three classes of affinity: an uncharacterized low-affinity complex (K_d = 145 nM) and complexes of intermediate (K_d = 250 pM) and high (K_d = 40 pM) affinity. The latter two consist of the IL-7Rα alone and complexes of IL-7Rα and γc, respectively. High-affinity (79 pM) and low-affinity (16 nM) IL-7R were identified on the murine stromal cell line IxN/2b. Treatment of these cells with anti-γc antibody reduced the affinity to 255 pM without affecting the low-affinity receptor and the combination of anti-γc plus anti IL-7Rα eliminated the high-affinity receptor altogether while the low-affinity binding was not affected. These findings suggest the existence of a γc-independent IL-7Rα-containing IL-7 receptor complex (Sugamura *et al.*, 1996). The nature of the low-affinity complex is unknown but may involve a yet to be identified receptor component. The IL-7Rα chain is also part of the receptor for murine thymus-derived lymphopoietin 1 (TSLP-1). In the functional TSLP-1R complex IL-7Rα does not seem to pair with γc but with another yet to be described receptor (Levin *et al.*, 1999). The affinity of this complex for TSLP-1 is unknown.

Cell types and tissues expressing the receptor

See **Table 1** (for references in reviews, see Akashi *et al.*, 1998; DiSanto and Rodewald, 1998; Maeurer *et al.*, 1998).

Release of soluble receptors

cDNA clones have been isolated that potentially encode a soluble receptor. These clones lack the sequences in exon 6 encoding the transmembrane domain (Pleiman *et al.*, 1991). It has not been

Table 1 Cell types and tissues that express the IL-7 receptor

Fetal NK cell/dendritic cell precursors in fetal lymph nodes
Common T/NK/B lymphoid cell precursors
Cryptopatch-associated lymphoid precursors
Developing T cells
All subsets of CD3− CD4− CD8− thymocytes
A fraction of CD4+ CD8+ DP cells
SP CD4+ or CD8+ thymocytes
TCR$\gamma\delta$ cells
Thymic NK-1.1+ T cells
Developing B cells
All pre-pro, pro-, and pre-B cell stages
Mature T cells
Bone marrow-derived macrophages
Malignant cell types
Colorectal cancer cells
Renal cancer cells
Cutaneous T cell lymphomas
Human intestinal epithelial cells

reported whether a soluble receptor protein is indeed secreted by cells expressing it. The function of the soluble receptor is unknown.

SIGNAL TRANSDUCTION

Associated or intrinsic kinases

The IL-7R complex requires both the α chain and γc for relaying signals. These two chains do not have intrinsic catalytic activity, but associate with tyrosine phosphokinases that are instrumental in transducing signals triggered by binding of IL-7 to the receptor complex. The src kinase p59fyn has been shown to associate with the receptor complex following activation with IL-7 in a human pre-B cell line (Seckinger and Fougereau, 1994). Association of p59fyn with the receptor was also found in T cells (Venkitaraman and Cowling, 1992). In both T cells and thymocytes p56lck has been shown to associate with IL-7Rα. The role of p56lck and p59fyn in IL-7 signaling is unknown. Fyn is certainly not essential in IL-7R signaling since Fyn−/− mice have no reduction in the size of the thymus and numbers of T and B cells in the periphery are normal (Appleby *et al.*, 1992), in contrast to IL-7Rα−/−mice. In addition, the phenotype of p56lckα−/− mice is distinct from that of IL-7R−/− mice (Molina *et al.*, 1992). Thus, these src kinases do not seem to be important in IL-7R signaling. However, an attenuating role of these enzymes in certain aspects of IL-7Rα signaling cannot be excluded.

Tyrosine kinases belonging to the family of Janus kinases are essential for IL-7R signaling. Two Janus kinases, JAK1 and JAK3, associate with the complex. JAK1 binds to IL-7Rα and JAK3 to γc (Leonard and O'Shea, 1998). Both kinases become phosphorylated when IL-7 binds to its receptor, but the extent of phosphorylation of JAK3 is much higher than that of JAK1. JAK3−/− mice have a phenotype similar to that of IL-7Rα−/− mice (Nosaka *et al.*, 1995; Thomis *et al.*, 1995), strongly suggesting that JAK3 activity is required for IL-7R-mediated signaling during T and B cell development. In contrast to JAK3, which is specifically expressed in lymphoid cells, JAK1 is ubiquitously expressed. Nonetheless, analysis of JAK1−/− mice revealed a dedicated role of JAK1 in T and B cell development (Rodig *et al.*, 1998). These mice were born but all pups died within 24 hours after birth. The thymus of newborn JAK1−/− mice was 260-fold reduced in size as compared with heterozygous controls. CD4+ CD8+ double positive (DP) cells were present, but the CD4− CD8− double negative (DN) compartment was increased. No further analysis was performed and it is therefore unknown whether the DN cells in the thymus of these mice were stromal cells or T cell precursors. B cells were strongly reduced in numbers. Importantly, whereas wild-type fetal liver cells form colonies in IL-3 and IL-7, JAK1−/− fetal liver cells formed colonies in IL-3 but not in IL-7 (Rodig *et al.*, 1998). These findings demonstrate that JAK1 is an essential constituent of IL-7R signaling.

Another kinase that associates with the IL-7R complex is phosphatidyl 3-kinase (PI-3 kinase). JAK3 appears also to control PI-3 kinase activity (Sharfe *et al.*, 1995). PI-3 kinase phosphorylates the D3 position of the inositol group of phosphoinositide lipids to generate phosphatidylinositol-3 phosphate (PtdIns(3)P), PtdIns(3,4)P_2 and PtdIns(3,4)P_3 (Kapeller and Cantley, 1994). The latter two products are important for cellular proliferation. The enzyme consists of a 85 kDa adapter and a 110 kDa catalytic unit. Upon ligand binding the p85 unit is tyrosine phosphorylated and associates with many receptors, including the IL-7Rα chain (Venkitaraman and Cowling, 1994). Stimulation of human thymocytes with IL-7 was shown to activate both isoforms of p85, α and β, in human thymocytes (Dadi *et al.*, 1994). Interestingly, phosphorylation of p85 required JAK3, which was shown to associate with p85 (Sharfe *et al.*, 1995). In a later study these authors presented evidence that two pools of PI-3 kinase are activated by IL-7: one is associated with the IL-7Rα chain and another with insulin receptor substrate 1 (IRS-1) and IRS-2 (Sharfe and Roifman, 1997). PI-3 kinase appears to be important for survival/proliferation of both T and B cell precursors. Using human IL-7Rα mutants transfected into murine pro-B cell lines, Venkitaraman and Cowling demonstrated association of PI-3 kinase with the IL-7Rα chain through an SH2 domain recognition motif (YXXM) spanning residue Tyr449 in its cytoplasmic domain (Venkitaraman and Cowling, 1994). Interestingly, this residue was found to be critical for IL-7-mediated proliferation of murine pre-B cells. Induction of IgH VDJ gene recombination, however, did not require Tyr449 residue (Corcoran *et al.*, 1996). These findings indicate that IL-7Rα-transduced signals can control both survival and/or proliferation and differentiation through distinct pathways. More recently it was demonstrated that PI-3 kinase also mediates survival and proliferation of human T cell precursors. Pallard *et al.* (1999) observed that the Tyr449 in the cytoplasmic domain of the IL-7R is important for survival and expansion of human thymocytes. A dominant negative mutant of p85 that binds to the receptor but fails to interact with p100, was introduced into human T cell precursors by retrovirus-mediated gene transfer. This mutant was found to inhibit survival and expansion of the T cell precursors in a fetal thymic organ culture but did not inhibit differentiation of these cells. The study of Pallard coworkers left unresolved which isoform, p85α or β, is recruited by IL-7 binding. These two isoforms are encoded by different genes. Two recent studies failed to provide evidence for a role of p85α in T cell development as the size of the thymus and the distribution of thymocyte subsets was unaffected in p85α−/− mice (Fruman *et al.*, 1999; Suzuki *et al.*, 1999). The observation that the level of p85β in these mice is elevated compared with wild-type mice strongly suggests a compensatory role of p85β in T cell development of p85α−/− mice.

Cytoplasmic signaling cascades

Recently it was demonstrated that protein kinase B (PKB) can be activated by IL-7 (Pallard *et al.*, 1999). PKB seems to play a central role in PI-3 kinase-mediated protection against apoptosis in a wide range of cell types (Coffer *et al.*, 1998). As activation of PKB as well as of PI-3 kinase requires residue Tyr449, it is very likely that in human T cell precursors PKB is a downstream effector of PI-3 kinase as well (Pallard *et al.*, 1999). The substrates of IL-7-activated PKB are not yet known.

Two sets of observations suggest a role for the p38 MAP kinase in IL-7-mediated proliferation of mature T cells. In the first place activation of a murine T cell line by IL-7 resulted in phosphorylation of p38 MAP kinase and secondly the proliferative response of human T cells was inhibited by the highly selective p38 MAP kinase inhibitor SB203580 (Crawley *et al.*, 1997). It is unknown which signaling pathway activates p38 MAP kinase; p21ras is not activated by IL-7 and therefore does not act in this pathway. The same report also documented activation of the stress-activated protein kinase (SAP kinase)/Jun-N-terminal kinase (JUN kinase) by IL-7, but there is no evidence that activation of this enzyme is important for the proliferative response induced by IL-7 (Crawley *et al.*, 1997). The p38 inhibitor SB203580 influenced negative selection through the TCR$\alpha\beta$ in a fetal thymic organ culture (FTOC) but did not affect thymic cellularity (Sugawara *et al.*, 1998). Since blocking the IL-7/IL-7R interaction resulted in a strong reduction of thymic cellularity and SB203580 did not, it is unlikely that p38 MAP kinase is involved in IL-7-mediated control of early T cell development in the thymus.

A potential constituent of the IL-7R signaling is the serine/threonine kinase pim-1 as pro/pre B cells of pim-1−/− mice do not proliferate in response to IL-7 (Domen *et al.*, 1993). Evidence for a direct link between pim-1 and the IL-7R complex is lacking, however, since the B and T cell phenotype of pim-1-deficient mice is comparable to that of wild-type mice. It is possible that the related kinases pim-2 and pim-3 compensate the pim-1 deficiency, but this has yet to be verified.

DOWNSTREAM GENE ACTIVATION

Transcription factors activated

In general JAKs phosphorylate the receptor chains, providing docking sites for SH2 domains of STATs (Leonard and O'Shea, 1998). STATs are recruited to the phosphorylated sites of the receptor and become phosphorylated. The phosphorylated STATs can dimerize, translocate to the nucleus and stimulate expression of cytokine-inducible genes. Two studies documented activation of STAT1 by IL-7 (Zeng *et al.*, 1994; van der Plas *et al.*, 1996). The phenotype of

STAT1-deficient mice, however, makes it unlikely that STAT1 plays an essential role in IL-7R signaling *in vivo* as the thymus of these mice is normal and T and B cell development proceeds undisturbed (Meraz *et al.*, 1996). STAT3 and STAT5 have been shown to be activated by IL-7 in several cell types (Leonard and O'Shea, 1998). It is unlikely that STAT3 is involved in IL-7R signaling *in vivo*. STAT3 deficiency results in embryonic lethality preventing analysis of the role of STAT3 in T cell development in conventional STAT3−/− mice. Recently, mice were generated with a STAT3 deficiency specifically in the T cell lineage by conditional gene targeting using the Cre-loxP system (Takeda *et al.*, 1998). Floxed-STAT3 mice were mated with transgenic mice with Cre recombinase under control of the T cell-specific Lck promoter. Although STAT3 was not expressed in the thymus of these mice, the cellularity of the thymus and distribution of thymic subsets was the same as in control mice (Takeda *et al.*, 1998). Thymocytes of these mice responded normally to IL-7 *in vitro*. Together, these findings indicate that STAT3 is not involved in IL-7-mediated control of T cell development.

IL-7 activates STAT5 in human PBMCs and thymocytes (Lin *et al.*, 1995; Pallard *et al.*, 1999). STAT5 was originally identified as a prolactin-induced mammary gland transcription factor. Two STAT5 genes, 5a and 5b, encode proteins that are approximately 95% identical in amino acid sequence. STAT5a and STAT5b differ in their C-terminal transactivation domains and exert relatively specific actions. STAT5-deficient mice display immunological defects. STAT5a−/− mice have a reduced expansion of peripheral T cells which is associated with a diminished IL-2-mediated induction of the IL-2Rα chain. STAT5b-deficient mice exhibit a reduced expansion of peripheral NK cells and a slightly reduced cellularity of the thymus (Imada *et al.*, 1998). In contrast, mice deficient for both STAT5a and STAT5b were reported to have no major decrease in thymic cellularity (Moriggl *et al.*, 1999). In addition, the distributions of CD4+, CD8+SP and DP cells were normal, although the number of peripheral T cells reduced progressively when these mice aged (Moriggl *et al.*, 1999). The mechanism of this reduction is not clear but it is likely that an inability of STAT5-deficient T cells to respond to IL-2 is a contributing factor. The reported findings on the phenotype of mice deficient for either STAT5a and STAT5b or both argues against a role for STAT5 isoforms in IL-7-mediated control of T cell development. However, Pallard *et al.* (1999) reported that overexpression of dominant negative mutant of STAT5b in human T cell precursors disrupted their development into mature T cells in an *in vitro* FTOC. The mutant used by Pallard and coworkers lacks the C-terminal transactivating domain and inhibits transactivation of wild-type STAT5 and one would expect that overexpression of dominant negative STAT5b has the same effect as STAT5 deficiency.

There is presently no explanation for these discordant effects of the absence of STAT5a and STAT5b and of overexpression of dominant negative STAT5b on T cell development. It is possible that there are species differences in the requirement of STAT5 in murine and human T cell development. Another possibility is that the effect of dominant negative STAT5b on development of human T cell precursors is a consequence of a gain of function of the mutant resulting in activities beyond those of inhibiting transcription mediated by full-length wild-type STAT5a or STAT5b.

Mel-18 is a mammalian homolog of *Drosophila melanogaster* Polycomb group genes. Mice deficient for this gene show a *severe combined immunodeficiency* phenotype with a strongly reduced thymic size (Akasaka *et al.*, 1997). Thymocytes of these mice fail to respond to IL-7, suggesting an involvement of mel-18 in cell cycle progression in response to the interaction of IL-7 and the IL-7R complex.

Genes induced

Presently there are no genes known that are proven to be direct targets of IL-7R-signaling.

BIOLOGICAL CONSEQUENCES OF ACTIVATING OR INHIBITING RECEPTOR AND PATHOPHYSIOLOGY

Unique biological effects of activating the receptors

The effects of activation of the IL-7R with IL-7 are discussed in the chapter on IL-7. Disrupting IL-7/IL-7R interactions also affect human T cell development. Antibodies to the human IL-7Rα strongly inhibited development of fetal liver and thymic T cell precursors in a hybrid human/mouse fetal thymic organ culture (Plum *et al.*, 1996; Pallard *et al.*, 1999). Development of the human T cell precursors was arrested at the CD3− CD4− CD8− stage.

Phenotypes of receptor knockouts and receptor overexpression mice

IL-7Rα-deficient mice were reported by Peschon *et al.* (1994). These mice had severely reduced numbers of T and B cells. NK cell numbers were not affected by the IL-7Rα deficiency and also development of myeloid cells was normal. Another IL-7Rα−/−mouse was generated by Ikuta and collaborators (Maki *et al.*, 1996). In these mice exon 2 was targeted rather than the exon 3 in the mice generated by Peschon and coworkers. The phenotype of these latter mice was similar to that of exon 3-targeted mice although some differences were noted.

Peschon *et al.* (1998) reported that the T cell deficiency in IL-7R(exon 3)−/− mice was variable. While all IL-7Rα-deficient mice demonstrated a reduction in thymic cellularity, the level of the reduction varied. In around 65% of the mice the size of the thymus was only 0.1% of the wild type, while the thymus of the other mice was between 1 and 10% of the wild type. In the mice with <0.1%, T cell development was arrested at the CD3− CD4− CD8− triple negative (TN) CD44+ CD25− stage with a few TN CD44+ CD25+ cells (**Figure 1**), while in mice with a less severe phenotype all thymic subsets including DP cells were present in normal proportions.

IL-7−/− mice have a phenotype similar to that of the 'less severe' type of IL-7Rα−/− mice. This suggests that another IL-7Rα ligand, for example TSLP-1, is involved in T cell development. Support for this notion comes from the observations that the phenotype of γc- and IL-7-deficient mice is similar to that of the 'mild' IL-7Rα−/− mice. As murine TSLP-1 interacts with the IL-7Rα chain in a γc-independent manner (Levin *et al.*, 1999), a role of TSLP-1 in T cell development could explain why T cell defect in the majority of the IL-7Rα (exon 3)-deficient mice is more severe than that in the γc−/− mice. A possible role of TSLP-1 in T cell development has yet to be determined.

It is puzzling that the severe phenotype was not found in another, independently generated mouse strain (Maki *et al.*, 1996). In this mouse strain exon 2 was targeted, in contrast to exon 3 that was targeted in the mice created by Peschon and coworkers. It is difficult to imagine that the differences in targeting can be responsible for the phenotypic differences between these two mouse strains. However it is clear that the variability in T cell phenotype in different IL-7Rα−/− mice is a complicating factor in the interpretation of the effect of IL-7Rα deficiency. As discussed in the chapter on IL-7 IL-7/IL-7R interactions can be important for cell survival, proliferation, and induction of TCRγ gene rearrangement in T cell precursors.

The IL-7R plays a crucial role in B cell development in the mouse. In IL-7Rα−/− mice B cell development is inhibited at the transition of pre-pro B cells (fraction A) to pro-B cells (**Figure 2**). Inhibition of B cell development in IL-7Rα−/− mice

Figure 1 A model of cellular stages in T cell development. The model is adapted from Shortman and Wu (1996) and DiSanto and Rodewald (1998). The stages where IL-7 and IL-7R deficiencies affect T cell development are indicated. *One of the IL-7Rα−/− strains that was reported by Peschon *et al.* (1994) showed variability in their thymic phenotype. Sixty percent had a block in the transition of CD44+CD25− to CD44+CD25+ cells whereas the other 40% had a phenotype comparable to that of the IL-7−/− and γc−/− mice. Another IL-7Rα−/− strain had a phenotype identical to that of IL-7−/− and γc−/− mice (Maki *et al.*, 1996).

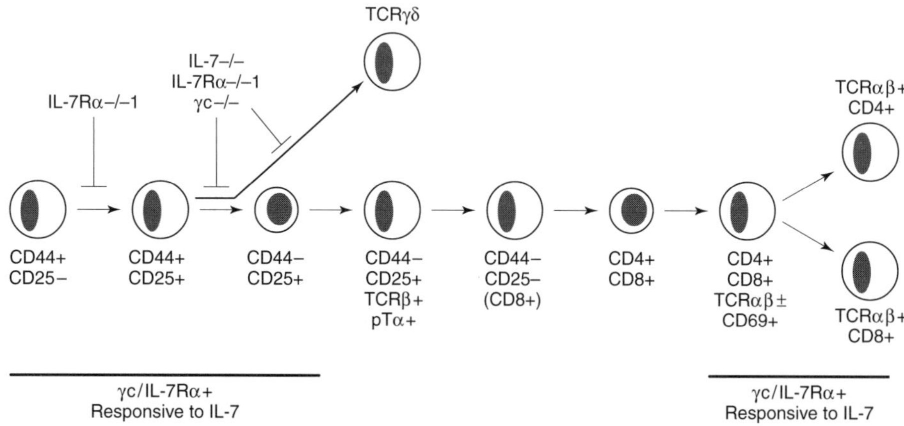

Figure 2 A model of early stages in murine B cell development. The model is based on observations made by Li *et al.* (1996). The effects of IL-7 on various B cell precursor populations are based on information that is summarized in the chapter and in a review by Maeurer *et al.* (1998).

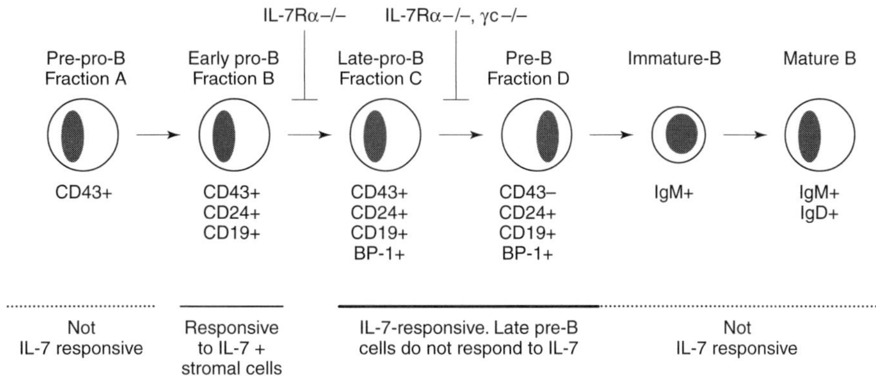

(Peschon *et al.*, 1998) at an earlier stage than in IL-7−/− mice (von Freeden-Jeffry *et al.*, 1998). It is possible that the alternative IL-7R ligand TSLP-1 plays a role in transition of pre-pro-B cells to pro-B cells and IL-7 in transition of pro-B into pre-B cells. Venkitaraman and coworkers have shown that signals through the IL-7Rα chain are essential for the induction of IgH gene rearrangements and are distinct from those that induce proliferation (Corcoran *et al.*, 1996). Development of IL-7Rα−/− B cells could be restored by means of retroviral transduction of the human IL-7Rα. Mutational analysis revealed that a single amino acid, Tyr449 (see Signal transduction) was required for IL-7-induced proliferation of the pre-B cells but not for IgH rearrangement. A later analysis of this group revealed that IL-7 signaling promotes accessibility of distal V regions to the recombination machinery (Corcoran *et al.*, 1998). The mechanism is yet unknown but it is of note that IL-7Rα-deficient pro-B cells have strongly reduced levels of the transcription factor Pax-5, which is important for IgH V-DJ recombination (Nutt *et al.*, 1998).

A more detailed discussion on the mechanism by which the IL-7/IL-7Rα interaction controls development of various T and B cell subsets can be found in the IL-7 chapter.

Interestingly, the IL-7R is critical for the formation of lymphoid organs in the intestine as Peyer's patches are absent in mice deficient for IL-7Rα and γc (Adachi *et al.*, 1998). One of the earliest events in the formation of Peyer's patches is the formation of an anlage with VCAM-1+ spots. Clusters of IL-7Rα+ cells are formed simultaneously. No VCAM-1+ spots and clusters of IL-7Rα+ cells were present in IL-7Rα−/− mice. IL-7Rα+ clusters were also absent in JAK3−/− mice although some IL-7Rα+ cells were found scattered over the intestine. As discussed in the Signal transduction section, JAK3 is a tyrosine kinase essential for IL-7R signaling and it is therefore clear that a signal through the IL-7Rα/γc complex which involves JAK3 is required for the formation of Peyer's patches. The observation that γc is required for formation of Peyer's patches makes it unlikely that the IL-7Rα ligand, TSLP-1, is the factor critical for generation of this organ as TSLP-1 is thought to act independently of the γc chain (Levin *et al.*, 1999).

Human abnormalities

Two patients have been described who have IL-7Rα deficiencies (Puel *et al.*, 1998). These *severe combined immunodeficiency* (SCID) patients presented with a profound lack of T cells but with normal numbers of B and NK cells. In one patient the disease was caused by two mutations preventing expression of a functional IL-7Rα; one allele had a splice-junction mutation at the 3′ end of intron 4 on one allele and a premature stop codon (Trp217 → stop codon) in exon 5 of the other allele. The disease-causing mutation of the second patient is not known. The extracellular domain contained two amino acid changes, but the IL-7Rα chain derived from this patient bound IL-7 and could transmit signals leading to activation of STAT5. Since the expression of the IL-7Rα chain in the second patient was strongly diminished, the defect in this patient is perhaps localized in the promoter region.

With regard to T and B cell development, similar defects were noted in patients with genetic alterations in the γc chain (Noguchi *et al.*, 1993) and JAK3 (Russell *et al.*, 1995). The finding that normal B cells were present in these patients underscores the dispensability of the IL-7/IL-7R system for development of B cells in humans. Therefore mice deficient for JAK3, γc, or IL-7Rα are not accurate models for humans with mutations in these genes.

THERAPEUTIC UTILITY

Effect of treatment with soluble receptor domain

Although transcripts for soluble IL-7Rα receptors have been found, there is no information on the function of soluble receptors. It is possible that soluble receptors play a regulatory role.

References

Adachi, S., Yoshida, H., Honda, K., Maki, K., Saijo, K., Ikuta, K., Saito, T., and Nishikawa, S. I. (1998). Essential role of IL-7 receptor alpha in the formation of Peyer's patch anlage. *Int. Immunol.* **10**, 1–6.

Akasaka, T., Tsuji, K., Kawahira, H., Kanno, M., Harigaya, K., Hu, L., Ebihara, Y., Nakahata, T., Tetsu, O., Taniguchi, M., and Koseki, H. (1997). The role of mel-18, a mammalian Polycomb group gene, during IL-7-dependent proliferation of lymphocyte precursors. *Immunity* **7**, 135–146.

Akashi, K., Kondo, M., and Weissman, I. L. (1998). Role of interleukin-7 in T-cell development from hematopoietic stem cells. *Immunol. Rev.* **165**, 13–28.

Appleby, M. W., Gross, J. A., Cooke, M. P., Levin, S. D., Qian, X., and Perlmutter, R. M. (1992). Defective T cell receptor signaling in mice lacking the thymic isoform of p59fyn. *Cell* **70**, 751–763.

Coffer, P. J., Jin, J., and Woodgett, J. R. (1998). Protein kinase B (c-Akt): a multifunctional mediator of phosphatidylinositol 3-kinase activation. *Biochem. J.* **335**, 1–13.

Corcoran, A. E., Smart, F. M., Cowling, R. J., Crompton, T., Owen, M. J., and Venkitaraman, A. R. (1996). The interleukin-7 receptor alpha chain transmits distinct signals for proliferation and differentiation during B lymphopoiesis. *EMBO J.* **15**, 1924–1932.

Corcoran, A. E., Riddell, A., Krooshoop, D., and Venkitaraman, A. R. (1998). Impaired immunoglobulin gene rearrangement in mice lacking the IL-7 receptor. *Nature* **391**, 904–907.

Crawley, J. B., Rawlinson, L., Lali, F. V., Page, T. H., Saklatvala, J., and Foxwell, B. M. (1997). T cell proliferation in response to interleukins 2 and 7 requires p38MAP kinase activation. *J. Biol. Chem.* **272**, 15023–15027.

Dadi, H., Ke, S., and Roifman, C. M. (1994). Activation of phosphatidylinositol-3 kinase by ligation of the interleukin-7 receptor is dependent on protein tyrosine kinase activity. *Blood* **84**, 1579–1586.

DiSanto, J. P., and Rodewald, H. R. (1998). *In vivo* roles of receptor tyrosine kinases and cytokine receptors in early thymocyte development. *Curr. Opin. Immunol.* **10**, 196–207.

Domen, J., van der Lugt, N. M., Acton, D., Laird, P. W., Linders, K., and Berns, A. (1993). Pim-1 levels determine the size of early B lymphoid compartments in bone marrow. *J. Exp. Med.* **178**, 1665–1673.

Fruman, D. A., Snapper, S. B., Yballe, C. M., Davidson, L., Yu, J. Y., Alt, F. W., and Cantley, L. C. (1999). Impaired B cell development and proliferation in absence of phosphoinositide 3-kinase p85alpha. *Science* **283**, 393–397.

Goodwin, R. G., Friend, D., Ziegler, S. F., Jerzy, R., Falk, B. A., Gimpel, S., Cosman, D., Dower, S. K., March, C. J., Namen, A. E., and Park, L. S. (1990). Cloning of the human and murine interleukin-7 receptors: demonstration of a soluble form and homology to a new receptor superfamily. *Cell* **60**, 941–951.

Imada, K., Bloom, E. T., Nakajima, H., Horvath, A. J., Udy, G. B., Davey, H. W., and Leonard, W. J. (1998). Stat5b is essential for natural killer cell-mediated proliferation and cytolytic activity. *J. Exp. Med.* **188**, 2067–2074.

Kapeller, R., and Cantley, L. C. (1994). Phosphatidylinositol 3-kinase. *Bioessays* **16**, 565–576.

Leonard, W. J., and O'Shea, J. J. (1998). Jaks and STATs: biological implications. *Annu. Rev. Immunol.* **16**, 293–322.

Levin, S. D., Koelling, R. M., Friend, S. L., Isaksen, D. E., Ziegler, S. F., Perlmutter, R. M., and Farr, A. G. (1999). Thymic stromal lymphopoietin: a cytokine that promotes the development of IgM+ B cells *in vitro* and signals via a novel mechanism. *J. Immunol.* **162**, 677–683.

Li, Y. S., Wasserman, R., Hayakawa, K., and Hardy, R. R. (1996). Identification of the earliest B lineage stage in mouse bone marrow. *Immunity* **5**, 527–535.

Lin, J. X., Migone, T. S., Tsang, M., Friedmann, M., Weatherbee, J. A., Zhou, L., Yamauchi, A., Bloom, E. T., Mietz, J., John, S., and Leonard, W. J. (1995). The role of shared receptor motifs and common Stat proteins in the generation of cytokine pleiotropy and redundancy by IL-2, IL-4, IL-7, IL-13, and IL-15. *Immunity* **2**, 331–339.

Lynch, M., Baker, E., Park, L. S., Sutherland, G. R., and Goodwin, R. G. (1992). The interleukin-7 receptor gene is at 5p13. *Hum. Genet.* **89**, 566–568.

Maeurer, M. J., Edington, H. D., and Lotze, M. T. (1998). In "The Cytokine Handbook" (eds. A.W. Thompson), Interleukin-7, pp. 229–269. Academic Press, London.

Maki, K., Sunaga, S., Komagata, Y., Kodaira, Y., Mabuchi, A., Karasuyama, H., Yokomuro, K., Miyazaki, J. I., and Ikuta, K. (1996). Interleukin 7 receptor-deficient mice lack gammadelta T cells. *Proc. Natl Acad. Sci. USA* **93**, 7172–7177.

Maraskovsky, E., Teepe, M., Morrissey, P. J., Braddy, S., Miller, R. E., Lynch, D. H., and Peschon, J. J. (1996). Impaired survival and proliferation in IL-7 receptor-deficient peripheral T cells. *J. Immunol.* **157**, 5315–5323.

Meraz, M. A., White, J. M., Sheehan, K. C., Bach, E. A., Rodig, S. J., Dighe, A. S., Kaplan, D. H., Riley, J. K., Greenlund, A. C., Campbell, D., Carver-Moore, K., DuBois, R. N., Clark, R., Aguet, M., and Schreiber, R. D. (1996). Targeted disruption of the Stat1 gene in mice reveals unexpected physiologic specificity in the JAK-STAT signaling pathway. *Cell* **84**, 431–442.

Molina, T. J., Kishihara, K., Siderovski, D. P., van, Ewijk, W., Narendran, A., Timms, E., Wakeham, A., Paige, C. J., Hartmann, K. U., Veillette, A., Davidson, D., and Mak, T. W. (1992). Profound block in thymocyte development in mice lacking p56lck. *Nature* **357**, 161–164.

Moriggl, R., Topham, D. J., Teglund, S., Sexl, V., McKay, C., Wang, D., Hoffmeyer, A., van Deursen, J., Sangster, M. Y., Bunting, K. D., Grosveld, G. C., and Ihle, J. N. (1999). Stat5 is required for IL-2-induced cell cycle progression of peripheral T cells. *Immunity* **10**, 249–259.

Noguchi, M., Yi, H., Rosenblatt, H. M., Filipovich, A. H., Adelstein, S., Modi, W. S., McBride, O. W., and Leonard, W. J. (1993). Interleukin-2 receptor gamma chain mutation results in X-linked severe combined immunodeficiency in humans. *Cell* **73**, 147–157.

Nosaka, T., van, D. J., Tripp, R. A., Thierfelder, W. E., Witthuhn, B. A., McMickle, A. P., Doherty, P. C., Grosveld, G. C., and Ihle, J. N. (1995). Defective lymphoid development in mice lacking Jak3. *Science* **270**, 800–802.

Nutt, S. L., Morrison, A. M., Dorfler, P., Rolink, A., and Busslinger, M. (1998). Identification of BSAP (Pax-5) target genes in early B-cell development by loss- and gain-of-function experiments. *EMBO J.* **17**, 2319–2333.

Pallard, C., Stegmann, A. P., van Kleffens, T., Smart, F., Venkitaraman, A., and Spits, H. (1999). Distinct roles of the phosphatidylinositol 3-kinase and STAT5 pathways in IL-7-mediated development of human thymocyte precursors. *Immunity* **10**, 525–535.

Park, L. S., Friend, D. J., Schmierer, A. E., Dower, S. K., and Namen, A. E. (1990). Murine interleukin 7 (IL-7) receptor. Characterization on an IL-7-dependent cell line. *J. Exp. Med.* **171**, 1073–1089.

Peschon, J. J., Morrissey, P. J., Grabstein, K. H., Ramsdell, F. J., Maraskovsky, E., Gliniak, B. C., Park, L. S., Ziegler, S. F., Williams, D. E., Ware, C. B., Meyer, J. D., and Davison, B. L. (1994). Early lymphocyte expansion is severely impaired in interleukin 7 receptor-deficient mice. *J. Exp. Med.* **180**, 1955–1960.

Peschon, J. J., Gliniak, B. C., Morissey, P., and Maraskowsky, E. (1998). In "Cytokine Knockouts" (eds. S.Durum and K. Muegge), Lymphoid development and function in IL-7R-deficient mice, pp. 37-52. Humana Press, Totaway, NJ.

Pleiman, C. M., Gimpel, S. D., Park, L. S., Harada, H., Taniguchi, T., and Ziegler, S. F. (1991). Organization of the murine and human interleukin-7 receptor genes: two mRNAs generated by differential splicing and presence of a type I- interferon-inducible promoter. *Mol. Cell. Biol.* **11**, 3052–3059.

Plum, J., De Smedt, M., Leclercq, G., Verhasselt, B., and Vandekerckhove, B. (1996). Interleukin-7 is a critical growth factor in early human T-cell development. *Blood* **88**, 4239–4245.

Puel, A., Ziegler, S. F., Buckley, R. H., and Leonard, W. J. (1998). Defective IL7R expression in T(−)B(+)NK(+) severe combined immunodeficiency. *Nature Genet.* **20**, 394–397.

Rodig, S. J., Meraz, M. A., White, J. M., Lampe, P. A., Riley, J. K., Arthur, C. D., King, K. L., Sheehan, K. C., Yin, L., Pennica, D., Johnson, E. M. Jr., and Schreiber, R. D. (1998). Disruption of the Jak1 gene demonstrates obligatory and nonredundant roles of the Jaks in cytokine-induced biologic responses. *Cell* **93**, 373–383.

Russell, S. M., Tayebi, N., Nakajima, H., Riedy, M. C., Roberts, J. L., Aman, M. J., Migone, T. S., Noguchi, M., Markert, M. L., Buckley, R. H., O'Shea, J. J., and Leonard, W. J. (1995). Mutation of Jak3 in a patient with SCID: essential role of Jak3 in lymphoid development. *Science* **270**, 797–800.

Seckinger, P., and Fougereau, M. (1994). Activation of src family kinases in human pre-B cells by IL-7. *J. Immunol.* **153**, 97–109.

Sharfe, N., and Roifman, C. M. (1997). Differential association of phosphatidylinositol 3-kinase with insulin receptor substrate (IRS)-1 and IRS-2 in human thymocytes in response to IL-7. *J. Immunol.* **159**, 1107–1114.

Sharfe, N., Dadi, H. K., and Roifman, C. M. (1995). JAK3 protein tyrosine kinase mediates interleukin-7-induced activation of phosphatidylinositol-3′ kinase. *Blood* **86**, 2077–2085.

Shortman, K., and Wu, L. (1996). Early T lymphocyte progenitors. *Annu. Rev. Immunol.* **14**, 29–47.

Sugamura, K., Asao, H., Kondo, M., Tanaka, N., Ishii, N., Ohbo, K., Nakamura, M., and Takeshita, T. (1996). The interleukin-2 receptor gamma chain: its role in the multiple cytokine receptor complexes and T cell development in XSCID. *Annu. Rev. Immunol.* **14**, 179–205.

Sugawara, T., Moriguchi, T., Nishida, E., and Takahama, Y. (1998). Differential roles of ERK and p38 MAP kinase pathways in positive and negative selection of T lymphocytes. *Immunity* **9**, 565–574.

Suzuki, H., Terauchi, Y., Fujiwara, M., Aizawa, S., Yazaki, Y., Kadowaki, T., and Koyasu, S. (1999). Xid-like immunodeficiency in mice with disruption of the p85alpha subunit of phosphoinositide 3-kinase. *Science* **283**, 390–392.

Takeda, K., Kaisho, T., Yoshida, N., Takeda, J., Kishimoto, T., and Akira, S. (1998). Stat3 activation is responsible for IL-6-dependent T cell proliferation through preventing apoptosis: generation and characterization of T cell-specific Stat3-deficient mice. *J. Immunol.* **161**, 4652–4660.

Thomis, D. C., Gurniak, C. B., Tivol, E., Sharpe, A. H., and Berg, L. J. (1995). Defects in B lymphocyte maturation and T lymphocyte activation in mice lacking Jak3. *Science* **270**, 794–797.

van der Plas, D. C., Smiers, F., Pouwels, K., Hoefsloot, L. H., Lowenberg, B., and Touw, I. P. (1996). Interleukin-7 signaling in human B cell precursor acute lymphoblastic leukemia cells and murine BAF3 cells involves activation of STAT1 and STAT5 mediated via the interleukin-7 receptor alpha chain. *Leukemia* **10**, 1317–1325.

Venkitaraman, A. R., and Cowling, R. J. (1992). Interleukin 7 receptor functions by recruiting the tyrosine kinase p59fyn through a segment of its cytoplasmic tail. *Proc. Natl Acad. Sci. USA* **89**, 12083–12087.

Venkitaraman, A. R., and Cowling, R. J. (1994). Interleukin-7 induces the association of phosphatidylinositol 3-kinase with the alpha chain of the interleukin-7 receptor. *Eur. J. Immunol.* **24**, 2168–2174.

von Freeden-Jeffry, U., Moore, T. A., Zlotnik, A., and Murray, R. (1998). In "Cytokine Knockouts" (eds. S. Durum and K. Muegge), IL-7 knockout mice and the generation of lymphocytes, pp. 21–36. Humana Press, Totaway, NJ.

Zeng, Y. X., Takahashi, H., Shibata, M., and Hirokawa, K. (1994). Jak-3 Janus kinase is involved in interleukin 7 signal pathway. *FEBS Lett.* **353**, 289–293.

IL-9 Receptor

Jean-Christophe Renauld*

Ludwig Institute for Cancer Research, Catholic University of Louvain, 74 Avenue Hippocrate, Brussels, B-1200, Belgium

*corresponding author tel: 32-2-764-7464, fax: 32-2-762-9405, e-mail: renauld@licr.ucl.ac.be

DOI: 10.1006/rwcy.2000.14004.

SUMMARY

High-affinity binding of IL-9 to target cells is mediated by a heterodimer consisting of the IL-9 receptor α chain and the IL-2 receptor γ chain, also called γc (common γ); both chains belong to the hematopoietin receptor superfamily. The IL-9 receptor α chain is sufficient to confer high-affinity binding but both chains are needed for signal transduction. The IL-9 receptor α is associated with JAK1 tyrosine kinase, and the IL-2 receptor γ chain with JAK3. Upon IL-9 binding, both kinases are activated and the IL-9 receptor γ is phosphorylated on a single tyrosine. This phosphorylated residue is used as a docking site for STAT1, STAT3, and STAT5 transcription factors. Activation of these transcription factors is considered to be critical for all IL-9 activities, because a mutant on this single tyrosine residue completely lost its activity. The human IL-9R gene is located on the long arm pseudoautosomal region of the X and Y chromosomes, a region which has been linked to *asthma*.

BACKGROUND

Discovery

The mouse IL-9R cDNA was identified by expression cloning in COS cells (Renauld *et al.*, 1992).

The human IL-9 receptor cDNA was isolated by crosshybridization with a mouse probe (Renauld *et al.*, 1992).

Structure

The IL-9 receptor (IL-9R) was found to interact with the γ chain of the IL-2R, which is required for signal transduction but not for IL-9 binding, since an antibody directed against this molecule completely inhibits the activity of IL-9 without affecting the K_d of IL-9 binding (Kimura *et al.*, 1995). The fact that this molecule, now called γc, is shared by IL-2R, IL-4R, IL-7R, IL-9R, and IL-15R could explain the overlapping activities observed for these T cell growth factors.

Main activities and pathophysiological roles

The IL-9R is believed to mediate all the biological activities of IL-9 on T cells, B cells, mast cells, eosinophils, and hematopoietic progenitors, particularly for the development of *asthma*, for which a genetic linkage has been reported (Holroyd *et al.*, 1998).

GENE

The human genome contains at least four IL-9R pseudogenes with $\sim$90% homology with the IL-9R gene (Kermouni *et al.*, 1995; Vermeesch *et al.*, 1997).

Accession numbers

Human IL-9R gene: L39064
Human IL-9R pseudogenes: L39063, L39062

Sequence

The mouse IL-9R gene is composed of nine exons and eight introns, sharing many characteristics with other genes encoding cytokine receptors (Renauld *et al.*, unpublished data).

The human IL-9R gene is composed of 11 exons and 10 introns, stretching over $\sim$17 kb.

A frequent alternative splicing of the human gene generates an intriguing heterogeneity in the 5′ untranslated region of the mRNA and introduces some short open reading frames that might represent an additional level in the regulation of IL-9R translation, as suggested for many genes involved in cell growth (Kozak, 1991; Kermouni *et al.*, 1995). More recently, another splice variant was identified that contained an in-frame deletion of a single residue of the extracellular domain and lacked the ability to bind IL-9 (Grasso *et al.*, 1998).

Analysis of the 5′ flanking region revealed multiple transcription initiation sites as well as potential binding motifs for AP-1, AP-2, AP-3, SP-1, and NFκB, although this region lacks a TATA box (Kermouni *et al.*, 1995).

Although the IL-9R pseudogenes are similar to the IL-9R gene ($\sim$90% identity), none of these copies encodes a functional receptor: none of them contain sequences homologous to the 5′ flanking region or exon 1 of the IL-9R gene and the remaining open reading frames have been inactivated by various point mutations and deletions (Kermouni *et al.*, 1995).

Chromosome location and linkages

In the mouse, the IL-9R gene is a single copy gene located on chromosome 11 (Vermeesch *et al.*, 1997).

The human IL-9R gene is located in the subtelomeric region of chromosomes X and Y. IL-9R was thus the first gene to be identified in the long arm pseudoautosomal region and turned out to be a unique tool to study this particular region of the genome. Using a polymorphism in the coding region of this gene, Vermeesch and colleagues showed that IL-9R is expressed both from X and Y and escapes X inactivation (Vermeesch *et al.*, 1997). Interestingly, a genetic linkage has been reported between this region and *asthma* or *bronchial hyperresponsiveness*, suggesting that different alleles of the IL-9R gene affect allergic responses (Holroyd *et al.*, 1998).

The four IL-9R pseudogenes are located in the subtelomeric regions of chromosomes 9q, 10p, 16p, and 18p (Kermouni *et al.*, 1995).

PROTEIN

Accession numbers

SwissProt:
Mouse: Q01114
Human: Q01113

Description of protein

The murine IL-9R contains 468 amino acids, including an extracellular domain, composed of 233 amino acids (Renauld *et al.*, 1992) that shows the typical features of the hematopoietin receptor superfamily, namely four conserved cysteines and a WSEWS motif, located a few residues upstream from the transmembrane domain (Bazan, 1990). The human IL-9R cDNA encodes a 522 amino acid protein.

Relevant homologies and species differences

The human IL-9R protein shows a 53% identity with the mouse IL-9R. The extracellular region is particularly conserved with 67% identity, while the cytoplasmic domain is significantly larger in the human receptor (231 versus 177 residues) (Renauld *et al.*, 1992).

The juxtamembrane region of the cytoplasmic tail of the IL-9R contains a Pro-X-Pro sequence preceded by a cluster of hydrophobic residues, which partially fits a consensus motif shared by many cytokine receptors (IL-4R, IL-7R, IL-3R, EPOR, IL-2Rβ, G-CSFR) (Murakami *et al.*, 1991). Downstream from this Pro-X-Pro motif, a striking homology was observed with the β chain of the IL-2R and with the erythropoietin receptor. As a result, for the first 33 amino acids of the cytoplasmic domain, a 40% identity was noticed between the human IL-9R and the IL-2Rβ. This homologous region probably explains why IL-9, like other cytokines such as IL-2, induces JAK1 and JAK3 phosphorylation (Russell *et al.*, 1994; Yin *et al.*, 1994; Demoulin *et al.*, 1996).

Affinity for ligand(s)

A variety of mouse hematopoietic cells express high-affinity receptors for IL-9 ($K_d \sim 100$ pM) (Druez *et al.*, 1990). The γ chain of the IL-2R, which associates with the IL-9R, is required for signal transduction but not for IL-9 high-affinity binding (Kimura *et al.*, 1995).

Cell types and tissues expressing the receptor

This issue has been poorly investigated so far. A variety of mouse hematopoietic cells, including T cells, mast cells, and macrophages, express high-affinity receptors for IL-9 (Druez *et al.*, 1990). More recently,

IL-9R was found to be expressed preferentially by peritoneal B-1 lymphocytes.

Release of soluble receptors

Soluble receptors have never been reported at the protein level. As observed for many members of the hematopoietin receptor superfamily, IL-9R mRNA has been identified that lack the sequences encoding the transmembrane and cytoplasmic domains, as a result of alternative splicing (Renauld *et al.*, 1992). However, the frequency of these mRNA seems quite low and it is not yet clear whether they really encode a soluble IL-9-binding protein.

SIGNAL TRANSDUCTION

Associated or intrinsic kinases

IL-9R interacts with the γ chain of the IL-2R, which is required for signal transduction and is shared by IL-2R, IL-4R, IL-7R, IL-9R, and IL-15R (Demoulin and Renauld, 1998). So far, the only function of γc seems to recruit the tyrosine kinase JAK3, while IL-9R is associated with JAK1. Upon IL-9 binding, both JAK1 and JAK3 become phosphorylated and catalytically active. These kinases are likely to be responsible for IL-9R phosphorylation on one of its five tyrosine residues.

Cytoplasmic signaling cascades

This single phosphorylated residue acts as a docking site for STAT1, STAT3, and STAT5 – three transcription factors that, after phosphorylation by the JAK kinases associated to the receptor, form hetero- or homodimers and migrate to the nucleus (Demoulin *et al.*, 1996; Bauer *et al.*, 1998).

IL-9 does not seem to induce or enhance the phosphorylation of the serine/threonine kinases Raf-1 or MAP kinases in the Mo7E leukemia cell line (Miyazawa *et al.*, 1992), or in mouse lymphoid cells (Grasso *et al.*, unpublished).

More clearly established is the activation by IL-9 of an adaptor protein called 4PS/IRS2, a feature shared with IL-4 signal transduction, where this pathway was shown to be critical for growth regulation (Keegan *et al.*, 1994; Yin *et al.*, 1995; Demoulin *et al.*, 1996). Phosphorylation of 4PS/IRS2 is not dependent on the phosphorylation of the IL-9R, contrasting with the IL-4 system in which 4PS/IRS2 associates with the IL-4R through a phosphotyrosine residue. Preliminary observations suggest that 4PS/IRS2 and JAK1 activation require the same region of the IL-9R (Demoulin *et al.*, unpublished data) and these two molecules were shown to be associated in response to IL-9 (Yin *et al.*, 1995). Taken together, these observations raise the possibility that, upon IL-9 activation, 4PS/IRS2 becomes phosphorylated by interacting directly with the JAK1 tyrosine kinase. After phosphorylation, 4PS/IRS2 binds the SH2 domain of various signaling proteins including the p85 subunit of the phosphatidylinositol 3-kinase (Demoulin *et al.*, unpublished data).

DOWNSTREAM GENE ACTIVATION

Transcription factors activated

STAT1, STAT3, and STAT5 (Demoulin *et al.*, 1996).

Genes induced

Granzyme A, granzyme B, mouse mast cell proteases (MMCP), Ly6A/E, L-selectin, IL-6, adseverin (Robbens *et al.*, 1998), Bcl-3 (Richard *et al.*, 1999), M-ras (Louahed *et al.*, 1999).

Promoter regions involved

GAS site for Ly6A/E (Demoulin *et al.*, 1999).

BIOLOGICAL CONSEQUENCES OF ACTIVATING OR INHIBITING RECEPTOR AND PATHOPHYSIOLOGY

Unique biological effects of activating the receptors

See chapter on IL-9 activities.

References

Bauer, J., Liu, K., You, Y., Lai, S., and Goldsmith, M. (1998). Heteromerization of the γc chain with the interleukin-9 receptor α subunit leads to STAT activation and prevention of apoptosis. *J. Biol. Chem.* **273**, 9255–9260.

Bazan, F. (1990). Structural design and molecular evolution of a cytokine receptor superfamily. *Proc. Natl Acad. Sci. USA* **87**, 6934–6938.

Demoulin, J.-B., and Renauld, J.-C. (1998). Signalling by cytokines interacting with the interleukin-2 receptor γ chain. *Cytokines Mol. Ther.* **4**, 243–256.

Demoulin, J.-B., Uyttenhove, C., Van Roost, E., de Lestré, B., Donckers, D., Van Snick, J., and Renauld, J.-C. (1996). A single tyrosine of the interleukin-9 (IL-9) receptor is required for STAT activation, antiapoptotic activity, and growth regulation by IL-9. *Mol. Cell. Biol.* **16**, 4710–4716.

Demoulin, J.-B., Maisin, D., and Renauld, J.-C. (1999). Ly-6A/E induction by interleukin-6 and interleukin-9 in T cells. *Eur. Cytokine Netw.* **10**, 49–56.

Druez, C., Coulie, P., Uyttenhove, C., and Van Snick, J. (1990). Functional and biochemical characterization of mouse P40/IL-9 receptors. *J. Immunol.* **145**, 2494–2499.

Grasso, L., Huang, M., Sullivan, C. D., Messler, C. J., Kiser, M. B., Dragwa, C. R., Holroyd, K. J., Renauld, J.-C., Levitt, R. C., and Nicolaides, N. C. (1998). Molecular analysis of human interleukin-9 receptor transcripts in peripheral blood mononuclear cells: identification of a splice variant encoding for a non-functional cell surface receptor. *J. Biol. Chem.* **273**, 24016–24024.

Holroyd, K. J., Martinati, L. C., Trabetti, E., Scherpbier, T., Eleff, S. M., Boner, A. L., Pignatti, P. F., Kiser, M. B., Dragwa, C. R., Hubbard, F., Sullivan, C. D., Grasso, L., Messler, C. J., Huang, M., Hu, Y., Nicolaides, N. C., Buetow, K. H., and Levitt, R. C. (1998). Asthma and bronchial hyperresponsiveness linked to the XY long arm pseudoautosomal region. *Genomics* **52**, 233–235.

Keegan, A., Nelms, K., Wang, L. M., Pierce, J., and Paul, W. (1994). Interleukin-4 receptor: signaling mechanisms. *Immunol. Today* **15**, 423–432.

Kermouni, A., Van Roost, E., Arden, K. C., Vermeesch, J. R., Weiss, S., Godelaine, D., Flint, J., Lurquin, C., Szikora, J.-P., Higgs, D. R., Maryunen, P., and Renauld, J.-C. (1995). The IL-9 receptor gene: genomic structure, chromosomal localization in the pseudoautosomal region of the long arm of the sex chromosomes and identification of IL-9R pseudogenes at 9qter, 10pter, 16pter and 18pter. *Genomics* **29**, 371–382.

Kimura, Y., Takeshita, T., Kondo, M., Ishii, N., Nakamura, M., Van Snick, J., and Sugamura, K. (1995). Sharing of the IL-2 receptor γ chain with the functional IL-9 receptor complex. *Int. Immunol.* **7**, 115–120.

Kozak, M. (1991). An analysis of vertebrate mRNA sequences: intimations of translational control. *J. Cell Biol.* **115**, 887–903.

Louahed, J., Grasso, L., De Smet, C., Van Roost, E., Wildmann, C., Nicolaides, N. C., Levitt, R. C., and Renauld, J. C. (1999). Interleukin-9-induced expression of M-Ras/R-Ras3 oncogene in T helper clones. *Blood* **94**, 1701–1710.

Miyazawa, K., Hendrie, P., Kim, Y.-J., Mantel, C., Yang, Y.-C., Se Kwon, B., and Broxmeyer, H. (1992). Recombinant human interleukin-9 induces protein tyrosine phosphorylation and synergizes with steel factor to stimulate proliferation of the human factor-dependent cell line, MO7e. *Blood* **80**, 1685–1692.

Murakami, M., Narazaki, M., Hibi, M., Yawata, H., Yasukawa, K., Hamaguchi, M., Taga, T., and Kishimoto, T. (1991). Critical cytoplasmic region of the interleukin-6 signal transducer gp130 is conserved in the cytokine receptor family. *Proc. Natl Acad. Sci. USA* **88**, 11349–11353.

Renauld, J.-C., Druez, C., Kermouni, A., Houssiau, F., Uyttenhove, C., Van Roost, E., and Van Snick, J. (1992). Expression cloning of the murine and human interleukin 9 receptor cDNAs. *Proc. Natl Acad. Sci. USA* **89**, 5690–5694.

Richard, M., Louahed, J., Demoulin, J. B., and Renauld, J. C. (1999). Interleukin-9 regulates NF-κB activity through BCL3 gene induction. *Blood* **93**, 4318–4327.

Robbens, J., Louahed, J., De Pestel, K., Van Colen, I., Ampe, C., Vandekerckhove, J., and Renauld, J.-C. (1998). Murine adseverin (D5), a novel member of the gelsolin family, and murine adseverin are induced by interleukin-9 in T-helper lymphocytes. *Mol. Cell. Biol.* **18**, 4589–4596.

Russell, S., Johnston, J., Noguchi, M., Kawamura, M., Bacon, C., Friedmann, M., Berg, M., McVicar, D., Witthuhn, B., Silvennoinen, O., Goldman, A., Schmalstieg, F., Ihle, J., O'Shea, J., and Leonard, W. (1994). Interaction of IL-2Rβ and γc chains with Jak1 and Jak3: implications for XSCID and XCID. *Science* **266**, 1042–1045.

Vermeesch, J. R., Petit, P., Kermouni, A., Renauld, J.-C., Van Den Berghe, H., and Marynen, P. (1997). The IL-9 receptor gene, located in the Xq/Yq pseudoautosomal region, has an autosomal origin and escapes X inactivation. *Hum. Mol. Genet.* **6**, 1–8.

Yin, T., Lik-Shing Tsang, M., and Yang, Y.-C. (1994). JAK1 kinase forms complexes with interleukin-4 receptor and 4PS/insulin receptor substrate-1-like protein and is activated by interleukin-4 and interleukin-9 in T lymphocytes. *J. Biol. Chem.* **269**, 26614–26617.

Yin, T., Keller, S., Quelle, F., Witthuhn, B., Lik-Shing Tsang, M., Lienhard, G., Ihle, J., and Yang, Y.-C. (1995). Interleukin-9 induces tyrosine phosphorylation of insulin receptor substrate-1 via JAK tyrosine kinases. *J. Biol. Chem.* **270**, 20497–20502.

IL-10 Receptor

Rene de Waal Malefyt*

Department of Molecular Biology, DNAX Research Institute, 901 California Avenue, Palo Alto, CA 94304-1104, USA

*corresponding author tel: 650-496-1164, fax: 650-496-1200, e-mail: rene@dnax.org

DOI: 10.1006/rwcy.2000.14005.

SUMMARY

IL-10 interacts with its tetrameric receptor complex consisting of two IL-10Rα and two IL-10Rβ chains resulting in the phosphorylation and activation of JAK1 and TYK2 kinases, which in turn phosphorylate two tyrosine residues in the intracytoplasmic parts of the IL-10Rα chains that form docking sites for STAT3. Binding of STAT3 results in phosphorylation by JAK1 and TYK2 kinases, homo- or heterodimerization and translocation to the nucleus where it binds to the promoters of IL-10-responsive genes such as the FcγRI and activates transcription. All IL-10-mediated responses are dependent on activation of STAT3, but the anti-inflammatory actions of IL-10 require additional sequences in the distal intracytoplasmic part of the IL-10Rα chain.

BACKGROUND

Discovery

IL-10 mediates its biological effects by interacting with specific cell surface receptors. The IL-10 receptor complex consists of at least two separate receptor chains, a ligand-binding IL-10Rα chain and a IL-10Rβ chain, which is essential for signal transduction. Radiolabeled IL-10 and FLAG epitope-tagged IL-10 were initially used to detect specific receptors on IL-10-responsive cells and to enrich for cells expressing high numbers of receptors (Tan *et al.*, 1993). Expression cloning strategies led to the isolation of cDNAs encoding murine and human IL-10Rα-binding proteins (Ho *et al.*, 1993; Liu *et al.*, 1994). Subsequently, an orphan class II cytokine receptor was identified as the IL-10Rβ chain by functional complementation studies in combination with the IL-10Rα chain (Kotenko *et al.*, 1997).

Structure

Both IL-10Rα and IL-10Rβ belong to the class II cytokine receptor family, together with the receptors for IFNα and IFNγ. Crystallization of human IL-10 or viral IL-10 to soluble extracellular IL-10Rα chains has indicated that the stoichiometry of the IL-10/sIL-10Rα complex contains two IL-10 dimers binding to four shIL-10Rα monomers (Hoover *et al.*, 1999), confirming results from previous gel filtration experiments (Tan *et al.*, 1995). A model of IL-10 complexed with its soluble receptor at high resolution, based on topological similarity between IL-10 and IFNγ is also available (Zdanov *et al.*, 1996).

Main activities and pathophysiological roles

The main activity for the IL-10 receptor is to bind IL-10 and initiate the transduction of a signaling cascade, which leads to the modification of biological responses. There have been no specific pathophysiological roles described for the IL-10 receptor complex itself. The association of IL-10 with disease is described in the IL-10 chapter.

GENE

Accession numbers

Mouse IL-10Rα: L12120
Human IL-10Rα: U00672, NM001558

Mouse IL-10Rβ: U53696
Human IL-10Rβ: NM000628

Sequence

See **Figure 1**.

Chromosome location and linkages

Mouse IL-10Rα: 9
Human IL-10Rα: 11q23.3
Mouse IL-10Rβ: 16
Human IL-10Rβ: 21q22.1–21q22.2

The structure of the IL-10Rβ loci is conserved among human, mouse, and chicken but not in fish (Reboul *et al.*, 1999).

PROTEIN

Accession numbers

Mouse IL-10Rα: AAA16156.1
Human IL-10Rα: NP001549.1
Mouse IL-10Rβ: AAC53062
Human IL-10Rβ: NP000619

Description of protein

The mIL-10Rα and hIL-10Rα polypeptides are composed of 576 and 578 amino acid residues respectively, whereas mIL-10Rβ and hIL-10Rβ chains consist of 349 and 325 amino acid residues respectively. IL-10Rα chains are expressed as 90–120 kDa proteins and IL-10Rβ chains have a molecular weight of ~40 kDa. IL-10Rα chains contain four (mouse) to six (human) *N*-linked glycosylation sites which are used. Positive cells express only a few hundred receptor complexes per cell.

Relevant homologies and species differences

Mouse and human IL-10Rα chains are ~60% identical and 73% similar. Mouse and human IL-10Rβ chains are 69% identical at the amino acid level. The first extracellular immunoglobulin-like binding domain of the hIL-10Rα lacks the intracellular disulfide bridge and the intracytoplasmic tail lacks one tyrosine residue as compared to the mIL-10Rα, although neither of these differences seem functionally significant.

Affinity for ligand(s)

Mouse and human IL-10 receptor complexes bind their ligand with high affinity ($K_d \sim$ 35–200 pM). Human IL-10 interacts with mIL-10R complex, but mouse IL-10 does not bind to the hIL-10R complex, which is consistent with the observed species specificities of these cytokines. The affinity and specific activity of human IL-10 is ~10-fold lower on cells expressing mixed receptor complexes consisting of hIL-10Rα and mIL-10Rβ chains as compared to mIL-10R$\alpha\beta$ complexes. EBV-derived viral IL-10 (vIL-10) bound human and mouse IL-10Rα transfectants with at least a 1000-fold lower affinity than their corresponding cellular ligand/receptor combinations. However, vIL-10 could activate cells through mIL-10R or hIL-10R with a specific activity identical to the cellular cytokine depending on the cell type and in agreement with the limited biological spectrum of vIL-10 as compared to cellular IL-10. Recently, it has been shown that a single amino acid substitution at position 87 (I-A) is responsible for the reduced immunostimulatory activity of viral IL-10 (Ding *et al.*, 2000). All responses to vIL-10 are dependent on the expression of the IL-10Rα as indicated by a blocking anti-hIL-10Rα monoclonal antibody (Liu *et al.*, 1997). CmvIL-10, a viral homolog of IL-10 identified in human cytomegalovirus ORF UL111a, also binds to the hIL-10R complex and induces signal transduction events and biological activities (Kotenko *et al.*, 2000). Interestingly, unlike EBV-derived vIL-10, which is 84% homologous to hIL-10, cmvIL-10 displays only 27% homology to hIL-10.

Cell types and tissues expressing the receptor

IL-10Rα chains are expressed on cells of hematopoietic origin including T cells, B cells, monocytes, macrophages, dendritic cells, NK cells, mast cells, and various hematopoietic progenitors. In addition, tumor cells of hematopoietic origin, such as myeloma cells, B CLL, microglia, etc. have been described to express the IL-10Rα The IL-10Rα is present on epithelial cells from murine small and large intestine, on human

Figure 1 Nucleotide sequences for the mouse and human IL-10Rα genes and the mouse and human IL-10Rβ genes.

Mouse IL-10Ra:

```
   1 CCATTGTGCT GGAAAGCAGG ACGCGCCGGC CGGAGGCGTA AAGGCCGGCT CCAGTGGACG
  61 ATGCCGCTGT GCGCCCAGGA TGTTGTCGCG TTTGCTCCCA TTCCTCGTCA CGATCTCCAG
 121 CCTGAGCCTA GAATTCATTG CATACGGGAC AGAACTGCCA AGCCCTTCCT ATGTGTGGTT
 181 TGAAGCCAGA TTTTTCCAGC ACATCCTCCA CTGGAAACCT ATCCCAAACC AGTCTGAGAG
 241 CACCTACTAT GAAGTGGCCC TCAAACAGTA CGGAAACTCA ACCTGGAATG ACATCCATAT
 301 CTGTAGAAAG GCTCAGGCAT TGTCCTGTGA TCTCACAACG TTCACCCTGG ATCTGTATCA
 361 CCGAAGCTAT GGCTACCGGG CCAGAGTCCG GGCAGTGGAC AACAGTCAGT ACTCCAACTG
 421 GACCACCACT GAGACTCGCT TCACAGTGGA TGAAGTGATT CTGACAGTGG ATAGCGTGAC
 481 TCTGAAAGCA ATGGACGGCA TCATCTATGG GACAATCCAT CCCCCCAGGC CCACGATAAC
 541 CCCTGCAGGG GATGAGTACG AACAAGTCTT CAAGGATCTC CGAGTTTACA AGATTTCCAT
 601 CCGGAAGTTC TCAGAACTAA AGAATGCAAC CAAGAGAGTG AAACAGGAAA CCTTCACCCT
 661 CACGGTCCCC ATAGGGGTGA GAAAGTTTTG TGTCAAGGTG CTGCCCCGCT TGGAATCCCG
 721 AATTAACAAG GCAGAGTGGT CGGAGGAGCA GTGTTTACTT ATCACGACGG AGCAGTATTT
 781 CACTGTGACC AACCTGAGCA TCTTAGTCAT ATCTATGCTG CTATTCTGTG GAATCCTGGT
 841 CTGTCTGGTT CTCCAGTGGT ACATCCGGCA CCCGGGGAAG TTGCCTACAG TCCTGGTCTT
 901 CAAGAAGCCT CACGACTTCT TCCCAGCCAA CCCTCTCTGC CCAGAAACTC CCGATGCCAT
 961 TCACATCGTG GACCTGGAGG TTTTCCCAAA GGTGTCACTA GAGCTGAGAG ACTCAGTCCT
1021 GCATGGCAGC ACCGACAGTG GCTTTGGCAG TGGTAAACCA TCACTTCAGA CTGAAGAGTC
1081 CCAATTCCTC CTCCCTGGCT CCCACCCCCA GATACAGGGG ACTCTGGGAA AAGAAGAGTC
1141 TCCAGGGCTA CAGGCCACCT GTGGGGACAA CACGGACAGT GGGATCTGCC TGCAGGAGCC
1201 CGGCTTACAC TCCAGCATGG GGCCCGCCTG GAAGCAGCAG CTTGGATATA CCCATCAGGA
1261 CCAGGATGAC AGTGACGTTA ACCTAGTCCA GAACTCTCCA GGGCAGCCTA AGTACACACA
1321 GGATGCATCT GCCTTGGGCC ATGTCTGTCT CCTAGAACCT AAAGCCCCTG AGGAGAAAGA
1381 CCAAGTCATG GTGACATTCC AGGGCTACCA GAAACAGACC AGATGGAAGG CAGAGGCAGC
1441 AGGCCCAGCA GAATGCTTGG ACGAAGAGAT TCCCTTGACA GATGCCTTTG ATCCTGAACT
1501 TGGGGTACAC CTGCAGGATG ATTTGGCTTG GCCTCCACCA GCTCTGGCCG CAGGTTATTT
1561 GAAACAGGAG TCTCAAGGGA TGGCTTCTGC TCCACCAGGG ACACCAAGTA GACAGTGGAA
1621 TCAACTGACC GAAGAGTGGT CACTCCTGGG TGTGGTTAGC TGTGAAGATC TAAGCATAGA
1681 AAGTTGGAGG TTTGCCCATA AACTTGACCC TCTGGACTGT GGGGCAGCCC CTGGTGGCCT
1741 CCTGGATAGC CTTGGCTCTA ACCTGGTCAC CCTGCCGTTG ATCTCCAGCC TGCAGGTAGA
1801 AGAATGACAG CGGCTAAGAG TTATTTGTAT TCCAGCCATG CCTGCTCCCC TCCCTGTACC
1861 TGGGAGGCTC AGGAGTCAAA GAAATATGTG GGTCCTTTTC TGCAGACCTA CTGTGACCAG
1921 CTAGCCAGGC TCCACGGGGC AAGGAAAGGC CATCTTGATA CACGAGTGTC AGGTACATGA
1981 GAGGTTGTGG CTAGTCTGCT GAGTGAGGGT CTGTAGATAC CAGCAGAGCT GAGCAGGATT
2041 GACAGAGACC TCCTCATGCC TCAGGGCTGG CTCCTACACT GGAAGGACCT GTGTTTGGGT
2101 GTAACCTCAG GGCTTTCTGG ATGTGGTAAG ACTGTAGGTC TGAAGTCAGC TGAGCCTGGA
2161 TGTCTGCGGA GGTGTTGGAG TGGCTAGCCT GCTACAGGAT AAAGGGAAGG CTCAAGAGAT
2221 AGAAGGGCAG AGCATGAGCC AGGTTTAATT TTGTCCTGTA GAGATGGTCC CCAGCCAGGA
2281 TGGGTTACTT GTGGCTGGGA GATCTTGGGG TATACACCAC CCTGAATGAT CAGCCAGTCA
2341 ATTCAGAGCT GTGTGGCAAA AGGGACTGAG ACCCAGAATT TCTGTTCCTC TTGTGAGGTG
2401 TCTCTGCTAC CCATCTGCAG ACAGACATCT TCATCTTTTT ACTATGGCTG TGTCCCCTGA
2461 ATTACCAGCA GTGGCCAAGC CATTACTCCC TGCTGCTCAC TGTTGTGACG TCAGACCAGA
2521 CCAGACGCTG TCTGTCTGTG TTAGTACACT ACCCTTTAGG TGGCCTTTGG GCTTGAGCAC
2581 TGGCCCAGGC TTAGGACTTA TGTCTGCTTT TGCTGCTAAT CTCTAACTGC AGACCCAGAG
2641 AACAGGGTGC TGGGCTGACA CCTCCGTGTT CAGCTGTGTG ACCTCCGACC AGCAGCTTCC
2701 TCAGGGGACT AAAATAATGA CTAGGTCATT CAGAAGTCCC TCATGCTGAA TGTTAACCAA
2761 GGTGCCCCTG GGGTGATAGT TTAGGTCCTG CAACCTCTGG GTTGGAAGGA AGTGGACTAC
2821 GGAAGCCATC TGTCCCCCTG GGGAGCTTCC ACCTCATGCC AGTGTTTCAG AGATCTTGTG
2881 GGAGCCTAGG GCCTTGTGCC AAGGGAGCTG CTAGTCCCTG GGGTCTAGGG CTGGTCCCTG
2941 CCTCCCTATA CTGCGTTTGA GACCTGTCTT CAAATGGAGG CAGTTTGCAG CCCCTAAGCA
3001 AGGATGCTGA GAGAAGCAGC AAGGCTGCTG ATCCCTGAGC CCAGAGTTTC TCTGAAGCTT
3061 TCCAAATACA GACTGTGTGA CGGGGTGAGG CCAGCCATGA ACTTTGGCAT CCTGCCGAGA
3121 AGGTCATGAC CCTAATCTGG TACGAGAGCT CCTTCTGGAA CTGGGCAAGC TCTTTGAGAC
3181 CCCCCTGGAA CCTTTATTTA TTTATTTGCT CACTTATTTA TTGAGGAAGC AGCGTGGCAC
3241 AGGCGCAAGG CTCTGGGTCT CTCAGGAGGT CTAGATTTGC CTGCCCTGTT TCTAGCTGTG
3301 TGACCTTGGG CAAGTCACGT TTCCTCGTGG AGCCTCAGTT TTCCTGTCTG TATGCAAAGC
3361 TTGGAAATTG AAATGTACCT GACGTGCTCC ATCCCTAGGA GTGCTGAGTC CCACTGAGAA
3421 AGCGGGCACA GACGCCTCAA ATGGAACCAC AAGTGGTGTG TGTTTTCATC CTAATAAAAA
3481 GTCAGGTGTT TTGTGGA
```

Figure 1 (*Continued*)

Human IL-10Ra:

```
   1 AAAGAGCTGG AGGCGCGCAG GCCGGCTCCG CTCCGGCCCC GGACGATGCG GCGCGCCCAG
  61 GATGCTGCCG TGCCTCGTAG TGCTGCTGGC GGCGCTCCTC AGCCTCCGTC TTGGCTCAGA
 121 CGCTCATGGG ACAGAGCTGC CCAGCCCTCC GTCTGTGTGG TTTGAAGCAG AATTTTTCCA
 181 CCACATCCTC CACTGGACAC CCATCCCAAA TCAGTCTGAA AGTACCTGCT ATGAAGTGGC
 241 GCTCCTGAGG TATGGAATAG AGTCCTGGAA CTCCATCTCC AACTGTAGCC AGACCCTGTC
 301 CTATGACCTT ACCGCAGTGA CCTTGGACCT GTACCACAGC AATGGCTACC GGGCCAGAGT
 361 GCGGGCTGTG GACGGCAGCC GGCACTCCAA CTGGACCGTC ACCAACACCC GCTTCTCTGT
 421 GGATGAAGTG ACTCTGACAG TTGGCAGTGT GAACCTAGAG ATCCACAATG GCTTCATCCT
 481 CGGGAAGATT CAGCTACCCA GGCCCAAGAT GGCCCCCGCG AATGACACAT ATGAAAGCAT
 541 CTTCAGTCAC TTCCGAGAGT ATGAGATTGC CATTCGCAAG GTGCCGGGAA ACTTCACGTT
 601 CACACACAAG AAAGTAAAAC ATGAAAACTT CAGCCTCCTA ACCTCTGGAG AAGTGGGAGA
 661 GTTCTGTGTC CAGGTGAAAC CATCTGTCGC TTCCCGAAGT AACAAGGGGA TGTGGTCTAA
 721 AGAGGAGTGC ATCTCCCTCA CCAGGCAGTA TTTCACCGTG ACCAACGTCA TCATCTTCTT
 781 TGCCTTTGTC CTGCTGCTCT CCGGAGCCCT CGCCTACTGC CTGGCCCTCC AGCTGTATGT
 841 GCGGCGCCGA AAGAAGCTAC CCAGTGTCCT GCTCTTCAAG AAGCCCAGCC CCTTCATCTT
 901 CATCAGCCAG CGTCCCTCCC CAGAGACCCA AGACACCATC CACCCGCTTG ATGAGGAGGC
 961 CTTTTTGAAG GTGTCCCCAG AGCTGAAGAA CTTGGACCTG CACGGCAGCA CAGACAGTGG
1021 CTTTGGCAGC ACCAAGCCAT CCCTGCAGAC TGAAGAGCCC CAGTTCCTCC TCCCTGACCC
1081 TCACCCCCAG GCTGACAGAA CGCTGGGAAA CGGGGAGCCC CCTGTGCTGG GGGACAGCTG
1141 CAGTAGTGGC AGCAGCAATA GCACAGACAG CGGGATCTGC CTGCAGGAGC CCAGCCTGAG
1201 CCCCAGCACA GGGCCCACCT GGGAGCAACA GGTGGGGAGC AACAGCAGGG GCCAGGATGA
1261 CAGTGGCATT GACTTAGTTC AAAACTCTGA GGGCCGGGCT GGGGACACAC AGGGTGGCTC
1321 GGCCTTGGGC CACCACAGTC CCCCGGAGCC TGAGGTGCCT GGGGAAGAAG ACCCAGCTGC
1381 TGTGGCATTC CAGGGTTACC TGAGGCAGAC CAGATGTGCT GAAGAGAAGG CAACCAAGAC
1441 AGGCTGCCTG GAGGAAGAAT CGCCCTTGAC AGATGGCCTT GGCCCCAAAT TCGGGAGATG
1501 CCTGGTTGAT GAGGCAGGCT TGCATCCACC AGCCCTGGCC AAGGGCTATT TGAAACAGGA
1561 TCCTCTAGAA ATGACTCTGG CTTCCTCAGG GGCCCCAACG GGACAGTGGA ACCAGCCCAC
1621 TGAGGAATGG TCACTCCTGG CCTTGAGCAG CTGCAGTGAC CTGGGAATAT CTGACTGGAG
1681 CTTTGCCCAT GACCTTGCCC CTCTAGGCTG TGTGGCAGCC CCAGGTGGTC TCCTGGGCAG
1741 CTTTAACTCA GACCTGGTCA CCCTGCCCCT CATCTCTAGC CTGCAGTCAA GTGAGTGACT
1801 CGGGCTGAGA GGCTGCTTTT GATTTTAGCC ATGCCTGCTC CTCTGCCTGG ACCAGGAGGA
1861 GGGCCCTGGG GCAGAAGTTA GGCACGAGGC AGTCTGGGCA CTTTTCTGCA AGTCCACTGG
1921 GGCTGGCCCA GCCAGGCTGC AGGGCTGGTC AGGGTGTCTG GGGCAGGAGG AGGCCAACTC
1981 ACTGAACTAG TGCAGGGTAT GTGGGTGGCA CTGACCTGTT CTGTTGACTG GGGCCCTGCA
2041 GACTCTGGCA GAGCTGAGAA GGGCAGGGAC CTTCTCCCTC CTAGGAACTC TTTCCTGTAT
2101 CATAAAGGAT TATTTGCTCA GGGGAACCAT GGGGCTTTCT GGAGTTGTGG TGAGGCCACC
2161 AGGCTGAAGT CAGCTCAGAC CCAGACCTCC CTGCTTAGGC CACTCGAGCA TCAGAGCTTC
2221 CAGCAGGAGG AAGGGCTGTA GGAATGGAAG CTTCAGGGCC TTGCTGCTGG GGTCATTTTT
2281 AGGGGAAAAA GGAGGATATG ATGGTCACAT GGGGAACCTC CCCTCATCGG GCCTCTGGGG
2341 CAGGAAGCTT GTCACTGGAA GATCTTAAGG TATATATTTT CTGGACACTC AAACACATCA
2401 TAATGGATTC ACTGAGGGGA GACAAAGGGA GCCGAGACCC TGGATGGGGC TTCCAGCTCA
2461 GAACCCATCC CTCTGGTGGG TACCTCTGGC ACCCATCTGC AAATATCTCC CTCTCTCCAA
2521 CAAATGGAGT AGCATCCCCC TGGGGCACTT GCTGAGGCCA AGCCACTCAC ATCCTCACTT
2581 TGCTGCCCCA CCATCTTGCT GACAACTTCC AGAGAAGCCA TGGTTTTTTG TATTGGTCAT
2641 AACTCAGCCC TTTGGGCGGC CTCTGGGCTT GGGCACCAGC TCATGCCAGC CCCAGAGGGT
2701 CAGGGTTGGA GGCCTGTGCT TGTGTTTGCT GCTAATGTCC AGCTACAGAC CCAGAGGATA
2761 AGCCACTGGG CACTGGGCTG GGGTCCCTGC CTTGTTGGTG TTCAGCTGTG TGATTTTGGA
2821 CTAGCCACTT GTCAGAGGGC CTCAATCTCC CATCTGTGAA ATAAGGACTC CACCTTTAGG
2881 GGACCCTCCA TGTTTGCTGG GTATTAGCCA AGCTGGTCCT GGGAGAATGC AGATACTGTC
2941 CGTGGACTAC CAAGCTGGCT TGTTTCTTAT GCCAGAGGCT AACAGATCCA ATGGGAGTCC
3001 ATGGTGTCAT GCCAAGACAG TATCAGACAC AGCCCCAGAA GGGGGCATTA TGGGCCCTGC
3061 CTCCCCATAG GCCATTTGGA CTCTGCCTTC AAACAAAGGC AGTTCAGTCC ACAGGCATGG
3121 AAGCTGTGAG GGGACAGGCC TGTGCGTGCC ATCCAGAGTC ATCTCAGCCC TGCCTTTCTC
3181 TGGAGCATTC TGAAAACAGA TATTCTGGCC CAGGGAATCC AGCCATGACC CCCACCCCTC
3241 TGCCAAAGTA CTCTTAGGTG CCAGTCTGGT AACTGAACTC CCTCTGGAGG CAGGCTTGAG
3301 GGAGGATTCC TCAGGGTTCC CTTGAAAGCT TTATTTATTT ATTTTGTTCA TTTATTTATT
3361 GGAGAGGCAG CATTGCACAG TGAAAGAATT CTGGATATCT CAGGAGCCCC GAAATTCTAG
3421 CTCTGACTTT GCTGTTTCCA GTGGTATGAC CTTGGAGAAG TCACTTATCC TCTTGGAGCC
3481 TCAGTTTCCT CATCTGCAGA ATAATGACTG ACTTGTCTAA TTCATAGGGA TGTGAGGTTC
3541 TGCTGAGGAA ATGGGTATGA ATGTGCCTTG AACACAAAGC TCTGTCAATA AGTGATACAT
3601 GTTTTTTATT CCAATAAATT GTCAAGACCA CA
```

Figure 1 (*Continued*)

Mouse IL-10Rb:

```
   1 ACATGGCCCC GTGCGTGGCG GGCTGGCTGG GTGGCTTCCT TCTGGTGCCA GCTCTAGGAA
  61 TGATTCCACC CCCTGAGAAG GTCAGAATGA ATTCAGTTAA TTTCAAGAAC ATTCTACAGT
 121 GGGAGGTACC TGCTTTCCCC AAAACGAACC TGACTTTCAC AGCTCAGTAT GAAAGTTACA
 181 GGTCTTTCCA AGATCACTGC AAGCGCACTG CCTCGACTCA GTGCGACTTC TCTCATCTTT
 241 CTAAATACGG AGACTACACT GTGAGAGTCA GGGCTGAATT GGCGGATGAA CATTCGGAGT
 301 GGGTCAATGT CACCTTCTGC CCCGTGGAAG ACACCATCAT TGGACCTCCT GAGATGCAGA
 361 TAGAATCCCT TGCTGAGTCT TTACACCTGC GTTTCTCAGC CCCACAAATT GAGAATGAGC
 421 CTGAGACGTG GACCTTGAAG AACATTTATG ACTCATGGGC TTACAGAGTG CAATACTGGA
 481 AAAATGGGAC TAATGAGAAG TTTCAAGTTG TGTCTCCGTA CGACTCTGAG GTCCTCCGGA
 541 ACCTGGAGCC GTGGACAACT TACTGCATTC AAGTTCAAGG GTTTCTTCTC GACCAGAACA
 601 GAACAGGAGA GTGGAGTGAA CCCATCTGTG AACGGACAGG CAATGACGAA ATAACCCCTT
 661 CCTGGATTGT GGCCATCATC CTCATAGTCT CCGTCCTGGT GGTCTTCCTC TTCCTCCTGG
 721 GCTGCTTTGT CGTGCTGTGG CTCATTTATA AGAAGACCAA GCATACCTTC CGTTCTGGGA
 781 CGTCTCTTCC ACAGCACCTG AAGGAGTTTC TGGGCCACCC CCATCACAGC ACGTTTCTGC
 841 TGTTCTCCTT CCCTCCCCCC GAGGAGGCCG AGGTGTTCGA CAAACTAAGC ATCATCAGCG
 901 AAGAGTCTGA AGGCAGCAAG CAGAGTCCTG AAGACAACTG TGCCTCAGAG CCCCCGTCTG
 961 ATCCAGGGCC TCGGGAGCTG GAGTCCAAGG ATGAAGCTCC CTCACCTCCA CACGATGACC
1021 CCAAACTGCT CACGTCGACC TCAGAAGTAT GACCAGAGAG CCACCTGAAA AAACTCCAAA
1081 TCTAGAACTT CCTGATGCTG CACTGGTACA CACAACCAAA GAGCTAGGTT TTAAACACTC
1141 TACTTGGGAA TTTGCTGCCA TATAAAGACT AATAATTTAG GGACTGAGGG TGTAGCTCAG
1201 TGACTAGAGC TCTTACTTGG CACACATGAA GTCCTAGCTT CGATCCCCAA CACCATATAA
1261 ACCAGGGATG GGGGCACCTA CCTATAAGCC CAGCACTTTG GAGGTAGAGG CAGGAGGATC
1321 ACAGTCATCT TGAACTATAC AGGGAGTTCA AGGCCAAGCT GGACTAGAGA CCCTGTCTAA
1381 GAGAGAGAGA GAGAACTTAT ATATTTTATG GCCACTGAAT GTAATTTGAG CCCTTTGTGC
1441 TCACTAAAAC AAGGATCACA TTTAACTTGT GACAAACAAA AATATTTTAA ATGGGGGGGG
1501 GGGCATGGAA ACACTATGAA ATTATAAGAA TGCCTATAGA CCACCCGCAT CTCAAAAGTG
1561 GTTGGCCCCA TGCGGGACAG ACATGAACAT TTTGGATTCC CAAGGAGCAA AGAGATTTCC
1621 TTCCTTACCT GTGTGTTTTG TATTAATATT AGTGTTCTGT AAATATTCTA
```

Human IL-10Rb:

```
  1 ATGGCGTGGA GTCTTGGGAG CTGGCTGGGT GGCTGCCTGC TGGTGTCAGC ATTGGGAATG
 61 GTACCACCTC CCGAAAATGT CAGAATGAAT TCTGTTAATT TCAAGAACAT TCTACAGTGG
121 GAGTCACCTG CTTTTGCCAA AGGGAACCTG ACTTTCACAG CTCAGTACCT AAGTTATAGG
181 ATATTCCAAG ATAAATGCAT GAATACTACC TTGACGGAAT GTGATTTCTC AAGTCTTTCC
241 AAGTATGGTG ACCACACCTT GAGAGTCAGG GCTGAATTTG CAGATGAGCA TTCAGACTGG
301 GTAAACATCA CCTTCTGTCC TGTGGATGAC ACCATTATTG GACCCCCTGG AATGCAAGTA
361 GAAGTACTTG ATGATTCTTT ACATATGCGT TTCTTAGCCC CTAAAATTGA GAATGAATAC
421 GAAACTTGGA CTATGAAGAA TGTGTATAAC TCATGGACTT ATAATGTGCA ATACTGGAAA
481 AACGGTACTG ATGAAAAGTT TCAAATTACT CCCCAGTATG ACTTTGAGGT CCTCAGAAAC
541 CTGGAGCCAT GGACAACTTA TTGTGTTCAA GTTCGAGGGT TTCTTCCTGA TCGGAACAAA
601 GCTGGGGAAT GGAGTGAGCC TGTCTGTGAG CAAACAACCC ATGACGAAAC GGTCCCCTCC
661 TGGATGGTGG CCGTCATCCT CATGGCCTCG GTCTTCATGG TCTGCCTGGC ACTCCTCGGC
721 TGCTTCTCCT TGCTGTGGTG CGTTTACAAG AAGACAAAGT ACGCCTTCTC CCCTAGGAAT
781 TCTCTTCCAC AGCACCTGAA AGAGGTAGGT AGGATGGAGT GA
```

epidermal cells and keratinocytes and can be induced on fibroblasts (Michel *et al.*, 1997; Denning *et al.*, 2000).

IL-10Rβ chain seems to be expressed ubiquitously.

Regulation of receptor expression

IL-10Rα expression is induced in fibroblasts by LPS activation (Weber-Nordt *et al.*, 1994). In contrast, the expression of IL-10Rα mRNA by human T cell clones was downregulated following activation (Liu *et al.*, 1994).

Release of soluble receptors

No data has been published on the production and release of soluble IL-10 receptors in normal or disease states.

SIGNAL TRANSDUCTION

Associated or intrinsic kinases

Neither the IL-10Rα nor the IL-10Rβ chain possesses an intrinsic kinase activity. However, IL-10 treatment of cells induces phosphorylation of the IL-10Rα-associated JAK1 and the IL-10Rβ-associated TYK2 kinases (Finbloom and Winestock, 1995; Ho *et al.*, 1995).

Cytoplasmic signaling cascades

Phosphorylation of JAK1 (Janus kinase 1) and TYK2 kinases by IL-10 binding to the receptor complex activates their kinase activity and leads to phosphorylation of two tyrosine residues, which are part of the cytokine receptor box 3 motif, (Y446 and Y496 in hIL-10Rα and Y427 and Y477 in mIL-10Rα) in the intracytoplasmic tail of the IL-10Rα chain. Phosphorylation of at least one tyrosine residue is sufficient for a biological response. These phosphorylated tyrosine residues then serve as docking sites for the latent transcription factor STAT3 (signal transducer and activator of transcription 3) which in turn becomes phosphorylated by the activated JAK1 and TYK2 kinases. Phosphorylated STAT molecules subsequently form homo- or heterodimers and translocate from the cytoplasm to the nucleus, where they bind with high affinity to STAT-binding elements (SBE) in the promoters of IL-10-responsive genes (Lai *et al.*, 1996; Wehinger *et al.*, 1996). IL-10 can also activate STAT5 and/or only STAT1 DNA-binding activity in some cell types (Weber-Nordt *et al.*, 1996a; Zocchia *et al.*, 1997). Whereas STAT3 is directly recruited to the IL-10Rα chain, activation of STAT1 and STAT5 may occur through other mechanisms (Weber-Nordt *et al.*, 1996b). The two membrane-distal tyrosine molecules in the intracytoplasmic tail of the IL-10Rα chain and STAT3 are essential for the antiproliferative and developmental actions of IL-10, but an additonal C-terminal sequence which contains at least one functionally critical serine residue is required for the anti-inflammatory effects of IL-10 such as inhibition of TNFα production by monocytes/macrophages (O'Farrell *et al.*, 1998; Riley *et al.*, 1999). Another membrane-proximal region of the IL-10Rα chain has been implicated as a functional domain involved in regative regulation (Ho *et al.*, 1995).

IL-10 antagonizes the induction of some IFNα- or IFNγ-inducible genes on human monocytes by preventing the IFN-induced phosphorylation of STAT1 (Ito *et al.*, 1999). This may be mediated by the rapid STAT3-dependent induction of SOCS3 (suppressor of cytokine signaling 3). SOCS proteins are members of a family of molecules that interfere with JAK/STAT signal transduction pathways. Alternatively, differences in DNA-binding activity and composition of IL-10-induced STAT1 and IFN-induced STAT1 complexes may play a role in the differential responses to IL-10 and IFNγ (Yamaoka *et al.*, 1999).

Interestingly, in contrast to IFNγ, IL-10 failed to upregulate FcγRI expression and GRR (IFNγ response region)- or SIE (serum-inducible element)-binding activity in human neutrophils, despite the presence of a functional IL-10R complex, and STAT1 and STAT3 expression by these cells (Bovolenta *et al.*, 1998). IL-10 did induce SOCS3 expression in neutrophils, indicating that activation of STAT1 or STAT3 phosphorylation is not required for SOCS3 induction in these cells as it is in monocytes (Cassatella *et al.*, 1999).

Other signaling pathways that have been implicated in biological responses to IL-10 include NFκB, phosphatidylinositol 3-kinase, p70 S6 kinase, and MAP kinase cascades, but none of these have been directly linked to molecules that interact with the IL-10 receptor (De Waal Malefyt and Moore, 1998).

DOWNSTREAM GENE ACTIVATION

Transcription factors activated

IL-10 activates STAT1, STAT3, and in some cell types STAT5 DNA-binding activity and transcriptional activation (Finbloom and Winestock, 1995; Ho *et al.*, 1995; Lai *et al.*, 1996; Weber Nordt *et al.*, 1996a; Wehinger *et al.*, 1996).

Genes induced

IL-10 induces expression of FcγRI, TIMP-1 (tissue inhibitor of metalloproteinases 1), MCP-1, Ilinck (IL-10-induced chemokine), CCR5 on monocytes, and enhances IL-1Ra and soluble p55 and p75 TNFαR. On mouse mast cells it induces expression of mast cell-specific proteases mMCP-1 and mMCP-2, whereas on mouse B cells it enhanced expression of MHC class II. On human B cells it enhanced expression of Bcl-2 and the high-affinity IL-2R. A protein homologous to SNAP23 (synaptosomal-associated protein of 23 kDa) was induced by IL-10 in OTT1 cells (Morikawa *et al.*, 1998). IL-10 inhibited the proliferation of bone marrow-derived macrophages and of J774 cells by STAT3-dependent induction of the cyclin-dependent kinase inhibitor p19INK4D, which acts on the interactions between cdk's 4 and 6 and D cyclins, and the STAT3-independent induction of p21CIP1, which has been shown to inhibit cyclins A or E-associated cdk-2 activity.

Promoter regions involved

IL-10-activated STAT1, STAT3, and STAT5 are able to bind to the GRR (IFNγ response region) of the FcγRI gene, the SIE (serum-inducible element) of the c-*fos* promoter and the PRL-STAT (prolactin STAT) consensus sequence of the β-casein gene (Wehinger *et al.*, 1996). In addition, STAT3 bound the IL-6 response element in hepatoma cells transfected with the IL-10R (Lai *et al.*, 1996) and could enhance the expression of hsp90α (heat shock protein) and hsp90β promoters in these cells and in peripheral blood mononuclear cells (Ripley, 1999). Finally, IL-10 activated two STAT3-binding sites in the proximal p19INK4D promoter (O'Farrell *et al.*, 2000).

IL-10 downregulated IFNγ-induced ICAM1 transcription in human monocytes by preventing IFNγ-induced binding activity at the NFκB site of the TNFα-responsive NFκB/CEBP composite element in the ICAM1 promoter (Song *et al.*, 1997).

BIOLOGICAL CONSEQUENCES OF ACTIVATING OR INHIBITING RECEPTOR AND PATHOPHYSIOLOGY

Unique biological effects of activating the receptors

IL-10 binding to its receptor complex induces the modulation of a wide variety of immunological and biological responses as described in the IL-10 chapter. However, the most prominent activity of IL-10 is probably its immunosuppressive and anti-inflammatory actions on monocytes (De Waal Malefyt and Moore, 1998).

Phenotypes of receptor knockouts and receptor overexpression mice

An IL-10Rβ chain knockout mouse has been described. These mice did not show abnormalities at birth and developed normally until 12 weeks of age, at which time a majority developed a *chronic colitis* and *splenomegaly*, reminiscent of the pathology and phenotype of IL-10 knockout mice (Spencer *et al.*, 1998). Cells from IL-10Rβ−/− mice responded normally to type I and type II interferons, but did not respond to IL-10.

Human abnormalities

No human abnormalities related to the IL-10R complex have been described to date.

THERAPEUTIC UTILITY

Effect of treatment with soluble receptor domain

ShIL-10R are able to neutralize the effects of IL-10 in proliferation and differentation assays (Tan *et al.*, 1995).

Effects of inhibitors (antibodies) to receptors

Anti-IL-10R antibodies that neutralize the biological activity of mIL-10 or hIL-10 have been described (Ho *et al.*, 1995; Liu *et al.*, 1998). Although *in vivo* studies in mice are ongoing, it can be predicted that anti-IL-10R monoclonal antibodies, like anti-IL-10 monoclonal antibodies, are able to enhance acquired cellular immune reponses against intracellular pathogens, bacteria, and tumor cells (De Waal Malefyt and Moore, 1998).

References

Bovolenta, C., Gasperini, S., McDonald, P. P., and Cassatella, M. A. (1998). High affinity receptor for IgG (Fc gamma RI/CD64) gene and STAT protein binding to the IFN-gamma response region (GRR) are regulated differentially in human neutrophils and monocytes by IL-10. *J. Immunol.* **160**, 911–919.

Cassatella, M. A., Gasperini, S., Bovolenta, C., Calzetti, F., Vollebregt, M., Scapini, P., Marchi, M., Suzuki, R., Suzuki, A., and Yoshimura, A. (1999). Interleukin-10 (IL-10) selectively enhances CIS3/SOCS3 mRNA expression in human neutrophils: evidence for an IL-10-induced pathway that is independent of STAT protein activation. *Blood* **94**, 2880–2889.

De Waal Malefyt, R., and Moore, K. W. (1998). In "The Cytokine Handbook, 3rd edn" (ed A. Thomson), Interleukin-10, pp. 333–364. Academic Press, London.

Denning, T. L., Campbell, N. A., Song, F., Garofalo, R. P., Klimpel, G. R., Reyes, V. E., and Ernst, P. B. (2000). Expression of IL-10 receptors on epithelial cells from the murine small and large intestine. *Int. Immunol.* **12**, 133–139.

Ding, Y., Qin, L., Kotenko, S. V., Pestka, S., and Bromberg, J. S. (2000). A single amino acid determines the immunostimulatory activity of interleukin 10. *J. Exp. Med.* **191**, 213–224.

Finbloom, D. S., and Winestock, K. D. (1995). IL-10 induces the tyrosine phosphorylation of tyk2 and Jak1 and the differential assembly of STAT1 alpha and STAT3 complexes in human T cells and monocytes. *J Immunol* **155**, 1079–1090.

Ho, A. S., Liu, Y., Khan, T. A., Hsu, D. H., Bazan, J. F., and Moore, K. W. (1993). A receptor for interleukin 10 is related to interferon receptors. *Proc. Natl Acad. Sci. USA* **90**, 11267–11271.

Ho, A. S., Wei, S. H., Mui, A. L., Miyajima, A., and Moore, K. W. (1995). Functional regions of the mouse interleukin-10 receptor cytoplasmic domain. *Mol. Cell Biol.* **15**, 5043–5053.

Hoover, D. M., Schalk-Hihi, C., Chou, C. C., Menon, S., Wlodawer, A., and Zdanov, A. (1999). Purification of receptor complexes of interleukin-10 stoichiometry and the importance of deglycosylation in their crystallization. *Eur. J. Biochem.* **262**, 134–141.

Ito, S., Ansari, P., Sakatsume, M, Dickensheets, H., Vazquez, N., Donnelly, R. P., Larner, A. C., and Finbloom, D. S. (1999). Interleukin-10 inhibits expression of both interferon alpha- and interferon gamma-induced genes by suppressing tyrosine phosphorylation of STAT1. *Blood* **93**, 1456–1463.

Kotenko, S. V., Krause, C. D., Izotova, L. S., Pollack, B. P., Wu, W., and Pestka, S. (1997). Identification and functional characterization of a second chain of the interleukin-10 receptor complex. *EMBO J.* **16**, 5894–5903.

Kotenko, S. V., Saccani, S., Izotova, L. S., Mirochnitchenko, O. V., and Pestka, S. (2000). Human cytomegalovirus harbors its own unique IL-10 homolog (cmvIL-10). *Proc. Natl Acad. Sci. USA* **97**, 1695–700.

Lai, C. F., Ripperger, J., Morella, K. K., Jurlander, J., Hawley, T. S., Carson, W. E., Kordula, T., Caligiuri, M. A., Hawley, R. G., Fey, G. H., and Baumann, H. (1996). Receptors for interleukin (IL)-10 and IL-6-type cytokines use similar signaling mechanisms for inducing transcription through IL-6 response elements. *J. Biol. Chem.* **271**, 13968–13975.

Liu, Y., Wei, S. H., Ho, A. S., de Waal Malefyt, R., and Moore, K. W. (1994). Expression cloning and characterization of a human IL-10 receptor. *J. Immunol.* **152**, 1821–1829.

Liu, Y., de Waal Malefyt, R., Briere, F., Parham, C., Bridon, J. M., Banchereau, J., Moore, K. W., and Xu, J. (1997). The EBV IL-10 homologue is a selective agonist with impaired binding to the IL-10 receptor. *J. Immunol.* **158**, 604–613.

Michel, G., Mirmohammadsadegh, A., Olasz, E., Jarzebska-Deussen, B., Muschen, A., Kemeny, L., Abts, H. F., and Ruzicka, T. (1997). Demonstration and functional analysis of IL-10 receptors in human epidermal cells: decreased expression in psoriatic skin, down-modulation by IL-8, and up-regulation by an antipsoriatic glucocorticosteroid in normal cultured keratinocytes. *J. Immunol.* **159**, 6291–6297.

Morikawa, Y., Nishida, H., Misawa, K., Nosaka, T., Miyajima, A., Senba, E., and Kitamura, T. (1998). Induction of synaptosomal-associated protein-23 kD (SNAP-23) by various cytokines. *Blood* **92**, 129–135.

O'Farrell, A.-M., Liu, Y., Moore, K. W., and Mui, A. L. (1998). IL-10 inhibits macrophage activation and proliferation by distinct signaling mechanisms: evidence for Stat3-dependent and -independent pathways. *EMBO J.* **17**, 1006–1018.

O'Farrell, A.-M., Parry, D. A., Zindy, F., Roussel, M. F., Lees, E., Moore, K. W., and Mui, A. L.-F. (2000). Stat3-dependent induction of p19INK4D by IL-10 contributes to inhibition of macrophage proliferation. *J. Immunol.* **164**, 4607–4615.

Reboul, J., Gardiner, K., Monneron, D., Uze, G., and Lutfalla, G. (1999). Comparative genomic analysis of the interferon/interleukin-10 receptor gene cluster. *Genome Res.* **9**, 242–250.

Riley, J. K., Takeda, K., Akira, S., and Schreiber, R. D. (1999). Interleukin-10 receptor signaling through the JAK-STAT pathway. Requirement for two distinct receptor-derived signals for anti-inflammatory action. *J. Biol. Chem.* **274**, 16513–16521.

Ripley, B. J., Stephanou, A., Isenberg, D. A., and Latchman, D. S. (1999). Interleukin-10 activates heat-shock protein 90beta gene expression. *Immunology* **97**, 226–231.

Spencer, S. D., Di Marco, F., Hooley, J., Pitts-Meek, S., Bauer, M., Ryan, A. M., Sordat, B., Gibbs, V. C., and Aguet, M. (1998). The orphan receptor CRF2–4 is an essential subunit of the interleukin 10 receptor. *J. Exp. Med.* **187**, 571–578.

Song, S., Ling-Hu, H., Roebuck, K. A., Rabbi, M. F., Donnelly, R. P., and Finnegan, A. (1997). Interleukin-10 inhibits interferon-gamma-induced intercellular adhesion molecule-1 gene transcription in human monocytes. *Blood* **89**, 4461–4469.

Tan, J. C., Indelicato, S. R., Narula, S. K., Zavodny, P. J., and Chou, C. C. (1993). Characterization of interleukin-10 receptors on human and mouse cells. *J. Biol. Chem.* **268**, 21053–21059.

Tan, J. C., Braun, S., Rong, H., DiGiacomo, R., Dolphin, E., Baldwin, S., Narula, S. K., Zavodny, P. J., and Chou, C. C. (1995). Characterization of recombinant extracellular domain of human interleukin-10 receptor. *J. Biol. Chem.* **270**, 12906–12911.

Weber-Nordt, R. M., Meraz, M. A., and Schreiber, R. D. (1994). Lipopolysaccharide-dependent induction of IL-10 receptor expression on murine fibroblasts. *J. Immunol.* **153**, 3734–3744.

Weber-Nordt, R. M., Egen, C., Wehinger, J., Ludwig, W., Gouilleux-Gruart, V., Mertelsmann, R., and Finke, J. (1996a). Constitutive activation of STAT proteins in primary lymphoid and myeloid leukemia cells and in Epstein–Barr virus (EBV)-related lymphoma cell lines. *Blood* **88**, 809–816.

Weber-Nordt, R. M., Riley, J. K., Greenlund, A. C., Moore, K. W., Darnell, J. E., and Schreiber, R. D. (1996b). Stat3 recruitment by two distinct ligand-induced, tyrosine-phosphorylated docking sites in the interleukin-10 receptor intracellular domain. *J. Biol. Chem.* **271**, 27954–28961.

Wehinger, J., Gouilleux, F., Groner, B., Finke, J., Mertelsmann, R., and Weber-Nordt, R. M. (1996). IL-10 induces DNA binding activity of three STAT proteins (Stat1, Stat3, and Stat5) and their distinct combinatorial assembly in the promoters of selected genes. *FEBS Lett.* **394**, 365–370.

Yamaoka, K., Otsuka, T., Niiro, H., Nakashima, H., Tanaka, Y., Nagano, S., Ogami, E., Niho, Y., Hamasaki, N., and Izuhara, K. (1999). Selective DNA-binding activity of interleukin-10-stimulated STAT molecules in human monocytes. *J. Interferon Cytokine Res.* **19**, 679–685.

Zdanov, A., Schalk-Hihi, C., and Wlodawer, A. (1996). Crystal structure of human interleukin-10 at 1.6 A resolution and a model of a complex with its soluble receptor. *Protein Sci.* **5**, 1955–1962.

Zocchia, C., Spiga, G., Rabin, S. J., Grekova, M., Richert, J., Chernyshev, O., Colton, C., and Mocchetti, I. (1997). Biological activity of interleukin-10 in the central nervous system. *Neurochem Int.* **30**, 433–439.

IL-12 Receptor

Clemens Esche, Michael R. Shurin and Michael T. Lotze*

Biological Therapeutics Program, University of Pittsburgh Cancer Institute, Pittsburgh, PA 15213, USA

*corresponding author tel: 412-624-9375, fax: 412-624-1172, e-mail: lotzemt@msx.upmc.edu

DOI: 10.1006/rwcy.2000.14006.

SUMMARY

The functional high-affinity IL-12 receptor is composed of at least two β-type receptor subunits, each independently exhibiting a low affinity for IL-12. Both subunits are members of the cytokine receptor superfamily. IL-12 p40 interacts primarily with IL-12Rβ1, while IL-12 p35 interacts primarily with the signal-transducing β2 subunit. IL-12 signaling involves activation of the receptor-associated tyrosine kinases JAK2 and TYK2 and downstream tyrosine phosphorylation of the transcription factors STAT3 and STAT4 (the central signal transducer of IL-12). The IL-12R is expressed primarily on activated T and NK cells. Expression of the β2 subunit determines TH1 development since the β2 component is selectively downregulated on TH2 cells. Inhibiting IL-12 responsiveness represents an experimental therapy for TH1-driven inflammatory and autoimmune diseases.

BACKGROUND

Discovery

A cDNA encoding a type I transmembrane protein representing a low-affinity component of the functional IL-12 receptor was identified on PHA-activated human PBMCs (Chua *et al.*, 1994). Receptors with this type of cytoplasmic region have been classified as β-type cytokine receptors (Stahl and Yancopoulos, 1993). Consequently, the new protein was termed IL-12Rβ (Chua *et al.*, 1994). The corresponding murine cDNA was isolated using crosshybridization (Chua *et al.*, 1995). More recently, an additional β-type IL-12R protein was identified and designated IL-12Rβ2 (Presky *et al.*, 1996), while the previously reported subunit was reclassified as IL-12Rβ1 (Presky *et al.*, 1996). A recent report suggests the existence of a third chain (Kawashima *et al.*, 1998).

Alternative names

IL-12Rβ1 was originally termed IL-12Rβ (Presky *et al.*, 1996).

Structure

The functional high-affinity IL-12R is a heterodimer consisting of a β1 and a β2 chain (Gately *et al.*, 1998).

Main activities and pathophysiological roles

The IL-12R is differentially expressed on T cell subsets. Naïve T cells do not express the IL-12R, while antigen stimulation induces expression of both IL-12R subunits. Cells developing along the TH1 pathway continue to express both receptor components, whereas TH2 cells selectively lose the signal-transducing component IL-12Rβ2 during differentiation. Consequently, expression levels of the β2 subunit represent a therapeutic target for the redirection of ongoing T cell responses. See also Main activities and pathophysiological roles in the IL-12 chapter.

GENE

Accession numbers

GenBank:
Human cDNA: U03187 (β1), U64198 (β2)
Murine cDNA: U23922 (β1), U64199.1 (β2)

Chromosome location and linkages

Human IL-12Rβ1 localizes to a region in chromosome 19 at band p13.1, while human IL-12Rβ2 localizes to chromosome 1 at band p31.2 (Yamamoto *et al.*, 1997).

PROTEIN

Accession numbers

GenBank:
Human IL-12R: AAA21340.1 (β1), AAB36675.1 (β2)
Murine IL-12R: AAA87457.1 (β1), AAB36676.1 (β2)
SwissProt:
Human protein: P42701 (β1)

Sequence

See **Figure 1**. Also see Chua *et al.* (1994) for human β1, Chua *et al.* (1995) for murine β1, and Presky *et al.* (1996) for β2.

Description of protein

The β1 subunit is a 662 amino acid type I transmembrane protein with an extracellular domain of 516 amino acids and a cytoplasmic domain of 91 amino acids lacking tyrosine residues (Chua *et al.*, 1994). In contrast, the β2 subunit, which consists of 862 amino acids, has three tyrosine residues in the cytoplasmic domain (Presky *et al.*, 1996).

Relevant homologies and species differences

Both IL-12R subunits are homologous to gp130 (a receptor subunit for IL-6), G-CSF, IL-11, oncostatin M, CNTF and LIF receptors (Chua *et al.*, 1994; Presky *et al.*, 1996; Trinchieri, 1998). Human and murine IL-12Rβ1 demonstrate a 54% amino acid homology (Chua *et al.*, 1995). Human and murine

Figure 1 Sequencesfor IL-12Rβ1 and IL-12Rβ2.

```
Human IL-12Rβ1
  1 MEPLVTWVVP LLFLFLLSRQ GAACRTSECC FQDPPYPDAD SGSASGPRDL RCYRISSDRY
 61 ECSWQYEGPT AGVSHFLRCC LSSGRCCYFA AGSATRLQFS DQAGVSVLYT VTLWVESWAR
121 NQTEKSPEVT LQLYNSVKYE PPLGDIKVSK LAGQLRMEWE TPDNQVGAEV QFRHRTPSSP
181 WKLGDCGPQD DDTESCLCPL EMNVAQEFQL RRRQLGSQGS SWSKWSSPVC VPPENPPQPQ
241 VRFSVEQLGQ DGRRRLTLKE QPTQLELPEG CQGLAPGTEV TYRLQLHMLS CPCKAKATRT
301 LHLGKMPYLS GAAYNVAVIS SNQFGPGLNQ TWHIPADTHT EPVALNISVG TNGTTMYWPA
361 RAQSMTYCIE WQPVGQDGGL ATCSLTAPQD PDPAGMATYS WSRESGAMGQ EKCYYITIFA
421 SAHPEKLTLW STVLSTYHFG GNASAAGTPH HVSVKNHSLD SVSVDWAPSL LSTCPGVLKE
481 YVVRCRDEDS KQVSEHPVQP TETQVTLSGL RAGVAYTVQV RADTAWLRGV WSQPQRFSIE
541 VQVSDWLIFF ASLGSFLSIL LVGVLGYLGL NRAARHLCPP LPTPCASSAI EFPGGKETWQ
601 WINPVDFQEE ASLQEALVVE MSWDKGERTE PLEKTELPEG APELALDTEL SLEDGDRCKA
661 KM

Human IL-12Rβ2
  1 MAHTFRGCSL AFMFIITWLL IKAKIDACKR GDVTVKPSHV ILLGSTVNIT CSLKPRQGCF
 61 HYSRRNKLIL YKFDRRINFH HGHSLNSQVT GLPLGTTLFV CKLACINSDE IQICGAEIFV
121 GVAPEQPQNL SCIQKGEQGT VACTWERGRD THLYTEYTLQ LSGPKNLTWQ KQCKDIYCDY
181 LDFGINLTPE SPESNFTAKV TAVNSLGSSS SLPSTFTFLD IVRPLPPWDI RIKFQKASVS
241 RCTLYWRDEG LVLLNRLRYR PSNSRLWNMV NVTKAKGRHD LLDLKPFTEY EFQISSKLHL
301 YKGSWSDWSE SLRAQTPEEE PTGMLDVWYM KRHIDYSRQQ ISLFWKNLSV SEARGKILHY
361 QVTLQELTGG KAMTQNITGH TSWTTVIPRT GNWAVAVSAA NSKGSSLPTR INIMNLCEAG
421 LLAPRQVSAN SEGMDNILVT WQPPRKDPSA VQEYVVEWRE LHPGGDTQVP LNWLRSRPYN
481 VSALISENIK SYICYEIRVY ALSGDQGGCS SILGNSKHKA PLSGPHINAI TEEKGSILIS
541 WNSIPVQEQM GCLLHYRIYW KERDSNSQPQ LCEIPYRVSQ NSHPINSLQP RVTYVLWMTA
601 LTAAGESSHG NEREFCLQGK ANWMAFVAPS ICIAIIMVGI FSTHYFQQKV FVLLAALRPQ
661 WCSREIPDPA NSTCAKKYPI AEEKTQLPLD RLLIDWPTPE DPEPLVISEV LHQVTPVFRH
721 PPCSNWPQRE KGIQGHQASE KDMMHSASSP PPPRALQAES RQLVDLYKVL ESRGSDPKPE
781 NPACPWTVLP AGDLPTHDGY LPSNIDDLPS HEAPLADSLE ELEPQHISLS VFPSSSLHPL
841 TFSCGDKLTL DQLKMRCDSL ML
```

IL-12Rβ2 proteins demonstrate a 68% amino acid sequence identity (Presky *et al.*, 1996). Although similar with respect to their molecular properties, significant differences exist between mice and humans. Ba/F3 cells expressing murine IL-12Rβ1 exhibited two binding affinities for ^{125}I-mouse IL-12: 50 pM and 470 pM (Chua *et al.*, 1995), while Ba/F3 cells transfected with human IL-12Rβ1 bound ^{125}I-human IL-12 only with very low affinity (> 50 nM) (Presky *et al.*, 1996). These differences should be considered when evaluating potential therapeutic IL-12 antagonists directed against IL-12R in murine models of human disease. Furthermore, expression of IL-12Rβ2 mRNA is regulated differentially in mice and humans. In mice, IFNγ treatment of early developing TH2 cells maintains IL-12Rβ2 expression (Szabo *et al.*, 1997). In humans, IFNα induces expression of IL-12Rβ2 during *in vitro* T cell differentiation (Rogge *et al.*, 1997).

Affinity for ligand(s)

Analysis of steady-state binding data of IL-12 by Scatchard analysis identified a single binding site on PHA-activated human lymphoblasts with an equilibrium dissociation constant of 100–600 pmol/L and 1000–9000 sites per cell (Chizzonite *et al.*, 1992). More recently, three classes of IL-12-binding sites were identified on PHA-activated human T lymphoblasts: high-affinity (K_d = 5–20 pM) (100–1000 sites/cell), intermediate-affinity (K_d = 50–200 pM) (200–1000 sites/cell), and low-affinity (K_d = 2–6 nM) (1000–5000 sites/cell) (Chua *et al.*, 1994; Stern *et al.*, 1997). Both IL-12Rβ1 and IL-12Rβ2 bind IL-12 with only low affinity when transfected into COS-7 cells. However, coexpression of both human subunits results in high- (K_d = 55 pM) and low-affinity IL-12-binding sites (K_d = 8 nM) and a receptor complex capable of signaling (Presky *et al.*, 1996). Dissociation constants on concanavalin (Con A)-stimulated murine splenocytes were determined to be 40 pM (100–500 sites/cell), 200 pM (600–800 sites/cell) and 7 nM (1000–2000 sites/cell) (Stern *et al.*, 1997).

Cell types and tissues expressing the receptor

Presence was detected mostly on activated T cells and NK cells (Desai *et al.*, 1992). Dendritic cells express a single class of high-affinity IL-12R (Grohmann *et al.*, 1998). In addition, IL-12Rβ1 was detected on human B cell lines and was upregulated on activated human peripheral blood or tonsillar B lymphoblasts that, however, failed to bind measurable amounts of IL-12 (Benjamin *et al.*, 1996; Wu *et al.*, 1996).

TH1 cells express both the β1 and β2 subunits, whereas TH2 cells express only the β1 subunit due to selective downregulation of β2 (Rogge *et al.*, 1997; Szabo *et al.*, 1997). Consequently, only TH1 cells are capable of signaling in response to IL-12. Murine B10.D2 T cells express IL-12Rβ2 at sufficient levels to allow functional responses to IL-12, whereas Balb/c T cells express low levels of β2 and demonstrate limited functional IL-12-induced STAT4 phosphorylation (Guler *et al.*, 1997).

Regulation of receptor expression

IL-4 rapidly extinguishes IL-12R β2 chains *in vitro* (Szabo *et al.*, 1997; Wu *et al.*, 1997a) and *in vivo* (Himmelrich *et al.*, 1998), while IFNγ maintains the ability to signal through IL-12Rβ2 both *in vitro* (Szabo *et al.*, 1997) and *in vivo* (Himmelrich *et al.*, 1998). Thus, IFNγ and IL-4 promote the development of TH1 and TH2 responses by differential regulation of IL-12 receptor expression (Gollob *et al.*, 1997). TGFβ and IL-10 prevent TH1 responses by nearly completely suppressing expression of high-affinity IL-12-binding sites, whereas expression of IL-12Rβ1 is only partially inhibited (Wu *et al.*, 1997a). Cholera toxin inhibits the expression of both β1 and β2 chains (Braun *et al.*, 1999). T cells from patients with *Sezary syndrome* express little or no message for the IL-12Rβ2 subunit and exhibit highly reduced levels of STAT4 (Showe *et al.*, 1999). Prostaglandin E_2 (PGE_2) and dexamethasone inhibit IL-12Rβ1 expression and mRNA for IL-12Rβ2 (Wu *et al.*, 1998). In contrast, IL-2, IL-7, IL-15, phytohemagglutinin (PHA) and anti-CD3 monoclonal antibodies induce IL-12Rβ1 expression (Wu *et al.*, 1997a). The natural killer T (NKT) cell ligand α-galactosylceramide (α-GalCer) induces mRNA for both IL-12Rβ1 and IL-12Rβ2 (Kitamura *et al.*, 1999). Upregulation of the receptor correlates with the ability of the cells to proliferate in response to IL-12 (Chizzonite *et al.*, 1992; Desai *et al.*, 1992).

SIGNAL TRANSDUCTION

Associated or intrinsic kinases

IL-12 receptor stimulation activates the receptor-associated tyrosine kinases JAK2 and TYK2 (Bacon *et al.*, 1995a). In addition, an isoform of MAP kinase is phosphorylated (Pignata *et al.*, 1994). IL-12 also induces tyrosine phosphorylation of the src family lck tyrosine kinase (Pignata *et al.*, 1995).

Cytoplasmic signaling cascades

IL-12 binding to its receptor on T and NK cells induces simultaneous activation of two members of the Janus kinase (JAK) family of protein tyrosine kinases, JAK2 and TYK2 (Bacon *et al.*, 1995a). TYK2 interacts with the $\beta 1$ subunit of the IL-12 receptor and JAK2 interacts with the $\beta 2$ subunit (Zou *et al.*, 1997). Both JAK2 and TYK2 also transduce signals via other cytokine receptors including type 1 (Velazquez *et al.*, 1992) and type 2 IFN (Watling *et al.*, 1993), IL-3 (Silvennoinen *et al.*, 1993), IL-6 (Stahl *et al.*, 1994), and GM-CSF (Quelle *et al.*, 1994). NOS2-derived nitric oxide is a prerequisite for IL-12-induced activation of TYK2 in NK cells (expressing NOS2) but not in T cells (lacking NOS2) (Diefenbach *et al.*, 1999).

JAK kinases phosphorylate the IL-12 receptor on tyrosines located in the intracellular domain. The phosphorylated regions are binding sites for transcription factors termed signal transducers and activators of transcription (STAT). IL-12 binding to its receptor results in activation of three members of the STAT family. STAT1 can dimerize with either STAT3 or STAT4 and STAT4 can dimerize with STAT3 (Jacobson *et al.*, 1995; Yu *et al.*, 1996). Only STAT4, but neither STAT1 nor STAT3 is activated directly through the IL-12 receptor (Naeger *et al.*, 1999). Only IL-12 and IFNα induce tyrosine phosphorylation and DNA binding of STAT4 (Bacon *et al.*, 1995b; Cho *et al.*, 1996). This signaling pathway is therefore relatively specific to IL-12. Deletion of the STAT4 gene in knockout mice results in defective responses specific to IL-12 (Thierfelder *et al.*, 1996; Kaplan *et al.*, 1996). Tyrosine phosphorylation of STAT proteins induces their dimerization and subsequent translocation to the nucleus where they bind to related DNA sequences and regulate transcription (Darnell *et al.*, 1994). Signaling induced by IL-12 varies with the particular cell evaluated.

DOWNSTREAM GENE ACTIVATION

Transcription factors activated

STAT1, STAT3, STAT4 (Trinchieri, 1998), NFκB and c-Jun (Zhang *et al.*, 1999) are all activated by IL-12R.

Genes induced

IL-12 induces upregulation of IFNγ (with costimulus of IL-18), the IL-18R (Xu *et al.*, 1998), the IL-12R$\beta 2$, IRF-1 (Coccia *et al.*, 1999), ERM (Ouyang *et al.*, 1999) and other, yet to be identified genes.

Promoter regions involved

Early reports suggested that IL-12 alone does not activate IFNγ promoter activity (Jacobson *et al.*, 1995). However, c-Jun represents at least one transcription factor induced by IL-12 that binds to the IFNγ promoter and can stimulate transcription (Zhang *et al.*, 1999). In addition, IL-12 strongly induces IFNγ promoter activity in the presence of costimulatory signals provided by soluble antibodies to CD3 and CD28 (Barbulescu *et al.*, 1998). Both STAT4 and AP-1 are required for IFNγ promoter activation (Barbulescu *et al.*, 1998). Activated STAT4 binds to the GAS element in the IRF-1 promoter (Coccia *et al.*, 1999). The precise mechanism by which STAT proteins mediate transcriptional activation remain to be elucidated (Hoey and Grusby, 1999).

BIOLOGICAL CONSEQUENCES OF ACTIVATING OR INHIBITING RECEPTOR AND PATHOPHYSIOLOGY

Unique biological effects of activating the receptors

Antibodies that activate the receptor have not been identified.

Phenotypes of receptor knockouts and receptor overexpression mice

IL-12R$\beta 1$-deficient (IL-12R$\beta 1-/-$) mice are grossly indistinguishable from their control littermates (Wu *et al.*, 1997b). However, splenocytes from IL-12R$-/-$ are deficient in all IL-12-induced biologic activities including proliferation, IFNγ secretion, TH1 development and enhancement of NK lytic activity (Wu *et al.*, 1997b). In addition, IL-12R$\beta 1-/-$ mice exhibit a severe defect in their ability to generate IFNγ-producing TH1 cells, whereas generation of TH2 cells is moderately enhanced (Wu *et al.*, 1997b).

Human abnormalities

Recessive mutations in the gene encoding the IL-12Rβ1 subunit have been identified in seven unrelated individuals with idiopathic mycobacterial and *Salmonella* infections (Altare *et al.*, 1998; de Jong *et al.*, 1998). Their cells were deficient in IL-12R signaling and IFNγ production and their remaining T cell responses were independent of endogenous IL-12. Thus, the lack of IL-12Rβ1 expression results in a human immunodeficiency and demonstrates the essential role of IL-12 in resistance to infections due to intracellular bacteria.

THERAPEUTIC UTILITY

Effects of inhibitors (antibodies) to receptors

Monoclonal antibodies against IL-12Rβ1 inhibit IL-12-induced proliferation of activated T cells, IL-12-induced secretion of IFNγ by resting PBMCs, and IL-12-mediated lymphokine-activated killer cell activation (Wu *et al.*, 1996). Thus, the IL-12Rβ1 chain appears to be an essential component of the functional IL-12R on both T and NK cells.

References

Altare, F., Durandy, A., Lammas, D., Emile, J. F., Lamhamedi, S., Le Deist, F., Drysdale, P., Jouanguy, E., Doffinger, R., Bernaudin, F., Jeppsson, O., Gollob, J. A., Meinl, E., Segal, A. W., Fischer, A., Kumararatne, D., and Casanova, J. L. (1998). Impairment of mycobacterial immunity in human interleukin-12 receptor deficiency. *Science* **280**, 1432–1435.

Bacon, C. M., McVicar, D. W., Ortaldo, J. R., Rees, R. C., O'Shea, J. J., and Johnston, J. A. (1995a). Interleukin 12 (IL-12) induces tyrosine phosphorylation of JAK2 and TYK2: differential use of Janus family tyrosine kinases by IL-2 and IL-12. *J. Exp. Med.* **181**, 399–404.

Bacon, C. M., Petricoin, E. F., 3rd, Ortaldo, J. R., Rees, R. C., Larner, A. C., Johnston, J. A., and O'Shea, J. J. (1995b). Interleukin 12 induces tyrosine phosphorylation and activation of STAT4 in human lymphocytes. *Proc. Natl Acad. Sci. USA* **92**, 7307–7311.

Barbulescu, K., Becker, C., Schlaak, J. F., Schmitt, E., Meyer zum Buschenfelde, K. H., and Neurath, M. F. (1998). IL-12 and IL-18 differentially regulate the transcriptional activity of the human IFN-gamma promoter in primary CD4+ T lymphocytes. *J. Immunol.* **160**, 3642–3647.

Benjamin, D., Sharma, V., Kubin, M., Klein, J. L., Sartori, A., Holliday, J., and Trinchieri, G. (1996). IL-12 expression in AIDS-related lymphoma B cell lines. *J. Immunol.* **156**, 1626–1637.

Braun, M. C., He, J., Wu, C. Y., and Kelsall, B. L. (1999). Cholera toxin suppresses interleukin (IL)-12 production and IL-12 receptor beta1 and beta2 chain expression. *J. Exp. Med.* **189**, 541–552.

Chizzonite, R., Truitt, T., Desai, B. B., Nunes, P., Podlaski, F. J., Stern, A. S., and Gately, M. K. (1992). IL-12 receptor. I. Characterization of the receptor on phytohemagglutinin-activated human lymphoblasts. *J. Immunol.* **148**, 3117–3124.

Cho, S. S., Bacon, C. M., Sudarshan, C., Rees, R. C., Finbloom, D., Pine, R., and O'Shea, J. J. (1996). Activation of STAT4 by IL-12 and IFN-alpha: evidence for the involvement of ligand-induced tyrosine and serine phosphorylation. *J. Immunol.* **157**, 4781–4789.

Chua, A. O., Chizzonite, R., Desai, B. B., Truitt, T. P., Nunes, P., Minetti, L. J., Warrier, R. R., Presky, D. H., Levine, J. F., Gately, M. K., and Gubler, U. (1994). Expression cloning of a human IL-12 receptor component. A new member of the cytokine receptor superfamily with strong homology to gp130. *J. Immunol.* **153**, 128–136.

Chua, A. O., Wilkinson, V. L., Presky, D. H., and Gubler, U. (1995). Cloning and characterization of a mouse IL-12 receptor-beta component. *J. Immunol.* **155**, 4286–4294.

Coccia, E. M., Passini, N., Battistini, A., Pini, C., Sinigaglia, F., and Rogge, L. (1999). Interleukin-12 induces expression of interferon regulatory factor-1 via signal transducer and activator of transcription-4 in human T helper type 1 cells. *J. Biol. Chem.* **274**, 6698–6703.

Darnell Jr., J. E., Kerr, I. M., and Stark, G. R. (1994). Jak-STAT pathways and transcriptional activation in response to IFNs and other extracellular signaling proteins. *Science* **264**, 1415–1421.

de Jong, R., Altare, F., Haagen, I. A., Elferink, D. G., Boer, T., van Breda Vriesman, P. J., Kabel, P. J., Draaisma, J. M., van Dissel, J. T., Kroon, F. P., Casanova, J. L., and Ottenhoff, T. H. (1998). Severe mycobacterial and Salmonella infections in interleukin-12 receptor-deficient patients. *Science* **280**, 1435–1438.

Desai, B. B., Quinn, P. M., Wolitzky, A. G., Mongini, P. K., Chizzonite, R., and Gately, M. K. (1992). IL-12 receptor. II. Distribution and regulation of receptor expression. *J. Immunol.* **148**, 3125–3132.

Diefenbach, A., Schindler, H., Rollinghoff, M., Yokoyama, W. M., and Bogdan, C. (1999). Requirement for type 2 NO synthase for IL-12 signaling in innate immunity [published erratum appears in Science 1999, 284(5421), 1776]. *Science* **284**, 951–955.

Gately, M. K., Renzetti, L. M., Magram, J., Stern, A. S., Adorini, L., Gubler, U., and Presky, D. H. (1998). The interleukin-12/interleukin-12-receptor system: role in normal and pathologic immune responses. *Annu. Rev. Immunol.* **16**, 495–521.

Gollob, J. A., Kawasaki, H., and Ritz, J. (1997). Interferon-gamma and interleukin-4 regulate T cell interleukin-12 responsiveness through the differential modulation of high-affinity interleukin-12 receptor expression. *Eur. J. Immunol.* **27**, 647–652.

Grohmann, U., Belladonna, M. L., Bianchi, R., Orabona, C., Ayroldi, E., Fioretti, M. C., and Puccetti, P. (1998). IL-12 acts directly on DC to promote nuclear localization of NF-kappaB and primes DC for IL-12 production. *Immunity* **9**, 315–323.

Guler, M. L., Jacobson, N. G., Gubler, U., and Murphy, K. M. (1997). T cell genetic background determines maintenance of IL-12 signaling: effects on BALB/c and B10.D2 T helper cell type 1 phenotype development. *J. Immunol.* **159**, 1767–1774.

Himmelrich, H., Parra-Lopez, C., Tacchini-Cottier, F., Louis, J. A., and Launois, P. (1998). The IL-4 rapidly produced in BALB/c mice after infection with Leishmania major down-regulates IL-12 receptor beta 2-chain expression on CD4+ T cells resulting in a state of unresponsiveness to IL-12. *J. Immunol.* **161**, 6156–6163.

Hoey, T., and Grusby, M. J. (1999). STATs as mediators of cytokine-induced responses. *Adv. Immunol.* **71**, 145–162.

Jacobson, N. G., Szabo, S. J., Weber-Nordt, R. M., Zhong, Z., Schreiber, R. D., Darnell, J.E., Jr., and Murphy, K. M. (1995). Interleukin 12 signaling in T helper type 1 (Th1) cells involves tyrosine phosphorylation of signal transducer and activator of transcription (Stat)3 and Stat4. *J. Exp. Med.* **181**, 1755–1762.

Kaplan, M. H., Sun, Y. L., Hoey, T., and Grusby, M. J. (1996). Impaired IL-12 responses and enhanced development of TH2 cells in Stat4-deficient mice. *Nature* **382**, 174–177.

Kawashima, T., Kawasaki, H., Kitamura, T., Nojima, Y., and Morimoto, C. (1998). Interleukin-12 induces tyrosine phosphorylation of an 85-kDa protein associated with the interleukin-12 receptor beta 1 subunit. *Cell Immunol.* **186**, 39–44.

Kitamura, H., Iwakabe, K., Yahata, T., Nishimura, S., Ohta, A., Ohmi, Y., Sato, M., Takeda, K., Okumura, K., Van Kaer, L., Kawano, T., Taniguchi, M., and Nishimura, T. (1999). The natural killer T (NKT) cell ligand alpha-galactosylceramide demonstrates its immunopotentiating effect by inducing interleukin (IL)-12 production by dendritic cells and IL-12 receptor expression on NKT cells. *J. Exp. Med.* **189**, 1121–1128.

Naeger, L. K., McKinney, J., Salvekar, A., and Hoey, T. (1999). Identification of a STAT4 binding site in the interleukin-12 receptor required for signaling. *J. Biol. Chem.* **274**, 1875–1878.

Ouyang, W., Jacobson, N. G., Bhattacharya, D., Gorham, J. D., Fenoglio, D., Sha, W. C., Murphy, T. L., and Murphy, K. M. (1999). The Ets transcription factor ERM is Th1-specific and induced by IL-12 through a Stat4-dependent pathway. *Proc. Natl Acad. Sci. USA* **96**, 3888–3893.

Pignata, C., Sanghera, J. S., Cossette, L., Pelech, S. L., and Ritz, J. (1994). Interleukin-12 induces tyrosine phosphorylation and activation of 44-kD mitogen-activated protein kinase in human T cells. *Blood* **83**, 184–190.

Pignata, C., Prasad, K. V., Hallek, M., Druker, B., Rudd, C. E., Robertson, M. J., and Ritz, J. (1995). Phosphorylation of src family lck tyrosine kinase following interleukin-12 activation of human natural killer cells. *Cell. Immunol.* **165**, 211–216.

Presky, D. H., Yang, H., Minetti, L. J., Chua, A. O., Nabavi, N., Wu, C. Y., Gately, M. K., and Gubler, U. (1996). A functional interleukin 12 receptor complex is composed of two beta-type cytokine receptor subunits. *Proc. Natl Acad. Sci. USA* **93**, 14002–14007.

Quelle, F. W., Sato, N., Witthuhn, B. A., Inhorn, R. C., Eder, M., Miyajima, A., Griffin, J. D., and Ihle, J. N. (1994). JAK2 associates with the beta c chain of the receptor for granulocyte-macrophage colony-stimulating factor, and its activation requires the membrane-proximal region. *Mol. Cell. Biol.* **14**, 4335–4341.

Rogge, L., Barberis-Maino, L., Biffi, M., Passini, N., Presky, D. H., Gubler, U., and Sinigaglia, F. (1997). Selective expression of an interleukin-12 receptor component by human T helper 1 cells. *J. Exp. Med.* **185**, 825–831.

Showe, L. C., Fox, F. E., Williams, D., Au, K., Niu, Z., and Rook, A. H. (1999). Depressed IL-12-mediated signal transduction in T cells from patients with Sezary syndrome is associated with the absence of IL-12 receptor beta 2 mRNA and highly reduced levels of STAT4. *J. Immunol.* **163**, 4073–4079.

Silvennoinen, O., Witthuhn, B. A., Quelle, F. W., Cleveland, J. L., Yi, T., and Ihle, J. N. (1993). Structure of the murine Jak2 protein-tyrosine kinase and its role in interleukin 3 signal transduction. *Proc. Natl Acad. Sci. USA* **90**, 8429–8433.

Stahl, N., and Yancopoulos, G. D. (1993). The alphas, betas, and kinases of cytokine receptor complexes. *Cell* **74**, 587–590.

Stahl, N., Boulton, T. G., Farruggella, T., Ip, N. Y., Davis, S., Witthuhn, B. A., Quelle, F. W., Silvennoinen, O., Barbieri, G., Pellegrini, S., Ihle, J. N., and Yancopoulos, G. D. (1994). Association and activation of Jak-Tyk kinases by CNTF-LIF-OSM-IL-6 beta receptor components. *Science* **263**, 92–95.

Stern, A. S., Gubler, U., Presky, D. H., and Magram, J. (1997). Structural and functional aspects of the IL-12 receptor complex. *Chem. Immunol.* **68**, 23–37.

Szabo, S. J., Dighe, A. S., Gubler, U., and Murphy, K. M. (1997). Regulation of the interleukin (IL)-12R beta 2 subunit expression in developing T helper 1 (Th1) and TH2 cells. *J. Exp. Med.* **185**, 817–824.

Thierfelder, W. E., van Deursen, J. M., Yamamoto, K., Tripp, R. A., Sarawar, S. R., Carson, R. T., Sangster, M. Y., Vignali, D. A., Doherty, P. C., Grosveld, G. C., and Ihle, J. N. (1996). Requirement for Stat4 in interleukin-12-mediated responses of natural killer and T cells. *Nature* **382**, 171–174.

Trinchieri, G. (1998). Interleukin-12: a cytokine at the interface of inflammation and immunity. *Adv. Immunol.* **70**, 83–243.

Velazquez, L., Fellous, M., Stark, G. R., and Pellegrini, S. (1992). A protein tyrosine kinase in the interferon alpha/beta signaling pathway. *Cell* **70**, 313–322.

Watling, D., Guschin, D., Muller, M., Silvennoinen, O., Witthuhn, B. A., Quelle, F. W., Rogers, N. C., Schindler, C., Stark, G. R., Ihle, J. N., and Kerr, I. M. (1993). Complementation by the protein tyrosine kinase JAK2 of a mutant cell line defective in the interferon-gamma signal transduction pathway [see comments]. *Nature* **366**, 166–170.

Wu, C. Y., Warrier, R. R., Carvajal, D. M., Chua, A. O., Minetti, L. J., Chizzonite, R., Mongini, P. K., Stern, A. S., Gubler, U., Presky, D. H., and Gately, M. K. (1996). Biological function and distribution of human interleukin-12 receptor beta chain. *Eur. J. Immunol.* **26**, 345–350.

Wu, C., Warrier, R. R., Wang, X., Presky, D. H., and Gately, M. K. (1997a). Regulation of interleukin-12 receptor beta1 chain expression and interleukin-12 binding by human peripheral blood mononuclear cells. *Eur. J. Immunol.* **27**, 147–154.

Wu, C., Ferrante, J., Gately, M. K., and Magram, J. (1997b). Characterization of IL-12 receptor beta1 chain (IL-12Rbeta1)-deficient mice: IL-12Rbeta1 is an essential component of the functional mouse IL-12 receptor. *J. Immunol.* **159**, 1658–1665.

Wu, C. Y., Wang, K., McDyer, J. F., and Seder, R. A. (1998). Prostaglandin E2 and dexamethasone inhibit IL-12 receptor expression and IL-12 responsiveness. *J. Immunol.* **161**, 2723–2730.

Xu, D., Chan, W. L., Leung, B. P., Hunter, D., Schulz, K., Carter, R. W., McInnes, I. B., Robinson, J. H., and Liew, F. Y. (1998). Selective expression and functions of interleukin 18 receptor on T helper (Th) type 1 but not Th2 cells. *J. Exp. Med.* **188**, 1485–1492.

Yamamoto, K., Kobayashi, H., Miura, O., Hirosawa, S., and Miyasaka, N. (1997). Assignment of IL12RB1 and IL12RB2, interleukin-12 receptor beta 1 and beta 2 chains, to human chromosome 19 band p13.1 and chromosome 1 band p31.2, respectively, by *in situ* hybridization. *Cytogenet. Cell Genet.* **77**, 257–258.

Yu, C. R., Lin, J. X., Fink, D. W., Akira, S., Bloom, E. T., and Yamauchi, A. (1996). Differential utilization of Janus kinase-signal transducer activator of transcription signaling pathways in the stimulation of human natural killer cells by IL-2, IL-12, and IFN-alpha. *J. Immunol.* **157**, 126–137.

Zhang, F., Nakamura, T., and Aune, T. M. (1999). TCR and IL-12 receptor signals cooperate to activate an individual response element in the IFN-gamma promoter in effector Th cells. *J. Immunol.* **163**, 728–735.

Zou, J., Presky, D. H., Wu, C. Y., and Gubler, U. (1997). Differential associations between the cytoplasmic regions of the interleukin-12 receptor subunits beta1 and beta2 and JAK kinases. *J. Biol. Chem.* **272**, 6073–6077.

LICENSED PRODUCTS

IL-12R has not been licensed.
R&D Systems (www.rndsystems.com), Sigma (www.sigma-aldrich.com).
Antibodies:
Flow cytometry: PharMingen (www.pharmingen.com)

ACKNOWLEDGEMENTS

This work has been supported in part by the 1999 Advanced Polymer Systems Research Fellowship of the Dermatology Foundation to CE, NIH Grant CA 80126 to MRS, and NIH Grants CA 68067 and CA 73743 to MTL. We apologize to all colleagues whose contributions have not been directly referenced in this chapter.

IL-13 Receptor

David J. Matthews and Andrew N.J. McKenzie*

MRC Laboratory of Molecular Biology, Hills Road, Cambridge, CB2 2QH, UK

*corresponding author tel: 44 1223 402377, fax: 44 1223 412178, e-mail: anm@mrc-lmb.cam.ac.uk

DOI: 10.1006/rwcy.2000.14007.

SUMMARY

There are two membrane-bound IL-13-binding proteins, IL-13Rα1 and IL-13Rα2. The IL-13Rα1 protein has a moderate affinity for IL-13 but requires the presence of IL-4Rα (CD124) to form a high-affinity receptor complex. IL-4Rα alone does not bind IL-13. Signaling through the IL-13Rα1/IL-4Rα complex activates STAT6 and IRS-1/2, and IL-13Rα1 appears to recruit a JAK kinase to the activated receptor. In contrast, IL-13Rα2 alone is a high-affinity receptor for IL-13 but there is little evidence that this protein has signal transduction properties. The function of IL-13Rα2 is still unclear, however, it has been proposed that it may act as an IL-13 antagonist. Both receptor chains have a wide tissue distribution, although notably, the IL-13Rα1 receptor does not appear to be expressed on the surface of T cells.

BACKGROUND

Discovery

Two IL-13-binding chains have been identified following the cloning of IL-13. Using an expression cloning approach, Caput *et al.* (1996) identified IL-13Rα2, a membrane-bound human protein with a high affinity for IL-13 but no apparent capacity for signal transduction. In contrast, Hilton *et al.* (1996) obtained a second type of IL-13 receptor by screening a mouse genomic library using redundant oligonucleotides to the WSXWS motif (a conserved motif found in all class I cytokine receptors). This approach identified IL-13Rα1, a low-affinity IL-13 binding chain, that requires the presence of the IL-4Rα chain (CD124) in order to form a high-affinity IL-13 receptor. As a consequence, the IL-4Rα/IL-13Rα1 complex was also identified as a functional IL-4 receptor (the type II IL-4 receptor) (Callard *et al.*, 1996). The relationship between the IL-13Rα1 and IL-13Rα2 is not clear but there is evidence that IL-13Rα2 may act as an IL-13 antagonist. The γc chain, the promiscuous receptor chain found in the IL-2, IL-4, IL-7, IL-9, and IL-15 receptors, does not appear to be a functional component of the IL-13 receptor (Matthews *et al.*, 1995).

Alternative names

A number of groups have cloned both IL-13 receptors in humans (Aman *et al.*, 1996; Caput *et al.*, 1996; Miloux *et al.*, 1996; Gauchat *et al.*, 1997) and in mice. The IL-13Rα1 has also been termed γ' and the IL-13Rα2 has also been called IL-13Rα'.

Structure

Both IL-13Rα1 and IL-13Rα2 are members of the class I cytokine receptor family and have a similar structure to the IL-5Rα chain and contain a WSXWS motif. The main structural difference between the two IL-13 receptors is that the IL-13Rα1 chain has a longer intracellular domain than IL-13Rα2.

Main activities and pathophysiological roles

Signals generated by the IL-13Rα1/IL-4Rα receptor complex promote TH2-driven immunological responses important in *parasitic worm infections* and *atopy* (McKenzie *et al.*, 1998a, 1998b). In humans, the signals generated by the IL-13 receptor promote the production of IgE (Punnonen *et al.*, 1993) but inhibit the production of TH1 proinflammatory mediators such as TNFα (Manna and Aggarwal, 1998).

GENE

Accession numbers

Human IL-13Rα1: Y09328
Mouse IL-13Rα1: S80963

Human IL-13Rα2: X95302
Mouse IL-13Rα2: U65747

Sequence

See **Figure 1**.

Figure 1 Nucleotide sequences for human and mouse IL-13Rα1 and IL-13Rα2.

hIL-13Rα1

```
1       CGGGTAATTT TTTCAAAGTA AACGCTTCGG GCCCCGCGGG ACACTCAGCT AAGAGCCCGG
61      CCGGGCTCCG AGGCGAGAGG CTGCATGGAG TGGCCGGCGC GGCTCTGCGG GCTGTGGGCG
121     CTGCTGCTCT GCGCCGGCGG CGGGGGCGGG GGCGGGGGCG CCGCGCCTAC GGAAACTCAG
181     CCACCTGTGA CAAATTTGAG TGTCTCTGTT GAAAACCTCT GCACAGTAAT ATGGACATGG
241     AATCCACCCG AGGGAGCCAG CTCAAATTGT AGTCTATGGT ATTTTAGTCA TTTTGGCGAC
301     AAACAAGATA AGAAAATAGC TCCGGAAACT CGTCGTTCAA TAGAAGTACC CCTGAATGAG
361     AGGATTTGTC TGCAAGTGGG GTCCCAGTGT AGCACCAATG AGAGTGAGAA GCCTAGCATT
421     TTGGTTGAAA AATGCATCTC ACCCCCAGAA GGTGATCCTG AGTCTGCTGT GATTGAGCTT
481     CAATGCATTT GGCACAACCT GAGCTACATG AAGTGTTCTT GGCTCCCTGG AAGGAATACC
541     AGTCCCGACA CTAACTATAC TCTCTACTAT TGGCACAGAA GCCTGGAAAA AATTCATCAA
601     TGTGAAAACA TCTTTAGAGA AGGCCAATAC TTTGGTTGTT CCTTTGATCT GACCAAAGTG
661     AAGGATTCCA GTTTTGAACA ACACAGTGTC CAAATAATGG TCAAGGATAA TGCAGGAAAA
721     ATTAAACCAT CCTTCAATAT AGTGCCTTTA ACTTCCCGTG TGAAACCTGA TCCTCCACAT
781     ATTAAAAACC TCTCCTTCCA CAATGATGAC CTATATGTGC AATGGGAGAA TCCACAGAAT
841     TTTATTAGCA GATGCCTATT TTATGAAGTA GAAGTCAATA ACAGCCAAAC TGAGACACAT
901     AATGTTTTCT ACGTCCAAGA GGCTAAATGT GAGAATCCAG AATTTGAGAG AAATGTGGAG
961     AATACATCTT GTTTCATGGT CCCTGGTGTT CTTCCTGATA CTTTGAACAC AGTCAGAATA
1021    AGAGTCAAAA CAAATAAGTT ATGCTATGAG GATGACAAAC TCTGGAGTAA TTGGAGCCAA
1081    GAAATGAGTA TAGGTAAGAA GCGCAATTCC ACACTCTACA TAACCATGTT ACTCATTGTT
1141    CCAGTCATCG TCGCAGATGC AATCATAGTA CTCCTGCTTT ACCTAAAAAG GCTCAAGATT
1201    ATTATATTCC CTCCAATTCC TGATCCTGGC AAGATTTTTA AAGAAATGTT TGGAGACCAG
1261    AATGATGATA CTCTGCACTG GAAGAAGTAC GACATCTATG AGAAGCAAAC CAAGGAGGAA
1321    ACCGACTCTG TAGTGCTGAT AGAAAACCTG AAGAAAGCCT CTCAGTGATG GAGATAATTT
1381    ATTTTTACCT TCACTGTGAC CTTGAGAAGA TTCTTCCCAT TCTCCATTTG TTATCTGGGA
1441    ACTTATTAGA TGGAAACTGA AACTACTGCA CCATTTAAAA ACAGGCAGCT CATAAGAGCC
1501    ACAGGTCTTT ATGTTGAGTC GCTAGCAAGA ACAAGAAAAG TTTTAAAGAA AGATGTTGCT
1561    TACTATGAGT GG
```

hIL-13Rα2

```
1       GTAAGAACAC TCTCGTGAGT CTAACGGTCT TCCGGATGAA GGCTATTTGA AGTCGCCATA
61      ACCTGGTCAG AAGTGTGCCT GTCGGCGGGG AGAGAGGCAA TATCAAGGTT TTAAATCTCG
121     GAGAAATGGC TTTCGTTTGC TTGGCTATCG GATGCTTATA TACCTTTCTG ATAAGCACAA
181     CATTTGGCTG TACTTCATCT TCAGACACCG AGATAAAAGT TAACCCTCCT CAGGATTTTG
241     AGATAGTGGA TCCCGGATAC TTAGGTTATC TCTATTTGCA ATGGCAACCC CCACTGTCTC
301     TGGATCATTT TAAGGAATGC ACAGTGGAAT ATGAACTAAA ATACCGAAAC ATTGGTAGTG
361     AAACATGGAA GACCATCATT ACTAAGAATC TACATTACAA AGATGGGTTT GATCTTAACA
421     AGGGCATTGA AGCGAAGATA CACACGCTTT TACCATGGCA ATGCACAAAT GGATCAGAAG
481     TTCAAAGTTC CTGGGCAGAA ACTACTTATT GGATATCACC ACAAGGAATT CCAGAAACTA
541     AAGTTCAGGA TATGGATTGC GTATATTACA ATTGGCAATA TTTACTCTGT TCTTGGAAAC
601     CTGGCATAGG TGTACTTCTT GATACCAATT ACAACTTGTT TTACTGGTAT GAGGGCTTGG
661     ATCATGCATT ACAGTGTGTT GATTACATCA AGGCTGATGG ACAAAATATA GGATGCAGAT
721     TTCCCTATTT GGAGGCATCA GACTATAAAG ATTTCTATAT TTGTGTTAAT GGATCATCAG
781     AGAACAAGCC TATCAGATCC AGTTATTTCA CTTTTCAGCT TCAAAATATA GTTAAACCTT
841     TGCCGCCAGT CTATCTTACT TTTACTCGGG AGAGTTCATG TGAAATTAAG CTGAAATGGA
901     GCATACCTTT GGGACCTATT CCAGCAAGGT GTTTTGATTA TGAAATTGAG ATCAGAGAAG
961     ATGATACTAC CTTGGTGACT GCTACAGTTG AAAATGAAAC ATACACCTTG AAAACAACAA
1021    ATGAAACCCG ACAATTATGC TTTGTAGTAA GAAGCAAAGT GAATATTTAT TGCTCAGATG
1081    ACGGAATTTG GAGTGAGTGG AGTGATAAAC AATGCTGGGA AGGTGAAGAC CTATCGAAGA
1141    AAACTTTGCT ACGTTTCTGG CTACCATTTG GTTTCATCTT AATATTAGTT ATATTTGTAA
1201    CCGGTCTGCT TTTGCGTAAG CCAAACACCT ACCCAAAAAT GATTCCAGAA TTTTTCTGTG
1261    ATACATGAAG ACTTTCCATA TCAAGAGACA TGGTATTGAC TCAACAGTTT CCAGTCATGG
1321    CCAAATGTTC AATATGAGTC TCAATAAACT GAATTTTTCT TGCGAATGTT GAAAAA
```

Figure 1(b) (*Contd.*)

mIL-13Rα1

```
1       TGAAAAGATA GAATAAATGG CCTCGTGCCG AATTCGGCAC GAGCCGAGGC GAGGGCCTGC
61      ATGGCGCGGC CAGCGCTGCT GGGCGAGCTG TTGGTGCTGC TACTGTGGAC CGCCACCGTG
121     GGCCAAGTTG CCGCGGCCAC AGAAGTTCAG CCACCTGTGA CGAATTTGAG CGTCTCTGTC
181     GAAAATCTCT GCACGATAAT ATGGACGTGG AGTCCTCCTG AAGGAGCCAG TCCAAATTGC
241     ACTCTCAGAT ATTTTAGTCA CTTTGATGAC CAACAGGATA AGAAAATTGC TCCAGAAACT
301     CATCGTAAAG AGGAATTACC CCTGGATGAG AAAATCTGTC TGCAGGTGGG CTCTCAGTGT
361     AGTGCCAATG AAAGTGAGAA GCCTAGCCCT TTGGTGAAAA AGTGCATCTC ACCCCCTGAA
421     GGTGATCCTG AGTCCGCTGT GACTGAGCTC AAGTGCATTT GGCATAACCT GAGCTATATG
481     AAGTGTTCCT GGCTCCCTGG AAGGAATACA AGCCCTGACA CACACTATAC TCTGTACTAT
541     TGGTACAGCA GCCTGGAGAA AAGTCGTCAA TGTGAAAACA TCTATAGAGA AGGTCAACAC
601     ATTGCTTGTT CCTTTAAATT GACTAAAGTG GAACCTAGTT TTGAACATCA GAACGTTCAA
661     ATAATGGTCA AGGATAATGC TGGGAAAATT AGGCCATCCT GCAAAATAGT GTCTTTAACT
721     TCCTATGTGA AACCTGATCC TCCACATATT AAACATCTTC TCCTCAAAAA TGGTGCCTTA
781     TTAGTGCAGT GGAAGAATCC ACAAAATTTT AGAAGCAGAT GCTTAACTTA TGAAGTGGAG
841     GTCAATAATA CTCAAACCGA CCGACATAAT ATTTTAGAGG TTGAAGAGGA CAAATGCCAG
901     AATTCCGAAT CTGATAGAAA CATGGAGGGT ACAAGTTGTT TCCAACTCCC TGGTGTTCTT
961     GCCGACGCTG TCTACACAGT CAGAGTAAGA GTCAAAACAA ACAAGTTATG CTTTGATGAC
1021    AACAAACTGT GGAGTGATTG GAGTGAAGCA CAGAGTATAG GTAAGGAGCA AAACTCCACC
1081    TTCTACACCA CCATGTTACT CACCATTCCA GTCTTTGTCG CAGTGGCAGT CATAATCCTC
1141    CTTTTTTACC TGAAAAGGCT TAAGATCATT ATATTTCCTC CAATTCCTGA TCCTGGCAAG
1201    ATTTTTAAAG AAATGTTTGG AGACCAGAAT GATGATACCC TGCACTGGAA GAAGTATGAC
1261    ATCTATGAGA AACAATCCAA AGAAGAAACG GATTCTGTAG TGCTGATAGA AAACCTGAAG
1321    AAAGCAGCTC CTTGATGGGG AGAAGTGATT TCTTTCTTGC CTTCAATGTG ACCCTGTGAA
1381    GATTTATTGC ATTCTCCATT TGTTATCTGG GGGACTTGTT AAATAGAAAC TGAAACTACT
1441    CTTGAAAAAC AGGCAGCTCC TAAGAGCCAC AGGTCTTGAT GTGACTTTTG CATTGAAAAC
1501    CCAAACCCAA AGGAGCTCCT TCCAAGAAAA GCAAGAGTTC TTCTCGTTCC TTGTTCCAAT
1561    CCCTAAAAGC AGATGTTTTG CCAAATCCCC AAACTAGAGG ACAAAGACAA GGGGACAATG
1621    ACCATCAATT CATCTAATCA GGAATTGTGA TGGCTTCCTA AGGAATCTCT GCTTGCTCTG
```

mIL-13Rα2

```
1       GGCACGAGGG AGAGGAGGAG GGAAAGATAG AAAGAGAGAG AGAAAGATTG CTTGCTACCC
61      CTGAACAGTG ACCTCTCTCA AGACAGTGCT TTGCTCTTCA CGTATAAGGA AGGAAAACAG
121     TAGAGATTCA ATTTAGTGTC TAATGTGGAA AGGAGGACAA AGAGGTCTTG TGATAACTGC
181     CTGTGATAAT ACATTTCTTG AGAAACCATA TTATTGAGTA GAGCTTTCAG CACACTAAAT
241     CCTGGAGAAA TGGCTTTTGT GCATATCAGA TGCTTGTGTT TCATTCTTCT TTGTACAATA
301     ACTGGCTATT CTTTGGAGAT AAAAGTTAAT CCTCCTCAGG ATTTTGAAAT ATTGGATCCT
361     GGATTACTTG GTTATCTCTA TTTGCAATGG AAACCTCCTG TGGTTATAGA AAAATTTAAG
421     GGCTGTACAC TAGAATATGA GTTAAAATAC CGAAATGTTG ATAGCGACAG CTGGAAGACT
481     ATAATTACTA GGAATCTAAT TTACAAGGAT GGGTTTGATC TTAATAAAGG CATTGAAGGA
541     AAGATACGTA CGCATTTGTC AGAGCATTGT ACAAATGGAT CAGAAGTACA AAGTCCATGG
601     ATAGAAGCTT CTTATGGGAT ATCAGATGAA GGAAGTTTGG AAACTAAAAT TCAGGACATG
661     AAGTGTATAT ATTATAACTG GCAGTATTTG GTCTGCTCTT GGAAACCTGG CAAGACAGTA
721     TATTCTGATA CCAACTATAC CATGTTTTTC TGGTATGAGG GCTTGGATCA TGCCTTACAG
781     TGTGCTGATT ACCTCCAGCA TGATGAAAAA AATGTTGGAT GCAAACTGTC CAACTTGGAC
841     TCATCAGACT ATAAAGATTT TTTTATCTGT GTTAATGGAT CTTCAAAGTT GGAACCCATC
901     AGATCCAGCT ATACAGTTTT TCAACTTCAA AATATAGTTA AACCATTGCC ACCAGAATTC
961     CTTCATATTA GTGTGGAGAA TTCCATTGAT ATTAGAATGA AATGGAGCAC ACCTGGAGGA
1021    CCCATTCCAC CAAGGTGTTA CACTTATGAA ATTGTGATCC GAGAAGACGA TATTTCCTGG
1081    GAGTCTGCCA CAGACAAAAA CGATATGAAG TTGAAGAGGA GAGCAAATGA AAGTGAAGAC
1141    CTATGCTTTT TTGTAAGATG TAAGGTCAAT ATATATTGTG CAGATGATGG AATTTGGAGC
1201    GAATGGAGTG AAGAGGAATG TTGGGAAGGT TACACAGGGC CAGACTCAAA GATTATTTTC
1261    ATAGTACCAG TTTGTCTTTT CTTTATATTC CTTTTGTTAC TTCTTTGCCT TATTGTGGAG
1321    AAGGAAGAAC CTGAACCCAC ATTGAGCCTC CATGTGGATC TGAACAAAGA AGTGTGTGCT
1381    TATGAAGATA CCCTCTGTTA AACCACCAAT TTCTTGACAT AGAGCCAGCC AGCAGGAGTC
1441    ATATTAAACT CAATTTCTCT TAAAATTTCG AATACATCTT CTTGAAAATC AGTGTTTGTC
1501    CTAATAGTGT TGGGTTTTTG ACTAAAGTGC TGGATATATA TCTCCAAAAA AAAAAAAAAA
1561    AAAAAAA
```

Chromosome location and linkages

The location of the huIL-13Rα1 gene is at present unknown. The huIL-13Rα2 gene is located on the X chromosome in the region Xq24 (Guo *et al.*, 1997). Both murine genes map to the X chromosome (Donaldson *et al.*, 1998). Murine IL-13Rα1 maps to the region DXMit85: 3.8 ± 2.1 cM (mIL-13Rα1,

Agtr2) 3.8 ± 2.1 cM: DXMit49. Murine IL-13Rα2 maps to the region DXMit4: 6.4 ± 2.5 cM (mIL-13Rα2, DXMit34) 7.9 ± 2.9 cM: DXMit120.

PROTEIN

Accession numbers

Human IL-13Rα1: P78552
Mouse IL-13Rα1: O09030
Human IL-13Rα2: Q14627
Mouse IL-13Rα2: 3483094

Sequence

See **Figure 2**.

Description of protein

Both IL-13Rα1 and IL-13Rα2 proteins contain an N-terminal immunoglobulin-like domain followed by a class I cytokine receptor region including a WSXWS motif near the transmembrane domain. The human and mouse IL-13Rα1 proteins have a cytoplasmic region capable of signal transduction and contain Box 1 and Box 2 motifs. Although the IL-4Rα chain has no measurable ability to bind IL-13 (Zurawski *et al.*, 1995), it does, however, act as an affinity converter for IL-13Rα1, increasing its affinity for IL-13 by 100-fold (Hilton *et al.*, 1996). The role of IL-13Rα2 is unclear, although it has a very short intracellular domain and it does not appear to be involved in signal transduction.

The IL-13 receptor displays a high degree of complexity in its relationship with its ligands and with the IL-4 receptor (Callard *et al.*, 1996). IL-13

Figure 2 Amino acid sequences for human and mouse IL-13Rα1 and IL-13Rα2. Putative signal peptides are shown in bold.

huIL-13Rα1

1 **MEWPARLCGL WALLLCAGGG GGGGG**AAPTE TQPPVTNLSV SVENLCTVIW TWNPPEGASS
61 NCSLWYFSHF GDKQDKKIAP ETRRSIEVPL NERICLQVGS QCSTNESEKP SILVEKCISP
121 PEGDPESAVT ELQCIWHNLS YMKCSWLPGR NTSPDTNYTL YYWHRSLEKI HQCENIFREG
181 QYFGCSFDLT KVKDSSFEQH SVQIMVKDNA GKIKPSFNIV PLTSRVKPDP PHIKNLSFHN
241 DDLYVQWENP QNFISRCLFY EVEVNNSQTE THNVFYVQEA KCENPEFERN VENTSCFMVP
301 GVLPDTLNTV RIRVKTNKLC YEDDKLWSNW SQEMSIGKKR NSTLYITMLL IVPVIVAGAI
361 IVLLLYLKRL KIIIFPPIPD PGKIFKEMFG DQNDDTLHWK KYDIYEKQTK EETDSVVLIE
421 NLKKASQ

huIL-13Rα2

1 **MAFVCLAIGC LYTFLISTTF GCTSSSD**TEI KVNPPQDFEI VDPGYLGYLY LQWQPPLSLD
61 HFKECTVEYE LKYRNIGSET WKTIITKNLH YKDGFDLNKG IEAKIHTLLP WQCTNGSEVQ
121 SSWAETTYWI SPQGIPETKV QDMDCVYYNW QYLLCSWKPG IGVLLDTNYN LFYWYEGLDH
181 ALQCVDYIKA DGQNIGCRFP YLEASDYKDF YICVNGSSEN KPIRSSYFTF QLQNIVKPLP
241 PVYLTFTRES SCEIKLKWSI PLGPIPARCF DYEIEIREDD TTLVTATVEN ETYTLKTTNE
301 TRQLCFVVRS KVNIYCSDDG IWSEWSDKQC WEGEDLSKKT LLRFWLPFGF ILILVIFVTG
361 LLLRKPNTYP KMIPEFFCDT

mIL-13Rα1

1 **MARPALLGEL LVLLLWTATV GQVAAAT**EVQ PPVTNLSVSV ENLCTIIWTW SPPEGASPNC
61 TLRYFSHFDD QQDKKIAPET HRKEELPLDE KICLQVGSQC SANESEKPSP LVKKCISPPE
121 GDPESAVTEL KCIWHNLSYM KCSWLPGRNT SPDTHYTLYY WYSSLEKSRQ CENIYREGQH
181 IACSFKLTKV EPSFEHQNVQ IMVKDNAGKI RPSCKIVSLT SYVKPDPPHI KHLLLKNGAL
241 LVQWKNPQNF RSRCLTYEVE VNNTQTDRHN ILEVEEDKCQ NSESDRNMEG TSCFQLPGVL
301 ADAVYTVRVR VKTNKLCFDD NKLWSDWSEA QSIGKEQNST FYTTMLLTIP VFVAVAVIIL
361 LFYLKRLKII IFPPIPDPGK IFKEMFGDQN DDTLHWKKYD IYEKQSKEET DSVVLIENLK
421 KAAP

mIL-13Rα2

1 **MAFVHIRCLC FILLCTITGY S**LEIKVNPPQ DFEILDPGLL GYLYLQWKPP VVIEKFKGCT
61 LEYELKYRNV DSDSWKTIIT RNLIYKDGFD LNKGIEGKIR THLSEHCTNG SEVQSPWIEA
121 SYGISDEGSL ETKIQDMKCI YYNWQYLVCS WKPGKTVYSD TNYTMFFWYE GLDHALQCAD
181 YLQHDEKNVG CKLSNLDSSD YKDFFICVNG SSKLEPIRSS YTVFQLQNIV KPLPPEFLHI
241 SVENSIDIRM KWSTPGGPIP PRCYTYEIVI REDDISWESA TDKNDMKLKR RANESEDLCF
301 FVRCKVNIYC ADDGIWSEWS EEECWEGYTG PDSKIIFIVP VCLFFIFLLL LLCLIVEKEE
361 PEPTLSLHVD LNKEVCAYED TLC

and IL-4 can both cross-compete for binding to the IL-13 receptor complex (Hilton *et al.*, 1996; Miloux *et al.*, 1996). In addition, the γc, a component of the type 1 IL-4R, appears to compete for the IL-4Rα chain and inhibit IL-13 binding by sequestering the IL-4Rα chain (Orchansky *et al.*, 1997; Kuznetsov and Puri, 1999). This complex relationship may be important in hematopoietic cells where both the type 1 IL-4 receptor and the IL-13 receptors are coexpressed.

Relevant homologies and species differences

See **Table 1**. It is also of note that both human and mouse IL-13 receptors share approximately 25% identity to the IL-5Rα chain.

Affinity for ligand(s)

See **Table 2**.

Table 1 Percentage of shared amino acid identity between the IL-13-binding chains

	huIL-13Rα1	huIL-13Rα2	mIL-13Rα1	mIL-13Rα2
huIL-13Rα1	100%	27%	74%	26%
huIL-13Rα2	27%	100%	25%	59%
mIL-13Rα1	74%	25%	100%	29%
mIL-13Rα2	26%	59%	29%	100%

Cell types and tissues expressing the receptor

See **Table 3**.

Regulation of receptor expression

Studies on mature human B cells found that potent B cell activators, such as the antibodies anti-μ or anti-CD40, upregulated IL-13Rα1 mRNA expression, especially when they were used to co-stimulate B cells (Graber *et al.*, 1998; Ogata *et al.*, 1998; Ford *et al.*, 1999). In addition, the expression of hIL-13Rα1 mRNA on human peripheral T cells was shown to be downregulated after stimulation by either anti-CD3 plus anti-CD28 or anti-CD3 plus PMA (Gauchat *et al.*, 1997). The activation of human monocytes by IL-13 results in the downregulation of IL-13Rα1 (Graber *et al.*, 1998).

Release of soluble receptors

In humans, soluble IL-13Rα1 has been detected in T cell supernatants (Graber *et al.*, 1998). In mice, a soluble receptor is present in serum and urine which binds IL-13 with high affinity. Purification and subsequent partial sequencing of the protein indicate that it is the soluble form of mIL-13Rα2 protein (Zhang *et al.*, 1997).

SIGNAL TRANSDUCTION

Associated or intrinsic kinases

Although the IL-13Rα1/IL-4Rα does not have an intrinsic kinase domain, it does associate with

Table 2 Summary of the physical properties of the IL-13-binding chains: number of potential glycosylation sites and approximate affinities of IL-13 receptors for IL-13

	Human		Mouse		
	IL-13Rα1	IL-13Rα2	IL-13Rα1	IL-13Rα2	sIL-13Rα2
M_r	70,000	70,000	60,000	70,000	40,000
Mature peptide	401 aa	354 aa	398 aa	362 aa	n.k.
Precursor peptide	427 aa	380 aa	424 aa	383 aa	n.k.
Glycosylation sites	10	4	4	4	n.k.
Affinity K_d	4 nM	450 pM	2–10 nM	250 pM	35 pM
+IL-4Rα	30 pM	n.c.	75 pM	n.c.	–

n.c., no change; n.k., not known; sIL-13Rα2, soluble IL-13Rα2.

Table 3 Summary of cell types shown to express IL-13 receptor messenger RNA

Tissue/cells	hIL-13Rα1	mIL-13Rα1	hIL-13Rα2	mIL-13Rα2
Brain	yes	not detected	yes	yes
Spleen	yes	yes	yes	yes
Liver	yes	yes	yes	yes
Fetal liver	yes	unknown	yes	unknown
Thymus	yes	yes	yes	unknown
Heart	yes	yes	yes	unknown
Lung	yes	yes	yes	unknown
Kidney	yes	yes	unknown	not detected
Testis	yes	yes	yes	not detected
Stomach	yes	yes	yes	unknown
Skin	yes	yes	yes	unknown
Appendix	yes	unknown	yes	unknown
PBC	yes	unknown	yes	unknown
Bone marrow	yes	no	yes	unknown
Skeletal muscle	yes	no	yes	unknown
Colon	yes	yes	unknown	unknown
Small intestine	yes	unknown	unknown	unknown
Ovary	yes	unknown	unknown	unknown
Prostate	yes	unknown	unknown	unknown
Pancreas	yes	unknown	yes	unknown
B cells	yes	unknown	yes	unknown
T cells	yes	unknown	yes	unknown
Endothelial cells	yes	unknown	yes	unknown

members of the JAK kinase family. The IL-13 receptor has been shown to phosphorylate JAK2 and TYK2 in fibroblasts (Murata *et al.*, 1998), JAK1, TYK2 in B9 cells (Welham *et al.*, 1995), JAK1 in TF-1 cells (Keegan *et al.*, 1995), JAK1, JAK2 and TYK2 in monocytes (Roy and Cathcart, 1998) and *colon carcinoma* cell lines (Murata *et al.*, 1996). Partial deletion analysis of the cytoplasmic domain of IL-13Rα1 cells has noted that the terminal 38 amino acids were not necessary for proliferation or for the tyrosine phosphorylation of JAK1 or TYK2 in FD5 cells (Orchansky *et al.*, 1999). There is no evidence that JAK3 kinase is involved in IL-13 signal transduction (Keegan *et al.*, 1995).

Cytoplasmic signaling cascades

The IL-13R initiates a JAK/STAT signaling cascade resulting in the activation of STAT6 (Lin *et al.*, 1995) and STAT3 (Orchansky *et al.*, 1999) and which may be downregulated by SOCS proteins (Starr *et al.*, 1997). In addition, IL-13 also induces the phosphorylation of IRS-1/2 and IL-4Rα (Keegan *et al.*, 1995) and activates PI-3 kinase (Wright *et al.*, 1999). In human monocytes, IL-13 has been described as inducing cAMP production, PLCγ1 activation, phosphoinositol metabolism, and mobilization of intracellular calcium stores (Sozzani *et al.*, 1998). The tyrosine phosphatase SHP-1 has been implicated in the negative regulation of IL-13 signal transduction (Haque *et al.*, 1998).

DOWNSTREAM GENE ACTIVATION

Transcription factors activated

STAT6 is activated upon IL-13 receptor activation and is an important mediator of IL-13 function

(Lin *et al.*, 1995; Palmer-Crocker *et al.*, 1996). STAT3 has also been reported to be activated by IL-13 (Orchansky *et al.*, 1999). The transcription factors c-fos, c-jun, and c-myc have also been shown to be upregulated by IL-13 (Doucet *et al.*, 1998).

Genes induced

Notable genes induced or upregulated by IL-13 include CD23 (Punnonen *et al.*, 1993), MHC class II, and in human B cells surface IgM (Zurawski and de Vries, 1994). VCAM-1 is found to be upregulated in fibroblasts (Doucet *et al.*, 1998) and endothelial cells (Kotowicz *et al.*, 1996).

Promoter regions involved

A STAT6 response element has been described in the IL-4 promoter GTCTGATTTCAGGAACAA-TTTTA (Curiel *et al.*, 1997).

BIOLOGICAL CONSEQUENCES OF ACTIVATING OR INHIBITING RECEPTOR AND PATHOPHYSIOLOGY

Unique biological effects of activating the receptors

In humans, only IL-4 and IL-13 have been reported to induce antibody class-switching to IgE (Punnonen *et al.*, 1993), which is a major mediator of *allergic responses*. Therefore, IL-13 and its receptor are likely to have an important role in the pathology of *atopy*. In mice, IL-13 has been demonstrated to play a unique role in inducing the rapid expulsion of certain nematode worms, suggesting an important role for IL-13 in gut immunology (McKenzie *et al.*, 1998b).

Phenotypes of receptor knockouts and receptor overexpression mice

As yet, neither IL-13R molecules have been knocked out. However, mice deficient in both the IL-4Rα chain (Barner *et al.*, 1998) and STAT6 (Takeda *et al.*, 1996) have been generated and these mice have impaired TH2 cell development and fail to expel nematode infections efficiently (Urban *et al.*, 1998). These results are in concordance with findings for IL-13-deficient mice (McKenzie *et al.*, 1998b).

Human abnormalities

There have not been cases described where either IL-13 receptor has been directly implicated in a human abnormality; however, in certain cases some polymorphisms in the IL-4Rα chain have been shown to be associated with *atopy* (Hershey *et al.*, 1997; Mitsuyasu *et al.*, 1998). The IL-13R has been found to be overexpressed and a marker for human gliomas (Debinski *et al.*, 1999a,b).

THERAPEUTIC UTILITY

Effect of treatment with soluble receptor domain

There are no published work describing the effects of using the soluble IL-13 receptor for therapeutic purposes in humans. In mice, treatment with a recombinant soluble IL-13Rα2-Fc fusion protein impaired the expulsion of nematode worms (Urban *et al.*, 1998). A soluble form of the mIL-13Rα1 has also been administered to mice and found to increase the production of IgG2a and IgG2b in germinal center B cells (Poudrier *et al.*, 1999).

Effects of inhibitors (antibodies) to receptors

There is no published works describing the effects of using anti-IL-13 receptor antibodies for therapeutic purposes. However, the mutant form of human IL-4, Y124D, is a potent antagonist of both IL-4 and IL-13 (Kruse *et al.*, 1992; Aversa *et al.*, 1993; Matthews *et al.*, 1997) and may have therapeutic potential. Some antibodies to the IL-4Rα chain have been shown to inhibit B cell responses to both IL-4 and IL-13 (Zurawski *et al.*, 1995; Matthews *et al.*, 1997).

References

Aman, J., Tayebi, N., Obiri, N., Puri, R., Modi, W., and Leonard, W. (1996). cDNA cloning and characterisation of the human interleukin 13 receptor α chain. *J. Biol. Chem.* **271**, 29265–29270.

Aversa, G., Punnonen, J., Cocks, B., de Waal Malefyt, R., Vega, F. J., Zurawski, S., Zurawski, G., and de Vries, J. (1993). An interleukin-4 (IL-4) mutant protein inhibits both IL-4 and IL-13-induced human immunoglobulin G4 (IgG4) and IgE synthesis and B cell proliferation: support for a common component shared by IL-4 and IL-13 receptors. *J. Exp. Med.* **178**, 2213–2218.

Barner, M., Mohrs, M., Brombacher, F., and Kopf, M. (1998). Differences between IL-4Rα-deficient and IL-4-deficient mice reveal a role for IL-13 in the regulation of Th2 responses. *Curr. Biol.* **8**, 669–672.

Callard, R. E., Matthews, D. J., and Hibbert, L. (1996). IL-4 and IL-13 receptors: are they one and the same? *Immunol. Today* **17**, 108–110.

Caput, D., Laurent, P., Kaghad, M., Lelias, J.-M., Lefort, S., Vita, N., and Ferrara, P. (1996). Cloning and characterisation of a specific interleukin (IL)-13 binding protein structurally related to the IL-5 receptor α chain. *J. Biol. Chem.* **271**, 16921–16926.

Curiel, R., Lahesmaa, R., Subleski, J., Cippitelli, M., Kirken, R., Young, H., and Ghosh, P. (1997). Identification of a Stat6-responsive element in the promoter of the human interleukin-4 gene. *Eur. J. Immunol.* **27**, 1982–1987.

Debinski, W., Gibo, D. M., Hulet, S. W., Connor, J. R., and Gillespie, G. Y. (1999a). Receptor for interleukin 13 is a marker and therapeutic target for human high-grade gliomas. *Clin. Cancer Res.* **5**, 985–990.

Debinski, W., Gibo, D. M., Slagle, B., Powers, S. K., and Gillespie, G. Y. (1999b). Receptor for interleukin 13 is abundantly and specifically over-expressed in patients with glioblastoma multiforme. *Int. J. Oncol.* **15**, 481–486.

Donaldson, D. D., Whitters, M. J., Fitz, L. J., Neben, T. Y., Finnerty, H., Henderson, S. L., O'Hara, R. M. J., Beier, D. R., Turner, K. J., Wood, C. R., and Collins, M. (1998). The murine IL-13 receptor alpha 2: molecular cloning, characterization, and comparison with murine IL-13 receptor alpha 1. *J. Immunol.* **161**, 2317–2324.

Doucet, C., Brounty-Boye, D., Pottin-Clemenceau, C., Jasmin, C., Canonica, G. W., and Azzarone, B. (1998). IL-4 and IL-13 specifically increase adhesion molecule and inflammatory cytokine expression in human lung fibroblasts. *Int. Immunol.* **10**, 1421–1433.

Ford, D., Sheehan, C., Girasole, C., Priester, R., Kouttab, N., Tigges, J., King, T. C., Luciani, A., Morgan, J. W., and Maizel, A. L. (1999). The human B cell response to IL-13 is dependent on cellular phenotype as well as mode of activation. *J. Immunol.* **163**, 3185–3193.

Gauchat, J.-F., Schlagenhauf, E., Feng, N.-P., Moser, R., Yamage, M., Jeanin, P., Alouani, S., Elson, G., Notarangelo, D., Wells, T., Eugster, H.-P., and Bonnefoy, J.-Y. (1997). A novel 4 kb interleukin-13 receptor α mRNA expressed in human B, T, and endothelial cells encoding an alternate type-II interleukin-4/interleukin-13 receptor. *Eur. J. Immunol.* **27**, 971–978.

Graber, P., Gretener, D., Herren, S., Aubry, J. P., Elson, G., Poudrier, J., Lecoanet-Henchoz, S., Alouani, S., Losberger, C., Bonnefoy, J. Y., Kosco-Vilbois, M. H., and Gauchat, J. F. (1998). The distribution of IL-13 receptor $\alpha 1$ expression on B cells, T cells and monocytes and its regulation by IL-13 and IL-4. *Eur. J. Immunol.* **28**, 4286–4298.

Guo, J., Apiou, F., Mellerin, M. P., Lebeau, B., Jaques, Y., and Minvielle, S. (1997). Chromosome mapping and expression of the human interleukin-13 receptor. *Genomics* **42**, 141–145.

Haque, S. J., Harbor, P., Tabrizi, M., Yi, T., and Williams, B. R. (1998). Protein-tyrosine phosphatase Shp-1 is a negative regulator of IL-4- and IL-13-dependent signal transduction. *J. Biol. Chem.* **273**, 33893–33896.

Hershey, G. K., Friedrich, M. F., Esswein, L. A., Thomas, M. L., and Chatila, T. A. (1997). The association of atopy with a gain-of-function mutation in the alpha subunit of the interleukin-4 receptor. *N. Engl. J. Med.* **337**, 1720–1725.

Hilton, D., Zhang, J.-G., Metcalf, D., Alexander, W., Nicola, N., and Wilson, T. (1996). Cloning and characterisation of a binding subunit of the interleukin 13 receptor that is also a component of the interleukin 4 receptor. *Proc. Natl Acad. Sci. USA* **93**, 497–501.

Keegan, A., Johnston, J., Tortolani, P., McReynolds, L., Kinzer, C., O'Shea, J., and Paul, W. (1995). Similarities and differences in signal transduction by interleukin 4 and interleukin 13: analysis of janus kinase activation. *Proc. Natl Acad. Sci. USA* **92**, 7681–7685.

Kotowicz, N., Callard, R., Friedrich, K., Matthews, D., and Klein, N. (1996). Biological activity of IL-4 and IL-13 on human endothelial cells: functional evidence that both cytokines act through the same receptor. *Int. Immunol.* **8**, 1915–1925.

Kruse, N., Tony, H., and Sebald, W. (1992). Conversion of human interleukin-4 into a high affinity antagonist by a single amino acid replacement. *EMBO J.* **11**, 3237–3244.

Kuznetsov, V. A., and Puri, R. K. (1999). Kinetic analysis of high affinity forms of interleukin (IL)-13 receptors: suppression of IL-13 binding by IL-2 receptor gamma chain. *Biophys. J.* **77**, 154–172.

Lin, J.-X., Migone, T.-S., Tsang, M., Friedmann, M., Weatherbee, J., Zhou, L., Yamauchi, A., Bloom, E., Meitz, J., John, S., and Leonard, W. (1995). The role of shared receptor motifs and common Stat proteins in the generation of cytokine pleiotropy and redundancy by IL-2, IL-4, IL-7, IL-13 and IL-5. *Immunity* **2**, 331–339.

Manna, S. K., and Aggarwal, B. B. (1998). IL-13 suppresses TNF-induced activation of nuclear factor-κB, activation protein-1, and apoptosis. *J. Immunol.* **161**, 2863–2872.

Matthews, D., Clark, P., Herbert, J., Morgan, G., Armitage, R., Kinnon, C., Minty, A., Grabstein, K., Caput, D., Ferrara, P., and Callard, R. (1995). Function of the interleukin-2 (IL-2) receptor gamma-chain in biologic responses of X-linked severe combined immunodeficient B cells to IL-2, IL-4, IL-13 and IL-15. *Blood* **85**, 38–42.

Matthews, D. J., Hibbert, L., Friedrich, K., Minty, A., and Callard, R. E. (1997). X-SCID B cell responses to interleukin-4 and interleukin-13 are mediated by a receptor complex that includes the interleukin-4 receptor alpha chain (p140) but not the gamma c chain. *Eur. J. Immunol.* **27**, 116–121.

McKenzie, G., Bancroft, A., Grencis, R., and Mckenzie, A. (1998a). A distinct role for interleukin-13 in Th2-cell-mediated immune responses. *Curr. Biol.* **8**, 339–342.

McKenzie, G. J., Emson, C. L., Bell, S. E., Anderson, S., Fallon, P., Zurawski, G., Murray, R., and McKenzie, A. N. J. (1998b). Impaired development of Th2 cells in IL-13-deficient mice. *Immunity* **9**, 423–432.

Miloux, B., Laurent, P., Bonnin, O., Lupker, J., Caput, D., Vita, N., and Ferrara, P. (1996). Cloning of the human IL-13R$\alpha 1$ chain and reconstitution with the IL-4Rα of a functional IL-4/IL-13 receptor complex. *FEBS Lett.* **401**, 163–166.

Mitsuyasu, H., Izuhara, K., Mao, X. O., Gao, P. S., Arinobu, Y., Enomoto, T., Kawai, M., Saski, S., Dake, Y., Hamasaki, N., Shirakawa, T., and Hopkin, J. M. (1998). IleVal variant of IL4R alpha upregulates IgE synthesis and associates with atopic asthma. *Nat. Genet.* **2**, 119–120.

Murata, T., Noguchi, P., and Puri, R. (1996). IL-13 induces phosphorylation and activation of JAK2 janus kinase in human colon carcinoma cell lines: similarities between IL-4 and IL-13 signalling. *J. Immunol.* **156**, 2972–2978.

Murata, T., Husain, S. R., Mohri, H., and Puri, R. K. (1998). Two different IL-13 receptor chains are expressed in normal human skin fibroblasts, and IL-13 mediate signal transduction through a common pathway. *Int. Immunol.* **8**, 1103–1110.

Ogata, H., Ford, D., Kouttab, N., King, T. C., Vita, N., Minty, A., Stoeckler, J., Morgan, D., Girasole, C., Morgan, W. J., and Maizel, A. L. (1998). Regulation of interleukin-13 receptor constituents on mature human B lymphocytes. *J. Biol. Chem.* **279**, 9864–9871.

Orchansky, P., Ayres, S., Hilton, D., and Schraeder, J. (1997). An interleukin (IL)-13 receptor lacking the cytoplasmic domain fails to transduce IL-13-induced signals and inhibits responses to IL-4. *J. Biol. Chem.* **272**, 22940–22947.

Orchansky, P. L., Kwan, R., Lee, F., and Schrader, J. W. (1999). Characterization of the cytoplasmic domain of interleukin-13 receptor-alpha. *J. Biol. Chem.* **274**, 20818–20825.

Palmer-Crocker, R., Hughes, C., and Pober, J. (1996). IL-4 and IL-13 activate the JAK2 tyrosine kinase and Stat6 in cultured human vascular endothelial cells through a common pathway that does not involve the γc chain. *J. Clin. Invest.* **98**, 604–609.

Poudrier, J., Graber, P., Herren, S., Gretener, D., Elson, G., Berney, C., Gauchat, J. F., and Kosco-Vilbois, M. H. (1999). A soluble form of IL-13 receptor alpha 1 promotes IgG2a and IgG2b production by murine germinal center B cells. *J. Immunol.* **163**, 1153–1161.

Punnonen, J., Aversa, G., Cocks, B., Mckenzie, A., Menon, S., Zurawski, G., de Waal Malefyt, R., and de Vries, J. (1993). Interleukin-13 induces interleukin-4-independent IgG4 and IgE synthesis and CD23 expression by human B-cells. *Proc. Natl Acad. Sci. USA* **90**, 3730–3734.

Roy, B., and Cathcart, M. K. (1998). Induction of 15-lipoxygenase expression by IL-13 requires tyrosine phosphorylation of JAK2 and TYK2 in human monocytes. *J. Biol. Chem.* **273**, 32023–32029.

Sozzani, P., Hasan, L., Seguelas, M. H., Caput, D., Ferrara, P., Pipy, B., and Cambon, C. (1998). Il-13 induces tyrosine phosphorylation of phospholipase C gamma-1 following IRS-2 association in human monocytes: relationship with the inhibitory effect of IL-13 on ROI production. *Biochem. Biophy. Res. Commun.* **244**, 665–670.

Starr, R., Willson, T., Viney, E., Murray, L., Rayner, J., Jenkins, B., Gonda, T., Alexander, W., Metcalf, D., Nicola, N., and Hilton, D. (1997). A family of cytokine-inducible inhibitors of signalling. *Nature* **387**, 917–921.

Takeda, K., Kamanaka, M., Tanaka, T., Kishimoto, T., and Akira, S. (1996). Impaired IL-13-mediated functions of macrophages in STAT6-deficient mice. *J. Immunol.* **157**, 3220–3222.

Urban, J., Noben-Trauth, N., Donaldson, D., Madden, K., Morris, S., Collins, M., and Finkelman, F. (1998). IL-13, IL-4R and Stat6 are required for the expulsion of the gastrointestinal nematode parasite *Nippostrongylus brasiliensis*. *Immunity* **8**, 255–264.

Welham, M. J., Learmonth, L., Bone, H., and Schrader, J. W. (1995). Interleukin-13 signal transduction in lymphohemopoietic cells. Similarities and differences in signal transduction with interleukin-4 and insulin. *J. Biol. Chem.* **270**, 12286–12296.

Wright, K., Kolios, G., Westwick, J., and Ward, S. G. (1999). Cytokine-induced apoptosis in epithelial HT-29 cells is independent of nitric oxide formation. Evidence for an interleukin-13-driven phosphatidylinositol 3-kinase-dependent survival mechanism. *J. Biol. Chem.* **274**, 17193–17201.

Zhang, J., Hilton, D., Willson, T., McFarlane, C., Roberts, B., Moritz, R., Simpson, R., Alexander, W., Metcalf, D., and Nicola, N. (1997). Identification, purification and characterisation of a soluble interleukin (IL)-13-binding protein. Evidence that it is distinct from the cloned IL-13 receptor and IL-4 receptor alpha chains. *J. Biol. Chem.* **272**, 9474–9480.

Zurawski, G., and de Vries, J. (1994). Interleukin 13, an interleukin 4-like cytokine that acts on monocytes and B-cells but not on T-cells. *Immunol. Today* **15**, 19–26.

Zurawski, S., Chomarat, P., Djossou, O., Bidaud, C., Mckenzie, A., Miossec, P., Banchereau, J., and Zurawski, G. (1995). The primary binding subunit of the human interleukin-4 receptor is also a component of the interleukin-13 receptor. *J. Biol. Chem.* **270**, 13869–13878.

IL-15 Receptor

Thomas A. Waldmann* and Yutaka Tagaya

Metabolism Branch, National Cancer Institute, NIH Building 10, Room 4N115, 10 Center Drive, MSC 1374, Bethesda, MD 20892-1374, USA

*corresponding author tel: 301-496-6653, fax: 301-496-9956, e-mail: tawald@helix.nih.gov

DOI: 10.1006/rwcy.2000.14008.

SUMMARY

IL-15 uses two distinct receptor and signaling pathways. In T and NK cells, the type 1 IL-15 receptor includes the IL-2/15Rβ subunit shared with IL-2, the γc subunit shared with IL-2, IL-4, IL-7, and IL-9, as well as an IL-15-specific receptor subunit IL-15Rα. Thus type 1 receptor uses a JAK1/JAK3 and STAT3/STAT5 signaling system. Mast cells respond to IL-15 with a type 2 receptor system that does not share elements with IL-2R but uses a novel 60–65 kDa IL-15RX subunit. This type 2 receptor signaling involves JAK2/STAT5 activation. In addition to the other functional activities in immune and nonimmune cells, signaling through the type 1 IL-15 receptor plays a pivotal role in the development, survival, and activation of NK cells.

BACKGROUND

Discovery

IL-15 is a 14–15 kDa member of the four α helix bundle family of cytokines. IL-15 utilizes two distinct receptor and signaling pathways (**Figure 1**). In T and NK cells the type 1 IL-15 receptor includes the γc shared with IL-2, IL-4, IL-7, IL-9, and the IL-2/15Rβ subunit shared with IL-2. Furthermore, it involves an IL-15-specific receptor subunit, IL-15Rα. In contrast, mast cells respond to IL-15 using a receptor system that does not share elements with the IL-2R system but involves a novel 60–65 kDa IL-15RX subunit. In mast cells IL-15 signaling involves JAK2 and STAT5 activation rather than the JAK1/JAK3 and STAT3/STAT5 that are used by IL-15 in activated T and NK cells.

The IL-2/15Rβ chain and the γc chain were defined as part of an analysis of the IL-2 receptor system. Using radiolabeled IL-2 in crosslinking studies Tsudo *et al.* (1986) and Sharon *et al.* (1986) discovered IL-2Rβ, a 70–75 kDa binding subunit. Grabstein *et al.* (1994) and Bamford *et al.* (1994) used IL-2Rβ-specific antibodies to demonstrate that IL-15 requires this subunit for its action in T and NK cells. Takeshita *et al.* (1992) defined and cloned the γc receptor and showed that it was a component of the IL-2 receptor. It was subsequently demonstrated that IL-2Rγ (now termed common gamma or γc) is not only an essential element of high- and intermediate-affinity receptors for IL-2, but is also required for the actions of IL-4, IL-7, and IL-9 (Kondo *et al.*, 1993; Noguchi *et al.*, 1993). Giri *et al.* (1994) demonstrated in cells transfected with IL-2R subunits that γc as well as IL-2Rβ are required for IL-15 binding and signaling. IL-15 does not use the α subunit of the IL-2 receptor. However, a novel IL-15-specific binding protein termed IL-15Rα was identified and its cDNA cloned by Giri *et al.* (1995). Thus in T and NK cells the IL-15

Figure 1 IL-15 receptors.

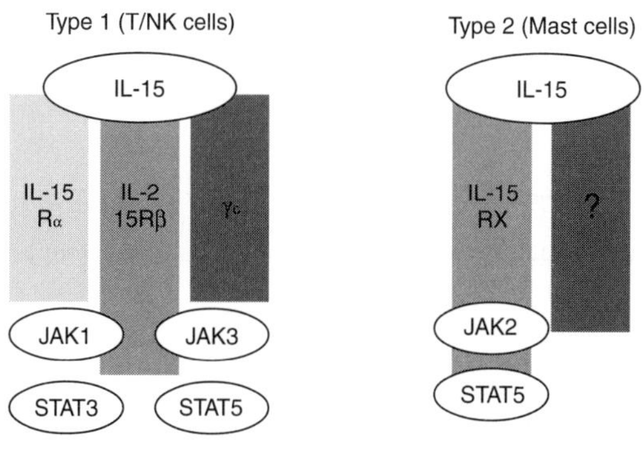

receptor was shown to involve an IL-15-specific subunit IL-15Rα, the IL-2/15Rβ subunit shared with IL-2, and the γc subunit shared with IL-2, IL-4, IL-7, and IL-9. Tagaya *et al.* (1996a,b) demonstrated that IL-15 uses a distinct type 2 receptor/signal transduction pathway in mast cells. Mast cells were shown to respond to IL-15 with a receptor system that does not share elements with the IL-2R but uses a novel 60–65 kDa IL-15RX subunit.

Alternative names

The β chain shared by IL-2 and IL-15 is frequently called IL-2Rβ. It is also referred to as IL-2/15Rβ. The common gamma chain, γc, was initially termed IL-2Rγ prior to the demonstration of its use by other cytokines. The IL-15-specific receptor in T and NK cells is termed IL-15Rα, whereas the chain involved in the type 2 receptor of mast cells is IL-15RX.

Structure

IL-2/15Rβ and γc are members of the cytokine I superfamily of receptors that contain four conserved cystines and the canonical WSXWS (Trp-Ser-X-Trp-Ser) motif. IL-15Rα is a type 1 membrane protein that is not a member of the cytokine receptor superfamily. However, a comparison of IL-2Rα and IL-15Rα revealed the shared presence of a conserved motif known as the GP-1 motif or a SUSHI domain (Giri *et al.*, 1995). The structure of the mast cell type 2 IL-15RX receptor subunit has not been defined.

Main activities and pathophysiological roles

Through its action on its receptor, IL-15 stimulates the proliferation of activated CD4−CD8−, CD4+ CD8+, CD4+ and CD8+ cells, and dendritic epidermal T cells (Burton *et al.*, 1994; Grabstein *et al.*, 1994; Edelbaum *et al.*, 1995; Garcia *et al.*, 1998). It was recently reported that IL-15 preferentially propagates CD8 memory T cells (Zhang *et al.*, 1998). Although IL-15 does not have an effect on resting B cells it induces proliferation and immunoglobulin synthesis by B cells costimulated by PMA or by an immobilized antibody to immunoglobulin M (Armitage *et al.*, 1995).

One of the most critical functions of IL-15 acting through the type 1 receptor is a pivotal role in the development, survival, and activation of NK cells. IL-15 acting through the type 2 receptor stimulates mast cells proliferation. IL-15 also acts on skeletal muscle (Quinn *et al.*, 1997), endothelial cells, and microglia (Lee *et al.*, 1996; Hanisch *et al.*, 1997).

GENE

Accession numbers

GenBank:
Human IL-15Rα: NM_002189
Human IL-2/15Rβ: NM_000878
Human γc: NM_000206
Mouse IL-15Rα: U22339
Mouse IL-2/15Rβ: M28052
Mouse γc: U21795

Sequence

Nucleotide sequences for the IL-15R components can be found in GenBank. The accession numbers are shown above.

Chromosome location and linkages

Human α: 10 p15-14; β: 22q11.2-q12; γc: Xq13
Mouse α: 2; β: 15: γc: X

The chromosomal localization of IL-15RX has not been defined.

Relevant linkages

IL-15Rα and IL-2Rα are linked in humans on chromosome 10 p15-14 and in mice on chromosome 2 (Anderson *et al.*, 1995).

PROTEIN

Accession numbers

Human IL-15Rα: NP_002180
Human IL-2/15Rβ: NP_000869
Human γc: NP_000197
Mouse IL-15Rα: AAC52240
Mouse IL-2/15Rβ: AAA39283
Mouse γc: AAA64279

Sequence

See **Figure 2**.

Description of protein

IL-15Rα is a type 1 membrane protein with a predicted signal peptide of 32 amino acids, a 173 amino acid extracellular domain, a single membrane-spanning region of 21 amino acids, and a 37 amino acid cytoplasmic domain (mouse IL-15Rα). In contrast to IL-2/15Rβ and γc, IL-15Rα is not a member of the cytokine receptor superfamily. However, IL-15Rα contains a motif known as a GP-1 motif or a SUSHI domain (Giri *et al.*, 1995). The human IL-2/15Rβ mRNA encodes a primary translation product of 551 amino acids (Hatakeyama *et al.*, 1989). The receptor contains a 26 amino acid signal peptide and a mature human IL-2/15Rβ is composed of 525 amino acids with an extracellular segment of 214 amino acids, a hydrophobic transmembrane stretch of 25 amino acids, and a 286 amino acid cytoplasmic domain. The human γc cDNA contains an open reading frame encoding a 369 amino acid residue polypeptide (Takeshita *et al.*, 1992). This protein contains a 22 amino acid signal peptide, a 233 amino acid extracellular domain, a 28 amino acid hydrophobic transmembrane domain, and an 86 amino acid terminal cytoplasmic domain. IL-2/15Rβ and γc are members of the hematopoietin or cytokine superfamily of receptors that contain four conserved cystines and canonical WSXWS (Trp-Ser-X-Trp-Ser) motif. IL-15RX is a 60–65 kDa receptor whose structure has not been defined.

Relevant homologies and species differences

IL-2/15Rβ and γc are members of the hematopoietin or cytokine superfamily of receptors. IL-15Rα shares with IL-2Rα the presence of the GP-1 or SUSHI domain motif. Furthermore, the IL-2Rα and IL-15Rα genes have a similar intron–exon organization. Moreover, they are closely linked on both human (10p-15-14) and murine genomes (chromosome 2) (Anderson *et al.*, 1995).

Affinity for ligand(s)

IL-15Rα binds IL-15 with a very high affinity (dissociation constant $K_d = 10^{-11}$ M) (Anderson *et al.*, 1995; Giri *et al.*, 1995). This affinity was not dramatically altered by the simultaneous presence of IL-2/15Rβ or γc. IL-2/15Rβ and γc acting together in the absence of IL-15Rα bind IL-15 with an intermediate affinity (approximately $K_d = 10^{-9}$ M). IL-15RX binds IL-15 with an intermediate affinity ($K_d = 10^{-9}$ M) (Tagaya *et al.*, 1996b).

Cell types and tissues expressing the receptor

IL-15Rα has a wide cellular distribution. Its expression is observed in T cells, B cells, macrophages, and

Figure 2 The amino acid sequences for human IL-15Rα (Anderson *et al.*, 1995) and mouse IL-15Rα (Giri *et al.*, 1995). The transmembrane domain is underlined.

Human IL-15Rα

```
-30   MAPRRARGCR TLGLPALLLL LLLRPPATRG
ITCPPPMSVE HADIWVKSYS LYSRERYICN SGFKRKAGTS SLTECVLNKA
51    TNVAHWTTPS LKCIRDPALV HQRPAPPSTV TTAGVTPQPE SLSPSGKEPA
ASSPSSNNTA ATTAAIVPGS QLMPSKSPST GTTEISSHES SHGTPSQTTA
KNWELTASAS HQPPGVYPQG HSDTTVAIST STVLLCGLSA VSLLACYLKS
RQTPPLASVE MEAMEALPVT WGTSSRDEDL ENCSHHL
```

Murine IL-15Rα

```
-32   MASPQLRGYG VQAIPVLLLL LLLLLLPLRV TP
1     GTTCPPPVSI EHADIRVKNY SVNSRERYVC NSGFKRKAGT STLIECVINK
NTNVAHWTTP SLKCIRDPSL AHYSPVPTVV TPKVTSQPES PSPSAKEPEA
AFSPKSDTAM TTETAIMPGS RLTPSQTTSA GTTGTGSHKS SRAPSLAATM
TLEPTASTSL RITEISPHSS KMTKVAISTS VLLVGAGVVM AFLAWYIKSR
201   QPSQPCRVEV ETMETVPMTV RASSKEDEDT GA
```

in thymic stroma cells and bone marrow stroma cells (Anderson *et al.*, 1995). In addition, IL-15Rα mRNA is widespread in such tissues as liver, heart, spleen, lung, skeletal muscle, and activated vascular endothelial cells (Giri *et al.*, 1995). IL-15Rα mRNA is increased in T cells after addition of IL-2, an anti-CD3 antibody or phorbol-myristate acetate (PMA) (Giri *et al.*, 1995). Furthermore, IL-15Rα expression is augmented in macrophage cell lines after treatment with IFNγ. IL-2/15Rβ is constitutively expressed by NK cells, monocytes, and resting CD8 cells but is not expressed by resting CD4 cells although it is inducible in such cells. The common gamma chain is expressed by most hematopoietic cells.

Regulation of receptor expression

IL-2/15Rβ: Promoter/enhancer region contains putative binding sites for Ets-1, GABP, SP-1 and Egr-1 (Lin and Leonard, 1997).

γc: The γc gene has a constitutive activation promoter that contains an Ets-binding site.

IL-15Rα: The 5′ regulatory region of this gene is not defined.

Release of soluble receptors

In contrast to the release of IL-2Rα, there is little release of IL-2/15Rβ or γc. Although levels have not been quantitated, it has been suggested that IL-15Rα is released from the cell surface and may act to inhibit IL-15 action.

SIGNAL TRANSDUCTION

Associated or intrinsic kinases

The type 1 IL-15 receptors in T and NK cells, like most cytokine receptors, do not possess intrinsic protein tyrosine kinase (PTK) domains, yet receptor stimulation invokes rapid tyrosine phosphorylation of intracellular proteins including the receptors themselves. In T and NK cells, IL-15 activates JAK1 and JAK3 of the tyrosine kinase family members (Witthuhn *et al.*, 1994; Johnston *et al.*, 1995). Furthermore, the addition of IL-15 to such receptor-expressing T cells led to the tyrosine phosphorylation and nuclear translocation of STAT 3 and STAT5 (Johnston *et al.*, 1995; Lin *et al.*, 1995). The IL-15-signaling pathway in T cells also involves the phosphorylation of the Src-related cytoplasmic tyrosine kinases p56lck and p72syk, the induction of the expression of the Bcl-2 anti-apoptotic protein and the stimulation of the Ras/Raf/MAP kinase pathway leading to fos/jun activation (Miyazaki *et al.*, 1995).

Mast cells respond to IL-15 with a type 2 receptor system that uses a novel 60–65 kDa IL-15RX subunit. This type 2 receptor involves JAK2/STAT5 activation rather than the JAK1/JAK3 and STAT3/STAT5 system used by the type 1 receptor in T/NK cells.

DOWNSTREAM GENE ACTIVATION

Transcription factors activated

IL-15 through the type 1 receptor activates jun/fos AP-1 complex. It also activates STAT3, STAT5a, and STAT5b transcription factors (Lin *et al.*, 1995).

The type 2 receptor signaling activates STAT5a and b molecules.

Genes induced

The genes induced by IL-15R include IL-2Rα (Treiber-Held *et al.*, 1996), CC chemokines and receptors (Perera *et al.*, 1999), bcl-2/bcl-X_L anti-apoptotic genes, caspase 8/FLICE (Perera and Waldmann, 1998), Pim-1, CIS/SIS/SOCS family member proteins, and c-myc.

Promoter regions involved

The promoter regions involved are the STAT5 consensus sequence and AP-1 sites.

BIOLOGICAL CONSEQUENCES OF ACTIVATING OR INHIBITING RECEPTOR AND PATHOPHYSIOLOGY

Unique biological effects of activating the receptors

For more detail on the biological effects of IL-15 receptor, see the review by Waldmann and Tagaya (1999).

IL-15 acting through the type 1 receptor stimulates the proliferation of activated CD4−CD8−, CD4+

CD8+, CD4+, and CD8+ cells (Burton *et al.*, 1994; Grabstein *et al.*, 1994; Edelbaum *et al.*, 1995; Zhang *et al.*, 1998; Garcia *et al.*, 1998). IL-15 also has an effect on activated but not resting T cells, inducing proliferation and immunoglobulin synthesis in cells costimulated by PMA or by an immobilized antibody to IgM (Armitage *et al.*, 1995). IL-15 may be an essential factor for the development of NK cells (Carson *et al.*, 1994). NK cells are absent in mice made deficient in elements required for IL-15 action, including IL-2/15Rβ (Suzuki *et al.*, 1997), γc (Cao *et al.*, 1995; DiSanto *et al.*, 1995), IRF-1−/− (Ogasawara *et al.*, 1998; Ohteki *et al.*, 1998), JAK3 (Biron *et al.*, 1989; Russell *et al.*, 1995; Macchi *et al.*, 1995), or STAT5a/b (Imada *et al.*, 1998; Teglund *et al.*, 1998). Furthermore, IL-15 is effective in inducing bone marrow progenitor differentiation into NK cells (Mrózek *et al.*, 1996; Cavazzana-Calvo *et al.*, 1996). In a similar way, the addition of IL-15 to immature postnatal thymocytes or to fetal thymic organ cultures led to the development of NK cells (Mingari *et al.*, 1997).

IL-15 also has unique functions on nonlymphoid cells. Acting through the type 2 receptor it stimulates mast cell proliferation (Tagaya *et al.*, 1996a,b). Although the type of receptor has not been defined, IL-15 also has actions on muscle, inducing skeletal muscle fiber hypertrophy (Quinn *et al.*, 1997), vascular endothelial cells promoting angiogenesis (Angiolillo *et al.*, 1997), and on brain microglia and astrocytes (Lee *et al.*, 1996; Hanisch *et al.*, 1997).

Phenotypes of receptor knockouts and receptor overexpression mice

IL-15Rα-null (IL-15Rα−/−) mice are markedly lymphopenic despite grossly normal T and B lymphocyte development (Lodolce *et al.*, 1998). This lymphopenia is due to decreased proliferation and decreased homing of IL-15Rα−/− lymphocytes to peripheral lymph nodes. These mice are also deficient in NK cells, natural killer T cells, CD8+ lymphocytes, and TCRγ/δ-intraepithelial lymphocytes. In addition, memory phenotype CD8+ T cells are selectively reduced in number.

Mice lacking the IL-2/15Rβ chain are deficient in functions mediated by either IL-2 or IL-15 since this receptor is shared by these two cytokines. Mice lacking this receptor lack NK cells. Furthermore, they manifest spontaneously activated T cells, increased differentiation of B cells into plasma cells, and high serum concentrations of immunoglobulins IgG and IgE as well as autoantibodies that cause *hemolytic anemia* (Suzuki *et al.*, 1997). These animals manifested marked infiltrative granulopoiesis and died after about 12 weeks. The γc chain is shared by IL-2, IL-4, IL-7, IL-9, and IL-15; thus mice deficient in this chain lack the ability to respond to all of these cytokines. Mice made deficient in this cytokine or its membrane-proximal signaling element, JAK3, manifest *severe combined immunodeficiency disease* with a virtual absence of NK cells markedly deficient T cell numbers and function and abnormalities of B cell function, presumably due to the lack of IL-7 function in early T/B cell development.

Human abnormalities

No patients with deficiency of IL-15Rα, IL-2/15Rβ, or IL-15RX have been defined. Patients deficient in γc manifest *X chromosome-linked severe combined immunodeficiency disease* (X-SCID) (Schorle *et al.*, 1991; Noguchi *et al.*, 1993). Patients with this disorder have a dramatic reduction in T and NK cells but have at least normal numbers of B cells that are functionally abnormal. Deficiency of JAK3 in humans yields an autosomal disorder with the same phenotype.

THERAPEUTIC UTILITY

Effect of treatment with soluble receptor domain

The injection of an IL-15 antagonist with the soluble form of IL-15Rα into DBS/1 mice suppressed their development of *collagen-induced arthritis* (Ruchatz *et al.*, 1998).

Effects of inhibitors (antibodies) to receptors

An IL-15 receptor antagonist has been produced by mutating glutamine residues within the C-terminus of IL-15 to aspartic acid, completely inhibiting IL-15-triggered cell proliferation (Kim *et al.*, 1998). This IL-15 mutant protein markedly attenuated antigen-specific delayed hypersensitivity responses in Balb/c mice and enhanced the acceptance of islet cell allografts. An antibody (Mikβ1) directed toward IL-2/IL-15Rβ inhibits the actions of IL-15 but not those mediated by IL-2 through the high-affinity IL-2 receptor. A humanized version of this antibody prolonged renal allograft survival in cynomolgus monkeys (Tinubu *et al.*, 1994).

The clinical application of new therapeutic agents that target IL-15 or the receptor and signaling elements shared by IL-15 and other T cell stimulatory cytokines may provide a new perspective for the treatment of *tropical spastic paraparesis HTLV-I-associated myelopathy* (TSP/HAM), *rheumatoid arthritis*, and *inflammatory bowel disease* where abnormalities of IL-15 expression have been demonstrated.

References

Anderson, D. M., Kumaki, S., Ahdieh, M., Bertles, J., Tometsko, M., Loomis, A., Giri, J., Copeland, G., Gilbert, D. J., Jenkins, N. A., Valentine, V., Shapiro, D. N., Morris, S. W., Park, L. S., and Cosman, D. (1995). Functional characterization of the human interleukin-15 receptor α chain and close linkage of IL15RA and IL2RA genes. *J. Biol. Chem.* **270**, 29862–29869.

Angiolillo, A. L., Kanegane, H., Sgadari, C., Reaman, G. H., and Tosato, G. (1997). Interleukin-15 promotes angiogenesis *in vivo*. *Biochem. Biophys. Res. Commun.* **233**, 231–237.

Armitage, R. J., Macduff, B. M., Eisenman, J., Paxton, R., and Grabstein, K. H. (1995). IL-15 has stimulatory activity for the induction of B cell proliferation and differentiation. *J. Immunol.* **154**, 483–490.

Bamford, R. N., Grant, A. J., Burton, J. D., Peters, C., Kurys, G., Goldman, C. K., Brennan, J., Roessler, E., and Waldmann, T. A. (1994). The interleukin (IL) 2 receptor β chain is shared by IL-2 and a cytokine, provisionally designated IL-T, that stimulates T-cell proliferation and the induction of lymphokine-activated killer cells. *Proc. Natl Acad. Sci. USA* **91**, 4940–4944.

Biron, C. A., Byron, K. S., and Sullivan, J. L. (1989). Severe herpesvirus infections in an adolescent without natural killer cells [see comments]. *N. Engl J. Med.* **320**, 1731–1735.

Burton, J. D., Bamford, R. N., Peters, C., Grant, A. J., Kurys, G., Goldman, C. K., Brennan, J., Roessler, E., and Waldmann, T. A. (1994). A lymphokine, provisionally designated interleukin T and produced by a human adult T-cell leukemia line, stimulates T-cell proliferation and the induction of lymphokine-activated killer cells. *Proc. Natl Acad. Sci. USA* **91**, 4935–4939.

Cao, X., Shores, E. W., Hu-Li, J., Anver, M. R., Kelsall, B. L., Russell, S. M., Drago, J., Noguchi, M., Grinberg, A., Bloom, E. T., Paul, W. E., Katz, S. I., Love, P. E., and Leonard, W. J. (1995). Defective lymphoid development in mice lacking expression of the common cytokine receptor γ chain. *Immunity* **2**, 223–238.

Carson, W. E., Giri, J. G., Lindemann, M. J., Linett, M. L., Ahdieh, M., Paxton, R., Anderson, D., Eisenmann, J., Grabstein, K., and Caligiuri, M. A. (1994). Interleukin (IL) 15 is a novel cytokine that activates human natural killer cells via components of the IL-2 receptor. *J. Exp. Med.* **180**, 1395–1403.

Cavazzana-Calvo, M., Hacein-Bey, S., de Saint Basile, G., De Coene, C., Selz, F., Le Deist, F., and Fischer, A. (1996). Role of interleukin-2 (IL-2), IL-7, and IL-15 in natural killer cell differentiation from cord blood hematopoietic progenitor cells and from γc transduced severe combined immunodeficiency X1 bone marrow cells. *Blood* **88**, 3901–3909.

DiSanto, J. P., Muller, W., Guy-Grand, D., Fischer, A., and Rajewsky, K. (1995). Lymphoid development in mice with a targeted deletion of the interleukin 2 receptor γ chain. *Proc. Natl Acad. Sci. USA* **92**, 377–381.

Edelbaum, D., Mohamadzadeh, M., Bergstresser, P. R., Sugamura, K., and Takashima, A. (1995). Interleukin (IL)-15 promotes the growth of murine epidermal gamma delta T cells by a mechanism involving the beta- and gamma c-chains of the IL-2 receptor. *J. Invest. Dermatol.* **105**, 837–843.

Garcia, V. E., Jullien, D., Song, M., Uyemura, K., Shuai, K., Morita, C. T., and Modlin, R. L. (1998). IL-15 enhances the response of human gamma delta T cells to nonpeptide microbial antigens. *J. Immunol.* **160**, 4322–4329.

Giri, J. G., Ahdieh, M., Eisenman, J., Shanebeck, K., Grabstein, K., Kumaki, S., Namen, A., Park, L. S., Cosman, D., and Anderson, D. (1994). Utilization of the beta and gamma chains of the IL-2 receptor by the novel cytokine IL-15. *EMBO J.* **13**, 2822–2830.

Giri, J. G., Kumaki, S., Ahdieh, M., Friend, D. J., Loomis, A., Shanebeck, K., DuBose, R., Cosman, D., Park, L. S., and Anderson, D. M. (1995). Identification and cloning of a novel IL-15 binding protein that is structurally related to the alpha chain of the IL-2 receptor. *EMBO J.* **14**, 3654–3663.

Grabstein, K. H., Eisenman, J., Shanebeck, K., Rauch, C., Srinivasan, S., Fung, V., Beers, C., Richardson, J., Schoenborn, M. A., Ahdieh, M., Johnson, L., Alderson, M. R., Watson, J. D., Anderson, D. M., and Giri, J. G. (1994). Cloning of a T cell growth factor that interacts with the β chain of the interleukin-2 receptor. *Science* **264**, 965–968.

Hanisch, U. K., Lyons, S. A., Prinz, M., Nolte, C., Weber, J. R., Kettenmann, H., and Kirchhoff, F. (1997). Mouse brain microglia express interleukin-15 and its multimeric receptor complex functionally coupled to Janus kinase activity. *J. Biol. Chem.* **272**, 28853–28860.

Hatakeyama, M., Tsudo, M., Minamoto, S., Kono, T., Doi, T., Miyata, T., Miyasaka, M., and Taniguchi, T. (1989). Interleukin-2 receptor beta chain gene: generation of three receptor forms by cloned human alpha and beta chain cDNA's. *Science* **244**, 551–556.

Imada, K., Bloom, E. T., Nakajima, H., Horvath-Arcidiacono, J. A., Udy, G. B., Davey, H. W., and Leonard, W. J. (1998). Stat5b is essential for natural killer cell-mediated proliferation and cytolytic activity. *J. Exp. Med.* **188**, 2067–2074.

Johnston, J. A., Bacon, C. M., Finbloom, D. S., Rees, R. C., Kaplan, D., Shibuya, K., Ortaldo, J. R., Gupta, S., Chen, Y. Q., Giri, J. D., and O'Shea, J. J. (1995). Tyrosine phosphorylation and activation of STAT5, STAT3, and Janus kinases by interleukins 2 and 15. *Proc. Natl Acad. Sci. USA* **92**, 8705–8709.

Kim, Y. S., Maslinski, W., Zheng, X. X., Stevens, A. C., Li, X. C., Tesch, G. H., Kelley, V. R., and Strom, T. B. (1998). Targeting the IL-15 receptor with an antagonist IL-15 mutant/Fc γ2a protein blocks delayed-type hypersensitivity. *J. Immunol.* **160**, 5742–5748.

Kondo, M., Takeshita, T., Ishii, N., Nakamura, M., Watanabe, S., Arai, K., and Sugamura, K. (1993). Sharing of the interleukin-2 (IL-2) receptor γ chain between receptors for IL-2 and IL-4. *Science* **262**, 1874–1877.

Lee, Y. B., Satoh, J., Walker, D. G., and Kim, S. U. (1996). Interleukin-15 gene expression in human astrocytes and microglia in culture. *Neuroreport* **7**, 1062–1066.

Lin, J. X., and Leonard, W. J. (1997). Signaling from the IL-2 receptor to the nucleus. *Cytokine Growth Factor Rev.* **8**, 313–332.

Lin, J. X., Migone, T. S., Tsang, M., Friedmann, M., Weatherbee, J. A., Zhou, L., Yamauchi, A., Bloom, E. T., Mietz, J., John, S., and Leonard, W. J. (1995). The role of shared receptor motifs and common Stat proteins in the generation of cytokine pleiotropy and redundancy by IL-2, IL-4, IL-7, IL-13, and IL-15. *Immunity* **2**, 331–339.

Lodolce, J. P., Boone, D. L., Chai, S., Swain, R. E., Dassopoulos, T., Trettin, S., and Ma, A. (1998). IL-15 receptor maintains lymphoid homeostasis by supporting lymphocyte homing and proliferation. *Immunity* **9**, 669–676.

Macchi, P., Villa, A., Gillani, S., Sacco, M. G., Frattini, A., Porta, F., Ugazio, A. G., Johnston, J. A., Candotti, F., O'Shea, J. J., Vezzoni, P., and Notarangelo, L. D. (1995). Mutations of Jak-3 gene in patients with autosomal severe combined immune deficiency (SCID). *Nature* **377**, 65–68.

Mingari, M. C., Vitale, C., Cantoni, C., Bellomo, R., Ponte, M., Schiavetti, F., Bertone, S., Moretta, A., and Moretta, L. (1997). Interleukin-15-induced maturation of human natural killer cells from early thymic precursors: selective expression of CD94/NKG2-A as the only HLA class I-specific inhibitory receptor. *Eur J. Immunol.* **27**, 1374–1380.

Miyazaki, T., Liu, Z. J., Kawahara, A., Minami, Y., Yamada, K., Tsujimoto, Y., Barsoumian, E. L., Permutter, R. M., and Taniguchi, T. (1995). Three distinct IL-2 signaling pathways mediated by bcl-2, c-myc, and lck cooperate in hematopoietic cell proliferation. *Cell* **81**, 223–231.

Mrózek, E., Anderson, P., and Caligiuri, M. A. (1996). Role of interleukin-15 in the development of human CD56+ natural killer cells from CD34+ hematopoietic progenitor cells. *Blood* **87**, 2632–2640.

Noguchi, M., Nakamura, Y., Russell, S. M., Ziegler, S. F., Tsang, M., Cao, X., and Leonard, W. J. (1993). Interleukin-2 receptor γ chain: a functional component of the interleukin-7 receptor. *Science* **262**, 1877–1880.

Ogasawara, K., Hida, S., Azimi, N., Tagaya, Y., Sato, T., Yokochi-Fukuda, T., Waldmann, T. A., Taniguchi, T., and Taki, S. (1998). Requirement for IRF-1 in the microenvironment supporting development of natural killer cells. *Nature* **391**, 700–703.

Ohteki, T., Yoshida, H., Matsuyama, T., Duncan, G. S., Mak, T. W., and Ohashi, P. S. (1998). The transcription factor interferon regulatory factor 1 (IRF-1) is important during the maturation of natural killer 1.1+ T cell receptor- alpha/beta+ (NK1+ T) cells, natural killer cells, and intestinal intraepithelial T cells. *J. Exp. Med.* **187**, 967–972.

Perera, L. P., and Waldmann, T. A. (1998). Activation of human monocytes induces differential resistance to apoptosis with rapid down regulation of caspase-8/FLICE. *Proc. Natl Acad. Sci. USA* **95**, 14308–14313.

Perera, L. P., Goldman, C. K., and Waldmann, T. A. (1999). IL-15 induces the expression of chemokines and their receptors in T lymphocytes. *J. Immunol.* **165**, 2606–2612.

Quinn, L. S., Haugk, K. L., and Damon, S. E. (1997). Interleukin-15 stimulates C2 skeletal myoblast differentiation. *Biochem. Biophys. Res. Commun.* **239**, 6–10.

Ruchatz, H., Leung, B. P., Wei, X. Q., McInnes, I. B., and Liew, F. Y. (1998). Soluble IL-15 receptor alpha-chain administration prevents murine collagen-induced arthritis: a role for IL-15 in development of antigen- induced immunopathology. *J. Immunol.* **160**, 5654–5660.

Russell, S. M., Tayebi, N., Nakajima, H., Riedy, M. C., Roberts, J. L., Aman, M. J., Migone, T. S., Noguchi, M., Markert, M. L., Buckley, R. H., O'Shea, J. J., and Leonard, W. J. (1995). Mutation of Jak3 in a patient with SCID: essential role of Jak3 in lymphoid development. *Science* **270**, 797–800.

Schorle, H., Holtschke, T., Hunig, T., Schimpl, A., and Horak, I. (1991). Development and function of T cells in mice rendered interleukin-2 deficient by gene targeting. *Nature* **352**, 621–624.

Sharon, M., Klausner, R. D., Cullen, B. R., Chizzonite, R., and Leonard, W. J. (1986). Novel interleukin-2 receptor subunit detected by cross-linking under high-affinity conditions. *Science* **234**, 859–863.

Suzuki, H., Duncan, G. S., Takimoto, H., and Mak, T. W. (1997). Abnormal development of intestinal intraepithelial lymphocytes and peripheral natural killer cells in mice lacking the IL-2 receptor beta chain. *J. Exp. Med.* **185**, 499–505.

Tagaya, Y., Bamford, R. N., DeFilippis, A. P., and Waldmann, T. A. (1996a). IL-15: a pleiotropic cytokine with diverse receptor/signaling pathways whose expression is controlled at multiple levels. *Immunity* **4**, 329–336.

Tagaya, Y., Burton, J. D., Miyamoto, Y., and Waldmann, T. A. (1996b). Identification of a novel receptor/signal transduction pathway for IL-15/T in mast cells. *EMBO J.* **15**, 4928–4939.

Takeshita, T., Asao, H., Ohtani, K., Ishii, N., Kumaki, S., Tanaka, N., Munakata, H., Nakamura, M., and Sugamura, K. (1992). Cloning of the γ chain of the human IL-2 receptor. *Science* **257**, 379–382.

Teglund, S., McKay, C., Schuetz, E., van Deursen, J. M., Stravopodis, D., Wang, D., Brown, M., Bodner, S., Grosveld, G., and Ihle, J. N. (1998). Stat5a and Stat5b proteins have essential and nonessential, or redundant, roles in cytokine responses. *Cell* **93**, 841–850.

Tinubu, S. A., Hakimi, J., Kondas, J. A., Bailon, P., Familletti, P. C., Spence, C., Crittenden, M. D., Parenteau, G. L., Dirbas, F. M., Tsudo, M., Bacher J. D., Kasten-Sportes, C., Martinucci J. L., Goldman, C. K., Clark, R. E., and Waldmann, T. A. (1994). Humanized antibody directed to the IL-2 receptor β-chain prolongs primate cardiac allograft survival. *J. Immunol.* **153**, 4330–4338.

Treiber-Held, S., Stewart, D. M., Kurman, C. C., and Nelson, D. L. (1996). IL-15 induces the release of soluble IL-2R-α from human peripheral blood mononuclear cells. *Clin. Immunol. Immunopathol.* **79**, 71–78.

Tsudo, M., Kozak, R. W., Goldman, C. K., and Waldmann, T. A. (1986). Demonstration of a non-Tac peptide that binds interleukin 2: a potential participant in a multichain interleukin 2 receptor complex. *Proc. Natl Acad. Sci. USA* **83**, 9694–9698.

Waldmann, T. A., and Tagaya, Y. (1999). The multifaceted regulation of interleukin-15 expression and the role of this cytokine in NK cell differentiation and host response to intracellular pathogens. *Annu. Rev. Immunol.* **17**, 19–49.

Witthuhn, B. A., Silvennoinen, O., Miura, O., Lai, K. S., Cwik, C., Liu, E. T., and Ihle, J. N. (1994). Involvement of the Jak-3 Janus kinase in signalling by interleukins 2 and 4 in lymphoid and myeloid cells. *Nature* **370**, 153–157.

Zhang, X., Sun, S., Hwang, I., Tough, D. F., and Sprent, J. (1998). Potent and selective stimulation of memory-phenotype CD8+ T cells *in vivo* by IL-15. *Immunity* **8**, 591–599.

LICENSED PRODUCTS

A polyclonal antibody against human and murine IL-15Rα peptide may be obtained from Santa Cruz Biotechnology Inc., Santa Cruz, CA, USA.

Antibodies against human IL-2/15Rβ molecule
Mikβ2 and β3 monoclonal antibodies are available from PharMingen (San Diego, CA, USA).

Polyclonal antibodies against IL-2/15β peptides are available from Santa Cruz Biotechnology (Santa Cruz, CA, USA).

Antibodies against murine IL-2/15Rβ molecule
Two monoclonal anti-murine IL-2/15Rβ antibodies may be obtained from PharMingen. Santa Cruz carries anti-mouse IL-2/15Rβ polyclonal IgGs that are raised against synthetic peptides from this molecule.

Antibodies against human γc molecule
A monoclonal anti-human γc antibody is available from PharMingen.

Antibodies against murine γc molecule
Monoclonal antibodies recognizing the murine γc molecule are available from PharMingen.
Polyclonal anti-mouse γc peptides are available from Santa Cruz Biotechnology.

IL-16 Receptor (CD4)

David M. Center*, Hardy Kornfeld and William W. Cruikshank

Pulmonary and Critical Care Division, Department of Medicine,
Boston University School of Medicine, 715 Albany Street, R-304, Boston, MA 02118, USA

*corresponding author tel: 617-638-4860, fax: 617-536-8093, e-mail: dcenter@bupula.bu.edu

DOI: 10.1006/rwcy.2000.14009.

SUMMARY

Interleukin 16 receptor (CD4) is a 62 kDa cell surface receptor that defines a major MHC class II-restricted T cell phenotype. It is linked intracellularly to an src family tyrosine kinase in T cells, p56lck, which is responsible for most, if not all of its signals. It is an essential coreceptor for T cell receptor (TCR) activation, facilitating antigen–MHC class II interaction with the TCR. When bound in isolation of the TCR, there are epitope-specific signals transduced through CD4 which result in a chemotactic response and competence growth factor-like activity resulting in IL-2- and IL-15-sensitive proliferation with simultaneous inhibition of responsiveness to antigens. An identical CD4 molecule exists on eosinophils, monocytes, and pro-B cells, all of which lack p56lck; thus the src kinase responsible for the earliest part of the signal transduction cascade is not known. In eosinophils and monocytes CD4 is a chemotactic factor receptor; and in pro-B cells it transduces a signal that induces differentiation to pre-B cells and RAG1 and RAG2 expression. IL-16 is a soluble ligand for CD4 of immune cell origin and therefore could mediate a sentinel role for CD4. If bound in the presence of antigen presented by MHC class II, CD4 permits T cell receptor activation and progression of antigen-dependent 'immune responses' while ligation of CD4 in the absence of antigen primes CD4 cells for an 'inflammatory phenotype' which is antigen-independent. The CD4 cell can therefore respond either as an immune or an inflammatory cell, but the two responses are mutually exclusive; as determined by CD4-mediated signals. Furthermore, the presence of CD4 on non-T cells (eosinophils, monocytes, etc.) implies non 'immune' non-TCR-dependent functions for CD4 signal transduction. In that regard, IL-16 and a number of other synthetic or virally encoded CD4 ligands induce chemotactic responses, cell progression and activation. It is unknown whether other Ig family receptors with homology to CD4 (e.g. LAG-3) also serve as receptors for IL-16.

BACKGROUND

All of the existing *in vitro* information regarding IL-16 functions suggests that the cell surface expression of CD4 (IL-16R) is required for transmission of signals. We will briefly review the structure and function of CD4 as an immune modulator in this chapter, but will concentrate mainly on the evidence that IL-16 interacts with CD4 in a ligand-receptor relationship. While it is possible that a coreceptor may be required for full expression of IL-16-mediated functions, none has been identified to date.

Discovery

CD4 was discovered by a number of investigators in the late 1970s when monoclonal antibody technology first became available. Its cell surface presence identified a major class of CD3+ (T) cells, which were subsequently identified functionally as MHC class II restricted 'helper' T cells (Greenstein *et al.*, 1984; Gay *et al.*, 1987, 1988).

Alternative names

The names of a variety of monoclonal antibodies generated against T cell membrane molecules recognized CD4. In the earliest work, human CD4 was known by the antibody designations, the most prevalent of which were OKT4(a–f) and Leu-3(a,b), shortened to T4 before the CD classification became popular. Mouse CD4 was known as L3T4.

Structure

CD4 is a 64 kDa immunoglobulin superfamily member expressed as a cell surface differentiation marker (Maddon *et al.*, 1985) associated with recognition of antigen in concert with MHC class II-bearing cells (Greenstein *et al.*, 1984; Gay *et al.*, 1987, 1988; Sleckman *et al.*, 1987; Eichmann *et al.*, 1989; Wang *et al.*, 1990; Fleury *et al.*, 1991; Sakihama *et al.*, 1995a).

Main activities and pathophysiological roles

On T cells, CD4 acts as an intercellular adhesion molecule with nonvariable regions of MHC class II to augment antigen activation of the T cell receptor (Doyle and Strominger, 1987; Anderson *et al.*, 1988; Eichmann *et al.*, 1989; Janeway *et al.*, 1987; reviewed in Janeway, 1992; Sakihama *et al.*, 1995b) and provides a physical noncovalent link to p56lck (Rudd *et al.*, 1988; Veillette *et al.*, 1988; Shaw *et al.*, 1989), whose function is essential in antigen-driven proliferation and interleukin 2 (IL-2) synthesis (Sleckman *et al.*, 1987; Rivas *et al.*, 1988; Glaichenhaus *et al.*, 1991; Collins *et al.*, 1992).

CD4 has also been identified as the major functional binding receptor for *HIV*-1 (Dalgleish *et al.*, 1984; Klatzman *et al.*, 1984; Maddon *et al.*, 1986; Lasky *et al.*, 1987; Berger *et al.*, 1988; Ryu *et al.*, 1990), at a site distinct from MHC class II binding (Lamarre *et al.*, 1989) but in the same immunoglobulin domain (Rosenstein *et al.*, 1990). The internalization of HIV-1 requires binding to a seven membrane spanning chemokine receptor family member (Dragic *et al.*, 1996; Keng *et al.*, 1997). In that regard, there is substantial information about the signal transduction pathways that are initiated following *HIV*-1 gp120 binding to CD4 (Diamond *et al.*, 1988; Wahl *et al.*, 1989; Cruikshank *et al.*, 1990; Oyaizu *et al.*, 1990; Chirmule *et al.*, 1990, 1995; Clouse *et al.*, 1991; Juszczak *et al.*, 1991; Jabado *et al.*, 1994; Sakihama *et al.*, 1995b; Wang *et al.*, 1998; Kornfeld *et al.*, 1998; reviewed in Capon and Ward, 1991), which will be compared with the IL-16 signal transduction events discussed below.

In addition, CD4 has been identified as a receptor for gp17, a secreted human seminal plasma glycoprotein identical to gross cystic disease fluid 15 protein and prolactin-inducible protein, which is present in malignant *breast tumors* (Autiero *et al.*, 1991, 1995, 1997). gp17 binds to the D1D2 region of CD4 and competitively displaces *HIV*-1 binding. There is no homology to IL-16. Little is known about the functions and signal transduction events following gp17 binding to CD4 and thus it will not be discussed further in this chapter.

GENE

Accession numbers

See Maddon *et al.* (1985).

Sequence

See Maddon *et al.* (1985) and Stewart *et al.* (1986).

PROTEIN

Accession numbers

See Maddon *et al.* (1985).

Description of protein

CD4 is a 62 kDa protein is expressed on the cell surface of subset of T cells associated with MHC class II molecules (Doyle and Strominger, 1987). There are four immunoglobulin-like domains defined by three disulfide-linked loop structures named D1 through D4 from N- to C-terminus (Maddon *et al.*, 1985). The D1 region is associated with MHC binding and contains the cell surface adhesion domain for HIV-1 (Doyle and Strominger, 1987). There is a single transmembrane region and a short (35 amino acid) intracellular region which is essential for transduction of direct CD4-mediated signals (Ledbetter *et al.*, 1988; Sleckman *et al.*, 1988; Veillette *et al.*, 1988, 1989; Cruikshank *et al.*, 1991; Glaichenhaus *et al.*, 1991; Juszczak *et al.*, 1991; Collins *et al.*, 1992; Ryan *et al.*, 1995). However, augmentation of T cell receptor function does not require the kinase activity of p56lck (Xu and Littman, 1993) and certain MHC-dependent coreceptor functions can occur through binding to the extracellular domain of CD4 in the absence of the intracytoplasmic domain. The intracellular portion of CD4 has a unique relationship with the src kinase p56lck in that cysteines at 420 and 422 are essential for noncovalent but extremely tight affinity.

Mutations of either cysteine results in loss of association of intracellular p56lck (Shaw *et al.*, 1989, 1990; Turner *et al.*, 1990). The relationship between CD4 and lck is unusual also in its stoichiometry. As much as 30% of all cellular lck is associated with CD4 in T cells (Veillette *et al.*, 1988).

The D1/D2 region was crystallized in 1990 (Kwong *et al.*, 1990; Wang *et al.*, 1990). The crystal structure revealed two unusual features for an immunoglobulin superfamily member: first, D1 contains an extra pair of β sheets which comprise a lateral extension of the molecule involved in MHC class II binding; and second, the last β sheet of D1 is shared with the first sheet in D2. This creates a linear, rigid structure between the two domains, making the N-terminus of the molecule an extended rod-like structure.

The D3/D4 region (of rat) CD4 has also been crystallized (Davis *et al.*, 1990). An interesting feature of the D4 domain is that it contains a touch point at the C-terminus of the third disulfide loop which permits spontaneous dimerization of the entire D3/D4 complex. The D3 domains then splay outwards in the shape of butterfly wings. Presumably, the N-terminal extension of D3 to D2/D1 creates a long rod, which when dimerized would be in the shape of a 'V'. The dimerization point is thought to be essential for approximating intracellular CD4-associated p56lck (auto)crossphosphorylation (Konig *et al.*, 1995; Sakihama *et al.*, 1995a,b). Other immunoglobulin family receptors share this characteristic (reviewed in Klemm *et al.*, 1988); dimerization or 'crosslinking' of Ig-like receptors is essential for signal transduction for surface Ig on mast cells (Metzger, 1992) and B cells (reviewed in Cambier *et al.*, 1994).

Relevant homologies and species differences

IL-16 proteins of various species have very high structural homology at the primary amino acid level (Keane *et al.*, 1998). Of interest, however, is conservation of IL-16-induced, CD4+ T cell-related functions among species. Keane *et al.* (1998) showed that rat, mouse, and human CD4+ T cells all migrate to all three species of IL-16, implying a substantial degree of homology at the level of the receptor, CD4. This is interesting because, in general, monoclonal antibodies raised against species-specific CD4s do not recognize the CD4 of other species. However, there is substantial interspecies homology in CD4 in three regions (Maddon *et al.*, 1987), the intracellular tail, the N-terminal 130 amino acids and a region which encompasses the fourth Ig domain (third disulfide linked loop). In T cells, the function of the intracytoplasmic tail has been studied in great detail and includes the noncovalent link with an src kinase family member, p56lck (Veillette *et al.*, 1989; Rudd *et al.*, 1988; Sleckman *et al.*, 1987, 1988, 1992). The N-terminal 130 amino acids comprise the invariant MHC-binding region and the *HIV*-1-binding region (Berger *et al.*, 1988; Camerini and Seed, 1990), which are separate and distinct (Lamarre *et al.*, 1989). The HIV-1-binding domain overlaps with the gp17-binding domain (Autiero *et al.*, 1995). The fourth Ig domain codes for a region reported to be essential for CD4–CD4 dimerization in the presence of MHC (see above). Dimerization or multimerization appears essential for clustering of intracellular kinases required for induction of signals both in concert with the intracellular kinases of the TCR and for CD4 signaling independent of the TCR (Sakihama *et al.*, 1995a).

All available data suggest that the D3/D4 region is responsible for IL-16 functions. In that regard, recombinant soluble CD4 (rsCD4) inhibits IL-16 chemotactic and activating functions (Cruikshank *et al.*, 1994). rsCD4 D1/D2 does not inhibit IL-16 functions, while rsCD4 D3/D4 has inhibitory activity identical to that of intact rsCD4. These findings could be interpreted to show that dimerization of CD4 is essential for IL-16 function and that the rsD3/D4 constructs are merely inhibiting dimer formation, not IL-16 interaction. While this is likely in part true, T cells that express CD4 mutated in the D4 region lose IL-16 activity while retaining HIV-1 gp120-dependent signals and functions and GST-linked CD4 directly binds this region of CD4 (Liu *et al.*, 1999).

Affinity for ligand(s)

The affinity for MHC appears low (Doyle and Strominger, 1987) and cannot be accurately determined in the context of k_d as the relationship between the two is one of cell to cell interaction not true soluble ligand–receptor. On T cells, there are about 100,000 CD4 molecules per cell. Monocytes express much lower numbers, in the range of 1 to 10,000 (Stewart *et al.*, 1986), while certain monocyte cell lines (e.g. U239) express high numbers of CD4 molecules per cell approaching the numbers seen on T cells. The number of CD4 molecules expressed on eosinophils has been estimated at less than 1000 (Lucey *et al.*, 1990); however this low number is fully capable of transducing a chemotactic signal to *HIV*-1 gp120, anti-CD4 antibodies and IL-16 (Rand *et al.*, 1991). There are no accurate affinity measurements of IL-16 for CD4. It is known to bind to CD4 as recombinant forms of the two coprecipitates (Cruikshank *et al.*, 1994) from solution. The dose-response kinetics of recombinant IL-16 reveal a peak chemotactic T cell response in the range of 10^{-10} M, however all formal binding experiments have resulted in loss of bioactivity, with iodination of the protein making calculation of a k_d value impossible.

Cell types and tissues expressing the receptor

The major cell type which expresses CD4 is the CD4+ T cell. It is expressed on monocytes and macrophages (Stewart *et al.*, 1986), dendritic cells, basophils, eosinophils (Lucey *et al.*, 1990), pro-B cells (Li *et al.*, 1996; Szabo *et al.*, 1998) and certain cells in the CNS. Since only T cells express T cell receptors, it is likely that CD4 has alternate functions in addition to acting as an intracellular adhesion molecule to augment MHC–antigen/TCR binding to antigen-presenting cells. Some of these potential functions are described under Signal transduction.

Regulation of receptor expression

CD4 expression is regulated during T cell development through stages of thymic development. The CD4−CD8− double negative cell becomes double positive before differentiating into single positive stage (Zuniga-Pflucker *et al.*, 1989, 1991; Fleury *et al.*, 1991; Robey *et al.*, 1991). Following full expression of CD4 on mature T cells modulation occurs following antigen-induced T cell activation in concert with the TCR (Saizawa *et al.*, 1987; Kupfer *et al.*, 1987; Anderson *et al.*, 1988). *HIV*-1 binding to CD4 and its coreceptors results in marked and prolonged downregulation of CD4 from the cell surface (Salmon *et al.*, 1988) which can last up to one week in culture (Theodore *et al.*, 1996) and phorbol esters modulate CD4 from the surface of T cells (Hoxie *et al.*, 1986). Pro-B cells lose CD4 when they pass to the pre-B cell stage (Li *et al.*, 1996).

Much less is known about regulation of CD4 on non-T cells. Isolation of monocytes on plastic results in downmodulation of CD4 which is re-expressed over time *in vitro* (Cruikshank *et al.*, 1987). *In vitro*, eosinophils appear to require supplemental growth factors for expression of CD4 (Lucey *et al.*, 1990) but the level of expression on these cells is so low that actual measurements of regulation have been hard to make.

Release of soluble receptors

CD4 is, in general, not thought to be released as a soluble receptor complex. Recombinant soluble CD4 has been utilized *in vitro* (Smith *et al.*, 1987; Traunecker *et al.*, 1988) and in short trials *in vivo* (Watanabe *et al.*, 1989; Kahn *et al.*, 1990; Schooley *et al.*, 1990) as a means to block *HIV*-1 entry into CD4+ T cells. This strategy, while effective *in vitro*, has not been shown to be effective *in vivo*.

SIGNAL TRANSDUCTION

Associated or intrinsic kinases

In T cells, CD4 is noncovalently linked to the src family member p56lck (Marth *et al.*, 1985; Rudd *et al.*, 1988; Shaw *et al.*, 1989). The stoichiometry of this reaction is unusual in that up to 30% of all cellular lck can be associated with CD4, and an equal percentage of CD4 molecules when activated are associated with an lck molecule. In this regard, lck is unusual among src family members. It is myristolated at the N-terminus, providing an anchor to the inner leaflet of the plasma membrane. Furthermore, it contains a unique sequence in the N-terminus which permits noncovalent binding to the intracytoplasmic tail of CD4 at CD4Cys420 and CD4Cys422 (Shaw *et al.*, 1990). Much of our knowledge of signaling through CD4 independent of the TCR comes from work with *HIV*-1 gp120 and with crosslinked anti-CD4 antibodies. These studies can be divided into two categories: direct signaling through CD4 by HIV-1 gp120 or anti-CD4 antibodies and the subsequent indirect effects this interaction has on TCR signaling. In fact, the distinction between indirect and direct signals is semantic as CD4 probably plays a sentinel role (see Janeway, 1992). When bound by antigen in the context of MHC class II, it is an essential coreceptor for TCR signals and functions. When bound in isolation of the TCR it transmits signals that simultaneously induce inflammatory competence (chemotaxis, IL-2 responsiveness, cytokine secretion) while preventing antigen activation. Thus, CD4 is an essential switch between CD4+ T cell-mediated immune and inflammatory responses and permits a single CD4+ T cell to participate in either response.

The signals transmitted by CD4 appear to depend upon the state of oligomerization of the ligand and the epitope bound by the ligand. In that regard, Fab fragments of anti-CD4 antibodies have never been demonstrated to induce any CD4-mediated signals. The major known direct consequence of divalent anti-CD4 antibody interaction with CD4 appears to be a chemotactic response (Cruikshank *et al.*, 1987). A 'negative signal' is also transduced by multiple CD4 antibodies. The chemotactic response is complex, but appears to be completely dependent in T cells on the association of CD4 with lck (Ryan *et al.*, 1995). However, like CD4-dependent T cell activation (Xu and Littman, 1993) the lck kinase activity is unimportant for chemotaxis as herbimycin does not inhibit the response and T cells that express only chimeric CD4 linked to lck lacking the enzymatic domain migrate normally to anti-CD4 antibodies. Mutations in CD4 that eliminate lck binding do not signal chemotactic responses (Ryan *et al.*, 1995). Therefore the association of lck with CD4, but not the enzymatic activity of lck is required for a migratory signal, suggesting that lck has some adapter role for signal transduction (reviewed in Pawson and Schlessinger, 1993). Cells that express CD4–lck chimeric molecules that lack the SH2 adapter portion of lck have a marked decrease in chemotactic response. The full chemotactic response to anti-CD4 antibodies is associated with activation of PI-3 kinase; and likewise the chemotactic response is inhibited by Wortmannin. Last, the chemotactic response is inhibited by all inhibitors of protein kinase C (PKC), suggesting that this kinase pathway is essential in the motile response (Parada *et al.*, 1996). Divalent anti-CD4 antibodies do not activate the lck-associated tyrosine kinase; this phenomena is observed only following crosslinking CD4 antibodies. Crosslinking CD4, however, does not result in augmentation of the motile response (Ryan *et al.*, 1995). As noted, not all CD4 epitopes transduce identical signals. Certain anti-CD4 antibodies are inefficient in inducing NF-AT (Baldari *et al.*, 1995) or in inhibiting CD95 expression by antigen; no anti-CD4 antibodies have been demonstrated to inhibit *HIV*-1 transcription which presumably occurs through induction of a repressor element (Baier *et al.*, 1995; Mackewicz *et al.*, 1996; Maciaszek *et al.*, 1997; Viglianti *et al.*, 1997; Zhou *et al.*, 1997).

The IL-16-induced chemotactic response following interaction with CD4 has been studied in some detail (Ryan *et al.*, 1995). CD4-negative mouse T cells do not respond to IL-16; while cells expressing wild-type CD4 have a marked chemotactic response. Cells expressing mutated CD4 that does not associate with lck fail to migrate in response to IL-16. Herbimycin does not inhibit this response, and cells expressing CD4–lck mutants that lack the kinase domain migrate normally, while those that lack the SH2 adapter region fail to migrate (Ryan *et al.*, 1995; Cruikshank *et al.*, 1996a).

While there is no lck in monocytes nor eosinophils, both these cell types express the identical CD4 molecule observed in T cells and both generate a motile response to IL-16 and the other CD4 ligands, gp120 and CD4 antibodies (Cruikshank *et al.*, 1987; Rand *et al.*, 1991). In that regard, Natke and Ryan (personal communication) have recently demonstrated that in the lck-negative THP monocytoid cell line, CD4 associates with lyn to induce subsequent downstream signals including phosphorylation of Syk. As predicted by the lack of sequence homology in the N-terminus of lyn to lck, lyn does not associate with the same stoichiometry to CD4. The best calculation is that less than 5% of cellular lyn is associated with CD4; and that a similar percentage of CD4 molecules have a lyn molecule associated. In these studies, unlike the T cell, herbimycin did inhibit the chemotactic response, suggesting that activation of the lyn kinase may be essential in downstream cascades associated with monocyte motility, but not T cell motility.

Cytoplasmic signaling cascades

In T cells, the direct interaction of IL-16 with CD4 results in activation of p56lck kinase activity and autophosphorylation of lck within minutes (Ryan *et al.*, 1995). A subsequent rise in intracellular Ca^{2+} is observed and unlike the sequence observed following activation of seven membrane spanning G protein-coupled receptors a peak in intracellular inositol trisphosphate (IP_3) follow several minutes later (Cruikshank *et al.*, 1991). The rises in Ca^{2+} and IP_3 are, however, dependent upon the association of CD4 with lck as these signals are not observed in T cells that express only mutant CD4 that does not associate with lck; or with CD4-lck chimeras that lack adapter domains (Cruikshank *et al.*, 1991). IL-16 interaction with CD4 results in translocation of PKC from cytosol and the chemotactic response is inhibited by PKC inhibitors (Parada *et al.*, 1996).

Identical direct signals are observed following gp120 interaction with CD4 (Kornfeld *et al.*, 1988) which have more recently been demonstrated to be dependent upon the presence of an appropriate chemokine coreceptor (Dragic *et al.*, 1996; Weissman *et al.*, 1997). Thus, there is a rise in intracellular Ca^{2+} and IP_3 following gp120 stimulation of T cells. In addition, gp120 induces activation of p56lck, NFκB, NF-AT, and AP-1 (Chirmule *et al.*, 1995). Subsequent antigen activation is markedly inhibited (Oyaizu *et al.*, 1990). If one follows the argument that viruses use molecular mimicry to bind and enter cells, this latter study might imply that there is a chemokine-like coreceptor for IL-16; however none has been identified.

The 'indirect' effects of CD4 ligation have been known for a much longer time than the potential direct effects. In that regard, Bank and Chess (1985) showed that CD4 antibody would subsequently inhibit activation through the TCR. This observation has been repeated following many activation signals in dozens of subsequent reports utilizing CD4 antibodies (Gutstein *et al.*, 1986; Haque *et al.*, 1987; Qin *et al.*, 1987; Janeway *et al.*, 1987; Neudorf *et al.*, 1990; Jabado *et al.*, 1994) and *HIV*-1 gp120 (Shalaby *et al.*, 1987; Diamond *et al.*, 1988; Salmon *et al.*, 1988; Chirmule *et al.*, 1990; Juszczak *et al.*, 1991; Cefai *et al.*, 1992). The studies are, in general, similar in that the signals measured and functions initiated by TCR activation are inhibited by prior binding of gp120 or CD4 antibodies. The studies differ in that gp120 markedly modulates CD4 from the cell surface, while this is not observed with CD4 antibodies nor IL-16 and as noted above (Baldari *et al.*, 1995) certain CD4 epitopes appear not to result in signals that inhibit CD95 expression.

As predicted by this synopsis, ligation of CD4 by IL-16 will transiently inhibit TCR activation. In that regard, IL-16 incubation with T cells for one hour inhibits subsequent CD3-mediated IL-2 synthesis, IL-2R expression, CD40L expression, CD95 expression, proliferation and Ca^{2+} signals (Cruikshank *et al.*, 1996a). The mixed leukocyte reaction (MLR) is likewise inhibited (Theodore *et al.*, 1996). IL-16-stimulated CD4+ T cells are thus protected from antigen-induced cell death (Idziorek *et al.*, 1998). Interestingly, IL-16 activation of CD4 can result in a 'complete activation signal' in certain T cell lymphomas (Cruikshank *et al.*, 1996b). In those studies, IL-16 but not CD4 antibodies induced CD4-dependent proliferation. Neither the signals required nor the abnormalities in cell cycle control that permit cell proliferation in these cell lines are known.

In CD4+ macrophages, IL-16 activates the SAP (stress-activated protein) kinase pathway although no correlations have been made to IL-16-induced monocyte functions (Krautwald, 1998). IL-16-induced phosphorylation of SEK-1 results in activation of the SAP kinase p46 and p54 protein and to the phosphorylation of c-Jun and p38 MAP kinase. Several downstream functions of CD4 signaling can be inferred from IL-16 and gp120 interaction with monocytes. IL-16 induces a chemotactic response and upregulation of MHC class II molecules in an IFNγ-independent fashion (Cruikshank *et al.*, 1987). *HIV*-1 gp120 has been demonstrated to induce TNFα, IL-1, and arachidonic acid metabolites (Merrill *et al.*, 1989; Wahl *et al.*, 1989) and most recently, Wang *et al.* (1998) have shown that HIV-1 gp120 inhibits subsequent chemotactic responses to chemokines by downregulation of the chemokine receptor. Anti-CD4 antibodies are chemotactic for monocytes (Cruikshank *et al.*, 1987), a response which has recently been demonstrated to be dependent upon CD4 association with the tyrosine kinase lyn and the kinase enzymatic activity of lyn (Natke, personal communication).

DOWNSTREAM GENE ACTIVATION

Transcription factors activated

A number of studies have reported the direct activation of AP-1, NF-AT, and NFκB by *HIV*-1 gp120 and subsequent inhibition of TCR-mediated activation of these transcription factors by pretreatment with gp120 (Chirmule *et al.*, 1990, 1995). Studies with IL-16 have demonstrated similar findings. There is direct activation of AP-1, NF-AT, and NFκB by IL-16 (Parada, personal communication). In addition preincubation of CD4+ T cells with IL-16 results in subsequent inhibition of PMA and CD3-mediated NFκB activation. These latter observations may lend some insights into the mechanisms by which IL-16 regulates CD3-induced IL-2R, CD95, and CD40L expression. In addition, IL-16 induces a repressor which binds to the core enhancer of the *HIV*-1 LTR (Maciaszek *et al.*, 1997). The identity of this repressor element is not known. This finding, however, demonstrated that IL-16 inhibits HIV-1 replication not by inhibiting binding and viral entry like the chemokines, but by a CD4-dependent signaling event.

Genes induced

Both IL-16 (Cruikshank *et al.*, 1987; Parada *et al.*, 1998) and *HIV*-1 gp120 (Kornfeld *et al.*, 1988) activation of CD4 result in induction of IL-2Rα and upregulation of cell surface expression of the mature functional protein. IL-2Rβ is similarly induced by IL-16 (Parada *et al.*, 1998), although no studies have been performed with HIV-1 gp120. In monocytes, IL-1 and TNFα synthesis are induced by HIV-1 gp120 (Merrill *et al.*, 1989; Wahl *et al.*, 1989), as is synthesis of prostaglandin E_2 (PGE_2), implying induction of cyclooxygenase activity, but it is not clear if this latter response is transcriptionally regulated.

Promoter regions involved

CD4 is constitutively expressed on mature peripheral T cells and other cells. Its expression during thymic education has been recently reviewed in detail by Mak (1995).

BIOLOGICAL CONSEQUENCES OF ACTIVATING OR INHIBITING RECEPTOR AND PATHOPHYSIOLOGY

Unique biological effects of activating the receptors

As noted above, the unique aspect of activating CD4 in isolation of the TCR results in subsequent inhibition of TCR-mediated responses. Several groups have attempted to exploit this phenomenon by using CD4 antibodies as immunosuppressive agents. Although *HIV*-1 gp120 is a much more potent immunosuppressive than CD4 antibodies, it has not been exploited for these purposes to date.

Phenotypes of receptor knockouts and receptor overexpression mice

Since deletion of CD4 markedly influences development of the CD4+ T cell itself, it is almost impossible to dissect the role of CD4 as a signaling molecule from the function of CD4+ T cells themselves. CD4−/− mice have, as expected, marked deficiencies in MHC-mediated immune responses including impaired clearance of certain parasites (Rottenberg *et al.*, 1993), persistent viremia and poor clearance of certain viruses (Battegay *et al.*, 1994), poor immune responses to MHC class II-restricted antigens (Rahemtulla *et al.*, 1994; reviewed in Mak, 1995), diminished contact sensitivity, altered and dampened autoimmune responses (Koh *et al.*, 1995). Of interest, overexpression of CD4 in CD4- mice results in susceptibility to staphylococcal superantigen-induced toxicity resembling *toxic shock syndrome* in humans (Yeung *et al.*, 1996). Studies of the responses to IL-16 in CD4−/− mice have not been published.

Human abnormalities

There is a CD4 phenotype that lacks the OKT4-binding domain originally described at the beginning of the *HIV*-1 infection era. These patients were reported to have profoundly reversed CD4/CD8 ratios, but are completely normal otherwise. Thus, this region of CD4 is not likely to impart much significant function for MHC binding or other functions of CD4 (Lederman and Chess, 1987).

THERAPEUTIC UTILITY

Effect of treatment with soluble receptor domain

Soluble receptors have been demonstrated to inhibit *HIV*-1 binding and internalization *in vitro* (Deen *et al.*, 1988; Byrn *et al.*, 1989). However, despite beneficial effects in simians there have been disappointing results when used *in vivo* in humans.

Effects of inhibitors (antibodies) to receptors

Anti-CD4 antibodies have been used in clinical trials in *asthma* and *rheumatoid arthritis* based upon their ability selectively to delete certain populations of CD4+ T cells (Chace *et al.*, 1994). Both are diseases in which IL-16 has been demonstrated to be present in affected organs; although the therapeutic effects of the antibodies are likely due to the ability of these antibodies to inhibit CD4+ T cell functions in general. The antibodies are not in current therapeutic trials for either disease. However, an immunosuppressive humanized CD4 monoclonal antibody is being tested for acute and chronic solid *organ transplant rejection.*

References

Anderson, P., Blue, M. L., and Schlossman, S. F. (1988). Comodulation of CD3 and CD4. Evidence for a specific association between CD4 and approximately 5% of the CD3:T cell receptor complexes on helper T lymphocytes. *J. Immunol.* **140**, 1732.

Autiero, M., Abrescia, P., and Guardiola, J. (1991). Interaction of seminal plasma proteins with cell surface antigens: presence of a CD4-binding glycoprotein in human seminal plasma. *Exp. Cell Res.* **197**, 268.

Autiero, M., Cammarota, G., Friedlein, A., Zulauf, M., Chiapeta, G., Dragone, V., and Guardiola, J. (1995). A 17 kDa CD4-binding glycoprotein present in human seminal plasma and in breast tumor cells. *Eur. J. Immunol.* **25**, 1461.

Autiero, M., Bouchier, C., Basmaciogullari, S., Zaborski, P., El Marhomy, S., Martin, M., Guardiola, J., and Piatier-Tonneau, D. (1997). Isolation from a human seminal vesicle library of the cDNA for gp17, a CD4 binding factor. *Immunogenetics* **46**, 345.

Baier, M., Werner, A., Bannert, N., Metzner, K., and Kurth, R. (1995). HIV suppression by interleukin-16. *Nature* **378**, 563.

Baldari, C. T., Milia, E., Di Somma, M. M., Baldoni, F., Balitutti, S., and Telford, J. L. (1995). Distinct signalling properties identify different CD4 epitopes. *Eur. J. Immunol.* **25**, 1843.

Bank, I., and Chess, L. (1985). Perturbation of the T4 molecule transmits a negative signal to T cells. *J. Exp. Med.* **162**, 1294–1299.

Battegay, M., Moskophidis, D., Rahemtulla, A., Hengartner, H., Mak, T. W., and Zinkernagel, R. M. (1994). Enhanced establishment of a virus carrier state in adult CD4+ T-cell-deficient mice. *J. Virol.* **68**, 4700.

Berger, E., Fuerst, T., and Moss, B. (1988). A soluble recombinant polypeptide comprising the amino-terminal half of the extracellular region of the CD4 molecule contains an active binding site for human immunodeficiency virus. *Proc. Natl Acad. Sci. USA* **85**, 2357–2361.

Byrn, R., Sekigawa, I., Chamow, S., Johnson, J., Gregory, T., Capon, D., and Groopman, J. (1989). Characterization of *in vitro* inhibition of human immunodeficiency virus by purified recombinant CD4. *J. Virol.* **63**, 4370–4375.

Cambier, J. C., Pleiman, C. M., and Clark, M. R. (1994). Signal transduction by the B cell antigen receptor and its coreceptors. *Annu. Rev. Immunol.* **12**, 457–486.

Camerini, D., and Seed, B. (1990). A CD4 domain important for HIV-mediated syncytium formation lies outside the virus binding site. *Cell* **60**, 747–754.

Capon, D. J., and Ward, R. H. R. (1991). The CD4-gp120 Interaction and AIDS Pathogenesis. *Annu. Rev. Immunol.* **11**, 649.

Cefai, D., Ferrer, M., Serpente, N., Idziorek, T., Dautry-Varsat, A., Debre, P., and Bismuth, G. (1992). Internalization of HIV glycoprotein gp120 is associated with down modulation of membrane CD4 and p56lck together with impairment of T cell activation. *J. Immunol.* **149**, 285–291.

Chace, J. H., Cowdery, J. S., and Field, E. H. (1994). Effect of anti-CD4 on CD4 subsets: I. Anti-CD4 preferentially deletes resting, naive CD4 cells and spares activated CD4 cells. *J. Immunol.* **152**, 405–412.

Chirmule, N., Kalyanaraman, V. S., Oyaizu, N., Slade, H. B., and Pahwa, S. (1990). Inhibition of functional properties of tetanus antigen-specific T-cell clones by envelope glycoprotein gp120 of human immunodeficiency virus. *Blood* **75**, 152–162.

Chirmule, N., Goonewardena, H., Pahwa, S., Pasieka, R., Kalyanaraman, V. S., and Pahwa, S. (1995). HIV-1 envelope glycoprotein induces activation of activated protein-1 in CD4+ T cells. *J. Biol. Chem.* **270**, 19364–19369.

Clouse, K. A., Cosentino, L. M., Weih, K. A., Pyle, S. W., Robbins, P. B., Hochstein, H. D., Natarajan, V., and Farrar, W. L. (1991). The HIV-1 gp120 envelope protein has the intrinsic capacity to stimulate monokine secretion. *J. Immunol.* **147**, 2892–2901.

Collins, T. L., Uniyal, S., Shin, J., Stominger, J. L., Mittler, R. S., and Burakoff, S. J. (1992). p56^{lck} association with CD4 is required for the interaction between CD4 and the TCR/CD3 complex and for optimal antigen stimulation. *J. Immunol.* **148**, 2159.

Cruikshank, W. W., Berman, J. S., Theodore, A. C., Bernardo, J., and Center, D. M. (1987). Lymphokine activation of T4+ lymphocytes and monocytes. *J. Immunol.* **138**, 3817.

Cruikshank, W. W., Center, D. M., Pyle, S. W., and Kornfeld, H. (1990). Biologic activities of HIV-1 envelope glycoprotein: the effects of crosslinking. *Biomed. Pharmacother.* **44**, 5.

Cruikshank, W. W., Greenstein, J. L., Theodore, A. C., and Center, D. M. (1991). Lymphocyte chemoattractant factor (LCF) induces CD4-dependent intracytoplasmic signalling in lymphocytes. *J. Immunol.* **146**, 2928–2934.

Cruikshank, W. W., Center, D. M., Nisar, N. Natke, B., Theodore, A. C., and Kornfeld, H. (1994). Molecular and functional analysis of a lymphocyte chemoattractant factor: association of biologic function with CD4 Expression. *Proc. Natl Acad. Sci. USA* **91**, 5109–5113.

Cruikshank, W. W., Kornfeld, H., Berman, J., Chupp, G., Keane, J., and Center, D. (1996a). Biologic activities of interleukin 16. *Nature* **382**, 501–502.

Cruikshank, W. W., Fine, G., Taylor, K., and Center, D. M. (1996b). Autocrine and paracrine growth regulation of CD4+ cell lines by IL-16. *FASEB J.* **10**, A1485.

Dalgleish, A. G., Beverly, P. C., Clapham, P. R., Crawford, D. H., Greaves, M. F., and Weiss, R. A. (1984). The CD4 (T4) antigen is an essential component of the receptor for the AIDS retrovirus. *Nature* **312**, 763–767.

Davis, S. J., Brady, R., Barclay, N., Harlos, K., Dodson, G., and Williams, A. (1990). Crystallization of a soluble form of the rat T-cell surface glycoprotein CD4 complexed with Fab from the W3/25 monoclonal antibody. *J. Mol. Biol.* **213**, 7–10.

Deen, K., McDougal, S., Inacker, R., Folena-Wasserman, G., Arthos, J., Rosenberg, J., Maddon, P., Axel, R., and Sweet, R. (1988). A soluble form of CD4 (T4) protein inhibits AIDS virus infection. *Nature* **331**, 82–84.

Diamond, D., Sleckman, B., Gregory, T., Lasky, L., Greenstein, J., and Burakoff, S. (1988). Inhibition of CD4+ T cell function by the HIV envelope protein, gp120. *J. Immunol.* **141**, 3715–3717.

Doyle, C., and Strominger, J. L. (1987). Interaction between CD4 and class II MHC molecules mediates cell adhesions. *Nature* **330**, 256.

Dragic, T., Litwin, V., Allaway, G., Martin, S. R., Huang, Y., Nagashima, K. A., Cayanan, C., Maddon, P. J., Koup, R. A., Moore, J. P., and Paxton, W. A. (1996). HIV-1 entry into CD4+ cells is mediated by the chemokine receptor CC-CKR-5. *Nature* **381**, 667–673.

Eichmann, K., Boyce, N. W., Schmidt-Ullrich, R., Jonsson, J.-I. (1989). Distinct functions of CD8 (CD4) are utilized at different stages of T-lymphocyte differentiation. *Immunol. Rev.* **109**, 40.

Fleury, S. G., Croteau, G., and Sekaly, R. P. (1991). CD4 and CD8 recognition of class II and class I molecules of the major histocompatibility complex. *Semin. Immunol.* **3**, 177.

Gay, D., Maddon, P., Sekaly, R., Talle, M. A., Godfrey, M., Long, E., Goldstein, G., Chess, L., Axel, R., Kappler, J., and Marrack, P. (1987). Functional interaction between human T-cell protein CD4 and the major histocompatibility complex HLA-DR antigen. *Nature* **328**, 626.

Gay, D., Brus, S., Pasternak, J., Kappler, J., and Marrack, P. (1988). The T cell accessory molecule CD4 recognizes a monomorphic determinant on isolated Ia. *Proc. Natl Acad. Sci. USA* **85**, 5629.

Glaichenhaus, N., Shastri, N., Littman, D. R., and Turner, J. M. (1991). Requirement for association of p56^{lck} with CD4 in antigen-specific signal transduction. *Cell* **64**, 511–520.

Greenstein, J. L., Kappler, J., Marrack, P., and Burakoff, S. J. (1984). The role of L3T4 in recognition of Ia by a cytotoxic, H-2Dd-specific T cell hybridoma. *J. Exp. Med.* **159**, 1213.

Gutstein, N. L., Seaman, W. E., Scott, J. H., and Wofsy, D. (1986). Induction of immune tolerance by administration of monoclonal antibody to L3T4. *J. Immunol.* **137**, 1127–1132.

Haque, S., Saizawa, K., Rojo, J., and Janeway, C. A. (1987). The influence of valence on the functional activities of monoclonal anti-L3T4 antibodies: Discrimination of signalling from other effects. *J. Immunol.* **139**, 3207.

Hoxie, J. A., Matthews, D. M., Callahan, K. J., Cassel, D. L., and Cooper, R. A. (1986). Transient modulation and internalization of T4 antigen induced by phorbol esters. *J. Immunol.* **140**, 786–795.

Idziorek, T., Khalife, J., Billaut-Mulot, O., Hermann, E., Aumercier, M., Mouton, Y., Capron, A., and Bahr, G. M. (1998). Recombinant human IL-16 inhibits HIV-1 replication and protects against activation-induced cell death (AICD). *Clin. Exp. Immunol.* **112**, 84–91.

Jabado, N., Le Deist, F., Fischer, A., and Hivroz, C. (1994). Interaction of HIV gp120 and anti-CD4 antibodies with the CD4 molecule on himan CD4+ T cells inhibits the binding activity of NF-AT, NF-κB and AP-1, three nuclear factors regulating interleukin-2 gene enhancer activity. *Eur. J. Immunol.* **24**, 2646.

Janeway, C. A. (1992). The T cell receptor as a multicomponent signalling machine: CD4/CD8 coreceptors and CD45 in T cell activation. *Annu. Rev. Immunol.* **10**, 645–674.

Janeway, C.A. Jr., Haque, S., Smith, L. A., and Saizawa, K. (1987). The role of the murine L3T4 molecule in T cell activation: differential effects of anti-L3T4 on activation on monoclonal anti-receptor antibodies. *J. Mol. Cell. Immunol.* **3**, 121.

Juszczak, R. J., Howard, T., Truneh, A., Culp, J., and Kassis, S. (1991). Effect of human immunodeficiency virus gp120 glycoprotein on the association of the protein tyrosine kinase $p56^{lck}$ with CD4 in human T lymphocytes. *J. Biol. Chem.* **266**, 11176.

Kahn, J. O., Allan, J. D., Hodges, T. L., Kaplan, L. D., Arri, C. J., Fitch, H. F., Izu, A. E., Mordenti, J., Sherwin, S., Groopman, J. E., and Volberding, P. (1990). The safety and pharmacokinetics of recombinant soluble CD4 (rCD4) in subjects with the acquired immunodeficiency syndrome (AIDS) and AIDS-related complex. A Phase I study. *Ann. Intern. Med.* **112**, 254–261.

Keane, J., Nicoll, J., Wu, D. M. H., Kim, S., Cruikshank, W. W., Brazer, W., Natke, B., Center, D. M., and Kornfeld, H. (1998). Conservation of structure and function between murine and human interleukin 16. *J. Immunol.* **160**, 5945–5954.

Keng, H., Unutmaz, D., Kewal-Ramani, V. N., and Littman, D. R. (1997). Expression cloning of new receptors used by simian and human immunodeficiency viruses. *Nature* **388**, 296–300.

Klatzman, D., Champagne, E., Chamaret, S., Gruest, J., Guetard, D., Hercend, T., Gluckman, J. C., Montagnier, L. (1984). T-lymphocyte T4 molecule behaves as the receptor for human retrovirus LAV. *Nature* **312**, 767–768.

Klemm, J. D., Schreiber, S. L., and Crabtree, G. R. (1998). Dimerization as a regulatory mechanism in signal transduction. *Annu. Rev. Immunol.* **16**, 569.

Koh, D. R., Ho, A., Rahemtulla, A., Fung-Leung, W. P., Giesser, H., and Mak, T. W. (1995). Murine lupus in MRL/lpr mice lacking CD4 or CD8 T cells. *Eur. J. Immunol.* **25**, 2558.

Konig, R., Shen, X., and Germain, R. N. (1995). Involvement of both major histocompatibility complex class II α and β chains in CD4 function indicates a role for ordered oligomerization in T cell activation. *J. Exp. Med.* **182**, 779–787.

Kornfeld, H., Cruikshank, W. W., Pyle, S. W., Berman, J. S., and Center, D. M. (1988). Lymphocyte activation by HIV-1 envelope glycoprotein. *Nature* **335**, 445–448.

Krautwald, S. (1998). IL-16 activates the SAPK signaling pathway in CD4+ macrophages. *J. Immunol.* **160**, 5874–5879.

Kupfer, A., Singer, S. J., Janeway, C. A., and Swain, S. L. (1987). Coclustering of CD4 (L3T4) molecule with the T cell receptor is induced by specific direct interaction of helper T cells and antigen presenting cells. *Proc. Natl Acad. Sci. USA* **84**, 5888.

Kwong, P. D., Ryu, S.-E., Hendrickson, W., Axel, R., Sweet, R., Folena-Wasserman, G., Hensley, P., and Sweet, R. (1990). Molecular characteristics of recombinant human CD4 as deduced from polymorphic crystals. *Proc. Natl Acad. Sci. USA* **87**, 6423–6427.

Lamarre, D., Ashkenazi, A., Fleury, S., Smith, D., Sekaly, R.-P., and Capon, D. J. (1989). The MHC-binding and gp120-binding functions of CD4 are separable. *Science* **245**, 743–746.

Lasky, L., Nakamura, G., Smith, D. H., Fennie, C., Shimasaki, C., Patzer, E., Berman, P., Gregory, T., and Capon, D. J. (1987). Delineation of a region of the human immunodeficiency virus type 1 gp120 glycoprotein critical for interaction with the CD4 receptor. *Cell* **50**, 975–985.

Ledbetter, J. A., June, C. H., Rabinovitch, P. S., Grossman, A., Tsu, T. T., and Imboden J. B. (1988). Signal transduction through CD4 receptors: stimulatory vs. inhibitory activity is regulated by CD4 proximity to the CD3/ T cell receptor. *Eur. J. Immunol.* **18**, 525.

Lederman, S., and Chess, L. (1987). The role of T4 in immune function and dysfunction. *Ann. Inst. Pasteur Immunol.* **138**, 158.

Li, Y. S. R., Waserman, R., Hayakawa, K., and Hardy, R. (1996). Identification of the earliest B leneage stages in mouse bone marrow. *Immunity* **5**, 527.

Liu, Y., Cruikshank, W. W., O'Loughlin, T., O'Reilly, P., Center, D. M., and Kornfeld, H. (1999). Identification of a CD4 domain required for interleukin-16 binding and lymphocyte activation. *J. Biol. Chem.* **274**, 23387.

Lucey, D. R., Dorsky, D. I., Nicholson-Weller, A., and Weller, P. F. (1990). Human eosinophils express CD4 protein and bind human immunodeficiency virus 1 gp120. *J. Exp. Med.* **169**, 327.

Maciaszek, J. W., Parada, N. A., Cruikshank, W. W., Center, D. M., Kornfeld, H., and Viglianti, G. A. (1997). Interleukin-16 represses HIV-1 promoter activity. *J. Immunol.* **158**, 5–8.

Mackewicz, C., Levy, J., Cruikshank, W., Kornfeld, H., and Center, D. (1996). Role of IL-16 in HIV replication. *Nature* **383**, 488–489.

Maddon, P.J. Littman, R. R., Godfrey, M., Maddon, D. E., Chess L., and Axel, R. (1985). The isolation and nucleotide sequence of a cDNA encoding the T cell surface protein T4: A new member of the immunoglobulin gene family. *Cell* **42**, 93.

Maddon, P. J., Dalgleish, A. G., McDougal, J. S., Clapham, P. R., Weiss, R. A., and Axel, R. (1986). The T4 gene encodes the AIDS virus receptor and is expressed in the immune system and the brain. *Cell* **47**, 333–348.

Maddon, P. J., Molineaux, S. M., Maddon, D. E., Zimmerman, K. A., Godfrey, M., Alt, F. W., Chess, L., and Axel, R. (1987). Structure and expression of the human and mouse T4 genes. *Proc. Natl Acad. Sci. USA* **84**, 9155.

Mak, T. W. (1995). Gaining insights into the ontogeny and activation of T cells through the use of gene-targeted mutant mice. *J. Inflammation* **45**, 79.

Marth, J. D., Peet, E. G., Krebs, E. G., and Perlmutter, R. M. (1985). A lymphocyte specific protein tyrosine kinase is rearranged and overexpressed in murine T cell lymphoma LSTRA. *Cell* **43**, 393.

Merrill, J. E., Koyanagi, Y., and Chen, I. S. Y. (1989). Interleukin-1 and tumor necrosis factor α can be induced from mononuclear phagocytes by human immunodeficiency virus type 1 binding to the CD4 receptor. *J. Virol.* **63**, 4404–4408.

Metzger, H. (1992). Transmembrane signaling: The joy of aggregation. *J. Immunol.* **149**, 1477.

Neudorf, N., Jones, M., McCarthy, B., Harmony, J., and Choi, E. (1990). The CD4 molecule transmits biochemical information important in the regulation of T lymphocyte activity. *Cell. Immun.* **125**, 301–314.

Oyaizu, N., Chirmule, N., Kalyanaraman, V., Hall, W., Good, R., and Pahwa, S. (1990). Human immunodeficiency virus type 1 envelope glycoprotein gp120 produces immune defects in CD4+ lymphocytes by inhibiting interleukin 2 mRNA. *Proc. Natl Acad. Sci. USA* **87**, 2379–2383.

Parada, N. A., Ryan, T. C., Danis, H., Cruikshank, W. W., and Center, D. M. (1996). IL-16 and other CD4 ligand-induced migration is dependent upon protein kinase C. *Cell. Immunol.* **168**, 100–106.

Parada, N. A., Cruikshank, W. W., Kornfeld, H., and Center, D. M. (1998). Synergistic activation of CD4+ T cells by interleukin 16 and interleukin 2. *J. Immunol.* **160**, 2115–2120.

Pawson, T., and Schlessinger, J. (1993). SH2 and SH3 domains. *Curr. Biol.* **3**, 434.

Qin, S., Cobbold, S., Tighe, H., Benjamin, R., and Waldmann, H. (1987). CD4 monoclonal antibody pairs for immunosuppression and tolerance induction. *Eur. J. Immunol.* **17**, 1159–1165.

Rahemtulla, A., Kundig, T. M., Narendran, A., Bachmann, M. F., Julius, M., Paige, C. J., Ohashi, P. S., Zinkernagel, R. M., and Mak, T. (1994). Class II major histocompatibility complex-restricted T cell function in CD4-deficient mice. *Eur. J. Immunol.* **24**, 2213.

Rand, T., Cruikshank, W. W., Center, D. M., and Weller, P. F. (1991). CD4-mediated stimulation of human eosinophils: lymphocyte chemoattractant factor and other CD4-binding ligands elicit eosinophil migration. *J. Exp. Med.* **173**, 1521–1528.

Rivas, A., Takada, S., Koide, J., Sonderstrup-McDevitt, G., and Engleman, E. G. (1988). CD4 molecules are associated with the antigen receptor complex on activated but not resting T cells. *J. Immunol.* **140**, 1337–1345.

Robey, E. A., Fowlkes, B. J., Gordon, J. W., Kioussis, D., von Boehmer, H., Ramsdell, F., and Axel, R. (1991). Thymic selection in CD8 transgenic mice supports an instructive model for commitment to a CD4 or CD8 lineage. *Cell* **64**, 99.

Rosenstein, Y., Burakoff, S. J., and Herrmann, S. H. (1990). HIV-gp120 can block CD4-class II MHC-mediated adhesion. *J. Immunol.* **144**, 526–531.

Rottenberg, M. E., Bakhiet, M., Olsson, T., Kristensson, K., Mak, T. W., Wigzell, H., and Orn, A. (1993). Differential susceptibilities of mice genomically deleted of CD4 and CD8 to infections with *Trypanosoma cruzi* or *Trypanosoma brucei*. *Infect. Immun.* **61**, 5129.

Rudd, C. E., Trevillyan, J. M., Dasgupta, J. V., Wong, L. L., and Schlossman, S. F. (1988). The CD4 antigen is complexed in detergent lystates to a protein tyrosine kinase (pp58) from human T lymphocytes. *Proc. Natl Acad. Sci. USA* **85**, 5190.

Ryan, T., Cruikshank, W. W., and Center, D. M. (1995). Activation of CD4 associated p56 lck by the lymphocyte chemoattractant factor. Dissociation of kinase enzymatic activity with chemotactic response. *J. Biol. Chem.* **270**, 17081–17086.

Ryu, S.-E., Kowng, P. D., Truneh, A., Porter, T. G., Arthos, J., Rosenberg, M., Dai, X., Xuong, N.-H., Axel, R., Sweet, R. W., and Hendrickson, W. A. (1990). Crystal structure of an HIV-binding recombinant fragment of human CD4. *Nature* **348**, 419.

Saizawa, K., Rojo, J., and Jane, C.A. Jr. (1987). Evidence for a physical association of CD4 and the CD4:α:β T cell receptor. *Nature* **328**, 260.

Sakihama, T., Smolyar, A., and Reinherz, E. (1995a). Molecular recognition of antigen involves lattice formation between CD4, MHC class II and TCR molecules. *Immunol. Today* **16**, 581–587.

Sakihama, T., Smolyar, A., and Reinherz, E. (1995b). Oligomerization of CD4 is required for stable binding to class II major histocompatibility complex proteins but not for interaction with human immunodeficiency virus gp120. *Proc. Natl Acad. Sci. USA* **92**, 6444–6448.

Salmon, P., Olivier, R., Riviere, Y., Brisson, E., Gluckman, J. C., Kieny, M.P. Montagnier, L., and Klatzman, D. (1988). Loss of CD4 membrane expression and CD4 mRNA during acute human immunodeficiency virus replication. *J. Exp. Med.* **168**, 1953–1969.

Schooley, R. T., Merigan, T. C., Gaut, P., Hirsch, M., Holodniy, M., Flynn, T., Liu, S., Byington, B. S., Henochowicz, S., Gubish, E., Spriggs, D., Kufe, D., Schindler, J., Dawson, A., Thomas, D., Hanson, D. G., Letwin, B., Liu, T., Gulinello, J., Kennedy, S., Fisher, R., and Ho, D. (1990). Recombinant soluble CD4 therapy in patients with the acquired immunodeficiency syndrome (AIDS) and AIDS-related complex. *Ann. Intern. Med.* **112**, 247–253.

Shalaby, M. R., Krowka, J. F., Gregory, T. J., Hirabayashi, S., McCabe, S., Kaufman, D., Stites, D., and Ammann, A. (1987). The effects of human immunodeficiency virus recombinant envelope glycoprotein on immune cell functions *in vitro*. *Cell. Immun.* **110**, 140–148.

Shaw, A. S., Amrein, K. E., Hannond, C., Stern, D. F., Sefteo, B. M., and Rose, J. K. (1989). The cytoplasmic domain of CD4 interacts with the tyrosine protein kinase, p56lck through its unique amino-terminal domain. *Cell* **59**, 627.

Shaw, A. S., Chalupny, J., Whitney, J. A., Hammond, C., Amrein, K. E., Kavathas, P., Sefton, B. M., and Rose, J. K. (1990). Short related sequences in the cytoplasmic domains of CD4 and CD8 mediate binding to the amino-terminal domain of the p56lck tyrosine protein kinase. *Mol. Cell. Biol.* **10**, 1853.

Sleckman, B., Peterson, A., Jones, W. K., Foran, J. A., Greenstein, J. L., Seed, B., and Burakoff, S. (1987). Expression and function of CD4 in murine T-cell hybridoma. *Nature* **328**, 351.

Sleckman, B. P., Peterson, A., Foran, J. A., Gorga, J. C., Kara, C. J., Strominger, J. L., Burakoff, S. J., and Greenstein, J. L. (1988). Functional analysis of a cytoplasmic domain-deleted mutant of the CD4 molecule. *J. Immunol.* **3**, 645–651.

Sleckman, B. P., Shin, J., Igras, V. E., Collins, T. L., Stominger, J. L., and Burakoff, S. J. (1992). Disruption of the CD4-p56^{lck} complex is required for rapid internalization of CD4. *Proc. Natl Acad. Sci. USA* **89**, 7566–7570.

Smith, D., Byrn, R., Marsters, S., Gregory, T., Groopman, J., and Capon, D. (1987). Blocking of HIV-1 infectivity by a soluble, secreted form of the CD4 antigen. *Science* **328**, 1704–1707.

Stewart, S. J., Fujimoto, J., and Levy, R. (1986). Human T lymphocytes and monocytes bear the same Leu 3(T4). antigen. *J. Immunol.* **136**, 3773–3778.

Szabo, P., Kesheng, Z., Kirman, I., Le Maoult, J., Dyall, R., Cruikshank, W., and Weksler, M. E. (1998). Maturation of B cell precursors is impaired in thymic-deprived nude and old mice. *J. Immunol.* **161**, 2248–2253.

Theodore, A. C., Center, D. M., Cruikshank, W. W., and Beer, D. J. (1986). A human T-T cell hybridoma-derived lymphocyte chemoattractant factor. *Cell. Immunol.* **98**, 411.

Theodore, A. C., Center, D. M., Nicoll, J., Fine, G., Kornfeld, H., and Cruikshank, W. W. (1996). The CD4 ligand interleukin 16 inhibits the mixed lymphocyte reaction. *J. Immunol.* **157**, 1958–1964.

Traunecker, A., Luke, W., and Karjalainen, K. (1988). Soluble CD4 molecules neutralize human immunodeficiency virus type 1. *Nature* **331**, 84–86.

Turner, J. M., Brodshy, M. H., Irving, B. A., Levin, S. D., Perlmutter, R. M., and Littman, D. R. (1990). Interaction of the unique N-terminal region of tyrosine kinase p56lck with cytoplasmic domains of CD4 and CD8 is mediated by cysteine motifs. *Cell* **60**, 755.

Veillette, A. Bookman, M. A., Horak, E. M., and Bolen, J. B. (1988). The CD4 and CD8 T cell surface antigens are associated with the internal membrane tyrosine protein kinase p56lck. *Cell* **55**, 301.

Veillette, A., Bookman, M. A., Horak, E. M., Samelson, L. E., and Bolen, J. B. (1989). Signal transduction through the CD4 receptor involves the activation of the internal membrane tyrosine-protein kinase p56lck. *Nature* **338**, 257.

Viglianti, G. A., Parada, N. A., Maciaszek, J. W., Kornfeld, H., Center, D. M., and Cruikshank, W. W. (1997). IL-16 anti-HIV-1 therapy. *Nature Med.* **3**, 938.

Wahl, L., Corcoran, M. L., and Pyle, S. W. (1989). Human immunodeficiency virus glycoprotein (gp120) induction of monocyte arachidonic acid metabolites and interleukin-1. *Proc. Natl Acad. Sci. USA* **86**, 621–625.

Wang, J., Yan, Y., Garrett, T. P. J., Liu, J., Rodgers, D. W., Garlick, R. L., Rarr, G. E., Husain, Y., Reinherz, E. L., and Harrison, S. C. (1990). Atomic structure of a fragment of human CD4 containing two immunoglobulin like domains. *Nature* **348**, 411.

Wang, J. M., Ueda, H., Howaard, O. M. A., Grimm, M. C., Chertov, O., Gong, X., Gong, W., Resau, J. J., Broder, C. C., Evans, G., Arthur, L. O., Ruscetti, F. W., and Oppenheim, J. J. (1998). HIV-1 envelope p120 inhibits the monocyte response to chemokines through CD4 signal dependent chemokine receptor down regulation. *J. Immunol.* **161**, 4309.

Watanabe, M., Reimann, K. A., DeLong, P. A., Liu, T., Fisher, R. A., and Letvin, N. L. (1989). Effect of recombinant soluble CD4 in rhesus monkeys infected with simian immunodeficiency virus of macaques. *Nature* **337**, 267–270.

Weissman, D., Rabin, R. L., Arthos, J., Rubbert, A., Dybul, M., Swofford, R., Venkatesan, S., Farber, J. M., and Fauci, A. S. (1997). Macrophage-tropic HIV and SIV envelope proteins induce a signal through the CCR5 chemokine receptor. *Nature* **389**, 981–985.

Xu, H., and Littman, D. R. (1993). A kinase-independent function of lck in potentiating antigen-specific T cell activation. *Cell* **74**, 633.

Yeung, R. S., Penninger, J. M., Kundig, T., Khoo, W., Ohashi, P. S., Kroemer, G., and Mak, T. W. (1996). Human CD4 and human major histocompatibility complex class II (DQ6) transgenic mice: supersensitivity to superantigen-induced septic shock. *Eur. J. Immunol.* **26**, 1074–1082.

Zhou, P., Goldstein, S., Devadas, K., Tewari, D., and Notkins, A. L. (1997). Human CD4+ cells transfected with IL-16 cDNA are resistant to HIV-1 infection: inhibition of mRNA expression. *Nature Med.* **3**, 659–664.

Zuniga-Pflucker, J. C., McCarthy, S. A., Weston, M., Longo, D. L., Singer, A., and Kruisbeek, A. M. (1989). Role of CD4 in thymocyte selection and maturation. *J. Exp. Med.* **169**, 2085.

Zuniga-Pflucker, J. C., Jones, L. A., Chin, L. T., and Kruisbeek, A. M. (1991). CD4 and CD8 act as co-receptors during thymic selection of the T cell repertoire. *Semin. Immun.* **3**, 167.

LICENSED PRODUCTS

Multiple monoclonal anti-CD4 antibodies and recombinant soluble CD4 products are available.

IL-17 Receptor

Serge Lebecque*, François Fossiez and Elizabeth Bates

Laboratory for Immunological Research, Schering-Plough, 27 Chemin des Peupliers, Dardilly, 69572, France

*corresponding author tel: (33) 4 72 17 27 00, fax: (33) 4 78 35 47 50, e-mail: serge.lebecque@spcorp.com

DOI: 10.1006/rwcy.2000.14010.

SUMMARY

A ubiquitously expressed, large (864 aa) mouse membrane glycoprotein has been cloned after its binding to soluble vIL-17–Fc fusion protein. This receptor, which also binds CTLA-8, is unrelated to previously identified cytokine receptor families. The cDNA encoding a human homolog of the mIL-17R has been isolated by crosshybridization. Like its mouse counterpart, the human IL-17R (866 aa) is also ubiquitously expressed. Monoclonal antibodies against the hIL-17R block the secretion by fibroblasts of IL-6 induced by hIL-17–Fc. Binding studies, dose–response, and cellular restriction of IL-17 biological activities suggest the existence of another, as yet unidentified, high-affinity IL-17R chain.

BACKGROUND

Discovery

A mouse cDNA encoding an IL-17-binding protein has been cloned using a chimeric protein comprising a portion of the Fc region of human IgG1 followed by amino acids 19–151 of vIL-17 (Yao *et al.*, 1995a). The vIL-17–Fc fusion protein specifically bound the murine thymoma cell line EL4, from which a cDNA library was screened to isolate the gene encoding the binding molecule. When expressed in mammalian cells, the putative IL-17 receptor was shown to bind vIL-17 as well as mIL-17–Fc fusion proteins. Using the mouse cDNA as a probe, the human homolog was cloned from a human T cell library (Yao *et al.*, 1997).

GENE

Accession numbers

Human IL-17R mRNA: U58917
Mouse IL-17R mRNA: U31993

Sequence

See **Figure 1**.

Chromosome location and linkages

The chromosomal localization of IL-17R is not homologous between mouse (chromosome 6, between *Raf1* and *CD4*, in the vicinity of the recently described macrophage-restricted group II C-type lectin *mcl* gene locus (Yao *et al.*, 1995b; Balch *et al.*, 1998) and human (chromosome 22, between markers F8VWFP and D22S420) (Yao *et al.*, 1997).

PROTEIN

Accession numbers

Human IL-17R: U58917
Mouse IL-17R: U31993

Sequence

See **Figure 2**.

Figure 1 Human and mouse IL-17R mRNA nucleotide sequences. Start codon is in bold; stop codon is underlined.

Human IL-17R mRNA

GGGGCCGAGCCCTCCGCGACGCCACCCGGGCC**ATG**GGGGCCGCACGCAGCCCGCCGTCCGCTGTCCCGGGGCCCCTGCTGGGGCTGCTCCTGCTGCT
CCTGGGCGTGCTGGCCCCGGGTGGCGCCTCCCTGCGACTCCTGGACCACCGGGCGCTGGTCTGCTCCCAGCCGGGGCTAAACTGCACGGTCAAGAAT
AGTACCTGCCTGGATGACAGCTGGATTCACCCTCGAAACCTGACCCCCTCCTCCCCAAAGGACCTGCAGATCCAGCTGCACTTTGCCCACACCCAAC
AAGGAGACCTGTTCCCCGTGGCTCACATCGAATGGACACTGCAGACAGACGCCAGCATCCTGTACCTCGAGGGTGCAGAGTTATCTGTCCTGCAGCT
GAACACCAATGAACGTTTGTGCGTCAGGTTTGAGTTTCTGTCCAAACTGAGGCATCACCACAGGCGGTGGCGTTTTACCTTCAGCCACTTTGTGGTT
GACCCTGACCAGGAATATGAGGTGACCGTTCACCACCTGCCCAAGCCCATCCCTGATGGGGACCCAAACCACCAGTCCAAGAATTTCCTTGTGCCTG
ACTGTGAGCACGCCAGGATGAAGGTAACCACGCCATGCATGAGCTCAGGCAGCCTGTGGGACCCCAACATCACCGTGGAGACCCTGGAGGCCCACCA
GCTGCGTGTGAGCTTCACCCTGTGGAACGAATCTACCCATTACCAGATCCTGCTGACCAGTTTTCCGCACATGGAGAACCACAGTTGCTTTGAGCAC
ATGCACCACATACCTGCGCCCAGACCAGAAGAGTTCCACCAGCGATCCAACGTCACACTCACTCTACGCAACCTTAAAGGGTGCTGTCGCCACCAAG
TGCAGATCCAGCCCTTCTTCAGCAGCTGCCTCAATGACTGCCTCAGACACTCCGCGACTGTTTCCTGCCCAGAAATGCCAGACACTCCAGAACCAAT
TCCGGACTACATGCCCCTGTGGGTGTACTGGTTCATCACGGGCATCTCCATCCTGCTGGTGGGCTCCGTCATCCTGCTCATCGTCTGCATGACCTGG
AGGCTAGCTGGGCCTGGAAGTGAAAAATACAGTGATGACACCAAATACACCGATGGCCTGCCTGCGGCTGACCTGATCCCCCCACCGCTGAAGCCCA
GGAAGGTCTGGATCATCTACTCAGCCGACCACCCCCTCTACGTGGACGTGGTCCTGAAATTCGCCCAGTTCCTGCTCACCGCCTGCGGCACGGAAGT
GGCCCTGGACCTGCTGGAAGAGCAGGCCATCTCGGAGGCAGGAGTCATGACCTGGGTGGGCCGTCAGAAGCAGGAGATGGTGGAGAGCAACTCTAAG
ATCATCGTCCTGTGCTCCCGCGGCACGCGCGCCAAGTGGCAGGCGCTCCTGGGCCGGGGGGCGCCTGTGCGGCTGCGCTGCGACCACGGAAAGCCCG
TGGGGGACCTGTTCACTGCAGCCATGAACATGATCCTCCCGGACTTCAAGAGGCCAGCCTGCTTCGGCACCTACGTAGTCTGCTACTTCAGCGAGGT
CAGCTGTGACGGCGACGTCCCCGACCTGTTCGGCGCGGCGCCGCGGTACCCGCTCATGGACAGGTTCGAGGAGGTGTACTTCCGCATCCAGGACCTG
GAGATGTTCCAGCCGGGCCGCATGCACCGCGTAGGGGAGCTGTCGGGGGACAACTACCTGCGGAGCCCGGGCGGCAGGCAGCTCCGCGCCGCCCTGG
ACAGGTTCCGGGACTGGCAGGTCCGCTGTCCCGACTGGTTCGAATGTGAGAACCTCTACTCAGCAGATGACCAGGATGCCCCGTCCCTGGACGAAGA
GGTGTTTGAGGAGCCACTGCTGCCTCCGGGAACCGGCATCGTGAAGCGGGCGCCCCTGGTGCGCGAGCCTGGCTCCCAGGCCTGCCTGGCCATAGAC
CCGCTGGTCGGGGAGGAAGGAGGAGCAGCAGTGGCAAAGCTGGAACCTCACCTGCAGCCCCGGGGTCAGCCAGCGCCGCAGCCCCTCCACACCCTGG
TGCTCGCCGCAGAGGAGGGGGCCCTGGTGGCCGCGGTGGAGCCTGGGCCCCTGGCTGACGGTGCCGCAGTCCGGCTGGCACTGGCGGGGGAGGGCGA
GGCCTGCCCGCTGCTGGGCAGCCCGGGCGCTGGGCGAAATAGCGTCCTCTTCCTCCCCGTGGACCCCGAGGACTCGCCCCTTGGCAGCAGCACCCCC
ATGGCGTCTCCTGACCTCCTTCCAGAGGACGTGAGGGAGCACCTCGAAGGCTTGATGCTCTCGCTCTTCGAGCAGAGTCTGAGCTGCCAGGCCCAGG
GGGGCTGCAGTAGACCCGCCATGGTCCTCACAGACCCACACACGCCCTACGAGGAGGAGCAGCGGCAGTCAGTGCAGTCTGACCAGGGCTACATCTC
CAGGAGCTCCCCGCAGCCCCCCGAGGGACTCACGGAAATGGAGGAAGAGGAGGAAGAGGAGCAGGACCCAGGGAAGCCGGCCCTGCCACTCTCTCCC
GAGGACCTGGAGAGCCTGAGGAGCCTCCAGCGGCAGCTGCTTTTCCGCCAGCTGCAGAAGAACTCGGGCTGGGACACGATGGGGTCAGAGTCAGAGG
GGCCCAGTGCATGAGGGCGGCTCCCCAGGGACCGCCCAGATCCCAGCTTTGAGAGAGGAGTGTGTGTGCACGTATTCATCTGTGTGTACATGTCTGC
ATGTGTATATGTTCGTGTGTGAAATGTAGGCTTTAAAATGTAAATGTCTGGATTTTAATCCCAGGCATCCCTCCTAACTTTTCTTTGTGCAGCGGTC
TGGTTATCGTCTATCCCCAGGGGAATCCACACAGCCCGCTCCCAGGAGCTAATGGTAGAGCGTCCTTGAGGCTCCATTATTCGTTCATTCAGCATTT
ATTGTGCACCTACTATGTGGCGGGCATTTGGGATACCAAGATAAATTGCATGCGGCATGGCCCCAGCCATGAAGGAACTTAACCGCTAGTGCCGAGG
ACACGTTAAACGAACAGGATGGGCCGGGCACGGTGGCTCACGCCTGTAATCCCAGCACACTGGGAGGCCGAGGCAGGTGGATCACTCTGAGGTCAGG
AGTTTGAGCCAGCCTG

Mouse IL-17R mRNA

GTCGACTGGAACGAGACGACCTGCTGCCGACGAGCGCCAGTCCTCGGCCGGGAAAGCCATCGCGGGCCCTCGCTGTCGCGCGGAGCCAGCTGCGAGC
GCTCCGCGACCGGGCCGAGGGCT**ATG**GCGATTCGGCGCTGCTGGCCACGGGTCGTCCCCGGGCCCGCGCTGGGATGGCTGCTTCTGCTGCTGAACGT
TCTGGCCCCGGGCCGCGCCTCCCCGCGCCTCCTCGACTTCCCGGCTCCGGTCTGCGCGCAGGAGGGGCTGAGCTGCAGAGTCAAGAATAGTACTTGT
CTGGATGACAGCTGGATCCACCCCAAAAACCTGACCCCGTCTTCCCCAAAAAACATCTATATCAATCTTAGTGTTTCCTCTACCCAGCACGGAGAAT
TAGTCCCTGTGTTGCATGTTGAGTGGACCCTGCAGACAGATGCCAGCATCCTGTACCTCGAGGGTGCAGAGCTGTCCGTCCTGCAGCTGAACACCAA
TGAGCGGCTGTGTGTCAAGTTCCAGTTTCTGTCCATGCTGCAGCATCACCGTAAGCGGTGGCGGTTTTCCTTCAGCCACTTTGTGGTAGATCCTGGC
CAGGAGTATGAAGTGACTGTTCACCACCTGCCGAAGCCCATCCCTGATGGGGACCCAAACCACAAATCCAAGATCATCTTTGTGCCTGACTGTGAGG
ACAGCAAGATGAAGATGACTACCTCATGCGTGAGCTCAGGCAGCCTTTGGGATCCCAACATCACTGTGGAGACCTTGGACACACAGCATCTGCGAGT
GGACTTCACCCTGTGGAATGAATCCACCCCCTACCAGGTCCTGCTGGAAAGTTTCTCCGACTCAGAGAACCACAGCTGCTTTGATGTCGTTAAACAA
ATATTTGCGCCCAGGCAAGAAGAATTCCATCAGCGAGCTAATGTCACATTCACTCTAAGCAAGTTTCACTGGTGCTGCCATCACCACGTGCAGGTCC
AGCCCTTCTTCAGCAGCTGCCTAAATGACTGTTTGAGACACGCTGTGACTGTGCCCTGCCCAGTAATCTCAAATACCACAGTTCCCAAGCCAGTTGC
AGACTACATTCCCCTGTGGGTGTATGGCCTCATCACACTCATCGCCATTCTGCTGGTGGGATCTGTCATCGTGCTGATCATCTGTATGACCTGGAGG
CTTTCTGGCGCCGATCAAGAGAAACATGGTGATGACTCCAAAATCAATGGCATCTTGCCCGTAGCAGACCTGACTCCCCCACCCCTGAGGCCCAGGA
AGGTCTGGATCGTCTACTCGGCCGACCACCCCCTCTATGTGGAGGTGGTCCTAAAGTTCGCCCAGTTCCTGATCACTGCCTGTGGCACTGAAGTAGC
CCTTGACCTCCTGGAAGAGCAGGTTATCTCTGAGGTGGGGGTCATGACCTGGGTGAGCCGACAGAAGCAGGAGATGGTGGAGAGCAACTCCAAAATC
ATCATCCTGTGTTCCCGAGGCACCCAAGCAAAGTGGAAAGCTATCTTGGGTTGGGCTGAGCCTGCTGTCCAGCTACGGTGTGACCACTGGAAGCCTG
CTGGGGACCTTTTCACTGCAGCCATGAACATGATCCTGCCAGACTTCAAGAGGCCAGCCTGCTTCGGCACCTACGTTGTTTGCTACTTCAGTGGCAT
CTGTAGTGAGAGGGATGTCCCCGACCTCTTCAACATCACCTCCAGGTACCCACTCATGGACAGATTTGAGGAGGTTTACTTCCGGATCCAGGACCTG
GAGATGTTTGAACCCGGCCGGATGCACCATGTCAGAGAGCTCACAGGGGACAATTACCTGCAGAGCCCTAGTGGCCGGCAGCTCAAGGAGGCTGTGC
TTAGGTTCCAGGAGTGGCAAACCCAGTGCCCCGACTGGTTCGAGCGTGAGAACCTCTGCTTAGCTGATGGCCAAGATCTTCCCTCCCTGGATGAAGA
AGTGTTTGAAGACCCACTGCTGCCACCAGGGGGAGGAATTGTCAAACAGCAGCCCCTGGTGCGGGAACTCCCATCTGACGGCTGCCTTGTGGTAGAT
GTCTGTGTCAGTGAGGAAGAAAGTAGAATGGCAAAGCTGGACCCTCAGCTATGGCCACAGAGAGAGCTAGTGGCTCACACCCTCCAAAGCATGGTGC
TGCCAGCAGAGCAGGTCCCTGCAGCTCATGTGGTGGAGCCTCTCCATCTCCCAGACGGCAGTGGAGCAGCTGCCCAGCTGCCCATGACAGAGGACAG
CGAGGCTTGCCCGCTGCTGGGGGTCCAGAGGAACAGCATCCTTTGCCTCCCCGTGGACTCAGATGACTTGCCACTCTGTAGCACCCCAATGATGTCA
CCTGACCACCTCCAAGGCGATGCAAGAGAGCAGCTAGAAAGCCTAATGCTCTCGGTGCTGCAGCAGAGCCTGAGTGGACAGCCCCTGGAGAGCTGGC
CGAGGCCAGAGGTGGTCCTCGAGGGCTGCACACCCTCTGAGGAGGAGCAGCGGCAGTCGGTGCAGTCGGACCAGGGCTACATCTCCAGGAGCTCCCC
GCAGCCCCCCGAGTGGCTCACGGAGGAGGAAGAGCTAGAACTGGGTGAGCCCGTTGAGTCTCTCTCTCCTGAGGAACTACGGAGCCTGAGGAAGCTC
CAGAGGCAGCTTTTCTTCTGGGAGCTCGAGAAGAACCCTGGCTGGAACAGCTTGGAGCCACGGAGACCCACCCCAGAAGAGCAGAATCCCTCCTAGG
CCTCCTGAGCCTGCTACTTAAGAGGGTGTATATTGTACTCTGTGTGTGCGTGCGTGTGTGTGTGTGTGTGTGTGTGTGTGCGTGTGTGTGTGTGT
GTGTGTGTGTGTGTGTGTGTAGTGCCCGGCTTAGAAATGTGAACATCTGAATCTGACATAGTGTTGTATACCTGAAGTCCCAGCACTTGGGAACTGA
GACTTGATGATCTCCTGAAGCCAGGTGTTCAGGGCCAGTGTGAAAACATAGCAAGACCTCAGAGAAATCAATGCAGACATCTTGGTACTGATCCCTA
AACACACCCCTTTCCCTGATAACCCGACATGAGCATCTGGTCATCATTGCACAAGAATCCACAGCCCGTTCCCAGAGCTCATAGCCAAGTGTGTTGC
TCATTCCTTGAATATTTATTCTGTACCTACTATTCATCAGACATTTGGAATTCAAAAACAAGTTACATGACACAGCCTTAGCCACTAAGAAGCTTAA
AATTCGGTAAGGATGTAAAATTAGCCAGGATGAATAGAGGGCTGCTGCCCTGGCTGCAGAAGAGCAGGTCGTCTCGTTCCAGTCGAC

Figure 2 Amino acid sequences for human and mouse IL-17R.

```
Human IL-17R
MGAARSPPSA VPGPLLGLLL LLLGVLAPGG ASLRLLDHRA LVCSQPGLNC
TVKNSTCLDD SWIHPRNLTP SSPKDLQIQL HFAHTQQGDL FPVAHIEWTL
QTDASILYLE GAELSVLQLN TNERLCVRFE FLSKLRHHHR RWRFTFSHFV
VDPDQEYEVT VHHLPKPIPD GDPNHQSKNF LVPDCEHARM KVTTPCMSSG
SLWDPNITVE TLEAHQLRVS FTLWNESTHY QILLTSFPHM ENHSCFEHMH
HIPAPRPEEF HQRSNVTLTL RNLKGCCRHQ VQIQPFFSSC LNDCLRHSAT
VSCPEMPDTP EPIPDYMPLW VYWFITGISI LLVGSVILLI VCMTWRLAGP
GSEKYSDDTK YTDGLPAADL IPPPLKPRKV WIIYSADHPL YVDVVLKFAQ
FLLTACGTEV ALDLLEEQAI SEAGVMTWVG RQKQEMVESN SKIIVLCSRG
TRAKWQALLG RGAPVRLRCD HGKPVGDLFT AAMNMILPDF KRPACFGTYV
VCYFSEVSCD GDVPDLFGAA PRYPLMDRFE EVYFRIQDLE MFQPGRMHRV
GELSGDNYLR SPGGRQLRAA LDRFRDWQVR CPDWFECENL YSADDQDAPS
LDEEVFEEPL LPPGTGIVKR APLVREPGSQ ACLAIDPLVG EEGGAAVAKL
EPHLQPRGQP APQPLHTLVL AAEEGALVAA VEPGPLADGA AVRLALAGEG
EACPLLGSPG AGRNSVLFLP VDPEDSPLGS STPMASPDLL PEDVREHLEG
LMLSLFEQSL SCQAQGGCSR PAMVLTDPHT PYEEEQRQSV QSDQGYISRS
SPQPPEGLTE MEEEEEEEQD PGKPALPLSP EDLESLRSLQ RQLLFRQLQK
NSGWDTMGSE SEGPSA

Mouse IL-17R
MAIRRCWPRV VPGPALGWLL LLLNVLAPGR ASPRLLDFPA PVCAQEGLSC
RVKNSTCLDD SWIHPKNLTP SSPKNIYINL SVSSTQHGEL VPVLHVEWTL
QTDASILYLE GAELSVLQLN TNERLCVKFQ FLSMLQHHRK RWRFSFSHFV
VDPGQEYEVT VHHLPKPIPD GDPNHKSKII FVPDCEDSKM KMTTSCVSSG
SLWDPNITVE TLDTQHLRVD FTLWNESTPY QVLLESFSDS ENHSCFDVVK
QIFAPRQEEF HQRANVTFTL SKFHWCCHHH VQVQPFFSSC LNDCLRHAVT
VPCPVISNTT VPKPVADYIP LWVYGLITLI AILLVGSVIV LIICMTWRLS
GADQEKHGDD SKINGILPVA DLTPPPLRPR KVWIVYSADH PLYVEVVLKF
AQFLITACGT EVALDLLEEQ VISEVGVMTW VSRQKQEMVE SNSKIIILCS
RGTQAKWKAI LGWAEPAVQL RCDHWKPAGD LFTAAMNMIL PDFKRPACFG
TYVVCYFSGI CSERDVPDLF NITSRYPLMD RFEEVYFRIQ DLEMFEPGRM
HHVRELTGDN YLQSPSGRQL KEAVLRFQEW QTQCPDWFER ENLCLADGQD
LPSLDEEVFE DPLLPPGGGI VKQQPLVREL PSDGCLVVDV CVSEEESRMA
KLDPQLWPQR ELVAHTLQSM VLPAEQVPAA HVVEPLHLPD GSGAAAQLPM
TEDSEACPLL GVQRNSILCL PVDSDDLPLC STPMMSPDHL QGDAREQLES
LMLSVLQQSL SGQPLESWPR PEVVLEGCTP SEEEQRQSVQ SDQGYISRSS
PQPPEWLTEE EELELGEPVE SLSPEELRSL RKLQRQLFFW ELEKNPGWNS
LEPRRPTPEE QNPS
```

Description of protein

The predicted human receptor is a type I membrane glycoprotein with a 293 amino acid extracellular domain, a 21 amino acid transmembrane region, and 525 amino acid cytoplasmic tail. Human and mouse IL-17 receptors appear highly conserved as they share 82% amino acid similarity and 69% identity, and six potential *N*-linked glycosylation sites. Comparison of both human and mouse IL-17R sequences with public databases revealed no significant homology with known nucleotide and protein sequences. IL-17R does not include structural features of the immunoglobulin superfamily nor of the TNF receptor family. The extracellular domain does not contain the WSXWS motif found in hematopoietin receptor family members (Cosman, 1993). However, a relatively large proportion of acidic (16%) and proline (9%) residues is shared with other growth factor receptors, and a segment (TPPPLRPRKVW) located close to the mIL-17R transmembrane domain is highly conserved among cytokine receptors. While no homology with tyrosine kinase catalytic domains could be found in the large cytoplasmic tail, two acidic regions and a serine-rich region are also present in the IL-2Rβ chain, the IL-4R, and the G-CSFR. The molecular mass of the molecule immunoprecipitated using the m202 anti-hIL-17R antibody is larger (128–132 kDa) than predicted from the hIL-17R cDNA sequence. Expression of hIL-17R in the presence of tunicamycin demonstrated that the *N*-linked glycosylation sites of the extracellular domain are indeed utilized.

Affinity for ligand(s)

Direct binding assays revealed a relatively weak affinity for hIL-17 (K_a values in the range of 2×10^7 to 2×10^8), which was unexpected from the low concentrations of IL-17 needed for most half-maximal biological activities (between 2 and 50 ng/mL). This discrepancy suggests that an as yet unidentified high-affinity converting subunit might be present on IL-17-responsive cells (Yao *et al.*, 1997).

Cell types and tissues expressing the receptor

Human IL-17R mRNA is constitutively and widely expressed, as it was detected by PCR in total PBMCs, in NK cells, in Raji B cell line, in myelomonocytic THP1 cell line, in lung epithelial cell lines, and in the embryonal kidney line 293. Cell surface expression was confirmed by flow cytometric analysis using hIL-17–Fc (Yao *et al.*, 1997). Likewise, analysis by northern blot of the tissue distribution of mIL-17R showed that a single band of approximatively 3.7 kb is present in all tested tissues, with strong signals observed in spleen and kidney. Moderate signals are observed in lung and liver, and weaker signals in brain, testes, heart and skeletal muscles. Likewise, northern blot and RT-PCR detected the mIL-17R mRNA in every cell line tested (including fetal liver epithelial cells, rat intestinal epithelial cells, fibroblasts, muscle cells, mast cells, splenic B cells, pre-B cells, triple negative thymocytes, T cell thymoma, and T cell clones), confirming the ubiquitous expression of this message (Yao *et al.*, 1995b).

The human, mouse, rat, and viral IL-17 proteins can induce IL-6 secretion by mouse stromal cells, indicating that they all recognize the mouse receptor (Kennedy *et al.*, 1996; Yao *et al.*, 1995a).

SIGNAL TRANSDUCTION

Associated or intrinsic kinases

IL-17 induces NFκB protein–DNA complexes, consisting of p65/p50 heterodimers in mouse 3T3 fibroblasts and in the rat intestinal epithelial cell line IEC-6 (Yao *et al.*, 1995a; Awane *et al.*, 1999). IL-17 regulates the activities of extracellular signal-regulated kinase ERK1, ERK2, c-Jun N-terminal kinase (JNK), and p38 mitogen-activated protein (MAP) kinases in IEC-6 cells and in chondrocytes (Shalom-Barak *et al.*, 1998; Awane *et al.*, 1999). Whereas the IL-17-mediated activation of ERK and MAP kinases was mediated through Ras, JNK activation was dependent on functional TRAF6 (but not TRAF2). The correlation between JNK inhibition by dexamethasone and the inhibitory effect of the latter on the response of chondrocytes to IL-17 suggests that JNK is more central to the inflammatory response than the other MAP kinases (Shalom-Barak *et al.*, 1998).

Cytoplasmic signaling cascades

These data suggest that IL-17R uses signaling mediators involved in the IL-1R-signaling pathway and that NFκB-inducing kinase can serve as the common mediator in the NFκB signaling cascades triggered by IL-17, TNFα, and IL-1β (Awane *et al.*, 1999).

As the analysis of the predicted structure of the cloned IL-17R revealed no homology with the catalytic domain of previously described tyrosine kinase-containing or -associated cytokine and growth factor receptors (Yao *et al.*, 1995a), the finding that the protein tyrosine kinase inhibitor genistein did not affect IL-17-induced stimulation of G-CSF production by mouse 3T3 fibroblasts was not unexpected (Cai *et al.*, 1998). In contrast, Subramaniam and coworkers have reported that the early signaling events triggered by hIL-17 in human U937 monocytic leukemia cells involve rapid tyrosine phosphorylation of several cellular proteins including Raf-1 serine/threonine kinase and several members of the JAK and STAT protein families (JAK1, 2, and 3, TYK2, and STAT1, 2, 3, and 4) (Subramaniam *et al.*, 1999a, 1999b). Furthermore, the IL-17-dependent stimulation of IL-1α and TNFα production by human macrophages was completely or partially blocked by PKA-specific or nonspecific tyrosine kinase inhibitor, and by PKC or MAP kinase inhibitors, respectively (Jovanovic *et al.*, 1998).

In conclusion, the signaling pathway of IL-17R in stromal cells seems to be shared in part with IL-1R. In contrast, analysis of the biological activities of IL-17 on monocytes/macrophages shows differences between mouse and human and suggests a role for JAK and STAT family members in the latter. It remains unclear, however, how the mouse and human IL-17Rs, which are highly conserved, would signal through two different pathways in two different cell types.

DOWNSTREAM GENE ACTIVATION

Transcription factors activated

The NFκB activation by IL-17 is in line with the activation of the secretion of several cytokines and chemokines (like IL-1β, IL-6, IL-8, MIP-2, MCP-1) known to be under the transcriptional control of this factor.

BIOLOGICAL CONSEQUENCES OF ACTIVATING OR INHIBITING RECEPTOR AND PATHOPHYSIOLOGY

No obvious phenotype was observed after disruption of the IL-17R in mice, although experiments addressing functional alterations have not yet been reported (Spriggs, 1997). Activation of mIL-17R by injecting either rhIL-17 or using adenovirus-mediated gene transfer of the murine IL-17 cDNA targeted to liver resulted in a marked acute neutrophilia, profound stimulation of splenic hematopoiesis, and IL-6-dependent protection against lethal challenge with *E. coli*. There are no data currently available regarding either the consequences of IL-17R dysregulated expression or abnormalities in humans.

THERAPEUTIC UTILITY

Effect of treatment with soluble receptor domain

The haematopoietic effect of IL-17, and in particular its ability to indirectly induce acute neutrophilia, suggest a potential therapeutic anti-infectious use in the context of immunosuppression or during bone marrow recovery. On the other hand, IL-17R might represent a target for therapeutic inhibition by blocking antibodies, by soluble receptor or by small receptor antagonists in T-dependent autoimmune diseases such as *rheumatoid arthritis* and *multiple sclerosis*, in chronic inflammatory conditions of the lung (*chronic pulmonary obstructive disease*, *asthma*) or the skin (*psoriasis* and *atopic dermatitis*), in *organ graft rejection*. Preclinical data also suggest that blocking IL-17R may reduce paracrine IL-6 secretion which acts as a growth factor for some *cancers*.

References

Awane, M., Andres, P. G., Li, D. J., and Reinecker, H. C. (1999). NF-kappa B-inducing kinase is a common mediator of IL-17-, TNF-alpha-, and IL-1 beta-induced chemokine promoter activation in intestinal epithelial cells. *J. Immunol.* **162**, 5337–5344.

Balch, S. G., McKnight, A. J., Seldin, M. F., and Gordon, S. (1998). Cloning of a novel C-type lectin expressed by murine macrophages. *J. Biol. Chem.* **273**, 18656–18664.

Cai, X. Y., Gommoll, C. P., Jr., Justice, L., Narula, S. K., and Fine, J. S. (1998). Regulation of granulocyte colony-stimulating factor gene expression by interleukin-17. *Immunol. Lett.* **62**, 51–58.

Cosman, D. (1993). The hematopoietin receptor superfamily. *Cytokine* **5**, 95–106.

Jovanovic, D. V., Di Battista, J. A., Martel-Pelletier, J., Jolicoeur, F. C., He, Y., Zhang, M., Mineau, F., and Pelletier, J. P. (1998). IL-17 stimulates the production and expression of proinflammatory cytokines, IL-beta and TNF-alpha, by human macrophages. *J. Immunol.* **160**, 3513–3521.

Kennedy, J., Rossi, D. L., Zurawski, S. M., Vega, F., Jr., Kastelein, R. A., Wagner, J. L., Hannum, C. H., and Zlotnik, A. (1996). Mouse IL-17: a cytokine preferentially expressed by alpha beta TCR+ CD4− CD8− T cells. *J. Interferon Cytokine Res.* **16**, 611–617.

Shalom-Barak, T., Quach, J., and Lotz, M. (1998). Interleukin-17-induced gene expression in articular chondrocytes is associated with activation of mitogen-activated protein kinases and NF-kappaB. *J. Biol. Chem.* **273**, 27467–27473.

Spriggs, M. K. (1997). Interleukin-17 and its receptor. *J. Clin. Immunol.* **17**, 366–369.

Subramaniam, S. V., Cooper, R. S., and Adunyah, S. E. (1999a). Evidence for the involvement of JAK/STAT pathway in the signaling mechanism of interleukin-17. *Biochem. Biophys. Res. Commun.* **262**, 14–19.

Subramaniam, S. V., Pearson, L. L., and Adunyah, S. E. (1999b). Interleukin-17 induces rapid tyrosine phosphorylation and activation of raf-1 kinase in human monocytic progenitor cell line U937. *Biochem. Biophys. Res. Commun.* **259**, 172–177.

Yao, Z., Fanslow, W. C., Seldin, M. F., Rousseau, A. M., Painter, S. L., Comeau, M. R., Cohen, J. I., and Spriggs, M. K. (1995a). *Herpesvirus Saimiri* encodes a new cytokine, IL-17, which binds to a novel cytokine receptor. *Immunity* **3**, 811–821.

Yao, Z., Painter, S. L., Fanslow, W. C., Ulrich, D., Macduff, B. M., Spriggs, M. K., and Armitage, R. J. (1995b). Human IL-17: a novel cytokine derived from T cells. *J. Immunol.* **155**, 5483–5466.

Yao, Z., Spriggs, M. K., Derry, J. M., Strockbine, L., Park, L. S., VandenBos, T., Zappone, J. D., Painter, S. L., and Armitage, R. J. (1997). Molecular characterization of the human interleukin (IL)-17 receptor. *Cytokine* **9**, 794–800.

LICENSED PRODUCTS

R&D Systems

Human IL-17:

Recombinant hIL-17 expressed in *E. coli* (catalog no.: 317-IL-050)

Polyclonal anti-hIL-17 goat antiserum (catalog no.: AF-317-NA)

Monoclonal anti-hIL-17 mouse IgG2b antibody (catalog no.: MAB317)

Quantitative hIL-17 colorimetric sandwich ELISA (catalog no.: D1700)

Mouse IL-17:

Recombinant mIL-17 expressed in *E. coli* (catalog no.: 421-ML-025)

Polyclonal anti-mIL-17 goat antiserum (catalog no.: AF-421-NA)

Monoclonal anti-mIL-17 mouse IgG2a antibody, selected to neutralize the bioactivity of mIL-17 (catalog no.: MAB421)

Monoclonal anti-mIL-17 mouse IgG2a antibody, selected as a capture antibody in mouse IL-17 sandwich ELISAs (catalog no.: MAB721)

Quantitative mIL-17 colorimetric sandwich ELISA (catalog no.: D1700)

Biosource

Quantitative hIL-17 colorimetric solid phase ELISA (catalog no.: KAC1591, KAC1592)

Pharmingen

Monoclonal anti-mIL-17 mouse IgG1 rat antibody, selected as a capture antibody in mouse IL-17 sandwich ELISAs (catalog no.: 23290D, purified 23291A/D, PE-conjugated 23295A)

Monoclonal anti-mIL-17 mouse IgG1 rat antibody, biotinnylated and selected as a detection antibody in mouse IL-17 sandwich ELISAs (catalog no.: 23282D)

ACKNOWLEDGEMENTS

The authors wish to thank first P. Golstein and E. Rouvier for sharing the early CTLA-8 data. The members of LIR and of DNAX are aknowledged for their contribution to the identification and the determination of IL-17 functions: J. Abrams, S. Ait-Yahia, J. Banchereau, E. Bates, F. Bazan, J.C. Bories,

F. Brière, C. Caux, P. Chomarat, B. Das Mahapatra, O. Djossou, L. Flores-Romo, C. Gaillard, E. Garcia, P. Garrone, D. Gorman, C. H. Hannum, R. Kastelein, J. Kennedy, P. Krishna, C. Maat, K. Moore, R. Murray, C. Perone, J.J. Pin, S. Saeland, A. Zlotnik, G. Zurawsky, and S. Zurawsky. They also want to thank J. Chiller, D. de Groote, P. Miossec, S. Narula, J.F. Nicolas, M. Spriggs, E. Tartour, and C. Von Kooten for exchange of unpublished informations and discussions. The editorial assistance of S. Bourdarel has been greatly appreciated.

Prolactin Receptor

Vincent Goffin* and Paul A. Kelly

INSERM Unit 344, Faculty of Medicine Necker, 156 rue de Vaugirard, Paris, Cedex 15, 75730, France

*corresponding author tel: +33-1-40615616, fax: +33-1-43060443, e-mail: goffin@necker.fr

DOI: 10.1006/rwcy.2000.14012.

SUMMARY

The prolactin receptor (PRLR) exists in all vertebrates in many isoforms, soluble or membrane-bound. It is expressed in a wide variety of tissues, and is responsible for the transmission of almost 300 distinct functions of its ligand, prolactin. This hormone activates the prolactin receptor by inducing its homodimerization, the first step required for triggering signaling cascades. No other membrane chain is required for signaling. Isoforms of prolactin receptor that vary in the length and sequence of the cytoplasmic tail have been identified; These variations account for the different signaling properties of the different isoforms. The JAK/STAT pathway (and mainly JAK2/STAT5) is the major signaling cascade, though many other proteins are also activated such as the MAP kinase pathway. Target genes are multiple, and confer to prolactin its numerous properties, such as lactogenic actions (milk protein gene induction) and mitogenic or antiapoptotic effects. Although no genetic abnormality has been linked to the PRLR gene, the involvement of this receptor in *breast cancer* has been proposed.

BACKGROUND

Discovery

The prolactin receptor (PRLR) was identified in the early 1970s as a specific, high-affinity, saturable membrane-anchored protein, binding both prolactin and human growth hormone (Posner *et al.*, 1974). It was first identified in liver, mammary gland, and reproductive organs from mammals, and later in many other tissues throughout all vertebrates, including fish. The original cloning of the cDNA encoding the rat PRLR (short isoform) was reported in the late 1980s (Boutin *et al.*, 1988).

Alternative names

The prolactin receptor is also referred to as the 'lactogen' receptor, or as the receptor for lactogenic or luteotropic hormones, two other names of prolactin. These names can be misleading, however, since the term 'lactogenic' refers to the biological activity of milk production, whereas the prolactin receptor is responsible for the transduction of many nonlactogenic signals, e.g. mitogenic. In view of the wide spectrum of biological activities of prolactin, 'prolactin receptor' is probably the most appropriate name for this protein.

Structure

Like all class 1 cytokine receptors, the PRLR is a single-pass transmembrane chain, with the N-terminus outside the cell. In contrast to many cytokine receptors, for example the IL-2 receptor or IL-6 receptor families, the active form of the PRLR does not appear to involve any chain other than the PRLR itself. Depending on the species, there are many isoforms of the PRLR. These only vary in the length of the cytoplasmic domains, and hence in their signaling properties. In mammals, the overall length of the PRLR varies from $\sim$200 amino acids for the soluble binding protein up to $\sim$600 residues for the long membrane isoforms. Within the cytokine receptor superfamily, the growth hormone receptor (GHR) is undoubtedly the closest member to PRLR, with respect to its structure, its signaling properties, and its ligand.

Main activities and pathophysiological roles

Prolactin was originally isolated based on its ability to stimulate mammary development and lactation in rabbits, and soon thereafter to stimulate the production of crop milk in pigeons (Stricker and Grueter, 1928; Riddle *et al.*, 1933). Prolactin was also shown to be luteotropic, that is to promote the formation and action of the corpus luteum (Astwood, 1941). Subsequently, a number of additional activities have been associated with this hormone in various vertebrate species. In the now classical reviews by Nicoll and Bern (Nicoll and Bern, 1972; Nicoll, 1974), 85 different biological functions of prolactin were subdivided into five broad categories related to reproduction, osmoregulation, growth, integument, and synergism with steroids. Since the publication of these first reviews, numerous other biological functions of prolactin have been identified. Bole-Feysot *et al.* (1998) listed up to 300 functions or molecules activated by the PRLR, organized into the following categories: water and electrolyte balance, growth and development, endocrinology and metabolism, brain and behavior, reproduction, and finally immunoregulation and protection. Thus, despite the fact that prolactin remains historically linked to its actions in lactation and reproduction, the biological role of this hormone can no longer be restricted to these functions.

Finally, although most circulating prolactin is of pituitary origin, emphasis has been given within the last few years to locally produced, nonpituitary prolactin. The wide distribution of the PRLR and the increasing number of tissues identified as prolactin sources (Ben-Jonathan *et al.*, 1996) probably explains the unusually large number of functions of this hormone. Surprisingly, there is, to date, no disease known to be caused by a mutation of the PRLR, or even of its ligand. *Hyperprolactinemia* due to pituitary adenoma leads to reproductive disorders, especially in women, which can be treated either by drugs reducing prolactin secretion, or by surgical hypophysectomy.

GENE

Accession numbers

The cDNAs encoding the PRLR have been cloned and sequenced in several vertebrates.
GenBank:
Human PRLR cDNA: M31661 (long isoform)
Bovine PRLR cDNA: L02549 (long isoform)
Rat PRLR cDNA: M57668 (long isoform), M74152 (intermediate (or Nb2) isoform), M19304 (short isoform)
Mouse PRLR cDNA: X73372 (long isoform), M22957/M22958/M22959 (short isoforms)
Chicken PRLR cDNA: D13154 (long isoform)
Fish PRLR cDNA: L34783 (long isoform)
The original cloning of the short isoform PRLR was performed from rat liver (Boutin *et al.*, 1988), then long isoforms were identified from mammary gland and ovarian cells (Boutin *et al.*, 1989).

Sequence

The coding sequence of the rat PRL receptor long isoform is given in **Figure 1** (Shirota *et al.*, 1990).

The gene encoding the PRLR is unique, but several PRLR mRNAs resulting from alternative splicing are observed in (almost) all species studied thus far. Depending on the tissue and/or species considered, these various PRLR transcripts encode identical or different mature proteins. In humans, at least three mRNAs have been identified (2.8, 3.5, and 7.3 kb) that probably encode the sole long isoform; nevertheless, there is at least one study which reports the existence of a C-terminally truncated PRLR isoform in breast tumors (Clevenger *et al.*, 1995). Conversely, in rodents, different mRNAs encode various isoforms differing in the length and composition of their cytoplasmic tail. These isoforms are referred to as short, intermediate, or long PRLR with respect to their size. In mouse, seven transcripts encode four different PRLR isoforms (one long, three short) (Buck *et al.*, 1992). In addition to mRNAs encoding these membrane-anchored PRLR isoforms, alternatively spliced mRNA encoding soluble prolactin-binding protein (PRLbp) have been reported, such as in humans (Fuh and Wells, 1995).

Chromosome location and linkages

The gene encoding the human PRLR is located on chromosome 5 (p13-14) and contains at least 10 exons for an overall length > 100 kb (Arden *et al.*, 1990). In the mouse, the PRLR gene is > 120 kb and is located within a cluster of cytokine receptor loci on chromosome 15 (p12-13) (Gearing *et al.*, 1993). The genomic organization of the mouse PRLR gene has recently been deciphered (Ormandy *et al.*, 1998). It contains 13 exons, nine of which are shared by all PRLR isoforms, the last four encoding the C-terminal tail of the long isoform (exon 10) and the three short isoforms named PRLR S1 (exon 12), S2 (exon 11),

Figure 1 Coding sequence of the rat long isoform of the PRLR. The sequence encoding the signal peptide is underlined and the sequence encoding the transmembrane domain is underlined and in italic.

-19+1

ATGCCATCTGCACTTGCTTTCGTCCTACTTGTTCTCAACATCAGCCTCCTGAAGGGACAG

TCACCACCAGGGAAACCTGAGATCCACAAATGTCGCTCTCCTGACAAGGAAACATTCACC

TGCTGGTGGAATCCTGGGACAGATGGAGGACTTCCTACCAATTATTCACTGACTTACAGC

AAAGAAGGAGAGAAAACCACCTACGAATGTCCAGACTACAAAACCAGTGGCCCCAACTCC

TGCTTCTTTAGCAAGCAGTACACTTCCATCTGGAAAATATATATCATCACAGTAAATGCC

ACGAACCAAATGGGAAGCAGTTCCTCGGATCCACTTTATGTGGATGTGACTTACATCGTT

GAGCCAGAGCCTCCTCGGAACCTGACATTAGAAGTAAAACAGCTAAAAGACAAAAAAACA

TATCTGTGGGTAAAATGGTCCCCACCCACCATAACTGATGTGAAAACTGGTTGGTTTACA

ATGGAATATGAAATTCGATTAAAGCCTGAAGAAGCAGAAGAGTGGGAGATCCATTTTACA

GGTCATCAAACACAGTTTAAAGTTTTTGACCTATATCCAGGGCAAAAGTATCTTGTCCAG

ACTCGCTGCAAGCCAGACCATGGATACTGGAGTAGATGGAGCCAGGAGAGTTCCGTTGAA

ATGCCAAATGACTTCACCTTGAAGGAC*ACAACCGTGTGGATCATTGTGGCCATTCTCTCT*

*GCTGTCATCTGTTTGATTATGGTCTGGGCAGTGGCTTTG*AAGGGCTATAGCATGATGACC

TGCATCTTTCCACCAGTTCCTGGGCCAAAAATAAAAGGATTTGATACCCATCTGCTGGAG

AAGGGCAAGTCTGAAGAGCTGCTGAGTGCCTTGGGGTGCCAAGACTTTCCCCCTACTTCT

GACTGTGAGGACTTGCTGGTGGAGTTCTTAGAAGTTGATGACAATGAGGACGAGCGGCTA

ATGCCATCCCATTCCAAAGAGTATCCAGGTCAAGGTGTTAAGCCCACACACCTAGATCCC

GACAGTGACTCTGGTCACGGAAGCTATGACAGCCATTCTCTTTTATCTGAAAAGTGTGAG

GAACCCCAGGCCTACCCCCCTACTTTGCACATCCCTGAGATCACTGAGAAGCCAGAGAAT

CCTGAAGCAAATATTCCTCCCACCGTGGACCCCCAAAGCACCAACCCCAATTTTCATGTA

GATGCACCCAAATCTTCAACATGGCCATTACTGCCTGGCCAACACATGCCCAGATCTCCT

TACCACAGTGTTGCTGATGTGTGCAAGCTAGCCGGAAGTCCTGTGAATACACTGGACTCT

TTCTTGGACAAAGCAGAGGAAAATGTTCTAAAGTTGTCTAAAGCCCTTGAGACTGGAGAG

GAAGAAGTGGCTAAGCAAAAAGGGGCAAAAAGCTTCCCTTCTGACAAACAAAACACACCT

TGGCCGCTGCTCCAGGAGAAAAGCCCCACTGTCTATGTTAAACCCCCAGATTATGTGGAG

ATTCACAAAGTCAACAAAGATGGAGTGCTATCATTATTCCCCAAGCAGAGAGAAAACAAC

CAGACAGAGAAGCCTGGGGTTCCTGAAACCAGTAAGGAGTATGCCAAGGTGTCTGGCATT

ATGGATAACAATATCCTCGTATTAGTGCCAGACTCACGAGCCCAGAACACAGCGTTGCTC

GAGGAATCAGCCAAGAAGGCTCCACCATCGTTTGAAGCTGACCAATCTGAGAAAGATCTG

GCCAGCTTCACTGCAACCTCAAGCAACCGCAGACTCCAACTGGGTAGGCTGGATTACCTG

GATCCTACGTGCTTCATGCACTCCTTTCAC**TGA**

Stop

and S3 (exon 13), the latter being homologous to the unique rat short isoform. Five different promoter regions have also been identified in the 5′ UTR of the mouse PRLR gene (Ormandy *et al.*, 1998). In the rat, three promoters have been identified and tissue-specific usage has been demonstrated for two of them (Moldrup *et al.*, 1996; Hu *et al.*, 1998). Interestingly, at least in mouse, the genomic organization (coding sequences of exons) closely parallels the functional/folding domains of the mature proteins; the two disulfides are encoded by exons 4 and 5, the WS motif by exon 7, the transmembrane domain by exon 8 and the Box-1 by exon 9 (Ormandy *et al.*, 1998).

PROTEIN

Accession numbers

SwissProt accession numbers for the PRLR precursors are listed in **Table 1**.

Sequence

The sequence of the rat long isoform of the PRLR is given in **Figure 2**.

Description of protein

The PRLR protein is a glycosylated, single-pass transmembrane protein with the N-terminus in the extracellular space. It is synthesized as a precursor including a signal peptide of ~19 to ~24 amino acids. Within each species, different PRLR isoforms can be observed that have strictly identical extracellular (ligand-binding) domains and differ only by the length of their cytoplasmic tail (**Figure 3**). For example, the rat PRLR isoforms contain 291 (short), 393 (intermediate), or 591 (long) amino acids, and are identical until residue 260 (after Box-1). In SDS-PAGE (western blots), the long PRLR migrates with an apparent size of 90–95 kDa due to posttranslational modifications, whereas the rat short PRLR appears as a doublet at 42–44 kDa. Of the other cytokine receptors, the growth hormone receptor (GHR) is closest to the PRLR in terms of protein structure, signaling properties, and ligands.

Table 1 Accession numbers (SwissProt) of PRLR precursors (including signal peptides)

Species	Accession number	Isoform	Precursor	Mature
Human	P16471	Long	622 aa	598 aa
Bovine	Q28172	Long	581 aa	557 aa
Rat	P05710	Long	610 aa	591 aa
Mouse	Q08501	Long	608 aa	589 aa
Fish (tilapia)	Q91513	Long	630 aa	606 aa

Figure 2 The sequence of the rat long isoform of the PRLR. The signal peptide is underlined, the transmembrane domain is underlined and in italic.

-19+1
MPSALAFVLLVLNISLLKGQSPPGKPEIHKCRSPDKETFTCWWNPGTDGGLPTNYSLTYS
KEGEKTTYECPDYKTSGPNSCFFSKQYTSIWKIYIITVNATNQMGSSSSDPLYVDVTYIV
EPEPPRNLTLEVKQLKDKKTYLWVKWSPPTITDVKTGWFTMEYEIRLKPEEAEEWEIHFT
GHQTQFKVFDLYPGQKYLVQTRCKPDHGYWSRWSQESSVEMPNDFTLKD*TTVWIIVAILS*
*AVICLIMVWAVAL*KGYSMMTCIFPPVPGPKIKGFDTHLLEKGKSEELLSALGCQDFPPTS
DCEDLLVEFLEVDDNEDERLMPSHSKEYPGQGVKPTHLDPDSDSGHGSYDSHSLLSEKCE
EPQAYPPTLHIPEITEKPENPEANIPPTVDPQSTNPNFHVDAPKSSTWPLLPGQHMPRSP
YHSVADVCKLAGSPVNTLDSFLDKAEENVLKLSKALETGEEEVAKQKGAKSFPSDKQNTP
WPLLQEKSPTVYVKPPDYVEIHKVNKDGVLSLFPKQRENNQTEKPGVPETSKEYAKVSGI
MDNNILVLVPDSRAQNTALLEESAKKAPPSFEADQSEKDLASFTATSSNRRLQLGRLDYL
DPTCFMHSFH
591

Figure 3 Schematic representation of soluble (human) and membrane (rat) isoforms of the PRLR. Although the mechanism of PRLbp generation remains unclear (alternative splicing or proteolysis or both), an mRNA encoding a soluble PRLbp of 206 amino acids has been isolated in the human breast cancer cell line BT-474. In a given species, all forms have an identical extracellular ligand-binding domain. Subdomain D1 contains two pairs of disulfide-bonded cysteines (C–C) and subdomain D2 the WS motif (green box), two characteristic features of the cytokine receptor superfamily. Box-1 (orange box) is found in the cytoplasmic domain of all membrane isoforms. In rat, the intermediate PRLR (only found in Nb2 cells) differs from the long isoform by a 198 amino acid deletion in the cytoplasmic domain (amino acids 323–520). Otherwise, the short PRLR is identical to both other isoforms up to residue 260, after which its sequence differs (light blue box). Cytoplasmic tyrosine residues are indicated. (Full colour figure may be viewed online.)

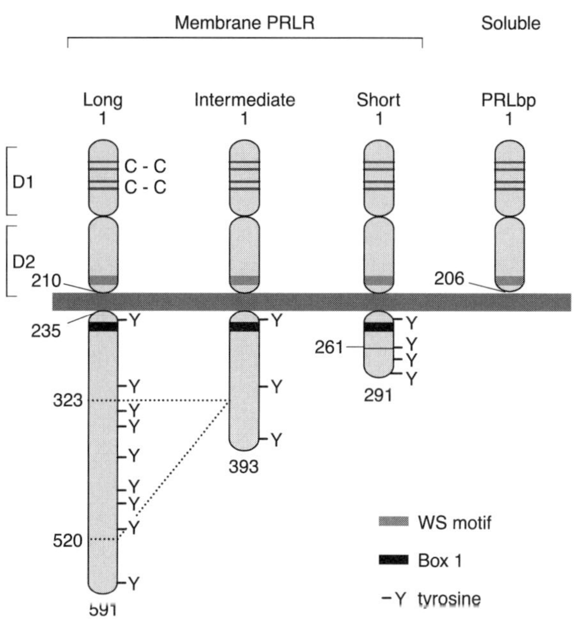

The extracellular domain of the PRLR is a typical cytokine extracellular domain, composed of a sequence of ~210 amino acids referred to as the cytokine receptor homology (CRH) region. In contrast to many other cytokine receptors, there is no additional domain. The PRLR extracellular domain is divided into two subdomains of ~100 amino acids (referred to as D1 and D2), both showing analogies with the fibronectin type III module (Kelly *et al.*, 1991).

Two highly conserved features are found in all extracellular domains of PRLR isoforms: the first is two pairs of disulfide-linked cysteines in the N-terminal subdomain D1 (Cys12–Cys22 and Cys51–Cys62 in human PRLR), and the second, a characteristic feature of cytokine receptors, is the 'WS motif' found in the membrane proximal region of subdomain D2. These features are required to obtain fully active receptors: mutation of cysteines leads to misfolded proteins with impaired ligand-binding properties, whereas mutation of the WS domain alters cell trafficking of the receptor (impaired export to cell surface). In addition to these features, two tryptophans (Trp72 and Trp139 in human PRLR) conserved in the PRLR and in the closely related growth hormone receptor are presumably important for binding prolactin.

The transmembrane domain is 24 amino acids long (residues 211–234 in human PRLR). The possible involvement of this region (or of any crucial amino acid within this domain) in the functional activity of the receptor is unknown.

The cytoplasmic domain is the only region which distinguishes PRLR isoforms. The cytoplasmic domain can be very short (57 amino acids in the short rat PRLR), and attains 357 amino acids in the long PRLR. The intracellular domain is devoid of any intrinsic enzymatic (tyrosine kinase) activity. Two regions, called Box-1 and Box-2, are conserved features (Bole-Feysot *et al.*, 1998). Box-1 is a membrane-proximal region composed of eight amino acids highly enriched in prolines and hydrophobic residues (amino acids 243–250 in rat PRLR). Due to the particular structural properties of proline residues, the conserved P–X–P (X = any amino acid) motif within Box-1 is assumed to adopt a consensus folding specifically recognized by transducing molecules. The second consensus region, Box-2, is much less conserved than Box-1 and consists of a succession of hydrophobic, negatively charged then positively charged residues (amino acids 288–298 in rat). While Box-1 is conserved in all membrane PRLR isoforms, Box-2 is not found in short isoforms.

Finally, two motifs involved in receptor internalization, namely a dileucine and a predicted β turn, have been identified in the rat short PRLR. It is noteworthy that apart from Box-1, no consensus folding domain (SH2, SH3, PTB, WW, etc.) has been identified within the cytoplasmic domain of the PRLR.

The three-dimensional structure of a genetically engineered human PRLR extracellular domain has been determined by crystallographic analysis (Somers *et al.*, 1994) (**Figure 4**). Each D1 and D2 subdomain folds in seven β strands forming a sandwich of two antiparallel β sheets. Both subdomains are linked by a small four-residue polypeptide. This structure is the conformational paradigm of the CRH domain (Bazan, 1990) and is shared by the growth hormone receptor, the EPO receptor, and the IFNγ receptor.

Figure 4 Ribbon representation of the three-dimensional X-ray structure of a monomer of the human PRLR extracellular domain. The extracellular domain folds in a β sandwich formed by two antiparallel β sheets (see text). N- and C-terminal ends are indicated by N and C, respectively. This figure was kindly provided by Dr A. M. de Vos.

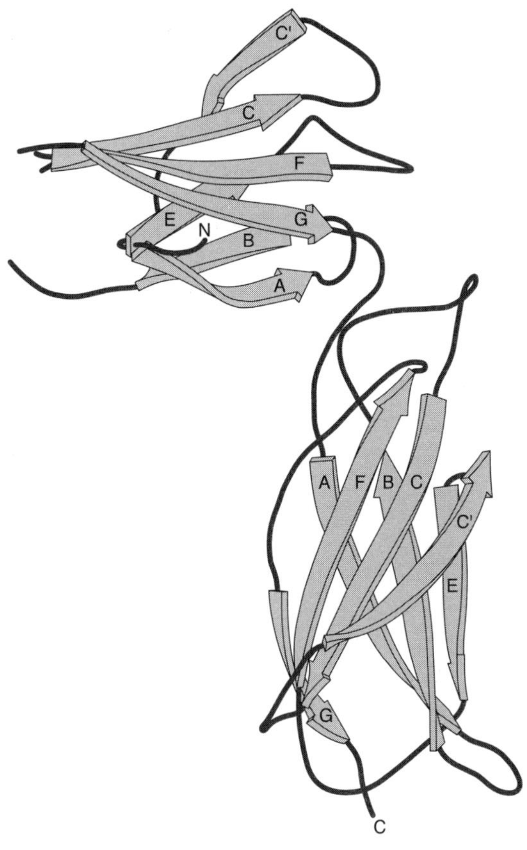

To the best of our knowledge, no structural data have been reported yet for the cytoplasmic domain of any cytokine receptor, including the PRLR.

Relevant homologies and species differences

The variability of PRL receptors has more to do with the existence of various isoforms within a given species than with any interspecies differences. However, two atypical cases can be cited. First, avian PRLRs are particular in that their extracellular domain is duplicated and contains two highly homologous CRH domains; the additional N-terminal module seems, however, to have no functional role (Gao *et al.*, 1996). Second, in cervine and bovine PRLR, the C-terminal tail is truncated by ~35 residues, including the last tyrosine which was shown to be functionally important in other mammalian PRLRs (see Signal transduction section). In addition to these species-specific variations, it has been observed that the *N*-linked glycosylation sites found in the PRLR extracellular domain are not strictly conserved in all species (Buteau *et al.*, 1998).

Affinity for ligand(s)

The PRL receptor binds to at least three types of ligands: prolactin, primate growth hormone, and placental lactogen (PL) which is synthesized by mammal placenta and is thus not found in lower vertebrates. These ligands belong to a hormone family termed the PRL/GH/PL family (Goffin *et al.*, 1996b). Although growth hormone binds to its specific receptor, growth hormone from primates (human, monkeys) is able to bind to the PRLR as well. There is currently no specific receptor identified for PLs, which binds to the PRLR and/or the growth hormone receptor, depending on the species considered (Gertler *et al.*, 1997).

At least two parameters can modulate the reported affinities of the PRLR for its ligands. First, it is usually observed that the soluble binding protein (PRLbp) has a higher affinity (~10 times) than the membrane-bound PRLR for a given ligand (Postel-Vinay *et al.*, 1991), and the length of cytoplasmic tail also influences the overall affinity, although to a lesser extent (Ali *et al.*, 1991). Second, the affinity of the PRLR will vary depending on the type and species of origin of the ligand considered. For example, the affinity of human growth hormone, but not of PRL, for the PRLbp is modulated by 8000-fold depending on the zinc concentration (Cunningham *et al.*, 1990), an effect which is explained by the fact that two amino acids within human growth hormone coordinate one zinc ion together with two residues of the human PRLbp (Somers *et al.*, 1994). In conclusion, depending on these parameters and cross-species variation, the affinity of the PRLR for its ligands is usually in the range of $K_d = 10^{-9}$ to 10^{-10} M.

The binding of these ligands to the PRLR is the first step of receptor activation. Several studies have shown indirectly that the PRLR is activated by dimerization (Goffin *et al.*, 1996b) (**Figure 5**), which involves two regions (so-called binding sites 1 and 2) of the ligands, each interacting with one molecule of PRLR. Even though crystallographic analysis of the PRL/PRLbp complex is lacking, it is anticipated from the three-dimensional structure of the closely-related hGH/hGHBP complex (De Vos *et al.*, 1992) that both binding sites interact with virtually overlapping epitopes within the receptor. Thus far, no accessory

Figure 5 PRLR activation by PRL-induced dimerization. Hormone binding to PRLR is sequential. First, the hormone interacts with the receptor through its binding site 1, forming an inactive $H_1:R_1$ complex. Then, the hormone binds to a second receptor through its site 2, which leads to receptor homodimerization and formation of an active $H_1:2R_2$ complex. Hormone analogs whose binding site 2 is sterically blocked are unable to induce receptor homodimerization and are thus inactive; since they still bind to the receptor through site 1, they behave as antagonists of wild-type hormones.

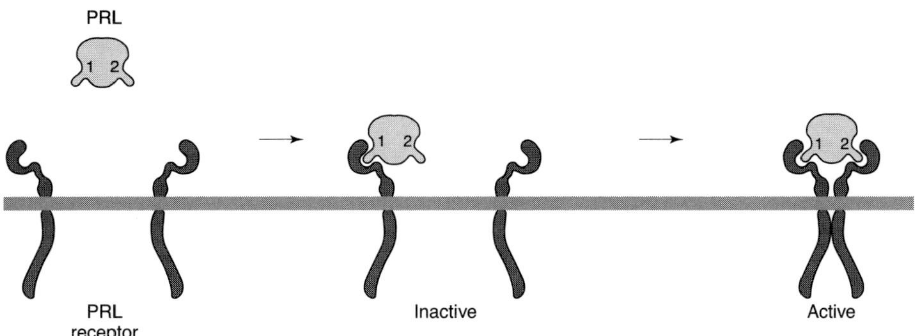

membrane protein has been shown to be required for PRLR signaling.

Cell types and tissues expressing the receptor

PRL receptors have been identified in a wide range of cells and tissues. In addition to the previously known PRL targets, such as mammary gland or reproductive organs, many other organs have been found to express the PRLR. An exhaustive list is provided in **Table 2** (Bole-Feysot *et al.*, 1998).

With respect to *in vitro* cell cultures, the Nb2 rat lymphoma is one of the preferred cell systems used to investigate PRLR-related events. First, this cell line expresses 12,000 PRL receptors per cell, which allows easy study of PRLR signaling (see below). Second, the proliferation of Nb2 cells is induced by very small amounts of lactogens (< 1 ng/mL) which makes this cell line the most widely used bioassay for quantifying the lactogen content of physiologic fluids or of cell-conditioned media, or for estimating the biological properties of mutated proteins of the PRL/GFH/PL family. In addition to the Nb2 cell line, mouse mammary epithelial cells such as HC-11 or human mammary tumor cell lines such as T-47D or MCF-7 also express detectable amounts of PRLR.

Regulation of receptor expression

The level of expression of the PRLR varies from ~ 10 to ~ 2000 fmol/mg of membrane protein. The expression of short and long forms of receptor have been shown to vary as a function of the stage of the estrous cycle, pregnancy, and lactation. However, the hormonal regulation is different depending on the target organ, and due to the extremely wide distribution of the PRLR, it is currently difficult and premature to propose a general overview of the regulation of PRLR expression. Since no recent review on this theme is available, it is probably best to do a MedLine search for the organ of interest.

In the mammary gland, where the long form is predominant, PRLR levels increase upon lactation, and in breast tumor cells its regulation is cross-regulated with that of sex steroid hormone receptors (Ormandy *et al.*, 1997b). In rat liver, the short PRLR isoform is predominant and is increased by estrogens. In the prostate, PRLR levels are increased by testosterone, and decreased by estrogens. In the mouse ovary, PRLR expression also varies during pregnancy (Clarke *et al.*, 1993), and in the human endometrium, the PRLR expression is upregulated during the secretory phase of the menstrual cycle (Jabbour *et al.*, 1998). Prolactin is able to both up- and downregulate its receptor, the latter process probably being due to an acceleration of internalization of hormone/receptor complexes. The effect of prolactin on its receptor is also a function of the hormone concentration and time of exposure of the tissue. Growth hormone is also able to upregulate the PRLR (Kelly *et al.*, 1991).

Release of soluble receptors

Soluble receptors are naturally observed, but whether they result from alternative splicing or membrane

Table 2 Distribution of prolactin-binding sites in vertebrates

Central nervous system	Brain
	Cortex
	Hippocampus
	Choroid plexus
	Striatum
	Cochlear duct
	Corpus callosum
	Hypothalamus
	Astrocytes
	Glial cells
	Retina
	Olfactory system
Ganglia	
Pituitary	Anterior lobe
	Intermediate lobe
Adrenal cortex	
Skin	Epidermis
	Hair follicle
	Sweat glands
Bone tissue	Chondrocytes
	Cartilage
	Osteoblasts
Gills (fishes and larval amphibians)	
Lung	
Heart	Cardiac muscle
	Atria
Skeletal muscle	
Adipocytes (birds)	
Brown adipose tissue	
Liver	Hepatocytes
	Kupffer cells
Submandicular gland	
Submaxillary gland	
Pancreas	Islet of Langerhans
Gastrointestinal tract	Esophagus
	Stomach
	Intestine
	Duodenum
	Jejunum
	Ileum
	Colon

Table 2 (*Continued*)

	Crop sac (birds)
Kidney	Cortex
Bladder (fishes, reptiles, amphibians)	
Lymphoid tissue	Spleen
	Thymus
	Nurse cells
	Epithelial cells
	Lymphocytes
	T
	B
	Macrophages
	Ganglia
	Intestinal cells
Reproductive system (female)	Ovary
	Ova
	Granulosa cells
	Thecal cells
	Corpus luteum (luteal cells)
	Oviduct
	Mammary gland
	Epithelial cells
	Milk
	Tumors
	Uterus (endometrium)
	Placenta
	Amnion
Reproductive system (male)	Testis
	Germ cells
	Spermatozoa
	Leydig cells
	Sertoli cells
	Epididymis
	Seminal vesicle
	Prostate

receptor proteolysis or both is still to be clearly established (Postel-Vinay *et al.*, 1991). To our knowledge, no protease has been identified, and few studies report the existence of alternatively spliced mRNAs encoding a PRLbp (Fuh and Wells, 1995). As for the GHBP, some species-specificity might be observed in the mechanism of binding protein release. Soluble isoforms are identical to the extracellular, ligand-binding domain of the PRLR, with minor differences regarding truncation of the last few C-terminal amino acids. The physiological role of binding proteins remains unclear. It is usually assumed that the formation of PRL/PRLbp complexes enhances the half-life of the hormone in blood circulation. Moreover, since activation of the PRLR occurs by ligand-induced dimerization, such complexes may interfere with signaling properties of membrane-bound PRLR if receptor heterodimers (one membrane-bound, one soluble) were to be formed (Lesueur *et al.*, 1993).

SIGNAL TRANSDUCTION

All of the actions mediated by the PRLR result from its interaction with any of the natural ligands, which leads to receptor dimerization and activation of cascades in the intracellular space. Nb2 cells are a rat lymphoma (pre-T cell line) which expresses 12,000 PRLRs per cell and proliferates in the presence of very low amounts of lactogens (Gout *et al.*, 1980). Although the PRLR expressed in Nb2 cells is a mutated form of the long PRLR (so-called intermediate isoform), lacking 198 amino acids within the cytoplasmic domain (Ali *et al.*, 1991) (Figure 3), this cell line remains one of the preferred models to study prolactin actions with respect to signal-transducing molecules that are activated. It is noteworthy that there is currently no clear picture whether the mutation of the Nb2 PRLR has any incidence on signaling cascades that are triggered by this truncated PRLR compared with the long isoform. In the following section, only a brief overview of PRLR signaling is proposed (for further discussions and references, see Hennighausen *et al.*, 1997; Bole-Feysot *et al.*, 1998; Clevenger *et al.*, 1998; Yu-Lee *et al.*, 1998).

Associated or intrinsic kinases

The PRLR is devoid of any intrinsic enzymatic activity. In 1994, JAK2 was identified as the Janus kinase associated with the PRLR (Rui *et al.*, 1994). Although the involvement of JAK1 has also been proposed in the particular context of mouse lymphoid BAF/3 cells transfected with the PRLR (Dusanter-Fourt *et al.*, 1994), and in addition, the possible interaction between JAK2 and JAK3 in avian PRLR signaling has been suggested (Gao *et al.*, 1996), JAK2 is unambiguously the major PRLR-associated Janus kinase. JAK2 is constitutively associated with the PRLR, i.e. its recruitment is not induced by ligand binding, contrary to what has been observed for the growth hormone receptor. The PRLR–JAK2 interaction requires the Box-1 domain, which does not preclude the involvement of additional residues towards the C-terminus. Since Box-1 is conserved in all PRLR isoforms, signaling differences between the different isoforms are anticipated not to be directed by the ability to associate/activate the kinase. The most C-terminal proline of Box-1 (Pro250 in rat PRLR) is critical for association with and subsequent activation of JAK2. Whether the interaction between JAK2 and the PRLR is direct or involves an intermediate protein (adapter) is currently unknown. To the best of our knowledge, no mutational study aimed at mapping the region of JAK2 interacting with the PRLR has been reported yet, although investigation of the structural features required for the formation of other cytokine receptors/JAK complexes suggest the involvement of the N-terminal region of the kinase in the interaction with cytokine receptors.

Other kinases have also been reported to be associated with/activated by the PRLR in the rat T lymphoma Nb2 cell line, such as the serine/threonine kinase Raf-1 (Clevenger *et al.*, 1994) and the Src tyrosine kinase family member, Fyn (Clevenger and Medaglia, 1994). Association of the PRLR with Src, the prototype member of the Src kinase family, has also been reported after prolactin stimulation in lactating rat hepatocytes (Berlanga *et al.*, 1995). Prolactin induces a rapid tyrosine phosphorylation of the 85 kDa subunit of the phosphatidylinositol (PI) 3'-kinase which, with IRS-1, appears to associate with the PRLR in a prolactin-dependent manner. For these numerous kinases, however, the sites of association on the receptor, their upstream and/or downstream effectors, as well as the biological functions/pathways in which they are involved are poorly understood.

Cytoplasmic signaling cascades

As the paradigm of cytokine receptor signaling, the JAK/STAT signaling pathway is the most widely described cascade for the PRLR (**Figure 6**). Activation of JAK2 by the PRLR occurs very rapidly after hormonal stimulation (within 1 minute), suggesting that this Janus kinase occupies a central and very upstream role in the activation of several signaling pathways of the PRLR. As expected, PRLR mutants unable to associate with JAK2, such as Box-1-deleted mutants, are unable to trigger tyrosine phosphorylation cascades and downstream activation of PRLR-responsive reporter genes.

Figure 6 Schematic representation of the PRLR receptor signaling pathways. Long and short isoforms of rat PRLR are represented. PRLR activates STAT1, STAT3 and, mainly, STAT5. Interaction of STAT5 with the glucocorticoid receptor (GR) has been reported. Whether the short PRLR isoform activates the STAT pathway is currently unknown. PRAP seems to interact preferentially with the short PRLR. The MAP kinase pathway involves the Shc, Grb2, Sos, Ras, Raf cascade and is presumably activated by both PRLR isoforms. Connections between the JAK/STAT and MAP kinase pathways have been suggested. Interactions between receptors and Src kinases (e.g. Fyn), SHP2, IRS-1, PI-3 kinase and other transducing molecules remain unclear.

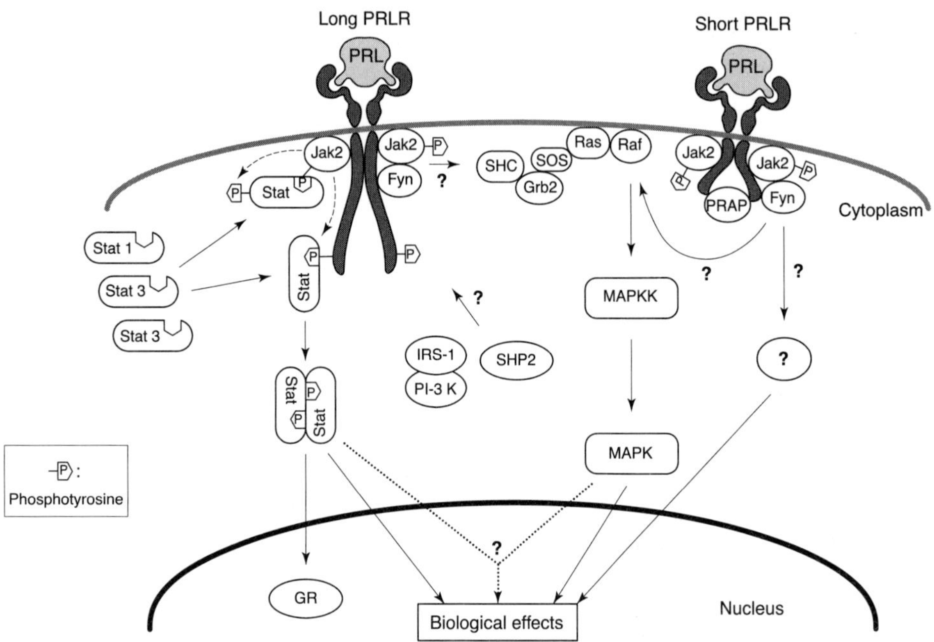

Heterodimerization of the short and the intermediate PRLR cytoplasmic tails produces complexes unable to stimulate JAK2 autophosphorylation (Chang *et al.*, 1998), whereas both can associate with and activate the kinase in the context of their respective wild-type receptor. This observation might be of importance in the physiological context since in species known to express different PRLR isoforms (e.g. rat), all tissues express both isoforms, although in varying ratios. In transfected cells, the short PRLR functions as a dominant negative isoform, inhibiting the activation of milk protein gene transcription by the receptor complex through heterodimerization (Perrot-Applanat *et al.*, 1997).

All rat PRLR isoforms are able to activate JAK2, which in turn phosphorylates the receptor on tyrosines, with the exception of the short isoform which does not undergo tyrosine phosphorylation in spite of the presence of four tyrosines in its cytoplasmic domain. In rat intermediate and long PRLR isoforms, the most C-terminal tyrosine is one major target of JAK2. This is not an absolute rule, however, since activation of the IRF-1 promoter (see below) requires other tyrosines as well (Wang *et al.*, 1997; Yu-Lee *et al.*, 1998), and several studies suggest that tyrosines other than the C-terminal can be phosphorylated in the long PRLR as well (Pezet *et al.*, 1997; Bole-Feysot *et al.*, 1998; Mayr *et al.*, 1998). Accordingly, the natural C-terminal truncation of the cytoplasmic tail (including the last tyrosine) in bovine and cervine PRLR is not detrimental to receptor activity, indicating that alternative intracellular tyrosines can be used (Jabbour *et al.*, 1996). To the best of our knowledge, no correlation between tyrosines that are preferentially phosphorylated by JAK kinases and their surrounding amino acids has been identified. Despite the possible redundancy of tyrosine phosphorylation within the long PRLR isoforms, the most C-terminal tyrosine appears critical for stimulating reporter genes containing milk protein gene promoters and for cell proliferation. Mutation of the C-terminal tyrosine within the intracellular domain of a single intermediate PRLR (Nb2 form) involved in a dimerized complex strongly decreases prolactin-induced proliferation of BA/F3 cells (Chang *et al.*, 1998). Thus, not only is this residue important, but it also needs to be present on both chains of the PRLR dimer.

The PRLR phosphorylated tyrosines serve as docking sites for STAT proteins (signal transducer and activator of transcription). Accordingly, it is noteworthy that the (in)ability of the short PRLR isoform (which is not phosphorylated on tyrosine) to activate STAT proteins remains to be definitely established. Three members of the STAT family have been thus far identified as transducer molecules of the PRLR (long and intermediate): STAT1, STAT3, and, mainly, STAT5. STAT5, initially referred to as mammary gland factor (MGF) was identified from sheep mammary gland, the major target tissue of prolactin (induction of milk protein gene expression). From mutational studies, the phosphorylated C-terminal tyrosine is a good candidate for recruiting STAT5 (Figure 6), even though involvement of other tyrosines cannot be disregarded (Bole-Feysot *et al.*, 1998; Mayr *et al.*, 1998).

Two homologous genes encoding STAT5 (namely A and B) have been identified, and both can be activated by prolactin, although with different kinetics (Kirken *et al.*, 1997). In addition to being tyrosine phosphorylated by JAK2, STAT5A/B are also phosphorylated on serine, but the kinase involved remains to be identified (Kirken *et al.*, 1997). A recent report suggests that two pathways are involved in serine phosphorylation of STAT5, one that is prolactin-activated and MAP kinase inhibitor-sensitive, and one that is constitutively activated and MAP kinase inhibitor-insensitive (Yamashita *et al.*, 1998). Finally, Yu-Lee and coworkers have shown recently that STAT5A and STAT5B exert an inhibitory effect on PRL-inducible IRF-1 promoter activity, and these authors have proposed this inhibition to involve squelching by STAT5 of a factor that STAT1 requires to stimulate the IRF-1 promoter (Luo and Yu-Lee, 1997; Yu-Lee, 1997).

STAT1 and STAT3 are also activated by the PRLR. The region(s) of the PRLR required for activation of these STATs are less well documented. In the Nb2 PRLR, the C-terminal and middle tyrosines (Tyr382 and Tyr309) are proposed to bind STAT 1 (Wang *et al.*, 1997). In the context of the growth hormone receptor, it has been hypothesized that phosphotyrosine(s) of JAK2 could also bind to STAT3, in agreement with the presence of the consensus STAT3-binding sites in the kinase. Although such an interaction does not preclude the possible occurrence of interactions with the receptor, this hypothesis remains to be tested in the context of STAT activation by the PRLR.

The recent discovery of a family of proteins downregulating the activation of the JAK/STAT pathway has greatly helped our understanding of how these activated (phosphorylated) proteins come back to their steady state after prolactin (and cytokine in general) stimulation. These proteins are named SOCS (suppressor of cytokine signaling) or CIS (cytokine-inducible SH2), and JAB (JAK2-binding proteins) and contain at least five members (CIS-1, -2, -3, -4, and JAB). A recent study demonstrates that PRLR-activation of a STAT5-responsive reporter gene in transfected 293 cells is abolished by transfected CIS-3 and JAB (Helman *et al.*, 1998). Interestingly, SOCS genes are target genes of the JAK/STAT pathway, which provides a direct negative regulation of this cascade (see below).

The JAK/STAT is undoubtedly the major cascade triggered by the PRLR, as emphasized by the poor understanding of PRLR signaling before the Janus kinase family was discovered in the early 1990s. However, many other signaling proteins were found to be activated by the PRLR. The well-known MAP kinase pathway involves the Shc/SOS/Grb2/Ras/Raf/MAP kinase cascade and this pathway has been demonstrated to be activated by the PRLR, including the short isoform, in various cell systems (Piccoletti *et al.*, 1994; Das and Vonderhaar, 1996; Erwin *et al.*, 1996). Whether activation of the MAP cascade requires JAK2, Fyn, Src or any other pathway is currently unknown. Although the JAK/STAT and the MAP kinase cascades were initially regarded as independent pathways, recent data rather suggest that these pathways are interconnected (Chida *et al.*, 1998). Finally, other proteins that are activated by the PRLR can be listed, despite the fact that the cascade they are involved in remain poorly deciphered: the phosphatase SHP-2, the adapters IRS-1, IRS-2, and IRS-3, the PRLR-associated protein PRAP recently identified as a 17β-hydroxysteroid dehydrogenase/17-ketosteroid reductase (Nokelainen *et al.*, 1998), the adapter Cbl, vav, and proteins linked to apoptotic pathway including Bax, Bcl-2 and Bag-1 (see referencecs in Bole-Feysot *et al.*, 1998). Currently, however, these molecules are not considered as major signaling events triggered by the PRLR. Crosstalk with the tyrosine kinase (noncytokine) receptor for EGF has also been reported (Lange *et al.*, 1998; Quijano and Sheffield, 1998).

DOWNSTREAM GENE ACTIVATION

Transcription factors activated

When activated, STAT factors translocate to the nucleus, where they transactivate target gene promoters by binding to consensus DNA sequences. The interaction of STATs with other transcription factors has been reported. For example, the glucocorticoid receptor interacts with STAT5 and positively modulates gene transactivation (Stocklin *et al.*, 1996, 1997). Interaction between STATs activated by the PRLR and other nuclear receptors, such as the progesterone receptor, has been demonstrated (Richer *et al.*, 1998) or, as for the estrogen receptor, is anticipated from functional interference observed *in vitro* (e.g. transfected cells) but awaits experimental confirmation. Interaction between STAT1 and CBP300 (CREB-binding protein), which itself does not bind DNA but interferes with transcription factors through protein–protein interactions, has also been shown to enhance IRF-1 promoter activity. Co-operative interaction between NF1 and STAT5 on WAP gene expression has also been described (Li and Rosen, 1995; see Chida *et al.*, 1998 and references therein).

Genes induced

Due to the extremely wide spectrum of prolactin activities, an exhaustive listing of genes activated is beyond the scope of this chapter (Horseman and Yu-Lee, 1994; Bole-Feysot *et al.*, 1998). In the mammary gland, milk protein genes are obvious targets (caseins, lactoglobulin, whey acidic protein, etc.), and full activation of these genes requires other hormones (insulin, glucocorticoid) and extracellular matrix components. In liver, prolactin stimulates the transcription of early growth-related genes such as c-*fos*, c-*jun*, c-*myc*, c-*src*, IGF-1, and mid- and late G1 genes such as ornithine decarboxylase (ODC), heat shock proteins (hsp), or gfi-1. In Nb2 cells, prolactin activates expression of most of these genes, plus those of the transcription factor IRF-1, Bax, Bcl-2, and many cyclins. In a more general fashion, members of the CIS protein family, the downregulators of the JAK/STAT pathways, are target genes of PRLR signaling; however, the cell-specificity of this induction remains to be investigated (Helman *et al.*, 1998).

Promoter regions involved

Consensus DNA motifs specifically recognized by STAT complexes have been identified in the promoters of target genes. The motif termed GAS (for gamma interferon-activated sequence) was defined using STAT homodimers and consists of a palindromic sequence TTTCxxxGAAA. This motif is found in the promoters of all STAT target genes. The specificity of the interaction between a particular STAT and a GAS motif found in a given target promoter has been proposed to depend, at least in part, on the center core nucleotide(s) (Ihle, 1996). Whether the synergism of STATs with other transcription factors involves DNA binding of the latter is not a general rule, and remains controversial for the glucocorticoid receptor (Lechner *et al.*, 1997; Stocklin *et al.*, 1997).

The activation of identical STAT proteins by different cytokine receptors questions the mechanisms by which specificity of signaling pathways is achieved in response to a particular hormonal stimulation. Although several cytokines (e.g. EPO, GM-CSF, GH, PRL, IL-2, IL-3, and IL-5) activate the DNA-binding ability of STAT5 and/or transactivate the β-casein luciferase reporter gene *in vitro*, β-casein is only found in mammary epithelial cells and in cytotoxic T cells. Moreover, in mammary gland, activation of this gene promoter requires other lactogenic hormones such as insulin and glucocorticoid. This suggests that different STAT combinations and/or involvement of other signal transducers/transcription factors direct the specificity of the final response. This question has been recently investigated by Miyajima and co-workers who proposed that the activation of STAT5 involves complex interactions with other signaling pathways and requires integration of opposing signals from Ras (Chida *et al.*, 1998).

BIOLOGICAL CONSEQUENCES OF ACTIVATING OR INHIBITING RECEPTOR AND PATHOPHYSIOLOGY

Unique biological effects of activating the receptors

As emphasized by the phenotypes of prolactin knockout and PRLR knockout mice, milk production and reproductive properties are the functions that can obviously not be taken over by other hormones or cytokines, despite the likelihood that cytokines display some functional redundancy. Besides these actions, the extremely wide spectrum of prolactin activities must probably be regarded as a panel of functions that are modulated by, rather than unique to prolactin or its receptor.

Phenotypes of receptor knockouts and receptor overexpression mice

Knockout mouse models for the PRL gene (Horseman *et al.*, 1997) and for the PRLR gene (Ormandy *et al.*, 1997a) have been developed. Both exhibit the same general pattern of reproductive deficiencies. Several additional phenotypes have been reported for the receptor knockout. This may be related to the fact that knockout of the PRLR abolishes all activities naturally depending on this receptor, while knockout of the ligand prolactin does not prevent other potential ligands from binding and activating the receptor. Neither of these knockouts is lethal, indicating that the PRLR is not an absolute requirement for fetal development.

The main phenotypes of PRLR knockout mice are linked to the sterility of double negative females, due to a failure of embryo implantation. Along with their impaired ability to lactate, these two phenotypes correlate well with the historically known functions of prolactin, referred to as the 'lactogenic and luteotropic hormone'. Interestingly, heterozygous females are not sterile, but also show impaired lactation, which can be restored to normal level by successive pregnancies. Other phenotypes have recently been evaluated. Maternal behavior of double negative females towards foster pups is altered (Bridges *et al.*, 1985; Lucas *et al.*, 1998). Bone formation is also impaired (Clement-Lacroix *et al.*, 1999), which was unexpected since the potential role, even indirect, of prolactin on bone cells had been ignored earlier. Despite studies on hypophysectomized animals indicating an immunomodulatory role of prolactin (Nagy and Berczi, 1991), no immunological phenotype was observed in either PRLR or PRL knockout mice. This suggests that PRL- or PRLR-deficient mice can probably compensate for the lack of receptor functions via redundancy of (an)other cytokine(s).

Human abnormalities

To date, no disease has been linked to any genetic abnormality of the PRLR. This would suggest either that mutations have no detectable effect *in vivo*, or that such mutations might be lethal and, thereby, never detected. PRLR knockout mice are viable, so it is clear that the PRLR is not essential for survival, at least in this species. On the other hand, since there are important reproductive effects in females, this could explain the lack of genetic transmission.

Hyperprolactinemia caused by hypersecretion of prolactin by pituitary adenomas usually leads to various endocrine effects, with consequences on reproductive functions, especially in women. These patients are treated with dopamine agonists, the natural inhibitor of prolactin secretion, or alternatively they often undergo surgical removal of the adenoma.

Prolactin is believed to be one of the factors favoring the proliferation of some tumors, such as *breast cancer* or *prostate cancer*. Mammary tumors express more PRLR than normal tissue (Touraine *et al.*, 1998), and the local secretion of mammary prolactin has been suggested to stimulate cell proliferation in an autocrine/paracrine manner (Ginsburg and Vonderhaar, 1995). If this is true, prolactin could be carcinogenic in the mammary gland, which is clearly established in animal models of prolactin hypersecretion (Wennbo *et al.*, 1997) but remains to be demonstrated in humans.

Finally, prolactin has been shown to be increased and to affect a number of autoimmune states (Bole-Feysot *et al.*, 1998), such as *systemic lupus erythematosus*, acute *experimental allergic encephalomyelitis*, *rheumatoid arthritis*, *adjuvant arthritis*, and *graft-versus-host disease*, for example as a marker of rejection in heart transplantation. Prolactin has also been suggested to be involved in the etiology of *cystic fibrosis*, although the precise mechanism remains unclear.

THERAPEUTIC UTILITY

Effect of treatment with soluble receptor domain

To the best of our knowledge, no disease has been tentatively treated with soluble prolactin-binding protein. The binding protein could, however, bind human growth hormone, and thus have an effect on the half-life and potentially the biological activity of growth hormone as well.

Effects of inhibitors (antibodies) to receptors

In vivo, we are not aware of any clinical trials involving anti-PRLR or anti-PRL antibodies. Disorders linked to prolactin synthesis are treated by lowering its secretion by dopamine. *In vitro* studies have shown that anti-human PRL antibodies can prevent proliferation of *breast cancer* cell lines induced by locally produced prolactin (Ginsburg and Vonderhaar, 1995), which might identify non-pituitary prolactin as a new target in breast cancer therapy. Since regulation of extrapituitary PRL is probably not controlled by dopamine, alternative strategies to dopamine agonists must be investigated to prevent the putative autocrine/paracrine effect of prolactin in mammary tumors. Prolactin antagonists have been generated by engineering prolactin mutants with highly impaired ability to induce PRLR dimerization (i.e. activation) (Goffin *et al.*, 1996a, 1996b; Bole-Feysot *et al.*, 1998). Such prolactin variants might be of use in future to counteract the proliferative effect of prolactin in breast tumors.

References

Ali, S., Pellegrini, I., Edery, M., Lesueur, L., Paly, J., Djiane, J., and Kelly, P. A. (1991). A prolactin-dependent immune cell line (Nb2) expresses a mutant form of prolactin receptor. *J. Biol. Chem.* **266**, 20110–20117.

Arden, K. C., Boutin, J. M., Djiane, J., Kelly, P. A., and Cavenee, W. K. (1990). The receptors for prolactin and growth hormone are localized in the same region of human chromosome 5. *Cytogenet. Cell Genet.* **53**, 161–165.

Astwood, E. B. (1941). The regulation of corpus luteum function by hypophysial luteotrophin. *Endocrinology* **29**, 309–319.

Bazan, J. F. (1990). Structural design of molecular evolution of a cytokine receptor superfamily. *Proc. Natl Acad. Sci. USA* **87**, 6934–6938.

Ben-Jonathan, N., Mershon, J. L., Allen, D. L., Steinmetz, R. W. (1996). Extrapituitary prolactin: distribution, regulation, functions, and clinical aspects. *Endocr. Rev.* **17**, 639–669.

Berlanga, J. J., Fresno, V., Martin-Perez, J., and Garcia-Ruiz, J. P. (1995). Prolactin receptor is associated with c-*src* kinase in rat liver. *Mol. Endocrinol.* **9**, 1461–1467.

Bole-Feysot, C., Goffin, V., Edery, M., Binart, N., and Kelly, P. A. (1998). Prolactin and its receptor: actions, signal transduction pathways and phenotypes observed in prolactin receptor knockout mice. *Endocr. Rev.* **19**, 225–268.

Boutin, J. M., Jolicoeur, C., Okamura, H., Gagnon, J., Edery, M., Shirota, M., Banville, D., Dusanter-Fourt, I., Djiane, J., and Kelly, P. A. (1988). Cloning and expression of the rat prolactin receptor, a member of the growth hormone/prolactin receptor gene family. *Cell* **53**, 69–77.

Boutin, J. M., Edery, M., Shirota, M., Jolicoeur, C., Lesueur, L., Ali, S., Gould, D., Djiane, J., and Kelly, P. A. (1989). Identification of a cDNA encoding a long form of prolactin receptor in human hepatoma and breast cancer cells. *Mol. Endocrinol.* **3**, 1455–1461.

Bridges, R. S., Dibiase, R., Loundes, D. D., and Doherty, P. C. (1985). Prolactin stimulation of maternal behavior in female rats. *Science* **227**, 782–784.

Buck, K., Vanek, M., Groner, B., and Ball, R. K. (1992). Multiple forms of prolactin receptor messenger ribonucleic acid are specifically expressed and regulated in murine tissues and the mammary cell line HC11. *Endocrinology* **130**, 1108–1114.

Buteau, H., Pezet, A., Ferrag, F., Perrot-Applanat, M., Kelly, P. A., and Edery, M. (1998). *N*-Glycosylation of the prolactin receptor is not required for activation of gene transcription but is crucial for its cell surface targeting. *Mol. Endocrinol.* **12**, 544–555.

Chang, W. P., Ye, Y., and Clevenger, C. V. (1998). Stoichiometric structure–function analysis of the prolactin receptor signaling domain by receptor chimeras. *Mol. Cell Biol.* **18**, 896–905.

Chida, D., Wakao, H., Yoshimura, A., and Miyajima, A. (1998). Transcriptional regulation of the beta-casein gene by cytokines: cross-talk between STAT5 and other signaling molecules. *Mol. Endocrinol.* **12**, 1792–1806.

Clarke, D. L., Arey, B. J., and Linzer, D. H. (1993). Prolactin receptor messenger ribonucleic acid expression in the ovary during rat estrous cycle. *Endocrinology* **133**, 2594–2603.

Clement-Lacroix, P., Ormandy, C., Lepescheux, L., Ammann, P., Damotte, D., Goffin, V., Bouchard, B., Amling, M., Gaillard-Kelly, M., Binart, N., Baron, R., and Kelly, P. A. (1999). Osteoblasts are a new target for prolactin: analysis of bone formation in prolactin receptor knockout mice. *Endocrinology* **140**, 96–105.

Clevenger, C. V., and Medaglia, M. V. (1994). The protein tyrosine kinase p59fyn is associated with prolactin (PRL) receptor and is activated by PRL stimulation of T-lymphocytes. *Mol. Endocrinol.* **8**, 674–681.

Clevenger, C. V., Torigoe, T., and Reed, J. C. (1994). Prolactin induces rapid phosphorylation and activation of prolactin receptor-associated RAF-1 kinase in a T-cell line. *J. Biol. Chem.* **269**, 5559–5565.

Clevenger, C. V., Chang, W. P., Ngo, W., Pasha, T. M., Montone, K. T., and Tomaszewski, J. E. (1995). Expression of prolactin and prolactin receptor in human breast carcinoma. *Am. J. Pathol.* **146**, 695–705.

Clevenger, C. V., Freier, D. O., and Kline, J. B. (1998). Prolactin receptor signal transduction in cells of the immune system. *J. Endocrinol.* **157**, 187–197.

Cunningham, B. C., Bass, S., Fuh, G., and Wells, J. A. (1990). Zinc mediation of the binding of human growth hormone to the human prolactin receptor. *Science* **250**, 1709–1712.

Das, R., and Vonderhaar, B. K. (1996). Involvement of SHC, GRB2, SOS and RAS in prolactin signal transduction in mammary epithelial cells. *Oncogene* **13**, 1139–1145.

De Vos, A. M., Ultsch, M., and Kossiakoff, A. A. (1992). Human growth hormone and extracellular domain of its receptor: crystal structure of the complex. *Science* **255**, 306–312.

Dusanter-Fourt, I., Muller, O., Ziemiecki, A., Mayeux, P., Drucker, B., Djiane, J., Wilks, A., Harper, A. G., Fischer, S., and Gisselbrecht, S. (1994). Identification of Jak protein tyrosine kinases as signaling molecules for prolactin, Functional analysis of prolactin receptor and prolactin-erythropoietin receptor chimera expressed in lymphoid cells. *EMBO J.* **13**, 2583–2591.

Erwin, R. A., Kirken, R. A., Malabarba, M. G., Farrar, W. L., and Rui, H. (1996). Prolactin activates ras via signaling proteins Shc, growth factor receptor bound 2 and Son of Sevenless. *Endocrinology* **136**, 3512–3518.

Fuh, G., and Wells, J. A. (1995). Prolactin receptor antagonists that inhibit the growth of breast cancer cell lines. *J. Biol. Chem.* **270**, 13133–13137.

Gao, J., Hughes, J. P., Auperin, B., Buteau, H., Edery, M., Zhuang, H., Wojchowski, D. M., and Horseman, N. D. (1996). Interactions among janus kinases and the prolactin (PRL) receptor in the regulation of a PRL response element. *Mol. Endocrinol.* **10**, 847–856.

Gearing, D. P., Druck, T., Huebner, K., Overhauser, J., Gilbert, D. J., Copeland, N. G., and Jenkins, N. A. (1993). The leukemia inhibitory factor receptor (LIFR) gene is located within a cluster of cytokine receptor loci on mouse chromosome 15 and human chromosome 5p12-p13. *Genomics* **18**, 148–150.

Gertler, A., Grosclaude, J., and Djiane, J. (1997). Interaction of lactogenic hormones with prolactin receptors. *Ann. N.Y. Acad. Sci.* **839**, 177–181.

Ginsburg, E., and Vonderhaar, B. K. (1995). Prolactin synthesis and secretion by human breast cancer cells. *Cancer Res.* **55**, 2591–2595.

Goffin, V., Kinet, S., Ferrag, F., Binart, N., Martial, J. A., and Kelly, P. A. (1996a). Antagonistic properties of human prolactin analogs that show paradoxical agonistic activity in the Nb2 bioassay. *J. Biol. Chem.* **271**, 16573–16579.

Goffin, V., Shiverick, K. T., Kelly, P. A., and Martial, J. A. (1996b). Sequence–function relationships within the expanding family of prolactin, growth hormone, placental lactogen and related proteins in mammals. *Endocr. Rev.* **17**, 385–410.

Gout, P. W., Beer, C. T., and Noble, R. L. (1980). Prolactin-stimulated growth of cell cultures established from malignant Nb rat lymphomas. *Cancer Res.* **40**, 2433–2436.

Helman, D., Sandowski, Y., Cohen, Y., Matsumoto, A., Yoshimura, A., Merchav, S., and Gertler, A. (1998). Cytokine-inducible SH2 protein (CIS3) and JAK2 binding protein (JAB) abolish prolactin receptor-mediated STAT5 signaling. *FEBS Lett.* **441**, 287–291.

Hennighausen, L., Robinson, G. W., Wagner, K. U., and Liu, W. (1997). Prolactin signaling in mammary gland development. *J. Biol. Chem.* **272**, 7567–7569.

Horseman, N. D., and Yu-Lee, L. Y. (1994). Transcriptional regulation by the helix bundle peptide hormones: growth hormone, prolactin, and hematopoietic cytokines. *Endocr. Rev.* **15**, 627–649.

Horseman, N. D., Zhao, W., Montecino-Rodriguez, E., Tanaka, M., Nakashima, K., Engle, S. J., Smith, F., Markoff, E., and Dorshkind, K. (1997). Defective mammopoiesis, but normal hematopoiesis, in mice with a targeted disruption of the prolactin gene. *EMBO J.* **16**, 6926–6935.

Hu, Z. Z., Zhuang, L., Meng, J., and Dufau, M. L. (1998). Transcriptional regulation of the generic promoter III of the rat prolactin receptor gene by C/EBPbeta and Sp1. *J. Biol. Chem.* **273**, 26225–26235.

Ihle, J. N. (1996). STATs: signal transducers and activators of transcription. *Cell* **84**, 331–334.

Jabbour, H. N., Clarke, L. A., Boddy, S., Pezet, A., Edery, M., and Kelly, P. A. (1996). Cloning, sequencing and functional analysis of a truncated cDNA encoding red deer prolactin receptor: an alternative tyrosine residue mediates β-casein promoter activation. *Mol. Cell. Endocrinol.* **123**, 17–26.

Jabbour, H. N., Critchley, H. O., and Boddy, S. C. (1998). Expression of functional prolactin receptors in nonpregnant human endometrium: Janus kinase-2, signal transducer and activator of transcription-1 (STAT1), and STAT5 proteins are phosphorylated after stimulation with prolactin. *J. Clin. Endocrinol. Metab.* **83**, 2545–2553.

Kelly, P. A., Djiane, J., Postel-Vinay, M. C., and Edery, M. (1991). The prolactin/growth hormone receptor family. *Endocr. Rev.* **12**, 235–251.

Kirken, R. A., Malabarba, M. G., Xu, J., Liu, X., Farrar, W. L., Hennighausen, L., Larner, A. C., Grimley, P. M., and Rui, H. (1997). Prolactin stimulates serine/tyrosine phosphorylation and formation of heterocomplexes of multiple STAT5 isoforms in Nb2 lymphocytes. *J. Biol. Chem.* **272**, 14098–14103.

Lange, C. A., Richer, J. K., Shen, T., and Horwitz, K. B. (1998). Convergence of progesterone and epidermal growth factor signaling in breast cancer. Potentiation of mitogen-activated protein kinase pathways. *J. Biol. Chem.* **273**, 31308–31316.

Lechner, J., Welte, T., Tomasi, J. K., Bruno, P., Cairns, C., Gustafsson, J., and Doppler, W. (1997). Promoter-dependent synergy between glucocorticoid receptor and STAT5 in the activation of beta-casein gene transcription. *J. Biol. Chem.* **272**, 20954–20960.

Lesueur, L., Edery, M., Paly, J., Kelly, P. A., and Djiane, J. (1993). Roles of the extracellular and cytoplasmic domains of the prolactin receptor in signal transduction to milk protein genes. *Mol. Endocrinol.* **7**, 1178–1184.

Li, S., and Rosen, J. M. (1995). Nuclear factor I and mammary gland factor (STAT5) play a critical role in regulating rat whey acidic protein gene expression in transgenic mice. *Mol. Cell. Biol.* **15**, 2063–2070.

Lucas, B. K., Ormandy, C., Binart, N., Bridges, R. S., and Kelly, P. A. (1998). Null mutation of prolactin receptor gene produces a defect in maternal behavior. *Endocrinology* **139**, 4102–4107.

Luo, G., and Yu-Lee, L. Y. (1997). Transcriptional inhibition by STAT5: differential activities at growth-related vs differentiation-specific promoters. *J. Biol. Chem.* **272**, 26841–26849.

Mayr, S., Welte, T., Windegger, M., Lechner, J., May, P., Heinrich, P. C., Horn, F., and Doppler, W. (1998). Selective coupling of STAT factors to the mouse prolactin receptor. *Eur. J. Biochem.* **258**, 784–793.

Moldrup, A., Ormandy, C., Nagano, M., Murthy, K., Banville, D., Tronche, F., and Kelly, P. A. (1996). Differential promoter usage in prolactin receptor gene expression: hepatocyte nuclear factor 4 binds to and activates the promoter preferentially active in the liver. *Mol. Endocrinol.* **10**, 661–671.

Nagy, E., and Berczi, I. (1991). Hypophysectomized rats depend on residual prolactin for survival. *Endocrinology* **128**, 2776–2784.

Nicoll, C. S. (1974). In "Handbook of Physiology: Endocrinology IV" (ed. E. Knobil and W. Sawyer), Physiological actions of prolactin, pp. 253–292. Williams and Wilkins, Baltimore.

Nicoll, C. S., and Bern, H. (1972). In "Lactogenic Hormone" (ed. G. Wolstenholme and J. Knight), On the actions of PRL among the vertebrates: is there a common denominator?, pp. 299–327. Churchill Livingstone, London.

Nokelainen, P., Peltoketo, H., Vihko, R., and Vihko, P. (1998). Expression cloning of a novel estrogenic mouse 17 beta-hydroxysteroid dehydrogenase/17-ketosteroid reductase (m17HSD7), previously described as a prolactin receptor-associated protein (PRAP) in rat. *Mol. Endocrinol.* **12**, 1048–1059.

Ormandy, C. J., Camus, A., Barra, J., Damotte, D., Lucas, B. K., Buteau, H., Edery, M., Brousse, N., Babinet, C., Binart, N., and Kelly, P. A. (1997a). Null mutation of the prolactin receptor gene produces multiple reproductive defects in the mouse. *Genes Dev.* **11**, 167–178.

Ormandy, C. J., Hall, R. E., Manning, D. L., Robertson, J. F., Blamey, R. W., Kelly, P. A., Nicholson, R. I., and Sutherland, R. L. (1997b). Coexpression and cross-regulation of the prolactin receptor and sex steroid hormone receptors in breast cancer. *J. Clin. Endocrinol. Metab.* **82**, 3692–3699.

Ormandy, C. J., Binart, N., Helloco, C., and Kelly, P. A. (1998). Mouse prolactin receptor gene: genomic organization reveals alternative promoter usage and generation of isoforms via alternative 3′-exon splicing. *DNA Cell Biol.* **17**, 761–770.

Perrot-Applanat, M., Gualillo, O., Pezet, A., Vincent, V., Edery, M., and Kelly, P. A. (1997). Dominant negative and cooperative effects of mutant forms of prolactin receptor. *Mol. Endocrinol.* **11**, 1020–1032.

Pezet, A., Ferrag, F., Kelly, P. A., and Edery, M. (1997). Tyrosine docking sites of the rat prolactin receptor required for association and activation of STAT5. *J. Biol. Chem.* **272**, 25043–25050.

Piccoletti, R., Maroni, P., Bendinelli, P., and Bernelli-Zazzera, A. (1994). Rapid stimulation of mitogen-activated protein kinase of rat liver by prolactin. *Biochem. J.* **303**, 429–433.

Posner, B. I., Kelly, P. A., Shiu, R. P., and Friesen, H. G. (1974). Studies of insulin, growth hormone and prolactin binding: tissue distribution, species variation and characterization. *Endocrinology* **95**, 521–531.

Postel-Vinay, M. C., Belair, L., Kayser, C., Kelly, P. A., and Djiane, J. (1991). Identification of prolactin and growth hormone binding proteins in milk. *Proc. Natl Acad. Sci. USA* **88**, 6687–6690.

Quijano, V. J. J., and Sheffield, L. G. (1998). Prolactin decreases epidermal growth factor receptor kinase activity via a phosphorylation-dependent mechanism. *J. Biol. Chem.* **273**, 1200–1207.

Richer, J. K., Lange, C. A., Manning, N. G., Owen, G., Powell, R., and Horwitz, K. B. (1998). Convergence of progesterone with growth factor and cytokine signaling in breast cancer.Progesterone receptors regulate signal transducers and activators of transcription expression and activity. *J. Biol. Chem.* **273**, 31317–31326.

Riddle, O., Bates, R. W., and Dykshorn, S. W. (1933). The preparation, identification and assay of prolactin – a hormone of the anterior pituitary. *Am. J. Physiol.* **105**, 191–216.

Rui, H., Kirken, R. A., and Farrar, W. L. (1994). Activation of receptor-associated tyrosine kinase JAK2 by prolactin. *J. Biol. Chem.* **269**, 5364–5368.

Shirota, M., Banville, D., Ali, S., Jolicoeur, C., Boutin, J. M., Edery, M., Djiane, J., and Kelly, P. A. (1990). Expression of two forms of prolactin receptor in rat ovary and liver. *Mol. Endocrinol.* **4**, 1136–1143.

Somers, W., Ultsch, M., De Vos, A. M., and Kossiakoff, A. A. (1994). The X-ray structure of the growth hormone-prolactin receptor complex. *Nature* **372**, 478–481.

Stocklin, E., Wissler, M., Gouilleux, F., and Groner, B. (1996). Functional interactions between STAT5 and the glucocorticoid receptor. *Nature* **383**, 726–728.

Stocklin, E., Wissler, M., Moriggl, R., and Groner, B. (1997). Specific DNA binding of STAT5, but not glucocorticoid receptor is required for their functional cooperation in the regulation of gene transcription. *Mol. Cell. Biol.* **11**, 6708–6716.

Stricker, P., and Grueter, R. (1928). Action du lobe antérieur de l'hypophyse sur la montée laiteuse. *C.R. Soc. Biol.* **99**, 1978–1980.

Touraine, P., Martini, J. F., Zafrani, B., Durand, J. C., Labaille, F., Malet, C., Nicolas, A., Trivin, C., Postel-Vinay, M. C., Kuttenn, F., and Kelly, P. A. (1998). Increased expression of prolactin receptor gene assessed by quantitative polymerase chain reaction in human breast tumors versus normal breast tissues. *J. Clin. Endocrinol. Metab.* **83**, 667–674.

Wang, Y. F., O'Neal, K. D., and Yu-Lee, L. Y. (1997). Multiple prolactin receptor cytoplasmic residues and STAT1 mediate prolactin signaling to the IRF-1 promoter. *Mol. Endocrinol.* **11**, 1353–1364.

Wennbo, H., Gebre-Medhin, M., Gritli-Linde, A., Ohlsson, C., Isaksson, O. G., and Tornell, J. (1997). Activation of the prolactin receptor but not the growth hormone receptor is important for induction of mammary tumors in transgenic mice. *J. Clin. Invest.* **100**, 2744–2751.

Yamashita, H., Xu, J., Erwin, R. A., Farrar, W. L., Kirken, R. A., and Rui, H. (1998). Differential control of the phosphorylation state of proline-juxtaposed serine residues Ser725 of STAT5a and Ser730 of STAT5b in prolactin-sensitive cells. *J. Biol. Chem.* **273**, 30218–30224.

Yu-Lee, L. Y. (1997). Molecular actions of actions of prolactin in the immune system. *Proc. Soc. Exp. Biol. Med.* **215**, 35–51.

Yu-Lee, L. Y., Luo, G., Book, M. L., and Morris, S. M. (1998). Lactogenic hormone signal transduction. *Biol. Reprod.* **58**, 295–301.

LICENSED PRODUCTS

1. Kelly, P.A., and Djiane, J. (1991) cDNA encoding human prolactin receptor. US patent number 4,992,378
2. Kelly, P.A., Edery, M., Prunet, M., and Sandra, O. (1994). cDNA encoding fish prolactin receptor. European patent number 94,10535

ACKNOWLEDGEMENTS

The authors are grateful to Dr A.M. de Vos for providing the figure of the three-dimensional structure of the prolactin receptor extracellular domain (Figure 4). They also thank Dr N. Binart for helpful discussions and Dr A. Pezet for help with figures.

IL-1 Family Receptors

IL-1 Receptor Family

Luke A. J. O'Neill[1,*] and Steve K. Dower[2]

[1]Department of Biochemistry & Biotechnology, Trinity College Dublin, Dublin, Ireland

[2]Division of Molecular & Genetic Medicine, University of Sheffield, Royal Hallamshire Hospital, Sheffield, S10 2JF, UK

*corresponding author tel: 353 1 6082439, fax: 353 1 6772400, e-mail: laoneill@tcd.ie

DOI: 10.1006/rwcy.2000.02005.

SUMMARY

Interleukin-1 (IL-1) is a potent proinflammatory cytokine which induces the expression of immune and inflammatory genes in target cells. Its effects are mediated via the type I IL-1 receptor (IL-1RI), in a complex with the IL-1 receptor accessory protein (IL-1R AcP). Both of these proteins are highly homologous but only IL-1R1 is capable of binding ligand. In addition, there is a type II IL-1 receptor (IL-1RII), which does not signal, but which instead is shed from cells, acting as a decoy.

Along with these receptors, there is a growing superfamily of proteins which share homologies with IL-1RI. These include receptors that, similar to IL-1RI, have immunoglobulin domains extracellularly and a conserved cytosolic sequence. The receptors for IL-18 (formerly termed IL-1RrP) and its accessory protein AcPL, and T1/ST2, whose ligand is still unknown, have these features. In addition there are receptors that, although they have the conserved cytosolic domain, are different extracellularly, having leucine-rich repeats. The founder member here is the *Drosophila* protein Toll, and the conserved cytosolic domain has as a result been named the Toll–IL-1R (TIR) domain. Other members with leucine-rich repeats extracellularly include six Toll-like receptors (TLRs). LPS from gram-negative bacteria appears to be able to signal using TLR-4, while products from gram-positive bacteria use TLR-2. In addition, a number of plant proteins have been described which are similar to Toll, including N protein from tobacco, L6 protein from flax, and Rpp5 from *Arabidopsis*, which participate in disease resistence in plants. The final member of the superfamily is MyD88, which is cytosolic and contains a TIR domain and a death domain. MyD88 is a critical adapter for IL-1R, IL-18R, and TLR-4, and may therefore be the adapter for the entire superfamily. This expanding receptor superfamily therefore represents an ancient signaling system which participates in innate immunity and inflammation.

INTRODUCTION

IL-1 is one of the most intensively studied cytokines. This is partly for historical reasons – IL-1 being one of the first cytokines to be discovered – but is also because of its pleiotropic effects (Dinarello, 1996). Virtually every cell type responds to IL-1, the major response being to acquire an inflamed phenotype, i.e. the enhanced expression of a range of genes which encode proteins with roles in inflammation and immunity. Current estimates indicate that up to 200 genes change in response to IL-1 in diverse cell types, although this number is bound to increase, given the latest microarray technologies, where thousands of genes can be monitored simultaneously.

The IL-1 ligand family comprises three members, IL-1α, IL-1β, and the IL-1 receptor antagonist (IL-1Ra). The earliest observations in relation to a possible receptor for IL-1 indicated that both IL-1α and IL-1β bound to the same receptor, which had a molecular weight based on crosslinking analysis of 80 kDa (Dower *et al.*, 1986a; Bird and Saklatvala, 1996). IL-1Ra is unique in that it is the only known endogenous receptor antagonist to a cytokine, acting to block binding of either type of IL-1 (Eisenberg *et al.*, 1990). The effects of IL-1 are mediated by the type I IL-1 receptor complex, comprising the type I receptor (IL-1RI) and its accessory protein, IL-1R AcP, which is unable to bind IL-1 but acts as the second chain in the receptor complex and is essential for signaling (Stylianou *et al*., 1992; Sims *et al*., 1993; Greenfeeder *et al*., 1995).

Apart from these two proteins, a steadily growing superfamily of homologous receptors has become apparent. These are shown in **Figure 1**. The superfamily includes the adapter protein MyD88, another essential component in the receptor complex (Wesche *et al*., 1997a; Burns *et al*., 1998), linking the receptor via a probable homotypic interaction to downstream signaling components, which include two IL-1 receptor associated kinases, IRAK and IRAK-2 (Martin *et al*., 1994; Cao *et al*., 1996; Muzio *et al*., 1997). Other members include the type II IL-1 receptor (which is shed from cells and acts as a decoy) (McMahon *et al*., 1991; Colotta *et al*., 1993), the IL-18 receptor complex (comprising IL-1Rrp1 (Parnet *et al*., 1996; Torigoe *et al*., 1997) and AcPL (Born *et al*., 1998)) and a number of receptors homologous to the *Drosophila* protein Toll, and termed TLRs (Rock *et al*., 1998; Takeuchi *et al*., 1999). It appears that TLRs may act as receptors for bacterial products such as LPS, TLR-4 being a strong contender for the much-sought-after LPS signaling receptor (Poltorak *et al*., 1998; Chow *et al*., 1999).

Figure 1 The major members of the IL-1 receptor/Toll-like receptor superfamily. All members contain a conserved cytosolic region termed the TIR domain (Toll–IL-1 receptor domain). The superfamily can be divided into two subgroups. The immunoglobulin subgroup members all contain immunoglobulin domains extracellularly and include receptors and accessory proteins for IL-1 and IL-18. Several members have no known ligand (e.g., T1/ST2, IL-1RAPL). The leucine-rich repeat subgroup members have leucine-rich repeats and include the signaling receptors for LPS (TLR4) and various plant proteins. Many of the members of this subgroup also have no known ligands. Finally, MyD88 is exclusively cytosolic and is a signaling adapter for IL-1RI, IL-18R, TLR-2, and TLR-4, if not the entire family.

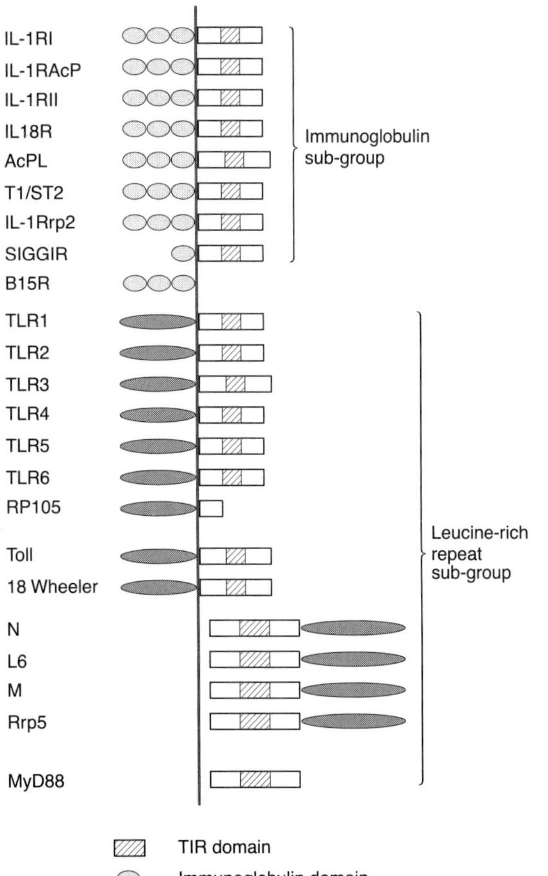

In addition, there are several plant members of the family, all involved in disease resistance in plants (Hammond-Kosack and Jones, 1996). This growing list of proteins, all of which share a common signaling sequence, is intriguing, and suggests that the IL-1 receptor system is very ancient and is used in signaling in a wide range of species, mainly in the context of defense and response to injury. The importance of IL-1 and its receptor system in disease is underscored by the recent clinical efficacy of IL-1Ra in rheumatoid arthritis (Bresnihan *et al*., 1998). The rapidly expanding list of receptors in the family represents a challenge common to all other branches of cytokine biology at present, given the deluge of information emerging from genome sequencing. The common theme in the IL-1 system, however, is that of inflammation and defense, the signaling region of the family obviously representing a highly efficient machine which, when triggered, leads to the induction of a bank of genes important for the elimination of infection and subsequent repair. The overactivation of this system appears to contribute to the pathogenesis of inflammatory diseases, for example, rheumatoid arthritis. The relationship of the IL-1/TLR family to innate and acquired immunity is also discussed in the chapter on Proinflammatory Cytokines.

DEFINING THE IL-1R/TLR FAMILY

Based on sequence homologies, it is possible to divide the IL-1R/TLR superfamily into two groups, as shown in Figure 1. The first consists of the founder member of the entire family (being the first one whose gene sequence was determined), IL-1RI (Sims *et al.*, 1988). The members of this group all contain extracellular (immunoglobulin domains) and intracellular (the IL-1R/TLR signaling domain, also termed the TIR domain) homologies. Apart from IL-1RI, this group contains the aforementioned IL-1R AcP (Greenfeeder *et al.*, 1995), IL-1RII (McMahon *et al.*, 1991) and its vaccinia virus homolog B15R (Alcami and Smith, 1992) (which are included here because of their three immunoglobulin domains), and IL-18R (Torigoe *et al.*, 1997) and its accessory protein AcPL (accessory protein-like) (Born *et al.*, 1998). In addition, there is IL-1RrP2 (Lovenberg *et al.*, 1996), T1/ST2 (Mitcham *et al.*, 1996), single Ig IL-1R-related molecule (SIGIRR) (Thomassen *et al.*, 1999), and IL-1 receptor accessory protein-like (IL-1RAPL) (Carrie *et al.*, 1999), whose ligands have yet to be discovered. SIGIRR, as its name suggests, only contains a single immunoglobulin-like domain, unlike the other group members. IL-1RAPL is expressed in the hippocampus, and when mutated it is responsible for one form of X-linked mental retardation (Carrie *et al.*, 1999). It may be a novel accessory protein.

The members of the second group also have the TIR domain, but are quite different extracellularly, having a series of leucine-rich repeats. The founder member if this group is Toll, as mentioned above (Hashimoto *et al.*, 1988; Gay and Keith, 1991). There are three further *Drosophila* members, termed 18-Wheeler (Eldon *et al.*, 1994), MstProx (Sims and Dower, 1994), and Tehao. The putative ligand for Toll is Spatzle (Morisato and Anderson, 1994), with ligands for the other two proteins as yet unknown. The six mammalian TLRs so far described are also in this subgroup (Rock *et al.*, 1998; Takeuchi *et al.*, 1999), as are the plant proteins (Whitham *et al.*, 1994; Hammond-Kozack and Jones, 1996; Parker *et al.*, 1997), which, although exclusively cytosolic, have leucine-rich repeats and are therefore more closely related to Toll than to IL-1RI.

The final protein in the family, MyD88, is an exclusively cytosolic protein which does not belong to either group, since it only has the region of homology which defines the entire family (i.e. the TIR domain) (Wesche *et al.*, 1997a; Burns *et al.*, 1998).

The degree of sequence similarity in each of these groups is quite striking. **Figure 2a–c** shows alignments of the superfamily, with **Table 1** giving the GenBank accession numbers of all of the sequences aligned. Figure 2a demonstrates an alignment of the extracellular regions of superfamily members in the immunoglobulin subgroup. From the consensus sequence, the conserved cysteines and bulky hydrophobic tryptophans, characteristic of immunoglobulin domains, can be seen. Figure 2b demonstrates an alignment of members in the leucine-rich repeat subgroup. From its consensus sequence it can seen that the conserved residues are mainly leucines. Decorin and platelet protein 5 are included as examples of other leucine-rich repeat proteins which will be detected in a BLAST search using TLRs. Figure 2c demonstrates an alignment of the TIR domain in all family members. The degree of cross-species – and, indeed, cross-kingdom – similarity is remarkable. Included in this alignment are sequences from human, mouse, rat, hamster, *Drosophila melanogaster*, *Caenorhabditis elegans* and, particularly surprisingly, a protein from the bacterium *Streptomyces coelicolor*. Such conservation supports the view that the TIR domain is very ancient, and may have arisen prior to the divergence of prokaryotes from eukaryotes. As can been seen from this alignment, there are three boxes within the sequences which align, termed box 1, box 2, and box 3.

In 1992, prior to the discovery of most of the superfamily members, biochemical data on the residues important for receptor function were described (Heguy *et al.*, 1992). Arg431 in box 2, and Phe 513 and Trp514 in box 3 of IL-1RI are essential for receptor function (Heguy *et al.*, 1992). In addition, more recently it was shown that Pro712 in Box2 of TLR-4 was required for signaling since, in the LPS-resistent C3H/HeJ mouse, this amino acid is mutated. Since TLR-4 is the signaling receptor for LPS, it can be inferred that this mutation leads to a defective receptor (Poltarak *et al.*, 1998). These features will be discussed below.

As yet, the structure of the TIR domain has not been elucidated, either from data obtained by X-ray crystallography or modeling.

IL-1RI

Prior to the cloning of IL-1RI, Scatchard analysis using radiolabeled IL-1 revealed a number of interesting features. First, the receptor had a high affinity for its ligands, with K_d values in the subnanomolar range (Dower *et al.*, 1986b). Second, IL-1 was active on cells far below the K_d of the putative receptor, with cells being responsive with receptors numbering less than 10 per cell (Dower *et al.*, 1986b). This implied that a large degree of amplification must occur during signaling. Finally, it appeared that IL-1, once bound to its receptor, was rapidly internalized, the receptor possibly acting simply as an internalization vector (Bird and Saklatvala, 1987).

The gene for IL-1RI was cloned in 1988 and was identified as the 80 kDa protein found by crosslinking

Figure 2 Sequence alignments for the IL-1 receptor/Toll-like receptor superfamily. Following database searches, sequence alignments were performed on members of the superfamily. (a) Alignment of the extracellular domains of the immunoglobulin subgroup of the superfamily. Twelve members are shown, including the predicted amino acid sequence of the vaccinia gene B15R, which is homologous to IL-1RII. The consensus sequence indicates key conserved cysteines characteristic of immunoglobulin domains. (b) Alignment of the extracellular domains of the leucine-rich repeat subgroup of the superfamily. Thirteen members are shown, including the human Toll-like receptors and rp105, which is homologous to the family extracellularly. Conserved leucines are indicated in the consensus sequence. (c) Alignment of the TIR domain of the superfamily. Members of both subgroups are included. Three conserved boxes are indicated, along with a consensus sequence. The function of these boxes is as yet unknown, although mutations in amino acids in boxes 2 and 3 abolish receptor function (see text for details). Family members from *C. elegans* and *S. coelicolor*, which have not been described before, are also shown.

(a)

```
               11                                                                                                    110
  huIL18Racp   LLWTYSTRSEEEFVL..FCD LPEPQKSHFCHRNRLSPKQV PEHLPFMGS.NDLSDVQWYQ QPSNGDPLEDIR...KSYPH IIQDKCTLHFLTPGVNNSGS
  muIL18Racp   LLWTYSARGAENFVL..FCD LQELQEQKFSHASQLSPTQS PAHKPCSGSQKDLSDVQWYM QPRSGSPLEEIS...RNSPH .MQSEGMLHILAPQTNSIWS
     huil18r   ~~~~~TAESCTSR.....PH ITVVEGEPF.YLKHCSC... ....SL.AHEIETTTKSWY. ..KSSGSQEHVELNPRSSSR IALHDCVLEFWPVELNDTGS
     muIl18R   ~~~~~ASKSCIHR.....SQ IHVVEGEPF.YLKPCGI... ....SAPVHRNETATMRWF. ..KGSASHEYRELNNRSSPR VTFHDHTLEFWPVEMEDEGT
        B15R   ~~QTFNAPECIDKG.QYFAS FMELENEPV.IL.PCPQI.. NTL.S...SGYNILDILWE. .KRGADNDRIIPID.NGS.N M......LILNPT.QSDSGI
     huIL1R2   AAHTGAARSCRFRG.RHYKR EFRLEGEPV.AL.RCPQV.P YWLWA...SVSPRINLTWH. .KN..DSARTVPGE.EET.R MWAQDGALWLLPALQEDSGT
    huil1rec   ~~~~LEADKCKERE.EKI.. ILVSSANEI.DVRPCP.LNP NE.HK...G.....TITWY. .KD..DSKTPVSTE.QAS.R IHQHKEKLWFVPAKVEDSGH
     muil1r1   ~~~~LEIDVCTEYP.NQI.. VLFLSVNEI.DIRKCP.LTP NKMHG...D.....TIIWY. .KN..DSKTPISAD.RDS.R IHQQNEHLWFVPAKVEDSGY
     chil1r1   ~~~LFSAEECV.IC.NYF.. VLVGEPTAI.S...CPVITL PMLHS...D....YNLTWY. .RN..GSNMPITTE.RRA.R IHQRKGLLWFIPAALEDSGL
   ratil1rrp   ~~~~~~AGNCTDVY.MHH.. EMISEGQPF.PFN.CTY..P PVTNG...A....VNLTWH. .RT..PSKSPISIN.RHV.R IHQDQSWILFLPLALEDSGI
     il1racp   ~~~SHASERCDDWGLDTMRQ IQVFEDEPA.RI.KCPLFEH FLKYNYSTAHSSGLTLIWYW TRQDRDLEEPINFRLPEN.R ISKEKDVLWFRPTLLNDTGN
       must2   ~~~~~~~~~~~~~~~~~~~Y LTVTEGSKS.SWG...LENE ALIVRCPQRGRSTYPVEWYY S....DTNESIPTQKR.N.R IFVSRDRLKFLPARVEDSGI
   Consensus   ------A--C---------- ----E---------C----- -----------------WY- -----DS---I----R---R I------LWF-P---EDSG-

               111                                                                                                   210
  huIL18Racp   YICRPKMIKSPYDVACCVKM ILEVKPQTNAS....CEYSA S..HKQDL.LLGSTGSISCP SLSCQSDAQS...PAVTWYK NGK..LLSVER.......SN
  muIL18Racp   YICRPR.IRSPQDMACCIKT VLEVKPQRNVS....CGNTA Q..DEQVL.LLGSTGSIHCP SLSCQSDVQS...PEMTWYK DGR..LLPEHK.......KN
     huil18r   YFFQMKNYT.....QKWKLN VIRRNKHSCFTERQVT.... ....SKIVEV.KKFFQITCE N..SYYQTLVN...STSLYK NCKKLLLENN........KN
     muIl18R   YISQVGNDR.....RNWTLN VTKRNKHSCFSDKLVT.... ....SRDVEV.NKSLHITCK N..PNYEELIQ...DTWLYK NCKEI....S........KT
        B15R   YICITTNETYCDMMS.LNLT IVS......VSESNIDL... ISYP.QIVNER.STGEMVCP NIN.AF.IASNVNADIIWSG HRRL...RNKR.LKQR.TPG
     huIL1R2   YVCTTRNASYCDKMS.IELR VFE......NTDAFLPF... ISYP.QILTLS.TSGVLVCP DLS.EF.TRDKTDVKIQWYK DSLLLDKDNEKFLSVRGTTH
    huil1rec   YYCVVRNSSYCLRIK.ISAK FVE.......NEPNLCYNAQ AIFK.QKLPVA.GDGGLVCP YME.FFKNENNELPKLQWYK DCKP.LLLDN..IHFSGVKD
     muil1r1   YYCIVRNSTYCLKTK.VTVT VLE.......NDPGLCYSTQ ATFP.QRLHIA.GDGSLVCP YVS.YFKDENNELPEVQWYK NCKP.LLLDN..VSFFGVKD
     chil1r1   YECEVRSLNRS.KQKIINLK VFK.......NDNGLCFNGE MKYD.QIVKSA.NAGKIICP DLE.NFKDEDNINPEIHWYK ECKSGFLEDKRLVLAEG.EN
   ratil1rrp   YQCVIKDAHSCYRI.AINLT VFRKHWCDSSNEES.SINSS DEYQ.QWLPIG.KSGSLTC. HL..YFP.ESCVLDSIKWYK GCE.EIKVSKKFCP.TGTK.
     il1racp   YTCMLRNTTYCSKV.AFPLE VVQK.......DS..CFNSA MRFPVHKMYIEHGIHKITCP NVDGYFP..SSVKPSVTWYK GC..TEIVDFHNVLPEGMNL
       must2   YACVIRSPN.LNKTGYLNVT IHKKPPSCNIPD.YLMYST. .......VRGSDKNFKITCP TIDLY.....NWTAPVQWFK NCKA..LQEPRF...RAHRS
   Consensus   Y-C--RN-----------L- V--------------C---- -----Q--------G-I-CP -----F-----------WYK -CK---L------------

               211                                                                                                   310
  huIL18Racp   RIVVDEVYDYHQGTYVCDYT QSDTVSSWTVRAVVQVRTIV GD.TKLK.PDILDPVEDTL. EVELGKPLTISCKARF..GF ERVFNPVIKW.....YIKDS
  muIL18Racp   PIEMADIYVFNQGLYVCDYT QSDNVSSWTVRAVVKVRTIG KD.INVK.PEILDPITDTL. DVELGKPLTLPCRVQF..GF QRLSKPVIKW.....YVKES
     huil18r   PTIKKNAEFEDQGYYSCVHF LHHNGKLFNITKTFN.ITIV EDRSNI.VPVLLGPKLNHVA .VELGKNVRLNCSALLNE.. ....EDVIYWMFGE...ENG
     muIl18R   PRILKDAEFGDEGYYSCVFS VHHNGTRYNITKTVN.ITVI EGRSKV.TPAILGPKCEKVG .VELGKDVELNCSASLNK.. ....DDLFYWSIRK...EDS
        B15R   IITIEDVRKNDAGYYTCVLE YIYGGKTYNVTRIVK.LEVR DK...I.IPSTMQLP.DGI. VTSIGSNLTIACRVSL...R PPTTDADVFWISNGMYYEED
     huIL1R2   LL.VHDVALEDAGYYRCVLT FAHEGQQYNITRSIE.LRIK KKKEET.IPVIIS.PLKTI. SASLGSRLTIPCKVFLGTGT PLTT..MLWWTANDTHI.ES
    huil1rec   RLIVMNVAEKHRGNYTCHAS YTYLGKQYPITRVIEFITLE ENK.PTR.PVIVSPANETM. EVDLGSQIQLICNV...TGQ ...LSDIAYWKWNGSVIDED
     muil1r1   KLLVRNVAEEHRGDYICRMS YTFRGKQYPVTRVIQFITID ENK.RDR.PVILSPRNETI. EADPGSMIQLICNV...TGQ ...FSDLVYWKWNGSEIEWN
     chil1r1   AILILNVTIQDKGNYTCRMV YTYMGKQYNVSRTMN.LEVK ESPLKMR.PEFIYPNNNTI. EVELGSHVVMECNVS..SGV ...YGLLPYWQVNDEDVDSF
   ratil1rrp   .LLVNNIDVEDSGSYACSAR LTHLGRIFTVRNYI.AVNTK EVGSGGRIPNITYPKNNSI. EVQLGSTLIVDCNIT.DTKE ...NTNLRCWRVNNTLVDDY
     il1racp   SFFIPLVS..NNGNYTCVVT YPENGRLFHLTRTV.TVKVV GSPKDALPPQIYSPNDRVVY EKEPGEELVIPCKVYFSF.. IMDSHNEVWWTIDGKKPDDV
       must2   YLFIDNVTHDDEGDYTCQFT HAENGTNYIVTAT.RSFTVE EKGF.SMFPVITNPPYNHTM EVEIGKPASIACSACFGKGS HFLAD..VLWQINKTVVGNF
   Consensus   ------V---D-G-Y-C--- ----G--Y-VT------T-- E-------P-I--P------ EVELG------C-V----G- ---------W--N-------
```

Figure 2 *(Continued)*

(b)

```
            1                                                                                                    100
hupgp5      .QPFPC.............. ...PPACKCVFRDAAQCSGG DVAR......ISALGLPTNL TH.ILLFGMGR........G VLQSQSFSGMTV.LQRLMIS
hudecorin   DEASGI.............. ...GPEVP.DDRD.FEPSLG PVCP......FRC...QCHL RV.VQCSDLGL........D KV.PKDLPPDTT.L..LDLQ
hutlr1      .................... .........SEES..EFLVD R.SKN.GLIH.....VPKDL SQKTTILNIS.QNYISELWT ....SDILSLS.KLRILIIS
hutlr6      MTK.DKEPIVKSFHFVCLMI IIVGTRIQFSDGN..EFAVD K.SKR.GLIH.....VPKDL PLKTKVLDMS.QNYIAELQV ....SDMSFLS.ELTVLRLS
hutlr2      LSK.EESSNQASLS..C... ..........DRN..G.ICK G.SSG.SLNS.....IPSGL TEAVKSLDLS.NNRITYISN ....SDLQRCV.NLQALVLT
hutlr4      L................G.. .................SY. ........S........... .....F..FS.F...P.... ............ELQVLDLS
rp105       MAF.DVSCFFWVVLFSAGCK VITSWDQMCIEKEA.NKTYN ..CENLGLSE.....IPDTL PNTTEFLEFS.FNFLPTIHN ....RTFSRLM.NLTFLDLT
hutlr3      ....KCTVSHEVADCSHLKL TQVPDDLPTNITVL.NLTHN Q.LRRLPAAN.....FTRYS QLTSLDVGFN..TISKLE.P ....ELCQKLP.MLKVLNLQ
hutlr5      IP..SCSFDGRIAFYRFCNL TQVPQVLNTTERLL..LSFN Y.IRTVTASS.....FPFLE QLQLLELGSQ.YTPLTID.K ....EAFRNLP.NLRILDLG
wheeler     MP..ATSSIITIIAVAACLL LLVAD.AHAQQQ.C.NWQYG .....LTTMD.....I.RCS .VRALESGTG..SPLDLQVA ....EAAGRLD.LQCSQELL
drotoll2    ML.....TYLPVVWLFFALL VL....RSATGQII.PL... P.TFCLGLS.......PQCT CAA......E.GNVVRFHCP ....DEYAML......LEVS
dmmstprox   M...KVRNYIPNKNFTA..L I.......CTHKNC.NF... ..IRGINGLE.....CPKLC QCLYIIDDLE.LN.ID.... .........CS.NLGLL...
drotoll     MSRLKAASELALLVIILQLL QWPGSEASFGRDACSEMSID GLCQCAPIMSEYEIICPANA ENPTFRLTIQPKDYVQIMCN LTDTTDYQQLPKKLRIGEVD
Consensus   M------------------L -----------------S-- -------L--------P--L ------L--S--N------- -----D---L---L--L-LS

            101                                                                                                  200
hupgp5      DSHISAV....APGTFSDLI KLKT.....LRLSRNKITH. .L............PGALLD KMVLLEQLFL.DHNALR... ......GIDQNMFQKL....
hudecorin   NNKITEI....KDGDFKNLK NLHA.....LILVNNKISK. .VS...........PGA.FT PLVKLERLYL.SKNQLK... ......ELPEKM.PK.....
hutlr1      HNTIQ..YL..DISVFKFNQ EL.....EYLDLSHNKLVK. .ISC.....HPTV....... NLKHLD...L.SFNAFDALP ICKEFGNM.SQL.KFL.GLS
hutlr6      HNRIQ..LL..DLSVFKFNQ DL.....EYLDLSHNQLQK. .ISC.....HPIV....... SFRHLD...L.SFNDFKALP ICKEFGNL.SQL.NFL.GLS
hutlr2      SNGIN..TI..EEDSFSSLG SL.....EHLDLSYNYLSN. .LSS..SWFKPLS....... SLTFLN...L.LGNPYKTLG ETSLFSHL.TKL.QIL.RVG
hutlr4      RCEIQ..TI..EDGAYQSLS HL.....STLILTGNPIQ.. SLAL..GAFSGLS....... SLQKLV...AVETN.LASLE NFP.IGHL.KTL.KEL.NVA
rp105       RCQIN..WI..HEDTFQSHH QL.....STLVLTGNPLI.. FMAE..TSLNGPK....... SLKHLF...LIQTG.ISNLE FIP.VHNL.ENL.ESL.YLG
hutlr3      HNELS..QL..SDKTFAFCT NL.....TELHLMSNSIQK. .IKN..NPFVKQK....... NLITLD...L.SHNGLSSTK LGTQV.QL.ENL.QEL.LLS
hutlr5      SSKIY..FL..HPDAFQGLF HL.....FELRLYFCGLSDA VLKD..GYFRNLK....... ALTRLD...L.SKNQIRSLY LHPSFGKL.NSL.KSI.DFS
wheeler     HA..S..EL..APGLFRQLQ KL.....SELRIDACKLQR. .VPP..NAFEGLM....... SLKRLT...LESHNAV.... WGP..GK...TL.E....LH
drotoll2    EPGAS...L..YMSYYAS.T EL.....QWL.......PRF NIS................. SLVKI......EFDAYIFWP E.KFLSDL...L.KTL.GVQ
dmmstprox   ..QIP..PL..PIPIY...G DV.....K.LNFSNNSLSQ. .LPT..MTLPGYK....... LVKRLD...V.SRNRLTNL. ...SINHLPAKL.DYL.DVS
drotoll     RVQMRRCMLPGHTPIASILL YLGIVSPTTLIFESDNLGMN ITRQHLDRLHGLKRFRFTTR RLTHIPANLLTDMRNLSHLE LRANIEEMPSHLFDDLENLE
Consensus   -N-I----L------F--L- -L-------L-LS-N-L--- -LS-----F-GLK------- SL--LD---L-S-N-L--L- -----G-L---L---L--LS
            201                                                                                                  300
hupgp5      .................... VNLQELALNQNQLDFLP... .................... .................... ...............ASLFT
hudecorin   .................... .TLQELRAHENEITKVR... .................... .................... ...............KVTFN
hutlr1      T..TH.LEKSSV.LPIAHL. .NISKVLLVLGE.TYGEKED PEGLQDFNTESLHIV.F... PTNKEFHFI..LDVSVK..T VANLELSN......IKCV..
hutlr6      A..MK.LQKLDL.LPIAHL. .HLSYILLDLRN.YYIKENE TESLQILNAKTLHLV.FH.. PTSL.FAIQ..VNISVN..T LGCLQLTN......IK....
hutlr2      N..MDTFTKIQR.KDFAGL. .TFLEE.LEIDA.SDLQSYE PKSLKSIQNVS.HLI.LH.. ..MKQHILL..LEIFVD..V TSSVEC........LE....
hutlr4      HNLIQSFKLPEY.FSNL.T. .NLEHLDLSSNKIQSIYCTD LRVLHQM..PLLNL.SLD.. LSLNPMNFIQ.PGAFKEIRL HKLTLRNNFDSLNVMKTCIQ
rp105       SNHISSIKFPKD.FP.A.R. .NLKVLDFQNNAIHYISRED MRSLEQ....AINL.SLN.. FNGNNVKGIE.LGAF.DSTV FQ.SL..NFGGTPNLSVIFN
hutlr3      NNKIQALKSEEL.DIFANS. .SLKKLELSSNQIKEFSPGC FHAIGRLFGLFLNNVQLG.. PSLTEKLCLELANTSIRNLS LSNSQLSTTSNTTFLGLKWT
hutlr5      SNQIFLVCEHEL.EPLQGK. .TLSFFSLAANSL..YS... ..RVSVDWGKCMN....... P..FRNMVLEIVDVSGNGWT VDIT..GNFSNA..IS.KSQ
wheeler     GQSFQGLKE........... ..LSELHLGDNNIRQLPEGV WCSMPSL..QLLNLTQ.... .NRIRS..AEFLGFS.EKLC AG.SALSN.ANGAVSG..GS
drotoll2    TVKTIIFRDRTL.ETVVTR. .DV....LNSGN.GYMETSQ PENI.....TTWHFGSV... PGLKKFKFFSHV...PE... LQESIFHGFDTLRDL.....
dmmstprox   FNEI................ .....INMGNDVIKYL.... .RTVP.IFKQTGNQWTIH.. CDDKP.....LLNFFRHLKL I..IRMKS....AEMKPMFL
drotoll     SIEFGSNKLRQMPRGIFGKM PKLKQLNLWSNQLHNLTKHD FEGATSVLGIDIHDNGIEQL PHDVFAHLTNVTDINLSANL FRSLPQGLFDHNKHLNEVRL
Consensus   -N-I---K--------A--- --L--L-L--N-I-YL---- ---L-------LNL------ P----------L--S----- -------NF-----LK----
```

Figure 2 *(Continued)*

(b)

```
              301                                                                                                   400
hupgp5        NLENLKL............. .................... .................... ..........LDLSGNNLTH LPKGLLGAQAKLERLLLHSN
hudecorin     GLNQMIV............. .................... .................... ..........IEL.GTN... .P..............LKSS
hutlr1        .LEDNKCSYF.LSILAKLQT NPKLS.NLTL.....NNIE. ..TTWNSFI..RILQLVWH. TTVWYFSISNVKL....QGQ LDFRDFD....YSGTSLKAL
hutlr6        .LNDDNCQVF.IKFLSELTR GPTLL.NFTL.....NHIE. ..TTWKCLV..RVFQFLWP. KPVEYLNIYNLTI....IES IREEDFT....YSKTTLKAL
hutlr2        .LRDTDLDTFHFSELSTGET NSLIK.KFTF.....RNVK. ..ITDESLF..QVMKLLNQI SGLLELEFDDCTL..NGVGN FRASDND..RVIDPGKVETL

hutlr4        GLAGLEV..HRL.VLGE... ...........FRNEGNLE. ..KFDKSAL..EGLCNLTIE EFRLAYLDYYLD.......D IID.LFNCLTNVSSFSLVSV
rp105         GLQNSTT..QSL.WLGT... ...........FEDIDD.E. ..DISSAML..KGLCEMSVE SLNLQ..EHRFS.......D ISSTTFQCFTQLQELDLTAT
hutlr3        NLTMLDLSYNNLNVVGND.S FAWLP.QLEYFFLEYNNIQ. ..HLFSHSL..HGLFNVRYL NLKRSFTKQSISL..ASLPK IDDFSFQWLKCLEHLNMEDN
hutlr5        AFSLI.LAH...HIMG.A.G FGF.H.NIK....DPD..Q. ..NTFA.GL...ARSSVRHL DLSHGFV...FSL..NS..R V....FETLKDLKVLNLAYN
wheeler       ELQTLDVSFNELRSLPDAWG ASRLR.RLQTLSLQHNNIS. ..TLAPNAL..AGLSSLRVL NISYN......HL..VSLP. ..SEAFAGNKELRELHLQGN
drotoll2      HLS.VNVTTLPGNMLSTV.. NGTLK.TLTI...ESPGIV. ..SFGNPLL..RELQQLRNL SLAL...IHPFH...ERDKQ LQPHFFGSMTNLEEVRLASA
dmmstprox     H.SLTELPKGFLKFLGKHFI WLGVR.KQEYYLI...NEE. ..QLLQSMH..RKLNNLNTI MSIYKYMEW...L..HRKLI FVNREYD.LFYIRQ..MAAP
drotoll       MNNRVPLATLPSRLFANQPE LQILRLRAELQSLPGDLFEH STQITNISLGDNLLKTLPAT LLEHQVNLLSLDLSNNRLTH LPDSLFAHTTNLTDLRLEDN
Consensus     -L----L----L--LG---- ---L------------NIE- --------L----L--L--- -L---------L-------- -----F-----L--L-L--N

              401                                                                                                   500
hupgp5        RLVSLDSGLLNSLGALTELQ FHRN...........HIRSI .................... ....APGAFDRLPNLSSLTL SRNHLAFLPSALFLHSHNLT
hutlrdecorin  ...GIENGAFQGMKKLSYIR IADT...........NITSI .................... .....PQGLP..PSLTELHL DGNKISRVDAA........S
hutlr1        SIHQV.VSDVFGF.PQSYI. .....YEIFSNM...NIKNF T.....VS..GTR..MVH.M LCPSK...IS...PFLHLDF SNNLL...............
hutlr6        TIEHI.TNQVFLF.SQTAL. .....YTVFSEM...NIMML T.....IS..DTP..FIH.M LCPHA...PS...TFKFLNF TQNVF...............
hutlr2        TIRRLHIPRFYLFYDLSTL. .....YSL.TE....RVKRI T.....VE..NSKVFLVP.C LLSQH...LK...SLEYLDL SENLM...............
hutlr4        TIERVKDFS.YNFG.WQHLE L.VN..CKFGQFPTLKLKSL KR.LTFTS..N.KG...G.N AFSEV.....DLPSLEFLDL SRNGL.........SFKGCC
rp105         HLKGLPSGM.KGLNLLKKLV LSVNHFDQLCQISAANFPSL TH.LYIRG..NVKKLHLG.V GCLE......KLGNLQTLDL SHNDI.........EASDCC
hutlr3        DIPGIKSNMFTGLINLKYLS LS.NSFTSLRTLTNETFVSL AHSPLHIL..NLTKNKIS.K IESDA...FSWLGHLEVLDL GLNEI.........G.QELT
hutlr5        KINKIADEAFYGLDNLQVLN L...SYNLLGELCSSNFYGL P.KVAYI...DLQKNHIA.I IQDQT...FKFLEKLQTLDL RDN..............ALT
wheeler       DLYELPKGLLHRLEQLLVLD LS.GNQLTSHHVDNSTFAGL IR..LIVL..NLSNNALT.R IGSKT...FKELYFLQILDM RNNSI.........G.H.IE
drotoll2      T.SSVNRSMFKGTNKLQLIK MNGND.DLM.ELPGEIF..L DQV.......N......... ..............LKTLDL SCNAI.........VT..LH
dmmstprox     CPHKCE..CCYSRDSL.ILK IDCRNKFVY......NFPDI .....VAR..NSR......L MRKQN...MS..SPME.LHL SKNNI.........SNITIA
drotoll       LLTGISGDIFSNLGNLVTLV MSRNRLRTIDSRAFVSTNGL RHLHLDHNDIDLQQPLLDIM LQTQINSPFGYMHGLLTLNL RNNSIIFVYNDWKNTMLQLR
Consensus     -I--I----F-GL--L--L- L--N-----------NF-SL ---------N---------- --------F--L--L--LDL S-N-I-------------L-
                501                                                                                                 600
hupgp5        ................LLTL FENPLA.ELPGVLFGEMGGL QELWLNR...TQLRTL...P AAAFRNLSRLRYLGVT.... ..................LS
hudecorin     ................LKGL ..NNLA.KL.GLSFNSISAV D....NG...S.LANT...P .....HLREL.HLD...... ...................N
hutlr1        ........TDTV..FE.... ..N...CG......HL.TEL ETLILQ...MNQLKELS.KI AEMTTQMKSLQQLDISQ... .....NSVSYDEKKGDCSWT
hutlr6        ........TDSI..FE.... ..K...CS......TL.VKL ETLILQ...KNGLKDLF.KV GLMTKDMPSLEILDVSW... .....NSLESGRHKENCTWV
hutlr2        ........VEEY..LK.... ..NSA.CE......DAWPSL QTLILR...QNHLASLE.KT GETLLTLKNLTNIDISK... .....NSFHS..MPETCQWP
hutlr4        S..QSDFGTTSL..KYL.DL SFNGV.ITMSSN.FLGLEQL EHLDFQ...HSNLK.QM.SE FSVFLSLRNLIYLDISH... .....THTRVAFNGIFNGL.
rp105         S..LQLKNLSHL..QTL.NL SHNEP.LGLQSQAFKECPQL ELLDLA...FTRLH.IN.AP QSPFQNLHFLQVLNLTY... .....CFLDTSNQHLLAGL.
hutlr3        G..QEWRGLENI..FEI.YL SYNKY.LQLTRNSFALVPSL QRLMLR...RVALKNVD.SS PSPFQPLRNLTILDLSN... .....NNIANINDDMLEGLE
hutlr5        T..IHF..IPSI..PDI.FL SGNKL.VTLPK..INLTANL IHLSEN...R..LENLD.IL YFLLR.VPHLQILILNQ... .....NRFSSCSGDQTPSEN
wheeler       E..GAFLPLYNL..HTL.NL AENRL.HTLDNRIFNGLYVL TKLTLN...NNLVSIVE.SQ ..AFRNCSDLKELDLSS... .....NQLTEV.PEAVQDLS
drotoll2      E..DVFKGLGNL..TLL.DL SKNRL.TNLSSTIFAPLTSL NVLRLN...KNSLTAMS... PSVFQDVVSLNYIEMVN... .....TQFYGAT..LLMNYE
dmmstprox     M..LP....KEL..RFL.DL RFNNL.VTLDDKV......L SYL..K...KNSIKTKL.SG NPWNCDCKSRSVLSILR... .....DH.....EPL..EYD
drotoll       ELDLSYNNISSLGYEDLAFL SQNRLHVNMTHNKIRRIALP EDVHLGEGYNNNLVHVDLND NPLVCDCTILWFIQLVRGVH KPQYSRQFKLRTDRLVCSQP
Consensus     ----------L----L--L- S-N-L---L----F-----L E-L-LN----N-LK-L---- ---F--L--L--LDLS---- -----N------------L-
```

Figure 2 (*Continued*)

(b)

```
            601                                                                                                    700
   hupgp5   PRL................ ...........SALPQGAFQ GLGELQVLALHSNGLTALPD G.LLRGLGKLRQVSLRRNRL ..RALP.RALFRNLSSLESV
hudecorin   NKL................ ...........TRVPGGLAE H.KYIQVVYLHNNNISVVGS S.DFCPPGH....NTKKASY ..SGV...SLFSN.....PV
   hutlr1   KSLL...SLNMSSNIL..T. ..DTIFR...C..LP..... ..PRIKVLDLHSNKIKSIPK QV.VK.LEALQELNVAFNSL T..DLPG.C..GSFSSLSVL
   hutlr6   ESIV...VLNLSSNML..T. ..DSVFR...C..LP..... ..PRIKVLDLHSNKIKSVPK QV.VK.LEALQELNVAFNSL T..DLPG.C..GSFSSLSVL
   hutlr2   EKMK...YLNLSSTRI.... ..HSVTG...C..IP..... ..KTLEILDVSNNNLNLFS. ...LN.LPQLKELYISRNKL M..TLPD.A..SLLPMLLVL
   hutlr4   SSLE...VLKMAGNS..... .FQENFL.......PDIFTE .LRNLTFLDLSQCQLEQLSP TA.FNSLSSLQVLNMSHNNF F..SLDT.FPYKCLNSLQVL
    rp105   PVLR...HLNLKGNH..... .FQDGTI...T..KTNLLQT .VGSLEVLILSSCGLLSIDQ QA.FHSLGKMSHVDLSHNSL TCDSIDS.LSH..LKGI.YL
   hutlr3   .KLE...ILDLQHNNL..AR LWKHANP...G..GPIYFLK GLSHLHILNLESNGFDEIPV EV.FKDLFELKIIDLGLNNL ..NTLPA.SVFNNQVSLKSL
   hutlr5   PSLE...QLFLGENML..QL AWE.TEL...C..WDV.F.E GLSHLQVLYLNHNYLNSLPP GV.FSHLTALRGLSLNSNRL ..TVL...SHNDLPANLEIL
  wheeler   .MLK...TLDLGENQI..SE .FKN.N.........TF.R NLNQLTGLRLIDNRIGNITV GM.FQDLPRLSVLNLAKNRI ..QSIER.GAFDKNTEIEAI
 drotoll2   AVVC...TNDEACQYK..SA EWQCDPR...C..I.CWVQR SVGSL.IVDCRGTSLEELPD LP.RTTLLS.TVLKVGNNSL T..SLPTVSEHSGYANVSGL
dmmstprox   VTLK...RCNISPT...... ..DCPDV...C..VCCLDNL TWPSF.IVDCRGEGLLQMPS ...LSS..RVTYVDLRNNNL TALS......QKNRSSIE..
  drotoll   NVLEGTPVRQIEPQTLICPL DFSDDPRERKCPRGCNCHVR TYDKALVINCHSGNLTHVPR LPNLHKNMQLMELHLENNTL LR..LPSANTPG.YESVTSL
Consensus   --L-----LNL--N-L---- ----------C---P----- -L--L-VLDLHSN-L---P- ---F--L--L--LNL--NSL T--SLP---------SL--L

            701                                                                                                    800
   hupgp5   QLDHNQLETLPGDVFGALPR .LTEVLL............. .................... ..........G.HN...... ....................
hudecorin   Q..YWEIQ..P.STFRCVYV .RSAIQL............. .................... ..........GNYK...... ....................
   hutlr1   IIDHNSVSHPSADFFQSC.Q KMRSIKAGDNPFQCTCE.LG EFVKNIDQVSSEVLEGWPDS YKCDYPESYRGTL..L.... ....KD.......FHMSELS
   hutlr6   IIDHNSVSHPSADFFQSC.Q KMRSIKAGDNPFQCTCE.LR EFVKNIDQVSSEVLEGWPDS YKCDYPESYRGSP..L.... ....KD.......FHMSE..
   hutlr2   KISRNAITTFSKEQLDSF.H TLKTLEAGGNNFICSCE.FL SFTQE.QQALAKVLIDWPAN YLCDSPSHVRGQQ..V.... ....QD.......VRLSVSE
   hutlr4   DYSLNHIMTSKKQELQHFPS SLAFLNLTQNDFACTCEHQS .FLQWIKDQRQL.LVE.VER MECATPSDKQGMP..VLSLN IT.CQ...............
    rp105   NLAANSINIISPR.LLPILS QQSTINLSHNPLDCTCSNIH .FLTWYKENLHK.LEG.SEE TTCANPPSLRGVK...LS.. ....D...............
   hutlr3   NLQKNLITSVEKKVFGPAFR NLTELDMRFNPFDCTCESIA WFVNWINETHTN.IPELSSH YLCNTPPHYHGFP..VRLFD TSSCKDSAPFEL.FFMINTS
   hutlr5   DISRNQLLAPNPDV....FV SLSVLDITHNKFICECE.LS TFINWLNHTNVT.IAGPPAD IYCVYPDSLSGVS..LFSLS TEGC.DEE..EV.L....KS
  wheeler   RLDKNFLTDIN.GIFA.TLA SLLWLNLSEN.......HL. ..V.WFDYAF...I...PSN L..KWLD.IHGNY..IEALG .NYYKLQE..EI...RVTT.
 drotoll2   FLSDNNLTSLGSG..DQLPD NLTHLDVRGNQIQSLSEEFL LFLQEPNNTMTLSLSGNP.. ITCGC.ESL..SL..LFFVR TNPQRVRDIADI.V.CTKQK
dmmstprox   ....N..............R SLK.LHLLDNPWSCSCNDIE K.INFMKSVSSSIV.DFTE. IKCS.....NGEK..LVSI. ....................
  drotoll   HLAGNNLTSID...VDQLPT NLTHLDISWNHLQMLNATVL GFLNRTMKWRSVKLSGNPWM CDCTAKPLLLFTQDNFERIG DRNEMMCVNAEMPTRMVELS
Consensus   -L--N--T------F----- -L--L-L--NPF-CTCE--- -F--W--------L-G-P-- --C--P-S--G----L---- -----D-------------
```

Figure 2 *(Continued)*

(c)

```
                 BOX 1                                           BOX 2
              31                                                                                                       130
huIL18Racp   .QTLGDKKDFDAFVSYAKWS SFPSEATSSLSEEHLALSLF PDVLENKYGYSLCLLERDVA PGGVYAE.DIVSIIKR.SRR GIFILSPNYVN........G
muIL18RacP   DETLGDKKEFDAFVSYSNWS SPETDAVGSLSEEHLALNLF PEVLEDTYGYRLCLLDRDVT PGGVYAD.DIVSIIKK.SRR GIFILSPSYLN........G
huIL18R      DETLTDGKTYDAFVSYLK.. ...ECRPENGEEHTFAVEIL PRVLEKHFGYKLCIFERDVV PGGAVVD.EIHSLIEK.SRR LIIVLSKSYMS........N
muIL18R      DETLTDGKTYDAFVSYLK.. ...ECHPENKEEYTFAVETL PRVLEKQFGYKLCIFERDVV PGGAVVE.EIHSLIEK.SRR LIIVLSQSYLT........N
muil1racp    DETILDGKEYDIYVSYAR.. ........NVEEEEFVLLTL RGVLENEFGYKLCIFDRDSL PGGIVTD.ETLSFIQK.SRR LLVVLSPNYVL........Q
muil1r1      PSKASDGKTYDAYILYPK.. TLGEGS..FSDLDTFVFKLL PEVLEGQFGYKLFIYGRDDY VGEDTIE.VTNENVKK.SRR LIIILVRDMGG........F
ratil1r      PRKASDGRTYDAYVLYPK.. TYGEGS..FAYLDTFVFKLL PEVLEGQFGYKLFICGRDDY VGEDTIE.VTNENVKR.SRR LIIILVRDMGS........F
huil1r1      PIKASDGKTYDAYILYPK.. TVGEGS..TSDCDIFVFKVL PEVLEKQCGYKLFIYGRDDY VGEDIVE.VINENVKK.SRR LIIILVRETSG........F
chil1r1      GKKVSDGKIYDAYVLYPK.. .NRESC..LYSSDIFALKIL PEVLERQCGYNLFIFGRNDL AGEAVID.VTDEKIHQ.SRR VIIILVPEPSC........Y
ratil1rrp   AQAPDDEKLYDAYVLYPK.. YPRESQ..GHDVDTLVLKIL PEVLEKQCGYKLFIFGRDEF PGQAVAS.VIDENIKL.CRR LMVLVAPETSS........F
ratst2       .KTQNDGKLYDAYIIYPR.. VFRGSAAGTGSVEYFVHYTL PDVLENKCGYKLCIYGRDLL PGQDAAT.VVESSIQN.SRR QVFVLAPHM.M........H
humyd88      LGHMPE..RFDAFICYCPSD .......IQFVQE.MIRQLE QTNYR....LKLCVSDRDVL PGTCV.WSIASELIEKRCRR MVVVVSDDYLQ........S
mumyd88      LGQTPE..LFDAFICYCPND .......IEFVQE.MIRQLE QTDYR....LKLCVSDRDVL PGTCV.WSIASELIEKRCRR MVVVVSDDYLQ........S
hutlr3       RQTE..QFEYAAYIIHAYKD .......KDWV....WEHFS SMEKEDQS.LKFCLEERDFE AGVFELEAIV.NSIKR.SRK IIFVITHHLLK........D
hutlr5       QGTEPDMYKYDAYLCFSSKD .......FTWVQNALLKHLD T.QYSDQNRFNLCFEERDFV PGENRIANIQ.DAIWN.SRK IVCLVSRHFLR........D
hutlr1       LEELQRNLQFHAFISYSGHD .......SFWVKNELLPNLE ......KEGMQICLHERNFV PGKSIVENII.TCIEK.SYK SIFVLSPNFVQ........S
hutlr6       LEELQRNLQFHAFISYSEHD .......SAWVKSELVPYLE ......KEDIQICLHERNFV PGKSIVENII.NCIEK.SYK SIFVLSPNFVQ........S
hutlr2       ....SRNICYDAFVSYSERD .......AYWVENLMVQELE ....NFNPPFKLCLHKRDFI PGKWIIDNII.DSIEK.SHK TVFVLSENFVK........S
hutlr4       .....ENI.YDAFVIYSSQD .......EDWVRNELVKNLE ....EGVPPFQLCLHYRDFI PGVAIAANIIHEGFHK.SRK VIVVVSQHFIQ........S
drotoll      L.DKDK..KFDAFISYSHKD .......QSFIEDYLVPQLE ....HGPQKFQLCVHERDWL VGGHIPENIM.RSVAD.SRR TIIVLSQNFIK........S
drotoll2     L.DKDK..TYDAFISYSHKD .......EELISK.LLPKLE ....SGPHPFRLCLHDRDWL VGDCIPEQIV.RTVDD.SKR VIIVLSQHFID........S
dromstprox   L.DKDK..RFDAFLAFTHKD .......EALLEEF.VDRLE ....RGRPRFQLCFYLRDWL AGESIPDCIG.QSIKD.SRR IIVLMTENFMN........S
wheeler      FEDAGK..LYDAIILHSEKD .......YEFVCRNIAAELE ....HGRPPFRLCIQQRD.L PPQASHLQLV.EGARA.SRK IILVLTRNLLA........T
arabrpp1     ....SRIWKHQVFPSF.H.. .GADVR......KTILSHIL .ESFRRK.GIDPFIDN.NIE RSKSIGHELK.EAIK.GSKI AIVLLSKNYAS........S
arabrrp5     ....SGRRRYDVFPSF.S.. .GVDVR......KTFLSHLI .EALDGK.SINTFIDH.GIE RSRTIAPELI.SAIR.EARI SIVIFSKNYAS........S
ngene        ....SSRWSYDVFLSF.R.. .GEDTR......KTFTSHLY .EVLNDK.GIKTFQDDKRLE YGATIPGELC.KAIE.ESQF AIVVFSENYAT........S
L6           ....PSV.EYDVFLSF.R.. .GPDTR......KQFTDFLY .HFLCYY.KIHTFRDDDELR KGKEIGPNLL.RAID.QSKI YVPIISSGYAD........S
CelegTran     .......KQIDVFISYRR.. .STGNQLA.SLIKVLLQL.. .RGYR......VFIDVDKLY AG.KFDSSLL.KNIQ.AAKH FILVLTPNSLDRLLNDDNCE
strepprot    .......RRIDAFVSYSR.. .AADSRLAPSIQRGLARLAK .KWYRTR.ALNVFRDQTDLS ASHALGASIE.RALA.DARF FVLLASPAAAE........S
CelegEST     ........SYHAFVSYSKKD ...........EKMVIDQL. CRPLEDE.DYQLCLLHRD.. .GPTYCSN............ .LXAISDELI.....AQMDS
Consensus    -----D-K-YDAF-SYS--D --------------V--L- P-VLE---GYKLCI--RD-- PG--I-E-II-E-I-K-SRR -I-VLSPNY---------S
```

Figure 2 (*Continued*)

(c)

```
                                                                                                  BOX 3
             131                                                                                    220
huIL18Racp   PSI...FELQAAVNLALDDQ .TLKLILIKFCYFQE..... .PESLPHLVKKALRVLPTVT WRG...LKSVPPNSR.FWAK MRYHMPVKNS
muIL18RacP   PRV...FELQAAVNLALVDQ .TLKLILIKFCSFQE..... .PESLPYLVKKALRVLPTVT WKG...LKSVHASSR.FWTQ IRYHMPVKNS
huIL18R      E.VR..YELESGLHEALVER .KIKIILIEFTPVTDFT... ...FLPQSLKLLKS.HRVLK WKAD...KSLSYNSR.FWKN LLYLMPAKTV
muIL18R      G.AR..RELESGLHEALVER .KIKIILIEFTPASNIT... ...FLPPSLKLLKS.YRVLK WRAD....SPSMNSR.FWKN LVYLMPAKAV
il1racp      G.TQALLELKAGL.ENMASR GNINVILVQYKAVKDMK... ...V..KELKRAKTVLTVIK WKGE...KSKYPQGR.FWKQ LQVAMPVKK.
muil1r1      SWLGQSSEEQIAIYNALIQE G.IKIVLLELEKIQDYE.K. ....MPDSIQFIKQKHGVIC WSGDFQERPQSAKTR.FWKN LRYQMPAQRR
ratil1r      SCLGQSSEEQIAIYDALIRE G.IKIILLELEKIQDYE.K. ....MPESIQFIKQKHGAIC WSGDFKERPQSAKTR.FWKN LRYQMPAQRR
huil1rec     SWLGGSSEEQIAMYNALVQD G.IKVVLLELEKIQDYE.K. ....MPESIKFIKQKHGAIR WSGDFTQGPQSAKTR.FWKN VRYHMPVQRR
chil1r       GILEDASEKHLAVYNALIQD G.IKIILIELEKIEDYA.N. ....MPESIKYVKQKYGAIR WTGDFSERSHSASTR.FWKK VRYHMPSRKH
ratil1rrp2   SFLKNLTEEQIAVYNALVQD G.MKVILIELERVKDYS.T. ....MPESIQYIRQKHGAIQ WDGDFTEQAQCAKTK.FWKK VRYHMP....
ratst2       S.KEFAYEQEIALHSALIQN N.SKVILIEMEPMGEAS.RL QLGDLQDSLQHLVKMQGTIK WREDHVADKQSLSSK.FWKH VRYQMPVPKR
humyd88      KECD..FQTKFAL..SLSPG AHQKRLI....PIKYKAMKK EFPSILRFI.....TV..CD YTNPCTK.SW......FWTR LAKAL.....
mumyd88      KECD..FQTKFAL..SLSPG VQQKRLI....PIKYKAMKK DFPSILRFI.....TI..CD YTNPCTK.SW......FWTR LAKAL.....
hutlr3       PLC.KRFKVHHAVQQAIEQN LDS.IILVFLEEIPDYKL.N HALCLRRGM.FKSHCI..LN WPVQKERIGA......FRHK L..QVALGSK
hutlr5       GWCLEAFS..YAQGRCL.SD LNSALIMVVVGSLSQYQLMK HQ.SI.RGF.VQKQQY..LR WPEDLQDVGW......FLHK LSQQILKKEK
hutlr1       EWC..HYELYFAHHNLFHEG .SNSLILILLEPIPQYSIPS SYHKLKSLMA..RRTY..LE WPKEKSKRGL......FWAN LRAAI.....
hutlr6       EWC..HYELYFAHHNLFHEG .SNNLILILLEPIPQNSIPN KYHKLKALMT..QRTY..LQ WPKEKSKRGL......FWAN IRAAF.....
hutlr2       EWC..KYELDFSHFRLFEEN .NDAAILILLEPIEKKAIPQ RFCKLRKIMN..TKTY..LE WPMDEAQREG......FWVN LRAAI.....
hutlr4       RWC..IFEYEIAQTWQFLSS .RAGIIFIVLQKVEKTLLRQ QV.ELYRLLS..RNTY..LE WEDSVLGRHI......FWRR LRKALLDGK.
drotoll      EWA..RLEFRAAHRSALNEG .RSRIIVIIYSDIGDVE.KL D.EELKAYLKM..NTY..LK WGDP.....W......FWDK LRFALPHRRP
drotoll2     VWA..RMEFRIAYQATLQDK .RERIIIILYRELEHMN.GI D.SELRAYLKL..NTY..LK WGDP.....L......FWSK LYYAMPHNRR
dromstprox   TWG..RLEFRLALHATSRDR .CKRLIVVLYPNVKNFD.SL D.SELRTYMAF..NTY..LE RSHP.....N......FWNK LIYSMPHT..
wheeler      EWN..RIEFRNAFHESLRGL .AQKLVIIEETSV.SAE.AE DVAELSPYLKSVPSNR..LL TCDR.....Y......FWEK LRYAIP....
arabrpp1     SWCLDELAEIMKCR.ELLG. ...QIVMTIFYEVDPTDIK. .KQT..GEFG..KAFTKTCK .GKTK...EYVERWRKALED VATIAGYHS.
arabrrp5     TWCLNELVEIHKCF.NDLG. ...QMVIPVFYDVDPSEVR. .KQT..GEFG..KVFEKTCE VSKDKQPGDQKQRWVQALTD IANIAGEDL.
ngene        RWCLNELVKIMECK.TRFK. ...QTVIPIFYDVDPSHVR. .NQK..ESFA..KAFEE..H ETKYKDDVEGIQRWRIALNE AANLKGSCD.
L6           KWCLMELAEIVRRQEEDPR. ...RIILPIFYMVDPSDVR. .HQT..GCYK..KAFRK..H ANKF..DGETIQNWKDALKK VGDL......
celegans     DWVHKELKCAFEHQKNIIPI FDTAFEFPTKEDQIPNDIRM ITKY..NGVKWVHDYQDACM AKVVRFITGELNRTTPTTKE MPSISRKTTQ
strepprot    KWVAKEIE..F.WQRN..RT SDT.F.LVALTD...GTIRW DDEA..GDFDW..SVTDA.. ..LPRSLSGYF.EAEPLWQD L.TWSRDADR
d70209       SQC................. ....LILVLTKHFVENE... .................... .................WKT LQ........
Consensus    -WC----E---A---AL--- ---KIILI-LE-I------- ----L----K--K--Y--L- W--D---------R-FWKK LRYAMP----
```

Table 1 Species and GenBank accession numbers for sequences used in the alignments shown in Figure 2

Sequence name	Species	Accession number
huIL18Racp	*Homo sapiens*	AAC72196
muIL18RacP	*Mus musculus*	AAC72197
huIL18R	*Homo sapiens*	U43672
muIL18R	*Mus musculus*	U43673
muil1racp	*Mus musculus*	U43673
muil1r1	*Mus musculus*	M20658
ratil1r	*Rattus norvegicus*	M95578
huil1r1	*Homo sapiens*	X16896
chil1r1	*Gallus gallus*	M81846
ratil1rrp	*Rattus norvegicus*	U49066
ratst2	*Rattus norvegicus*	U04317
humyd88	*Homo sapiens*	U84408
mumyd88	*Mus musculus*	U84409
hutlr3	*Homo sapiens*	U88879
hutlr5	*Homo sapiens*	AF051151
hutlr1	*Homo sapiens*	D13637
hutlr6	*Homo sapiens*	NM_006068
hutlr2	*Homo sapiens*	U88878
hutlr4	*Homo sapiens*	U88880
drotoll	*Drosphila melanogaster*	P08953
drotoll2	*Drosphila melanogaster*	AF140019
dromstprox	*Drosphila melanogaster*	U42425
wheeler	*Drosphila melanogaster*	L23171
arabrpp1	*Arabdopsis thaliana*	AF098963
arabrrp5	*Arabdopsis thaliana*	CAB46048
ngene	*Nicotiana glutinosa*	A54810
L6	*Linum usitatissimum*	AAD25976
CelegTran	*Caenorhabditis elegans*	Z49936
Strepprot	*Streptomyces coelicolor*	AL021411
CelegEST	*Caenorhabditis elegans*	D70209
B15R	Vaccinia virus	A19577
HuIL1R2	*Homo sapiens*	U64094
hupgp5	*Homo sapiens*	Z23091
hudecorin	*Homo sapiens*	P07585
rp105	*Homo sapiens*	D83597

analysis (Sims *et al.*, 1988). Its sequence revealed three immunoglobulin domains extracellularly. The cytosolic region was unique at that time, and therefore gave no clues as to a possible signal transduction mechanism. In 1992, site-directed mutagenesis studies revealed that six amino acids, conserved in the sequences of those receptors which had been cloned (human, mouse, and chicken IL-1RI and Toll) were found to be essential for signaling, namely Arg431, Lys515, Arg518, Phe513, Trp514, and Tyr519 (Heguy *et al.*, 1992). In addition, Pro521 was also required for maximal signaling capacity. Phe513 and Trp514 are in the conserved box 3 in the TIR domain of IL-1RI, as described above and shown in Figure 2c. The precise role of each of these amino acids is still undetermined. Presumably they are required for correct recognition of the signaling domain by signaling proteins. Given that they also occur in IL-1R AcP and MyD88, it is possible that the homotypic association of this region is required for signaling. Such a trimerization (i.e. between TIR domains in IL-1RI, IL-1R AcP, and MyD88) also occurs in the p55 TNF receptor system, in which trimeric TNF brings together three death domains in each p55 TNFR.

One amino acid, Tyr479, has been suggested to interact directly with a signaling protein. It has been suggested that p85 from PI-3 kinase associates with this Tyr, when phosphorylated, thereby recruiting p110, the catalytic subunit of PI-3 kinase, to the membrane (Marmiroli *et al.*, 1998). Tyr479 occurs in a motif recognized by p85, Tyr-Glu-X-Met. Another group has recently shown that p85 associates with IL-1R AcP rather than IL-1RI (Sizemore *et al.*, 1999). IL-1R AcP does not, however, contain a homologous Tyr. In that study the role of PI-3 kinase was to induce the phosphorylation of the p65 subunit of the transcription factor NFκB.

The importance of IL-1RI for biological effects of IL-1 can be seen from a number of studies. Antibodies to the IL-1RI have been shown to block IL-1 responses in several cell systems. In addition, IL-1RI knockout mice have revealed that this receptor mediates the induction of such IL-1 responses as increased IL-6 production and fever (Labow *et al.*, 1997). In addition, the acute-phase response, delayed-type hypersensitivity, and the ability to combat infection by *Listeria monocytogenes* were all impaired (Labow *et al.*, 1997). There was also an enhanced TH2-like response in such mice, which was suggested to be due to enhanced IL-4 and IL-10 production, implying that IL-1 limits the production of these cytokines. These results all indicate the importance of IL-1RI for inflammation and infection.

There has also been progress on the structural basis for IL-1RI recognition by IL-1. In 1997 the crystal structures of IL-1RI in a complex with IL-1β or IL-1Ra were solved to 2.7 Å resolution (Schreuder *et al.*, 1997; Vigers *et al.*, 1997). The structures revealed that immunoglobulin domains 1 and 2 are tightly linked, but that domain 3 is separate and connected by a flexible linker. IL-1β contacts the receptor via two regions, one of these interacting with the first two domains and the other with domain 3. The binding of IL-1β actually induces a conformational change, which results in domain 3 moving, such that the receptor actually wraps around the ligand (Vigers *et al.*, 1997). IL-1Ra, unlike IL-1β, does not induce a movement in domain 3 (Schreuder *et al.*, 1997). This may be because IL-1Ra only contacts IL-1RI at domains 1 and 2. Such a movement may allow interaction between IL-1RI and IL-1R AcP, which then triggers the signal. Using antibodies, two hydrophilic domains have been identified in IL-1R AcP, which are critical for signaling and hence possible interaction with IL-1RI (Yoon and Dinarello, 1998). A full account of the molecular changes which occur upon binding IL-1 in IL-1RI and IL-1R AcP extracellularly, and indeed intracellularly, will await the full elucidation of all the structures involved, both in the absence and presence of ligand.

IL-1RII

Crosslinking analysis had also revealed that on certain cell types, e.g. 70Z/3 pre-B cells, a second complex could be detected with a molecular weight of about 60 kDa, which could not be immunoprecipitated with anti-IL-1RI antibodies (Matsushima *et al.*, 1986; Horuk *et al.*, 1987). This was subsequently cloned and named IL-1RII (McMahon *et al.*, 1991). Extracellularly, it was highly homologous to IL-1RI, having three immunoglobulin domains. Intracellularly, however, it proved very different, having only a short cytoplasmic tail of 29 amino acids. In terms of expression, the expression of IL-1RII was more restricted than that of IL-1RI, being highly expressed on lymphoid and myeloid cells (McMahon *et al.*, 1991). Certain cell types express both receptors, and even in cells where IL-1RII appears to predominate, low levels of IL-1RI can be detected (Stylianou *et al.*, 1992).

Clear evidence then emerged that IL-1RI was responsible for generating the signal, with IL-1RII being unable to signal (Stylianou *et al.*, 1992; Sims *et al.*, 1993). Finally, it was shown that IL-1RII is actually shed from cells and acts as a decoy receptor, preventing IL-1 from binding to IL-1RI (Colotta *et al.*, 1993). This provides another means (along with IL-1Ra) of limiting IL-1 action. The shedding of IL-1RII from neutrophils can be induced by glucocorticoids and IL-4, which may form part of their anti-inflammatory effects.

Another interesting feature in relation to inhibitory effects of IL-1RII is its ability to interact with and thereby limit the availabilty of IL-1R AcP (Lang *et al.*, 1998). This may prevent IL-1RI from recruiting IL-1R AcP, which would inhibit IL-1 signaling. Also of note is the observation that IL-1RII has a very low affinity for IL-1Ra (Symons *et al.*, 1995), which is consistent with the role of both proteins as inhibitors of IL-1 binding to IL-1RI.

A search of the database with IL-1RII revealed a homolog in several viruses of the pox family, most notably vaccinia virus, whose B15R encodes a protein very similar to IL-1RII (Alcami and Smith, 1992; Spriggs *et al.*, 1992), and which, if deleted from the vaccinia genome, gives rise to a much more virulent virus. This indicates that the role of B15R in vaccinia is to limit the damage caused by the overproduction of IL-1 during infection. Of particular note was the observation that virus lacking B15R is much more pyrogenic than wild-type virus (Alcami and Smith, 1996), providing further proof that IL-1 is a key pyrogenic cytokine, and in the case of viral infection probably the most important endogenous pyrogen.

IL-1R ACCESSORY PROTEIN

The discovery of IL-1RI and IL-1RII appears to account for all of the binding capacity of cells for IL-1. Early experiments also indicated that IL-1RI alone appeared capable of driving IL-1 signals (Curtis *et al.*, 1989). This was based on the reconstitution of signaling in Chinese hamster ovary cells by transfection of IL-1RI. It was then found that an antibody which did not recognize IL-1RI or IL-1RII was able to block IL-1 signals in cells without affecting IL-1 binding (Greenfeeder *et al.*, 1995). The target for this antibody was found to be IL-1R AcP. As mentioned above, this is the second chain of the IL-1R complex, which does not bind IL-1 but which is essential for signaling (Greenfeeder *et al.*, 1995; Wesche *et al.*, 1997b).

IL-1R AcP is widely expressed and on the whole its expression correlates with that for IL-1RI. The exception to this is in brain, where in rat they have been found not to colocalize (Liu *et al.*, 1996), although whether this is a problem with different detection sensitivities requires further analysis. Apart from its role in signaling, IL-1R AcP also appears to increase the affinity of IL-1RI for ligand by about 5-fold (Greenfeeder *et al.*, 1995). In spite of this, it is still not clear whether IL-1RI and IL-1R AcP are in a complex prior to the addition of ligand. As discussed above, IL-1 may cause a conformational change in IL-1RI which recruits IL-1R AcP to the complex. Another more likely scenario is that the addition of ligand increases the affinity of IL-1RI for IL-1R AcP, which prior to ligand is in a low-affinity complex. It is also unclear whether IL-1RI and IL-1R AcP interact with the same signaling components. Even though both have a TIR domain, there are differences in sequence, which might lead to the recruitment of different proteins during signaling. Indeed, it has been suggested from immunoprecipitation studies that the IL-1 receptor-associated kinase (IRAK) is recruited via IL-1R AcP, with IRAK-2 being recruited to IL-1RI (Huang *et al.*, 1997; Muzio *et al.*, 1997; Volpe *et al.*, 1997). Both also appear to be recruited via MyD88 (Muzio *et al.*, 1997; Wesche *et al.*, 1997a; Burns *et al.*, 1998). Whether IL-1R AcP will have a subtly different role from IL-1RI in signaling will await further clarification. A plausible scenario is that the interaction between IL-1RI and IL-1R AcP creates a novel surface which allows MyD88 to be recruited in a homotypic manner.

IL-18 RECEPTOR COMPLEX

In 1996 a homology-based search of the database uncovered two additional members of the IL-1R family, which were termed IL-1 receptor-related proteins 1 and 2 (IL-1Rrp1 and 2) (Lovenberg *et al.*, 1996; Parnet *et al.*, 1996). At the time these had no known ligands but chimeric receptors comprising extracellular IL-1RI and intracellular IL-1Rrp1 could signal in response to IL-1 (Parnet *et al.*, 1996). It was subsequently shown that IL-1Rrp1 is the receptor for IL-18 (Torigoe *et al.*, 1997), a cytokine first described as IFNγ-inducing factor, which had a strikingly similar structure to IL-1 (Bazan *et al.*, 1996). Similar structural features are therefore likely to form the basis for interactions between IL-1/IL-1RI and IL-18/IL-1Rrp1. IL-18R has replaced IL-1Rrp1 as the name of the IL-18 receptor.

Similarities between the two cytokine systems also extend to signaling pathways, with IL-18 activating identical pathways to IL-1 (Kojima *et al.*, 1998). In addition, MyD88 knockout mice are unresponsive to both IL-1 and IL-18 (Adachi *et al.*, 1998), confirming the importance of MyD88 as the key adapter for TIR domain-containing receptors. Cells prepared from IRAK knockout mice are similarly unresponsive to IL-1 or IL-18 (Kanakaraj *et al.*, 1999), further emphasizing the importance of IRAK in signaling by the family. The lack of IRAK could obviously not be compensated for by IRAK-2, or indeed the recently described third IRAK, termed IRAKm, which appears to be exclusively expressed in myeloid cells (Wesche *et al.*, 1999).

IL-18R is detectable in lung, spleen, heart, testis, peripheral blood T cells, and NK cells (Torigoe *et al.*, 1997). It is notably absent from brain, although again this may be a detection problem. IL-18R is particularly associated with natural killer cell and TH1 cell activation. An intriguing observation concerns the relative roles of IL-1 and IL-18 in helper T cell function. IL-18 appears preferentially to augment TH1 cell effector responses, whereas IL-1, although capable of potentiating IL-12-induced TH1 cell development, is also capable of augmenting TH2 cell function (Robinson *et al.*, 1997). This latter result is inconsistent with the data from IL-1RI knockout mice, which have an enhanced TH2 response (Labow *et al.*, 1997). This is discussed further below. A role for IL-18 in TH2 cell function is consistent with the lack of IL-18R expression on TH2 cells (Xu *et al.*, 1998). A role for IL-18R in TH1 cells is borne out from studies with IL-18R knockout mice (Hoshino *et al.*, 1999a). These mice had TH1 cells which were unresponsive to IL-18 and defective cytolytic responses in their NK cells. In addition, TH1 cell development was impaired. Similar phenotypes were observed in MyD88 and IRAK knockout mice, which again points to the importance of these signaling proteins in IL-18R action.

In 1998 the cloning of the accessory protein for IL-18R was reported (Born *et al.*, 1988). This was again discovered from a homology search and, because of its homology particularly with the IL-1R AcP, was named AcPL, for accessory protein-like. Its expression appears to be more restricted than IL-18R, in that it is not detectable in heart or testis by northern blot analysis. It seems highly likely at this stage that AcPL will function in an exactly analogous manner to IL-1R AcP in IL-1 signaling.

T1/ST2

In 1989 two separate groups reported on the cloning of what was then the only other protein with a similar sequence to IL-1RI (Klemenz *et al.*, 1989; Tominaga, 1989). This was named T1 or ST2, and was cloned as a delayed early-response gene induced in response to proliferative signals. It was characterized as a secreted protein whose sequence suggested a soluble IL-1 receptor, with a transmembrane form subsequently being characterized (Yanagisawa *et al.*, 1993). The homology with IL-1RI in the transmembrane form extends throughout the molecule: T1/ST2 contains immunoglobulin domains extracellularly and the TIR domain intracellularly. Soluble and membrane versions are expressed in different tissues. This is because different promoters lead to different polyadenylation sites and from there to different forms of mRNA (Bergers *et al.*, 1994). The soluble form is mainly produced by fibroblasts and possibly mast cells, whereas the membrane-bound form is predominantly expressed on hematopoietic cells (Bergers *et al.*, 1994). However, it has also been suggested that in hematopoietic cells, transmembrane T1/ST2 is expressed constitutively, while the soluble form is induced upon stimulation with proinflammatory agents, including IL-1. Soluble T1/ST2 is in fact strongly induced during inflammation *in vivo* (Kumar *et al.*, 1997). A model for the roles of soluble and membrane-bound T1/ST2 has been proposed whereby the putative ligand when bound to the receptor prevents proliferation, with soluble receptor acting to block this effect (Gayle *et al.*, 1996). This may explain the high levels of soluble T1/ST2 correlating with proliferation.

The ligand for T1/ST2 has yet to be found. A putative ligand was described in 1996 which could bind T1/ST2 (Gayle *et al.*, 1996). This ligand was unrelated to IL-1 and could not generate a signal. The TIR domain in T1/ST2 is capable of signaling, as demonstrated in studies with chimeras between extracellular IL-1RI and intracellular T1/ST2 (Mitcham *et al.*, 1996).

Recently, a role for T1/ST2 in TH2 cell function has been suggested. Differential display analysis of TH1 or TH2 cells revealed T1/ST2 to be strongly expressed in TH2 cells (Lohning *et al.*, 1988). Its expression on TH2 cells was independent of IL-4, IL-5, or IL-10, making it the only TH2 marker that is IL-4-independent. T1/ST2 was found to play a critical role in TH2 cell function in that a neutralizing antibody to T1/ST2 attenuated TH2-driven responses *in vivo*. The role of soluble T1/ST2 may therefore be to limit TH2 responses. The uncovering of the ligand for T1/ST2 on TH1/TH2 will therefore be of great significance for TH1/TH2 biology (see chapter on TH1/TH2 Interleukins).

Two reports have recently appeared describing T1/ST2-deficient transgenic mice. Confusingly, one report showed no impairment in TH2 cell development or function (Hoshino *et al.*, 1999b), while the other showed impaired TH2 cell effector function (Townsend *et al.*, 2000). These opposing results are unexplained at present.

IL-1RI, IL-18R, AND T1/ST2 IN TH1 AND TH2 CELL FUNCTION

From studies on T1/ST2, IL-18R, and IL-1R it is clear that these members of the IL-1R/TLR family have divergent roles in T helper subtype activation. From studies into the development of TH1 or TH2 responses it can be inferred that IL-1RI, IL-18R, and

Figure 3 The signaling pathway to NFκB activated by IL-1 in mammalian cells, and that to Dorsal activated by Spatzle in *Drosophila melanogaster* is highly homologous. The ligands in each case (IL-1 and Spatzle) are very different, since the extracellular domains of the receptors to which they bind differ. The cytosolic regions of the receptors have TIR domains, however. In spite of this homology, the known adapters utilized differ, in that MyD88 is the adapter for IL-1RI, while that for Toll is Tube. A mammalian homolog of Tube has not been described. IL-1 receptor-associated kinase (IRAK) is highly homologous to Pelle, however, with Cactus being an I-κB homolog and Dorsal, similar to p50 and p65 in mammals, being a Rel family member. The identity of the proteins linking Pelle to phosphorylation of cactus is not known. The pathway in *Drosophila* shown is involved in the establishment of dorsoventral polarity. In adult flies a similar pathway is used in response to infection, with Dif, another Rel family member, being involved.

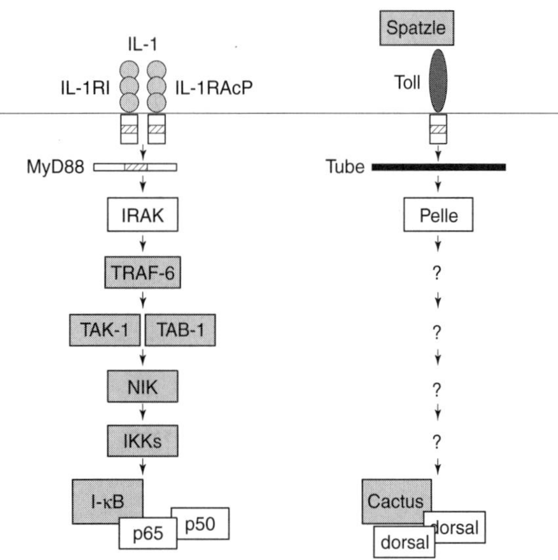

T1/ST2 are expressed on TH0 cells. IL-1 acts to potentiate IL-12-driven development of TH1 cells (Robinson *et al.*, 1997). IL-18R but not IL-1R is expressed on TH1 cells, where IL-18 acts to promote TH1 responses in combination with IL-12. IL-1RI on the other hand is expressed on TH2 cells where it induces IL-4 production and proliferation. T1/ST2, however, is only expressed on TH2 cells and is required for TH2 cell function (Lohning *et al.*, 1998). It therefore appears that IL-1 can affect both cell types, with IL-18 affecting TH1 cells and the as-yet undiscovered ligand for T1/ST2 affecting TH2 cells. Such observations are in part based on certain strains of mice and a definitive account of this process in humans is as yet lacking.

Furthermore, studies with transgenic mice indicate that the role of IL-1 is to promote TH1 responses in spite of its *in vitro* effect on TH2 cells (Labow *et al.*, 1997). In addition, T1/ST2 signaling would be expected to be impaired in MyD88 and IRAK knockout mice. TH1 responses are impaired in such mice but the status of TH2 responses was not determined (Adachi *et al.*, 1998; Kanakaraj *et al.*, 1999). It would be of interest to determine such responses, which, if impaired, could be explained in terms of a lack of T1/ST2 signaling. The fine balance of IL-1, IL-18, and the T1/ST2 ligand may, therefore, have a key role in driving TH1 or TH2 responses (see chapter on TH1/TH2 Interleukins).

TOLL

Subsequent to T1/ST2 being described as a homolog of IL-1R, an important homology was described in 1991 between the *Drosophila melanogaster* protein Toll and IL-1RI (Gay and Keith, 1991). This represented the first description of the TIR domain, although the term TIR has only recently been coined. The homology in the cytosolic regions of Toll and IL-1R defines the receptor family, as shown in Figure 1 and **Figure 3**. Extracellularly, Toll is unlike IL-1RI, in that it does not have immunoglobulin domains. Instead, Toll has leucine-rich repeats (Hashimoto *et al.*, 1988). As a result, the ligand for Toll is quite different: it is a protein termed Spatzle (Morisato and Anderson, 1994), although direct binding of Spatzle to Toll has not been described. Spatzle has a cysteine knot structure, which also occurs in members of the NGF family (Mizuguchi *et al.*, 1998). In addition, the structure of Toll has been modeled and, intriguingly, it resembles platelet glycoprotein 1b (Keith and Gay, 1990), the receptor for von Willebrand factor. In a further similarity, Spatzle, like von Willebrand factor, is generated as a result of a protease cascade, activated in *Drosophila* during development.

Toll was first described as a maternal-effect gene involved in the development of dorsoventral polarity in the developing *Drosophila* embryo (Hashimoto *et al.*, 1988). The homology between Toll and IL-1RI was particularly intriguing given the target for both pathways (reviewed in O'Neill and Green, 1998). The homologs which occur in both pathways are shown in Figure 3. Toll activates two proteins, termed Tube and Pelle, which in turn lead to the release of a protein cactus from the transcription factor Dorsal. Dorsal is the *Drosophila* homolog of NFκB, while cactus is the IκB homolog. Given that the NFκB activation is central to IL-1 action, the TIR domain may have evolved to activate NFκB. Also of note is the homology between IRAK and Pelle, pointing to further conservation in both pathways. A homolog of Tube has not yet been found and it is possible that MyD88 acts as a functional homolog in mammals. In adult *Drosophila* Toll has an additional role in the response to infection. Here, Toll activates another NFκB family member, Dif, which regulates the expression of antifungal peptides such as drosomycin (Lemaitre *et al.*, 1997; Meng *et al.*, 1999). It therefore appeared that the Toll and IL-1 systems were conserved across evolution, being activated during infection. This view is further strengthened by the discovery of a human Toll-like receptor, described below.

18-Wheeler, MstProx, and Tehao

Three further proteins homologous to Toll have been described in *Drosophila*: 18-Wheeler, MstProx, and Tehao (Eldon *et al.*, 1994; Sims and Dower, 1994). 18-Wheeler is again involved in development but in the adult is also involved in infection, inducing the production of antibacterial peptides such as attacin, again via the activation of Dif (Williams *et al.*, 1998). This is a potentially interesting difference in that Toll responds to fungal pathogens while 18-Wheeler responds to bacteria (Lemaitre *et al.*, 1997; Williams *et al.*, 1997). Such subtlety may also be apparent in the mammalian TLRs, described below. The role of MstProx and Tehao are as yet unknown. The fact that similar systems are being used in development and immunity possibly indicates the efficiency of this signaling system in the regulation of gene expression.

Human Toll-like receptors

In 1998, five proteins were described in humans which were homologs of Toll (Rock *et al.*, 1998). Similar to IL-1RI, they all had conserved cytosolic regions, but extracellularly they resembled Toll, having leucine-rich repeats. They were termed human TLRs 1–5. More recently, a sixth TLR, termed TLR-6, has been described (Takeuchi *et al.*, 1999). The sequence of TLR-1 had been described previously and termed rsc786 (Sims and Dower, 1994). In addition, at virtually the same time a human Toll (termed hToll) had been described, which was identical to TLR-4 (Medzhitov *et al.*, 1997). The description of this family and its homology to Toll indicated that TLRs might be important participants in innate immunity since it had long been postulated that human homologs of proteins involved in defense in lower organisms would be found. TLR-4 was shown to be a potent upregulator of B7.1 expression in T cells, and also inflammatory cytokine production (including IL-1) when expressed as a fusion protein with CD2, which targeted it to the membrane (Medzhitov *et al.*, 1997). These responses are part of the initial innate response to infection. It therefore appeared that the TLR family, using the TIR domain to signal, was a key participant in innate immunity and therefore inflammation.

There are differences between the expression patterns of the TLRs (Rock *et al.*, 1998; Takeuchi *et al.*, 1999). TLR-1 is ubiquitously expressed and at a higher level compared to other TLRs. TLR-2 is expressed in brain, heart, muscle, and lung. TLR-3 is expressed in a similar pattern to TLR-2, but with high expression in the pancreas and placenta. In contrast, TLR-4 and TLR-5 are less well-expressed, with placenta and peripheral blood leukocytes showing high-level expression for TLR-4, and ovary and monocytes highly expressing TLR-5. TLR-6 is expressed in spleen, thymus, ovary, and lung (Takeuchi *et al.*, 1999). In addition, certain TLRs have different-length mRNA transcripts in certain tissues, indicating further complexity (Rock *et al.*, 1998).

A protein termed RP105 can also be included in the TLR subfamily. This was described in 1995 – before the TLR family was described – and was shown to be homologous to Toll extracellularly (Miyake *et al.*, 1995). It has a short cytoplasmic sequence lacking the TIR domain. It is expressed predominantly on B cells and when crosslinked drives proliferation (Chan *et al.*, 1998).

TLR-2 and TLR-4 as signaling receptors for bacterial products

The most immediate question which arises concerning TLRs is the nature of the ligands which bind them. The work on TLR-4 that demonstrated that it could signal, involved expressing it as a membrane-localized fusion protein (Medzhitov *et al.*, 1997). The breakthrough on the identity of a possible ligand came from work on gram-negative LPS. In 1998 it was demonstrated by two separate groups that in human cells TLR-2 was required for LPS signaling (Kirschning *et al.*, 1998; Yang *et al.*, 1998). LPS had been known for some time to induce signals very similar to IL-1, and there had been suspicions that the effect of LPS may have been mediated via IL-1 production. It was also known that CD14 was required for LPS action (Wright *et al.*, 1990), although since CD14 was a glycophosphatidylinositol-anchored protein, it was suspected that a second chain was needed to generate the signal. This second chain was postulated to be TLR-2 since it was possible fully to reconstitute LPS responses in 293 cells cotransfected with CD14 and TLR-2 (Kirschning *et al.*, 1998; Yang *et al.*, 1998).

There is also evidence that TLR-4 is required for LPS signaling. This has come from studies into LPS-resistant mice. The genetic mutation in the C3H/HeJ mouse has long been sought after and was found to be in the gene for TLR-4 (Poltarok *et al.*, 1998). A single point mutation at Pro712 in box 2 of the TIR domain was found in the TLR-4 gene in these mice, which rendered TLR-4 unable to signal. Similarly, in the C57BL/10ScCr mouse, which is also LPS-resistant, there is a null mutation in the TLR-4 gene (Poltorak *et al.*, 1998).

It therefore appears that TLR-2 and/or TLR-4 are signaling receptors for LPS. This further expands the utility of the IL-1/TLR family of proteins, and clearly emphasizes their importance for innate immunity. There are some unanswered questions in relation to LPS and the TLRs. TLR-2 is intact in both C3H/HeJ and C57BL/10ScCr mice and yet they are still unresponsive to LPS (Poltorak *et al.*, 1998). Could there be differences between humans and mice in relation to which TLR is responsible for LPS action?. It has been known for some time that different animals respond differently to LPS mimetics, with certain lipid A analogs acting as LPS antagonists in humans but agonists in mice and hamsters (Golenbock *et al.*, 1991). This could be due to species differences in terms of which TLR is responding to LPS. Alternatively, perhaps a dimer of TLR-2 and -4 provides the optimal signal transducer for LPS. It has also been shown recently that Chinese hamster ovary cells are responsive to LPS when transfected with CD14, but that TLR-2 is not required for this response since the TLR-2 gene has a deletion which results in a protein without a cytosolic region (Heine *et al.*, 1999); indeed, this feature is also present in the hamster genome. TLR-4 may therefore be more important in hamsters, as in mice.

Further complexity has recently been indicated, demonstrating that another protein, termed MD-2, is required for TLR-4 responsiveness, indicating that all of the components required for LPS responses may not have been uncovered (Shimazu *et al.*, 1999). Another question concerns whether LPS binds to either TLR and the precise role of CD14. If TLRs are functionally analogous to Toll, they might be expected to have endogenous ligands, although no Spatzle homolog, if indeed this is the ligand for Toll, has yet been found in mammals. It is possible that they might function as adhesion molecules, as has been suggested for Toll (Keith and Gay, 1990), possibly aggregating homo- or heterotypically. Another

Figure 4 IL-1, IL-18, and LPS: receptors and signaling. The IL-1 receptor complex comprises IL-1RI and IL-1R AcP, with IL-1RI undergoing a conformational change upon ligand binding. A similar conformational change may occur in IL-18R, which has its own accessory protein (AcPL or IL-18RAcP). LPS acts via LBP, CD14, MD-2, and TLR-4, with TLR-4 acting as the signaling component. The TIR domains (hatched box), probably through homotypic interactions, recruit MyD88 via its TIR domain, which then mediates the signals common to all three stimuli via IRAK, IRAK-2, and TRAF-6. This leads to the activation of NFκB and the MAP kinases p38, JNK, or p42/p44, ultimately driving proinflammatory gene expression. The signaling pathways are reviewed by O'Neill and Greene (1998). The role of NFκB-inducing kinase (NIK) has recently been disputed (Shinkura *et al.*, 1999).

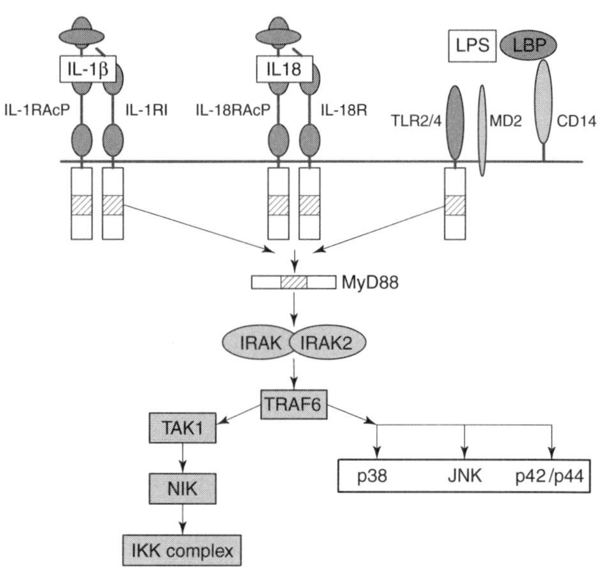

question concerns specificity in the recognition of different TLRs by microbial products. As mentioned above, it is possible that, as in *Drosophila*, different TLRs may recognize different pathogens. It has been shown that peptidoglycan and lipoteichoic acid, both important determinants of the innate response to gram-positive bacteria, also appear to signal via TLR-2 (Schwander *et al.*, 1999). This indicates that TLR-2 may not be specific for LPS. It has also been shown that TLR-2 shows different sensitivities to LPS from different bacteria, possibly pointing to specificity in the response (Yang *et al.*, 1998).

Finally, it has been shown that TLR-2 internalizes in response to products from gram-positive bacteria and yeast, and is recruited to the phagosome during phagocytosis. This study also demonstrates that TLR-2 does not respond to LPS, which requires TLR-4, which in turn is unresponsive to gram-positive products (Underhill *et al.*, 1999). It would appear, therefore, that a consensus is emerging whereby it is TLR-4 which is required for responses to LPS, with TLR-2 mediating the response to products from gram-positive bacteria and yeast. Intern-alization would also appear to be critical for their effects. This area is likely to be most fruitful for our understanding of host defense against microbial pathogens.

TLR signaling

With regard to signaling, given the presence of the TIR domain it was expected that TLRs would signal in the same way as IL-1R. This indeed appears to be the case with TLR-2 and TLR-4 utilizing MyD88, IRAK, IRAK-2, and TRAF-6 in the pathway to NFκB (Muzio *et al.*, 1998; Yang *et al.*, 1999; Zhang *et al.*, 1999). It can therefore be concluded that IL-1, IL-18, and LPS all use the same signaling components in their mechanism of action, as shown in **Figure 4**. Early studies with TLR-1 (then called rsc786) revealed that a chimera between extracellular IL-1RI and intracellular TLR-1 was unable to signal, at least in terms of activation of the IL-8 promoter (Mitcham *et al.*, 1996), possibly indicating that there may be different signaling roles for different TLRs. The area of TLRs has therefore opened up new avenues of research which are likely to yield important results for our understanding of immunity and inflammation.

PLANT MEMBERS OF THE FAMILY

The IL-1R/TLR family also has plant members. In 1994 the tobacco mosaic virus resistance gene N was cloned from tobacco and, intriguingly, was found to be homologous to IL-1RI and Toll (Whitham *et al.*, 1994). N protein was found to be more related to Toll in that, along with the TIR domain, it also had four leucine-rich repeats. The role of N protein in tobacco is to activate a hypersensitive response in the plant during infection, which leads to necrosis of infected plant tissue, thereby halting pathogen growth and spread. The observation that it contained a TIR domain indicated that this family of proteins is likely to be very ancient, arising in the common unicellular ancestor to plants and animals, if not before, and having a role in defense. It is also likely that the Toll subgroup of the family arose first, with immunoglobulin domains replacing leucine-rich repeats in the IL-1RI subgroup at a later stage. Subsequently several other plant proteins involved in disease resistance were described, including L6 in flax and Rpp5 in *Arabidopsis* (Hammond-Kosack and Jones, 1996; Parker *et al.*, 1997). As more genome information is emerging from plant DNA sequencing, this is the fastest-growing group in the family. The ligands for the plant receptors are wholly unknown. Since they are exclusively cytosolic, the ligands must also be cytosolic and could conceivably be derived from intracellular infectious agents.

Apart from receptors, disease resistance in plants also involves homologs of IRAK. These include pto and fen in tomato (Jia *et al.*, 1997). This implies that signaling components are also conserved between species, further emphasizing the evolutionary conservation of this system. In terms of downstream consequences NFκB family members have yet to be described in plants, although p38 MAP kinases are strongly conserved in all species examined (Caffrey *et al.*, 1999), implying that, similar to IRAK, pto and the other plant homologs may be involved in p38 activation.

The conclusion one can draw at this stage is that the IL-1/TLR receptor system and certain of the signaling proteins activated by it are pangenomic.

MYD88

The final member of the family, MyD88, like the plant members, is exclusively cytosolic but, unlike the plant members, the homology is only apparent in the TIR domain: It lacks immunoglobulin domains and leucine-rich repeats. MyD88 was first cloned as an IL-6-inducible protein expressed during myeloid differentiation, which had no known function (Lord *et al.*, 1990). Subsequently, its C-terminal domain was shown to be homologous to Toll, its N-terminus intriguingly containing a death domain (Burns *et al.*, 1998), implying that some of the protein–protein interactions in IL-1 signaling may be similar to those in signaling by the TNF receptor family. IRAK is also purported to have a death domain. Based on these structural features, experiments were performed to determine whether MyD88 would dimerize (Burns *et al.*, 1998). This was shown to be the case, via both the death domains and the TIR domains. In addition it was shown that MyD88 coprecipitates with the IL-1 receptor complex, but only when IL-1RI was liganded, implying that IL-1RI/IL-1R AcP acts to recruit MyD88 to the complex (Wesche *et al.*, 1997a; Burns *et al.*, 1998). Subsequent experiments clearly indicated that MyD88 then recruits IRAK to the complex, MyD88 therefore acting in an analogous manner to TRADD in TNF signaling and Tube in Toll signaling (Muzio *et al.*, 1997; Wesche *et al.*, 1997a; Burns *et al.*, 1998). The simplest scenario is therefore for MyD88 to be the adapter for recruitment of both IRAKs. As stated above and as shown in Figure 4, the importance of MyD88 in IL-1 and IL-18 signaling can be seen in the MyD88 knockout mouse, which is unable to respond to either cytokine (Adachi *et al.*, 1998). Further studies have shown that *in vivo*, these mice are unresponsive to LPS (Kawai *et al.*, 1999). *In vitro*, however, LPS still activates NFκB and JNK, although these responses are delayed. This implies that MyD88 is critical for *in vivo* responses to LPS, but that another adapter may compensate in the cells examined in culture.

CONCLUSIONS AND PERSPECTIVES

With increasing information emerging from DNA sequencing, it is clear that the IL-1R/TLR family, and some of the signaling proteins used by the family, is growing rapidly. The role of the family at this stage appears to be in host defense and response to injury, with family members providing the receptors for the key cytokines IL-1 and IL-18, and important pathogen-derived factors such as LPS. The remarkable conservation across species, both in receptors and signaling components, points to the efficiency of this system as a rapid responder to infection and injury, which probably arose in early eukaryotes if not prokaryotes. A number of questions can be posed at this stage. These include:

1. What are the ligands for the orphan receptors in the family? Finding the ligand for ST2 would be an important goal for understanding the control of TH2 cells. Similarly, determining the ligands for TLR family members would be of great use in efforts to understand host–pathogen interactions and also in the area of adjuvants.
2. Are there other family members yet to be described? It is distinctly possible that many more receptors will be described, particularly in plants, whose ligands will be novel cytokines or molecules from pathogens.
3. IL-1RI and IL-1R AcP, IL-18R and AcPL are not necessarily coexpressed in all tissues, indicating that other receptors for IL-1 and IL-18 may exist. This may be particularly relevant in brain, where such coexpression is particularly absent.
4. Are there other MyD88-like molecules, some of which might be inhibitory in an analogous way to the recently described suppressor of death domains proteins?

Future work into this receptor family will yield much information, not only on the molecular basis to aspects of immunity and inflammation, but also on the evolution of the inflammatory and immune responses.

References

Adachi, O., Kawai, T., Takeda, K., Matsumoto, M., Tsutsui, H., Sakagami, M., Nakanishi, K., and Akira, S. (1998). Targetted disruption of the MyD88 gene results in loss of IL-1 and IL-18-mediated function. *Immunity* **9**, 143–150.

Alcami, A., and Smith, G. L. (1992). A soluble receptor for interleukin-1beta encoded by Vaccinia virus: a novel mechanism of virus modulation of the host response to infection. *Cell* **71**, 153–162.

Alcami, A., and Smith, G. L. (1996). A mechanism for the inhibition of fever by a virus. *Proc. Natl Acad. Sci. USA* **93**, 11029–11034.

Bazan, J. F., Timmins, J. C., and Kastelein, R. A. (1996). A newly defined interleukin-1? *Nature* **379**, 591–592.

Bergers, G., Reikerstorfer, A., Braselmann, S., Graninger, P., and Busslinger, M. (1994). Alternative promoter usage of the Fos-responsive gene Fit-1 generates mRNA isoforms coding for either secreted or membrane-bound proteins related to the IL-1 receptor. *EMBO J.* **13**, 1176–1188.

Bird, T. A., and Saklatvala, J. (1986). Identification of a common class of high affinity receptors for both types of porcine interleukin-1 on connective tissue cells. *Nature* **324**, 263–266.

Bird, T. A., and Saklatvala, J. (1987). Studies on the fate of receptor-bound ^{125}I-interleukin 1 beta in porcine synovial fibroblasts. *J. Immunol.* **139**, 92–97.

Born, T. L., Thomassen, E., Bird, T. A., and Sims, J. E. (1998). Cloning of a novel receptor subunit, AcPL, required for interleukin-18 signaling. *J. Biol. Chem.* **273**, 29445–29450.

Bresnihan, B., Alvaro-Gracia, J. M., Cobby, M., Doherty, M., Domljan, Z., Emery, P., Nuki, G., Pavelka, K., Rau, R., Rozman, B., Watt, I., Williams, B., Aitchison, R., McCabe, D., and Musikic, P. (1998). Treatment of rheumatoid arthritis with recombinant human interleukin-1 receptor antagonist. *Arthr. Rheum.* **41**, 2196–2204.

Burns, K., Martinon, F., Esslinger, C., Pahl, H., Schneider, P., Bodmer, J. L., Di Marco, F., French, L., and Tschopp, J. (1998). MyD88, an adapter protein involved in interleukin-1 signaling. *J. Biol. Chem.* **273**, 12203–12209.

Caffrey, D., O'Neill, L. A. J., and Shields, D. (1999). Evidence for co-duplication in MAP kinase pathways in multiple species. *J. Mol. Evol.* **49**, 567–582.

Cao, Z., Henzel, W. J., and Gao, X. (1996). IRAK: a kinase association with the interleukin-1 receptor. *Science* **271**, 1128–1131.

Carrie, A., Jun, L., Bienvenu, T., Vinet, M. C., McDonell, N., Couvert, P., Zemni, R., Cardona, A., Van Buggenhout, G., Frints, S., Hamel, B., Moraine, C., Ropers, H. H., Strom, T., Howell, G. R., Whittaker, A., Ross, M. T., Kahn, A., Fryns, J. P., Beldjord, C., Marynen, P., and Chelly, J. (1999). A new member of the IL-1 receptor family highly expressed in hippocampus and involved in X-linked mental retardation. *Nature Genet.* **23**, 25–31.

Chan, V., Mecklenbrauker, I., Su, I., Texido, G., Leitges, M., Carsetti, R., Lowell, C. A., Rajewsky, K., Miyake, K., and Tarahovsky, A. (1998). The molecular mechanism of B cell activation by toll-like receptor RP105. *J. Exp. Med.* **188**, 93–101.

Chow, J. C., Young, D. W., Golenbock, D. T., Christ, W. J., and Gusovsky, F. (1999). Toll-like receptor-4 mediates LPS-induced signal transduction. *J. Biol. Chem.* **274**, 10689–10692.

Colotta, F., Re, F., Muzio, M., Bertini, R., Polentarutti, N., Sironi, M., Giri, J. G., Dower, S. K., Sims, J. E., and Mantovani, A. (1993). Interleukin 1 type II receptor: a decoy target for IL-1 that is regulated by IL-4. *Science* **261**, 472–475.

Curtis, B. M., Gallis, B., Overell, R. W., McMahan, C. J., DeRoos, P., Ireland, R., Eisenman, J., Dower, S. K., and Sims, J. E. (1989). T-cell interleukin 1 receptor cDNA expressed in Chinese hamster ovary cells regulates functional responses to interleukin 1. *Proc. Natl Acad. Sci. USA* **86**, 3045–3049.

Dinarello, C. A. (1996). Biologic basis for interleukin-1 in disease. *Blood* **87**, 2095–2147.

Dower, S. K., Kronheim, S. R., Hopp, T. P., Cantrell, M., Deeley, M., Gillis, S., Henney, C. S., and Urdal, D. L. (1986a). The cell surface receptors for interleukin-1α and interleukin-1β are identical. *Nature* **324**, 266–270.

Dower, S. K., Call, S. M., Gillis, S., and Urdal, D. L. (1986b). Similarities between the interleukin-1 receptors on a murine T-lymphoma cell line and on a murine fibroblast cell line. *Proc. Natl Acad. Sci. USA* **83**, 1060–1066.

Eisenberg, S. P., Evans, R. J., Arend, W. P., Verderber, E., Brewer, M. T., Hannum, C. H., and Thompson, R. C. (1990). Primary structure and functional expression from complimentary DNA of a human interleukin-1 receptor antagonist. *Nature* **343**, 341–343.

Eldon, E., Kooyer, S., D'Evelyn, D., Duman, M., Lawinger, P., Botas, J., and Bellen, H. (1994). The *Drosophila* 18-Wheeler is required for morphogenesis and has striking similarities to Toll. *Development* **120**, 885–895.

Gay, N. J., and Keith, F. (1991). *Drosophila* Toll and IL-1 receptor. *Nature* **351**, 355–356.

Gayle, M. A., Slack, J., Bonnert, T., Renshaw, B. R., Sonoda, G., Taguchi, T., Testa, J. R., Dower, S. K., and Sims, J. E. (1996). Cloning of a putative ligand for the T1/ST2 receptor. *J. Biol. Chem.* **271**, 5784–5789.

Golenbock, D. T., Hampton, R. Y., Qureshi, N., Takayama, K., and Raetz, C. R. (1991). Lipid A-like molecules that antagonise the effects of endotoxin on human monocytes. *J. Biol. Chem.* **266**, 19490–19497.

Greenfeeder, S. A., Nunes, P., Kwee, L., Labow, M., Chizzonite, R. A., and Ju, G. (1995). Molecular cloning and characterisation of a second subunit of the interleukin-1 receptor complex. *J. Biol. Chem.* **270**, 13757–13765.

Hammond-Kosack, K. E., and Jones, J. D. (1996). Resistance gene-dependent plant defense responses. *Plant Cell.* **8**, 773–791.

Hashimoto, C., Hudson, K. L., and Anderson, K. V. (1988). *Cell* **52**, 269–279.

Heguy, A., Baldari, C. T., Macchia, G., Telford, J. L., and Melli, M. (1992). Amino acids conserved in interleukin-1 receptors (IL-1Rs) and the *Drosophila* Toll protein are essential for IL-1R signal transduction. *J. Biol. Chem.* **267**, 2605–2609.

Heine, H., Kirschning, C. J., Lien, E., Monks, B. G., Rothe, M., and Golenbock, D. T. (1999). Cells that carry a null allele for Toll-like receptor 2 are capable of responding to endotoxin. *J. Immunol.* **162**, 6971–6975.

Horuk, R., Huang, J. J., Covington, M., and Newton, R. C. (1987). A biochemical and kinetic analysis of the interleukin-1 receptor: evidence for differences in the molecular properties of IL-1 receptors. *J. Biol. Chem.* **262**, 16275–16282.

Hoshino, K., Tsutsui, H., Kawai, T., Takeda, K., Nakanishi, K., Takeda, Y., and Akira, S. (1999a). Generation of IL-18 receptor-deficient mice: evidence for IL-1 receptor related protein as an essential IL-18 binding receptor. *J. Immunol.* **162**, 5041–5044.

Hoshino, K., Kashiwmura, S., Kuribayashi, K., Kodama, T., Tsujimura, T., Nakanishi, K., Matsuyama, T., Takeda, K., and Akira, S. (1999b). The absence of interleukin-1 receptor-related T1/ST2 does not affact T helper cell type 2 development and its effector function. *J. Exp. Med.* **190**, 1541–1547.

Huang, J., Gao, X., Li, S., and Cao, Z. (1997). Recruitment of IRAK to the interleukin 1 receptor complex requires interleukin 1 receptor accessory protein. *Proc. Natl Acad. Sci. USA* **94**, 12829–12832.

Jia, Y., Loh, Y. T., Zhou, J., and Martin, G. B. (1997). Alleles of pto and fen occur in bacterial speck-susceptible and fenthion-insensitive tomato cultivars and encode active protein kinases. *Plant Cell* **9**, 61–73.

Kanakaraj, P., Ngo, K., Wu, Y., Angulo, A., Ghazal, P., Harris, C. A., Siekierka, J. J., Peterson, P. A., and Fung-Leung, W. P. (1999). Defective IL-18-mediated natural killer and T helper cell type 1 responses in IL-1 receptor-associated kinase deficient mice. *J. Exp. Med.* **189**, 1129–1138.

Keith, F. J., and Gay, N. J. (1990). The *Drosophila* membrane receptor Toll can function to promoter cellular adhesion. *EMBO J.* **9**, 4299–4396.

Kirschning, C. J., Wesche, H., Ayres, T. M., and Rothe, M. (1998). Human Toll-like receptor 2 confers responsiveness to lipopolysaccharide. *J. Exp. Med.* **188**, 2091–2097.

Klemenz, R., Hoffmann, S., and Werenskiold, A. K. (1989). Serum and oncoprotein mediated induction of a gene with sequence similarity to the gene encoding carcinoembryonic antigen. *Proc. Natl Acad. Sci. USA* **86**, 5708–5712.

Kojima, H., Takeuchi, M., Ohta, T., Nishida, Y., Arai, N., Ikeda, M., Ikegami, H., and Kurimoto, M. (1998). Interleukin-18 activates IRAK-TRAF-6 pathway in mouse EL-4 cells. *Biochem. Biophys. Res. Commun.* **244**, 183–186.

Kumar, S., Tzimas, M. N., Griswold, D. E., and Young, P. R. (1997). Expression of ST2, an interleukin-1 receptor homolog, is induced by proinflammatory stimuli. *Biochem. Biophys. Res. Commun.* **235**, 474–478.

Labow, M., Shuster, D., Zetterstrom, M., Nunes, P., Terry, R., Cullinan, E. B., Bartfai, T., Solorzano, C., Moldawer, L. L., Chizzonite, R., and McIntyre, K. W. (1997). Absence of IL-1 signaling and reduced inflammatory response in IL-1 type I receptor-deficient mice. *J. Immunol.* **159**, 2452–2461.

Lang, D., Knop, J., Wesche, H., Raffetseder, U., Kurrle, R., Boraschi, D., and Martin, M. U. (1998). The type II IL-1 receptor interacts with the IL-1 receptor accessory protein: a novel mechanism of regulation of IL-1 responsiveness. *J. Immunol.* **161**, 6871–6877.

Lemaitre, B., Reichart, J. M., and Hoffmann, J. A. (1997). *Drosophila* host defense: differential induction of antimicrobial peptide genes after infection by various classes of microorganisms. *Proc. Natl Acad. Sci. USA* **94**, 14614–14619.

Liu, C., Chalmers, D., Maki, R., and De Souza, E. B. (1996). Rat homolog of mouse interleukin-1 receptor accessory protein:cloning, localization and modulation studies. *J. Neuroimmunol.* **66**, 41–48.

Lohning, M., Stroemann, A., Coyle, A. J., Grogan, J. L., Lin, S., Gutierrez-Ramos, J. C., Levinson, D., Radbruch, A., and Kamradt, T. (1998). T1/ST2 is preferentially expressed on murine TH2 cells, independent of interleukin 4, interleukin 5 and interleukin 10, and important for TH2 effector function. *Proc. Natl Acad. Sci. USA* **95**, 6930–6935.

Lord, K. A., Hoffman-Liebermann, B., and Liebermann, D. A. (1990). Nucleotide sequence and expression of a cDNA encoding MyD88, a novel myeloid differentiation primary response gene induced by IL6. *Oncogene* **5**, 1095–1097.

Lovenberg, T. W., Crowe, P. D., Liu, C., Chalmers, D. T., Liu, X. J., Liaw, C., Clevenger, W., Oltersdorf, T., De Souza, E. B., and Maki, R. A. (1996). Cloning of a cDNA encoding a novel interleukin-1 receptor related protein (IL 1R-rp2). *J. Neuroimmunol.* **70**, 113–122.

McMahon, C. J., Slack, J. L., Mosely, B., Cosman, D., Lupton, S. D., Brunton, L. L., Grubin, C. E., Wignall, J. M., Jenkins, N. A., Brannan, C. I., Copeland, N. G., Huebner, K., Croce, C. M., Cannizzaro, L. A., Benjamin, D., Dower, S. K., Spriggs, M. G., and Sims, J. E. (1991). A novel IL-1 receptor cloned from B cells by mammalian expression is expressed in many cell types. *EMBO J.* **10**, 2821–2832.

Marmiroli, S., Bavelloni, A., Faenza, I., Sirri, A., Ognibene, A., Cenni, V., Tsukada, J., Koyama, Y., Ruzzene, M., Ferri, A., Auron, P. E., Toker, A., and Maraldi, N. M. (1998). Phosphatidylinositol 3-kinase is recruited to a specific site in the activated IL-1 receptor I. *FEBS Lett.* **438**, 49–54.

Martin, M., Bol, G. F., Eriksson, A., Resch, K., and Brigelius-Flohe, R. (1994). Interleukin-1-induced activation of a protein kinase co-precipitating with the type I interleukin-1 receptor in T cell. *Eur. J. Immunol.* **24**, 1566–1571.

Matsushima, K., Akahoshi, T., Yamada, M., Furutani, Y., and Oppenheim, J. J. (1986). Properties of specific IL-1 receptor on human Epstein Barr-transformed B lymphocytes: identity of the receptors for IL-1alpha and IL-1beta. *J. Immunol.* **136**, 4496–4503.

Medzhitov, R., Preston-Hurlburt, P., and Janeway, C. (1997). A human homologue of the *Drosophila* Toll protein signals activation of adaptive immunity. *Nature* **388**, 394–397.

Meng, X., Khanuja, B. S., and Ip, Y. T. (1999). Toll receptor-mediated *Drosophila* immune response requires Dif, an NF-kappaB factor. *Genes Dev.* **13**, 792–797.

Mitcham, J. L., Parnet, P., Bonnert, T., Garka, K., Gerhart, M. J., Slack, J., Gayle, M. A., Dower, S. K., and Sims, J. E. (1996). T1/ST2 signaling establishes it as a member of an expanding interleukin-1 receptor family. *J. Biol. Chem.* **271**, 5777–5783.

Miyake, K., Yamashita, M., Ogata, T., and Kimoto, M. (1995). RP105, a novel B cell surface molecule implicated in B cell activation, is a member of the leucine-rich repeat protein family. *J. Immunol.* **154**, 3333–3340.

Mizuguchi, K., Parker, J. S., Blundell, T. L., and Gay, N. J. (1998). Getting knotted: a model for the structure and activation of Spatzle. *Trends Biochem. Sci.* **23**, 239–242.

Morisato, D., and Anderson, K. V. (1994). The spatzle gene encodes a component of the extracellular signaling pathway establishing the dorsal–ventral pattern in the *Drosophila* embryo. *Cell* **76**, 677–684.

Muzio, M., Ni, J., Feng, P., and Dixit, V. M. (1997). IRAK (pelle) family member IRAK-2 and MyD88 as proximal mediators of IL-1 signaling. *Science* **278**, 1612–1615.

Muzio, M., Natoli, G., Saccani, S., Levrero, M., and Mantovani, A. (1998). The human toll signaling pathway: divergence of nuclear factor kappaB and JNK/SAPK activation upstream of tumor necrosis factor receptor-associated factor 6 (TRAF6). *J. Exp. Med.* **187**, 2097–2101.

O'Neill, L. A. J., and Greene, C. (1998). Signal transduction pathways activated by the IL-1 receptor family: ancient signaling machinery in mammals, insects and plants. *J. Leuk. Biol.* **63**, 650–657.

Parker, J. E., Coleman, M. J., Szabo, V., Frost, L. N., Schmidt, R., van der Biezen, E. A., Moores, T., Dean, C., Daniels, M. J., and Jones, J. D. (1997). The arabidopsis downy mildew resistance gene RPP5 shares similarity to the toll and interleukin-1 receptors with N and L6. *Plant Cell* **9**, 879–894.

Parnet, P., Garka, K. E., Bonnert, T. E., Dower, S. K., and Sims, J. E. (1996). IL-1RrP is a novel receptor-like molecule similar to the type I interleukin-1 receptor and its homologs T1/ST2 and IL-1RAcP. *J. Biol. Chem.* **271**, 3967–3970.

Poltorak, A., He, X., Smirnova, I., Lie, M. Y., Huffel, C. V., Du, X., Birdwell, D., Alejos, E., Silva, M., Galanos, C., Freudenberg, M., Ricciardi-Castagnoli, P., Layton, B., and Beutler, B. (1998). Defective LPS signaling in CSH/HeJ and C57BL/10ScCr mice: mutations in TLR-4 gene. *Science* **282**, 2085–2088.

Robinson, D., Shibuya, K., Mui, A., Zonin, F., Murphy, E., Sana, T., Hartley, S. B., Menon, S., Kastelein, R., Bazan, F., and O'Garra, A. (1997). IGIF does not drive TH1 development but synergises with IL-12 for interferon gamma production and activated IRAK and NF-kappaB. *Immunity* **7**, 571–581.

Rock, F. L., Hardiman, G., Timans, J. C., Kastelein, R. A., and Bazan, F. (1998). A family of human receptors structurally related to *Drosophila* Toll. *Proc. Natl Acad. Sci. USA* **95**, 588–593.

Schreuder, H., Tarfif, C., Trump-Kallmeyer, S., Soffientini, A., Sarubbi, E., Akeson, A., Bowlin, T., Yanofsky, S., and Barrett, R. W. (1997). A new cytokine receptor binding mode revealed by the crystal structure of the IL-1 receptor with an antagonist. *Nature* **386**, 194–200.

Schwander, R., Dziarski, R., Wesche, H., Rothe, M., and Kirschning, C. J. (1999). Peptidoglycan and lipoteichoic acid-induced cell activation is mediated by Toll-like receptor 2. *J. Biol. Chem.* **274**, 17406–17409.

Shimazu, R., Akashi, S., Ogata, H., Nagai, Y., Fukudome, K., Miyake, K., and Kimoto, M. (1999). MD-2, a molecule that confers lipopolysaccharaide responsiveness on toll-like receptor 4. *J. Exp. Med.* **189**, 1777–1782.

Shinkura, R., Kitada, K., Matsuda, F., Tashiro, K., Ikuta, K., Suzuki, M., Kogishi, K., Serikawa, T., and Honjo, T. (1999). Alymphoplasia is caused by a point mutation in the mouse gene encoding Nf-kappa B-inducing kinase. *Nature Genet.* **22**, 74–77.

Sims, J. E., and Dower, S. K. (1994). Interleukin-1 receptors. *Eur. Cytok. Netw.* **5**, 539–546.

Sims, J. E., March, C. J., Cosman, D., Widmer, M. B., MacDonald, H. R., MacMahon, C. J., Grubin, C. E., Wignall, J. M., Jackson, J. L., Call, S. M., Friend, D., Alpert, A. R., Gillis, S., Urdal, D. L., and Dower, S. K. (1988). cDNA expression cloning of the IL-1 receptor, a member of the immunoglobulin superfamily. *Science* **241**, 585–589.

Sims, J. E., Gayle, M. A., Slack, J. L., Alderson, M. R., Bird, T. A., Giri, J. G., Colotta, F., Re, F., Mantovani, A., Shanebeck, K., Grabstein, K. H., and Dower, S. K. (1993). Interleukin 1 signaling occurs exclusively via the type I receptor. *Proc. Natl Acad. Sci. USA* **90**, 6155–6161.

Sizemore, N., Leung, S., and Stark, G. R. (1999). Activation of phosphatidlyinositol 3-kinase in response to interleukin-1 leads to phosphorylation and activation of the NF-kappaB p65/RelA subunit. *Mol. Cell Biol.* **19**, 4798–4805.

Spriggs, M. K., Hruby, D. E., Maliszewski, C. R., Pickup, D. J., Sims, J. E., Buller, R. M. L., and Van Slyke, J. (1992). Vaccinia and cowpox viruses encode a novel secreted interleukin-1 binding protein. *Cell* **71**, 145–153.

Stylianou, E., O'Neill, L. A. J., Edbrooke, M. R.., Woo, P., and Saklatvala, J. (1992). Interleukin-1 induces NFκB through its type I and not type II receptor. *J. Biol. Chem.* **270**, 13757–13765.

Symons, J. A., Young, P. R., and Duff, G. W. (1995). Soluble type II IL-1 receptor binds and blocks processing of IL-1 beta precursor and loses affinity for IL-1 receptor antagonist. *Proc. Natl Acad. Sci. USA* **92**, 1714–1718.

Takeuchi, O., Kawai, T., Sanjo, H., Copeland, N. G., Gilbert, D. J., Jenkins, N. A., Takeda, K., and Akira, S.

(1999). TLR-6: a novel member of an expanding toll-like receptor family. *Gene* **29**, 59–65.

Thomassen, E., Renshaw, B. R., and Sims, J. E. (1999). Identification and characterisation of SIGIRR, a molecule representing a novel sub-type of the IL-1R superfamily. *Cytokine* **11**, 389–399.

Tominaga, S. (1989). A putative protein of a growth specific cDNA from BALB/c 3T3 cells is highly similar to the extracellular portion of mouse interleukin-1 receptor. *FEBS Lett.* **258**, 301–304.

Torigoe, K., Ushio, S., Okura, T., Kobayashi, S., Taniai, M., Kunikata, T., Murakami, T., Sanou, O., Kojima, H., Fugii, M., Ohta, T., Ikeda, M., Ikegami, H., and Kurimoto, M. (1997). Purification and characterization of the human interleukin-18 receptor. *J. Biol. Chem.* **272**, 25737–25742.

Townsend, M. J., Fallon, P. G., Matthews, D. J., Jolin, H. E., and McKenzie, A. N. J. (2000). T1/ST2-deficient mice demonstrate the importance of T1/ST2 in developing promary T helper cell type 2 responses. *J. Exp. Med.* **191**, 1069–1075.

Underhill, D. M., Ozinsky, A., Hajjar, A. M., Stevens, A., Wilson, C. B., Bassetti, M., and Aderam, A. (1999). The Toll-like receptor 2 is recruited to macrophage phagosomes and discriminates between pathogens. *Nature* **401**, 811–815.

Vigers, G. P., Anderson, L. J., Caffes, P., and Brandhuber, B. J. (1997). Crystal structure of the type-I interleukin-1 receptor complexed with interleukin-1beta. *Nature* **386**, 190–194.

Volpe, F., Clatworthy. J., Kaptein, A., Maschera, B., Griffin, A. M., and Ray, K. (1997). The IL-1 receptor accessory protein is responsible for the recruitment of the interleukin-1 receptor associated kinase to the IL1/IL1 receptor I complex. *FEBS Lett.* **419**, 41–44.

Wesche, H., Henzel, W. J., Shillinglaw, W., Li, S., and Cao, Z. (1997a). MyD88: an adapter that recruits IRAK to the IL-1 receptor complex. *Immunity* **7**, 837–847.

Wesche, H., Korherr, C., Kracht, M., Falk, W., Resch, K., and Martin, M. U. (1997b). The interleukin-1 receptor accessory protein in essential for IL-1-induced activation of interleukin-1 receptor associated kinase and stress-activated protein kinases. *J. Biol. Chem.* **272**, 7727–7731.

Wesche, H., Gao, X., Li, X., Kirschning, C. J., Stark, G. R., and Cao, Z. (1999). IRAK-M is a novel member of the pelle/interleukin-1 receptor-associated kinase (IRAK) family. *J. Biol. Chem.* **274**, 19403–19410.

Whitham, S., Dinesh-Kumar, S. P., Choi, D., Hehl, R., Corr, C., and Baker, B. (1994). The product of the tobacco mosaic virus resistence gene N: similarity to Toll and the interleukin-1 receptor. *Cell* **78**, 1101–1115.

Williams, M. J., Rodriguez, A., Kimbrell, D. A., and Eldon, E. D. (1998). The 18-wheeler mutation reveals complex antibacterial gene regulation in *Drosophila* host defense. *EMBO J.* **16**, 6120–6130.

Wright, S. D., Ramos, R. A., Tobias, P. S., Ulevitch, R. J., and Mathison, J. C. (1990). CD14, a receptor for complexes of LPS and LPS binding protein. *Science* **249**, 1431–1434.

Xu, D., Chan, W. L., Leung, B. P., Hunter, D., Schulz, K., Carter, R. W., McInnes, I. B., Robinson, J. H., and Liew, F. Y. (1998). Selective expression and functions of interleukin 18 receptor on T helper type 1 but not TH2 cells. *J. Exp. Med.* **188**, 1485–1492.

Yanagisawa, K., Takagi, T., Tsukamoto, T., Tetsuka, T., and Tominaga, S. (1993). Presence of a novel primary response gene ST2L, encoding a product highly similar to the interleukin 1 receptor type 1. *FEBS Lett.* **318**, 83–87.

Yang, R. B., Mark, M. R., Gray, A., Huang, A., Xie, M. H., Zhang, M., Goddard, A., Wood, W. I., Gurney, A. L., and Godowski, P. J. (1998). Toll-like receptor-2 mediates lipopolysaccharide-induced cellular signaling. *Nature* **395**, 284–288.

Yang, R. B., Mark, M. R., Gurney, A. L., and Godowski, P. J. (1999). Signaling events induced by lipopolysaccharide-activated toll-like receptor 2. *J. Immunol.* **163**, 639–643.

Yoon, D. Y., and Dinarello, C. A. (1998). Antibodies to domains II and III of the IL-1 receptor accessory protein inhibit IL-1beta activity but not binding: regulation of IL-1 responses is via type I receptor, not the accessory protein. *J. Immunol.* **160**, 3170–3179.

Zhang, F. X., Kirschning, C. J., Mancinelli, R., Xu, X. P., Jin, Y., Faure, E., Mantovani, A., Rothe, M., Muzio, M., and Arditi, M. (1999). Bacterial lipopolysaccharide activates nuclear factor-kappaB through interleukin-1 signaling mediators in cultured human dermal endothelial cells and mononuclear phagocytes. *J. Biol. Chem.* **274**, 7611–7614.

IL-1 Receptor Type I

Charles A. Dinarello*

Department of Infectious Diseases, University of Colorado Health Sciences Center, 4200 East Ninth Avenue, B168, Denver, CO 80262, USA

*corresponding author tel: 303-315-3589, fax: 303-315-8054, e-mail: cdinare333@aol.com

DOI: 10.1006/rwcy.2000.15001.

SUMMARY

The IL-1 receptor type I is the ligand-binding chain of the IL-1 heterodimer complex. It is a three-domain Ig-like extracellular receptor with a cytoplasmic domain containing the Toll protein-like sequences. The IL-1 R type I does not function without the second chain of the dimer, namely the IL-1R accessory protein. Although the IL-1R accessory protein chain does contain similar extracellular domains as the type I, it does not bind IL-1 in solution. However, the IL-1R accessory protein, together with the IL-1RI, form a complex with the IL-1 ligand (IL-1α or IL-1β) with a high affinity.

Soluble forms of the IL-1RI, produced by proteolytic cleavage, are found in the circulation of healthy humans and in elevated levels during disease. The soluble IL-1RI has an unusually high binding constant to the IL-1Ra and hence with preferentially bind this antagonist member of the IL-1 family. Although in animals administration of soluble IL-1RI has reduced the severity of disease, in humans with *rheumatoid arthritis* this method of neutralizing IL-1 has not been successful because it binds IL-1Ra before it binds IL-1α or IL-1β.

BACKGROUND

Discovery

The molecular cloning of the IL-1 receptor type I (IL-1RI) was first made in the mouse (Sims *et al.*, 1988). Earlier, several groups had described an 80 kDa glycoprotein on T cells and fibroblasts. This, in retrospect, was the IL-1RI (Mosley *et al.*, 1987; Bird *et al.*, 1987; Chin *et al.*, 1988; Qwarnstrom *et al.*, 1988; Scapigliati *et al.*, 1989; Savage *et al.*, 1989).

Structure

The extracellular domains of the IL-1RI are typical of an Ig-like receptor of the fibroblast growth factor family. The 80 kDa glycoprotein has three Ig-like domains in the extracellular segment, a transmembrane domain, and a cytoplasmic domain containing the Toll-like protein of *Drosophila* (Gay and Keith, 1991).

Main activities and pathophysiological roles

When IL-1 binds to the cell membrane of IL-1RI, it recruits the IL-1R accessory protein (IL-1R AcP). The heterodimer of the IL-1RI and the IL-1R AcP initiate signal transduction.. A soluble form of this receptor has been described that acts to reduce the biological effects of IL-1 (Dower *et al.*, 1989).

GENE

Accession numbers

The gene for the IL-1RI has been reported (Sims *et al.*, 1995).

Sequence

The sequence of the human IL-1RI gene can be found in Sims *et al.*, (1995).

PROTEIN

Description of protein

The extracellular domains of the IL-1RI form a typical Ig-like receptor. It is in the same family as the fibroblast growth factor family. The glycoprotein has three Ig-like domains in the extracellular segment, a transmembrane domain and a cytoplasmic domain (Sims *et al.*, 1988). Glycosylation of IL-1RI appears to be necessary for optimal activity (Mancilla *et al.*, 1992). In fact, IL-1RI is heavily glycosylated and blocking the glycosylation sites reduces the binding of IL-1. Surface expression of this receptor is likely on most IL-1-responsive cells as biological activity of IL-1 is a better assessment of receptor expression than ligand binding to cell surfaces (Rosoff *et al.*, 1988). Failure to show specific and saturable IL-1 binding is often due to the low numbers of surface IL-1RI on primary cells. In cell lines, the number of IL-1RI can reach 5000 per cell, but primary cells usually express less than 200 receptors per cell. In some primary cells there are less than 50 per cell (Shirakawa *et al.*, 1987) and IL-1 signal transduction has been observed in cells expressing less than 10 type I receptors per cell (Stylianou *et al.*, 1992). The low number of IL-1RI on cells and the discrepancy between binding affinities and biological activities can be explain by the increased binding affinity of IL-1 in the complex with the IL-1R AcP (Greenfeder *et al.*, 1995).

IL-1RI has a single transmembrane segment and a cytoplasmic domain. Using specific neutralizing antibodies, IL-1RI but not IL-1RII is the primary signal transducing receptor. Antisense oligonucleotides directed against IL-1RI block IL-1 activities *in vitro* and *in vivo* (Burch and Mahan, 1991). The cytoplasmic domain of IL-1RI has no apparent intrinsic tyrosine kinase activity but when IL-1 binds to only a few receptors, the remaining unoccupied receptors appear to undergo phosphorylation (Gallis *et al.*, 1989), probably by a member of the MAP kinase family. Interestingly, the cytosolic domain of IL-1RI has a 45% amino acid homology with the cytosolic domain of the *Drosophila Toll* gene (Gay and Keith, 1991). Toll is a transmembrane protein acting like a receptor. Gene organization and amino acid homology suggests that the IL-1RI and the cytosolic Toll are derived from a common ancestor and trigger similar signals (Guida *et al.*, 1992; Heguy *et al.*, 1992). Because of the finding of the Toll protein in the cytoplasmic domain of other receptors (IL-18, for example), IL-1RI belongs to an entire family of receptors characterized by their common cytoplasmic Toll domains. This family is called Toll-like receptors (TLR) and is discussed in detail in the chapter on the IL-1 receptors.

Relevant homologies and species differences

There is considerable homology between the IL-1RI and IL-1R AcP in the extracellular and cytoplasmic domains but only in the extracellular domains of IL-1RII in all species. The extracellular domains of the IL-1RI share homologies to IL-1R-related protein (Parnet *et al.*, 1996) and the ST2 receptor (Gayle *et al.*, 1996).

Affinity for ligand(s)

Although the type I receptor binds IL-1α or IL-1β and transmits a signal upon complex formation with the IL-1R AcP, of the three members of the IL-1 family, IL-1β has the lowest affinity for the cell-bound form of IL-1RI (500 pM–1 nM). The greatest binding affinity of the three IL-1 ligands for the IL-1RI is that of IL-1Ra. In fact, the off-rate is slow and binding of IL-1Ra to IL-1RI is nearly irreversible. After IL-1Ra, IL-1α binds to IL-1RI with affinities ranging from 100 to 300 pM. By comparison, IL-1β binds more avidly to the nonsignal transducing type II receptor (100 pM). IL-1β binding to the soluble form of the IL-1RI is lower compared to the cell-bound receptor.

Differences exist in the binding affinity, association and dissociation rates of the mature forms of each member of the IL-1 family to cell-bound IL-1RI and soluble IL-1RI receptors. In some cells, there is a discrepancy between the dissociation constant of either form of IL-1 (usually 200–300 pM) and concentrations of IL-1 which can elicit a biological response (10–100 fM). In cells expressing large amounts of IL-1R AcP, the high-affinity binding of the IL-1R/IL-1R AcP complex may explain why two classes of binding have been observed. Human recombinant 17 kDa IL-1α binds to cell surface and soluble type I receptors with approximately the same affinity (100–300 pM). If the binding of IL-1Ra is examined, the affinity is even higher than that of IL-1α. IL-1Ra avidly binds to the surface type I receptor (50–100 pM). Although

IL-1Ra binds less so the soluble form of this receptor, it is, nevertheless, a high-affinity binding.

Cell types and tissues expressing the receptor

IL-1RI is found predominantly on endothelial cells, smooth muscle cells, epithelial cells, hepatocytes, fibroblasts, keratinocytes, epidermal dendritic cells, and T lymphocytes.

Regulation of receptor expression

The genomic organization of the human type I receptor reveals three distinct transcription initiation sites contained in three separate segments of the first exon which is distributed over 29 kb of the gene (Sims *et al.*, 1995). Each part of this first exon is thought to possess a separate promoter which functions independently in different cells (Sims *et al.*, 1995; Ye *et al.*, 1996). Despite evidence that type I receptor gene expression can be upregulated *in vitro* (Koch *et al.*, 1992; Ye *et al.*, 1992), the most proximal (5′) promoter region lacks a TATA or CAAT box (Ye *et al.*, 1996). In fact, this promoter region for the human IL-1RI shares striking similarity to those of housekeeping genes rather than highly regulated genes. The transcription initiation start site contains nearly the same motif as that for the TdT gene (Ye *et al.*, 1993). There is a G+C-rich segment (75%) following the transcription initiation site of exon I which accounts for considerable secondary RNA structure. Low numbers of surface IL-1RI may, in fact, be due to multiple secondary RNA structures which reduce optimal translation of the mRNA (Ye *et al.*, 1996).

Surface expression of IL-1RI clearly impacts upon the biological response to IL-1. Similar to IL-1β, cells can express high steady-state levels of mRNA for IL-1RI but low levels of the protein. This may be due the amount of secondary structure in each of the polyadenylated RNA species. Studies on IL-1R surface expression have mostly used binding of labeled ligands rather than assessment of surface receptor density using specific antibodies. Nevertheless, phorbol esters, PGE_2, dexamethasone, epidermal growth factor, IL-2, and IL-4 increase surface expression of IL-1RI. In cells which synthesize PGE_2, IL-1 upregulates its own receptor via PGE_2; however, when PGE_2 synthesis is inhibited, IL-1 downregulates IL-1RI in the same cells (Takii *et al.*, 1992, 1994; Pronost *et al.*, 1993). TGFβ and IL-1 downregulate surface expression of IL-1RI on T cells (Dubois *et al.*, 1990; Silvera *et al.*, 1994). In the case of TH2 lymphocytes, IL-1 downregulates IL-1RI surface expression and this is associated with a decrease in mRNA half-life (Ye *et al.*, 1992). Therefore, despite the housekeeping nature of its promoters, IL-1RI is regulated in the context of inflammation and immune responses.

Using *in situ* hybridization, IL-1RI gene expression can be detected in the endocrine pancreas, cardiac endothelium, epidermis, hair follicles, uterine serosa, developing oocytes and granulosa cells from ruptured ovarian follicles of healthy mice (Deyerle *et al.*, 1992). In contrast, the spleen and thymus are largely negative. IL-1RI has also been localized to specific hippocampal areas of the normal mouse brain, adrenal and testis (Cunningham *et al.*, 1991). There is an uneven distribution with localization in the hippocampus, the anterior lobe of the pituitary gland, and the dentate gyrus (Deyerle *et al.*, 1992). The importance of constitutive gene expression for IL-1RI in ontogeny and homeostasis is unclear since mice deficient in this gene exhibit a grossly normal phenotype (Labow *et al.*, 1997).

Release of soluble receptors

In contrast to the IL-1RII in which LPS induces a rapid shedding of the extracellular domain, LPS induced IL-1RI and also surface IL-1R AcP expression (Penton-Rol *et al.*, 1999). In addition to the major 5 kb IL-1RI mRNA the 2.4 kb IL-1RI transcripts were observed. The 2.4 kb inducible band includes incompletely spliced, polyadenylated transcripts which may encode for truncated receptors. Thus, LPS has divergent effects on IL-1 receptors by increasing soluble IL-1RII but stimulating expression of IL-1RI and IL-1R AcP.

Soluble receptors for IL-1RI have been found in a variety of disease states. Soluble type II IL-1R are found in the circulation and urine of healthy subjects and in inflammatory synovial and other pathologic body fluids. In healthy humans, the circulating levels of IL-1sRI and IL-1sRII are 50–100 pM. The rank of affinities for the IL-1RI soluble receptors are remarkably different for each of the three IL-1 molecules. The rank for the three IL-1 ligands binding to soluble IL-1RI is IL-1Ra > IL-1α > IL-1β. Elevated levels of soluble IL-1RI are found in the circulation of patients with sepsis, in the synovial fluid of patients with active *rheumatoid arthritis* (Arend *et al.*, 1994) and in patients with *hairy cell leukemia* (Barak *et al.*, 1996). In patients undergoing aorta resection, crossclamping of the aorta results in significant ischemia and a dramatic release of soluble IL-1R I (Pruitt *et al.*, 1996).

Similar to IL-1RI and IL-1RII, a soluble form of the IL-1R AcP exists but this form appears to result from an RNA splice donor/acceptor site resulting in a truncated protein ending before the transmembrane region. Unlike the soluble forms of the IL-1RI and IL-1RII, the soluble IL-1R AcP is not formed by proteolytic cleavage of the full-length accessory protein. It is unclear how soluble IL-1R AcP mRNA is expressed compared to the cell-bound protein. Furthermore, since the IL-1R AcP does not bind IL-1 itself (Greenfeder *et al.*, 1995), the effect of the soluble IL-1R AcP on the binding of IL-1 remains unclear.

SIGNAL TRANSDUCTION

Associated or intrinsic kinases

Hopp reported a detailed sequence and structural comparison of the cytosolic segment of IL-1RI with the *ras*-family of GTPases (Hopp, 1995). In this analysis, the known amino acids residues for GTP binding and hydrolysis by the GTPase family were found to align with residues in the cytoplasmic domain of the IL-1RI. In addition, Rac, a member of the Rho family of GTPases, was also present in the binding and hydrolytic domains of the IL-1RI cytosolic domains. These observations are consistent with the observations that GTP analogs undergo a rapid hydrolysis when membrane preparations of IL-1RI are incubated with IL-1. Amino acid sequences in the cytosolic domain of the IL-1R AcP also align with the same binding and hydrolytic regions of the GTPases (T. P. Hopp, personal communication). A protein similar to G protein-activating protein has been identified which associates with the cytosolic domain of the IL-1RI (Michum and Sims, 1995). This finding is consistent with the hypothesis that an early event in IL-1R signaling involves dimerization of the two cytosolic domains, activation of putative GTP-binding sites on the cytosolic domains, binding of a G protein, hydrolysis of GTP and activation of a phospholipase. It then follows that hydrolysis of phospholipids generates diacylglycerol or phosphatidic acids.

Characteristics of the Cytoplasmic Domain of the IL-1RI

The cytoplasmic domain of the IL-1RI does not contain a consensus sequence for intrinsic tyrosine phosphorylation but deletion mutants of the receptor reveal specific functions of some domains. There are four nuclear localization sequences which share homology with the glucocorticoid receptor. Three amino acids (Arg431, Lys515, and Arg518), also found in the Toll protein, are essential for IL-1-induced IL-2 production (Heguy *et al.*, 1992). However, deletion of a segment containing these amino acids did not affect IL-1-induced IL-8 (Kuno *et al.*, 1993). There are also two cytoplasmic domains in the IL-1RI which share homology with the IL-6-signaling gp130 receptor. When these regions are deleted, there is a loss of IL-1-induced IL-8 production (Kuno *et al.*, 1993).

The C-terminal 30 amino acids of the IL-1RI can be deleted without affecting biological activity (Croston *et al.*, 1995). Two independent studies have focused on the area between amino acids 513 and 529. Amino acids 508–521 contain sites required for the activation of NFκB. In one study, deletion of this segment abolished IL-1-induced IL-8 expression and in another study, specific mutations of amino acids 513 and 520 to alanine prevented IL-1-driven E-selectin promoter activity. This area is also present in the Toll protein domain associated with NFκB translocation and previously shown to be part of the IL-1 signaling mechanism. This area (513–520) is also responsible for activating a kinase which associates with the receptor. Termed IL-1RI-associated kinase, this kinase phosphorylates a 100 kDa substrate. Others have reported a serine/threonine kinase that coprecipitates with the IL-1RI (Martin *et al.*, 1994). Amino acid sequence comparisons of the cytosolic domain of the IL-1RI have revealed similarities with a protein kinase C (PKC) acceptor site. Because PKC activators usually do not mimic IL-1-induced responses, the significance of this observation is unclear.

Recruitment of MyD88 and IL-1 Receptor Activating Kinase

An event that may be linked to the binding of G proteins to the IL-1 receptor complex is the recruitment of the cytosolic protein called MyD88. This small protein has many of the characteristics of cytoplasmic domains of receptors but lacks any known extracellular or transmembrane structure. It is unclear exactly how this protein functions, since it does not have any known kinase activity. However, it may assist in the binding of IRAK to the complex and hence has been said to function as an adapter molecule. The binding of IRAK to the IL-1R complex appears to be a critical step in the activation of NFκB (Croston *et al.*, 1995). The IL-1R AcP is essential for the recruitment and activation of IRAK (Huang *et al.*, 1997; Wesche *et al.*, 1997). In fact, deletion of specific amino acids in the IL-1R AcP cytoplasmic domain results in loss of IRAK association (Wesche *et al.*,

1997). In addition, the intracellular adapter molecule termed MyD88 appears to dock to the complex, allowing IRAK to become phosphorylated (Croston *et al.*, 1995; Cao, 1998). IRAK then dissociates from the IL-1R complex and associates with TNF receptor-associated factor 6 (TRAF6) (Cao *et al.*, 1996). TRAF6 then phosphorylates NIK (Malinin *et al.*, 1997) and NIK phosphorylates the inhibitory κB kinases (IKK-1 and IKK-2) (DiDonato *et al.*, 1997). Once phosphorylated, IκB is rapidly degraded by a ubiquitin pathway, liberating NFκB which translocates to the nucleus for gene transcription.

In mice deficient in TRAF6, there is no IL-1 signaling in thymocytes and the phenotype exhibits severe *osteopetrosis* and defective formation of osteoclasts (Lomaga *et al.*, 1999).

IL-1 and IL-18 Signaling are Similar

A review of the biology of IL-18 has been published (Dinarello, 1999). There are both high (0.4 nM) and low (40 nM) affinity binding sites for IL-18 in murine primary T cells (Yoshimoto *et al.*, 1998). Clearly there are two chains to the IL-18 receptor complex. Similar to the nomenclature for other heterodimeric receptors, there is an α chain which functions as a ligand-binding chain and a β chain which is the signaling chain. For IL-18, the ligand-binding chain was isolated and found to be a member of the IL-1R family. Hence the IL-18Rα had been previous cloned and called the IL-1R-related protein IL-1Rrp (Parnet *et al.*, 1996). This receptor remained an orphan receptor until the work by Torigoe *et al.*, (1997). The binding of a second chain would explain the higher affinity. This 'other' receptor chain has been identified as the 'accessory protein-like' (AcPL) receptor (Born *et al.*, 1998), and is related to the IL-1 accessory protein identified by Greenfeder *et al.* (1995). Similar to the IL-1 accessory protein, the IL-18 AcPL does not bind to its ligand but rather binds to the complex formed by IL-18 with the IL-18Rα chain (Born *et al.*, 1998). This heterodimeric is probably the high-affinity complex.

Therefore, it is likely that the IL-18 receptor complex is comprised of a ligand-binding chain of IL-1Rrp (IL-18Rα) with the second chain being the non-IL-18 binding AcPL (IL-18Rα) (Born *et al.*, 1998). An IL-18-binding protein (IL-18BP) has been purified from human urine and cloned (Novick *et al.*, 1999). But a transmembrane form of the IL-18BP has not been isolated to date despite murine cDNA cloning and human genomic analyses (Novick *et al.*, 1999). Thus, the IL-18BP is a third gene product and appears to be the 'decoy' receptor of IL-18, similar to the IL-1R type II which is the decoy receptor for IL-1.

IRAK also associates with the IL-18R complex (Robinson *et al.*, 1997; Kojima *et al.*, 1998). This was demonstrated using IL-12-stimulated T cells followed by immunoprecipitation with anti-IL-18R or anti-IRAK (Kojima *et al.*, 1998). Furthermore, IL-18-triggered cells also recruited TRAF6 (Kojima *et al.*, 1998). Like IL-1 signaling, MyD88 has a role in IL-18 signaling. MyD88-deficient mice do not produce acute phase proteins and have diminished cytokine responses. Recently, TH1-developing cells from MyD88-deficient mice were unresponsive to IL-18-induced activation of NFκB and c-Jun N-terminal kinase (JNK) (Adachi *et al.*, 1998). Thus, MyD88 is an essential component in the signaling cascade that follows IL-1 receptor as well as IL-18 receptor binding. It appears that the cascade of sequential recruitment of MyD88, IRAK, and TRAF6 followed by the activation of NIK and degradation of IκBK and release of NFκB are nearly identical for IL-1 as well as for IL-18. Indeed, in cells transfected with IL-18Rα (IL-1Rrp) and then stimulated with IL-18, translocation of NFκB takes is observed using electromobility shift assay (Torigoe *et al.*, 1997). In U1 macrophages, which already express the gene for IL-1Rrp, there is translocation of NFκB and stimulation of *HIV*-1 production (Shapiro *et al.*, 1998a).

Cytoplasmic signaling cascades

Signal transduction of IL-1 depends on the formation of a heterodimer between IL-1RI and IL-1R AcP (Greenfeder *et al.*, 1995). This interaction recruits MyD88, a cytoplasmic adapter molecule. This is followed by recruitment of the IL-1R-activating kinase (IRAK) (Croston *et al.*, 1995; Cao *et al.*, 1996; Huang *et al.*, 1997). Antibodies to IL-1RI block IL-1 activity. Although IL-1R AcP does not bind IL-1, antibodies to IL-1R AcP also prevent IL-1 activity (Yoon and Dinarello, 1998). Both the extracellular domain of the IL-1R AcP and its cytoplasmic segment share homology with the IL-1RI. There is a perfectly conserved protein kinase C acceptor site in both cytoplasmic domains, although agents activating protein kinase C do not mimic IL-1 signal transduction. Limited sequence homology of the gp130 cytoplasmic domain with those of IL-1RI (Kuno and Matsushima, 1994) and IL-1R AcP suggest that complex formation of the IL-1R/IL-1/IL-1R AcP transduces a signal similar to that observed with ligands which cause the dimerization of gp130. In fact, deletion of the gp130-shared sequences from the IL-1RI cytoplasmic domain results in a reduced response to IL-1 (Kuno and Matsushima, 1994). IL-1 shares some prominent biological properties with gp130 ligands; for example, fever, hematopoietic stem cell activation, and the stimulation of the hypothalamic-pituitary-adrenal axis are common to IL-1 and IL-6. Other biological activities of IL-1 and IL-6 are distinctly antagonistic.

High levels of IL-1R AcP are expressed in mouse and human brain tissue. The discovery and function of the IL-1R AcP has placed IL-1 receptor biology and signaling mechanisms into the same arena as other cytokines and growth factors. The IL-1R AcP also explains previous studies describing low- and high-binding affinities of IL-1 to various cells. Like other models of two-chain receptors, IL-1 binds first to the IL-1RI with a low affinity. Although there is no direct evidence, a structural change may take place in IL-1 allowing for docking of IL-1R AcP to the IL-1RI/IL-1 complex. Once IL-1RI/IL-1 binds to IL-1R AcP, a high-affinity binding is observed. Antibodies to the type I receptor and to the IL-1R AcP block IL-1 binding and activity. Therefore, IL-1 may bind to the type I receptor with a low affinity, causing a structural change in the ligand followed by recognition by the IL-1R AcP.

Within a few minutes following binding to cells, IL-1 induces several biochemical events (reviewed by Rossi, 1993; Kuno and Matsushima, 1994; Brooks and Mizel, 1994; Martin and Falk, 1997). It remains unclear which is the most 'upstream' triggering event or whether several occur at the same time. No sequential order or cascade has been identified but several signaling events appear to be taking place during the first 2–5 minutes. Some of the biochemical changes associated with signal transduction are probably cell specific. Within 2 minutes, hydrolysis of GTP, phosphatidylcholine, phosphatidylserine, or phosphatidylethanolamine (Rosoff, 1989) and release of ceramide by neutral (Schutze *et al.*, 1994), not acidic, sphingomyelinase (Andrieu *et al.*, 1994) have been reported. In general, multiple protein phosphorylations and activation of phosphatases can be observed within 5 minutes (Bomalaski *et al.*, 1992) and some are thought to be initiated by the release of lipid mediators. The release of ceramide has attracted attention as a possible early signaling event (Kolesnick and Golde, 1994). Phosphorylation of PLA_2-activating protein also occurs within the first few minutes (Gronich *et al.*, 1994) which would lead to a rapid release of arachidonic acid. Multiple and similar signaling events have also been reported for TNF.

Of special consideration to IL-1 signal transduction is the unusual discrepancy between the low number of receptors (< 10 in some cells) and the low concentrations of IL-1 that can induce a biological response. This latter observation, however, may be clarified in studies on high-affinity binding with the IL-1R AcP complex. A rather extensive 'amplification' step(s) takes place following the initial postreceptor binding event. The most likely mechanism for signal amplification is multiple and sequential phosphorylations (or dephosphorylations) of kinases which result in nuclear translocation of transcription factors and activation of proteins participating in translation of mRNA. IL-1RI is phosphorylated following IL-1 binding. It is unknown whether the IL-1R AcP is phosphorylated during receptor complex formation. In primary cells, the number of IL-1R type I receptors is very low (< 100 per cell) and a biological response occurs when only as few as 2–3% of them are occupied (Gallis *et al.*, 1989; Ye *et al.*, 1992). In IL-1-responsive cells, one assumes that there is constitutive expression of the IL-1R AcP.

With few exceptions, there is general agreement that IL-1 does not stimulate hydrolysis of phosphatidylinositol nor an increase in intracellular calcium. Without a clear increase in intracellular calcium, early postreceptor binding events nevertheless include hydrolysis of a GTP with no associated increase in adenyl cyclase, activation of adenyl cyclase (Minzel, 1990; Munoz *et al.*, 1990), hydrolysis of phospholipids (Rosoff *et al.*, 1988; Kester *et al.*, 1989), release of ceramide (Mathias *et al.*, 1993), and release of arachidonic acid from phopholipids via cytosolic phospholipase A_2 (PLA_2) following its activation by PLA_2-activating protein (Gronich *et al.*, 1994). Some IL-1-signaling events are prominent in different cells. Postreceptor signaling mechanisms may therefore provide cellular specificity. For example, in some cells, IL-1 is a growth factor and signaling is associated with serine/threonine phosphorylation of the MAP kinase p42/44 in mesangial cells (Huwiler and Pfeilschifter, 1994). The MAP p38 kinase, another member of the MAP kinase family, is phosphorylated in fibroblasts (Freshney *et al.*, 1994), as is the p54α MAP kinase in hepatocytes (Kuno *et al.*, 1993).

Activation of MAP Kinases Following IL-1 Receptor Binding

Multiple phosphorylations take place during the first 15 minutes following IL-1 receptor binding. Most consistently, IL-1 activates protein kinases which phosphorylate serine and threonine residues, targets of the MAP kinase family. An early study reported an IL-1-induced serine/threonine phosphorylation of a 65 kDa protein clearly unrelated to those phosphorylated via PKC (Matsushima *et al.*, 1987). As reviewed by O'Neill (1995), prior to IL-1 activation of serine/threonine kinases, IL-1 receptor binding results in the phosphorylation of tyrosine residues (Freshney *et al.*, 1994; Kracht *et al.*, 1994). Tyrosine phosphorylation induced by IL-1 is probably due to activation of MAP kinase kinase which then phosphorylates tyrosine and threonine on MAP kinases.

Following activation of MAP kinases, there are phosphorylations on serine and threonine residues of the epidermal growth factor receptor, heat shock protein p27, myelin basic protein and serine 56 and 156 of β casein, each of which has been observed in IL-1-stimulated cells (Bird *et al.*, 1991; Guesdon *et al.*, 1993). TNF also activates these kinases. There are at least three families of MAP kinases. The p42/44 MAP kinase family is associated with signal transduction by growth factors including Ras/Raf-1 signal pathways. In rat mesangial cells, IL-1 activates the p42/44 MAP kinase within 10 minutes and also increases *de novo* synthesis of p42 (Huwiler and Pfeilschifter, 1994).

p38 MAP Kinase Activation

The stress-activated protein kinase (SAPK), which is molecularly identified as Jun N-terminal kinase (JNK), is phosphorylated in cells stimulated with IL-1 (Stylianou and Saklatvala, 1998). In addition to p42/44, two members of the MAP kinase family (p38 and p54) have been identified as part of an IL-1 phosphorylation pathway and are responsible for phosphorylating hsp 27 (Freshney *et al.*, 1994; Kracht *et al.*, 1994). In rabbit primary liver cells, IL-1 selectively activates JNK without apparent activation of p38 or p42 p38 MAP kinases (Finch *et al.*, 1997). These MAP kinases are highly conserved proteins, homologous to the *HOG-1* stress gene in yeasts. In fact, when *HOG-1* is deleted, yeasts fail to grow in hyperosmotic conditions; however, the mammalian gene coding for the IL-1-inducible p38 MAP kinase (Kracht *et al.*, 1994) can reconstitute the ability of the yeast to grow in hyperosmotic conditions (Galcheva-Gargova *et al.*, 1994). In cells stimulated with hyperosmolar NaCl, LPS, IL-1, or TNF, indistinguishable phosphorylation of the p38 MAP kinase takes place (Han *et al.*, 1994). In human monocytes exposed to hyperosmolar NaCl (375–425 mosmol/L), IL-8, IL-1β, IL-1α, and TNFα gene expression and synthesis takes place, which is indistinguishable from that induced by LPS or IL-1 (Shapiro and Dinarello, 1995, 1997). Thus, the MAP p38 kinase pathways involved in IL-1, TNF, and LPS signal transductions share certain elements that are related to the primitive stress-induced pathway. The dependency of Rho members of the GTPase family for IL-1-induced activation of p38 MAP kinases has been demonstrated (Zhang *et al.*, 1995). This latter observation links the intrinsic GTPase domains of IL-1RI and IL-1R AcP with activation of the p38 MAP kinase.

Inhibition of p38 MAP Kinase

The target for pyridinyl imidazole compounds has been identified as a homolog of the *HOG-1* family (Lee *et al.*, 1994); its sequence is identical to that of the p38 MAP kinase-activating protein 2 (Han *et al.*, 1995). Inhibition of the p38 MAP kinase is highly specific for reducing LPS- and IL-1-induced cytokines (Lee *et al.*, 1994). IL-1-induced expression of *HIV*-1 is suppressed by specific inhibition using pyridinyl imidazole compounds (Shapiro *et al.*, 1998b). As expected, this class of imidazoles also prevents the downstream phosphorylation of hsp 27 (Cuenda *et al.*, 1995). Compounds of this class appear to be highly specific for inhibition of the p38 MAP kinase in that there was no inhibition of 12 other kinases. Using one of these compounds, both hyperosmotic NaCl- and IL-1α-induced IL-8 synthesis was inhibited (Shapiro and Dinarello, 1995). It has been proposed that MAP kinase-activating protein 2 is one of the substrates for the p38 MAP kinases and that MAP kinase-activating protein 2 is the kinase which phosphorylates hsp 27 (Cuenda *et al.*, 1995).

DOWNSTREAM GENE ACTIVATION

Transcription factors activated

NFκB and AP-1 (Muegge *et al.*, 1989, 1993).

Genes induced

See chapter on IL-1β.

BIOLOGICAL CONSEQUENCES OF ACTIVATING OR INHIBITING RECEPTOR AND PATHOPHYSIOLOGY

Unique biological effects of activating the receptors

See Signal transduction.

Phenotypes of receptor knockouts and receptor overexpression mice

Mice deficient in IL-1RI have been generated and show no abnormal phenotype in health and exhibit normal homeostasis, similar to that observed in IL-1β- or IL-1α-deficient mice (Horai *et al.*, 1998; Zheng *et al.*, 1995) but distinctly different from mice deficient in IL-1Ra (Hirsch *et al.*, 1996). Mice deficient in IL-1RI do not exhibit significant disruption of reproduction aside from a somewhat reduced litter size (Abbondanzo *et al.*, 1996). In some laboratories, however, the body weights of the IL-1RI-deficient mice were 30% less than wild-type, whereas the TNFRp55-deficient mice weighed 30% more than wild-type mice of equivalent age (Vargas *et al.*, 1996). Although IL-1α is constitutively expressed in the skin, the barrier function of skin remains intact in mice deficient in IL-1RI (Man *et al.*, 1999). Similarly, mice deficient in IL-1R AcP appear normal but have no responses to IL-1 *in vivo* (Cullinan *et al.*, 1998). However, cells deficient in IL-1R AcP have normal binding of IL-1α and IL-1Ra (binding to the IL-1RI being intact) but a 70% reduction in binding of IL-1β (Cullinan *et al.*, 1998). In these cells, there is no biological response to IL-1α despite binding of IL-1α. The results suggest that IL-1R AcP and not IL-1RI is required for IL-1β binding and biological response to IL-1.

Mice injected with LPS have been studied. IL-1RI-deficient mice exhibit the same decrease in hepatic lipase as do wild-type mice. However, injection of LPS directly into the eye of mice deficient in IL-1RI reveals a decrease in the number of infiltrating leukocytes whereas there was no decrease in mice deficient in both TNF receptors (Rosenbaum *et al.*, 1998). IL-1RI-deficient mice failed to respond to IL-1 in a variety of assays, including IL-1-induced IL-6 and E-selectin expression and IL-1-induced fever (Labow *et al.*, 1997). Similar to IL-1β-deficient mice, IL-1RI-deficient mice had a reduced acute phase response to turpentine. Also similar to IL-1β-deficient mice (Shornick *et al.*, 1996), IL-1RI-deficient mice had a reduced delayed-type hypersensitivity response and were highly susceptible to infection by *Listeria monocytogenes*.

Endotoxin-treated wild-type animals showed extensive ultrastructural endothelial damage with most sections showing only denuded endothelium on the luminal surface. IL-1RI-deficient mice injected with endotoxin showed complete maintenance of endothelial structural integrity compared to endotoxin-treated wild-type with ultrastructural morphology appearing identical to those given saline vehicle (Sutton *et al.*, 1997). Also, no apparent correlation was observed be-tween serum IL-6 concentrations or serum nitric oxide levels and aortic endothelial damage. These studies confirm observations of endothelial cell protection with IL-1Ra and suggest that IL-1 contributes significantly to sepsis-induced endothelial damage.

Calvariae and humeri of IL-1RI-deficient mice were normal with respect to trabecular bone volume, osteoclast number, osteoclast surface, growth plate widths, and cortical thickness (Vargas *et al.*, 1996). Parathyroid hormone or TNFα induced IL-6 mRNA and protein levels in IL-1RI-deficient bone marrow stromal cells whereas IL-1α had no effect. These findings demonstrate that normal bone development in mice can occur in the absence of IL-1RI (Vargas *et al.*, 1996).

IL-1RI-deficient mice have normal serum Ig levels and generate primary and secondary antibody responses equivalent to those of wild-type mice. Similar to IL-1β-deficient mice (Fantuzzi *et al.*, 1996), in response to a lethal challenge with D-galactosamine plus LPS or high-dose LPS, IL-1RI-deficient mice are equivalent to wild-type mice. In addition, acute phase protein mRNA induction in IL-1RI-deficient and wild-type mice are similar (Glaccum *et al.*, 1997).

In mice infected with *Leishmania* or following immunization with keyhole limpet hemocyanin, IL-1RI deficiency resulted in a greater production of IL-4 and IL-10 compared to wild-type mice (Satoskar *et al.*, 1998). In addition, wild-type mice produced more IFNγ compared to IL-1RI-deficient mice. These results suggest that IL-1 functions in these models to downregulate the TH2 response, particularly IL-4.

Human abnormalities

A case of a cortisol-secreting adrenal adenoma causing *Cushing's syndrome* in a 62-year-old woman has been described. The patient exhibited the classic clinical and laboratory findings of Cushing's syndrome which abated once the tumor was removed. Examination of the tissue revealed high expression of IL-1RI (Willenberg *et al.*, 1998). Moreover, unlike normal adrenal cells, the tumor did not respond to corticotropin-induced cortisol production but rather responded to IL-1β stimulation of cortisol production. In contrast to the patient's tumor, other adrenal tumors responded to corticotropin-induced cortisol production but not IL-1β. There was abundant expression of the IL-1RI in the patient's tumor but not in other tumors. It was concluded that the unique expression of IL-1RI in this tumor and induction of cortisol by IL-1β resulted in the pathological disease.

THERAPEUTIC UTILITY

Effect of treatment with soluble receptor domain

In Animal Studies

The extracellular domain of the type I receptor has been used in several models of inflammatory and autoimmune disease. Administration of murine IL-1sRI to mice has increased the survival of heterotopic heart allografts and reduced the hyperplasic lymph node response to allogeneic cells (Fanslow *et al.*, 1990). In a rat model of *antigen-induced arthritis*, local instillation of the murine soluble IL-1RI reduced joint swelling and tissue destruction (Dower *et al.*, 1994). When a dose of soluble receptor (1 μg) was instilled into the contralateral, unaffected joint, a reduction in the degree of tissue damage was observed in the affected joint. These data suggest that the amount of soluble IL-1RI given in the normal, contralateral joint was acting systemically. In a model of *experimental autoimmune encephalitis*, the soluble IL-1RI reduced the severity of this disease (Jacobs *et al.*, 1991). Administration of soluble IL-1RI to animals has also been reported to reduce the physiologic response to LPS, acute lung injury, and delayed-type hypersensitivity (reviewed in Dower *et al.*, 1994).

However, there are also data suggesting that exogenous administration of soluble IL-1RI may be harmful. In mice, an i.v. injection of soluble IL-RI alone induced a rapid release of circulating IL-1α, but not of TNFα or IL-1β (Netea *et al.*, 1999). The soluble receptor did not interfere with the IL-1α assay. Treatment of mice with soluble IL-1RI improved the survival during a lethal infection with *Candida albicans*.

In the accelerated model of *autoimmune diabetes* induced by cyclophosphamide in the nonobese diabetic (NOD) mouse, repeated injections with soluble IL-1R protected NOD mice from *insulin-dependent diabetes mellitus* in a dose-dependent fashion; the incidence of the diabetes was 53.3% in the mice treated with 0.2 mg/kg and only 6.7% in mice treated with 2 mg/kg. However, none of the doses of the soluble IL-1RI reduced the extent of *insulitis* in NOD mice. Splenic lymphoid cells from NOD mice treated with 2 mg/kg soluble IL-1R for 5 consecutive days showed a normal distribution of mononuclear cell subsets and maintained their capacity to secrete IFNγ and IL-2 (Nicoletti *et al.*, 1994).

In Humans

Recombinant human soluble IL-1RI has been administered intravenously to healthy humans in a phase I trial without side-effects or changes in physiologic, hematologic, or endocrinologic parameters. Thus, similar to infusions of IL-1Ra, soluble IL-1RI appears safe and reinforces the conclusion that IL-1 does not have a role in homeostasis in humans.

Volunteers have been injected with LPS and pretreated with soluble IL-1RI. The basis for these studies are that in animal models, blocking IL-1 with IL-1Ra has reduced the severity of the response (reviewed in Dinarello, 1996). Pretreatment of subjects with 10 mg/kg of IL-1Ra prior to intravenous endotoxin resulted in a statistically significant but modest decrease (40%) in circulating neutrophils (Granowitz *et al.*, 1993). Volunteers were also pretreated with soluble IL-1R type I or placebo and then challenged with endotoxin. There were no effects on fever or systemic symptoms noted. Although there was a decrease in the level of circulating IL-1β compared to placebo-treated volunteers, there was also a decrease in the level of circulating IL-1Ra ($p < 0.001$) due to complexing of the soluble receptor to endogenous IL-1Ra (Preas *et al.*, 1996). This was dose-dependent and resulted in a 43-fold decrease in endotoxin-induced IL-1Ra. High doses of soluble IL-1R type I were also associated with higher levels of circulating TNFα and IL-8 as well as cell-associated IL-1β (Preas *et al.*, 1996). These results support the concept that soluble IL-1R type I binds endogenous IL-1Ra and reduces the biological effectiveness of this natural IL-1 receptor antagonist in inhibiting IL-1. As discussed below, patients with *rheumatoid arthritis* treated with soluble IL-1R type I do not exhibit improved clinical outcome and the mechanism is probably the binding of endogenous IL-1Ra with a reduction in its biological role.

Soluble IL-1R type was administered subcutaneously to 23 patients with active *rheumatoid arthritis* in a randomized, double-blind, two-center study. Patients received subcutaneous doses of the receptor at 25, 250, 500, or 1000 μg/m^2/day or placebo for 28 consecutive days. Although four of eight patients receiving 1000 μg/m^2/day showed improvement in at least one measure of disease activity, only one of these four patients exhibited clinical improvement (Drevlow *et al.*, 1996). Similar to the placebo-treated patients, lower doses of the receptor did not result in any improvement by acceptable criteria. Despite this lack of clinical or objective improvement in disease activity, cell surface monocyte IL-1α expression in all patients receiving the soluble IL-1R type I was significantly reduced. Other parameters of altered immune function in common in patients with rheumatoid arthritis also showed reduction. One possible explanation for the lack of clinical response despite efficacy in suppressing immune responses could be the inhibition of endogenous IL-1Ra. This was observed in volunteers receiving soluble IL-1R type I before challenge by endotoxin (Preas *et al.*, 1996).

A phase I trial of soluble IL-1RI was conducted in patients with relapsed and refractory *acute myeloid leukemia*. Soluble IL-1RI was well tolerated. Serum levels of IL-1β, IL-6, and TNFα did not change. Circulating levels of soluble IL-1RI were elevated 360- and 25-fold after i.v. and s.c. administration, respectively. There were no complete, partial or minor responses to treatment (Bernstein *et al.*, 1999).

Human soluble IL-1RI inhibits *HIV*-1 expression in acutely infected primary PBMCs and in the chronically infected promonocytic cell line, U1. A phase I/II trial was conducted in HIV-1-infected individuals with CD4 T cell counts $< 300/\mu L$. Twelve evaluable patients were enrolled at three increasing doses of the soluble receptor injected three times a week for 8 weeks, followed by a 4-week observation period (Takebe *et al.*, 1998). Seven patients reported improvements in one or more symptoms, including weight gain (3), improved energy level (4), decreased diarrhea (1), decreased night sweats (1), improvement in *psoriatic arthritis* (1), and improvement in a non-specific chronic diffuse skin rash (1). No consistent trends in absolute CD4 counts or percentages, quantitative HIV-1 cultures, or serum p24 antigen, β_2-microglobulin, or triglyceride levels were observed.

The goal of any anti-IL-1 strategy is to prevent IL-1 binding to surface receptors. Using soluble receptors to block IL-1 activity in disease is similar to using neutralizing antibodies against IL-1 and distinct from using receptor blockade with IL-1Ra. Since the molar concentrations of circulating IL-1 in disease are relatively low, pharmacologic administration of soluble IL-1RI to reach a 100-fold molar excess of the soluble receptor over that of IL-1 is feasible. The human trial of soluble IL-1RI in delayed-hypersensitivity reactions supports the notion that low doses (100 µg/patient) can have anti-inflammatory effects. The fusion of two chains of extracellular domains of the type IL-1RI to the Fc portion of immunoglobulin enhances the binding of IL-1 over that of monomeric soluble IL-RI (Pitti *et al.*, 1994) and may have a greater plasma half-life compared to that of the monomeric form. However, as shown in the study of soluble IL-1RI in *rheumatoid arthritis*, binding of the endogenous IL-1Ra was associated with a worsening of the disease. In contrast to neutralizing IL-1 itself, the goal of receptor blockade requires the condition of blocking all unoccupied IL-1 surface receptors since triggering only a few evokes a response. Receptor blockade is a formidable task and the large amounts of IL-1Ra required to reduce disease activity contributes to this conclusion. The potential disadvantage for using soluble IL-1 receptor type I therapy is the possibility that these receptors will either prolong the clearance of IL-1 or bind the natural IL-1Ra.

Host Defense Impairment with IL-1RI Blockade

Mice given anti-IL-1RI antibody are more vulnerable to lethal infection with *Listeria*, an obligate intracellular bacterium (Havell *et al.*, 1992). Similar findings have been reported for TNFR blockade.

Why do Humans Respond to IL-1?

Unlike other cytokine receptors, in cells expressing both IL-1 type I and type II receptors, there is competition to bind IL-1 first. This competition between signaling and nonsignaling receptors for the same ligand appears unique to cytokine receptors, although it exists for atrial natriuretic factor receptors (Leitman *et al.*, 1986). Since the type II receptor is more likely to bind to IL-1β than IL-1α, this can result in a diminished response to IL-1β. The soluble forms of IL-1RI and IL-1RII circulate in healthy humans at molar concentrations which are 10- to 50-fold greater than those of IL-1β measured in septic patients and 100-fold greater than the concentration of IL-1β following intravenously administration (Crown *et al.*, 1991). Why do humans have a systemic response to an infusion of IL-1α (Smith *et al.*, 1993) or IL-1β? One concludes that binding of IL-1 to the soluble forms of IL-1R types I and II exhibits a slow 'on' rate compared to the cell IL-1RI.

In addition, naturally occurring neutralizing antibodies to IL-1α are present in many subjects and probably reduce the activity of IL-1α. Despite the portfolio of soluble receptors and naturally occurring antibodies, IL-1 produced during disease does, in fact, trigger the type I receptor since in animals and humans, blocking receptors or neutralizing IL-1 ameliorates disease. These findings underscore the high functional level of only a few IL-1 type I receptors. They also imply that the postreceptor triggering events are greatly amplified. It seems reasonable to conclude that treating disease based on blocking IL-1R needs to take into account the efficiency of so few type I receptors initiating a biological event.

Effects of inhibitors (antibodies) to receptors

Antibodies to the mouse IL-1RI block IL-1 responses and reduce the severity of disease (Gershenwald *et al.*, 1990). In mice overexpressing TNFα a *rheumatoid arthritis*-like disease develops, and treatment with blocking antibodies to the IL-1RI prevents the onset of disease (Probert *et al.*, 1995). However, IL-1RI-blocking antibodies have not been tested in humans to date.

References

Abbondanzo, S. J., Cullinan, E. B., McIntyre, K., Labow, M. A., and Stewart, C. L. (1996). Reproduction in mice lacking a functional type 1 IL-1 receptor. *Endocrinology* **137**, 3598–3601.

Adachi, O., Kawai, T., Takeda, K., Matsumoto, M., Tsutsui, H., Sakagami, M., Nakanishi, K., and Akira, S. (1998). Targeted disruption of the MyD88 gene results in loss of IL-1- and IL-18-mediated function. *Immunity* **9**, 143–150.

Andrieu, N., Salvayre, R., and Levade, T. (1994). Evidence against involvement of the acid lysosomal shingomyelinase in the tumor necrosis factor and interleukin-1-induced sphigomyelin cycle and cell proliferation in human fibroblasts. *Biochem. J.* **303**, 341–345.

Arend, W. P., Malyak, M., Smith, M. F., Whisenand, T. D., Slack, J. L., Sims, J. E., Giri, J. G., and Dower, S. K. (1994). Binding of IL-1α, IL-1β, and IL-1 receptor antagonist by soluble IL-1 receptors and levels of soluble IL-1 receptors in synovial fluids. *J. Immunol.* **153**, 4766–4774.

Barak, V., Vannier, E., Nisman, B., Pollack, A., and Dinarello, C. A. (1996). Cytokines and their soluble receptors as markers for hairy cell leukemia. *Eur. Cytokine Netw.* **7**, 536 (abstract).

Bernstein, S. H., Fay, J., Frankel, S., Christiansen, N., Baer, M. R., Jacobs, C., Blosch, C., Hanna, R., and Herzig, G. (1999). A phase I study of recombinant human soluble interleukin-1 receptor (rhu IL-1R) in patients with relapsed and refractory acute myeloid leukemia. *Cancer Chemother. Pharmacol.* **43**, 141–144.

Bird, T. A., Gearing, A. J., and Saklatvala, J. (1987). Murine interleukin-1 receptor: differences in binding properties between fibroblastic and thymoma cells and evidence for a two-chain receptor model. *FEBS Lett.* **225**, 21–26.

Bird, T. A., Sleath, P. R., de Roos, P. C., Dower, S. K., and Virca, G. D. (1991). Interleukin-1 represents a new modality for the activation of extracellular signal-related kinases/microtubule-associated protein-2 kinases. *J. Biol. Chem.* **266**, 22661–22670.

Bomalaski, J. S., Steiner, M. R., Simon, P. L., and Clark, M. A. (1992). IL-1 increases phospholipase A_2 activity, expression of phospholipase A_2-activating protein, and release of linoleic acid from the murine T helper cell line EL-4. *J. Immunol.* **148**, 155–160.

Born, T. L., Thomassen, E., Bird, T. A., and Sims, J. E. (1998). Cloning of a novel receptor subunit, AcPL, required for interleukin-18 signaling. *J. Biol. Chem.* **273**, 29445–29450.

Brooks, J. W., and Mizel, S. B. (1994). Interleukin-1 signal transduction. *Eur. Cytokine Netw.* **5**, 547–561.

Burch, R. M., and Mahan, L. C. (1991). Oligonucleotides antisense to the interleukin-1 receptor mRNA block the effects of interleukin-1 in cultured murine and human fibroblasts and in mice. *J. Clin. Invest.* **88**, 1190–1196.

Cao, Z. (1998). Signal transduction of interleukin-1. *Eur. Cytokine Netw.* **9**, 378 (abstract).

Cao, Z., Xiong, J., Takeuchi, M., Kurama, T., and Goeddel, D. V. (1996). Interleukin-1 receptor activating kinase. *Nature* **383**, 443–446.

Chin, J., Rupp, E., Cameron, P. M., MacNaul, K. L., Lotke, P. A., Tocci, M. J., Schmidt, J. A., and Bayne, E. K. (1988). Identification of a high-affinity receptor for interleukin-1α and interleukin-1β on cultured human rheumatoid synovial cells. *J. Clin. Invest.* **82**, 420–426.

Croston, G. E., Cao, Z., and Goeddel, D. V. (1995). NFκB activation by interleukin-1 requires an IL-1 receptor-associated protein kinase activity. *J. Biol. Chem.* **270**, 16514–16517.

Crown, J., Jakubowski, A., Kemeny, N., Gordon, M., Gasparetto, C., Wong, G., Toner, G., Meisenberg, B., Botet, J., Applewhite, J., Sinha, S., Moore, M., Kelsen, D., Buhles, W., and Gabrilove, J. (1991). A phase I trial of recombinant human interleukin-1β alone and in combination with myelosuppressive doses of 5-fluoruracil in patients with gastrointestinal cancer. *Blood* **78**, 1420–1427.

Cuenda, A., Rouse, J., Doza, Y. N., Meier, R., Cohen, P., Gallagher, T. F., Young, P. R., and Lee, J. C. (1995). SB 203580 is a specific inhibitor of a MAP kinase homologue which is stimulated by stresses and interleukin-1. *FEBS Lett.* **364**, 229–233.

Cullinan, E. B., Kwee, L., Nunes, P., Shuster, D. J., Ju, G., McIntyre, K. W., Chizzonite, R. A., and Labow, M. A. (1998). IL-1 receptor accessory protein is an essential component of the IL-1 receptor. *J. Immunol.* **161**, 5614–5620.

Cunningham, E. T., Wada, E., Carter, D. B., Tracey, D. E., Battey, J. F., and DeSouza, E. B. (1991). Localization of interleukin-1 receptor messenger RNA in murine hippocampus. *Endocrinology* **128**, 2666–2668.

Deyerle, K. L., Sims, J. E., Dower, S. K., and Bothwell, M. A. (1992). Pattern of IL-1 receptor gene expression suggests a role in noninflammatory processes. *J. Immunol.* **149**, 1657–1665.

DiDonato, J. A., Hayakawa, M., Rothwarf, D. M., Zandi, E., and Karin, M. (1997). A cytokine-responsive I kappaB kinase that activates the transcription factor NF-kappaB. *Nature* **388**, 548–554.

Dinarello, C. A. (1996). Biological basis for interleukin-1 in disease. *Blood* **87**, 2095–2147.

Dinarello, C. A. (1999). IL-18: A Th1-inducing, proinflammatory cytokine and new member of the IL-1 family. *J. Allergy Clin. Immunol.* **103**, 11–24.

Dower, S. K., Wignall, J. M., Schooley, K., McMahan, C. J., Jackson, J. L., Prickett, K. S., Lupton, S., Cosman, D., and Sims, J. E. (1989). Retention of ligand binding activity by the extracellular domain of the IL-1 receptor. *J. Immunol.* **142**, 4314–4320.

Dower, S. K., Fanslow, W., Jacobs, C., Waugh, S., Sims, J. E., and Widmer, M. B. (1994). Interleukin-1 antagonists. *Ther. Immunol.* **1**, 113–122.

Drevlow, B. E., Lovis, R., Haag, M. A., Sinacore, J. M., Jacobs, C., Blosche, C., Landay, A., Moreland, L. W., and Pope, R. M. (1996). Recombinant human interleukin-1 receptor type I in the treatment of patients with active rheumatoid arthritis. *Arthritis Rheum.* **39**, 257–265.

Dubois, C. M., Ruscetti, F. W., Palaszynski, E. W., Falk, L. A., Oppenheim, J. J., and Keller, J. R. (1990). Transforming growth factor β is a potent inhibitor of interleukin 1 (IL-1) receptor expression: proposed mechanism of inhibition of IL-1 action. *J. Exp. Med.* **172**, 737–744.

Fanslow, W. C., Sims, J. E., Sassenfeld, H., Morrissey, P. J., Gillis, S., Dower, S. K., and Widmer, M. B. (1990). Regulation of alloreactivity *in vivo* by a soluble form of the interleukin-1 receptor. *Science* **248**, 739–742.

Fantuzzi, G., Zheng, H., Faggioni, R., Benigni, F., Ghezzi, P., Sipe, J. D., Shaw, A. R., and Dinarello, C. A. (1996). Effect of endotoxin in IL-1β-deficient mice. *J. Immunol.* **157**, 291–296.

Finch, A., Holland, P., Cooper, J., Saklatvala, J., and Kracht, M. (1997). Selective activation of JNK/SAPK by interleukin-1 in rabbit liver is mediated by MKK7. *FEBS Lett.* **418**, 144–148.

Freshney, N. W., Rawlinson, L., Guesdon, F., Jones, E., Cowley, S., Hsuan, J., and Saklatvala, J. (1994). Interleukin-1 activates a novel protein cascade that results in the phosphorylation of hsp27. *Cell* **78**, 1039–1049.

Galcheva-Gargova, Z., Dérijard, B., Wu, I.-H., and Davis, R. J. (1994). An osmosensing signal transduction pathway in mammalian cells. *Science* **265**, 806–809.

Gallis, B., Prickett, K. S., Jackson, J., Slack, J., Schooley, K., Sims, J. E., and Dower, S. K. (1989). IL-1 induces rapid phosphorylation of the IL-1 receptor. *J. Immunol.* **143**, 3235–3240.

Gay, N. J., and Keith, F. J. (1991). Drosophila Toll and IL-1 receptor. *Nature* **351**, 355–356.

Gayle, M. A., Slack, J. L., Bonnert, T. P., Renshaw, B. R., Sonoda, G., Taguchi, T., Testa, J. R., Dower, S. K., and Sims, J. E. (1996). Cloning of a putative ligand for the T1/ST2 receptor. *J. Biol. Chem.* **271**, 5784–5789.

Gershenwald, J. E., Fong, Y. M., Fahey, T. J., Calvano, S. E., Chizzonite, R., Kilian, P. L., Lowry, S. F., and Moldawer, L. L. (1990). Interleukin 1 receptor blockade attenuates the host inflammatory response. *Proc. Natl Acad. Sci. USA* **87**, 4966–4970.

Glaccum, M. B., Stocking, K. L., Charrier, K., Smith, J. L., Willis, C. R., Maliszewski, C., Livingston, D. J., Peschon, J. J., and Morrissey, P. J. (1997). Phenotypic and functional characterization of mice that lack the type I receptor for IL-1. *J. Immunol.* **159**, 3364–3371.

Granowitz, E. V., Porat, R., Mier, J. W., Orencole, S. F., Callahan, M. V., Cannon, J. G., Lynch, E. A., Ye, K., Poutsiaka, D. D., Vannier, E., Shapiro, L., Pribble, J. P., Stiles, D. M., Catalano, M. A., Wolff, S. M., and Dinarello, C. A. (1993). Hematological and immunomodulatory effects of an interleukin-1 receptor antagonist coinfusion during low-dose endotoxemia in healthy humans. *Blood* **82**, 2985–2990.

Greenfeder, S. A., Nunes, P., Kwee, L., Labow, M., Chizzonite, R. A., and Ju, G. (1995). Molecular cloning and characterization of a second subunit of the interleukin-1 receptor complex. *J. Biol. Chem.* **270**, 13757–13765.

Gronich, J., Konieczkowski, M., Gelb, M. H., Nemenoff, R. A., and Sedor, J. R. (1994). Interleukin-1α causes a rapid activation of cytosolic phospholipase A_2 by phosphorylation in rat mesangial cells. *J. Clin. Invest.* **93**, 1224–1233.

Guesdon, F., Freshney, N., Waller, R. J., Rawlinson, L., and Saklatvala, J. (1993). Interleukin 1 and tumor necrosis factor stimulate two novel protein kinases that phosphorylate the heat shock protein hsp27 and beta-casein. *J. Biol. Chem.* **268**, 4236–4243.

Guida, S., Heguy, A., and Melli, M. (1992). The chicken IL-1 receptor: differential evolution of the cytoplasmic and extracellular domains. *Gene* **111**, 239–243.

Han, J., Lee, J.-D., Bibbs, L., and Ulevitch, R. J. (1994). A MAP kinase targeted by endotoxin and hyperosmolarity in mammalian cells. *Science* **265**, 808–811.

Han, J., Richter, B., Li, Z., Kravchenko, V. V., and Ulevitch, R. J. (1995). Molecular cloning of human p38 MAP kinase. *Biochim. Biophys. Acta* **1265**, 224–227.

Havell, E. A., Moldawer, L. L., Helfgott, D., Kilian, P. L., and Sehgal, P. B. (1992). Type I IL-1 receptor blockade exacerbates murine listeriosis. *J. Immunol.* **148**, 1486–1492.

Heguy, A., Baldari, C. T., Macchia, G., Telford, J. L., and Melli, M. (1992). Amino acids conserved in interleukin-1 receptors and the *Drosophila* Toll protein are essential for IL-1R signal transduction. *J. Biol. Chem.* **267**, 2605–2609.

Hirsch, E., Irikura, V. M., Paul, S. M., and Hirsh, D. (1996). Functions of interleukin-1 receptor antagonist in gene knockout and overproducing mice. *Proc. Natl Acad. Sci. USA* **93**, 11008–11013.

Hopp, T. P. (1995). Evidence from sequence information that the interleukin-1 receptor is a transmembrane GTPase. *Protein Sci.* **4**, 1851–1859.

Horai, R., Asano, M., Sudo, K., Kanuka, H., Suzuki, M., Nishihara, M., Takahashi, M., and Iwakura, Y. (1998). Production of mice deficient in genes for interleukin (IL)-1α, IL-1β, IL-1α/β, and IL-1 receptor antagonist shows that IL-1β is crucial in turpentine-induced fever development and glucocorticoid secretion. *J. Exp. Med.* **187**, 1463–1475.

Huang, J., Gao, X., Li, S., and Cao, Z. (1997). Recruitment of IRAK to the interleukin 1 receptor complex requires interleukin 1 receptor accessory protein. *Proc. Natl Acad. Sci. USA* **94**, 12829–12832.

Huwiler, A., and Pfeilschifter, J. (1994). Interleukin-1 stimulates *de novo* synthesis of mitogen-activated protein kinase in glomerular mesangial cells. *FEBS Lett.* **350**, 135–138.

Jacobs, C. A., Baker, P. E., Roux, E. R., Picha, K. S., Toivola, B., Waugh, S., and Kennedy, M. K. (1991). Experimental autoimmune encephalomyelitis is exacerbated by IL-1α and suppressed by soluble IL-1 receptor. *J. Immunol.* **146**, 2983–2989.

Kester, M., Siomonson, M. S., Mene, P., and Sedor, J. R. (1989). Interleukin-1 generate transmembrane signals from phospholipids through novel pathways in cultured rat mesangial cells. *J. Clin. Invest.* **83**, 718–723.

Koch, K.-C., Ye, K., Clark, B. D., and Dinarello, C. A. (1992). Interleukin 4 (IL) 4 up-regulates gene and surface IL-1 receptor type I in murine T helper type 2 cells. *Eur. J. Immunol.* **22**, 153–157.

Kojima, H., Takeuchi, M., Ohta, T., Nishida, Y., Arai, N., Ikeda, M., Ikegami, H., and Kurimoto, M. (1998). Interleukin-18 activates the IRAK-TRAF6 pathway in mouse EL-4 cells. *Biochem. Biophys. Res. Commun.* **244**, 183–186.

Kolesnick, R., and Golde, D. W. (1994). The sphingomyelin pathway in tumor necrosis factor and interleukin-1 signalling. *Cell* **77**, 325–328.

Kracht, M., Truong, O., Totty, N. F., Shiroo, M., and Saklatvala, J. (1994). Interleukin-1α activates two forms of p54α mitogen-activated protein kinase in rabbit liver. *J. Exp. Med.* **180**, 2017–2027.

Kuno, K., and Matsushima, K. (1994). The IL-1 receptor signaling pathway. *J. Leukocyte Biol.* **56**, 542–547.

Kuno, K., Okamoto, S., Hirose, K., Murakami, S., and Matsushima, K. (1993). Structure and function of the intracellular portion of the mouse interleukin-1 receptor (Type I). *J. Biol. Chem.* **268**, 13510–13518.

Labow, M., Shuster, D., Zetterstrom, M., Nunes, P., Terry, R., Cullinan, E. B., Bartfai, T., Solorzano, C., Moldawer, L. L., Chizzonite, R., and McIntyre, K. W. (1997). Absence of IL-1 signaling and reduced inflammatory response in IL-1 type I receptor-deficient mice. *J. Immunol.* **159**, 2452–2461.

Lee, J. C., Laydon, J. T., McDonnell, P. C., Gallagher, T. F., Kumar, S., Green, D., McNulty, D., Blumenthal, M. J., Heys, J. R., Landvatter, S. W., Strickler, J. E., Mc Laughlin, M. M., Slemens, I. R., Fisher, S. M., Livi, G. P., White, J. R., Adams, J. L., and Young, P. R. (1994). A protein kinase involved in the regulation of inflammatory cytokine biosynthesis. *Nature* **372**, 739–747.

Leitman, D. C., Andersen, J. W., Kuno, T., Kamisaki, Y., Chang, J., and Murad, F. (1986). Identification of multiple binding sites for atrial natriuretic factor by affinity cross-linking in cultured endothelial cells. *J. Biol. Chem.* **261**, 11650–11656.

Lomaga, M. A., Yeh, W. C., Sarosi, I., Duncan, G. S., Furlonger, C., Ho, A., Morony, S., Capparelli, C., Van, G., Kaufman, S., van der Heiden, A., Itie, A., Wakeham, A., Khoo, W., Sasaki, T., Cao, Z., Penninger, J. M., Paige, C. J., Lacey, D. L., Dunstan, C. R., Boyle, W. J., Goeddel, D. V., and Mak, T. W. (1999). TRAF6 deficiency results in osteopetrosis and defective interleukin-1, CD40, and LPS signaling. *Genes Dev.* **13**, 1015–1024.

Malinin, N. L., Boldin, M. P., Kovalenko, A. V., and Wallach, D. (1997). MAP3K-related kinase involved in NF-kappaB induction by TNF, CD95 and IL-1. *Nature* **385**, 540–544.

Man, M. Q., Wood, L., Elias, P. M., and Feingold, K. R. (1999). Cutaneous barrier repair and pathophysiology following barrier disruption in IL-1 and TNF type I receptor deficient mice. *Exp. Dermatol.* **8**, 261–266.

Mancilla, J., Ikejima, I., and Dinarello, C. A. (1992). Glycosylation of the interleukin-1 receptor type I is required for optimal binding of interleukin-1. *Lymphokine Cytokine Res.* **11**, 197–205.

Martin, M. U., and Falk, W. (1997). The interleukin-1 receptor complex and interleukin-1 signal transduction. *Eur. Cytokine Netw.* **8**, 5–17.

Martin, M., Bol, G. F., Eriksson, A., Resch, K., and Brigelius-Flohe, R. (1994). Interleukin-1-induced activation of a protein kinase co-precipitating with the type I interleukin-1 receptor in T-cells. *Eur. J. Immunol.* **24**, 1566–1571.

Mathias, S., Younes, A., Kan, C.-C., Orlow, I., Joseph, C., and Kolesnick, R. N. (1993). Activation of the sphingomyelin signaling pathway in intact EL4 cells and in a cell-free system by IL-1β. *Science* **259**, 519–522.

Matsushima, K., Kobayashi, Y., Copeland, T. D., Akahoshi, T., and Oppenheim, J. J. (1987). Phosphorylation of a cytosolic 65-kDa protein induced by interleukin-1 in glucocorticoid pretreated normal human peripheral blood mononuclear leukocytes. *J. Immunol.* **139**, 3367–3374.

Mitchum, J. L., and Sims, J. E. (1995). IIP1: a novel human that interacts with the IL-1 receptor. *Cytokine* **7**, 595 (abstract).

Mizel, S. B. (1990). Cyclic AMP and interleukin-1 signal transduction. *Immunol. Today* **11**, 390–391.

Mosley, B., Urdal, D. L., Prickett, K. S., Larsen, A., Cosman, D., Conlon, P. J., Gillis, S., and Dower, S. K. (1987). The interleukin-1 receptor binds the human interleukin-1α precursor but not the interleukin-1β precusor. *J. Biol. Chem.* **262**, 2941–2944.

Muegge, K., Williams, T. M., Kant, J., Karin, M., Chiu, R., Schmidt, A., Siebenlist, U., Young, H. A., and Durum, S. K. (1989). Interleukin-1 costimulatory activity on the interleukin-2 promoter via AP-1. *Science* **246**, 249–251.

Muegge, K., Vila, M., Gusella, G. L., Musso, T., Herrlich, P., Stein, B., and Durum, S. K. (1993). IL-1 induction of the c-jun promoter. *Proc. Natl Acad. Sci. USA* **90**, 7054–7058.

Munoz, E., Beutner, U., Zubiaga, A., and Huber, B. T. (1990). IL-1 activates two separate signal transduction pathways in T helper type II cells. *J. Immunol.* **144**, 964–969.

Netea, M. G., Kullberg, B. J., Boerman, O. C., Verschueren, I., Dinarello, C. A., and Van der Meer, J. W. (1999). Soluble murine IL-1 receptor type I induces release of constitutive IL-1 alpha. *J. Immunol.* **162**, 4876–4881.

Nicoletti, F., Di Marco, R., Barcellini, W., Magro, G., Schorlemmer, H. U., Kurrle, R., Lunetta, M., Grasso, S., Zaccone, P., and Meroni, P. (1994). Protection from experimental autoimmune diabetes in the non-obese diabetic mouse with soluble interleukin-1 receptor. *Eur. J. Immunol.* **24**, 1843–1847.

Novick, D., Kim, S.-H., Fantuzzi, G., Reznikov, L., Dinarello, C. A., and Rubinstein, M. (1999). Interleukin-18 binding protein: a novel modulator of the Th1 cytokine response. *Immunity* **10**, 127–136.

O'Neill, L. A. J. (1995). Towards and understanding of the signal transduction pathways for interleukin-1. *Biochim. Biophys. Acta* **1266**, 31–44.

O'Neill, L. A. J., and Greene, C. (1998). Signal transduction pathways activated by the IL-1 receptor family: ancient signaling machinery in mammals, insects, and plants. *J. Leukocyte Biol.* **63**, 650–657.

Parnet, P., Garka, K. E., Bonnert, T. P., Dower, S. K., and Sims, J. E. (1996). IL-1Rrp is a novel receptor-like molecule similar to the type I interleukin-1 receptor and its homologues T1/ST2 and IL-1R AcP. *J. Biol. Chem.* **271**, 3967–3970.

Penton-Rol, G., Orlando, S., Polentarutti, N., Bernasconi, S., Muzio, M., Introna, M., and Mantovani, A. (1999). Bacterial lipopolysaccharide causes rapid shedding, followed by inhibition of mRNA expression, of the IL-1 type II receptor, with concomitant up-regulation of the type I receptor and induction of incompletely spliced transcripts. *J. Immunol.* **162**, 2931–2938.

Pitti, R. M., Marsters, S. A., Haak-Frendscho, M., Osaka, G. C., Mordenti, J., Chamow, S. M., and Ashkenazi, A. (1994). Molecular and biological properties of an interleukin-1 receptor immunoadhesin. *Mol. Immunol.* **31**, 1345–1351.

Preas, H.L., II, Reda, D., Tropea, M., Vandivier, R. W., Banks, S. M., Agosti, J. M., and Suffredini, A. F. (1996). Effects of recombinant soluble type I interleukin-1 receptor on human inflammatory responses to endotoxin. *Blood* **88**, 2465–2472.

Probert, L., Plows, D., Kontogeorgos, G., and Kollias, G. (1995). The type I interleukin-1 receptor acts in series with tumor necrosis factor (TNF) to induce arthritis in TNF-transgenic mice. *Eur. J. Immunol.* **25**, 1794–1797.

Pronost, S., Delecouillerie, G., Redini, F., Paolozzi, L., Vivien, D., Galera, P., Loyau, G., and Pujol, J. P. (1993). Interleukin-1 and naproxen down-regulate the expression of IL-1 receptors in cultured human rheumatoid synovial cells. *Agents Actions* **39**, 213–217.

Pruitt, J. H., Welborn, M. B., Edwards, P. D., Harward, T. R., Seeger, J. W., Martin, T. D., Smith, C., Kenney, J. A., Wesdorp, R. I., Meijer, S., Cuesta, M. A., Abouhanze, A., Copeland, E. M., Giri, J., Sims, J. E., Moldawer, L. L., and Oldenburg, H. S. (1996). Increased soluble interleukin-1 type II receptor concentrations in postoperative patients and in patients with sepsis syndrome. *Blood* **87**, 3282–3288.

Qwarnstrom, E. E., Page, R. C., Gillis, S., and Dower, S. K. (1988). Binding, internalization, and intracellular localization of interleukin-1 beta in human diploid fibroblasts. *J. Biol. Chem.* **263**, 8261–8269.

Robinson, D., Shibuya, K., Mui, A., Zonin, F., Murphy, E., Sana, T., Hartley, S. B., Menon, S., Kastelein, R., Bazan, F., and O'Garra, A. (1997). IGIF does not drive Th1 development but synergizes with IL-12 for interferon-γ production and activates IRAK and NFκB. *Immunity* **7**, 571–581.

Rosenbaum, J. T., Han, Y. B., Park, J. M., Kennedy, M., and Planck, S. R. (1998). Tumor necrosis factor-alpha is not essential in endotoxin induced eye inflammation: studies in cytokine receptor deficient mice. *J. Rheumatol.* **25**, 2408–2416.

Rosoff, P. M. (1989). Characterization of the interleukin-1-stimulated phospholipase C activity in human T lymphocytes. *Lymphokine Res.* **8**, 407–413.

Rosoff, P. M., Savage, N., and Dinarello, C. A. (1988). Interleukin-1 stimulates diacylglycerol production in T lymphocytes by a novel mechanism. *Cell* **54**, 73–81.

Rossi, B. (1993). IL-1 transduction signals. *Eur. Cytokine Netw.* **4**, 181–187.

Satoskar, A. R., Okano, M., Connaughton, S., Raisanen-Sokolwski, A., David, J. R., and Labow, M. (1998).

Enhanced Th2-like responses in IL-1 type 1 receptor-deficient mice. *Eur. J. Immunol.* **28**, 2066–2074.

Savage, N., Puren, A. J., Orencole, S. F., Ikejima, T., Clark, B. D., and Dinarello, C. A. (1989). Studies on IL-1 receptors on D10S T-helper cells: demonstration of two molecularly and antigenically distinct IL-1 binding proteins. *Cytokine* **1**, 23–25.

Scapigliati, G., Ghiara, P., Bartalini, A., Taglibue, A., and Boraschi, D. (1989). Differential binding of IL-1α and IL-1β to receptors on B and T cells. *FEBS Lett.* **243**, 394–398.

Schutze, S., Machleidt, T., and Kronke, M. (1994). The role of diacylglycerol and ceramide in tumor necrosis factor and interleukin-1 signal transduction. *J. Leukocyte Biol.* **56**, 533–541.

Shapiro, L., and Dinarello, C. A. (1995). Osmotic regulation of cytokine synthesis *in vitro*. *Proc. Natl Acad. Sci. USA* **92**, 12230–12234.

Shapiro, L., and Dinarello, C. A. (1997). Cytokine expression during osmotic stress. *Exp. Cell Res.* **231**, 354–362.

Shapiro, L., Puren, A. J., Barton, H. A., Novick, D., Peskind, R. L., Su, M.S.-S., Gu, Y., and Dinarello, C. A. (1998a). Interleukin-18 stimulates HIV type 1 in monocytic cells. *Proc. Natl Acad. Sci. USA* **95**, 12550–12555.

Shapiro, L., Heidenreich, K. A., Meintzer, M. K., and Dinarello, C. A. (1998b). Role of p38 mitogen-activated protein kinase in HIV type 1 production *in vitro*. *Proc. Natl Acad. Sci. USA* **95**, 7422–7426.

Shirakawa, F., Tanaka, Y., Ota, T., Suzuki, H., Eto, S., and Yamashita, U. (1987). Expression of interleukin-1 receptors on human peripheral T-cells. *J. Immunol.* **138**, 4243–4248.

Shornick, L. P., De Togni, P., Mariathasan, S., Goellner, J., Strauss-Schoenberger, J., Karr, R. W., Ferguson, T. A., and Chaplin, D. D. (1996). Mice deficient in IL-1beta manifest impaired contact hypersensitivity to trinitrochlorobenzone. *J. Exp. Med.* **183**, 1427–1436.

Silvera, M. R., Sempowski, G. D., and Phipps, R. P. (1994). Expression opf TGFβ isoforms by Thy-1$^+$ and Thy-1$^-$ pulmonary fibroblast subsets: evidence for TGFβ as a regulator of IL-dependent stimulation of IL-6. *Lymphokine Cytokine Res.* **13**, 277–285.

Sims, J. E., March, C. J., Cosman, D., Widmer, M. B., MacDonald, H. R., McMahan, C. J., Grubin, C. E., Wignall, J. M., Jackson, J. L., and Call, S. M. (1988). cDNA expression cloning of the IL-1 receptor, a member of the immunoglobulin superfamily. *Science* **241**, 585–589.

Sims, J. E., Painter, S. L., and Gow, I. R. (1995). Genomic organization of the type I and type II IL-1 receptors. *Cytokine* **7**, 483–490.

Smith, J. W., Longo, D., Alford, W. G., Janik, J. E., Sharfman, W. H., Gause, B. L., Curti, B. D., Creekmore, S. P., Holmlund, J. T., Fenton, R. G., Sznol, M., Miller, L. L., Shimzu, M., Oppenheim, J. J., Fiem, S. J., Hursey, J. C., Powers, G. C., and Urba, W. J. (1993). The effects of treatment with interleukin-1α on platelet recovery after high-dose carboplatin. *N. Engl. J. Med.* **328**, 756–761.

Stylianou, E., and Saklatvala, J. (1998). Interleukin-1. *Int. J. Biochem. Cell Biol.* **30**, 1075–1079.

Stylianou, E., O'Neill, L. A. J., Rawlinson, L., Edbrooke, M. R., Woo, P., and Saklatvala, J. (1992). Interleukin-1 induces NFkB through its type I but not type II receptor in lymphocytes. *J. Biol. Chem.* **267**, 15836–15841.

Sutton, E. T., Norman, J. G., Newton, C. A., Hellermann, G. R., and Richards, I. S. (1997). Endothelial structural integrity is maintained during endotoxic shock in an interleukin-1 type 1 receptor knockout mouse. *Shock* **7**, 105–110.

Takebe, N., Paredes, J., Pino, M. C., Lownsbury, W. H., Agosti, J., and Krown, S. E. (1998). Phase I/II trial of the type I soluble recombinant human interleukin-1 receptor in HIV-1-infected patients. *J. Interferon Cytokine Res.* **18**, 321–326.

Takii, T., Akahoshi, T., Kato, K., Hayashi, H., Marunouchi, T., and Onozaki, K. (1992). Interleukin-1 upregulates transcription of its own receptor in a human fibroblast cell line TIG-1. Role of endogenous PGE_2 and cAMP. *Eur. J. Immunol.* **22**, 1221–1227.

Takii, T., Hayashi, H., Marunouchi, T., and Onozaki, K. (1994). Interleukin-1 downregulates type I interleukin-1 receptor mRNA expression in a human fibroblast cell line TIG-1 in the absence of prostaglandin E(2) synthesis. *Lymphokine Cytokine Res.* **13**, 213–219.

Torigoe, K., Ushio, S., Okura, T., Kobayashi, S., Taniai, M., Kunikate, T., Murakami, T., Sanou, O., Kojima, H., Fuji, M., Ohta, T., Ikeda, M., Ikegami, H., and Kurimoto, M. (1997). Purification and characterization of the human interleukin-18 receptor. *J. Biol. Chem.* **272**, 25737–25742.

Vargas, S. J., Naprta, A., Glaccum, M., Lee, S. K., Kalinowski, J., and Lorenzo, J. A. (1996). Interleukin-6 expression and histomorphometry of bones from mice deficient in receptors for interleukin-1 or tumor necrosis factor. *J. Bone Miner. Res.* **11**, 1736–1744.

Wesche, H., Korherr, C., Kracht, M., Falk, W., Resch, K., and Martin, M. U. (1997). The interleukin-1 receptor accessory protein is essential for IL-1-induced activation of interleukin-1 receptor-associated kinase (IRAK) and stress-activated protein kinases (SAP kinases). *J. Biol. Chem.* **272**, 7727–7731.

Willenberg, H. S., Stratakis, C. A., Marx, C., Ehrhart-Bornstein, M., Chrousos, G. P., and Bornstein, S. R. (1998). Aberrent interleukin-1 receptors in a cortisol-secreting adrenal adenoma causing Cushing's syndrome. *N. Engl. J. Med.* **339**, 27–31.

Ye, K., Dinarello, C. A., and Clark, B. D. (1993). Identification of the promoter region of the human interleukin 1 type I receptor gene: multiple initiation sites, high G+C content, and constitutive expression. *Proc. Natl Acad. Sci. USA* **90**, 2295–2299.

Ye, K., Koch, K.-C., Clark, B. D., and Dinarello, C. A. (1992). Interleukin-1 down regulates gene and surface expression of interleukin-1 receptor type I by destabilizing its mRNA whereas interleukin-2 increases its expression. *Immunology* **75**, 427–434.

Ye, K., Vannier, E., Clark, B. D., Sims, J. E., and Dinarello, C. A. (1996). Three distinct promoters direct transcription of different 5′ untranslated regions of the human interleukin 1 type I receptor: a possible mechanism for control of translation. *Cytokine* **8**, 421–429.

Yoon, D. Y., and Dinarello, C. A. (1998). Antibodies to domains II and III of the IL-1 receptor accessory protein inhibit IL-1β activity but not binding: regulation of IL-1 responses is via type I receptor, not the accessory protein. *J. Immunol.* **160**, 3170–3179.

Yoshimoto, T., Takeda, K., Tanaka, T., Ohkusu, K., Kashiwamura, S., Okamura, H., Akira, S., and Nakanishi, K. (1998). IL-12 upregulates IL-18 receptor expression on T cells, Th1 cells and B cells: synergism with IL-18 for IFNγ production. *J. Immunol.* **161**, 3400–3407.

Zhang, S., Han, J., Sells, M. A., Chernoff, J., Knaus, U. G., Ulevitch, R. J., and Bokoch, G. M. (1995). Rho family GTPases regulate p38 mitogen-activated protein kinase through the downstream mediator Pak1. *J. Biol. Chem.* **270**, 23934–23936.

Zheng, H., Fletcher, D., Kozak, W., Jiang, M., Hofmann, K., Conn, C. C., Siszynski, D., Grabiec, C., Trumbauer, M. A., Shaw, A. R., Kostura, M. J., Stevens, K., Rosen, H., North, R. J., Chen, H. Y., Tocci, M. J., Kluger, M. J., and Van der Ploeg, L. H. T. (1995). Resistance to fever induction and impaired acute-phase response in interleukin-1β deficient mice. *Immunity* **3**, 9–19.

Poxvirus IL-1β Receptor Homologs

Grant McFadden[1,*] and Richard Moyer[2]

[1]The John P. Robarts Research Institute and Department of Microbiology and Immunology, The University of Western Ontario, 1400 Western Road, London, Ontario, N6G 2V4, Canada

[2]Department of Molecular Genetics and Microbiology, University of Florida College of Medicine, PO Box 100266, Gainesville, FL 32610-0266, USA

*corresponding author tel: (519)663-3184, fax: (519)663-3847, e-mail: mcfadden@rri.on.ca

DOI: 10.1006/rwcy.2000.14013.

SUMMARY

The IL-1β receptor homologs of poxviruses were the second examples discovered, following the poxvirus TNF receptor homologs, of 'viroceptors' or virus-encoded receptor mimics that function to bind and sequester cellular ligands away from their cognate cellular receptors. The prototypic member of this family, B15R of vaccinia virus, is a secreted member of the Ig superfamily, with highest sequence similarity to the ligand-binding domain of the cellular type II IL-1 receptor. The protein plays an important role in modulating the inflammatory cascade, particularly the fever response, following virus infection.

BACKGROUND

Discovery

Routine sequencing of *vaccinia virus* Western Reserve (WR) revealed two open reading frames (B15R and B18R) with significant homology to type I IL-1 receptors (Smith *et al.*, 1991). Subsequent publication of sequences from murine and human type II IL-1 receptors revealed that the B15R ORF has far more significant homology (McMahan *et al.*, 1991) than that of B18R. The B15R gene encodes a 326 amino acid protein and is found in the vaccinia virus WR, Lister, and Ankara strains. A nearly identical gene is found in *cowpox virus*. A nonfunctional protein is encoded by vaccinia Copenhagen, *ectromelia virus*, and *variola virus* (*smallpox*). The corresponding variola ORF is repeatedly disrupted by frameshifts and translation termination codons (Shchelkunov *et al.*, 1993; Massung *et al.*, 1994).

Alternative names

IL-1β soluble receptor, ORF B15R (the vaccinia WR designation used throughout this discussion), ORF B16R (vaccinia Copenhagen), cowpox virus (B14R).

Structure

The ORF is typical of poxvirus genes, contiguous and with no introns.

Main activities and pathophysiological roles

The protein inhibits murine and human IL-1β, but not IL-1α or IL-1Ra proteins (Alcamí and Smith, 1992, 1996). Affinities for human IL-1β are high ($K_d \sim 230$ pM) which when coupled to very high levels of expression as compared with cellular receptors, lead to effective blocking of IL-1β binding to receptor-bearing cells and IL-1β-mediated activities *in vitro* (Alcamí and Smith, 1992; Spriggs *et al.*, 1992).

GENE

Accession numbers

Vaccinia virus WR: X56121, D01018
Vaccinia virus Ankara: U94848
Cowpox virus: M95202

Sequence

In vaccinia WR, the gene is located 174 kb from the left end of the ~190 kb genome in the *Hin*dIII B fragment, the rightmost terminal *Hin*dIII fragment of the viral genome. Transcription is in the rightward direction.

The protein is synthesized and secreted 'late' in infection, subsequent to DNA replication (Alcamí and Smith, 1992).

PROTEIN

Accession numbers

Vaccinia virus WR: 222697
Vaccinia virus Ankara: 2772816
Cowpox virus: 418180

Sequence

See **Figure 1**.

Description of protein

The protein consists of 326 amino acids (see Figure 1). There is an ~21 amino acid N-terminal secretory signal sequence which is cleaved to generate the mature protein. The protein is characterized by the presence of three Ig domains which comprise most of the mature protein (Figure 1). All three Ig domains contain a pair of conserved cysteine residues which form intradomain disulfide bridges. These residues are present at positions 48 and 99, 143 and 194, and 242 and 309. There is no transmembrane sequence nor a cytoplasmic domain (Smith *et al.*, 1991). While IL-1β levels are controlled extracellularly by this protein (ORF B15R), IL-1β levels are also controlled intracellularly in many orthopoxvirus strains by the SPI-2/crmA protein (a serpin) of vaccinia/cowpox virus. In vaccinia WR, the SPI-2 ORF is designated B13R. The B13R protein inhibits the proteinase interleukin-1β convertase (ICE) and prevents activation of precursor IL-1β (Ray *et al.*, 1992).

Relevant homologies and species differences

All active B15R proteins expressed by various orthopoxviruses are virtually identical. The most significant homology to a nonviral protein is to the rat type II IL-1β receptor (Bristulf *et al.*, 1994) (**Figure 2**). Cleavage of an N-terminal signal precedes secretion into the infected cell media.

Figure 1 Sequence of the vaccinia WR IL-1β-secreted soluble receptor precursor. The signal sequence of approximately 20 amino acids is cleaved to generate the mature protein. The three immunoglobulin domains which dominate the protein are indicated in bold and are separated by small intervening peptides. The active proteins synthesized by vaccinia Ankara and cowpox virus are virtually identical.

1 MSILPVIFLS IFFYSSFVQT FNAPECIDKG **QYFASFMELE**
51 INTLSSGYNI LDILWEKRGA DNDRIIPIDN GSNMLILNPT
QSDSGIYICI
101 TTNETYCDMM SLNLTIVSVS **ESNIDLISYP QIVNERSTGE**
MVCPNINAFI
151 ASNVNADIIW SGHRRLRNKR LKQRTPGIIT IEDVRKNDAG
YYTCVLEYIY
201 GGKTYNVTRI VKLEVRDKII **PSTMQLPDGI VTSIGSNLTI**
ACRVSLRPPT
251 TDADVFWISN GMYYEEDDGD GNGRISVANK IYMTDKRRVI
TSRLNINPVK
301 EEDATTFTCM AFTIPSISKT VTVSIT

Figure 2 A 'GAP' alignment of the B15R 326 amino acid ORF with a portion of the rat 416 amino acid type II IL-1 receptor precursor. The proteins are 39% similar and 29% identical through the regions compared. The conserved pairs of cysteine residues, conserved within the three immunoglobulin domains of both proteins are shown in bold.

```
VV     1 ................MSILPVIFLSIFFYSSFVQTFNAPECIDKGQYFA  34
                          .:.    .|    .|        :|  :|. |
RAT    1 MFILLVLVTGVSAFTTPAVVHTGRVSESPVTSEKHPVLGDDCWFRGRDFK  50
                  .         .         .         .         .
      35 SFMELENEPVILPCPQINTLSSGYNILDILWEKRGADNDRIIPID.....  79
         | : || |||:| || :    .  .   :|   : .|. .:|| |
      51 SELRLEGEPVVLRCPLVPHSDTSSSSRSLLTWSK.SDSSQLIPGDEPRMW  99
                  .         .         .         .         .
      80 .NGSNMLILNPTQSDSGIYICITTNETYCDMMSLNLTIVSVSESNIDLIS 128
              : :|   | ||| |||   | .:|: ||| | :   .|..  |:|
     100 VKDDTLWVLPAVQQDSGTYICTFRNASHCEQMSLELKVFKNTEASFPLVS 149
                  .         .         .         .         .
     129 YPQIVNERSTGEMVCPNINAFIASNVNADIIWSGHRRL....RNKRLKQR 174
         | ||    ||| :|||.:  ||.|    .| |     |      | |
     150 YLQISALSSTGLLVCPDLKEFISSRTDGKIQWYKGSILLDKGNKKFLSAG 199
                  .         .         .         .         .
     175 TPGIITIEDVRKNDAGYYTCVLEYIYGGKTYNVTRIVKLEVR...DKIIP 221
          |  : | .    ||||| ||: : | || ||:|| :.| |:     .||
     200 DPTRLLISNTSMGDAGYYRCVMTFTYEGKEYNITRNIELRVKGITTEPIP 249
                  .         .         .         .         .
     222 STMQLPDGIVTSIGSNLTIACRVSLRPPT.TDADVFWISNGMYYEEDDGD 270
            . :  |  |:|| | : |:| |   | ..  |.|..|  :
     250 VIISPLETIPASLGSRLIVPCKVFLGTGTSSNTIVWWMANSTFISVAYPR 299
                  .         .         .         .         .
     271 GNGRISVANKIYMTDKRRVITSRLNINPVKEEDATTFTCMAFTIPSISKT 320
         |   .  ..     |.  |  | :   ||:|   | |.|      |
     300 GRVTEGLHHQYSENDENYVEVSLIFDPVTKEDLNTDFKCVATNPRSFQSL 349
                  .         .         .         .         .
     321 VTVSIT............................................ 326
         |
     350 HTTVKEVSSTFSWGIALAPLSLIILVVGAIWIRRRCKRQAGKTYGLTKLP 399
```

BIOLOGICAL CONSEQUENCES OF ACTIVATING OR INHIBITING RECEPTOR AND PATHOPHYSIOLOGY

The role of this protein was initially investigated by two groups using different model infections of mice. In one case, based on intracranial injection, it was reported that *vaccinia virus* WR deleted for the B15R ORF was attenuated (Spriggs *et al.*, 1992). However, when mice were infected intranasally, deletion of the B15R ORF had the opposite effect, i.e. the infection was more virulent, the animals showing increased weight loss and a more generally severe illness (Alcamí and Smith, 1992). It was proposed that the increased illness was due to increased levels of IL-1β. Consistent with this prediction is the fact that IL-1β knockout mice are resistant to the induction of fever by chemical irritants (Kozak *et al.*, 1995; Zheng *et al.*, 1995).

The specific link of IL-1β presence to fever suppression in *vaccinia virus* infections was proven when it was subsequently found that mice infected intranasally with wild-type vaccinia do not generate a fever despite signs of a severe infection and up to 30% loss of body weight. By contrast, animals infected with vaccinia deleted for the B15R ORF developed a significant fever which lasted up to 5 days (Alcamí and Smith, 1996). Hence, it appears that intracellular IL-1β is mainly responsible for the control of fever in vaccinia virus-infected animals, despite the presence of the viral encoded intracellular B13R ORF gene product (SPI-2/crmA) which prevents ICE (caspase 3) activation of proIL-1β to active IL-1β.

References

Alcamí, A., and Smith, G. L. (1992). A soluble receptor for interleukin-1β encoded by vaccinia virus: A novel mechanism of virus modulation of the host response to infection. *Cell* **71**, 153–167.

Alcamí, A., and Smith, G. L. (1996). A mechanism for the inhibition of fever by a virus. *Proc. Natl Acad. Sci. USA* **93**, 11029–11034.

Bristulf, J., Gatti, S., Malinowsky, D., Bjork, L., Sundgren, A. K., and Bartfai, T. (1994). Interleukin-1 stimulates the expression of type I and type II interleukin-1 receptors in the rat insulinoma cell line Rinm5F, sequencing a rat type II interleukin-1 receptor cDNA. *Eur. Cytokine Netw.* **5**, 319–330.

Kozak, W., Zheng, H., Conn, C. A., Soszynski, D., van der Ploeg, L. H., and Kluger, M. J. (1995). Thermal and behavioral effects of lipopolysaccharide and influenza in interleukin-1 beta-deficient mice. *Am. J. Physiol.* **269**, R969–R977.

McMahan, C. J., Slack, J. L., Mosley, B., Cosman, D., Lupton, S. D., Bruton, L. L., Grubin, C. E., Wignall, J. M., Jenkins, N. A., and Brannan, C. I. (1991). A novel IL-1 receptor, cloned from B cells by mammalian expression, is expressed in many cell types. *EMBO J.* **10**, 2821–2832.

Massung, R. F., Liu, L. I., Qi, J., Knight, J. C., Yuran, T. E., Kerlavage, A. R., Parsons, J. M., Venter, J.C., and Esposito J. J. (1994). Analysis of the complete genome of smallpox variola major virus strain Bangladesh-1975. *Virology* **201**, 215–240.

Ray, C. A., Black, R. A., Kronheim, S. R., Greenstreet, T. A., Sleath, P. R., Salvesen, G. S., and Pickup, D. J. (1992). Viral inhibition of inflammation: cowpox virus encodes an inhibitor of the interleukin-1β converting enzyme. *Cell* **69**, 597–604.

Shchelkunov, S. N., Blinov, V. M., and Sandakhchiev, L. S. (1993). Genes of variola and vaccinia viruses necessary to overcome the host protective mechanism. *FEBS Lett.* **319**, 80–83.

Smith, G., Chan, Y. S., and Howard, S. T. (1991). Two vaccinia virus proteins structurally related to the interleukin-1 receptor and the immunoglobulin superfamily. *J. Gen. Virol.* **72**, 511–518.

Spriggs, M. K., Hruby, D. E., Maliszewski, C. R., Pickup, D. J., Sims, J. E., Buller, R. M., and VanSlyke, J. (1992). Vaccinia and cowpox viruses encode a novel secreted interleukin-1 binding protein. *Cell* **71**, 145–152.

Zheng, H., Fltecher, D., Kozak, W., Jiang, M., Hofmann, K. J., Conn, C. A., Soszynski, D., Grabiec, C., Trumbauer, M. E., and Shaw, A. (1995). Resistance to fever induction and impaired acute-phase response in interleukin-1 beta-deficient mice. *Immunity* **3**, 9–19.

IL-18 Receptor

Haruki Okamura*, Hiroko Tsutsui, Sin-ichiro Kashiwamura, Tomohiro Yoshimoto and Kenji Nakanishi

Hyogo College of Medicine, Nishinomiya, Japan

* corresponding author tel: 81-0798-45-6744, fax: 81-0798-45-6746, e-mail: haruoka@hyo-med.ac.jp

DOI: 10.1006/rwcy.2000.15002.

SUMMARY

Much interest has been focused on the proinflammatory cytokines, such as TNFα, IL-1α, IL-1β, and IL-18, because these cytokines have critical roles in the immune and inflammatory reactions. Of these cytokines, IL-1 and IL-18 have common structural features, and indeed, their receptors form a family. They also share parts of a common signal transducing system, comprising IL-1 receptor-associated kinase (IRAK), MyD88, and TNF receptor-associated factor (TRAF-6), and activation of the transcription factor NFκB. Although IL-1R type I and IL-18Rα (IL-1Rrp) have a high homology, the receptors for each cytokine are expressed on different types of cells. For example, expression of IL-18Rα (IL-18R), but not IL-1R type I, on T cells is upregulated by IL-12, a powerful differentiation factor for TH1. IL-18Rα is now accepted as one of the surface markers for TH1 cells. IL-18, like IL-1, is expressed in a variety of tissues other than immune organs, and the expression of IL-18R in these tissues needs to be examined in relationship to its pathophysiological roles.

BACKGROUND

Discovery

Human IL-18 receptor has been purified and characterized using monoclonal antibody which specifically inhibits the binding of IL-18 to *Hodgkin's disease* cell line and inhibits its actions (Torigoe *et al.*, 1997). Its amino acid sequence completely matched that of human IL-1 receptor-related protein (IL-1Rrp), whose DNA sequence had been published as an orphan receptor belonging to the IL-1 receptor family (Parnet *et al.*, 1996). A novel member of the IL-1R family, with the highly conserved IL-1R hallmark domains, was recently cloned (Born *et al.*, 1998). Based on its homology to IL-1R accessory protein (AcP), it is termed as an IL-1R AcP-like protein (AcPL). Like IL-1R AcP, which does not directly bind to IL-1 but stabilizes IL-1R conformation, AcPL as well as IL-1Rrp is required to mediate signaling by IL-18. Recently, a soluble IL-18-binding protein (IL-18BP) has also been cloned (Novick *et al.*, 1999). This protein was purified from human urine by chromatography of IL-18 beads, sequenced, and cloned. IL-18BP directly binds to IL-18 and inhibits the IFNγ-inducing action of IL-18.

Alternative names

The IL-18-binding component of IL-18R has been identified as IL-1Rrp (Parnet *et al.*, 1996; Torigoe *et al.*, 1997). It was proposed that IL-18, once designated as an IFNγ inducing factor (IGIF), might be renamed as IL-1γ because of its similarities in secondary structure to IL-1 and because, like IL-1, it lacks a signal peptide in its sequence. However, IL-1α and IL-1β utilize the same receptor, whereas IL-18 does not bind to or share their receptor. Therefore, it seems better to call IL-1Rrp the IL-18 receptor. There is also a proposal that IL-1Rrp should be renamed as IL-18Rα and AcPL as IL-18Rβ, but general agreement on this has not been attained.

Structure

The IL-1 receptor consists of an IL-1-binding component, IL-1R type I, and IL-1 signaling component, IL-1R accessory protein (AcP). Similarly, the IL-18 receptor is composed of a ligand-binding subunit

(IL-18Rα/IL-1Rrp) and a signaling component (IL-18Rβ/AcPL).

Main activities and pathophysiological roles

IL-18R mediates the signal of IL-18 and activates signal-transducing molecules such as IRAK and TRAF-6, resulting in the translocation of NFκB to the nucleus. The pathophysiological roles of IL-18 remain to be clarified. Since IL-18R is expressed on TH1 cells, but not on TH2 cells (Yoshimoto *et al.*, 1998; Xu *et al.*, 1998), IL-18R may be an indicator of TH1 and be concerned with TH1-dominant tissue injury (Rothe *et al.*, 1997).

GENE

Accession numbers

GenBank:
Human IL-18Rα (IL-1Rrp): U43672
Murine IL-18Rα (IL-1Rrp): U43673
Human IL-18Rβ (AcPL): AF077346
Murine IL-18Rβ (AcPL): AF077347
Human IL-18BP (IL-18-binding protein): AF110799
Murine IL-18BP: AA498857

Sequence: Chromosome location and linkages

Human IL-18Rα (IL-1Rrp) maps to chromosome 2q13-21, where genes encoding IL-1α, IL-1β, IL-1Ra, and IL-1 receptors (type I, a binding component, and type II, a decoy receptor) are located (Nolan *et al.*, 1998; Aizawa *et al.*, 1999). On the other hand, human IL-18 maps to chromosome 11q22, closely linked to DRD2. Therefore, the genes for the IL-18 ligand and the IL-18 receptor map to different chromosomes. The gene for IL-18BP, a recently cloned soluble IL-18 decoy receptor, is localized on chromosome 11q13 (Novick *et al.*, 1999).

PROTEIN

Relevant homologies and species differences

Murine IL-18R (IL-1Rrp) is homologous to human IL-18R (IL-1Rrp) by 65% in overall amino acid sequence. It is related to murine accessory protein (31% homology), to murine T1/ST2 (30% homology), and to murine IL-1R type I (27% homology). The cytoplasmic domains have slightly greater sequence homology (36–44%) than the extracellular portions (20–27%) (Parnet *et al.*, 1996).

Human and murine AcPL share 65% identity. AcPL shows homology of 25% to IL-1R type I, 27% to IL-1 AcP, and 26% to IL-18R (IL-1Rrp), respectively (Born *et al.*, 1998).

Murine IL-18BP is 65.7% identical at the amino acid level to human IL-18BP. There is no apparent homology with any other cytokine or cytokine receptor, when examined by the homology research. Yet, the Ig domain of IL-18BP is homologous to the third Ig domain of IL-1R type II. In addition, IL-18BP is significantly homologous to the putative proteins encoded by several pox viruses (Novick *et al.*, 1999).

Affinity for ligand(s)

Scatchard plot analysis on the affinity of IL-18 binding to L428 (human *Hodgkin's disease* cell line) suggested a single class of binding site for IL-18 on these cells. The apparent K_d value was 18.5 nM with 18,000 binding sites/cells (Torigoe *et al.*, 1997). COS-1 cells transfected with hIL-1Rrp cDNA express low-affinity hIL-18R (49,000/cell, K_d 46 nM) by the same analysis using [^{125}I]hIL-18, indicating that IL-18R is composed of IL-18Rα/IL-1Rrp. Although IL-18Rβ/AcPL does not bind to IL-18 directly, coexpression of IL-18Rα/IL-1Rrp and IL-18Rβ/AcPL is required for IL-18 responsiveness in terms of NFκB activation and c-Jun N-terminal kinase (JNK) activation (Born *et al.*, 1998). Furthermore, IL-18Rα-deficient mice reveal that IL-18Rα is an essential component of IL-18R for both IL-18 binding and exertion of IL-18-induced signal transduction (Hoshino *et al.*, 1999). Splenocytes from IL-18Rα-deficient mice do not produce IFNγ or upregulate NK cell functions in response to IL-18. Neither do they bind to IL-18 or cause activation of NFκB or c-Jun-terminal kinase after stimulation with IL-18. However, at this time, it is uncertain whether high-affinity IL-18R is composed of both IL-18Rα/IL-1Rrp and IL-18Rβ/AcPL.

Cell types and tissues expressing the receptor

IL-18Rα/IL-1Rrp mRNA is detectable in various organs of mice including thymus, spleen, liver, lung, intestine, colon, placenta, prostate, and heart. However, it is not detectable in the brain, kidney, skeletal muscle, and pancreas (Parnet *et al.*, 1996). This is interesting because IL-18 mRNA is strongly expressed in the pancreas, skeletal muscle, and kidney, where expression of IL-18Rα/IL-1Rrp mRNA is very low (Ushio *et al.*, 1996). In addition to above organs, weak expression of IL-18Rα/IL-1Rrp mRNA is also observed in the testis and ovary.

Thus, IL-18Rα/IL-1Rrp mRNA is widely distributed, but not universally. However, it needs to be considered whether the reported IL-18Rα mRNA expression in these organs is due to the possible contamination of blood cells. It is also important to determine which cell types express IL-18R in these IL-18Rα-expressing tissues. This will help to elucidate the physiological roles of IL-18 in these tissues, particularly in nonimmune organs.

The expression pattern of IL-18Rβ/AcPL is closely similar to that of IL-18Rα/IL-1Rrp. It is expressed in lung, spleen, leukocytes, and colon, but not detected in heart, brain, kidney, and muscle (Born *et al.*, 1998).

Resting murine T cells express no IL-18R mRNA and protein, whereas T cells stimulated with IL-12 alone or cultured with antigen plus IL-12 to induce TH1 cells begin to express IL 18Rα/IL-1Rrp mRNA and protein (Xu *et al.*, 1998; Yoshimoto *et al.*, 1998). IL-18Rα is not expressed on TH2 cells.

NK cells constitutively express IL-18Rα, as well as IL-12Rβ, and both IL-18 and IL-12 augment NK activity independently of each other (Hyodo *et al.*, 1999). However, as for induction of IFNγ, stimulation of NK cells solely with IL-12 or IL-18 induces only a trace amount of IFNγ, while the combined stimulation induces several hundred-fold amount of IFNγ.

Although coexpression of IL-18Rα/IL-1Rrp and IL-18Rβ/AcPL is required for signaling by IL-18 (Born *et al.*, 1998), it remains to be clarified whether both IL-18Rα and IL-18Rβ are expressed simultaneously on IL-18-responsive cells, such as TH1 cells and NK cells.

Regulation of receptor expression

IL-18, in combination with IL-12, induces TH1 clones and T cells to produce IFNγ (Ahn *et al.*, 1997; Yoshimoto *et al.*, 1998; Xu *et al.*, 1998; Murphy, 1998). Naïve T cells do not produce IFNγ in response to IL-18. Murine T cells that had been stimulated with IL-12 exhibit dose-dependent proliferation and IFNγ production in response to IL-18, suggesting that IL-12 renders naïve T cells responsive to IL-18 through induction of IL-18R. T cells stimulated with IL-12 have the capacity to specifically bind IL-18. The shape of the Scatchard plot obtained from our initial binding study is consistent with the presence of both high-affinity and low-affinity IL-18-binding sites. Measurement of [^{125}I]IL-18 bound to IL-12-stimulated T cells revealed that they express 405 high-affinity IL-18R (K_d 430 pM) and 5500 low-affinity IL-18R (K_d 31.4 nM) on each cell (Yoshimoto *et al.*, 1998). While IL-18R on T cells was suggested to be induced by IL-12, that on NK cells was shown to be constitutively expressed (Kunikata *et al.*, 1998; Hyodo *et al.*, 1999).

Release of soluble receptors

A soluble IL-18 receptor (an IL-18-binding protein: IL-18BP) was purified and characterized from human urine (Novick *et al.*, 1999) and from the sera of the mice sequentially administered with *Propionibacterium acnes* and LPS (Aizawa *et al.*, 1999). IL-18BP (38 kDa protein) is constitutively expressed in the spleen, and belongs to the immunoglobulin superfamily. No exon coding for a transmembrane domain was found in an 8.3 kb genomic sequence. IL-18BP can bind to IL-18 in high affinity, suggesting that IL-18BP may physiologically play a soluble decoy receptor, functionally similar to the membrane-associated IL-1R type II. Indeed, IL-18BP abolishes the induction of IFNγ production by IL-18 both *in vitro* and *in vivo*. No significant homology was found between IL-18BP and IL-18Rα/IL-1Rrp or IL-18Rβ/AcPL. However, the Ig domain of IL-18BP is homologous to the third Ig domain of the decoy receptor of IL-1, IL-1R type II (Novick *et al.*, 1999). Interestingly, IL-18BP is highly homologous to proteins derived from several Pox viruses (Novick *et al.*, 1999). In fact, Molluscum contagiosum virus, a common human poxvirus, encodes a family of proteins with homology to IL-18BP (Xiang and Moss, 1999). The proteins in this family have high-affinity binding activity to IL-18 and have capacity to inhibit both IL-18 binding and IL-18-induced IFNγ production. Since IL-18 plays an important role in host defense against viral infection, these proteins may act as an important escape tool from attack by the host immune system.

SIGNAL TRANSDUCTION

All the member of IL-1R family are involved in the response to infections, and their cytoplasmic regions contain six distinct regions which are well conserved (O'Neill and Green, 1998). These homologies seem to be important for interactions with signal transducing proteins. IL-18R does not activate the JAK/STAT signaling pathway, but it shares IL-1R-associated kinase (IRAK) and TRAF-6 with IL-1, resulting in nuclear translocation of NFκB (Matsumoto *et al.*, 1997; Robinson *et al.*, 1997; Kojima *et al.*, 1998). The IFNγ gene promoter has consensus sequences for NFκB, cyclosporin A-sensitive NF-AT-binding site, intronic enhancer region (C3), and STAT4, which is essential for IL-12 signaling (Xu *et al.*, 1996; Sica *et al.*, 1997). These regions are involved in the regulation of IFNγ gene expression and p50 and p65 NFκB subunits specifically bind to NFκB and C3 sites. The synergism between IL-12 and IL-18 for IFNγ in T cells might be partly caused by the combined activation of these distinct IFNγ-inducing signaling. MyD88, an adapter molecule in IL-1 signaling (Muzio *et al.*, 1997), also participates in signaling by IL-18 (Adachi *et al.*, 1998). The targeted disruption of MyD88 gene results in the destruction of signaling by both IL-1 and IL-18 (Adachi *et al.*, 1998). T cells from mice targeted for MyD88 gene are defective in proliferative response as well as induction of acute phase proteins and cytokines in response to IL-1. Increases in IFNγ production and NK cell activity in response to IL-18 are also abrogated. Furthermore, IL-18-induced activation of NFκB and c-Jun N-terminal kinase (JNK) is blocked in MyD88−/− mice. Moreover, IRAK-deficient mice have been shown to display an impaired response in their splenocytes to stimulation with IL-18 as well as IL-1 (Kanakaray *et al.*, 1999; Thomas *et al.*, 1999). Taken together, a MyD88-IRAK-TRAF6-NFκB line seems to be a major signaling pathway for IL-18 as well as IL-1. IL-18 has also been suggested to directly activate another transcription factor, AP-1, required for IFNγ gene expression (Barbulescu *et al.*, 1998).

There is a second signal pathway of IL-18. IL-18 has been suggested to activate MAP kinase. Protein tyrosine kinase, as well as the src kinase Lck, is activated in TH1 cells stimulated with IL-18 (Tsuji-Takayama *et al.*, 1997). Since MAP kinase pathways are considered to be involved in cell growth, IL-18 may exert a proliferating activity on T and NK cells through this pathway (Tomura *et al.*, 1998). These observations have been reviewed (Dinarello, 1998).

References

Adachi, O., Kawai, T., Matsumoto, M., Tsutsui, H., Sakagami, M., Nakanishi, K., and Akira, S. (1998). Targeted disruption of the MyD88 gene results in loss of IL-1-and IL-18 mediated function. *Immunity* **9**, 143–150.

Ahn, H.-J., Maruo, S., Tomura, M., Mu, J., Hamaoka, T., Nakanishi, K., Clark, S., Kurimoto, M., Okamura, H., and Fujiwara, H. (1997). A mechanism underlying synergy between IL-12 and IFN-γ-inducing factor in enhanced production of IFN-γ. *J. Immunol.* **159**, 2125–2131.

Aizawa, Y., Akita, K., Taniai, M., Torigoe, K., Mori, T., Nishida, Y., Ushio, S., Nukada, Y., Tanimoto, T., Ikegami, H., Ikeda, M., and Kurimoto, M. (1999). Cloning and expression of interleukin-18 binding protein. *FEBS Lett.* **445**, 338–342.

Barbulescu, K., Becker, C., Schlaak, J. F., Schmitt, E., Meyer zum Büschenfelde, K. H., and Neurath, M. F. (1998). IL-12 and IL-18 differentially regulate the transcriptional activity of the human IFN-gamma promoter in primary CD4+ T lymphocytes. *J. Immunol.* **160**, 3642–3647.

Born, T. L., Thomasson, E., Bird, T. A., and Sims, J. E. (1998). Cloning of a novel receptor subunit, AcPL, required for interleukin-18 signaling. *J. Biol. Chem.* **273**, 29445–29450.

Dinarello, C. A. (1998). Interleukin-1, interleukin-1 receptor and interleukin-1 receptor antagonist. *Int. Rev. Immunol.* **16**, 457–499.

Hoshino, K., Tsutsui, H., Kawai, T., Takeda, K., Nakanishi, K., Takeda, Y., and Akira, S. (1999). Cutting edge: generation of IL-18 receptor-deficient mice: evidence for IL-1 receptor-related protein as an essential IL-18 binding receptor. *J. Immunol.* **162**, 5041–5044.

Hyodo, Y., Matsui, K., Hayashi, N., Tsutsui, H., Kashiwamura, S.-I., Yamauchi, H., Hiroishi, K., Takeda, K., Tagawa, Y., Iwakura, Y., Kayagaki, N., Kurimoto, M., Okamura, H., Hada, T., Yagita, H., Akira, S., Nakanishi, K., and Higashino, K. (1999). Interleukin 18 upregulates perforin-mediated NK activity without increasing perforin mRNA expression by binding to constitutively expressed IL-18R. *J. Immunol.* **162**, 1662–1668.

Kanakaray, P., Ngo, K., Wu, Y., Angulo, A., Ghazal, P., Harris, C. A., Siekierka, J. J., Peterson, P. A., and Fung-Leung, W.-P. (1999). Defective interleukin (IL)-18-mediated natural killer and T helper cell type 1 responses in IL-1 receptor-associated kinase (IRAK)-deficient mice. *J. Exp. Med.* **189**, 1129–1138.

Kojima, H., Takeuchi, M., Ohta, T., Nishida, Y., Arai, N., Ikeda, M., Ikegami, H., and Kurimoto, M. (1998). Interleukin-18 activates the IRAK-TRAF6 pathway in mouse EL-4 cells. *Biochem. Biophys. Res. Commun.* **244**, 183–186.

Kunikata,T., Torigoe, K., Usio, S., Okura, T., Ushio, C., Yamauchi, H., Ikeda, M., Ikegami, H., and Kurimoto, M. (1998). Constitutive and induced IL-18 receptor expression by various peripheral blood cell subsets as determined by anti-hIL-18R monoclonal antibody. *Cell. Immunol.* **189**, 135–143.

Matsumoto, S., Tuji-Takayama, K., Aizawa, Y., Koide, K., Takeuchi, M., Ohta, T., and Kurimoto, M. (1997). Interleukin-18 activates NF-κB in murine T helper type 1 cells. *Biochem. Biophys. Res. Commun.* **234**, 454–457.

Murphy, K. M. (1998). T lymphocyte differentiation in the periphery. *Curr. Opin. Immunol.* **10**, 226–232.

Muzio, M., Ni, J., Feng, P., and Dixit, M. V. (1997). IRAK (Pelle) family member IRAK-2 and MyD88 as proximal mediators of IL-1 signaling. *Science* **278**, 1612–1615.

O'Neill, L. A., and Green, C. (1998). Signal transduction pathways activated by the IL-1 receptor family: ancient signaling machinery in mammals, insects, and plants. *J. Leukoc. Biol.* **63**, 650–657.

Nolan, K. F., Greaves, D. R., and Waldmann, H. (1998). The human interleukin 18 gene IL-18 maps to 11q22.2-q22.3, closely linked to the DRD2 gene locus and distinct from mapped IDDM loci. *Genomics* **51**, 161–163.

Novick, D., Kim, S.-H., Fantuzzi, G., Reznikov, L. L., Dinarello, C. A., and Rubinstein, M. (1999). Interleukin-18 binding protein: a novel modulator of the Th1 cytokine response. *Immunity* **10**, 127–136.

Parnet, P., Garka, K. E., Bonner, T. P., Dower, S. K., and Sims, J. E. (1996). IL-1Rrp is a novel receptor-like molecule similar to the type I interleukin-1 receptor and its homologues T1/ST2 and IL-1R AcP. *J. Biol. Chem.* **271**, 3967–3970.

Robinson, D., Shibuya, K., Mui, A., Zoni, F., Murphy, E., Sana, T., Hartley, S. B., Menon, S., Kastelein, R., Bazan, F., and O'Garra, A. (1997). IGIF does not drive Th1 development but synergizes with IL-12 for interferon-γ production and activates IRAK and NFκB. *Immunity* **7**, 71–581.

Rothe, H., Hibino, T., Itoh, Y., Kolb, H., and Martin, S. (1997). Systemic production of interferon-gamma inducing factor (IGIF) versus local IFN-gamma expression involved in the development of Th1 insulitis in NOD mice. *J. Autoimmun.* **10**, 251–256.

Sica, A., Dorman, L., Viggiano, V., Cippitelli, M., Ghosh, P., Rice, N., and Young, H. A. (1997). Interaction of NF-κB and NFAT with the Interferon-γ promoter. *J. Biol. Chem.* **272**, 30412–30420.

Thomas, J. A., Allen, J. L., Tsen, M., Dubnicoff, T., Nanao, J., Liao, X. C., Cao, Z., and Wasserman, S. A. (1999). Impaired cytokine signaling in mice lacking the IL-1 receptor-associated kinase. *J. Immunol.* **163**, 978–984.

Tomura, M., Zhou, X. Y., Maruo, S., Ahn, H. J., Hamaoka, T., Okamura, H., Nakanishi, K., Tanimoto, T., Kurimoto, M., and Fujiwara, H. (1998). A critical role for IL-18 in the proliferation and activation of NK1.1+ $CD3^-$ cells. *J. Immunol.* **160**, 4738–4746.

Torigoe, K., Ushio, S., Okura, T., Kobayashi, S., Taniai, M., Kunikata, T., Murakami, T., Sanou, O., Kojima, H., Fujii, M., Ohta, T., Ikeda, M., Ikegami, H., and Kurimoto, M. (1997). Purification and characterization of human interleukin-18 receptor. *J. Biol. Chem.* **272**, 25737–25742.

Tsuji-Takayama, K., Matsumoto, S., Koide, K., Takeuchi, M., Ikeda, M., Ohta, T., and Kurimoto, M. (1997). Interleukin-18 induces activation and association of p56lck and MAPK in a murine TH1 clone. *Biochem. Biophys. Res. Commmun.* **237**, 126–130.

Ushio, S., Namba, M., Okura, T., Hattori, K., Nukada, Y., Akita, K., Tanabe, F., Konishi, K., Micallef, M., Fujii, M., Torigoe, K., Tanimoto, T., Fukuda, S., Ikede, M., Okamura, H., and Kurimoto, M. (1996). Cloning of the cDNA for human IFN-γ-inducing factor, expression in *Escherichia coli*, and studies on the biologic activities of the protein. *J. Immunol.* **156**, 4274–4279.

Xiang, Y., and Moss, B. (1999). IL-18 binding and inhibition of interferon γ induction by human poxvirus-encoded proteins. *Proc. Natl Acad. Sci. USA* **96**, 11537–11542.

Xu, B. D., Chan, W. L., Leung, B. P., Hunter, D., Schulz, K., Carter, R. W., McInnes, I. B., Robinson, J. H., and Liew, F. Y. (1998). Selective expression and functions of interleukin 18 receptor on T helper (Th) Type 1 but not Th2 cells. *J. Exp. Med.* **188**, 1485–1492.

Xu, X., Sun, Y.-L., and Hoey, T. (1996). Cooperative DNA binding and sequence-selective recognition conferred by the STAT4 amino-terminal domain. *Science* **273**, 794–797.

Yoshimoto, T., Takeda, K., Tanaka, T., Ohkusu, K., Kashiwamura, S.-I., Okamura, H., and Nakanishi, K. (1998). IL-12 upregulates IL-18 receptor expression on T cells, Th1 cells, and B cells: Synergism with IL-18 for IFN-γ production. *J. Immunol.* **161**, 3400–3407.

IL-1 Receptor Type II

Charles A. Dinarello*

Department of Infectious Diseases, University of Colorado Health Sciences Center, 4200 East Ninth Avenue, B168, Denver, CO 80262, USA

*corresponding author tel: 303-315-3589, fax: 303-315-8054, e-mail: cdinare333@aol.com

DOI: 10.1006/rwcy.2000.15003.

SUMMARY

The extracellular domains of type II IL-1R are structurally related to those of the type I IL-1R; however, the type II IL-1R is a decoy receptor in that the primary ligand, IL-1β, preferentially binds to this receptor rather than the signaling receptor. As such, in the presence of increasing expression of the type II receptor on the cell surface, less IL-1 signaling takes place. This is because the type II IL-1R lacks a cytoplasmic domain capable of cell signal transduction. The type II IL-1R also exists in a shed form as a soluble receptor. The soluble receptor has a high affinity to bind IL-1β over that of IL-1α or IL-1Ra. The soluble IL-1R type II is ideally suited for clinical use because it has a high affinity for IL-1β and a low affinity for IL-1Ra.

BACKGROUND

Discovery

The discovery of the IL-1 receptor type II (IL-1RII) was made by in 1991 (McMahon *et al.*, 1991). Others also contributed to the discovery of this receptor (Symons *et al.*, 1991, 1993). The ability of IL-1β to preferentially bind to B cells is also probably the binding to the type II receptor (Scapigliati *et al.*, 1989; Ghiara *et al.*, 1991).

Alternative names

Another name for the IL-1RII is the IL-1 'decoy' receptor (Colotta *et al.*, 1993, 1994).

Structure

The extracellular domains of the IL-1RII are typical of an Ig-like receptor in the fibroblast growth factor family. The glycoprotein has three Ig-like domains in the extracellular segment, a transmembrane domain, and a short cytoplasmic domain (McMahon *et al.*, 1991).

Main activities and pathophysiological roles

When IL-1 binds to the cell membrane, IL-1RII does not signal. A soluble form of this receptor has been described to act to reduce the biological effects of IL-1β (Dower *et al.*, 1994).

GENE

Accession numbers

The gene for the IL-1RII has been reported (Sims *et al.*, 1995).

PROTEIN

Description of protein

IL-1RII has three Ig-like domains in the extracellular segment, a transmembrane domain, and a short cytoplasmic domain (McMahon *et al.*, 1991). The transmembrane segment is linked to a short cytoplasmic domain. In the rat, this cytoplasmic domain is longer (Bristulf *et al.*, 1994) but still does not signal. In the human and mouse, IL-1RII has a short cytosolic domain consisting of 29 amino acids; in the rat, there are an additional six charged amino acids (Bristulf *et al.*, 1994).

Relevant homologies and species differences

There is considerable homology between the IL-1 receptor type I and IL-1RII in all species. Limited homology between the IL-18-binding protein (Novick *et al.*, 1999) and the IL-1RII exists. Vaccinia and cowpox virus genes encode for a protein with a high amino acid homology to the type II receptor and this protein binds IL-1β (Alcami and Smith, 1992; Spriggs *et al.*, 1992). These viruses also code for IL-18-binding protein-like molecules.

Affinity for ligand(s)

The cell-bound IL-1RII does not appear to form a complex with the IL-1R type I receptor (Slack *et al.*, 1993; Dower *et al.*, 1994) nor does it transduce a signal (Sims *et al.*, 1993, 1994). The rank for the three IL-1 ligands (IL-1α, IL-1β, and IL-1Ra) binding to IL-1RII is IL-1β > IL-1α > IL-1Ra (Arend *et al.*, 1994; Dower *et al.*, 1994; Sims *et al.*, 1994). In some cells, there is a discrepancy between the dissociation constant of either form of IL-1 (usually 200–300 pM) and concentrations of IL-1 which can elicit a biological response (10–100 fM). In cells expressing large amounts of IL-1R AcP, the high-affinity binding of the IL-1R/IL-1R AcP complex may explain why two classes of binding have been observed. Human recombinant 17 kDa IL-1α binds to cell surface and soluble type I receptors with approximately the same affinity (200–300 pM); however, binding to surface and soluble type II receptors is nearly 100-fold less (30 and 10 nM, respectively).

Of the three members of the IL-1 family, IL-1β has the lowest affinity for the cell-bound form of IL-1RI (500 pM^{-1} nM). By comparison, IL-1β binds more avidly to the nonsignal transducing type II receptor (100 pM). IL-1β binding to the soluble form of the IL-1RI is lower than that to the cell-bound receptor. However, the most dramatic differences in IL-1β binding can be seen at the level of the soluble form of the type II receptor. Of the three ligands, the most avid binding is that of mature IL-1β (500 pM). IL-1β binding to the soluble IL-1RII is nearly irreversible due to a long dissociation rate (2 hours) (Arend *et al.*, 1994; Dower *et al.*, 1994; Symons *et al.*, 1994). Moreover, precursor IL-1β also preferentially binds to the soluble form of IL-1RII (Symons *et al.*, 1991, 1993).

Cell types and tissues expressing the receptor

The primary cells expressing the IL-1RII are monocytes, macrophages, neutrophils, B lymphocytes, myelomonocytic leukemia cells, and hairy cell leukemic cells.

Regulation of receptor expression

Increased surface expression of IL-1RII reduces the biological response to IL-1 (Colotta *et al.*, 1993, 1994). Gene expression of the IL-1RII is under the control of two promoters, each of which controls the usage of a divided first exon (exon 1A or 1B) (Vannier *et al.*, 1995). Early studies using B cells, monocytes, or bone marrow cells (type II receptor-bearing cells) demonstrated that hematopoietic growth factors, dexamethasone, and PGE_2 increase the number of IL-1-binding sites. Surface expression of IL-1RII is upregulated on neutrophils exposed to dexamethasone and IL-4 (Colotta *et al.*, 1993) and on monocytes or B cell lines exposed to dexamethasone (Dower *et al.*, 1994). These observations have been confirmed using gene expression in different cell lines (Vannier *et al.*, 1995). A transcription factor called PU.1, which is present in cells of hematopoietic origin, is required for expression of IL-1RII. In patients with bacterial sepsis, elevated IL-1RII expression has been observed on neutrophils (Fasano *et al.*, 1991). Although IL-1 itself downregulates gene and surface expression of IL-1RI, IL-1 upregulates gene and surface expression of the IL-1RII on an insulinoma cell line (Bristulf *et al.*, 1994).

Release of soluble receptors

Unlike soluble TNF receptors, it is unknown whether the soluble form of IL-1RII acts as a carrier for IL-1β and prolongs its half-life in the circulation. It is likely that as cell-bound IL-1RII increases, there is a comparable increase in soluble forms (Giri *et al.*, 1990). Similar to soluble receptors for TNF, the extracellular domains of the type II IL-1R are found as 'soluble' molecules in the circulation and urine of healthy subjects and in inflammatory synovial and other pathologic body fluids (Symons *et al.*, 1993, 1994; Arend *et al.*, 1994; Sims *et al.*, 1994; Barak *et al.*, 1996). In healthy humans, the circulating concentrations of the soluble IL-1RII are 100–200 pM. Elevated levels of the soluble IL-1RII are found in the circulation of patients with *sepsis* (Giri *et al.*, 1994) and in the synovial fluid of patients with active *rheumatoid arthritis* (Arend *et al.*, 1994) and in patients with *hairy cell leukemia* (Barak *et al.*, 1996). In patients undergoing aorta resection, crossclamping of the aorta results in significant ischemia and a dramatic release of soluble IL-1RII (Pruitt *et al.*, 1996). High-dose IL-2 therapy induces soluble IL-1RII (Orencole *et al.*, 1995).

LPS causes rapid shedding of the IL-1 type II. This effect of LPS is reduced by inhibition of metalloprotease. Following LPS, monocytes exhibited a reduction in steady-state mRNA levels (Penton-Rol *et al.*, 1999). Chemoattractants also cause a rapid release of the extracellular domain of the IL-1RII from myelomonocytic cells within a few minutes following exposure (Mantovani *et al.*, 1998). This induction of release of the decoy receptor suggests an early event in inflammation to limit the cascade. Inhibitors of matrix metalloproteases such as hydroxamic acid inhibit the release of the extracellular domain of IL-1RII (Orlando *et al.*, 1997). These protease inhibitors also reduced the slow release of soluble IL-1RII from monocytes and neutrophils and from cells stimulated with dexamethasone, TNF, chemoattractants, or phorbol myristate acetate (PMA). Inhibitors of other protease classes did not affect release. Inhibitors of serine proteases increased the molecular size of the released form of IL-1RII from 45 to 60 kDa (Orlando *et al.*, 1997).

SIGNAL TRANSDUCTION

Associated or intrinsic kinases

Because the IL-1RII has no significant cytoplasmic domain, there are no kinases intrinsic to the receptor. See reveiw of Martin and Falk (1997).

Cytoplasmic signaling cascades

The IL-1RII does not signal and serves only to bind IL-1 (preferentially IL-1β) and prevent signaling by IL-1 binding to the type I receptor (Colotta *et al.*, 1994).

BIOLOGICAL CONSEQUENCES OF ACTIVATING OR INHIBITING RECEPTOR AND PATHOPHYSIOLOGY

Unique biological effects of activating the receptors

The type II receptor appears to act as 'decoy' molecule, particularly for IL-1β. The receptor binds IL-1β tightly, thus preventing binding to the signal-transducing type I receptor (Colotta *et al.*, 1993). It is the lack of a signal-transducing cytosolic domain that makes the type II receptor a functionally negative receptor. For example, when the extracellular portion of the type II receptor is fused to the cytoplasmic domain of the type I receptor, a biological signal occurs (Heguy *et al.*, 1993). The extracellular portion of the type II receptor is found in body fluids, where it is termed IL-1 soluble receptor type II. A proteolytic cleavage of the extracellular domain of the IL-1RII from the cell surface is the source of the soluble receptor.

Phenotypes of receptor knockouts and receptor overexpression mice

Constructs encoding IL-1RII were transfected into U937 cells. Gene transfer resulted in receptor numbers (approximately 10^3/cell) of the same order of magnitude as that found in normal myelomonocytic cells. Transfer of IL-1RII reduced responsiveness to IL-1 (Penton-Rol *et al.*, 1997). Mice overexpressing IL-1RII or deficient in IL-1RII are not reported to date.

THERAPEUTIC UTILITY

Effect of treatment with soluble receptor domain

Because soluble IL-1RII binds IL-1β so avidly, a considerable therapeutic use is likely.

Effects of inhibitors (antibodies) to receptors

In general, antibodies to IL-1RI block IL-1-mediated activities *in vitro* and *in vivo*, whereas antibodies specific for the IL-1RII have no effect. An antibody (ALVA°42) which recognizes α and β subunits of HLA-DR (Gayle *et al.*, 1994) also binds to cells expressing type II receptors (Ghiara *et al.*, 1991). The ability of this antibody to inhibit IL-1-mediated effects *in vivo* may be due to inhibition of IL-1-induced production of IL-1. For example, anti-HLA-DR monoclonal antibodies stimulate the production of IL-1β by macrophages and enhance (or suppress) IL-1β induced by either superantigens or LPS.

References

Alcami, A., and Smith, G. L. (1992). A soluble receptor for interleukin-1β encoded by vaccinia virus: a novel mechanism of virus modulation of the host response to infection. *Cell* **71**, 153–167.

Arend, W. P., Malyak, M., Smith, M. F., Whisenand, T. D., Slack, J. L., Sims, J. E., Giri, J. G., and Dower, S. K. (1994). Binding of IL-1α, IL-1β, and IL-1 receptor antagonist by soluble IL-1 receptors and levels of soluble IL-1 receptors in synovial fluids. *J. Immunol.* **153**, 4766–4774.

Barak, V., Vannier, E., Nisman, B., Pollack, A., Dinarello, C. A. (1996). Cytokines and their soluble receptors as markers for hairy cell leukemia. *Eur. Cytokine Netw.* **7**, 536 (abstract).

Bristulf, J., Gatti, S., Malinowsky, D., Bjork, L., Sundgren, A. K., and Bartfai, T. (1994). Interleukin-1 stimulates the expression of type I and type II interleukin-1 receptors in the rat insulinoma cell line Rinm5F; sequencing a rat type II interleukin-1 receptor cDNA. *Eur. Cyokine Netw.* **5**, 319–330.

Colotta, F., Re, F., Muzio, M., Bertini, R., Polentarutti, N., Sironi, M., Giri, J., Dower, S. K., Sims, J. E., and Mantovani, A. (1993). Interleukin-1 type II receptor: a decoy target for IL-1 that is regulated by IL-4. *Science* **261**, 472–475.

Colotta, F., Dower, S. K., Sims, J. E., and Mantovani, A. (1994). The type II 'decoy' receptor: a novel regulatory pathway for interleukin-1. *Immunol. Today* **15**, 562–566.

Dower, S. K., Fanslow, W., Jacobs, C., Waugh, S., Sims, J. E., and Widmer, M. B. (1994). Interleukin-1 antagonists. *Ther. Immunol.* **1**, 113–122.

Fasano, M. B., Cousart, S., Neal, S., and McCall, C. E. (1991). Increased expression of the interleukin-1 receptor on blood neutrophils of humans with sepsis syndrome. *J. Clin. Invest.* **88**, 1452–1459.

Gayle, M. A., Sims, J. E., Dower, S. K., and Slack, J. L. (1994). Monoclonal antibody 1994–01 (also known as ALVA 42) reported to recognize type II IL-1 receptor is specific for HLA-DR alpha and beta chains. *Cytokine* **6**, 83–86.

Ghiara, P., Armellini, D., Scapigliati, G., Nuti, S., Nucci, D., Bugnoli, M., Censini, S., Villa, L., Tagliabue, A., Bossu, P., and Boraschi, D. (1991). Biological role of the IL-1 receptor type II as defined by a monoclonal antibody. *Cytokine* **3**, 473 (abstract).

Giri, J., Newton, R. C., and Horuk, R. (1990). Identification of soluble interleukin-1 binding protein in cell-free supernatants. *J. Biol. Chem.* **265**, 17416–17419.

Giri, J. G., Wells, J., Dower, S. K., McCall, C. E., Guzman, R. N., Slack, J., Bird, T. A., Shanebeck, K., Grabstein, K. H., Sims, J. E., and Alderson, M. R. (1994). Elevated levels of shed type II IL-1 receptor in sepsis. *J. Immunol.* **153**, 5802–5813.

Heguy A., Baldari, C. T., Censini, S., Ghiara, P., and Telford, J. L. (1993). A chimeric type II/I interleukin-1 receptor can mediate interleukin-1 induction of gene expression in T cells. *J. Biol. Chem.* **268**, 10490–10494.

McMahon, C. J., Slack, J. L., Mosley, B., Cosman, D., Lupton, S. D., Brunton, L. L., Grubin, C. E., Wignall, J. M., Jenkins, N. A., Brannan, C. I., Copeland, N. G., Huebner, K., Croce, C. M., Cannizzaro, L. A., Benjamin, D., Dower, S., Spriggs, M. K., and Sims, J. E. (1991). A novel IL-1 receptor cloned form B cells by mammalian expression is expressed in many cell types. *EMBO J.* **10**, 2821–2832.

Mantovani, A., Muzio, M., Ghezzi, P., Colotta, C., and Introna, M. (1998). Regulation of inhibitory pathways of the interleukin-1 system. *Ann. NY Acad. Sci.* **840**, 338–351.

Martin, M. U., and Falk, W. (1997). The interleukin-1 receptor complex and interleukin-1 signal transduction. *Eur. Cytokine Netw.* **8**, 5–17.

Novick D., Kim, S.-H., Fantuzzi, G., Reznikov, L., Dinarello, C. A., and Rubinstein, M. (1999). Interleukin-18 binding protein: a novel modulator of the Th1 cytokine response. *Immunity* **10**, 127–136.

Orencole, S. F., Fantuzzi, G., Vannier, E., and Dinarello, C. A. (1995). Circulating levels of IL-1 soluble receptors in health and after endotoxin or IL-2. *Cytokine* **7**, 642.

Orlando S., Sironi, M., Bianchi, G., Drummond, A. H., Boraschi, D., Yabes, D., and Mantovani, A. (1997). Role of metalloproteases in the release of the IL-1 type II decoy receptor. *J. Biol. Chem.* **272**, 31764–31769.

Penton–Rol, G., Polentarutti, N., Sironi, M., Saccani, S., Introna, M., and Mantovani, A. (1997). Gene transfer-mediated expression of physiological numbers of the type II decoy receptor in a myelomonocytic cellular context dampens the response to interleukin-1. *Eur. Cytokine Netw.* **8**, 265–269.

Penton-Rol, G., Orlando, S., Polentarutti, N., Bernasconi, S., Muzio, M., Introna, M., and Mantovani, A. (1999). Bacterial lipopolysaccharide causes rapid shedding, followed by inhibition of mRNA expression, of the IL-1 type II receptor, with conscomitant up-regulation of the type I receptor and induction of incompletely spliced transcripts. *J. Immunol.* **162**, 2931–2938.

Pruitt, J. H., Welborn, M. B., Edwards, P. D., Harward, T. R., Seeger, J. W., Martin, T. D., Smith, C., Kenney, J. A., Wesdorp, R. I., Meijer, S., Cuesta, M. A., Abouhanze, A., Copeland, E. M., Giri, J., Sims, J. E., Moldawer, L. L., and Oldenburg, H. S. (1996). Increased soluble interleukin-1 type II receptor concentrations in postoperative patients and in patients with sepsis syndrome. *Blood* **87**, 3282–3288.

Scapigliati, G., Ghiara, P., Bartalini, A., Taglibue, A., and Boraschi, D. (1989). Differential binding of IL-1α and IL-1β to receptors on B and T cells. *FEBS Lett.* **243**, 394–398.

Sims, J. E., Gayle, M. A., Slack, J. L., Alderson, M. R., Bird, T. A., Giri, J. G., Colotta, F., Re, F., Mantovani, A., Shanebeck, K., Grabstein, K. H., and Dower, S. K. (1993). Interleukin-1 signaling occurs exclusively via the type I receptor. *Proc. Natl Acad. Sci. USA* **90**, 6155–6159.

Sims, J. E., Giri, J. G., and Dower, S. K. (1994). The two interleukin-1 receptors play different roles in IL-1 activities. *Clin. Immunol. Immunopathol.* **72**, 9–14.

Sims, J. E., Painter, S. L., and Gow, I. R. (1995). Genomic organization of the type I and type II IL-1 receptors. *Cytokine* **7**, 483–490.

Slack, J., McMahan, C. J., Waugh, S., Schooley, K., Spriggs, M. K., Sims, J. E., and Dower, S. K. (1993). Independent binding of interleukin-1 alpha and interleukin-1 beta to type I and type II interleukin-1 receptors. *J. Biol. Chem.* **268**, 2513–2524.

Spriggs, M. K., Hruby, D. E., Maliszewski, C. R., Pickup, D. J., Sims, J. E., Buller, R. M., and VanSlyke, J. (1992). Vaccinia and cowpox viruses encode a novel secreted interleukin-1 binding protein. *Cell* **71**, 145–152.

Symons, J. A., Eastgate, J. A., and Duff, G. W. (1991). Purification and characterization of a novel soluble receptor for interleukin-1. *J. Exp. Med.* **174**, 1251–1254.

Symons, J. A., Young, P. A., and Duff, G. W. (1993). The soluble interleukin-1 receptor: ligand binding properties and mechanisms of release. *Lymphokine Cytokine Res.* **12**, 381.

Symons, J. A., Young, P. A., and Duff, G. W. (1994). Differential release and ligand binding of type II IL-1 receptors. *Cytokine* **6**, 555 (abstract).

Vannier, E., Ye, K., Fenton, M. J., Sims, J. E., and Dinarello, C. A. (1995). IL-1 receptor type II gene expression: a role for PU.1. *Cytokine* **7**.

TNF Family Receptors

TNF Receptors

Bharat B. Aggarwal[1,*], Ajoy Samanta[1] and Marc Feldmann[2]

[1]Cytokine Research Laboratory, Department of Bioimmunotherapy, The University of Texas M.D. Anderson Cancer Center, 1515 Holcombe Boulevard, PO Box 143, Houston, TX 77030, USA

[2]Kennedy Institute of Rheumatology, 1 Aspenlea Road, Hammersmith, London, W6 8LH, UK

* corresponding author tel: 713-792-3503, fax: 713-794-1613, e-mail: aggarwal@utmdacc.mda.uth.tmc.edu

DOI: 10.1006/rwcy.2000.16001.

SUMMARY

The TNF receptors, commonly referred to as type I and type II with a molecular mass of 55–60 kDa and 75–80 kDa, respectively, share a cysteine-rich extracellular domain and a distinct transmembrane domain. Type I receptor is expressed in all cell types, whereas type II is expressed only by cells of the immune system and on endothelial cells. Most TNF signals are mediated through the type I receptor; the precise role of type II receptor is still unclear. The cytoplasmic domains of both receptors lack any enzymatic activity. The death domain present in the cytoplasmic portion of the type I receptor is known to recruit at least 20 different proteins to form a cascade leading to activation of various cellular responses including apoptosis, nuclear transcription factor NFκB, and c-Jun N-terminal kinase. Soluble forms of both types of TNF receptors, consisting of an extracellular domain, have been identified in *in vitro* cell culture conditioned media and in serum, urine, synovial fluids, and cerebral spinal fluids of patients with various diseases. The soluble form of the type II receptor has been approved for human use in *rheumatoid arthritis*.

BACKGROUND

Discovery

That there are specific high-affinity cell surface receptors for TNFα were first discovered in 1985 (Aggarwal *et al.*, 1985). By crosslinking the ligand with the receptor through reversible and irreversible crosslinkers and by immunoaffinity chromatography, the TNF receptor was isolated as a protein with an approximate molecular mass of 70 kDa (Stauber *et al.*, 1988 and references therein). In 1990, the cDNA for two different TNF receptors with a predicted molecular mass of 55–60 kDa or 75–80 kDa were cloned and thus referred to as p60 (or p55) and p80 (or p75) (Loetscher *et al.*, 1990; Schall *et al.*, 1990; Smith *et al.*, 1990).

Alternative names

The smaller TNF receptor is referred to as p60, p55, type I, or CD120a receptor and the larger form as p80, p70, p75, type II receptor, or CD120b.

Structure

Both TNF receptors are type II transmembrane proteins consisting of an extracellular domain (ECD), a transmembrane domain (TMD), and an intracellular domain (ICD) (Figure 1). The ECDs of both receptors contain four well conserved cysteine-rich domains (CRDs). The amino acid sequences of the ICD of the two receptors are quite dissimilar and lack any intrinsic enzymatic activity.

Main activities and pathophysiological roles

In general, most TNF proinflammatory activities are mediated through the p60 receptor; the role of the p80 receptor is less clear (Tartaglia and Goeddel, 1992; Gruss and Dower, 1995). The p60 receptor has been shown to activate apoptosis, nuclear transcription factor NFκB, and c-Jun N-terminal kinase (JNK) (Chainy *et al.*, 1996 and references therein). In addition, the p60 receptor has been implicated in

TNF-mediated inflammation, viral replication, and protection against bacterial and fungal infections. In selective studies, the p80 receptor has been shown to mediate cytotoxicity and primary thymocyte and T cell proliferation (Tartaglia *et al.*, 1993a). The overexpression of p80 receptor can also cause activation of NFκB and JNK (Haridas *et al.*, 1998 and references therein).

GENE

Accession numbers

p60 receptor: M75866
p80 receptor: S63368

Sequence

The p60 and p80 TNF receptors are each encoded by a single gene. The p60 receptor is encoded by at least three exons. The gene structure for the human p80 receptor, elucidated by Santee and Owen-Schaub (1996), spans approximately 43 kbp, consists of 10 exons (ranging in length from 34 bp to 2.5 kbp) and nine introns ranging from 343 bp to 19 kbp. Consensus elements for transcription factors present in the promoter region include T cell factor 1 (TCF-1), Ikaros, AP-1, CK-2, IL-6 receptor E (IL-6RE), ISRE, GAS, NFκB, and SP-1 in the 5′-flanking region. The unusual $(GATA)_n$ and (GAA)(GGA) repeats were found within intron 1.

Chromosome location and linkages

The gene for the human p60 receptor is located on chromosome 12p13 and the p80 receptor gene is on chromosome 1p36.

PROTEIN

Accession numbers

p60 receptor: M75866
p80 receptor: S63368

Description of protein

The p60 receptor has 426 amino acid residues consisting of an ECD of 182 amino acids, a TMD of 21 amino acids, and an ICD of 221 amino acids. From this the predicted molecular mass of this receptor is about 47.5 kDa. Since the apparent molecular mass of the p60 receptor is between 55 and 60 kDa, the difference most probably is due to three potential *N*-linked glycosylation sites present in the ECDs of the receptors. The ECD of p60 receptor has a net charge opposite to that of the TNF, suggesting electrostatic interaction.

The p80 receptor is a 46 kDa protein, and it consists of 439 amino acid residues with an ECD of 235 amino acids, a TMD of 30 amino acid residues, and an ICD of 174 amino acids. This receptor is also glycosylated (Tartaglia and Goeddel, 1992; Gruss and Dower, 1995).

The ECDs of both p60 and p80 receptors contain four cysteine-rich domain repeats, each consisting of six cysteine residues. The ICDs of the two receptors, however, are completely distinct, indicating distinct signaling pathways. The most striking feature of the ICD of the p60 receptor is a region of approximately 80 amino acid residues near the C-terminus called the death domain (DD) because of its importance in TNF-mediated cell death (Tartaglia *et al.*, 1993b). This region is homologous to Fas, death receptor (DR)-3, DR4, DR5 and DR6, all the receptors implicated in cell death (Ashkenazi and Dixit, 1998). The p80 receptor, in contrast, lacks a DD but contains a serine-rich region that undergoes phosphorylation in a ligand-independent manner (Pennica *et al.*, 1992; Darnay *et al.*, 1994a; Beyaert *et al.*, 1995).

The crystal structure of the p60 receptor bound to TNFβ has been solved. This structure predicts that one ligand trimer brings three receptor chains together to form a complex. The receptor binds in three grooves in the ligand trimer formed by the subunit interfaces, so each receptor makes contact with two subunits (Banner *et al.*, 1993).

Initially it was found that the TNF receptor binds TNF with the same affinity as it binds to lymphotoxin (LT) (Aggarwal *et al.*, 1985). Later it was found that both p60 and p80 forms of the TNF receptors can bind both LT and TNF with comparable affinities (for references see Tartaglia and Goeddel, 1992).

Relevant homologies and species differences

The cDNA of mouse TNF receptor has also been cloned (Barrett *et al.*, 1991; Lewis *et al.*, 1991). The sequence homologies show that the human p60 and p80 receptors are 64% and 62%, respectively, identical to the corresponding mouse receptor. The p60 receptor, however, is most conserved in the ECD

(70%) whereas the p80 is conserved in the ICD (73%). This may explain why human p60 receptor binds both human and murine TNF whereas human p80 receptor binds only human and not mouse TNF (Lewis *et al.*, 1991). The ECD of murine p60 and p80 are 28% identical to each other. The p60 and p80 form of the TNF receptors are also homologous to several other members of the TNF receptor superfamily characterized by the presence of cysteine-rich domains in their ECDs. These include Fas, DR3, DR4, DR5, DR6, NGF (31%), RANK, CD40 (40%), CD27, CD30, Ox40, and 4-1BB (for references see Gruss and Dower, 1995; Ashkenazi and Dixit, 1998). The major area of homology between these receptors, which is in their ECD, may range from 25 to 30%.

In addition, several viral open reading frames (ORF) have been found to encode for soluble TNF receptor-like molecules. This includes SFV-T2 in Shope fibrosarcoma virus and Va53 or SaIF19R in vaccinia virus, MYX-T2, G4R, CrmB, and CrmD.

Affinity for ligand(s)

Both types of TNF receptor have a high affinity for TNF, in the range of 0.1–1 nM. Most cells exhibit a receptor density of around 1000 sites/cell but in some it is as high as 5000 sites/cell. A recombinant human TNF has been engineered that binds either the p60 or p80 form of the TNF receptors. TNF mutated at R32W and S86T binds to the p60 and that mutated at D143N and A145R exclusively binds to the p80 forms, thus suggesting that p80 receptor binds at the C-terminal of the cytokine whereas p60 binds more towards the N-terminus (Loetscher *et al.*, 1993; Van Ostade *et al.*, 1994; Haridas *et al.*, 1998).

Cell types and tissues expressing the receptor

The p60 form of the TNF receptor is expressed by all cell types examined to date. The p80 receptor, in contrast, appears to be expressed by the cells of the immune system and hematopoietic cells such as macrophages, neutrophils, lymphocytes (B cells and T cells), thymocytes, and mast cells. Endothelial cells, cardiac myocytes, and prostate cells have also been shown to express the p80 receptor.

Regulation of receptor expression

When the TNF receptor was first discovered, it was found to be upregulated by IFNγ (Aggarwal *et al.*, 1985). Since then several agents have been found to regulate TNF receptors, including IFNα, IFNβ, IL-2, IL-4, and phorbol ester (**Table 1**; for references see Aggarwal and Natarajan, 1996). Interestingly, TNF can also both upregulate and downregulate its own receptors in a cell type-specific manner.

Table 1 Agents that can regulate TNF receptors in different cells

Protein kinase C activators	Phorbol esters
Protein kinase C inhibitors	Staurosporine
Cytokines	IFN$\alpha\beta\gamma$
	IL-2
	IL-4
	IL-6
	IL-8
	TNF
	Thyroid-stimulating hormone
	GM-CSF
Microtubule depolymerizing agents	Nocodazole
	Vincristine
	Vinblastine
	Colchicine
	Podophyllotoxin
	P-Lumicolchicine
Protein kinase A activators	Dibutyryl cAMP
	Forskolin
Others	Hydrogen peroxide
	Retinal
	Butyrate
	Glucocorticoids
	Lipopolysaccharide
	Taxol
	fMLP
	Complement (C5a)
	Calcium ionophore
	Platelet-activating factor
	Leukotriene B_4
	Lectins
	Sulfasalazine
	Okadaic acid
	Iodoacetic acid
	Pervanadate

So far, there is no report of differential regulation by one type of TNF receptor over another. There is also very little known about the regulation of TNF receptor at the transcription, translation, or post-translation levels.

Release of soluble receptors

The TNF receptor was first purified in its soluble form (Seckinger *et al.*, 1989, 1990; Kohno *et al.*, 1990; Nophar *et al.*, 1990; Gatanaga *et al.*, 1990; Lantz *et al.*, 1990). Studies during the last decade have shown that both the p60 and p80 forms of the TNF receptor are released in cell culture-conditioned medium in response to a variety of stimuli. In addition, the soluble forms of both receptors have been detected *in vivo*, in serum, synovial fluids, cerebral spinal fluids, ovarian ascites and urine (Engelmann *et al.*, 1989; Seckinger *et al.*, 1989). The levels of these receptors appear to be elevated when a person is attacked by a pathological condition (Aderka *et al.*, 1993; Deloron *et al.*, 1994). In general, there are higher levels of soluble p80 receptors (2–4 ng/mL) than the soluble p60 receptors in normal sera. These receptors appear to be able to bind TNF with almost the same affinity as the transmembrane receptor. How these receptors are released is not fully understood. Metalloprotease inhibitors block both TNF and TNFR release with similar IC_{50} values by blocking. TNFα-converting enzyme (TACE), the enzyme that causes the release of TNF, is also involved in the release of the receptor. Analyses of cells lacking this metalloproteinase-disintegrin revealed an expanded role for TACE in the processing of other cell surface proteins, including a TNF receptor, the L-selectin adhesion molecule, and TGFα. The phenotype of mice lacking TACE suggests an essential role for soluble TGFα in normal development (Peschon *et al.*, 1998).

SIGNAL TRANSDUCTION

Associated or intrinsic kinases

The ICD of both the p60 and p80 receptor lacks homology to the catalytic domain of either Tyr or Ser/Thr-specific protein kinases or to nucleotide-binding proteins. The cytoplasmic domains of both the p60 and p80 receptors have been shown to bind to distinct serine/threonine kinases and cause the phosphorylation of the receptor (Darnay *et al.*, 1994a, 1994b, 1995; Beyaert *et al.*, 1995). In the case of the p60 receptor, the DD recruits a protein called TRADD which binds to TRAF2. TRAF2 has been shown to recruit a serine/threonine kinase called NIK that binds to IKKα and IKKβ (for references see Wallach *et al.*, 1997) (see also The TNF Ligand and TNF/NGF Receptor Families).

Cytoplasmic signaling cascades

A series of proteins have been identified that bind to the p60 receptor, leading to various cellular responses (**Table 2**; for references see Tewari and Dixit, 1996; Darnay and Aggarwal, 1997; Singh and Aggarwal, 1998). Three major activities assigned to the p60 receptor include activation of NFκB, JNK, and apoptosis (**Figure 1**). As indicated above, the DD present in the ICD of p60 receptor binds to TRADD, which binds to TRAF2, which in turn binds to NIK (for references see Wallach *et al.*, 1997) (see also The TNF Ligand and TNF/NGF Receptor Families). NIK then binds to activate IKKα and IKKβ, which cause the phosphorylation of IκB-α, leading to its degradation through a ubiquitin-dependent pathway, thus releasing in the cytoplasm the p50/p65 and other heterodimers of NFκB, which is then translocated to the nucleus to bind to the DNA, resulting finally in NFκB-dependent gene activation (**Figure 2**).

The cytoplasmic protein TRADD has also been found to recruit FADD, which in turn binds to FLICE, and the latter activates a family of aspartate-specific cysteine proteases, called caspases, which are responsible for inducing apoptosis or cell death (for references see Tewari and Dixit, 1996; Darnay and Aggarwal, 1997; Salvesen and Dixit, 1997; Singh and Aggarwal, 1998). Although it has been shown that TRAF2 is needed for JNK activation through the p60

Figure 1 Architecture of the p60 and p80 forms of the TNF receptors and their function.

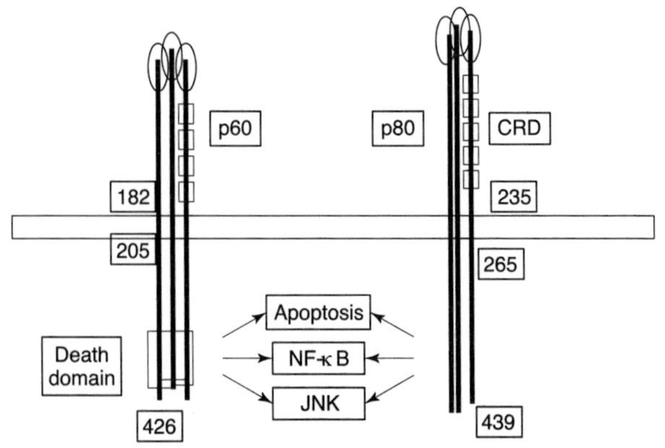

Table 2 Proteins known to interact with the cytoplasmic domain of the TNF receptors

Protein	Interacting partner
p60 receptor	
TNF receptor-associated death domain protein (TRADD)	p60 receptor
Sentrin	p60 receptor
Factor-associated with N-Smase activation (FAN)	p60 receptor
MAP kinase-activated death domain protein (MADD)	p60 receptor
TNF receptor-associated protein (TRAP1)	p60 receptor
TNF receptor-associated protein (TRAP2)	p60 receptor
p60 TNF receptor-associated kinase 60-TRAK	p60 receptor
BRE (brain and reproductive organ expression)	p60 receptor
55.11 protein	p60 receptor
TNF receptor-associated factor (TRAF2)	TRADD
Receptor-interacting protein (RIP)	TRAF2
NFκB-inducing kinase (NIK)	TRAF2
Apoptosis signal-regulating kinase 1 (ASK1)	TRAF2
Germinal center kinase (GCK)	TRAF2
TRAF2-interacting protein (I-TRAF/TANK)	TRAF2
Cellular inhibitor of apoptosis (cIAP1)	TRAF2
Cellular inhibitor of apoptosis (cIAP2)	TRAF2
A20 zinc finger protein	TRAF2
Silencer of death domains (SODD)	p60 receptor
Fas-associated death domain protein (FADD/Mort 1)	TRADD
FADD-like ICE (FLICE/MACH)	FADD
FLICE-interacting protein (I-FLICE/CASH/FLIP/MRIT)	FLICE
p80 receptor	
TNF receptor-associated factor (TRAF2)	TRAF1
TNF receptor-associated factor (TRAF1)	TRAF2
Cellular inhibitor of apoptosis (cIAP1)	TRAF2
Cellular inhibitor of apoptosis (cIAP2)	TRAF2
p80-TRAK	p80 and p60 receptor

receptor, how TRAF2 induces JNK activation is not clear (Figure 2). Recent evidence indicates that apoptosis signal-regulating kinase 1 (ASK1) interacts with TRAF2 and activates JNK (Nishitoh *et al.*, 1998). ASK1 is a MAPKKK that activates SEK1/JNK and MKK6/p38 signaling cascades. Another study indicates that interaction of TRAF2 with germinal center kinase (GCK) leads to activation of JNK (Yuasa *et al.*, 1998). Yuasa's group also showed that the interaction of TRAF2 with RIP leads to activation of both JNK and p38 MAPK.

Some studies indicate that the p80 receptor can also activate apoptosis, JNK and NFκB (Haridas *et al.*, 1998). The p80 receptor lacks the DD. Exactly how the p80 receptor mediates these responses is not fully understood, but its ICD is known to bind to TRAF2 directly, which can activate NFκB and JNK through a similar cascade used by the p60 receptor (see chapters on TNFα and The TNF Ligand and TNF/NGF Receptor Families).

In addition to these responses, the p60 receptor is known to activate a family of kinases, the mitogen-

Figure 2 Signaling cascade leading to activation of NFκB, JNK, and apoptosis.

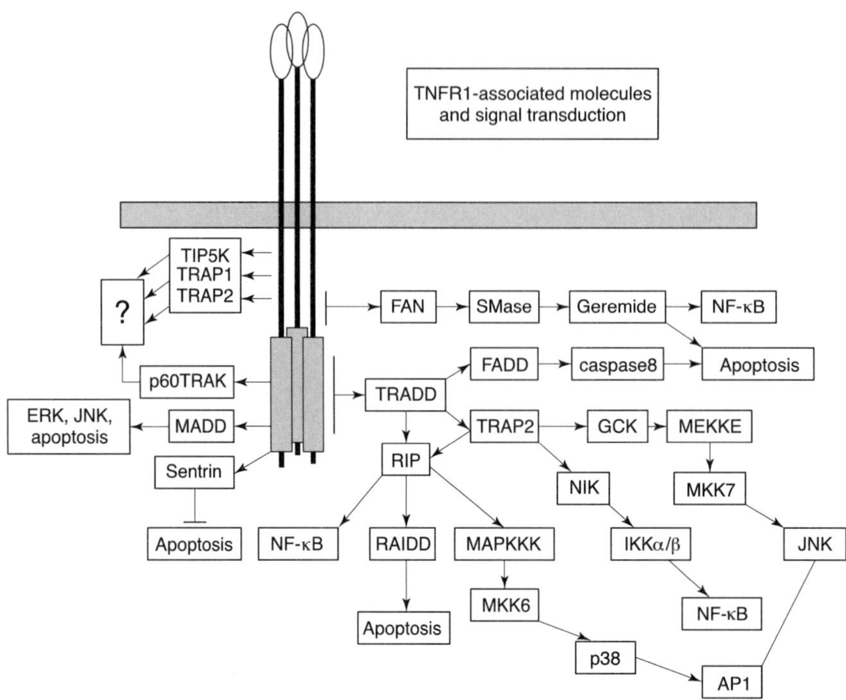

activated protein kinases (MAPK) as outlined in **Figure 3**. In addition, the p60 receptor has also been shown to activate various other kinases including germinal center kinase (GCK) (Pombo *et al.*, 1995), apoptosis signal-regulating kinase (ASK1) (Nishitoh *et al.*, 1998) and p21-activated kinase (PAK) (**Table 3**). Besides kinases, the p60 receptor activates acidic and neutral sphingomyelinases to induce the release of ceramide which is involved in downstream signaling (Kim *et al.*, 1991 and references therein). This receptor is also known to activate various phospholipases leading to the release of arachidonic acid and its metabolites, the prostaglandins. The p60 receptor is also a potent inducer of reactive oxygen intermediates. The signals mediated through each type of receptor are listed in **Table 4**.

DOWNSTREAM GENE ACTIVATION

Transcription factors activated

TNF activates various transcription factors (**Table 5**). Perhaps most important of these is NFκB, which is responsible for many of the inflammatory effects of the TNF. TNF is the most potent activator of this transcription factor described to date. The activation occurs within 5 minutes and with as little as 1 pM TNF (Chaturvedi *et al.*, 1994). How the p60 receptor activates this transcription factor is shown in Figure 2. The activation of another transcription factor, AP-1, by the p60 receptor is mediated through JNK (Brenner *et al.*, 1989).

Genes induced

The TNF receptors mediate the induction of a wide variety of genes that are involved in autoimmunity, inflammation, viral replication, cell proliferation, tumorigenesis, and tumor metastasis. Some of these genes are listed in Table 5. This includes genes for inflammatory mediators, acute phase response proteins, cytokines, receptors, members of the MHC, enzymes, oncogenic proteins, cell adhesion molecules, and growth regulatory molecules (for references see Aggarwal and Natarajan, 1996).

Promoter regions involved

The promoter of the inflammatory genes activated by the TNF receptor contains NFκB-binding sites. Other genes also contain AP-1, SP-1, or c-*myc*-binding sites needed for TNF receptor-mediated activation.

Figure 3 Activation of MAPK pathways by TNF receptors.

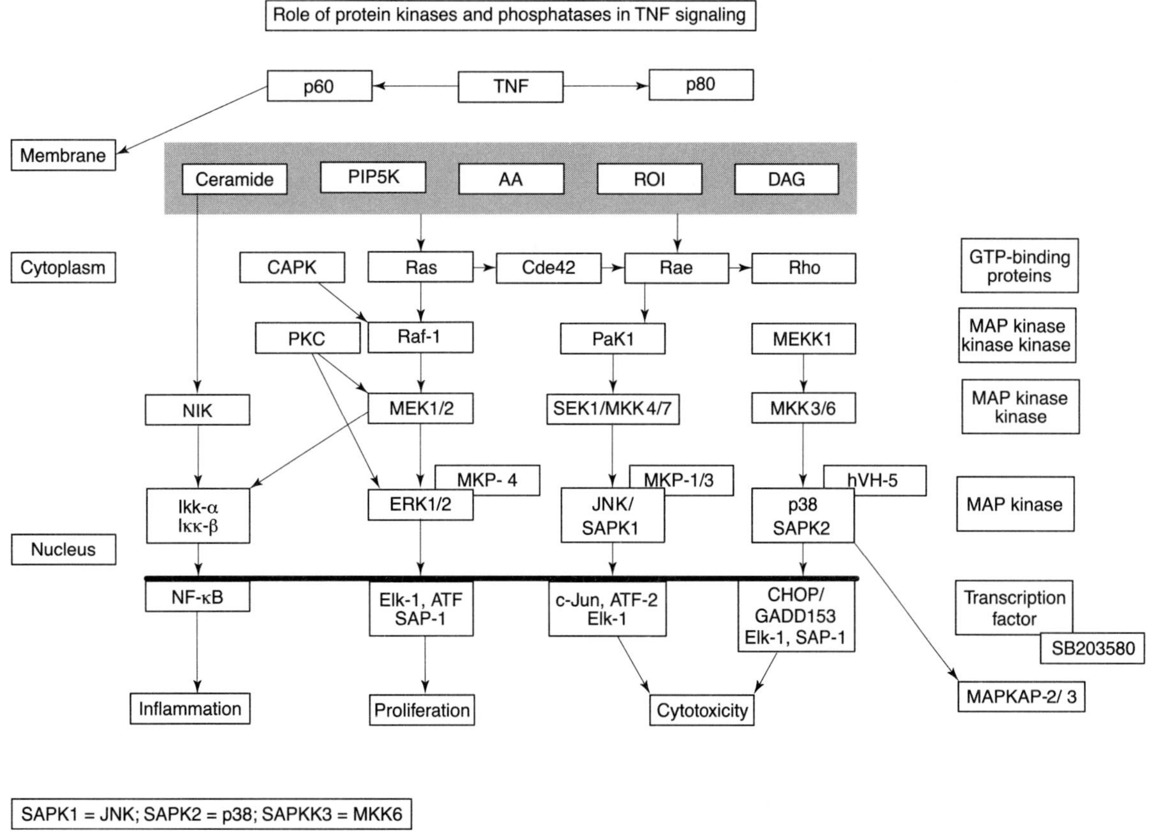

BIOLOGICAL CONSEQUENCES OF ACTIVATING OR INHIBITING RECEPTOR AND PATHOPHYSIOLOGY

Unique biological effects of activating the receptors

As indicated above, most activities of TNF are mediated through the p60 receptor (see Table 4). There are some reports which suggest that proliferation of thymocyte and circulating T lymphocytes and inhibition of early hematopoiesis are mediated through the p80 receptor (Tartaglia *et al.*, 1993a; Jacobsen *et al.*, 1994). It has been proposed that the p80 receptor may concentrate the ligand to be passed on to the p60 receptor. There are also reports suggesting that the p80 receptor is unique in mediating signaling initiated by the transmembrane form of TNF (Grell *et al.*, 1995). In contrast to the effects of TNF and p60 receptor-specific TNF mutein, p80 receptor-specific TNF mutein given to baboons failed to induce inflammation or shock, failed to induce hemodynamic changes, and failed to induce the plasma appearance of IL-6 and IL-8 (Van Zee *et al.*, 1994; Welborn *et al.*, 1996). It also increased baboon thymocyte proliferation *in vitro*. In addition, local skin necrosis and tissue neutrophil infiltration was seen after subcutaneous administration only of the TNF and TNF-p60 mutein, not with the p80-specific mutein. A similar role of the p80 receptor was discovered by administering receptor antibodies (Sheehan *et al.*, 1995).

Phenotypes of receptor knockouts and receptor overexpression mice

The deletion of genes for p60 or p80 receptor have quite distinct effects in mice (**Table 6**). Mice deficient in p60 receptor are resistant to endotoxin-induced shock but are susceptible to infection by *Listeria monocytogenes* (Pfeffer *et al.*, 1993). The p60 receptor

Table 3 Kinases activated by TNF receptors

P60 receptor	p42/44 Mitogen-activated protein kinase (MAPK or Erk)
	MAPK kinase (MAPKK or MEK)
	MAPKK kinase (MAPKKK or MEKK; MKK4, MKK7)
	p38 MAPK
	c-Jun N-terminal protein kinase (JNK or SAPK)
	NFκB-inducing kinase (NIK)
	IκBa kinase (IKK-α)
	IκBa kinase (IKK-β)
	Germinal center kinase (GCK)
	Apoptosis signal-regulating kinase 1 (ASK1)
	Receptor interacting protein (RIP)
	p21-activated protein kinase (PAK)
	hsp27 and β-casein kinase (65 kDa)
	c-raf-1 kinase
	Protein kinase C
	Casein kinase 2
	pp90rsk
	Cyclin-dependent kinase (CDK) 2
	Ceramide-activated protein kinase (CAPK)
	HPK/GCK-like (HGK)
	TGFβ-activated kinase (TAK1)
p80 receptor	c-Jun N-terminal kinase (JNK)

Both NIK and ASK1 belong to the MAPKKK family. All kinases are Ser/Thr kinases except MEK, which is a dual specificity kinase and phosphorylates at Ser-Thr and Tyr residues. For relationships between these kinases see Figure 2.

Table 4 Signals transmitted through each type of the TNF receptors

p60 receptor	Induction of NFκB activation
	Induction of c-fos
	Stimulation of protein kinase C
	Stimulation of sphingomyelinase
	Stimulation of phospholipase A_2
	Production of diacylglycerol
	Production of ceramide
	Induction of IL-6
	Induction of Mn superoxide dismutase mRNA
	Prostaglandin E_2 synthesis
	Induction of IL-2 receptor
	HLA class I and II Ag expression
	Antiproliferation/cytotoxicity/ apoptosis
	Growth stimulation
	Endothelial cell adhesion molecules
	Generation of lymphocyte-activated killer (LAK) cells
	Proliferation of natural killer (NK) cells
	Antiviral activities
p80 receptor	Induction of NFκB
	Proliferation of thymocytes
	Induction of IL-6
	Generation of NK and LAK cells
	DNA fragmentation
	Antiproliferation/cytotoxicity/ apoptosis

was also found to control early *graft-versus-host disease* (Speiser *et al.*, 1997). These results also suggest that the p60 receptor plays an important role in *septic shock* and in protection from bacterial infection. Another study indicated that such mice are also resistant to TNF-mediated toxicity (Rothe *et al.*, 1993). Interestingly, however, deletion of the gene for the p80 receptor in mice also decreased the sensitivity to TNF (Erickson *et al.*, 1994). In addition, p80 receptor-deleted mice had normal T cell development though they exhibited depressed Langerhans cell migration and reduced contact hypersensitivity (Wang *et al.*, 1997).

There are limited studies on the effects of TNF receptor transgenes *in vivo*. It was found that the production of the human p80 receptor in transgenic mice results in a severe inflammatory syndrome involving mainly the pancreas, liver, kidney, and lung, and characterized by constitutively increased NFκB activity in the peripheral blood mononuclear cell compartment (Douni and Kollias, 1998).

Mice have been generated in which the genes for the signaling proteins through which p60 receptor mediates its effects (Table 3) are deleted. The effect of these gene deletions on p60 signaling has been examined (see Table 6). For instance the deletion of TRAF2, FADD, FLICE, caspase 9, and the caspase 3

Table 5 Genes and transcription factors activated and induced by TNF

Transcription factors	SP-1
	c-fos
	c-jun/AP-1
	NFκB
	IRF-1
Oncogenes	c-*myc*
	c-*abl*
	c-*sis*
	Egr-1
	p53
Cytokines	TNF
	IL-1
	IL-6
	IL-8
	IFN
	MCP-1
	NCF
	MCAF
	G-CSF
	GM-CSF
	M-CSF
	HB-EGF
	Activin A
	Basic FGF
	PDGF
	TGFα
	NGF
Receptors	fMLPR
	IL-1R
	IL-2Rα
	EGFR
	IL-6R (gp130)
	IFNγR
	PTHR
Cell adhesion molecules	VCAM-1
	ICAM-1
	ELAM-1
Inflammatory mediators	Collagenase
	Stromelysin
	Tissue factor
	α1 acid glycoprotein

Table 5 (*Continued*)

Transcription factors	SP-1
	Haptoglobin
	C3 complement
	TNF-stimulated gene 14
MHC proteins	MHC class I and II
Viruses	HIV
Enzymes	Acyl-CoA synthetase
	Stearoyl-CoA desaturase 1
Other proteins	Plasminogen activator inhibitor I and II
	MnSOD
	2′,5′-oligoadenylate synthetase
	Glycoprotein Ib
	HMG-CoA reductase
	B12
	GLUT-1
	B94
	Phosphodiesterase
	M1, M2
	PCNA
	Endothelin 1
	A20
	Cyclin 1
	Secretory component
	CD4 surface antigen
	Dimethylnitrosamine
	Cystic fibrosis transmembrane
	Conductance regulator gene
	VP16

gene was lethal in mouse (Yeh *et al.*, 1997, 1998; Woo *et al.*, 1998; Varfolomeev *et al.*, 1998; Hakem *et al.*, 1998), whereas deletion of Apaf-1 was not (Yoshida *et al.*, 1998). Embryonic lethality suggests a critical role of FADD and FLICE in development. This role must be independent of TNF receptor as deletion of either p60 or p80 has no effect on the survival of the animals. Cells derived from FLICE or FADD knockout animals could be activated for NFκB and JNK but not apoptosis, suggesting a critical role of FADD and FLICE in TNF-induced apoptosis.

Table 6 Effect of gene deletion for TNF receptors and receptor-associated proteins on the mouse phenotypes

Deletion	Phenotype
p60	Resistant to *endotoxin-induced shock*
	Susceptible to infection by *Listeria monocytogenes*
	Resistant to TNF-mediated toxicity
	Controls early *graft-versus-host disease*
	Controls induction of *autoimmune heart disease*
	Mediates deletion of peripheral cytotoxic T lymphocytes *in vivo*
	Expression of adhesion molecules and leukocyte organ infiltration
p80	Decreased the sensitivity to TNF
	Normal T cell development
	Depressed Langerhans cell migration
	Reduced contact hypersensitivity
TRAF2	Appeared normal at birth but became progressively runted and died prematurely
	Atrophy of the thymus and spleen and depletion of B cell precursors
	Elevated serum TNF levels
	Normal NFκB activation
	Severe reduction in TNF-induced JNK activation
	Increased sensitivity to TNF-induced cell death
FADD/Mort-1	Lethal, mice do not survive beyond day 11.5 of embryogenesis
	Signs of *cardiac failure* and *abdominal hemorrhage*
	Cells are resistant to p60 receptor-mediated apoptosis
FLICE/caspase 8	Lethal *in utero*
	Impaired heart muscle development and congested accumulation of erythrocytes
	Low recovery of hematopoietic colony-forming cells
	TNF-activated NFκB and JNK but not apoptosis
Caspase 3	Reduced viability of animals
	Defective neuronal apoptosis and neurological defects
	Show cell type-specific inihibition of DNA fragmentation
Caspase 9	Embryonically lethal
	Defective brain development with decreased apoptosis
	Cells are resistant to apoptosis induction by various agents
	Normal TNF-induced apoptosis in some cells but not in others
Apaf-1	Animals are viable but exhibit craniofacial abnormalities with hyperproliferation of neuronal cells
	Cells are resistant to apoptotic stimuli but not to Fas
	Impaired processing of caspase 2, 3, and 8

Human abnormalities

There are several familial disorders characterized by periodic fever. The most common is the recessive *familial mediterranean fever* (FMF) which is caused by a polymorphism of pyrin, a protein expressed predominantly in leukocytes homologous to nuclear factors (The International FMF Consortium, 1997).

Familial hibernian fever (FHF) is a dominant disease, found in several large Irish/Scottish families, which has longer inflammatory attacks than FMF, and responds less well to colchicine. FHF has been studied by genetic analysis, and the susceptibility mapped to the distal part of chromosome 12p, an area encoding CD4, LAG-3, CD27, complement genes CIr and CIs, and the p55 TNFR (McDermott *et al.*, 1998). As during attacks the patients have low levels of soluble p55 TNFR compared to controls, this latter gene was analysed in seven families. In each family, a single nucleotide substitution was found, most in the first cysteine-rich domain of the extracellular region, some in the second cysteine-rich domain of the extracellular region (McDermott *et al.*, 1999).

In the affected individuals, soluble p55 TNFR levels were half those of normals (500 pg against 1 ng/mL), whereas levels of p75 soluble TNFR were normal. The affected individuals have increased levels of membrane-bound p55 TNFR, with a similar binding affinity for TNF. Shedding of p55 TNFR after activation, e.g. with PMA, was reduced, with more membrane TNFR and less in the supernatant. These studies confirm that the p55 TNFR is a major receptor for inflammation, and that its regulation by shedding is critical to the regulation of inflammation.

Regulatory Roles of TNF Receptor Cleavage

TNF receptors signal after they have been aggregated, usually by the agonist, TNF, or experimentally by antibodies. Thus the density of TNFR is important in facilitating signaling, and TNFR cleavage will reduce TNFR signaling by reducing TNFR density (Chan and Aggarwal, 1994). Moreover, the soluble TNFR can bind TNF and so can act as a competitor to the cell surface receptors (Engelmann *et al.*, 1989; Seckinger *et al.*, 1989; Olsson *et al.*, 1989). The concentration of soluble receptors in the serum in normal healthy individuals is high, about 1 ng/mL of p55 and 2–4 ng/mL of the p75 receptor. In inflammatory diseases these levels are increased significantly and this has been well documented in *rheumatoid arthritis*, with levels increased in synovial fluid several fold higher than in the serum, where the levels are also augmented (Cope, 1992; Roux-Lombard, 1993).

It is likely that a major function of soluble TNFR is to augment the clearance of TNF from the serum. This concept is supported by studies in volunteer patients or primates given TNF i.v., who had transient rises in serum TNF and subsequent rises in serum TNFR with rapid clearance of the serum TNF (Van Zee *et al.*, 1992; Bemelmans *et al.*, 1993). Furthermore, studies in nephrectomized mice have supported this conclusion.

THERAPEUTIC UTILITY

See chapter on TNF, where the effects of TNF blockade especially in *rheumatoid arthritis* and *Crohns' disease* are discussed. So far, two TNF-blocking agents have been licensed for use: Enbrel™, a p75 dimeric Fc fusion protein (Weinblatt *et al.*, 1999) and Remicade™, a chimeric anti-TNFα antibody. More are in clinical trials. The second aspect is the use of soluble TNFR to monitor inflammatory disease activity. It has been found that in a wide spectrum of diseases, such as *systemic lupus erythematosus* (Aderka *et al.*, 1993), *rheumatoid arthritis* (e.g. Cope *et al.*, 1992; Roux-Lombard *et al.*, 1993) and *cancer* (e.g. Aderka *et al.*, 1991), the levels of soluble TNFR are augmented as disease activity gets worse, and so it is a useful marker of disease activity.

References

Aderka, D., Engelmann, D., Hornick, V., Skornick, Y., Levo, Y., Wallach, D., and Kushtai, G. (1991). Increased serum levels of soluble receptors for tumour necrosis factor in cancer patients. *Cancer Res.* **51**, 5602–5607.

Aderka, D., Wysenbeek, A., Engelmann, H., Cope, A. P., Brennan, F., Molad, Y., Hornick, V., Levo, Y., Maini, R. N., Feldmann, M., and Wallach, D. (1993). Correlation between serum levels of soluble tumour necrosis factor receptor and disease activity in systemic lupus erythematosus. *Arthritis Rheum.* **36**, 1111–1120.

Aggarwal, B. B., and Natarajan, K. (1996). Tumor necrosis factors: Developments during last decade. *Eur. Cytokine Netw.* **7**, 93–124.

Aggarwal, B. B., Eessalu, T. E., and Hass, P. E. (1985). Characterization of receptor for human tumor necrosis factor and their regulation by γ-interferon. *Nature* **318**, 665–667.

Ashkenazi, A., and Dixit, V. M. (1998). Death receptors: signaling and modulation. *Science* **281**, 1305–1308.

Barrett, K., Taylor-Fishwick, D. A., Cope, A. P., Kissonerghis, A. M., Gray, P. W., Feldmann, M., and Foxwell, B. M. (1991). Cloning, expression and cross-linking analysis of the murine p55 tumor necrosis factor receptor. *Eur. J. Immunol.* **21**, 1649–1656.

Banner, D. W., D'Arcy, A., Janes, W., Gentz, R., Schoenfeld, H. J., Broger, C., Loetscher, H., and Lesslauer, W. (1993). Crystal structure of the soluble human 55 kd TNF receptor-human TNF beta complex: implications for TNF receptor activation. *Cell* **73**, 431–445.

Bemelmans, M. H. A., Gouma, D. J., and Buurman, W. A. (1993). Influence of nephrectomy on TNF receptor clearance in a murine model. *J. Immunol.* **150**, 2007–2017.

Beyaert, R., Vanhaesebroeck, B., Declercq, W., Van Lint, J., Vandenabeele, P., Agostinis, P., Vandenheede, J. R., and Fiers, W. (1995). Casein kinase-1 phosphorylates the p75 tumor necrosis factor receptor and negatively regulates tumor necrosis factor signaling for apoptosis. *J. Biol. Chem.* **270**, 23293–23299.

Brenner, D. A., O'Hara, M., Angel, P., Chojkier, M., and Karin, M. (1989). Prolonged activation of jun and collagenase genes by tumour necrosis factor-alpha. *Nature* **337**, 661–663.

Cope, A. P., Aderka, D., Doherty, M., Engelmann, H., Gibbons, D., Jones, A. C., Brennan, F. M., Maini, R. N., Wallach, D., and Feldmann, M. (1992). Increased levels of soluble tumor necrosis factor receptors in the sera and synovial fluid of patients with rheumatic diseases. *Arthritis Rheum.* **35**, 1160–1169.

Chan, H., and Aggarwal, B. B. (1994). Role of tumor necrosis factor receptors in the activation of nuclear factor κB in human histiocytic lymphoma U-937 cells. *J. Biol. Chem.* **269**, 31424–31429.

Chainy, G. B. N., Singh, S., Raju, U., and Aggarwal, B. B. (1996). Differential activation of the nuclear factor-κB by TNF muteins-specific for the p60 and p80 TNF receptors *J. Immunol.* **157**, 2410–2417.

Chaturvedi, M., LaPushin, R., and Aggarwal, B. B. (1994). Tumor necrosis factor and lymphotoxin: qualitative and quantitative differences in the mediation of early and late cellular responses. *J. Biol. Chem.* **269**, 14575–14583.

Darnay, B. G., and Aggarwal, B. B. (1997). Early events in TNF signaling: a story of associations and dissociations. *J. Leukocyte Biol.* **61**, 559–566.

Darnay, B. G., Reddy, S. A. G., and Aggarwal, B. B. (1994a). Physical and functional association of a serine-threonine protein kinase to the cytoplasmic domain of the p80 form of the human tumor necrosis factor receptor in human histiocytic lymphoma U-937 cells. *J. Biol. Chem.* **269**, 19687–19690.

Darnay, B. G., Reddy, S. A. G., and Aggarwal, B. B. (1994b). Identification of a protein kinase associated with the cytoplsmic domain of the p60 tumor necrosis factor receptor. *J. Biol. Chem.* **269**, 20299–20304.

Darnay, B. G., Singh, S., Chaturvedi, M. M., and Aggarwal, B. B. (1995). The p60 tumor necrosis factor (TNF) receptor-associated kinase (TRAK) binds residues 344-397 within the cytoplasmic domain involved in TNF signaling. *J. Biol. Chem.* **270**, 14867–14870.

Deloron, P., Roux Lombard, P., Ringwald, P., Wallon, M., Niyongabo, T., Aubry, P., Dayer, J. M., and Peyron, F. (1994). Plasma levels of TNF-alpha soluble receptors correlate with outcome in human falciparum malaria. *Eur. Cytokine Netw.* **5**, 331–336.

Douni, E., and Kollias, G. A. (1998). Critical role of the p75 tumor necrosis factor receptor (p75TNF-R) in organ inflammation independent of TNF, lymphotoxin alpha, or the p55TNF-R. *J. Exp. Med.* **188**, 1343–52.

Engelmann, H., Aderka, D., Rubinstein, M., Rotman, D., and Wallach, D. (1989). A tumor necrosis factor-binding protein purified to homogeneity from human urine protects cells from tumor necrosis factor toxicity. *J. Biol. Chem.* **264**, 11974–11980.

Erickson, S. L., de Sauvage, F. J., Kikly, K., Carver-Moore, K., Pitts-Meek, S., Gillett, N., Sheehan, K. C., Schreiber, R. D., and Goeddel, D. V. (1994). Moore MW. Decreased sensitivity to tumour-necrosis factor but normal T-cell development in TNF receptor-2-deficient mice. *Nature* **372**, 560–563.

Gatanaga, T., Hwang, C. D., Kohr, W., Cappuccini, F., Lucci, J. A. 3rd., Jeffes, E. W., Lentz, R., Tomich, J., Yamamoto, R. S., and Granger, G. A. (1990). Purification and characterization of an inhibitor (soluble tumor necrosis factor receptor) for tumor necrosis factor and lymphotoxin obtained from the serum ultrafiltrates of human cancer patients. *Proc. Natl Acad. Sci. USA* **87**, 8781–8784.

Grell, M., Douni, E., Wajant, H., Lohden, M., Clauss, M., Maxeiner, B., Georgopoulos, S., Lesslauer, W., Kollias, G., Pfizenmaier, K., and Scheurich, P. (1995). The transmembrane form of tumor necrosis factor is the prime activating ligand of the 80 kDa tumor necrosis factor receptor. *Cell* **83**, 793–802.

Gruss, H. J., and Dower, S. K. (1995). Tumor necrosis factor ligand superfamily: involvement in the pathology of malignant lymphomas. *Blood* **85**, 3378–404.

Hakem, R., Hakem, A., Duncan, G. S., Henderson, J. T., Woo, M., Soengas, M. S., Elia, A., de la Pompa, J. L., Kagi, D., Khoo, W., Potter, J., Yoshida, R., Kaufman, S. A., Lowe, S. W., Penninger, J. M., and Mak, T. W. (1998). Differential requirement for caspase 9 in apoptotic pathways *in vivo*. *Cell* **94**, 339–352.

Haridas, V., Darnay, B., Natarajan, K., Heller, R., and Aggarwal, B. B. (1998). Overexpression of the p80 TNF receptor leads to TNF-dependent apoptosis, NF-κB activation, and c-Jun kinase activation. *J. Immunol.* **160**, 3152–3162.

Jacobsen, F. W., Rothe, M., Rusten, L., Goeddel, D. V., Smeland, E. B., Veiby, O. P., Slordal, L., and Jacobsen, S. E. (1994). Role of the 75-kDa tumor necrosis factor receptor: inhibition of early hematopoiesis. *Proc. Natl Acad. Sci. USA* **91**, 10695–10699.

Kohno, T., Brewer, M. T., Baker, S. L., Schwaltz, P. E., King, M. W., Hale, K. K., Squires, C. H., Thompson, R. C., and Vannice, J. L. (1990). Second tumor necrosis factor receptor gene product can shed a naturally occurring tumor necrosis factor inhibitor. *Proc. Natl Acad. Sci. USA* **87**, 8331.

Kim, M.-Y., Linardic, C., Obeid, L., and Hannun, Y. (1991). Identification of sphingomyelin turnover as an effector mechanism for the action of tumor necrosis factor alfa and gamma interferon: specific role in cell differentiation. *J. Biol. Chem.* **266**, 484.

Lantz, M., Gullberg, U., Nilsson, E., and Olsson, I. (1990). Characterization *in vitro* of a human tumor necrosis factor-binding protein. A soluble form of a tumor necrosis factor receptor. *J. Clin. Invest.* **86**, 1396–1342.

Lewis, M., Tartaglia, L. A., Lee, A., Bennett, G. L., Rice, G. C., Wong, G. H., Chen, E. Y., and Goeddel, D. V. (1991). Cloning and expression of cDNAs for two distinct murine tumor necrosis factor receptors demonstrate one receptor is species specific. *Proc. Natl Acad. Sci. USA* **88**, 2830–2834.

Loetscher H. Y., Pan, C. E., Lahm, H. W., Gentz, R., Brockhaus, M., Tabuchi, H., and Lesslauer, W. (1990). Molecular cloning and expression of the human 55 kd tumor necrosis factor receptor. *Cell* **61**, 351.

Loetscher, H., Stueber, D., Banner, D., Mackay, F., and Lesslauer, W. (1993). Human tumor necrosis factor alpha (TNF alpha) mutants with exclusive specificity for the 55-kDa or 75-kDa TNF receptors. *J. Biol. Chem.* **268**, 26350–26357.

McDermott, M. F., Ogunkolade, B. W., McDermott, E. M., Jones, L. C., Wan, Y., Quane, K. A., McCarthy, J., Phelan, M., Molloy, M. G., and Powell, R. J. (1998). Linkage of familial Hibernian fever to chromosome 12p13. *Am. J. Hum. Genet.* **62**, 1446–1451.

McDermott, M. F., Aksentijevich, I., Galon, J., McDermott, E. M., Ogunkolade, B. W., Centola, M., Mansfield, E., Gadina, M., Karenko, L., Pettersson, T., McCarthy, J., Frucht, D. M., Aringer, M., Torosyan, Y., Teppo, A.-M., Wilson, M., Karaarslan, H. M., Wan, Y., Todd, I., Wood, G., Schlimgen, R., Kumarajeewa, T. R., Cooper, S. P., Amos, C. I., Mulley, J., Quane, K. A., Molloy, M. G., Ranki, A., Powell, R. J., Hitman, G. A., O'Shea, J. J., and Kastner, D. L. (1999). Germline mutations in the extracellular domains of the 55 kDa TNF receptor, TNFR1, define a family of dominantly inherited autoinflammatory syndromes. *Cell* **97**, 133–144.

Nishitoh, H., Saitoh, M., Mochida, Y., Takeda, K., Nakano, H., Rothe, M., Miyazono, K., and Ichijo, H. (1998). ASK1 is essential for JNK/SAPK activation by TRAF2. *Mol. Cell* **2**, 389–395.

Nophar, Y., Kemper, O., Brakebusch, C., Englemann, H., Zwang, R., Aderka, D., Holtmann, H., and Wallach, D. (1990). Soluble forms of tumor necrosis factor receptors (TNF-Rs). The cDNA for the type I TNF-R, cloned using amino acid sequence data of its soluble form, encodes both the cell surface and a soluble form of the receptor. *EMBO J.* **9**, 3269–3278.

Olsson, I., Lantz, S., Nilsson, E., Peetre, C., Thysell, H., Grubb, A., and Adolf, G. (1989). Isolation and characterization of a tumor necrosis factor binding protein from urine. *Eur. J. Haematol.* **42**, 270–275.

Pennica, D., Lam, V. T., Mize, N. K., Weber, R. F., Lewis, M., Fendly, B. M., Lipari, M. T., and Goeddel, D. V. (1992). Biochemical properties of the 75-kDa tumor necrosis factor receptor: characterization of ligand binding, internalization, and receptor phosphorylation. *J. Biol. Chem.* **267**, 21172–21178.

Peschon, J. J., Slack, J. L., Reddy, P., Stocking, K. L., Sunnarborg, S. W., Lee, D. C., Russell, W. E., Castner, B. J., Johnson, R. S., Fitzner, J. N., Boyce, R. W., Nelson, N., Kozlosky, C. J., Wolfson, M. F., Rauch, C. T., Cerretti, D. P., Paxton, R. J., March, C. J., and Black, R. A. (1998). An essential role for ectodomain shedding in mammalian development. *Science* **282**, 1281–1284.

Pfeffer, K., Matsuyama, T., Kundig, T. M., Wakeham. A., Kishihara, K., Shahinian, A., Wiegmann, K., Ohashi, P. S., Kronke, M., and Mak, T. W. (1993). Mice deficient for the 55 kd tumor necrosis factor receptor are resistant to endotoxin shock, yet succumb to *L. monocytogenes* infection. *Cell* **73**, 457.

Pombo, C. M., Kehrl, J. H., Sanchez, I., Katz, P., Avruch, J., Zon, L. I., Woodgett, J. R., Force, T., and Kyriakis, J. M. (1995). Activation of the SAPK pathway by the human STE20 homologue germinal centre kinase. *Nature* **377**, 750–754.

Rothe, J., Lesslauer, W., Lotscher, H., Lang, Y., Koebel, P., Kontgen, F., Althage, A., Zinkernagel, R., Steinmetz, M., and Bluethmann, H. (1993). Mice lacking the tumour necrosis factor receptor 1 are resistant to TNF-mediated toxicity but highly susceptible to infection by *Listeria monocytogenes*. *Nature* **364**, 798–802.

Roux-Lombard, P., Punzi, L., Hasler, F., Bas, S., Todesco, S., Gallati, H., Guerne, P. A., and Dayer, J. M. (1993). Soluble tumor necrosis factor receptors in human inflammatory synovial fluids. *Arthritis Rheum.* **36**, 485–489.

Salvesen, G. S., and Dixit, V. M. (1997). Caspases: intracellular signaling by proteolysis. *Cell* **91**, 443–446.

Santee, S. M., and Owen-Schaub, L. B. (1996). Human tumor necrosis factor receptor p75/80 (CD120b) gene structure and promoter characterization. *J. Biol. Chem.* **271**, 21151–21159.

Seckinger, P., Isaaz, S., and Dayer, J. M. (1989). Purification and biologic characterization of a specific tumor necrosis factor alpha inhibitor. *J. Biol. Chem.* **264**, 11966–11973.

Seckinger, P., Zhang, J. H., Hauptmann, B., and Dayer, J. M. (1990). Characterization of a tumor necrosis factor alpha (TNF-alpha) inhibitor: evidence of immunological cross-reactivity with the TNF receptor. *Proc. Natl Acad. Sci. USA* **87**, 5188–5192.

Schall, T. J., Lewis, M., Koller, K. J., Lee, A., Rice, G. C., Wong, H. W., Gatanaga, T., Granger, G. A., Lentz, R., Raab, H., Kohr, W. J., and Goeddel, D. V. (1990). Molecular cloning and expression of a receptor for human tumor necrosis factor. *Cell* **61**, 361.

Sheehan, K. C., Pinckard, J. K., Arthur, C. D., Dehner, L. P., Goeddel, D. V., and Schreiber, R. D. (1995). Monoclonal antibodies specific for murine p55 and p75 tumor necrosis factor receptors: identification of a novel *in vivo* role for p75. *J. Exp. Med.* **181**, 607–617.

Singh, A., Ni, J., and Aggarwal, B. B. (1998). Death domain receptors and their role in cell demise. *J. Interferon Cytokine Res.* **18**, 439–450.

Smith, C. A., Davis, T., Anderson, D., Solam, L., Beckmann, M. P., Jerzy, R., Dower, S. K., Cosman, D., and Goodwin, R. G. (1990). A receptor for tumor necrosis factor defines an unusual family of cellular and viral proteins. *Science* **248**, 1019.

Speiser, D. E., Bachmann, M. F., Frick, T. W., McKall-Faienza, K., Griffiths, E., Pfeffer, K., Mak, T. W., and Ohashi, P. S. (1997). TNF receptor p55 controls early acute graft-versus-host disease. *J. Immunol.* **158**, 5185–5190.

Stauber, G. B., Aiyer, R. A., and Aggarwal, B. B. (1988). Human tumor necrosis factor receptor: purification by immunoaffinity chromatography and initial characterization. *J. Biol. Chem.* **263**, 19098–19104.

Tartaglia, L. A., and Goeddel, D. V. (1992). Two TNF receptors. *Immunol. Today* **13**, 151–153.

Tartaglia, L. A., Goeddel, D. V., Reynolds, C., Figari, I. S., Weber, R. F., Fendly, B. M., and Palladino, M.A. Jr. (1993a). Stimulation of human T-cell proliferation by specific activation of the 75-kDa tumor necrosis factor receptor. *J. Immunol.* **151**, 4637–4641.

Tartaglia, L. A., Ayres, T. M., Wong, G. H., and Goeddel, D. V. (1993b). A novel domain within the 55 kd TNF receptor signals cell death. *Cell* **74**, 845–853.

Tewari, M., and Dixit, V. M. (1996). Recent advances in tumor necrosis factor and CD40 signaling. *Curr. Opin. Genet. Dev.* **6**, 39–44.

The International FMF Consortium.(1997). Ancient missense mutations in a new member of the RoRet gene family are likely to cause familial Mediterranean fever. *Cell* **90**, 797–807.

Van Ostade, X., Tavernier, J., and Fiers, W. (1994). Structure-activity studies of human tumour necrosis factors. *Protein Eng.* **7**, 5–22.

Van Zee, K. J., Kohno, T., Fischer, E., Rock, C. S., Moldawer, L. L., and Lowry, S. F. (1992). Tumor necrosis factor soluble receptors circulate during experimental and clinical inflammation and can protect against excessive tumor necrosis factor alpha *in vitro* and *in vivo*. *Proc. Natl Acad. Sci. USA* **89**, 4845–4849.

Van Zee, K. J., Stackpole, S. A., Montegut, W. J., Rogy, M. A., Calvano, S. E., Hsu, K. C., Chao, M., Meschter, C. L., Loetscher, H., Stuber, D., Ettlin, U., Wipf, B., Lesslauer, W., Lowry, S. F., and Moldawer, L. L. (1994). A human tumor necrosis factor (TNF) alpha mutant that binds exclusively to the p55 TNF receptor produces toxicity in the baboon. *J. Exp. Med.* **179**, 1185–1191.

Varfolomeev, E. E., Schuchmann, M., Luria, V., Chiannilkulchai, N., Beckmann, J. S., Mett, I. L., Rebrikov, D., Brodianski, V. M., Kemper, O. C., Kollet, O., Lapidot, T., Soffer, D., Sobe, T., Avraham, K. B., Goncharov, T., Holtmann, H., Lonai, P., and Wallach, D. (1998). Targeted disruption of the mouse Caspase 8 gene ablates cell death induction by the TNF receptors, Fas/Apo1, and DR3 and is lethal prenatally. *Immunity* **9**, 267–276.

Wallach, D., Boldin, M., Varfolomeev, E., Beyaert, R., Vandenabeele, P., and Fiers, W. (1997). Cell death induction by receptors of the TNF family: towards a molecular understanding. *FEBS Lett.* **410**, 96–106.

Wang, B., Fujisawa, H., Zhuang, L., Kondo, S., Shivji, G. M., Kim, C. S., Mak, T. W., and Sauder, D. N. (1997). Depressed Langerhans cell migration and reduced contact hypersensitivity response in mice lacking TNF receptor p75. *J. Immunol.* **159**, 6148–6155.

Weinblatt, M. E., Kremer, J. M., Bankhurst, A. D., Bulpitt, K. J., Fleischmann, R. M., Fox R. I., Jackson, C. G., Lange, M., and Burge, D. J. (1999). A trial of Etanercept, a recombinant tumor necrosis factor receptor:Fc fusion protein, in patients with rheumatoid arthritis receiving methotrexate. *N. Engl. J. Med.* **340**, 253–259.

Welborn, M. B. 3rd., Van Zee, K., Edwards, P. D., Pruitt, J. H., Kaibara, A., Vauthey, J. N., Rogy, M., Castleman, W. L., Lowry, S. F., Kenney, J. S., Suber, D., Ettlin, U., Wiof, B., Loetscher, H., Copeland, E. M., Lesslauer, W., and Moldawer, L. L. (1996). A human tumor necrosis factor p75 receptor agonist stimulates *in vitro* T cell proliferation but does not produce inflammation or shock in the baboon. *J. Exp. Med.* **184**, 165–171.

Woo, M., Hakem, R., Soengas, M. S., Duncan, G. S., Shahinian, A., Kagi, D., Hakem, A., McCurrach, M., Khoo, W., Kaufman, S. A., Senaldi, G., Howard, T., Lowe, S. W., and Mak, T. W. (1998). Essential contribution of caspase 3/CPP32 to apoptosis and its associated nuclear changes. *Genes Dev.* **12**, 806–819.

Yuasa, T., Ohno, S., Kehrl, J. H., and Kyriakis, J. M. (1998). Tumor necrosis factor signaling to stress-activated protein kinase (SAPK)/Jun NH2-terminal kinase (JNK) and p38. Germinal center kinase couples TRAF2 to mitogen-activated protein kinase/ERK kinase kinase 1 and SAPK while receptor interacting protein associates with a mitogen-activated protein kinase kinase kinase upstream of MKK6 and p38. *J. Biol. Chem.* **273**, 22681–22692.

Yeh, W. C., Shahinian, A., Speiser, D., Kraunus, J., Billia, F., Wakeham, A., de la Pompa, J. L., Ferrick, D., Hum, B., Iscove, N., Ohashi, P., Rothe, M., Goeddel, D. V., and Mak, T. W. (1997). Early lethality, functional NF-kappaB activation, and increased sensitivity to TNF-induced cell death in TRAF2-deficient mice. *Immunity* **7**, 715–725.

Yeh, W. C., Pompa, J. L., McCurrach, M. E., Shu, H. B., Elia, A. J., Shahinian, A., Ng, M., Wakeham, A., Khoo, W., Mitchell, K., El-Deiry, W. S., Lowe, S. W., Goeddel, D. V., and Mak, T. W. (1998). FADD: essential for embryo development and signaling from some, but not all, inducers of apoptosis. *Science* **279**, 1954–1958.

Yoshida, H., Kong, Y. Y., Yoshida, R., Elia, A. J., Hakem, A., Hakem, R., Penninger, J. M., and Mak, T. W. (1998). Apaf1 is required for mitochondrial pathways of apoptosis and brain development. *Cell* **94**, 739–750.

LICENSED PRODUCTS

Etanercept (Enbrel^{TM}), a dimeric fusion protein of p75 TNFR coupled to IgG1 Fc (Immunex/AHP).
TNF p60 receptor antibodies supplied by StressGen Biotechnology Corp. (www.stressgen.com/reagents)
TNF p80 receptor antibodies
Soluble p60 receptor
Soluble p80 receptor
Enbrel (soluble p80 ECD fused to Fc) supplied by Immunex Corporation for use in advanced rheumatoid arthritis

Poxvirus TNF Receptor Homologs

Grant McFadden[1,*] and Richard Moyer[2]

[1]The John P. Robarts Research Institute and Department of Microbiology and Immunology, The University of Western Ontario, 1400 Western Road, London, Ontario, N6G 2V4, Canada

[2]Department of Molecular Genetics and Microbiology, University of Florida College of Medicine, PO Box 100266, Gainesville, FL 32610-0266, USA

*corresponding author tel: (519)663-3184, fax: (519)663-3847, e-mail: mcfadden@rri.on.ca

DOI: 10.1006/rwcy.2000.14011.

SUMMARY

Following the cloning and sequencing of the cellular TNF receptors in the early 1990s, it was noticed that several virus genes in the database shared remarkable sequence similarity to the external cysteine-rich domain (CRD) repeat sequences that comprised the binding sites for TNF. Unlike the cellular receptors, however, the viral versions were predicted to be secreted glycoproteins that could scavenge host TNF prior to binding with its cognate cellular receptors. These viral genes, first identified in several poxviruses, were the first documented examples of 'viroceptors', or virus-encoded receptor homologs that function to bind and sequester cellular ligands away from the appropriate cellular receptors.

BACKGROUND

Discovery

The first known example of a virus-encoded homolog of a cellular cytokine receptor was the S-T2 gene of *Shope fibroma virus* (SFV) (Smith *et al.*, 1990). In 1987, the SFV T2 open reading frame, the second gene from the terminus, was sequenced and shown to be expressed as a virus early gene (Macaulay *et al.*, 1987; Upton *et al.*, 1987). When the cellular type I and type II tumor necrosis factor receptors (TNFR) were cloned in 1990, it was realized that the S-T2 gene belonged to the TNFR superfamily and we now know that this family includes not only TNFRI and TNFRII, but also other important immunomodulators such as Fas, CD40, CD30, CD27, Ox40, and 4-1BB (Smith *et al.*, 1990, 1994). S-T2 protein was subsequently expressed from transfected COS cells and shown to bind TNF (Smith *et al.*, 1991). A related TNFR homolog, designated M-T2, was also found in *myxoma virus* and was shown in 1991 to be an important virulence determinant for myxoma pathogenesis (Upton *et al.*, 1991).

Multiple TNFR homologs (designated crmB, crmC, crmD) have also been found in other poxviruses, such as *cowpox virus*, *ectromelia virus*, and *variola virus* (Shchelkunov *et al.*, 1993; Hu *et al.*, 1994; Smith *et al.*, 1996; Loparev *et al.*, 1998). The vaccinia virus genome contains two discontinuous, and thus nonfunctional, fragments of TNFR, but this virus is highly attenuated by passage outside a true vertebrate host (Goebel *et al.*, 1990; Howard *et al.*, 1991). Table 1 summarizes the different poxviral and cellular TNFR family members. All the poxvirus TNFR homologs possess 3 or 4 cysteine-rich domains (CRDs) at the N-terminus, which constitute the binding domain for TNF (see Figure 1).

Alternative names

M-T2, S-T2, vTNF-R, crmB, crmC, crmD.

Structure

The structural features of poxvirus TNFR homologs are reviewed elsewhere (Sedger and McFadden, 1996; McFadden *et al.*, 1997; Xu *et al.*, 1998). The secreted viral proteins generally have an N-terminal signal sequence followed by three or four copies of CRDs that define their membership in the TNFR superfamily, and a C-terminal extension that is unrelated to other cellular genes. The C-terminal domains of the viral proteins are generally not hydrophobic, allowing for protein secretion of at least some of the members from infected cells.

Main activities and pathophysiological roles

The viral TNFR homologs are believed to be secreted scavengers of host TNF, and function as viroceptors to block the biological activities of host TNF and lymphotoxin prior to receptor engagement (Sedger and McFadden, 1996; McFadden *et al.*, 1997; Xu *et al.*, 1998). In the case of the M-T2 protein from myxoma, a second activity has been described of inhibiting apoptosis in virus-infected lymphocytes in a TNF-independent fashion (Macen *et al.*, 1996; Schreiber *et al.*, 1997).

Table 1 Accession numbers for poxvirus-encoded TNFR homologs: genes

Poxvirus	Gene	GenBank
Myxoma virus	M-T2	M95181
Shope fibroma virus	S-T2	A23727
Variola virus Bangladesh 1975[a]	G2R	L22579
Cowpox virus	crmB	L08906
Cowpox virus	crmC	Y15035
Cowpox virus	crmD[b]	U87234
Monkeypox virus[c]	crmB	U87842

[a]G2R/crmB homologs have been sequenced from other variola virus strains, including Garcia 1966 (Brazilian alastrim, minor), Harvey 1944, Butler 1952, Congo 1970, Somalia 1977, Sierra Leone 1968, Chimp 9-2, and Chimp 9-4.

[b]Four strains of ectromelia virus also encode crmD homologs.

[c]CrmB genes from monkeypox virus strains Zaire 1996, Sierra Leone 1970, Nigeria 1971, Benin 1978, Zaire 1979, CW-N1, CV1, Zaire 1970, WMP, and UTC have been sequenced and reported in GenBank. Truncated and presumably nonfunctional homologs of crmB are found in rabbitpox, buffalopox, taterapox, and most strains of vaccinia virus.

GENE

Accession numbers

See **Table 1**.

PROTEIN

Accession numbers

See **Table 2**.

Table 2 Accession numbers for poxvirus-encoded TNFR homologs: proteins

Poxvirus	Amino acids	PID
Myxoma virus	326	332310
Shope fibroma virus	325	139627
Variola virus Bangladesh 1975	349	439102
Cowpox virus (CrmB)	355	333519
Cowpox virus (CrmC)	186	3097018
Cowpox virus (CrmD)	320	2738033
Monkeypox virus	349	2738063

Sequence

See **Figure 1**.

Description of protein

In early studies examining the biological activity of S-T2, the secreted viral protein was independently expressed and found to bind TNF in ligand blot assays. Characterization of the binding properties of M-T2 to TNF have also revealed that the cytotoxicity of rabbit TNF, but not the human or murine ligands, could be completely inhibited by addition of purified M-T2 protein. Hence M-T2 binds and inhibits TNF in a distinctly species-specific manner. On the other hand, the cowpox TNF-R homolog, crmC, inhibits murine TNF and lymphotoxin while the expressed variola version inhibits the human ligands.

Figure 1 Amino acid alignment of representative members of the poxvirus TNF receptor superfamily.

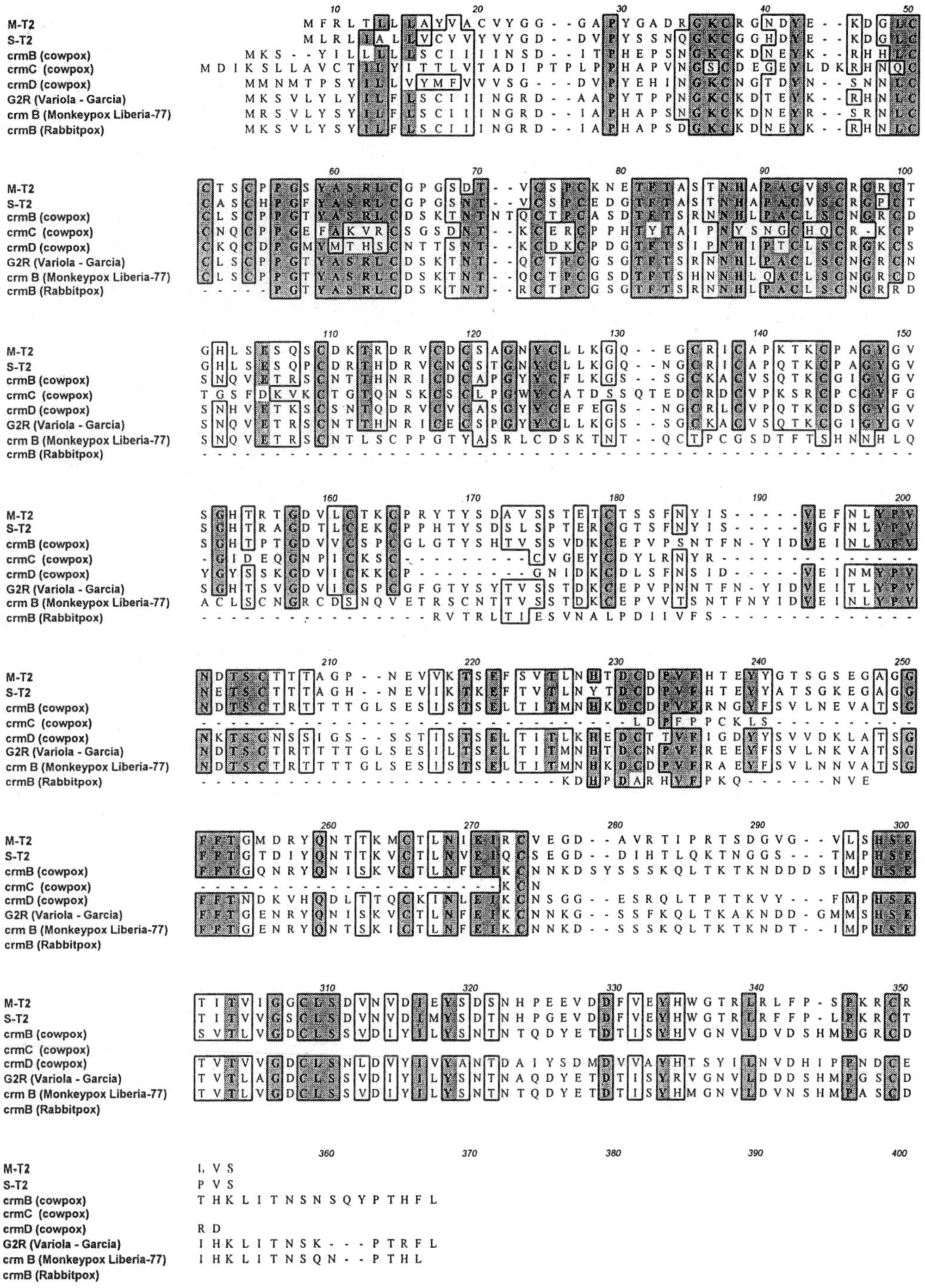

Relevant homologies and species differences

Although sequence identity of the extracellular ligand-binding domain of the different TNF receptor family members is not high, amino acid sequence alignments reveal striking conservation in the position of conserved cysteine residues (Figure 1). Similarly, the ligands that engage these receptors can also be described in a superfamily of structurally related proteins, whose predicted secondary structures are virtually superimposable. These include TNF, lymphotoxin α and β, CD40L, CD30L,

CD27L, FasL, and TRAIL. Following binding to their respective ligands, cellular TNF receptor family members oligomerize and mediate important immunomodulatory functions, including lymphocyte activation and proliferation, B cell costimulation and, under certain circumstances, cell death or apoptosis.

Affinity for ligand(s)

Scatchard analysis revealed that M-T2 binds rabbit TNF with an affinity in the range of K_d 170-195 pM. This is comparable to that measured for TNF binding to cellular receptors. The Scatchard assays also indicated that binding of M-T2 to murine TNF occurs with an affinity that is an order of magnitude lower (K_d 1.7 nM), whereas binding to human TNF could not be demonstrated at all in these assays. The crmC protein from cowpox was shown to bind TNF, but not lymphotoxin α, suggesting that ligand specificities might vary substantially between the poxvirus proteins (Smith *et al.*, 1996). The crmD protein of cowpox and ectromelia binds both TNF and lymphotoxin α (Loparev *et al.*, 1998).

Regulation of receptor expression

All of the poxvirus TNFR homologs are early genes, except for crmC of cowpox virus, which is expressed at late times post infection.

Release of soluble receptors

Most poxvirus TNFR homologs are secreted as soluble glycoproteins in monomeric, dimeric, or higher-ordered oligomeric forms (Schreiber *et al.*, 1996; Loparev *et al.*, 1998). The dimer of M-T2 is a more potent inhibitor of TNF cytolysis than is the monomeric form, presumably by blocking ligand-induced dimerization of the receptor (Schreiber *et al.*, 1996).

BIOLOGICAL CONSEQUENCES OF ACTIVATING OR INHIBITING RECEPTOR AND PATHOPHYSIOLOGY

Phenotypes of receptor knockouts and receptor overexpression mice

In order to evaluate how M-T2 contributes to viral virulence, a recombinant *myxoma virus* was constructed in which both copies of the M-T2 ORF present in the virus genome were disrupted by insertion of a dominant selectable marker. This myxoma T2-minus virus (vMyxT2$^-$) does not express detectable T2 protein, but replicates normally within infected fibroblasts *in vitro*. When the vMyxT2$^-$ virus was used to infect susceptible European rabbits, virus disease was found to be significantly attenuated (Upton *et al.*, 1991).

An important aspect of *Leporipoxvirus* infection and pathogenesis is the ability of these viruses to productively infect lymphocytes and spread to secondary sites via the lymphatics. Although both myxoma and the vMyxT2$^-$ viruses can replicate normally in infected fibroblasts *in vitro* the M-T2 knockout virus cannot productively infect CD4+ rabbit T cells. Replication of vMyxT2$^-$ in these cells is abortive, due to a rapid induction of DNA fragmentation and morphological alterations of cell death by apoptosis (Macen *et al.*, 1996). Deletion analysis revealed that only the first two CRDs of M-T2 are required to prevent apoptosis of infected lymphocytes (Schreiber *et al.*, 1997), whereas the first three CRDs are necessary for M-T2 to bind and inhibit TNF (Schreiber and McFadden, 1996).

References

Goebel, S. J., Johnson, G. P., Perkus, M. E., Davis, S. W., Winslow, J. P., and Paoletti, E. (1990). The complete DNA sequence of vaccinia virus. *Virology* **179**, 247–266.

Howard, S. T., Chan, Y. C., and Smith, G. L. (1991). Vaccinia virus homologues of the Shope fibroma virus inverted terminal repeat proteins and a discontinuous ORF related to the tumour necrosis factor receptor family. *Virology* **180**, 633–647.

Hu, F.-Q., Smith, C. A., and Pickup, D. J. (1994). Cowpox virus contains two copies of an early gene encoding a soluble secreted form of the Type II TNF receptor. *Virology* **204**, 343–356.

Loparev, V. N., Parsons, J. M., Knight, J. C., Panus, J. F., Ray, C. A., Buller, R. M. L., Pickup, D. J., and Esposito, J. J. (1998). A third distinct tumor necrosis factor receptor of orthopoxviruses. *Proc. Natl Acad. Sci. USA* **95**, 3786–3791.

Macaulay, C., Upton, C., and McFadden, G. (1987). Tumorigenic poxviruses: Transcriptional mapping of the terminal inverted repeats of Shope fibroma virus. *Virology* **158**, 381–393.

Macen, J. L., Graham, K. A., Lee, S. F., Schreiber, M., Boshkov, L. K., and McFadden, G. (1996). Expression of the myxoma virus tumor necrosis factor receptor homologue (T2) and M11L genes is required to prevent virus-induced apoptosis in infected rabbit T lymphocytes. *Virology* **218**, 232–237.

McFadden, G., Schreiber, M., and Sedger, L. (1997). Myxoma T2 protein as a model for poxvirus encoded TNF receptor homologs. *J. Neuroimmunol.* **72**, 119–126.

Schreiber, M., and McFadden, G. (1996). Mutational analysis of the ligand binding domain of M-T2 protein, the tumor necrosis factor receptor homologue of myxoma virus. *J. Immunol.* **157**, 4486–4495.

Schreiber, M., Rajarathnam, K., and McFadden, G. (1996). Myxoma virus T2 protein, a tumor necrosis factor (TNF) receptor homolog, is secreted as a monomer and dimer that each bind rabbit TNFα, but the dimer is a more potent TNF inhibitor. *J. Biol. Chem.* **271**, 13333–13341.

Schreiber, M., Sedger, L., and McFadden, G. (1997). Distinct domains of M-T2, the myxoma virus TNF receptor homolog, mediate extracellular TNF binding and intracellular apoptosis inhibition. *J. Virol.* **71**, 2171–2181.

Sedger, L., and McFadden, G. (1996). M-T2: A poxvirus TNF receptor homologue with dual activities. *Immunol. Cell Biol.* **74**, 538–545.

Shchelkunov, S. N., Blinov, V. M., and Sandakhchiev, L. S. (1993). Genes of variola and vaccinia viruses necessary to overcome the host protective mechanism. *FEBS Lett.* **319**, 80–83.

Smith, C. A., Davis, T., Anderson, D., Solam, L., Beckman, M. P., Jerzy, R., Dower, S. K., Cosman, D., and Goodwin, R. G. (1990). A receptor for tumor necrosis factor defines an unusual family of cellular and viral proteins. *Science* **248**, 1019–1023.

Smith, C. A., Davis, T., Wignall, J. M., Din, W. S., Farrah, T., Upton, C., McFadden, G., and Goodwin, R. G. (1991). T2 open reading frame from Shope fibroma virus encodes a soluble form of the TNF receptor. *Biochem. Biophys. Res. Commun.* **176**, 335–342.

Smith, C. A., Farrah, T., and Goodwin, R. G. (1994). The TNF receptor superfamily of cellular and viral proteins: activation, costimulation and death. *Cell* **76**, 959–962.

Smith, C. A., Hu, F.-Q., Smith, T. D., Richards, C. L., Smolak, P., Goodwin, R. G., and Pickup, D. J. (1996). Cowpox virus genome encodes a second soluble homologue of cellular TNF receptors, distinct from CrmB, that binds TNF but not LTα. *Virology* **223**, 132–147.

Upton, C., DeLange, A. M., and McFadden, G. (1987). Tumorigenic poxviruses: Genomic organization and DNA sequence of the telomeric region of the Shope fibroma virus genome. *Virology* **160**, 20–30.

Upton, C., Macen, J. L., Schreiber, M., and McFadden, G. (1991). Myxoma virus expresses a secreted protein with homology to the tumor necrosis factor receptor gene family that contributes to viral virulence. *Virology* **184**, 370–382.

Xu, X.-M., Nash, P., and McFadden, G. (1998). Myxoma virus expresses a TNF receptor homolog with two distinct functions. *Virus Genes* (in press).

Lymphotoxin β Receptor

Nancy Ruddle[1] **and Carl F. Ware**[2,*]

[1]Department of Epidemiology and Public Health Immunology, Yale University School of Medicine, 815 LEPH, New Haven, CT 06520-8034, USA

[2]Division of Molecular Immunology, LaJolla Institute for Allergy and Immunology, 10355 Science Center Drive, San Diego, CA 92121, USA

*corresponding author tel: 858-678-4660, fax: 858-558-3595, e-mail: carl_ware@liai.org

DOI: 10.1006/rwcy.2000.16003.

SUMMARY

The LTβ receptor (LTβR), a member of the TNFR superfamily, functions as an essential element in the organization of lymphoid tissue and initiation of innate and immune defenses. The LTβR is a type 1 transmembrane glycoprotein with four cysteine-rich motifs in the ectodomain involved in binding to ligands, LTα1β2 and LIGHT. LTβR gene maps near TNFRI and CD27 on chromosome 12p13. LTβR signaling induces NFκB and Jnk/AP-1 transcription factors and activates a slow apoptotic death in certain adenocarcinoma cell lines. The LTβR cytoplasmic tail binds TRAF2, TRAF3, and TRAF5, but not TRAF6 to propagate signaling. Mice deficient in LTβR lack peripheral lymphoid organs and have disorganized splenic microarchitecture, similar to mice deficient in LTα or LTβ genes. LTβR is expressed on stromal tissue and thus receives signals from activated lymphocytes that mediate tissue organization. The LTβR–LTα1β2 cytokine system has unique roles, but functions in some cases with TNFRI and II, and HVEM cytokine systems as an integrated network that orchestrates multiple developmental processes and immune responses.

BACKGROUND

Discovery

The lymphotoxin β receptor (LTβR) was initially identified as a transcript in somatic cell hybrids expressing genes on chromosome 12p and recognized as a cysteine-rich TNFR-like domain (Baens *et al.*, 1993). The ligand specificity for LTα1β2 was defined by Crowe *et al.* (1994) and provided key evidence that explained the unusual phenotype of LTα-deficient mice and proved that surface lymphotoxin was unique and had a biological role distinct from that of TNF and LTα. Recently, Mauri *et al.* (1998) showed that LTβR also recognized another closely related ligand, LIGHT.

Alternative names

LTβR was previously referred to as TNF receptor-related protein (TNFRrp). For the purposes of cataloging human genes, the Human Gene Nomenclature Committee, HUGO (http://www.gene.ucl.ac.uk/users/hester/tnftop.html), has assigned members of the TNF ligand and receptor superfamilies numerical indicators. LTβR is TNFRSF3.

Structure

The LTβR is a type 1 single transmembrane glycoprotein of 435 amino acids that contains a cysteine-rich extracellular domain, which defines it as a member of the TNFR superfamily (**Figure 1**) (Ware *et al.*, 1998). LTβR maps to chromosome 12p13 near TNFRI and CD27. The LTβR binds three known ligands: LTα1β2 and LIGHT with high affinity, and the LTα2β1 complex with low affinity. The LTβR

Figure 1 Main features of the lymphotoxin β receptor.

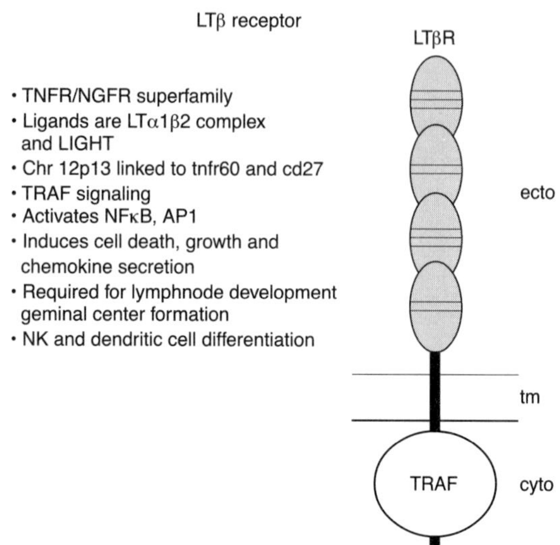

signals via members of the TRAF (TNF receptor-associated factors) family of zinc ring finger proteins that mediate activation of NFκB, AP-1 transcription factors, and cell death. The mouse LTβR is highly homologous to the human version (68% identity).

Figure 2 The immediate LT/TNF family. Depicted are the individual cytokine systems that utilize the common set of receptors which defines members of the immediate LT family. Arrows indicate high-affinity ligand–receptor interactions, the dashed line indicates the low-affinity binding of LTα to HVEM. HSVgD is the envelope glycoprotein gD of herpes simplex virus. DcR3 is the same as OPG2 or TR6 and exists as a soluble product that also binds Fas ligand. Not pictured is LT$\alpha 2\beta 1$ as the functional significance is not known, however it binds TNFRI and TNFRII with high affinity and LTβR with low affinity.

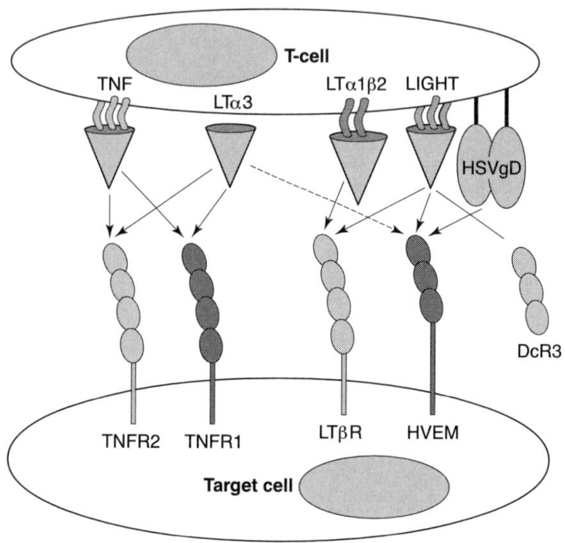

Main activities and pathophysiological roles

The LTβR is one of five receptors that interact with a common group of ligands including LTα, TNF, LTβ, and LIGHT, which form individual cytokine systems, and together define the immediate TNF/LT family (**Figure 2**) (Smith *et al.*, 1994; Ware *et al.*, 1995; Wallach *et al.*, 1999). The LTβR is now proven to be the critical signaling molecule for the LT-$\alpha\beta$ complex. It is required for the formation of secondary lymphoid tissue, lymph nodes (LN) and Peyer's patches (PP) (Fu and Chaplin, 1999). The LTβR contributes to the signaling for organization of T and B cells in the spleen, and germinal center formation during immune responses. LTα- and LTβ-deficient mice also exhibit the loss of progenitor cells required for NK and NK-T cell development and maturation. The LTβR is expressed on stromal cells, while the ligands are produced by activated lymphocytes. This unidirectional signaling from lymphocyte to stroma is thought to organize the architecture of the tissue to convene efficient immune responses. The physiologic signaling by LTβR activated by its other ligand, LIGHT is just beginning to be explored.

GENE

Accession numbers

Mouse LTβR cDNA: U29173, L38423 (for gene organization see Force *et al.*, 1996)
Human LTβR cDNA: L04270

Chromosome location and linkages

The gene encoding mouse LTβR contains 10 exons spanning 6.9 kb and maps to chromosome 6 closely linked to genes for TNFRI (CD120a) and CD27 (**Figure 3**). Approximately 1.5 cM separates *ltbr* and *tnfr1*. This region exhibits conserved synteny with human chromosome 12p13. The human gene encoding LTβR spans ~9 kb with an intron/exon arrangement very similar to that of TNFRI. The promoter region lacks TATA and CAAT sequences and resembles that of a housekeeping-like gene, and is

Figure 3 Genetic organization of the murine LTβ receptor. (a) Chromosomal map location of *Ltbr* and *Tnfr1* on chromosome 6. The black circle represents the centromere. The relative position of each locus pair is illustrated by vertical hash marks. The recombinational differences between loci and standard errors are shown in parentheses and conserved synteny with the human map locations for the indicated genes are shown below. (b) Exon organization of *Ltbr*. Each exon is represented as a black rectangle. (c) The mLTβR cDNA represented as a compilation of its exons. The number of each exon is given above and the amino acid position located at the intron/exon junction is given below denoted by a vertical hash mark. (d) Schematic of the mLTβR protein. L, leader sequence; D1–4, the ligand-binding region including the cysteine-rich repeats; TM, transmembrane domain; CYTO, the cytoplasmic signaling domain. Reproduced with permission from Force *et al.* (1996).

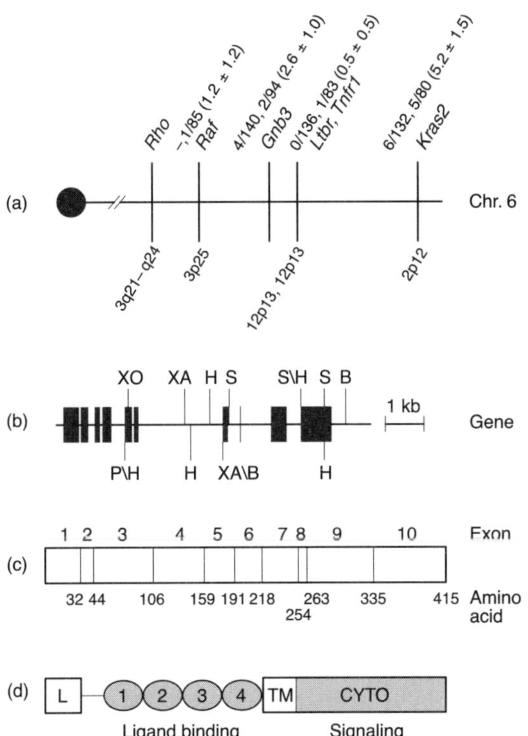

similar to the promoter found in *tnfr1* and consistent with the constitutive expression of this protein.

Disease linkage has not been identified for *ltbr*; however, a spontaneous fever syndrome has been mapped to 12p13, although initial analysis indicates mutations that elevate expression of TNFRI causing an inflammation-based disease.

PROTEIN

Accession numbers

Mouse LTβR cDNA: U29173, L38423
Human LTβR cDNA is L04270

Sequence

See **Figure 4**.

Description of protein

LTβR is translated from a 2.2 kb mRNA with a theoretical mass of 46.7 kDa. The observed mass of 61 kDa indicates that the two potential *N*-glycosylation sites are probably utilized. Analysis of the human LTβR cDNA sequence encodes a predicted 435 amino acid protein sharing 41% and 46% homology with TNFRI and TNFRII, respectively. LTβR has a ligand-binding ectodomain with four cysteine-rich pseudo repeats followed by a short proline-rich membrane proximal region, characteristic of the TNFR superfamily. The ligand binding domain of LTβR has characteristics of both TNFRI and TNFRII in the positioning of the cysteine residues and can be readily modeled on the TNFRI structure (Guex and Peitsch, 1997). Similarity with TNFRI is found in the first and second domains. In contrast, the equivalent of loop 1 in domain 3 of LTβR is dramatically shortened, having only five residues, and in this regard is highly similar to domain 3 of TNFRII. The fourth domain of LTβR resembles TNFRII more closely than TNFRI. An additional similarity between LTβR and TNFRII is the proline-rich region proximal to the membrane-spanning sequence, suggested the formation of a 'stalk' extending from the cell surface.

The cytoplasmic portion of LTβR shares limited homology with other members of the family by sequence analysis but is grouped functionally with those other members that bind to TRAF signaling molecules, such as CD40, CD30, and CD27.

Relevant homologies and species differences

Mouse and human LTβRs are homologous (68% identity) over the entire length of the sequence, and are similar to many other receptors in this family. The mouse LTβR lacks 20 amino acids overall, and 12 of those are at the C-terminal tail.

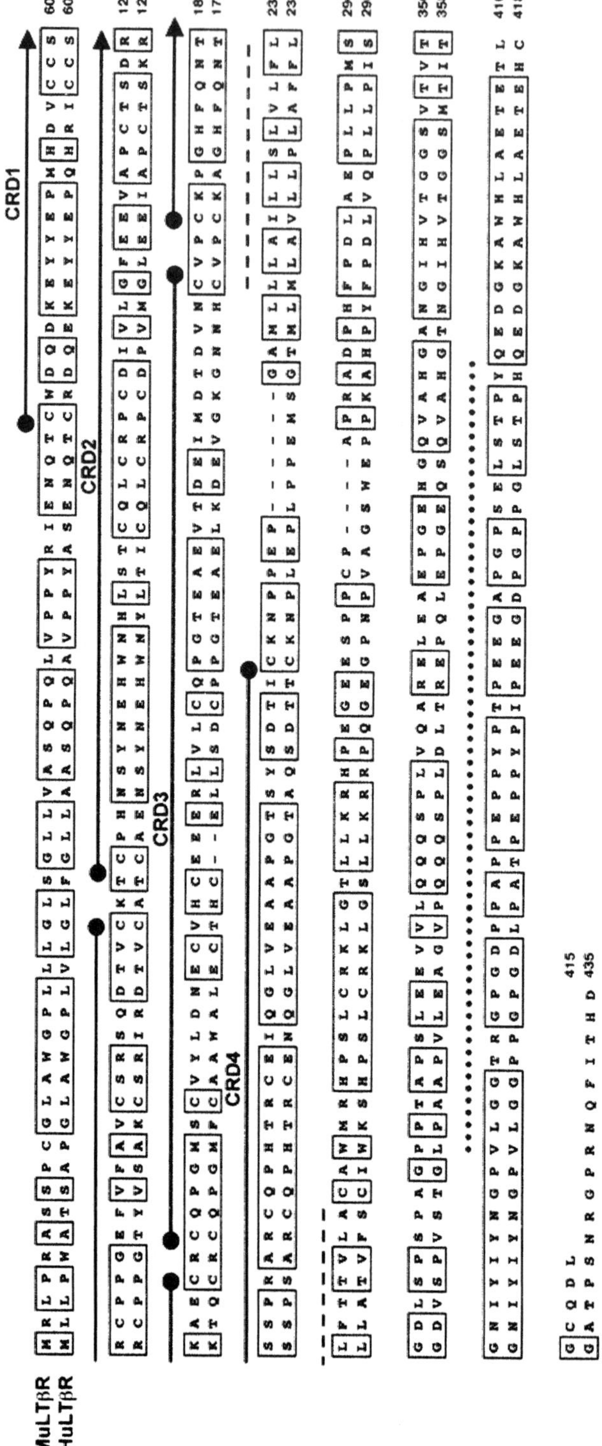

Figure 4 Sequence of the LTβR. Sequences were aligned with ClustalW program. The cysteine-rich domains (CRD) are denoted by solid bars above the sequence. The transmembrane region is identified by a dashed line and the TRAF-binding region in the cytoplasmic domain by a stippled line.

Affinity for ligand(s)

LTβR binds membrane forms of LTα1β2 and LIGHT with high affinity, and also to their recombinant soluble forms, but only weakly to LTα2β1 (**Table 1**). It cannot be excluded that relatively weak interactions between receptor and ligand in the soluble phase may be highly significant in the context of membrane localization where the avidity may be increased substantially.

Cell types and tissues expressing the receptor

The LTβR is expressed on stromal cells in lymphoid tissue such as thymus and spleen and absent on lymphocytes as determined by immunohistochemistry (Murphy *et al.*, 1998). The LTβR is expressed on normal dermal fibroblasts, normal bronchial airway epithelial cells, but absent on human peripheral blood derived mononuclear cells including T and B lymphocytes. Most adherent cell lines including FDC-1, follicular dendritic cell line, U937 promyelomonocytic cell line, HT-29 colon adenocarcinoma, HeLa cervical carcinoma line, and the HEK 293 embryonic kidney cells express LTβR as measured by cell surface staining (Murphy *et al.*, 1998).

Release of soluble receptors

Unlike TNFRI or TNFRII, no evidence indicates that LTβR is shed or has alternate spliced forms that create soluble versions.

SIGNAL TRANSDUCTION

Associated or intrinsic kinases

The binding of the trivalent ligand, which induces an ordered aggregation or 'clustering' of receptors, initiates signal transduction by TNFR family members. Antireceptor antibodies mimic receptor clustering, and for many receptors mere overexpression can lead to activation of signal transduction pathways. The LTβR, as well as all the other related TNFRs, has no intrinsic kinase or other enzymatic activities encoded by its cytosolic domain. Rather, signal transduction occurs through the recruitment of cytosolic adapters, which directly or indirectly confer enzymatic activity to the receptor. Recent findings indicate that the mutation in the alymphoplasia (*aly*) mutant mice, which lack lymph nodes and PPs, is due to NFκB-inducing kinase (NIK). This result indicates that NIK is likely to be required for LTβR-specific signaling (Shinkura *et al.*, 1999).

Cytoplasmic signaling cascades

The LTβR activates gene transcription and can activate apoptosis, although it is not a death domain receptor, which interact directly with adapters for caspases. The LTβR utilizes select members of the TRAF family (Arch *et al.*, 1998) to propagate signals received via LTα1β2 and LIGHT (VanArsdale *et al.*, 1997) (**Figure 5**). TRAF2, 3, and 5 are known to bind directly to the cytosolic domain of the LTβR. The

Table 1 Receptor specificity of the LT/TNF cytokine system

Ligand	Form[a]	Binding receptor	Affinity ($\sim K_d$ = nM)
LTα	Soluble	TNFRI and TNFRII	High (0.1)[b]
LTα1β2	Membrane	LTβR	High (0.5)
LTα2β1	Membrane	TNFRI and TNFRII	High (0.5)
	Soluble	LTβR	Low (20)
LIGHT	Membrane	LTβR	High (1)
	Soluble		High (0.5)
TNF	Soluble	TNFRII and TNFRI	High (0.1)[b]

[a]Membrane ligands used to measure binding avidity were expressed on Tn5 insect cells infected with recombinant baculoviruses or by transfection of 293T cells with LTα or LTβ cDNAs. Soluble ligands were purified proteins that were radioiodinated with ^{125}I. With ligands expressed on cells binding specificity was performed using graded amounts of LTβR:Fc or TNFRI:Fc fusion proteins and detection of bound fraction by flow cytometry. With soluble ligands, competitive binding assays were performed with receptor:Fc proteins bound to microplate wells and [^{125}I]LTα1β2. Soluble LIGHT LTβR interactions were determined with Plasmon resonance (Biacore).

[b]Values derived by Scatchard analysis for ligands bound to mammalian cells.

Figure 5 Signal transduction pathways for the LTβR. This cartoon depicts the LTβR clustered by its trivalent ligands, LTα1β2, or LIGHT, which recruits TRAF adapter molecules. TRAF2 and 5 are involved in gene activation through NFκB or AP-1 transcription factors. The link between TRAF3 and the caspase-dependent apoptotic pathway is unknown.

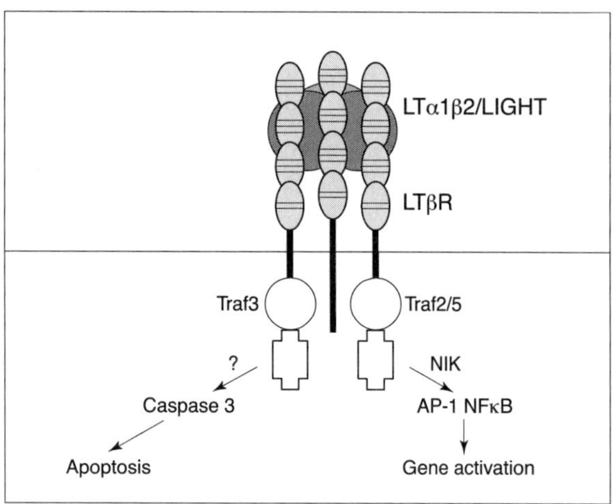

TRAF-binding region is located within a proline-rich sequence between residues 345 and 396 (Force *et al.*, 2000). This is also a region that binds to the hepatitis C virus core protein (Matsumoto *et al.*, 1997). TRAF2 and 5 are potent inducers of NFκB and AP-1 transcription factors. TRAF3, however, appears to be involved in signaling cell death as dominant negative mutants of TRAF3 inhibit LTβR signaling of cell death, but not NFκB activation (Force *et al.*, 1997). Thus the LTβR signaling pathways bifurcate and appear to be independently controlled (Figure 5). NIK may mediate some signaling events for the LTβR. One likely scenario envisions TRAF2 or 5, both of which interact directly with NIK, acting as the adapters coupling LTβR to IκB kinases essential for activation of NFκB. However, the NIK phenotype in *aly* mice is not recapitulated by knockout of TRAF2 or 5, suggesting either a redundancy of function or that an alternate mechanism exists for LTβR to activate NFκB. At present, it is not understood how TRAF3 may be connected to the death pathway. It is possible that TRAF3 acts as an adapter that connects LTβR to TRADD, which in turn interacts with FADD thus coupling directly to caspase 8, which activates caspase 3, the executioner caspase.

DOWNSTREAM GENE ACTIVATION

Transcription factors activated

The ability of LTβR to activate both NFκB and AP-1 transcription factors suggests that signaling will elicit inflammation and cellular stress responses. The plethora of genes activated via these transcription factors makes it difficult to predict specific responses. Cellular responses will also be controlled in part by the strength and duration of LTβR-generated signals, as well as within the context of the tissue and cell type. LTβR can activate expression of adhesion molecules such as ICAM-1 in follicular dendritic cells (FDCs) and is presumably involved in forming FDC clusters, which are absent in LTβR knockout mice. The induction of adhesion molecules by the LTβR is weak compared to that by TNF, especially in other cells types.

In vivo, MAdCAM expression by FDCs is suppressed by treatment with LTβR:Fc fusion protein. LTβR also induces chemokine expression which attracts leukocytes and other cells. Intriguingly, mice deficient in the chemokine receptor BLR1 (Forster *et al.*, 1996) partially resembles the phenotype of LTβR−/− mice.

Genes induced

Because NFκB and AP-1 transcription factors are activated by LTβR, it is expected that a subset of the genes regulated by these factors will be induced (Whitmarsh and Davis, 1996).

Promoter regions involved

The NFκB and AP-1 transcriptional regulation systems are well studied and found in numerous genes (see reviews by Whitmarsh and Davis, 1996; Ghosh *et al.*, 1998).

BIOLOGICAL CONSEQUENCES OF ACTIVATING OR INHIBITING RECEPTOR AND PATHOPHYSIOLOGY

The distinct receptor specificity of LTα1β2 complex for the LTβR hinted that new functions would be associated with this cytokine–receptor system. The production of LT-deficient mice has revealed molecular insight into fundamental processes central to the immune response (De Togni *et al.*, 1994; Banks *et al.*, 1995; Futterer *et al.*, 1998). LTα-deficient mice have no detectable popliteal, inguinal, para-aortic, mesenteric, axillary, or cervical lymph nodes, whereas LTβ-deficient mice have mesenteric and cervical lymph nodes (Koni *et al.*, 1997; Alimzhanov *et al.*, 1997).

Since mice deficient in TNF have a normal lymph node structure (Pasparakis *et al.*, 1996), LTα and TNF are not functionally redundant molecules. Recent evidence has shown that NK and NK-T cells are absent in LT-deficient mice (Iizuka *et al.*, 1999). Interestingly, no easily detectable defect in T lymphocyte number or function could be ascertained by traditional assays. However lack of NK-T cells and no circulating $\alpha_E\beta_7$+ T cells indicates that at least these subsets are abnormal. There was, however, an increase in circulating B cells in the LTα-deficient animals. Bone marrow cells obtained from LTα-deficient mice were able to home to both spleen and lymph nodes of SCID or lethally irradiated wild-type mice, indicating no intrinsic defect in the hematopoietic compartment. In contrast, normal bone marrow was unable to reconstitute lymph nodes in irradiated LTα-deficient mice, indicating that lymph node organogenesis is developmentally fixed.

Since mice deficient for the TNF receptors TNFRI and TNFRII, or TNF had no defects in lymph node organogenesis, it was hypothesized that an interaction between membrane-bound LT$\alpha\beta$ and LTβR located on the stromal elements might play a role in lymph node genesis (Crowe *et al.*, 1994). Three approaches were taken, two using an LTβR:Fc fusion protein to neutralize ligands either expressed as a transgene (Ettinger *et al.*, 1996) or injected into pregnant mice for introduction into embryonic circulation (Rennert *et al.*, 1996), and the other by interruption of the LTβ and LTβR locus (Koni *et al.*, 1997; Alimzhanov *et al.*, 1997; Futterer *et al.*, 1998). The LTβR:Fc protein does not block soluble, trimeric LTα binding with TNFRI, however it does block the activity of LIGHT. LTβ-deficient mice and mice that were treated *in utero* with LTβR:Fc fusion protein have no histologically detectable inguinal and popliteal lymph nodes nor PPs. As this matches the phenotype of LTα and LTβ mice it is unlikely that LIGHT plays a role here.

Temporal development of lymph nodes could be determined by administering the LTβR:Fc fusion protein at different days of gestation; the full effect of LTβR:Fc must be administered prior to day 12 of gestation. By this staging, PPs were the latest in sequence to develop. By contrast the mice with LTβR:Fc transgene had normal lymph node development, but PPs were reduced or absent, probably due to low neutralizing levels of LTβR:Fc until late and/or the mutated Fc region did not cross the placental barrier.

LTβR−/− mice, and LTα- and LTβ-deficient mice, display disorganized splenic architecture. The other shared phenotypes of the LTβR−/− and LTα−/− mice was the absence of colon-associated lymphoid tissues and all lymph nodes. Other phenotypes observed in the LTβR−/− mice included loss of $\alpha_E\beta_7^{high}$ integrin+ T cells and loss of marginal zones, T/B cell segregation, and follicular dendritic cell networks in the spleen. Peanut agglutinin+ cells were aberrantly detectable around central arterioles. In contrast to TNF receptor p55−/− mice, antibody affinity maturation was impaired. Since LTβR−/− mice exhibit distinct defects when compared to LTα−/− and LTβ−/− mice, for example, affinity maturation of the antihapten Ig response occurs almost unimpaired in TNFRI−/− mice and is not markedly impaired in LTα−/− mice, it implies that other ligands may be able to activate LTβR. Here LIGHT becomes a lead candidate (Mauri *et al.*, 1998). Nonetheless, the LTβR is essential for generation of lymphoid tissues.

In addition to the LT system regulating lymph node genesis, several genes have been identified that also play a role in this process. Mice deficient for the early hematopoietic- and lymphocyte-restricted transcription factor Ikaros, lack lymph nodes and PPs. Mice that developed a spontaneous autosomal recessive mutation, termed *aly* (alymphoplasia), also lack lymph nodes and this defect is attributable to a mutation in NIK, which links the LTβR to activation of NFκB (Shinkura *et al.*, 1999).

LTβR:Fc injected into pregnant mice recapitulated the NK cell defect observed in LTα−/− mice. The lack of functional NK cells has serious consequences, including the inability to reject some tumors (Iizuka *et al.*, 1999; Ito *et al.*, 1999).

Unique biological effects of activating the receptors

Although *in vitro* the LTβR shares significant overlap with TNFRI, *in vivo* studies clearly indicate these receptors play distinct and nonredundant functions.

Phenotypes of receptor knockouts and receptor overexpression mice

See **Table 2**.

Table 2 Phenotypes of mice deficient in members of the immediate LT/TNF family

	Lymph nodes							Spleen	Thymus
Gene:	mes	cer	axi	ing	pop	PP	cec		
LTβR−/−	−	−	−	−	−	−	−	+	+
TNFRI−/−	+	+	+	+	+	±	−	+	+
LTα−/−	±	−	−	−	−	−	−	+	+
LTβ−/−	+	+	−	−	−	−	+	+	+
TNF−/−	+	+	+	+	+	+	+	+	+

	Splenic architecture						Germinal centers	
Gene:	T & B zones	MZ sinuses	Moma-1+ Mφ	Sialo-adhesion	MAd-CAM-1	MZB cells	PNA^+ clusters	FDC network
LTβR−/−	−	−	−	−	−	−	−	−
TNFRI−/−	+	+	+	+	−	+	−	−
LTα−/−	−	−	−	−	−	−	−	−
LTβ−/−	+	−	−	−	−	−	+	−
TNF−/−	+	+	+	+	−	+	+	−

	Immune response				
	Isotype switching		Affinity maturation		$\alpha_E\beta_7$ T cells
Gene:	sRBC	Hapten-Ag	Low Ag	High Ag	
LTβR−/−	−	+	−	+	
TNFRI−/−	−	+	+	+	+
LTα−/−	−	+	−	+	−
LTβ−/−	−	+	+	+	−
TNF−/−	−	+	−	+	+

mes, mesenteric; cer, cervical; axi, axillary; ing, inguinal; pop, popliteal; PP, Peyer's patches; cec, cecal; MZ, marginal zone; MAdCAM-1, mesenteric addressin cell adhesion molecule-1; PNA, peanut agglutinin; FDC, follicular dendritic cell; sRBC, sheep red blood cell; $\alpha_E\beta_7$, integrin; Ag, antigen.

Human abnormalities

None described to date.

THERAPEUTIC UTILITY

Effect of treatment with soluble receptor domain

The LTβR:Fc fusion protein is a potent neutralizing agent for LTα1β2 and LIGHT. Mackay *et al.* (1998) investigated the action of this inhibitor in two independent rodent models of *colitis*. LTβR:Fc attenuated the development of both the clinical and histological manifestations of the disease and was as efficacious as antibody to TNF. This lead to the conclusion that LT pathway plays a role in the development of colitis and represents a potential novel intervention point for the treatment of *inflammatory bowel disease*.

Effects of inhibitors (antibodies) to receptors

Anti-LTβR monoclonal antibodies have been shown to induce death of certain *adenocarcinomas* and also show an effect on growth of tumors *in vivo* (Browning *et al.*, 1996), suggesting the possibility of their use in antitumor therapy.

References

Alimzhanov, M. B., Kuprash, D. V., Kosco-Vilbois, M. H., Luz, A., Turetskaya, R. L., Tarakhovsky, A., Rajewsky, K., Nedospasov, S. A., and Pfeffer, K. (1997). Abnormal development of secondary lymphoid tissues in lymphotoxin β-deficient mice. *Proc. Natl Acad. Sci. USA* **94**, 9302–9307.

Arch, R., Gedrich, R., and Thompson, C. (1998). Tumor necrosis factor receptor-associated factors (TRAFs) – a family of adapter proteins that regulates life and death. *Genes Dev.* **12**, 2821–2830.

Baens, M., Chaffanet, M., Cassiman, J. J., van den Berghe, H., and Marynen, P. (1993). Construction and evaluation of a hncDNA library of human 12p transcribed sequences derived from a somatic cell hybrid. *Genomics* **16**, 214–218.

Banks, T. A., Rouse, B. T., Kerley, M. K., Blair, P. J., Godfrey, V. L., Kuklin, N. A., Bouley, D. M., Thomas, J., Kanangat, S., and Mucenski, M. L. (1995). Lymphotoxin-α-deficient mice. Effects on secondary lymphoid organ development and humoral immune responsiveness. *J. Immunol.* **155**, 1685–1693.

Browning, J. L., Miatkowski, K., Sizing, I., Griffiths, D. A., Zafari, M., Benjamin, C. D., Meier, W., and Mackay, F. (1996). Signalling through the lymphotoxin-β receptor induces the death of some adenocarcinoma tumor lines. *J. Exp. Med.* **183**, 867–878.

Chaplin, D., and Fu, Y.-X. (19??). Cytokine regulation of secondary lymphoid organ development. *Curr. Opin. Immunol.* **10**, 289–297.

Crowe, P. D., VanArsdale, T. L., Walter, B. N., Ware, C. F., Hession, C., Ehrenfels, B., Browning, J. L., Din, W. S., Goodwin, R. G., and Smith, C. A. (1994). A lymphotoxin-beta-specific receptor. *Science* **264**, 707–710.

De Togni, P., Goellner, J., Ruddle, N. H., Streeter, P. R., Fick, A., Mariathasan, S., Smith, S. C., Carlson, R., Shornick, L. P., Strauss-Schoenberger, J., Russell, J. H., Karr, R., and Chaplin, D. D. (1994). Abnormal development of peripheral lymphoid organs in mice deficient in lymphotoxin. *Science* **264**, 703–706.

Ettinger, R., Browning, J. L., Michie, S. A., van Ewijk, W., and McDevitt, H. O. (1996). Disrupted splenic architecture, but normal lymphnode development in mice expressing a soluble lymphotoxin-β receptor-IgG1 chimeric fusion protein. *Proc. Natl Acad. Sci.* **93**, 13102–13107.

Force, W. R., Walter, B. N., Hession, C., Tizard, R., Kozak, C. A., Browning, J. L., and Ware, C. F. (1996). Mouse lymphotoxin-β receptor. Molecular genetics, ligand binding, and expression. *J.Immunol.* **155**, 5280–5288.

Force, W. R., Cheung, T. C., and Ware, C. F. (1997). Dominant negative mutants of TRAF3 reveal an important role for the coiled coil domains in cell death signaling by the lymphotoxin-β receptor (LTbR). *J. Biol. Chem.* **272**, 30835–30840.

Force, W. R., Glass, A. A., Benedict, C. A., Cheung, T. C., Lama, J., and Ware, C. F. (2000). Discrete signaling regions in the lymphotoxin-β receptor for TRAF binding, subcellular localization and activation of cell death and NFκB pathways. *J. Biol. Chem.* **275**, 11121–11129.

Forster, R., Mattis, A. E., Kremmer, E., Wolf, E., Brem, G., and Lipp, M. (1996). A putative chemokine receptor, BLR1, directs B cell migration to defined lymphoid organs and specific anatomic compartments of the spleen. *Cell* **87**, 1037–1047.

Fu, Y. X., and Chaplin, D. D. (1999). Development and maturation of secondary lymphoid tissues. *Annu. Rev. Immunol.* **17**, 399–433.

Futterer, A., Mink, K., Luz, A., Kosco-Vilbois, M. H., and Pfeffer, K. (1998). The lymphotoxin beta receptor controls organigenesis and affinity maturation in peripheral lymphoid tissues. *Immunity* **9**, 59–70.

Ghosh, S., May, M., and Kopp, E. (1998). NF-kB and REL proteins: Evolutionarily conserved mediators of immune responses. *Annu. Rev. Immunol.* **16**, 225–260.

Guex, N., and Peitsch, M. C. (1997). SWISS-MODEL and the Swiss-Pdb Viewer: An environment for comparative protein modelling. *Electrophoresis* **18**, 2714–2723.

Iizuka, K., Chaplin, D. D., Wang, Y., Wu, Q., Pegg, L. E., Yokoyama, W. M., and Fu, Y. X. (1999). Requirement for membrane lymphotoxin in natural killer cell development. *Proc. Natl Acad. Sci. USA* **96**, 6336–6340.

Ito, D., Back, T. C., Shakhov, A. N., Wiltrout, R. H., and Nedospasov, S. A. (1999). Mice with a targeted mutation in lymphotoxin-alpha exhibit enhanced tumor growth and metastasis: impaired NK cell development and recruitment. *Immunology* **163**, 2809–2815.

Koni, P. A., Sacca, R., Lawton, P., Browning, J. L., Ruddle, N. H., and Flavell, R. A. (1997). Distinct roles in lymphoid organogenesis for lymphotoxins α and β revealed in lymphotoxin β-deficient mice. *Immunity* **6**, 491–500.

Mackay, F., Browning, J. L., Lawton, P., Shah, S. A., Comiskey, M., Bhan, A. K., Mizoguchi, E., Terhorst, C., and Simpson, S. J. (1998). Both the lymphotoxin and tumor necrosis factor pathways are involved in experimental murine models of colitis. *Gastroenterology* **115**, 1464–1475.

Matsumoto, M., Hsieh, T. Y., Zhu, N. L., VanArsdale, T., Hwang, S. B., Jeng, K. S., Gorbalenya, A. E., Lo, S. Y., Ou, J. H., Ware, C. F., and Lai, M. M. C. (1997). Hepatitis C virus core protein interacts with the cytoplasmic tail of lymphotoxin-β receptor. *J. Virol.* **71**, 1301–1309.

Mauri, D. N., Ebner, R., Montgomery, R. I., Kochel, K. D., Cheung, T. C., Yu, G.-L., Ruben, S., Murphy, M., Eisenbery, R. J., Cohen, G. H., Spear, P. G., and Ware, C. F. (1998). LIGHT, a new member of the TNF superfamily and lymphotoxin α are ligands for herpesvirus entry mediator. *Immunity* **8**, 21–30.

Murphy, M., Walter, B. N., Pike-Nobile, L., Fanger, N. A., Guyre, P. M., Browning, J. L., Ware, C. F., and Epstein, L. B. (1998). Expression of the lymphotoxin beta receptor on follicular stromal cells in human lymphoid tissues. *Cell Death Differ.* **6**, 497–505.

Pasparakis, M., Alexopoulou, L., Episkopou, V., and Kollias, G. (1996). Immune and inflammatory responses in TNFα-deficient mice: a critical requirement for TNFα in the formation of primary B cell follicles, follicular dendritic cell networks and germinal centers, and in the maturation of the humoral immune response. *J. Exp. Med.* **184**, 1397–1411.

Rennert, P. D., Browning, J. L., Mebius, R., Mackay, F., and Hochman, P. S. (1996). Surface lymphotoxin α/β complex is required for the development of peripheral lymphoid organs. *J. Exp. Med.* **184**, 1999–2006.

Shinkura, R., Kitada, K., Matsuda, F., Tashiro, K., Ikuta, K., Suzuki, M., Kogishi, K., Serikawa, T., and Honjo, T. (1999). Alymphoplasia is caused by a point mutation in the mouse gene encoding Nf-kappa B-inducing kinase. *Nature Genet.* **22**, 74–77.

Smith, C. A., Farrah, T., and Goodwin, R. G. (1994). The TNF receptor superfamily of cellular and viral proteins: activation, costimulation, and death. *Cell* **76**, 959–962.

VanArsdale, T. L., VanArsdale, S. L., Force, W. R., Walter, B. N., Mosialos, G., Kieff, E., Reed, J. C., and Ware, C. F. (1997). Lymphotoxin-β receptor signaling complex: role of tumor necrosis factor receptor-associated factor 3 recruitment in cell death and activation of nuclear factor kB. *Proc. Natl Acad. Sci. USA* **94**, 2460–2465.

Wallach, D., Varfolomeev, E. E., Malinin, N. L., Goltsev, Y. V., Kovalenko, A. V., and Boldin, M. P. (1999). Tumor necrosis factor receptor and Fas signaling mechanisms. *Annu. Rev. Immunol.* **17**, 331–367.

Ware, C. F., VanArsdale, T. L., Crowe, P. D., and Browning, J. L. (1995). In "Pathways for Cytolysis" (ed G. M. Griffiths and J. Tschopp) The ligands and receptors of the lymphotoxin system, pp. 175–218. Springer-Verlag, Basel.

Ware, C. F., Santee, S., and Glass, A. (1998). In "The Cytokine Handbook" (ed A. Thompson) Tumor necrosis factor-related ligands and receptors, pp. 549–592. Academic Press, San Diego.

Whitmarsh, A., and Davis, R. (1996). Transcription factor AP-1 regulation by mitogen-activated protein kinase signal transduction pathways. *J. Mol. Med.* **74**, 589–607.

Fas

Shigekazu Nagata*

Osaka University Medical School, 2-2 Yamada-oka, Suita, Osaka, 565-0871, Japan

*corresponding author tel: 81-6-6879-3310, fax: 81-6-6879-3319, e-mail: nagata@genetic.med.osaka-u.ac-jp

DOI: 10.1006/rwcy.2000.16004.

SUMMARY

Fas is a type I membrane protein belonging to the TNF/NGF receptor family, and is expressed in various tissues and cell lines. Binding of FasL or agonistic anti-Fas antibody to Fas causes apoptosis in Fas-bearing cells. That is, the Fas engagement recruits pro-caspase 8 through an adapter molecule called FADD. Oligomerization of pro-caspase 8 induces processing of pro-caspase 8 to active forms. The caspase 8 then activates other caspases in the downstream of the caspase cascade. These caspases then cleave various cellular substrates to cause morphological changes of the cells, and to degrade chromosomal DNA. The Fas-mediated apoptosis is involved to kill the activated T cells and B cells, thus downregulating the immune reaction. The loss-of-function mutations in Fas in human and mouse cause lymphoproliferation, and accelerate *autoimmune diseases*.

BACKGROUND

Discovery

Fas was discovered as a cell surface antigen that can be recognized by a cytotoxic monoclonal antibody (anti-Fas or anti-Apo1 antibody) against human cell-surface protein (Trauth *et al.*, 1989; Yonehara *et al.*, 1989). The human Fas cDNA was identified from human leukemia cell line by expression cloning using the anti-Fas antibody (Itoh *et al.*, 1991), or by using the amino acid sequence information of the purified protein (Oehm *et al.*, 1992). The mouse Fas cDNA was subsequently isolated from mouse cDNA library by cross-hybridization with human cDNA (Watanabe-Fukunaga *et al.*, 1992b).

Alternative names

CD95, Apo1.

Structure

Human Fas is a type I membrane protein with apparent molecular mass of 48 kDa (Itoh *et al.*, 1991; Oehm *et al.*, 1992).

Main activities and pathophysiological roles

Fas transduces an apoptotic signal into cells upon engagement by Fas ligand (FasL) or agonistic anti-Fas antibody (Itoh *et al.*, 1991; Oehm *et al.*, 1992; Nagata and Golstein, 1995; Nagata, 1997).

GENE

Accession numbers

Human Fas: M67454 (Itoh *et al.*, 1991).
Mouse Fas: M83649 (Watanabe-Fukunaga *et al.*, 1992b).

Sequence

See **Figure 1**.

Figure 1 Nucleotide sequence for human Fas.

```
1    CACGCTTCTG GGGAGTGAGG GAAGCGGTTT ACGAGTGACT TGGCTGGAGC CTCAGGGGCG
61   GGCACTGGCA CGGAACACAC CCTGAGGCCA CCCCTTGCTG CCCAGGCGGA GCTGCCTCTT
121  CTCCCGCGGG TTGGTGGACC CGCTCACTAC GGAGTTCGGG AAGCTCTTTC ACTTCCCACC
181  ATTGCTCAAC AACCATGCTG GGCATCTGGA CCCTCCTACC TCTGGTTCTT ACGTCTGTTG
241  CTAGATTATC GTCCAAAAGT GTTAATGCCC AAGTGACTGA CATCAACTCC AAGGGATTGG
301  AATTGAGGAA GACTGTTACT ACAGTTGAGA CTCAGAACTT GGAAGGCCTG CATCATGATG
361  GCCAATTCTG CCATAAGCCC TGTCCTCCAG GTGAAAGGAA ACCTAGGGAC TGCACAGTCA
421  ATGGGGATGA ACCAGACTGC GTGCCCTGCC AAGAAGGGAA GGAGTACACA GACAAAGCCC
481  ATTTTTCTTC CAAATGCAGA AGATGTAGAT TGTGTGATGA AGGACATGGC TTAGAAGTGG
541  AAATAAACTG CACCCGGACC CAGAATACCA AGTGCAGATG TAAACCAAAC TTTTTTTGTA
601  ACTCTACTGT ATGTGAACAC TGTGACCCTT GCACCAAATG TGAACATGGA ATCATCAAGG
661  AATGCACACT CACCAGCAAC ACCAAGTGCA AAGAGGAAGG ATCCAGATCT AACTTGGGGT
721  GGCTTTGTCT TCTTCTTTTG CCAATTCCAC TAATTGTTTG GGTGAAGAGA AAGGAAGTAC
781  AGAAAACATG CAGAAAGCAC AGAAAGGAAA ACCAAGGTTC TCATGAATCT CCAACCTTAA
841  ATCCTGAAAC AGTGGCAATA AATTTATCTG ATGTTGACTT GAGTAAATAT ATCACCACTA
901  TTCCTGGAGT CATGACACTA AGTCAAGTTA AAGGCTTTGT TCGAAAGAAT GGTGTCAATG
961  AAGCCAAAAT AGATGAGATC AAGAATGACA ATGTCCAAGA CACAGCAGAA CAGAAAGTTC
1021 AACTGCTTCG TAATTGGCAT CAACTTCATG GAAAGAAAGA AGCGTATGAC ACATTGATTA
1081 AAGATCTCAA AAAAGCCAAT CTTTGTACTC TTGCAGAGAA AATTCAGACT ATCATCCTCA
1141 AGGACATTAC TAGTGACTCA GAAAATTCAA ACTTCAGAAA TGAAATCCAA AGCTTGGTCT
1201 AGAGTGAAAA ACAACAAATT CAGTTCTGAG TATATGCAAT TAGTGTTTGA AAAGATTCTT
1261 AATAGCTGGC TGTAAATACT GCTTGGTTTT TTACTGGGTA CATTTTATCA TTTATTAGCG
1321 CTGAAGAGCC AACATATTTG TAGATTTTTA ATATCTCATG ATTCTGCCTC CAAGGATGTT
1381 TAAAATCTAG TTGGGAAAAC AAACTTCATC AAGAGTAAAT GCAGTGGCAT GCTAAGTACC
1441 CAAATAGGAG TGTATGCAGA GGATGAAAGA TTAAGATTAT GCTCTGGCAT CTAACATATG
1501 ATTCTGTAGT ATGAATGTAA TCAGTGTATG TTAGTACAAA TGTCTATCCA CAGGCTAACC
1561 CCACTCTATG AATCAATAGA AGAAGCTATG ACCTTTTGCT GAAATATCAG TTACTGAACA
1621 GGCAGGCCAC TTTGCCTCTA AATTACCTCT GATAATTCTA GAGATTTTAC CATATTTCTA
1681 AACTTTGTTT ATAACTCTGA GAAGATCATA TTTATGTAAA GTATATGTAT TTGAGTGCAG
1741 AATTTAAATA AGGCTCTACC TCAAAGACCT TTGCACAGTT TATTGGTGTC ATATTATACA
1801 ATATTTCAAT TGTGAATTCA CATAGAAAAC ATTAAATTAT AATGTTTGAC TATTATATAT
1861 GTGTATGCAT TTTACTGGCT CAAAACTACC TACTTCTTTC TCAGGCATCA AAAGCATTTT
1921 GAGCAGGAGA GTATTACTAG AGCTTTGCCA CCTCTCCATT TTTGCCTTGG TGCTCATCTT
1981 AATGGCCTAA TGCACCCCCA AACATGGAAA TATCACCAAA AAATACTTAA TAGTCCACCA
2041 AAAGGCAAGA CTGCCCTTAG AAATTCTAGC CTGGTTTGGA GATACTAACT GCTCTCAGAG
2101 AAAGTAGCTT TGTGACATGT CATGAACCCA TGTTTGCAAT CAAAGATGAT AAAATAGATT
2161 CTTATTTTTC CCCCACCCCC GAAAATGTTC AATAATGTCC CATGTAAAAC CTGCTACAAA
2221 TGGCAGCTTA TACATAGCAA TGGTAAAATC ATCATCTGGA TTTAGGAATT GCTCTTGTCA
2261 TACCCTCAAG TTTCTAAGAT TTAAGATTCT CCTTACTACT ATCCTACGTT TAAATATCTT
2341 TGAAAGTTTG TATTAAATGT GAATTTTAAG AAATAATATT TATATTTCTG TAAATGTAAA
2401 CTGTGAAGAT AGTTATAAAC TGAAGCAGAT ACCTGGAACC ACCTAAAGAA CTTCCATTTA
2461 TGGAGGATTT TTTTGCCCCT TGTGTTTGGA ATTATAAAAT ATAGGTAAAA GTACGTAATT
2521 AAATAATGTT TTTG
```

Chromosome location and linkages

Human chromosome 10q24.1 (Inazawa *et al.*, 1992) Mouse chromosome 11 (Watanabe-Fukunaga *et al.*, 1992b) linked to *lpr* mutation.

PROTEIN

Accession numbers

Human Fas: PID g105364, g182410 (Itoh *et al.*, 1991) Mouse Fas: PID g346698, g193226 (Watanabe-Fukunaga *et al.*, 1992b).

Description of protein

Human Fas is comprised of 319 amino acids with a signal sequence of 15 amino acids at the N-terminus (Itoh *et al.*, 1991; Oehm *et al.*, 1992). A single transmembrane domain of 17 amino acids divides the molecule into the extracellular region of 157 amino acids, and the cytoplasmic region of 145 amino acids. The extracellular region can be further divided into three cysteine-rich subregions.

Relevant homologies and species differences

The extracellular region of Fas has a homology (25–30% homology) with that of other TNF/NGF receptor family members (Itoh *et al.*, 1991; Oehm *et al.*, 1992; Smith *et al.*, 1994). The cytoplasmic region of Fas contains a domain of about 80 amino acids that shows a homology with the corresponding region of the TNF type I receptor (Itoh and Nagata, 1993; Tartaglia *et al.*, 1993) and the death receptors DR3 and DR4 (Ashkenazi and Dixit, 1998). The domain is called the 'death domain'. The amino acid sequence of mouse Fas has an identity of 49.3% with human Fas (Watanabe-Fukunaga *et al.*, 1992b). There is no species-specificity among human, mouse, and rat (Takahashi *et al.*, 1994).

Affinity for ligand(s)

Fas ligand binds to Fas with a K_d of about 1.0 nM.

Cell types and tissues expressing the receptor

Fas is ubiquitously expressed in various tissues and cells (Watanabe-Fukunaga *et al.*, 1992b; Leithäuser *et al.*, 1993). In particular, Fas is expressed in the thymus, activated T cells, hepatocytes, and heart. Lymphomas of T cell and B cell origin express a high level of Fas, constitutively (Falk *et al.*, 1992; Debatin *et al.*, 1994).

Regulation of receptor expression

Expression of Fas in mature T cells is upregulated by activation with phorbol myristate acetate and ionomycin. A cis-regulatory element of NFκB at positions −295 to −286 in the 5′ flanking region of human Fas chromosomal gene is responsible for the induced expression of Fas in T cells (Chan *et al.*, 1999). Oncogene p53 upregulates the expression of Fas (Owen-Schaub *et al.*, 1995). p53-responsive elements in the intron 1 and 5′ promoter flanking region of human Fas gene seem to be responsible for the p53-dependent expression of Fas (Muller *et al.*, 1999).

Release of soluble receptors

The soluble Fas can be produced by alternative splicing (Cascino *et al.*, 1995; Hughes and Crispe, 1995; Papoff *et al.*, 1996). Patients with nonhematopoietic malignancies or ATL (*adult T cell leukemia*) have a high level of soluble Fas in the serum (Sugawara *et al.*, 1997). There is a soluble decoy receptor (DcR3) to which FasL binds with an affinity similar to that with which it binds the authentic Fas receptor (Pitti *et al.*, 1999). The DcR3 gene is amplified in about half of human *lung cancer* and *colon cancer* cells.

SIGNAL TRANSDUCTION

Associated or intrinsic kinases

The cytoplasmic region of Fas and its signaling molecule, FADD, is constitutively phosphorylated (Kennedy and Budd, 1998). Several kinases including RIP (receptor-interacting protein) and p59fyn were reported to interact with the cytoplasmic region of the Fas receptor (Stanger *et al.*, 1995; Atkinson *et al.*, 1996). However, the cells deficient in RIP kinase are potent to transduce the apoptotic signal, suggesting that RIP is not essential for the Fas-mediated apoptotic signal transduction (Ting *et al.*, 1996).

Cytoplasmic signaling cascades

A cascade of caspases (cysteine proteases) is activated by Fas engagement (Enari *et al.*, 1996; Nagata, 1997; Tewari and Dixit, 1995) (**Figure 2**). Binding of Fas ligand or agonistic anti-Fas antibody to Fas activates Fas. Fas ligand is a trimer, and the agonistic anti-Fas antibodies are IgM (immunoglobulin pentamer), or IgG3 that tends to aggregate. The $F(ab')_2$ fragment of anti-Apo1 antibody or other isotypes does not activate Fas (Dhein *et al.*, 1992), suggesting that Fas must be trimerized or aggregated to the higher order to transduce the signal (Takahashi *et al.*, 1996). Trimerization of Fas by FasL or agonistic anti-Fas or Apo1 antibody causes recruitment of caspase 8 to Fas through an adapter called FADD (Boldin *et al.*, 1995, 1996; Chinnaiyan *et al.*, 1995; Muzio *et al.*, 1996). This activates caspase 8, which then sequentially activates other caspases such as caspases 3, 6, and 7 in the downstream of the caspase cascade (Hirata *et al.*, 1998; Kawahara *et al.*, 1998). Caspases thus activated would cleave various cellular substrates to progress the apoptotic program (Martin and Green, 1995; Nagata, 1997). One of the caspase substrates is ICAD (inhibitor of caspase-activated DNase)/DFF-45 (DNA fragmentation factor 45) (Liu *et al.*, 1997; Sakahira *et al.*, 1998). The cleavage of ICAD by caspase 3 leads to the activation of a DNase (CAD), which causes the DNA fragmentation seen in most apoptotic cells (Enari *et al.*, 1998). This signal transduction cascade was confirmed by knocking-out the respective signaling molecules. That is, the cells lacking FADD or caspase 8 do not undergo apoptosis induced by Fas activation (Varfolomeev *et al.*, 1998; Yeh *et al.*, 1998; Zhang *et al.*, 1998a). The cells lacking caspase 3 or ICAD do not show DNA fragmentation induced by Fas activation (Woo *et al.*, 1998; Zhang *et al.*, 1998b).

Figure 2 Fas-induced apoptosis. Binding of Fas ligand to Fas induces trimerization of Fas, which recruits pro-caspases 8 via an adapter called FADD/MORT1, and induces its processing to the mature form. In one signaling pathway, caspase 8 directly activates caspase 3 downstream of the caspase cascade. In another pathway, caspase 8 cleaves Bid, a proapoptotic member of the Bcl-2 family. The cleaved Bid then enters mitochondria to cause release of cytochrome C, which activates caspase 9 together with Apaf-1. The caspase 9 then activates caspase 3. Caspase 3, thus activated, cleaves various cellular substrates to induce morphological changes of cells and nuclei. One of the caspase 3 substrates is ICAD/DFF-45, which is complexed with CAD, a specific DNase. Cleavage of ICAD by caspase 3 releases CAD from ICAD, and CAD then enters nuclei to cause DNA fragmentation.

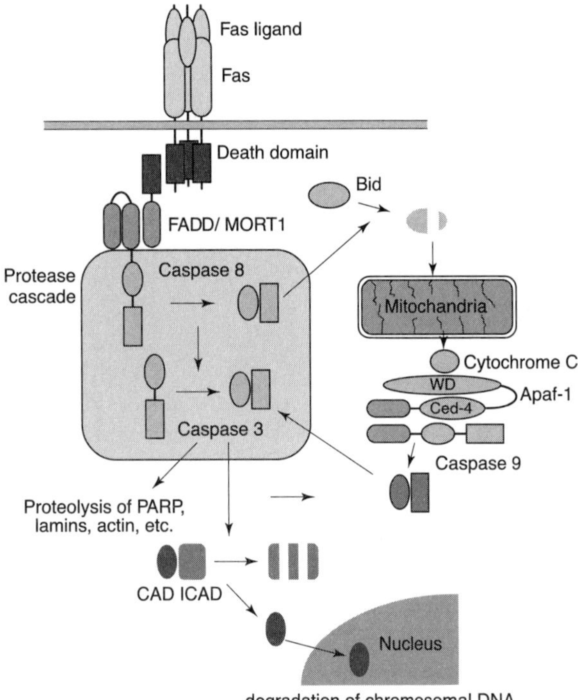

Many other signaling cascades initiated by the activated Fas leading to apoptosis have been proposed. For example, ceramide was postulated to mediate the Fas signal by activating the Ras/MAP kinase pathway (Cifone *et al.*, 1994; Gulbin *et al.*, 1995; Tepper *et al.*, 1995). However, recent analysis suggests that ceramide is not activated during Fas-mediated apoptosis, and ceramide is probably not involved in this signaling pathway (Hofmann and Dixit, 1998; Hsu *et al.*, 1998; Watts *et al.*, 1997). Other pathways proposed involving DAXX, JNK, or Bid (Brenner *et al.*, 1997; Yang *et al.*, 1997; Li *et al.*, 1998; Luo *et al.*, 1998) also need to be confirmed.

DOWNSTREAM GENE ACTIVATION

Transcription factors activated

Transcription factors are not activated by Fas engagement, although NFκB was reported to be activated in some cell lines (Rensing *et al.*, 1995).

Genes induced

The Fas activation quickly leads to cleavage of chromosomal DNA (Itoh *et al.*, 1991), and does not induce gene expression.

BIOLOGICAL CONSEQUENCES OF ACTIVATING OR INHIBITING RECEPTOR AND PATHOPHYSIOLOGY

Unique biological effects of activating the receptors

Activation of Fas causes apoptotic cell death in most cases (Nagata, 1997). It may also cause necrotic cell death in certain conditions (Vercammen *et al.*, 1998).

Phenotypes of receptor knockouts and receptor overexpression mice

Mouse mutation *lpr* (lymphoproliferation) is a loss-of-function mutation of Fas (Watanabe-Fukunaga *et al.*, 1992a) (**Figure 3**). In one allele (*lpr*), an early transposable element (ETn) is inserted in intron 2 of the Fas chromosomal gene (Adachi *et al.*, 1993). The Fas transcript prematurely terminates in intron 2. However, the inhibition of expression is not complete, as demonstrated by the presence of full-length Fas mRNA, albeit at a low level, suggesting that *lpr* is a leaky mutation. In the other allele (lpr^{cg}), a point mutation that causes replacement of an amino acid from isoleucine to asparagine is introduced in the death domain of the Fas cytoplasmic region, which abolishes the ability of Fas to transduce the apoptotic signal (Watanabe-Fukunaga *et al.*, 1992a). Mice carrying the *lpr* mutation develop *lymphadenopathy* and *splenomegaly*, produce autoantibodies, and suffer

Figure 3 Mutations of Fas in *lpr*-mice and in human patients with Canale–Smith syndrome. (a) A mutation in the Fas gene in *lpr* mice. The structure of mouse Fas chromosomal gene is schematically shown. In *lpr* mice an early transposable element (ETn) carrying poly(A) addition signal on a long terminal repeat (LTR) is inserted in intron 2 of the gene. This causes premature termination of Fas transcript. (b) Point mutations in the Fas death domain in human Canale–Smith syndrome and lpr^{cg} mice. The amino acid sequences of the death domain of human and mouse Fas are aligned. The amino acid replacements in patients of Canale–Smith syndrome are indicated on the upper line, while the mutation in lpr^{cg} mice is shown in the lower line.

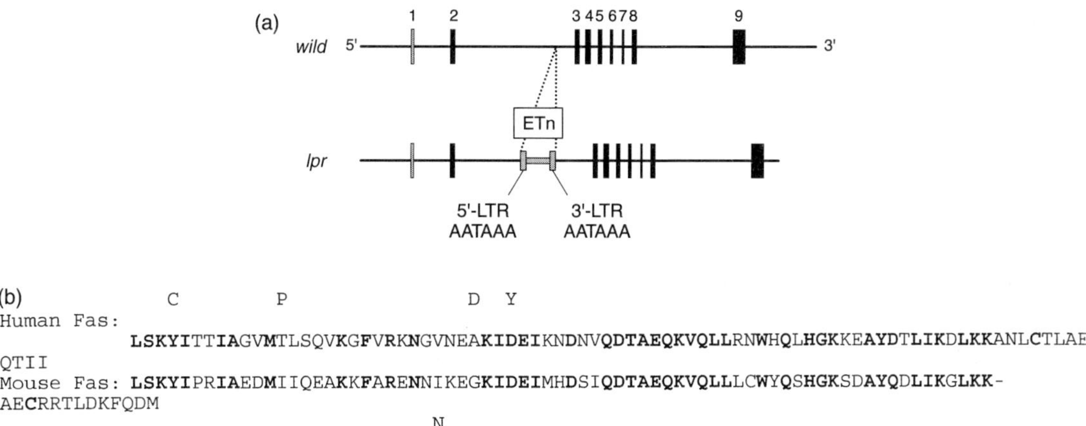

from *autoimmune diseases* such as *nephritis* and *arthritis* (Cohen and Eisenberg, 1991). Fas knockout mice were also established, and the mice shows more accelerated and pronounced lymphadenopathy and splenomegaly than *lpr* mice (Adachi *et al.*, 1995).

Human abnormalities

Human patients with *Canale–Smith syndrome* or *autoimmune lymphoproliferative syndrome* show phenotypes (lymphadenopathy, splenomegaly, and autoimmune diseases) similar to those in *lpr* mice, and carry loss-of-function mutations in the Fas gene (Fisher *et al.*, 1995; Rieux-Laucat *et al.*, 1995; Nagata, 1998). Mutations in the Fas death domain in this syndrome are shown in Figure 3b.

THERAPEUTIC UTILITY

Effect of treatment with soluble receptor domain

Administration of soluble Fas (the extracellular region of Fas fused to the Fc region of human immunoglobulin) blocks development of CTL-induced *hepatitis* in a mouse model system (Kondo *et al.*, 1997).

References

Adachi, M., Watanabe-Fukunaga, R., and Nagata, S. (1993). Aberrant transcription caused by the insertion of an early transposable element in an intron of the Fas antigen gene of *lpr* mice. *Proc. Natl Acad. Sci. USA* **90**, 1756–1760.

Adachi, M., Suematsu, S., Kondo, T., Ogasawara, J., Tanaka, T., Yoshida, N., and Nagata, S. (1995). Targeted mutation in the Fas gene causes hyperplasia in the peripheral lymphoid organs and liver. *Nature Genet.* **11**, 294–300.

Ashkenazi, A., and Dixit, V. M. (1998). Death receptors: signaling and modulation. *Science* **281**, 1305–1308.

Atkinson, E., Ostergaard, H., Kane, K., Pinkoski, M., Caputo, A., Olszowy, M., and Bleackley, R. (1996). A physical interaction between the cell death protein Fas and the tyrosine kinase $p59^{fynT}$. *J. Biol. Chem.* **271**, 5968–5971.

Boldin, M. P., Varfolomeev, E. E., Pancer, Z., Mett, I. L., Camonis, J. H., and Wallach, D. (1995). A novel protein that interacts with the death domain of Fas/APO1 contains a sequence motif related to the death domain. *J. Biol. Chem.* **270**, 7795–7798.

Boldin, M. P., Goncharov, T. M., Goltsev, Y. V., and Wallach, D. (1996). Involvement of MACH, a novel MORT1/FADD-interacting protease, in Fas/APO-1- and TNF receptor-induced cell death. *Cell* **85**, 803–815.

Brenner, B., Koppenhoefer, U., Weinstock, C., Linderkamp, O., Lang, F., and Gulbins, E. (1997). Fas- or ceramide-induced apoptosis is mediated by a Rac1-regulated activation of Jun N-terminal kinase/p38 kinases and GADD153. *J. Biol. Chem.* **272**, 22173–22181.

Cascino, I., Fiucci, G., Papoff, G., and Ruberti, G. (1995). Three functional soluble forms of the human apoptosis-inducing Fas molecule are produced by alternative splicing. *J. Immunol.* **154**, 2706–2713.

Chan, H., Bartos, D. P., and Owen-Schaub, L. B. (1999). Activation-dependent transcriptional regulation of the human fas promoter requires NF-kappaB p50–p65 recruitment. *Mol. Cell. Biol.* **19**, 2098–2108.

Chinnaiyan, A. M., O'Rourke, K., Tewari, M., and Dixit, V. M. (1995). FADD, a novel death domain-containing protein, interacts with the death domain of Fas and initiates apoptosis. *Cell* **81**, 505–512.

Cifone, M. G., De, M. R., Roncaioli, P., Rippo, M. R., Azuma, M., Lanier, L. L., Santoni, A., and Testi, R. (1994). Apoptotic signaling through CD95 (Fas/Apo-1) activates an acidic sphingomyelinase. *J. Exp. Med.* **180**, 1547–1552.

Cohen, P. L., and Eisenberg, R. A. (1991). *Lpr* and *gld*: single gene models of systemic autoimmunity and lymphoproliferative disease. *Annu. Rev. Immunol.* **9**, 243–269.

Debatin, K.-M., Fahrig-Faissner, A., Enenkel-Stoodt, S., Kreuz, W., Benner, A., and Krammer, P. H. (1994). High expression of APO-1 (CD95) on T lymphocytes from human immunodeficiency virus-1-infected children. *Blood* **83**, 3101–3103.

Dhein, J., Daniel, P. T., Trauth, B. C., Oehm, A., Möller, P., and Krammer, P. H. (1992). Induction of apoptosis by monoclonal antibody anti-APO-1 class switch variants is dependent on cross-linking of APO-1 cell surface antigens. *J. Immunol.* **149**, 3166–3173.

Enari, M., Talanian, R. V., Wong, W. W., and Nagata, S. (1996). Sequential activation of ICE-like and CPP32-like proteases during Fas-mediated apoptosis. *Nature* **380**, 723–726.

Enari, M., Sakahira, H., Yokoyama, H., Okawa, K., Iwamatsu, A., and Nagata, S. (1998). A caspase-activated DNase that degrades DNA during apoptosis and its inhibitor ICAD. *Nature* **391**, 43–50.

Falk, M. H., Trauth, B. C., Debatin, K.-M., Klas, C., Gregory, C. D., Rickinson, A. B., Calender, A., Lenoir, G. M., Ellwart, J. W., Krammer, P. H., and Bornkamm, G. W. (1992). Expression of the APO-1 antigen in Burkitt lymphoma cell lines correlates with a shift towards a lymphoblastoid phenotype. *Blood* **79**, 3300–3306.

Fisher, G. H., Rosenberg, F. J., Straus, S. E., Dale, J. K., Middelton, L. A., Lin, A. Y., Strober, W., Lenardo, M. J., and Puck, J. M. (1995). Dominant interfering Fas gene mutations impair apoptosis in a human autoimmune lymphoproliferative syndrome. *Cell* **81**, 935–946.

Gulbin, E., Bissonnette, R., Mahboubi, A., Martin, S., Nishioka, W., Brunner, T., Baier, G., Baier-Bitterkich, G., Byrd, C., Lang, F., Kolesnick, R., Altman, A., and Green, D. (1995). Fas-induced apoptosis is mediated via a ceramide-initiated ras signaling pathway. *Immunity* **2**, 341–351.

Hirata, H., Takahashi, A., Kobayashi, S., Yonehara, S., Sawai, H., Okazaki, T., Yamamoto, K., and Sasada, M. (1998). Caspases are activated in a branched protease cascade and control distinct downstream processes in Fas-induced apoptosis. *J. Exp. Med.* **187**, 587–600.

Hofmann, K., and Dixit, V. M. (1998). Ceramide in apoptosis-does it really matter? *Trends Biochem. Sci.* **23**, 374–377.

Hsu, S.-C., Wu, C.-C., Luh, T.-Y., Chou, C.-K., Han, S.-H., and Lai, M.-Z. (1998). Apoptosis signal of Fas is not mediated by ceramide. *Blood* **91**, 2658–2663.

Hughes, D. P., and Crispe, I. N. (1995). A naturally occurring soluble isoform of murine Fas generated by alternative splicing. *J. Exp. Med.* **182**, 1395–1401.

Inazawa, J., Itoh, N., Abe, T., and Nagata, S. (1992). Assignment of the human Fas antigen gene (FAS) to 10q24.1. *Genomics* **14**, 821–822.

Itoh, N., and Nagata, S. (1993). A novel protein domain required for apoptosis: mutational analysis of human Fas antigen. *J. Biol. Chem.* **268**, 10932–10937.

Itoh, N., Yonehara, S., Ishii, A., Yonehara, M., Mizushima, S., Sameshima, M., Hase, A., Seto, Y., and Nagata, S. (1991). The polypeptide encoded by the cDNA for human cell surface antigen Fas can mediate apoptosis. *Cell* **66**, 233–243.

Kawahara, A., Enari, M., Talanian, R. V., Wong, W. W., and Nagata, S. (1998). Fas-induced DNA fragmentation and proteolysis of nuclear proteins. *Genes Cells* **3**, 297–306.

Kennedy, N. J., and Budd, R. C. (1998). Phosphorylation of FADD/MORT1 and Fas by kinases that associate with the membrane-proximal cytoplasmic domain of Fas. *J. Immunol.* **160**, 4881–4888.

Kondo, T., Suda, T., Fukuyama, H., Adachi, M., and Nagata, S. (1997). Essential roles of the Fas ligand in the development of hepatitis. *Nature Med.* **3**, 409–413.

Leithäuser, F., Dhein, J., Mechtersheimer, G., Koretz, K., Brüderlein, S., Henne, C., Schmidt, A., Debatin, K.-M., Krammer, P. H., and Möller, P. (1993). Constitutive and induced expression of APO-1, a new member of the nerve growth factor/tumor necrosis factor receptor superfamily, in normal and neoplastic cells. *Lab. Invest.* **69**, 415–429.

Li, H., Zhu, H., Xu, C. J., and Yuan, J. (1998). Cleavage of BID by caspase 8 mediates the mitochondrial damage in the Fas pathway of apoptosis. *Cell* **94**, 491–501.

Liu, X., Zou, H., Slaughter, C., and Wang, X. (1997). DFF, a heterodimeric protein that functions downstream of caspase-3 to trigger DNA fragmentation during apoptosis. *Cell* **89**, 175–184.

Luo, X., Budihardjo, I., Zou, H., Slaughter, C., and Wang, X. (1998). Bid, a Bcl2 interacting protein, mediates cytochrome c release from mitochondria in response to activation of cell surface death receptors. *Cell* **94**, 481–490.

Martin, S., and Green, D. (1995). Protease activation during apoptosis: death by a thousand cuts. *Cell* **82**, 349–352.

Muller, M., Wilder, S., Bannasch, D., Israeli, D., Lehlbach, K., Li-Weber, M., Friedman, S. L., Galle, P. R., Stremmel, W., Oren, M., and Krammer, P. H. (1999). p53 activates the CD95 (APO-1/Fas) gene in response to DNA damage by anticancer drugs. *J. Exp. Med.* **188**, 2033–2045.

Muzio, M., Chinnaiyan, A. M., Kischkel, F. C., O'Rourke, K., Shevchenko, A., Ni, J., Scaffidi, C., Bretz, J. D., Zhang, M., Gentz, R., Mann, M., Krammer, P. H., Peter, M. E., and Dixit, V. M. (1996). FLICE, a novel FADD-homologous ICE/CED-3-like protease, is recruited to the CD95 (Fas/APO-1) death-inducing signaling complex. *Cell* **85**, 817–827.

Nagata, S. (1997). Apoptosis by death factor. *Cell* **88**, 355–365.

Nagata, S. (1998). Human autoimmune lymphoproliferative syndrome, a defect in the apoptosis-inducing Fas receptor: A lesson from the mouse model. *J. Hum. Genet.* **43**, 2–8.

Nagata, S., and Golstein, P. (1995). The Fas death factor. *Science* **267**, 1449–1456.

Oehm, A., Behrmann, I., Falk, W., Pawlita, M., Maier, G., Klas, C., Li-Weber, M., Richards, S., Dhein, J., Trauth, B. C., Ponstingl, H., and Krammer, P. H. (1992). Purification and molecular cloning of the APO-1 cell surface antigen, a member of the tumor necrosis factor/nerve growth factor receptor superfamily: sequence identity with the Fas antigen. *J. Biol. Chem.* **267**, 10709–10715.

Ogasawara, J., Watanabe-Fukunaga, R., Adachi, M., Matsuzawa, A., Kasugai, T., Kitamura, Y., Itoh, N., Suda, T., and Nagata, S. (1993). Lethal effect of the anti-Fas antibody in mice. *Nature* **364**, 806–809.

Owen-Schaub, L. B., Zhang, W., Cusack, J. C., Angelo, L. S., Santee, S. M., Fujiwara, T., Roth, J. A., Deisseroth, A. B., Zhang, W. W., and Kruzel, E. (1995). Wild-type human p53 and a temperature-sensitive mutant induce Fas/APO-1 expression. *Mol. Cell. Biol.* **15**, 3032–3040.

Papoff, G., Cascino, I., Eramo, A., Starace, G., Lynch, D. H., and Ruberti, G. (1996). An N-terminal domain shared by Fas/Apo-1 (CD95) soluble variants prevents cell death *in vitro*. *J. Immunol.* **156**, 4622–4630.

Pitti, R. M., Marsters, S. A., Lawrence, D. A., Roy, M., Kishkel, F. C., Dowd, P., Huang, A., Donahue, C. J., Sherwood, S. W., Baldwin, D. T., Godowski, P. J., Wood, W. I., Gurney, A. L., Hillan, K. J., Cohen, R. L., Goddard, A. D., Botstein, D., and Ashkenazi, A. (1999). Genomic amplification of a decoy receptor for Fas ligand in lung and colon cancer. *Nature* **396**, 699–703.

Rensing, E. A., Hess, S., Ziegler, H. H., Riethmuller, G., and Engelmann, H. (1995). Fas/Apo-1 activates nuclear factor kappa B and induces interleukin-6 production. *J. Inflamm.* **45**, 161–174.

Rieux-Laucat, F., Le Deist, F., Hivroz, C., Roberts, I. A. G., Denatin, K. M., Fischer, A., and de Villarty, J. P. (1995). Mutations in Fas associated with human lymphoproliferative syndrome and autoimmunity. *Science* **268**, 1347–1349.

Sakahira, H., Enari, M., and Nagata, S. (1998). Cleavage of CAD inhibitor in CAD activation and DNA degradation during apoptosis. *Nature* **391**, 96–99.

Smith, C. A., Farrah, T., and Goodwin, R. G. (1994). The TNF receptor superfamily of cellular and viral proteins: activation, costimulation, and death. *Cell* **76**, 959–962.

Stanger, B. Z., Leder, P., Lee, T.-H., Kim, E., and Seed, B. (1995). RIP: A novel protein containing a death domain that interacts with FAS/APO-1 (CD95) in yeast and causes cell death. *Cell* **81**, 513–523.

Sugawara, K., Yamada, Y., Hiragata, Y., Matsuo, Y., Tsuruda, K., Tomonaga, M., Maeda, T., Atogami, S., Tsukasaki, K., and Kamihira, S. (1997). Soluble and membrane isoforms of Fas/CD95 in fresh adult T-cell leukemia (ATL) cells and ATL-cell lines. *Int. J. Cancer* **72**, 128–132.

Takahashi, T., Tanaka, M., Inazawa, J., Abe, T., Suda, T., and Nagata, S. (1994). Human Fas ligand; gene structure, chromosomal location and species specificity. *Int. Immunol.* **6**, 1567–1574.

Takahashi, T., Tanaka, M., Ogasawara, J., Suda, T., Murakami, H., and Nagata, S. (1996). Swapping between Fas and G-CSF receptor. *J. Biol. Chem.* **271**, 17555–17560.

Tartaglia, L. A., Ayres, T. M., Wong, G. H. W., and Goeddel, D. V. (1993). A novel domain within the 55 kd TNF receptor signals cell death. *Cell* **74**, 845–853.

Tepper, C. G., Jayadev, S., Liu, B., Bielawska, A., Wolff, R., Yonehara, S., Hannun, Y. A., and Seldin, M. F. (1995). Role for ceramide as an endogenous mediator of Fas-induced cytotoxicity. *Proc. Natl Acad. Sci. USA* **92**, 8443–8447.

Tewari, M., and Dixit, V. M. (1995). Fas- and tumor necrosis factor-induced apoptosis is inhibited by the poxvirus crmA gene product. *J. Biol. Chem.* **270**, 3255–3260.

Ting, A. T., Pimentel, M. F., and Seed, B. (1996). RIP mediates tumor necrosis factor receptor 1 activation of NF-kappaB but not Fas/APO-1-initiated apoptosis. *EMBO J.* **15**, 6189–6196.

Trauth, B. C., Klas, C., Peters, A. M. J., Matzuku, S., Möller, P., Falk, W., Debatin, K.-M., and Krammer, P. H. (1989). Monoclonal antibody-mediated tumor regression by induction of apoptosis. *Science* **245**, 301–305.

Varfolomeev, E. E., Schuchmann, M., Luria, V., Chiannikulchai, N., Beckmann, J. S., Mett, I. L., Rebrikov, D., Brodianski, V. M., Kemper, O. C., Kollet, O., Lapidot, T., Soffer, D., Sobe, T., Avraham, K. B., Goncharov, T., Holtmann, H., Lonai, P., and Wallach, D. (1998). Target disruption of the mouse caspase 8 gene abalates cell death induction by the TNF receptors, Fas/Apo1, and DR3 and is lethal prenatally. *Immunity* **9**, 267–276.

Vercammen, D., Brouckaert, G., Denecker, G., Van de Craen, M., Declercq, W., Fiers, W., and Vandenabeele, P. (1998). Dual signaling of the Fas receptor: initiation of both apoptotic and necrotic cell death pathways. *J. Exp. Med.* **188**, 919–930.

Watanabe-Fukunaga, R., Brannan, C. I., Copeland, N. G., Jenkins, N. A., and Nagata, S. (1992a). Lymphoproliferation disorder in mice explained by defects in Fas antigen that mediates apoptosis. *Nature* **356**, 314–317.

Watanabe-Fukunaga, R., Brannan, C. I., Itoh, N., Yonehara, S., Copeland, N. G., Jenkins, N. A., and Nagata, S. (1992b). The cDNA structure, expression, and chromosomal assignment of the mouse Fas antigen. *J. Immunol.* **148**, 1274–1279.

Watts, J. D., Gu, M., Polverino, A. J., Patterson, S. D., and Aebersold, R. (1997). Fas-induced apoptosis of T cells occurs independently of ceramide generation. *Proc. Natl Acad. Sci. USA* **94**, 7292–7296.

Woo, M., Hakem, R., Soengas, M. S., Duncan, G. S., Shahinian, A., Kagi, D., Hakem, A., McCrrach, M., Khoo, W., Kaufman, S. A., Senaldi, G., Howard, T., Lowe, S. W., and Mak, T. W. (1998). Essential contribution of caspase 3/CPP32 to apoptosis and its associated nuclear changes. *Genes Dev.* **12**, 806–819.

Yang, X., Khosravi-Far, R., Chang, H. Y., and Baltimore, D. (1997). Daxx, a novel Fas-binding protein that activates JNK and apoptosis. *Cell* **89**, 1067–1076.

Yeh, W.-C., Pompa, J. L. D. L., McCurrach, M. E., Shu, H.-B., Elia, A. J., Shahinian, A., Ng, M., Wakeham, A., Khoo, W., Mitchell, K., El-Deiry, W. S., Lowe, S. W., Goeddel, D. V., and Mak, T. W. (1998). FADD: essential for embryo development and signaling from some, but not all, inducers of apoptosis. *Science* **279**, 1954–1958.

Yonehara, S., Ishii, A., and Yonehara, M. (1989). A cell-killing monoclonal antibody (anti-Fas) to a cell surface antigen co-downregulated with the receptor of tumor necrosis factor. *J. Exp. Med.* **169**, 1747–1756.

Zhang, J., Cado, D., Chen, A., Kabra, N. H., and Winoto, A. (1998a). Fas-mediated apoptosis and activation-induced T-cell proliferation are defective in mice lacking FADD/Mort1. *Nature* **392**, 296–300.

Zhang, J., Liu, X., Scherer, D. C., Kaer, L. v., Wang, X., and Xu, M. (1998b). Resistance to DNA fragmentation and chromatin condensation in mice lacking the DNA fragmentation factor 45. *Proc. Natl Acad. Sci. USA* **95**, 12480–12485.

LICENSED PRODUCTS

Monoclonal antibodies (clones CH-11, UB-2, and ZB-4) against human Fas are available from Medical & Biological Laboratories (MBL), Nagoya, Japan. The CH-11 antibody works as an agonist (Yonehara *et al.*, 1989), while the ZB-4 antibody works as an antagonist.

Monoclonal antibody against mouse Fas (clone Jo2) (Ogasawara *et al.*, 1993) is available form PharMingen (San Diego, CA, USA). The Jo2 antibody works as an agonist.

CD40

Edward A. Clark*

Primate Center, University of Washington, Box 357330, Seattle, WA 98195-0001, USA

*corresponding author tel: 206-543-8706, fax: 206-685-0305, e-mail: eclark@bart.rprc.washington.edu

DOI: 10.1006/rwcy.2000.16005.

SUMMARY

CD40 is a surface receptor within the TNF receptor family expressed on B lymphocytes, dendritic cells, activated monocytes, endothelial cells, and epithelial cells. The ligand for CD40, CD40L, is expressed on activated T cells, and CD40L+ T cells provide the classic 'T cell help' by inducing via CD40 B cell proliferation and differentiation. The CD40L/CD40 signaling pathway also plays an important role in regulating T cell activation by antigen-presenting cells. Engaging CD40 on antigen-presenting cells such as dendritic cells induces them to express more costimulatory receptors such as CD80 and CD86, to produce cytokines such as IL-12, and to effectively stimulate CD8 effector T cells. Thus, the CD40 receptor plays a key role in the development of T cell-dependent responses to foreign antigens and pathogens.

BACKGROUND

The MedLine database circa April 1999 listed over 1800 references where CD40 is mentioned, with over 380 citations for CD40 in 1998 alone. Thus, it is impossible to cover all aspects of CD40 in a short chapter. In this chapter, as well as briefly summarizing key features about CD40, I describe how various cytokines interact with or are regulated by the CD40 pathway. The reader can readily access the chapter for a particular cytokine with which CD40 interacts to further explore the relationship. In this way, it should be possible to begin to develop a sense of the network of relationships which CD40 has with other signaling elements.

Discovery

CD40 was defined in the Third International CD Workshop using mAb G28-5 and cloned using this mAb (Stamenkovic *et al.*, 1989).

Initial studies of CD40 function were principally done with B lymphocytes. Ligation of CD40 on B cells has a number of consequences depending on the stage of the B cell. Ligating CD40 on B cells can induce B cell proliferation, prevent B cell antigen receptor (BCR)-induced cell death, promote isotype class switching or trigger a program that makes the B cell more susceptible to CD95/Fas-mediated cell death. CD40 receptor engagement in general must integrate with signals delivered via the BCR to regulate B cell responses. It is important to recognize that the combination of signal transduction pathways activated via CD40 may be unique at different stages of B cell differentiation, since specific signaling molecules depend, for example, on which sets of kinases, phosphatases, G proteins, adapters, inhibitors are being expressed. A range of CD40-regulated effects are described below in the individual cytokine/chemokine sections.

Alternative names

CD40 is also called CDw40 and Bp50.

Main activities and pathophysiological roles

The major known functions of CD40 are to promote B lymphocyte survival, proliferation and differentiation, and to promote inflammatory responses in other antigen-presenting cells. There are a number of reviews on CD40 which provide more detail and a more linear historical background (e.g. Clark, 1990; Clark and Ledbetter, 1994; Banchereau *et al.*, 1994; van Kooten and Banchereau, 1997). For discussion of the CD40 signaling pathway see Kehry (1996) and Craxton *et al.* (1999). The chapter on CD40L emphasizes the expression, regulation, modulation, and clinical studies with this molecule.

CD40 and IL-2

IL-2 plays a key role in the regulation of CD40L. In general, IL-2 and CD40L, unlike IL-4 and CD40L, do not work well alone to stimulate B cells. However, once CD40-activated B cells are stimulated with IL-10, they express high-affinity IL-2 receptors and then respond to IL-2 and both proliferate and secrete IgM and IgG (Fluckiger *et al.*, 1993). This suggests that IL-10 may make CD40-activated B cells more responsive to T cell-dependent B cell maturation signals such as IL-2. CD40 ligation can also upregulate the expression of IL-2 receptors on B cells (Grabstein *et al.*, 1993), so there may be some circumstances when IL-10 is not required for IL-2 and CD40L to act together (e.g. Armitage *et al.*, 1993), particularly in the presence of follicular dendritic cells (FDCs) (Grouard *et al.*, 1995). Memory B cell in particular may be induced to differentiate into plasma cells following exposure to IL-2 plus CD40L (Arpin *et al.*, 1997).

Dendritic cells, which can express CD40L (Pinchuk *et al.*, 1996), have recently been found to promote CD40-dependent B cell differentiation with IL-2 (Fayette *et al.*, 1998). It has been reported that CD40L can stimulate T cells to produce IL-2 (Fanslow *et al.*, 1994). However, it is not clear if this is due to a direct effect on T cells or via activation of small numbers of CD40+ DCs in the cultures.

CD40 and IL-4

IL-4 and CD40L are major partners during immune responses and in particular work together to promote B cell proliferation and differentiation. More than 450 studies have involved IL-4 and CD40, so I will just mention a few here. Early on it was recognized that IL-4 upregulates CD40 expression on B cells and along with CD40 ligation stimulates proliferation of activated B cells (e.g. Gordon *et al.*, 1987; Clark *et al.*, 1989). In B lymphocytes, CD40 ligation and IL-4 act together to induce isotype class switching, particularly to IgE (Shapira *et al.*, 1992; Kimata *et al.*, 1992; Fujita *et al.*, 1995; Warren and Berton, 1995; Hasbold *et al.*, 1998). CD40 plus IL-4-induced isotype class switch can be blocked by signaling B cells through CD30 (Cerutti *et al.*, 1998), while CD40 plus IL-4-induced IgE production is augmented by CD27L (Nagumo *et al.*, 1998).

The DNA-dependent protein kinase, the catalytic subunit associated with the Ku70/Ku86 heterodimer, may play a role in switch recombination. Resting B cells express relatively little Ku, but the combination of CD40 ligation and IL-4, but not either stimulus alone, strongly upregulates Ku expression (Zelazowski *et al.*, 1997).

Underscoring this coordinated signaling activity of IL-4 and CD40L is the finding that both IL-4 signaling and CD40 ligation are required for the activation of NF-AT in B cells, a key transcription factor required for lymphocyte proliferation (Choi *et al.*, 1994). Similarly, although CD40 ligation alone can activate NFκB in human B cells (Berberich *et al.*, 1994), the combination of CD40 ligation and IL-4 induces a very strong activation (Jeppson *et al.*, 1998). The CD40 and IL-4 signaling pathways induce a number of other phenotypes together, including changing B cell morphology (Davey *et al.*, 1998). Follicular mantle (FM) B cells in particular can be induced to survive and proliferate in response to IL-4 and CD40 ligation (Holder *et al.*, 1991). Germinal center (GC) B cells can also be induced to become IgG1-producing antibody-forming cells by IL-4 plus CD40 mAb (e.g. Baba *et al.*, 1997).

Other noteworthy findings are as follows. (a) IL-4 along with IL-10 can block the ability of CD40 ligation to induce the expression of inflammatory cytokines in monocytes (see Suttles *et al.*, 1999). (b) By contrast, in renal epithelial cells, IL-4 has the opposite effect and upregulates CD40-induced expression of RANTES (Deckers *et al.*, 1998). (c) Retinoic acid can inhibit the dual effects on CD40 ligation plus IL-4 on B cells (Worm *et al.*, 1998a). (d) Although IL-4 can block CD95-induced apoptosis of B cells, in *non-Hodgkin's lymphoma*, CD40 stimulation and IL-4 synergize to upregulate CD95 (Plumas *et al.*, 1998).

CD40 and IL-5

A number of studies have found that IL-5 is particularly effective at enhancing B cell maturation induced by the combination of CD40 and IL-4 signaling described above (Armitage *et al.*, 1993; Maliszewski *et al.*, 1993; Baba *et al.*, 1997), but unlike the CD40 signaling pathway, IL-5 signaling in B cells requires the Btk kinase (Baba *et al.*, 1997). B cell differentiation induced by CD40 ligation with IL-4 and IL-5 also requires the p50 NFκB subunit (Snapper *et al.*, 1996).

CD40 and IL-7

Not much has been published showing interplay between CD40 and IL-7. IL-7 can augment the costimulatory effect of IL-3 and CD40 ligation on early B lineage cells (Saeland *et al.*, 1993); the outcome of stimulation with IL-3/IL-7 with anti-CD40 may depend on the stage of differentiation of the B cell precursor since IL-3 plus IL-7-induced proliferation of pro-B cells is blocked via CD40 stimulation (Larson and LeBien, 1994). IL-7 does not seem to affect CD40 signaling in mature B cells (e.g. Jeannin *et al.*, 1998a) but may regulate a step in germinal center B cell development (Hikida *et al.*, 1998).

CD40 and IL-10

Ligating CD40 on B cells upregulates IL-10 protein and mRNA expression (Burdin *et al.*, 1995; Zan *et al.*, 1998; Aicher *et al.*, 1999), and this endogenously produced IL-10 is essential for CD40-induced B cell differentiation (Burdin *et al.*, 1995). CD40 crosslinking also induces the CD14+ subpopulation of DCs to make IL-10 but not a CD1a+ DC subset (de Saint Vis *et al.*, 1998; Aicher *et al.*, 1999). IL-10 can promote the ability of CD40 ligation to induce expression of IL-13 receptors (Billard *et al.*, 1997) and to make B cells class switch to IgG and produce IgG (Briere *et al.*, 1994; Nonoyama *et al.*, 1993; Malisan *et al.*, 1996). CD40-activated B cells can be induced to proliferate (Rousset *et al.*, 1992) or to become plasma cells by IL-10 (Rousset *et al.*, 1995), and together with IL-2, IL-10 induces human CD40-activated memory B cells to become plasma cells (Arpin *et al.*, 1997).

Some effects of IL-10 are countered by CD40 ligation: for example, IL-10 can block the ability of LPS to induce B cell proliferation but this inhibitory effect is overridden by CD40 crosslinking (Marcelletti, 1996). And the ability of IL-10 to downregulate IgE production by B cells is prevented by CD40 ligation (Jeannin *et al.*, 1998b). Conversely, IL-10 can mitigate certain effects induced via CD40 such as CD40-induced IL-12 production by DCs (Hino and Nariuchi, 1996; Kelsall *et al.*, 1996; Koch *et al.*, 1996); IL-10 treatment makes DCs less responsive to CD40 ligation so that they produce less IL-12 (Buelens *et al.*, 1997), and in TH1 T cell/dendritic cell cultures, IL-10, but not IL-4, inhibits the production of IL-12 (Ria *et al.*, 1998). IL-10 also blocks the ability of CD40L to block DC apoptosis (Ludewig *et al.*, 1995).

IL-10 plus IL-4 can also block CD40-induced production of inflammatory cytokines in monocytes (Suttles *et al.*, 1999) and, in contrast to CD40 ligation, potentiate the maturation of the DC1 subpopulation of DCs (Rissoan *et al.*, 1999). However, in general, IL-10 is more effective in blocking the inflammatory effects of LPS than of CD40 ligation (Buelens *et al.*, 1997).

CD40 and IL-12

The key relationship between CD40 and IL-12 is that ligating CD40 on antigen-presenting cells (APCs) induces expression of IL-12. T cells promote IL-12 production in monocytes via CD40L/CD40-dependent interaction (Shu *et al.*, 1995; Hino and Nariuchi, 1996; Kato *et al.*, 1996). Ligating CD40 on DCs also rapidly induces increases in IL-12 mRNA and protein and this requires a p38 MAPK-dependent pathway (Koch *et al.*, 1996; Cella *et al.*, 1996; Aicher *et al.*, 1999). This IL-12 appears to play a key role in DC-regulated differentiation of B cells. Whether or not CD40 can induce B cells to make IL-12 is controversial (see Aicher *et al.*, 1999); the varying results may depend on the stage of the B cell. The induction of IL-12 by DCs according to one study does not require IFNγ but does require T cell–DC contact and CD40L/CD40 signaling (Ria *et al.*, 1998). Another study (Snijders *et al.*, 1998) claims that two signals are required for IL-12 production: one via CD40, the other via IFNγ.

CD40-induced IL-12 production appears to play an important role in the resistance to certain infections requiring TH1 cell immunity. Both CD40L- and CD40-deficient mice, unlike wild-type littermates, are susceptible to infection by *Leishmania major* and make significantly less IL-12 than controls (Campbell *et al.*, 1996; Kamanaka *et al.*, 1996). Treatment with IL-12 can prevent disease progression. Consistent with these findings, actual blockade of CD40L/CD40 *in vivo* prevents IL-12 production and the generation of TH1 cells (Stuber *et al.*, 1996). CD40 is required for antigen-induced production of IL-12 but not IL-12 induced in response to bacteria (DeKruyff *et al.*, 1997; Maruo *et al.*, 1997). CD40-induced IL-12 may be so essential in resistance to certain infections that pathogens target this pathway: *measles* virus-infected DCs become defective in producing IL-12 (Fugier-Vivier *et al.*, 1997), suggesting a means by which the virus induces immunosuppression in its host.

CD40 and IL-13

IL-13 shares a common receptor and a number of properties with IL-4, so it is not surprising that like IL-4, IL-13 promotes CD40-induced B cell proliferation and maturation (Aversa *et al.*, 1993; Cocks *et al.*, 1993). In general, the effect of IL-13 on CD40-activated B cells is weaker than that of IL-4. IL-13 and CD40L also work together to block spontaneous B cell apoptosis (Lomo *et al.*, 1997). IL-13 can also promote CD40-induced production of RANTES in epithelial cells (Deckers *et al.*, 1998).

CD40 activation can upregulate IL-13 receptors (Billard *et al.*, 1997), which thus enhance the ability of IL-13 to affect CD40 signaling.

CD40 and IL-15 or IL-17

IL-15 is a monocyte-derived cytokine that shares several biologic activities with IL-2. It can upregulate CD40 expression on monocytes (Avice *et al.*, 1998). IL-17 can promote increased expression of CD40 on DCs (Antonysamy *et al.*, 1999), and may be costimulatory with CD40 ligation on epithelial cells (Fossiez *et al.*, 1998).

CD40 and TNFα

TNFα and CD40L are both members of the TNF receptor family. TNFα upregulates expression of CD40 on thymic epithelial cells (Galy and Spits, 1992) or airway smooth muscle cells (Lazaar *et al.*, 1998). Although TNFα cannot augment CD40-induced B cell proliferation, it can augment proliferation induced by the combination of CD40L and IL-4 (Armitage *et al.*, 1993). CD40 ligation of monocytes (Alderson *et al.*, 1993; Caux *et al.*, 1994), neonatal thymic $\gamma\delta$ T cells (Ramsdell *et al.*, 1994), basal epithelial cells (Peguet Navarro *et al.*, 1997), or B cells (Boussiotis *et al.*, 1994; Burdin *et al.*, 1995) can promote production of TNFα. One study suggests that the TNFα produced after ligating CD40 on B cells is required for B cell proliferation (Boussiotis *et al.*, 1994). The regulation of TNFα production via CD40 may contribute to inflammatory disease processes; for example, Sekine *et al.* (1998) found that CD40 may be essential for the production of TNFα by *rheumatoid arthritis* synovial monocytes.

CD40 crosslinking not only makes cells susceptible to CD95-mediated cell death (see below), but also makes epithelial cells more susceptible to TNFα-induced death (Eliopoulos *et al.*, 1996). CD40L and TNFα have many similar effects on B cells and DCs, such as blocking BCR-induced death (Park *et al.*, 1996), sensitizing B cells to CD95-induced death (Lens *et al.*, 1996), or inducing DC maturation (Cella *et al.*, 1996). However, CD40 ligation, unlike TNFα, is a potent activator of IL-12 in DCs (Cella *et al.*, 1996; DeKruyff *et al.*, 1997).

CD40 and IL-1

IL-1α can upregulate CD40 expression on thymic epithelial cells and then work in concert with CD40 to induce expression of GM-CSF (Galy and Spits, 1992). Ligating CD40 on monocytes but not DCs induces expression of IL-1α and IL-1β (Caux *et al.*, 1994; Wagner *et al.*, 1994; Kiener *et al.*, 1995), while B cells can only be induced to make IL-1β after CD40 stimulation (Burdin *et al.*, 1995). The activation of IL-1β in monocytes requires the ERK pathway (Suttles *et al.*, 1999). Interestingly, CD40 ligation of vascular smooth muscle cells or endothelial cells not only induces increases in IL-1β precursor, but also activates the IL-1β-converting enzyme (ICE or caspase 1), which in turn produces the mature IL-1β (Schönbeck *et al.*, 1997). CD40L can also induce fibroblasts to produce IL-1α (Cao *et al.*, 1998). The reason why CD40 ligation preferentially upregulates IL-1α rather than IL-1β in various cell types remains obscure.

CD40 and IL-6

Ligating CD40 on B cells induces IL-6 expression and IL-6 signaling induces CD40 to be phosphorylated (Clark and Shu, 1990). CD40 also induces IL-6 expression in myeloma cells which may act as an autocrine growth signal (Westendorf *et al.*, 1994; Tong *et al.*, 1994; Urashima *et al.*, 1995). Naïve B cells but not germinal center B cells can be induced to make IL-6 (Burdin *et al.*, 1996), and IL-4 promotes the CD40-induced expression of IL-6 (e.g. Jeppson *et al.*, 1998). The IL-6 produced after CD40 ligation appears to be necessary for B cells to isotype class switch and make IgE (e.g. Bjorck *et al.*, 1998). In monocytes, ligating CD40 is also a potent costimulus for increasing expression of IL-6 (Alderson *et al.*, 1993). Fibroblasts, epithelial cells, and endothelial cells can also be induced to make IL-6 after CD40 ligation (Hess *et al.*, 1995; Yellin *et al.*, 1995; D'echanet *et al.*, 1997; Eliopoulos *et al.*, 1997), and so can airway smooth muscle cells (Lazaar *et al.*, 1998). Thus, CD40 ligation of a variety of cell types may promote inflammatory responses.

CD40 and Lymphotoxin (LT) α/LTβ

Ligating CD40 on B cells induces mRNA expression of LTα but not LTβ (Worm and Geha, 1994; Boussiotis *et al.*, 1994). LTα, like CD27L, can promote CD40 and IL-4-induced B cell proliferation and IgE production (Worm *et al.*, 1998b).

CD40 and IL-18

IL-18, originally known as IFNγ-inducing factor, as in other cells can promote CD40-stimulated B cells to produce IFNγ (Yoshimoto *et al.*, 1997).

CD40 and Chemokines

CD40 stimulation of human monocytes, DCs, or epithelial cells induces changes that enhance and prolong inflammatory responses, including the production of chemokines. Caux *et al.* (1994) found that ligating CD40 on DCs induces them to produce mainly TNFα, IL-8, and MIP-1α, while the same stimulus induced monocytes to make mainly IL-1α, IL-1β, IL-6, IL-8, IL-10, TNFα, and MIP-1α, a result confirmed by Kiener *et al.* (1995). Ligating CD40 on monocytes or macrophages in particular is effective at inducing expression of inflammatory cytokines and chemokines such as MIP-1α, MIP-1β, and RANTES (Caux *et al.*, 1994; Kornbluth *et al.*, 1998). Ligating CD40 on renal epithelial cells also upregulates expression of MCP-1, IL-8, and RANTES (van Kooten *et al.*, 1997; Deckers *et al.*, 1998), and CD40 crosslinking also induces expression of IL-8 in basal epithelial cells (Peguet-Navarro *et al.*, 1997).

CD40-induced RANTES production can be further enhanced by IL-4 or IL-13 (Deckers *et al.*, 1998). This is potentially interesting since unlike IL-8, which inhibits IgE production by B cells induced via CD40 plus IL-4 (Kimata *et al.*, 1992), RANTES or MIP-1α enhance IgE production (Kimata *et al.*, 1996). Thus, a context whereby CD40 selectively induces one chemokine or another (e.g. more or less IL-4) could help to tip the balance toward or against an allergic response.

CD40 and IL-3

IL-3 and CD40 ligation together promote the proliferation of B cell precursors derived from human fetal bone marrow (Saeland *et al.*, 1993; Larson and LeBien, 1994). Similarly, IL-3 and CD40 ligation together can induce acute lymphocytic leukemia cells to proliferate (Planken *et al.*, 1996). IL-3 can also induce CD40 mRNA and protein expression in monocytes and also can promote CD40L-induced expression of IL-6 and TNFα by monocytes (Alderson *et al.*, 1993). Lymphoid-like DCs prevented from dying in culture by IL-3 can then be induced to differentiate by CD40L (Grouard *et al.*, 1997; Strobl *et al.*, 1998).

CD40 and IFNγ

IFNγ upregulates CD40 expression on B cells and epithelial cells (Stamenkovic *et al.*, 1989), monocytes (Alderson *et al.*, 1993), fibroblasts (Yellin *et al.*, 1995) and keratinocytes (Gaspari *et al.*, 1996). IFNγ also potentiates signaling via CD40 such as CD40-mediated protection from programmed cell death (Johnson-L'eger *et al.*, 1997). A key function of CD40 on DCs to upregulate IL-12 production (see CD40 and IL-12) may require the initial and sustained expression of IFNγ (Hilkens *et al.*, 1997), underscoring how TH1 cytokines and CD40 may function together. Blockade of CD40L/CD40 interactions or IL-12 may inhibit the expression of IFNγ by T cells (Schultze *et al.*, 1999). However, TH1 regulation may be complex since maturation of the so-called DC1 subset of DCs required for activating TH1 T cells is blocked by IFNα and CD40 ligation (Rissoan *et al.*, 1999).

CD40 and IFNα

IFNα can induce apoptosis of B cells which is blocked by CD40 ligation (Yanase *et al.*, 1998).

CD40 and TGFβ

Ligating CD40 on B cells induces production of endogenous TGFβ, which is essential for isotype class switching to IgA to occur (Zan *et al.*, 1998). And TGFβ and IL-10 then in turn induce CD40-activated B cells to secrete IgA (Defrance *et al.*, 1992). In microglial cells TGFβ inhibits IFNγ-induced CD40 protein and mRNA expression by enhancing degradation of CD40 mRNA (Nguyen *et al.*, 1998). Similarly, TGFβ decreases CD40 expression on macrophages and thereby reduces their ability to be induced to make IL-12 (Takeuchi *et al.*, 1998). TGFβ can also inhibit CD40-induced B cell proliferation of *chronic lymphocytic leukemia* cells (Lotz *et al.*, 1994). TGFβ can induce apoptosis in some B cells, and CD40 ligation neutralizes this apoptotic signal, apparently by activation of the NFκB pathway (Arsura *et al.*, 1996).

CD40 and Other Surface Receptors

Ligating CD40 on B cells upregulates expression of CD54/ICAM-1 (Barrett *et al.*, 1991) and CD80 (Ranheim and Kipps, 1993). CD40 stimulation also upregulates CD80 and CD86 on DCs (Caux *et al.*, 1994; Pinchuk *et al.*, 1994). Ligating CD40 on monocyte-derived DCs also induces expression of a functional Ox40L (Ohshima *et al.*, 1997).

CD40 and CD95

CD40 and CD95/Fas, which are both members of the TNF receptor family, have an important and complex relationship for regulating the fate of B lymphocytes and other cells (Craxton *et al.*, 1999). Unlike BCR-stimulated B cells, CD40L-stimulated B cells are very sensitive to CD95-mediated cell death (Rothstein *et al.*, 1995). Although CD40 ligation can protect low-density CD95+ B cells from CD95-induced death (Cleary *et al.*, 1995), it also can strongly upregulate

CD95 expression on resting B cells and make these B cells more susceptible to Fas-mediated death (Garrone *et al.*, 1995; Schattner *et al.*, 1995). Whether or not B cells are induced to divide or die by CD40L and CD95L on activated T cells depends on whether they have been acutely triggered via the BCR (Rathmell *et al.*, 1996): if the BCR pathway is activated, then the CD40/CD95 pathway promotes proliferation; otherwise CD40L then CD95L leads to B cell death. Adding further to the complexity is the fact that in the same cell that CD40 ligation blocks BCR-induced death, it induces expression of CD95 and susceptibility to CD95-induced death (Lens *et al.*, 1996). Thus, again there is a fine balancing act between BCR, CD40, and CD95 signals and final outcomes.

CD40 and CD95 also interact in other cells: although CD40 signaling upregulates CD95 expression in DCs, it can block CD95-induced DC death (Bjorck *et al.*, 1997; Koppi *et al.*, 1997). This suggests that there may be a complex set of CD40/CD95 interactions in DCs as in B cells. Likewise, CD40 crosslinking protects *bladder carcinomas* from Fas-mediated death (Jakobson *et al.*, 1998).

CD40 and CD27

CD27, another member of the TNF receptor family, is expressed on T cells and memory B cells (Klein *et al.*, 1998). Ligating CD40 on B cells and B leukemic cells downregulates CD27 and upregulates its ligand, CD70 (Ranheim *et al.*, 1995). Signaling via CD40 or CD27 can promote immunoglobulin production in B cells, but the signaling is not synergistic (Jacquot *et al.*, 1997).

GENE

Accession numbers

GenBank:
Human CD40 (Stamenkovic *et al.*, 1989): P25942
Mouse CD40 (Torres and Clark, 1992): M83312

Sequence

See **Figure 1**.

PROTEIN

Description of protein

The CD40 receptor is a 45–50 kDa type I transmembrane glycoprotein member of the TNF receptor superfamily (Stamenkovic *et al.*, 1989; Banchereau *et al.*, 1994).

Cell types and tissues expressing the receptor

CD40 is expressed on B cells, follicular dendritic cells (FDCs), dendritic cells (DCs), activated monocytes, macrophages, endothelial cells, epithelial cells, and vascular smooth muscle cells (e.g. Schönbeck *et al.*, 1997).

SIGNAL TRANSDUCTION

Ligating CD40 rapidly activates the NFκB pathway in a number of cell types such as B cells (Lalmanach-Girard *et al.*, 1993; Berberich *et al.*, 1994), fibroblasts (Hess *et al.*, 1995), and endothelial cells (Karmann *et al.*, 1996); activation of NFκB correlates with CD40-induced antibody secretion and upregulation of CD54/ICAM-1 (Hsing *et al.*, 1997; Lee *et al.*, 1999). CD40 activation of NFκB also helps to mediate CD40 rescue of BCR-induced cell death (Schauer *et al.*, 1996).

Ligating CD40 also activates in B cells and DCs members of the MAP family of kinases including JNK/SAPK (Sakata *et al.*, 1995; Berberich *et al.*, 1996; Li *et al.*, 1996; Sutherland *et al.*, 1996; Aicher *et al.*, 1999), and p38 MAPK (Sutherland *et al.*, 1996; Salmon *et al.*, 1997; Grammer *et al.*, 1998; Craxton *et al.*, 1998; Aicher *et al.*, 1999). JNK is also induced via CD40 in endothelial cells (Karmann *et al.*, 1996). Whether or not ERK1/2 are activated via CD40 appears to depend on the cell type: in resting human B cells or the WEHI 231 cells, CD40 ligation may not activate ERK2 (Sakata *et al.*, 1996; Berberich *et al.*, 1996; Sutherland *et al.*, 1996) but in some B cells, DCs, and monocytes it does (Li *et al.*, 1996; Purkerson and Parker, 1998; Aicher *et al.*, 1999; Suttles *et al.*, 1999). In monocytes, p38 MAPK and JNK are not activated via CD40 (Suttles *et al.*, 1999). Thus, the set of MAP family kinases induced via CD40 is cell type-specific and perhaps stage-specific.

Associated or intrinsic kinases

Although CD40 has been found to associate with the kinase JAK3, JAK3 does not appear to be required for a number of CD40-induced phenotypes such as B cell proliferation or isotype class switching (Jabara *et al.*, 1998). However, B cell proliferation induced via CD40 requires the phosphoinositide 3-kinase (PI3-K) p85α subunit (Fruman *et al.*, 1999).

Figure 1 Amino acid sequences for human CD40 (Stamenkovic *et al.*, 1989) and mouse CD40 (Torres and Clark, 1992).

Gene

Human CD40 (Stamenkovic *et al.*, 1989); Genbank Assession number _______

```
MVRLPLQCVLWGCLLTAVHPEPPTACREKQYLINSQCCSLCQPGQKLVSDCTEF
TETECLPCGESEFLDTWNRETHCHQHKYCDPNLGLRVQQKGTSETDTICTCEEG
WHCTSEACESCVLHRSCSPGFGVKQIATGVSDTICEPCPVGFFSNVSSAFEKCHP
WTSCETKDLVVQQAGTNKTDVVCGPQDRLRALVVIPIIFGWFAILLVLVFIKKVA
KKPTNKAPHPKQEPQEINFPDDLPGSNTAAPVQETLHGCQPVTQEDGKESRISVQ
ERQ
```

Mouse CD40 (Torres and Clark, 1992); Genebank Assession number M83312

```
MVSLPRLCALWGCLLTAVHLGQCVTCSDKQYLHDGQCCDLCQPGSRLTSHCTA
LEKTQCHPCDSGEFSAQWNREIRCHQHRHCEPNQGLRVKKEGTAESDTVCTCKE
GQHCTSKDCEACAQHTPCIPGFGVMEMATETTDTVCHPCPVGFFSNQSSLFEKC
YPWTSCEDKNLEVLQKGTSQTNVICGLKSRMRALLVIPVVMGILITIFGVFLYIKV
VKKPKDNEMLPPAARRQDPQEMEDYPGHNTAAPVQETLHGCQPVTQEDGKESI
SVQERQVTDSIALRPLV
```

Cytoplasmic signaling cascades

Receptor-mediated CD40 signaling is initiated by binding of CD40 ligand (CD40L or CD154) to its receptor, CD40. Normally CD40 must be at least dimerized in order to transmit a signal.

References

Aicher, A., Shu, G., Magaletti, D., Mulvania, T., Pezzutto, A., Craxton, A., and Clark, E. A. (1999). Differential role for p38 MAPK in regulating CD40-induced gene expression in human dendritic cells and B cells. *J. Immunol.* **163**, 5786–5795.

Alderson, M. R., Armitage, R. J., Tough, T. W., Strockbine, L., Fanslow, W. C., and Spriggs, M. K. (1993). CD40 expression by human monocytes: regulation by cytokines and activation of monocytes by the ligand for CD40. *J. Exp. Med.* **178**, 669–674.

Antonysamy, M. A., Fanslow, W. C., Fu, F., Li, W., Qian, S., Troutt, A. B., and Thomson, A. W. (1999). Evidence for a role of IL-17 in organ allograft rejection: IL-17 promotes the functional differentiation of dendritic cell progenitors. *J. Immunol.* **162**, 577–584.

Armitage, R. J., Macduff, B. M., Spriggs, M. K., and Fanslow, W. C. (1993). Human B cell proliferation and Ig secretion induced by recombinant CD40 ligand are modulated by soluble cytokines. *J. Immunol.* **150**, 3671–3680.

Arpin, C., Banchereau, J., and Liu, Y.-J. (1997). Memory B cells are biased towards terminal differentiation: a strategy that may prevent repertoire freezing. *J. Exp. Med.* **186**, 931–940.

Arsura, M., Wu, M., and Sonenshein, G. E. (1996). TGF beta 1 inhibits NF-kappa B/Rel activity inducing apoptosis of B cells: transcriptional activation of I kappa B alpha. *Immunity* **5**, 31–40.

Aversa, G., Punnonen, J., Cocks, B. G., de Waal Malefyt, R., Vega Jr, F., Zurawski, S. M., Zurawski, G., and de Vries, J. E. (1993). An interleukin 4 (IL-4) mutant protein inhibits both IL-4 or IL-13-induced human immunoglobulin G4 (IgG4) and IgE synthesis and B cell proliferation: support for a common component shared by IL-4 and IL-13 receptors. *J. Exp. Med.* **178**, 2213–2218.

Avice, M. N., Demeure, C. E., Delespesse, G., Rubio, M., Armant, M., and Sarfati, M. (1998). IL-15 promotes IL-12 production by human monocytes via T cell-dependent contact and may contribute to IL-12-mediated IFN-gamma secretion by CD4+ T cells in the absence of TCR ligation. *J. Immunol.* **161**, 3408–3415.

Baba, M., Kikuchi, Y., Mori, S., Kimoto, H., Inui, S., Sakaguchi, N., Inoue, J., Yamamoto, T., Takemori, T., Howard, M., and Takatsu, K. (1997). Mouse germinal center B cells with the xid mutation retain responsiveness to antimouse CD40 antibodies but diminish IL-5 responsiveness. *Int. Immunol.* **9**, 1463–1473.

Banchereau, J., Bazan, F., Blanchard, D., Briere, F., Galizzi, J. P., van Kooten, C., Liu, Y. J., Rousset, F., and Saeland, S. (1994). The CD40 antigen and its ligand. *Annu. Rev. Immunol.* **12**, 881–922.

Barrett, T. B., Shu, G., and Clark, E. A. (1991). CD40 signaling activates CD11a/CD18-(LFA-1)-mediated adhesion in B cells. *J. Immunol.* **146**, 1722–1729.

Berberich, I., Shu, G. L., and Clark, E. A. (1994). Cross-linking CD40 on B cells rapidly activates nuclear factor-kappa B. *J. Immunol.* **153**, 4357–4366.

Berberich, I., Shu, G., Siebelt, F., Woodgett, J. R., Kyriakis, J. M., and Clark, E. A. (1996). Cross-linking CD40 on B cells preferentially induces stress-activated protein kinases rather than mitogen-activated protein kinases. *EMBO J.* **15**, 92–101.

Billard, C., Caput, D., Vita, N., Ferrara, P., Orrico, M., Gaulard, P., Boumsell, L., Bensussan, A., and Farcet, J. P. (1997). Interleukin-13 responsiveness and interleukin-13 receptor expression in non-Hodgkin's lymphoma and reactive lymph node B cells. Modulation by CD40 activation. *Eur. Cytokine Netw.* **8**, 19–27.

Bjorck, P., Banchereau, J., and Flores-Romo, L. (1997). CD40 ligation counteracts Fas-induced apoptosis of human dendritic cells. *Int. Immunol.* **9**, 365–372.

Bjorck, P., Larsson, S., Andang, M., Ahrlund-Richter, L., and Paulie, S. (1998). IL-6 antisense oligonucleotides inhibit IgE production in IL-4 and anti-CD40-stimulated human B-lymphocytes. *Immunol. Lett.* **61**, 1–5.

Boussiotis, V. A., Nadler, L. M., Strominger, J. L., and Goldfeld, A. E. (1994). Tumor necrosis factor alpha is an autocrine growth factor for normal human B cells. *Proc. Natl Acad. Sci. USA* **91**, 7007–7011.

Briere, F., Servet Delprat, C., Bridon, J. M., Saint Remy, J. M., and Banchereau, J. (1994). Interleukin 10 induces naive surface immunoglobulin D+ (IgD+) B cells to secrete IgG1 and IgG3. *J. Exp. Med.* **179**, 757–762.

Buelens, C., Verhasselt, V., De Groote, D., Thielemans, K., Goldman, M., and Willems, F. (1997). Interleukin-10 prevents the generation of dendritic cells from human peripheral blood mononuclear cells cultured with interleukin-4 and granulocyte/macrophage-colony-stimulating factor. *Eur. J. Immunol.* **27**, 756–762.

Burdin, N., Van Kooten, C., Galibert, L., Abrams, J. S., Wijdenes, J., Banchereau, J., and Rousset, F. (1995). Endogenous IL-6 and IL-10 contribute to the differentiation of CD40-activated human B lymphocytes. *J. Immunol.* **154**, 2533–2544.

Burdin, N., Galibert, L., Garrone, P., Durand, I., Banchereau, J., and Rousset, F. (1996). Inability to produce IL-6 is a functional feature of human germinal center B lymphocytes. *J. Immunol.* **156**, 4107–4113.

Campbell, K. A., Ovendale, P. J., Kennedy, M. K., Fanslow, W. C., Reed, S. G., and Maliszewski, C. R. (1996). CD40 ligand is required for protective cell-mediated immunity to *Leishmania major*. *Immunity* **4**, 283–289.

Cao, H. J., Wang, H. S., Zhang, Y., Lin, H. Y., Phipps, R. P., and Smith, T. J. (1998). Activation of human orbital fibroblasts through CD40 engagement results in a dramatic induction of hyaluronan synthesis and prostaglandin endoperoxide H synthase-2 expression. Insights into potential pathogenic mechanisms of thyroid-associated ophthalmopathy. *J. Biol. Chem.* **273**, 29615–29625.

Caux, C., Massacrier, C., Vanbervliet, B., Dubois, B., Van Kooten, C., Durand, I., and Banchereau, J. (1994). Activation of human dendritic cells through CD40 cross-linking. *J. Exp. Med.* **180**, 1263–1272.

Cella, M., Scheidegger, D., Palmer Lehmann, K., Lane, P., Lanzavecchia, A., and Alber, G. (1996). Ligation of CD40 on dendritic cells triggers production of high levels of interleukin-12 and enhances T cell stimulatory capacity: T–T help via APC activation. *J. Exp. Med.* **184**, 747–752.

Cerutti, A., Schaffer, A., Shah, S., Zan, H., Liou, H. C., Goodwin, R. G., and Casali, P. (1998). CD30 is a CD40-inducible molecule that negatively regulates CD40-mediated immunoglobulin class switching in non-antigen-selected human B cells. *Immunity* **9**, 247–256.

Choi, M. S., Brines, R. D., Holman, M. J., and Klaus, G. G. (1994). Induction of NF-AT in normal B lymphocytes by anti-immunoglobulin or CD40 ligand in conjunction with IL-4. *Immunity* **1**, 179–187.

Clark, E. A. (1990). CD40: A cytokine receptor in search of a ligand. *Tissue Antigens* **35**, 33–36.

Clark, E. A., and Ledbetter, J. A. (1994). How B and T cells talk to each other. *Nature* **367**, 425–428.

Clark, E. A., and Shu, G. (1990). Association between IL-6 and CD40 signaling.IL-6 induces phosphorylation of CD40 receptors. *J. Immunol.* **145**, 1400–1406.

Clark, E. A., Shu, G. L., Luscher, B., Draves, K. E., Banchereau, J., Ledbetter, J. A., and Valentine, M. A. (1989). Activation of human B cells. Comparison of the signal transduced by IL-4 to four different competence signals. *J. Immunol.* **143**, 3873–3880.

Cleary, A. M., Fortune, S. M., Yellin, M. J., Chess, L., and Lederman, S. (1995). Opposing roles of CD95 (Fas/APO-1) and CD40 in the death and rescue of human low density tonsillar B cells. *J. Immunol.* **155**, 3329–33237.

Cocks, B. G., de Waal Malefyt, R., Galizzi, J. P., de Vries, J. E., and Aversa, G. (1993). IL-13 induces proliferation and differentiation of human B cells activated by the CD40 ligand. *Int. Immunol.* **5**, 657–663.

Craxton, A., Shu, G., Graves, J. D., Saklatvala, J., Krebs, E. G., and Clark, E. A. (1998). p38 MAPK is required for CD40-induced gene expression and proliferation in B lymphocytes. *J. Immunol.* **161**, 3225–3236.

Craxton, A., Otipoby, K., Jiang, A., and Clark, E. A. (1999). Signal transduction pathways which regulate the fate of B lymphocytes. *Adv. Immunol.* **73**, 79–152.

Davey, E. J., Thyberg, J., Conrad, D. H., and Severinson, E. (1998). Regulation of cell morphology in B lymphocytes by IL-4: evidence for induced cytoskeletal changes. *J. Immunol.* **160**, 5366–5373.

D'echanet, J., Grosset, C., Taupin, J. L., Merville, P., Banchereau, J., Ripoche, J., and Moreau, J. F. (1997). CD40 ligand stimulates proinflammatory cytokine production by human endothelial cells. *J. Immunol.* **159**, 5640–5647.

Deckers, J. G., De Haij, S., van der Woude, F. J., van der Kooij, S. W., Daha, M. R., and van Kooten, C. (1998). IL-4 and IL-13 augment cytokine- and CD40-induced RANTES production by human renal tubular epithelial cells *in vitro*. *J. Am. Soc. Nephrol.* **9**, 1187–1193.

Defrance, T., Vanbervliet, B., Briere, F., Durand, I., Rousset, F., and Banchereau, J. (1992). Interleukin 10 and transforming growth factor beta cooperate to induce anti-CD40-activated naive human B cells to secrete immunoglobulin A. *J. Exp. Med.* **175**, 671–682.

DeKruyff, R. H., Gieni, R. S., and Umetsu, D. T. (1997). Antigen-driven but not lipopolysaccharide-driven IL-12 production in macrophages requires triggering of CD40. *J. Immunol.* **158**, 359–366.

Eliopoulos, A. G., Dawson, C. W., Mosialos, G., Floettmann, J. E., Rowe, M., Armitage, R. J., Dawson, J., Zapata, J. M., Kerr, D. J., Wakelam, M. J., Reed, J. C., Kieff, E., and Young, L. S. (1996). CD40-induced growth inhibition in epithelial cells is mimicked by Epstein–Barr Virus-encoded LMP1: involvement of TRAF3 as a common mediator. *Oncogene* **13**, 2243–2254.

Eliopoulos, A. G., Stack, M., Dawson, C. W., Kaye, K. M., Hodgkin, L., Sihota, S., Rowe, M., and Young, L. S. (1997). Epstein–Barr virus-encoded LMP1 and CD40 mediate IL-6 production in epithelial cells via an NF-kappaB pathway involving TNF receptor-associated factors. *Oncogene* **14**, 2899–2916.

Fanslow, W. C., Clifford, K. N., Seaman, M., Alderson, M. R., Spriggs, M. K., Armitage, R. J., and Ramsdell, F. (1994). Recombinant CD40 ligand exerts potent biologic effects on T cells. *J. Immunol.* **152**, 4262–4269.

Fayette, J., Durand, I., Bridon, J. M., Arpin, C., Dubois, B., Caux, C., Liu, Y. J., Banchereau, J., and Briere, F. (1998). Dendritic cells enhance the differentiation of naive B cells into plasma cells *in vitro*. *Scand. J. Immunol.* **48**, 563–570.

Fluckiger, A. C., Garrone, P., Durand, I., Galizzi, J. P., and Banchereau, J. (1993). Interleukin 10 (IL-10) upregulates functional high affinity IL-2 receptors on normal and leukemic B lymphocytes. *J. Exp. Med.* **178**, 1473–1481.

Fossiez, F., Banchereau, J., Murray, R., Van Kooten, C., Garrone, P., and Lebecque, S. (1998). Interleukin-17. *Int. Rev. Immunol.* **16**, 541–551.

Fruman, D. A., Snapper, S. B., Yballe, C. M., Davidson, L., Yu, J. Y., Alt, F. W., and Cantley, L. C. (1999). Impaired B cell development and proliferation in absence of phosphoinositide 3-kinase p85alpha. *Science* **283**, 393–397.

Fugier-Vivier, I., Servet-Delprat, C., Rivailler, P., Rissoan, M. C., Liu, Y. J., and Rabourdin-Combe, C. (1997). Measles virus suppresses cell-mediated immunity by interfering with the survival and functions of dendritic and T cells. *J. Exp. Med.* **186**, 813–823.

Fujita, K., Jumper, M. D., Meek, K., and Lipsky, P. E. (1995). Evidence for a CD40 response element, distinct from the IL-4 response element, in the germline epsilon promoter. *Int. Immunol.* **7**, 1529–1533.

Galy, A. H., and Spits, H. (1992). CD40 is functionally expressed on human thymic epithelial cells. *J. Immunol.* **149**, 775–782.

Garrone, P., Neidhardt, E. M., Garcia, E., Galibert, L., van Kooten, C., and Banchereau, J. (1995). Fas ligation induces apoptosis of CD40-activated human B lymphocytes. *J. Exp. Med.* **182**, 1265–1273.

Gaspari, A. A., Sempowski, G. D., Chess, P., Gish, J., and Phipps, R. P. (1996). Human epidermal keratinocytes are induced to secrete interleukin-6 and co-stimulate T lymphocyte proliferation by a CD40-dependent mechanism. *Eur. J. Immunol.* **26**, 1371–1377.

Gordon, J., Millsum, M. J., Guy, G. R., and Ledbetter, J. A. (1987). Synergistic interaction between interleukin 4 and anti-Bp50 (CDw40) revealed in a novel B cell restimulation assay. *Eur. J. Immunol.* **17**, 1535–1538.

Grabstein, K. H., Maliszewski, C. R., Shanebeck, K., Sato, T. A., Spriggs, M. K., Fanslow, W. C., and Armitage, R. J. (1993). The regulation of T cell-dependent antibody formation *in vitro* by CD40 ligand and IL-2. *J. Immunol.* **150**, 3141–3147.

Grammer, A. C., Swantek, J. L., McFarland, R. D., Miura, Y., Geppert, T., and Lipsky, P. E. (1998). TNF receptor-associated factor-3 signaling mediates activation of p38 and Jun N-terminal kinase, cytokine secretion, and Ig production following ligation of CD40 on human B cells. *J. Immunol.* **161**, 1183–1193.

Grouard, G., de Bouteiller, O., Banchereau, J., and Liu, Y. J. (1995). Human follicular dendritic cells enhance cytokine-dependent growth and differentiation of CD40-activated B cells. *J. Immunol.* **155**, 3345–3352.

Grouard, G., Rissoan, M. C., Filgueira, L., Durand, I., Banchereau, J., and Liu, Y. J. (1997). The enigmatic plasmacytoid T cells develop into dendritic cells with interleukin (IL)-3 and CD40-ligand. *J. Exp. Med.* **185**, 1101–1111.

Hasbold, J., Lyons, A. B., Kehry, M. R., and Hodgkin, P. D. (1998). Cell division number regulates IgG1 and IgE switching of B cells following stimulation by CD40 ligand and IL-4. *Eur. J. Immunol.* **28**, 1040–1051.

Hess, S., Rensing Ehl, A., Schwabe, R., Bufler, P., and Engelmann, H. (1995). CD40 function in nonhematopoietic cells. Nuclear factor kappa B mobilization and induction of IL-6 production. *J. Immunol.* **155**, 4588–4595.

Hikida, M., Nakayama, Y., Yamashita, Y., Kumazawa, Y., Nishikawa, S. I., and Ohmori, H. (1998). Expression of recombination activating genes in germinal center B cells: involvement of interleukin 7 (IL-7) and the IL-7 receptor. *J. Exp. Med.* **188**, 365–372.

Hilkens, C. M., Kalinski, P., de Boer, M., and Kapsenberg, M. L. (1997). Human dendritic cells require exogenous interleukin-12-inducing factors to direct the development of naive T-helper cells toward the Th1 phenotype. *Blood* **90**, 1920–1926.

Hino, A., and Nariuchi, H. (1996). Negative feedback mechanism suppresses interleukin-12 production by antigen-presenting cells interacting with T helper 2 cells. *Eur. J. Immunol.* **26**, 623–628.

Holder, M. J., Liu, Y.-J., Defrance, T., Flores-Romo, L., MacLennan, I. C., and Gordon, J. (1991). Growth factor requirements for the stimulation of germinal center B cells: evidence for an IL-2-dependent pathway of development. *Int. Immunol.* **3**, 1243–1251.

Hsing, Y., Hostager, B. S., and Bishop, G. A. (1997). Characterization of CD40 signaling determinants regulating nuclear factor-kappa B activation in B lymphocytes. *J. Immunol.* **159**, 4898–4906.

Jabara, H. H., Buckley, R. H., Roberts, J. L., Lefranc, G., Loiselet, J., Khalil, G., and Geha, R. S. (1998). Role of JAK3 in CD40-mediated signaling. *Blood* **92**, 2435–2440.

Jacquot, S., Kobata, T., Iwata, S., Morimoto, C., and Schlossman, S. F. (1997). CD154/CD40 and CD70/CD27 interactions have different and sequential functions in T cell-dependent B cell responses: enhancement of plasma cell differentiation by CD27 signaling. *J. Immunol.* **159**, 2652–2657.

Jakobson, E., Jonsson, G., Bjorck, P., and Paulie, S. (1998). Stimulation of CD40 in human bladder carcinoma cells inhibits anti-Fas/APO-1 (CD95)-induced apoptosis. *Int. J. Cancer* **77**, 849–853.

Jeppson, J. D., Patel, H. R., Sakata, N., Domenico, J., Terada, N., and Gelfand, E. W. (1998). Requirement for dual signals by anti-CD40 and IL-4 for the induction of nuclear factor-kappa B, IL-6, and IgE in human B lymphocytes. *J. Immunol.* **161**, 1738–1742.

Jeannin, P., Delneste, Y., Lecoanet Henchoz, S., Gretener, D., and Bonnefoy, J. Y. (1998a). Interleukin-7 (IL-7) enhances class switching to IgE and IgG4 in the presence of T cells via IL-9 and sCD23. *Blood* **91**, 1355–1361.

Jeannin, P., Lecoanet, S., Delneste, Y., Gauchat, J. F., and Bonnefoy, J. Y. (1998b). IgE versus IgG4 production can be differentially regulated by IL-10. *J. Immunol.* **160**, 3555–3561.

Johnson L'eger, C., Hasbold, J., Holman, M., and Klaus, G. G. (1997). The effects of IFN-gamma on CD40-mediated activation of B cells from X-linked immunodeficient or normal mice. *J. Immunol.* **159**, 1150–1159.

Kamanaka, M., Yu, P., Yasui, T., Yoshida, K., Kawabe, T., Horii, T., Kishimoto, T., and Kikutani, H. (1996). Protective role of CD40 in *Leishmania major* infection at two distinct phases of cell-mediated immunity. *Immunity* **4**, 275–281.

Karmann, K., Min, W., Fanslow, W. C., and Pober, J. S. (1996). Activation and homologous desensitization of human endothelial cells by CD40 ligand, tumor necrosis factor, and interleukin 1. *J. Exp. Med.* **184**, 173–182.

Kato, T., Hakamada, R., Yamane, H., and Nariuchi, H. (1996). Induction of IL-12 p40 messenger RNA expression and IL-12 production of macrophages via CD40–CD40 ligand interaction. *J. Immunol.* **156**, 3932–3938.

Kehry, M. R. (1996). CD40-mediated signaling in B cells. Balancing cell survival, growth, and death. *J. Immunol.* **156**, 2345–2348.

Kelsall, B. L., Stuber, E., Neurath, M., and Strober, W. (1996). Interleukin-12 production by dendritic cells. The role of CD40–CD40L interactions in Th1 T-cell responses. *Ann. N.Y. Acad. Sci.* **795**, 116–126.

Kiener, P. A., Moran Davis, P., Rankin, B. M., Wahl, A. F., Aruffo, A., and Hollenbaugh, D. (1995). Stimulation of CD40 with purified soluble gp39 induces proinflammatory responses in human monocytes. *J. Immunol.* **155**, 4917–4925.

Kimata, H., Yoshida, A., Ishioka, C., Lindley, I., and Mikawa, H. (1992). Interleukin 8 (IL-8) selectively inhibits immunoglobulin

E production induced by IL-4 in human B cells. *J. Exp. Med.* **176**, 1227–1231.

Kimata, H., Yoshida, A., Ishioka, C., Fujimoto, M., Lindley, I., and Furusho, K. (1996). RANTES and macrophage inflammatory protein 1 alpha selectively enhance immunoglobulin (IgE) and IgG4 production by human B cells. *J. Exp. Med.* **183**, 2397–2402.

Klein, U., Rajewsky, K., and Kuppers, R. (1998). Human immunoglobulin (Ig)M+IgD+ peripheral blood B cells expressing the CD27 cell surface antigen carry somatically mutated variable region genes: CD27 as a general marker for somatically mutated (memory) B cells. *J. Exp. Med.* **188**, 1679–1689.

Koch, F., Stanzl, U., Jennewein, P., Janke, K., Heufler, C., Kampgen, E., Romani, N., and Schuler, G. (1996). High level IL-12 production by murine dendritic cells: upregulation via MHC class II and CD40 molecules and downregulation by IL-4 and IL-10 *J. Exp. Med.* **184**, 741–746.

Koppi, T. A., Tough-Bement, T., Lewinsohn, D. M., Lynch, D. H., and Alderson, M. R. (1997). CD40 ligand inhibits Fas/CD95-mediated apoptosis of human blood-derived dendritic cells. *Eur. J. Immunol.* **27**, 3161–3165.

Kornbluth, R. S., Kee, K., and Richman, D. D. (1998). CD40 ligand (CD154) stimulation of macrophages to produce HIV-1-suppressive beta-chemokines. *Proc. Natl Acad. Sci. USA* **95**, 5205–5210.

Lalmanach-Girard, A. C., Chiles, T. C., Parker, D. C., and Rothstein, T. L. (1993). T cell-dependent induction of NF-kappa B in B cells. *J. Exp. Med.* **177**, 1215–1219.

Larson, A. W., and LeBien, T. W. (1994). Cross-linking CD40 on human B cell precursors inhibits or enhances growth depending on the stage of development and the IL costimulus. *J. Immunol.* **153**, 584–594.

Lazaar, A. L., Amrani, Y., Hsu, J., Panettieri Jr, R. A., Fanslow, W. C., Albelda, S. M., and Pure, E. (1998). CD40-mediated signal transduction in human airway smooth muscle. *J. Immunol.* **161**, 3120–3127.

Lee, H. H., Dempsey, P. W., Parks, T. P., Zhu, X., Baltimore, D., and Cheng, G. (1999). Specificities of CD40 signaling: involvement of TRAF2 in CD40-induced NF-kappaB activation and intercellular adhesion molecule-1 up-regulation. *Proc. Natl Acad. Sci. USA.* **96**, 1421–1426.

Lens, S. M., Tesselaar, K., den Drijver, B. F., van Oers, M. H., and van Lier, R. A. (1996). A dual role for both CD40-ligand and TNF-alpha in controlling human B cell death. *J. Immunol.* **156**, 507–514.

Li, Y. Y., Baccam, M., Waters, S. B., Pessin, J. E., Bishop, G. A., and Koretzky, G. A. (1996). CD40 ligation results in protein kinase C-independent activation of ERK and JNK in resting murine splenic B cells. *J. Immunol.* **157**, 1440–1447.

Lomo, J., Blomhoff, H. K., Jacobsen, S. E., Krajewski, S., Reed, J. C., and Smeland, E. B. (1997). Interleukin-13 in combination with CD40 ligand potently inhibits apoptosis in human B lymphocytes: upregulation of Bcl-xL and Mcl-1. *Blood* **89**, 4415–4424.

Lotz, M., Ranheim, E., and Kipps, T. J. (1994). Transforming growth factor beta as endogenous growth inhibitor of chronic lymphocytic leukemia B cells. *J. Exp. Med.* **179**, 999–1004.

Ludewig, B., Graf, D., Gelderblom, H. R., Becker, Y., Kroczek, R. A., and Pauli, G. (1995). Spontaneous apoptosis of dendritic cells is efficiently inhibited by TRAP (CD40-ligand) and TNF-alpha, but strongly enhanced by interleukin-10. *Eur. J. Immunol.* **25**, 1943–1950.

Malisan, F., Briere, F., Bridon, J. M., Harindranath, N., Mills, F. C., Max, E. E., Banchereau, J., and Martinez-Valdez, H. (1996). Interleukin-10 induces immunoglobulin G isotype switch recombination in human CD40-activated naive B lymphocytes. *J. Exp. Med.* **183**, 937–947.

Maliszewski, C. R., Grabstein, K., Fanslow, W. C., Armitage, R., Spriggs, M. K., and Sato, T. A. (1993). Recombinant CD40 ligand stimulation of murine B cell growth and differentiation: cooperative effects of cytokines. *Eur. J. Immunol.* **23**, 1044–1049.

Marcelletti, J. F. (1996). IL-10 inhibits lipopolysaccharide-induced murine B cell proliferation and cross-linking of surface antigen receptors or ligation of CD40 restores the response. *J. Immunol.* **157**, 3323–3333.

Maruo, S., Oh-hora, M., Ahn, H. J., Ono, S., Wysocka, M., Kaneko, Y., Yagita, H., Okumura, K., Kikutani, H., Kishimoto, T., Kobayashi, M., Hamaoka, T., Trinchieri, G., and Fujiwara, H. (1997). B cells regulate CD40 ligand-induced IL-12 production in antigen-presenting cells (APC) during T cell/APC interactions. *J. Immunol.* **158**, 120–126.

Nagumo, H., Agematsu, K., Shinozaki, K., Hokibara, S., Ito, S., Takamoto, M., Nikaido, T., Yasui, K., Uehara, Y., Yachie, A., and Komiyama, A. (1998). CD27/CD70 interaction augments IgE secretion by promoting the differentiation of memory B cells into plasma cells. *J. Immunol.* **161**, 6496–6502.

Nguyen, V. T., Walker, W. S., and Benveniste, E. N. (1998). Post-transcriptional inhibition of CD40 gene expression in microglia by transforming growth factor-beta. *Eur. J. Immunol.* **28**, 2537–2548.

Nonoyama, S., Hollenbaugh, D., Aruffo, A., Ledbetter, J. A., and Ochs, H. D. (1993). B cell activation via CD40 is required for specific antibody production by antigen-stimulated human B cells. *J. Exp. Med.* **178**, 1097–1102.

Ohshima, Y., Tanaka, Y., Tozawa, H., Takahashi, Y., Maliszewski, C., and Delespesse, G. (1997). Expression and function of OX40 ligand on human dendritic cells. *J. Immunol.* **159**, 3838–3848.

Park, E., Kalunta, C. I., Nguyen, T. T., Wang, C. L., Chen, F. S., Lin, C. K., Kaptein, J. S., and Lad, P. M. (1996). TNF-alpha inhibits anti-IgM-mediated apoptosis in Ramos cells. *Exp. Cell. Res.* **226**, 1–10.

Peguet-Navarro, J., Dalbiez-Gauthier, C., Moulon, C., Berthier, O., Reano, A., Gaucherand, M., Banchereau, J., Rousset, F., and Schmitt, D. (1997). CD40 ligation of human keratinocytes inhibits their proliferation and induces their differentiation. *J. Immunol.* **158**, 144–152.

Pinchuk, L. M., Polacino, P. S., Agy, M. B., Klaus, S. J., and Clark, E. A. (1994). Role of CD40 and CD80 accessory cell molecules in dendritic cell-dependent HIV-1 infection. *Immunity* **1**, 317–325.

Pinchuk, L. M., Klaus, S. J., Magaletti, D. M., Pinchuk, G. V., Norsen, J. P., and Clark, E. A. (1996). Functional CD40 ligand (CD40L) expressed by human blood dendritic cells is upregulated by CD40 ligation. *J. Immunol.* **157**, 4363–4370.

Planken, E. V., Dijkstra, N. H., Bakkus, M., Willemze, R., and Kluin Nelemans, J. C. (1996). Proliferation of precursor B-lineage acute lymphoblastic leukaemia by activating the CD40 antigen. *Br. J. Haematol.* **95**, 319–326.

Plumas, J., Jacob, M. C., Chaperot, L., Molens, J. P., Sotto, J. J., and Bensa, J. C. (1998). Tumor B cells from non-Hodgkin's lymphoma are resistant to CD95 (Fas/Apo-1)-mediated apoptosis. *Blood* **91**, 2875–2885.

Purkerson, J. M., and Parker, D. C. (1998). Differential coupling of membrane Ig and CD40 to the extracellularly regulated kinase signaling pathway. *J. Immunol.* **160**, 2121–2129.

Ramsdell, F., Seaman, M. S., Clifford, K. N., and Fanslow, W. C. (1994). CD40 ligand acts as a costimulatory signal for neonatal thymic gamma delta T cells. *J. Immunol.* **152**, 2190–2197.

Ranheim, E. A., and Kipps, T. J. (1993). Activated T cells induce expression of B7/BB1 on normal or leukemic B cells through a CD40-dependent signal. *J. Exp. Med.* **177**, 925–935.

Ranheim, E. A., Cantwell, M. J., and Kipps, T. J. (1995). Expression of CD27 and its ligand, CD70, on chronic lymphocytic leukemia B cells. *Blood* **85**, 3556–3565.

Rathmell, J. C., Townsend, S. E., Xu, J. C., Flavell, R. A., and Goodnow, C. C. (1996). Expansion or elimination of B cells *in vivo*: dual roles for CD40- and Fas (CD95)-ligands modulated by the B cell antigen receptor. *Cell* **87**, 319–329.

Ria, F., Penna, G., and Adorini, L. (1998). Th1 cells induce and Th2 inhibit antigen-dependent IL-12 secretion by dendritic cells. *Eur. J. Immunol.* **28**, 2003–2016.

Rissoan, M. C., Soumelis, V., Kadowaki, N., Grouard, G., Briere, F., de-Waal Malefyt, R., and Liu, Y. J. (1999). Reciprocal control of T helper cell and dendritic cell differentiation. *Science* **283**, 1183–1186.

Rothstein, T. L., Wang, J. K., Panka, D. J., Foote, L. C., Wang, Z., Stanger, B., Cui, H., Ju, S. T., and Marshak-Rothstein, A. (1995). Protection against Fas-dependent Th1-mediated apoptosis by antigen receptor engagement in B cells. *Nature* **374**, 163–165.

Rousset, F., Garcia, E., Defrance, T., Peronne, C., Vezzio, N., Hsu, D. H., Kastelein, R., Moore, K. W., and Banchereau, J. (1992). Interleukin 10 is a potent growth and differentiation factor for activated human B lymphocytes. *Proc. Natl Acad. Sci. USA* **89**, 1890–1893.

Rousset, F., Peyrol, S., Garcia, E., Vezzio, N., Andujar, M., Grimaud, J. A., and Banchereau, J. (1995). Long-term cultured CD40-activated B lymphocytes differentiate into plasma cells in response to IL-10 but not IL-4. *Int. Immunol.* **7**, 1243–1253.

Saeland, S., Duvert, V., Moreau, I., and Banchereau, J. (1993). Human B cell precursors proliferate and express CD23 after CD40 ligation. *J. Exp. Med.* **178**, 113–120.

de Saint Vis, B., Fugier Vivier, I., Massacrier, C., Gaillard, C., Vanbervliet, B., Ait Yahia, S., Banchereau, J., Liu, Y. J., Lebecque, S., and Caux, C. (1998). The cytokine profile expressed by human dendritic cells is dependent on cell subtype and mode of activation. *J. Immunol.* **160**, 1666–1676.

Sakata, N., Patel, H. R., Terada, N., Aruffo, A., Johnson, G. L., and Gelfand, E. W. (1995). Selective activation of c-Jun kinase mitogen-activated protein kinase by CD40 on human B cells. *J. Biol. Chem.* **270**, 30823–30828.

Salmon, R. A., Foltz, I. N., Young, P. R., and Schrader, J. W. (1997). The p38 mitogen-activated protein kinase is activated by ligation of the T or B lymphocyte antigen receptors, Fas or CD40, but suppression of kinase activity does not inhibit apoptosis induced by antigen receptors. *J. Immunol.* **159**, 5309–5317.

Schattner, E. J., Elkon, K. B., Yoo, D. H., Tumang, J., Krammer, P. H., Crow, M. K., and Friedman, S. M. (1995). CD40 ligation induces Apo-1/Fas expression on human B lymphocytes and facilitates apoptosis through the Apo-1/Fas pathway. *J. Exp. Med.* **182**, 1557–1565.

Schauer, S. L., Wang, Z., Sonenshein, G. E., and Rothstein, T. L. (1996). Maintenance of nuclear factor-kappa B/Rel and c-myc expression during CD40 ligand rescue of WEHI 231 early B cells from receptor-mediated apoptosis through modulation of I kappa B proteins. *J. Immunol.* **157**, 81–86.

Schönbeck, U., Mach, F., Bonnefoy, J. Y., Loppnow, H., Flad, H. D., and Libby, P. (1997). Ligation of CD40 activates interleukin 1beta-converting enzyme (caspase-1) activity in vascular smooth muscle and endothelial cells and promotes elaboration of active interleukin 1beta. *J. Biol. Chem.* **272**, 19569–19574.

Schultze, J. L., Michalak, S., Lowne, J., Wong, A., Gilleece, M. H., Gribben, J. G., and Nadler, L. M. (1999). Human non-germinal center B cell interleukin (IL)-12 production is primarily regulated by T cell signals CD40 ligand, interferon gamma, and IL-10: role of B cells in the maintenance of T cell responses. *J. Exp. Med.* **189**, 1–12.

Sekine, C., Yagita, H., Miyasaka, N., and Okumura, K. (1998). Expression and function of CD40 in rheumatoid arthritis synovium. *J. Rheumatol.* **25**, 1048–1053.

Shapira, S. K., Vercelli, D., Jabara, H. H., Fu, S. M., and Geha, R. S. (1992). Molecular analysis of the induction of immunoglobulin E synthesis in human B cells by interleukin 4 and engagement of CD40 antigen. *J. Exp. Med.* **175**, 289–292.

Shu, U., Kiniwa, M., Wu, C. Y., Maliszewski, C., Vezzio, N., Hakimi, J., Gately, M., and Delespesse, G. (1995). Activated T cells induce interleukin-12 production by monocytes via CD40–CD40 ligand interaction. *Eur. J. Immunol.* **25**, 1125–1128.

Snapper, C. M., Zelazowski, P., Rosas, F. R., Kehry, M. R., Tian, M., Baltimore, D., and Sha, W. C. (1996). B cells from p50/NF-kappa B knockout mice have selective defects in proliferation, differentiation, germ-line CH transcription, and Ig class switching. *J. Immunol.* **156**, 183–191.

Snijders, A., Kalinski, P., Hilkens, C. M., and Kapsenberg, M. L. (1998). High-level IL-12 production by human dendritic cells requires two signals. *Int. Immunol.* **10**, 1593–1598.

Stamenkovic, I., Clark, E. A., and Seed, B. (1989). A B-lymphocyte activation molecule related to the nerve growth factor receptor and induced by cytokines in carcinomas. *EMBO J.* **8**, 1403–1410.

Strobl, H., Scheinecker, C., Riedl, E., Csmarits, B., Bello Fernandez, C., Pickl, W. F., Majdic, O., and Knapp, W. (1998). Identification of CD68+lin- peripheral blood cells with dendritic precursor characteristics. *J. Immunol.* **161**, 740–748.

Stuber, E., Strober, W., and Neurath, M. (1996). Blocking the CD40L–CD40 interaction *in vivo* specifically prevents the priming of T helper 1 cells through the inhibition of interleukin 12 secretion. *J. Exp. Med.* **183**, 693–698.

Sutherland, C. L., Heath, A. W., Pelech, S. L., Young, P. R., and Gold, M. R. (1996). Differential activation of the ERK, JNK, and p38 mitogen-activated protein kinases by CD40 and the B cell antigen receptor. *J. Immunol.* **157**, 3381–3390.

Suttles, J., Milhorn, D. M., Miller, R. W., Poe, J. C., Wahl, L. M., and Stout, R. D. (1999). CD40 signaling of monocyte inflammatory cytokine synthesis through an ERK1/2-dependent pathway. A target of interleukin (il)-4 and il-10 anti-inflammatory action. *J. Biol. Chem.* **274**, 5835–5842.

Takeuchi, M., Alard, P., and Streilein, J. W. (1998). TGF-beta promotes immune deviation by altering accessory signals of antigen-presenting cells. *J. Immunol.* **160**, 1589–1597.

Tong, A. W., Zhang, B. Q., Mues, G., Solano, M., Hanson, T., and Stone, M. J. (1994). Anti-CD40 antibody binding modulates human multiple myeloma clonogenicity *in vitro*. *Blood*. **84**, 3026–3033.

Torres, R. M., and Clark, E. A. (1992). Differential increase of an alternatively polyadenylated mRNA species of murine CD40 upon B lymphocyte activation. *J. Immunol.* **148**, 620–626.

Urashima, M., Chauhan, D., Uchiyama, H., Freeman, G. J., and Anderson, K. C. (1995). CD40 ligand triggered interleukin-6 secretion in multiple myeloma. *Blood*. **85**, 1903–1912.

van Kooten, C., and Banchereau, J. (1997). Functions of CD40 on B cells, dendritic cells and other cells. *Curr. Opin. Immunol.* **9**, 330–337.

van Kooten, C., Gerritsma, J. S., Paape, M. E., van Es, L. A., Banchereau, J., and Daha, M. R. (1997). Possible role for

CD40–CD40L in the regulation of interstitial infiltration in the kidney. *Kidney Int.* **51**, 711–721.

Wagner Jr, D. H., Stout, R. D., and Suttles, J. (1994). Role of the CD40–CD40 ligand interaction in CD4+ T cell contact-dependent activation of monocyte interleukin-1 synthesis. *Eur. J. Immunol.* **24**, 3148–3154.

Warren, W. D., and Berton, M. T. (1995). Induction of germ-line gamma 1 and epsilon Ig gene expression in murine B cells. IL-4 and the CD40 ligand-CD40 interaction provide distinct but synergistic signals. *J. Immunol.* **155**, 5637–5646.

Westendorf, J. J., Ahmann, G. J., Armitage, R. J., Spriggs, M. K., Lust, J. A., Greipp, P. R., Katzmann, J. A., and Jelinek, D. F. (1994). CD40 expression in malignant plasma cells. Role in stimulation of autocrine IL-6 secretion by a human myeloma cell line. *J. Immunol.* **152**, 117–128.

Worm, M., and Geha, R. S. (1994). CD40 ligation induces lymphotoxin alpha gene expression in human B cells. *Int. Immunol.* **6**, 1883–1890.

Worm, M., Krah, J. M., Manz, R. A., and Henz, B. M. (1998a). Retinoic acid inhibits CD40+ interleukin-4-mediated IgE production *in vitro*. *Blood.* **92**, 1713–1720.

Worm, M., Ebermayer, K., and Henz, B. M. (1998b). Lymphotoxin-alpha is an important autocrine factor for CD40+ interleukin-4-mediated B-cell activation in normal and atopic donors. *Immunology* **94**, 395–402.

Yanase, N., Takada, E., Yoshihama, I., Ikegami, H., and Mizuguchi, J. (1998). Participation of Bax-alpha in IFN-alpha-mediated apoptosis in Daudi B lymphoma cells. *J. Interferon Cytokine Res.* **18**, 855–861.

Yellin, M. J., Winikoff, S., Fortune, S. M., Baum. D., Crow, M. K., Lederman, S., and Chess, L. (1995). Ligation of CD40 on fibroblasts induces CD54 (ICAM-1) and CD106 (VCAM-1) up-regulation and IL-6 production and proliferation. *J. Leukoc. Biol.* **58**, 209–216.

Yoshimoto, T., Okamura, H., Tagawa, Y. I., Iwakura, Y., and Nakanishi, K. (1997). Interleukin 18 together with interleukin 12 inhibits IgE production by induction of interferon-gamma production from activated B cells. *Proc. Natl Acad. Sci. USA* **94**, 3948–3953.

Zan, H., Cerutti, A., Dramitinos, P., Schaffer, A., and Casali, P. (1998). CD40 engagement triggers switching to IgA1 and IgA2 in human B cells through induction of endogenous TGF-beta: evidence for TGF-beta but not IL-10-dependent direct S mu → S alpha and sequential S mu → S gamma, S gamma → S alpha DNA recombination. *J. Immunol.* **161**, 5217–5225.

Zelazowski, P., Max, E. E., Kehry, M. R., and Snapper, C. M. (1997). Regulation of Ku expression in normal murine B cells by stimuli that promote switch recombination. *J. Immunol.* **159**, 2559–2562.

CD30

Francesco Annunziato[1], Paola Romagnani[2], Carmelo Mavilia[1], Gianni Pizzolo[3], Harald Stein[4] and Sergio Romagnani[1,*]

[1]Department of Internal Medicine, Section of Immunoallergology and Respiratory Diseases, University of Florence, Florence, Italy

[2]Department of Physiopathology, Endocrinology Unit, University of Florence, Florence, Italy

[3]Department of Clinical and Experimental Medicine, Section of Hematology, University of Verona, Verona, Italy

[4]Institute of Pathology, UK Benjamin Franklin, Freie University of Berlin, Berlin, Germany

*corresponding author tel: 39-055-413663, fax: 39-055-412867, e-mail: s.romagnani@dfc.unifi.it

DOI: 10.1006/rwcy.2000.16006.

SUMMARY

CD30 is a member of the TNF receptor family that is expressed by a subset of activated T cells (both CD4+ and CD8+) and B cells and is constitutively present in decidual and exocrine pancreatic cells. In activated T cells, CD30 appears to be preferentially associated with type 2 T helper (TH2) responses, since its expression is at least partly dependent on the presence of IL-4. Moreover, CD30 is present in a variety of tumors, including *Hodgkin's lymphoma* and *non-Hodgkin's lymphoma*, and *embryonal carcinomas*. A soluble form of CD30 (sCD30) is released in the serum of patients with Hodgkin's lymphoma and other CD30+ tumors, as well as in inflammatory conditions characterized by strong B cell or TH0/TH2 cell activation. The measurement of sCD30 in biological fluids may represent a good marker of disease activity and/or prognosis in such conditions. *In vivo* CD30-targeting in patients with refractory CD30+ tumors is also being attempted.

BACKGROUND

Discovery

CD30 was discovered by using a monoclonal antibody (Ki-1), which was raised against a *Hodgkin's disease* (HD)-derived cell line (Schwab *et al.*, 1982). This antibody recognized a molecule selectively expressed by Hodgkin's and Reed–Sternberg (H-RS) cells in tissues involved in HD. Ki-1, as well as other subsequently raised anti-CD30 antibodies, also reacted with a small population of large cells preferentially localized to reactive lymphoid tissue around B cell follicles (Stein *et al.*, 1982) and medullary areas of human thymus (Stein *et al.*, 1985; Romagnani *et al.*, 1998a).

Alternative names

The original name of CD30 was Ki-1, after the name of the first antibody used for its identification.

Structure

CD30 is a membrane glycoprotein consisting of two chains with an apparent molecular weight of 120 and 105 kDa, belonging to the TNFR superfamily. The extracellular portion of CD30 is proteolytically cleaved to produce an 88 kDa soluble form of the molecule (sCD30).

Main activities and pathophysiological roles

The physiological roles of membrane and soluble forms of CD30 are still unknown, although these molecules probably play an important regulatory role in both the function and interactions of normal lymphoid cells, as well as of other cell types. The physiological role of CD30 ligand (CD30L) is also presently unknown.

GENE

Accession numbers

GenBank:
CD30: U25416

Sequence

See **Figure 1**.

The complete nucleotide sequence of human cDNA for CD30 is 3630 bp with a G/C content equal to 62%. The opening reading frame (ORF) extends from nucleotide 231 to nucleotide 2015. The ATG initiation codon is flanked by sequences which are in agreement with the consensus sequences.

Chromosome location and linkages

The human CD30 gene is located on the short arm of chromosome 1 at position 36 (1p36). The ORF consists of two similar domains sharing 77% homology (nucleotides 381–472 and 906–1270). The 5′ untranslated leader sequence contains short ORFs and is not required for gene expression. The ORF is followed by an untranslated sequence with a short palindromic sequence extending from nucleotide 2867 to 2888. There are two polyadenylation sites in the 3′ untranslated region preceded by the unusual poly(A) signal sequences TGTAAA and AATAAT, respectively. By northern blot analysis of poly(A)+ RNA from various human cell lines it is possible to detect a major RNA species of about 3.8 kb and, after longer exposure time, a minor species of about 2.6 kb (Durkop *et al.*, 1992).

Recently, cDNAs for two novel CD30 mRNAs of 2.3 kb have been identified and cloned (GenBank accession number D86042), which are induced by TPA in the myeloid leukemia cell line HL-60 (Horie *et al.*, 1996). They were transcribed from the intronic region just upstream of the exon coding the transmembrane domain of CD30 protein. The shorter cDNA had a deletion of 54 nucleotides (nucleotide position from 1479 to 1532 of the CD30 cDNA), corresponding to the 3′ region of the transmembrane domain of CD30, which was probably caused by the alternative splicing (DDBJ, accession number D86042). Translation of this transcript appeared to start from the internal methionine codon at nucleotide position 289 that corresponds to that of 1612 in the CD30 cDNA, and encoded a protein of 132 amino acid residues, which corresponds exactly to the C-terminal cytoplasmic domain of CD30 (Horie *et al.*, 1996).

PROTEIN

Accession numbers

SwissProt:
CD30: P28908

Sequence

See **Figure 2**.

Description of protein

The deduced amino acid sequence to the ORF of the cDNA predict of 595 amino acids is a polypeptide with a molecular weight of about 64 kDa. The primary structure indicates that the protein traverses the cellular membrane and is, therefore, composed of an 18 residue leader peptide, an extracellular domain of 365 residues, a single transmembrane domain of 24 residues, and a cytoplasmic domain of 188 residues (Durkop *et al.*, 1992).

The full-length murine CD30 cDNA encodes a protein of 498 amino acids, consisting of an 18 amino acid residue leader peptide, 263 extracellular domain amino acid, a 27 amino acid transmembrane domain, and a 190 amino acid intracellular domain. The calculated molecular mass of the mature protein core is 52 kDa. Murine CD30 is 97 amino acid shorter than its human homolog, because of a large 90 amino acid deletion in the extracellular region and the loss of another 7 amino acids throughout the sequence. The extracellular domain of human CD30 has proved to be homologous to that of other TNF receptor (TNFR) superfamily members, which are characterized by the content of some (usually 3 or 4) cysteine-rich motifs of about 40 residues in the extracellular part of the molecule. In particular, the extracellular part of CD30, which is regarded as the putative ligand-binding domain, can be divided by a hinge sequence of about 60 amino acids that may be derived from the central region of another TNFR motif. Potential phosphorylation sites for the tyrosine kinase are present only in the extracellular domain, whereas sites for serine/threonine kinase are localized in the intracellular domain (Durkop *et al.*, 1992).

Figure 1 Nucleotide sequence for CD30.

Sequence

```
   1 ATACGGGAGA ACTAAGGCTG AAACCTCGGA GGAACAACCA CTTTTGAAGT GACTTCGCGG
  61 CGTGCGTTGG GTGCGGACTA GGTGGCCCCG GCGGGAGTGT GCTGGAGCCT GAAGTCCACG
 121 CGCGCGGCTG AGAACCGCCG GGACCGCACG TGGGCGCCGC GCGCTTCCCC CGCTTCCCAG
 181 GTGGGCGCCG GCCGCCAGGC CACCTCACGT CCGGCCCCGG GGATGCGCGT CCTCCTCGCC
 241 GCGCTGGGAC TGCTGTTCCT GGGGGCGCTA CGAGCCTTCC CACAGGATCG ACCCTTCGAG
 301 GACACCTGTC ATGGAAACCC CAGCCACTAC TATGACAAGG CTGTCAGGAG GTGCTGTTAC
 361 CGCTGCCCCA TGGGGCTGTT CCCGACACAG CAGTGCCCAC AGAGGCCTAC TGACTGCAGG
 421 AAGCAGTGTG AGCCTGACTA CTACCTGGAT GAGGCCGACC GCTGTACAGC CTGCGTGACT
 481 TGTTCTCGAG ATGACCTCGT GGAGAAGACG CCGTGTGCAT GGAACTCCTC CCGTGTCTGC
 541 GAATGTCGAC CCGGCATGTT CTGTTCCACG TCTGCCGTCA ACTCCTGTGC CCGCTGCTTC
 601 TTCCATTCTG TCTGTCCGGC AGGGATGATT GTCAAGTTCC CAGGCACGGC GCAGAAGAAC
 661 ACGGTCTGTG AGCCGGCTTC CCCAGGGGTC AGCCCTGCCT GTGCCAGCCC AGAGAACTGC
 721 AAGGAACCCT CCAGTGGCAC CATCCCCCAG GCCAAGCCCA CCCCGGTGTC CCCAGCAACC
 781 TCCAGTGCCA GCACCATGCC TGTAAGAGGG GGCACCCGCC TCGCCCAGGA AGCTGCTTCT
 841 AAACTGACGA GGGCTCCCGA C~CTCCCTCC TCTGTGGGAA GGCCTAGTTC AGATCCAGGT
 901 CTGTCCCCAA CACAGCCATG CCCAGAGGGG TCTGGTGATT GCAGAAAGCA GTGTGAGCCC
 961 GACTACTACC TGGACGAGGC CGGCCGCTGC ACAGCCTGCG TGAGCTGTTC TCGAGATGAC
1021 CTTGTGGAGA AGACGCCATG TGCATGGAAC TCCTCCCGCA CCTGCGAATG TCGACCTGGC
1081 ATGATCTGTG CCACATCAGC CACCAACTCC TGTGCCCGCT GTGTCCCCTA CCCAATCTGT
1141 GCAGCAGAGA CGGTCACCAA GCCCCAGGAT ATGGCTGAGA AGGACACCAC CTTTGAGGCG
1201 CCACCCCTGG GGACCCAGCC GGACTGCAAC CCCACCCCAG AGAATGGCGA GGCGCCTGCC
1261 AGCACCAGCC CCACTCAGAG CTTGCTGGTG GACTCCCAGG CCAGTAAGAC GCTGCCCATC
1321 CCAACCAGCG CTCCCGTCGC TCTCTCCTCC ACGGGGAAGC CCGTTCTGGA TGCAGGGCCA
1381 GTGCTCTTCT GGGTGATCCT GGTGTTGGTT GTGGTGGTCG GCTCCAGCGC CTTCCTCCTG
1441 TGCCACCGGA GGGCCTGCAG GAAGCGAATT CGGCAGAAGC TCCACCTGTG CTACCCGGTC
1501 CAGACCTCCC AGCCCAAGCT AGAGCTTGTG GATTCCAGAC CCAGGAGGAG CTCAACGCAG
1561 CTGAGGAGTG GTGCGTCGGT GACAGAACCC GTCGCGGAAG AGCGAGGGTT AATGAGCCAG
1621 CCACTGATGG AGACCTGCCA CAGCGTGGGG GCAGCCTACC TGGAGAGCCT GCCGCTGCAG
1681 GATGCCAGCC CGGCCGGGGG CCCCTCGTCC CCCAGGGACC TTCCTGAGCC CCGGGTGTCC
1741 ACGGAGCACA CCAATAACAA GATTGAGAAA ATCTACATCA TGAAGGCTGA CACCGTGATC
1801 GTGGGGACCG TGAAGGCTGA GCTGCCGGAG GGCCGGGGCC TGGCGGGGCC AGCAGAGCCC
1861 GAGTTGGAGG AGGAGCTGGA GGCGGACCAT ACCCCCCACT ACCCCGAGCA GGAGACAGAA
1921 CCGCCTCTGG GCAGCTGCAG CGATGTCATG CTCTCAGTGG AAGAGGAAGG GAAAGAAGAC
1981 CCCTTGCCCA CAGCTGCCTC TGGAAAGTGA GGCCTGGGCT GGGCTGGGGC TAGGAGGGCA
2041 GCAGGGTGGC CTCTGGGAGG CCAGGATGGC ACTGTTGGCA CCGAGGTTGG GGGCAGAGGC
2101 CCATCTGGCC TGAACTGAGG CTCCAGCATC TAGTGGTGGA CCGGCCGGTC ACTGCAGGGG
2161 TCTGGTGGTC TCTGCTTGCA TCCCCAACTT AGCTGTCCCC TGACCCAGAG CCTAGGGGAT
2221 CCGGGGCTTG TACAGAAGAG ACAGTCCAAG GGGACTGGAT CCCAGCAGTG ATGTTGGTTG
2281 AGGCAGCAAA CAGATGGCAG GATGGGCACT GCCGAGAACA GCATTGGTCC CAGAGCCCTG
2341 GGCATCAGAC CTTAACCACC AGGCCCACAG CCCAGCGAGG GAGAGGTCGT GAGGCCAGCT
2401 CCCGGGGCCC CTGTAACCCT ACTCTCCTCT CTCCCTGGAC CTCAGAGGTG ACACCCATTG
2461 GGCCCTTCCG GCATGCCCCC AGTTACTGTA AATGTGGCCC CCAGTGGGCA TGGAGCCAGT
2521 GCCTGTGGTT GTTTCTCCAG AGTCAAAAGG GAAGTCGAGG GATGGGGCGT CGTCAGCTGG
2581 CACTGTCTCT GCTGCAGCGG CCACACTGTA CTCTGCACTG GTGTGAGGGC CCCTGCCTGG
2641 ACTGTGGGAC CCTCCTGGTG CTGCCCACCT TCCCTGTCCT GTAGCCCCCT CGGTGGGCCC
2701 AGGGCCTAGG GGCCCAGGAT CAAGTCACTC ATCTCAGAAT GTCCCCACCA ATCCCCGCCA
2761 CAGCAGGCGC CTCGGGTCCC AGATGTCTGC AGCCCTCAGC AGCTGCAGAC CGCCCCTCAC
2821 CAACCCAGAG AACCTGCTTT ACTTTGCCCA GGGACTTCCT CCCCATGTGA ACATGGGGAA
2881 CTTCGGGCCC TGCCTGGAGT CCTTGACCGC TCTCTGTGGG CCCCACCCAC TCTGTCCTGG
2941 GAAATGAAGA AGCATCTTCC TTAGGTCTGC CCTGCTTGCA AATCCACTAG CACCGACCCC
3001 ACCACCTGGT TCCGGCTCTG CACGCTTTGG GGTGTGGATG TCGAGAGGCA CCACGGCCTC
3061 ACCCAGGCAT CTGCTTTACT CTGGACCATA GGAAACAAGA CCGTTTGGAG GTTTCATCAG
3121 GATTTTGGGT TTTTCACATT TCACGCTAAG GAGTAGTGGC CCTGACTTCC GGTCGGCTGG
3181 CCAGCTGACT CCCTAGGGCC TTCAGACGTG TATGCAAATG AGTGATGGAT AAGGATGAGT
3241 CTTGGAGTTG CGGGCAGCCT GGAGACTCGT GGACTTACCG CCTGGAGGCA GGCCCGGGAA
3301 GGCTGCTGTT TACTCATCGG GCAGCCACGT GCTCTCTGGA GGAAGTGATA GTTTCTGAAA
3361 CCGCTCAGAT GTTTTGGGGA AAGTTGGAGA AGCCGTGGCC TTGCGAGAGG TGGTTACACC
3421 AGAACCTGGA CATTGGCCAG AAGAAGCTTA AGTGGGCAGA CACTGTTTGC CCAGTGTTTG
3481 TGCAAGGATG GAGTGGGTGT CTCTGCATCA CCCACAGCCG CAGCTGTAAG GCACGCTGGA
3541 AGGCACACGC CTGCCAGGCA GGGCAGTCTG GCGCCCATGA TGGGAGGGAT TGACATGTTT
3601 CAACAAAATA ATGCACTTCC TTAAAAAAAA
```

Figure 2 Amino acid sequence for CD30.

```
Peptide Sequence of CD30

MRVLLAALGL LFLGALRAFP QDRPFEDTCH GNPSHYYDKA VRRCCYRCPM
GLFTQQCPQR PTDCRKQCEP DYYLDEADRC TACVTCSRDD LVEKTPCAWN
SSRVCECRPG MFCSTSAVNS CARCFFHSVC PAGMIVKFPG TAQKNTVCEP
ASPGVSPACA SPENCKEPSS GTIPQAKPTP VSPATSSAST MPVRGGTRLA
QEAASKLTRA PDSPSSVGRP SSDPGLSPTQ PCPEGSGDCR KQCEPDYYLD
EAGRCTACVS CSRDDLVEKT PCAWNSSRTC BCRPGMICAT SATNSCARCV
PYPICAAETV TKPQDMAEKD TTFEAPPLGQ PDCNPTPENG EAPASTSPTQ
SLLVDSQASK TLPIPTSAPV ALSSTGKPVL DAGPVLFWVI LVLVVVVGSS
AFLLCHRRAC RKRIRQKLHL CYPVQTSQPK LELVDSRPRS STQLRSGASV
TEPVAEERGL MSQPLMETCH SVGAAYLESL PLQDASPAGG PSSPRDLPEP
RVSTEHTNNK IEKIYIMKAD TVIVGTVKAE LPEGRGLAGP AEPELEEELA
DHTPHYPEQE TEPPLGSCSD VMLSVEEEGK EDPLPTAASG K
```

The 132 amino acid CD30 variant protein (CD30v) had a calculated molecular mass of 14,087. Thus, the predicted CD30v protein retains most of the cytoplasmic region, but lacks both extracellular and transmembrane domains. This C-terminal cytoplasmic portion includes a highly conserved domain between human and rat CD30 with potential phosphorylation sites for PKC (amino acid positions 529–531 and 593–595) and for CK2 (amino acid positions 571–574 and 578–581) (Horie *et al.*, 1996).

Cell types and tissues expressing the receptor

Following the discovery of CD30 expression by Hodgkin's and Reed–Sternberg (H-RS) cells present in tissues involved by *Hodgkin's disease* (HD) (Schwab *et al.*, 1982) (**Figure 3a**), CD30 expression was also observed on a variable proportion of human cells in *lymphomatoid papulosis*, *angio-immunoblastic lymphoadenopathy*, *peripheral T cell lymphomas*, and on all cells in a new category of *non-Hodgkin's lymphomas* (so-called *Ki-1 lymphomas*), more appropriately referred to as *anaplastic large cell lymphomas* (ALCL) (Stein *et al.*, 1985) (**Table 1**). A small population of CD30+ cells was also found to be localized in reactive lymphoid tissue around B cell follicles (Stein *et al.*, 1982) (Figure 3b), and in the medullary areas of postnatal thymus (Romagnani *et al.*, 1998a) (Figure 3d).

By contrast, CD30 was not expressed on resting mature circulating T or B cells, but it could be induced on either cell types by phytohemagglutinin (PHA) or *Staphylococcus aureus* transformation, or by infection with human T leukemia viruses (*HTLV-I* and *HTLV-II*) or *Epstein–Barr virus* (EBV), suggesting that CD30+ lymphomatous cells may represent the malignant transformation of activated lymphoid cells of either T cell or, less commonly, B cell type (Stein *et al.*, 1985). CD30 expression was also observed on a proportion of auto- or alloactivated peripheral blood T cells (Andreesen *et al.*, 1984), whereas anti-CD3-induced mitogenesis was found to be restricted to a small subset of activated CD45RO+ T cells (Ellis *et al.*, 1993). Subsequently, Alzona *et al.* (1994) showed that CD30 was apparently a marker of a subset of memory T cells producing IFNγ and IL-5 and providing B cell helper function. Based on this finding, as well as on the demonstration that CD30 expression was upregulated by IL-12, a powerful type 1 T helper (TH1) cell inducer (Manetti *et al.*, 1993), the same authors suggested that CD30 expression was mainly associated with TH1 function (Alzona *et al.*, 1995). However, by examining CD30 expression on CD4+ and CD8+ human T cell clones with an established profile of cytokine production (TH1 and TC1, TH0 and TC0, TH2 and TC2), we found that all TH2 or TC2 clones and the majority of TH0 or TC0 clones expressed CD30 after activation, whereas TH1 and TC1 clones showed poor and transient, or no, CD30 expression (Manetti *et al.*, 1994; Del Prete *et al.*, 1995a). The preferential expression of CD30 on established TH2 clones was also confirmed by Bengtsson *et al.* (1995) and by Hamann *et al.* (1996), although both these groups questioned the possibility that CD30 expression may discriminate between human TH1- and TH2-type T cells.

CD30 expression is not limited to human cells of the lymphoid system. It has been found on tumors of nonlymphoid origin, such as *embryonal carcinomas* (Pera *et al.*, 1997; Pallesen and Hamilton-Dutoit, 1988), *seminomas* (Hittmair *et al.*, 1996), *mesenchymal tumors* (Mechtersheimer and Moller, 1990), as well as some *gastric plasmocytomas* (Moller *et al.*, 1989), and *histiocytic lymphomas* (van der Putte *et al.*, 1988). Moreover, high CD30 expression was found in normal nonlymphoid tissues, such as human decidual cells (Figure 3c) and exocrine pancreatic cells (Ito *et al.*, 1994). With the exception of embryonal carcinomas, the above mentioned CD30 expression on seminomas, mesenchymal tumors of histiocytic lymphomas could not be confirmed in subsequent studies (Mechterheimer and Stein, unpublished results). These discrepancies probably depend on the fact that a number of weak immunohistological staining patterns reported in the early studies with anti-CD30 antibodies turned out to be nonreproducible following antigen retrieval improvements by boiling the paraffin sections instead of enzymatic digestion. More recently, a truncated variant form of CD30 (CD30v), which had only the cytoplasmic domain of CD30, was found to be expressed also in alveolar macrophages (Horie *et al.*, 1996).

Murine CD30 has been characterized only recently and much less is known about its distribution. Murine

Figure 3 Detection by immunohistochemistry of CD30 expression in human tissues. (a) CD30-positive Reed–Sternberg cells in the lymph node from a patient with Hodgkin's disease. (b) Large CD30-positive lymphoid cells surrounding a B cell follicle in the lymph node from a subject with nonspecific reactive lymphoadenitis. (c) CD30-positive epithelial cells in the decidua. (d) CD30-positive cells in the medulla of postnatal thymus. From Romagnani *et al.* (1998a).

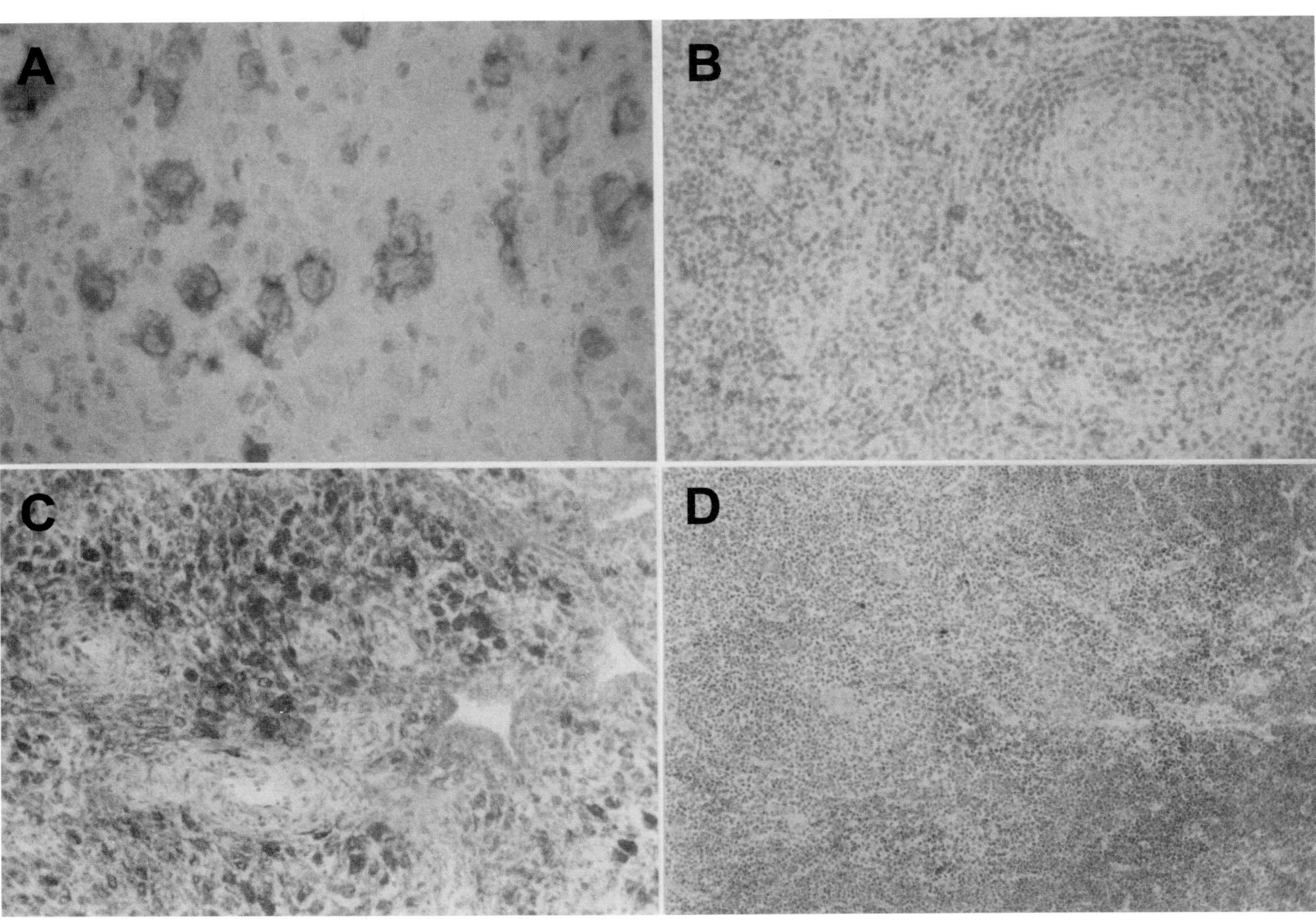

Table 1 Cells expressing CD30

Cell or tissue type	Reference
Human normal cells	
Activated macrophages	Horie *et al.*, 1996
Activated T lymphocytes	Stein *et al.*, 1982
CD4+ (TH0/TH2)	Del Prete *et al.*, 1995a
CD8+ (TC0, TC2)	Manetti *et al.*, 1994
Medullary IL-4R-positive thymocytes	Romagnani *et al.*, 1998a
Activated B lymphocytes	Stein *et al.*, 1985
Decidual cells	Ito *et al.*, 1994
Exocrine pancreatic cells	Ito *et al.*, 1994
Human neoplasias	
Hodgkin's lymphomas	Schwab *et al.*, 1982
Anaplastic large cell lymphomas	Stein *et al.*, 1985
Peripheral T cell lymphomas	Stein *et al.*, 1985
Lymphomatoid papulosis	Stein *et al.*, 1985
Angioimmunoblastic lymphoadenopathy	Stein *et al.*, 1985
Embryonal carcinomas	Pera *et al.*, 1997
Murine normal cells	
Activated B lymphocytes	Shanebek *et al.*, 1995
Activated T lymphocytes	Nakamura *et al.*, 1997
Thymus (northern blot)	Bowen *et al.*, 1996

CD30 mRNA was detected in the thymus but not in lung, brain, kidney, liver, spleen, or bone marrow. Pokeweed mitogen and Con A stimulation induced CD30 expression in splenocytes (Bowen *et al.*, 1996; Nakamura *et al.*, 1997), whereas unactivated and LPS-stimulated spleen cells did not express CD30 mRNA levels (Bowen *et al.*, 1996).

Regulation of receptor expression

Experiments performed in both mice and humans have shown that CD30 expression by activated naïve T cells is strongly dependent on the presence of IL-4, which favors the development of TH2 cells, whereas it is downregulated by IFNγ (Annunziato *et al.*, 1997; Nakamura *et al.*, 1997). In murine T cells, CD30 was expressed after activation by a majority of CD8 cells and a minority of CD4 cells. Stimulation of CD8 cells with anti-CD30 resulted in IL-5 production without IFNγ production (Bowen *et al.*, 1996). The IL-4-dependence of CD30 in activated T cells was clearly demonstrated in IL-4 knockout mice, in which CD30 expression was poor or absent (Nakamura *et al.*, 1997; Barner *et al.*, 1997; Gilfillan *et al.*, 1998), but could be restored by the addition of exogenous IL-4 (Nakamura *et al.*, 1997; Gilfillan *et al.*, 1998). Surface CD30 expression by activated T cells was even more severely affected in IL-4R knockouts than in IL-4-knockouts, suggesting that IL-4R was more important than IL-4 for CD30 expression (Barner *et al.*, 1998). Activated T cells from CD28 knockout mice were also unable to express CD30, but surprisingly, this ability was reconstituted by the addition of exogenous IL-4 (Gilfillan *et al.*, 1998). However, CD28 signaling did not upregulate CD30 expression solely through the augmentation of IL-4 production because IL-4-deficient T cells stimulated with anti-CD3 and anti-CD28 expressed CD30 (Gilfillan *et al.*, 1998). Taken together, these findings indicate that CD30 expression depends on at least three signals: (a) TCR triggering, (b) CD28 activation, and (c) IL-4 interaction with its receptor. Thus, the reason for preferential CD30 expression in TH0/TH2 and TC0/TC2 cells (Manetti *et al.*, 1994; Del Prete *et al.*, 1995a) can be explained on the basis of its partial dependence on the interaction of IL-4 (which is produced by these cell types) with their IL-4R (which is also upregulated by IL-4) (Romagnani *et al.*, 1998b) (Figure 4).

Both activation with *Staphylococcus aureus* (a T cell-independent mitogen) and infection with EBV result in CD30 expression by human B cells (Stein *et al.*, 1985). More recently, it was found that CD30 expression on B cells could also be induced by the interaction of CD40 with its ligand (CD40L), and was downregulated by B cell receptor coengagement and/or exposure of B cells to IL-6 and IL-12 (Cerutti *et al.*, 1998).

The reason why CD30 expression is highly expressed on decidual cells in the pregnant uterus and endometrium with decidual change in the secretory phase is also unclear (Ito *et al.*, 1994). However, a possible hormonal regulation can be suggested. Indeed, at least *in vitro*, progesterone was found to be capable of upregulating both IL-4 production and CD30 expression by established human T cell clones (Piccinni *et al.*, 1995). CD30 is absent from normal monocytes even after *in vitro* activation, but it has been reported that CD30 is expressed by macrophage-type cells developed during subsequent *in vitro* differentiation on Teflon membranes, as well as on some cell lines derived from *histiocytic malignancies* (Andreesen *et al.*, 1989). However, recent studies suggest that this positive CD30 staining is not specific since it is mediated by a Fc receptor-like binding site with affinities for murine IgG3, but not IgG1 monoclonal antibodies (Durkop *et al.*, unpublished

Figure 4 Detection by flow cytometry of CD30 in TH2-polarized activated human naïve and memory T cells. (a) IL-4, IFNγ, and CD30 expression by IL-12-conditioned (right part) and IL-4-conditioned (left part) naïve CD4+ T lymphocytes; (b) IL-4, IFNγ, and CD30 expression by streptokinase (SK)- (right part) and *Dermatophagoides pteronyssinus* group 1 (Der p1)- (left part) specific short-term T cell lines. The black areas of the histograms express the reactivity with anti-CD30 mAb; the open areas express the reactivity with the isotype control mAb.

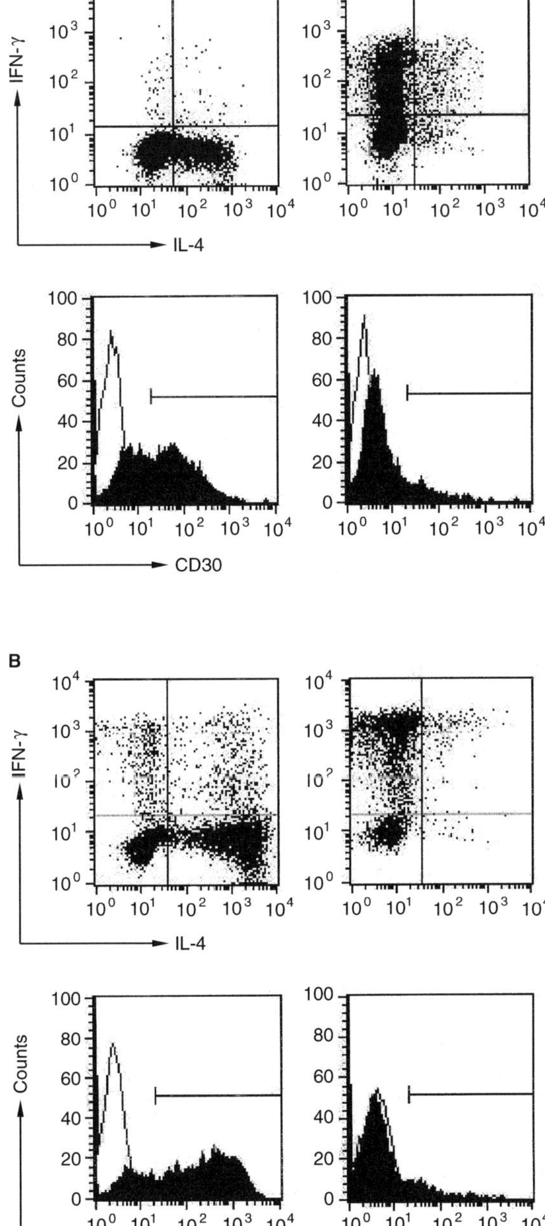

results). In addition, the re-evaluation of Andreesen's histiocytic malignancies could not confirm their true histiocytic nature, since they lacked lysozyme expression (Durkop *et al.*, unpublished results).

Release of soluble receptors

In 1989, Josimovic-Alasevic *et al.* (1989) developed an ELISA assay which was able specifically to reveal the presence of an 88 kDa soluble form of the CD30 molecule (sCD30) in culture supernatants of CD30+ cell lines, as well as in serum samples collected from patients with CD30-expressing lymphomas (*Hodgkin's disease*, *anaplastic large cell lymphomas*, HTLV-1-related *adult T cell leukemia* (ATLL), and AILD (*angioimmunoblastic lymphoadenopathy with dysproteinemia*)-like T cell lymphomas). Subsequent studies confirmed and extended these findings (Pizzolo *et al.*, 1990a, 1990b; Gause *et al.*, 1991; Nadali *et al.*, 1994, 1995, 1998; Zinzani *et al.*, 1998). As a whole, sCD30 serum levels in patients with HD and ALCL appeared to be a reliable tumor marker (Figure 5). In particular, in HD patients elevated sCD30 levels at diagnosis represented the strongest predictor of poor outcome. Also, in testicular *embryonal carcinomas* the cellular expression of CD30 was associated with increased serum levels of sCD30, possibly representing a new serological marker for this neoplastic condition (Latza *et al.*, 1995).

The extension of sCD30 determination to a large range of conditions showed that sCD30 can also frequently be detected in the serum and other biologic fluids of subjects with infections, autoimmune disorders, allergic disorders or even in normal subjects particularly during early childhood. Among infections, high levels of sCD30 have been found in the serum of subjects with *infectious mononucleosis* (Pfreundschuh *et al.*, 1990; Vinante *et al.*, 1994), *measles* (Del Prete *et al.*, 1995b), *HIV* (Pizzolo *et al.*, 1994, 1997; Rizzardi *et al.*, 1996; Sabin *et al.*, 1997), *hepatitis B virus* (HBV) (Fattovich *et al.*, 1996), and *hepatitis C virus* (HCV) (Woitas *et al.*, 1997) infections. While it is probable that in infectious mononucleosis the high serum sCD30 levels simply reflect the high turnover of EBV-infected B cells and/or EBV-specific activated CD8+ T cells, the role of high sCD30 levels in HIV infection is more complex and reveals important pathophysiological consequences (Del Prete *et al.*, 1995b), which are discussed below. The demonstration that CD30 expression preferentially associated with TH2 rather than TH1 responses suggests that high levels of sCD30 in HBV and HCV possibly reflect concomitant production of TH2-type cytokines (IL-4 and IL-10) (Reiser *et al.*, 1997) in

Figure 5 Event free survival probability of 303 patients with Hodgkin's disease according to sCD30 serum levels at diagnosis. Thick lines indicate last follow-up. Vertical bars indicate 95% CI. From Nadali *et al.* (1998).

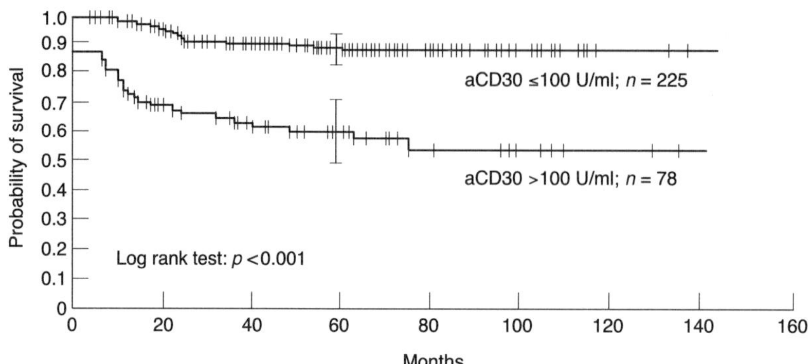

addition to IFNγ during viral infection, thus providing explanation for the chronicity of hepatitis (Woitas *et al.*, 1997).

Likewise, the elevated levels of sCD30 in *systemic lupus erythematosus* (Caligaris Cappio *et al.*, 1995), *systemic sclerosis* (Giacomelli *et al.*, 1997; Mavilia *et al.*, 1997), *primary biliary cirrhosis* (Krams *et al.*, 1996), *bullous pemphigoid* (De Pita *et al.*, 1997), have been interpreted as a consequence of TH2 cell predominance in these disorders. The same hypothesis has been formulated to explain the presence of CD30+ cells and/or the increased levels of sCD30 in the serum of patients with *ulcerative colitis* (Elewaut *et al.*, 1998; Giacomelli *et al.*, 1998), burn-associated *Candida albicans infection* (Kobayashi *et al.*, 1998), *periodontal diseases* (Cury *et al.*, 1998; Gemmell and Seymour, 1998), or during clinical remission in *multiple sclerosis* (a TH1-dominated disorder) (McMillan *et al.*, 1998), as well as the absence of CD30+ cells and/or sCD30 in the serum of patients TH1-dominated disorders, such as *Crohn's disease* (Elewaut *et al.*, 1998; Giacomelli *et al.*, 1998), *Helicobacter pylori-induced peptic ulcer* (D'Elios *et al.*, 1997; Bamford *et al.*, 1998). Finally, skin-homing CD30+ cells and/or high levels of soluble CD30 have been found in *atopic dermatitis* (Piletta *et al.*, 1996; Dummer *et al.*, 1997; Bengtsson *et al.*, 1997), or other atopic disorders (Leonard *et al.*, 1997; Nogueira *et al.*, 1998), whereas neither high levels of sCD30 nor skin-infiltrating CD30+ T cells were found in patients with contact dermatitis (Dummer *et al.*, 1998).

It should be mentioned, however, that CD30+ cells and/or high levels of sCD30 have also been reported in a proportion of patients with diseases in which TH1 responses should be predominant, such as *tuberculosis* (Munk *et al.*, 1997), *rheumatoid arthritis* (Gerli *et al.*, 1995), *Hashimoto's thyroiditis* (Okumura *et al.*, 1997), *Wegener's granulomatosis* (Wang *et al.*, 1997). These findings probably indicate that even in these diseases (or at least in some patients suffering from these diseases) IL-4 is produced and it may be responsible for the elevated CD30 expression. Recently, we found that many T cell clones generated from pleural exudates of patients with *acute tuberculosis* produce high amounts of IL-4 in addition to IFNγ, which then shift to a clearcut TH1 phenotype after successful treatment and recovery. Likewise, in the bronchial biopsy of one patient with *Wegener's granulomatosis*, we recently found a clearcut TH2 cell infiltration, characterized by the presence of CD3+ CDR30+ T cells showing high IL-4, but no IFNγ, mRNA expression and CCR3+ eosinophils. In this respect, it is of interest that elevated sCD30 levels were found in children with *Omenn's syndrome* (a TH2-dominated condition) (Chilosi *et al.*, 1996), but also in many normal children aged less than 5 years, which appeared to decrease with aging, just when the number of IFNγ-producing cells showed a progressive increase (Krampera *et al.*, unpublished results).

SIGNAL TRANSDUCTION

The CD30 cytoplasmic tail interacts with TNFR-associated factors (TRAFs), which have been shown to transduce signals mediated by TNFRII and CD40. TRAF2 also plays an important role in CD30 crosslinking-induced NFκB activation (see Transcription factors activated) (Lee *et al.*, 1996a). The same region of CD30 interacts with TRAF1, suggesting that TRAF1 and/or TRAF2 play an important role in cell death in addition to their previously identified roles in cell proliferation (Lee *et al.*, 1996b; Duckett *et al.*, 1997). Two other TNFR-associated factors, TRAF1 or TRAF3, and TRAF5 were also found to react with a CD30-binding site (Gedrich *et al.*, 1996; Aizawa *et al.*, 1997). TRAF5 has recently been cloned and sequenced (Mizushima *et al.*, 1998). Binding of TRAF2 to the cytoplasmic domain of CD30 resulted in the rapid depletion of TRAF2 and the associated protein TRAF1 by proteolysis, suggesting a model in which CD30 limits its own ability to transduce cell survival signals through signal-coupled depletion of TRAF2 (Duckett and Thompson, 1997). The CD30 induction of *HIV* gene transcription also appears to be mediated by TRAF2 (Tstitsikov *et al.*, 1997). By contrast, the latent membrane protein 1 (LMP-1) of *Epstein–Barr virus* (EBV) that contributes to the immortalizing activity of EBV in primary human B lymphocytes, associates with TRAF1, TRAF2, and TRAF3, even if much more avidly with TRAF3 (Sandeberg *et al.*, 1997). A TRAF-interacting protein (TRIP) can associate with TRAF2 and inhibits the TRAF2-mediated NFκB activation, thus acting as a receptor-proximal regulator (Lee *et al.*, 1997). The TRAF-binding motifs of CD30 intracellular tail have recently been characterized. The more N-terminal motif, 558PHYPEQET565, binds TRAF2 and TRAF3, while the more C-terminal motif, 576MLSVEEEG583, binds TRAF1 and TRAF2 (Boucher *et al.*, 1997). However, each of the three CD30 domains (D1, D2, and D3) alone can be sufficient to induce NFκB activation despite the fact that only two of them contain binding sites for TRAF proteins, suggesting involvement of a still unknown TRAF protein(s) in the signal transduction pathway of CD30 (Horie *et al.*, 1998). A diagram showing the putative role of the various CD30-associated factors in the signal transduction is depicted in **Figure 6**.

It should also be noted that CD30 may not only act as receptor for its ligand (CD30L), but also deliver reverse signaling by using CD30L as a receptor (Wiley *et al.*, 1996). However, both the physiologic meaning and the mechanisms that regulate this phenomenon are unknown.

DOWNSTREAM GENE ACTIVATION

Transcription factors activated

CD30 crosslinking leads to activation of NFκB at levels comparable to those induced by TNFα. This was independently shown by both northern blot analysis and by examining CAT activity in ACH-2 cells transfected with the HIV LTR-CAT construct (Biswas *et al.*, 1995) and in HD-derived CD30+ line L540, as well as in human T cell clones showing a TH0/TH2 cytokine

Figure 6 A flow diagram showing the relationship of various CD30 receptor-associated factors and their putative role in the signal transduction.

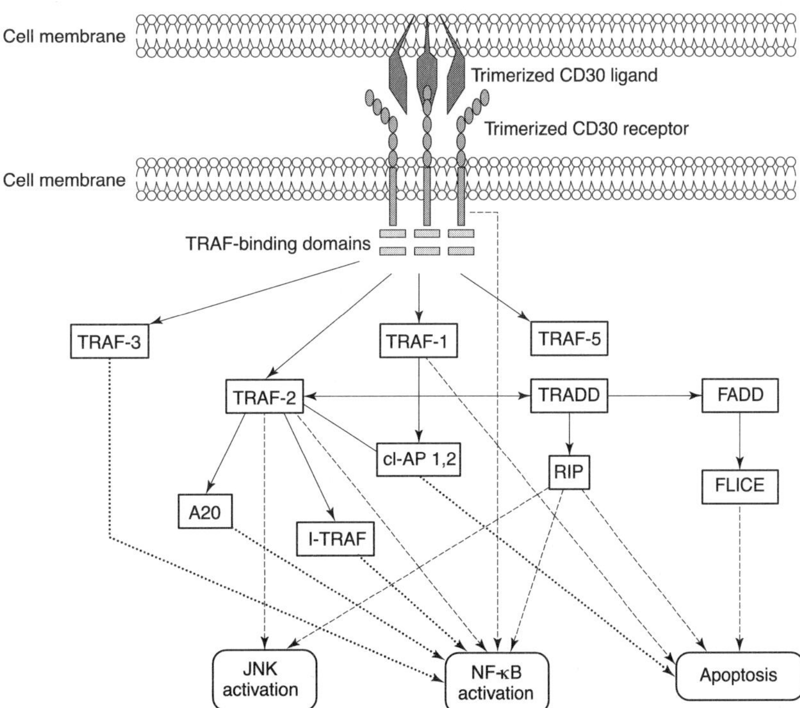

profile (McDonald *et al.*, 1995). Both p50 and p65, but not p52 or c-Rel, proteins appeared to be involved in the inducible formation of nuclear NFκB complexes in CD30-crosslinked L540 cells and T cell clones (McDonald *et al.*, 1995; Gruss *et al.*, 1996a).

BIOLOGICAL CONSEQUENCES OF ACTIVATING OR INHIBITING RECEPTOR AND PATHOPHYSIOLOGY

Unique biological effects of activating the receptors

The effects of CD30 activation have been investigated by mimicking the activity of its ligand (CD30L) with agonistic anti-CD30 antibodies (Gruss *et al.*, 1994, 1995, 1995c; McDonald *et al.*, 1995; Gruss *et al.*, 1996b), a CD30 fusion protein (Telford *et al.*, 1997), CD30L-expressing fixed CD8+ T-cell clones (Maggi *et al.*, 1995), or an insolubilized recombinant form of CD30L (Powell *et al.*, 1998). Based on these experimental models, CD30 activation *in vitro* may result in either T cell proliferation or T cell death, depending on the cell type receiving the signal and perhaps also on the microenvironmental conditions in which the signal is provided (Gruss *et al.*, 1994). Opposite effects of the CD30L are not due to CD30 mutations and do not correlate with differences in calcium mobilization after CD30 crosslinking (Jung *et al.*, 1994). On the other side, it is clear that CD30L provides an important signal for the proliferation of, and possibly also immunoglobulin production by, both murine and human B lymphocytes (Gruss *et al.*, 1994; Shanebek *et al.*, 1995).

There is strong evidence to suggest that CD30-mediated signaling plays a critical pathophysiological role in the regulation of the growth of malignant lymphomas, particularly *Hodgkin's disease*, *large and anaplastic T cell lymphoma* (LCAL), *Burkitt's lymphoma*, and CD30+ *cutaneous T cell lymphoma* (Gruss *et al.*, 1994, 1995; Tian *et al.*, 1995; Lee *et al.*, 1996b). In Hodgkin's disease (HD), CD30L enhances cytokine production, such as IL-6 and TNFα, as well as surface intercellular adhesion molecule 1 (ICAM-1/CD54) expression and shedding, by H-RS cells (Gruss *et al.*, 1994). CD30L transduction in H-RS cells involves activation of a tyrosine kinase and of a mitogen-activated protein kinase (Wendtner *et al.*, 1995). Of interest is the demonstration of high CD30L expression by eosinophils present in HD-involved lymph nodes, as well as its ability to transduce proliferative signals on CD30+ target cells, including H-RS cells (Gruss *et al.*, 1995). These data suggest that eosinophils may contribute to the deregulated network of CD30L/CD30-mediated interactive signals between H-RS cells and surrounding reactive cells in HD-involved tissues (Pinto *et al.*, 1996).

In lymphoma patients with CD30+ tumors, CD30L expression was limited to a few B cells and was reduced in those showing high levels of sCD30, suggesting that enhanced production of sCD30 by these tumors may provide a mechanism to escape the apoptosis-inducing activity of CD30L (Younes *et al.*, 1997). A direct growth-inhibitory effect, or even cytolytic cell death, was induced by CD30L on CD30+ *large and anaplastic T cell lymphoma* (LCAL). The CD30L/CD30 interaction seems to play some role in the autocrine regulation of *embryonal carcinoma* stem cells (Pera *et al.*, 1998).

CD30L/CD30 interactions have also been suggested to play a role in the progression of *HIV* infection. High levels of sCD30 were found in the sera of HIV-infected individuals, which appeared to act as an independent predictor of unfavorable prognosis for the development of full-blown disease (Pizzolo *et al.*, 1994). Subsequent reports confirmed the elevated levels of sCD30 in the serum of HIV-infected patients (Rizzardi *et al.*, 1996; Pizzolo *et al.*, 1997; Sabin *et al.*, 1997). Of interest, very high values were observed during acute primary HIV-1 infection. In this phase of the disease a multivariate logistic regression analysis showed that sCD30 was, together with TNFα levels, the best predictor of outcome, independently of CD4+ T cell count (Rizzardi *et al.*, 1996). *In vitro* studies have clearly shown that CD30 crosslinking strongly enhances *HIV* replication in chronically HIV-infected T cell lines (Biswas *et al.*, 1995). This effect was found to be mediated by NFκB activation, which in turn activates the viral LTR (Biswas *et al.*, 1995). Accordingly, both agonistic anti-CD30 antibodies and CD30L-expressing, glutaraldehyde-fixed, human CD8+ T cell clones significantly enhanced HIV expression in human CD4+ T cells obtained from HIV-infected individuals (Maggi *et al.*, 1995). Taken together, these data suggest that activation of CD30 expression in HIV-infected CD4+ T cells, which mainly occurs during TH2 responses, may allow interaction of these cells with CD30L-expressing cells and, therefore, favor HIV replication.

Recently, an important role for CD30 in the protection against *autoimmune disorders* has been reported (Kurts *et al.*, 1999). Indeed, in the absence of CD30 signaling, CD8+ T cells reactive with pancreatic β cells, gained the ability to proliferate extensively upon secondary encounter with antigen on pancreatic target tissue. This suggests that CD30 signaling can limit the proliferative potential of autoreactive CD8+ effector T cells, thus providing an important mechanism of peripheral tolerance which protects the body against autoimmunity (Kurts *et al.*, 1999).

Phenotypes of receptor knockouts and receptor overexpression mice

CD30 knockout mice exhibited elevated number of thymocytes. Moreover, activation-induced death of thymocytes from these animals after CD30 cross-linking was impaired both *in vitro* and *in vivo*. Breeding the CD30 mutation separately into $\alpha\beta$TCR or $\gamma\delta$TCR transgenic mice revealed a gross defect in negative but not positive selection (Amakawa *et al.*, 1996). More recently, CD30 knockouts were also tested for their ability to mount T cell effector responses. TH2 subset differentiation was normal in CD30-deficient mice infected with *Nippostrongylus brasiliensis*. Moreover, serum IgE and IgG1 responses, and lung eosinophilia were unaltered, demonstrating that TH2 differentiation and effector responses were not dependent on CD30 (Barner *et al.*, 1997). Experiments with CD30 transgenic mice have been recently reported. CD30 transgenic thymocytes are induced to undergo apoptosis upon crosslinking of CD30. CD30-mediated apoptosis in these animals required caspase 1 and caspase 3 and was not associated with the activation of NFκB or c-Jun, but was totally prevented by Bcl-2. These findings support the concept that CD30 may act as a costimulatory molecule in thymic negative selection (Chiarle *et al.*, 1999).

Human abnormalities

Abnormalities of CD30 expression and sCD30 release have been described in the preceding appropriate paragraphs. Taken together, all these data suggest that CD30 expression is usually not expressed in human tissues under physiologic normal conditions, except in thymocytes during thymus development, pancreatic exocrine cells, and decidual cells in the pregnant uterus and endometrium with decidual change in the secretory phase. The appearance of CD30+ cells in other sites and/or the release of sCD30 in older children and adults may result from the neoplastic transformation of some cell types (T cells, B cells, embryonic stem cells). In addition, these changes may reflect B cell infection by EBV infection or T cell infection by HTLV or effector responses characterized by IL-4 production, alone (TH2), or associated with IFNγ production (TH0), due to infections, autoimmune disorders, allergic disorders, or other chronic inflammatory disorders (Romagnani, 1997; D'Elios *et al.*, 1997; Romagnani *et al.*, 1998b; Horie and Watanabe, 1998). For example, remarkable and consistent CD30 expression was found in CD4+ T cells present in the skin perivascular infiltrates of patients with *Omenn's syndrome* (Chilosi *et al.*, 1996) (**Figure 7a**), chronic *graft-versus-host disease* (Figure 7b) (which is considered to be a TH2-dominated disorder) (Ushiyama *et al.*, 1995), as well as of patients with *progressive systemic sclerosis* (Figure 7c), which also expressed high levels of IL-4, but no IFNγ mRNA, as detected by *in situ* hybridization (Figure 7d) (Mavilia *et al.*, 1997). The preferential association of CD30 expression with TH0/TH2 effector responses is consistent with the results obtained in IL-4 knockout animal models, showing the IL-4-dependence of CD30 expression (see Regulation of receptor expression). Thus, since IL-4 generated by T cells is rapidly transported, and usually does not accumulate in sufficient concentrations to be easily demonstrated in tissues by immunohistochemistry or in biological fluids with ELISA assays, *in vivo* detection of CD30 expression or sCD30 release can be considered as useful surrogate markers for the identification of immune responses characterized by IL-4 production (D'Elios *et al.*, 1997; Romagnani, 1997).

THERAPEUTIC UTILITY

Preliminary studies clearly demonstrated that anti-CD30 immunotoxins specifically inhibited protein synthesis by *Hodgkin's disease* target cell lines (Tazzari *et al.*, 1992) and displayed a powerful *in vivo* antitumor effect in SCID mice bearing human HD (Engert *et al.*, 1990) and ALCL (Pasqualucci *et al.*, 1993). On this basis and following the demonstration that *in vivo* injection of the anti-CD30 Ber-H2 monoclonal antibody was able optimally to target CD30-expressing tumor cells, anti-CD30/saponin immunotoxin was administered to patients with advanced HD refractory to conventional therapies (Falini *et al.*, 1992a, 1992b). The observed remarkable, although transient, regression of tumor masses suggested that this therapeutic approach could possibly have a role in CD30+ neoplasias.

Subsequent studies showed cell growth inhibition *in vitro* on human ALCL xenografts of CD30+ lymphoma cells following anti-CD30 antibodies, provided that these antibodies recognized the ligand-binding site (Tian *et al.*, 1995) or by using a chimeric anti-CD30 antibody that mediates MHC and

Figure 7 Detection by immunohistochemistry of CD30-positive T cells in target tissue of subjects with TH2-dominated pathological conditions. (a) CD30-positive T cells in the skin of a child with Omenn's syndrome (courtesy of M. Chilosi, University of Verona); (b) CD30-positive T cells in the skin of a patient with chronic graft-versus-host disease (from D'Elios *et al.*, 1997); (c) CD30-positive T cells in the skin of a patient with systemic sclerosis; (d) *in situ* hybridization for IL-4 mRNA in the skin of the same patient (dark field); (e) IL-4 mRNA expression (bright field) in the skin of another patient with systemic sclerosis; (f) absence of IFNγ mRNA expression in a consecutive section of the same skin specimen (from Mavilia *et al.*, 1997).

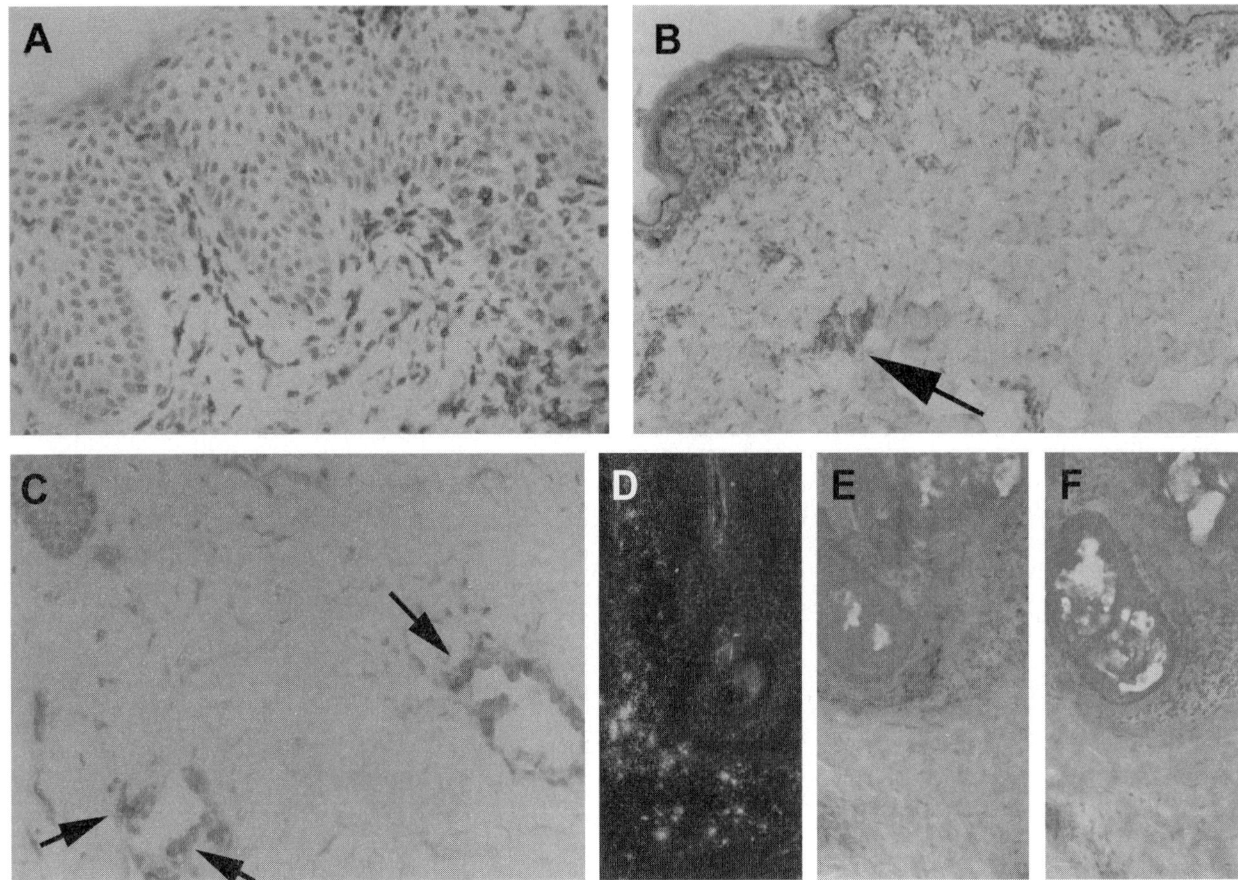

TCR/CD3-zeta-independent T-cell activation (Hombach *et al.*, 1998), or an anti-CD25/CD39 or anti-CD30/anti-saporin bispecific antibody (Sforzini *et al.*, 1998). More importantly, *in vivo* treatment in 15 patients with refractory *Hodgkin's disease* has recently been attempted by using an anti-CD16/CD30 bispecific antibody, which resulted in one complete remission, one partial remission, three minor responses and two disease stabilizations (Hartmann *et al.*, 1998). These preliminary results encourage further clinical trials with these novel immunotherapeutic approaches.

References

Aizawa, S., Nakano, H., Ishida, T., Horie, R., Nagai, M., Ito, K., Yagita, H., Okumura, K., Inoue, J., and Watanabe, T. (1997). Tumor necrosis factor receptor-associated factor (TRAF) 5 and TRAF2 are involved in CD30-mediated NF-κB activation. *J. Biol. Chem.* **272**, 2042–2045.

Alzona, M., Jack, H.-M., Fisher, R. I., and Ellis, T. M. (1994). CD30 defines a subset of activated human T-cells that produce IFN-γ and IL-5 and exhibit enhanced B-cell helper activity. *J. Immunol.* **153**, 2861–2867.

Alzona, M., Jack, H.-M., Fisher, R. I., and Ellis, T. M. (1995). IL-12 activates IFN-γ production through the preferential activation of CD30+ T cells. *J. Immunol.* **154**, 9–16.

Amakawa, R., Hakem, A., Kundig, T. M., Matsuyama, T., Simard, J. J. L., Timms, E., Wakeham, A., Mittruecker, H.-W., Griesser, H., Takimoto, H., Schmits, R., Shahinian, A., Ohashi, P. S., Penninger, J. F., and Mak, T. W. (1996). Impaired negative selection of T cells in Hodgkin's disease antigen CD30-deficient mice. *Cell* **84**, 551–562.

Andreesen, R., Osterholz, J., Lohr, G. W., and Bross, K. J. (1984). A hodgkin-specific antigen is expressed on a subset of auto- and allo-activated T (helper) lymphoblasts. *Blood* **63**, 1299–1302.

Andreesen, R., Brugger, W., Lohr, G. W., and Bross, K. J. (1989). Human macrophages can express the Hodgkin's cell-associated antigen K-1 (CD30). *Am. J. Pathol.* **134**, 187–192.

Annunziato, F., Manetti, R., Cosmi, L., Galli, G., Heusser, C. H., Romagnani, S., and Maggi, E. (1997). Opposite regulatory effect of IL-4 and IFN-γ on CD30 and LAG-3 expression by activated human naive T cells. *Eur. J. Immunol.* **27**, 2239–2243.

Bamford, K. B., Fan, X., Crowe, S. E., Leary, J. F., Gourley, W. K., Luthra, G. K., Brooks, E. G., Graham, D. Y., Reyes, V. E., and Ernst, P. B. (1998). Lymphocytes in the human gastric mucosa during *Helicobacter pylori* have a T helper cell 1 phenotype. *Gastroenterology* **114**, 482–492.

Barner, M., Kopf, M., and Lefrang, K. (1997). CD30 is a specific marker for TH2 cells but is not required for their development. Basel Institute for Immunology Annual Report. *Research Report* **54.**

Barner, M., Mohrs, M., Brombacher, F., and Kopf, M. (1998). Differences between IL-4R alpha-deficient and IL-4-deficient mice reveal a role for IL-13 in the regulation of TH2 responses. *Curr. Biol.* **8**, 669–672.

Bengtsson, A., Johansson, C., Linder Tengvall, M., Hallden, G., van der Ploeg, I., and Scheynius, A. (1995). Not only TH2 cells but also Th1 and Th0 cells express CD30 after activation. *J. Leukoc. Biol.* **58**, 683–689.

Bengtsson, A., Holm, L., Back, O., Fransson, J., and Scheynius, A. (1997). Elevated serum levels of soluble CD30 in patients with atopic dermatitis. *Clin. Exp. Immunol.* **109**, 533–537.

Biswas, P., Smith, C. A., Goletti, D., Hardy, E. C., Jackson, R. W., and Fauci, A. S. (1995). Cross-linking of CD30 induces HIV expression in chronically infected T cells. *Immunity* **2**, 587–596.

Boucher, L. M., Marengere, L. E., Lu, Y., Thukral, S., and Mak, T. W. (1997). Binding sites of cytoplasmic effectors TRAF1, 2, and 3 on CD30 and other members of the TNF receptor superfamily. *Biochem. Biophys. Res. Commun.* **233**, 592–600.

Bowen, M. A., Lee, R. K., Miragliotta, G., Nam, S. Y., and Podack, E. R. (1996). Structure and expression of murine CD30 and its role in cytokine production. *J. Immunol.* **156**, 442–449.

Caligaris-Cappio, F., Bertero, M. T., Converso, M., Stacchini, A., Vinante, F., Romagnani, S., and Pizzolo, G. (1995). Circulating levels of soluble CD30, a marker of cells producing Th2-type cytokines, are increased in patients with systemic lupus erythematosus and correlate with disease activity. *Clin. Exp. Rheumatol.* **13**, 339–343.

Cerutti, A., Schaffer, A., Shah, S., Zan, H., Liou, H. C., Goodwin, R. G., and Casali, P. (1998). CD30 is a CD40-inducible molecule that negatively regulates CD40-mediated immunoglobulin class switching in non-antigen-selected human B cells. *Immunity* **9**, 247–256.

Chiarle, R., Podda, A., Prolla, G., Podack, E. R., Jeanette Thorbecke, G., and Inghirami, G. (1999). CD30 overexpression enhances negative selection in the thymus and mediates programmed cell death via a Bcl-2 sensitive pathway. *J. Immunol.* **163**, 194–205.

Chilosi, M., Facchetti, F., Notarangelo, L. D., Romagnani, S., Del Prete, G.-F., Almerigogna, F., De Carli, M., and Pizzolo, G. (1996). CD30 cell expression and abnormal soluble CD30 serum accumulation in Omenn's syndrome. Evidence for a Th2-mediated condition. *Eur. J. Immunol.* **26**, 329–334.

Cury, V. C., Sette, P. S., da Silva, J. V., de Araujo, V. C., and Gomez, R. S. (1998). Immunohistochemical study of apical periodontal cysts. *J. Endod.* **24**, 36–37.

D'Elios, M. M., Romagnani, P., Scaletti, C., Annunziato, F., Manghetti, M., Mavilia, C., Parronchi, P., Pupilli, C., Pizzolo, G., Maggi, E., Del Prete, G.-F., and Romagnani, S. (1997). *In vivo* CD30 expression in human diseases with predominant activation of Th2-like T cells. *J. Leukoc. Biol.* **61**, 539–544.

Del Prete, G.-F., De Carli, M., Almerigogna, F., Daniel, C. K., D'Elios, M. M., Zancuoghi, G., Vinante, F., Pizzolo, G., and Romagnani, S. (1995a). Preferential expression of CD30 by human CD4+ T cells producing Th2-type cytokines. *FASEB J.* **9**, 81–86.

Del Prete, G.-F., Maggi, E., Pizzolo, G., and Romagnani S. (1995b). CD30, Th2 cytokines and HIV infection: a complex and fascinating link. *Immunol. Today* **16**, 76–80.

Del Prete, G.-F., De Carli, M., D'Elios, M. M., Daniel, K. C., Almerigogna, F., Alderson, M., Smith, C. A., Thomas, E., and Romagnani, S. (1995c). CD30-mediated signalling promotes the development of human T helper type 2-like T cells. *J. Exp. Med.* **182**, 1655–1661.

De Pita, O., Frezzolini, A., Cianchini, G., Ruffelli, M., Teofoli, P., and Puddu, P., T-helper 2 involvement in the pathogenesis of bullous pemphigoid: role of soluble CD30 (sCD30) (1997). *Arch. Dermatol. Res.* **289**, 667–670.

Duckett, C. S., and Thompson, C. B. (1997). CD30-dependent degradation of TRAF2: implications for negative regulation of TRAF signalling and the control of cell survival. *Genes Dev.* **11**, 2810–2821.

Duckett, C. S., Gedrich, R. W., Gilfillan, M. C., and Thompson, C. B. (1997). Induction of nuclear factor κB by the CD30 receptor is mediated by TRAF1 and TRAF2. *Mol. Cell. Biol.* **17**, 1535–1542.

Dummer, W., Brocker, E. B., and Bastian, B. C. (1997). Elevated serum levels of soluble CD30 are associated with atopic dermatitis, but not with respiratory atopic disorders and allergic contact dermatitis. *Br. J. Dermatol.* **137**, 185–187.

Dummer, W., Rose, C., and Brocker, E. B. (1998). Expression of CD30 on T helper cells in the inflammatory infiltrate of acute atopic dermatitis but not of allergic contact dermatitis. *Arch. Dermatol. Res.* **90**, 598–602.

Durkop, H., Latza, U., Hummel, M., Eitelbach, F., Seed, B., and Stein, H. (1992). Molecular cloning and expression of a new member of the nerve growth factor receptor family that is characteristic for Hodgkin's disease. *Cell* **68**, 421–427.

Elewaut, D., De Keyser, F., Cuvelier, C., Lazarovits, A. I., Mielants, H., Verbruggen, G., Sas, S., Devos, M., and Veys, E. M. (1998). Distinctive activated cellular subsets in colon from patients with Crohn's disease and ulcerative colitis. *Scand. J. Gastroenterol.* **33**, 743–748.

Ellis, T. M., Simms, P. E., Slivnick, D. J., Jack, H.-M., and Fisher, R. I. (1993). CD30 is a signal-transducing molecule that defines a subset of human activated CD45RO+ T cells. *J. Immunol.* **151**, 2380–2389.

Engert, A., Martin, G., Pfreundschuh, M., Amlot, P., Hsu, S. M., Diehl, V., and Thorpe, P. (1990). Antitumor effects of ricin A chain immunotoxins prepared from intact antibodies and Fab′ fragments on solid human Hodgkin's disease tumors in mice. *Cancer Res.* **50**, 2929–2935.

Falini, B., Flenghi, L., Fedeli, L., Bros, M. K., Bonino, C., Stein, H., Bigerna, B., Barbabietola, G., Venturi, S., Aversa, F., Pizzolo, G., Bartoli, A., Pileri, S., Sabattini, E., Palumbo, R., and Martelli, M. F. (1992a). *In vivo* targeting of Hodgkin's and Reed–Sternberg cells of Hodgkin's disease with monoclonal antibody Ber-H2.Immunoscintigraphic and immunohistological evidence. *Br. J. Haematol.* **82**, 38–45.

Falini, B., Bolognesi, A., Flenghi, L., Tazzari, P. L., Bros, M. K., Stein, H., Dürkop, H., Aversa, F., Corneli, P., Pizzolo, G., Barbabietola, G., Sabattini, E., Pileri, S., Martelli, M. F., and Stirpa, F. (1992b). Response of refractory Hodgkin's disease to therapy with anti-CD30 monoclonal antibody linked to saporin (BER-H2/SO6) immunotoxin. *Lancet* **339**, 1195–1196.

Fattovich, G., Vinante, F., Giustina, G., Morosato, L., Alberti, A., Ruol, A., and Pizzolo, G. (1996). Serum levels of soluble CD30 in chronic hepatitis B virus infection. *Clin. Exp. Immunol.* **103**, 105–110.

Gause, A., Pohl, C., Tschiersch, A., Da Costa, L., Jung, W., Diehl, V., Hasenclever, D., and Pfreundschuh, M. (1991). Clinical significance of soluble CD30 antigen in the sera of patients with untreated Hodgkin's disease. *Blood* **77**, 1983–1988.

Gedrich, R. W., Gilfillan, M. C., Duckett, C. S., Van Dongen, J. L., and Thompson, C. B. (1996). CD30 contains two binding sites with different specificities for members of the tumor necrosis factor receptor-associated factor family of signal transducing proteins. *J. Biol. Chem.* **271**, 12852–12858.

Gemmell, E., and Seymour, G. J. (1998). Cytokine profiles of cells extracted from humans with periodontal disease. *J. Dent. Res.* **77**, 16–26.

Gerli, R., Muscat, C., Bistoni, O., Falini, B., Tomassini, C., Agea, E., Tognellini, R., Biagini, P., and Bertotto, A. (1995). High levels of the soluble form of CD30 molecule in rheumatoid arthritis (RA) are expression of CD30+ T cell involvement in the inflammed fluids. *Clin. Exp. Immunol.* **102**, 547–550.

Giacomelli, R., Cipriani, P., Lattanzio, R., Di Franco, M., Locanto, M., Parzanese, I., Passacantando, A., Ciocci, A., and Tonietti, G. (1997). Circulating levels of soluble CD30 are increased in patients with systemic sclerosis (SSc) and correlate with serological and clinical features of the disease. *Clin. Exp. Immunol.* **108**, 42–46.

Giacomelli, R., Passacantando, A., Parzanese, I., Vernia, P., Klidara, N., Cucinelli, F., Lattanzio, R., Santori, A., Cipriani, P., Caprilli, R., and Tonietti, G. (1998). Serum levels of soluble CD30 are increased in ulcerative colitis (UC) but not in Crohn's disease. *Clin. Exp. Immunol.* **111**, 532–535.

Gilfillan, M. C., Noel, P. J., Podack, E. R., Reiner, S. L., and Thompson, C. B. (1998). Expression of the costimulatory receptor CD30 is regulated by both CD28 and cytokines. *J. Immunol.* **160**, 2180–2187.

Gruss, H., Boiani, N., Williams, D. E., Armitage, R. J., Smith, C. A., and Goodwin, R. G. (1994). Pleiotropic effects of the CD30 ligand on CD30-expressing cells and lymphoma cell lines. *Blood* **83**, 2045–2056.

Gruss, H.-J., Ulrich, D., Braddy, S., Armitage, R. J., and Dower, S. K. (1995). Recombinant CD30 ligand and CD40 ligand share common biological activities on Hodgkin and Reed–Sternberg cells. *Eur. J. Immunol.* **25**, 2083–2089.

Gruss, H.-J., Ulrich, D., Dower, S. K., Herrmann, F., and Brach, M. A. (1996a). Activation of Hodgkin cells via the CD30 receptor induced autocrine secretion of interleukin-6 engaging the NF-κB transcription factor. *Blood* **87**, 2443–2449.

Gruss, H.-J., Scheffrahn, I., Hubinger, G., Duyster, J., and Hermann, F. (1996b). The CD30 ligand and CD40 ligand regulate CD54 surface expression and release of its soluble form by cultured Hodgkin and Reed–Sternberg cells. *Leukemia* **10**, 829–835.

Hamann, D., Hilkens, C. M. U., Grogan, J. L., Lens, S. M. A., Kapsenberg, M. L., Yazdanbakash, M., and van Lier, R. A. W. (1996). CD30 expression does not discriminate between human Th1- and Th2-type T cells. *J. Immunol.* **156**, 1387–1391.

Hartmann, F., Renner, C., Jung, W., and Pfreundschuh, M. (1998). Anti-CD16/CD30 bispecific antibodies as possible treatment for refractory Hodgkin's disease. *Leuk. Lymphoma* **31**, 385–392.

Hittmair, A., Rogatsch, H., Hobisch, A., Mikuz, G., and Feichtinger, H. (1996). CD30 expression in seminoma. *Hum. Pathol.* **27**, 1166–1171.

Hombach, A., Heuser, C., Sircar, R., Tillman, T., Diehl, V., Pohl, C., and Abken, H. (1998). An anti-CD30 chimeric receptor that mediates CD3-zeta-independent T-cell activation against Hodgkin's lymphoma cells in the presence of soluble CD30). *Cancer Res.* **58**, 1116–1119.

Horie, R., and Watanabe, T. (1998). CD30: expression and function in health and disease. *Semin. Immunol.* **10**, 457–470.

Horie, R., Ito, K., Tatekawi, M., Nagai, M., Aizawa, S., Higashihara, M., Ishida, T., Inoue, J., Takizawa, H., and Watanabe, T. (1996). A variant CD30 protein lacking extracellular and transmembrane domains is induced in HL-60 by tetradecanoylphorbol acetate and is expressed in alveolar macrophages. *Blood* **88**, 2422–2432.

Horie, R., Aizawa, S., Nagai, M., Ito, K., Higashihara, M., Ishida, T., Inoue, J., and Watanabe, T. (1998). A novel domain in the CD30 cytoplasmic tail mediates NFκB activation. *Int. Immunol.* **10**, 203–210.

Ito, K., Watanabe, T., Horie, R., Shiota, M., Kawamura, S., and Mori, S. (1994). High expression of the CD30 molecule in human decidual cells. *Am. J. Pathol.* **145**, 276–280.

Josimovic-Alasevic, O., Durkop, H., Schwarting, R., Backe, E., Stein, H., and Diamantstein, T. (1989). Ki-1 (CD30) antigen is released by Ki-1 positive tumor cells *in vitro* and *in vivo*. I. Partial characterization of soluble Ki-1 antigen and detection of the antigen in cell culture supernatants and in serum by an enzyme-linked immunosorbent assay. *Eur. J. Immunol.* **19**, 157–162.

Jung, W., Krueger, S., Renner, C., Gause, A., Sahin, U., Trumper, L., and Pfreundschuh, M. (1994). Opposite effects of the CD30 ligand are not due to CD30 mutations: results from cDNA cloning and sequence comparison of the CD30 antigen from different sources. *Mol. Immunol.* **31**, 1329–1334.

Krams, S. M., Cao, S., Hayashi, M., Villanueva, J. C., and Martinez, O. M. (1996). Elevations in IFNγ, IL-5, and IL-10 in patients with the autoimmune disease primary biliary cirrhosis: association with autoantibodies and soluble CD30. *Clin. Immunol. Immunopathol.* **80**, 311–320.

Kobayashi, M., Kobayashi, H., Herndon, D. N., Pollard, R. B., and Suzuki, F. (1998). Burn-associated *Candida albicans* infection caused by CD30+ type T cells. *J. Leukoc. Biol.* **63**, 723–731.

Kurts, C., Carbone, F. R., Krummel, M. F., Koch, K. M., Miller, J. F. A. P., and Heath, W. R. (1999). Signalling through CD30 protects against autoimmune diabetes mediated by CD8 T cells. *Nature* **398**, 341–344.

Latza, U., Foss, H.-D., Dürkop, H., Eitelbach, F., Dieckmann, K.-P., Loy, W., Unger, M., Pizzolo, G., and Stein, H. (1995). CD30 antigen in embryonal carcinoma and embryogenesis and release of the soluble molecule. *Am. J. Pathol.* **146**, 463–471.

Lee, S. Y., Lee, S. Y., Kandala, G., Liou, M. L., Liou, H. C., and Choi, Y. (1996a). CD30/TNF receptor-associated factor interaction: NF-κB activation and binding specificity. *Proc. Natl Acad. Sci. USA* **93**, 9699–9703.

Lee, S. Y., Park, C. G., and Choi, Y. (1996b). T cell receptor-dependent cell death of T cell hybridomas mediated by the CD30 cytoplasmic domain in association with tumor necrosis factor receptor-associated factors. *J. Exp. Med.* **183**, 669–674.

Lee, S. Y., Lee, S. Y., and Choi, Y. (1997). TRAF-interacting protein (TRIP): a novel component of the tumor necrosis factor receptor (TNFR)- and CD30-TRAF signalling complexes that inhibits TRAF2-mediated NF-κB activation. *J. Exp. Med.* **185**, 1275–1285.

Leonard, C., Tormey, V., Burke, C., and Poulter, L. W. (1997). Allergen-induced cytokine production in atopic disease and its relationship to disease severity. *Am. J. Respir. Cell. Mol. Biol.* **17**, 368–375.

McDonald, P., Cassatella, M., Bald, A., Maggi, E., Romagnani, S., Gruss, H.-J., and Pizzolo, G. (1995). CD30 ligation induces nuclear factor-κ B activation in human T cell lines. *Eur. J. Immunol.* **25**, 2870–2876.

McMillan, S. A., McDonell, G. V., Douglas, J. P., Droogan, A. G., and Hawkins, S. A. (1998). Elevated serum and CSF levels of soluble CD30 during clinical remission in multiple sclerosis. *Neurology* **51**, 1156–1160.

Maggi, E., Annunziato, F., Manetti, R., Biagiotti, R., Giudizi, M.-G., Ravina, A., Almerigogna, F., Boiani, M., Alderson, M., and Romagnani, S. (1995). Activation of HIV expression by CD30 triggering in CD4+ T cells from HIV-infected individuals. *Immunity* **3**, 1–5.

Manetti, R., Parronchi, P., Giudizi, M. G., Piccinni, M. P., Maggi, E., Trinchieri, G., and Romagnani, S. (1993). Natural killer stimulatory factor (NKSF/IL-12) induces Th1-type specific immune responses and inhibits the development of IL-4-producing Th cells. *J. Exp. Med.* **177**, 1199–1204.

Manetti, R., Annunziato, F., Biagiotti, R., Giudizi, M.-G., Piccinni, M.-P., Giannarini, L., Sampognaro, S., Parronchi, P., Vinante, F., Pizzolo, G., Maggi, E., and Romagnani S. (1994). CD30 expression by CD8+ T cells producing type 2 helper cytokines. Evidence for large numbers of CD8+CD30+ T cell clones in human immunodeficiency virus infection. *J. Exp. Med.* **180**, 2407–2411.

Mavilia, C., Scaletti, C., Romagnani, P., Carossino, A. M., Pignone, A., Emmi, L., Pupilli, C., Pizzolo, G., Maggi, E., and Romagnani, S. (1997). Type 2 helper T (Th2) cell predominance and high CD30 expression in systemic sclerosis. *Am. J. Pathol.* **151**, 1751–1758.

Mechtersheimer, G., and Moller, P. (1990). Expression of Ki-1 antigen (CD30) in mesenchymal tumors. *Cancer* **66**, 1732–1737.

Mizushima, S., Fujita, M., Ishida, T., Azuma, S., Kato, K., Hirai, M., Otsuka, M., Yamamoto, T., and Inoue, J. (1998). Cloning and characterization of a cDNA encoding the human homolog of tumor necrosis factor receptor-associated factor 5 (TRAF5). *Gene* **207**, 135–140.

Moller, P., Matthaei-Maurer, D. U., and Moldenhauer, G. (1989). CD30(Ki-1) antigen expression in a subset of gastric mucosal plasma cells and in a primary gastric plasmacytoma. *Am. J. Clin. Pathol.* **91**, 18–23.

Munk, M. E., Kern, P., and Kaufmann, S. H. (1997). Human CD30+ cells are induced by *Mycobacterium tuberculosis* and present in tuberculosis lesions. *Int. Immunol.* **9**, 713–720.

Nadali, G., Vinante, F., Ambrosetti, A., Todeschini, G., Veneri, D., Zanotti, R., Meneghini, V., Ricetti, M. M., Benedetti, F., Vassanelli, A., Perona, G., Chilosi, M., Menestrina, F., Fiacchini, M., Stein, H., and Pizzolo, G. (1994). Serum levels of soluble CD30 are elevated in the majority of untreated patients with Hodgkin's disease and correlate with clinical features and prognosis. *J. Clin. Oncol.* **12**, 793–797.

Nadali, G., Vinante, F., Stein, H., Todeschini, G., Tecchio, C., Morosato, L., Chilosi, M., Menestrina, F., Kinney, M. C., Greer, J. P., Latza, U., Perona, G., and Pizzolo, G. (1995). Serum levels of the soluble form of CD30 molecule as a tumor marker in CD30+ anaplastic large cell lymphoma. *J. Clin. Oncol.* **13**, 1355–1360.

Nadali, G., Tavecchia, L., Zanolin, E., Bonfante, V., Viavini, S., Camerini, E., Musto, P., Di Renzo, N., Carotenuto, M., Chilosi, M., Krampera, M., and Pizzolo, G. (1998). Serum level of the soluble form of the CD30 molecule identifies patients with Hodgkin's disease at high risk of unfavorable outcome. *Blood* **91**, 3011–3016.

Nakamura, T., Lee, R. K., Nam, S. Y., Al-Ramadi, B. K., Koni, P. A., Bottomly, K., Podack, E. R., and Flavell, R. A. (1997). Reciprocal regulation of CD30 expression on CD4+ T cells by IL-4 and IFNγ. *J. Immunol.* **158**, 2090–2098.

Nogueira, J. M., Pinto, P. L., Loureiro, V., Prates, S., Gaspar, A., Almeida, M. M., and Pinto, J. E. (1998). Soluble CD30, dehydroepiandrosterone in atopic and non atopic children. *Allerg. Immunol.* **30**, 3–8.

Okumura, M., Hidaka, Y., Kuroda, S., Takeoka, K., Tada, H., and Amino, N. (1997). Increased serum concentrations of soluble CD30 in patients with Graves' disease and Hashimoto's thyroiditis. *J. Clin. Endocrinol. Metab.* **82**, 1757–1760.

Pallesen, G., and Hamilton-Dutoit, S. J. (1988). Ki-1 (CD30) antigen is regularly expressed by tumor cells of embryonal carcinoma. *Am. J. Pathol.* **133**, 446–450.

Pasqualucci, L., Wasik, M., Teicher, B., Martelli, M. F., Falini, B., and Kadin, M. E. (1993). *In vivo* antitumor activity of Ber-H2 (CD30)/saporin immunotoxin against human Ki-1+ anaplastic large cell lymphoma in SCID mice. *Blood* **82** (Suppl), 137a.

Pera, M. F., Bennett, W., and Cerretti, D. P. (1997). Expression of CD30 and CD30L in cultured cell lines from human germ-cell tumors. *Lab. Invest.* **76**, 497–504.

Pera, M. F., Bennett, W., and Cerretti, D. P. (1998). CD30 and its ligand: possible role in regulation of teratoma stem cells. *APMIS* **106**, 169–172.

Pfreundschuh, M., Pohl, C., Berenbeck, C., Schroeder, J., Jung, W., Schmits, R., Tschiersch, A., Diehl, V., and Gause, A. (1990). Detection of a soluble form of the CD30 antigen in sera of patients with lymphoma, adult T-cell leukemia and infectious mononucleosis. *Int. J. Cancer* **45**, 869–874.

Piccinni, M.-P., Giudizi, M.-G., Biagiotti, R., Giannarini, L., Sampognaro, S., Parronchi, P., Manetti, R., Annunziato, F., Livi, C., Romagnani, S., and Maggi E. (1995). Progesterone favors the development of human T helper cells producing Th2-type cytokines and promotes both IL-4 production and membrane CD30 expression in established Th1 cell clones. *J. Immunol.* **155**, 128–133.

Piletta, P. A., Wirth, S., Hommel, L., Saurat, J. H., and Hauser, C. (1996). Circulating skin-homing T cells in atopic dermatitis. *Arch. Dermatol.* **132**, 1171–1176.

Pinto, A., Aldinucci, D., Gloghini, A., Zagonel, V., Degan, M., Improta, S., Juzbasic, S., Todesco, M., Perin, V., Gattei, V., Herrman, F., Gruss, H.-J., and Carbone, A. (1996). Human eosinophils express functional CD30 ligand and stimulate proliferation of a Hodgkin's disease cell line. *Blood* **88**, 3299–3305.

Pizzolo, G., Vinante, F., Chilosi, M., Dallenbach, F., Josimovic-Alasevic, O., Diamantstein, T., and Stein, H. (1990a). Serum levels of soluble CD30 molecule (Ki-1 antigen) in Hodgkin's disease: relationship with disease activity and clinical stage. *Br. J. Haematol.* **75**, 282–284.

Pizzolo, G., Stein, H., Josimovic-Alasevic, O., Vinante, F., Zanotti, R., Chilosi, M., Feller, A. C., and Diamanstein, T. (1990b). Increased serum levels of soluble IL-2 receptor, CD30 and CD8 molecules, and γ-interferon in angioimmunoblastic lymphoadenopathy: possible pathogenic role of immunoactivation mechanisms. *Br. J. Haematol.* **75**, 485–488.

Pizzolo, G., Vinante, F., Morosato, L., Nadali, G., Chilosi, M., Gandini, G., Sinicco, A., Raiteri, R., Semenzato, G., Stein, H., and Perona, G. (1994). High serum level of the soluble form of CD30 molecule in the early phase of HIV-1 infection as an independent predictor of progression to AIDS. *AIDS* **8**, 741–745.

Pizzolo, G., Vinante, F., Nadali, G., Krampera, M., Morosato, L., Chilosi, M., Raiteri, R., and Sinicco, A. (1997). High serum level of soluble CD30 in acute primary HIV-1 infection. *Clin. Exp. Immunol.* **108**, 251–253.

Powell, I. F., Li, T., Jack, H. M., and Ellis, T. M. (1998). Construction and expression of a soluble form of human CD30 ligand with functional activity. *J. Leukoc. Biol.* **63**, 752–757.

Reiser, M., Marousis C. G., Nelson, D. R., Lauer, G., Gonzalez-Peralta, R. P., Davis, G. L., and Lau, J. Y. (1997). Serum interleukin 4 and interleukin 10 levels in patients with chronic hepatitis C virus infection. *J. Hepatol.* **26**, 471–478.

Rizzardi, G. P., Barcellini, W., Tambussi, G., Lillo, F., Malnati, M., Perrin, L., and Lazzarin, A. (1996). Plasma levels of soluble CD30, tumor necrosis factor (TNF)-alpha and TNF receptors during primary HIV-1 infection: correlation with HIV-1 RNA and the clinical outcome. *AIDS* **10**, 45–50.

Romagnani, P., Annunziato, F., Manetti, R., Mavilia, C., Lasagni, L., Manuelli, C., Vannelli, G. B., Vanini, V., Maggi, E., Pupilli, C., and Romagnani, S. (1998a). High CD30 ligand expression by epithelial cells and Hassal's corpuscles in the medulla of human thymus. *Blood* **91**, 3323–3332.

Romagnani, P., Annunziato, F., and Romagnani, S. (1998b). Pleiotropic biologic functions of CD30/CD30L. Does it contribute to negative selection in thymus? *The Immunologist* **6**, 137–141.

Romagnani, S. (1997). The Th1/Th2 paradigm. *Immunol. Today* **18**, 263–266.

Sabin, C. A., Bofill, M., Phillips, A. N., Elford, J., Janossy, G., and Lee, C. A. (1997). Relation between soluble CD30 levels measured soon after HIV seroconversion and disease progression in men with hemophilia. *J. Acquir. Immune Defic. Syndr. Hum. Retrovirol.* **16**, 279–283.

Sandeberg, M., Hammerschmidt, W., and Sugden, B. (1997). Characterization of LMP-1's association with TRAF1, TRAF2, and TRAF3. *J. Virol.* **71**, 4649–4656.

Shanebeck, K. D., Maiszewski, C. R., Kennedy, M. K., Picha, K. S., Smith, C. A., Goodwin, R. G., and Grabstein, K. H. (1995). Regulation of murine B cell growth and differentiation by CD30 ligand. *Eur. J. Immunol.* **25**, 2147–2153.

Schwab, U., Stein, H., Gerdes, J., Lemke, H., Kirchner, H., Schaadt, M., and Diehl, V. (1982). Production of a monoclonal antibody specific for Hodgkin and Sternberg–Reed cells of Hodgkin's disease and a subset of normal lymphoid cells. *Nature* **299**, 65–67.

Sforzini, S., de Totero, D., Gaggero, A., Ippoliti, R., Glennie, M. J., Canevari, S., Stein, H., and Ferrini, S. (1998). Targeting of saporin to Hodgkin's lymphoma cells by anti-CD30 and anti-CD25 bispecific antibodies. *Br. J. Haematol.* **102**, 1061–1068.

Stein, H., Gerdes, J., Schwab, U., Lemke, H., Mason, D. Y., Ziegler, A., Schienle, W., and Diehl, V. (1982). Identification of Hodgkin and Sternberg-Reed cells as a unique cell type derived from a newly-detected small-cell population. *Int. J. Cancer* **30**, 445–459.

Stein, H., Mason, D. Y., Gerdes, J., O'Connor, N., Wainscoat, J., Pallesen, G., Gatter, K., Falini, B., Delson, G., Lemke, H., Schwarting, R., and Lennert, K. (1985). The expression of Hodgkin's disease associated antigen Ki-1 in reactive and neoplastic lymphoid tissue: evidence that the Reed Sternberg cells and histiocytic malignancies are derived from activated lymphoid cells. *Blood* **66**, 848–858.

Tazzari, P. L., Bolognesi, A., de Totero, D., Falini, B., Lemoli, R. M., Soria, M. R., Gobbi, M., Stein, H., Flenghi, L., Martelli, M. F., and Stirpe, F. (1992). Ber-H2 (anti-CD30)-saporin immunotoxin: A new tool for the treatment of Hodgkin's disease and CD30+ lymphoma. *In vitro* evaluation. *Br. J. Haematol.* **81**, 203–211.

Telford, W. G., Nam, S. Y., Podack, E. R., and Miller, R. A. (1997). CD30-regulated apoptosis in murine CD8 T cells after cessation of TCR signals. *Cell Immunol.* **182**, 125–136.

Tian, Z. G., Longo, D. L., Funahoshi, S., Asai, O., Ferris, D. K., Widmer, M., and Murphy, W. J. (1995). *In vivo* antitumor effects of unconjugated CD30 monoclonal antibodies on human anaplastic large-cell lymphoma xenografts. *Cancer Res.* **55**, 5335–5341.

Tsitsikov, E. N., Wright, D. A., and Geha, R. S. (1997). CD30 induction of human immunodeficiency virus gene transcription is mediated by TRAF2. *Proc. Natl Acad. Sci. USA* **94**, 1390–1395.

Ushiyama, C., Hirano, T., Miyajima, H., Okumura, K., Ovary, Z., and Hashimoto, H. (1995). Anti-IL-4 antibody prevents graft-versus-host disease in mice after bone marrow transplantation. **154**, 2687–2696.

van der Putte, S. C., Toonstra, J., van Wichen, D. F., van Unnik, J. A., and van Vloten, W. A. (1988). The expression of the Hodgkin's disease-associated antigen Ki-1 in cutaneous infiltrates. *Acta Derm. Venereol.* **68**, 202–208.

Vinante, F., Morosato, L., Siviero, F., Nadali, G., Rigo, A., Veneri, D., de Sabata, D., Vincenzi, C., Chilosi, M., Semenzato, G., and Pizzolo, G. (1994). Soluble forms of p55-IL-2Ra, CD8, and CD30 molecules as markers of lymphoid cell activation in infectious mononucleosis. *Haematologica* **79**, 413–419.

Wang, G., Hansen, H., Tatsis, E., Csernok, E., Lemke, H., and Gross, W. L. (1997). High plasma levels of the soluble form of CD30 activation molecule reflect disease activity in patients with Wegener's granulomatosis. *Am. J. Med.* **102**, 517–523.

Wendtner, C. M., Schmitt, B., Gruss, H. J., Druker, B. J., Emmerich, B., Goodwin, R. G., and Hallek, M. (1995). CD30 ligand signal transduction involves activation of a tyrosine kinase and of mitogen-activated protein kinase in a Hodgkin's lymphoma cell line. *Cancer Res.* **55**, 4157–4161.

Wiley, S. R., Goodwin, R. G., and Smith, C. A. (1996). Reverse signalling via CD30 ligand. *J. Immunol.* **157**, 3635–3639.

Woitas, R. P., Lechmann, M., Jung, G., Kaiser, R., Sauerbruch, T., and Spengler, U. (1997). CD30 induction and cytokine profiles in hepatitis C virus core-specific peripheral blood T lymphocytes. *J. Immunol.* **159**, 1012–1018.

Younes, A., Consoli, U., Snell, V., Clodi, K., Kliche, K. O., Palmer, J. L., Gruss, H.-J., Armitage, R. J., Thomas, E. K., Cabanillas, F., and Andreeff, M. (1997). CD30 ligand in lymphoma patients with CD30+ tumors. *J. Clin. Oncol.* **15**, 3355–3362.

Zinzani, P. L., Pileri, S., Bendandi, M., Buzzi, M., Sabatini, E., Ascani, S., Gherlinzoni, F., Magagnoli, M., Albertini, P., and Tura, S. (1998). Clinical implications of serum levels of soluble CD30 in 70 adult anaplastic large-cell lymphoma patients. *J. Clin. Oncol.* **16**, 1532–1537.

4-1BB

Dass S. Vinay[1] and Byoung S. Kwon[2,3,*]

[1]Department of General Surgery, University of Michigan Medical School, 1516 MSRB I, 1150 West Medical Center Drive, Ann Arbor, MI 48109, USA

[2]The Immunomodulation Research Center, University of Ulsan, Ulsan, Korea

[3]Department of Ophthalmology, LSUMC, 2020 Gravier Street Suite B, New Orleans, LA 70112, USA

*corresponding author tel: 504-412-1200 ext 1379, fax: 504-412-1315, e-mail: bkwon@lsumc.edu

DOI: 10.1006/rwcy.2000.16007.

SUMMARY

4-1BB (CD137) is a member of the tumor necrosis factor (TNF) receptor superfamily expressed on activated CD4+, CD8+, as well as on NK cells (Vinay and Kwon, 1998). 4-1BB binds to 4-1BB ligand (4-1BBL) found on activated B cells, macrophages, and dendritic cells (Alderson *et al.*, 1994; Goodwin *et al.*, 1993; Pollok *et al.*, 1994; DeBenedette *et al.*, 1997). Data accumulated thus far show that signals generated through 4-1BB induce T cell activation and cytokine modulation (Vinay and Kwon, 1998). These observations are further substantiated by the findings that 4-1BBL-expressing cell lines or soluble 4-1BBL induce T cell proliferation and IL-2 secretion (Goodwin *et al.*, 1993; DeBenedette *et al.*, 1995, 1997; Hurtado *et al.*, 1995; Saoulli *et al.*, 1998). Also, the finding that addition of 4-1BB fusion proteins block T cell proliferation mediated by 4-1BBL (Goodwin *et al.*, 1993; DeBenedette *et al.*, 1995, 1997; Hurtado *et al.*, 1995; Saoulli *et al.*, 1998) further support a strong role for 4-1BB in T cell regulation. Evidence gathered also points out that antibodies to 4-1BB possess an increasing ability to activate and control CD8 responses (Shuford *et al.*, 1997; Melero *et al.*, 1997). This fact is further evidenced by recent findings using 4-1BBL-deficient mice (DeBenedette *et al.*, 1999; Tan *et al.*, 1999). Interestingly, literature also points out a role for 4-1BB in TH1 (Kim *et al.*, 1998) and TH2 (Chu *et al.*, 1997) and TC1 and TC2 development (Vinay and Kwon, 1999). Although, CD28 is considered as central to the T cell immune responses (Chambers and Allison, 1997), evidence available demonstrates that during T cell activation, especially when stimulus through TCR is acute, 4-1BB can replace CD28 in the costimulation of naïve T cells (Watts and DeBenedette, 1999).

The molecular basis underlying 4-1BB-mediated T cell activation is beginning to be appreciated. Although exhaustive studies are lacking, the available data show that this T cell-activating antigen transmits signals via interactions with the TNF receptor-associated (TRAF) family of molecules, such as TRAF1, TRAF2, and TRAF3, and activates NFκB (Saoulli *et al.*, 1998; Arch and Thompson, 1998; Jang *et al.*, 1998).

BACKGROUND

Optimal induction of T cell activation is incomplete without cognate interaction between T cells and antigen-presenting cell (APC)-derived cell surface molecules (Schwarz, 1990). A number of APC-derived cell surface determinants have been shown to possess the ability to potentiate/desensitize T cell effector functions. Although it is considered that B7/CD28 is central to this pathway, studies of CD28-deficient mice (Shahnian *et al.*, 1993) have shown that this may not be the limiting factor in immune regulation. Recent studies indicate that several ligand/receptor pairs, beyond the B7/CD28 pathway, also have the ability to initiate and propel the ongoing immune reaction. Among these are three members of the expanding TNFR family: 4-1BB, Ox40, and GITR that may be involved in T cell function in the post CD28 phase.

Discovery

The T cell-activating antigen (Ag) 4-1BB was initially discovered in screens for receptors on activated mouse lymphocytes (Kwon and Weissman, 1989).

Alternative names

4-1BB was originally named 'induced by lymphocyte activation' (ILA) in humans and 4-1BB in the mouse (Pollok *et al.*, 1993; Schwarz *et al.*, 1993). Both ILA and 4-1BB recently received the human leukocyte differentiation antigen (HLDA) nomenclature CD137.

Structure

4-1BB is a 30 kDa glycoprotein and exists both as a monomer and a 55 kDa dimer on the T cell surface. The entire gene spans approximately 13 kb of mouse chromosome 4.

Main activities and pathophysiological roles

4-1BB is shown to transmit costimulatory signals to T cells in synergy with anti-CD3. Functional studies reveal that antibodies to the 4-1BB molecule can increase *graft-versus-host disease*, accelerate the rejection of cardiac allografts and skin transplants, and eradicate established tumors. 4-1BB is also known to play an important role in *HIV* and *Hodgkin's disease* (see Vinay and Kwon, 1998 for a review).

GENE

Accession numbers

U022567

Sequence

The 4-1BB gene consists of 10 exons and 9 introns in which there are two exons for 5′UTR and 8 for the coding region (**Figure 1**). Two types of UTR sequences were identified in the genomic sequence and found to be separated by an intron ~2.5 kb in length. The cysteine-rich extracellular domain is divided into six exons. The signal sequence, transmembrane region,

Figure 1 Organization of the 4-1BB gene. Exons are shown as boxes; UTRs are displayed as red boxes while the protein-coding regions are green boxes (Roman numerals). Lengths of exons and introns are shown in Arabic numerals (base pairs). (Full colour figure may be viewed online.)

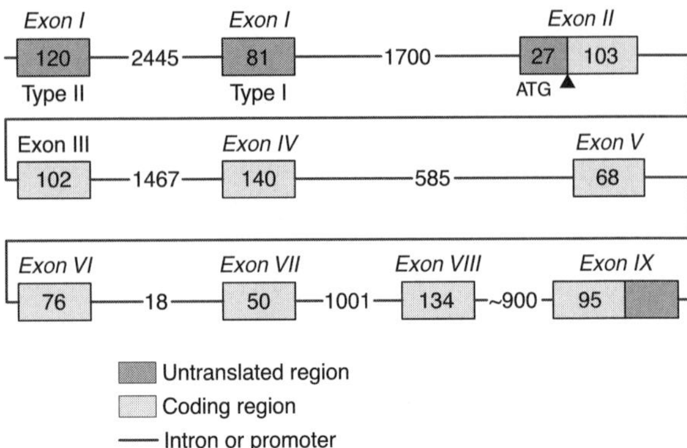

and serine, threonine, proline (STP)-rich region immediately outside the transmembrane domain are contained in separate exons. Finally, the cytoplasmic domain that contains the p56lck-binding site is located in the last exon of the gene. The exon/intron boundaries are assigned by comparing the 4-1BB cDNA sequence with the genomic sequence. In the flanking sequence of the type I 5′ UTR, no TATA box-related elements were found. Instead, they are very good matches of the consensus TPA-responsive element (AP-1) at positions −18 to −10, and of the NFκB-binding sequence at positions −49 to −39. Upstream of these elements, this region contains a potential ets-binding site at positions −169 to −162, an activator protein 2 (AP-2)-binding site at positions −460 to −453, a typical CAAT element at positions −498 to −494, and an SP-1-binding site at positions −522 to −516.

The 5' flanking region of the type II 5′ UTR contains a TATA-related element at positions −28 to −23. Two potential ets-binding sites appear at positions −15 to −8 and −139 to −132, two potential AP-2-binding sites at positions −89 to −82 and −331, and a very good match of an AP-1-binding site at positions −311 to −302.

PROTEIN

Accession numbers

Murine 4-1BB: J04492
Human 4-1BB: U03397
Mouse 4-1BB ligand: L15435
Human 4-1BB ligand: U03398

Description of protein

The nucleotide sequence of murine 4-1BB revealed a single open reading frame which codes for a polypeptide of 256 amino acids with a calculated mass of 27,587 Da. The predicted protein contained an unusually large number (23) of cysteines. This region comprises four potential TNFR motifs, of which the first is partial and the third distinct from those of the TNFR.

Following this ligand-binding domain is a stretch of amino acids (residues 140–185), in which almost 30% of the amino acids are serine and threonines; potential sites for *O*-linked glycosylation. Amino acids 186–211 constitute the hydrophobic transmembrane domain followed by the stop-transfer sequence containing several basic residues. The C-terminal part of the cytoplasmic domain contains two short runs of three and four acidic residues, respectively, and a sequence of five glycines followed by a tyrosine.

The human homolog of 4-1BB (h4-1BB) contains 255 amino acids with two potential *N*-linked glycosylation sites. The molecular weight of its protein backbone is calculated to be 27 kDa (Zhou *et al.*, 1995) and is 60% identical to mouse 4-1BB. In the cytoplasmic domain, five regions of amino acid sequences are conserved between mouse and human, indicating that these residues might be important in 4-1BB function.

Affinity for ligand(s)

Although the h4-1BB ligand was detected in both T and B cells of human peripheral blood, the ligand was preferentially expressed in primary B cells and B cell lines. Daudi, a B cell lymphoma, was one of the B cell lines that carried a higher number of ligands. Scatchard analysis showed that the $K_d = 1.4 \times 10^9$ M and the number of ligands in Daudi cells was 4.2×10^3. Primary B cells when stimulated with pokeweed mitogen showed enhanced ligand expression receptor binding. On the other hand, the ligand for murine 4-1BB (m4-1BBL) can be found at low levels on T cell lines (nonactivated and anti-CD3-activated), pre-B cell lines and a few immature macrophage cell lines. Also, the binding analysis with 4-1BB AP (a fusion protein consisting of the extracellular domain of 4-1BB fused to human placental alkaline phosphatase) exhibited no binding to glial tumor cell line, HeLa cells, or COS cells. On the other hand, the anti-IgM-activated primary B cells compared with anti-CD3-activated primary T cells showed higher binding of 4-1BB AP. Scatchard analysis indicated that the A 20 B cell lymphoma expressed 3680 binding sites per cell with a K_d of 1.86 nM. Western analysis showed that 4-1BBL has a molecular mass of approximately 18–25 kDa.

Cell types and tissues expressing the receptor

The 4-1BB molecule is present on activated T cells and monocytes.

Regulation of receptor expression

The 4-1BB is not detected (< 3%) on resting T cells and T cell lines. However, when the T cells, in the presence of APCs, stimulated with a variety of agonists (plate-bound anti-CD3, Con A, PHA, IL-2, IL-4, anti-CD28, PMA, ionomycin alone or in combinations) upregulates and maintain its expression (Pollok *et al.*, 1993). Upon activation, 4-1BB mRNA is detected within hours and cell surface expression can be detected between 10 and 16 hours. The expression peaks at 64 hours and is maintained until 120 hours after single stimulation (Hurtado *et al.*, 1995). Interestingly, expression of 4-1BB is detectable on T cells derived from individuals with pathological conditions (Michel *et al.*, 1998) indicating that its expression is indeed activation dependent.

Release of soluble receptors

One of the interesting aspects of 4-1BB in immune regulation is its ability to occur in soluble form in tissue microenvironment. The 4-1BB shares this feature with certain other receptor forms like TNFR, NGFR, CD27, CD30, and CD95. Evidence documented till now suggests that the 4-1BB molecule is secreted in soluble form in sera and lymphocyte secretions in patients with *rheumatoid arthritis* (for details, see Michel *et al.*, 1998). The significance of such a phenomenon remains to be explored.

SIGNAL TRANSDUCTION

Associated or intrinsic kinases

Though not studied in detail, the signal transduction pathways chosen by 4-1BB is beginning to be appreciated. Studies carried out thus far suggest that no apparent kinase activity is observed with 4-1BB, suggesting that 4-1BB-associated molecules may be involved in 4-1BB-mediated signal transduction.

Cytoplasmic signaling cascades

Available data indicate that, like its relatives in the TNFR superfamily, signals by 4-1BB are relayed through TNF receptor-associated factors (TRAFs). Studies from our lab found that TRAF1, TRAF2, and TRAF3 all interact with the cytoplasmic domain of 4-1BB and mutation analysis showed the involvement of the runs of acidic residues in the cytoplasmic domain of 4-1BB (Jang *et al.*, 1998). Jang *et al.* (1998) and Arch and Thompson (1998) demonstrated that 4-1BB crosslinking induces activation of NFκB and is known to be inhibited by dominant negative TRAF2 and NFκB-inducing kinase (NIK). Our laboratory also identified a putative binding site for the T cell-specific protein tyrosine kinase p56lck, in the cytoplasmic domain (Kim *et al.*, 1993). The latter observation suggests that 4-1BB may be linked to the src tyrosine kinase family signaling pathway, but this has not been investigated further as yet. A recent study showed a role for apoptosis signal regulatory kinase (ASK-1) and c-Jun N-terminal kinase (JNK)/ stress-activating protein kinase (SAPK) in 4-1BB-mediated T cell activation (Cannons *et al.*, 1999).

BIOLOGICAL CONSEQUENCES OF ACTIVATING OR INHIBITING RECEPTOR AND PATHOPHYSIOLOGY

Following introduction of agonistic antibodies to 4-1BB to activate quiescent T cells it is generally 4–5 days before a response is noted (Kim *et al.*, 1998), possibly because T cells require 2–3 days to bring out optimal cell surface expression of 4-1BB (Pollok *et al.*, 1993).

Once crosslinked either by anti-4-1BB or 4-1BB ligand (4-1BBL), the 4-1BB molecule transmits a distinct and potent costimulatory signal leading to the activation and differentiation of CD4+ and CD8+ cells in the context of anti-CD3 as first signal, resulting in the secretion of cytokines (IL-2, IL-4, and IFNγ) (Chu *et al.*, 1997). Stimulation through 4-1BB leading to T cell activation appears to work in both a CD28-dependent and -independent manner (DeBenedette *et al.*, 1997). The capacity of 4-1BB to prevent activation-induced cell death (AICD) (Hurtado *et al.*, 1997) and to rescue and sustain the ongoing immune reactions in the apparent absence of the CD28 molecule (Kim *et al.*, 1998) is well documented in literature. It is interesting to note that 4-1BB-mediated signaling lies in its ability to potentiate the cytolytic capacity of IL-2-activated intestinal intraepithelial cells (IELs) (Zhou *et al.*, 1994).

Our laboratory recently provided evidence that 4-1BB may be involved in the pathogenesis of *AIDS*. The level of 4-1BB expression and the percentage of 4-1BB-expressing T cells were higher in HIV-1-positive individuals than in HIV-1-negative controls (Wang *et al.*, 1998). In addition, crosslinking 4-1BB with agonistic monoclonal antibody enhanced *HIV*-1 replication. Administration of monoclonal antibodies to the 4-1BB molecule can completely eradicate established tumors (Melero *et al.*, 1997). Shuford and colleagues (1997) demonstrated that *in vivo* administration of such monoclonal antibodies enhances *graft-versus-host disease* (GVHD), and increases the rapidity of cardiac allograft and skin transplant rejection. The role played by 4-1BB in *Hodgkin's disease* (HD) suggests that eosinophils bearing ligands for 4-1BB may act as an important element in the pathology of HD (Gruss *et al.*, 1996).

Phenotypes of receptor knockouts and receptor overexpression mice

Our laboratory has recently generated mice deficient in the 4-1BB molecule. Unpublished data from our laboratory point out that the absence of 4-1BB leads to dysregulated lymphocyte activation and abnormalities in the myeloid progenitor pool. These mice, under nonimmunized conditions, tend to secrete increased amounts of IgG2a, IgA; increased antigen-specific (T-dependent) IgG1, IgG2a, and offer no interference in the IgG class-switch.

References

Alderson, M. R., Smith, C. A., Tough, T. W., Davis-Smith, T., Armitage, R. J., Falk, B., Roux, E., Baker, E., Sutherland, G. R., and Din, W. S. (1994). Molecular and biological characterization of human 4-1BB and its ligand. *Eur. J. Immunol.* **24**, 2219–2227.

Arch, R. H., and Thompson, C. B. (1998). 4-1BB and OX-40 are members of tumor necrosis factor (TNF)-nerve growth factor receptor subfamily that bind TNF receptor-associated factors and activate nuclear factor κB. *Mol. Cell. Biol.* **18**, 558–565.

Cannons, J. L., Hoeflich, K. P., Woodgett, J. R., and Watts, T. H. (1999). Role of the stress kinase pathway in signaling via the T cell costimulatory receptor 4-1BB. *J. Immunol.* **163**, 2990–2998.

Chambers, C. A., and Allison, J. P. (1997). Co-stimulation in T cell reponses. *Curr. Opin. Immunol.* **9**, 396–404.

Chu, N. R., DeBenedette, M. A., Stiernholm, B. J., Barber, B. H., and Watts, T. H. (1997). Role of IL-12 and 4-1BBL in cytokine production by $CD28^+$ $CD28^-$ T cells. *J. Immunol.* **158**, 3081–3089.

DeBenedette, M. A., Chu, N. R., Pollok, K. E., Hurtado, J. C., Wade, W. F., Kwon, B. S., and Watts, T. H. (1995). Role of 4-1BB ligand in costimulation of T lymphocyte growth and its upregulation on M12 B lymphomas by cAMP. *J. Exp. Med.* **181**, 985–992.

DeBenedette, M. A., Shahnian A., Mak, T. W., and Watts, T. H. (1997). Costimulation of $CD28^-$ T lymphocytes by 4-1BB ligand. *J. Immunol.* **158**, 551–559.

DeBenedette, M. A., Wen, T., Bachmann, M. F., Ohashi, P. S., Barber, B. H., Stocking, K. L., Peschon, J. J., and Watts, T. H. (1999). Analysis of 4-1BB ligand (4-1BBL)-deficient mice and of mice lacking both 4-1BBL and CD28 reveals a role for 4-1BBL in skin allograft rejection and in the cytotoxic T cell response to influenza virus. *J. Immunol.* **163**, 4833–4841.

Goodwin, R. G., Din, W. S, Davis-Smith, T., Anderson, D. M., Gimpel, S. D., Sato, T. A., Maliszewski, C. R., Brannan, C. I., Copeland, N. G., Jenkins, N. A. *et al.* (1993). Molecular cloning of a ligand for the inducible T cell gene 4-1BB: a member of an emerging family of cytokines with homology to tumor necrosis factor. *Eur. J. Immunol.* **23**, 2631–2641.

Gruss, H. J., Duyster, J., and Herrmann, F. (1996). Structural and biological features of the TNF ligand superfamilies: interactive signals in the pathobiology of Hodgkin's disease. *Ann. Oncol.* **4**, 19–26.

Hurtado J. C., Kim, S. H., Pollok, K. E., Lee, Z. E., and Kwon, B. S. (1995). Potential role of 4-1BB in T cell activation. Comparison with the costimulatory molecule CD28. *J. Immunol.* **155**, 3360–3367.

Hurtado, J. C., Kim, Y.-J., and Kwon, B. S. (1997). Signals generated through 4-1BB are costimulatory to previously activated splenic T cells and inhibit activation-induced cell death. *J. Immunol.* **158**, 2600–2609.

Jang, I.-K., Lee, Z. H., Kim, Y.-J., Kim, S. H., and Kwon. B. S. (1998). Human 4-1BB signals are mediated by TRAF2 and activate nuclear factor-κB (NF-κB). *Biophys. Biochem. Acta* **242**, 613–620.

Kim, Y.-J., Pollok, K. E., Zhou, Z., Shaw, A., Bohlen, J., Fraser, M., and Kwon, B. S. (1993). A T cell antigen 4-1BB associates with the protein tyrosine kinase $p56^{lck}$. *J. Immunol.* **151**, 1255–1262.

Kim, Y.-J., Kim, S. H., Mantel, P., and Kwon, B. S. (1998). Human 4-1BB regulates CD28 costimulation to promote Th1 cell responses. *Eur. J. Immunol.* **28**, 881–890.

Kwon, B. S., and Weissman, S. M. (1989). cDNA sequences of two inducible T-cell genes. *Proc. Natl Acad. Sci. USA* **86**, 1963–1967.

Melero, I., Shuford, W. W., Newsby, S. A., Aruffo, A., Ledbetter, K. E., Hellstrom, R., Mittler, R. S., and Chen, L. (1997). Monoclonal antibodies against 4-1BB T cell activation molecule eradicate established tumors. *Nature Med.* **3**, 682–685.

Michel, J., Langstein, J., Hofstadter, F., and Schwarz, H. (1998). A soluble form of CD137 (ILA/4-1BB), a member of TNF receptor family, is released by activated lymphocytes and is detectable in sera of patients with rheumatoid arthritis. *Eur. J. Immunol.* **28**, 290–295.

Pollok, K. E., Kim, Y.-J., Zhou, Z., Hurtado, J. C., Kim, K.-K., Pickard, R. T., and Kwon, B. S. (1993). The inducible T cell antigen 4-1BB: analysis of expression and function. *J. Immunol.* **150**, 771–781.

Pollok, K. E., Kim, Y.-J., Hurtado, J. C., Zhou, Z., Kim, K. K., and Kwon, B. S. (1994). 4-1BB T-cell antigen binds to mature B cells and macrophages, and costimulates anti-mu-primed splenic B cells. *Eur. J. Immunol.* **24**, 367–374.

Saoulli, K., Lee, S. Y., Cannons, J. L., Yeh, W. C., Santana, A., Goldstein, M. D., Bangia, N., DeBenedette, M. A., Mak, T. W., Choi, Y., and Watts, T. H. (1998). CD28-independent, TRAF2-dependent costimulation of resting T cells by 4-1BB ligand. *J. Exp Med.* **187**, 1849–1862.

Schwarz, R. H. (1990). A cell culture model for T lymphocyte clonal anergy. *Science* **248**, 1349–1356.

Schwarz H., Tuckwell, J., and Lotz, M. (1993). A receptor-induced by lymphocyte activation (ILA): a new member of the human nerve growth factor/tumor necrosis factor receptor family. *Gene* **134**, 295–298.

Shahnian, A., Pieffer, K., Lee, K. P., Kundig, T. M., Kishihara, K., Wakeham, A., Kawai, K., Ohashi, P. M., Thompson, C. B., and Mak, T. W. (1993). Differential T cell costimulatory requirements in CD28-deficient mice. *Science* **261**, 609–612.

Shuford, W. W., Brown, T. J., Emswiler, J., Raecho, H., Larsen, C. P., Pearson, T. C., Ledbetter, J. A., Aruffo, A., and Mittler, R. S. (1997). 4-1BB costimulatory signals preferentially induce $CD8^+$ T cell proliferation and leads to the amplification of *in vivo* cytotoxic T cell responses. *J. Exp. Med.* **186**, 47–55.

Tan, J. T., Whitmire, J. K., Ahmed, R., Pearson, T. C., and Larsen, C. P. (1999). 4-1BB ligand, a member of the TNF family, is important for the generation of antiviral CD8 T cell responses. *J. Immunol.* **163**, 4859–4868.

Vinay, D. S., and Kwon, B. S. (1998). Role of 4-1BB in immune responses. *Semin. Immunol.* **10**, 481–489.

Vinay, D. S., and Kwon, B. S. (1999). Differential expression and costimulatory effect of 4-1BB (CD137) and CD28 molecules on cytokine-induced murine CD8(+) Tc1 and Tc2 cells. *Cell. Immunol.* **192**, 63–71.

Wang , S., Kim, Y.-J., Bick, C., Kim, S. H., and Kwon, B. S. (1998). The potential roles of 4-1BB costimulation in HIV type I infection. *AIDS Res. Hum. Retrovirus* **14**, 223–231.

Watts, T. H., and DeBenedette, M. A. (1999). T cell co-stimulatory molecules other than CD28. *Curr. Opin. Immunol.* **11**, 286–293.

Zhou, Z., Pollok, K. E., Kim, Y.-J., and Kwon, B. S. (1994). Functional analysis of T cell antigen 4-1BB in activated intestinal intra-epithelial T lymphocytes. *Immunol. Lett.* **41**, 177–184.

Zhou, Z., Kim, S. H., Hurtado, J. C., Lee, Z. H., Kim, K. K., Pollok, K. E., and Kwon, B. S. (1995). Characterization of human homologue of 4-1BB and its ligand. *Immunol. Lett.* **45**, 67–73.

ACKNOWLEDGEMENTS

SRC funds to IRC from the Korean Ministry of Science and Technology and NIH Grants (AI28125 and DE12156) are greatly appreciated.

RANK

William J. Boyle*

Department of Cell Biology, Amgen, Inc., One Amgen Center Drive, Thousand Oaks, CA 91320-1799, USA

*corresponding author tel: 805-447-4304, fax: 805-447-1982, e-mail: bboyle@amgen.com

DOI: 10.1006/rwcy.2000.16008.

SUMMARY

RANK is a novel member of the TNFR superfamily that is capable of activating the NFκB and JNK pathways during cell survival. It was originally cloned from a dendritic cell cDNA library suggesting a role in modulating dendritic cell function(s) during regulation of immune responses. RANK is one of the largest TNFR-related proteins, characterized as having four cysteine-rich repeats motifs in its extracellular domain, and a long intracellular signaling domain. The intracellular domain is now known to bind to members of the TRAF family of signal transducers, which mediate the activation of NFκB and JNK following ligand binding.

RANK is expressed at very high levels on osteoclast precursors, and is the hematopoietic cell surface protein that mediates the osteoclastogenic effects of RANKL. Since RANKL induction by calciotropic hormones and pro-resorptive cytokines is known to regulate bone density and calcium metabolism, RANK is implicated as the osteoclast receptor that integrates these humoral signals during physiologic conditions and during disease.

BACKGROUND

Discovery

The TNFR-related protein RANK (receptor activator of NFκB) was first identified as an interesting EST sequence obtained from a dendritic cell cDNA library (Anderson *et al.*, 1997). It is now known to be an important component of the OPG–RANKL–RANK axis involved in regulating bone and immune homeostasis. The full-length cDNA encoded a novel TNFR-related protein, and was predicted to be involved in regulating dendritic cell function(s) during adaptive immune responses. Functional analysis of the receptor in mammalian cell lines indicated that, like other TNFR family members, RANK was capable of activating the transcription factor NFκB. A ligand for RANK (RANKL) was expression cloned from a T cell line library using a soluble receptor fusion dimer. RANK was also identified by two independent groups as the receptor that regulated osteoclastogenesis mediated by osteoclast differentiation factor (ODF) or osteoprotegerin ligand (OPGL) (Nakagawa *et al.*, 1998; Hsu *et al.*, 1999). Interestingly, RANKL/ODF/OPGL/TRANCE was also identified as a T cell protein whose expression was controlled by calcineurin-regulated transcription factors during activation (Wong *et al.*, 1997).

Thus, the various experimental approaches used to identify both RANK and RANKL strongly implicated a role for this receptor in regulating T cell and dendritic cell interactions. In addition, cell biological and molecular genetic analysis in mouse has revealed an essential function in regulating osteoclast differentiation and activation during bone remodeling and metabolism (see Suda *et al.*, 1999, for review).

Alternative names

ODFR, osteoclast differentiation factor receptor (Nakagawa *et al.*, 1999).
ODAR, osteoclast differentiation and activation receptor (Hsu *et al.*, 1999).
TRANCE-R, TRANCE receptor (Wong *et al.*, 1998).

Structure

The human RANK polypeptide is a 616 amino acid type I transmembrane protein that is functionally divided into two regions (Anderson *et al.*, 1997). The N-terminal region of the protein is the extracellular ligand-binding domain of the protein that is displayed on the cell surface. The C-terminal region of the receptor is the cytoplasmic, signal-transducing portion that effects cellular metabolic functions in response to activation by ligand.

Main activities and pathophysiological roles

Activation of RANK by overexpression in transfected cells, or by treatment of receptor-bearing cells *in vitro* with soluble RANKL (a.k.a. TRANCE/ODF/OPGL), has been shown to stimulate signal transduction leading to the activation of NFκB (Anderson *et al.*, 1997; Darney *et al.*, 1999, Hsu *et al.*, 1999). The mechanism for this is described below, but is known to involve cytoplasmic factors that belong to the TRAF family of proteins. RANK activation has also been shown to rapidly stimulate Jun N-terminal kinase (JNK/SPK), leading to activation of AP-1 related transcription factors (Wong *et al.*, 1997; Galibert *et al.*, 1998; Hsu *et al.*, 1999). Our current view of RANK is that it is involved in both immune and bone homeostasis.

RANK was identified as a dendritic surface receptor, indicating a potential role in modulating dendritic cell differentiation and survival. Soluble RANKL can act as a costimulatory factor during antigen presentation during *in vitro* culture (Anderson *et al.*, 1997), and impacts the survival of dendritic cells during *in vitro* culture (Wong *et al.*, 1997). The effects mediated by RANKL on dendritic cell function are dependent on native folding of this cytokine, and can be blocked by the addition of a soluble RANK-Ig fusion protein.

RANK has also been identified as the intrinsic hematopoietic cell surface determinant required for the differentiation and activation of the osteoclast (Nakagawa *et al.*, 1998; Hsu *et al.*, 1999). RANK is expressed at very high levels on osteoclast progenitor cells (Lacey *et al.*, 1998; Hsu *et al.*, 1999). Addition of soluble RANKL to bone marrow cultures in the presence of CSF-1 (M-CSF) stimulates osteoclastogenesis and the activation of mature osteoclasts to resorb bone via binding to RANK (Hsu *et al.*, 1999; Burgess *et al.*, 1999). These effects can be blocked by the addition of either OPG or soluble RANK-Ig. Recombinant soluble RANK-Ig can block osteoclastogenesis and activation when administered to mice, leading to increases in bone mass (Hsu *et al.*, 1999).

GENE

Accession numbers

Human and mouse cDNA sequences:
AF018253
AF019046
AF019047
AF019048

Sequence

See **Figure 1**.

PROTEIN

Accession numbers

Human RANK protein: AF018253

Sequence

See **Figure 2**.

Description of protein

Human RANK is a TNFR-related, type I transmembrane protein of 616 amino acid residues (**Figure 3**), whereas the mouse polypeptide is 625 residues. A hydrophobic signal peptide of 29 amino acids located at the extreme N-terminus, and is removed during synthesis. The C-terminal extracellular domain is composed of four tandem cysteine-rich pseudo repeat sequences (I–IV) that are characteristic of this family (Smith *et al.*, 1994), followed by a stalk of about 17 residues. The pseudo repeat sequences are involved in ligand binding, and are most closely related to domains I–IV of CD40, a TNFR-related protein important for B cell maturation and activation.

A hydrophobic sequence of about 23 amino acids forms a membrane-spanning domain and helps to define the extracellular domain of this receptor. The C-terminal 383 amino acid residues forms the cytoplasmic region or RANK, and is one of the longest cytoplasmic domains of receptors in this class. This region of the protein has been shown to bind to a class of cytoplasmic tumor necrosis factor receptor-associated factors (TRAFs), which are involved in ligand-induced signal transduction leading to the activation of NFκB and JNK/SAPK pathways (Arch *et al.*, 1998).

Figure 1 Nucleotide sequence for human RANK.

```
1981  CGATCGGTAC AGTCGAGGAA GACCACCCGG CATTCTCTGC CCACTTTGCC TTCCAGGAAA
2041  TGGGCTTTTC AGGAAGTGAA TTGATGAGGA CTGTCCCCAT GCCCACGGAT GCTCAGCAGC
2101  CCGCCGCACT GGGGCAGATG TCTCCCCTGC CACTCCTCAA ACTCGCAGCA GTAATTTGTG
2161  GCACTATGAC AGCTATTTTT ATGACTATCC TGTTCTGTGG GGGGGGGTC TATGTTTTCC
2221  CCCCATATTT GTATTCCTTT TCATAACTTT TCTTGATATC TTTCCTCCCT CTTTTTTAAT
2281  GTAAAGGTTT TCTCAAAAAT TCTCCTAAAG GTGAGGGTCT CTTTCTTTTC TCTTTTCCTT
2341  TTTTTTTTCT TTTTTTGGCA ACCTGGCTCT GGCCCAGGCT AGAGTGCAGT GGTGCGATTA
2401  TAGCCCGGTG CAGCCTCTAA CTCCTGGGCT CAAGCAATCC AAGTGATCCT CCCACCTCAA
2461  CCTTCGGAGT AGCTGGGATC ACAGCTGCAG GCCACGCCCA GCTTCCTCCC CCCGACTCCC
2521  CCCCCCCAGA GACACGGTCC CACCATGTTA CCCAGCCTGG TCTCAAACTC CCCAGCTAAA
2581  GCAGTCCTCC AGCCTCGGCC TCCCAAAGTA CTGGGATTAC AGGCGTGAGC CCCCACGCTG
2641  GCCTGCTTTA CGTATTTTCT TTTGTGCCCC TGCTCACAGT GTTTTAGAGA TGGCTTTCCC
2701  AGTGTGTGTT CATTGTAAAC ACTTTTGGGA AAGGGCTAAA CATGTGAGGC CTGGAGATAG
2761  TTGCTAAGTT GCTAGGAACA TGTGGTGGGA CTTTCATATT CTGAAAAATG TTCTATATTC
2821  TCATTTTTCT AAAAGAAAGA AAAAAGGAAA CCCGATTTAT TTCTCCTGAA TCTTTTTAAG
2881  TTTGTGTCGT TCCTTAAGCA GAACTAAGCT CAGTATGTGA CCTTACCCGC TAGGTGGTTA
2941  ATTTATCCAT GCTGGCAGAG GCACTCAGGT ACTTGGTAAG CAAATTTCTA AAACTCCAAG
3001  TTGCTGCAGC TTGGCATTCT TCTTATTCTA GAGGTCTCTC TGGAAAAGAT GGAGAAAATG
3061  AACAGGACAT GGGGCTCCTG GAAAGAAAGG GCCCGGGAAG TTCAAGGAAG AATAAAGTTG
3121  AAATTTTAAA AAAAAA
```

Figure 2 Amino acid sequence for human RANK protein. The hydrophobic signal peptide and transmembrane domain are underlined.

RANK protein

Accession number and sequence

Human RANK protein: AF018253

```
1     MAPRARRRRP LFALLLLCAL LARLQVALQI APPCTSEKHY EHLGRCCNKC EPGKYMSSKC
61    TTTSDSVCLP CGPDEYLDSW NEEDKCLLHK VCDTGKALVA VVAGNSTTPR RCACTAGYHW
121   SQDCECCRRN TECAPGLGAQ HPLQLNKDTV CKPCLAGYFS DAFSSTDKCR PWTNCTFLGK
181   RVEHHGTEKS DAVCSSSLPA RKPPNEPHVY LPGLIILLLF ASVALVAAII FGVCYRKKGK
241   ALTANLWHWI NEACGRLSGD KESSGDSCVS THTANFGQQG ACEGVLLLTL EEKTFPEDMC
301   YPDQGGVCQG TCVGGGPYAQ GEDARMLSLV SKTEIEEDSF RQMPTEDEYM DRPSQPTDQL
361   LFLTEPGSKS TPPFSEPLEV GENDSLSQCF TGTQSTVGSE SCNCTEPLCR TDWTPMSSEN
421   YLQKEVDSGH CPHWAASPSP NWADVCTGCR NPPGEDCEPL VGSPKRGPLP QCAYGMGLPP
481   EEEASRTEAR DQPEDGADGR LPSSARAGAG SGSSPGGQSP ASGNVTGNSN STFISSGQVM
541   NFKGDIIVVY VSQTSQEGAA AAAEPMGRPV QEETLARRDS FAGNGPRFPD PCGGPEGLRE
601   PEKASRPVQE QGGAKA
```

Relevant homologies and species differences

The mouse and human RANK cDNAs have been cloned and sequenced, and their protein products compared (Anderson *et al.*, 1997). The human and mouse proteins are about 85% identical, with sequence gaps due to the fact that the murine protein is longer. All of the cysteine residues located in the full-length mature proteins (residues 29–194) are conserved

Figure 3 The primary structure of the RANK polypeptide. The extracellular domain contains a short hydrophobic signal peptide (black box), four tandem cysteine-rich pseudo repeat sequences (I–IV), and a short stock region. A hydrophobic transmembrane domain anchors the RANK protein to the cell membrane. The cytoplasmic region contains at least three functional TRAF-binding sites indicated by the hatched bars. The amino acid coordinates for the various RANK structural features are listed below.

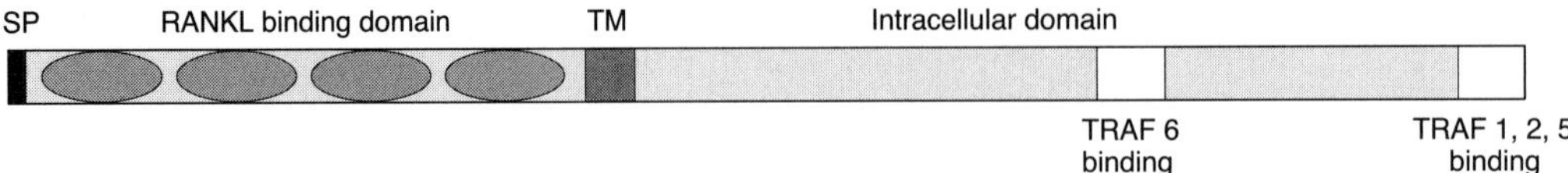

and in identical positions. RANK is most closely related to the TNFR-related protein CD40, and is about 40% similar in the ligand-binding domain.

Affinity for ligand(s)

Various lines of evidence suggest that RANKL is the major, if not sole, ligand for RANK. Mice in which either the RANK (Dougall *et al.*, 1999; Li *et al.*, 2000) or RANKL (OPGL) (Kong *et al.*, 1999a) genes have been disrupted by homologous recombination have near identical phenotypes with respect to bone metabolism and lymph node organogenesis. Biochemical analysis of surface proteins found on primary marrow-derived osteoclast progenitors indicates that RANK is the only protein present on these cells that interacts with soluble RANKL (Hsu *et al.*, 1999; Nakagawa *et al.*, 1998). Soluble RANK-Ig fusion protein (sRANK-Ig) and RANKL binding coefficients have been determined by BIAcore and solution binding, and found to be in the range of $\sim 3 \times 10^{-9}$ M (Hsu *et al.*, 1999). Osteoprotegerin, a secreted RANKL neutralizing receptor, binds to RANKL with approximately 10–100 times higher affinity ($\sim 3^{-10} \times 10^{-11}$) using these same assays. In support of this difference in affinity for the same ligand, OPG and sRANK-Ig have similar differences in EC_{50} values for neutralizing RANKL-induced osteoclastogenesis *in vitro* and *in vivo* (Hsu *et al.*, 1999).

Cell types and tissues expressing the receptor

See **Table 1**.

Table 1 Cell types and tissues expressing the receptor

Organs and tissues	Cartilaginous bone primordia during fetal development
	Intestinal epithelium
	Kidney
	Liver
	Bone and growth plate cartilage
	Lymph node
	Spleen
	Thymus
	Heart
Cell types	Hypertrophic chondrocytes
	Osteoclast progenitors
	Myeloid precursors
	B cells, activated
	Intestinal epithelium (small and large)
	Dendritic cells
	Foreskin fibroblasts
Cell lines	RAW 264.7 (murine macrophage)
	KG-1 (human myeloid leukemia)
	K562 (human erythroleukemia)
	LIM 1863 (human colorectal carcinoma)
	MP-1 (human lymphoblastoid)
	A-172 (human glioblastoma)
	W1-26 (human lung fibroblast, SV-40 transformed)

Regulation of receptor expression

Little is actually known about the regulation of RANK expression in cells and tissues. Like many receptors, regulation of its activity is via ligand binding, and not necessarily at the transcriptional level. Cell types bearing this receptor are likely to acquire expression during development and lineage allocation, such as in the osteoclast. RANK expression is detected early during hematopoietic development from stem cells, and is found on early myeloblast cell lines such as KG-1 and K562 (Anderson *et al.*, 1997; Lacey *et al.*, 1998). RANK expression on the surface of hematopoietic precursor cells is the key determinant that typifies the osteoclast progenitor (Lacey *et al.*, 1998; Hsu *et al.*, 1999; Li *et al.*, 2000). Anderson *et al.* (1997) have identified cells that express high levels of RANK mRNA, but do not express surface protein that crossreacts with anti-RANK antibodies, suggesting that posttranscriptional mechanisms regulate surface localization of the receptor. Interestingly, activation by the cytokines IL-4 and TGFβ1 induce surface expression in these cells.

Osteoclast progenitors express very high levels of RANK mRNA and surface receptor (Lacey *et al.*, 1998; Hsu *et al.*, 1999). During differentiation into mature osteoclasts, RANK expression is downregulated, although the receptor is still present and able to bind and respond to soluble RANKL (Burgess *et al.*, 1999).

The mouse RANK transcript is expressed in the cartilaginous primordia of bone during embryonic development, then later is expressed in the intestine, kidney, lung, and bone (Hsu *et al.*, 1999). *In situ* hybridization of embryonic and adult mouse bone indicates that RANK is expressed at sites of robust bone resorption and remodeling, such as in the growth plate cartilage region. In addition, RANK is expressed on hypertrophic chondrocytes, suggesting a role in regulating growth plate physiology. In support of this concept, mice deficient in RANKL have developmental alterations in chondrocyte development, and have abnormal-looking growth plate cartilage (Kong *et al.*, 1999a).

Release of soluble receptors

The mature RANK polypeptide has not been shown to be cleaved from the cell surface, releasing a soluble ectodomain. However, as mentioned above, posttranscriptional regulation of RANK is thought to occur, and this could involve cleavage of the extracellular domain. If this occurs, this could provide a mechanism for modulating the effects of RANKL during immune and bone homeostasis. Osteoprotegerin is a known secreted inhibitor of RANKL.

SIGNAL TRANSDUCTION

The primary structure of RANK suggests that it mediates the effects of RANKL in a fashion similar to that of other TNRF-related proteins when they engage their cognate ligands: induction of cellular metabolic processes via the activation of signal transducing cytoplasmic factors. RANK contains no death domain motifs, and is not believed to play a role in mediating apoptosis. In contrast, its structure is more related to receptors such as CD40, which tend to promote cell survival and can stimulate differentiation. RANK was first characterized as a dendritic cell surface protein proposed to function in T cell and dendritic cell interactions during the immune response. Wong *et al.* (1997) first reported that RANKL (TRANCE) could stimulate JNK/SAPK activity in treated cells, suggesting that RANK was involved in stimulating cell activation mechanisms. RANK itself was first characterized as a receptor-like protein capable of stimulating the transcription factor NFκB (Anderson *et al.*, 1997). These data suggested that two major cellular signaling pathways, JNK/SAPK and NFκB, were regulated by activation of this receptor. Activation of RANK on osteoclast progenitor cells leads to the rapid stimulation of JNK activity, and subsequently the induction of osteoclast-specific gene expression (Lacey *et al.*, 1998; Hsu *et al.*, 1999). Apparently, NFκB is not activated during this process, but is already constitutively active at this stage of osteoclast precursor development. However, RANK present on mature osteoclasts responds to ligand treatment by induction of both JNK/SAPK and NFκB activity, and these functions both appear to be critical for osteoclast survival and activation (Jimi *et al.*, 1999).

The RANK cytoplasmic domain is about 383 amino acids in, and contains no obvious structural features that imply a signaling mechanism. Other members of the TNFR family are known to mediate signal transduction via cytoplasmic factors belong to the TRAF family of proteins (Arch *et al.*, 1998). RANK presumably mediates the activation of the transcription factor NFκB and JNK/SAPK activity by coupling with this class of cytoplasmic factors. The TRAF family members 1, 2, 3, 5, and 6 were all found to interact with the cytoplasmic domain of RANK *in vitro* (Galibert *et al.*, 1998; Wong *et al.*, 1998; Darney *et al.*, 1999; Hsu *et al.*, 1999) (**Figure 4**). Putative TRAF-binding sites have been found at several locations within the RANK cytoplasmic domain of about 5–6 amino acids in length (Darney *et al.*, 1999). In cultured cells, TRAF2, 5, and 6 interactions with the RANK cytoplasmic domain have been detected (Galibert *et al.*, 1998; Darney *et al.*, 1999; Hsu *et al.*, 1999). Both TRAF2 and TRAF5 have been shown to bind within the same sites at the very C-terminus of the protein, while TRAF6 binds to a two potential juxtaposed sites lying between amino acid residues 340 to 358 (Galibert *et al.*, 1998; Hsu *et al.*, 1999). Yeast two-hybrid interaction screening

Figure 4 RANK signaling pathway. Illustration of the RANK type I transmembrane protein on the cell surface in relation to cytoplasmic factors involved in RANK signal transduction. RANKL binding to RANK induces the aggregation of TRAF2, 5, and 6, and subsequently the activation of NFκB and JNK/SAPK. TRAF6 has been implicated as the major signal transducing TRAF protein involved in regulating osteoclastogenesis. During RANK-L-induced osteoclastogenesis, JNK/SAPK is rapidly induced, followed by induction of osteoclast specific gene expression (TRAP, cathepsin K, calcitonin receptor, and $\alpha v\beta 3$).

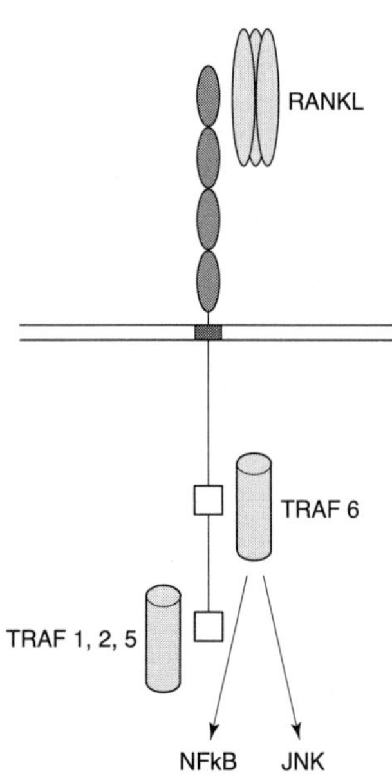

of osteoclast precursor cDNA libraries identified TRAF6 as the major TRAF-related protein that binds to the RANK cytoplasmic domain in these cells. This same internal region of the RANK cytoplasmic domain was also found to be required for the induction of NFκB and JNK activity in osteoclast precursors during the induction of osteoclast-specific gene expression, suggesting that TRAF6 is an important signal transducer during induction of osteoclastogenesis (Hsu *et al.*, 1999). Mice deficient in TRAF6 have recently been analyzed, and found to have an osteopetrotic phenotype (*osteosclerosis* and failure in tooth eruption), confirming a biological role for this protein in osteoclast function and in bone metabolism (Lomaga *et al.*, 1999). *In vivo*, mature osteoclasts are observed, but ultrastructural analysis indicates that there is a defect in the ability of osteoclasts lacking TRAF6 to adhere to bone surfaces, and formation of normal resorption lacunae.

DOWNSTREAM GENE ACTIVATION

Transcription factors activated

Activation of RANK on osteoclast progenitor cells leads to the rapid stimulation of JNK activity, and subsequently the expression of several genes that typify the osteoclast lineage. These genes include tartrate-resistant acid phosphatase (TRAP), the calcitonin receptor, cathepsin K, the integrin $\alpha v\beta 3$, and the c-src proto-oncogene (Lacey *et al.*, 1998; Hsu *et al.*, 1999). The promoter regions of these genes are currently being characterized to identify the transcription factors that mediate RANK regulation of osteoclast development.

BIOLOGICAL CONSEQUENCES OF ACTIVATING OR INHIBITING RECEPTOR AND PATHOPHYSIOLOGY

Unique biological effects of activating the receptors

The following *in vitro* and *in vivo* biological effects have been reported as a direct or indirect consequence of activating RANK on receptor-bearing cells following stimulation with RANKL:

- Activation of osteoclast differentiation (Lacey *et al.*, 1998; Nakagawa *et al.*, 1998; Hsu *et al.*, 1999).
- Induction of osteoclast-specific gene expression (Lacey *et al.*, 1998; Hsu *et al.*, 1999).
- Induction of mature osteoclast survival (Jimi *et al.*, 1999).
- Activation of osteoclast-mediated bone resorption (Lacey *et al.*, 1998; Burgess *et al.*, 1999).
- Regulation of lymph node organogenesis (Kong *et al.*, 1999a; Dougall *et al.*, 1999; Li *et al.*, 2000).
- Stimulation of alloreactive T cell proliferation (Anderson *et al.*, 1997; Wong *et al.*, 1997).

- Regulation of chondrocyte development (Kong *et al.*, 1999a; Li *et al.*, 2000).
- Regulation of calcium metabolism (Li *et al.*, 2000).
- Mediating the effects of calciotropic hormones and proresorptive cytokines on bone metabolism (Li *et al.*, 2000).

Phenotypes of receptor knockouts and receptor overexpression mice

As one would predict from the effects of transgenically delivered OPG, sRANK-Ig expressed in transgenic mice blocks RANKL-induced osteoclastogenesis, leading to increases in bone mass similar to native OPG (Hsu *et al.*, 1999). sRANK-Ig transgenic mice have severe *osteopetrosis*, characterized by a defect in bone resorption and lack mature osteoclasts. Normal osteoclast progenitors are present in these mice, and osteoclastogenesis from these cells can be demonstrated *in vitro*.

Mice deficient in RANK are also osteopetrotic, and lack osteoclasts (Dougall *et al.*, 1999; Li *et al.*, 2000), a situation similar to that previously observed for the RANKL (OPGL)−/− mice (Kong *et al.*, 1999a). These results confirm that a major physiological role of the OPG/RANKL/RANK axis is to regulate bone metabolism and density. Like RANKL−/− mice, RANK−/− mice have a developmental defect in lymph node organogenesis that is not observed in sRANK-Ig transgenic animals (Dougall *et al.*, 1999; Li *et al.*, 2000). Interestingly, the defect in T cell development seen in RANKL−/− mice is not seen in RANK−/− mice. T cells have recently been shown to produced copious amounts RANKL, and can play a role in the pathophysiology of bone loss during experimental inflammation models (Kong *et al.*, 1999b). RANKL may be required for T cell activities during these processes either as a trophic factor or as a transmembrane signaling receptor.

Human abnormalities

To date, there have been no reports of any structural abnormalities in the RANK gene or protein that provide any obvious links to the pathophysiology of human disease. The human RANK gene has been localized to chromosome 18q22.1 (Anderson *et al.*, 1997). This same region has been shown to harbor heritable mutations that effect susceptibility to *Paget's disease* of the bone, and *familial expansile exostosis*, both of which are diseases that affect normal bone metabolism.

THERAPEUTIC UTILITY

Effect of treatment with soluble receptor domain

sRANK-Ig fusion protein has similar biological activity to OPG, although it is somewhat less potent (Hsu *et al.*, 1999). Administration of sRANK-Ig into mice blocks osteoclastogenesis and inhibits bone resorption, and acts as an antiresorptive agent in disease models characterized by pathological increases in osteoclast activity. See the chapter on OPG for a description for its potential uses in the treatment of human diseases.

Effects of inhibitors (antibodies) to receptors

There have been no reports of small molecules that inhibit RANK bioactivity. The analysis of RANK knockout mice provides insight into the effects that inhibition of the RANK signaling pathway would be likely to produce *in vivo* (see above). Polyclonal antibodies to RANK have been isolated, and have been shown to stimulate osteoclast development *in vitro* (Nakagawa *et al.*, 1998; Hsu *et al.*, 1999). Yasuda *et al.* (1998) have prepared Fab fragments of this polyclonal antibody, and have shown that it blocks receptor activation by RANKL during osteoclastogenesis *in vitro*.

References

Anderson, D. M., Maraskovsky, E., Billingsley, W. L., Dougall, W. C., Tometsko, M. E., Roux, E. R., Teepe, M. C., DuBose, R. F., Cosman, D., and Galibert, L. (1997). A homologue of the TNF receptor and its ligand enhance T-cell growth and dendritic-cell function. *Nature* **390**, 175–179.

Arch, R. H., Gedroch, R. W., and Thompson, C. B. (1998). Tumor necrosis factor receptor-associated factors (TRAFs) – a family of adapter proteins that regulates life and death. *Genes Dev.* **12**, 2821–2830.

Burgess, T. L., Qian, Y.-X., Kaufman, S., Ring, B. D., Van, G., Capparelli, C., Kelley, M., Hsu, H., Boyle, W. J., Dunstan, C. R., Hu, S., and Lacey, D. L. (1999). The ligand for osteoprotegerin (OPGL) directly activates mature osteoclasts. *J. Cell Biol.* **145**, 527–538.

Darney, B. G., Ni, J., Moore, P. A., and Aggarwal, B. B. (1999). Activation of NFkB by RANK requires tumor necrosis factor receptor associated factor (TRAF) 6 and NF-kB-inducing kinase. *J. Biol. Chem.* **274**, 7724–7731.

Dougall, W. C., Glaccum, M., Charrier, K., Rohrbach, K., Brasel, K., De-Smedt, T., Daro, E., Smith, J., Tometsko, M. E., Maliszewski, C. R., Armstrong, A., Shen, V., Bain, S., Cosman, D., Anderson, D., Morrissey, P. J.,

Peschon, J. J., and Schuh, J. (1999). RANK is essential for osteoclast and lymph node development. *Genes Dev.* **13**, 2412–2424.

Galibert, L., Tometsko, M. E., Anderson, D. M., Cosman, D., and Dougall, W. C. (1998). The involvement of multiple tumor necrosis factor receptor (TNFR)-associated factors in the signaling mechanisms of receptor activator of NF-κB, a member of the TNFR superfamily. *J. Biol. Chem.* **273**, 34120–34127.

Hsu, H., Lacey, D. L., Dunstan, C. R., Solovyev, I., Colombero, A., Timms, E., Tan, H. L., Elliott, G., Kelley, M. J., Sarosi, I., Wang, L., Xia, X. Z., Elliott, R., Chiu, L., Black, T., Scully, S., Capparelli, C., Morony, S., Shimamoto, G., Bass, M. B., and Boyle, W. J. (1999). Tumor necrosis factor receptor family member RANK mediates osteoclast differentiation and activation induced by osteoprotegerin ligand. *Proc. Natl Acad. Sci. USA* **96**, 3540–3545.

Jimi, E., Akiyama, S., Tsurukai, T., Okahashi, N., Kobayashi, K., Udagawa, N., Nishihashi, T., Takahashi, N., and Suda, T. (1999). Osteoclast differentiation factor acts as a multifunctional regulator in murine osteoclast differentiation and function. *J. Immunol.* **163**, 434–442.

Kong, Y.-Y., Yoshida, H., Sarosi, I., Tan, H. L., Timms, E., Capparelli, C., Morony, S., Oliveira dos Santos, A. J., Van, G., Itie, A., Khoo, W., Wakeham, A., Dunstan, C. R., Lacey, D. L., Mak, T. W., Boyle, W. J., and Penninger, J. M. (1999a). OPGL is a key regulator of osteoclastogenesis, lymphocyte development and lymph-node organogenesis. *Nature* **397**, 315–323.

Kong, Y.-Y., Feige, U., Sarosi, I., Bolon, B., Tafuri, A., Morony, S., Capparelli, C., Li, J., Elliott, R., McCabe, S., Wong, T., Campagnuolo, G., Moran, E., Bogoch, E. R., Van, G., Nguyen, L. T., Ohashi, P. S., Lacey, D. L., Fish, E., Boyle, W. J., and Penninger, J. M. (1999b). Activated T cells regulate bone loss and joint destruction in adjuvant arthritis through osteoprotegerin ligand. *Nature* **402**, 304–309.

Lacey, D. L., Timms, E., Tan, H.-L., Kelley, M. J., Dunstan, C. R., Burgess, T., Elliott, R., Colombero, A., Elliott, G., Scully, S., Hsu, H., Sullivan, J., Hawkins, N., Davy, E., Capparelli, C., Eli, A., Qian, Y. X., Kaufman, S., Sarosi, I., Shalhoub, V., Senaldi, G., Guo, J., Delaney, J., and Boyle, W. J. (1998). Osteoprotegerin ligand is a cytokine that regulates osteoclast differentiation and activation. *Cell* **93**, 165–176.

Li, J., Sarosi, I., Yan, X. Q., Morony, S., Capparelli, C., Tan, H. L., McCabe, S., Elliott, R., Scully, S., Van, G., Kaufman, S., Juan, S. C., Sun, Y., Tarpley, J., Martin, L., Christensen, K., McCabe, J., Kostenuik, P., Hsu, H., Fletcher, F., Dunstan, C. R., Lacey, D. L., and Boyle, W. J. (2000). RANK is the intrinsic hematopoietic cell surface receptor that controls osteoclastogenesis and regulation of bone mass and calcium metabolism. *Proc. Natl Acad. Sci. USA* **97**, 1566–1571.

Lomaga, M. A., Yeh, W.-C., Sarosi, I. *et al.* (1999). TRAF 6 deficiency results in osteopetrosis and defective interleukin-1, CD40, and LPS signaling. *Genes Dev.* **13**, 1015–1024.

Nakagawa, N., Kinosaki, M., Yamaguchi, K., Shima, N., Yasuda, H., Yano, K., Morinaga, T., and Higashio, K. (1998). RANK is the essential signaling receptor for osteoclast differentiation factor in osteoclastogenesis. *Biochem. Biophys. Res. Commun.* **253**, 395–400.

Smith, C. A., Farrah, T., and Goodwin, R. G. (1994). The TNF receptor superfamily of cellular and viral proteins: activation, co-stimulation, and death. *Cell* **76**, 959–962.

Suda, T., Takahashi, N., Udagawa, N., Jimi, E., Gillespie, M. T., and Martin, T. J. (1999). Modulation of osteoclast differentiation and function by the new members of the tumor necrosis factor receptor and ligand families. *Endocr. Rev.* **20**, 345–357.

Wong, B. R., Rho, J., Arron, J. *et al.* (1997). TRANCE is a novel ligand of the Tumor Necrosis Factor Receptor family that activates c-Jun N-terminal kinase in T cells. *J. Biol. Chem.* **272**, 25190–25194.

Wong, B. R., Josien, R., Lee, S. Y., Vologodskaia, M., Steinman, R. M., and Choi, Y. (1998). The TRAF family of signal transducers mediates NF-kB activation by the TRANCE receptor. *J. Biol. Chem.* **273**, 28355–28359.

Yasuda, H., Shima, N., Nakagawa, N., Yamaguchi, K., Kinosaki, M., Mochizuki, S., Tomoyasu, A., Yano, K., Goto, M., Murakami, A., Tsuda, E., Morinaga, T., Higashio, K., Udagawa, N., Takahashi, N., and Suda, T. (1998). Osteoclast differentiation factor is a ligand for osteoprotegerin/osteoclastogenesis-inhibitory factor and is identical to TRANCE/RANKL. *Proc. Natl Acad. Sci. USA* **95**, 3597–3602.

LICENSED PRODUCTS

RANK recombinant protein and antibodies for research

Alexis Biochemicals, 6181 Cornerstone Court East, Suites 102-104, San Diego, CA 92121, USA

R&D Systems, Inc., 614 McKinley Place N.E., Minneapolis, MN 55413, USA

Santa Cruz Biotechnology, Inc., 2161 Delaware Avenue, Santa Cruz, CA 95060, USA

Osteoprotegerin

William J. Boyle*

Department of Cell Biology, Amgen, Inc., One Amgen Center Drive,
Thousand Oaks, CA 91320-1799, USA

*corresponding author tel: 805-447-4304, fax: 805-447-1982, e-mail: bboyle@amgen.com

DOI: 10.1006/rwcy.2000.16009

SUMMARY

Osteoprotegerin (OPG) is a new member of the TNFR superfamily that functions solely as a secreted cytokine antagonist or decoy receptor. The structure and function of OPG is highly conserved during evolution, involves the binding to, and neutralization of RANKL during the regulation of bone metabolism. OPG is secreted by expressing cells, and can bind tightly to extracellular matrix proteins via a C-terminal heparin-binding domain.

Overexpression of OPG, or systemic administration of recombinant protein, blocks osteoclast differentiation and activation in mice, rats, primates, and humans. The resulting effect is to negatively regulate bone resorption, which leads to increases in bone mass. Loss of OPG results in early onset *osteoporosis* in mice, leading to brittle bone structure and accumulation of long bone fractures. OPG recombinant protein can block the pathological bone loss that occurs following estrogen loss, during adjuvant-induced *arthritis*, and during the growth of *lytic tumors*, indicating a possible use in the treatment of osteopenic disorders characterized by increased osteoclast activity.

BACKGROUND

Discovery

Osteoprotegerin (OPG) is a recently described protein that appears to be a major factor affecting bone metabolism (Simonet *et al.*, 1997). It is a member of the tumor necrosis factor receptor superfamily, with special properties. OPG is a secreted protein, and not a traditional surface membrane-associated receptor that transmits signals regulating cell functions. In this regard it is unlike most other TNFR family members. The primary structure of OPG suggests that it can act in the extracellular milieu as a receptor-like protein that binds to, and regulates, the activity of a TNF-related ligand (see Suda *et al.*, 1999 for review). The term 'osteoprotegerin' simply refers to a protein that protects or guards against the destruction of bone. Based on *in vitro* and *in vivo* analysis, OPG has been found to negatively regulate osteoclastogenesis and osteoclast activation, and in doing so, to control bone remodeling and density.

Osteoprotegerin was first identified as a novel expressed sequence tag (EST) from a cDNA library prepared from fetal rat intestine, independent of any perceived biological function, i.e. via genomics (Adams *et al.*, 1992). Computational analysis of this sequence revealed that it encoded a homolog of the TNFR superfamily of receptors. Using this sequence information, full-length rat, mouse, and human cDNA clones were isolated, and the amino acid sequence of its protein product was deduced. OPG was also purified from the conditioned media produced by human fibroblasts as a factor that inhibited osteoclast development from hematopoietic precursors (Tsuda *et al.*, 1997).

Alternative names

OCIF, osteoclastogenesis inhibitory factor (Yasuda *et al.*, 1998a)
TR1, TNF receptor-like protein 1 (Kwon *et al.*, 1998)
FDCR-1, follicular dendritic cell receptor 1 (Yun *et al.*, 1998)

Structure

OPG is a secreted TNFR-related protein with two functional domains (Figure 3). The N-terminal half of OPG contains four tandem cysteine-rich repeat sequences that are characteristic of the ligand-binding domains of all members of this family. The C-terminal half of the protein has no distinct homologies to any known proteins, and is involved in OPG homodimerization following biosynthesis. The C-terminal domain binds to heparin, and may be involved in localization to extracellular matrix.

Main activities and pathophysiological roles

OPG plays a central role in regulating skeletal metabolism by inhibiting differentiation and activation of the osteoclast. Bone remodeling and homeostasis is an essential function that regulates skeletal integrity throughout adult life in higher vertebrates and mammals (Suda *et al.*, 1992; Roodman, 1996). The maintenance of skeletal mass is controlled by the activities of specialized cells within the bone that have seemingly antagonistic activities: bone synthesis and bone resorption. Osteoblastic cells of mesenchymal origin synthesize and deposit bone matrix and increase bone mass. Osteoclastic cells are large, multinucleated phagocytes of hematopoietic origin that resorb both mature and newly synthesized bone upon activation. Bone synthesis and resorption processes are highly coordinated, and are controlled by osteotropic and calciotropic hormones during physiological and pathological conditions (Rodan and Martin, 1981; Martin and Udagawa, 1998). Osteoclast recruitment and activation can be accelerated in certain disease processes, leading to inappropriate increases in bone resorption, and subsequently a net loss of bone mass. Increased osteoclast activity is associated with bone loss in several disorders, including primary *osteoporosis*, *Paget's disease* of the bone, *hypercalcemia* of malignancy, and *rheumatoid arthritis*. Understanding the biology of the osteoclast is an important field of study, and carries with it implications that may lead to the development of effective therapies to treat bone loss in humans.

The ability of OPG to regulate bone resorption and remodeling *in vivo* implicates the protein as a key determinant in the regulating bone mass and calcium metabolism. OPG does not act alone, it effects a pathway of hematopoietic cell development that is dependent on a key ligand/receptor interaction found to be essential for osteoclast maturation and activation. OPG has been useful as a reagent to probe the nature of this regulatory pathway, and combined biochemical, cell biological, and molecular genetic analysis in mice has provided a clear mechanism-based view of how osteoclastogenesis is regulated.

Osteoclasts are specialized macrophage cells that develop from hematopoietic precursors in the bone marrow and spleen. They stem from a common monocyte/macrophage precursor that can give rise to several mature macrophage cell types, including avelolar macrophages and hepatic Küpffer cells. At the time OPG was first identified and characterized, little was actually known about the various steps involved in the commitment to the osteoclast lineage, and the processes that control osteoclast maturation (osteoclastogenesis). However, using *in vitro* culture systems, mature osteoclasts could be formed from bone marrow and spleen cells (Udagawa *et al.*, 1989). The polypeptide factor M-CSF (CSF-1), along with stromal cells, vitamin D3, and dexamethasone are required for osteoclastogenesis to occur in this *in vitro* systems.

A vitamin D3-dependent bone marrow and stromal cell co-culture system was used to determine the effects of OPG on osteoclast development (Udagawa *et al.*, 1989; Lacey *et al.*, 1995). This assay system was used to purify the native human OPG polypeptide to homogeneity from conditioned media produced by a normal human diploid lung fibroblast cell line (Tsuda *et al.*, 1997). The protein product that was purified included both homodimeric and monomeric OPG polypeptide chains, similar to the expression product characterized in the circulation of OPG transgenic mice (Simonet *et al.*, 1997). Native OPG and recombinant OPG protein has been shown to block *in vitro* osteoclastogenesis in a dose-dependent manner, with a half-maximal effective concentration in the range of 1–2 ng/mL. OPG blocks the formation of large multinucleated osteoclasts that express tartrate-resistant acid phosphatase (TRAP) activity, the calcitonin receptor, and the integrin $\alpha v\beta 3$, all markers of the mature osteoclast (Figure 5). Only the N-terminal half of the protein, which resembles a ligand-binding domain, is required for biological activity in this *in vitro* system (Simonet *et al.*, 1997; Yamaguchi *et al.*, 1998).

GENE

Accession numbers

Rat OPG cDNA: U94331
Murine OPG cDNA: U94331, E15271
Human OPG cDNA: U94332, AB013898
Mouse OPG gene: AB013899S

Sequence

See **Figure 1**.

PROTEIN

Accession numbers

Rat OPG protein: AAB53707
Murine OPG protein: AAB53708, AB013903
Human OPG protein: AAB53709, AB002146

Sequence

See **Figure 2**.

Description of protein

The long open reading frame of the OPG OPG/OCIF encodes a 401 residue polypeptide with two functional regions. The N-terminal half of OPG (residues 22–185) consists of four tandem cysteine-rich repeat sequences, and bears striking homology to the ligand-binding domain of all TNFR family members (Smith *et al.*, 1994). The C-terminal half of the protein (residues 186–401) bears no obvious homologies to other known proteins, but appears to contain minimal sequence identity to a portion of the death domain motif found on certain intracellular proteins (Simonet *et al.*, 1997; Yamaguchi *et al.*, 1998). This region is involved in OPG homodimerization and extracellular matrix binding. At the extreme N-terminus of the protein lies a functional 21 amino acid hydrophobic

Figure 1 Nucleotide sequence for OPG.

```
Human OPG cDNA Sequence:
   1 GTATATATAA CGTGATGAGC GTACGGGTGC GGAGACGCAC CGGAGCGCTC GCCCAGCCGC
  61 CGYCTCCAAG CCCCTGAGGT TTCCGGGGAC CACAATGAAC AAGTTGCTGT GCTGCGCGCT
 121 CGTGTTTCTG GACATCTCCA TTAAGTGGAC CACCCAGGAA ACGTTTCCTC CAAAGTACCT
 181 TCATTATGAC GAAGAAACCT CTCATCAGCT GTTGTGTGAC AAATGTCCTC CTGGTACCTA
 241 CCTAAAACAA CACTGTACAG CAAAGTGGAA GACCGTGTGC GCCCCTTGCC CTGACCACTA
 301 CTACACAGAC AGCTGGCACA CCAGTGACGA GTGTCTATAC TGCAGCCCCG TGTGCAAGGA
 361 GCTGCAGTAC GTCAAGCAGG AGTGCAATCG CACCCACAAC CGCGTGTGCG AATGCAAGGA
 421 AGGGCGCTAC CTTGAGATAG AGTTCTGCTT GAAACATAGG AGCTGCCCTC CTGGATTTGG
 481 AGTGGTGCAA GCTGGAACCC CAGAGCGAAA TACAGTTTGC AAAAGATGTC CAGATGGGTT
 541 CTTCTCAAAT GAGACGTCAT CTAAAGCACC CTGTAGAAAA CACACAAATT GCAGTGTCTT
 601 TGGTCTCCTG CTAACTCAGA AAGGAAATGC AACACACGAC AACATATGTT CCGGAAACAG
 661 TGAATCAACT CAAAAATGTG GAATAGATGT TACCCTGTGT GAGGAGGCAT TCTTCAGGTT
 721 TGCTGTTCCT ACAAAGTTTA CGCCTAACTG GCTTAGTGTC TTGGTAGACA ATTTGCCTGG
 781 CACCAAAGTA AACGCAGAGA GTGTAGAGAG GATAAAACGG CAACACAGCT CACAAGAACA
 841 GACTTTCCAG CTGCTGAAGT TATGGAAACA TCAAAACAAA GCCCAAGATA TAGTCAAGAA
 901 GATCATCCAA GATATTGACC TCTGTGAAAA CAGCGTGCAG CGGCACATTG GACATGCTAA
 961 CCTCACCTTC GAGCAGCTTC GTAGCTTGAT GGAAAGCTTA CCGGGAAAGA AAGTGGGAGC
1021 AGAAGACATT GAAAAAACAA TAAACCCATC CAAACCCACT CACCACATCC TGAAGCTGCT
1081 CAGTTTGTGG CGAATAAAAA ATGGCGACCA AGACACCTTG AAGGGCCTAA TGCACGCACT
1141 AAAGCACTCA AAGACGTACC ACTTTCCCAA AACTGTCACT CAGAGTCTAA AGAAGACCAT
1201 CAGGTTCCTT CACAGCTTCA CAATGTACAA ATTGTATCAG AAGTTATTTT TAGAAATGAT
1261 AGGTAACCAG GTCCAATCAG TAAAAATAAG CTGCTTATAA CTGGAAATGG CCATTGAGCT
1321 GTTTCCTCAC AATTGGCGAG ATCCCATGGA TGATAA
```

Figure 2 Amino acid sequence for OPG.

```
Human OPG protein sequence:
  1   MNKLLCCALV FLDISIKWTT QETFPPKYLH YDEETSHQLL CDKCPPGTYL KQHCTAKWKT
 61   CAPCPDHYY  TDSWHTSDEC LYCSPVCKEL QYVKQECNRT HNRVCECKEG RYLEIEFCLK
121   HRSCPPGFGV VQAGTPERNT VCKRCPDGFF SNETSSKAPC RKHTNCSVFG LLLTQKGNAT
181   HDNICSGNSE STQKCGIDVT LCEEAFFRFA VPTKFTPNWL SVLVDNLPGT KVNAESVERI
241   KRQHSSQEQT FQLLKLWKHQ NKAQDIVKKI IQDIDLCENS VQRHIGHANL TFEQLRSLME
301   SLPGKKVGAE DIEKTIKACK PSDQILKLLS LWRIKNGDQD TLKGLMHALK HSKTYHFPKT
361   VTQSLKKTIR FLHSFTMYKL YQKLFLEMIG NQVQSVKISC L
```

signal peptide that directs secretion and is cleaved during biosynthesis.

Further analysis of the OPG amino acid sequence failed to reveal potential any hydrophobic transmembrane-spanning domain such as that seen in other members of the TNFR family, and was an early indication that it might be a secreted protein. Pulse-chase studies in mammalian cell lines that overexpress the murine OPG cDNA revealed that it is indeed a secreted protein that is proteolytically processed after synthesis, modified by glycosylation, and exported into the conditioned media. Analysis of the biosynthetic product under nonreducing conditions also reveals that the primary OPG polypeptide self associates during secretion, and is secreted from the cell both as a disulfide-linked dimer, and as a monomer (Simonet *et al.*, 1997; Tsuda *et al.*, 1997).

Structural analysis of the C-terminal region indicates that cysteine residue at position 400 is critical for the formation of secreted homodimer (Yamaguchi *et al.*, 1998). Dimerization and oligomerization of TNFR-related proteins are associated with high affinity for ligand. Formation of a secreted dimer is a unique property for members of the TNFR superfamily, and suggests that OPG is biosynthesized as a high-affinity cytokine antagonist.

Structure and function analyses of OPG reveals that the protein region required for biological activity resides within the N-terminal 185 amino acids, which harbors the four tandem cysteine-rich repeat sequences (**Figure 3**). All four domains are required for biological activity, as well as each of the 18 individual cysteine residues that stabilize the tertiary structure of these motifs. This region of TNFR-related proteins comprises the ligand-binding domain of members to this family, implying that OPG activity is mediated via binding to a TNF-related protein. The C-terminal domain of the protein can be deleted, and OPG molecules retain biological activity *in vitro* and *in vivo*, further suggesting that OPG bioactivity involves binding to a TNF-related cytokine.

Figure 3 Structure of rat OPG mRNA and protein. The rat OPG mRNA is a 2.4 kb transcript and is represented by the thin line. The 401 amino acid long open reading frame encoding the rat OPG product is represented by the coded box. The first 21 amino acids (black) indicate the position of the signal peptide sequence. The N-terminal half of OPG contains four tandem cysteine-rich repeat sequences (shaped ellipses). The C-terminal dimerization domain is indicated by a gray box. The coding frame begins at a methionine codon (AUG), and terminates at a stop codon (TAG) following leucine 401. C185 represents the last cysteine residue of the TNFR homology domain. The mouse and human OPG cDNA sequences also encode 401 amino acid polypeptides that are approximately 89–94% identical to the rat protein shown here. Below is a illustration of the mature secreted form of the OPG homodimer, beginning at amino acid reisdue 22 and ending at 401. The last disulfide residue of the cysteine-rich domain IV is labeled as Cys185.

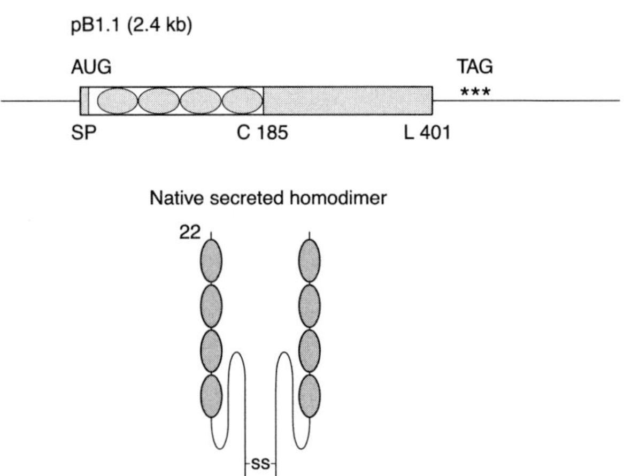

Relevant homologies and species differences

The rat, mouse, and human OPG cDNAs have been cloned and sequenced, and their protein products compared. OPG sequences are highly conserved during evolution at the DNA and protein level. The rat and mouse proteins are about 94% identical, whereas the mouse and human proteins are about 89% identical, without sequence gaps. All of the cysteine residues located in the full-length mature proteins (residues 22–401) are conserved and in identical positions.

Affinity for ligand(s)

The biological activity of OPG strongly suggests that it negatively regulates a cytokine that stimulates osteoclast maturation and activation. It had been known for many years that osteoblastic cells could be stimulated by calciotropic agents to produce a factor that stimulates osteoclast development and activation (Suda *et al.*, 1999). The ligand for OPG was a likely candidate for this factor, but had not yet been identified and its biological activities remained elusive. Using OPG as a probe, a putative OPG ligand was expression cloned from cDNA libraries made from osteoblastic stromal cells induced with vitamin D3 (Yasuda *et al.*, 1998b), or from the murine myelomonocytic cell line 32D (Lacey *et al.*, 1998). Both clones encoded an identical 316 amino acid mouse protein that was termed osteoclast differentiation factor (ODF) and OPG ligand (OPGL) (see RANKL). The protein product was predicted to be a type II transmembrane surface protein with clear structural motifs found in TNF family members.

OPG ligand (RANKL) can be cleaved from the surface of expressing cells (Lacey *et al.*, 1998) and released in a soluble, biologically active form. It is not yet clear whether the soluble form of OPG ligand plays a role in regulating osteoclast function(s) *in vivo*, although recombinant forms of soluble of the protein have potent bioactivity when administered to animals (Lacey *et al.*, 1998). The cDNA encoding ODF/OPGL had previously been isolated by two different approaches that suggested a role for this cytokine in regulation of immune responses. The first report of this sequence was the differential cloning of a T cell protein induced by calcineurin-regulated transcription factors, and named TRANCE (TNF related activation-induced cytokine) (Wong *et al.*, 1998). It was subsequently also identified as RANKL, a ligand for a novel TNFR-related protein RANK (receptor activator of NFκB), a dendritic cell surface protein (Anderson *et al.*, 1997). Activated T cells and some T cell lines express TRANCE/RANKL, suggesting a role in costimulatory processes during antigen processing by regulating dendritic cell survival (Wong *et al.*, 1997; Anderson *et al.*, 1997).

OPG has also been shown to bind to the TNF-related cytotoxin TRAIL (Emery *et al.*, 1998). The coaddition of TRAIL and OPG during *in vitro* osteoclast-forming assays indicates that recombinant TRAIL can block OPG activity. This suggests that TRAIL may regulate OPG function during bone metabolism *in vivo*. Administration of recombinant TRAIL protein into mice at relatively high doses (25 mg/kg) apparently does not affect bone resorption or remodeling, as one would expect if TRAIL protein interacted with OPG *in vivo* (Walczak *et al.*, 1999). TRAIL exists as a membrane-bound protein on expressing cells *in vivo*, and OPG interaction with native TRAIL may regulate important cellular functions.

Several lines of evidence provide definitive evidence of the role of ODF/OPGL/TRANCE/RANKL as the key factor that regulates osteoclastogenesis (referred herein as RANKL for continuity). First, recombinant RANKL binds specifically and with high affinity to biologically active forms of OPG, but not inactive OPG mutants (Yasuda *et al.*, 1998; Lacey *et al.*, 1998). Second, either native cell surface expressed RANKL, or recombinant soluble RANKL acts as a potent osteoclast differentiation factor *in vitro* using either murine or human hematopoietic precursors. Osteoclastogenesis *in vitro* is dependent on the presence of CSF-1, and is characterized by induction of the expression of key markers that typify the osteoclast cell lineage. This includes the calcitonin receptor, TRAP, cathepsin K, and the integrin $\alpha v\beta 3$, as well as the formation of large multinucleated cells that are capable or resorbing bone. RANKL not only stimulates osteoclast differentiation, but also mature osteoclast activation (Fuller *et al.*, 1998; Burgess *et al.*, 1999) and survival (Jimi *et al.*, 1999) *in vitro* and *in vivo*. Third, using fluorescenated RANKL as a probe, all of the hematopoietic progenitors capable of giving rise to osteoclasts during *in vitro* culture could be isolated and purified from bone marrow cells by flow cytometry (Lacey *et al.*, 1998). Finally, expression of RANKL in osteoblasts was induced by agents that increase osteoclast development *in vitro*, such as vitamin D3, IL-1 and IL-11, PGE_2, and PTH (Yasuda *et al.*, 1998b). Interestingly, some of these agents also appear to simultaneously downregulate or inhibit OPG expression in osteoblasts (Vidal *et al.*, 1998a). This raises the possibility that coordinate regulation of OPG and OPGL expression by osteotropic and calciotropic hormones may be a mechanism used to couple osteoblast and osteoclast development and activation during bone remodeling. Thus, the combination of cell biological data measuring osteoclast differentiation and activation activity, its ability to tightly bind to OPG, and its relatedness to TNF family members all indicated that OPGL is a critical factor that regulates osteoclastogenesis.

OPG binds to soluble RANKL in the 30–100 pM range using protein–protein interaction assays in solution, and in the 1 nM range using BIAcore analysis. OPG/TRAIL interactions have been measured in the 1–3 nM range in solution assays.

Cell types and tissues expressing the receptor

See **Table 1**.

Regulation of receptor expression

The mouse OPG transcript is expressed in the cartilaginous primordia of bone during embryonic development, then later is expressed in the intestine, kidney, lung, and bone (Simonet *et al.*, 1997). In humans, OPG expression in tissues is detected at relatively high levels in the lung, kidney, intestine, spleen, thymus, and heart (Simonet *et al.*, 1997; Yasuda *et al.*, 1998a). OPG expression can be

Table 1 Cell types and tissues expressing the receptor

Organs and tissues	Cartilaginous bone primordia during fetal development
	Intestinal epithelium
	Kidney
	Liver
	Lung
	Thyroid
	Bone and growth plate cartilage
	Lymph node
	Spleen
	Thymus
	Heart
Cell types	Hypertrophic chondrocytes
	Osteoblastic stromal cells
	Osteoblasts
	T cells (subset undefined)
	B cells
	Intestinal epithelium (small and large)
	M cells
	Follicular dendritic cells
Cell lines	CTLL-2 (IL-2-dependent T cell line)
	IMR-90 (human diploid lung fibroblasts)
	LIM 1863 (human colorectal carcinoma)
	MG63 (human osteosarcoma)

detected in osteoblastic stromal cells, and is found to be regulated by cytokines, growth factors, and steroid hormones, suggesting a role as a factor that regulates osteoblast and osteoclast coupling during bone development and remodeling (Yasuda *et al.*, 1998b; Vidal *et al.*, 1998b). Members of the TGFβ superfamily, such as TGFβ1 and BMP-2 can induce expression of OPG (Hofbauer *et al.*, 1998, Takai *et al.*, 1998; Horwood *et al.*, 1998). Calciotropic agents known to induce bone resorption, such as vitamin D3, PGE_2, and hydrocortisone, can also downregulate OPG expression in osteoblasts (Vidal *et al.*, 1998a). OPG transcripts have also been detected in dendritic cells and osteoblastic sarcomas (Yun *et al.*, 1998; Vidal *et al.*, 1998b). Its expression in some lymphoid cells is upregulated by the ligation of CD40 receptor, a TNFR protein that is closely related to OPG (Yun *et al.*, 1998). Cytokines such as TNFα and TNFβ, as well as interleukins IL-1, upregulate the production of OPG in an osteosarcoma cell line and in osteoblast-like stromal cells (Vidal *et al.*, 1998b; Brandstrom *et al.*, 1998), indicating a possible link between immune system responses and the regulation of bone metabolism.

Release of soluble receptors

OPG is a secreted protein and is not a cell-associated, transmembrane signaling receptor. Analysis of the OPG gene has not revealed any alternatively spliced forms that contain a transmembrane domain or cell-anchoring motifs.

SIGNAL TRANSDUCTION

Associated or intrinsic kinases

OPG is a cytokine antagonist that can block RANKL binding to RANK, and induction of signal transduction. See the chapter on RANK for discussion on cytoplasmic factors and signaling from this receptor.

DOWNSTREAM GENE ACTIVATION

Transcription factors activated

OPG, RANKL, and RANK have all been implicated as the key extracellular proteins that coordinately interact to regulate osteoclast differentiation and activation. Each of these proteins has been evaluated in mouse molecular genetic models, and their interrelated functions during osteoclastogenesis have been confirmed. A mechanistic model for regulation of osteoclastogenesis can be derived that invokes binding of these proteins in a selective fashion to produce negative or positive regulation of bone mass (**Figure 4**). RANKL is a pathway agonist that activates the RANK receptor on osteoclast precursors, or on mature osteoclasts, resulting in bone resorption. OPG is a cytokine antagonist that acts as a negative regulator of osteoclast development and activation by sequestering its ligand. The resulting effect *in vivo* is to block bone resorption, leading to the accumulation of bone mass. Differential expression of OPG and RANKL by the osteoblast is under the control of a diverse collection of hormones and cytokines that influence the rate of bone remodeling,

Figure 4 Molecular mechanism for the regulation of osteoclast differentiation and activation via OPGL and OPG on osteoclast precursors. OPGL (RANKL) is represented as a transmembrane and soluble protein that binds to and activates the osteoclast receptor RANK, stimulating signal transduction leading to the induction of osteoclast specific gene expression. OPG is a secreted neutralizing or decoy receptor that functions by binding to and sequestering OPGL, thus rendering it unable to activate RANK and thereby preventing the induction of osteoclastogenesis. The membrane-bound arrowhead-shaped complex represents the putative OPGL convertase activity detected in human 293 fibroblasts that liberates soluble OPGL. TRAF 2, 5, 6 are TNFR-associated cytoplasmic factors involved in signal transduction. TRAP, tartrate-resistant acid phosphatase; CTR, calcitonin receptor; $\alpha v\beta 3$, osteoclast-specific integrin $\alpha v\beta 3$; CatK, osteoclast-specific protease cathepsin K

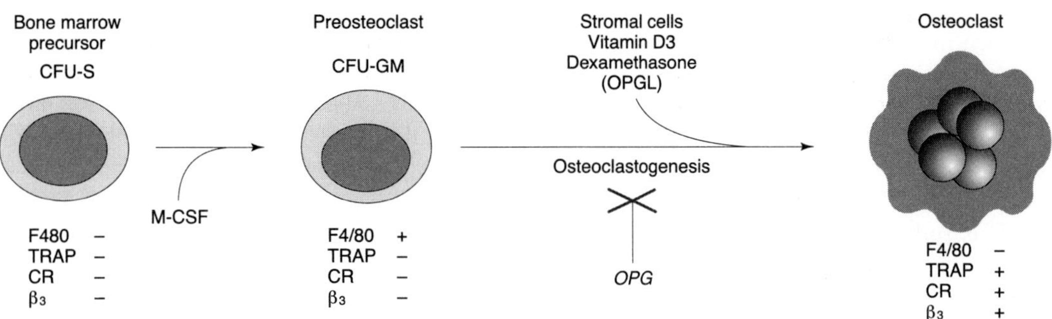

and can control the numbers and activity of osteoclasts. At the level of the bone, RANKL would appear to control the coupling of osteoblast and osteoclast functions to coordinate bone remodeling, while OPG can uncouple this process and block bone remodeling. There is still much to earn about this pathway, and exactly how RANKL/RANK interactions can produce changes in gene expression patterns in hematopoietic progenitors to result in the osteoclast cell phenotype. Further study of this pathway should provide insights as to how bone density is regulated by cytokines and hormones during normal development and disease.

BIOLOGICAL CONSEQUENCES OF ACTIVATING OR INHIBITING RECEPTOR AND PATHOPHYSIOLOGY

Unique biological effects of activating the receptors

Our current view of OPG is that it is not activated like traditional receptors that are cell bound. Since OPG is a secreted protein, its activity is regulated at the level of OPG expression and secretion. Overexpression of OPG blocks osteoclast-mediated bone resorption, and leads to increases in bone mass (see below).

Phenotypes of receptor knockouts and receptor overexpression mice

The biological activity of OPG was determined *in vivo* by analyzing the effects of systemic administration of OPG via transgenic delivery of the rat and mouse cDNA into mouse (Simonet *et al.*, 1997). This system employed the use of the ApoE promoter and hepatocyte control element to produce high levels of OPG protein in the liver beginning around day one of postnatal development, and continuing throughout adulthood. Overexpression of OPG in this way is designed to mimic the chronic systemic administration of protein, and therefore would perturb any OPG-regulated process and lead to an identifiable phenotype.

Transgenic OPG founder mice expressing varying levels of OPG were found to be healthy, and appeared of normal size and weight. They had no abnormalities in blood cell levels or in serum chemistry. They were able to breed, and readily gave rise to transgenic mouse lines. However, they all had a noticeable phenotype that correlated in severity with the level of exogenous OPG measured in their circulation. OPG transgenics had increased bone density (*osteopetrosis*), resulting from the accumulation of newly synthesized bone. The bones were of normal length and shape, suggesting that bone growth and modeling was not affected by excess OPG. The increase in bone mass detected in transgenic animal was restricted to the endosteal regions of the trabecular or marrow-containing bones. Within the endosteal region, there were few if any TRAP-positive multinucleated osteoclasts observed, suggesting a defect in bone resorption.

Comparison of bone sections obtained from OPG transgenic mice and control littermates indeed showed that OPG transgenics lacked mature osteoclasts within the endosteal regions of the affected bones. Normal osteoclast progenitor cells can be detected in the spleens of these mice using *in vitro* osteoclast formation assays, and immunohistochemistry of the spleen indicated normal expression of the myelomonocytic cell surface marker F4/80 (Austyn and Gordon, 1981). F4/80 is a surface antigen associated with macrophage and osteoclast precursors, and its abundant expression in the spleens of these animals indicates normal development in the myeloid cell lineages. This indicates that the bone phenotype seen in OPG transgenic phenotype is not due to alterations in hematopoiesis leading to the production of the osteoclast progenitor, yet rather these mice harbor a defect in the latter stages of osteoclast maturation. The OPG transgenic mice still produce and harbor osteoclast progenitor cells that could differentiate *in vitro* after being removed from exogenous OPG exposure *in vivo*. This means that hematopoiesis leading to the production of osteoclast precursors is unaffected by OPG, and these precursor cells are blocked in their ability to differentiate along the terminal osteoclast maturation pathway (**Figure 5**). From this work, we now know that OPG impacts a crucial regulatory step during osteoclast development, and that further study of this novel protein will lead to a better understanding of this process, and physiological cues that help to regulate bone density.

The murine and human OPG gene has been isolated and partially characterized (Morinaga *et al.*, 1998; Bucay *et al.*, 1998). Using the murine gene, OPG knockout mice have been generated and analyzed to further characterize the function of OPG (Bucay *et al.*, 1998; Mizuno *et al.*, 1998). OPG−/− mice were born live, and found primarily to have early onset *osteoporosis* that increases in severity during aging. These mice have extremely fragile bones, and multiple long bone fractures can be readily observed upon X-ray examination of mice maintained with normal handling. Histological examination of the bones of OPG−/− mice shows a dramatic decrease in trabecular and cortical bone mass. The cortical region of the long bones degenerates over time into a trabecularized structure with woven matrix, indicating a very weak form of bone in these animals. Interestingly, the heterozygous OPG−/+ mice also have decreased bone mass, at a level intermediate to that of +/+ and −/− mice (Bucay *et al.*, 1998). If the OPG gene dosing levels seen in OPG transgenic, wild-type, and OPG−/+ and −/− mice are compared, it can readily be concluded that the level of OPG directly determines the level of bone mass, implying that OPG is a secreted molecule that regulates bone density.

Figure 5 OPG effect on osteoclastogenesis *in vitro*. Diagram depicting the hypothetical action of OPG on osteoclast development *in vitro* using the bone marrow and stromal cell co-culture system. Bone marrow precursor cells treated with M-CSF (CSF-1) give rise to bone marrow macrophage cells capable of differentiating into mature osteoclasts. Co-culture of bone marrow macrophages with ST-2 stromal cells in the presence of 1,25 dihydroxyvitamin D3 and dexamethasone differentiate into mature osteoclasts, Prostaglandin E_2 (PGE_2) potentiates osteoclast formation in this system. OPG blocks osteoclast development at the stage which requires stromal cells, vitamin D3, and dexamethasone. F4/80, a monocyte/macrophage surface marker; TRAP, tartrate-resistant acid phosphatase; CR, calcitonin receptor; $\alpha v\beta 3$, osteoclast specific integrin $\alpha v\beta 3$; CFU-S; colony-forming unit stem cells; CFU-GM, colony-forming unit granulocyte/macrophage.

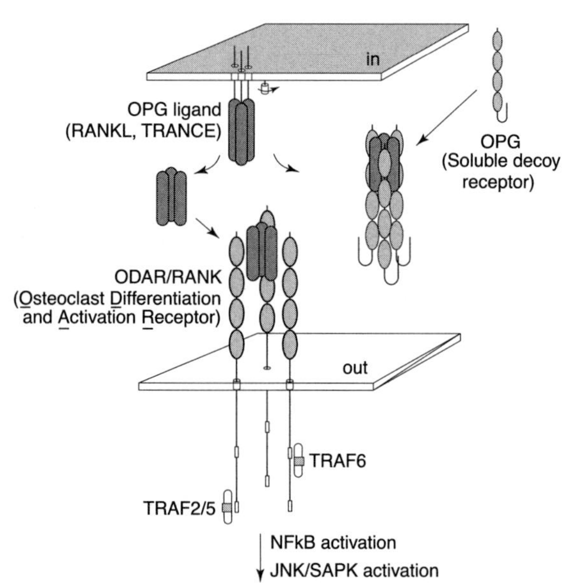

OPG−/− mice have also been shown to have calcifications in the vessel walls of major arteries in the heart and kidney (Bucay *et al.*, 1998). The underlying mechanism of this effect is not known, but may be related to secondary events that occur following losses in bone mineral density. The knockout of the OPG actually provided a phenotype in mice that is diametrically opposite that of OPG overexpression, reinforcing the concept that it acts to regulate bone mass as a negative regulator of osteoclast activity.

Human abnormalities

To date, there have been no reports of any structural abnormalities in the OPG gene or protein that provide any obvious links to the pathophysiology of human disease.

THERAPEUTIC UTILITY

Effect of treatment with soluble receptor domain

The early hypothesis that OPG functions as a secreted osteoclast inhibitor *in vivo* was tested by administration of recombinant OPG in neonate animals. Young growing mice were injected with recombinant murine OPG protein daily for 3–7 days, and its effects on bone were measured by radiography and quantitative histology (Simonet *et al.*, 1997; Yasuda *et al.*, 1998). In growing mice, newly synthesized bone is actively generated at the growth plates of the long bones, particularly at the proximal tibial metaphysis. Injection of OPG leads to the rapid accumulation of bone at this site, coincident with the decreased numbers of TRAP-positive osteoclasts. Microscopic analysis of this area shows the accumulation of newly synthesized bone-encased cartilage, or osteoid. This is the same histologic picture as that seen in OPG transgenic mice, and indicates that the phenotypic effects seen in transgenics can be recreated in normal mice when given intravenous recombinant protein. The effects of recombinant OPG in young growing mice were similar to those seen in mice treated with bisphosphonates (Sietsema *et al.*, 1989), small molecule antiresorptive therapeutics useful for the treatment of osteoporosis. The effects of OPG and the bisphosphonate pamidronate were tested in this *in vivo* model, and both compounds were found to lead to equivalent changes in the accumulation of bone mass (Simonet *et al.*, 1997). Thus, the biological activity of OPG on osteoclast maturation acts like an antiresorptive agent in normal animals.

OPG can also block bone resorption in a disease model characterized by increased osteoclast numbers and/or activity. A rat model for *osteoporosis* associated with the loss of estrogen was used to compare the effects of OPG with those of pamidronate. In this model, ovairectomy leads to the rapid loss of bone mass due to increased osteoclast activity (Kalu *et al.*, 1989). OPG and pamidronate both blocked increases in osteoclast numbers and activity, and prevented bone loss seen in the untreated ovariectomized rat controls. OPG has also been analyzed in the thyroparathyroidectomized rat model (Yamamoto *et al.*, 1998). In this model, serum calcium levels are raised by the exogenous addition of either parathyroid hormone or vitamin D3 via induction of osteoclast-mediated bone resorption. OPG can both block the induction of serum hypercalcemia in a rapid manner and reverse it once established, suggesting an immediate effect on osteoclast function and/or survival.

These data indicate that the OPG bioactivity translates into a protective agent that may be useful for the treatment of *osteopenic disorders* characterized by excessive bone loss due to elevated osteoclast numbers or activity, and may have important implications for the clinic. OPG is currently being tested in phase I/II clinical trials in the United States, and single-dose injections have been shown to suppress surrogate bone resorption markers in normal postmenopausal women (Bekker *et al.*, 1999).

Effects of inhibitors (antibodies) to receptors

There have been no reports of molecules that inhibit OPG bioactivity. The analysis of OPG knockout mice provides insight into the effects that OPG inhibition would probably produce *in vivo* (see above). OPG neutralizing antibodies have been isolated, and have been shown to block OPG effects on osteoclast development *in vitro* (Yasuda *et al.*, 1998a).

References

Adams, M. D., Dubnick, M., Kerlavage, A. R., Moreno, R., Kelley, J. M., Utterback, T. R., Nagle, J. W., Fields, C., and Venter, J. C. (1992). Sequence identification of 2,375 human brain genes. *Nature* **355**, 632–634.

Anderson, D. M., Maraskovsky, E., Billingsley, W. L., Dougall, W. C., Tometsko, M. E., Roux, E. R., Teepe, M. C., DuBose, R. F., Cosman, D., and Galibert, L. (1997). A homologue of the TNF receptor and its ligand enhance T-cell growth and dendritic-cell function. *Nature* **390**, 175–179.

Austyn, J. M., and Gordon, S. (1981). F4/80, a monoclonal antibody directed specifically against the mouse macrophage. *Eur. J. Immunol.* **11**, 805–815.

Bekker, P. J., Holloway, D., Nakanishi, A., Arrighi, H. M., and Dunstan, C. R. (1999). Osteoprotegerin (OPG) has potent and sustained anti-resorptive activity in postmenopausal women. *J. Bone Miner. Res.* **14**, (Suppl. 1), S180 (Abstract 1190).

Brandstrom, H., Jonsson, K. B., Vidal, O., Ljunghall, S., Ohlsson, C., and Ljunggren, O. (1998). Tumor necrosis factor-alpha and -beta upregulate the levels of osteoprotegerin mRNA in human osteosarcoma MG-63 cells. *Biochem. Biophys. Res. Commun.* **248**, 454–457.

Bucay, N., Sarosi, I., Dunstan, C. R., Morony, S., Tarpley, J., Capparelli, C., Scully, S., Tan, H. L., Xu, W., Lacey, D. L., Boyle, W. J., and Simonet, W. S. (1998). Osteoprotegerin-deficient mice develop early onset osteoporosis and arterial calcifications. *Genes Dev.* **12**, 1260–1268.

Burgess, T. L., Qian, Y.-X., Kaufman, S., Ring, B. D., Van, G., Capparelli, C., Kelley, M., Hsu, H., Boyle, W. J., Dunstan, C. R., Hu, S., and Lacey, D. L. (1999). The ligand for osteoprotegerin (OPGL) directly activates mature osteoclasts. *J. Cell Biol.* **145**, 527–538.

Emery, J. G., McDonnell, P., Burke, M. B., Deen, K. C., Lyn, S., Silverman, C., Dul, E., Appelbaum, E. R., Eichman, C., DiPrinzio, R., Dodds, R. A., James, I. E., Rosenberg, M., Lee, J. C., and Young, P. R. (1998). Osteoprotegerin is a receptor for the cytotoxic ligand TRAIL. *J. Biol. Chem* **273**, 14363–14367.

Fuller, K., Wong, B., Fox, S., Choi, Y., and Chambers, T. J. (1998). TRANCE is necessary and sufficient for osteoblast-mediated activation of bone resorption in osteoclasts. *J. Exp. Med.* **188**, 977–1001.

Hofbauer, L. C., Dunstan, C. R., Spelsberg, T. C., Riggs, B. L., and Khosla, S. (1998). Osteoprotegerin production by human osteoblast lineage cells is stimulated by vitamin D, bone morphogenetic protein-2, and cytokines. *Biochem. Biophys. Res. Commun.* **250**, 776–781.

Horwood, N. J., Elliott, J., Martin, T. J., and Gillespie, M. T. (1998). Osteotropic agents regulate the expression of osteoclast differentiation factor and osteoprotegerin in osteoblastic stromal cells. *Endocrinology* **139**, 4743–4746.

Jimi, E., Akiyama, S., Tsurukai, T., Okahashi, N., Kobayashi, K., Udagawa, N., Nishihashi, T., Takahashi, N., and Suda, T. (1999). Osteoclast differentiation factor acts as a multifunctional regulator in murine osteoclast differentiation and function. *J. Immunol.* **163**, 434–442.

Kalu, D. N., Liu, C. C., Hardin, R. R., and Hollis, B. W. (1989). The aged rat model of ovarian deficiency bone loss. *Endocrinology* **124**, 7–16.

Kwon, B. S., Wang, S., Udagawa, N., Haridas, V., Lee, Z. H., Kim, K. K., Oh, K. O., Greene, J., Li, Y., Su, J., Gentz, R., Aggarwal, B. B., and Ni, J. (1998). TR1, a new member of the tumor necrosis factor superfamily, induces fibroblast proliferation and inhibits osteoclastogenesis and bone resorption. *FASEB J.* **12**, 845–854.

Lacey, D. L., Erdmann, J. M., Teitelbaum, S. L., Tan, H.-L., Ohara, J., and Shioi, A. (1995). Interleukin 4, interferon-γ, and prostaglandin E impact the osteoclastic cell-forming potential of murine bone marrow macrophages. *Endocrinology* **136**, 2367–2376.

Lacey, D. L., Timms, E., Tan, H.-L., Kelley, M. J., Dunstan, C. R., Burgess, T., Elliott, R., Colombero, A., Elliott, G., Scully, S., Hsu, H., Sullivan, J., Hawkins, N., Davy, E., Capparelli, C., Eli, A., Qian, Y. X., Kaufman, S., Sarosi, I., Shalhoub, V., Senaldi, G., Guo, J., Delaney, J., and Boyle, W. J. (1998). Osteoprotegerin ligand is a cytokine that regulates osteoclast differentiation and activation. *Cell* **93**, 165–176.

Martin, T. J., and Udagawa, N. (1998). Hormonal regulation of osteoclast function. *Endrocrinol. Metab.* **9**, 6–12.

Mizuno, A., Amizuka, N., Irie, K., Murakami, A., Fujise, N., Kanno, T., Sato, Y., Nakagawa, N., Yasuda, H., Mochizuki, S., Gomibuchi, T., Yano, K., Shima, N., Washida, N., Tsuda, E., Morinaga, T., Higashio, K., and Ozawa, H. (1998). Severe osteoporosis in mice lacking osteoclastogenesis inhibitory factor/osteoprotegerin. *Biochem. Biophys. Res. Commun.* **237**, 610–615.

Morinaga, T., Nakagawa, N., Yasuda, H., Tsuda, E., and Higashio, K. (1998). Cloning and characterization of the gene encoding human osteoprotegerin/osteoclastogenesis inhibitory factor. *Eur. J. Biochem.* **254**, 685–691.

Rodan, G. A., and Martin, T. J. (1981). Role of osteoblasts in hormonal control of bone resorption – A hypothesis. *Calcif. Tissue Int.* **95**, 13361–13362.

Roodman, G. D. (1996). Advances in bone biology: the osteoclast. *Endocr. Rev.* **17**, 308–332.

Sietsema, W. K., Ebetino, F. H., Salvagno, A. M., and Bevan, J. A. (1989). Antiresorptive dose-response relationships across three generations of bisphosphonates. *Drugs Under Exp. Clin. Res.* **15**, 389–396.

Simonet, W. S., Lacey, D. L., Dunstan, C. R., Kelley, M., Chang, M. S., Luethy, R., Nguyen, H. Q., Wooden, S., Bennett, L., Boone, T., Shimamoto, G., DeRose, M., Elliott, R., Colombero, A., Tan, H. L., Trail, G., Sullivan, J., Davy, E., Bucay, N., Renshaw-Gegg, L., Hughes, T. M., Hill, D., Pattison, W., Campbell, P., and Boyle, W. J. (1997). Osteoprotegerin: A novel secreted protein involved in the regulation of bone density. *Cell* **89**, 309–319.

Smith, C. A., Farrah, T., and Goodwin, R. G. (1994). The TNF receptor superfamily of cellular and viral proteins: activation, co-stimulation, and death. *Cell* **76**, 959–962.

Suda, T., Takahashi, N, and Martin, T. J. (1992). Modulation of osteoclast differentiation. *Endocr. Rev.* **13**, 66–80.

Suda, T., Takahashi, N., Udagawa, N., Jimi, E., Gillespie, M. T., and Martin, T. J. (1999). Modulation of osteoclast differentiation and function by the new members of the tumor necrosis factor receptor and ligand families. *Endocr. Rev.* **20**, 345–57.

Takai, H., Kanematsu, M., Yano, K., Tsuda, E., Higashio, K., Ikeda, K., Watanabe, K., and Yamada, Y. (1998). Transforming growth factor-beta stimulates the production of osteoprotegerin/osteoclastogenesis inhibitory factor by bone marrow stromal cells. *J. Biol. Chem.* **273**, 27091–27096.

Tsuda, E., Goto, M., Mochizuki, S.-I., Yano, K., Kobayashi, F., Morinaga, T., and Hgiashio, K. (1997). Isolation of a novel cytokine from human fibroblasts that specifically inhibits osteoclastogenesis. *Biochem. Biophys. Res. Commun.* **234**, 137–142.

Udagawa, N., Takahashi, N., Akatsu, T., Sasaki, T., Tamaguchi, A., Kodama, H., Martin, T. J., and Suda, T. (1989). The bone marrow-derived stromal cells line MC3T3/PA6 and ST2 support osteoclast-like cell differentiation in cocultures with mouse spleen cells. *Endocrinology* **125**, 1805–1813.

Vidal, O. N., Brandstrom, H., Jonsson, K. B., and Ohlsson, C. (1998a). Osteoprotegerin mRNA is expressed in primary human osteoblast-like cells: down-regulation by glucorticoids. *J. Endocrinol.* **159**, 191–195.

Vidal, O. N., Sjogren, K., Eriksson, B. I., Ljunggren, O., and Ohlsson, C. (1998b). Osteoprotegerin mRNA is increased by interleukin-1 alpha in the human osteosarcoma cell line MG-63 and in human osteoblast-like cells. *Biochem. Biophys. Res. Commun.* **248**, 696–700.

Walczak, H., Miller, R. E., Ariail, K., Gliniak, B., Griffith, T. S., Kubin, M., Chin, W., Jones, J., Woodward, A., Le, T., Smith, C., Smolak, P., Goodwin, R. G., Rauch, C. T., Schuh, J. C., and Lynch, D. H. (1999). Tumoricidal activity of tumor necrosis factor-related apoptosis-inducing ligand *in vivo*. *Nature Med.* **5**, 157–163.

Wong, B. R., Josien, R., Lee, S. Y., Sauter, B., Li, H. L., Steinman, R. M., and Choi, Y. (1997). TRANCE (tumor necrosis factor (TNF)-related activation-induced cytokine), a new

TNF family member predominantly expressed in T cells, is a dendritic cell-specific survival factor. *J. Exp. Med.* **186**, 2075–2080.

Yamaguchi, K., Kinosaki, M., Goto, M., Kobayashi, F., Tsuda, E., Morinaga, T., and Higashio, K. (1998). Characterization of structural domains of human osteoclastogenesis inhibitory factor (OCIF)/OPG. *J. Biol. Chem.* **273**, 5117–7123.

Yamamoto, M., Murakami, T., Nishikawa, M. E., Tsuda, E., Mochizuki, S., Higashio, K., Akatsu, T., Motoyoshi, K., and Nagata, N. (1998). Hypocalcemic effect of osteoclastogenesis inhibitory factor/osteoprotegerin in the thyroparathryoidectomized rat. *Endocrinology* **139**, 4012–4015.

Yasuda, H., Shima, N., Nakagawa, N., Mochizuki, S. I., Yano, K., Fujise, N., Sato, Y., Goto, M., Yamaguchi, K., Kuriyama, M., Kanno, T., Murakami, A., Tsuda, E., Morinaga, T., and Higashio, K. (1998a). Identity of osteoclastogenesis inhibitory factor (OCIF) and osteoprotegerin (OPG): a mechanism by which OPG inhibits osteoclastogenesis *in vitro*. *Endocrinology* **139**, 1329–1337.

Yasuda, H., Shima, N., Nakagawa, N., Yamaguchi, K., Kinosaki, M., Mochizuki, S., Tomoyasu, A., Yano, K., Goto, M., Murakami, A., Tsuda, E., Morinaga, T., Higashio, K., Udagawa, N., Takahashi, N., and Suda, T. (1998b). Osteoclast differentiation factor is a ligand for osteoprotegerin/osteoclastogenesis-inhibitory factor and is identical to TRANCE/RANKL. *Proc. Natl Acad. Sci. USA* **95**, 3597–3602.

Yun, T. J., Chaudhary, P. M., Shu, G. L., Frazer, J. K., Ewings, M. K., Schwartz, S. M., Pascual, V., Hood, L. E., and Clark, E. A. (1998). OPG/FDCR-1, a TNF receptor family member, is expressed in lymphoid cells and is up-regulated by ligating CD40. *J. Immunol.* **161**, 6133–6121.

LICENSED PRODUCTS

OPG recombinant protein and antibodies for research

Alexis Biochemicals, 6181 Cornerstone Court East, Suites 102–104, San Diego, CA 92121, USA

R&D Systems, Inc., 614 McKinley Place N.E., Minneapolis, MN 55413, USA

Santa Cruz Biotechnology, Inc., 2161 Delaware Avenue, Santa Cruz, CA 95060, USA

CD27

W. A. M. Loenen*

Department of Medical Microbiology, University Hospital Maastricht, PO Box 5800, Maastricht, 6202AZ, The Netherlands

*corresponding author tel: (31)43 387 6642/6644, fax: (31)43 387 6643, e-mail: Wil.Loenen@medmic.unimaas.nl

DOI: 10.1006/rwcy.2000.16010.

SUMMARY

An important role for the CD27/CD70 ligand pair is emerging with respect to lymphocyte maturation in the periphery, while the putative role of CD27 and CD70 in thymic selection awaits further research. Triggering of the CD27/CD70 route affects levels of proliferation and differentiation of different lymphocyte subpopulations upon antigenic stimulation in different ways, and accumulating data support the notion that the receptor/ligand pair controls proliferation and facilitates maturation processes without interfering with programmed details of individual cells. It appears to control the crosstalk of the different subsets with each other both within the T, B, and NK cell lineages as well as between them. Its emerging role in synergy with CD28 and CD40 routes overlaps the immune response with respect to the T helper-mediated cascade of T and B cells, but allows extension of this help to the NK cells as well. On the other hand it does not (as far as current data go) appear to affect the inflammatory response involving nonlymphoid cells, as reported for CD28 and CD40, or directly influence the interaction with antigen-presenting cells.

BACKGROUND

Discovery

CD27 was discovered over a decade ago as a human glycoprotein on resting peripheral blood T cells and medullary thymocytes, and recognized by the monoclonal antibodies (mAbs) VIT14, S152, OKT18A, and CLB-9F4. Its restricted expression, the effect of mAbs, as well as an activation-related second protein form indicated a specific role for CD27 in the T cell-mediated immune response. Cloning of the cDNA identified CD27 as related to nerve growth factor receptor (NGFR) and Bp50 (CD40), the third member of a new group of receptors, currently known as the NGFR/TNFR supergene family, and involved in control of the immune cascade. The human ligand CD27L proved identical to the lymphoma antigen CDw70, recognized by the KI-24 mAb and highly expressed on Reed–Sternberg cells, as well as other lymphoid malignancies, including *Hodgkin's lymphoma* and *non-Hodgkin lymphoma*.

Structure

CD27 is a type I disulfide-linked glycoprotein and member of the NGFR/TNFR gene family. The ligand-binding domain contains two and a half of the characteristic cysteine-rich TNFR repeats, predicted to resemble an elongated structure. This region is followed by a membrane-proximal heavily *O*-glycosylated stalk-like domain and a short cytoplasmic domain. The activation-related monomeric soluble CD27 protein comprises the complete extracellular domain, is also found associated with T cell membranes and is most likely cleaved from the surface receptor (Camerini *et al.*, 1991; Loenen *et al.*, 1992a).

There are two *N*-linked carbohydrates in the N-terminal ligand-binding domain, and extensive *O*-linked glycosylation in the serine/threonine/proline-rich membrane-proximal extracellular part, increasing the M_r by 25 kDa. CD27 is phosphorylated on serine residues in resting cells, and becomes hyperphosphorylated after T cell activation. A consensus site for PKC in the human protein is not conserved in mouse. Recently a novel extracellular posttranslational modification, i.e. attachment of ADP-ribose to arginine residues, implicated in cytotoxic T cell function and homing, has been reported for CD27, which may regulate CD27 function *in vivo* (Okamoto *et al.*, 1998).

Main activities and pathophysiological roles

CD27 is emerging as a receptor that coordinates and controls interactions including both naïve and memory lymphocyte subsets during the expansion of T, B, and NK cells upon immune activation (**Figure 1**) (Gravestein and Borst, 1998; Lens *et al.*, 1998; Loenen, 1998). CD27 plays a role in cytokine and antibody production and is involved in the generation of cytotoxic T cells, antibody-producing B cells, memory cells, and the cytolytic activity of NK cells (**Figure 2**). Expression of CD27 is a hallmark for memory B lymphocytes, the double positive CD27+CD70+ cells comprising recently activated memory B cells (Klein *et al.*, 1998; Nagumo *et al.*, 1998; Tangye *et al.*, 1998). Levels of soluble CD27 in body fluids correlate with disease status in patients with autoimmune or malignant disease and certain viral infections (Hintzen *et al.*, 1991; Portegies *et al.*, 1993; Van Oers *et al.*, 1993; Kersten *et al.*, 1996), as reported for CD30 and TNFR (Aderka *et al.*, 1991; Cope *et al.*, 1992). Studies with CD27 mAb implicate a role for CD27 in thymocyte selection. *In vitro* experiments demonstrate bidirectional signaling by both CD27 and CD70, suggesting mutual involvement in expansion of the response, as reported for CD27 relatives, such as CD40, Fas, CD30, Ox40, and 4-1BB (Horie and Watanabe, 1998; Moulian and Berrih-Aknin, 1998; Toes *et al.*, 1998; Vogel and Noelle, 1998; Weinberg *et al.*, 1998; Vinay and Kwon, 1998). CD27 and CD70 activity on B cell lymphomas and lines may balance an autocrine loop and determine decisions of growth or apoptosis (Loenen, 1998). Evidence is emerging for enhancement of the *in vivo* T cell-mediated antitumor response via

Figure 1 Potential crosstalk of CD27+ lymphocytes with memory T and B cells. Interactions between the CD70+ memory CD4+CD45RO+ T cell and activated B cell memory population with resting T and NK cells, activated T and NK cells, as well as resting and activated memory B cells, would allow extensive crosstalk between subsets. It should be noted that expression of the CD70 ligand CD27 strongly increases after activation of human cells (for simplicity reasons not included in the figure), concomitant with the generation of soluble CD27. The role of soluble CD27, which can also be found associated with T cell membranes, is not clear, as is coexpression on NK cells of CD70 and CD27.

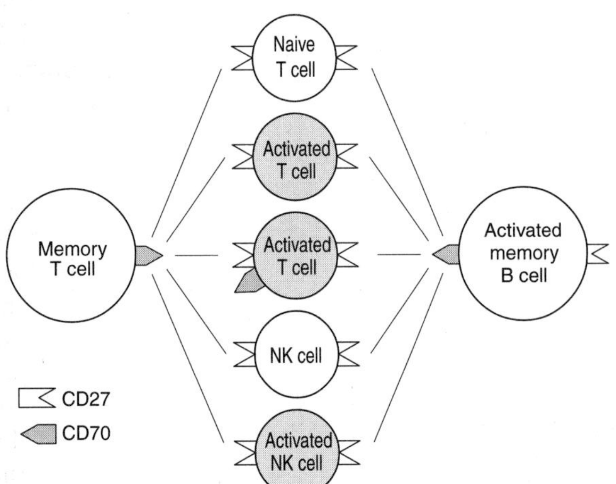

Figure 2 Potential interactions of naïve CD27+ T cells with lymphocytes expressing CD70. Interactions with CD27+ T cells would allow recruitment of the naïve lymphocytes by both resting memory CD4+CD45RO+ T cells and activated memory B cells, as well as activated T and NK cells. It should be noted that expression of CD27 strongly increases after activation of human cells (for simplicity reasons not included in the figure), concomitant with the generation of soluble CD27. The role of soluble CD27, which can also be found associated with T cell membranes, is not clear, as is coexpression on NK cells of CD70 and CD27.

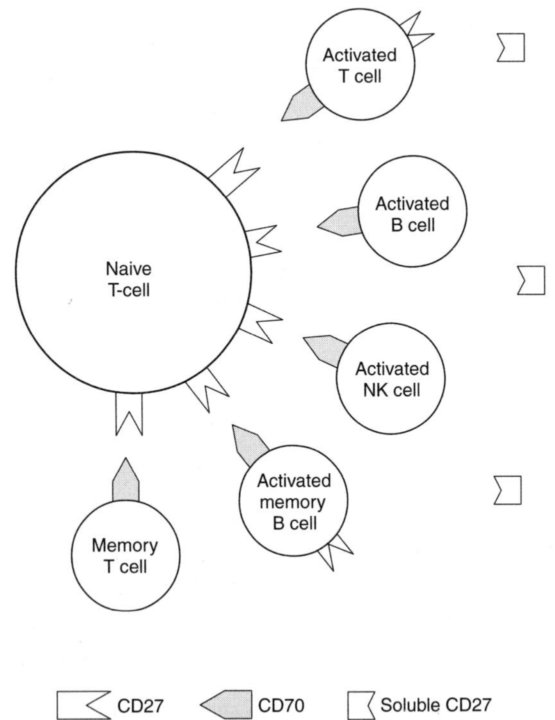

manipulations of the CD27/CD70 route, presumably via the synergistic effects with CD28/B7 and CD40/CD40L (CD154) reported *in vitro* (Couderc *et al.*, 1998; Nieland *et al.*, 1998; Stuhler *et al.*, 1999).

In contrast to most of its TNFR relatives, expression of CD27 appears to be (1) lymphocyte-restricted, (2) present on resting as well as activated T, B, and NK cells, (3) involved in crosstalk between all lymphocyte subsets, including both naïve and memory cells, and (4) a disulfide-linked homodimeric surface protein.

GENE

Accession numbers

EMBL/GenBank:
Human cDNA: M63928 (Camerini *et al.*, 1991)
Human gene: L24493, L24494 (Loenen *et al.*, 1992b)
Mouse cDNA: L24495 (Gravestein *et al.*, 1993)

Chromosome location and linkages

Human 12p13.

PROTEIN

Accession numbers

Human CD27: A46517 (PIR), P26842 (SwissProt)
Murine CD27: A49053 (PIR), P41272 (SwissProt)

Sequence

See **Figure 3**.

Description of protein

The human protein is a type I transmembrane protein of 260 amino acids, expressed as a disulfide-linked homodimer with subunits of 55 kDa (Camerini *et al.*, 1991). A backbone of 29 kDa carries two *N*-linked sugars in the N-terminal domain (80 residues), and is heavily *O*-glycosylated in the membrane-proximal extracellular serine/threonine/proline-rich region (70 residues). The N-terminal ligand-binding domain contains a large number of cysteine residues in the specific repeated pattern characteristic of the NGFR/TNFR supergene family, whose structure presumably resembles TNFR, an elongated structure held together with disulfide bridges, and lacking α helices or β sheets. The C-terminal part of the intracellular domain of 48 residues shares limited homology with 4-1BB and GITR, and is involved in binding members of the TRAF (TNFR-associated factors) family of signal transducers (Gravestein and Borst, 1998) and an apparently TRAF-unrelated death domain-containing protein called Siva, that may play a role *in vivo* in ischemia (Padanilam *et al.*, 1999). An activation-related 32 kDa monomeric protein form of human CD27 is a soluble receptor, also found associated with T cell membranes, most likely generated from the surface receptor via a proteolytic event that does not require receptor internalization (Loenen *et al.*, 1992a).

The murine receptor is a 45 kDa protein with nearly 80% identity to the human protein in the N-terminal ligand binding domain and the C-terminal

Figure 3 Amino acid sequence of human CD27 (Camerini *et al.*, 1991) and mouse CD27 (Gravestein *et al.*, 1993).

Human CD27

```
  1 MARPHPWWLC VLGTLVGLSA TPAPKSCPER HYWAQGKLCC QMCEPGTFLV
 51 KDCDQHRKAA QCDPCIPGVS FSPDHHTRPH CESCRHCNSG LLVRNCTITA
101 NAECACRNGW QCRDKECTEC DPLPNPSLTA RSSQALSPHP QPTHLPYVSE
151 MLEARTAGHM QTLADFRQLP ARTLSTHWPP QRSLCSSDFI RILVIFSGMF
201 LVFTLAGALF LHQRRKYRSN KGESPVEPAE PCRYSCPREE EGSTIPIQED
251 YRKPEPACSP
```

Murine CD27

```
  1 MAWPPPYWLC MLGTLVGLSA TLAPNSCPDK HYWTGGGLCC RMCEPGTFFV
 51 KDCEQDRTAA QCDPCIPGTS FSPDYHTRPH CESCRHCNSG FLIRNCTVTA
101 NAECSCSKNW QCRDQECTEC DPPLNPALTR QPSETPSPQP PPTHLPHGTE
151 KPSWPLHRQL PNSTVYSQRS SHRPLCSSDC IRIFVTFSSM FLIFVLGAIL
201 FFHQRRNHGP NEDRQAVPEE PCPYSCPREE EGSAIPIQED YRKPEPAFYP
```

cytoplasmic region. Overall homology is about 65% (Gravestein *et al.*, 1993).

Relevant homologies and species differences

A ligand-binding cysteine-rich domain is characteristic of the TNFR superfamily, which includes NGFR, TNFR-I, -II and -III, CD30, 4-1BB (CD137), Ox40, GITR, LT-β receptor, Herpes virus entry receptor (HVEM/ATAR), RANK, TR2, CD95 (Fas/Apo-1), Apo-3, and TRAIL receptors. The cytoplasmic C-terminus, the region involved in signal transduction, is conserved between human and murine CD27, 4-1BB, and GITR (Gravestein and Borst, 1998; Lens *et al.*, 1998).

Homology in the ligand-binding domain and C-terminal region exists between human and murine CD27 and allows cross-species recognition: human CD27 and CD70 can stimulate murine cells bearing CD70 or CD27, and vice versa. A major species difference of unknown consequence is CD27 tissue expression on resting T cells and thymocytes. CD27 is expressed in murine thymus from the early DP cell stage onwards, while in human expression is confined to mature medullary thymocytes. Expression of murine CD27 is high on thymocytes and resting T cells, while in human it is relatively low, but strongly enhanced upon activation.

Affinity for ligand(s)

Scatchard analysis of soluble CD27-Fc binding to the EBV-transformed B-cell line MP-1 shows a biphasic curve with low and high binding affinities with reported K_d values of $1.83 \times 10^8\,M^{-1}$ and $1.58 \times 10^9\,M^{-1}$, while that of recombinant sCD27 binding to 3T3 cells, transfected with the CD70 cDNA, gave values of $1.14 \times 10^8\,M^{-1}$ and $1.25 \times 10^9\,M^{-1}$ (Goodwin *et al.*, 1993; Agematsu *et al.*, 1994).

Cell types and tissues expressing the receptor

CD27 appears to be confined to lymphoid cells. Expression on human thymocytes appears after CD69, but can be induced on earlier cells. Outside the thymus, CD27-negative lymphocytes are often found in skin and lung, and hardly in lymphoid tissue, probably comprising the CD4+CD45RO+ T cells. This helper subset carries organ-specific homing receptors, and produces IL-4, IL-5, or IFNγ. Within the CD8+ subset, CD27-negative cells are CD45RA+CD95L(FasL)+ cytotoxic T cells, producing granzyme A/B and perforin, and these cells can kill *in vitro* without prestimulation. CD27 on NK cells has been reported to synergize with IL-2 or IL-12 in NK-mediated lysis. The data are summarized in **Table 1**.

Regulation of receptor expression

In humans, CD27 is upregulated after T, B and NK cell activation upon antigenic triggering and leads to high CD27 surface expression as well as soluble receptor (Lens *et al.*, 1998). Enhanced expression appears to be transient on all cells analyzed, with concomitant expression of its ligand CD70 on different subsets at different time points after activation. CD27 expression is downregulated on cells after prolonged antigenic stimulation. Expression on human thymocytes can be induced and/or upregulated by lectins, or phorbol ester plus ionomycin on both early and more mature cells. CD27 is absent on final stage effector cells, a situation parallel to that of CD40 on B cells from early stages onwards but for the effector plasma B cells. The data are summarized in **Table 2**.

Release of soluble receptors

Soluble CD27 is released in an activation-related manner. *In vivo* levels in serum and urine in general mark immune activation and correlate with tumor load in patients with *leukemias* and *lymphomas* (Van Oers *et al.*, 1993; Lens *et al.*, 1998). High soluble CD27 levels have also been reported in synovial fluid of *rheumatoid arthirtis* patients and cerebrospinal fluid of *multiple sclerosis* patients (Lens *et al.*, 1998).

SIGNAL TRANSDUCTION

Associated or intrinsic kinases

CD27 lacks intrinsic kinase activity.

Cytoplasmic signaling cascades

The C-terminal region associates with members of the TRAF (TNFR-associated factor) family, TRAF2 and TRAF5, that link the receptor to the NIK and

Table 1 Lymphocyte subsets expressing the CD27 surface receptor

	Positive subset	Negative subset	Comments
Periphery			
T cells	Naïve CD4+ and CD8+.	Effector and memory	Memory cells appear to reside within the CD70+ subset
	Both CD45RA+	CD4+ CD45RO+	
	75–90%	CD8+CD45RA+	
B cells	Memory cells	All other B cells	Hallmark for memory cells
	IgM+/IgD+IgM+		Activated memory B cells are
	5–30%	70–95%	CD27+CD70+
NK cells	30–40% CD56+	60–70% CD56+	Mainly CD16+, reciprocal high-level expression with ligand?
Thymus			
Thymocytes	CD1+CD3+CD69+	CD1+CD3+CD69+	Human medullary cells; in mouse both in cortex andmedulla (low level on highly immature DP)
	CD45RA−/CD45RO+	CD45RA−/CD45RO+	
	Mainly $CD4^{dull}$	Mainly CD4+	Ligand on medullary epithelial cells
	$CD8\alpha+\beta+$ and later stages	$CD8\alpha^{dull}$ and earlier stages	Thymic emigrants are CD45RA+

Table 2 Expression of CD27 and CD70 on T and B lymphocytes after activation

	CD27	CD70
T cells		
Resting cells	+	–
Activated cells	++	+
Memory cells	–	+
Activated memory cells	–	+
B cells		
Resting cells	–	–
Activated cells	–	+
Memory cells	+	+
Activated memory cells	+	+

JNK signaling pathways involving the transcription factors NFκB and Jun (**Figure 4**). This involves sequences in the C-terminal domain that have also been identified as important for interactions of its relatives CD40 and CD30 (EEEG) with TRAF1 and TRAF2, and of CD30 and TNFRII (PIQED) with TRAF1, TRAF2, and TRAF3 (**Figure 5**) (Gravestein and Borst, 1998).

DOWNSTREAM GENE ACTIVATION

Transcription factors activated

NFκB and c-Jun (Figure 4).

BIOLOGICAL CONSEQUENCES OF ACTIVATING OR INHIBITING RECEPTOR AND PATHOPHYSIOLOGY

Unique biological effects of activating the receptors

CD27 has a general stimulatory effect on lymphocyte function and appears to facilitate maturation of all lymphocyte subsets involved in the antigenic response. Its emerging unique role in this response within the TNFR family appears to be its ability for crosstalk between memory and naïve cells of the different lineages, including the natural killer response as well as that of the antigen-specific T and B cells. The potential interactions between memory and naïve cells are shown in Figure 1. Addition of CD70 to the model discussed by Toes

Figure 4 Signal transduction by CD27 involves TRAF/MAP kinase routes to transcription factors Jun and NFκB. Like its relatives, CD27 uses signal-transducing molecules (TRAF2 and TRAF5) that are members of the TNFR-associated factor (TRAF) family, and MAP kinase, that activate the transcription factors Jun (via phosphorylation by Jun kinase, JNK) and NFκB (via inactivation of IκB by NIK, which leads to translocation of NFκB from the cytosol to the nucleus). In addition, association of CD27 with an unrelated death domain-containing protein, called Siva, has been reported, which is not included in this figure.

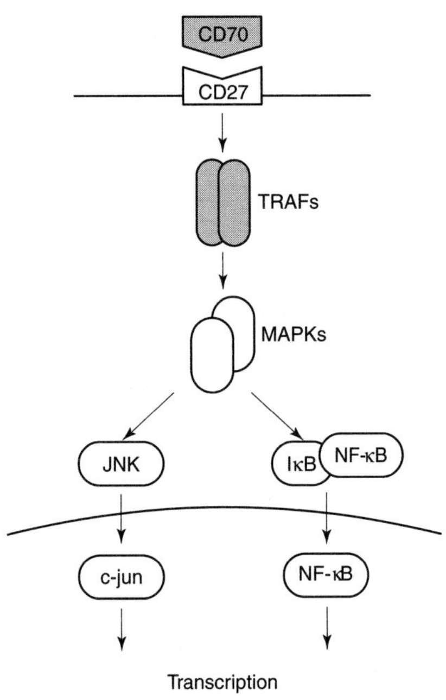

Figure 5 Segments of human and mouse CD27 showing the EEEG and PIQED sequences important for interactions with TRAF1, TRAF2, and TRAF3.

Human CD27
PAEPCRYSCPR**EEEG**STI**PIQED**YRKPEPACSP

Mouse CD27
PEEPCPYSCPR**EEEG**SAI**PIQED**YRKPEPAFYP

et al. (1998) on the involvement of CD40 in the interactions between helper and cytotoxic T cells with dendritic cells is shown in Figure 5 of the chapter on CD70.

Phenotypes of receptor knockouts and receptor overexpression mice

Preliminary data on CD27−/− mice show no major alterations (Gravestein, 1998).

Human abnormalities

In *X-linked hyper-IgM syndrome*, in which patients' class switching is impaired due to a defect in the CD40L gene, the IgD-CD27+ B-cell memory population has been reported greatly decreased, supporting its putative role *in vivo* in the generation of memory B cells and development of plasma cells (Agematsu *et al.*, 1998).

References

Aderka, D., Englemann, H., Hornik, V., Skornick, Y., Levo, Y., Wallach, D., and Kushtai, G. (1991). Increased serum levels of soluble receptors for tumor necrosis factor in cancer patients. *Cancer Res.* **51**, 5602–5607.

Agematsu, K., Kobata, T., Sugita, K., Freeman, G. J., Beckmann, M. P., Schlossman, S. F., and Morimoto, C. (1994). Role of CD27 in T cell immune response. Analysis by recombinant soluble CD27. *J. Immunol.* **153**, 1421–1429.

Agematsu, K., Nagumo, H., Shinozaki, K., Hokibara, S., Terada, K., Kawamura, N., Toba, T., Nonoyama, S., Ochs, H. D., and Komiyama, A. (1998). Absence of IgD-CD27+ memory B cell population in X-linked hyper-IgM syndrome. *J. Clin. Invest.* **102**, 853–860.

Camerini, D., Walz, G., Loenen, W. A. M., Borst, J., and Seed, B. (1991). The T cell activation antigen CD27 is a member of the nerve growth factor/tumor necrosis factor receptor gene family. *J. Immunol.* **147**, 3165–3169.

Cope, A. P., Aderka, D., Doherty, M., Engelmann, H., Gibbons, D., Jones, A. C., Brennan, F. M., Maini, R. N., Wallach, D., and Feldmann, M. (1992). Increased levels of soluble tumor necrosis factor receptors in the sera and synovial fluid of patients with rheumatic diseases. *Arthritis Rheum.* **35**, 1160–1169.

Couderc, B., Zitvogel, L., Douin-Echinard, V., Djennane, L., Tahara, H., Favre, G., Lotze, M. T., and Robbins, P. D. (1998). Enhancement of antitumor immunity by expression of CD70 (CD27 ligand) or CD154 (CD40 ligand) costimulatory molecules in tumor cells. *Cancer Gene Ther.* **5**, 163–175.

Goodwin, R. G., Alderson, M. R., Smith, C. A., Armitage, R. J., VandenBos, T., Jerzy, R., Tough, T. W., Schoenborn, M. A., Davis-Smith, T., Hennen, K., Falk, B., Cosman, D., Baker, E., Sutherland, G. R., Grabstein, K. H., Farrah, T., Giri, J. G., and Beckmann, M. P. (1993). Molecular and biological characterisation of a ligand for CD27 defines a new family of cytokines with homology to tumor necrosis factor. *Cell* **73**, 447–456.

Gravestein, L. A. (1998). Molecular and functional characterisation of murine CD27. PhD thesis, University of Amsterdam, NL.

Gravestein, L. A., and Borst, J. (1998). Tumor necrosis factor receptor family members in the immune system. *Semin. Immunol.* **10**, 423–434.

Gravestein, L. A., Blom, B., Nolten, L. A., De Vries, E., Van der Horst, G., Ossendorp, F., Borst, J., and Loenen, W. A. M. (1993). Cloning and expression of murine CD27: comparison with 4-1BB, another lymphocyte-specific member of the nerve growth factor receptor family. *Eur. J. Immunol.* **23**, 943–950.

Hintzen, R. Q., Van Lier, R. A. W., Kuijpers, K. C., Baars, P. A., Schaasberg, W., Lucas, C. J., and Polman, C. H. (1991). Elevated levels of a soluble form of the T cell activation antigen CD27 in cerebrospinal fluid of multiple sclerosis patients. *J. Neuroimmunol.* **35**, 211–216.

Horie, R., and Watanabe, T. (1998). CD30: expression and function in health and disease. *Semin. Immunol.* **10**, 4457–4470.

Kersten, M. J., Evers, L. M., Dellemijn, P. L. I., Portegies, P., Hintzen, R. Q., Van Lier, R. A. W., Von dem Borne, A. E. G. Kr., and Van Oers, R. H. J. (1996). Elevation of cerebrospinal fluid soluble CD27 levels in patients with meningeal localisation of lymphoid malignancies. *Blood* **87**, 1985–1889.

Klein, U., Rajewski, K., and Kuppers, R. (1998). Human immunoglobulin (Ig)M+IgD+ peripheral blood B cells expressing the CD27 cell surface antigen carry somatically mutated variable region genes: CD27 as a general marker for somatically mutated (memory) B cells. *J. Exp. Med.* **188**, 1679–1689.

Lens, S. M. A., Tesselaar, K., Van Oers, M. H. J., and Van Lier, R. A. W. (1998). Control of lymphocyte function through CD27–CD70 interactions. *Semin. Immunol.* **10**, 491–499.

Loenen, W. A. M. (1998). Editorial overview: CD27 and (TNFR) relatives in the immune system: their role in health and disease. *Semin. Immunol.* **10**, 417–422.

Loenen, W. A. M., De Vries, E., Gravestein, L. A., Hintzen, R. Q., Van-Lier, R. A., and Borst, J. (1992a). The CD27 membrane receptor, a lymphocyte-specific member of the nerve growth factor receptor family, gives rise to a soluble form by protein processing that does not involve receptor endocytosis. *Eur. J. Immunol.* **22**, 447–455.

Loenen, W. A. M., Gravestein, L. A., Beumer, S., Melief, C. J. M., Hagemeijer, A., and Borst, J. (1992b). Genomic organization and chromosomal localization of the human CD27 gene. *J. Immunol.* **149**, 3937–3943.

Moulian, N., and Berrih-Aknin, S. (1998). Fas/Apo-1/CD95 in health and autoimmune disease: thymic and peripheral aspects. *Semin. Immunol.* **10**, 449–456.

Nagumo, H., Agematsu, K., Shinozaki, K., Hokibara, S., Ito, S., Takamoto, M., Nikaido, T., Yasui, K., Uehara, Y., Yachie, A., Komiyama, A. (1998). CD27/CD70 interaction augments IgE secretion by promoting the differentiation of memory B cells into plasma cells. *J. Immunol.* **161**, 6496–6502.

Nieland, J. D., Graus, Y. F., Dortmans, Y. E., Kremers, B. L. J. M., and Kruisbeek, A. M. (1998). CD40 and CD70 co-stimulate a potent *in vivo* antitumor T cell response. *J. Immunother.* **21**, 225–236.

Okamoto, S., Azhipa, O., Yu, Y., Russo, E., and Dennert, G. (1998). Expression of ADP-ribosyltransferase on normal T lymphocytes and effects of nicotinamide adenine nucleotide on their function. *J. Immunol.* **160**, 4190–4198.

Padanilam, B. J., Lewington, A. J. P., and Hammerman, M. R. (1998). Expression of CD27 and ischemia/reperfusion-induced expression of its ligand Siva in rat kidneys. *Kidney Intern.* **54**, 1967–1975.

Portegies, P., Godfried, M. H., Hintzen, R. Q., Stam, J., Van der Poll, T., Bakker, M., Van Dventer, S. J. H., Van Lier, R. A. W., and Goudsmit, J. (1993). Low levels of specific T cell activation marker CD27 accompanied by elevated levels of markers for non-specific immune activation in the cerebrospinal fluid of patients with AIDS dementia complex. *J. Neuroimmunol.* **48**, 241–248.

Stuhler, G., Zobywalski, A., Grunebach, F., Brossaert, P., Reichardt, V. L., Barth, H., Stevanovich, S., Brugger, W., Kanz, L., and Schlossman, S. F. (1999). Immune regulatory loops determine productive interactions within human T-lymphocyte-dendritic cell clusters. *Proc. Natl Acad. Sci. USA* **96**, 1532–1535.

Tangye, S. G., Liu, Y. J., Aversa, G., Phillips, J., and De Vries, J. E. (1998). Identification of functional human splenic memory B cells by expression of CD148 and CD27. *J. Exp. Med.* **188**, 1691–1703.

Toes, R. E. M., Schoenberger, S. P., Van der Voort, E. I. H., Offringa, R., and Melief, C. J. M. (1998). CD40–CD40 Ligand interactions and their role in cytotoxic T lymphocyte priming and anti-tumor immunity. *Semin. Immunol.* **10**, 443–448.

Van Oers, M. H. J., Pals, S. T., Evers, L. M., Van der Schoot, C. E., Koopman, G., Bonfrer, J. M. G., Hintzen, R. Q., Von dem Borne, A. E. G. Kr., and Van Lier, R. A. W. (1993). Expression and release of CD27 in human B-cell malignancies. *Blood* **82**, 3430–3436.

Vinay, D. S., and Kwon, B. S. (1998). Role of 4-1BB in immune responses. *Semin. Immunol.* **10**, 4481–4490.

Vogel, L. A., and Noelle, R. J. (1998). CD40 and its crucial role as a member of the TNFR family. *Semin. Immunol.* **10**, 435–442.

Weinberg, A. D., Vella, A. T., and Croft, M. (1998). OX-40: life beyond the effector T cell stage. *Semin. Immunol.* **10**, 471–480.

HVEM

Carl F. Ware*

Division of Molecular Immunology, La Jolla Institute for Allergy and Immunology, 10355 Science Center Drive, San Diego, CA 92121, USA

*corresponding author tel: 858-678-4660, fax: 858-558-3595, e-mail: carl_ware@liai.org

DOI: 10.1006/rwcy.2000.16011.

SUMMARY

Herpesvirus entry mediator type A (HVEM, also known as HveA) is related to the receptors for tumor necrosis factor. HVEM is a single transmembrane glycoprotein with a canonical cysteine-rich extracellular domain that binds to two cellular ligands, LIGHT and lymphotoxin α, which are related to TNF. HVEM shares ligand specificity with the LTβ receptor and the two receptors for TNF, which also bind LTα. HVEM binds a virus-encoded ligand, the envelope glycoprotein D (gD) of *Herpes simplex virus* (HSV) types 1 and 2, and serves as one of several entry routes used by HSV to infect cells. Envelope gD can sterically inhibit the binding of membrane LIGHT to HVEM, acting as a virokine, potentially disrupting signal transduction. The cytoplasmic tail of HVEM binds TRAF2 and TRAF5 adapter proteins, which propagate signals leading to the activation of NFκB and AP-1 transcription factors, which control many genes involved in inflammation and immune responses.

HVEM is prominently expressed on cells in the immune system, particularly T cells and dendritic cells and activation of HVEM can provide a costimulator function for T cells. HVEM targets HSV to activated T cells inducing Fas ligand, which can lead to fratricide of bystander lymphocytes. HSV can also infect dendritic cells and block their maturation. Together, HVEM-mediated entry of HSV may lead to localized immune suppression. LIGHT can interfere with HSV entry, thus potentially acting as a virus deterrent. The UL144 gene in human *cytomegalovirus* (β-herpesvirus) is a structural homolog of HVEM, suggesting a long evolutionary history between the TNF superfamily and herpesviruses.

BACKGROUND

Discovery

Herpesvirus entry mediator (HVEM) was discovered by Spear and colleagues as a factor that allowed *Herpes simplex virus* (HSV-1 and -2) to infect Chinese hamster ovary cells (CHO) (Montgomery *et al.*, 1996). CHO cells are normally resistant to infection by HSV, and thus, transfer of a cDNA library into CHO cells was used to select genes that allowed virus entry and subsequent expression of a virus-encoded marker gene (β-galactosidase). Sequencing of one such clone revealed a membrane glycoprotein with a cysteine-rich extracellular domain that showed significant homology to TNFR superfamily. This assay led to the discovery of several additional entry factors for HSV-1 (Geraghty *et al.*, 1998). In addition, HVEM was identified in several databases as a TNFR homolog (Marsters *et al.*, 1997; Kwon *et al.*, 1997; Hsu *et al.*, 1997).

Table 1 Cellular entry factors for herpes simplex virus

Gene	Other names[a]	Gene family[b]	Virus attachment[c]	Tissue[d]
HveA	HVEM, ATAR,TR2	TNF receptor	gD HSV1&2	Lymphoid
HveB	Prr2	Ig fold	gD HSV2	Epithelial
HveC	Prr1	Ig fold	gD HSV1&2	Neuronal
Pvr-HveD	Pvr	Ig fold	PRV and BHV1	Unknown

[a]HVEM, herpesvirus entry mediator; ATAR, another TRAF-associated receptor; TR2, TNFR-like receptor 2; Pvr, poliovirus receptor; Prr, poliovirus receptor-related protein.

[b]Structural relationship with: TNF receptor superfamily or immunoglobulin superfamily (Ig fold).

[c]HSV, herpes simplex virus; PRV, pseudorabiesvirus; BHV1, bovineherpesvirus 1.

[d]Tissue type within which this entry protein may serve as an HSV entry route.

Alternative names

Herpesvirus entry mediator (HVEM) is the original designation for this protein, but TR2 and ATAR are also found in the literature. Recently, Spear suggested redesignation as HveA, based on a nomenclature scheme that groups it with other HSV entry factors (Geraghty *et al.*, 1998) (**Table 1**). This scheme groups functionally similar, but structurally distinct, proteins that bind envelope glycoprotein D (gD) of HSV. Additionally, for the purposes of cataloging the large numbers of human genes, HVEM is also designated TNFRSF14 (TNF receptor superfamily 14) by the Human Gene Nomenclature Committee, HUGO (www.gene.ucl.ac.uk/nomenclature).

Figure 1 Significant features of herpesvirus entry mediator (HVEM).

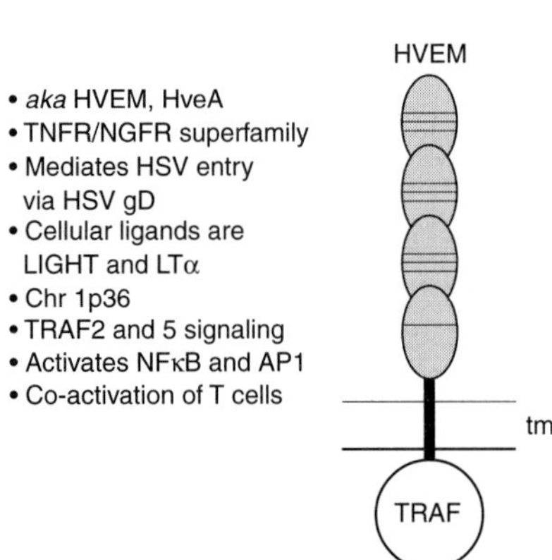

Herpesvirus entry mediator
HVEM
• aka HVEM, HveA
• TNFR/NGFR superfamily
• Mediates HSV entry via HSV gD
• Cellular ligands are LIGHT and LTα
• Chr 1p36
• TRAF2 and 5 signaling
• Activates NFκB and AP1
• Co-activation of T cells
tm
TRAF

Structure

HVEM is a type 1 single transmembrane glycoprotein of 283 amino acids that contains a cysteine-rich extracellular domain, which is homologous to members of the TNFR superfamily (**Figure 1**). The HVEM gene maps to chromosome 1p36. HVEM binds three known ligands: LIGHT, LTα, and envelope glycoprotein D (gD) of *Herpes simplex virus*. The relatively short cytosolic signaling domain binds to select members of the TRAF (TNFR-associated factor) family of signaling proteins that can activate NFκB and AP-1 transcription factors. The mouse HVEM homolog is 276 amino acids long and shares 44% overall identity.

Main activities and pathophysiological roles

HVEM is linked to the biology of herpesvirus as an entry factor and as a signaling receptor for the lymphotoxin-like cytokine LIGHT (Mauri *et al.*, 1998). HVEM is considered an integral component of the immediate TNF family because it also binds the LTα homotrimer, and LIGHT binds to the LTβR, the receptor for the LTα1β2 heterotrimer (**Figure 2**). Gene deletion or transgenic systems have not yet defined the role of HVEM in immune physiology, however, it is well established by these systems that the LTβR is an essential regulator of lymphoid organogenesis and splenic architecture (Futterer *et al.*, 1998) as well as NK-T cell differentiation (Fu and Chaplin, 1999). These results, and the close structural homology and shared ligands with HVEM, entice speculation that this receptor may participate in similar physiologic processes. However, the prominent expression of HVEM on T and B lymphocytes (LTβR is absent on lymphocytes), and the ability of HVEM to activate NFκB and AP-1 transcription factors hints that HVEM is intimately involved in the cascade of cytokines orchestrating lymphocyte differentiation and effector activity. Tissue culture models suggest that HVEM may serve as a costimulator for lymphocyte activation and cytokine secretion (Kwon *et al.*, 1997).

HVEM is likely to play a significant role in the pathophysiology of HSV infection as an entry factor. Viral entry is a complex process involving several viral envelope glycoproteins and several cellular components (Spear *et al.*, 1992). Virus attachment to the cell surface is mediated via the binding of virus envelope glycoprotein C (gC) to cell surface sulfated proteoglycans. Virus fusion with the cell membrane (entry) requires HSV gD, gB, gH, gL acting in concert with one of several distinct cellular receptors. Of the several entry routes that HSV can utilize (Geraghty *et al.*, 1998), HVEM may serve as a major entry route for a subset of T lymphocytes (Montgomery *et al.*, 1996). Interestingly, contact between HSV-infected fibroblasts and CTLs or NK cells rapidly inactivates their cytolytic capacity (Posavad *et al.*, 1994), suggesting that HSV may use the HVEM entry route as a mechanism of immune suppression. HSV-1 entry is only dependent on the ligand-binding and transmembrane domains, as deletion of the cytoplasmic tail of HVEM does not interfere with entry. Soluble HVEM inhibits HSV infection in many cell types by binding to gD in the envelope of the virion (Geraghty *et al.*, 1998). The binding site on gD for HVEM topographically overlaps with the site recognized by HveC as they compete with each other to bind gD (Whitbeck *et al.*, 1997). HVEM also plays a role in virus-induced cell fusion (Terry-Allison *et al.*, 1998) and promotes cell-to-cell spread of virus (Roller and Rauch, 1998).

Figure 2 HVEM is a member of the immediate TNF/LT$\alpha\beta$ family. Arrows indicate the receptor–ligand binding specificity of the various members. Solid lines indicate high-affinity interactions; dashed line refers to weak interactions.

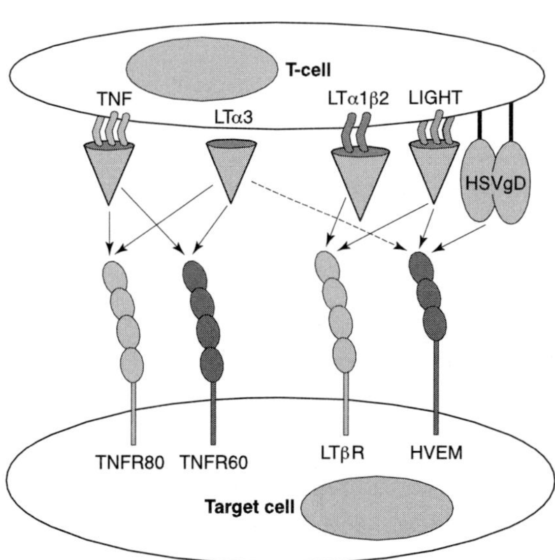

As a possible viral deterrent, LIGHT interferes with the entry of HSV-1 into CHO cells transfected with HVEM (Mauri *et al.*, 1998). HSV envelope gD also causes a similar interference phenomenon (Johnson and Spear, 1989), indicating that coexpression of HVEM with either ligand inhibits virus entry. This could occur by direct competition with the binding of virion gD, or by the ability of ligands to inhibit HVEM expression at the cell surface. HSV-1 gD directly and specifically competes with HVEM binding to membrane-anchored LIGHT. Thus, at the membrane contact region between the lymphocyte and virus-infected cell, gD could block the initiation of LIGHT/HVEM signaling pathways and subsequent cellular responses.

An additional molecular link between herpesvirus and the LT/TNF cytokine family was revealed by the discovery of a structural homolog of HVEM that is encoded in clinical isolates of human *cytomegalovirus*. The UL144 open reading frame (ORF) encodes a single transmembrane glycoprotein with 46% identity to the cysteine-rich extracellular (ecto) domain of HVEM (Benedict *et al.*, 1999). UL144 is expressed early after infection of fibroblasts, however it is retained intracellularly. A YRTL motif in the highly conserved cytoplasmic tail of UL144 contributes to its subcellular distribution consistent with a site of action in clathrin-coated vesicles. This, together with the findings that no known ligand of the TNF family binds UL144, suggests that its mechanism of action is distinct from other known virus immune-evasion genes. Specific antibodies to UL144 can be detected in the serum of a subset of HCMV seropositive individuals infected with *HIV*, indicating that this virus gene product is relevant to the human immune response.

GENE

Accession numbers

Human HVEM cDNA: U70321

Chromosome location and linkages

The HVEM gene is located at chromosome 1p36.22–36.3, which is close to several other TNFR genes including TNFR2, 4-1BB, CD30, Ox40, and DR3/TRAMP (Kwon *et al.*, 1997). The genomic organization has not been published at this time. Possible polymorphisms in the coding regions of HVEM cDNA between different isolates (U81232) are not apparent in corrected sequences.

PROTEIN

Description of protein

HVEM is a type I single transmembrane glycoprotein of 283 amino acids (276 for mouse HVEM) encoded by a major mRNA transcript of 1.8 kb. HVEM is defined as a member of the TNF receptor superfamily by virtue of its extracellular cysteine-rich (20 residues) domain (CRD) (**Figure 3**). For many TNFR-related proteins, each CRD contains six cysteines that form three disulfide bonds creating loops linked around a core sequence of CxxCxxC that is highly conserved. This basic motif is repeated from two to six times depending upon the individual receptor. The crystal structure of TNFR1 reveals an elongated molecule, where the three disulfide bonds in each CRD resemble the rungs of a ladder (Banner *et al.*, 1993).

For HVEM, the first two CRDs correspond exactly to the pattern displayed by TNFR1 and can be readily modeled on the TNFR1 structure (Guex and Peitsch, 1997). However, the spacing of remaining cysteines has a complex pattern that deviates substantially from TNFR1, making the demarcation

Figure 3 Sequence and domain structure of human HVEM. The predicted protein sequence for HVEM reveals a type 1 transmembrane glycoprotein. The repeated cysteine-rich domains (CRD) are predicted by Profilescan (www.isrec.isb-sib.ch/software/PFSCAN_form.html) although the division for CRD3 and 4 is arbitrary. The disulfide bond pattern (1–2, 3–5, 4–6) for CRD1 and 2 is based on TNFR1 and is shown by connecting lines. Demarcation of transmembrane domain and cytoplasmic tail is by PSORTII (www.psort.nibb.ac.jp:8800/). The TRAF-binding region was deduced by mutation (Hsu *et al.*, 1997).

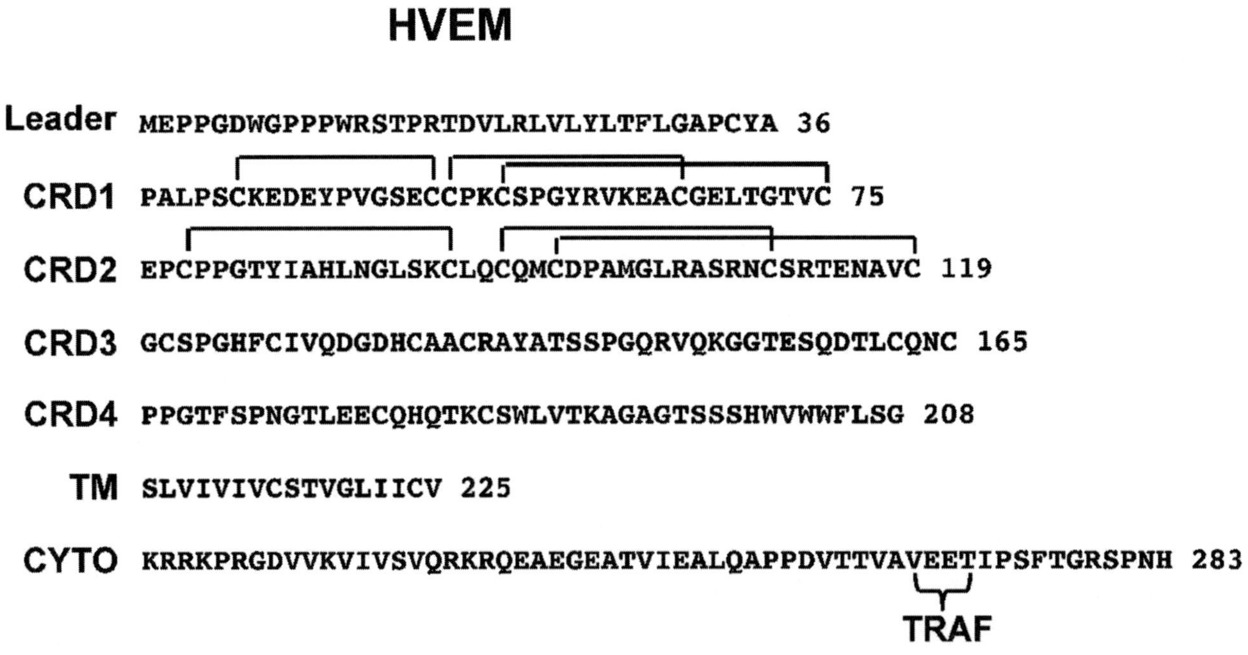

HVEM

Leader	MEPPGDWGPPPWRSTPRTDVLRLVLYLTFLGAPCYA 36
CRD1	PALPSCKEDEYPVGSECCPKCSPGYRVKEACGELTGTVC 75
CRD2	EPCPPGTYIAHLNGLSKCLQCQMCDPAMGLRASRNCSRTENAVC 119
CRD3	GCSPGHFCIVQDGDHCAACRAYATSSPGQRVQKGGTESQDTLCQNC 165
CRD4	PPGTFSPNGTLEECQHQTKCSWLVTKAGAGTSSSHWVWWFLSG 208
TM	SLVIVIVCSTVGLIICV 225
CYTO	KRRKPRGDVVKVIVSVQRKRQEAEGEATVIEALQAPPDVTTVAVEETIPSFTGRSPNH 283

TRAF

of CRD3 and 4 in HVEM difficult. Motif profiling programs do not recognize this pattern as a TNFR motif, unless low stringent criteria are used.

The HVEM ectodomain (truncated at His200) under native conditions behaves as a dimer of $\sim$40 kDa, and treatment with reducing agents does not alter the dimeric structure, indicating that all cysteines are probably involved in intramolecular disulfide bonds (Whitbeck *et al.*, 1997). The ectodomain of HVEM has two *N*-linked glycosylation sites and exhibits size heterogeneity on SDS-PAGE that is reduced by treatment with endoglycosidase F and *O*-glycanase (**Table 2**).

Relevant homologies and species differences

HVEM is most closely related in overall sequence homology to TNFR80, CD40, and LTβR, and this homology is primarily reflected in the ectodomain (**Figure 4**). HVEM also shows significant homology (46%) to the UL144 ORF encoded by human *cytomegalovirus* (**Figure 5**). The species homology for mouse and human HVEM is only 41% (**Figure 6**) significantly less than for related receptors in this system, which range from 63 to 68% or higher (Hsu *et al.*, 1997) (**Table 3**). Given that mice do not have a natural α-herpesvirus equivalent, it is tempting to speculate that HSV may have served as the strong selective pressure driving this divergence.

Affinity for ligand(s)

HVEM binds three distinct ligands: LIGHT, LTα (Mauri *et al.*, 1998), and HSV-encoded gD (Whitbeck *et al.*, 1997). LIGHT and LTα are members of the superfamily of TNF-related ligands. LIGHT is a type II membrane protein, whereas LTα is naturally produced as a soluble protein unless assembled as a complex with LTβ (Browning *et al.*, 1993). These proteins are structurally related and both form trimers. By contrast, HSV gD is a type I membrane protein that readily forms dimers. HSV gD is found in the virion envelope and is also expressed on the surface of infected cells (Cohen *et al.*, 1988). No structural relationship is observed between gD and members of the TNF ligand superfamily.

HVEM exhibits an avidity binding constant, as a dimeric HVEM-Fc molecule, of $K_d = 0.5$ nM with a soluble form of recombinant LIGHT; the estimated

Table 2 Some physical properties of HVEM

Property	Human	Mouse	
mRNA (kb)	1.8 and 3.8	nd	
Amino acids	283	276	
Molecular mass (kDa)[a]			
Sequence	30.4	30.3	
Observed	32/42	nd	
N-Glycosylation[b]	2 (110; 173)	2 (184; 197)	
Signal cleavage[c]	36–37	38–39	
Transmembrane[c]	209–225	214–230	
Alternate forms	nd	Deletion 101–158	CRD2/3
		Deletion 61–84	CRD1
		Truncation 158	soluble protein
		Truncation 101	soluble protein

[a]Calculated from sequence; observed by SDS-PAGE for *in vitro* translated cDNA and estimated from a partial cDNA expressed in CHO cells.

[b]Number of sites and their amino acid position.

[c]Determined using PSORTII analysis (www.psort.nibb.ac.jp:8800/)

Figure 4 Alignment of HVEM with other TNFR superfamily members. The cysteine-rich domain 2 of several TNFR superfamily members were aligned with ClustalW (PAM250 series, MacVector). The boxed areas show identical regions.

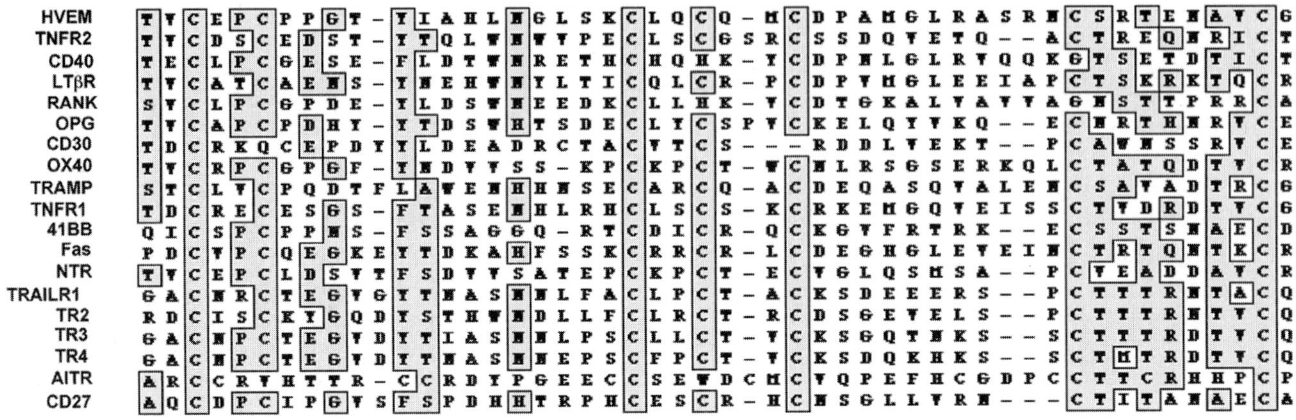

```
HVEM     T V C E P C P P G T - Y I A H L N G L S K C L Q C Q - M C D P A M G L R A S R N C S R T E N A V C G
TNFR2    T V C D S C E D S T - Y T Q L W N W V P E C L S C G S R C S S D Q V E T Q - - A C T R E Q N R I C T
CD40     T E C L P C G E S E - F L D T W N R E T H C H Q H K - Y C D P N L G L R V Q Q K G T S E T D T I C T
LTβR     T V C A T C A E N S - Y N E H W N Y L T I C Q L C R - P C D P V M G L E E I A P C T S K R K T Q C R
RANK     S V C L P C G P D E - Y L D S W N E E D K C L L H K - V C D T G K A L V A V V A G N S T T P R R C A
OPG      T V C A P C P D H Y - Y T D S W H T S D E C L Y C S P V C K E L Q Y V K Q - - E C N R T H N R V C E
CD30     T D C R K Q C E P D Y Y L D E A D R C T A C V T C S - - - R D D L V E K T - - P C A W N S S R V C E
OX40     T V C R P C G P G F - Y N D V V S S - K P C K P C T - W C N L R S G S E R K Q L C T A T Q D T V C R
TRAMP    S T C L V C P Q D T F L A W E N H H N S E C A R C Q - A C D E Q A S Q V A L E N C S A V A D T R C G
TNFR1    T D C R E C E S G S - F T A S E N H L R H C L S C S - K C R K E M G Q V E I S S C T V D R D T V C G
41BB     Q I C S P C P P N S - F S S A G G Q - R T C D I C R - Q C K G V F R T R K - - E C S S T S N A E C D
Fas      P D C V P C Q E G K E Y T D K A H F S S K C R R C R - L C D E G H G L E V E I N C T R T Q N T K C R
NTR      T V C E P C L D S V T F S D V V S A T E P C K P C T - E C V G L Q S M S A - - P C V E A D D A V C R
TRAILR1  G A C N R C T E G V G Y T N A S N N L F A C L P C T - A C K S D E E E R S - - P C T T T R N T A C Q
TR2      R D C I S C K Y G Q D Y S T H W N D L L F C L R C T - R C D S G E V E L S - - P C T T T R N T V C Q
TR3      G A C N P C T E G V D Y T I A S N N L P S C L L C T - V C K S G Q T N K S - - S C T T T R D T V C Q
TR4      G A C N P C T E G V D Y T N A S N N E P S C F P C T - V C K S D Q K H K S - - S C T M T R D T V C Q
AITR     A R C C R V H T T R - C C R D Y P G E E C C S E W D C M C V Q P E F H C G D P C C T T C R H H P C P
CD27     A Q C D P C I P G V S F S P D H H T R P H C E S C R - H C N S G L L V R N - - - C T I T A N A E C A
```

affinity for a single site was 44 nM (Harrop *et al.*, 1998) (**Table 4**). Given that HVEM forms dimers, binding to LIGHT is predicted to be of relatively high affinity on the cell surface. Competition binding assays using HVEM-Fc and membrane-bound LIGHT showed half maximum binding at 10–30 nM for HVEM-Fc, whereas LTα behaved as a competitive binding ligand at 50–70 nM. This suggests that LTα-HVEM interaction is relatively weak compared to LTα binding to TNFR ($K_d = \sim 0.1$ nM) (Mauri *et al.*, 1998). In this same format soluble recombinant gD exhibited half-maximal binding at ~ 1 µM in competition with HVEM-Fc binding to membrane LIGHT. A mutant of gD, gDΔ290–299t, competes with LIGHT binding to HVEM-Fc at $\sim$10-fold higher affinity than wild-type gD. Interestingly, HSV-1 virions that express the gDΔ290–299t mutant are inactive. Hence, from the point of view of the virus, too high an affinity between gD and HVEM is deleterious to infection. The molar stoichiometry of gD binding HVEM is 2 : 1, whereas LIGHT or LTα are trimers and can bind up to three HVEM molecules.

Cell types and tissues expressing the receptor

HVEM exhibits relatively broad tissue distribution, but is prominent in lymphoid tissues, including spleen, peripheral blood, and thymus (Montgomery *et al.*, 1996) (**Table 5**). At the cellular level, T and B lymphocytes express HVEM prominently, a feature that distinguishes HVEM from the LTβR. The level of HVEM detected by FACS is in the same range as TNFR2 (CD120b or 75–80 kDa TNFR) on activated T cells, which is on the order of 5×10^3 receptors per cell.

Regulation of receptor expression

HVEM mRNA and protein is constitutively expressed in freshly isolated T and B lymphocytes from peripheral blood, and levels increase following activation (Kwon *et al.*, 1997). In cell lines of myelomonocytic lineage, such as U937, mRNA levels are induced following activation with phorbol ester, or in HL60 with dimethyl sulfoxide; TNF can also induce expression in the osteosarcoma line MG63.

Release of soluble receptors

In the mouse, four alternately spliced forms of HVEM have been identified that result in deletions in the ectodomain or truncate the C-terminus, creating soluble receptors (Hsu *et al.*, 1997) (Table 2). No evidence has been reported that these forms are expressed or functional. Additionally, no evidence has been obtained indicating that HVEM is proteolytically cleaved (shed) from the cell surface.

Figure 5 Sequence homology of UL144 with the TNFR superfamily. (a) Deduced amino acid sequence of UL144-F. The predicted signal cleavage site is denoted by the arrow, *N*-linked glycosylation sites by (*), the transmembrane region is underlined and the putative disulfide bonds in CRD2 modeled after those in TNFR60 are connected by bars. (b) Multiple alignment of UL144 sequences (ClustalW, PAM series) from Fiala (F), Toledo (T) and two other low passages isolates, LU and ME. The outlined and shaded regions identify identical and conserved residues. The bars denote positions of CRD1 and 2, the transmembrane (tm) and cytoplasmic (cyto) domains. (c) Pairwise alignment of CRD1 and 2 of HVEM and UL144F where the boxed and shaded regions show identity and conservative substitutions. (d) Homology between UL144 and members of the TNF superfamily. Alignments between the extracellular domains (ecto) or the second cysteine-rich domains of UL144 and the indicated TNFR were made with ClustalW and the percentage of identical residues was determined.

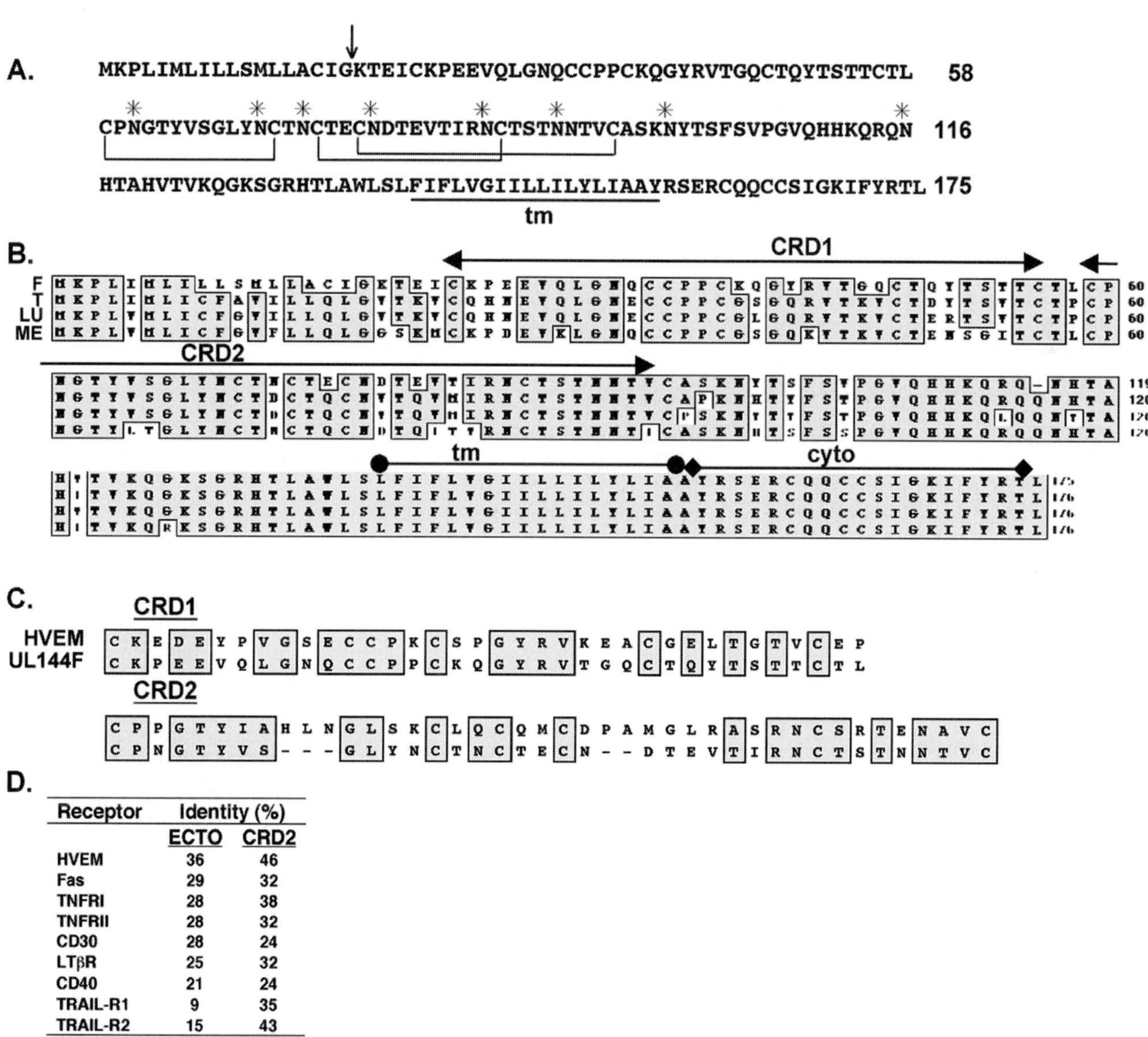

Figure 6 Human and mouse HVEM. Alignment of protein sequence of human and mouse sequences were performed using pairwise alignment (Pam 250 series Macvector) of the (a) ecto and (b) cytoplasmic domains. Shaded areas show sequence identity.

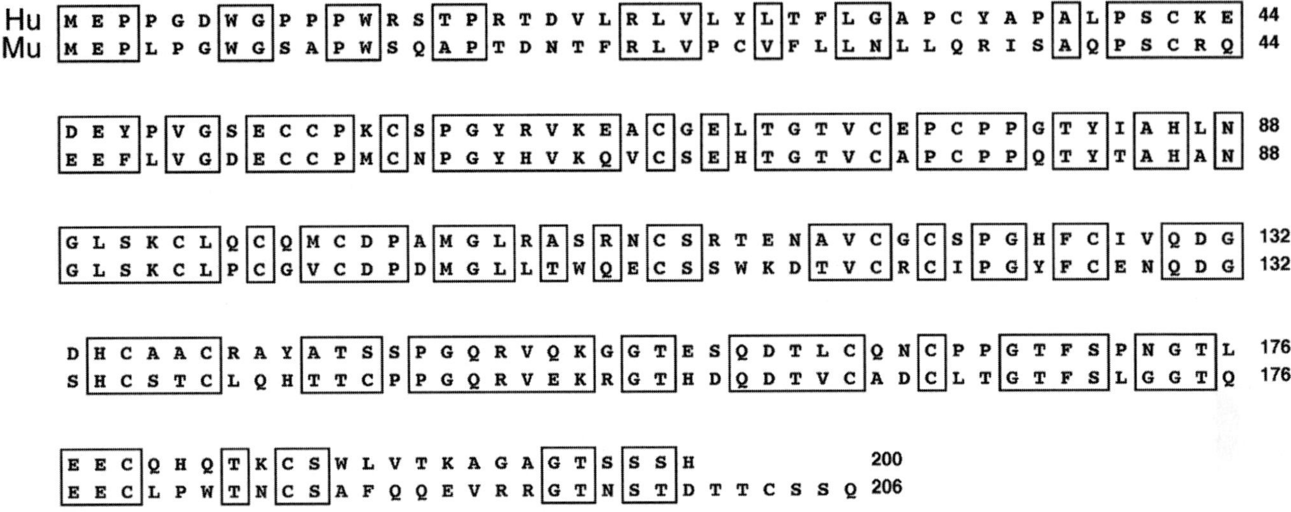

B. HVEM cytoplasmic domain

Hu K R R K P R G D V V K V I V S V Q R K R Q E A E G E A T V I E A L Q A P P D V T T V A
Mu T R R H L H T S S V A K E L E P F Q Q E Q Q E - - - - - - - - N T I R F P - - V T E V G

V E E T I P S F T G R S P N H
F A E T E E E T A S N

Table 3 Species homology in the TNF superfamily

	% identity human and mouse	
	Overall	Ectodomain[a]
Receptors		
HVEM	41	51
LTβR	68	69
TNFR1	63	70
Ligands		
LIGHT	76	83
LTβ	60	75
LTα	72	79

[a]Ectodomain for the receptors includes leader up to the transmembrane segment. The ligand ectodomain is defined as the receptor-binding region involved in trimer formation.

Table 4 Affinity of HVEM for cellular and viral ligands

Receptor[a]	Ligand affinity ($\sim K_d$ nM)			
	LIGHT[b]	LTα[c]	gD[d]	Super gD[e]
HVEM-Fc (FACS)	1	50	1000	200
HVEM-Fc (PRS)	0.4	nd	nd	nd
HVEM (206t)	nd	nd	1000	200

[a]Different forms of HVEM used in affinity measurements: as a Fc fusion protein of the ectodomain or as a soluble ectodomain truncated at residue 206. Plasmon resonance (BIACORE) or flow cytometry was used to calculate half-maximum binding concentrations.

[b]FACS analysis used stably transfected lines expressing membrane LIGHT.

[c] LTα-binding affinity was determined by competition with HVEM-Fc binding to membrane LIGHT.

[d]Recombinant soluble form of gD.

[e]Recombinant soluble form of the deletion mutant of gDΔ290–299t.

Table 5 Tissue and cellular expression patterns of HVEM

Tissue	Human[a]	Mouse[b]
Adult	Strong: spleen, PBL, thymus, heart, lung, liver, kidney	Strong: bone marrow, thymus, spleen, small intestine, lung, heart, cecum, colon
	Weak: prostate, small intestine, colon, brain, prostate, skeletal muscle	Weak: kidney, skin, heart, liver, stomach
Fetal	Lung, liver, kidney	Lung, liver, kidney
Cell lines[a,c]		
Lymphocytes	II-23 CD4+ T cell hybridoma	
	Jurkat CD4+	
	Raji B lymphoblastoid	
Peripheral blood:	CD4+ resting T cells	
	CD8+ resting T cells	
	B cells (CD19+)	
Monocytic	U937	
	HL60	
	THP1	
Nonlymphoid	HT29 colon[c]	
	HeLa carcinoma[c]	

[a]Determined by northern blots.
[b]Determined by RNase protection assay.
[c]Determined by flow cytometry.

SIGNAL TRANSDUCTION

Associated or intrinsic kinases

The binding of a trivalent ligand, which induces an ordered aggregation or 'clustering' of receptors, initiates signal transduction by TNFR family members. Antireceptor antibodies, or mere overexpression, can induce receptor clustering and lead to activation of signal transduction pathways. HVEM, as well as all the other related TNF receptors, have no intrinsic kinase or other enzymatic activities encoded by their cytosolic domain (Figure 3). Rather, signal transduction occurs through the recruitment of cytosolic adapters, which directly or indirectly confer enzymatic activity to the receptor. Initial characterization of HVEM signaling pathways indicates that the cytoplasmic domain interacts directly with members of the TNFR-associated factor (TRAF) family of zinc RING finger proteins (Arch *et al.*, 1998) (**Figure 7**).

Our understanding of the mechanisms and pathways involved in HVEM signaling is at an early stage, however, reasonable predictions can be drawn from the fact that HVEM interacts with a subset of the TRAF adapter molecules (Hsu *et al.*, 1997; Marsters *et al.*, 1997). TRAF2 and 5 have been shown to bind HVEM directly using yeast two-hybrid analysis or direct binding assays with GST fusion proteins. The binding of TRAF1 and 3 to HVEM is equivocal as two groups differ in their results. Yeast two-hybrid analysis showed strong interactions between HVEM with TRAF2 and 5, but not 1 or 3 (Hsu *et al.*, 1997). In contrast, GST fusion protein binding analysis using extracts from TRAFs overexpressed in HEK293 cells showed TRAF1 and 3, as well as 2 and 5, but not TRAF6 bound to HVEM (Marsters *et al.*, 1997). TRAF1 binding could occur indirectly in the latter model because of the propensity of TRAF1 to form heterocomplexes with TRAF2. Regardless, TRAF3 binding to HVEM is weak when compared to that to LTβR. Mutation of glutamic acid at position 271

Figure 7 Schematic diagram of the HVEM-signaling pathway. HVEM utilizes TRAF2 and TRAF5 in initiating signaling pathways. NFκB induces gene expression resulting in proinflammatory and antiapoptotic activities. The JNK/AP-1 pathway induces stress responses, including cytokine secretion.

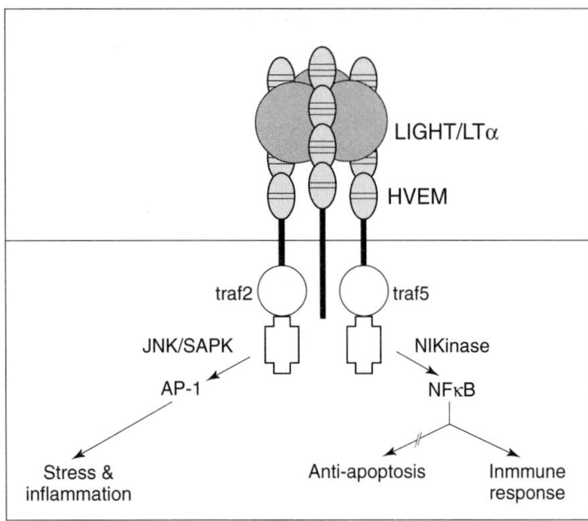

(E271) to alanine in HVEM specifically ablates TRAF2 and 5 binding in yeast two-hybrid analysis (Hsu *et al.*, 1997). The V269EET region is homologous to the TRAF2 and 5-binding region in CD40 (Figure 3). Recent results indicate that HVEM is located in a distinct subcellular compartment separate from LTβR and cannot recruit TRAF3, which may account for the inability of HVEM to signal death.

The interaction of HVEM with TRAF2 or 5 is expected to initiate activation of both NFκB and c-jun N-terminal kinase (JNK)/AP-1 pathways. HVEM overexpression in HEK caused activation of the p65/p50 heterocomplex of κB as measured by gel shift and reporter (pELAM-luciferase) assays, especially when cotransfected with TRAF5 (Hsu *et al.*, 1997). TRAF2 signaling was less robust in this model. Overexpression of HVEM in 293 cells activated JNK as measured by phosphorylation of c-Jun-GST fusion protein on serine 63 (Marsters *et al.*, 1997). Electrophoretic mobility shift assays showed that AP-1 nucleotide binding activity induced by HVEM was supershifted by an antibody to JunD. Overexpression of TRAF2 or TRAF5 has similar effect on AP-1. TRAF2 is known to be essential for TNFR1 activation of the JNK/AP-1 pathway in mice, suggesting that TRAF2 may be utilized by HVEM to activate this pathway (Yeh and Ohashi, 1997; Lee *et al.*, 1997).

Cytoplasmic signaling cascades

The TNFR superfamily is divided into two major groups based on the functional properties of their cytoplasmic tails: the death domain receptors, such as Fas and TNFR1, and the TRAF-binding receptors, such as LTβR, CD40, and HVEM among others. The TRAF family of signal transducers consists of six members. TRAF2, 5, and 6 are involved directly in activating NFκB and JNK pathways, which are prominent for the induction of genes involved in inflammation and stress responses (Arch *et al.*, 1998). TRAF3 is involved in regulating cell death for LTβR (Force *et al.*, 1997), CD40, and CD30, whereas roles for TRAF1 and 4 have not clearly emerged.

Three distinct structural motifs are found in the TRAF molecule. Located at the N-terminus, most TRAF proteins have a zinc RING finger and multiple zinc fingers (TRAF1 lacks a RING motif), a centrally positioned coiled coil region, and a C-terminal (TRAF) domain that binds directly to the receptor's cytoplasmic domain. TRAF2, 5, and 6 act as adapter molecules that recruit kinases and other components to their receptors. As exemplified by TRAF2, several distinct proteins interact with this adapter including NIK and ASK1. The NFκB-inducing kinase (NIK) is a serine kinase involved in activating IKK-1 and 2, which in turn phosphorylate IκB (inhibitor of κB) causing its degradation and the release the transcriptionally active form of NFκB. The interaction of TRAF2 with ASK1, a MAP3 kinase, activates the JNK-dependent cascade that in turn activates the AP-1 transcription factor. TRAF5 and 6 exhibit several redundant activities with TRAF2, including NIK and ASK1 activation. Mice deleted of TRAF2 reveal absolute dependence of this adapter for JNK activation, but not NFκB activation; these mice show a perinatal demise indicating the importance of this protein in development. TRAF5 knockout mice show no deficit in signaling, suggesting it may function redundantly with TRAF2. For a comprehensive review of these signaling pathways see the chapter on The TNF Ligand and TNF/NGF Receptor Families.

DOWNSTREAM GENE ACTIVATION

Transcription factors activated

Genes specifically induced by HVEM signaling have not been identified. The ability of HVEM to activate both NFκB and AP-1 transcription factors suggests that HVEM signaling will elicit components involved in inflammation and cellular stress responses. However, the plethora of genes activated via these transcription factors makes it difficult to predict responses specific to HVEM. Cellular responses will also be controlled in part by the strength and duration of HVEM signals, as well as by the responding tissue and cell types. Here, the fact that naïve T cells express HVEM but lack TNFR suggests that HVEM signaling could activate genes involved during the early phases of T cell activation, such as chemokines or adhesion molecules.

Genes induced

Because NFκB and AP-1 transcription factors are activated by HVEM, it is expected that a subset of the genes regulated by these factors will be induced (Whitmarsh and Davis, 1996).

Promoter regions involved

Genetic elements controlling HVEM gene expression have not been analyzed.

BIOLOGICAL CONSEQUENCES OF ACTIVATING OR INHIBITING RECEPTOR AND PATHOPHYSIOLOGY

Unique biological effects of activating the receptors

The unique feature of HVEM is its ability to serve as an entry factor for HSV-1 and 2.

Human abnormalities

No known human genetic disease is linked to the HVEM gene. However, the ability of HVEM to serve as an entry factor for *Herpes simplex virus* argues strongly that this molecule will play some role in the initial or recurrent infections associated with this α-herpesvirus. There are two major types of HSV that differ genetically and immunologically. HSV-1 is associated with primary infection of oropharyngeal mucosal surfaces. HSV-2 is associated with infection of genitalia. Primary infection typically resolves in 2–3 weeks and stimulates strong cellular and humoral immunity, which is specific for subtypes and strains of HSV. HSV-1 is widespread in the human population and is typically acquired in childhood by a large portion of the population. However, both types persist in a latent phase in sensory ganglion that innervate the area of the primary site of infection. Sporadically, the virus re-emerges to cause a vesicular lesion supplied by the neuron harboring latent virus. The reactivation of HSV-1 is associated with exposure to UV light, stress, menstruation, or chemotherapy, processes that may compromise local and systemic immunity. Recurrent infections typically resolve quickly (< 1 week), however in immune compromised individuals control of recurrent HSV lesions is poor. The lesions can engross a large area, and virus can disseminate to other tissues.

THERAPEUTIC UTILITY

Effect of treatment with soluble receptor domain

In tissue culture systems soluble HVEM can inhibit infection of cells by HSV-1 (Whitbeck *et al.*, 1997). This effect is completely dependent on the type of gD protein expressed by the virion. Soluble HVEM is thought to act by steric hindrance of gD in the virion membrane, resulting in reduced attachment of gD to the cellular entry factors HveA or HveC that are expressed on the surface of the target cell. It is not known whether HVEM-Fc administered *in vivo* can block primary or recurrent HSV infections.

HVEM-Fc inhibits the action of its ligands, LIGHT and LTα, irrespective of the receptors that these ligands engage, and thus could be used to modulate immune responses that are dependent upon these cytokines. In tissue culture models, HVEM-Fc blocks LIGHT-induced growth arrest of HT29 cells and shows modest inhibition of cellular proliferation induced in mixed lymphocyte cultures (Harrop *et al.*, 1998).

Effects of inhibitors (antibodies) to receptors

Antibodies to HVEM inhibit entry of HSV-1 in a subset of T cells (Montgomery *et al.*, 1996). No evidence is available on whether anti-HVEM antibodies suppress infection *in vivo*. Recent studies (Sarrias *et al.*, 1999) have isolated peptides that bind HVEM. These were identified by affinity selection from phage-display libraries. One peptide inhibits the binding of gD to HVEM and can block infection of HVEM-transfected CHO cells by HSV-1. This exciting observation suggests that small molecules may prove useful in blocking virus entry pathways.

References

Arch, R., Gedrich, R., and Thompson, C. (1998). Tumor necrosis factor receptor-associated factors (TRAFs)–a family of adaptor proteins that regulates life and death. *Genes Dev.* **12**, 2821–2830.

Banner, D. W., D'Arcy, A., Janes, W., Gentz, R., Schoenfeld, H. J., Broger, C., Loetscher, H., and Lesslauer, W. (1993). Crystal structure of the soluble human 55 kd TNF receptor-human TNF beta complex: implications for TNF receptor activation. *Cell* **73**, 431–445.

Benedict, C., Butrovich, K., Lurain, N., Corbeil, J., Rooney, I., Schneider, P., Tschopp, J., and Ware, C. (1999). Cutting edge: A novel viral TNF receptor superfamily member in virulent strains. *J. Immunol.* **162**, 6967–6970.

Browning, J. L., Ngam-ek, A., Lawton, P., DeMarinis, J., Tizard, R., Chow, E. P., Hession, C., O'Brine-Greco, B., Foley, S. F., and Ware, C. F. (1993). Lymphotoxin β, a novel member of the TNF family that forms a heteromeric complex with lymphotoxin on the cell surface. *Cell* **72**, 847–856.

Cohen, G. H., Wilcox, W. C., Sodora, D. L., Long, D., Levin, J. Z., and Eisenberg, R. J. (1988). Expression of herpes simplex virus type 1 glycoprotein D deletion mutants in mammalian cells. *J. Virol.* **62**, 1932–1940.

Force, W. R., Cheung, T. C., and Ware, C. F. (1997). Dominant negative mutants of TRAF3 reveal an important role for the coiled coil domains in cell death signaling by the lymphotoxin-β receptor (LTβR). *J. Biol. Chem.* **272**, 30835–30840.

Fu, Y.-X., and Chaplin, D. (1999). Development and maturation of secondary lymphoid tissues. *Annu. Rev. Immunol.* **17**, 399–433.

Futterer, A., Mink, K., Luz, A., Kosco-Vilbois, M. H., and Pfeffer, K. (1998). The lymphotoxin beta receptor controls organigenesis and affinity maturation in peripheral lymphoid tissues. *Immunity* **9**, 59–70.

Geraghty, R. J., Krummenacher, C., Cohen, G. H., Eisenberg, R. J., and Spear, P. G. (1998). Entry of alphaherpesviruses mediated by poliovirus receptor-related protein 1 and poliovirus receptor. *Science* **280**, 1618–1620.

Guex, N., and Peitsch, M. C. (1997). SWISS-MODEL and the Swiss-Pdb Viewer: an environment for comparative protein modelling. *Electrophoresis* **18**, 2714–2723.

Harrop, J. A., McDonnell, P. C., Brigham-Burke, M., Lyn, S. D., Minton, J., Tan, K. B., Dede, K., Spampanato, J., Silverman, C., Hensley, P., DiPrinzio, R., Emery, J. G., Deen, K., Eichman, C., Chabot-Fletcher, M., Truneh, A., and Young, P. R. (1998). Herpesvirus entry mediator ligand (HVEM-L), a novel ligand for HVEM/TR2, stimulates proliferation of T cells and inhibits HT29 cell growth. *J. Biol. Chem.* **273**, 27548–27556.

Hsu, H., Solovyev, I., Colobero, A., Elliott, R., Kelley, M., and Boyle, W. J. (1997). ATAR, a novel tumor necrosis factor receptor family member, signals through TRAF2 and TRAF5. *J. Biol. Chem.* **272**, 13471–13474.

Johnson, R. M., and Spear, P. G. (1989). Herpes simplex virus glycoprotein D mediates interference with herpes simplex virus infection. *J. Virol.* **63**, 819–827.

Kwon, B. S., Tan, K. B., Ni, J., Lee, K. O., Kim, K. K., Kim, Y. J., Wang, S., Gentz, R., Yu, G. L., Harrop, J., Lyn, S. D., Silverman, C., Porter, T. G., Truneh, A., and Young, P. R. (1997). A newly identified member of the tumor necrosis factor receptor superfamily with a wide tissue distribution and involvement in lymphocyte activation. *J. Biol. Chem.* **272**, 14272–14276.

Lee, S., Reichlin, A., Santana, A., Sokol, K., Nussenzweig, M., and Choi Y (1997). TRAF2 is essential for JNK but not NF-kappaB activation and regulates lymphocyte proliferation and survival. *Immunity* **7**, 703–713.

Marsters, S. A., Ayres, T. M., Skuatch, M., Gray, C. L., Rothe, M. L., and Ashkenazi, A. (1997). Herpesvirus entry mediator, a member of the tumor necrosis factor receptor family (TNFR), interacts with members of the TNFR-associated factor family and activates the transcription factors NF-κB and AP-1. *J. Biol. Chem.* **272**, 14029–14032.

Mauri, D. N., Ebner, R., Montgomery, R. I., Kochel, K. D., Cheung, T. C., Yu, G.-L., Ruben, S., Murphy, M., Eisenbery, R. J., Cohen, G. H., Spear, P. G., and Ware, C. F. (1998). LIGHT, a new member of the TNF superfamily and lymphotoxin α are ligands for herpesvirus entry mediator. *Immunity* **8**, 21–30.

Montgomery, R. I., Warner, M. S., Lum, B., and Spear, P. G. (1996). Herpes simplex virus 1 entry into cells mediated by a novel member of the TNF/NGF receptor family. *Cell* **87**, 427–436.

Posavad, C. M., Newton, J. J., and Rosenthal, K. L. (1994). Infection and inhibition of human cytotoxic T lymphocytes by herpes simplex virus. *J. Virol.* **68**, 4072–4074.

Roller, R., and Rauch, D. (1998). Herpesvirus entry mediator HVEM mediates cell-cell spread in BHK(TK-) cell clones. *J.Virol.* **72**, 1411–1417.

Sarrias, M., Whitbeck, J., Rooney, I., Spruce, L., Kay, B., Montgomery, R., Spear, P., Ware, C., Eisenberg, R., Cohen, G., and Lambris, J. (1999). Inhibition of herpes simplex virus gD and lymphotoxin-alpha binding to HveA by peptide antagonists. *J. Virol.* **73**, 5681–5687.

Spear, P., Shieh, M., Herold, B., WuDunn, D., and Koshy, T. (1992). Heparan sulfate glycosaminoglycans as primary cell surface receptors for herpes simplex virus. *Adv. Exp. Med. Biol.* **313**, 341–353.

Terry-Allison, T., Montgomery, R., Whitbeck, J., Xu, R., Cohen, G., Eisenberg, R., and Spear, P. (1998). HveA (herpesvirus entry mediator A), a co-receptor for herpes simplex virus entry, also participates in virus-induced cell fusion. *J. Virol.* **72**, 5802–5810.

Whitbeck, J. C., Peng, C., Lou, H., Xu, R., Willis, S. H., Ponce de Leon, M., Peng, T., Nicola, A. V., Montgomery, R. I., Warner, M. S., Soulika, A., Spruce, L., Moor, W. T., Lambris, J. D., Spear, P. G., Cohen, G. H., and Eisenberg, R. J. (1997). Glycoprotein D of herpes simplex virus binds directly to HVEM, a mediator of HSV entry. *J. Virol.* **71**, 6083–6093.

Whitmarsh, A., and Davis, R. (1996). Transcription factor AP-1 regulation by mitogen-activated protein kinase signal transduction pathways. *J. Mol. Med.* **74**, 589–607.

Yeh, W. C., and Ohashi, P. (1997). Early lethality, functional NF-kappaB activation, and increased sensitivity to TNF-induced cell death in TRAF2-deficient mice. *Immunity* **7**, 715–725.

DR4, DR5, DcR1, DcR2

Avi Ashkenazi*

Department of Molecular Oncology, Genentech, Inc., 1 DNA Way, South San Francisco, CA 94080-4918, USA

*corresponding author tel: 650-225-1853, fax: 650-225-6443, e-mail: aa@gene.com

DOI: 10.1006/rwcy.2000.16012.

SUMMARY

Researchers have identified five receptors that bind to the death ligand Apo2L/TRAIL, of which four are cell-associated. These four receptors contain closely related extracellular cysteine-rich domains, and belong to the TNFR gene superfamily; their encoding human genes map to chromosome 8p21-22. DR4 and DR5 have a cytoplasmic death domain that signals apoptosis. DcR1 is a GPI-linked protein that lacks a cytoplasmic region. DcR2 has a cytoplasmic tail that contains a truncated, nonfunctional death domain. Upon overexpression, DcR1 and DcR2 inhibit apoptosis induction by Apo2L/TRAIL, indicating that they can act as decoys that compete for ligand binding to DR4 and DR5. Whereas many tumor cell lines express DR4 and/or DR5, few express significant levels of DcR1 or DcR2. Activation of T cells suppresses DcR1 expression completely, which is consistent with the implication of Apo2L/TRAIL in activation-induced apoptosis of T cells. Activation of p53 in some tumor cell lines induces DR5 and DcR1 mRNA expression, which suggests a possible connection between p53 and receptors for Apo2L/TRAIL. Apo2L/TRAIL induces apoptosis in a wide variety of cancer cell lines by engaging DR4 and DR5, but it is not cytotoxic towards most normal cell types studied so far. Thus, it may be possible to treat cancer effectively by targeting tumor DR4 and DR5 with Apo2L/TRAIL, without significant toxicity to normal tissues.

BACKGROUND

Apoptosis is an evolutionarily conserved cell-suicide program that eliminates unneeded or damaged cells during development and homeostasis of metazoans (Jacobson *et al.*, 1997). A variety of extracellular and intracellular cues keep the cell's apoptotic machinery in check. In the event of loss-of-survival stimuli from its environment or in case of irreparable internal damage, the cell ignites its apoptotic caspase machinery, leading to the apoptotic death and engulfment of the cell by neighboring cells (Thornberry and Lazebnik, 1998). In higher metazoans, an additional mechanism of apoptosis-initiation has evolved, called 'instructive apoptosis', which enables the organism actively to direct individual cells to commit suicide. In instructive apoptosis, specific 'death' ligands deliver the suicide instructions to target cells through binding to cell surface 'death' receptors (Nagata, 1997; Ashkenazi and Dixit, 1998).

The death ligands that are known to date belong to the tumor necrosis factor (TNF) gene superfamily (Gruss and Dower, 1995); their cognate death receptors belong to the TNF receptor (TNFR) gene superfamily (Smith *et al.*, 1994). Upon ligation, death receptors rapidly engage the cell's death caspase machinery, leading to an apoptotic demise (Ashkenazi and Dixit, 1998). Death receptors share a homologous region of about 70 amino acids in their cytoplasmic tail, dubbed the 'death domain' (Itoh and Nagata, 1993; Tartaglia *et al.*, 1993), which transmits the apoptosis signal. The best characterized death ligands are Fas ligand (FasL) and TNF, and their respective receptors, Fas (also called Apo1 or CD95) and TNFR1 (Nagata, 1997; Ashkenazi and Dixit, 1998). Through a homophilic interaction, the death domain of Fas recruits a death domain-containing adapter molecule called FADD (or MORT1) into a signaling complex (Boldin *et al.*, 1995; Chinnaiyan *et al.*, 1995). FADD in turn recruits the apoptotic protease caspase 8 into the complex, through homophilic interaction of so-called death effector domains which are found in FADD and in caspase 8 (Boldin *et al.*,

1996; Muzio *et al.*, 1996). The juxtaposition of caspase 8 molecules in the complex leads to their autocatalytic activation (Muzio *et al.*, 1998); caspase 8 then activates downstream effector caspases to execute the apoptosis program. TNFR1 activates apoptosis through FADD and caspase 8 as well, except that its death domain recruits FADD indirectly, through a FADD-related, death domain-containing adapter called TRADD (Hsu *et al.*, 1995). TRADD is a platform adapter that links TNFR1 to additional signaling pathways, which leads to activation of the transcription factors NFκB or AP-1 (Ashkenazi and Dixit, 1998).

Recent work has identified a novel death ligand, called Apo2 ligand (Apo2L) (Pitti *et al.*, 1996) or TNF-related apoptosis-inducing ligand (TRAIL) (Wiley *et al.*, 1995), which has significant sequence homology to FasL and TNF. Early studies with Apo2L/TRAIL indicated that this ligand induces apoptosis independently of Fas or TNFR1. Subsequently, four related yet distinct receptors that bind to Apo2L/TRAIL were identified: two are death receptors that signal apoptosis, whereas the other two are decoys that bind the ligand but do not transmit an apoptosis signal (Ashkenazi and Dixit, 1998, 1999). Apo2L/TRAIL also interacts with a soluble member of the TNFR superfamily called osteoprotegerin (OPG) (Simonet *et al.*, 1997; Emery *et al.*, 1998); however, OPG is much less homologous to DR4, DR5, DcR1, and DcR2 than the latter four are to one another.

Discovery

Death receptor 4 (DR4) was discovered by searching expressed sequence tag (EST) DNA databases for ESTs that have homology to the death domain of TNFR1 (Pan *et al.*, 1997). A human EST with TNFR homology facilitated isolation of a full-length cDNA that encoded a previously unknown member of the TNFR superfamily. An Fc-fusion protein (immunoadhesin) based upon the extracellular domain (ECD) of DR4 bound to soluble Apo2L/TRAIL and blocked apoptosis induction by the ligand, which identified DR4 as a specific receptor for Apo2L/TRAIL.

DR5 was discovered independently by several groups, through different approaches. Sheridan *et al.* (1997) identified DR5 before the DR4 sequence was available, on the basis of an EST that showed homology to the death domains of TNFR1, Fas, and DR3. Screaton *et al.* (1997a) discovered DR5 on the basis of ESTs that showed homology to the cysteine-rich domains (CRDs) found in the ECDs of TNFR family members, together with ESTs that showed homology to death domains. Several other groups (Pan and Dixit, 1997; McFarlane *et al.*, 1997; Schneider *et al.*, 1997; Chaudhary *et al.*, 1997) cloned DR5 on the basis of ESTs that showed homology to DR4. Walczak *et al.* (1997) used ligand-based affinity purification to isolate the DR5 protein. Wu *et al.* (1997) isolated DR5 as a gene that is induced by activation of the tumor suppressor p53 protein.

Decoy receptor 1 (DcR1) was discovered in a yeast-based screen for signal peptide-containing cDNAs (Sheridan *et al.*, 1997), and by EST-based approaches (Pan and Dixit, 1997; Schneider *et al.*, 1997; Degli-Esposti *et al.*, 1997a; Mongkolsapaya *et al.*, 1998).

DcR2 was discovered by screening cDNA libraries with DcR1-based probes (Marsters *et al.*, 1997; Degli-Esposti *et al.*, 1997b) or by searching DNA databases for ESTs with homology to previously known Apo2L/TRAIL receptors (Pan *et al.*, 1998).

Alternative names

DR4: TNFRSF10A; TRAIL-R1
DR5: TNFRSF10B; Apo2; TRICK2; TRAIL-R2; KILLER
DcR1: TNFRSF10C; TRID; TRAIL-R3; LIT
DcR2: TNFRSF10D; TRAIL-R4; TRUNDD

Structure

DR4 and DR5 are similar, type 1 transmembrane proteins (**Figure 1**). They each contain three CRDs in their extracellular portion and a death domain in their cytoplasmic region. DcR1 is a glycosyl phosphatidylinositol (GPI)-linked cell surface protein; it contains two CRDs and lacks a cytoplasmic region (Figure 1). DcR2 has a type 1 transmembrane topology similar

Figure 1 Schematic structure of cellular death and decoy receptors that bind to Apo2L/TRAIL. S, signal sequence; CRD, cysteine-rich domain; TM, transmembrane region; DD, death domain; TD, truncated death domain. Red vertical lines indicate cysteines.

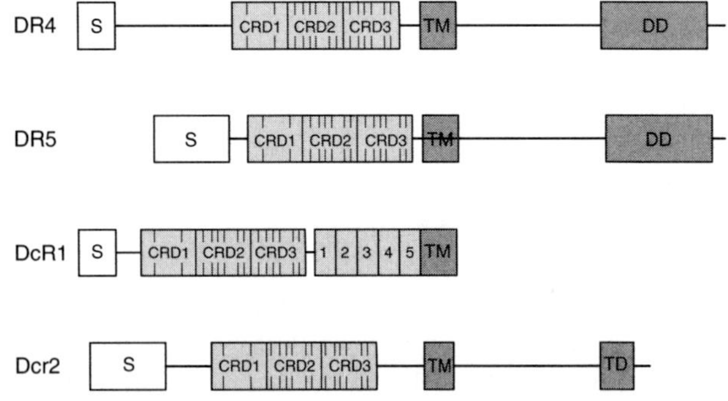

to that of DR4 and DR5, with two extracellular CRDs; however, in the cytoplasmic region, DcR2 has a truncated death domain which is much shorter than the death domains of DR4, DR5, or other death receptors (Figure 1).

Main activities and pathophysiological roles

Like Fas and TNFR1, DR4 (Pan *et al.*, 1997) and DR5 (Sheridan *et al.*, 1997; Pan and Dixit, 1997; McFarlane *et al.*, 1997; Schenider *et al.*, 1997; Chaudhary *et al.*, 1997; Screaton *et al.*, 1997b) initiate apoptosis upon transient transfection. Overexpressed DR4 and DR5 also can activate the transcription factor NFκB (Sheridan *et al.*, 1997; Schneider *et al.*, 1997; Chaudhary *et al.*, 1997); however, as compared with TNF, Apo2L/TRAIL activates this pathway only at very high doses (Sheridan *et al.*, 1997; Ashkenazi *et al.*, 1999). Hence, the physiological relevance of NFB-induction by overexpressed DR4 or DR5 remains unclear.

The absence of a cytoplasmic region in DcR1 suggests that this receptor does not transduce signals. Indeed, ectopic expression of DcR1 in cells did not induce apoptosis (Sheridan *et al.*, 1997; Pan and Dixit, 1997; MacFarlane *et al.*, 1997; Schneider *et al.*, 1997; Degli-Esposti *et al.*, 1997a; Mongkolsapaya *et al.*, 1998). Rather, transfection of Apo2L/TRAIL-sensitive cells with DcR1 markedly reduced sensitivity to apoptosis-induction by the ligand (Sheridan *et al.*, 1997; Pan and Dixit, 1997; Mongkolsapaya *et al.*, 1998). Further, enzymatic treatment of DcR1-expressing cells with phosphatidylinositol phospholipase C (PI-PLC), which cleaves GPI moieties, resulted in marked sensitization to Apo2L/TRAIL-induced apoptosis (Sheridan *et al.*, 1997). Thus, DcR1 appears able to function as a decoy that may compete with DR4 and DR5 for binding of Apo2L/TRAIL.

DcR2 has a truncated cytoplasmic death domain that is about one-third the length of a canonical death domain. Four out of six death domain amino acid positions that are critical for apoptosis signaling and for NFκB activation by TNFR1 (Tartaglia *et al.*, 1993), one of which corresponds to the inactivating *lpr* mutation in mouse Fas (Nagata, 1997), are absent in DcR2. DcR2 overexpression in cells did not induce apoptosis (Marsters *et al.*, 1997; Degli-Esposti *et al.*, 1997b; Pan *et al.*, 1998). Instead, transfection with DcR2 inhibited apoptosis induction by Apo2L/TRAIL (Marsters *et al.*, 1997; Degli-Esposti *et al.*, 1997b; Pan *et al.*, 1998). Deletion of the DcR2 cytoplasmic region did not abrogate the inhibitory activity of this receptor, indicating that signaling by DcR2 is not required for this activity (Sheridan *et al.*, 1997). Thus, DcR2 appears also to be able to act as a decoy that may compete with DR4 and DR5 for binding to Apo2L/TRAIL. To date, no pathophysiological roles of these four Apo2L/TRAIL receptors have been found.

GENE

Accession numbers

DR4: HSU90875
DR5: AF012535 (411 aa); AF016849 (440 aa)
DcR1: AF012536
DcR2: AF029761

Sequence

See **Figure 2**.

Chromosome location and linkages

All four receptors map to human chromosome 8p21-22 (Marsters *et al.*, 1997; Degli-Esposti *et al.*, 1997b).
Nearest markers:
DR4: D8S2127
DR5: DS481
DcR1: WI6536
DcR2: SHGC33989

PROTEIN

Accession numbers

DR4: P_W64483
DR5: AF012535_1 (411 aa); AF016849_1 (440 aa)
DcR1: AF012536_1
DcR2: AF029761_1

Sequence

See **Figure 3**.

Description of protein

DR4 is a 468 amino acid polypeptide (Pan *et al.*, 1997). The N-terminus contains a signal sequence, which is predicted to be 23 residues long, followed by an ECD that contains three CRDs (CRD1-3, with 2, 6, and 6 cysteines, respectively), a transmembrane region, and a cytoplasmic region that contains a canonical death domain.

DR5 has two apparent alternate splicing forms, which are 411 and 440 amino acids long; the longer form has 29 extra residues in the C-terminal portion of the ECD. N-terminal sequencing of the mature protein indicates a 53 amino acid signal sequence (Sheridan *et al.*, 1997). The ECD contains three CRDs, which have the same cysteine organization as DR4. The CRDs are followed by a transmembrane

Figure 2 Nucleotide sequences for DR4, DR5 (411 aa), DR5 (440 aa), DcR1, and DcR2.

DR4

```
ATGGCGCCACCACCAGCTAGAGTACATCTAGGTGCGTTCCTGGCAGTGACTCCGAATCCCGGGAGCGCAGCGAGTGGGACAGAGGCAGCCGCGGCCACACCCA
GCAAAGTGTGGGGCTCTTCCGCGGGGAGGATTGAACCACGAGGCGGGGGCCGAGGAGCGCTCCCTACCTCCATGGGACAGCACGGACCCAGTGCCCGGGCCCG
GGCAGGGCGCGCCCCAGGACCCAGGCCGGCGCGGGAAGCCAGCCCTCGGCTCCGGGTCCACAAGACCTTCAAGTTTGTCGTCGTCGGGGTCCTGCTGCAGGTC
GTACCTAGCTCAGCTGCAACCATCAAACTTCATGATCAATCAATTGGCACACAGCAATGGGAACATAGCCCTTTGGGAGAGTTGTGTCCACCAGGATCTCATA
GATCAGAACGTCCTGGAGCCTGTAACCGGTGCACAGAGGGTGTGGGTTACACCAATGCTTCCAACAATTTGTTTGCTTGCCTCCCATGTACAGCTTGTAAATC
AGATGAAGAAGAGAGAAGTCCCTGCACCACGACCAGGAACACAGCATGTCAGTGCAAACCAGGAACTTTCCGGAATGACAATTCTGCTGAGATGTGCCGGAAG
TGCAGCACAGGGTGCCCCAGAGGGATGGTCAAGGTCAAGGATTGTACGCCCTGGAGTGACATCGAGTGTGTCCACAAAGAATCAGGCAATGGACATAATATAT
GGGTGATTTTGGTTGTGACTTTGGTTGTTCCGTTGCTGTTGGTGGCTGTGCTGATTGTCTGTTGTTGCATCGGCTCAGGTTGTGGAGGGGACCCCAAGTGCAT
GGACAGGGTGTGTTTCTGGCGCTTGGGTCTCCTACGAGGGCCTGGGGCTGAGGACAATGCTCACAACGAGATTCTGAGCAACGCAGACTCGCTGTCCACTTTC
GTCTCTGAGCAGCAAATGGAAAGCCAGGAGCCGGCAGATTTGACAGGTGTCACTGTACAGTCCCCAGGGGAGGCACAGTGTCTGCTGGGACCGGCAGAAGCTG
AAGGGTCTCAGAGGAGGAGGCTGCTGGTTCCAGCAAATGGTGCTGACCCCACTGAGACTCTGATGCTGTTCTTTGACAAGTTTGCAAACATCGTGCCCTTTGA
CTCCTGGGACCAGCTCATGAGGCAGCTGGACCTCACGAAAAATGAGATCGATGTGGTCAGAGCTGGTACAGCAGGCCCAGGGGATGCCTTGTATGCAATGCTG
ATGAAATGGGTCAACAAAACTGGACGGAACGCCTCGATCCACACCCTGCTGGATGCCTTGGAGAGGATGGAAGAGAGACATGCAAAAGAGAAGATTCAGGACC
TCTTGGTGGACTCTGGAAAGTTCATCTACTTAGAAGATGGCACAGGCTCTGCCGTGTCCTTGGAGTGA
```

DR5 (411 aa)

```
CCCACGCGTCCGCATAAATCAGCACGCGGCCGGAGAACCCCGCAATCTCTGCGCCCACAAAATACACCGACGATGCCCGATCTACTTTAAGGGCTGAAACCCA
CGGGCCTGAGAGACTATAAGAGCGTTCCCTACCGCCATGGAACAACGGGGACAGAACGCCCCGGCCGCTTCGGGGGCCCGGAAAAGGCACGGCCCAGGACCCA
GGGAGGCGCGGGGAGCCAGGCCTGGGCTCCGGGTCCCCAAGACCCTTGTGCTCGTTGTCGCCGCGGTCCTGCTGTTGGTCTCAGCTGAGTCTGCTCTGATCAC
CCAACAAGACCTAGCTCCCCAGCAGAGAGCGGCCCCACAACAAAAGAGGTCCAGCCCCTCAGAGGGATTGTGTCCACCTGGACACCATATCTCAGAAGACGGT
AGAGATTGCATCTCCTGCAAATATGGACAGGACTATAGCACTCACTGGAATGACCTCCTTTTCTGCTTGCGCTGCACCAGGTGTGATTCAGGTGAAGTGGAGC
TAAGTCCCTGCACCACGACCAGAAACACAGTGTGTCAGTGCGAAGAAGGCACCTTCCGGGAAGAAGATTCTCCTGAGATGTGCCGGAAGTGCCGCACAGGGTG
TCCCAGAGGGATGGTCAAGGTCGGTGATTGTACACCCTGGAGTGACATCGAATGTGTCCACAAAGAATCAGGCATCATCATAGGAGTCACAGTTGCAGCCGTA
GTCTTGATTGTGGCTGTGTTTGTTTGCAAGTCTTTACTGTGGAAGAAAGTCCTTCCTTACCTGAAAGGCATCTGCTCAGGTGGTGGTGGGGACCCTGAGCGTG
TGGACAGAAGCTCACAACGACCTGGGGCTGAGGACAATGTCCTCAATGAGATCGTGAGTATCTTGCAGCCCACCCAGGTCCCTGAGCAGGAAATGGAAGTCCA
GGAGCCAGCAGAGCCAACAGGTGTCAACATGTTGTCCCCCGGGGAGTCAGAGCATCTGCTGGAACCGGCAGAAGCTGAAAGGTCTCAGAGGAGGAGGCTGCTG
GTTCCAGCAAATGAAGGTGATCCCACTGAGACTCTGAGACAGTGCTTCGATGACTTTGCAGACTTGGTGCCCTTTGACTCCTGGGAGCCGCTCATGAGGAAGT
TGGGCCTCATGGACAATGAGATAAAGGTGGCTAAAGCTGAGGCAGCGGGCCACAGGGACACCTTGTACACGATGCTGATAAAGTGGGTCAACAAAACCGGGCG
AGATGCCTCTGTCCACACCCTGCTGGATGCCTTGGAGACGCTGGGAGAGAGACTTGCCAAGCAGAAGATTGAGGACCACTTGTTGAGCTCTGGAAAGTTCATG
TATCTAGAAGGTAATGCAGACTCTGCCTTGTCCTAAGTGTGATTCTCTTCAGGAAGTGAGACCTTCCCTGGTTTACCTTTTTTCTGGAAAAAGCCCAACTGGA
CTCCAGTCAGTAGGAAAGTGCCACAATTGTCACATGACCGGTACTGGAAGAAACTCTCCCATCCAACATCACCCAGTGGATGGAACATCCTGTAACTTTTCAC
TGCACTTGGCATTATTTTTATAAGCTGAATGTGATAATAAGGACACTATGGAAATGTCTGGATCATTCCGTTTGTGCGTACTTTGAGATTTGGTTTGGGATGT
CATTGTTTTCACAGCACTTTTTTATCCTAATGTAAATGCTTTATTTATTTATTTGGGCTACATTGTAAGATCCATCTACAAAAAAAAAAAAAAAAAAAAAAAGG
GCGGCCGCGACTCTAGAGTCGACCTGCAGAAGCTTGGCCGCCATGGCC
```

DR5 (440 aa)

```
GAATTCGCGGCACCGCTCATAAATCAGCACGCGGCCGGAGAACCCCGCAATCTTTGCGCCCACAAAATACACCGACGATGCCCGATCTACTTTAAGGGCTGAA
ACCCACGGGCCTGAGAGACTATAAGAGCGTTCCCTACCGCCATGGAACAACGGGGACAGAACGCCCCGGCCGCTTCGGGGGCCCGGAAAAGGCACGGCCCAGG
ACCCAGGGAGGCGCGGGGAGCCAGGCCTGGGCCCCGGGTCCCCAAGACCCTTGTGCTCGTTGTCGCCGCGGTCCTGCTGTTGGTCTCAGCTGAGTCTGCTCTG
ATCACCCAACAAGACCTAGCTCCCCAGCAGAGAGCGGCCCCACAACAAAAGAGGTCCAGCCCCTCAGAGGGATTGTGTCCACCTGGACACCATATCTCAGAAG
ACGGTAGAGATTGCATCTCCTGCAAATATGGACAGGACTATAGCACTCACTGGAATGACCTCCTTTTCTGCTTGCGCTGCACCAGGTGTGATTCAGGTGAAGT
GGAGCTAAGTCCGTGCACCACGACCAGAAACACAGTGTGTCAGTGCGAAGAAGGCACCTTCCGGGAAGAAGATTCTCCTGAGATGTGCCGGAAGTGCCGCACA
GGGTGTCCCAGAGGGATGGTCAAGGTCGGTGATTGTACACCCTGGAGTGACATCGAATGTGTCCACAAAGAATCAGGTACAAAGCACAGTGGGGAAGCCCCAG
CTGTGGAGGAGACGGTGACCTCCAGCCCAGGGACTCCTGCCTCTCCCTGTTCTCTCTCAGGCATCATCATAGGAGTCACAGTTGCAGCCGTAGTCTTGATTGT
GGCTGTGTTTGTTTGCAAGTCTTTACTGTGGAAGAAAGTCCTTCCTTACCTGAAAGGCATCTGCTCAGGTGGTGGTGGGGACCCTGAGCGTGTGGACAGAAGC
TCACAACGACCTGGGGCTGAGGACAATGTCCTCAATGAGATCGTGAGTATCTTGCAGCCCACCCAGGTCCCTGAGCAGGAAATGGAAGTCCAGGAGCCAGCAG
AGCCAACAGGTGTCAACATGTTGTCCCCCGGGGAGTCAGAGCATCTGCTGGAACCGGCAGAAGCTGAAAGGTCTCAGAGGAGGAGGCTGCTGGTTCCAGCAAA
TGAAGGTGATCCCACTGAGACTCTGAGACAGTGCTTCGATGACTTTGCAGACTTGGTGCCCTTTGACTCCTGGGAGCCGCTCATGAGGAAGTTGGGCCTCATG
GACAATGAGATAAAGGTGGCTAAAGCTGAGGCAGCGGGCCACAGGGACACCTTGTACACGATGCTGATAAAGTGGGTCAACAAAACCGGGCGAGATGCCTCTG
TCCACACCCTGCTGGATGCCTTGGAGACGCTGGGAGAGAGACTTGCCAAGCAGAAGATTGAGGACCACTTGTTGAGCTCTGGAAAGTTCATGTATCTAGAAGG
TAATGCAGACTCTGCCATGTCCTAAGTGTGATTCTCTTCAGGAAGTGAGACCTTCCCTGGTTTACCTTTTTTCTGGAAAAAGCCCAACTGGACTCCAGTCAGT
AGGAAAGTGCCACAATTGTCACATGACCGGTACTGGAAGAAACTCTCCCATCCAACATCACCCAGTGGATGGAACATCCTGTAACTTTTCACTGCACTTGGCA
TTATTTTTATAAGCTGAATGTGATAATAAGGACACTATGGAAATGTCTGGATCATTCCGTTTGTGCGTACTTTGAGATTTGGTTTGGGATGTCATTGTTTTCA
CAGCACTTTTTTATCCTAATGTAAATGCTTTATTTATTTATTTGGGCTACATTGTAAGATCCAGCAGGTCGTCTCGTTTCAAGATCTGTTTAAACTAGTTAGC
TAGGC
```

Figure 2 (*Continued*)

DcR1

```
GCTGTGGGAACCTCTCCACGCGCACGAACTCAGCCAACGATTTCTGATAGATTTTTGGGAGTTTGACCAGAGATGCAAGGGGTGAAGGAGCGCTTCCTACCGT
TAGGGAACTCTGGGGACAGAGCGCCCCGGCCGCCTGATGGCCGAGGCAGGGTGCGACCCAGGACCCAGGACGGCGTCGGGAACCATACCATGGCCCGGATCCC
CAAGACCCTAAAGTTCGTCGTCGTCATCGTCGCGGTCCTGCTGCCAGTCCTAGCTTACTCTGCCACCACTGCCCGGCAGGAGGAAGTTCCCCAGCAGACAGTG
GCCCCACAGCAACAGAGGCACAGCTTCAAGGGGGAGGAGTGTCCAGCAGGATCTCATAGATCAGAACATACTGGAGCCTGTAACCCGTGCACAGAGGGTGTGG
ATTACACCAACGCTTCCAACAATGAACCTTCTTGCTTCCCATGTACAGTTTGTAAATCAGATCAAAAACATAAAAGTTCCTGCACCATGACCAGAGACACAGT
GTGTCAGTGTAAAGAAGGCACCTTCCGGAATGAAAACTCCCCAGAGATGTGCCGGAAGTGTAGCAGGTGCCCTAGTGGGGAAGTCCAAGTCAGTAATTGTACG
TCCTGGGATGATATCCAGTGTGTTGAAGAATTTGGTGCCAATGCCACTGTGGAAACCCCAGCTGCTGAAGAGACAATGAACACCAGCCCGGGGACTCCTGCCC
CAGCTGCTGAAGAGACAATGAACACCAGCCCAGGGACTCCTGCCCCAGCTGCTGAAGAGACAATGACCACCAGCCCGGGGACTCCTGCCCCAGCTGCTGAAGA
GACAATGACCACCAGCCCGGGGACTCCTGCCCCAGCTGCTGAAGAGACAATGACCACCAGCCCGGGGACTCCTGCCTCTTCTCATTACCTCTCATGCACCATC
GTAGGGATCATAGTTCTAATTGTGCTTCTGATTGTGTTTGTTTGAAAGACTTCACTGTGGAAGAAATTCCTTCCTTACCTGAAAGGTTCAGGTAGGCGCTGGC
TGAGGGCGGGGGGCGCTGGACACTCTCTGCCCTGCCTCCCTCTGCTGTGTTCCCACAGACAGAAACGCCTGCCCCTGCCCCAAAAAAAAAAAAAAAAAAAAAA
AAAAAAAAAAAAAAAAAAAAAAAAAAAAAAAAAAAAAAAAAAAAAAA
```

DcR2

```
CGAGAACCTTTGCACGCGCACAAACTACGGGGACGATTTCTGATTGATTTTTGGCGCTTTCGATCCACCCTCCTCCCTTCTCATGGGACTTTGGGGACAAAGC
GTCCCGACCGCCTCGAGCGCTCGAGCAGGGCGCTATCCAGGAGCCAGGACAGCGTCGGGAACCAGACCATGGCTCCTGGACCCCAAGATCCTTAAGTTCGTCG
TCTTCATCGTCGCGGTTCTGCTGCCGGTCCGGGTTGACTCTGCCACCATCCCCCGGCAGGACGAAGTTCCCCAGCAGACAGTGGCCCCACAGCAGCAGCCTCA
AGGAGGAGGAGTGTCCAGCAGGATCTCATAGATCAGAATATACTGGAGCCTGTAACCCGTGCACAGAGGGTGTGGATTACACCATTGCTTCCAACAATTTGCC
TTCTTGCCTGCTATGTACAGTTTGTAAATCAGGTCAAACAAATAAAAGTTCCTGTACCACGACCAGAGACACCGTGTGTCAGTGTGAAAAAGGAAGCTTCCAG
GATAAAAACTCCCCTGAGATGTGCCGGACGTGTAGAACAGGGTGTCCCAGAGGGATGGTCAAGGTCAGTAATTGTACGCAGCGGAGGAGACAGTGACCACCAT
CCTGGGGATGCTTGCCTCTCCCTATCACTACCTTATCATCATAGTGGTTTTAGTCATCATTTTAGCTGTGGTTGTGGTTGGCTTTTCATGTCGGAAGAAATTC
ATTTCTTACCTCAAAGGCATCTGCTCAGGTGGTGGAGGAGGTCCCGAACGTGTGCACAGAGTCCTTTTCCGGCGGCGTTCATGTCCTTCACGAGTTCCTGGGG
CGGAGGACAATGCCCGCAACGAGACCCTGAGTAACAGATACTTGCAGCCCACCCAGGTCTCTGAGCAGGAAATCCAAGGTCAGGAGCTGGCAGAGCTAACAGG
TGTGACTGTAGAGTCGCCAGAGGAGCCACAGCGTCTGCTGGAACAGGCAGAAGCTGAAGGGTGTCAGAGGAGGAGGCTGCTGGTTCCAGTGAATGACGCTGAC
TCCGCTGACATCAGCACCTTGCTGGATGCCTCGGCAACACTGGAAGAAGGACATGCAAAGGAAACAATTCAGGACCAACTGGTGGGCTCCGAAAAGCTCTTTT
ATGAAGAAGATGAGGCAGGCTCTGCTACGTCCTGCCTGTGAAAGAATCTCTTCAGGAAACCAGAGCTTCCCTCATTTACCTTTTCTCCTACAAAGGGAAGCAG
CCTGGAAGAAACAGTCCAGTACTTGACCCATGCCCCAACAAACTCTACTATCCAATATGGGGCAGCTTACCAATGGTCCTAGAACTTTGTTAACGCACTTGGA
GTAATTTTTATGAAATACTGCGTGTGATAAGCAAACGGGAGAAATTTATATCAGATTCTTGGCTGCATAGTTATACGATTGTGTATTAAGGGTCGTTTTAGGC
CACATGCGGTGGCTCATGCCTGTAATCCCAGCACTTTGATAGGCTGAGGCAGGTGGATTGCTTGAGCTCGGGAGTTTGAGACCAGCCTCATCAACACAGTGAA
ACTCCATCTCAATTTAAAAAGAAAAAAAGTGGTTTTAGGATGTCATTCTTTGCAGTTCTTCATCATGAGACAAGTCTTTTTTTCTGCTTCTTATATTGCAAGC
TCCATCTCT
```

domain and a cytoplasmic region that contains a death domain. The DR5 sequence does not predict any *N*-linked glycosylation sites.

The DcR1 protein is 259 amino acids long. N-terminal sequencing of the mature polypeptide indicates that the signal sequence is 25 amino acids long (Sheridan *et al.*, 1997). A longer form of the protein with an extended N-terminus has also been reported (McFarlane *et al.*, 1997). The protein contains three CRDs with the same cysteine arrangement as DR4 and DR5, followed by five sequence pseudorepeats, each 15 amino acids long and rich in threonine, alanine, proline, and glutamic acid (TAPE repeats). These repeats may assume an elongated, rod-like structure (Schneider *et al.*, 1997). Downstream of the TAPE repeats is a hydrophobic C-terminal region, which is not followed by charged amino acids, unlike type 1 transmembrane proteins. The hydrophobic C-terminus is reminiscent of the C-termini of proteins that are tethered to the cell surface by a GPI anchor. Binding of Apo2L/TRAIL to DcR1-expressing cells can be reduced substantially by treatment of the cells with PI-PLC, an enzyme that cleaves the GPI moiety, supporting the notion that DcR1 is a GPI-linked molecule (Sheridan *et al.*, 1997; Degli-Esposti *et al.*, 1997a). DcR1 has five potential *N*-linked glycosylation sites: one in the first CRD, and the other four in the first four TAPE repeats.

DcR2 is a 386 amino acid protein. N-terminal sequence analysis of the mature polypeptide indicates a signal sequence of 55 residues (Marsters *et al.*, 1997). There are apparently three allelic variants of DcR2 that differ at amino acid 35, or 310, or both. The ECD contains three CRDs with the same cysteine arrangement as DR4, DR5, and DcR1. The CRDs are followed by a transmembrane domain and a cytoplasmic region. The intracellular portion contains a truncated death domain, which is about one-third the length of a canonical death domain (Itoh and Nagata, 1993; Tartaglia *et al.*, 1993). The ECD of DcR2 contains three potential *N*-linked glycosylation sites.

Relevant homologies and species differences

The ECD of each of the four Apo2L/TRAIL receptors contains three CRDs, with 2, 6, and 6

Figure 3 Amino acid sequences for DR4, DR5 (411 aa), DR5 (440 aa), DcR1, and DcR2.

DR4

```
MAPPPARVHLGAFLAVTPNPGSAASGTEAAAATPSKVWGSSAGRIEPRGGGRGALPTSMGQHGPSARARAGRAPGPRPAREASPRLRVHKTFKFVVVGVLLQV
VPSSAATIKLHDQSIGTQQWEHSPLGELCPPGSHRSERPGACNRCTEGVGYTNASNNLFACLPCTACKSDEEERSPCTTTRNTACQCKPGTFRNDNSAEMCRK
CSTGCPRGMVKVKDCTPWSDIECVHKESGNGHNIWVILVVTLVVPLLLVAVLIVCCCIGSGCGGDPKCMDRVCFWRLGLLRGPGAEDNAHNEILSNADSLSTF
VSEQQMESQEPADLTGVTVQSPGEAQCLLGPAEAEGSQRRRLLVPANGADPTETLMLFFDKFANIVPFDSWDQLMRQLDLTKNEIDVVRAGTAGPGDALYAML
MKWVNKTGRNASIHTLLDALERMEERHAKEKIQDLLVDSGKFIYLEDGTGSAVSLE
```

DR5 (411 aa)

```
MEQRGQNAPAASGARKRHGPGPREARGARPGLRVPKTLVLVVAAVLLLVSAESALITQQDLAPQQRAAPQQKRSSPSEGLCPPGHHISEDGRDCISCKYGQDY
STHWNDLLFCLRCTRCDSGEVELSPCTTTRNTVCQCEEGTFREEDSPEMCRKCRTGCPRGMVKVGDCTPWSDIECVHKESGIIIGVTVAAVVLIVAVFVCKSL
LWKKVLPYLKGICSGGGGDPERVDRSSQRPGAEDNVLNEIVSILQPTQVPEQEMEVQEPAEPTGVNMLSPGESEHLLEPAEAERSQRRRLLVPANEGDPTETL
RQCFDDFADLVPFDSWEPLMRKLGLMDNEIKVAKAEAAGHRDTLYTMLIKWVNKTGRDASVHTLLDALETLGERLAKQKIEDHLLSSGKFMYLEGNADSALS
```

DR5 (440 aa)

```
MEQRGQNAPAASGARKRHGPGPREARGARPGPRVPKTLVLVVAAVLLLVSAESALITQQDLAPQQRAAPQQKRSSPSEGLCPPGHHISEDGRDCISCKYGQDY
STHWNDLLFCLRCTRCDSGEVELSPCTTTRNTVCQCEEGTFREEDSPEMCRKCRTGCPRGMVKVGDCTPWSDIECVHKESGTKHSGEAPAVEETVTSSPGTPA
SPCSLSGIIIGVTVAAVVLIVAVFVCKSLLWKKVLPYLKGICSGGGGDPERVDRSSQRPGAEDNVLNEIVSILQPTQVPEQEMEVQEPAEPTGVNMLSPGESE
HLLEPAEAERSQRRRLLVPANEGDPTETLRQCFDDFADLVPFDSWEPLMRKLGLMDNEIKVAKAEAAGHRDTLYTMLIKWVNKTGRDASVHTLLDALETLGER
LAKQKIEDHLLSSGKFMYLEGNADSAMS
```

DcR1

```
MARIPKTLKFVVVIVAVLLPVLAYSATTARQEEVPQQTVAPQQQRHSFKGEECPAGSHRSEHTGACNPCTEGVDYTNASNNEPSCFPCTVCKSDQKHKSSCTM
TRDTVCQCKEGTFRNENSPEMCRKCSRCPSGEVQVSNCTSWDDIQCVEEFGANATVETPAAEETMNTSPGTPAPAAEETMNTSPGTPAPAAEETMTTSPGTPA
PAAEETMTTSPGTPAPAAEETMTTSPGTPASSHYLSCTIVGIIVLIVLLIVFV
```

DcR2

```
MGLWGQSVPTASSARAGRYPGARTASGTRPWLLDPKILKFVVFIVAVLLPVRVDSATIPRQDEVPQQTVAPQQQRRSLKEEECPAGSHRSEYTGACNPCTEGV
DYTIASNNLPSCLLCTVCKSGQTNKSSCTTTRDTVCQCEKGSFQDKNSPEMCRTCRTGCPRGMVKVSNCTPRSDIKCKNESAASSTGKTPAAEETVTTILGML
ASPYHYLIIIVVLVIILAVVVVGFSCRKKFISYLKGICSGGGGGPERVHRVLFRRRSCPSRVPGAEDNARNETLSNRYLQPTQVSEQEIQGQELAELTGVTVE
SPEEPQRLLEQAEAEGCQRRRLLVPVNDADSADISTLLDASATLEEGHAKETIQDQLVGSEKLFYEEDEAGSATSCL
```

cysteines respectively in CRD1, 2, and 3. Other known mammalian TNFR family members contain 3–6 CRDs, whereas the chicken death receptor CAR1 has two CRDs (Brojatsch *et al.*, 1996). The ECD sequences of the four Apo2L/TRAIL receptors are related more closely to each other (~60–65% identity) than to other human TNFR family members (~20–25% identity). In addition, the cytoplasmic death domains of DR4 and DR5 are related more closely to each other (~65%) than to the death domains of other death receptors (~15–30%). A murine receptor for Apo2L/TRAIL was identified recently and reported to be an ortholog of human DR5 (Wu *et al.*, 1999). The full-length protein, CRDs, and death domain of mouse DR5 show 36, 40, and 56% sequence identity to the corresponding regions of human DR5, and 32, 42, and 59% sequence identity to the corresponding regions of human DR4.

Affinity for ligand(s)

Studies on binding of a soluble leucine-zipper fusion protein based on Apo2L/TRAIL to DR4- or DR5-transfected cells or to DcR1- or DcR2-based immunoadhesins (Degli-Esposti *et al.*, 1997a, 1997b) indicate two classes of binding sites for each of the four receptors, with K_d values ranging from 0.04 to 0.09 nM for the high-affinity sites and 0.4 to 39 nM for the low-affinity sites.

Cell types and tissues expressing the receptor

Northern blot analysis indicated expression of three major DR4 mRNA transcripts of 2.6, 4.6, and 7.2 kb in several human tissues, including spleen, peripheral blood leukocytes (PBLs), small intestine, thymus, and activated T cells (Pan *et al.*, 1997). In addition, DR4 transcripts were detected by northern hybridization in K562 erythroleukemia cells, MCF7 breast carcinoma cells (Pan *et al.*, 1997), and by RT-PCR in 8/12 human glioma cell lines (Rieger *et al.*, 1998), 4/8 melanoma cell lines (Griffith *et al.*, 1998), and 11/11 breast carcinoma cell lines (Keane *et al.*, 1999).

Northern blot analysis indicated expression of one major DR5 mRNA transcript of ~4.5 kB in human fetal kidney, liver, and lung, and in many adult tissues, including PBL, colon, small intestine, ovary, testis, prostate, thymus, spleen, pancreas, kidney, skeletal muscle, liver, lung, and heart (Sheridan *et al.*, 1997; Pan and Dixit, 1997; Walczak *et al.*, 1997).

RT-PCR analysis indicated that resting PBL, CD4+ and CD8+ T cells, and B cells from peripheral blood express DR5 mRNA (Mongkolsapaya *et al.*, 1998). Northern analysis revealed that several cancer cell lines express the DR5 mRNA, including HL-60 promyelocytic leukemia, HeLa S3 hystiocytic lymphoma, K562 erythroleukemia, MOLT-4 T cell leukemia, Raji B cell lymphoma, SW480 colon carcinoma, A549 lung carcinoma, and G361 melanoma (Sheridan *et al.*, 1997; Pan and Dixit, 1997). In addition, RT-PCR analysis indicated DR5 mRNA expression in 11/12 glioma cell lines (Rieger *et al.*, 1998), 8/8 melanoma cell lines (Griffith *et al.*, 1998), and 11/11 breast carcinoma cell lines (Keane *et al.*, 1999).

Northern analysis revealed four DcR1 mRNA transcripts of approximately 1.5, 3.5, 4.5, and 7.5 kB in human tissues; the 1.5 kB transcript, which corresponds in size to the DcR1 cDNA, appeared to be the most abundant (Sheridan *et al.*, 1997; Pan and Dixit, 1997). DcR1 message was present in several tissues including heart, placenta, lung, liver, skeletal muscle, kidney, pancreas, spleen, PBL, and bone marrow; of note, DcR1 mRNA was not detectable in the HL-60, HeLa S3, K562, MOLT-4, Raji, SW480, A549, or G361 cancer cell lines, which showed abundant DR5 mRNA expression (Sheridan *et al.*, 1997; Pan and Dixit, 1997). RT-PCR analysis indicated DcR1 mRNA expression in resting PBL, in resting CD4+ and CD8+ T cells, and in B cells, but not in phytohemagglutinin (PHA)-activated PBL (Mongkolsapaya *et al.*, 1998). In addition, RT-PCR analysis indicated DcR1 mRNA expression in 4/12 glioma cell lines (Rieger *et al.*, 1998), 1/8 melanoma cell lines (Griffth *et al.*, 1998), and 3/11 breast carcinoma cell lines (Keane *et al.*, 1999).

Northern analysis indicated one predominant DcR2 mRNA transcript of approximately 4 kb in several human tissues including fetal kidney, liver, and lung, and adult PBL, colon, small intestine, ovary, testis, prostate, thymus, spleen, pancreas, kidney, liver, lung, placenta, and heart (Degli-Esposti *et al.*, 1997a; Marsters *et al.*, 1997; Pan *et al.*, 1998). In addition, DcR2 mRNA was detected in the A549, SW480, and HeLa S3 cell lines. RT-PCR analysis indicated DcR2 mRNA expression in 2/12 glioma cell lines (Rieger *et al.*, 1998), and in 1/8 melanoma cell lines (Griffith *et al.*, 1998).

In conclusion, these mRNA expression data indicate that DR4 and DR5 transcripts are expressed in many normal tissues; both death receptors, particularly DR5, are expressed also in the majority of tumor cell lines studied so far. DcR1 and DcR2 transcripts also are expressed in many normal tissues, with DcR2 being more widely detected; however, only a small fraction of the tumor cell lines studied so far express significant amounts of DcR1 or DcR2 mRNA.

Regulation of receptor expression

Activation of the p53 tumor-suppressor induced DR5 mRNA, but not DR4 or DcR1 mRNA in certain tumor cell lines (Wu *et al.*, 1997; Sheikh *et al.*, 1998). TNF also induced DR5 mRNA expression in certain tumor cell lines, independently of their p53 status (Sheikh *et al.*, 1998). Stimulation of monocytes with interferon γ, which induced expression of Apo2L/TRAIL mRNA, induced a concomitant downregulation of DR5 expression, perhaps as a mechanism of reducing sensitivity of the Apo2L/TRAIL-producing cells to the ligand (Griffith *et al.*, 1999).

DcR1 mRNA expression in resting PBLs was readily detectable by RT-PCR; treatment with PHA reduced DcR1 message to below the level of detection (Mongkolsapaya *et al.*, 1998). In addition, DcR1 mRNA was found to be upregulated in certain tumor cell lines that have a wild-type p53 gene upon DNA damage and p53 activation (Sheikh *et al.*, 1999).

SIGNAL TRANSDUCTION

Apoptosis induction by Apo2L/TRAIL requires caspase activity. Ectopic expression of a dominant-negative version of the FADD adapter protein at levels that readily inhibited CD95-induced apoptosis did not block apoptosis induction by Apo2L/TRAIL (Marsters *et al.*, 1996), suggesting that a FADD-independent pathway can link this ligand to caspase activation. Overexpression of DR4 or DR5 triggers caspase-dependent apoptosis; however, there are conflicting reports on the effect of transfection with dominant-negative FADD on this response: in some studies no effect was observed (Pan *et al.*, 1997; Pan and Dixit, 1997; Sheridan *et al.*, 1997), whereas in other studies inhibition was seen (Chaudhary *et al.*, 1997; Scheider *et al.*, 1997; Walczak *et al.*, 1997). Published results also disagree with respect to the ability of DR4 and DR5 to bind to known adapters: some experiments show no such interaction (McFarlane *et al.*, 1997; Pan *et al.*, 1997; Pan and Dixit, 1997), whereas others show binding of DR4 and DR5 to TRADD, FADD, TRAF2, and RIP (Chaudhary *et al.*, 1997; Schneider *et al.*, 1997). In addition, apoptosis-induction by Apo2L/TRAIL can be inhibited by overexpression of c-FLIP, which intervenes with caspase activation by death receptors

Figure 4 Potential signaling pathways that operate downstream of DR4 and DR5. Question marks indicate implication on the basis of overexpression studies.

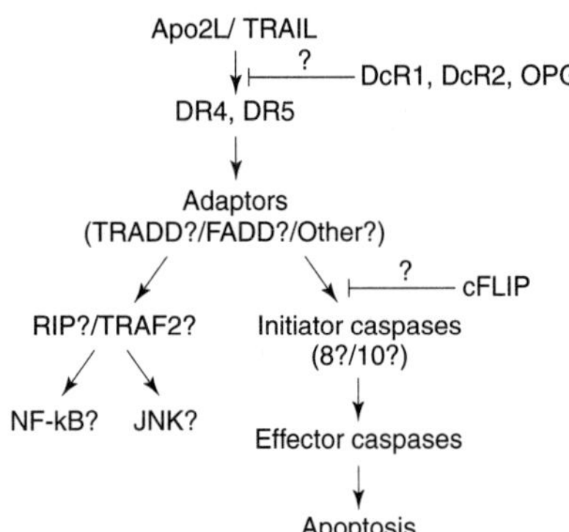

(Thorne *et al.*, 1997). Because the latter interactions were observed in cotransfection experiments, it is possible that the abnormally high levels of receptors and adapters led to promiscuous homophilic association between related domains that do not interact physiologically. Importantly, cells from FADD-deficient mice, which are resistant to apoptosis induction by CD95, TNFR1, and DR3, show full responsiveness to DR4, suggesting that a FADD-independent pathway can couple Apo2L/TRAIL to caspases (Yeh *et al.*, 1998). Signal transduction pathways that may operate downstream of DR4 and DR5 are shown schematically in **Figure 4**.

BIOLOGICAL CONSEQUENCES OF ACTIVATING OR INHIBITING RECEPTOR AND PATHOPHYSIOLOGY

Unique biological effects of activating the receptors

Overexpression of DR4 or DR5, or activation of these receptors by Apo2L/TRAIL, triggers apoptosis in a variety of tumor cell lines, but not in most normal cell types studied to date. One exception may be cultured human astrocytes, which showed sensitivity to Apo2L/TRAIL in one study (Walczak *et al.*, 1999) but not in another study (Ashkenazi *et al.*, 1999). Upon overexpression, DR4 and DR5 can also induce activation of NFκB (Chaudhary *et al.*, 1997; Schneider *et al.*, 1997; Sheridan *et al.*, 1997). Because ligand activation of this response requires 100–1000-fold higher concentrations of Apo2L/TRAIL than of TNF (Ashkenazi *et al.*, 1999), the physiological significance of NFκB activation by DR4 and DR5 remains uncertain.

Overexpression of DcR2 activated NFκB in one study (Degli-Esposti *et al.*, 1997b), but not in another study (Marsters *et al.*, 1997). As noted above, five out of six amino acid positions in the death domain, each of which is required for NFB-signaling by TNFR1, are absent in DcR2. Hence, if DcR2 activates NFB, then it must do so by engaging a signaling pathway which is distinct from the pathway engaged by TNFR1. Whether the ligand itself stimulates NFB through DcR2 is yet to be investigated. In any event, a truncated DcR2 molecule that lacks most of the cytoplasmic sequence inhibited apoptosis-induction by Apo2L/TRAIL similarly to the full-length DcR2 molecule, indicating that signaling by DcR2 is not required for its inhibitory action.

Human abnormalities

Two alterations in the DR5 genes in *head and neck cancers* have been identified, including a truncating loss-of-function mutation in the death domain (Pai *et al.*, 1998).

THERAPEUTIC UTILITY

Targeting DR4 and DR5 with Apo2L/TRAIL may be a useful approach to the treatment of human *cancer* (Ashkenazi *et al.*, 1999; Walczak *et al.*, 1999).

References

Ashkenazi, A., and Dixit, V. M. (1998). Death receptors: signaling and modulation. *Science* **281**, 1305–1308.

Ashkenazi, A., and Dixit, V. M. (1999). Apoptosis control by death and decoy receptors. *Curr. Opin. Cell Biol.* **11**, 255–260.

Ashkenazi, A., Pai, R., Fong, S., Leung, S., Lawrence, D., Marsters, S., Blackie, C., Chang, L., McMurtrey, A., Hebert, A., DeForge, L., Khoumenis, I., Lewis, D., Harris, L., Bussiere, J., Koeppen, H., Shahrokh, Z., and Schwall, R. (1999). Safety and anti-tumor activity of recombinant soluble Apo2 ligand. *J. Clin. Invest.* **104**, 155–162.

Boldin, M. P., Varfolomeev, E. E., Pancer, Z., Mett, I. L., Cmonis, J. H., and Wallach, D. (1995). A novel protein that interacts with the death domain of Fas/APO1 contains a sequence motif related to the death domain. *J. Biol. Chem.* **270**, 387–391.

Boldin, M., Goncharov, T., Goltsev, Y., and Wallach, D. (1996). Involvement of MACH, a novel MORT1/FADD-interacting protease, in FAS/APO-1-and TNF receptor-induced cell death. *Cell* **85**, 803–815.

Brojatsch, J., Naughton, J., Rolls, M. M., Zingler, K., and Young, J. A. T. (1996). CAR1, a TNFR-related protein, is a cellular receptor for cytopathic avian leukosis-sarcoma viruses and mediates apoptosis. *Cell* **87**, 845–855.

Chaudhary, P. M., Eby, M., Jasmin, A., Bookwalter, A., Murray, J., and Hood, L. (1997). Death receptor 5, a new member of the TNFR family, and DR4 induce FADD-dependent apoptosis and activate the NF-κB pathway. *Immunity* **7**, 821–830.

Chinnaiyan, A. M., O'Rourke, K., Tewari, M., and Dixit, V. M. (1995). FADD, a novel death domain-containing protein, interacts with the death domain of Fas and initiates apoptosis. *Cell* **81**, 505–512.

Degli-Esposti, M., Smolak, P. J., Walczak, H., Waugh, J., Huang, C. P., Dubose, R. F., Goodwin, R. G., and Smith, C. A. (1997a). Cloning and characterization of TRAIL-R3, a novel member of the emerging TRAIL receptor family. *J. Exp. Med.* **186**, 1165–1170.

Degli-Esposti, M. A., Dougall, W. C., Smolak, P. J., Waugh, J. Y., Smith, C. A., and Goodwin, R. G. (1997b). The novel receptor TRAIL-R4 induces NF-κB and protects aginst TRAIL-mediated apoptosis, yet retains an incomplete death domain. *Immunity* **7**, 813–820.

Emery, J. G., McDonell, P., Burke, M. C., Deen, K. C., Lyn, S., Silverman, C., Dul, E., Appelbaum, E. R., Eichman, C., DiPrinzio, R., Dodds, R. A., James, I. E., Rosenberg, M., Lee, J. C., and Young, P. R. (1998). Osteoprotegerin is a receptor for the cytotoxic ligand TRAIL. *J. Biol. Chem.* **273**, 14363–14367.

Griffith, T. S., Chin, W. A., Jackson, G. C., Lynch, D. H., and Kubin, M. Z. (1998). Intracellular regulation of TRAIL-induced apoptosis in human melanoma cells. *J. Immunol.* **161**, 2833–2840.

Griffith, T., Wiley, S., Kubin, M., Sedger, L., Maliszewski, C., and Fanger, N. (1999). Monocyte-mediated tumoricidal activity via the tumor necrosis factor-related cytokine, TRAL. *J. Exp. Med.* **189**, 1343–1353.

Gruss, H. J., and Dower, S. K. (1995). Tumor necrosis factor ligand superfamily: involvement in the pathology of malignant lymphomas. *Blood* **85**, 3378–3404.

Hsu, H., Xiong, J., and Goeddel, D. (1995). The TNF receptor-associated protein TRADD signals cell death and NF-κB activation. *Cell* **81**, 495–504.

Itoh, N., and Nagata, S. (1993). A novel protein domain required for apoptosis: mutational analysis of human Fas antigen. *J. Biol. Chem.* **268**, 10932–10937.

Jacobson, M. D., Weil, M., and Raff, M. C. (1997). Programmed cell death in animal development. *Cell* **88**, 347–354.

Keane, M. M., Ettenberg, S. A., Nau, M. M., Russel, E., and Lipkowitz, S. (1999). Chemotherapy augments TRAIL-induced apoptosis in breast cell lines. *Cancer Res.* **59**, 734–741.

McFarlane, M., Ahmad, M., Srinivasula, S. M., Fernandes-Alnemri, T., Cohen, G. M., and Alnemri, E. S. (1997). Identification and molecular cloning of two novel receptors for the cytotoxic ligand TRAIL. *J. Biol. Chem.* **272**, 25417–25420.

Marsters, S., Pitti, R., Donahue, C., Ruppert, S., Bauer, K., and Ashkenazi, A. (1996). Activation of apoptosis by Apo-2 ligand is independent of FADD but blocked by CrmA. *Curr. Biol.* **6**, 750–752.

Marsters, S. A., Sheridan, J. P., Pitti, R. M., Huang, A., Skubatch, M., Baldwin, D., Yuan, J., Gurney, A., Goddard, A. D., Godowski, P., and Ashkenazi, A. (1997). A novel receptor for Apo2L/TRAIL contains a truncated death domain. *Curr. Biol.* **7**, 1003–1006.

Mongkolsapaya, J., Cowper, A., Xu, X. N., Morris, G., McMichael, A., Bell, J. I., and Screaton, G. R. (1998). Lymphocyte inhibitor of TRAIL: a new receptor protecting lymphocytes from the death ligand TRAIL. *J. Immunol.* **160**, 3–6.

Muzio, M., Chinnaiyan, A., Kischkel, F., O'Rourke, K., Shevchenko, A., Ni, J., Scaffidi, C., Bretz, J., Zhang, M., Gentz, R., Mann, M., Krammer, P., Peter, M., and Dixit, V. (1996). FLICE, a novel FADD-homologous ICE/CED-3-like protease, is recruited to the CD95 (Fas/APO1) death-inducing signaling complex. *Cell* **85**, 817–827.

Muzio, M., Stockwell, B. R., Stennicke, H. R., Salvesen, G. S., and Dixit, V. M. (1998). An induced proximity model for caspase 8 activation. *J. Biol. Chem.* **273**, 2926–2930.

Nagata, S. (1997). Apoptosis by death factor. *Cell* **88**, 355–365.

Pai, S. I., Wu, G. S., Ozoren, N., Wu, L., Jen, J., Siransky, D., and El-Deiry, W. S. (1998). Rare loss-of-function mutation of a death receptor gene in head and neck cancer. *Cancer Res.* **58**, 3513–3518.

Pan, G., and Dixit, V. M. (1997). An antagonist decoy receptor and a new death domain-containing receptor for TRAIL. *Science* **277**, 815–818.

Pan, G., O'Rourke, K., Chinnaiyan, A. M., Gentz, R., Ebner, R., Ni, J., and Dixit, V. M. (1997). The receptor for the cytotoxic ligand TRAIL. *Science* **276**, 111–113.

Pan, G., Ni, J., Yu, G. L., Wei, Y. F., and Dixit, V. M. (1998). TRUNDD, a new member of the TRAIL receptor family that antagonizes TRAIL signaling. *FEBS Lett.* **424**, 41–45.

Pitti, R. M., Marsters, S. A., Ruppert, S., Donahue, C. J., Moore, A., and Ashkenazi, A. (1996). Induction of apoptosis by Apo-2 Ligand, a new member of the tumor necrosis factor receptor family. *J. Biol. Chem.* **271**, 12697–12690.

Rieger, J., Naumann, U., Glaser, T., Ashkenazi, A., and Weller, M. (1998). Apo2 ligand: a novel lethal weapon against malignant glioma? *FEBS Lett.* **427**, 124–128.

Schneider, P., Bodmer, J. L., Thome, M., Hofmann, K., Hohller, N., and Tschopp, J. (1997). Characterization of two receptors for TRAIL. *FEBS Lett.* **416**, 329–334.

Screaton, G., Xu, X., Olsen, A., Cowper, A., Tan, R., McMichael, A., and Bell, J. (1997a). LARD: a new lymphoid-specific death domain containing receptor regulated by alternative pre-mRNA splicing. *Proc. Natl Acad. Sci. USA* **94**, 4615–4619.

Screaton, G. R., Mongkolsapaya, J., Xu, X. N., Cowper, A. E., McMichael, A. J., and Bell, J. I. (1997b). TRICK2, a new alternatively spliced receptor that transduces the cytotoxic signal from TRAIL. *Curr. Biol.* **7**, 693–696.

Sheikh, M., Burns, T. F., Huang, Y., Wu, G. S., Amundson, S., Brooks, K. S., and Fornace Jr, A. J., El-Deiry, W. S. (1998). p53-dependent and -independent regulation of the death receptor KILLER/DR5 gene expression in response to genotoxic stress and tumor necrosis factor alpha. *Cancer Res.* **58**, 1593–1598.

Sheikh, M., Huang, Y., Fernandez-Salas, E., El-Deiry, W., Friess, H., Amundson, S., Yin, J., Meltzer, S., Holbrook, N., and Fornace, A. J. (1999). The antiapoptotic decoy receptor TRID/TRAIL-R3 is a p53-regulated DNA damage-inducible gene that is overexpressed in primary tumors of the gastrointestinal tract. *Oncogene* **18**, 4153–4159.

Sheridan, J. P., Marsters, S. A., Pitti, R. M., Gurney, A., Skubatch, M., Baldwin, D., Ramakrishnan, L., Gray, C., Baker, K., Wood, W. I., Goddard, A. D., Godowski, P., and Ashkenazi, A. (1997). Control of TRAIL-induced apoptosis by a family of signaling and decoy receptors. *Science* **277**, 818–821.

Simonet, W. S., Lacey, D. L., Dunstan, C. R., Kelley, M., Chang, M. S., Luthy, R., Nguyen, H. Q., Wooden, S.,

Bennet, L., Boone, T., Shimamoto, G., DeRose, M., Elliot, R., Colombero, A., Tan, H. L., Trail, G., Sullivan, J., Davy, E., Bucay, N., and Boyle, W. J. (1997). Osteoprotegerin: a novel secreted protein involved in the regulation of bone density. *Cell* **89**, 309–319.

Smith, C. A., Farrah, T., and Goodwin, R. G. (1994). The TNF receptor superfamily of cellular and viral proteins: activation, costimulation, and death. *Cell* **76**, 959–962.

Tartaglia, L. A., Ayers, T. M., Wong, G. H. W., and Goeddel, D. V. (1993). A novel domain within the 55 kd TNF receptor signals cell death. *Cell* **74**, 845–853.

Thome, M., Schneider, P., Hofmann, K., Fickenscher, H., Meinl, E., Neipel, F., Mattmann, C., Burns, K., Bodmer, J., Schroter, M., Scaffidi, C., Krammer, P., Peter, M., and Tschopp, J. (1997). Viral FLICE-inhibitory proteins (FLIPs) prevent apoptosis induced by death receptors. *Nature* **386**, 517–521.

Thornberry, N. A., and Lazebnik, Y. (1998). Caspases: enemies within. *Science* **281**, 1312–1316.

Walczak, H., Degli-Esposti, M. A., Johnson, R. S., Smolak, P. J., Waugh, J. Y., Boiani, N., Timour, M. S., Gerhart, M. J., Schooley, K. A., Smith, C. A., Goodwin, R. G., and Rauch, C. T. (1997). TRAIL-R2: a novel apoptosis-mediating receptor for TRAIL. *EMBO J.* **16**, 5386–5397.

Walczak, H., Miller, R. E., Ariail, K., Gliniak, B., Griffith, T. S., Kubin, M., Chin, W., Jones, J., Woodward, A., Le, T., Smith, C., Smolak, P., Goodwin, R. G., Rauch, C. T., Schuh, J. C. L., and Lynch, D. H. (1999). Tumoricidal activity of tumor necrosis factor-related apoptosis-inducing ligand *in vivo. Nature Med.* **5**, 157–163.

Wiley, S. R., Schooley, K., Smolak, P. J., Din, W. S., Huang, C. P., Nicholl, J. K., Sutherland, G. R., Davis-Smith, T., Rauch, C., Smith, C. A., and Goodwin, R. G. (1995). Identification and characterization of a new member of the TNF family that induces apoptosis. *Immunity* **3**, 673–682.

Wu, G. S., Burns, T. F., McDonald, E. R., Jiang, W., Meng, R., Krantz, I. D., Kao, G., Gan, D. D., Zhou, J. Y., Muschel, R., Hamilton, S. R., Spinner, N. B., Markowitz, S., Wu, G., El-Deiry, W. (1997). KILLER/DR5 is a DNA damage-inducible p53-regulated death receptor gene. *Nature Genet.* **17**, 141–143.

Wu, G., Burns, T., Zhan, Y., Alnemri, E., and El-Deiry, W. (1999). Molecular cloning and functional analysis of the mouse homologue of the KILLER/DR5 tumor necrosis factor-related apoptosis-inducing ligand (TRAIL) death receptor. *Cancer Res.* **59**, 2770–2775.

Yeh, W. C., de la Pompa, J. L., McCurrach, M. E., Shu, H. B., Elia, A. J., Shahinian, A., Ng, M., Wakeham, A., Khoo, W., Mitchel, K., El-Deiry, W. S., Lowe, S. W., Goeddel, D. V., and Mak, T. W. (1998). FADD: essential for embryo development and signaling from some, but not all, inducers of apoptosis. *Science* **279**, 1954–1958.

DcR3

Avi Ashkenazi*

Department of Molecular Oncology, Genentech Inc., 1 DNA Way, South San Francisco, CA 94080-4918, USA

*corresponding author tel: 650-225-1853, fax: 650-225-6443, e-mail: aa@gene.com

DOI: 10.1006/rwcy.2000.16013.

SUMMARY

DcR3 is a novel secreted member of the TNFR family. Its closest relative is OPG, to which it shows 31% protein sequence homology. *In vitro* studies show that DcR3-Fc binds to FasL with an affinity that is comparable to that of Fas-Fc; binding of DcR3-Fc and Fas-Fc to FasL is mutually exclusive, which suggests that DcR3 is a decoy receptor that competes with Fas for FasL binding. DcR3-Fc blocks apoptosis induction by FasL in several cell-based assays. DcR3 binds also to the TNF family member LIGHT, and DcR3-Fc inhibits LIGHT-induced apoptosis in *colon cancer* cells. The DcR3 gene appears to be amplified frequently in *cancer*, and DcR3 mRNA is often detectable in malignant tissue. Thus, DcR3 may play a pathologic role in cancer, perhaps through its interaction with FasL and/or LIGHT.

BACKGROUND

Discovery

Apoptosis is a physiological cell-suicide mechanism that enables metazoans to eliminate individual cells that threaten the organism's survival. Fas ligand (FasL, also called CD95 or Apo1 ligand) regulates mainly three types of apoptosis: (a) activation-induced cell death (AICD) of mature T lymphocytes; (b) elimination of inflammatory cells from immune-privileged sites; and (c) immune-cytotoxic killing of damaged cells (Nagata, 1997). T cell AICD helps shut down the host's immune response once an infection has been cleared. Repeated stimulation of the T cell receptor (TCR) by antigen induces expression of FasL and Fas on the surface of T helper cells; subsequently, FasL engages Fas and triggers apoptosis in the activated lymphocytes, leading to their elimination. Immune-privileged sites are tissues such as the eye, brain, or testis, in which any inflammatory immune response perturbs function; cells in immune-privileged sites constitutively express FasL, and eliminate infiltrating leukocytes that express Fas through Fas-dependent apoptosis. Immune-cytotoxic cells such as natural killer (NK) cells and cytotoxic T lymphocytes eliminate cells that have been damaged by viral or bacterial infection or by oncogenic transformation. This elimination occurs through two major mechanisms: one pathway involves release of perforin and granzymes; an alternative pathway involves expression of FasL and apoptosis-induction through Fas on target cells (Nagata, 1997; Moretta, 1997).

FasL belongs to a family of type 2 transmembrane proteins that are structurally related to tumor necrosis factor (TNF); its receptor, Fas, belongs to a family of type 1 transmembrane proteins that are structurally similar to TNF receptors (TNFRs) (Nagata, 1997). TNFR family members contain homologous extracellular cysteine-rich domains (CRDs) (Smith *et al.*, 1994). A subset of the TNFR family, including TNFRI, Fas, death receptor 3 (DR3), DR4, and DR5, contain a related cytoplasmic region dubbed 'death domain' (Itoh and Nagata, 1993; Tartaglia *et al.*, 1993). The death domain of Fas mediates apoptosis signaling in response to FasL binding, by activating the apoptotic protease caspase 8 through the adapter molecule FADD/Mort1 (Ashkenazi and Dixit, 1998). Until the recent identification of decoy receptor 3 (DcR3) (Pitti *et al.*, 1998), Fas was the only receptor which was known to bind to FasL.

DcR3 was discovered by a search of expressed sequence tag (EST) databases for ESTs that showed sequence homology to members of the TNF receptor (TNFR) gene superfamily (Pitti *et al.*, 1998). A set of overlapping ESTs with homology to TNFR was identified, and was used to isolate a full-length cDNA from human fetal lung. The cDNA encoded a previously unknown polypeptide that resembles the TNFR family, which was named DcR3 based upon functional studies.

Alternative names

TNFRSF6B; TR6; OPG2.

Structure

The N-terminal sequence of DcR3 contains a typical secretion signal, followed by four tandem cysteine-rich domains (CRD), which are characteristic of the TNFR family. Unlike most TNFR family members, DcR3 lacks an apparent transmembrane sequence, suggesting that it is a soluble, rather than a membrane-associated molecule. This was confirmed by expressing a recombinant, histidine-tagged form of DcR3 in mammalian cells: the protein was secreted into the cell culture medium (Pitti *et al.*, 1998). DcR3 has one potential N-linked glycosylation site (Asn173).

Main activities and pathophysiological roles

DcR3 binds to FasL (Pitti *et al.*, 1998). FasL plays a key role in regulating the immune response; however, how the function of FasL is controlled is not fully understood. One mechanism involves the molecule cFLIP, which modulates the apoptosis signal transduction pathway downstream of Fas (Tschopp *et al.*, 1998). A second mechanism involves proteolytic shedding of FasL from the cell surface (Tanaka *et al.*, 1998). DcR3 acts as a decoy for FasL; hence, it may be involved in extracellular regulation of FasL activity. The closest known relative of DcR3, osteoprotegerin (OPG), appears to act as a decoy for the TNF family members OPGL (Simonet *et al.*, 1997), as well as for Apo2L/TRAIL (Emery *et al.*, 1998). Thus, DcR3 and OPG define a novel subset of TNFR homologs that may function as secreted decoys to modulate specific TNF-related ligands. A similar mechanism is used by certain Pox viruses, which produce soluble TNFR homologs, presumably to inhibit the host's antiviral immune response (Smith *et al.*, 1994).

The frequent amplification of the DcR3 gene in lung and colon tumors (Pitti *et al.*, 1998) raises the possibility that DcR3 may have a pathological role in *cancer*.

GENE

Accession numbers

AF104419.

Sequence

See **Figure 1**.

Figure 1 Nucleotide sequence of the DcR3 gene.

```
cDNA  Sequence
TCCGCAGGCGGACCGGGGGCAAAGGAGGTGGCATGTCGGTCAGGCACAGCAGGGTCCTGTGTCCGCGCTGAGCCGCGCTCTCCCTGCTC
CAGCAAGGACCATGAGGGCGCTGGAGGGGCCAGGCCTGTCGCTGCTGTGCCTGGTGTTGGCGCTGCCTGCCCTGCTGCCGGTGCCGGCT
GTACGCGGAGTGGCAGAAACACCCACCTACCCCTGGCGGACGCAGAGACAGGGGAGCGGCTGGTGTGCGCCCAGTGCCCCCCAGGCAC
CTTTGTGCAGCGGCCGTGCCGCCGAGACAGCCCCACGACGTGTGGCCCGTGTCCACCGCGCCACTACACGCAGTTCTGGAACTACCTGG
AGCGCTGCCGCTACTGCAACGTCCTCTGCGGGGAGCGTGAGGAGGAGGCACGGGCTTGCCACGCCACCCACAACCGTGCCTGCCGCTGC
CGCACCGGCTTCTTCGCGCACGCTGGTTTCTGCTTGGAGCACGCATCGTGTCCACCTGGTGCCGGCGTGATTGCCCCGGGCACCCCCAG
CCAGAACACGCAGTGCCAGCCGTGCCCCCCAGGCACCTTCTCAGCCAGCAGCTCCAGCTCAGAGCAGTGCCAGCCCCACCGCAACTGCA
CGGCCCTGGGCCTGGCCCTCAATGTGCCAGGCTCTTCCTCCCATGACACCCTGTGCACCAGCTGCACTGGCTTCCCCCTCAGCACCAGG
GTACCAGGAGCTGAGGAGTGTGAGCGTGCCGTCATCGACTTTGTGGCTTTCCAGGACATCCCATCAAGAGGCTGCAGCGGCTGCTGCAG
GCCCTCGAGGCCCCGGAGGGCTGGGGTCCGACACCAAGGGCGGGCCGCGCGGCCTTGCAGCTGAAGCTGCGTCGGCGGCTCACGGAGCT
CCTGGGGGCGCAGGACGGGGCGCTGCTGGTGCGGCTGCTGCAGGCGCTGCGCGTGGCCAGGATGCCCGGGCTGGAGCGGAGCGTCCGTG
AGCGCTTCCTCCCTGTGCACTGATCCTGGCCCCCTCTTATTTATTCTACATCCTTGGCACCCCACTTGCACTGAAAGAGGCTTTTTTTT
AAATAGAAGAAATGAGGTTTCTTAAAAAAAAAAAAAAAAAAAAAAAAAAA
```

Chromosome location and linkages

Human chromosome 20q13. Nearest linked marker: AFM218xe7 (Pitti *et al.*, 1998).

PROTEIN

Accession numbers

AF104419_1

Sequence

See **Figure 2**.

Description of protein

The DcR3 polypeptide is 300 amino acids long; the mature secreted protein is predicted to be 277 amino acids long. Immunoprecipitation of native DcR3 from conditioned media of the HCT116 colon carcinoma cell line, which shows about 4-fold amplification of the DcR3 gene, reveals a relative molecular mass of aproximately 37 kDa (**Figure 3**).

Relevant homologies and species differences

There is one other soluble, secreted TNFR homolog known, namely, OPG (Simonet *et al.*, 1997). DcR3 shows sequence similarity in particular to OPG (31%) (**Figure 4**), and TNFRII (29%), and relatively less homology to Fas (17%). All the cysteines in the four CRDs of DcR3 and OPG are conserved; however, the C-terminal portion of DcR3 is 101 residues shorter. So far, only the human DcR3 gene has been reported.

Affinity for ligand(s)

A recombinant DcR3-Fc fusion protein (immunoadhesin) was used to study ligand interactions (Pitti *et al.*, 1998). DcR3-Fc bound specifically to human 293 cells transfected with full-length FasL (Suda *et al.*, 1993), but not to cells transfected with TNF (Pennica *et al.*, 1984); Apo2L/TRAIL (Wiley *et al.*, 1995; Pitti *et al.*, 1996); Apo3L/TWEAK (Chicheportiche *et al.*, 1997; Marsters *et al.*, 1998); or TRANCE/RANKL/OPGL (Anderson *et al.*, 1997; Wong *et al.*, 1997; Lacey *et al.*, 1998). DcR3-Fc coimmunoprecipitated purified soluble FasL, or shed soluble FasL from cells transfected with full-length FasL (Pitti *et al.*, 1998). Size-exclusion chromatography confirmed that DcR3-Fc and soluble FasL formed a stable complex. Equilibrium binding analysis indicated that DcR3-Fc bound to soluble FasL with a K_d of 0.8 nM, compared to a K_d of 1.1 nM for Fas-Fc. DcR3-Fc blocked binding of soluble FasL to Fas-Fc, suggesting that the binding sites for the two receptors on FasL overlap. Recent work (Yu *et al.*, 1999) confirms the binding of DcR3 to FasL, and demonstrates that DcR3 also binds to LIGHT, a TNF family member that binds two other receptors: HVEM and LTβR (Mauri *et al.*, 1998).

Cell types and tissues expressing the receptor

Northern blot analysis indicated expression of a predominant 1.2 kb DcR3 mRNA transcript in human fetal lung, brain, and liver, and in human adult spleen, colon, and lung (Pitti *et al.*, 1998), as well as in human umbilical vein endothelial cells and in phorbol ester-stimulated Jurkat T cells (Yu *et al.*, 1999). In addition, a relatively high DcR3 mRNA level was observed in the human colon carcinoma cell line SW480 (Pitti *et al.*, 1998). DcR3 protein was detected in conditioned medium of HCT116 human colon carcinoma cells, which overexpress the DcR3 mRNA, but not in human colon carcinoma Colo205 cells, which do not overexpress DcR3 (Figure 3). DcR3 mRNA was also detected in a number of primary tumor specimens by *in situ* hybridization; expression of DcR3 message was detected specifically over malignant areas of the tumors (Pitti *et al.*, 1998).

Figure 2 Amino acid sequence of the DcR3 protein.

Protein Sequence

MRALEGPGLSLLCLVLALPALLPVPAVRGVAETPTYPWRDAETGERLVCAQCPPGTFVQRPCRRDSPTTCGPCPPRHYTQFWNYLERCR

YCNVLCGEREEEARACHATHNRACRCRTGFFAHAGFCLEHASCPPGAGVIAPGTPSQNTQCQPCPPGTFSASSSSSEQCQPHRNCTALG

LALNVPGSSSHDTLCTSCTGFPLSTRVPGAEECERAVIDFVAFQDISIKRLQRLLQALEAPEGWGPTPRAGRAALQLKLRRRLTELLGA

QDGALLVRLLQALRVARMPGLERSVRERFLPVH

Figure 3 Migration of native DcR3 protein on SDS-PAGE. Human HCT116 or COLO205 cell lines were cultured for 4 days. Conditioned media were collected, and subjected to immunoprecipitation and western blot analysis with mouse anti-human DcR3 monoclonal antibody.

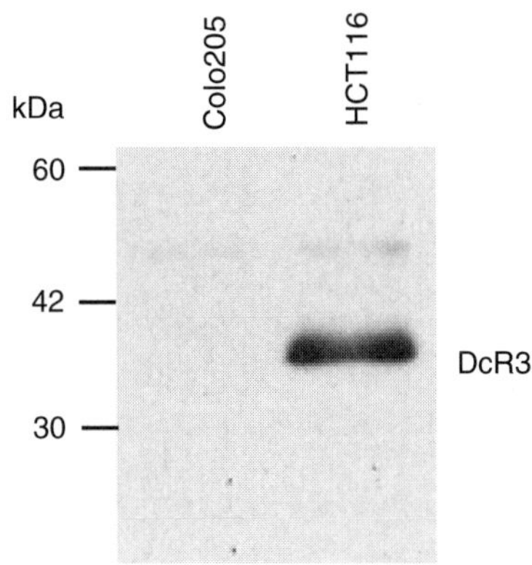

Figure 4 Schematic representation of the DcR3 and OPG proteins. The cysteine-rich domains (CRDs) are indicated by ovals; cysteines are indicated by vertical lines.

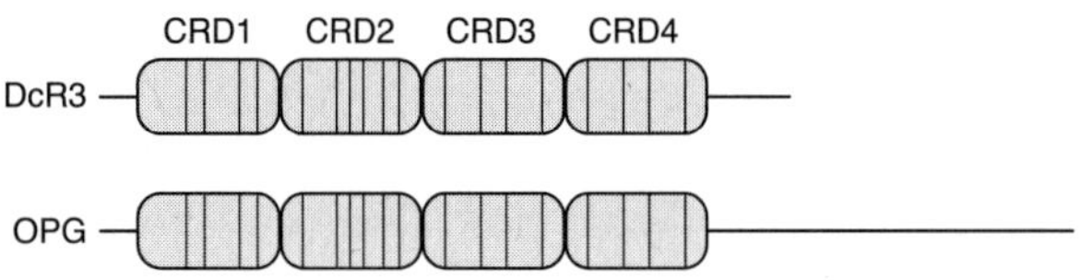

Regulation of receptor expression

There is frequent overexpression of DcR3 in certain types of *cancer*.

Release of soluble receptors

DcR3 is a secreted soluble protein. A membrane-associated form of the protein has not been found.

BIOLOGICAL CONSEQUENCES OF ACTIVATING OR INHIBITING RECEPTOR AND PATHOPHYSIOLOGY

Human abnormalities

The DcR3 gene is amplified frequently in certain types of human *cancer* (Pitti *et al.*, 1998). Eight of 18 primary *lung tumors*, and 9 of 17 primary *colon tumors* showed genomic amplification of DcR3, ranging from 2- to 18-fold. Epicenter analysis in the nine colon tumors that had DcR3 amplification showed that DcR3 was amplified significantly, in contrast to neighboring chromosomal regions; this result is consistent with the possibility that amplification of DcR3 is functionally relevant to tumorigenesis. Analysis of DcR3 mRNA expression in primary tumor tissue sections by *in situ* hybridization revealed clear DcR3 expression in 2/2 *colon tumors*, 1/4 *lung tumors*, 2/5 *breast tumors*, and 1/1 *gastric tumor*. DcR3 message was typically localized to malignant areas of the tumors, but was absent from surrounding stromal tissue, indicating tumor-specific expression.

There are several modes through which DcR3 overproduction could potentially provide growth/survival advantages to tumors. (a) Because immune-cytotoxic cells use FasL (among other mechanisms) to trigger apoptosis in cancer cells, tumors that amplify DcR3 may use the decoy to evade FasL-dependent immune-surveillence (Pitti *et al.*, 1998). (b) Recent evidence suggests that FasL is inducibly expressed in several tissues; if tumor cells induce FasL expression in nearby stromal cells, then DcR3 might serve to protect tumor cells against apoptosis-induction by stromal FasL (Green, 1988). (c) DNA-damaging agents can induce FasL expression in tumor cells; DcR3 amplification might protect tumor cells against chemotherapy-induced, FasL-dependent apoptosis (Green, 1988). (d) DcR3 might contribute directly to tumorigenesis, perhaps by cooperation with oncogenes such as Myc (Green, 1988).

THERAPEUTIC UTILITY

Effect of treatment with soluble receptor domain

Three cell-based assay systems were used to investigate whether binding of DcR3 inhibits FasL activity (Pitti *et al.*, 1998). First, apoptosis-induction by transient transfection of full-length FasL in HeLa cells, which express Fas, was studied. FasL induced apoptosis in about 25% of the cells; Fc-tagged DcR3 or Fas completely blocked this effect. Second, AICD of peripheral blood T cells, a process that involves endogenous FasL (Nagata, 1997), was examined. Consistent with previous results (Dhein *et al.*, 1995), T cell receptor engagement with anti-CD3 antibody increased the level of apoptosis in IL-2-stimulated CD4+ T cells by 2-fold. Fc-tagged DcR3 or Fas blocked this effect substantially. Third, the killing of

Fas-expressing target tumor cells by peripheral blood NK cells, which involves FasL (Arase *et al.*, 1995; Medvedev *et al.*, 1997), was investigated. NK cells triggered significant cell death in Jurkat T leukemia cells; Fc-tagged DcR3 or Fas inhibited target cell killing substantially, whereas control IgG did not. These results indicated that binding of DcR3 inhibits FasL activity. DcR3-Fc also was shown to inhibit apoptosis-induction in human colon carcinoma H29 cells by LIGHT or LIGHT plus IFNγ (Yu *et al.*, 1999). Hence, DcR3-immunoadhesins may be useful for treating pathologic conditions that involve excessive activity of FasL and/or LIGHT.

Effects of inhibitors (antibodies) to receptors

Because DcR3 overexpression appears to be functionally relevant to tumorigenesis, antibodies that block the interaction of DcR3 with FasL and LIGHT and/or augment clearance of DcR3 from the blood circulation might be useful for treating *cancers* that overproduce DcR3.

References

Anderson, D. M., Maraskovsk, E., Billingsley, W. L., Dougall, W. C., Tometsko, M. E., Roux, E. R., Teepe, M. C., DuBose, R. F., Cosman, D., and Galibert, L. (1997). A homolog of the TNF receptor and its ligand enhance T-cell growth and dendritic-cell function. *Nature* **390**, 175–179.

Arase, H., Arase, N., and Saito, T. (1995). Fas-mediated cytotoxicity by freshly isolated natural killer cells. *J. Exp. Med.* **181**, 1235–1238.

Ashkenazi, A., and Dixit, V. M. (1998). Death receptors: signaling and modulation. *Science* **281**, 1305–1308.

Chicheportiche, Y., Bourdon, P. R., Xu, H., Hsu, Y., Scott, H., Hession, C., and Browning, J. L. (1997). TWEAK, a new secreted ligand in the TNF family that weakly induces apoptosis. *J. Biol. Chem.* **272**, 32401–32410.

Dhein, J., Walczak, H., Baumler, C., Debatin, K. M., and Krammer, P. H. (1995). Autocrine T-cell suicide mediated by Apo1/(Fas/CD95). *Nature* **373**, 438–441.

Emery, J. G., McDonell, P., Burke, M. C., Deen, K. C., Lyn, S., Silverman, C., Dul, E., Appelbaum, E. R., Eichman, C., DiPrinzio, R., Dodds, R. A., James, I. E., Rosenberg, M., Lee, J. C., and Young, P. R. (1998). Osteoprotegerin is a receptor for the cytotoxic ligand TRAIL. *J. Biol. Chem.* **273**, 14363–14367.

Green, D. (1988). Death deciever. *Nature* **396**, 629–630.

Itoh, N., and Nagata, S. (1993). A novel protein domain required for apoptosis: mutational analysis of human Fas antigen. *J. Biol. Chem.* **268**, 10932–10937.

Lacey, D. L., Timms, E., Tan, H. L., Kelley, M. J., Dunstan, C. R., Burgess, T., Elliot, R., and Boyle, W. J. (1998). Osteoprotegerin ligand is a cytokine that regulates osteoclast differentiation and activation. *Cell* **93**, 165–176.

Marsters, S. A., Sheridan, J. P., Pitti, R. M., Brush, J., Goddard, A., and Ashkenazi, A. (1998). Identification of a ligand for the death-domain-containing receptor Apo3. *Curr. Biol.* **8**, 525–528.

Mauri, D., Ebner, R., Montgomery, R. I., Kochel, K. D., Cheung, T. C., Guo-Liang, Y., Ruben, S., Murphy, M., Eisenberg, R. J., Cohen, G. H., Spear, P. G., and Ware, C. F. (1998). LIGHT, a new member of the TNF superfamily, and lymphotoxin alpha are ligands for Herpesvirus entry mediator. *Immunity* **8**, 21–30.

Medvedev, A. E., Johnsen, A. C., Haux, J., Steinkjer, B., Egeberg, K., Lynch, D. H., Sundan, A., and Espevik, T. (1997). Regulation of Fas and Fas ligand expression in NK cells by cytokines and the involvement of Fas ligand in NK/LAK cell-mediated cytotoxicity. *Cytokine* **9**, 394–404.

Moretta, A. (1997). Molecular mechanisms in cell-mediated cytotoxicity. *Cell* **90**, 13–18.

Nagata, S. (1997). Apoptosis by death factor. *Cell* **88**, 355–365.

Pennica, D., Nedwin, G. E., Hayflick, J. S., Seeburg, P. H., Derynck, R., Palladino, M. A., Kohr, W. J., Aggarwal, B. B., and Goeddel, D. V. (1984). Human tumour necrosis factor: precursor structure, expression and homology to lymphotoxin. *Nature* **312**, 724–729.

Pitti, R. M., Marsters, S. A., Ruppert, S., Donahue, C. J., Moore, A., and Ashkenazi, A. (1996). Induction of apoptosis by Apo-2 ligand, a new member of the tumor necrosis factor receptor family. *J. Biol. Chem.* **271**, 12697–12690.

Pitti, R., Marsters, S. A., Lawrence, D. A., Roy, M., Kischkel, F. C., Dowd, P., Huang, A., Donahue, C. J., Sherwood, S. W., Baldwin, D. T., Godowski, P. J., Wood, W. I., Gurney, A. L., Hillan, K. J., Cohen, R. L., Goddard, A. D., and Botstein, D., A, A. (1998). Genomic amplification of a decoy receptor for Fas ligand in lung and colon cancer. *Nature* **396**, 699–703.

Simonet, W. S., Lacey, D. L., Dunstan, C. R., Kelley, M., Chang, M. S., Luthy, R., Nguyen, H. Q., Wooden, S., Bennet, L., Boone, T., Shimamoto, G., DeRose, M., Elliot, R., Colombero, A., Tan, H. L., Trail, G., Sullivan, J., Davy, E., Bucay, N., and Boyle, W. J. (1997). Osteoprotegerin: a novel secreted protein involved in the regulation of bone density. *Cell* **89**, 309–319.

Smith, C. A., Farrah, T., and Goodwin, R. G. (1994). The TNF receptor superfamily of cellular and viral proteins: activation, costimulation, and death. *Cell* **76**, 959–962.

Suda, T., Takahashi, T., Golstein, P., and Nagata, S. (1993). Molecular cloning and expression of the Fas ligand, a novel member of the tumor necrosis factor family. *Cell* **75**, 1169–1178.

Tanaka, M., Itai, T., Adachi, M., and Nagata, S. (1998). Downregulation of Fas ligand by shedding. *Nature Med.* **4**, 31–36.

Tartaglia, L. A., Ayers, T. M., Wong, G. H. W., and Goeddel, D. V. (1993). A novel domain within the 55 kd TNF receptor signals cell death. *Cell* **74**, 845–853.

Tschopp, J., Irmler, M. a., and Thome, M. (1998). Inhibition of Fas death signals by FLIPs. *Curr. Opin. Immunol.* **10**, 552–558.

Wiley, S. R., Schooley, K., Smolak, P. J., Din, W. S., Huang, C. P., Nicholl, J. K., Sutherland, G. R., Davis-Smith, T., Rauch, C., Smith, C. A., and Goodwin, R. G. (1995). Identification and characterization of a new member of the TNF family that induces apoptosis. *Immunity* **3**, 673–682.

Wong, B. R., Rho, J., Arron, J., Robinson, E., Orlinick, J., Chao, M., Kalachikov, S., Cayani, E., Bartlet, F. S., Frankel, W. N., Lee, S. W., and Choi, Y. (1997). TRANCE is a novel ligand of the TNFR family that activates c-Jun-N-terminal kinase in T cells. *J. Biol. Chem.* **272**, 25190–25194.

Yu, K., Kwon, B., Ni, J., Zhai, Y., Ebner, R., and Kwon, B. (1999). A newly identified member of tumor necrosis factor receptor superfamily (TR6) suppresses LIGHT-mediated apoptosis. *J. Biol. Chem.* **274**, 13733–13736.

Ox40

Dass S. Vinay[1] and Byoung S. Kwon[2,3,*]

[1]Department of General Surgery, University of Michigan Medical School, 1516 MSRB I, 1150 West Medical Center Drive, Ann Arbor, MI 48109, USA

[2]The Immunomodulation Research Center, University of Ulsan, Ulsan, Korea

[3]Department of Ophthalmology, LSUMC, 2020 Gravier Street Suite B, New Orleans, LA 70112, USA

*corresponding author tel: 504-412-1200 ex 1379, fax: 504-412-1315, e-mail: bkwon@lsumc.edu

DOI: 10.1006/rwcy.2000.16014.

SUMMARY

Ox40 is a member of TNF receptor superfamily and is found on activated T cells (Watts and DeBenedette, 1999) but there is a difference in their expression patterns on T cell subsets. While *in vitro* studies showed the expression peaks at 48 hours following activation on both CD4+ and CD8+ T cells, the *vivo* experiments, on the other hand, demonstrate a selective expression on CD4+ T cells upon immunizing with adjuvant or under specific clinical conditions (Weinberg, 1998; Weinberg *et al.*, 1998; Higgins *et al.*, 1999). Ox40 is an activation-dependent molecule with a broad tissue distribution, and has been localized on activated T cells (Baum *et al.*, 1994), B cells (Calderhead *et al.*, 1993), dendritic cells (Ohshima *et al.*, 1997), and vascular endothelial cells (Imura *et al.*, 1996). Data available indicate that both Ox40–Ox40L interactions and anti-Ox40–anti-CD3 combinations provide costimulation to T cells, resulting in elevated cytokine secretion (Weinberg *et al.*, 1998). The authenticity of these findings was further substantiated by the findings that in Ox40-deficient mice the T helper responses were greatly diminished, while the B cell and CTL responses remain unaffected (Kopf *et al.*, 1999). Also, studies with Ox40–Ig fusion protein demonstrate decreased T cell responses under the conditions tested (Weinberg *et al.*, 1999). Ox40 has also been shown to play an important role in various disease conditions (Kaleeba *et al.*, 1998).

Detailed functional outcomes of Ox40-mediated immune reactions are beginning to be appreciated. Reports suggest that ligation of Ox40 promotes the differentiation of naïve CD4+ T cells into TH2 cells, producing IL-4 *in vitro* (Ohshima *et al.*, 1998; Flynn *et al.*, 1998) indicating that Ox40 may be involved in the development of TH1 and TH2 *in vivo*. Interestingly, Ox40 expression on CD4+ T cells and Ox40L expression on CD11c+ dendritic cells (DC) in popliteal lymph nodes of *Leishmania major*-infected mice has been noted. This indicates a crucial role of Ox40–Ox40L interaction in the development of TH2 cells *in vivo*, possibly through T cell–DC interaction in the draining lymph nodes (Akibaa *et al.*, 2000).

In addition, Ox40-positive T cells have been noted *in vivo* in several inflammatory states, including *experimental autoimmune encephalomyelitis*, *rheumatoid arthritis*, and *graft-versus-host disease* (Weinberg *et al.*, 1996a, 1996b; Tittle *et al.*, 1997). The clinical importance of Ox40-expressing T cells has also been highlighted in experimental autoimmune encephalomyelitis in which disease prevention was achieved by selective depletion with an immunotoxin-conjugated antibody to Ox40 (Weinberg *et al.*, 1996a, 1996b). Besides its role in costimulation of CD4 cells, evolving results point out that Ox40/Ox40L interactions are closely involved in effector functions as well. For example, Ox40L crosslinking has been reported to support B cell stimulation and antibody production (Stuber *et al.*, 1995) and elevated dendritic cell effector functions (Ohshima *et al.*, 1997) and interfering with this association can lead to the inhibition of both primary and secondary IgG responses (Stuber and Strober, 1996). Also, the Ox40L-associated B lymphocyte and dendritic cell activation appear to depend on the strength of CD40 signaling, thereby leading to the assumption that Ox40/Ox40L association may initially appear during later phase responses after engagement of one or more other ligand–receptor pairs (Gramaglia *et al.*, 1998).

Members of TNF family are known to bring about intricate signaling mechanisms in interacting cells. However, the signal transductory pathways traversed by Ox40 in regulating T cell-dependent immune regulation are still at an infant stage. The few available data suggest that Ox40, like some of the other TNFR superfamily members, is closely linked to the TRAF family of signaling molecules, especially TRAF2, TRAF3, and TRAF5, resulting in NFκB activation (Arch and Thompson, 1998; Kawamata *et al.*, 1998).

BACKGROUND

Discovery

The Ox40 molecule was originally described as a cell surface antigen found on activated rat T cells (Paterson *et al.*, 1987).

Alternative names

The human homolog of Ox40 is called ACT35 antigen.

Structure

Northern analysis showed that the Ox40 cDNA reacted with a single mRNA of ~1.4 kb present in both Con A-activated thymocytes and lymph node cells. The predicted molecular weight of the monomeric protein, based on the amino acid sequence alone, is ~49,000.

GENE

Accession numbers

Murine Ox40: X85214
Human Ox40: X75962

Sequence

The cDNA of 1.2 kb contains a translational site, ATC, surrounded by residues consistent with the consensus sequence found around initiator ATGs. This is followed by an open reading frame (ORF) coding for 271 amino acids. The 3′ noncoding sequence contains a typical polyadenylation site 33 residues from a poly(A) tail of more than 80 residues (see Mallet *et al.*, 1990).

PROTEIN

Accession numbers

The sequence data are available from EMBL/GenBank/DDBJ under accession number X75962 (human Ox40), Z21674 (murine Ox40), X79929 (human Ox40 ligand), and U12763 (murine Ox40 ligand).

Sequence

The initiator methionine is followed by a sequence containing a high proportion of hydrophobic residues typical of a signal sequence (**Figure 1**). The most likely signal cleavage point, predicted on the basis of other signal sequence cleavage sites (von Heijine, 1986), is after residue 19 which could give a mature protein of 252 amino acids. There is a single putative transmembrane domain of 25 predominantly hydrophobic amino acids, dividing the protein into an extracellular domain of 191 amino acids and a cytoplasmic region of 36 amino acids.

The extracellular part contains a cysteine-rich region in which 18 out of 140 residues are cysteines and, adjacent to the membrane, a 46 amino acid hinge-like region which includes 25 residues that are threonine, serine, or proline. This latter type of sequence is often glycosylated with *O*-linked sugars. This seems likely to be the case for Ox40 as the apparent M_r on SDS-PAGE is 47–51 kDa (Paterson *et al.*, 1987) but the M_r predicted for the mature protein from the cDNA sequence is 27,777 and there are only two potential sites for *N*-linked glycosylation. One of these is NCTP and may not be glycosylated as the sequence NTTP and NCTP in an Ig and the leukocyte common antigen, respectively, have been shown not to be glycosylated (Barclay *et al.*, 1987) presumably due to the juxtaposition of the proline residue and the NXS/T sequence.

Relevant homologies and species differences

The murine Ox40 was cloned from mouse TH cell line TH2 D.10. Comparison of the mouse sequence with the rat revealed greater than 90% homology between

Figure 1 Homology between the amino acid sequences of human, mouse, and rat species. Identical amino acids are shown in red. (Full colour figure may be viewed online.)

```
1     MCVGARRLGRGPCAALLLLGLGLS-TVTGLHCVGDTYPSN    0x40-Human
1     MYV----WVQQP-TALLLLGLTLGVTARRLNCVKHTYPSG    0x40-Mouse
1     MYV----WVQQP-TAFLLLGLSLGVTVK-LNCVKDTYPSG    0x40-Rat

40    DRCCHECRPGNGMVSRCSRSQNTVCRPCGPGFYNDVVSSK    0x40-Human
36    HKCCRECQPGHGMVSRCDHTRDTLCHPCETGFYNEAVNYD    0x40-Mouse
35    HKCCRECQPGHGMVSRCDHTRDTVCHPCEPGFYNEAVNYD    Ox40-Rat

80    PCKPCTWCNLRSGSERKQLCTATQDTVCRCRAGRQPLD--    Ox40-Human
76    TCKQCTQCNHRSGSELKQNCTPTQDTVCRCRPGTQPRQDS    0x40-Mouse
75    TCKQCTQCNHRSGSELKQNTPTTEDTVCQCRPGTQPRQDS    Ox40-Rat

118   SYKPGVDCAPCPPGHFSPGDNQACKPWTNCTLAGKHTLQP    0x40-Human
116   GYKLGVDCVPCPPGHFSPGNNQACKPWTNCTLSGKQTRHP    0x40-Mouse
115   SHKLGVDCVPCPPGHFSPGSNQACKPWTNCTLSGKQIRHP    0x40-Rat

158   ASNSSDAICEDRDPPATQPQETQGPPARPITVQPTEAWPR    0x40-Human
156   ASDSLDAVCEDRSLLATLLWETQRPTFRPTTVQSTTVWPR    0x40-Mouse
155   ASNSLDTVCEDRSLLATLLWETQRTTFRPTTVPSTTVWPR    0x40-Rat

198   TSQGPSTRPVEVPGGRAVAAILGLGLVLGLLGPLAILLAL    0x40-Human
196   TSELPSPPTLVTPEGPAFAVLLGLG--LGLLAPLTVLLAL    0x40-Mouse
195   TSQLPSTPTLVAPEGPAFAVILGLG--LGLLAPLTVLLAL    0x40-Rat

238   YLLRRDQRLPPDAHKPPGGSFRTPIQEEQADAHSTLAKI     0x40-Human
234   YLLRKAWRLP-NTPKPCWGNSFRTPIQEEHTDAHFTLAK-    0x40-Mouse
233   YLLRKAWRSP-NTPKPCWGNSFRTPIQEEQTDTHFTLAKI    0x40-Rat
```

the two sequences at both the DNA and protein level. Northern analysis found that, as in the rat, Ox40 expression appears to be confined to activated T cells. The most significant difference between rat and mouse Ox40 sequences is the change of the valine to alanine at position 22, insertion of an arginine residue, and the conversion of the neighboring lysine to an arginine at positions 23 and 24 in the proposed mouse sequence. All of the 19 cysteines, including the 18 found in the predicted extracellular domain, are conserved between the rat and mouse. Ox40 has two sites within the putative extracellular domain for *N*-linked glycosylation as well as a serine/threonine/proline-rich region near the proposed transmembrane domain, which may contain sites for *O*-linked glycosylation. The intracellular domain possesses two putative protein kinase C phosphorylation sites (S/T-X-K/R), but no other obvious intracellular signaling domain (see Calderhead *et al.*, 1993).

A search of the protein databases using the amino acid sequence of the Ox40 shows sequence similarities in the N-terminal cysteine-rich region of both NGFR and CD40. The cytoplasmic regions of the Ox40, NGFR and CD40 proteins are dissimilar in length, spanning 36, 152, and 62 amino acids, respectively. In contrast to CD40 and NGFR, both of which have four repeating sequence homology units that usually contain six cysteine residues, Ox40 contains only three clear repeats as domain 3 is much shorter than the rest.

A search of the nucleic acid databases with the ACT35 cDNA sequence revealed a good score with the rat Ox40 (65% identity) and the murine Ox40 (68% identity). However, no conservation exists in the 3′ UTR of the rat Ox40 and the ACT35 sequences. The human and rat sequences show an overall identity of 63%. The sequence conservation is more pronounced in the cysteine-rich potential ligand-binding domain than in the intracellular portion, including all 16 conserved cysteine residues. The first putative *N*-linked glycosylation site in the human sequence corresponds to the most probable *N*-glycosylation site found in the rat Ox40 antigen. Furthermore, the potential PKC phosphorylation site is conserved in human, rat and mouse. Of the other members of the NGFR/TNFR family, best scores were obtained with the human CD40 (24% identity) and the murine 4-1BB (22% identity) (for details, see Latza *et al.*, 1994).

Cell types and tissues expressing the receptor

Murine Ox40 DNA probe recognizes an mRNA species of approximately 1.4 kb that is expressed in splenic T cells as well as thymus and spleen. Ox40 could not be localized in heart, lung, brain, kidney or liver. The level of Ox40 greatly increased after activation with plate-bound anti-CD3 (Calderhead *et al.*, 1993). The expression of Ox40 is restricted to activated T lymphocytes (CD4+ and CD8+ cells).

Regulation of receptor expression

The transcription and translation of the Ox40 genes are induced after primary activation of T cells by engagement of the TCR by peptide/MHC complexes and costimulatory signals or by mitogenic stimulation of the cells (Al-Shamkhani *et al.*, 1996).

SIGNAL TRANSDUCTION

Utilizing the yeast two-hybrid system with the cytoplasmic domain of Ox40 as baits, TRAF2 and 3 were found to interact with the Ox40 molecule. This was further confirmed by co-immunoprecipitation studies. In contrast to 4-1BB, a single TRAF-binding domain was identified in the cytoplasmic tail of Ox40, leading to the activation of NFκB (Arch and Thompson, 1998).

BIOLOGICAL CONSEQUENCES OF ACTIVATING OR INHIBITING RECEPTOR AND PATHOPHYSIOLOGY

Unique biological effects of activating the receptors

Signaling through Ox40 has been shown to regulate T cell-dependent immune responses. Human Ox40 has been shown to promote adhesion of T lymphocytes to vascular endothelial cells (Imura *et al.*, 1996). Recent experiments suggest that the expression of Ox40 on T cells specific for myelin basic protein in *experimental autoimmune encephalomyelitis* (EAE). Experimental data also show that an Ox40 immunotoxin led to specific depletion of these autoreactive lymphocytes and to amelioration of EAE in rats (Weinberg *et al.*, 1996). In addition, OX40-mediated signaling also augments the primary B cell proliferation leading to enhanced humoral responses by binding to its ligand on B cells (Stuber and Strober, 1996).

References

Akibaa, H., Miyahira, Y., Atsuta, M., Takeda, K., Noharaa, C., Toshiro, F., Matsuda, F., Aoki, T., Yagita, H., and Okumura, K. (2000). Critical contribution of Ox-40 ligand to T helper cell type 2 differentiation in experimental leishmaniasis. *J. Exp. Med.* **191**, 375–380.

Al-Shamkhani, A., Birkeland, M. L., Puklavec, M., Brown, M. H., James, W., and Barclay, A. N. (1996). OX40 is differentially expressed on activated rat and mouse T cells and is the sole receptor for the OX40 ligand. *Eur. J. Immunol.* **26**, 1695–1699.

Arch, R. H., and Thompson, C. B. (1998). 4-1BB and OX-40 are members of tumor necrosis factor (TNF)-nerve growth factor receptor subfamily that bind TNF receptor-associated factors and activate nuclear factor κB. *Mol. Cell. Biol.* **18**, 558–565.

Barclay, A. N., Jackson, D. J., Willis, A. C., and Williams, A. F. (1987). Lymphocyte-specific heterogeneity in the rat leukocyte common antigen (T220) is due to differences in polypeptide sequences near the NH_2-terminus. *EMBO J.* **6**, 1259–1264.

Baum, P. R., Gayle, R. B., Ramsdell, F., Srinivasan, S., Sorensen, R. A., Watson, M. L., Seldin, M. F., Baker, E., Sutherland, G. R., and Clifford, K.N. *et al.* (1994). Molecular characterization of murine and human Ox-40/Ox-40 ligand systems: Identification of a human Ox-40 ligand as the HTLV-1-regulated protein gp34. *EMBO J.* **13**, 3992–4001.

Calderhead, D. M., Buhlmann, J. E., van den Eertwegh, A. J. M., Claassen, E., Noelle, R. J., and Fell, H. P. (1993). Cloning of mouse Ox40: A T cell activation marker that may mediate T-B collaborations. *J. Immunol.* **151**, 5261–5271.

Flynn, S., Toellner, K. M., Raykundalia, C., Goodall, M., and Lane, P. (1998). CD4 T cell cytokine differentiation: the B cell activation molecule, Ox-40 ligand, instructs CD4 T cells to express interleukin 4 and upregulates expression of the chemokine receptor, Blr-1. *J. Exp. Med.* **188**, 297–304.

Gramaglia, I., Andrew, D., Weinberg, A. D., Michael Lemon, M., and Croft, M. (1998). Ox-40 ligand: a potent costimulatory molecule for sustaining primary CD4 T cell responses. *J. Immunol.* **161**, 6510–6517.

Higgins, L. M., McDonald, S. A., Whittle, N., Crockett, N., Shields, J. G., and MacDonald, T. T. (1999). Regulation of T cell activation *in vitro* and *in vivo* by targeting the Ox-40–Ox-40 ligand interaction: Amelioration of ongoing inflammatory bowel disease with an Ox-40–IgG fusion protein, but not with an Ox-40 ligand–IgG fusion protein. *J. Immunol.* **162**, 486–493.

Imura, A., Hori, T., Imada, K., Ishikawa, T., Tanaka, Y., Maeda, M., Imamura, S., and Uchiyama, T. (1996). The human OX40/gp34 system directly mediates adhesion of activated T cells to vascular endothelial cells. *J. Exp. Med.* **183**, 2185–2195.

Kaleeba, J. A., Offner, H., Vandenbark, A. A., Lublinski, A., and Weinberg, A. D. (1998). Ox-40 receptor provides a potent costimulatory signal capable of inducing encephalitogenicity in myelin-specific CD4+ T cells. *Int. Immunol.* **10**, 453–461.

Kawamata, S., Hori, T., Imura, A., Takaori-Kondo, A., and Uchiyama, T. (1998). Activation of Ox-40 signal transduction pathways lead to tumor necrosis factor receptor-associated factor (TRAF)-2 and -5 mediated NF-kappaB activation. *J. Biol. Chem.* **273**, 5808–5814.

Kopf, M., Ruedl, C., Schmitz, N., Gallimore, A., Lefrang, K., Ecabert, B., Odermatt, B., and Bachmann, M. F. (1999). Ox-40-deficient mice are defective in Th cell proliferation but are competent in generating B cell and CTL responses after virus infection. *Immunity*. **11**, 699–708.

Latza, U., Durkop, H., Schnittger, S., Ringeling, J., Eitelbach, F., Hummel, M., Fonatsch, C., and Stein, H. (1994). The human OX40 homologue: cDNA structure, expression and chromosomal assignment of the ACT35 antigen. *Eur. J. Immunol.* **24**, 677–683.

Mallet, S., Fossum, S., and Barclay, A. N. (1990). Characterization of the MRC OX-40 antigen of CD4 positive T lymphocytes – a molecule related to nerve growth factor receptor. *EMBO J.* **9**, 1063–1068.

Ohshima, Y., Tanaka, Y., Tozawa, H., Takahashi, Y., Maliszewski, C., and Delespesse, G. (1997). Expression and function of Ox-40 ligand on human dendritic cells. *J. Immunol.* **189**, 3838–3848.

Ohshima, Y., Yang, L. P., Uchiyama, T., Tanaka, Y., Baum, P., Sergerie, M., Hermann, P., and Delespesse, G. (1998). Ox-40 costimulation enhances interleukin-4 (IL-4) expression at priming and promotes the differentiation of naive human CD4 (+) T cells into high IL-4-producing effectors. *Blood* **92**, 3338–3345.

Paterson, D. J., Jefferies, D. J., Green, J. R., Brandon, M. R., Corthesy, P., Puklavec, M., and Williams, A. F. (1987). Antigens of activated rat T lymphocytes including a molecule of ~50,000 Mr detected only on CD4+ T blasts. *Mol. Immunol.* **24**, 1281–1290.

Stuber, E., and Strober, W. (1996). The T cell-B cell interactions via OX40/OX40L is necessary for the T cell-dependent humoral immune response. *J. Exp. Med.* **183**, 979–989.

Stuber, E., Neurath, M., Calderhead, D., Fell, H. P., and Strober, W. (1995). Cross-linking of Ox-40 ligand, a member of the TNF/NGF cytokine family, induces proliferation and differentiation in murine splenic B cells. *Immunity* **2**, 507–521.

Tittle, T. V., Weinberg, A. D., Steinkeler, C. N., and Maziarz, R. T. (1997). Expression of the T-cell activation antigen, Ox-40, identifies alloreactive T cells in acute graft-versus-host disease. *Blood* **89**, 4652–4658.

von Heinije, G. (1986). A new method for predicting signal sequence cleavage sites. *Nucleic Acids Res.* **14**, 4683–4690.

Watts, T. H., and DeBenedette, M. A. (1999). T cell costimulatory molecules other than CD28. *Curr. Opin. Immunol.* **11**, 286–293.

Weinberg, A. D. (1998). Antibodies to Ox-40 (CD134) can identify and eliminate autoreactive T cells: implications for human autoimmune disease. *Mol. Med. Today*. **4**, 76–83.

Weinberg, A. D., Lemon, M., Jones, A. J., Vainiene, M., Celnik, B., Buenafe, A. C., Culbertson, N., Bakke, A., Vandenbark, A. A., and Offner. H. (1996a). Ox-40 antibody enhances for autoantigen specific Vb8.2+ T cells within the spinal cord of Lewis rats with autoimmune encephalomyelitis. *J. Neurosci. Res.* **43**, 42–49.

Weinberg, A. D., Bourdette, D. N., Sullivan, T. J., Lemon, M., Wallin, J. J., Maziarz, R., Davey, M., Palida, F., Godfrey, W., Engleman, E., Fulton, R. J., Offner, H., and Vandenbark, A. A. (1996b). Selective depletion of myelin-reactive T cells with the anti-Ox-40 antibody ameliorates autoimmune encephalomyelitis. *Nature Med.* **2**, 183–189.

Wienberg, A. D., Vella, A. T., and Croft, M. (1998). Ox-40: Life beyond the effector T cell stage. *Semin. Immunol.* **10**, 471–480.

Weinberg, A. D., Wegmann, K. W., Funutake, C., and Whitham, R. H. (1999). Blocking Ox-40/Ox-40 ligand interaction *in vitro* and *in vivo* leads to decreased T-cell function and amelioration of EAE. *J. Immunol.* **162**, 1818–1826.

ACKNOWLEDGEMENTS

SRC funds to IRC from the Korean Ministry of Science and Technology and NIH Grants (AI28125 and DE12156) are greatly appreciated.

GIT Receptor

Dass S. Vinay[1] and Byoung S. Kwon[2,3,*]

[1]Department of General Surgery, University of Michigan Medical School, 1516 MSRB I, 1150 West Medical Center Drive, Ann Arbor, MI 48109, USA

[2]The Immunomodulation Research Center, University of Ulsan, Ulsan, Korea

[3]Department of Ophthalmology, LSUMC, 2020 Gravier Street Suite B, New Orleans, LA 70112, USA

*corresponding author tel: 504-412-1200 ex 1379, fax: 504-412-1315, e-mail: bkwon@lsumc.edu

DOI: 10.1006/rwcy.2000.16015.

SUMMARY

The TNF and TNF receptor gene superfamilies control a variety of distinct physiological functions such as cell proliferation, differentiation, and survival, etc. A newly emerging member this family with strong role in T cell homeostasis is GITR (glucocorticoid-induced tumor necrosis factor receptor) (Nocentini *et al.*, 1998). A majority of glucocorticoid hormones are known to induce apoptosis (Nocentini *et al.*, 1998). It is surprising to note that these hormones also protect the cells from undergoing apoptosis under the influence of discreet stimuli (Riccardi *et al.*, 1999). These events are thought to involve the participation of GITR and GILZ genes by controlling events like activation of NFκB and expression of Fas/FasL molecules (Riccardi *et al.*, 1999). Recently, a human homolog of murine GITR was discovered (Kwon *et al.*, 1999; Gurney *et al.*, 1999). The human receptor was called AITR (activation-induced TNFR member) (Kwon *et al.*, 1999) and hGITR (Gurney *et al.*, 1999). Its ligand was cloned and called AITRL (Kwon *et al.*, 1999) and hGITRL (Gurney *et al.*, 1999). Within the cytoplasmic domain, the AITR shares a striking homology with 4-1BB and CD27 (Kwon *et al.*, 1999). AITR associates with TRAF1 (TNF receptor-associated factor 1), TRAF2, and TRAF3, and induces NFκB activation via TRAF2 (Kwon *et al.*, 1999). AITRL was expressed in endothelial cells (Kwon *et al.*, 1999). Expression of GITR appears to be activation dependent as stimulation of T lymphocytes by anti-CD3 mAb, Con A, or phorbol 12-myristate 13-acetate plus Ca-ionophore treatment readily upregulates its levels (Nocentini *et al.*, 1997).

BACKGROUND

Optimal induction of T cell activation is incomplete without the cognate interaction between T cells and antigen presenting cell (APC)-derived cell surface molecules (Schwarz, 1990). A number of APC-derived cell surface determinants have been shown to possess the ability to potentiate/desensitize T cell effector functions. Although it is considered that B7/CD28 is central to this pathway, studies of CD28-deficient mice (Shahnian *et al.*, 1993) have shown that this may not be the limiting factor in immune regulation. Recent studies indicate that several ligand/receptor pairs, beyond the B7/CD28 pathway, also have the ability to initiate and propel the ongoing immune reaction. Among these are three members of the expanding TNFR family: 4-1BB, Ox40, and GITR.

Discovery

The glucocorticoid-induced tumor necrosis factor receptor (GITR) family-related gene was cloned first from dexamethasone-treated murine T cell hybridoma (3DO) cells by differential display technique.

Alternative names

Published data from our laboratory provide evidence that a novel 25 kDa protein named activation-inducible protein of the TNF receptor (AITR) is the human homolog of the murine GITR (Kwon *et al.*, 1999). The AITR has 55% identity with murine GITR at the amino acid level. It is shown to activate by transducing signals through TRAF2-mediated mechanism. The expression of AITR is inducible by PMA and ionomycin, anti-CD3 plus anti-CD28 monoclonal antibodies. It is detected as a 1.25 kb mRNA in lymph nodes, PBLs and weakly in the spleen and colorectal adenocarcinoma cell line (SW 480) (Kwon *et al.*, 1999).

Structure

GITR is a 228 amino acid type I transmembrane protein characterized by three cysteine pseudorepeats in the extracellular domain. It is similar to 4-1BB in the intracellular domain. The full-length GITR cDNA revealed a 1005 bp long sequence. Northern blot analysis suggested that GITR mRNA is about 1.1 kb long.

Main activities and pathophysiological roles

Evidence accumulated thus far suggests that GITR expression confers resistance to TCR/CD3-induced apoptosis of transfected T cells. This resistance is independent of Fas triggering. However, modulation of GITR expression does not seem to modify the sensitivity to apoptotic stimuli other than TCR triggering (Nocentini *et al.*, 1997).

GENE

Accession numbers

GenBank:
Murine GITR: U82534

Sequence

Full cDNA from a T lymphocyte cDNA library revealed a 1005 bp long sequence. Northern analysis of GITR mRNA was found to be 1.1 kb long. Nucleotide sequencing of the three cDNA clones showed the presence of a single 684 bp ORF beginning at nucleotide position 46 and extending to a TGA termination codon at position 730. The putative initiation codon at position 46 is surrounded by a sequence (AGCACTATGG). The termination codon is followed by a 3′ UTR of 276 bp. A canonical polyadenylylation signal is present from 18 bp 5′ to the poly (A) tail.

PROTEIN

Sequence

See **Figure 1**.

Description of protein

The molecule putatively encoded by the GITR mRNA is a cysteine-rich protein of 228 amino acids. The two hydrophobic regions are present, probably representing the signal peptide and the transmembrane domain. A cleavage site for the signal peptide can be found between GLY (at position −1) and Gln (at position 1), despite the unusual presence of Asp at position −3. The transmembrane domain is located between positions 135 and 157 of the mature protein. Given this, GITR can be categorized as a type I membrane protein. The molecular weight of the predicted native protein is 25,334, consistent with that obtained by *in vitro* translation of the cloned cDNA. The predicted molecular weight of the putative mature protein before further posttranslational modifications is 23,321.

Relevant homologies and species differences

The GITR amino acid sequence exhibits marked homologies with 4-1BB, a member of the TNF/NGFR family. The GITR cytoplasmic domain spans amino acids 158–209 of the mature protein. It has a striking homology with the cytoplasmic domains of murine and human 4-1BB and CD27 (Figure 1) but does not show any significant homology with other members of the TNF/NGFR family (Gravestein *et al.*, 1993). This similarity defines a new intracellular motif that could identify a subfamily of the TNF/NGFR family, including GITR, 4-1BB, and CD27.

Cell types and tissues expressing the receptor

GITR is not detectable in freshly derived lymphoid tissues (including thymocytes, spleen, and lymph node T cells), liver, kidney, and brain and T cell hybridoma 3D0. However, low levels of GITR mRNA were detected by competitive RT-PCR in T cell hybridoma, thymocytes, spleen, and lymph node T cells.

Figure 1 (a) Deduced amino acid sequence of AITR. The potential signal sequence is indicated as bold characters and the transmembrane region is indicated in boxes. (b) Comparison of the amino acid sequence of AITR with the murine GITR. Bold Cs within three cysteine pseudorepeat motifs indicate cysteine residues found in the extracellular domain of the TNFR superfamily members. Conserved acidic amino acid clusters of the cytoplasmic domain are indicated in boxes.

A

```
  1   MAQHGAMGAFRALCGLALLCALSLGQRPTGGPGCGPGRLLLGTGTDARCCRVHTTRCCRD
 61   YPGEECCSEWDCMCVQPEFHCGDPCCTTCRHHPCPPGQGVQSQGKFSFGFQCIDCASGTF
121   SGGHEGHCKPWTDCTQFGFLTVFPGNKTHVAVCVPGSPPAEPLGWLTVVLLAVAACVLLL
181   TSAQLGLHIWQLRKTQLLLEVPPSTEDARSCQFPEEERGERSAEEKGRLGDLWV
```

B

```
                                               cysteine pseudorepeat #1
AITR  MAQHGAMGAFRALCGLALLCALSLGQRPT-GGPGCGPGRLLLGTGTDARCCRVHTTRCCRD    60
GITR  M------GAWAMLYGVSMLCVLDLGQPSVVEEPGCGPGKVQNGSGNNTRCCSLYA------    48
                               cysteine pseudorepeat #2
AITR  YPGEECCSEWDCMCVQPEFHCGDPCCTTCRHHPCPPGQGVQSQGKFSFGFQCIDCASGTF   120
GITR  -PGKEDCPKERCICVTPEYHCGDPQCKICKHYPCQPGQRVESQGDIVFGFRCVACAMGTF   109
      cysteine pseudorepeat #3
AITR  SGGHEGHCKPWTDCTQFGFLTVFPGNKTHVAVCVPGSPPAEPLGWLTVVLLAVAACVLLL   180
GITR  SAGRDGHCRLWTNCSQFGFLTMFPGNKTHNAVCIPEPLPTEQYGHLTVIFLVMAACIFFL   169

AITR  TSAQLGLHIWQLRKTQL-------LLEVPPSTEDARSCQFPEEERGERSAEEKGRLGDLWV   234
GITR  TTVQLGLHIWQLRRQHMCPRETQPFAEVQLSAEDACSFQFREEERGEQT-EEKCHLGGRWP   229
```

Regulation of receptor expression

GITR expression in T cells was found to increase 4- to 8-fold upon treatment with immobilized anti-CD3, Con A, PMA+ Ca^{2+} ionophore. However, the induction kinetics was slow (no increase before 6 hours) (Nocentini *et al.*, 1997).

BIOLOGICAL CONSEQUENCES OF ACTIVATING OR INHIBITING RECEPTOR AND PATHOPHYSIOLOGY

Unique biological effects of activating the receptors

To date, known GITR-induced biological effects are restricted to offering resistance to TCR/CD3-induced apoptosis. The protection toward TCR/CD3-induced apoptosis appears to be specific, as other apoptotic signals (Fas triggering, dexamethasone treatment, and UV irradiation) were not modulated by GITR transfection, indicating that GITR is a new member of TNF/NGFR family involved in the regulation of TCR-mediated cell death.

References

Gravestein, L. A., Blom, B., Nolten, L. A., de Vries, E., van der Horst, G., Ossendorp, F., Borst, J., and Loenen, W. A. M. (1993). Cloning and expression of murine CD27: comparison with 4-1BB, another lymphocyte-specific member of the nerve growth factor receptor family. *Eur. J. Immunol.* **23**, 943–950.

Gurney, A. L., Marsters, S. A., Huang, R. M., Pitti, R. M., Mark, D. T., Baldwin, D. T., Gray, A. M., Dowd, A. D., Brush, A. D., Heldens, A. D., Schow, A. D., Goddard, A. D., Wood, W. I., Baker, K. P., Godowski, P. J., and Ashkenazi, A. (1999). Identification of a new member of the tumor necrosis factor family and its receptor, a human ortholog of mouse GITR. *Curr. Biol.* **9**, 215–218.

Kwon, B., Yu, K. Y., Ni, J., Yu, G. L., Jang, I.-K., Kim, Y.-J., Xing, L., Lium D., Wang, S. X., and Kwon, B. S. (1999). Identification of a novel activation-inducible protein of the

tumor necrosis factor receptor superfamily and its ligand. *J. Biol. Chem.* **274**, 6056–6061.
Nocentini, G., Giunchi, L., Ronchetti, S., Krausz, L. T., Bartoli, A., Moaca, R., Migliorati, G., and Riccardi, C. (1997). A new member of the tumor necrosis factor/nerve growth factor receptor family inhibits T cell receptor-induced apoptosis. *Proc. Natl Acad. Sci. USA* **94**, 6216–6221.
Nocentini, G., Giunchi, L., Ronchetti, S., Bartoli, A., Migliorati, G., and Riccardi, C. (1998). Glucocorticoids: regulation of gene expression and apoptosis. *J. Chemother.* **10**, 187–191.
Riccardi, C., Cifone, M. G., and Migliorati, G. (1999). Glucocorticoid hormone-induced modulation of gene expression and regulation of T-cell death: role of GITR and GILZ, two dexamethasone-induced genes. *Cell Death Differ.* **6**, 1182–1189.
Schwarz, R. H. (1990). A cell culture model for T lymphocyte clonal anergy. *Science* **248**, 1349–1356.
Shahnian, A., Pieffer, K., Lee, K. P., Kundig, T. M., Kishihara, K., Wakeham, A., Kawai, K., Ohashi, P. M., Thompson, C. B., and Mak, T. W. (1993). Differential T cell costimulatory requirements in CD28-deficient mice. *Science* **261**, 609–612.

ACKNOWLEDGEMENTS

Authors are grateful to Dr Byoung Suk Kwon for sharing certain unpublished research data on the human homolog of murine GITR, the AITR.

SRC funds to IRC from the Korean Ministry of Science and Technology and NIH Grants (AI28125 and DE12156) are greatly appreciated.

IL-6 Family Receptors

IL-6 Receptor

Masahiko Hibi* **and Toshio Hirano**

Division of Molecular Oncology, Department of Oncology, Biomedical Research Center, Osaka University Graduate School of Medicine(C7), 2-2, Yamada-oka, Suita, Osaka, 565, Japan

*corresponding author tel: 81-6-879-3880, fax: 81-6-879-3889, e-mail: hirano@molonc.med.osaka-u.ac.jp

DOI: 10.1006/rwcy.2000.17001.

SUMMARY

Interleukin 6 (IL-6) is a pleiotropic cytokine that regulates immune reaction, hematopoiesis, and differentiation of the nervous system. The receptor for IL-6 (IL-6R) consists of two chains, namely IL-6Rα and gp130. Both IL-6Rα and gp130 belong to the type I cytokine receptor superfamily. IL-6Rα is the binding component specific to IL-6. In contrast, gp130 transmits signals not only of IL-6 but also of IL-6-related cytokines such as leukemia-inhibitory factor (LIF), ciliary neurotropic factor (CNTF), oncostatin M (OSM), IL-11, cardiotropin 1 (CT-1), and possibly neurotrophin-1/B cell-stimulating factor 3 (NNT-1/BSF-3). The ligand's binding to the receptor leads to homo- and heterodimerization of gp130, resulting in the activation of gp130-associated JAKs (JAK1, JAK2, and TYK2) and subsequently tyrosine phosphorylation of gp130. The tyrosine-phosphorylated gp130 further transmits signals by recruiting SH2 domain-containing signaling molecules such as the protein tyrosine phosphatase SHP-2 and STAT1 and STAT3 (signal transducers and activators of transcription). We describe the roles of these signal transduction pathways in the biological responses of the IL-6 family cytokines.

BACKGROUND

Discovery

IL-6 is a pleiotropic cytokine that acts on various cells. Signals of IL-6 are transmitted into the inside of cells through the IL-6R complex that exists on the cell surface. Initially, ligand-binding assays using isotope-labeled IL-6 revealed the existence of high-affinity binding receptors expressed on the surface of cells which can respond to IL-6 (Taga *et al.*, 1987). The human cDNA for IL-6Rα (originally IL-6R) was cloned by an expression-cloning strategy (Yamasaki *et al.*, 1988). Later, IL-6Rα turned out to be a member of cytokine receptor family (type I cytokine receptor superfamily, hematopoietin receptor family), which includes receptors for many interleukins, colony-stimulating factors, and hormones (e.g. prolactin, growth hormone, and leptin) (Bazan, 1990).

Expression of IL-6Rα in cells constitutes the low-affinity binding sites, but IL-6Rα alone does not create either the high-affinity binding sites or signals that elicit biological functions of IL-6, suggesting the existence of another component of the IL-6R complex (Taga *et al.*, 1989) in a manner which is called receptor conversion. It was shown that in all the IL-6-responsive cells, IL-6Rα associates with a transmembrane glycoprotein, molecular weight 130 kDa in the human myeloma cell line U266, in response to IL-6 (Taga *et al.*, 1989). Thus it was named gp130. Furthermore, a complex of IL-6 and soluble IL-6Rα was shown to bind the extracellular domain of gp130 and transmit signals (Taga *et al.*, 1989). Molecular cloning of gp130 further revealed that it also belongs to the type I cytokine receptor superfamily. Expression of gp130 was reported to generate the high-affinity IL-6 binding sites, and render a certain cell line to receive the signal of IL-6 and soluble IL-6R (Hibi *et al.*, 1990).

Thereafter it was found that gp130 transmits signals not only of IL-6 but also of IL-6-related cytokines such as LIF, CNTF, OSM, IL-11, and CT-1, and possibly NNT-1/BSF-3 (Senaldi *et al.*, 1999) which share the biological function with IL-6. gp130 forms a complex with α subunits of IL-6R, IL-11R, CNTF receptor, and possibly CT-1 receptor, and LIF receptor β or OSM receptor β to act as receptors for the IL-6-related cytokines (Hirano *et al.*, 1997; Heinrich *et al.*, 1998).

Alternative names

IL-6 receptor α: B cell stimulatory factor-2, BSF-2 receptor, IL-6 receptor, CD126.
gp130: IL-6 signal transducer, IL-6 receptor β chain, CD130.

Structure

The IL-6 receptor system is composed of IL-6Rα and gp130 (**Figure 1**). A reconstitution experiment of the soluble IL-6R system suggested that the IL-6 receptor consists of two molecules of IL-6, IL-6Rα, and gp130 (Paonessa *et al.*, 1995). Alternatively, it was proposed that the IL-6 receptor consists of one molecule of IL-6 and IL-6Rα, and two molecules of gp130 (Grotzinger *et al.*, 1997).

The molecular weight of IL-6Rα is 80 kDa (the molecular weight predicted from the amino acid sequence is 49.9 kDa) (Yamasaki *et al.*, 1988; Hirata *et al.*, 1989; Taga *et al.*, 1989). IL-6Rα is a type I transmembrane protein. The precursor of human IL-6Rα contains 468 amino acids and its putative mature form consists of 449 amino acids. It is composed of an extracellular region of 339 amino acids, a membrane-spanning region of 28 amino acids, and a cytoplasmic region of 82 amino acids. It has five potential *N*-glycosylation sites. IL-6Rα belongs to the type I cytokine receptor superfamily. It has one cytokine-binding module (CBM) composed of two fibronectin type III-like (FNIII) domains, in which the N-terminal domain contains a set of four conserved cysteine residues and the C-terminal one contains a WSXWS motif, conserved in the type I cytokine receptor superfamily. It was reported that the C-terminal FNIII domain in the CBM is sufficient for IL-6 binding but not for the association with gp130 (Ozbek *et al.*, 1998). In addition to the CRM, IL-6Rα contains an immunoglobulin-like domain at the N-terminus. The structure and functions of the cytoplasmic domain of IL-6Rα remain to be clarified.

The molecular weight of gp130 is 130–160 kDa (Taga *et al.*, 1989; Hibi *et al.*, 1990), the variation in the molecular weight probably being due to the glycosylation. Deglycosylated gp130 displays 100 kDa, which is consistent with the predicted molecular weight from the amino acid sequence (101 kDa) (Taga *et al.*, 1989). gp130 is also a type I transmembrane protein. The precursor of human gp130 contains 918 amino acids and its putative mature form consists of 896 amino acids. It is composed of an extracellular region of 597 amino acids, a membrane-spanning region of 22 amino acids, and a cytoplasmic region of 277 amino acids. gp130 is also a member of the type I cytokine receptor superfamily. It has a immunoglobulin-like domain at the N-terminus, which is followed by a CBM containing two FNIII domains. gp130 has three additional FNIII domains following the CBM, and thus contains five FNIII domains in the extracellular region.

Figure 1 Structure of the IL-6 receptor system. IL-6Rα and gp130 belong to the cytokine receptor type superfamily. The extracellular domains of both IL-6Rα and gp130 contain an immunoglobulin-like (Ig) domain and a cytokine-binding module (CBM). In addition, gp130 has three repeats of fibronectin type III (FNIII) domains in the extracellular domain. gp130 has regions (box 1 and box 2) conserved with the family and six tyrosines in its cytoplasmic domain (numbers indicate locations of tyrosine in human gp130). Box 1 and box 2 are involved in the association and activation of JAKs (JAK1, JAK2, and TYK2). Phosphorylation of the Y759 and the four tyrosines (Y767, Y814, Y904, and Y915) in the C-terminal are involved in the phosphorylation and activation of SHP-2 and STAT3. The phosphorylation of Y904 and Y915 was shown to be involved in the phosphorylation and activation of STAT1.

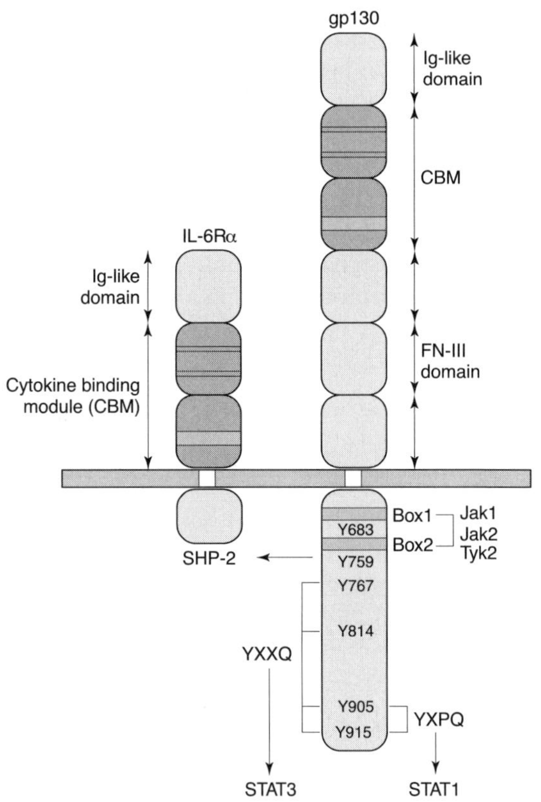

The three-dimensional structure of the CBM of gp130 has been analyzed (Bravo *et al.*, 1998). Structurally, it is closely related to the CBMs of growth hormone receptor (de Vos *et al.*, 1992) and erythropoietin receptor. The cytoplasmic region of gp130 has several distinct domains. The membrane-proximal region of gp130 has a similar sequence to those of all members of the type I cytokine receptor superfamily. The conserved region was separated by a short nonconserved amino acid stretch and thus these domains are named box 1 and box 2 domains (Murakami *et al.*, 1991). The box 1 domain is a proline-rich amino acid stretch and box 2 contains a cluster of hydrophobic amino acids followed by acidic amino acid residues. The box 1 and box 2 regions constitute a docking site for Janus tyrosine kinases (JAK1, JAK2, and TYK2). gp130 has a region which is homologous to G-CSF receptor in the middle of the cytoplasmic domain (amino acids 772–787) and this region is termed box 3 (Fukunaga *et al.*, 1991). In addition to box 1, box 2, and box 3, gp130 contains six tyrosines, some of which have been shown to be important for signal transduction.

Main activities and pathophysiological roles

IL-6Rα and gp130 transmit all the biological functions of IL-6. gp130 is also required for the biological functions of the IL-6-related cytokines: LIF, CNTF, IL-11, OSM, and CT-1.

GENE

Accession numbers

Human IL-6Rα: X12830, M20566
Mouse IL-6Rα: X51975, X51976
Human gp130: M57230
Mouse gp130: X62646, M83336
Rat gp130: M92340
Chick gp130: AJ011688, GGA011688
Genome:
IL-6Rα: chromosomal location not determined.
Human gp130: chromosome 5 (a pseudogene exists in chromosome 17).

Sequence

See **Figure 2**.

Figure 2 Nucleotide sequences for human IL-6 receptor α and gp130.

Human IL-6 receptor a

```
GGCGGTCCCCTGTTCTCCCCGCTCAGGTGCGGCGCTGTGGCAGGAAGCCA
CCGCCAGGGGACAAGAGGGGCGAGTCCACGCCGCGACACCGTCCTTCGGT
CCCCCTCGGTCGGCCGGTGCGCGGGGCTGTTGCGCCATCCGCTCCGGCTT
GGGGGAGCCAGCCGGCCACGCGCCCCGACAACGCGGTAGGCGAGGCCGAA
TCGTAACCGCACCCTGGGACGGCCCAGAGACGCTCCAGCGCGAGTTCCTC
AGCATTGGCGTGGGACCCTGCCGGGTCTCTGCGAGGTCGCGCTCAAGGAG
AAATGTTTTCCTGCGTTGCCAGGACCGTCCGCCGCTCTGAGTCATGTGCG
TTTACAAAAGGACGCAACGGTCCTGGCAGGCGGCGAGACTCAGTACACGC
AGTGGGAAGTCGCACTGACACTGAGCCGGGCCAGAGGGAGAGGAGCCGAG
TCACCCTTCAGCGTGACTGTGACTCGGCCCGGTCTCCCTCTCCTCGGCTC
CGCGGCGCGGGGCCGAGGGACTCGCAGTGTGTGTAGAGAGCCGGGCTCCT
GCGCCGCGCCCCGGCTCCCTGAGCGTCACACACATCTCTCGGCCCGAGGA
GCGGATGGGGGCTGCCCCCGGGGCCTGAGCCCGCCTGCCCGCCCACCGCC
CGCCTACCCCCGACGGGGGCCCCGGACTCGGGCGGACGGGCGGGTGGCGG
CCGCCCCGCCCCTGCCACCCCTGCCGCCCGGTTCCCATTAGCCTGTCCGC
GGCGGGGCGGGGACGGTGGGGACGGCGGGCCAAGGGTAATCGGACAGGCG
CTCTGCGGGACCATGGAGTGGTAGCCGAGGAGGAAGCATGCTGGCCGTCG
GAGACGCCCTGGTACCTCACCATCGGCTCCTCCTTCGTACGACCGGCAGC
GCTGCGCGCTGCTGGCTGCCCTGCTGGCCGCGCCGGGAGCGGCGCTGGCC
CGACGCGCGACGACCGACGGGACGACCGGCGCGGCCCTCGCCGCGACCGG
CCAAGGCGCTGCCCTGCGCAGGAGGTGGCAAGAGGCGTGCTGACCAGTCT
GGTTCCGCGACGGACGCGTCCTCCACCGTTCTCCGCACGACTGGTCAGA
GCCAGGAGACAGCGTGACTCTGACCTGCCCGGGGGTAGAGCCGGAAGACA
CGGTCCTCTGTCGCACTGAGACTGGACGGGCCCCCATCTCGGCCTTCTGT
ATGCCACTGTTCACTGGGTGCTCAGGAAGCCGGCTGCAGGCTCCCACCCC
TACGGTGACAAGTGACCCACGAGTCCTTCGGCCGACGTCCGAGGGTGGGG
AGCAGATGGGCTGGCATGGGAAGGAGGCTGCTGCTGAGGTCGGTGCAGCT
TCGTCTACCCGACCGTACCCTTCCTCCGACGACGACTCCAGCCACGTCGA
CCACGACTCTGGAAACTATTCATGCTACCGGGCCGGCCGCCCAGCTGGGA
GGTGCTGAGACCTTTGATAAGTACGATGGCCCGGCCGGCGGGTCGACCCT
CTGTGCACTTGCTGGTGGATGTTCCCCCCGAGGAGCCCCAGCTCTCCTGC
GACACGTGAACGACCACCTACAAGGGGGGCTCCTCGGGGTCGAGAGGACG
TTCCGGAAGAGCCCCCTCAGCAATGTTGTTTGTGAGTGGGGTCCTCGGAG
AAGGCCTTCTCGGGGGAGTCGTTACAACAAACACTCACCCCAGGAGCCTC
CACCCCATCCCTGACGACAAAGGCTGTGCTCTTGGTGAGGAAGTTTCAGA
GTGGGGTAGGGACTGCTGTTTCCGACACGAGAACCACTCCTTCAAAGTCT
ACAGTCCGGCCGAAGACTTCCAGGAGCCGTGCCAGTATTCCCAGGAGTCC
TGTCAGGCCGGCTTCTGAAGGTCCTCGGCACGGTCATAAGGGTCCTCAGG
CAGAAGTTCTCCTGCCAGTTAGCAGTCCCGGAGGGAGACAGCTCTTTCTA
GTCTTCAAGAGGACGGTCAATCGTCAGGGCCTCCCTCTGTCGAGAAAGAT
CATAGTGTCCATGTGCGTCGCCAGTAGTGTCGGGAGCAAGTTCAGCAAAA
GTATCACAGGTACACGCAGCGGTCATCACAGCCCTCGTTCAAGTCGTTTT
CTCAAACCTTTCAGGGTTGTGGAATCTTGCAGCCTGATCCGCCTGCCAAC
GAGTTTGGAAAGTCCCAACACCTTAGAACGTCGGACTAGGCGGACGGTTG
ATCACAGTCACTGCCGTGGCCAGAAACCCCCGCTGGCTCAGTGTCACCTG
TAGTGTCAGTGACGGCACCGGTCTTTGGGGGCGACCGAGTCACAGTGGAC
GCAAGACCCCCACTCCTGGAACTCATCTTTCTACAGACTACGGTTTGAGC
CGTTCTGGGGGTGAGGACCTTGAGTAGAAAGATGTCTGATGCCAAACTCG
TCAGATATCGGGCTGAACGGTCAAAGACATTCACAACATGGATGGTCAAG
AGTCTATAGCCCGACTTGCCAGTTTCTGTAAGTGTTGTACCTACCAGTTC
GACCTCCAGCATCACTGTGTCATCCACGACGCCTGGAGCGGCCTGAGGCA
CTGGAGGTCGTAGTGACACAGTAGGTGCTGCGGACCTCGCCGGACTCCGT
CGTGGTGCAGCTTCGTGCCCAGGAGGAGTTCGGGCAAGGCGAGTGGAGCG
GCACCACGTCGAAGCACGGGTCCTCCTCAAGCCCGTTCCGCTCACCTCGC
AGTGGAGCCCGGAGGCCATGGGCACGCCTTGGACAGAATCCAGGAGTCCT
TCACCTCGGGCCTCCGGTACCCGTGCGGAACCTGTCTTAGGTCCTCAGGA
CCAGCTGAGAACGAGGTGTCCACCCCCATGCAGGCACTTACTACTAATAA
GGTCGACTCTTGCTCCACAGGTGGGGGTACGTCCGTGAATGATGATTATT
AGACGATGATAATATTCTCTTCAGAGATTCTGCAAATGCGACAAGCCTCC
```

Figure 2 (*Continued*)

```
TCTGCTACTATTATAAGAGAAGTCTCTAAGACGTTTACGCTGTTCGGAGG
CAGTGCAAGATTCTTCTTCAGTACCACTGCCCACATTCCTGGTTGCTGGA
GTCACGTTCTAAGAAGAAGTCATGGTGACGGGTGTAAGGACCAACGACCT
GGGAGCCTGGCCTTCGGAACGCTCCTCTGCATTGCCATTGTTCTGAGGTT
CCCTCGGACCGGAAGCCTTGCGAGGAGACGTAACGGTAACAAGACTCCAA
CAAGAAGACGTGGAAGCTGCGGGCTCTGAAGGAAGGCAAGACAAGCATGC
GTTCTTCTGCACCTTCGACGCCCGAGACTTCCTTCCGTTCTGTTCGTACG
ATCCGCCGTACTCTTTGGGGCAGCTGGTCCCGGAGAGGCCTCGACCCACC
TAGGCGGCATGAGAAACCCCGTCGACCAGGGCCTCTCCGGAGCTGGGTGG
CCAGTGCTTGTTCCTCTCATCTCCCCACCGGTGTCCCCCAGCAGCCTGGG
GGTCACGAACAAGGAGAGTAGAGGGGTGGCCACAGGGGGTCGTCGGACCC
GTCTGACAATACCTCGAGCCACAACCGACCAGATGCCAGGGACCCACGGA
CAGACTGTTATGGAGCTCGGTGTTGGCTGGTCTACGGTCCCTGGGTGCCT
GCCCTTATGACATCAGCAATACAGACTACTTCTTCCCCAGATAGCTGGCT
CGGGAATACTGTAGTCGTTATGTCTGATGAAGAAGGGGTCTATCGACCGA
GGGTGGCACCAGCAGCCTGGACCCTGTGGATGACAAAACACAAACGGGCT
CCCACCGTGGTCGTCGGACCTGGGACACCTACTGTTTTGTGTTTGCCCGA
CAGCAAAAGATGCTTCTCACTGCCATGCCAGCTTATCTCAGGGGTGTGCG
GTCGTTTTCTACGAAGAGTGACGGTACGGTCGAATAGAGTCCCCACACGC
GCCTTTGGCTTCACGGAAGAGCCTTGCGGAAGGTTCTACGCCAGGGGAAA
CGGAAACCGAAGTGCCTTCTCGGAACGCCTTCCAAGATGCGGTCCCCTTT
ATCAGCCTGCTCCAGCTGTTCAGCTGGTTGAGGTTTCAAACCTCCCTTTC
TAGTCGGACGAGGTCGACAAGTCGACCAACTCCAAAGTTTGGAGGGAAAG
CAAATGCCCAGCTTAAAGGGGTTAGAGTGAACTTGGGCCACTGTGAAGAG
GTTTACGGGTCGAATTTCCCCAATCTCACTTGAACCCGGTGACACTTCTC
AACCATATCAAGACTCTTTGGACACTCACACGGACACTCAAAAGCTGGGC
TTGGTATAGTTCTGAGAAACCTGTGAGTGTGCCTGTGAGTTTTCGACCCG
AGGTTGGTGGGGGCCTCGGTGTGGAGAAGCGGCTGGCAGCCCACCCCTCA
TCCAACCACCCCCGGAGCCACACCTCTTCGCCGACCGTCGGGTGGGGAGT
ACACCTCTGCACAAGCTGCACCCTCAGGCAGGTGGGATGGATTTCCAGCC
TGTGGAGACGTGTTCGACGTGGGAGTCCGTCCACCCTACCTAAAGGTCGG
AAAGCCTCCTCCAGCCGCCATGCTCCTGGCCCACTGCATCGTTTCATCTT
TTTCGGAGGAGGTCGGCGGTACGAGGACCGGGTGACGTAGCAAAGTAGAA
CCAACTCAAACTCTTAAAACCCAAGTGCCCTTAGCAAATTCTGTTTTTCT
GGTTGAGTTTGAGAATTTTGGGTTCACGGGAATCGTTTAAGACAAAAAGA
AGGCCTGGGGACGGCTTTTACTTAAACGCCAAGGCCTGGGGGAAGAAGCT
TCCGGACCCCTGCCGAAAATGAATTTGCGGTTCCGGACCCCCTTCTTCGA
CTCTCCTCCCTTTCTTCCCTACAGTTCAAAAACAGCTGAGGGTGAGTGGG
GAGAGGAGGGAAAGAAGGGATGTCAAGTTTTTGTCGACTCCCACTCACCC
TGAATAATACAGTATGTCAGGGCCTGGTCGTTTTCAACAGAATTATAATT
ACTTATTATGTCATACAGTCCCGGACCAGCAAAAGTTGTCTTAATATTAA
AGTTCCTCATTAGCAGTTTTGCCTAAATGTGAATGATGATCCTAGGCATT
TCAAGGAGTAATCGTCAAAACGGATTTACACTTACTACTAGGATCCGTAA
TGCTGAATACAGAGGCAACTGCATTGGCTTTGGGTTGCAGGACCTCAGGT
ACGACTTATGTCTCCGTTGACGTAACCGAAACCCAACGTCCTGGAGTCCA
GAGAAGCAGAGGAAGGAGAGGAGAGGGGCACAGGGTCTCTACCATCCCCT
CTCTTCGTCTCCTTCCTCTCCTCTCCCCGTGTCCCAGAGATGGTAGGGGA
GTAGAGTGGGAGCTGAGTGGGGGATCACAGCCTCTGAAAACCAATGTTCT
CATCTCACCCTCGACTCACCCCCTAGTGTCGGAGACTTTTGGTTACAAGA
CTCTTCTCCACCTCCCACAAAGGAGAGCTAGCAGCAGGGAGGGCTTCTGC
GAGAAGAGGTGGAGGGTGTTTCCTCTCGATCGTCGTCCCTCCCGAAGACG
CATTTCTGAGATCAAAACGGTTTTACTGCAGCTTTGTTTGTTGTCAGCTG
GTAAAGACTCTAGTTTTGCCAAAATGACGTCGAAACAAACAACAGTCGAC
AACCTGGGTAACTAGGGAAGATAATATTAAGGAAGACAATGTGAAAAGAA
TTGGACCCATTGATCCCTTCTATTATAATTCCTTCTGTTACACTTTTCTT
AAATGAGCCTGGCAAGAATGCGTTTAAACTTGGTTTTTAAAAAACTGCTG
TTTACTCGGACCGTTCTTACGCAAATTTGAACCAAAAATTTTTTGACGAC
ACTGTTTTCTCTTGAGAGGGTGGAATATCCAATATTCGCTGTGTCAGCAT
TGACAAAAGAGAACTCTCCCACCTTATAGGTTATAAGCGACACAGTCGTA
AGAAGTAACTTACTTAGGTGTGGGGGAAGCACCATAACTTTGTTTAGCCC
TCTTCATTGAATGAATCCACACCCCCTTCGTGGTATTGAAACAAATCGGG
AAAACCAAGTCAAGTGAAAAAGGAGGAAGAGAAAAAATATTTTCCTGCCA
TTTTGGTTCAGTTCACTTTTTCCTCCTTCTCTTTTTTATAAAAGGACGGT
GGCATGGAGGCCCACGCACTTCGGGAGGTCGAGGCAGGAGGATCACTTGA
CCGTACCTCCGGGTGCGTGAAGCCCTCCAGCTCCGTCCTCCTAGTGAACT
GTCCAGAAGTTTGAGATCAGCCTGGGCAATGTGATAAAACCCCATCTCTA
CAGGTCTTCAAACTCTAGTCGGACCCGTTACACTATTTTGGGGTAGAGAT
CAAAAAGCATAAAAATTAGCCAAGTGTGGTAGAGTGTGCCTGAAGTCCCA
GTTTTTCGTATTTTTAATCGGTTCACACCATCTCACACGGACTTCAGGGT
GATACTTGGGGGGCTGAGGTGGGAGGATCTCTTGAGCCTGGGAGGTCAAG
CTATGAACCCCCCGACTCCACCCTCCTAGAGAACTCGGACCCTCCAGTTC
GCTGCAGTGAGCCGAGATTGCACCACTGCACTCCAGCCTGGGGTGACAGA
CGACGTCACTCGGCTCTAACGTGGTGACGTGAGGTCGGACCCCACTGTCT
GCAAGTGAGACCCTGTCTCCGTTCACTCTGGGACAGAG
```

Figure 2 (*Continued*)

Human gp130

```
GAGCAGCCAAAAGGCCCGCGGAGTCGCGCTGGGCCGCCCCGGCGCAGCTG
CTCGTCGGTTTTCCGGGCGCCTCAGCGCGACCCGGCGGGGCCGCGTCGAC
AACCGGGGGCCGCGCCTGCCAGGCCGACGGGTCTGGCCCAGCCTGGCGCC
TTGGCCCCCGGCGCGGACGGTCCGGCTGCCCAGACCGGGTCGGACCGCGG
AAGGGGTTCGTGCGCTGTGGAGACGCGGAGGGTCGAGGCGGCGCGGCCTG
TTCCCCAAGCACGCGACACCTCTGCGCCTCCCAGCTCCGCCGCGCCGGAC
AGTGAAACCCAATGGAAAAAGCATGACATTTAGAAGTAGAAGACTTAGCT
TCACTTTGGGTTACCTTTTTCGTACTGTAAATCTTCATCTTCTGAATCGA
TCAAATCCCTACTCCTTCACTTACTAATTTTGTGATTTGGAAATATCCGC
AGTTTAGGGATGAGGAAGTGAATGATTAAAACACTAAACCTTTATAGGCG
GCAAGATGTTGACGTTGCAGACTTGGGTAGTGCAAGCCTTGTTTATTTTC
CGTTCTACAACTGCAACGTCTGAACCCATCACGTTCGGAACAAATAAAAG
CTCACCACTGAATCTACAGGTGAACTTCTAGATCCATGTGGTTATATCAG
GAGTGGTGACTTAGATGTCCACTTGAAGATCTAGGTACACCAATATAGTC
TCCTGAATCTCCAGTTGTACAACTTCATTCTAATTTCACTGCAGTTTGTG
AGGACTTAGAGGTCAACATGTTGAAGTAAGATTAAAGTGACGTCAAACAC
TGCTAAAGGAAAAATGTATGGATTATTTTCATGTAAATGCTAATTACATT
ACGATTTCCTTTTTACATACCTAATAAAAGTACATTTACGATTAATGTAA
GTCTGGAAAACAAACCATTTTACTATTCCTAAGGAGCAATATACTATCAT
CAGACCTTTTGTTTGGTAAAATGATAAGGATTCCTCGTTATATGATAGTA
AAACAGAACAGCATCCAGTGTCACCTTTACAGATATAGCTTCATTAAATA
TTTGTCTTGTCGTAGGTCACAGTGGAAATGTCTATATCGAAGTAATTTAT
TTCAGCTCACTTGCAACATTCTTACATTCGGACAGCTTGAACAGAATGTT
AAGTCGAGTGAACGTTGTAAGAATGTAAGCCTGTCGAACTTGTCTTACAA
TATGGAATCACAATAATTTCAGGCTTGCCTCCAGAAAAACCTAAAAATTT
ATACCTTAGTGTTATTAAAGTCCGAACGGAGGTCTTTTTGGATTTTTAAA
GAGTTGCATTGTGAACGAGGGGAAGAAAATGAGGTGTGAGTGGGATGGTG
CTCAACGTAACACTTGCTCCCCTTCTTTTACTCCACACTCACCCTACCAC
GAAGGGAAACACACTTGGAGACAAACTTCACTTTAAAATCTGAATGGGCA
CTTCCCTTTGTGTGAACCTCTGTTTGAAGTGAAATTTTAGACTTACCCGT
ACACACAAGTTTGCTGATTGCAAAGCAAAACGTGACACCCCCACCTCATG
TGTGTGTTCAAACGACTAACGTTTCGTTTTGCACTGTGGGGGTGGAGTAC
CACTGTTGATTATTCTACTGTGTATTTTGTCAACATTGAAGTCTGGGTAG
GTGACAACTAATAAGATGACACATAAAACAGTTGTAACTTCAGACCCATC
AAGCAGAGAATGCCCTTGGGAAGGTTACATCAGATCATATCAATTTTGAT
TTCGTCTCTTACGGGAACCCTTCCAATGTAGTCTAGTATAGTTAAAACTA
CCTGTATATAAAGTGAAGCCCAATCCGCCACATAATTTATCAGTGATCAA
GGACATATATTTCACTTCGGGTTAGGCGGTGTATTAAATAGTCACTAGTT
CTCAGAGGAACTGTCTAGTATCTTAAAATTGACATGGACCAACCCAAGTA
GAGTCTCCTTGACAGATCATAGAATTTTAACTGTACCTGGTTGGGTTCAT
TTAAGAGTGTTATAATACTAAAATATAACATTCAATATAGGACCAAAGAT
AATTCTCACAATATTATGATTTTATATTGTAAGTTATATCCTGGTTTCTA
GCCTCAACTTGGAGCCAGATTCCTCCTGAAGACACAGCATCCACCCGATC
CGGAGTTGAACCTCGGTCTAAGGAGGACTTCTGTGTCGTAGGTGGGCTAG
TTCATTCACTGTCCAAGACCTTAAACCTTTTACAGAATATGTGTTTAGGA
AAGTAAGTGACAGGTTCTGGAATTTGGAAAATGTCTTATACACAAATCCT
TTCGCTGTATGAAGGAAGATGGTAAGGGATACTGGAGTGACTGGAGTGAA
AAGCGACATACTTCCTTCTACCATTCCCTATGACCTCACTGACCTCACTT
GAAGCAAGTGGGATCACCTATGAAGATAGACCATCTAAAGCACCAAGTTT
CTTCGTTCACCCTAGTGGATACTTCTATCTGGTAGATTTCGTGGTTCAAA
CTGGTATAAAATAGATCCATCCCATACTCAAGGCTACAGAACTGTACAAC
GACCATATTTTATCTAGGTAGGGTATGAGTTCCGATGTCTTGACATGTTG
TCGTGTGGAAGACATTGCCTCCTTTTGAAGCCAATGGAAAAATCTTGGAT
AGCACACCTTCTGTAACGGAGGAAAACTTCGGTTACCTTTTTAGAACCTA
TATGAAGTGACTCTCACAAGATGGAAATCACATTTACAAAATTACACAGT
ATACTTCACTGAGAGTGTTCTACCTTTAGTGTAAATGTTTTAATGTGTCA
TAATGCCACAAAACTGACAGTAAATCTCACAAATGATCGCTATCTAGCAA
ATTACGGTGTTTTGACTGTCATTTAGAGTGTTTACTAGCGATAGATCGTT
CCCTAACAGTAAGAAATCTTGTTGGCAAATCAGATGCAGCTGTTTTAACT
GGGATTGTCATTCTTTAGAACAACCGTTTAGTCTACGTCGACAAAATTGA
ATCCCTGCCTGTGACTTTCAAGCTACTCACCCTGTAATGGATCTTAAAGC
TAGGGACGGACACTGAAAGTTCGATGAGTGGGACATTACCTAGAATTTCG
ATTCCCCAAAGATAACATGCTTTGGGTGGAATGGACTACTCCAAGGGAAT
```

Figure 2 (*Continued*)

```
TAAGGGGTTTCTATTGTACGAAACCCACCTTACCTGATGAGGTTCCCTTA
CTGTAAAGAAATATATACTTGAGTGGTGTGTGTTATCAGATAAAGCACCC
GACATTTCTTTATATATGAACTCACCACACACAATAGTCTATTTCGTGGG
TGTATCACAGACTGGCAACAAGAAGATGGTACCGTGCATCGCACCTATTT
ACATAGTGTCTGACCGTTGTTCTTCTACCATGGCACGTAGCGTGGATAAA
AAGAGGGAACTTAGCAGAGAGCAAATGCTATTTGATAACAGTTACTCCAG
TTCTCCCTTGAATCGTCTCTCGTTTACGATAAACTATTGTCAATGAGGTC
TATATGCTGATGGACCAGGAAGCCCTGAATCCATAAAGGCATACCTTAAA
ATATACGACTACCTGGTCCTTCGGGACTTAGGTATTTCCGTATGGAATTT
CAAGCTCCACCTTCCAAAGGACCTACTGTTCGGACAAAAAAAGTAGGGAA
GTTCGAGGTGGAAGGTTTCCTGGATGACAAGCCTGTTTTTTTCATCCCTT
AAACGAAGCTGTCTTAGAGTGGGACCAACTTCCTGTTGATGTTCAGAATG
TTTGCTTCGACAGAATCTCACCCTGGTTGAAGGACAACTACAAGTCTTAC
GATTTATCAGAAATTATACTATATTTTATAGAACCATCATTGGAAATGAA
CTAAATAGTCTTTAATATGATATAAAATATCTTGGTAGTAACCTTTACTT
ACTGCTGTGAATGTGGATTCTTCCCACACAGAATATACATTGTCCTCTTT
TGACGACACTTACACCTAAGAAGGGTGTGTCTTATATGTAACAGGAGAAA
GACTAGTGACACATTGTACATGGTACGAATGGCAGCATACACAGATGAAG
CTGATCACTGTGTAACATGTACCATGCTTACCGTCGTATGTGTCTACTTC
GTGGGAAGGATGGTCCAGAATTCACTTTTACTACCCCAAAGTTTGCTCAA
CACCCTTCCTACCAGGTCTTAAGTGAAAATGATGGGGTTTCAAACGAGTT
GGAGAAATTGAAGCCATAGTCGTGCCTGTTTGCTTAGCATTCCTATTGAC
CCTCTTTAACTTCGGTATCAGCACGGACAAACGAATCGTAAGGATAACTG
AACTCTTCTGGGAGTGCTGTTCTGCTTTAATAAGCGAGACCTAATTAAAA
TTGAGAAGACCCTCACGACAAGACGAAATTATTCGCTCTGGATTAATTTT
AACACATCTGGCCTAATGTTCCAGATCCTTCAAAGAGTCATATTGCCCAG
TTGTGTAGACCGGATTACAAGGTCTAGGAAGTTTCTCAGTATAACGGGTC
TGGTCACCTCACACTCCTCCAAGGCACAATTTTAATTCAAAAGATCAAAT
ACCAGTGGAGTGTGAGGAGGTTCCGTGTTAAAATTAAGTTTTCTAGTTTA
GTATTCAGATGGCAATTTCACTGATGTAAGTGTTGTGGAAATAGAAGCAA
CATAAGTCTACCGTTAAAGTGACTACATTCACAACACCTTTATCTTCGTT
ATGACAAAAAGCCTTTTCCAGAAGATCTGAAATCATTGGACCTGTTCAAA
TACTGTTTTTCGGAAAAGGTCTTCTAGACTTTAGTAACCTGGACAAGTTT
AAGGAAAAAATTAATACTGAAGGACACAGCAGTGGTATTGGGGGGTCTTC
TTCCTTTTTTAATTATGACTTCCTGTGTCGTCACCATAACCCCCCAGAAG
ATGCATGTCATCTTCTAGGCCAAGCATTTCTAGCAGTGATGAAAATGAAT
TACGTACAGTAGAAGATCCGGTTCGTAAAGATCGTCACTACTTTTACTTA
CTTCACAAAACACTTCGAGCACTGTCCAGTATTCTACCGTGGTACACAGT
GAAGTGTTTTGTGAAGCTCGTGACAGGTCATAAGATGGCACCATGTGTCA
GGCTACAGACACCAAGTTCCGTCAGTCCAAGTCTTCTCAAGATCCGAGTC
CCGATGTCTGTGGTTCAAGGCAGTCAGGTTCAGAAGAGTTCTAGGCTCAG
TACCCAGCCCTTGTTAGATTCAGAGGAGCGGCCAGAAGATCTACAATTAG
ATGGGTCGGGAACAATCTAAGTCTCCTCGCCGGTCTTCTAGATGTTAATC
TAGATCATGTAGATGGCGGTGATGGTATTTTGCCCAGGCAACAGTACTTC
ATCTAGTACATCTACCGCCACTACCATAAAACGGGTCCGTTGTCATGAAG
AAACAGAACTGCAGTCAGCATGAATCCAGTCCAGATATTTCACATTTTGA
TTTGTCTTGACGTCAGTCGTACTTAGGTCAGGTCTATAAAGTGTAAAACT
AAGGTCAAAGCAAGTTTCATCAGTCAATGAGGAAGATTTTGTTAGACTTA
TTCCAGTTTCGTTCAAAGTAGTCAGTTACTCCTTCTAAAACAATCTGAAT
AACAGCAGATTTCAGATCATATTTCACAATCCTGTGGATCTGGGCAAATG
TTGTCGTCTAAAGTCTAGTATAAAGTGTTAGGACACCTAGACCCGTTTAC
AAAATGTTTCAGGAAGTTTCTGCAGCAGATGCTTTTGGTCCAGGTACTGA
TTTTACAAAGTCCTTCAAAGACGTCGTCTACGAAAACCAGGTCCATGACT
GGGACAAGTAGAAAGATTTGAAACAGTTGGCATGGAGGCTGCGACTGATG
CCCTGTTCATCTTTCTAAACTTTGTCAACCGTACCTCCGACGCTGACTAC
AAGGCATGCCTAAAAGTTACTTACCACAGACTGTACGGCAAGGCGGCTAC
TTCCGTACGGATTTTCAATGAATGGTGTCTGACATGCCGTTCCGCCGATG
ATGCCTCAGTGAAGGACTAGTAGTTCCTGCTACAACTTCAGCAGTACCTA
TACGGAGTCACTTCCTGATCATCAAGGACGATGTTGAAGTCGTCATGGAT
TAAAGTAAAGCTAAAATGATTTTATCTGTGAATTC
ATTTCATTTCGATTTTACTAAAATAGACACTTAAG
```

PROTEIN

Accession numbers

Human IL-6Rα: 49726, 33846
Mouse IL-6Rα: 52693
Rat IL-6Rα: 111882
Human gp130: 729833, 106982, 186354
Mouse gp130: 729834, 2137360, 840817, 193592
Rat gp130: 729835, 348455

Sequence

See **Figure 3**.

Description of protein

See **Table 1**.

Relevant homologies and species differences

The overall homology between human and mouse IL-6Rα is 69% and 54% at the DNA and protein level, respectively (Yamasaki *et al.*, 1988; Sugita *et al.*, 1990).

The overall homology between human and mouse gp130 is 76.6% and 76.8% at the DNA and protein level, respectively (Hibi *et al.*, 1990; Saito *et al.*, 1992). The homologies in the extracellular and cytoplasmic regions between human and mouse gp130 are 72% and 85%, respectively. The transmembrane domains of human and mouse gp130 are identical. The homology between human and chick gp130 is 74% at the amino acid level (Geissen *et al.*, 1998).

Affinity for ligand(s)

Low-affinity binding sites for IL-6: $K_d = 1–6$ nM.
High-affinity binding sites for IL-6: $K_d = 40–70$ pM.

The low-affinity binding site for IL-6 is composed of IL-6Rα alone (Hibi *et al.*, 1990). The high-affinity binding site for IL-6 is composed of both IL-6Rα and gp130, although gp130 does not have the ability to bind IL-6 by itself (Taga *et al.*, 1989; Hibi *et al.*, 1990).

Cell types and tissues expressing the receptor

IL-6Rα is expressed in all the cell lines that respond to IL-6. These include T cell lymphoma and B cell

Figure 3 Amino acid sequences for human and mouse IL-6 receptor α and human and mouse gp130. Signal peptides are underlined; transmembrane regions are in bold.

Human IL-6 receptor α

MLAVGCALLAALLAAPGAALAPRRCPAQEVARGVLTSLPGDSVT
LTCPGVEPEDNATVHWVLRKPAAGSHPSRWAGMGRRLLLRSVQLHDSGNYSCYRAGRP
AGTVHLLVDVPPEEPQLSCFRKSPLSNVVCEWGPRSTPSLTTKAVLLVRKFQNSPAED
FQEPCQYSQESQKFSCQLAVPEGDSSFYIVSMCVASSVGSKFSKTQTFQGCGILQPDP
PANITVTAVARNPRWLSVTWQDPHSWNSSFYRLRFELRYRAERSKTFTTWMVKDLQHH
CVIHDAWSGLRHVVQLRAQEEFGQGEWSEWSPEAMGTPWTESRSPPAENEVSTPMQAL
TTNKDDDNILFRDSANATSLPVQD**SSSVPLPTFLVAGGSLAFGTLLCIAIVL**RFKKTW
KLRALKEGKTSMHPPYSLGQLVPERPRPTPVLVPLISPPVSPSSLGSDNTSSHNRPDA
RDPRSPYDISNTDYFFPR

Mouse IL-6 receptor α

MLTVGCTLLVALLAAPAVALVLGSCRALEVANGTVTSLPGATVT
LICPGKEAAGNVTIHWVYSGSQNREWTTTGNTLVLRDVQLSDTGDYLCSLNDHLVGTV
PLLVDVPPEEPKLSCFRKNPLVNAICEWRPSSTPSPTTKAVLFAKKINTTNGKSDFQV
PCQYSQQLKSFSCQVEILEGDKVYHIVSLCVANSVGSKSSHNEAFHSLKMVQPDPPAN
LVVSAIPGRPRWLKVSWQHPETWDPSYYLLQFQLRYRPVWSKEFTVLLLPVAQYQCVI
HDALRGVKHVVQVRGKEELDLGQWSEWSPEVTGTPWIAEPRTTPAGILWNPTQVSVED
SANHEDQYESSTEATSVLAPVQE**SSSMSLPTFLVAGGSLAFGLLLCVFIIL**RLKQKWK
SEAEKESKTTSPPPPPYSLGPLKPTFLLVPLLTPHSSGSDNTVNHSCLGVRDAQSPYD
NSNRDYLFPR

Human gp130

MLTLQTWVVQALFIFLTTESTGELLDPCGYISPESPVVQLHSNF
TAVCVLKEKCMDYFHVNANYIVWKTNHFTIPKEQYTIINRTASSVTFTDIASLNIQLT
CNILTFGQLEQNVYGITIISGLPPEKPKNLSCIVNEGKKMRCEWDGGRETHLETNFTL
KSEWATHKFADCKAKRDTPTSCTVDYSTVYFVNIEVWVEAENALGKVTSDHINFDPVY
KVKPNPPHNLSVINSEELSSILKLTWTNPSIKSVIILKYNIQYRTKDASTWSQIPPED
TASTRSSFTVQDLKPFTEYVFRIRCMKEDGKGYWSDWSEEASGITYEDRPSKAPSFWY
KIDPSHTQGYRTVQLVWKTLPPFEANGKILDYEVTLTRWKSHLQNYTVNATKLTVNLT
NDRYLATLTVRNLVGKSDAAVLTIPACDFQATHPVMDLKAFPKDNMLWVEWTTPRESV
KKYILEWCVLSDKAPCITDWQQEDGTVHRTYLRGNLAESKCYLITVTPVYADGPGSPE
SIKAYLKQAPPSKGPTVRTKKVGKNEAVLEWDQLPVDVQNGFIRNYTIFYRTIIGNET
AVNVDSSHTEYTLSSLTSDTLYMVRMAAYTDEGGKDGPEFTFTTPKFAQGEIE**AIVVP**
VCLAFLLTTLLGVLFCFNKRDLIKKHIWPNVPDPSKSHIAQWSPHTPPRHNFNSKDQM
YSDGNFTDVSVVEIEANDKKPFPEDLKSLDLFKKEKINTEGHSSGIGGSSCMSSSRPS
ISSSDENESSQNTSSTVQYSTVVHSGYRHQVPSVQVFSRSESTQPLLDSEERPEDLQL
VDHVDGGDGILPRQQYFKQNCSQHESSPDISHFERSKQVSSVNEEDFVRLKQQISDHI
SQSCGSGQMKMFQEVSAADAFGPGTEGQVERFETVGMEAATDEGMPKSYLPQTVRQGG
YMPQ

Mouse gp130

MSAPRIWLAQALLFFLTTESIGQLLEPCGYIYPEFPVVQRGSNF
TAICVLKEACLQHYYVNASYIVWKTNHAAVPREQVTVINRTTSSVTFTDVVLPSVQLT
CNILSFGQIEQNVYGVTMLSGFPPDKPTNLTCIVNEGKNMLCQWDPGRETYLETNYTL
KSEWATEKFPDCQSKHGTSCMVSYMPTYYVNIEVWVEAENALGKVSSESINFDPVDKV
KPTPPYNLSVTNSEELSSILKLSWVSSGLGGLLDLKSDIQYRTKDASTWIQVPLEDTM
SPRTSFTVQDLKPFTEYVFRIRSIKDSGKGYWSDWSEEASGTTYEDRPSRPPSFWYKT
NPSHGQEYRSVRLIWKALPLSEANGKILDYEVILTQSKSVSQTYTVTGTELTVNLTND
RYVASLAARNKVGKSAAAVLTIPSPHVTAAYSVVNLKAFPKDNLLWVEWTPPPKPVSK
YILEWCVLSENAPCVEDWQQEDATVNRTHLRGRLLESKCYQITVTPVFATGPGGSESL
KAYLKQAAPARGPTVRTKKVGKNEAVLAWDQIPVDDQNGFIRNYSISYRTSVGKEMVV
HVDSSHTEYTLSSLSSDTLYMVRMAAYTDEGGKDGPEFTFTTPKFAQGEIE**AIVVPVC**
LAFLLTTLLGVLFCFNKRDLIKKHIWPNVPDPSKSHIAQWSPHTPPRHNFNSKDQMYS
DGNFTDVSVVEIEANNKKPCPDDLKSVDLFKKEKVSTEGHSSGIGGSSCMSSSRPSIS
SNEENESAQSTASTVEYSTVVHSGYRHQVPSVQVFSRSESTQPLLDSEERPEDLQLVD
SVDGGDEILPRQPYFKQNCSQPEACPEISHFERSNQVLSGNEEDFVRLKQQQVSDHIS
QPYGSEQRRLFQEGSTADALGTGADGQMERFESVGMETTIDEEIPKSYLPQTVRQGGY
MPQ

Table 1 Composition of human IL-6 receptor α and gp130

	Human IL-6Rα	Human gp130
Composition (amino acids)		
Precursor	468	918
Mature form	449	896
Extracellular domain	339	597
Transmembrane domain	28	22
Cytoplasmic domain	82	277
Molecular weight (kDa)		
Predicted	49.9	101
Observed	80	130–160

lymphoma, multiple myeloma (plasmacytoma), hepatoma, hepatocellular carcinoma, and glioma cell lines (Taga *et al.*, 1987; Sugita *et al.*, 1990). IL-6Rα is expressed in spleen, liver, lung, and thymus strongly in mice (Sugita *et al.*, 1990). The size of mRNA for human and mouse IL-6Rα is about 5.5 kb (Sugita *et al.*, 1990). In a mouse plasmacytoma line P3U1, the cytoplasmic domain of IL-6Rα was replaced by a part of LTR intracisternal A particle gene and the mRNA size of the aberrant receptor is 1.8 kb (Sugita *et al.*, 1990).

gp130 is ubiquitously expressed (Hibi *et al.*, 1990; Saito *et al.*, 1992). Two mRNAs for gp130 are detected by northern blotting. The size of gp130 mRNAs are 7 kb and 10 kb. The relative ratio of 7 kb and 10 kb gp130 mRNA varies among tissues: in heart, spleen, and lung; expression of 10 kb mRNA was much lower than 7 kb mRNA (Saito *et al.*, 1992).

Release of soluble receptors

Both IL-6Rα and gp130 have soluble forms: 30–70 ng/mL of soluble IL-6Rα is found in human serum (Honda *et al.*, 1992; Muller-Newen *et al.*, 1996). Soluble IL-6 Rα was reported to be elevated in serum of *multiple myeloma* (Gaillard *et al.*, 1993), *juvenile chronic arthritis* (Keul *et al.*, 1998), *HIV* (Honda *et al.*, 1992), and *Graves' disease* (Salvi *et al.*, 1996). Although soluble IL-6Rα was generated by shedding *in vitro* (Mullberg *et al.*, 1993), it is possible that it is generated by alternative splicing (Lust *et al.*, 1992; Muller-Newen *et al.*, 1996). It was reported that a complex of soluble IL-6Rα and IL-6 has agonistic effects on gp130 (Taga *et al.*, 1989). The complex of IL-6 and soluble IL-6Rα can act on cells expressing only gp130 and such a mechanism, which is called receptor conversion, may generate functional diversity of cytokines *in vivo* (Hirano *et al.*, 1997). In fact, double transgenic mice expressing human IL-6 and soluble IL-6Rα showed *myocardial hypertrophy* (Hirota *et al.*, 1995) and extramedullary expansion of hematopoietic progenitors (Peters *et al.*, 1997; Schirmacher *et al.*, 1998). Furthermore, the fusion protein consisting of IL-6 and the extracellular domain of IL-6Rα linked together by a flexible peptide chain was shown to stimulate hematopoietic progenitor cells effectively (Fischer *et al.*, 1997). However, it is not yet clear whether soluble IL-6Rα acts positively on gp130 signaling under physiological conditions. Soluble gp130 was also found in human serum at a concentration of 300 ng/mL (Narazaki *et al.*, 1993). Soluble gp130 was reported to be generated by alternative splicing (Diamant *et al.*, 1997).

SIGNAL TRANSDUCTION

Associated or intrinsic kinases

Janus kinases, JAK1, JAK2, and TYK2 constitutively associate with gp130 (Lutticken *et al.*, 1994; Narazaki *et al.*, 1994; Stahl *et al.*, 1994). The structure of Janus kinases is shown in **Figure 4**. Janus kinases contain a tyrosine kinase domain in their C-termini. Adjacent to the kinase domain, they have a kinase-like domain, which is similar to the authentic kinase domain but lacks several amino acids conserved among the kinase domains (Ihle, 1995). Upon stimulation of IL-6-related cytokines, these Janus kinases are tyrosine phosphorylated and become activated, resulting in tyrosine phosphorylation of the cytoplasmic domain of gp130. The box 1/box 2 domain in the membrane-proximal regions of the cytoplasmic domain is involved in the interaction between gp130 and the JAKs. Mutations of the box 1 region reduces the interaction between gp130 and JAKs (Tanner *et al.*, 1995), and abrogates ligand-dependent activation of JAK1 and JAK2 (Narazaki *et al.*, 1994) and downstream signaling (Murakami *et al.*, 1991; Lai *et al.*, 1995a). Truncations of the box 2 region also reduced the interaction between gp130 and JAK2, and abrogated the biological functions mediated by gp130 (Murakami *et al.*, 1991; Lai *et al.*, 1995a). However, JAKs were reported to interact with gp130 lacking the box 2 region when they are overexpressed (Lai *et al.*, 1995a; Tanner *et al.*, 1995). These indicate that the box 1 region is required and sufficient for the JAK binding to some extent, and the box 2 region

Figure 4 Structure of JAK and STAT. A JAK contains seven domains (JAK homology: JH1–7) conserved among JAKs. The JH1 and JH2 domains correspond to the tyrosine kinase and kinase-like (pseudokinase) domains, respectively. A STAT contains an SH2 domain, an SH3-like domain (functional significance unknown). Y (tyrosine) indicates a phosphoacceptor site for JAKs. S indicates a serine residue, the phosphorylation of which is involved in the activation of STATs.

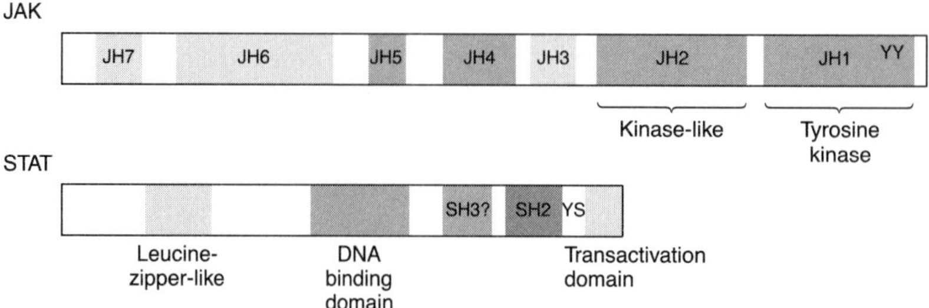

may support the interaction or should be necessary for the activation of the JAKs. Although JAK1, JAK2, and TYK2 interact with and are activated by gp130, it is not clear which JAKs are involved in downstream signaling and biological functions and to what extent.

It was demonstrated that tyrosine phosphorylation of gp130 and STAT3 induced by soluble IL-6Rα and IL-6 was significantly reduced in a JAK1-deficient fibrosarcoma cell line but not in a JAK2-deficient cell line (Guschin *et al.*, 1995). This is also consistent with data from knockout experiments. Several types of cells derived from JAK1-deficient mice did not respond to the IL-6-related cytokines (Rodig *et al.*, 1998), but JAK2-deficient fibroblasts and embryonic stem (ES) cells did respond to IL-6 and LIF, respectively (Neubauer *et al.*, 1998; Parganas *et al.*, 1998). Furthermore, the embryonic lethal phenotype of JAK1-deficient mice resembled that of gp130-deficient mice (Yoshida *et al.*, 1996; Rodig *et al.*, 1998). Further analysis will be required to reveal which JAK is involved in individual biological responses elicited by gp130.

In addition to Janus kinase, several tyrosine kinases were shown to interact with gp130 directly or through the JAKs in certain cells. Hck, a Src-family tyrosine kinase, was reported to associate with gp130, and was activated in response to LIF in ES cells (Ernst *et al.*, 1994). Tec and Btk, which are distant members of the Src-family tyrosine kinases and contain a pleckstrin homology (PH) domain, were reported to associate with gp130 through the JAKs and tyrosine phosphorylated upon the stimulation of gp130 (Matsuda *et al.*, 1995b; Takahashi-Tezuka *et al.*, 1997). Fes was also reported to interact with and was activated by gp130 (Matsuda *et al.*, 1995a). It was shown that erbB2 interacts with gp130 and the activity of erbB2 is necessary for ERK/MAP kinase activation in *prostate carcinoma* cells (Qiu *et al.*, 1998). The biological significance of these tyrosine kinases remains to be elucidated.

Cytoplasmic signaling cascades

The current model of gp130-mediated signal transduction is illustrated in **Figure 5** (also see review by Hirano *et al.*, 1997). Once the ligand is bound to the receptor complex, gp130 becomes a homodimer, and gp130-associated JAKs come close and transphosphorylate each other. It is believed that transphosphorylation of JAKs activates its kinase activity. In the case of JAK2 and Tyk2, phosphorylation of the two tyrosines in the activation loop of JAKs was shown to be required for tyrosine kinase activity (Gauzzi *et al.*, 1996; Feng *et al.*, 1997). The gp130-associated JAKs, once activated by the transphosphorylation, phosphorylate tyrosines in the cytoplasmic domain of gp130. gp130 contains six tyrosines in the cytoplasmic region (Y683, Y759, Y767, Y814, Y905, and Y915 in human gp130; the numbers are indicated amino acids from the translational initiation site). It is not known which tyrosines are preferentially phosphorylated on stimulation. However, phosphorylated tyrosines on gp130 are known to recruit signal transducing molecules such as SHP-2 and STATs (described below), resulting in the tyrosine phosphorylation of these molecules by the gp130-associated JAKs. Experiments of truncation and point mutations in

Figure 5 Model of gp130-mediated signal transduction.

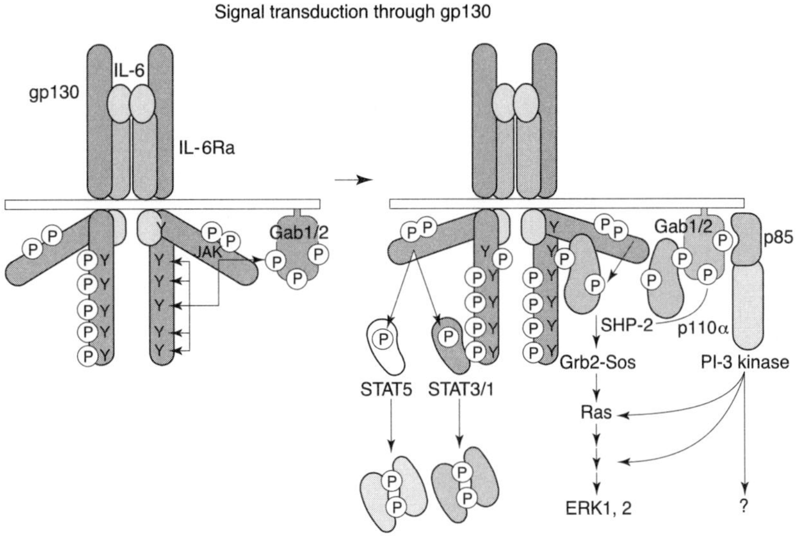

gp130 revealed the roles of individual tyrosine for downstream signaling. Y683 is not significantly phosphorylated since a truncated mutant gp130, which possesses the membrane-proximal 68 amino acid containing only Y683, could not be tyrosyl phosphorylated in spite of the tyrosine phosphorylation of JAKs (Fukada *et al.*, 1996). Phosphorylation of Y759 was shown to be required for tyrosine phosphorylation of SHP-2, a protein tyrosine phosphatase (Stahl *et al.*, 1995; Fukada *et al.*, 1996; Yamanaka *et al.*, 1996). The four tyrosines (Y767, Y814, Y905, and Y915) in the C-terminus have a glutamine at position +3 of tyrosines (YXXQ) and phosphorylation of any one of these tyrosines has been shown to be required for tyrosine phosphorylation and activation of STAT3 (Stahl *et al.*, 1995; Fukada *et al.*, 1996; Gerhartz *et al.*, 1996; Yamanaka *et al.*, 1996).

The two tyrosines in the carboxy-end (Y905 and Y915) among these four tyrosines have a proline at position +2 of tyrosines (YXPQ) and their phosphorylation was further reported to be necessary for tyrosine phosphorylation of STAT1 (Gerhartz *et al.*, 1996). In addition to SHP-2, STATs, and tyrosine kinases, there are several proteins that are reported to be tyrosyl phosphorylated on stimulation of gp130. These include Shc (Boulton *et al.*, 1994; Kumar *et al.*, 1994), IRS-1 (Burfoot *et al.*, 1997), Gab1 (Takahashi-Tezuka *et al.*, 1998), and Gab2 (Nishida *et al.*, 1999). Tyrosine phosphorylation of STAT5, Gab1, and Gab2 does not require tyrosine phosphorylation of gp130 (Lai *et al.*, 1995b; Fujitani *et al.*, 1997; Takahashi-Tezuka *et al.*, 1998; Nishida *et al.*, 1999).

Figure 6 Structure of SHP-2, Gab1, and Gab2. SHP-2 contains two SH2 and a phosphatase domains in its N- and C-terminal, respectively. Y (tyrosine) indicates possible phosphorylation sites. Both Gab1 and Gab2 have a pleckstrin homology (PH) domain, a c-Met-binding domain (MBD), and tyrosine-based motifs for SHP-2 and p85 PI-3 kinase. The region around the MBD contains the proline-rich PXXP motifs, which are known to bind the SH3 domains. The white bars indicate predicted tyrosine-phosphorylated sites, the surrounding sequences of which fit with the consensus-binding sequences for Grb2, Crk, and PLCγ.

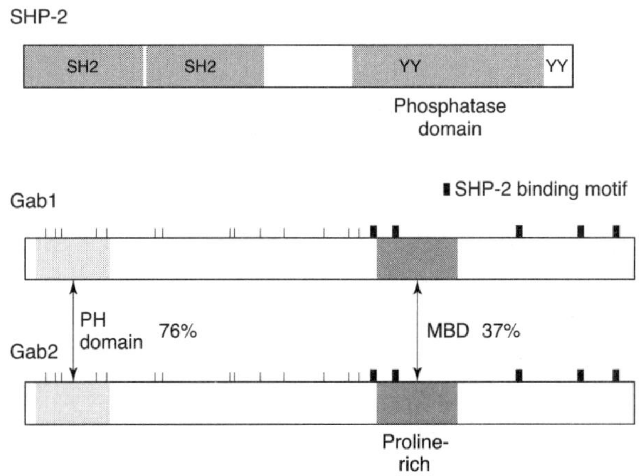

SHP-2 is a protein tyrosine phosphatase bearing two SH2 domains in the N-terminal region and a phosphatase domain in the C-terminal region (**Figure 6**) (Adachi *et al.*, 1996). In comparison to SHP-1, which is a negative regulator for signaling of a cytokine receptor (Klingmuller *et al.*, 1995) and B cell antigen receptor (Blery *et al.*, 1998; Maeda *et al.*, 1998), SHP-2 is a positive regulator of signaling from receptor tyrosine kinases and cytokine receptors (Tonks and Neel, 1996; Neel and Tonks, 1997). SHP-2 is tyrosyl phosphorylated on stimulation of gp130 (Stahl *et al.*, 1995; Fukada *et al.*, 1996). SHP-2 contains three to four YXNX (depending on splicing variants) motifs, which is the consensus sequence for Grb2 binding, in its C-terminal region (Feng *et al.*, 1993; Vogel *et al.*, 1993). Upon stimulation, tyrosine phosphorylated SHP-2 was reported to interact with Grb2 (Bennett *et al.*, 1994; Li *et al.*, 1994; Fukada *et al.*, 1996). Since Grb2 constitutively interacts with Sos, the GDP–GTP exchange factor for Ras, SHP-2 may act as an adapter molecule mediating gp130 signaling to the Ras pathway. The mutation of Y759 to a phenylalanine abrogated tyrosine phosphorylation of SHP-2, the interaction between SHP-2 and Grb2, and activation of an ERK/MAP kinase, a major target of the Ras pathway, suggest a role for SHP-2 in the Ras pathway (Fukada *et al.*, 1996). In addition to the possible adapter function of SHP-2, the catalytic activity of SHP-2 was shown to be necessary for transmitting signals to the ERK/MAP kinases in the case of receptor tyrosine kinases such as receptors for insulin, EGF, and FGF (Noguchi *et al.*, 1994; Tang *et al.*, 1995; Bennett *et al.*, 1996). In gp130 signaling, SHP-2 was shown to interact with Gab1 or Gab2 upon stimulation (Takahashi-Tezuka *et al.*, 1998; Nishida *et al.*, 1999). Tyrosyl phosphorylated Gab1 and Gab2 were shown to be dephosphorylated by SHP-2 *in vitro* (Nishida *et al.*, 1999). Gab1 and Gab2 are adapter proteins which display homology to *Drosophila* Dos (daughter of sevenless) (Figure 6) (Herbst *et al.*, 1996; Raabe *et al.*, 1996), a substrate for corkscrew, a *Drosophila* homolog of SHP-2 (Perkins *et al.*, 1992). They contain a PH domain, and proline-rich sequences including c-Met-binding domain (MBD), and tyrosine-based motifs that bind to SH2 domains when they are phosphorylated (Holgado-Madruga *et al.*, 1996; Weidner *et al.*, 1996).

Tyrosine phosphorylation of Gab1 and Gab2 was independent of tyrosine phosphorylation of gp130 (Takahashi-Tezuka *et al.*, 1998; Nishida *et al.*, 1999). They are likely phosphorylated by the gp130-associated JAKs, since overexpression of JAKs induced their phosphorylation. Gab1 and Gab2 interacted with not only SHP-2 but also the p85 PI-3 kinase on stimulation. Overexpression of Gab1 or Gab2 was reported to enhance the gp130-dependent activation of ERK/MAP kinases, suggesting that Gab family proteins act upstream of ERK/MAP kinases (Takahashi-Tezuka *et al.*, 1998; Nishida *et al.*, 1999). The mutation of Y759 disrupted the interaction between SHP-2 and Gab1, and Gab1 and PI-3 kinases, suggesting that Gab1-mediating signals are dependent on Y759 of gp130. Gab1-dependent ERK activation was shown to be inhibited by a dominant negative p85, wortmannin, an inhibitor for PI-3 kinases, or a dominant negative Ras (Takahashi-Tezuka *et al.*, 1998). These results suggest that Gab family proteins mediate signals from SHP-2 to PI-3 kinases to activate ERK/MAP kinases. Further analysis will be required to clarify this point.

A negative role of SHP-2 in gp130 signaling has also been reported. The mutation of Y759 was shown to prolong the activation of a STAT3-mediated pathway (Kim *et al.*, 1998). Since gp130 and JAK1 but not STAT3 could be substrates for SHP-2 *in vitro* (Nishida *et al.*, 1999), SHP-2 may distinguish gp130-mediated signals as a feedback mechanism.

STATs are transcription factors containing an SH2 domain (Figure 4) (Darnell *et al.*, 1994; Ihle and Kerr, 1995). Upon stimulation, STAT1 and STAT3 are recruited on gp130 and tyrosyl phosphorylated by the gp130-associated JAKs (Akira *et al.*, 1994; Zhong *et al.*, 1994; Guschin *et al.*, 1995). Tyrosine 701 of STAT1 and tyrosine 705 of STAT3 are phosphoacceptor sites by JAKs. Phosphorylated STAT1 and STAT3 form homodimers (STAT1/STAT1, STAT3/STAT3) or a heterodimer (STAT1/STAT3) through the interaction between the phosphotyrosines and the SH2 domains. It was also reported that STAT5 (STAT5a and STAT5b) is tyrosine phosphorylated independently of tyrosine phosphorylation of gp130 (Lai *et al.*, 1995a; Fujitani *et al.*, 1997). Tyrosine phosphorylation of STAT5 by JAKs was shown to be mediated in part by a direct interaction between STAT5 and the JH2 domain of JAKs (Fujitani *et al.*, 1997). In addition to tyrosine phosphorylation of STATs, serine phosphorylation of STAT1 and STAT3 was required for full transcriptional activity of the molecules (Boulton *et al.*, 1995; Nakajima *et al.*, 1995; Wen *et al.*, 1995). Serine 727 of both STAT1 and STAT3 is responsible for the phosphorylation site and their mutations were shown to reduce STATs-dependent transcription (Wen and Darnell, 1997). Serine 727 is located prior to proline and fits a part of the consensus sequence of phosphoacceptor sites by ERK/MAP kinase. In fact, serine 727 could be phosphorylated by ERK/MAP kinase *in vitro* (Wen *et al.*, 1995). Upon growth hormone stimulation, STAT3 is phosphorylated dependent on ERK/MAP kinases, whereas serine phosphorylation of STAT3 induced by IL-6 is independent of ERK/MAP kinase (Chung *et al.*, 1997; Ng and Cantrell, 1997). Several groups reported that the gp130-mediated serine phosphorylation of STAT3 was inhibited by a protein kinase inhibitor H7 (Boulton *et al.*, 1995; Nakajima *et al.*, 1995). The responsible kinase remains to be elucidated.

After dimers are formed, they enter the nucleus. It has been reported that nuclear transport of STAT1 is mediated by the importin-α family member NPI-1 and it depends on tyrosine phosphorylation of STAT1 (Sekimoto *et al.*, 1997) and the GTPase activity of Ran (Sekimoto *et al.*, 1996). After dimerization and nuclear translocation, STATs bind the specific sequence of DNA through their DNA-binding domains located in the middle of the molecules (amino acids 300–500) (Horvath *et al.*, 1995), and activate a series of transcriptions as described below.

The biological functions of the individual signal transduction cascades have been analyzed using mutant gp130 and dominant negative forms of STAT3.

1. Mouse leukemia cell line M1 is differentiated from macrophages on gp130 stimulation, and the differentiation depends on the activity of STAT3 (Nakajima *et al.*, 1996; Yamanaka *et al.*, 1996). The STAT3-mediated pathway was shown to be involved in downregulation of c-myc and c-myb, and upregulation of p19INK4D (Nakajima *et al.*, 1996; Yamanaka *et al.*, 1996; Narimatsu *et al.*, 1997).
2. Mouse proB cell line BAF-B03 proliferates in response to gp130 stimulation. Mutations of tyrosine 759 disrupted a G_2-M cell cycle transition. A mutation of YXXQ motifs or expression of dominant negative forms of STAT3 rendered cells susceptible to apoptosis and inhibited induction of *bcl-2*, suggesting that STAT3 mediates anti-apoptosis in part through expression of *bcl-2*. It was also reported that gp130-mediated G_1–S cell cycle transition in BAF-B03 cells depends on the activity of STAT3 (Fukada *et al.*, 1996, 1998). STAT3 controls the G_1–S transition through upregulation of cyclin D2, D3, A, and cdc25A, and downregulation of p21 and p27 in BAF-B03 cells (Fukada *et al.*, 1998).
3. Differentiation of PC12 cells to neuron-like cells in response to IL-6 depends on the tyrosine 759-mediated Ras/MAP kinase pathway (Ihara *et al.*, 1997). Intriguingly, STAT3 was shown to regulate gp130-mediated PC12 differentiation negatively (Ihara *et al.*, 1997).
4. LIF maintains the pluripotential activity of mouse embryonic stem cells. It has been reported that this activity is dependent on the activity of STAT3 (Niwa *et al.*, 1998).
5. Differentiation from cerebral cortical precursor cells to type I astrocytes by CNTF was reported to depend on the STAT3-mediated pathway (Bonni *et al.*, 1997).

DOWNSTREAM GENE ACTIVATION

Transcription factors activated

STATs: STAT3, STAT1, and STAT5

Dimers of STATs bind the specific DNA element (TTN5AA) (Seidel *et al.*, 1995). STAT1/STAT3 dimer preferentially binds the TTCN3GAA sequence, which presents in the promoter of acute-phase protein genes (Wegenka *et al.*, 1993; Seidel *et al.*, 1995). STATs binds DNA cooperatively with other transcription factors. Promoter regions of *junB*, *IRF-1*, and *Stat3* genes contain both a STAT-binding site and a CRE-like element (Kojima *et al.*, 1996; Ichiba *et al.*, 1998), and these genes were shown to be regulated by STAT3 and unidentified transcription factors binding the CRE-like element. Functional interaction of STAT3 and other transcription factors was also reported, including C/EBPβ (Schumann *et al.*, 1996), NF-IL6 (Stephanou *et al.*, 1998), NFκB (Brown *et al.*, 1995), AP-1 (Schumann *et al.*, 1996; Korzus *et al.*, 1997; Symes *et al.*, 1997), and glucocorticoid receptor (Zhang *et al.*, 1997). STAT3 has an alternatively splicing product, named STAT3β. The C-terminal 55 amino acids of STAT3 were replaced by five unrelated amino acids in STAT3β (Schaefer *et al.*, 1995; Caldenhoven *et al.*, 1996). In certain cell lines, STAT3β was reported to be a constitutive activator (Sasse *et al.*, 1997). Furthermore, it was shown that STAT3 interacts with c-Jun and synergistically activated gene expression (Schaefer *et al.*, 1995). However, it was also reported that STAT3β acts as a dominant negative protein against STAT3 in a certain condition (Caldenhoven *et al.*, 1996). The physiological roles of STAT3β remain to be elucidated.

The mechanism by which STATs activate transcription is not documented clearly. Transcriptional coactivators CREB-binding protein (CBP) and p300 have been shown to interact with STAT1 and STAT2 and to be involved in STAT-dependent transcription (Bhattacharya *et al.*, 1996; Zhang *et al.*, 1996b; Horvai *et al.*, 1997). Both the N-terminal and C-terminal domains of STAT1 interact with CBP/p300 (Zhang *et al.*, 1996b). CBP and p300 act as bridging factors linking transcription factors to general transcriptional machinery (Chakravarti *et al.*, 1996). It is also reported that CBP/p300 or their associated factors possess histone acetyl transferase activity and regulate chromatin structure for gene regulation (Ogryzko *et al.*, 1996).

The Ras/MAP Kinase Pathway and Others

There are few reports describing transcription factors other than STATs in gp130 signaling. As described above, however, Ras/MAP kinase pathways are activated on tyrosine phosphorylation of the tyrosine 759 of gp130 (Fukada *et al.*, 1996). Therefore, transcription factors controlled by ERK/MAP kinases should work to some extent. For instance, IL-6 induces the expression of *egr-1* in the mouse leukemia cell line M1. The expression of *egr-1* was dependent on tyrosine 759 (Yamanaka *et al.*, 1996) and the promoter of *egr-1* contains several serum-responsive elements (SREs) (Qureshi *et al.*, 1991), suggesting that gp130-mediated signals activate ERK/MAP kinases, induce phosphorylation of Elk-1, a binding partner for serum-responsive factor (SRF), by ERKs and activate SRE-dependent gene expression. Transcription factors downstream of the other signaling cascades such as PI-3 kinase in gp130 signaling have not been reported.

Genes induced

Acute-phase Protein Genes

C-reactive protein (Zhang *et al.*, 1996a), α_1-antichymotrypsin (Kordula *et al.*, 1998), α_2-macroglobulin (Wegenka *et al.*, 1993), lipopolysaccharide-binding protein (Schumann *et al.*, 1996), tissue inhibitor of metalloproteinases (Bugno *et al.*, 1995).

Immediate Early Genes

junB (Nakajima and Wall, 1991), *c-fos* (Hill and Treisman, 1995), *c-myc* (Fukada *et al.*, 1996; Kiuchi *et al.*, 1999), *egr-1* (Yamanaka *et al.*, 1996), IRF-1 (Harroch *et al.*, 1994).

Cell Cycle Regulators

p19IND4D (Narimatsu *et al.*, 1997), p18INK4C (Morse *et al.*, 1997), p21WAF/CIP (Morse *et al.*, 1997), cyclin D_2, cyline D3, cyclin A, cdc25A (Fukada *et al.*, 1998).

Others

C/EBPδ (Yamada *et al.*, 1997), interstitial collagenase (Korzus *et al.*, 1997), vasoactive intestinal peptide (Symes *et al.*, 1997), proopiomelanocortin (Ray *et al.*, 1996), hsp90 (Stephanou *et al.*, 1998), bcl-2 (Fukada *et al.*, 1996), bcl-x (Fujio *et al.*, 1997), gp130 (Saito *et al.*, 1992; O'Brien and Manolagas, 1997), STAT3 (Ichiba *et al.*, 1998), SOCS/JAB/SSI-1 (Endo *et al.*, 1997; Naka *et al.*, 1997; Starr *et al.*, 1997).

Promoter regions involved

Promoter regions of acute-phase protein genes contain two types of IL-6-responsive elements (Nakajima *et al.*, 1995; Hirano *et al.*, 1997). Type I IL-6-responsive element is a binding site for NF-IL6 (C/EBPβ) and other members of the C/EBP transcription factor family. Type II IL-6-responsive element, also termed acute-phase responsive element (APRE), is a binding site for STAT3.

As described above, promoters of junB, IRF-1, and STAT3 contain both a STAT3-binding and a CRE-like site in close proximity (Kojima *et al.*, 1996; Ichiba *et al.*, 1998). In the case of the junB promoter (termed JRE-IL-6), an unidentified 36 kDa protein forms a complex with STAT3 and binds this element (Kojima *et al.*, 1996).

The promoter region of *c-fos* contain several elements such as Sis-inducible element (SIE), SRE, and cyclic AMP-responsive element (CRE). Of these, SIE is a STAT-binding element and SRE is a binding site for a complex of SRF and a tertiary complex factor (TCF: Elk1 and SAP1/SAP2). SIE and SRE are likely targets for STATs and Ras/MAP kinase pathways activated by gp130.

The promoter of *c-myc* contains an E2F site, which is known to be regulated by E2F/DP family transcription factors and the retinoblastoma gene product RB. The E2F site in the promoter overlaps with a STAT3-binding site. Full activation of the c-*myc* gene by gp130 stimulation was shown to depend on STAT3, since a mutation of the YXXQ motif or dominant negative STAT3s inhibited gp130-mediated c-*myc* induction (Kiuchi *et al.*, 1999).

SOCS/JAB/SSI-1 is an inhibitor for JAKs (Endo *et al.*, 1997; Naka *et al.*, 1997; Starr *et al.*, 1997). It has an SH2 domain which directly binds tyrosyl phosphorylated JAK and inhibits the activity of JAKs. It is shown to be induced by gp130 stimulation through the STAT3-mediated pathway.

BIOLOGICAL CONSEQUENCES OF ACTIVATING OR INHIBITING RECEPTOR AND PATHOPHYSIOLOGY

Phenotypes of receptor knockouts and receptor overexpression mice

Knockout of IL-6Rα has not yet been reported.

Embryos homozygous for the gp130 disruption die between 12.5 days postcoitum (dpc) and term

(Yoshida *et al.*, 1996). The gp130-deficient embryos of later stages (16.5 dpc or later) displayed *cardiac hypertrophy*. The embryos have reduced numbers of pluripotential and committed hematopoietic progenitors in the liver, and differentiated lineages of T cells in thymus. The development of erythroid lineage cells was also impaired in some of the knockout embryos.

Postnatal inactivation of gp130 gene in mice was also performed by the Cre/loxP-mediated conditional targeting method (Betz *et al.*, 1998). The conditional gp130 mutants exhibited a thinning of the left and right ventricular wall of the heart. They displayed a reduction in thrombocyte numbers in the peripheral blood together with an increased number of circulating leukocytes. Recovery from *thrombocytopenia* was also impaired in the mutants. Red blood cells were normal in the mutant mice. The number of hematopoietic precursors (CFU-S, GM-CFU) were reduced in bone marrow to some extent but less than the conventional gp130 mutants. T cell content was reduced in the conditional gp130 mutants, but B cell development appeared to be normal. Increased susceptibility to viral (*vesicular stomatitis virus* and *vaccinia virus*) and bacterial infections (*Listeria monocytogenes*) was also observed in the conditional gp130 mutants.

With regard to the immune fraction, phenotypes of the gp130 mutants were similar to those of IL-6-deficient mice, indicating that IL-6 is the principal member of the gp130-dependent cytokine family in the immune system. The conditional gp130 mutants exhibited liver abnormalities: a decreased content of binucleated hepatocytes, increased number of Kupffer cells, a low content of endoplasmic reticulum, an increase in lipid vacuoles, an increase in laminar bodies, a widening of the Disse space and intercellular space between hepatocytes, dense clustering of mitochondria, reduction and morphological alteration of microvilli, and a shift in glycogen particles from the rosette α to the monoparticulated β form. Consistent with the morphological abnormalities of liver, the acute-phase reaction in the liver was also perturbed in the mutants. The gp130 mutants also displayed an age-dependent reduction of elastic fibers in the lungs, followed by the development of *emphysema*.

Expression of a Dominant Negative Form of gp130

Transgenic expression of a dominant negative form of gp130, which is a truncated form containing only 63 amino acids with box 1 and box 2 of the membrane-proximal cytoplasmic region, was performed (Kumanogoh *et al.*, 1997). The transgenic mice exhibited defects in antigen-specific antibody production of most immunoglobulin isotypes other than IgM after immunization with 2,4-dinitrophenol-conjugated ovalbumin.

Expression of Antisense gp130 RNA in Chick Embryos

When antisense gp130 RNA was introduced into chick sympathetic neurons through *retroviral infection*, a strong reduction in the number of VIP-expressing cells, but not of choline acetyltransferase (ChAT)- or tyrosine hydroxylase (a marker for adrenergic neuron)-expressing cells, was reported (Geissen *et al.*, 1998). This indicates that gp130 signaling is involved in the control of VIP expression during the development of cholinergic sympathetic neurons.

Overexpression of IL-6 and Soluble IL-6Rα

Transgenic mice overexpressing IL-6 and IL-6Rα exhibited *hepatocellular hyperplasia* (Schirmacher *et al.*, 1998), *plasmacytoma* formation (Schirmacher *et al.*, 1998), extramedullar hematopoiesis (Peters *et al.*, 1997; Schirmacher *et al.*, 1998), and *cardiac hypertrophy* (Hirota *et al.*, 1995).

THERAPEUTIC UTILITY

Effects of inhibitors (antibodies) to receptors

A humanized monoclonal antibody (PM-1) against human IL-6Rα was reported to inhibit IL-6 binding to the IL-6R (Hirata *et al.*, 1989). It was reported that humanized PM-1 had a therapeutic effect in *rheumatoid arthritis* (Yoshizaki *et al.*, 1998).

References

Adachi, M., Fischer, E. H., Ihle, J., Imai, K., Jirik, F., Neel, B., Pawson, T., Shen, S., Thomas, M., Ullrich, A., and Zhao, Z. (1996). Mammalian SH2-containing protein tyrosine phosphatases [letter]. *Cell* **85**, 15.

Akira, S., Nishio, Y., Inoue, M., Wang, X. J., Wei, S., Matsusaka, T., Yoshida, K., Sudo, T., Naruto, M., and Kishimoto, T. (1994). Molecular cloning of APRF, a novel IFN-stimulated gene factor 3 p91-related transcription factor involved in the gp130-mediated signaling pathway. *Cell* **77**, 63–71.

Bazan, J. F. (1990). Structural design and molecular evolution of a cytokine receptor superfamily. *Proc. Natl Acad. Sci. USA* **87**, 6934–6938.

Bennett, A. M., Tang, T. L., Sugimoto, S., Walsh, C. T., and Neel, B. G. (1994). Protein-tyrosine-phosphatase SHPTP2 couples platelet-derived growth factor receptor beta to Ras. *Proc. Natl Acad. Sci. USA* **91**, 7335–7339.

Bennett, A. M., Hausdorff, S. F., O'Reilly, A. M., Freeman, R. M., and Neel, B. G. (1996). Multiple requirements for SHPTP2 in epidermal growth factor-mediated cell cycle progression. *Mol. Cell Biol.* **16**, 1189–1202.

Betz, U. A. K., Bloch, W., van den Broek, M., Yoshida, K., Taga, T., Kishimoto, T., Addicks, K., Rajewsky, K., and Muller, W. (1998). Postnatally induced inactivation of gp130 in mice results in neurological, cardiac, hematopoietic, immunological, hepatic, and pulmonary defects. *J. Exp. Med.* **188**, 1955–1965.

Bhattacharya, S., Eckner, R., Grossman, S., Oldread, E., Arany, Z., D'Andrea, A., and Livingston, D. M. (1996). Cooperation of Stat2 and p300/CBP in signalling induced by interferon-alpha. *Nature* **383**, 344–347.

Blery, M., Kubagawa, H., Chen, C. C., Vely, F., Cooper, M. D., and Vivier, E. (1998). The paired Ig-like receptor PIR-B is an inhibitory receptor that recruits the protein-tyrosine phosphatase SHP-1. *Proc. Natl Acad. Sci. USA* **95**, 2446.

Bonni, A., Sun, Y., Nadal-Vicens, M., Bhatt, A., Frank, D. A., Rozovsky, I., Stahl, N., Yancopoulos, G. D., and Greenberg, M. E. (1997). Regulation of gliogenesis in the central nervous system by the JAK-STAT signaling pathway. *Science* **278**, 477–483.

Boulton, T. G., Stahl, N., and Yancopoulos, G. D. (1994). Ciliary neurotrophic factor/leukemia inhibitory factor/interleukin 6/oncostatin M family of cytokines induces tyrosine phosphorylation of a common set of proteins overlapping those induced by other cytokines and growth factors. *J. Biol. Chem.* **269**, 11648–11655.

Boulton, T. G., Zhong, Z., Wen, Z., Darnell, J.E. Jr., Stahl, N., and Yancopoulos, G. D. (1995). STAT3 activation by cytokines utilizing gp130 and related transducers involves a secondary modification requiring an H7-sensitive kinase. *Proc. Natl Acad. Sci. USA* **92**, 6915–6919.

Bravo, J., Staunton, D., Heath, J. K., and Jones, E. Y. (1998). Crystal structure of a cytokine-binding region of gp130. *EMBO J.* **17**, 1665–1674.

Brown, R. T., Ades, I. Z., and Nordan, R. P. (1995). An acute phase response factor/NF-kappa B site downstream of the junB gene that mediates responsiveness to interleukin-6 in a murine plasmacytoma. *J. Biol. Chem.* **270**, 31129–31135.

Bugno, M., Graeve, L., Gatsios, P., Koj, A., Heinrich, P. C., Travis, J., and Kordula, T. (1995). Identification of the interleukin-6/oncostatin M response element in the rat tissue inhibitor of metalloproteinases-1 (TIMP-1) promoter. *Nucl. Acids Res.* **23**, 5041–5047.

Burfoot, M. S., Rogers, N. C., Watling, D., Smith, J. M., Pons, S., Paonessaw, G., Pellegrini, S., White, M. F., and Kerr, I. M. (1997). Janus kinase-dependent activation of insulin receptor substrate 1 in response to interleukin-4, oncostatin M, and the interferons. *J. Biol. Chem.* **272**, 24183–24190.

Caldenhoven, E., van Dijk, T. B., Solari, R., Armstrong, J., Raaijmakers, J. A. M., Lammers, J. W. J., Koenderman, L., and de Groot, R. P. (1996). STAT3beta, a splice variant of transcription factor STAT3, is a dominant negative regulator of transcription. *J. Biol. Chem.* **271**, 13221–13227.

Chakravarti, D., LaMorte, V. J., Nelson, M. C., Nakajima, T., Schulman, I. G., Juguilon, H., Montminy, M., and Evans, R. M. (1996). Role of CBP/P300 in nuclear receptor signalling [see comments]. *Nature* **383**, 99–103.

Chung, J., Uchida, E., Grammer, T. C., and Blenis, J. (1997). STAT3 serine phosphorylation by ERK-dependent and -independent pathways negatively modulates its tyrosine phosphorylation. *Mol. Cell Biol.* **17**, 6508–6516.

Darnell, J.E. Jr., Kerr, I. M., and Stark, G. R. (1994). JAK-STAT pathways and transcriptional activation in response to IFNs and other extracellular signaling proteins. *Science* **264**, 1415–1421.

de Vos, A. M., Ultsch, M., and Kossiakoff, A. A. (1992). Human growth hormone and extracellular domain of its receptor: crystal structure of the complex. *Science* **255**, 306–312.

Diamant, M., Rieneck, K., Mechti, N., Zhang, X. G., Svenson, M., Bendtzen, K., and Klein, B. (1997). Cloning and expression of an alternatively spliced mRNA encoding a soluble form of the human interleukin-6 signal transducer gp130. *FEBS Lett.* **412**, 379–384.

Endo, T. A., Masuhara, M., Yokouchi, M., Suzuki, R., Sakamoto, H., Mitsui, K., Matsumoto, A., Tanimura, S., Ohtsubo, M., Misawa, H., Miyazaki, T., Leonor, N., Taniguchi, T., Fujita, T., Kanakura, Y., Komiya, S., and Yoshimura, A. (1997). A new protein containing an SH2 domain that inhibits JAK kinases. *Nature* **387**, 921–924.

Ernst, M., Gearing, D. P., and Dunn, A. R. (1994). Functional and biochemical association of Hck with the LIF/IL-6 receptor signal transducing subunit gp130 in embryonic stem cells. *EMBO J.* **13**, 1574–1584.

Feng, G. S., Hui, C. C., and Pawson, T. (1993). SH2-containing phosphotyrosine phosphatase as a target of protein-tyrosine kinases. *Science* **259**, 1607.

Feng, J., Witthuhn, B. A., Matsuda, T., Kohlhuber, F., Kerr, I. M., and Ihle, J. N. (1997). Activation of JAK2 catalytic activity requires phosphorylation of Y1007 in the kinase activation loop. *Mol. Cell Biol.* **17**, 2497–2501.

Fischer, M., Goldschmitt, J., Peschel, C., Brakenhoff, J. P., Kallen, K. J., Wollmer, A., Grotzinger, J., and Rose-John, S. (1997). I. A bioactive designer cytokine for human hematopoietic progenitor cell expansion. *Nature Biotechnol.* **15**, 142–145.

Fujio, Y., Kunisada, K., Hirota, H., Yamauchi-Takihara, K., and Kishimoto, T. (1997). Signals through gp130 upregulate bcl-x gene expression via STAT1-binding cis-element in cardiac myocytes. *J. Clin. Invest.* **99**, 2898–2905.

Fujitani, Y., Hibi, M., Fukada, T., Takahashi-Tezuka, M., Yoshida, H., Yamaguchi, T., Sugiyama, K., Yamanaka, Y., Nakajima, K., and Hirano, T. (1997). An alternative pathway for STAT activation that is mediated by the direct interaction between JAK and STAT. *Oncogene* **14**, 751–761.

Fukada, T., Hibi, M., Yamanaka, Y., Takahashi-Tezuka, M., Fujitani, Y., Yamaguchi, T., Nakajima, K., and Hirano, T. (1996). Two signals are necessary for cell proliferation induced by a cytokine receptor gp130: involvement of STAT3 in anti-apoptosis. *Immunity* **5**, 449–460.

Fukada, T., Ohtani, T., Yoshida, Y., Shirogane, T., Nishida, K., Nakajima, K., Hibi, M., and Hirano, T. (1998). STAT3 orchestrates contradictory signals in cytokine-induced G1 to S cell-cycle transition. *EMBO J.* **17**, 6670–6677.

Fukunaga, R., Ishizaka-Ikeda, E., Pan, C. X., Seto, Y., and Nagata, S. (1991). Functional domains of the granulocyte colony-stimulating factor receptor. *EMBO J.* **10**, 2855–2865.

Gaillard, J. P., Bataille, R., Brailly, H., Zuber, C., Yasukawa, K., Attal, M., Maruo, N., Taga, T., Kishimoto, T., and Klein, B. (1993). Increased and highly stable levels of functional soluble interleukin-6 receptor in sera of patients with monoclonal gammopathy. *Eur. J. Immunol.* **23**, 820–824.

Gauzzi, M. C., Velazquez, L., McKendry, R., Mogensen, K. E., Fellous, M., and Pellegrini, S. (1996). Interferon-alpha-dependent activation of Tyk2 requires phosphorylation of positive regulatory tyrosines by another kinase. *J. Biol. Chem.* **271**, 20494–20500.

Geissen, M., Heller, S., Pennica, D., Ernsberger, U., and Rohrer, H. (1998). The specification of sympathetic neurotransmitter phenotype depends on gp130 cytokine receptor signaling. *Development* **125**, 4791–4801.

Gerhartz, C., Heesel, B., Sasse, J., Hemmann, U., Landgraf, C., Schneider-Mergener, J., Horn, F., Heinrich, P. C., and Graeve, L. (1996). Differential activation of acute phase response factor/STAT3 and STAT1 via the cytoplasmic domain of the interleukin 6 signal transducer gp130. I. Definition of a novel phosphotyrosine motif mediating STAT1 activation. *J. Biol. Chem.* **271**, 12991–12998.

Grotzinger, J., Kurapkat, G., Wollmer, A., Kalai, M., and Rose-John, S. (1997). The family of the IL-6-type cytokines: specificity and promiscuity of the receptor complexes. *Proteins* **27**, 96–109.

Guschin, D., Rogers, N., Briscoe, J., Witthuhn, B., Watling, D., Horn, F., Pellegrini, S., Yasukawa, K., Heinrich, P., Stark, G. R., Ihle, J. N., and Kerr, I. M. (1995). A major role for the protein tyrosine kinase JAK1 in the JAK/STAT signal transduction pathway in response to interleukin-6. *EMBO J.* **14**, 1421–1429.

Harroch, S., Revel, M., and Chebath, J. (1994). Induction by interleukin-6 of interferon regulatory factor 1 (IRF-1) gene expression through the palindromic interferon response element pIRE and cell type-dependent control of IRF-1 binding to DNA. *EMBO J.* **13**, 1942–1949.

Heinrich, P. C., Behrmann, I., Muller-Newen, G., Schaper, F., and Graeve, L. (1998). Interleukin-6-type cytokine signalling through the gp130/JAK/STAT pathway. *Biochem. J.* **334**, 297–314.

Herbst, R., Carroll, P. M., Allard, J. D., Schilling, J., Raabe, T., and Simon, M. A. (1996). Daughter of sevenless is a substrate of the phosphotyrosine phosphatase Corkscrew and functions during sevenless signaling. *Cell* **85**, 899–909.

Hibi, M., Murakami, M., Saito, M., Hirano, T., Taga, T., and Kishimoto, T. (1990). Molecular cloning and expression of an IL-6 signal transducer, gp130. *Cell* **63**, 1149–1157.

Hill, C. S., and Treisman, R. (1995). Differential activation of c-fos promoter elements by serum, lysophosphatidic acid, G proteins and polypeptide growth factors. *EMBO J.* **14**, 5037–5047.

Hirano, T., Nakajima, K., and Hibi, M. (1997). Signaling mechanisms through gp130: a model of the cytokine system. *Cytokine Growth Factor Rev.* **8**, 241–252.

Hirata, Y., Taga, T., Hibi, M., Nakano, N., Hirano, T., and Kishimoto, T. (1989). Characterization of IL-6 receptor expression by monoclonal and polyclonal antibodies. *J. Immunol.* **143**, 2900–2906.

Hirota, H., Yoshida, K., Kishimoto, T., and Taga, T. (1995). Continuous activation of gp130, a signal-transducing receptor component for interleukin 6-related cytokines, causes myocardial hypertrophy in mice. *Proc. Natl Acad. Sci. USA* **92**, 4862–4866.

Holgado-Madruga, M., Emlet, D. R., Moscatello, D. K., Godwin, A. K., and Wong, A. J. (1996). A Grb2-associated docking protein in EGF- and insulin-receptor signalling. *Nature* **379**, 560–564.

Honda, M., Yamamoto, S., Cheng, M., Yasukawa, K., Suzuki, H., Saito, T., Osugi, Y., Tokunaga, T., and Kishimoto, T. (1992). Human soluble IL-6 receptor: its detection and enhanced release by HIV infection. *J. Immunol.* **148**, 2175–2180.

Horvai, A. E., Xu, L., Korzus, E., Brard, G., Kalafus, D., Mullen, T. M., Rose, D. W., Rosenfeld, M. G., and Glass, C. K. (1997). Nuclear integration of JAK/STAT and Ras/AP-1 signaling by CBP and p300. *Proc. Natl Acad. Sci. USA* **94**, 1074–1079.

Horvath, C. M., Wen, Z., and Darnell Jr, J. E. (1995). A STAT protein domain that determines DNA sequence recognition suggests a novel DNA-binding domain. *Genes Dev.* **9**, 984–994.

Ichiba, M., Nakajima, K., Yamanaka, Y., Kiuchi, N., and Hirano, T. (1998). Autoregulation of the Stat3 gene through cóoperation with a cAMP-responsive element-binding protein. *J. Biol. Chem.* **273**, 6132–6138.

Ihara, S., Nakajima, K., Fukada, T., Hibi, M., Nagata, S., Hirano, T., and Fukui, Y. (1997). Dual control of neurite outgrowth by STAT3 and MAP kinase in PC12 cells stimulated with interleukin-6. *EMBO J.* **16**, 5345–5352.

Ihle, J. N. (1995). The Janus protein tyrosine kinase family and its role in cytokine signaling. *Adv. Immunol.* **60**, 1–35.

Ihle, J. N., and Kerr, I. M. (1995). Jaks and Stats in signaling by the cytokine receptor superfamily. *Trends Genet.* **11**, 69–74.

Keul, R., Heinrich, P. C., Muller-newen, G., Muller, K., and Woo, P. (1998). A possible role for soluble IL-6 receptor in the pathogenesis of systemic onset juvenile chronic arthritis. *Cytokine* **10**, 729–734.

Kim, H., Hawley, T. S., Hawley, R. G., and Baumann, H. (1998). Protein tyrosine phosphatase 2 (SHP-2) moderates signaling by gp130 but is not required for the induction of acute-phase plasma protein genes in hepatic cells. *Mol. Cell Biol.* **18**, 1525–1533.

Kiuchi, N., Nakajima, K., Ichiba, M., Fukada, T., Marimatsu, M., Mizuno, K., Hibi, M., and Hirano, T. (1999). STAT3 is required for the gp130-mediated full activation of c-myc gene. *J. Exp. Med.* **189**, 63–73.

Klingmuller, U., Lorenz, U., Cantley, L. C., Neel, B. G., and Lodish, H. F. (1995). Specific recruitment of SH-PTP1 to the erythropoietin receptor causes inactivation of JAK2 and termination of proliferative signals. *Cell* **80**, 729–738.

Kojima, H., Nakajima, K., and Hirano, T. (1996). IL-6-inducible complexes on an IL-6 response element of the junB promoter contain Stat3 and 36 kDa CRE-like site binding protein(s). *Oncogene* **12**, 547–554.

Kordula, T., Rydel, R. E., Brigham, E. F., Horn, F., Heinrich, P. C., and Travis, J. (1998). Oncostatin M and the interleukin-6 and soluble interleukin-6 receptor complex regulate alpha1-antichymotrypsin expression in human cortical astrocytes. *J. Biol. Chem.* **273**, 4112–4118.

Korzus, E., Nagase, H., Rydell, R., and Travis, J. (1997). The mitogen-activated protein kinase and JAK-STAT signaling pathways are required for an oncostatin M-responsive element-mediated activation of matrix metalloproteinase 1 gene expression. *J. Biol. Chem.* **272**, 1188–1196.

Kumanogoh, A., Marukawa, S., Kumanogoh, T., Hirota, H., Yoshida, K., Lee, I. S., Yasui, T., Taga, T., and Kishimoto, T. (1997). Impairment of antigen-specific antibody production in transgenic mice expressing a dominant-negative form of gp130. *Proc. Natl Acad. Sci. USA* **94**, 2478–2482.

Kumar, G., Gupta, S., Wang, S., and Nel, A. E. (1994). Involvement of Janus kinases, p52shc, Raf-1, and MEK-1 in the IL-6-induced mitogen-activated protein kinase cascade of a growth-responsive B cell line. *J. Immunol.* **153**, 4436–4447.

Lai, C. F., Ripperger, J., Morella, K. K., Wang, Y., Gearing, D. P., Fey, G. H., and Baumann, H. (1995a). Separate signaling mechanisms are involved in the control of STAT protein activation and gene regulation via the interleukin 6 response element by the box 3 motif of gp130. *J. Biol. Chem.* **270**, 14847–14850.

Lai, C. F., Ripperger, J., Morella, K. K., Wang, Y., Gearing, D. P., Horseman, N. D., Campos, S. P., Fey, G. H., and Baumann, H. (1995b). STAT3 and STAT5B are targets of two different signal pathways activated by hematopoietin receptors and control transcription via separate cytokine response elements. *J. Biol. Chem.* **270**, 23254–23257.

Li, W., Nishimura, R., Kashishian, A., Batzer, A. G., Kim, W. J., Cooper, J. A., and Schlessinger, J. (1994). A new function for a phosphotyrosine phosphatase: linking GRB2-Sos to a receptor tyrosine kinase. *Mol. Cell Biol.* **14**, 509–517.

Lust, J. A., Donovan, K. A., Kline, M. P., Greipp, P. R., Kyle, R. A., and Maihle, N. J. (1992). Isolation of an mRNA encoding a soluble form of the human interleukin-6 receptor. *Cytokine* **4**, 96–100.

Lutticken, C., Wegenka, U. M., Yuan, J., Buschmann, J., Schindler C., Ziemiecki, A., Harpur, A. G., Wilks, A. F., Yasukawa, K., Taga, T., Kishimoto, T., Barbieri, G., Pellegrini, S., Sendtner, M., Heinrich, P. C., and Horn, F. (1994). Association of transcription factor APRF and protein kinase Jak1 with the interleukin-6 signal transducer gp130. *Science* **263**, 89–92.

Maeda, A., Kurosaki, M., Ono, M., Takai, T., and Kurosaki, T. (1998). Requirement of SH2-containing protein tyrosine phosphatases SHP-1 and SHP-2 for paired immunoglobulin-like receptor B (PIR-B)-mediated inhibitory signal. *J. Exp. Med.* **187**, 1355.

Matsuda, T., Fukada, T., Takahashi-Tezuka, M., Okuyama, Y., Fujitani, Y., Hanazono, Y., Hirai, H., and Hirano, T. (1995a). Activation of Fes tyrosine kinase by gp130, an interleukin-6 family cytokine signal transducer, and their association. *J. Biol. Chem.* **270**, 11037–11039.

Matsuda, T., Takahashi-Tezuka, M., Fukada, T., Okuyama, Y., Fujitani, Y., Tsukada, S., Mano, H., Hirai, H., Witte, O. N., and Hirano, T. (1995b). Association and activation of Btk and Tec tyrosine kinases by gp130, a signal transducer of the interleukin-6 family of cytokines. *Blood* **85**, 627–633.

Morse, L., Chen, D., Franklin, D., Xiong, Y., and Chen-Kiang, S. (1997). Induction of cell cycle arrest and B cell terminal differentiation by CDK inhibitor p18(INK4c) and IL-6. *Immunity* **6**, 47–56.

Mullberg, J., Schooltink, H., Stoyan, T., Gunther, M., Graeve, L., Buse, G., Mackiewicz, A., Heinrich, P. C., and Rose-John, S. (1993). The soluble interleukin-6 receptor is generated by shedding. *Eur. J. Immunol.* **23**, 473–480.

Muller-Newen, G., Kohne, C., Keul, R., Hemmann, U., Muller-Esterl, W., Wijdenes, J., Brakenhoff, J. P., Hart, M. H., and Heinrich, P. C. (1996). Purification and characterization of the soluble interleukin-6 receptor from human plasma and identification of an isoform generated through alternative splicing. *Eur. J. Biochem.* **236**, 837–842.

Murakami, M., Narazaki, M., Hibi, M., Yawata, H., Yasukawa, K., Hamaguchi, M., Taga, T., and Kishimoto, T. (1991). Critical cytoplasmic region of the interleukin 6 signal transducer gp130 is conserved in the cytokine receptor family. *Proc. Natl Acad. Sci. USA* **88**, 11349–11353.

Naka, T., Narazaki, M., Hirata, M., Matsumoto, T., Minamoto, S., Aono, A., Nishimoto, N., Kajita, T., Taga, T., Yoshizaki, K., Akira, S., and Kishimoto, T. (1997). Structure and function of a new STAT-induced STAT inhibitor. *Nature* **387**, 924–929.

Nakajima, K., and Wall, R. (1991). Interleukin-6 signals activating junB and TIS11 gene transcription in a B-cell hybridoma. *Mol. Cell Biol.* **11**, 1409–1418.

Nakajima, K., Matsuda, T., Fujitani, Y., Kojima, H., Yamanaka, Y., Nakae, K., Takeda, T., and Hirano, T. (1995). Signal transduction through IL-6 receptor: involvement of multiple protein kinases, stat factors, and a novel H7-sensitive pathway. *Ann. NY Acad. Sci.* **762**, 55–70.

Nakajima, K., Yamanaka, Y., Nakae, K., Kojima, H., Ichiba, M., Kiuchi, N., Kitaoka, T., Fukada, T., Hibi, M., and Hirano, T. (1996). A central role for Stat3 in IL-6-induced regulation of growth and differentiation in M1 leukemia cells. *EMBO J.* **15**, 3651–3658.

Narazaki, M., Yasukawa, K., Saito, T., Ohsugi, Y., Fukui, H., Koishihara, Y., Yancopoulos, G. D., Taga, T., and Kishimoto, T. (1993). Soluble forms of the interleukin-6 signal-transducing receptor component gp130 in human serum possessing a potential to inhibit signals through membrane-anchored gp130. *Blood* **82**, 1120–1126.

Narazaki, M., Witthuhn, B. A., Yoshida, K., Silvennoinen, O., Yasukawa, K., Ihle, J. N., Kishimoto, T., and Taga, T. (1994). Activation of JAK2 kinase mediated by the interleukin 6 signal transducer gp130. *Proc. Natl Acad. Sci. USA* **91**, 2285–9.

Narimatsu, M., Nakajima, K., Ichiba, M., and Hirano, T. (1997). Association of Stat3-dependent transcriptional activation of p19INK4D with IL-6-induced growth arrest. *Biochem. Biophys. Res. Commun.* **238**, 764–768.

Neel, B. G., and Tonks, N. K. (1997). Protein tyrosine phosphatases in signal transduction. *Curr. Opin. Cell Biol.* **9**, 193–204.

Neubauer, H., Cumano, A., Muller, M., Wu, H., Huffstadt, U., and Pfeffer, K. (1998). Jak2 deficiency defines an essential developmental checkpoint in definitive hematopoiesis. *Cell* **93**, 397–409.

Ng, J., and Cantrell, D. (1997). STAT3 is a serine kinase target in T lymphocytes. Interleukin 2 and T cell antigen receptor signals converge upon serine 727. *J. Biol. Chem.* **272**, 24542–24549.

Nishida, K., Yoshida, Y., Itoh, M., Fukada, T., Ohtani, T., Shirogane, T., Atsumi, T., Takahashi-Tezuka, M., Ishihara, K., Hibi, M., and Hirano, T. (1999). Gab-family adapter proteins act downstream of cytokine and growth factor receptors and T- and B-cell antigen receptors. *Blood* **93**, 1809–1816.

Niwa, H., Burdon, T., Chambers, I., and Smith, A. (1998). Self-renewal of pluripotent embryonic stem cells is mediated via activation of STAT3. *Genes Dev.* **12**, 2048–2060.

Noguchi, T., Matozaki, T., Horita, K., Fujioka, Y., and Kasuga, M. (1994). Role of SH-PTP2, a protein-tyrosine phosphatase with Src homology 2 domains, in insulin-stimulated Ras activation. *Mol. Cell Biol.* **14**, 6674–6682.

O'Brien, C. A., and Manolagas, S. C. (1997). Isolation and characterization of the human gp130 promoter. Regulation by STATS. *J. Biol. Chem.* **272**, 15003–15010.

Ogryzko, V. V., Schiltz, R. L., Russanova, V., Howard, B. H., and Nakatani, Y. (1996). The transcriptional coactivators p300 and CBP are histone acetyltransferases. *Cell* **87**, 953–959.

Ozbek, S., Grotzinger, J., Krebs, B., Fischer, M., Wollmer, A., Jostock, T., Mullberg, J., and Rose-John, S. (1998). The membrane proximal cytokine receptor domain of the human interleukin-6 receptor is sufficient for ligand binding but not for gp130 association. *J. Biol. Chem.* **273**, 21374–21379.

Paonessa, G., Graziani, R., De Serio, A., Savino, R., Ciapponi, L., Lahm, A., Salvati, A. L., Toniatti, C., and Ciliberto, G. (1995). Two distinct and independent sites on IL-6 trigger gp 130 dimer formation and signalling. *EMBO J.* **14**, 1942–1951.

Parganas, E., Wang, D., Stravopodis, D., Topham, D. J., Marine, J. C., Teglund, S., Vanin, E. F., Bodner, S., Colamonici, O. R., van Deursen, J. M., Grosveld, G., and Ihle, J. N. (1998). Jak2 is essential for signaling through a variety of cytokine receptors. *Cell* **93**, 385–395.

Perkins, L. A., Larsen, I., and Perrimon, N. (1992). corkscrew encodes a putative protein tyrosine phosphatase that functions to transduce the terminal signal from the receptor tyrosine kinase torso. *Cell* **70**, 225–236.

Peters, M., Schirmacher, P., Goldschmitt, J., Odenthal, M., Peschel, C., Fattori, E., Ciliberto, G., Dienes, H. P., Meyer zum Buschenfelde, K. H., and Rose-John, S. (1997).

Extramedullary expansion of hematopoietic progenitor cells in interleukin (IL)-6-sIL-6R double transgenic mice. *J. Exp. Med.* **185**, 755–766.

Qiu, Y., Ravi, L., and Kung, H. J. (1998). Requirement of ErbB2 for signalling by interleukin-6 in prostate carcinoma cells. *Nature* **393**, 83–85.

Qureshi, S. A., Cao, X. M., Sukhatme, V. P., and Foster, D. A. (1991). v-Src activates mitogen-responsive transcription factor Egr-1 via serum response elements. *J. Biol. Chem.* **266**, 10802–10806.

Raabe, T., Riesgo-Escovar, J., Liu, X., Bausenwein, B. S., Deak, P., Maroy, P., and Hafen, E. (1996). DOS, a novel pleckstrin homology domain-containing protein required for signal transduction between sevenless and Ras1 in *Drosophila*. *Cell* **85**, 911–920.

Ray, D. W., Ren, S. G., and Melmed, S. (1996). Leukemia inhibitory factor (LIF) stimulates proopiomelanocortin (POMC) expression in a corticotroph cell line. Role of STAT pathway. *J. Clin. Invest.* **97**, 1852–1859.

Rodig, S. J., Meraz, M. A., White, J. M., Lampe, P. A., Riley, J. K., Arthur, C. D., King, K. L., Sheehan, K. C., Yin, L., Pennica, D., Johnson, E. M., Jr., and Schreiber, R. D. (1998). Disruption of the Jak1 gene demonstrates obligatory and nonredundant roles of the Jaks in cytokine-induced biologic responses. *Cell* **93**, 373–383.

Saito, M., Yoshida, K., Hibi, M., Taga, T., and Kishimoto, T. (1992). Molecular cloning of a murine IL-6 receptor-associated signal transducer, gp130, and its regulated expression *in vivo*. *J. Immunol.* **148**, 4066–4071.

Salvi, M., Girasole, G., Pedrazzoni, M., Passeri, M., Giuliani, N., Minelli, R., Braverman, L. E., and Roti, E. (1996). Increased serum concentrations of interleukin-6 (IL-6) and soluble IL-6 receptor in patients with Graves' disease [see comments]. *J. Clin. Endocrinol. Metab.* **81**, 2976–2979.

Sasse, J., Hemmann, U., Schwartz, C., Schniertshauer, U., Heesel, B., Landgraf, C., Schneider-Mergener, J., Heinrich, P. C., and Horn, F. (1997). Mutational analysis of acute-phase response factor/Stat3 activation and dimerization. *Mol. Cell Biol.* **17**, 4677–4686.

Schaefer, T. S., Sanders, L. K., and Nathans, D. (1995). Cooperative transcriptional activity of Jun and Stat3 beta, a short form of Stat3. *Proc. Natl Acad. Sci. USA* **92**, 9097–9101.

Schirmacher, P., Peters, M., Ciliberto, G., Blessing, M., Lotz, J., Meyer zum Buschenfelde, K. H., and Rose-John, S. (1998). Hepatocellular hyperplasia, plasmacytoma formation, and extramedullary hematopoiesis in interleukin (IL)-6/soluble IL-6 receptor double-transgenic mice. *Am. J. Pathol.* **153**, 639–648.

Schumann, R. R., Kirschning, C. J., Unbehaun, A., Aberle, H. P., Knope, H. P., Lamping, N., Ulevitch, R. J., and Herrmann, F. (1996). The lipopolysaccharide-binding protein is a secretory class 1 acute-phase protein whose gene is transcriptionally activated by APRF/STAT/3 and other cytokine-inducible nuclear proteins. *Mol. Cell Biol.* **16**, 3490–3503.

Seidel, H. M., Milocco, L. H., Lamb, P., Darnell Jr, J. E. Stein, R. B., and Rosen, J. (1995). Spacing of palindromic half sites as a determinant of selective STAT (signal transducers and activators of transcription) DNA binding and transcriptional activity. *Proc. Natl Acad. Sci. USA* **92**, 3041–3245.

Sekimoto, T., Nakajima, K., Tachibana, T., Hirano, T., and Yoneda, Y. (1996). Interferon-gamma-dependent nuclear import of Stat1 is mediated by the GTPase activity of Ran/TC4. *J. Biol. Chem.* **271**, 31017–31020.

Sekimoto, T., Imamoto, N., Nakajima, K., Hirano, T., and Yoneda, Y. (1997). Extracellular signal-dependent nuclear import of Stat1 is mediated by nuclear pore-targeting complex formation with NPI-1, but not Rch1. *EMBO J.* **16**, 7067–7077.

Senaldi, G., Varnum, B. C., Sarmiento, U., Starnes, C., Lile, J., Scully, S., Guo, J., Elliott, G., McNinch, J., Shaklee, C. L., Freeman, D., Manu, F., Simonet, W. S., Boone, T., and Chang, M. S. (1999). Novel neurotrophin-1/B cell-stimulating factor-3: a cytokine of the IL-6 family. *Proc. Natl Acad. Sci. USA* **96**, 11458–11463.

Stahl, N., Boulton, T. G., Farruggella, T., Ip, N. Y., Davis, S., Witthuhn, B. A., Quelle, F. W., Silvennoinen, O., Barbieri, G., Pellegrini, S., Ihle, J. N., and Yancopoulos, G. D. (1994). Association and activation of Jak-Tyk kinases by CNTF-LIF-OSM-IL-6 beta receptor components. *Science* **263**, 92–95.

Stahl, N., Farruggella, T. J., Boulton, T. G., Zhong, Z., Darnell Jr, J. E., and Yancopoulos, G. D. (1995). Choice of STATs and other substrates specified by modular tyrosine-based motifs in cytokine receptors. *Science* **267**, 1349–1353.

Starr, R., Willson, T. A., Viney, E. M., Murray, L. J., Rayner, J. R., Jenkins, B. J., Gonda, T. J., Alexander, W. S., Metcalf, D., Nicola, N. A., and Hilton, D. J. (1997). A family of cytokine-inducible inhibitors of signalling. *Nature* **387**, 917–921.

Stephanou, A., Isenberg, D. A., Akira, S., Kishimoto, T., and Latchman, D. S. (1998). The nuclear factor interleukin-6 (NF-IL6) and signal transducer and activator of transcription-3 (STAT-3) signalling pathways co-operate to mediate the activation of the hsp90beta gene by interleukin-6 but have opposite effects on its inducibility by heat shock. *Biochem. J.* **330**, 189–195.

Sugita, T., Totsuka, T., Saito, M., Yamasaki, K., Taga, T., Hirano, T., and Kishimoto, T. (1990). Functional murine interleukin 6 receptor with the intracisternal A particle gene product at its cytoplasmic domain. Its possible role in plasmacytomagenesis. *J. Exp. Med.* **171**, 2001–2009.

Symes, A., Gearan, T., Eby, J., and Fink, J. S. (1997). Integration of Jak-Stat and AP-1 signaling pathways at the vasoactive intestinal peptide cytokine response element regulates ciliary neurotrophic factor-dependent transcription. *J. Biol. Chem.* **272**, 9648–9654.

Taga, T., Kawanishi, Y., Hardy, R. R., Hirano, T., and Kishimoto, T. (1987). Receptors for B cell stimulatory factor 2. Quantitation, specificity, distribution, and regulation of their expression. *J. Exp. Med.* **166**, 967–981.

Taga, T., Hibi, M., Hirata, Y., Yamasaki, K., Yasukawa, K., Matsuda, T., Hirano, T., and Kishimoto, T. (1989). Intereukin-6 triggers the association of its receptor with a possible signal transducer, gp130. *Cell* **58**, 573–581.

Takahashi-Tezuka, M., Hibi, M., Fujitani, Y., Fukada, T., Yamaguchi, T., and Hirano, T. (1997). Tec tyrosine kinase links the cytokine receptors to PI-3 kinase probably through JAK. *Oncogene* **14**, 2273–2282.

Takahashi-Tezuka, M., Yoshida, Y., Fukada, T., Ohtani, T., Yamanaka, Y., Nishida, K., Nakajima, K., Hibi, M., and Hirano, T. (1998). Gab1 acts as an adapter molecule linking the cytokine receptor gp130 to ERK mitogen-activated protein kinase. *Mol. Cell Biol.* **18**, 4109–4117.

Tang, T. L., Freeman Jr, R. M., O'Reilly, A. M., Neel, B. G., and Sokol, S. Y. (1995). The SH2-containing protein-tyrosine phosphatase SH-PTP2 is required upstream of MAP kinase for early *Xenopus* development. *Cell* **80**, 473–483.

Tanner, J. W., Chen, W., Young, R. L., Longmore, G. D., and Shaw, A. S. (1995). The conserved box 1 motif of cytokine receptors is required for association with JAK kinases. *J. Biol. Chem.* **270**, 6523–6530.

Tonks, N. K., and Neel, B. G. (1996). From form to function: signaling by protein tyrosine phosphatases [see comments]. *Cell* **87**, 365–368.

Vogel, W., Lammers, R., Huang, J., and Ullrich, A. (1993). Activation of a phosphotyrosine phosphatase by tyrosine phosphorylation. *Science* **259**, 1611.

Wegenka, U. M., Buschmann, J., Lutticken, C., Heinrich, P. C., and Horn, F. (1993). Acute-phase response factor, a nuclear factor binding to acute-phase response elements, is rapidly activated by interleukin-6 at the posttranslational level. *Mol. Cell Biol.* **13**, 276–288.

Weidner, K. M., Di Cesare, S., Sachs, M., Brinkmann, V., Behrens, J., and Birchmeier, W. (1996). Interaction between Gab1 and the c-Met receptor tyrosine kinase is responsible for epithelial morphogenesis. *Nature* **384**, 173–176.

Wen, Z., and Darnell Jr, J. E. (1997). Mapping of Stat3 serine phosphorylation to a single residue (727) and evidence that serine phosphorylation has no influence on DNA binding of Stat1 and Stat3. *Nucl. Acids Res.* **25**, 2062–2067.

Wen, Z., Zhong, Z., and Darnell Jr, J. E. (1995). Maximal activation of transcription by Stat1 and Stat3 requires both tyrosine and serine phosphorylation. *Cell* **82**, 241–250.

Yamada, T., Tobita, K., Osada, S., Nishihara, T., and Imagawa, M. (1997). CCAAT/enhancer-binding protein delta gene expression is mediated by APRF/STAT3. *J. Biochem. (Tokyo)* **121**, 731–738.

Yamanaka, Y., Nakajima, K., Fukada, T., Hibi, M., and Hirano, T. (1996). Differentiation and growth arrest signals are generated through the cytoplasmic region of gp130 that is essential for Stat3 activation. *EMBO J.* **15**, 1557–1565.

Yamasaki, K., Taga, T., Hirata, Y., Yawata, H., Kawanishi, Y., Seed, B., Taniguchi, T., Hirano, T., and Kishimoto, T. (1988). Cloning and expression of the human interleukin-6 (BSF-2/IFN beta 2) receptor. *Science* **241**, 825–828.

Yoshida, K., Taga, T., Saito, M., Suematsu, S., Kumanogoh, A., Tanaka, T., Fujiwara, H., Hirata, M., Yamagami, T., Nakahata, T., Hirabayashi, T., Yoneda, Y., Tanaka, K., Wang, W. Z., Mori, C., Shiota, K., Yoshida, N., and Kishimoto, T. (1996). Targeted disruption of gp130, a common signal transducer for the interleukin 6 family of cytokines, leads to myocardial and hematological disorders. *Proc. Natl Acad. Sci. USA* **93**, 407–411.

Yoshizaki, K., Nishimoto, N., Mihara, M., and Kishimoto, T. (1998). Therapy of rheumatoid arthritis by blocking IL-6 signal transduction with a humanized anti-IL-6 receptor antibody. *Springer Semin. Immunopathol.* **20**, 247–259.

Zhang, D., Sun, M., Samols, D., and Kushner, I. (1996a). STAT3 participates in transcriptional activation of the C-reactive protein gene by interleukin-6. *J. Biol. Chem.* **271**, 9503–9509.

Zhang, J. J., Vinkemeier, U., Gu, W., Chakravarti, D., Horvath, C. M., and Darnell Jr, J. E. (1996b). Two contact regions between Stat1 and CBP/p300 in interferon gamma signaling. *Proc. Natl Acad. Sci. USA* **93**, 15092–15096.

Zhang, Z. X., Jones, S., Hagood, J. S., Fuentes, N. L., and Fuller, G. M. (1997). STAT3 acts as a co-activator of gluocorticoid receptor signaling. *J. Biol. Chem.* **272**, 30607–30610.

Zhong, Z., Wen, Z., and Darnell Jr, J. E. (1994). Stat3: a STAT family member activated by tyrosine phosphorylation in response to epidermal growth factor and interleukin-6. *Science* **264**, 95–98.

IL-11 Receptor

James Keith*

Genetics Institute, One Burtt Road, Andover, MA 01810, USA

*corresponding author tel: 978-247-1372, fax: 978-247-1333, e-mail: jkeith@genetics.com

DOI: 10.1006/rwcy.2000.17004.

SUMMARY

The cDNAs encoding the genes for the murine and human interleukin 11 receptor α (IL-11Rα) proteins were reported between 1989 and 1994. The IL-11 high-affinity binding receptor complex is composed of a low-affinity IL-11 ligand-binding α chain (mIL-11Rα) and the signal-transducing subunit, gp130, that is shared with the other members of the IL-6 cytokine family: IL-6; leukemia inhibitory factor (LIF); oncostatin M; cardiotropin 1 (CT-1); and ciliary neurotropic factor (CNTF). The mIL-11Rα is composed of an extracellular domain, a transmembrane domain, and a cytoplasmic tail. The hIL-11Rα protein exists in two isoforms, with one lacking the cytoplasmic domain. The extracellular regions of the receptor from both species exhibit structural features typical of hematopoietic receptors such as proline residues preceding each 100 amino acid subdomain, a motif of four conserved cysteines, and one tryptophan residue, a series of polar and hydrophobic amino acid residues, and the WSXWS domain between the cysteine and transmembrane domain. IL-11 binding to the IL-11Rα alone occurs with low affinity ($K_d \sim 10$ nM) and apparently does not transduce an intracellular signal, while a high-affinity receptor complex capable of transducing a signal ($K_d \sim 400$–800 pM) is produced with coexpression of IL-11Rα and gp130. The exact stoichiometry of the high-affinity IL-11R complex is unknown. IL-11 and other IL-6 family cytokines, in a complex with their specific α receptors, interact with gp130 and, as a consequence, activate the Janus kinase/signal transducer and activator of transcription (JAK/STAT3) signaling pathway in their target cells. IL-11Rα and gp130 are widely distributed throughout the body, and activation of the IL-11 signaling pathway results in effects on a variety of cells, including hematopoietic, bone, and immune cells. Biological functions essential to IL-11 have been investigated through the generation of mice with a null mutation of the mIL-11Rα gene. Although the mice with the null mutation were healthy and had normal peripheral white blood cell, hematocrit, and platelet levels, female mice were infertile because of defective decidualization. Therefore, IL-11 as well as LIF appears to play a critical role during reproduction.

BACKGROUND

Discovery

The cDNA for the murine IL-11 receptor α chain (mIL-11Rα) of the receptor complex for IL-11 was published in 1994 (Hilton *et al.*, 1994), and the human IL-11 receptor α chain (hIL-11Rα) was reported over the next 2 years (Nandurkar *et al.*, 1996; Cherel *et al.*, 1995, 1996; Van Leuven *et al.* (1996).

Alternative names

A murine gene coding for protein Etl2 was first described in 1989 (Gossler *et al.*, 1989). Molecular cloning of its cDNA described an orphan receptor of type I cytokines (Neuhaus *et al.*, 1994). Simultaneously, publication of a murine cDNA (NR-1) revealed that NR-1 and Etl2 coded for an identical protein, mIL-11Rα (Hilton *et al.*, 1994).

Structure

The IL-11 high-affinity binding receptor complex is composed of a low-affinity IL-11 ligand-binding α chain and the signal transducing subunit, gp130, that is shared with other members of the IL-6 cytokine family: IL-6; LIF; oncostatin M; CT-1; and CNTF.

mIL-11Rα is composed of an extracellular domain, a transmembrane domain, and a cytoplasmic tail. The receptor exhibits 24% and 22% amino acid homology with the IL-6Rα chain and CNTF receptor (CNTFR) α chain, respectively. The extracellular region exhibits structural features typical of hematopoietic receptors such as proline residues preceding each 100 amino acid subdomain, a motif of four conserved cysteines and one tryptophan residue, a series of polar and hydrophobic amino acid residues, and the WSXWS domain between the cysteine and transmembrane domain. IL-11 binding to IL-11Rα alone occurs with low affinity ($K_d \sim 10$ nM) and apparently does not transduce an intracellular signal. A high-affinity receptor complex capable of transducing a signal ($K_d \sim 400$–800 pM) is produced with coexpression of IL-11Rα and gp130 (Nandurkar *et al.*, 1996).

The exact stoichiometry of the high-affinity IL-11R complex is unknown. Evidence from *in vitro* solution-phase binding assays indicates that the high-affinity IL-6R consists as a hexameric complex of two IL-6 molecules, two IL-6Rα chains, and a gp130 homodimer. A similar hexameric complex has also been proposed for the CNTFR complex, with the exception of a heterodimer of gp130 and the LIFRα chain replacing the gp130 homodimer. Similar studies examining the IL-11R complex indicate a dimer of IL-11 and IL-11Rα forms in the presence of gp130, resulting in a pentameric complex.

However, homodimerization of gp130 or heterodimerization of gp130 and LIFR did not occur (Neddermann *et al.*, 1996). These studies may point to an unidentified IL-11Rβ chain being involved in IL-11 signal transduction. The IL-11Rα chain does not have to be membrane-bound to elicit a biological effect because soluble forms of the α chain receptor have been generated that can bind IL-11 and activate gp130 (Baumann *et al.*, 1996; Neddermann *et al.*, 1996; Curtis *et al.*, 1997).

GENE

The mIL-11Rα gene includes 14 exons with alternative use of the first two exons regulated in a developmental fashion (Nandurkar *et al.*, 1997b). The gene contains two loci (1 and 2), while locus 2 is restricted to only some mouse strains. Two alternatively spliced exons (1a and 1b) encode the 5′ untranslated region (5′ UTR) of the murine locus 1. Northern analysis was also used to examine the human gene expression, and its chromosomal location was determined by fluorescence *in situ* hybridization. The presence of exon(s) encoding the 5′ UTR and mapping of transcription initiation sites was determined by reverse transcriptase polymerase chain reaction and 5′ rapid amplification of cDNA ends (5′ RACE) techniques. The human locus was 10 kb and contained 14 exons. Two alternatively spliced first exons (1a and 1b), encoding the 5′ UTR, shared 76 and 73% nucleotide identity with murine exons 1a and 1b. Multiple transcription start sites were seen for human exon 1a. The promoter regions of both human exons 1a and 1b did not display a canonical TATA box. A predominant 1.8 kb transcript for the hIL-11Rα was present in heart, brain, skeletal muscle, lymph nodes, thymus, appendix, pancreas, and fetal liver. The hIL-11Rα gene was localized to chromosome 9p13. The hIL-11Rα gene was highly related to locus 1 of the murine gene, but there was no evidence of a second hIL-11Rα locus.

A second mIL-11Rα locus has been identified (IL-11Rα2) adjacent to the IL-11Rα1 gene (Robb *et al.*, 1997). It shares 99% sequence identity with IL-11Rα1 in the coding exons but contains differences in 5′ UTR (Bilinski *et al.*, 1996, 1998; Robb *et al.*, 1997). The mIL-11Rα1 gene is expressed at relatively low levels in several tissues, including bone marrow, spleen, thymus, lung, bladder, heart, brain, kidney, muscle, salivary gland, small and large intestine, ovary, testis, and uterus. Primary cell types such as macrophages, osteoblasts, and osteoclasts also express the murine IL-11Rα1 gene (Romas *et al.*, 1996; Trepicchio *et al.*, 1997) while expression of the IL-11Rα2 gene is restricted to the testis, lymph node, and thymus (Robb *et al.*, 1997).

The hIL-11Rα chain shares 85% nucleotide identity and 84% amino acid identity with the murine gene (Van Leuven *et al.*, 1996; Nandurkar *et al.*, 1997a). hIL-11Rα was cloned and its structure analyzed. The gene is composed of 13 exons comprising nearly 10 kb of DNA that was completely sequenced. The intron–exon boundaries were determined based on the mouse Etl2 and IL-11R cDNAs that were recently cloned. The protein sequence predicted by the human gene was over 83% identical with its murine counterpart, with very strict conservation of functionally important domains and signatures. Fluorescence *in situ* hybridization confirmed that the gene was located on human chromosome 9p13, syntenic with the mouse etl2 gene on chromosome 4.

Accession numbers

mIL-11Rα mRNA: U14412 (Hilton *et al.*, 1994), U69491 (Robb *et al.*, 1997)
hIL-11Rα mRNA: U32324
hIL-11Rα full cDNA: U32323

Sequence

The sequence for mIL-11Rα (Hilton *et al.*, 1994) is seen in **Figure 1**. The exon–intron boundary sequences of hIL-11Rα (Van Leuven *et al.*, 1996) can be seen in **Figure 2**.

PROTEIN

Accession numbers

mIL-11Rα1: AAA53248
mIL-11Rα2: AAC53114
hIL-11Rα (422 aa): NP_004503
hIL-1Rα (388 aa, lacking the cytoplasmic region): CAA86570

Description of protein

The IL-11 high-affinity binding receptor complex is composed of a low-affinity IL-11 ligand-binding α chain and the signal-transducing subunit, gp130, that is shared with the other members of the IL-6 cytokine family: IL-6; LIF; oncostatin M; CT-1; and CNTF (Yoshida *et al.*, 1996). The 432 amino acid mIL-11Rα is composed of an extracellular domain, a transmembrane domain, and a cytoplasmic tail. The extracellular region exhibits structural features typical of hematopoietic receptors such as proline residues preceding each 100 amino acid subdomain, a motif of four conserved cysteines and one tryptophan residue, a series of polar and hydrophobic amino acid residues, and the WSXWS domain between the cysteine and transmembrane domain.

Two isoforms of the hIL-11Rα have been identified and they differ in the structure of their cytoplasmic domains. One isoform, a 422 amino acid protein, has a short cytoplasmic domain similar to IL-6Rα and mIL-11Rα. The second isoform, a 388 amino acid protein, lacks the cytoplasmic domain and is similar to human CNTFR. **Figure 3** shows a comparison of the full-length mIL-11Rα and hIL-11Rα proteins (Van Leuven *et al.*, 1996).

Relevant homologies and species differences

hIL-11Rα exhibits 24% and 22% amino acid homology with the IL-6Rα chain and CNTFRα chain, respectively. The structural similarities between hIL-11Rα and CNTFR suggest that they may have evolved from a common ancestor. This idea is supported by the fact that the CNTFR and the hIL-11Rα genes are both found in the same band on chromosome 9 (Cherel *et al.*, 1996). mIL-11Rα and hIL-11Rα are over 83% identical (Van Leuven *et al.*, 1996).

Affinity for ligand(s)

The mature human and murine IL-11 ligand proteins share 88% homology at the amino acid level, while the human and nonhuman primate proteins share 94% homology. Amino acid residues 59 (methionine), 41 (lysine), and 98 (lysine) are critical for function (receptor binding and signaling) of the protein, and these residues are completely conserved in the mouse, nonhuman primate, and human proteins (Czupryn *et al.*, 1995a). Specific alkylation of a single methionine residue, Met59, produces a 25-fold reduction of *in vitro* biological activity of rhIL-11 on mouse plasmacytoma cells. Modification of the N-terminal amino group and partial labeling of two lysines, Lys41 and Lys98, causes a 3-fold decrease in activity. Removal of the last four C-terminal residues reduces rhIL-11 activity 25-fold, whereas removal of eight or more amino acids results in an inactive molecule. Using the four helix bundle model, Met59, Lys41, and Lys98 are located on the surface of the molecule; it is postulated that Met58 and the C-terminus of rhIL-11 are involved in the primary receptor-binding site (site I), whereas Lys41 and Lys98 may be a part of binding site II.

Biological activities of the C-terminal deletion mutants of human IL-11 have also been analyzed (Miyadai *et al.*, 1996). Removal of only one amino acid residue (leucine) from the C-terminus caused nearly an 80% loss of its biological activity. This shows the importance of the C-terminus of human IL-11 in terms of conserving the biological activity.

IL-11 binding to hIL-11Rα alone occurs with low affinity ($K_d \sim 10\,\text{nM}$) and apparently does not transduce an intracellular signal. A high-affinity receptor complex capable of transducing a signal ($K_d \sim 400$–$800\,\text{pM}$) is produced with coexpression of IL-11Rα and gp130 (Nandurkar *et al.*, 1996).

Cell types and tissues expressing the receptor

The expression of murine IL-11 ligand and mIL-11Rα in adult mouse tissues, in embryos, and during development of embryonic stem (ES) cells into cystic

Figure 1 Structure, nucleotide sequence, and predicted amino acids of the mIL-11Rα gene. From Hilton *et al.* (1994), with permission.

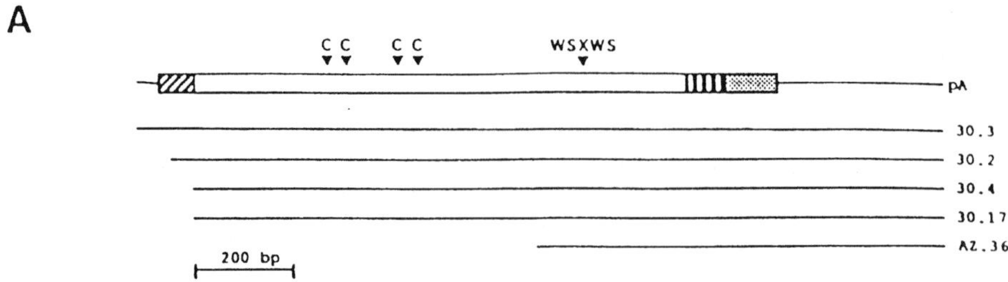

B

```
 - 44                                            gagagggtgagggcggaggccgctggcggcggctgccgcagaag

    1  ATG AGC AGC AGC TGC TCA GGG CTG ACC AGG GTC CTG GTG GCC GTG GCT ACA GCC CTG GTG
        M   S   S   S   C   S   G   L   T   R   V   L   V   A   V   A   T   A   L   V     20

   61  TCT TCC TCC TCC CCC TGC CCC CAA GCT TGG GGT CCT CCA GGG GTC CAG TAT GGA CAA CCT
        S   S   S   S   P   C   P   Q   A   W   G   P   P   G   V   Q   Y   G   Q   P     40

  121  GGC AGG CCC GTG ATG CTG TGC TGC CCC GGA GTG AGT GCT GGG ACT CCA GTG TCC TGG TTT
        G   R   P   V   M   L   C   C   P   G   V   S   A   G   T   P   V   S   W   F     60

  181  CGG GAT GGA GAT TCA AGG CTG CTC CAG GGA CCT GAC TCT GGG TTA GGA CAC AGA CTG GTC
        R   D   G   D   S   R   L   L   Q   G   P   D   S   G   L   G   H   R   L   V     80

  241  TTG GCC CAG GTG GAC AGC CCT GAT GAA GGC ACT TAT GTC TGC CAG ACC CTG GAT GGT GTA
        L   A   Q   V   D   S   P   D   E   G   T   Y   V   C   Q   T   L   D   G   V    100

  301  TCA GGG GGC ATG GTG ACC CTG AAG CTG GGC TTT CCC CCA GCA CGT CCT GAA GTC TCC TGC
        S   G   G   M   V   T   L   K   L   G   F   P   P   A   R   P   E   V   S   C    120

  361  CAA GCG GTA GAC TAT GAA AAC TTC TCC TGT ACT TGG AGT CCA GGC CAG GTC AGC GGT TTG
        Q   A   V   D   Y   E   N   F   S   C   T   W   S   P   G   Q   V   S   G   L    140

  421  CCC ACC CGC TAC CTT ACT TCC TAC AGG AAG AAG ACG CTG CCA GGA GCT GAG AGT CAG AGG
        P   T   R   Y   L   T   S   Y   R   K   K   T   L   P   G   A   E   S   Q   R    160

  481  GAA AGT CCA TCC ACC GGG CCT TGG CCG TGT CCA CAG GAC CCT CTG GAG GCC TCC CGA TGT
        E   S   P   S   T   G   P   W   P   C   P   Q   D   P   L   E   A   S   R   C    180

  541  GTG GTC CAT GGG GCA GAG TTC TGG AGT GAG TAC CGG ATC AAT GTG ACC GAG GTG AAC CCA
        V   V   H   G   A   E   F   W   S   E   Y   R   I   N   V   T   E   V   N   P    200

  601  CTG GGT GCC AGC ACG TGC CTA CTG GAT GTG AGA TTA CAG AGC ATC TTG CGT CCT GAT CCA
        L   G   A   S   T   C   L   L   D   V   R   L   Q   S   I   L   R   P   D   P    220

  661  CCC CAA GGA CTG CGG GTG GAA TCC GTA CCT GGT TAC CCG AGA CGC CTG CAT GCC AGC TGG
        P   Q   G   L   R   V   E   S   V   P   G   Y   P   R   R   L   H   A   S   W    240

  721  ACA TAC CCT GCC TCC TGG CGT CGC CAA CCC CAC TTT CTG CTC AAG TTC CGG TTG CAA TAC
        T   Y   P   A   S   W   R   R   Q   P   H   F   L   L   K   F   R   L   Q   Y    260

  781  CGA CCA GCA CAG CAT CCA GCC TGG TCC ACG GTG GAG CCC ATT GGC TTG GAG GAA GTG ATA
        R   P   A   Q   H   P   A   W   S   T   V   E   P   I   G   L   E   E   V   I    280

  841  ACA GAT GCT GTG GCT GGG CTG CCA CAC GCG GTA CGA GTC AGT GCC AGG GAC TTT CTG GAT
        T   D   A   V   A   G   L   P   H   A   V   R   V   S   A   R   D   F   L   D    300

  901  GCT GGC ACC TGG AGC GCC TGG AGC CCA GAG GCC TGG GGT ACT CCT AGC ACT GGT CCC CTG
        A   G   T   W   S   A   W   S   P   E   A   W   G   T   P   S   T   G   P   L    320

  961  CAG GAT GAG ATA CCT GAT TGG AGC CAG GGA CAT GGA CAG CAG CTA GAG GCA GTA GTA GCT
        Q   D   E   I   P   D   W   S   Q   G   H   G   Q   Q   L   E   A   V   V   A    340

 1021  CAG GAG GAC AGC CCG GCT CCT GCA AGG CCT TCC TTG CAG CCG GAC CCA AGG CCA CTT GAT
        Q   E   D   S   P   A   P   A   R   P   S   L   Q   P   D   P   R   P   L   D    360

 1081  CAC AGG GAC CCC TTG GAG CAA GTA GCT GTG TTA GCG TCT CTG GGA ATC TTC TCT TGC CTT
        H   R   D   P   L   E   Q   V   A   V   L   A   S   L   G   I   F   S   C   L    380

 1141  GGC CTG GCT GTT GGA GCT CTG GCA CTG GGG CTC TGG CTG AGG CTG AGA CGG AGT GGG AAG
        G   L   A   V   G   A   L   A   L   G   L   W   L   R   L   R   R   S   G   K    400

 1201  GAT GGA CCG CAA AAA CCT GGG CTC TTG GCA CCC ATG ATC CCG GTG GAA AAG CTT CCA GGA
        D   G   P   Q   K   P   G   L   L   A   P   M   I   P   V   E   K   L   P   G    420

 1261  ATT CCA AAC CTG CAG AGG ACC CCA GAG AAC TTC AGC TGA tttcatctgtaacccggtcagactggg
        I   P   N   L   Q   R   T   P   E   N   F   S   *

 1327  ggcagaaagaggcggggcagtggatccctgtggatggaggtctcagctgaaagtctgagctcttttctttgacacctat
 1405  actccaaacttgctgccggctgaaggctgtctggacttccgatgtcctgaggtggaagtccacctgaggaatgtgtaca
 1483  gaagtctgtgttcctgtgatcgtgtgtgtatgtgagacagggagcaaaagttctctgcatgtgtgtacagatgattgga
 1561  gagtgtgtgcggtcttgggcttggcccttctgggaagtgtgaagagttgaaataaaagagacggaagttttggaaaaa
 1639  aaaaaaaaaaaaaaaaaa
```

Figure 2 Exon–intron boundary sequences of the hIL-11Rα gene. Intron sequences are in lower case letters; exon sequences in upper case letters. From Van Leuven *et al.* (1996), with permission.

	Position							
	cDNA	Genomic						
1	1–61	802–862	ggagacgggg	GCTGTAGCTG	[61 bp]	GATCACCGAG	gtagggtggg	[1.21 kb]
2	61–161	2073–2172	ctccccacag	ATGAGCAGCA	[100 bp]	GGCCCCCCAG	gtgagaagaa	[0.29 kb]
3	162–222	2463–2523	ccctccacag	GGGTCCAGTA	[61 bp]	TGACTGCCGG	gtaagtgccc	[1.07 kb]
4	223–392	3597–3766	tcacttccag	GGACCCAGTG	[170 bp]	CAGCTGGGCT	gtgagttggg	[0.12 kb]
5	393–507	3893–4007	ctgcctctag	ACCCTCCAGC	[115 bp]	CCTCCTACAG	gtgtgtgtgt	[0.15 kb]
6	508–540	4161–4193	tccccaccag	GAAGAAGACA	[33 bp]	ATAGCCAGAG	gtaggacgtg	[0.08 kb]
7	541–707	4279–4445	tcccttctag	GAGGAGTCCA	[167 bp]	CAGAGCATCT	gtgagtaccc	[0.93 kb]
8	708–871	5376–5539	atgcccttag	TGCGCCCTGA	[164 bp]	CTGGTCCACG	gtgaggcctg	[1.05 kb]
9	872–1013	6595–6736	ttacccccag	GTGGAGCCAG	[142 bp]	CCGAGCACTG	gtgagagaca	[0.37 kb]
10	1014–1133	7110–7229	cttcctttag	GGACCATACC	[120 bp]	CGGCTACTTG	gtgagcttgg	[0.11 kb]
11	1134–1230	7340–7436	tccccctcag	ATCACAGGGA	[97 bp]	TGGGGCTCTG	gtaagtgact	[0.25 kb]
12	1231–1313	7690–7772	tatgccccag	GCTGAGGCTG	[83 bp]	AGGCGTCCAG	gtgagtagga	[0.54 kb]
13	1314	8319–8701	cttcttctag	GAGCTCCAAA	[383 bp]	(AATAAA)		

Note. The size of each exon and its boundaries are tabulated. The position in the sequence as shown in column 2 is based on that of the murine cDNA (Neuhaus *et al.*, 1994) and is included for ease of comparison. The actual position of the exons in the human genomic sequence contig as determined in the present work are given in column 3. In the sequences of the exon–intron boundaries, intron sequences are in lowercase and exon sequences in uppercase. Note that the size of exon 13 is based on the position of the polyadenylation signal, as indicated (AATAAA), and not on the actual polyadenylation site (see text for details).

Figure 3 Alignment of the predicted protein chains of hIL-11Rα and mIL-11Rα genes. From Van Leuven *et al.* (1996), with permission.

```
Ex.2                                ↓ Ex.3               ↓ Ex.4
MSSSCSGLSRVLVAVATALVSASSPCPQAWGPPGVQYGQPGRSVKLCCPGVTAGDPVSWF     60
::::::::.:::::::::::: :::::::::::::::::::: : ::::::.::: ::::
MSSSCSGLTRVLVAVATALVSSSSPCPQAWGPPGVQYGQPGRPVMLCCPGVSAGTPVSWF     60

                                      ↓ Ex.5
RDGEPKLLQGPDSGLGHELVLAQADSTDEGTYICQTLDGALGGTVTLQLGYPPARPVVSC    120
:::. .::::::::::: ::::: :: :::::.::::::  :: ::: ::.::::: :::
RDGDSRLLQGPDSGLGHRLVLAQVDSPDEGTYVCQTLDGVSGGMVTLKLGFPPARPEVSC    120

     ▼                   ↓ Ex.6      ↓ Ex.7
QAADYENFSCTWSPSQISGLPTRYLTSYRKKTVLGADSQRRSPSTGPWPCPQDPLGAARC    180
:: ::::::::::: :.:::::::::::::::. ::.::: :::::::::::::: : ::
QAVDYENFSCTWSPGQVSGLPTRYLTSYRKKTLPGAESQRESPSTGPWPCPQDPLEASRC    180

             ▼                 ↓ Ex.8
VVHGAEFWSQYRINVTEVNPLGASTRLLDVSLQSILRPDPPQGLRVESVPGYPRRLRASW    240
::::::::: ::::::::::::::: :::: ::::::::::::::::::::::::: :::
VVHGAEFWSEYRINVTEVNPLGASTCLLDVRLQSILRPDPPQGLRVESVPGYPRRLHASW    240

                          ↓ Ex.9
TYPASWPCQPHFLLKFRLQYRPAQHPAWSTVEPAGLEEVITDAVAGLPHAVRVSARDFLD    300
::::::  ::::::::::::::::::::::::: ::::::::::::::::::::::::::
TYPASWRRQPHFLLKFRLQYRPAQHPAWSTVEPIGLEEVITDAVAGLPHAVRVSARDFLD    300

              ↓ Ex.10
AGTWSTWSPEAWGTPSTGTIPKEIPAWGQLHTQPEVE---PQVDSPAPPRPSLQPHPRLL    357
::::: :::::::::::: . .::: : : : :  .:    : ::::: :::::: :: :
AGTWSAWSPEAWGTPSTGPLQDEIPDWSQGHGQ-QLEAVVAQEDSPAPARPSLQPDPRPL    359

 ↓ Ex.11                        ↓ Ex.12
DHRDSVEQVAVLASLGILSFLGLVAGALALGLWLRLRRGGKDGSPKPGFLASVIPVDRRP    417
:::: .::::::::::: : :::  ::::::::::::: ::::  ::: ::  :::.. :
DHRDPLEQVAVLASLGIFSCLGLAVGALALGLWLRLRRSGKDGPQKPGLLAPMIPVEKLP    419

 ↓ Ex.13
GAPNL                                                           422
: :::
GIPNLQRTPENFS                                                   432
```

embryoid bodies *in vitro* has been examined by RNase protection assays (Davidson *et al.*, 1997). The testis showed a high level of IL-11 gene expression, and a much lower level of expression was detected in the lung, stomach, small intestine, and large intestine. Expression of the IL-11 ligand was not detected between day 10.5 and day 18.5 postcoitum of embryonic development or in differentiating ES cells *in vitro*. However, mIL-11Rα was expressed in all adult tissues examined, during embryonic development, and in totipotent and differentiating ES cells.

Murine megakaryocytes (MKs) are direct targets of rhIL-11 since they expressed functional mIL-11Rα (Weich *et al.*, 1997). Exposure of purified bone marrow MKs to rhIL-11 enhanced phosphorylation of both its signal transduction subunit, gp130, and evoked the transcription factor STAT3, showing a direct activation of receptor signaling by the cytokine. Consistent with the lack of effect of rhIL-11 on human platelets *in vivo*, hIL-11Rα mRNA and protein were not detected in isolated human platelets. These data indicate that rhIL-11 acts directly on MKs and MK progenitors but not on platelets.

Because rhIL-11 has immunomodulating activities in several animal models and biologic activity in patients with inflammation (Dorner *et al.*, 1997), the relative expression of mIL-11Rα has been determined in immune cells (Trepicchio and Dorner, 1998). RNase protection assays revealed the receptor component in murine blood marrow cells, peritoneal macrophages, and spleen cells, all known to be responsive to rhIL-11 treatment. However, only very low levels of hIL-11Rα were detectable by RT-PCR in human neutrophils (Bozza *et al.*, 1998). Not surprisingly, rhIL-11 treatment had no effect on neutrophil function.

Regulation of receptor expression

Primary osteoblasts constitutively expressed mRNAs for mIL-11Rα and gp130 (Romas *et al.*, 1996). Osteotropic factors did not modulate mIL-11Rα mRNA at 24 hours, but steady-state gp130 mRNA expression in osteoblasts was upregulated by $1\alpha,25(OH)_2D_3$, PTH, or IL-1. In co-cultures, formation of multinucleated osteoclast-like cells (OCLs) in response to IL-11, or IL-6 with its soluble IL-6 receptor, was dose-dependently inhibited by rat monoclonal antimouse gp130 antibody. Addition of anti-gp130 antibody abolished OCL formation induced by IL-1, and partially inhibited OCL formation induced by PGE_2, PTH, or $1\alpha,25(OH)_2D_3$. During osteoclast formation in marrow cultures, a sequential relationship existed between the expression of calcitonin receptor mRNA and mIL-11Rα mRNA. Osteoblasts as well as OCLs expressed transcripts for mIL-11Rα, as indicated by RT-PCR analysis and *in situ* hybridization. These results suggest a central role of gp130-coupled cytokines, especially IL-11, in osteoclast development. Since osteoblasts and mature osteoclasts expressed mIL-11Rα mRNA, both bone-forming and bone-resorbing cells are potential targets of IL-11.

Release of soluble receptors

Although a transcript of the hIL-11Rα gene has been detected which codes for a protein without the cytoplasmic domain (Cherel *et al.*, 1995), which could lead to speculation concerning a soluble receptor, circulating soluble IL-11Rα has not been detected in animals or humans.

SIGNAL TRANSDUCTION

The activities of murine (Barton *et al.*, 1999) and human (Czupryn *et al.*, 1995a, 1995b) IL-11 ligand mutants in receptor-binding and cell proliferation assays have been used to characterize the critical residues involved in the binding of murine and human IL-11 to both IL-11R and gp130. The location of these residues, as predicted from structural studies and a model of IL-11, suggest that murine and human IL-11 have three distinct receptor-binding sites, structurally and functionally analogous receptor-binding sites I, II, and III of IL-6. These data support the concept that IL-11 signals through the formation of a hexameric receptor complex and suggests that site III is a common feature of cytokines that signal through gp130. The signaling mechanisms through gp130 have been elegantly reviewed by Hirano *et al.* (1997).

Both IL-6 and IL-11, in a complex with their specific α receptors, interact with gp130 and, as a consequence, activate the Janus kinase/signal transducer and activator of transcription (JAK/STAT) signaling pathway in their target cells. However, it is not clear whether gp130 is bound to these cytokines and their specific α receptor subunits through identical or different epitopes. Dahmen *et al.* (1998) studied the interaction of IL-11 and IL-11R with human gp130 using a soluble hIL-11Rα, expressed in baculovirus-infected insect cells. Coprecipitation binding assays revealed that IL-11 and IL-6 compete for binding to gp130. Then, deletion and point mutations of gp130 were used to show that IL-11–IL-11R and IL-6–IL-6R recognize overlapping binding motifs on gp130.

Transfection of IL-3-dependent immortalized hematopoietic cells, Ba/F3 cells, with the two isoforms of the hIL-IIRα (α_1 full length or α_2 lacking the cytoplasmic domain) in combination with human gp130 was performed (Lebeau *et al.*, 1997). These cells were stimulated with similar efficiencies and proliferated with superimposable dose–response curves to IL-11, showing that the intracellular domain of IL-11Rα has no significant effect on IL-11 ligand binding and signaling.

Associated or intrinsic kinases

The initial characterization of the biochemical nature of mIL-11Rα and possible signal transduction pathways mediated by IL-11 in 3T3-L1 mouse preadipocytes was performed in the Yang laboratory (Yin *et al.*, 1992). IL-11 strongly inhibited lipoprotein lipase activity and adipogenesis in 3T3-L1 cells, and the suppression of lipoprotein lipase activity by IL-11 was controlled at the posttranscriptional level. Scatchard plot analysis according to specific binding data revealed the existence of a single class of high-affinity IL-11R with a K_d of 3.49×10^{-10} M and a receptor density of 5140 sites/cell on 3T3-L1 cells. Affinity crosslinking studies with [^{125}I]IL-11 indicated that IL-11R consists of a single polypeptide chain of 151 kDa in size. Studies of the role of protein tyrosine phosphorylation in the IL-11R-linked signal transduction pathways revealed that IL-11R ligation rapidly and transiently stimulated tyrosine phosphorylation of 152, 94, 47, and 44 kDa proteins. The effect is specific for IL-11, as neutralizing antibody to IL-11 blocked the IL-11-induced tyrosine phosphorylation.

Subsequently, treatment of 3T3-L1 cells with IL-11, IL-6, LIF, and oncostatin M was shown to induce overlapping but distinct patterns of tyrosine phosphorylation and activation-indistinguishable primary response genes (Yin *et al.*, 1994). It was demonstrated that IL-11, IL-6, LIF, and oncostatin M trigger the activation of mitogen-activated protein kinases and the 85–92 kDa ribosomal S6 protein kinase (pp90rsk). Preincubation of cells with a tyrosine kinase inhibitor herbimycin A, but not with a serine/threonine kinase inhibitor H7, blocked activation of mitogen-activated protein kinases and pp90rsk. H7, but not herbimycin A, and inhibited pp90rsk activity in the *in vitro* kinase assays, suggesting that pp90rsk is one of the potential candidates for the H7-sensitive protein kinases, which is critical for the activation of primary response genes by these cytokines.

A 130 kDa tyrosine-phosphorylated protein induced by IL-11 in 3T3-L1 cells was identified as JAK2 tyrosine kinase (Yin and Yang, 1994). The *in vitro* kinase activity of JAK2 is greatly enhanced following stimulation with IL-11 in 3T3-L1 cells and TF-1 cells, and JAK2 physically associated with the signal transducer gp130. Similar results were observed following stimulation with IL-6, LIF, and oncostatin M.

Cytoplasmic signaling cascades

The effects of IL-11 on [^{3}H]phosphatidic acid (PA) formation in [^{3}H]arachidonic acid (AA) prelabeled quiescent mouse 3T3-L1 cells has been studied (Siddiqui and Yang, 1995). The result of this study suggested that one of the cellular signaling mechanisms of IL-11 in 3T3-L1 cells involves the activation of phospholipase D to produce the second messenger PA. The increased level of PA then enhanced tyrosine phosphorylation of p44 and p47, which belong to the members of the mitogen-activated protein kinase family, and thus transduced some of the mitogenic signals of IL-11 in this cell line.

Addition of IL-11 to 3T3-L1 cells resulted in an increase in the tyrosine phosphorylation of Syp (Fuhrer *et al.*, 1995). Syp was inducibly associated with both gp130 and JAK2. A phosphopeptide containing the sequence for a potential Syp-binding site (YXXV) was used to compete with the associations of Syp with gp130 and JAK2. The phosphopeptide reduced the Syp association with both gp130 and JAK2. Syp had multiple interactions in IL-11 signal transduction. In addition to the IL-11-induced tyrosine phosphorylation of Syp, Syp coprecipitated with gp130, JAK2, and other tyrosine-phosphorylated proteins in response to IL-11.

IL-11 promotes the formation of the active GTP-bound form of Ras, suggesting that IL-11 actions may be transduced in part through the Ras/mitogen-activated protein kinase signaling pathway (Wang *et al.*, 1995). The association of tyrosine phosphoproteins with Grb2, an adapter protein, may serve as a key intermediate for Ras activation. These phosphotyrosine-containing proteins have been subsequently identified to be JAK2, Fyn, and Syp. JAK2 and Fyn are transiently associated with Grb2 upon stimulation with IL-11, suggesting that JAK2 and Fyn may be involved in transducing signals from the IL-11R gp130 to the Ras system through Grb2.

When the IL-11-responsive cell line 3T3-L1 (mouse preadipocytes) was tested for the presence of Src family protein tyrosine kinases (PTKs), only p62yes, p59fyn, and p60src were found (Fuhrer and Yang, 1996). Immune complex kinase reactions using the artificial substrate enolase showed that IL-11 activated p62yes and p60src. IL-11-stimulated cells also exhibited increased phosphatidylinositol 3-kinase (PI-3 kinase) activity. The increased activity of PI-3 kinase was found to be associated with tyrosine-phosphorylated proteins and p62yes, after IL-11 treatment. Immunoprecipitation studies with anti-PI-3 kinase revealed that p62yes associated with PI-3 kinase in response to IL-11. Spencer and Adunyah (1997) studied the possible involvement of PKC in the IL-11-signaling pathway. IL-11 stimulated rapid PKC activation and markedly induced cytosolic to particulate (membrane) association of α and β PKC isoforms, suggesting that PKC may be involved in the IL-11-signaling cascade.

Dependent on the cell type and the condition of that cell, whether damaged or normal, the effect of IL-11R activation may vary. The consequences of gp130 signaling and the subsequent activation of STAT3 and STAT1 are shown in **Figure 4** and **Figure 5** (Hirano *et al.*, 1997).

DOWNSTREAM GENE ACTIVATION

Transcription factors activated

Analysis of rhIL-11 effects on transcription factors that activate proinflammatory cytokines revealed that the level of LPS-induced NFκB-binding activity in the nucleus of rhIL-11-treated peritoneal macrophages was significantly reduced (Trepicchio *et al.*, 1997). The block to NFκB nuclear translocation was related to the ability of rhIL-11 to maintain or increase protein levels of the inhibitors of NFκB, IκB-α, and IκB-β following LPS treatment. Treatment of LPS-stimulated macrophages resulted in significant elevation of IκB-α and IκB-β mRNA levels. These results suggest that the antiinflammatory activity of rhIL-11 is mediated in part by inhibition of NFκB-dependent transcriptional activation and demonstrate for the first time the regulation of IκB-β by an anti-inflammatory cytokine.

Genes induced

Depending on cell type, a variety of early-response genes are induced, including *tis8*, *tis11*, *tis21*, and *junB* (Du and Williams, 1997). Of course, the acute-phase proteins are activated, as is the case with IL-6, but the potency of IL-11 appears weak, compared to IL-6 (Fukuda and Sassa, 1993).

Promoter regions involved

The specific promoter regions of various genes activated specifically by IL-11-evoked gp130 signaling remain to be defined.

Figure 4 Cytoplasmic topology of gp130 signal transduction pathways. From Hirano *et al.* (1997), with permission.

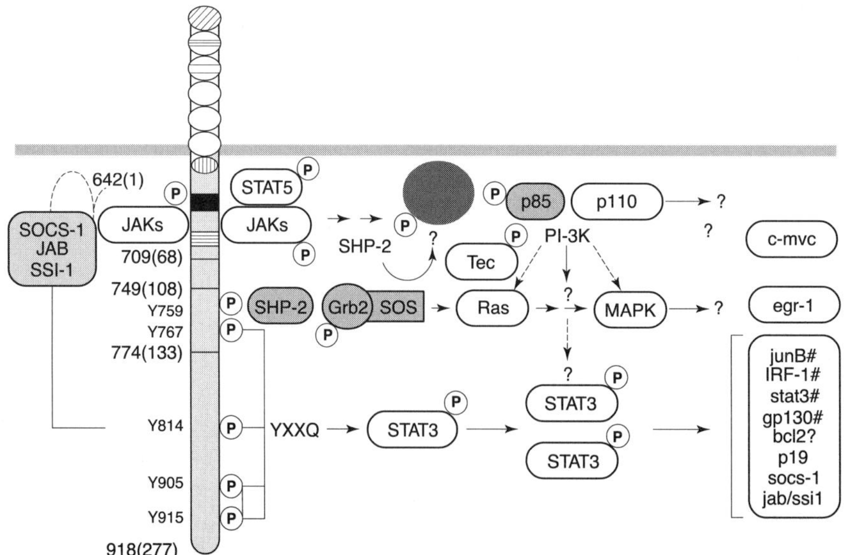

Figure 5 Gp130 signal transduction pathways regulating cell growth/differentiation and death. Depending on the state of the target cell, IL-11-evoked signaling may produce cell cycle arrest or cell cycle progression, with anti-apoptotic activity. From Hirano *et al.* (1997), with permission.

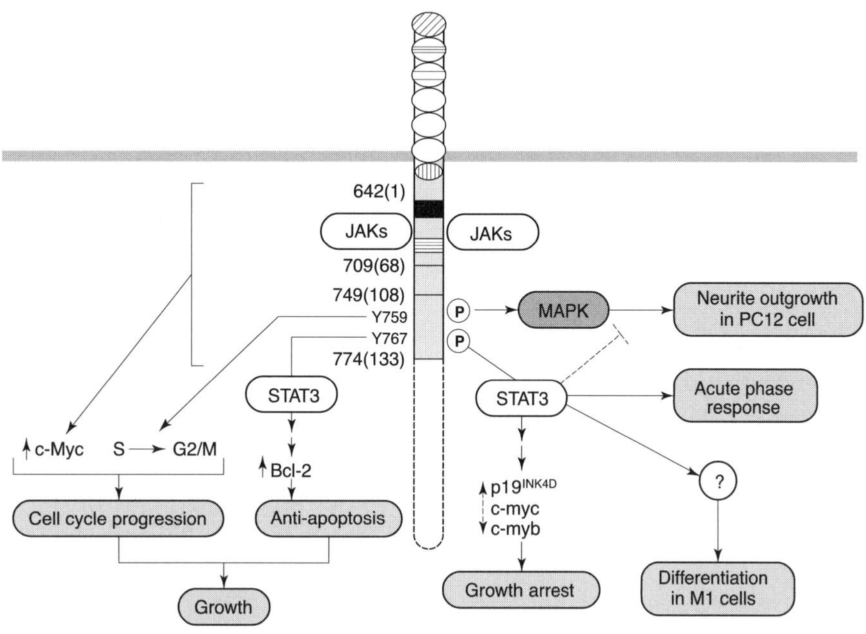

BIOLOGICAL CONSEQUENCES OF ACTIVATING OR INHIBITING RECEPTOR AND PATHOPHYSIOLOGY

Unique biological effects of activating the receptors

IL-11 has multiple activities that continue to be characterized. Initially, *in vitro* and *in vivo* studies demonstrated a hematopoietic activity for IL-11, largely manifest as thrombopoiesis in humans, and rhIL-11 has been developed and recently approved by the Food and Drug Administration for use in the prevention of *severe thrombocytopenia* occurring after *cancer* chemotherapy.

Studies of cells and tissues from other organ systems indicate that IL-11 has: activity in protection and restoration of the gastrointestinal mucosa, major effects as an immunomodulating agent, and activity in bone metabolism. Developmental investigations in mice indicate a widespread distribution of IL-11 expression in the embryo, but of great interest is the finding that IL-11 signaling is an absolute requirement for normal development of placentation and survival to birth.

Phenotypes of receptor knockouts and receptor overexpression mice

Murine studies of the IL-11Rα chain knockout mouse indicate that activity of IL-11 is an absolute requirement for successful reproductive function (Robb *et al.*, 1998). Blastocysts were able to implant initially, but then decidualization did not occur and pregnancy failed. The secondary decidual zone did not develop, and the space normally occupied by mesometrial decidua was filled with trophoblast giant cells at 6.5 days. Most embryos were dead by 7.5 days, and no viable embryos were present after 10.5 days.

THERAPEUTIC UTILITY

Effect of treatment with soluble receptor domain

The influence of cardiotropin (CT-1) and IL-11 on newborn rat dorsal root ganglion neuron survival *in vitro* has been reported (Their *et al.*, 1999). Mouse CT-1 showed prominent trophic effects that were comparable to those of CNTF and LIF. Mouse IL-11 alone did not enhance neuronal survival, but soluble mouse IL-11Rα rendered neurons sensitive to IL-11. Surprisingly, soluble IL-11Rα even had slight neurotropic effects by itself. These results suggest that CT-1 and IL-11 might also be involved in the regulation of sensory neuron survival.

Curtis *et al.* (1997) expressed a soluble *N*-glycosylated form of the murine IL-11Rα chain (sIL-11R) and examined signaling in cells expressing the gp130 molecule. In the presence of gp130 but not the transmembrane IL-11R, the sIL-11R mediated IL-11-dependent differentiation of M1 leukemic cells and proliferation in the IL-3-dependent hematopoietic cell line Ba/F3 cells. Early intracellular events stimulated by sIL-11R, including phosphorylation of gp130, STAT3, and SHP-2, were similar to signaling through the transmembrane IL-11R. IL-11 bound to sIL-11R with low affinity ($K_d = 10$–$50\,nM$). Binding of sIL-11R to gp130 was IL-11-dependent with intermediate affinity ($K_d = 1.5$–$3.0\,nM$). However, the concentration of IL-11 required for signaling through the sIL-11R was 10- to 20-fold greater than that required for cells expressing transmembrane IL-11R and gp130 in the absence of sIL-11R.

Pflanz *et al.* (1999) have shown that a recombinant fusion protein of a fragment of the hIL-11Pα ectodomain linked to human IL-11 acts as a superagonist on cells expressing gp130 but lacking membrane-bound hIL-11Pα. It induces acute-phase protein synthesis in hepatoma cells and efficiently promotes proliferation of Ba/F3 cells stably transfected with gp130. In these bioassays, fusion protein was 50 times more potent than the combination of IL-11 and soluble hIL-11Pα.

Effects of inhibitors (antibodies) to receptors

Although these studies suggest a similar structural motif that is involved with IL-11 ligand and receptor complex interactions, other studies indicate that they are not identical. IL-6 mutants have been produced that function as IL-6R antagonists (IL-6Ra) (Sun *et al.*, 1997). These mutants had substitutions that increased their affinity with IL-6R and abolished one of the two sites of interaction with gp130. The IL-6Ra with one mutated binding site to gp130 inhibited IL-11 activity. It did not affect the interactions of CNTF, LIF, and oncostatin M, even when used at a very high concentration, suggesting that the interaction of one gp130 chain with IL-6R/IL-6R complexes further inhibited the dimerization of gp130 induced by IL-11/IL-11R, but not its heterodimerization with LIFR. Gu *et al.* (1996) used human plasmacytoma cell lines, completely dependent on the addition of one of the IL-6 family of cytokines for their growth, to produce anti-gp130 monoclonal antibodies specifically inhibiting one of these five cytokines without affecting the biological activity of the others. Administration of specific anti-gp130 antibodies to dogs has also been demonstrated to be a potent inhibitor of the IL-6-induced acute-phase response, thereby blocking IL-6-mediated increments in fibrinogen, C-reactive protein, and platelet count (Harrison *et al.*, 1996).

sIL-11R was capable of antagonizing the activity of IL-11 when tested on cells expressing transmembrane IL-11R and gp130 (Curtis *et al.*, 1997). These data support the contention that the observed IL-11 antagonism by the sIL-11R may depend on limiting numbers of gp130 molecules on cells already expressing transmembrane IL-11R.

A stromal protein, designated restrictin-P, subsequently shown to be identical to activin A, specifically kills plasma-like cells by competitively antagonizing the proliferation-inducing effects of IL-6 and IL-11 (Brosh *et al.*, 1995). Interestingly, the competition-binding assay indicated that activin A did not interfere with the binding of IL-6 to its receptor on plasma-like cells, but suggested that it may act by intervening in the signal transduction pathway of IL-6. In B9 cells, addition of IL-6 and activin A was followed by sustained overexpression of the *junB* gene until cell death occurred. This is in contrast to transient expression of *junB* typically evoked by IL-6 and IL-11.

References

Barton, V. A., Hudson, K. R., and Heath, J. K. (1999). Identification of three distinct receptor binding sites of murine interleukin-11. *J. Biol. Chem.* **274**, 5755–5761.

Baumann, H., Wang, Y., Morella, K. K., Lai, C. F., Dams, H., Hilton, D. J., Hawley, R. G., and Mackiewicz, A. (1996). Complex of the soluble IL-11 receptor and IL-11 acts as IL-6-type cytokine in hepatic and nonhepatic cells. *Immunology* **157**, 284–290.

Bilinski, P., Hall, M. A., Neuhaus, H., Gissel, C., Heath, J. K., and Gossler, A. (1996). Two differentially expressed interleukin-11 receptor genes in the mouse genome. *Biochem J.* **320**, 359–363.

Bilinski, P., Roopenian, D., and Gossler, A. (1998). Maternal IL-11Rα function is required for normal decidua and fetoplacental development in mice. *Genes Dev.* **12**, 2234–2243.

Bozza, M., Kyvelos, D., Trepicchio, W. L., Collins, M., Klempner, M. S., and Dorner, A. J. (1998). Recombinant human interleukin-11 does not affect functions of purified human neutrophils *in vitro*. *J. Interferon Cytokine Res.* **18**, 889–895.

Brosh, N., Sternberg, D., Honigwachs-Sha'anani, J., Lee, B. C., Shav-Tal, Y., Tzehoval, E., Shulman, L. M., Toledo, J., Hacham, Y., Carmi, P., Jiang, W., Sasse, J., Horn, F., Burstein, Y., and Zipori, D. (1995). The plasmacytoma growth inhibitor restrictin-P is an antagonist of interleukin 6 and interleukin 11. Identification as a stroma-derived activin A. *J. Biol. Chem.* **270**, 29594–29600.

Cherel, M., Sorel, M., Lebeau, B., Dubois, S., Moreau, J. F., Bataille, R., Minvielle, S., and Jacques, Y. (1995). Molecular cloning of two isoforms of a receptor for the human hematopoietic cytokine interleukin-11. *Blood* **86**, 2534–2540.

Cherel, M., Sorel, M., Apiou, F., Lebeau, B., Dubois, S., Jacques, Y., and Minvielle, S. (1996). The human interleukin-11 receptor alpha gene (IL11RA): genomic organization and chromosome mapping. *Genomics* **32**, 49–53.

Curtis, D. J., Hilton, D. J., Roberts, B., Murray, L., Nicola, N., and Begley, C. G. (1997). Recombinant soluble interleukin-11 (IL-11) receptor alpha-chain can act as an IL-11 antagonist. *Blood* **90**, 4403–4412.

Czupryn, M. J., McCoy, J. M., and Scoble, H. A. (1995a). Structure–function relationships in human interleukin-11. Identification of regions involved in activity by chemical modification and site-directed mutagenesis. *J. Biol. Chem.* **270**, 978–985.

Czupryn, M., Bennett, F., Dube, J., Grant, K., Scoble, H., Sookdeo, H., and McCoy, J. M. (1995b). Alanine-scanning mutagenesis of human interleukin-11: identification of regions important for biological activity. *Ann. NY Acad. Sci.* **762**, 152–164.

Dahmen, H., Horsten, U., Kuster, A., Jacques, Y., Minvielle, S., Kerr, I. M., Ciliberto, G., Paonessa, G., Heinrich, P. C., and Muller-Newen, G. (1998). Activation of the signal transducer gp130 by interleukin-11 and interleukin-6 is mediated by similar molecular interactions. *Biochem. J.* **331**, 695–702.

Davidson, A. J., Freeman, S. A., Crosier, K. E., Wood, C. R., and Crosier, P. S. (1997). Expression of murine interleukin 11 and its receptor alpha-chain in adult and embryonic tissues. *Stem Cells* **15**, 119–124.

Dorner, A. J., Goldman, S., and Keith, JR. (1997). Interleukin-11: biological activity and clinical studies. *BioDrugs* **8**, 418–429.

Du, X., and Williams, D. A. (1997). Interleukin-11: review of molecular, cell biology, and clinical use. *Blood* **89**, 3897–3908.

Fuhrer, D. K., and Yang, Y. C. (1996). Activation of Src-family protein tyrosine kinases and phosphatidylinositol 3-kinase in 3T3-L1 mouse preadipocytes by interleukin-11. *Exp. Hematol.* **24**, 195–203.

Fuhrer, D. K., Feng, G. S., and Yang, Y. C. (1995). Syp associates with gp130 and Janus kinase 2 in response to interleukin-11 in 3T3-L1 mouse preadipocytes. *J. Biol. Chem.* **270**, 24826–24830.

Fukuda, Y., and Sassa, S. (1993). Effect of interleukin-11 on the levels of mRNAs encoding heme oxygenase and haptoglobin in human HepG2 hepatoma cells. *Biochem. Biophys. Res. Commun.* **193**, 297–302.

Gossler, A., Joyner, A. L., Rossant, J., and Skarnes, W. C. (1989). Mouse embryonic stem cells and reporter constructs to detect developmentally regulated genes. *Science* **244**, 463–465.

Gu, Z. J., Wijdenes, J., Zhang, X. G., Hallet, M. M., Clement, C., and Klein, B. (1996). Anti-gp130 transducer monoclonal antibodies specifically inhibiting ciliary neurotrophic factor, interleukin-6, interleukin-11, leukemia inhibitory factor or oncostatin M. *J. Immunol. Methods* **190**, 21–27.

Harrison, P., Downs, T., Friese, P., Wolf, R., George, J. N., and Burstein, S. A. (1996). Inhibition of the acute-phase response *in vivo* by anti-gp130 monoclonal antibodies. *Br. J. Haematol.* **95**, 443–451.

Hilton, D. J., Hilton, A. A., Raicevic, A., Rakar, S., Harrison-Smith, M., Gough, N. M., Begley, C. G., Metcalf, D., Nicola, N. A., and Wilson, T. A. (1994). Cloning of a murine IL-11 receptor α-chain; requirement for gp130 for high affinity binding and signal transduction. *EMBO J.* **13**, 4765–4775.

Hirano, T., Nakajima, K., and Hibi, M. (1997). Signaling mechanisms through gp130: a model of the cytokine system. *Cytokine Growth Factor Rev.* **8**, 241–252.

Lebeau, B., Montero Julian, F. A., Wijdenes, J., Muller-Newen, G., Dahmen, H., Cherel, M., Heinrich, P. C., Brailly, H., Hallet, M. M., Godard, A., Minvielle, S., and Jacques, Y. (1997). Reconstitution of two isoforms of the human interleukin-11 receptor and comparison of their functional properties. *FEBS Lett.* **407**, 41–47.

Miyadai, K., Ohsumi, J., Yoshimura, C., Kawashima, I., and Ito, Y. (1996). Importance of the carboxy-terminus of human interleukin-11 in conserving its biological activity. *Biosci. Biotechnol. Biochem.* **60**, 541–542.

Nandurkar, H. H., Hilton, D. J., Nathan, P., Willson, T., Nicola, N., and Begley, C. G. (1996). The human IL-11 receptor requires gp130 for signaling: demonstration by molecular cloning of the receptor. *Oncogene* **12**, 585–593.

Nandurkar, H. H., Robb, L., Tarlinton, D., Barnett, L., Kontgen, F., and Begley, C. G. (1997a). Adult mice with targeted mutation of the interleukin-11 receptor (IL11Ra) display normal hematopoiesis. *Blood* **90**, 2148–2159.

Nandurkar, H. H., Robb, L., Nicholl, J. K., Hilton, D. J., Sutherland, G. R., and Begley, C. G. (1997b). The gene for the human interleukin-11 receptor alpha chain locus is highly homologous to the murine gene and contains alternatively spliced first exons. *Int. J. Biochem. Cell Biol.* **29**, 753–766.

Neddermann, P., Graziani, R., Ciliberto, G., and Paonessa, G. (1996). Functional expression of soluble human interleukin-11 (IL-11) receptor alpha and stoichiometry of *in vitro* IL-11 receptor complexes with gp130. *J. Biol. Chem.* **271**, 30986–30991.

Neuhaus, H., Bettenhausen, B., Bilinski, P., Simon-Chazottes, D., Guenet, J. L., and Gossler, A. (1994). Etl2, a novel putative type-1 cytokine receptor expressed during mouse embryogenesis at high levels in skin and cells with skeletogenic potential. *Dev. Biol.* **166**, 531–542.

Pflanz, S., Tacken, I., Grotzinger, J., Jacques, Y., Dahmen, H., Heinrich, P. C., and Muller-Newen, G. (1999). A fusion protein of interleukin-11 and soluble interleukin-11 receptor acts as a superagonist on cells expressing gp130. *FEBS Lett.* **450**, 117–122.

Robb, L., Hilton, D. J., Brook-Carter, P. T., and Begley, C. G. (1997). Identification of a second murine interleukin-11 receptor alpha-chain gene (IL11Ra2) with a restricted pattern of expression. *Genomics* **40**, 387–394.

Robb, L., Li, R., Hartley, L., Nandurkar, H. H., Koentgen, F., and Begley, C. G. (1998). Infertility in female mice lacking the receptor for interleukin 11 is due to a defective uterine response to implantation. *Nature Med.* **4**, 303–308.

Romas, E., Udagawa, N., Zhou, H., Tamura, T., Saito, M., Taga, T., Hilton, D. J., Suda, T., Ng, K. W., and Martin, T. J. (1996). The role of gp130-mediated signals in osteoclast development: regulation of interleukin 11 production by osteoblasts and distribution of its receptor in bone marrow cultures. *J. Exp. Med.* **183**, 2581–2591.

Siddiqui, R. A., and Yang, Y. C. (1997). Interleukin-11 induces phosphatidic acid formation and activates MAP kinase in mouse 3T3-L1 cells. *Cell Signal* **7**, 247–259.

Spencer, G. C., and Adunyah, S. E. (1995). Interleukin-11 induces rapid PKC activation and cytosolic to particulate translocation of alpha and beta PKC isoforms in human erythroleukemia K562 cells. *Biochem. Biophys. Res. Commun.* **232**, 61–64.

Sun, R. X., Gennaro, C., Rocco, S., Gu, Z. J., and Klein, B. (1997). Interleukin-6 receptor antagonists inhibit interleukin-11 biological activity. *Eur. Cytokine Netw.* **8**, 51–56.

Their, M., Hall, M., Heath, J. K., Pennica, D., and Weis, J. (1999). Trophic effects of cardiotrophin-1 and interleukin-11 on rat dorsal root ganglion neurons *in vitro*. *Brain Res. Mol. Brain Res.* **64**, 80–84.

Trepicchio, W. L., Wang, L., Bozza, M., and Dorner, A. J. (1997). IL-11 regulates macrophage effector function through the inhibition of nuclear factor-kappaB. *J. Immunol.* **159**, 5661–5670.

Trepicchio, W. L., and Dorner, A. J. (1998). Interleukin-11 A gp130 cytokine. *Ann. NY Acad. Sci.* **856**, 12–21.

Van Leuven, F., Stas, L., Hillicker, C., Miyake, Y., Bilinski, P., and Gossler, A. (1996). Molecular cloning and characterization of the human interleukin-11 receptor α-chain gene, IL11RA, located on chromosome 9p13. *Genomics* **31**, 65–70.

Wang, X. Y., Fuhrer, D. K., Marshall, M. S., and Yang, Y. C. (1995). Interleukin-11 induces complex formation of Grb2, Fyn, and JAK2 in 3T3L1 cells. *J. Biol. Chem.* **270**, 27999–28002.

Weich, N. S., Wang, A., Fitzgerald, M., Neben, T. Y., Donaldson, D., Giannott, J., Yetz-Aldape, J., Leven, R. M., and Turner, K. J. (1997). Recombinant human interleukin-11 directly promotes megakaryocytopoiesis *in vitro*. *Blood* **90**, 3893–3902.

Yin, T., Miyazawa, K., and Yang, Y. C. (1992). Characterization of interleukin-11 receptor and protein tyrosine phosphorylation induced by interleukin-11 in mouse 3T3-L1 cells. *J. Biol. Chem.* **267**, 8347–8351.

Yin, T., and Yang, Y. C. (1994). Mitogen-activated protein kinases and ribosomal S6 protein kinases are involved in signaling pathways shared by interleukin-11, interleukin-6, leukemia inhibitory factor, and oncostatin M in mouse 3T3-L1 cells. *J. Biol. Chem.* **269**, 3731–3738.

Yin, T., Yasukawa, K., Taga, T., Kishimoto, T., and Yang, Y. C. (1994). Identification of a 130-kilodalton tyrosine-phosphorylated protein induced by interleukin-11 as JAK2 tyrosine kinase, which associates with gp130 signal transducer. *Exp. Hematol.* **22**, 467–472.

Yoshida, K., Taga, T., Saito, M., Suematsu, S., Kumanogoh, A., Tanaka, T., Fujiwara, H., Hirata, M., Yamagami, T., Nakahata, T., Hirabayashi, T., Yoneda, Y., Tanaka, K., Wang, W. Z., Mori, C., Shiota, K., Yoshida, N., and Kishimoto, T. (1996). Targeted disruption of gp130, a common signal transducer for the interleukin 6 family of cytokines, leads to myocardial and hematological disorders. *Proc. Natl Acad. Sci. USA* **93**, 407–411.

OSM Receptor

Timothy M. Rose* **and A. Gregory Bruce**

Department of Pathobiology, School of Public Health and Community Medicine, University of Washington, Box 357238, Seattle, WA 98195, USA

*corresponding author tel: 206 616 2084, fax: 206 543 3873, e-mail: trose@u.washington.edu

DOI: 10.1006/rwcy.2000.17005.

SUMMARY

Oncostatin M (OSM), a member of the IL-6 family of cytokines, interacts with low-affinity receptor subunit monomers and high-affinity heterodimeric receptor complexes composed of members of the class I cytokine receptor family. Important species-specific differences in receptor binding have been identified. OSM binds directly with low-affinity to gp130, which was originally characterized as the signal transducer subunit within the high-affinity IL-6 receptor complex.

BACKGROUND

Discovery

High- and low-affinity receptors for oncostatin M were originally detected using binding assays on a wide variety of cell types (Linsley *et al.*, 1989; Horn *et al.*, 1990). Crosslinking studies using ^{125}I-labeled OSM revealed a major binding protein of approximately 160 kDa (Linsley *et al.*, 1989). The nature and identity of these receptors first came to light after the discovery that OSM was structurally and functionally related to leukemia-inhibitory factor (LIF) Rose and Bruce, 1991), and that OSM shared with LIF the ability to bind the high-affinity LIF receptor (Bruce *et al.*, 1992a; Gearing *et al.*, 1992a). The high-affinity LIF receptor is a heterodimeric complex of two receptor subunits, the LIFα receptor (LIFRα) which binds LIF directly at low affinity (Gearing *et al.*, 1991) and gp130, a molecule previously shown to be the signaling subunit of the high-affinity interleukin 6 (IL-6) receptor complex (Hibi *et al.*, 1990). Molecular characterization of LIFRα and gp130 revealed a close similarity between the two and to the members of a newly described cytokine receptor family which includes the IL-6α receptor (IL-6Rα) (Bazan, 1990). Ligand binding to this class of receptors is characterized by low-affinity binding to an α receptor subunit which is converted to high affinity by further association of an additional receptor component or components involved in signal transduction, usually refered to as converting or β receptor subunits. Binding studies demonstrated that gp130 is the low-affinity receptor for OSM (Gearing *et al.*, 1992a; Liu *et al.*, 1992b). Thus, while gp130 is the β affinity-converting receptor subunit for the high-affinity receptor complexes for LIF, IL-6 and other members of the IL-6 family of cytokines, it is the α receptor subunit for OSM within the high-affinity OSM receptor complex, and, in this context, will be referred to as OSMRα(gp130).

Previous studies noted the presence of a high-affinity receptor which was specific to OSM and did not bind LIF (Bruce *et al.*, 1992b). While the shared LIF/OSM receptor described above is composed of OSMRα(gp130) and LIFRα, the OSM-specific receptor is composed of OSMRα(gp130) and a previously undescribed receptor subunit closely related to OSMRα(gp130), LIFRα, and other members of the cytokine receptor family. This receptor subunit converted the receptor binding of OSM to high affinity in the absence of LIFRα and was thus termed the OSMβ receptor (OSMRβ) (Mosley *et al.*, 1996). An important note is that studies in mice have shown that murine OSM binds only the OSM-specific receptor with high affinity, as discussed further below.

Alternative names

The low-affinity OSMα receptor gp130, herein designated as OSMRα(gp130), has also been called the IL-6 signal transducer or the IL-6 receptor β

chain (Hibi *et al.*, 1990). The LIF α receptor, which acts as a β affinity-converting receptor subunit for the LIF/OSM shared receptor in humans, is variously designated as LIFR, LIFRα, LIFRβ, or differentiation-stimulating factor receptor (Gearing *et al.*, 1991; Gearing *et al.*, 1992b). The β affinity-converting receptor for the high-affinity OSM-specific receptor is termed OSMRβ or OSM receptor β subunit in humans (Mosley *et al.*, 1996). In the murine system, only one high-affinity receptor complex exists for murine OSM, which consists of the murine homologs of OSMRα(gp130) and OSMRβ (Ichihara *et al.*, 1997; Lindberg *et al.*, 1998). Binding and crosslinking studies in heterologous cells suggest that murine OSM binds separately to both subunits of this receptor complex at low affinity, although the relative affinities are unknown at this time. As such, the murine homolog of OSMRβ is referred to as a specific receptor for OSM to denote the fact that it binds murine OSM directly at low affinity (Lindberg *et al.*, 1998). However, the murine homolog of gp130, which also binds OSM directly at low affinity, has not been termed as such in the literature.

Structure

Molecular cloning studies of OSMRα(gp130), OSMRβ, and LIFRα have demonstrated a structural relationship between the three receptor subunits which categorizes them as members of the class I cytokine receptor family (Mosley *et al.*, 1996; Taga, 1996; Lindberg *et al.*, 1998). Members of this family contain conserved hematopoietin domains of approximately 200 amino acids in the extracellular portion of the receptor (Bazan, 1990). Within this domain are positioned four conserved cysteine residues and a WSXWS motif, where X is any

Table 1 OSMRα(gp130), OSMRβ, and LIFRα gene sequences

Accession	Species	Source	Type	Size (bp)	Reference
OSMRα(gp130)					
M57230	Human	Placenta	mRNA, complete	3085	Hibi *et al.*, 1990
S80479	Human	Embryo	Alternate splice IF-1[a] mRNA, partial	150	Sharkey *et al.*, 1995
U58146	Human	Blood cells	Alternate splice IF-2[a] mRNA, partial	153	Diamant *et al.*, 1997
X62646	Mouse	Macrophage	mRNA, complete	2995	Saito *et al.*, 1992
M92340	Rat	Liver	mRNA, complete	3053	Wang *et al.*, 1992
OSMRβ					
U60805	Human	Placenta/bone marrow/fibroblast	mRNA composite, complete	4171	Mosley *et al.*, 1996
AB015978	Mouse		mRNA, complete	4026	Unpublished
AF058805	Mouse	Skeletal muscle	mRNA, complete	4792	Lindberg *et al.*, 1998
LIFRα					
X61615	Human	Placenta	mRNA, complete	3591	Gearing *et al.*, 1991
U78628	Human	Placenta	Alternate 5′ noncoding exon	224	Unpublished
Not deposited	Human	Liver	3 alternate spliced isoforms		Tomida, 1997
U78104	Human	Placenta	Promoter and partial exon 1	4935	Unpublished
S83362	Human	Placenta	5′ region and exon 1	1350	Tomida and Gotoh, 1996
AF018079	Human		Alternate promoter (nonplacental)	681	Unpublished

[a]IF = isoform designation used within this chapter.

amino acid. Additionally, three fibronectin type III modules which are considered to function as ligand-binding pockets are positioned proximal to the transmembrane-spanning domain.

Main activities and pathophysiological roles

OSM is a pleiotropic cytokine which regulates cell growth and differentiation in a wide variety of biological systems, including hematopoiesis, neurogenesis, and osteogenesis (Bruce *et al.*, 1992b). However, the elaboration of the biological activities of OSM has been confounded by the presence of different OSM receptor signaling systems in humans and mice. In humans, OSM signals through two different receptors complexes: the LIF/OSM shared receptor (Gearing and Bruce, 1992), which shares high-affinity binding with LIF, an evolutionarily related protein with structural similarity to OSM (Rose and Bruce, 1991; Rose *et al.*, 1993), and the OSM-specific receptor, which binds OSM uniquely (Bruce *et al.*, 1992a). In mice, OSM signals only through the murine homolog of the OSM-specific receptor (Ichihara *et al.*, 1997; Lindberg *et al.*, 1998).

To confuse matters, human OSM, used historically for *in vitro* and *in vivo* studies in mice, binds uniquely to the murine LIF receptor and thus exhibits only the biological activities of LIF in mice and not those of OSM (Ichihara *et al.*, 1997; Lindberg *et al.*, 1998). Therefore, the biological activities for OSM are derived from signaling through two different receptors and overlap those of LIF in human but not murine systems. Receptor utilization of OSM in other species has not yet been well defined. As such, the literature on OSM should be reviewed with careful consideration of these findings.

GENE

Accession numbers

See **Table 1**.

Sequence

The complete mRNA coding sequences for the membrane-bound forms of human and mouse OSMRα(gp130), OSMRβ, and LIFRα have been determined (**Figure 1**, **Figure 2**, and **Figure 3**; Table 1). In addition, alternately spliced mRNAs have been detected for OSMRα(gp130) and LIFRα (Table 1) which produce different translated products that correspond to soluble forms of the receptor subunits (**Figure 4** and **Figure 5**). An alternate splice of a 5′ noncoding exon of the human LIFRα has also been identified (Table 1). The gene for human LIFRα spans more than 70 kilobases and contains 20 exons (Tomida and Gotoh, 1996).

PROTEIN

Accession numbers

See **Table 2**.

Description of protein

A general comparison of the different OSM membrane-bound receptor subunits encoded by the mRNAs for OSMRα(gp130) (Figure 1), OSMRβ (Figure 2), and LIFRα (Figure 3) is shown in **Figure 6** and **Table 3**. All contain three fibronectin type III repeats proximal to a hydrophobic transmembrane domain. In addition, all have a hydrophobic signal sequence at the N-terminus and a C-terminal cytoplasmic domain (200–300 amino acids). Conserved hematopoietin domains ($\sim$200 amino acids) containing four positionally conserved cysteine residues in the N-terminal region and a WSXWS motif in the C-terminal region are found in all three receptor subunits. OSMRβ and LIFR have additional variant hematopoietin domains, with a domain lacking the N-terminal cysteine residues in OSMRβ and a domain lacking one pair of conserved cysteine residues in LIFRα. In the C-terminal cytoplasmic domain of each receptor subunit are conserved sequences corresponding to the box 1, box 2, and box 3 motifs involved in signal transduction (Murakami *et al.*, 1991; Baumann *et al.*, 1994).

Relevant homologies and species differences

OSMRα(gp130), OSMRβ, and LIFRα are related to each other and to other members of the hematopoietin receptor family. OSMRβ shares closest similarity to the LIFR, with a 32% amino acid identity, while OSMRα(gp130) is less similar (Mosley *et al.*, 1996). Structurally, OSMRβ and LIFRα are very similar, with the exception that the LIFRα contains two intact hematopoietin domains, whereas the OSMRβ has an N-terminal truncated domain lacking the conserved cysteine residues. OSMRβ is unique among the hematopoietin receptors in this regard, since all other receptors have domains with both the conserved cysteine residues and the WSXWS motif. OSMRα(gp130) contains only one hematopoietin domain but contains an immunoglobulin (Ig)-like domain at its N-terminus (Bazan, 1990).

Figure 1 Nucleotide and encoded amino acid sequence of the transmembrane form of human OSMRα(gp130). The hydrophobic signal sequence and transmembrane-spanning domains are shown in bold and the WSXWS hematopoietin motif is boxed. Exon splice junctions yielding alternately spliced mRNAs are indicated, using the exon numbering of the human LIFRα gene (Tomida and Gotoh, 1996).

```
GAGCAGCCAAAAGGCCCGCGGAGTCGCGCTGGGCCGCCCCGGCGCAGCTGAACCGGGGGC    60
CGCGCCTGCCAGGCCGACGGGTCTGGCCCAGCCTGGCGCCAAGGGGTTCGTGCGCTGTGG   120
AGACGCGGAGGGTCGAGGCGGCGCGGCCTGAGTGAAACCCAATGGAAAAAGCATGACATT   180
TAGAAGTAGAAGACTTAGCTTCAAATCCCTACTCCTTCACTTACTAATTTTGTGATTTGG   240

               M  L  T  L  Q  T  W  V  V  Q  A  L  F  I  F
AAATATCCGCGCAAGATGTTGACGTTGCAGACTTGGGTAGTGCAAGCCTTGTTTATTTTC   300

L  T  T  E  S  T  G  E  L  L  D  P  C  G  Y  I  S  P  E  S
CTCACCACTGAATCTACAGGTGAACTTCTAGATCCATGTGGTTATATCAGTCCTGAATCT   360

P  V  V  Q  L  H  S  N  F  T  A  V  C  V  L  K  E  K  C  M
CCAGTTGTACAACTTCATTCTAATTTCACTGCAGTTTGTGTGCTAAAGGAAAAATGTATG   420

D  Y  F  H  V  N  A  N  Y  I  V  W  K  T  N  H  F  T  I  P
GATTATTTTCATGTAAATGCTAATTACATTGTCTGGAAAACAAACCATTTTACTATTCCT   480

K  E  Q  Y  T  I  I  N  R  T  A  S  S  V  T  F  T  D  I  A
AAGGAGCAATATACTATCATAAACAGAACAGCATCCAGTGTCACCTTTACAGATATAGCT   540

S  L  N  I  Q  L  T  C  N  I  L  T  F  G  Q  L  E  Q  N  V
TCATTAAATATTCAGCTCACTTGCAACATTCTTACATTCGGACAGCTTGAACAGAATGTT   600

Y  G  I  T  I  I  S  G  L  P  P  E  K  P  K  N  L  S  C  I
TATGGAATCACAATAATTTCAGGCTTGCCTCCAGAAAAACCTAAAAATTTGAGTTGCATT   660

V  N  E  G  K  K  M  R  C  E  W  D  G  G  R  E  T  H  L  E
GTGAACGAGGGGAAGAAAATGAGGTGTGAGTGGGATGGTGGAAGGGAAACACACTTGGAG   720

T  N  F  T  L  K  S  E  W  A  T  H  K  F  A  D  C  K  A  K
ACAAACTTCACTTTAAAATCTGAATGGGCAACACACAAGTTTGCTGATTGCAAAGCAAAA   780

R  D  T  P  T  S  C  T  V  D  Y  S  T  V  Y  F  V  N  I  E
CGTGACACCCCCACCTCATGCACTGTTGATTATTCTACTGTGTATTTTGTCAACATTGAA   840

V  W  V  E  A  E  N  A  L  G  K  V  T  S  D  H  I  N  F  D
GTCTGGGTAGAAGCAGAGAATGCCCTTGGGAAGGTTACATCAGATCATATCAATTTTGAT   900

P  V  Y  K  V  K  P  N  P  P  H  N  L  S  V  I  N  S  E  E
CCTGTATATAAAGTGAAGCCCAATCCGCCACATAATTTATCAGTGATCAACTCAGAGGAA   960

L  S  S  I  L  K  L  T  W  T  N  P  S  I  K  S  V  I  I  L
CTGTCTAGTATCTTAAAATTGACATGGACCAACCCAAGTATTAAGAGTGTTATAATACTA  1020

K  Y  N  I  Q  Y  R  T  K  D  A  S  T  W  S  Q  I  P  P  E
AAATATAACATTCAATATAGGACCAAAGATGCCTCAACTTGGAGCCAGATTCCTCCTGAA  1080

D  T  A  S  T  R  S  S  F  T  V  Q  D  L  K  P  F  T  E  Y
GACACAGCATCCACCCGATCTTCATTCACTGTCCAAGACCTTAAACCTTTTACAGAATAT  1140

V  F  R  I  R  C  M  K  E  D  G  K  G  Y [W  S  D  W  S] E
GTGTTTAGGATTCGCTGTATGAAGGAAGATGGTAAGGGATACTGGAGTGACTGGAGTGAA  1200

E  A  S  G  I  T  Y  E  D  R  P  S  K  A  P  S  F  W  Y  K
GAAGCAAGTGGGATCACCTATGAAGATAGACCATCTAAAGCACCAAGTTTCTGGTATAAA  1260

I  D  P  S  H  T  Q  G  Y  R  T  V  Q  L  V  W  K  T  L  P
ATAGATCCATCCCATACTCAAGGCTACAGAACTGTACAACTCGTGTGGAAGACATTGCCT  1320

P  F  E  A  N  G  K  I  L  D  Y  E  V  T  L  T  R  W  K  S
CCTTTTGAAGCCAATGGAAAAATCTTGGATTATGAAGTGACTCTCACAAGATGGAAATCA  1380
```

Figure 1 (*Continued*)

```
H  L  Q  N  Y  T  V  N  A  T  K  L  T  V  N  L  T  N  D  R
CATTTACAAAATTACACAGTTAATGCCACAAAACTGACAGTAAATCTCACAAATGATCGC    1440

Y  L  A  T  L  T  V  R  N  L  V  G  K  S  D  A  A  V  L  T
TATCTAGCAACCCTAACAGTAAGAAATCTTGTTGGCAAATCAGATGCAGCTGTTTTAACT    1500

I  P  A  C  D  F  Q  A  T  H  P  V  M  D  L  K  A  F  P  K
ATCCCTGCCTGTGACTTTCAAGCTACTCACCCTGTAATGGATCTTAAAGCATTCCCCAAA    1560

D  N  M  L  W  V  E  W  T  T  P  R  E  S  V  K  K  Y  I  L
GATAACATGCTTTGGGTGGAATGGACTACTCCAAGGGAATCTGTAAAGAAATATATACTT    1620

E  W  C  V  L  S  D  K  A  P  C  I  T  D  W  Q  Q  E  D  G
GAGTGGTGTGTGTTATCAGATAAAGCACCCTGTATCACAGACTGGCAACAAGAAGATGGT    1680

T  V  H  R  T  Y  L  R  G  N  L  A  E  S  K  C  Y  L  I  T
ACCGTGCATCGCACCTATTTAAGAGGGAACTTAGCAGAGAGCAAATGCTATTTGATAACA    1740

V  T  P  V  Y  A  D  G  P  G  S  P  E  S  I  K  A  Y  L  K
GTTACTCCAGTATATGCTGATGGACCAGGAAGCCCTGAATCCATAAAGGCATACCTTAAA    1800

Q  A  P  P  S  K  G  P  T  V  R  T  K  K  V  G  K  N  E  A
CAAGCTCCACCTTCCAAAGGACCTACTGTTCGGACAAAAAAAGTAGGGAAAAACGAAGCT    1860

V  L  E  W  D  Q  L  P  V  D  V  Q  N  G  F  I  R  N  Y  T
GTCTTAGAGTGGGACCAACTTCCTGTTGATGTTCAGAATGGATTTATCAGAAATTATACT    1920

I  F  Y  R  T  I  I  G  N  E  T  A  V  N  V  D  S  S  H  T
ATATTTTATAGAACCATCATTGGAAATGAAACTGCTGTGAATGTGGATTCTTCCCACACA    1980

E  Y  T  L  S  S  L  T  S  D  T  L  Y  M  V  R  M  A  A  Y
GAATATACATTGTCCTCTTTGACTAGTGACACATTGTACATGGTACGAATGGCAGCATAC    2040

                                            exon 17> ↓<exon 18
T  D  E  G  G  K  D  G  P  E  F  T  F  T  T  P  K  F  A  Q
ACAGATGAAGGTGGGAAGGATGGTCCAGAATTCACTTTTACTACCCCAAAGTTTGCTCAA    2100

G  E  I  E  A  I  V  V  P  V  C  L  A  F  L  L  T  T  L  L
GGAGAAATTGAAGCCATAGTCGTGCCTGTTTGCTTAGCATTCCTATTGACAACTCTTCTG    2160

                    exon 18> ↓<exon 19
G  V  L  F  C  F  N  K  R  D  L  I  K  K  H  I  W  P  N  V
GGAGTGCTGTTCTGCTTTAATAAGCGAGACCTAATTAAAAAACACATCTGGCCTAATGTT    2220

P  D  P  S  K  S  H  I  A  Q  W  S  P  H  T  P  P  R  H  N
CCAGATCCTTCAAAGAGTCATATTGCCCAGTGGTCACCTCACACTCCTCCAAGGCACAAT    2280

F  N  S  K  D  Q  M  Y  S  D  G  N  F  T  D  V  S  V  V  E
TTTAATTCAAAAGATCAAATGTATTCAGATGGCAATTTCACTGATGTAAGTGTTGTGGAA    2340

I  E  A  N  D  K  K  P  F  P  E  D  L  K  S  L  D  L  F  K
ATAGAAGCAAATGACAAAAAGCCTTTTCCAGAAGATCTGAAATCATTGGACCTGTTCAAA    2400

K  E  K  I  N  T  E  G  H  S  S  G  I  G  G  S  S  C  M  S
AAGGAAAAAATTAATACTGAAGGACACAGCAGTGGTATTGGGGGGTCTTCATGCATGTCA    2460

S  S  R  P  S  I  S  S  S  D  E  N  E  S  S  Q  N  T  S  S
TCTTCTAGGCCAAGCATTTCTAGCAGTGATGAAAATGAATCTTCACAAAACACTTCGAGC    2520

T  V  Q  Y  S  T  V  V  H  S  G  Y  R  H  Q  V  P  S  V  Q
ACTGTCCAGTATTCTACCGTGGTACACAGTGGCTACAGACACCAAGTTCCGTCAGTCCAA    2580
```

Comparison of the human and murine OSM receptor subunits demonstrates a close similarity between the OSMRα(gp130) (76% amino acid identity) and OSMRβ (55% amino acid identity) homologs (Lindberg *et al.*, 1998). Although the mouse OSMRβ has the same structural domains as the human protein, it contains variant sequences in the WSXWS motifs present in the hematopoietin domains with a WGNWS sequence in the N-terminal truncated domain and a WSDWT motif in the second complete domain. Whereas the human LIFRα forms part of the LIF/OSM shared receptor complex with OSMRα(gp130), the murine homolog of LIFRα does not participate in binding or signaling of mouse OSM (Ichihara *et al.*, 1997; Lindberg *et al.*, 1998). Many studies examining the biological function of

Figure 1 *(Continued)*

```
V   F   S   R   S   E   S   T   Q   P   L   L   D   S   E   E   R   P   E   D
GTCTTCTCAAGATCCGAGTCTACCCAGCCCTTGTTAGATTCAGAGGAGCGGCCAGAAGAT   2640

L   Q   L   V   D   H   V   D   G   G   D   G   I   L   P   R   Q   Q   Y   F
CTACAATTAGTAGATCATGTAGATGGCGGTGATGGTATTTTGCCCAGGCAACAGTACTTC   2700

K   Q   N   C   S   Q   H   E   S   S   P   D   I   S   H   F   E   R   S   K
AAACAGAACTGCAGTCAGCATGAATCCAGTCCAGATATTTCACATTTTGAAAGGTCAAAG   2760

Q   V   S   S   V   N   E   E   D   F   V   R   L   K   Q   Q   I   S   D   H
CAAGTTTCATCAGTCAATGAGGAAGATTTTGTTAGACTTAAACAGCAGATTTCAGATCAT   2820

I   S   Q   S   C   G   S   G   Q   M   K   M   F   Q   E   V   S   A   A   D
ATTTCACAATCCTGTGGATCTGGGCAAATGAAAATGTTTCAGGAAGTTTCTGCAGCAGAT   2880

A   F   G   P   G   T   E   G   Q   V   E   R   F   E   T   V   G   M   E   A
GCTTTTGGTCCAGGTACTGAGGGACAAGTAGAAAGATTTGAAACAGTTGGCATGGAGGCT   2940

A   T   D   E   G   M   P   K   S   Y   L   P   Q   T   V   R   Q   G   G   Y
GCGACTGATGAAGGCATGCCTAAAAGTTACTTACCACAGACTGTACGGCAAGGCGGCTAC   3000

M   P   Q
ATGCCTCAGTGAAGGACTAGTAGTTCCTGCTACAACTTCAGCAGTACCTATAAAGTAAAG   3060
CTAAAATGATTTTATCTGTGAATTC                                      3085
```

OSM in mice have used human OSM which only mimics mouse LIF by binding and signaling uniquely through the mouse LIF receptor complex (Ichihara *et al.*, 1997; Lindberg *et al.*, 1998).

Affinity for ligand(s)

A summary of low-affinity direct binding for individual receptor subunits is shown in **Table 4**. Direct binding of OSMRα(gp130) to OSM at low affinity has been detected in both human and murine systems (Linsley *et al.*, 1989; Ichihara *et al.*, 1997; Lindberg *et al.*, 1998). Although OSMRβ can bind OSM directly in the murine system (low-affinity; Lindberg *et al.*, 1998), no evidence for binding is seen in the human system (Mosley, 1997). LIFRα binds only LIF directly (low-affinity) and not OSM (Gearing and Bruce, 1992).

A summary of binding to high-affinity receptor complexes is shown in **Table 5**. The shared LIF/OSM receptor complex composed of OSMRα(gp130) and LIFRα binds both OSM and LIF with high affinity in humans (Gearing and Bruce, 1992; Bruce *et al.*, 1992a). However, the murine homolog of the LIF/OSM receptor complex binds murine and human LIF, as well as human OSM, but does not bind murine OSM (Ichihara *et al.*, 1997). Therefore, in murine cells, human OSM mimics the activities of LIF and does not display the activities of murine OSM. Murine OSM binds with high affinity only to the murine OSM-specific receptor (Ichihara *et al.*, 1997), which is composed of the murine homologs of OSMRα(gp130) and OSMRβ (Lindberg *et al.*, 1998).

Cell types and tissues expressing the receptor

The OSMRα(gp130) receptor is ubiquitously expressed on a wide variety of cell types and tissues (Saito *et al.*, 1992). Distinctive patterns of expression have been demonstrated in the brain (Watanabe *et al.*, 1996). Alternately spliced products are found in embryonic tissues (Sharkey *et al.*, 1995) and in blood mononuclear cells (Diamant *et al.*, 1997).

OSMRβ receptor mRNA is detected in mouse heart, brain, spleen, lung, liver, skeletal muscle, and kidney tissue, but not in testis (Lindberg *et al.*, 1998). Human LIFRα is expressed in a variety of cell tissues, including the oocytes, preimplantation embryos and the placenta (Gearing *et al.*, 1991; Kojima *et al.*, 1995; van Eijk *et al.*, 1996). Alternately spliced mRNAs encoding soluble human LIFRα have been detected in liver, placenta, and choriocarcinoma cells (Tomida, 1997). Studies on bone marrow stromal/osteoblastic cells have shown the presence of OSMRα(gp130), OSMRβ, and LIFRα (Bellido *et al.*, 1996).

A number of studies have determined sites of expression of mouse LIFRα but the exact correlation with the human situation is not clear, since mouse LIFRα, unlike human LIFRα, does not participate in OSM signaling. A comparison of the expression of the different human OSM receptor subunits, derived from Mosley *et al.* (1996) is shown in **Table 6**.

Regulation of receptor expression

Of the OSM receptor subunits, only the promoter region for the hLIFRα has been reported. The region upstream of the transcriptional start site for LIFRα has a consensus TATA motif 30 bp upstream of the initiation site and several potential regulatory elements, including AP-2-, SP-1-, and NF-IL6-binding sites (Tomida and Gotoh, 1996). An alternate promoter in the LIFRα gene with an upstream enhancer which is active in placental tissues has also been characterized (Wang and Melmed, 1997).

Figure 2 Nucleotide and encoded amino acid sequence of the transmembrane form of human OSMRβ. The hydrophobic signal sequence and transmembrane-spanning domains are shown in bold and the WSXWS hematopoietin motif is boxed.

```
GGGCCGCCTCTGCACGTCCGCCCCGGAGCCCGCACCCGCGCCCCACGCGCCGCCGAGGAC    60
TCGGCCCGGCTCGTGGAGCCCTTCGCCCGCGGCGTGAGTACCCCCGACCCGCCCGTCCCC   120
GCTCTGCTCGCGCCCTGCCGCTGCGCCGCCCTCGGTGGCTTTTCCGACGGGCGAGCCCCG   180
TGCTGTGCGGGAAAGAATCCGACAACTTCGCAGCCCATCCCGGCTGGACGCGACCGGGAG   240
TGCAGCAGCCCGTTCCCCTCCTCGGTGCCGCCTCTGCCCAGCGTTTGCTTGGCTGGGCTA   300
CCACCTGCGCTCGGACGGCGCTCGGAGGGTCCTCGCCCCCGGCCTGCCTACCTGAAAACC   360

        M  A  L  F  A  V  F  Q  T  T  F  F  L  T  L  L  S  L
AGAACTGATGGCTCTATTTGCAGTCTTTCAGACAACATTCTTCTTAACATTGCTGTCCTT   420

 R  T  Y  Q  S  E  V  L  A  E  R  L  P  L  T  P  V  S  L  K
GAGGACTTACCAGAGTGAAGTCTTGGCTGAACGTTTACCATTGACTCCTGTATCACTTAA   480

 V  S  T  N  S  T  R  Q  S  L  H  L  Q  W  T  V  H  N  L  P
AGTTTCCACCAATTCTACGCGTCAGAGTTTGCACTTACAATGGACTGTCCACAACCTTCC   540

 Y  H  Q  E  L  K  M  V  F  Q  I  Q  I  S  R  I  E  T  S  N
TTATCATCAGGAATTGAAAATGGTATTTCAGATCCAGATCAGTAGGATTGAAACATCCAA   600

 V  I  W  V  G  N  Y  S  T  T  V  K  W  N  Q  V  L  H  W  S
TGTCATCTGGGTGGGGAATTACAGCACCACTGTGAAGTGGAACCAGGTTCTGCATTGGAG   660

 W  E  S  E  L  P  L  E  C  A  T  H  F  V  R  I  K  S  L  V
CTGGGAATCTGAGCTCCCTTTGGAATGTGCCACACACTTTGTAAGAATAAAGAGTTTGGT   720

 D  D  A  K  F  P  E  P  N  F [W  S  N  W  S] S  W  E  E  V
GGACGATGCCAAGTTCCCTGAGCCAAATTTCTGGAGCAACTGGAGTTCCTGGGAGGAAGT   780

 S  V  Q  D  S  T  G  Q  D  I  L  F  V  F  P  K  D  K  L  V
CAGTGTACAAGATTCTACTGGACAGGATATATTGTTCGTTTTCCCTAAAGATAAGCTGGT   840

 E  E  G  T  N  V  T  I  C  Y  V  S  R  N  I  Q  N  N  V  S
GGAAGAAGGCACCAATGTTACCATTTGTTACGTTTCTAGGAACATTCAAAATAATGTATC   900

 C  Y  L  E  G  K  Q  I  H  G  E  Q  L  D  P  H  V  T  A  F
CTGTTATTTGGAAGGGAAACAGATTCATGGAGAACAACTTGATCCACATGTAACTGCATT   960

 N  L  N  S  V  P  F  I  R  N  K  G  T  N  I  Y  C  E  A  S
CAACTTGAATAGTGTGCCTTTCATTAGGAATAAAGGGACAAATATCTATTGTGAGGCAAG  1020

 Q  G  N  V  S  E  G  M  K  G  I  V  L  F  V  S  K  V  L  E
TCAAGGAAATGTCAGTGAAGGCATGAAAGGCATCGTTCTTTTTGTCTCAAAAGTACTTGA  1080

 E  P  K  D  F  S  C  E  T  E  D  F  K  T  L  H  C  T  W  D
GGAGCCCAAGGACTTTTCTTGTGAAACCGAGGACTTCAAGACTTTGCACTGTACTTGGGA  1140

 P  G  T  D  T  A  L  G  W  S  K  Q  P  S  Q  S  Y  T  L  F
TCCTGGGACGGACACTGCCTTGGGGTGGTCTAAACAACCTTCCCAAAGCTACACTTTATT  1200

 E  S  F  S  G  E  K  K  L  C  T  H  K  N  W  C  N  W  Q  I
TGAATCATTTTCTGGGGAAAAGAAACTTTGTACACACAAAAACTGGTGTAATTGGCAAAT  1260

 T  Q  D  S  Q  E  T  Y  N  F  T  L  I  A  E  N  Y  L  R  K
AACTCAAGACTCACAAGAAACCTATAACTTCACACTCATAGCTGAAAATTACTTAAGGAA  1320

 R  S  V  N  I  L  F  N  L  T  H  R  V  Y  L  M  N  P  F  S
GAGAAGTGTCAATATCCTTTTTAACCTGACTCATCGAGTTTATTTAATGAATCCTTTTAG  1380

 V  N  F  E  N  V  N  A  T  N  A  I  M  T  W  K  V  H  S  I
TGTCAACTTTGAAAATGTAAATGCCACAAATGCCATCATGACCTGGAAGGTGCACTCCAT  1440
```

Figure 2 *(Continued)*

```
 R  N  N  F  T  Y  L  C  Q  I  E  L  H  G  E  G  K  M  M  Q
AAGGAATAATTTCACATATTTGTGTCAGATTGAACTCCATGGTGAAGGAAAAATGATGCA    1500

 Y  N  V  S  I  K  V  N  G  E  Y  F  L  S  E  L  E  P  A  T
ATACAATGTTTCCATCAAGGTGAACGGTGAGTACTTCTTAAGTGAACTGGAACCTGCCAC    1560

 E  Y  M  A  R  V  R  C  A  D  A  S  H  F  W  K  W  S  E  W
AGAGTACATGGCGCGAGTACGGTGTGCTGATGCCAGCCACTTCTGGAAATGGAGTGAATG    1620

 S  G  Q  N  F  T  T  L  E  A  A  P  S  E  A  P  D  V  W  R
GAGTGGTCAGAACTTCACCACACTTGAAGCTGCTCCCTCAGAGGCCCCTGATGTCTGGAG    1680

 I  V  S  L  E  P  G  N  H  T  V  T  L  F  W  K  P  L  S  K
AATTGTGAGCTTGGAGCCAGGAAATCATACTGTGACCTTATTCTGGAAGCCATTATCAAA    1740

 L  H  A  N  G  K  I  L  F  Y  N  V  V  V  E  N  L  D  K  P
ACTGCATGCCAATGGAAAGATCCTGTTCTATAATGTAGTTGTAGAAAACCTAGACAAACC    1800

 S  S  S  E  L  H  S  I  P  A  P  A  N  S  T  K  L  I  L  D
ATCCAGTTCAGAGCTCCATTCCATTCCAGCACCAGCCAACAGCACAAAACTAATCCTTGA    1860

 R  C  S  Y  Q  I  C  V  I  A  N  N  S  V  G  A  S  P  A  S
CAGGTGTTCCTACCAAATCTGCGTCATAGCCAACAACAGTGTGGGTGCTTCTCCTGCTTC    1920

 V  I  V  I  S  A  D  P  E  N  K  E  V  E  E  E  R  I  A  G
TGTAATAGTCATCTCTGCAGACCCCGAAAACAAAGAGGTTGAGGAAGAAAGAATTGCAGG    1980

 T  E  G  G  F  S  L  S  W  K  P  Q  P  G  D  V  I  G  Y  V
CACAGAGGGTGGATTCTCTCTGTCTTGGAAACCCCAACCTGGAGATGTTATAGGCTATGT    2040

 V  D  W  C  D  H  T  Q  D  V  L  G  D  F  Q  W  K  N  V  G
TGTGGACTGGTGTGACCATACCCAGGATGTGCTCGGTGATTTCCAGTGGAAGAATGTAGG    2100

 P  N  T  T  S  T  V  I  S  T  D  A  F  R  P  G  V  R  Y  D
TCCCAATACCACAAGCACAGTCATTAGCACAGATGCTTTTAGGCCAGGAGTTCGATATGA    2160

 F  R  I  Y  G  L  S  T  K  R  I  A  C  L  L  E  K  K  T  G
CTTCAGAATTTATGGGTTATCTACAAAAAGGATTGCTTGTTTATTAGAGAAAAAAACAGG    2220

 Y  S  Q  E  L  A  P  S  D  N  P  H  V  L  V  D  T  L  T  S
ATACTCTCAGGAACTTGCTCCTTCAGACAACCCTCACGTGCTGGTGGATACATTGACATC    2280

 H  S  F  T  L  S  W  K  D  Y  S  T  E  S  Q  P  G  F  I  Q
CCACTCCTTCACTCTGAGTTGGAAAGATTACTCTACTGAATCTCAACCTGGTTTTATACA    2340

 G  Y  H  V  Y  L  K  S  K  A  R  Q  C  H  P  R  F  E  K  A
AGGGTACCATGTCTATCTGAAATCCAAGGCGAGGCAGTGCCACCCACGATTTGAAAAGGC    2400

 V  L  S  D  G  S  E  C  C  K  Y  K  I  D  N  P  E  E  K  A
AGTTCTTTCAGATGGTTCAGAATGTTGCAAATACAAAATTGACAACCCGGAAGAAAAGGC    2460

 L  I  V  D  N  L  K  P  E  S  F  Y  E  F  F  I  T  P  F  T
ATTGATTGTGGACAACCTAAAGCCAGAATCCTTCTATGAGTTTTTCATCACTCCATTCAC    2520

 S  A  G  E  G  P  S  A  T  F  T  K  V  T  T  P  D  E  H  S
TAGTGCTGGTGAAGGCCCCAGTGCTACGTTCACGAAGGTCACGACTCCGGATGAACACTC    2580

 S  M  L  I  H  I  L  L  P  M  V  F  C  V  L  L  I  M  V  M
CTCGATGCTGATTCATATCCTACTGCCCATGGTTTTCTGCGTCTTGCTCATCATGGTCAT    2640

 C  Y  L  K  S  Q  W  I  K  E  T  C  Y  P  D  I  P  D  P  Y
GTGCTACTTGAAAAGTCAGTGGATCAAGGAGACCTGTTATCCTGACATCCCTGACCCTTA    2700
```

Studies on lung-derived epithelial cells have shown that mRNA levels of OSMRα(gp130) and OSMRβ are upregulated by OSM (Cichy *et al.*, 1998).

Release of soluble receptors

Alternately spliced mRNAs encoding two different soluble forms of OSMRα(gp130) have been identified (Sharkey *et al.*, 1995; Diamant *et al.*, 1997) (Figure 4). In addition, alternately spliced mRNAs encoding three different soluble forms of human LIFRα have been detected in adult liver (Figure 5) (Tomida, 1997). Interestingly, some of the soluble forms encode new cysteine residues in the C-terminal domain (Figure 4 and Figure 5), suggesting the possibility of lipid linkages to membranes, as is found with the receptor

Figure 2 (*Continued*)

```
 K  S  S  I  L  S  L  I  K  F  K  E  N  P  H  L  I  I  M  N
CAAGAGCAGCATCCTGTCATTAATAAAATTCAAGGAGAACCCTCACCTAATAATAATGAA    2760

 V  S  D  C  I  P  D  A  I  E  V  V  S  K  P  E  G  T  K  I
TGTCAGTGACTGTATCCCAGATGCTATTGAAGTTGTAAGCAAGCCAGAAGGGACAAAGAT    2820

 Q  F  L  G  T  R  K  S  L  T  E  T  E  L  T  K  P  N  Y  L
ACAGTTCCTAGGCACTAGGAAGTCACTCACAGAAACCGAGTTGACTAAGCCTAACTACCT    2880

 Y  L  L  P  T  E  K  N  H  S  G  P  G  P  C  I  C  F  E  N
TTATCTCCTTCCAACAGAAAAGAATCACTCTGGCCCTGGCCCCTGCATCTGTTTTGAGAA    2940

 L  T  Y  N  Q  A  A  S  D  S  G  S  C  G  H  V  P  V  S  P
CTTGACCTATAACCAGGCAGCTTCTGACTCTGGCTCTTGTGGCCATGTTCCAGTATCCCC    3000

 K  A  P  S  M  L  G  L  M  T  S  P  E  N  V  L  K  A  L  E
AAAAGCCCCAAGTATGCTGGGACTAATGACCTCACCTGAAAATGTACTAAAGGCACTAGA    3060

 K  N  Y  M  N  S  L  G  E  I  P  A  G  E  T  S  L  N  Y  V
AAAAAACTACATGAACTCCCTGGGAGAAATCCCAGCTGGAGAAACAAGTTTGAATTATGT    3120

 S  Q  L  A  S  P  M  F  G  D  K  D  S  L  P  T  N  P  V  E
GTCCCAGTTGGCTTCACCCATGTTTGGAGACAAGGACAGTCTCCCAACAAACCCAGTAGA    3180

 A  P  H  C  S  E  Y  K  M  Q  M  A  V  S  L  R  L  A  L  P
GGCACCACACTGTTCAGAGTATAAAATGCAAATGGCAGTCTCCCTGCGTCTTGCCTTGCC    3240

 P  P  T  E  N  S  S  L  S  S  I  T  L  L  D  P  G  E  H  Y
TCCCCCGACCGAGAATAGCAGCCTCTCCTCAATTACCCTTTTAGATCCAGGTGAACACTA    3300

 C  *
CTGCTAACCAGCATGCCGATTTCATACCTTATGCTACACAGACATTAAGAAGAGCAGAGC    3360
TGGCACCCTGTCATCACCAGTGGCCTTGGTCCTTAATCCCAGTACAATTTGCAGGTCTGG    3420
TTTATATAAGACCACTACAGTCTGGCTAGGTTAAAGGCCAGAGGCTATGGAACTTAACAC    3480
TCCCCATTGGAGCAAGCTTGCCCTAGAGACGGCAGGATCATGGGAGCATGCTTACCTTCT    3540
GCTGTTTGTTCCAGGCTCACCTTTAGAACAGGAGACTTGAGCTTGACCTAAGGATATGCA    3600
TTAACCACTCTACAGACTCCCACTCAGTACTGTACAGGGTGGCTGTGGTCCTAGAAGTTC    3660
AGTTTTTACTGAGGAAATATTTCCATTAACAGCAATTATTATATTGAAGGCTTTAATAAA    3720
GGCCACAGGAGACATTACTATAGCATAGATTGTCAAATGTAAATTTACTGAGCGTGTTTT    3780
ATAAAAAACTCACAGGTGTTTGAGGCCAAAACAGATTTTAGACTTACCTTGAACGGATAA    3840
GAATCTATAGTTCACTGACACAGTAAAATTAACTCTGTGGGTGGGGGCGGGGGGCATAGC    3900
TCTAATCTAATATATAAAATGTGTGATGAATCAACAAGATTTCCACAATTCTTCTGTCAA    3960
GCTTACTACAGTGAAAGAATGGGATTGGCAAGTAACTTCTGACTTACTGTCAGTTGTACT    4020
TCTGCTCCATAGACATCAGTATTCTGCCATCATTTTTGATGACTACCTCAGAACATAAAA    4080
AGGAACGTATATCACATAATTCCAGTCACAGTTTTTGGTTCCTCTTTTCTTTCAAGAACT    4140
ATATATAAATGACCTGTTTTCACGCGGCCGC                                 4171
```

for ciliary neurotropic factor (CNTFRα), which contains a glycosylphosphatidylinositol anchor at a C-terminal cysteine residue. Soluble forms of OSMRα(gp130) (50 and 100 kDa) and LIFRα have been detected in normal human serum, plasma, and urine (Narzaki *et al.*, 1993; Zhang *et al.*, 1998). Soluble murine OSMRα(gp130) has been detected in the ascitic fluid of tumor-bearing mice (Matsuda and Hirano, 1994).

SIGNAL TRANSDUCTION

Associated or intrinsic kinases

The OSM receptor subunits OSMRα(gp130), OSMRβ, and LIFRα all contain cytoplasmic domains with critical tyrosine residues involved in signaling. However, these molecules contain no intrinsic kinase activity and are dependent upon members of the JAK (Janus-activated kinase) family of constitutively associated kinases (JAK1, JAK2, JAK3, TYK2) for phosphorylation and subsequent signal transduction (Stahl *et al.*, 1994; reviewed in Nakashima and Taga, 1998). Activation of the JAK kinases does not explain all downstream signaling events, and other pathways involving the Src family tyrosine kinases, Ras, mitogen-activated protein kinases (MAPK), phosphatidylinositol 3-kinase (PI-3 kinase) are also implicated in cytokine signaling (Schiemann *et al.*, 1997; reviewed in Hirano *et al.*, 1997). Signal transduction by OSM in endothelial cells has been shown to involve activation of the p62yes tyrosine kinase (Schieven *et al.*, 1992). Studies have suggested that the OSM-specific receptor signal transduction pathway utilizes the MAPK activation more than the LIF/OSM shared receptor

(Amaral *et al.*, 1993; Thoma *et al.*, 1994). OSM activates Raf-1 which leads to the ultimate activation of MAPK. This requires the expression of STAT1 and is mediated through a JAK1-dependent pathway (Stancato *et al.*, 1997, 1998). Phosphorylation of a 250 kDa protein is apparently a specific consequence of OSM signaling through the OSM-specific receptor in A375 cells which involves the JAK1, JAK2, and TYK2 tyrosine kinases (Auguste *et al.*, 1997).

Figure 3 Nucleotide and encoded amino acid sequence of the transmembrane form of human LIFRα. The hydrophobic signal sequence and transmembrane-spanning domains are shown in bold and the WSXWS hematopoietin motif is boxed. Exon splice junctions yielding alternately spliced mRNAs are indicated, using the exon numbering of the human LIFRα gene (Tomida and Gotoh, 1996).

```
AGATCTTGGAACGAGACGACCTGCTCTCTCTCCCAGAACGTGTCTCTGCTGCAAGGCACC     60
GGGCCCTTTCGCTCTGCAGAACTGCACTTGCAAGACCATTATCAACTCCTAATCCCAGCT    120

                                                          M
CAGAAAGGGAGCCTCTGCGACTCATTCATCGCCCTCCAGGACTGACTGCATTGCACAGAT    180

 M  D  I  Y  V  C  L  K  R  P  S  W  M  V  D  N  K  R  M  R
GATGGATATTTACGTATGTTTGAAACGACCATCCTGGATGGTGGACAATAAAAGAATGAG    240

 T  A  S  N  F  Q  W  L  L  S  T  F  I  L  L  Y  L  M  N  Q
GACTGCTTCAAATTTCCAGTGGCTGTTATCAACATTTATTCTTCTATATCTAATGAATCA    300

 V  N  S  Q  K  K  G  A  P  H  D  L  K  C  V  T  N  N  L  Q
AGTAAATAGCCAGAAAAAGGGGGCTCCTCATGATTTGAAGTGTGTAACTAACAATTTGCA    360

 V  W  N  C  S  W  K  A  P  S  G  T  G  R  G  T  D  Y  E  V
AGTGTGGAACTGTTCTTGGAAAGCACCCTCTGGAACAGGCCGTGGTACTGATTATGAAGT    420

 C  I  E  N  R  S  R  S  C  Y  Q  L  E  K  T  S  I  K  I  P
TTGCATTGAAAACAGGTCCCGTTCTTGTTATCAGTTGGAGAAAACCAGTATTAAAATTCC    480

 A  L  S  H  G  D  Y  E  I  T  I  N  S  L  H  D  F  G  S  S
AGCTCTTTCACATGGTGATTATGAAATAACAATAAATTCTCTACATGATTTTGGAAGTTC    540

 T  S  K  F  T  L  N  E  Q  N  V  S  L  I  P  D  T  P  E  I
TACAAGTAAATTCACACTAAATGAACAAAACGTTTCCTTAATTCCAGATACTCCAGAGAT    600

 L  N  L  S  A  D  F  S  T  S  T  L  Y  L  K  W  N  D  R  G
CTTGAATTTGTCTGCTGATTTCTCAACCTCTACATTATACCTAAAGTGGAACGACAGGGG    660

 S  V  F  P  H  R  S  N  V  I  W  E  I  K  V  L  R  K  E  S
TTCAGTTTTTCCACACCGCTCAAATGTTATCTGGGAAATTAAAGTTCTACGTAAAGAGAG    720

 M  E  L  V  K  L  V  T  H  N  T  T  L  N  G  K  D  T  L  H
TATGGAGCTCGTAAAATTAGTGACCCACAACACAACTCTGAATGGCAAAGATACACTTCA    780

 H  W  S  W  A  S  D  M  P  L  E  C  A  I  H  F  V  E  I  R
TCACTGGAGTTGGGCCTCAGATATGCCCTTGGAATGTGCCATTCATTTTGTGGAAATTAG    840

 C  Y  I  D  N  L  H  F  S  G  L  E  E  W  S  D  W  S  P  V
ATGCTACATTGACAATCTTCATTTTTCTGGTCTCGAAGAGTGGAGTGACTGGAGCCCTGT    900

 K  N  I  S  W  I  P  D  S  Q  T  K  V  F  P  Q  D  K  V  I
GAAGAACATTTCTTGGATACCTGATTCTCAGACTAAGGTTTTTCCTCAAGATAAAGTGAT    960

 L  V  G  S  D  I  T  F  C  C  V  S  Q  E  K  V  L  S  A  L
ACTTGTAGGCTCAGACATAACATTTTGTTGTGTGAGTCAAGAAAAAGTGTTATCAGCACT   1020

 I  G  H  T  N  C  P  L  I  H  L  D  G  E  N  V  A  I  K  I
GATTGGCCATACAAACTGCCCCTTGATCCATCTTGATGGGGAAAATGTTGCAATCAAGAT   1080

 R  N  I  S  V  S  A  S  S  G  T  N  V  V  F  T  T  E  D  N
TCGTAATATTTCTGTTTCTGCAAGTAGTGGAACAAATGTAGTTTTTACAACCGAAGATAA   1140

 I  F  G  T  V  I  F  A  G  Y  P  P  D  T  P  Q  Q  L  N  C
CATATTTGGAACCGTTATTTTTGCTGGATATCCACCAGATACTCCTCAACAACTGAATTG   1200

 E  T  H  D  L  K  E  I  I  C  S  W  N  P  G  R  V  T  A  L
TGAGACACATGATTTAAAAGAAATTATATGTAGTTGGAATCCAGGAAGGGTGACAGCGTT   1260

 V  G  P  R  A  T  S  Y  T  L  V  E  S  F  S  G  K  Y  V  R
GGTGGGCCCACGTGCTACAAGCTACACTTTAGTTGAAAGTTTTTCAGGAAAATATGTTAG   1320
```

Figure 3 (*Continued*)

```
 L  K  R  A  E  A  P  T  N  E  S  Y  Q  L  L  F  Q  M  L  P
ACTTAAAAGAGCTGAAGCACCTACAAACGAAAGCTATCAATTATTATTTCAAATGCTTCC   1380

 N  Q  E  I  Y  N  F  T  L  N  A  H  N  P  L  G  R  S  Q  S
AAATCAAGAAATATATAATTTTACTTTGAATGCTCACAATCCGCTGGGTCGATCACAATC   1440

 T  I  L  V  N  I  T  E  K  V  Y  P  H  T  P  T  S  F  K  V
AACAATTTTAGTTAATATAACTGAAAAAGTTTATCCCCATACTCCTACTTCATTCAAAGT   1500

 K  D  I  N  S  T  A  V  K  L  S  W  H  L  P  G  N  F  A  K
GAAGGATATTAATTCAACAGCTGTTAAACTTTCTTGGCATTTACCAGGCAACTTTGCAAA   1560

 I  N  F  L  C  E  I  E  I  K  K  S  N  S  V  Q  E  Q  R  N
GATTAATTTTTTATGTGAAATTGAAATTAAGAAATCTAATTCAGTACAAGAGCAGCGGAA   1620

 V  T  I  K  G  V  E  N  S  S  Y  L  V  A  L  D  K  L  N  P
TGTCACAATCAAAGGAGTAGAAAATTCAAGTTATCTTGTTGCTCTGGACAAGTTAAATCC   1680

 Y  T  L  Y  T  F  R  I  R  C  S  T  E  T  F  W  K [W  S  K]
ATACACTCTATATACTTTTCGGATTCGTTGTTCTACTGAAACTTTCTGGAAATGGAGCAA   1740

[W  S] N  K  K  Q  H  L  T  T  E  A  S  P  S  K  G  P  D  T
ATGGAGCAATAAAAAACAACATTTAACAACAGAAGCCAGTCCTTCAAAGGGGCCTGATAC   1800

 W  R  E  W  S  S  D  G  K  N  L  I  I  Y  W  K  P  L  P  I
TTGGAGAGAGTGGAGTTCTGATGGAAAAAATTTAATAATCTATTGGAAGCCTTTACCCAT   1860

 N  E  A  N  G  K  I  L  S  Y  N  V  S  C  S  S  D  E  E  T
TAATGAAGCTAATGGAAAAATACTTTCCTACAATGTATCGTGTTCATCAGATGAGGAAAC   1920

 Q  S  L  S  E  I  P  D  P  Q  H  K  A  E  I  R  L  D  K  N
ACAGTCCCTTTCTGAAATCCCTGATCCTCAGCACAAAGCAGAGATACGACTTGATAAGAA   1980

 D  Y  I  I  S  V  V  A  K  N  S  V  G  S  S  P  P  S  K  I
TGACTACATCATCAGCGTAGTGGCTAAAAATTCTGTGGGCTCATCACCACCTTCCAAAAT   2040

 A  S  M  E  I  P  N  D  D  L  K  I  E  Q  V  V  G  M  G  K
AGCGAGTATGGAAATTCCAAATGATGATCTCAAAATAGAACAAGTTGTTGGGATGGGAAA   2100

 G  I  L  L  T  W  H  Y  D  P  N  M  T  C  D  Y  V  I  K  W
GGGGATTCTCCTCACCTGGCATTACGACCCCAACATGACTTGCGACTACGTCATTAAGTG   2160

 C  N  S  S  R  S  E  P  C  L  M  D  W  R  K  V  P  S  N  S
GTGTAACTCGTCTCGGTCGGAACCATGCCTTATGGACTGGAGAAAAGTTCCCTCAAACAG   2220

                exon 14> ↓<exon 15
 T  E  T  V  I  E  S  D  E  F  R  P  G  I  R  Y  N  F  F  L
CACTGAAACTGTAATAGAATCTGATGAGTTTCGACCAGGTATAAGATATAATTTTTTCCT   2280

 Y  G  C  R  N  Q  G  Y  Q  L  L  R  S  M  I  G  Y  I  E  E
GTATGGATGCAGAAATCAAGGATATCAATTATTACGCTCCATGATTGGATATATAGAAGA   2340

 L  A  P  I  V  A  P  N  F  T  V  E  D  T  S  A  D  S  I  L
ATTGGCTCCCATTGTTGCACCAAATTTTACTGTTGAGGATACTTCTGCAGATTCGATATT   2400
```

Cytoplasmic signaling cascades

Signal transduction through OSMRα(gp130) has mainly been studied in the context of IL-6 activation, which has become a model for the cytokine system. Binding of ligand to its receptor induces dimerization of OSMRα(gp130) which leads to activation of members of the JAK family of tyrosine kinases (reviewed in Heinrich *et al.*, 1998) and subsequent phosphorylation of members of the STAT (signal transducer and activator of transcription) family of transcriptional activators, including STAT1, STAT3, and STAT5 (Darnell *et al.*, 1994; Schindler and Darnell, 1995). Phosphorylated STATs dimerize and are translocated to the nucleus where they activate expression of genes containing STAT-recognition sites.

DOWNSTREAM GENE ACTIVATION

Transcription factors activated

Signaling through OSM-specific and LIF/OSM shared receptors activates the DNA-binding activity

Figure 3 (*Continued*)

```
 V  K  W  E  D  I  P  V  E  E  L  R  G  F  L  R  G  Y  L  F
AGTAAAATGGGAAGACATTCCTGTGGAAGAACTTAGAGGCTTTTTAAGAGGATATTTGTT   2460

                                             exon 16> ↓<exon 17
 Y  F  G  K  G  E  R  D  T  S  K  M  R  V  L  E  S  G  R  S
TTACTTTGGAAAAGGAGAAAGAGACACATCTAAGATGAGGGTTTTAGAATCAGGTCGTTC   2520

 D  I  K  V  K  N  I  T  D  I  S  Q  K  T  L  R  I  A  D  L
TGACATAAAAGTTAAGAATATTACTGACATATCCCAGAAGACACTGAGAATTGCTGATCT   2580

 Q  G  K  T  S  Y  H  L  V  L  R  A  Y  T  D  G  G  V  G  P
TCAAGGTAAAACAAGTTACCACCTGGTCTTGCGAGCCTATACAGATGGTGGAGTGGGCCC   2640

                        exon 17> ↓<exon 18
 E  K  S  M  Y  V  V  T  K  E  N  S  V  G  L  I  I  A  I  L
GGAGAAGAGTATGTATGTGGTGACAAAGGAAAATTCTGTGGGATTAATTATTGCCATTCT   2700

 I  P  V  A  V  A  V  I  V  G  V  V  T  S  I  L  C  Y  R  K
CATCCCAGTGGCAGTGGCTGTCATTGTTGGAGTGGTGACAAGTATCCTTTGCTATCGGAA   2760

exon 18> ↓<exon 19
 R  E  W  I  K  E  T  F  Y  P  D  I  P  N  P  E  N  C  K  A
ACGAGAATGGATTAAAGAAACCTTCTACCCTGATATTCCAAATCCAGAAAACTGTAAAGC   2820

                      exon 19> ↓<exon 20
 L  Q  F  Q  K  S  V  C  E  G  S  S  A  L  K  T  L  E  M  N
ATTACAGTTTCAAAAGAGTGTCTGTGAGGGAAGCAGTGCTCTTAAAACATTGGAAATGAA   2880

 P  C  T  P  N  N  V  E  V  L  E  T  R  S  A  F  P  K  I  E
TCCTTGTACCCCAAATAATGTTGAGGTTCTGGAAACTCGATCAGCATTTCCTAAAATAGA   2940

 D  T  E  I  I  S  P  V  A  E  R  P  E  D  R  S  D  A  E  P
AGATACAGAAATAATTTCCCCAGTAGCTGAGCGTCCTGAAGATCGCTCTGATGCAGAGCC   3000

 E  N  H  V  V  V  S  Y  C  P  P  I  I  E  E  E  I  P  N  P
TGAAAACCATGTGGTTGTGTCCTATTGTCCACCCATCATTGAGGAAGAAATACCAAACCC   3060

 A  A  D  E  A  G  G  T  A  Q  V  I  Y  I  D  V  Q  S  M  Y
AGCCGCAGATGAAGCTGGAGGGACTGCACAGGTTATTTACATTGATGTTCAGTCGATGTA   3120

 Q  P  Q  A  K  P  E  E  E  Q  E  N  D  P  V  G  G  A  G  Y
TCAGCCTCAAGCAAAACCAGAAGAAGAACAAGAAAATGACCCTGTAGGAGGGGCAGGCTA   3180

 K  P  Q  M  H  L  P  I  N  S  T  V  E  D  I  A  A  E  E  D
TAAGCCACAGATGCACCTCCCCATTAATTCTACTGTGGAAGATATAGCTGCAGAAGAGGA   3240

 L  D  K  T  A  G  Y  R  P  Q  A  N  V  N  T  W  N  L  V  S
CTTAGATAAAACTGCGGGTTACAGACCTCAGGCCAATGTAAATACATGGAATTTAGTGTC   3300

 P  D  S  P  R  S  I  D  S  N  S  E  I  V  S  F  G  S  P  C
TCCAGACTCTCCTAGATCCATAGACAGCAACAGTGAGATTGTCTCATTTGGAAGTCCATG   3360

 S  I  N  S  R  Q  F  L  I  P  P  K  D  E  D  S  P  K  S  N
CTCCATTAATTCCCGACAATTTTTGATTCCTCCTAAAGATGAAGACTCTCCTAAATCTAA   3420

 G  G  G  W  S  F  T  N  F  F  Q  N  K  P  N  D  *
TGGAGGAGGGTGGTCCTTTACAAACTTTTTTCAGAACAAACCAAACGATTAACAGTGTCA   3480
CCGTGTCACTTCAGTCAGCCATCTCAATAAGCTCTTACTGCTAGTGTTGCTACATCAGCA   3540
CTGGGCATTCTTGGAGGGATCCTGTGAAGTATTGTTAGGAGGTGAACTTCA            3591
```

of STAT1, STAT3, and STAT5b (Auguste *et al.*, 1997; Kuropatwinski *et al.*, 1997; Stephens *et al.*, 1998). Although many similarities are seen with the activation by the IL-6 receptor, an increase in the activation of STAT5 over that seen with IL-6 suggests that differences in biological activity could result from differential activation of the various STATs (Kuropatwinski *et al.*, 1997). In addition to tyrosine phosphorylation of STATs, phosphorylation on serine residues is also important, especially for binding to low-affinity sites where homodimerization of the STATs is essential (reviewed in Hirano *et al.*, 1997).

Genes induced

Table 7 summarizes the gene expression induced by activation of the OSM receptors.

Figure 4 Comparison of the alternately spliced isoforms of human OSMRα(gp130). (a) The hydrophobic signal peptide (SP) sequence and transmembrane (TM) domain are blue. In the extracellular (EC) domain, the hematopoietin domains with conserved cysteine residues and WSXWS motifs are shown. The fibronectin (FN) type III repeats are colored green. The C-terminal alternately spliced domains are indicated with striped boxes and the presence of a new cysteine residue in these domains is indicated with a C. The intracellular (IC) domain of the membrane form is shown. (b) The exon origins of the alternately spliced mRNAs are shown for the extracellular membrane-form (ECM) and the two putative soluble isoforms (IF-1 and IF-2). Variations in splicing results in the use of different reading frames within the same exon. (c) The encoded amino acid sequence for the C-terminus of each form is shown. The positions of the splice junctions are indicated by bold type and underlining using the exon numbering of the human LIFRα gene (Tomida and Gotoh, 1996). The C-terminal cysteine residues (C) are boxed. (Full colour figure can be viewed online.)

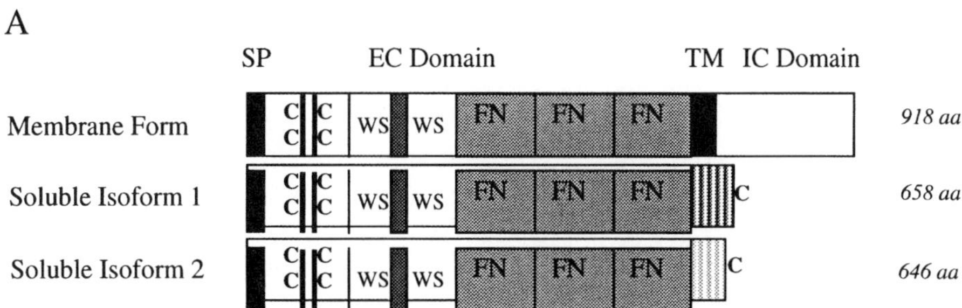

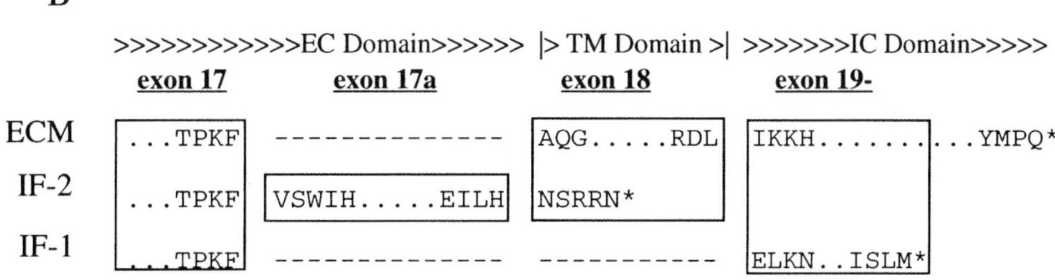

C

ECM ...TFTTPKF**A**QGEIEAIVVPVCLAFLLTTLLGVLFCFNKRD**L**IKKHIWPNVPDPSKSHI
AQWSPHTPPRHNFNSKDQMYSDGNFTDVSVVEIEANDKKPFPEDLKSLDLFKKEKINTEG
HSSGIGGSSCMSSSRPSISSSDENESSQNTSSTVQYSTVVHSGYRHQVPSVQVFSRSEST
QPLLDSEERPEDLQLVDHVDGGDGILPRQQYFKQNCSQHESSPDISHFERSKQVSSVNEE
DFVRLKQQISDHISQSCGSGQMKMFQEVSAADAFGPGTEGQVERFETVGMEAATDEGMPK
SYLPQTVRQGGYMPQ*

IF-2 ...TFTTPKF**v**swihygfftwlescqrgpyslveeilh**n**srrn*

IF-1 ...TFTTPKF**e**lkntsglmfqilqrvilpsghltllqgtiliqkikciqmaislm*

Promoter regions involved

STAT-recognition sites, including the types I and II IL-6 response elements (IL-6RE), have been identified in a variety of genes induced by IL-6. Subsequently, genes activated by OSM have also been shown to contain these recognition sites (Kordula *et al.*, 1998). In addition, OSM-responsive elements have been detected in other genes induced by OSM, including tissue inhibitor of matrix metalloproteinase 1 (TIMP-1) and matrix metalloproteinase 1 (MMP1) (Korzus *et al.*, 1997).

Figure 5 Comparison of the alternately spliced isoforms of LIFRα. (a) The hydrophobic signal peptide (SP) sequence and transmembrane (TM) domain are blue. In the extracellular (EC) domain, the hematopoietin domains with conserved cysteine residues and WSXWS motifs are shown. The fibronectin (FN) type III repeats are colored green. The C-terminal alternately spliced domains are indicated with striped boxes and the presence of a new cysteine residue in these domains is indicated with a C. The intracellular (IC) domain of the membrane form is shown. (b) The exon origins of the alternately spliced mRNAs are shown for the extracellular membrane-form (ECM) and the three putative soluble isoforms (IF-1, IF-2, and IF-3). The exon numbering is derived from that of the human LIFRα (Tomida and Gotoh, 1996). The amino acids which are not capitalized are derived from the adjacent intron. (c) The encoded amino acid sequence for the C-terminus of each form is shown. The positions of the splice junctions are indicated by bold type and underlining. The C-terminal cysteine residue (C) is boxed.

A

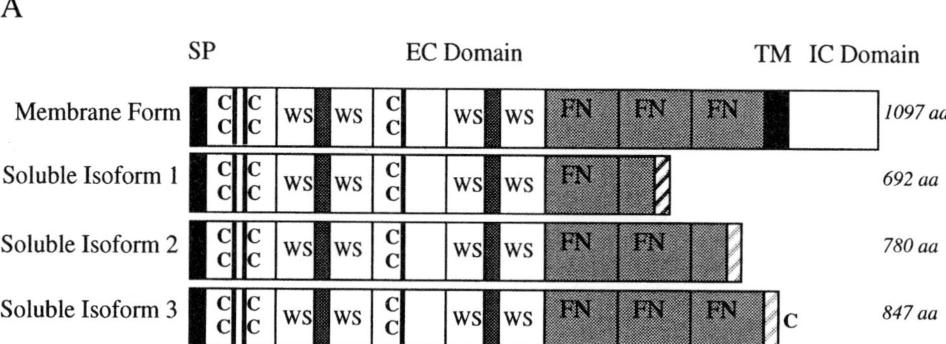

B

>>>>>>>>>>>>>>>>>>>>>> EC Domain >>>>>>>>>>>>>>>> | >TM and IC Domain>

	exon 1-14	**exon 15-16**	**exon 17**	**exon 18-20**
ECM	...VIES	DEFR...VLES	GRSD...TKEN	SVGL...GGEL*
IF-1	...VIESgkmn*			
IF-2	...VIES	DEFR...VLESge*		
IF-3	...VIES	DEFR...VLES	GRSD...TKENckfqvavevlsilha*	

C

ECM ...VIES**D**EFRPGIRYNFFLYGCRNQGYQLLRSMIGYIEELAPIVAPNFTVEDTSADSIL
VKWEDIPVEELRGFLRGYLFYFGKGERDTSKMRVLES**G**RSDIKVKNITDISQKTLRIADL
QGKTSYHLVLRAYTDGGVGPEKSMYVVTKEN**S**VGLIIAILIPVAVAVIVGVVTSILCYRK
REWIKETFYPDIPNPENCKALQFQKSVCEGSSALKTLEMNPCTPNNVEVLETRSAFPKIE
DTEIISPVAERPEDRSDAEPENHVVVSYCPPIIEEEIPNPAADEAGGTAQVIYIDVQSMY
QPQAKPEEEQENDPVGGAGYKPQMHLPINSTVEDIAAEEDLDKTAGYRPQANVNTWNLVS
PDSPRSIDSNSEIVSFGSPCSINSRQFLIPPKDEDSPKSNGGGWSFTNFFQNKPND*

IF-1 ...VIESgkmn*

IF-2 ...VIES**D**EFRPGIRYNFFLYGCRNQGYQLLRSMIGYIEELAPIVAPNFTVEDTSADSIL
VKWEDIPVEELRGFLRGYLFYFGKGERDTSKMRVLESge*

IF-3 ...VIES**D**EFRPGIRYNFFLYGCRNQGYQLLRSMIGYIEELAPIVAPNFTVEDTSADSIL
VKWEDIPVEELRGFLRGYLFYFGKGERDTSKMRVLES**G**RSDIKVKNITDISQKTLRIADL
QGKTSYHLVLRAYTDGGVGPEKSMYVVTKEN[c]kfqvavevlsilha*

Table 2 OSMRα(gp130), OSMRβ, and LIFRα protein sequences

Accession	Species	Source	Type	Size (amino acids)	Reference
OSMRα(gp130)					
106982	Human	Placenta	Complete	918	Hibi *et al.*, 1990
1246098	Human	Embryo	Alternate splice soluble IF-1[a], partial	49	Sharkey *et al.*, 1995
2253598	Human	Blood cells	Alternate splice soluble IF-2[a], partial	47	Diamant *et al.*, 1997
3660079	Human		Binding domain A, partial	214	Bravo *et al.*, 1998
3660080	Human		Binding domain B, partial	215	Bravo *et al.*, 1998
2137360	Mouse	Macrophage	Complete	917	Saito *et al.*, 1992
729835	Rat	Liver	Complete	918	Wang *et al.*, 1992
OSMRβ					
1794211	Human	Placenta/bone marrow/fibroblast	Complete (composite)	979	Mosley *et al.*, 1998
3721860	Mouse		Complete	970	Unpublished
3153816	Mouse	Muscle	Complete	971	Lindbergh *et al.*, 1998
LIFRα					
1170784	Human	Placenta	Complete	1097	Gearing *et al.*, 1991
258656	Human		Complete	1078	Gearing *et al.*, 1992b

[a]IF = isoform designation used within this chapter.

Figure 6 Comparison of the membrane-bound forms of the different OSM receptor subunits. The membrane-spanning domains are colored blue and the relative positions of the receptors with respect to the cellular membrane are indicated. The hematopoietin domains with conserved cysteine residues and the WSXWS motifs are shown. The fibronectin type III repeats (FN) are colored green. (Full colour figure can be viewed online.)

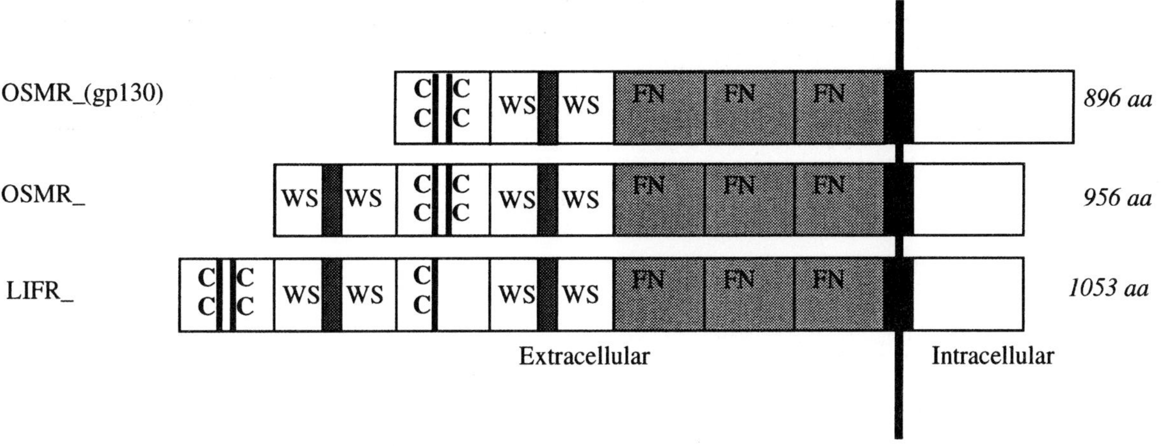

Table 3 Protein properties of OSM receptor subunits

	OSMRα(gp130)				OSMRβ		LIFRα				
	Memb.		Soluble			Memb.		Memb.	Soluble		
	Human	Mouse	Human		Mouse	Human	Mouse	Human	Human		
			IF1	IF2					IF1	IF2	IF3
Number of amino acids											
Precursor	918	917	658	646		979	970	1097	692	780	847
Signal peptide domain	22	22	22	22		23	23	44	44	44	44
Mature protein	896	895	636	624		956	947	1053	648	736	803
Extracellular domain	597	595				716	712	789			
Transmembrane domain	22	22				22	21	25			
Cytoplasmic domain	277	278				218	214	221			
***N*-Glycosylation sites**	14	12	11	10		20	21	20	17	19	20

Memb., membrane-bound form; IF = isoform designation for this chapter.

Table 4 Low-affinity (direct) receptor subunit–ligand binding

	OSM		LIF	
	Human	Murine	Human	Murine
OSMRα(gp130)				
Human	Yes[a,b]	No[c]	Yes[d], No[e]	ND
Murine	Yes[f]	Yes[c,g]	ND	No[g]
OSMRβ				
Human	No[e]	ND	No[e]	ND
Murine	No[c]	Yes[c]	ND	ND
LIFRα				
Human	No[e]	ND	Yes[e]	ND
Murine	No[h]	No[g]	Yes[d,h]	Yes[i]

[a]Modrell *et al.*, 1994; [b]Gearing *et al.*, 1992b; [c]Lindberg *et al.*, 1998; [d]Zhang *et al.*, 1997; [e]Mosley *et al.*, 1996; [f]Liu *et al.*, 1994; [g]Ichihara *et al.*, 1997; [h]Gearing *et al.*, 1992a; [i]Gearing and Bruce, 1992; ND, no data.

Table 5 High-affinity receptor complex–ligand binding

	OSM		LIF	
	Human	Murine	Human	Murine
OSMRα(gp130)/LIFRα				
Human	Yes [a]	No [b]	Yes [a]	ND
Murine	Yes [b,c,d]	No [b,e]	Yes [b,c]	Yes [b,e]
OSMRα(gp130)/OSMRβ				
Human	Yes [a,c,d]	ND	No [a,c]	ND
Murine	No [b]	Yes [b,e]	No [b]	No [b,e]

[a]Mosley *et al.*, 1996; [b]Lindberg *et al.*, 1998; Gearing *et al.*, 1992a; [d]Bruce *et al.*, 1992a; [e]Ichihara *et al.*, 1997; ND, no data.

Table 6 Expression profile of the different OSM receptor subunits

Tissue type	Cell line	Relative mRNA expression level		
		OSMRα(gp130)	OSMRβ	LIFRα
Primary cells				
Bone marrow		0	0	0
Brain (fetal)		25	5	68
Monocyte		15	0	0
Muscle (smooth)		224	129	57
Peripheral blood T cells		61	0	0
Placenta		124	107	142
Skin		44	46	84
Tonsil B cells		0	0	0
Tonsil T cells		92	0	0
Pre-B cell lines	JM-1	0	0	0
	Nalm 6	0	0	0
B cell lines	CESS	4	1	0
	Raji	2	0	0
Endothelial cell lines				
Umbilical vein	HUVE	53	17	39
Uterine, mesodermal tumor	SK-UT-1	14	25	0
Epithelial cell lines				
Cervical carcinoma	HeLa	91	52	38
Epidermal carcinoma	KB	81	43	93
Fibroblast cell lines				
Foreskin	HFF	114	94	41
Lung, adult	LL97A	116	116	79
Lung, embryonic	WI26 VA4	32	36	7
Liver				
Hepatocarcinoma	HepG2	21	47	18
	Hep3B	15	0	0
	SK Hep	30	36	17
Monocytic cell lines				
Acute monocytic leukemia	HL-60	39	0	0
	THP-1	37	6	2
	U937	5	0	0
Neural cell lines				
Astrocytoma	CCF STT G1	49	28	41
Glioblastoma	A172	52	82	83
Medulloblastoma	Daoy	29	30	26
Neuroblastoma	SK-N-SH	25	72	67
T cell lines	clone 22	8	0	0
	Jurkat	8	1	0

Table 6 (*Continued*)

Tissue type	Cell line	Relative mRNA expression level		
		OSMRα(gp130)	OSMRβ	LIFRα
Other				
Bone marrow stromal	IMTLH	42	42	7
Leiosarcoma	SK-LMS	70	23	5
Megakaryocyte	Mo7E	8	0	0
Melanoma, malignant	A375	59	47	73
Pancreatic tumor	HPT	35	29	8
Placental choriocarcinoma	JAR	22	0	26
Promonocyte	TF-1	7	0	0
Rhabdomyosarcoma	A673	17	22	38

Modified with data from Mosley *et al.* (1996).

Table 7 Gene expression induced by activation of the OSM receptors

Affected gene	Cell type	Species	Reference
α_1-Antichymotrypsin	HepG2 cells	Human	Richards *et al.*, 1992
α_1-Antichymotrypsin	Astrocytes	Human	Kordula *et al.*, 1998
α_1-Proteinase inhibitor	Epithelial	Human	Sallenave *et al.*, 1997
Basic fibroblast growth factor	Endothelial	Bovine	Wijelath *et al.*, 1997
EGR-1, c-*jun*, c-*myc*	Fibroblasts	Human	Liu *et al.*, 1992a
Haptoglobin	HepG2 cells	Human	Richards *et al.*, 1992
IL-6	Endothelial	Human	Brown *et al.*, 1991
Matrix metalloproteinase 1	Fibroblasts	Human	Korzus *et al.*, 1997
P21 kinase inhibitor	Osteoblasts	Human	Bellido *et al.*, 1998
P-Selectin	Endothelial	Human	Yao *et al.*, 1996
TIMP-1 (tissue inhibitor of metalloproteinase 1)	Cartilage	Human	Nemoto *et al.*, 1996
TIMP-1	Synovial	Human	Gatsios *et al.*, 1996
TIMP-1	Fibroblasts	Mouse	Richards *et al.*, 1997
TIMP-1	Fibroblasts	Human	Korzus *et al.*, 1997
Urokinase-type plasminogen activator	Fibroblasts	Human	Hamilton *et al.*, 1991

BIOLOGICAL CONSEQUENCES OF ACTIVATING OR INHIBITING RECEPTOR AND PATHOPHYSIOLOGY

Unique biological effects of activating the receptors

Although most of the biological effects of activating the OSM receptors are similar to other receptor complexes with shared receptor subunits, some effects appear to be specific to the OSM-specific receptors. OSM-specific receptor appears specifically to inhibit the growth of normal and malignant mammary epithelial cells (Liu *et al.*, 1998), stimulate expression of the tissue inhibitor of metalloproteinase 1 (TIMP-1) (Richards *et al.*, 1997), and induce the synthesis of α_1-antichymotrypsin and α_1-antiproteinase inhibitor (Cichy *et al.*, 1998; Kordula *et al.*, 1998).

Phenotypes of receptor knockouts and receptor overexpression mice

Both receptor knockouts and receptor overexpression in mice have been used to study the biological effects of OSMRα(gp130). Because of the ubiquitous use of OSMRα(gp130) as a signaling subunit in the receptor complexes for the entire IL-6 family of cytokines, the elimination of this gene has dire consequences in embryonic development, including hematological disorders, hypoplasia of the myocardium, structural and function defects in the placenta and reduction of bone mass (Yoshida *et al.*, 1996; reviewed in Nakashima and Taga, 1998). Inducible inactivation of OSMRα(gp130) postnatally results in neurological, cardiac, hematopoietic, immunological, hepatic, and pulmonary defects in mice (Betz *et al.*, 1998). Mice expressing a dominant-negative form of OSMRα(gp130) with a truncated signaling domain demonstrated the necessity for OSMRα(gp130) in antigen-specific antibody production (Kumanogoh *et al.*, 1997).

Targeted disruption of the LIFRα in mice causes placental, skeletal, neural, and metabolic defects and results in perinatal death (Ware *et al.*, 1995; Koblar *et al.*, 1998). Since OSM utilizes receptor complexes containing LIFRα in humans, but not in mice, generalizing the murine knockout studies to the functions of OSM-induced LIFRα signaling through the LIF/OSM shared receptor in humans should be done with some caution. The biological effects of OSMRβ knockouts or overexpressors are not yet known.

THERAPEUTIC UTILITY

Effect of treatment with soluble receptor domain

Although soluble forms of OSMRα(gp130) and LIFRα act as antagonists of members of the IL-6 family of cytokines *in vitro* (Layton *et al.*, 1992; Yamaguchi-Yamamoto *et al.*, 1993; Montero-Julian *et al.*, 1997), their function *in vivo* is unknown.

Effects of inhibitors (antibodies) to receptors

Monoclonal antibodies to OSMRα(gp130) have been derived that specifically inhibit the growth of OSM-dependent cell lines (Liu *et al.*, 1992b; Taga *et al.*, 1992; Gu *et al.*, 1996). Although monoclonal antibodies to LIFRα have been derived that specifically block the biological activity of LIF (Pitard *et al.*, 1997), it is unknown whether they also block the activity of OSM binding to the LIF/OSM shared receptor in humans. Mutants of LIF have been derived which antagonize OSM signaling through the LIF/OSM shared receptor (Vernallis *et al.*, 1997).

References

Amaral, M. C., Miles, S., Kumar, G., and Nel, A. E. (1993). Oncostatin M stimulates tyrosine protein phosphorylation in parallel with the activation of p42MAPK/ERK-2 in Kaposi's cells. Evidence that this pathway is important in Kaposi cell growth. *J. Clin. Invest.* **92**, 848–857.

Auguste, P., Guillet, C., Fourcin, M., Olivier, C., Veziers, J., Poouplard-Barthelaix, A., and Gascan, H. (1997). Signaling of type II oncostatin M receptor. *J. Biol. Chem.* **272**, 15760–15764.

Baumann, H., Symes, A. J., Comeau, M. R., Morella, K. K., Wang, Y., Friend, D., Ziegler, S. F., Fink, J. S., and Gearing, D. P. (1994). Multiple regions within the cytoplasmic domains of the leukemia inhibitory factor receptor and gp130 cooperate in signal transduction in hepatic and neuronal cells. *Mol. Cell. Biol.* **14**, 138–146.

Bazan, J. F. (1990). Structural design and molecular evolution of a cytokine receptor superfamily. *Proc. Natl Acad. Sci. USA* **87**, 6934–6938.

Bellido, T., Stahl, N., Farruggella, T. J., Borba, V., Yancopoulos, G. D., and Manolagas, S. C. (1996). Detection of receptors for interleukin-6, interleukin-11, leukemia inhibitory factor, oncostatin M, and ciliary neurotrophic factor in bone marrow stromal/osteoblastic cells. *J. Clin. Invest.* **97**, 431–437.

Bellido, T., O'Brian, C. A., Roberson, P. K., and Manolagas, S. C. (1998). Transcriptional activation of the $p21^{WAF1,CIP1,SDI1}$ gene by interleukin-6 type cytokines. *J. Biol. Chem.* **273**, 21137–21144.

Betz, U. A. K., Bloch, W., van den Broek, M., Yoshida, K., Taga, T., Kishimoto, T., Addicks, K., Rajewsky, K., and Muller, W. (1998). Postnatally induced inactivation of gp130 in mice results in neurological, cardiac, hematopoietic, immunological, hepatic, and pulmonary defects. *J. Exp. Med.* **188**, 1955–1965.

Bravo, J., Staunton, D., Heath, J. K., and Jones, E. Y. (1998). Crystal structure of a cytokine-binding region of gp130. *EMBO J.* **17**, 1665–1674.

Brown, T. J., Rowe, J. M., Liu, J., and Shoyab, M. (1991). Regulation of IL-6 expression by oncostatin M. *J. Immunol.* **147**, 2175–2180.

Bruce, A. G., Hoggatt, I. H., and Rose, T. M. (1992a). Oncostatin M is a differentiation factor for myeloid leukemia cells. *J. Immunol.* **149**, 1271–1275.

Bruce, A. G., Linsley, P. S., and Rose, T. M. (1992b). Oncostatin M. *Progr. Growth Factor Res.* **4**, 157–170.

Cichy, J., Rose-John, S., and Pure, E. (1998). Regulation of the type II oncostatin M receptor expression in lung-derived epithelial cells. *FEBS Lett.* **429**, 412–416.

Darnell Jr, J. E., Kerr, I. M., and Stark, G. R. (1994). Jak-STAT pathways and transcriptional activation in response to IFNs and other extracellular signaling proteins. *Science* **264**, 1415–1421.

Diamant, M., Rieneck, K., Mechti, N., Zhang, X.-G., Svenson, M., Bendtzen, K., and Klein, B. (1997). Cloning and expression of an alternatively spliced mRNA encoding a soluble form of the human interleukin-6 signal transducer gp130. *FEBS Lett.* **412**, 379–384.

Gatsios, P., Haubeck, H.-D., Van de Leur, E., Frisch, W., Apte, S. S., Greiling, H., Heinrich, P. C., and Graeve, L. (1996). Oncostatin M differentially regulates tissue inhibitors of metalloproteinases TIMP-1 and TIMP-3 gene expression in human synovial lining cells. *Eur. J. Biochem.* **241**, 56–63.

Gearing, D. P., and Bruce, A. G. (1992). Oncostatin M binds the high affinity leukemia inhibitory factor receptor. *New Biol.* **4**, 61–65.

Gearing, D. P., Thut, C. J., VandeBos, T., Gimpel, S. D., Delaney, P. B., King, J., Price, V., Cosman, D., and Beckmann, M. P. (1991). Leukemia inhibitory factor receptor is structurally related to the IL-6 signal transducer, gp130. *Embo J.* **10**, 2839–2848.

Gearing, D. P., Comeau, M. R., Griend, D. J., Gimpel, S. D., Thut, C. H., McGourty, J., Brasher, K. K., King, J. A., Gillis, S., Mosley, B., Ziegler, S. F., and Cosman, D. (1992a). The IL-6 signal transducer, gp130: an oncostatin M receptor and affinity converter for the LIF receptor. *Science* **255**, 1434–1437.

Gearing, D. P., VandenBos, T., Beckmann, M. P., Thut, C. J., Comeau, M. R., Mosley, B., and Ziegler, S. F. (1992b). Reconstitution of high affinity leukaemia inhibitory factor (LIF) receptors in haemopoietic cells transfected with the cloned human LIF receptor. *Ciba Found. Symp.*. **167**, 245–255.

Gu, Z. J., Wijdenes, J., Zhang, X. G., Hallet, M. M., Clement, C., and Klein, B. (1996). Anti-gp130 transducer monoclonal antibodies specifically inhibiting ciliary neurotrophic factor, interleukin-6, interleukin-11, leukemia inhibitory factor or oncostatin M. *J. Immunol. Methods* **190**, 21–27.

Hamilton, J. A., Leizer, T., Piccoli, D. S., Royston, K. M., Butler, D. M., and Croatto, M. (1991). Oncostatin M stimulates urokinase-type plasminogen activator activity in human synovial fibroblasts. *Biochem. Biophys. Res. Commun.* **180**, 652–659.

Heinrich, P. C., Behrmann, I., Muller-Newen, G., Schaper, F., and Graeve, L. (1998). Interleukin-6-type cytokine signalling through the gp130/Jak/STAT pathway. *Biochem. J.* **334**, 297–314.

Hibi, M., Murakami, M., Saito, M., Hirano, T., Taga, T., and Kishimoto, T. (1990). Molecular cloning and expression of an IL-6 signal transducer, gp130. *Cell* **63**, 1149–1157.

Hirano, T., Nakajima, K., and Hibi, M. (1997). Signaling mechanisms through gp130: a model of the cytokine system. *Cytokine Growth Factor Rev.* **8**, 241–252.

Horn, D., Fitzpatrick, W. C., Gompper, P. T., Ochs, V., Bolton-Hanson, M., Zarling, J., Malik, N., Todaro, G. J., and Linsley, P. S. (1990). Regulation of cell growth by recombinant oncostatin M. *Growth Factors* **22**, 157–165.

Ichihara, M., Hara, T., Kim, H., Murate, T., and Miyajima, A. (1997). Oncostatin M and leukemia inhibitory factor do not use the same functional receptor in mice. *Blood* **90**, 165–173.

Koblar, S. A., Turnley, A. M., Classon, B. J., Reid, K. L., Ware, C. B., Cheema, S. S., Murphy, M., and Bartlett, P. F. (1998). Neural precursor differentiation into astrocytes requires signaling through the leukemia inhibitory factor receptor. *Proc. Natl Acad. Sci. USA* **95**, 3178–3181.

Kojima, K., Kanzaki, H., Iwai, M., Hatayama, H., Fujimoto, M., Narukawa, S., Higuchi, T., Kaneko, Y., Mori, T., and Fujita, J. (1995). Expression of leukaemia inhibitory factor (LIF) receptor in human placenta: a possible role for LIF in the growth and differentiation of trophoblasts. *Hum. Reprod.* **10**, 1907–1911.

Kordula, T., Rydel, R. E., Brigham, E. F., Horn, F., Heinrich, P. C., and Travis, J. (1998). Oncostatin M and the interleukin-6 and soluble interleukin-6 receptor complex regulate α_1-antichymotrypsin expression in human cortical astrocytes. *J. Biol. Chem.* **273**, 4112–4118.

Korzus, E., Nagase, H., Rydell, R., and Travis, J. (1997). The mitogen-activated protein kinase and JAK-STAT signaling pathways are required for an oncostatin M-responsive element-mediated activation of matrix metalloproteinase 1 gene expression. *J. Biol. Chem.* **272**, 1188–1196.

Kumanogoh, A., Marukawa, S., Kumanogoh, T., Hirota, H., Yoshida, K., Lee, I.-S., Yasui, T., Yoshida, K., Taga, T., and Kishimoto, T. (1997). Impairment of antigen-specific antibody production in transgenic mice expressing a dominant-negative form of gp130. *Proc. Natl Acad. Sci. USA* **94**, 2478–2482.

Kuropatwinski, K. K., Imus, C. D., Gearing, D., Baumann, H., and Mosley, B. (1997). Influence of subunit combinations on signaling by receptors for oncostatin M, leukemia inhibitory factor and interleukin-6. *J. Biol. Chem.* **272**, 15135–15144.

Layton, M. J., Cross, B. A., Metcalf, D., Ward, L. D., Simpson, R. J., and Nicola, N. A. (1992). A major binding protein for leukemia inhibitory factor in normal mouse serum: identification as a soluble form of the cellular receptor. *Proc. Natl Acad. Sci. USA* **89**, 8616–8620.

Lindberg, R. A., Juan, T. S.-C., Welcher, A. A., Sun, Y., Cupples, R., Guthrie, B., and Fletcher, F. A. (1998). Cloning and characterization of a specific receptor for mouse oncostatin M. *Mol. Cell. Biol.* **18**, 3357–3367.

Linsley, P. S., Hanson, M. B., Horn, D., Malik, N., Kallestad, J. C., Ochs, V., Zarling, J. L., and Shoyab, M. (1989). Identification and characterization of cellular receptors for the growth regulator, oncostatin M. *J. Biol. Chem.* **264**, 4282–4289.

Liu, J., Clegg, C. H., and Shoyab, M. (1992a). Regulation of EGR-1, c-*jun*, and c-*myc* gene expression by oncostatin M. *Cell Growth Diff.* **3**, 307–313.

Liu, J., Modrell, B., Aruffo, A., Marken, J. S., Taga, T., Yasukawa, K., Murakami, M., Kishimoto, T., and Shoyab, M. (1992b). Interleukin-6 signal transducer gp130 mediates oncostatin M signaling. *J. Biol. Chem.* **267**, 16763–16766.

Liu, J., Modrell, B., Aruffo, A., Scharnowske, S., and Shoyab, M. (1994). Interactions between oncostatin M and the IL-6 signal transducer, gp130. *Cytokine* **6**, 272–278.

Liu, J., Hadjokas, N., Mosley, B., Estrov, Z., Spence, M. J., and Vestal, R. E. (1998). Oncostatin M-specific receptor expression and function in regulating cell proliferation of normal and malignant mammary epithelial cells. *Cytokine* **10**, 295–302.

Matsuda, T., and Hirano, T. (1994). Establishment of the ELISA for murine soluble gp130, a signal transducer for the IL-6 family cytokine, and its detection in the ascitic fluids of tumor-bearing mice. *Biochem. Biophys. Res. Commun.* **202**, 637–642.

Modrell, B., Liu, J., Miller, H., and Shoyab, M. (1994). LIF and OM directly interact with a soluble form of gp130, the IL-6 receptor signal transducing subunit. *Growth Factors* **11**, 81–91.

Montero-Julian, F. A., Brailly, H., Sautes, C., Joyeux, I., Dorval, T., Mosseri, V., Yasukawa, K., Wijdenes, J., Adler, A., Gorin, I., Fridman, W. H., and Tartour, E. (1997). Characterization of soluble gp130 released by melanoma cell lines. A polyvalent antagonist of cytokines from the interleukin 6 family. *Clin. Cancer Res.* **3**, 1443–1451.

Mosley, B., Imus, C. D., Friend, D., Boiani, N., Thoma, B., Park, L. S., and Cosman, D. (1996). Dual oncostatin M (OSM) receptors. *J. Biol. Chem.* **271**, 32635–32643.

Murakami, M., Narazaki, M., Masahiko, H., Yawata, H., Yasukawa, K., Hamaguchi, M., Taga, T., and Kishimoto, T. (1991). Critical cytoplasmic region of the interleukin 6 signal transducer gp130 is conserved in the cytokine receptor family. *Proc. Natl Acad. Sci. USA* **88**, 11349–11353.

Nakashima, K., and Taga, T. (1998). gp130 and the IL-6 family of cytokines: signaling mechanisms and thrombopoietic activities. *Semin. Hematol.* **35**, 210–221.

Narzaki, M., Yasukawa, K., Saito, T., Ohsugi, Y., Fukui, H., Koishihara, Y., Yancopoulos, G. D., Taga, T., and Kishimoto, T. (1993). Soluble forms of the interleukin-6 signal-transducing receptor component gp130 in human serum possessing a potential to inhibit signals through membrane-anchored gp130. *Blood* **82**, 1120–1126.

Nemoto, O., Yamada, H., Mukaida, M., and Shimmei, M. (1996). Stimulation of TIMP-1 production by oncostatin M in human articular cartilage. *Arthr. Rheum.* **39**, 560–566.

Pitard, V., Taupin, J. L., Miossec, V., Blanchard, F., Cransac, M., Jollet, I., Vernallis, A., Hudson, K., Godard, A., Jacques, Y., and Moreau, J. F. (1997). Production and characterization of monoclonal antibodies against the leukemia inhibitory factor low affinity receptor, gp190. *J. Immunol. Methods* **205**, 177–190.

Richards, C. D., Brown, T. J., Shoyab, M., Baumann, H., and Gauldie, J. (1992). Recombinant oncostatin M stimulates the production of acute phase proteins in HepG2 cells and rat primary hepatocytes *in vitro*. *J. Immunol.* **148**, 1731–1736.

Richards, C. D., Kerr, C., Tanaka, M., Hara, T., Miyajima, A., Pennica, D., Botelho, F., and Langdon, C. M. (1997). Regulation of tissue inhibitor of metalloproteinase-1 in fibroblasts and acute phase proteins in hepatocytes *in vitro* by mouse oncostatin M, cardiotrophin-1, and IL-6. *J. Immunol.* **159**, 2431–2437.

Rose, T. M., and Bruce, A. G. (1991). Oncostatin M is a member of a cytokine family that includes leukemia-inhibitory factor, granulocyte colony-stimulating factor, and interleukin 6. *Proc. Natl Acad. Sci. USA* **88**, 8641–8645.

Rose, T. M., Lagrou, M. J., and Fransson, I. (1993). The genes for oncostatin M (OSM) and leukemia inhibitory factor (LIF) are tightly linked on human chromosome 22. *Genomics* **17**, 136–140.

Saito, M., Yoshida, K., Hibi, M., Taga, T., and Kishimoto, T. (1992). Molecular cloning of a murine IL-6 receptor-associated signal transducer, gp130, and its regulated expression *in vivo*. *J. Immunol.* **148**, 4066–4071.

Sallenave, J.-M., Tremblay, G. M., Gauldie, J., and Richards, C. D. (1997). Oncostatin M, but not interleukin-6 or leukemia inhibitory factor, stimulates expressions of alpha1-proteinase inhibitor in A549 human alveolar epithelial cells. *J. Interferon Cytokine Res.* **17**, 337–346.

Schiemann, W. P., Bartoe, J. L., and Nathanson, N. M. (1997). Box 3-independent signaling mechanisms are involved in leukemia inhibitory factor receptor α- and gp130-mediated stimulation of mitogen-activated protein kinase. *J. Biol. Chem.* **272**, 16631–16636.

Schieven, G. L., Kallestad, J. C., Brown, T. J., Ledbetter, J. A., and Linsley, P. S. (1992). Oncostatin M induces tyrosine phosphorylation in endothelial cells and activation of $p62^{yes}$ tyrosine kinase. *J. Immunol.* **149**, 1676–1682.

Schindler, C., and Darnell Jr, J. E. (1995). Transcriptional responses to polypeptide ligands: the JAK-STAT pathway. *Annu. Rev. Biochem.* **64**, 621–651.

Sharkey, A. M., Dellow, K., Blayney, M., Mcnamee, M., Charnock-Jones, S., and Smith, S. K. (1995). Stage-specific expression of cytokine and receptor messenger ribonucleic acids in human preimplantation embryos. *Biol. Reprod.* **53**, 974–981.

Stahl, N., Boulton, T. G., Farruggella, T., Ip, N. Y., Davis, S., Witthuhn, B. A., Quelle, F. W., Silvennoinen, O,, Barbieri, G., Pellegrini, S., Ihle, J. N., and Yancopoulos, G. D. (1994). Association and activation of Jak-Tyk kinases by CNTF-LIF-OSM-IL-6 beta receptor components. *Science* **263**, 92–95.

Stancato, L. F., Sakatsume, M., David, M., Dent, P., Dong, F., Petricoin, E. F., Krolewski, J. J., Silvennoinen, O., Saharinen, P., Pierce, J., Marshall, C. J., Sturgill, T., Finbloom, D. S., and Larner, A. C. (1997). Beta interferon and oncostatin M activate Raf-1 and mitogen-activated protein kinase through a JAK1-dependent pathway. *Mol. Cell. Biol.* **17**, 3833–3840.

Stancato, L. F., Yu, C.-R., Petricoin, E.F. III, and Larner, A. C. (1998). Activation of Raf-1 by interferon γ and oncostatin M requires the expression of the Stat1 transcription factor. *J. Biol. Chem.* **273**, 18701–18704.

Stephens, J. M., Lumpkin, S. J., and Fishman, J. B. (1998). Activation of signal transducers and activators of transcription 1 and 3 by leukemia inhibitory factor, oncostatin M, and interferon-γ in adipocytes. *J. Biol. Chem.* **273**, 31408–31416.

Taga, T. (1996). gp130, a shared signal transducing receptor component for hematopoietic and neuropoietic cytokines. *J. Neurochem.* **67**, 1–10.

Taga, T., Narazaki, M., Yasukawa, K., Saito, T., Miki, D., Hamaguchi, M., Davis, S., Shoyab, M., Yancopoulos, G. D., and Kishimoto, T. (1992). Functional inhibition of hematopoietic and neurotrophic cytokines by blocking the interleukin 6 signal transducer gp130. *Proc. Natl Acad. Sci. USA* **89**, 10998–11001.

Thoma, B., Bird, T. A., Friend, D. J., Gearing, D. P., and Dower, S. K. (1994). Oncostatin M and leukemia inhibitory factor trigger overlapping and different signals through partially shared receptor complexes. *J. Biol. Chem.* **269**, 6215–6222.

Tomida, M. (1997). Presence of mRNAs encoding the soluble D-factor/LIF receptor in human choriocarcinoma cells and production of the soluble receptor. *Biochem. Biophys. Res. Commun.* **232**, 427–431.

Tomida, M., and Gotoh, O. (1996). Structure of the gene encoding the human differentiation-stimulating factor/leukemia inhibitory factor receptor. *J. Biochem.* (*Tokyo*) **120**, 201–205.

Van Eijk, M. J., Mandelbaum, J., Salat-Baroux, J., Belaisch-Allart, J., Plachot, M., Junca, A. M., and Mummery, C. L. (1996). Expression of leukaemia inhibitory factor receptor subunits LIFR beta and gp130 in human oocytes and preimplantation embryos. *Mol. Hum. Reprod.* **2**, 355–360.

Vernallis, A. B., Hudson, K. R., and Heath, J. K. (1997). An antagonist for the leukemia inhibitory factor receptor inhibits leukemia inhibitory factor, cardiotrophin-1, ciliary neurotrophic factor, and oncostatin M. *J. Biol. Chem.* **272**, 26947–26952.

Wang, Z., and Melmed, S. (1997). Identification of an upstream enhancer within a functional promoter of the human leukemia inhibitory factor receptor gene and its alternative promoter usage. *J. Biol. Chem.* **272**, 27957–27965.

Wang, Y., Nesbitt, J. E., Fuentes, N. L., and Fuller, G. M. (1992). Molecular cloning and characterization of the rat liver IL-6 signal transducing molecule, gp130. *Genomics* **14**, 666–672.

Ware, C. B., Horowitz, M. C., Renshaw, B. R., Hunt, J. S., Liggitt, D., Koblar, S. A., Gliniak, B. C., McKenna, H. J., Papayannopoulou, T., Thoma, B., Cheng, L., Donovan, P. J., Pescon, J. J., Bartlett, P. F., Willis, C. R., Wright, B. D.,

Carpenter, M. K., Davison, B. L., and Gearing, D. P. (1995). Targeted disruption of the low-affinity leukemia inhibitory factor receptor gene causes placental, skeletal, neural and metabolic defects and results in perinatal death. *Development* **121**, 1283–1299.

Watanabe, D., Yoshimura, R., Khalil, M., Yoshida, K., Kishimoto, T., Taga, T., and Kiyama, H. (1996). Characteristic localization of gp130 (the signal transducing receptor component used in common for IL-6/IL-11/CNTF/LIF/OSM) in the rat brain. *Eur. J. Neurosci.* **8**, 1630–1640.

Wijelath, E. S., Carlsen, B., Cole, T., Chen, J., Kothari, S., and Hammond, W. P. (1997). Oncostatin M induces basic fibroblast growth factor expression in endothelial cells and promotes endothelial cell proliferation, migration and spindle morphology. *J. Cell Sci.* **110**, 871–879.

Yamaguchi-Yamamoto, Y., Tomida, M., and Hozumi, M. (1993). Pregnancy associated increase in differentiation-stimulating factor (D-factor)/leukemia inhibitory factor (LIF)-binding substance(s) in mouse serum. *Leuk. Res.* **17**, 515–522.

Yao, L., Pan, J., Setiadi, H., Patel, K. D., and McEver, R. P. (1996). Interleukin-4 or oncostatin M induces a prolonged increase in P-selectin mRNA and protein in human endothelial cells. *J. Exp. Med.* **184**, 81–92.

Yoshida, K., Taga, T., Saito, M., Suematsu, S., Kumanogoh, A., Tanaka, T., Fujiwara, H., Hirata, M., Yamagami, T., Nakahata, T., Hirabayashi, T., Yoneda, Y., Tanaka, K., Wang, W.-Z., Mori, C., Shiota, K., Yoshida, N., and Kishimoto, T. (1996). Targeted disruption of gp130, a common signal transducer for the interleukin 6 family of cytokines, leads to myocardial and hematological disorders. *Proc. Natl Acad. Sci. USA* **93**, 407–411.

Zhang, J.-G., Owczarek, C. M., Ward, L. D., Howlett, G., Fabri, L. J., Roberts, B. A., and Nicola, N. A. (1997). Evidence for the formation of a heterotrimeric complex of leukaemia inhibitory factor with its receptor subunits in solution. *Biochem. J.* **325**, 693–700.

Zhang, J.-G., Zhang, Y., Owczarek, C. M., Ward, L. D., Moritz, R. L., Simpson, R. J., Yasukawa, K., and Nicola, N. A. (1998). Identification and characterization of two distinct truncated forms of gp130 and a soluble form of leukemia inhibitory factor receptor α-chain in normal human urine and plasma. *J. Biol. Chem.* **273**, 10798–10805.

CT-1 Receptor

Kenneth R. Chien*

Department of Medicine, Center for Molecular Genetics, and the American Heart Association-Bugher Foundation, Center for Molecular Biology, University of California, San Diego School of Medicine, 9500 Gilman Drive, La Jolla, CA 92093, USA

*corresponding author tel: 619 534 6835, fax: 619 534 8081, e-mail: kchien@ucsd.edu

DOI: 10.1006/rwcy.2000.17006.

SUMMARY

The CT-1 ligand receptor system has not been fully elucidated. This notion should be advanced by the molecular cloning of the CT-1-specific receptor.

BACKGROUND

Discovery

Cardiotropin 1, CT-1, signals via heterodimerization of the gp130 (IL-6 receptor) and LIF receptors (LIFR) (Pennica *et al.*, 1995; Wollert *et al.*, 1996). The subsequent activation events reveal that CT-1 induces and utilizes JAK1-, JAK2-, and TYK2-associated tyrosine kinases, which are in turn relayed by signaling via the STAT3 transcription factor (Robledo *et al.*, 1997). Crosslinking of iodinated CT-1 to the cell surface led to the identification of a third α component in addition to gp130 and LIFR, with an apparent molecular mass of 80 kDa. On motor neurons, CT-1 action was inhibited by phosphatidylinositol-specific phospholipase C, suggesting that CT-1 may act through a GPI-linked component. Since no binding of CT-1 to ciliary neurotropic factor receptor (CNTFR; GPI-linked receptor) was detected, CT-1 may use a novel cytokine receptor α subunit (Pennica *et al.*, 1996). Thus, a CT-1-specific receptor subunit has been proposed, but thus far not verified by molecular cloning of the corresponding cDNA.

Alternative names

gp130, LIFR.

Structure

See chapters on gp130 (IL-6 receptor) and LIFR.

Main activities and pathophysiological roles

See chapter on CT-1.

GENE

Accession numbers

See chapters on gp130 (IL-6 receptor) and LIFR. A CT-1-specific receptor gene has not been cloned.

Sequence

See chapters on gp130 (IL-6 receptor) and LIFR.

Chromosome location and linkages

See chapters on gp130 (IL-6 receptor) and LIFR.

PROTEIN

Accession numbers

See chapters on gp130 (IL-6 receptor) and LIFR.

Description of protein

See chapters on gp130 (IL-6 receptor) and LIFR.

Crosslinking of iodinated CT-1 to the cell surface led to the identification of a third component in addition to gp130 and LIFR, with an apparent molecular mass of 80 kDa. Removal of *N*-linked carbohydrates from the protein backbone of the third component resulted in a protein of 45 kDa. On motor neurons, CT-1 action was inhibited by phosphatidylinositol-specific phospholipase C, suggesting that CT-1 may act through a GPI-linked component (Pennica *et al.*, 1996) (see chapter on CNTF receptor).

Relevant homologies and species differences

See chapters on gp130 (IL-6 receptor) and LIFR.

Affinity for ligand(s)

Binding studies employing purified soluble gp130 and LIFR molecules demonstrate that CT-1 binds to the LIFR with about the same affinity as LIF but fails to bind to gp130 alone; however, the addition of gp130 enhanced the binding of CT-1 to the LIFR (Pennica *et al.*, 1995). Therefore CT-1 receptor binding involves an initial low-affinity interaction with the LIFR, followed by the recruitment of gp130 into a high-affinity heterotrimeric complex (Pennica *et al.*, 1995).

Cell types and tissues expressing the receptor

See the chapters on gp130 (IL-6 receptor) and LIFR.

Functional studies and receptor-binding studies in cultured cardiomyocytes and the mouse myeloid leukemia cell line M1 have revealed that CT-1 signals through the gp130/LIFR heterodimer in cardiomyocytes, most likely without a further requirement for a specific receptor (Pennica *et al.*, 1995, 1996; Wollert *et al.*, 1996). By contrast, CT-1 signaling in neuronal cells may require an additional CT-1-specific receptor (Pennica *et al.*, 1996; Robledo *et al.*, 1997).

Regulation of receptor expression

See chapters on gp130 (IL-6 receptor) and LIFR.

Release of soluble receptors

See chapters on gp130 (IL-6 receptor) and LIFR.

SIGNAL TRANSDUCTION

Associated or intrinsic kinases

CT-1 induces the rapid tyrosine phosphorylation of gp130 and LIFR, leading to the activation of JAK1, JAK2, and TYK2 kinase, as reported for the other members of the IL-6 cytokine family (Wollert *et al.*, 1996; Robledo *et al.*, 1997).

Cytoplasmic signaling cascades

See section on *In vitro* activities in the chapter on CT-1.

DOWNSTREAM GENE ACTIVATION

Transcription factors activated

See section on *In vitro* activities in the chapter on CT-1.

Genes induced

See section on *In vitro* activities in the chapter on CT-1.

Promoter regions involved

See the chapters on gp130 (IL-6 receptor) and LIFR.

BIOLOGICAL CONSEQUENCES OF ACTIVATING OR INHIBITING RECEPTOR AND PATHOPHYSIOLOGY

Unique biological effects of activating the receptors

See section on *In vivo* biological activities of ligands in animal models in the chapter on CT-1.

References

Pennica, D., Shaw, K. J., Swanson, T. A., Moore, M. W., Shelton, D. L., Zioncheck, K. A., Rosenthal, A., Taga, T., Paoni, N. F., and Wood, W. I. (1995). Cardiotrophin-1. Biological activities and binding to the leukemia inhibitory factor receptor/gp130 signaling complex. *J. Biol. Chem.* **270**, 10915–10922.

Pennica, D., Arce, V., Swanson, T. A., Vejsada, R., Pollock, R. A., Armanini, M., Dudley, K., Phillips, H. S., Rosenthal, A., Kato, A. C., and Henderson, C. E. (1996). Cardiotrophin-1, a cytokine present in embryonic muscle, supports long-term survival of spinal motoneurons. *Neuron* **17**, 63–74.

Robledo, O., Fourcin, M., Chevalier, S., Guillet, C., Auguste, P., Pouplard-Barthelaix, A., Pennica, D., and Gascan, H. (1997). Signaling of the cardiotrophin-1 receptor. Evidence for a third receptor component. *J. Biol. Chem.* **272**, 4855–4863.

Wollert, K. C., Taga, T., Saito, M., Narazaki, M., Kishimoto, T., Glembotski, C. C., Vernallis, A. B., Heath, J. K., Pennica, D., Wood, W. I., and Chien, K. R. (1996). Cardiotrophin-1 activates a distinct form of cardiac muscle cell hypertrophy. Assembly of sarcomeric units in series VIA gp130/leukemia inhibitory factor receptor-dependent pathways. *J. Biol. Chem.* **271**, 9535–9545.

LICENSED PRODUCTS

See the chapters on gp130 (IL-6 Receptor) and LIF Receptor.

Receptors for Chronic Inflammatory Mediators

IFNγ Receptor

Vijay Shankaran and Robert D. Schreiber*

Center for Immunology and Department of Pathology, Washington University School of Medicine, 660 South Euclid Avenue, Mailstop 8118, St. Louis, MO 63108, USA

*corresponding author tel: 314-362-8747, fax: 314-747-4888, e-mail: schreiber@immunology.wustl.edu

DOI: 10.1006/rwcy.2000.18001.

SUMMARY

The IFNγ receptor consists of two subunits, IFNGR1 and IFNGR2, and is expressed nearly ubiquitously on all cell surfaces. The 90 kDa IFNGR1 glycoprotein is predominantly responsible for mediating high-affinity, species-specific ligand binding, ligand trafficking, and signal transduction. The 62 kDa IFNGR2 glycoprotein plays a minor role in ligand binding but is required for signaling. Activated IFNγ receptors utilize the JAK/STAT signaling pathway and specifically JAK1, JAK2, and STAT1 for mediating many IFNγ–dependent effects on cells. A large number of immediate–early genes have been identified that are regulated through an IFNγ- and STAT1-dependent mechanism. These genes contain common promoter elements known as GAS sites which function as target sites for activated STAT1 homodimers.

Interference with the IFNγ-dependent JAK/STAT signaling pathway (i.e. through the use of blocking antibodies or gene deletion) renders the host exquisitely sensitive to infection by a variety of microbial pathogens and certain viruses and also increases host susceptibility to tumors. Recently, patients suffering from rare mycobacterial infections have been identified who have mutations in either IFNGR1 or IFNGR2. Thus the physiologic relevance of signaling through the IFNγ receptor has been unequivocally established.

BACKGROUND

Discovery

The IFNγ receptor was initially characterized in the early 1980s in radioligand-binding studies conducted in several laboratories on a variety of different cell types (Anderson *et al.*, 1982; Celada *et al.*, 1984, 1985; reviewed in Farrar and Schreiber, 1993). These experiments showed that most primary and cultured cells expressed a moderate level (250–25,000 sites/cell) of high-affinity ($K_a = 10^9$–10^{10} M^{-1}) binding sites for IFNγ. The interaction of IFNγ with its receptor was not inhibited by other interferon classes, which explained the basis for the biologic specificity of IFNγ. In addition, human and murine IFNγ bound to their respective receptors in a strictly species-specific manner and thereby induced biologic responses only in species-matched cells. The latter observation proved to be critical in defining the subunits of the functionally active IFNγ receptor and in determining the structure–function relationships operative within each subunit.

A major step forward in defining the subunit composition of IFNγ receptors came from key genetic experiments conducted in 1987 (Jung *et al.*, 1987). These studies employed a family of stable murine–human somatic cell hybrids that contained the full complement of murine chromosomes and a random assortment of human chromosomes. All hybrids that contained human chromosome 6 bound human IFNγ with high affinity – an observation that was later explained by the presence of the human IFNγ receptor α chain gene on this chromosome (Pfizenmaier *et al.*, 1988). However, biologic responsiveness to human IFNγ was found only in hybrids that contained both human chromosomes 6 and 21. These observations, together with similar studies using hamster–murine somatic cell hybrids, led to the hypothesis that functionally active human or murine IFNγ receptors consist of two (or more) species-matched subunits (Jung *et al.*, 1987; Hibino *et al.*, 1991). The first is the receptor subunit responsible for binding ligand in a species-specific manner. The second is a species-matched subunit that is required for induction of biologic responses.

This concept was further refined by independent reports in 1987–88 of the purification of the ligand-binding component of the human IFNγ receptor (Aguet and Merlin, 1987; Novick *et al.*, 1987; Calderon *et al.*, 1988) and the subsequent cloning of its gene (Aguet *et al.*, 1988). This event was followed one year later by the isolation of the gene encoding the murine homolog (Gray *et al.*, 1989; Hemmi *et al.*, 1989; Kumar *et al.*, 1989; Munro and Maniatis, 1989; Cofano *et al.*, 1990). When the ligand-binding chains of the human or murine IFNγ receptor were expressed at high levels in murine or human cells, respectively, the transfected cells bound human or murine ligand in a manner that was identical to endogenous receptors expressed on homologous cells. However, treatment of the transfected cells with heterologous ligand failed to effect induction of cellular responses. In contrast, when the human IFNγ-binding protein was expressed in murine cells that also contained human chromosome 21, these cells not only bound the human ligand but also responded to it (Fischer *et al.*, 1990a; Jung *et al.*, 1990; Farrar *et al.*, 1991). These observations thus added significant support to the concept that functionally active IFNγ receptors require a second, species-specific subunit. Definitive proof of this concept came in 1994 when the second subunit of both the human and murine IFNγ receptors were simultaneously identified by two independent laboratories using complementation cloning approaches (Hemmi *et al.*, 1994; Soh *et al.*, 1994). Thus, functionally active IFNγ receptors are now known to consist of two species-matched chains: a 90 kDa polypeptide responsible for mediating high-affinity ligand binding, ligand trafficking through the cell and signal transduction, and a distinct 62 kDa subunit which is responsible for signaling (**Figure 1**).

Alternative names

The nomenclature for the IFNγ receptor subunits was recently established by the investigators in the field. Currently, the ligand-binding component of the IFNγ receptor (originally denoted the IFNγ receptor α chain or IFNγR1) is now known as IFNGR1 or CDw119. The second subunit, originally designated the IFNγ receptor β chain, accessory factor-1 (AF-1) or IFNγR2, is now known as IFNGR2.

Structure

Functionally active IFNγ receptors are comprised of two species-matched polypeptide chains (Figure 1 and **Table 1**). IFNGR1 is a 90 kDa glycoprotein that plays important roles in mediating ligand binding, ligand trafficking through the cell, and signal transduction. IFNGR2 is a 62 kDa glycoprotein that plays only a minor role in ligand binding but which is required in an obligatory manner for signaling. In unstimulated cells the two receptor subunits are not tightly preassociated with one another. Association is induced upon exposure of the cell to ligand and this association initiates the signal transduction process.

Figure 1 Polypeptide chain structure of the human IFNγ receptor. The IFNγ receptor consists of two species-matched polypeptide subunits. IFNGR1 is a 90 kDa glycoprotein required for ligand binding, trafficking of ligand through the cell, and signal transduction. IFNGR2 is a 62 kDa glycoprotein required predominantly for signaling. The intracellular domain of IFNGR1 contains two functionally critical sequences: first, an LPKS sequence that participates in the binding of JAK1 to IFNGR1, and second, a YDKPH sequence that, when phosphorylated, forms the docking site on the activated receptor for STAT1. The intracellular domain of IFNGR2 contains a 12 amino acid sequence required for JAK2 association.

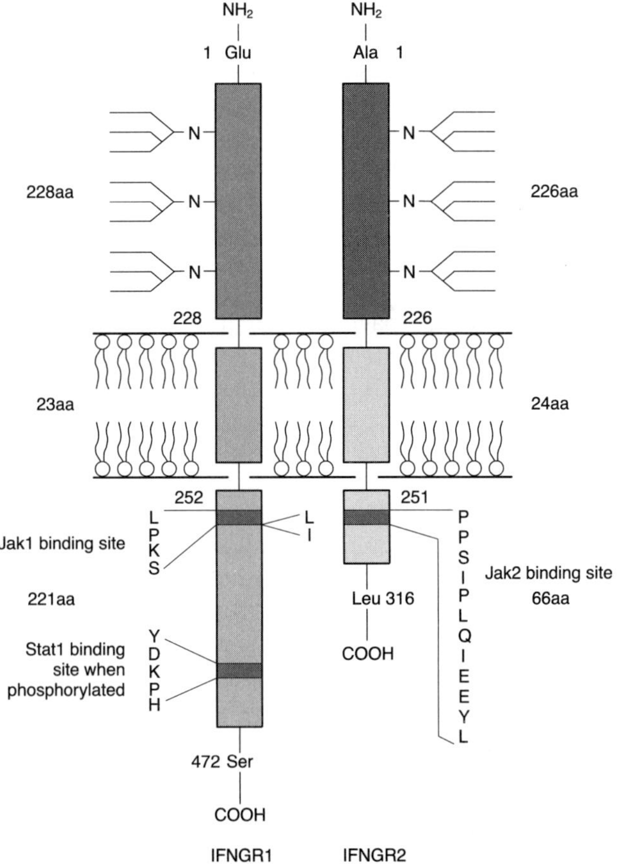

Table 1 Properties of the IFNγ receptor subunits

Property	IFNGR1		IFNGR2	
	Human	Murine	Human	Murine
Primary sequence				
Signal peptide	17 aa	26 aa	21 aa	18 aa
Mature form	472 aa	451 aa	316 aa	314 aa
Homology		52%		58%
Chromosomal localization	6	10	21	16
Domain structure				
Extracellular	228 aa	228 aa	226 aa	224 aa
Transmembrane	23 aa	23 aa	24 aa	24 aa
Intracellular	221 aa	200 aa	66 aa	66 aa
Potential *N*-linked glycosylation sites	5	5	6	6
Predicted M_r (kDa)	52.5	49.8	34.8	35.6
Mature protein M_r (kDa)	90	90	61–67	60–65
Conserved intracellular tyrosines	5		3	

Main activities and pathophysiological roles

Analysis of IFNγ receptor and IFNα/β receptor signaling led to the discovery of a novel signaling pathway known as the JAK/STAT signaling pathway (reviewed in Darnell *et al.*, 1994; Ihle *et al.*, 1995; Schindler and Darnell, 1995; Leonard and O'Shea, 1998). This work has had far-reaching effects in understanding cytokine receptor signaling in general because this signal transduction pathway was subsequently found to participate in the development of many biologic responses induced by a wide variety of different cytokines. The study of IFN receptor signaling led to the identification of two classes of signaling proteins that comprised this pathway. One was a family of latent cytosolic transcription factors that eventually became known as STAT proteins (abbreviation of signal transducers and activators of transcription) (Fu *et al.*, 1992; Schindler *et al.*, 1992). The other was a family of structurally distinct protein tyrosine kinases known as Janus family kinases or JAKs (Darnell *et al.*, 1994). The unique feature of the JAK/STAT signaling pathway was that receptor ligation resulted in the activation of specific subsets of Janus family protein tyrosine kinases (i.e. the JAKs) that subsequently tyrosine phosphorylated and activated cytosolic STAT proteins. The phosphorylated STAT proteins then dimerized and translocated directly from the membrane to the nucleus and effected transcriptional activation of specific target genes. The events linking receptor ligation with signal transduction were ultimately defined when IFNγ was found to effect the tyrosine phosphorylation of the IFNγ receptor α chain, thereby forming a specific docking site on the activated receptor for a particular STAT family member, namely STAT1 (Greenlund *et al.*, 1994). Subsequent work generalized this finding by showing that other STAT family members were recruited to their distinct cytokine receptors by a similar mechanism.

GENE

Accession numbers

Human IFNGR1: J03143
Mouse IFNGR1: M26711, M28233, M25764, M28995
Human IFNGR2: U05877
Mouse IFNGR2: S69336

Sequence

See **Figure 2**.

Figure 2 Nucleotide sequences for human and mouse IFNGR1.

Human IFNGR1

GAATTCCGCAGGCGCTCGGGGTTGGAGCCAGCGACCGTCGGTAGCAGCA
TGGCTCTCCTCTTTCTCCTACCCCTTGTCATGCAGGGTGTGAGCAGGGCTG
AGATGGGCACCGCGGATCTGGGGCCGTCCTCAGTGCCTACACCAACTAA
TGTTACAATTGAATCCTATAACATGAACCCTATCGTATATTGGGAGTACC
AGATCATGCCACAGGTCCCTGTTTTTACCGTAGAGGTAAAGAACTATGGT
GTTAAGAATTCAGAATGGATTGATGCCTGCATCAATATTTCTCATCATTAT
TGTAATATTTCTGATCATGTTGGTGATCCATCAAATTCTCTTTGGGTCAGA
GTTAAAGCCAGGGTTGGACAAAAAGAATCTGCCTATGCAAAGTCAGAAG
AATTTGCTGTATGCCGAGATGGAAAAATTGGACCACCTAAACTGGATATC
AGAAAGGAGGAGAAGCAAATCATGATTGACATATTTCACCCTTCAGTTTT
TGTAAATGGAGACGAGCAGGAAGTCGATTATGATCCCGAAACTACCTGT
TACATTAGGGTGTACAATGTGTATGTGAGAATGAACGGAAGTGAGATCC
AGTATAAAATACTCACGCAGAAGGAAGATGATTGTGACGAGATTCAGTG
CCAGTTAGCGATTCCAGTATCCTCACTGAATTCTCAGTACTGTGTTTCAGC

AGAAGGAGTCTTACATGTGTGGGGTGTTACAACTGAAAAGTCAAAAGAA
GTTTGTATTACCATTTTCAATAGCAGTATAAAAGGTTCTCTTTGGATTCCA
GTTGTTGCTGCTTTACTACTCTTTCTAGTGCTTAGCCTGGTATTCATCTGTT
TTTATATTAAGAAAATTAATCCATTGAAGGAAAAAAGCATAATATTACCC
AAGTCCTTGATCTCTGTGGTAAGAAGTGCTACTTTAGAGACAAAACCTGA
ATCAAAATATGTATCACTCATCACGTCATACCAGCCATTTTCCTTAGAAA
AGGAGGTGGTCTGTGAAGAGCCGTTGTCTCCAGCAACAGTTCCAGGCAT
GCATACCGAAGACAATCCAGGAAAAGTGGAACATACAGAAGAACTTTCT
AGTATAACAGAAGTGGTGACTACTGAAGAAAATATTCCTGACGTGGTCC
CGGGCAGCCATCTGACTCCAATAGAGAGAGAGAGTTCTTCACCTTTAAGT
AGTAACCAGTCTGAACCTGGCAGCATCGCTTTAAACTCGTATCACTCCAG
AAATTGTTCTGAGAGTGATCACTCCAGAAATGGTTTTGATACTGATTCCA
GCTGTCTGGAATCACATAGCTCCTTATCTGACTCAGAATTTCCCCCAAAT
AATAAAGGTGAAATAAAAACAGAAGGACAAGAGCTCATAACCGTAATA
AAAGCCCCCACCTCCTTTGGTTATGATAAACCACATGTGCTAGTGGATCT
ACTTGTGGATGATAGCGGTAAAGAGTCCTTGATTGGTTATAGACCAACAG
AAGATTCCAAAGAATTTTCATGAGATCAGCTAAGTTGCACCAACTTTGAA
GTCTGATTTTCCTGGACAGTTTTCTGCTTTAATTTCATGAAAAGATTATGA
TCTCAGAAATTGTATCTTAGTTGGTATCAACCAAATGGAGTGACTTAGTG
TACATGAAAGCGTAAAGAGGATGTGTGGCATTTTCACTTTTGGCTTGTAA
AGTACAGACTTTTTTTTTTTTTAAACAAAAAAAGCATTGTAACTTATGAA
CCTTTACATCCAGATAGGTTACCAGTAACGGAACATATCCAGTACTCCTG
GTTCCTAGGTGAGCAGGTGATGCCCCAGGGACCTTTGTAGCCACTTCACT
TTTTTTCTTTTCTCTGCCTTGGTATAGCATATGTGTTTTGTAAGTTTATGCAT
ACAGTAATTTTAAGTAATTTCAGAAGAAATTCTCGAAGCTTTCAAAATT

Figure 2 (*Continued*)

GGACTTAAAATCTAATTCAAACTAATAGAATTAATGGAATATGTAAATA
GAAACGTGTATATTTTTTATGAAACATTACAGTTAGAGATTTTTAAATAA
AGAATTTTAAAACTC

Mouse IFNGR1

CGGCAGGCCGCTTGCGGACTTGGCGACTAGTCTGCGGCGGACGTGACGC
CAAGGCCAGGCCACGGGCAGCGCGGGTCCCCTGTCAGAGGTGTCCCTCG
CGCAGGAATGGGCCCGCAGGCGGCAGCTGGCAGGATGATTCTGCTGGTG
GTCCTGATGCTGTCTGCGAAGGTCGGGAGTGGAGCTTTGACGAGCACTGA
GGATCCTGAGCCTCCCTCGGTGCCTGTACCGACGAATGTTCTAATTAAGT
CTTATAACTTGAACCCTGTCGTATGCTGGGAATACCAGAACATGTCACAG
ACTCCTATTTTTACTGTACAGGTAAAGGTGTATTCGGGTTCCTGGACTGAT
TCCTGCACCAACATTTCTGATCATTGTTGTAATATCTATGAACAAATTATG
TATCCTGATGTATCTGCCTGGGCCAGAGTTAAAGCTAAGGTTGGACAAAA
AGAATCTGACTATGCACGGTCAAAAGAGTTCCTTATGTGCCTAAAGGGA
AAGGTCGGGCCCCCTGGCCTGGAGATCAGGAGGAAGAAGGAAGAACAG
CTCTCCGTCCTCGTATTTCACCCTGAAGTCGTTGTGAATGGAGAGAGCCA
GGGAACCATGTTTGGTGACGGGAGCACCTGTTACACATTCGACTATACTG
TGTATGTGGAGCATAACCGGAGTGGGGAGATCCTACATACGAAACATAC
GGTCGAAAAAGAAGAGTGTAATGAGACTCTGTGTGAGTTAAACATCTCA
GTATCCACACTGGATTCCAGATATTGTATTTCAGTAGACGGAATCTCATCT
TTCTGGCAAGTTAGAACAGAAAAATCGAAAGACGTCTGTATCCCTCCTTT
CCATGATGACAGAAAGGATTCAATTTGGATTCTGGTGGTTGCTCCTCTTAC
CGTCTTTACAGTAGTTATCCTGGTATTTGCGTATTGGTATACTAAGAAGAA
TTCATTCAAGAGAAAAAGCATAATGTTACCTAAGTCCTTGCTCTCTGTGG

Chromosome location and linkages

Human IFNGR1 is encoded by a 30 kb gene located on the long arm of chromosome 6 (Table 1) (Pfizenmaier *et al.*, 1988). The murine homolog is a 22 kb gene present on chromosome 10 (Mariano *et al.*, 1987). Both genes consist of seven exons. Exons 1–5 encode the receptor extracellular domain; exon 6 encodes a small portion of the membrane-proximal region of the extracellular domain and the transmembrane domain; and exon 7 encodes the entire intracellular domain. Transcription of the human and murine IFNGR1 genes gives rise to mRNA transcripts of 2.3 kb (Farrar and Schreiber, 1993).

The human IFNGR2 gene has been localized to chromosome 21q22.1 (Cook *et al.*, 1994; Soh *et al.*, 1994). The murine homolog resides on chromosome 16 (Table 1) (Hibino *et al.*, 1991). These syntenic chromosomal regions also contain the genes of several other IFN receptor family members, including the subunits of the IFNα/β receptor (IFNAR1 and IFNAR2) and the nonligand-binding subunit of the IL-10 receptor, originally denoted CRF2-4 (Lutfalla *et al.*, 1993; Cook *et al.*, 1994).

At the present time, structural data are only available for the mouse IFNGR2 gene. This 17 kb gene appears to consist of seven exons. Transcriptional activation of the IFNGR2 gene results in the generation of an mRNA transcript of 1.8 kb in human cells or 2 kb in mouse cells (Hemmi *et al.*, 1994; Soh *et al.*, 1994).

Figure 2 (*Continued*)

TAAAAAGTGCCACGTTAGAGACAAAACCTGAATCGAAGTATTCACTTGT
CACACCGCACCAGCCAGCTGTCCTAGAGAGTGAGACGGTGATCTGTGAA
GAGCCCCTGTCCACAGTGACAGCTCCAGACAGCCCCGAAGCAGCAGAAC
AGGAAGAACTTTCAAAAGAAACAAAGGCTCTGGAGGCTGGAGGAAGCA
CGTCTGCCATGACCCCAGACAGCCCTCCAACTCCGACACAAAGACGCAG
CTTTTCCCTGTTAAGTAGTAACCAGTCAGGCCCTTGTAGCCTCACCGCCTA
TCACTCCCGAAACGGCTCTGACAGTGGCCTCGTGGGATCGGGCAGCTCC
ATATCGGACTTGGAATCTCTCCCAAACAACAACTCAGAAACAAAGATGG
CAGAGCACGACCCTCCACCCGTGAGAAAGGCCCCCATGGCCTCCGGTTA
TGACAAACCGCACATGTTGGTGGACGTGCTTGTGGATGTTGGGGGGAAG
GAGTCTCTCATGGGGTATAGACTCACAGGAGAGGCCCAGGAGCTGTCCT
AAGGTCTCCCGAGGCCTGCTGGTGGTAAAGAAACTGACCTTTTAGGCAGT
TTTTCTGCATTGATTTCATGAAAGAAGCTATACATTAGCTAATACTAACCA
CATAGAATATCAGACTTAGATACGTGAATAAGGATCCTGTGGGCACTGCT
GGGTCCACTCTGCAAATGCCAAGACTATCAAAGGAACGTATTGTCGCTTC
TGGCTCCTTCCCAGGTGGGCTAGCATCTGTGAGTTTGCCTCGGCTAGCCTT
GCTTCCTACAGCCGCCACTGCTCCTCCACCCTGATCATCTCACAGGACAG
GGTGGACCGGGTTTTTTTTTTTTTCACACACCTTTGTATATGTAAGTTCATG
TATATAATATGTTTACATGTTTCACTTTGAACTGAAAGCTACTCAAAGCCA
GCCGTAAGTCTATGGTAGAATGTGATGGAACATGTTGGTGGAAGCTTGTA
CAATAGAACACATTGGTGGGAGCTTGTACATACTTTTTTATGGAGCATTA
CTTACGATTTTTTAAGTAAAATGTTTTGAAACCAAAAAAAAAA

PROTEIN

Accession numbers

Human IFNGR1: P15260
Mouse IFNGR1: P15261
Human IFNGR2: P38484
Mouse IFNGR2: A49947

Sequence

See **Figure 3**.

Description of protein

The human and murine IFNGR1 proteins are organized in a similar manner and are symmetrically oriented around a single transmembrane domain (Figure 1 and Table 1) (Aguet *et al.*, 1988; Gray *et al.*, 1989; Hemmi *et al.*, 1989; Kumar *et al.*, 1989; Munro and Maniatis, 1989; Cofano *et al.*, 1990). The mature proteins consist of 472 and 451 amino acids, respectively, and have predicted molecular masses of 52.5 and 49.8 kDa. Both proteins are symmetrically oriented around single 23 amino acid transmembrane domains. Each possesses a 228 amino acid extracellular domain that contains 10 cysteine residues and five potential *N*-linked glycosylation sites. Based on biosynthetic labeling experiments, all five glycosylation sites appear to be occupied (Hershey and Schreiber, 1989; Mao *et al.*, 1989; Fischer *et al.*, 1990b) and *N*-linked oligosaccharides contribute approximately 25 kDa to the apparent molecular mass of the fully mature protein. The size of the fully mature IFNGR1 derived from different cell types ranges from 85 to 105 kDa depending on the extent of

Figure 3 Amino acid sequences for human and mouse IFNGR1 and IFNGR2.

Human IFNGR1

MALLFLLPLVMQGVSRAEMGTADLGPSSVPTPTNVTIESYNMNPIVYWEYQI
MPQVPVFTVEVKNYGVKNSEWIDACINISHHYCNISDHVGDPSNSLWVRVKA
RVGQKESAYAKSEEFAVCRDGKIGPPKLDIRKEEKQIMIDIFHPSVFVNGDEQE
VDYDPETTCYIRVYNVYVRMNGSEIQYKILTQKEDDCDEIQCQLAIPVSSLNSQ
YCVSAEGVLHVWGVTTEKSKEVCITIFNSSIKGSLWIPVVAALLLFLVLSLVFICF
YIKKINPLKEKSIILPKSLISVVRSATLETKPESKYVSLITSYQPFSLEKEVVCEEPLS
PATVPGMHTEDNPGKVEHTEELSSITEVVTTEENIPDVVPGSHLTPIERESSSPLS
SNQSEPGSIALNSYHSRNCSESDHSRNGFDTDSSCLESHSSLSDSEFPPNNKGEI
KTEGQELITVIKAPTSFGYDKPHVLVDLLVDDSGKESLIGYRPTEDSKEFS

Mouse IFNGR1

MGPQAAAGRMILLVVLMLSAKVGSGALTSTEDPEPPSVPVPTNVLIKSYNLNP
VVCWEYQNMSQTPIFTVQVKVYSGSWTDSCTNISDHCCNIYGQIMYPDVSAW
ARVKAKVGQKESDYARSKEFLMCLKGKVGPPGLEIRRKKEEQLSVLVFHPEVV
VNGESQGTMFGDGSTCYTFDYTVYVEHNRSGEILHTKHTVEKEECNETLCEL
NISVSTLDSRYCISVDGISSFWQVRTEKSKDVCIPPFHDDRKDSIWILVVAPLTVF
TVVILVFAYWYTKKNSFKRKSIMLPKSLLSVVKSATLETKPESKYSLVTPHQPA
VLESETVICEEPLSTVTAPDSPEAAEQEELSKETKALEAGGSTSAMTPDSPPTPT

cell-specific glycosylation (Hershey and Schreiber, 1989; Mao *et al.*, 1989; Fischer *et al.*, 1990b).

The human and murine IFNGR2 proteins are also structurally similar to one another (Figure 1 and Table 1) (Hemmi *et al.*, 1994; Soh *et al.*, 1994). The mature human protein consists of a 226 amino acid extracellular domain, a 24 amino acid transmembrane domain, and a 66 amino acid intracellular domain. It has a predicted molecular mass of 34.8 kDa. The mature murine equivalent has an extracellular domain that is two amino acids shorter. The predicted molecular mass of murine IFNGR2 is 35.6 kDa. The human and murine proteins contain five and six *N*-linked glycosylation sites, respectively, and both are heavily glycosylated. Glycosylation contributes to the significant size heterogeneity seen in the mature proteins, even when derived from the same cell. The fully mature human protein displays M_r values that range from 61 to 67 kDa, while mature forms of the mouse equivalent display M_r values of 60–65 kDa (Bach *et al.*, 1995).

Relevant homologies and species differences

Examination of the extracellular domains of both IFNGR1 and IFNGR2 reveal that they belong to a small family of cytokine receptors known as the class 2 cytokine receptors. The other members of this protein family include the two subunits of the IFNα/β receptor (IFNAR1 and IFNAR2), the two subunits of the IL-10 receptor, and tissue factor (Bazan, 1990). The members of this protein family share a similarly organized 210 amino acid-binding domain which contains conserved cysteine pairs at both N- and C-termini. The class 2 receptor family appears to be

Figure 3 (*Continued*)

QRRSFSLLSSNQSGPCSLTAYHSRNGSDSGLVGSGSSISDLESLPNNNSETKMAE

HDPPPVRKAPMASGYDKPHMLVDVLVDVGGKESLMGYRLTGEAQELS

Human IFNGR2

MRPTLLWSLLLLLGVFAAAAAAPPDPLSQLPAPQHPKIRLYNAEQVLSWEPV

ALSNSTRPVVYRVQFKYTDSKWFTADIMSIGVNCTQITATECDFTAASPSAGFP

MDFNVTLRLRAELGALHSAWVTMPWFQHYRNVTVGPPENIEVTPGEGSLIIR

FSSPFDIADTSTAFFCYYVHYWEKGGIQQVKGPFRSNSISLDNLKPSRVYCLQV

QAQLLWNKSNIFRVGHLSNISCYETMADASTELQQVILISVGTFSLLSVLAGAC

FFLVLKYRGLIKYWFHTPPSIPLQIEEYLKDPTQPILEALDKDSSPKDDVWDSVSI

ISFPEKEQEDVLQTL

Mouse IFNGR2

MRPLPLWLPSLLLCGLGAAASSPDSFSQLAAPLNPRLHLYNDEQILTWEPSPSS

NDPRPVVYQVEYSFIDGSWHRLLEPNCTDITETKCDLTGGGRLKLFPHPFTVFL

RVRAKRGNLTSKWVGLEPFQHYENVTVGPPKNISVTPGKGSLVIHFSPPFDVF

HGATFQYLVHYWEKSETQQEQVEGPFKSNSIVLGNLKPYRVYCLQTEAQLILK

NKKIRPHGLLSNVSCHETTANASARLQQVILIPLGIFALLLGLTGACFTLFLKY

QSRVKYWFQAPPNIPEQIEEYLKDPDQFILEVLDKDGSPKEDSWDSVSIISSPEK

ERDDVLQTP

only distantly related structurally to the much larger class 1 cytokine receptor family.

IFNGR1 and IFNGR2 display a strict species-specificity in their ability to bind and respond to ligand, i.e. the human IFNγ receptor binds and responds only to human and not to murine IFNγ while the murine IFNγ receptor binds and responds only to the murine ligand (reviewed in Farrar and Schreiber, 1993). The sequence identity between human and murine IFNGR1 is only 52% overall (50% identity between the extracellular domains and 55% identity between the intracellular domains). Human and murine IFNGR2 exhibit 58% identity overall but show greater identity (73%) within their intracellular domains (Hemmi *et al.*, 1994; Soh *et al.*, 1994). The species-dependent sequence differences within the receptor subunits are the reason for the species-restricted interactions between the receptor subunits with ligand and with one another. In contrast, the functionally critical regions within the intracellular domains of each subunit have been preserved between mice and humans and thus no species-specificity is observed in the interactions of the IFNγ receptor subunits with the intracellular proteins required for signal transduction.

Affinity for ligand(s)

Immunochemical and radioligand binding experiments on intact cells indicate that the IFNγ receptor binds ligand with a single high affinity (K_a) of 10^9–$10^{10}\,M^{-1}$ (Farrar and Schreiber, 1993). The majority of this affinity is derived from the interaction of

IFNGR1 with the ligand which is characterized by a K_a of 10^8–$10^9\,M^{-1}$. Using ligand-binding assays, sucrose density gradient ultracentrifugation, and HPLC gel filtration chromatography, a soluble form of the IFNGR1 extracellular domain (sECD) was found to form stable complexes in free solution with IFNγ that consisted of 1 mole ligand and 2 mole soluble receptor (Fountoulakis *et al.*, 1992; Greenlund *et al.*, 1993). Formation of the 2:1 (receptor–ligand) complex was also demonstrated on cell surfaces using either chemical crosslinking or immunochemical approaches. The crystal structure of the IFNγ–IFNγ receptor complex has been solved and confirms the structure predicted by the biochemical approaches (Walter *et al.*, 1995).

Deletional mutagenesis analysis of the soluble IFNGR1 extracellular domain showed that the majority of the sECD (residues 6–227) was required for expression of ligand-binding activity (Fountoulakis *et al.*, 1991). However, by exchanging corresponding regions between the human and murine IFNGR1 extracellular domains, several important internal sequences were identified throughout the extracellular domain that contributed to the species-specificity of the ligand-binding process (Axelrod *et al.*, 1994). Moreover, this study also revealed the presence of distinct regions within IFNGR1 that played an obligate role in biologic response induction, but not in ligand binding. One explanation for the latter observation is that the functionally important sequences may contribute to the interaction between the IFNGR1 and IFNGR2 subunits.

The contribution of IFNGR2 to the ligand-binding process has also been established (Marsters *et al.*, 1995). Using an experimental system where the two human IFNγ receptor subunits were expressed either individually or together in murine fibroblasts, no direct interaction was detected between human IFNγ and human IFNGR2. However, when human IFNGR2 was expressed in murine cells that also expressed human IFNGR1, IFNγ binding was increased 4-fold over that observed on cells that expressed the human IFNGR1 chain alone. Thus, one function of IFNGR2 is to stabilize the complex formed between ligand and the IFNGR1 subunit.

Cell types and tissues expressing the receptor

On the basis of immunochemical, radioligand binding, and molecular genetic analyses, IFNGR1 is ubiquitously expressed on nearly all cells (except erythrocytes) (Farrar and Schreiber, 1993). Even platelets express IFNγ receptors at a level of 300 receptors/cell. Considering the large number of platelets in the circulation (3×10^8/mL), it is possible that this cell plays an important role in transporting IFNγ through the circulatory system. It is noteworthy that when receptor expression in different tissues is analyzed at either the mRNA or protein levels, the highest expression is observed in tissues which are not generally considered to have primary immunologic functions. Specifically, skin, nerve, and syncytial trophoblasts of the placenta express levels of IFNγ receptor that are often 10–100 times that observed in spleen or on hematopoietic cells. Expression patterns of the IFNGR2 gene generally follow those of the IFNGR1 gene, except in T lymphocytes.

Regulation of receptor expression

In most cell types, IFNGR1 expression is constitutive and unregulated. The 5′ flanking regions of the IFNGR1 gene contains a GC-rich region with no TATA box like that of promoters for noninducible housekeeping genes (Pfizenmaier *et al.*, 1988; Merlin *et al.*, 1989). This observation suggests that expression of IFNGR1 is not regulated by external stimuli, a result that has been largely confirmed experimentally. Nevertheless, expression of the fully mature IFNGR1 protein at the plasma membrane varies widely between tissues (250–25,000 sites/cell). However there does not appear to be a direct correlation between the extent of receptor α chain expression and the magnitude of IFNγ-induced responses in cells (Farrar and Schreiber, 1993). Following IFNγ receptor ligation, the IFNGR1-ligand complex is internalized and enters an acidified compartment. Within this compartment, the complex dissociates and free IFNγ is trafficked to the lysosome where it is degraded. In many cells, such as fibroblasts and macrophages, the uncoupled IFNGR1 enters a large intracellular pool of IFNGR1 subunits and eventually recycles back to the cell surface. In most cells, the size of the intracellular pool is approximately 2–4 times that of the receptors expressed at the cell surface (Anderson *et al.*, 1983; Celada and Schreiber, 1987; Finbloom, 1988; Hershey and Schreiber, 1989; Farrar *et al.*, 1991).

Little is known about the intracellular trafficking of IFNGR2. However, in certain cells (such as T lymphocytes), IFNGR2 expression is altered in a stimulus-linked manner. Several potential binding sites for a variety of externally regulated activated transcription factors have been identified within the 5′ flanking region of the IFNGR2 gene (Ebensperger *et al.*, 1996). This observation suggested that transcription of the IFNGR2 gene may be tightly regulated. In fact, this hypothesis has been supported experimentally by the observation of differential expression of IFNGR2 on distinct murine helper T cell subsets. Based on the observation that murine CD4+ T helper cell subsets differed in their ability to respond to IFNγ (Gajewski and Fitch, 1988), two independent groups demonstrated in 1995 that the IFNγ-unresponsive state seen exclusively in TH1 cells was due to a lack of cellular expression of IFNGR2 (Bach *et al.*, 1995; Pernis *et al.*, 1995). Unresponsiveness was shown to be a result of IFNγ-dependent IFNGR2 downregulation and was not linked to T cell differentiation (Bach *et al.*, 1995). Thus, mouse TH1 cells, which produce IFNγ, downregulate expression of IFNGR2, become IFNγ-unresponsive, and thus circumvent the growth-inhibitory effects of the cytokine that they produce. In contrast, TH2 cells, which do not produce IFNγ, express IFNGR2 and remain IFNγ-responsive. However, IFNGR2 downregulation could also be induced in murine TH2 cells, and also in human peripheral blood T cells, upon exposure to IFNγ (Bach *et al.*, 1995; Sakatsume and Finbloom, 1996). Interestingly, ligand-induced IFNGR2 downregulation did not occur in certain fibroblast cell lines. Thus, IFNγ appears to regulate expression of IFNGR2 on certain cell types and thereby determines the ability of these cells to respond to subsequent exposure to IFNγ. Recently, treatment of T cells with phorbol esters or with CD3 antibodies has been shown to effect induction of IFNGR2 mRNA (Sakatsume and Finbloom, 1996). Taken together, these results demonstrate that IFNGR2 expression can be regulated either negatively or positively in a stimulus-specific manner.

SIGNAL TRANSDUCTION

Associated or intrinsic kinases

Like all members of the class 1 and class 2 cytokine receptor families, the intracellular domains of the IFNγ receptor subunits do not express intrinsic kinase activity. However, distinct sequences within each intracellular domain function as the specific binding sites on the receptor for two members of the Janus family of protein tyrosine kinases. IFNGR1 associates with JAK1 and IFNGR2 associates with JAK2.

The association of the IFNγ receptor subunits with distinct Janus kinases was revealed by three lines of evidence. First, using human fibrosarcoma cells that were mutated and selected for lack of responsiveness to both type I IFN and IFNγ, two of the Janus kinases – JAK1 and JAK2 – were found to be required for induction of IFNγ-dependent biologic responses (Pellegrini *et al.*, 1989; McKendry *et al.*, 1991; Müller *et al.*, 1993; Watling *et al.*, 1993). U1A cells that lack TYK2 did not respond to IFNα but responded normally to IFNγ while γ1A cells that lack JAK2 responded to IFNγ but did not respond to IFNγ. In contrast, U4A cells that lack JAK1 responded to neither type I IFN nor to IFNγ. Responsiveness was restored in each cell line following transfection with plasmids encoding each of the missing JAKs. Thus IFNγ signaling, at least in this human fibrosarcoma line, required the concomitant presence of JAK1 and JAK2, while IFNα/β signaling required the concomitant presence of JAK1 and TYK2.

Second, using a combination of deletion and substitution mutagenesis approaches, distinct sequences within the intracellular domains of IFNGR1 and IFNGR2 were identified that acted as specific binding sites on the receptor subunits for JAK1 and JAK2 (reviewed in Farrar and Schreiber, 1993; Bach *et al.*, 1997). In IFNGR1, a membrane proximal Leu-Pro-Lys-Ser (LPKS) sequence residing at positions 266–269 in the intracellular domain was found to function as the constitutive binding site for JAK1 (Farrar *et al.*, 1991; Greenlund *et al.*, 1994). Replacement of this sequence with four alanine residues or substitution of alanine for proline at position 267 produced mutant IFNGR1 proteins that first, did not associate with JAK1; second, did not support ligand-induced activation of protein tyrosine kinase activity; and third, did not promote development of IFNγ-dependent cellular responses (Kaplan *et al.*, 1996). Coprecipitation studies coupled with biologic function analyses performed on cells treated with either buffer or IFNγ revealed that, whereas inactive JAK1 constitutively associates with the IFNγ receptor α chain, its activation is ligand-dependent. Similar studies conducted on the IFNGR2 subunit demonstrated that a 12 amino acid sequence (263-PPSIPLQIEEYL-274) located 13 amino acids away from the membrane in IFNGR2 functioned as the major site of attachment for JAK2 (Bach *et al.*, 1996). Mutation within this region produced an IFNGR2 that was unable to interact with JAK2 and failed to support IFNγ-dependent biologic response induction.

Third, disruption of the genes encoding JAK1 or JAK2 in mice led to IFNγ unresponsiveness in primary cells derived from these animals (Neubauer *et al.*, 1998; Paraganas *et al.*, 1998; Rodig *et al.*, 1998). In contrast, cells derived from JAK3−/− mice and cells lacking TYK2 responded normally to IFNγ.

The role of STAT1 in IFNγ signaling was also revealed by three experimental approaches. First, IFNγ treatment of cells was shown to result in the selective activation of only one member of the STAT protein family, namely STAT1 (Fu *et al.*, 1992; Schindler *et al.*, 1992; Shuai *et al.*, 1992). Following addition of IFNγ to receptor-bearing cells, STAT1 was rapidly (30 seconds to 5 minutes) phosphorylated on a single tyrosine residue residing at position 701, forming a homodimer that translocated to the nucleus where it associated with the promoter regions of IFNγ-inducible genes and promoted gene transcription. IFNγ-induced tyrosine phosphorylation of STAT1 required the presence in the cell of both JAK1 and JAK2, a result that directly linked the JAKs to the STATs.

Second, structure–function analyses performed on IFNGR1 revealed the presence of a functionally critical five amino acid region located at positions 440–444 near the C-terminus that contains the residues Tyr-Asp-Lys-Pro-His (YDKPH) (Farrar *et al.*, 1992). The tyrosine residue at position 440 in human IFNGR1 (Tyr420 in murine IFNGR1) was found to be a major substrate site for the activated JAKs (Greenlund *et al.*, 1994; Igarashi *et al.*, 1994). Once phosphorylated, the Y(PO_4)DKPH sequence functioned as the attachment site on the activated receptor for STAT1. Mutational analysis of these five residues demonstrated that only Tyr440, Asp441, and His444 were functionally important. IFNGR1 harboring alanine substitutions at any of these three residues failed to activate STAT1 and failed to induce IFNγ-dependent cellular responses. Moreover, STAT1 bound to peptides containing the Y(PO_4)DKPH sequence (but not to unphosphorylated forms of the peptides) in a specific and high-affinity manner and phosphopeptides containing this specific sequence were able to inhibit IFNγ-dependent STAT1 activation in a broken cell experimental system (Greenlund *et al.*, 1995). Using antibodies specific for different regions of the STAT1 protein and STAT1/STAT2 chimeric proteins, the SH2 domain of STAT1 was shown to be responsible for mediating association with the phosphorylated receptor sequence (Greenlund *et al.*, 1995; Heim *et al.*, 1995).

Third, cells from mice lacking the STAT1 gene failed to respond to IFNγ (and also failed to respond to IFNα/β) and the mice themselves were highly susceptible to infection with a wide variety of microbial pathogens and viruses (Durbin *et al.*, 1996; Meraz *et al.*, 1996). In addition, STAT1-deficient mice were more sensitive than their wild-type counterparts to tumor formation induced by chemical carcinogens (Kaplan *et al.*, 1998). However, STAT1-deficient mice showed no other defects that could be attributed to signaling through other cytokine receptors and mice lacking other STAT proteins (i.e. STAT2, 3, 4, 5a, 5b, and 6) responded normally to IFNγ. Thus, under physiologic conditions STAT1 plays a dedicated role in promoting IFN signaling and most IFN-dependent biologic responses require the participation of STAT1.

Cytoplasmic signaling cascades

Based on the results discussed above, it has been possible to formulate a relatively complete model of the IFNγ receptor signaling process (**Figure 4**) (Bach *et al.*, 1997). In unstimulated cells, the IFNγ receptor subunits are not preassociated with each other but rather associate through their intracellular domains with inactive forms of specific Janus family kinases. JAK1 associates with the IFNGR1 subunit by binding to the 266-LPKS-269 sequence and JAK2 constitutively associates with the 263-PPSIPLQIE-EYL-274 sequence in IFNGR2. Addition of IFNγ, a homodimeric ligand, to the cells induces the rapid dimerization of IFNGR1 subunits, thereby forming a site that is recognized, in a species-specific manner, by the extracellular domains of IFNGR2 subunits. The ligand-induced assembly of the complete receptor complex containing two IFNGR1 and two IFNGR2 subunits brings into close juxtaposition the intracellular domains of these proteins together with the inactive JAK enzymes that they carry. In this complex, JAK1 and JAK2 transactivate one another and then phosphorylate the functionally critical Tyr440 residue on IFNGR1, thereby forming a paired set of STAT1 docking sites on the activated receptor. Two STAT1 molecules then associate with the paired docking sites, are brought into close proximity with receptor-associated, activated JAK enzymes, and are activated by phosphorylation of the STAT1 Tyr701 residue. Tyrosine-phosphorylated STAT1 molecules dissociate from their receptor tether and form homodimeric complexes. The activated STAT1 complex is then phosphorylated on a specific C-terminal serine residue (Ser723) (Wen *et al.*, 1995). Recent reports suggest that the serine phosphorylation is mediated by an as yet undefined enzyme with MAP kinase-like specificity (David *et al.*, 1995; Wen *et al.*, 1995). Activated STAT1 translocates to the nucleus and, after binding to a specific sequence in the promoter region of immediate-early IFNγ-inducible genes, effects gene transcription.

This signaling pathway is controlled by five distinct mechanisms. First, because the pathway is driven by

Figure 4 Signaling through the IFNγ receptor. The details of this process are described in the text. In unstimulated cells, inactive forms of JAK1 and JAK2 associate in a constitutive manner with the IFNGR1 and IFNGR2 receptor subunits, respectively. Upon ligand addition, the subunits oligomerize, leading to the transactivation of JAK1 and JAK2. The activated kinases then phosphorylate Y440 of IFNGR1, forming a paired set of docking sites on the activated receptor for STAT1. Two STAT1 proteins bind to the receptor, become tyrosine and serine phosphorylated and then dimerize, translocate to the nucleus, and initiate IFNγ-dependent gene transcription. This process is regulated by phosphatases, kinase inhibitors, STAT inhibitors, and by degradation and/or transcriptional silencing of the receptor subunits and signaling proteins. The sites of regulation in the pathway are noted.

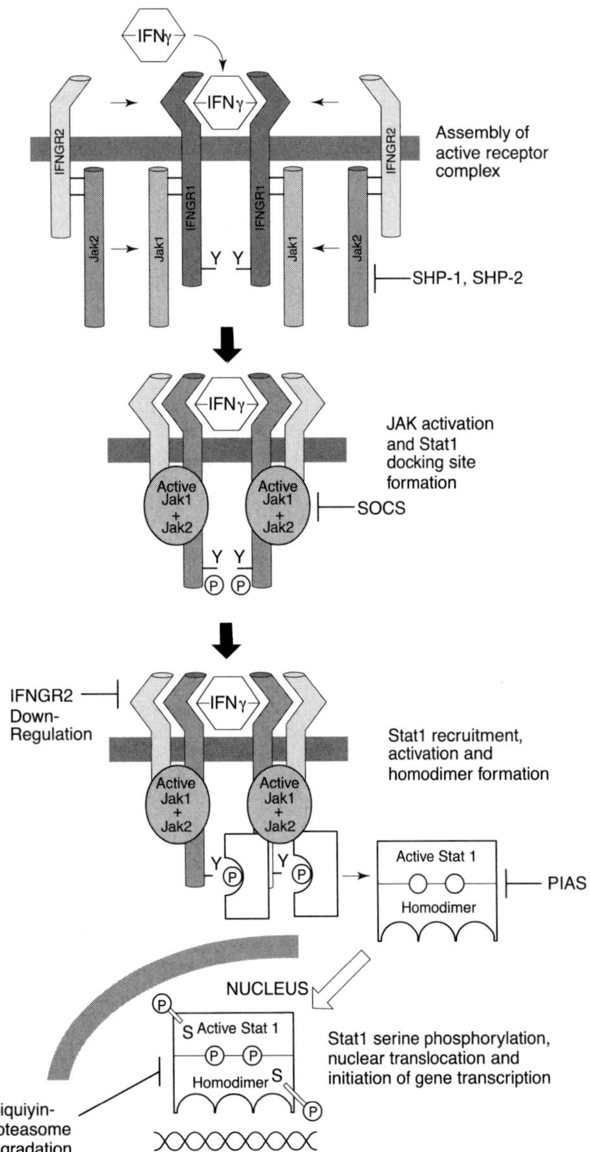

tyrosine phosphorylation, protein tyrosine phosphatases were considered obvious candidates capable of regulating IFNγ receptor signaling. In fact, there is some evidence that suggests that both SHP-1 and SHP-2 may play a role in pathway regulation (You and Zhao, 1997; You *et al.*, 1999). Second, a family of proteins was identified that regulate the activation of the Janus kinases. This family is known as suppressors of cytokine signaling (SOCS), JAK-binding proteins (JABs), or STAT-induced STAT inhibitors (SSIs) (Endo *et al.*, 1997; Naka *et al.*, 1997; Starr *et al.*, 1997). These proteins bind to activated JAKs and inhibit their catalytic activity. There are eight members of this family that participate in regulating the JAK/STAT signaling pathway and another 12 members whose function remains unknown (Hilton *et al.*, 1998).

Interestingly, mice lacking SOCS-1 die within 3 weeks of age of a massive inflammatory disease (Naka *et al.*, 1998; Starr *et al.*, 1998). This disease is prevented if the SOCS-1 deficiency is bred into an IFNγ-deficient background (Alexander *et al.*, 1999). Thus, SOCS-1 appears to be important in regulating signaling through the IFNγ receptor pathway. The third family of proteins, known as proteins that inhibit activated STATs (PIAS), act downstream of the receptor and function by binding to activated STAT complexes and inhibiting their capacity to bind DNA (Chung *et al.*, 1997; Liu *et al.*, 1998). Two family members have been characterized thus far. One of these, termed PIAS1, binds selectively to STAT1 homodimers and thus is likely to play an important role in controlling IFNγ receptor signaling. The fourth proposed mechanism is one in which components of the IFNγ signaling pathway are marked for degradation by the proteasome via ubiquination. Whereas one group showed that an inhibitor of the proteasome (MG132) stabilized phosphorylated STAT1 in IFNγ-treated HeLa cells, another group showed that STAT1 molecules quantitatively cycle between the cytosol and nucleus as nonphosphorylated and phosphorylated proteins, respectively (Haspel *et al.*, 1996; Kim and Maniatis, 1996). Thus, it remains unclear how much this pathway contributes to signaling regulation under physiologic conditions. Finally, as discussed above, IFNγ receptor signaling can be regulated in certain cells by mechanisms involving expression of the receptor IFNGR2 subunit (Bach *et al.*, 1995; Pernis *et al.*, 1995).

Thus, IFNγ signaling is an ordered, affinity-driven, and highly regulated process that derives its specificity from first, the specific binding of a particular STAT protein, i.e. STAT1, to a defined, ligand-induced but transiently expressed docking site on the activated receptor, and second, the ability of the

STAT1 homodimer specifically to activate IFNγ-induced genes.

DOWNSTREAM GENE ACTIVATION

Genes induced

The rapid signaling of the JAK/STAT pathway makes it an ideal system to regulate the activation of immediate-early genes and thus provides the host with a rapid mechanism to respond to an infectious agent. In fact, over the years it has been possible to identify a large number of IFNγ-stimulated genes that are induced rapidly (within 15–30 minutes) after IFNγ treatment of cells, and whose transcription does not depend on new protein synthesis (Kerr and Stark, 1991; Lewin *et al.*, 1991).

Examples of IFNγ-induced immediate-early genes include interferon regulatory factor 1 (IRF-1), guanylate-binding protein 1 (GBP-1), and the type I Fcγ receptor (FcγRI), which encode proteins that participate in inflammatory and immune responses. Several IFNγ-regulated intermediate genes have also been identified. These genes are induced within 6–8 hours of stimulation and require additional protein synthesis for transcriptional activation to occur. Examples of these genes include those that encode class I and class II major histocompatibility (MHC) proteins that play a central role in determining adaptive immune responses. Studies utilizing STAT1-, JAK1-, or JAK2-deficient mice have shown that these three signaling proteins play an obligate role in activating most IFNγ-inducible genes (Durbin *et al.*, 1996; Meraz *et al.*, 1996; Neubauer *et al.*, 1998; Paraganas *et al.*, 1998). More than 100 IFNγ regulated genes have been identified. This subject has recently been reviewed elsewhere and the reader is referred to several excellent reviews on the subject (Kerr and Stark, 1991; Boehm *et al.*, 1997; Der *et al.*, 1998).

Promoter regions involved

The promoter regions of IFNγ-stimulated immediate-early genes contain GAS elements (IFNγ-activated sequence) which act as the binding sites for activated STAT1 homodimers. When occupied, these *cis*-acting elements collaborate with basal transcriptional factors and promote gene transcription. The GAS site is a 9 nucleotide sequence with a consensus motif of TTNCNNNAA. For STAT1 dimers, the most common sequence recognized is TTCC(G > C)GGAA.

Two major pathways account for induction of IFNγ-induced intermediate genes. These pathways are driven by the transcription factors IRF-1 and CIITA (Tanaka and Taniguchi, 1992; Mach *et al.*, 1996). IRF-1 is itself an immediate-early gene that gives rise to a protein that functions by binding directly to DNA sequences in the promoters of IFN-induced genes. In contrast, CIITA is an intermediate gene induced by a collaboration between STAT1, IRF-1, and a ubiquitous transcription factor (USF-1), and functions by binding to other transcription factors that make contact with DNA sequences within the X, X2, and Y boxes of the promoters in MHC class II genes.

BIOLOGICAL CONSEQUENCES OF ACTIVATING OR INHIBITING RECEPTOR AND PATHOPHYSIOLOGY

Phenotypes of receptor knockouts and receptor overexpression mice

The physiologic consequences of global *in vivo* inactivation of the IFNγ signaling pathway in mice were originally uncovered using neutralizing monoclonal antibodies specific for IFNγ (Buchmeier and Schreiber, 1985). However, the physiologic role of IFNγ has been more fully elucidated using mice with engineered disruptions in either the IFNγ structural gene (Dalton *et al.*, 1993), the genes encoding the two IFNγ receptor subunits (IFNGR1 and IFNGR2) (Huang *et al.*, 1993; Kamijo *et al.*, 1993; Lu *et al.*, 1998), or the genes encoding the three signaling proteins required by IFNγ for biologic response induction, i.e. STAT1 (Durbin *et al.*, 1996; Meraz *et al.*, 1996), JAK1 (Rodig *et al.*, 1998), and JAK2 (Neubauer *et al.*, 1998; Paraganas *et al.*, 1998). As a group, these mice display a greatly impaired ability to resist infection by a variety of microbial pathogens, including *Listeria monocytogenes*, *L. major*, and several mycobacterial species, including *Mycobacterium bovis* and *M. avium*. Moreover, mice that cannot respond to IFNγ display enhanced tumor development when challenged with chemical carcinogens (Dighe *et al.*, 1994; Kaplan *et al.*, 1998). Importantly, mice lacking either IFNGR1 or IFNGR2 are able to mount a curative response to many viruses, while mice lacking the IFNα/β receptor or STAT1, and cells of mice deficient in JAK1 are not. These results thus demonstrate that under physiologic conditions the majority of the antiviral effects of the IFN system are largely mediated via type I interferons (Müller *et al.*, 1994). Mice have also been generated which overexpress a truncated dominant-negative mutant form of IFNGR1 in either the T cell or macrophage compartments and thereby represent mice that display a tissue-targeted IFNγ insensitivity (Dighe *et al.*, 1995). These types of mice will be useful in defining the tissue-specific actions of IFNγ in protective host responses against pathogens and neoplastic cells.

Human abnormalities

Two research groups initially identified children from three unrelated families with inactivating mutations of IFNGR1 who manifest a severe susceptibility to weakly pathogenic mycobacterial species (Jouanguy *et al.*, 1996; Newport *et al.*, 1996). Genetic analysis of these patients' families revealed that susceptibility to *atypical mycobacterial infection* was inherited in an autosomal recessive manner. Sequence analysis of the patients' IFNGR1 alleles identified missense mutations in genetic regions coding for the extracellular domain of the IFNGR1 polypeptide, leading to the production of truncated receptor proteins that were not retained at the cell surface. The clinical syndromes of these patients were similar. In one study, a group of related Maltese children were identified that showed extreme susceptibility to infection with *M. fortuitum*, *M. avium*, and *M. chelonei* (Newport *et al.*, 1996). In another study, a Tunisian child was identified with disseminated *M. bovis* infection following bacillus Calmette-Guérin (BCG) vaccination (Jouanguy *et al.*, 1996). A third study identified a child of distinct ancestry who had a similar immunocompromised phenotype (Pierre-Audigier *et al.*, 1997). Biopsies from these patients revealed the presence of multibacillary, poorly defined granulomas, which contained scattered macrophages, but lacked epithelioid and giant cells and surrounding lymphocytes. Importantly, these patients showed enhanced susceptibility to mycobacteria and occasionally to salmonella but not to other typical bacteria or other common microbial pathogens or fungi. Moreover, in all three kindred, the patients were able to mount antibody and/or curative responses to several different viruses.

Subsequently, a number of patients with other IFNγ receptor mutations have been identified. A child with severe, disseminated infections due to *M. fortuitum* and *M. avium* was identified who lacked mutations in the IFNGR1 gene (Dorman and Holland, 1998). This child had a homozygous mutation in the IFNGR2 gene. The mutation resulted in a premature stop codon in the extracellular domain-encoding region, and led to the production of IFNGR2 proteins that were not expressed at the cell surface. The clinical and histopathological phenotype of this patient closely resembled that of patients lacking expression of the IFNGR1 chain.

Other patients have been identified who develop less severe mycobacterial disease than the children described above. Upon analysis of their IFNγ receptor subunit genes, some of the patients were found to display distinct mutations in the IFNGR1 gene, leading to reduced but not ablated receptor function. Two patients were homozygous for a point mutation in the extracellular domain-encoding region of the IFNGR1 subunit gene that produced an isoleucine-to-threonine amino acid substitution. The mutant receptors were found to require 100- to 1000-fold higher concentrations of IFNγ than normal receptors in order to activate STAT1 (Jouanguy *et al.*, 1997). These patients responded to high-dose IFNγ therapy.

Another set of 19 patients from 12 unrelated families were found to inherit partial IFNγ insensitivity in an autosomal dominant manner (Jouanguy *et al.*, 1999). All of these patients were found to be heterozygous for a wild-type IFNGR1 allele and an IFNGR1 allele with a frameshift mutation that produced an IFNGR1 protein that lacked most of the intracellular domain, including the JAK1- and STAT1-binding sites. Interestingly, in this group of patients, there were 12 independent mutation events at a single site, defining a small deletion hotspot in the IFNGR1 gene. The truncated receptor chain accumulated on cell surfaces and was shown to act in a dominant negative manner to inhibit IFNγ responses in cells. The definition of the molecular basis for this defect was facilitated by the observation that these patients phenotypically resembled IFNγ-insensitive cells and transgenic mice which overexpressed a genetically engineered truncated IFNGR1 subunit (Dighe *et al.*, 1993, 1994). In all of these patients, defects in IFNγ responsiveness were partial, and cells from the patients retained some degree of sensitivity to IFNγ. This phenotype correlated with the milder infections in these patients, and their positive responses to exogenous IFNγ therapy.

References

Aguet, M., and Merlin, G. (1987). Purification of human gamma interferon receptors by sequential affinity chromatography on immobilized monoclonal anti-receptor antibodies and human gamma interferon. *J. Exp. Med.* **165**, 988–999.

Aguet, M., Dembic, Z., and Merlin, G. (1988). Molecular cloning and expression of the human interferon-γ receptor. *Cell* **55**, 273–280.

Alexander, W. S., Starr, R., Fenner, J. E., Scott, C. L., Handman, E., Sprigg, N. S., Corbin, J. E., Cornish, A. L., Darwiche, R., Owczarek, C. M., Kay, T. W. H., Nicola, N. A., Herzog, P. J., Metcalf, D., and Hilton, D. J. (1999). SOCS1 is a critical inhibitor of interferon γ signaling and prevents the potentially fatal neonatal actions of this cytokine. *Cell* **98**, 597–608.

Anderson, P., Yip, Y. K., and Vilcek, J. (1982). Specific binding of 125I-human interferon-gamma to high affinity receptors on human fibroblasts. *J. Biol. Chem.* **257**, 11301–11304.

Anderson, P., Yip, Y. K., and Vilcek, J. (1983). Human interferon-gamma is internalized and degraded by cultured fibroblasts. *J. Biol. Chem.* **258**, 6497–6502.

Axelrod, A., Gibbs, V. C., and Goeddel, D. V. (1994). The interferon γ receptor extracellular domain. Non-identical requirements for ligand binding and signaling. *J. Biol. Chem.* **269**, 15533–15539.

Bach, E. A., Szabo, S. J., Dighe, A. S., Ashkenazi, A., Aguet, M., Murphy, K. M., and Schreiber, R. D. (1995). Ligand-induced autoregulation of IFN-γ receptor β chain expression in T helper cell subsets. *Science* **270**, 1215–1218.

Bach, E. A., Tanner, J. W., Marsters, S. A., Ashkenazi, A., Aguet, M., Shaw, A. S., and Schreiber, R. D. (1996). Ligand-induced assembly and activation of the gamma interferon receptor in intact cells. *Mol. Cell. Biol.* **16**, 3214–3221.

Bach, E. A., Aguet, M., and Schreiber, R. D. (1997). The IFNγ receptor: a paradigm for cytokine receptor signaling. *Annu. Rev. Immunol.* **15**, 563–591.

Bazan, J. F. (1990). Structural design and molecular evolution of a cytokine receptor superfamily. *Proc. Natl Acad. Sci. USA* **87**, 6934–6938.

Boehm, U., Klamp, T., Groot, M., and Howard, J. C. (1997). Cellular responses to interferon-γ. *Annu. Rev. Immunol.* **15**, 749–795.

Buchmeier, N. A., and Schreiber, R. D. (1985). Requirement of endogenous interferon-gamma production for resolution of *Listeria monocytogenes* infection. *Proc. Natl Acad. Sci. USA* **82**, 7404–7408.

Calderon, J., Sheehan, K. C. F., Chance, C., Thomas, M. L., and Schreiber, R. D. (1988). Purification and characterization of the human interferon-gamma receptor from placenta. *Proc. Natl Acad. Sci. USA* **85**, 4837–4841.

Celada, A., and Schreiber, R. D. (1987). Internalization and degradation of receptor-bound interferon-gamma by murine macrophages. Demonstration of receptor recycling. *J. Immunol.* **139**, 147–153.

Celada, A., Gray, P. W., Rinderknecht, E., and Schreiber, R. D. (1984). Evidence for a gamma-interferon receptor that regulates macrophage tumoricidal activity. *J. Exp. Med.* **160**, 55–74.

Celada, A., Allen, R., Esparza, I., Gray, P. W., and Schreiber, R. D. (1985). Demonstration and partial characterization of the interferon-gamma receptor on human mononuclear phagocytes. *J. Clin. Invest.* **76**, 2196–2205.

Chung, C. D., Liao, J. Y., Liu, B., Rao, X. P., Jay, P., Berta, P., and Shuai, K. (1997). Specific inhibition of STAT3 signal transduction by PIAS3. *Science* **278**, 1803–1805.

Cofano, F., Moore, S. K., Tanaka, S., Yuhki, N., Landolfo, S., and Applella, E. (1990). Affinity purification, peptide analysis, and cDNA sequence of the mouse interferon-γ receptor. *J. Biol. Chem.* **265**, 4064–4071.

Cook, J. R., Emanuel, S. L., Donnelly, R. J., Soh, J., Mariano, T. M., Schwartz, B., Rhee, S., and Pestka, S. (1994). Sublocalization of the human interferon-gamma receptor accessory factor gene and characterization of accessory factor activity by yeast artificial chromosomal fragmentation. *J. Biol. Chem.* **269**, 7013–7018.

Dalton, D. K., Pitts-meek, S., Keshav, S., Figari, I. S., and Bradley, A. Stewart, T. A. (1993). Multiple defects of immune function in mice with disrupted interferon-γ genes. *Science* **259**, 1739–1742.

Darnell Jr., J. E., Kerr, I. M., and Stark, G. R. (1994). JAK-STAT pathways and transcriptional activation in response to IFNs and other extracellular signaling proteins. *Science* **264**, 1415–1421.

David, M., Petricoin III, E., Bejamin, C., Pine, R., Weber, M. J., and Larner, A. C. (1995). Requirement of MAP kinase (ERK2) activity in interferon α- and interferon β-stimulated gene expression through STAT proteins. *Science* **269**, 1721–1723.

Der, S. D., Zhou, A., Williams, B. R., and Silverman, R. H. (1998). Identification of genes differentially regulated by interferon alpha, beta, or gamma using oligonucleotide arrays. *Proc. Natl Acad. Sci. USA* **95**, 15623–15628.

Dighe, A. S., Farrar, M. A., and Schreiber, R. D. (1993). Inhibition of cellular responsiveness to interferon-γ (IFNγ) induced by overexpression of inactive forms of the IFNγ receptor. *J. Biol. Chem.* **268**, 10645–10653.

Dighe, A. S., Richards, E., Old, L. J., and Schreiber, R. D. (1994). Enhanced *in vivo* growth and resistance to rejection of tumor cells expressing dominant negative IFNγ receptors. *Immunity* **1**, 447–456.

Dighe, A. S., Campbell, D., Hsieh, C.-S., Clarke, S., Greaves, D. R., Gordon, S., Murphy, K. M., and Schreiber, R. D. (1995). Tissue specific targeting of cytokine unresponsiveness in transgenic mice. *Immunity* **3**, 657–666.

Dorman, S. E., and Holland, S. M. (1998). Mutation in the signal-transducing chain of the interferon-gamma receptor and susceptibility to mycobacterial infection. *J. Clin. Invest.* **101**, 2364–2369.

Durbin, J. E., Hackenmiller, R., Simon, M. C., and Levy, D. E. (1996). Targeted disruption of the mouse *Stat1* gene results in compromised innate immunity to viral infection. *Cell* **84**, 443–450.

Ebensperger, C., Rhee, S., Muthukumaran, G., Lembo, D., Donnelly, R., Pestka, S., and Dembic, Z. (1996). Genomic organization and promoter analysis of the gene *ifngr2* encoding the second chain of the mouse interferon-γ receptor. *Scand. J. Immunol.* **44**, 599–606.

Endo, T. A., Masuhara, M., Yokouchi, M., Suzuki, R., Sakamoto, H., Mitsui, K., Matsumoto, A., Tanimura, S., Ohtsubo, M., Misawa, H., Miyazaki, T., Leonor, N., Taniguchi, T., Fujita, T., Kanakura, Y., Komiya, S., and Yoshimura, A. (1997). A new protein containing an SH2 domain that inhibits JAK kinases. *Nature* **387**, 921–924.

Farrar, M. A., and Schreiber, R. D. (1993). The molecular cell biology of interferon-γ and its receptor. *Annu. Rev. Immunol.* **11**, 571–611.

Farrar, M. A., Fernandez-Luna, J., and Schreiber, R. D. (1991). Identification of two regions within the cytoplasmic domain of the human interferon-gamma receptor required for function. *J. Biol. Chem.* **266**, 19626–19635.

Farrar, M. A., Campbell, J. D., and Schreiber, R. D. (1992). Identification of a functionally important sequence motif in the carboxy terminus of the interferon-γ receptor. *Proc. Natl Acad. Sci. USA* **89**, 11706–11710.

Finbloom, D. S. (1988). Internalization and degradation of human recombinant interferon-gamma in the human histocytic lymphoma cell line, U937, relationship to Fc receptor enhancement and anti-proliferation. *Clin. Immunol. Immunopathol.* **47**, 93–105.

Fischer, T., Rehm, A., Aguet, M., and Pfizenmaier, K. (1990a). Human chromosome 21 is necessary and sufficient to confer human IFNγ responsiveness to somatic cell hybrids expressing the cloned human IFNγ receptor gene. *Cytokine* **2**, 157–161.

Fischer, T., Thoma, B., Scheurich, P., and Pfizenmaier, K. (1990b). Glycosylation of the human interferon-gamma receptor. *N*-linked carbohydrates contribute to structural heterogeneity and are required for ligand binding. *J. Biol. Chem.* **265**, 1710–1717.

Fountoulakis, M., Lahm, H.-W., Maris, A., Friedlein, A., Manneberg, M., Stueber, D., and Garotta, G. (1991). A 25-kDa stretch of the extracellular domain of the human interferon γ receptor is required for full ligand binding capacity. *J. Biol. Chem.* **266**, 14970–14977.

Fountoulakis, M., Zulauf, M., Lustig, A., and Garotta, G. (1992). Stoichiometry of interaction between inferferon-γ and its receptor. *Eur. J. Biochem.* **208**, 781–787.

Fu, X.-Y., Schindler, C., Improta, T., Aebersold, R., and Darnell Jr., J. E. (1992). The proteins of ISGF-3, the interferon α-induced transcriptional activator, define a gene family involved in signal transduction. *Proc. Natl Acad. Sci. USA* **89**, 7840–7843.

Gajewski, T. F., and Fitch, F. W. (1988). Anti-proliferative effect of IFN-gamma in immune regulation. I. IFN-gamma inhibits the proliferation of Th2 but not Th1 murine helper T lymphocyte clones. *J. Immunol.* **140**, 4245–4252.

Gray, P. W., Leong, S., Fennie, E. H., Farrar, M. A., Pingel, J. T., Fernandez-Luna, J., and Schreiber, R. D. (1989). Cloning and expression of the cDNA for the murine interferon gamma receptor. *Proc. Natl Acad. Sci. USA* **86**, 8497–8501.

Greenlund, A. C., Schreiber, R. D., Goeddel, D. V., and Pennica, D. (1993). Interferon-γ induces receptor dimerization in solution and on cells. *J. Biol. Chem.* **268**, 18103–18110.

Greenlund, A. C., Farrar, M. A., Viviano, B. L., and Schreiber, R. D. (1994). Ligand induced IFNγ receptor phosphorylation couples the receptor to its signal transduction system (p91). *EMBO J.* **13**, 1591–1600.

Greenlund, A. C., Morales, M. O., Viviano, B. L., Yan, H., Krolewski, J., and Schreiber, R. D. (1995). STAT recruitment by tyrosine-phosphorylated cytokine receptors: an ordered reversible affinity-driven process. *Immunity* **2**, 677–687.

Haspel, R. L., Salditt-Georgieff, M., and Darnell, J. E. (1996). The rapid inactivation of nuclear tyrosine phosphorylated Stat1 depends upon a protein tyrosine phosphatase. *EMBO J.* **15**, 6262–6268.

Heim, M. H., Kerr, I. M., Stark, G. R., and Darnell, J.E. Jr. (1995). Contribution of STAT SH2 groups to specific interferon signaling by the Jak-STAT pathway. *Science* **267**, 1347–1349.

Hemmi, S., Peghini, P., Metzler, M., Merlin, G., Dembic, Z., and Aguet, M. (1989). Cloning of murine interferon gamma receptor cDNA: expression in human cells mediates high-affinity binding but is not sufficient to confer sensitivity to murine interferon gamma. *Proc. Natl Acad. Sci. USA* **86**, 9901–9905.

Hemmi, S., Bohni, R., Stark, G., DiMarco, F., and Aguet, M. (1994). A novel member of the interferon receptor family complements functionality of the murine interferon γ receptor in human cells. *Cell* **76**, 803–810.

Hershey, G. K., and Schreiber, R. D. (1989). Biosynthetic analysis of the human interferon-gamma receptor. Identification of *N*-linked glycosylation intermediates. *J. Biol. Chem.* **264**, 11981–11988.

Hibino, Y., Mariano, T. M., Kumar, C. S., Kozak, C. A., and Pestka, S. (1991). Expression and reconstitution of a biologically active mouse interferon gamma receptor in hamster cells. Chromo-somal location of an accessory factor. *J. Biol. Chem.* **266**, 6948–6951.

Hilton, D. J., Richardson, R. T., Alexander, W. S., Viney, E. M., Willson, T. A., Sprigg, N. S., Starr, R., Nicholson, S. E., Metcalf, D., and Nicola, N. A. (1998). Twenty proteins containing a C-terminal SOCS box form five structural classes. *Proc. Natl Acad. Sci. USA* **95**, 114–119.

Huang, S., Hendriks, W., Althage, A., Hemmi, S., Bluethmann, H., Kamijo, R., Vilcek, J., Zinkernagel, R., and Aguet, M. (1993). Immune response in mice that lack the interferon-γ receptor. *Science* **259**, 1742–1745.

Igarashi, K., Garotta, G., Ozmen, L., Ziemieckl, A., Wilks, A. F., Harpur, A. G., Larner, A. C., and Finbloom, D. S. (1994). Interferon-γ induces tyrosine phosphorylation of interferon-γ receptor and regulated association of protein tyrosine kinases, Jak1 and Jak2 with its receptor. *J. Biol. Chem.* **269**, 14333–14336.

Ihle, J. N., Witthuhn, B. A., Quelle, F. W., Yamamoto, K., and Silvennoinen, O. (1995). Signaling through the hematopoietic cytokine receptors. *Annu. Rev. Immunol.* **13**, 369–398.

Jouanguy, E., Altare, F., Lamhamedi, S., Revy, P., Newport, M., Levin, M., Blanche, S., Fischer, A., and Casanova, J-L. (1996). Interferon-γ-receptor deficiency in an infant with fatal bacille Calmette-Guerin infection. *N. Engl. J. Med.* **335**, 1956–1961.

Jouanguy, E., Lamhamedi-Cherradi, S., Altare, F., Fondaneche, M., Tuerlinckx, D., Blanche, S., Emile, J., Gaillard, J., Schreiber, R., Levin, M., Fischer, A., Hivroz, C., and Casanova, J. (1997). Partial interferon-gamma receptor 1 deficiency in a child with tuberculoid bacillus calmette-guerin infection and a sibling with clinical tuberculosis. *J. Clin. Invest.* **100**, 2658–2664.

Jouanguy, E., Lamhamedi-Cherradi, S., Lammas, D., Dorman, S. E., Fondaneche, M. C., Dupuis, S., Doffinger, R., Altare, F., Girdlestone, J., Emile, J. F., Ducoulombier, H., Edgar, D., Clarke, J., Oxelius, V. A., Brai, M., Novelli, V., Heyne, K., Fischer, A., Holland, S. M., Kumararatne, D. S., Schreiber, R. D., and Casanova, J. L. (1999). A human IFNGR1 small deletion hotspot associated with dominant susceptibility to mycobacterial infection. *Nature Genet.* **21**, 370–378.

Jung, V., Rashidbaigi, A., Jones, C., Tischfield, J. A., Shows, T. B., and Pestka, S. (1987). Human chromosomes 6 and 21 are required for sensitivity to human interferon gamma. *Proc. Natl Acad. Sci. USA* **84**, 4151–4155.

Jung, V., Jones, C., Kumar, C. S., Stefanos, S., O'Connell, S., and Pestka, S. (1990). Expression and reconstitution of a biologically active human IFNγ receptor in hamster cells. *J. Biol. Chem.* **265**, 1827–1830.

Kamijo, R., Le, J., Shapiro, D., Havell, E. A., Huang, S., Aguet, M., Bosland, M., and Vilcek, J. (1993). Mice that lack the interferon-γ receptor have profoundly altered responses to infection with Bacillus Calmette-Guérin and subsequent challenge with lipopolysaccharide. *J. Exp. Med.* **178**, 1435–1440.

Kaplan, D. H., Greenlund, A. C., Tanner, W. J., Shaw, A. S., and Schreiber, R. D. (1996). Identification of an interferon-γ receptor α chain sequence required for JAK-1 binding. *J. Biol. Chem.* **271**, 9–12.

Kaplan, D. H., Shankaran, V., Dighe, A. S., Stockert, E., Aguet, M., Old, L. J., and Schreiber, R. D. (1998). Demonstration of an interferon γ-dependent tumor surveillance system in immunocompetent mice. *Proc. Natl Acad. Sci. USA* **95**, 7556–7561.

Kerr, I. M., and Stark, G. R. (1991). The control of interferon-inducible gene expression. *FEBS Lett.* **285**, 194–198.

Kim, T. K., and Maniatis, T. (1996). Regulation of interferon-gamma-activated Stat1 by the ubiquitin-proteasome pathway. *Science* **273**, 1717–1719.

Kumar, C. S., Muthukumaran, G., Frost, L. J., Noe, M., Ahn, Y. H., Mariano, T. M., and Pestka, S. (1989). Molecular characterization of the murine interferon-γ receptor cDNA. *J. Biol. Chem.* **264**, 17939–17946.

Leonard, W. J., and O'Shea, J. J. (1998). JAKs and STATs: biological implications. *Annu. Rev. Immunol.* **16**, 293–322.

Lewin, A. R., Reid, L. E., McMahon, M., Stark, G. R., and Kerr, I. M. (1991). Molecular analysis of a human interferon-inducible gene family. *Eur. J. Biochem.* **199**, 417–423.

Liu, B., Liao, J., Rao, X., Kushner, S., Chung, C. D., Chang, D. D., and Shuai, K. (1998). Inhibition of Stat1-mediated gene activation by PIAS1. *Proc. Natl Acad. Sci. USA* **95**, 10626–10631.

Lu, B., Ebensperger, C., Dembic, Z., Wang, Y., Kvatyuk, M., Lu, T., Coffman, R. L., Pestka, S., and Rothman, P. B. (1998). Targeted disruption of the interferon-gamma receptor 2 gene results in severe immune defects in mice. *Proc. Natl Acad. Sci. USA* **95**, 8233–8238.

Lutfalla, G., Gardiner, K., and Uzé, G. (1993). A new member of the cytokine receptor gene family maps on chromosome 21 at less than 35 kb from IFNAR. *Genomics* **16**, 366–373.

Mach, B., Steimle, V., Martinez-Soria, E., and Reith, W. (1996). Regulation of MHC class II genes: lessons from a disease. *Annu. Rev. Immunol.* **14**, 301–331.

Mao, C., Aguet, M., and Merlin, G. (1989). Molecular characterization of the human interferon-gamma receptor: analysis of polymorphism and glycosylation. *J. Interferon Res.* **9**, 659–669.

Mariano, T. M., Kozak, C. A., Langer, J. A., and Pestka, S. (1987). The mouse immune interferon receptor gene is located on chromosome 10. *J. Biol. Chem.* **262**, 5812–5814.

Marsters, S., Pennica, D., Bach, E., Schreiber, R. D., and Ashkenazi, A. (1995). Interferon γ signals via a high-affinity multisubunit receptor complex that contains two types of polypeptide chain. *Proc. Natl Acad. Sci. USA* **92**, 5401–5405.

McKendry, R., John, J., Flavell, D., Müller, M., Kerr, I. M., and Stark, G. R. (1991). High-frequency mutagenesis of human cells and characterization of a mutant unresponsive to both α and γ interferons. *Proc. Natl Acad. Sci. USA* **88**, 11455–11459.

Meraz, M. A., White, J. M., Sheehan, K. C. F., Bach, E. A., Rodig, S. J., Dighe, A. S., Kaplan, D. H., Riley, J. K., Greenlund, A. C., Campbell, D., Carver-Moore, K., DuBois, R. N., Clark, R., Aguet, M., and Schreiber, R. D. (1996). Targeted disruption of the STAT1 gene in mice reveals unexpected physiologic specificity in the Jak-STAT signaling pathway. *Cell* **84**, 431–442.

Merlin, G., Van der Leede, B.-J., Aguet, M., and Dembic, Z. (1989). The human interferon gamma receptor gene. *J. Interferon Res.* **9**, 89.

Müller, M., Briscoe, J., Laxton, C., Guschin, D., Ziemiecki, A., Silvennoinen, O., Harpur, A. G., Barbier, G., Witthuhn, B. A., Schindler, C., Pellegrini, S., Wilks, A. F., Ihle, J. N., Stark, G. R., and Kerr, I. M. (1993). The protein tyrosine kinase JAK1 complements a mutant cell line defective in the interferon-α/β and -γ signal transduction pathways. *Nature* **366**, 129–135.

Müller, U., Steinhoff, U., Reis, L. F. L., Hemmi, S., Pavlovic, J., Zinkernagel, R. M., and Aguet, M. (1994). Functional role of type I and type II interferons in antiviral defense. *Science* **264**, 1918–1921.

Munro, S., and Maniatis, T. (1989). Expression and cloning of the murine interferon-γ receptor cDNA. *Proc. Natl Acad. Sci. USA* **86**, 9248–9252.

Naka, T., Narazaki, M., Hirata, M., Matsumoto, T., Minamoto, S., Aono, A., Nishimoto, N., Kajita, T., Taga, T., Yoshizaki, K., Akira, S., and Kishimoto, T. (1997). Structure and function of a new STAT-induced STAT inhibitor. *Nature* **387**, 924–928.

Naka, T., Matsumoto, T., Narazaki, M., Fujimoto, M., Morita, Y., Ohsawa, Y., Saito, H., Nagasawa, T., Uchiyama, Y., and Kishimoto, T. (1998). Accelerated apoptosis of lymphocytes by augmented induction of Bax in SSI-1 (STAT-induced STAT inhibitor-1) deficient mice. *Proc. Natl Acad. Sci. USA* **95**, 15577–15582.

Neubauer, H., Cumano, A., Muller, M., Wu, H., Huffstadt, U., and Pfeffer, K. (1998). Jak2 deficiency defines an essential developmental checkpoint in definitive hematopoiesis. *Cell* **93**, 397–409.

Newport, M. J., Huxley, C. M., Huston, S., Hawrylowicz, C. M., Oostra, B. A., Williamson, R., and Levin, M. (1996). A mutation in the interferon-γ-receptor gene and susceptibility to mycobacterial infection. *N. Engl. J. Med.* **335**, 1941–1949.

Novick, D., Orchansky, P., Revel, M., and Rubinstein, M. (1987). The human interferon-gamma receptor. Purification, characterization, and preparation of antibodies. *J. Biol. Chem.* **262**, 8483–8487.

Paraganas, E., Wang, D., Stravopodis, D., Topham, D. J., Vanin, E. F., Bodner, S., Colamonici, O. R., van Deursen, J. M., Grosveld, G., and Ihle, J. N. (1998). Jak2 is essential for signaling through a variety of cytokine receptors. *Cell* **93**, 385–395.

Pellegrini, S., John, J., Shearer, M., Kerr, I. M., and Stark, G. R. (1989). Use of a selectable marker regulated by alpha interferon to obtain mutations in the signaling pathway. *Mol. Cell Biol.* **9**, 4605–4612.

Pernis, A., Gupta, S., Gollob, K. J., Garfein, E., Coffman, R. L., Schindler, C., and Rothman, P. (1995). Lack of interferon-γ receptor β chain and the prevention of interferon-γ signaling in TH1 cells. *Science* **269**, 245–247.

Pfizenmaier, K., Wiegmann, K., Scheurich, P., Krönke, M., Merlin, G., Aguet, M., Knowles, B. B., and Ucer, U. (1988). High affinity human IFN-gamma-binding capacity is encoded by a single receptor gene located in proximity to c-ras on human chromosome region 6q16 to 6q22. *J. Immunol.* **141**, 856–860.

Pierre-Audigier, C., Jouanguy, E., Lamhamedi, S., Altare, F., Rauzier, J., Vincent, V., Canioni, D., Emile, J.-F., Fischer, A., Blanche, S., Gaillard, J.-L., and Casanova, J.-L. (1997). Fatal disseminated *Mycobacterium smegmatis* infection in a child with inherited interferon γ receptor deficiency. *Clin. Infect. Dis.* **24**, 982–984.

Rodig, S. J., Meraz, M. A., White, J. M., Lampe, P. A., Riley, J. K., Arthur, C. D., King, K. L., Sheehan, K. C. F., Yin, L., Pennica, D., Johnson Jr., E. M., and Schreiber, R. D. (1998). Disruption of the *Jak1* gene demonstrates obligatory and nonredundant roles of the Jaks in cytokine-induced biologic responses. *Cell* **93**, 373–383.

Sakatsume, M., and Finbloom, D. S. (1996). Modulation of the expression of the IFN-γ receptor β-chain controls responsiveness to IFN-γ in human peripheral blood T cells. *J. Immunol.* **156**, 4160–4166.

Schindler, C., and Darnell Jr., J. E. (1995). Transcriptional responses to polypeptide ligands: the JAK-STAT pathway. *Annu. Rev. Biochem.* **64**, 621–651.

Schindler, C., Fu, X.-Y., Improta, T., Aebersold, R., and Darnell Jr., J. E. (1992). Proteins of transcription factor ISGF-3: one gene encodes the 91- and 84-kDa ISGF-3 proteins that are activated by interferon α. *Proc. Natl Acad. Sci. USA* **89**, 78360–7839.

Shuai, K., Schindler, C., Prezioso, V. R., and Darnell Jr., J. E. (1992). Activation of transcription by IFN-γ: tyrosine phosphorylation of a 91-kD DNA binding protein. *Science* **258**, 1808–1812.

Soh, J., Donnelly, R. O., Kotenko, S., Mariano, T. M., Cook, J. R., Wang, N., Emanuel, S., Schwartz, B., Miki, T., and Pestka, S. (1994). Identification and sequence of an accessory factor required for activation of the human interferon-γ receptor. *Cell* **76**, 793–802.

Starr, R., Wilson, T. A., Viney, E. M., Murray, L. J. L., Rayner, J. R., Jenkins, B. J., Gonda, T. J., Alexander, W. S., Metcalf, D., Nicola, N. A., and Hilton, D. J. (1997). A family of cytokine-inducible inhibitors of signalling. *Nature* **387**, 917–920.

Starr, R., Metcalf, D., Elefanty, A. G., Brysha, M., Willson, T. A., Nicola, N. A., Hilton, D. J., and Alexander, W. S. (1998). Liver degeneration and lymphoid deficiencies in mice lacking suppressor of cytokine signaling-1. *Proc. Natl Acad. Sci. USA* **95**, 14395–14399.

Tanaka, N., and Taniguchi, T. (1992). Cytokine gene regulation: regulatory cis-elements and DNA binding factors involved in the interferon system. *Adv. Immunol.* **52**, 263–281.

Walter, M. R., Windsor, W. T., Nagabhushan, T. L., Lundell, D. J., Lunn, C. A., Zauodny, P. J., and Narula, S. W. (1995). Crystal structure of a complex between interferon-γ and its soluble high-affinity receptor. *Nature* **376**, 230–235.

Watling, D., Guschin, D., Müller, M., Silvennoinen, O., Witthuhn, B. A., Quelle, F. W., Rogers, N. C., Schindler, C., Stark, G. R., Ihle, J. N., and Kerr, I. M. (1993). Complementation by the protein tyrosine kinase JAK2 of a mutant cell line defective in the interferon-γ signal transduction pathway. *Nature* **366**, 166–170.

Wen, Z., Zhong, Z., and Darnell Jr., J. E. (1995). Maximal activation of transcription by Stat1 and Stat3 requires both tyrosine and serine phosphorylation. *Cell* **82**, 241–250.

You, M., and Zhao, Z. (1997). Positive effects of SH2 domain-containing tyrosine phosphatase SHP-1 on epidermal growth factor- and interferon-gamma-stimulated activation of STAT transcription factors in HeLa cells. *J. Biol. Chem.* **272**, 23376–23381.

You, M., Yu, D., and Feng, G. (1999). SHP-2 tyrosine phosphatase functions as a negative regulator of the interferon-stimulated JAK/STAT pathway. *Mol. Cell. Biol.* **19**, 2416–2424.

LICENSED PRODUCTS

Antibodies against IFNGR1

Human IFNGR1

Monoclonal IgG

Mouse anti-human IFNγ receptor α chain (anti-human IFNGR1; blocking)	BD Pharmingen
	Catalog number: GIR-208
	Concentration: 500 μg/mL
Mouse anti-human IFNγ receptor α chain (anti-human IFNGR2; nonblocking)	BD Pharmingen
	Catalog number: GIR-94
	Concentration: 500 μg/mL

Murine IFNGR1

Rat antimouse IFNγ receptor α chain (antimurine IFNGR1; blocking)	BD Pharmingen
	Catalog number: 09811A
	Concentration: 500 μg/mL
Hamster antimouse IFNγ receptor α chain (antimurine IFNGR2; nonblocking)	BD Pharmingen
	Catalog number: 2E2
	Concentration: 500 μg/mL

Polyclonal IgG

Rabbit polyclonal IgG	Santa Cruz Biotechnology
	Catalog number: sc-703
	Concentration: 200 μg/mL
Rabbit polyclonal IgG	Santa Cruz Biotechnology
	Catalog number: sc-702
	Concentration: 200 μg/mL

Human and murine IFNGR1

Rabbit polyclonal IgG	Santa Cruz Biotechnology
	Catalog number: sc-700
	Concentration: 200 μg/mL

Antibodies against IFNGR2

Human IFNGR2

Monoclonal IgG

Hamster anti-human IFNγ receptor β chain (anti-human IFNGR2)	BD Pharmingen
	Catalog number: 2HUB-159
	Concentration: 500 μg/mL

Polyclonal IgG

Rabbit polyclonal IgG	Santa Cruz Biotechnology
	Catalog number: sc-971
	Concentration: 200 μg/mL
Rabbit polyclonal IgG	Santa Cruz Biotechnology
	Catalog number: sc-970
	Concentration: 200 μg/mL

Murine IFNGR2	
Hamster antimurine IFNγ receptor β chain (antimurine IFNGR2)	BD Pharmingen
	Catalog number: MOB-47
	Concentration: 500 μg/mL
Rabbit polyclonal IgG	Santa Cruz Biotechnology
	Catalog number: sc-973
	Concentration: 200 μg/mL
Rabbit polyclonal IgG	Santa Cruz Biotechnology
	Catalog number: sc-972
	Concentration: 200 μg/mL

Poxvirus IFNγ Receptor Homologs

Grant McFadden[1,*] and Richard Moyer[2]

[1]The John P. Robarts Research Institute and Department of Microbiology and Immunology, The University of Western Ontario, 1400 Western Road, London, Ontario, N6G 2V4, Canada

[2]Department of Molecular Genetics and Microbiology, University of Florida College of Medicine, PO Box 100266, Gainesville, FL 32610-0266, USA

*corresponding author tel: (519) 663-3184, fax: (519) 663-3847, e-mail: mcfadden@rri.on.ca

DOI: 10.1006/rwcy.2000.14014.

SUMMARY

The first example discovered of a viroceptor (virus-encoded receptor homolog) targeted against the interferon family was the M-T7 gene product of *myxoma virus*, a secreted glycoprotein of 37 kDa that shares sequence similarity to the external ligand-binding domain of the cellular IFNγ receptor. Later, M-T7 was shown to be closely related to a variety of homologous genes in many other poxviruses that share sequence homology to the same domain of the cellular receptors. Studies have shown that these poxvirus IFNγ receptors bind and inhibit IFNγ in a relatively species-specific fashion, with the highest affinity generally being to the ligand derived from the host species of the particular poxvirus.

BACKGROUND

Discovery

The first example of a virus-encoded IFNγ receptor was uncovered when DNA sequencing studies revealed that *myxoma virus* encodes a 263 amino acid protein possessing considerable homology with the cellular IFNγR (Upton *et al.*, 1992). This 37 kDa protein is the major protein secreted from cells infected with myxoma virus (strain Lausanne) and is designated M-T7 because it is encoded by the seventh open reading from the viral DNA terminus (Upton *et al.*, 1992). Database analysis revealed that the conserved amino acids mapped to the extracellular ligand-binding domain of the mammalian IFNγRs. Particularly noteworthy was the cysteine residue spacing important for ligand interaction (Figure 1).

Due to the availability of virus genome sequence information, soluble IFNγR homologs have since been discovered in a variety of other poxviruses (Mossman *et al.*, 1995a; Alcamí and Smith, 1996a,b). The T7 open reading frame of a related leporipoxvirus, *Shope fibroma virus* (SFV), encodes a 265 residue soluble IFNγR homolog, called S-T7 (Upton and McFadden, 1986; Upton *et al.*, 1992). *Vaccinia virus* strains Copenhagen and Western Reserve (WR) also encode related 272 amino acid homologs from their respective B8R open reading frames (Goebel *et al.*, 1990; Howard *et al.*, 1991). By probing *ectromelia virus* DNA fragments with the vaccinia virus B8R gene, an ectromelia virus IFNγR homolog was discovered, and subsequent cloning and sequencing revealed an open reading frame encoding a 266 residue protein, whose location within the ectromelia virus genome remains to be determined (Mossman *et al.*, 1995b). Two strains of *variola virus*, the causative agent of *smallpox* (strains Bangladesh 1975 and India 1967), have been completely sequenced (Massung *et al.*, 1993a; Shchelkunov *et al.*, 1993) and found to encode related soluble IFNγRs, designated B8R and B9R, respectively. Both proteins are predicted to contain 266 amino acids, of which 265 are identical. Finally, swinepox virus, a member of the suipoxvirus genus, encodes a 274 amino acid

IFNγR homolog from the C6L open reading frame (Massung *et al.*, 1993b). Tanapox virus, a member of the yatapoxvirus genus, secretes a 38 kDa glycoprotein which binds and inhibits the effects of human IFNγ, IL-2, and IL-5 (Essani *et al.*, 1994), but its sequence remains to be reported.

Alternative names

vIFNγR.

Structure

All the poxvirus IFNγR homologs are secreted proteins with sequence homology to the ligand-binding domain of the cellular IFNγ receptors. All contain from 6 to 8 of the conserved cysteines that define the folding domains of this family (Figure 1). The family has been extensively reviewed (Upton and McFadden, 1994; Mossman *et al.*, 1995a, 1996a; Alcamí and Smith, 1996b; Smith, 1996).

Main activities and pathophysiological roles

IFNγ-binding proteins from supernatants infected with 10 different *vaccinia virus* strains, as well as cowpox, rabbitpox, camelpox, buffalopox, elephantpox, and tanapox viruses have been identified by crosslinking studies (Alcamí and Smith, 1995; Mossman *et al.*, 1995b). The secreted versions of all poxvirus IFNγR homologs possess apparent molecular masses of 37–43 kDa (Alcamí and Smith, 1995; Mossman *et al.*, 1995b). To date, all poxviruses tested express an IFNγ-binding protein, indicating the universal importance of IFNγ as a critical ligand in the orchestration of an effective immune response against virus infection.

While all of the sequenced poxviral IFNγR homologs have been found to exhibit significant homology with the mammalian IFNγR ligand-binding domain, the extent of overall sequence similarity with either the human or murine IFNγRs is relatively low (20–25% identity). While the percentage identity between IFNγR homologs from different poxvirus genera is equally low, the homology between IFNγR proteins within the same poxvirus genus is high. The leporipoxvirus homologs, myxoma virus M-T7 and SFV S-T7, are 68% identical and among the orthopoxvirus homologs encoded by vaccinia, ectromelia, and variola viruses there is greater than 90% identity. An important feature of all poxviral IFNγR homologs is the strong conservation of the cysteine residues shown to be critical for proper folding and functioning of the corresponding mammalian receptors (Figure 1). An intriguing observation, however, is that the orthopoxvirus receptor homologs lack the corresponding first two cysteine residues which form the lone disulfide bond within domain 1 of the mammalian receptors (Figure 1). This observation is particularly unexpected since the domain 1 disulfide bond is the only bond whose removal completely abolishes activity of the mammalian receptors (Stuber *et al.*, 1993). The significance of this important missing disulfide bond with respect to the function of the orthopoxvirus IFNγR homologs remains to be determined.

GENE

Accession numbers

Myxoma virus (M-T7): M81919
Shope fibroma virus (S-T7): M14899
Vaccinia virus Western Reserve (B8R): M58056
Vaccinia virus Copenhagen (B8R): M35027
Swinepox virus (C6L): L22013
Variola virus major (B8R): L22579
Variola virus minor (H9R): U18339
Cowpox virus (B7R): Y15035
Ectromelia virus: U19584

PROTEIN

Accession numbers

Myxoma virus (M-T7): 332308
Vaccinia virus Western Reserve (B8R): 335312
Vaccinia virus Copenhagen (B8R): 335553
Swinepox virus (C6L): 418186
Variola virus major (B8R): 439086
Variola virus minor (H9R): 885773
Cowpox virus (B7R): 3097028
Ectromelia virus: 761722

Sequence

See **Figure 1**.

Description of protein

All the poxvirus IFNγR homologs are secreted from virus-infected cells as 35–45 kDa glycoproteins that

Figure 1 Amino acid sequence alignment of the poxviral IFNγR homologs: myxoma virus M-M-T7, Shope fibroma virus S-M-T7, swinepox virus SPV C6L, vaccinia virus B8R (VV-B8R), and *variola virus* Bangladesh 1975 B8R (VAR-B8R), and the cellular murine (muIFNR) and human (huIFNR) IFNγ receptor chains. Boxes indicate amino acid identity among all proteins, while (*) denotes amino acids which are conserved in at least four of the seven proteins. The disulfide-forming cysteine residues conserved between the mammalian and viral proteins are both boxed and numbered 1–8. The arrow indicates the location of the N-terminal residue, determined by sequencing, of the mature secreted myxoma M-T7 protein, while the predicted transmembrane domains of the mammalian IFNγRs are underlined. The full lengths of the human and murine IFNγRs are 489 and 477 amino acids, respectively.

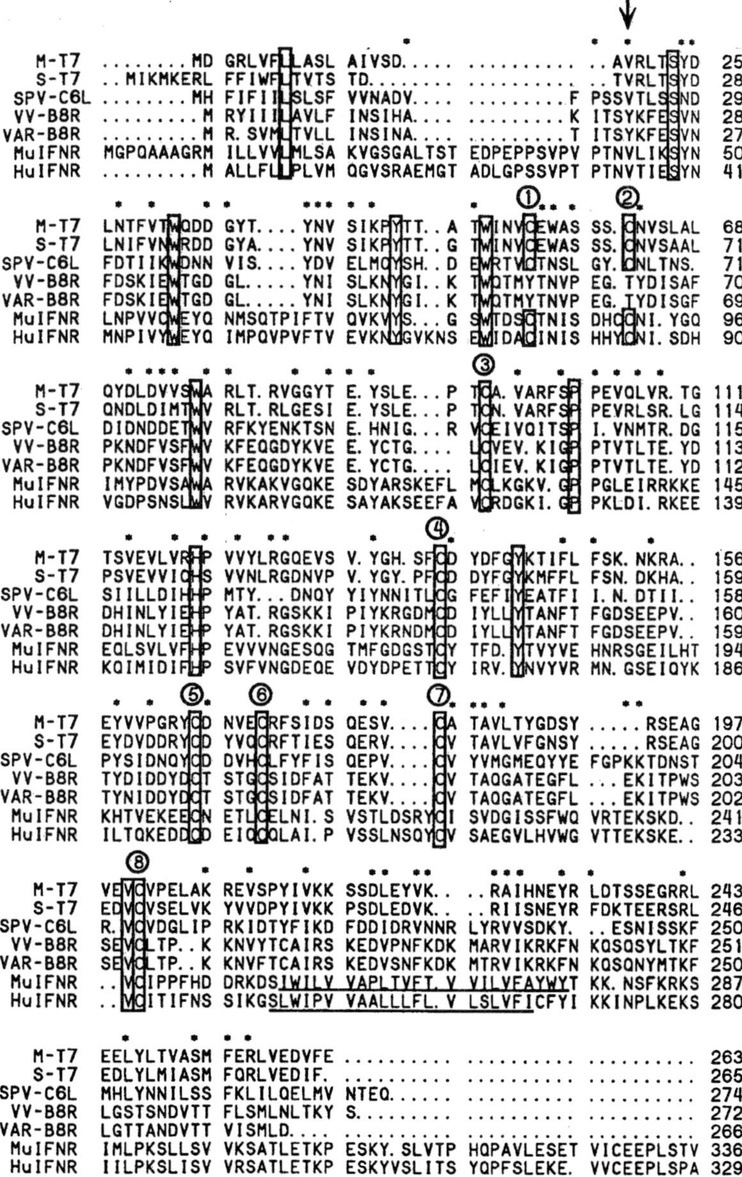

form soluble inhibitory complexes with IFNγ, with variable species specifications. Thus, all the downstream signaling events triggered by IFNγ-dependent receptor stimulators are blocked at a stage prior to receptor engagement.

Studies on the *myxoma virus* M-T7 protein have illustrated the significant role played by viral IFNγR homologs during a productive infection. M-T7 is expressed early during infection from a strong poxviral promoter (Mossman *et al.*, 1995c), an important property considering the requirement to counteract the potent activities of IFNγ. M-T7 binds rabbit IFNγ with an affinity comparable to that demonstrated by the mammalian soluble receptors and their cognate ligands (10^{-9} M) (Mossman *et al.*, 1995c). M-T7 is secreted from myxoma virus-infected cells at levels which exceed 10^7 molecules per cell per hour at early times of infection (Mossman *et al.*, 1995c), which constitutes a vast excess over the 10^2–10^4 cellular receptors typically expressed at the surface of most mammalian cells.

Relevant homologies and species differences

See Figure 1.

Affinity for ligand(s)

IFNγ interacts exclusively with receptors from the same species (Farrar and Schreiber, 1993), and since the ligand specificity of individual receptor homologs has the potential to shed light on virus evolutionary history, determining the species specificity of poxviral IFNγR homologs has particular relevance. For example, *myxoma virus* is one of the few poxviruses whose natural host, the South American bush rabbit or tapeti (*Silvilagus brasiliensis*), has been well established (Fenner and Ratcliffe, 1965; Fenner and Ross, 1994). M-T7 demonstrates strict species specificity for rabbit IFNγ (Mossman *et al.*, 1995e), while the IFNγR homologs from members of the orthopoxvirus genus, such as vaccinia, exhibit much broader species specificity for IFNγ inhibition (Alcamí and Smith, 1995).

Unlike myxoma virus, the natural host of most orthopoxviruses, particularly *vaccinia virus*, remains obscure. In chemical cross-linking assays, the vaccinia virus IFNγR homolog has been found capable of binding a variety of radiolabeled ligands, including human, bovine, rabbit, and rat IFNγ (Alcamí and Smith, 1995; Mossman *et al.*, 1995b). In one study, the *vaccinia virus* receptor homolog was shown to bind murine IFNγ with only low affinity (Mossman *et al.*, 1995b) while another study failed to observe any detectable binding (Alcamí and Smith, 1995). Broad ligand specificity was also demonstrated by IFNγR homologs encoded by cowpox, rabbitpox, camelpox, buffalopox and elephantpox viruses, whose receptor homologs bind human, bovine, and rat IFNγ with comparable affinities (Alcamí and Smith, 1995). Only camelpox and elephantpox viruses express proteins that showed weak binding of murine IFNγ. Interestingly, *ectromelia virus* displayed equal binding affinities for human, rabbit, and murine IFNγ (Mossman *et al.*, 1995b).

The IFNγR homologs of *variola virus* (strains major and minor) were expressed in bacterial vectors and shown to be specific inhibitors of human, but not murine, IFNγ (Seregin *et al.*, 1996). Interestingly, the vaccinia B8R protein is far more promiscuous than other poxviral IFNγR homologs, and can inhibit even chicken IFNγ (Puehler *et al.*, 1998). Very recently, the myxoma M-T7 protein was also shown to exhibit a second unpredicted property of binding to the glycosamino-glycan domain of a broad spectrum of chemokines, a property not shared by the IFNγR homologs from other poxviruses (Lalani *et al.*, 1997), but its sequence remains to be determined.

Regulation of receptor expression

All the poxvirus IFNγR homologs are expressed from early viral promoters, which function independently from the infected cell type.

BIOLOGICAL CONSEQUENCES OF ACTIVATING OR INHIBITING RECEPTOR AND PATHOPHYSIOLOGY

Phenotypes of receptor knockouts and receptor overexpression mice

Myxoma virus pathogenesis is an attractive poxvirus system to study virus virulence due to the established nature of both its natural host and the disease syndrome it causes. In the European rabbit (*Oryctolagus cuniculus*), myxoma virus induces a rapidly lethal infection known as *myxomatosis* (Fenner and Myers, 1978; Fenner and Ratcliffe, 1965). The disseminating nature of the infection, coupled with the extensive immunosuppression that is associated with both cellular immune defects and cytokine dysregulations, leads to death of the infected rabbits within 2 weeks (Strayer, 1989; McFadden, 1994).

Deletion of the M-T7 gene from the myxoma virus genome dramatically attenuated the virus and confirmed the importance of this protein during viral pathogenesis (Mossman *et al.*, 1996b).

References

Alcamí, A., and Smith, G. L. (1995). Vaccinia, cowpox and camelpox viruses encode soluble gamma interferon receptors with novel broad species specificity. *J. Virol.* **69**, 4633–4639.

Alcamí, A., and Smith, G. L. (1996a). Receptors for gamma-interferon encoded by poxviruses: Implications for the unknown origin of vaccinia virus. *Trends Microbiol.* **4**, 321–326.

Alcamí, A., and Smith, G. L. (1996b). Soluble interferon-γ receptors encoded by poxviruses. *Comp. Immunol. Microbiol. Infect. Dis.* **19**, 305–317.

Essani, K., Chalasani, S., Eversole, R., Beuving, L., and Birmingham, L. (1994). Multiple anti-cytokine activities secreted from tanapox virus-infected cells. *Microbial Pathogenesis* **17**, 347–353.

Farrar, M. A., and Schreiber, R. D. (1993). The molecular cell biology of interferon-γ and its receptor. *Annu. Rev. Immunol.* **11**, 571–611.

Fenner, F., and Myers, K. (1978). In "Myxoma Virus and Myxomatosis in Retrospect: The First Quarter Century of a New Disease." Academic Press, New York.

Fenner, F., and Ratcliffe, F. N. (1965). In "Myxomatosis." Cambridge University Press, Cambridge.

Fenner, F., and Ross, J. (1994). In "The European Rabbit, the History and Biology of a Successful Colonizer" (ed. G. V. Thompson and C. M. King), Myxomatosis, pp. 205–239. Oxford University Press, Oxford.

Goebel, S. J., Johnson, G. P., Perkus, M. E., Davis, S. W., Winslow, J. P., and Paoletti, E. (1990). The complete DNA sequence of vaccinia virus. *Virology* **179**, 247–266.

Howard, S. T., Chan, Y. C., and Smith, G. L. (1991). Vaccinia virus homologues of the Shope fibroma virus inverted terminal repeat proteins and a discontinuous ORF related to the tumour necrosis factor receptor family. *Virology* **180**, 633–647.

Lalani, A. S., Graham, K., Mossman, K., Rajarathnam, K., Clark-Lewis, I., Kelvin, D., and McFadden, G. (1997). The purified myxoma virus IFN-γ receptor homolog, M-T7, interacts with the heparin binding domains of chemokines. *J. Virol.* **71**, 4356–4363.

McFadden, G. (1994). "Rabbit, Hare, Squirrel and Swine Poxviruses." Academic Press, San Diego.

Massung, R. F., Esposito, J. J., Liu, L. I., Qi, J., Utterback, T. R., Knight, J. C., Aubin, L., Yuran, T. E., Parsons, J. M., Loparev, V. N., Selivanov, N. A., Cavallaro, K. F., Kerlavage, A. R., Mahy, B. W. J., and Venter, J. C. (1993a). Potential virulence determinants in terminal regions of variola smallpox virus genome. *Nature* **366**, 748–751.

Massung, R. F., Jayarama, V., and Moyer, R. W. (1993b). DNA sequence analysis of conserved and unique regions of swinepox virus: identification of genetic elements supporting phenotypic observations including a novel G protein-coupled receptor homologue. *Virology* **197**, 511–528.

Mossman, K., Barry, M., and McFadden, G. (1995a). In "Viroceptors, Virokines and Related Immune Modulators" (ed. G. McFadden), Interferon-γ receptors encoded by poxviruses, pp. 41–54. R.G. Landes & Co., Austin, TX.

Mossman, K., Upton, C., Buller, R. M., and McFadden, G. (1995b). Species specificity of ectromelia virus and vaccinia virus interferon-γ binding proteins. *Virology* **208**, 762–769.

Mossman, K., Upton, C., and McFadden, G. (1995c). The myxoma virus soluble interferon-γ receptor homolog, M-T7, inhibits interferon-γ in a species-specific manner. *J. Biol. Chem.* **270**, 3031–3038.

Mossman, K., Upton, C., and McFadden, G. (1995e). The myxoma virus-soluble interferon-γ receptor homolog, M-T7, inhibits interferon-γ in a species specific manner. *J. Biol. Chem.* **270**, 3031–3038.

Mossman, K., Barry, M., and McFadden, G. (1996a). In "Gamma Interferon: A Pleiotropic Cytokine with Antiviral Activity" (ed. G. Karupiah), Regulation of interferon-γ gene expression and extracellular ligand function by immunomodulatory viral proteins, pp. 175–188. R.G. Landes & Co., Austin, TX.

Mossman, K., Nation, P., Macen, J., Garbutt, M., Lucas, A., and McFadden, G. (1996b). Myxoma virus M-T7, a secreted homolog of the interferon-γ receptor, is a critical virulence factor for the development of myxomatosis in European rabbits. *Virology* **215**, 17–30.

Puehler, F., Weining, K. C., Symons, J. A., Smith, G. L., and Staeheli, P. (1998). Vaccinia virus-encoded cytokine receptor binds and neutralizes chicken interferon-gamma. *Virology* **248**, 231–240.

Seregin, S. V., Babkina, I. N., Nesterov, A. E., Sinakov, A. N., and Shchelkunov, S. N. (1996). Comparative studies of gamma-interferon receptor-like proteins of variola major and variola minor viruses. *FEBS Lett.* **382**, 79–83.

Shchelkunov, S. N., Blinov, V. M., and Sandakhchiev, L. S. (1993). Genes of variola and vaccinia viruses necessary to overcome the host protective mechanism. *FEBS Lett.* **319**, 80–83.

Smith, G. L. (1996). Virus proteins that bind cytokines, chemokines or interferons. *Curr. Opin. Immunol.* **8**, 467–471.

Strayer, D. (1989). In "Virus-Induced Immunosuppression" (ed. S. Specter, M. Bendinelli, and H. Friedman), Poxviruses, pp. 173–192. Plenum Press, New York.

Stuber, D., Friedlein, A., Fountoulakis, M., Lahm, H. W., and Garotta, G. (1993). Alignment of disulfide bonds of the extracellular domain of the interferon γ receptor and investigation of their role in biological activity. *Biochemistry* **32**, 2423–2430.

Upton, C., and McFadden, G. (1986). Tumorigenic poxviruses: analysis of viral DNA sequences implicated in the tumorigenicity of Shope fibroma virus and malignant rabbit virus. *Virology* **152**, 308–321.

Upton, C., and McFadden, G. (1994). In "Methods in Molecular Genetics"Detection of viral homologs of cellular interferon γ receptors, pp. 383–390. Academic Press, London.

Upton, C., Mossman, K., and McFadden, G. (1992). Encoding of a homolog of the interferon-γ receptor by myxoma virus. *Science* **258**, 1369–1372.

IFNα/β Receptor

S. Jaharul Haque[1,2] and B. R. G. Williams[1,*]

[1]Department of Cancer Biology, Lerner Research Institute, Cleveland Clinic Foundation, 9500 Euclid Avenue, Cleveland, OH 44195, USA

[2]Department of Pulmonary and Critical Care Medicine, Cleveland Clinic Foundation, 9500 Euclid Avenue, Cleveland, OH 44195, USA

*corresponding author tel: 216-445-9652, fax: 216-444-3164, e-mail: williab@ccf.org

DOI: 10.1006/rwcy.2000.18002.

SUMMARY

Interferons (IFNs) comprise a family of secreted proteins produced by a variety of vertebrate cells in response to viral and other microbial infections, and function as antiviral, antiproliferative, and immunomodulatory agents. IFNs bind to specific receptors on the surface of target cells and activate multiple intracellular signaling cascades including the JAK/STAT pathway which in turn activates the transcription of IFN-stimulated genes. Type I IFNs, including α, β, and ω, bind to the same receptor that comprises two transmembrane subunits (IFNAR-1 and IFNAR-2) that are classified as type II cytokine receptors based on their amino acid sequence and structural features. These two proteins function in a species-specific fashion. IFNAR-1 and IFNAR-2 physically associate with TYK2 and JAK1 respectively that catalyze the tyrosyl phosphorylation of IFN-signaling proteins including JAK1, TYK2, IFNAR-1, IFNAR-2, STAT1, STAT2, and STAT3. Mice lacking the IFNAR-1 are completely unresponsive to type I IFNs and unable to combat viral infection.

BACKGROUND

Interferons (IFNs) belong to the cytokine superfamily of secreted proteins that were originally identified as antiviral agents in 1957 by Isaacs and Lindenmann. Subsequent studies revealed that IFNs exert pleiotropic biological effects, all mediated through the activation of specific receptors on the cell surface (Pestka *et al.*, 1987; Sen and Lengyel, 1992; Darnell *et al.*, 1994; Uze *et al.*, 1995; Domanski and Colamonici, 1996; Haque and Williams, 1998; Stark *et al.*, 1998). IFNs are divided into two types: human type I IFNs include 13 nonallelic isoforms of IFNα, one IFNβ, and one IFNω (Weissmann and Weber, 1986; Pestka *et al.*, 1987; Sen and Lengyel, 1992; Haque and Williams, 1998). All type I IFNs compete for binding to the same receptor on the cell surface, known as type I IFN receptor or IFNα/β receptor (Aguet *et al.*, 1984; Merlin *et al.*, 1985; Pestka *et al.*, 1987; Uze *et al.*, 1995; Haque and Williams, 1998). IFNγ binds to a distinct transmembrane receptor (Aguet *et al.*, 1984; Pestka *et al.*, 1987; Uze *et al.*, 1995; Haque and Williams, 1998; Merlin *et al.*, 1985).

IFNs exhibit a high level of species-specificity with a few exceptions (Stwert, 1979; Uze *et al.*, 1990; Novick *et al.*, 1994; Domanski and Colamonici, 1996). Binding of IFNs to their receptors initiates signals that are transmitted from cell surface to the nucleus, resulting in the rapid induction of a number of IFN-stimulated genes (ISGs) in the absence of *de novo* protein synthesis (Sen and Lengyel, 1992; Darnell *et al.*, 1994; Levy, 1995; Ihle, 1996; Haque and Williams, 1998). Each type of IFN induces a distinct set of genes with a certain degree of overlap (Sen and Lengyel, 1992; Darnell *et al.*, 1994; Haque and Williams, 1998; Stark *et al.*, 1998). Most of the IFNα/β-responsive genes contain an enhancer element, termed the IFN-stimulated response element (ISRE) that binds to the transcription factor IFN-stimulated gene factor 3 (ISGF3), which is induced through an IFNα/β-dependent activation of the Janus kinase/signal transducer and activator of transcription (JAK/STAT) signal transduction pathway (Darnell *et al.*, 1994; Levy, 1995; Ihle, 1996; Haque and Williams, 1998; Stark *et al.*, 1998). The SH2 domain-containing proteins STAT1 and STAT2 are phosphorylated at unique tyrosine residues by a pair of activated JAKs that are physically associated with the IFN receptor subunits (Darnell *et al.*, 1994; Stark *et al.*, 1998). This results in an SH2 domain–phosphotyrosine interaction-mediated formation of a STAT1/STAT2 heterodimer that migrates to the nucleus and binds with a member of the IFN regulatory factor (IRF) family protein p48 (ISFG3-γ) to form the functional trimeric complex ISGF3 that binds to an ISRE to activate transcription (Darnell *et al.*, 1994; Stark *et al.*, 1998).

A second set of genes, including major histocompatibility complexes, is induced as a delayed response that requires new protein synthesis (Pestka *et al.*, 1987; Sen and Lengyel, 1992; Haque and Williams, 1998; Stark *et al.*, 1998). The mechanisms of this delayed response to IFNs are not well understood, but in some cases IFNs may regulate gene expression at the posttranscriptional level (Stark *et al.*, 1998). The proteins encoded by the IFN-regulated genes mediate multiple biological activities attributed to IFNs (Stark *et al.*, 1998).

Discovery

Early investigations using radiolabeled recombinant type I IFNs revealed the cell surface expression of both low-affinity and high-affinity receptors for type I IFNs (Aguet and Blanchard, 1981; Branca and Baglioni, 1981; Joshi *et al.*, 1982; Hannigan *et al.*, 1983, 1984, 1986; Aguet *et al.*, 1984; Merlin *et al.*, 1985; Raziuddin and Gupta, 1985; Zhang *et al.*, 1986; Pestka *et al.*, 1987; Vanden Broecke and Pfeffer, 1988; Sen and Lengyel, 1992; Faltynek *et al.*, 1993). Studies on binding and crosslinking of radiolabeled IFNs to the cell surface suggested the existence of the multisubunit structure for the type I IFN receptors (Joshi *et al.*, 1982; Aguet *et al.*, 1984; Merlin *et al.*, 1985; Raziuddin and Gupta, 1985; Hannigan *et al.*, 1986; Zhang *et al.*, 1986; Pestka *et al.*, 1987; Vanden Broecke and Pfeffer, 1988; Sen and Lengyel, 1992; Faltynek *et al.*, 1993). The existence of the multisubunit structure was later demonstrated using specific monoclonal antibodies to receptor subunits (Colamonici *et al.*, 1990, 1992).

The first receptor subunit to be cloned was the human IFNAR-1 (formerly known as IFN receptor α chain) (Uze *et al.*, 1990). The cDNA encoding the human IFNAR-1 was isolated by an expression cloning strategy utilizing the species-specific recognition property of type I IFNs. This protein conferred resistance to vesicular stomatitis virus replication in mouse cells in the presence of human IFNα8 (Uze *et al.*, 1990). However, when the human IFNAR-1 cDNA was expressed in mouse cells, the protein could only bind to IFNα8 at 37°C, but not to the other type I IFNs of human origin (Uze *et al.*, 1990). The human IFNα8 was later found to possess some binding affinity for mouse receptor (Domanski and Colamonici, 1996). Therefore, IFNAR-1 was not sufficient to confer responsiveness to all the type I IFNs, indicating that another species-specific ligand-binding receptor subunit was required for signal transduction (Uze *et al.*, 1990; Domanski and Colamonici, 1996).

The purification of human IFNα-binding protein from urine and its amino acid sequence analyses led to the isolation of an 1.5 kb cDNA for a novel subunit of type I IFN receptor encoding 331 amino acids (Novick *et al.*, 1994). However, this novel protein of 331 amino acids termed IFNAR-2B (β_S subunit) was not a functional receptor subunit. A longer form of human IFNAR-2 encoding 515 amino acids, termed IFNAR-2C (β_L subunit) was identified as the universal ligand-binding subunit of the type I IFN receptor (Zhang *et al.*, 1986; Bazan, 1990; Colamonici *et al.*, 1990, 1992; Domanski *et al.*, 1995; Lutfalla *et al.*, 1995).

Alternative names

The receptor for the type I IFNs is also referred to as type I IFNR, IFNR, IFNαR, and IFNα/βR (Uze *et al.*, 1995; Domanski and Colamonici, 1996). The first cloned subunit of the type I IFN receptor is IFNAR-1, which was initially termed IFNAR by Uze *et al.* (1990). It is identical to IFNRα of the Colamonici group (Uze *et al.* 1995). The second subunit of type I IFN receptor, originally cloned by Novick *et al.* (1994), is IFNAR-2, which is also known as IFNRβ (Domanski *et al.*, 1995). This subunit has three isoforms: IFNAR-2A, IFNAR-2B (also termed IFNRβ_S), and IFNAR-2C, which is also known as IFNRβ_L (Domanski *et al.*, 1995; Lutfalla *et al.* 1995; Domanski and Colamonici, 1996).

Structure

Cytokine receptors are transmembrane proteins with a single membrane-spanning region. Based on the amino acid sequence and structural features, cytokine receptors are divided into two classes (Bazan, 1990; Kishimoto *et al.*, 1994; Heldin, 1995; Haque and Williams, 1998). While the majority of the cytokine receptors fall into class I, the receptors for IFNα/β, IFNγ, and IL-10 as well as the tissue factor (a membrane receptor for the coagulation protease factor VII) belong to the class II cytokine receptor family (Bazan, 1990).

The cytokine receptors contain one or two characteristic external domain structures (D200) consisting of two homologous subdomains (SD100) of ~100 amino acids, each of which adopts an immunoglobulin-like fold with seven β strands (s1–s7) organized into two β sheets (Bazan, 1990). While the D200 module of class I receptors contains a set of four conserved cysteine residues and a WSXWS motif, the class II cytokine receptors share one cysteine pair with class I and contain an additional conserved cysteine pair, and several conserved proline and tryptophan residues, but lack the WSXWS motif (Bazan, 1990).

The INFAR-1 chain of the receptor contains two D200 modules, while IFNAR-2 has one (Bazan, 1990; Gaboriaud *et al.*, 1990; Thoreau *et al.*, 1991). Human IFNα2 binds to the human IFNAR-1 with low affinity, while it exhibits moderate binding affinity towards the bovine homolog of IFNAR-1. This has facilitated the analysis of IFNα binding to IFNAR-1 (Rehberg *et al.*, 1982; Zoon *et al.*, 1982; Lundgren and Langer, 1997; Goldman *et al.*, 1998). Binding of human IFNα2 to bovine/human IFNAR-1 chimeras reveals that bovine SD2 and SD3 contain residues necessary but not sufficient for moderate-affinity ligand binding; SD1 and SD4 also contribute either directly or indirectly to ligand binding (Goldman *et al.*, 1998). Mapping of epitopes for neutralizing monoclonal antibodies to IFNAR-1 also implicates direct roles for amino acid residues in all four subdomains of the IFNAR-1 molecule (Eid and Tovey, 1995; Goldman *et al.*, 1998; Lu *et al.*, 1998).

The human IFNAR-2 binds to type I IFNs with moderate affinity (2–8 nmol/L; Novick *et al.*, 1994; Domanski and Colamonici, 1996; Lewerenz *et al.*, 1998). Mutant IFNAR-2 proteins generated by alanine substitutions of selected amino acids in two loop regions (amino acids 71–75 in s3–s4 and amino acids 103–106 in s5–s6) in the N-terminal subdomain confer complete resistance to IFNα but not IFNβ binding when expressed in IFNAR-2-null human cell line U5A (Lewerenz *et al.*, 1998). In contrast, unlike IFNα, IFNβ binding to IFNAR-2 is sensitive to an alanine substitution of Try127 located in the intersubdomain link of the IFNAR-2 molecule (Lewerenz *et al.*, 1998). Cytokines in general contain four main α helices labeled A through D connected by two long loops (AB and CD) and a short loop (BC) with an up-up-down-down arrangement (Uze *et al.*, 1995; Mitsui and Senda, 1997). IFNs having an additional helix (E) instead of CD loop constitute a subclass of cytokine with an up-up-down-up-down arrangement (Uze *et al.*, 1995; Mitsui and Senda, 1997). At the amino acid levels, different IFN α subtypes are ~80% homologous, while the homology between the α subtypes and the β subtype is about 35% (Pestka *et al.*, 1987). The AB loop, C helix, and D helix are important for functional IFNα binding to its receptors (Runkel *et al.*, 1998). Mutational studies reveal that substitutions of amino acids that are highly conserved in the AB loop and D helix reduce the binding of IFNα compared with that of IFNβ, suggesting a differential interaction of α and β IFNs with their receptor (Runkel *et al.*, 1998). It is evident that both IFNα and IFNβ need to make physical contact with IFNAR-1 and IFNAR-2 for intracellular signal transduction, although IFNAR-1 seems to engage IFNβ quite differently than IFNα (Novick *et al.*, 1994; Cohen *et al.*, 1995; Domanski *et al.*, 1995; Lutfalla *et al.*, 1995; Cutrone and Langer, 1997; Karpusas *et al.*, 1997; Rani *et al.*, 1996). This suggests that there are some functional differences between these two subtypes of IFNs at the level of receptor recognition.

The three-dimensional structure of the cytoplasmic domain of cytokine receptors is not currently available. Most of the information on structure–function relationship has been derived from the receptor mutagenesis studies. Cytoplasmic domains of cytokine receptors are of variable lengths (Bazan, 1990; Uze *et al.*, 1990; Kishimoto *et al.*, 1994; Novick *et al.*, 1994; Ihle, 1995; Ihle and Kerr, 1995; Ihle *et al.*, 1995; Domanski and Colamonici, 1996; Haque and Williams, 1998; Leonard and O'Shea, 1998; Stark *et al.*, 1998). Many cytokine receptors contain two membrane proximal loosely conserved motifs, Box 1 and Box 2 (Bazan, 1990; Kishimoto *et al.*, 1994; Ihle, 1995; Ihle and Kerr, 1995; Ihle *et al.*, 1995; Leonard and O'Shea, 1998).

Main activities and pathophysiological roles

Protein tyrosine phosphorylation is a key reaction in the activation of cytokine and growth factor receptors (Velazquez *et al.*, 1992; Müller *et al.*, 1993; Watling *et al.*, 1993; Kishimoto *et al.*, 1994; Heldin, 1995; Ihle, 1995, 1996; Ihle and Kerr, 1995; Ihle *et al.*, 1995; Krowlewski, 1995; Levy, 1995; Domanski and Colamonici, 1996; Haque and Williams, 1998; Leonard and O'Shea, 1998; Stark *et al.*, 1998). Unlike most growth factor receptors, the cytokine receptors do not possess any cytoplasmic tyrosine kinase domain, rather they constitutively associate with members of Janus family tyrosine kinases that provide tyrosine kinase activity necessary for receptor activation and subsequent signal transduction

(Velazquez *et al.*, 1992; Müller *et al.*, 1993; Watling *et al.*, 1993; Kishimoto *et al.*, 1994; Ihle, 1995; Ihle and Kerr, 1995; Ihle *et al.*, 1995; Domanski and Colamonici, 1996; Haque and Williams, 1998; Leonard and O'Shea, 1998; Stark *et al.*, 1998;). After being activated by JAK-mediated tyrosine phosphorylation cytokine receptor subunits recruit a number of downstream signaling components through protein–protein interactions (Velazquez *et al.*, 1992; Müller *et al.*, 1993; Watling *et al.*, 1993; Kishimoto *et al.*, 1994; Ihle, 1995; Ihle and Kerr, 1995; Ihle *et al.*, 1995; Domanski and Colamonici, 1996; Haque and Williams, 1998; Leonard and O'Shea, 1998; Stark *et al.*, 1998).

The type I IFN receptor subunits IFNAR-1 and IFNAR-2 constitutively associate with the Janus kinases TYK2 and JAK1 respectively (Velazquez *et al.*, 1992; Müller *et al.*, 1993; Watling *et al.*, 1993; Domanski and Colamonici, 1996; Haque and Williams, 1998; Stark *et al.*, 1998). TYK2 was the first member of the JAK family to be identified as an essential component of IFNα signaling by the use of somatic cell genetic studies and subsequent investigations established the role of other JAKs in cytokine signaling (Velazquez *et al.*, 1992).

The C-terminal kinase domain (JH1) of JAKs shares sequence homology with the catalytic domains of other protein tyrosine kinases within the defined conserved (Ihle, 1995; Ihle and Kerr, 1995; Ihle *et al.*, 1995; Krowlewski, 1995; Wilks, 1995; Hanks *et al.*, 1988; Leonard and O'Shea, 1998). Adjacent to the kinase domain JAKs have a pseudokinase (i.e. kinase-related) domain (JH2) that has a number of sequence motifs characteristic of a catalytic domain but is missing some conserved amino acids including the catalytic aspartic acid (Ihle *et al.*, 1995; Leonard and O'Shea, 1998). The precise function of the pseudokinase domain is not clear yet. The N-terminal half of the JAKs contains five regions (JH3–7) that share sequence homology among the Janus family members. In contrast, however, the extreme N-terminal region of each JAK protein is unique, and may confer the specificity in binding to the membrane-proximal regions of cytokine receptors (Ihle, 1995; Ihle *et al.*, 1995; Leonard and O'Shea, 1998).

TYK2-binding domain in IFNAR-1 protein has been mapped to a $\sim$33 amino acids membrane-proximal region that comprises the Box 1 and Box 2 motifs (Domanski and Colamonici, 1996; Yan *et al.*, 1996a). Almost all phylogenetically conserved residues in this region of IFNAR-1 are essential for TYK2 recognition (Domanski and Colamonici, 1996; Yan *et al.*, 1996a). In contrast to other cytokine receptors, the Box 1 motif, which is not well conserved in IFNAR-1, does not play a critical role in this protein–protein interaction (Domanski and Colamonici, 1996). The three amino acid residues Ile504, Leu505, and Glu506 in IFNAR-1 are essential for TYK2 binding (Colamonici *et al.*, 1994a, 1994b; Domanski and Colamonici, 1996; Yan *et al.*, 1996a).

In vitro binding studies suggest that the N-terminal $\sim$600 amino acids containing JH7 to JH3 comprise the major binding site of TYK2 to the IFNAR-1 protein (Yan *et al.*, 1998). Glutathione *S*-transferase-TYK2-JH3 or JH6 domain physically interacts *in vitro* with the IFNAR-1 protein, suggesting that JH3 and JH6 are the major sites of interaction of TYK2 with IFNAR-1 (Yan *et al.*, 1998). A truncated TYK2 protein containing amino acids 1–601 can function *in vivo* as a dominant negative mutant kinase inhibiting IFNα-dependent JAK/STAT signaling (Yan *et al.*, 1998). Further mutagenesis studies have revealed that JH7 and a part of the JH6 domain containing amino acids 22–221 are essential for TYK2 binding to IFNAR-1 (Richter *et al.*, 1998).

Tyr466 and, to some extent, Tyr481 on human IFNAR-1 are phosphorylated by TYK2 and phospho-Tyr466 serves as a docking site for STAT2 (Krishnan *et al.*, 1996, 1998; Yan *et al.*, 1996b; Li *et al.*, 1997; Richter *et al.*, 1998; Nadeau *et al.*, 1999). The SH2-containing protein tyrosine phosphatase SHP-2 preassociates with IFNAR-1 and is phosphorylated in response to IFNα/β (David *et al.*, 1995a). In transient transfection assays a dominant negative mutant SHP-2 inhibits IFNα/β-induced expression of an ISRE-driven reporter gene, suggesting that SHP-2 may function as a positive regulator of type IFN signaling (David *et al.*, 1995a). In contrast, SHP-1 negatively controls IFNα/β signaling. Bone marrow-derived macrophages from viable moth-eaten mice (expressing mutant SHP-1 with substantially reduced phosphatase activity) exhibit enhanced IFNα/β signaling compared with normal littermate control (David *et al.*, 1995a). SHP-1 physically associates with IFNAR-1 in a ligand-regulated fashion (David *et al.*, 1995a). Mutation of tyrosines to phenylalanines at positions 527 and 538 of IFNAR-1 enhances IFNα signaling, suggesting that IFNα/β signaling is negatively regulated through this region of IFNAR-1 (Gibbs *et al.*, 1996).

The cytoplasmic domains of IFNAR-2B and IFNAR-2C are divergent after the membrane-proximal 15 amino acids. Human IFNAR-2B contains two tyrosine residues (269 and 321) and no characteristic motifs of cytokine receptors and it does not interact with JAK1 (Domanski *et al.*, 1995; Lutfalla *et al.*, 1995; Domanski and Colamonici, 1996). Like IFNAR-2A, IFNAR-2B probably functions as decoy receptor for type I IFNs. In contrast, the cytoplasmic domain of human IFNAR-2C contains a Box 1 motif and seven tyrosine residues (269, 306, 316, 337, 411, and 512) and a number of acidic residues (Domanski *et al.*, 1995; Lutfalla *et al.*, 1995; Domanski and Colamonici, 1996). The JAK1-interaction domain of IFNAR-2 has been mapped to amino acids 300–346 (Domanski *et al.*, 1996). The Box 1 motif in IFNAR-2 plays a minor role in JAK1 binding (Domanski *et al.*, 1995; Lutfalla *et al.*, 1995; Domanski and Colamonici, 1996).

It has been demonstrated by indirect approaches using a JAK1–JAK2 chimeric protein that the JH3–JH7 of JAK1 is involved in the physical association with IFNAR-2 to elicit biological responses of type I IFNs (Kohlhuber *et al.*, 1997). Both STAT1 and STAT2 physically associate with the IFNAR-2C protein in unstimulated cells, but the biological significance of the interaction is not clear (Stancato *et al.*, 1996; Stark *et al.*, 1998). Murine cells expressing human IFNAR-2C truncated at amino acid 417 show a marked decrease in IFNβ-mediated antiviral response without any defect in JAK/STAT signaling by both IFNα and IFNβ (Platanias *et al.*, 1998). Protein tyrosine phosphatase activity associated with the distal region of IFNAR-2C has been implicated in the negative regulation of the growth-inhibitory action of type I IFNs (Platanias *et al.*, 1998).

The mitogen-activated protein kinase (MAPK) is activated by IFNα/β treatment of cells and ERK2 (p42-MAPK) associates with IFNAR-1 both *in vitro* and *in vivo* (David *et al.*, 1995b). Binding of type I IFNs to receptor causes tyrosine phosphorylation of insulin receptor substrate 1 and subsequent activation of the phosphatidylinositol 3-kinase pathway, which requires functional JAK1 and TYK2 proteins (Pfeffer *et al.*, 1997). However, the contribution of the PI-3 kinase pathway to the biological outcome of type I IFN action is not yet known (Wang *et al.*, 1997). STAT3 is activated by type I IFNs and may serve as an adapter for the recruitment of PI-3 kinase in Daudi cells (Burfoot *et al.*, 1997). Stimulation of cells with IFNα causes a JAK1-dependent phosphorylation of cytosolic phospholipase A_2 ($cPLA_2$) and pretreatment of cells with inhibitors of cPLA2 inhibits the activation of ISGF3 but not GAF (STAT1 homodimer) formation (Flati *et al.*, 1996). Moreover, JAK1 and cPLA2 can be co-immunoprecipitated from IFN-treated cell lysate (Flati *et al.*, 1996). Type I IFNs also activate p38 mitogen-activated kinase (p38 MAPK) and this is essential for $cPLA_2$-dependent ISGF3 formation (Goh *et al.*, 1999; Uddin *et al.*, 1999).

IFNAR-1, IFNAR-2C, JAK1, and TYK2 are known to form the functional type I IFN receptor complex (Darnell *et al.*, 1994; Uze *et al.*, 1995; Domanski and Colamonici, 1996; Haque and Williams, 1998; Stark *et al.*, 1998). However, differences have been observed between cell signaling by IFNα and IFNβ. Interestingly, U1A cells which are completely defective in response to recombinant IFNα1, IF-α2, or a mixture of natural IFNαs, retain a partial response to IFNβ (Pellegrini *et al.*, 1989). But none of the mutant cell lines lacking JAK1, STAT1, or STAT2 exhibits any IFNβ response, suggesting that TYK2 activity is not absolutely required for IFNβ signaling (Stark *et al.*, 1998). U1A cells express low levels of IFNAR-1 which is not stably maintained at the plasma membrane. Therefore, in U1A cells IFNβ likely signals through the IFNAR-2 dimer (Lewerenz *et al.*, 1998). Reconstitutions of U1A cells with mutant TYK2 proteins reveal that TYK2 N-terminal domain is required to maintain functional IFNAR-1, JH2 (pseudokinase) domain is necessary for high-affinity ligand binding and JH1(kinase) domain phosphorylates IFNAR-1 which is dispensable for JAK/STAT signaling by IFNα or IFNβ (Pellegrini *et al.*, 1989; Velazquez *et al.*, 1992; Richter *et al.*, 1998; Rani *et al.*, 1999). Recently it has been demonstrated that the IFNAR-1 and TYK2 expression is coordinately regulated in cells (Gauzzi *et al.*, 1997). This is consistent with the model that IFNβ interacts with the receptor in a manner that is different from that of IFNα (Pellegrini *et al.*, 1989; Velazquez *et al.*, 1992; Lewerenz *et al.*, 1998; Stark *et al.*, 1998; Rani *et al.*, 1999).

IFNα2 and IFNβ may require distinct cytoplasmic regions of IFNAR-2C to elicit an antiviral response. IFNAR-2C is phosphorylated at comparable levels in response to IFNα and IFNβ; however, the IFNAR1/IFNAR-2C complex can be immunoprecipitated only from the IFNβ-treated but not IFNα-treated cells (Croze *et al.*, 1996; Platanias *et al.*, 1996). Identification of the β-R1 gene which shows a selective induction by IFNβ compared with IFNα by Rani *et al.* has provided further evidence that IFNβ may use a distinct pathway for cell signaling (Rani *et al.*, 1996).

IFNβ shares only 35% identity with the IFNα subtypes, and appears to engage the α/β receptor differently. This can result in the activation of selective subsets of genes by IFNβ (Der *et al.*, 1998), even though it binds and activates the same receptor. Apparently distinctive structural differences are transmitted through the membrane to the cytoplasmic domains of the receptors that then mediate a differential response. Mutant cell lines lacking TYK2 which are completely defective in their response to IFNα2 still respond to IFNβ or α8, albeit with reduced activity. In fact, using these and other mutant cell lines reconstituted with different mutant proteins, three distinct modes of type I interactions with receptor subunits have been discerned: IFNα with IFNAR-1 and IFNAR-2, IFNβ with IFNAR-1, and IFNAR-2 and IFNβ with IFNAR-2 only (Lewerenz *et al.*, 1998). Thus IFNα and β signal differently through their receptors because they interact with the receptor in different ways. There is a tight correlation between receptor occupancy and the transcriptional response to IFN (Hannigan and Williams, 1986). The degree of receptor occupancy is a rate-limiting step in determining the transcriptional response to IFN which is transient and is accompanied by the downregulation of receptors on the cell surface. Downregulation of type I IFN receptors is also seen *in vivo* on peripheral blood lymphocytes following IFN therapy (Lau *et al.*, 1986). This is a temporary state and receptors reappear on the surface over 24–48 hours.

GENE

Accession numbers

Human IFNAR-1 cDNA: J03171
Human IFNAR-1 gene: X60459
Mouse IFNAR-1 cDNA: M89641
Bovine IFNAR-1 cDNA: L06320
Sheep IFNAR-1 cDNA: U65978
Human IFNAR-2.2 cDNA: L41942
Mouse IFNAR-2 cDNA: Y09813
Mouse IFNAR-2B cDNA: Y09864
Mouse IFNAR-2C cDNA: Y09865
Sheep IFNAR-2 cDNA: U65979

Chromosome location and linkages

The human IFNAR-1 gene comprises 11 exons covering a 32.9 kb region on chromosome 21q22.1 (Lutfalla *et al.*, 1990, 1992). The human IFNAR-2 gene is also located on human chromosome 21q22.1 (Lutfalla *et al.*, 1995). The murine IFNAR gene is located on chromosome 16.

The genes encoding two other members of the class II cytokine receptors, namely IFNGR-2 and CRF2–4, are also mapped to chromosome 21q22.1 within a ~300 kb span (Hertzog *et al.*, 1994; Rubinstein *et al.*, 1998). IFNGR-2 is the second chain of IFNγ receptor and recently the orphan receptor CRF2–4 has been shown to encode the second chain of IL-10 receptor (Spencer *et al.*, 1998; Reboul *et al.*, 1999).

PROTEIN

Accession numbers

Bovine IFNAR-2: Q95141
Human IFNAR-2: P48551
Sheep IFNAR-2: Q95207
Human IFNAR-1: P17181
Sheep IFNAR-1: Q28589, Q95206
Bovine IFNAR-1: Q04790
Mouse IFNAR-1: P33896

Description of protein

See Structure.

Both *N*- and *O*-linked glycosylation of the IFNAR-1 has been demonstrated (Novick *et al.*, 1992; Platanias *et al.*, 1993; Constantinescu *et al.*, 1995; Ling *et al.*, 1995).

Relevant homologies and species differences

The receptor subunits are species-specific (Uze *et al.*, 1990; Domanski and Colamonici, 1996; Stark *et al.*, 1998).

Affinity for ligand(s)

See Structure.

Cell types and tissues expressing the receptor

The IFNAR subunits are expressed in all tissues. Soluble IFNAR-1 and IFNAR-2 are present in human body fluids (Novick *et al.*, 1992, 1994).

Regulation of receptor expression

Recent study shows that TYK2 protein can regulate the expression and ligand-binding activity of IFNAR-1 protein (Gauzzi *et al.*, 1997). Binding studies with iodinated IFα2 and Scatchard plot analysis reveal that monocyte differentiation to macrophages results in a 3- to 4-fold increase in cell surface receptors with no change in their affinity (Eantuzzi *et al.*, 1997).

BIOLOGICAL CONSEQUENCES OF ACTIVATING OR INHIBITING RECEPTOR AND PATHOPHYSIOLOGY

Phenotypes of receptor knockouts and receptor overexpression mice

Mice lacking the IFNAR-1 were completely unresponsive to type I IFNs. These animals showed no overt abnormalities but were unable to cope with viral infection (Muller *et al.*, 1994; Steinhoff *et al.*, 1995; Van den Broek *et al.*, 1995a, 1995b). Despite compelling evidence for modulation of cell proliferation and differentiation by type I IFNs, there were no gross signs of abnormal fetal development or morphological changes in adult IFNAR-1-deficient mice (Hwang *et al.*, 1995). However, abnormalities of hematopoietic cells were detected in these animals.

References

Aguet, M., and Blanchard, B. (1981). High affinity binding of ^{125}I-labeled mouse interferon to a specific cell surface receptor. II. Analysis of binding properties. *Virology* **115**, 249–261.

Aguet, M., Grobke, M., and Dreiding, P. (1984). Various human interferon alpha subclasses cross-react with common receptors: the binding affinities correlate with their specific biological activities. *Virology* **132**, 211–216.

Bazan, J. F. (1990). Structural design and molecular evolution of a cytokine receptor superfamily. *Proc. Natl Acad. Sci. USA* **876**, 934–938.

Branca, A. A., and Baglioni, C. (1981). Evidence that type I and type II interferons have different receptors. *Nature* **294**, 768–770.

Burfoot, M. S., Rogers, N. C., Watling, D., Smith, J. M., Pons, S., Paonessaw, G., Pellegrini, S., White, M. F., and Kerr, I. M. (1997). Janus kinase-dependent activation of insulin receptor substrate 1 in response to interleukin-4, oncostatin M, and the interferons. *J. Biol. Chem.* **272**, 24183–24190.

Cohen, B., Novick, D., Barak, S., and Rubinstein, M. (1995). Ligand-induced association of the type-I interferon receptor components. *Mol. Cell Biol.* **15**, 4208–4214.

Colamonici, O. R., D'Alessandro, F., Diaz, M. O., Gregory, S. A., Neckers, L. M., and Nordan, R. (1990). Characterization of three monoclonal antibodies that recognize the IFNα2 receptor. *Proc. Natl Acad. Sci. USA* **87**, 7230–7234.

Colamonici, O. R., Pfeffer, L. M., D'Alessandro, F., Platanias, L. C., Gregory, S. A., Rosolen, A., Nordan, R., Cruciani, R. A., and Diaz, M. O. (1992). Multichain structure of the IFN-alpha receptor on hematopoietic cells. *J. Immunol.* **148**, 2126–2132.

Colamonici, O., Yan, H., Domanski, P., Handa, R., Smalley, D., Mullersman, J., Witte, M., Krishnan, K., and Krolewski, J. (1994a). Direct binding to and tyrosine phosphorylation of the alpha subunit of the type I interferon receptor by p135tyk2 tyrosine kinase. *Mol. Cell Biol.* **14**, 8133–8142.

Colamonici, O. R., Uyttendaele, H., Domanski, P., Yan, H., and Krolewski, J. J. (1994b). p135tyk2, an interferon-alpha-activated tyrosine kinase, is physically associated with an interferon-alpha receptor. *J. Biol. Chem.* **269**, 3518–3522.

Constantinescu, S. N., Croze, E., Murti, A., Wang, C., Basu, L., Hollander, D., Russell-Harde, D., Betts, M., Garcia-Martinez, V., Mullersman, J. E., and Pfeffer, L. M. (1995). Expression and signaling specificity of the IFNAR chain of the type I interferon receptor complex. *Proc. Natl Acad. Sci. USA* **92**, 10487–10491.

Croze, E., Russell-Harde, D., Wagner, T. C., Pu, H., Pfeffer, L. M., and Perez, H. D. (1996). The human type I interferon receptor. Identification of the interferon beta-specific receptor-associated phosphoprotein. *J. Biol. Chem.* **271**, 33165–33168.

Cutrone, E. C., and Langer, J. A. (1997). Contributions of cloned type I interferon receptor subunits to differential ligand binding. *FEBS Lett.* **404**, 197–202.

Darnell Jr., J.E., Kerr, I. M., and Stark, G. R. (1994). Jak-Stat pathways and transcriptional activation in response to IFNs and other extracellular proteins. *Science* **264**, 1415–1421.

David, M., Chen, H. E., Goelz, S., Larner, A. C., and Neel, B. G. (1995a). Differential regulation of the alpha/beta interferon-stimulated Jak/Stat pathway by the SH2 domain-containing tyrosine phosphatase SHPTP1. *Mol. Cell Biol.* **15**, 7050–7058.

David, M., Petricoin III, E., Benjamin, C., Pine, R., Weber, M. J., and Larner, A. C. (1995b). Requirement for MAP kinase (ERK2) activity in interferon alpha- and interferon beta-stimulated gene expression through STAT proteins. *Science* **269**, 1721–1723.

Der, S. D., Zhou, A., Williams, B. R., and Silverman, R. H. (1998). Identification of genes differentially regulated by interferon alpha, beta, or gamma using oligonucleotide arrays. *Proc. Natl Acad. Sci. USA* **95**, 15623–15628.

Domanski, P., and Colamonici, O. R. (1996). The type I interferon receptor. The long and short of it. *Cytokine Growth Factor Rev.* **7**, 143–151.

Domanski, P., Witte, M., Kellum, M., Rubinstein, M., Hackett, R., Pitha, P., and Colamonici, O. R. (1995). Cloning and expression of a long form of the beta subunit of the interferon alpha beta receptor that is required for signaling. *J. Biol. Chem.* **270**, 21606–21611.

Eantuzzi, L., Eid, P., Malorni, W., Rainaldi, G., Gauzzi, M. C., Pellegrini, S., Belardelli, F., and Gessani, S. (1997). Post-translational up-regulation of the cell surface-associated alpha component of the human type I interferon receptor during differentiation of peripheral blood monocytes: role in the biological response to type I interferon. *Eur. J. Immunol.* **27**, 1075–1081.

Eid, P., and Tovey, M. G. (1995). Characterization of a domain of a human type I interferon receptor protein involved in ligand binding. *J. Interferon Cytokine Res.* **15**, 205–211.

Faltynek, C. R., Branca, A. A., McCandless, S., and Baglioni, C. (1993). Characterization of an interferon receptor on human lymphoblastoid cells. *Proc. Natl Acad. Sci. USA* **80**, 3269–3273.

Flati, V., Haque, S. J., and Williams, B. R. (1996). Interferon-alpha-induced phosphorylation and activation of cytosolic phospholipase A_2 is required for the formation of interferon-stimulated gene factor three. *EMBO J.* **15**, 1566–1571.

Gaboriaud, C., Uze, G., Lutfalla, G., and Mogensen, K. (1990). Hydrophobic cluster analysis reveals duplication in the external structure of human alpha-interferon receptor and homology with gamma-interferon receptor external domain. *FEBS Lett.* **269**, 1–3.

Gauzzi, M. C., Barbieri, G., Richter, M. F., Uze, G., Ling, L., Fellous, M., and Pellegrini, S. (1997). The amino-terminal region of TYK2 sustains the level of interferon alpha receptor 1, a component of the interferon. *Proc. Natl Acad. Sci. USA* **94**, 11839–11844.

Gibbs, V. C., Takahashi, M., Aguet, M., and Chuntharapai, A. (1996). A negative regulatory region in the intracellular domain of the human interferon-alpha receptor. *J. Biol. Chem.* **271**, 28710–28716.

Goh, K. C., Haque, S. J., and Williams, B. R. (1999). p38 MAP kinase is required for STAT1 serine phosphorylation and transcriptional activation induced by interferons. *EMBO J.* **18**, 5601–5608.

Goldman, L. A., Cutrone, E. C., Dang, A., Hao, X., Lim, J., and Langer, J. A. (1998). Mapping human interferon-alpha (IFNα2) binding determinants of the type I interferon receptor subunit IFNAR-1 with human/bovine IFNAR-1 chimeras. *Biochemistry* **37**, 13003–13010.

Hanks, S. K., Quinn, A. M., and Hunter, T. (1988). The protein kinase family: conserved features and deduced phylogeny of the catalytic domains. *Science* **241**, 42–52.

Hannigan, G., and Williams, B. R. G. (1986). Transcriptional regulation of interferon-responsive genes is closely linked to interferon receptor occupancy. *EMBO J.* **5**, 1607–1613.

Hannigan, G. E., Gewert, D. R., Fish, E. N., Read, S. E., and Williams, B. R. G. (1983). Differential binding of human interferon-alpha subtypes to receptors on lymphoblastoid cells. *Biochem. Biophys. Res. Commun.* **110**, 537–544.

Hannigan, G. E., Gewert, D. R., and Williams, B. R. G. (1984). Characterization of interferon receptor binding to interferon sensitive and resistant human lymphoblastoid cells. *J. Biol. Chem.* **259**, 9456–9460.

Hannigan, G. E., Lau, A. S., and Williams, B. R. G. (1986). Differential human interferon alpha receptor expression on proliferating and non-proliferating cells. *Eur. J. Biochem.* **157**, 187–193.

Haque, S. J., and Williams, B. R. G. (1998). Signal transduction in the interferon system. *Semin. Oncol.* **25**, (Suppl. 1), 14–22.

Heldin, C. H. (1995). Dimerization of cell surface receptors in signal transduction. *Cell* **80**, 213–223.

Hertzog, P. J., Hwang, S. Y., Holland, K. A., Tymms, M. J., Iannello, R., and Kola, I. (1994). A gene on human chromosome 21 located in the region 21q22.2 to 21q22.3 encodes a factor necessary for signal transduction and antiviral response to type I interferons. *J. Biol. Chem.* **269**, 14088–14093.

Hwang, S. Y., Hertzog, P. J., Holland, K. A., Sumarsono, S. H., Tymms, M. J., Hamilton, J. A., Whitty, G., Bertoncello, I., and Kola, I. (1995). A null mutation in the gene encoding a type I interferon receptor component eliminates antiproliferative and antiviral responses to interferons alpha and beta and alters macrophage responses. *Proc. Natl Acad. Sci. USA* **92**, 11284–11288.

Ihle, J. N. (1995). Cytokine receptor signalling. *Nature* **377**, 591–594.

Ihle, J. N. (1996). STATs: signal transducers and activators of transcription. *Cell* **84**, 331–334.

Ihle, J. N., and Kerr, I. M. (1995). Jaks and Stats in signaling by the cytokine receptor superfamily. *Trends Genet.* **11**, 69–74.

Ihle, J. N., Witthuhn, B. A., Quelle, F. W., Yamamoto, K., and Silvennoinen, O. (1995). Signaling through the hematopoietic cytokine receptors. *Annu. Rev. Immunol.* **13**, 369–398.

Isaacs, A., and Lindenmann, J. (1957). Virus interference II. Some properties of interferon. *Proc. R. Soc. Lond. B Biol. Sci.* **147**, 268–273.

Joshi, A. R., Sarkar, F. H., and Gupta, S. L. (1982). Interferon receptors. Cross-linking to human leukocytes interferon alpha-2 to its receptor on human cell. *J. Biol. Chem.* **257**, 13884–13887.

Karpusas, M., Nolte, M., Benton, C. B., Meier, W., Lipscomb, W. N., and Goelz, S. (1997). The crystal structure of human interferon beta at 2.2-A resolution. *Proc. Natl Acad. Sci. USA* **94**, 11813–11818.

Kishimoto, T., Taga, T., and Akira, S. (1994). Cytokine signal transduction. *Cell* **76**, 253–262.

Kohlhuber, F., Rogers, N. C., Watling, D., Feng, J., Guschin, D., Briscoe, J., Witthuhn, B. A., Kotenko, S. V., Pestka, S., Stark, G. R., Ihle, J. N., and Kerr, I. M. (1997). A JAK1/JAK2 chimera can sustain alpha and gamma interferon responses. *Mol. Cell Biol.* **17**, 695–706.

Krishnan, K., Yan, H., Lim, J. T., and Krolewski, J. J. (1996). Dimerization of a chimeric CD4-interferon-alpha receptor reconstitutes the signaling events preceding STAT phosphorylation. *Oncogene* **13**, 125–133.

Krishnan, K., Singh, B., and Krolewski, J. J. (1998). Identification of amino acid residues critical for the Src-homology 2 domain-dependent docking of Stat2 to the interferon alpha receptor. *J. Biol. Chem.* **273**, 19495–19501.

Krowlewski, J. J. (1995). In "The Protein Kinase Facts Book, Protein-Tyrosine Kinases" (ed G. Hardie and S. Hanks), Interferon-α regulated PTK (vertebrate)(p135tyk2) pp. 114–116. Academic Press, London.

Lau, A. S., Hannigan, G. E., Freedman, M. H., and Williams, B. R. G. (1986). Regulation of interferon receptor expression in human blood lymphocytes *in vitro* and during interferon therapy. *J. Clin. Invest.* **77**, 1632–1638.

Leonard, W. J., and O'Shea, J. J. (1998). Jaks and STATs: biological implications. *Annu. Rev. Immunol.* **16**, 293–322.

Levy, D. E. (1995). Interferon induction of gene expression through the Jak-Stat pathway. *Semin. Virol.* **6**, 181–189.

Lewerenz, M., Mogensen, K. E., and Uze, G. (1998). Shared receptor components but distinct complexes for alpha and beta interferons. *J. Mol. Biol.* **282**, 585–599.

Li, X., Leung, S., Kerr, I. M., and Stark, G. R. (1997). Functional subdomains of STAT2 required for preassociation with the alpha interferon receptor and for signaling. *Mol. Cell Biol.* **17**, 2048–2056.

Ling, L. E., Zafari, M., Reardon, D., Brickelmeier, M., Goelz, S. E., and Benjamin, C. D. (1995). Human type I interferon receptor, IFNAR, is a heavily glycosylated 120–130 kD membrane protein. *Interferon Cytokine Res.* **15**, 55–61.

Lu, J., Chuntharapai, A., Beck, J., Bass, S., Ow, A., De Vos, A. M., Gibbs, V., and Kim, K. J. (1998). Structure–function study of the extracellular domain of the human IFNalpha receptor (hIFNAR1) using blocking monoclonal antibodies: the role of domains 1 and 2. *J. Immunol.* **160**, 1782–1788.

Lundgren, E., and Langer, J. A. (1997). Nomenclature of interferon receptors and interferon-delta. *Interferon Cytokine Res.* **17**, 431–432.

Lutfalla, G., Roeckel, N., Mogensen, K. E., Mattel, M.-G., and Uze, G. (1990). Assignment of the human interferon alpha gene to chromosome 21q22.1 by *in situ* hybridization. *J. IFN Res.* **10**, 515–517.

Lutfalla, G., Gardiner, K., Proudhon, D., Vielh, E., and Uze, G. (1992). The structure of the human interferon alpha/beta receptor gene. *J. Biol. Chem.* **267**, 2802–2809.

Lutfalla, G., Holland, S. J., Cinato, E., Monneron, D., Reboul, J., Rogers, N. C., Smith, J. M., Stark, G. R., Gardiner, K., Mogensen, K. E., Kerr, I. M., and Uze, G. (1995). Mutant U5A cells are complemented by an interferon-alpha beta receptor subunit generated by alternative processing of a new member of a cytokine receptor gene cluster. *EMBO J.* **14**, 5100–5108.

Merlin, G., Falcoff, E., and Aguet, M. (1985). ^{125}-labeled human interferon alpha, beta and gamma: Comparative receptor binding data. *J. Gen. Virol.* **66**, 1149–1152.

Mitsui, Y., and Senda, T. (1997). Elucidation of the basic three-dimensional structure of type I interferons and its functional and evolutionary implications. *J. Interferon Cytokine Res.* **17**, 319–326.

Müller, M., Briscoe, J., Laxton, C., Guschin, D., Ziemiecki, A., Silvennoinen, O., Harpur, A. G., Barbieri, G., Witthuhn, B. A., Schindler, C., Pellegrini, S., Wilks, A. F., Ihle, J. N., Stark, G. R., and Kerr, I. M. (1993). The protein tyrosine kinase JAK1 complements defects in interferon-alpha/beta and -gamma signal transduction. *Nature* **366**, 129–135.

Muller, U., Steinhoff, U., Reis, L. F., Hemmi, S., Pavlovic, J., Zinkernagel, R. M., and Aguet, M. (1994). Functional role of type I and type II interferons in antiviral defense. *Science* **264**, 1918–1921.

Nadeau, O. W., Domanski, P., Usacheva, A., Uddin, S., Platanias, L. C., Pitha, P., Raz, R., Levy, D., Majchrzak, B., Fish, E., and Colamonici, O. R. (1999). The proximal tyrosines of the cytoplasmic domain of the beta chain of the type I interferon receptor are essential for signal transducer and activator of transcription (Stat) 2 activation. Evidence that two Stat2 sites are required to reach a threshold of interferon alpha-induced Stat2 tyrosine phosphorylation that allows normal formation of interferon-stimulated gene factor 3. *J. Biol. Chem.* **274**, 4045–4052.

Novick, D., Cohen, B., and Rubinstein, M. (1992). Soluble interferon-alpha receptor molecules are present in body fluids. *FEBS Lett.* **314**, 445–448.

Novick, D., Cohen, B., and Rubinstein, M. (1994). The human interferon α/β receptor: characterization and cloning. *Cell* **77**, 391–400.

Pellegrini, S., John, J., Shearer, M., Kerr, I. M., and Stark, G. R. (1989). Use of a selectable marker regulated by alpha interferon to obtain mutations in the signaling pathway. *Mol. Cell. Biol.* **9**, 4605–4612.

Pestka, S., Langer, J. A., Zoon, K. C., and Samuel, C. (1987). Interferons and their actions. *Annu. Rev. Biochem.* **56**, 727–777.

Pfeffer, L. M., Mullersman, J. E., Pfeffer, S. R., Murti, A., Shi, W., and Yang, C. H. (1997). STAT3 as an adapter to couple phosphatidylinositol 3-kinase to the IFNAR1 chain of the type I interferon receptor. *Science* **276**, 1418–1420.

Platanias, L. C., Pfeffer, L. M., Cruciani, R., and Colamonici, O. R. (1993). Characterization of the alpha subunit of the IFNalpha receptor. Evidence of *N*- and *O*-linked glycosylation and association with other surface proteins. *J. Immunol.* **150**, 3382–3388.

Platanias, L. C., Uddin, S., Domanski, P., and Colamonici, O. R. (1996). Differences in interferon alpha and beta signaling. Interferon beta selectively induces the interaction of the alpha and beta subunits of the type I interferon receptor. *J. Biol. Chem.* **271**, 23630–23633.

Platanias, L. C., Domanski, P., Nadeau, O. W., Yi, T., Uddin, S., Fish, E., Neel, B. G., and Colamonici, O. R. (1998). Identification of a domain in the beta subunit of the type I interferon (IFN) receptor that exhibits a negative regulatory effect in the growth inhibitory action of type I IFNs. *J. Biol. Chem.* **273**, 5577–5581.

Rani, M. R. S., Foster, G. R., Leung, S., Leaman, D., Stark, G. R., and Ransohoff, R. M. (1996). Characterization of beta-R1, a gene that is selectively induced by interferon beta (IFNbeta) compared with IFNalpha. *J. Biol. Chem.* **271**, 22878–22884.

Rani, M. R., Gauzzi, C., Pellegrini, S., Fish, E. N., Wei, T., and Ransohoff, R. M. (1999). Induction of beta-R1/I-TAC by interferon-beta requires catalytically active TYK2. *J. Biol. Chem.* **274**, 1891–1897.

Raziuddin, A., and Gupta, S. L. (1985). In "The 2-5A System: Molecular and Clinical Aspects of Interferon-regulated Pathway" (ed B. R. G. Williams and R. H. Silvermann), Interferon receptors pp. 219–226. New York.

Reboul, J., Gardiner, K., Monneron, D., Uze, G., and Lutfalla, G. (1999). Comparative genomic analysis of the interferon/interleukin-10 receptor gene cluster. *Genome Res.* **9**, 242–250.

Rehberg, E., Kelder, B., Hoal, E. G., and Pestka, S. (1982). Specific molecular activities of recombinant and hybrid leukocyte interferons. *J. Biol. Chem.* **257**, 11497–11502.

Richter, M. F., Dumenil, G., Uze, G., Fellous, M., and Pellegrini, S. (1998). Specific contribution of Tyk2 JH regions to the binding and the expression of the interferon alpha/beta receptor component IFNAR1. *J. Biol. Chem.* **273**, 24723–24729.

Rubinstein, M., Dinarello, C. A., Oppenheim, J. J., and Hertzog, P. (1998). Recent advances in cytokines, cytokine receptors and signal transduction. *Cytokine Growth Factor Rev.* **9**, 175–181.

Runkel, L., Pfeffer, L., Lewerenz, M., Monneron, D., Yang, C. H., Murti, A., Pellegrini, S., Goelz, S., Uze, G., and Mogensen, K. (1998). Differences in activity between alpha and beta type I interferons explored by mutational analysis. *J. Biol. Chem.* **273**, 8003–8008.

Sen, G. C., and Lengyel, P. (1992). The interferon system: a bird's eye view of biochemistry. *J. Biol. Chem.* **267**, 5017–5020.

Spencer, S. D., Di Marco, F., Hooley, J., Pitts-Meek, S., Bauer, M., Ryan, A. M., Sordat, B., Gibbs, V. C., and Aguet, M. (1998). The orphan receptor CRF2–4 is an essential subunit of the interleukin 10 receptor. *J. Exp. Med.* **187**, 571–578.

Stancato, L. F., David, M., Carter-Su, C., Larner, A. C., and Pratt, W. B. (1996). Preassociation of STAT1 with STAT2 and STAT3 in separate signaling complexes prior to cytokine stimulation. *J. Biol. Chem.* **271**, 4134–4137.

Stark, G. R., Kerr, I. M., Williams, B. R. G., Silverman, R. H., and Schreiber, R. D. (1998). How cells respond to interferons. *Annu. Rev. Biochem.* **67**, 227–264.

Steinhoff, U., Muller, U., Schertler, A., Hengartner, H., Aguet, M., and Zinkernagel, R. M. (1995). Antiviral protection by vesicular stomatitis virus-specific antibodies in alpha/beta interferon receptor-deficient mice. *J. Virol.* **69**, 2153–2158.

Stewart II, W.E. (1979). In "The Interferon System" Springer-Verlag, Vienna.

Thoreau, E., Petridou, B., Kelly, P. A., Djiane, J., and Mornon, J. P. (1991). Structural symmetry of the extracellular domain of the cytokine/growth hormone/prolactin receptor family and interferon receptors revealed by hydrophobic cluster analysis. *FEBS Lett.* **282**, 26–31.

Uddin, S., Majchrzak, B., Woodson, J., Arunkumar, P., Alsayed, Y., Pine, R., Young, P. R., Fish, E. N., and Platanias, L. C. (1999). Activation of the p38 mitogen-activated protein kinase by type I interferons. *J. Biol. Chem.* **274**, 30127–30131.

Uze, G., Lutfalla, G., and Mogensen, K. E. (1995). α and β Interferons and their receptor and their friends and relations. *J. Interferon Cytokine Res.* **15**, 3–26.

Uze, G., Lutfalla, G., and Gresser, I. (1990). Genetic transfer of a functional human interferon α receptor into mouse cells: cloning and expression of its cDNA. *Cell* **60**, 225–234.

Van den Broek, M. F., Muller, U., Huang, S., Aguet, M., and Zinkernagel, R. M. (1995a). Antiviral defense in mice lacking both alpha/beta and gamma interferon receptors. *J. Virol.* **69**, 4792–4796.

Van den Broek, M. F., Muller, U., Huang, S., Zinkernagel, R. M., and Aguet, M. (1995b). Immune defence in mice lacking type I and/or type II interferon receptors. *Immunol. Rev.* **148**, 5–18.

Vanden Broecke, C., and Pfeffer, L. M. (1988). Characterization of interferon-alpha binding sites on human cell lines. *J. IFN Res.* **8**, 803–811.

Velazquez, L., Fellous, M., Stark, G. R., and Pellegrini, S. (1992). A protein tyrosine kinase in the interferon alpha/beta signaling pathway. *Cell* **70**, 313–322.

Wang, H. Y., Zamorano, J., Yoerkie, J. L., Paul, W. E., and Keegan, A. D. (1997). The IL-4-induced tyrosine phosphorylation of the insulin receptor substrate is dependent on JAK1 expression in human fibrosarcoma cells. *J. Immunol.* **158**, 1037–1040.

Watling, D., Guschin, D., Muller, M., Silvennoinen, O., Witthuhn, B. A., Quelle, F. W., Rogers, N. C., Schindler, C., Stark, G. R., Ihle, J. N., and Kerr, I. M. (1993). Complementation by the protein tyrosine kinase JAK2 of a mutant cell line defective in the interferon-gamma signal transduction pathway. *Nature* **366**, 166–170.

Weissmann, C., and Weber, H. (1986). The interferon genes. *Proc. Nucl. Acids Res. Mol. Biol.* **33**, 251–300.

Wilks, A. (1995). In "The Protein Kinase Facts Book, Protein-Tyrosine Kinases" (ed G. Hardie and S. Hanks), Janus kinases 1 and 2 (vertebrates) (just another kinase 1 and 2) pp. 117–119. Academic Press, London.

Yan, H., Krishnan, K., Greenlund, A. C., Gupta, S., Lim, J. T., Schreiber, R. D., Schindler, C. W., and Krolewski, J. J. (1996a). Phosphorylated interferon-alpha receptor 1 subunit (IFNαR1) acts as a docking site for the latent form of the 113 kDa STAT2 protein. *EMBO J.* **15**, 1064–1074.

Yan, H., Krishnan, K., Lim, J. T., Contillo, L. G., and Krolewski, J. J. (1996b). Molecular characterization of an alpha interferon receptor 1 subunit (IFNαR1) domain required for TYK2 binding and signal transduction. *Mol. Cell Biol.* **16**, 2074–2082.

Yan, H., Piazza, F., Krishnan, K., Pine, R., and Krolewski, J. J. (1998). Definition of the interferon-alpha receptor-binding domain on the TYK2 kinase. *J. Biol. Chem.* **273**, 4046–4051.

Zhang, Z. Q., Fournier, A., and Tan, Y. H. (1986). The isolation of human beta interferon receptor by wheat germ lectin affinity and immunosorbent column chromatographies. *J. Biol. Chem.* **261**, 8017–8022.

Zoon, K., Zur Nedden, D., and Arnheiter, H. (1982). Specific binding of human alpha interferon to a high affinity cell surface binding site on bovine kidney cells. *J. Biol. Chem.* **257**, 4695–4697.

Poxvirus IFNα/β Receptor Homologs

Grant McFadden[1,*] and Richard Moyer[2]

[1]The John P. Robarts Research Institute and Department of Microbiology and Immunology, The University of Western Ontario, 1400 Western Road, London, Ontario, N6G 2V4, Canada

[2]Department of Molecular Genetics and Microbiology, University of Florida College of Medicine, PO Box 100266, Gainesville, FL 32610-0266, USA

*corresponding author tel: (519)663-3184, fax: (519)663-3847, e-mail: mcfadden@rri.on.ca

DOI: 10.1006/rwcy.2000.14015.

SUMMARY

The discovery of extracellular poxvirus inhibitors of type I interferons in 1995 was made by direct inhibition studies, rather than by sequence homology analysis. In fact, the *vaccinia virus* (strain Western Reserve) prototype of this family, B18R, is more closely related to members of the Ig superfamily than to the cellular type I interferon receptors, at least in terms of overall similarity scores. Nevertheless, B18R binds and inhibits cellular type I interferons with a relatively broad species-specificity. The protein is expressed both as a secreted and cell surface glycoprotein and plays an important role in the pathogenesis of virus infection. In fact, the safest vaccinia strains used historically for *smallpox* vaccinations either did not express this gene, or the variant protein expressed by the strain was relatively inefficient in terms of type I interferon inhibition.

BACKGROUND

Discovery

Poxvirus control of interferons (IFNs), like that of IL-1β, is both intracellular and extracellular. In the orthopoxviruses, intracellular control of IFNs is mediated by genes E3L and K3L (vaccinia). The E3L gene, an RNA-binding protein, functions to prevent PKR activation by double-stranded RNA (Chang *et al.*, 1992) whereas the K3L gene acts to prevent phosphorylation of eIF2α by PKR (Beattie *et al.*, 1991; Davies *et al.*, 1992).

The action of extracellular type I interferons (IFNα/β) is inhibited by the secreted IFNα/β receptor homolog encoded by the B18R ORF (vaccinia). The B18R ORF was initially identified by routine sequencing (Smith and Chan, 1991) and was believed to be involved in the regulation of IL-1β. The role of this ORF in IFNα/β control was unanticipated because, unlike the cellular type IFN receptors which belong to the type II cytokine receptor superfamily and contain type 3 fibronectin domain repeats (Farrar and Schreiber, 1993), the B18R protein is a member of the Ig superfamily (Smith and Chan, 1991). An independent screening of proteins secreted from *vaccinia virus*-infected cells led to the identification of an approximately 65 kDa protein which bound to human type I interferon (Symons *et al.*, 1995). The protein was rapidly identified as the B18R gene product. The protein is not expressed in all orthopoxviruses, and is notably absent from both the Lister and Ankara strains of vaccinia. The Wyeth strain contains the gene but expresses a protein which exhibits atypically low avidity for type I interferons (Symons *et al.*, 1995). The gene is also present in *ectromelia virus* (Colamonici *et al.*, 1995). The protein is also known as the orthopoxvirus S antigen (Morikawa and Ueda, 1993).

Alternative names

B18R ORF: (vaccinia WR, used as reference in this article) orthopoxvirus S antigen.

Structure

The ORF, typical of poxvirus genes, is contiguous and contains no introns.

Main activities and pathophysiological roles

The soluble B18R protein binds type I IFNs with high avidity to prevent IFN engagement with the appropriate cellular receptors (Symons *et al.*, 1995), and thereby serves to block signal transduction and subsequent phosphorylation of JAK kinases (Colamonici *et al.*, 1995). The protein binds many human IFNs (IFNα1, IFNα2, IFNα7, IFNα8, IFNβ). Binding affinities are high, with K_d values of $\sim$1–4 nM (Colamonici *et al.*, 1995; Symons *et al.*, 1995). The B18R protein also binds a novel type of IFNα (Vancova *et al.*, 1998) but not IFNγ (Colamonici *et al.*, 1995). Type I IFN binding exhibits a broad specificity that includes bovine, rabbit, and rat. Binding affinities to mouse type I IFNs are considerably lower.

It is important to note that the interaction of this protein with type I IFNs is probably different from that for cellular type I IFN receptors. Binding of type I IFNs to cellular receptors is blocked by monoclonal antibodies to the N-terminal but not the C-terminal portion of the molecule. However, antibodies against both the C- and N-terminal portion of the B18R protein inhibit IFN binding (Liptáková *et al.*, 1997).

GENE

Accession numbers

Nucleic acid accession numbers:
Vaccinia virus WR: X56122
Vaccinia virus Copenhagen: D01019
Cowpox virus: Y15035
Variola virus Garcia: X72086
Variola virus Somalia: U18341
Variola virus Bangladesh: L22579

Sequence

In vaccinia virus, the gene is located approximately 176 kb from the left end of the $\sim$190 kb genome within the *Hin*dIII B fragment, the rightmost terminal *Hin*dIII fragment of the viral genome. Transcription is in the rightward direction. The protein is expressed early, prior to viral DNA synthesis.

PROTEIN

Accession numbers

Vaccinia virus WR: 62238, 514205
Vaccinia virus Copenhagen: 222699
Cowpox virus: 3097038
Variola virus Garcia: 1150677
Variola virus Somalia: 439095
Variola virus India: 516438, 457079

Sequence

See **Figure 1**.

Description of protein

The complete vaccinia virus open reading frame predicts a product of 351 amino acids (see Figure 1). The mature protein is formed following cleavage of the N-terminal secretory signal sequence. Although the protein exhibits significant homology to the IgSF repeats as indicated, these repeats themselves have homology to the fibronectin repeats typical of cellular type I IFN receptors. Thus, the B18R ORF exhibits significant homology to the mouse, human, and bovine α subunits of the type I IFN receptor (Colamonici *et al.*, 1995), particularly within the binding domains. Although classified as a secreted protein, a certain amount of the active protein is also found bound to the cell surface (Colamonici *et al.*, 1995).

Relevant homologies and species differences

The *vaccinia virus* protein is 93% identical to the corresponding ORF in *cowpox virus*, and 88–89% identical to the *variola virus* ORFs. The protein binds to type I IFNs in solution, which serves to prevent IFN binding to appropriate cellular receptors (Symons *et al.*, 1995), subsequent signal transduction and phosphorylation of the JAK kinases (Colamonici *et al.*, 1995). The protein can rebind to cells and under certain conditions be expressed at the surface of

Figure 1 The sequence of the B18R ORF precursor protein from vaccinia virus WR. A predicted cleavage site is present between amino acids 19 and 20 of the N-terminal secretory signal to generate the mature, secreted protein. The predicted molecular weight ($\sim$41 kDa), differs from the observed molecular weight of $\sim$65 kDa because of glycosylation. Putative glycosylation sites are located at amino acids 117–119, 261–263, 269–271, 321–324. Shown in boldface are the three IgSF repeat domains, which comprise most of the mature protein (Smith and Chan, 1991).

1 MTMKMMVHIY FVSLLLLLFH SYAIDIENEI TEFFNKMRDT LPAKDSKWLN **PACMFGGTMN**

61 DIAALGEPFS AKCPPIEDSL LSHRYKDYVV KWERLEKNRR RQVSNKRVKH GDLWIANYTS

121 KFSNRRYLCT VTTKNGDCVQ GIVRSHIRKP **PSCIPKTYEL GTHDKYGIDL YCGILYAKHY**

181 NNITWYKDNK EINIDDIKYS QTGKELIIHN PELEDSGRYD CYVHYDDVRI KNDIVVSRCK

241 ILTVIPSQDH RFKLI**LDPKI NVTIGEPANI TCTAVSTSLL IDDVLIEWEN PSGWLIGFDF**

301 DVYSVLTSRG GITEATLYFE NVTEEYIGNT YKCRGHNYYF EKTLTTTVVL E

infected cells via proposed type II membrane topology (Morikawa and Ueda, 1993). Whether free in solution, or bound to the surface of cells, the protein is believed to afford protection against extracellular type I IFNs.

BIOLOGICAL CONSEQUENCES OF ACTIVATING OR INHIBITING RECEPTOR AND PATHOPHYSIOLOGY

Both type I and type II interferons are known to be important in controlling poxvirus infections. Indeed as pointed out in the Discovery section, IFN is controlled by poxviruses in both an intracellular as well as extracellular fashion. Treatment of *ectromelia virus*-infected mice with neutralizing antibodies to either type I or type II IFN prevents normal clearance of the virus in infected mice (Karupiah *et al.*, 1993).

The role of the B18R ORF in virulence was tested by deleting the gene from *vaccinia virus* and performing intranasal infections of mice. An attenuation was noted and signs of clinical illness were reduced in animals infected with the mutant virus. The lethal dose of virus was some 100-fold less than for wild-type virus. Growth and viremia was measured by the assay for virus in lungs and brains respectively of infected animals. At selected doses, titres of virus in the lungs were drastically reduced in mice (undetectable in some animals) lacking the B18R gene and absent from the brains. In contrast, for animals infected with wild-type virus, virus was readily detected in both lungs and brain (Symons *et al.*, 1995). It is noteworthy that these effects are observed despite the fact that affinity of murine type I IFNs for this protein is poor compared with the binding of human, rabbit, and rat IFNs. These results clearly implicate a role of the type I IFNs in controlling orthopoxvirus infections.

References

Beattie, E., Tartaglia, J., and Paoletti, E. (1991). Vaccinia virus-encoded eIF-2α homolog abrogates the antiviral effect of interferon. *Virology* **183**, 419–422.

Chang, H.-W., Watson, J. C., and Jacobs, B. L. (1992). The E3L gene of vaccinia virus encodes an inhibitor of the interferon-induced, double-stranded RNA-dependent protein kinase. *Proc. Natl Acad. Sci. USA* **89**, 4825–4829.

Colamonici, O. R., Domanski, P., Sweitzer, S. M., Larner, A., and Buller, R. M. L. (1995). Vaccinia virus B18R gene encodes a type I interferon-binding protein that blocks interferon α transmembrane signaling. *J. Biol. Chem.* **270**, 15974–15978.

Davies, M. V., Eroy-Stein, O., Jagus, R., Moss, B., and Kaufman, R. J. (1992). The vaccinia virus K3L gene product potentiates translation by inhibiting double-stranded-RNA-activated protein kinase and phosphorylation of the alpha subunit of eukaryotic initiation factor 2. *J. Virol.* **66**, 1943–1950.

Farrar, M. A., and Schreiber, R. D. (1993). The molecular cell biology of interferon-γ and its receptor. *Annu. Rev. Immunol.* **11**, 571–611.

Karupiah, G., Fredrickson, T. N., Holmes, K. L., Khairallah, L. H., and Buller, R. M. L. (1993). Importance of inteferons in recovery from mousepox. *J. Virol.* **67**, 4214–4226.

Liptáková, H., Kontsekova, E., Alcami, A., Smith, G. L., and Kontsek, P. (1997). Analysis of an interaction between the soluble vaccinia virus-coded Type I interferon (IFN)-receptor and human IFN-α1 and IFN-α2. *Virology* **232**, 86–90.

Morikawa, S., and Ueda, Y. (1993). Characterization of vaccinia surface antigen expressed by recombinant baculovirus. *Virology* **193**, 753–761.

Smith, G., and Chan, Y. S. (1991). Two vaccinia virus proteins structurally related to the interleukin-1 receptor and the immunoglobulin superfamily. *J. Gen. Virol.* **72**, 511–518.

Symons, J. A., Alcamí, A., and Smith, G. L. (1995). Vaccinia virus encodes a soluble type I interferon receptor of novel structure and broad species specificity. *Cell* **81**, 551–560.

Vancova, I., La Bonnadiere, C., and Kontsek, P. (1998). Vaccinia virus protein B18R inhibits the activity and cellular binding of the novel type interferon-delta. *J. Gen. Virol.* **79**, 1647–1649.

Osteopontin Receptor

Gerard J. Nau*

Infectious Disease Unit, Massachusetts General Hospital, Boston, MA 02114, USA

*corresponding author tel: 617-726-3812, fax: 617-726-7416, e-mail: nau@wi.mit.edu

DOI: 10.1006/rwcy.2000.18003.

SUMMARY

The two classes of osteopontin (OPN) receptors identified so far, CD44 and integrins, each have an extensive literature independent of OPN. Multiple activities have been attributed to these receptors, from development to leukocyte and lymphocyte homing and activation. Further investigation is required first to characterize better what biological activities OPN possesses, and then to determine which receptors are responsible for these activities. There are promising therapeutic agents under development for inhibiting integrin function. Some of these biological effects may be the result of inhibiting OPN activity. The opportunities for modulation of CD44 activity and for specific OPN antagonists have yet to be pursued.

BACKGROUND

Discovery

Several receptors for OPN have been identified. Broadly, these include CD44 and various integrins (Uede *et al.*, 1997). It would be impractical to summarize all of the data related to these receptor classes. For the purposes of this chapter, general references and specifics regarding the interaction of these receptors and OPN will be reviewed.

CD44 was identified on the surface of human leukocytes, T cells, and thymocytes by a monoclonal antibody raised against human lymph node lymphocytes (Dalchau *et al.*, 1980). The first indication that CD44 was relevant to OPN physiology came with the discovery of an interaction between these two molecules on cells transfected with CD44 (Weber *et al.*, 1996).

Integrins have been studied for nearly two decades. Early work identified major cell surface antigens on several cell types, such as platelets, fibroblasts, and leukocytes, and the interaction of these receptors with extracellular matrix proteins (Phillips *et al.*, 1980; Jennings and Phillips, 1982; Horwitz *et al.*, 1985; Pytela *et al.*, 1985a,b; Springer *et al.*, 1985). Great interest in integrins was generated when the first OPN sequence became available. This sequence identified an RGD integrin-binding motif within the molecule (Oldberg *et al.*, 1986).

Alternative names

After the specific monoclonal antibody was isolated, CD44 was referred to as human leukocyte-common antigen (Dalchau *et al.*, 1980). The receptor is also known as the polymorphic glycoprotein Pgp-1 antigen and the gp90 Hermes antigen (Goldstein *et al.*, 1989; Stamenkovic *et al.*, 1989).

The term integrin was first proposed by Tamkun and colleagues after cloning a cDNA that encoded a protein believed to link the extracellular matrix and intracellular actin (Tamkun *et al.*, 1986). Alternative names for this chicken integrin include 140K complex, CSAT antigen, and JG22 antigen (Hynes, 1987). The integrin appellation has been applied to the many $\alpha\beta$ heterodimers that have been identified. Alternative names have been used to identify these receptors, particularly with regards to the extracellular matrix to which they are associated. For example, the $\alpha_v\beta_3$ integrin is known as the vitronectin receptor (Pytela *et al.*, 1985b). Other alternative names reflect the cell from which the molecule was identified. For example, the $\alpha_L\beta_2$ molecule on lymphocytes was termed leukocyte function-associated (LFA-1) and the $\alpha_M\beta_2$ on macrophages was identified as Mac-1, Mo-1, p150,95 (Hynes, 1987). The $\alpha_{IIb}\beta_3$ structure on platelets is known as GPIIb-IIIa (Hynes, 1992).

Structure

Upon cloning the gene, Stamenkovic *et al.* found that CD44 is a transmembrane glycoprotein of 341 amino acids (Stamenkovic *et al.*, 1989). Goldstein and colleagues cloned a cDNA encoding a transmembrane protein of 294 amino acids (Goldstein *et al.*, 1989). It is now known that the protein exists in several isoforms that are generated by the selection of 10 out of 20 exons through alternative RNA splicing. These variant exons are called v1–v10. There is one standard isoform that is present on erythrocytes, leukocytes, and in the brain (Borland *et al.*, 1998). The extracellular domain is large and is heavily glycosylated (Borland *et al.*, 1998).

Integrins are transmembrane, heterodimeric receptors composed of α and β subunits. In mammals, there are 17 α chains, 8 β chains, and 22 distinct heterodimers that have been identified (Kumar, 1998). There is a large extracellular portion comprised of three repeats of a cysteine-rich domain in the β chain. The short transmembrane domain is followed by an intracellular domain of variable length (Kumar, 1998).

Main activities and pathophysiological roles

CD44 function has been the subject of much study (Borland *et al.*, 1998). The receptor is known to be involved in lymphocyte homing, allowing binding to mucosal high endothelial venules (Duijvestijn and Hamann, 1989). Homing of prothymocytes is also dependent on CD44. CD44 binds hyaluronic acid (HA) in the extracellular matrix, fragments of HA, fibronectin, and collagen (Kincade *et al.*, 1997). Small HA fragments, acting through CD44 (McKee *et al.*, 1996), can induce expression of nitric oxide synthase 2 (NOS2; McKee *et al.*, 1997). When coupled with IFNγ, the CD44–HA interaction can enhance the expression of some chemokines (Horton *et al.*, 1998). CD44 has also been shown to present the chemokine MIP-1β (Tanaka *et al.*, 1993). Finally, many tumors express altered isoforms and glycosylation of CD44 which presumably enhances metastatic potential (Borland *et al.*, 1998).

Integrins are known to mediate cell-to-cell and cell-to-extracellular matrix (ECM) interactions. Many, though not all, integrins bind to ligands with an RGD motif; some show strict specificity to their ligands (Kumar, 1998). Multiple second messenger pathways are known to be activated upon integrin receptor ligation. These pathways include phospholipase activity, phosphatidylinositol turnover, increases in intracellular calcium, and intracellular alkylanization. The cytoskeleton reorganizes after engagement of integrins and is believed to be related to cell adhesion, migration, and possibly phagocytosis (Hynes, 1987). In *Glanzmann's thrombasthenia*, there is a defect in platelet aggregation manifested by *hemorrhagic symptoms*. This is attributable to a deficiency of the $\alpha_{IIb}\beta_3$, GPIIb-IIIa integrin (George *et al.*, 1990). Deficiencies in Mac-1, LFA-1, or p150,95, which share a common β_2 subunit, lead to *leukocyte adhesion deficiency syndrome* characterized by granulocytosis, impaired leukocyte adhesion, and increased susceptibility to pyogenic infections (Anderson and Springer, 1987).

GENE

Accession numbers

A search of the GenBank nucleic acid database for CD44 yields 221 listings. The cloning of human CD44 was published simultaneously by two groups and entered in M25078 and M24915 (Goldstein *et al.*, 1989; Stamenkovic *et al.*, 1989). Goldstein and colleagues used a λgt11 expression library derived from a human B lymphoblastoid line that expressed high levels of gp90^{Hermes} (Goldstein *et al.*, 1989). Stamenkovic *et al.* used transient transfection of COS cells and antibody panning on dishes to select for positive clones (Stamenkovic *et al.*, 1989). Pgp-1, the murine homolog of the human Hermes antigen, was cloned from a cDNA library derived from PU5-1.8 cells, accession number M30655 (Zhou *et al.*, 1989)

A search for integrins in GenBank yields an even larger number of entries: 2136. The first β chain was cloned in 1986, accession number M14049, using a λgt11 expression library constructed from chicken fibroblasts and a polyclonal antisera that recognized the chicken fibroblast 140 kDa complex (Tamkun *et al.*, 1986). The human fibronectin receptor α chain was also cloned with a λgt11 screen, accession number M13918 (Argraves *et al.*, 1986). Particularly relevant for OPN binding, the human α_v chain was cloned by a similar approach, accession number M13918 (Suzuki *et al.*, 1986).

PROTEIN

Accession numbers

CD44

AAA36138 (Goldstein *et al.*, 1989)
AAA35674 (Stamenkovic *et al.*, 1989)
AAA39922 (Zhou *et al.*, 1989)

Integrins

AAA48926 (Tamkun *et al.*, 1986)
AAA52467 (Argraves *et al.*, 1986)
AAA52467 (Suzuki *et al.*, 1986)

Description of protein

As outlined in the background section, both of the receptor classes for OPN, i.e. CD44 and integrins, are transmembrane glycoproteins. The standard form of CD44, CD44s (Borland *et al.*, 1998), is composed of a single polypeptide with a 248 amino acid extracellular domain, a 21 amino acid transmembrane domain, and a 72 amino acid intracellular domain. CD44 isoforms are created by alternative splicing of 10 out of 20 exons during RNA processing. CD44s is extensively glycosylated, accounting for the molecular weight difference between the predicted polypeptide mass of 40 kDa and the 80–95 kDa observed on SDS-PAGE (Borland *et al.*, 1998). These glycosylation sites include *N*-linked (primarily in the N-terminus cartilage link domain) and *O*-linked moieties. The murine CD44 is similar in structure and has six possible *N*-linked and six possible *O*-linked sites (Zhou *et al.*, 1989).

In contrast to CD44, integrins are $\alpha\beta$ heterodimers. The α subunits range between 120 and 180 kDa and the β chains between 90 and 110 kDa (Hynes, 1992). The β chains span the cellular membrane and have a cysteine-rich motif that is repeated four times in the extracellular domain. The α chains are also transmembrane proteins, some of which are post-translationally cleaved, yielding subunits that are linked by disulfide bonds (Hynes, 1987).

Relevant homologies and species differences

CD44 is highly conserved between the murine and human sequences (Zhou *et al.*, 1989). The 155 amino acids on the N-terminus of the receptor have 89% identity between the two species and the 105 amino acids on the C-terminus have 86% identity (Zhou *et al.*, 1989). Interestingly, there is an area of significant divergence in the extracellular domain near the plasma membrane; this area shares only 42% identity between the species, which is likely the result of the splice variations (Zhou *et al.*, 1989).

Integrins are homologous between *Drosophila* and vertebrates (Hynes, 1992). Hynes has proposed that these similarities allowed for the development of multicellular organisms and that the differences permitted the development of variety and specificity of binding (Hynes, 1992; Borland *et al.*, 1998).

Affinity for ligand(s)

CD44 is known to bind several ligands, including HA, HA fragments, fibronectin, laminin, and collagen type I (Borland *et al.*, 1998). OPN was shown to bind CD44 transfected into a murine embryonic cell line; however, a dissociation constant was not calculated (Weber *et al.*, 1996).

Integrins have a range of ligands where some receptors have single ligands and others have multiple ligands (Hynes, 1987, 1992). Of the 20 or more heterodimers identified, the $\alpha_v\beta_3$ complex appears to be the most promiscuous, binding at least seven ligands (Hynes, 1992). The K_d values of the fibronectin receptor for fibronectin and laminin are 10^{-6} and 2×10^{-6} M, respectively (Horwitz *et al.*, 1985). OPN binding to integrins has largely been demonstrated by the use of blocking antibodies. **Table 1** summarizes the specific integrin heterodimers that are believed to bind OPN. One report identified a binding constant of 2.35×10^{-10} M of radiolabeled OPN on P388D1 cells (Nasu *et al.*, 1995). However, it is unclear to which integrin OPN was binding.

Cell types and tissues expressing the receptor

The tissue distribution of CD44 is broad. CD44 is expressed on T lymphocytes and B lymphocytes, thymocytes, granulocytes, and dendritic cells (Dalchau *et al.*, 1980; Weiss *et al.*, 1997). The murine homolog is also known to be present on macrophages and 3T3 cells (Hughes and August, 1982; Hughes *et al.*, 1983). Immunostaining of frozen tissue showed thymic medulla, white pulp in spleen, and lymph node cortex express this molecule (Dalchau *et al.*, 1980).

Integrins also have broad cell and tissue distribution. Some integrins, however, are cell-specific. For example, the $\alpha_{IIb}\beta_3$ integrin is expressed only on platelets. The β_2 subfamily is found on leukocytes (e.g. $\alpha_L\beta_2$ (LFA-1) on lymphoid and various myeloid cells and $\alpha_M\beta_2$ (Mac-1) on macrophages; Hynes, 1987).

Regulation of receptor expression

CD44 is constitutively expressed. Modulation of receptor expression occurs when normal resting lymphocytes are activated, resulting in a transient shift

Table 1 Integrin heterodimers that bind OPN

Integrin heterodimer	Cell type	Assay	Reference
$\alpha_v\beta_3$	Osteosarcoma cells	Adhesion	Oldberg *et al.*, 1986
	Smooth muscle	Adhesion/migration	Liaw *et al.*, 1994, 1995; Yue *et al.*, 1994
	Fibroblasts	Adhesion	van Dijk *et al.*, 1993[a]
	Platelets	Adhesion	Bennett *et al.*, 1997
$\alpha_v\beta_5$	Smooth muscle	Adhesion	Liaw *et al.*, 1995
$\beta_v\beta_1$	Smooth muscle	Adhesion	Liaw *et al.*, 1995
$\alpha_9\beta_1$	Melanoma lines	Adhesion	Smith *et al.*, 1996
$\alpha_4\beta_1$	HL-60 promyelocytic, Ramos lymphoblastoid	Adhesion	Bayless *et al.*, 1998
$\alpha_8\beta_1$	K562 line	Adhesion	Denda *et al.*, 1998
α_4-, α_5-	P388D1 cells	Adhesion, haptotaxis	Nasu *et al.*, 1995

[a]Not dependent on the presence of an RGD sequence.

from CD44s to splice variants (Ponta *et al.*, 1998). Langerhans cells and dendritic cells increase their expression of certain isoforms after activation (Weiss *et al.*, 1997).

The regulation of integrin expression appears to be involved in development (Hynes, 1987). More important in the regulation of integrin function, however, is the concept of inside-to-out signaling (Hynes, 1992). Both the $\alpha_{IIb}\beta_3$ and the β_2 subfamily integrins exhibit this mechanism of control; activation of platelets and leukocytes is required to permit integrin-mediated signaling/events. This is believed to occur through a conformational change in the integrin because extracellular activating antibodies can achieve the same end (Hynes, 1992).

Release of soluble receptors

The release of soluble CD44 (sCD44) has been well-documented and a molecular basis for this phenomenon has been identified (Yu and Toole, 1996). sCD44 from *non-Hodgkin's lymphoma* cells blocks cell interaction with hyaluronate and inhibits lymphocyte binding to high endothelial venules (Ristamaki *et al.*, 1997). This suggests that sCD44 could inhibit immune responses against the tumor cells.

Naturally occurring soluble integrins have not been described. A soluble recombinant molecule has been produced in which the heterodimers were joined with fos- and jun-binding domains (Eble *et al.*, 1998).

SIGNAL TRANSDUCTION

Little is known about signal transduction from CD44 (Lesley and Hyman, 1998). In contrast, much information has accumulated on the signaling pathways associated with integrins; excellent reviews are available on this topic (Howe *et al.*, 1998; Kumar, 1998; Dedhar, 1999).

Associated or intrinsic kinases

One report indicates that monoclonal antibody ligation of CD44 on T cells leads to ZAP-70 tyrosine phosphorylation (Taher *et al.*, 1996). This was attributed to increased p56lck activity. In addition, CD44 was found to be physically associated with p56lck (Taher *et al.*, 1996). However, CD44 lacks a CXCP sequence necessary for p56lck binding (Lesley and Hyman, 1998). An adapter molecule may be required for this signaling pathway.

Crosslinking or clustering of integrins leads to the formation of focal adhesions that activate focal adhesion kinase (FAK), a Rho-GTPase phenomenon (Kumar, 1998; Dedhar, 1999). Integrin ligation appears to inhibit the serine-threonine kinase that is associated with the β chain, integrin-linked kinase (ILK) (Hannigan *et al.*, 1996). Importantly, integrin engagement that leads to FAK activation can initiate the MAP kinase cascade (Howe *et al.*, 1998). The adapter protein Shc can associate with the α chain via caveolin, which can also activate the MAPK (Howe *et al.*, 1998). There may also be a Ras-independent mechanism for integrin activation of the MAPK (Howe *et al.*, 1998).

Regarding integrins that bind OPN, phosphatidylinositol 3-kinase (PI-3 kinase) and Src kinase, in addition to FAK, are found associated with $\alpha_v\beta_3$ (Hruska *et al.*, 1995). OPN signaling increases FAK activity (Hruska *et al.*, 1995). OPN also appears to activate Src kinase, at least in melanoma cells (Chellaiah *et al.*, 1996). Not surprisingly, OPN treatment of cells has been found to increase tyrosine phosphorylation (Lopez *et al.*, 1995).

Cytoplasmic signaling cascades

Ligation of CD44 by HA leads to a rise in intracellular Ca^{2+} in mouse T lymphoma cells (BW5147) (Bourguignon *et al.*, 1993). Increased intracellular Ca^{2+} by either HA or calcium ionophore leads to patching and capping of CD44 in these same cells (Bourguignon *et al.*, 1993). This effect has not been demonstrated with OPN.

Several papers have evaluated potential signaling cascades induced by receptor–OPN ligation. OPN binding to $\alpha_v\beta_3$ results in immediate signals in osteoclasts (Miyauchi *et al.*, 1991). These include a reduction of cytosolic Ca^{2+} through a calmodulin-dependent activation of a Ca^{2+}-ATPase (Miyauchi *et al.*, 1991). Because peptides had a similar effect as whole protein, the authors argue that immobilization is not required (Miyauchi *et al.*, 1991). OPN binding to $\alpha_v\beta_3$ also initiates phophatidylinositol turnover, including the production of phosphatidylinositol triphosphate (Hruska *et al.*, 1995). This is due to increased activity of PI-3 kinase (Hruska *et al.*, 1995). An important consequence of this kinase's activity is to inhibit apoptosis (Kumar, 1998), though this has yet to be demonstrated with OPN as the integrin ligand. Other intracellular signals generated by integrin ligation have been identified in other systems. These include tyrosine phosphorylation, cytoplasmic alkalinization, and increases in intracellular Ca^{2+} (Hynes, 1992).

DOWNSTREAM GENE ACTIVATION

To date, studies of gene activation induced by OPN–receptor interactions have not been done. Gene expression changes after receptor activation have been studied for integrins, but not for CD44.

Transcription factors activated

It is known that integrin receptor ligation activates several different transcription factors. For example, NFκB is activated in the monocytic line THP-1 when the cells are cultured on fibronectin (Rosales and Juliano, 1996). Shear stress on endothelial cells triggers MAP kinase activity, culminating in the activation of AP-1, SP-1, Elk-C, and NFκB (Shyy and Chien, 1997). Integrins also function as costimulatory receptors in T lymphocytes. For example, fibronectin binding to $\alpha_5\beta_1$, the very late activation antigen-5 (VLA-5) receptor, induces AP-1 activity in human peripheral blood T cells (Yamada *et al.*, 1991). T cells treated with immobilized anti-CD3 antibody and fibronectin have increased IL-2 production, IL-2 receptor expression, and proliferation compared to cells without exposure to fibronectin (Yamada *et al.*, 1991).

Genes induced

Several studies have examined the gene expression changes after integrin ligation. In one analysis, cross-linking of the fibronectin receptor increased the expression of collagenase and stromelysin (Werb *et al.*, 1989). Granulocyte–macrophage colony-stimulating factor (GM-CSF) mRNA increases after macrophages adhere to fibronectin (Thorens *et al.*, 1987). Monocyte adherence to fibronectin increases the expression of TNFα and CSF-1 (Eierman *et al.*, 1989). Another report demonstrated that macrophages cultured on collagen or fibronectin-coated plastic increase their expression of IL-8 (Standiford *et al.*, 1991). Using two-dimensional gel electrophoresis and a subtractive cloning strategy, Danen *et al.* identified 12 novel genes whose expression is increased in human salivary gland cells cultured on fibronectin or collagen (Danen *et al.*, 1998). Presumably, these gene expression changes are the result of signaling through integrins, though it is unknown if OPN is capable of eliciting any of these responses.

BIOLOGICAL CONSEQUENCES OF ACTIVATING OR INHIBITING RECEPTOR AND PATHOPHYSIOLOGY

The details of OPN's biological effects have been summarized in the OPN chapter. What follows here is an overview of some biological effects of integrins and CD44.

Unique biological effects of activating the receptors

CD44 is intimately involved in immune cell physiology. CD44 is an important receptor for lymphocyte homing to high endothelial venules (HEV) of lymphoid tissue (Duijvestijn and Hamann, 1989). Langerhans and dendritic cell homing to T cell zones of lymph nodes also relies on variant CD44 isoforms (Weiss *et al.*, 1997). Structures similar to HEV are associated with *chronic inflammation* and are likely to participate in lymphocyte recruitment to these areas (Duijvestijn and Hamann, 1989). In at least one instance, CD44 was shown to immobilize a chemokine, MIP-1β, thereby augmenting adhesion *in vitro* (Tanaka *et al.*, 1993). Several studies have shown that anti-CD44 monoclonal antibodies enhance T cell activation *in vitro*. For example, the proliferation of human peripheral blood T cells in response to anti-CD2 or anti-CD3 antibodies increases when anti-CD44 antibodies are included in the culture (Denning *et al.*, 1990). In contrast, Rothman and colleagues have generated an anti-CD44 monoclonal that inhibits activation induced by anti-CD3 antibodies (Rothman *et al.*, 1991). CD44 can mediate redirected lysis by NK cells (Sconocchia *et al.*, 1994, 1997). CD44 ligation participates in the activation of monocytes (Webb *et al.*, 1990), which may account for the monocyte-dependence of T cell proliferation augmented by anti-CD44 antibodies (Denning *et al.*, 1990). Related to macrophage activation, HA fragments induce the expression of NOS2 after binding CD44 (McKee *et al.*, 1997).

Integrins have significant effects on cellular adhesion, cell viability, and cell proliferation (for example, see reviews by Howe *et al.*, 1998, and Hynes, 1992). Endothelial cells (Meredith *et al.*, 1993) and epithelial cells (Frisch and Francis, 1994) separated from their extracellular matrix undergo apoptosis; in this context, it has specifically been named anoikis (Frisch and Francis, 1994; Frisch and Ruoslahti, 1997). The inhibition of anoikis depends on integrin activation of phosphoinositide 3-OH kinase and activation of protein kinase B/Akt (Khwaja *et al.*, 1997). Failure to undergo anoikis after detachment is characteristic of some tumors, particularly those with activated Src or Ras (Kumar, 1998). Integrin–ligand binding also appears to potentiate the effects of growth factors, presumably through activation of the growth factor receptor (Schneller *et al.*, 1997; Howe *et al.*, 1998). Schneller *et al.* found that $\alpha_v\beta_3$ integrins were associated with PDGF receptors and that $\alpha_v\beta_3$ ligation by vitronectin increased the mitogenic and chemotactic effects of PDGF (Schneller *et al.*, 1997). Antibodies that block the $\alpha_v\beta_3$ integrin inhibit angiogenesis induced by FGF in the chicken chorioallantoic membrane (Brooks *et al.*, 1994). These findings are of particular interest because $\alpha_v\beta_3$ is a receptor for OPN.

A substantial literature has accumulated regarding integrins and leukocyte function. Leukocyte recruitment to sites of inflammation relies on the interplay between selectins on endothelial cells and their respective ligands (Springer, 1994). However, the model of leukocyte rolling, tight adhesion, and transendothelial migration also involves the interaction of leukocytes with integrin ligands: intercellular adhesion molecule (ICAM), vascular cell adhesion molecule 1 (VCAM-1), and mucosal addressin cell adhesion molecule 1 (MAdCAM-1) (Springer, 1994). Moreover, there are critical intercellular interactions between leukocytes involving ICAMs and the leukointegrins of the β_2 family: CD11a (α_L), CD11b (α_M), CD11c (α_{x2}), CD11d (α_{d2}), and CD18 (β_2) (Gahmberg, 1997). These interactions are important for T cell activation, cytolysis, neutrophil function, and NK cell function (Gahmberg, 1997). Divalent cations are required and specialized outside-to-in and inside-to-out activation occurs (Gahmberg, 1997; Hynes, 1992). As mentioned previously, inside-to-out activation refers to enhanced ligand binding after the activation of the integrin (Hynes, 1992); chemokines play an important role in this process (Gahmberg, 1997; Springer, 1994). The α_4 family of integrins are also expressed by T cells, B cells, and monocytes. An example of this group is very late antigen 4 (VLA-4), $\alpha_4\beta_1$, a receptor for VCAM-1 (Springer, 1994).

Phenotypes of receptor knockouts and receptor overexpression mice

In spite of the multiple functions attributed to CD44, the phenotype of the null mouse was relatively modest (Schmits *et al.*, 1997). The null animals generated fewer hematopoietic precursors, measured as colony-forming units in blood and spleen, after systemic administration of G-CSF. T and B cell function, CTL activity, and delayed-type hypersensitivity reactions appeared unchanged. Relevant to OPN physiology, CD44-null mice had enhanced granuloma burdens after i.v. injection of heat-killed *Corynebacterium parvum*. Transfection of CD44 to fibroblasts from these animals increased binding to HA. Finally, CD44− fibroblasts transformed with SV40 large T antigen generated large tumors in syngeneic (CD44+) animals compared to lines derived from CD44+ fibroblasts (Schmits *et al.*, 1997).

Several studies of integrin-null animals have demonstrated profound phenotypes. Attesting to the importance of integrins in development, numerous integrin knockout strains have embryonic lethal phenotypes with morphogenic and placental defects (Yang *et al.*, 1993, 1995; Stephens *et al.*, 1995; Fassler and Meyer, 1995; Kreidberg *et al.*, 1996; Bader *et al.*, 1998). Some integrin-deficient animals do not have lethal phenotypes. These strains have demonstrated the function of integrins in adhesion and in normal tissue architecture (Dowling *et al.*, 1996; Gardner *et al.*, 1996; van der Neut *et al.*, 1996).

In contrast to the anti-angiogenic effects of anti-antibodies, embryos lacking α_v chains have normal vessel growth (Bader *et al.*, 1998). In an effort to reconcile this discrepancy, Bader and colleagues point out there may be functional redundancy among integrins during development. Alternatively, model systems where $\alpha_v\beta_3$ antibodies are effective at blocking angiogenesis may be overly dependent on this integrin (Bader *et al.*, 1998). This difference, though unexplained, is not moot because anti-$\alpha_v\beta_3$ reagents may be used clinically (see Therapeutic utility).

Several groups have generated mouse strains deficient in the leukointegrins. Studies of these animals have reiterated the importance of these integrins in *acute inflammation* and immune function. Neutrophils from CD11b-null mice have diminished adhesion to endothelium and, unexpectedly, a reduced rate of apoptosis induced by phagocytosis (Coxon *et al.*, 1996). In addition, these animals have impaired mast cell development and mast cell function in a model of *acute peritonitis* (Rosenkranz *et al.*, 1998). Another CD11b-null mouse shows reduced neutrophil adherence to fibrinogen-coated disks but normal neutrophil accumulation (Lu *et al.*, 1997). A neutrophil migration defect was expected, but CD11a appears to compensate for the CD11b deficiency (Lu *et al.*, 1997). Mice expressing a low amount of β_2, the common β chain of the leukointegrins, have fewer neutrophils accumulate in the peritoneum after administration of thioglycollate and delayed rejection of cardiac tissue transplants (Wilson *et al.*, 1993). Mice deficient in ICAM-1, an important ligand for CD11a/CD18 (LFA-1) and CD11b/CD18 (Mac-1), have leukocytosis with diminished T cell function in mixed lymphocyte reactions, reduced contact hypersensitivity, and a reduced sensitivity to LPS (Xu *et al.*, 1994).

Human abnormalities

There has been one report of a human who was deficient in some form of CD44 (Parsons *et al.*, 1994). CD44 was exclusively absent from red blood cells in a patient with *anemia*; lymphocytes, granulocytes, and monocytes had normal levels of expression (Parsons *et al.*, 1994). Bone marrow sampling revealed abnormal erythropoiesis with erythroblast accumulation in G_1 and G_2 phases of the cell cycle (Wickramasinghe *et al.*, 1991). This implicates at least one CD44 isoform in hematopoiesis.

There are two well-described clinical syndromes that are the direct result of integrin defects. The case of a boy with recurrent infections and diminished neutrophil adhesion (Crowley *et al.*, 1980) led to similar observations in other patients (Arnaout *et al.*, 1982; Bowen *et al.*, 1982; Dana *et al.*, 1984; Kohl *et al.*, 1984). Clinically, these patients had *recurrent bacterial infections*, *granulocytosis*, *aggressive periodontitis*, *poor wound healing* and delayed umbilical cord separation. Phagocytosis and neutrophil recruitment, grossly observed as the inability to form pus, was also impaired (Anderson and Springer, 1987). This constellation of findings was labeled *leukocyte adhesion deficiency syndrome*. Analysis of these patients revealed the absence of some plasma membrane integrins and a defect in β_2 expression, though α chains were present within the cell (Springer *et al.*, 1984). This observation led to the understanding that coexpression of both the α and the β chains are required for the surface expression of integrins (Springer *et al.*, 1984). Ultimately, the defect was characterized on the molecular level: mutant β mRNAs were identified and the protein products failed to associate with the α subunits (Kishimoto *et al.*, 1987).

The second integrin defect is manifested as *Glanzmann's thrombasthenia*. This hemorrhagic disorder is characterized by *cutaneous purpura* and *mucosal hemorrhage* with a prolonged bleeding time and normal platelet numbers. However, the platelets fail to aggregate and form a platelet plug (George *et al.*, 1990). A defect was found on platelet membrane glycoproteins (Nurden and Caen, 1974), later identified to be GPIIb/IIIa (Phillips and Agin, 1977). Ultimately this was found to be the $\alpha_{IIb}\beta_3$ integrin (George *et al.*, 1990). The absence of this integrin reduces platelet adhesion to fibrinogen, fibronectin, and von Willebrand factor (Ruggeri *et al.*, 1982).

THERAPEUTIC UTILITY

Effect of treatment with soluble receptor domain

Although soluble CD44 molecules have been defined (Yu and Toole, 1996) and are capable of blocking lymphocyte adhesion to HEV (Ristamaki *et al.*, 1997), this has not been exploited clinically. Recombinant $\alpha_3\beta_1$ dimers have also been described (Eble *et al.*, 1998), but these have not been used therapeutically.

Effects of inhibitors (antibodies) to receptors

Integrins make a critical contribution to inflammation, which makes them possible therapeutic targets. Antibodies, peptides, and peptidomimetics to block VLA-4 are being developed as anti-inflammatory drugs (Lin and Castro, 1998). A CD11b/CD18 (Mac-1) antagonist reduces inflammation in an animal model of *colitis* (Meenan *et al.*, 1996).

Because of the importance of the $\alpha_{IIb}\beta_3$ integrin in platelet aggregation, the therapeutic potential, and the financial possibilities, this platelet integrin has been a prime target for inhibition during *acute coronary syndromes*. Several glycoprotein inhibitors, namely tirofiban, lamifiban, and eptifibatide, have been studied in clinical trials with favorable results (Alexander and Harrington, 1998). Similar results of blocking platelet aggregation have been observed with antibody inhibitors and peptide inhibitors (Domanovits *et al.*, 1998; Mousa *et al.*, 1998).

Several potential products are being developed as antagonists of $\alpha_v\beta_3$. A synthetic peptide antagonist inhibits bone resorption and *osteoporosis* in oopherectomized rats (Engleman *et al.*, 1997). Presumably this results from disrupting osteoclast $\alpha_v\beta_3$ adhesion to bone OPN. A cyclic peptidomimetic antagonist of $\alpha_v\beta_3$ reduces restenosis after stent placement in a coronary artery (Srivatsa *et al.*, 1997). The OPN–$\alpha_v\beta_3$ interaction is believed to be critical in the pathophysiology of restenosis (Hirota *et al.*, 1993; Panda *et al.*, 1997). A Fab product under development, abciximab, should be particularly enticing for interventional cardiologists because it inhibits both the platelet $\alpha_{IIb}\beta_3$ integrin and $\alpha_v\beta_3$ (Tam *et al.*, 1998). Antibodies and a cyclic peptide antagonist of $\alpha_v\beta_3$ reduce neovascularization, attesting to a role for $\alpha_v\beta_3$ in vessel formation (Brooks *et al.*, 1994; Drake *et al.*, 1995; Hammes *et al.*, 1996).

Finally, the study of snake venom led to the discovery of an entirely new class of integrin antagonists, the disintegrins (Gould *et al.*, 1990). These are short peptides that contain an RGD integrin-binding motif with an extremely high affinity for integrins (Gould *et al.*, 1990). Further analysis revealed that these molecules are truncated from larger proteins (Paine *et al.*, 1992). Full-length proteins contain a disintegrin domain and a zinc metalloprotease domain, the hemorrhagin domain (Paine *et al.*, 1992). Ultimately similar proteins that are named a disintegrin and metalloprotease (ADAM) were disovered in many species, including humans (Black and White, 1998). These are physiologically important molecules; for example, an ADAM controls the release of TNFα (Black *et al.*, 1997; Moss *et al.*, 1997).

Venom proteins are known for their inhibition of platelet function (Gould *et al.*, 1990) and their activation of the coagulation cascade (Paine *et al.*, 1992). The integrin-binding function is responsible for the block of platelet function (Huang *et al.*, 1989); this integrin-binding property is being harnessed to study other cell biological and clinical questions. The disintegrin contortrostatin can bind $\alpha_v\beta_3$ and inhibit osteoclast attachment (Mercer *et al.*, 1998). Echistatin also binds $\alpha_v\beta_3$ and inhibits M-CSF-directed cell migration and the formation of multinucleated osteoclasts (Nakamura *et al.*, 1998). The disintegrin accutin is an $\alpha_v\beta_3$ antagonist that blocks angiogenesis by inducing endothelial apoptosis (Yeh *et al.*, 1998). There are numerous reports using disintegrins to block platelet aggregation. For example, bitistatin has been used to block $\alpha_{IIb}\beta_3$, thereby preventing platelet loss during extracorporeal membrane oxygenation (Shigeta *et al.*, 1992). Disintegrins may be useful anti-inflammatory agents (Schluesener, 1998).

Thus, there is great promise for innovative applications of integrin antagonists. These applications include reducing inflammation, blocking platelet function, particularly during *acute cardiac thromboses*, blocking *restenosis* after angioplasty, and inhibiting vessel growth. The latter may be applicable to conditions such as *diabetic retinopathy* (Hammes *et al.*, 1996) and *neovascularization* associated with *cancer* (Yeh *et al.*, 1998).

References

Alexander, J. H., and Harrington, R. A. (1998). Recent antiplatelet drug trials in the acute coronary syndromes. Clinical interpretation of PRISM, PRISM-PLUS, PARAGON A and PURSUIT. *Drugs* **56**, 965–976.

Anderson, D. C., and Springer, T. A. (1987). Leukocyte adhesion deficiency: an inherited defect in the Mac-1, LFA-1, and p150,95 glycoproteins. *Annu. Rev. Med.* **38**, 175–194.

Argraves, W. S., Pytela, R., Suzuki, S., Millan, J. L., Pierschbacher, M. D., and Ruoslahti, E. (1986). cDNA sequences from the alpha subunit of the fibronectin receptor predict a transmembrane domain and a short cytoplasmic peptide. *J. Biol. Chem.* **261**, 12922–12924.

Arnaout, M. A., Pitt, J., Cohen, H. J., Melamed, J., Rosen, F. S., and Colten, H. R. (1982). Deficiency of a granulocyte-membrane glycoprotein (gp150) in a boy with recurrent bacterial infections. *N. Engl. J. Med.* **306**, 693–699.

Bader, B. L., Rayburn, H., Crowley, D., and Hynes, R. O. (1998). Extensive vasculogenesis, angiogenesis, and organogenesis precede lethality in mice lacking all alpha v integrins. *Cell* **95**, 507–519.

Bayless, K. J., Meininger, G. A., Scholtz, J. M., and Davis, G. E. (1998). Osteopontin is a ligand for the $\text{alpha}_4\text{beta}_1$ integrin. *J. Cell Sci.* **111**, 1165–1174.

Bennett, J. S., Chan, C., Vilaire, G., Mousa, S. A., and DeGrado, W. F. (1997). Agonist-activated $\text{alpha}_v\text{beta}_3$ on platelets and lymphocytes binds to the matrix protein osteopontin. *J. Biol. Chem.* **272**, 8137–8140.

Black, R. A., and White, J. M. (1998). ADAMs: focus on the protease domain. *Curr. Opin. Cell Biol.* **10**, 654–659.

Black, R. A., Rauch, C. T., Kozlosky, C. J., Peschon, J. J., Slack, J. L., Wolfson, M. F., Castner, B. J., Stocking, K. L., Reddy, P., Srinivasan, S., Nelson, N., Boiani, N., Schooley, K. A., Gerhart, M., Davis, R., Fitzner, J. N., Johnson, R. S., Paxton, R. J., March, C. J., and Cerretti, D. P. (1997). A metalloproteinase disintegrin that releases tumour-necrosis factor-alpha from cells. *Nature* **385**, 729–733.

Borland, G., Ross, J. A., and Guy, K. (1998). Forms and functions of CD44. *Immunology* **93**, 139–148.

Bourguignon, L. Y., Lokeshwar, V. B., Chen, X., and Kerrick, W. G. (1993). Hyaluronic acid-induced lymphocyte signal transduction and HA receptor (GP85/CD44)-cytoskeleton interaction. *J. Immunol.* **151**, 6634–6644.

Bowen, T. J., Ochs, H. D., Altman, L. C., Price, T. H., Van Epps, D. E., Brautigan, D. L., Rosin, R. E., Perkins, W. D., Babior, B. M., Klebanoff, S. J., and Wedgwood, R. J. (1982). Severe recurrent bacterial infections associated with defective adherence and chemotaxis in two patients with neutrophils deficient in a cell-associated glycoprotein. *J. Pediatr.* **101**, 932–940.

Brooks, P. C., Clark, R. A., and Cheresh, D. A. (1994). Requirement of vascular integrin alpha v beta 3 for angiogenesis. *Science* **264**, 569–571.

Chellaiah, M., Fitzgerald, C., Filardo, E. J., Cheresh, D. A., and Hruska, K. A. (1996). Osteopontin activation of c-src in human melanoma cells requires the cytoplasmic domain of the integrin alpha v-subunit. *Endocrinology* **137**, 2432–2440.

Coxon, A., Rieu, P., Barkalow, F. J., Askari, S., Sharpe, A. H., von Andrian, U. H., Arnaout, M. A., and Mayadas, T. N. (1996). A novel role for the beta 2 integrin CD11b/CD18 in neutrophil apoptosis: a homeostatic mechanism in inflammation. *Immunity* **5**, 653–666.

Crowley, C. A., Curnutte, J. T., Rosin, R. E., Andre-Schwartz, J., Gallin, J. I., Klempner, M., Snyderman, R., Southwick, F. S., Stossel, T. P., and Babior, B. M. (1980). An inherited abnormality of neutrophil adhesion. Its genetic transmission and its association with a missing protein. *N. Engl. J. Med.* **302**, 1163–1168.

Dalchau, R., Kirkley, J., and Fabre, J. W. (1980). Monoclonal antibody to a human leukocyte-specific membrane glycoprotein probably homologous to the leukocyte-common (L-C) antigen of the rat. *Eur. J. Immunol.* **10**, 737–744.

Dana, N., Todd, R. F. D., Pitt, J., Springer, T. A., and Arnaout, M. A. (1984). Deficiency of a surface membrane glycoprotein (Mo1) in man. *J. Clin. Invest.* **73**, 153–159.

Danen, E. H., Lafrenie, R. M., Miyamoto, S., and Yamada, K. M. (1998). Integrin signaling: cytoskeletal complexes, MAP kinase activation, and regulation of gene expression. *Cell Adhes. Commun.* **6**, 217–224.

Dedhar, S. (1999). Integrins and signal transduction. *Curr. Opin. Hematol.* **6**, 37–43.

Denda, S., Reichardt, L. F., and Muller, U. (1998). Identification of osteopontin as a novel ligand for the integrin alpha_8 beta_1 and potential roles for this integrin–ligand interaction in kidney morphogenesis. *Mol. Biol. Cell* **9**, 1425–1435.

Denning, S. M., Le, P. T., Singer, K. H., and Haynes, B. F. (1990). Antibodies against the CD44 p80, lymphocyte homing receptor molecule augment human peripheral blood T cell activation. *J. Immunol.* **144**, 7–15.

Domanovits, H., Nikfardjam, M., Janata, K., Hornykewycz, S., Maurer, G., Laggner, A. N., and Huber, K. (1998). Restoration of coronary blood flow by single bolus injection of the GPIIb/IIIa receptor antagonist c7E3 Fab in a patient with acute myocardial infarction of recent onset. *Clin. Cardiol.* **21**, 525–528.

Dowling, J., Yu, Q. C., and Fuchs, E. (1996). Beta4 integrin is required for hemidesmosome formation, cell adhesion and cell survival. *J. Cell Biol.* **134**, 559–572.

Drake, C. J., Cheresh, D. A., and Little, C. D. (1995). An antagonist of integrin alpha v beta 3 prevents maturation of blood vessels during embryonic neovascularization. *J. Cell Sci.* **108**, 2655–2661.

Duijvestijn, A., and Hamann, A. (1989). Mechanisms and regulation of lymphocyte migration. *Immunol. Today* **10**, 23–28.

Eble, J. A., Wucherpfennig, K. W., Gauthier, L., Dersch, P., Krukonis, E., Isberg, R. R., and Hemler, M. E. (1998). Recombinant soluble human alpha 3 beta 1 integrin: purification, processing, regulation, and specific binding to laminin-5 and invasin in a mutually exclusive manner. *Biochemistry* **37**, 10945–10955.

Eierman, D. F., Johnson, C. E., and Haskill, J. S. (1989). Human monocyte inflammatory mediator gene expression is selectively regulated by adherence substrates. *J. Immunol.* **142**, 1970–1976.

Engleman, V. W., Nickols, G. A., Ross, F. P., Horton, M. A., Griggs, D. W., Settle, S. L., Ruminski, P. G., and Teitelbaum, S. L. (1997). A peptidomimetic antagonist of the alpha(v)beta3 integrin inhibits bone resorption *in vitro* and prevents osteoporosis *in vivo*. *J. Clin. Invest.* **99**, 2284–2292.

Fassler, R., and Meyer, M. (1995). Consequences of lack of beta 1 integrin gene expression in mice. *Genes Dev.* **9**, 1896–1908.

Frisch, S. M., and Francis, H. (1994). Disruption of epithelial cell-matrix interactions induces apoptosis. *J. Cell Biol.* **124**, 619–626.

Frisch, S. M., and Ruoslahti, E. (1997). Integrins and anoikis. *Curr. Opin. Cell Biol.* **9**, 701–706.

Gahmberg, C. G. (1997). Leukocyte adhesion: CD11/CD18 integrins and intercellular adhesion molecules. *Curr. Opin. Cell Biol.* **9**, 643–650.

Gardner, H., Kreidberg, J., Koteliansky, V., and Jaenisch, R. (1996). Deletion of integrin alpha 1 by homologous recombination permits normal murine development but gives rise to a specific deficit in cell adhesion. *Dev. Biol.* **175**, 301–313.

George, J. N., Caen, J. P., and Nurden, A. T. (1990). Glanzmann's thrombasthenia: the spectrum of clinical disease. *Blood* **75**, 1383–1395.

Goldstein, L. A., Zhou, D. F., Picker, L. J., Minty, C. N., Bargatze, R. F., Ding, J. F., and Butcher, E. C. (1989). A human lymphocyte homing receptor, the hermes antigen, is related to cartilage proteoglycan core and link proteins. *Cell* **56**, 1063–1072.

Gould, R. J., Polokoff, M. A., Friedman, P. A., Huang, T. F., Holt, J. C., Cook, J. J., and Niewiarowski, S. (1990). Disintegrins: a family of integrin inhibitory proteins from viper venoms. *Proc. Soc. Exp. Biol. Med.* **195**, 168–171.

Hammes, H. P., Brownlee, M., Jonczyk, A., Sutter, A., and Preissner, K. T. (1996). Subcutaneous injection of a cyclic peptide antagonist of vitronectin receptor-type integrins inhibits retinal neovascularization. *Nature Med.* **2**, 529–533.

Hannigan, G. E., Leung-Hagesteijn, C., Fitz-Gibbon, L., Coppolino, M. G., Radeva, G., Filmus, J., Bell, J. C., and Dedhar, S. (1996). Regulation of cell adhesion and anchorage-dependent growth by a new beta 1-integrin-linked protein kinase. *Nature* **379**, 91–96.

Hirota, S., Imakita, M., Kohri, K., Ito, A., Morii, E., Adachi, S., Kim, H. M., Kitamura, Y., Yutani, C., and Nomura, S. (1993). Expression of osteopontin messenger RNA by macrophages in atherosclerotic plaques. A possible association with calcification. *Am. J. Pathol.* **143**, 1003–1008.

Horton, M. R., McKee, C. M., Bao, C., Liao, F., Farber, J. M., Hodge-DuFour, J., Pure, E., Oliver, B. L., Wright, T. M., and Noble, P. W. (1998). Hyaluronan fragments synergize with interferon-gamma to induce the C-X-C chemokines mig and interferon-inducible protein-10 in mouse macrophages. *J. Biol. Chem.* **273**, 35088–35094.

Horwitz, A., Duggan, K., Greggs, R., Decker, C., and Buck, C. (1985). The cell substrate attachment (CSAT) antigen has properties of a receptor for laminin and fibronectin. *J. Cell Biol.* **101**, 2134–2144.

Howe, A., Aplin, A. E., Alahari, S. K., and Juliano, R. L. (1998). Integrin signaling and cell growth control. *Curr. Opin. Cell Biol.* **10**, 220–231.

Hruska, K. A., Rolnick, F., Huskey, M., Alvarez, U., and Cheresh, D. (1995). Engagement of the osteoclast integrin alpha v beta 3 by osteopontin stimulates phosphatidylinositol 3-hydroxyl kinase activity. *Endocrinology* **136**, 2984–2992.

Huang, T. F., Holt, J. C., Kirby, E. P., and Niewiarowski, S. (1989). Trigramin: primary structure and its inhibition of von Willebrand factor binding to glycoprotein IIb/IIIa complex on human platelets. *Biochemistry* **28**, 661–666.

Hughes, E. N., and August, J. T. (1982). Murine cell surface glycoproteins. Identification, purification, and characterization of a major glycosylated component of 110,000 daltons by use of a monoclonal antibody. *J. Biol. Chem.* **257**, 3970–3977.

Hughes, E. N., Colombatti, A., and August, J. T. (1983). Murine cell surface glycoproteins. Purification of the polymorphic Pgp-1 antigen and analysis of its expression on macrophages and other myeloid cells. *J. Biol. Chem.* **258**, 1014–1021.

Hynes, R. O. (1987). Integrins: a family of cell surface receptors. *Cell* **48**, 549–554.

Hynes, R. O. (1992). Integrins: versatility, modulation, and signaling in cell adhesion. *Cell* **69**, 11–25.

Jennings, L. K., and Phillips, D. R. (1982). Purification of glycoproteins IIb and III from human platelet plasma membranes and characterization of a calcium-dependent glycoprotein IIb-III complex. *J. Biol. Chem.* **257**, 10458–10466.

Khwaja, A., Rodriguez-Viciana, P., Wennstrom, S., Warne, P. H., and Downward, J. (1997). Matrix adhesion and Ras transformation both activate a phosphoinositide 3-OH kinase and protein kinase B/Akt cellular survival pathway. *EMBO J.* **16**, 2783–2793.

Kincade, P. W., Zheng, Z., Katoh, S., and Hanson, L. (1997). The importance of cellular environment to function of the CD44 matrix receptor. *Curr. Opin. Cell Biol.* **9**, 635–642.

Kishimoto, T. K., Hollander, N., Roberts, T. M., Anderson, D. C., and Springer, T. A. (1987). Heterogeneous mutations in the beta subunit common to the LFA-1, Mac-1, and p150,95 glycoproteins cause leukocyte adhesion deficiency. *Cell* **50**, 193–202.

Kohl, S., Springer, T. A., Schmalstieg, F. C., Loo, L. S., and Anderson, D. C. (1984). Defective natural killer cytotoxicity and polymorphonuclear leukocyte antibody-dependent cellular cytotoxicity in patients with LFA-1/OKM-1 deficiency. *J. Immunol.* **133**, 2972–2978.

Kreidberg, J. A., Donovan, M. J., Goldstein, S. L., Rennke, H., Shepherd, K., Jones, R. C., and Jaenisch, R. (1996). Alpha 3 beta 1 integrin has a crucial role in kidney and lung organogenesis. *Development* **122**, 3537–3547.

Kumar, C. C. (1998). Signaling by integrin receptors. *Oncogene* **17**, 1365–1373.

Lesley, J., and Hyman, R. (1998). CD44 structure and function. *Front Biosci.* **3**, D616–D630.

Liaw, L., Almeida, M., Hart, C. E., Schwartz, S. M., and Giachelli, C. M. (1994). Osteopontin promotes vascular cell adhesion and spreading and is chemotactic for smooth muscle cells *in vitro*. *Circ. Res.* **74**, 214–224.

Liaw, L., Skinner, M. P., Raines, E. W., Ross, R., Cheresh, D. A., Schwartz, S. M., and Giachelli, C. M. (1995). The adhesive and migratory effects of osteopontin are mediated via distinct cell surface integrins. Role of alpha v beta 3 in smooth muscle cell migration to osteopontin *in vitro*. *J. Clin. Invest.* **95**, 713–724.

Lin, K. C., and Castro, A. C. (1998). Very late antigen 4 (VLA4) antagonists as anti-inflammatory agents. *Curr. Opin. Chem. Biol.* **2**, 453–457.

Lopez, C. A., Davis, R. L., Mou, K., and Denhardt, D. T. (1995). Activation of a signal transduction pathway by osteopontin. *Ann. NY Acad. Sci.* **760**, 324–326.

Lu, H., Smith, C. W., Perrard, J., Bullard, D., Tang, L., Shappell, S. B., Entman, M. L., Beaudet, A. L., and Ballantyne, C. M. (1997). LFA-1 is sufficient in mediating neutrophil emigration in Mac-1-deficient mice. *J. Clin. Invest.* **99**, 1340–1350.

McKee, C. M., Penno, M. B., Cowman, M., Burdick, M. D., Strieter, R. M., Bao, C., and Noble, P. W. (1996). Hyaluronan (HA) fragments induce chemokine gene expression in alveolar macrophages. The role of HA size and CD44. *J. Clin. Invest.* **98**, 2403–2413.

McKee, C. M., Lowenstein, C. J., Horton, M. R., Wu, J., Bao, C., Chin, B. Y., Choi, A. M., and Noble, P. W. (1997). Hyaluronan fragments induce nitric-oxide synthase in murine macrophages through a nuclear factor kappaB-dependent mechanism. *J. Biol. Chem.* **272**, 8013–8018.

Meenan, J., Hommes, D. W., Mevissen, M., Dijkhuizen, S., Soule, H., Moyle, M., Buller, H. R., ten Kate, F. W., Tytgat, G. N., and van Deventer, S. J. (1996). Attenuation of the inflammatory response in an animal colitis model by neutrophil inhibitory factor, a novel beta 2-integrin antagonist. *Scand. J. Gastroenterol.* **31**, 786–791.

Mercer, B., Markland, F., and Minkin, C. (1998). Contortrostatin, a homodimeric snake venom disintegrin, is a potent inhibitor of osteoclast attachment. *J. Bone Miner. Res.* **13**, 409–414.

Meredith Jr, J. E., Fazeli, B., and Schwartz, M. A. (1993). The extracellular matrix as a cell survival factor. *Mol. Biol. Cell* **4**, 953–961.

Miyauchi, A., Alvarez, J., Greenfield, E. M., Teti, A., Grano, M., Colucci, S., Zambonin-Zallone, A., Ross, F. P., Teitelbaum, S. L., Cheresh, D., and Hruska, K. A. (1991). Recognition of osteopontin and related peptides by an alpha v beta 3 integrin stimulates immediate cell signals in osteoclasts. *J. Biol. Chem.* **266**, 20369–20374.

Moss, M. L., Jin, S. L., Milla, M. E., Bickett, D. M., Burkhart, W., Carter, H. L., Chen, W. J., Clay, W. C., Didsbury, J. R., Hassler, D., Hoffman, C. R., Kost, T. A., Lambert, M. H., Leesnitzer, M. A., McCauley, P., McGeehan, G., Mitchell, J., Moyer, M., Pahel, G., Rocque, W., Overton, L. K., Schoenen, F., Seaton, T., Su, J. L., Warner, J., Willard, D., and Becherer, J. D. (1997). Cloning of a disintegrin metalloproteinase that processes precursor tumour-necrosis factor-alpha [published erratum appears in *Nature* (1997) **386**, 738]. *Nature* **385**, 733–736.

Mousa, S. A., Forsythe, M., Wityak, J., Bozarth, J., and Mu, D. X. (1998). Intravenous and oral antiplatelet/antithrombotic efficacy and specificity of XR300, a novel nonpeptide platelet GPIIb/IIIa antagonist. *J. Cardiovasc. Pharmacol.* **31**, 441–448.

Nakamura, I., Tanaka, H., Rodan, G. A., and Duong, L. T. (1998). Echistatin inhibits the migration of murine prefusion osteoclasts and the formation of multinucleated osteoclast-like cells. *Endocrinology* **139**, 5182–5193.

Nasu, K., Ishida, T., Setoguchi, M., Higuchi, Y., Akizuki, S., and Yamamoto, S. (1995). Expression of wild-type and mutated rabbit osteopontin in *Escherichia coli*, and their effects on adhesion and migration of P388D1 cells. *Biochem. J.* **307**, 257–265.

Nurden, A. T., and Caen, J. P. (1974). An abnormal platelet glycoprotein pattern in three cases of Glanzmann's thrombasthenia. *Br. J. Haematol.* **28**, 253–260.

Oldberg, A., Franzen, A., and Heinegard, D. (1986). Cloning and sequence analysis of rat bone sialoprotein (osteopontin) cDNA reveals an Arg-Gly-Asp cell-binding sequence. *Proc. Natl Acad. Sci. USA* **83**, 8819–8823.

Paine, M. J., Desmond, H. P., Theakston, R. D., and Crampton, J. M. (1992). Purification, cloning, and molecular characterization of a high molecular weight hemorrhagic metalloprotease, jararhagin, from *Bothrops jararaca* venom. Insights into the disintegrin gene family. *J. Biol. Chem.* **267**, 22869–22876.

Panda, D., Kundu, G. C., Lee, B. I., Peri, A., Fohl, D., Chackalaparampil, I., Mukherjee, B. B., Li, X. D., Mukherjee, D. C., Seides, S., Rosenberg, J., Stark, K., and Mukherjee, A. B. (1997). Potential roles of osteopontin and alpha(v)betainf 3 integrin in the development of coronary artery restenosis after angioplasty. *Proc. Natl Acad. Sci. USA* **94**, 9308–9313.

Parsons, S. F., Jones, J., Anstee, D. J., Judson, P. A., Gardner, B., Wiener, E., Poole, J., Illum, N., and Wickramasinghe, S. N. (1994). A novel form of congenital dyserythropoietic anemia associated with deficiency of erythroid CD44 and a unique blood group phenotype [In(a-b-), Co(a-b-)]. *Blood* **83**, 860–868.

Phillips, D. R., and Agin, P. P. (1977). Platelet membrane defects in Glanzmann's thrombasthenia. Evidence for decreased amounts of two major glycoproteins. *J. Clin. Invest.* **60**, 535–545.

Phillips, D. R., Jennings, L. K., and Edwards, H. H. (1980). Identification of membrane proteins mediating the interaction of human platelets. *J. Cell Biol.* **86**, 77–86.

Ponta, H., Wainwright, D., and Herrlich, P. (1998). The CD44 protein family. *Int. J. Biochem. Cell Biol.* **30**, 299–305.

Pytela, R., Pierschbacher, M. D., and Ruoslahti, E. (1985a). A 125/115-kDa cell surface receptor specific for vitronectin interacts with the arginine-glycine-aspartic acid adhesion sequence derived from fibronectin. *Proc. Natl Acad. Sci. USA* **82**, 5766–5770.

Pytela, R., Pierschbacher, M. D., and Ruoslahti, E. (1985b). Identification and isolation of a 140 kd cell surface glycoprotein with properties expected of a fibronectin receptor. *Cell* **40**, 191–198.

Ristamaki, R., Joensuu, H., Gron-Virta, K., Salmi, M., and Jalkanen, S. (1997). Origin and function of circulating CD44 in non-Hodgkin's lymphoma. *J. Immunol.* **158**, 3000–3008.

Rosales, C., and Juliano, R. (1996). Integrin signaling to NF-kappa B in monocytic leukemia cells is blocked by activated oncogenes. *Cancer Res.* **56**, 2302–2305.

Rosenkranz, A. R., Coxon, A., Maurer, M., Gurish, M. F., Austen, K. F., Friend, D. S., Galli, S. J., and Mayadas, T. N. (1998). Impaired mast cell development and innate immunity in Mac-1 (CD11b/CD18, CR3)-deficient mice. *J. Immunol.* **161**, 6463–6467.

Rothman, B. L., Blue, M. L., Kelley, K. A., Wunderlich, D., Mierz, D. V., and Aune, T. M. (1991). Human T cell activation by OKT3 is inhibited by a monoclonal antibody to CD44. *J. Immunol.* **147**, 2493–2499.

Ruggeri, Z. M., Bader, R., and de Marco, L. (1982). Glanzmann thrombasthenia: deficient binding of von Willebrand factor to thrombin-stimulated platelets. *Proc. Natl Acad. Sci. USA* **79**, 6038–6041.

Schluesener, H. J. (1998). The disintegrin domain of ADAM 8 enhances protection against rat experimental autoimmune encephalomyelitis, neuritis and uveitis by a polyvalent autoantigen vaccine. *J. Neuroimmunol.* **87**, 197–202.

Schmits, R., Filmus, J., Gerwin, N., Senaldi, G., Kiefer, F., Kundig, T., Wakeham, A., Shahinian, A., Catzavelos, C., Rak, J., Furlonger, C., Zakarian, A., Simard, J. J., Ohashi, P. S., Paige, C. J., Gutierrez-Ramos, J. C., and Mak, T. W. (1997). CD44 regulates hematopoietic progenitor distribution, granuloma formation, and tumorigenicity. *Blood* **90**, 2217–2233.

Schneller, M., Vuori, K., and Ruoslahti, E. (1997). Alphavbeta3 integrin associates with activated insulin and PDGFbeta receptors and potentiates the biological activity of PDGF. *EMBO J.* **16**, 5600–5607.

Sconocchia, G., Titus, J. A., and Segal, D. M. (1994). CD44 is a cytotoxic triggering molecule in human peripheral blood NK cells. *J. Immunol.* **153**, 5473–5481.

Sconocchia, G., Titus, J. A., and Segal, D. M. (1997). Signaling pathways regulating CD44-dependent cytolysis in natural killer cells. *Blood* **90**, 716–725.

Shigeta, O., Gluszko, P., Downing, S. W., Lu, W., Niewiarowski, S., and Edmunds Jr, L. H. (1992). Protection of platelets during long-term extracorporeal membrane oxygenation in sheep with a single dose of a disintegrin. *Circulation* **86**, II398–II404.

Shyy, J. Y., and Chien, S. (1997). Role of integrins in cellular responses to mechanical stress and adhesion. *Curr. Opin. Cell Biol.* **9**, 707–713.

Smith, L. L., Cheung, H. K., Ling, L. E., Chen, J., Sheppard, D., Pytela, R., and Giachelli, C. M. (1996). Osteopontin N-terminal domain contains a cryptic adhesive sequence recognized by alpha9beta1 integrin. *J. Biol. Chem.* **271**, 28485–28491.

Springer, T. A. (1994). Traffic signals for lymphocyte recirculation and leukocyte emigration: the multistep paradigm. *Cell* **76**, 301–314.

Springer, T. A., Thompson, W. S., Miller, L. J., Schmalstieg, F. C., and Anderson, D. C. (1984). Inherited deficiency of the Mac-1, LFA-1, p150,95 glycoprotein family and its molecular basis. *J. Exp. Med.* **160**, 1901–1918.

Springer, T. A., Teplow, D. B., and Dreyer, W. J. (1985). Sequence homology of the LFA-1 and Mac-1 leukocyte adhesion glycoproteins and unexpected relation to leukocyte interferon. *Nature* **314**, 540–542.

Srivatsa, S. S., Fitzpatrick, L. A., Tsao, P. W., Reilly, T. M., Holmes Jr, D. R., Schwartz, R. S., and Mousa, S. A. (1997). Selective alpha v beta 3 integrin blockade potently limits neointimal hyperplasia and lumen stenosis following deep coronary arterial stent injury: evidence for the functional importance of integrin alpha v beta 3 and osteopontin expression during neointima formation. *Cardiovasc. Res.* **36**, 408–428.

Stamenkovic, I., Amiot, M., Pesando, J. M., and Seed, B. (1989). A lymphocyte molecule implicated in lymph node homing is a member of the cartilage link protein family. *Cell* **56**, 1057–1062.

Standiford, T. J., Kunkel, S. L., Kasahara, K., Milia, M. J., Rolfe, M. W., and Strieter, R. M. (1991). Interleukin-8 gene expression from human alveolar macrophages: the role of adherence. *Am. J. Respir. Cell Mol. Biol.* **5**, 579–585.

Stephens, L. E., Sutherland, A. E., Klimanskaya, I. V., Andrieux, A., Meneses, J., Pedersen, R. A., and Damsky, C. H. (1995). Deletion of beta 1 integrins in mice results in inner cell mass failure and peri-implantation lethality. *Genes Dev.* **9**, 1883–1895.

Suzuki, S., Argraves, W. S., Pytela, R., Arai, H., Krusius, T., Pierschbacher, M. D., and Ruoslahti, E. (1986). cDNA and amino acid sequences of the cell adhesion protein receptor recognizing vitronectin reveal a transmembrane domain and homologies with other adhesion protein receptors. *Proc. Natl Acad. Sci. USA* **83**, 8614–8618.

Taher, T. E., Smit, L., Griffioen, A. W., Schilder-Tol, E. J., Borst, J., and Pals, S. T. (1996). Signaling through CD44 is mediated by tyrosine kinases. Association with p56lck in T lymphocytes. *J. Biol. Chem.* **271**, 2863–2867.

Tam, S. H., Sassoli, P. M., Jordan, R. E., and Nakada, M. T. (1998). Abciximab (ReoPro, chimeric 7E3 Fab) demonstrates equivalent affinity and functional blockade of glycoprotein IIb/IIIa and alpha(v)beta3 integrins. *Circulation* **98**, 1085–1091.

Tamkun, J. W., DeSimone, D. W., Fonda, D., Patel, R. S., Buck, C., Horwitz, A. F., and Hynes, R. O. (1986). Structure of integrin, a glycoprotein involved in the transmembrane linkage between fibronectin and actin. *Cell* **46**, 271–282.

Tanaka, Y., Adams, D. H., Hubscher, S., Hirano, H., Siebenlist, U., and Shaw, S. (1993). T-cell adhesion induced by proteoglycan-immobilized cytokine MIP-1 beta [see comments]. *Nature* **361**, 79–82.

Thorens, B., Mermod, J. J., and Vassalli, P. (1987). Phagocytosis and inflammatory stimuli induce GM-CSF mRNA in macrophages through posttranscriptional regulation. *Cell* **48**, 671–679.

Uede, T., Katagiri, Y., Iizuka, J., and Murakami, M. (1997). Osteopontin, a coordinator of host defense system: a cytokine or an extracellular adhesive protein? *Microbiol. Immunol.* **41**, 641–648.

van der Neut, R., Krimpenfort, P., Calafat, J., Niessen, C. M., and Sonnenberg, A. (1996). Epithelial detachment due to absence of hemidesmosomes in integrin beta 4 null mice. *Nature Genet.* **13**, 366–369.

van Dijk, S., D'Errico, J. A., Somerman, M. J., Farach-Carson, M. C., and Butler, W. T. (1993). Evidence that a non-RGD domain in rat osteopontin is involved in cell attachment. *J. Bone Miner. Res.* **8**, 1499–1506.

Webb, D. S., Shimizu, Y., Van Seventer, G. A., Shaw, S., and Gerrard, T. L. (1990). LFA-3, CD44, and CD45: physiologic triggers of human monocyte TNF and IL-1 release. *Science* **249**, 1295–1297.

Weber, G. F., Ashkar, S., Glimcher, M. J., and Cantor, H. (1996). Receptor–ligand interaction between CD44 and osteopontin (Eta-1). *Science* **271**, 509–512.

Weiss, J. M., Sleeman, J., Renkl, A. C., Dittmar, H., Termeer, C. C., Taxis, S., Howells, N., Hofmann, M., Kohler, G., Schopf, E., Ponta, H., Herrlich, P., and Simon, J. C. (1997). An essential role for CD44 variant isoforms in epidermal Langerhans cell and blood dendritic cell function. *J. Cell Biol.* **137**, 1137–1147.

Werb, Z., Tremble, P. M., Behrendtsen, O., Crowley, E., and Damsky, C. H. (1989). Signal transduction through the fibronectin receptor induces collagenase and stromelysin gene expression. *J. Cell Biol.* **109**, 877–889.

Wickramasinghe, S. N., Illum, N., and Wimberley, P. D. (1991). Congenital dyserythropoietic anaemia with novel intra-erythroblastic and intra-erythrocytic inclusions. *Br. J. Haematol.* **79**, 322–330.

Wilson, R. W., Ballantyne, C. M., Smith, C. W., Montgomery, C., Bradley, A., O'Brien, W. E., and Beaudet, A. L. (1993). Gene targeting yields a CD18-mutant mouse for study of inflammation. *J. Immunol.* **151**, 1571–1578.

Xu, H., Gonzalo, J. A., St Pierre, Y., Williams, I. R., Kupper, T. S., Cotran, R. S., Springer, T. A., and Gutierrez-Ramos, J. C. (1994). Leukocytosis and resistance to septic shock in intercellular adhesion molecule 1-deficient mice. *J. Exp. Med.* **180**, 95–109.

Yamada, A., Nikaido, T., Nojima, Y., Schlossman, S. F., and Morimoto, C. (1991). Activation of human CD4 T lymphocytes. Interaction of fibronectin with VLA-5 receptor on CD4 cells induces the AP-1 transcription factor. *J. Immunol.* **146**, 53–56.

Yang, J. T., Rayburn, H., and Hynes, R. O. (1993). Embryonic mesodermal defects in alpha 5 integrin-deficient mice. *Development* **119**, 1093–1105.

Yang, J. T., Rayburn, H., and Hynes, R. O. (1995). Cell adhesion events mediated by alpha 4 integrins are essential in placental and cardiac development. *Development* **121**, 549–560.

Yeh, C. H., Peng, H. C., and Huang, T. F. (1998). Accutin, a new disintegrin, inhibits angiogenesis *in vitro* and *in vivo* by acting as integrin alphavbeta3 antagonist and inducing apoptosis. *Blood* **92**, 3268–3276.

Yu, Q., and Toole, B. P. (1996). A new alternatively spliced exon between v9 and v10 provides a molecular basis for synthesis of soluble CD44. *J. Biol. Chem.* **271**, 20603–20607.

Yue, T. L., McKenna, P. J., Ohlstein, E. H., Farach-Carson, M. C., Butler, W. T., Johanson, K., McDevitt, P., Feuerstein, G. Z., and Stadel, J. M. (1994). Osteopontin-stimulated vascular smooth muscle cell migration is mediated by beta 3 integrin. *Exp. Cell Res.* **214**, 459–464.

Zhou, D. F., Ding, J. F., Picker, L. J., Bargatze, R. F., Butcher, E. C., and Goeddel, D. V. (1989). Molecular cloning and expression of Pgp-1. The mouse homolog of the human H-CAM (Hermes) lymphocyte homing receptor. *J. Immunol.* **143**, 3390–3395.

LICENSED PRODUCTS

Multiple vendors of immunologic reagents have products related to CD44 and integrins. For example, Pharmingen (San Diego, CA) and Harlan Bioproducts (Indianapolis, IN) have multiple antibodies directed to these antigens. Sigma-Aldrich (St Louis, MO) sells extracellular matrix proteins, RGD-containing peptides, and several disintegrins.

TGFβ Receptors

Peter ten Dijke[1,*] and Carl-Henrik Heldin[2]

[1]Division of Cellular Biochemistry, The Netherlands Cancer Institute, Plesmanlaan 121, Amsterdam, 1066 CX, The Netherlands

[2]Ludwig Institute for Cancer Research, Box 595, Uppsala, S-75124, Sweden

*corresponding author tel: 31-20-5121979, fax: 31-20-5121989, e-mail: ptdijke@nki.nl

DOI: 10.1006/rwcy.2000.18005.

SUMMARY

TGFβ is part of a large family of structurally related secreted signaling proteins, which also includes activins and bone morphogenetic proteins (BMPs), that have an important role in intracellular communication during development and tissue homeostasis. TGFβ regulates the proliferation, differentiation, migration, and apoptosis of many different cell types through interaction with different receptor types: TGFβ type I receptor (TβR-I) and TGFβ type II receptor (TβR-II) are required for TGFβ signal transduction, whereas TGFβ type III receptor (TβR-III) and endoglin appear to have a more indirect role in signal transduction, and may act as accessory receptors.

TβR-I and TβR-II were found to encode serine/threonine kinases. TGFβ-mediated receptor activation involves formation of heteromeric oligomers of two distinct sequentially acting kinases; the constitutively active TβR-II kinase phosphorylates and activates TβR-I. Recently, genetic studies in *Drosophila* and *Caenorhabditis elegans* have led to the identification of downstream effector molecules of the TGFβ receptor complex, known as Smad proteins. Smads relay the signal from the membrane to the nucleus; Smad2 and Smad3 transiently interact with and become phosphorylated by the activated TβR-I, and form heteromeric complexes with Smad4 that translocate to the nucleus, where they act as transcriptional regulators of target genes.

BACKGROUND

Discovery

TGFβ receptors were initially identified by affinity crosslinking of radiolabeled TGFβ1 to cell surface proteins on TGFβ-responsive cells (Massagué *et al.*, 1990). On most cells TGFβ binds to three distinct receptor types, termed TβR-I (65–70 kDa), TβR-II (85–110 kDa), and TβR-III (approximately 300 kDa). Endothelial cells often lack expression of TβR-III, but express a protein structurally related to TβR-III, termed endoglin.

The TβR-II receptor cDNA was identified through an expression cloning strategy using [^{125}I]TGFβ1 to screen transfected COS cells (Lin *et al.*, 1992). The protein sequence predicted from the cDNA revealed high sequence similarity towards the previously identified activin type II receptor (ActR-II), a transmembrane serine/threonine kinase (Mathews and Vale, 1991). TβR-I cDNA was cloned through a polymerase chain reaction (PCR)-mediated approach using PCR primers based upon short stretches of high sequence similarity between ActR-II and a related transmembrane serine/threonine kinase from *C. elegans*, termed Daf-1 (Franzén *et al.*, 1993).

TβR-III cDNA was identified through an expression-cloning strategy using transfected COS cells with iodinated TGFβ1 as a probe (Wang *et al.*, 1991), and by purification on a TGFβ affinity column, protein sequence analysis, and subsequent cDNA cloning (López-Casillas *et al.*, 1991). Endoglin was originally identified by a monoclonal antibody against a cell surface protein present on the pre-B leukemic cell line HOON (Gougos and Letarte, 1990). TβR-III and endoglin are structurally related transmembrane proteins with short intracellular regions.

Alternative names

TβR-I is also known as activin receptor-like kinase (ALK)-5. Alternate names for TβR-III and endoglin are β-glycan and CD105, respectively.

Structure

The overall structures of TβR-I and TβR-II are similar: small cysteine-rich extracellular parts, single transmembrane regions and intracellular parts that contain serine/threonine kinase domains (**Figure 1a**). The extracellular domain of TβR-I is shorter than that of TβR-II. The extracellular domains of human TβR-I and TβR-II contain one and three glycosylation sites, respectively. The kinase domains contain two short stretches at analogous positions with no significant similarity to kinases. The C-terminal extension distal to the kinase domain is longer in TβR-II than in TβR-I. TβR-I, but not TβR-II, has in its intracellular juxtamembrane sequence a region which is rich in serine and glycine residues, termed the Gly-Ser (GS) domain. Both receptors are present in the plasma membrane as homo-oligomeric complexes, possibly homodimers.

TβR-III is a transmembrane proteoglycan which migrates as a 300 kDa component in SDS gel electrophoresis. The core protein is only 120 kDa but the large glycosaminoglycan (GAG) chains, that are rich in chondroitin sulfate and heparan sulfate (six putative *N*-linked glycosylation sites and two GAG attachment sites in the extracellular domain), slows down the migration. TβR-III is present as a homo-oligomer, possibly a homodimer, and has a short intracellular domain (43 amino acid residues) that does not reveal any enzymatic activity but contains many serine and threonine residues (42%). Several transmembrane proteins show homology to a short segment in extracellular domain of TβR-III, including major zymogen granule membrane glycoprotein, sperm receptors Zp2 and Zp3, and urinary protein uromodulin (Bork and Sander, 1992).

Endoglin is a disulfide-linked dimeric transmembrane glycoprotein of 180 kDa (four putative *N*-linked glycosylation sites in the extracellular domain), which is structurally related to TβR-III. Comparison of the two proteins reveals that, in addition to the transmembrane and the short intracellular domain, there are two stretches in the extracellular domain that show substantial sequence similarity (Figure 1b). An additional splice variant of endoglin, termed S-endoglin, has been reported with 14 amino acids in the cytoplasmic domain instead of 46 amino acids of L-endoglin. The intracellular domain of endoglin is constitutively phosphorylated on serine residues.

Figure 1 Structure of TGFβ receptors. (a) A schematic illustration of the structure of human TβR-I and TβR-II. (b) A schematic illustration of structure of human endoglin and TβR-III. The number and localization of intrachain disulfide bonds are unknown.

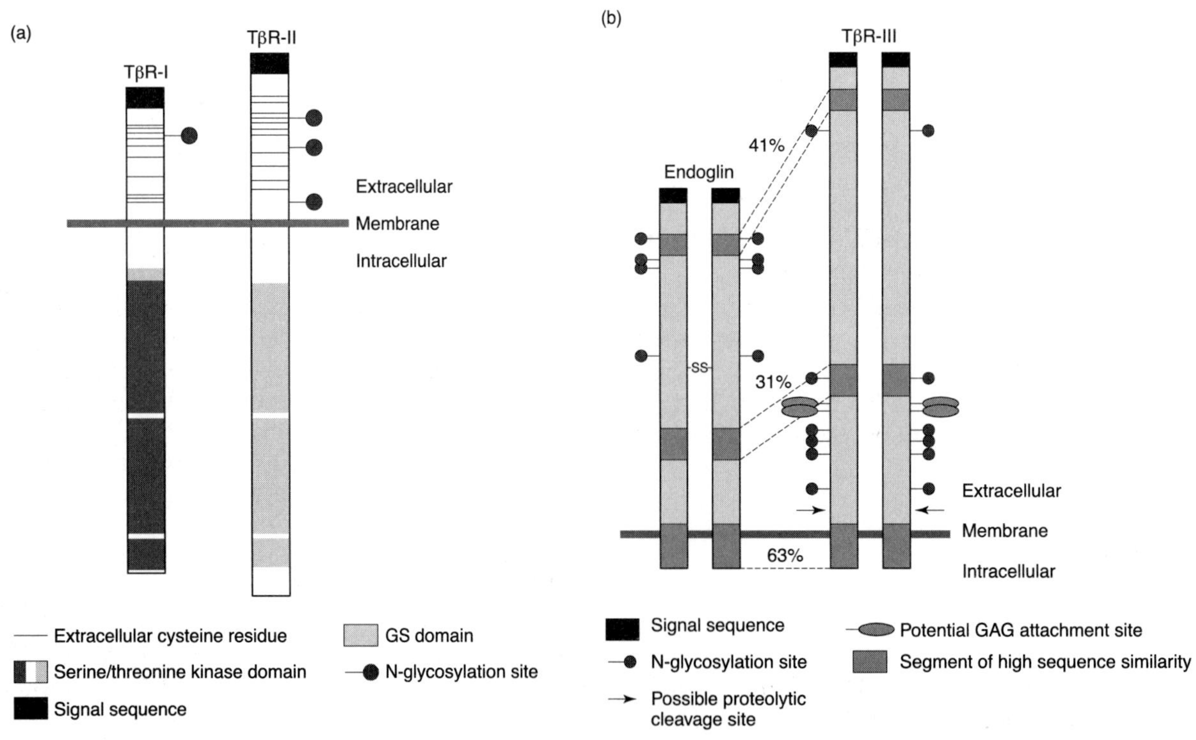

Main activities and pathophysiological roles

TGFβ-mediated signaling through its receptors leads to inhibition of proliferation in many different cell types, including epithelial, endothelial, and hematopoietic cells. For some mesenchymal cells, however, TGFβ is a mitogenic factor. TGFβ is also a potent inducer of extracellular matrix formation and an important regulator of immune responses (Letterio and Roberts, 1998). Inappropriate functioning of TGFβ receptors has been implicated in several pathological conditions, including carcinogenesis, *rheumatoid arthritis*, and *fibrosis*.

GENE

Accession numbers

TβR-I: Human, bovine U97485; rat L26110; mouse D25540; chicken D14460
TβR-II: Human M85079; mouse S69407; rat L09653; chicken L18784
TβR-IIB (alternative splice variant): Human D28131; mouse D32072
TβR-III: Human L07594; pig L07595; rat M80784; mouse AF039601; chicken L01121
Endoglin: Human U37439; pig Z23142; mouse S69407

Sequence

See **Figure 2**.

Chromosome location and linkages

TβR-I maps to human chromosome 9q22, mouse chromosome 4, and bovine chromosome 8.
TβR-II maps to human chromosome 3p22 and mouse chromosome distal 9. Human tumor cells often have a loss of 3p. TβR-II gene is a candidate tumor suppressor gene in this region.
TβR-III maps to human chromosome 1p32-p33.
Endoglin maps to human chromosome 9q33-q34 and mouse chromosome 2.

PROTEIN

Accession numbers

TβR-I: Human A49432; bovine U97485; rat P36897; mouse D25540; chicken D14460
TβR-II: Human P37173; pig P38551; mouse S69114; rat P38438; chicken L48784
TβR-IIB (alternative splice variant): Human D28131; mouse Q62312
TβR-III: Human Q03167; pig P35054; rat P26342; mouse AF039601; chicken 511843
Endoglin: Human S50831; pig P37176; mouse Q63961

Sequence

See **Figure 3**.

Relevant homologies and species differences

TβR-II and, in particular, TβR-I are highly conserved among different species. The overall sequence similarity of human, rat, and mouse TβR-III is approximately 80%. The putative dibasic proteolytic cleavage site in TβR-III and the two potential GAG attachment sites are conserved in human, rat, and mouse. The overall sequence similarity between human and porcine endoglin is 67%. For both TβR-III and endoglin the sequence similarity in the transmembrane and intracellular domains is higher than in the extracellular domains. Human endoglin contains an RGD sequence motif; however, this motif is not found in porcine endoglin.

Affinity for ligand(s)

TβR-I and TβR-II are specific for TGFβ. Upon overexpression of receptors in COS cells, TGFβ1 was found to complex with TβR-II and with each one of the seven known type I receptors. However, in nontransfected cells antibodies against TβR-I, but not antibodies against other type I receptors, were able to immunoprecipitate a crosslinked TGFβ receptor complex from a large variety of TGFβ-responsive cell lines (ten Dijke *et al.*, 1994). Whereas TβR-II can bind ligand by itself, TβR-I requires TβR-II for TGFβ1 binding. TβR-II binds TGFβ1 and TGFβ3 with higher affinity (K_d 5–50 pM) than TGFβ2. However, TGFβ2 can bind to a heteromeric complex of TβR-I and TβR-II.

TβR-III binds TGFβ2 with slightly higher affinity than TGFβ1 and TGFβ3 (K_d 50–200 pM); other TGFβ superfamily ligands do not show appreciable binding to TβR-III. TβR-III facilitates the binding of TGFβ to TβR-II. This is particularly important for TGFβ2, which has only low intrinsic affinity for

TβR-II. TGFβ binding occurs through the core protein part of TβR-III and GAG chains are not required. Two binding sites for TGFβ are present in TβR-III (Kaname and Ruoslahti, 1996; López-Casillas *et al.*, 1994). In addition, TβR-III has been shown to bind fibroblast growth factor through its heparan sulfate proteoglycan chains (Andres *et al.*, 1992). Endoglin binds TGFβ1 and TGFβ3 with higher affinity than TGFβ2. The K_d for the binding of TGFβ1 to endoglin is approximately 50 pM. Interestingly, endoglin was recently shown also to interact with activin A, BMP-2, and BMP-7, in the presence of the appropriate type I or type II receptors (Barbara *et al.*, 1999).

Cell types and tissues expressing the receptor

Most cell types express TβR-I and TβR-II. TβR-III is expressed broadly, but not in many endothelial cells and certain types of hematopoietic, epithelial cells, and myoblasts. Endoglin is selectively expressed in vascular endothelial cells and certain hematopoietic cells.

Figure 2 Nucleotide sequences for human TβR-I, TβR-II, TβR-III, and endoglin.

Sequence

Human TβR-I

```
GGCGAGGCGAGGTTTGCTGGGGTGAGGCAGCGGCGCGGCCGGGCCGGGCCGGGCCACAGGCGGTGGCGGCGGGACCATGGAGGCGGCGGTCGCTGCTCCGCGT
CCCCGGCTGCTCCTCCTCGTGCTGGCGGCGGCGGCGGCGGCGGCGGCGGCGCTGCTCCCGGGGGCGACGGCGTTACAGTGTTTCTGCCACCTCTGTACAAAAG
ACAATTTTACTTGTGTGACAGATGGGCTCTGCTTTGTCTCTGTCACAGAGACCACAGACAAAGTTATACACAACAGCATGTGTATAGCTGAAATTGACTTAAT
TCCTCGAGATAGGCCGTTTGTATGTGCACCCTCTTCAAAAACTGGGTCTGTGACTACAACATATTGCTGCAATCAGGACCATTGCAATAAAATAGAACTTCCA
ACTACTGTAAAGTCATCACCTGGCCTTGGTCCTGTGGAACTGGCAGCTGTCATTGCTGGACCAGTGTGCTTCGTCTGCATCTCACTCATGTTGATGGTCTATA
TCTGCCACAACCGCACTGTCATTCACCATCGAGTGCCAAATGAAGAGGACCCTTCATTAGATCGCCCTTTTATTTCAGAGGGTACTACGTTGAAAGACTTAAT
TTATGATATGACAACGTCAGGTTCTGGCTCAGGTTTACCATTGCTTGTTCAGAGAACAATTGCGAGAACTATTGTGTTACAAGAAAGCATTGGCAAAGGTCGA
TTTGGAGAAGTTTGGAGAGGAAAGTGGCGGGGAGAAGAAGTTGCTGTTAAGATATTCTCCTCTAGAGAAGAACGTTCGTGGTTCCGTGAGGCAGAGATTTATC
AAACTGTAATGTTACGTCATGAAAACATCCTGGGATTTATAGCAGCAGACAATAAAGACAATGGTACTTGGACTCAGCTCTGGTTGGTGTCAGATTATCATGA
GCATGGATCCCTTTTTGATTACTTAAACAGATACACAGTTACTGTGGAAGGAATGATAAAACTTGCTCTGTCCACGGCGAGCGGTCTTGCCCATCTTCACATG
GAGATTGTTGGTACCCAAGGAAAGCCAGCCATTGCTCATAGAGATTTGAAATCAAAGAATATCTTGGTAAAGAAGAATGGAACTTGCTGTATTGCAGACTTAG
GACTGGCAGTAAGACATGATTCAGCCACAGATACCATTGATATTGCTCCAAACCACAGAGTGGGAACAAAAAGGTACATGGCCCCTGAAGTTCTCGATGATTC
CATAAATATGAAACATTTTGAATCCTTCAAACGTGCTGACATCTATGCAATGGGCTTAGTATTCTGGGAAATTGCTCGACGATGTTCCATTGGTGGAATTCAT
GAAGATTACCAACTGCCTTATTATGATCTTGTACCTTCTGACCCATCAGTTGAAGAAATGAGAAAAGTTGTTTGTGAACAGAAGTTAAGGCCAAATATCCCAA
ACAGATGGCAGAGCTGTGAAGCCTTGAGAGTAATGGCTAAAATTATGAGAGAATGTTGGTATGCCAATGGAGCAGCTAGGCTTACAGCATTGCGGATTAAGAA
AACATTATCGCAACTCAGTCAACAGGAAGGCATCAAAATGTAATTCTACAGCTTTGCCTGAACTCTCCTTTTTTCTTCAGATCTGCTCCTGGGTTTTAATTTG
GGAGGTCAGTTGTTCTACCTCACTGAGAGGGAACAGAAGGATATTGCTTCCTTTTGCAGCAGTGTAATAAAGTCAATTAAAAACTTCCCAGGATTTCTTTGGA
CCCAGGAAACAGCCATGTGGGTCCTTTCTGTGCACTATGAACGCTTCTTTCCCAGGACAGAAAATGTGTAGTCTACCTTTATTTTTTATTAACAAAACTTGTT
TTTTAAAAAGATGATTGCTGGTCTTAACTTTAGGTAACTCTGCTGTGCTGGAGATCATCTTTAAGGGCAAAGGAGTTGGATTGCTGAATTACAATGAAACATG
TCTTATTACTAAAGAAAGTGATTTACTCCTGGTTAGTACATTCTCAGAGGATTCTGAACCACTAGAGTTTCCTTGATTCAGACTTTGAATGTACTGTTCTATA
GTTTTTCAGGATCTTAAAACTAACACTTATAAAACTCTTATCTTGAGTCTAAAAATGACCTCATATAGTAGTGAGGAACATAATTCATGCAATTGTATTTTGT
ATACTATTATTGTTCTTTCACTTATTCAGAACATTACATGCCTTCAAAATGGGTTGTACTATACCAGTAAGTGCCACTTCTGTGTCTTTCTAATGGAAATGAG
TAGATTGCTGAAAGTCTCTATGTTAAAACCTATAGTGTTT
```

Human TβR-II

```
GTTGGCGAGGAGTTTCCTGTTTCCCCCGCAGCGCTGAGTTGAAGTTGAGTGAGTCACTCGCGCGCACGGAGCGACGACACCCCCGCGCGTGCACCCGCTCGGG
ACAGGAGCCGGACTCCTGTGCAGCTTCCCTCGGCCGCCGGGGGCCTCCCCGCGCCTCGCCGGCCTCCAGGCCCCTCCTGGCTGGCGAGCGGGCGCCACATCTG
GCCCGCACATCTGCGCTGCCGGCCCGGCGCGGGGTCCGGAGAGGGCGCGGCGCGGAGCGCAGCCAGGGGTCCGGGAAGGCGCCGTCCGTGCGCTGGGGGCTCG
GTCTATGACGAGCAGCGGGGTCTGCCATGGGTCGGGGGCTGCTCAGGGGCCTGTGGCCGCTGCACATCGTCCTGTGGACGCGTATCGCCAGCACGATCCCACC
GCACGTTCAGAAGTCGGTTAATAACGACATGATAGTCACTGACAACAACGGTGCAGTCAAGTTTCCACAACTGTGTAAATTTTGTGATGTGAGATTTTCCACC
TGTGACAACCAGAAATCCTGCATGAGCAACTGCAGCATCACCTCCATCTGTGAGAAGCCACAGGAAGTCTGTGTGGCTGTATGGAGAAAGAATGACGAGAACA
TAACACTAGAGACAGTTTGCCATGACCCCAAGCTCCCCTACCATGACTTTATTCTGGAAGATGCTGCTTCTCCAAAGTGCATTATGAAGGAAAAAAAAAAGCC
TGGTGAGACTTTCTTCATGTGTTCCTGTAGCTCTGATGAGTGCAATGACAACATCATCTTCTCAGAAGAATATAACACCAGCAATCCTGACTTGTTGCTAGTC
ATATTTCAAGTGACAGGCATCAGCCTCCTGCCACCACTGGGAGTTGCCATATCTGTATCATCATCTTCTACTGCTACCGCGTTAACCGGCAGCAGAAGCTGAG
TTCAACCTGGGAAACCGGCAAGACGCGGAAGCTCATGGAGTTCAGCGAGCACTGTGCCATCATCCTGGAAGATGACCGCTCTGACATCAGCTCCACGTGTGCC
AACAACATCAACCACAACACAGAGCTGCTGCCCATTGAGCTGGACACCCTGGTGGGAAAGGTCGCTTTGCTGAGGTCTATAAGGCCAAGCTGAAGCAGAACA
CTTCAGAGCAGTTTGAGACAGTGGCAGTCAAGATCTTTCCCTATGAGGAGTATGCCTCTTGGAAGACAGAGAAGGACATCTTCTCAGACATCAATCTGAAGCA
TGAGAACATACTCCAGTTCCTGACGGCTGAGGAGCGGAAGACGGAGTTGGGGAAACAATACTGGCTGATCACCGCCTTCCACGCCAAGGGCAACCTACAGGAG
TACCTGACGCGGCATGTCATCAGCTGGGAGGACCTGCGCAAGCTGGGCAGCTCCCTCGCCCGGGGGATTGCTCACCTCCACAGTGATCACACTCCATGTGGGA
GGCCCAAGATGCCCATCGTGCACAGGGACCTCAAGAGCTCCAATATCCTCGTGAAGAACGACCTAACCTGCTGCCTGTGTGACTTTGGGCTTTCCCTGCGTCT
GGACCCTACTCTGTCTGTGGATGACCTGGCTAACAGTGGGCAGGTGGGAACTGCAAGATACATGGCTCCAGAAGTCCTAGAATCCAGGATGAATTTGGAGAAT
GCTGAGTCCTTCAAGCAGACCGATGTCTACTCCATGGCTCTGGTGCTCTGGGAAATGACATCTCGCTGTAATGCAGTGGGAGAAGTAAAAGATTATGAGCCTC
CATTTGGTTCCAAGGTGCGGGAGCACCCCTGTGTCGAAAGCATGAAGGACAACGTGTTGAGAGATCGAGGGCGACCAGAAATTCCCAGCTTCTGGCTCAACCA
CCAGGGCATCCAGATGGTGTGTGAGACGTTGACTGAGTGCTGGGACCACGACCCAGAGGCCCGTCTCACAGCCCAGTGTGTGGCAGAACGCTTCAGTGAGCTG
GAGCATCTGGACAGGCTCTCGGGGAGGAGCTGCTCGGAGGAGAAGATTCCTGAAGACGGCTCCCTAAACACTACCAAATAGCTCTTATGGGGCAGGCTGGGCA
TGTCCAAAGAGGCTGCCCCTCTCACCAAA
```

Figure 2 (*Continued*)

```
Human TβR-III
TCTTTAAGATTTGTAGCTACTAAGAAAGAAAGGAGCTTTTTTTCCTTGGGCCTTCAAACTGAAAGAACCGCATGAGCCTGACGGCGCATGGTCTTAACATCAG
GCTGTGCAGGAAGAAGCTATCTGCAGATGGATGCCAGCACACACAAGGAAGCAGAGCTCTGGCAACATTGAGTCAAAGCAAGGACACAACATCAGAGGGACGG
CAGAGAATCCTTGTGTGTAGTCTTTGGTGGCAGTTTGAAAATTGCAAGGAGGGACTTTAAGACTACTTCTGATTTGCAAAGATGGTCTGTGCTCCGAGCAGGC
TAAAGTGACTGGACGAGACGCACTGTTGGAGAAATAAAAATGACTTCCCATTATGTGATTGCCATCTTTGCCCTGATGAGCTTCTGTTTAGCCACTGCAGGTC
CAGAGCCTGGTGCACTGTGTGAACTGTCACCTGTCAGTGCCTCCCATCCTGTCCAGGCCTTGATGGAGAGCTTCACTGTTTTGTCAGGCTGTGCCAGCAGAGG
CACAACTGGGCTGCCACAGGAGGTGCATGTCCTGAATCTCGCACTGCGCCAGGGGCCTGGCCAGCTACAGAGAGAGGTCACACTTCACCTGAATCCCATCTCC
TCAGTCCACATCCACCACAAGTCTGTTGTGTTCCTGCTCAACTCCCCACACCCCCTGGTGTGGCATCTGAAGACAGAGAGACTTGCCACTGGGGTCTCCAGAC
TGTTTTTGGTGTCTGAGGGTTCTGTGGTCCAGTTTTCATCAGCAAACTTCTCCTTGACAGCAGAAACAGAAGAAAGGAACTTCCCCCATGGAAATGAACATCT
GTTAAATTGGGCCCGAAAAGAGTATGGAGCAGTTACTTCATTCACCGAACTCAAGATAGCAAGAAACATTTATATTAAAGTGGGGGAAGATCAAGTGTTCCCT
CCAAAGTGCAACATAGGGAAGAATTTTCTCTCACTCAATTACCTTGCTGAGTACCTTCAACCCAAAGCAGCAGAAGGGTGTGTGATGTCCAGCCAGCCCCAGA
ATGAGGAAGTACACATCATCGAGCTAATCACCCCCAACTCTAACCCCTACAGTGCTTTCCAGGTGGATATAACAATTGATATAAGACCTTCTCAAGAGGATCT
TGAAGTGGTCAAAAATCTCATCCTGATCTTGAAGTGCAAAAAGTCTGTCAACTGGGTGATCAAATCTTTTGATGTTAAGGGAAGCCTGAAAATTATTGCTCCT
AACAGTATTGGCTTTGGAAAAGAGAGTGAAAGATCTATGACAATGACCAAATCAATAAGAGATGACATTCCTTCAACCCAAGGGAATCTGGTGAAGTGGGCTT
TGGACAATGGCTATAGTCCAATAACTTCATACACAATGGCTCCTGTGGCAATAGTATTTCATCTTCGGCTTGAAAATAATGAGGAGATGGGAGATGAGGAAGT
CCACACTATTCCTCCTGAGCTACGGATCCTGCTGGACCCTGGTGCCCTGCCTGCCCTGCAGAACCCGCCCATCCGGGGAGGGGAAGGCCAAAATGGAGGCCTT
CCGTTTCCTTTCCCAGATATTTCCAGGAGAGTCTGGAATGAAGAGGGAGAAGATGGGCTCCCTCGGCCAAAGGACCCTGTCATTCCCAGCATACAACTGTTTC
CTGGTCTCAGAGAGCCAGAAGAGGTGCAAGGGAGCGTGGATATTGCCCTGTCTGTCAAATGTGACAATGAGAAGATGATCGTGGCTGTAGAAAAAGATTCTTT
TCAGGCCAGTGGCTACTCGGGGATGGACGTCACCCTGTTGGATCCTACCTGCAAGGCCAAGATGAATGGCACACACTTTGTTTTGGAGTCTCCTCTGAATGGC
TGCGGTACTCGGCCCCGGTGGTCAGCCCTTGATGGTGTGGTCTACTATAACTCCATTGTGATACAGGTTCCAGCCCTTGGGGACAGTAGTGGTTGGCCAGATG
GTTATGAAGATCTGGAGTCAGGTGATAATGGATTTCCGGGAGATATGGATGAAGGAGATGCTTCCCTGTTCACCCGACCTGAAATCGTGGTGTTTAATTGCAG
CCTTCAGCAGGTGAGGAACCCCAGCAGCTTCCAGGAACAGCCCCACGGAAACATCACCTTCAACATGGAGCTATACAACACTGACCTCTTTTTGGTGCCCTCC
CAGGGCGTCTTCTCTGTGCCAGAGAATGGACACGTTTATGTTGAGGTATCTGTTACTAAGGCTGAACAAGAACTGGGATTTGCCATCCAAACGTGCTTTATCT
CTCCATATTCGAACCCTGATAGGATGTCTCATTACACCATTATTGAGAATATTTGTCCTAAAGATGAATCTGTGAAATTCTACAGTCCCAAGAGAGTGCACTT
CCCTATCCCGCAAGCTGACATGGATAAGAAGCGATTCAGCTTTGTCTTCAAGCCTGTCTTCAACACCTCACTGCTCTTTCTACAGTGTGAGCTGACGCTGTGT
ACGAAGATGGAGAAGCACCCCCAGAAGTTGCCTAAGTGTGTGCCTCCTGACGAAGCCTGCACCTCGCTGGACGCCTCGATAATCTGGGCCATGATGCAGAATA
AGAAGACGTTCACCAAGCCCCTTGCTGTGATCCACCATGAAGCAGAATCTAAAGAAAAAGGTCCAAGCATGAAGGAACCAAATCCAATTTCTCCACCAATTTT
CCATGGTCTGGACACCCTAACCGTGATGGGCATTGCGTTTGCAGCCTTTGTGATCGGAGCACTCCTGACGGGGGCCTTGTGGTACATCTATTCTCACACAGGG
GAGACAGCAGGAAGGCAGCAAGTCCCCACCTCCCCGCCAGCCTCGGAAAACAGCAGTGCTGCCCACAGCATCGGCAGCACGCAGAGCACGCCTTGCTCCAGCA
GCAGCACGGCCTAGCCCAACCCAGGCCCAACCCGGCCCAACCCAGCCCAGCCCAGCTCAGCTCAGCTACTCCAAGGGCAGGACCAATGGCTGAGCCTCGTGTC
CAGACTCAGAGGGCTGGATTTTGGTTCCCTTGTAAAGACAGAGTGAATTTCAGTATAAAGATCACCCGTTGTATTCACCCCACACCCAGGGCTAGTATAAACA
TGACCCTGGGCTTCTGTACCACACTAGAATTCATGTGAGAAAGCTAAAATGGTGGTCTTCTCCACCAGCCCCTCACAGGCTTGGGGGTTTCTATGTGAAACA
CATGCCAGTTTTTAAAATGCTGCTTTGTCCAGGTGAGAACATCCATAATTTGGGGCCCTGAGTTTTACCCAGACTCAAGGAGTTGGTAAAGGGTTAATAGCCA
GATAGTAGAACCAGTGAGGAGATGCGGCCAAAGATTCTTTATATCTGAACCAAGATGTAAAACAAGAAATGCTTTGAGGCTTTCTAAGCGATCCTCCTGTCTA
ATTTGCACCTTTGTCTGGATGCACTCTTCTGACCTTGCTGCCACAACCTGTGGGGTCTGATGTGTCCCAAGATGGGTGCTGCCCTCAGGGACTGCACCCTGAC
AAGTGTTAAGGCAACATTCCTTGCTTGTGCCCTGGGCCAAAACCAATGCTGATGACCTTATCAGCTTCCTGTTTCTTCCCATACTGCATACACCACTGCAAAA
TGTCTTAATGCAAATTTTGTATTTCTTACAGGCCTACAGAAATTGAAAATGACCAAAATCAGGAACCACAGATTTGTGCCCATTCCTAATATTTTGTTCTGCA
AATTAATGTATAATTTGAGGTGAAATTCAGTTATAAAGTCAAGGACGAATTTGCACAGTGATATATTTCTATGTGTATGCAAGTACAAGTATATAATATGTCA
CCTGGCACATTCATTTTCTCAGTTGAAGAAGAGAAAATTTGAAAATGTCCTTATGCTTTTAGAGTTGCAACTTAAGTATATTTGGTAGGGTGAGTGTTTCCAC
TCAAAATATGTCAACTTAAAAAAAAATAGGCCCTTTCATAAAAACCAAACTGTAGCAAGATGCAAATGCATGGCAAATCTGTCGGTCTCCAGTTGGTTATCTG
AATAGTGTCACCAATTCCACCAAGACAGTGCTGAGATTGGAAAGGGCACTCATTTGGATTGCCTTACTTCTCTTGCCTTAAATATATCCCATATATTTAATAT
GTCAAAAAGGGCTTGAGGTGAATTTCATTAAATGGAATAATATGATGCCACTTTGCAGCTAAAATAAGCTCAGTGATACCTCCTTGTT

Human endoglin
CCCTCTAGGTGGACAGTCCTAGCAACCATGGCTCAATNNCAGGCCTGGCTGTGATGAGCCCGTTTGCTGCAAGAGGAGACTGAGGTTCAGAGAAGTCGAGGGT
CCATGGCTCAGCAGAGCTGGCACCAAACCCACATGGGCCAGCACAACAGGGTAGGGGATGGGGCAGGGGCAGAGTGGCAGTGCTGATGGCGTCGGCCCTCTCT
AGGTTGCACAAGCAAAGGCCTCGTCCTGCCCGCCGTGCTGGGCATCACCTTTGGTGCCTTCCTCATCGGGGCCCTGCTCACTGCTGCACTCTGGTACATCTAC
TCGCACACGCGTGAGTACCCCAGGCCCCCACAGTGAGCATGCCGGGCCCCTCCATCCACCCGGGGGAGCCCAGTGAAGCCTCTGAGGGATTGAGGGGCCCTGG
CAGGACCCTGACCTCCGCCCCTGCCCCCGCTCCCGCTCCCAGGTTCCCCCAGCAAGCGGGAGCCCGTGGTGGCGGTGGCTGCCCCGGCCTCCTCGGAGAGCAG
CAGCACCAACCACAGCATCGGGAGCACCCAGAGCACCCCCTGCTCCACCAGCAGCATGGCATAGCCCCGGCCCCCCGCGCTCGCCCAGCAGGAGAGACTGAGC
AGCCGCCAGCTGGGAGCACTGGTGTGAACTCACCCTGGGAGCCAGTCCTCCACTCGACCCAGAATGGAGCCTGCTCTCCGCGCCTACCCTTCCCGCCTCCCTC
TCAGAGGCCTGCTGCCAGTGCAGCCACTGGCTTGGAACACCTTGGGGTCCCTCCACCCCACAGAACCTTCAACCCAGTGGGTCTGGGATATGGCTGCCCAGGA
GACAGACCACTTGCCACGCTGTTGTAAAAACCCAAGTCCCTGTCATTTGAACCTGGATCCAGCACTGGTGAACTGAGCTGGGCAGGAAGGGAGAACTTGAAAC
AGATTCAGGCCAGCCCAGCCAGGCCAACAGCACCTCCCCGCTGGGAAGAGAAGAGGGCCCAGCCCAGAGCCACCTGGATCTATCCCTGCGGCCTCCACACCTG
AACTTGCCTAACTAACTGGCAGGGGAGACAGGAGCCTAGCGGAGCCCAGCCTGGGAGCCCAGAGGGTGGCAAGAACAGTGGGCGTTGGGAGCCTAGCTCCTGC
CACATGGAGCCCCCTCTGCCGGTCGGGCAGCCAGCAGAGGGGGAGTAGCCAAGCTGCTTGTCCTGGGCCTGCCCCTGTGTATTCACCACCAATAAATCAGACC
ATGAAACCTGAAA
```

Regulation of receptor expression

The expression of TGFβ receptors on the surface of cultured cells is regulated by external stimuli as well as by the growth conditions. Thus, TGFβ1 has been shown to upregulate mRNA and protein for TβR-I and -II in a pancreatic cancer cell line (Kleeff and Korc, 1998). Moreover, TβR-II levels were found to be lowered on microvascular endothelial cells grown in three-dimensional collagen gels, compared to cells grown on a solid support (Sankar *et al.*, 1996). The

Figure 3 Amino acid sequences for human TβR-I, TβR-II, TβR-III, and endoglin.

Description of protein

Human TβR-I

MEAAVAAPRPRLLLLVLAAAAAAAAALLPGATALQCFCHLCTKDNFTCVTDGLCFVSVTETTDKVIHNSMCIAEIDLIPRDRPFVCAPSSKTGSVTTTYCCNQ
DHCNKIELPTTVKSSPGLGPVELAAVIAGPVCFVCISLMLMVYICHNRTVIHHRVPNEEDPSLDRPFISEGTTLKDLIYDMTTSGSGSGLPLLVQRTIARTIV
LQESIGKGRFGEVWRGKWRGEEVAVKIFSSREERSWFREAEIYQTVMLRHENILGFIAADNKDNGTWTQLWLVSDYHEHGSLFDYLNRYTVTVEGMIKLALST
ASGLAHLHMEIVGTQGKPAIAHRDLKSKNILVKKNGTCCIADLGLAVRHDSATDTIDIAPNHRVGTKRYMAPEVLDDSINMKHFESFKRADIYAMGLVFWEIA
RRCSIGGIHEDYQLPYYDLVPSDPSVEEMRKVVCEQKLRPNIPNRWQSCEALRVMAKIMRECWYANGAARLTALRIKKTLSQLSQQEGIKM

Human TβR-II

MGRGLLRGLWPLHIVLWTRIASTIPPHVQKSVNNDMIVTDNNGAVKFPQLCKFCDVRFSTCDNQKSCMSNCSITSICEKPQEVCVAVWRKNDENITLETVCHD
PKLPYHDFILEDAASPKCIMKEKKKPGETFFMCSCSSDECNDNIIFSEEYNTSNPDLLLVIFQVTGISLLPPLGVAISVIIIFYCYRVNRQQKLSSTWETGKT
RKLMEFSEHCAIILEDDRSDISSTCANNINHNTELLPIELDTLVGKGRFAEVYKAKLKQNTSEQFETVAVKIFPYEEYASWKTEKDIFSDINLKHENILQFLT
AEERKTELGKQYWLITAFHAKGNLQEYLTRHVISWEDLRKLGSSLARGIAHLHSDHTPCGRPKMPIVHRDLKSSNILVKNDLTCCLCDFGLSLRLDPTLSVDD
LANSGQVGTARYMAPEVLESRMNLENAESFKQTDVYSMALVLWEMTSRCNAVGEVKDYEPPFGSKVREHPCVESMKDNVLRDRGRPEIPSFWLNHQGIQMVCE
TLTECWDHDPEARLTAQCVAERFSELEHLDRLSGRSCSEEKIPEDGSLNTTK

Human TβR-III

MTSHYVIAIFALMSFCLATAGPEPGALCELSPVSASHPVQALMESFTVLSGCASRGTTGLPQEVHVLNLALRQGPGQLQREVTLHLNPISSVHIHHKSVVFLL
NSPHPLVWHLKTERLATGVSRLFLVSEGSVVQFSSANFSLTAETEERNFPHGNEHLLNWARKEYGAVTSFTELKIARNIYIKVGEDQVFPPKCNIGKNFLSLN
YLAEYLQPKAAEGCVMSSQPQNEEVHIIELITPNSNPYSAFQVDITIDIRPSQEDLEVVKNLILILKCKKSVNWVIKSFDVKGSLKIIAPNSIGFGKESERSM
TMTKSIRDDIPSTQGNLVKWALDNGYSPITSYTMAPVAIVFHLRLENNEEMGDEEVHTIPPELRILLDPGALPALQNPPIRGGEGQNGGLPFPFPDISRRVWN
EEGEDGLPRPKDPVIPSIQLFPGLREPEEVQGSVDIALSVKCDNEKMIVAVEKDSFQASGYSGMDVTLLDPTCKAKMNGTHFVLESPLNGCGTRPRWSALDGV
VYYNSIVIQVPALGDSSGWPDGYEDLESGDNGFPGDMDEGDASLFTRPEIVVFNCSLQQVRNPSSFQEQPHGNITFNMELYNTDLFLVPSQGVFSVPENGHVY
VEVSVTKAEQELGFAIQTCFISPYSNPDRMSHYTIIENICPKDESVKFYSPKRVHFPIPQADMDKKRFSFVFKPVFNTSLLFLQCELTLCTKMEKHPQKLPKC
VPPDEACTSLDASIIWAMMQNKKTFTKPLAVIHHEAESKEKGPSMKEPNPISPPIFHGLDTLTVMGIAFAAFVIGALLTGALWYIYSHTGETAGRQQVPTSPP
ASENSSAAHSIGSTQSTPCSSSSTA

Human Endoglin

MDRGTLPLAVALLLASCSLSPTSLAETVHCDLQPVGPERGEVTYTTSQVSKGCVAQAPNAILEVHVLFLEFPTGPSQLELTLQASKQNGTWPREVLLVLSVNS
SVFLHLQALGIPLHLAYNSSLVTFQEPPGVNTTELPSFPKTQILEWAAERGPITSAAELNDPQSILLRLGQAQGSLSFCMLEASQDMGRTLEWRPRTPALVRG
CHLEGVAGHKEAHILRVLPGHSAGPRTVTVKVELSCAPGDLDAVLILQGPPYVSWLIDANHNMQIWTTGEYSFKIFPEKNIRGFKLPDTPQGLLGEARMLNAS
IVASFVELPLASIVSLHASSCGGRLQTSPAPIQTTPPKDTCSPELLMSLIQTKCADDAMTLVLKKELVAHLKCTITGLTFWDPSCEAEDRGDKFVLRSAYSSC
GMQVSASMISNEAVVNILSSSSPQRKKVHCLNMDSLSFQLGLYLSPHFLQASNTIEPGQQSFVQVRVSPSVSEFLLQLDSCHLDLGPEGGTVELIQGRAAKGN
CVSLLSPSPEGDPRFSFLLHFYTVPIPKTGTLSCTVALRPKTGSQDQEVHRTVFMRLNIISPDLSGCTSKGLVLPAVLGITFGAFLIGALLTAALWYIYSHTR
SPSKREPVVAVAAPASSESSSTNHSIGSTQSTPCSTSSMA

decreased expression of TβR-II correlated with a decreased growth-inhibitory response to TGFβ. However, the matrix response was retained, suggesting that different responses require different thresholds of signaling. TGFβ receptor levels have also been found to be modulated *in vivo*; TβR-I, -II as well as -III are increased in hepatocytes after partial hepatectomy (Nishikawa *et al.*, 1998). The notion that TGFβ receptors are suppressed in the normal liver is further supported by the observations that receptors are upregulated on hepatocytes (Nishikawa *et al.*, 1998) and fat-storing cells (Friedman *et al.*, 1994) after cells have been explanted into tissue culture. As for other receptors, TGFβ receptors are downregulated at the cell surface in response to ligand-induced internalization (Koli and Arteaga, 1997).

The promoters of TβR-I and TβR-II have been cloned and characterized. The TβR-I promoter lacks a TATA box, but is GC-rich and contains several putative SP-1-binding sites. TGFβ1 was found to upregulate TβR-I transcription. PEBP2/CBFa and AP-1 were identified as possible *cis*-acting elements in the rat TβR-I promoter (Ji *et al.*, 1998). The human TβR-II was found to contain multiple positive and negative regulatory elements in addition to the core promoter. A novel ETS-related transcription factor (ERT) was found to be a major transcription factor regulating transcription of the human TβR-II gene (Choi *et al.*, 1998).

The promoter for human endoglin was characterized (Rius *et al.*, 1998); it was found to lack consensus TATA and CAAT boxes, but contained consensus motifs for multiple transcription factors, among them SP-1 and AP-2. An ETS-binding site at position −68 was found to be important for basal activity of the promoter. TGFβ1 was found to activate the endoglin promoter.

Release of soluble receptors

A cDNA encoding a soluble TβR-I form was isolated from a rat kidney cDNA library, and reported to bind

TGFβ1 in the presence of TβR-II (Choi, 1999). No naturally occurring soluble forms of TβR-II have been reported.

Soluble forms of TβR-III have been detected in the conditioned media of various cell types. Proteolytic cleavage occurs at a dibasic site, Lys-Lys744 in human TβR-III. Soluble TβR-III acts as an antagonist of TGFβ bioactivity by preventing TGFβ binding to TβR-I/TβR-II heteromeric complex (López-Casillas *et al.*, 1994). Endoglin is mutated in *hereditary hemorrhagic telangiectasia* (McAllister *et al.*, 1994). Mutants of endoglin were found to be expressed intracellularly; no dominant negative effect has been demonstrated for these mutated receptors (Pece *et al.*, 1997).

SIGNAL TRANSDUCTION

Associated or intrinsic kinases

TβR-I and TβR-II have intrinsic kinase domains. Although these domains show some of the structural characteristics of tyrosine kinase domains (Huse *et al.*, 1999), the receptors were found mainly to autophosphorylate on serine and threonine residues *in vitro* and *in vivo*. This is consistent with the analysis of the amino acid residues in kinase subdomains VI and VIII, which predicts TβR-I and TβR-II to be serine/threonine kinases.

Phosphorylation sites have been mapped in TβR-II. Interestingly, phosphorylation of Ser213 and Ser409 are required for TβR-II activity, whereas phosphorylation of Ser416 inhibits TβR-II activity (Luo and Lodish, 1997). TβR-II has also been shown to autophosphorylate on Tyr259, Tyr336, and Tyr424 upon overexpression (Lawler *et al.*, 1997), the physiological significance of which, however, remains unknown.

TβR-I is phosphorylated by TβR-II kinase at several residues in the GS domain (Thr185, Thr186, Ser187, Ser189, and Ser191). In addition, TβR-I is phosphorylated at Ser165, which is located N-terminally of the GS domain; phosphorylation of this residue was found to modulate TGFβ signaling responses (Souchelnytskyi *et al.*, 1997).

A constitutively active TβR-I mutant, TβR-I/T204D, has been described (Wieser *et al.*, 1995); this mutant receptor signals in the absence of ligand and TβR-II. Thr204 has not been identified as a phosphorylation site by TβR-II kinase. Presumably the replacement of Thr204 by Asp causes a conformational change in the receptor similar to that induced by the phosphorylation of residues in GS domain by TβR-II kinase.

A kinase defective mutant of TβR-II is a phosphoprotein, indicating that TβR-II, in addition to undergoing autophosphorylation, is a substrate for other kinases.

Cytoplasmic signaling cascades

Recent genetic and biochemical studies have established an intracellular pathway for TGFβ from the cell membrane to the nucleus (Massagué, 1998). Upon TGFβ-induced heteromeric complex formation of TβR-I and TβR-II (most likely consisting of a heterotetramer of two TβR-Is and two TβR-IIs) and subsequent phosphorylation of TβR-I in the GS domain by the constitutively active TβR-II kinase, Smad2 and Smad3 transiently interact with and become phosphorylated in their C-terminal SXS motifs by the activated TβR-I. Recruitment of Smad2 and Smad3 to the TGFβ receptor complex is achieved through SARA, a FYVE zinc finger domain containing protein that interacts with both the TGFβ receptor complex and Smad2 or Smad3 (Tsukazaki *et al.*, 1998) (**Figure 4**). Phosphorylated Smad2 and Smad3 assemble with common mediator Smad4 into heteromeric complexes that translocate into the nucleus, where they regulate, in combination with other transcription factors, the transcription of target genes (**Figure 5**) (Derynck *et al.*, 1998).

In addition to the TGFβ/Smad pathway, recent studies point to the existence of other (parallel) pathways. TGFβ was found to activate TAK-1, a serine/threonine kinase of the MAP kinase family (Yamaguchi *et al.*, 1995). In addition, members of the Rac or Ras families of small GTP-binding proteins have been implicated in intracellular TGFβ signaling (Atfi *et al.*, 1997). In certain cell types, certain MAP kinases, including extracellular signal-regulated kinases (ERK)1 and 2 and stress-activated protein kinase (SAPK)/Jun-N-terminal kinase (JNK), have been found to be activated by TGFβ (Frey and Mulder, 1997).

DOWNSTREAM GENE ACTIVATION

Transcription factors activated

Smad proteins are nuclear effectors for TGFβ receptors. Smad2, Smad3, and Smad4 have conserved N- and C-terminal domains, known as MAD homology (MH)1 and MH2 domains, respectively. The MH2 domain has transcriptional activation activity

Figure 4 Mechanism of activation of TGFβ receptors. Current model for transmembrane signaling of TGFβ through TGFβ receptor complex is depicted. (a) TGFβ initially binds to TβR-III, which facilitates the binding to TβR-II. (b) Subsequently, TGFβ bound to TβR-II recruits TβR-I into the complex. (c) Thereafter, TβR-I is phosphorylated by the constitutively active TβR-II kinase. Activated TβR-I propagates the signal downstream through phosphorylation of downstream targets, including Smad2 and Smad3.

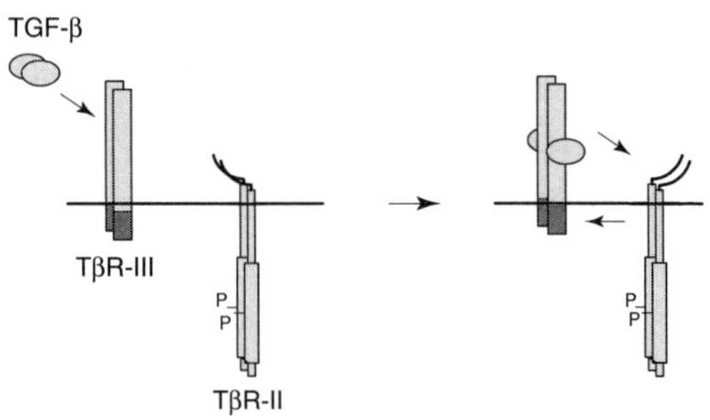

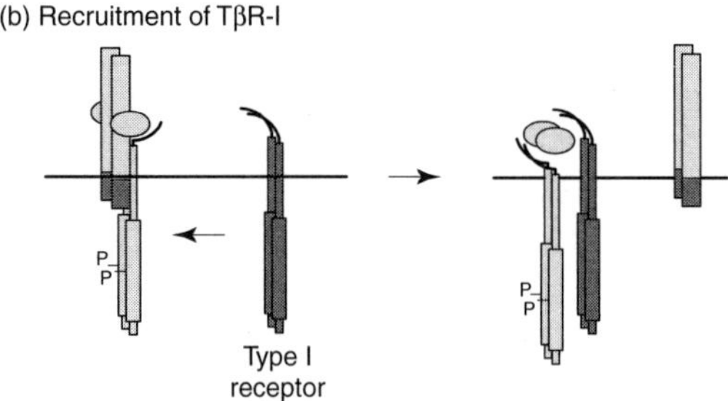

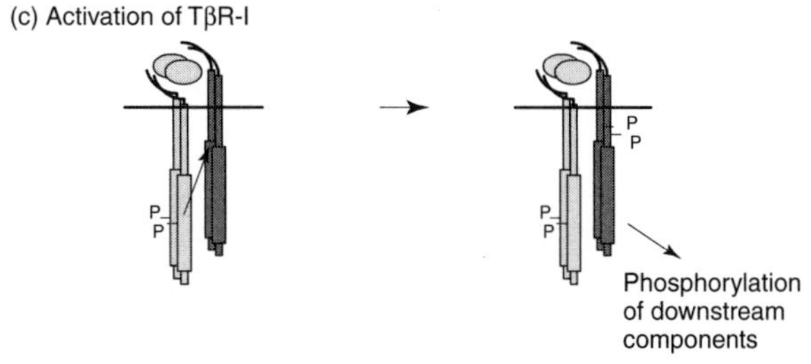

(Liu *et al.*, 1996), and was found to interact with transcriptional coactivators, i.e. p300/CBP for Smad2 and Smad3 (Feng *et al.*, 1998; Janknecht *et al.*, 1998) and MSG1 for Smad4 (Shioda *et al.*, 1998). MH1 domains of Smad3 and Smad4 have been shown to have an intrinsic sequence-specific DNA binding activity (Dennler *et al.*, 1998). Smad2, in contrast to Smad3, does not appear to bind DNA directly. Smads have been shown to interact with other transcription factors; Smad2 and Smad3 interact with FAST (Chen *et al.*, 1997a), and Smad3 with c-Jun and c-Fos (Zhang *et al.*, 1998).

Genes induced

TGFβ elicits its multifunctional effects through transcriptional responses on a large variety of target genes. TGFβ receptor activation potently induces the transcription of extracellular matrix genes, like plasminogen activator inhibitor 1, fibronectin, and collagen. TGFβ also induces its own expression and the transcription of cyclin-dependent kinase (CDK) inhibitors p15 and p21.

Promoter regions involved

Smad-binding elements have been identified in many target genes for TGFβ, including PAI-1 (Dennler *et al.*, 1998), JunB (Jonk *et al.*, 1998), and collagen VII (Vindevoghel *et al.*, 1998). The presence of AP-1 transcription binding sites in TGFβ-responsive promoters has been demonstrated to be important, e.g. PAI-1 and TGFβ1. In the case of p15 and p21, there appears to be an involvement of SP-1 (Moustakas and Kardassis, 1998).

BIOLOGICAL CONSEQUENCES OF ACTIVATING OR INHIBITING RECEPTOR AND PATHOPHYSIOLOGY

Unique biological effects of activating the receptors

TGFβ isoforms induce a multitude of cellular effects, which are context-dependent and vary depending on which other signals the target cell is exposed to. TGFβs are part of a large superfamily with overlapping biological effects. Individual members of the family have unique functions during the development, probably mainly because of their specific expression patterns.

Phenotypes of receptor knockouts and receptor overexpression mice

Mice deficient in TβR-II were found have defects in yolk sac hematopoiesis and vasculogenesis, resulting in an embryonal lethality around 10.5 days of gestation (Oshima *et al.*, 1996). This phenotype is strikingly similar to that of homozygous TGFβ1 null mice (Dickson *et al.*, 1995).

Figure 5 The TGFβ/Smad pathway. (a) Upon TGFβ-mediated heteromeric complex formation of TβR-I and TβR-II and activation of TβR-I by constitutively active TβR-II kinase, Smad2, and Smad3 interact with, and become phosphorylated by TβR-I kinase. Smad2 and Smad3 are recruited to the TGFβ receptor complex through SARA. (b) Subsequently, Smad2 and Smad3 assemble in heteromeric complexes with Smad4 that translocate into the nucleus. (c) Thereafter, the nuclear heteromeric complex, in combination with other transcription factors, regulates the transcription of target genes.

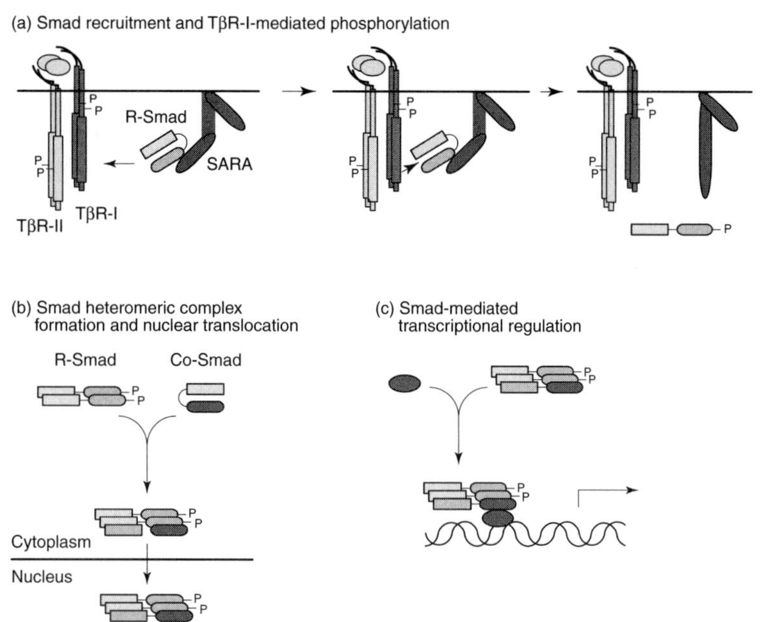

The effect of overexpression of dominant negative TβR-II mutant (lacking the intracellular domain) in transgenic mice has been investigated in different cellular/tissue contexts. Expression of dominant negative TβR-II in epidermis induced a hyperplastic and hyperkeratotic epidermis, and led to a thickened and wrinkled skin (Wang *et al.*, 1997). The results suggest an important role for TGFβ-induced growth inhibition of keratinocytes *in vivo* and maintenance of epidermal homeostasis. In another report, overexpression of dominant negative TβR-II in the basal cell compartment reduced the TGFβ responsiveness of transgenic keratinocytes to TGFβ, and increased the *carcinoma* frequency and accelerated development of carcinomas (Amendt *et al.*, 1998). In addition, challenging transgenic mice overexpressing dominant-negative TβR-II in mammary glands with the carcinogen 7,12-dimethylbenz-[α]-anthracene revealed an enhanced tumorigenesis in the mammary gland (Böttinger *et al.*, 1997a). Expression of dominant negative TβR-II under control of metallothionein 1 (MT1) promoter in transgenic mice led to inhibition of TβR-II-induced signaling in select epithelial cells, and revealed the essential role of TGFβ in maintaining epithelial homeostasis and the differentiated phenotype in the exocrine pancreas (Böttinger *et al.*, 1997b).

Human abnormalities

Tumor cells have been shown to escape from the potent growth-inhibitory effects of TGFβ by decreasing the receptor expression or by inactivating mutations in the TGFβ receptor genes. TβR-II is frequently mutated in an inherited form of *colon cancer* with a microsatellite instability phenotype (Markowitz *et al.*, 1995); the inactivating mutations in TβR-II results in an inability of the cell to respond to TGFβ, and was shown to contribute to cancer progression. In addition, a microsatellite instability in the TβR-II gene has been reported in atherosclerotic and restenotic vascular cells, providing a mechanistic explanation for the inability of TGFβ to inhibit their cell growth (McCaffrey *et al.*, 1995). Loss of functional TβR-I expression was found to correlate with resistance to TGFβ1-mediated growth inhibition in *chronic lymphocytic leukemia* (DeCoteau *et al.*, 1997).

THERAPEUTIC UTILITY

Effect of treatment with soluble receptor domain

The extracellular domain of TβR-II has been expressed in mammalian cells, either alone (Lin *et al.*, 1995) or as fusion with Fc region of human immunoglobulin (Komesli *et al.*, 1998). It was found capable of binding TGFβ1 and TGFβ3, but with lower affinity than wild-type TβR-II; TGFβ2 was not bound. Accordingly, the soluble extracellular domain of TβR-II was found to inhibit TGFβ1 and TGFβ3, but not TGFβ2 bioactivity. The soluble domain of TβR-III was also found to inhibit TGFβ bioactivity.

Effects of inhibitors (antibodies) to receptors

An antibody against the extracellular domain of TβR-II was found to block effectively TGFβ binding to TβR-I without affecting binding to TβR-II or TβR-III, indicating a selective interference of TβR-II/TβR-I heteromeric complex formation. Interestingly, the antibodies were found to inhibit TGFβ-induced PAI-1 transcriptional response without affecting the TGFβ-induced growth inhibition (Hall *et al.*, 1996).

The immunophilin FKBP12 interacts with TβR-I, and inhibits TβR-II-mediated TβR-I phosphorylation. It may function to prevent inappropriate signaling by spontaneous complex formation between TβR-I and TβR-II (Chen *et al.*, 1997b).

Smad7 interacts with activated TβR-I and was shown to prevent the receptor interaction and phosphorylation of Smad2 and Smad3 (Hayashi *et al.*, 1997; Nakao *et al.*, 1997). Smad7 thus functions as a intracellular inhibitor of TGFβ signaling.

References

Amendt, C., Schirmacher, P., Weber, H., and Blessing, M. (1998). Expression of a dominant negative type II TGFβ receptor in mouse skin results in an increase in carcinoma incidence and an acceleration of carcinoma development. *Oncogene* **17**, 25–34.

Andres, J. L., DeFalcis, D., Noda, M., and Massagué, J. (1992). Binding of two growth factor families to separate domains of the proteoglycan betaglycan. *J. Biol. Chem.* **267**, 5927–5930.

Atfi, A., Djelloul, S., Chastre, E., Davis, R., and Gespach, C. (1997). Evidence for a role of Rho-like GTPases and stress-activated protein kinase/c-Jun N-terminal kinase (SAPK/JNK) in transforming growth factor β-mediated signaling. *J. Biol. Chem.* **272**, 1429–1432.

Barbara, N. P., Wrana, J. L., and Letarte, M. (1999). Endoglin is an accessory protein that ineracts with the signaling receptor complex of multiple members of the transforming growth factor-β superfamily. *J. Biol. Chem.* **274**, 584–594.

Bork, P., and Sander, C. (1992). A large domain common to sperm receptors (Zp2 and Zp3) and TGFβ type III receptor. *FEBS Lett.* **300**, 237–240.

Böttinger, E. P., Jakubczak, J. L., Haines, D. C., Bagnall, K., and Wakefield, L. M. (1997a). Transgenic mice overexpressing a dominant-negative mutant type II transforming growth factor β receptor show enhanced tumorigenesis in the mammary gland and lung in response to the carcinogen 7,12-dimethylbenz-[α]-anthracene. *Cancer Res.* **57**, 5564–5570.

Böttinger, E. P., Jakubczak, J. L., Roberts, I. S. D., Mumy, M., Hemmati, P., Bagnall, K., Merlino, G., and Wakefield, L. M. (1997b). Expression of a dominant-negative mutant TGFβ type II receptor in transgenic mice reveals essential roles for TGFβ in regulation of growth and differentiation in the exocrine pancreas. *EMBO J.* **16**, 2621–2633.

Chen, X., Weisberg, E., Fridmacher, V., Watanabe, M., Naco, G., and Whitman, M. (1997a). Smad4 and FAST-1 in the assembly of activin-responsive factor. *Nature* **389**, 85–89.

Chen, Y. G., Liu, F., and Massagué, J. (1997b). Mechanism of TGFβ receptor inhibition by FKBP12. *EMBO J.* **16**, 3866–3876.

Choi, M. E. (1999). Cloning and characterization of a naturally occurring soluble form of TGFβ type I receptor. *Am. J. Physiol.* **276**, F88–F95.

Choi, S. G., Yi, Y., Kim, Y. S., Kato, M., Chang, J., Chung, H. W., Hahm, K. B., Yang, H. K., Rhee, H. H., Bang, Y. J., and Kim, S. J. (1998). A novel ets-related transcription factor, ERT/ESX/ESE-1, regulates expression of the transforming growth factor-β type II receptor. *J. Biol. Chem.* **273**, 110–117.

DeCoteau, J. F., Knaus, P. I., Yankelev, H., Reis, M. D., Lowsky, R., Lodish, H. F., and Kadin, M. E. (1997). Loss of functional cell surface transforming growth factor β (TGFβ) type I receptor correlates with insensitivity to TGFβ in chronic lymphocytic leukemia. *Proc. Natl Acad. Sci. USA* **94**, 5877–5881.

Dennler, S., Itoh, S., Vivien, D., ten Dijke, P., Huet, S., and Gauthier, J.-M. (1998). Direct binding of Smad3 and Smad4 to critical TGFβ-inducible elements in the promoter of human plasminogen activator inhibitor-type 1 gene. *EMBO J.* **17**, 3091–3100.

Derynck, R., Zhang, Y., and Feng, X.-H. (1998). Smads: transcriptional activators of TGFβ responses. *Cell* **95**, 737–740.

Dickson, M. C., Martin, J. S., Cousins, F. M., Kulkarni, A. B., Karlsson, S., and Akhurst, R. J. (1995). Defective haematopoiesis and vasculogenesis in transforming growth factor-β1 knock out mice. *Development* **121**, 1845–1854.

Feng, X. H., Zhang, Y., Wu, R. Y., and Derynck, R. (1998). The tumor suppressor Smad4/DPC4 and transcriptional adaptor CBP/p300 are coactivators for Smad3 in TGFβ-induced transcriptional activation. *Genes Dev.* **12**, 2153–2163.

Franzén, P., ten Dijke, P., Ichijo, H., Yamashita, H., Schulz, P., Heldin, C.-H., and Miyazono, K. (1993). Cloning of a TGFβ type I receptor that forms a heteromeric complex with the TGFβ type II receptor. *Cell* **75**, 681–692.

Frey, R. S., and Mulder, K. M. (1997). Involvement of extracellular signal-regulated kinase 2 and stress-activated protein kinase Jun N-terminal kinase activation by transforming growth factor β in the negative growth control of breast cancer cells. *Cancer Res.* **57**, 628–633.

Friedman, S. L., Yamasaki, G., and Wong, L. (1994). Modulation of transforming growth factor β receptors of rat lipocytes during the hepatic wound healing response. Enhanced binding and reduced gene expression accompany cellular activation in culture and *in vivo*. *J. Biol. Chem.* **269**, 10551–10558.

Gougos, A., and Letarte, M. (1990). Primary structure of endoglin, an RGD-containing glycoprotein of human endothelial cells. *J. Biol. Chem.* **265**, 8361–8364.

Hall, F. L., Benya, P. D., Padilla, S. R., Carbonaro-Hall, D., Williams, R., Buckley, S., and Warburton, D. (1996). Transforming growth factor-β type-II receptor signalling: intrinsic/associated casein kinase activity, receptor interactions and functional effects of blocking antibodies. *Biochem. J.* **316**, 303–310.

Hayashi, H., Abdollah, S., Qiu, Y., Cai, J., Xu, Y. Y., Grinnell, B. W., Richardson, M. A., Topper, J. N., Gimbrone, M. A. J., Wrana, J. L., and Falb, D. (1997). The MAD-related protein Smad7 associates with the TGFβ receptor and functions as an antagonist of TGFβ signaling. *Cell* **89**, 1165–1173.

Huse, M., Chen, Y. G., Massagué, J., and Kuriyan, J. (1999). Crystal structure of the cytoplasmic domain of the type I TGF β receptor in complex with FKBP12. *Cell* **96**, 425–436.

Janknecht, R., Wells, N. J., and Hunter, T. (1998). TGFβ-stimulated cooperation of smad proteins with the coactivators CBP/p300. *Genes Dev.* **12**, 2114–2119.

Ji, C. H., Casinghino, S., Chang, D. J., Chen, Y., Javed, A., Ito, Y., Hiebert, S. W., Lian, J. B., Stein, G. S., McCarthy, T. L., and Centrella, M. (1998). CBFa(AML/PEBP2)-related elements in the TGFβ type I receptor promoter and expression with osteoblast differentiation. *J. Cell Biochem.* **69**, 353–363.

Jonk, L. J. C., Itoh, S., Heldin, C.-H., ten Dijke, P., and Kruijer, W. (1998). Identification and functional characterization of a Smad binding element (SBE) in the *JunB* promoter that acts as a transforming growth factor-β, activin, and bone morphogenetic protein-inducible enhancer. *J. Biol. Chem.* **273**, 21145–21152.

Kaname, S., and Ruoslahti, E. (1996). Betaglycan has multiple binding sites for transforming growth factor-β_1. *Biochem. J.* **315**, 815–820.

Kleeff, J., and Korc, M. (1998). Up-regulation of transforming growth factor (TGF)-β receptors by TGFβ_1 in COLO-357 cells. *J. Biol. Chem.* **273**, 7495–7500.

Koli, K. M., and Arteaga, C. L. (1997). Processing of the transforming growth factor β type I and II receptors. Biosynthesis and ligand-induced regulation. *J. Biol. Chem.* **272**, 6423–6427.

Komesli, S., Vivien, D., and Dutartre, P. (1998). Chimeric extracellular domain type II transforming growth factor (TGF)-β receptor fused to the Fc region of human immunoglobulin as a TGFβ antagonist. *Eur. J. Biochem.* **254**, 505–513.

Lawler, S., Feng, X. H., Chen, R. H., Maruoka, E. M., Turck, C. W., Griswold-Prenner, I., and Derynck, R. (1997). The type II transforming growth factor-β receptor autophosphorylates not only on serine and threonine but also on tyrosine residues. *J. Biol. Chem.* **272**, 14850–14859.

Letterio, J. J., and Roberts, A. B. (1998). Regulation of immune responses by TGFβ. *Annu. Rev. Immunol.* **16**, 137–161.

Lin, H. Y., Wang, X.-F., Ng-Eaton, E., Weinberg, R. A., and Lodish, H. F. (1992). Expression cloning of the TGFβ type II receptor, a functional transmembrane serine/threonine kinase. *Cell* **68**, 775–785.

Lin, H. Y., Moustakas, A., Knaus, P., Wells, R. G., Henis, Y. I., and Lodish, H. F. (1995). The soluble exoplasmic domain of the type II transforming growth factor (TGF)-β receptor. A heterogeneously glycosylated protein with high affinity and selectivity for TGFβ ligands. *J. Biol. Chem.* **270**, 2747–2754.

Liu, C. B., Wallace, K., Shi, C. Z., Heyner, S., Komm, B., and Haddad, J. G. (1996). Post-transcriptional stimulation of transforming growth factor β_1 mRNA by TGFβ_1 treatment of transformed human osteoblasts. *J. Bone Miner. Res.* **11**, 211–217.

López-Casillas, F., Cheifetz, S., Doody, J., Andres, J. L., Lane, W. S., and Massagué, J. (1991). Structure and expression of the membrane proteoglycan betaglycan, a component of the TGFb receptor system. *Cell* **67**, 785–795.

López-Casillas, F., Payne, H. M., Andres, J. L., and Massagué, J. (1994). Betaglycan can act as a dual modulator of TGFβ access to signaling receptors: mapping of ligand binding and GAG attachment sites. *J. Cell Biol.* **124**, 557–568.

Luo, K. X., and Lodish, H. F. (1997). Positive and negative regulation of type II TGFβ receptor signal transduction by autophosphorylation on multiple serine residues. *EMBO J.* **16**, 1970–1981.

Markowitz, S., Wang, J., Myeroff, L., Parsons, R., Sun, L., Lutterbaugh, J., Fan, R. S., Zborowska, E., Kinzler, K. W., Vogelstein, B., Brattain, M., and Willson, J. K. V. (1995). Inactivation of the type II TGFβ receptor in colon cancer cells with microsatellite instability. *Science* **268**, 1336–1338.

Massagué, J. (1998). TGFβ signal transduction. *Annu. Rev. Biochem.* **67**, 753–791.

Massagué, J., Cheifetz, S., Boyd, F. T., and Andres, J. L. (1990). TGFβ receptors and TGFβ binding proteoglycans: recent progress in identifying their functional properties. *Ann. NY Acad. Sci.* **593**, 59–72.

Mathews, L. S., and Vale, W. W. (1991). Expression cloning of an activin receptor, a predicted transmembrane serine kinase. *Cell* **65**, 973–982.

McAllister, K. A., Grogg, K. M., Johnson, D. W., Gallione, C. J., Baldwin, M. A., Jackson, C. E., Helmbold, E. A., Markel, D. S., McKinnon, W. C., Murrell, J., McCormick, M. K., Pericak-Vance, M. A., Heutink, P., Oostra, B. A., Haitjema, T., Westerman, C. J. J., Porteous, M. E., Guttmacher, A. E., Letarte, M., and Marchuk, D. A. (1994). Endoglin, a TGFβ binding protein of endothelial cells, is the gene for hereditary haemorrhagic telangiectasia type 1. *Nature Genet.* **8**, 345–351.

McCaffrey, T. A., Consigli, S., Du, B., Falcone, D. J., Sanborn, T. A., Spokojny, A. M., and Bush Jr, H. L. (1995). Decreased type II/type I TGFβ receptor ratio in cells derived from human atherosclerotic lesions. Conversion from an antiproliferative to profibrotic response to TGFβ_1. *J. Clin. Invest.* **96**, 2667–2675.

Moustakas, A., and Kardassis, D. (1998). Regulation of the human p21/WAF1/Cip1 promoter in hepatic cells by functional interactions between Sp1 and Smad family members. *Proc. Natl Acad. Sci. USA* **95**, 6733–6738.

Nakao, A., Afrakhte, M., Morén, A., Nakayama, T., Christian, J. L., Heuchel, R., Itoh, S., Kawabata, M., Heldin, N.-E., Heldin, C.-H., and ten Dijke, P. (1997). Identification of Smad7, a TGFβ-inducible antagonist of TGFβ signalling. *Nature* **389**, 631–635.

Nishikawa, Y., Wang, M., and Carr, B. I. (1998). Changes in TGFβ receptors of rat hepatocytes during primary culture and liver regeneration: Increased expression of TGFβ receptors associated with increased sensitivity to TGFβ-mediated growth inhibition. *J. Cell Physiol.* **176**, 612–623.

Oshima, M., Oshima, H., and Taketo, M. M. (1996). TGFβ receptor type II deficiency results in defects of yolk sac hematopoiesis and vasculogenesis. *Dev. Biol.* **179**, 297–302.

Pece, N., Vera, S., Cymerman, U., White, R. I. J., Wrana, J. L., and Letarte, M. (1997). Mutant endoglin in hereditary hemorrhagic telangiectasia type 1 is transiently expressed intracellularly and is not a dominant negative. *J. Clin. Invest.* **100**, 2568–2579.

Rius, C., Smith, J. D., Almendro, N., Langa, C., Botella, L. M., Marchuk, D. A., Vary, C. P., and Bernabeu, C. (1998). Cloning of the promoter region of human endoglin, the target gene for hereditary hemorrhagic telangiectasia type 1. *Blood* **92**, 4677–4690.

Sankar, S., Mahooti-Brooks, N., Bensen, L., McCarthy, T. L., Centrella, M., and Madri, J. A. (1996). Modulation of transforming growth factor β receptor levels on microvascular endothelial cells during *in vitro* angiogenesis. *J. Clin. Invest.* **97**, 1436–1446.

Shioda, T., Lechleider, R. J., Dunwoodie, S. L., Li, H., Yahata, T., de Caestecker, M. P., Fenner, M. H., Roberts, A. B., and Isselbacher, K. J. (1998). Transcriptional activating activity of Smad4: roles of SMAD hetero-oligomerization and enhancement by an associating transactivator. *Proc. Natl Acad. Sci. USA* **95**, 9785–9790.

Souchelnytskyi, S., Tamaki, K., Engström, U., Wernstedt, C., ten Dijke, P., and Heldin, C.-H. (1997). Phosphorylation of Ser465 and Ser467 in the C terminus of Smad2 mediates interaction with Smad4 and is required for transforming growth factor-β signaling. *J. Biol. Chem.* **272**, 28107–28115.

ten Dijke, P., Yamashita, H., Ichijo, H., Franzén, P., Laiho, M., Miyazono, K., and Heldin, C.-H. (1994). Characterization of type I receptors for transforming growth factor-β and activin. *Science* **264**, 101–104.

Tsukazaki, T., Chiang, T. A., Davison, A. F., Attisano, L., and Wrana, J. L. (1998). SARA, a FYVE domain protein that recruits Smad2 to the TGFβ receptor. *Cell* **95**, 779–791.

Vindevoghel, L., Lechleider, R. J., Kon, A., de Caestecker, M. P., Uitto, J., Roberts, A. B., and Mauviel, A. (1998). SMAD3/4-dependent transcriptional activation of the human type VII collagen gene (COL7A1) promoter by transforming growth factor β. *Proc. Natl Acad. Sci. USA* **95**, 14769–14774.

Wang, X.-F., Lin, H. Y., Ng-Eaton, E., Downward, J., Lodish, H. F., and Weinberg, R. A. (1991). Expression cloning and characterization of the TGFβ type III receptor. *Cell* **67**, 797–805.

Wang, X. J., Greenhalgh, D. A., Bickenbach, J. R., Jiang, A. B., Bundman, D. S., Krieg, T., Derynck, R., and Roop, D. R. (1997). Expression of a dominant-negative type II transforming growth factor beta (TGFβ) receptor in the epidermis of transgenic mice blocks TGFβ-mediated growth inhibition. *Proc. Natl Acad. Sci. USA* **94**, 2386–2391.

Wieser, R., Wrana, J. L., and Massagué, J. (1995). GS domain mutations that constitutively activate TβR-I, the downstream signaling component in the TGFβ receptor complex. *EMBO J.* **14**, 2199–2208.

Yamaguchi, K., Shirakabe, K., Shibuya, H., Irie, K., Oishi, I., Ueno, N., Taniguchi, T., Nishida, E., and Matsumoto, K. (1995). Identification of a member of the MAPKKK family as a potential mediator of TGFβ signal transduction. *Science* **270**, 2008–2011.

Zhang, Y., Feng, X. H., and Derynck, R. (1998). Smad3 and Smad4 cooperate with c-Jun/c-Fos to mediate TGFβ-induced transcription. *Nature* **394**, 909–913.

LICENSED PRODUCTS

Antibodies against TβR-I

Rabbit polyclonal IgG against TβR-I	Santa Cruz Biotechnology, 2161 Delaware Avenue, Santa Cruz, CA 95060, USA Catalogue number: sc-399 Concentration: 100 µg/mL; epitope corresponding to amino acids 482–501 of TβR-I Antibody works in western blotting, immunoprecipitation, and immunohistochemistry
Rabbit polyclonal IgG against TβR-I	Santa Cruz Biotechnology, 2161 Delaware Avenue, Santa Cruz, CA 95060, USA Catalogue number: sc-398 Concentration: 100 µg/mL; epitope corresponding to amino acids 158–179 of TβR-I Antibody works in western blotting, immunoprecipitation, and immunohistochemistry

Antibodies against TβR-II

Rabbit polyclonal IgG against TβR-II	Santa Cruz Biotechnology, 2161 Delaware Avenue, Santa Cruz, CA 95060, USA

	Catalogue number: sc-220 Concentration: 100 µg/mL; epitope corresponding to amino acids 550–565 of TβR-II Antibody works in western blotting, immunoprecipitation, and immunohistochemistry
Rabbit polyclonal IgG against TβR-II	Santa Cruz Biotechnology, 2161 Delaware Avenue, Santa Cruz, CA 95060, USA Catalogue number: sc-400 Concentration: 100 µg/mL; epitope corresponding to amino acids 246–266 of TβR-II Antibody works in western blotting, immunoprecipitation, and immunohistochemistry
Rabbit polyclonal IgG against TβR-II	Santa Cruz Biotechnology, 2161 Delaware Avenue, Santa Cruz, CA 95060, USA Catalogue number: sc-1700 Concentration: 100 µg/mL; epitope corresponding to amino acids 1–567 of TβR-II Antibody works in western blotting and immunohistochemistry
Mouse monoclonal (IgG1 isotype) against TβR-II	Transduction Laboratories, 133 Venture Ct, Suite 5, Lexington, KY 40511-2624, USA Catalogue number: T42520 Pack size 50 µg or 150 µg; concentration: 250 µg/mL Antibody works in western blotting (1:2500 dilution).
Antibodies against TβR-III	
Goat polyclonal IgG against TβR-III	Santa Cruz Biotechnology, 2161 Delaware Avenue, Santa Cruz, CA 95060, USA Catalogue number: sc-6199 Concentration: 200 mg/mL; epitope corresponding to amino acids 830–849 Antibody works in western blotting and immunohistochemistry
Mouse monoclonal (IgG isotype) against TβR-III	Calbiochem, PO Box 12087, La Jolla, CA 92039-2087, USA Catalogue number: GR19-Q Pack size: 100 µg Antibody works in immunohistochemistry (frozen and paraffin sections).
Antibodies against endoglin	
Mouse monoclonal (IgG1 isotype) against endoglin	Biodesign International, 105 York Street, Kennebunk, ME 04043, USA Catalogue number: P61016M Pack size: 1 mL supernatant (100–200 tests) Antibody works in immunohistochemistry (1 : 10–1 : 2 preferably in phosphate-buffered saline). The antibody is not useful for paraffin sections.
Mouse monoclonal (IgM isotype) against endoglin	Biodesign International, 105 York Street, Kennebunk, ME 04043, USA Catalogue number: K54418M Concentration upon delivery: 200 µg/2 mL
Mouse monoclonal (IgG3 isotype) against endoglin	Biodesign International, 105 York Street, Kennebunk, ME 04043, USA

	Catalogue number: P42226M Pack size: 0.2 mg freeze-dried powder Antibody can be used in flow cytometry ($2\,mg/5 \times 10^5$ cells/test)
Soluble TβR-II protein	
Recombinant human TGFβ soluble type II receptor	R&D Systems, 614 McKinley Place NE, Minneapolis, MN 55413, USA Catalogue number: 241-R2-025 Pack size: 25 µg; lyophilized with BSA as a carrier protein; purity over 97% ED_{50} on TGFβ1-induced growth inhibition of mouse T cell line: 200–500 ng/mL
Recombinant extracellular domain of human TGFβ sRII (159 amino acids) Fc chimera	R&D Systems, 614 McKinley Place NE, Minneapolis, MN 55413, USA Catalogue number: 341-BR-050 Pack size: 50 µg; lyophilized carrier free; purity over 97% ED_{50} on TGFβ1-induced growth inhibition of mouse T cell line: 5–15 ng/mL
Soluble TβR-III protein	
Recombinant human TGFβ soluble type III receptor	R&D Systems, 614 McKinley Place NE, Minneapolis, MN 55413, USA Catalogue number: 242-R3-100 Pack size: 100 mg; lyophilized with BSA as a carrier protein; purity over 97% ED_{50} on TGFβ2-induced growth inhibition of mouse T cell line: 20–50 ng/mL

BMP Receptor

A. Hari Reddi*

Department of Orthopaedic Surgery, Center for Tissue Regeneration and Repair, University of California, Davis, 4635 Second Avenue, Room 2000, Sacramento, CA 95817, USA

*corresponding author tel: 916-734-5749, fax: 916-734-5750, e-mail: ahreddi@ucdavis.edu

DOI: 10.1006/rwcy.2000.18006.

SUMMARY

Bone morphogenetic proteins (BMPs) are osteotrophic and osteoinductive cytokines. BMPs are pleiotropic cytokines with actions not only on bone and cartilage but also on brain, teeth, skin, heart, lung, kidney, and a host of other tissues. They act on chemotaxis, mitosis, differentiation, cell survival and cell death. The biological actions of BMPs are mediated by binding to specific BMP receptors types I and II. There is a collaboration between the receptors and downstream substrates such as receptor-regulated R-Smads 1, 5, and 8. The phosphorylated R-Smads partner with Co-Smad, Smad4, and enter the nucleus to initiate the transcription of BMP-responsive genes. Smad6 inhibits the activity of type I BMP receptor protein kinase. Thus, the BMP receptors provide a finely regulated homeostatic system with checks and balances.

BACKGROUND

It is axiomatic in biology that critical cytokines, growth factors, and morphogens bind to specific receptors and act in collaboration to initiate signal transduction to activate effector genes, culminating in biological action. In certain instances the biological actions may translate into therapeutic actions with clinical applications. Certainly, this is the case with members of the bone morphogenetic protein family (BMPs) and their cognate receptors and downstream effectors. The BMP family is a burgeoning area of signaling molecules with wide-ranging biological actions. However, although there are over 15 members of the family in mammals, they convergence on a paradoxically small number of receptors.

In addition to the BMPs found in mammals, there are a number of homologs found in nematodes, *Drosophila*, sea urchins, and in nonmammalian vertebrates. BMPs are distantly related to the TGFβ family, and it is customary to discuss them together. However, in this article the focus is exclusively on BMP receptor signaling in mammals, with emphasis on the conserved pathways between mammals and *Drosophila*.

Discovery

The early experiments demonstrated the binding of radio-iodinated BMP-4 to cellular proteins (Paralkar *et al.*, 1991). Specific interactions between BMP-4 and BMP-7 and BMP receptors were discovered by ten Dijke *et al.* (1994).

Members of the BMP family bind to BMP type I and II receptors (Heldin *et al.*, 1997; Kawabata *et al.*, 1998; Reddi, 1998). The BMPs play a pivotal role in heteroligomerization of the BMP type IA and B receptors with BMP type II receptor. Type IA BMP receptor (earlier known as ALK-3) and type IB BMP receptor (earlier known as ALK-6) are phosphorylated by type II receptor (Kawabata *et al.*, 1998).

During the course of work on TGFβ type I receptor, seven different activin receptor-like kinases (ALKs) were cloned and called ALK-1 through ALK-7 (ten Dijke *et al.*, 1994; Kang and Reddi, 1996). Certain type I receptors for TGFβ and BMPs interact with an immunophilin called FKBP-12 (Wang *et al.*, 1994).

Main activities and pathophysiological roles

BMPs are multifunctional morphogens involved in pattern formation, mesoderm induction, differentiation, regeneration, and the inevitable cell death and related apoptosis (Reddi, 1998; Bhatia *et al.*, 1999). Unlike TGFβ, which causes fibrosis of diverse tissues, BMPs are critical for embryonic development and recapitulation of developmental stages during regeneration and tissue repair (Reddi, 1998). In amphibians such as *Xenopus*, BMPs induce ventral mesoderm derivatives, while activin induces dorsal mesoderm. Disruption of the BMP signals may induce neural lineages by default. Thus BMP signaling mechanisms play a key morphogenetic role.

The general scheme of the BMP signaling pathway includes interaction and binding to BMP type I and II receptors that function as serine/threonine kinases (Heldin *et al.*, 1997; Kawabata *et al.*, 1998; Reddi, 1998). The resulting phosphorylated acceptor proteins are called Smad1, Smad5, and Smad8. These receptor-regulated Smads (R-Smads) interact intimately with plasma membrane proteins, including SARA (Smad anchor for receptor activation). The phosphorylation of R-Smads enables them to seek their common functional partner Smads, the so-called Co-Smads. Smads are trimeric molecules forming oligomeric cytoplasmic complexes that are translocated into the nucleus. In the nucleus, Smads interact with specific DNA sequences directly or indirectly via Smad-interacting proteins (SIPs). The biological functions of the R-Smad/Co-Smad partnership is regulated by inhibitory Smads (I-Smads), such as Smad6 and Smad7. The inhibitory Smads are generally resident in the nucleus and BMPs appear to translocate them into cytoplasm to competitively inhibit receptor-mediated phosphorylation of R-Smads.

GENE

Little is known at present.

PROTEIN

Little is known at present.

SIGNAL TRANSDUCTION

Associated or intrinsic kinases

The interaction of the BMP family ligand with the type I and II BMP receptor serine/threonine kinases results in activation of type I receptor kinase. The type I and II receptors for BMPs appear to interact with the ligand concurrently. The activated type I BMP receptor phosphorylates the BMP receptor-regulated R-Smads called Smad1, Smad5, and Smad8. It is noteworthy to recall that TGFβ and activin receptors phosphorylate Smad2 and Smad3. The receptor-regulated Smads are in this respect pathway-restricted. However, irrespective of whether the BMP or TGFβ/activin pathway is involved, the various R-Smads interact with a common Smad4 (Co-Smad) and form oligomeric complexes which translocate into the nucleus, possibly due to a conformational change (Heldin *et al.*, 1997; Attisano and Wrana, 1998; Reddi, 1998).

Smads are an evolutionarily conserved family of molecules from nematodes (Savage *et al.*, 1996). They have a molecular mass between 40 and 65 kDa. The three functional classes of Smads are: (1) The pathway-restricted receptor-regulated Smads (R-Smads), of which Smad1, Smad5, and Smad8 are specific for the BMP signaling pathway (Tamaki *et al.*, 1998; Nishimura *et al.*, 1998); (2) the Co-Smads, of which the primary member is Smad4, also known as dpc4 (deleted in pancreatic carcinoma 4), a tumor suppressor gene (Hahn *et al.*, 1996; Zawel *et al.*, 1998); (3) the inhibitory Smads (I-Smads, also called anti-Smads), as exemplified by Smad6 and Smad7. Smad1 is also activated by BMP-7 type I receptor via ALK-2, an activin type I receptor (Yamamoto *et al.*, 1997; Macias-Silva *et al.*, 1998).

Smads have conserved N- and C-terminal domains called MAD (Mothers against dpp in *Drosophila*) homology domain (MH) 1 and 2, respectively. A linker domain with a proline-rich sequence is found between the MH1 and MH2 domains. The 'C'-terminal SSXS motif is the critical phosphorylation site in R-Smads (Souchelnytskyi *et al.*, 1998). The MH2 domain of Smad is the effector domain involved in oligomerization and DNA interaction.

The MH1 domain consists of about 130 amino acids and is implicated in DNA binding. The proline-rich linker region contains a PY motif with potential to interact with the WW domain. Emerging data from several groups have pointed to Smads as the key signal mediators of the BMP family. Smad1 and Smad5 are the signaling substrates for BMP type I receptor kinase.

Cytoplasmic signaling cascades

Although considerable evidence has accumulated in favor of Smad pathways, it is possible that BMPs may act via other signaling cascades, including Jun and MAP kinases (Adachi-Yamada *et al.*, 1999; Ulloa *et al.*, 1999; Liberati *et al.*, 1999).

DOWNSTREAM GENE ACTIVATION

Transcription factors activated

The precise nature of the BMP-induced promoter sequences are not clear. Smad1 interacts with DNA-binding proteins (Shi *et al.*, 1999). BMPs induce a homeobox gene related to human Hox 11 gene and called Tlx-2 (Tang *et al.*, 1998).

R-Smads are phosphorylated by type I BMP receptors. The potential role of the plasma membrane protein called Smad anchor for receptor activation (SARA) in BMP signaling is not clear, although it has been suggested that it might play a role in recruitment of Smad2 and Smad3 to TGFβ receptors. Upon translocation into the nucleus, Smad1 and Smad5 in partnership with Smad4 activate downstream genes. Although BMP-response genes are known to be activated by BMPs few studies have implicated BMPs in the activation of specific transcription factors such as AP-1, c-bfa1, and c-fos. Although mice devoid of core-binding factor A1 (CBFA1) lack bone, the role of BMP-4 or BMP-7 in the regulation of CBFA1 is not clear. Equal attention should also be paid to transcriptional repressors. The recent work of Verschueren *et al.* (2000) has implicated Smad-interacting protein 1 (SIP-1), a zinc finger-containing homeodomain protein that binds to R-Smads. SIP-1 interacts with the brachyury gene in *Xenopus*. It is probable that SIP-1 functions as a transcriptional repressor and R-Smads may overcome this repression by interaction with SIP-1 (Verschueren *et al.*, 2000).

Both Smad6 and Smad7 are induced by BMPs and are translocated from nucleus to cytoplasm. In the cytoplasm Smad6 and Smad7 may inhibit type I receptor kinase-mediated phosphorylation of Smad1, Smad5, and Smad8 (Imamura *et al.*, 1997; Nakao *et al.*, 1997; Kawabata *et al.*, 1998; Bhushan *et al.*, 1998).

BIOLOGICAL CONSEQUENCES OF ACTIVATING OR INHIBITING RECEPTOR AND PATHOPHYSIOLOGY

The most important action of the BMP family of ligands is new bone formation *in vivo*. Despite the advances of the past two years there is a gap in our knowledge of the processes between the entry of the oligomeric R-Smads – Smad1, Smad5, and Smad8 – and Co-Smad4 into the nucleus and the attendant gene expression and the lineage of bone differentiation. What cascade of genes is induced and/or repressed to set into motion the train of events leading to the developmental cascade of bone and cartilage formation? Although a flash of excitement was heralded by the discovery of CBFA1 in bone formation, the connection between the Smad pathway and BMP signaling is still shrouded in mystery. Perhaps BMP-binding proteins such as noggin and chordin, which interact with BMP-4 with the same affinity as the type I and II serine/threonine kinase BMP receptors, may hold some secrets. However, the nuclear gene activation needs considerable scrutiny and is the next big step in the BMP receptor pathway.

THERAPEUTIC UTILITY

There is much expectation and hope that BMPs can be used with BMP receptors and the signaling substrate Smads in preclinical and clinical situations. In this regard it is noteworthy that upon transfection into mesenchymal cells, adenovirus vectors with Smad1 and Smad5 confer an osteoblastic phenotype as assessed by alkaline phosphatase activity (Fujii and Miyazono, personal communication). This raises the possibility that Smad1 and Smad5 may initiate the lower threshold of the biological response and that additional stimulation by BMP ligand might complete the progression of the cascade. However, additional experiments *in vivo* are needed to further extend the conceptual framework for the action of R-Smads.

The critical role of R-Smad5 and Co-Smad4 is illustrated by embryonic lethality and other developmental defects. Smad4 knockout mice die by embryonic day 7.5 *in utero* due to gastrulation defects (Sirard *et al.*, 1998; Yang *et al.*, 1998). Smad5 knockout results in impaired angiogenesis with decreased smooth muscle cells and attendant vascular pathology (Yang *et al.*, 1999). It is noteworthy that angiogenesis and vascular invasion into hypertrophic cartilage is a prerequisite for optimal bone formation and modeling by osteoclasts.

References

Adachi-Yamada, T., Nakamura, M., Irie, K., Tomoyasu, Y., Sano, Y., Mori, E., Goto, S., Ueno, N., Nishida, Y., and Matsumoto, K. (1999). p38 mitogen-activated protein kinase can be involved in transforming growth factor β superfamily signal transduction in *Drosophila* wing morphogenesis. *Mol. Cell Biol.* **19**, 2322–2329.

Attisano, L., and Wrana, J. L. (1998). Mads and Smads in TGFβ signaling. *Curr. Opin. Cell Biol.* **10**, 188–194.

Bhatia, M., Bonnet, D., Wu, D., Murdoch, B., Wrana, J., Gallacher, L., and Dick, J. E. (1999). Bone morphogenetic proteins regulate the developmental program of human hematopoietic stem cells. *J. Exp. Med.* **189**, 1139–1148.

Bhushan, A., Chen, Y., and Vale, W. (1998). Smad7 inhibits mesoderm formation and promotes neural cell fate in *Xenopus* embryos. *Dev. Biol.* **200**, 260–268.

Hahn, S. A., Shutte, M., Hoque, A. T. M. S., Moskaluk, C. A., da Costa, L. T., Rozenblum, E., Weinstein, C. L., Fischer, A., Yeo, C. J., Hruban, R. H., and Kern, S. E. (1996). DPC4, a candidate tumor suppressor gene at human chromosome 18q21.1. *Science* **271**, 350–353.

Heldin, C.-H., Miyazono, K., and ten Dijke, P. (1997). TGF-β signaling from cell membrane to nucleus through SMAD proteins. *Nature* **390**, 465–471.

Imamura, T., Takase, M., Nishihara, A., Oeda, E., Hanai, J.-I., Kawabata, M., and Miyazono, K. (1997). Smad6 inhibits signaling by the TGF-β superfamily. *Nature* **389**, 622–626.

Kang, Y., and Reddi, A. H. (1996). Identification and cloning of novel type I serine/threonine kinase receptor of the TGFβ/BMP superfamily in rat prostate. *Biochem. Mol. Biol. Int.* **40**, 993–1001.

Kawabata, M., Imamura, T., and Miyazono, K. (1998). Signal transduction by bone morphogenetic proteins. *Cytokine Growth Factor Rev.* **9**, 49–61.

Liberati, N. T., Datto, M. B., Frederick, J. P., Shen, X., Wong, C., Rougier-Chapman, E. M., and Wang, X.-F. (1999). Smads bind directly to the Jun family of AP-1 transcription factors. *Proc. Natl Acad. Sci. USA* **96**, 4844–4849.

Macias-Silva, M., Hoodless, P. A., Tang, S. J., Buchwald, M., and Wrana, J. L. (1998). Specific activation of Smad1 signaling pathways by the BMP7 type I receptor, ALK2. *J. Biol. Chem.* **273**, 25628–25636.

Nakao, A., Afrakhte, M., Morén, A., Nakayama, T., Christian, J. L., Heuchel, R., Itoh, S., Kawabata, M., Heldin, N.-E., Heldin, C.-H., and ten Dijke, P. (1997). Identification of Smad7, a TGFβ-inducible antagonist of TGF-β signaling. *Nature* **389**, 631–635.

Nishimura, R., Kato, Y., Chen, D., Harris, S. E., Mundy, G. R., and Yoneda, T. (1998). Smad5 and DPC4 are key molecules in mediating BMP-2-induced osteoblastic differentiation of the pluripotent mesenchymal precursor cell line C2C12. *J. Biol. Chem.* **273**, 1872–1879.

Paralkar, V. M., Hammonds, R. G., and Reddi, A. H. (1991). Identification and characterization of cellular binding proteins (receptors) for recombinant human bone morphogenetic protein 2B. *Proc. Natl Acad. Sci. USA* **88**, 3397–3401.

Reddi, A. H. (1998). Role of morphogenetic proteins in skeletal tissue engineering and regeneration. *Nature Biotechnol.* **16**, 247–252.

Savage, C., Das, P., Finelli, A. L., Townsend, S. R., Sun, C. Y., Baird, S. E., and Padgett, R. W. (1996). *Caenorhabditis elegans* genes sma-2, sma-3, and sma-4 define a conserved family of transforming growth factor β pathway components. *Proc. Natl Acad. Sci. USA* **93**, 790–794.

Shi, X., Yang, X., Chen, D., Chang, D., and Cao, X. (1999). Smad1 interacts with homeobox DNA-binding proteins in bone morphogenetic protein signaling. *J. Biol. Chem.* **274**, 13711–13717.

Sirard, C., de la Pompa, J. L., Elia, A., Itie, A., Mirtsos, C., Cheung, A., Hahn, S., Wakeham, A., Schwartz, L., Kern, S. E., Rossant, J., and Mak, T. W. (1998). The tumor suppressor gene Smad4/Dpc4 is required for gastrulation and later for anterior development of the mouse embryo. *Genes Dev.* **12**, 107–119.

Souchelnytskyi, S., Nakayama, T., Nakao, A., Morén, A., Heldin, C.-H., Christian, J. L., and ten Dijke, P. (1998). Physical and functional interaction of murine and *Xenopus* Smad7 with bone morphogenetic protein receptors and transforming growth factor-β receptors. *J. Biol. Chem.* **273**, 25364–25370.

Tamaki, K., Souchelnytskyi, S., Itoh, S., Nakao, A., Sampath, K., Heldin, C.-H, and ten Dijke, P. (1998). Intracellular signaling of osteogenic protein-1 through Smad5 activation. *J. Cell. Physiol.* **177**, 355–363.

Tang, S. J., Hoodless, P. A., Lu, Z., Breitman, M. L., McInnes, R. R., Wrana, J. L., and Buchwald, M. (1998). The Tlx-2 homeobox gene is a downstream target of BMP signaling and is required for mouse mesoderm development. *Development* **125**, 1877–1887.

ten Dijke, P., Yamashita, H., Samptah, T. K., Reddi, A. H., Estevez, M., Riddle, D. L., Ichijo, H., Heldin, C.-H, and Miyazono, K. (1994). Identification of type I receptors for osteogenic protein-1 and bone morphogenetic protein-4. *J. Biol. Chem.* **269**, 16985–16988.

Ulloa, L., Doody, J., and Massagu, J. (1999). Inhibition of transforming growth factor-β/SMAD signaling by the interferon-γ/STAT pathway. *Nature* **397**, 710–713.

Verschueren, K., Remacle, J. E., Collart, C., Kraft, H., Baker, B. S., Tylzanowski, P., Nelles, L., Wuytens, G., Su, M.-T., Bodmer, R., Smith, J. C., and Huylebroeck, D. (2000). SIP1, a novel zinc finger/homeodomain repressor, interacts with Smad proteins and binds to 5′-CACCT sequences in candidate target genes. *J. Biol. Chem.* (in press).

Wang, T., Donahoe, P. K., and Zervos, A. S. (1994). Specific interaction of type I receptors of the TGF-β family with the immunophilin FKBP-12. *Science* **265**, 674–676.

Yamamoto, N., Akiyama, S., Katagiri, T., Namiki, M., Kurokawa, T., and Suda, T. (1997). Smad1 and Smad5 act downstream of intracellular signalings of BMP-2 that inhibits myogenic differentiation and induces osteoblast differentiation in C2C12 myoblasts. *Biochem. Biophys. Res. Commun.* **238**, 574–580.

Yang, X., Li, C., Xu, X., and Deng, C. (1998). The tumor suppressor SMAD4/DPC4 is essential for epiblast proliferation and mesoderm induction in mice. *Proc. Natl Acad. Sci. USA* **95**, 3667–3672.

Yang, X., Castilla, L. H., Xu, X., Li, C., Gotay, J., Weinstein, M., Liu, P. P., and Deng, C. X. (1999). Angiogenesis defects and mesenchymal apoptosis in mice lacking SMAD5. *Development* **126**, 1571–1580.

Zawel, L., Le Dai, J., Buckhaults, P., Zhou, S., Kinzler, K. W., Vogelstein, B., and Kern, S. E. (1998). Human Smad3 and Smad4 are sequence-specific transcription activators. *Mol. Cell* **1**, 611–617.

Hematopoietic Receptors

Hematopoietic Receptor Family

Atsushi Miyajima*

Institute of Molecular and Cellular Biosciences, The University of Tokyo, 1-1-1 Yayoi, Tokyo, Bunkyo-Ku, 113-0032, Japan

*corresponding author tel: +81-3-5800-3551, fax: +81-3-4800-3550, e-mail: miyajima@hgc.ems.u-tokyo.ac.jp

DOI: 10.1006/rwcy.2000.02010.

SUMMARY

The functional high-affinity receptors for IL-3, IL-5, and GM-CSF are heterodimeric receptors consisting of a cytokine-specific α subunit and the common β subunit, βc, which is shared by the three cytokines. In the mouse, the IL-3-specific β_{IL3} subunit, homologous to βc, is present and forms a high-affinity receptor only with the IL-3Rα. These subunits are members of the type I cytokine receptor family.

As the β subunits play a major role in signaling, the three cytokines exhibit similar biological functions on the same target cells. The JAK2 tyrosine kinase is associated with both β subunits and is activated upon cytokine stimulation. The membrane-proximal region of βc is responsible for the activation of JAK2 and STAT5 as well as induction of c-myc. The signals induced by this region are required for cell cycle progression and DNA synthesis. The activation of the Ras pathway requires the distal region of βc and is involved in suppression of apoptosis. As IL-3, IL-5, and GM-CSF are produced mainly from activated T cells and mast cells, it has been hypothesized that these cytokines play a major role in inductive hematopoiesis associated with immune and inflammatory reactions. However, mice lacking the entire functions of IL-3, IL-5, and GM-CSF are viable and no significant defects other than a reduced level of eosinophils and some defects in mast cell development were found, even in mice infected with bacteria and parasites. Thus, there appears to be a compensatory mechanism to overcome any defect in the three cytokines.

INTRODUCTION

IL-3, IL-5, and GM-CSF are related to each other genetically, structurally, and functionally, and stimulate development of hematopoietic cells. IL-3 and GM-CSF act on a broad range of hematopoietic progenitors and thereby exhibit a number of different biological activities, whereas IL-5-responsive cells are restricted to eosinophils, basophils, and some mouse B cells. IL-3, GM-CSF, and IL-5 exhibit almost identical biological activities when they act on the same target cells such as eosinophilic progenitors. These cytokines are mainly produced by activated T cells and mast cells, but the production by bone marrow stroma cells is minimal. Thus, it has been hypothesized that these cytokines play a major role in hematopoiesis in an emergency situation such as inflammation (Arai *et al*., 1990).

The common activities of the three cytokines are now explained by the structures of the receptors. The receptors for IL-3, IL-5, and GM-CSF are composed of a cytokine-specific α subunit and a common β subunit, βc (Miyajima *et al*., 1993). As the βc subunit plays a critical role in signal transduction, similar signals are transmitted through the β subunit, and the α subunits provide specificity to cytokines. This chapter describes the structure, expression, and functions of the IL-3, IL-5, and GM-CSF receptors.

STRUCTURE OF THE IL-3, IL-5, AND GM-CSF RECEPTORS

The high-affinity receptors for IL-3, GM-CSF, and IL-5 consist of a cytokine-specific α subunit and a common β subunit, βc (Miyajima *et al.*, 1993) (**Figure 1**). Both are members of the class I cytokine receptor superfamily. The α subunits are glycoproteins of 60–70 kDa with a small cytoplasmic domain of about 50 amino acid residues and bind their specific ligand with low affinity. The shared βc is a glycoprotein of 120–130 kDa with two repeats of a conserved motif of the class I cytokine receptors in the extracellular domain. Its large cytoplasmic domain contains two motifs known as Box-1 and Box-2 which are conserved in members of the class I cytokine receptors. The βc does not bind any cytokine by itself, but is required for the formation of a high-affinity receptor with any of the three α subunits, IL-3 receptor α (IL-3Rα), GM-CSF receptor α (GM-CSFRα), and IL-5 receptor α (IL-5Rα). IL-3Rα and GM-CSFRα genes are colocalized in the human pseudoautosomal region of the sex chromosomes (Gough *et al.*, 1990), whereas the IL-5Rα gene is on human chromosome 3p25–p26 (Isobe *et al.*, 1992).

In the mouse, there are two homologous β subunits: βc (also known as AIC2B) and the IL-3-specific β subunit, β_{IL3} (also known as AIC2A) (Itoh *et al.*, 1990; Gorman *et al.*, 1990), which binds IL-3 with low affinity and forms a high-affinity receptor with IL-3Rα only (Hara and Miyajima, 1992) (Figure 1). No functional differences have been found between the two high-affinity mIL-3Rs which consist of either βc or β_{IL3}. The two β subunit genes are tightly linked on mouse chromosome 15 (Gorman *et al.*, 1992) and are placed in a head-to-head configuration (Hannemann *et al.*, 1995). The human βc gene is on chromosome 22q12.2–13.1 (Shen *et al.*, 1992) and there is no evidence that β_{IL3} exists in the human.

Like receptor tyrosine kinases, multimerization of the receptor subunits is a key step for activation of the class I cytokine receptors. Although the cytoplasmic domains of both α and βc are required

Figure 1 Receptors for IL-3, GM-CSF, and IL-5. The high-affinity receptors for IL-3, GM-CSF, and IL-5 consist of a cytokine-specific α subunit and the common β subunit, βc. There is the IL-3-specific β subunit β_{IL3} in the mouse but not in the human.

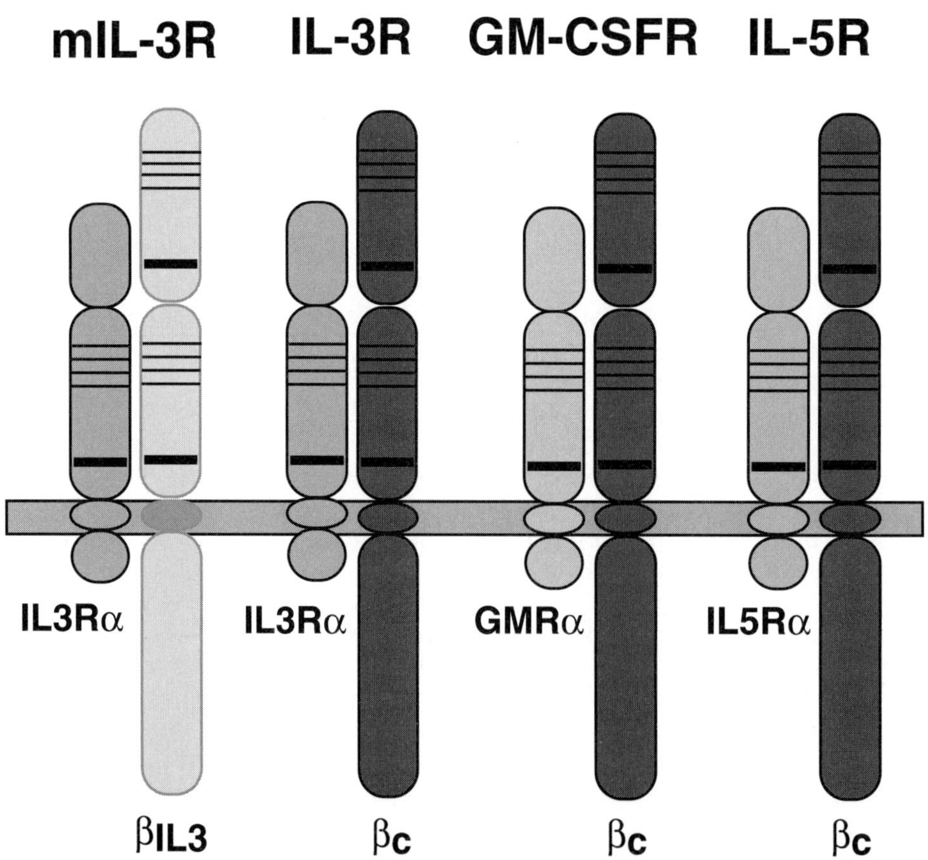

for signaling, the βc cytoplasmic domain plays a major role in signal transduction. This is supported by experiments using chimeric receptors that are forced to form a dimer, e.g. the chimeric receptor which consists of the βc cytoplasmic domain and the extracellular domain of homodimerizing EPOR induced full growth signals in response to EPO (Sakamaki *et al.*, 1993), whereas similar chimeric receptors with a cytoplasmic domain of the α subunits failed to induce any signals (Eder *et al.*, 1994; Muto *et al.*, 1995). Thus the α subunits provide the cytokine specificity and βc induces common signals. This model clearly explains the functional overlap between the three cytokines.

Alanine substitution mutagenesis of βc identified the amino acid residues of βc (Y365–I368) essential for binding to GM-CSF (Woodcock *et al.*, 1994). Evidence was presented that the GM-CSF receptor exists as a preformed complex that can be activated by GM-CSF, IL-3, and IL-5 (Woodcock *et al.*, 1997) and the IL-3 and GM-CSF receptors undergo covalent dimerization of the respective α subunit with βc in the presence of cognate cytokine. Cys86, Cys91, and Cys96 of βc are involved in the covalent dimerization (Stomski *et al.*, 1998). IL-3 was shown to induce disulfide-linked dimerization between IL-3Rα and βc, which is required for receptor activation but not for high-affinity binding (Stomski *et al.*, 1996). It was also reported that βc forms an inactive dimer in the absence of cognate ligand and is activated by binding of a cytokine (Muto *et al.*, 1996). These results suggest that the activated receptor complex may be a multimeric complex.

Polymerase chain reaction (PCR)-based random mutagenesis of the βc subunit led to identification of several mutations that result in constitutive dimerization of βc (Jenkins *et al.*, 1995; Jenkins *et al.*, 1998). Mutations in the extracellular domain of βc are clustered in the membrane-proximal domain (domain 4). Two mutations in the transmembrane domain were found and one of them (V449E) is similar to the constitutive active mutant of the Neu/ErbB oncogene. Interestingly, mutations at two positions (R461C,H and H544R) in the cytoplasmic domain also result in constitutive activity. Whether these constitutive active βc form either a dimer or a more complex structure remains unknown.

EXPRESSION OF THE RECEPTOR SUBUNITS

Expression of α and β subunits of the IL-3/GM-CSF/IL-5 receptors is mainly restricted to hematopoietic cells (Sato *et al.*, 1993a), while some expression is also found in nonhematopoietic tissues such as testis, placenta, and brain (Morikawa *et al.*, 1996; Korpelainen *et al.*, 1993; Gearing *et al.*, 1989). βc and β_{IL3} are expressed in various myeloid progenitor cells, macrophages, mast cells, CD5-positive B cells, and some endothelial cells, but not in erythroblasts, mature T cells, and fibroblasts (Gorman *et al.*, 1990). Expression of the α subunits is more restricted to cytokine-responsive cells: IL-3Rα, but not GM-CSFRα and IL-5Rα, is expressed in mast cells and multipotential progenitor cells that form the CFU mix. In contrast, IL-5Rα is predominantly expressed in eosinophils and a subset of B cells, and IL-5 is a major cytokine for eosinophils, but not for other hematopoietic cells (Takatsu *et al.*, 1994).

IL-3 and GM-CSF exhibit a broad spectrum of biological functions, while IL-5 function is restricted to mainly eosinophils (Arai *et al.*, 1990). The functional differences between the three cytokines may be due to the restricted expression of the α subunits. Alternatively, each α subunit may play an active role in inducing cytokine-specific functions. These possibilities were tested by generating a transgenic mouse strain that expresses IL-5Rα ubiquitously by the constitutive promoter of the phosphoglycerokinase 1 (PGK) gene (Takagi *et al.*, 1995). Bone marrow cells of the transgenic mice formed colonies of various lineages and mixed colonies in response to IL-5 in a manner similar to IL-3, indicating that the limited activity of IL-5 is mainly due to the restricted expression of IL-5Rα and that IL-5Rα is functionally equivalent to IL-3Rα. The results further suggest that hematopoietic cells have their own differentiation program which is not affected by the α subunits. However, it should be noted that the cytoplasmic domains of the α subunits are required for signaling.

MICE DEVOID OF THE IL-3, GM-CSF, AND IL-5 RECEPTORS

As there are two IL-3 receptors in the mouse (Miyajima *et al.*, 1993), either β_{IL3} or βc may be dispensable for IL-3 function, whereas βc is crucial for IL-5 and GM-CSF. To test the role of each β subunit, mice devoid of either one of the β subunits were generated (Nishinakamura *et al.*, 1995). As expected, the β_{IL3}-deficient mice showed no apparent phenotype and no hematological defect was found. In contrast, bone marrow cells of the βc-deficient mice did not form any colonies in the presence of either GM-CSF or IL-5, while IL-3-induced colony formation was normal, indicating that βc is essential for the function of GM-CSF and IL-5. A significant reduction of the number of eosinophils in the peripheral blood was noticed in the βc-deficient mouse, consistent with the idea that IL-5 is the major cytokine for eosinophil development (Takatsu *et al.*, 1994). This is consistent with the observation that IL-5Rα knockout mice exhibited only a basal level of eosinophils. In addition, IL-5Rα knockout mice showed decreased numbers of B-1 cells concomitant with low serum concentrations of IgM and IgG3 (Yoshida *et al.*, 1996).

The βc-deficient mice exhibited lung abnormalities, including accumulation of proteinous material in the alveolar spaces and peribronchovascular lymphocytic infiltration. These observations are probably attributable to the deficiency of GM-CSF function as the same phenotype was observed in the GM-CSF-deficient mice (Dranoff *et al.*, 1994). Alveolar macrophages may play a role in lung homeostasis by clearing surfactant and other debris from the alveolar space and the function of these macrophages may be impaired in such mice.

The apparently normal hematopoiesis in the βc-deficient mice may be due to the presence of the functional IL-3 receptor consisting of β_{IL3}. As IL-3 knockout mice were generated and were apparently normal, the IL-3 ligand knockout mouse was crossed with the βc knockout mouse to generate a mouse line lacking the entire function of IL-3, GM-CSF, and IL-5 (Nishinakamura *et al.*, 1996). Interestingly, the mice developed normally and were fertile. Hematopoiesis in these mice was similar to the βc knockout. While the eosinophil number was reduced and lung disease developed in the double knockout mice, the severity was not changed compared to that of βc mutant mice. Thus, the entire function of IL-3, GM-CSF, and IL-5 is dispensable for hematopoiesis in normal life.

Since a major source of these cytokines is activated T cells and basal levels of these cytokines are almost negligible in the normal bone marrow, it is possible that the major role of these cytokines is to promote hematopoiesis in an emergency situation such as inflammation (Arai *et al.*, 1990). To address the question whether these cytokines are essential for emergency hematopoiesis, the mutant mice were infected with parasites and bacteria. IL-5Rα knockout mice showed sensitivity to infection with *Angiostrongylus cantonensis* (Sugaya *et al.*, 1997). While IL-3 was shown not to be required for the generation of mast cells and basophils, it was recently shown to contribute to increases in the numbers of tissue mast cells, enhanced basophil production and immunity in mice infected with the nematode *Stronglyoides venezuelensis* (Lantz *et al.*, 1998). However, the susceptibility to infection appears to be rather restricted to particular cases, and the double knockout mice lacking both βc and IL-3 were completely resistant to infection by *Listeria monocytogenes* (Nishinakamura *et al.*, 1996). Thus, it appears that IL-3, GM-CSF, and IL-5 are largely dispensable even in emergency situations and the defects can be compensated by other unknown mechanism.

Consistent with the notion that IL-3 is dispensable, naturally occurring mice with IL-3 hyporesponsiveness were found. It was known that several mouse strains such as A/J show hyporesponsiveness to IL-3. Molecular genetic analysis revealed that the defect is due to a small deletion (four base pairs) in the branch point in intron 7 of the IL-3Rα gene. As branch points are required for proper splicing, the A/J mouse produces mostly aberrant IL-3Rα mRNA (Ichihara *et al.*, 1995). Curiously, the same mutation was found in 10 out of 27 laboratory mouse strains we analyzed. These mouse strains include A/J, C58J, A/WySnJ, A/HeY, RF/J, AKR/J, SM/J, BUB/BnJ, CE/J, and NZB/BINJ (Hara *et al.*, 1995).

SIGNAL TRANSDUCTION

IL-3, GM-CSF, and IL-5 induce rapid tyrosine phosphorylation of various cellular proteins including the β subunit, phosphatidylinositol 3-kinase, Vav, Shc, and PTP-1D (Miyajima *et al.*, 1993; Itoh *et al.*, 1996). While members of the Src family of tyrosine kinases, such as Lyn and Fyn, as well as Btk, a member of another tyrosine kinase family, were initially shown to be involved in signaling in certain cell types (Torigoe *et al.*, 1992; Li *et al.*, 1995), the roles of these kinases in IL-3/GM-CSF signaling still remain unknown. In contrast, JAK kinases are now believed to play a major role in cytokine signaling (Ihle, 1995), and JAK2 was found to bind to the β subunits. The conserved motif among various cytokine receptors known as Box-1 is present in the membrane-proximal region of the βc subunit (between 455 and 544) and is sufficient for activation of JAK2 (Quelle *et al.*, 1994), followed by activation of signal transducer and activator of transcription 5 (STAT5) (**Figure 2**). STAT5 activation leads to the induction of various genes such as Pim-1, Id-1, CIS, and OSM (Yoshimura *et al.*, 1995, 1996; Mui *et al.*, 1996). This region is also responsible for induction of several cytokine-inducible genes, including c-myc. The signals derived from the membrane-proximal region are important for DNA replication and cell cycle progression.

The more distal portion (544–626) of βc is required for activation of Ras, Raf, MAP kinase and PI-3 kinase as well as induction of c-fos and c-jun (Sato *et al.*, 1993b; Itoh *et al.*, 1996). Analysis of signaling

Figure 2 Signal transduction pathways from the IL-3/IL-5/GM-CSF receptors. JAK2 kinase is associated with the β subunit and is activated by cytokine binding. The activated JAK2 induces various signals. Ras activated by SOS in turn activates the Raf/MAP kinase pathway and also PI-3 kinase. PI-3 kinase can be activated directly by the receptor as well. The activated STAT5 translocates to the nucleus and induces various genes, including CIS, a negative feedback regulator which turns off the STAT5 pathway. STAM1 appears to link JAK2 and induction of c-myc. Cell proliferation requires signals for suppression of cell death and for cell cycle progression; antiapoptotic signals are mainly delivered by the Ras pathway and signals for cell cycle progression are derived from the membrane-proximal region of the β subunit.

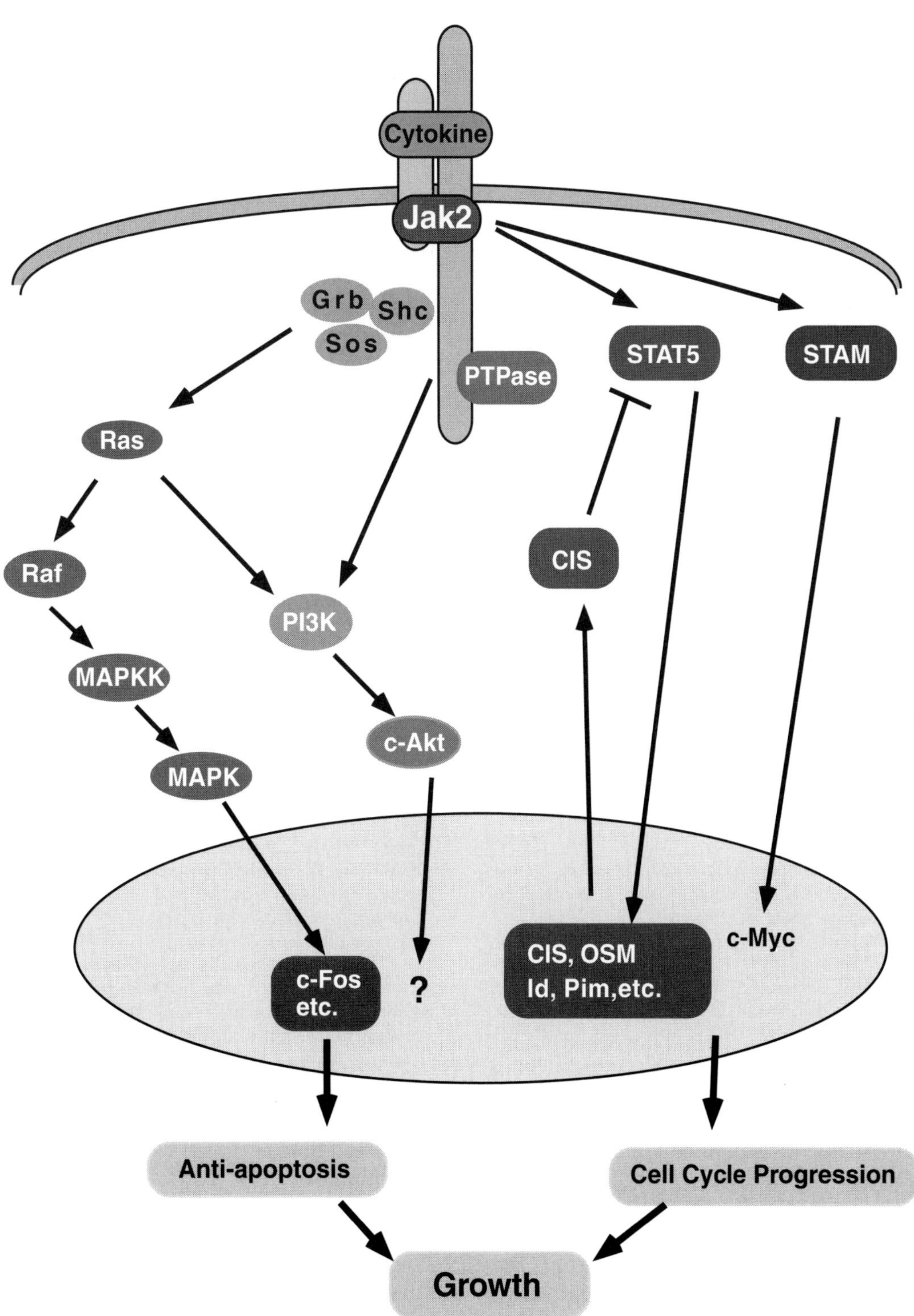

by mutant β subunits with a C-terminal truncation as well as substitution mutants in which tyrosine residues are substituted with phenylalanine within the β cytoplasmic domain suggests that multiple regions mediate Ras activation (Sato *et al.*, 1993b; Itoh *et al.*, 1996): Shc is an adapter molecule that is bound to the tyrosine phosphorylated receptor and recruits SOS, a Ras-guanine nucleotide exchange factor, to membranes where Ras activation takes place. Phosphorylation of Shc by GM-CSF requires the tyrosine residue 577 of βc. A protein tyrosine phosphatase, PTP-1D, is also known to serve as an adapter to recruit SOS, and its tyrosine phosphorylation is mediated by Tyr577 as well as other tyrosine residues downstream (Itoh *et al.*, 1996). Thus, activation of Ras appears to be mediated by multiple pathways. Activation of Ras results in suppression of apoptosis of hematopoietic cells (Kinoshita *et al.*, 1995). While Ras activates Raf kinase as well as PI-3 kinase, both pathways are involved in suppression of apoptosis (Kinoshita *et al.*, 1997). Long-term cell proliferation requires signals from the membrane-proximal region of the β subunit as well as the antiapoptotic function mediated by Ras (Kinoshita *et al.*, 1995).

A further C-terminal portion (763 to the C-terminus) appears to be involved in negative regulation, as the deletion of this region rather enhances signaling such as tyrosine phosphorylation of the β subunit and Shc (Sato *et al.*, 1993b). A tyrosine phosphatase may bind to this region and negatively regulate the signaling. The best candidate of this negative regulator is PTP-1C (HCP) as it has been shown to bind to the β subunit and is also known to be a negative regulator of cytokine signaling (Yi *et al.*, 1993).

References

Arai, K., Lee, F., Miyajima, A., Miyatake, S., Arai N., and Yokota, T. (1990). Cytokines: coordinators of immune and inflamatory responses. *Annu. Rev. Biochem.* **59**, 783–836.

Dranoff, G., Crawford, A. D., Sadelain, M., Ream, B., Rashid, A., Bronson, R. T., Dickersin, G.R, Bachurski, C. J., Mark, E. L., Whitsett J. A., and Mulligan R. C. (1994). Involvement of granulocyte-macrophage colony-stimulating factor in pulmonary homeostasis. *Science* **264**, 713–716.

Eder, M., Ernst, T. J., Ganser, A., Jubinsky, P. T., Inhorn, P., Hoelzer, D., and Griffin, J. D. (1994). A low affinity chimeric human alpha/beta-granulocyte-macrophage colony-stimulating factor receptor induces ligand-dependent proliferation in a murine cell line. *J. Biol. Chem.* **269**, 30173–30180.

Gearing, D. P., King, J. A., Gough, N. M., and Nicola, N. A. (1989). Expression cloning of a receptor for human granulocyte-macrophage colony-stimulating factor. *EMBO J.* **8**, 3667–3676.

Gorman, D., Itoh, M. N., Kitamura, T., Schreurs, J., Yonehara, S., Yahara, I., Arai, K., and Miyajima, A. (1990). Cloning and expression of a gene encoding an interleukin 3 receptor-like protein: identification of another member of the cytokine receptor gene family. *Proc. Natl Acad. Sci. USA* **87**, 5459–5463.

Gorman, D., Itoh, M. N., Jenkins, N. A., Gilbert, D. J., Copeland, N. G., and Miyajima, A. (1992). Chromosomal localization and organization of the murine genes encoding the β subunits (AIC2A and AIC2B) of the interleukin 3, granulocyte/macrophage colony-stimulating factor and interleukin 5 receptors. *J. Biol. Chem.* **267**, 15842–15848.

Gough, N. M., Gearing, D. P., Nicola, N. A., Baker, E.., Pritchard, M., Callen, D. F., and Sutherland, G. R. (1990). Localization of the human GM-CSF receptor gene to the X-Y pseudoautosomal region. *Nature* **345**, 734–736.

Hannemann, J., Hara, T., Kawai, M., Miyajima, A., Ostertag, W., and Stocking, C. (1995). Sequential mutations in the interleukin-3 (IL3)/granulocyte-macrophage colony-stimulating factor/IL5 receptor beta-subunit genes are necessary for the complete conversion to growth autonomy mediated by a truncated beta C subunit. *Mol. Cell Biol.* **15**, 2402–2412.

Hara, T., and Miyajima A. (1992). Two distinct functional high affinity receptors for mouse IL-3. *EMBO J.* **10**, 1875–1884.

Hara, T., Ichihara, M., Takagi, M., and Miyajima, A. (1995). Interleukin-3 (IL-3) poor-responsive inbred mouse strains carry the identical deletion of a branch point in the IL-3 receptor alpha subunit gene. *Blood* **85**, 2331–2336.

Ichihara, M., Hara, T., Takagi, M., Cho, L. C., Gorman D. M., and Miyajima, A. (1995). Impaired interleukin-3 (IL-3) response of the A/J mouse is caused by a branch point deletion in the IL-3 receptor alpha subunit gene. *EMBO J.* **14**, 939–950.

Ihle, J. N. (1995). Cytokine receptor signalling. *Nature* **377**, 591–594.

Isobe, M., Kumura, Y., Murata, Y., Takaki, S., Tominaga, A., Takatsu, K., and Ogita, Z. (1992). Localization of the gene encoding the alpha subunit of human interleukin-5 receptor (IL5RA) to chromosome region 3p24–3p26. *Genomics* **14**, 755–758.

Itoh, N., Yonehara, S., Schreurs,, Gorman, J., Maruyama, K., Ishii, A., Yahara, I., Arai, K., and Miyajima, A. (1990). Cloning of an interleukin-3 receptor: a member of a distinct receptor gene family. *Science* **247**, 324–327.

Itoh, T., Muto, A., Watanabe, S., Miyajima, A., Yokoa, T., and Arai, K. (1996). Granulocyte-macrophage colony-stimulating factor provoles RAS activation and transcription of c-fos through different modes of signaling. *J. Biol. Chem.* **271**, 7587–7592.

Jenkins, B., D'Andrea, J. R., and Gonda, T. J. (1995). Activating point mutations in the common beta subunit of the human GM-CSF, IL-3 and IL-5 receptors suggest the involvement of beta subunit dimerization and cell type-specific molecules in signalling. *EMBO J.* **14**, 4276–4287.

Jenkins, B., Blake, T., and Gonda, T. (1998). Saturation mutagenesis of the beta subunit of the human granulocyte-macrophage colony-stimulating factor receptor shows clustering of constitutive mutations, activation of ERK MAP kinase and STAT pathways, and differential beta subunit tyrosine phosphorylation. *Blood* **92**, 1989–2002.

Kinoshita, T., Yokota, T., Arai, K., and Miyajima, A. (1995). Suppression of apoptotic death in hematopoietic cells by signalling through the IL-3/GM-CSF receptors. *EMBO J.* **14**, 266–275.

Kinoshita, T., Shirouzu, M., Kamiya, A., Hashimoto, K., Yokoyama, S., and Miyajima, A. (1997). Raf/MAPK- and rapamycin-sensitive pathways mediate the anti-apoptotic function of p21Ras in IL-3 dependent hematopoietic cells. *Oncogene* **15**, 619–627.

Korpelainen, E. I., Gamble, J. R., Smith, W. B., Goodall, G. J., Qiyu, S., Woodcock, J. M., Dottore, M., Vadas, M. A., and

Lopez, A. F. (1993). The receptor for interleukin 3 is selectively induced in human endothelial cells by tumor necrosis factor alpha and potentiates interleukin 8 secretion and neutrophil transmigration. *Proc. Natl Acad. Sci. USA* **90**, 11137–11141.

Lantz, C., Boesiger, J., Song, C., Mach, N., Kobayashi, T., Mulligan, R., Nawa, Y., Dranoff, G., and Galli, S. (1998). Role for interleukin-3 in mast-cell and basophil development and in immunity to parasites. *Nature* **392**, 90–93.

Li, T., Tsukada, S., Satterthwaite, A., Havlik, M. H., Park, H., Takatsu, K., and Witte, O. N. (1995). Activation of Bruton's tyrosine kinase (BTK) by a point mutation in its pleckstrin homology (PH) domain. *Immunity* **2**, 451–460.

Miyajima, A., Mui, A. L., Ogorochi, T., and Sakamaki, K. (1993). Receptors for granulocyte-macrophage colony-stimulating factor, interleukin-3, and interleukin-5. *Blood* **82**, 1960–1974.

Morikawa, Y., Tohya, K., Hara, T., Kitamura, T., and Miyajima, A. (1996). Expression of the IL-3 receptor in testis. *Biochem. Biophys. Res. Commun.* 107–112.

Mui, A.L-F., Wakao, H., Kinoshita, T., Kitamura, T., and Miyajima, A. (1996). Suppression of interleukin-3-induced gene expression by a C-terminal truncated Stat5: role of Stat5 in proliferation. *EMBO J.* **15**, 2425–2433.

Muto, A., Watanabe, S., Itoh, T., Miyajima, A., Yokota, T., and Arai, K. (1995). Roles of the cytoplasmic domains of the alpha and beta subunits of human granulocyte-macrophage colony-stimulating factor receptor. *J. Allergy Clin. Immunol.* **96**, 1100–1114.

Muto, A., Watanabe, S., Miyajima, A., Yokota, T., and Arai, K. (1996). The beta subunit of human granulocyte-macrophage colony-stimulating factor receptor forms a homodimer and is activated via association with the alpha subunit. *J. Exp. Med.* **183**, 1911–1916.

Nishinakamura, R., Nakayama, N., Hirabayashi, Y., Inoue, T., Aud, D., McNeil, T., Azuma, S., Yoshida, S., Toyoda, Y., Arai, K., Miyajima, A., and Murray, R. (1995). Mice deficient for the IL-3/GM-CSF/IL-5 beta c receptor exhibit lung pathology and impaired immune response, while beta IL3 receptor-deficient mice are normal. *Immunity* **2**, 211–222.

Nishinakamura, R., Miyajima, A., Mee, P. J., Tybulewicz, V. L. J., and Murray, R. (1996). Hematopoiesis in mice lacking the entire granulocyte macrophage-colony stimulating factor/interleukin-3/interleukin-5 functions. *Blood* **88**, 2458–2464.

Quelle, F. W., Sato, N., Witthuhn, B. A., Inhorn, R. C., Eder, M., Miyajima, A., Griffin, J. D., and Ihle, J. N. (1994). JAK2 associates with the βc chain of the receptor for granulocyte-macrophage colony stimulating factor, and its activation requires the membrane-proximal region. *Mol. Cell Biol.* **14**, 4335–4341.

Sakamaki, K., Wang, H. M., Miyajima, I., Kitamura, T., Todokoro, K., Harada, N., and Miyajima, A. (1993). Ligand-dependent activation of chimeric receptors with the cytoplasmic domain of the interleukin-3 receptor beta subunit (beta IL3). *J. Biol. Chem.* **268**, 15833–15839.

Sato, N., Caux, C., Kitamura, T., Watanabe, Y., Arai, K., Banchereau, J., and Miyajima, A. (1993a). Expression and factor-dependent modulation of the interleukin-3 receptor subunits on human hematopoietic cells. *Blood* **82**, 752–761.

Sato, N., Sakamaki, K., Terada, N., Arai, K., and Miyajima, A. (1993b). Signal transduction by the high-affinity GM-CSF receptor: two distinct cytoplasmic regions of the common beta subunit responsible for different signaling. *EMBO J.* **12**, 4181–4189.

Shen, Y., Baker, E., Callen, D. F., Sutherland, G. R., Willson, T. A., Rakar, S., and Gough, N. M. (1992). Localization of the human GM-CSF receptor β chain gene (CSF2RB) to chromosome 22q12.2–q13.1. *Cytogenet Cell Genet.* **61**, 175–177.

Stomski, F., Sun, Q., Bagley, C., Woodcock, J., Goodall, G., Andrews, R., Berndt, M., and Lopez, A. (1996). Human interleukin-3 (IL-3) induces disulfide-linked IL-3 receptor alpha- and beta-chain heterodimerization, which is required for receptor activation but not high-affinity binding. *Mol. Cell Biol.* **16**, 3035–3046.

Stomski, F., Woodcock, J., Zacharakis, B., Bagley, C., Sun, Q., and Lopez, A. (1998). Identification of a Cys motif in the common beta chain of the interleukin 3, granulocyte-macrophage colony-stimulating factor, and interleukin 5 receptors essential for disulfide-linked receptor heterodimerization and activation of all three receptors. *J. Biol. Chem.* **273**, 1192–1199.

Sugaya, H., Aoki, M., Yoshida, T., Takatsu, K., and Yoshimura, K. (1997). Eosinophilia and intracranial worm recovery in interleukin-5 transgenic and interleukin-5 receptor alpha chain-knockout mice infected with Angiostrongylus cantonensis. *Parasitol. Res.* **83**, 583–590.

Takagi, M., Hara, T., Ichihara, M., Takatsu, K., and Miyajima, A. (1995). Multi-colony stimulating activity of interleukin 5 (IL-5) on hematopoietic progenitors from transgenic mice that express IL-5 receptor alpha subunit constitutively. *J. Exp. Med.* **181**, 889–899.

Takatsu, K., Takaki, S., and Hitoshi, Y. (1994). Interleukin-5 and its receptor: implications in the immune system and inflammation. *Adv. Immunol.* **57**, 145–190.

Torigoe, T., O'Conner, R., Santoli, D., and Reed, J. C. (1992). Interleukin-3 regulates the activity of the Lyn protein-tyrosine kinase in myeloid-committed leukemic cell lines. *Blood* **80**, 617–624.

Woodcock, J. M., Zacharakis, B., Plaetinck, G., Bagley, C. J., Qiyu, S., Hersuc, T. R., Tavernie, R. J., and Lopez, A. F. (1994). Three residues in the common β chain of the human GM-CSF, IL-3 and IL-5 receptors are essential for GM-CSF and IL-5 high affinity binding but not IL-3 high affinity binding and interact with Glu21 of GM-CSF. *EMBO J.* **13**, 5176–5185.

Woodcock, J., McClure, B., Stomski, F., Elliott, M., Bagley, C., and Lopez, A. (1997). The human granulocyte-macrophage colony-stimulating factor (GM-CSF) receptor exists as a preformed receptor complex that can be activated by GM-CSF, interleukin-3, or interleukin-5. *Blood* **90**, 3005–3017.

Yi, T., Mui, A. L., Krystal, G., and Ihle, J. N. (1993). Hematopoietic cell phosphatase associates with the interleukin-3 (IL-3) receptor beta chain and down-regulates IL-3-induced tyrosine phosphorylation and mitogenesis. *Mol. Cell Biol.* **13**, 7577–7586.

Yoshida, T., Ikuta, K., Sugaya, H., Maki, K., Takagi, M., Kanazawa, H., Sunaga, S., Kinashi, T., Yoshimura, K., Miyazaki, J., Takaki, S., and Takatsu, K. (1996). Defective B-1 cell development and impaired immunity against *Angiostrongylus cantonensis* in IL-5R alpha-deficient mice. *Immunity* **4**, 483–494.

Yoshimura, A., Ohkubo, T., Kiguchi, T., Jenkins, N. A., Gilbert, D. J., Copeland, N. G., Hara, T., and Miyajima, A. (1995). A novel cytokine-inducible gene CIS encodes an SH2-containing protein that binds to tyrosine-phosphorylated interleukin 3 and erythropoietin receptors. *EMBO J.* **14**, 2816–2826.

Yoshimura, A., Ichihara, M., Kinjyo, I., Moriyama, M., Copeland, N. G., Gilbert, D. J., Jenkins, N. A., Hara, T., and Miyajima, A. (1996). Mouse oncostatin M: an immediate early gene induced by multiple cytokines through the JAK-STAT5 pathway. *EMBO J.* **15**, 1055–1063.

IL-3 Receptor

John W. Schrader*

Department of Medicine, Biomedical Research Center, University of British Columbia, 2222 Health Science Mall, Vancouver, British Columbia, Canada V6T 1Z3

*corresponding author tel: 604-822-7810, fax: 604-822-7815, e-mail: john@brc.ubc.ca

DOI: 10.1006/rwcy.2000.20001.

SUMMARY

The receptor for IL-3 is made up of two subunits, both members of the cytokine receptor superfamily. The smaller α subunit has a low affinity for IL-3 and the complex of IL-3 and the α subunit binds to the β subunit to form a high-affinity signaling complex. The β subunit is shared with the receptors for GM-CSF and IL-5, hence its name, β common (βc), and alone, has no affinity for any of these three cytokines. In the mouse, however, there is an additional, IL-3-specific β subunit, which is encoded by a duplication of the βc gene and which has itself a low affinity for IL-3. Signaling downstream of the receptor complex involves activation of JAK2 kinase, tyrosine phosphorylation of βc and the initiation of signaling paths regulating growth, survival, and differentiation.

BACKGROUND

Discovery

The first IL-3-binding protein cloned was the murine AIC2A protein (Itoh *et al.*, 1990). This bound IL-3 with low affinity and was homologous with a closely related molecule, AIC2B (Gorman *et al.*, 1990). AIC2A is the result of the recent duplication in the mouse of the AIC2B gene, which is the murine ortholog of the β common chain (βc). βc is shared between the receptors for IL-3, GM-CSF, and IL-5 and interacts with complexes of GM-CSF, IL-3, or IL-5 with their respective low-affinity α subunits. The α subunit of the IL-3 receptor was subsequently discovered (Kitamura *et al.*, 1991). It was shown to bind IL-3 with low affinity and then to interact with β common in the human, or either AIC2A or AIC2B in the mouse, to generate a high-affinity IL-3 receptor.

Alternative names

AIC2A and AIC2B in the mouse refer respectively to the murine product of the duplicated β common gene that binds IL-3 with low affinity, and the ortholog of β common in the human which can interact with the α chains of the IL-3, GM-CSF or IL-5 receptors.

Structure

The three-dimensional structures of the α and β subunits of the IL-3 receptor are not yet available. In the case of both the α and β subunits, the deduced protein sequences of the extracellular domains predict structures that are homologous to those of other receptors in this family, for which structures are available such as the receptors for growth hormone, prolactin, and erythropoietin.

Main activities and pathophysiological roles

The only known activity of the α unit of the IL-3 receptor is to bind IL-3. As noted above, the β common chain is able to interact with complexes of the IL-3, GM-CSF, and IL-5 receptors, each with their respective α subunits. There is no evidence that complexes of the IL-3 receptor α and β subunits exist in the absence of IL-3. In contrast, in the case of the related GM-CSF receptor, there is evidence that the α and β subunits of the receptor are associated before the ligand binds.

GENE

Accession numbers

Mouse α subunit: X64534
Human α subunit: M74782
Human βc receptor subunit: A39255
Mouse βc receptor subunit: P26955 (*AIC2B*), AAA39295 (*AIC2A*)

Sequence

The murine IL-3 receptor α subunit gene spans 10 kb and includes 12 exons. The general genomic structure is similar to that of other genes in the cytokine receptor family, consistent with their common phylogenetic origin (Miyajima *et al.*, 1995). The promoter of the IL-3 receptor α subunit gene in the mouse has potential sites for GATA, Ets, c-myb, p1, and AP-2.

Chromosome location and linkages

In the mouse, the gene for the IL-3 receptor α subunit is on chromosome 14, whereas that encoding the corresponding α subunit for the GM-CSF receptor is on chromosome 19. In the human, the genes for the α subunits of both the IL-3 receptor and the GM-CSF receptor are tightly linked and are on the pseudoautosomal regions of the X (Xp22.3) and Y (Yp11.3) chromosomes.

PROTEIN

Accession numbers

Mouse α subunit: X64534
Human α subunit: M74782
Human βc receptor subunit: A39255
Mouse βc receptor subunit: P26955 (AIC2B), AAA39295 (AIC2A)

Description of protein

The α subunit of the IL-3 receptor is a member of the cytokine receptor family. Its deduced protein sequence has 378 amino acids in the human and has the characteristic distribution of cysteines, and the WSXWS motif characteristic of receptors of the cytokine family. The cytoplasmic domain of the human IL-3 receptor α subunit is relatively short and contains no tyrosines (the murine ortholog has one) but does contain a proline-rich motif ('Box 1' motif) characteristic of the cytoplasmic domain of receptors of this family.

The extracellular domains of βc (or AIC2A) again show the characteristic features of the members of the cytokine receptor family but are larger than that of the α subunit as a result of a duplication of the basic cytokine receptor domain. The cytoplasmic domains of βc or AIC2A have multiple tyrosines, many of which have been shown by mutational studies to be relevant to receptor function and to be phosphorylated following binding of IL-3.

Relevant homologies and species differences

As noted only in the mouse, are there two β subunits of the IL-3 receptor, the additional β chain AIC2A having arisen by duplication of the βc gene, and evolved the capacity to itself bind IL-3.

Cell types and tissues expressing the receptor

The IL-3 receptor α subunit is expressed on the broad range of cells of hematopoietic origin upon which it is alive. It is lacking on mature T and B lymphocytes and neutrophils but present on most other nucleated cells derived from the pluripotential hematopoietic stem cells. There is some evidence for expression of the IL-3 receptor on endothelial cells. The βc subunit is expressed more widely, being also present on neutrophils.

Regulation of receptor expression

Binding of its ligand results in internalization and downregulation of the levels of the IL-3 receptor on the cell surface. There is some evidence for upregulation of expression of mRNA encoding the IL-3 receptor α subunit by IL-3. A genetic mutation leads to decreased expression of the IL-3 receptor α subunit and thus diminished responses to IL-3 in certain strains of mice (e.g. A/J) (Ichihara *et al.*, 1995; Leslie *et al.*, 1996).

SIGNAL TRANSDUCTION

Associated or intrinsic kinases

JAK2 kinase appears to be associated with the active ligand-bound complex of the IL-3 receptor α subunit and βc or AIC2A. There is some evidence suggesting that JAK2 kinase may associate constitutively with the βc. The cytoplasmic domain of the IL-3 receptor α subunit is absolutely required for IL-3-mediated signaling and one hypothesis is that the cytoplasmic domain of the IL-3 receptor α subunit is involved in recruiting JAK kinase to the heterodimeric, ligand-induced complex of the α and β subunits.

Cytoplasmic signaling cascades

An early event is activation of the JAK2 protein tyrosine kinase. As noted above, this may result from apposition of two JAK2 kinase molecules brought together by ligand-induced approximation of the α and β subunits. In keeping with this notion, there is evidence that the formation of simple heterodimers of the cytoplasmic domains of the α and β subunits is both sufficient and necessary for mitogenesis (Orban *et al.*, 1999). However, higher order complexes involving multiple β chains may also be formed. The cytoplasmic domain of βc or AIC2A have multiple tyrosines, many of which have been shown by mutational studies to be relevant to receptor function and to be phosphorylated following binding of IL-3. These phosphotyrosine residues serve as docking sites. The PTB domain of the adapter protein Shc links the activated receptor via Grb2 to the Ras exchange factor mSOS1. Another potential link to the Ras pathway is provided by the binding to βc via its SH2 domain of the tyrosine phosphatase SHP2, which itself becomes tyrosine phosphorylated and serves as a further docking site for complexes of Grb2 and mSOS1. The translocation of mSOS1 to the membrane brings it into proximity with its substrates, the small GTPases of the Ras family. IL-3 binding results in activation of M-Ras, a 29 kDa relative of p21ras and probably also of p21ras, although this has not been demonstrated formally.

Activation of the Ras family leads to activation of the MAP kinase family through cascades of serine/threonine kinases. Activation of p21ras leads to activation of Raf-1, and in turn MEK1 and MEK2 and then the ERK1 and ERK2 kinases. One of the substrates of the ERK kinases is Stathmin, a protein involved in regulation of the stability of microtubules. Others include transcription factors such as c-Myc or Elk-1. IL-3 also activates members of the other two families of the MAP kinase superfamily, the p38MAP kinases, and the JNK/stress-activated kinases.

At the head of another pathway activated by phosphorylation of the activated IL-3 receptor complex is the lipid kinase PI-3 kinase. The SH2 domains of the p85 subunit of PI-3 kinase bind to phosphotyrosines on SHP2 and IRS2, another of the proteins that bind to the receptor complex via a PTB domain and themselves become phosphorylated on tyrosine. This approximation of PI-3 kinase with its lipid substrates in the membrane results in increased levels of the products of its action. These phospholipids in turn activate the PH domains of two enzymes PDK1 and PDK2, which phosphorylate and activate two serine/threonine protein kinases, PKB and p70 S6 kinase. PI-3 kinase activity is absolutely required for IL-3-induced increases in levels of c-Myc mRNA, but p70 S6 kinase activity is not, suggesting that PKB activity is required for IL-3-induced upregulation of c-Myc.

Also docking onto the activated IL-3 receptor complex are molecules with counter-regulatory activity. These include the tyrosine phosphotase SHP2, and the lipid phosphatase SHIP1, which dephosphorylates products of PI-3 kinase action.

DOWNSTREAM GENE ACTIVATION

Transcription factors activated

IL-3-induced activation of tyrosine kinases leads to activation of STAT5a and STAT5b transcription factors. These cytoplasmic factors dock onto phosphotyrosines on the β subunit and themselves become phosphorylated on tyrosine by JAK2 kinase. This results in dimerization and translocation to the nucleus, where they bind promoters with GAS elements.

BIOLOGICAL CONSEQUENCES OF ACTIVATING OR INHIBITING RECEPTOR AND PATHOPHYSIOLOGY

Unique biological effects of activating the receptors

There is little evidence for any unique biological effects downstream of the IL-3 receptor. Many of its best-studied effects, such as stimulation of growth and survival, involve paths activated by many receptors such as the Ras/MAP kinases, PI-3 kinase, Bcl-2, etc. STAT5 is also activated downstream of other cytokine receptors (e.g. those for erythropoietin or IL-2) and any unique effects are likely to result from combinatorial and quantitative balances of signals also involved in signaling from other receptors.

Phenotypes of receptor knockouts and receptor overexpression mice

The knockout of the βc gene results in a phenotype similar to that seen in the GM-CSF knockout.

References

Gorman, D. M., Itoh, N., Kitamura, T., Schreurs, J., Yonehara, S., Yahara, I., Arai, K., and Miyajima, A. (1990). Cloning and expression of a gene encoding an interleukin 3 receptor-like protein: identification of another member of the cytokine receptor gene family. *Proc. Natl Acad. Sci. USA* **87**, 5459–5463.

Ichihara, M., Hara, T., Takagi, M., Cho, L. C., Gorman, D. M., and Miyajima, A. (1995). Impaired interleukin-3 (IL-3) response of the A/J mouse is caused by a branch point deletion in the IL-3 receptor alpha subunit gene. *EMBO J.* **14**, 939–950.

Itoh, N., Yonehara, S., Schreurs, J., Gorman, D. M., Maruyama, K., Ishii, A., Yahara, I., Arai, K., and Miyajima, A. (1990). Cloning of an interleukin-3 receptor gene: a member of a distinct receptor gene family. *Science* **247**, 324–327.

Kitamura, T., Sato, N., Arai, K., and Miyajima, A. (1991). Expression cloning of the human IL-3 receptor cDNA reveals a shared beta subunit for the human IL-3 and GM-CSF receptors. *Cell* **66**, 1165–1174.

Leslie, K. B., Jalbert, S., Orban, P., Welham, M., Duronio, V., and Schrader, J. W. (1996). Genetic basis of hypo-responsiveness of A/J mice to interleukin-3. *Blood* **87**, 3186–3194.

Miyajima, I., Levitt, L., Hara, T., Bedell, M. A., Copeland, N. G., Jenkins, N. A., and Miyajima, A. (1995). The murine interleukin-3 receptor alpha subunit gene: chromosomal localization, genomic structure, and promoter function. *Blood* **85**, 1246–1253.

Orban, P. C., Levings, M. K., and Schrader, J. W. (1999). Heterodimerization of the alpha and beta chains of the interleukin-3 (IL-3) receptor is necessary and sufficient for IL-3-induced mitogenesis. *Blood* **94**, 1614–1622.

IL-5 Receptor

Christopher J. Bagley[1,2,*], Jan Tavernier[3], Joanna M. Woodcock[4,2] and Angel F. Lopez[4,2]

[1]Protein Laboratory, Hanson Centre for Cancer Research, Frome Road, Adelaide, SA, 5000, Australia

[2]Human Immunology, Institute of Medical and Veterinary Science, Frome Road, Adelaide, SA, 5000, Australia

[3]Department of Medical Protein Research, Flanders' Interuniversity Institute for Biotechnology, University of Ghent, Ghent, Belgium

[4]Cytokine Receptor Laboratory, Hanson Centre for Cancer Research, Frome Road, Adelaide, SA, 5000, Australia

*corresponding author tel: +61-8-82223714, fax: +61-8-82324092, e-mail: chris.bagley@imvs.sa.gov.au

DOI: 10.1006/rwcy.2000.20002.

SUMMARY

The receptor for IL-5 is comprised of two chains: an α chain that binds IL-5 with moderate affinity but alone is unable to mediate signaling, and a β chain that represents the major signaling component of the receptor. The β chain is unable to bind IL-5 alone, but when expressed together with the IL-5 receptor α chain (IL-5Rα) a high-affinity receptor is formed ($K_d = 0.15$ nM). The β chain is shared with the related receptors for granulocyte–macrophage colony-stimulating factor (GM-CSF) and IL-3 and many of the events in IL-5-induced signal transduction are paralleled in the GM-CSF and IL-3 receptors. The IL-5Rα is expressed on a restricted range of cell types, principally eosinophils, basophils, and their immediate precursors confining the actions of IL-5 to these cells. Thus, the IL-5 receptor is considered to be a target in the development of treatments for allergic inflammatory diseases such as *asthma*.

BACKGROUND

Discovery

The two components of the IL-5R were first identified in crosslinking studies (Mita *et al.*, 1989). The cDNA for IL-5Rα was cloned by Takaki *et al.* (1990) and by Tavernier *et al.* (1991) who also identified the β chain of the GM-CSF receptor (βc, cloned by Hayashida *et al.*, 1990) as a component of the IL-5R.

Alternative names

IL-5R α chain: IL-5Rα, Cdw125.

IL-5R β chain: IL-5Rβ, Cdw131, common β chain of the GM-CSF, IL-3 and IL-5 receptors, βc.

Structure

The IL-5 receptor consists of two chains, denoted α and β. The α subunit has specific ligand-binding activity whereas the β subunit does not bind IL-5 by itself but enhances the binding of IL-5 and provides the major determinants of signaling capacity of the receptor. The β subunit is also a component of the high-affinity receptors for GM-CSF and IL-3.

Main activities and pathophysiological roles

Stimulation of the IL-5 receptor *in vivo* results in the stimulation of production of eosinophils and basophils which exacerbates *allergic inflammation*.

GENE

Accession numbers

GenBank:
Human IL-5Rα: M75914, M96651, M96652, X61176, X61177, X61178, X62156
Human IL-5Rβ: M59941, M38275

Sequence

Both the murine IL-5Rα gene (Imamura *et al.*, 1994) and IL-5Rβ gene (Gorman *et al.*, 1992) have been mapped.

PROTEIN

Accession numbers

See **Table 1**.

Description of protein

The nucleotide sequence of the human IL-5Rα cDNA predicts a polypeptide of 420 amino acids. It is characterized by a 20 residue N-terminal signal peptide, followed by a 322 amino acid extracellular domain, a membrane anchor spanning 20 residues, and a 58 amino acid cytoplasmic tail. The predicted molecular mass for the α chain is 45.5 kDa, indicating that *N*-linked glycosylation of one or more of the six potential *N*-glycosylation sites (and perhaps *O*-glycosylation) contributes to the apparent molecular mass of 60 kDa. The biological role of glycosylation remains unresolved since deglycosylation appears to lead to loss of IL-5 binding (Johanson *et al.*, 1995) although IL-5Rα produced in *Escherichia coli* has been reported to bind IL-5 with near normal affinity (Monahan *et al.*, 1997).

The extracellular part of the protein folds into three fibronectin type III structural domains, each having a seven β sheet scaffold. The juxtamembrane domain contains a canonical Trp-Ser-Xaa-Trp-Ser motif (WSXWS box) and forms together with the central domain, which itself is characterized by four conserved cysteines, a so-called cytokine receptor module (CRM) (Goodall *et al.*, 1993) also described as a cytokine-binding domain (CBD). The N-terminal domain is more related to the WSXWS box domain (Tuypens *et al.*, 1992), and contains two Cys residues, which may form an intradomain disulfide bond (Cornelis *et al.*, 1995a). With the exception of a Box-1 motif (Pro-Pro-Xaa-Pro), no consensus sequences are found in the cytoplasmic domain. Ligand recognition involves residues in the hinge between the central and membrane-proximal domains as well as a region in the N-terminal domain (Cornelis *et al.*, 1995b).

The nucleotide sequence of the human IL-5Rβ cDNA predicts a polypeptide of 897 amino acids. The protein is characterized by a 16 residue N-terminal signal peptide, followed by a 422 amino acid extracellular domain, a membrane anchor spanning 22 residues, and a 437 amino acid cytoplasmic tail. The predicted molecular mass for the β chain is 97.3 kDa, indicating that *N*-linked glycosylation and probably *O*-glycosylation contribute to the apparent molecular mass of approximately 130 kDa. The extracellular part of the protein folds into four fibronectin type III structural domains grouped into two CRMs (Bagley *et al.*, 1997). The membrane-proximal CRM contains the main determinants for interaction with IL-5, principally in the loops between the B and C and F and G strands of domain four (Woodcock *et al.*, 1994, 1996). The cytoplasmic domain contains a Box-1 motif (Pro-Pro-Xaa-Pro) responsible for

Table 1 Protein accession numbers

	ID (Swiss)	AC (Swiss)	AC (GenPept)
Human IL-5Rα	IL5R_HUMAN	Q01344	
Mouse IL-5Rα	IL5R_MOUSE	P21183	
Guinea pig IL-5Rα			U55215
Human βc	CYRB_HUMAN	P32927	
Mouse βc	CYRB_MOUSE	P26955	
Guinea pig βc			U94688
Rat IL-3-specific (or common) β chain		Q64146	

interaction with JAK kinases, eight tyrosine residues that are susceptible to phosphorylation, and a recently reported phosphoserine-containing site that binds 14-3-3 proteins (Stomski *et al.*, 1999).

Relevant homologies and species differences

Both chains of the IL-5R are members of the cytokine receptor family. The IL-5Rα is most closely related to the GM-CSFRα and IL-3Rα and more distantly related to the IL-13Rα2. The two CRMs of βc are not closely related to each other, nor to other members of the cytokine receptor family.

Apart from the immediate membrane-proximal region, the cytoplasmic domain of βc is not significantly related to those of other cytokine receptors.

Human and murine IL-5R α chains are 68% identical (extracellular 71%; intracellular 58%). Human and murine common (GM-CSF and IL-5 receptor) β chains are 56% identical (extracellular 57%; intracellular 55%).

Affinity for ligand(s)

The IL-5Rα subunit is ligand-specific and binds with intermediate affinity in humans ($K_d = \sim 0.5$–1 nM) and low affinity in the mouse ($K_d = \sim 5$–10 nM). Upon association with the β subunit, similar high-affinity binding ($K_d = 150$ pM) is observed for both species. The β subunit does not have any detectable affinity for IL-5 by itself (Takaki *et al.*, 1990, 1991; Devos *et al.*, 1991; Tavernier *et al.*, 1991; Murata *et al.*, 1992). The dissociation rate of mIL-5 from the α/β complex is considerably slower ($t_{1/2} > 1$ hour) than from the low-affinity α-binding site ($t_{1/2} < 2$ min) (Devos *et al.*, 1991). IL-5 binds to its receptor with unidirectional species-specificity: mIL-5 binds with comparable affinities to both murine and human α subunits, but hIL-5 displays 100-fold lower binding affinity for the murine α chain, compared with its human counterpart, a cross-species pattern that is mirrored in their biological activities (Lopez *et al.*, 1986).

Cell types and tissues expressing the receptor

Expression of the IL-5R in humans is most prominent on eosinophils and basophils (Lopez *et al.*, 1991). Its appearance on multipotential myeloid progenitors is critical for their development towards the eosinophilic lineage. It remains expressed on mature cells, but the expression level is controlled at different levels. In the mouse, in addition to these cell types, the IL-5Rα is also found prominently on B cells belonging to the B1 lineage (Hitoshi *et al.*, 1990). Although there has been no report of direct measurement of IL-5R on human B cells, IL-5 appears to augment terminal differentiation of mitogen-stimulated B cells in some cases.

Regulation of receptor expression

Two promoters, P1 and P2, have been identified in the hIL-5Rα gene (Sun *et al.*, 1995; Zhang *et al.*, 1997). P1 and P2 precede the first and the second exon respectively, and show no significant sequence similarities. P1 is myeloid and eosinophil lineage-specific, whilst the use of P2 is restricted to eosinophilic HL60-C15 cells. It is at present unclear whether differential use and regulation of both promoters occur during eosinophilic differentiation.

The P1 promoter contains multiple consensus binding sites for AP-1, C/EBP, GATA, and PU.1. In addition, the region between −432 and −398 was shown to contain a unique *cis*-element (EOS1), necessary and sufficient for the expression in eosinophilic cell lines (Sun *et al.*, 1995). The putative myeloid- or eosinophil-specific binding factor(s) has not been identified so far. Involvement of the adjacent AP-1 element at position −440 to −432 in the expression in eosinophilic HL-60 cells has been demonstrated (Baltus *et al.*, 1998), suggesting cooperation between the cognate binding transcription factors. Supershift analysis experiments showed the presence of cJun, CREB, and CREM in the AP-1-binding complexes.

The P2 promoter sequence shows the presence of AP-1, C/EBP, GATA, CLEO (IL-5), and its consensus binding sites. A unique functional motif was identified between positions −19 and −14. It is involved in the binding of a hitherto unidentified eosinophilic HL-60-C15-specific transcription factor(s). Alternatively, it may also serve as a non-canonical TATA box, since such a motif is lacking in the P2 promoter.

Rapid downregulation (maximum inhibition within 2 hours) of the hIL-5Rα mRNA is induced in peripheral blood eosinophils upon treatment with IL-3, IL-5, or GM-CSF. In contrast, similar treatment leads to upregulation of the IL-3Rα, GM-CSFRα, and βc mRNA. The mechanisms involved were shown to be promoter activation and reduced mRNA degradation, respectively, indicating differential regulation (Wang *et al.*, 1998). It is at present unclear whether the downmodulation occurs for P1, P2, or both.

The promoter of the mIL-5Rα contains consensus binding sites for AP-1, GATA, NF-IL-6, NFκB, and

SP-1. Little is known about transcription factors controlling cell type-specific expression of the mIL-5Rα chain.

Release of soluble receptors

Human eosinophils express through alternative splicing different transcripts from the same IL-5Rα locus (Tavernier *et al.*, 1992; Tuypens *et al.*, 1992). As a result, in addition to the membrane-anchored receptor, two soluble isoforms can be produced. One of these soluble variants is the predominant (>90%) transcript detected in eosinophilic HL-60-C15 cells and in eosinophils obtained from cord blood cultures. Variable isoform mRNA expression has been observed in eosinophils purified from peripheral blood. The forced expiratory volume in 1 second in patients with mild *asthma* has been reported to be inversely correlated with the expression of the membrane-anchored isoform and directly correlated with the soluble isoform in eosinophils in endobronchial biopsies (Yasruel *et al.*, 1997).

In mouse B cells, transcripts encoding secreted variants are also generated through alternative splicing. In contrast to humans, there is no evidence for a similar soluble variant-specific exon. Rather, these soluble variant-specific transcripts are formed by skipping of the membrane anchor exon (Takaki *et al.*, 1990; Tavernier *et al.*, 1992; Imamura *et al.*, 1994).

The soluble hIL-5Rα isoform has antagonistic properties *in vitro*. It binds one IL-5 dimer in solution. It inhibits various IL-5 activities, including induced tyrosine phosphorylation of JAK2 and βc, proliferation of IL-5-dependent cell lines, and eosinophilic differentiation and survival (Tavernier *et al.*, 1991; Monahan *et al.*, 1997), suggesting a role in the regulation of *eosinophilia in vivo*. No IL-5-potentiating effects have been observed in *in vitro* assays, underscoring its anti-inflammatory potential. So far, however, neither translation of the message encoding this soluble variant *in vitro* in eosinophils, nor circulating soluble hIL-5Rα *in vivo* has been reported. One possible explanation might be the thermolability of this soluble receptor. Alternatively, this splice regulation could merely serve a regulatory function driving transcription into a nonproductive pathway, reducing the expression level of the membrane-associated receptor. Soluble murine IL-5Rα also has antagonistic properties *in vitro*, albeit to a lesser degree than its human counterpart, consistent with its lower affinity for IL-5.

SIGNAL TRANSDUCTION

Signal transduction via the IL-5 receptor involves ligand binding and receptor dimerization, requirements shared by other cytokine receptors (Bagley *et al.*, 1997). The structural elements utilized by the IL-5 receptor α and β chains to bind IL-5 have been discussed above. Following binding of IL-5, dimerization of the IL-5 receptor ensues, a process which shares certain events with the cytokine receptor superfamily at large but which has some features more limited to the IL-5, GM-CSF, and IL-3 subfamily of receptors.

Dimerization of the IL-5 receptor involves the association of the IL-5 receptor α chain with βc. This takes place by noncovalent as well as covalent means. The covalent linkage of IL-5Rα and βc is probably the most functionally relevant one as it is associated with tyrosine phosphorylation of the receptor (Stomski *et al.*, 1998). The cysteines involved are Cys86 and Cys91 in the most N-terminal domain of βc (Stomski *et al.*, 1998). These cysteines interact with a Cys in the IL-5 receptor α chain which has not yet been determined, however, since this α chain has an odd number of cysteines and all cysteines except Cys86 appear to form intramolecular bonds (based on alignment with other members of the cytokine receptor superfamily); Cys86 is the prime candidate. The related GM-CSF and IL-3 receptor α chains also exhibit an odd number of cysteines consistent with them also forming high-order complexes with βc.

Both the IL-5Rα and βc subunits are required for signaling. Deletion of the cytoplasmic domains of either chain leads to complete loss of signaling, without altered ligand binding (Sakamaki *et al.*, 1992; Takaki *et al.*, 1994; Cornelis *et al.*, 1995b). IL-5 binding leads to the rapid tyrosine phosphorylation of a multitude of cytoplasmic proteins. Mutations at position 13 of IL-5 cause diminished or abrogated activation of the βc chain and may yield variants with antagonistic activity (Tavernier *et al.*, 1995; Bagley *et al.*, 1999). An E to K substitution at residue 13 leads to loss of detectable phosphorylation of βc in eosinophils. Yet, whilst being deficient in inducing TF-1 proliferation and eosinophil adhesion, this IL-5 mutein still retains the capacity to support eosinophil survival, indicating that the different signaling pathways and functional responses can be segregated (McKinnon *et al.*, 1997). The expression level of the IL-5Rα subunit may control this agonist/antagonist balance (van Ostade *et al.*, 1999).

Associated or intrinsic kinases

Neither subunit of the IL-5 receptor possesses intrinsic tyrosine kinase activity.

Studies in cell lines showed association of JAK2 and JAK1 (or JAK2) with the IL-5Rα and βc subunits, respectively, and rapid tyrosine phosphorylation upon IL-5-binding (Ogata *et al.*, 1998). This activation critically depends on the presence of intact, membrane-proximal proline-rich motifs (Box-1 motif) in both chains (Quelle *et al.*, 1994; Takaki *et al.*, 1994).

Cytoplasmic signaling cascades

In addition to the JAK kinases that associate directly with IL-5R, other kinases such as the Src family kinases Lyn and Fyn, and the Bruton tyrosine kinase, Btk, are activated by IL-5, suggesting a cascade of tyrosine phosphorylation events. Btk has been implicated in IL-5 signaling in B cells only. Mutations in the *btk* gene lead to B cell deficiencies in humans (*X-linked agammaglobulinemia*) and mice (*X-linked immunodeficiency*) (Hitoshi *et al.*, 1993; Koike *et al.*, 1995). Btk functions in concert with the Src family kinases Lyn and Fyn (Cheng *et al.*, 1994; Appleby *et al.*, 1995).

Multiple tyrosine residues become phosphorylated on the βc subunit and provide docking sites for signaling molecules. Both STAT1 (Pazdrak *et al.*, 1995; van der Bruggen *et al.*, 1995) and STAT5 (Mui *et al.*, 1995) can become phosphorylated and activated by IL-5 treatment. STAT1 appears to be the major STAT activated by IL-5 in eosinophils. In the case of STAT5, a high degree of redundancy in recruitment sites has been reported (van Dijk *et al.*, 1997).

Via rapid recruitment and phosphorylation of the adapters Shc or SHP2, IL-5 signaling can be coupled to the Ras pathway (Pazdrak *et al.*, 1997). Downstream effector molecules of Ras include PI-3 kinase and MAP kinase. PI-3 kinase plays a critical role in the induction by IL-5 of a chemokinetic response in bone marrow eosinophils (Palframan *et al.*, 1998). The downstream targets of PI-3 kinase remain unclear. Activation of MAP kinase is required for induction of members of the AP-1 family, including the c-fos and c-jun proto-oncogenes. Members of this family are involved in myeloid differentiation (Foletta *et al.*, 1998).

Ras activation has also been implicated in suppression of apoptosis of eosinophils by IL-5. The βc not only undergoes tyrosine phosphorylation but is also phosphorylated on serine residues such as Ser585. This allows β to bind to the 14-3-3ζ adapter protein and presumably associate with other molecules (Stomski *et al.*, 1999). The biological significance of this is being determined.

DOWNSTREAM GENE ACTIVATION

Transcription factors activated

STAT1 (Pazdrak *et al.*, 1995; van der Bruggen *et al.*, 1995) and STAT5, c-fos, c-myc (Mui *et al.*, 1995), NF-AT (Jinquan *et al.*, 1999).

Genes induced

Activation of STAT5 leads to rapid induction of immediate early response genes, including Cis, OSM, Id, *pim-1*, c-*fos* (Mui *et al.*, 1995).

Other genes induced include c-*myc*, VEGF (Horiuchi and Weller, 1997), β_2 integrin (Palframan *et al.*, 1998), PAF (Kishimoto *et al.*, 1996), Bcl-xL (Dibbert *et al.*, 1998) and Bcl2 (Ochiai *et al.*, 1997; Dewson *et al.*, 1999). The expression of the IL-3Rα, GM-CSFRα, and βc chains is induced, whereas that of the IL-5Rα is repressed (Wang *et al.*, 1998).

BIOLOGICAL CONSEQUENCES OF ACTIVATING OR INHIBITING RECEPTOR AND PATHOPHYSIOLOGY

Unique biological effects of activating the receptors

Although the IL-5 receptor exhibits the same biological activities as the GM-CSF and IL-3 receptors in a given cell type by virtue of their sharing the β subunit, its unique biological effect stems from the spectrum of cells types that express the receptor. Thus, in contrast to the GM-CSF receptor and to a lesser degree the IL-3 receptor, which are more widely distributed, the expression of the IL-5 receptor is largely restricted to eosinophils and basophils in humans (and to B cells also in the mouse). Stimulation of the IL-5 receptor *in vivo* thus results in selective stimulation of eosinophil and basophil production which exacerbates allergic inflammation.

Phenotypes of receptor knockouts and receptor overexpression mice

Disruption of the IL-5Rα gene in mice leads to decreased levels of IgA in mucosal secretions (Hiroi *et al.*, 1999) and an inability to induce eosinophilia in response to *parasitic infection* (Sugaya *et al.*, 1997). Overexpression of the IL-5Rα does not lead to increased levels of eosinophils or B cells, although they exhibit some enhancement of sensitivity to IL-5 (Sugaya *et al.*, 1997).

Mice that are deficient in the β chain of the IL-5 receptor are apparently normal except for a profound decrease in the number of eosinophils and a *pulmonary alveolar proteinosis* (PAP)-like disease (Nishinakamura *et al.*, 1995). Given that the β subunit is shared with the GM-CSF and IL-3 receptors, it is not clear which defect is specific for the IL-5 receptor. The reduction of eosinophils is likely to be so since a similar phenotype is seen in IL-5 knockouts; however, PAP is also seen in GM-CSF but not IL-5 knockout mice.

THERAPEUTIC UTILITY

Effect of treatment with soluble receptor domain

Soluble IL-5Rα inhibits the ability of IL-5 to promote the survival, proliferation, and activation of eosinophils and basophils.

Effects of inhibitors (antibodies) to receptors

Inhibition of the IL-5 receptor is being tried as an alternative to inhibiting IL-5 itself for the treatment of allergic inflammation such as *asthma*. One approach relies on blocking the specific α chain of the IL-5 receptor by constructing IL-5 mutants defective in interacting with the β chain only. This can be achieved by modifying IL-5 itself by mutating Glu13, a residue that is conserved in position in the tertiary structure and function in GM-CSF and IL-3. Substitution of Glu13 by a Gln, Arg, or Lys results in an IL-5 molecule that behaves as a specific IL-5 antagonist. However, the E13K mutant is still able to support eosinophil survival. A second approach involves blocking the β subunit, an approach which has the additional therapeutical advantage of blocking stimulation by IL-5 but also the stimulation by GM-CSF and IL-3, of eosinophils and basophils, the major inflammatory cell types in allergy. The antibody BION-1 has recently been developed and this binds to the B-C loop of the fourth domain of βc and blocks the production, survival, and activation of eosinophils in response to IL-5, GM-CSF, and IL-3 (Sun *et al.*, 1999).

References

Appleby, M. W., Kerner, J. D., Chien, S., Maliszewski, C. R., Bondadaa, S., and Perlmutter, R. M. (1995). Involvement of p59fynT in interleukin-5 receptor signaling. *J. Exp. Med.* **182**, 811–820.

Bagley, C. J., Woodcock, J. M., Stomski, F. C., and Lopez, A. F. (1997). The structural and functional basis of cytokine receptor activation: lessons from the common beta subunit of the granulocyte-macrophage colony-stimulating factor, interleukin-3 (IL-3), and IL-5 receptors. *Blood* **89**, 1471–1482.

Bagley, C. J., Woodcock, J. M., Stomski, F. C., and López, A. F. (1999). In "Interleukin-5, From Molecule to Drug Target for Asthma" (ed. C. J. Sanderson), The structural basis for interleukin-5 receptor assembly pp. 189–203. Marcel Dekker, New York.

Baltus, B., van Dijk, T. B., Caldenhoven, E., Zanders, E., Raaijmakers, J. A., Lammers, J. W., Koenderman, L., and de Groot, R. P. (1998). An AP-1 site in the promoter of the human IL-5R alpha gene is necessary for promoter activity in eosinophilic HL60 cells. *FEBS Lett.* **434**, 251–254.

Cheng, G., Ye, Z. S., and Baltimore, D. (1994). Binding of Bruton's tyrosine kinase to Fyn, Lyn, or Hck through a Src homology 3 domain-mediated interaction. *Proc. Natl Acad. Sci. USA* **91**, 8152–8155.

Cornelis, S., Plaetinck, G., Devos, R., Van der Heyden, J., Tavernier, J., Sanderson, C., Guisez, Y., and Fiers, W. (1995a). Detailed analysis of the IL-5–IL-5R alpha interaction: characterization of crucial residues on the ligand and the receptor. *EMBO J.* **14**, 3395–3402.

Cornelis, S., Fache, I., Van der Heyden, J., Guisez, Y., Tavernier, J., Devos, R., Fiers, W., and Plaetinck, G. (1995b). Characterization of critical residues in the cytoplasmic domain of the human interleukin-5 receptor alpha chain required for growth signal transduction. *Eur. J. Immunol.* **25**, 1857–1864.

Devos, R., Plaetinck, G., Van der Heyden, J., Cornelis, S., Vandekerckhove, J., Fiers, W., and Tavernier, J. (1991). Molecular basis of a high affinity murine interleukin-5 receptor. *EMBO J.* **10**, 2133–2137.

Dewson, G., Walsh, G. M., and Wardlaw, A. J. (1999). Expression of Bcl-2 and its homologues in human eosinophils. Modulation by interleukin-5. *Am. J. Respir. Cell Mol. Biol.* **20**, 720–728.

Dibbert, B., Daigle, I., Braun, D., Schranz, C., Weber, M., Blaser, K., Zangemeister-Wittke, U., Akbar, A. N., and Simon, H. U. (1998). Role for Bcl-xL in delayed eosinophil apoptosis mediated by granulocyte-macrophage colony-stimulating factor and interleukin-5. *Blood* **92**, 778–783.

Foletta, V. C., Segal, D. H., and Cohen, D. R. (1998). Transcriptional regulation in the immune system: all roads lead to AP-1. *J. Leukoc. Biol.* **63**, 139–152.

Goodall, G. J., Bagley, C. J., Vadas, M. A., and Lopez, A. F. (1993). A model for the interaction of the GM-CSF, IL-3 and IL-5 receptors with their ligands. *Growth Factors* **8**, 87–97.

Gorman, D. M., Itoh, N., Jenkins, N. A., Gilbert, D. J., Copeland, N. G., and Miyajima, A. (1992). Chromosomal localization and organization of the murine genes encoding the beta subunits (AIC2A and AIC2B) of the interleukin 3, granulocyte/macrophage colony-stimulating factor, and interleukin 5 receptors. *J. Biol. Chem.* **267**, 15842–15848.

Hayashida, K., Kitamura, T., Gorman, D. M., Arai, K., Yokota, T., and Miyajima, A. (1990). Molecular cloning of a second subunit of the receptor for human granulocyte-macrophage colony-stimulating factor (GM-CSF): reconstitution of a high-affinity GM-CSF receptor. *Proc. Natl Acad. Sci. USA* **87**, 9655–9659.

Hiroi, T., Yanagita, M., Iijima, H., Iwatani, K., Yoshida, T., Takatsu, K., and Kiyono, H. (1999). Deficiency of IL-5 receptor alpha-chain selectively influences the development of the common mucosal immune system independent IgA-producing B-1 cell in mucosa-associated tissues. *J. Immunol.* **162**, 821–828.

Hitoshi, Y., Yamaguchi, N., Mita, S., Sonoda, E., Takaki, S., Tominaga, A., and Takatsu, K. (1990). Distribution of IL-5 receptor-positive B cells. Expression of IL-5 receptor on Ly-1(CD5)+ B cells. *J. Immunol.* **144**, 4218–4225.

Hitoshi, Y., Sonoda, E., Kikuchi, Y., Yonehara, S., Nakauchi, H., and Takatsu, K. (1993). IL-5 receptor positive B cells, but not eosinophils, are functionally and numerically influenced in mice carrying the X-linked immune defect. *Int. Immunol.* **5**, 1183–1190.

Horiuchi, T., and Weller, P. F. (1997). Expression of vascular endothelial growth factor by human eosinophils: upregulation by granulocyte macrophage colony-stimulating factor and interleukin-5. *Am. J. Respir. Cell Mol. Biol.* **17**, 70–77.

Imamura, F., Takaki, S., Akagi, K., Ando, M., Yamamura, K. I., Takatsu, K., and Tominaga, A. (1994). The murine interleukin-5 receptor alpha-subunit gene: characterization of the gene structure and chromosome mapping DNA. *Cell. Biol.* **13**, 283–292.

Jinquan, T., Quan, S., Jacobi, H. H., Reimert, C. M., Millner, A., Hansen, J. B., Thygesen, C., Ryder, L. P., Madsen, H. O., Malling, H. J., and Poulsen, L. K. (1999). Cutting edge: expression of the NF of activated T cells in eosinophils: regulation by IL-4 and IL-5. *J. Immunol.* **163**, 21–24.

Johanson, K., Appelbaum, E., Doyle, M., Hensley, P., Zhao, B., Abdel-Meguid, S. S., Young, P., Cook, R., Carr, S., and Matico, R. (1995). Binding interactions of human interleukin 5 with its receptor alpha subunit. Large scale production, structural, and functional studies of *Drosophila*-expressed recombinant proteins. *J. Biol. Chem.* **270**, 9459–9471.

Kishimoto, S., Shimadzu, W., Izumi, T., Shimizu, T., Fukuda, T., Makino, S., Sugiura, T., and Waku, K. (1996). Regulation by IL-5 of expression of functional platelet-activating factor receptors on human eosinophils. *J. Immunol.* **157**, 4126–4132.

Koike, M., Kikuchi, Y., Tominaga, A., Takaki, S., Akagi, K., Miyazaki, J., Yamamura, K., and Takatsu, K. (1995). Defective IL-5-receptor-mediated signaling in B cells of X-linked immunodeficient mice. *Int. Immunol.* **7**, 21–30.

Lopez, A. F., Begley, C. G., Williamson, D. J., Warren, D. J., Vadas, M. A., and Sanderson, C. J. (1986). Murine eosinophil differentiation factor. An eosinophil-specific colony-stimulating factor with activity for human cells. *J. Exp. Med.* **163**, 1085–1099.

Lopez, A. F., Vadas, M. A., Woodcock, J. M., Milton, S. E., Lewis, A., Elliott, M. J., Gillis, D., Ireland, R., Olwell, E., and Park, L. S. (1991). Interleukin-5, interleukin-3, and granulocyte-macrophage colony-stimulating factor cross-compete for binding to cell surface receptors on human eosinophils. *J. Biol. Chem.* **266**, 2474–2477.

McKinnon, M., Page, K., Uings, I. J., Banks, M., Fattah, D., Proudfoot, A. E., Graber, P., Arod, C., Fish, R., Wells, T. N., and Solari, R. (1997). An interleukin 5 mutant distinguishes between two functional responses in human eosinophils. *J. Exp. Med.* **186**, 121–129.

Mita, S., Tominaga, A., Hitoshi, Y., Sakamoto, K., Honjo, T., Akagi, M., Kikuchi, Y., Yamaguchi, N., and Takatsu, K. (1989). Characterization of high-affinity receptors for interleukin 5 on interleukin 5-dependent cell lines. *Proc. Natl Acad. Sci. USA* **86**, 2311–2315.

Monahan, J., Siegel, N., Keith, R., Caparon, M., Christine, L., Compton, R., Cusik, S., Hirsch, J., Huynh, M., Devine, C., Polazzi, J., Rangwala, S., Tsai, B., and Portanova, J. (1997). Attenuation of IL-5-mediated signal transduction, eosinophil survival, and inflammatory mediator release by a soluble human IL-5 receptor. *J. Immunol.* **159**, 4024–4034.

Mui, A. L., Wakao, H., O'Farrell, A. M., Harada, N., and Miyajima, A. (1995). Interleukin-3, granulocyte-macrophage colony stimulating factor and interleukin-5 transduce signals through two STAT5 homologs. *EMBO J.* **14**, 1166–1175.

Murata, Y., Takaki, S., Migita, M., Kikuchi, Y., Tominaga, A., and Takatsu, K. (1992). Molecular cloning and expression of the human interleukin 5 receptor. *J. Exp. Med.* **175**, 341–351.

Nishinakamura, R., Nakayama, N., Hirabayashi, Y., Inoue, T., Aud, D., McNeil, T., Azuma, S., Yoshida, S., Toyoda, Y., Arai, K., Miyajima, A., and Murray, R. (1995). Mice deficient for the IL-3/GM-CSF/IL-5 beta c receptor exhibit lung pathology and impaired immune response, while beta IL3 receptor-deficient mice are normal. *Immunity* **2**, 211–222.

Ochiai, K., Kagami, M., Matsumura, R., and Tomioka, H. (1997). IL-5 but not interferon-gamma (IFN-gamma) inhibits eosinophil apoptosis by up-regulation of bcl-2 expression. *Clin. Exp. Immunol.* **107**, 198–204.

Ogata, N., Kouro, T., Yamada, A., Koike, M., Hanai, N., Ishikawa, T., and Takatsu, K. (1998). JAK2 and JAK1 constitutively associate with an interleukin-5 (IL-5) receptor a and bc subunit, respectively, and are activated upon IL-5 stimulation. *Blood* **91**, 2264–2271.

Pazdrak, K., Stafford, S., and Alam, R. (1995). The activation of the Jak-STAT 1 signaling pathway by IL-5 in eosinophils. *J. Immunol.* **155**, 397–402.

Pazdrak, K., Adachi, T., and Alam, R. (1997). Src homology 2 protein tyrosine phosphatase (SHPTP2)/Src homology 2 phosphatase 2 (SHP2) tyrosine phosphatase is a positive regulator of the interleukin 5 receptor signal transduction pathways leading to the prolongation of eosinophil survival. *J. Exp. Med.* **186**, 561–568.

Palframan, R. T., Collins, P. D., Severs, N. J., Rothery, S., Williams, T. J., and Rankin, S. M. (1998). Mechanisms of acute eosinophil mobilization from the bone marrow stimulated by interleukin 5: the role of specific adhesion molecules and phosphatidylinositol 3-kinase. *J. Exp. Med.* **188**, 1621–1632.

Quelle, F. W., Sato, N., Witthuhn, B. A., Inhorn, R. C., Eder, M., Miyajima, A., Griffin, J. D., and Ihle, J. N. (1994). JAK2 associates with the beta c chain of the receptor for granulocyte-macrophage colony-stimulating factor, and its activation requires the membrane-proximal region. *Mol. Cell. Biol.* **14**, 4335–4341.

Sakamaki, K., Miyajima, I., Kitamura, T., and Miyajima, A. (1992). Critical cytoplasmic domains of the common beta subunit of the human GM-CSF, IL-3 and IL-5 receptors for growth, signal transduction and tyrosine phosphorylation. *EMBO J.* **11**, 3541–3549.

Stomski, F. C., Woodcock, J. M., Zacharakis, B., Bagley, C. J., Sun, Q., and Lopez, A. F. (1998). Identification of a Cys motif in the common beta chain of the interleukin 3, granulocyte-macrophage colony-stimulating factor, and interleukin 5 receptors essential for disulfide-linked receptor heterodimerization and activation of all three receptors. *J. Biol. Chem.* **273**, 1192–1199.

Stomski, F. C., Dottore, M., Winnall, W., Guthridge, M. A., Woodcock, J., Bagley, C. J., Thomas, D. T., Andrews, R. K., Berndt, M. C., and Lopez, A. F. (1999). Identification of a 14-3-3 binding sequence in the common beta chain of the granulocyte-macrophage colony-stimulating factor (GM-CSF), interleukin-3 (IL-3), and IL-5 receptors that is serine-phosphorylated by GM-CSF. *Blood* **94**, 1933–1942.

Sugaya, H., Aoki, M., Yoshida, T., Takatsu, K., and Yoshimura, K. (1997). Eosinophilia and intracranial worm recovery in interleukin-5 transgenic and interleukin-5 receptor alpha chain-knockout mice infected with *Angiostrongylus cantonensis*. *Parasitol. Res.* **83**, 583–590.

Sun, Z., Yergeau, D. A., Tuypens, T., Tavernier, J., Paul, C. C., Baumann, M. A., Tenen, D. G., and Ackerman, S. J. (1995). Identification and characterization of a functional promoter region in the human eosinophil IL-5 receptor alpha subunit gene. *J. Biol. Chem.* **270**, 1462–1471.

Sun, Q., Jones, K., McClure, B., Cambareri, B., Zacharakis, B., Iversen, P. O., Stomski, F., Woodcock, J. M., Bagley, C. J., D'Andrea, R., and Lopez, A. F. (1999). Simultaneous antagonism of IL-5, GM-CSF and IL-3 stimulation of human eosinophils by targetting the common cytokine binding site of their receptors. *Blood* **94**, 1943–1951.

Takaki, S., Tominaga, A., Hitoshi, Y., Mita, S., Sonoda, E., Yamaguchi, N., and Takatsu, K. (1990). Molecular cloning and expression of the murine interleukin-5 receptor. *EMBO J.* **9**, 4367–4374.

Takaki, S., Mita, S., Kitamura, T., Yonehara, S., Yamaguchi, N., Tominaga, A., Miyajima, A., and Takatsu, K. (1991). Identification of the second subunit of the murine interleukin-5 receptor: interleukin-3 receptor-like protein, AIC2B is a component of the high affinity interleukin-5 receptor. *EMBO J.* **10**, 2833–2838.

Takaki, S., Kanazawa, H., Shiiba, M., and Takatsu, K. (1994). A critical cytoplasmic domain of the interleukin-5 (IL-5) receptor alpha chain and its function in IL-5-mediated growth signal transduction. *Mol. Cell. Biol.* **14**, 7404–7413.

Tavernier, J., Devos, R., Cornelis, S., Tuypens, T., Van der Heyden, J., Fiers, W., and Plaetinck, G. (1991). A human high affinity interleukin-5 receptor (IL5R) is composed of an IL5-specific alpha chain and a beta chain shared with the receptor for GM-CSF. *Cell* **66**, 1175–1184.

Tavernier, J., Tuypens, T., Plaetinck, G., Verhee, A., Fiers, W., and Devos, R. (1992). Molecular basis of the membrane-anchored and two soluble isoforms of the human interleukin 5 receptor alpha subunit. *Proc. Natl Acad. Sci. USA* **89**, 7041–7045.

Tavernier, J., Tuypens, T., Verhee, A., Plaetinck, G., Devos, R., Van der Heyden, J., Guisez, Y., and Oefner, C. (1995). Identification of receptor-binding domains on human interleukin 5 and design of an interleukin 5-derived receptor antagonist. *Proc. Natl Acad. Sci. USA* **92**, 5194–5198.

Tuypens, T., Plaetinck, G., Baker, E., Sutherland, G., Brusselle, G., Fiers, W., Devos, R., and Tavernier, J. (1992). Organization and chromosomal localization of the human interleukin 5 receptor alpha-chain gene. *Eur. Cytokine Netw.* **3**, 451–459.

van der Bruggen, T., Caldenhoven, E., Kanters, D., Coffer, P., Raaijmakers, J. A., Lammers, J. W., and Koenderman, L. (1995). Interleukin-5 signaling in human eosinophils involves JAK2 tyrosine kinase and Stat1 alpha. *Blood* **85**, 1442–1448.

van Dijk, T. B., Caldenhoven, E., Raaijmakers, J. A., Lammers, J. W., Koenderman, L., and de Groot, R. P. (1997). Multiple tyrosine residues in the intracellular domain of the common beta subunit of the interleukin 5 receptor are involved in activation of STAT5. *FEBS Lett.* **412**, 161–164.

van Ostade, X., Van der Heyden, J., Verhee, A., Vandekerckhove, J., and Tavernier, J. (1999). The cell surface expression level of the human interleukin-5 receptor alpha subunit determines the agonistic/antagonistic balance of the human interleukin-5 E13Q mutein. *Eur. J. Biochem.* **259**, 954–960.

Wang, P., Wu, P., Cheewatrakoolpong, B., Myers, J. G., Egan, R. W., and Billah, M. M. (1998). Selective inhibition of IL-5 receptor alpha-chain gene transcription by IL-5, IL-3, and granulocyte-macrophage colony-stimulating factor in human blood eosinophils. *J. Immunol.* **160**, 4427–4432.

Woodcock, J. M., Zacharakis, B., Plaetinck, G., Bagley, C. J., Sun, Q., Hercus, T. R., Tavernier, J., and Lopez, A. F. (1994). Three residues in the common β chain of the human GM-CSF, IL-3 and IL-5 receptors are essential for GM-CSF and IL-5 but not IL-3 high affinity binding and interact with Glu^{21} of GM-CSF. *EMBO J.* **13**, 5176–5185.

Woodcock, J. M., Zacharakis, B., Plaetinck, G., Bagley, C. J., Sun, Q., Hercus, T. R., Tavernier, J., and Lopez, A. F. (1996). A single tyrosine residue in the membrane proximal domain of the GM-CSF, IL-3 and IL-5 receptor common β chain is necessary and sufficient for high affinity binding and signalling by all three ligands. *J. Biol. Chem.* **271**, 25999–26006.

Yasruel, Z., Humbert, M., Kotsimbos, T. C., Ploysongsang, Y., Minshall, E., Durham, S., Pfister, R., Menz, G., Tavernier, J., Kay, A. B., and Hamid, Q. (1997). Membrane-bound and soluble alpha IL-5 receptor mRNA in the bronchial mucosa of atopic and nonatopic asthmatics. *Am. J. Respir. Crit. Care Med.* **155**, 1413–1418.

Zhang, J., Kuvelkar, R., Cheewatrakoolpong, B., Williams, S., Egan, R. W., and Billah, M. M. (1997). Evidence for multiple promoters of the human IL-5 receptor alpha subunit gene: a novel 6-base pair element determines cell-specific promoter function. *J. Immunol.* **159**, 5412–5421.

LICENSED PRODUCTS

See **Table 2**.

Table 2 Suppliers of anti-IL-5Rα and anti-IL-5Rβ antibodies

Type	Class	Clone/ID	Supplier
Anti-IL-5Rα antibodies			
Polyclonal (goat)	IgG	AF-253-NA	R&D Systems
Monoclonal	IgG1	A14	Pharmingen
Anti-IL-5Rβ antibodies			
Monoclonal	IgG1	MAB1008	Chemicon International
Monoclonal	IgG1	202325	Stratagene
Monoclonal	IgG2b	S16	Santa Cruz
Monoclonal	IgG1k	AR-1635	Maine Biotechnology Services
Monoclonal	IgG1	3D7	Pharmingen
Monoclonal	IgG1	4F3	Amrad

SCF Receptor

Diana Linnekin[1,*] and Jonathan R. Keller[2]

[1]Basic Research Laboratory, Division of Basic Sciences, National Cancer Institute–Frederick Cancer Research & Development Center, Frederick, MD 21702, USA

[2]Intramural Research and Support Program, SAIC Frederick, National Cancer Institute–Frederick Cancer Research & Development Center, Frederick, MD 21702, USA

*corresponding author tel: 301-846-5631 fax: 301-846-6641, e-mail: dlinnekin@mail.ncifcrf.gov

DOI: 10.1006/rwcy.2000.20003.

SUMMARY

The stem cell factor receptor (SCFR) is a member of the receptor tyrosine kinase superfamily. It is expressed on a variety of cell types, including hematopoietic cells, germ cells, neural cells, melanocytes, and the interstitial cells of Cajal in the intestine. The absence of functional murine SCFR is lethal *in utero*. Mutations resulting in reduced expression, or function, of the SCFR lead to abnormalities in hematopoiesis, pigmentation, and reproduction in mice. Inappropriate expression or activation of the SCFR is associated with a variety of diseases, including *mastocytosis* in humans. Further, downregulation of the SCFR may play a role in *metastatic melanoma*. The importance of the SCFR in both normal and pathophysiology has led to extensive study of its mechanism of action. Recent work indicates that the SCFR activates multiple signal transduction pathways, including the Ras/Raf/MAP kinase cascade, the JAK/STAT pathway, phosphatidylinositol 3-kinase, and Src family members. This review will provide an overview of the structure, biology, and mechanism of action of the SCFR.

BACKGROUND

Discovery

As early as 1929, naturally occurring mutations in mice resulting in profound impairment in pigmentation, reproduction, and blood cell development were observed by geneticists. Subsequent work found this phenotype resulted from alterations in either the white spotting (W) or steel (Sl) locus. In 1979, Elizabeth Russell proposed that the gene product of the white spotting locus was a receptor and that the gene product of the steel locus was its cognate ligand (Russell, 1979). In 1986, v-kit, an oncogene isolated from Hardy–Zuckerman feline sarcoma virus, was cloned and characterized (Besmer *et al.*, 1986). Within 2 years both the human and murine forms of the proto-oncogene c-Kit had been cloned (Yarden *et al.*, 1987; Qiu *et al.*, 1988). c-Kit was subsequently mapped to the white spotting locus and in 1990 the gene product of the steel locus was cloned and found to be the ligand for c-Kit (Chabot *et al.*, 1988; Geissler *et al.*, 1988; Huang *et al.*, 1990; Zsebo *et al.*, 1990). Names for the ligand include steel factor, stem cell factor, c-Kit ligand, and mast cell growth factor.

Alternative names

The stem cell factor receptor (SCFR) is also known as c-Kit and has been designated CD117.

Structure

The SCFR is a receptor tyrosine kinase (RTK) and is related to the receptors for platelet-derived growth factor (PDGF) and colony-stimulating factor 1 (CSF-1). The full-length murine protein consists of 975 amino acids and the human protein is 976 amino acids (Yarden *et al.*, 1987; Qiu *et al.*, 1988). The extracellular domain consists of approximately 500 amino acids and is followed by a series of hydrophobic amino acids that span the cell membrane. The intracellular region is slightly less than 400 amino acids long and consists of a 30 amino acid juxtamembrane region, a catalytic domain divided into two regions by an insert of 77 amino acids and a 50 amino acid carboxyl tail (Figure 1). An alternately spliced form of the SCFR, termed c-KitA in the literature, contains a four-codon insert in the extracellular domain (Reith *et al.*, 1991). In addition, a 24 kDa truncated form of the SCFR consisting of a portion of the second catalytic domain and the carboxyl tail is expressed in spermatids (Rossi *et al.*, 1992).

Main activities and pathophysiological roles

Early studies of white spotting and steel mice, as well as subsequent work since the cloning of SCF and the SCFR, have shown that SCF is critical for the survival and development of stem cells involved in hematopoiesis, pigmentation, and reproduction (Galli *et al.*, 1994; Broudy, 1997; Lyman and Jacobsen, 1998; Ashman, 1999). More recently, the SCFR has been found to be important in development of the interstitial cells of Cajal in the intestine, and in certain learning functions in the hippocampal region of the brain (Huizinga *et al.*, 1995; Motro *et al.*, 1996). A comprehensive overview of SCF and its receptor in relation to all of the physiological target tissues was written in 1994 by Galli and coworkers. The following section will summarize the early literature and highlight recent findings relating to the activities and pathophysiological roles of the SCFR in different organ systems.

Reviews of the role of the SCFR in hematopoiesis have recently been published (Galli *et al.*, 1994; Broudy, 1997; Lyman and Jacobsen, 1998; Ashman, 1999). The role of the SCFR in the ontogeny of murine hematopoietic progenitor cells is illustrated by the dramatic anemias found in steel and white spotting mice. These animals are characterized by *macrocytic anemia*, decreases in myeloid and erythroid progenitors, and mast cell deficiencies (Russell, 1979). The ablation of CFU-S, CFU-GM, CFU-IL-3 and CFU-M in adult mice injected with ACK-2, an antibody that blocks the murine SCFR, demonstrates the important role of the SCFR in erythro- and myelopoiesis in adult mice as well (Ogawa *et al.*, 1991).

Expression and function of the SCFR have been examined extensively in relation to *hematopoietic disease* in humans. One disease in which the SCFR likely plays a role is *mastocytosis*. Reviews dealing with the SCFR and mastocytosis have recently been published (Tsujimura, 1996; Födinger and Mannhaller, 1997; Pignon, 1997; Vliagoftis *et al.*, 1997). In brief, Metcalfe and coworkers found that patients with mastocytosis with associated hematological disorder have a mutation in the SCFR (substitution of aspartic acid 816 with valine, D816V) that results in constitutive activation (Nagata *et al.*, 1995). A patient with aggressive systemic mastocytosis with the identical mutation in mast cells was also reported by Longley *et al.* (1996). In all patients described to date, the mutation in the SCFR is somatic. Interestingly, mutation of the analogous amino acid in the RTKs Ret and Met results in constitutive activation and is associated with *multiple endocrine neoplasia type 2B* and *papillary renal carcinomas*, respectively (Hofstra *et al.*, 1994; Schmidt *et al.*, 1997). Since these initial reports, a number of groups have reported the D816V mutation in patients with different forms of mastocytosis (Afonja *et al.*, 1998; Büttner *et al.*, 1998; Worobec *et al.*, 1998; Longley *et al.*, 1999). Recently, other gain-of-function mutations in the SCFR have been identified in patients with mastocytosis. These include mutations at codons 820 and 560 (Pignon *et al.*, 1997; Büttner *et al.*, 1998). In dogs with mastocytosis, mutations in the juxtamembrane region of the SCFR have also been found (Ma *et al.*, 1999). A summary of c-Kit mutations found in patients with mastocytosis is given in **Table 1**.

The relationship of the SCFR to *leukemia* and *lymphoma* in humans has also been studied rigorously. These data have recently been reviewed (Galli *et al.*, 1994; Hassan and Zander, 1996; Broudy, 1997; Sperling *et al.*, 1997; Escribano *et al.*, 1998; Lyman and Jacobsen, 1998). In summary, the SCFR is expressed on blast cells from most patients with *acute myeloid leukemia* (AML), although it is unknown if this plays a role in the onset and progression of the disease. The frequency of SCFR expression on cells from patients with *chronic myelogenous leukemia* (CML) is lower; however, studies in cell lines suggest that p210bcr/abl binds and activates the SCFR (Hallek *et al.*, 1996). In addition, several proteins phosphorylated in response to SCF are constitutively phosphorylated in CML cells (Wisniewski *et al.*, 1996). These include Dok, a 62 kDa protein, and the adapter protein CRKL (Carpino *et al.*, 1997; Sattler *et al.*, 1997; Yamanashi and Baltimore, 1997; Sattler and Salgia, 1998). In *acute lymphoid leukemia* (ALL), the SCFR is expressed on malignant lymphoid cells from T-ALL patients while SCFR expression on cells from patients with B-ALL is quite rare. The SCFR may also be a marker for certain types of *biphenotypic leukemias*.

The importance of SCFR in development of melanocytes is highlighted by the abnormal pigmentation found when expression is reduced, or the kinase activity is impaired. In both mice and pigs,

Table 1 Activating mutations in the SCFR associated with diseases

Region	Amino acid residue(s)	Disease[a]	Species	Reference
Extracellular domain	D52N	Myeloproliferative disorder	Human	Kimura *et al.*, 1997
Juxtamembrane	Deletion 559, 560	GIST	Human	Hirota *et al.*, 1998
Juxtamembrane	Deletion 551–555	GIST	Human	
Juxtamembrane	V559D	GIST	Human	
Juxtamembrane	K550I, Deletion 551–555	GIST	Human	
Juxtamembrane	Deletion 550–558	GIST	Human	
Juxtamembrane	Deletion 551–554, V555L	GIST	Human	Moskaluk *et al.*, 1999
Juxtamembrane	Deletion 556–560	GIST	Human	
Juxtamembrane	Q556H, Deletion 557–572	GIST	Human	
Juxtamembrane	Deletion 560	GIST	Human	
Juxtamembrane	Deletion 557–575	GIST	Human	
Juxtamembrane	Deletion 570–576	GIST	Human	
Juxtamembrane	V560D	GIST	Human	
Juxtamembrane	12 codon insert between 574 and 575	GIST	Human	
Juxtamembrane	Deletion 556–557	Mastocytoma	Canine	Ma *et al.*, 1999
Juxtamembrane	Deletion 558	Mastocytoma	Canine	
Juxtamembrane	W556R	Mastocytoma	Canine	
Juxtamembrane	L575P	Mastocytoma	Canine	
Juxtamembrane	V560G	Mastocytosis	Human	Büttner *et al.*, 1998
Catalytic domain	D816V, D816Y, D816F	Mastocytosis	Human	Nagata *et al.*,1995; Longley *et al.*, 1996; Afonja *et al.*, 1998; Worobec *et al.*, 1998; Büttner *et al.*, 1998; Longley *et al.*, 1999
Catalytic domain	D820G	Aggressive mast cell disease	Human	Pignon *et al.*, 1997

[a]GIST designates gastrointestinal stromal cell tumor.

abnormalities in the SCFR are associated with a characteristic pattern of white spots occurring predominantly on the ventral trunk, head, tail, and feet (Russell, 1979; Marklund *et al.*, 1998). Similarly, in humans, *autosomal dominant piebaldism* results from mutations in the SCFR (Fleischman *et al.*, 1991; Giebel and Spritz, 1991; Spritz, 1994, 1997; Boissy and Nordlund, 1997). Expression of functional SCFR appears to be critical at several points during the development and maturation of melanocytes (Nishikawa *et al.*, 1991; Okura *et al.*, 1995; Wehrle-Haller and Weston, 1995). This includes proliferation of melanoblasts in the mesoderm leading to entry into the epidermal layers during embryogenesis. In addition, the SCFR plays a role in the replenishment of melanocytes involved in hair pigmentation in adults. The important role of SCFR in melanocyte development has led investigators to examine involvement in *melanoma*. While SCFR has been reported in some melanoma tissue, as well as in melanoma-derived cell lines, many investigators have found that SCFR is either absent, or expressed at reduced levels, as compared with normal melanocytes (Funasaka *et al.*, 1992; Lassam and Bickford, 1992; Gutman *et al.*, 1994; Mattei *et al.*, 1994; Ohashi *et al.*, 1996; Montone *et al.*, 1997). Huang and coworkers found that infection of a highly metastatic melanoma cell line with a retroviral expression vector containing the SCFR reduced metastatic potential (Huang *et al.*, 1996). They also found that SCF induced apoptosis of these cells. Thus, downregulation of the SCFR in melanoma may promote expansion of transformed cells due to escape from SCF-induced apoptosis. Recently, a strong correlation between loss of expression of the AP-2 transcription factor and downregulation of the SCFR in melanoma-derived tissues has been demonstrated (Huang *et al.*, 1998). The data examining the SCFR in relation to *metastatic melanoma* have recently been reviewed (Lu and Kerbel, 1994; Luca and Bar-Eli, 1998).

The receptor for SCF can be detected in primordial germ cells (PGCs) as early as day 7 of embryonic development. SCF acts as a viability factor for PGCs and also induces limited increases in proliferation (Donovan, 1994). Similar to hematopoietic cells, SCF in combination with other growth factors elicits synergistic increases in proliferation of PGCs. In developing sperm, the SCFR has been detected in type A spermatogonia as well as primary spermatocytes (Morrison-Graham and Yoshiko, 1993; Loveland and Schlatt, 1997). A truncated form of the SCFR, consisting of the second portion of the catalytic domain, as well as the C-terminal tail, is expressed in spermatids (Rossi *et al.*, 1992). Interestingly, injection of this form of SCFR into mouse oocytes results in parthenogenesis (Sette *et al.*, 1997). The SCFR has also been detected in Leydig cells in the testicular interstitium. The role of SCFR in the development of sperm in adults was recently reviewed by Loveland and Schlatt (1997). The importance of SCFR in male fertility is illustrated by male sterility in both white spotting and steel mice. In addition, SCFR may also be a marker for certain forms of *testicular cancer* (Rukstalis, 1996; Donovan *et al.*, 1998). In an experimental model, nearly 100% of transgenic mice expressing *papillomavirus type 16* (HPV16) develop testicular cancer (Kondoh *et al.*, 1995; Li *et al.*, 1996). Introduction of either the steel dickie mutation encoding soluble SCF, or white spotting mutants, with impaired SCFR kinase activity, dramatically reduces tumor volume in these animals (Kondoh *et al.*, 1995; Li *et al.*, 1996).

In postembryonic females, SCFR is expressed on both oocytes and theca cells, while the SCF ligand is expressed on granulosa cells (Driancourt and Thuel, 1998). SCF interaction with SCFR plays a role in early follicular development, formation of preantrum follicules and ovarian follicular maturation (Yoshida *et al.*, 1997). In contrast, formation and survival of primordial follicles, antral follicular development, ovulation, and follicular luteinization do not require the SCFR. In addition, activation of the SCFR on oocytes by SCF on granulosa cells may play a role in meiotic arrest of oocytes. Further, downregulation of SCF by gonadotropins may relieve meiotic arrest (Laitinen *et al.*, 1995; Ismail *et al.*, 1996, 1997). Both SCF and the SCFR are presently being examined for potential involvement in premature *ovarian failure* in humans (Bondy *et al.*, 1998).

In the gastrointestinal tract, the SCFR is expressed on the interstitial cells of Cajal (ICC) (Torihashi *et al.*, 1995). These cells play a critical role in the pacemaker activity of the gut. Interestingly, in white spotting viable mice, ICC cells are absent and pacemaker activity is dramatically reduced (Huizinga *et al.*, 1995). In addition, these animals may be more susceptible to *necrotizing enterocolitis* (Yamataka *et al.*, 1998). Humans with *piebaldism* resulting from mutations in the SCFR have also been reported to have defects in intestinal function (Giebel and Spritz, 1991). Many *gastrointestinal stromal tumors* (GISTs) express a constitutively active form of the SCFR resulting from mutations in the juxtamembrane region (Table 1). The association of mutations in c-Kit with GISTs has recently been reviewed (Kitamura *et al.*, 1998; Chan, 1999). As described above, another constitutively active SCFR mutant, D816V, is found in patients with *mastocytosis* and associated hematological disorders. Interestingly, some of these patients also have gastrointestinal problems.

Another organ system where the SCFR is expressed extensively is the central nervous system (CNS). Detailed descriptions of SCFR localization in the adult CNS have previously been published and reviewed (Hirota *et al.*, 1992; Morii *et al.*, 1992; Hamel and Westphal, 1997; Zhang and Fedoroff, 1997). Heterozygote steel dickie mice have normal long-term potentiation but impaired configural learning (Motro *et al.*, 1996). This demonstrates an important role for signaling of membrane-bound SCF in some aspects of brain function. With regard to *cancers of the CNS*, the SCFR has been observed in some, but not all, neuroblastoma and glioblastoma cell lines (Hamel and Westphal, 1997). SCFR is expressed on Nf1-deficient Schwann cell lines and may play a role in Schwann cell hyperplasia observed in neurofibromin-deficient patients (Badache *et al.*, 1998). Transgenic mice expressing portions of the SCFR promoter fused to the large T antigen of SV40 develop *neuroendocrine tumors* at high frequencies (Bossé *et al.*, 1997). These findings suggest CNS-specific regulatory sequences in the 5′ region of the SCFR gene. The SCFR may also be involved in nonmalignant diseases of the CNS. A recent study by He and coworkers (1997) found that infection of cultured fetal brain cells with *HIV-1* induced expression of the SCFR. This was associated with high levels of apoptosis in astrocytes, suggesting that inappropriate expression of the SCFR in the CNS could be involved in neurological disorders associated with *AIDS*.

The Kit gene product has also been associated with other forms of *cancer*. The v-Kit oncogene is a component of Hardy-Zuckerman strain of feline sarcoma virus (Besmer *et al.*, 1986). Approximately 70% of all *small cell lung carcinomas* (SCLC) coexpress the SCFR and SCF inappropriately (Hibi *et al.*, 1991; Sekido *et al.*, 1991; Plummer *et al.*, 1993; Rygaard *et al.*, 1993). Thus, an autocrine loop may play a role in the etiology of this disease (Krystal *et al.*, 1996). Similarly, in one study, the SCFR and SCF were coexpressed in 9 of 11 breast tumors, and in 7 of 13 cell lines derived from breast tumors (Hines *et al.*, 1995). Downregulation of the SCFR in the epithelial cells of both *breast cancer* and *thyroid cancer* tissue has also been reported (Natali *et al.*, 1995; Chui *et al.*, 1996). Thus, similar to its actions on melanocytes, SCF may play a role in promoting apoptosis in these tissues and downregulation of SCFR may promote transformation. In cell lines derived from *colon carcinoma* tissue, truncated forms of the SCFR encoding a portion of the second catalytic domain have been identified (Toyota *et al.*, 1994; Takaoka *et al.*, 1997). What, if any, role of these forms of the SCFR play in colon carcinoma is unknown. The SCFR has also been found in cell lines derived from a variety of other types of cancers (Natali *et al.*, 1992; Turner *et al.*, 1992; Matsuda *et al.*, 1993; Inoue *et al.*, 1994; Ridings *et al.*, 1995; DiPaola *et al.*, 1997).

GENE

Accession numbers

The cDNA sequence for the human SCFR (GenBank X06182) was reported in 1987 and for the murine SCFR (GenBank Y00864) in 1988 (Yarden *et al.*, 1987; Qiu *et al.*, 1988). The genomic sequence for both murine and human SCFR have also been reported (André *et al.*, 1992; Giebel *et al.*, 1992; Gokkel *et al.*, 1992; Vandenbark *et al.*, 1992; Yasuda *et al.*, 1993). The GenBank accession number for the complete coding sequence of the human SCFR is U63834 and for portions of the murine gene, including the 5′ regulatory region, is X65998. The human SCFR gene is located on chromosome 4 and the murine gene is on chromosome 5 (Yarden *et al.*, 1987).

The c-Kit gene is over 70 kb in size and has 21 exons (André *et al.*, 1992; Giebel *et al.*, 1992; Gokkel *et al.*, 1992; Vandenbark *et al.*, 1992; Yasuda *et al.*, 1993). The organization is highly conserved between the mouse and human genome. The c-Kit transcript is 5.2 kb and alternately spliced forms have been reported.

PROTEIN

Accession numbers

GenBank:
Human SCFR: 1817733
Murine SCFR: 125473

Description of protein

The SCFR belongs to the type III family of receptor tyrosine kinases (Yarden and Ullrich, 1988). A model depicting the structural domains of the SCFR is shown in **Figure 1**. The extracellular domain is organized into a series of five Ig-like regions. The first three Ig-like regions bind SCF while the fourth may play a role in either forming or stabilizing SCFR homodimers (Lev *et al.*, 1992a; Blechman *et al.*, 1993a, 1995; Lev *et al.*, 1993; Blechman *et al.*, 1993b). The role of the fifth Ig-like region is unknown.

The cytoplasmic domain of the SCFR can be divided into five regions. The 30 amino acids distal to the

Figure 1 Structural domains of the SCFR. The transmembrane domain divides the SCFR into an intracellular and extracellular region. The intracellular region consists of five immunoglobulin-like domains. The first three bind SCF, the fourth domain may play a role in forming or stabilizing receptor dimers, and the function of the fifth is unknown. The 30 amino acids distal to the transmembrane domain form the juxtamembrane region. The catalytic domain is divided into two by a kinase insert and the last 50 amino acids comprise the kinase tail.

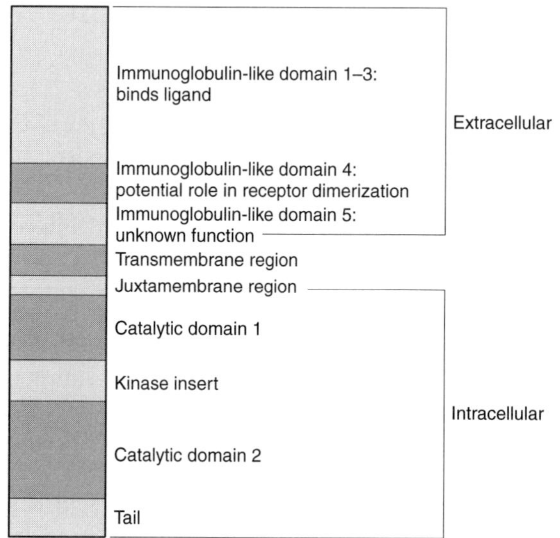

transmembrane region comprises the juxtamembrane region. The catalytic domain is divided by a 77 amino acid insert (Figure 1). The first catalytic domain contains the ATP-binding region while the second catalytic domain has a number of possible autophosphorylation sites. Lastly, there is a carboxyl-tail of approximately 50 amino acids.

Relevant homologies and species differences

The sequences of the human and murine SCFR were published in 1987 and 1988 respectively (Yarden *et al.*, 1987; Qiu *et al.*, 1988). The amino acid homology is 81.9%. Although the human and murine forms of the SCFR are relatively similar, there are significant differences in ligand specificity. Both murine and human SCFR bind and are activated by murine SCF. In contrast, human SCF binds the murine receptor with extremely low affinity and does not have significant biological activity on murine cells. Other species for which the SCFR sequence is known include rat, bird, goat, pig, cow, cat, and dog (Tsujimura *et al.*, 1991; Sasaki *et al.*, 1993; Kubota *et al.*, 1994; Zhang and Anthony, 1994; Herbst *et al.*, 1995a; Tanaka *et al.*, 1997; Ma *et al.*, 1999).

Affinity for ligand(s)

SCF binds the SCFR with an affinity of 50–200 pM in both hematopoietic and nonhematopoietic target tissue (Abkowitz *et al.*, 1992; Turner *et al.*, 1992).

Cell types and tissues expressing the receptor

The embryonic expression patterns of the SCFR have previously been published (Matsui *et al.*, 1990; Orr-Urtreger *et al.*, 1990; Keshet *et al.*, 1991; Motro *et al.*, 1991). The following section will summarize expression of the SCFR in adult tissues.

SCFR expression in hematopoietic cells was recently reviewed by Lyman and Jacobsen (1998). In brief, the SCFR is expressed on stem cells, pluripotential progenitor cells and more differentiated progenitor cells including those leading to the myeloid, lymphoid, erythroid and megakaryocytic lineages. SCFR expression levels vary with the differentiation of murine marrow cells. Early progenitors can be divided into $SCFR^{dull}$ or $SCFR^{bright}$ populations and in general, SCFR expression declines as progenitor cells differentiate. Certain terminally differentiated lineages do express the SCFR. These include mast cells, activated platelets and a subset of NK cells.

SCFR is also expressed in numerous cell lines isolated from hematopoietic cells. **Table 2** is a summary of some of these lines, the species of origin, the hematopoietic lineage, if known, and whether the cells require hematopoietic growth factors for propagation in culture.

In the postnatal female, SCFR is expressed in both oocytes and theca cells (Draincourt and Truel, 1998). In the postnatal male, full-length SCFR is expressed in type A spermatogonia as well as primary spermatocytes (Morrison-Graham and Yoshiko, 1993; Loveland and Schlatt, 1997). In addition, a truncated form has been found in spermatids. The SCFR is also expressed in the testicular interstitium in the Leydig cells.

SCFR is expressed throughout the CNS. Descriptions of its localization have previously been published (Hirota *et al.*, 1992; Morii *et al.*, 1992; Hamel and Westphal, 1997; Zhang and Fedoroff, 1997). Other tissues that express the SCFR in adults are the interstitial cells of Cajal in the intestine, the oval and bile epithelial cells in the liver, alveolar, ductal, and epithelial cells in the breast and the glandular epithelial cells in the parotid, dermal sweat, and salivary glands as well as the esophagus (Lammie *et al.*, 1994; Torihashi *et al.*, 1995; Fujio *et al.*, 1996). SCFR has also been found in human endometrium and placenta during pregnancy (Saito *et al.*, 1994; Kauma *et al.*, 1996).

Regulation of receptor expression

Surface expression of the SCFR is regulated through multiple mechanisms. Similar to other receptors, ligand-induced activation rapidly induces internalization (Yee *et al.*, 1993, 1994a; Miyazawa *et al.*, 1994; Shimizu *et al.*, 1996; Gommerman *et al.*, 1997; Broudy *et al.*, 1998). While both soluble and membrane-bound SCF induce internalization of the SCFR, the kinetics are more rapid in response to the soluble form (Miyazawa *et al.*, 1995). One mechanism mediating internalization involves interaction of the SCFR with clathrin, incorporation into clathrin-coated pits and subsequent degradation, in part, by lysosomal proteases (Gommerman *et al.*, 1997). A second mechanism mediating downregulation of activated SCFR is through rapid ubiquination and degradation though the proteasome proteolytic pathway (Miyazawa *et al.*, 1994).

Table 2 Hematopoietic cell lines that express the SCFR

Designation	Species	Lineage	Growth factor-dependence
HEL	Human	Erythroid	No
LAMA	Human	Erythroid	No
K562YO	Human	Erythroid	No
OCIM1	Human	Erythroid	No
OCIM2	Human	Erythroid	No
JK-1	Human	Erythroid	No
HMC1	Human	Mast cell	No
CMK	Human	Megakaryoblastic	No
DAMI	Human	Megakaryoblastic	No
MEG-01	Human	Megakaryoblastic	No
Mo7e	Human	Megakaryoblastic	Yes
HML-2	Human	Megakaryoblastic	Yes
MB-02	Human	Erythroid	Yes
TF-1[a]	Human	Erythroid	Yes
UT-7	Human	Erythroid	Yes
F-36P	Human	Erythroid	Yes
GF-D8	Human	Myeloid	Yes
AML193[a]	Human	Myeloid	Yes
P815	Mouse	Mastocytoma	No
FMA3	Mouse	Mastocytoma	No
SKT6	Mouse	Erythroid	No
B6M	Mouse	Mast cell	Yes
MC-9	Mouse	Mast cell	Yes
R6-X	Mouse	Mast cell/meg.	Yes
FDC-P1	Mouse	Progenitor	Yes
FDCP-Mix A4	Mouse	Progenitor	Yes
B6SUtA1	Mouse	Myeloid	Yes
NSF-60	Mouse	Myeloid	Yes
HCD-57	Mouse	Erythroid	Yes

[a]Clones of these cell lines have been reported that do not express the SCFR.

Expression of the SCFR can also be regulated by other cytokines and growth factors (Galli *et al.*, 1994; Broudy *et al.*, 1998; Lyman and Jacobsen, 1998). In brief, treatment of hematopoietic progenitor cells, as well as cell lines, with TGFβ, TNFα, or IL-4 reduces expression of the SCFR (Sillaber *et al.*, 1991; Khoury *et al.*, 1994; Nilsson *et al.*, 1994; Rusten *et al.*, 1994; Heinrich *et al.*, 1995; Jacobsen *et al.*, 1995). One mechanism mediating TGFβ-induced downregulation of the SCFR is reduction in mRNA stability (de Vos *et al.*, 1993; Dubois *et al.*, 1994; Heinrich *et al.*, 1995). IFNγ also downregulates expression of SCFR on erythroid progenitor cells (Taniguchi *et al.*, 1997). There are varied reports on the effects of IL-3 on SCFR expression. One group found that IL-3 and GM-CSF had no effect on expression of SCFR mRNA in the megakaryoblastic cell line Mo7e (Hu *et al.*, 1994). Treatment of mast cells, a mast cell line, or a myeloid cell line with IL-3 downregulated SCFR mRNA as well as protein levels (Welham and

Schrader, 1991). Similar results were found after treatment of a mast cell line with EPO or GM-CSF, but not IL-4.

Release of soluble receptors

Another mechanism for regulation of cell surface expression of SCFR is the release of soluble SCFR from the cell surface through proteolytic cleavage and release of the extracellular domain. Soluble SCFR is generated by umbilical endothelial cells, hematopoietic cell lines, and mast cells (Brizzi *et al.*, 1994a; Turner *et al.*, 1995). Further, soluble SCFR has been detected at relatively high concentrations in human plasma (Wypych *et al.*, 1995). Treatment of cells with either phorbol myristate acetate or calcium ionophore results in release of soluble SCFR into the supernatant (Asano *et al.*, 1993; Yee *et al.*, 1993). While the biological role of soluble SCFR is unknown, it is capable of binding SCF and antagonizing SCF-mediated responses.

SIGNAL TRANSDUCTION

Associated or intrinsic kinases

The SCFR is a receptor tyrosine kinase. Binding of SCF induces rapid increases in homodimerization of the receptor followed by increases in the SCFR autophosphorylation activity in multiple lineages of cells including hematopoietic, melanocytes, and spermatozoa (Blume-Jensen *et al.*, 1991; Funasaka *et al.*, 1992; Lev *et al.*, 1992a, 1992b; Philo *et al.*, 1996; Lemmon *et al.*, 1997; Feng *et al.*, 1998). The critical role of the SCFR kinase activity in responses mediated by SCF is illustrated by the severe phenotype of white spotting mice expressing SCFR lacking kinase activity. In addition, gain-of-function mutations in the SCFR lead to oncogenic forms of the receptor and have also been found in patients with *gastrointestinal stromal tumors* (GISTs) and *mastocytosis* (Nagata *et al.*, 1995; Longley *et al.*, 1996; Hirota *et al.*, 1998; Moskaluk *et al.*, 1999).

Cytoplasmic signaling cascades

Similar to other receptor tyrosine kinases, SCFR activates multiple signal transduction components that couple to a variety of biochemical pathways. Among these are phosphatidylinositol 3-kinase (PI-3 kinase), Src family members, the JAK/STAT pathway and the Ras/Raf/MAP kinase cascade. A model of signaling components and pathways activated by SCF is shown in **Figure 2**. A review of signaling pathways activated by SCF in hematopoietic cells has recently been published (Linnekin, 1999). The following summarizes each of these pathways in relation to SCF signaling.

PI-3 kinase is a heterodimer composed of an 85 kDa regulatory subunit and a 110 kDa catalytic subunit involved in the signaling pathways of a variety of growth factors and cytokines (Shepherd *et al.*, 1998). Studies in both transfected fibroblasts, as well as hematopoietic cells, have shown that SCF induces increases in SCFR-associated PI-3 kinase activity. SCF also induces increases in tyrosine phosphorylation of the 85 kDa subunit of PI-3 kinase (Lev *et al.*, 1991; Reith *et al.*, 1991; Rottapel *et al.*, 1991). SCFR mutants lacking kinase activity do not activate PI-3 kinase in response to SCF, while cells expressing a constitutively active SCFR mutant have increased levels of receptor-associated PI-3 kinase activity (Rottapel *et al.*, 1991). Thus, activation of PI-3 kinase by the SCFR correlates directly with SCFR catalytic activity.

Tyrosine 719 of murine SCFR (Y721 of human SCFR) interacts with the SH2 domain of PI-3 kinase.

Figure 2 SCF activates multiple signaling pathways. Ligand-induced increases in tyrosine phosphorylation of the SCFR recruit multiple signaling components to the receptor complex. These include Grb2, Shc, and SHP2, proteins involved in regulation of the Ras/Raf/MAP kinase cascade. The 85 kDa regulatory subunit of PI-3 kinase also binds phosphorylated SCFR and results in activation of the catalytic subunit of PI-3 kinase. Src family members, JAK2, Tec, PLCγ and STAT1 also associate with the SCFR and are activated by SCF.

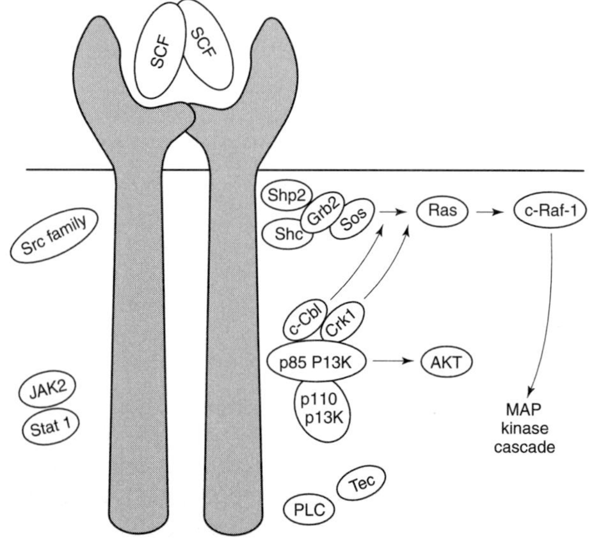

Studies of SCFR mutants containing phenyalanine at residue 719 (Y719F) suggest that PI-3 kinase has multiple roles in SCF-mediated responses. These include adhesion to fibronectin, as well as membrane ruffling and actin assembly in mast cells (Serve *et al.*, 1995; Vosseller *et al.*, 1997). Partial inhibition of SCF-mediated proliferation and survival have also been observed in murine mast cells expressing Y719F SCFR (Serve *et al.*, 1995; Vosseller *et al.*, 1997). In DA-1 cells, a murine myeloid cell line, PI-3 kinase plays a role in trafficking of the SCFR through endocytic pathways (Gommerman *et al.*, 1997). In the myeloid progenitor cell line FDC-P1, PI-3 kinase may also play a role in differentiation mediated by the SCFR (Kubota *et al.*, 1998). Lastly, SCF-induced activation of the serine threonine kinase Akt is impaired in fibroblast cell lines expressing Y721F human SCFR. This correlates with a partial impairment in capacity of SCF to prevent apoptosis (Blume-Jensen *et al.*, 1998). Thus, the role of PI-3 kinase in SCF-mediated responses depends on cell lineage.

SCF also activates the JAK/STAT pathway (Linnekin *et al.*, 1997a). Studies from two groups have shown that JAK2 associates with the SCFR and that SCF activates JAK2 (Brizzi *et al.*, 1994b; Weiler *et al.*, 1996). Studies with antisense oligonucleotides suggest that JAK2 is required for maximal proliferation in response to SCF. SCF also induces association of STAT1 with the SCFR, increases in STAT1 tyrosine phosphorylation and increases in STAT1 DNA-binding activity in multiple cell lines (DeBerry *et al.*, 1997). In addition to STAT1, SCF also activates STAT5, and induces serine phosphorylation of STAT3 (Gotoh *et al.*, 1996; Ryan *et al.*, 1997). In contrast to these findings, several investigators have reported that SCF has no effect on JAK or STAT family members (Tang *et al.*, 1994; O'Farrell *et al.*, 1996; Jacobs-Helber *et al.*, 1997; Joneja *et al.*, 1997; Pearson *et al.*, 1998). It remains unclear if these differences relate to technical parameters or possibly to lineage-specific differences in cell lines.

Another family of signaling components activated in response to SCF are Src family members. SCF induces association of Lyn with phosphorylated tyrosine residues in the juxtamembrane region of the SCFR, as well as increases in both Lyn tyrosine phosphorylation and kinase activity (Linnekin *et al.*, 1997b). Inhibition of Lyn activity with PP1, a Src-family inhibitor, or reduction in Lyn protein with antisense oligonucleotides, inhibits SCF-induced proliferation. In cells lacking Lyn, SCF activates other Src family members. Further, SCF induces increases in Fyn activity in mast cells transfected with murine SCFR. *In vitro*, Fyn binds a phosphopeptide sequence spanning tyrosines 568 and 570 of human SCFR, while interaction of murine SCFR with the SH2 domain of Fyn requires tyrosine 567 (Price *et al.*, 1997; Timokhina *et al.*, 1998). Expression of Y567F murine SCF in SCFR-deficient murine mast cells inhibits activation of Fyn in response to SCF as well as partially inhibiting SCF-induced proliferation (Timokhina *et al.*, 1998).

Tec, a Src-like kinase related to Btk and Akt, is constitutively associated with the SCFR (Tang *et al.*, 1994). Stimulation with SCF also results in increases in tyrosine phosphorylation of Tec, as well as increases in its catalytic activity. The role of Tec in SCF-mediated responses is presently unknown.

The Ras/Raf/MAP kinase cascade is a signaling pathway activated in response to many growth factors (Madhani *et al.*, 1998). SCF induces association of the SCFR with Grb2, Grap, SHP2, and Shc (Cutler *et al.*, 1993; Matsuguchi *et al.*, 1994; Tauchi *et al.*, 1994a, 1994b; Feng *et al.*, 1996). This in turn leads to colocalization of Sos and Ras and subsequent activation of Ras. RasGAP, Nf1, SHIP, and Vav may also play a role in modulating Ras activity in response to SCF (Miyazawa *et al.*, 1991; Alai *et al.*, 1992; Matsuguchi *et al.*, 1995; Liu *et al.*, 1997; Zhang *et al.*, 1998). Activation of Ras results in activation of c-Raf-1, a serine threonine kinase critical in SCF-mediated responses (Brennscheidt *et al.*, 1994; Muszynski *et al.*, 1995). SCF also induces increases in tyrosine phosphorylation and kinase activity of MEK1, MEK2, Erk1, and Erk2 (Hallek *et al.*, 1992; Okuda *et al.*, 1992; Tsai *et al.*, 1993; Jacobs-Helber *et al.*, 1997; Joneja *et al.*, 1997; Pearson *et al.*, 1998; Suie *et al.*, 1998). Downstream of the Erks, the S6 kinase pp90rsk is activated (Tsai *et al.*, 1993). In melanocytes, one substrate of the Erks is the microphthalmia transcription factor (MITF). Serine phosphorylation of MITF results in interaction with the transcriptional coactivator p300/CBP and transactivation (Hemesath *et al.*, 1998; Price *et al.*, 1998).

In addition to its effects on the classical MAP kinase pathway, SCF also activates JNK, a MAP kinase involved in stress responses (Folz and Schrader, 1997). Recent studies by Besmer and coworkers indicate that activation of JNK is mediated by convergence of PI-3 kinase and Src signaling pathways on the GTPase Rac1 (Timokhina *et al.*, 1998).

SCF induces tyrosine phosphorylation of phospholipase Cγ (PLCγ), as well as a weak association between the SH2 domain of PLCγ and tyrosine 936 of the human SCFR (Lev *et al.*, 1991; Reith *et al.*, 1991; Rottapel *et al.*, 1991; Hallek *et al.*, 1992; Blume-Jensen *et al.*, 1994; Herbst *et al.*, 1995b). The truncated form of the SCFR expressed in spermatids contains tyrosine 936. Interestingly, injection of mouse oocytes with this form of the SCFR activates

PLCγ in oocytes and induces parthenogenesis (Sette *et al.*, 1997, 1998). Although PLCγ is a substrate for the SCFR, it is unlikely that SCF activates the classical PLCγ pathway. SCF does not induce increases in inositol trisphosphate (IP3) levels and increases in diacylglycerol (DAG) are likely mediated by phospholipase D (Lev *et al.*, 1991; Koike *et al.*, 1993; Kozawa *et al.*, 1997).

Possible negative regulators of SCFR signaling are SHP1, SOCS, SHIP, Chk, and PKC isoforms (Blume-Jenson *et al.*, 1993, 1994, 1995; Lorenz *et al.*, 1996; Paulson *et al.*, 1996; Grgurevich *et al.*, 1997; Huber *et al.*, 1998; Kozlowski *et al.*, 1998; De Sepulveda *et al.*, 1999). SHP2 may also be a negative regulator of SCFR signaling in more differentiated lineages (Kozlowski *et al.*, 1998). **Figure 3** summarizes which pathways are likely targeted by these signaling components.

In summary, SCF activates multiple signal transduction components, including PI-3 kinase, Src family members, the Ras/Raf/MAP kinase cascade and the JAK/STAT pathway (Figure 2 and Figure 3). Activation of many of these pathways is dependent on interaction of an upstream signaling component with specific sites on the SCFR. **Table 3** summarizes sites on the SCFR that interact with these signaling components as well as some of the known structure-function relationships.

Figure 3 Summary of putative negative regulators of SCF-mediated signaling pathways. Src family members are negatively regulated by Chk, PKC isoforms, and SHP1. SOCS and SHP1 probably regulate SCF-induced activation of JAK family members. The Ras/Raf/MAP kinase cascade can be regulated by SHIP1, SHP1, SHP2, and PKC isoforms, while PI-3 kinase activation can be regulated by PKC isoforms and SHIP.

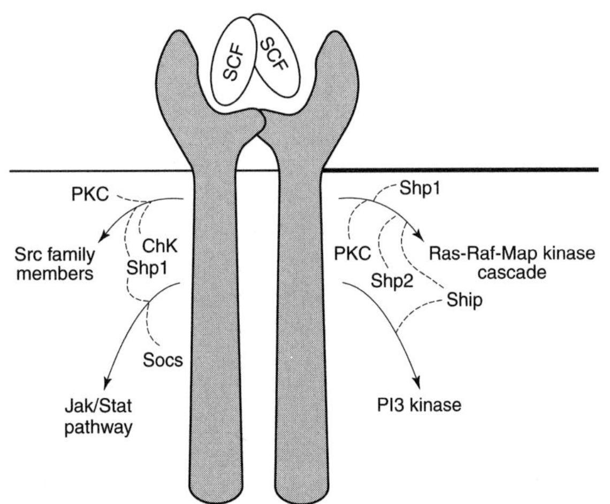

Downstream of the signaling pathways described above, pathways involved in mediating cell survival and cell cycle progression are activated in response to SCF. Although SCF prevents apoptosis in hematopoietic cells, in some epithelial cells it may be pro-apoptotic. Thus, the effects of SCF on cell survival are dependent on cellular context. With regards to the antiapoptotic effects of SCF, PI-3 kinase clearly plays a role. One kinase downstream of PI-3 kinase and involved in SCF-mediated survival is Akt. SCF-induced activation of Akt in fibroblasts expressing wild-type human SCFR resulted in interaction of Akt and Bad, and serine phosphorylation of Bad (Blume-Jensen *et al.*, 1998). SCF inhibits p53-mediated apoptosis in *erythroleukemia* cell lines (Abrahamson *et al.*, 1995; Lin and Benchimol, 1995; Lee, 1998). Further, there is a reduced rate of apoptosis in SCF-deprived, bone marrow-derived mast cells from p53-deficient mice as compared with animals expressing normal levels of p53 (Yee *et al.*, 1994b).

Signaling pathways involved in SCF-induced cell cycle progression have also been examined in recent years. SCF induces events classically associated with the G_1/S transition, such as increases in Cdk2 activity and phosphorylation of Rb (Mantel *et al.*, 1996). SCF also induces increases in expression of CaM-BP68, a calmodulin-binding protein likely involved in the G_1/S transition of the cell cycle (Reddy and Quesenberry, 1996). One group has suggested that synergistic responses induced by SCF in combination with GM-CSF are mediated by alterations in expression of the Cdk inhibitors p21cip-1 and p27kip-1 (Mantel *et al.*, 1996).

DOWNSTREAM GENE ACTIVATION

Transcription factors activated

SCF induces increases in both the expression and activity of a variety of transcription factors. Several groups have reported SCF-induced increases in tyrosine phosphorylation and DNA-binding activity of STAT family members, including STAT1 and STAT5 (DeBerry *et al.*, 1997; Ryan *et al.*, 1997). SCF also induces increases in serine phosphorylation of STAT3 (Gotoh *et al.*, 1996).

As discussed above, one transcription factor studied rather extensively in relation to SCF signaling is MITF. SCF induces serine phosphorylation of MITF and interaction with the coactivator p300/CBP. This complex binds DNA and activates DNA transcription (Hemesath *et al.*, 1998; Price *et al.*, 1998).

Table 3 Signaling components that associate with specific regions of the SCFR and potential function

Protein	Region	Tyrosine residue(s) (human/murine sequence)	Function	Cell line	Reference
Lyn	Juxta	Within 544–577 (human)	Proliferation	Mo7e	Linnekin *et al.*, 1997b
Fyn	Juxta	Y568, Y570 (human)	–	–	Price *et al.*, 1997
Fyn	Juxta	Y567 (murine)	Partial effect survival	Mast cells	Timokhina *et al.*, 1998
Fyn	Juxta	Y567 (murine)	Partial effect proliferation	Mast cells	Timokhina *et al.*, 1998
Shc	Juxta	Y568, Y570 (human)	–	–	Price *et al.*, 1997
SHP1	Juxta	Y569 (murine)	Negative regulator of proliferation	BaF3	Kozlowski *et al.*, 1998
SHP2	Juxta	Y567 (murine)	Negative regulator of proliferation	BaF3	Kozlowski *et al.*, 1998
Chk	Juxta	Y568, Y570 (human)	–	–	Price *et al.*, 1997
p85 P-I3 kinase	Insert	Y719 (murine)	Adhesion	Mast cells	Serve *et al.*, 1995
	Insert	Y719 (murine)	Partial effect survival, proliferation	Mast cells	Serve *et al.*, 1995
	Insert	Y719 (murine)	c-Kit trafficking	DA1 cells	Gommerman *et al.*, 1997
	Insert	Y721 (human)	Differentiation	FDC-P1 cells	Kubota *et al.*, 1998
	Insert	Y721 (human)	Survival	Fibroblasts	Blume-Jenson *et al.*, 1998
Unknown	Kinase II	Y821 (murine)	Survival	Mast cells	Serve *et al.*, 1995
	Kinase II	Y821 (murine)	Proliferation	Mast cells	Serve *et al.*, 1995
PLCγ	Tail	Y936 (human)	Parthenogenesis	Oocytes	Herbst *et al.*, 1995b; Sette *et al.*, 1997

Transcription factors induced by SCF include a number of immediate early-response genes. SCF induces increases in expression of c-*fos*, *junB*, *egr-1* and c-*myc* (Horie and Broxmeyer, 1993; Serve *et al.*, 1995; O'Farrell *et al.*, 1996). Further, synergistic induction of these genes may play a role in the capacity of SCF to synergize with other growth factors (Horie and Broxmeyer, 1993). SCF also induces increases in expression of SCL and c-Myb. As described below, both of these transcription factors play a role in maintaining expression of the SCFR (Ratajczak *et al.*, 1992; Miller *et al.*, 1994; Melotti and Calabretta, 1996; Krosl *et al.*, 1998).

Genes induced

Several classes of signaling molecules are induced in response to SCF. Csk-homologous kinase (Chk) is a kinase involved in downregulation of Src family kinase activity. SCF induces increases in Chk expression that peak 8 hours after stimulation and are maintained for 24 hours (Grgruvevich *et al.*, 1997). Studies with phosphopeptides suggest that, similar to Lyn, Chk associates with phosphorylated tyrosine residues in the juxtamembrane region of the SCFR (Price *et al.*, 1997). Similarly, SOCS, an inhibitor of JAK family tyrosine kinases, is also induced in response to SCF (De Sepulveda *et al.*, 1999). Another signaling molecule induced after stimulation of Mo7e cells with SCF in combination with IFNγ is the serine/threonine kinase Pim1 (Yip-Schneider *et al.*, 1995). Pim1 was also induced by IL-3 and SCF treatment of the mast cell line B6M (O'Farrell *et al.*, 1996). In contrast, in FDC2 cells, SCF as a single factor induced expression of Pim1 (Joneja *et al.*, 1997). Possible explanations for the differences in these findings could be differences in either cell lineage or the differentiation state of these cell lines. The role of Pim1 in responses mediated by SCF alone or in combination with other factors remains to be defined.

Recently, Keller and coworkers found that SCF in combination with IL-3 induces D3 in lineage-negative hematopoietic progenitor cells (Weiler *et al.*, 1999). Although the function of this protein is not yet known, expression correlates with myeloid differentiation in both cell lines and normal progenitor cells.

Lastly, SCF also induces a number of immediate early-response genes such as c-*fos*, *junB*, *egr-1*, and c-*myc* (Horie and Broxmeyer, 1993; Serve *et al.*, 1995; O'Farrell *et al.*, 1996).

Promoter regions involved

The promoter region for both the murine and human SCFR has been cloned and partially characterized. Analysis of the sequence suggests multiple transcription factors regulate expression of the SCFR in different tissues. Included among these are AP-2, SP-1, Myb/Ets-family members and SCL. Early studies of the SCFR promoter were reviewed in 1994 by Galli and coworkers. This section will review recent data relating to regulation of the SCFR promoter.

One transcription factor regulating SCFR expression in mast cells and melanocytes is MITF, a member of the basic-helix-loop-helix-leucine zipper family (bHLH-Zip) Similar to white spotting mice, mice deficient in MITF have reduced numbers of mast cells, as well as defects in pigmentation. Interestingly, the SCFR is not expressed in either cultured mast cells or melanocytes from these mice, but is expressed at normal levels in erythroid progenitors and germ cells (Isozaki *et al.*, 1994). These findings suggest tissue-specific elements are involved in MITF regulation of SCFR expression. While the SCFR promoter has three CANNTG motifs that could serve as binding sites for bHLH-Zip transcription factors, only one binds MITF (Tsujimura *et al.*, 1996a).

Another bHLH-Zip transcription factor with potential importance in the regulation of SCFR expression is SCL, also known as TAL1. Inhibition of SCL expression by transfecting cells with SCL cDNA in the antisense orientation reduces expression of SCFR protein (Krosl *et al.*, 1998). Similar results are obtained when cells are transfected with a dominant negative SCL construct. The induction of SCL by SCF during erythroid differentiation suggests a positive feedback loop for SCFR expression (Miller *et al.*, 1994).

The c-Kit promoter also contains multiple AP-2 binding sites (Giebel *et al.*, 1992; Yamamoto *et al.*, 1993; Yasuda *et al.*, 1993). AP-2 binds the SCFR promoter and upregulates SCFR expression. Downregulation of AP-2 in melanoma cells results in downregulation of the SCFR and may contribute to an increase in metastatic potential (Huang *et al.*, 1998).

Other transcription factors with binding sites in the c-Kit promoter region are SP-1, Myb, and Ets (Ratajczak *et al.*, 1992; Melloti and Calabretta, 1996; Park *et al.*, 1998; Ratajczak *et al.*, 1998).

BIOLOGICAL CONSEQUENCES OF ACTIVATING OR INHIBITING RECEPTOR AND PATHOPHYSIOLOGY

Unique biological effects of activating the receptors

The transforming capabilities of the SCFR are illustrated by the oncogenic potential of v-Kit. Structurally, v-Kit lacks the extracellular region and the 50 amino acid C-terminus of the SCFR (Qiu *et al.*, 1988; Herbst *et al.*, 1995a). In addition, there is a five amino acid insertion in the carboxyl tail, a deletion of two amino acids in the juxtamembrane region and one point mutation in the kinase insert region. As expected, v-Kit has constitutive tyrosine kinase activity *in vitro* (Majumder *et al.*, 1990).

Other constitutively active forms of the SCFR have been identified and are associated with cellular transformation. One of the best characterized gain-of-function mutations in the human SCFR is substitution of valine for aspartic acid 816 (aspartic acid 814 of murine SCFR). This mutant has been identified in factor-independent mastocytoma lines derived from humans (HMC1), mice (P-815) and rats (RBL-2H3) (Furitsu *et al.*, 1993; Tsujimura *et al.*, 1994). Expression of D816V SCFR in factor-dependent cell lines results in factor-independent cell lines that produce tumors in mice (Kitayama *et al.*, 1995, 1996; Piao and Bernstein, 1996; Ferrao *et al.*, 1997). As discussed above, this mutation is also found in patients with *mastocytosis* with associated hematological disorders, as well as in patients with other forms of mastocytosis (Nagata *et al.*, 1995; Longley *et al.*, 1996, 1999; Afonja *et al.*, 1998; Büttner *et al.*, 1998; Worobec *et al.*, 1998). The D816V SCFR mutant is constitutively phosphorylated on tyrosine residues and the catalytic activity is increased as compared with wild-type SCFR (Furitsu *et al.*, 1993; Tsujimura *et al.*, 1994; Kitayama *et al.*, 1995; Piao and Bernstein, 1996; Piao *et al.*, 1996). D816V SCFR may also have alterations in substrate specificity that result in autophosphorylation of different sites (Piao *et al.*, 1996). In addition, accelerated degradation of SHP1, a negative regulator SCFR signaling, has also been reported. Although D816V SCFR is constitutively active, it does not form dimers through the extracellular domain (Kitayama *et al.*, 1995). In contrast, this form of the SCFR may dimerize through regions of the intracellular domain (Lam *et al.*, 1999; Tsujimura *et al.*, 1999).

Several gain-of-function mutations of the SCFR have been detected in the juxtamembrane region. Substitution of glycine for valine 559 of murine SCFR results in constitutive dimerization, activation, and increases in oncogenic potential (Kitayama *et al.*, 1995, 1996; Tsujimura, 1996; Tsujimura *et al.*, 1997). Deletion of seven amino acids in the juxtamembrane region also generates a constitutively active form of the SCFR that spontaneously dimerizes (Tsujimura *et al.*, 1996b). In addition, a series of mutations in the SCFR juxtamembrane region resulting in constitutive kinase activity have been found in some patients with *gastrointestinal stromal tumors*, as well as dogs with *mastocytosis* (Hirota *et al.*, 1998; Ma *et al.*, 1999; Moskaluk *et al.*, 1999). Table 1 summarizes activating mutations found in the SCFR and diseases with which they are associated.

Phenotypes of receptor knockouts and receptor overexpression mice

Studies of mice with a variety of mutations in the white spotting locus have demonstrated the importance of the SCFR in hematopoiesis, melanocyte, and germ cell development, and in intestine and brain function (Russell, 1979; Galli *et al.*, 1994; Huizinga *et al.*, 1995; Motro *et al.*, 1996). The molecular basis of the known white spotting mutants and the corresponding phenotypes have previously been described (Nocka *et al.*, 1990; Reith *et al.*, 1991; Fleischman, 1993; Morrison-Graham and Takahashu, 1993). In summary, the animals die *in utero* at mid-gestation. In addition, heterozygotic animals expressing kinase-defective SCFR also have severe phenotypes. This dominant negative effect is the result of dimerization of wild-type and kinase inactive SCFR and subsequent impairment in signaling.

Human abnormalities

Mutations in the SCFR have been detected in patients with *autosomal dominant piebaldism*. These individuals have aberrations in pigmentation on the abdomen, extremities, and forehead. Interestingly, the distribution of the pigmentation defects is similar to that of the white spotting mice. The mutations in several patients have been identified and occur predominantly in the SCFR catalytic domain. Alterations in amino acids 561, 583, 584, 642, and 664 have all been described (Spritz, 1994). Although no abnormalities in hematopoiesis have been reported, all of the patients examined were heterozygotes (Spritz, 1992). Many of the heterozygotic white spotting mice reported to date also have normal hematopoiesis. As discussed previously, activated forms of the SCFR have been found in patients with *gastrointestinal stromal tumors* and *mastocytosis* (Nagata *et al.*, 1995; Longley *et al.*, 1996, 1999; Hirota *et al.*, 1998; Moskaluk *et al.*, 1999).

THERAPEUTIC UTILITY

Effect of treatment with soluble receptor domain

As discussed above, SCF binds the SCFR extracellular domain and induces dimerization. The soluble form of the SCFR binds SCF and antagonizes SCF-induced increases in autophosphorylation as well as activation of downstream signaling pathways (Lev *et al.*, 1992a). This results in inhibition of SCF-mediated responses (Lev *et al.*, 1992a, 1993). To date this has not been utilized clinically.

Effects of inhibitors (antibodies) to receptors

AG1296 is a quinoxaline tyrphostin compound that inhibits the kinase activity of the PDGF and SCF receptors (Kovalenko *et al.*, 1994). Because a high percentage of SCLC cells inappropriately express both SCF and its receptor, Krystal and coworkers examined the effect of AG1296 on the growth of SCLC cell lines *in vitro* (Krystal *et al.*, 1997). This drug inhibited SCLC growth as well as induced apoptosis. These results suggest that an autocrine loop mediates survival and growth of some SCLC cells. Another drug that is reported to inhibit SCFR-mediated signaling is AG776 (Wessely *et al.*, 1997). Treatment of avian erythroid progenitor cells with this drug decreased SCF-mediated survival (Wessely *et al.*, 1997). To date, there are no published reports relating to the use of either of these drugs in patients.

In humans, antibodies blocking the binding of SCF to its receptor have not been used *in vivo*. However, recent studies *in vitro* suggest that the SCFR may be utilized to transfer genes into hematopoietic progenitor cells. Keller and coworkers demonstrated targeted transfection of hematopoietic cell lines expressing the SCFR with a molecular conjugate vector containing SCF (Schwarzenberger *et al.*, 1996). SCF has also been incorporated into retroviral vectors and utilized successfully for targeted gene transfer *in vitro* (Yajima *et al.*, 1998). Lastly, a plasmid containing IL-3 was recently conjugated to an antibody specific for the SCFR and transiently transfected into a hematopoietic cell line as well as CD34-positive progenitor cells (Chapel *et al.*, 1999). Thus, the SCFR may be an ideal target for gene transfer into hematopoietic cells.

References

Abkowitz, J. L., Broudy, V. C., Bennett, L. G., Zsebo, K. M., and Martin, F. H. (1992). Absence of abnormalities of *c-kit* or its ligand in two patients with diamond-blackfan anemia. *Blood* **79**, 25–28.

Abrahamson, J. L. A., Lee, J. M., and Bernstein, A. (1995). Regulation of p53-mediated apoptosis and cell cycle arrest by Steel factor. *Mol. Cell. Biol.* **15**, 6953–6960.

Afonja, O., Amorosi, E., Ashman, L., and Takeshita, K. (1998). Multilineage involvement and erythropoietin-'independent' erythroid progenitor cells in a patient with systemic mastocytosis. *Ann. Hematol.* **77**, 183–186.

Alai, M., Mui, A. L. -F., Cutler, R. L., Bustelo, S. R., Barbacid, M., and Krystal, G. (1992). Steel factor stimulates the tyrosine phosphorylation of the proto-oncogene product, $p95^{vav}$, in human hemopoietic cells. *J. Biol. Chem.* **267**, 18021–18025.

André, C., Martin, E., Cornu, F., Hu, W.-X., Wang, X.-P., and Galibert, F. (1992). Genomic organization of the human c-kit gene: evolution of the receptor tyrosine kinase subclass III. *Oncogene* **7**, 685–691.

Asano, Y., Brach, M. A., Ahlers, A., de Vos, S., Butterfield, J. H., Ashman, L. K., Valent, P., Gruss, H.-J., and Herrmann, F. (1993). Phorbol ester 12-*O*-tetradecanoylphorbol-13-acetate down-regulates expression of the c-*kit* proto-oncogene product. *J. Immunol.* **151**, 2345–2354.

Ashman, L. (1999). The biology of stem cell factor and its receptor c-Kit. *Int. J. Biochem. Cell Biol.* **31**, 1037–1051.

Badache, A., Muja, N., and De Vries, G. H. (1998). Expression of Kit in neurofibromin-deficient human Schwann cells: role in Schwann cell hyperplasia associated with type 1 neurofibromatosis. *Oncogene* **17**, 795–800.

Besmer, P., Murphy, J. E., George, P. C., Qiu, F., Bergold, P. J., Lederman, L., Snyder Jr. H. W., Brodeur, D., Zuckerman, E. E., and Hardy, W. D. (1986). A new acute transforming feline retrovirus and relationship of its oncogene v-*kit* with the protein kinase gene family. *Nature* **320**, 415–421.

Blechman, J. M., Lev, S., Brizzi, M. F., Leitner, O., Pegoraro, L., Givol, D., and Yarden, Y. (1993a). Soluble c-kit proteins and antireceptor monoclonal antibodies confine the binding site of the stem cell factor. *J. Biol. Chem.* **268**, 4399–4406.

Blechman, J. M., Lev, S., Givol, D., and Yarden, Y. (1993b). Structure–function analyses of the kit receptor for the steel factor. *Stem Cells* **11**, 12–21.

Blechman, J. M., Lev, S., Barg, J., Eisenstein, M., Vaks, B., Vogel, Z., Givol, D., and Yarden, Y. (1995). The fourth immunoglobulin domain of the stem cell factor receptor couples ligand binding to signal transduction. *Cell* **80**, 103–113.

Blume-Jensen, P., Claesson-Welsh, L., Siegbahn, A., Zsebo, K. M., Westermark, B., and Heldin, C.-H. (1991). Activation of the human c-*kit* product by ligand-induced dimerization mediates circular actin reorganization and chemotaxis. *EMBO J.* **10**, 4121–4128.

Blume-Jensen, P., Siegbahn, A., Stabel, S., Heldin, C.-H., and Rönnstrand, L. (1993). Increased Kit/SCF receptor induced mitogenicity but abolished cell motility after inhibition of protein kinase C. *EMBO J.* **12**, 4199–4209.

Blume-Jensen, P., Rönnstrand, L., Gout, I., Waterfield, M. D., and Heldin, C.-H. (1994). Modulation of Kit/stem cell factor receptor-induced signaling by protein kinase C. *J. Biol. Chem.* **269**, 21793–21802.

Blume-Jensen, P., Wernstedt, C., Heldin, C.-H., and Rönnstrand, L. (1995). Identification of the major phosphorylation sites for protein kinase C in Kit/stem cell factor receptor *in vitro* and in intact cells. *J. Biol. Chem.* **270**, 14192–14200.

Blume-Jensen, P., Janknecht, R., and Hunter, T. (1998). The Kit receptor promotes cell survival via activation of PI 3-kinase and subsequent Akt-mediated phosphorylation of Bad on Ser136. *Curr. Biol.* **8**, 779–782.

Boissy, R. E., and Nordlund, J. J. (1997). Molecular basis of congenital hypopigmentary disorders in humans: a review. *Pigment Cell Res.* **10**, 12–24.

Bondy, C. A., Nelson, L. M., and Kalantaridou, S. N. (1998). The genetic origins of ovarian failure. *J. Women's Health* **7**, 1225–1229.

Bossé, P., Bernex, F., De Sepulveda, P., Salaün, P., Panthier, J.-J. (1997). Multiple neuroendocrine tumours in transgenic mice induced by c-*kit*-SV40 T antigen fusion genes. *Oncogene* **14**, 2661–2670.

Brennscheidt, U., Riedel, D., Kölch, W., Bonifer, R., Brach, M. A., Ahlers, A., Mertelsmann, R. H., and Herrmann, F. (1994). Raf-1 is a necessary component of the mitogenic response of the human megakaryoblastic leukemia cell line MO7 to human stem cell factor, granulocyte-macrophage colony-stimulating factor, interleukin 3, and interleukin 9. *Cell Growth Differ.* **5**, 367–372.

Brizzi, M. F., Blechman, J. M., Cavalloni, G., Givol, D., Yarden, Y., and Pegoraro, L. (1994a). Protein kinase C-dependent release of a functional whole extracellular domain of the mast cell growth factor (MGF) receptor by MGF-dependent human myeloid cells. *Oncogene* **9**, 1583–1589.

Brizzi, M. F., Zini, M. G., Aronica, M. G., Blechman, J. M., Yarden, Y., and Pegoraro, L. (1994b). Convergence of signaling by interleukin-3, granulocyte-macrophage colony-stimulating factor, and mast cell growth factor on JAK2 tyrosine kinase. *J. Biol. Chem.* **269**, 31680–31684.

Broudy, V. C. (1997). Stem cell factor and hematopoiesis. *Blood* **90**, 1345–1364.

Broudy, V. C., Lin, N. L., Bühring, H.-J., Komatsu, N., and Kavanagh, T. J. (1998). Analysis of *c-kit* receptor dimerization by fluorescence resonance energy transfer. *Blood* **91**, 898–906.

Büttner, C., Henz, B. M., Welker, P., Sepp, N. T., and Grabbe, J. (1998). Identification of activating c-kit mutations in adult-, but not in childhood-onset indolent mastocytosis: a possible explanation for divergent clinical behavior. *J. Invest. Dermatol.* **111**, 1227–1231.

Carpino, N., Wisniewski, D., Strife, A., Marshak, D., Kobayashi, R., Stillman, B., and Clarkson, B. (1997). $p62^{dok}$: a constitutively tyrosine-phosphorylated, GAP-associated protein in chronic myelogenous leukemia progenitor cells. *Cell* **88**, 197–204.

Chabot, B., Stephenson, D. A., Chapman, V. M., Besmer, P., and Bernstein, A. (1988). The proto-oncogene *c-kit* encoding a transmembrane tyrosine kinase receptor maps to the mouse *W* locus. *Nature* **335**, 88–89.

Chan, J. K. (1999). Mesenchymal tumors of the gastrointestinal tract: a paradise for acronyms (STUMP, GIST, GANT, and now GIPACT), implication of c-kit in genesis, and yet another of the many emerging roles of the interstitial cell of Cajal in the pathogenesis of gastrointestinal diseases? *Adv. Anat. Pathol.* **6**, 19–40.

Chapel, A., Poncet, P., Neildez-Nguyen, T. M. A., Vétillard, J., Brouard, N., Goupy, C., Chavanel, G., Hirsch, F., and Thierry, D. (1999). Targeted transfection of the *IL-3* gene into primary human hematopoietic progenitor cells through the c-kit receptor. *Exp. Hematol.* **27**, 250–258.

Chui, X., Egami, H., Yamashita, J., Kurizaki, T., Ohmachi, H., Yamamoto, S., and Ogawa, M. (1996). Immunohistochemical expression of the c-kit proto-oncogene product in human malignant and non-malignant breast tissues. *Br. J. Cancer* **73**, 1233–1236.

Cutler, R. L., Liu, L., Damen, J. E., and Krystal, G. (1993). Multiple cytokines induce the tyrosine phosphorylation of Shc and its association with Grb2 in hemopoietic cells. *J. Biol. Chem.* **268**, 21463–21465.

DeBerry, C., Mou, S., and Linnekin, D. (1997). Stat1 associates with c-kit and is activated in response to stem cell factor. *Biochem. J.* **327**, 73–80.

De Sepulveda, P., Okkenhaug, K., La Rose, J., Hawley, R. G., Dubreuil, P., and Rottapel, R. (1999). Socs1 binds to multiple signalling proteins and suppresses Steel factor-dependent proliferation. *EMBO J.* **18**, 904–915.

de Vos, S., Brach, M. A., Asano, Y., Ludwig, W.-D., Beffelheim, P., Gruss, H.-J., and Herrmann, F. (1993). Transforming growth factor-β1 interferes with the proliferation-inducing activity of stem cell factor in myelogenous leukemia blasts through functional down-regulation of the c-*kit* proto-oncogene product. *Cancer Res.* **53**, 3638–3642.

DiPaola, R. S., Kuczynski, W. I., Onodera, K., Ratajczak, M. Z., Hijiya, N., Moore, J., and Gewirtz, A. M. (1997). Evidence for a functional kit receptor in melanoma, breast, and lung carcinoma cells. *Cancer Gene Ther.* **4**, 176–182.

Donovan, P. J. (1994). Growth factor regulation of mouse primordial germ cell development. *Curr. Topics Dev. Biol.* **29**, 189–225.

Donovan, P. J., De Miguel, M., Cheng, L., and Resnick, J. L. (1998). Primordial germ cells, stem cells and testicular cancer. *APMIS* **106**, 134–141.

Driancourt, M.-A., and Thuel, B. (1998). Control of oocyte growth and maturation by follicular cells and molecules present in follicular fluid. *Reprod. Nutr. Dev.* **38**, 345–362.

Dubois, C. M., Ruscetti, F. W., Stankova, J., and Keller, J. R. (1994). Transforming growth factor-beta regulates c-kit message stability and cell-surface protein expression in hematopoietic progenitors. *Blood* **83**, 3138–3145.

Escribano, L., Ocqueteau, M., Almeida, J., Orfao, A., and San Miguel, J. F. (1998). Expression of the c-kit (CD117) molecule in normal and malignant hematopoiesis. *Leuk. Lymph.* **30**, 459–466.

Feng, G.-S., Ouyang, Y.-B., Hu, D.-P., Shi, Z-Q., Gentz, R., and Ni, J. (1996). Grap is a novel SH3-SH2-SH3 adaptor protein that couples tyrosine kinases to the ras pathway. *J. Biol. Chem.* **271**, 12129–12132.

Feng, H., Sandlow, J. I., and Sandra, A. (1998). The C-*kit* receptor and its possible signaling transduction pathway in mouse spermatozoa. *Mol. Reprod. Dev.* **49**, 317–326.

Ferrao, P., Gonda, T. J., and Ashman, L. K. (1997). Expression of constitutively activated human c-kit in myb transformed early myeloid cells leads to factor independence, histiocytic differentiation, and tumorigenicity. *Blood* **90**, 4539–4552.

Fleischman, R. A. (1993). From white spots to stem cells: the role of the Kit receptor in mammalian development. *Trends Genet.* **9**, 285–290.

Fleischman, R. A., Saltman, D. L., Stastny, V., and Zneimer, S. (1991). Deletion of the c-*kit* protooncogene in the human developmental defect piebald trait. *Proc. Natl Acad. Sci. USA* **88**, 10885–10889.

Födinger, M., and Mannhalter, C. (1997). Molecular genetics and development of mast cells: implications for molecular medicine. *Mol. Med. Today* **3**, 131–137.

Foltz, I. N., and Schrader, J. W. (1997). Activation of the stress-activated protein kinases by multiple hematopoietic growth factors with the exception of interleukin-4. *Blood* **89**, 3092–3096.

Fujio, K., Hu, Z., Evarts, R. P., Marsden, E. R., Niu, C.-H., and Thorgeirsson, S. S. (1996). Coexpression of stem cell factor and c-*kit* in embryonic and adult liver. *Exp. Cell Res.* **224**, 243–250.

Funasaka, Y., Boulton, T., Cobb, M., Yarden, Y., Fan, B., Lyman, S. D., Williams, D. E., Anderson, D. M., Zakut, R., Mishima, Y., and Halaban, R. (1992). c-Kit-kinase induces a cascade of protein tyrosine phosphorylation in normal human melanocytes in response to mast cell growth factor and stimulates mitogen-activated protein kinase but is down-regulated in melanomas. *Mol. Biol. Cell* **3**, 197–209.

Furitsu, T., Tsujimura, T., Tono, T., Ikeda, H., Kitayama, H., Koshimizu, U., Sugahara, H., Butterfield, J. H., Ashman, L. K., Kanayama, Y., Matsuzawa, Y., Kitamura, Y., and Kanakura, Y. (1993). Identification of mutations in the coding sequence of the proto-oncogene c-*kit* in a human mast cell leukemia cell line causing ligand-independent activation of c-*kit* product. *J. Clin. Invest.* **92**, 1736–1744.

Galli, S. J., Zsebo, K. M., and Geissler, E. N. (1994). The kit ligand, stem cell factor. *Adv. Immunol.* **55**, 1–97.

Geissler, E. N., Ryan, M. A., and Housman, D. E. (1988). The dominant-white spotting *(W)* locus of the mouse encodes the *c-kit* proto-oncogene. *Cell* **55**, 185–192.

Giebel, L. B., and Spritz, R. A. (1991). Mutation of the *KIT* (mast/stem cell growth factor receptor) protooncogene in human piebaldism. *Proc. Natl Acad. Sci. USA* **88**, 8696–8699.

Giebel, L. B., Strunk, K. M., Holmes, S. A., and Spritz, R. A. (1992). Organization and nucleotide sequence of the human KIT (mast/stem cell growth factor receptor) proto-oncogene. *Oncogene* **7**, 2207–2217.

Gokkel, E., Grossman, Z., Ramot, B., Yarden, Y., Rechavi, G., and Givol, D. (1992). Structural organization of the murine c-*kit* proto-oncogene. *Oncogene* **7**, 1423–1429.

Gommerman, J. L., Rottapel, R., and Berger, S. A. (1997). Phosphatidylinositol 3-kinase and Ca^{2+} influx dependence for ligand-stimulated internalization of the c-Kit receptor. *J. Biol. Chem.* **272**, 30519–30525.

Gotoh, A., Takahira, H., Mantel, C., Litz-Jackson, S., Boswell, H. S., and Broxmeyer, H. E. (1996). Steel factor induces serine phosphorylation of Stat3 in human growth factor-dependent myeloid cell lines. *Blood* **88**, 138–145.

Grgurevich, S., Linnekin, D., Musso, T., Zhang, X., Modi, W., Varesio, L., Ruscetti, F. W., Ortaldo, J. R., and McVicar, D. W. (1997). The Csk-like proteins Lsk, Hyl, and Matk represent the same *C*sk *h*omologous *k*inase (Chk) and are regulated by stem cell factor in the megakaryoblastic cell line M07e. *Growth Factors* **14**, 103–115.

Gutman, M., Singh, R. K., Radinsky, R., and Bar-Eli, M. (1994). Intertumoral heterogeneity of receptor-tyrosine kinases expression in human melanoma cell lines with different metastatic capabilities. *Anticancer Res.* **14**, 1759–1766.

Hallek, M., Druker, B., Lepisto, E. M., Wood, K. W., Ernst, T. J., and Griffin, J. D. (1992). Granulocyte-macrophage colony-stimulating factor and steel factor induce phosphorylation of both unique and overlapping signal transduction intermediates in a human factor-dependent hematopoietic cell line. *J. Cell Physiol.* **153**, 176–186.

Hallek, M., Danhauser-Riedl, S., Herbst, R., Warmuth, M., Winkler, A., Kolb, H.-J., Druker, B., Griffin, J. D., Emmerich, B., and Ullrich, A. (1996). Interaction of the receptor tyrosine kinase p145$^{c\text{-}kit}$ with the p210$^{bcr/abl}$ kinase in myeloid cells. *Br. J. Haematol.* **94**, 5–16.

Hamel, W., and Westphal, M. (1997). The road less travelled: c-kit and stem cell factor. *J. Neuro-Oncol.* **35**, 327–333.

Hassan, H. T., and Zander, A. (1996). Stem cell factor as a survival and growth factor in human normal and malignant hematopoiesis. *Acta Haematol.* **95**, 257–262.

He, J., DeCastro, C. M., Vandenbark, G. R., Busciglioi, J., and Gabuzda, D. (1997). Astrocyte apoptosis induced by HIV-1 transactivation of the c-*kit* protooncogene. *Proc. Natl Acad. Sci. USA* **94**, 3954–3959.

Heinrich, M. C., Dooley, D. C., and Keeble, W. W. (1995). Transforming growth factor $\beta 1$ inhibits expression of the gene products for Steel factor and its receptor (c-kit). *Blood* **85**, 1769–1780.

Hemesath, T. J., Price, E. R., Takemoto, C., Badalian T., and Fisher, D. E. (1998). MAP kinase links the transcription factor microphthalmia to c-Kit signalling in melanocytes. *Nature* **391**, 298–301.

Herbst, R., Munemitsu, S., and Ullrich, A. (1995a). Oncogenic activation of v-*kit* involves deletion of a putative tyrosine-substrate interaction site. *Oncogene* **10**, 369–379.

Herbst, R., Shearman, M. S., Jallal, B., Schlessinger, J., and Ullrich, A. (1995b). Formation of signal transfer complexes between stem cell and platelet-derived growth factor receptors and SH2 domain proteins *in vitro*. *Biochemistry* **34**, 5971–5979.

Hibi, K., Takahashi, T., Sekido, Y., Ueda, R., Hida, T., Ariyoshi, Y., Takagi, H., and Takahashi, T. (1991). Coexpression of the stem cell factor and the c-*kit* genes in small-cell lung cancer. *Oncogene* **6**, 2291–2296.

Hines, S. J., Organ, C., Kornstein, M. J., and Krystal, G. W. (1995). Coexpression of the c-*kit* and stem cell factor genes in breast carcinomas. *Cell Growth Differ.* **6**, 769–779.

Hirota, S., Ito, A., Morii, E., Wanaka, A., Tohyama, M., Kitamura, Y., and Nomura, S. (1992). Localization of mRNA for c-*kit* receptor and its ligand in the brain of adult rats: an analysis using *in situ* hybridization histochemistry. *Mol. Brain Res.* **15**, 47–54.

Hirota, S., Isozaki, K., Moriyama, Y., Hashimoto, K., Nishida, T., Ishiguro, S., Kawano, K., Hanada, M., Kurata, A., Takeda, M., Tunio, G. M., Matsuzawa, Y., Kanakura, Y., Shinomura, Y., and Kitamura, Y. (1998). Gain-of-function mutations of c-*kit* in human gastrointestinal stromal tumors. *Science* **279**, 577–580.

Hofstra, R. M., Landsvater, R. M., Ceccherini, I., Stulp, R. P., Stelwagen, T., Luo, Y., Pasini, B., Hoppener, J. W., van Amstel, H. K., Romeo, G., Lips, C. J. M., and Buys, C.H. C. M. (1994). A mutation in the RET proto-oncogene associated with multiple endocrine neoplasia type 2B and sporadic medullary thyroid carcinoma. *Nature* **367**, 375–376.

Horie, M., and Broxmeyer, H. E. (1993). Involvement of immediate-early gene expression in the synergistic effects of steel factor in combination with granulocyte-macrophage colony-stimulating factor or interleukin-3 on proliferation of a human factor-dependent cell line. *J. Biol. Chem.* **268**, 968–973.

Hu, Z.-B., Ma, W., Uphoff, C. C., Quentmeier, H., and Drexler, H. G. (1994). *C-kit* expression in human megakaryoblastic leukemia cell lines. *Blood* **83**, 2133–2144.

Huang, E., Nocka, K., Beier, D. R., Chu, T.-Y., Buck, J., Lahm, H.-W., Wellner, D., Leder, P., and Besmer, P. (1990). The hematopoietic growth factor KL is encoded by the *Sl* locus and is the ligand of the c-*kit* receptor, the gene product of the *W* locus. *Cell* **63**, 225–233.

Huang, S., Luca, M., Gutman, M., McConkey, D. J., Langley, K. E., Lyman, S. D., and Bar-Eli, M. (1996). Enforced c-KIT expression renders highly metastatic human melanoma cells susceptible to stem cell factor-induced apoptosis and inhibits their tumorigenic and metastatic potential. *Oncogene* **13**, 2339–2347.

Huang, S., Jean, D., Luca, M., Tainsky, M. A., and Bar-Eli, M. (1998). Loss of AP-2 results in downregulation of c-KIT and enhancement of melanoma tumorigenicity and metastasis. *EMBO J.* **17**, 4358–4369.

Huber, M., Helgason, C. D., Scheid, M. P., Duronio, V., Humphries, R. K., and Krystal, G. (1998). Targeted disruption of SHIP leads to steel factor-induced degranulation of mast cells. *EMBO J.* **17**, 7311–7319.

Huizinga, J. D., Thuneberg, L., Klüppel, M., Malysz, J., Mikkelsen, H. B., and Bernstein, A. (1995). *W/kit* gene required for interstitial cells of Cajal and for intestinal pacemaker activity. *Nature* **373**, 347–349.

Inoue, M., Kyo, S., Fujita, M., Enomoto, T., and Kondoh, G. (1994). Coexpression of the c-*kit* receptor and the stem cell factor in gynecological tumors. *Cancer Res.* **54**, 3049–3053.

Ismail, R. S., Okawara, Y., Fryer, J. N., and Vanderhyden, B. C. (1996). Hormonal regulation of the ligand for c-*kit* in the rat ovary and its effects on spontaneous oocyte meiotic maturation. *Mol. Reprod. Dev.* **43**, 458–469.

Ismail, R. S., Dubé, M., and Vanderhyden, B. C. (1997). Hormonally regulated expression and alternative splicing of kit ligand may regulate kit-induced inhibition of meiosis in rat oocytes. *Dev. Biol.* **184**, 333–342.

Isozaki, K., Tsujimura, T., Nomura, S., Morii, E., Koshimizu, U., Nishimune, Y., and Kitamura, Y. (1994). Cell type-specific deficiency of c-*kit* gene expression in mutant mice of *mi/mi* genotype. *Am. J. Pathol.* **145**, 827–836.

Jacobs-Helber, S. M., Penta, K., Sun, Z., Lawson, A., and Sawyer, S. T. (1997). Distinct signaling from stem cell factor and erythropoietin in HCD57 cells. *J. Biol. Chem.* **272**, 6850–6853.

Jacobsen, F. W., Dubois, C. M., Rusten, L. S., Veiby, O. P., and Jacobsen, S. E. (1995). Inhibition of stem cell factor-induced proliferation of primitive murine hematopoietic progenitor cells signaled through the 75-kilodalton tumor necrosis factor receptor. Regulation of c-kit and p53 expression. *J. Immunol.* **154**, 3732–3741.

Joneja, B., Chen, H.-C., Seshasayee, D., Wrentmore, A. L., and Wojchowski, D. M. (1997). Mechanisms of stem cell factor and erythropoietin proliferative co-signaling in FDC2-ER cells. *Blood* **90**, 3533–3545.

Kauma, S., Huff, T., Krystal, G., Ryan, J., Takacs, P., and Turner, T. (1996). The expression of stem cell factor and its receptor, c-kit in human endometrium and placental tissues during pregnancy. *J. Clin. Endocrinol. Metab.* **81**, 1261–1266.

Keshet, E., Lyman, S. D., Williams, D. E., Anderson, D. M., Jenkins, N. A., Copeland, N. G., and Parada, L. F. (1991). Embryonic RNA expression patterns of the c-*kit* receptor and its cognate ligand suggest multiple functional roles in mouse development. *EMBO J.* **10**, 2425–2435.

Khoury, E., Andre, C., Pontvert-Delucq, S., Drenou, B., Baillou, C., Guigon, M., Najman, A., and Lemoine, F. M. (1994). Tumor necrosis factor α (TNFα) downregulates c-*kit* proto-oncogene product expression in normal and acute myeloid leukemia CD34$^+$ cells via p55 TNFα receptors. *Blood* **84**, 2506–2514.

Kimura, A., Nakata, Y., Katoh, O., and Hyodo, H. (1997). c-kit point mutation in patients with myeloproliferative disorders. *Leuk. Lymph.* **25**, 281–287.

Kitamura, Y., Hirota, S., and Nishida, T. (1998). Molecular pathology of c-*kit* proto-oncogene and development of gastrointestinal stromal tumors. *Ann. Chir. Gyn.* **87**, 282–286.

Kitayama, H., Kanakura, Y., Furitsu, T., Tsujimura, T., Oritani, K., Ikeda, H., Sugahara, H., Mitsui, H., Kanayama, Y., Kitamura, Y., and Matsuzawa, Y. (1995). Constitutively activating mutations of c-*kit* receptor tyrosine kinase confer factor-independent growth and tumorigenicity of factor-dependent hematopoietic cell lines. *Blood* **85**, 790–798.

Kitayama, H., Tsujimura, T., Matsumura, I., Oritani, K., Ikeda, H., Ishikawa, J., Okabe, M., Suzuki, M., Yamamura, K., Matsuzawa, Y., Kitamura, Y., and Kanakura, Y. (1996). Neoplastic transformation of normal hematopoietic cells by constitutively activating mutations of *c-kit* receptor tyrosine kinase. *Blood* **88**, 995–1004.

Koike, T., Hirai, K., Morita, Y., and Nozawa, Y. (1993). Stem cell factor-induced signal transduction in rat mast cells. *J. Immunol.* **151**, 359–366.

Kondoh, G., Hayasaka, N., Li, Q., Nishimune, Y., and Hakura, A. (1995). An *in vivo* model for receptor tyrosine kinase autocrine-paracrine activation: auto-stimulated *KIT* receptor acts as a tumor promoting factor in papillomavirus-induced tumorigenesis. *Oncogene* **10**, 341–347.

Kovalenko, M., Gazit, A., Bohmer, A., Rorsman, C., Ronnstrand, L., Heldin, C. H., Waltenberger, J., Bohmer, F. D., and Levitzki, A. (1994). Selective platelet-derived growth factor receptor kinase blockers reverse sis-transformation. *Cancer Res.* **54**, 6106–6114.

Kozawa, O., Blume-Jensen, P., Heldin, C.-H., and Rönnstrand, L. (1997). Involvement of phosphatidylinositol 3′-kinase in stem-cell-factor-induced phospholipase D activation and arachidonic acid release. *Eur. J. Biochem.* **248**, 149–155.

Kozlowski, M., Larose, L., Lee, F., Le, D. M., Rottapel, R., and Siminovitch, K. A. (1998). SHP-1 binds and negatively modulates the c-Kit receptor by interaction with tyrosine 569 in the c-Kit juxtamembrane domain. *Mol. Cell. Biol.* **18**, 2089–2099.

Krosl, G., He, G., Lefrancois, M., Charron, F., Roméo, P.-H., Jolecoeur, P., Kirsch, I. R., Nemer, M., and Hoang, T. (1998). Transcription factor SCL is required for c-kit expression and c-Kit function in hemopoietic cells. *J. Exp. Med.* **188**, 439–450.

Krystal, G. W., Hines, S. J., and Organ, C. P. (1996). Autocrine growth of small cell lung cancer mediated by coexpression of c-*kit* and stem cell factor. *Cancer Res.* **56**, 370–376.

Krystal, G. W., Carlson, P., and Litz, J. (1997). Induction of apoptosis and inhibition of small cell lung cancer growth by the quinoxalin tyrphostins. *Cancer Res.* **57**, 2203–2208.

Kubota, T., Hikona, H., Sasaki, E., and Sakurai, M. (1994). Sequence of a bovine c-kit proto-oncogene cDNA. *Gene* **141**, 305–306.

Kubota, Y., Angelotti, T., Niederfellner, G., Herbst, R., and Ullrich, A. (1998). Activation of phosphatidylinositol 3-kinase is necessary for differentiation of FDC-P1 cells following stimulation of type III receptor tyrosine kinases. *Cell Growth Differ.* **9**, 247–256.

Laitinen, M., Rutanen, E.-M., and Ritvos, O. (1995). Expression of c-*kit* ligand messenger ribonucleic acids in human ovaries and regulation of their steady state levels by gonadotropins in cultured granulosa-luteal cells. *Endocrinology* **136**, 4407–4414.

Lam, L. P. Y., Chow, R. Y. K., and Berger, S. A. (1999). A transforming mutation enhances the activity of the c-Kit soluble tyrosine kinase domain. *Biochem. J.* **338**, 131–138.

Lammie, A., Drobnjak, M., Gerald, W., Saad, A., Cote, R., and Cordon-Cardo, C. (1994). Expression of c-kit and kit ligand proteins in normal human tissues. *J. Histochem. Cytochem.* **42**, 1417–1425.

Lassam, N., and Bickford, S. (1992). Loss of c-*kit* expression in cultured melanoma cells. *Oncogene* **7**, 51–56.

Lee, J. M. (1998). Inhibition of p53-dependent apoptosis by the KIT tyrosine kinase: regulation of mitochondrial permeability transition and reactive oxygen species generation. *Oncogene* **17**, 1653–1662.

Lemmon, M. A., Pinchasi, D., Zhou, M., Lax, I., and Schlessinger, J. (1997). Kit receptor dimerization is driven by bivalent binding of stem cell factor. *J. Biol. Chem.* **272**, 6311–6317.

Lev, S., Givol, D., and Yarden, Y. (1991). A specific combination of substrates is involved in signal transduction by the *kit*-encoded receptor. *EMBO J.* **10**, 647–654.

Lev, S., Yarden, Y., and Givol, D. (1992a). A recombinant ectodomain of the receptor for the stem cell factor (SCF) retains ligand-induced receptor dimerization and antagonizes SCF-stimulated cellular responses. *J. Biol. Chem.* **267**, 10866–10873.

Lev, S., Yarden, Y., and Givol, D. (1992b). Dimerization and activation of the kit receptor by monovalent and bivalent binding of the stem cell factor. *J. Biol. Chem.* **267**, 15970–15977.

Lev, S., Blechman, J., Nishikawa, S.-I., Givol, D., and Yarden, Y. (1993). Interspecies molecular chimeras of kit help define the binding site of the stem cell factor. *Mol. Cell. Biol.* **13**, 2224–2234.

Li, Q., Kondoh, G., Inafuku, S., Nishimune, Y., and Hukura, A. (1996). Abrogation of c-*kit*/*Steel factor*-dependent tumorigenesis by kinase defective mutants of the c-*kit* receptor: c-*kit* kinase defective mutants as candidate tools for cancer gene therapy. *Cancer Res.* **56**, 4343–4346.

Lin, Y., and Benchimol, S. (1995). Cytokines inhibit p53-mediated apoptosis but not p53-mediated G_1 arrest. *Mol. Cell. Biol.* **15**, 6045–6054.

Linnekin, D. (1999). Early signaling pathways activated by c-Kit in hematopoietic cells. *Int. J. Biochem. Cell Biol.*, **31**, 1053–1074.

Linnekin, D., Mou, S., Deberry, C. S., Weiler, S. R., Keller, J. R., Ruscetti, F. W., and Longo, D. L. (1997a). Stem cell factor, the JAK-STAT pathway and signal transduction. *Leuk. Lymphoma.* **27**, 439–444.

Linnekin, D., DeBerry, C. S., and Mou, S. (1997b). Lyn associates with the juxtamembrane region of c-Kit and is activated by stem cell factor in hematopoietic cell lines and normal progenitor cells. *J. Biol. Chem.* **272**, 27450–27455.

Liu, Y.-C., Kawagishi, M., Kameda, R., and Ohashi, H. (1993). Characterization of a fusion protein composed of the extracellular domain of *c-kit* and the Fc region of human IgG expressed in a baculovirus system. *Biochem. Biophys. Res. Commun.* **197**, 1094–1102.

Liu, L., Damen, J. E., Ware, M., Hughes, M., and Krystal, G. (1997). SHIP, a new player in cytokine-induced signalling. *Leukemia* **11**, 181–184.

Longley, B. J., Tyrrell, L., Lu, S.-Z., Ma, Y.-S., Langley, K., Ding, T., Duffy, T., Jacobs, P., Tang, L. H., and Modlin, I. (1996). Somatic c-*KIT* activating mutation in urticaria pigmentosa and aggressive mastocytosis: establishment of clonality in a human mast cell neoplasm. *Nature Genet.* **12**, 312–314.

Longley Jr., B. J., Metcalfe, D. D., Tharp, M., Wang, X., Tyrrell, L., Lu, S.-Z., Heitjan, D., and Ma, Y. (1999). Activating and dominant inactivating c-*KIT* catalytic domain mutations in distinct clinical forms of human mastocytosis. *Proc. Natl Acad. Sci. USA.* **96**, 1609–1614.

Lorenz, U., Bergemann, A. D., Steinberg, H. N., Flanagan, J. G., Li, X., Galli, S. J., and Neel, B. G. (1996). Genetic analysis reveals cell type-specific regulation of receptor tyrosine kinase c-Kit by the protein tyrosine phosphatase SHP1. *J. Exp. Med.* **184**, 1111–1126.

Loveland, K. L., and Schlatt, S. (1997). Stem cell factor and c-kit in the mammalian testis: lessons originating from Mother Nature's gene knockouts. *J. Endocrinol.* **153**, 337–344.

Lu, C., and Kerbel, R. S. (1994). Cytokines, growth factors and the loss of negative growth controls in the progression of human cutaneous malignant melanoma. *Curr. Opin. Oncol.* **6**, 212–220.

Luca, M. R., and Bar-Eli, M. (1998). Molecular changes in human melanoma metastasis. *Histol. Histopathol.* **13**, 1225–1231.

Lyman, S. D., and Jacobsen, S. E. W. (1998). C-*kit* ligand and Flt3 ligand: stem/progenitor cell factors with overlapping yet distinct activities. *Blood* **91**, 1101–1134.

Ma, Y., Longley, B. J., Wang, X., Blount, J. L., Langley, K., and Caughey, G. H. (1999). Clustering of activating mutations in *c-KIT*'s juxtamembrane coding region in canine mast cell neoplasms. *J. Invest. Dermatol.* **112**, 165–170.

Madhani, H. D., and Fink, G. R. (1998). The riddle of MAP kinase signaling specificity. *TIG* **14**, 151–155.

Majumder, S., Ray, P., and Besmer, P. (1990). Tyrosine protein kinase activity of the HZ4-feline sarcoma virus P80$^{gag-kit}$-transforming protein. *Oncogene Res.* **5**, 329–335.

Mantel, C., Luo, Z., Canfield, J., Braun, S., Deng, C., and Broxmeyer, H. E. (1996). Involvement of p21^{cip-1} in the maintenance of stem/progenitor cells *in vivo*. *Blood* **88**, 3710–3719.

Marklund, S., Kijas, J., Rodriguez-Martinez, H., Rönnstrand, L., Funa, K., Moller, M., Lange, D., Edfors-Lilja, I., and Andersson, L. (1998). Molecular basis for the dominant white phenotype in the domestic pig. *Genome Res.* **8**, 826–833.

Matsuda, R., Takahashi, T., Nakamura, S., Sekido, Y., Nishida, K., Seto, M., Seito, T., Sugiura, T., Ariyoshi, Y., Takahashi, T., and Ueda, R. (1993). Expression of the c-*kit* protein in human solid tumors and in corresponding fetal and adult normal tissues. *Am. J. Pathol.* **142**, 339–346.

Matsuguchi, T., Salgia, R., Hallek, M., Eder, M., Druker, B., Ernst, T. J., and Griffin, J. D. (1994). Shc phosphorylation in myeloid cells is regulated by granulocyte macrophage colony-stimulating factor, interleukin-3, and steel factor and is constitutively increased by p210$^{BCR/ABL}$. *J. Biol. Chem.* **269**, 5016–5021.

Matsuguchi, T., Inhorn, R. C., Carlesso, N., Xu, G., Druker, B., and Griffin, J. D. (1995). Tyrosine phosphorylation of p95^{Vav} in myeloid cells is regulated by GM-CSF, IL-3 and Steel factor and is constitutively increased by p210$^{BCR/ABL}$. *EMBO J.* **14**, 257–265.

Matsui, Y., Zsebo, K. M., and Hogan, B. L. (1990). Embryonic expression of a haematopoietic growth factor encoded by the Sl locus and the ligand for c-kit. *Nature* **347**, 667–669.

Mattei, S., Colombo, M. P., Melani, C., Silvani, A., Parmiani, G., and Herlyn, M. (1994). Expression of cytokine/growth factors and their receptors in human melanoma and melanocytes. *Int. J. Cancer* **56**, 853–857.

Melloti, P., and Calabretta, B. (1996). The transcription factors c-*myb* and GATA-2 act independently in the regulation of normal hematopoiesis. *Proc. Natl Acad. Sci. USA* **93**, 5313–5318.

Miller, B. A., Floros, J., Cheung, J. Y., Wojchowski, D. M., Bell, L., Begley, C. G., Elwood, N. J., Kreider, J., and Christian, C. (1994). Steel factor affects SCL expression during normal erythroid differentiation. *Blood* **84**, 2971–2976.

Miyazawa, K., Hendrie, P. C., Mantel, C., Wood, K., Ashman, L. K., and Broxmeyer, H. E. (1991). Comparative analysis of signaling pathways between mast cell growth factor (*c-kit* ligand) and granulocyte-macrophage colony-stimulating factor in a human factor-dependent myeloid cell line involves phosphorylation of Raf-1, GTPase-activating protein and mitogen-activated protein kinase. *Exp. Hematol.* **19**, 1110–1123.

Miyazawa, K., Toyama, K., Gotoh, A., Hendrie, P. C., Mantel, C., and Broxmeyer, H. E. (1994). Ligand-dependent polyubiquitination of c-*kit* gene product: a possible mechanism of receptor down modulation in M07e cells. *Blood* **83**, 137–145.

Miyazawa, K., Williams, D. A., Gotoh, A., Nishimaki, J., Broxmeyer, H. E., and Toyama, K. (1995). Membrane-bound steel factor induces more persistent tyrosine kinase activation and longer life span of c-*kit* gene-encoded protein than its soluble form. *Blood* **85**, 641–649.

Montone, K. T., van Belle, P., Elenitsas, R., and Elder, D. E. (1997). Proto-oncogene c-*kit* expression in malignant melanoma: protein loss with tumor progression. *Mod. Pathol.* **10**, 939–944.

Morii, E., Hirota, S., Kim, H.-M., Mikoshiba, K., Nishimune, Y., Kitamura, Y., and Nomura, S. (1992). Spatial expression of genes encoding c-*kit* receptors and their ligands in mouse cerebellum as revealed by *in situ* hybridization. *Dev. Brain Res.* **65**, 123–126.

Morrison-Graham, K., and Takahashi, Y. (1993). Steel factor and c-Kit receptor: from mutants to a growth factor system. *BioEssays* **15**, 77–83.

Moskaluk, C. A., Tian, Q., Marshall, C. R., Rumpel, C. A., Franquemont, D. W., and Frierson Jr., H. F. (1999). Mutations of c-*kit* JM domain are found in a minority of human gastrointestinal stromal tumors. *Oncogene* **18**, 1897–1902.

Motro, B., van der Kooy, D., Rossant, J., Reith, A., and Bernstein, A. (1991). Contiguous patterns of c-*kit* and *steel* expression: analysis of mutations at the *W* and *Sl* loci. *Development* **113**, 1207–1221.

Motro, B., Wojtowicz, J. M., Bernstein, A., and van der Kooy, D. (1996). Steel mutant mice are deficient in hippocampal learning but not long-term potentiation. *Proc. Natl Acad. Sci. USA* **93**, 1808–1813.

Muszynski, K. W., Ruscetti, F. W., Heidecker, G., Rapp, U., Troppmair, J., Gooya, J. M., and Keller, J. R. (1995). Raf-1 protein is required for growth factor-induced proliferation of hematopoietic cells. *J. Exp. Med.* **181**, 2189–2199.

Nagata, H., Worobec, A. S., Oh, C. K., Chowdhury, B. A., Tannenbaum, S., Suzuki, Y., and Metcalfe, D. D. (1995). Identification of a point mutation in the catalytic domain of the protooncogene c-*kit* in peripheral blood mononuclear cells of patients who have mastocytosis with an associated hematologic disorder. *Proc. Natl Acad. Sci. USA* **92**, 10560–10564.

Natali, P. G., Nicotra, M. R., Sures, I., Santoro, E., Bigotti, A., and Ullrich, A. (1992). Expression of c-*kit* receptor in normal and transformed human nonlymphoid tissues. *Cancer Res.* **52**, 6139–6143.

Natali, P. G., Berlingieri, M. T., Nicotra, M. R., Fusco, A., Santoro, E., Bigotti, A., and Vecchio, G. (1995). Transformation of thyroid epithelium is associated with loss of c-*kit* receptor. *Cancer Res.* **55**, 1787–1791.

Nilsson, G., Miettinen, U., Ishizaka, T., Ashman, L. K., Irani, A. M., and Schwartz, L. B. (1994). Interleukin-4 inhibits the expression of Kit and tryptase during stem cell factor-dependent development of human mast cells from fetal liver cells. *Blood* **84**, 1519–1527.

Nishikawa, S., Kusakabe, M., Yoshinaga, K., Ogawa, M., Hayashi, S.-I., Kunisada, T., Era, T., Sakakura, T., and Nishikawa, S.-I. (1991). *In utero* manipulation of coat color formation by a monoclonal anti-*c-kit* antibody: two distinct waves of *c-kit*-dependency during melanocyte development. *EMBO J.* **10**, 2111–2118.

Nocka, K., Tan, J. C., Chiu, E., Chu, T. Y., Ray, P., Traktman, P., and Besmer, P. (1990). Molecular bases of dominant negative and loss of function mutations at the murine c-*kit*/white spotting locus: W^{37}, W^{V}, W^{41} and *W*. *EMBO J.* **9**, 1805–1813.

O'Farrell, A.-M., Ichihara, M., Mui, A.L.-F., and Miyajima, A. (1996). Signaling pathways activated in a unique mast cell line where interleukin-3 supports survival and stem cell factor is required for a proliferative response. *Blood* **87**, 3655–3665.

Ogawa, M., Matsuzaki, Y., Nishikawa, S., Hayashi, S.-I., Kunisada, T., Sudo, T., Kina, T., Nakuchi, H., and Nishikawa, S.-I. (1991). Expression and function of c-*kit* in hemopoietic progenitor cells. *J. Exp. Med.* **174**, 63–71.

Ohashi, A., Funasaka, Y., Ueda, M., and Ichihashi, M. (1996). c-KIT receptor expression in cutaneous malignant melanoma and benign melanocytic naevi. *Melanoma Res.* **6**, 25–30.

Okuda, K., Sanghera, J. S., Pelech, S. L., Kanakura, Y., Hallek, M., Griffin, J. D., and Druker, B. J. (1992). Granulocyte-macrophage colony-stimulating factor, interleukin-3, and Steel factor induce rapid tyrosine phosphorylation of p42 and p44 MAP kinase. *Blood* **79**, 2880–2887.

Okura, M., Maeda, H., Nishikawa, S., and Mizoguchi, M. (1995). Effects of monoclonal anti-c-kit antibody (ACK2) on melanocytes in newborn mice. *J. Invest. Dermatol.* **105**, 322–328.

Orr-Urtreger, A., Avivi, A., Zimmer, Y., Givol, D., Yarden, Y., and Lonai, P. (1990). Developmental expression of c-*kit*, a proto-oncogene encoded by the *W* locus. *Development* **109**, 911–923.

Park, G. H., Plummer III, H. K., and Krystal, G. W. (1998). Selective Sp1 binding is critical for maximal activity of the human c-*kit* promoter. *Blood* **92**, 4138–4149.

Paulson, R. F., Vesely, S., Siminovitch, K. A., and Bernstein, A. (1996). Signalling by the *W/Kit* receptor tyrosine kinase is negatively regulated *in vivo* by the protein tyrosine phosphatase Shp1. *Nature Genet.* **13**, 309–315.

Pearson, M. A., O'Farrell, A.-M., Dexter, T. M., Whetton, A. D., Owen-Lynch, P. J., and Heyworth, C. M. (1998). Investigation of the molecular mechanisms underlying growth factor synergy: the role of ERK 2 activation in synergy. *Growth Factors* **15**, 293–306.

Philo, J. S., Wen, J., Wypych, J., Schwartz, M. G., Mendiaz, E. A., and Langley, K. E. (1996). Human stem cell factor dimer forms a complex with two molecules of the extracellular domain of its receptor, kit. *J. Biol. Chem.* **271**, 6895–6902.

Piao, X., and Bernstein, A. (1996). A point mutation in the catalytic domain of c-*kit* induces growth factor independence, tumorigenicity, and differentiation of mast cells. *Blood* **87**, 3117–3123.

Piao, X., Paulson, R., van der Geer, P., Pawson, T., and Bernstein, A. (1996). Oncogenic mutation in the Kit receptor tyrosine kinase alters substrate specificity and induces degradation of the protein tyrosine phosphatase SHP-1. *Proc. Natl Acad. Sci. USA* **93**, 14665–14669.

Pignon, J.-M. (1997). C-kit mutations and mast cell disorders: a model of activating mutations of growth factor receptors. *Hematol. Cell Ther.* **39**, 114–116.

Pignon, J. M., Giraudier, S., Duquesnoy, P., Jouault, H., Imbert, M., Vainchenker, W., Vernant, J. P., and Tulliez, M. (1997). A new c-kit mutation in a case of aggressive mast cell disease. *Br. J. Haematol.* **96**, 374–376.

Plummer III, H., Catlett, J., Leftwich, J., Armstrong, B., Carlson, P., Huff, T., and Krystal, G. (1993). c-*myc* expression correlates with suppression of c-*kit* protooncogene expression in small cell lung cancer cell lines. *Cancer Res.* **53**, 4337–4342.

Price, D. J., Rivnay, B., Fu, Y., Jiang, S., Avraham, S., and Avraham, H. (1997). Direct association of Csk homologous kinase (CHK) with the diphosphorylated site $Tyr^{568/570}$ of the activated c-KIT. *J. Biol. Chem.* **272**, 5915–5920.

Price, E. R., Ding, H.-F., Badalian, T., Bhattachhary, S., Takemoto, C., Yao, T.-P., Hemesath, T. J., and Fisher, D. E. (1998). Lineage-specific signaling in melanocytes. *J. Biol. Chem.* **273**, 17983–17986.

Qiu, F., Ray, P., Brown, K., Barker, P. E., Jhanwar, S., Ruddle, F. H., and Besmer, P. (1988). Primary structure of c-*kit*: relationship with the CSF-1/PDGF receptor kinase family – oncogenic activation of v-*kit* involves deletion of extracellular domain and C terminus. *EMBO J.* **7**, 1003–1011.

Ratajczak, M. Z., Luger, S. M., DeRiel, K., Abrahm, J., Calabretta, B., and Gewirtz, A. M. (1992). Role of the *KIT* protooncogene in normal and malignant human hematopoiesis. *Proc. Natl Acad. Sci. USA* **89**, 1710–1714.

Ratajczak, M. Z., Perrotti, D., Melotti, P., Powzaniuk, M., Calabretta, B., Onodera, K., Kregenow, D. A., Machalinski, B., and Gewirtz, A. M. (1998). *Myb* and *Ets* proteins are candidate regulators of c-*kit* expression in human hematopoietic cells. *Blood* **91**, 1934–1946.

Reddy, G. P., and Quesenberry, P. J. (1996). Stem cell factor enhances interleukin-3 dependent induction of 68-kD calmodulin-binding protein and thymidine kinase activity in NFS-60 cells. *Blood* **87**, 3195–3202.

Reith, A. D., Ellis, C., Lyman, S. D., Anderson, D. M., Williams, D. E., Bernstein, A., and Pawson, T. (1991). Signal transduction by normal isoforms and *W* mutant variants of the kit receptor tyrosine kinase. *EMBO J.* **10**, 2451–2459.

Ridings, J., Macardle, P. J., Byard, R. W., Skinner, J., and Zola, H. (1995). Cytokine receptor expression by solid tumours. *Ther. Immunol.* **2**, 67–76.

Rossi, P., Marziali, G., Albanesi, C., Charlesworth, A., Geremia, R., and Sorrentino, V. (1992). A novel *c-kit* transcript, potentially encoding a truncated receptor, originates within a *kit* gene intron in mouse spermatids. *Dev. Biol.* **152**, 203–207.

Rottapel, R., Reedijk, M., Williams, D. E., Lyman, S. D., Anderson, D. M., Pawson, T., and Bernstein, A. (1991). The *Steel*/*W* transduction pathway: kit autophosphorylation and its association with a unique subset of cytoplasmic signaling proteins is induced by the steel factor. *Mol. Cell. Biol.* **11**, 3043–3051.

Rukstalis, D. B. (1996). Molecular mechanisms of testicular carcinogenesis. *World J. Urol.* **14**, 347–352.

Russell, E. S. (1979). Hereditary anemias of the mouse: a review for geneticists. *Adv. Genet.* **20**, 357–459.

Rusten, L. S., Smeland, E. B., Jacobsen, F. W., Lien, E., Lesslauer, W., Loetscher, H., Dubois, C. M., and Jacobsen, S. E. (1994). Tumor necrosis factor-alpha inhibits stem cell factor-induced proliferation of human bone marrow progenitor cells *in vitro*. Role of p55 and p75 tumor necrosis factor receptors. *J. Clin. Invest.* **94**, 165–172.

Ryan, J. J., Huang, H., McReynolds, L. J., Shelburne, C., Hu-Li, J., Huff, T. F., and Paul, W. E. (1997). Stem cell factor activates STAT-5 DNA binding in IL-3-derived bone marrow mast cells. *Exp. Hematol.* **25**, 357–362.

Rygaard, K., Nakamura, T., and Spang-Thomsen, M. (1993). Expression of the proto-oncogenes c-*met* and c-*kit* and their ligands, hepatocyte growth factor/scatter factor and stem cell factor in SCLC cell lines and xenografts. *Br. J. Cancer* **67**, 137–146.

Saito, S., Enomoto, M., Sakakura, S., Ishii, Y., Sudo, T., and Ichijo, M. (1994). Localization of stem cell factor (SCF) and c-kit mRNA in human placental tissue and biological effects of

SCF on DNA synthesis in primary cultured cytotrophoblasts. *Biochem. Biophys. Res. Commun.* **205**, 1762–1769.

Sasaki, E., Okamura, H., Chikamune T., Kanai, Y., Watanabe, M., Naito, M., and Sakurai, M. (1993). Cloning and expression of the chicken c-kit proto-oncogene. *Gene* **128**, 257–261.

Sattler, M., and Salgia, R. (1998). Role of the adapter protein CRKL in signal transduction of normal hematopoietic and BCR/ABL-transformed cells. *Leukemia* **12**, 637–644.

Sattler, M., Salgia, R., Shrikhande, G., Verma, S., Pisick, E., Prasad, K. V. S., and Griffin, J. D. (1997). Steel factor induces tyrosine phosphorylation of CRKL and binding of CRKL to a complex containing c-Kit, phosphatidylinositol 3-kinase, and p120^{CBL}. *J. Biol. Chem.* **272**, 10248–10253.

Schmidt, L., Duh, F.-M., Chen, F., Kishida, T., Glenn, G., Choyke, P., Scherer, S. W., Zhuang, Z., Lubensky, I., Dean, M., Allikmets, R., Chidambaram, A., Bergerheim, U. R., Feltis, J. T., Casadevall, C., Zamarron, A., Bernues, M., Richard, S., Lips, C. J. M., Walther, M. M., Tsui, L.-C., Geil, L., Orcutt, M. L., Stackhouse, T., Lipan, J., Slife, L., Brauch, H., Decker, J., Niehans, G., Hughson, M. D., Moch, H., Storkel, S., Lerman, M. I., Linehan, W. M., and Zbar, B. (1997). Germline and somatic mutations in the tyrosine kinase domain of the *MET* proto-oncogene in papillary renal carcinomas. *Nature Genet.* **16**, 68–73.

Schwarzenberger, P., Spence, S. E., Gooya, J. M., Michiel, D., Curiel, D. T., Ruscetti, F. W., and Keller, J. R. (1996). Targeted gene transfer to human hematopoietic progenitor cell lines through the c-kit receptor. *Blood* **87**, 472–478.

Sekido, Y., Obata, Y., Ueda, R., Hida, T., Suyama, M., Shimokata, K., Ariyoshi, Y., and Takahashi, T. (1991). Preferential expression of c-*kit* protooncogene transcripts in small cell lung cancer. *Cancer Res.* **51**, 2416–2419.

Serve, H., Yee, N. S., Stella, G., Sepp-Lorenzino, L., Tan, J. C., and Besmer, P. (1995). Differential roles of PI3-kinase and Kit tyrosine 821 in Kit receptor-mediated proliferation, survival and cell adhesion in mast cells. *EMBO J.* **14**, 473–483.

Sette, C., Bevilacqua, A., Bianchini, A., Mangia, F., Geremia, R., and Rossi, P. (1997). Parthenogenetic activation of mouse eggs by microinjection of a truncated c-kit tyrosine kinase present in spermatozoa. *Development* **124**, 2267–2274.

Sette, C., Bevilacqua, A., Geremia, R., and Rossi, P. (1998). Involvement of phospholipase Cγ1 in mouse egg activation induced by a truncated form of the c-kit tyrosine kinase present in spermatozoa. *J. Cell Biol.* **142**, 1063–1074.

Shepherd, P. R., Withers, D. J., and Siddle, K. (1998). Phosphoinositide 3-kinase: the key switch mechanism in insulin signalling. *Biochem. J.* **333**, 471–490.

Shimizu, Y., Ashman, L. K., Du, Z., and Schwartz, L. B. (1996). Internalization of kit together with stem cell factor on human fetal liver-derived mast cells. *J. Immunol.* **156**, 3443–3449.

Sillaber, C., Strobl, H., Bevec, D., Ashman, L. K., Butterfield, J. H., Lechner, K., Maurer, D., Bettelheim, P., and Valent, P. (1991). IL-4 regulates c-kit proto-oncogene product expression in human mast and myeloid progenitor cells. *J. Immunol.* **147**, 4224–4228.

Sperling, C., Schwartz, S., Büchner, T., Thiel, E., Ludwig, W.-D. (1997). Expression of the stem cell factor receptor c-kit (CD117) in acute leukemias. *Haematologica* **82**, 617–621.

Spritz, R. A. (1992). Lack of apparent hematologic abnormalities in human patients with c-*kit* (stem cell factor receptor) gene mutations. *Blood* **79**, 2497–2499.

Spritz, R. A. (1994). Molecular basis of human piebaldism. *J. Invest. Derm.* **103**, 1375–1405.

Spritz, R. A. (1997). Piebaldism, Waardenburg syndrome, and related disorders of melanocyte development. *Semin. Cutaneous Med. Surg.* **16**, 15–23.

Sui, X., Krantz, S. B., You, M., and Zhao, Z. (1998). Synergistic activation of MAP kinase (ERK1/2) by erythropoietin and stem cell factor is essential for expanded erythropoiesis. *Blood* **92**, 1142–1149.

Takaoka, A., Toyota, M., Hinoda, Y., Itoh, F., Mita, H., Kakiuchi, H., Adachi, M., and Imai, K. (1997). Expression and identification of aberrant c-*kit* transcripts in human cancer cells. *Cancer Lett.* **115**, 257–261.

Tanaka, S., Yanagisawa, N., Tojo, H., Kim, Y. J., Tsujimura, T., Kitamura, Y., Sawasaki, T., and Tachi, C. (1997). Molecular cloning of cDNA encoding the c-kit receptor of Shiba goats and a novel alanine insertion specific to goats and sheep in the kinase insert region. *Biochim. Biophys. Acta* **1352**, 151–155.

Tang, B., Mano, H., Yi, T., and Ihle, J. N. (1994). Tec kinase associates with c-kit and is tyrosine phosphorylated and activated following stem cell factor binding. *Mol. Cell. Biol.* **14**, 8432–8437.

Taniguchi, S., Dai, C. H., Price, J. O., and Krantz, S. B. (1997). Interferon gamma downregulates stem cell factor and erythropoietin receptors but not insulin-like growth factor-I receptors in human erythroid colony-forming cells. *Blood* **90**, 2244–2252.

Tauchi, T., Boswell, H. S., Leibowitz, D., and Broxmeyer, H. E. (1994a). Coupling between p210bcr-abl and Shc and Grb2 adaptor proteins in hematopoietic cells permits growth factor receptor-independent link to Ras activation pathway. *J. Exp. Med.* **179**, 167–175.

Tauchi, T., Feng, G.-S., Marshall, M. S., Shen, R., Mantel, C., Pawson, T., and Broxmeyer, H. E. (1994b). The ubiquitously expressed Syp phosphatase interacts with c-*kit* and Grb2 in hematopoietic cells. *J. Biol. Chem.* **269**, 25206–25211.

Timokhina, I., Kissel, H., Stella, G., and Besmer, P. (1998). Kit signaling through PI 3-kinase and Src kinase pathways: an essential role for Rac1 and JNK activation in mast cell proliferation. *EMBO J.* **17**, 6250–6262.

Torihashi, S., Ward, S. M., Nishikawa, S.-I., Nishi, K., Kobayashi, S., and Sanders, K. M. (1995). C-*kit*-dependent development of interstitial cells and electrical activity in the murine gastrointestinal tract. *Cell Tissue Res.* **280**, 97–111.

Toyota, M., Hinoda, Y., Itoh, F., Takaoka, A., Imai, K., and Yachi, A. (1994). Complementary DNA cloning and characterization of truncated form of c-*kit* in human colon carcinoma cells. *Cancer Res.* **54**, 273–275.

Tsai, M., Chen, R.-H., Tam, S.-Y., Blenis, J., and Galli, S. J. (1993). Activation of MAP kinases, pp90^{rsk} and pp70-S6 kinases in mouse mast cells by signaling through the c-*kit* receptor tyrosine kinase or FcRI: rapamycin inhibits activation of pp70-S6 kinase and proliferation in mouse mast cells. *Eur. J. Immunol.* **23**, 3286–3291.

Tsujimura, T. (1996). Role of c-*kit* receptor tyrosine kinase in the development, survival and neoplastic transformation of mast cells. *Pathol. Int.* **46**, 933–938.

Tsujimura, T., Hirota, S., Nomura, S., Niwa, Y., Yamazaki, M., Tono, T., Morii, E., Kim, H. M., Kondo, K., and Nishimune, Y. (1991). Characterization of Ws mutant allele of rats: a 12-base deletion in tyrosine kinase domain of c-kit gene. *Blood* **78**, 1942–1946.

Tsujimura, T., Furitsu, T., Morimoto, M., Isozaki, K., Nomura, S., Matsuzawa, Y., Kitamura, Y., and Kanakura, Y. (1994). Ligand-independent activation of c-*kit* receptor tyrosine kinase in a murine mastocytoma cell line P-815 generated by a point mutation. *Blood* **83**, 2619–2626.

Tsujimura, T., Morii, E., Nozaki, M., Hashimoto, K., Moriyama, Y., Takebayashi, K., Kondo, T., Kanakura, Y., and Kitamura, Y. (1996a). Involvement of transcription factor encoded by the *mi* locus in the expression of c-*kit* receptor tyrosine kinase in cultured mast cells of mice. *Blood* **88**, 1225–1233.

Tsujimura, T., Morimoto, M., Hashimoto, K., Moriyama, Y., Kitayama, H., Matsuzawa, Y., Kitamura, Y., and Kanakura, Y. (1996b). Constitutive activation of c-*kit* in FMA3 murine mastocytoma cells caused by deletion of seven amino acids at the juxtamembrane domain. *Blood* **87**, 273–283.

Tsujimura, T., Kanakura, Y., and Kitamura, Y. (1997). Mechanisms of constitutive activation of c-*kit* receptor tyrosine kinase. *Leukemia* **11**, 396–398.

Tsujimura, T., Hashimoto, K., Kitayama, H., Ikeda, H., Sugahara, H., Matsumura, I., Kaisho, T., Terada, N., Kitamura, Y., and Kanakura, Y. (1999). Activating mutation in the catalytic domain of c-*kit* elicits hematopoietic transformation by receptor self-association not at the ligand-induced dimerization site. *Blood* **93**, 1319–1329.

Turner, A. M., Zsebo, K. M., Martin, F., Jacobsen, F. W., Bennett, L. G., and Broudy, V. C. (1992). Nonhematopoietic tumor cell lines express stem cell factor and display c-*kit* receptors. *Blood* **80**, 374–381.

Turner, A. M., Bennett, L. G., Lin, N. L., Wypych, J., Bartley, T. D., Hunt, R. W., Atkins, H. L., Langley, K. E., Parker, V., Martin, F., and Broudy, V. C. (1995). Identification and characterization of a soluble c-*kit* receptor produced by human hematopoietic cell lines. *Blood* **85**, 2052–2058.

Vandenbark, G. R., deCastro, C. M., Taylor, H., Dew-Knight, S., and Kaufman, R. E. (1992). Cloning and structural analysis of the human c-*kit* gene. *Oncogene* **7**, 1259–1266.

Vliagoftis, H., Worobec, A. S., and Metcalfe, D. D. (1997). Updates on cells and cytokines: the protooncogene c-*kit* and c-*kit* ligand in human disease. *J. Allergy Clin. Immunol.* **100**, 435–440.

Vosseller, K., Stella, G., Yee, N. S., and Besmer, P. (1997). c-Kit receptor signaling through its phosphatidylinositide-3′-kinase-binding site and protein kinase C: role in mast cell enhancement of degranulation, adhesion, and membrane ruffling. *Mol. Biol. Cell.* **8**, 909–922.

Wehrle-Haller, B., and Weston, J. A. (1995). Soluble and cell-bound forms of steel factor activity play distinct roles in melanocyte precursor dispersal and survival on the lateral neural crest migration pathway. *Development* **121**, 731 742.

Weiler, S. R., Mou, S., DeBerry, C. S., Keller, J. R., Ruscetti, F. W., Ferris, D. K., Longo, D. L., and Linnekin, D. (1996). JAK2 is associated with the c-kit proto-oncogene product and is phosphorylated in response to stem cell factor. *Blood* **87**, 3688–3693.

Weiler, S. R., Gooya, J. M., Ortiz, M., Tsai, S., Collins, S. J., and Keller, J. R. (1999). D3: a gene induced during myeloid cell differentiation of Lin^{lo} $c\text{-}Kit^{+}$ $Sca\text{-}1^{+}$ progenitor cells. *Blood* **93**, 527–536.

Welham, M. J., and Schrader, J. W. (1991). Modulation of c-*kit* mRNA and protein by hemopoietic growth factors. *Mol. Cell. Biol.* **11**, 2901–2904.

Wessely, O., Mellitzer, G., von Lindern, M., Levitzki, A., Gazit, A., Ischenko, I., Hayman, M. J., and Beug, H. (1997). Distinct roles of the receptor tyrosine kinases c-ErbB and c-Kit in regulating the balance between erythroid cell proliferation and differentiation. *Cell Growth Differ.* **8**, 481–493.

Wisniewski, D., Strife, A., Berman, E., and Clarkson, B. (1996). c-*kit* ligand stimulates tyrosine phosphorylation of a similar pattern of phosphotyrosyl proteins in primary primitive normal hematopoietic progenitors that are constitutively phosphorylated in comparable primitive progenitors in chronic phase chronic myelogenous leukemia. *Leukemia* **10**, 229–237.

Worobec, A. S., Semere, T., Nagata, H., and Metcalfe, D. D. (1998). Clinical correlates of the presence of the Asp816Val c-*kit* mutation in the peripheral blood mononuclear cells of patients with mastocytosis. *Cancer* **83**, 2120–2129.

Wypych, J., Bennett, L. G., Schwartz, M. G., Clogston, C. L., Lu, H. S., Broudy, V. C., Bartley, T. D., Parker, V. P., and Langley, K. E. (1995). Soluble kit receptor in human serum. *Blood* **85**, 66–73.

Yajima, T., Kanda, T., Yoshiike, K., and Kitamura, Y. (1998). Retroviral vector targeting human cells via c-Kit-stem cell factor interaction. *Hum. Gene Ther.* **9**, 779–787.

Yamamoto, K., Tojo, A., Aoki, N., and Shibuya, M. (1993). Characterization of the promoter region of the human c-kit proto-oncogene. *Jpn J. Cancer Res.* **84**, 1136–1144.

Yamanashi, Y., and Baltimore, D. (1997). Identification of the Abl- and rasGAP-associated 62 kDa protein as a docking protein, Dok. *Cell* **88**, 205–211.

Yamataka, A., Yamataka, T., Lane, G. J., Kobayashi, H., Sueyoshi, N., and Miyano, T. (1998). Necrotizing enterocolitis and C-KIT. *J. Pediatr. Surg.* **33**, 1682–1685.

Yarden, Y., and Ullrich, A. (1988). Molecular analysis of signal transduction by growth factors. *Biochemistry* **27**, 3113–3119.

Yarden, Y., Kuang, W.-J., Yang-Feng, T., Coussens, L., Munemitsu, S., Dull, T. J., Chen, E., Schlessinger, J., Francke, U., and Ullrich, A. (1987). Human proto-oncogene c-*kit*: a new cell surface receptor tyrosine kinase for an unidentified ligand. *EMBO J.* **6**, 3341–3351.

Yasuda, H., Galli, S. J., and Geissler, E. N. (1993). Cloning and functional analysis of the mouse c-kit promoter. *Biochem. Biophys. Res. Commun.* **191**, 893–901.

Yee, N. S., Langen, H., and Besmer, P. (1993). Mechanism of kit ligand, phorbol ester, and calcium-induced down-regulation of c-*kit* receptors in mast cells. *J. Biol. Chem.* **268**, 14289–14301.

Yee, N. S., Hsiau, C.-W. M., Serve, H., Vosseller, K., and Besmer, P. (1994a). Mechanism of down-regulation of c-*kit* receptor. *J. Biol. Chem.* **269**, 31991–31998.

Yee, N. S., Paek, I., and Besmer, P. (1994b). Role of *kit*-ligand in proliferation and suppression of apoptosis in mast cells: basis for radiosensitivity of *White Spotting* and *Steel* mutant mice. *J. Exp. Med.* **179**, 1777–1787.

Yip-Schneider, M. T., Horie, M., and Broxmeyer, H. E. (1995). Transcriptional induction of *pim*-1 protein kinase gene expression by interferon γ and posttranscriptional effects on costimulation with steel factor. *Blood* **85**, 3494–3502.

Yoshida, H., Takakura, N., Kataoka, H., Kunisada, T., Okamura, H., and Nishikawa, S.-I. (1997). Stepwise requirement of c-kit tyrosine kinase in mouse ovarian follicle development. *Dev. Biol.* **184**, 122–137.

Zhang, Z., and Anthony, R. V. (1994). Porcine stem cell factor/c-kit ligand: its molecular cloning and localization within the uterus. *Biol. Reprod.* **50**, 95–102.

Zhang, S.-C., and Fedoroff, S. (1997). Cellular localization of stem cell factor and c-kit receptor in the mouse nervous system. *J. Neurosci. Res.* **47**, 1–15.

Zhang, B. Y.-Y., Vik, T. A., Ryder, J. W., Srour, E. F., Jacks, T., Shannon, K., and Clapp, D. W. (1998). *Nf1* regulates hematopoietic progenitor cell growth and Ras signaling in response to multiple cytokines. *J. Exp. Med.* **187**, 1893–1902.

Zsebo, K. M., Williams, D. A., Geissler, E. N., Broudy, V. C., Martin, F. H., Atkins, H. L., Hsu, R.-Y., Birkett, N. C., Okino, K. H., Murdock, D. C., Jacobsen, F. W.,

Langley, K. E., Smith, K. A., Takeishi, T., Cattanach, B. M., Galli, S. J., and Suggs, S. V. (1990). Stem cell factor is encoded at the *Sl* locus of the mouse and is the ligand for the c-*kit* tyrosine kinase receptor. *Cell* **63**, 213–224.

LICENSED PRODUCTS

A variety of companies sell antisera specific for either human or murine SCFR. These products, the sources and some known uses are summarized in **Table 4**.

A recombinant form of the SCFR extracellular domain is sold by R&D Systems. This product has been used to antagonize SCF-mediated proliferative responses.

An inhibitor of the SCFR kinase activity, AG1296, is sold by Calbiochem.

ACKNOWLEDGEMENTS

The authors would like to thank Dr Douglas Lowy, Dr RuJu Chian, Sherry Mou, and Bridget O'Laughlin for critical review of this manuscript.

Table 4 Commercially available antisera specific for the SCFR

Species specificity	Monoclonal/ polyclonal	Epitope	Designation	Vendor	Use
Mouse	M	Binding site	ACK2	Gibco-BRL-LTI	FC, IB, N
Mouse	M	ECD	2B8	Pharmingen	FC
Mouse	M	ECD	ACK45	Pharmingen	FC, N
Mouse	P	C-tail	M-14	Santa Cruz	IP, IB
Mouse/human	P	C-tail	–	UBI	IP, IB
Mouse/human	P	C-tail	C-19	Santa Cruz	IP, IB
Human	M	Binding site	3D6	Boehringer-Mannheim	IP, IB, FC, N
Human	M	Binding site	K44.2	Sigma	IP, FC
Human	M	ECD	YB5.B8	Pharmingen	FC
Human	M	ECD	95C3	Coulter-Immunotech	FC
Human	M	ECD	17F11	Coulter-Immunotech	FC
Human	M	ECD	104D2D1	oulter-Immunotech	FC
Human	P	ECD	–	R&D	IB
Human	P	C-tail	C-14	Santa Cruz	IP, IB
Human	P	C-tail	–	Oncogene	IF

M, monoclonal; P, polyclonal; ECD, extracellular domain; C-tail, carboxyl tail; IB, immunoblotting; IP, immunoprecipitation; FC, flow cytometry; N, neutralizing.

GM-CSF Receptor

Nicos A. Nicola*

Division of Cancer and Haematology, The Walter and Eliza Hall Institute of Medical Research, P.O. Royal Melbourne Hospital, Parkville, Victoria 3050, Australia

*corresponding author tel: 61-3-9345-2526, fax: 61-3-9345-2616, e-mail: nicola@wehi.edu.au

DOI: 10.1006/rwcy.2000.20004.

SUMMARY

The GM-CSF receptor is expressed primarily on myeloid cells and consists of two separate subunits, both of which belong to the type I cytokine receptor family. The α subunit binds GM-CSF with high specificity but low affinity while the β chain is shared with the IL-3 and IL-5 receptors and converts each of these low-affinity interactions into one of high affinity. The α chain has only a short cytoplasmic tail which is nevertheless essential for intracellular signaling but the longer cytoplasmic tail of the β chain appears to transmit most of the biological signals. Known signaling pathways include activation of the JAK2/STAT5 and MAP kinase pathways and induction of cell survival, proliferation, and functional activation.

BACKGROUND

Discovery

The GM-CSF receptor was first defined using binding of radioactive GM-CSF to mouse and human bone marrow cells and cell lines (Park *et al.*, 1986a, 1986b; Walker and Burgess, 1985). These studies revealed receptor heterogeneity by demonstrating the existence of both high- and low-affinity binding sites. A specific GM-CSF-binding receptor was cloned from a human placental cDNA expression library in 1989 (Gearing *et al.*, 1989) and shown to constitute only the low-affinity binding site. It has subsequently been termed the GM-CSFRα chain. Meanwhile, cloning of the single human equivalent of mouse IL-3 receptor subunits defined by inhibitory antibodies (AIC2A and AIC2B) revealed that the human molecule (KH97) did not bind GM-CSF by itself but could generate high-affinity binding in the presence of GM-CSFRα (Hayashida *et al.*, 1990). This chain, now termed the common β chain, was shown to be a shared affinity-converting chain for GM-CSF, IL-3 and IL-5 receptors each of which contains a unique ligand-specific α chain (Kitamura *et al.*, 1991; Tavernier *et al.*, 1991).

Alternative names

The GM-CSF receptor α chain is also called CSF-2RA and, in the CD nomenclature, CDw116. The common β chain of this receptor was also known as AIC2B in the mouse and as KH97 in human.

Structure

The high-affinity GM-CSF receptor consists of a ligand-specific, low-affinity α chain and a shared high-affinity converting common β chain. Although both chains are required for intracellular signaling, most signaling molecules have been mapped to the β chain. Both receptor chains belong to the hematopoietin or cytokine type I receptor family defined by the presence of an extracellular 200 amino acid motif called the hematopoietin domain or cytokine receptor homology domain. This domain consists of two halves, each of which is similar to a fibronectin type III repeat and contains seven β strands. The stoichiometry of the complex is still unknown but may consist of two each of the α and β chains.

Main activities and pathophysiological roles

These are as shown for GM-CSF and, for the common β chain, also as shown for IL-3 and IL-5. In one patient with *pulmonary alveolar proteinosis*, a point mutation in the common β chain gene (C to A at nucleotide position 1135 leads to Pro602 to Thr alteration) results in reduced expression and function of the protein (Dirksen *et al.*, 1997) and this phenotype can also be created in mice with loss of the common β gene (Nishinakamura *et al.*, 1995; Robb *et al.*, 1995).

GENE

The GM-CSF receptor α chain gene (*CSF2RA*) occurs in the pseudoautosomal region of the X and Y chromosomes at Xp22.32 in humans (Gough *et al.*, 1990; Rappold *et al.*, 1992) and on chromosome 19 (51.00 cM) in the mouse (Disteche *et al.*, 1992). The common β chain gene (*CSF2RB*) is on chromosome 22q12.2-13.1 in humans (Shen *et al.*, 1992) and on chromosome 15 in the mouse (Gorman *et al.*, 1992). In humans the GM-CSF and IL-3 receptor α chains are closely linked (within 190 kb) in the pseudoautosomal regions of the X and Y chromosomes (Kremer *et al.*, 1993) and in the mouse the common β chain and the IL-3-specific β chain (*AIC2A*) are closely linked to each other and to c-*sis* (within 250 kb) (Gorman *et al.*, 1992). The human α chain gene consists of 13 exons spanning about 44 kb (Nakagawa *et al.*, 1994) while the mouse β chain gene consists of 14 exons spanning about 28 kb (Gorman *et al.*, 1992). In both cases the intron/exon boundaries are well conserved and suggest a common evolutionary origin for this entire class of receptors.

Accession numbers

GenBank:
Human α chain cDNA (main membrane form): X17648 (Gearing *et al.*, 1989)
Human α chain cDNA (minor membrane form): M64445 (Crosier *et al.*, 1991), L29349 (Hu *et al.*, 1994)
Human soluble forms (by alternate splicing): L29348 (Hu *et al.*, 1994), M73832 (Raines *et al.*, 1991)
Mouse α chain: M85078 (Park *et al.*, 1992)
Human β chain cDNA: M59941, M38275 (Hayashida *et al.*, 1990)
Mouse β chain cDNA: M93429 (Gorman *et al.*, 1992)

PROTEIN

Accession numbers

Human α chain: SwissProt P15509 and other isoforms TrEMBL O00207, Q16564, Q14431
Mouse α chain: TrEMBL Q00941
Human β chain: SwissProt P32927
Mouse β chain: SwissProt P26955

Sequence

See **Figure 1**.

Description of protein

The human α chain is a type I membrane glycoprotein of 400 amino acids (including the leader sequence). The extracellular domain extends from amino acids 23 to 320 and consists of about 100 amino acids preceding a 200 amino acid domain that is conserved in all type I cytokine receptors called the cytokine receptor homology domain (CRHD). This domain contains two conserved disulfide bonds (126–136, 165–178) and the conserved WSXWS sequence (WSSWS in the α chain) and is predicted to have the structure of two seven β-stranded fibronectin III modules separated by a hinge region. There are 11 potential *N*-glycosylation sites (at amino acids 46, 54, 99, 123, 135, 182, 195, 223, 229, 272, and 305) in the extracellular domain and these appear to be necessary for obtaining the correct structure of the receptor for ligand binding (Ding *et al.*, 1995; Shibuya *et al.*, 1991). Glycosylation probably explains the difference between the molecular mass of the native protein (80 kDa) and the predicted mass (44 kDa). The short cytoplasmic domain of 55 amino acids contains the conserved box 1 and box 2 elements characteristic of class I cytokine receptors. Two alternate forms of this transmembrane receptor have been described to arise from alternate splicing. In one the last 25 amino acids are replaced by the sequence EMGPQRHHR-CGWNLYPTPGPSPGSGSSPRLGSESSL (Crosier *et al.*, 1991) while in the other the last 85 amino acids are replaced by the sequence DDHLGGIHP-RGRERLPRRGLDREGNYLRPRGCRNGMDIS-ASATRGNCFLDDAVNLYIIFYVFI (Hu *et al.*, 1994). Alternative splicing is also known to generate several different soluble forms of the human GM-CSF receptor α chain. In one, the last 130 amino acids are replaced by the sequence VVLTTG-TSALCTFMCS (Hu *et al.*, 1994) while in another the last 83 amino acids are replaced by the sequence

Figure 1 Amino acid sequence for human GM-CSF receptor α chain and β chain. Leader signal sequence is underlined. Transmembrane region is bold and underlined.

Human GM-CSF receptor α chain:
MLLLVTSLLL CELPHPAFLL IPEKSDLRTV APASSLNVRF DSRTMNLSWD CQENTTFSKC
FLTDKKNRVV EPRLSNNECS CTFREICLHE GVTFEVHVNT SQRGFQQKLL YPNSGREGTA
AQNFSCFTYN ADLMNCTWAR GPTAPRDVQY FLYIRNSKRR REIRCPYYIQ DSGTHVGCHL
DNLSGLTSRN YFLVNGTSRE IGIQFFDSLL DTKKIERFNP PSNVTVTCNT THCLVRWKQP
RTYQKLSYLD FQYQLDVHRK NTQPGTENLL INVSGDLENR YNFPSSEPRA KHSVKIRAAD
VRILNWSSWS EAIEFGSDDG **NLGSVYIYVL LIVGTLVCGI VLGFL**FKRFL RIQRLFPPVP
QIKDKLNDNH EVEDEIIWEE FTPEEGKGYR EEVLTVKEIT

Human GM-CSF receptor β chain:
MVLAQGLLSM ALLALCWERS LAGAEERIPL QTLRCYNDYT SHIRVRWADT QDAQRLVNVT
LIRRVNEDLL EPVSCDLSDD MPWSACPHPR CVPRRCVIPC QSFVVTDVDY FSFQPDRPLG
TRLTVTLTQH VQPPEPRDLQ ISTDQDHFLL TWSVALGSPQ SHWLSPGDLE FEVVYKRLQD
SWEDAAILLS NTSQATLGPE HLMPSSTYVA RVRTRLAPGS RLSGRPSKWS PEVCWDSQPG
DEAQPQNLEC FFDGAAVLSC SWEVRKEVAS SVSFGLFYKP SPDAGEEECS PVLREGLGSL
HTRHHCQIPV PDPATHGQYI VSVQPRRAEK HIKSSVNIQM APPSLNVTKD GDSYSLTWET
MKMRYEHIDH TFEIQYRKDT ATWKDSKTET LQNAHSMALP ALEPSTRYWA RVRVRTSRTC
YNGIWSEWSE ARSWDTESVL PMW**VLALIVI FLRIAVLLAL** RFCGIYGYRL RRKWEEKIPN
PSKSHLFQNG SAELWPPGSM SAFTSGSPPH QGPWGSRFPE LEGVFPVGFG DSEVSPLTIE
DPKHVCDPPS GPDTTPAASD LPTEQPPSPQ PGPPAASHTP EKQASSFDFN GPYLGPPHSR
SLPDILGQPE PPQEGGSQKS PPPGSLEYLC LPAGGQVQLV PLAQAMGPGQ AVEVERRPSQ
GAAGSPSKES GGGPAPPALG PRVGGQDQKD SPVAIPMSSG DTEDPGVASG YVSSADLVFT
PNSGASSVSL VPSLGLPSDQ TPSLCPGLAS GPPGAPGPVK SGFEGYVELP PIEGRSPRSP
RNNPVPPEAK SPVLNPCERP ADVSPTSPQP EGLLVLQQVG DYCFLPGLGP GPLSLRSKPS
SPGPGPEIKN LDQAFQVKKP PGQAVPQVPV IQLFKALKQQ DYLSLPPWEV NKPGEVC

LGYS-GCSRGFHRSKTN (Ashworth and Kraft, 1990; Raines *et al.*, 1991).

The human common β chain (897 amino acids including the leader sequence) is also a type I membrane glycoprotein that belongs to the class I cytokine receptor family. However it contains two copies of the CRHD in the extracellular region (427 amino acids) and also displays a much longer cytoplasmic domain (437 amino acids). It contains three potential *N*-glycosylation sites (positions 58, 191, and 346) but appears to be more lightly glycosylated than the α chain (Shibuya *et al.*, 1991). The predicted molecular mass of the mature protein is 95 kDa and the observed molecular weight of 120 kDa suggests that some glycosylation does occur.

By analogy with the site II binding site on the growth hormone receptor, the binding site on the common β chain (βc) for GM-CSF (especially at Glu21) consists of the predicted B′-C′ and F′-G′ loops of the membrane proximal hematopoietin domain. In the former, His367 as well as Tyr365 and Ile368 are the most critical residues (Lock *et al.*, 1994; Woodcock *et al.*, 1994) while in the latter Tyr421 is critical (Woodcock *et al.*, 1996). In the α chain Arg280 in the predicted F′-G′ loop of the hematopoietin domain also appears to be critical for binding GM-CSF, probably through a charge–charge interaction with Asp112 (Rajotte *et al.*, 1997).

The stoichiometry of the signaling receptor complex is not known although it has been proposed to be GM-CSF$_2\alpha_2\beta_2$. There is some evidence that β homodimers exist prior to ligand binding but become tyrosine phosphorylated only after association with the complex of GM-CSF with the receptor α chain (Muto *et al.*, 1996). On the other hand there is also evidence that α/βc heterodimers also pre-exist before ligand association (Woodcock *et al.*, 1997) and that two α chains are present in the receptor complex (Lia *et al.*, 1996). Finally, there is also some evidence that GM-CSF-induced covalent disulfide bond formation between α and β chains (involving Cys86 and Cys91 in the latter) may be necessary for productive signaling (Stomski *et al.*, 1998).

Relevant homologies and species differences

The overall structure of the GM-CSF receptor complex is the same in mice and humans. However, there is no crossreactivity in the binding specificity of the two species. Nevertheless, it is known that the mouse common β chain can form an active signaling complex with the human GM-CSF receptor α chain despite its inability to convert the low-affinity binding reaction to one of high affinity. Consequently, significantly higher concentrations of human GM-CSF are required for signaling in this artificial situation (Metcalf *et al.*, 1990).

Affinity for ligand(s)

The human GM-CSF receptor α chain binds GM-CSF specifically but with a low-affinity equilibrium dissociation constant (K_d) of 1–5 nM at 4°C (Gearing *et al.*, 1989). Binding of GM-CSF to the common β chain is undetectable but the complex of α and β chains binds GM-CSF with high affinity (K_d = 10–100 pM at 4°C) (Hayashida *et al.*, 1990). In some cells intermediate affinity receptors have also been described (K_d = 300–800 pM) but these also appear to be constituted from α and β chains (Wheadon *et al.*, 1997). Essentially identical results were obtained for mouse GM-CSF receptors (Park *et al.*, 1992; Walker and Burgess, 1985).

Cell types and tissues expressing the receptor

In mice and humans GM-CSF receptors have been detected on neutrophils, macrophages and eosinophils and their precursors (DiPersio *et al.*, 1988) as well as myeloid dendritic cells (Yamada *et al.*, 1997). There have been conflicting reports on whether or not functional GM-CSF receptors are expressed on human endothelial cells (Bussolino *et al.*, 1993; Yong *et al.*, 1991). The low-affinity α chain of the GM-CSF receptor has been detected in placental trophoblasts (Gearing *et al.*, 1989; Hampson *et al.*, 1993). GM-CSF receptors have also been detected on a range of myeloid cell lines, as well as hematopoietic and nonhematopoietic tumor cells (Baldwin *et al.*, 1989; Hirsch *et al.*, 1995; Rokhlin *et al.*, 1996; Crosier *et al.*, 1997; Rivas *et al.*, 1998).

Regulation of receptor expression

The α chain gene contains binding sites for the ets-like transcription factor PU1 at nucleotides −53 to −41 and studies *in vitro* and *in vivo* (in PU1 knockout mice) have demonstrated an essential role for this transcription factor in the induction of α chain expression (Hohaus *et al.*, 1995; Anderson *et al.*, 1998). In addition a C/EBPα-binding CCAAT site at nucleotides −70 to −54 is required for both positive and negative transcriptional regulation (Hohaus *et al.*, 1995).

The βc gene contains putative PU1- and GATA-1-binding sites upstream of a conserved transcriptional initiation site but regulation of expression of this gene in mice and humans is poorly understood (Gorman *et al.*, 1992).

Release of soluble receptors

Soluble forms of the α chain arise from alternate splicing but little is known of the regulation of this process or secretion or *in vivo* half-lives.

SIGNAL TRANSDUCTION

Associated or intrinsic kinases

The GM-CSF receptor has no intrinsic tyrosine kinase activity. The best-studied tyrosine kinase associated with GM-CSF receptor signaling is the cytoplasmic kinase JAK2. JAK2 binds to a proline-rich sequence in βc (box 1) proximal to the transmembrane domain but does not appear to bind to the α chain (Quelle *et al.*, 1994). Only the N-terminal 290 amino acids but not the kinase or kinase-like domains of JAK2 are required for binding (Zhao *et al.*, 1995). Despite the lack of binding to JAK2, the cytoplasmic domain of the α chain is necessary for JAK2 phosphorylation, activation, and signaling (Doyle and Gasson, 1998). It may be that the α chain cytoplasmic domain is necessary to aggregate βc chains so that two JAK2 molecules can cross-activate each other or that the α chain binds a distinct kinase needed for JAK2 activation. The importance of JAK2 kinase activity for most GM-CSF-dependent signaling functions has been demonstrated by experimental mutation of the JAK2-binding site on βc or by the use of dominant-negative forms of JAK2 (Watanabe *et al.*, 1996). Several other cytoplasmic tyrosine kinases such as fps/fes, lyn, fyn, yes and hck have been shown to be activated by GM-CSF signaling and in some cases to be associated with βc or to be necessary for GM-CSF signaling events (Corey *et al.*, 1993; Hanazono *et al.*, 1993; Linnekin *et al.*, 1994; Li and Chen, 1995; Li *et al.*, 1995; Brizzi *et al.*, 1996; Wei *et al.*, 1996; Yousefi *et al.*, 1996; Park *et al.*, 1998).

Cytoplasmic signaling cascades

Receptor activation leads to phosphorylation of βc cytoplasmic tyrosine residues and of cytoplasmic latent transcription factors STAT5a and STAT5b

(Mui *et al.*, 1995). In one study neither GM-CSF-dependent STAT5 phosphorylation nor proliferation were dependent on receptor tyrosine phosphorylation (Okuda *et al.*, 1997) but in another study all tyrosine phosphorylation sites on the receptor were required for STAT5 phosphorylation and for optimal proliferation (Itoh *et al.*, 1998). STAT5a isoforms appear to be preferentially activated by GM-CSF (Rosen *et al.*, 1996) and gene deletion studies of STAT5a in mice suggested that this isoform is most critical for a full proliferative response to GM-CSF (Feldman *et al.*, 1997). After phosphorylation, STAT5 proteins form homo- or heterodimers (STAT5a:5a, STAT5a:5b) by binding of the SH2 domain of one STAT molecule to the tyrosine phosphate site of another. They are then translocated to the nucleus and activate transcription of genes with STAT-response elements (see below).

A second link to transcriptional activation is through activation of the Ras/MAP kinase pathway. This occurs through binding of the PTB domain of the adapter protein SHC to phosphorylated tyrosine 577 (numbering from the mature protein) or 593 (numbering from the initiating methionine) on the βc cytoplasmic domain (Pratt *et al.*, 1996; Okuda *et al.*, 1997). SHC becomes tyrosine phosphorylated and binds to GRB2, which in turn associates with membrane-associated SOS, a guanine nucleotide exchanger that activates the small GTPase Ras. Ras activates the cytoplasmic serine/threonine kinase Raf-1 which in turn sets up a serine/threonine kinase cascade through MEK and MAP kinase that ultimately phosphorylates and activates nuclear transcription factors that induce fos expression. GM-CSF also activates jun kinase (JNK) activation through a process that involves both the box 1 region and multiple tyrosine residues in the cytoplasmic domain of βc (Liu *et al.*, 1997).

The tyrosine phosphatase SHP-2 binds to any of three phosphotyrosine residues on activated βc (577, 612, 695 mature protein or 593, 628, 712 full-length protein) (Okuda *et al.*, 1997; Itoh *et al.*, 1998) but the consequences of activating SHP-2 are unknown. SHP-1 also binds to βc, although the site is unknown, and it appears to act as a negative regulator at least of proliferative signaling since moth-eaten mouse macrophages (which lack SHP1 activity) are hyperresponsive to GM-CSF (Jiao *et al.*, 1997).

GM-CSF action leads to tyrosine phosphorylation of phosphatidylinositol 3-kinase and activation of its activity (conversion of $PI(4,5)P_2$ to $PI(3,4,5)P_3$). Distal regions of βc (residues 626–763 of the mature protein) are required for the binding and activation of this kinase (Corey *et al.*, 1993; Sato *et al.*, 1993; Jucker and Feldman, 1996).

In terms of functional effects, the GM-CSF receptor appears to consist of several discrete regions. Initiation of cell proliferation requires only the membrane proximal region (box 1 and box 2) of βc which correlates with activation of JAK2 and transcriptional activation of myc and pim-1 (Sato *et al.*, 1993; Watanabe *et al.*, 1996; Smith *et al.*, 1997). Downstream regions up to residue 643 of the mature protein are required for long-term cell survival even in the presence of serum (Inhorn *et al.*, 1995; Kinoshita *et al.*, 1995; Smith *et al.*, 1997; Chao *et al.*, 1998) while further downstream regions including tyrosine 750 (mature protein) and Ras activation enhance cell survival but are not required in the presence of serum (Inhorn *et al.*, 1995; Kinoshita *et al.*, 1995). Paradoxically, residues up to 643 were also required for clonal suppression accompanying cell differentiation in myeloid leukemia cell lines, while downstream regions to residue 799 were required for the full spectrum of the differentiated phenotype (Smith *et al.*, 1997; Matsuguchi *et al.*, 1998). Finally, the region between residues 799 and the C-terminus negatively regulates proliferative signaling by the GM-CSF receptor (Smith *et al.*, 1997).

The α chain cytoplasmic domain is required for all aspects of GM-CSF signaling, presumably by initiating tyrosine phosphorylation and activation of βc (Muto *et al.*, 1995; Doyle and Gasson, 1998). However, if βc cytoplasmic domains are artificially dimerized in the absence of α chain cytoplasmic domains by the use of chimeric receptors, signaling is intact so the role of the α chain appears to be to activate the signaling capacity of βc (Muto *et al.*, 1995; Patel *et al.*, 1996). It has been suggested that the α chain alone is sufficient to signal GM-CSF-mediated increases in glucose transport (Ding *et al.*, 1994) but this could not be achieved in cells from mice which express only the α chain (Scott *et al.*, 1998).

DOWNSTREAM GENE ACTIVATION

Transcription factors activated

GM-CSF induces fos and jun components of the AP-1 or serum response transcriptional complexes. It also induces phosphorylation and activation of STAT5a and STAT5b with occasional reports of activation of STAT1 and STAT3. GM-CSF also induces transcriptional activity of E2F probably by dissociation of the p107 inhibitor.

Genes induced

GM-CSF induces transcription of the AP-1 transcription factors fos (Sato *et al.*, 1993; Watanabe *et al.*, 1993) and jun (Liu *et al.*, 1997) in most cell types that display GM-CSF receptors, Fc receptor $\gamma 1$ (Rosen *et al.*, 1996), fms (the M-CSF receptor) (Helftenbein *et al.*, 1996), and scavenger receptor A (SRA) (Guidez *et al.*, 1998) in macrophages, and mcl-1 and A1 (pro-survival bcl-2-related genes) (Chao *et al.*, 1998; Feldman *et al.*, 1997), early growth response 1 gene (egr-1) (Watanabe *et al.*, 1997), myc and pim-1 (Sato *et al.*, 1993; Watanabe *et al.*, 1995) in myeloid progenitor cells. GM-CSF also induces expression of a family of negative regulators of cytokine signaling including CIS, SOCS-1, SOCS-2, and SOCS-3 (Yoshimura *et al.*, 1995; Starr *et al.*, 1997).

Promoter regions involved

Induction of myc and pim-1 gene expression by GM-CSF requires only the box 1 region of the receptor although downstream regions enhance this induction. The *cis*-acting P2 promoter region of the myc gene appears to mediate the induction and the effect of GM-CSF is somehow to allow dissociation of p107 (a pocket protein transcriptional inhibitor) from the E2F transcriptional activator (Watanabe *et al.*, 1995).

Induction of fos and egr-1 gene expression requires the concerted action of both the MAP kinase and STAT pathways acting on the serum response element (SRE) and STAT-inducible elements (SIE), respectively (Watanabe *et al.*, 1997).

Induction of the mcl-1 gene by GM-CSF requires a region between −197 to −69 (Chao *et al.*, 1998) while induction of the SRA gene requires an enhancer about 4 kb upstream of the initiation site (Guidez *et al.*, 1998).

BIOLOGICAL CONSEQUENCES OF ACTIVATING OR INHIBITING RECEPTOR AND PATHOPHYSIOLOGY

Unique biological effects of activating the receptors

The unique activities are those associated with GM-CSF and include the maintenance of surfactant clearance from the lung as well as resistance to some infections.

Phenotypes of receptor knockouts and receptor overexpression mice

Mice in which the βc gene has been deleted have a phenotype expected for a combined GM-CSF and IL-5 knockout with no effect on IL-3 function because of the existence of an alternate IL-3-specific β chain. Thus the mice display lymphoid infiltration in the lungs with a pulmonary alveolar proteinosis-like disease in which surfactant is not cleared and accumulates in the alveoli. In addition, eosinophil numbers in the blood and bone marrow are very low and the mice fail to mount an eosinophilic response to parasitic infections (Nishinakamura *et al.*, 1995; Robb *et al.*, 1995). There are no reports of the phenotype of GM-CSF receptor α chain knockouts.

Transgenic mice engineered to express human GM-CSF receptor α and β chains off constitutive promoters showed that progenitor cells from most hematopoietic lineages (including erythroid cells, megakaryocytes, mast cells, blast cells, and NK cells) responded to human GM-CSF with proliferation and differentiation. This suggested that the hematopoietic specificity of GM-CSF is determined by where receptors are expressed rather than by the signaling capacity of the receptor itself (Nishijima *et al.*, 1995, 1997b). Injection of human GM-CSF into these mice resulted in multilineage proliferation, depletion of the bone marrow and expansion of extramedullary hematopoiesis in the spleen and liver. The thymus, however, was shrunken and although GM-CSF stimulated proliferation of most T cell subsets it inhibited the formation and failed to stimulate the proliferation of CD4+ CD8+ T cells (Nishijima *et al.*, 1997a; Yasuda *et al.*, 1997).

Human abnormalities

Defective expression or mutation of βc has been associated with some cases of *pulmonary alveolar proteinosis* (PAP) (Dirksen *et al.*, 1997). It has also been reported that some pediatric patients with *acute myeloid leukemia* associated with respiratory distress or pulmonary alveolar proteinosis-like disease have defective expression of βc and of the α chain (Dirksen *et al.*, 1998). While some experimental mutations of βc have been shown to lead to *myelodysplasia* (Jenkins *et al.*, 1995, 1998; McCormack and Gonda, 1997), analyses of *leukemia* patients have generally failed to reveal any mutations in the α or βc chains of the GM-CSF receptor (Brown *et al.*, 1994; Wagner *et al.*, 1994; Decker *et al.*, 1995; Freeburn *et al.*, 1996, 1997, 1998).

THERAPEUTIC UTILITY

Effect of treatment with soluble receptor domain

The soluble α and β chains are expected to be inhibitors of GM-CSF action but no therapeutic studies have yet been performed.

Effects of inhibitors (antibodies) to receptors

Antagonistic antibodies to both the α and β chains have been described that inhibit binding of GM-CSF and inhibit biological activity (Nicola *et al.*, 1993; Takaki *et al.*, 1991) but, again, no therapeutic studies have been performed.

References

Anderson, K. L., Smith, K. A., Conners, K., McKercher, S. R., Maki, R. A., and Torbett, B. E. (1998). Myeloid development is selectively disrupted in PU.1 null mice. *Blood* **91**, 3702–3710.

Ashworth, A., and Kraft, A. (1990). Cloning of a potentially soluble receptor for human GM-CSF. *Nucleic Acids Res.* **18**, 7178.

Baldwin, G. C., Gasson, J. C., Kaufman, S. E., Quan, S. G., Williams, R. E., Avalos, B. R., Gazdar, A. F., Golde, D. W., and DiPersio, J. F. (1989). Nonhematopoietic tumor cells express functional GM-CSF receptors. *Blood* **73**, 1033–1037.

Brizzi, M. F., Aronica, M. G., Rosso, A., Bagnara, G. P., Yarden, Y., and Pegoraro, L. (1996). Granulocyte-macrophage colony-stimulating factor stimulates JAK2 signaling pathway and rapidly activates p93fes, STAT1 p91, and STAT3 p92 in polymorphonuclear leukocytes. *J. Biol. Chem.* **271**, 3562–3567.

Brown, M. A., Harrison-Smith, M., DeLuca, E., Begley, C. G., and Gough, N. M. (1994). No evidence for GM-CSF receptor alpha chain gene mutation in AML-M2 leukemias which have lost a sex chromosome. *Leukemia* **8**, 1774–1779.

Bussolino, F., Colotta, F., Bocchietto, E., Guglielmetti, A., and Mantovani, A. (1993). Recent developments in the cell biology of granulocyte-macrophage colony-stimulating factor and granulocyte colony-stimulating factor: activities on endothelial cells. *Int. J. Clin. Lab. Res.* **23**, 8–12.

Chao, J. R., Wang, J. M., Lee, S. F., Peng, H. W., Lin, Y. H., Chou, C. H., Li, J. C., Huang, H. M., Chou, C. K., Kuo, M. L., Yen, J. J., and Yang-Yen, H. F. (1998). mcl-1 is an immediate-early gene activated by the granulocyte-macrophage colony-stimulating factor (GM-CSF) signaling pathway and is one component of the GM-CSF viability response. *Mol. Cell. Biol.* **18**, 4883–4898.

Corey, S., Eguinoa, A., Puyana-Theall, K., Bolen, J. B., Cantley, L., Mollinedo, F., Jackson, T. R., Hawkins, P. T., and Stephens, L. R. (1993). Granulocyte macrophage-colony stimulating factor stimulates both association and activation of phosphoinositide 3OH-kinase and src-related tyrosine kinase(s) in human myeloid derived cells. *EMBO J.* **12**, 2681–2690.

Crosier, K. E., Wong, G. G., Mathey-Prevot, B., Nathan, D. G., and Sieff, C. A. (1991). A functional isoform of the human granulocyte/macrophage colony-stimulating factor receptor has an unusual cytoplasmic domain. *Proc. Natl Acad. Sci. USA* **88**, 7744–7748.

Crosier, K. E., Hall, L. R., Vitas, M. R., and Crosier, P. S. (1997). Expression and functional analysis of two isoforms of the human GM-CSF receptor alpha chain in myeloid development and leukaemia. *Br. J. Haematol.* **98**, 540–548.

Decker, J., Fiedler, W., Samalecos, A., and Hossfeld, D. K. (1995). Absence of point mutations in the extracellular domain of the alpha subunit of the GM-CSF receptor in a series of patients with acute myeloid leukemia (AML). *Leukemia* **9**, 185–188.

Ding, D. X., Rivas, C. I., Heaney, M. L., Raines, M. A., Vera, J. C., and Golde, D. W. (1994). The alpha subunit of the human granulocyte-macrophage colony-stimulating factor receptor signals for glucose transport via a phosphorylation-independent pathway. *Proc. Natl Acad. Sci. USA* **91**, 2537–2541.

Ding, D. X., Vera, J. C., Heaney, M. L., and Golde, D. W. (1995). N-glycosylation of the human granulocyte-macrophage colony-stimulating factor receptor alpha subunit is essential for ligand binding and signal transduction. *J. Biol. Chem.* **270**, 24580–24584.

DiPersio, J., Billing, P., Kaufman, S., Eghtesady, P., Williams, R. E., and Gasson, J. C. (1988). Characterization of the human granulocyte-macrophage colony-stimulating factor receptor. *J. Biol. Chem.* **263**, 1834–1841.

Dirksen, U., Nishinakamura, R., Groneck, P., Hattenhorst, U., Nogee, L., Murray, R., and Burdach, S. (1997). Human pulmonary alveolar proteinosis associated with a defect in GM-CSF/IL-3/IL-5 receptor common beta chain expression. *J. Clin. Invest.* **100**, 2211–2217.

Dirksen, U., Hattenhorst, U., Schneider, P., Schroten, H., Gobel, U., Bocking, A., Muller, K. M., Murray, R., and Burdach, S. (1998). Defective expression of granulocyte-macrophage colony-stimulating factor/interleukin-3/interleukin-5 receptor common beta chain in children with acute myeloid leukemia associated with respiratory failure. *Blood* **92**, 1097–1103.

Disteche, C. M., Brannan, C. I., Larsen, A., Adler, D. A., Schorderet, D. F., Gearing, D., Copeland, N. G., Jenkins, N. A., and Park, L. S. (1992). The human pseudoautosomal GM-CSF receptor alpha subunit gene is autosomal in mouse. *Nature Genet.* **1**, 333–336.

Doyle, S. E., and Gasson, J. C. (1998). Characterization of the role of the human granulocyte-macrophage colony-stimulating factor receptor alpha subunit in the activation of JAK2 and STAT5. *Blood* **92**, 867–876.

Feldman, G. M., Rosenthal, L. A., Liu, X., Hayes, M. P., Wynshaw-Boris, A., Leonard, W. J., Hennighausen, L., and Finbloom, D. S. (1997). STAT5A-deficient mice demonstrate a defect in granulocyte-macrophage colony-stimulating factor-induced proliferation and gene expression. *Blood* **90**, 1768–1776.

Freeburn, R. W., Gale, R. E., Wagner, H. M., and Linch, D. C. (1996). The beta subunit common to the GM-CSF, IL-3 and IL-5 receptors is highly polymorphic but pathogenic point mutations in patients with acute myeloid leukaemia (AML) are rare. *Leukemia* **10**, 123–129.

Freeburn, R. W., Gale, R. E., Wagner, H. M., and Linch, D. C. (1997). Analysis of the coding sequence for the GM-CSF receptor alpha and beta chains in patients with juvenile chronic myeloid leukemia (JCML). *Exp. Hematol.* **25**, 306–311.

Freeburn, R. W., Gale, R. E., and Linch, D. C. (1998). Activating point mutations in the betaC chain of the GM-CSF, IL-3 and IL-5 receptors are not a major contributory factor in the pathogenesis of acute myeloid leukaemia. *Br. J. Haematol.* **103**, 66–71.

Gearing, D. P., King, J. A., Gough, N. M., and Nicola, N. A. (1989). Expression cloning of a receptor for human granulocyte-macrophage colony-stimulating factor. *EMBO J.* **8**, 3667–3676.

Gorman, D. M., Itoh, N., Jenkins, N. A., Gilbert, D. J., Copeland, N. G., and Miyajima, A. (1992). Chromosomal localization and organization of the murine genes encoding the beta subunits (AIC2A and AIC2B) of the interleukin 3, granulocyte/macrophage colony-stimulating factor, and interleukin 5 receptors. *J. Biol. Chem.* **267**, 15842–15848.

Gough, N. M., Gearing, D. P., Nicola, N. A., Baker, E., Pritchard, M., Callen, D. F., and Sutherland, G. R. (1990). Localization of the human GM-CSF receptor gene to the X-Y pseudoautosomal region. *Nature* **345**, 734–736.

Guidez, F., Li, A. C., Horvai, A., Welch, J. S., and Glass, C. K. (1998). Differential utilization of Ras signaling pathways by macrophage colony-stimulating factor (CSF) and granulocyte-macrophage CSF receptors during macrophage differentiation. *Mol. Cell. Biol.* **18**, 3851–3861.

Hampson, J., McLaughlin, P. J., and Johnson, P. M. (1993). Low-affinity receptors for tumour necrosis factor-alpha, interferon-gamma and granulocyte-macrophage colony-stimulating factor are expressed on human placental syncytiotrophoblast. *Immunology* **79**, 485–490.

Hanazono, Y., Chiba, S., Sasaki, K., Mano, H., Miyajima, A., Arai, K., Yazaki, Y., and Hirai, H. (1993). c-fps/fes protein-tyrosine kinase is implicated in a signaling pathway triggered by granulocyte-macrophage colony-stimulating factor and interleukin-3. *EMBO J.* **12**, 1641–1646.

Hayashida, K., Kitamura, T., Gorman, D. M., Arai, K., Yokota, T., and Miyajima, A. (1990). Molecular cloning of a second subunit of the receptor for human granulocyte-macrophage colony-stimulating factor (GM-CSF): reconstitution of a high-affinity GM-CSF receptor. *Proc. Natl Acad. Sci. USA* **87**, 9655–9659.

Helftenbein, G., Krusekopf, K., Just, U., Cross, M., Ostertag, W., Niemann, H., and Tamura, T. (1996). Transcriptional regulation of the c-fms proto-oncogene mediated by granulocyte/macrophage colony-stimulating factor (GM-CSF) in murine cell lines. *Oncogene* **12**, 931–935.

Hirsch, T., Eggstein, S., Frank, S., Farthmann, E. H., and von Specht, B. U. (1995). Expression of GM-CSF and a functional GM-CSF receptor in the human colon carcinoma cell line SW403. *Biochem. Biophys. Res. Commun.* **217**, 138–143.

Hohaus, S., Petrovick, M. S., Voso, M. T., Sun, Z., Zhang, D. E., and Tenen, D. G. (1995). PU.1 (Spi-1) and C/EBP alpha regulate expression of the granulocyte-macrophage colony-stimulating factor receptor alpha gene. *Mol. Cell. Biol.* **15**, 5830–5845.

Hu, X., Emanuel, P. D., and Zuckerman, K. S. (1994). Cloning and sequencing of the cDNAs encoding two alternative splicing-derived variants of the alpha subunit of the granulocyte-macrophage colony-stimulating factor receptor. *Biochim. Biophys. Acta* **1223**, 306–308.

Inhorn, R. C., Carlesso, N., Durstin, M., Frank, D. A., and Griffin, J. D. (1995). Identification of a viability domain in the granulocyte/macrophage colony-stimulating factor receptor beta-chain involving tyrosine-750. *Proc. Natl Acad. Sci. USA* **92**, 8665–8669.

Itoh, T., Liu, R., Yokota, T., Arai, K. I., and Watanabe, S. (1998). Definition of the role of tyrosine residues of the common beta subunit regulating multiple signaling pathways of granulocyte-macrophage colony-stimulating factor receptor. *Mol. Cell. Biol.* **18**, 742–752.

Jenkins, B. J., D'Andrea, R., and Gonda, T. J. (1995). Activating point mutations in the common beta subunit of the human GM-CSF, IL-3 and IL-5 receptors suggest the involvement of beta subunit dimerization and cell type-specific molecules in signalling. *EMBO J.* **14**, 4276–4287.

Jenkins, B. J., Blake, T. J., and Gonda, T. J. (1998). Saturation mutagenesis of the beta subunit of the human granulocyte-macrophage colony-stimulating factor receptor shows clustering of constitutive mutations, activation of ERK MAP kinase and STAT pathways, and differential beta subunit tyrosine phosphorylation. *Blood* **92**, 1989–2002.

Jiao, H., Yang, W., Berrada, K., Tabrizi, M., Shultz, L., and Yi, T. (1997). Macrophages from motheaten and viable motheaten mutant mice show increased proliferative responses to GM-CSF: detection of potential HCP substrates in GM-CSF signal transduction. *Exp. Hematol.* **25**, 592–600.

Jucker, M., and Feldman, R. A. (1996). Novel adapter proteins that link the human GM-CSF receptor to the phosphatidylinositol 3-kinase and Shc/Grb2/ras signaling pathways. *Curr. Top. Microbiol. Immunol.* **211**, 67–75.

Kinoshita, T., Yokota, T., Arai, K., and Miyajima, A. (1995). Suppression of apoptotic death in hematopoietic cells by signalling through the IL-3/GM-CSF receptors. *EMBO J.* **14**, 266–275.

Kitamura, T., Sato, N., Arai, K., and Miyajima, A. (1991). Expression cloning of the human IL-3 receptor cDNA reveals a shared beta subunit for the human IL-3 and GM-CSF receptors. *Cell* **66**, 1165–1174.

Kremer, E., Baker, E., D'Andrea, R. J., Slim, R., Phillips, H., Moretti, P. A., Lopez, A. F., Petit, C., Vadas, M. A., and Sutherland, G.R. *et al.* (1993). A cytokine receptor gene cluster in the X-Y pseudoautosomal region? *Blood* **82**, 22–28.

Li, Y., and Chen, B. (1995). Differential regulation of fyn-associated protein tyrosine kinase activity by macrophage colony-stimulating factor (M-CSF) and granulocyte-macrophage colony-stimulating factor (GM-CSF). *J. Leukoc. Biol.* **57**, 484–490.

Li, Y., Shen, B. F., Karanes, C., Sensenbrenner, L., and Chen, B. (1995). Association between Lyn protein tyrosine kinase (p53/56lyn) and the beta subunit of the granulocyte-macrophage colony-stimulating factor (GM-CSF) receptors in a GM-CSF-dependent human megakaryocytic leukemia cell line (M-07e). *J. Immunol.* **155**, 2165–2174.

Lia, F., Rajotte, D., Clark, S. C., and Hoang, T. (1996). A dominant negative granulocyte-macrophage colony-stimulating factor receptor alpha chain reveals the multimeric structure of the receptor complex. *J. Biol. Chem.* **271**, 28287–28293.

Linnekin, D., Howard, O. M., Park, L., Farrar, W., Ferris, D., and Longo, D. L. (1994). Hck expression correlates with granulocyte-macrophage colony-stimulating factor-induced proliferation in HL-60 cells. *Blood* **84**, 94–103.

Liu, R., Itoh, T., Arai, K., and Watanabe, S. (1997). Activation of c-Jun N-terminal kinase by human granulocyte macrophage-colony stimulating factor in BA/F3 cells. *Biochem. Biophys. Res. Commun.* **234**, 611–615.

Lock, P., Metcalf, D., and Nicola, N. A. (1994). Histidine-367 of the human common beta chain of the receptor is critical for high-affinity binding of human granulocyte-macrophage colony-stimulating factor. *Proc. Natl Acad. Sci. USA* **91**, 252–256.

McCormack, M. P., and Gonda, T. J. (1997). Expression of activated mutants of the human interleukin-3/interleukin-5/granulocyte-macrophage colony-stimulating factor receptor common beta subunit in primary hematopoietic cells induces factor-independent proliferation and differentiation. *Blood* **90**, 1471–1481.

Matsuguchi, T., Lilly, M. B., and Kraft, A. S. (1998). Cytoplasmic domains of the human granulocyte-macrophage colony-stimulating factor (GM-CSF) receptor beta chain (hbetac) responsible for human GM-CSF-induced myeloid cell differentiation. *J. Biol. Chem.* **273**, 19411–19418.

Metcalf, D., Nicola, N. A., Gearing, D. P., and Gough, N. M. (1990). Low-affinity placenta-derived receptors for human granulocyte-macrophage colony-stimulating factor can deliver a proliferative signal to murine hemopoietic cells. *Proc. Natl Acad. Sci. USA* **87**, 4670–4674.

Mui, A. L., Wakao, H., O'Farrell, A. M., Harada, N., and Miyajima, A. (1995). Interleukin-3, granulocyte-macrophage colony stimulating factor and interleukin-5 transduce signals through two STAT5 homologs. *EMBO J.* **14**, 1166–1175.

Muto, A., Watanabe, S., Itoh, T., Miyajima, A., Yokota, T., and Arai, K. (1995). Roles of the cytoplasmic domains of the alpha and beta subunits of human granulocyte-macrophage colony-stimulating factor receptor. *J. Allergy Clin. Immunol.* **96**, 1100–1114.

Muto, A., Watanabe, S., Miyajima, A., Yokota, T., and Arai, K. (1996). The beta subunit of human granulocyte-macrophage colony-stimulating factor receptor forms a homodimer and is activated via association with the alpha subunit. *J. Exp. Med.* **183**, 1911–1916.

Nakagawa, Y., Kosugi, H., Miyajima, A., Arai, K., and Yiokota, T. (1994). Structure of the gene encoding the alpha subunit of the human granulocyte-macrophage colony stimulating factor receptor. Implications for the evolution of the cytokine receptor superfamily. *J. Biol. Chem.* **269**, 10905–10912.

Nicola, N. A., Wycherley, K., Boyd, A. W., Layton, J. E., Cary, D., and Metcalf, D. (1993). Neutralizing and nonneutralizing monoclonal antibodies to the human granulocyte-macrophage colony-stimulating factor receptor alpha-chain. *Blood* **82**, 1724–1731.

Nishijima, I., Nakahata, T., Hirabayashi, Y., Inoue, T., Kurata, H., Miyajima, A., Hayashi, N., Iwakura, Y., Arai, K., and Yokota, T. (1995). A human GM-CSF receptor expressed in transgenic mice stimulates proliferation and differentiation of hemopoietic progenitors to all lineages in response to human GM-CSF. *Mol. Biol. Cell* **6**, 497–508.

Nishijima, I., Nakahata, T., Watanabe, S., Tsuji, K., Tanaka, I., Hirabayashi, Y., Inoue, T., and Arai, K. (1997a). Hematopoietic and lymphopoietic responses in human granulocyte-macrophage colony-stimulating factor (GM-CSF) receptor transgenic mice injected with human GM-CSF. *Blood* **90**, 1031–1038.

Nishijima, I., Watanabe, S., Nakahata, T., and Arai, K. (1997b). Human granulocyte-macrophage colony-stimulating factor (hGM-CSF)-dependent *in vitro* and *in vivo* proliferation and differentiation of all hematopoietic progenitor cells in hGM-CSF receptor transgenic mice. *J. Allergy Clin. Immunol.* **100**, S79–86.

Nishinakamura, R., Nakayama, N., Hirabayashi, Y., Inoue, T., Aud, D., McNeil, T., Azuma, S., Yoshida, S., Toyoda, Y., Arai, K., Miyajima, A., and Murray, R. (1995). Mice deficient for the IL-3/GM-CSF/IL-5 beta c receptor exhibit lung pathology and impaired immune response, while beta IL3 receptor-deficient mice are normal. *Immunity* **2**, 211–222.

Okuda, K., Smith, L., Griffin, J. D., and Foster, R. (1997). Signaling functions of the tyrosine residues in the beta c chain of the granulocyte-macrophage colony-stimulating factor receptor. *Blood* **90**, 4759–4766.

Park, L. S., Friend, D., Gillis, S., and Urdal, D. L. (1986a). Characterization of the cell surface receptor for granulocyte-macrophage colony-stimulating factor. *J. Biol. Chem.* **261**, 4177–4183.

Park, L. S., Friend, D., Gillis, S., and Urdal, D. L. (1986b). Characterization of the cell surface receptor for human granulocyte/macrophage colony-stimulating factor. *J. Exp. Med.* **164**, 251–262.

Park, L. S., Martin, U., Sorensen, R., Luhr, S., Morrissey, P. J., Cosman, D., and Larsen, A. (1992). Cloning of the low-affinity murine granulocyte-macrophage colony-stimulating factor receptor and reconstitution of a high-affinity receptor complex. *Proc. Natl Acad. Sci. USA* **89**, 4295–4299.

Park, W. Y., Ahn, J. H., Feldman, R. A., and Seo, J. S. (1998). c-Fes tyrosine kinase binds to and activates STAT3 after granulocyte-macrophage colony-stimulating factor stimulation. *Cancer Lett.* **129**, 29–37.

Patel, N., Herrman, J. M., Timans, J. C., and Kastelein, R. A. (1996). Functional replacement of cytokine receptor extracellular domains by leucine zippers. *J. Biol. Chem.* **271**, 30386–30391.

Pratt, J. C., Weiss, M., Sieff, C. A., Shoelson, S. E., Burakoff, S. J., and Ravichandran, K. S. (1996). Evidence for a physical association between the Shc-PTB domain and the beta c chain of the granulocyte-macrophage colony-stimulating factor receptor. *J. Biol. Chem.* **271**, 12137–12140.

Quelle, F. W., Sato, N., Witthuhn, B. A., Inhorn, R. C., Eder, M., Miyajima, A., Griffin, J. D., and Ihle, J. N. (1994). JAK2 associates with the beta c chain of the receptor for granulocyte-macrophage colony-stimulating factor, and its activation requires the membrane-proximal region. *Mol. Cell. Biol.* **14**, 4335–4341.

Raines, M. A., Liu, L., Quan, S. G., Joe, V., DiPersio, J. F., and Golde, D. W. (1991). Identification and molecular cloning of a soluble human granulocyte-macrophage colony-stimulating factor receptor. *Proc. Natl Acad. Sci. USA* **88**, 8203–8207.

Rajotte, D., Cadieux, C., Haman, A., Wilkes, B. C., Clark, S. C., Hercus, T., Woodcock, J. A., Lopez, A., and Hoang, T. (1997). Crucial role of the residue R280 at the F′-G′ loop of the human granulocyte/macrophage colony-stimulating factor receptor alpha chain for ligand recognition. *J. Exp. Med.* **185**, 1939–1950.

Rappold, G., Willson, T. A., Henke, A., and Gough, N. M. (1992). Arrangement and localization of the human GM-CSF receptor alpha chain gene CSF2RA within the X-Y pseudoautosomal region. *Genomics* **14**, 455–461.

Rivas, C. I., Vera, J. C., Delgado Lopez, F., Heaney, M. L., Guaiquil, V. H., Zhang, R. H., Scher, H. I., Concha, II, Nualart, F., Cordon-Cardo, C., and Golde, D. W. (1998). Expression of granulocyte-macrophage colony-stimulating factor receptors in human prostate cancer. *Blood* **91**, 1037–1043.

Robb, L., Drinkwater, C. C., Metcalf, D., Li, R., Kontgen, F., Nicola, N. A., and Begley, C. G. (1995). Hematopoietic and lung abnormalities in mice with a null mutation of the common beta subunit of the receptors for granulocyte-macrophage colony-stimulating factor and interleukins 3 and 5. *Proc. Natl Acad. Sci. USA* **92**, 9565–9569.

Rokhlin, O. W., Griebling, T. L., Karassina, N. V., Raines, M. A., and Cohen, M. B. (1996). Human prostate carcinoma cell lines secrete GM-CSF and express GM-CSF-receptor on their cell surface. *Anticancer Res.* **16**, 557–563.

Rosen, R. L., Winestock, K. D., Chen, G., Liu, X., Hennighausen, L., and Finbloom, D. S. (1996). Granulocyte-macrophage colony-stimulating factor preferentially activates the 94-kD STAT5A and an 80-kD STAT5A isoform in human peripheral blood monocytes. *Blood* **88**, 1206–1214.

Sato, N., Sakamaki, K., Terada, N., Arai, K., and Miyajima, A. (1993). Signal transduction by the high-affinity GM-CSF receptor: two distinct cytoplasmic regions of the common beta subunit responsible for different signaling. *EMBO J.* **12**, 4181–4189.

Scott, C. L., Hughes, D. A., Cary, D., Nicola, N. A., Begley, C. G., and Robb, L. (1998). Functional analysis of mature hematopoietic cells from mice lacking the beta chain of the granulocyte-macrophage colony-stimulating factor receptor. *Blood* **92**, 4119–4127.

Shen, Y., Baker, E., Callen, D. F., Sutherland, G. R., Willson, T. A., Rakar, S., and Gough, N. M. (1992). Localization of the human GM-CSF receptor beta chain gene (CSF2RB) to chromosome 22q12.2-q13.1. *Cytogenet. Cell Genet.* **61**, 175–177.

Shibuya, K., Chiba, S., Miyagawa, K., Kitamura, T., Miyazono, K., and Takaku, F. (1991). Structural and functional analyses of glycosylation on the distinct molecules of human GM-CSF receptors. *Eur. J. Biochem.* **198**, 659–666.

Smith, A., Metcalf, D., and Nicola, N. A. (1997). Cytoplasmic domains of the common beta-chain of the GM-CSF/IL-3/IL-5 receptors that are required for inducing differentiation or clonal suppression in myeloid leukaemic cell lines. *EMBO J.* **16**, 451–464.

Starr, R., Willson, T. A., Viney, E. M., Murray, L. J., Rayner, J. R., Jenkins, B. J., Gonda, T. J., Alexander, W. S., Metcalf, D., Nicola, N. A., and Hilton, D. J. (1997). A family of cytokine-inducible inhibitors of signalling. *Nature* **387**, 917–921.

Stomski, F. C., Woodcock, J. M., Zacharakis, B., Bagley, C. J., Sun, Q., and Lopez, A. F. (1998). Identification of a Cys motif in the common beta chain of the interleukin 3, granulocyte-macrophage colony-stimulating factor, and interleukin 5 receptors essential for disulfide-linked receptor heterodimerization and activation of all three receptors. *J. Biol. Chem.* **273**, 1192–1199.

Takaki, S., Mita, S., Kitamura, T., Yonehara, S., Yamaguchi, N., Tominaga, A., Miyajima, A., and Takatsu, K. (1991). Identification of the second subunit of the murine interleukin-5 receptor: interleukin-3 receptor-like protein, AIC2B is a component of the high affinity interleukin-5 receptor. *EMBO J.* **10**, 2833–2838.

Tavernier, J., Devos, R., Cornelis, S., Tuypens, T., Van der Heyden, J., Fiers, W., and Plaetinck, G. (1991). A human high affinity interleukin-5 receptor (IL5R) is composed of an IL5-specific alpha chain and a beta chain shared with the receptor for GM-CSF. *Cell* **66**, 1175–1184.

Wagner, H. M., Gale, R. E., Freeburn, R. W., Devereux, S., and Linch, D. C. (1994). Analysis of mutations in the GM-CSF receptor alpha coding sequence in patients with acute myeloid leukaemia and haematologically normal individuals by RT-PCR-SSCP. *Leukemia* **8**, 1527–1532.

Walker, F., and Burgess, A. W. (1985). Specific binding of radioiodinated granulocyte-macrophage colony-stimulating factor to hemopoietic cells. *EMBO J.* **4**, 933–939.

Watanabe, S., Muto, A., Yokota, T., Miyajima, A., and Arai, K. (1993). Differential regulation of early response genes and cell proliferation through the human granulocyte macrophage colony-stimulating factor receptor: selective activation of the c-fos promoter by genistein. *Mol. Biol. Cell* **4**, 983–992.

Watanabe, S., Ishida, S., Koike, K., and Arai, K. (1995). Characterization of cis-regulatory elements of the c-myc promoter responding to human GM-CSF or mouse interleukin 3 in mouse proB cell line BA/F3 cells expressing the human GM-CSF receptor. *Mol. Biol. Cell* **6**, 627–636.

Watanabe, S., Itoh, T., and Arai, K. (1996). JAK2 is essential for activation of c-fos and c-myc promoters and cell proliferation through the human granulocyte-macrophage colony-stimulating factor receptor in BA/F3 cells. *J. Biol. Chem.* **271**, 12681–12686.

Watanabe, S., Kubota, H., Sakamoto, K. M., and Arai, K. (1997). Characterization of cis-acting sequences and trans-acting signals regulating early growth response 1 and c-fos promoters through the granulocyte-macrophage colony-stimulating factor receptor in BA/F3 cells. *Blood* **89**, 1197–1206.

Wei, S., Liu, J. H., Epling-Burnette, P. K., Gamero, A. M., Ussery, D., Pearson, E. W., Elkabani, M. E., Diaz, J. I., and Djeu, J. Y. (1996). Critical role of Lyn kinase in inhibition of neutrophil apoptosis by granulocyte-macrophage colony-stimulating factor. *J. Immunol.* **157**, 5155–5162.

Wheadon, H., Devereux, S., Khwaja, A., and Linch, D. C. (1997). Granulocyte-macrophage colony stimulating factor receptor alpha and beta chain complexes can form both high and intermediate affinity functional receptors. *Br. J. Haematol.* **98**, 809–818.

Woodcock, J. M., Zacharakis, B., Plaetinck, G., Bagley, C. J., Qiyu, S., Hercus, T. R., Tavernier, J., and Lopez, A. F. (1994). Three residues in the common beta chain of the human GM-CSF, IL-3 and IL-5 receptors are essential for GM-CSF and IL-5 but not IL-3 high affinity binding and interact with Glu21 of GM-CSF. *EMBO J.* **13**, 5176–5185.

Woodcock, J. M., Bagley, C. J., Zacharakis, B., and Lopez, A. F. (1996). A single tyrosine residue in the membrane-proximal domain of the granulocyte-macrophage colony-stimulating factor, interleukin (IL)-3, and IL-5 receptor common beta-chain is necessary and sufficient for high affinity binding and signaling by all three ligands. *J. Biol. Chem.* **271**, 25999–26006.

Woodcock, J. M., McClure, B. J., Stomski, F. C., Elliott, M. J., Bagley, C. J., and Lopez, A. F. (1997). The human granulocyte-macrophage colony-stimulating factor (GM-CSF) receptor exists as a preformed receptor complex that can be activated by GM-CSF, interleukin-3, or interleukin-5. *Blood* **90**, 3005–3017.

Yamada, K., Yamakawa, M., Imai, Y., and Tsukamoto, M. (1997). Expression of cytokine receptors on follicular dendritic cells. *Blood* **90**, 4832–4841.

Yasuda, Y., Nishijima, I., Watanabe, S., Arai, K., Zlotnik, A., and Moore, T. A. (1997). Human granulocyte-macrophage colony-stimulating factor (hGM-CSF) induces inhibition of intrathymic T-cell development in hGM-CSF receptor transgenic mice. *Blood* **89**, 1349–1356.

Yong, K., Cohen, H., Khwaja, A., Jones, H. M., and Linch, D. C. (1991). Lack of effect of granulocyte-macrophage and granulocyte colony-stimulating factors on cultured human endothelial cells [see comments]. *Blood* **77**, 1675–1680.

Yoshimura, A., Ohkubo, T., Kiguchi, T., Jenkins, N. A., Gilbert, D. J., Copeland, N. G., Hara, T., and Miyajima, A. (1995). A novel cytokine-inducible gene CIS encodes an SH2-containing protein that binds to tyrosine-phosphorylated interleukin 3 and erythropoietin receptors. *EMBO J.* **14**, 2816–2826.

Yousefi, S., Hoessli, D. C., Blaser, K., Mills, G. B., and Simon, H. U. (1996). Requirement of Lyn and Syk tyrosine kinases for the prevention of apoptosis by cytokines in human eosinophils. *J. Exp. Med.* **183**, 1407–1414.

Zhao, Y., Wagner, F., Frank, S. J., and Kraft, A. S. (1995). The amino-terminal portion of the JAK2 protein kinase is necessary for binding and phosphorylation of the granulocyte-macrophage colony-stimulating factor receptor beta c chain. *J. Biol. Chem.* **270**, 13814–13818.

G-CSF Receptor

Shigekazu Nagata*

Department of Genetics, Osaka University Medical School, 2-2 Yamada-oka Suita, Osaka, 565-0871, Japan

*corresponding author tel: +81-6-6879-3310, fax: 81-6-6879-3319, e-mail: nagata@genetic.med.osaka-u.ac.jp

DOI: 10.1006/rwcy.2000.20006.

SUMMARY

The G-CSF receptor is a type I membrane protein which belongs to the cytokine receptor superfamily, and is specifically expressed in mature neutrophils and neutrophilic precursors. Binding of G-CSF to G-CSFR induces its dimerization. The dimerized receptor transduces growth and differentiation signals which are mediated by the N-terminal and C-terminal halves of the G-CSFR cytoplasmic region. G-CSF activates JAK family kinases, which cause tyrosine phosphorylation of STAT family transcription factors. STATs then stimulate proliferation, and induce differentiation of neutrophilic precursor cells. Genetic defects in the G-CSFR gene cause severe *neutropenia* (*Kostmann's syndrome*).

BACKGROUND

Discovery

The receptor for G-CSF was discovered as a membrane protein expressed in myeloid *leukemia* cells or neutrophilic granulocytes to which [^{125}I]-labeled G-CSF binds (Nicola and Metcalf, 1984). Murine G-CSFR was purified from the membrane fraction of mouse myeloid leukemia NFS60 cells that respond to G-CSF for proliferation (Fukunaga *et al.*, 1990b), and its cDNA was isolated from an NFS60 cDNA library by expression cloning (Fukunaga *et al.*, 1990a). The human counterpart was subsequently identified from a human placenta cDNA library by crosshybridization with mouse cDNA (Fukunaga *et al.*, 1990c), or from a human neutrophil cDNA library by expression cloning (Larsen *et al.*, 1990).

Alternative names

G-CSFR is also known as CD114.

Structure

G-CSFR is a type I membrane protein of about 130 kDa.

Main activities and pathophysiological roles

G-CSFR transduces proliferation and differentiation signals into neutrophilic precursor cells, thus mediating the action of G-CSF (production of neutrophils, granulopoiesis) (Fukunaga *et al.*, 1991).

GENE

Accession numbers

Human G-CSFR: M59818 (Fukunaga *et al.*, 1990c; Larsen *et al.*, 1990)
Murine G-CSFR: M32699 (Fukunaga *et al.*, 1990a)

Sequence

See **Figure 1**.

Figure 1 Nucleotide sequence for the human G-CSFR gene.

```
Human G-CSFR
1    GAAGCTGGAC TGCAGCTGGT TTCAGGAACT TCTCTTGACG AGAAGAGAGA CCAAGGAGGC
61   CAAGCAGGGG CTGGGCCAGA GGTGCCAACA TGGGGAAACT GAGGCTCGGC TCGGAAAGGT
121  GAAGTAACTT GTCCAAGATC ACAAAGCTGG TGAACATCAA GTTGGTGCTA TGGCAAGGCT
181  GGGAAACTGC AGCCTGACTT GGGCTGCCCT GATCATCCTG CTGCTCCCCG GAAGTCTGGA
241  GGAGTGCGGG CACATCAGTG TCTCAGCCCC CATCGTCCAC CTGGGGGATC CCATCACAGC
301  CTCCTGCATC ATCAAGCAGA ACTGCAGCCA TCTGGACCCG GAGCCACAGA TTCTGTGGAG
361  ACTGGGAGCA GAGCTTCAGC CCGGGGGCAG GCAGCAGCGT CTGTCTGATG GGACCCAGGA
421  ATCTATCATC ACCCTGCCCC ACCTCAACCA CACTCAGGCC TTTCTCTCCT GCTGCCTGAA
481  CTGGGGCAAC AGCCTGCAGA TCCTGGACCA GGTTGAGCTG CGCGCAGGCT ACCCTCCAGC
541  CATACCCCAC AACCTCTCCT GCCTCATGAA CCTCACAACC AGCAGCCTCA TCTGCCAGTG
601  GGAGCCAGGA CCTGAGACCC ACCTACCCAC CAGCTTCACT CTGAAGAGTT TCAAGAGCCG
661  GGGCAACTGT CAGACCCAAG GGGACTCCAT CCTGGACTGC GTGCCCAAGG ACGGGCAGAG
721  CCACTGCTGC ATCCCACGCA AACACCTGCT GTTGTACCAG AATATGGGCA TCTGGGTGCA
781  GGCAGAGAAT GCGCTGGGGA CCAGCATGTC CCCACAACTG TGTCTTGATC CCATGGATGT
841  TGTGAAACTG GAGCCCCCCA TGCTGCGGAC CATGGACCCC AGCCCTGAAG CGGCCCCTCC
901  CCAGGCAGGC TGCCTACAGC TGTGCTGGGA GCCATGGCAG CCAGGCCTGC ACATAAATCA
961  GAAGTGTGAG CTGCGCCACA AGCCGCAGCG TGGAGAAGCC AGCTGGGCAC TGGTGGGCCC
1021 CCTCCCCTTG GAGGCCCTTC AGTATGAGCT CTGCGGGCTC CTCCCAGCCA CGGCCTACAC
1081 CCTGCAGATA CGCTGCATCC GCTGGCCCCT GCCTGGCCAC TGGAGCGACT GGAGCCCCAG
1141 CCTGGAGCTG AGAACTACCG AACGGGCCCC CACTGTCAGA CTGGACACAT GGTGGCGGCA
1201 GAGGCAGCTG GACCCCAGGA CAGTGCAGCT GTTCTGGAAG CCAGTGCCCC TGGAGGAAGA
1261 CAGCGGACGG ATCCAAGGTT ATGTGGTTTC TTGGAGACCC TCAGGCCAGG CTGGGGCCAT
1321 CCTGCCCCTC TGCAACACCA CAGAGCTCAG CTGCACCTTC CACCTGCCTT CAGAAGCCCA
1381 GGAGGTGGCC CTTGTGGCCT ATAACTCAGC CGGGACCTCT CGCCCCACCC CGGTGGTCTT
1441 CTCAGAAAGC AGAGGCCCAG CTCTGACCAG ACTCCATGCC ATGGCCCGAG ACCCTCACAG
1501 CCTCTGGGTA GGCTGGGAGC CCCCCAATCC ATGGCCTCAG GGCTATGTGA TTGAGTGGGG
1561 CCTGGGCCCC CCCAGCGCGA GCAATAGCAA CAAGACCTGG AGGATGGAAC AGAATGGGAG
1621 AGCCACGGGG TTTCTGCTGA AGGAGAACAT CAGGCCCTTT CAGCTCTATG AGATCATCGT
1681 GACTCCCTTG TACCAGGACA CCATGGGACC CTCCCAGCAT GTCTATGCCT ACTCTCAAGA
1741 AATGGCTCCC TCCCATGCCC CAGAGCTGCA TCTAAAGCAC ATTGGCAAGA CCTGGGCACA
1801 GCTGGAGTGG GTGCCTGAGC CCCCTGAGCT GGGGAAGAGC CCCCTTACCC ACTACACCAT
1861 CTTCTGGACC AACGCTCAGA ACCAGTCCTT CTCCGCCATC CTGAATGCCT CCTCCCGTGG
1921 CTTTGTCCTC CATGGCCTGG AGCCCGCCAG TCTGTATCAC ATCCACCTCA TGGCTGCCAG
1981 CCAGGCTGGG GCCACCAACA GTACAGTCCT CACCCTGATG ACCTTGACCC CAGAGGGGTC
2041 GGAGCTACAC ATCATCCTGG GCCTGTTCGG CCTCCTGCTG TTGCTCACCT GCCTCTGTGG
2101 AACTGCCTGG CTCTGTTGCA GCCCCAACAG GAAGAATCCC CTCTGGCCAA GTGTCCCAGA
2161 CCCAGCTCAC AGCAGCCTGG GCTCCTGGGT GCCCACAATC ATGGAGGAGG ATGCCTTCCA
2221 GCTGCCCGGC CTTGGCACGC CACCCATCAC CAAGCTCACA GTGCTGGAGG AGGATGAAAA
2281 GAAGCCGGTG CCCTGGGAGT CCCATAACAG CTCAGAGACC TGTGGCCTCC CCACTCTGGT
2341 CCAGACCTAT GTGCTCCAGG GGGACCCAAG AGCAGTTTCC ACCCAGCCCC AATCCCAGTC
2401 TGGCACCAGC GATCAGGTCC TTTATGGGCA GCTGCTGGGC AGCCCCACAA GCCCAGGGCC
2461 AGGGCACTAT CTCCGCTGTG ACTCCACTCA GCCCCTCTTG GCGGGCCTCA CCCCCAGCCC
2521 CAAGTCCTAT GAGAACCTCT GGTTCCAGGC CAGCCCCTTG GGGACCCTGG TAACCCCAGC
2581 CCCAAGCCAG GAGGACGACT GTGTCTTTGG GCCACTGCTC AACTTCCCCC TCCTGCAGGG
2641 GATCCGGGTC CATGGGATGG AGGCGCTGGG GAGCTTCTAG GGCTTCCTGG GGTTCCCTTC
2701 TTGGGCCTGC CTCTTAAAGG CCTGAGCTAG CTGGAGAAGA GGGGAGGGTC CATAAGCCCA
2761 TGACTAAAAA CTACCCCAGC CCAGGCTCTC ACCATCTCCA GTCACCAGCA TCTCCCTCTC
2821 CTCCCAATCT CCATAGGCTG GGCCTCCCAG GCGATCTGCA TACTTTAAGG ACCAGATCAT
2881 GCTCCATCCA GCCCCACCCA ATGGCCTTTT GTGCTTGTTT CCTATAACTT CAGTATTGTA
2941 AAC
```

Chromosome location and linkages

G-CSFR is on chromosome 1p35-34.3 in humans (Inazawa *et al.*, 1991) and on chromosome 4 in the mouse (Ito *et al.*, 1994).

PROTEIN

Accession numbers

Human G-CSFR: Q99062 (Fukunaga *et al.*, 1990c; Larsen *et al.*, 1990)

Mouse G-CSFR: P40223 (Fukunaga *et al.*, 1990a)

Sequence

See **Figure 2**.

Description of protein

Human G-CSFR comprises 836 amino acids with a signal sequence of 23 amino acids at the N-terminus (Fukunaga *et al.*, 1990b). A single transmembrane

Figure 2 Amino acid sequence for the human G-CSF receptor. Leader sequence is underlined and transmembrane domain is in bold and underlined.

```
MARLGNCSLT WAALIILLLP GSLEECGHIS VSAPIVHLGD PITASCIIKQ NCSHLDPEPQ
ILWRLGAELQ PGGRQQRLSD GTQESIITLP HLNHTQAFLS CCLNWGNSLQ ILDQVELRAG
YPPAIPHNLS CLMNLTTSSL ICQWEPGPET HLPTSFTLKS FKSRGNCQTQ GDSILDCVPK
DGQSHCCIPR KHLLLYQNMG IWVQAENALG TSMSPQLCLD PMDVVKLEPP MLRTMDPSPE
AAPPQAGCLQ LCWEPWQPGL HINQKCELRH KPQRGEASWA LVGPLPLEAL QYELCGLLPA
TAYTLQIRCI RWPLPGHWSD WSPSLELRTT ERAPTVRLDT WWRQRQLDPR TVQLFWKPVP
LEEDSGRIQG YVVSWRPSGQ AGAILPLCNT TELSCTFHLP SEAQEVALVA YNSAGTSRPT
PVVFSESRGP ALTRLHAMAR DPHSLWVGWE PPNPWPQGYV IEWGLGPPSA SNSNKTWRME
QNGRATGFLL KENIRPFQLY EIIVTPLYQD TMGPSQHVYA YSQEMAPSHA PELHLKHIGK
TWAQLEWVPE PPELGKSPLT HYTIFWTNAQ NQSFSAILNA SSRGFVLHGL EPASLYHIHL
MAASQAGATN STVLTLMTLT PEGSELHIIL GLFGLLLLLT CLCGTAWLCC SPNRKNPLWP
SVPDPAHSSL GSWVPTIMEE DAFQLPGLGT PPITKLTVLE EDEKKPVPWE SHNSSETCGL
PTLVQTYVLQ GDPRAVSTQP QSQSGTSDQV LYGQLLGSPT SPGPGHYLRC DSTQPLLAGL
TPSPKSYENL WFQASPLGTL VTPAPSQEDD CVFGPLLNFP LLQGIRVHGM EALGSF
```

domain of 26 amino acids divides the molecule into the extracellular region of 604 amino acids, and the cytoplasmic region of 183 amino acids.

Relevant homologies and species differences

Human and mouse G-CSF receptors have a homology of 62.5%, and there is no species specificity between human and mouse G-CSFRs (Nicola *et al.*, 1985; Fukunaga *et al.*, 1990c). The overall structure of G-CSFR is similar to that of gp130 (the IL-6 receptor signal transducer) (Hibi *et al.*, 1990) and LIF receptor (Gearing *et al.*, 1991), and leptin receptor (Tartaglia *et al.*, 1995). The ligand-binding domain (CRH domain, amino acids from 97 to 308 in mouse G-CSFR) in the extracellular region of G-CSFR (Fukunaga *et al.*, 1991; Layton *et al.*, 1997b) shows similarity to the corresponding region of other cytokine receptors (Bazan, 1990a,b; Fukunaga *et al.*, 1990a,c). Two regions (box 1 and box 2) of the membrane-proximal region of G-CSFR are related to the corresponding regions of other cytokine receptors, while another region (box 3) is related to the corresponding region of gp130 (Fukunaga *et al.*, 1991; Murakami *et al.*, 1991).

Affinity for ligand(s)

G-CSF binds to G-CSFR with K_d of about 100 pM (Fukunaga *et al.*, 1990b; Nicola and Metcalf, 1984).

Cell types and tissues expressing the receptor

G-CSFR is expressed in neutrophils, neutrophilic precursors in the bone marrow, myeloid leukemia cells, and placenta (Fukunaga *et al.*, 1990a,c; Nicola and Metcalf, 1984).

Regulation of receptor expression

Two regions in the promoter seem to be important for G-CSF receptor gene expression. C/EBPα binds to a region located at −49 from the transcription initiation site (Smith *et al.*, 1996), which is essential for the expression of G-CSFR (Zhang *et al.*, 1997). The other two *cis*-regulatory elements are located at +36 and +43, to which the ets family member PU.1. binds. Mutation of these sites reduces the promoter activity by 75% (Smith *et al.*, 1996). However, mice lacking PU.1 can express G-CSFR (Olson *et al.*, 1995), suggesting that this element may not be essential for the expression of the G-CSFR gene.

Release of soluble receptors

An alternatively spliced mRNA coding for a protein lacking the transmembrane region can be detected in human U937 cells (Fukunaga *et al.*, 1990c). However, the existence of soluble G-CSFR in human serum has not yet been reported.

SIGNAL TRANSDUCTION

Associated or intrinsic kinases

Binding of G-CSF to its receptor causes dimerization or oligomerization of the receptor (Ishizaka-Ikeda *et al.*, 1993). It is reported that at a low concentration of G-CSF, an asymmetric 2:1 receptor–ligand complex is formed, while at high ligand concentrations it is converted to 2:2 or 4:4 complex (Hiraoka *et al.*, 1995; Horan *et al.*, 1996). A 76 amino acid stretch proximal to the transmembrane domain, containing the box 1 and box 2 motifs, is essential for transducing growth signaling (Fukunaga *et al.*,

1993; Ziegler *et al.*, 1993) while both the N- and C-terminal domains of the cytoplasmic region are indispensable for transducing differentiation signals (Dong *et al.*, 1993; Fukunaga *et al.*, 1993). The C-terminal domain of the cytoplasmic region of the mouse G-CSFR contains four tyrosine residues (Tyr703, Tyr728, Tyr743, and Tyr763), which are phosphorylated upon stimulation by G-CSF (Pan *et al.*, 1993), and seems to be involved in different aspects of signal transduction (de Koning *et al.*, 1996b; Yoshikawa *et al.*, 1995). JAK1 is constitutively associated with the G-CSFR and becomes activated by the binding of G-CSF to the receptor (Nicholson *et al.*, 1994). G-CSF also activates JAK2 and TYK2 (Shimoda *et al.*, 1994; Tian *et al.*, 1996; Avalos *et al.*, 1997). The membrane-proximal region containing the box 1 and 2 motifs is required for the activation of the JAK kinases (Nicholson *et al.*, 1995).

In addition to the JAK family kinases, an src-related protein tyrosine kinase, Lyn, and a non-src-related Syk tyrosine kinase have been reported to be associated with the G-CSFR and activated upon G-CSF stimulation (Corey *et al.*, 1994). The requirement of Lyn kinase in the G-CSFR-mediated proliferation signal was demonstrated with a reconstitution system using a Lyn-deficient chicken B cell line (Corey *et al.*, 1998). On the other hand, irradiated mice reconstituted with Syk-deficient fetal liver show no gross perturbations in G-CSF responsiveness, suggesting no requirement of Syk for G-CSF signaling (Turner *et al.*, 1995). Several other tyrosine kinases such as Tec, a cytoplasmic src-related protein kinase, and p72sak tyrosine kinase, are tyrosine-phosphorylated and specifically activated by G-CSF (Matsuda *et al.*, 1995; Miyazato *et al.*, 1996). However, their physiological roles in G-CSF-induced signal transduction is unknown.

Cytoplasmic signaling cascades

The JAK kinases strongly activate STAT3 and weakly activate STAT1, which leads to the formation of STAT1 and STAT3 homodimeric and heterodimeric complexes (Tian *et al.*, 1994). Activation of other STAT proteins such as STAT5 and a novel STAT-like protein, STAT G in G-CSF-stimulated neutrophils has also been reported (Tweardy *et al.*, 1995; Nicholson *et al.*, 1996; Tian *et al.*, 1996). The activation of STAT3, but not of either STAT1 or STAT5, requires the membrane-distal region of the G-CSFR which carries Tyr703 (de Koning *et al.*, 1996b; Tian *et al.*, 1996). The surrounding sequence of Tyr703 is YXXQ, which fits to the consensus sequence for the STAT3-docking site found in gp130 (Stahl *et al.*, 1995). Although the tyrosine phosphorylation of the G-CSF receptor is not an absolute requirement for STAT3 activation (Cleveland *et al.*, 1989; de Koning *et al.*, 1996a; Nicholson *et al.*, 1996; Welte *et al.*, 1987), it is possible that its phosphorylation increases the affinity of STAT3 for this docking site. The activated, i.e. phosphorylated STAT3, seems to be released from the receptor by forming a homo- or heterodimer with STAT1, and is translocated into the nucleus (Shimozaki *et al.*, 1997). At high concentration of G-CSF, G-CSF activates STAT5 through the box 1 and box 2 region of G-CSFR, which is responsible for G-CSF-induced proliferation signals (Dong *et al.*, 1998).

The Ras/MAP kinase pathway is another signaling cascade activated by G-CSF. In the proB cell line BAF-B03, G-CSF activates Ras and MAP kinase (Bashey *et al.*, 1994; Nicholson *et al.*, 1995). Various molecules such as the Shc, Grb2, and Syp adaptors, and the vav guanine nucleotide exchanger are phosphorylated by G-CSF in various cells responding to G-CSF (de Koning *et al.*, 1996b), which are responsible for activation of the Ras/MAP kinase pathway. The activation of Ras requires the membrane-proximal region of G-CSFR (Barge *et al.*, 1996), as well as the membrane-distal region, specifically the domain containing Tyr763 (Duronio *et al.*, 1992; de Koning *et al.*, 1996b). It is likely that the JAK family kinases, activated through the membrane-proximal region of the receptor, phosphorylate Tyr763 of the receptor, to which adaptor molecules are recruited to activate the Ras/MAP kinase pathway. The kinds of genes activated by the signal from the Ras/MAP kinases are not elucidated yet.

In addition to tyrosine kinases and Ras/MAP kinase, G-CSF seems to regulate turnover of phosphatidylinositol. Upon binding of G-CSF to the receptor, phosphatidylinositol 3-kinase (PI-3 kinase) is recruited to the region containing Tyr-1 (amino acids 682–715), and it blocks apoptosis leading to cell survival (Hunter and Avalos, 1998). On the other hand, SH2-containing inositol phosphatase (SHIP) binds to the membrane-distal region of the receptor and, together with Shc, downregulates proliferation signals.

DOWNSTREAM GENE ACTIVATION

Transcription factors activated

As described above, G-CSF activates STAT1 and STAT3. c-rel, a proto-oncogene belonging to the

NFκB family, was also shown to be activated by G-CSF (Druker *et al.*, 1994; Avalos *et al.*, 1995). Although the box 1 motif in the membrane-proximal region of the receptor is required for NFκB activation, it is not clear what roles NFκB plays in G-CSF-mediated signaling. One possibility is that NFκB activation by G-CSF induces anti-apoptotic signals, as found in the IL-1 and TNF systems (Liu *et al.*, 1996b).

Genes induced

G-CSF induces in myeloid cells the expression of various neutrophil-specific genes such as myeloperoxidase (MPO) and neutrophilic elastase (Fukunaga *et al.*, 1993; Morishita *et al.*, 1987). Transcription of c-fos oncogene is upregulated in myeloid cells by treatment with G-CSF (Gonda and Metcalf, 1984), while the transcription of other oncogenes such as c-myc and c-myb is suppressed by the same treatment (Gonda and Metcalf, 1984; Shimozaki *et al.*, 1997).

Promoter regions involved

A DNA fragment of about 800 bp in the 5′ flanking region of human myeloperoxidase gene responds to G-CSF for gene activation in a myeloid cell-specific manner (Suzow and Friedman, 1993; Orita *et al.*, 1997). Several transcription factors, such as NF-Y, PEBP/CBP, and MyNF-1, were suggested to be involved in neutrophil-specific expression of myeloperoxidase (Suzow and Friedman, 1993; Nuchprayoon *et al.*, 1994; Orita *et al.*, 1997). However, the precise mechanism for the G-CSF-induced activation of neutrophil-specific genes is not elucidated yet.

BIOLOGICAL CONSEQUENCES OF ACTIVATING OR INHIBITING RECEPTOR AND PATHOPHYSIOLOGY

Unique biological effects of activating the receptors

Granulopoiesis (production of neutrophilic granulocytes).

Phenotypes of receptor knockouts and receptor overexpression mice

Similar to the G-CSF-null mice, the mice lacking G-CSFR show chronic *neutropenia*: the peripheral blood neutrophil level is 20–30% of those of wild-type mice (Liu *et al.*, 1996a). The number of neutrophilic precursor cells is also reduced in G-CSFR-null mice, confirming an involvement of G-CSF in the initial stage of neutrophilic development. The residual neutrophils in G-CSFR-null mice rapidly undergo apoptosis.

Human abnormalities

Severe congenital *neutropenia* (*Kostmann's syndrome*) is characterized by profound neutropenia and a maturation arrest of marrow progenitor cells at the promyelocyte–myelocyte stage. Somatic point mutations in one allele of the G-CSF receptor gene have been identified in some patients with severe congenital neutropenia (Dong *et al.*, 1994, 1995, 1997).

References

Avalos, B. R., Hunter, M. G., Parker, J. M., Ceselski, S. K., Druker, B. J., Corey, S. J., and Mehta, V. B. (1995). Point mutations in the conserved box 1 region inactivate the human granulocyte colony-stimulating factor receptor for growth signal transduction and tyrosine phosphorylation of p75c-rel. *Blood* **8535**, 3117–3126.

Avalos, B. R., Parker, J. M., Ware, D. A., Hunter, M. G., Sibert, K. A., and Druker, B. J. (1997). Dissociation of the Jak kinase pathway from G-CSF receptor signaling in neutrophils. *Exp. Hematol.* **25**, 160–168.

Barge, R. M., de Koning, J. P., Pouwels, K., Dong, F., Lowenberg, B., and Touw, I. P. (1996). Tryptophan 650 of human granulocyte colony-stimulating factor (G-CSF) receptor, implicated in the activation of JAK2, is also required for G-CSF-mediated activation of signaling complexes of the p21ras route. *Blood* **87**, 2148–2153.

Bashey, A., Healy, L., and Marshall, C. J. (1994). Proliferative but not nonproliferative responses to granulocyte colony-stimulating factor are associated with rapid activation of the p21ras/MAP kinase signalling pathway. *Blood* **83**, 949–957.

Bazan, J. F. (1990a). Haemopoietic receptors and helical cytokines. *Immunol. Today* **11**, 350–354.

Bazan, J. F. (1990b). Structural design and molecular evolution of a cytokine receptor superfamily. *Proc. Natl Acad. Sci. USA* **87**, 6934–6938.

Cleveland, J. L., Dean, M., Rosenberg, N., Wang, J. Y., and Rapp, U. R. (1989). Tyrosine kinase oncogenes abrogate interleukin-3 dependence of murine myeloid cells through signaling pathways c-myc: conditional regulation of c-myc transcription by temperature-sensitive v-abl. *Mol. Cell. Biol.* **264**, 11699–11705.

Corey, S. J., Burkhardt, A. L., Bolen, J. B., Geahlen, R. L., Tkatch, L. S., and Tweardy, D. J. (1994). Granulocyte colony-stimulating factor receptor signaling involves the formation of a three-component complex with Lyn and Syk protein-tyrosine kinases. *Proc. Natl Acad. Sci. USA* **91**, 4683–4687.

Corey, S. J., Dombrosky-Ferlan, P. M., Zuo, S., Krohn, E., Donnenberg, A. D., Zorich, P., Romero, G., Takata, M., and Kurosaki, T. (1998). Requirement of Src kinase Lyn for induction of DNA synthesis by granulocyte colony-stimulating factor. *J. Biol. Chem.* **273**, 3230–3235.

de Koning, J. P., Dong, F., Smith, L., Schelen, A. M., Barge, R. M., van der Plas, D. C., Hoefsloot, L. H., Lowenberg, B., and Touw, I. P. (1996a). The membrane-distal cytoplasmic region of human granulocyte colony-stimulating factor receptor is required for STAT3 but not STAT1 homodimer formation. *Blood* **87**, 1335–1342.

de Koning, J. P., Schelen, A. M., Dong, F., van Buitenen, C., Burgering, B. M., Bos, J. L., Lowenberg, B., and Touw, I. P. (1996b). Specific involvement of tyrosine 764 of human granulocyte colony-stimulating factor receptor in signal transduction mediated by p145/Shc/GRB2 or p90/GRB2 complexes. *Blood* **87**, 132–140.

Dong, F., van Buitenen, C., Pouwels, K., Hoefsloot, L. H., Löwenberg, B., and Touw, I. P. (1993). Distinct cytoplasmic regions of the human granulocyte colony-stimulating factor receptor involved in induction of proliferation and maturation. *Mol. Cell. Biol.* **13**, 7774–7781.

Dong, F., Hoefsloot, L. H., Schelen, A. M., Broeders, C. A., Meijer, Y., Veerman, A. J., Touw, I. P., and Lowenberg, B. (1994). Identification of a nonsense mutation in the granulocyte-colony-stimulating factor receptor in severe congenital neutropenia. *Proc. Natl Acad. Sci. USA* **91**, 4480–4484.

Dong, F., Brynes, R. K., Tidow, N., Welte, K., Lowenberg, B., and Touw, I. P. (1995). Mutations in the gene for the granulocyte colony-stimulating-factor receptor in patients with acute myeloid leukemia preceded by severe congenital neutropenia. *N. Engl. J. Med.* **333**, 487–493.

Dong, F., Dale, D. C., Bonilla, M. A., Freedman, M., Fasth, A., Neijens, H. J., Palmblad, J., Briars, G. L., Carlsson, G., Veerman, A. J., Welte, K., Lowenberg, B., and Touw, I. P. (1997). Mutations in the granulocyte colony-stimulating factor receptor gene in patients with severe congenital neutropenia. *Leukemia* **11**, 120–125.

Dong, F., Liu, X., de, K. J., Touw, I. P., Henninghausen, L., Larner, A., and Grimley, P. M. (1998). Stimulation of Stat5 by granulocyte colony-stimulating factor (G-CSF) is modulated by two distinct cytoplasmic regions of the G-CSF receptor. *J. Immunol.* **161**, 6503–6509.

Druker, B. J., Neumann, M., Okuda, K., Franza, B. J., and Griffin, J. D. (1994). rel is rapidly tyrosine-phosphorylated following granulocyte-colony stimulating factor treatment of human neutrophils. *J. Biol. Chem.* **269**, 5387–5390.

Duronio, V., Clark-Lewis, I., Federsppiel, B., Wieler, J., and Schrader, J. W. (1992). Tyrosine phosphorylation of receptor b subunits and common substrates in response to interleukin-3 and granulocyte-macrophage colony-stimulating factor. *J. Biol. Chem.* **267**, 21856–21863.

Fukunaga, R., Ishizaka-Ikeda, E., Seto, Y., and Nagata, S. (1990a). Expression cloning of a receptor for murine granulocyte colony-stimulating factor. *Cell* **61**, 341–350.

Fukunaga, R., Seto, Y., Mizushima, S., and Nagata, S. (1990b). Three different mRNAs encoding human granulocyte colony-stimulating factor receptor. *Proc. Natl Acad. Sci. USA* **87**, 8702–8706.

Fukunaga, R., Ishizaka-Ikeda, E., and Nagata, S. (1990c). Purification and characterization of the receptor for murine granulocyte colony-stimulating factor. *J. Biol. Chem.* **265**, 14008–14015.

Fukunaga, R., Ishizaka-Ikeda, E., Pan, C.-X., Seto, Y., and Nagata, S. (1991). Functional domains of the granulocyte colony-stimulating factor receptor. *EMBO J.* **10**, 2855–2865.

Fukunaga, R., Ishizaka-Ikeda, E., and Nagata, S. (1993). Growth and differentiation signals mediated by two distinct regions in the cytoplasmic domain of G-CSF receptor. *Cell* **14**, 1079–1087.

Gearing, D. P., Thut, C. J., VandeBos, T., Gimpel, S. D., Delaney, P. B., King, J., Price, V., Cosman, D., and Beckmann, M. P. (1991). Leukemia inhibitory factor receptor is structurally related to the IL-6 signal transducer, gp130. *EMBO J.* **10**, 2839–2348.

Gonda, T. J., and Metcalf, D. (1984). Expression of *myb*, *myc* and *fos* proto-oncogenes during the differentiation of a murine myeloid leukemia. *Nature* **310**, 249–251.

Hibi, M., Murakami, M., Saito, M., Hirano, T., Taga, T., and Kishimoto, T. (1990). Molecular cloning and expression of an IL-6 signal transducer, gp130. *Cell* **63**, 1149–1157.

Hiraoka, O., Anaguchi, H., Asakura, A., and Ota, Y. (1995). Requirement for the immunoglobulin-like domain of granulocyte colony-stimulating factor receptor in formation of a 2 : 1 receptor-ligand complex. *J. Biol. Chem.* **270**, 25928–25934.

Horan, T., Wen, J., Narhi, L., Parker, V., Garcia, A., Arakawa, T., and Philo, J. (1996). Dimerization of the extracellular domain of granulocyte-colony stimulating factor receptor by ligand binding: a monovalent ligand induces 2:2 complexes. *Biochemistry* **35**, 4886–4896.

Hunter, M. G., and Avalos, B. R. (1998). Phosphatidylinositol 3 kinase and SH2-containing inositol phosphatase (SHIP) are recruited by distinct positive and negative growth-regulatory domains in the granulocyte colony-stimulating factor receptor. *J. Immunol.* **160**, 4979–4987.

Inazawa, J., Fukunaga, R., Seto, Y., Nakagawa, H., Misawa, S., Abe, T., and Nagata, S. (1991). Assignment of the human granulocyte colony-stimulating factor receptor gene (CSF3R) to chromosome 1 at region p35–p34.3. *Genomics* **10**, 1075–1078.

Ishizaka-Ikeda, E., Fukunaga, R., Wood, W. I., Goeddel, D. V., and Nagata, S. (1993). Signal transduction mediated by growth hormone receptor and its chimeric molecules with the granulocyte colony-stimulating factor receptor. *Proc. Natl Acad. Sci. USA* **90**, 123–127.

Ito, Y., Seto, Y., Brannan, C. I., Copeland, N. G., Jenkins, N. A., Fukunaga, R., and Nagata, S. (1994). Structural analysis of the functional gene and pseudogene encoding the murine granulocyte colony-stimulating-factor receptor. *Eur. J. Biochem.* **220**, 881–891.

Larsen, A., Davis, T., Curtis, B. M., Gimpel, S., Sims, J. E., Cosman, D., Park, L., Sorensen, E., March, C. J., and Smith, C. A. (1990). Expression cloning of a human granulocyte colony-stimulating factor receptor: a structural mosaic of hematopoietin receptor, immunoglobulin, and fibronectin domains. *J. Exp. Med.* **172**, 1559–1570.

Layton, J. E., Iaria, J., and Nicholson, S. E. (1997a). Neutralising antibodies to the granulocyte colony-stimulating factor receptor recognise both the immunoglobulin-like domain and the cytokine receptor homologous domain. *Growth Factors* **14**, 117–130.

Layton, J. E., Iaria, J., Smith, D. K., and Treutlein, H. R. (1997b). Identification of a ligand-binding site on the granulocyte colony-stimulating factor receptor by molecular modeling and mutagenesis. *J. Biol. Chem.* **272**, 29735–29741.

Liu, F., Wu, H. Y., Wesselschmidt, R., Kornaga, T., and Link, D. C. (1996a). Impaired production and increased apoptosis of neutrophils in granulocyte colony-stimulating factor receptor-deficient mice. *Immunity* **5**, 491–501.

Liu, Z.-G., Hsu, H., Goeddel, D., and Karin, M. (1996b). Dissection of TNF receptor 1 effector functions: JNK activation is not linked to apoptosis while NF-κB activation prevents cell death. *Cell* **87**, 565–576.

Matsuda, T., Takahashi, T. M., Fukada, T., Okuyama, Y., Fujitani, Y., Tsukada, S., Mano, H., Hirai, H., Witte, O. N., and Hirano, T. (1995). Association and activation of Btk and Tec tyrosine kinases by gp130, a signal transducer of the interleukin-6 family of cytokines. *Blood* **85**, 627–633.

Miyazato, A., Yamashita, Y., Hatake, K., Miura, Y., Ozawa, K., and Mano, H. (1996). Tec protein tyrosine kinase is involved in the signaling mechanism of granulocyte colony-stimulating factor receptor. *Cell Growth Differ.* **7**, 1135–1139.

Morishita, K., Tsuchiya, M., Asano, S., Kaziro, Y., and Nagata, S. (1987). Chromosomal gene structure of human myeloperoxidase and regulation of its expression by granulocyte colony-stimulating factor. *J. Biol. Chem.* **262**, 3844–3851.

Murakami, M., Narazaki, M., Hibi, M., Yawata, H., Yasukawa, K., Hamaguchi, M., Taga, T., and Kishimoto, T. (1991). Critical cytoplasmic region of the interleukin 6 signal transducer gp130 is conserved in the cytokine receptor family. *Proc. Natl Acad. Sci. USA* **88**, 11349–11353.

Nicholson, S. E., Oates, A. C., Harpur, A. G., Ziemiecki, A., Wilks, A. F., and Layton, J. E. (1994). Tyrosine kinase JAK1 is associated with the granulocyte-colony-stimulating factor receptor and both become tyrosine-phosphorylated after receptor activation. *Proc. Natl Acad. Sci. USA* **91**, 2985–2988.

Nicholson, S. E., Novak, U., Ziegler, S. F., and Layton, J. E. (1995). District regions of the granulocyte colony-stimulating factor receptor are required for tyrosine phosphorylation of the signaling molecules JAK2, Stat3, and p42, p44MAPK. *Blood* **86**, 3698–3704.

Nicholson, S. E., Starr, R., Novak, U., Hilton, D. J., and Layton, J. E. (1996). Tyrosine residues in the granulocyte colony-stimulating factor (G-CSF) receptor mediate G-CSF-induced differentiation of murine myeloid leukemic (M1) cells. *J. Biol. Chem.* **271**, 26947–26953.

Nicola, N. A., and Metcalf, D. (1984). Binding of the differentiation-inducer, granulocyte colony-stimulating factor, to responsive but not unresponsive leukemic cell lines. *Proc. Natl Acad. Sci. USA* **81**, 8765–8769.

Nicola, N. A., Begley, C. G., and Metcalf, D. (1985). Identification of the human analogue of a regulator that induces differentiation in murine leukaemic cells. *Nature* **314**, 625–628.

Nuchprayoon, I., Meyers, S., Scott, L. M., Suzow, J., Hiebert, S., and Friedman, A. D. (1994). PEBP2/CBF, the murine homolog of the human myeloid AML1 and PEBP2 beta/CBF beta proto-oncoproteins, regulates the murine myeloperoxidase, and neutrophil elastase genes in immature myeloid cells. *Mol. Cell. Biol.* **14**, 5558–5568.

Olson, M. C., Scott, E. W., Hack, A. A., Su, G. H., Tenen, D. G., Singh, H., and Simon, M. C. (1995). PU.1 is not essential for early myeloid gene expression but is required for terminal myeloid differentiation. *Immunity* **3**, 703–714.

Orita, T., Shimozaki, K., Murakami, H., and Nagata, S. (1997). Binding of NF-Y transcription factor to one of cis-elements in the myeloperoxidase gene promoter that responds to G-CSF. *J. Biol. Chem.* **272**, 23216–23223.

Pan, C.-X., Fukunaga, R., Yonehara, S., and Nagata, S. (1993). Uni-directional cross-phosphorylation between the G-CSF and IL-3 receptors. *J. Biol. Chem.* **268**, 25818–25823.

Shimoda, K., Iwasaki, H., Okamura, S., Ohno, Y., Kubota, A., Arimura, F., Otsuka, T., and Niho, Y. (1994). G-CSF induces tyrosine phosphorylation of the JAK2 protein in the human myeloid G-CSF responsive and proliferative cells, but not in mature neutrophils. *Biochem. Biophys. Res. Commun.* **203**, 922–928.

Shimozaki, K., Nakajima, K., Hirano, T., and Nagata, S. (1997). Involvement of STAT3 in the granulocyte colony stimulating factor-induced differentiation of myeloid cells. *J. Biol. Chem.* **272**, 25184–25189.

Smith, L. T., Hohaus, S., Gonzalez, D. A., Dziennis, S. E., and Tenen, D. G. (1996). PU.1 (Spi-1) and C/EBP alpha regulate the granulocyte colony-stimulating factor receptor promoter in myeloid cells. *Blood* **88**, 1234–1247.

Stahl, N., Farruggella, T. J., Boulton, T. G., Zhong, Z., Darnell Jr., J. E., and Yancopoulos, G. D. (1995). Choice of STATs and other substrates specified by modular tyrosine-based motifs in cytokine receptors. *Science* **267**, 1349–1353.

Suzow, J., and Friedman, A. D. (1993). The murine myeloperoxidase promoter contains several functional elements, one of which binds a cell type-restricted transcription factor, myeloid nuclear factor 1 (MyNF1). *Mol. Cell. Biol.* **13**, 2141–2151.

Tartaglia, L. A., Dembski, M., Weng, X., Deng, N., Culpepper, J., Devos, R., Richards, G. J., Campfield, L. A., Clark, F. T., and Deeds, J. (1995). Identification and expression cloning of a leptin receptor, OB-R. *Cell* **83**, 1263–1271.

Tian, S. S., Lamb, P., Seidel, H. M., Stein, R. B., and Rosen, J. (1994). Rapid activation of the STAT3 transcription factor by granulocyte colony-stimulating factor. *Blood* **84**, 1760–1764.

Tian, S. S., Tapley, P., Sincich, C., Stein, R. B., Rosen, J., and Lamb, P. (1996). Multiple signaling pathways induced by granulocyte colony-stimulating factor involving activation of JAKs, STAT5, and/or STAT3 are required for regulation of three distinct classes of immediate early genes. *Blood* **88**, 4435–4444.

Turner, M., Mee, P. J., Costello, P. S., Williams, O., Price, A. A., Duddy, L. P., Furlong, M. T., Geahlen, R. L., and Tybulewicz, V. L. (1995). Perinatal lethality and blocked B-cell development in mice lacking the tyrosine kinase Syk. *Nature* **378**, 298–302.

Tweardy, D. J., Wright, T. M., Ziegler, S. F., Baumann, H., Chakraborty, A., White, S. M., Dyer, K. F., and Rubin, K. A. (1995). Granulocyte colony-stimulating factor rapidly activates a distinct STAT-like protein in normal myeloid cells. *Blood* **86**, 4409–4416.

Welte, K., Bonilla, M. A., Gabrilove, J. L., Gillio, A. P., Potter, G. K., Moore, M. A. S., O'Reilly, R. J., Boone, T. C., and Souza, L. M. (1987). Recombinant human granulocyte-colony stimulating factor: *In vitro* and *in vivo* effects on myelopoiesis. *Blood Cells* **13**, 17–30.

Yoshikawa, A., Murakami, H., and Nagata, S. (1995). Distrinct signal transduction through the tyrosine-containing domains of the granulocyte colony-stimulating factor receptor. *EMBO J.* **14**, 5288–5296.

Zhang, D. E., Zhang, P., Wang, N. D., Hetherington, C. J., Darlington, G. J., and Tenen, D. G. (1997). Absence of granulocyte colony-stimulating factor signaling and neutrophil development in CCAAT enhancer binding protein alpha-deficient mice. *Proc. Natl Acad. Sci. USA* **94**, 569–574.

Ziegler, S. F., Bird, T. A., Morella, K. K., Mosley, B., Gearing, D. P., and Baumann, H. (1993). Distinct regions of the human granulocyte-colony-stimulating factor receptor cytoplasmic domain are required for proliferation and gene induction. *Mol. Cell. Biol.* **13**, 2384–2390.

LICENSED PRODUCTS

Mouse anti-human G-CSFR monoclonal antibody (clone LMM 741) (Layton *et al.*, 1997a; Nicholson *et al.*, 1994) is available from PharMingen (San Diego).

TPO Receptor

David J. Kuter* **and Junzhi Li**

Hematology Unit, COX 640, Massachusetts General Hospital, 100 Blossom Street, Boston, MA 02114, USA

*corresponding author tel: 617-726-8743, fax: 617-724-3166, e-mail: kuter.david@MGH.harvard.edu

DOI: 10.1006/rwcy.2000.20008.

SUMMARY

The thrombopoietin receptor, c-*mpl*, is a member of the hematopoietin/cytokine receptor superfamily and consists of two duplicated hematopoietin receptor domains (HRD). The proximal HRD is responsible for receptor dimerization and consequent signal transduction and the distal HRD may regulate its action. The thrombopoietin receptor is found on many hematopoietic progenitors but is primarily found on megakaryocytes and platelets. The platelet receptor is of high binding affinity (120–200 pM) but low surface density (30–60 sites per cell). Receptor binding by thrombopoietin results in the promotion of cell growth/cellular differentiation, the prevention of apoptosis, and the internalization and degradation of the receptor–ligand complex. Loss of the distal HRD by fusion with the murine leukemia virus env protein results in unregulated expression of c-*mpl* and a myeloproliferative syndrome in mice. Mice deficient in the thrombopoietin receptor produce 10–15% as many megakaryocytes and platelets as wild-type mice but also have 25–35% the normal number of erythroid and myeloid progenitors despite normal peripheral red and white blood cell counts. No human abnormalities of c-*mpl* have yet been described and therapeutic use of the soluble receptor has not yet been demonstrated.

BACKGROUND

Discovery

In 1986 Wendling *et al.* reported that a murine retrovirus, the *myeloproliferative leukemia virus* (MPLV), caused an acute hematological disorder in most mice strains that was characterized by multilineage proliferation and differentiation. Cells in this disease became growth factor-independent and showed autonomous growth *in vitro*. By sequencing the DNA encoding the envelope gene of the MPLV a novel oncogene called v-*mpl* was found that was in phase with two parts of the Friend murine leukemia virus envelope gene. v-*mpl* showed close structural analogies with the hematopoietin receptor superfamily but was not related to any known receptor. In infected cells the *env-mpl* fusion gene produced a truncated transmembrane receptor for a then unidentified hematopoietic growth factor (Souryi *et al.*, 1990).

When the full-length cellular homolog, c-*mpl*, was cloned in 1992 (Vigon *et al.*, 1992) it was found to be a novel hematopoietin receptor that was present in a large number of hematopoietic cells and most megakaryocytic cell lines (Vigon *et al.*, 1993b,c). Subsequently it was shown that c-*mpl* was present in purified populations of platelets and megakaryocytes and that antisense oligonucleotides against c-*mpl* inhibited megakaryocyte growth (Methia *et al.*, 1993). This close association between c-*mpl* and megakaryocyte growth led to the use of recombinant c-Mpl protein on an affinity column to purify its ligand (Bartley *et al.*, 1994; de Sauvage *et al.*, 1994; Kato *et al.*, 1995). It is now known that the c-Mpl ligand is indeed the hematopoietic growth factor long called thrombopoietin (Kelemen *et al.*, 1958). Final proof of this association was provided by experiments in mice in which the c-*mpl* genes were eliminated; such animals had less than 10% of the normal number of platelets and megakaryocytes (Gurney *et al.*, 1994).

Alternative names

The terms thrombopoietin (TPO) receptor (TPOR), c-Mpl, c-Mpl receptor, Mpl, and Mpl receptor are used interchangeably.

Structure

The TPO receptor is a member of the hematopoietin/cytokine receptor superfamily, which is characterized by a highly conserved HRD in the extracellular region (Vigon *et al.*, 1992; Drachman and Kaushansky, 1995). Although the HRD of different members of this family shows only about 20–30% amino acid sequence homology, each HRD contains two fibronectin type III (FNIII)-like domains, each of which consists of seven β strands. Two to four amino acid residues link the two FNIII-like domains. The N-terminus of the first FNIII-like domain contains four highly conserved cysteine residues which form two disulfide bonds between β strands A/B and C′/E, respectively. Substitution or deletion of any of these four cysteine residues abolishes receptor binding of the ligand, presumably due to disruption of proper conformation at the ligand binding site. At the C-terminus of the second FNIII-like domain, a 5 amino acid sequence, termed the WSXWS motif, is highly conserved in the hematopoietin/cytokine receptor superfamily. Although the WSXWS motif is located away from the ligand binding site in the three-dimensional structural model, mutation of this motif also results in inactivation of ligand binding.

Along with a few other members of the hematopoietin/cytokine receptor superfamily, like the leukemia-inhibitory factor receptor and the β chain of the IL-3 receptor, the TPO receptor has two duplicated HRDs (referred to hereafter as the membrane-proximal HRD and the membrane-distal HRD). In addition, the TPO receptor is unique in having a 50 amino acid insertion between β strands C′ and E in the second FNIII-like domain of the membrane-distal HRD.

Structure–function analysis has demonstrated that the membrane proximal HRD is responsible for the dimerization and consequent signal transduction that occurs upon ligand binding (Alexander *et al.*, 1995). Whether both proximal and distal HRD can bind ligand is unclear. Recent data (Sabath *et al.*, 1999) using mutants of c-*mpl* suggest that the distal HRD binds thrombopoietin and is responsible for thrombopoietin-dependent growth. Furthermore, in the absence of ligand the distal HRD seems to inhibit receptor activation; cells transfected with a c-*mpl* construct lacking the distal HRD displayed thrombopoietin-independent growth, just like v-*mpl*, and did not bind thrombopoietin.

Four isoforms of the TPO receptor mRNA have been identified by cDNA sequencing; all of them arise from alternative splicing of c-*mpl* RNA (Vigon *et al.*, 1992; Kiladjian *et al.*, 1997; Li *et al.*, 2000). Of these, the P isoform is the only one demonstrated to be a functional receptor. After cleavage of the signal peptide, the mature wild-type TPO receptor (P form) is composed of an extracellular domain of 463 amino acids, a transmembrane domain of 22 amino acids, and intracellular domain of 122 amino acids. The biological and/or physiological role of the other three alternative RNA splicing isoforms is completely unclear. The K isoform lacks much of the cytoplasmic domain and presumably would lack the ability to initiate signal transduction pathways. A third RNA isoform lacks the transmembrane region and would be predicted to be a soluble receptor (Skoda *et al.*, 1993). The existence of the protein products from these two RNA isoforms has not been demonstrated *in vivo*. The fourth, called c-*mpl-del*, lacks 72 bp in the extracellular region of c-*mpl* and arises as a consequence of alternative RNA splicing between exons 8 and 9 (Kiladjian *et al.*, 1997; Li *et al.*, 2000). c-*mpl-del* mRNA and protein are expressed in platelets, megakaryocytes, and CD34 progenitor cells but the protein does not seem to be transported to the cell surface and hence it lacks biological activity.

Main activities and pathophysiological roles

Like all other members of the hematopoietin/cytokine receptor superfamily, the extracellular domain of c-Mpl must provide two essential functions in order to permit TPO signal transduction: a high-affinity binding site for TPO and a dimerization site upon ligand binding. As a model system for analyzing the TPO receptor, platelets have a single class of binding sites (approximately 50–60 per platelet) for TPO with binding affinity (K_d) of 100–200 pM (Li *et al.*, 1996, 1999b; Broudy *et al.*, 1997; Fielder *et al.*, 1997).

Receptor binding initiates at least three major events in cells. The first is initiation of signal transduction pathways that promote cell growth and/or cellular differentiation. The second is the activation of signal transduction pathways that prevent apoptosis. The third is internalization and degradation of the receptor–ligand complex for regulating the circulating TPO level (Kuter and Rosenberg, 1995; Fielder *et al.*, 1996; Kuter, 1996a; Broudy *et al.*, 1997; Fielder *et al.*, 1997; Li *et al.*, 1999).

The consequences of receptor binding vary with the cell type. For megakaryocyte colony-forming cells (Meg-CFC), receptor activation inhibits apoptosis and increases the rate of mitosis. For early megakaryocytes, receptor binding inhibits apoptosis, stimulates megakaryocyte endomitosis, and promotes cytoplasmic maturation (Kuter, 1996a, 1997). This results in an increase in the number, size, and ploidy of bone marrow megakaryocytes. For late, mature megakaryocytes receptor binding may actually inhibit platelet shedding (Choi *et al.*, 1996; Nagahisi *et al.*, 1996; Sheridan *et al.*, 1997). For platelets, receptor binding produces a decreased threshold for platelet activation (Chen *et al.*, 1995; Harker *et al.*, 1996; Kubota *et al.*, 1996; Montrucchio *et al.*, 1996). For platelets and possibly megakaryocytes, receptor binding results in internalization and degradation of thrombopoietin. This is the major mechanism by which circulating thrombopoietin levels are regulated (Kuter, 1996a,b, 1997; Li *et al.*, 1999b). For other early, nonmegakaryocyte bone marrow progenitors and the pluripotential stem cell, receptor activation prevents apoptosis and modestly increases the progenitor number. This pluripotential effect is most strikingly illustrated in mice in which c-*mpl* has been eliminated. Although the circulating red and white blood cell numbers were not altered, the number of erythroid and myeloid progenitor cells was reduced to 20–30% of normal (Carver-Moore *et al.*, 1996). Finally, there is no TPO effect (and probably no TPO receptor) on the more mature nonmegakaryocyte precursors in the bone marrow.

At the present time there are no human disorders directly attributable to changes in c-Mpl structure or expression. The initial description of the murine v-*mpl-env* fusion protein in the MPLV-infected mice was associated with a myeloproliferative syndrome in which all lineages were stimulated. Deletion of all but the 43 most membrane proximal amino acids of the extracellular domain was responsible for this oncogenic effect. The structural basis for this oncogene activation has recently been explored (Sabath *et al.*, 1999). The intact c-*mpl* cDNA was compared with a c-*mpl* construct lacking the distal HRD. Upon transfection of the wild-type c-*mpl* cDNA into Baf3 cells, the cells exhibited thrombopoietin-dependent growth and bound thrombopoietin. However, the cells transfected with the c-*mpl* construct lacking the distal HRD displayed thrombopoietin-independent growth, just like v-*mpl*, and did not bind thrombopoietin. These results suggest that the distal HRD imposes some inhibitory effect on the proximal HRD, binds thrombopoietin, and is responsible for thrombopoietin dependence; absence of the distal HRD confers thrombopoietin-independent growth.

In *polycythemia vera* (Moliterno *et al.*, 1998) and in *essential thrombocythemia* (Li *et al.*, 1996) the amount of thrombopoietin receptor on platelets seems to be reduced. Whether this plays any pathophysiological role is unclear. Platelets from patients with polycythemia vera failed to undergo tyrosine phosphorylation upon exposure to thrombopoietin and had reduced levels of c-Mpl on their surface. Platelets from patients with essential thrombocythemia had a normal binding affinity for thrombopoietin but had only 5.6 receptors/platelet, compared with the 56 receptors/platelet found on normal platelets (Li *et al.*, 1996).

The thrombopoietin receptor is found on cells and cell lines from many nonlymphoid hematological malignancies but not on cells from purely lymphoid malignancies or nonhematopoietic solid tumors (Columbyova *et al.*, 1995; Drexler and Quentmeier, 1996; Graf *et al.*, 1996; Matsumura *et al.*, 1996; Quentmeier *et al.*, 1996; Bredoux *et al.*, 1997; Hirai *et al.*, 1997). There is a suggestion that the presence of c-*mpl* in *acute myeloid leukemia* cells confers a poor prognosis (Vigon *et al.*, 1993a; Wetzler *et al.*, 1997).

GENE

Accession numbers

The accession numbers of the GenBank database for both human genomic DNA and cDNA of the TPO receptor are available from National Center for Biotechnology Information (NCBI: www.ncbi.org). These are U68159 (exons 1–6), U68160 (exons 7 and 8), U68161 (exons 9 and 10), and U68162 (exons 11 and 12) for the genomic DNA, M90102 and M90103 for the P and K forms of the cDNA, respectively. The accession numbers for mouse genomic and cDNA are Z22657, X73677, and Z22649, respectively.

Sequence

See **Figure 1** for the human genomic DNA sequence.

PROTEIN

Accession numbers

The accession numbers for cDNA-derived P and K forms of the human TPO receptor protein are AAA69971 and AAA69972, respectively. The accession number for the genomic DNA-derived TPO

Figure 1 Human genomic DNA sequence.

```
   1 CAAGGGGGCA GGGTAAGGAG TGTGAGCCAT CTCCAATCTG AGCAAACAGG ATAAAAGTAC
  61 TGAAGACCAT TGCTTCTCCA ACTTTAAGTT ACATTCAAAC CATCTGGGGA TTTTGGCAAA
 121 ATGCAGGTTC AACTTTAATA GGCCTGGGGT GGGGCCTAAG ATTCTGCATT TCTAATAAGT
 181 GCCCACCTAA TGCTGTTGCT CCTGGTCTAC AGACCACATT TTGAGTAGCA AGGGATTAGA
 241 GGCAGCCGTG GAGGCCACCC ATAATGGAGA GGGCATTTCC AGTTCCAGGG AGAATGTGTT
 301 TATTCATGCG CAAAGTCACA GAACTATTAA AGTGGAAATA CCCACCATTT GGCATGGTTG
 361 TGGGGCAGAG AGTTATTTCT CAGCAGATTA GCCTCCCAAG AGCATAGCAC AGCATGGCTG
 421 TTATGCAGGT GGGTTCTGAA GGTTGAACCT CCTGAGCTTT ATGTCTTAGT ATGTAATCTT
 481 GGGCAAATCT TTAACTTCTC TGTATCTCCG TTTCCTCATC TCTATAATGG AAATAACAAG
 541 ACTATTGTGC GGATTGTCTA AGTGAATGGA TATAAAGTGC TTAACACACG GCAGCATAGT
 601 AAGTGCTCAG TAAAAGTTGC TATTCTGATG CTATTATACC ATTCTATAAT CGTTAGCACA
 661 GATACAGAGG CTGAGTTGCC TTTTGGTACA CGGTCAAATA TACAACCCCC ATCTCCCTCA
 721 CACCAAAAGG CCTGTGTCCT CTCTTTCAAA TTCCTCCTTC CCTCCTGCCC ACTCCCCCTC
 781 CCCTGGCCCC TGGCCCCAGT GTGGTCTGGA TGGGCCCCAG AGGGGCAGGG ACAGGGACAG
 841 GACGTGGGGC TGTATCTGAC AGGAACCTGA GGGGCTGGCC TGGGAGGGGA TTGGGGCCCA
 901 GCTTCCTGAA GGGAGGATGG GCTAAGGCAG GCACACAGTG GCGGAGAAGA TGCCCTCCTG
 961 GGCCCTCTTC ATGGTCACCT CCTGCCTCCT CCTGGCCCCT CAAAACCTGG CCCAAGTCAG
1021 CAGCCAAGGT GAGGTGCACA GAGGGTGGAG ATCACCTATG CCCCAGGAAG AGGGAGCCCT
1081 GGGAGGTGAT GCAGGGCCCC CGGAGGGGAG GTAGAGTAAG AGGCTCTCCT GCTGGTCCCC
1141 TCCCCTTCCA CATAAACATG CCTGGGAGGA CCCAGGGCCA ACTCACCAGC TGTTCCTTAG
1201 ATGTCTCCTT GCTGGCATCA GACTCAGAGC CCCTGAAGTG TTTCTCCCGA ACATTTGAGG
1261 ACCTCACTTG CTTCTGGGAT GAGGAAGAGG CAGCGCCCAG TGGGACATAC CAGCTGCTGT
1321 ATGCCTACCC GCGGTAGGTG CTGGACTGTG CCCCACTCCC CATGTATCTG TCCCTGCACT
1381 TAGCTGAGTC CCACTCCAGC AGCTTTCCTG CCTGTCCGAG GACCACTCTG AATACAAGCC
1441 CTAAGTACCT ACTTTTTGAA CAGATGTGCT TTGGATGTAT GTGGGCCCCA GCTCCAGCCC
1501 TACATAACCC CTAATCCCAC CTATCCCAGG CAGTGAGAAG AAAAATGGCA GTACTAGAGA
1561 CAGAAGTTGG GCATGGGCCC AGGTCTGGGT CCTCAGGCGT CCGCATGGTG GCTGTGTAGG
1621 AGGGACCTCT TCTATGCCAA CAGGGAGAAG CCCCGTGCTT GCCCCCTGAG TTCCCAGAGC
1681 ATGCCCCACT TTGGAACCCG ATACGTGTGC CAGTTTCCAG ACCAGGAGGA AGTGCGTCTC
1741 TTCTTTCCGC TGCACCTCTG GGTGAAGAAT GTGTTCCTAA ACCAGACTCG GACTCAGCGA
1801 GTCCTCTTTG TGGACAGTGT AGGTAAGAGC CATCCTCCTG TCACCCTGCC CCCTCCACTT
1861 GCTGCCCCCA GTCCAGCTCC CGGAATCAGA CTTGCCTGTG CCCTTCCAGC TCAGCACGGA
1921 TACCCCTATT ACCAGACCCC CTGAGGCACC CCAAGACTCC CTTGTCATTC CTCCCAGCCT
1981 TGGCATCTAG AGCTAGAAAT GCCCAACAGA TCATTCACAC CCTGTNNCTC CCATCGTCAA
2041 CAAATCATTT ATTCATCCAT TCAAGAGTTA CAGAATACCT ACTGTGTGCC AGACACTGGG
2101 CTAGGTGATG GGGATATGGA GAGAAAAAGG GATCCCTGCT TACAAGAAGC ACTGAGTCTA
2161 GCAAGCCATC ATACTGCAAA GAAGTCACTG TTCTCATGAG AGAAGAACAG AATCAGGAGC
2221 CTCCTCAGCC TGCCCTCCAG GAGAATGACA GCCTACAATT TTTGAATCCA GAAGCTGCCC
2281 CAATTCAGCC CCCAGCACCC CTCTCTGCAG TCCAGAGGCT GAGCCATAGA CTGTGGTACT
2341 CAGAGTTCTG ATGTGCCCTG TCTTGCCCTC AGGCCTGCCG GCTCCCCCCA GTATCATCAA
2401 GGCCATGGGT GGGAGCCAGC CAGGGGAACT TCAGATCAGC TGGGAGGAGC CAGCTCCAGA
2461 AATCAGTGAT TTCCTGAGGT ACGAACTCCG CTATGGCCCC AGAGATCCCA AGAACTCCAC
2521 TGGTCCCACG GTCATACAGC TGATTGCCAC AGAAACCTGC TGCCCTGCTC TGCAGAGGCC
2581 TCACTCAGCC TCTGCTCTGG ACCAGTCTCC ATGTGCTCAG CCCACAATGC CCTGGCAAGA
2641 TGGACCAAAG CAGACCTCCC CAAGTAGAGA AGTATGCTGA CCTTCTTCTG CCCCACCTCT
2701 TATCTCCTAC CTTCAATCTT GCCCCAGGAA AGGACAGACC ATACTTTGGG ATTCCAGAC
2761 CTAGGTTCGT CCTTGGAGAG TTAGTATAGG CTCAGATATG AGCACGCCTA CTTAGGGGCT
2821 TCCTACTTTG GGTCATTCCC ACTGACAAAA GCAAGGCTTT CAGGCCTCCA AATTAATGGA
2881 GATTTCGCAA CAAAACCCTG AACACCACCT GAAACCCACC AACTTAGCCC CTGGTCTGTG
2941 TGCATATCTA TCCAGCAAGG AGCTGAGCTC CTCTCCACAG AGATGCTGTG CAAATATAAG
3001 GGTTGGAGGC TCTCTCAGCT GACAGGCAGA CCTAGATTGT GAAGCTGGGA TTTTCCTCCC
3061 AAGGCTTCAG CTCTGACAGC AGAGGGTGGA AGCTGCCTCA TCTCAGGACT CCAGCCTGGC
3121 AACTCCTACT GGCTGCAGCT GCGCAGCGAA CCTGATGGGA TCTCCCTCGG TGGCTCCTGG
3181 GGATCCTGGT CCCTCCCTGT GACTGTGGAC CTGCCTGGAG ATGCAGGTGA GTCAACAAAG
3241 GAATAGGGAG ATGGGGAGGA GATAAAAGAA TATCTCTAGG GAAGCCTGGG CTAGATCTGA
3301 AGCTCTGGGA ACCATGGTCC TTCCTGATGA TCTCGAACTT GCCACTGGAC AGGAACTATG
3361 TTCAGGGAAA GAAGAGAGAA TAGGAGTCAA TGTTCTAAGT TATATGTGTA GAAATTATCT
3421 GAAATCTGAA CACCCTATAC AGTAGGGGCA CACGGGCCCT GATGGGACTT ACTTCTTTGA
3481 CTTTAGTGGC ACTTGGACTG CAATGCTTTA CCTTGGACCT GAAGAATGTT ACCTGTCAAT
3541 GGCAGCAACA GGACCATGCT AGCTCCCAAG GCTTCTTCTA CCACAGCAGG GCACGGTGCT
3601 GCCCCAGAGA CAGGTGAGAG CTGAACTGCT GATTGAGGTT GGTGTCATGG GAGTGAGCCA
3661 CAATCTTGCA GAAAAAAAGA AGAGAGTGTT CTTGGTCCTC TTCACTCTCC TTCCTTTGTC
3721 TCCAAACCAT ACAGCTTTCA ATGCTCTCTT ATCTATTCTG TCATCCTCCA ATCGATATTT
3781 TATCTTCTCA ACCCCTCTCT GTTCCCATTG GGAATGCTTT GGTTTAGTTT AGCCTTAATC
3841 ATGTCACTGG GACCATTGCA ACATCCTACA GTCTAATTCA CCTCCAATGT ATTCCCCATG
3901 CTGCTAACAG AGTGATCTTT GTTAAGCTCA AACTGTGCAT GACCTT
```

Figure 1 (*Continued*)

Exons 7 and 8:

```
   1 AAAATGCAAA AACAAACGAA CAAATGAACA AAGGAAAAAA CAAAAAGAAA ACATCATGGC
  61 ATAAAGACAG AGGGAAAGCT CTGGAGGCCC CTGGAAGGAA GGAGTTAAAT GATAATCCCT
 121 GCAGGCCATC GTTCTTGTAG GATGGGAAGC CTTGGGATTA GTCTCTGAGG CAGGCCTGAT
 181 TCAATGACTC TGTGGGGCTG GGTCTTAGGT ACCCCATCTG GGAGAACTGC GAAGAGGAAG
 241 AGAAAACAAA TCCAGGACTA CAGACCCCAC AGTTCTCTCG CTGCCACTTC AAGTCACGAA
 301 ATGACAGCAT TATTCACATC CTTGTGGAGG TGACCACAGC CCCGGGTACT GTTCACAGCT
 361 ACCTGGGCTC CCCTTTCTGG ATCCACCAGG CTGGTAAGAA CTTTCTTCCT CATTCTTCCA
 421 CATAGTTCCC ACCCCCACTG AATCTGACCC TGTGCCCAGG ATCCCCAACT CTGACCCTTC
 481 TGACCGATGG CTCTGGTGGC ACAATGCCTT GTGCACAGAA GGACTTAAGC TGCTCCCTGC
 541 TGACATCCCT GTAGTGCGCC TCCCCACCCC AAACTTGCAC TGGAGGGAGA TCTCCAGTGG
 601 GCATCTGGAA TTGGAGTGGC AGCACCCATC GTCCTGGGCA GCCCAAGAGA CCTGTTATCA
 661 ACTCCGATAC ACAGGAGAAG GCCATCAGGA CTGGAAGGTA TGGTCAAGCA ACAAATGCCC
 721 ACAGACCTCA CTACGCAGGG GATCCCTGGG GTTGGCCATG CCTGTTAGCA GGAGTGAAAG
 781 TGTCTATGTA TCCAGTCTCT GCTAGTATAT CTATGTTTAT CAGATTCAAC TGGTATCTTA
 841 GTCTCTCTTG GGCATCTGAT ATTTCTGGCT ACTCCTACTT AACATTTCTC TTCCCTTTGT
 901 CTTTGGGATA TCAGTCTTCC CCGATTTTCC TCCTACCTCT CTGACTGAAC TTTCTCACTC
 961 TCCTCTCCTG GCTCCTCCTT TCCAGCCTTC CACAAAATGC TGTGCTTCCT AGAGTCTCAT
1021 CCTAGGCTCT TTTCTCCTCT GATAAAATAA ATATGTTCTC CCCAGGTGAG TGCTTTAACT
1081 CCATGGTTTC AATGTCCATA CATTAACAAT CGCAATTTTA CCATTCTAAC TCCCATATGG
1141 AAAGGAATTT ACAAACCACA GGCAAATTAC CAAAATGTTT GAAGTCTGTT TCCTCATTGG
1201 TGAAATGAGG ATGAAGCTAT CTACCTTATG GAGTTATAGT AAAGATTAAA TACTATATCT
1261 AAATACCTAG CATACTGCCT GGCATAAGGA AGGTGCTCAA TATACATTAC TTAAATATTA
1321 GTTTGTCTAC TTTTATTCTC ATTTGCCACA CATATATACA ACTATGTACA ACATCTTATG
1381 GTT
```

Exons 9 and 10:

```
   1 CCTTGCACTA ATTAATTGTT GTGATTATTT ACCTGTTTGG ATGTCTTTCT CATTAGATAT
  61 GTAAGAGACG GTGTTTTATT GCTTTCTCCC CAGTGCCTAA CCCAATGTTT GGAGTATCAC
 121 ACAGTTCATA AATGTTTGCT GCCTTCCTGC TTGTACACAT GATGCTTGTA CCCCAGCGCC
 181 TAACGCAAAT CTGGCATCCT CTGCAGCATG AGTATTATTT GTGGCATGGT CTGTGTGGCT
 241 CTGAATATAT CTGTTTCTGG GGGTGTCCAT CGCCCGACCT GGAGTTGTGA GAACACCCGG
 301 TAGGGTGTGC GTGTACCGGG ATCCCTGCCG AGGGTGTACC TGGGTGTTGG TGTTAGGATA
 361 CGTAGCTCTC TGAGGTGAGG TCTGTGTCTC AGGAGGTGGG CCTGGCACGT TTCTCTCGGG
 421 CAGAGCGTGA TCCCGTTAAG GAGGCTCTCG GTTAGGGCGC TCTATCCTGT TGCTGGGAAG
 481 CGTGTTTTCT GCCGGTGGGG CTCTTTGTGG GAATCTCCGA CCGCCTGGGG ATTCGGAGCT
 541 GCAGGATTTG GGTCAAACAG ACGCTGGGCT ATCGAAGCCC CGACGCCGGG CCACCGCACG
 601 CTTCTTTGCT CAGGTGCTGG AGCCGCCTCT CGGGGCCCGA GGAGGGACCC TGGAGCTGCG
 661 CCCGCGATCT CGCTACCGTT TACAGCTGCG CGCCAGGCTC AACGGCCCCA CCTACCAAGG
 721 TCCCTGGAGC TCGTGGTCGG ACCCAACTAG GGTGGAGACC GCCACCGAGA CCGGTGAGGC
 781 AAGCCCCGGC CGCACCAAAG CCGCACAGCG CCTGCGCAGG GACTGGGCGC CGGGTGCGAG
 841 TGGGGCGGGG CTCGGAGAGG GGCGAGNNGG GGCGNGGAGA GGGCGGGGCC CTGACCTTGC
 901 GGGCCGACGC TGCGCAGGTG CCCGCAGTCC CAGGGGCGGC GAGGGGCGGG GCCAGAGTAG
 961 CCCCTCCCTG GATGAGGGCG GGGCTCCGGC CCGGGTGGGC CGAAGTCTGA CCCTTTTTGT
1021 CTCCTAGCCT GGATCTCCTT GGTGACCGCT CTGCATCTAG TGCTGGGCCT CAGCGCCGTC
1081 CTGGGCCTGC TGCTGCTGAG GTGGCAGTTT CCTGCACACT ACAGGTACCG CCCCCGCCAG
1141 GCAGGAGACT GGCGGTGGAC CAGGTGGAGC CGAACCGGTG TAAACAGGCA TTCTTGGTTC
1201 GCTCTGTGAC CCCAGATCTC CGTCCACCGC CCGTGCGCAC CTACGGCTTC GCACTTCCTG
1261 CACGTCACCT CTGGGACTCG CCGCGGCTCC TTACACTCTA ACACGCCCAC TATACCGCCC
1321 ACCTCGAACA GCCCCGCCTC CTGCTGCTCA CCTCGGCGAC TAGGCCACCG TCCACCCTTC
1381 AGCCAAACTG CCCACTCCAC CCCCATCCAA TCTGCCGTCA GTCCCACCTC CTAAACCTAG
1441 TCCAAACAAT GGCCCCCTTT CTCTAGCCCT ACAGACACGA CCTGACTCAC TGAAACAGAA
1501 CCTCTTGCCC CTCCGATCTA GCTACACAGC CACCAGAATA ATCTTTCTAA AATGCGCACA
1561 TGAACCTATC AATTCCCTCC TTAAATCCTT TCAATAGCTC CCAGGATATA GTCCAGGCTC
1621 CCTATGATGA CAAGCGAGGC TCTGCAAAAT CGGCCCCTGC CAGCAAGCAC TTCAGCCTGT
1681 TCTACCCTGT GCATCAGCCT TTCCAAACTG GTTTTCTCCA CAAGTCAGGT TTTACTTCTC
1741 AGTGCCTTTG CACATAACGT CCGCTCTGCC TGAAATAGCC TCCCT
```

Figure 1 (*Continued*)

```
Exons 11 and 12:
   1 CTGCAGATAA AGAGGGAGTA CTGTGTTTCA GGCAAGGTAG TCAGGGATGA CATTTGAACA
  61 GAGGCCTACA ATGAGGTGAG GGAATGAATC CAGTGTTCCA GACAGAACAA ACATTGGAGT
 121 GTTTGAGGAA ACTGCAGCAC CCTTTGTTGA ACCTGCCCAC CACTGACTTC CCTCCCCTAC
 181 CTGAGGTTTT CTTCTTACTC TACTCCCCTG CCCAACTTTA TTGCCTTGAC CATGCTCTTA
 241 TATCTCCAAG CCTTACTCCC CGGCCTCACT GCTCCCCCAC CCCAACTGCT GTCCTCCTCC
 301 CTGCCAATCC ACTGCCATGG CTCAGTCTGC TTCTCTTCCT TCTCCCCCAG GAGACTGAGG
 361 CATGCCCTGT GGCCCTCACT TCCAGACCTG CACCGGGTCC TAGGCCAGTA CCTTAGGGAC
 421 ACTGCAGCCC TGAGCCCGGT GAGTGTGCTT CCCTCCCCTG TGCCCACCAC CAACCCTGCC
 481 TGGTACTGGA TCCTTGCCCC AACAATACAA CTTGTTCAAG GTCCTTGCCC TACCAGTGTG
 541 CCATCCCCAG GCACTACCCC AGCACTACCC CAGCCCTTCT CCTTCCTGTA CAGTCCAGCC
 601 CCTCCTCCCA CAGGATCTGC TTTAATCCAG CGCCTCTCCT CATCTCTCCC AGCCCAAGGC
 661 CACAGTCTCA GATACCTGTG AAGAAGTGGA ACCCAGCCTC CTTGAAATCC TCCCCAAGTC
 721 CTCAGAGAGG ACTCCTTTGC CCCTGTGTTC CTCCCAGGCC CAGATGGACT ACCGAAGATT
 781 GCAGCCTTCT TGCCTGGGGA CCATGCCCCT GTCTGTGTGC CCACCCATGG CTGAGTCAGG
 841 GTCCTGCTGT ACCACCCACA TTGCCAACCA TTCCTACCTA CCACTAAGCT ATTGGCAGCA
 901 GCCTTGAGGA CAGGCTCCTC ACTCCCAGTT CCCTGGACAG AGCTAAACTC TCGAGACTTC
 961 TCTGTGAACT TCCCTACCCT ACCCCCACAA CACAAGCACC CCAGACCTCA CCTCCATCCC
1021 CCTCTGTCTG CCCTCACAAT TAGGCTTCAT TGCACTGATC TTACTCTACT GCTGCTGACA
1081 TAAAACCAGG ACCCTTTCTC CACAGGCAGG CTCATTTCAC TAAGCTCCTC CTTTACTTCC
1141 TCTCTCCTCT TTGATGTCAA ACGCCTTGAA AACAAGCCTC CACTTCCCCA CACTTCCCAT
1201 TTACTCTTGA GACTACTTCA ATTAGTTCCC CTACTACACT TTGCTAGTGA ACTGCCCAGG
1261 CAAAGTGCAC CTCAAATCTT CTAATTCCAA GATCCAATAG GATCTCGTTA ATCATCAGTT
1321 CCTTTGATCT CGCTGTAAGA TTTGTCAAGG CTGACTACTC ACTTCTCCTT TAAATTCTTT
1381 CCTACCTTGG TCCTGCCTCT TTGAGTATAT TAGTAGGTTT TTTTTATTTG TTTGAGACAG
1441 GGTCTCACTC TGTCACCCAG GCTGCAGTGC AATGGCGCGA TCTCAGCTCA CTGCAACCTC
1501 CACCTCCGGG TTCAAGCGAT TCTTGTGCCT CGGCCTCCCT AGTAGCTGGG ATTACAGGCG
1561 CACACCACCA CACACAGCTA ATTTTTTTTT TTTTTTTTTT TTTTTTTTTT TTTTTTAGAC
1621 GGAGCCTTGC CTGTTGCCAG ACTGGAGTGC AGTGGCACGA TCTCGGCTCA CTGCAACCTC
1681 TGCCTCCCGG GTTCAAGCCA TTCTGCCTCA GCCTCCCAAG TAGCTGGGAG TACAGCGTCT
1741 GCCACCATGC CTAATTTTTT TCTATTTTTA GGAGAGACCG GTTTTCACCA CGTTGGCCAG
1801 GATGGTCTCG ATATCTGATC TCGTGATCCG CCTGCCTCTG CCTCCCAAAG TGCTGGGATT
1861 ACAGGTGTGA CCCACTGCGC ACAGCCCCAG CTAATTTTCA TATTTTTAGT AGAGACAGGG
1921 TTTTGCCATG TTGCCCAGGC TGGTCTTGAA CTCCTAACCT CGGGTGATCC ACCCACCTTG
1981 GCCTCCCAAA GTGTTAGGAT TACAGGCATG AGCCACTGCG CCCGGCTGAG TGTACTAGTA
2041 GTTAAGAGAA TAAACTAGAT CTAGAATCAG AGCTGGATTC AATTCCTGTC CTTCACATTT
2101 ACTAGCTGTG CAACCTTGGG CACATAACTT AATGTCTTTG AGCCTTAGTT TTTTCATCTG
2161 TAAAACAGGG ATAATAACAG CACCCCATAG AGTTGTGACG AGGATTGAGA TAATCTAAGT
2221 AAAGCACAGT CCCTAGGACA TAGTAAATGA TTCATATATC CGAACTACTG TTATAATTAT
2281 TCCTTCTTAC TCTCCTCTTC TAGCATTTCT TCCAATTATT ACAGTCCTTC AAGATTCCAT
2341 TTCTTAACAG TCTCCAATCC CATCTATTCT CTGCCTTTAC TATATGTTGA CCATTCCAAA
2401 GTTCTTATCT CTAGCTCAGA CATCTACTAC AGCACTGTGA TGCTTTATGC AACTAACTGT
2461 TTACATATCT GTCCCCTGCT ACTAGATTGT GAGCTCCTTG AGGGAAAGGA ACATGATTTA
2521 TTTGTCCTTT TCCCCCAGCA CCTAGAGTAG TGCTTGGTGC ATGATAGTAG GCCTTCAATA
2581 AATTTTTTCT AAATGAATGA AATCTTTCTG GACAAGGTTA CCTTAATCCC ATGCCAGCTG
2641 TCCTCATCCT CTAAACCGTT CATAGGCAGC AGTAAATGTG CCAGACAGTT CTGATTATGG
2701 ACTCTCCTAA CACATTCTAA TTCTCAAACA AATGGTTTTG TTGTCATTGT TTTTNCTGGC
2761 TCCAAATATA CCCACTGTAT CAGTCAGGGT CTTGGAAGAA ATAGATGACA CACTCAAATT
2821 GGGTAATATG ATGACAGTGT AATAAACT
```

receptor protein is AAB08424. Those for the mouse TPO receptor are CAA80365 and CAA52031.

Description of protein

Human full-length cDNAs encode 635 and 579 amino acid residues for the P and K forms of the TPO receptor, respectively (Vigon *et al.*, 1992). The P and K forms share identical amino acid sequence from residues 1 to 522 at the N-terminus, but are completely different after residue 523. The P and K forms result from alternative TPO receptor mRNA splicing at exon 10–11 (Vigon *et al.*, 1992). The first 25 amino acid residues at the N-terminus encode signal peptides. A hydrophobic region from residues 492 to 513 in the transmembrane region separates the extracellular domain residues 26–491 from the intracellular

Figure 1 (*Continued*)

Human cDNA sequence for TPO receptor P form:

```
   1 ATGCCCTCCT GGGCCCTCTT CATGGTCACC TCCTGCCTCC TCCTGGCCCC TCAAAACCTG
  61 GCCCAAGTCA GCAGCCAAGA TGTCTCCTTG CTGGCATCAG ACTCAGAGCC CCTGAAGTGT
 121 TTCTCCCGAA CATTTGAGGA CCTCACTTGC TTCTGGGATG AGGAAGAGGC AGCGCCCAGT
 181 GGGACATACC AGCTGCTGTA TGCCTACCCG CGGGAGAAGC CCCGTGCTTG CCCCCTGAGT
 241 TCCCAGAGCA TGCCCCACTT TGGAACCCGA TACGTGTGCC AGTTTCCAGA CCAGGAGGAA
 301 GTGCGTCTCT TCTTTCCGCT GCACCTCTGG GTGAAGAATG TGTTCCTAAA CCAGACTCGG
 361 ACTCAGCGAG TCCTCTTTGT GGACAGTGTA GGCCTGCCGG CTCCCCCCAG TATCATCAAG
 421 GCCATGGGTG GGAGCCAGCC AGGGGAACTT CAGATCAGCT GGGAGGAGCC AGCTCCAGAA
 481 ATCAGTGATT TCCTGAGGTA CGAACTCCGC TATGGCCCCA GAGATCCCAA GAACTCCACT
 541 GGTCCCACGG TCATACAGCT GATTGCCACA GAAACCTGCT GCCCTGCTCT GCAGAGGCCT
 601 CACTCAGCCT CTGCTCTGGA CCAGTCTCCA TGTGCTCAGC CCACAATGCC CTGGCAAGAT
 661 GGACCAAAGC AGACCTCCCC AAGTAGAGAA GCTTCAGCTC TGACAGCAGA GGGTGGAAGC
 721 TGCCTCATCT CAGGACTCCA GCCTGGCAAC TCCTACTGGC TGCAGCTGCG CAGCGAACCT
 781 GATGGGATCT CCCTCGGTGG CTCCTGGGGA TCCTGGTCCC TCCCTGTGAC TGTGGACCTG
 841 CCTGGAGATG CAGTGGCACT TGGACTGCAA TGCTTTACCT TGGACCTGAA GAATGTTACC
 901 TGTCAATGGC AGCAACAGGA CCATGCTAGC TCCCAAGGCT TCTTCTACCA CAGCAGGGCA
 961 CGGTGCTGCC CCAGAGACAG GTACCCCATC TGGGAGAACT GCGAAGAGGA AGAGAAAACA
1021 AATCCAGGAC TACAGACCCC ACAGTTCTCT CGCTGCCACT TCAAGTCACG AAATGACAGC
1081 ATTATTCACA TCCTTGTGGA GGTGACCACA GCCCCGGGTA CTGTTCACAG CTACCTGGGC
1141 TCCCCTTTCT GGATCCACCA GGCTGTGCGC CTCCCCACCC CAAACTTGCA CTGGAGGGAG
1201 ATCTCCAGTG GGCATCTGGA ATTGGAGTGG CAGCACCCAT CGTCCTGGGC AGCCCAAGAG
1261 ACCTGTTATC AACTCCGATA CACAGGAGAA GGCCATCAGG ACTGGAAGGT GCTGGAGCCG
1321 CCTCTCGGGG CCCGAGGAGG GACCCTGGAG CTGCGCCCGC GATCTCGCTA CCGTTTACAG
1381 CTGCGCGCCA GGCTCAACGG CCCCACCTAC CAAGGTCCCT GGAGCTCGTG GTCGGACCCA
1441 ACTAGGGTGG AGACCGCCAC CGAGACCGCC TGGATCTCCT TGGTGACCGC TCTGCATCTA
1501 GTGCTGGGCC TCAGCGCCGT CCTGGGCCTG CTGCTGCTGA GGTGGCAGTT TCCTGCACAC
1561 TACAGGAGAC TGAGGCATGC CCTGTGGCCC TCACTTCCAG ACCTGCACCG GGTCCTAGGC
1621 CAGTACCTTA GGGACACTGC AGCCCTGAGC CCGCCCAAGG CCACAGTCTC AGATACCTGT
1681 GAAGAAGTGG AACCCAGCCT CCTTGAAATC CTCCCCAAGT CCTCAGAGAG GACTCCTTTG
1741 CCCCTGTGTT CCTCCCAGGC CCAGATGGAC TACCGAAGAT TGCAGCCTTC TTGCCTGGGG
1801 ACCATGCCCC TGTCTGTGTG CCCACCCATG GCTGAGTCAG GGTCCTGCTG TACCACCCAC
1861 ATTGCCAACC ATTCCTACCT ACCACTAAGC TATTGGCAGC AGCCTTGA
```

Human cDNA sequence for TPO receptor K form:

```
   1 ATGCCCTCCT GGGCCCTCTT CATGGTCACC TCCTGCCTCC TCCTGGCCCC TCAAAACCTG
  61 GCCCAAGTCA GCAGCCAAGA TGTCTCCTTG CTGGCATCAG ACTCAGAGCC CCTGAAGTGT
 121 TTCTCCCGAA CATTTGAGGA CCTCACTTGC TTCTGGGATG AGGAAGAGGC AGCGCCCAGT
 181 GGGACATACC AGCTGCTGTA TGCCTACCCG CGGGAGAAGC CCCGTGCTTG CCCCCTGAGT
 241 TCCCAGAGCA TGCCCCACTT TGGAACCCGA TACGTGTGCC AGTTTCCAGA CCAGGAGGAA
 301 GTGCGTCTCT TCTTTCCGCT GCACCTCTGG GTGAAGAATG TGTTCCTAAA CCAGACTCGG
 361 ACTCAGCGAG TCCTCTTTGT GGACAGTGTA GGCCTGCCGG CTCCCCCCAG TATCATCAAG
 421 GCCATGGGTG GGAGCCAGCC AGGGGAACTT CAGATCAGCT GGGAGGAGCC AGCTCCAGAA
 481 ATCAGTGATT TCCTGAGGTA CGAACTCCGC TATGGCCCCA GAGATCCCAA GAACTCCACT
 541 GGTCCCACGG TCATACAGCT GATTGCCACA GAAACCTGCT GCCCTGCTCT GCAGAGGCCT
 601 CACTCAGCCT CTGCTCTGGA CCAGTCTCCA TGTGCTCAGC CCACAATGCC CTGGCAAGAT
 661 GGACCAAAGC AGACCTCCCC AAGTAGAGAA GCTTCAGCTC TGACAGCAGA GGGTGGAAGC
 721 TGCCTCATCT CAGGACTCCA GCCTGGCAAC TCCTACTGGC TGCAGCTGCG CAGCGAACCT
 781 GATGGGATCT CCCTCGGTGG CTCCTGGGGA TCCTGGTCCC TCCCTGTGAC TGTGGACCTG
 841 CCTGGAGATG CAGTGGCACT TGGACTGCAA TGCTTTACCT TGGACCTGAA GAATGTTACC
 901 TGTCAATGGC AGCAACAGGA CCATGCTAGC TCCCAAGGCT TCTTCTACCA CAGCAGGGCA
 961 CGGTGCTGCC CCAGAGACAG GTACCCCATC TGGGAGAACT GCGAAGAGGA AGAGAAAACA
1021 AATCCAGGAC TACAGACCCC ACAGTTCTCT CGCTGCCACT TCAAGTCACG AAATGACAGC
1081 ATTATTCACA TCCTTGTGGA GGTGACCACA GCCCCGGGTA CTGTTCACAG CTACCTGGGC
1141 TCCCCTTTCT GGATCCACCA GGCTGTGCGC CTCCCCACCC CAAACTTGCA CTGGAGGGAG
1201 ATCTCCAGTG GGCATCTGGA ATTGGAGTGG CAGCACCCAT CGTCCTGGGC AGCCCAAGAG
1261 ACCTGTTATC AACTCCGATA CACAGGAGAA GGCCATCAGG ACTGGAAGGT GCTGGAGCCG
1321 CCTCTCGGGG CCCGAGGAGG GACCCTGGAG CTGCGCCCGC GATCTCGCTA CCGTTTACAG
1381 CTGCGCGCCA GGCTCAACGG CCCCACCTAC CAAGGTCCCT GGAGCTCGTG GTCGGACCCA
1441 ACTAGGGTGG AGACCGCCAC CGAGACCGCC TGGATCTCCT TGGTGACCGC TCTGCATCTA
1501 GTGCTGGGCC TCAGCGCCGT CCTGGGCCTG CTGCTGCTGA GGTGGCAGTT TCCTGCACAC
1561 TACAGGTACC GCCCCCGCCA GGCAGGAGAC TGGCGGTGGA CCAGGTGGAG CCGAACGTGT
1621 AAACAGGCAT TCTTGGTTCG CTCTGTGACC CCAGATCTCC GTCCACCGCC CGTGCGCACC
1681 TACGGCTTCG CACTTCCTGC ACGTCACCTC TGGGACTCGC CGCGGCTCCT TACACTCTAA
```

Figure 1 (*Continued*)

```
Mouse cDNA for TPO receptor:
   1 CCTCTTCATG GTCACCTCCT GCCTCCTCTT GGCCCTTCCA AACCAGGCAC AAGTCACCAG
  61 CCAAGATGTC TTCTTGCTGG CCTTGGGCAC AGAGCCCCTG AACTGCTTCT CCCAAACATT
 121 TGAGGACCTC ACCTGCTTCT GGGATGAGGA AGAGGCAGCA CCCAGTGGGA CATACCAGCT
 181 GCTGTATGCC TACCGAGGAG AGAAGCCCCG TGCATGCCCC CTGTATTCCC AGAGTGTGCC
 241 CACCTTTGGA ACCCGGTATG TGTGCCAGTT TCCAGCCCAG GTAGAAGTGC GCCTCTTCTT
 301 TCCGCTGCAC CTCTGGGTGA AGAATGTGTC CCTCAACCAG ACTTTGATCC AGCGGGTGCT
 361 GTTTGTGGAT AGTGTGGGCC TGCCAGCTCC CCCCAGGGTC ATCAAGGCCA GGGGTGGGAG
 421 CCAACCAGGG GAACTTCAGA TCCACTGGGA GGCCCCTGCT CCTGAAATCA GTGACTTCCT
 481 GAGGCATGAA CTCCGCTATG GCCCCACGGA TTCCAGCAAC GCCACTGCCC CCTCCGTCAT
 541 TCAGCTGCTC TCCACAGAAA CCTGCTGCCC CACTTTGTGG ATGCCGAACC CAGTCCCTGT
 601 TCTTGACCAG CCTCCGTGTG TTCATCCGAC AGCATCCCAA CCGCATGGAC CAGTGAGGAC
 661 CTCCCCAGCT GGAGAAGCTC CATTTCTGAC AGTGAAGGGT GGAAGCTGTC TCGTCTCAGG
 721 CCTCCAGGCT AGCAAATCCT ACTGGCTCCA GCTACGCAGC CAACCCGACG GGGTCTCCCT
 781 TCGTGGCTCC TGGGGACCCT GGTCCTTCCC TGTGACTGTG GATCTTCCAG GAGATGCAGT
 841 GACAATTGGA CTTCAGTGCT TTACCTTGGA TCTGAAGATG GTCACCTGCC AGTGGCAGCA
 901 ACAAGACCGC ACTAGCTCCC AAGGCTTCTT CCGTCACAGC AGGACGAGGT GCTGCCCCAC
 961 AGACAGGGAC CCCACCTGGG AGAAATGTGA AGAGGAGGAA CCGCGTCCAG GATCACAGCC
1021 CGCTCTCGTC TCCCGCTGCC ACTTCAAGTC ACGAAATGAC AGTGTTATTC ACATCCTTGT
1081 AGAGGTGACC ACAGCGCAAG GTGCCGTTCA CAGCTACCTG GGCTCCCCTT TTTGGATCCA
1141 CCAGGCTGTG CTCCTTCCCA CCCCGAGCCT GCACTGGAGG GAGGTCTCAA GTGGAAGGCT
1201 GGAGTTGGAG TGGCAGCACC AGTCATCTTG GGCAGCTCAA GAGACCTGCT ACCAGCTCCG
1261 GTACACGGGA GAAGGCCGTG AGGACTGGAA GGTGCTGGAG CCATCTCTCG GTGCCCGGGG
1321 AGGGACCCTA GAGCTGCGCC CCCGAGCTCG CTACAGCTTG CAGCTGCGTG CCAGGCTCAA
1381 CGGCCCCACC TACCAAGGTC CCTGGAGCGC CTGGTCTCCC CCAGCTAGGG TGTCCACGGG
1441 CTCCGAGACT GCTTGGATCA CCTTGGTGAC TGCTCTGCTC CTGGTGCTGA GCCTCAGTGC
1501 CCTTCTGGGC CTACTGCTGC TAAAGTGGCA ATTTCCTGCG CACTACAGGA GACTGAGGCA
1561 TGCTTTGTGG CCCTCGCTTC CAGACCTACA CCGGGTCCTA GGCCAGTACC TCAGAGACAC
1621 TGCAGCCCTA AGTCCTTCTA AGGCCACGGT TACCGATAGC TGTGAAGAAG TGGAACCCAG
1681 CCTCCTGGAA ATCCTCCCTA AATCCTCAGA GAGCACTCCT TTACCTCTGT GTCCCTCCCA
1741 ACCTCAGATG GACTACAGAG GACTGCAACC TTGCCTGCGG ACCATGCCCC TGTCTGTGTG
1801 TCCACCCATG GCTGAGACGG GGTCCTGCTG CACCACACAC ATTGCCAACC ACTCCTACCT
1861 ACCACTAAGC TATTGGCAGC AGCCCTGAAG GCAGTCCCCA TGCTACTGCA GACCTATACA
1921 TTCCTACACA CTACCTTATC CATCCTCAAC ACCATCCATT CTGTTGCCAC CCCACTCCCC
1981 CTCTGGCTTT ATAACACTGA TCACTCCAAG ATGGCTGCTC ACAAATCCAG AGCTCTGTCT
2041 CTGCAG
```

domain residues 514–635. There are four potential *N*-linked glycosylation sites in the extracellular domain located at residues 117, 178, 298, and 358, respectively. The extracellular domain contains a duplicated HRD which is highly conserved in the hematopoietin/cytokine receptor superfamily. Each HRD contains two highly conserved disulfide bonds and two FNIII-like domains, each of which consists of seven β strands. A 5 amino acid motif called WSXWS is found at the C-terminus of the second FNIII-like domain in both HRDs of the TPO receptor – a finding which is characteristic of the hematopoietin/cytokine receptor family. In addition, the TPO receptor is unique in having a 50 amino acid insertion in the second FNIII-like domain of the membrane-distal HRD. A novel variant TPO receptor called c-Mpl-del in which 24 amino acid residues are deleted in the membrane-proximal HRD has been described (Li *et al.*, 2000). Functional analysis of c-Mpl-del protein demonstrated that it was expressed but did not transport to the cell surface and was therefore not functional.

The intracellular domain contains two short membrane-proximal motifs, box 1 and box 2, which are conserved in the hematopoietin/cytokine superfamily. Deletion of either box 1 or box 2 results in a defective TPO receptor (Gurney *et al.*, 1995). The membrane-distal region of the intracellular domain shares no homology with other cytokine receptors. Although there is no enzymatic motif found in the intracellular domain, it does contain five tyrosine residues which are conserved in both human and mouse and which play a critical role in the TPO receptor-mediated signal transduction pathways.

Relevant homologies and species differences

The genes for the human and mouse TPO receptors have been cloned and extensively studied. Comparison of the amino acid sequences of these TPO receptors reveals 81% amino acid identity (Foster *et al.*, 1994). Further comparison of the intracellular domains shows that there is 91% identity (Vigon *et al.*, 1993c), suggesting that the TPO-induced signaling pathways should be very similar in these two species.

Affinity for ligand(s)

Most information about TPO receptor binding affinity for their ligands has been obtained from experiments on human platelets (Broudy *et al.*, 1997; Fielder *et al.*, 1997; Li *et al.*, 1999). Scatchard analysis has shown that platelets contain a single class of binding sites for TPO with affinity constant (K_d) ranging from 120 to 200 pM. Each platelet contains 30–60 binding sites on the surface. After binding to platelets, c-Mpl ligands are internalized within 60 minutes and then degraded. Although megakaryocytes and their progenitor cells have been experimentally demonstrated to have TPO receptors on the cell surface (Broudy *et al.*, 1997; Solar *et al.*, 1998), the precise quantification of their number and binding affinity has not been done.

Cell types and tissues expressing the receptor

The thrombopoietin receptor can be directly demonstrated or its presence inferred on early bone marrow hematopoietic progenitor cells of all lineages as well as on the pluripotential stem cell (Carver-Moore *et al.*, 1996; Rasko *et al.*, 1997; Ratajczak *et al.*, 1997; Tanimukai *et al.*, 1997; Yoshida *et al.*, 1997; Yamada *et al.*, 1998; Solar *et al.*, 1998). It is present on common early erythroid/megakaryocyte progenitors (Kaushansky *et al.*, 1995, 1996; Kobayashi *et al.*, 1995; Carver-Moore *et al.*, 1996; Drexler and Quentmeier, 1996; Graf *et al.*, 1996; Kieran *et al.*, 1996; Bonsi *et al.*, 1997; Drexler *et al.*, 1997; Goncalves *et al.*, 1997; Higuchi *et al.*, 1997; Komatsu *et al.*, 1997; Miura *et al.*, 1998). It is present on late megakaryocyte progenitors but not on late erythroid progenitors. It is not present on mature erythrocytes, neutrophils, eosinophils, basophils, or lymphocytes but it is present on platelets (Sasaki *et al.*, 1995a). c-Mpl has been reported to be present on monocytes (but contamination by associated platelets could not be discounted; Sasaki *et al.*, 1995a). The receptor is found on cells and cell lines from many nonlymphoid hematological malignancies but not on cells from purely lymphoid malignancies or nonhematopoietic solid tumors (Columbyova *et al.*, 1995; Drexler *et al.*, 1996; Graf *et al.*, 1996; Matsumura *et al.*, 1996; Quentmeier *et al.*, 1996; Bredoux *et al.*, 1997; Hirai *et al.*, 1997).

Regulation of receptor expression

There is no clinical or experimental situation in which regulation of c-Mpl has been convincingly demonstrated. A 200 bp portion of the c-*mpl* promoter has been demonstrated to be adequate for high-level expression of the gene in cell lines (Deveaux *et al.*, 1996). GATA-1 and two members of the Ets family (Ets-1 and Fli-1) have specific binding sites and all three seem necessary for high-level expression in tissue culture cells.

As mentioned previously, platelets from patients with *essential thrombocythemia* or *polycythemia vera* have reduced amounts of c-Mpl on the cell surface (Li *et al.*, 1996; Moliterno *et al.*, 1998). Whether this is due to altered expression or ligand-induced downregulation of the receptor is unclear.

Release of soluble receptors

mRNA that lacks the c-*mpl* transmembrane region is found in cells that express c-*mpl* but the existence of its protein product has not been convincingly demonstrated. Like other alternatively spliced transcripts, the protein product may not be efficiently translated or exported from the endoplasmic reticulum (Li *et al.*, 2000). In whole plasma and serum a c-Mpl protein that is not associated with cells can be demonstrated but it is unclear whether this is the predicted soluble receptor or simply the extracellular portion of the full-length, wild-type platelet receptor that has been cleaved *in vivo* or during sample preparation. In unpublished studies, the circulating level of c-Mpl protein not associated with cells is directly proportional to the platelet count.

SIGNAL TRANSDUCTION

Associated or intrinsic kinases

As predicted by its cDNA structure, the TPO receptor does not contain any known intrinsic kinase activity – a finding that is a characteristic feature of the hematopoietin/cytokine receptor family. Therefore its signal transduction following ligand binding and dimerization must utilize cytoplasmic enzymes such as nonreceptor protein tyrosine kinases and other signaling molecules. These molecules are able to associate by protein–protein interaction with the TPO receptor to form a signaling complex. Indeed, upon exposure to TPO, a number of tyrosine kinases and signaling adapter molecules undergo tyrosine phosphorylation.

Cytoplasmic signaling cascades

Signal transduction starts with c-Mpl ligand binding to c-Mpl on the cell surface. Upon dimerization, the receptors transduce the ligand-specific signal by transient and covalent chemical modifications of intracellular signaling molecules. These molecules are either enzymatic proteins (e.g. kinases or phosphatases) or nonenzymatic adapter proteins. In most cases they contain multiple domains for protein–protein interaction, such as the Src-homology 2 domain (SH2) which binds to phosphotyrosine. One commonly seen modification is tyrosine phosphorylation. Tyrosine phosphorylation of these molecules could change their molecular conformation, leading to regulation of either enzymatic activity or affinity for interaction with other bio-macromolecules, or both. Frequently, tyrosine phosphorylation also induces translocation of the signaling molecule in the cells. Finally, the ligand-initiated signaling cascades usually affect cellular proliferation, differentiation/maturation, and/or apoptosis.

Treatment of cells with TPO induces tyrosine phosphorylation not only on the intracellular domain of the TPO receptor but also on a number of protein tyrosine kinases and signaling adaptor molecules. Although the biological specificity of the signaling pathways for TPO is still unsolved, there are at least three signaling cascades through which TPO may mediate its effects.

Like most hematopoietic cytokines, TPO stimulates the tyrosine phosphorylation of two members of JAK kinase family, JAK2 (125 kDa) (Drachman *et al.*, 1995; Miyakawa *et al.*, 1995) and TYK2 (130 kDa) (Ezumi *et al.*, 1995; Sattler *et al.*, 1995), as well as two members of the STAT (signal transducer and activator of transcription) family, STAT5 (92 kDa) and STAT3 (90 kDa) (Miyakawa *et al.*, 1996). STAT family proteins are well-known cytoplasmic substrates for the JAK family tyrosine kinases. Upon JAK2 phosphorylation and activation, STAT molecules are phosphorylated on their tyrosine residues and multimerized by protein–protein interactions via binding of the phosphotyrosine of one molecule to the SH2 domain of another. Then the STAT signaling complexes translocate into the nucleus to modulate the transcription of their downstream genes. Although evidence suggests that JAK2 constitutively associates with the TPO receptor via box 1 and box 2 motifs (Mu *et al.*, 1995; Drachman and Kaushansky, 1997), the role of the JAK2/STAT5 signaling pathway has been challenged by the observation that a c-*mpl* deletion mutant was able to convey a proliferative signal without activation of JAK2/TYK2 kinases or phosphorylation of STAT5 (Dorsch *et al.*, 1997).

TPO-induced phosphorylation of Shc (60 kDa) is another hallmark signaling pathway (Drachman *et al.*, 1995; Miyakawa *et al.*, 1995; Mu *et al.*, 1995). Shc is an adapter protein containing multiple protein interaction domains, including SH2 and SH3 domains, and plays important roles in activation of the Ras signaling pathway in various systems. Shc phosphorylation and association with the TPO receptor are independent of JAK2 activation, but dependent on TPO receptor phosphorylation on the tyrosine-599 of the intracellular domain, which provides a docking site for association with the SH2 domain of Shc (Alexander *et al.*, 1996a; Drachman and Kaushansky, 1997). It is still unclear which tyrosine kinase contributes to Shc phosphorylation. However, it appears that the signaling complexes between the TPO receptor and Shc probably further activate the Ras signaling pathway. Indeed, a growing body of evidence strongly supports this model. Firstly, a dominant negative Ras mutant inhibited TPO-induced megakaryocytic differentiation and reduced cell proliferation (Matsumura *et al.*, 1998). Secondly, TPO treatment induced phosphorylation of MAPK (Mu *et al.*, 1995; Rouyez *et al.*, 1997), Raf1 (Dorsch *et al.*, 1997), and Sos (Sasaki *et al.*, 1995b), key components of the Ras signaling pathway. Thirdly, the expression of c-*fos* and c-*myc* was upregulated by TPO treatment (Dorsch *et al.*, 1997; Kunitama *et al.*, 1997). Furthermore, the TPO-initiated Shc/Ras/Raf/MAPK pathway is primarily responsible for induction of megakaryocytic differentiation (Alexander *et al.*, 1996a; Dorsch *et al.*, 1997; Rouyez *et al.*, 1997).

TPO-induced activation of phosphatidylinositol-3′ (PI-3) kinase (85 kDa, regulatory subunit) is a third important signaling pathway (Chen *et al.*, 1995; Dorsch *et al.*, 1997; Sattler *et al.*, 1997; Zauli *et al.*, 1997). The activation of PI-3 kinase is independent of either JAK2 activation or Shc phosphorylation (Dorsch *et al.*, 1997), but is associated with TPO-induced tyrosine phosphorylation of Vav (95 kDa) and Cbl (120 kDa; Oda *et al.*, 1996; Sattler *et al.*, 1997). These two proto-oncogene products are expressed primarily in multiple lineages of hematopoietic cells, including megakaryocytes and platelets. As demonstrated in human B cells, it is more likely that PI-3 kinase activation is a downstream event of Vav phosphorylation. Protein kinase C is also associated with PI-3 kinase activation (Kunitama *et al.*, 1997; Hong *et al.*, 1998). Two biological roles of TPO-induced Vav/PI-3 signaling pathway have been implicated. One is to provide a mitogenic signal to cells (Dorsch *et al.*, 1997). The other is to increase the adherence of cells to the extracellular matrix by activation of integrin on the cell surface (Chen *et al.*, 1995; Cui *et al.*, 1997; Gotoh *et al.*, 1997; Zauli *et al.*, 1997). Addition of the PI-3 kinase inhibitor, wortmanin, abolished both biological effects of PI-3 kinase.

DOWNSTREAM GENE ACTIVATION

Transcription factors activated

TPO exerts its wide range of biological activities by initiation of several critical signaling pathways that lead to activation of a group of transcription factors. These activated transcription factors come from two basic sources. One is the conversion of preexisting, inactive transcription factors into active ones by covalent chemical modification. The best example of this is TPO-induced activation of STAT3/STAT5. A second is the synthesis of new transcription molecules by induction of gene expression. The best example of this is TPO-induced upregulation of c-*fos* and c-*myc*. However, no matter what their source, these TPO-induced and activated transcription factors are also used by many other hematopoietic cytokines. For example, the JAK/STAT signaling pathway is used by almost all hematopoietic cytokines.

As soon as the TPO receptor was discovered, TPO receptor-mediated induction of megakaryocyte-specific genes expressed in both CD34 cell cultures and c-*mpl*-transfected hematopoietic progenitor cell lines was described (Morita *et al.*, 1996). For example, treatment of c-*mpl*-transfected UT7 cells with TPO could increase the expression of GPIIb/IIIa (Porteu *et al.*, 1996). This TPO effect was associated with increased expression and binding activity of PU.1/Spi-1, a member of the Ets transcription factor family (Doubeikovski *et al.*, 1997). Overexpression of PU.1/Spi-1 enhanced GPIIb promoter activity, suggesting that PU.1/Spi-1 was a TPO-inducible transcription factor responsible for regulation of megakaryocyte-specific gene expression. Further analysis of the GPIIb gene promoter by transfection of a series of the promoter fragments indicated that the TPO-responsive *cis* element was within an enhancer region of the promoter, which included GATA- and Ets-binding sites. PU.1/Spi-1 was strongly and preferentially bound to this region.

Another protein, named nucleosome assembly protein (NAP), is reportedly upregulated in hematopoietic cells by TPO (Cataldo *et al.*, 1999). Although this protein is thought to affect megakaryocyte polyploidation, the mechanism for regulation of its gene expression by TPO is unclear.

BIOLOGICAL CONSEQUENCES OF ACTIVATING OR INHIBITING RECEPTOR AND PATHOPHYSIOLOGY

Unique biological effects of activating the receptors

TPO receptor binding initiates at least three major events in cells. The first is initiation of signal transduction pathways that promote cell growth and/or cellular differentiation. The second is the activation of signal transduction pathways that prevent apoptosis. The third is internalization and degradation of the receptor–ligand complex (Fielder *et al.*, 1996, 1997; Broudy *et al.*, 1997; Li *et al.*, 1999). All of these have been discussed above.

Phenotypes of receptor knockouts and receptor overexpression mice

Homozygous c-*mpl* knockout mice had platelet and megakaryocyte numbers that were 10–15% of normal but had normal numbers of white and red blood cells (Gurney *et al.*, 1994; Alexander *et al.*, 1996b). They had a compensatory rise in the circulating level of thrombopoietin. Analysis of the bone marrow of these mice (Alexander *et al.*, 1996b; Carver-Moore *et al.*, 1996) showed that while the megakaryocyte progenitor cell (Meg-CFC) levels were about 5% of normal, the erythroid (E-CFC) and myeloid (G-CFC) progenitor cell levels were also reduced to 20–30% of normal. Heterozygous knockout animals were entirely normal.

c-*mpl* is the only member of the hematopoietin receptor family to be identified as a proto-oncogene. TPO and c-Mpl provide a very potent hematopoietic stimulus and excess *in vivo* activation of either produces serious consequences. Ligand-dependent activation of c-Mpl via overexpression of TPO causes increased platelet reactivity (Montrucchio *et al.*, 1996; Fontenay-Roupie *et al.*, 1998), *myelofibrosis* (Yan *et al.*, 1995, 1996; Villeval *et al.*, 1997; Abina *et al.*, 1998), a *myeloproliferative syndrome* (Villeval *et al.*, 1997), and *osteosclerosis* (Yan *et al.*, 1996; Villeval *et al.*, 1997). As described in the introduction to this review, expression in mice of the fusion protein *env-mpl* made by the *myeloproliferative leukemia virus* (MPLV) produced a hematological disorder characterized by multilineage proliferation and differentiation (Wendling *et al.*, 1986; Wendling and Tambourin, 1991). Cells in this disease became growth factor-independent and showed autonomous growth *in vitro*. Furthermore MPLV could directly transform committed and multipotential hematopoietic progenitor cells *in vitro* leading to the creation of factor-independent immortalized lines that could differentiate spontaneously (Wendling and Tambourin, 1991). However, overexpression of c-*mpl* in mice using a murine retroviral vector produced *hepatosplenomegaly*, massive expansion of erythroblasts, no early leukocytosis or thrombocytosis, *thrombocytopenia* at late stage, and death within 9–12 weeks (Cocault *et al.*, 1996). The enormous stimulation of erythroblasts and the lack of a multilineage myeloproliferative syndrome in these mice differ from the effects seen in either the TPO or v-*mpl* overexpression mice.

Human abnormalities

No human abnormalities of c-Mpl have yet been described. c-*mpl* gene structure is normal in patients with *essential thrombocythemia* (Li *et al.*, 1996) and other myeloproliferative disorders. The amount of c-Mpl on platelets in *polycythemia vera* and essential thrombocythemia are markedly reduced (Li *et al.*, 1996; Moliterno *et al.*, 1998). The pathophysiological significance of this is unknown.

The receptor is found on cells from many nonlymphoid hematological malignancies but not on cells from purely lymphoid malignancies or non-hematopoietic solid tumors (Columbyova *et al.*, 1995; Drexler *et al.*, 1996; Graf *et al.*, 1996; Matsumura *et al.*, 1996; Quentmeier *et al.*, 1996; Bredoux *et al.*, 1997; Hirai *et al.*, 1997). There is a suggestion that the presence of c-*mpl* in *acute myeloid leukemia* cells confers a poor prognosis (Vigon *et al.*, 1993a; Wetzler *et al.*, 1997).

THERAPEUTIC UTILITY

Effect of treatment with soluble receptor domain

The effect of infusing soluble murine c-Mpl (sMpl) into mice has produced strikingly variable results. Presumably infusion of the soluble receptor would bind the available TPO in the circulation, lower its concentration, reduce the stimulatory effect on megakaryocyte growth, and decrease platelet production. A similar experiment had been previously performed in rats in which the infusion of excess platelets (containing c-Mpl) was followed by a reduction in TPO levels, megakaryocyte growth, and platelet production (Jackson *et al.*, 1984; Kuter and Rosenberg, 1990). Also, after transfusing platelets into thrombocytopenic patients with *aplastic anemia*, the elevated TPO levels were reduced.

It should be noted that all of these sMpl experiments have been done with the extracellular portion of recombinant murine c-Mpl and not with the 'native sMpl'. The latter is a naturally occurring c-*mpl* mRNA splice variant that is predicted to encode a protein containing both intracellular and extracellular regions of c-Mpl but lacking the transmembrane domain.

When thrombocytopenic irradiated mice were treated with multiple injections of the extracellular portion of recombinant murine c-Mpl, platelet recovery was delayed, suggesting that the injected sMpl blocked the interaction of endogenous TPO with the c-Mpl on megakaryocytes. The platelets produced in these sMpl-treated animals did not show a reduction in JAK2 protein in platelets – something which was observed in platelets from thrombocytopenic control animals and attributed to the action of the high concentration of endogenous TPO on the c-Mpl receptor (Nishiyama *et al.*, 1998).

Paradoxically, infusion of the sMpl into normal mice for 7 days caused a 50% increase in the platelet count but no change in the red or white blood cell number (Sheridan *et al.*, 1997); infusion of control 1% normal mouse serum or heat-inactivated sMpl had no effect. These results were interpreted as being due to removal of the normal inhibitory effect TPO has on the shedding of platelets from megakaryocytes (Sheridan *et al.*, 1997). It had previously been demonstrated that excess TPO suppressed *in vitro* proplatelet formation from megakaryocytes (alleviated by addition of sMpl) and that excess TPO suppressed *in vivo* platelet production (Choi *et al.*, 1996; Nagahisa *et al.*, 1996). Whether longer infusion of sMpl would eventually have reduced total megakaryocyte growth and platelet production was not studied.

Effects of inhibitors (antibodies) to receptors

Antibodies that activate the c-Mpl receptor have been made. A murine monoclonal antibody, BAH-1, made against human megakaryocyte cells binds the c-Mpl receptor. It stimulates Meg-CFC growth *in vitro* just like TPO and *in vivo* expands the number of megakaryocyte progenitor cells in myelosuppressed mice (Deng *et al.*, 1997). Such agonistic antibodies will probably have no advantage over recombinant TPO or TPO peptide mimetics in stimulating platelet production *in vivo* but could therapeutically serve to target cells expressing c-Mpl for the purpose of drug delivery or bone marrow purging.

Antibodies that may inhibit c-Mpl have also been described. One patient with persistent *hypomegakaryocytic thrombocytopenia* and elevated TPO levels was found to have antibody that was felt to bind to either c-Mpl or c-Mpl ligand (Nichol *et al.*, 1996).

Although TPO peptide mimetics have been developed that activate c-Mpl (Cwirla *et al.*, 1997), no competitive antagonist peptides have yet been published.

References

Abina, M. A., Tulliez, M., Lacout, C., Debili, N., Villeval, J. L., Pflumio, F., Wendling, F., Vainchenker, W., and Haddada, H. (1998). Major effects of TPO delivered by a single injection of a recombinant adenovirus on prevention of septicemia and anemia associated with myelosuppression in mice: risk of sustained expression inducing myelofibrosis due to immunosuppression. *Gene Ther.* **5**, 497–506.

Alexander, W. S., Metcalf, D., and Dunn, A. R. (1995). Point mutations within a dimer interface homology domain of c-Mpl induce constitutive receptor activity and tumorigenicity. *EMBO J.* **14**, 5569–5578.

Alexander, W. S., Maurer, A. B., Novak, U., and Harrison-Smith, M. (1996a). Tyrosine-599 of the c-Mpl receptor is required for Shc phosphorylation and the induction of cellular differentiation. *EMBO J.* **15**, 6531–6540.

Alexander, W. S., Roberts, A. W., Nicola, N. A., Li, R., and Metcalf, D. (1996b). Deficiencies in progenitor cells of multiple hematopoietic lineages and defective megakaryocytopoiesis in mice lacking the thrombopoietic receptor c-Mpl. *Blood* **87**, 2162–2170.

Bartley, T. D., Bogenberger, J., Hunt, P., Li, Y. S., Lu, H. S., Martin, F., Chang, M. S., Samal, B., Nichol, J. L., Swift, S., Johnson, M. J., Hsu, R. Y., Parker, V. P., Suggs, S., Skrine, J. D., Merewether, L. A., Clogston, C., Hsu, E., Hokom, M. M., Hornkohl, A., Choi, E., Pangelinan, M., Sun, Y., Mar, V., McNinch, J., Simonet, L., Jacobsen, F., Xie, C., Shutter, J., Chute, H., Basu, R., Selander, L., Trollinger, D., Sieu, L., Padilla, D., Trail, G., Elliott, G., Izumi, R., Covey, T., Crouse, J., Garcia, A., Xu, W., Del Castillo, J., Biron, J., Cole, S., Hu, M. C. T., Pacifici, R., Ponting, I., Saris, C., Wen, D., Yung, Y. P., Lin, H., and Bosselman, R. A. (1994). Identification and cloning of a megakaryocyte growth and development factor that is a ligand for the cytokine receptor Mpl. *Cell* **77**, 1117–1124.

Bonsi, L., Grossi, A., Strippoli, P., Tumietto, F., Tonelli, R., Vannucchi, A. M., Ronchi, A., Ottolenghi, S., Visconti, G., Avanzi, G. C., Pegoraro, L., and Bagnara, G. P. (1997). An erythroid and megakaryocytic common precursor cell line (B1647) expressing both c-mpl and erythropoietin receptor (Epo-R) proliferates and modifies globin chain synthesis in response to megakaryocyte growth and development factor (MGDF) but not to erythropoietin (Epo). *Br. J. Haematol.* **98**, 549–559.

Bredoux, C., Uphoff, C. C., and Drexler, H. G. (1997). Expression of tie receptor tyrosine kinase in human leukemia cell lines. *Leuk. Res.* **21**, 595–601.

Broudy, V. C., Lin, N. L., Sabath, D. F., Papayannopoulou, T., and Kaushansky, K. (1997). Human platelets display high-affinity receptors for thrombopoietin. *Blood* **89**, 1896–1904.

Carver-Moore, K., Broxmeyer, H. E., Luoh, S. M., Cooper, S., Peng, J., Burstein, S. A., Moore, M. W., and de Sauvage, F. J. (1996). Low levels of erythroid and myeloid progenitors in thrombopoietin-and c- mpl-deficient mice. *Blood* **88**, 803–808.

Cataldo, L. M., Zhang, Y., Lu, J., and Ravid, K. (1999). Rat NAP1: cDNA cloning and upregulation by mpl ligand. *Gene* **226**, 355–364.

Chen, J., Herceg-Harjacek, L., Groopman, J. E., and Grabarek, J. (1995). Regulation of platelet activation *in vitro* by the c-Mpl ligand, thrombopoietin. *Blood* **86**, 4054–4062.

Choi, E. S., Hokom, M. M., Chen, J. L., Skrine, J., Faust, J., Nichol, J., and Hunt, P. (1996). The role of megakaryocyte growth and development factor in terminal stages of thrombopoiesis. *Br. J. Haematol.* **95**, 227–233.

Cocault, L., Bouscary, D., Le Bousse Kerdiles, C., Clay, D., Picard, F., Gisselbrecht, S., and Souyri, M. (1996). Ectopic expression of murine TPO receptor (c-mpl) in mice is pathogenic and induces erythroblastic proliferation. *Blood* **88**, 1656–1665.

Columbyova, L., Loda, M., and Scadden, D. T. (1995). Thrombopoietin receptor expression in human cancer cell lines and primary tissues. *Cancer Res.* **55**, 3509–3512.

Cui, L., Ramsfjell, V., Borge, O. J., Veiby, O. P., Lok, S., and Jacobsen, S. E. (1997). Thrombopoietin promotes adhesion of primitive human hemopoietic cells to fibronectin and vascular cell adhesion molecule-1: role of activation of very late antigen (VLA)-4 and VLA-5. *J Immunol.* **159**, 1961–1969.

Cwirla, S. E., Balasubramanian, P., Duffin, D. J., Wagstrom, C. R., Gates, C. M., Singer, S. C., Davis, A. M., Tansik, R. L., Mattheakis, L. C., Boytos, C. M., Schatz, P. J., Baccanari, D. P., Wrighton, N. C., Barrett, R. W., and Dower, W. J. (1997). Peptide agonist of the thrombopoietin receptor as potent as the natural cytokine. *Science* **276**, 1696–1969.

de Sauvage, F. J., Hass, P. E., Spencer, S. D., Malloy, B. E., Gurney, A. L., Spencer, S. A., Darbonne, W. C., Henzel, W. J., Wong, S. C., and Kuang, W. J. (1994). Stimulation of megakaryocytopoiesis and thrombopoiesis by the c-Mpl ligand. *Nature* **369**, 533–538.

Deng, B., Banu, N., Eaton, D., Wang, J. F., Cavacini, L., and Avraham, H. (1997). An agonist murine monoclonal antibody to the human c-Mpl receptor stimulates megakaryocytopoiesis. *Blood* **90**, 55a.

Deveaux, S., Filipe, A., Lemarchandel, V., Ghysdael, J., Romeo, P. H., and Mignotte, V. (1996). Analysis of the thrombopoietin receptor (MPL) promoter implicates GATA and Ets proteins in the coregulation of megakaryocyte-specific genes. *Blood* **87**, 4678–4685.

Dorsch, M., Fan, P. D., Danial, N. N., Rothman, P. B., and Goff, S. P. (1997). The thrombopoietin receptor can mediate proliferation without activation of the Jak-STAT pathway. *J. Exp. Med.* **186**, 1947–1955.

Doubeikovski, A., Uzan, G., Doubeikovski, Z., Prandini, M. H., Porteu, F., Gisselbrecht, S., and Dusanter-Fourt, I. (1997). Thrombopoietin-induced expression of the glycoprotein IIb gene involves the transcription factor PU.1/Spi-1 in UT7-Mpl cells. *J. Biol. Chem.* **272**, 24300–24307.

Drachman, J. G., and Kaushansky, K. (1995). Structure and function of the cytokine receptor superfamily. *Curr. Opin. Hematol.* **2**, 22–28.

Drachman, J. G., and Kaushansky, K. (1997). Dissecting the thrombopoietin receptor: functional elements of the Mpl cytoplasmic domain. *Proc. Natl Acad. Sci. USA* **94**, 2350–2355.

Drachman, J. G., Griffin, J. D., and Kaushansky, K. (1995). The c-Mpl ligand (thrombopoietin) stimulates tyrosine phosphorylation of Jak2, Shc, and c-Mpl. *J. Biol. Chem.* **270**, 4979–4982.

Drexler, H. G., and Quentmeier, H. (1996). Use of human leukemia-lymphoma cell lines in hematological research: effects of thrombopoietin on human leukemia cell lines. *Hum. Cell.* **9**, 309–316.

Drexler, H. G., Zaborski, M., and Quentmeier, H. (1997). Thrombopoietin supports the continuous growth of cytokine-dependent human leukemia cell lines. *Leukemia* **11**, 541–551.

Ezumi, Y., Takayama, H., and Okuma, M. (1995). Thrombopoietin, c-Mpl ligand, induces tyrosine phosphorylation of Tyk2, JAK2, and STAT3, and enhances agonists-induced aggregation in platelets *in vitro*. *FEBS Lett.* **374**, 48–52.

Fielder, P. J., Gurney, A. L., Stefanich, E., Marian, M., Moore, M. W., Carver-Moore, K., and de Sauvage, F. J. (1996). Regulation of thrombopoietin levels by c-mpl-mediated binding to platelets. *Blood* **87**, 2154–2161.

Fielder, P. J., Hass, P., Nagel, M., Stefanich, E., Widmer, R., Bennett, G. L., Keller, G. A., de Sauvage, F. J., and Eaton, D. (1997). Human platelets as a model for the binding and degradation of thrombopoietin. *Blood* **89**, 2782–2788.

Fontenay-Roupie, M., Huret, G., Loza, J. P., Adda, R., Melle, J., Maclouf, J., Dreyfus, F., and Levy-Toledano, S. (1998). Thrombopoietin activates human platelets and induces tyrosine phosphorylation of p80/85 cortactin. *Thromb. Haemost.* **79**, 195–201.

Foster, D. C., Sprecher, C. A., Grant, F. J., Kramer, J. M., Kuijper, J. L., Holly, R. D., Whitmore, T. E., Heipel, M. D., Bell, L. A., Ching, A. F., McGrane, V., Hart, C., O'Hara, P. J., and Lok, S. (1994). Human thrombopoietin: gene structure, cDNA sequence, expression, and chromosomal localization. *Proc. Natl Acad. Sci. USA* **91**, 13023–13027.

Goncalves, F., Lacout, C., Villeval, J. L., Wendling, F., Vainchenker, W., and Dumenil, D. (1997). Thrombopoietin does not induce lineage-restricted commitment of Mpl-R expressing pluripotent progenitors but permits their complete erythroid and megakaryocytic differentiation. *Blood* **89**, 3544–3553.

Gotoh, A., Ritchie, A., Takahira, H., and Broxmeyer, H. E. (1997). Thrombopoietin and erythropoietin activate inside-out signaling of integrin and enhance adhesion to immobilized fibronectin in human growth-factor-dependent hematopoietic cells. *Ann. Hematol.* **75**, 207–213.

Graf, G., Dehmel, U., and Drexler, H. G. (1996). Expression of thrombopoietin and thrombopoietin receptor MPL in human leukemia-lymphoma and solid tumor cell lines. *Leuk. Res.* **20**, 831–838.

Gurney, A. L., Carver-Moore, K., de Sauvage, F. J., and Moore, M. W. (1994). Thrombocytopenia in c-mpl-deficient mice. *Science* **265**, 1445–1447.

Gurney, A. L., Wong, S. C., Henzel, W. J., and de Sauvage, F. J. (1995). Distinct regions of c-Mpl cytoplasmic domain are coupled to the JAK- STAT signal transduction pathway and Shc phosphorylation. *Proc. Natl Acad. Sci. USA* **92**, 5292–5296.

Harker, L. A., Marzec, U. M., Hunt, P., Kelly, A. B., Tomer, A., Cheung, E., Hanson, S. R., and Stead, R. B. (1996). Dose–response effects of pegylated human megakaryocyte growth and development factor on platelet production and function in nonhuman primates. *Blood* **88**, 511–521.

Higuchi, T., Koike, K., Sawai, N., and Koike, T. (1997). Proliferative and differentiative potential of thrombopoietin-responsive precursors: expression of megakaryocytic and erythroid lineages. *Exp. Hematol.* **25**, 463–470.

Hirai, H., Shimazaki, C., Yamagata, N., Goto, H., Inaba, T., Kikuta, T., Sumikuma, T., Sudo, Y., Ashihara, E., Fujita, N., Hibi, S., Imashuku, S., Ito, E., and Nakagawa, M. (1997). Effects of thrombopoietin (c-mpl ligand) on growth of blast cells from patients with transient abnormal myelopoiesis and acute myeloblastic leukemia. *Eur. J. Haematol.* **59**, 38–46.

Hong, Y., Dumenil, D., van der Loo, B., Goncalves, F., Vainchenker, W., and Erusalimsky, J. D. (1998). Protein kinase C mediates the mitogenic action of thrombopoietin in c- Mpl-expressing UT-7 cells. *Blood* **91**, 813–822.

Jackson, C. W., Brown, L. K., Somerville, B. C., Lyles, S. A., and Look, A. T. (1984). Two-color flow cytometric measurement of DNA distributions of rat megakaryocytes in unfixed, unfractionated marrow cell suspensions. *Blood* **63**, 768–778.

Kato, T., Ogami, K., Shimada, Y., Iwamatsu, A., Sohma, Y., Akahori, H., Horie, K., Kokubo, A., Kudo, Y., Maeda, E., Kobayashi, K., Ohashi, H., Ozawa, T., Inoue, H., Kawamura, K., and Miyazaki, H. (1995). Purification and characterization of thrombopoietin. *J. Biochem.* **119**, 229–236.

Kaushansky, K., Broudy, V. C., Grossmann, A., Humes, J., Lin, N., Ren, H. P., Bailey, M. C., Papayannopoulou, T., Forstrom, J. W., and Sprugel, K. H. (1995). Thrombopoietin expands erythroid progenitors, increases red cell production, and enhances erythroid recovery after myelosuppressive therapy. *J. Clin. Invest.* **96**, 1683–1687.

Kaushansky, K., Lin, N., Grossmann, A., Humes, J., Sprugel, K. H., and Broudy, V. C. (1996). Thrombopoietin expands erythroid, granulocyte-macrophage, and megakaryocytic progenitor cells in normal and myelosuppressed mice. *Exp. Hematol.* **24**, 265–269.

Kelemen, E., Cserhati, I., and Tanos, B. (1958). Demonstration and some properties of human thrombopoietin in thrombocythaemic sera. *Acta Haematol.* **20**, 350–355.

Kieran, M. W., Perkins, A. C., Orkin, S. H., and Zon, L. I. (1996). Thrombopoietin rescues *in vitro* erythroid colony formation from mouse embryos lacking the erythropoietin receptor. *Proc. Natl Acad. Sci. USA* **93**, 9126–9131.

Kiladjian, J. J., Elkassar, N., Hetet, G., Briere, J., Grandchamp, B., and Gardin, C. (1997). Study of the thrombopoietin receptor in essential thrombocythemia. *Leukemia* **11**, 1821–1826.

Kobayashi, M., Laver, J. H., Kato, T., Miyazaki, H., and Ogawa, M. (1995). Recombinant human thrombopoietin (Mpl ligand) enhances proliferation of erythroid progenitors. *Blood* **86**, 2494–2499.

Komatsu, N., Kirito, K., Shimizu, R., Kunitama, M., Yamada, M., Uchida, M., Takatoku, M., Eguchi, M., and Miura, Y. (1997). *In vitro* development of erythroid and megakaryocytic cells from a UT-7 subline, UT-7/GM. *Blood* **89**, 4021–4033.

Kubota, Y., Arai, T., Tanaka, T., Yamaoka, G., Kiuchi, H., Kajikawa, T., Kawanishi, K., Ohnishi, H., Yamaguchi, M., Takahara, J., and Irino, S. (1996). Thrombopoietin modulates platelet activation *in vitro* through protein-tyrosine phosphorylation. *Stem Cells* **14**, 439–444.

Kunitama, M., Shimizu, R., Yamada, M., Kato, T., Miyazaki, H., Okada, K., Miura, Y., and Komatsu, N. (1997). Protein kinase C and c-*myc* gene activation pathways in thrombopoietin signal transduction. *Biochem. Biophys. Res. Commun.* **231**, 290–294.

Kuter, D. J. (1996a). The physiology of platelet production. *Stem Cells* **14**, (Suppl 1), 88–101.

Kuter, D. J. (1996b). Thrombopoietin: biology, clinical applications, role in the donor setting. *J. Clin. Apheresis* **11**, 149–159.

Kuter, D. J. (1997). In "Thrombopoiesis and Thrombopoietins: Molecular, Cellular, Preclinical and Clinical Biology" (ed. D. J. Kuter, P. Hunt, W. Sheridan and D. Zucker-Franklin), The regulation of platelet production pp. 377–395. Humana, Totowa.

Kuter, D. J., and Rosenberg, R. D. (1990). Regulation of megakaryocyte ploidy *in vivo* in the rat. *Blood* **75**, 74–81.

Kuter, D. J., and Rosenberg, R. D. (1995). The reciprocal relationship of thrombopoietin (c-Mpl ligand) to changes in the platelet mass during busulfan-induced thrombocytopenia in the rabbit. *Blood* **85**, 2720–2730.

Li, J., Xia, Y., and Kuter, D. J. (1996). Analysis of the thrombopoietin receptor (MPL) on platelets from normal and essential thrombocythemic (ET) patients. *Blood* **88**, 545a.

Li, J., Xia, Y., and Kuter, D. (1999). Interaction of thrombopoietin with the platelet c-mpl receptor in plasma: binding, internalization, stability and pharmacodynamics. *Br. J. Haematol.* **106**, 345–356.

Li, J., Sabath, D. F., and Kuter, D. J. (2000). Cloning and functional characterization of a novel c-mpl variant expressed in human CD34 cells and platelets. *Cytokine* (in press).

Matsumura, I., Ikeda, H., and Kanakura, Y. (1996). The effects of thrombopoietin on the growth of acute myeloblastic leukemia cells. *Leuk. Lymphoma.* **23**, 533–538.

Matsumura, I., Nakajima, K., Wakao, H., Hattori, S., Hashimoto, K., Sugahara, H., Kato, T., Miyazaki, H., Hirano, T., and Kanakura, Y. (1998). Involvement of prolonged ras activation in thrombopoietin-induced megakaryocytic differentiation of a human factor-dependent hematopoietic cell line. *Mol. Cell. Biol.* **18**, 4282–4290.

Methia, N., Louache, F., Vainchenker, W., and Wendling, F. (1993). Oligodeoxynucleotides antisense to the proto-oncogene c-mpl specifically inhibit *in vitro* megakaryocytopoiesis. *Blood* **82**, 1395–1401.

Miura, Y., Kirito, K., and Komatsu, N. (1998). Regulation of both erythroid and megakaryocytic differentiation of a human leukemia cell line, UT-7. *Acta Haematol.* **99**, 180–184.

Miyakawa, Y., Oda, A., Druker, B. J., Kato, T., Miyazaki, H., Handa, M., and Ikeda, Y. (1995). Recombinant thrombopoietin induces rapid protein tyrosine phosphorylation of Janus kinase 2 and Shc in human blood platelets. *Blood* **86**, 23–27.

Miyakawa, Y., Oda, A., Druker, B. J., Miyazaki, H., Handa, M., Ohashi, H., and Ikeda, Y. (1996). Thrombopoietin induces tyrosine phosphorylation of Stat3 and Stat5 in human blood platelets. *Blood* **87**, 439–446.

Moliterno, A. R., Hankins, W. D., and Spivak, J. L. (1998). Impaired expression of the thrombopoietin receptor by platelets from patients with polycythemia vera. *N. Engl. J. Med.* **338**, 572–580.

Montrucchio, G., Brizzi, M. F., Calosso, G., Marengo, S., Pegoraro, L., and Camussi, G. (1996). Effects of recombinant human megakaryocyte growth and development factor on platelet activation. *Blood* **87**, 2762–2768.

Morita, H., Tahara, T., Matsumoto, A., Kato, T., Miyazaki, H., and Ohashi, H. (1996). Functional analysis of the cytoplasmic domain of the human Mpl receptor for tyrosine-phosphorylation of the signaling molecules, proliferation and differentiation. *FEBS Lett.* **395**, 228–234.

Mu, S. X., Xia, M., Elliott, G., Bogenberger, J., Swift, S., Bennett, L., Lappinga, D. L., Hecht, R., Lee, R., and Saris, C. J. (1995). Megakaryocyte growth and development factor and interleukin-3 induce patterns of protein-tyrosine phosphorylation that correlate with dominant differentiation over proliferation of mpl-transfected 32D cells. *Blood* **86**, 4532–4543.

Nagahisa, H., Nagata, Y., Ohnuki, T., Osada, M., Nagasawa, T., Abe, T., and Todokoro, K. (1996). Bone marrow stromal cells produce thrombopoietin and stimulate megakaryocyte growth and maturation but suppress proplatelet formation. *Blood* **87**, 1309–1316.

Nichol, J., Hornkohl, A., Best, D., Dunn, V., Rich, D., Hunt, P., and Yaffe, B. (1996). Acquired thrombocytopenia associated with an antibody that interferes with the Mpl receptor and/or its ligand:a case report. *Blood* **88**, 516a.

Nishiyama, U., Morita, H., Akahori, H., Kuwaki, T., Shimizu, E., Kato, T., Miyazaki, H., and Shimosaka, A. (1998). Markedly reduced reactivity of platelets from thrombocytopenic mice to recombinant murine megakaryocyte growth and development factor *in vitro*. *Blood* **92**, 30a.

Oda, A., Ozaki, K., Druker, B. J., Miyakawa, Y., Miyazaki, H., Handa, M., Morita, H., Ohashi, H., and Ikeda, Y. (1996). p120c-cbl is present in human blood platelets and is differentially involved in signaling by thrombopoietin and thrombin. *Blood* **88**, 1330–1338.

Porteu, F., Rouyez, M. C., Cocault, L., Benit, L., Charon, M., Picard, F., Gisselbrecht, S., Souyri, M., and Dusanter-Fourt, I. (1996). Functional regions of the mouse thrombopoietin receptor cytoplasmic domain: evidence for a critical region which is involved in differentiation and can be complemented by erythropoietin. *Mol. Cell. Biol.* **16**, 2473–2482.

Quentmeier, H., Zaborski, M., Graf, G., Ludwig, W. D., and Drexler, H. G. (1996). Expression of the receptor MPL and proliferative effects of its ligand thrombopoietin on human leukemia cells. *Leukemia* **10**, 297–310.

Rasko, J. E., O'Flaherty, E., and Begley, C. G. (1997). Mpl ligand (MGDF) alone and in combination with stem cell factor (SCF) promotes proliferation and survival of human megakaryocyte, erythroid and granulocyte/macrophage progenitors. *Stem Cells* **15**, 33–42.

Ratajczak, M. Z., Ratajczak, J., Marlicz, W., Pletcher, C.H. Jr., Machalinski, B., Moore, J., Hung, H., and Gewirtz, A. M. (1997). Recombinant human thrombopoietin (TPO) stimulates erythropoiesis by inhibiting erythroid progenitor cell apoptosis. *Br. J. Haematol.* **98**, 8–17.

Rouyez, M. C., Boucheron, C., Gisselbrecht, S., Dusanter-Fourt, I., and Porteu, F. (1997). Control of thrombopoietin-induced megakaryocytic differentiation by the mitogen-activated protein kinase pathway. *Mol. Cell. Biol.* **17**, 4991–5000.

Sabath, D. F., Kaushansky, K., and Broudy, V. C. (1999). Deletion of the extracellular membrane-distal cytokine receptor homology module of Mpl results in constitutive cell growth and loss of thrombopoietin binding. *Blood* **94**, 365–367.

Sasaki, A., Katoh, O., Kawaishi, K., Hyodo, H., Kimura, A., Satow, Y., and Kuramoto, A. (1995a). Expression of c-Mpl and c-Mpl ligand gene in hematopoietic cells of individuals with and without myeloproliferative disorders and leukemia cell lines. *Int. J. Hematol.* **62**, 217–223.

Sasaki, K., Odai, H., Hanazono, Y., Ueno, H., Ogawa, S., Langdon, W. Y., Tanaka, T., Miyagawa, K., Mitani, K., and Yazaki, Y. (1995b). TPO/c-mpl ligand induces tyrosine phosphorylation of multiple cellular proteins including proto-oncogene products, Vav and c-Cbl, and Ras signaling molecules. *Biochem. Biophys. Res. Commun.* **216**, 338–347.

Sattler, M., Durstin, M. A., Frank, D. A., Okuda, K., Kaushansky, K., Salgia, R., and Griffin, J. D. (1995). The thrombopoietin receptor c-MPL activates JAK2 and TYK2 tyrosine kinases. *Exp. Hematol.* **23**, 1040–1048.

Sattler, M., Salgia, R., Durstin, M. A., Prasad, K. V., and Griffin, J. D. (1997). Thrombopoietin induces activation of the phosphatidylinositol-3′ kinase pathway and formation of a complex containing p85PI3K and the protooncoprotein p120CBL. *J. Cell Physiol.* **171**, 28–33.

Sheridan, W. P., Choi, E., Toombs, C. F., Nichol, J., Fanucchi, M., and Baser, R. I. (1997). Biology of thrombopoiesis and the role of Mpl ligand in the production and function of platelets. *Platelets* **8**, 319–332.

Skoda, R. C., Seldin, D. C., Chiang, M. K., Peichel, C. L., Vogt, T. F., and Leder, P. (1993). Murine c-mpl: a member of the hematopoietic growth factor receptor superfamily that transduces a proliferative signal. *EMBO J.* **12**, 2645–2653.

Solar, G. P., Kerr, W. G., Zeigler, F. C., Hess, D., Donahue, C., de Sauvage, F. J., and Eaton, D. L. (1998). Role of c-mpl in early hematopoiesis. *Blood* **92**, 4–10.

Souryi, M., Vigon, I., Penciolelli, J.-F., Heard, J.-M., Tambourin, P., and Wendling, F. (1990). A putative truncated cytokine receptor gene transduced by the myeloproliferative leukemia virus immortalizes hematopoietic progenitors. *Cell* **63**, 1137–1147.

Tanimukai, S., Kimura, T., Sakabe, H., Ohmizono, Y., Kato, T., Miyazaki, H., Yamagishi, H., and Sonoda, Y. (1997). Recombinant human c-Mpl ligand (thrombopoietin) not only acts on megakaryocyte progenitors, but also on erythroid and multipotential progenitors *in vitro*. *Exp. Hematol.* **25**, 1025–1033.

Vigon, I., Mornon, J.-P., Cocault, L., Mitjavila, M.-T., Tambourin, P., Gisselbrecht, S., and Souyri, M. (1992). Molecular cloning and characterization of MPL, the human homolog of the v-mpl oncogene: Identification of a member of the hematopoietic growth factor receptor superfamily. *Proc. Natl Acad. Sci. USA* **89**, 5640–5644.

Vigon, I., Dreyfus, F., Melle, J., Viguie, F., Ribrag, V., Cocault, L., Souryi, M., and Gisselbrech, S. (1993a). Expression of the c-mpl proto-oncogene in human hematologic malignancies. *Blood* **82**, 877–883.

Vigon, I., Dreyfus, F., Melle, J., Viguie, F., Ribrag, V., Cocault, L., Souyri, M., and Gisselbrecht, S. (1993b). Expression of the c-mpl proto-oncogene in human hematologic malignancies. *Blood* **82**, 877–883.

Vigon, I., Florindo, C., Fichelson, S., Guenet, J. L., Mattei, M. G., Souyri, M., Cosman, D., and Gisselbrecht, S. (1993c). Characterization of the murine Mpl proto-oncogene, a member of the hematopoietic cytokine receptor family: molecular cloning, chromosomal location and evidence for a function in cell growth. *Oncogene* **8**, 2607–2615.

Villeval, J. L., Cohen-Solal, K., Tulliez, M., Giraudier, S., Guichard, J., Burstein, S. A., Cramer, E. M., Vainchenker, W., and Wendling, F. (1997). High thrombopoietin production by hematopoietic cells induces a fatal myeloproliferative syndrome in mice. *Blood* **90**, 4369–4383.

Wendling, F., and Tambourin, P. (1991). The oncogene V-MPL, a putative truncated cytokine receptor which immortalized hemtopoietic progenitors. *Nouv. Rev. Fr. Hematol.* **33**, 145–146.

Wendling, F., Varlet, P., Charon, M., and Tambourin, P. (1986). A retrovirus complex inducing an acute myeloproliferative leukemia disorder in mice. *Virology* **149**, 242–246.

Wetzler, M., Baer, M. R., Bernstein, S. H., Blumenson, L., Stewart, C., Barcos, M., Mrozek, K., Block, A. W., Herzig, G. P., and Bloomfield, C. D. (1997). Expression of c-mpl mRNA, the receptor for thrombopoietin, in acute myeloid leukemia blasts identifies a group of patients with poor response to intensive chemotherapy. *J. Clin. Oncol.* **15**, 2262–2268.

Yamada, M., Komatsu, N., Kirito, K., Kashii, Y., Tomizuka, H., Okada, K., Endo, T., Fukumaki, Y., Shinjo, K., Abe, K., and Miura, Y. (1998). Thrombopoietin supports *in vitro* erythroid differentiation via its specific receptor c-Mpl in a human leukemia cell line. *Cell Growth Differ.* **9**, 487–496.

Yan, X.-Q., Lacey, D., Fletcher, F., Hartley, C., McElroy, T., Sun, Y., Xia, M., Mu, S., Saris, C., Hill, D., Hawley, R. G., and McNiece, I. (1995). Chronic exposure to retroviral vector encoded MGDF (mpl-ligand) induces lineage-specific growth and differentiation of megakaryocytes in mice. *Blood* **86**, 4025–4033.

Yan, X. Q., Lacey, D., Hill, D., Chen, Y., Fletcher, F., Hawley, R. G., and McNiece, I. K. (1996). A model of myelofibrosis and osteosclerosis in mice induced by overexpressing thrombopoietin (mpl ligand): reversal of disease by bone marrow transplantation. *Blood* **88**, 402–409.

Yoshida, M., Tsuji, K., Ebihara, Y., Muraoka, K., Tanaka, R., Miyazaki, H., and Nakahata, T. (1997). Thrombopoietin alone stimulates the early proliferation and survival of human erythroid, myeloid and multipotential progenitors in serum-free culture. *Br. J. Haematol.* **98**, 254–264.

Zauli, G., Bassini, A., Vitale, M., Gibellini, D., Celeghini, C., Caramelli, E., Pierpaoli, S., Guidotti, L., and Capitani, S. (1997). Thrombopoietin enhances the alpha IIb beta 3-dependent adhesion of megakaryocytic cells to fibrinogen or fibronectin through PI 3 kinase. *Blood* **89**, 883–895.

CXC Chemokine Receptors

Chemokine Receptors: Overview

Philip M. Murphy*

Molecular Signaling Section, Laboratory of Host Defenses, National Institute of Allergy and Infectious Diseases, National Institutes of Health, Bethesda, MD 20892, USA

*corresponding author tel: 301-496-2877, fax: 301-402-4369, e-mail: pmm@nih.gov

DOI: 10.1006/rwcy.2000.02012.

INTRODUCTION

Chemokine receptors are defined by their ability to bind chemokines in a specific and saturable manner, and to transduce a cellular response. At the molecular level, this definition has been met by 16 human cell surface proteins (named CXCR1 to CXCR5, CCR1 to CCR9, XCR1, and CX3CR1), which together comprise the largest known structurally defined division of the rhodopsin superfamily of seven transmembrane domain, G protein-coupled receptors (GPCRs) (Murphy, 1994; Premack and Schall, 1996; Yoshie *et al.*, 1997; Locati and Murphy, 1999; Zlotnik *et al.*, 1999). In addition, four herpesvirus-encoded chemokine receptors (ORFs US28 of human cytomegalovirus, ECRF3 of herpesvirus saimiri, UL12 of HHV-6 and no. 74 of HHV-8/Kaposi's sarcoma herpesvirus (KSHV; also known as KSHV GPCR) (Isegawa *et al.*, 1998; Pease and Murphy, 1998), and two non-signaling mammalian chemokine-binding proteins (D6 and the Duffy antigen receptor for chemokines, DARC) (Horuk *et al.*, 1994; Nibbs *et al.*, 1997) have been described. Studied initially for their roles in leukocyte trafficking, chemokine receptors are now known to have multiple additional functions, including regulation of development of the cardiovascular, gastrointestinal, immune, and central nervous systems (Tachibana *et al.*, 1998; Zou *et al.*, 1998), and usage as cell entry factors by HIV-1 (Cocchi *et al.*, 1995; Feng *et al.*, 1996; Berger *et al.*, 1999) and *Plasmodium vivax* (Horuk *et al.*, 1993), the causative agents of AIDS and a form of malaria, respectively. The purpose of this chapter is to provide an overview of the shared and differential features of these molecules, as introduction to the chapters devoted to individual chemokine receptors.

CHEMOKINE RECEPTOR STRUCTURE

The deduced amino acid sequences of chemokine receptors have 25–80% identity, indicating a common ancestry. This has facilitated discovery of additional family members by crosshybridization of cDNAs and genes to DNA probes from known receptors (Murphy, 1996). The structural boundary which separates chemokine receptors as a group from other types of G protein-coupled receptors is not sharp, and they lack a single structural signature. However, several structural fea-tures common to known chemokine receptors are less common in other types of G protein-coupled receptors, including a length of 340–370 amino acids; a highly acidic N- terminal segment; the sequence DRYLAIVHA, or a minor variation thereof, in the second intracellular loop; a short basic third intracellular loop, and a cysteine in each of the four extracellular domains (Murphy, 1994). A novel sequence which contains all of these features is likely to represent a chemokine receptor.

The folded structure of chemokine receptors has not been determined; however a model can be constructed for the transmembrane helices based on the known structure of rhodopsin (Unger *et al.*, 1997). In addition, domain-specific antibodies have been used in some cases to establish the transmembrane topography of N- and C-termini and the loop regions, which are consistent with the rhodopsin model. Early biochemical crosslinking data were consistent with a monomer structure for neutrophil IL-8 receptors (Moser *et al.*, 1991) and Duffy (Neote *et al.*, 1993), whereas more recent data consistent with a dimer have been reported for CCR2 (Rodriguez-Frade *et al.*, 1999), CCR5 (Benkirane *et al.*, 1997), and CXCR4 (Lapham *et al.*, 1999). In the case of CCR2, a dimer has been implicated as the functional form of the receptor. This has further stimulated the unsettled debate as to whether chemokines bind to receptors as monomers or dimers. Although they exist as monomers at physiologic concentrations, most chemokines form dimers at high concentrations.

A major unanswered question is how chemokines bind to receptors. Mutagenesis studies support a Velcro-like interaction between ligand and receptor, in which multiple low-affinity binding sites integrate to produce an overall high-affinity binding energy (Ahuja *et al.*, 1996; Berson and Doms, 1998; Paavola *et al.*, 1998). There appear to be two classes of binding sites: the first for chemokine docking to receptor, the second for receptor triggering. Because of the seven transmembrane domains, the binding sites are not contiguous in the primary sequence, but instead are gathered together dirndl-like in the folded protein. The chemokine N-terminus is not usually important for docking but is typically critical for activation. Putative roles of the transmembrane domains include determination of the conformation of the extracellular domains, contact with the chemokine activation domain, and signal transduction. Additional functional domains include the C-terminus, which typically contains multiple serine and threonine residues, some of which may be phosphorylated upon activation and may act to desensitize the receptor (Ali *et al.*, 1999), domains in the second and third intracellular loops important for G protein-binding and activation (Damaj *et al.*, 1996), and a conserved consensus sequence for tyrosine sulfation in the N-terminus (Farzan *et al.*, 1999), which in the case of CCR5 must be utilized for HIV coreceptor activity.

CHEMOKINE RECEPTOR SPECIFICITY

Chemokine receptors can bind multiple chemokines, and vice versa; however the specificities are generally restricted by chemokine class (**Table 1** and **Table 2**).

Table 1 Chemokine receptor classification and specificity

Class	Subtype	Ligands
CC	CCR1	MIP-1α, RANTES, MCP-3
	CCR2[a]	MCP-1, MCP-3, MCP-4
	CCR3[a]	Eotaxin, eotaxin 2, RANTES, MCP-2, MCP-3, MCP-4, MCP-5
	CCR4	TARC, MDC
	CCR5[a]	MIP-1α, MIP-1β, RANTES, MCP-2
	CCR6	LARC
	CCR7	ELC, SLC
	CCR8[a]	I-309
	CCR9	TECK
	CMV US28[a]	MIP-1α, MIP-1β, RANTES, MCP-1, fractalkine
	HHV-6 UL12	MIP-1α, MIP-1β, RANTES, MCP-1
CXC	CXCR1	IL-8, GCP-2
	CXCR2	IL-8, GROα, GROβ, GROγ, NAP-2, ENA-78, GCP-2
	CXCR3	MIG, IP-10, I-TAC, eotaxin, SLC
	CXCR4[a]	SDF-1
	ECRF3	IL-8, GROα, NAP-2
CC/CXC	*KSHV GPCR*	IL-8, GROα, RANTES, MCP-1, I-309
C	XCR1	Lymphotactin
CX_3C	CX3CR1[a]	Fractalkine

Italicized receptors are viral in origin. All others are human.
[a]HIV-1 coreceptors.

Table 2 Chemokine-binding proteins with seven transmembrane domain motif, not yet demonstrated to signal

Molecule	Ligands
Duffy	IL-8, NAP-2, GROα, I-309, RANTES, MCP-1
D6	MIP-1α, RANTES, MCP-1

This provides for a simple nomenclature in which each receptor is named by the chemokine class it recognizes, followed by the letter R and an arabic numeral assigned by the order of discovery relative to other receptors. Thus, CCR1 is the first receptor specific for CC chemokines to be discovered. There are several known exceptions to this rule: CXCR3 also binds two CC chemokines, eotaxin and SLC, but apparently with lower affinity than its CXC ligands IP-10, MIG, and I-TAC (Soto *et al.*, 1998; Weng *et al.*, 1998); KSHV GPCR binds multiple CC and CXC chemokines with comparable affinity (Arvanitakis *et al.*, 1997); US28 binds multiple CC chemokines and the CX3C chemokine fractalkine (Kledal *et al.*, 1998); and Duffy binds multiple CC and CXC chemokines with comparable high affinity (Chaudhuri *et al.*, 1994).

Each receptor has a distinct specificity for chemokine ligands and leukocyte subsets; however, the specificities may overlap considerably. For example, CCR1 and CCR5 both bind MIP-1α and RANTES but can be distinguished by their specificities for MCP-3 and MIP-1β respectively. The structural basis of specificity is counterintuitive. For example, CCR2 is much more related structurally to CCR5 than is CCR1 (82% versus 56% amino acid identity), but has only one high-affinity ligand in common with CCR5, whereas CCR1 has several. Instead, CCR2 shares several ligands with CCR1, yet the two receptors are only 56% identical in amino acid sequence.

Promiscuous chemokine ligand–receptor relationships are common. As a result, defining the chemokine receptor responsible for stimulus–response coupling in primary cells is often not straightforward due to overlapping specificities of receptors for ligands and leukocytes, and a paucity of receptor subtype-selective blocking agents. Not only can distinct receptor subtypes specific for the same chemokine and the same function be coexpressed on the same cell, but also distinct chemokines acting at separate receptors coexpressed on the same cell can induce the same cellular response. Although antireceptor monoclonal antibodies and mice with targeted gene disruptions are now being used to resolve specificities *in vivo*, problems of interpretation persist due to the inequality of chemokine and chemokine receptor repertoires, tissue distribution, and biological usage among species. For example, IL-8 is found in human and rabbit, but not in mouse, and CXCR1 is expressed mainly in human neutrophils versus rat macrophages (Dunstan *et al.*, 1996). Fortunately, this situation appears to affect only a minority of chemokine receptors.

Chemokine receptors have been described on all leukocyte subsets studied (Luster, 1998; Mantovani *et al.*, 1998). Details of expression and regulation are complex, and in some cases controversial, and are beyond the scope of this overview; however, a few key patterns can be summarized. Cells of the granulocyte series appear to express a limited repertoire of chemokine receptors. Neutrophils express predominantly CXCR1 and CXCR2, whereas human eosinophils express mainly CCR3, the IL-8 receptors and, to a more variable extent, CCR1. CCR3 has also been reported on basophils and a small subset of TH2 T lymphocytes, consistent with a role in allergic inflammation. Monocytes and macrophages express a broader repertoire, which includes CXCR1, CXCR2, CXCR4, CCR1, CCR2, CCR5, CCR8, and CX3CR1. T lymphocytes express the complete chemokine receptor repertoire, but in an incomplete and differential manner on specific subsets. TH1 and TH2 cells have distinct receptor repertoires. Of note, CXCR3 and CCR4 have been reported to be markers of TH1 and TH2 cells, respectively, although this has been challenged. Little information has been reported for mature B cells, although it is clear that CXCR5 is expressed at high levels and is functionally important. Additional information on this subject with extensive lists of primary references can be found in the individual receptor chapters.

CHEMOKINE RECEPTOR SIGNALING

Aspects of signaling common to all known mammalian chemokine receptors include induction of calcium flux and chemotaxis, and marked inhibition of both by *Bordetella pertussis* toxin (Bokoch, 1995; Ward *et al.*, 1998). The latter reflects coupling of receptors in primary cells to G_i-type heterotrimeric G proteins, whose α subunits are covalently ADP-ribosylated by pertussis toxin, which inactivates the protein. There are several noteworthy exceptions to this. First, constitutive signaling by the viral receptor KSHV GPCR is completely insensitive to pertussis toxin; the presumptive G protein involved has not yet been identified (Arvanitakis *et al.*, 1997). Second, although CX3CR1 can signal in a conventional pertussis toxin-sensitive chemotactic pathway in response to its soluble chemokine ligand fractalkine, it can also function as a powerful cell–cell adhesion molecule by binding to membrane-bound fractalkine in a pertussis toxin-insensitive manner (Imai *et al.*, 1997). Third, inhibition of chemokine action in primary cells by pertussis toxin is often incomplete, perhaps reflecting coupling to other classes of G proteins. Consistent with this, several chemokine receptors, including CCR1, CXCR1, CXCR2, and CCR2, signal in a pertussis toxin-insensitive manner in cell lines cotransfected to express receptor and proteins from the G_q class, including G_{16} which is preferentially expressed in hematopoietic cells (Kuang *et al.*, 1996; Xie *et al.*, 1997).

Signaling by the IL-8 receptors, CXCR1 and CXCR2, has been studied most extensively, and includes additional common elements such as stimulation of phospholipase $C\beta2$ and inhibition of adenylyl cyclase (Hall *et al.*, 1999), and at least two differences: selective activation of phospholipase D and the NADPH oxidase by CXCR1 (Jones *et al.*, 1996). Activation of CCR1 and CCR2 has also been shown to inhibit adenylyl cyclase (Myers *et al.*, 1995), consistent with coupling to G_i. Activation of several chemokine receptors has been associated with protein tyrosine phosphorylation, including pyk-2 in the case of CCR5 (Davis *et al.*, 1997). IL-8 signaling to the Ras/Raf/MAP kinase/PI-3 kinase pathway has also been reported (Knall *et al.*, 1996). Unusual dual signaling pathways have been reported for RANTES in T cells. At low concentrations pertussis toxin-sensitive calcium flux occurs, whereas at high concentrations, threshold $\sim 1\ \mu M$, activation of ZAP70 in a pertussis toxin-insensitive manner is observed. The receptor mechanism for this latter phenomenon has not yet been defined (Bacon *et al.*, 1995).

Activation of chemokine receptors ultimately results in desensitization, which has been associated with phosphorylation of serines and threonines in the C tail, and clathrin-mediated endocytosis (Ali *et al.*, 1999; Oppermann *et al.*, 1999; Yang *et al.*, 1999). This process may be important for receptor resensitization and chemotaxis. Signaling is not required for HIV coreceptor activity by CCR5 or CXCR4 (Alkhatib *et al.*, 1997; Amara *et al.*, 1997; Farzan *et al.*, 1997), or its inhibition by cognate chemokines, or, as mentioned previously, for proadhesive activity by CX3CR1 (Imai *et al.*, 1997).

CHEMOKINE RECEPTOR FUNCTION IN HEALTH AND DISEASE

The main function shared by chemokines and chemokine receptors is leukocyte chemotaxis, which, together with differential expression, allows for orchestration of specific leukocyte trafficking *in vivo*. Despite the redundancy in organization, there is increasing evidence from gene knockout and immunologic neutralization experiments for substantial specificity in chemokine and chemokine receptor function *in vivo*, affecting three main areas: organ development, susceptibility to infection, and inflammation (**Table 3**) (Gerard, 1999). This implies that chemokines which share leukocyte specificities and receptors which share chemokine specificities may not always be expressed at equivalent levels and in the same temporal and spatial context *in vivo*, and there is increasing experimental evidence for this (Amichay *et al.*, 1996). Differential expression of combinations of chemokines or receptors could allow sequential action, directing leukocytes with high specificity to their *in vivo* targets (Foxman *et al.*, 1997). In this scenario, agents that neutralize single chemokines or block single chemokine receptor subtypes could be very effective at terminating the entire signaling relay, and may be useful therapeutically in diseases where chemokine-dependent inflammation contributes to pathology.

Chemokines and chemokine receptors can be loosely divided into three functional groups: immune, inflammatory, and an overlap group. Immune chemokine receptors, such as CXCR5 and CCR7, bind ligands that are constitutively expressed in a restricted manner and regulate basal leukocyte trafficking (Forster *et al.*, 1996; Gunn *et al.*, 1998a,b, 1999; Tang and Cyster, 1999; Saeki *et al.*, 1999). This group of receptors regulates organization of the lymphoid system and determines the migration and position of T cells, B cells, and dendritic cells within specific areas of organized lymphoid tissue. In contrast, inflammatory receptors, such as CXCR1, CXCR2, CCR2, and CCR3, regulate emergency leukocyte trafficking to tissue sites of inflammation by binding ligands whose expression is less spatially restricted than immune chemokines, and instead is

Table 3 Function of chemokine receptors *in vivo*: phenotypes associated with targeted gene disruptions in mice and naturally occurring inactivating mutations in humans

Receptor	Viable?	Development	Major phenotypes
Mouse CXCR2	Yes	Abnormal	Neutrophil and B cell expansion in blood, lymph nodes, spleen, and bone marrow
			Impaired neutrophil recruitment to i.p. thioglycollate
Mouse CXCR4	No	Abnormal	Ventricular septal defect
			Impaired B cell lymphopoiesis
			Impaired bone marrow myelopoiesis
			Defective cerebellar and gastric vascular development
Mouse CXCR5	Yes	Abnormal	Absent inguinal lymph nodes
			Absent or abnormal Peyer's patches
			Defective B cell trafficking and localization
Mouse CCR1	Yes	Normal	Impaired lung granuloma formation to *Schistosoma mansoni* eggs
			Reduced pancreatitis-induced pulmonary inflammation
			Increased susceptibility to *Aspergillus fumigatus*
			Abnormal TH1/TH2 cytokine balance in *S. mansoni* egg challenge
			Abnormal steady-state and induced trafficking and proliferation of myeloid progenitor cells
Mouse CCR2	Yes	Normal	Reduced monocyte recruitment after i.p. thioglycollate
			Reduced lung granuloma size to PPD challenge
			Abnormal TH1/TH2 cytokine balance in PPD challenge
			Increased susceptibility to *Listeria*
			Reduced atherogenesis
Mouse CCR5	Yes	Normal	Increased susceptibility to *Listeria*
			Increased susceptibility to LPS-induced endotoxemia
			Enhanced DTH reaction
			Increased humoral responses to T cell-dependent antigenic challenge
Human CCR5	Yes	NAD	Resistance to HIV-1 and AIDS
Human Duffy	Yes	NAD	Resistance to *Plasmodium vivax* form of malaria

highly temporally restricted by primary proinflammatory cytokines (e.g. IL-1 and TNF).

Chemokines and chemokine receptors also regulate hematopoiesis. Analysis of knockout mice has shown that CXCR4 is required for B cell lymphopoiesis and bone marrow myelopoiesis, and that CXCR2, CCR1, and CCR2 regulate hematopoietic progenitor cell growth and distribution (Broxmeyer *et al.*, 1996, 1999; Reid *et al.*, 1999). In particular, CXCR2 is a negative regulator of myeloid progenitors, which may explain in part the massive expansion of neutrophils in mice lacking this receptor (Cacalano *et al.*, 1994). Chemokines and chemokine receptors may also play a role in T lymphocyte differentiation into TH1 and TH2 phenotypes, as suggested by studies of CCR1 and CCR2 knockout mice in schistosome egg challenge of the lung; however the mechanism underlying this is not yet clear (Boring *et al.*, 1997; Gao *et al.*, 1997).

In addition to leukocytes, some chemokine receptors are also expressed on various other cell types, including erythrocytes, endothelial cells, neurons, and microglial cells of the brain (Hadley *et al.*, 1994; Horuk *et al.*, 1997; Gupta *et al.*, 1998). The biological significance of this is still undefined for most receptors, with the exception of CXCR4 (Tachibana *et al.*, 1998; Zou *et al.*, 1998). Consistent with expression in endothelial cells and neurons, genetic elimination of CXCR4 in mice causes defective neuronal cell migration in the cerebellum during development, a ventricular septal defect, and defective gastric vascular development. Another biologic process regulated by chemokines outside the hematopoietic system is angiogenesis. ELR-positive and -negative chemokines have been shown to have angiogenic and angiostatic activity, respectively, when injected in the rat cornea or overexpressed in animal models of cancer (Koch *et al.*, 1992; Strieter *et al.*, 1995). Whether this activity occurs physiologically is not yet known.

Apart from roles in development, host defense, and inflammation, certain chemokine receptors may also function paradoxically and pathologically as promicrobial factors, the result of exploitation or subversion by specific microorganisms (McFadden *et al.*, 1998; Pease and Murphy, 1998). The herpesvirus chemokine receptors mentioned earlier represent one mode of exploitation. The functions of these molecules are not yet understood, but possibilities include immune evasion through chemokine scavenging in the case of CMV US28 (Bodaghi *et al.*, 1998; Vieira *et al.*, 1998), and Kaposi's sarcoma tumorigenesis in the case of KSHV GPCR (Arvanitakis *et al.*, 1997). Better understood are the subset of human chemokine receptors which are exploited by HIV as coreceptors, and function with CD4 as target cell entry factors. Of these, CCR5 and CXCR4 appear to be the most important in pathogenesis, and have distinct specificity for two major classes of HIV viruses, defined by leukocyte cytotropism. CCR5 is essential for efficient person-to-person HIV transmission and may also regulate the rate of disease progression, as revealed by analysis of the naturally occurring inactive allele CCR5Δ32 (reviewed in Moore *et al.*, 1997; Doms and Peiper, 1997; Berger *et al.*, 1999). Similarly, study of a defective Duffy allele affecting the promoter has revealed the obligate usage of normal Duffy by the protozoan *Plasmodium vivax* as a receptor for erythrocyte entry in the pathogenesis of malaria (Horuk *et al.*, 1993; Horuk, 1994; Tournamille *et al.*, 1995).

Virally encoded chemokines have also been discovered, in several herpesviruses (e.g. HHV-8, mouse cytomegalovirus) and in molluscum contagiosum virus (MCV), a human poxvirus which causes the skin disease molluscum contagiosum. Interestingly, broad-spectrum chemokine receptor antagonist activity has been reported for several of these molecules, including vMIP-II of HHV-8 and MC148R of MCV (Kledal *et al.*, 1997; Damon *et al.*, 1998), which suggests a mechanism for immune evasion by these viruses, and, reciprocally, argues for the importance of chemokines in antiviral host defense. Furthermore, various orthopoxviruses encode two structurally unique classes of secreted, broad-spectrum chemokine scavengers, one of which also binds IFNγ (reviewed in McFadden *et al.*, 1998). Neither has structural homology to other proteins currently recorded in the public databases. To date, no naturally occurring mammalian chemokines have been identified that have chemokine receptor antagonist activity.

DEVELOPMENT OF CHEMOKINE RECEPTOR ANTAGONISTS

Because of their specificity for leukocyte subsets, chemokine receptors are logical targets for drug development in human diseases characterized by inflammation. Over the past decade, substantial progress has been made in identifying targets and disease associations, and more recently in developing blocking strategies (Baggiolini and Moser, 1997).

Animal models in which IL-8 has been neutralized immunologically have pointed to CXCR1 and CXCR2 as targets for diseases characterized by acute neutrophil-mediated inflammation, such as pustular psoriasis, glomerulonephritis, and ischemia–reperfusion injury, as may occur during angioplasty. Still, the available data do not discriminate specific roles of CXCR1 and CXCR2 in clinical disease. A small molecule antagonist specific for CXCR2 has been reported by White *et al.* (1998) from SmithKlein-Beecham, but not yet evaluated preclinically.

CCR3 is an attractive target in allergic inflammation and asthma because it appears to be the dominant chemokine receptor in human eosinophils, and is also expressed on basophils and a subset of TH2 lymphocytes. Consistent with this, genetic elimination of eotaxin, a major CCR3 ligand, in the mouse results in ~50% reduction in airway inflammation after ovalbumin sensitization and challenge (Rothenberg *et al.*, 1997). Extrapolation of this result to humans is restricted by the relative overexpression of CCR1 in mouse versus human eosinophils, the existence of other CCR3 ligands, and the potential for compensatory mechanisms in a nonconditional knockout. Direct CCR3 knockout will solve the second of these problems, but other strategies to determine the significance of CCR3 in disease are clearly needed.

Using specific neutralizing antibodies, Kennedy *et al.* have found that acute and relapsing components of experimental allergic encephalomyelitis in mice are regulated by MIP-1α and MCP-1, respectively, suggesting potential roles of CCR1 and/or CCR5 and CCR2 in the corresponding phases of multiple sclerosis in humans (Kennedy *et al.*, 1998). Selective small-molecule antagonists have been reported for all three of these receptors (Hesselgesser *et al.*, 1998; Baba *et al.*, 1999), but their effects in this and other disease models have not yet been published. Genetic knockouts in the mouse have revealed that CCR2 and its ligand MCP-1 contribute to the severity of atherosclerosis in dietary challenges (Boring *et al.*, 1998), providing justification for investigation of specific antagonists in this disease.

As alluded to above, direct proof of principle for the importance of specific chemokine receptors in clinical disease has been obtained only for CCR5 in AIDS and Duffy in vivax malaria. Indirect evidence has supported a role for CXCR4 in late stages of HIV infection (reviewed in Berger *et al.*, 1999). Specific blocking agents, including small-molecule antagonists (Schols *et al.*, 1997; Baba *et al.*, 1999) have been identified for these receptors, but not yet tested in animal models or clinical trials. Blocking only one HIV coreceptor may be hazardous in HIV-infected persons, since the virus may simply develop resistance or the capacity to use another coreceptor more efficiently. A promising result for exploiting HIV coreceptors in vaccine development was recently reported (LaCasse *et al.*, 1999), in which broad-spectrum neutralizing antibodies to HIV-1 were raised in mice immunized with CD4+ CCR5+ target cells crosslinked to HIV Env+ effector cells. Presumably, neutralizing epitopes are exposed transiently during coreceptor-promoted conformational changes in gp120 that occur transiently during the envelope–target cell membrane fusion reaction, and these are fixed by the crosslinking agent for stable and effective display to the immune system.

Another potentially useful direction in therapeutics is to develop the viral chemokine antagonists and scavengers reviewed above, which may be best suited as topical agents or single administration agents for acute inflammation, such as occurs after angioplasty.

CONCLUSION

Although additional chemokines and chemokine receptors will probably continue to be identified for some time, adding to what is already a system of enormous complexity, there is already enough known to discern important patterns of specificity and activity, as described in this overview. An area which remains relatively poorly developed and in which advances may have the greatest impact on further knowledge of the biology of the system is in chemokine receptor pharmacology. At present the system can be viewed as extremely rich in endogenous agonists, but equally poor in selective antagonists. Filling this void may facilitate biological experiments that cannot be done easily, or at all, using genetic and immunologic approaches currently in use, and lead to new treatments of human disease.

References

Ahuja, S. K., Lee, J. C., and Murphy, P. M. (1996). CXC chemokines bind to unique sets of selectivity determinants that can function independently and are broadly distributed on multiple domains of human interleukin-8 receptor B. Determinants of high affinity binding and receptor activation are distinct. *J. Biol. Chem.* **271**, 225–232.

Ali, H., Richardson, R. M., Haribabu, B., and Snyderman, R. (1999). Chemoattractant receptor cross-desensitization. *J. Biol. Chem.* **274**, 6027–6030.

Alkhatib, G., Locati, M., Kennedy, P. E., Murphy, P. M., and Berger, E. A. (1997). HIV-1 coreceptor activity of CCR5 and its inhibition by chemokines: independence from G protein signaling and importance of coreceptor downmodulation. *Virology* **234**, 340–348.

Amara, A., Gall, S. L., Schwartz, O., Salamero, J., Montes, M., Loetscher, P., Baggiolini, M., Virelizier, J. L., and Arenzana-Seisdedos, F. (1997). HIV coreceptor downregulation as antiviral principle: SDF-1alpha-dependent internalization of the chemokine receptor CXCR4 contributes to inhibition of HIV replication. *J. Exp. Med.* **186**, 139–146.

Amichay, D., Gazzinelli, R. T., Karupiah, G., Moench, T. R., Sher, A., and Farber, J. M. (1996). Genes for chemokines MuMig and Crg-2 are induced in protozoan and viral infections in response to IFN-gamma with patterns of tissue expression that suggest nonredundant roles *in vivo*. *J. Immunol.* **157**, 4511–4520.

Arvanitakis, L., Geras-Raaka, E., Varma, A., Gershengorn, M. C., and Cesarman, E. (1997). Human herpesvirus KSHV encodes a constitutively active G-protein-coupled receptor linked to cell proliferation. *Nature* **385**, 347–350.

Baba, M., Nishimura, O., Kanzaki, N., Okamoto, M., Sawada, H., Iizawa, Y., Shiraishi, M., Aramaki, Y., Okonogi, K., Ogawa, Y., Meguro, K., and Fujino, M. (1999). A small-molecule, nonpeptide CCR5 antagonist with highly potent and selective anti-HIV-1 activity. *Proc. Natl Acad. Sci. USA* **96**, 5698–5703.

Bacon, K. B., Premack, B. A., Gardner, P., and Schall, T. J. (1995). Activation of dual T cell signaling pathways by the chemokine RANTES. *Science* **269**, 1727–1730.

Baggiolini, M., and Moser, B. (1997). Blocking chemokine receptors. *J. Exp. Med.* **186**, 1189–1191.

Benkirane, M., Jin, D. Y., Chun, R. F., Koup, R. A., and Jeang, K. T. (1997). Mechanism of transdominant inhibition of CCR5-mediated HIV-1 infection by ccr5delta32. *J. Biol. Chem.* **272**, 30603–30606.

Berger, E. A., Murphy, P. M., and Farber, J. M. (1999). Chemokine receptors as HIV-1 coreceptors: roles in viral entry, tropism and disease. *Annu. Rev. Immunol.* **17**, 657–700.

Berson, J. F., and Doms, R. W. (1998). Structure–function studies of the HIV-1 coreceptors. *Semin. Immunol.* **10**, 237–248.

Bodaghi, B., Jones, T. R., Zipeto, D., Vita, C., Sun, L., Laurent, L., Arenzana-Seisdedos, F., Virelizier, J. L., and Michelson, S. (1998). Chemokine sequestration by viral chemoreceptors as a novel viral escape strategy: withdrawal of chemokines from the environment of cytomegalovirus-infected cells. *J. Exp. Med.* **188**, 855–866.

Bokoch, G. M. (1995). Chemoattractant signaling and leukocyte activation. *Blood* **86**, 1649–1660.

Boring, L., Gosling, J., Chensue, S. W., Kunkel, S. L., Farese, R.V. Jr., Broxmeyer, H. E., and Charo, I. F. (1997). Impaired monocyte migration and reduced type 1 (TH1) cytokine responses in CC chemokine receptor 2 knockout mice. *J. Clin. Invest.* **100**, 2552–2561.

Boring, L., Gosling, J., Cleary, M., and Charo, I. F. (1998). Decreased lesion formation in CCR2−/− mice reveals a role for chemokines in the initiation of atherosclerosis. *Nature* **394**, 894–897.

Broxmeyer, H. E., Cooper, S., Cacalano, G., Hague, N. L., Bailish, E., and Moore, M. W. (1996). Involvement of interleukin (IL) 8 receptor in negative regulation of myeloid progenitor cells *in vivo*: evidence from mice lacking the murine IL-8 receptor homologue. *J. Exp. Med.* **184**, 1825–1832.

Broxmeyer, H. E., Cooper, S., Hangoc, G., Gao, J. L., and Murphy, P. M. (1999). Dominant myelopoietic effector functions mediated by chemokine receptor CCR1. *J. Exp. Med.* **189**, 1987–1992.

Cacalano, G., Lee, J., Kikly, K., Ryan, A. M., Pitts-Meek, S., Hultgren, B., Wood, W. I., and Moore, M. W. (1994). Neutrophil and B cell expansion in mice that lack the murine IL-8 receptor homolog. *Science* **265**, 682–685.

Chaudhuri, A., Zbrzezna, V., Polyakova, J., Pogo, A. O., Hesselgesser, J., and Horuk, R. (1994). Expression of the Duffy antigen in K562 cells. Evidence that it is the human erythrocyte chemokine receptor. *J. Biol. Chem.* **269**, 7835–7838.

Cocchi, F., DeVico, A. L., Garzino-Demo, A., Arya, S. K., Gallo, R. C., and Lusso, P. (1995). Identification of RANTES, MIP-1α, and MIP-1β as the major HIV-suppressive factors produced by CD8+ T cells. *Science* **270**, 1811–1815.

Cyster, J. G. (1999). Chemokines and the homing of dendritic cells to the T cell areas of lymphoid organs. *J. Exp. Med.* **189**, 447–450.

Damaj, B. B., McColl, S. R., Neote, K., Songqing, N., Ogborn, K. T., Hebert, C. A., and Naccache, P. H. (1996). Identification of G-protein binding sites of the human interleukin-8 receptors by functional mapping of the intracellular loops. *FASEB J.* **10**, 1426–1434.

Damon, I., Murphy, P. M., and Moss, B. (1998). Broad spectrum chemokine antagonistic activity of a human poxvirus chemokine homolog. *Proc. Natl Acad. Sci. USA* **95**, 6403–6407.

Davis, C. B., Dikic, I., Unutmaz, D., Hill, C. M., Arthos, J., Siani, M. A., Thompson, D. A., Schlessinger, J., and Littman, D. R. (1997). Signal transduction due to HIV-1 envelope interactions with chemokine receptors CXCR4 or CCR5. *J. Exp. Med.* **186**, 1793–1798.

Doms, R. W., and Peiper, S. C. (1997). Unwelcomed guests with master keys: how HIV uses chemokine receptors for cellular entry. *Virology* **235**, 179–190.

Dunstan, C. A. N., Salafranca, M. N., Adhikari, S., Xia, Y., Feng, L., and Harrison, J. K. (1996). Identification of two rat genes orthologous to the human interleukin-8 receptors. *J. Biol. Chem.* **271**, 32770–32776.

Farzan, M., Choe, H., Martin, K. A., Sun, Y., Sidelko, M., Mackay, C. R., Gerard, N. P., Sodroski, J., and Gerard, C. (1997). HIV-1 entry and macrophage inflammatory protein-1beta-mediated signaling are independent functions of the chemokine receptor CCR5. *J. Biol. Chem.* **272**, 6854–6857.

Farzan, M., Mirzabekov, T., Kolchinsky, P., Wyatt, R., Cayabyab, M., Gerard, N. P., Gerard, C., Sodroski, J., and Choe, H. (1999). Tyrosine sulfation of the amino terminus of CCR5 facilitates HIV-1 entry. *Cell* **96**, 667–676.

Feng, Y., Broder, C. C., Kennedy, P. E., and Berger, E. (1996). HIV-1 entry co-factor: functional cDNA cloning of a seven-transmembrane G-protein coupled receptor. *Science* **272**, 872–877.

Forster, R., Mattis, A. E., Kremmer, E., Wolf, E., Brem, G., and Lipp, M. (1996). A putative chemokine receptor, BLR1, directs B cell migration to defined lymphoid organs and specific anatomic compartments of the spleen. *Cell* **87**, 1037–1047.

Foxman, E. F., Campbell, J. J., and Butcher, E. C. (1997). Multistep navigation and the combinatorial control of leukocyte chemotaxis. *J. Cell. Biol.* **139**, 349–360.

Gao, J. L., Wynn, T. A., Chang, Y., Lee, E. J., Broxmeyer, H. E., Cooper, S., Tiffany, H. L., Westphal, H., Kwon-Chung, J., and Murphy, P. M. (1997). Impaired host defense, hematopoiesis, granulomatous inflammation and type 1-type 2 cytokine balance in mice lacking CC chemokine receptor 1. *J. Exp. Med.* **185**, 1959–1968.

Gerard, C. (1999). In "Chemokines in Disease" (ed C. A. Hebert), Understanding chemokine biology through mouse genetics: riddles and answers, pp. 41–52. Humana, Totowa, NJ.

Gunn, M. D., Ngo, V. N., Ansel, K. M., Ekland, E. H., Cyster, J. G., and Williams, L. T. (1998a). A B-cell-homing chemokine made in lymphoid follicles activates Burkitt's lymphoma receptor-1. *Nature* **391**, 799–803.

Gunn, M. D., Tangemann, K., Tam, C., Cyster, J. G., Rosen, S. D., and Williams, L. T. (1998b). A chemokine expressed in lymphoid high endothelial venules promotes the adhesion and chemotaxis of naive T lymphocytes. *Proc. Natl Acad. Sci. USA* **95**, 258–263.

Gunn, M. D., Kyuwa, S., Tam, C., Kakiuchi, T., Matsuzawa, A., Williams, L. T., and Nakano, H. (1999). Mice lacking expression of secondary lymphoid organ chemokine have defects in lymphocyte homing and dendritic cell localization. *J. Exp. Med.* **189**, 451–460.

Gupta, S. K., Lysko, P. G., Pillarisetti, K., Ohlstein, E., and Stadel, J. M. (1998). Chemokine receptors in human endothelial cells. Functional expression of CXCR4 and its transcriptional regulation by inflammatory cytokines. *J. Biol. Chem.* **273**, 4282–4287.

Hadley, T. J., Lu, Z. H., Wasniowska, K., Martin, A. W., Peiper, S. C., Hesselgesser, J., and Horuk, R. (1994). Postcapillary venule endothelial cells in kidney express a multispecific chemokine receptor that is structurally and functionally identical to the erythroid isoform, which is the Duffy blood group antigen. *J. Clin. Invest.* **94**, 985–991.

Hall, D. A., Beresford, I. J., Browning, C., and Giles, H. (1999). Signalling by CXC-chemokine receptors 1 and 2 expressed in CHO cells: a comparison of calcium mobilization, inhibition of adenylyl cyclase and stimulation of GTPgammaS binding induced by IL-8 and GROalpha. *Br. J. Pharmacol.* **126**, 810–818.

Hesselgesser, J., Ng, H. P., Liang, M., Zheng, W., May, K., Bauman, J. G., Monahan, S., Islam, I., Wei, G. P., Ghannam, A., Taub, D. D., Rosser, M., Snider, R. M., Morrissey, M. M., Perez, H. D., and Horuk, R. (1998). Identification and characterization of small molecule functional antagonists of the CCR1 chemokine receptor. *J. Biol. Chem.* **273**, 15687–15692.

Horuk, R., Martin, A. W., Wang, Z., Schweitzer, L., Gerassimides, A., Guo, H., Lu, Z., Hesselgesser, J., Perez, H. D., Kim, J., Parker, J., Hadley, T. J., and Peiper, S. C. (1997). Expression of chemokine receptors by subsets of neurons in the central nervous system. *J. Immunol.* **158**, 2882–2890.

Horuk, R., Chitnis, C., Darbonne, W., Colby, T. J., Rybicki, A., Hadley, T. J., and Miller, L. H. (1993). The erythrocyte chemokine receptor is a receptor for the malarial parasite *Plasmodium vivax*. *Science* **261**, 1182–1184.

Horuk, R. (1994). The interleukin-8-receptor family: from chemokines to malaria. *Immunol. Today* **15**, 169–174.

Imai, T., Hieshima, K., Haskell, C., Baba, M., Nagira, M., Nishimura, M., Kakizaki, M., Takagi, S., Nomiyama, H., Schall, T. J., and Yoshie, O. (1997). Identification and molecular characterization of fractalkine receptor CX_3CR1, which mediates both leukocyte migration and adhesion. *Cell* **91**, 521–530.

Isegawa, Y., Ping, Z., Nakano, K., Sugimoto, N., and Yamanishi, K. (1998). Human herpesvirus 6 open reading frame U12 encodes a functional beta-chemokine receptor. *J. Virol.* **72**, 6104–6112.

Jones, S. A., Wolf, M., Qin, S., Mackay, C. R., and Baggiolini, M. (1996). Different functions for the interleukin 8 receptors (IL-8R) of human neutrophil leukocytes: NADPH oxidase and phospholipase D are activated through IL-8R1 but not IL-8R2. *Proc. Natl Acad. Sci. USA* **93**, 6682–6686.

Kennedy, K. J., Strieter, R. M., Kunkel, S. L., Lukacs, N. W., and Karpus, W. J. (1998). Acute and relapsing experimental autoimmune encephalomyelitis are regulated by differential expression of the CC chemokines macrophage inflammatory protein-1alpha and monocyte chemotactic protein-1. *J. Neuroimmunol.* **92**, 98–108.

Kledal, T. N., Rosenkilde, M. M., Coulin, F., Simmons, G., Johnsen, A. H., Alouani, S., Power, C. A., Luttichau, H. R., Gerstoft, J., Clapham, P. R., Clark-Lewis, I., Wells, T. N. C., and Schwartz, T. W. (1997). A broad-spectrum chemokine antagonist encoded by Kaposi's sarcoma-associated herpesvirus. *Science* **277**, 1656–1659.

Kledal, T. N., Rosenkilde, M. M., and Schwartz, T. W. (1998). Selective recognition of the membrane-bound CX3C chemokine, fractalkine, by the human cytomegalovirus-encoded broad-spectrum receptor US28. *FEBS Lett.* **441**, 209–214.

Knall, C., Young, S., Nick, J. A., Buhl, A. M., Worthen, G. S., and Johnson, G. L. (1996). Interleukin-8 regulation of the Ras/Raf/mitogen-activated protein kinase pathway in human neutrophils. *J. Biol. Chem.* **271**, 2832–2838.

Koch, A. E., Polverini, P. J., Kunkel, S. L., Harlow, L. A., DiPietro, L. A., Elner, V. M., Elner, S. G., and Strieter R. M. (1992). Interleukin-8 as a macrophage-derived mediator of angiogenesis. *Science* **258**, 1798–1801.

Kuang, Y., Wu, Y., Jiang, H., and Wu, D. (1996). Selective G protein coupling by C-C chemokine receptors. *J. Biol. Chem.* **271**, 3975–3978.

LaCasse, R. A., Follis, K. E., Trahey, M., Scarborough, J. D., Littman, D. R., and Nunberg, J. H. (1999). Fusion-competent vaccines: broad neutralization of primary isolates of HIV. *Science* **283**, 357–362.

Lapham, C. K., Zaitseva, M. B., Lee, S., Romanstseva, T., and Golding, H. (1999). Fusion of monocytes and macrophages with HIV-1 correlates with biochemical properties of CXCR4 and CCR5. *Nat. Med.* **5**, 303–308.

Locati, M., and Murphy, P. M. (1999). Chemokines and chemokine receptors: biology and clinical relevance in inflammation and AIDS. *Annu. Rev. Med.* **50**, 425–440.

Luster, A. D. (1998). Chemokines – chemotactic cytokines that mediate inflammation. *N. Engl. J. Med.* **338**, 436–445.

Mantovani, A., Allavena, P., Vecchi, A., and Sozzani, S. (1998). Chemokines and chemokine receptors during activation and deactivation of monocytes and dendritic cells and in amplification of TH1 versus TH2 responses. *Int. J. Clin. Lab. Res.* **28**, 77–82.

McFadden, G., Lalani, A., Everett, H., Nash, P., and Xu, X. (1998). Virus-encoded receptors for cytokines and chemokines. *Semin. Cell Dev. Biol.* **9**, 359–368.

Moore, J. P., Trkola, A., and Dragic, T. (1997). Co-receptors for HIV-1 entry. *Curr. Opin. Immunol.* **9**, 551–562.

Moser, B., Schumacher, C., von Tscharner, V., Clark-Lewis, I., and Baggiolini, M. (1991). Neutrophil-activating peptide 2 and gro/melanoma growth-stimulatory activity interact with neutrophil-activating peptide 1/interleukin 8 receptors on human neutrophils. *J. Biol. Chem.* **266**, 10666–10671.

Murphy, P. M. (1994). The molecular biology of leukocyte chemoattractant receptors. *Annu. Rev. Immunol.* **12**, 593–633.

Murphy, P. M. (1996). Chemokine receptors: cloning strategies. *Methods: Comp. Methods Enzymol.* **10**, 104–118.

Myers, S. J., Wong, L. M., and Charo, I. F. (1995). Signal transduction and ligand specificity of the human monocyte chemoattractant protein-1 receptor in transfected embryonic kidney cells. *J. Biol. Chem.* **270**, 5786–5792.

Neote, K., Darbonne, W., Ogez, J., Horuk, R., and Schall, T. J. (1993). Identification of a promiscuous inflammatory peptide receptor on the surface of red blood cells. *J. Biol. Chem.* **268**, 12247–12249.

Nibbs, R. J., Wylie, S. M., Yang, J., Landau, N. R., and Graham, G. J. (1997). Cloning and characterization of a novel promiscuous human beta-chemokine receptor D6. *J. Biol. Chem.* **272**, 32078–32083.

Oppermann, M., Mack, M., Proudfoot, A. E., and Olbrich, H. (1999). Differential effects of CC chemokines on CC chemokine receptor 5 (CCR5) phosphorylation and identification of phosphorylation sites on the CCR5 carboxyl terminus. *J. Biol. Chem.* **274**, 8875–8885.

Paavola, C. D., Hemmerich, S., Grunberger, D., Polsky, I., Bloom, A., Freedman, R., Mulkins, M., Bhakta, S., McCarley, D., Wiesent, L., Wong, B., Jarnagin, K., and Handel, T. M. (1998). Monomeric monocyte chemoattractant protein-1 (MCP-1) binds and activates the MCP-1 receptor CCR2B. *J. Biol. Chem.* **273**, 33157–33165.

Pease, J. E., and Murphy, P. M. (1998). Microbial corruption of the chemokine system: an expanding paradigm. *Semin. Immunol.* **10**, 169–178.

Premack, B. A., and Schall, T. J. (1996). Chemokine receptors: gateways to inflammation and infection. *Nat. Med.* **2**, 1174–1178.

Reid, S., Ritchie, A., Boring, L., Gosling, J., Cooper, S., Hangoc, G., Charo, I. F., and Broxmeyer, H. E. (1999). Enhanced myeloid progenitor cell cycling and apoptosis in mice lacking the chemokine receptor, CCR2. *Blood* **93**, 1524–1533.

Rodriguez-Frade, J. M., Vila-Coro, A. J., de Ana, A. M., Albar, J. P., Martinez, A. C., and Mellado, M. (1999). The chemokine monocyte chemoattractant protein-1 induces functional responses through dimerization of its receptor CCR2. *Proc. Natl Acad. Sci. USA* **96**, 3628–3633.

Rothenberg, M. E., MacLean, J. A., Pearlman, E., Luster, A. D., and Leder, P. (1997). Targeted disruption of the chemokine eotaxin partially reduces antigen-induced tissue eosinophilia. *J. Exp. Med.* **185**, 785–790.

Saeki, H., Moore, A. M., Brown, M. J., and Hwang, S. T. (1999). Secondary lymphoid-tissue chemokine (SLC) and CC chemokine receptor 7 (CCR7) participate in the emigration pathway of mature dendritic cells from the skin to regional lymph nodes. *J. Immunol.* **162**, 2472–2475.

Schols, D., Struyf, S., Van Damme, J., Este, J. A., Henson, G., and De Clercq, E. (1997). Inhibition of T-tropic HIV strains by selective antagonization of the chemokine receptor CXCR4. *J. Exp. Med.* **186**, 1383–1388.

Soto, H., Wang, W., Strieter, R. M., Copeland, N. G., Gilbert, D. J., Jenkins, N. A., Hedrick, J., and Zlotnik, A. (1998). The CC chemokine 6Ckine binds the CXC chemokine receptor CXCR3. *Proc. Natl Acad. Sci. USA* **95**, 8205–8210.

Strieter, R. M., Polverini, P. J., Arenberg, D. A., Walz, A., Opdenakker, G., Van Damme, J., and Kunkel, S. L. (1995). Role of C-X-C chemokines as regulators of angiogenesis in lung cancer. *J. Leukoc. Biol.* **57**, 752–762.

Tachibana, K., Hirota, S., Iizasa, H., Yoshida, H., Kawabata, K., Kataoka, Y., Kitamura, Y., Matsushima, K., Yoshida, N., Nishikawa, S., Kishimoto, T., and Nagasawa, T. (1998). The chemokine receptor CXCR4 is essential for vascularization of the gastrointestinal tract. *Nature* **393**, 591–594.

Tang, H. L., and Cyster, J. G. (1999). Chemokine up-regulation and activated T cell attraction by maturing dendritic cells. *Science* **284**, 819–822.

Tournamille, C., Colin, Y., Cartron, J. P., and Le Van Kim, C. (1995). Disruption of a GATA motif in the Duffy gene promoter abolishes erythroid gene expression in Duffy-negative individuals. *Nat. Genet.* **10**, 224–228.

Unger, V. M., Hargrave, P. A., Baldwin, J. M., and Schertler, G. F. (1997). Arrangement of rhodopsin transmembrane alpha-helices. *Nature* **389**, 203–206.

Vieira, J., Schall, T. J., Corey, L., and Geballe, A. P. (1998). Functional analysis of the human cytomegalovirus US28 gene by insertion mutagenesis with the green fluorescent protein gene. *J. Virol.* **72**, 8158–8165.

Ward, S. G., Bacon, K., and Westwick, J. (1998). Chemokines and T lymphocytes: more than an attraction. *Immunity* **9**, 1–11.

Weng, Y., Siciliano, S. J., Waldburger, K. E., Sirotina-Meisher, A., Staruch, M. J., Daugherty, B. L., Gould, S. L., Springer, M. S., and DeMartino, J. A. (1998). Binding and functional properties of recombinant and endogenous CXCR3 chemokine receptors. *J. Biol. Chem.* **273**, 18288–18291.

White, J. R., Lee, J. M., Young, P. R., Hertzberg, R. P., Jurewicz, A. J., Chaikin, M. A., Widdowson, K., Foley, J. J., Martin, L. D., Griswold, D. E., and Sarau, H. M. (1998). Identification of a potent, selective non-peptide CXCR2 antagonist that inhibits interleukin-8-induced neutrophil migration. *J. Biol. Chem.* **273**, 10095–10098.

Xie, W., Jiang, H., Wu, Y., and Wu, D. (1997). Two basic amino acids in the second inner loop of the interleukin-8 receptor are essential for Galpha16 coupling. *J. Biol. Chem.* **272**, 24948–24951.

Yang, W., Wang, D., and Richmond, A. (1999). Role of clathrin-mediated endocytosis in CXCR2 sequestration, resensitization, and signal transduction. *J. Biol. Chem.* **274**, 11328–11333.

Yoshie, O., Imai, T., and Nomiyama, H. (1997). Novel lymphocyte-specific CC chemokines and their receptors. *J. Leukoc. Biol.* **62**, 634–644.

Zlotnik, A., Morales, J., and Hedrick, J. A. (1999). Recent advances in chemokines and chemokine receptors. *Crit. Rev. Immunol.* **19**, 1–47.

Zou, Y. R., Kottmann, A. H., and Littman, D. R. (1998). Function of the chemokine receptor CXCR4 in haematopoiesis and in cerebellar development. *Nature* **393**, 595–598.

CXCR1 and CXCR2

Iris Roth* **and Caroline Hebert**

Department of Molecular Oncology, Genentech, 1 DNA Way South San Francisco, San Francisco, CA 94080, USA

*corresponding author tel: 650 225 2176, fax: 650 225 8221, e-mail: irisr@gene.com

DOI: 10.1006/rwcy.2000.21001.

SUMMARY

Chemokines induce cell migration and activation by binding to specific seven transmembrane (7TM) G protein-coupled cell surface receptors (GPCRs) on target cells (Murphy, 1994; Premack and Schall, 1996). Chemokine receptors, like all members of the GPCR superfamily, mediate signal transduction through G proteins. They have a single polypeptide chain and have 25–80% amino acid sequence identity among all family members. CXCR1 and CXCR2 were the first members of the chemokine receptor family to be cloned, and share a high degree of homology with receptors for C5a and formyl peptide. There is also a high level of homology between receptors from human, rabbit, rat, and mouse. Like the other chemokine receptor family members, CXCR1 and CXCR2 are expressed on a variety of cells and have a characteristically narrow ligand-binding profile, including a subgroup of human CXC chemokines defined by the conserved sequence motif glutamic acid-leucine-arginine (ELR). The human ELR-containing CXC chemokines IL-8, growth-related oncogene (GRO)α, GROβ, GROγ, neutrophil-activating peptide-2 (NAP-2), epithelial cell-derived neutrophil activating peptide 78 (ENA-78), and granulocyte chemoattractant protein 2 (GCP-2) are 40–90% identical in amino acid sequence. CXCR1 and CXCR2 are distinguished by their different selectivities for these chemokine ligands, with CXCR1 displaying a relatively narrow selectivity and preference for IL-8.

Both receptors are expressed on a wide variety of cell types, including neutrophils, monocytes, CD8+ T cells, mast cells, basophils, natural killer cells, keratinocytes, fibroblasts, neurons, endothelial cells, and melanocytes. They are distinguished by their differential affinities for a group of chemokine ligands. CXCR1 and CXCR2 play an important role in acute inflammation by transducing the signal for one ligand, IL-8. These receptors are attractive targets for the development of therapeutics for inflammatory disease, as GPCRs have proven to be good targets for small molecule antagonists. However, the relative contributions of each receptor to disease process have not yet been established, thus it is still not known which of the two must be targeted for drug development. Selective antagonists for CXCR1 and CXCR2 have begun to emerge, and these will be useful tools for further elucidating the role of each receptor in the pathophysiology of inflammatory diseases. Because the two receptors share a multitude of characteristics, including expression patterns and functional properties, they will be discussed together in this review. This chapter will include brief descriptions of their genetics, structure–function analysis, ligand-binding requirements, and signal transduction mechanisms, as well as expression patterns and roles in normal tissues as well as in disease states.

BACKGROUND

Discovery

The discovery of IL-8 marked the emergence of the chemokine field. Previously described chemotactic factors, including the formyl peptide *N*-formyl methionyl-leucyl-phenylalanine (fMLP), complement factor C5a, platelet-activating factor (PAF), and leukotriene B_4 (LTB_4), were promiscuous, acting on a variety of leukocyte subsets (Locati and Murphy, 1999). IL-8, produced by lipopolysaccharide (LPS)-stimulated monocytes, was active on neutrophils but not monocytes/macrophages. This chemoattractant was described by independent work in several laboratories and named neutrophil-activating factor (NAF)

(Walz *et al.*, 1987), monocyte-derived neutrophil-activating peptide (MONAP) (Schroder *et al.*, 1987), and monocyte-derived neutrophil chemotactic factor (MDNC) (Yoshimura *et al.*, 1987). It was subsequently shown that this factor is produced by various cell types and has multiple targets, thus it was renamed interleukin 8 (IL-8) (Balkwill and Burke, 1989; Larsen *et al.*, 1989). Lack of cross-desensitization suggested that this chemokine acts through selective receptors on neutrophils distinct from those for fMLP, C5a, PAF, and LTB_4 (Peveri *et al.*, 1988).

Samanta *et al.* first demonstrated specific binding of [^{125}I]IL-8 to the surface of human neutrophils (Samanta *et al.*, 1989). Scatchard analysis yielded a curvilinear plot, which the authors interpreted as representing a single binding affinity receptor. Chemical crosslinking of ligand–receptor complexes identified two IL-8-binding proteins, thought to be noncovalently associated subunits of this receptor. Subsequent binding studies using different preparations of IL-8 resulted in conflicting interpretations. Besemer *et al.* also observed nonlinear Scatchard plots, but concluded that their data were consistent with two populations of receptors (Besemer *et al.*, 1989). Others demonstrated a linear Scatchard plot and only one protein chemically crosslinked to IL-8 – results consistent with the presence of only one class of receptors on neutrophils (Grob *et al.*, 1990). The binding affinities ranged from 11 pM to 2 nM, with estimated receptor numbers ranging from 20,000 to 75,000 per cell.

The search for chemokine receptors began when several groups utilized expression cloning strategies to identify the human formyl peptide receptor (FPR) (Boulay *et al.*, 1990; Coats and Navarro, 1990; Murphy *et al.*, 1990). One such cDNA clone was capable of conferring specific fMLP binding to transfected mammalian COS cells (Boulay *et al.*, 1990). A putative rabbit homolog, F3R, was later cloned from a neutrophil cDNA library and shown to bind fMLP when expressed in *Xenopus* oocytes (Thomas *et al.*, 1990). However, F3R was only 26% homologous to the clone previously identified as the human fMLP receptor.

Unique clones for two distinct human IL-8 receptors were later independently described by Holmes and Murphy in the same issue of *Science* (Holmes *et al.*, 1991; Murphy and Tiffany, 1991). Using an expression cloning strategy in COS-7 cells, Holmes *et al.* isolated the clone for CXCR1 from a human neutrophil cDNA library. COS cells transfected with this cDNA specifically bound IL-8 with a K_d of 3.6 nM, within the range of 0.8–4 nM reported for binding to neutrophils. The cDNA contains a single long open reading frame, which encodes a 350 amino acid protein with seven hydrophobic domains. The IL-8 receptor described in this study is 79% identical to F3R at the amino acid level. Because F3R shares only a 26% identity with the human fMLP receptor, the authors suggested that this sequence might actually encode an IL-8 receptor.

In an accompanying paper, Murphy and Tiffany screened a differentiated HL-60 cDNA library to identify the human homolog of F3R (Murphy and Tiffany, 1991). One clone shared a 69% amino acid identity to F3R, yet demonstrated a pattern of expression more like that of IL-8-binding sites than *N*-formyl peptide-binding sites. When expressed in *Xenopus* oocytes, this clone was found to respond to IL-8, but not to fMLP. Because specific binding was not saturated at the highest concentration of [^{125}I]IL-8 tested, a dissociation constant could not be determined. Thus, the receptor encoded by this cDNA appeared to be a low-affinity IL-8 receptor (CXCR2). The structurally related ligands NAP-2 and melanoma growth-stimulatory activity (MGSA)/growth-related oncogene (GROα) induced calcium flux in transfected oocytes, though with decreased potency compared to IL-8. These results correlated with the effectiveness of these chemokines to compete with IL-8 for binding to neutrophils, as demonstrated by Moser *et al.* (1991).

Meanwhile, the original rabbit F3R clone was expressed in COS cells and found to confer saturatable specific binding of IL-8, not fMLP (Thomas *et al.*, 1991). IL-8 bound neutrophils and F3R-transfected cells with apparent K_d values of 1.2 and 1.4 nM, respectively. F3R antibodies specifically immunoprecipitated [^{125}I]IL-8 bound to the membranes of F3R-transfected COS cells as well as to neutrophils. With these results, Navarro's laboratory and others then correctly identified the F3R protein as the rabbit homolog of CXCR1 (Thomas *et al.*, 1991; Cerretti *et al.*, 1993a).

Alternative names

CXCR1: IL-8RA, IL-8R1, IL-8Rα
CXCR2: IL-8RB, IL-8R2, IL-8Rβ

The official names CXCR1 and CXCR2 were assigned by a consensus nomenclature agreement at the second Gordon Conference on Chemotactic Cytokines in Plymouth, NH, USA in 1996.

Structure

Human CXCR1 and CXCR2 have a single polypeptide chain 350, and 355 or 360 amino acids in length, respectively. The receptors share 76% amino acid identity to one another (Figure 3) (Holmes *et al.*, 1991; Murphy, 1994). As members of the GPCR superfamily, CXCR1 and CXCR2 are 7TM receptors that signal via G proteins (Murphy, 1994; Premack and Schall, 1996). Their structure can be inferred from homology to another family member, rhodopsin, which signals light sensitivity in retinal cones (Luo *et al.*, 1997). The seven hydrophobic regions of the receptors are embedded in the plasma membrane, giving it a serpentine appearance (**Figure 1**). The free N-terminal tail is extracellular, and the C-terminus is in the cytoplasm.

Main activities and pathophysiological roles

CXCR1 and CXCR2 mediate a diversity of chemokine functions in a variety of cell types. In leukocytes they are important in mediating antimicrobial host defenses. Both receptors act to induce chemotaxis and calcium flux in different leukocyte subsets. In neutrophils, receptor activation also stimulates the release of granule enzymes as well as the generation of superoxide in respiratory burst (Loetscher *et al.*, 1994; Hammond *et al.*, 1995; Jones *et al.*, 1996). In addition to their effects on immune cells, CXCR1 and/or CXCR2 may be important in regulating vasculogenesis and consequent tumor growth (Koch *et al.*, 1992; Strieter *et al.*, 1995). ELR-containing CXC chemokines signal via these receptors to induce endothelial cell chemotaxis *in vitro* as well as angiogenesis in various animal models (Koch *et al.*, 1992; Strieter *et al.*, 1995).

GENE

Accession numbers

Human CXCR1: L19592
Rat CXCR1-like: U71089
Gorilla CXCR1: X91110
Chimpanzee CXCR1: X91109
Human CXCR2: M99412
Mouse CXCR2: L23637
Rat CXCR2: U70988
Gorilla CXCR2: X91114
Rhesus CXCR2: X91116
Orangutan CXCR2: X91115
Chimpanzee CXCR2: X91113

Chromosome location and linkages

The human genes for CXCR1 and CXCR2 (*il8ra* and *il8rb*, respectively), along with one pseudogene (*il8rp*), have been mapped to chromosome 2q34–35, 20 kb apart (Ahuja *et al.*, 1992; Morris *et al.*, 1992; Lloyd *et al.*, 1993). The close clustering of the three genes, as well as their high degree of homology, suggests that they may have arisen by duplication of a common ancestral gene. The CXCR2 gene is approximately 12 kb in length while *il8ra* spans just 4 kb (Kelvin *et al.*,

Figure 1 Putative model of the proposed tertiary structure of CXCR1 and CXCR2. The transmembrane α helical domains are indicated by the tubular structures, numbered 1–7. Three extracellular loops are designated EL1-3. Two potential disulfide bridges, linking the extracellular N-terminal domain with the third extracellular loop, and the first and second extracellular loops, are shown in red. (Full colour figure can be viewed online.)

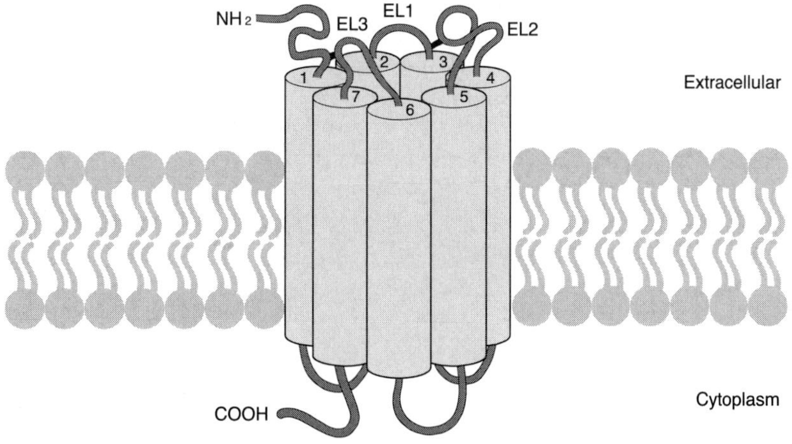

1993; Ahuja *et al.*, 1994). In both cases, the open reading frame (ORF) and 3′ untranslated regions are contained in a single exon, while the 5′ UTR is more complex.

There is one pseudogene, named *il8rp*, contained in the CXCR1/CXCR2 gene cluster. This locus has an 87% identity to CXCR2, but contains multiple frameshifts and point mutations introducing stop codons (Ahuja *et al.*, 1992; Mollereau *et al.*, 1993).

The 5′ UTR of *il8ra* resides in the first two of three exons (Ahuja *et al.*, 1994). Neutrophils contain two equally abundant mRNAs for CXCR1, 2.0 and 2.4 kb in length, the result of usage of two alternative polyadenylation signals (**Figure 2**). For *il8rb*, alternative splicing of 11 exons forms seven distinct mRNAs with one predominant splice variant, designated IL8RB3. Primer extension analysis identified two major transcriptional start sites for *il8ra* and 11 for *il8rb*. The promoters of both genes appeared to be very similar: a nonclassical TATA box and a GC-rich 5′ flanking region was identified immediately upstream of the transcription start site (Sprenger *et al.*, 1995). These minimal promoters were sufficient to induce strong constitutive activity when cloned upstream from a chloramphenicol acetyltransferase (CAT) reporter gene. A granulocyte colony-stimulating factor (G-CSF) responsive element was mapped within the first 118 nucleotides upstream of the transcription start site of *il8rb* (Sprenger *et al.*, 1995). Expression analyses of additional regulatory regions suggested that both promoters are negatively controlled by silencer elements. Recent work has identified promoter elements of *il8ra* and demonstrated that the Ets family transcription factor PU.1 is a major regulator for activation of the CXCR1 promoter (Wilkinson and Navarro, 1999).

PROTEIN

Accession numbers

Human CXCR1: P25024
Rabbit CXCR1: P21109
Rat CXCR1-like protein: P70612

Figure 2 Genomic organization of CXCR1 (a) and CXCR2 (b). Boxes, exons; lines connecting exons, introns; dotted lines, gaps in the sequence; H, sites for *Hin*dIII cleavage. Exons are numbered in the upper left of the corresponding box; introns are numbered in blue (CXCR1) or red (CXCR2) boxes, and their lengths are indicated in base pairs below the intron number. Splicing patterns are shown below each gene map, and the name for each mRNA form is given at the right of the splice pattern diagram. CXCR2 mRNA names are based on the number of the 5′-most exon. Horizontal lines for each mRNA indicate the corresponding exon in the gene, and internal tick marks indicate exon boundaries; dashed lines indicate splicing patterns. (Full colour figure can be viewed online.)

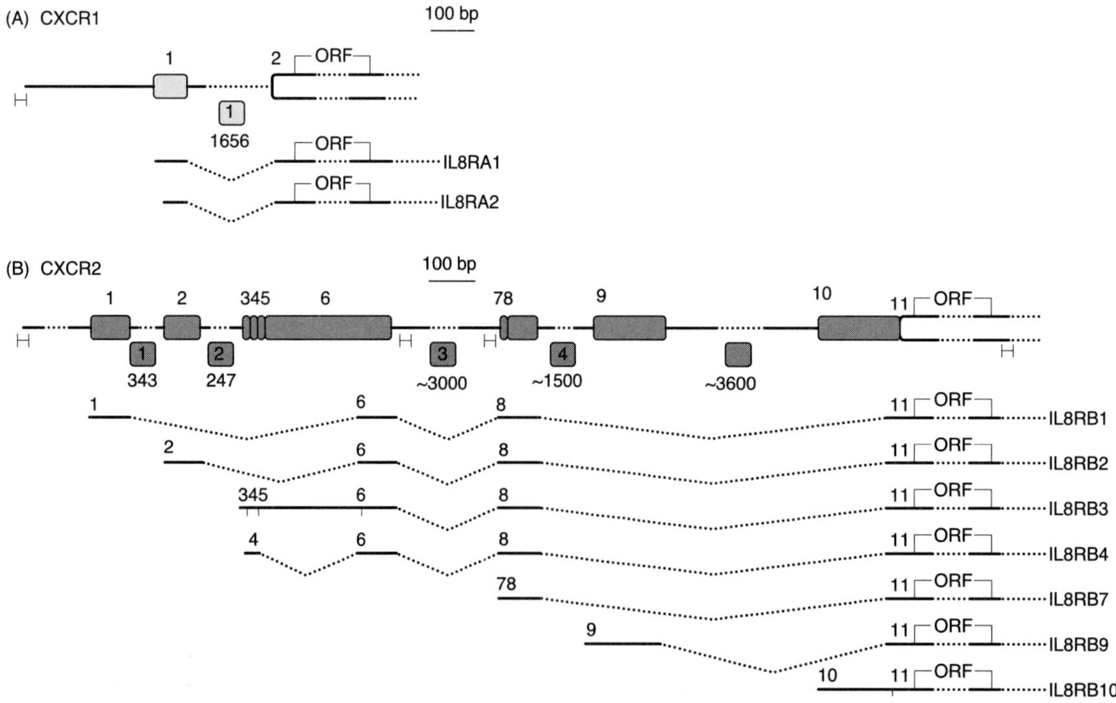

Gorilla CXCR1: P55919
Chimpanzee CXCR1: P55920
Human CXCR2: P25025
Mouse CXCR2: P35343
Rabbit CXCR2: P35344
Rat CXCR2: P35407
Gorilla CXCR2: Q28422
Rhesus CXCR2: Q28519
Chimpanzee CXCR2: Q28807
Bovine CXCR2: Q28003

Sequence

See **Figure 3**.

Description of protein

The human CXCR1 and CXCR2 proteins are highly homologous at the amino acid level (76%) (Figure 3). Both proteins also have a significant degree of homology to other chemokine receptors (25–30%). Residues important for ligand binding and receptor signaling have been elucidated using alanine scanning mutagenesis as well as the construction of receptor chimeras. Sequence differences between CXCR1 and CXCR2 appear to be clustered at the N-terminus, as well as in the second extracellular loop (Murphy, 1997). These regions may thus be important in conferring differential ligand specificity to each receptor.

The key residues of CXCR1 important for ligand binding were defined by extensive site-directed mutagenesis, in which extracellular acidic residues were replaced with alanine (**Figure 4**) (Hebert *et al.*, 1993; Leong *et al.*, 1994). While it was originally thought that acidic residues at the N-terminus interact with the basic ligand, IL-8, alanine substitutions at eight of the nine charged positions of the CXCR1 N-terminal domain had no effect on binding. Ligand binding was only affected when one residue, Asp11, was mutated to Ala. As predicted, IL-8 binding was retained when Asp11 was replaced with another acidic residue, such as Glu, or with Lys, found at position 11 in CXCR2 (Hebert *et al.*, 1993). These data suggest that the acidic nature of the N-terminal segment of the IL-8 receptors is not functionally important.

Additional residues in extracellular domains of CXCR1 were also shown to be important for IL-8 binding and subsequent signaling (Figure 4). Cells transformed with constructs containing alanine substitutions at residues Arg199 and Arg203, in the second extracellular loop, as well as Asp265, in the third extracellular loop, exhibited markedly diminished or absent IL-8 binding compared with wild-type CXCR1 (Leong *et al.*, 1994). These positions are also conserved in the sequence of CXCR2 and are thus unlikely to play a role in receptor subtype specificity. Ala substitution of each of another two residues, Glu275 and Arg280, both in the third extracellular loop of CXCR1, also resulted in a dramatic loss of IL-8 binding and signaling (Hebert *et al.*, 1993). These two residues are conserved in CXCR1 and CXCR2, as well as rabbit and mouse homologs of CXCR1. Glu275 and Arg280 are thought to interact with the positively charged Arg6 and Glu4 residues of IL-8, respectively. The Arg6 residue of IL-8 is itself critical for receptor binding: Ala or Lys substitution of this single residue causes a 1000-fold decrease in IL-8 binding affinity (Hebert *et al.*, 1991).

Site-directed mutagenesis studies demonstrated that binding occurs via determinants on multiple extracellular domains of CXCR1, in addition to the N-terminus. However, others have shown that an

Figure 3 Alignment of the sequences of human CXCR1 and CXCR2. Identical residues are marked by an asterisk. Conservative substitutions are marked by a dot.

Human CXCR1:

```
  1 MSNITDPQMW DFDDLNFTGM PPADEDYSPC MLETETLNKY VVIIAYALVF LLSLLGNSLV
 61 MLVILYSRVG RSVTDVYLLN LALADLLFAL TLPIWAASKV NGWIFGTFLC KVVSLLKEVN
121 FYSGILLLAC ISVDRYLAIV HATRTLTQKR HLVKFVCLGC WGLSMNLSLP FFLFRQAYHP
181 NNSSPVCYEV LGNDTAKWRM VLRILPHTFG FIVPLFVMLF CYGFTLRTLF KAHMGQKHRA
241 MRVIFAVVLI FLLCWLPYNL VLLADTLMRT QVIQETCERR NNIGRALDAT EILGFLHSCL
301 NPIIYAFIGQ NFRHGFLKIL AMHGLVSKEF LARHRVTSYT SSSVNVSSNL
```

Human CXCR2:

```
  1 MEDFNMESDS FEDFWKGEDL SNYSYSSTLP PFLLDAAPCE PESLEINKYF VVIIYALVFL
 61 LSLLGNSLVM LVILYSRVGR SVTDVYLLNL ALADLLFALT LPIWAASKVN GWIFGTFLCK
121 VVSLLKEVNF YSGILLLACI SVDRYLAIVH ATRTLTQKRY LVKFICLSIW GLSLLLALPV
181 LLFRRTVYSS NVSPACYEDM GNNTANWRML LRILPQSFGF IVPLLIMLFC YGFTLRTLFK
241 AHMGQKHRAM RVIFAVVLIF LLCWLPYNLV LLADTLMRTQ VIQETCERRN HIDRALDATE
301 ILGILHSCLN PLIYAFIGQK FRHGLLKILA IHGLISKDSL PKDSRPSFVG SSSGHTSTTL
```

Figure 3 (*Continued*)

```
                       10            20        30        40
hCXCR1         MSNITDPQMWDFDDL-NF---TGMPPADEDYSPCMLETETLNKYV
               * . .  . *  .** *.   . .**   * .**  *.  .***
hCXCR2    MEDFNMESDSFEDFWKGEDLSNYSYSSTLPPFLLDAAPCEPESLEINKYF
                  10        20        30        40        50

             50        60        70        80        90
hCXCR1    VIIAYALVFLLSLLGNSLVMLVILYSRVGRSVTDVYLLNLALADLLFALT
          *.* **********************************************
hCXCR2    VVIIYALVFLLSLLGNSLVMLVILYSRVGRSVTDVYLLNLALADLLFALT
                  60        70        80        90       100

            100       110       120       130       140
hCXCR1    LPIWAASKVNGWIFGTFLCKVVSLLKEVNFYSGILLLACISVDRYLAIVH
          **************************************************
hCXCR2    LPIWAASKVNGWIFGTFLCKVVSLLKEVNFYSGILLLACISVDRYLAIVH
                 110       120       130       140       150

            150       160       170       180       190
hCXCR1    ATRTLTQKRHLVKFVCLGCWGLSMNLSLPFFLFRQAYHPNNSSPVCYEVL
          ********* ****.**. ****. *.** .***..  ..* ** *** .
hCXCR2    ATRTLTQKRYLVKFICLSIWGLSLLLALPVLLFRRTVYSSNVSPACYEDM
                 160       170       180       190       200

            200       210       220       230       240
hCXCR1    GNDTAKWRMVLRILPHTFGFIVPLFVMLFCYGFTLRTLFKAHMGQKHRAM
          **.**.***.*****..*******..************************
hCXCR2    GNNTANWRMLLRILPQSFGFIVPLLIMLFCYGFTLRTLFKAHMGQKHRAM
                 210       220       230       240       250

            250       260       270       280       290
hCXCR1    RVIFAVVLIFLLCWLPYNLVLLADTLMRTQVIQETCERRNNIGRALDATE
          ****************************************.*.*******
hCXCR2    RVIFAVVLIFLLCWLPYNLVLLADTLMRTQVIQETCERRNHIDRALDATE
                 260       270       280       290       300

            300       310       320       330       340
hCXCR1    ILGFLHSCLNPIIYAFIGQNFRHGFLKILAMHGLVSKEFLARHRVTSYTS
          ***.*******.*******.****.*****.***.**. *...   *. .
hCXCR2    ILGILHSCLNPLIYAFIGQKFRHGLLKILAIHGLISKDSLPKDSRPSFVG
                 310       320       330       340       350

            350
hCXCR1    SSVNVSSNL
          **   .*
hCXCR2    SSSGHTSTTL
                 360
```

antibody that maps to the N-terminal region of CXCR1 blocks IL-8 binding (Chuntharapai *et al.*, 1994a). Similarly, an antibody that maps to the N-terminus of CXCR2 blocks binding of IL-8 and MGSA/GROα (Hoch *et al.*, 1996; Norgauer *et al.*, 1996). While antibodies to the N-terminus inhibit binding, peptides corresponding to the N-termini of each receptor only weakly inhibit IL-8 binding to

Figure 4 Model of the secondary structures of human CXCR1 and CXCR2. As deduced by alanine scanning mutagenesis, residues in CXCR1 that are important for receptor expression or for ligand binding are indicated in green (Hebert *et al.* 1993; Leong *et al.* 1994). Residues involved in G protein signaling are shown in blue (Damaj *et al.*, 1996b; Xie *et al.*, 1997). Important structural determinants conserved in both CXCR1 and CXCR2 are indicated in red. (Full colour figure can be viewed online.)

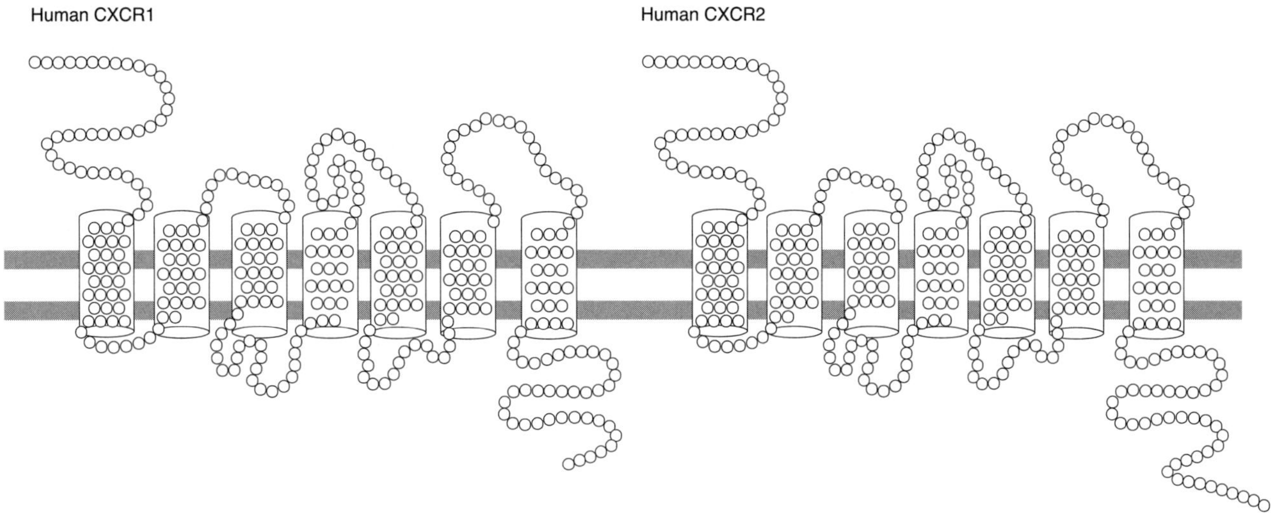

CXCR1 and CXCR2 ($K_i = 17–2.2\,\mu M$, and $K_i = 150\,\mu M$, respectively; Gayle *et al.*, 1993). In addition, a soluble peptide of the first 37 amino acids of CXCR1 is only capable of binding IL-8 with low affinity (Gayle *et al.*, 1993). Another peptide corresponding to residues 1–40 of CXCR1 was also shown to bind IL-8 with a dissociation constant of $170 \pm 50\,mM$ (Clubb *et al.*, 1994). These results are in agreement with data from alanine-scanning mutagenesis of CXCR1, demonstrating that multiple extracellular domains of the receptor are involved in high-affinity ligand interactions.

Using receptor chimeras, others have studied the selectivity determinants in CXCR1 and CXCR2 (LaRosa *et al.*, 1992; Gayle *et al.*, 1993; Ahuja *et al.*, 1996; Wu *et al.*, 1996). Experiments with human/rabbit chimeric receptors first implicated the N-terminal domains in determining ligand specificities (LaRosa *et al.*, 1992; Gayle *et al.*, 1993). Chimeras containing the CXCR1 N-terminus were selective for IL-8, whereas those carrying the N-terminal domain of CXCR2 bound both IL-8 and MGSA/GROα with high affinity. Further studies utilized chimeric forms of both human receptors, yet observations of individual groups using the same constructs did not agree. For example, a chimera composed of the CXCR1 N-terminus displayed a CXCR2 phenotype in one study: MGSA/GROα and NAP-2 bound with high affinity and induced a strong calcium flux (Ahuja *et al.*, 1996). The results of these studies by Ahuja *et al.* suggest that, in contrast to previous reports, the N-terminal segment of CXCR1 is not a dominant determinant of receptor subtype specificity. In another report, the same chimeric receptor had a CXCR1 phenotype: MGSA/GROα did not bind transfected cells, nor did it induce chemotaxis (Wu *et al.*, 1996). These observations reported by Wu *et al.* are more consistent with those previously reported by groups using the rabbit CXCR1 N-terminus in the chimera rather than the human CXCR1 (LaRosa *et al.*, 1992; Gayle *et al.*, 1993; Wu *et al.*, 1996).

More recently, Katancik *et al.* mapped the domains involved in binding of the ligands IL-8, NAP-2, and MGSA/GROα to CXCR2 using peptides corresponding to distinct regions of this receptor (Katancik *et al.*, 1997). Peptides representing the N-terminus and first extracellular loop of CXCR2 both inhibited IL-8 binding to the receptor. However, NAP-2 binding was inhibited only by the peptide corresponding to the first extracellular loop, and MGSA/GROα binding was inhibited by portions of the N-terminus. These results suggest that distinct extracellular domains of CXCR2 mediate the binding of each ligand to this receptor.

Additional residues in both CXCR1 and CXCR2 are important for maintaining correct tertiary structure of the receptors. Scanning mutagenesis demonstrated that amino acids Cys30 (in the N-terminus), Cys187 (in the second extracellular loop), and Cys277 (in the third extracellular loop) of CXCR1 are sensitive to mutation to alanine (Leong *et al.*, 1994). Cys30 and Cys277 are conserved in human CXCR1 and CXCR2, as well as rabbit and mouse IL-8 receptors, and may thus be critical for correct protein folding (Hoch *et al.*, 1996). These two cysteine residues likely form a disulfide bridge, bringing Asp11, Glu275, and Arg280 into close spatial proximity to one another. These three charged residues, each critical for ligand binding, thus form a binding domain in CXCR1 where they likely make contact with oppositely charged residues in IL-8 (Hebert *et al.*, 1993). An additional feature common to nearly all GPCRs includes Asp85, located in the second transmembrane domain of CXCR1. This residue is conserved in more than 90% of members of the GPCR superfamily, and is also thought to maintain correct tertiary structure of the receptor. An Asp residue in the second transmembrane domain is also conserved in almost all GPCRs and replacement of this residue leads to a loss of ligand binding (Probst *et al.*, 1992). When Asp85 of CXCR1 is replaced with Ala, the receptor is not expressed on the cell surface, thus abolishing binding of IL-8 (Hebert *et al.*, 1993).

The domains in CXCR1 and CXCR2 involving ligand binding may not be important for receptor signaling. A detailed study of chimeric forms of both human CXCR1 and CXCR2 demonstrated that IL-8, GROα, and NAP-2 signal via interactions with multiple receptor domains (Ahuja *et al.*, 1996). In this study, the authors demonstrated that all three ligands could elicit calcium flux via a CXCR2 chimeric receptor containing the N-terminus of human CC chemokine receptor 1 (CCR1), even though no IL-8 binding was detected. The structural determinants important for ligand binding and signaling have been further dissected using various blocking antibodies against the IL-8 receptors (Wu *et al.*, 1996). Studies showed that an antibody against CXCR1 can block IL-8-mediated functional responses, but is ineffective at blocking ligand binding. Together, these findings suggest that the structural requirements for IL-8 binding and signaling are distinct. This is consistent with the idea that conformational changes of the receptor secondary to ligand binding are required to elicit biological responses (Gilman, 1987).

Additional structural determinants are also important for receptor signaling (Figure 4). All chemokine receptors, including CXCR1 and CXCR2, contain a highly conserved DRYLAIVHA motif at the end of the third transmembrane domain (TM3) (Baggiolini *et al.*, 1997). As the DRY motif is necessary for G protein activation in other receptors, this domain may be important for signaling and biological activity of CXCR1 and CXCR2 (O'Dowd *et al.*, 1989). Additional motifs conserved among members of the chemokine receptor family are: TD(X)YLLNLA(X2)DLLF(X2) TLP(X)W in TM2, PLL(X)M(X2)CY in TM5, W(X)-PYN in TM6, and HCC(X)NP(X)IYAF(X) G(X2) FR in TM7.

A serine/threonine-rich C-terminal region is also a common feature of GPCRs, including CXCR1 and CXCR2 (Probst *et al.*, 1992). This region may be a target for phosphorylation by serine threonine kinases. Phosphorylation of these residues at the C-terminus of the β-adrenergic and rhodopsin receptors, both GPCR family members, leads to receptor desensitization (Lefkowitz *et al.*, 1990; Palczewski and Benovic, 1991).

A series of truncation mutants further defined key determinants important for receptor signaling. Because expression of CXCR2 requires an intact N-terminal tail, the role of this domain could not be determined (Gayle *et al.*, 1993; Ahuja *et al.*, 1996). Analysis of truncated forms of CXCR2 suggests that signaling requires the C-terminal sequence between amino acids 317 and 324 (Ben-Baruch *et al.*, 1995a). Cells transfected with truncated mutant receptors bound IL-8 with the same high affinity as those transfected with wild-type receptors, but were unable to chemotax in response to IL-8. In addition, the C-terminus-truncated mutant of CXCR2 is not phosphorylated by high doses of IL-8, as is the wild-type receptor (Ben-Baruch *et al.* 1997). Phosphorylation analysis of the wild-type receptor further demonstrates that IL-8 induced higher levels of phosphorylation than did NAP-2. The greater ability of IL-8 to induce receptor phosphorylation may thus also contribute to its more potent stimulation of chemotaxis compared with NAP-2.

Chemokine receptor structure, including that of CXCR1 and CXCR2, is unknown. However, a model can be constructed based on analogy with another GPCR, bacteriorhodopsin, whose crystal structure has been solved (Strader *et al.*, 1994; Luo *et al.*, 1997; Unger *et al.*, 1997; Lomize *et al.*, 1999). The hydrophobic regions of the receptor traverse the plasma membrane, separating extracellular and cytoplasmic loops of the protein and thus giving it a serpentine appearance (Figure 1). CXCR1 and CXCR2 are probably in the cell membrane with the N-terminal tail extracellular, and the C-terminus in the cytoplasm, with the seven hydrophobic domains embedded in the membrane as α helices. The resulting structure comprises a total of three extracellular and three intracellular loops connecting each of the seven transmembrane domains, and free N- and C-termini.

CXCR1 has five potential *N*-linked glycosylation sites (Strosberg, 1991). The molecular size predicted from the amino acid sequence is approximately 40 kDa (Holmes *et al.*, 1991). However, the receptor is glycosylated, resulting in an observed mass of 55–69 kDa (Samanta *et al.*, 1989). CXCR2 is also glycosylated at its one *N*-linked glycosylation site, to give a final mass of approximately 60 kDa (Horuk, 1994).

Relevant homologies and species differences

The human CXCR1 and CXCR2 proteins are highly homologous at the amino acid level (77%) with the highest homology over the membrane-spanning regions and significant divergence at both N- and C-termini (Kelvin *et al.*, 1993). Both proteins also have a significant degree of homology to other chemokine receptors: sequence alignment of the human receptors for the three neutrophil chemoattractants IL-8, fMLP, and C5a shows a 29–34% amino acid identity between these receptors.

Genes encoding CXCR1 and CXCR2 homologs from four nonhuman primates have been cloned and sequenced (Alvarez *et al.*, 1996). Both gorilla and chimpanzee CXCR1 homologs show 98–99% similarity to human CXCR1 (**Figure 5**). Rhesus and orang utan CXCR1 homologs are pseudogenes, in which a 2 bp insertion has generated a sequence with several stop codons. However, the CXCR2 genes from all four primate species are 95–99% identical to their human homolog.

Human and rabbit CXCR1 proteins share a 84% amino acid identity (Holmes *et al.*, 1991). The affinity profile and ligand specificity of rabbit CXCR1 is much like that of the human receptor: IL-8 ($K_d \sim$ 4 nM) $\ggg$ MGSA > NAP-2 (Prado *et al.*, 1994). No binding was observed with radiolabeled GROα or PF4 (Thomas *et al.*, 1994). IL-8 also stimulates calcium flux of transfected cells, while fMLP has no effect. Rabbit CXCR2 has an 80% amino acid identity to human CXCR2, 74% to rabbit CXCR1, and 73% to human CXCR1 (Prado *et al.*, 1994). Rabbit CXCR2 binds IL-8, NAP-2, and MGSA with apparent K_i values of 4, 120, and 320 nM, respectively. IL-8 induced calcium flux and desensitization in mammalian cells transfected with rabbit CXCR2.

Figure 5 Graphical representation of the sequence homology among CXCR1 and CXCR2 proteins from different animal species

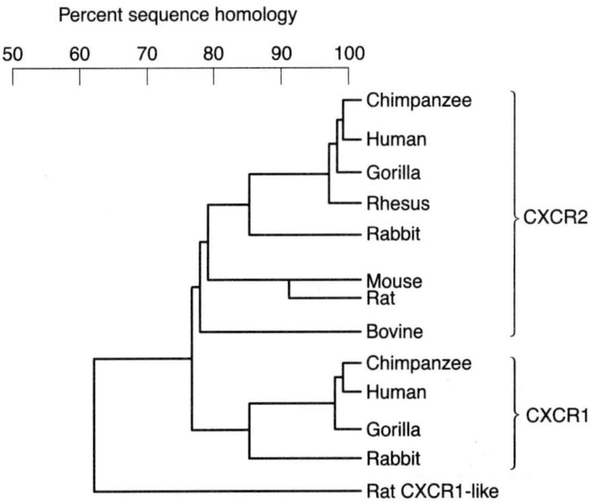

A mouse IL-8 receptor homolog was cloned by screening a cDNA library for sequences homologous to human and rabbit receptors (Bozic *et al.*, 1994; Harada *et al.*, 1994). The cDNA corresponding to human CXCR2 hybridizes strongly with two restriction fragments in mouse genomic DNA, suggesting that there are two candidate murine homologs of the human IL-8 receptors (Bozic *et al.*, 1994). However, only one murine IL-8 receptor-like protein has been isolated to date, most closely resembling human CXCR2 in its binding characteristics (Lee *et al.*, 1995). Mouse CXCR2 binds the murine counterpart of human GRO proteins, macrophage inflammatory protein 2 (MIP-2), with high affinity ($K_d \sim$ 5 nM), activating both calcium flux and a chemotactic response in neutrophils (Bozic *et al.*, 1994; Lee *et al.*, 1995). Another murine GRO homolog, KC, is approximately 10-fold less potent at triggering these responses. Mouse CXCR2 is 359 amino acids in length, and shares a 68 and 71% homology to human CXCR1 and CXCR2, respectively (Cerretti *et al.*, 1993b). The mRNA for this receptor is expressed in peritoneal neutrophils, and southern blotting analysis suggests that this is a single-copy gene (Cerretti *et al.*, 1993b; Harada *et al.*, 1994). A murine form of IL-8 has not been identified, but human IL-8 is an agonist for mouse CXCR2 ($K_d \sim$ 400 nM) (Bozic *et al.*, 1994, Suzuki *et al.*, 1994).

Two rat genes, CXCR1-like and CXCR2, have also been cloned (Dunstan *et al.*, 1996). Both genes are approximately 70% identical to their human homologs at the amino acid level (Dunstan *et al.*, 1996; Gobl *et al.*, 1997). Both rat genes also share a 65 and 86% identity, respectively, with murine CXCR2. Southern blot analysis indicates that rat CXCR1-like and CXCR2 are each single-copy genes. CXCR2 mRNA is detected in adult rat lung, spleen, and neutrophils. CXCR1-like mRNA is expressed in adult rat lung and primary rat macrophages, but not in neutrophils. Murine MIP-2 induces calcium flux in mammalian cells expressing the rat CXCR2 receptor, but not the CXCR1-like protein. Hybridization analysis of mouse genomic DNA suggests that the rat

CXCR1-like gene is homologous to the second candidate murine IL-8 receptor homolog (Bozic *et al.*, 1994; Dunstan *et al.*, 1996). Since the rat CXCR1-like protein is nearly equally homologous to both human CXCR1 and CXCR2 sequences, these unique rat and murine genes may represent a third class of receptors, distinct from CXCR1 and CXCR2.

Affinity for ligand(s)

IL-8 and GCP-2 are equipotent agonists at both CXCR1 and CXCR2, while the affinity for each receptor for other ligands is distinct. The rank order of potency of all agonists of CXCR1 and CXCR2 has been determined based on mean effective concentration values (EC_{50}) in calcium flux assays (Ahuja and Murphy, 1996; Baggiolini *et al.*, 1997; Van Damme *et al.*, 1997; Wuyts *et al.*, 1997, 1998; Wolf *et al.*, 1998). In the case of CXCR2, ligand potency is as follows: GROγ is the most potent (1 nM) > IL-8 (4 nM) ~ GCP-2 (3 nM) ~ GROα (5 nM) ~ GROβ (4 nM) ~ NAP-2 (7 nM) > ENA-78 (11 nM). For CXCR1, potency was far more selective, with IL-8 and GCP-2 being the most potent agonists (4 and 3 nM, respectively) $\ggg$ ENA-78 (40 nM) ~ NAP-2 (45 nM) GROα (63 nM) ~ GROγ (65 nM) $\gg$ GROβ. All six agonists for CXCR2 competed for high-affinity [^{125}I]IL-8, [^{125}I]GROα, [^{125}I]NAP-2, and [^{125}I]ENA-78 binding to CXCR2. GROα, GROβ, GROγ, NAP-2, and ENA-78 each competed weakly for high-affinity IL-8 binding to CXCR1. Although the affinity of CXCR1 for GROα and NAP-2 is almost 100-fold lower than that for IL-8, these ligands can act as low-potency agonists in chemotaxis assays (Loetscher *et al.*, 1994; Ahuja and Murphy, 1996). CXCR1 has also recently been shown selectively to bind the N-terminal cytokine module of human tyrosyl tRNA synthetase with an affinity similar to that of IL-8 (Wakasugi and Schimmel, 1999). This ELR-containing enzyme is secreted under apoptotic conditions and may be involved in inflammatory signaling by apoptotic cells.

Selective blocking antibodies to CXCR1 and CXCR2 have been used to determine the role of each receptor in mediating specific neutrophil functions. Responses such as chemotaxis, calcium flux, and the release of neutrophil granule enzymes appear to be independently mediated through both CXCR1 and CXCR2, while the respiratory burst and PLD activation depend exclusively on stimulation through CXCR1 (Loetscher *et al.*, 1994; Hammond *et al.*, 1995; Jones *et al.*, 1996). This is consistent with the ability of GROα and NAP-2 to induce calcium flux and degranulation, but not PLD activation, in human neutrophils (L'Heureux *et al.*, 1995). IL-8-induced neutrophil chemotaxis and priming were predominantly mediated by CXCR1, whereas priming by GROα and ENA-78 is mediated by CXCR2 (Hammond *et al.*, 1995; Green *et al.*, 1996).

The biological significance of the differential activity of CXCR1 and CXCR2 is unclear. For ligands other than IL-8, the two receptors may be involved in differential spatial transduction of chemotactic signals: At low chemokine concentrations, chemotaxis may be preferentially mediated by CXCR2. Near the site of inflammation, where chemokine concentrations are high and CXCR2 is unable to signal, CXCR1 might be more important. However, both receptors bind IL-8 with similar affinity, thus this mechanism would not be useful for IL-8 gradients. In this case, CXCR2 may mediate chemotaxis at more distal points, while CXCR1 induces respiratory burst at high chemokine concentrations in the inflammatory focus. In fact, the two receptor subtypes are differentially desensitized in human neutrophils. Higher concentrations of IL-8 are required for internalization of CXCR1 than CXCR2, thus downmodulating receptor expression on the cell surface (Chuntharapai and Kim, 1995; Sabroe *et al.*, 1997).

Cell types and tissues expressing the receptor

Using monoclonal antibodies specific for CXCR1 and CXCR2, their expression pattern on various peripheral blood leukocytes have been studied by flow cytometry (**Figure 6**) (Chuntharapai *et al.*, 1994b; Oin *et al.*, 1996). A wide range of donor variation in expression levels was observed. All neutrophils, all monocytes, and 5–25% of CD8+ T cells and CD56+ NK cells expressed both receptors (Chuntharapai *et al.*, 1994b). No CD20+ B cells or CD4+ T cells expressed CXCR1 or CXCR2. Neutrophils expressed the highest level of both CXCR1 and CXCR2 at an approximately equal ratio, while CXCR2 expression prevailed on other leukocyte subsets, including monocytes and lymphocytes.

Consistent with this report, Gerszten *et al.* recently demonstrated the prevalence of CXCR2 expression on elutriated monocytes (Gerszten *et al.*, 1999). Relative prevalence varied among donors, with CXCR1 expressed on 23–90% of monocytes, while 22–93% of monocytes expressed CXCR2. These receptors on monocytes are functional, as antibodies to both receptors inhibited IL-8-induced monocyte chemotaxis as well as calcium flux.

Figure 6 Double-color flow cytometric analysis of the distribution of CXCR1 and CXCR2 and human peripheral CD4+, CD8+, CD56+, and CD20+ B cells. Cells were stained simultaneously with a F-mAb to each receptor and a PE-mAb detecting each leukocyte subset.

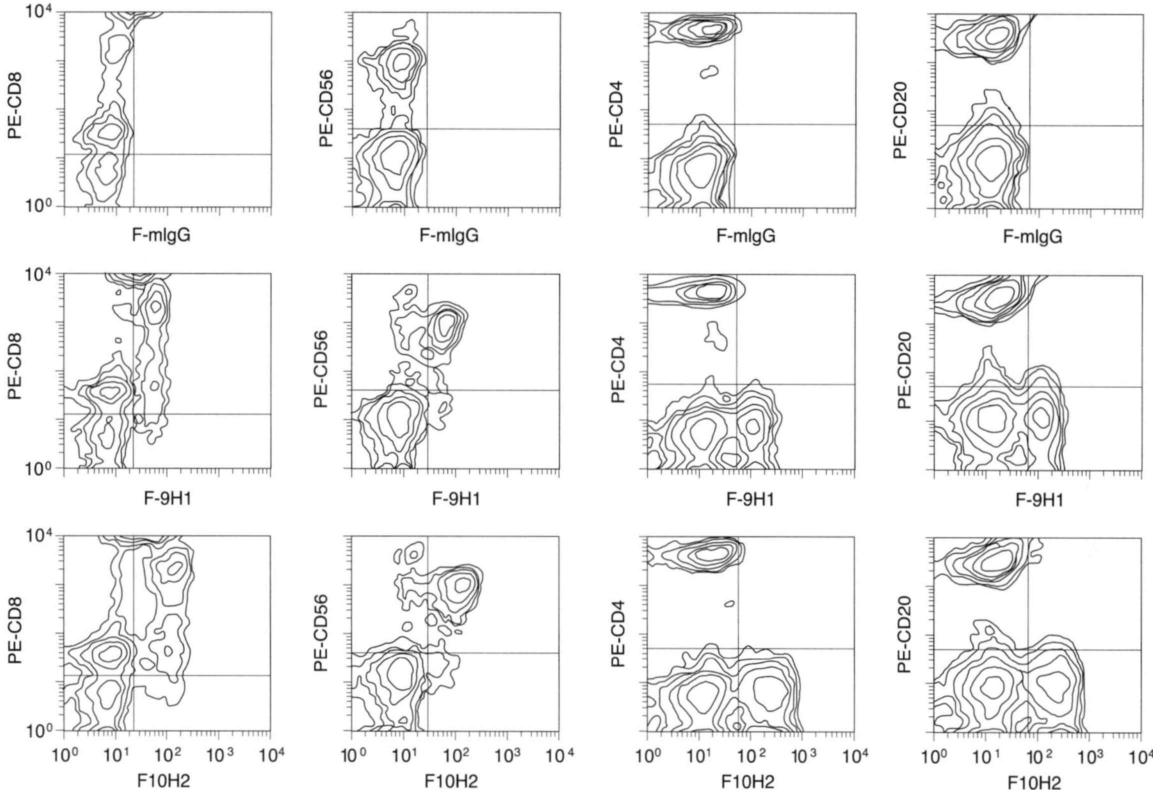

CXCR1 and CXCR2 are also expressed by certain subsets of lymphocytes. The mRNA for both receptors can be detected in T cells by RT-PCR, but not by northern blotting, suggesting that expression levels are quite low (Moser *et al.*, 1993; Xu *et al.*, 1995). Using monoclonal antibodies and flow cytometry, 7–42% of CD8+ T cells and 39–76% of CD56+ NK cells, but no CD20+ B cells or CD4+ T cells were shown to express CXCR2 (Chuntharapai *et al.*, 1994b). Expression of both CXCR1 and CXCR2 was also demonstrated on a subset of T cells that were both CD8+ and CD56+ (Oin *et al.*, 1996). Activation of T cells *in vitro* with anti-CD3 antibodies did not affect expression of these receptors.

CXCR1 and CXCR2 expression has also been demonstrated on the human leukemic mast cell line, HMC-1 (Lippert *et al.*, 1998; Nilsson *et al.*, 1999). The mRNA for both receptors was detected by RT-PCR, and flow cytometry demonstrated the presence of both proteins on the cell surface. The ligands IL-8, GROα, and NAP-2 all induced calcium flux, as well as chemotaxis of this cell line, demonstrating that the receptors are functionally active. Using postembedding immunoelectron microscopy, the expression of CXCR1 was shown on the cytoplasmic membrane of isolated human skin mast cells, whereas CXCR2 was found in mast cell-specific granules. However, using the RNAase protection assay, Nilsson *et al.* (1999) detected only CXCR2 mRNA in HMC-1 cells. Flow cytometry analysis also documented the surface expression of CXCR2. Both reports suggest that ligands for CXCR1 and/or CXCR2 may play an important role in mast cell recruitment during the initiation of an inflammatory response.

Controversial results have been published regarding the role of IL-8 in eosinophil activation and chemotaxis, particularly in allergic disease. Using a purified population of eosinophils, Petering *et al.* recently demonstrated that IL-8 does not induce chemotaxis of or calcium flux in these cells (Petering *et al.*, 1999). Furthermore, RT-PCR experiments showed that eosinophils do not express the mRNA for either CXCR1 or CXCR2. As little as 5% neutrophil contamination was sufficient to produce a measurable IL-8-induced calcium flux in an eosinophil preparation. This suggests that previous reports of eosinophil expression of IL-8 receptors may have been the result of experiments with impure populations of cells.

IL-8 has also been proposed as a stimulus for IgE-independent basophil activation. [^{125}I]IL-8 binding studies revealed approximately 9600 receptors with a mean K_d value of 0.15 nM on human basophils (Krieger *et al.*, 1992). NAP-2 weakly competed for IL-8 binding, and both chemokines led to a transient rise of cytosolic free calcium in these cells. GROα, β, and γ have also been shown to induce calcium flux and chemotaxis of human basophils (Geiser *et al.*, 1993). However, these studies only indirectly imply the presence of CXCR1 and/or CXCR2 on the cell surface. A more recent study demonstrated the expression of intermediate levels of CXCR1 and CXCR2 on basophils (Ochensberger *et al.*, 1999). However, receptor agonists did not stimulate cell functions such as cytokine and leukotriene release.

There is a lack of consensus regarding endothelial cell expression of either CXCR1 or CXCR2. The mRNA for GROα is upregulated when this chemokine is added to human umbilical vein endothelial cell (HUVEC) cultures *in vitro*, implying that these cells express the GROα receptor, CXCR2 (Wen *et al.*, 1989). More indirect evidence comes from experiments that show IL-8-induced albumin flux across HUVEC monolayers in the absence of neutrophils, an effect that is inhibited by cycloheximide (Biffl *et al.*, 1995). Additionally, IL-8 appears to be chemotactic for HUVECs in an *in vitro* chemotaxis assay (Koch *et al.*, 1992). IL-8 was also shown to bind to the surface of endothelial cells at different anatomical locations; however the identity of these binding sites is unknown (Rot *et al.*, 1996). Studies demonstrating direct expression of CXCR1 or CXCR2 on endothelial cells have been inconclusive. Schonbeck *et al.* found that IL-8 binds to both HUVEC and saphenous vein endothelial cells, and used RT-PCR techniques to demonstrate the expression of CXCR1 on these cells (Schonbeck *et al.*, 1995). Others were unable to show IL-8 binding, IL-8-induced calcium flux, or IL-8-induced endothelial cell proliferation using both large-vessel and microvascular endothelial cells (Petzelbauer *et al.*, 1995). A recent report by Murdoch *et al.* demonstrates endothelial cell expression of the mRNAs for CXCR1 and CXCR2, but low levels of surface expression of only CXCR1 protein (Murdoch *et al.*, 1999).

RT-PCR and immunocytochemical techniques have also detected CXCR1 and CXCR2 expression in normal melanocytes as well as melanoma cells (Moser *et al.*, 1993; Norgauer *et al.*, 1996). Keratinocytes have also been shown to possess binding sites for IL-8 (Kemeny *et al.*, 1994b). Receptors for IL-8 are also expressed throughout the central nervous system (reviewed in Hesselgesser and Horuk, 1999a; Mennicken *et al.*, 1999). While CXCR1 expression was not detected, CXCR2 is expressed at high levels by subsets of neurons in diverse regions of the brain and spinal cord (Horuk *et al.*, 1997).

Regulation of receptor expression

Bacterial LPS has been shown to downmodulate neutrophil CXCR1 and CXCR2 expression via the stimulation of tyrosine kinase activity (Khandaker *et al.*, 1998). However, others have demonstrated that serum-activated LPS acts to induce the surface expression of IL-8 receptors through *de novo* protein synthesis (Manna *et al.*, 1995). IL-8 receptor expression is also upregulated by formyl peptide via degranulation of the neutrophil secretory vesicle (Manna and Samanta, 1995). CXCR1 mRNA expression, IL-8 binding, and neutrophil chemotactic responses are upregulated by G-CSF as well (Lloyd *et al.*, 1995). Conversely, treatment of neutrophils with tumor necrosis factor α (TNFα) decreases cell surface CXCR1 and CXCR2 expression (Jawa *et al.*, 1999). Others have demonstrated that TNFα downregulates the expression of CXCR2, but CXCR1 levels are unaffected (Asagoe *et al.*, 1998). Transcription of the CXCR2 gene in human T lymphocytes is downregulated by incubating cells at 37°C, and is restored by co-culturing T cells with monocytes (Xu *et al.*, 1995).

CXCR1 and CXCR2 expression on the cell surface is also regulated endogenously by receptor internalization. More than 90% of [^{125}I]IL-8-bound neutrophil receptors is endocytosed within 10 minutes at 37°C, and the receptors are expressed on the cell surface 10 minutes later (Besemer *et al.*, 1989; Grob *et al.*, 1990; Samanta *et al.*, 1990). Inhibitory lysosomotropic agents, such as ammonium chloride, inhibit receptor internalization as well as chemotaxis, suggesting that chemotactic activity is dependent on receptor recycling (Samanta *et al.*, 1990).

The expression of all GPCRs, including the chemokine receptors CXCR1 and CXCR2, is strictly regulated by desensitization (Baggiolini *et al.*, 1994; Ben-Baruch *et al.*, 1995b). In homologous desensitization, exposure to high concentrations of ligand renders the cell unresponsive to the same ligand (Kelvin *et al.*, 1993; Ben-Baruch *et al.*, 1995b). In contrast, heterologous desensitization is a reversible loss of responsiveness to multiple ligands.

Homologous desensitization after stimulation with IL-8 or GROα is mediated by phosphorylation of serines and threonines in the C-terminus of CXCR1 and CXCR2, respectively (Baggiolini *et al.*, 1994; Mueller *et al.*, 1994; Richardson *et al.*, 1995; Schraufstatter *et al.*, 1998). Ser346 and Ser348 are the primary sites of phosphorylation of CXCR2 (Schraufstatter *et al.*, 1998). Alanine substitution at these residues prevents receptor desensitization after repeated exposure to IL-8; however, receptor internalization was unaffected. Heterologous desensitization also involves C-terminus phosphorylation, as phosphorylation-deficient mutant CXCR1 is resistant to cross-desensitization by receptors to fMLP, C5a, or PAF (Richardson *et al.*, 1995, 1998). Phosphorylation at these resides in the β-adrenergic and rhodopsin receptors, both GPCR family members, results in arrestin binding, thus sterically inhibiting the binding of G proteins to the receptor and preventing signaling (Zhao *et al.*, 1995; Lefkowitz, 1998). Receptor desensitization is likely an important mechanism in the detection of changes in ligand concentration, allowing cells to move toward a chemokine gradient.

SIGNAL TRANSDUCTION

Two groups of G proteins, the heterotrimeric G proteins and monomeric low molecular weight G proteins, mediate signal transduction through chemokine receptors. Like all members of the GPCR superfamily, CXCR1 and CXCR2 act as guanine nucleotide exchange factors for the heterotrimeric G proteins (**Figure 7**) (Kupper *et al.*, 1992; Strader *et al.*, 1994). Heterotrimeric G proteins associate with the intracellular domains of GPCRs when in their inactive, or guanosine diphosphate (GDP)-bound, state. Upon ligand binding, the GDP is exchanged for guanosine triphosphate (GTP), thus activating the G protein. The active G protein subsequently dissociates into its α and $\beta\gamma$ subunits to stimulate effector molecules.

In neutrophils, the free α subunit can activate both phospholipase C (PLC) β_1 and PLCβ_2, while the free $\beta\gamma$ subunit preferentially activates PLCβ_2 (Ben-Baruch *et al.*, 1995b). The activated PLCs catalyze the hydrolysis of the membrane phospholipid phosphatidylinositol-4,5-bisphosphate (PIP$_2$) to generate the second messengers inositol-1,4,5-trispohsphate (IP$_3$) and 1,2-diacylglycerol (DAG) (Baggiolini and Clark-Lewis, 1992; Baggiolini *et al.*, 1993). IP$_3$ diffuses into the cytosol and mobilizes intracellular calcium, and DAG activates protein kinase C (PKC; Thelen *et al.*, 1988; Murphy, 1994). IL-8-induced activation of PKC and the rise in cytosolic calcium further induce PLC and phospholipase D (PLD), which yield DAG and phosphatidic acid, respectively, and result in a positive feedback loop (Sozzani *et al.*, 1994a; Howard *et al.*, 1996). Others have been able to detect PLD activation only in IL-8-stimulated T lymphocytes, but not in neutrophils (Bacon *et al.*, 1995). Additional signaling pathways, such as the MAPK cascade and the serine/threonine and tyrosine kinases, are also activated following G protein activation (Grinstein and Furuya, 1992; Grinstein *et al.*, 1994; Thompson *et al.*, 1994; Jones *et al.*, 1995).

Experiments using neutrophils as well as transfected cells suggest that both CXCR1 and CXCR2 physically couple to many G protein subtypes, including Gα_{i2}, Gα_{i3}, Gα_{14}, Gα_{15}, and Gα_{16}, but not to Gα_q or Gα_{11} (Wu *et al.*, 1993; Damaj *et al.*, 1996a). Furthermore, Gα_{i2} peptides or antisera can effectively

Figure 7 Schematic representation of chemokine signal transduction. Abbreviations: DAG, diaglycerol; IP3, inositol-1,4,5-trisphosphate; PA, phosphatidic acid; PIP$_2$, phosphatidylinositol-4,5-bisphosphate, PKC, protein kinase C; PLC, phospholipase C; PLD, phospholipase D; [Ca^{2+}]$_i$, intracellular calcium.

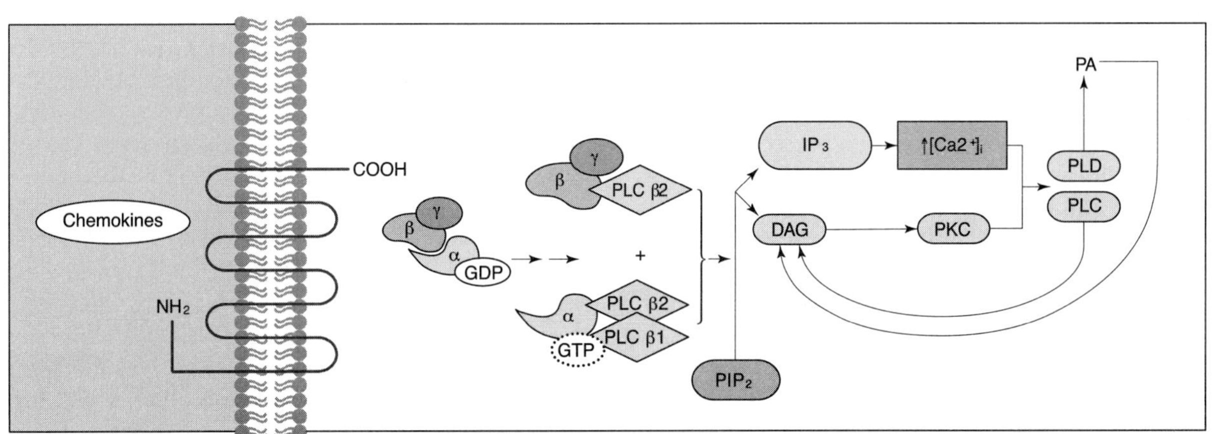

block calcium flux responses to IL-8 in cells expressing CXCR1 and CXCR2 (Thelen *et al.*, 1988; Damaj *et al.*, 1996a).

The structural determinants on CXCR1 and CXCR2 involved in mediating G protein signaling have been described (Damaj *et al.*, 1996b; Xie *et al.*, 1997). Using site-directed mutagenesis, key residues of CXCR1 were shown to be involved in the interaction of the receptor with $G\alpha_{i2}$ and $G\alpha_{16}$ (Figure 4). Activation of $G\alpha_{16}$ was abolished by alanine substitutions at both Lys158 and Arg159 in the second intracellular loop of CXCR1, suggesting that either of these two basic residues is sufficient for $G\alpha_{16}$ coupling (Xie *et al.*, 1997). Four residues in the second intracellular loop (Tyr136, Leu137, Ile139, and Val140) and one residue in the third intracellular loop (Met241) were shown to be crucial for $G\alpha_{i2}$ coupling and transduction of an IL-8-induced calcium flux (Damaj *et al.*, 1996). Other residues in the second and third intracellular loops were also found to affect CXCR1 signaling, but to a lesser extent.

An intrinsic GTPase activity regulates G protein signaling by hydrolyzing GTP to GDP. The resulting inactive GDP-bound G protein then reforms a complex with unoccupied receptors. GTPase-activating proteins known as regulators of G protein signaling (RGS) modulate this activity for heterotrimeric G proteins (Koelle, 1997). Studies with cotransfected HEK293 cells show that several RGS proteins modulate IL-8 signaling through CXCR2 (Bowman *et al.*, 1998).

In addition to signaling via the heterotrimeric G proteins, CXCR1 and CXCR2 signaling also activates monomeric, low molecular weight G proteins of the Ras and Rho families (Laudanna *et al.*, 1996). Rho proteins are usually involved in cell motility through regulation of actin-dependent processes such as membrane ruffling, pseudopod formation, and assembly of focal adhesion complexes. In mouse pre-B cells transfected with CXCR1, IL-8 signals via RhoA to stimulate adhesion to fibrinogen (Laudanna *et al.*, 1996).

Studies suggest that different neutrophil functions are activated by distinct signaling pathways. The assembly or activation of NADPH oxidase and the subsequent generation of superoxide anions in respiratory burst are associated with tyrosine phosphorylation and DAG–PLD interactions, respectively (Qualliotine-Mann *et al.*, 1993; Richard *et al.*, 1994). IL-8 stimulates the activation of PLD only at concentrations required for stimulating respiratory burst, not at the 10- to 100-fold lower concentrations that mediate chemotaxis (Sozzani *et al.*, 1994a).

Chemotaxis can be triggered by agonists for receptors coupled to the G_i subfamily of G proteins, but not by agonists for receptors coupled to G_s and G_q (Wu *et al.*, 1993). Furthermore, treatment of neutrophils with pertussis toxin inhibits IL-8-directed chemotaxis by preventing the activation of G_i. These observations suggest that chemotaxis is mediated by the specific pertussis-sensitive α subunit of G_i proteins. However, recent evidence demonstrates that chemotaxis does not require activation of $G\alpha_i$, and that the free $\beta\gamma$ subunit transmits the necessary chemotactic signal (Neptune *et al.*, 1999).

There is evidence that neutrophil chemotaxis results from PLC activation and subsequent calcium flux (Murphy, 1994). However, others have shown that cell migration does not require an increase in intracellular calcium (Sha'afi *et al.*, 1986; Sham *et al.*, 1993; Sozzani *et al.*, 1994b). More recent reports suggest that IL-8-induced neutrophil chemotaxis is dependent on phosphatidylinositol 3-kinase (PI-3 kinase) activity, as migration can be inhibited by wortmannin and the PI-3 kinase inhibitor LY294002 (Knall *et al.*, 1997). PI-3 kinase signaling further activates the Ras/Raf/extracellularly responsive kinase (ERK) pathway in neutrophils and in cells transfected with CXCR1 and CXCR2 (Knall *et al.*, 1996; Shyamala and Khoja, 1998). However, ERK activation does not play a role in mediating IL-8-directed chemotaxis, as the ERK kinase inhibitor, PD098059, has no effect on neutrophil migration (Knall *et al.*, 1997).

BIOLOGICAL CONSEQUENCES OF ACTIVATING OR INHIBITING RECEPTOR AND PATHOPHYSIOLOGY

Unique biological effects of activating the receptors

CXCR1 and CXCR2 mediate a diversity of chemokine functions in a variety of cell types. Both CXCR1 and CXCR2 induce chemotaxis and calcium flux of cells expressing these receptors. Both receptors also stimulate the release of granule enzymes in neutrophils. The generation of superoxide in the neutrophil respiratory burst is mediated exclusively by CXCR1 (Loetscher *et al.*, 1994; Hammond *et al.*, 1995; Jones *et al.*, 1996). CXCR1 and/or CXCR2 also mediate the angiogenic effects of ELR-containing CXC chemokines. These ligands induce endothelial cell proliferation and chemotaxis *in vitro* as well as angiogenesis in various animal models (Koch *et al.*, 1992; Strieter *et al.*, 1995).

CXCR1 and CXCR2 are important in mediating antimicrobial host defenses. Their function in directing leukocyte recruitment and activation leads to clearance of infective agents. In addition to their role in regulating immune responses, CXCR1 and/or CXCR2 are also involved in stimulating vasculogenesis. CXC chemokines are divided into two groups that either contain or lack the ELR motif at their N-terminus. ELR-containing chemokines are angiogenic, as shown by their effects on endothelial cell proliferation and chemotaxis *in vitro* as well as angiogenesis in animal models, including the rat cornea and tumor regression *in vivo* (Koch *et al.*, 1992; Strieter *et al.*, 1995; Arenberg *et al.*, 1997). Chemokines that lack the ELR motif inhibit the mitogenic effects of angiogenic chemokines, as well as the major angiogenic mediators vascular endothelial growth factor (VEGF) and fibroblast growth factor (FGF). ELR-containing CXC chemokines are ligands for CXCR1 and CXCR2. It is unclear which of these receptors are expressed on endothelial cells, and therefore which are responsible for the angiogenic effects.

Bioassays

A variety of assays are available to measure the biological activities of chemokine receptors and their ligands. Ligand-binding assays are initially useful for analyzing receptor-binding affinities. Cells expressing receptors are incubated with ^{125}I-labeled ligand, as well as varying concentrations of unlabeled ligand to compete for binding sites. After incubation, the cells are washed and γ emissions are counted. Data are then analyzed using Scatchard analysis to determine K_d values (Scatchard, 1949). More recently, the binding of nonradioactive, recombinant His-tagged ligands can be visualized by flow cytometry using monoclonal antibodies against the His tag (Wilken *et al.*, 1999). Following receptor binding, chemokine receptor activation is classically associated with changes in intracellular calcium concentration (Murphy, 1994). Intracellular calcium flux can be readily detected by monitoring the response of fluorescent calcium-binding dyes. Compounds such as Fura-2, Indo-1, and Fluo-3 (Molecular Probes, Eugene, OR) show changes in their spectral characteristics (fluorescence intensity or excitation or emission spectra) upon calcium binding (reviewed in Kao, 1994; McColl and Naccache, 1997). As the spectral characteristics of the three probes are unique, each may be useful under distinct experimental conditions or with different recording equipment, such as fluorescence microscopy, flow cytometry, and fluorescence spectroscopy.

Several assays measure the ability of chemokine ligands to stimulate activated primary neutrophils, cells which express both CXCR1 and CXCR2 (Baly *et al.*, 1997). As part of the host response to invading pathogens, neutrophils migrate to the site of infection, where they undergo oxidative burst then the release of degradative granular enzymes. The oxidative burst can be monitored *in vitro* in response to chemokine ligands (Baly *et al.*, 1997). One method involves the detection of hydrogen peroxide (H_2O_2), generated during respiratory burst. Neutrophils are incubated with dichlorofluorescein diacetate (DCFH-DA), a stable, nonfluorescent, nonpolar compound, and the DCFH-DA diffuses into the cells where it is deacetylated by intracellular esterases. The resulting nonfluorescent compound, DCFH, is polar, thus trapped within the cell and serves as a substrate for hydrogen peroxide. During the oxidative burst, hydrogen peroxide catalyzes the generation of the polar and highly fluorescent 2′,7′-dichlorofluorescein (DCF), which can be readily measured by flow cytometry.

Additional assays detect the enzymes elastase and β-glucuronidase, both released from intracellular azurophilic storage granules upon stimulation of activated neutrophils (Baly *et al.*, 1997). Enzyme concentrations are determined by measuring their proteolytic activity on synthetic substrates, such as MeOSuc-Ala-Ala-Pro-Val-PNA for elastase and 4-methylumbelliferyl-β-D-glucuronide for β-glucuronidase. The two enzymes differ in the time course of their release, their assay stability, and their susceptibility to interfering substances in the test samples. Following stimulation with IL-8, elastase release continues for 3 hours or more, while the release of β-glucuronidase is complete after 15 minutes. One benefit to measuring β-glucuronidase release is that enzymatic activity can be readily stopped by diluting samples in a high-pH glycine solution. This stable end-point reduces the risk of assay drift when large numbers of samples are assayed. Lastly, unlike that of β-glucuronidase, elastase activity is extremely sensitive to inhibition by plasma at concentrations as low as 0.05% (v/v). A recent report describes a method for measuring granule enzyme release in a 96-well microplate format (Tiberghien *et al.*, 1999).

Cell migration in response to chemokine ligands is also studied *in vitro*. Boyden first described the measurement of chemotaxis using membrane filters in 1962. Since then, a variety of methods have been used to assay chemotaxis with membrane filters (reviewed in Frevert *et al.*, 1998; Wilkinson, 1998). New assays for chemotaxis utilize a rapid fluorescence-based 96-well plate format. In this technique, cells expressing chemokine receptors are labeled with a fluorescent probe, such as calcein-AM (Molecular Probes, Eugene, OR) and added above a porous filter in a disposable 96-well plate chemotaxis chamber (ChemoTx, Neuro Probe, Gaithersburg, MD). Chemokine ligand is added below the filter, and cell migration into the lower chamber is measured using total fluorescence as a marker for cell number. This fluorescence end-point assay is more rapid and more sensitive, and is a significant advance over traditional chemotaxis assays.

Phenotypes of receptor knockouts and receptor overexpression mice

Disruption of the murine CXCR2 gene demonstrates that this expression of this chemokine receptor is essential for normal myeloid and lymphoid development. CXCR2 knockout mice fail to mobilize neutrophils in response to intraperitoneal thioglycollate *in vivo* (Cacalano *et al.*, 1994). Neutrophils isolated from these animals also fail to chemotax in response to KC or MIP-2 *in vitro*, indicating that CXCR2 is the dominant neutrophil receptor for these chemokines (Lee *et al.*, 1995). When raised in a specific pathogen-free environment, knockout mice have a massive expansion of neutrophils and B cells in bone marrow, lymphoid tissue, and blood (Cacalano *et al.*, 1994). However, neutrophil counts are not increased when mice are raised in a germ-free environment. This may be due to the fact that CXCR2 is a negative regulator of hematopoiesis (Broxmeyer *et al.*, 1996). Alternatively, Cacalano *et al.* have speculated that the inability to survey tissues properly and completely to eliminate external pathogens in the knockouts may result in the release of cytokines that stimulate neutrophil and B cell production (Cacalano *et al.*, 1994). Regardless, this defect does not appear to alter susceptibility to environmental or challenge pathogens. Another unexpected observation is an elevation of IL-4 and IgE plasma levels in these CXCR2 knockout mice.

Human abnormalities

Individuals without functional CXCR1 or CXCR2 have not been identified. However, a recent study using FACS analysis suggests that both receptors are downregulated in neutrophils from patients suffering from *pulmonary tuberculosis* and in those seropositive for *HIV-1* (Meddows-Taylor *et al.*, 1998). The reduced expression of IL-8 receptors also resulted in an impairment of both chemotaxis and calcium flux in response to IL-8. Exposure of neutrophils to hypoxia/reoxygenation also caused a downregulation of CXCR1 expression at both the mRNA and protein level (Grutkoski *et al.*, 1999). CXCR2 surface expression is downregulated by 50% on neutrophils from patients with *sepsis* (Cummings *et al.*, 1999). While neutrophil chemotaxis to IL-8 and the bacterial peptide fMLP were unaffected, cells from septic patients had a markedly suppressed migratory response to the CXCR2 ligands, ENA-78 and GROα.

Increased expression of epidermal IL-8 receptors has been observed in *psoriasis* and other inflammatory and hyperproliferative skin diseases (reviewed in Kemeny *et al.*, 1994b). For example, RT-PCR and immunohistochemical techniques have detected CXCR1 and CXCR2 expression on fibroblasts and smooth muscle cells of *burn lesions* (Nanney *et al.*, 1995). Using immunohistochemistry and *in situ* hybridization techniques, CXCR2 but not CXCR1 was localized to suprabasal lesional psoriatic keratinocytes (Kulke *et al.*, 1998). Keratinocyte IL-8 receptors are not expressed in normal skin, and thus may be useful targets in the treatment of *inflammatory skin diseases* (Schulz *et al.*, 1993; Kemeny *et al.*, 1994a; Kulke *et al.*, 1998).

Chemokines and their receptors, including CXCR1 and CXCR2, may be involved in pathologies of the central nervous system (reviewed in Hesselgesser and Horuk, 1999b). CXCR2 is widely expressed in the normal central nervous system, and also in neuritic plaques surrounding amyloid deposits in *Alzheimer's disease* (Horuk *et al.*, 1997; Xia *et al.*, 1997; Xia and Hyman, 1999). As the receptor for GROα, CXCR2 may also mediate this chemokine's effects on *melanoma* cell growth. CXCR2 expression on various melanoma cell lines was demonstrated by RT-PCR as well as by flow cytometry using specific anti-CXCR2 antibodies (Norgauer *et al.*, 1996). CXCR2 may be involved in syncytium formation in *HIV* infection and the direct cell-to-cell transfer of virus. The envelope protein of a T cell line-adapted HIV-2 strain has been shown to utilize CXCR2 to induce fusion. Although less efficient than CXCR4 and CCR3, both also used by the well-characterized HIV-1 strain, fusion with CXCR2 expressing cells was specific and was inhibited by antibodies against CXCR2 (Bron *et al.*, 1997).

THERAPEUTIC UTILITY

Antibody neutralization studies have demonstrated that IL-8 is a key mediator of neutrophil-mediated acute inflammation in the rabbit (Sekido *et al.*, 1993; Broaddus *et al.*, 1994; Folkesson *et al.*, 1995).

However, the selectivity of the two known rabbit IL-8 receptors for different rabbit chemokines has not been defined, thus their relative importance in inflammatory models is also not yet understood. Potential disease targets include many neutrophil-mediated inflammatory processes such as *ischemia–reperfusion* injury, *adult respiratory distress syndrome*, and certain forms of *glomerulonephritis* and *dermatitis*. Recent evidence suggests that IL-8 may also play a role in *Alzheimer's disease*, as well as in the growth of *tumors* via its angiogenic effects. Because of ligand redundancy, chemokine receptors may be better targets for development of therapeutic blocking agents. However, it is still not known whether it will be required to block both receptor subtypes, or whether inhibition of just one will suffice. Regardless, the abnormal phenotype of CXCR2 knockout mice calls into question the safety of therapeutically blocking these receptors.

Effects of inhibitors (antibodies) to receptors

Selective neutralizing monoclonal and polyclonal antibodies against CXCR1 and CXCR2 have been described (Hammond *et al.*, 1995; Green *et al.*, 1996; Jones *et al.*, 1996). N-terminal truncations of IL-8 and GROα are also effective antagonists of CXCR2 signaling (Baggiolini *et al.*, 1997). Similarly, N-terminally modified analogs of IL-8 block IL-8-induced calcium mobilization in cells transfected with CXCR1 or CXCR2 (Jones *et al.*, 1997). Analogs of GROα and PF4 bound only CXCR2 with high affinity, and blocked calcium flux at this receptor alone. These two analogs had no effect on IL-8-elicited superoxide generation and release of granule enzymes in neutrophils, as these functional responses are induced via CXCR1 alone.

As such proteins have some limitations as therapeutics, the search for small molecule antagonists of chemokine receptors has been an active area of pharmaceutical research in the past few years. A potent and selective nonpeptide small molecule inhibitor of CXCR2 was recently described (Ponath, 1998; White *et al.*, 1998). SB 225002 (*N*-(2-hydroxy-4-nitrophenyl)-N'-(2-bromophenyl)urea) is an antagonist of [^{125}I]IL-8 binding to CXCR2 with an $IC_{50} = 22\,nM$ and 150-fold selectivity over CXCR1. SB 225002 potently inhibits the chemotaxis of both rabbit and human neutrophils to IL-8 and GROα *in vitro*. *In vivo*, this molecule selectively blocks IL-8-induced neutrophil migration in rabbits, suggesting that CXCR2 plays an important role in this process. Using receptor-specific antibodies, others have shown that CXCR1 is the key receptor involved in neutrophil chemotaxis *in vitro* (Quan *et al.*, 1996). Regardless, this compound will be useful in defining the role of CXCR2 in neutrophil-mediated inflammatory diseases.

References

Ahuja, S. K., and Murphy, P. M. (1996). The CXC chemokines growth-regulated oncogene (GRO) alpha, GRObeta, GRO-gamma, neutrophil-activating peptide-2, and epithelial cell-derived neutrophil-activating peptide-78 are potent agonists for the type B, but not the type A, human interleukin-8 receptor. *J. Biol. Chem.* **271**, 20545–20550.

Ahuja, S. K., Ozcelik, T., Milatovitch, A., Francke, U., and Murphy, P. M. (1992). Molecular evolution of the human interleukin-8 receptor gene cluster. *Nature Genet.* **2**, 31–36.

Ahuja, S. K., Shetty, A., Tiffany, H. L., and Murphy, P. M. (1994). Comparison of the genomic organization and promoter function for human interleukin-8 receptors A and B. *J. Biol. Chem.* **269**, 26381–26389.

Ahuja, S. K., Lee, J. C., and Murphy, P. M. (1996). CXC chemokines bind to unique sets of selectivity determinants that can function independently and are broadly distributed on multiple domains of human interleukin-8 receptor B. Determinants of high affinity binding and receptor activation are distinct. *J. Biol. Chem.* **271**, 225–232.

Alvarez, V., Coto, E., Setien, F., Gonzalez, S., Gonzalez-Roces, S., and Lopez-Larrea, C. (1996). Characterization of interleukin-8 receptors in non-human primates. *Immunogenetics* **43**, 261–267.

Arenberg, D. A., Polverini, P. J., Kunkel, S. L., Shanafelt, A., and Strieter, R. M. (1997). *In vitro* and *in vivo* systems to assess role of C–X–C chemokines in regulation of angiogenesis. *Methods Enzymol.* **288**, 190–220.

Asagoe, K., Yamamoto, K., Takahashi, A., Suzuki, K., Maeda, A., Nohgawa, M., Harakawa, N., Takano, K., Mukaida, N., Matsushima, K., Okuma, M., and Sasada, M. (1998). Down-regulation of CXCR2 expression on human polymorphonuclear leukocytes by TNF-α. *J. Immunol.* **160**, 4518–4525.

Bacon, K. B., Flores-Romo, L., Life, P. F., Taub, D. D., Premack, B. A., Arkinstall, S. J., Wells, T. N., Schall, T. J., and Power, C. A. (1995). IL-8-induced signal transduction in T lymphocytes involves receptor- mediated activation of phospholipases C and D. *J. Immunol.* **154**, 3654–3666.

Baggiolini, M., and Clark-Lewis, I. (1992). Interleukin-8, a chemotactic and inflammatory cytokine. *FEBS Lett.* **307**, 97–101.

Baggiolini, M., Boulay, F., Badwey, J. A., and Curnutte, J. T. (1993). Activation of neutrophil leukocytes: chemoattractant receptors and respiratory burst. *Faseb J.* **7**, 1004–1010.

Baggiolini, M., Dewald, B., and Moser, B. (1994). Interleukin-8 and related chemotactic cytokines – CXC and CC chemokines. *Adv. Immunol.* **55**, 97–179.

Baggiolini, M., Dewald, B., and Moser, B. (1997). Human chemokines: an update. *Annu. Rev. Immunol.* **15**, 675–705.

Balkwill, F. R., and Burke, F. (1989). The cytokine network. *Immunol. Today* **10**, 299–304.

Baly, D., Gibson, U., Allison, D., and DeForge, L. (1997). Biological assays for C–X–C chemokines. *Methods Enzymol.* **287**, 69–88.

Ben-Baruch, A., Bengali, K. M., Biragyn, A., Johnston, J. J., Wang, J. M., Kim, J., Chuntharapai, A., Michiel, D. F., Oppenheim, J. J., and Kelvin, D. J. (1995a). Interleukin-8 receptor β. The role of the carboxyl terminus in signal transduction. *J. Biol. Chem.* **270**, 9121–9128.

Ben-Baruch, A., Michiel, D. F., and Oppenheim, J. J. (1995b). Signals and receptors involved in recruitment of inflammatory cells. *J. Biol. Chem.* **270**, 11703–11706.

Ben-Baruch, A., Grimm, M., Bengali, K., Evans, G. A., Chertov, O., Wang, J. M., Howard, O. M., Mukaida, N., Matsushima, K., and Oppenheim, J. J. (1997). The differential ability of IL-8 and neutrophil-activating peptide-2 to induce attenuation of chemotaxis is mediated by their divergent capabilities to phosphorylate CXCR2 (IL-8 receptor B). *J. Immunol.* **158**, 5927–5933.

Besemer, J., Hujber, A., and Kuhn, B. (1989). Specific binding, internalization, and degradation of human neutrophil activating factor by human polymorphonuclear leukocytes. *J. Biol. Chem.* **264**, 17409–17415.

Biffl, W. L., Moore, E. E., Moore, F. A., Carl, V. S., Franciose, R. J., and Banerjee, A. (1995). Interleukin-8 increases endothelial permeability independent of neutrophils. *J. Trauma* **39**, 98–103.

Boulay, F., Tardif, M., Brouchon, L., and Vignais, P. (1990). Synthesis and use of a novel N-formyl peptide derivative to isolate a human N-formyl peptide receptor cDNA. *Biochem. Biophys. Res. Commun.* **168**, 1103–1109.

Bowman, E. P., Campbell, J. J., Druey, K. M., Scheschonka, A., Kehrl, J. H., and Butcher, E. C. (1998). Regulation of chemotactic and proadhesive responses to chemoattractant receptors by RGS (regulator of G-protein signaling) family members. *J. Biol. Chem.* **273**, 28040–28048.

Boyden, S. V. (1962). The chemotactic effect of mixtures of antibody and antigen on polymorphonuclear leucocytes. *J. Exp. Med.* **115**, 453.

Bozic, C. R., Gerard, N. P., von Uexkull-Guldenband, C., Kolakowski, L. F., Jr., Conklyn, M. J., Breslow, R., Showell, H. J., and Gerard, C. (1994). The murine interleukin 8 type B receptor homologue and its ligands. Expression and biological characterization. *J. Biol. Chem.* **269**, 29355–29358.

Broaddus, V. C., Boylan, A. M., Hoeffel, J. M., Kim, K. J., Sadick, M., Chuntharapai, A., and Hebert, C. A. (1994). Neutralization of IL-8 inhibits neutrophil influx in a rabbit model of endotoxin-induced pleurisy. *J. Immunol.* **152**, 2960–2967.

Bron, R., Klasse, P. J., Wilkinson, D., Clapham, P. R., Pelchen-Matthews, A., Power, C., Wells, T. N., Kim, J., Peiper, S. C., Hoxie, J. A., and Marsh, M. (1997). Promiscuous use of CC and CXC chemokine receptors in cell-to-cell fusion mediated by a human immunodeficiency virus type 2 envelope protein. *J. Virol.* **71**, 8405–8415.

Broxmeyer, H. E., Cooper, S., Cacalano, G., Hague, N. L., Bailish, E., and Moore, M. W. (1996). Involvement of interleukin (IL) 8 receptor in negative regulation of myeloid progenitor cells *in vivo*: evidence from mice lacking the murine IL-8 receptor homologue. *J. Exp. Med.* **184**, 1825–1832.

Cacalano, G., Lee, J., Kikly, K., Ryan, A. M., Pitts-Meek, S., Hultgren, B., Wood, W. I., and Moore, M. W. (1994). Neutrophil and B cell expansion in mice that lack the murine IL-8 receptor homolog [see comments] [published erratum appears in *Science* 1995; **270**, 365]. *Science* **265**, 682–684.

Cerretti, D. P., Kozlosky, C. J., Vanden Bos, T., Nelson, N., Gearing, D. P., and Beckmann, M. P. (1993a). Molecular characterization of receptors for human interleukin-8, GRO/melanoma growth-stimulatory activity and neutrophil activating peptide-2. *Mol. Immunol.* **30**, 359–367.

Cerretti, D. P., Nelson, N., Kozlosky, C. J., Morrissey, P. J., Copeland, N. G., Gilbert, D. J., Jenkins, N. A., Dosik, J. K., and Mock, B. A. (1993b). The murine homologue of the human interleukin-8 receptor type B maps near the Ity-Lsh-Bcg disease resistance locus. *Genomics* **18**, 410–413.

Chuntharapai, A., and Kim, K. J. (1995). Regulation of the expression of IL-8 receptor A/B by IL-8: possible functions of each receptor. *J. Immunol.* **155**, 2587–2594.

Chuntharapai, A., Lee, J., Burnier, J., Wood, W. I., Hebert, C., and Kim, K. J. (1994a). Neutralizing monoclonal antibodies to human IL-8 receptor A map to the NH_2-terminal region of the receptor. *J. Immunol.* **152**, 1783–1789.

Chuntharapai, A., Lee, J., Hebert, C. A., and Kim, K. J. (1994b). Monoclonal antibodies detect different distribution patterns of IL-8 receptor A and IL-8 receptor B on human peripheral blood leukocytes. *J. Immunol.* **153**, 5682–5688.

Clubb, R. T., Omichinski, J. G., Clore, G. M., and Gronenborn, A. M. (1994). Mapping the binding surface of interleukin-8 complexes with an N-terminal fragment of the type 1 human interleukin-8 receptor. *FEBS Lett.* **338**, 93–97.

Coats Jr., W. D., and Navarro, J. (1990). Functional reconstitution of fMet-Leu-Phe receptor in *Xenopus laevis* oocytes. *J. Biol. Chem.* **265**, 5964–5966.

Cummings, C. J., Martin, T. R., Frevert, C. W., Quan, J. M., Wong, V. A., Mongovin, S. M., Hagen, T. R., Steinberg, K. P., and Goodman, R. B. (1999). Expression and function of the chemokine receptors CXCR1 and CXCR2 in sepsis. *J. Immunol.* **162**, 2341–2346.

Damaj, B. B., McColl, S. R., Mahana, W., Crouch, M. F., and Naccache, P. H. (1996a). Physical association of Gi2α with interleukin-8 receptors. *J. Biol. Chem.* **271**, 12783–12789.

Damaj, B. B., McColl, S. R., Neote, K., Songqing, N., Ogborn, K. T., Hebert, C. A., and Naccache, P. H. (1996b). Identification of G-protein binding sites of the human interleukin-8 receptors by functional mapping of the intracellular loops. *FASEB J.* **10**, 1426–1434.

Dunstan, C. A. N., Salafranca, M. N., Adhikari, S., Xia, Y., Feng, L., and Harrison, J. K. (1996). Identification of two rat genes orthologous to the human interleukin-8 receptors. *J. Biol. Chem.* **271**, 32770–32776.

Folkesson, H. G., Matthay, M. A., Hebert, C. A., and Broaddus, V. C. (1995). Acid aspiration-induced lung injury in rabbits is mediated by interleukin-8-dependent mechanisms. *J. Clin. Invest.* **96**, 107–116.

Frevert, C. W., Wong, V. A., Goodman, R. B., Goodwin, R., and Martin, T. R. (1998). Rapid fluorescence-based measurement of neutrophil migration *in vitro*. *J. Immunol. Methods* **213**, 41–52.

Gayle, R. B. D., Sleath, P. R., Srinivason, S., Birks, C. W., Weerawarna, K. S., Cerretti, D. P., Kozlosky, C. J., Nelson, N., Vanden Bos, T., and Beckmann, M. P. (1993). Importance of the amino terminus of the interleukin-8 receptor in ligand interactions. *J. Biol. Chem.* **268**, 7283–7289.

Geiser, T., Dewald, B., Ehrengruber, M. U., Clark-Lewis, I., and Baggiolini, M. (1993). The interleukin-8-related chemotactic cytokines GRO α, GRO β, and GRO γ activate human neutrophil and basophil leukocytes. *J. Biol. Chem.* **268**, 15419–15424.

Gerszten, R. E., Garcia-Zepeda, E. A., Lim, Y. C., Yoshida, M., Ding, H. A., Gimbrone, M. A., Jr., Luster, A. D., Luscinskas, F. W., and Rosenzweig, A. (1999). MCP-1 and IL-8 trigger firm adhesion of monocytes to vascular endothelium under flow conditions. *Nature* **398**, 718–723.

Gilman, A. G. (1987). G proteins: transducers of receptor-generated signals. *Annu. Rev. Biochem.* **56**, 615–649.

Gobl, A. E., Huang, M. R., Wang, S., Zhou, Y., and Oberg, K. (1997). Molecular cloning and characterization of a cDNA encoding the rat interleukin-8 receptor. *Biochim. Biophys. Acta* **1326**, 171–177.

Green, S. P., Chuntharapai, A., and Curnutte, J. T. (1996). Interleukin-8 (IL-8), melanoma growth-stimulatory activity, and neutrophil-activating peptide selectively mediate priming of the neutrophil NADPH oxidase through the type A or type B IL-8 receptor. *J. Biol. Chem.* **271**, 25400–25405.

Grinstein, S., and Furuya, W. (1992). Chemoattractant-induced tyrosine phosphorylation and activation of microtubule-associated protein kinase in human neutrophils. *J. Biol. Chem.* **267**, 18122–18125.

Grinstein, S., Butler, J. R., Furuya, W., L'Allemain, G., and Downey, G. P. (1994). Chemotactic peptides induce phosphorylation and activation of MEK-1 in human neutrophils. *J. Biol. Chem.* **269**, 19313–19320.

Grob, P. M., David, E., Warren, T. C., DeLeon, R. P., Farina, P. R., and Homon, C. A. (1990). Characterization of a receptor for human monocyte-derived neutrophil chemotactic factor/interleukin-8. *J. Biol. Chem.* **265**, 8311–8316.

Grutkoski, P. S., Graeber, C. T., D'Amico, R., Keeping, H., and Simms, H. H. (1999). Regulation of IL-8RA (CXCR1) expression in polymorphonuclear leukocytes by hypoxia/reoxygenation. *J. Leukoc. Biol.* **65**, 171–178.

Hammond, M. E., Lapointe, G. R., Feucht, P. H., Hilt, S., Gallegos, C. A., Gordon, C. A., Giedlin, M. A., Mullenbach, G., and Tekamp-Olson, P. (1995). IL-8 induces neutrophil chemotaxis predominantly via type I IL-8 receptors. *J. Immunol.* **155**, 1428–1433.

Harada, A., Kuno, K., Nomura, H., Mukaida, N., Murakami, S., and Matsushima, K. (1994). Cloning of a cDNA encoding a mouse homolog of the interleukin-8 receptor. *Gene* **142**, 297–300.

Hebert, C. A., Vitangcol, R. V., and Baker, J. B. (1991). Scanning mutagenesis of interleukin-8 identifies a cluster of residues required for receptor binding. *J. Biol. Chem.* **266**, 18989–18994.

Hebert, C. A., Chuntharapai, A., Smith, M., Colby, T., Kim, J., and Horuk, R. (1993). Partial functional mapping of the human interleukin-8 type A receptor. Identification of a major ligand binding domain [published erratum appears in *J. Biol. Chem.* 1994, **269**, 16520]. *J. Biol. Chem.* **268**, 18549–18553.

Hesselgesser, J., and Horuk, R. (1999a). Chemokine and chemokine receptor expression in the central nervous system. *J. Neurovirol.* **5**, 13–26.

Hesselgesser, J., and Horuk, R. (1999b). In "Chemokines in Disease: Biology and Clinical Research" (ed C. A. Hebert), Chemokines and chemokine receptors in the brain, pp. 295–312. Humana Press, Totowa, NJ.

Hoch, R. C., Schraufstatter, I. U., and Cochrane, C. G. (1996). *In vivo*, *in vitro*, and molecular aspects of interleukin-8 and the interleukin-8 receptors. *J. Lab. Clin. Med.* **128**, 134–145.

Holmes, W. E., Lee, J., Kuang, W. J., Rice, G. C., and Wood, W. I. (1991). Structure and functional expression of a human interleukin-8 receptor. *Science* **253**, 1278–1280.

Horuk, R. (1994). The interleukin-8-receptor family: from chemokines to malaria. *Immunol. Today* **15**, 169–174.

Horuk, R., Martin, A. W., Wang, Z., Schweitzer, L., Gerassimides, A., Guo, H., Lu, Z., Hesselgesser, J., Perez, H. D., Kim, J., Parker, J., Hadley, T. J., and Peiper, S. C. (1997). Expression of chemokine receptors by subsets of neurons in the central nervous system. *J. Immunol.* **158**, 2882–2890.

Howard, O. M., Ben-Baruch, A., and Oppenheim, J. J. (1996). Chemokines: progress toward identifying molecular targets for therapeutic agents. *Trends Biotechnol.* **14**, 46–51.

Jawa, R. S., Quaid, G. A., Williams, M. A., Cave, C. M., Robinson, C. T., Babcock, G. F., Lieberman, M. A., Witt, D., and Solomkin, J. S. (1999). Tumor necrosis factor α regulates CXC chemokine receptor expression and function. *Shock* **11**, 385–390.

Jones, S. A., Moser, B., and Thelen, M. (1995). A comparison of post-receptor signal transduction events in Jurkat cells transfected with either IL-8R1 or IL-8R2. Chemokine mediated activation of p42/p44 MAP-kinase (ERK-2). *FEBS Lett.* **364**, 211–214.

Jones, S. A., Wolf, M., Qin, S., Mackay, C. R., and Baggiolini, M. (1996). Different functions for the interleukin 8 receptors (IL-8R) of human neutrophil leukocytes: NADPH oxidase and phospholipase D are activated through IL-8R1 but not IL-8R2. *Proc. Natl Acad. Sci. U.S.A.* **93**, 6682–6686.

Jones, S. A., Dewald, B., Clark-Lewis, I., and Baggiolini, M. (1997). Chemokine antagonists that discriminate between interleukin-8 receptors. Selective blockers of CXCR2. *J. Biol. Chem.* **272**, 16166–16169.

Kao, J. P. (1994). Practical aspects of measuring [Ca^{2+}] with fluorescent indicators. *Methods Cell Biol.* **40**, 155–181.

Katancik, J. A., Sharma, A., Radel, S. J., and De Nardin, E. (1997). Mapping of the extracellular binding regions of the human interleukin-8 type B receptor. *Biochem. Biophys. Res. Commun.* **232**, 663–668.

Kelvin, D. J., Michiel, D. F., Johnston, J. A., Lloyd, A. R., Sprenger, H., Oppenheim, J. J., and Wang, J. M. (1993). Chemokines and serpentines: the molecular biology of chemokine receptors. *J. Leukoc. Biol.* **54**, 604–612.

Kemeny, L., Kenderessy, A. S., Olasz, E., Michel, G., Ruzicka, T., Farkas, B., and Dobozy, A. (1994a). The interleukin-8 receptor: a potential target for antipsoriatic therapy? *Eur. J. Pharmacol.* **258**, 269–272.

Kemeny, L., Ruzicka, T., Dobozy, A., and Michel, G. (1994b). Role of interleukin-8 receptor in skin. *Int. Arch. Allergy Immunol.* **104**, 317–322.

Khandaker, M. H., Xu, L., Rahimpour, R., Mitchell, G., DeVries, M. E., Pickering, J. G., Singhal, S. K., Feldman, R. D., and Kelvin, D. J. (1998). CXCR1 and CXCR2 are rapidly down-modulated by bacterial endotoxin through a unique agonist-independent, tyrosine kinase-dependent mechanism. *J. Immunol.* **161**, 1930–1938.

Knall, C., Young, S., Nick, J. A., Buhl, A. M., Worthen, G. S., and Johnson, G. L. (1996). Interleukin-8 regulation of the Ras/Raf/mitogen-activated protein kinase pathway in human neutrophils. *J. Biol. Chem.* **271**, 2832–2838.

Knall, C., Worthen, G. S., and Johnson, G. L. (1997). Interleukin 8-stimulated phosphatidylinositol-3-kinase activity regulates the migration of human neutrophils independent of extracellular signal-regulated kinase and p38 mitogen-activated protein kinases. *Proc. Natl Acad. Sci. U.S.A.* **94**, 3052–3057.

Koch, A. E., Polverini, P. J., Kunkel, S. L., Harlow, L. A., DiPietro, L. A., Elner, V. M., Elner, S. G., and Strieter, R. M. (1992). Interleukin-8 as a macrophage-derived mediator of angiogenesis [see comments]. *Science* **258**, 1798–1801.

Koelle, M. R. (1997). A new family of G-protein regulators – the RGS proteins. *Curr. Opin. Cell Biol.* **9**, 143–147.

Krieger, M., Brunner, T., Bischoff, S. C., von Tscharner, V., Walz, A., Moser, B., Baggiolini, M., and Dahinden, C. A. (1992). Activation of human basophils through the IL-8 receptor. *J. Immunol.* **149**, 2662–2667.

Kulke, R., Bornscheuer, E., Schluter, C., Bartels, J., Rowert, J., Sticherling, M., and Christophers, E. (1998). The CXC receptor 2 is overexpressed in psoriatic epidermis. *J. Invest. Dermatol.* **110**, 90–94.

Kupper, R. W., Dewald, B., Jakobs, K. H., Baggiolini, M., and Gierschik, P. (1992). G-protein activation by interleukin 8 and related cytokines in human neutrophil plasma membranes. *Biochem. J.* **282**, 429–434.

LaRosa, G. J., Thomas, K. M., Kaufmann, M. E., Mark, R., White, M., Taylor, L., Gray, G., Witt, D., and Navarro, J. (1992). Amino terminus of the interleukin-8 receptor is a major determinant of receptor subtype specificity. *J. Biol. Chem.* **267**, 25402–25406.

Laudanna, C., Campbell, J. J., and Butcher, E. C. (1996). Role of Rho in chemoattractant-activated leukocyte adhesion through integrins. *Science* **271**, 981–983.

Lee, J., Cacalano, G., Camerato, T., Toy, K., Moore, M. W., and Wood, W. I. (1995). Chemokine binding and activities mediated by the mouse IL-8 receptor. *J. Immunol.* **155**, 2158–2164.

Lefkowitz, R. J. (1998). G protein-coupled receptors. III. New roles for receptor kinases and β-arrestins in receptor signaling and desensitization. *J. Biol. Chem.* **273**, 18677–18680.

Lefkowitz, R. J., Hausdorff, W. P., and Caron, M. G. (1990). Role of phosphorylation in desensitization of the β-adrenoceptor. *Trends Pharmacol. Sci.* **11**, 190–194.

Leong, S. R., Kabakoff, R. C., and Hebert, C. A. (1994). Complete mutagenesis of the extracellular domain of interleukin-8 (IL- 8) type A receptor identifies charged residues mediating IL-8 binding and signal transduction. *J. Biol. Chem.* **269**, 19343–19348.

L'Heureux, G. P., Bourgoin, S., Jean, N., McColl, S. R., and Naccache, P. H. (1995). Diverging signal transduction pathways activated by interleukin-8 and related chemokines in human neutrophils: interleukin-8, but not NAP-2 or GRO α, stimulates phospholipase D activity. *Blood* **85**, 522–531.

Lippert, U., Artuc, M., Grutzkau, A., Moller, A., Kenderessy-Szabo, A., Schadendorf, D., Norgauer, J., Hartmann, K., Schweitzer-Stenner, R., Zuberbier, T., Henz, B. M., and Kruger-Krasagakes, S. (1998). Expression and functional activity of the IL-8 receptor type CXCR1 and CXCR2 on human mast cells. *J. Immunol.* **161**, 2600–2608.

Lloyd, A., Modi, W., Sprenger, H., Cevario, S., Oppenheim, J., and Kelvin, D. (1993). Assignment of genes for interleukin-8 receptors (IL8R) A and B to human chromosome band 2q35. *Cytogenet. Cell Genet.* **63**, 238–240.

Lloyd, A. R., Biragyn, A., Johnston, J. A., Taub, D. D., Xu, L., Michiel, D., Sprenger, H., Oppenheim, J. J., and Kelvin, D. J. (1995). Granulocyte-colony stimulating factor and lipopolysaccharide regulate the expression of interleukin 8 receptors on polymorphonuclear leukocytes. *J. Biol. Chem.* **270**, 28188–28192.

Locati, M., and Murphy, P. M. (1999). Chemokines and chemokine receptors: biology and clinical relevance in inflammation and AIDS. *Annu. Rev. Med.* **50**, 425–440.

Loetscher, P., Seitz, M., Clark-Lewis, I., Baggiolini, M., and Moser, B. (1994). Both interleukin-8 receptors independently mediate chemotaxis. Jurkat cells transfected with IL-8R1 or IL-8R2 migrate in response to IL-8, GRO α and NAP-2. *FEBS Lett.* **341**, 187–192.

Lomize, A. L., Pogozheva, I. D., and Mosberg, H. I. (1999). Structural organization of G-protein-coupled receptors. *J. Comput. Aided Mol. Des.* **13**, 325–353.

Luo, Z., Butcher, D. J., and Huang, Z. (1997). Molecular modeling of interleukin-8 receptor β and analysis of the receptor-ligand interaction. *Protein Eng.* **10**, 1039–1045.

Manna, S. K., and Samanta, A. K. (1995). Upregulation of interleukin-8 receptor in human polymorphonuclear neutrophils by formyl peptide and lipopolysaccharide. *FEBS Lett.* **367**, 117–121.

Manna, S. K., Bhattacharya, C., Gupta, S. K., and Samanta, A. K. (1995). Regulation of interleukin-8 receptor expression in human polymorphonuclear neutrophils. *Mol. Immunol.* **32**, 883–893.

McColl, S. R., and Naccache, P. H. (1997). Calcium mobilization assays. *Methods Enzymol.* **288**, 301–309.

Meddows-Taylor, S., Martin, D. J., and Tiemessen, C. T. (1998). Reduced expression of interleukin-8 receptors A and B on polymorphonuclear neutrophils from persons with human immunodeficiency virus type 1 disease and pulmonary tuberculosis. *J. Infect. Dis.* **177**, 921–930.

Mennicken, F., Maki, R., de Souza, E. B., and Quirion, R. (1999). Chemokines and chemokine receptors in the CNS: a possible role in neuroinflammation and patterning. *Trends Pharmacol. Sci.* **20**, 73–78.

Mollereau, C., Muscatelli, F., Mattei, M. G., Vassart, G., and Parmentier, M. (1993). The high-affinity interleukin 8 receptor gene (IL8RA) maps to the 2q33–q36 region of the human genome: cloning of a pseudogene (IL8RBP) for the low-affinity receptor. *Genomics* **16**, 248–251.

Morris, S. W., Nelson, N., Valentine, M. B., Shapiro, D. N., Look, A. T., Kozlosky, C. J., Beckmann, M. P., and Cerretti, D. P. (1992). Assignment of the genes encoding human interleukin-8 receptor types 1 and 2 and an interleukin-8 receptor pseudogene to chromosome 2q35. *Genomics* **14**, 685–691.

Moser, B., Schumacher, C., von Tscharner, V., Clark-Lewis, I., and Baggiolini, M. (1991). Neutrophil-activating peptide 2 and gro/melanoma growth-stimulatory activity interact with neutrophil-activating peptide 1/interleukin 8 receptors on human neutrophils. *J. Biol. Chem.* **266**, 10666–10671.

Moser, B., Barella, L., Mattei, S., Schumacher, C., Boulay, F., Colombo, M. P., and Baggiolini, M. (1993). Expression of transcripts for two interleukin 8 receptors in human phagocytes, lymphocytes and melanoma cells. *Biochem. J.* **294**, 285–292.

Mueller, S. G., Schraw, W. P., and Richmond, A. (1994). Melanoma growth stimulatory activity enhances the phosphorylation of the class II interleukin-8 receptor in non-hematopoietic cells. *J. Biol. Chem.* **269**, 1973–1980.

Murdoch, C., Monk, P. N., and Finn, A. (1999). Cxc chemokine receptor expression on human endothelial cells. *Cytokine* **11**, 704–712.

Murphy, P. M. (1994). The molecular biology of leukocyte chemoattractant receptors. *Annu. Rev. Immunol.* **12**, 593–633.

Murphy, P. M. (1997). Neutrophil receptors for interleukin-8 and related CXC chemokines. *Semin. Hematol.* **34**, 311–318.

Murphy, P. M., and Tiffany, H. L. (1991). Cloning of complementary DNA encoding a functional human interleukin-8 receptor. *Science* **253**, 1280–1283.

Murphy, P. M., Gallin, E. K., Tiffany, H. L., and Malech, H. L. (1990). The formyl peptide chemoattractant receptor is encoded by a 2 kilobase messenger RNA. Expression in *Xenopus* oocytes. *FEBS Lett.* **261**, 353–357.

Nanney, L. B., Mueller, S. G., Bueno, R., Peiper, S. C., and Richmond, A. (1995). Distributions of melanoma growth stimulatory activity of growth-regulated gene and the interleukin-8 receptor B in human wound repair. *Am. J. Pathol.* **147**, 1248–1260.

Neptune, E. R., Iiri, T., and Bourne, H. R. (1999). Galphai is not required for chemotaxis mediated by Gi-coupled receptors [published erratum appears in *J. Biol. Chem.* 1999 **274**, 7598]. *J. Biol. Chem.* **274**, 2824–2828.

Nilsson, G., Mikovits, J. A., Metcalfe, D. D., and Taub, D. D. (1999). Mast cell migratory response to interleukin-8 is mediated through interaction with chemokine receptor CXCR2/interleukin-8RB. *Blood* **93**, 2791–2797.

Norgauer, J., Metzner, B., and Schraufstatter, I. (1996). Expression and growth-promoting function of the IL-8 receptor β in human melanoma cells. *J. Immunol.* **156**, 1132–1137.

Ochensberger, B., Tassera, L., Bifrare, D., Rihs, S., and Dahinden, C. A. (1999). Regulation of cytokine expression and leukotriene formation in human basophils by growth factors, chemokines and chemotactic agonists. *Eur. J. Immunol.* **29**, 11–22.

O'Dowd, B. F., Lefkowitz, R. J., and Caron, M. G. (1989). Structure of the adrenergic and related receptors. *Annu. Rev. Neurosci.* **12**, 67–83.

Oin, S., LaRosa, G., Campbell, J. J., Smith-Heath, H., Kassam, N., Shi, X., Zeng, L., Buthcher, E. C., and Mackay, C. R. (1996). Expression of monocyte chemoattractant protein-1 and interleukin-8 receptors on subsets of T cells: correlation with transendothelial chemotactic potential. *Eur. J. Immunol.* **26**, 640–647.

Palczewski, K., and Benovic, J. L. (1991). G-protein-coupled receptor kinases. *Trends Biochem. Sci.* **16**, 387–391.

Petering, H., Gotze, O., Kimmig, D., Smolarski, R., Kapp, A., and Elsner, J. (1999). The biologic role of interleukin-8: functional analysis and expression of CXCR1 and CXCR2 on human eosinophils. *Blood* **93**, 694–702.

Petzelbauer, P., Watson, C. A., Pfau, S. E., and Pober, J. S. (1995). IL-8 and angiogenesis: evidence that human endothelial cells lack receptors and do not respond to IL-8 *in vitro*. *Cytokine* **7**, 267–272.

Peveri, P., Walz, A., Dewald, B., and Baggiolini, M. (1988). A novel neutrophil-activating factor produced by human mononuclear phagocytes. *J. Exp. Med.* **167**, 1547–1559.

Ponath, P. D. (1998). Chemokine receptor antagonists: novel therapeutics for inflammation and AIDS. *Exp. Opin. Invest. Drugs* **7**, 1–18.

Prado, G. N., Thomas, K. M., Suzuki, H., LaRosa, G. J., Wilkinson, N., Folco, E., and Navarro, J. (1994). Molecular characterization of a novel rabbit interleukin-8 receptor isotype. *J. Biol. Chem.* **269**, 12391–12394.

Premack, B. A., and Schall, T. J. (1996). Chemokine receptors: gateways to inflammation and infection. *Nat. Med.* **2**, 1174–1178.

Probst, W. C., Snyder, L. A., Schuster, D. I., Brosius, J., and Sealfon, S. C. (1992). Sequence alignment of the G-protein coupled receptor superfamily. *DNA Cell Biol.* **11**, 1–20.

Qualliotine-Mann, D., Agwu, D. E., Ellenburg, M. D., McCall, C. E., and McPhail, L. C. (1993). Phosphatidic acid and diacylglycerol synergize in a cell-free system for activation of NADPH oxidase from human neutrophils. *J. Biol. Chem.* **268**, 23843–23849.

Quan, J. M., Martin, T. R., Rosenberg, G. B., Foster, D. C., Whitmore, T., and Goodman, R. B. (1996). Antibodies against the N-terminus of IL-8 receptor A inhibit neutrophil chemotaxis. *Biochem. Biophys. Res. Commun.* **219**, 405–411.

Richard, S., Farrell, C. A., Shaw, A. S., Showell, H. J., and Connelly, P. A. (1994). C5a as a model for chemotactic factor-stimulated tyrosine phosphorylation in the human neutrophil. *J. Immunol.* **152**, 2479–2487.

Richardson, R. M., DuBose, R. A., Ali, H., Tomhave, E. D., Haribabu, B., and Snyderman, R. (1995). Regulation of human interleukin-8 receptor A: identification of a phosphorylation site involved in modulating receptor functions. *Biochemistry* **34**, 14193–14201.

Richardson, R. M., Ali, H., Pridgen, B. C., Haribabu, B., and Snyderman, R. (1998). Multiple signaling pathways of human interleukin-8 receptor A. Independent regulation by phosphorylation. *J. Biol. Chem.* **273**, 10690–10695.

Rot, A., Hub, E., Middleton, J., Pons, F., Rabeck, C., Thierer, K., Wintle, J., Wolff, B., Zsak, M., and Dukor, P. (1996). Some aspects of IL-8 pathophysiology. III: Chemokine interaction with endothelial cells. *J. Leukoc. Biol.* **59**, 39–44.

Sabroe, I., Williams, T. J., Hebert, C. A., and Collins, P. D. (1997). Chemoattractant cross-desensitization of the human neutrophil IL-8 receptor involves receptor internalization and differential receptor subtype regulation. *J. Immunol.* **158**, 1361–1369.

Samanta, A. K., Oppenheim, J. J., and Matsushima, K. (1989). Identification and characterization of specific receptors for monocyte-derived neutrophil chemotactic factor (MDNCF) on human neutrophils. *J. Exp. Med.* **169**, 1185–1189.

Samanta, A. K., Oppenheim, J. J., and Matsushima, K. (1990). Interleukin 8 (monocyte-derived neutrophil chemotactic factor) dynamically regulates its own receptor expression on human neutrophils. *J. Biol. Chem.* **265**, 183–189.

Scatchard, G. (1949). The attractions of proteins for small molecules and ions. *Ann. NY Acad. Sci.* **51**, 660–672.

Schonbeck, U., Brandt, E., Petersen, F., Flad, H. D., and Loppnow, H. (1995). IL-8 specifically binds to endothelial but not to smooth muscle cells. *J. Immunol.* **154**, 2375–2383.

Schraufstatter, I. U., Burger, M., Hoch, R. C., Oades, Z. G., and Takamori, H. (1998). Importance of the carboxy-terminus of the CXCR2 for signal transduction. *Biochem. Biophys. Res. Commun.* **244**, 243–248.

Schroder, J. M., Mrowietz, U., Morita, E., and Christophers, E. (1987). Purification and partial biochemical characterization of a human monocyte-derived, neutrophil-activating peptide that lacks interleukin 1 activity. *J. Immunol.* **139**, 3474–3483.

Schulz, B. S., Michel, G., Wagner, S., Suss, R., Beetz, A., Peter, R. U., Kemeny, L., and Ruzicka, T. (1993). Increased expression of epidermal IL-8 receptor in psoriasis. Down-regulation by FK-506 *in vitro*. *J. Immunol.* **151**, 4399–4406.

Sekido, N., Mukaida, N., Harada, A., Nakanishi, I., Watanabe, Y., and Matsushima, K. (1993). Prevention of lung reperfusion injury in rabbits by a monoclonal antibody against interleukin-8. *Nature* **365**, 654–657.

Sha'afi, R. I., Shefcyk, J., Yassin, R., Molski, T. F., Volpi, M., Naccache, P. H., White, J. R., and Feinstein, M. B, Becker, E. L. (1986). Is a rise in intracellular concentration of free calcium necessary or sufficient for stimulated cytoskeletal-associated actin? *J. Cell Biol.* **102**, 1459–1463.

Sham, R. L., Phatak, P. D., Ihne, T. P., Abboud, C. N., and Packman, C. H. (1993). Signal pathway regulation of interleukin-8-induced actin polymerization in neutrophils. *Blood* **82**, 2546–2551.

Shyamala, V., and Khoja, H. (1998). Interleukin-8 receptors R1 and R2 activate mitogen-activated protein kinases and induce c-fos, independent of Ras and Raf-1 in Chinese hamster ovary cells. *Biochemistry* **37**, 15918–15924.

Sozzani, S., Agwu, D. E., Ellenburg, M. D., Locati, M., Rieppi, M., Rojas, A., Mantovani, A., and McPhail, L. C. (1994a). Activation of phospholipase D by interleukin-8 in human neutrophils. *Blood* **84**, 3895–3901.

Sozzani, S., Zhou, D., Locati, M., Rieppi, M., Proost, P., Magazin, M., Vita, N., van Damme, J., and Mantovani, A. (1994b). Receptors and transduction pathways for monocyte chemotactic protein-2 and monocyte chemotactic protein-3. Similarities and differences with MCP-1. *J. Immunol.* **152**, 3615–3622.

Sprenger, H., Lloyd, A. R., and Kelvin, D. J. (1995). Promoter analysis of the human interleukin-8 receptor genes, IL-8RA and IL-8RB. *Immunobiology* **193**, 334–340.

Strader, C. D., Fong, T. M., Tota, M. R., Underwood, D., and Dixon, R. A. (1994). Structure and function of G protein-coupled receptors. *Annu. Rev. Biochem.* **63**, 101–132.

Strieter, R. M., Polverini, P. J., Arenberg, D. A., Walz, A., Opdenakker, G., Van Damme, J., and Kunkel, S. L. (1995). Role of C–X–C chemokines as regulators of angiogenesis in lung cancer. *J. Leukoc. Biol.* **57**, 752–762.

Strosberg, A. D. (1991). Structure/function relationship of proteins belonging to the family of receptors coupled to GTP-binding proteins. *Eur. J. Biochem.* **196**, 1–10.

Suzuki, H., Prado, G. N., Wilkinson, N., and Navarro, J. (1994). The N terminus of interleukin-8 (IL-8) receptor confers high affinity binding to human IL-8. *J. Biol. Chem.* **269**, 18263–18266.

Thelen, M., Peveri, P., Kernen, P., von Tscharner, V., Walz, A., and Baggiolini, M. (1988). Mechanism of neutrophil activation by NAF, a novel monocyte-derived peptide agonist. *Faseb J.* **2**, 2702–2706.

Thomas, K. M., Pyun, H. Y., and Navarro, J. (1990). Molecular cloning of the fMet-Leu-Phe receptor from neutrophils [published erratum appears in *J. Biol. Chem.* 1992, **267**, 13780]. *J. Biol. Chem.* **265**, 20061–20064.

Thomas, K. M., Taylor, L., and Navarro, J. (1991). The interleukin-8 receptor is encoded by a neutrophil-specific cDNA clone, F3R. *J. Biol. Chem.* **266**, 14839–14841.

Thomas, K. M., Taylor, L., Prado, G., Romero, J., Moser, B., Car, B., Walz, A., Baggiolini, M., and Navarro, J. (1994). Functional and ligand binding specificity of the rabbit neutrophil IL-8 receptor. *J. Immunol.* **152**, 2496–2500.

Thompson, H. L., Marshall, C. J., and Saklatvala, J. (1994). Characterization of two different forms of mitogen-activated protein kinase kinase induced in polymorphonuclear leukocytes following stimulation by N-formylmethionyl-leucyl-phenylalanine or granulocyte-macrophage colony-stimulating factor. *J. Biol. Chem.* **269**, 9486–9492.

Tiberghien, F., Didier, A., Bohbot, A., and Loor, F. (1999). The MultiScreen filtration system to measure chemoattractant-induced release of leukocyte granule enzymes by differentiated HL-60 cells or normal human monocytes. *J. Immunol. Methods* **223**, 63–75.

Unger, V. M., Hargrave, P. A., Baldwin, J. M., and Schertler, G. F. (1997). Arrangement of rhodopsin transmembrane α-helices. *Nature* **389**, 203–206.

Van Damme, J., Wuyts, A., Froyen, G., Van Coillie, E., Struyf, S., Billiau, A., Proost, P., Wang, J. M., and Opdenakker, G. (1997). Granulocyte chemotactic protein-2 and related CXC chemokines: from gene regulation to receptor usage. *J. Leukoc. Biol.* **62**, 563–569.

Wakasugi, K., and Schimmel, P. (1999). Highly differentiated motifs responsible for two cytokine activities of a split human tRNA synthetase. *J. Biol. Chem.* **274**, 23155–23159.

Walz, A., Peveri, P., Aschauer, H., and Baggiolini, M. (1987). Purification and amino acid sequencing of NAF, a novel neutrophil-activating factor produced by monocytes. *Biochem. Biophys. Res. Commun.* **149**, 755–761.

Wen, D. Z., Rowland, A., and Derynck, R. (1989). Expression and secretion of gro/MGSA by stimulated human endothelial cells. *EMBO J.* **8**, 1761–1766.

White, J. R., Lee, J. M., Young, P. R., Hertzberg, R. P., Jurewicz, A. J., Chaikin, M. A., Widdowson, K., Foley, J. J., Martin, L. D., Griswold, D. E., and Sarau, H. M. (1998). Identification of a potent, selective non-peptide CXCR2 antagonist that inhibits interleukin-8-induced neutrophil migration. *J. Biol. Chem.* **273**, 10095–10098.

Wilken, H. C., Rogge, S., Gotze, O., Werfel, T., and Zwirner, J. (1999). Specific detection by flow cytometry of histidine-tagged ligands bound to their receptors using a tag-specific monoclonal antibody. *J. Immunol. Methods* **226**, 139–145.

Wilkinson, P. C. (1998). Assays of leukocyte locomotion and chemotaxis. *J. Immunol. Methods* **216**, 139–153.

Wilkinson, N. C., and Navarro, J. (1999). PU. 1 regulates the CXCR1 promoter. *J. Biol. Chem.* **274**, 438–443.

Wolf, M., Delgado, M. B., Jones, S. A., Dewald, B., Clark-Lewis, I., and Baggiolini, M. (1998). Granulocyte chemotactic protein 2 acts via both IL-8 receptors, CXCR1 and CXCR2. *Eur J. Immunol.* **28**, 164–170.

Wu, D., LaRosa, G. J., and Simon, M. I. (1993). G protein-coupled signal transduction pathways for interleukin-8. *Science* **261**, 101–103.

Wu, L., Ruffing, N., Shi, X., Newman, W., Soler, D., Mackay, C. R., and Qin, S. (1996). Discrete steps in binding and signaling of interleukin-8 with its receptor. *J. Biol. Chem.* **271**, 31202–32109.

Wuyts, A., Van Osselaer, N., Haelens, A., Samson, I., Herdewijn, P., Ben-Baruch, A., Oppenheim, J. J., Proost, P., and Van Damme, J. (1997). Characterization of synthetic human granulocyte chemotactic protein 2: usage of chemokine receptors CXCR1 and CXCR2 and *in vivo* inflammatory properties. *Biochemistry* **36**, 2716–2723.

Wuyts, A., Proost, P., Lenaerts, J. P., Ben-Baruch, A., Van Damme, J., and Wang, J. M. (1998). Differential usage of the CXC chemokine receptors 1 and 2 by interleukin-8, granulocyte chemotactic protein-2 and epithelial-cell-derived neutrophil attractant-78. *Eur. J. Biochem.* **255**, 67–73.

Xia, M. Q., and Hyman, B. T. (1999). Chemokines/chemokine receptors in the central nervous system and Alzheimer's disease. *J. Neurovirol.* **5**, 32–41.

Xia, M., Qin, S., McNamara, M., Mackay, C., and Hyman, B. T. (1997). Interleukin-8 receptor B immunoreactivity in brain and neuritic plaques of Alzheimer's disease. *Am. J. Pathol.* **150**, 1267–1274.

Xie, W., Jiang, H., Wu, Y., and Wu, D. (1997). Two basic amino acids in the second inner loop of the interleukin-8 receptor are essential for Gα16 coupling. *J. Biol. Chem.* **272**, 24948–24951.

Xu, L., Kelvin, D. J., Ye, G. Q., Taub, D. D., Ben-Baruch, A., Oppenheim, J. J., and Wang, J. M. (1995). Modulation of IL-8 receptor expression on purified human T lymphocytes is associated with changed chemotactic responses to IL-8. *J. Leukoc. Biol.* **57**, 335–342.

Yoshimura, T., Matsushima, K., Tanaka, S., Robinson, E. A., Appella, E., Oppenheim, J. J., and Leonard, E. J. (1987). Purification of a human monocyte-derived neutrophil chemotactic factor that has peptide sequence similarity to other host defense cytokines. *Proc. Natl Acad. Sci. USA* **84**, 9233–9237.

Zhao, X., Palczewski, K., and Ohguro, H. (1995). Mechanism of rhodopsin phosphorylation. *Biophys. Chem.* **56**, 183–188.

CXCR3

Joshua Marion Farber[1,*] and Bernhard Moser[2]

[1]Laboratory of Clinical Investigation, National Institute of Allergy and Infectious Diseases, NIH, 10 Center Drive, Room 11N-228, MSC 1888, Bethesda, MD 0892-1888, USA

[2]Theodor-Kocher Institute, University of Bern, Freiestrasse 1, Bern, CH-3012, Switzerland

*corresponding author tel: 301-402-4910, fax: 301-402-0627, e-mail: joshua_farber@nih.gov

DOI: 10.1006/rwcy.2000.21003.

SUMMARY

CXCR3 is a seven transmembrane domain G protein-coupled receptor for the interferon-inducible CXC chemokines IP-10, MIG, and I-TAC. CXCR3 is expressed on T cells, B cells, and NK cells. Receptor expression and function on T cells are induced dramatically by T cell activation, and CXCR3 has been reported to be expressed preferentially on TH1 cell lines and clones. CXCR3 has been found on T cells in a variety of human inflammatory lesions including *rheumatoid arthritis* and *multiple sclerosis* and it is presumed that CXCR3 is involved in the recruitment of activated T cells to inflammatory sites.

BACKGROUND

Discovery

The gene for CXCR3 was first published as GPR9 with an incomplete sequence encoding a chemokine receptor-related protein (Marchese *et al.*, 1995), with subsequent independent identification of the complete cDNA sequence and activity as a receptor for IP-10 and human MIG (Loetscher *et al.*, 1996).

Alternative names

No alternative names are now used in the literature, although there are CXCR3 sequences in the database as GPR9 and CKR-L2.

Structure

No experimental data are available. From the primary structure, CXCR3 is predicted to be a seven transmembrane domain receptor of 368 amino acids.

Main activities and pathophysiological roles

CXCR3 is a receptor for the interferon-inducible CXC chemokines IP-10, MIG, and I-TAC (Loetscher *et al.*, 1996; Cole *et al.*, 1998). Murine CXCR3 has also been reported to signal in response to the CC chemokine SLC (6Ckine) (Soto *et al.*, 1998), although this has not been confirmed in transfected cells expressing the human receptor, where SLC (6Ckine) did not signal and did not compete with radiolabeled human MIG for binding. CXCR3 is expressed on subsets of peripheral blood T cells, B cells and NK cells (see Table 1). Although a chemotactic response to human MIG can be detected using fresh cells (Rabin *et al.*, 1999), CXCR3 expression and function are enhanced markedly following T cell activation (M. Loetscher *et al.*, 1996; Qin *et al.*, 1998; Rabin *et al.*, 1999) and receptor expression and activity have been reported to be highest on TH1 as compared with TH2 cell lines (Bonecchi *et al.*, 1998; Sallusto *et al.*, 1998). CXCR3 is presumed to mediate the trafficking of activated/effector T cells and NK cells to inflammatory sites in response to IP-10, MIG, and I-TAC. CXCR3 is active on tumor infiltrating lymphocyte lines (Liao *et al.*, 1995) and has been reported on a majority of *B cell chronic lymphocytic leukemias* (Jones *et al.*, 1998), on lymphocytes in the demyelinating lesions of *multiple sclerosis* (Sorensen *et al.*, 1999), in synovial fluid and in synovium from joints affected with *rheumatoid arthritis* (Qin *et al.*, 1998; P. Loetscher *et al.*, 1998), in inflammatory infiltrates in *ulcerative colitis* and *chronic vaginitis* (Qin *et al.*, 1998), in bronchoalveolar lavage fluid from patients with *sarcoidosis* (Agostini *et al.*, 1998), and in the *exanthem*-affected skin of macaques infected with SIVmac239 (Sasseville *et al.*, 1998) (Table 2).

GENE

Accession numbers

Human gene: U32674, Z79783
Human cDNA: X95876
Mouse cDNA: AF045146

Sequence

See **Figure 1**.

Chromosome location and linkages

Both mouse (Soto *et al.*, 1998) and human (M. Loetscher *et al.*, 1998) CXCR3 are located on the X chromosome.

PROTEIN

Accession numbers

Human: CAA65126
Mouse: AAC40163

Sequence

See **Figure 2**.

Description of protein

From the primary structure, CXCR3 is predicted to be a seven transmembrane domain receptor of 368 amino acids. Like other chemokine receptors, CXCR3 contains multiple acidic residues in the N-terminal domain, cysteine residues in the N-terminal domain and the third extracellular loop, a DRYL sequence following the predicted third transmembrane domain, and predicted sites for *N*-linked glycosylation in extracellular domains.

Figure 1 Nucleotide sequences for human and mouse CXCR3.

Human cDNA

```
CCAACCACAA GCACCAAAGC AGAGGGGCAG GCAGCACACC ACCCAGCAGC
CAGAGCACCA GCCCAGCCAT GGTCCTTGAG GTGAGTGACC ACCAAGTGCT
AAATGACGCC GAGGTTGCCG CCCTCCTGGA GAACTTCAGC TCTTCCTATG
ACTATGGAGA AAACGAGAGT GACTCGTGCT GTACCTCCCC GCCCTGCCCA
CAGGACTTCA GCCTGAACTT CGACCGGGCC TTCCTGCCAG CCCTCTACAG
CCTCCTCTTT CTGCTGGGGC TGCTGGGCAA CGGCGCGGTG GCAGCCGTGC
TGCTGAGCCG GCGGACAGCC CTGAGCAGCA CCGACACCTT CCTGCTCCAC
CTAGCTGTAG CAGACACGCT GCTGGTGCTG ACACTGCCGC TCTGGGCAGT
GGACGCTGCC GTCCAGTGGG TCTTTGGCTC TGGCCTCTGC AAAGTGGCAG
GTGCCCTCTT CAACATCAAC TTCTACGCAG GAGCCCTCCT GCTGGCCTGC
ATCAGCTTTG ACCGCTACCT GAACATAGTT CATGCCACCC AGCTCTACCG
CCGGGGGCCC CCGGCCCGCG TGACCCTCAC CTGCCTGGCT GTCTGGGGGC
TCTGCCTGCT TTTCGCCCTC CCAGACTTCA TCTTCCTGTC GGCCCACCAC
GACGAGCGCC TCAACGCCAC CCACTGCCAA TACAACTTCC CACAGGTGGG
CCGCACGGCT CTGCGGGTGC TGCAGCTGGT GGCTGGCTTT CTGCTGCCCC
TGCTGGTCAT GGCCTACTGC TATGCCCACA TCCTGGCCGT GCTGCTGGTT
TCCAGGGGCC AGCGGCGCCT GCGGGCCATG CGGCTGGTGG TGGTGGTCGT
GGTGGCCTTT GCCCTCTGCT GGACCCCCTA TCACCTGGTG GTGCTGGTGG
ACATCCTCAT GGACCTGGGC GCTTTGGCCC GCAACTGTGG CCGAGAAAGC
AGGGTAGACG TGGCCAAGTC GGTCACCTCA GGCCTGGGCT ACATGCACTG
CTGCCTCAAC CCGCTGCTCT ATGCCTTTGT AGGGGTCAAG TTCCGGGAGC
GGATGTGGAT GCTGCTCTTG CGCCTGGGCT GCCCCAACCA GAGAGGGCTC
CAGAGGCAGC CATCGTCTTC CCGCCGGGAT TCATCCTGGT CTGAGACCTC
AGAGGCCTCC TACTCGGGCT TGTGAGGCCG GAATCCGGGC TCCCCTTTCG
CCCACAGTCT GACTTCCCCG CATTCCAGGC TCCTCCCTCC CTCTGCCGGC
TCTGGCTCTC CCCAATATCC TCGCTCCCGG GACTCACTGG CAGCCCCAGC
ACCACCAGGT CTCCCGGGAA GCCACCCTCC CAGCTCTGAG GACTGCACCA
TTGCTGCTCC TTAGCTGCCA AGCCCCATCC TGCCGCCCGA GGTGGCTGCC
TGGAGCCCCA CTGCCCTTCT CATTTGGAAA CTAAAACTTC ATCTTCCCCA
AGTGCGGGGA GTACAAGGCA TGGCGTAGAG GGTGCTGCCC CATGAAGCCA
CAGCCCAGGC CTCCAGCTCA GCAGTGACTG TGGCCATGGT CCCCAAGACC
TCTATATTTG CTCTTTTATT TTTATGTCTA AAATCCTGCT TAAAACTTTT
CAATAAACAA GATCGTCAGG ACCAAAAAAA AAAAAAAAAA AAAAAAAAAA
AAAAAAAAAA AAAAAAAAAA
```

Mouse cDNA

```
ATGTACCTTG AGGTTAGTGA ACGTCAAGTG CTAGATGCCT CGGACTTTGC
CTTTCTTCTG GAAAACAGCA CCTCTCCCTA CGATTATGGG GAAAACGAGA
GCGACTTCTC TGACTCCCCG CCCTGCCCAC AGGATTTCAG CCTGAACTTT
GACAGAACCT TCCTGCCAGC CCTCTACAGC CTCCTCTTCT TGCTGGGGCT
GCTAGGCAAT GGGGCGGTGG CTGCTGTGCT ACTGAGTCAG CGCACTGCCC
TGAGCAGCAC GGACACCTTC CTGCTCCACC TGGCTGTAGC CGATGTTCTG
CTGGTGTTAA CTCTTCCATT GTGGGCAGTG GATGCTGCTG TCCAGTGGGT
TTTCGGCCCT GGCCTCTGCA AAGTGGCAGG CGCCTTGTTC AACATCAACT
TCTATGCAGG GGCCTTCCTG CTGGCTTGTA TAAGCTTCGA CAGATATCTG
AGCATAGTGC ACGCCACCCA GATCTACCGC AGGGACCCCC GGGTACGTGT
AGCCCTCACC TGCATAGTTG TATGGGGTCT CTGTCTGCTC TTTGCCCTCC
CAGATTTCAT CTACCTATCA GCCAACTACG ATCAGCGCCT CAATGCCACC
CATTGCCAGT ACAACTTCCC ACAGGTGGGT CGCACTGCTC TGCGTGTACT
GCAGCTAGTG GCTGGTTTCC TGCTGCCCCT TCTGGTCATG GCCTACTGCT
ATGCCCATAT CCTAGCTGTT CTGCTGGTCT CCAGAGGCCA GAGGCGTTTT
CGAGCTATGA GGCTAGTGGT AGTGGTGGTG GCAGCCTTTG CTGTCTGCTG
GACCCCCTAT CACCTGGTGG TGCTAGTGGA TATCCTCATG GATGTGGGAG
TTTTGGCCCG CAACTGTGGT CGAAAAAGCC ACGTGGATGT GGCCAAGTCA
GTCACCTCGG GCATGGGGTA CATGCACTGC TGCCTCAATC CGCTGCTCTA
TGCCTTTGTG GGAGTGAAGT TCAGAGAGAA AATGTGGATG TTGTTCACGC
GCCTGGGCCG CTCTGACCAG AGAGGGCCCC AGCGGCAGCC GTCATCTTCA
CGGAGAGAAT CATCCTGGTC TGAGACAACT GAGGCCTCCT ACCTGGGCTT
GTAATTCTGG ACTGGAACTG TAGCCTGCGC AGCCCAAGTC CTAACACACT
CCAAGTGCTT GTCCTCCTTG TAGTTGGGCT AGCTCGAACT TACCCGTAAC
TTTGCTGCCA GGATGCACTG ACAGCTCAGC ATATATCCAG GTCTCCTGAG
AATCAATTTC AGCAACAAGG ACAACACCAT TACTGTGCCT TAGCTGCCAT
GCCCTATCTT GCTGTTTTAG AACTAGCTGC CTGGAGCCCC ACCGCCCTAC
TAAATTAGCA AGTAGAACTC AGCCATCCCT GTGTGAGAAG AGGGAGAGGC
AAATAGCACA GAGGGCCAGG CGTTGTCAGC ACTGAATGTG CCCATCTCAG
TATCTCAATA TTTGCCCAAT TTTATTTCTA GAAACCTCAC TTAAACTTTC
AATAAACAAG GTAATGAGGG AAAAA
```

Figure 2 Comparison of human and mouse CXCR3. Numbers at the right indicate the positions of the residues at the end of each line. Solid backgrounds indicate identities between the two proteins. The dot indicates a single gap introduced in the mouse sequence for optimal alignment. Putative transmembrane domains are indicated. The alignment was created using the PileUp and PrettyBox programs of the Wisconsin Sequence Analysis Package, Genetics Computer Group, Madison, WI.

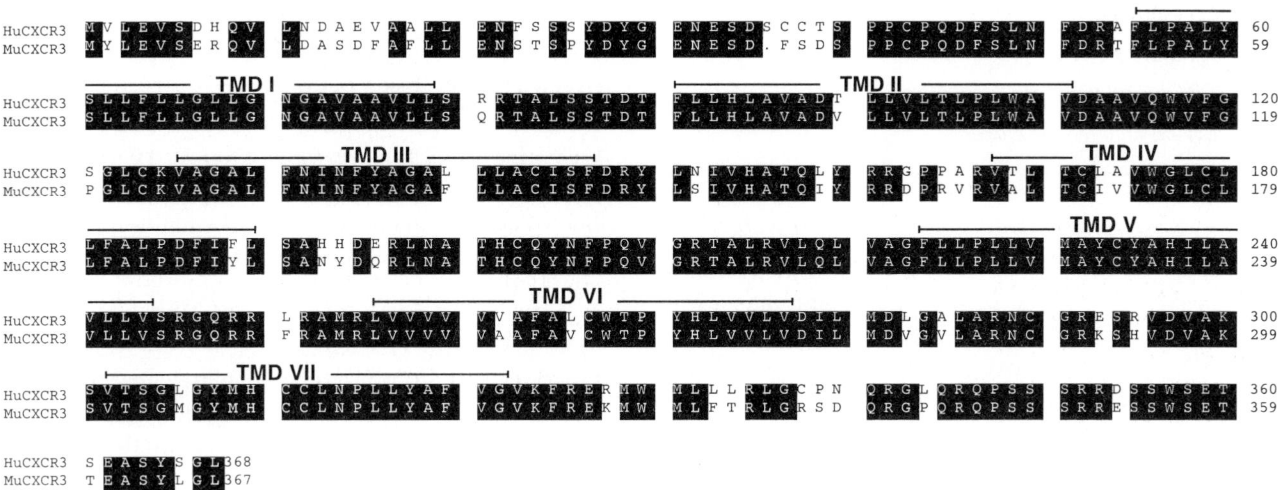

Relevant homologies and species differences

CXCR3 shows greatest similarity (approximately 40% identity) with the CXC receptors CXCR1, CXCR2, and CXCR5 (Zlotnik *et al.*, 1999). The mouse CXCR3 shows 87% sequence identity with the human receptor (Soto *et al.*, 1998), as shown in Figure 2. Mouse and human MIG and IP-10/-CRG-2 are active on both mouse and human receptors.

Affinity for ligand(s)

IP-10 has been reported to bind to CXCR3 on transfected cells with a K_d of 0.14–0.5 nM (Weng *et al.*, 1998; M. Loetscher *et al.*, 1998) and to activated T cells with a K_i of 0.04 nM. For human MIG, these numbers are 0.9–4.9 nM and 0.8 nM respectively. Using transfected HEK293 cells two binding sites were reported for I-TAC with K_i values of 0.3 and 36 nM from homologous displacement curves (Cole *et al.*, 1998). CXCR3 has also been reported to bind eotaxin and MCP-4 with K_i values of 60–70 nM, although these chemokines fail to produce a signal on CXCR3-transfected cells (Weng *et al.*, 1998).

Table 1 Cell type expression of CXCR3

Cell types	Subsets
CD4+ T cells	Memory, activated, TH1 preference
CD8+ T cells	Naïve, memory, activated
NK cells	
B cells	

Cell types and tissues expressing the receptor

CXCR3 mRNA was first reported in IL-2-activated T cells and NK cells (Loetscher *et al.*, 1996). Antibody to CXCR3 has revealed expression on subsets of peripheral blood T cells, B cells, and NK cells (Qin *et al.*, 1998). CXCR3 has been reported to be preferentially expressed on peripheral blood T cells that are CD45R0 high and CD45RA low as well as on most T cells that are positive for the activation markers CD25 and CD69 (Qin *et al.*, 1998) (*see* **Table 1**), although CXCR3 is also expressed and functional on freshly isolated, naïve CD8+ T cells (Rabin *et al.*, 1999). CXCR3 is active on tumor infiltrating lymphocyte lines (Liao *et al.*, 1995), has

Table 2 CXCR3 expression in disease

Tumor-infiltrating lymphocyte lines
Chronic lymphocytic leukemias (B cell)
Rheumatoid arthritis
Ulcerative colitis
Multiple sclerosis
SIV-associated exanthem
SIV-associated encephalitis

been reported on a majority of *B cell chronic lymphocytic leukemias* (Jones *et al.*, 1998), and has been found on lymphocytes in a variety of inflammatory infiltrates as noted above under Main activities and pathophysiological roles and in **Table 2**.

In the mouse, CXCR3 mRNA is expressed in spleen, lung, and heart, and CXCR3 sequences were detected in cDNA prepared from various populations of T cells, B cells, and endothelial cells (Soto *et al.*, 1998).

Regulation of receptor expression

CXCR3 can be upregulated by incubating T cells in IL-2 for 10–21 days (Loetscher *et al.*, 1996; Qin *et al.*, 1998; M. Loetscher *et al.*, 1998), with expression reportedly accelerated by treatment with PHA and downregulated in IL-2-treated cells by crosslinking CD3 (M. Loetscher *et al.*, 1998). CXCR3 expression and signaling can be upregulated on both naïve and memory T cells following 3 days of activation of PBMCs with OKT3 (Rabin *et al.*, 1999). CXCR3 has been reported to be preferentially expressed on TH1 T cell lines and clones as compared with TH2 T cells (Bonecchi *et al.*, 1998; Sallusto *et al.*, 1998), although CXCR3 is also expressed on activated TH2 T cells and the strength of the correlation with the TH1 phenotype has been questioned (P. Loetscher *et al.*, 1998; Annunziato *et al.*, 1998).

SIGNAL TRANSDUCTION

Cytoplasmic signaling cascades

On transfected cells, calcium flux and chemotaxis mediated through CXCR3 were blocked by pertussis toxin, indicating signaling through $G\alpha_i$ proteins (M. Loetscher *et al.*, 1998).

BIOLOGICAL CONSEQUENCES OF ACTIVATING OR INHIBITING RECEPTOR AND PATHOPHYSIOLOGY

Unique biological effects of activating the receptors

Activating CXCR3 with the ligands IP-10, human MIG, and I-TAC leads to calcium flux and chemotaxis in activated T cells, including tumor-infiltrating lymphocytes, and in CXCR3-transfected cells (Taub *et al.*, 1993; Liao *et al.*, 1995; Loetscher *et al.*, 1996; Cole *et al.*, 1998). In addition, CXCR3 rapidly activates IL-2-stimulated T cells to adhere to ICAM-1- and VCAM-1-coated plates and to human umbilical vein endothelial cells under conditions of flow (Piali *et al.*, 1998). IP-10, presumably through CXCR3, has been reported to enhance the production of IFNγ by antigen-activated lymphocytes (Gangur *et al.*, 1998).

THERAPEUTIC UTILITY

Effects of inhibitors (antibodies) to receptors

A monoclonal antibody against CXCR3, 1C6, has been described (Qin *et al.*, 1998) but no therapeutic effects have been reported. The antibody blocks binding of IP-10, but not human MIG, to CXCR3.

References

Agostini, C., Cassatella, M., Zambello, R., Trentin, L., Gasperini, S., Perin, A., Piazza, F., Siviero, M., Facco, M., Dziejman, M., Chilosi, M., Qin, S., Luster, A. D., and Semenzato, G. (1998). Involvement of the IP-10 chemokine in sarcoid granulomatous reactions. *J. Immunol.* **161**, 6413–6420.

Annunziato, F., Galli, G., Cosmi, L., Romagnani, P., Manetti, R., Maggi, E., and Romagnani, S. (1998). Molecules associated with human Th1 or Th2 cells. *Eur. Cytokine Netw.* **9**, 12–16.

Bonecchi, R., Bianchi, G., Bordignon, P. P., D'Ambrosio, D., Lang, R., Borsatti, A., Sozzani, S., Allavena, P., Gray, P. A., Mantovani, A., and Sinigaglia, F. (1998). Differential expression of chemokine receptors and chemotactic responsiveness of type 1 T helper cells (Th1s) and Th2s. *J. Exp. Med.* **187**, 129–134.

Cole, K. E., Strick, C. A., Paradis, T. J., Ogborne, K. T., Loetscher, M., Gladue, R. P., Lin, W., Boyd, J. G., Moser, B., Wood, D. E., Sahagan, B. G., and Neote, K. (1998). Interferon-inducible T cell alpha chemoattractant (I-TAC): a novel non-ELR CXC chemokine with potent activity on activated T cells through selective high affinity binding to CXCR3. *J. Exp. Med.* **187**, 2009–2021.

Gangur, V., Simons, F. E., and Hayglass, K. T. (1998). Human IP-10 selectively promotes dominance of polyclonally activated and environmental antigen-driven IFN-gamma over IL-4 responses. *FASEB J.* **12**, 705–713.

Jones, D., Shahsafaei, A., Ponath, P., and Dorfman, D. M. (1998). The chemokine receptor CXCR3 is expressed in a subset of peripheral blood B cells and is a marker of B-cell chronic lymphocytic leukemia. *Blood.* **92**, Suppl 1, 237A.

Liao, F., Rabin, R. L., Yannelli, J. R., Koniaris, L. G., Vanguri, P., and Farber, J. M. (1995). Human Mig chemokine: biochemical and functional characterization. *J. Exp. Med.* **182**, 1301–1314.

Loetscher, M., Gerber, B., Loetscher, P., Jones, S. A., Piali, L., Clark-Lewis, I., Baggiolini, M., and Moser, B. (1996). Chemokine receptor specific for IP10 and mig: structure, function, and expression in activated T-lymphocytes. *J. Exp. Med.* **184**, 963–969.

Loetscher, M., Loetscher, P., Brass, N., Meese, E., and Moser, B. (1998). Lymphocyte-specific chemokine receptor CXCR3: regulation, chemokine binding and gene localization. *Eur. J. Immunol.* **28**, 3696–3705.

Loetscher, P., Uguccioni, M., Bordoli, L., Baggiolini, M., Moser, B., Chizzolini, C., and Dayer, J. M. (1998). CCR5 is characteristic of Th1 lymphocytes. *Nature* **391**, 344–345.

Marchese, A., Heiber, M., Nguyen, T., Heng, H. H. Q., Saldivia, V. R., Cheng, R., Murphy, P. M., Tsui, L. C., Shi, X. M., Gregor, P., George, S. R., O'Dowd, B. F., and Docherty, J. M. (1995). Cloning and chromosomal mapping of three novel genes, GPR9, GPR10 and GPR14 encoding receptors related to interleukin 8, neuropeptide Y and somatostatin receptors. *Genomics* **29**, 335–344.

Piali, L., Weber, C., LaRosa, G., Mackay, C. R., Springer, T. A., Clark-Lewis, I., and Moser, B. (1998). The chemokine receptor CXCR3 mediates rapid and shear-resistant adhesion–induction of effector T lymphocytes by the chemokines IP10 and Mig. *Eur. J. Immunol.* **28**, 961–972.

Qin, S., Rottman, J. B., Myers, P., Kassam, N., Weinblatt, M., Loetscher, M., Koch, A. E., Moser, B., and Mackay, C. R. (1998). The chemokine receptors CXCR3 and CCR5 mark subsets of T cells associated with certain inflammatory reactions. *J. Clin. Invest.* **101**, 746–754.

Rabin, R. L., Park, M. K., Liao, F., Swofford, R., Stephany, D., and Farber, J. M. (1999). Chemokine receptor responses are achieved through regulation of both receptor expression and signaling. *J. Immunol.* **162**, 3840–3850.

Sallusto, F., Lenig, D., Mackay, C. R., and Lanzavecchia, A. (1998). Flexible programs of chemokine receptor expression on human polarized T helper 1 and 2 lymphocytes. *J. Exp. Med.* **187**, 875–883.

Sasseville, V. G., Rottman, J. B., Du, Z., Veazey, R., Knight, H. L., Caunt, D., Desrosiers, R. C., and Lackner, A. A. (1998). Characterization of the cutaneous exanthem in macaques infected with a Nef gene variant of SIVmac239. *J. Invest. Dermatol.* **110**, 894–901.

Sorensen, T. L., Tani, M., Jensen, J., Pierce, V., Lucchinetti, C., Folcik, V. A., Qin, S., Rottman, J., Sellebjerg, F., Strieter, R. M., Frederiksen, J. L., and Ransohoff, R. M. (1999). Expression of specific chemokines and chemokine receptors in the central nervous system of multiple sclerosis patients. *J. Clin. Invest.* **103**, 807–815.

Soto, H., Wang, W., Strieter, R. M., Copeland, N. G., Gilbert, D. J., Jenkins, N. A., Hedrick, J., and Zlotnik, A. (1998). The CC chemokine 6Ckine binds the CXC chemokine receptor CXCR3. *Proc. Natl Acad. Sci. USA* **95**, 8205–8210.

Taub, D. D., Lloyd, A. R., Conlon, K., Wang, J. M., Ortaldo, J. R., Harada, A., Matsushima, K., Kelvin, D. J., and Oppenheim, J. J. (1993). Recombinant human interferon-inducible protein 10 is a chemoattractant for human monocytes and T lymphocytes and promotes T cell adhesion to endothelial cells. *J. Exp. Med.* **177**, 1809–1814.

Weng, Y., Siciliano, S. J., Waldburger, K. E., Sirotina-Meisher, A., Staruch, M. J., Daugherty, B. L., Gould, S. L., Springer, M. S., and DeMartino, J. A. (1998). Binding and functional properties of recombinant and endogenous CXCR3 chemokine receptors. *J. Biol. Chem.* **273**, 18288–18291.

Zlotnik, A., Morales, J., and Hedrick, J. A. (1999). Recent advances in chemokines and chemokine receptors. *Crit. Rev. Immunol.* **19**, 1–47.

LICENSED PRODUCTS

R&D Systems: Anti-human CXCR3 mouse monoclonal IgG1, clone 49801.111, for flow cytometry.

PharMingen: Anti-human CXCR3 mouse monoclonal IgG1, clone 1C6, for flow cytometry, for immunohistochemistry and for neutralizing binding/signaling by IP-10.

CXCR4

Bernhard Moser*

Theodor-Kocher Institute, University of Bern, Freiestrasse 1, Bern, 3012, Switzerland

*corresponding author tel: 41-31-631-4157, fax: 41-31-631-3799, e-mail: bernhard.moser@tki.unibe.ch

DOI: 10.1006/rwcy.2000.21004.

SUMMARY

CXCR4 was cloned on the basis of structural homology with other chemokine receptors and on the basis of its coreceptor activity in an *HIV-1* fusion assay. The receptor selectively interacts with a single CXC chemokine, SDF-1. An unusual feature of this chemokine receptor is its broad range of cellular distribution with high expression on immature and mature hematopoietic cells as well as many tissue cells, including endothelial cells, neurons, and astrocytes. Freshly isolated peripheral blood T lymphocytes express high levels of CXCR4, most of which is stored in intracellular compartments, and cell culture under stimulatory conditions reduces CXCR4 gene expression. Short-term activation (e.g. via TCR-triggering) or exposure to SDF-1 leads to rapid CXCR4 phosphorylation and receptor internalization.

CXCR4 plays a critical role in embryogenesis, as demonstrated in CXCR4- or SDF-1-deficient mice which die at late embryonic stages and show severe defects in myelo- and B lymphopoiesis, brain and heart organogenesis, and blood vessel formation in the intestinal tract. The function of CXCR4 (and its ligand) in adult blood and tissue cells is not known but some reports suggest a role in the localization of hematopoietic progenitor cells in the bone marrow and the maturation of platelets during trans-endothelial migration of megakaryocytes. CXCR4 is one of the two major *HIV-1/2* coreceptors, characterizing CXCR4-dependent (X4) viruses which tend to appear at late stages in infected individuals. The envelope glycoproteins gp120 of X4 viruses act as a surrogate ligand for CXCR4 and mediate monocyte chemotaxis and apoptosis in CD8+ T lymphocytes (via macrophage activation) and neurons.

BACKGROUND

Discovery

Several groups cloned the cDNA for human CXCR4 in the early 1990s by hybridization screening of cDNA libraries with DNA probes corresponding to the rabbit IL-8 receptor (Loetscher *et al.*, 1994), the bovine neuropeptide Y receptor (Herzog *et al.*, 1993; Jazin *et al.*, 1993), or by PCR amplification using degenerate oligonucleotide primers corresponding to chemokine/chemoattractant receptors (Federsppiel *et al.*, 1993; Nomura *et al.*, 1993).

Alternative names

Prior to identification of SDF-1 as the selective ligand for this receptor, CXCR4 was an orphan receptor termed LESTR (Loetscher *et al.*, 1994), pBE1.3/HUMSTR (Federsppiel *et al.*, 1993), HM89 (Nomura *et al.*, 1993), hFB22 (Jazin *et al.*, 1993) and hL5R (Herzog *et al.*, 1993).

Structure

CXCR4 is a typical chemokine receptor with seven transmembrane-spanning domains which couples to heterotrimeric G proteins for signal transduction.

Main activities and pathophysiological roles

Signaling through CXCR4 induces chemotactic migration and adhesion of receptor-bearing cells to integrin ligands, as is typical for chemokine receptors. However, unlike other chemokine receptors, CXCR4 is expressed on all types of leukocytes and a role of this receptor in the steady-state turnover and homing of blood leukocytes is suggested. In contrast to chemokine receptors for inducible chemokines which are produced in inflammatory conditions, CXCR4 does not regulate leukocyte traffic to sites of infection and disease. In addition, CXCR4 is present on hematopoietic progenitor cells in the bone marrow and thymus and its ligand SDF-1 mobilizes CD34+ progenitor cells. CXCR4 and SDF-1 have additional functions that are unrelated to leukocyte chemotaxis. An essential role in embryogenesis was demonstrated in studies of mouse embryos that were genetically engineered to be deficient in SDF-1 or CXCR4. Deletion of either gene was lethal before or shortly after birth, and mutant embryos showed severe defects in B cell development and myelopoiesis, and had defective ventricular septum formation of the heart, impaired cerebellar development, and abnormal mesenteric blood vessel formation. CXCR4 is the only entry cofactor for syncytium-inducing *HIV* particles which typically emerge at the onset of *AIDS*, and rapid progression of this disease may be due to the unusually wide distribution of CXCR4 in blood and tissue cells. CXCR4 antagonists may represent a novel strategy for the treatment of patients with AIDS.

GENE

Accession numbers

The human LESTR/CXCR4 cDNA of 1645 bp has the GenBank/EMBL databank accession number X71635 (Loetscher *et al.*, 1994). Ortholog receptor cDNAs are described in chimpanzee (U89797), rhesus monkey (U73740), baboon (AF031089), African green monkey (AB015943), cow (M86739), and mouse (D87747) which express alternatively spliced mRNAs for two CXCR4 isoforms (Nagasawa *et al.*, 1996b; Heesen *et al.*, 1997; Moepps *et al.*, 1997), rat (U90610), cat (U63558), and fish (AB012310).

Sequences of the human and murine CXCR4 genes are deposited under Y14739, AJ224869, and X99581, respectively.

Chromosome location and linkages

The human and murine CXCR4 genes are localized on human chromosome 2q21 and mouse chromosome 1.

PROTEIN

Sequence

Protein sequence information can be retrieved from the GenBank/EMBL databank entries (see Gene: Accession numbers).

Description of protein

Human CXCR4 is a membrane-spanning glycoprotein of 352 amino acids (M_r 39,745). CXCR4 has seven transmembrane helices, which is typical for chemokine receptors (Loetscher *et al.*, 1994). CXCR4 shares with other chemokine receptors the DRYLAIVHA motif in the second intracellular loop and two Cys residues in the N-terminal region and the third extracellular loop, which may form an essential disulfide bond. Two potential *N*-glycosylation sites are located in the N-terminal region (Asn11-Tyr-Thr) and the second extracellular loop (Asn177-Val-Ser). CXCR4 shares several putative substrate motifs for protein kinase C with other chemokine receptors. The intracellular C-terminal region contains 18 Ser/Thr residues, some of which are involved in SDF-1-induced CXCR4 internalization through phosphorylation by receptor-specific kinases and/or Ser/Thr protein kinases (Amara *et al.*, 1997; Haribabu *et al.*, 1997).

Relevant homologies and species differences

Overall, amino acid sequence identity with other chemokine receptors is approximately 30%. Highest similarity is found in the transmembrane domains whereas the extracellular N-terminal region and the intracellular C-terminal region are less well conserved. CXCR4 couples to the class of G proteins which are sensitive to *Bordetella pertussis* toxin. CXCR4 is a member of the superfamily of G protein-coupled receptors which recognize ligands as varied as hormones, neurotransmitters, lipid derivatives, bacterial cell wall components, odorants, and light.

The sequences of human and African green monkey CXCR4 differ in only four amino acids, resulting in an amino acid sequence identity of 98.6%. Sequence conservation is further documented by 91.2% identity between human and murine CXCR4, and fish CXCR4 which shares 64.2% amino acids with the human receptor. The sequence of its ligand SDF-1 is equally well conserved, as documented by the single conservative amino acid difference between the human and mouse form.

Affinity for ligand(s)

The two isoforms SDF-1α and SDF-1β bind to CXCR4 with a K_d of 2–9 nM, as assessed in [^{125}I]-SDF-1 binding studies using CEM cells (Crump *et al.*, 1997; Loetscher *et al.*, 1998). Numerous structural variants of SDF-1 exist which bind to CXCR4 with lower affinity (Crump *et al.*, 1997; Heveker *et al.*, 1998; Loetscher *et al.*, 1998).

Cell types and tissues expressing the receptor

CXCR4 is widely expressed in blood and tissue cells, which is highly unusual for chemokine receptors (Table 1). The expression and function of CXCR4 in neutrophils is controversial. Some reports describe CXCR4 expression, by mRNA and protein analysis, and chemotactic responses to SDF-1 (Loetscher *et al.*, 1994; Oberlin *et al.*, 1996; Foerster *et al.*, 1998), whereas no activity was reported by others (Bleul *et al.*, 1996b, 1997; Hori *et al.*, 1998). Donor-to-donor variations in the neutrophil preparations may explain the observed differences. Alternatively, the antibodies that were used may have differed in their capacity to detect cell surface CXCR4. All reports agree that peripheral blood monocytes, T, and B lymphocytes are positive for CXCR4 and respond to SDF-1. In addition, CXCR4 is prominently expressed on hematopoietic progenitor cells and tissue cells, notably endothelial cells.

Regulation of receptor expression

The recognition of CXCR4 as an *HIV* entry coreceptor has mobilized many laboratories to investigate the regulation of its expression in CD4+ cells, notably T lymphocytes. Flow cytometry showed that only about 20% of freshly isolated blood lymphocytes expressed cell surface CXCR4 whereas the majority (> 80%) of these cells stained positive for intracellular CXCR4 (Bermejo *et al.*, 1998; Foerster *et al.*, 1998). Interestingly, short-term culture in the absence of T cell stimuli resulted in the relocation of CXCR4 to the cell surface. These findings are in striking contrast to chemokine receptors for inducible chemokines which are absent in resting T lymphocytes and are induced upon cell culture in the presence of IL-2 (Baggiolini, 1998; Moser *et al.*, 1998). Bermejo and colleagues did not find preferential expression of CXCR4 on certain T cell subsets (Bermejo *et al.*, 1998), whereas Bleul and colleagues (1997) reported predominant expression in resting naïve CD45RA+ T cells. Stimulation with PHA, IL-2, or a combination of anti-CD3 and anti-CD28 antibodies enhanced CXCR4 mRNA expression transiently, with peak values within days 3–6 of culture (Bleul *et al.*, 1997; Carroll *et al.*, 1997), mimicking the pattern of expression of the homing chemokine receptor CCR7 in T lymphocytes (Willimann *et al.*, 1998; Yoshida *et al.*, 1998). In agreement with these observations, promoter studies of human CXCR4 gene revealed considerable induction by IL-2 or antibodies to CD3 and CD28 (Moriuchi *et al.*, 1997). The marked base-level expression seen in resting circulating T lymphocytes may be attributed to the putative nuclear respiratory factor-binding site NRF-1. In contrast, responses to inducible chemokines and expression of corresponding chemokine receptors by cultured T cells are rapidly inhibited by anti-CD3 treatment (Loetscher *et al.*, 1996; Carroll *et al.*, 1997), indicating that signaling through the T cell receptor differentially regulates expression of the two classes of proinflammatory and homing types of chemokine receptors. IL-4 and TGFβ were also found to induce CXCR4 expression, whereas IFNγ was inhibitory, as shown for T cells, monocytes, dendritic cells, and endothelial cells (Gupta *et al.*, 1998; Jourdan *et al.*, 1998; Penton-Rol *et al.*, 1998; Zella *et al.*, 1998; Zoeteweij *et al.*, 1998). In addition, cultured endothelial cells express increased levels of CXCR4 after treatment with basic FGF, which is prevented by TNFα (Feil and Augustin, 1998).

In addition to regulation of gene expression by cytokines and other stimuli, the level of cell surface CXCR4 is rapidly modulated during T cell activation, as shown by short-term treatment (up to 120 minutes) with phytohemagglutinin, anti-CD3, or PMA which resulted in depletion of cell surface CXCR4 (Amara *et al.*, 1997; Haribabu *et al.*, 1997; Signoret *et al.*, 1997; Bermejo *et al.*, 1998; Foerster *et al.*, 1998; Jourdan *et al.*, 1998). This effect was probably due to Ser/The kinase-dependent receptor phosphorylation which induced receptor internalization through coated pits/vesicles and localization in endosomal compartments (Haribabu *et al.*, 1997; Signoret *et al.*, 1997). CXCR4 internalization was also rapidly achieved by addition of SDF-1, which was independent of signaling by *B. pertussis* toxin-sensitive G proteins, but involved the intracellular C-terminal region of the receptor (Amara *et al.*, 1997; Foerster *et al.*, 1998; Tarasova *et al.*, 1998). Removal of SDF-1 restored cell surface levels of CXCR4, indicating that local production of

Table 1 Expression of CXCR4 in blood and tissue cells

	RNA		Protein		References
	Northern[a]	*In situ*[b]	Cell surface[c]	Tissue[d]	
Mature blood cells					
Neutrophils	+		+		Loetscher *et al.*, 1994; Foerster *et al.*, 1998
Monocytes	+		+		Loetscher *et al.*, 1994; Bleul *et al.*, 1997; Foerster *et al.*, 1998; Hori *et al.*, 1998; Ostrowski *et al.*, 1998; Penton-Rol *et al.*, 1998
Macrophages			+	+	Vallat *et al.*, 1998; Zhang *et al.*, 1998
Dendritic cells	+		+		Granelli-Piperno *et al.*, 1996; Ayehunie *et al.*, 1997; Zoeteweij *et al.*, 1998
Microglia	+		+	+	Lavi *et al.*, 1997; Tanabe *et al.*, 1997; Vallat *et al.*, 1998
T lymphocytes	+		+		Loetscher *et al.*, 1994; Bleul *et al.*, 1997; Berkowitz *et al.*, 1998; Bermejo *et al.*, 1998; Foerster *et al.*, 1998; Hori *et al.*, 1998; Jourdan *et al.*, 1998; Ostrowski *et al.*, 1998; Zhang *et al.*, 1998;
B lymphocytes	+		+		Loetscher *et al.*, 1994; Bleul *et al.*, 1997, 1998; Foerster *et al.*, 1998; Hori *et al.*, 1998
Megakaryocytes/platelets			+		Hamada *et al.*, 1998
Hematopoietic cells					
CD34– cells			+		Deichmann *et al.*, 1997; Möhle, 1998
Pro-/pre-B cells	+		+		D'Apuzzo *et al.*, 1997
Thymocytes	+				Kitchen and Zack, 1997; Moepps *et al.*, 1997; Berkowitz *et al.*, 1998; Kim *et al.*, 1998; Zhang *et al.*, 1998
Tissue cells					
Endothelial cells	+	+	+	+	Gupta *et al.*, 1998; Tachibana *et al.*, 1998; Feil and Augustin, 1998; Volin *et al.*, 1998
Neurons	+			+	Hesselgesser *et al.*, 1997; Lavi *et al.*, 1997; Meucci *et al.*, 1998; Vallat *et al.*, 1998
Astrocytes	+		+		Tanabe *et al.*, 1997
Tissues					
Liver	+				Federsppiel *et al.*, 1993
Colon	+				Federsppiel *et al.*, 1993
Brain	+	+			Federsppiel *et al.*, 1993; Jazin, 1993; Lavi *et al.*, 1997; Moepps *et al.*, 1997; Zou *et al.*, 1998;
Heart	+				Federsppiel *et al.*, 1993
Thymus	+				Nagasawa *et al.*, 1996b; Moepps *et al.*, 1997
Bone marrow	+				Moepps *et al.*, 1997
Lymph node	+				Nagasawa *et al.*, 1996b; Moepps *et al.*, 1997
Spleen	+				Nagasawa *et al.*, 1996b; Moepps *et al.*, 1997

[a] CXCR4 mRNA in freshly isolated or cultured cells and tissues detected by northern blot or RT-PCR.

[b] CXCR4 mRNA expression in tissue sections analyzed by *in situ* hybridization.

[c] Flow cytometric detection of cell surface/intracellular CXCR4 protein.

[d] CXCR4 protein in tissue sections analyzed by immunohistochemistry.

this chemokine plays an important (negative-feedback) role in the regulation of T lymphocyte migration to SDF-1 and infection by CXCR4-tropic *HIV* particles. Finally, the envelope glycoprotein gp120 of X4 HIV-1 isolates was reported to inhibit chemokine responses in CD4-bearing cells through binding to and surface depletion of CD4 and CXCR4 (Madani *et al.*, 1998; Wang *et al.*, 1998).

Release of soluble receptors

There is no report of soluble, secreted CXCR4 which is probably due to its serpentine-like membrane insertion.

SIGNAL TRANSDUCTION

CXCR4, like all other chemokine receptors, couples to heterotrimeric G proteins (G_i or G_q subtype) and its activity is regulated by phosphorylation and subsequent internalization. SDF-1-mediated signaling leads to a range of cellular responses which can be divided into immediate responses (chemotactic migration involving cytoskeletal rearrangement and integrin-mediated adhesion) and long-term responses which depend on gene expression.

Associated or intrinsic kinases

Signal transduction elements that become activated during SDF-1-mediated CXCR4 signaling include focal adhesion components and diverse kinases which are critical for chemokine receptor inactivation, cell migration and gene expression (Davis *et al.*, 1997; Haribabu *et al.*, 1997; Bermejo *et al.*, 1998; Ganju *et al.*, 1998; Jourdan *et al.*, 1998). SDF-1 induces protein tyrosine phosphorylation of the related adhesion focal kinase RAFTK (also known as Pyk2), paxillin and the docking protein Crk which show enhanced association upon activation (Davis *et al.*, 1997; Ganju *et al.*, 1998). Similarly, phosphatidylinositol 3-kinase becomes rapidly phosphorylated, and its inhibition by wortmannin prevents paxillin phosphorylation and cell migration (Ganju *et al.*, 1998).

Cytoplasmic signaling cascades

Further downstream elements involved in transcriptional activation are p44/42 mitogen-activated protein (MAP) kinases Erk 1 and 2 which show enhanced phosphorylation upon SDF-1 stimulation (possibly through MAP kinase kinase 1 activation), whereas p38 MAP kinase and JNK remain unaffected (Ganju *et al.*, 1998; Jourdan *et al.*, 1998).

DOWNSTREAM GENE ACTIVATION

Transcription factors activated

CXCR4 signaling leads to activation of the nuclear transcriptional factor NFκB which is critically involved in gene expression during stimulation of lymphocytes and other inflammatory cells as well as *HIV* replication (Ganju *et al.*, 1998).

Genes induced

The effect of chemokines on gene expression is not well understood. Yet, induction of membrane-bound TNFα and TNF receptor type II expression during SDF-1-mediated apoptosis in T cells suggests that signaling through CXCR4 in macrophages and T cells resulted in expression of the corresponding genes (Herbein *et al.*, 1998).

BIOLOGICAL CONSEQUENCES OF ACTIVATING OR INHIBITING RECEPTOR AND PATHOPHYSIOLOGY

Unique biological effects of activating the receptors

CXCR4 mediates chemotactic migration and rapid/transient adhesion through integrin activation which are typical responses seen with chemokines in leukocytes. Of note, freshly isolated or short-term activated lymphocytes which do not respond to those chemokines known to be produced at sites of inflammation express high levels of CXCR4 and readily migrate to SDF-1 (Loetscher *et al.*, 1994; Bleul *et al.*, 1996a,b, 1997; Oberlin *et al.*, 1996; Bermejo *et al.*, 1998). An additional hallmark of CXCR4 is its wide expression in mature leukocytes present in blood and tissues (Table 1), which may be relevant for SDF-1-mediated localization of leukocytes at sites of steady-state turnover. The envelope glycoprotein gp120 of CXCR4-tropic *HIV* particles acts as a second type of CXCR4 ligand and induces CD4+ T cell activation.

CXCR4 may also be involved in the CD8+ T lymphocyte apoptosis generally observed during HIV infection (Davis *et al.*, 1997; Herbein *et al.*, 1998). This was confirmed in mixed cultures containing HIV-infected lymphocytes and macrophages in which the number of CD8+ T lymphocytes progressively diminished during viral propagation (Herbein *et al.*, 1998). CXCR4-tropic HIV particles could be replaced by soluble gp120 or SDF-1, and apoptosis was attributed to induction of two genes, one for the membrane-bound TNFα in macrophages and the other one for the TNF receptor type II in CD8+ T lymphocytes. Apoptosis in CD8+ T lymphocytes did not occur in the absence of macrophages. However, ligation of gp120 with CD4 and CXCR4 induced apoptosis in CD4+ T lymphocytes which was blocked by SDF-1, and in this case the mechanisms involved remain to be determined (Berndt *et al.*, 1998). A report describes the direct induction of apoptosis in a human neuronal cell line by SDF-1 via CXCR4 (Hesselgesser *et al.*, 1998), which is in contrast to another report demonstrating inhibition (rather than induction) of HIV envelope protein-mediated apoptosis in cultured hippocampal neurons by SDF-1 (Meucci *et al.*, 1998).

Numerous articles describe the expression of CXCR4 and activity of SDF-1 in myelo- and lymphopoietic progenitor cells (Nagasawa *et al.*, 1994; Aiuti *et al.*, 1997; D'Apuzzo *et al.*, 1997; Deichmann *et al.*, 1997; Kitchen and Zack, 1997; Kim *et al.*, 1998; Berkowitz *et al.*, 1998; Möhle *et al.*, 1998). Originally, SDF-1 was identified on the basis of its co-mitogenic activity together with IL-7 for pre-B cells but this function has not been investigated in great detail (Nagasawa *et al.*, 1994). Instead, SDF-1 was found to be a potent chemoattractant for early progenitor cells in the bone marrow, including CD34+ cells, and pre-/pro-B cells (Aiuti *et al.*, 1997; D'Apuzzo *et al.*, 1997). More developed B cell progenitors still expressed CXCR4 but lacked responsiveness, suggesting that SDF-1 acts on those hematopoietic progenitor cells which fully depend on contact with bone marrow stromal cells (D'Apuzzo *et al.*, 1997). A similar situation is found in the thymus where highest CXCR4 expression and marked responses to SDF-1 were seen in the most immature thymocyte subsets, the triple-negative ($\alpha\beta$TCR−CD4−CD8−) and the double-positive (CD4+CD8+) thymocytes (Kitchen and Zack, 1997; Berkowitz *et al.*, 1998; Kim *et al.*, 1998). Also here, SDF-1 may have a trapping function by preventing egress of immature thymocytes from cortex to medulla. SDF-1/CXCR4 may also play an important role in platelet formation. Mature megakaryocytes were shown to migrate in response to SDF-1 through cultured bone marrow endothelial cell monolayers, and this process resulted in the release of platelets which expressed CXCR4 (Hamada *et al.*, 1998).

Using an elegant cDNA cloning strategy, Feng and colleagues have identified CXCR4 (termed fusin by these authors) as entry cofactor, which complemented CD4 on target cells during fusion with syncytium-inducing *HIV-1* strains (Feng *et al.*, 1996). This finding demonstrated for the first time that chemokine receptors are critically involved in HIV infection. Today, it is clear that the distribution of CXCR4 and other chemokine receptors with cofactor function determines viral tropism and pathogenesis (Moore *et al.*, 1997; Cairns and D'Souza, 1998). SDF-1 prevents entry of CXCR4-tropic (X4) *HIV-1* strains through binding to CXCR4 on CD4+ T cells and consequent rapid internalization of ligand–receptor complexes (Bleul *et al.*, 1996a; Oberlin *et al.*, 1996; Amara *et al.*, 1997). In contrast to CCR5-tropic (R5) viruses which predominate at asymptomatic stages in infected individuals, X4 viruses typically emerge during onset of *AIDS* (Moore *et al.*, 1997; Cairns and D'Souza, 1998). Disease progression is accompanied by a change in coreceptor usage in emerging viruses from CCR5 to CXCR4, and the consequent rapid decline in immune functions may reflect the broad expression of CXCR4 in blood and tissue cells (Table 1). Of note, CXCR4 is also expressed in microglial cells, astrocytes, and neurons which may function as viral reservoirs during HIV infection and, in addition, may contribute to brain disorders (*dementia*) frequently observed in infected individuals (Hesselgesser *et al.*, 1997; Lavi *et al.*, 1997; Tanabe *et al.*, 1997; Vallat *et al.*, 1998). CXCR4 also functions as a CD4-independent receptor for *HIV-2* (Endres *et al.*, 1996). As with CCR5, much effort is being taken to develop synthetic inhibitors which selectively bind to CXCR4, and several reports describe the structure and function of such molecules (Baggiolini and Moser, 1997).

Phenotypes of receptor knockouts and receptor overexpression mice

Mice deficient in either SDF-1 (Nagasawa *et al.*, 1996a; Ma *et al.*, 1998) or CXCR4 (Ma *et al.*, 1998; Tachibana *et al.*, 1998; Zou *et al.*, 1998) have been generated by targeted disruption of the respective genes. Both gene knockout animals show identical phenotypes, indicating that SDF-1 is the primary physiologic ligand for CXCR4 and that other receptors for this chemokine do not exist. Whereas heterozygous (SDF-1+/−, CXCR4+/−) animals are healthy and fertile, homozygous disruption resulted in fetal mortality in SDF-1−/− and CXCR4−/− embryos. Mutant embryos were present at expected ratios up to day 15.5 of embryogenesis (E15.5), but half of the embryos were dead at E18.5 and neonates died within an hour after birth.

Phenotypic aberrations include defects in hematopoiesis, heart and brain development, and vascularization of the gastrointestinal tract. SDF-1−/− and CXCR4−/− embryos show severely reduced B lymphopoiesis and myelopoiesis in the bone marrow, which is in agreement with expression of CXCR4 in early hematopoietic progenitor cells and their chemotactic responses to SDF-1 (Aiuti *et al.*, 1997; D'Apuzzo *et al.*, 1997; Deichmann *et al.*, 1997). In the fetal liver, however, myelopoiesis was not affected and monocytes, macrophages, granulocytes, and megakaryocytes developed normally. Also, T lymphopoiesis was not affected in mutant embryos, indicating that thymocyte maturation does not depend on SDF-1/CXCR4. All mutants showed defects in the ventricular septum formation of the heart. In CXCR4−/− embryos all major blood vessels were present. However, blood vessel branching, notably in the vasculature of the gastrointestinal tract, was severely reduced (Tachibana *et al.*, 1998). This striking dependence on SDF-1/CXCR4 is supported by the recent evidence of CXCR4 expression in cultured endothelial cells and responses to SDF-1 (Feil and Augustin, 1998; Gupta *et al.*, 1998; Tachibana *et al.*, 1998; Volin *et al.*, 1998). Similarly, the expression of CXCR4 in the brain suggests that SDF-1 and its receptor may play a role in brain development (Hesselgesser *et al.*, 1997; Lavi *et al.*, 1997; Moepps *et al.*, 1997; Tanabe *et al.*, 1997; Vallat *et al.*, 1998) and this was confirmed by the observed defects in cerebellar anlage formation and abnormal granule cell localization in CXCR4−/− embryos (Zou *et al.*, 1998).

THERAPEUTIC UTILITY

Effects of inhibitors (antibodies) to receptors

Several low molecular weight compounds with selectivity for CXCR4 are described. The first is a heterocyclic bicyclam derivative, called AMD3100, which has been known for some time to inhibit the replication of T cell line-adapted (X4) *HIV-1* isolates. The mechanism of action remained unknown until De Clercq and colleagues demonstrated a selective interaction of AMD3100 with CXCR4 (Schols *et al.*, 1997). It diminished binding of a specific antibody to cell surface CXCR4, possibly through induction of receptor internalization, and completely blocked SDF-1-induced Ca^{2+} responses at 100 ng/mL. Although nontoxic at high concentrations in mice, poor oral bioavailability makes AMD3100 an unlikely candidate drug for use in clinical trials. Two other inhibitors of CXCR4 are small peptides which, like SDF-1-derived peptides, are presently not considered useful for the treatment of *AIDS*. T22 is an 18 amino acid peptide from the hemocyte debris of American horseshoe crab which prevents infection by X4 HIV-1 isolates through blockage of CXCR4 on CD4+ target cells (Murakami *et al.*, 1997). Similarly, the nona-D-arginine peptide ALX40-4C shows a high degree of selectivity for CXCR4 and blocked SDF-1-induced Ca^{2+} responses and infection by X4 HIV-1 strains at low micromolar concentrations (Doranz *et al.*, 1997). Finally, a distamycin derivative (NSC 651016) was found to block infection by both CXCR4- and CCR5-dependent HIV-1 strains, which was probably due to coreceptor downmodulation (Howard *et al.*, 1998). The use of monoclonal antibodies (mAbs) represents an alternative strategy for the treatment of HIV-infected individuals and several anti-CXCR4 mAbs with HIV-blocking activity are described. The first such mAb 12G5 was generated by Hoxie and colleagues who made it readily available to the scientific community (Endres *et al.*, 1996; McKnight *et al.*, 1997). Additional mAbs that block infection by X4 HIV-1 and *HIV-2* isolates through binding to CXCR4 are 2B11 (Foerster *et al.*, 1998), IVR7, and TSH123 (Hori *et al.*, 1998).

References

Amara, A., Gall, S. L., Schwartz, O., Salamero, J., Montes, M., Loetscher, P., Baggiolini, M., Virelizier, J. L., and Arenzana Seisdedos, F. (1997). HIV coreceptor downregulation as antiviral principle: SDF-1alpha-dependent internalization of the chemokine receptor CXCR4 contributes to inhibition of HIV replication. *J. Exp. Med.* **186**, 139–146.

Aiuti, A., Webb, I. J., Bleul, C., Springer, T., and Gutierrez-Ramos, J. C. (1997). The chemokine SDF-1 is a chemoattractant for human CD34+ hematopoietic progenitor cells and provides a new mechanism to explain the mobilization of CD34+ progenitors to peripheral blood. *J. Exp. Med.* **185**, 111–120.

Ayehunie, S., Garcia-Zepeda, E. A., Hoxie, J. A., Horuk, R., Kupper, T. S., Luster, A. D., and Ruprecht, R. M. (1997). Human immunodeficiency virus-1 entry into purified blood dendritic cells through CC and CXC chemokine coreceptors. *Blood* **90**, 1379–1386.

Baggiolini, M. (1998). Chemokines and leukocyte traffic. *Nature* **392**, 565–568.

Baggiolini, M., and Moser, B. (1997). Blocking chemokine receptors. *J. Exp. Med.* **186**, 1189–1191.

Berkowitz, R. D., Beckerman, K. P., Schall, T. J., and McCune, J. M. (1998). CXCR4 and CCR5 expression delineates targets for HIV-1 disruption of T cell differentiation. *J. Immunol.* **161**, 3702–3710.

Bermejo, M., Martin-Serrano, J., Oberlin, E., Pedraza, M. A., Serrano, A., Santiago, B., Caruz, A., Loetscher, P., Baggiolini, M., Arenzana-Seisdedos, F., and Alcami, J. (1998). Activation of blood T lymphocytes down-regulates CXCR4 expression and interferes with propagation of X4 HIV strains. *Eur. J. Immunol.* **28**, 3192–3204.

Berndt, C., Mopps, B., Angermuller, S., Gierschik, P., and Krammer, P. H. (1998). CXCR4 and CD4 mediate a rapid CD95-independent cell death in CD4(+) T cells. *Proc. Natl Acad. Sci. USA* **95**, 12556–12561.

Bleul, C. C., Farzan, M., Choe, H., Parolin, C., Clark-Lewis, I., Sodroski, J., and Springer, T. A. (1996a). The lymphocyte chemoattractant SDF-1 is a ligand for LESTR/fusin and blocks HIV-1 entry. *Nature* **382**, 829–833.

Bleul, C. C., Fuhlbrigge, R. C., Casasnovas, J. M., Aiuti, A., and Springer, T. A. (1996b). A highly efficacious lymphocyte chemoattractant, stromal cell-derived factor 1 (SDF-1). *J. Exp. Med.* **184**, 1101–1109.

Bleul, C. C., Wu, L., Hoxie, J. A., Springer, T. A., and Mackay, C. R. (1997). The HIV coreceptors CXCR4 and CCR5 are differentially expressed and regulated on human T lymphocytes. *Proc. Natl Acad. Sci. USA* **94**, 1925–1930.

Bleul, C. C., Schultze, J. L., and Springer, T. A. (1998). B lymphocyte chemotaxis regulated in association with microanatomic localization, differentiation state, and B cell receptor engagement. *J. Exp. Med.* **187**, 753–762.

Cairns, J. S., and D'Souza, M. P. (1998). Chemokines and HIV-1 second receptors: the therapeutic connection. *Nature Med.* **4**, 563–568.

Carroll, R. G., Riley, J. L., Levine, B. L., Feng, Y., Kaushal, S., Ritchey, D. W., Bernstein, W., Weislow, O. S., Brown, C. R., Berger, E. A., June, C. H., and St Louis, D. C. (1997). Differential regulation of HIV-1 fusion cofactor expression by CD28 costimulation of CD4+ T cells. *Science* **276**, 273–276.

Crump, M. P., Gong, J. H., Loetscher, P., Rajarathnam, K., Amara, A., Arenzana-Seisdedos, F., Virelizier, J. L., Baggiolini, M., Sykes, B. D., and Clark-Lewis, I. (1997). Solution structure and basis for functional activity of stromal cell-derived factor-1; dissociation of CXCR4 activation from binding and inhibition of HIV-1. *EMBO J.* **16**, 6996–7007.

D'Apuzzo, M., Rolink, A., Loetscher, M., Hoxie, J. A., Clark-Lewis, I., Melchers, F., Baggiolini, M., and Moser, B. (1997). The chemokine SDF-1, stromal cell-derived factor 1, attracts early stage B cell precursors via the chemokine receptor CXCR4. *Eur. J. Immunol.* **27**, 1788–1793.

Davis, C. B., Dikic, I., Unutmaz, D., Hill, C. M., Arthos, J., Siani, M. A., Thompson, D. A., Schlessinger, J., and Littman, D. R. (1997). Signal transduction due to HIV-1 envelope interactions with chemokine receptors CXCR4 or CCR5. *J. Exp. Med.* **186**, 1793–1798.

Deichmann, M., Kronenwett, R., and Haas, R. (1997). Expression of the human immunodeficiency virus type-1 coreceptors CXCR-4 (fusin, LESTR) and CKR-5 in CD34+ hematopoietic progenitor cells. *Blood* **89**, 3522–3528.

Doranz, B. J., Grovit-Ferbas, K., Sharron, M. P., Mao, S. H., Goetz, M. B., Daar, E. S., Doms, R. W., and O'Brien, W. A. (1997). A small-molecule inhibitor directed against the chemokine receptor CXCR4 prevents its use as an HIV-1 coreceptor. *J. Exp. Med.* **186**, 1395–1400.

Endres, M. J., Clapham, P. R., Marsh, M., Ahuja, M., Turner, J. D., McKnight, A., Thomas, J. F., Stoebenau-Haggarty, B., Choe, S., Vance, P. J., Wells, T. N., Power, C. A., Sutterwala, S. S., Doms, R. W., Landau, N. R., and Hoxie, J. A. (1996). CD4-independent infection by HIV-2 is mediated by fusin/CXCR4. *Cell* **87**, 745–756.

Federsppiel, B., Melhado, I. G., Duncan, A. M., Delaney, A., Schappert, K., Clark-Lewis, I., and Jirik, F. R. (1993). Molecular cloning of the cDNA and chromosomal localization of the gene for a putative seven-transmembrane segment (7-TMS) receptor isolated from human spleen. *Genomics* **16**, 707–712.

Feil, C., and Augustin, H. G. (1998). Endothelial cells differentially express functional CXC-chemokine receptor-4 (CXCR-4/fusin) under the control of autocrine activity and exogenous cytokines. *Biochem. Biophys. Res. Commun.* **247**, 38–45.

Feng, Y., Broder, C. C., Kennedy, P. E., and Berger, E. A. (1996). HIV-1 entry cofactor: functional cDNA cloning of a seven-transmembrane, G protein-coupled receptor. *Science* **272**, 872–877.

Foerster, R., Kremmer, E., Schubel, A., Breitfeld, D., Kleinschmidt, A., Nerl, C., Bernhardt, G., and Lipp, M. (1998). Intracellular and surface expression of the HIV-1 coreceptor CXCR4/fusin on various leukocyte subsets: rapid internalization and recycling upon activation. *J. Immunol.* **160**, 1522–1531.

Ganju, R. K., Brubaker, S. A., Meyer, J., Dutt, P., Yang, Y., Qin, S., Newman, W., and Groopman, J. E. (1998). The alpha-chemokine, stromal cell-derived factor-1alpha, binds to the transmembrane G-protein-coupled CXCR-4 receptor and activates multiple signal transduction pathways. *J. Biol. Chem.* **273**, 23169–23175.

Granelli-Piperno, A., Moser, B., Pope, M., Chen, D., Wei, Y., Isdell, F., O'Doherty, U., Paxton, W., Koup, R., Mojsov, S., Bhardwaj, N., Clark-Lewis, I., Baggiolini, M., and Steinman, R. M. (1996). Efficient interaction of HIV-1 with purified dendritic cells via multiple chemokine coreceptors. *J. Exp. Med.* **184**, 2433–2438.

Gupta, S. K., Lysko, P. G., Pillarisetti, K., Ohlstein, E., and Stadel, J. M. (1998). Chemokine receptors in human endothelial cells. Functional expression of CXCR4 and its transcriptional regulation by inflammatory cytokines. *J. Biol. Chem.* **273**, 4282–4287.

Hamada, T., Mohle, R., Hesselgesser, J., Hoxie, J., Nachman, R. L., Moore, M. A., and Rafii, S. (1998). Transendothelial migration of megakaryocytes in response to stromal cell-derived factor 1 (SDF-1) enhances platelet formation. *J. Exp. Med.* **188**, 539–548.

Haribabu, B., Richardson, R. M., Fisher, I., Sozzani, S., Peiper, S. C., Horuk, R., Ali, H., and Snyderman, R. (1997). Regulation of human chemokine receptors CXCR4. Role of phosphorylation in desensitization and internalization. *J. Biol. Chem.* **272**, 28726–28731.

Heesen, M., Berman, M. A., Hopken, U. E., Gerard, N. P., and Dorf, M. E. (1997). Alternate splicing of mouse fusin/CXC chemokine receptor-4: stromal cell-derived factor-1alpha is a ligand for both CXC chemokine receptor-4 isoforms. *J. Immunol.* **158**, 3561–3564.

Herbein, G., Mahlknecht, U., Batliwalla, F., Gregersen, P., Pappas, T., Butler, J., O'Brien, W. A., and Verdin, E. (1998). Apoptosis of CD8+ T cells is mediated by macrophages through interaction of HIV gp120 with chemokine receptor CXCR4. *Nature* **395**, 189–194.

Herzog, H., Hort, Y. J., Shine, J., and Selbie, L. A. (1993). Molecular cloning, characterization, and localization of the human homolog to the reported bovine NPY Y3 receptor: lack of NPY binding and activation. *DNA Cell Biol.* **12**, 465–471.

Hesselgesser, J., Halks-Miller, M., DelVecchio, V., Peiper, S. C., Hoxie, J., Kolson, D. L., Taub, D., and Horuk, R. (1997).

CD4-independent association between HIV-1 gp120 and CXCR4: functional chemokine receptors are expressed in human neurons. *Curr. Biol.* **7**, 112–121.

Hesselgesser, J., Taub, D., Baskar, P., Greenberg, M., Hoxie, J., Kolson, D. L., and Horuk, R. (1998). Neuronal apoptosis induced by HIV-1 gp120 and the chemokine SDF-1 alpha is mediated by the chemokine receptor CXCR4. *Curr. Biol.* **8**, 595–598.

Heveker, N., Montes, M., Germeroth, L., Amara, A., Trautmann, A., Alizon, M., and Schneider-Mergener, J. (1998). Dissociation of the signalling and antiviral properties of SDF-1-derived small peptides. *Curr. Biol.* **8**, 369–376.

Hori, T., Sakaida, H., Sato, A., Nakajima, T., Shida, H., Yoshie, O., and Uchiyama, T. (1998). Detection and delineation of CXCR-4 (fusin) as an entry and fusion cofactor for T cell-tropic HIV-1 by three different monoclonal antibodies. *J. Immunol.* **160**, 180–188.

Howard, O. M. Z., Korte, T., Tarasova, N. I., Grimm, M., Turpin, J. A., Rice, W. G., Michejda, C. J., Blumenthal, R., and Oppenheim, J. J. (1998). Small molecule inhibitor of HIV-1 cell fusion blocks chemokine receptor-mediated function. *J. Leukoc. Biol*, **64**, 6–13.

Jazin, E. E., Yoo, H., Blomqvist, A. G., Yee, F., Weng, G., Walker, M. W., Salon, J., Larhammar, D., and Wahlestedt, C. (1993). A proposed bovine neuropeptide Y (NPY) receptor cDNA clone, or its human homologue, confers neither NPY binding sites nor NPY responsiveness on transfected cells. *Regul. Pept.* **47**, 247–258.

Jourdan, P., Abbal, C., Nora, N., Hori, T., Uchiyama, T., Vendrell, J. P., Bousquet, J., Taylor, N., Pene, J., and Yssel, H. (1998). IL-4 induces functional cell-surface expression of CXCR4 on human T cells. *J. Immunol.* **160**, 4153–4157.

Kim, C. H., Pelus, L. M., White, J. R., and Broxmeyer, H. E. (1998). Differential chemotactic behavior of developing T cells in response to thymic chemokines. *Blood* **91**, 4434–4443.

Kitchen, S. G., and Zack, J. A. (1997). CXCR4 expression during lymphopoiesis: implications for human immunodeficiency virus type 1 infection of the thymus. *J. Virol.* **71**, 6928–6934.

Lavi, E., Strizki, J. M., Ulrich, A. M., Zhang, W., Fu, L., Wang, Q., O'Connor, M., Hoxie, J. A., and Gonzalez-Scarano, F. (1997). CXCR-4 (Fusin), a co-receptor for the type 1 human immunodeficiency virus (HIV-1), is expressed in the human brain in a variety of cell types, including microglia and neurons. *Am. J. Pathol.* **151**, 1035–1042.

Loetscher, M., Geiser, T., O'Reilly, T., Zwahlen, R., Baggiolini, M., and Moser, B. (1994). Cloning of a human seven-transmembrane domain receptor, LESTR, that is highly expressed in leukocytes. *J. Biol. Chem.* **269**, 232–237.

Loetscher, P., Seitz, M., Baggiolini, M., and Moser, B. (1996). Interleukin-2 regulates CC chemokine receptor expression and chemotactic responsiveness in T lymphocytes. *J. Exp. Med.* **184**, 569–577.

Loetscher, P., Gong, J. H., Dewald, B., Baggiolini, M., and Clark-Lewis, I. (1998). N-terminal peptides of stromal cell-derived factor-1 with CXC chemokine receptor 4 agonist and antagonist activities. *J. Biol. Chem.* **273**, 22279–22283.

Ma, Q., Jones, D., Borghesani, P. R., Segal, R. A., Nagasawa, T., Kishimoto, T., Bronson, R. T., and Springer, T. A. (1998). Impaired B-lymphopoiesis, myelopoiesis, and derailed cerebellar neuron migration in CXCR4- and SDF-1-deficient mice. *Proc. Natl Acad. Sci. USA* **95**, 9448–9453.

Madani, N., Kozak. S. L., Kavanaugh, M. P., and Kabat. D. (1998). gp120 envelope glycoproteins of human immunodeficiency viruses competitively antagonize signaling by co-receptors CXCR4 and CCR5. *Proc. Natl Acad. Sci. USA* **95**, 8005–8010.

McKnight, A., Wilkinson, D., Simmons, G., Talbot, S., Picard, L., Ahuja, M., Marsh, M., Hoxie, J. A., and Clapham, P. R. (1997). Inhibition of human immunodeficiency virus fusion by a monoclonal antibody to a coreceptor (CXCR4) is both cell type and virus strain dependent. *J. Virol.* **71**, 1692–1696.

Meucci, O., Fatatis, A., Simen, A. A., Bushell, T. J., Gray, P. W., and Miller, R. J. (1998). Chemokines regulate hippocampal neuronal signaling and gp120 neurotoxicity. *Proc. Natl Acad. Sci. USA* **95**, 14500–14505.

Möhle, R., Bautz, F., Rafii, S., Moore, M. A., Brugger, W., and Kanz, L. (1998). The chemokine receptor CXCR-4 is expressed on CD34+ hematopoietic progenitors and leukemic cells and mediates transendothelial migration induced by stromal cell-derived factor-1. *Blood* **91**, 4523–4530.

Moepps, B., Frodl, R., Rodewald, H. R., Baggiolini, M., and Gierschik, P. (1997). Two murine homologues of the human chemokine receptor CXCR4 mediating stromal cell-derived factor 1alpha activation of Gi2 are differentially expressed *in vivo*. *Eur. J. Immunol.* **27**, 2102–2112.

Moore, J. P., Trkola, A., and Dragic, T. (1997). Co-receptors for HIV-1 entry. *Curr. Opin. Immunol.* **9**, 551–562.

Moriuchi, M., Moriuchi, H., Turner, W., and Fauci, A. S. (1997). Cloning and analysis of the promoter region of CXCR4, a co-receptor for HIV-1 entry. *J. Immunol.* **159**, 4322–4329.

Moser, B., Loetscher, M., Piali, L., and Loetscher, P. (1998). Lymphocyte responses to chemokines. *Int. Rev. Immunol.* **16**, 323–344.

Murakami, T., Nakajima, T., Koyanagi, Y., Tachibana, K., Fujii, N., Tamamura, H., Yoshida, N., Waki, M., Matsumoto, A., Yoshie, O., Kishimoto, T., Yamamoto, N., and Nagasawa, T. (1997). A small molecule CXCR4 inhibitor that blocks T cell line-tropic HIV-1 infection. *J. Exp. Med.* **186**, 1389–1393.

Nagasawa, T., Kikutani, H., and Kishimoto, T. (1994). Molecular cloning and structure of a pre-B-cell growth-stimulating factor. *Proc. Natl Acad. Sci. USA* **91**, 2305–2309.

Nagasawa, T., Hirota, S., Tachibana, K., Takakura, N., Nishikawa, S., Kitamura, Y., Yoshida, N., Kikutani, H., and Kishimoto, T. (1996a). Defects of B-cell lymphopoiesis and bone-marrow myelopoiesis in mice lacking the CXC chemokine PBSF/SDF-1. *Nature* **382**, 635–638.

Nagasawa, T., Nakajima, T., Tachibana, K., Iizasa, H., Bleul, C. C., Yoshie, O., Matsushima, K., Yoshida, N., Springer, T. A., and Kishimoto, T. (1996b). Molecular cloning and characterization of a murine pre-B-cell growth-stimulating factor/stromal cell-derived factor 1 receptor, a murine homolog of the human immunodeficiency virus 1 entry coreceptor fusin. *Proc. Natl Acad. Sci. USA* **93**, 14726–14729.

Nomura, H., Nielsen, B. W., and Matsushima, K. (1993). Molecular cloning of cDNAs encoding a LD78 receptor and putative leukocyte chemotactic peptide receptors. *Int. Immunol.* **5**, 1239–1249.

Oberlin, E., Amara, A., Bachelerie, F., Bessia, C., Virelizier, J. L., Arenzana-Seisdedos, F., Schwartz, O., Heard, J. M., Clark-Lewis, I., Legler, D. F., Loetscher, M., Baggiolini, M., and Moser, B. (1996). The CXC chemokine SDF-1 is the ligand for LESTR/fusin and prevents infection by T-cell-line-adapted HIV-1. *Nature* **382**, 833–835.

Ostrowski, M. A., Justement, S. J., Catanzaro, A., Hallahan, C. A., Ehler, L. A., Mizell, S. B., Kumar, P. N., Mican, J. A., Chun, T. W., and Fauci, A. S. (1998). Expression of chemokine receptors CXCR4 and CCR5 in HIV-1-infected and uninfected individuals. *J. Immunol.* **161**, 3195–3201.

Penton-Rol, G., Polentarutti, N., Luini, W., Borsatti, A., Mancinelli, R., Sica, A., Sozzani, S., and Mantovani, A. (1998). Selective inhibition of expression of the chemokine receptor CCR2 in human monocytes by IFN-gamma. *J. Immunol.* **160**, 3869–3873.

Schols, D., Struyf, S., Van Damme, J., Este, J. A., Henson, G., and De Clercq, E. (1997). Inhibition of T-tropic HIV strains by selective antagonization of the chemokine receptor CXCR4. *J. Exp. Med.* **186**, 1383–1388.

Signoret, N., Oldridge, J., Pelchen-Matthews, A., Klasse, P. J., Tran, T., Brass, L. F., Rosenkilde, M. M., Schwartz, T. W., Holmes, W., Dallas, W., Luther, M. A., Wells, T. N., Hoxie, J. A., and Marsh, M. (1997). Phorbol esters and SDF-1 induce rapid endocytosis and down modulation of the chemokine receptor CXCR4. *J. Cell Biol.* **139**, 651–664.

Tachibana, K., Hirota, S., Iizasa, H., Yoshida, H., Kawabata, K., Kataoka, Y., Kitamura, Y., Matsushima, K., Yoshida, N., Nishikawa, S., Kishimoto, T., and Nagasawa, T. (1998). The chemokine receptor CXCR4 is essential for vascularization of the gastrointestinal tract. *Nature* **393**, 591–594.

Tanabe, S., Heesen, M., Yoshizawa, I., Berman, M. A., Luo, Y., Bleul, C. C., Springer, T. A., Okuda, K., Gerard, N., and Dorf, M. E. (1997). Functional expression of the CXC-chemokine receptor-4/fusin on mouse microglial cells and astrocytes. *J. Immunol.* **159**, 905–911.

Tarasova, N. I., Stauber, R. H., and Michejda, C. J. (1998). Spontaneous and ligand-induced trafficking of CXC-chemokine receptor 4. *J. Biol. Chem.* **273**, 15883–15886.

Vallat, A. V., De Girolami, U., He, J., Mhashilkar, A., Marasco, W., Shi, B., Gray, F., Bell, J., Keohane, C., Smith, T. W., and Gabuzda, D. (1998). Localization of HIV-1 co-receptors CCR5 and CXCR4 in the brain of children with AIDS. *Am. J. Pathol.* **152**, 167–178.

Volin, M. V., Joseph, L., Shockley, M. S., and Davies, P. F. (1998). Chemokine receptor CXCR4 expression in endothelium. *Biochem. Biophys. Res. Commun.* **242**, 46–53.

Wang, J. M., Ueda, H., Howard, O. M., Grimm, M. C., Chertov, O., Gong, X., Gong, W., Resau, J. H., Broder, C. C., Evans, G., Arthur, L. O., Ruscetti, F. W., and Oppenheim, J. J. (1998). HIV-1 envelope gp120 inhibits the monocyte response to chemokines through CD4 signal-dependent chemokine receptor down-regulation. *J. Immunol.* **161**, 4309–4317.

Willimann, K., Legler, D. F., Loetscher, M., Roos, R. S., Delgado, M. B., Clark-Lewis, I., Baggiolini, M., and Moser, B. (1998). The chemokine SLC is expressed in T cell areas of lymph nodes and mucosal lymphoid tissues and attracts activated T cells via CCR7. *Eur. J. Immunol.* **28**, 2025–2034.

Yoshida, R., Nagira, M., Kitaura, M., Imagawa, N., Imai, T., and Yoshie, O. (1998). Secondary lymphoid-tissue chemokine is a functional ligand for the CC chemokine receptor CCR7. *J. Biol. Chem.* **273**, 7118–7122.

Zella, D., Barabitskaja, O., Burns, J. M., Romerio, F., Dunn, D. E., Revello, M. G., Gerna, G., Reitz, M. S. Jr., Gallo, R. C., and Weichold, F. F. (1998). Interferon-gamma increases expression of chemokine receptors CCR1, CCR3, and CCR5, but not CXCR4 in monocytoid U937 cells. *Blood* **91**, 4444–4450.

Zhang, L., He, T., Talal, A., Wang, G., Frankel, S. S., and Ho, D. D. (1998). *In vivo* distribution of the human immunodeficiency virus/simian immunodeficiency virus coreceptors: CXCR4, CCR3, and CCR5. *J. Virol.* **72**, 5035–5045.

Zoeteweij, J. P., Golding, H., Mostowski, H., and Blauvelt, A. (1998). Cytokines regulate expression and function of the HIV coreceptor CXCR4 on human mature dendritic cells. *J. Immunol.* **161**, 3219–3223.

Zou, Y. R., Kottmann, A. H., Kuroda, M., Taniuchi, I., and Littman, D. R. (1998). Function of the chemokine receptor CXCR4 in haematopoiesis and in cerebellar development. *Nature* **393**, 595–599.

CXCR5

Reinhold Förster*

Molecular Tumor Genetics and Immunogenetics, Max-Delbruck-Center for Molecular Medicine, Robert Rossle Strasse 10, Berlin, 13092, Germany

* corresponding author tel: +49-30-9406-3330, fax: +49-30-9406-3884, e-mail: rfoerst@mdc-berlin.de

DOI: 10.1006/rwcy.2000.21005.

SUMMARY

The chemokine receptor CXCR5/BLR1 is expressed on mature B cells and a small subpopulation of T memory cells. Using gene targeting in mice, CXCR5 could be identified as the first chemokine receptor involved in homeostatic trafficking of lymphocytes to and within secondary lymphoid organs. CXCR5-deficient mice lack B cell follicles in spleen and Peyer's patches. As the chemokine BLC/BCA-1, a ligand for CXCR5, is expressed on follicular stroma cells, both molecules represent a receptor-ligand pair that mediates the formation of follicles in lymphoid organs. In addition, CXCR5-deficient mice lack defined lymph nodes, suggesting that this chemokine receptor also plays a pivotal role in lymphoid organ development.

BACKGROUND

Discovery

Chromosomal translocations causing deregulated expression of the proto-oncogene MYC had been delineated as an essential but not sufficient event during the development of a B cell neoplasia known as *Burkitt's lymphoma*. In order to identify additional factors involved in the pathogenesis of this disease, Dobner *et al.* (1992) compared the Burkitt's lymphoma cell line BL64 with the lymphoblastoid cell line IARC549 by means of subtractive hybridization. Using this approach, a novel G protein-coupled receptor was identified which was predicted to belong to the chemokine family as it contained the DRYLA motif at the end of the third transmembrane domain. Northern blot analysis revealed that this transcript was exclusively expressed in mature B cells and Burkitt's lymphoma cells and thus it was termed Burkitt's lymphoma receptor 1 (BLR1).

Alternative names

Burkitt's lymphoma receptor 1, BLR1, NLR, CXCR5.

Structure

CXCR5 is a member of the G protein-coupled seven transmembrane chemokine receptor family.

Main activities and pathophysiological roles

The restricted expression pattern of CXCR5 on mature B cells and a subpopulation of T memory cells suggested that this receptor might function as a regulator of B cell migration (Förster *et al.*, 1994). This view was recently confirmed by analyzing gene-targeted mice lacking CXCR5. These mice lack inguinal lymph nodes, possess none or only few aberrantly developed Peyer's patches and show altered primary and secondary lymphoid follicles in the spleen, but have normal mesenteric and peripheral lymph nodes. Upon adoptive transfer in wild-type animals, B cells isolated from CXCR5-mutant mice failed to migrate from the outer periarteriolar lymphoid sheath (PALS) to the B cell follicles (Förster *et al.*, 1996). These experiments identified CXCR5 as the first chemokine receptor directing the homing of B cells to lymphoid organs and defined anatomic compartments within. The chemokine BLC/BCA-1 is the only physiological ligand for CXCR5 identified so far.

Figure 1 Nucleotide sequence of the translated cDNA encoding human CXCR5. The codons for the first and last amino acid are underlined.

```
GCTGCCACCT CTCTAGAGGC ACCTGGCGGG GAGCCTCTCA ACATAAGACA GTGACCAGTC     60
TGGTGACTCA CAGCCGGCAC AGCCATGAAC TACCCGCTAA CGCTGGAAAT GGACCTCGAG    120
AACCTGGAGG ACCTGTTCTG GGAACTGGAC AGATTGGACA ACTATAACGA CACCTCCCTG    180
GTGGAAAATC ATCTCTGCCC TGCCACAGAG GGTCCCCTCA TGGCCTCCTT CAAGGCCGTG    240
TTCGTGCCCG TGGCCTACAG CCTCATCTTC CTCCTGGGCG TGATCGGCAA CGTCCTGGTG    300
CTGGTGATCC TGGAGCGGCA CCGGCAGACA CGCAGTTCCA CGGAGACCTT CCTGTTCCAC    360
CTGGCCGTGG CCGACCTCCT GCTGGTCTTC ATCTTGCCCT TTGCCGTGGC CGAGGGCTCT    420
GTGGGCTGGG TCCTGGGGAC CTTCCTCTGC AAAACTGTGA TTGCCCTGCA CAAAGTCAAC    480
TTCTACTGCA GCAGCCTGCT CCTGGCCTGC ATCGCCGTGG ACCGCTACCT GGCCATTGTC    540
CACGCCGTCC ATGCCTACCG CCACCGCCGC CTCCTCTCCA TCCACATCAC CTGTGGGACC    600
ATCTGGCTGG TGGGCTTCCT CCTTGCCTTG CCAGAGATTC TCTTCGCCAA AGTCAGCCAA    660
GGCCATCACA ACAACTCCCT GCCACGTTGC ACCTTCTCCC AAGAGAACCA AGCAGAAACG    720
CATGCCTGGT TCACCTCCCG ATTCCTCTAC CATGTGGCGG GATTCCTGCT GCCCATGCTG    780
GTGATGGGCT GGTGCTACGT GGGGGTAGTG CACAGGTTGC GCCAGGCCCA GCGGCGCCCT    840
CAGCGGCAGA AGGCAGTCAG GGTGGCCATC CTGGTGACAA GCATCTTCTT CCTCTGCTGG    900
TCACCCTACC ACATCGTCAT CTTCCTGGAC ACCCTGGCGA GGCTGAAGGC CGTGGACAAT    960
ACCTGCAAGC TGAATGGCTC TCTCCCCGTG GCCATCACCA TGTGTGAGTT CCTGGGCCTG   1020
GCCCACTGCT GCCTCAACCC CATGCTCTAC ACTTTCGCCG GCGTGAAGTT CCGCAGTGAC   1080
CTGTCGCGGC TCCTGACCAA GCTGGGCTGT ACCGGCCCTG CCTCCCTGTG CCAGCTCTTC   1140
CCTAGCTGGC GCAGGAGCAG TCTCTCTGAG TCAGAGAATG CCACCTCTCT CACCACGTTC   1200
TAGGTCCCAG TGTCCCCTTT TATTGCTGCT TTTCCTTGGG GCAGGCAGTG ATGCTGGATG   1260
```

GENE

Accession numbers

EMBL:
Human CXCR5 cDNA: X68149, S48709, S48717
Murine CXCR5 cDNA: X71788
Rat CXCR5 cDNA: X71463, S59748
Human CXCR5 promoter: X83755
Murine CXCR5 promoter: X83756

Figure 2 Schematic representation of putative binding sites of the human CXCR5 promoter. Those sites which were identified by mutational analysis to confirm cell- and differentiation-specific expression of CXCR5 are depicted in red. (Full colour figure can be viewed online.)

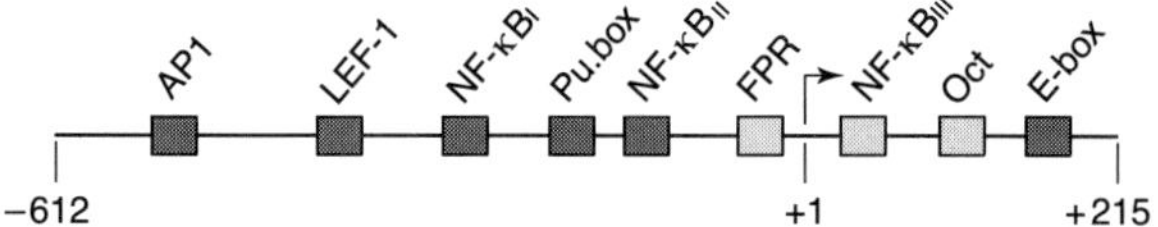

Sequence

See **Figure 1**. The cDNA sequence of human CXCR5 contains an open reading frame of 1116 bp coding for a 372 amino acid protein. Genomic analysis of the human CXCR5 gene shows that it consists of two exons separated by a 9 kb intron (Dobner *et al.*, 1992). The first exon encodes for the first 17 amino acids.

A similar situation has been found in the mouse. Murine CXCR5 consists of 374 amino acids. The first 19 amino acids are encoded about 12 kb upstream of the second exon (Kaiser *et al.*, 1993). In addition to the full-length transcript an alternative form of CXCR5, termed MDR15, has been reported. The two forms differ in the 5′ region, where the open reading frame of MDR15 is shorter by 45 codons. This cDNA would encode a 327 amino acid protein lacking almost the entire N-terminus of CXCR5 (Barella *et al.*, 1995).

A single specific start side of the human CXCR5 transcript could be mapped 267 bp upstream of the ATG start codon (Wolf *et al.*, 1998). Alignment of the nucleotide sequence of the human and murine CXCR5 promoter identified highly conserved regions for potential regulatory elements. They include an AP-1, a LEF-1, a purine-rich sequence (Pu box), a non-canonical octamer side (Oct), an E-box as well as three NFκB-binding motifs. Neither the human nor the murine promoter show CCAAT or TATA boxes or initiator-like sequences. Mutational analysis revealed that three essential elements confer cell type and differentiation-specific expression to B cells: the octamer motif (+157 with respect to the transcription start site), the NFκB$_{III}$ side (+44), and a functional promoter region (−36). **Figure 2** depicts the putative binding sites of the human CXCR5 promoter and functionally important binding sites. The importance of the NFκB- and Oct-binding sites was confirmed by *in vivo* studies using gene-targeted mice deficient either in both NFκB subunits p50 and p52, in Oct-2 or Bob1. The latter factor is known to act as a B cell-specific coactivator of octamer-binding factors. In all these mutated animals, expression of CXCR5 on

B cells was reduced or even totally absent in the case of p50/p52 double-mutant mice (Wolf *et al.*, 1998).

PROTEIN

Accession numbers

SwissProt:
Human CXCR5: P32302
Murine CXCR5: Q04683
Rat CXCR5: P34997

Sequence

See **Figure 3**.

Description of protein

CXCR5 shows all the characteristic features of chemokine receptors including seven hydrophobic transmembrane-spanning domains and the DRYLAIVHA motif at the end of transmembrane domain III. Murine and rat CXCR5 consist of 374 amino acids. The human homolog lacks two residues at positions 13 and 14 of the extracellular N-terminus and thus consists of 372 amino acids. Human and murine CXCR5 show an overall identity of 83%. The extracellular N-terminal domain of the receptor is less conserved between the two species, where an identity of 47% can be observed (Kaiser *et al.*, 1993). Murine and rat CXCR5 show an overall 92% amino acid identity (Kouba *et al.*, 1993). Human CXCR5 contains two potential *N*-linked glycosylation motifs at position Asn28 (N-terminus) and Asn196 (second extracellular domain) and three putative C-terminal phosphorylation sites at Ser354, Ser359, and Ser361 (see Figure 3) (Dobner *et al.*, 1992).

Relevant homologies and species differences

CXCR5 shows considerable homology to other members of the chemokine receptor family including CXCR2 (37% amino acid identity), CXCR1 (35%), CCR7 (33%), and CXCR4 (30%).

Affinity for ligand(s)

The chemokine BLC/BCA-1 has been recently identified as a ligand for CXCR5 (Gunn *et al.*, 1998; Legler *et al.*, 1998). Binding affinities have not yet been determined.

Cell types and tissues expressing the receptor

Using monoclonal antibodies directed against the N-terminus, CXCR5 was identified as the first lymphoid-specific member of the chemokine receptor family. In both human and murine peripheral blood cells, expression of CXCR5 is restricted to B cells (**Figure 4a**) and to minor subpopulations of CD4+ (approx. 15%; Figure 4b) and CD8+ (approx. 2–8%, Figure 4c) T cells. However, in contrast to most other chemokine receptors, CXCR5 is not expressed on monocytes or neutrophils (Figure 4d) (Förster *et al.*, 1994, 1996). Immunohistology on secondary lymphoid organs such as human tonsils and spleen demonstrated that all B cells found in the marginal and mantle zone express high levels of CXCR5, whereas most of the germinal center cells do not express this chemokine receptor (Förster *et al.*, 1994). After the murine and the rat homolog of CXCR5 had been identified, it was reported that in addition to the lymphoid system CXCR5 is also expressed in defined areas of the cerebrum and the cerebellum

Figure 3 Amino acid sequence of human CXCR5: Putative transmembrane domains are shaded and depicted by Roman numbers. The boundaries of the two exons (blue), putative sites for disulfide bonds (yellow), phosphorylation (green), and glycosylation (red) are shown. (Full colour figure can be viewed online.)

```
MNYPLTLEMDLENLEDLFWELDRLDNYNDTSLVENHLCPATEGPLMASFKAVFVPVAYSL   60
IFLLGVIGNVLVLVILERHRQTRSSTETFLFHLAVADLLLVFILPFAVAEGSVGWVLGTF  120
LCKTVIALHKVNFYCSSLLLACIAVDRYLAIVHAVHAYRHRRLLSIHITCGTIWLVGFLL  180
ALPEILFAKVSOGHHNNSLPRCTFSQENQAETHAWFTSRFLYHVAGFLLPMLVMGWCYVG  240
VVHRLRQAQRRPQRQKAVRVAILVTSIFFLCWSPYHIVIFLDTLARLKAVDNTCKLNGSL  300
PVAITMCEFLGLAHCCLNPMLYTFAGVKFRSDLSRLLTKLGCTGPASLCOLFPSWRRSSL  360
SESENATSLTIF
```

Figure 4 Conserved expression pattern of CXCR5 on human and murine white blood cells (WBC). Leukocytes were isolated from peripheral blood and stained with biotinylated monoclonal antibodies specific for human or murine CXCR5. Cells were stained with streptavidin-cychrome and FITC- or PE-labeled antibodies as indicated and analyzed by flow cytometry.

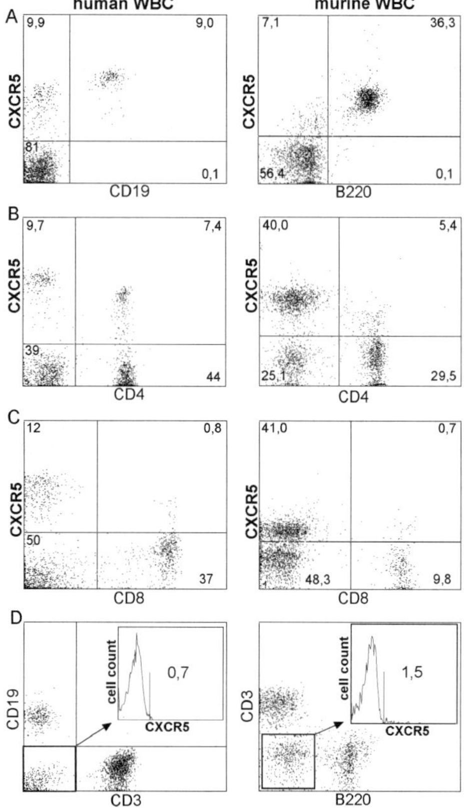

(Kaiser *et al.*, 1993; Kouba *et al.*, 1993). In contrast to the situation observed in the immune system, there are, at present, no data available concerning a possible role of CXCR5 in the central nervous system.

Organs, tissues, and cells expressing CXCR5 include spleen, lymph nodes, tonsil, brain, bone marrow, B cells, T memory cells, cerebrum, cerebellum, hippocampus, pituitary.

CXCR5-positive cell lines: BL64, BL21, JBL2, JI, LY66, LY91, Daudi, BL40, BL70, BL72, ESIII, Raji, BL60, BL90, BL99, BL106, LY67, WEHI 231, WEHI 279, 2PK3, A20.2J, M12-13, BFO.3.

Regulation of receptor expression

CXCR5 is primarily expressed on resting cells. Activation of B cells *in vitro* with anti-CD40 monoclonal antibodies and IL-4 leads to complete downregulation of surface CXCR5. The same has been observed on T cells after stimulation with anti-CD3 monoclonal antibody (Förster *et al.*, 1994). In contrast, activating naïve T cells with OX40 ligand induces expression of CXCR5 on these cells (Flynn *et al.*, 1998).

SIGNAL TRANSDUCTION

As shown for most members of the chemokine receptor family, pertussis toxin can inhibit B cell migration and Ca^{2+}-flux *in vitro*, suggesting that CXCR5 couples to $G\alpha_i$ proteins.

BIOLOGICAL CONSEQUENCES OF ACTIVATING OR INHIBITING RECEPTOR AND PATHOPHYSIOLOGY

Unique biological effects of activating the receptors

The chemokine BLC/BCA-1 is the only natural ligand for CXCR5 identified so far. Stimulating leukocytes *in vitro* with this chemokine induces B cell migration with high efficacy but also induces a weak chemotactic response in T cells and macrophages (Gunn *et al.*, 1998; Legler *et al.*, 1998).

Phenotypes of receptor knockouts and receptor overexpression mice

Data derived from CXCR5-gene targeted mice support the idea that CXCR5 is a major regulator of B cell migration (Förster *et al.*, 1996). CXCR5-deficiency prevents the formation of inguinal lymph nodes and severely affects the development of Peyer's patches (PPs). In 55% of all CXCR5−/− mice no PPs could be identified. In about 30% of the mice, small rudimentary structures could be identified resembling PPs, whereas in about 15% of the mice few regularly shaped PPs were found. Although outwardly normal, these PPs revealed severe histological alterations as no differentiation in B cell- and T cell-rich areas could be detected.

Further morphological and functional alterations in CXCR5-mutant mice can be found in the spleen. Compared with the situation in wild-type animals,

Figure 5 Transferred B cells from CXCR5-deficient mice fail to enter splenic B cell follicles. B cells (green) isolated from wild-type (left) or CXCR5-mutant (right) mice were intravenously transferred into wild-type recipients. After 5 hours wild-type B cells (left) had populated the T cell-rich zone (blue) and the B cell follicle (red). In contrast, B cells derived from CXCR5-deficient mice accumulated at the outer edge of the T cell zone but failed to migrate into the B cell follicle (right). (Full colour figure may be viewed online.)

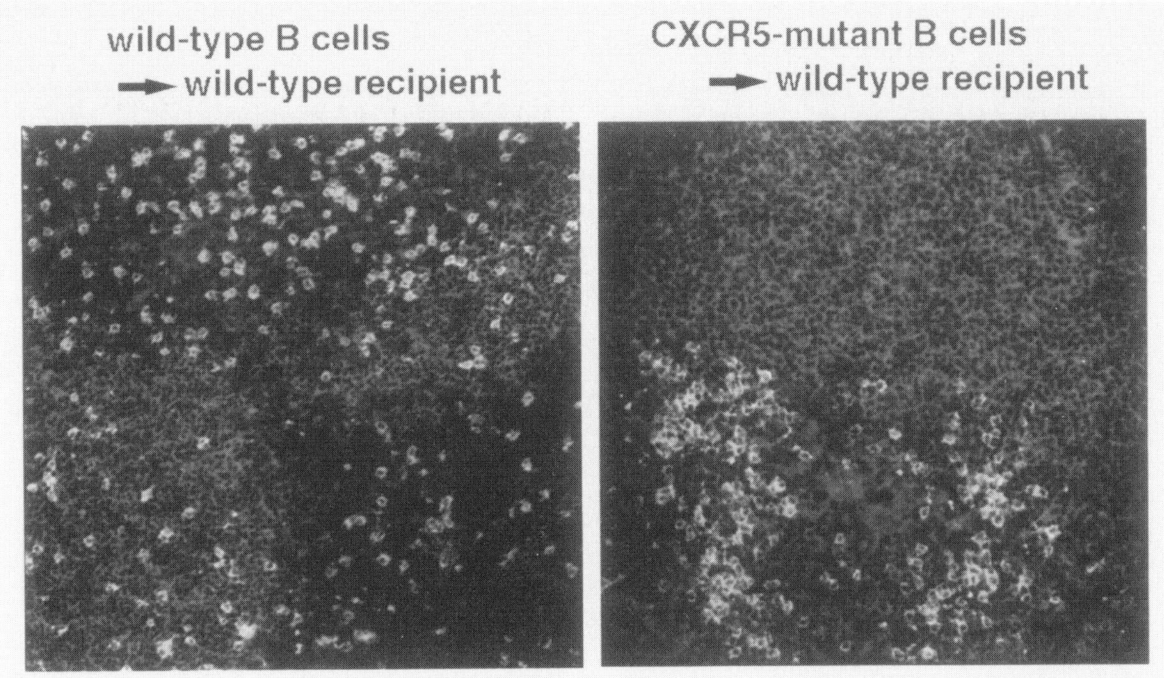

primary splenic follicles of CXCR5-deficient mice show three major differences: (1) the T cell zone is surrounded by a small rim of IgD+ IgM− cells, whereas very few cells can be found expressing both, IgM and IgD, (2) the marginal zone is enlarged, and (3) the T cell zone is situated centrally but not polarized in the follicle. After immunization with T cell-dependent antigens, the development of germinal centers is also impaired. Peanut agglutinin (PNA)-positive cells are not found in the B cell follicle but are scattered through the red pulp and within the T cell-rich zone around the central artery. Interestingly, mesenteric and peripheral lymph nodes do not show obvious differences in CXCR5-deficient mice when compared with wild-type animals. The aberrant follicular structure of spleen and PPs is caused by impaired B cell migration. Upon adoptive transfer, B cells isolated from CXCR5-mutant animals are not able to migrate to the B cell follicles in spleen (**Figure 5**) and PPs of wild-type recipients but accumulate at the outer edge of the T cell zone (Förster *et al.*, 1996). These data support the multistep model for lymphocyte migration and identify CXCR5 as the first chemokine receptor regulating microenvironmental homing.

Human abnormalities

Clinical studies demonstrate that CXCR5 is aberrantly expressed during the progression of the acquired immune deficiency syndrome (*AIDS*). As deregulated expression of CXCR5 on peripheral blood B and T cells can be observed even at the clinically latent stage, it has been suggested that this observation might reflect activating and remodeling processes leading to the destruction of lymphoid tissues by the human immunodeficiency virus, *HIV* (Förster *et al.*, 1997).

References

Barella, L., Loetscher, M., Tobler, A., Baggiolini, M., and Moser, B. (1995). Sequence variation of a novel heptahelical leucocyte receptor through alternative transcript formation. *Biochem. J.* **309**, 773–779.

Dobner, T., Wolf, I., Emrich, T., and Lipp, M. (1992). Differentiation-specific expression of a novel G protein-coupled receptor from Burkitt's lymphoma. *Eur. J. Immunol.* **22**, 2795–2799.

Flynn, S., Toellner, K. M., Raykundalia, C., Goodall, M., and Lane, P. (1998). CD4 T cell cytokine differentiation: the B cell activation molecule, OX40 ligand, instructs CD4 T cells to express interleukin 4 and upregulates expression of the chemokine receptor, Blr-1. *J. Exp. Med.* **188**, 297–304.

Förster, R., Emrich, T., Kremmer, E., and Lipp, M. (1994). Expression of the G-protein-coupled receptor BLR1 defines mature recirculating B cells and a subset of T memory helper cells. *Blood* **84**, 830–840.

Förster, R., Mattis, E. A., Kremmer, E., Wolf, E., Brem, G., and Lipp, M. (1996). A putative chemokine receptor, BLR1, directs B cell migration to defined lymphoid organs and specific anatomic compartments of the spleen. *Cell* **87**, 1037–1047.

Förster, R., Schweigard, G., Johann, S., Emrich, T., Kremmer, E., Nerl, C., and Lipp, M. (1997). Abnormal expression of the B-cell homing chemokine receptor BLR1 during the progression of acquired immunodeficiency syndrome. *Blood* **90**, 520–525.

Gunn, M. D., Ngo, V. N., Ansel, K. M., Ekland, E. H., Cyster, J. G., and Williams, L. T. (1998). A B-cell homing chemokine made in lymphoid follicles activates Burkitt's lymphoma receptor-1. *Nature* **391**, 799–803.

Kaiser, E., Förster, R., Wolf, I., Ebensperger, C., Kuehl, W. M., and Lipp, M. (1993). The G protein-coupled receptor BLR1 is involved in murine B cell differentiation and is also expressed in neuronal tissues. *Eur. J. Immunol.* **23**, 2532–2539.

Kouba, M., Vanetti, M., Wang, X., Schäfer, M., and Höllt, V. (1993). Cloning of a novel putative G-protein-coupled receptor (NLR) which is expressed in neuronal and lymphatic tissue. *FEBS Lett.* **321**, 173–178.

Legler, D. F., Loetscher, M., Roos, R. S., Clark-Lewis, I., Baggiolini, M., and Moser, B. (1998). B cell-attracting chemokine 1, a human CXC chemokine expressed in lymphoid tissues, selectively attracts B lymphocytes via BLR1/CXCR5. *J. Exp. Med.* **187**, 655–660.

Wolf, I., Pevzner, V., Kaiser, E., Bernhardt, G., Claudio, E., Siebenlist, U., Förster, R., and Lipp, M. (1998). Downstream activation of a TATA-less promoter by Oct-2, Bob1, and NFκB directs expression of the homing receptor BLR1 to mature B cells. *J. Biol. Chem.* **273**, 28831–28836.

LICENSED PRODUCTS

Monoclonal antibodies against human CXCR5 are distributed by R&D Systems.

CC, C, and CX3C Chemokine Receptors

CCR1

Richard Horuk*

Department of Immunology, Berlex Bioscience, 15049 San Pablo Avenue, Richmond, CA 94804, USA

*corresponding author tel: 510-669-4625, fax: 510-669-4244, e-mail: horuk@crl.com

DOI: 10.1006/rwcy.2000.22001.

SUMMARY

Chemokines belong to a large family of chemoattractant molecules involved in the directed migration of immune cells. They achieve their cellular effects by direct interaction with cell surface receptors. One such chemokine receptor is CCR1, which is particularly responsive to the CC chemokines RANTES and MIP-1α. This review seeks to highlight the biology, molecular biology, physiology, and pathophysiology of CCR1 and looks at the potential of this receptor as a therapeutic target in autoimmunity and inflammation.

BACKGROUND

Discovery

Although numerous reports had described specific effects of the chemokines RANTES and MIP-1α on T lymphocytes and monocytes, the identity of the putative receptor for these ligands was unknown (Schall, 1994). However, cloning of this receptor was aided by the fact that the primary sequences of the C5a, fMLP, and IL-8 receptors revealed domains which were conserved in receptors associated with cell motility, but not in other seven transmembrane-spanning receptors (Boulay *et al.*, 1990; Gerard and Gerard, 1991; Holmes *et al.*, 1991; Murphy and Tiffany, 1991). These similarities were exploited using PCR technology to obtain several orphan receptor cDNA clones which were then expressed and screened by receptor binding and functional assays. Using this homology hybridization cloning approach, the molecular cloning and functional expression of CCR1 was reported by two separate groups (Gao *et al.*, 1993; Neote *et al.*, 1993). The open reading frame for human CCR1 is on a single exon and predicts a protein of 355 amino acids (**Figure 1**).

The gene, cmkbr1, is located on human chromosome 3p21. The expressed human CCR1 was able to bind MIP-1α and RANTES with high affinity and physiological concentrations of both ligands induced an increase in intracellular Ca^{2+}. CCR1 was specific for these ligands and showed a poor response to MIP-1β and MCP-1. In addition to these ligands, CCR1 has been shown to respond with high affinity and to signal in response to a variety of other CC chemokines, including MCP-3, MPIF-1, leukotactin 1, and HCC-1.

Alternative names

CC CKR1, HM145, MIP-1α/R, MIP-1α/RANTES.

Structure

CCR1 is a putative seven transmembrane domain G protein-coupled receptor. For a more detailed discussion, see the section on Description of protein.

Main activities and pathophysiological roles

Assigning biological activities and elucidating pathophysiological roles for CCR1 is difficult for several reasons. First, several chemokines including MIP-1α, RANTES, MCP-3, leukotactin 1, and HCC-1 that bind with high affinity to CCR1 can also bind with high affinity and activate other chemokine receptors. RANTES, for example, can also bind to CCR3 and CCR5 and MIP-1α can also bind to CCR5. Second, there are no known commercially available neutralizing antibodies to CCR1. Consequently, current ideas regarding the physiologic and pathologic roles of

Figure 1 Proposed membrane topography of CCR1. Membrane spanning α helices are defined based on hydropathy analysis. CHO, potential *N*-linked glycosylation sites.

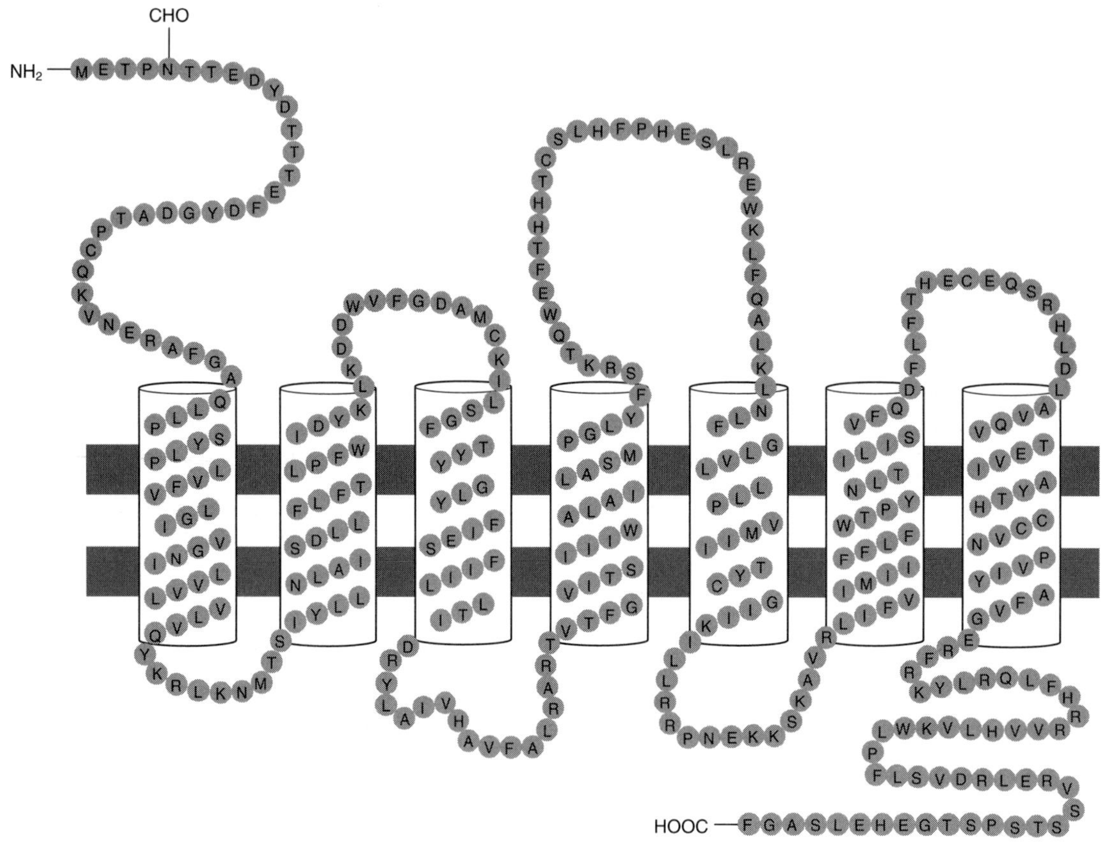

CCR1 come mainly from a consideration of the roles of its ligands in biology, recognizing that these ligands may also be acting upon other receptors. Knockout studies of CCR1 (see under the section on Phenotypes of receptor knockouts) have also provided some information of a rather limited nature, as have studies employing CCR1-transfected cell lines. For instance, MIP-1α and RANTES can induce predictable biological responses in CCR1-transfected cells – chemotaxis for example. While such information is useful, it does not convey a very clear picture of the role that CCR1 plays in the intact animal since, depending upon the physiological circumstances, the receptor may or may not even be expressed in its normal target cell.

Given these and other difficulties, a discussion of the potential pathophysiological roles of CCR1 (see the section on Unique biological effects of activating the receptors) is, of necessity, based upon a consideration of the roles of its ligands, which could of course also be acting upon other chemokine receptors. A clearer indication of the biological roles of CCR1 will come from access to CCR1 antagonists, a number of which are starting to become available (see the section on Effect of treatment with small molecule antagonists), which can be used in animal models of disease to elucidate directly the role of CCR1 in that disease. In this respect, further studies with CCR1 knockout mice will also be valuable.

GENE

Accession numbers

Human CCR1 was cloned independently by two separate groups (Gao *et al.*, 1993; Neote *et al.*, 1993). The accession number for the human mRNA sequence is L09230. So far, no description of the genomic sequence or promoter regulation of this receptor has been described. In addition to the human nucleotide sequence three other species of CCR1 have been described, mouse (Gao and Murphy, 1995; Post *et al.*, 1995), rat (Honda and Fujisawa, 1997), and rhesus monkey (Hauer *et al.*, unpublished results). The accession numbers for these sequences are U28404,

Figure 2 Nucleotide sequence of human CCR1.

```
   1 ggcttcccca ggactgttcc tgctccggct cttcaggctc cctgctttgt ccttttccac
  61 tgtccgcact gcatctgact cctgcagaga ccttgttctc ccacccgacc ttcctctctg
 121 tcctcccctc ccacctgccc ctcagttccc aggagactct tccggtgtaa ctctgatggc
 181 ctcctctggg tatgtcctcc aggcggagct ctccccctca actgagaact caagtcagct
 241 ggacttcgaa gatgtatgga attcttccta tggtgtgaat gattccttcc cagatggaga
 301 ctatgatgcc aacctggaag cagctgcccc ctgccactcc tgtaacctgc tggatgactc
 361 tgcactgccc ttcttcatcc tcaccagtgt cctgggtatc ctagctagca gcactgtcct
 421 cttcatgctt ttcagacctc tcttccgctg gcagctctgc cctggctggc ctgtcctggc
 481 acagctggct gtgggcagtg ccctcttcag cattgtggtg cccgtcttgg ccccagggct
 541 aggtagcact cgcagctctg ccctgtgtag cctgggctac tgtgtctggt atggctcagc
 601 ctttgcccag gctttgctgc tagggtgcca tgcctccctg ggccacagac tgggtgcagg
 661 ccaggtccca ggcctcaccc tggggctcac tgtgggaatt tggggagtgg ctgccctact
 721 gacactgcct gtcaccctgg ccagtggtgc ttctggtgga ctctgcaccc tgatatacag
 781 cacggagctg aaggctttgc aggccacaca cactgtagcc tgtcttgcca tctttgtctt
 841 gttgccattg ggtttgtttg gagccaaggg gctgaagaag gcattgggta tggggccagg
 901 cccctggatg aatatcctgt gggcctggtt tattttctgg tggcctcatg ggtggttct
 961 aggactggat ttcctggtga ggtccaagct gttgctgttg tcaacatgtc tggcccagca
1021 ggctctggac ctgctgctga acctggcaga agccctggca attttgcact gtgtggctac
1081 gcccctgctc ctcgccctat tctgccacca ggccacccgc accctcttgc cctctctgcc
1141 cctccctgaa ggatggtctt ctcatctgga caccсttgga agcaaatcct agttctcttc
1201 ccacctgtca acctgaatta aagtctacac tgcctttgtg
```

U29678, E13732, and AF 017282 respectively. The protein sequences are described below.

Sequence

The human CCR1 sequence has an open reading frame of 1065 bases, encoding a protein of 355 amino acids (**Figure 2**).

PROTEIN

Accession numbers

Human CCR1: SwissProt P32246, EMBL/GenBank L09230
Mouse CCR1: SwissProt P51675

Description of protein

The amino acid sequence of the human CCR1 protein shows several key features related to G protein-coupled receptors (GCPR) of the seven transmembrane-spanning receptor superfamily. For example, it has seven hydrophobic regions predicted to span the cell membrane, and cysteine residues in the first and the second extracellular loops that are implicated in forming a disulfide bond (Figure 1). However, certain features of the predicted CCR1 protein make it distinct from classical GCPRs. The C-terminus is relatively short and lacks cysteine residues involved in membrane anchorage via a palmitoylated moiety (O'Dowd *et al.*, 1989) and the hydrophobic membrane-spanning domains are relatively short – a feature consistent with other chemokine receptors (Horuk, 1994). There are three potential glycosylation sites in CCR1 (Figure 1); however, only the one in the N-terminus is likely to be glycosylated. There is also a consensus sequence for a protein kinase C phosphorylation site, at position 192, but this position is predicted to be extracellular.

Relevant homologies and species differences

The CCR1 protein from a number of species, including rhesus, mouse, and rat have been cloned, and they are aligned against human CCR1 in **Figure 3**. There is a high degree of sequence homology among all of these sequences: the human and rhesus are 87% identical and the rat and mouse proteins share 80% sequence identity with human CCR1. As expected, most of the salient features of the human CCR1 sequence are conserved and, where it exists, the variation is mainly confined to the N-terminus and the extracellular loops, regions which are likely involved in ligand binding. Human and mouse CCR1 proteins bind both human and mouse MIP-1α and RANTES with high affinity; however, human RANTES is 100-fold less potent than the MIP proteins in inducing

Figure 3 Alignment of the primary structures of the cloned human, rhesus, rat, and mouse CCR1. Conserved residues are boxed in orange. The solid overlines numbered with roman numerals depict the predicted 7TM spanning domains. (Full colour figure may be viewed online.)

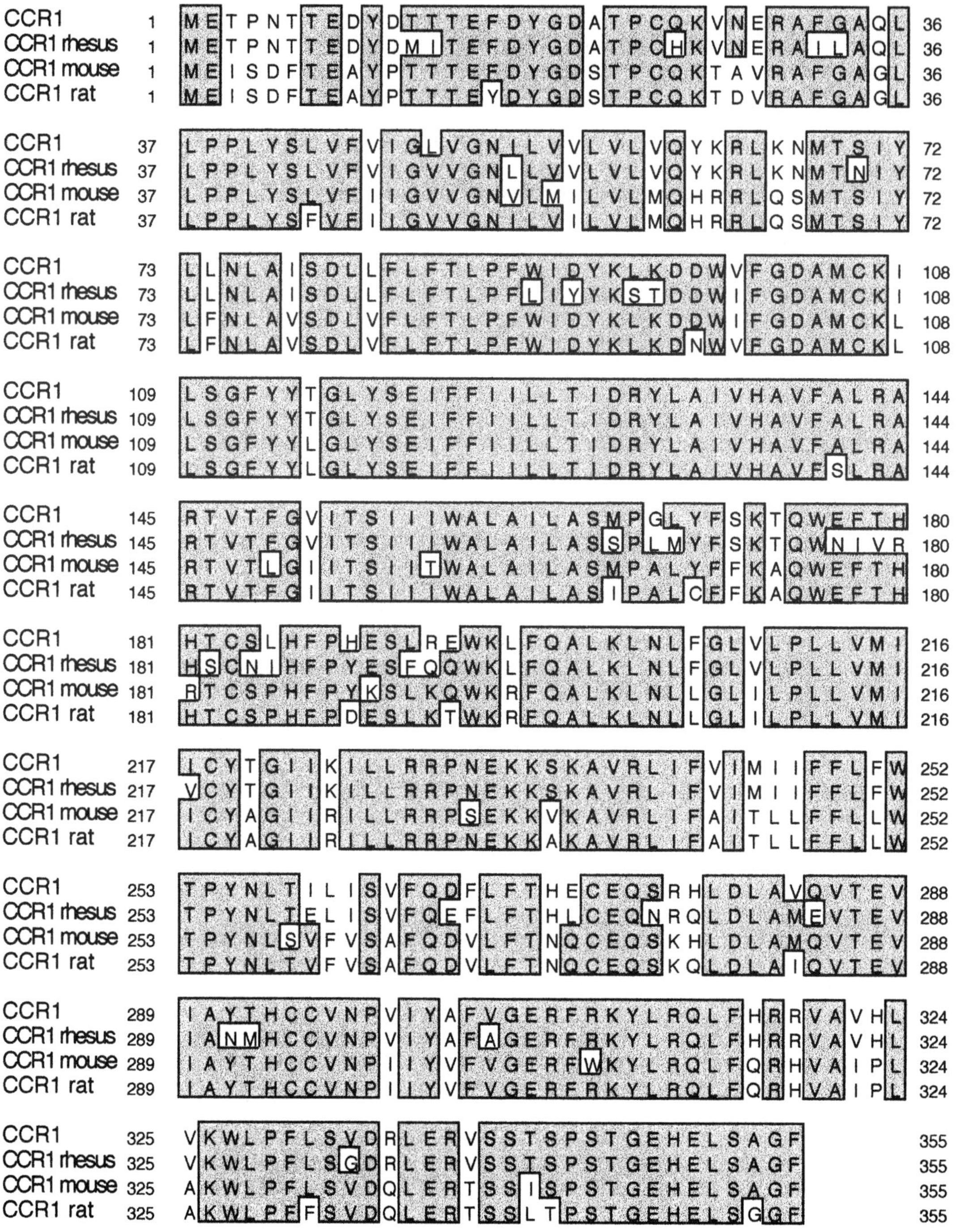

```
CCR1          1  METPNTTEDYDTTTEFDYGDATPCQKVNERAFGAQL  36
CCR1 rhesus   1  METPNTTEDYDMITEFDYGDATPCHKVNERAILAQL  36
CCR1 mouse    1  MEISDFTEAYPTTTEFDYGDSTPCQKTAVRAFGAGL  36
CCR1 rat      1  MEISDFTEAYPTTTEYDYGDSTPCQKTDVRAFGAGL  36

CCR1         37  LPPLYSLVFVIGLVGNILVVLVLVQYKRLKNMTSIY  72
CCR1 rhesus  37  LPPLYSLVFVIGVVGNLLVVLVLVQYKRLKNMTNIY  72
CCR1 mouse   37  LPPLYSLVFIIGVVGNVLMILVLMQHRRLQSMTSIY  72
CCR1 rat     37  LPPLYSFVFIIGVVGNILVILVLMQHRRLQSMTSIY  72

CCR1         73  LLNLAISDLLFLFTLPFWIDYKLKDDWVFGDAMCKI  108
CCR1 rhesus  73  LLNLAISDLLFLFTLPFLIYYKSTDDWIFGDAMCKI  108
CCR1 mouse   73  LFNLAVSDLVFLFTLPFWIDYKLKDDWIFGDAMCKL  108
CCR1 rat     73  LFNLAVSDLVFLFTLPFWIDYKLKDNWVFGDAMCKL  108

CCR1        109  LSGFYYTGLYSEIFFIILLTIDRYLAIVHAVFALRA  144
CCR1 rhesus 109  LSGFYYTGLYSEIFFIILLTIDRYLAIVHAVFALRA  144
CCR1 mouse  109  LSGFYYLGLYSEIFFIILLTIDRYLAIVHAVFALRA  144
CCR1 rat    109  LSGFYYLGLYSEIFFIILLTIDRYLAIVHAVFSLRA  144

CCR1        145  RTVTFGVITSIIIWALAILASMPGLYFSKTQWEFTH  180
CCR1 rhesus 145  RTVTFGVITSIIIWALAILASSPLMYFSKTQWNIVR  180
CCR1 mouse  145  RTVTLGIITSIITWALAILASMPALYFFKAQWEFTH  180
CCR1 rat    145  RTVTFGIITSIIIWALAILASIPALCFFKAQWEFTH  180

CCR1        181  HTCSLHFPHESLREWKLFQALKLNLFGLVLPLLVMI  216
CCR1 rhesus 181  HSCNIHFPYESFQQWKLFQALKLNLFGLVLPLLVMI  216
CCR1 mouse  181  RTCSPHFPYKSLKQWKRFQALKLNLLGLILPLLVMI  216
CCR1 rat    181  HTCSPHFPDESLKTWKRFQALKLNLLGLILPLLVMI  216

CCR1        217  ICYTGIIKILLRRPNEKKSKAVRLIFVIMIIFFLFW  252
CCR1 rhesus 217  VCYTGIIKILLRRPNEKKSKAVRLIFVIMIIFFLFW  252
CCR1 mouse  217  ICYAGIIRILLRRPSEKKVKAVRLIFAITLLFFLLW  252
CCR1 rat    217  ICYAGIIRILLRRPNEKKAKAVRLIFAITLLFFLLW  252

CCR1        253  TPYNLTILISVFQDFLFTHECEQSRHLDLAVQVTEV  288
CCR1 rhesus 253  TPYNLTELISVFQEFLFTHLCEQNRQLDLAMEVTEV  288
CCR1 mouse  253  TPYNLSVFVSAFQDVLFTNQCEQSKHLDLAMQVTEV  288
CCR1 rat    253  TPYNLTVFVSAFQDVLFTNQCEQSKQLDLAIQVTEV  288

CCR1        289  IAYTHCCVNPVIYAFVGERFRKYLRQLFHRRVAVHL  324
CCR1 rhesus 289  IANMHCCVNPVIYAFAGERFRKYLRQLFHRRVAVHL  324
CCR1 mouse  289  IAYTHCCVNPIIYVFVGERFWKYLRQLFQRHVAIPL  324
CCR1 rat    289  IAYTHCCVNPIIYVFVGERFRKYLRQLFQRHVAIPL  324

CCR1        325  VKWLPFLSVDRLERVSSTSPSTGEHELSAGF  355
CCR1 rhesus 325  VKWLPFLSGDRLERVSSTSPSTGEHELSAGF  355
CCR1 mouse  325  AKWLPFLSVDQLERTSSISPSTGEHELSAGF  355
CCR1 rat    325  AKWLPFFSVDQLERTSSLTPSTGEHELSGGF  355
```

Ca^{2+} transients (Neote *et al.*, 1993; Gao and Murphy, 1995).

The cloned human CCR1 protein also has a striking homology with an open reading frame in human *cytomegalovirus*, designated US28. There is around 60% amino acid identity in the presumed extracellular N-terminus prior to TM1, but only about 35% identity over the entire length of the predicted proteins. Receptor-binding studies showed that the protein encoded by US28 can bind CC chemokines with similar high affinity, but not CXC chemokines (Neote *et al.*, 1993). The US28 protein was shown to signal by producing a Ca^{2+} flux in response to RANTES and MIP-1α both in US28-transfected cells and in the context of virus-infected cells (Gao and Murphy, 1994; Billstrom *et al.*, 1998).

Affinity for ligand(s)

Human CCR1 has been shown to respond to a number of human CC chemokines in a variety of assays, including calcium mobilization, inhibition of adenylyl cyclase, increase in extracellular acidification, and chemotaxis. The range of chemokines that can signal through CCR1 is broad and includes MIP-1α, RANTES, MCP-2, MCP-3, MCP-4, leukotactin 1, MPIF-1, and HCC-1 (Neote *et al.*, 1993; Coulin *et al.*, 1997; Gong *et al.*, 1997; Sarau *et al.*, 1997; Youn *et al.*, 1997, 1998; Pardigol *et al.*, 1998). All of these ligands are potent agonists for human CCR1 (EC_{50} values lower than 100 nM). In addition, MIP-1α, RANTES, MCP-2, MCP-3, and leukotactin 1 have been shown to bind to human CCR1 with high affinity (K_i values lower than 50 nM: Neote *et al.*, 1993; Coulin *et al.*, 1997; Gong *et al.*, 1997; Sarau *et al.*, 1997). Human CCR1 is also able to bind human MIP-1β and MCP-1 with low affinity (greater than 100 nM) but neither ligand is able to signal (Neote *et al.*, 1993). Recently, a number of small molecule antagonists and peptide antagonists have been described.

Cell types and tissues expressing the receptor

A rabbit polyclonal antibody raised to the N-terminus of CCR1 demonstrated receptor expression in monocytes and lymphocytes but not in neutrophils (Su *et al.*, 1996). In contrast, mouse neutrophils express CCR1 (Gao *et al.*, 1997); the physiological basis for these differences is not known. CCR1 was also expressed on CD3+, CD4+, CD8+, and CD16+ peripheral blood lymphocytes but not on CD19+ B lymphocytes (Su *et al.*, 1996). Among CD4+ peripheral blood lymphocytes, CD45RO+ cells expressed a larger number of CCR1 compared with CD45RO−. In addition, CD34+ cells in human bone marrow as well as cord blood were uniformly stained with this antibody. CCR1 is also expressed on glycophorin A-positive erythroblasts in addition to lymphocytes and granulocytes (Su *et al.*, 1997). In addition, CCR1 RNA has been detected in human dendritic cells (Sozzani *et al.*, 1997) and the receptor also appears to be expressed in human eosinophils (Elsner *et al.*, 1997).

Regulation of receptor expression

A common feature of the chemokine receptors, including CCR1, is the presence of a serine/threonine-rich C-terminal region which is a target for phosphorylation (Horuk, 1994). Many G protein-coupled receptors are 'switched off' after agonist stimulation by the phosphorylation of C-terminal cytoplasmic residues. This process, termed desensitization, has been well documented for the β-adrenergic and rhodopsin receptors, and the receptor kinases involved in receptor regulation by these means have been characterized. For example, Mueller *et al.* (1994) have shown that CXCR2 is phosphorylated under basal conditions. MGSA treatment of cells transfected with CXCR2 resulted in a further increase in receptor phosphorylation on serine residues. The nature of the kinases involved in mediating these events is unknown, as are the physiological consequences of their actions. Similar data are available for other chemokine receptors, including CXCR4 (Haribabu *et al.*, 1997). The ligand for CXCR4, SDF-1, was shown to induce rapid phosphorylation and desensitization of CXCR4. SDF-1 as well as PMA induced rapid internalization of CXCR4. These studies indicate that signaling and internalization of chemokine receptors are regulated by receptor phosphorylation, although formal evidence that this is the case for CCR1 is lacking so far.

Programmed expression of chemokine receptors probably plays a very important role in ensuring that an appropriate immune response is mounted against a particular pathogen. For example, the selection of the appropriate effector T cells, i.e. TH1 or TH2 can be influenced by the expression of chemokine receptors which are regulated by cytokines. Recent studies have clearly demonstrated that chemokine receptor expression and association with TH1 and TH2 phenotypes is affected by cytokines present during polarization (Sallusto *et al.*, 1998). Thus, while TGFβ inhibits CCR3 expression, it enhances CCR4 and CCR7 expression, whereas IFNα inhibits CCR3 but upregulates CXCR3 and CCR1. These results demonstrate that flexible programs of chemokine receptor gene expression may control tissue-specific migration of effector T cells and play an important role in disease.

Chemokine receptor expression in immune cells can also be profoundly influenced during an infection by bacterial constituents such as lipopolysaccharide (LPS). In a recent study, LPS treatment of monocytes caused a rapid and drastic reduction of CCR1, CCR2, and CCR5 mRNA levels by reducing mRNA half-life. As expected, LPS-induced inhibition of chemokine receptor mRNA expression was associated with a reduction in binding and chemotactic responsiveness (Sica *et al.*, 1997). Cytokines like IFNγ, released in response to infection, have also been shown selectively and rapidly to inhibit expression of some chemokine receptors –CCR2, for example – but have no effect on others like CCR1, CCR3, CCR4, and CCR5 (Penton-Rol *et al.*, 1998).

SIGNAL TRANSDUCTION

Associated or intrinsic kinases

No associated or intrinsic kinases specific for CCR1 have so far been described. However, a family of important regulatory kinases called G protein-coupled receptor kinase (GRK) have been described (Sibley *et al.*, 1987; Lefkowitz, 1993). These kinases are involved in phosphorylation of target serine and threonine residues that are concentrated in the carboxyl tail of most G protein-coupled receptors. The phosphorylated receptor is then able to interact with a class of soluble proteins called arrestins (Wilson and Applebury, 1993), which mediate trafficking to clathrin-coated pits (endocytosis) and could also involve recruitment of c-Src, and thereby activation of the MAPK signaling pathway.

Cytoplasmic signaling cascades

The cytoplasmic signaling cascades that are set into action following ligand binding to chemokine recepors have received considerable attention primarily by examining the biological signals induced by specific chemokines. One of the best studied is the CC chemokine RANTES which binds with high affinity to both CCR1 and CCR5. Since there are no known CCR1 and CCR5 neutralizing antibodies commercially available and until recently no specific receptor antagonists for either receptor existed, it has proven to be incredibly difficult to assign particular intracellular signaling events that are induced by RANTES to either receptor with any certainty. In spite of these difficulties, it is known that ligation of chemokine receptors by specific high-affinity ligands induces a conformational change that leads to a dissociation of the receptor-associated heterotrimeric G proteins (guanine nucleotide-binding proteins) into α and $\beta\gamma$ subunits.

There is now considerable evidence that both α and $\beta\gamma$ subunits can function as second messengers in G protein-coupled receptor signaling. Furthermore, the identification of specific adapter proteins, G protein-induced switching, and demonstration of interaction with receptor tyrosine kinases has increased the diversity of signaling that occurs through these receptors (Lefkowitz, 1998). Although the amount of information regarding intracellular signaling for chemokine receptors is limited compared with that from well-studied receptors such as the β_2-adrenergic receptor, it is nevertheless likely that similarities in signaling pathways for these proteins exist and that information gathered from one class of this family of receptors can shed insight into the intracellular signaling mechanisms of another. For example, it was recently shown that protein kinase A-mediated phosphorylation of the β_2-adrenergic receptor could switch the G protein coupling in this receptor from G_s to G_i (Lefkowitz, 1998). In this context it is interesting that chemokine receptors like CCR1 have also been shown to couple to multiple G proteins, G_i and G_q, primarily via contact points in the third intracellular loop (Arai and Charo, 1996). These differences in G protein coupling could affect intracellular signaling, for example G_q-coupled receptors stimulate mitogen-activated protein kinase (MAPK) by increasing phosphatidylinositol turnover, leading to the activation of protein kinase C and resulting in activation of Raf (Faure *et al.*, 1994). In contrast, G_i-coupled receptors stimulate MAPK via $\beta\gamma$ subunits, which can be independent of their ability to stimulate phospholipase C in some cell lines but not in others (Hawes *et al.*, 1995; Della-Rocca *et al.*, 1997).

Recently, RANTES has been shown to stimulate the tyrosine phosphorylation and activation of the cytoplasmic signaling protein Pyk-2 (Davis *et al.*, 1997; Dikic and Schlessinger, 1998). Pyk-2 is known to be involved in the recruitment of the tyrosine kinase c-Src and it is tempting to speculate that activation of these proteins could provide a potential link between chemokine receptors and the MAPK pathway. Coupling of receptors to the c-Src kinase is known to require the interaction of specific SH3 domains within the kinase, with specific phosphotyrosine motifs in tyrosine receptor kinases such as the EGF receptor. Although these phosphotyrosine motifs are not present in CCR1 or CCR5, it has recently been shown that arrestin, which binds to phosphorylated serine residues on G protein-coupled receptors, can act as an adapter protein that mediates the specific targeting of the c-Src kinase through a poly-Pro-SH3 domain interaction. Whether CCR1 is involved in such signaling cascades remains to be seen; however, a negative regulatory role for the CCR1 ligand MIP-1α in the MAPK pathway has recently been demonstrated (Aronica *et al.*, 1997). In this report the authors showed that the chemokines

MIP-1α and IP-10 inhibited the combined stimulatory and synergistic effects of granulocyte-macrophage colony-stimulating factor and steel factor on MAPK activity by suppressing the phosphorylation of the eukaryotic initiation factor 4E and 4E-binding protein 1.

A role for RANTES in phosphatidylinositol 3-kinase (PI3K) activation was previously suggested, given that wortmannin, a PI3K inhibitor, could block the RANTES-induced increase in chemotaxis and polarization of T lymphocytes (Turner *et al.*, 1995). In line with these findings, direct evidence for a role of RANTES in PI3K activation was recently reported (Coffer *et al.*, 1998). In this study RANTES activation of PI3K was shown to activate protein kinase B (PKB), a downstream target of PI3K. The induction of PKB phosphorylation in human eosinophils was transiently induced on activation with the chemoattractants PAF, C5a, and RANTES.

In addition, recent evidence points to a role for RANTES in phospholipase D activation, probably mediated by $G\beta\gamma$ subunits which appear to require the GTP-binding proteins ARF and RhoA (Bacon *et al.*, 1998). It has been suggested that phosphatidic acid, the major metabolite of phospholipase D, can induce activation of the AP-1 transcription factor, perhaps pointing to a role for RANTES in gene regulation (Mollinedo *et al.*, 1994). In this context, RANTES has been shown to stimulate T lymphocyte cell proliferation and the upregulation of adhesion proteins (Bacon *et al.*, 1995). Furthermore, RANTES activation of ARF, which has been implicated in vesicle formation and retrograde transport of proteins from the Golgi to the endoplasmic reticulum (Franco *et al.*, 1995), may play a role in receptor downregulation. In line with these observations, ligand-induced chemokine receptors trafficking has been amply demonstrated and this pathway may provide one of a variety of ways of inducing receptor downregulation.

BIOLOGICAL CONSEQUENCES OF ACTIVATING OR INHIBITING RECEPTOR AND PATHOPHYSIOLOGY

Unique biological effects of activating the receptors

Role of CCR1 in Multiple Sclerosis

Multiple sclerosis is an autoimmune disease mediated by T and B lymphocytes, and macrophages. It results in extensive inflammation and demyelination of the white matter (Ebers, 1986). Although the mechanisms responsible for causing this immunologic damage in the CNS are still unknown, they are almost certainly mediated by infiltrating leukocytes. Initial interactions between invading T cells and monocytes in the CNS result in the production of cytokines such as TNF and IL-1. These cytokines induce a variety of effects that culminate in the recruitment of activated T cells and macrophages. It is likely that a chemotactic gradient of immobilized chemokines, possibly bound to sulfated glycans (Strieter *et al.*, 1989) on the subendothelial matrix (Huber *et al.*, 1991), guides the directed flow of these blood leukocytes across the endothelium into the CNS.

A variety of evidence implicates chemokines in multiple sclerosis. For instance, in an experimental *experimental allergic encephalitis* (EAE) model of *multiple sclerosis* in the mouse, mRNA levels for a number of chemokines, including KC, IP-10, MIP-1α, RANTES, MARC (murine MCP-3), and TCA-3 (murine I-309) are upregulated in the spinal cord during the course of disease (Godiska *et al.*, 1995). Chemokine transcript levels are induced several days prior to the onset of clinical disease and sustained throughout disease progression (Godiska *et al.*, 1995). Colocalization studies demonstrated that MIP-1α and RANTES were produced exclusively by infiltrating leukocytes (Glabinski *et al.*, 1997; Miyagishi *et al.*, 1997). Studies from another group have also demonstrated that the chemokines JE (murine MCP-1), RANTES, MIP-1α, IP-10, and KC are also upregulated in the spinal cord and brain during the acute stages and chronic relapse of murine EAE (Glabinski *et al.*, 1997).

Recent studies by Karpus and coworkers (Karpus *et al.*, 1995; Karpus and Kennedy, 1997) provide strong *in vivo* concept validation for a role of MIP-1α in a mouse EAE model of *multiple sclerosis*. These investigators were able to show that antibodies to MIP-1α prevented the development of both initial and relapsing paralytic disease as well as infiltration of mononuclear cells into the CNS. Treatment with MIP-1α antibody was also able to ameliorate the severity of ongoing clinical disease. These results led the authors to conclude that MIP-1α plays an important role in this T cell-mediated disease. In addition, Godiska *et al.* (1995) have shown an upregulation of mRNA for a number of chemokines, including MIP-1α, in the lesions and spinal fluid of SJL mice during the course of acute EAE.

Role of CCR1 in Rheumatoid Arthritis

There is accumulating evidence from a number of studies to implicate RANTES in the progression of *rheumatoid arthritis*. Rheumatoid arthritis is a chronic inflammatory disease characterized in part by a memory T lymphocyte and monocyte infiltrate (Rathanaswami *et al.*, 1993; Snowden *et al.*, 1994). This process is thought to be mediated by chemotactic factors released by inflamed tissues. Rheumatoid synovial fibroblasts upregulate RANTES mRNA in response to IL-1β, TNFα, and IFNγ.

Rathanswami *et al.* (1993) demonstrated by northern blot and ELISA that cultured synovial fibroblasts isolated from rheumatoid patients were capable of expressing and producing RANTES and other chemokines in response to IL-1β. Snowden *et al.* (1994) have used reverse transcriptase PCR to detect RANTES mRNA in four out of seven synovial tissue samples from *rheumatoid arthritis* patients. By contrast, *osteoarthritis* tissue does not express RANTES mRNA (Snowden *et al.*, 1994).

In addition to these studies we have recently obtained strong evidence implicating RANTES in the pathophysiology of *rheumatoid arthritis* (Barnes et *al.*, 1998). We were able to show in an *adjuvant-induced arthritis* (AIA) model in the rat that antibodies to RANTES greatly reduced the development of disease in animals induced for AIA. Polyclonal antibodies to either MIP-1α or KC were ineffective. Recently, Plater-Zyberk *et al.* (1997), using an altered form of RANTES, met-RANTES, which acts as a CCR1 antagonist, were able to show efficacy in an animal model of rheumatoid arthritis. In their studies delivery of the antagonist i.p. 3 times per week through day 21 resulted in the delay of onset and amelioration of collagen-induced arthritis in DBA/1 mice. These and other studies suggest that chemokines that activate CCR1 could play a major role in attracting T cells and monocytes into joints during onset of AIA, and makes CCR1 a logical candidate for involvement in the inflammatory and destructive processes culminating in rheumatoid arthritis.

Role of CCR1 in Organ Transplant Rejection

The classic signs of acute cellular rejection during organ transplantation includes the infiltration of mononuclear cells into the interstitium (Pattison *et al.*, 1994). This cellular infiltrate consists mainly of T cells and macrophages, cell types that express CCR1 and thus respond to RANTES. Several studies provide evidence for a role of RANTES in organ transplant rejection, particularly of the kidney. In a model of reperfusion injury in the rat, RANTES levels were increased over normal and remained high for more than a week, correlating with the peak of infiltrating macrophages (Takada *et al.*, 1997). RANTES protein was detected in infiltrating mononuclear cells, tubular epithelium, and vascular endothelium of renal allograft biopsy specimens from patients with *cyclosporin nephrotoxicity*, but not in normal kidney (Pattison *et al.*, 1994). A recent study suggests that RANTES may play a role in *graft atherosclerosis* (Pattison *et al.*, 1996). Increased levels of RANTES, both mRNA and protein, were detected in mononuclear cells, myofibroblasts, and endothelial cells of arteries undergoing accelerated atherosclerosis compared with normal coronary arteries. In summary, these studies strongly suggest that RANTES through activation of CCR1 receptors on mononuclear cells may play an important role in *allograft rejection*.

Role of CCR1 in Atherosclerosis

Atherosclerosis and *coronary artery disease* result from intimal thickening of the blood vessels due to localized accumulation of lipids, known as atheromas. Although the exact mechanism of atherosclerotic plaque formation remains unclear, it can be viewed as an inflammatory process involving macrophages and T lymphocytes. The presence of substantial numbers of T lymphocytes in the lesion and local and circulating autoantibodies to plaque components suggests that a specific immune response is operating. Expression of adhesion molecules and local secretion of chemokines help to recruit inflammatory cells to the lesion and CC chemokines in particular have been postulated to play a role in this process. Investigation of RANTES expression in transplant-associated accelerated atherosclerosis revealed an increased expression of the chemokine at both mRNA and protein levels in T cells, macrophages, myofibroblasts, and endothelial cells of arteries undergoing accelerated atherosclerosis but not in normal coronary arteries (Pattison *et al.*, 1996). Human vascular smooth muscle cells treated with IL-1 or TNF produce a number of chemokines, including RANTES. In contrast, very low amounts of RANTES (assessed by specific ELISA) are produced under basal conditions (Jordan *et al.*, 1997).

Other Indications

CCR1 may play a role in a variety of other autoimmune and inflammatory diseases including *asthma*, *atopic dermatitis*, and *endometriosis*. Evidence for a role in these diseases is limited primarily to the presence of the CCR1 ligands RANTES and MIP-1α. For example, RANTES has been shown to induce eosinophil activation and is a very potent eosinophil chemotactic agent (Chihara *et al.*, 1994), while MIP-1α has been shown to attract and activate eosinophils (Rot *et al.*, 1992), basophils, and mast cells (Alam *et al.*, 1994). These reports thus provide a potential link for CCR1 in asthma.

Along similar lines, several reports have shown elevated levels of RANTES in endometrial tissue, suggesting a potential role for this chemokine in the disease (Hornung *et al.*, 1997). The expression of RANTES mRNA in dermal tissue of patients with *atopic dermatitis* revealed that RANTES mRNA was present in the rashes of almost all of the patients (Yamada *et al.*, 1996).

Phenotypes of receptor knockouts and receptor overexpression mice

Further insight into the physiological role of CCR1 has been provided by targeted gene disruption studies (Gao *et al.*, 1997; Gerard *et al.*, 1997). These studies revealed that, although the distribution of leukocytes in the CCR1−/− mice was normal, the trafficking and proliferation of myeloid progenitor cells were disordered (Gerard *et al.*, 1997). In addition, neutrophils from CCR1−/− mice failed to migrate *in vitro* and mobilize into peripheral blood *in vivo* in response to MIP-1α. The CCR1−/− mice also exhibited a 40% reduction in the size of lung granulomas when injected with schistosome eggs compared with normal mice. This impaired immune response was also associated with an altered TH1/TH2 cytokine balance. In an independent study from another group, CCR1 gene deletion was reported to be associated with protection from *pulmonary inflammation* secondary to *acute pancreatitis* (Gerard *et al.*, 1997). This protection from lung injury was associated with decreased levels of TNF, indicating that the activation of the CCR1 is an early event in the systemic inflammatory response. These studies demonstrate that CCR1 has nonredundant functions in host defense and inflammation.

THERAPEUTIC UTILITY

Effects of inhibitors (antibodies) to receptors

Effect of Treatment with Protein Antagonists

Only one group has disclosed the use of a protein for inhibition of binding to CCR1. Glaxo have reported that a modified RANTES protein, called Met-RANTES, was an effective CCR1 antagonist (Proudfoot *et al.*, 1996). Met-RANTES had a K_i of 25 nM in competition binding studies with [^{125}I]RANTES and was shown to be a functional *in vitro* antagonist measured by inhibition of chemotaxis and calcium mobilization. Furthermore, it was recently shown to be effective *in vivo* as well, since treatment of animals with Met-RANTES in a *collagen-induced arthritis* animal model of *rheumatoid arthritis* delayed the onset and ameliorated disease in DBA/1 mice (Plater-Zyberk *et al.*, 1997).

The Glaxo group also recently described a derivative of RANTES that was created by chemical modification of the N-terminus, aminooxypentane (AOP)-RANTES (Simmons *et al.*, 1997). AOP-RANTES did not induce chemotaxis and was a subnanomolar antagonist of CCR5 function in monocytes. Thus, AOP-RANTES is a potent antagonist of CCR5 and also presumably of CCR1. It potently inhibited infection of macrophages and lymphocytes by *HIV-1*. Although these peptide antagonists of CCR1 are potent and are effective in animal models of disease, they suffer from poor metabolic stability and oral bioavailability, thereby limiting their therapeutic utility.

Effect of Treatment with Small Molecule Antagonists

The compound for which most biological information has been disclosed was reported by Berlex Biosciences (**Figure 4**) (Hesselgesser *et al.*, 1998). This compound, a member of the 4-hydroxypiperidine family, was shown to have high potency in a competition-binding assay (40nM). Its functional activity was determined by calcium mobilization, microphysiometry, and inhibition of chemotaxis. The most potent member of this class of compounds, 2-2-diphenyl-5-(4-chlorophenyl)-piperidin-lyl valeronitrite, dose-responsively inhibited the ability of MIP-1α to induce an increase in extracellular acidification and intracellular Ca^{2+} mobilization, demonstrating functional antagonism. When given alone the compound did not elicit any responses, indicating the absence of intrinsic agonist activity. Furthermore, the lead compound from this series of antagonists dose-responsively inhibited MIP-1α and RANTES induced migration in PBMCs. This last experiment also demonstrated functional selectivity against other chemokine receptors, since the compound had no effects on the migration of PBMCs stimulated with MIP-1β, MCP-1, or SDF-1α. These data demonstrate that the lead compound is a potent antagonist for CCR1 but has no effects on the related chemokine receptors CCR5, CXCR2, or CXCR4. Selectivity against other GPCRs was shown by screening against a panel of human GPCRs. Selectivity is important because of the vital roles that GPCRs play in regulating homeostasis. It was quite encouraging to note that, while the activity of the lead compound was

Figure 4 Structure of Berlex, Takeda, and Banyu CCR1 receptor antagonists.

1
Berlex
IC_{50} (MIP-1α) = 40 nM
IC_{50} (RANTES) = 60 nM

2
Takeda
IC_{50} (MIP-1α) = 5000 nM
IC_{50} (RANTES) = 6 nM

IC_{50} (MIP-1α) = 9 nM*
IC_{50} (RANTES) = 3 nM*

3
Takeda
IC_{50} (MIP-1α) = 50 nM
IC_{50} (RANTES) = 20 nM

IC_{50} (MIP-1α) = 8 nM*
IC_{50} (RANTES) = 5 nM*

*HEK-293 cell line

4
Banyu
IC_{50} (CCR1) = 1.9 nM
IC_{50} (CCR3) = 2.7 nM

5
Banyu
IC_{50} (CCR1) = 1.8 nM
IC_{50} (CCR3) = 0.7 nM

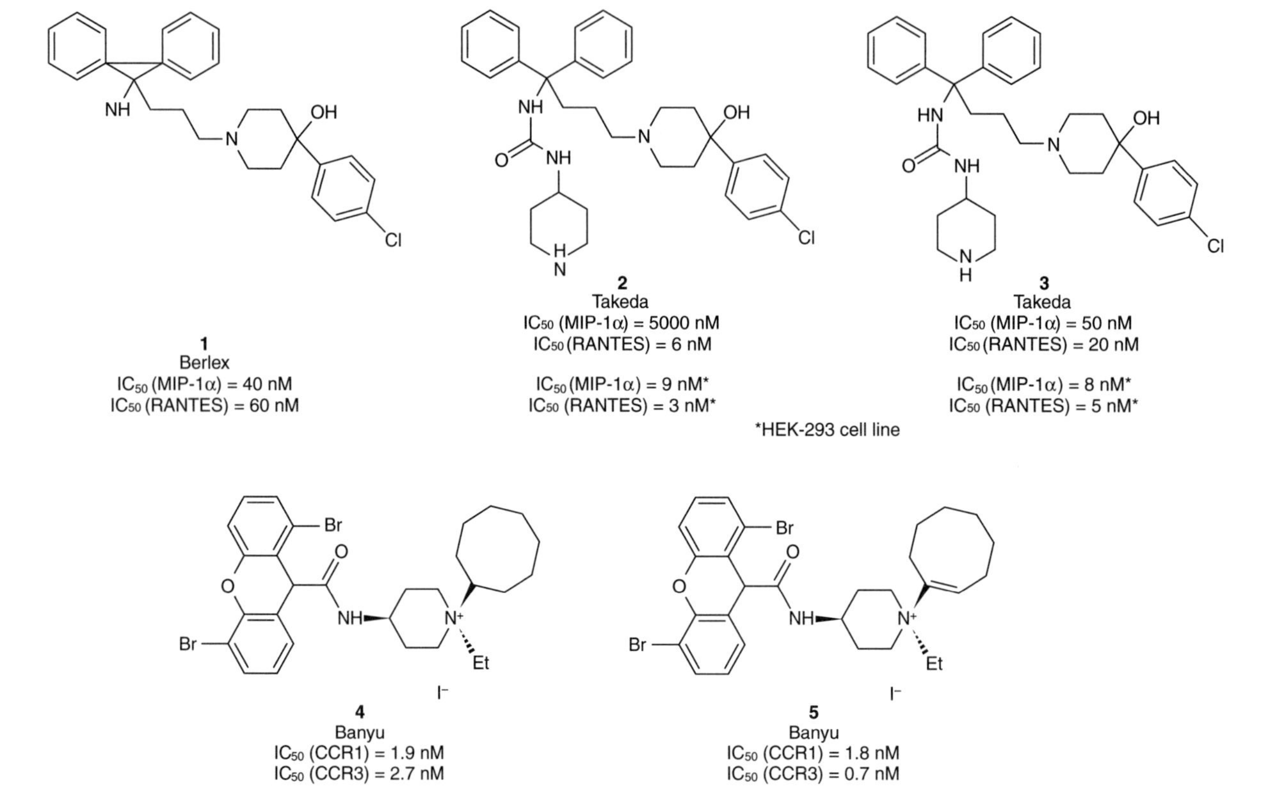

less than 40 nM for CCR1, it was more than 250-fold less active for most of the other receptors tested, including the related chemokine receptor CCR5, which binds RANTES and MIP-1α with high affinity. The only potential crossreactivity of the lead compound was with several biogenic amine neurotransmitter receptors, a result not surprising for a structure reminiscent of the typical neuroleptic or antidepressant structural motif (Cusack *et al.*, 1994; Richelson, 1996). No *in vivo* data have been reported yet on this compound.

A group of similar structures was disclosed by Takeda Chemical Industries as CCR1 receptor antagonists (Figure 4) (Kato *et al.*, 1997). The compounds reported share many structural features with the Berlex compound and were reported to be potent for inhibition of binding of RANTES to CCR1. However, these researchers reported that the compounds inhibited MIP-1α binding with lower potency. It should be noted that these researchers tested their compounds in a hamster CCR1-expressing cell line (CHO). When the compounds were screened using a human CCR1-expressing cell line (HEK), the potencies for inhibition of binding of MIP-1α and RANTES were almost identical (Horuk, unpublished data).

Figure 5 Structure of Merck CCR1 receptor antagonists.

6

7

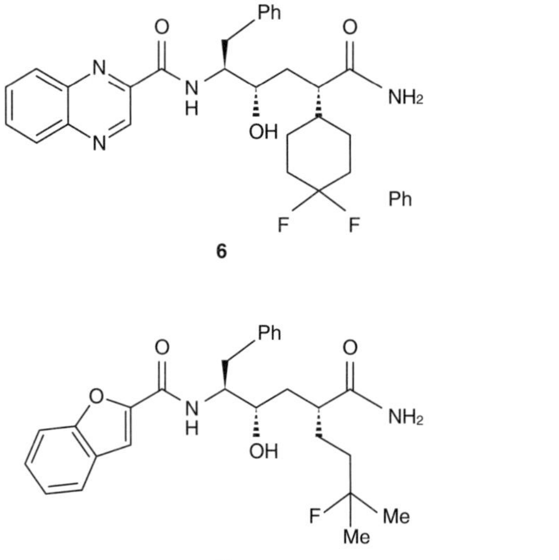

Banyu Pharmaceutical has recently reported small-molecule CCR1 antagonists (Figure 4) (Naya *et al.*, 1998). In this patent disclosure they report a group of tricyclic amides which inhibited receptor binding with an IC_{50} of 1.8 nM. These compounds were also reported to inhibit binding to CCR3 with a similar IC_{50} (1.7 nM), making them less specific than the Berlex compounds. In addition, the fact that these compounds are quaternary salts may further limit their therapeutic use, due to potential problems of oral absorption and rapid elimination.

Most recently Pfizer has disclosed a family of novel compounds which inhibit MIP-1α binding to CCR1 in a published patent (Brown *et al.*, 1998). Nineteen compounds were specifically claimed in this patent. One specific stereochemical alignment of substituents seemed preferred. The level to which these compounds antagonized binding to CCR1 was not disclosed.

References

Alam, R., Kumar, D., Anderson-Walters, D., and Forsythe, P. A. (1994). Macrophage inflammatory protein-1 alpha and monocyte chemoattractant peptide-1 elicit immediate and late cutaneous reactions and activate murine mast cells *in vivo*. *J. Immunol.* **152**, 1298–1303.

Arai, H., and Charo, I. F. (1996). Differential regulation of G-protein-mediated signaling by chemokine receptors. *J. Biol. Chem.* **271**, 21814–21819.

Aronica, S. M., Gingras, A. C., Sonenberg, N., Cooper, S., Hague, N., and Broxmeyer, H. E. (1997). Macrophage inflammatory protein-1alpha and interferon-inducible protein 10 inhibit synergistically induced growth factor stimulation of MAP kinase activity and suppress phosphorylation of eukaryotic initiation factor 4E and 4E binding protein 1. *Blood* **89**, 3582–3595.

Bacon, K. B., Premack, B. A., Gardner, P., and Schall, T. J. (1995). Activation of dual T cell signaling pathways by the chemokine RANTES. *Science* **269**, 1727–1730.

Bacon, K. B., Schall, T. J., and Dairaghi, D. J. (1998). RANTES activation of phospholipase D in Jurkat T cells: requirement of GTP-binding proteins ARF and RhoA. *J. Immunol.* **160**, 1894–1900.

Barnes, D. A., Tse, J., Kaufhold, M., Owen, M., Hesselgesser, J., Strieter, R., Horuk, R., and Perez, H. D. (1998). Polyclonal antibody directed against human RANTES ameliorates disease in the Lewis rat adjuvant-induced arthritis model. *J. Clin. Invest.* **101**, 2910–2919.

Billstrom, M. A., Johnson, G. L., Avdi, N. J., and Worthen, G. S. (1998). Intracellular signaling by the chemokine receptor US28 during human cytomegalovirus infection. *J. Virol.* **72**, 5535–5544.

Boulay, F., Tardif, M., Brouchon, L., and Vignais, P. (1990). The human N-formylpeptide receptor. Characterization of two cDNA isolates and evidence for a new subfamily of G-protein-coupled receptors. *Biochemistry* **29**, 11123–11133.

Brown, M. F., Kath, J. C., and Poss, C. S. (1998). Heteroarylhexanoic acid amide derivatives, their preparation and their use as selective inhibitors of MIP-1-alpha binding to its CCR1 receptor. World (PCT) patent W0-9838167.

Chihara, J., Hayashi, N., Kakazu, T., Yamamoto, T., Kurachi, D., and Nakajima, S. (1994). RANTES augments radical oxygen products from eosinophils. *Int. Arch. Allergy Immunol.* **104**, 52–58.

Coffer, P. J., Schweizer, R. C., Dubois, G. R., Maikoe, T., Lammers, J. W., and Koenderman, L. (1998). Analysis of signal transduction pathways in human eosinophils activated by chemoattractants and the T-helper 2-derived cytokines interleukin-4 and interleukin-5. *Blood* **91**, 2547–2557.

Coulin, F., Power, C. A., Alouani, S., Peitsch, M. C., Schroeder, J. M., Moshizuki, M., Clark-Lewis, I., and Wells, T. N. C. (1997). Characterisation of macrophage inflammatory protein-5 human CC cytokine-2, a member of the macrophage-inflammatory-protein family of chemokines. *Eur. J. Biochem.* **248**, 507–515.

Cusack, B., Nelson, A., and Richelson, E. (1994). Binding of antidepressants to human brain receptors: focus on newer generation compounds. *Psychopharmacology (Berlin)* **114**, 559–565.

Davis, C. B., Dikic, I., Unutmaz, D., Hill, C. M., Arthos, J., Siani, M. A., Thompson, D. A., Schlessinger, J., and Littman, D. R. (1997). Signal transduction due to HIV-1 envelope interactions with chemokine receptors CXCR4 or CCR5. *J. Exp. Med.* **186**, 1793–1798.

Della-Rocca, G. J., van Biesen, T., Daaka, Y., Luttrell, D. K., Luttrell, L. M., and Lefkowitz, R. J. (1997). Ras-dependent mitogen-activated protein kinase activation by G protein-coupled receptors. Convergence of Gi- and Gq-mediated pathways on calcium/calmodulin, Pyk2, and Src kinase. *J. Biol. Chem.* **272**, 19125–19132.

Dikic, I., and Schlessinger, J. (1998). Identification of a new Pyk2 isoform implicated in chemokine and antigen receptor signaling. *J. Biol. Chem.* **273**, 14301–14308.

Ebers, G. C. (1986). In "Diseases of the Nervous System" (ed A. K. Asbury, G. M. McKhann and W. I. McDonald), Multiple sclerosis and other demyelinating disease, pp. 1268–1281. Ardmore Medical Books, Philadelphia.

Elsner, J., Petering, H., Hochstetter, R., Kimmig, D., Wells, T. N., Kapp, A., and Proudfoot, A. E. (1997). The CC chemokine antagonist Met-RANTES inhibits eosinophil effector functions through the chemokine receptors CCR1 and CCR3. *Eur. J. Immunol.* **27**, 2892–2898.

Faure, M., Voyno-Yasenetskaya, T. A., and Bourne, H. R. (1994). cAMP and beta gamma subunits of heterotrimeric G proteins stimulate the mitogen-activated protein kinase pathway in COS-7 cells. *J. Biol. Chem.* **269**, 7851–7854.

Franco, M., Paris, S., and Chabre, M. (1995). The small G-protein ARF1GDP binds to the Gt beta gamma subunit of transducin, but not to Gt alpha GDP-Gt beta gamma. *FEBS Lett.* **362**, 286–290.

Gao, J. L., and Murphy, P. M. (1994). Human cytomegalovirus open reading frame US28 encodes a functional beta chemokine receptor. *J. Biol. Chem.* **269**, 28539–28542.

Gao, J. L., and Murphy, P. M. (1995). Cloning and differential tissue-specific expression of three mouse β chemokine receptor-like genes, including the gene for a functional macrophage inflammatory protein-1α receptor. *J. Biol. Chem.* **270**, 17494–17501.

Gao, J.-L., Kuhns, D. B., Tiffany, H. L., McDermott, D., Li, X., Francke, U., and Murphy, P. M. (1993). Structure and functional expression of the human macrophage inflammatory protein 1α/Rantes receptor. *J. Exp. Med.* **177**, 1421–1427.

Gao, J. L., Wynn, T. A., Chang, Y., Lee, E. J., Broxmeyer, H. E., Cooper, S., Tiffany, H. L., Westphal, H., Kwon-Chung, J., and Murphy, P. M. (1997). Impaired host defense, hematopoiesis, granulomatous inflammation and type 1–type 2 cytokine balance in mice lacking CC chemokine receptor 1. *J. Exp. Med.* **185**, 1959–1968.

Gerard, C., Frossard, J. L., Bhatia, M., Saluja, A., Gerard, N. P., Lu, B., and Steer, M. (1997). Targeted disruption of the

β-chemokine receptor CCR1 protects against pancreatitis-associated lung injury. *J. Clin. Invest.* **100**, 2022–2027.

Gerard, N. P., and Gerard, C. (1991). The chemotactic receptor for human C5a anaphylatoxin. *Nature* **349**, 614–617.

Glabinski, A. R., Tani, M., Strieter, R. M., Tuohy, V. K., and Ransohoff, R. M. (1997). Synchronous synthesis of α- and β-chemokines by cells of diverse lineage in the central nervous system of mice with relapses of chronic experimental autoimmune encephalomyelitis. *Am. J. Pathol.* **150**, 617–630.

Godiska, R., Chantry, D., Dietsch, G. N., and Gray, P. W. (1995). Chemokine expression in murine experimental allergic encephalomyelitis. *J. Neuroimmunol.* **58**, 167–176.

Gong, X., Gong, W., Kuhns, D. B., Ben-Baruch, A., Howard, O. M., and Wang, J. M. (1997). Monocyte chemotactic protein-2 (MCP-2) uses CCR1 and CCR2B as its functional receptors. *J. Biol Chem.* **272**, 11682–11685.

Haribabu, B., Richardson, R. M., Fisher, I., Sozzani, S., Peiper, S. C., Horuk, R., Ali, H., and Snyderman, R. (1997). Regulation of human chemokine receptors CXCR4. Role of phosphorylation in desensitization and internalization. *J. Biol. Chem.* **272**, 28726–28731.

Hawes, B. E., van Biesen, T., Koch, W. J., Luttrell, L. M., and Lefkowitz, R. J. (1995). Distinct pathways of Gi- and Gq-mediated mitogen-activated protein kinase activation. *J. Biol. Chem.* **270**, 17148–17153.

Hesselgesser, J., Ng, H. P., Liang, M., Zheng, W., May, K., Bauman, J. G., Monahan, S., Islam, I., Wei, G. P., Ghannam, A., Taub, D. D., Rosser, M., Snider, R. M., Morrissey, M. M., Perez, H. D., and Horuk, R. (1998). Identification and characterization of small molecule functional antagonists of the CCR1 chemokine receptor. *J. Biol. Chem.* **273**, 15687–15692.

Holmes, W. E., Lee, J., Kuang, W. J., Rice, G. C., and Wood, W. I. (1991). Structure and functional expression of a human interleukin-8 receptor. *Science* **253**, 1278–1280.

Honda, S., and Fujisawa, T. (1997). CC chemokine receptor protein its production and use. *Japan Patent JP*-2, 27599.

Hornung, D., Ryan, I. P., Chao, V. A., Vigne, J. L., Schriock, E. D., and Taylor, R. N. (1997). Immunolocalization and regulation of the chemokine RANTES in human endometrial and endometriosis tissues and cells. *J. Clin. Endocrinol. Metab.* **82**, 1621–1628.

Horuk, R. (1994). Molecular properties of the chemokine receptor family. *Trends Pharmacol. Sci.* **15**, 159–165.

Huber, A. R., Kunkel, S. L., Todd, R. F., and Weiss, S. J. (1991). Regulation of transendothelial neutrophil migration by endogenous interleukin-8. *Science* **254**, 99–102.

Jordan, N. J., Watson, M. L., Williams, R. J., Roach, A. G., Yoshimura, T., and Westwick, J. (1997). Chemokine production by human vascular smooth muscle cells: modulation by IL-13. *Br. J. Pharmacol.* **122**, 749–757.

Karpus, W. J., and Kennedy, K. J. (1997). MIP-1alpha and MCP-1 differentially regulate acute and relapsing autoimmune encephalomyelitis as well as TH1/TH2 lymphocyte differentiation. *J. Leukocyte Biol.* **62**, 681–687.

Karpus, W. J., Lukacs, N. W., McRae, B. L., Strieter, R. M., Kunkel, S. L., and Miller, S. D. (1995). An important role for the chemokine macrophage inflammatory protein-1α in the pathogenesis of the T cell-mediated autoimmune disease, experimental autoimmune encephalomyelitis. *J. Immunol.* **155**, 5003–5010.

Kato, K., Yamamoto, M., Honda, S., and Fujisawa, T. (1997). Use of heterocyclic derivatives as MIP-1α/RANTES antagonists useful for treating e.g. inflammatory and allergic diseases, arteriosclerosis, asthma and multiple sclerosis. World (PCT) Patent WO-9724325.

Lefkowitz, R. J. (1993). G protein-coupled kinases. *Cell* **74**, 409–412.

Lefkowitz, R. J. (1998). G protein-coupled receptors. III. New roles for receptor kinases and beta-arrestins in receptor signaling and desensitization. *J. Biol. Chem.* **273**, 18677–18680.

Miyagishi, R., Kikuchi, S., Takayama, C., Inoue, Y., and Tashiro, K. (1997). Identification of cell types producing RANTES, MIP-1 alpha and MIP-1 beta in rat experimental autoimmune encephalomyelitis by *in situ* hybridization. *J. Neuroimmunol.* **77**, 17–26.

Mollinedo, F., Gajate, C., and Flores, I. (1994). Involvement of phospholipase D in the activation of transcription factor AP-1 in human T lymphoid Jurkat cells. *J Immunol.* **153**, 2457–2469.

Mueller, S. G., Schraw, W. P., and Richmond, A. (1994). Melanoma growth stimulatory activity enhances the phosphorylation of the class II interleukin-8 receptor in non-haematopoietic cells. *J. Biol. Chem.* **269**, 1973–1980.

Murphy, P. M., and Tiffany, H. L. (1991). Cloning of complementary DNA encoding a functional human interleukin-8 receptor. *Science* **253**, 1280–1283.

Naya, A., Owada, Y., Saeki, T., Ohkawi, K., and Iwasawa, K. (1998). New fused compounds are chemokine antagonists useful for the treatment and prevention of acute and chronic inflammatory disorders, AIDS, cancer, ischemic reflow disorders and arteriosclerosis. World (PCT) patent W0-9804554.

Neote, K., DiGregorio, D., Mak, J. Y., Horuk, R., and Schall, T. J. (1993). Molecular cloning, functional expression, and signaling characteristics of a C-C chemokine receptor. *Cell* **72**, 415–425.

O'Dowd, B. F., Hnatowich, M., Caron, M. G., Lefkowitz, R. J., and Bouvier, M. (1989). Palmitoylation of the human beta 2-adrenergic receptor. Mutation of Cys341 in the carboxyl tail leads to an uncoupled nonpalmitoylated form of the receptor. *J. Biol. Chem.* **264**, 7564–7569.

Pardigol, A., Forssmann, U., Zucht, H. D., Loetscher, P., Schulz-Knappe, P., Baggiolini, M., Forssmann, W. G., and Magert, H. J. (1998). HCC-2, a human chemokine: gene structure, expression pattern, and biological activity. *Proc. Natl Acad. Sci. USA* **95**, 6308–6313.

Pattison, J., Nelson, P. J., Huie, P., Von, L. I., Farshid, G., Sibley, R. K., and Krensky, A. M. (1994). RANTES chemokine expression in cell-mediated transplant rejection of the kidney. *Lancet* **343**, 209–211.

Pattison, J. M., Nelson, P. J., Huie, P., Sibley, R. K., and Krensky, A. M. (1996). RANTES chemokine expression in transplant-associated accelerated atherosclerosis. *J. Heart Lung Transplant.* **15**, 1194–1199.

Penton-Rol, G., Polentarutti, N., Luini, W., Borsatti, A., Mancinelli, R., Sica, A., Sozzani, S., and Mantovani, A. (1998). Selective inhibition of expression of the chemokine receptor CCR2 in human monocytes by IFN-gamma. *J. Immunol.* **160**, 3869–3873.

Plater-Zyberk, C., Hoogewerf, A. J., Proudfoot, A. E. I., Power, C. A., and Wells, T. N. C. (1997). Effect of a CC chemokine receptor antagonist on collagen induced arthritis in DBA/1 mice. *Immunol. Lett.* **57**, 117–120.

Post, T. W., Bozic, C. R., Rothenberg, M. E., Luster, A. D., Gerard, N., and Gerard, C. (1995). Molecular characterization of two murine eosinophil β chemokine receptors. *J. Immunol.* **155**, 5299–5305.

Proudfoot, A. E. I., Power, C. A., Hoogewerf, A. J., Montjovent, M. O., Borlat, F., Offord, R. E., and Wells, T. N. C. (1996). Extension of recombinant human

RANTES by the retention of the initiating methionine produces a potent antagonist. *J. Biol. Chem.* **271**, 2599–2603.

Rathanaswami, P., Hachicha, M., Sadick, M., Schall, T. J., and McColl, S. R. (1993). Expression of the cytokine RANTES in human rheumatoid synovial fibroblasts. *J. Biol. Chem.* **268**, 5834–5839.

Richelson, E. (1996). Preclinical pharmacology of neuroleptics: focus on new generation compounds. *J. Clin. Psychiatry* **57** (Suppl 11), 4–11.

Rot, A., Krieger, M., Brunner, T., Bischoff, S. C., Schall, T. J., and Dahinden, C. A. (1992). RANTES and macrophage inflammatory protein 1 alpha induce the migration and activation of normal human eosinophil granulocytes. *J. Exp. Med.* **176**, 1489–1495.

Sallusto, F., Lenig, D., Mackay, C. R., and Lanzavecchia, A. (1998). Flexible programs of chemokine receptor expression on human polarized T helper 1 and 2 lymphocytes. *J. Exp. Med.* **187**, 875–883.

Sarau, H. M., Rush, J. A., Foley, J. J., Brawner, M. E., Schmidt, D. B., White, J. R., and Barnette, M. S. (1997). Characterization of functional chemokine receptors (CCR1 and CCR2) on EoL-3 cells: a model system to examine the role of chemokines in cell function. *J. Pharmacol. Exp. Ther.* **283**, 411–418.

Schall, T. (1994). In "The Cytokine Handbook, 2nd edn" (ed A. Thompson), The chemokines, pp. 419–460. Academic Press, San Diego.

Sibley, D. R., Benovic, J. L., Caron, M. G., and Lefkowitz, R. J. (1987). Regulation of transmembrane signaling by receptor phosphorylation. *Cell* **48**, 913–922.

Sica, A., Saccani, A., Borsatti, A., Power, C. A., Wells, T. N. C., Luini, W., Polentarutti, N., Sozzani, S., and Mantovani, A. (1997). Bacterial lipopolysaccharide rapidly inhibits expression of C-C chemokine receptors in human monocytes. *J. Exp. Med.* **185**, 969–974.

Simmons, G., Clapham, P. R., Picard, L., Offord, R. E., Rosenkilde, M. M., Schwartz, T. W., Buser, R., Wells, T. N. C., and Proudfoot, A. E. I. (1997). Potent inhibition of HIV-1 infectivity in macrophages and lymphocytes by a novel CCR5 antagonist. *Science* **276**, 276–279.

Snowden, N., Hajeer, A., Thomson, W., and Ollier, B. (1994). RANTES role in rheumatoid arthritis. *Lancet* **343**, 547–548.

Sozzani, S., Luini, W., Borsatti, A., Polentarutti, N., Zhou, D., Piemonti, L., D'Amico, G., Power, C. A., Wells, T. N., Gobbi, M., Allavena, P., and Mantovani, A. (1997). Receptor expression and responsiveness of human dendritic cells to a defined set of CC and CXC chemokines. *J. Immunol.* **159**, 1993–2000.

Strieter, R. M., Kunkel, S. L., Showell, H. J., Rennick, D. J., Phan, S. H., Ward, R. A., and Marks, R. M. (1989). Endothelial cell gene expression of a neutrophil chemotactic factor by TNF-α, LPS, and IL-1β. *Science* **243**, 1467–1469.

Su, S. B., Mukaida, N., Wang, J., Nomura, H., and Matsushima, K. (1996). Preparation of specific polyclonal antibodies to a C-C chemokine receptor, CCR1, and determination of CCR1 expression on various types of leukocytes. *J. Leukoc. Biol.* **60**, 658–666.

Su, S., Mukaida, N., Wang, J., Zhang, Y., Takami, A., Nakao, S., and Matsushima, K. (1997). Inhibition of immature erythroid progenitor cell proliferation by macrophage inflammatory protein-1alpha by interacting mainly with a C-C chemokine receptor, CCR1. *Blood* **90**, 605–611.

Takada, M., Nadeau, K. C., Shaw, G. D., Marquette, K. A., and Tilney, N. L. (1997). The cytokine-adhesion molecule cascade in ischemia/reperfusion injury of the rat kidney. Inhibition by a soluble P-selectin ligand. *J. Clin. Invest.* **99**, 2682–2690.

Turner, L., Ward, S. G., and Westwick, J. (1995). RANTES-activated human T lymphocytes: a role for phosphoinositide 3-kinase. *J. Immunol.* **155**, 2437–2444.

Wilson, C. J., and Applebury, M. L. (1993). Arresting G-protein coupled receptor activity. *Curr. Biol.* **3**, 683–686.

Yamada, H., Izutani, R., Chihara, J., Yudate, T., Matsukura, M., and Tezuka, T. (1996). RANTES mRNA expression in skin and colon of patients with atopic dermatitis. *Int. Arch. Allergy Immunol.* **111** (Suppl. 1), 19–21.

Youn, B.-S., Zhang, S. M., Lee, E. K., Park, D. H., Broxmeyer, H. E., Murphy, P. M., Locati, M., Pease, J. E., Kim, K. K., Antol, K., and Kwon, B. S. (1997). Molecular cloning of leukotactin 1: a novel human beta-chemokine, a chemoattractant for neutrophils, monocytes, and lymphocytes, and a potent agonist at CC chemokine receptors 1 and 3. *J. Immunol.* **159**, 5210–5205.

Youn, B. S., Zhang, S. M., Broxmeyer, H. E., Cooper, S., Antol, K., Fraser, M., and Kwon, B. S. (1998). Characterization of CKbeta8 and CKbeta8-1: two alternatively spliced forms of human beta-chemokine, chemoattractants for neutrophils, monocytes, and lymphocytes, and potent agonists at CC chemokine receptor 1. *Blood* **91**, 3118–3126.

CCR2

Jose Miguel Rodriguez-Frade, Mario Mellado and Carlos Martinez-A.*

Department of Immunology and Oncology, Centro Nacional de Biotecnología, CSIC, Campus de Cantoblanco, Madrid, 28049, Spain

*corresponding author tel: 34-915854559, fax: 34-913720493, e-mail: cmartineza@cnb.uam.es

DOI: 10.1006/rwcy.2000.22002.

SUMMARY

The name CCR2 refers to two alternatively spliced chemokine receptors: CCR2A and CCR2B. Although first identified as the specific, high-affinity receptor for MCP-1 present in monocytic cell lines, other chemokines have been shown to elicit responses through CCR2. CCR2 is expressed in monocytes, macrophages, T and B lymphocytes, natural killer (NK) cells, dendritic cells (DC), and in nonlymphoid cells and tissues. As one of the first chemokine receptors to be cloned and characterized, it has been studied extensively. Its implication in several pathologies, its structure–activity relationship, the signaling pathways activated, as well as its regulation have been analyzed in detail and it has served as a model for the study of other chemokine receptors.

BACKGROUND

Discovery

The presence of this receptor was inferred from the robust and specific responses of monocytes to MCP-1, which could not be accounted for by known chemokine receptors such as CCR1, CXCR1, or CXCR2.

It was first cloned from total RNA of a monocytic cell line (Mono Mac 6) by Charo *et al.* (1994), using degenerate primers corresponding to sequences of the second and third transmembrane domains of CCR1, CXCR1, and CXCR2 receptors, and PCR with Mono Mac 6 cDNA. CCR2 exists in two forms, CCR2A and CCR2B, which represent alternately spliced variants of a single gene. Shortly after the first description, Yamagami *et al.* (1994) cloned a functional MCP-1 receptor from THP-1 cells using similar strategies, corresponding to the CCR2B variant. Equivalent CCR2 receptors have been described in mice (Boring *et al.*, 1996; Kurihara and Bravo, 1996) and rats (Jiang *et al.*, 1998).

Alternative names

In addition to the accepted term CCR2, references to this receptor can be found in the literature as MCP-1R (for MCP-1 receptor), CC-CKR2, and multiple variations on these nomenclatures. It is worth noting that, in most cases, references made to CCR2 usually correspond to the CCR2B variant, since the CCR2A variant is poorly expressed on cell surface.

Structure

Like the rest of the chemokine receptors, both forms of the CCR2 are members of family 1 of the G protein-coupled receptors. CCR2 consists of seven membrane-spanning regions, which are linked by three extracellular and three intracellular loops, an extracellular N-terminal region and intracellular C-terminal region. The extracellular regions are believed to participate in ligand binding, whereas the intracellular regions act in concert to transduce biological signals. Of particular structural interest is the presence of extracellular disulfide bonds, required for maintaining the structural integrity necessary for ligand-binding and receptor activation. As is the case with most chemokine receptors, there are several potential *N*-glycosylation sites in the N-terminal region.

The conserved DRYLAIVHA motif in the second intracellular loop (amino acids 137–145) has an important role in signal transduction; a nonfunctional CCR2B receptor has been described, CCR2BY139F, in which tyrosine 139 is replaced by phenylalanine (Mellado *et al.*, 1998).

The two spliced forms of the CCR2 differ at the C-terminal end, resulting in a shorter 2B tail with a completely different amino acid sequence. Several putative Ser/Thr phosphorylation sites, involved in desensitization and internalization of the receptor, are present in this C-terminal region.

Main activities and pathophysiological roles

The principal function of CCR2 is considered to be mediating migration of different cell types in response to members of the monocyte chemoattractant proteins (MCP). Among the cells known to respond to a chemotactic stimulus through this receptor are monocytes, macrophages, basophils, dendritic cells, NK cells, and T and B lymphocytes. The response varies depending on the cell differentiation status and its degree of activation, however, which is reflected by the CCR2 expression on the surface of the cells. CCR2 mediates leukocyte adhesion and monocyte extravasation, and has been implicated in monocyte differentiation and homing of recently activated T cells.

The pathophysiological effects of CCR2 derive mainly from this function as a mediator of cell migration. It is well established that CCR2 has a major role in inflammatory diseases, including *asthma*, *chronic inflammation*, *experimental encephalomyelitis*, and *atherosclerosis* (Baggiolini, 1998; Gonzalo *et al.*, 1998; Lukacs *et al.*, 1999). In fact, CCR2, together with CCR3, may extensively support the recruitment and/or activation of the different populations found within the asthmatic airways. Although not yet well documented, it may be involved in those pathologies in which its principal ligand, MCP-1, plays a role, including *sepsis*, *tumor rejection*, *CNS trauma*, *multiple sclerosis*, bacterial and viral *meningitis*, *psoriasis*, and *glomerulonephritis* (Luster, 1998; Rollins, 1998).

CCR2 is not one of the main coreceptors for *HIV-1* strains; in fact, it appears to act as a coreceptor only for certain SIV-1 strains and dual-tropic (X4/R5) strains as 89.6. Population studies nevertheless indicate the influence of a mutation in the CCR2 receptor, CCR2V64I, in *AIDS* progression (Smith *et al.*, 1997). Individuals bearing this mutant CCR2, which is as functional as the wild-type receptor, show delayed HIV-1 infection through R5 and X4 strains. This has been linked to mutations in the promoter region of the CCR5 gene (Berger *et al.*, 1999) and to the fact that this receptor forms heterodimers with both CXCR4 and CCR5 (Mellado *et al.*, 1999), which in turn block HIV-1 infection.

GENE

Accession numbers

EMBL:
UO3882, UO3905, D29984, U80924, U95626

Sequence

See **Figure 1**.

Chromosome location and linkages

The gene (*CMKBR2*) (**Figure 2**) is clustered within 285 kb of chromosome 3p21 including CCR1 and CCR3. It comprises approximately 7 kb and consists of three exons and multiple polyadenylation sites. The first exon contains the 5′ UTR; the second exon contains the entire ORF of CCR2B, and the third exon contains the carboxy-tail and 3′ UTR of CCR2A. Strong consensus sequences for canonical 5′ donor and 3′ acceptor splice sites are found at all proposed intron/exon splice junctions (Wong *et al.*, 1997).

PROTEIN

Accession numbers

SwissProt:
Human: P41597
Mouse: P51683
Rat: O55193

Sequence

See **Figure 3**.

Human: Length: 374 (CCR2A) or 360 amino acids (CCR2B). Molecular weight: 41,914 Da (Charo *et al.*, 1994; Yamagami *et al.*, 1994)

Mouse: Length: 355 amino acids. Molecular weight: 40,901 Da (Kurihara and Bravo, 1996; Boring *et al.*, 1997)

Rat: Length: 373 amino acids. Molecular weight: 42,763 Da (Jiang *et al.*, 1998)

Description of protein

The protein consists of seven hydrophobic transmembrane domains linked by successive intracellular and

Figure 1 Nucleotide sequence for the CCR2 gene (2232 bp).

```
GGATTGAACA AGGACGCATT TCCCCAGTAC ATCCACAACA TGCTGTCCAC ATCTCGTTCT -60
CGGTTTATCA GAAATACCAA CGAGAGCGGT GAAGAAGTCA CCACCTTTTT TGATTATGAT 120
TACGGTGCTC CCTGTCATAA ATTTGACGTG AAGCAAATTG GGGCCCAACT CCTGCCTCCG 180
CTCTACTCGC TGGTGTTCAT CTTTGGTTTT GTGGGCAACA TGCTGGTCGT CCTCATCTTA 240
ATAAACTGCA AAAAGCTGAA GTGCTTGACT GACATTTACC TGCTCAACCT GGCCATCTCT 300
GATCTGCTTT TTCTTATTAC TCTCCCATTG TGGGCTCACT CTGCTGCAAA TGAGTGGGTC 360
TTTGGGAATG CAATGTGCAA ATTATTCACA GGGCTGTATC ACATCGGTTA TTTTGGCGGA 420
ATCTTCTTCA TCATCCTCCT GACAATCGAT AGATACCTGG CTATTGTCCA TGCTGTGTTT 480
GCTTTAAAAG CCAGGACGGT CACCTTTGGG GTGGTGACAA GTGTGATCAC CTGGTTGGTG 540
GCTGTGTTTG CTTCTGTCCC AGGAATCATC TTTACTAAAT GCCAGAAAGA AGATTCTGTT 600
TATGTCTGTG GCCCTTATTT TCCACGAGGA TGGAATAATT TCCACACAAT AATGAGGAAC 660
ATTTTGGGGC TGGTCCTGCC GCTGCTCATC ATGGTCATCT GCTACTCGGG AATCCTGAAA 720
ACCCTGCTTC GGTGTCGAAA CGAGAAGAAG AGGCATAGGG CAGTGAGAGT CATCTTCACC 780
ATCATGATTG TTTACTTTCT CTTCTGGACT CCCTATAACA TTGTCATTCT CCTGAACACC 840
TTCCAGGAAT TCTTCGGCCT GAGTAACTGT GAAAGCACCA GTCAACTGGA CCAAGCCACG 900
CAGGTGACAG AGACTCTTGG GATGACTCAC TGCTGCATCA ATCCCATCAT CTATGCCTTC 960
GTTGGGGAGA AGTTCAGAAG CCTTTTTCAC ATAGCTCTTG GCTGTAGGAT TGCCCCACTC 1020
CAAAAACCAG TGTGTGGAGG TCCAGGAGTG AGACCAGGAA AGAATGTGAA AGTGACTACA 1080
CAAGGACTCC TCGATGGTCG TGGAAAAGGA AAGTCAATTG GCAGAGCCCC TGAAGCCAGT 1140
CTTCAGGACA AAGAAGGAGC CTAGAGACAG AAATGACAGA TCTCTGCTTT GGAAATCACA 1200
CGTCTGGCTT CACAGATGTG TGATTCACAG TGTGAATCTT GGTGTCTACG TTACCAGGCA 1260
GGAAGGCTGA GAGGAGAGAG ACTCCAGCTG GGTTGGAAAA CAGTATTTTC CAAACTACCT 1320
TCCAGTTCCT CATTTTTGAA TACAGGCATA GAGTTCAGAC TTTTTTTAAA TAGTAAAAAT 1380
AAAATTAAAG CTGAAAACTG CAACTTGTAA ATGTGGTAAA GAGTTAGTTT GAGTTGCTAT 1440
CATGTCAAAC GTGAAAATGC TGTATTAGTC ACAGAGATAA TTCTAGCTTT GAGCTTAAGA 1500
ATTTTGAGCA GGTGGTATGT TTGGGAGACT GCTGAGTCAA CCCAATAGTT GTTGATTGGC 1560
AGGAGTTGGA AGTGTGTGAT CTGTGGGCAC ATTAGCCTAT GTGCATGCAG CATCTAAGTA 1620
ATGATGTCGT TTGAATCACA GTATACGCTC CATCGCTGTC ATCTCAGCTG GATCTCCATT 1680
CTCTCAGGCT TGCTGCCAAA AGCCTTTTGT GTTTTGTTTT GTATCATTAT GAAGTCATGC 1740
GTTTAATCAC ATTCGAGTGT TTCAGTGCTT CGCAGATGTC CTTGATGCTC ATATTGTTCC 1800
CTAATTTGCC AGTGGGAACT CCTAAATCAA ATTGGCTTCT AATCAAAGCT TTTAAACCCT 1860
ATTGGTAAAG AATGGAAGGT GGAGAAGCTC CCTGAAGTAA GCAAAGACTT TCCTCTTAGT 1920
CGAGCCAAGT TAAGAATGTT CTTATGTTGC CCAGTGTGTT TCTGATCTGA TGCAAGCAAG 1980
AAACACTGGG CTTCTAGAAC CAGGCAACTT GGGAACTAGA CTCCCAAGCT GGACTATGGC 2040
TCTACTTTCA GGCCACATGG CTAAAGAAGG TTTCAGAAAG AAGTGGGGAC AGAGCAGAAC 2100
TTTCACCTTC ATATATTTGT ATGATCCTAA TGAATGCATA AAATGTTAAG TTGATGGTGA 2160
TGAAATGTAA ATACTGTTTT TAACAACTAT GATTTGGAAA ATAAATCAAT GCTATAACTA 2220
TGTTGATAAA AG 2232
```

Figure 2 Structural organization of the CCR2 gene.

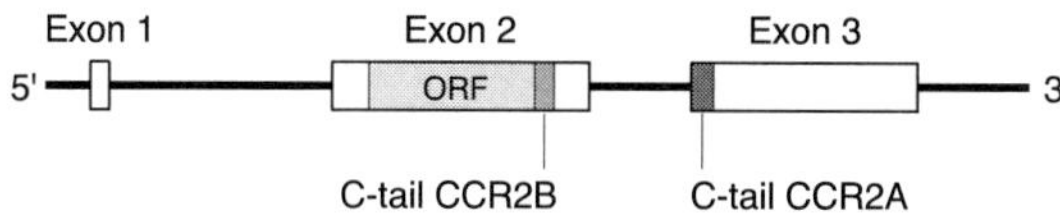

extracellular loops, an extracellular N-terminus, and a cytoplasmic C-terminal tail.

The conserved cysteine residues form disulfide bonds that link the extracellular loops and are required for maintaining the structural integrity necessary for ligand-binding and receptor activation.

The extracellular regions also contain putative *N*-linked glycosylation sites. The extracellular regions are involved in ligand-binding. Using chimeric CCR2/CCR1 or CCR2/CCR5 receptors (Samson *et al.*, 1997), it has been shown that the N-terminal region is required for high-affinity binding of MCP-1 (Monteclaro and Charo, 1996). Monoclonal antibodies raised against the N-terminal region of CCR2 (amino acids 24–38) and the third extracellular loop (amino acids 273–292), which have agonist and antagonist activities, indicate that the third loop is also involved in ligand-binding, whereas the N-terminus has a role in activating the receptor (Frade *et al.*, 1997a). The importance of the extracellular loops has been dissected in detail, and determinants present in the first extracellular loop support high-affinity binding (Asn104 and Glu105), whereas charged amino acid residues in this loop (His100) are critical for ligand-induced activation of G proteins, without affecting ligand-binding affinity (Han *et al.*, 1999a).

All together, these data indicate that the extracellular surface of the receptor acts not only to bind the appropriate ligands, but also to elicit the conformational changes needed for receptor activation. This

Figure 3 Amino acid sequences for human (CCR2A and CCR2B), mouse, and rat CCR2.

```
Human CCRA2
MLSTSRSRFI RNTNESGEEV TTFFDYDYGA PCHKFDVKQI GAQLLPPLYS
LVFIFGFVGN MLVVLILINC KKLKCLTDIY LLNLAISDLL FLITLPLWAH
SAANEWVFGN AMCKLFTGLY HIGYFGGIFF IILLTIDRYL AIVHAVFALK
ARTVTFGVVT SVITWLVAVF ASVPGIIFTK CQKEDSVYVC GPYFPRGWNN
FHTIMRNILG LVLPLLIMVI CYSGILKTLL RCRNEKKRHR AVRVIFTIMI
VYFLFWTPYN IVILLNTFQE FFGLSNCEST SQLDQATQVT ETLGMTHCCI
NPIIYAFVGE KFRSLFHIAL GCRIAPLQKP VCGGPGVRPG KNVKVTTQGL
LDGRGKGKSI GRAPEASLQD KEGA

Human CCR2B
MLSTSRSRFI RNTNESGEEV TTFFDYDYGA PCHKFDVKQI GAQLLPPLYS
LVFIFGFVGN MLVVLILINC KKLKCLTDIY LLNLAISDLL FLITLPLWAH
SAANEWVFGN AMCKLFTGLY HIGYFGGIFF IILLTIDRYL AIVHAVFALK
ARTVTFGVVT SVITWLVAVF ASVPGIIFTK CQKEDSVYVC GPYFPRGWNN
FHTIMRNILG LVLPLLIMVI CYSGILKTLL RCRNEKKRHR AVRVIFTIMI
VYFLFWTPYN IVILLNTFQE FFGLSNCEST SQLDQATQVT ETLGMTHCCI
NPIIYAFVGE KFRRYLSVFF RKHITKRFCK QCPVFYRETV DGVTSTNTPS
TGEQEVSAGL
              ETADRVSSTF TPSTGEQEVS VGL
Mouse
MEISDFTEAY PTTTEFDYGD STPCQKTAVR AFGAGLLPPL YSLVFIIGVV
GNVLMILVLM QHRRLQSMTS IYLFNLAVSD LVFLFTLPFW IDYKLKDDWI
FGDAMCKLLS GFYYLGLYSE IFFIILLTID RYLAIVHAVF ALRARTVTLG
IITSIITWAL AILASMPALY FFKAQWEFTH RTCSPHFPYK SLKQWKRFQA
LKLNLLGLIL PLLVMIICYA GIIRILLRRP SEKKVKAVRL IFAITLLFFL
LWTPYNLSVF VSAFQDVLFT NQCEQSKHLD LAMQVTEVIA YTHCCVNPII
YVFVGERFWK YLRQLFQRHV AIPLAKWLPF LSVDQLERTS SISPSTGEHE
LSAGF

Rat
MEDSNMLPQF IHGILSTSHS LFPRSIQELD EGATTPYDYD DGEPCHKTSV
KQIGAWILPP LYSLVFIFGF VGNMLVIIIL ISCKKLKSMT DIYLFNLAIS
DLLFLLTLPF WAHYAANEWV FGNIMCKLFT GLYHIGYFGG IFFIILLTID
RYLAIVHAVF ALKARTVTFG VITSVVTWVV AVFASLPGII FTKSEQEDDQ
HTCGPYFPTI WKNFQTIMRN ILSLILPLLV MVICYSGILH TLFRCRNEKK
RHRAVRLIFA IMIVYFLFWT PYNIVLFLTT FQEFLGMSNC VVDMHLDQAM
QVTETLGMTH CCVNPIIYAF VGEKFRRYLS IFFRKHIAKN LCKQCPVFYR
```

N-terminal region is also involved in the activity of CCR2 as a coreceptor for some *HIV-1* strains (Rucker *et al.*, 1996).

In accordance with their structural role, the seven transmembrane domains are the most highly conserved regions, not only among chemokine receptors, but in all GPCR, although several residues may be involved in ligand-binding or receptor activation.

The intracellular regions are mainly involved in coupling the CCR2 to signaling molecules. A critical role in signaling through GPCR has been assigned to the DRY motif at the junction of the third TM and the second intracellular loop. In the case of CCR2, the Y139 in this motif is critical, as its mutation results in a loss-of-function CCR2 (Mellado *et al.*, 1998). Study of chimeric CCR2/CCR1 receptors has identified the role of the third intracellular loop in G protein coupling (Arai and Charo, 1996). Other authors described differential coupling of CCR2A and CCR2B to G protein, thus assigning a role for the C-terminal region in G protein coupling (Kuang *et al.*, 1996).

Several tyrosine residues susceptible to phosphorylation are present in the CCR2B, mainly in the C-terminus. In contrast, CCR2A, which has a completely different C-terminal region, lacks most of these tyrosine residues. Comparison of the activity of these variants would help elucidate the role of these tyrosines in receptor function.

The cytoplasmic tails also contain serine and threonine residues susceptible to phosphorylation. The role of phosphorylation of these residues by GRK in receptor desensitization and internalization has been documented (Franci *et al.*, 1996; Aragay *et al.*, 1998). The differences between CCR2A and 2B must therefore influence receptor desensitization and cellular trafficking, although this question has not been fully addressed. The membrane-proximal region of the CCR2B C-terminus is involved in chemotaxis and signal transduction, although neither phosphorylation of Ser/Thr residues nor internalization is required for chemotaxis (Arai *et al.*, 1997). Differences have also been shown in the trafficking of CCR2A and 2B to the membrane surface. Experiments using mutant receptors show that there are cytoplasmic retention signals in the C-terminal region (amino acids 316–349); these signals result in clear surface expression of chimeras containing the CCR2B carboxy-tail, whereas mutant receptors with CCR2A C-terminal ends are found mainly in the cytoplasm (Wong *et al.*, 1997).

It has recently been suggested that chemokine receptors, including CCR2, can exist in multiple interconvertible states (Lee *et al.*, 1999). The CCR2 dimerizes following ligand stimulation, although the receptor regions involved in this process remain to be elucidated. In fact, dimerization is the first step in the initiation of the signaling cascade not only of CCR2 (Rodriguez-Frade *et al.*, 1999a), but also of CCR5 (Rodriguez-Frade *et al.*, 1999b) and CXCR4 (Vila-Coro *et al.*, 1999). *See* **Figure 4**.

Relevant homologies and species differences

CCR2 has the greatest amino acid similarity with CCR5 (71%), sharing nearly identical transmembrane domains with this receptor, as well as extra- and intracellular loops. The main differences, which influence ligand specificity, are located in the N-terminal extracellular domains. Significant similarities are also found with CCR1 (55%) and CCR3 (50%).

The murine CCR2 shows 80% similarity to the human CCR2B and 71% to human CCR2A.

Figure 4 CCR2 structure. Regions implicated in ligand binding, coupling of effector molecules and suggested targets for GRK and JAK are highlighted.

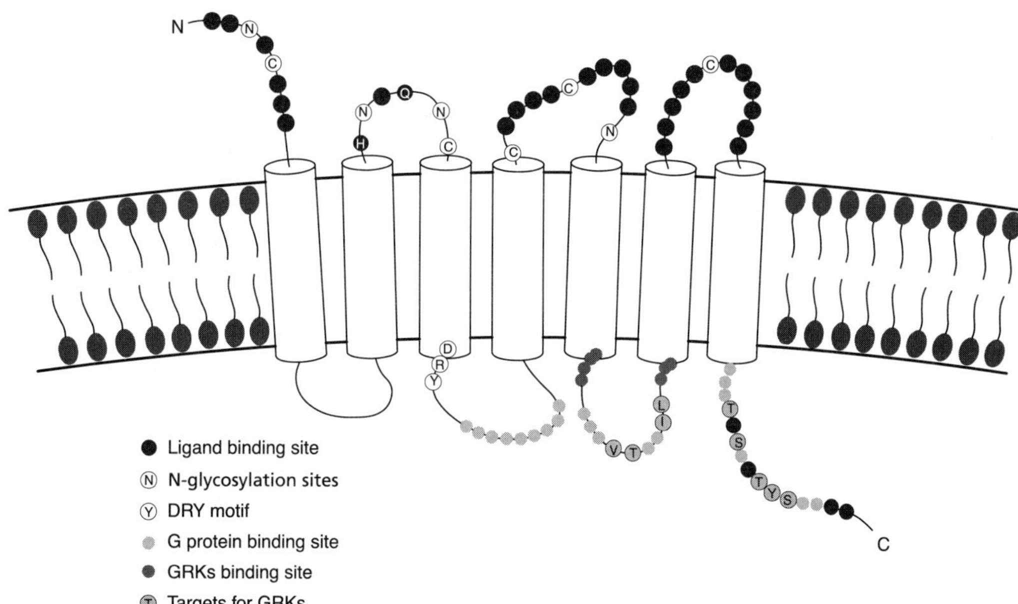

Only one CCR2 form has been described in mice which, in contrast to the human homologs, lacks potential *N*-glycosylation sites.

Affinity for ligand(s)

MCP-1: 0.5 ± 0.1 nM
MCP-2: 3.0 ± 1 nM
MCP-3: 5.0 nM
MCP-4: not determined
(Gong *et al.*, 1997)

Cell types and tissues expressing the receptor

CCR2 expression was first established in monocytes. Since then, many other cell types have been shown to respond to MCP-1 and to express CCR2. In most cases, however, receptor expression is not constitutive and is detected only following stimulation or under inflammatory conditions.

Constitutive CCR2 expression has been assessed in monocytes/macrophages and monocytic cell lines (Mono Mac 1 and 6, THP-1), although levels vary since its expression is regulated upon differentiation (Charo *et al.*, 1994; Frade *et al.*, 1997a; Fantuzzi *et al.*, 1999). Freshly isolated B cells from tonsils and peripheral blood also show constitutive CCR2 expression (Frade *et al.*, 1997b).

Peripheral blood T lymphocytes do not express CCR2 under resting conditions, although its expression is readily observed after activation and IL-2 stimulation (Loetscher *et al.*, 1996). Both TH1 and TH2 cells express CCR2. Memory T cells (CD45RO+) of the CD62L–CD26+ subset express CCR2 (Rabin *et al.*, 1999).

Other cells known to express CCR2, at least under certain conditions, include neutrophils (only under inflammatory conditions (Johnston *et al.*, 1999), elicited eosinophils (Lukacs *et al.*, 1999), natural killer cells (Polentarutti *et al.*, 1997; Sozzani *et al.*, 1997), immature dendritic cells (Sallusto *et al.*, 1998), and mast cells (Campbell *et al.*, 1999).

CCR2 mRNA expression has also been demonstrated in endothelial cells under inflammatory conditions (Weber *et al.*, 1999a) and in lung fibroblasts derived from *pulmonary granulomas* after cytokine treatment (Hogaboam *et al.*, 1999). Expression of CCR2 in astrocytes and cultured glia increased following induction of *experimental allergic encephalomyelitis* (Jiang *et al.*, 1998).

mRNA expression of CCR2 has been determined by northern blot in lung, kidney, heart, bone marrow, spleen, and thymus, but protein expression in the cell membrane is not well documented.

Regulation of receptor expression

The many functions of the chemokines, the redundancy observed among them in receptor usage, as well as their coordinated action described in several models, suggest that both chemokine and receptor expression must be tightly regulated. Several stimuli and pathophysiological conditions have been shown to up- or downregulate CCR2 expression.

IL-2 upregulates CCR2 expression in monocytes (Sozzani *et al.*, 1997), NK cells (Pollentaruti *et al.*, 1997), and T lymphocytes (Loetscher *et al.*, 1996). Other cytokines such as IL-4 and IL-15 upregulate CCR2 levels in different cell types (Hogaboam *et al.*, 1999; Perera *et al.*, 1999). Dexamethasone selectively upregulates CCR2 expression in monocytes by increasing mRNA half-life. Steroids devoid of glucocorticoid activity are inactive. This upregulation correlates with the chemotactic potential of the corresponding cells in response to CCR2 ligands. In cases of hypercholesterolemia, CCR2 upregulation is detected in monocytes, an effect mediated by plasma lipoproteins. LDL is responsible for CCR2 upregulation in both monocytes and monocytic cell lines, through LDL-derived cholesterol. In contrast, HDL reduces and even reverses the effect of LDL in monocytes (Han *et al.*, 1998, 1999b).

CCR2 is downregulated by LPS, making cells unresponsive to MCP-1 but not to other chemokines that activate other receptors, such as MIP-1α. This downregulation has been observed in PBLs, monocytes, macrophages (Zhou *et al.*, 1999), and NK cells (Sozzani *et al.*, 1997), and is mediated through a decrease in mRNA half-life. Similar downregulation of CCR2 has been described for other microbial agents (Sica *et al.*, 1997). TNFα downregulates CCR2 in Mono Mac 6 and THP 1 monocytic cell lines (Weber *et al.*, 1999b), an effect that is counteracted by OXLDL. TCR triggering with anti-CD3 antibodies downregulates CCR2 in memory/effector T cells (Sallusto *et al.*, 1998).

Diverse effects have been described for IFNγ in CCR2 regulation, including CCR2 upregulation in lung fibroblasts (Hogaboam *et al.*, 1999), downregulation in monocytes (Penton-Rol *et al.*, 1998), and no effect in neutrophils (Bonecchi *et al.*, 1999), suggesting cell-type specific receptor regulation.

SIGNAL TRANSDUCTION

Several biochemical events are stimulated by MCP-1, including inhibition of adenylate cyclase, activation of PLC, calcium flux, IP_3 generation and G protein-dependent mechanisms (Rollins, 1998; Pelchen-Matthews *et al.*, 1999).

Associated or intrinsic kinases

Although no intrinsic kinase activity has been documented, data in the literature support the association of several kinases with CCR2. The activation and association of members of the Janus kinase family to CCR2B have been described (Mellado *et al.*, 1998). It is also known that MCP-1 promotes activation of members of the GRK Ser/Thr kinase family (Franci *et al.*, 1996); these kinases associate to CCR2 to form a macromolecular complex that also involves β-arrestin (Aragay *et al.*, 1998).

Cytoplasmic signaling cascades

The current view of CCR2 signaling indicates that this receptor couples to multiple signaling cascades. Following ligand stimulation, the first events are conformational changes in the receptor, leading to receptor dimerization (Rodriguez-Frade *et al.*, 1999a). This dimerization allows JAK coupling to the CCR2, as well as receptor and JAK phosphorylation, which are needed for effective G protein coupling (Mellado *et al.*, 1998). The JAK pathway not only initiates G protein signaling, but also activates the STAT family of transcription factors. These events are not exclusive to the CCR2, but also occur in other chemokine receptors such as CCR5 and CXCR4 (Rodriguez-Frade *et al.*, 1999b; Vila-Coro *et al.*, 1999), as well as GPCR from other families, including the angiotensin II receptor (McWhinney *et al.*, 1997). Both JAK2 association and CCR2 tyrosine phosphorylation occur in the presence of pertussis toxin (PTX), indicating no G_i participation in this process. JAK2 dissociates from the receptor in a G_i-dependent manner, however, as it is blocked by PTX treatment. No MCP-1-induced, PTX-sensitive G protein-mediated physiological effects or $G\alpha_i$ association to CCR2 are observed after treatment of Mono Mac 1 cells with the JAK inhibitor tyrphostin B42. This is not the case when cells are treated with other tyrphostins, indicating that inhibition of JAK2 kinase activity abolishes the association and activation of the G proteins responsible for this response. This, and the JAK2 association with the CCR2 receptor in PTX-treated cells, implies that the first event after MCP-1 binding to CCR2 in Mono Mac 1 cells is JAK2 kinase association. The conformational changes promoted by both ligand interaction and tyrosine kinase association induce $G\alpha_i$ protein association to its binding site, which is probably located in the second intracellular loop, as is the case for the IL-8 receptor (Damaj *et al.*, 1996). Close examination of the intracellular domains of the different chemokine receptors reveals a highly conserved motif in the second intracellular loop (DRY[I/L]A[I/V]V[H/Q]A) (Murphy, 1994). This domain includes the residues critical for GPCR activation and is also present in other seven transmembrane receptors, such as the angiotensin II receptor, which is also associated with ligand-induced STAT activation (Marrero *et al.*, 1995). When signaling was analyzed through a receptor in which the CCR2BY-139 was replaced by phenylalanine, CCR2BY139F, MCP-1-triggered functional responses such as Ca^{2+} mobilization, cell migration, and CCR2B tyrosine phosphorylation were not elicited, although the receptor binds MCP-1 as well as does the wild-type receptor. This lack of response is due to abolition of JAK2 kinase association to the receptor, and thus of the G_i protein (Mellado *et al.*, 1998).

Most of these responses induced through the CCR2 can be inhibited by PTX treatment, indicating that members of the G_i protein family are the primary transduction partners associated with the receptors (Murphy, 1994; Frade *et al.*, 1997b). The physical association of $G\alpha_i$ to the CCR2 in response to MCP-1 stimulus in a human monocytic cell line (Mono Mac 1) has been described (Aragay *et al.*, 1998). Signaling studies of the CC chemokine receptors in transfected HEK-293 cells revealed potent, agonist-dependent inhibition of adenylyl cyclase and mobilization of intracellular calcium, consistent with receptor coupling to $G\alpha_i$ (Myers *et al.*, 1995). In these studies, the calcium response was not completely blocked by PTX, suggesting that chemokine receptors may couple to multiple G proteins, such as G_i, G_q, or G_{16}, depending on the chemokine receptor studied, and indicating that receptor–G protein pairings are cell type-specific (Al-Aoukaty *et al.*, 1996; Arai and Charo, 1996).

Following activation by the chemokine-triggered receptor, the heterotrimeric $G\alpha_i\beta\gamma$ protein dissociates into the $\beta\gamma$ subunit complex and the GTP-bound $G\alpha_i$ subunit, which remains associated to the receptor, probably through interaction with one or more regions of the intracellular loops (Arai and Charo, 1996). Both events, receptor association and subunit dissociation, initiate independent intracellular signaling responses by acting on different effector molecules.

$G\beta\gamma$ also acts as a docking protein, providing an interface for the GPCR, which would facilitate the interaction of GPCR in diverse signaling pathways. This is the case for the coupling of G protein-coupled receptor kinases (GRK; Wu *et al.*, 1998), in which $\beta\gamma$, but not heterotrimeric G protein or $G\alpha$, interacts with the third intracellular loop of the M2–M3 muscarinic receptors. The association of $\beta\gamma$ with the activated CCR2 allows the formation of a ternary complex with GRK2, required for effective receptor phosphorylation. It has been shown that MCP-1 induces both $\beta\gamma$ release and GRK2 association to the activated CCR2, allowing the formation of a macromolecular complex that also includes arrestin (Aragay *et al.*, 1998). The formation of such a complex is necessary to promote desensitization of the CCR2, and is the first step leading to its internalization. In fact, a CCR2 mutant in which Ser/Thr residues of the C-terminal region were replaced by alanines shows greatly reduced internalization (Arai *et al.*, 1997). The role of this region in chemotaxis is not conclusive, as although this mutant CCR2 supports chemokine-induced chemotaxis, other observations implicate the C-terminal region of CCR2 (amino acids 316–328) in chemotaxis.

Regarding the GRK2-mediated desensitization process, both the GRK2 catalytic activity as well as the Ser/Thr residues in the CCR2B C-terminal domain are critical. Coexpression of GRK2 and CCR2 blocks MCP-1-induced responses. When a CCR2 receptor mutant lacking Ser/Thr residues in the carboxyl tail was expressed, the MCP-1-induced signal was not inhibited by GRK2 coexpression (Franci *et al.*, 1996). This critical role of GRK kinase in CCR2 deactivation was also demonstrated when the CCR2 receptor was coexpressed with a dominant negative GRK2 mutant (Kong *et al.*, 1994); the cellular response to a second MCP-1 challenge was equivalent to the original response (Franci *et al.*, 1996).

The formation of phosphatidylinositol 3-kinase (PI-3 kinase) lipid products has also been observed after CCR2 activation, a process which is inhibited by PTX. MCP-1 activates at least two different PI-3 kinase isoforms, p85/p110 and $C2\alpha$. The biological consequences of this activation remain to be clarified, although PI-3 kinase activation has been implicated in a variety of cellular responses such as adhesion molecule upregulation (Shimizu and Hunt, 1996), superoxide release (Chung *et al.*, 1994), and chemotaxis (Turner *et al.*, 1995; Knall *et al.*, 1997; Turner *et al.*, 1998).

PTX-dependent phospholipase C (PLC) activation through MCP-1-activated CCR2B and 2A has been described (Kuang *et al.*, 1996).

GPCR signaling is linked to the mitogen-activated protein kinase (MAPK) cascade. MCP-1 binding to CCR2B promotes rapid, transient activation of MAPK. This process appears to be PTX-insensitive and PKC-dependent. Although its role remains to be clarified, several authors include this pathway among those required for chemotaxis (Yen *et al.*, 1997) (**Figure 5**).

Figure 5 Schematic representation of the signaling pathways activated through the CCR2. Ligand-binding to CCR2 induces conformational changes in the receptor, leading to its dimerization; this in turns disables JAK2 association and activation. JAK activation initiates the JAK/STAT pathway and enables efficient G protein coupling to CCR2. G proteins mediate signaling cascades through GRK and arrestins, leading to receptor desensitization and internalization. Other signaling molecules triggered through CCR2 include PI-3 kinase, PLC, and MAPK.

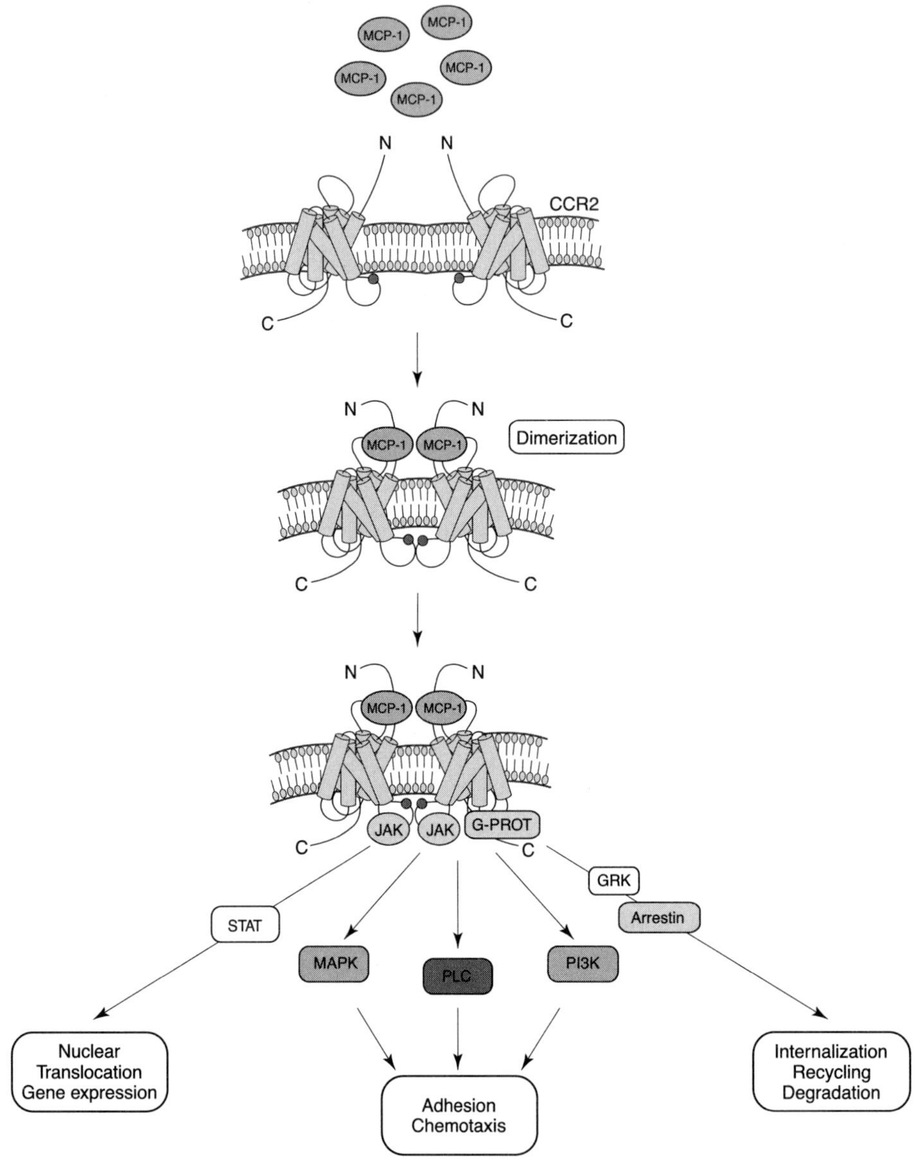

DOWNSTREAM GENE ACTIVATION

Transcription factors activated

JAK2/STAT5 pathway activation by MCP-1 has been reported (Mellado *et al.*, 1998), although other members of the JAK/STAT pathway may also be activated. The consequences of this activation remain to be elucidated. The fact that other chemokines, such as RANTES, activate transcription factors such as JNK through MAPK-related pathways, suggests that CCR2 may mediate activation of similar factors.

Promoter regions involved

1. Transcription initiation sites 26 bp downstream from the TATA box.
2. Consensus AATAAA polyadenylation signal sequences upstream of the 3′ exon 2 and 3 (Wong *et al.*, 1997), consistent with the presence of two different CCR2 transcripts.
3. Oct-1 binding sequences 36 bp upstream TATA box, important for transcriptional activation of CCR2.
4. Tandem CAAT/enhancer-binding protein (C/EBP) binding sequences at +50 and +77 within the 5′ UTR, which are essential for transcriptional activation and tissue specificity of CCR2.
5. Typical mammalian promoter consensus elements TATAA (−31 to −27) and CCAAT (−128 to −124).

It is interesting to note its different transcriptional activation compared with CCR5, although both receptors are supposed to share the same ancestral gene (Yamamoto *et al.*, 1999).

BIOLOGICAL CONSEQUENCES OF ACTIVATING OR INHIBITING RECEPTOR AND PATHOPHYSIOLOGY

Unique biological effects of activating the receptors

As for other chemokine receptors, one of the first consequences of CCR2 activation is a rapid, transient elevation of intracellular calcium that is PTX-sensitive. In accordance with G_i pathway activation, inhibition of cAMP production after CCR2 activation has also been described (Monteclaro and Charo, 1997). Another classical effect seen after CCR2 activation is the accumulation of inositol phosphate, related to PLC activation.

One of the most analyzed chemokine-related phenomena is chemotaxis. Chemotaxis is a key phenomenon both in cell movement and in the inflammatory response, and CCR2-mediated chemotaxis is well-documented (Rollins, 1998). It is implicated in the polarization of lymphocytes during migration, such that the leading cell edge develops cytoplasmic extensions, whereas the posterior part of the cell forms an appendage, the uropod. Following stimulation by molecules such as MCP-1, RANTES, IL-8, IL-15 or IL-2, the CCR2 is redistributed and located at the advancing front of migrating lymphocytes (Nieto *et al.*, 1997).

It has recently been demonstrated that NK cells contact target cells through the advancing front, a region in which CCR2 and CCR5 are concentrated; this suggests a role for these receptors during cytotoxic phenomena (Nieto *et al.*, 1998).

Phenotypes of receptor knockouts and receptor overexpression mice

While there are no references to mice overexpressing CCR2, several authors have developed CCR2-knockout mice. CCR2 deficiency has no observable effects on growth, development, or fertility. With respect to the immune response, the mice are not severely immunocompromised, with normal hematopoietic development including myeloid progenitors and development of lymphocyte subsets. The biological consequences nevertheless become clear when mice are challenged with proinflammatory stimuli. *In vivo*, it appears that CCR2 is highly specific for MCP-1, in accordance with the fact that MCP-1-knockout mice display a phenotype similar to that of the CCR2−/− mice (Rollins, 1996; Lu *et al.*, 1998).

Kuziel *et al.* (1997) reported that CCR2 is required for firm adhesion and diapedesis of leukocytes, which results in reduced macrophage accumulation at inflammation sites. Similar defects were observed in macrophages recovered from CCR2−/− mouse peritoneum after nonspecific inflammatory challenge (Kurihara *et al.*, 1997), whereas neutrophils and eosinophils are unimpaired, indicating the selectivity of CCR2 in eliciting such responses. This decrease in macrophage recruitment, which is not cholesterol-mediated, results in decreased lesion formation in CCR2−/− mice backcrossed with apoE null mice, which develop severe *atherosclerosis* (Boring *et al.*, 1998). These defects compromise the macrophage-dependent immunity to intracellular pathogens (Kurihara *et al.*, 1997).

Impaired monocyte and leukocyte migration has been observed in CCR2-knockout mice, as well as decreased IFNγ and IL-2 production following exposure to antigenic and nonantigenic challenge. It is not yet clear whether this decrease in IFNγ production is due to a direct effect of CCR2 on T cells, or to defects in CCR2-mediated trafficking to lymph nodes or spleen (Boring *et al.*, 1997).

Defects in macrophage recruitment and host defense, and enhanced myeloid progenitor cell cycle and apoptosis in bone marrow but not in spleen, have been described (Reid *et al.*, 1999). It is interesting to note that other myelosuppressive chemokines, acting through non-CCR2 receptors, do not have compensatory suppressive effects.

An altered airway hyperreactivity response, with decreased histamine levels following challenge, has also been described in CCR2−/− mice (Campbell *et al.*, 1999).

Human abnormalities

A polymorphism has been reported in CCR2, in which Val64 is replaced by Ile, giving the CCR2V64I (Smith *et al.*, 1997). This polymorphism occurs at an allele frequency of 10–25%, depending on the ethnic population. There is no evidence of functional difference between the CCR2V64I and the wild-type CCR2; however, this polymorphism is associated with a 2–4-year delay in progression to *AIDS* in HIV-1-infected individuals. More recently, an increased incidence of this polymorphism has been described in children with insulin-dependent *diabetes mellitus* (Szalai *et al.*, 1999), whereas the presence of the CCR2V64I allele confers a lower risk for development of *sarcoidosis* (Hizawa *et al.*, 1999). Whether or not these effects are directly related to CCR2 function remains to be elucidated.

THERAPEUTIC UTILITY

Effects of inhibitors (antibodies) to receptors

Antibodies against CCR2 have been developed by several groups (Frade *et al.*, 1997b; Sallusto *et al.*, 1998), some of which can block MCP-1-induced effects. This blockage results in reduced migration both in mouse models and in *in vitro* assays, as well as in reduced calcium mobilization and arachidonic acid release; some anti-CCR2 antibodies also neutralize *HIV-1* strains.

Modified MCP-1 with antagonist activity has also been described (Zhang *et al.*, 1996).

References

Al-Aoukaty, A., Schall, T. J., and Maghazachi, A. A. (1996). Differential coupling of CC chemokine receptors to multiple heterotrimeric G proteins in human interleukin-2-activated natural killer cells. *Blood* **87**, 4255–4260.

Aragay, A., Frade, J. M. R., Mellado, M., Serrano, A., Martínez-A. C., and Mayor, F. Jr. (1998). MCP-1-induced CCR2B receptor desensitization by the G protein-coupled receptor kinase-2. *Proc. Natl Acad. Sci. USA* **95**, 2985–2990.

Arai, H., and Charo, I. F. (1996). Differential regulation of G-protein-mediated signaling by chemokine receptors. *J. Biol. Chem.* **271**, 21814–21819.

Arai, H., Monteclaro, F. S., Tsou, C.-L., Franci, C., and Charo, I. F. (1997). Dissociation of chemotaxis from agonist-induced receptor internalization in a lymphocyte cell line transfected with CCR2B. Evidence that directed migration does not require rapid modulation of signaling at the receptor level. *J. Biol. Chem.* **272**, 25037–25042.

Baggiolini, M. (1998). Chemokines and leukocyte traffic. *Nature* **392**, 565–568.

Berger, E. A., Murphy, P. M., and Farber, J. M. (1999). Chemokine receptors as HIV-1 coreceptors: roles in viral entry, tropism, and disease. *Annu. Rev. Immunol.* **17**, 657–700.

Bonecchi, R., Polentarutti, N., Luini, W., Borsatti, A., Bernasconi, S., Locati, M., Power, C., Proudfoot, A., Wells, T. N. C., Mackay, C., Mantovani, A., and Sozzani, S. (1999). Up-regulation of CCR1 and CCR3 and induction of chemotaxis to CC chemokines by IFN-gamma in human neutrophils. *J. Immunol.* **162**, 474–479.

Boring, L., Gosling, J., Monteclaro, E. S., Lusis, A. J., Tsou, C.-L., and Charo, I. F. (1996). Molecular cloning and functional expression of murine JE (monocyte chemoattractant protein 1) and murine macrophage inflammatory protein 1 alpha receptors: evidence for two closely linked C-C receptors on chromosome 9. *J. Biol. Chem.* **271**, 7551–7558.

Boring, L., Gosling, J., Chensue, S. W., Kunkel, S. L., Farese Jr, R. V., Broxmeyer, H. E., and Charo, I. F. (1997). Impaired monocyte migration and reduced type 1 (TH1) cytokine responses in C-C chemokine receptor 2 knockout mice. *J. Clin. Invest.* **100**, 2552–2561.

Boring, L., Gosling, J., Cleray, M., and Charo, I. F. (1998). Decreased lesion formation in CCR2−/− mice reveals a role for chemokines in the initiation of atherosclerosis. *Nature* **394**, 894–897.

Campbell, E. M., Charo, I. F., Kunkel, S. L., Strieter, R. M., Boring, L., Gosling, J., and Lukacs, N. W. (1999). Monocyte chemoattractant protein-1 mediates cockroach allergen-induced bronchial hyperreactivity in normal but not CCR2−/− mice: the role of mast cells. *J. Immunol.* **163**, 2160–2167.

Charo, I. F., Myers, S. J., Herman, A., Franci, C., Connolly, A. J., and Coughlin, S. R. (1994). Molecular cloning and functional expression of two monocyte chemoattractant protein 1 receptors reveals alternative splicing of the carboxyl-terminal tails. *Proc. Natl Acad. Sci. USA* **91**, 2752–2756.

Chung, J., Grammer, T., Lemon, C., Kazlauskas, A., and Blenis, J. (1994). PDGF- and insulin-dependent pp70S6k activation mediated by phosphatidylinositol-3-OH kinase. *Nature* **370**, 71–73.

Damaj, B. B., McColl, S. R., Neote, K., Songquing, N., Ogborn, K. T., Hebert, C. A., and Naccache, P. H. (1996). Identification of G-protein binding sites of the human interleukin-8 receptors by functional mapping of the intracellular loops. *FASEB J.* **12**, 1426–1434.

Fantuzzi, L., Borghi, P., Ciolli, V., Pavlakis, G., Belardelli, F., and Gessani, S. (1999). Loss of CCR2 expression and functional response to monocyte chemotactic protein (MCP-1) during the differentiation of human monocytes: role of secreted MCP-1 in the regulation of the chemotactic response. *Blood* **94**, 875–883.

Frade, J. M. R., Llorente, M., Mellado, M., Alcami, J., Gutierrez-Ramos, J. C., Zaballos, A., del Real, G., and Martínez-A. C. (1997a). The amino terminal domain of the CCR2 chemokine receptor acts as coreceptor for HIV-1 infection. *J. Clin. Invest.* **100**, 497–502.

Frade, J. M. R., Mellado, M., del Real, G., Gutierrez-Ramos, J. C., Lind, P., and Martínez-A. C. (1997b). Characterization of the CCR2 chemokine receptor: functional CCR2 receptor expression in B cells. *J. Immunol.* **159**, 5576–5584.

Franci, C., Gosling, J., Tsou, C. L., Coughlin, S. R., and Charo, I. F. (1996). Phosphorylation by a G protein-coupled kinase inhibits signaling and promotes internalization of the monocyte chemoattractant protein-1 receptor. Critical role of carboxyl-tail serines/threonines in receptor function. *J. Immunol.* **157**, 5606–5612.

Gong, X., Gong, W., Kuhns, D. B., Ben-Baruch, A., Howard, O. M., and Wang, J. M. (1997). Monocyte chemotactic protein-2 (MCP-2) uses CCR1 and CCR2B as its functional receptors. *J. Biol. Chem.* **272**, 11682–11685.

Gonzalo, J. A., Lloyd, C., Wen, D., Albar, J. P., Wells, T. N. C., Proudfoot, A., Martínez-A. C., Dorf, M., Bjerke, T., Coyle, A. J., and Gutierrez-Ramos, J. C. (1998). The coordinated action of CC chemokines in the lung orchestrates allergic inflammation and airway hyperresponsiveness. *J. Exp. Med.* **188**, 157–167.

Han, K. H., Tangirala, R. K., Green, S. R., and Quehenberger, O. (1998). Chemokine receptor CCR2 expression and monocyte chemoattractant protein-1 mediated chemotaxis in human monocytes. A regulatory role for plasma LDL. *Arterioscler. Thromb. Vasc. Biol.* **18**, 1983–1991.

Han, K. H., Green, S. R., Tangirala, R. K., Tanaka, S., and Quehenberger, O. (1999a). Role of the first extracellular loop in the functional activation of CCR2. The first extracellular loop contains distinct domains necessary for both agonist binding and transmembrane signaling. *J. Biol. Chem.* **274**, 32055–32062.

Han, K. H., Han, K. O., Green, S. R., and Quehenberger, O. (1999b). Expression of the monocyte chemoattractant protein-1 receptor CCR2 is increased in hypercholesterolemia. Differential effects of plasma lipoproteins on monocyte function. *J. Lipid Res.* **40**, 1053–1063.

Hizawa, N., Yamaguchi, E., Furuya, K., Jinushi, E., Ito, A., and Kawakami, Y. (1999). The role of C-C chemokine receptor gene polymorphism V64I (CCR2-64I) in sarcoidosis in Japanese population. *Am. J. Respir. Crit. Care Med.* **159**, 2021–2023.

Hogaboam, C. M., Bone-Larson, C. L., Lipinski, S., Lukacs, N. W., Chensue, S. W., Strieter, R. M., and Kunkel, S. L. (1999). Differential monocyte chemoattractant protein-1 and chemokine receptor 2 expression by murine lung fibroblasts derived from TH1- and TH2-type pulmonary granuloma models. *J. Immunol.* **163**, 2193–2201.

Jiang, Y., Salafranca, M. N., Adhikari, S., Xia, Y., Feng, L., Sonntag, M. K., deFiebre, C. M., Pennell, N. A., Striet, W. J., and Harrison, J. K. (1998). Chemokine receptor expression in cultured glia and rat experimental allergic encephalomyelitis. *J. Neuroimmunol.* **86**, 1–12.

Johnston, B., Burns, A. R., Suematsu, M., Issekutz, T. B., Woodman, R. C., and Kubes, P. (1999). Chronic inflammation upregulates chemokine receptors and induces neutrophil migration to moncyte chemoattractant protein-1. *J. Clin. Invest.* **103**, 1269–1276.

Knall, C., Worthen, G. S., and Johnson, G. L. (1997). Interleukin 8-stimulated phosphatidylinositol-3-kinase activity regulates the migration of human neutrophils independent of extracellular signal-regulated kinase and p38 mitogen-activated protein kinases. *Proc. Natl Acad. Sci. USA* **94**, 3052–3057.

Kong, G., Penn, R., and Benovic, J. L. (1994). A β-adrenergic receptor kinase dominant negative mutant attenuates desensitization of the β_2-adrenergic receptor. *J. Biol. Chem.* **269**, 13084–13087.

Kuang, Y., Wu, Y., Jiang, H., and Wu, D. (1996). Selective G protein coupling by C-C chemokine receptors. *J. Biol. Chem.* **271**, 3975–3978.

Kurihara, T., and Bravo, R. (1996). Cloning and functional expression of mCCR2, a murine receptor for the C-C chemokines JE and FIC. *J. Biol. Chem.* **271**, 11603–11606.

Kurihara, T., Warr, G., Loy, J., and Bravo, R. (1997). Defects in macrophage recruitment and host defense in mice lacking the CCR2 chemokine receptor. *J. Exp. Med.* **186**, 1757–1762.

Kuziel, W. A., Morgan, S. J., Dawson, T. C., Griffin, S., Smithies, O., Ley, K., and Maeda, N. (1997). Severe reduction in leukocyte adhesion and monocyte extravasation in mice deficient in CC chemokine receptor 2. *Proc. Natl Acad. Sci. USA* **94**, 12053–12058.

Lee, B., Sharron, M., Blanpain, C., Doranz, B. J., Vakili, J., Setoh, P., Berg, E., Liu, G., Guy, H. R., Durell, S. R., Parmentier, M., Chang, C. N., Price, K., Tsang M., and Doms, R. W. (1999). Epitope mapping of CCR5 reveals multiple conformational states and distinct but overlapping structures involved in chemokine and coreceptor function. *J. Biol. Chem.* **274**, 9617–9626.

Loetscher, P., Seitz, M., Baggiolini, M., and Moser, B. (1996). Interleukin-2 regulates CC chemokine receptor expression and chemotactic responsiveness in T lymphocytes. *J. Exp. Med.* **184**, 569–577.

Lu, B., Rutledge, B. J., Gu, L., Fiorillo, J., Lukacs, N. W., Kunkel, S. L., North, R., Gerard, C., and Rollins, B. J. (1998). Abnormalities in monocyte recruitment and cytokine expression in monocyte chemoattractant protein-1 deficient mice. *J. Exp. Med.* **187**, 601–608.

Lukacs, N. W., Oliveira, S. H. P., and Hogaboam, C. M. (1999). Chemokines and asthma: redundancy of function or coordinated effort? *J. Clin. Invest.* **104**, 995–999.

Luster, A. D. (1998). Chemokines – chemotactic cytokines that mediate inflammation. *N. Engl. J. Med.* **338**, 436–445.

Marrero, M. B., Schieffer, B., Paxton, W. G., Eerdt, L. H., Berk, B. C., Delafontaine, P., and Bernstein, K. E. (1995). Direct stimulation of Jak/STAT pathway by the angiotensin II AT1 receptor. *Nature* **375**, 247–250.

McWhinney, C. D., Hunt, R. A., Conrad, K. M., Dostal, D. E., and Baker, K. M. (1997). The type I angiotensin II receptor couples to STAT1 and STAT3 activation through Jak2 kinase in neonatal rat cardiac myocytes. *J. Mol. Cell. Cardiol.* **29**, 2513–2524.

Mellado, M., Rodríguez-Frade, J. M., Aragay, A., del Real, G., Martin de Ana, A., Vila-Coro, A., Serrano, A., Mayor, F. Jr., and Martínez-A. C. (1998). The chemokine MCP-1 triggers JAK2 kinase activation and tyrosine phosphorylation of the CCR2B receptor. *J. Immunol.* **161**, 805–813.

Mellado, M., Rodríguez-Frade, J. M., Vila-Coro, A. J., Martín de Ana, A., and Martínez-A. C. (1999). Chemokine control of HIV-1 infection. *Nature* **400**, 723–724.

Monteclaro, F. S., and Charo, I. F. (1996). The amino terminal extracellular domain of the MCP-1 receptor, but not the RANTES/MIP-1α receptor, confers chemokine selectivity. Evidence for a two-step mechanism for MCP-1 receptor activation. *J. Biol. Chem.* **271**, 19084–19092.

Monteclaro, F. S., and Charo, I. F. (1997). The amino-terminal domain of CCR2 is both necessary and sufficient for high affinity binding of monocyte chemoattractant protein 1. *J. Biol. Chem.* **272**, 23186–23190.

Murphy, P. M. (1994). The molecular biology of leukocyte chemoattractant receptors. *Annu. Rev. Immunol.* **12**, 593–633.

Myers, S. J., Wong, L. M., and Charo, I. F. (1995). Signal transduction and ligand specificity of the human monocyte chemoattractant protein-1 receptor in transfected embryonic kidney cells. *J. Biol. Chem.* **270**, 5786–5792.

Nieto, M., Frade, J. M. R., Sancho, D., Mellado, M., Martínez-A. C., and Sánchez-Madrid, F. (1997). Polarization of the chemokine receptors to the leading edge during lymphocyte migration. *J. Exp. Med.* **186**, 153–158.

Nieto, M., Navarro, F., Perez-Villar, J. J., del Pozo, M. A., Gonzalez-Amaro, R., Mellado, M., Rodríguez-Frade, J. M., Martínez-A. C., López-Botet, M., and Sánchez-Madrid, F. (1998). Roles of chemokines and receptor polarization in NK-target cell interactions. *J. Immunol.* **161**, 3330–3339.

Pelchen-Matthews, A., Signoret, N., Klasse, P. J., Fraile-Ramos, A., and Marsh, M. (1999). Chemokine receptor trafficking and viral replication. *Immunol. Rev.* **168**, 33–49.

Penton-Rol, G., Polentarutti, N., Luini, W., Borsatti, A., Mancinelli, R., Sica, A., Sozzani, S., and Manyovani, A. (1998). Selective inhibition of expression of the chemokine receptor CCR2 in human monocytes by IFN-γ. *J. Immunol.* **160**, 3869–3873.

Perera, L. P., Goldman, C. K., and Waldmann, T. A. (1999). IL-15 induces the expression of chemokines and their receptors in T lymphocytes. *J. Immunol.* **162**, 2606–2612.

Polentarutti, N., Allavena, P., Bianchi, G., Giardina, G., Basile, A., Sozzani, S., Mantovani, A., and Introna, M. (1997). IL-2-regulated expression of the monocyte chemotactic protein-1 receptor (CCR2) in human NK cells: characterization of a predominant 3.4-kilobase transcript containing CCR2B and CCR2A sequences. *J. Immunol.* **158**, 2689–2694.

Rabin, R. L., Park, M. K., Liao, F., Swofford, R., Stephany, D., and Farber, J. M. (1999). Chemokine receptor responses on T cells are achieved through regulation of both receptor expression and signaling. *J. Immunol.* **162**, 3840–3850.

Reid, S., Ritchie, A., Boring, L., Gosling, J., Cooper, S., Hangoc, G., Charo, I. F., and Broxmeyer, H. E. (1999). Enhanced myeloid progenitor cell cycling and apoptosis in mice lacking the chemokine receptor, CCR2. *Blood* **93**, 1524–1533.

Rodriguez-Frade, J. M., Vila-Coro, A., Martín de Ana, A., Albar, J. P., Martínez-A. C., and Mellado, M. (1999a). Monocyte chemoattractant protein-1 induces functional responses through CCR2 dimerization. *Proc. Natl Acad. Sci. USA* **96**, 3628–3633.

Rodriguez-Frade, J. M., Vila-Coro, A., Martín de Ana, A., Nieto, M., Sánchez-Madrid, F., Proudfoot, A. E. I., Wells, T. N. C., Martínez-A. C., and Mellado, M. (1999b). Similarities and differences in RANTES- and (AOP)-RANTES-triggered signals: implications for chemotaxis. *J. Cell. Biol.* **144**, 755–765.

Rollins, B. J. (1996). Monocyte chemoattractant protein 1: a potential regulator of monocyte recruitment in inflammatory disease. *Mol. Med. Today* **2**, 198–204.

Rollins, B. J. (1998). Chemokines. *Blood* **90**, 909–928.

Rucker, J., Samson, M., Doranz, B. J., Libert, F., Berson, J. F., Yi, Y., Smyth, R. J., Collman, R. G., Broder, C. C., Vassart, G., Doms, R. W., and Parmentier, M. (1996). Regions in β-chemokine receptors CCR5 and CCR2b that determine HIV-1 cofactor specificity. *Cell* **87**, 437–446.

Sallusto, F., Schaerli, P., Loetscher, P., Schaniel, C., Lening, D., Mackay, C. R., Qin, S., and Lanzavecchia, A. (1998). Rapid and coordinated switch in chemokine receptor expression during dendritic cell maturation. *Eur. J. Immunol.* **28**, 2760–2769.

Samson, M., LaRosa, G., Libert, F., Paindavoine, P., Detheux, M., Vassart, G., and Parmentier, M. (1997). The second extracellular loop of CCR5 is the major determinant of ligand specificity. *J. Biol. Chem.* **272**, 24934–24941.

Shimizu, Y., and Hunt III, S. W. (1996). Regulating integrin-mediated adhesion: one more function for PI3-kinase? *Immunol. Today* **17**, 565–573.

Sica, A., Saccani, A., Borsatti, A., Power, C. A., Wells, T. N. C., Luini, W., Polentarutti, N., Sozzani, S., and Mantovani, A. (1997). Bacterial lipopolysaccharide rapdly inhibits expression of C-C chemokine receptors in human monocytes. *J. Exp. Med.* **185**, 969–974.

Smith, M. W., Dean, M., Carrington, M., Winkler, C., Huttley, G. A., Lomb, D. A., Goedert, J. J., O'Brien, T. R., Jacobson, L. P., Kaslow, R., Buchbinder, S., Vittinghoff, E., Vlahov, D., Hoots, K., Hilgartner, M. W., and O'Brien, S. J. (1997). Contrasting genetic influence of CCR2 and CCR5 variants on HIV-1 infection and disease progression. *Science* **277**, 959–965.

Sozzani, S., Introna, M., Bernasconi, S., Polentarutti, N., Cinque, P., Poli, G., Sica, A., and Mantovani, A. (1997). MCP-1 and CCR2 in HIV-1 infection: regulation of agonist and receptor expression. *J. Leukoc. Biol.* **62**, 30–33.

Szalai, C., Csaszar, A., Czinner, A., Szabo, T., Panczel, P., Madacsy, L., and Falus, A. (1999). Chemokine receptor CCR2 and CCR5 polymorphisms in children with insulin-dependent diabetes mellitus. *Pediatr. Res.* **46**, 82–84.

Turner, L., Ward, S. G., and Westwick, J. (1995). RANTES-activated human T lymphocytes. A role for phosphoinositide 3-kinase. *J. Immunol.* **155**, 2437–2444.

Turner, S. J., Domin, J., Waterfield, M. D., Ward, S. G., and Westwick, J. (1998). The CC chemokine monocyte chemotactic peptide-1 activates both the class I p85/p110 phosphatidylinositol 3-kinase and the class II PI3K-C2α. *J. Biol. Chem.* **273**, 25987–25995.

Vila-Coro, A. J., Rodríguez-Frade, J. M., Martín de Ana, A., Moreno-Ortíz, M. C., Martínez-A. C., and Mellado, M. (1999). The chemokine SDF-1α triggers CXCR4 receptor dimerization and activates the JAK/STAT pathway. *FASEB J.* **13**, 1699–1710.

Weber, K. S., Nelson, P. J., Grone, H. J., and Weber, C. (1999a). Expression of CCR2 by endothelial cells: implications for MCP-1 mediated wound injury repair and *in vivo* inflammatory activation of endothelium. *Arterioscler. Thromb. Vasc. Biol.* **19**, 2085–2093.

Weber, C., Draude, G., Weber, K. S., Wubert, J., Lorenz, R. L., and Weber, P. C. (1999b). Downregulation by tumor necrosis factor-alpha of monocyte CCR2 expression and monocyte chemotactic protein-1-induced transendothelial migration is antagonized by oxidized low-density lipoprotein: a potential mechanism of monocyte retention in atherosclerosis lesions. *J. Immunol.* **163**, 2160–2167.

Wong, L.-M., Myers, S. J., Tsou, C.-L., Gosling, J., Arai, H., and Charo, I. F. (1997). Organization and differential expression of the human monocyte chemoattractant protein 1 receptor gene. Evidence for the role of the carboxyl-terminal tail in receptor trafficking. *J. Biol. Chem.* **272**, 1038–1045.

Wu, G., Benovic, J. L., Hildebrandt, J. D., and Lanier, S. M. (1998). Receptor docking sites for G-protein $\beta\gamma$ subunits. Implications for signal regulation. *J. Biol. Chem.* **273**, 7197–7200.

Yamagami , S., Tokuda, Y., Ishii, K., Tanaka, H., and Endo, N. (1994). cDNA cloning and functional expression of a human monocyte chemoattractant protein 1 receptor. *Biochem. Biophys. Res. Commun.* **202**, 1156–1162.

Yamamoto, K., Takeshima, H., Hamada, K., Nakao, M., Kino, T., Nishi, T., Kochi, M., Kuratsu, J.-I., Yoshimura, T., and Ushio, Y. (1999). Cloning and functional characterization of the 5′ flanking region of the human monocyte chemoattractant protein-1 receptor (CCR2) gene. *J. Biol. Chem.* **274**, 4646–4654.

Yen, H., Zhang, Y., Penfold, S., and Rollins, B. J. (1997). MCP-1-mediated chemotaxis requires activation of non-overlapping signal transduction pathways. *Leukoc. Biol.* **61**, 529–532.

Zhang, Y., Ernst, C. A., and Rollins, B. J. (1996). MCP-1: structure/activity analysis. *Methods* **10**, 93–103.

Zhou, Y., Yang, Y., Warr, G., and Bravo, R. (1999). LPS down-regulates the expression of chemokine receptor CCR2 in mice and abolishes macrophage infiltration in acute inflammation. *J. Leukoc. Biol.* **65**, 265–269.

LICENSED PRODUCTS

Antibodies against CCR2, both monoclonal and polyclonal, are available from several suppliers, including R&D Systems (614 McKinley Place, Minneapolis, MN 55413, USA, http://www.rndsystems.com) and Santa Cruz (2161 Delaware Ave., Santa Cruz, CA 95060, USA, http://www.scbt.com). Among these are antibodies suitable for flow cytometry analysis, western blot and immunohistochemistry. Some antibodies have been derived using peptides from the CCR2 C-terminal tail; although they are only suitable for analysis of CCR2 in permeabilized or lysed cells, they offer the advantage of specific distinction between CCR2A and 2B variants. Note that some of the antibodies available may react differently with CCR2, depending on the cell line employed.

The corresponding recombinant ligands of human, rat, and murine origin are available from Peprotech (5 Crescent Ave., Rocky Hill, NJ 08553-0275, USA, http://www.peprotech.com), Pharmingen (10975 Torreyana Road, San Diego, CA 92121-1111, USA, http://www.pharmingen.com), Genzyme (One Kendall Square, Cambridge, MA 02139, USA, http://www.genzyme.com), and R&D Systems (614 McKinley Place, Minneapolis, MN 55413, USA, http://www.rndsystems.com).

Radiolabeled ligands are available from Amersham (http://www.apbiotech.com) and NEN (http://www.nenlifesci.com).

A multi-probe ribonuclease protection assay for mRNA analysis of chemokines and chemokine receptors (including CCR2) is available from Pharmingen.

CCR3

Charles R. Mackay*

The Garvan Institute of Medical Research, 384 Victoria Street, Darlinghurst (Sydney), New South Wales, 2010, Australia

*corresponding author tel: 61-2-92958405, fax: 61-2-92958404, e-mail: c.mackay@garvan.unsw.edu.au

DOI: 10.1006/rwcy.2000.22003.

SUMMARY

CCR3 is a β chemokine receptor, closely related to CCR1, CCR2, and CCR5. CCR3 is the principal chemokine receptor on eosinophils that mediates their migration to a variety of chemokines, including eotaxin, eotaxin 2, RANTES, MCP-2, MCP-3, and MCP-4. CCR3 is also expressed by other allergic leukocytes such as basophils, mast cells, and TH2 cells. Blocking CCR3 or CCR3 ligands has proven to be effective in animal models of *asthma*, for reducing inflammation and *bronchohyperreactivity*. For these reasons CCR3 is one of the promising targets for drug development for *allergic disease*.

BACKGROUND

Discovery

The discovery of CCR3 arose from studies aimed at characterizing the chemokines and chemokine receptors for eosinophil migration. Eosinophils are selectively recruited into certain inflammatory lesions as a result of IgE-mediated reactions, for instance in *rhinitis* and *allergic asthma*, and also in response to certain *parasitic infections* (Kay and Corrigan, 1992; Gleich *et al.*, 1993). This suggested the presence of a chemoattractant receptor on eosinophils that was expressed in a relatively restricted fashion, distinct from other receptors such as the C5a receptor which is also expressed on neutrophils. The chemokine eotaxin, first identified in guinea pigs (Jose *et al.*, 1994) and subsequently in humans (Ponath *et al.*, 1996a) and mouse (Gonzalo *et al.*, 1996), is unusual in that it is selectively chemotactic for eosinophils. Binding studies with ^{125}I-labeled eotaxin and peripheral blood eosinophils showed that eotaxin bound a receptor on eosinophils distinct from CCR1 or CCR2, as specific binding could not compete with MIP-1α or MCP-1 (Ponath *et al.*, 1996a). In addition, receptor cross-desensitization experiments using eosinophils suggested that at least one additional receptor existed that signals in response to RANTES and MCP-3 (Baggiolini and Dahinden, 1994; Dahinden *et al.*, 1994), but not MIP-1α.

Based on the structure of other chemokine receptors, novel seven transmembrane receptors were sought that were expressed in eosinophils, and bound eotaxin. In humans, CCR3 was identified as an eosinophilic β chemokine receptor that principally bound eotaxin and with lower affinity RANTES and MCP-3 (Daugherty *et al.*, 1996; Ponath *et al.*, 1996b). In mice, CCR3 was reported as a MIP-1α receptor (Post *et al.*, 1995), although murine CCR3 has subsequently been shown also to bind murine eotaxin with high affinity. A difference between mouse and human CCR3 is the ability of mouse CCR3 to bind MIP-1α.

Alternative names

The accepted nomenclature for this receptor is CCR3, although in the early papers this receptor was referred to as CCR-3, CKR-3, and CC CKR-3. This receptor is sometimes referred to as the eotaxin receptor since eotaxin binds with high affinity and fidelity to this receptor. Nevertheless, this is a misnomer since numerous chemokines bind CCR3.

Structure

CCR3 exhibits the seven transmembrane, G protein-coupled receptor structure typical of all of the chemokine receptors. It contains the motif DRYLAIVHA in the second intracellular loop, which is common to many members of the chemokine receptor family. CCR3 is somewhat unusual in that it does not contain sites for *N*-linked glycosylation.

Main activities and pathophysiological roles

CCR3 expression on eosinophils, basophils, and TH2 cells is the likely explanation for the selective recruitment of these cells to certain inflammatory sites. Its expression on eosinophils, in particular, may account for the striking accumulation of eosinophils at sites where eotaxin is produced, because eotaxin binds only to CCR3, and eosinophils are much more chemotactic in response to eotaxin than other CCR3-bearing cells such as basophils or TH2 cells. CCR3-eotaxin interactions may trigger eosinophil arrest during the multistep process of leukocyte binding to endothelium under flow conditions (Kitayama *et al.*, 1998), leading to selective eosinophil recruitment to a tissue. In addition, expression of eotaxin by epithelial cells in allergic airways (Ponath *et al.*, 1996; Ying *et al.*, 1997) might also lead to selective eosinophil migration to certain areas of a tissue. Epithelial cell destruction by eosinophils is one of the pathological hallmarks of *allergic asthma*. CCR3 has been the target of numerous drug development efforts, for the treatment of allergic disease, particularly *asthma*.

In addition to its role in chemoattraction, CCR3 activation also leads to mediator release by eosinophils and basophils (Uguccioni *et al.*, 1997). The importance of this aspect of CCR3 for allergic inflammation is not known. Although CCR3 is expressed at high levels on basophils, other chemokine receptors appear to be more important for triggering basophil mediator release, particularly CCR2 (Uguccioni *et al.*, 1997). All activities of eotaxin presumably occur through CCR3, and so reports that eotaxin stimulates the growth of myeloid cell progenitors and the differentiation of mast cells during embryonic development (Quackenbush *et al.*, 1997) implicates a role for CCR3 in these processes. CCR3 is also expressed by mature human mast cells *in vitro* and appears to be one of the principal chemokine receptors for these cells, although CCR3 appears not to function for chemotaxis (Ochi *et al.*, 1999). CCR3 ligands also enhance antigen-stimulated IL-4 production by basophils, which suggests that chemokines such as eotaxin amplify allergic inflammation through effects on cytokine production (Devouassoux *et al.*, 1999).

CCR3 also serves as a coreceptor for certain strains of *HIV-1* (Choe *et al.*, 1996), and may be one of the principal receptors in the brain for neurotropic strains of *HIV-1* (He *et al.*, 1997). CCR3 can interact with Envs from certain macrophage (M)-tropic strains (which also use CCR5), T cell line-tropic laboratory-adapted strains (which also use CXCR4), and certain dual-tropic primary isolates (which also use CCR2b, CCR5, and CXCR4). CCR3 expression on some cytotoxic T cells enables chemokines such as RANTES to stimulate anti-HIV cytotoxic activity (Hadida *et al.*, 1998). *Kaposi's sarcoma-associated herpesvirus* (KSHV or HHV-8) encodes chemokine-like proteins, including vMIP-II, which activate and chemoattract human eosinophils by way of CCR3 (Boshoff *et al.*, 1997).

GENE

Accession numbers

GenBank:
Human: U49727, U51241
Mouse: U29677
Guinea pig: AF060698

Sequence

See **Figure 1** for human CCR3.

PROTEIN

Accession numbers

SwissProt:
Human: P51677
Mouse: P51678

Sequence

See **Figure 2**.

Relevant homologies and species differences

CCR3 is most closely related to CCR1 (62% amino acid similarity in humans).

Figure 1 Gene sequence for human CCR3.

```
   1 AATCCTTTTC CTGGCACCTC TGATATCCTT TTGAAATTCA TGTTAAAGAA TCCCTAGGCT
  61 GCTATCACAT GTGGCATCTT TGTTGAGTAC ATGAATAAAT CAACTGGTGT GTTTTACGGA
 121 GGATGATTAT GCTTCATTGT GGGATTGTAT TTTTCTTCTT CTATCACAGG GAGAAGTGAA
 181 ATGACAACCT CACTAGATAC AGTTGAGACC TTTGGTACCA CATCCTACTA TGATGACGTG
 241 GGCCTGCTCT GTGAAAAAGC TGATACCAGA GCACTGATGG CCCAGTTTGT GCCCCCGCTG
 301 TACTCCCTGG TGTTCACTGT GGGCCTCTTG GGCAATGTGG TGGTGGTGAT GATCCTCATA
 361 AAATACAGGA GGCTCCGAAT TATGACCAAC ATCTACCTGC TCAACCTGGC CATTTCGGAC
 421 CTGCTCTTCC TCGTCACCCT TCCATTCTGG ATCCACTATG TCAGGGGGCA TAACTGGGTT
 481 TTTGGCCATG GCATGTGTAA GCTCCTCTCA GGGTTTTATC ACACAGGCTT GTACAGCGAG
 541 ATCTTTTTCA TAATCCTGCT GACAATCGAC AGGTACCTGG CCATTGTCCA TGCTGTGTTT
 601 GCCCTTCGAG CCCGGACTGT CACTTTTGGT GTCATCACCA GCATCGTCAC CTGGGGCCTG
 661 GCAGTGCTAG CAGCTCTTCC TGAATTTATC TTCTATGAGA CTGAAGAGTT GTTTGAAGAG
 721 ACTCTTTGCA GTGCTCTTTA CCCAGAGGAT ACAGTATATA GCTGGAGGCA TTTCCACACT
 781 CTGAGAATGA CCATCTTCTG TCTCGTTCTC CCTCTGCTCG TTATGGCCAT CTGCTACACA
 841 GGAATCATCA AAACGCTGCT GAGGTGCCCC AGTAAAAAAA AGTACAAGGC CATCCGGCTC
 901 ATTTTTGTCA TCATGGCGGT GTTTTTCATT TTCTGGACAC CCTACAATGT GGCTATCCTT
 961 CTCTCTTCCT ATCAATCCAT CTTATTTGGA AATGACTGTG AGCGGACGAA GCATCTGGAC
1021 CTGGTCATGC TGGTGACAGA GGTGATCGCC TACTCCCACT GCTGCATGAA CCCGGTGATC
1081 TACGCCTTTG TTGGAGAGAG GTTCCGGAAG TACCTGCGCC ACTTCTTCCA CAGGCACTTG
1141 CTCATGCACC TGGGCAGATA CATCCCATTC CTTCCTAGTG AGAAGCTGGA AAGAACCAGC
1201 TCTGTCTCTC CATCCACAGC AGAGCCGGAA CTCTCTATTG TGTTTTAGGT AGATGCAGAA
1261 AATTGCCTAA AGAGGAAGGA CCAAGGAGAT NAAGCAAACA CATTAAGCCT TCCACACTCA
1321 CCTCTAAAAC AGTCCTTCAA ACCTTCCAGT GCAACACTGA AGCTCTTAAG ACACTGAAAT
1381 ATACACACAG CAGTAGCAGT AGATGCATGT ACCCTAAGGT CATTACCACA GGCCAGGGCT
1441 GGGCAGCGTA CTCATCATCA ACCTAAAAAG CAGAGCTTTG CTTCTCTCTC TAAAATGAGT
1501 TACCTATATT TTAATGCACC TGAATGTTAG ATAGTTACTA TATGCCGCTA CAAAAAGGTA
1561 AAACTTTTTA TATTTTATAC ATTAACTTCA GCCAGCTATT ATATAAATAA AACATTTTCA
1621 CACAATACAA TAAGTTAACT ATTTTATTTT CTAATGTGCC TAGTTCTTTC CCTGCTTAAT
1681 GAAAAGCTT
```

Figure 2 Amino acid sequence for CCR3.

```
MTTSLDTVET FGTTSYYDDV GLLCEKADTR ALMAQFVPPL YSLVFTVGLL
GNVVVVMILI KYRRLRIMTN IYLLNLAISD LLFLVTLPFW IHYVRGHNWV
FGHGMCKLLS GFYHTGLYSE IFFIILLTID RYLAIVHAVF ALRARTVTFG
VITSIVTWGL AVLAALPEFI FYETEELFEE TLCSALYPED TVYSWRHFHT
LRMTIFCLVL PLLVMAICYT GIIKTLLRCP SKKKYKAIRL IFVIMAVFFI
FWTPYNVAIL LSSYQSILFG NDCERTKHLD LVMLVTEVIA YSHCCMNPVI
YAFVGERFRK YLRHFFHRHL LMHLGRYIPF LPSEKLERTS SVSPSTAEPE LSIVF
```

Affinity for ligand(s)

The known ligands for CCR3 are listed in **Table 1**. Of all the ligands, eotaxin has the highest affinity for CCR3, and produces the most potent biological effects on eosinophils.

Cell types and tissues expressing the receptor

CCR3 is the main chemokine receptor expressed on eosinophils. It is also expressed on other leukocytes involved in allergic reactions, such as basophils (Uguccioni *et al.*, 1997), mast cells (Ochi *et al.*, 1999), and TH2 cells (Gerber *et al.*, 1997; Sallusto *et al.*, 1997, 1998) (**Table 2**). As such, it is thought to be important for the development of *allergic* or *antiparasitic reactions*. CCR3 was considered at first to be relatively restricted in expression; however, a more detailed analysis of various cells revealed CCR3 expression or upregulation on other cell types (Table 2). For instance, some cells of the monocyte-macrophage lineage express CCR3, including microglia (Ghorpade *et al.*, 1998).

Regulation of receptor expression

CCR3 is expressed at high levels on eosinophils, and cytokines such as IL-5 do not affect receptor level or functional activity (Heath *et al.*, 1997). CCR3 can be

Table 1 CCR3 ligands and their receptor binding properties

Chemokine	K_d (nmol/L)	Other receptors bound	Principal references
Eotaxin	1.5		Ponath *et al.*, 1996b; Heath *et al.*, 1997
Eotaxin 2	ND		Forssmann *et al.*, 1997
RANTES	3.1	CCR1, CCR5	Ponath *et al.*, 1996b; Heath *et al.*, 1997
MCP-2	ND	CCR2, CCR5	Ponath *et al.*, 1996b; Heath *et al.*, 1997
MCP-3	2.7	CCR1, CCR2, CCR5	Ponath *et al.*, 1996b; Heath *et al.*, 1997
MCP-4	ND	CCR2	Garcia-Zepeda *et al.*, 1996; Heath *et al.*, 1997
MIP-5/Leukotactin 1	ND	CCR1	Coulin *et al.*, 1997; Youn *et al.*, 1997
MIP-1α[a]	ND	CCR1, CCR5	Post *et al.*, 1995
vMIP-II [b]	ND	CCR5	Boshoff *et al.*, 1997

[a]Mouse chemokine active on mouse CCR3. Human MIP-1α does not bind human CCR3.
[b]Chemokine-like protein encoded by *Kaposi's sarcoma-associated herpesvirus*.

Table 2 CCR3 expression pattern

Cell type	Expression		References
	Resting	Activated	
Eosinophils	+ + +[a]	+ + +	Heath *et al.*, 1997
Basophils	+ + +	+ + +	Uguccioni *et al.*, 1997
TH2 T cells	+	+/−	Sallusto *et al.*, 1997
Mast cells	+	?	Ochi *et al.*, 1999
Monocyte-macrophages	+	+	Ghorpade *et al.*, 1998
Microglia	+ +	+ +	Ghorpade *et al.*, 1998
Neutrophils	−	+	Bonecchi *et al.*, 1999

[a]Levels of CCR3 on eosinophils and basophils are approximately 5×10^4 sites per cell.

expressed on neutrophils after activation by IFNγ (Bonnechi *et al.*, 1999). CCR3 expression is regulated on TH2 cells, through the influence of cytokines such as IL-2 and IL-4 (Jinquan *et al.*, 1999). Anti-CD3 activation of TH2 cells may lead to CCR3 down-regulation, as it does for certain chemokine receptors on other cell types (Sallusto *et al.*, 1999).

SIGNAL TRANSDUCTION

Cytoplasmic signaling cascades

CCR3 is a G protein-coupled receptor, and is inhibited by pertussis toxin. There is currently no information on CCR3 signaling that distinguishes this receptor from other CC chemokine receptors.

BIOLOGICAL CONSEQUENCES OF ACTIVATING OR INHIBITING RECEPTOR AND PATHOPHYSIOLOGY

Unique biological effects of activating the receptors

CCR3 is principally a chemoattractant receptor, although, like other chemokine receptors, it functions in mediator release from basophils and eosinophils. Expression of CCR3 ligands in tissues is thought to be responsible for the accumulation of allergic leukocytes – eosinophils, basophils, and TH2 cells, in the tissue.

Phenotypes of receptor knockouts and receptor overexpression mice

The phenotype of CCR3 knockout mice has not yet been reported, although these mice have been generated (Gerard, personal communication) and their phenotype should be published in the near future.

THERAPEUTIC UTILITY

CCR3 is one of the most actively pursued targets in the pharmaceutical industry for *allergic inflammation*. The seven transmembrane structure of this receptor lends itself to inhibition by small molecule antagonists. In addition, an mAb, 7B11, has been produced that completely blocks ligand binding and functional activity of CCR3.

Effects of inhibitors (antibodies) to receptors

The best characterized antagonist of human CCR3 reported to date is the 7B11 mAb. This mAb is fully antagonistic in functional assays, including transendothelial chemotaxis and calcium release assays (Heath *et al.*, 1997). The responses of eosinophils from most individuals to eotaxin, RANTES, MCP-2, MCP-3, and MCP-4 is mediated entirely through CCR3, since in most individuals this anti-CCR3 mAb is able completely to block eosinophil responses to these chemokines (Heath *et al.*, 1997). At present this mAb is not commercially available, and inquiries should be made directly to Millennium Pharmaceuticals, Cambridge, MA, USA. A potentially useful CCR3 antagonist is Met-Ckβ7, which is highly specific and prevents signaling through CCR3 (Nibbs *et al.*, 2000).

Proof of principle for CCR3 *in vivo* has been difficult to establish, since mAbs to rodent CCR3 have been difficult to generate. A complicating factor for proof of principle studies in rodents is that mouse eosinophils also express CCR1, whereas in humans CCR1 expression on eosinophils is variable. However, blocking studies using antibodies to one of the ligands for CCR3, eotaxin, showed that anti-eotaxin polyclonal antibody was able substantially to reduce *allergic inflammation* and *broncho-hyperreactivity* in a mouse airway model (Gonzalo *et al.*, 1998). In addition, eotaxin-deficient mice have reduced eosinophil recruitment to the airways in lung allergic models (Rothenberg *et al.*, 1997). Met-RANTES, which blocks CCR1 and CCR3 (the two main eosinophilic chemokine receptors in the mouse; Elsner *et al.*, 1997), is also able to inhibit *airway inflammation* and *bronchoreactivity* in mouse models (Gonzalo *et al.*, 1998). Recently, an mAb to guinea pig CCR3 has been produced that blocks eosinophil migration to eotaxin injected *in vivo* (Sabroe *et al.*, 1998). Whether this mAb blocks eosinophil migration during an allergic reaction, or broncho-hyperreactivity in the guinea pig *asthma* model, has yet to be reported. A rat mAb to mouse CCR3 has been produced (Grimaldi *et al.*, 1999), which is nonblocking, although it does deplete eosinophils from the circulation in parasite-challenged mice.

References

Baggiolini, M., and Dahinden, C. A. (1994). CC chemokines in allergic inflammation. *Immunol. Today* **15**, 127–133.

Bonecchi, R., Polentarutti, N., Luini, W., Borsatti, A., Bernasconi, S., Locati, M., Power, C., Proudfoot, A., Wells, T. N., Mackay, C., Mantovani, A., and Sozzani, S. (1999). Up-regulation of CCR1 and CCR3 and induction of chemotaxis to CC chemokines by IFN-gamma in human neutrophils. *J. Immunol.* **162**, 474–479.

Boshoff, C., Endo, Y., Collins, P. D., Takeuchi, Y., Reeves, J. D., Schweickart, V. L., Siani, M. A., Sasaki, T., Williams, T. J., Gray, P. W., Moore, P. S., Chang, Y., and Weiss, R. A. (1997). Angiogenic and HIV-inhibitory functions of KSHV-encoded chemokines [see comments]. *Science* **278**, 290–294.

Choe, H., Farzan, M., Sun, Y., Sullivan, N., Rollins, B., Ponath, P. D., Wu, L., Mackay, C. R., LaRosa, G., Newman, W., Gerard, N., Gerard, C., and Sodroski, J. (1996). The beta-chemokine receptors CCR3 and CCR5 facilitate infection by primary HIV-1 isolates. *Cell* **85**, 1135–1148.

Coulin, F., Power, C. A., Alouani, S., Peitsch, M. C., Schroeder, J. M., Moshizuki, M., Clark-Lewis, I., and Wells, T. N. (1997). Characterisation of macrophage inflammatory protein-5/human CC cytokine-2, a member of the macrophage-inflammatory-protein family of chemokines. *Eur. J. Biochem.* **248**, 507–515.

Daugherty, B. L., Siciliano, S. J., DeMartino, J. A., Malkowitz, L., Sirotina, A., and Springer, M. S. (1996). Cloning, expression, and characterization of the human eosinophil eotaxin receptor. *J. Exp. Med.* **183**, 2349–2354.

Dahinden, C. A., Geiser, T., Brunner, T., von Tscharner, V., Caput, D., Ferrara, P., Minty, A., and Baggiolini, M. (1994). Monocyte chemotactic protein 3 is a most effective basophil- and eosinophil-activating chemokine. *J. Exp. Med.* **179**, 751–756.

Devouassoux, G., Metcalfe, D. D., and Prussin, C. (1999). Eotaxin potentiates antigen-dependent basophil IL-4 production. *J. Exp. Med.* **190**, 267–280.

Elsner, J., Petering, H., Hochstetter, R., Kimmig, D., Wells, T. N., Kapp, A., and Proudfoot, A. E. (1997). The CC chemokine antagonist Met-RANTES inhibits eosinophil effector functions through the chemokine receptors CCR1 and CCR3. *Eur. J. Immunol.* **27**, 2892–2898.

Forssmann, U., Uguccioni, M., Loetscher, P., Dahinden, C. A., Langen, H., Thelen, M., and Baggiolini, M. (1997). Eotaxin-2, a novel CC chemokine that is selective for the chemokine receptor CCR3, and acts like eotaxin on human eosinophil and basophil leukocytes. *J. Exp. Med.* **185**, 2171–2176.

Garcia-Zepeda, E. A., Combadiere, C., Rothenberg, M. E., Sarafi, M. N., Lavigne, F., Hamid, Q., Murphy, P. M., and Luster, A. D. (1996). Human monocyte chemoattractant protein (MCP)-4 is a novel CC chemokine with activities on monocytes, eosinophils, and basophils induced in allergic and nonallergic inflammation that signals through the CC chemokine receptors (CCR)-2 and -3. *J. Immunol.* **157**, 5613–5626.

Gerber, B. O., Zanni, M. P., Uguccioni, M., Loetscher, M., Mackay, C. R., Pichler, W. J., Yawalkar, N., Baggiolini, M., and Moser, B. (1997). Functional expression of the eotaxin receptor CCR3 in T lymphocytes co-localizing with eosinophils. *Curr. Biol.* **7**, 836–843.

Ghorpade, A., Xia, M. Q., Hyman, B. T., Persidsky, Y., Nukuna, A., Bock, P., Che, M., Limoges, J., Gendelman, H. E., and Mackay, C. R. (1998). Role of the beta-chemokine receptors CCR3 and CCR5 in human immunodeficiency virus type 1 infection of monocytes and microglia. *J. Virol.* **72**, 3351–3361.

Gleich, G. J., Adolphson, C. R., and Leiferman, K. M. (1993). The biology of the eosinophilic leukocyte. *Annu. Rev. Med.* **44**, 85–101.

Gonzalo, J.-A., Jia, G.-Q., Aguirre, V., Friend, D., Coyle, A. J., Jenkins, N. A., Lin, G.-S., Katz, H., Lichtman, A., Copeland, N., Kopf, M., and Gutierrez-Ramos, J.-C. (1996). Mouse eotaxin expression parallels eosinophil accumulation during lung allergic inflammation but is not restricted to a TH2-type response. *Immunity* **4**, 1–14.

Gonzalo, J. A., Lloyd, C. M., Wen, D., Albar, J. P., Wells, T. N., Proudfoot, A., Martinez, A. C., Dorf, M., Bjerke, T., Coyle, A. J., and Gutierrez-Ramos, J. C. (1998). The coordinated action of CC chemokines in the lung orchestrates allergic inflammation and airway hyperresponsiveness. *J. Exp. Med.* **188**, 157–167.

Grimaldi, J. C., Yu, N. X., Grunig, G., Seymour, B. W., Cottrez, F., Robinson, D. S., Hosken, N., Ferlin, W. G., Wu, X., Soto, H., O'Garra, A., Howard, M. C., and Coffman, R. L. (1999). Depletion of eosinophils in mice through the use of antibodies specific for C-C chemokine receptor 3 (CCR3). *J. Leukoc. Biol.* **65**, 846–853.

Hadida, F., Vieillard, V., Autran, B., Clark-Lewis, I., Baggiolini, M., and Debre, P. (1998). HIV-specific T cell cytotoxicity mediated by RANTES via the chemokine receptor CCR3. *J. Exp. Med.* **188**, 609–614.

He, J., Chen, Y., Farzan, M., Choe, H., Ohagen, A., Gartner, S., Busciglio, J., Yang, X., Hofmann, W., Newman, W., Mackay, C. R., Sodroski, J., and Gabuzda, D. (1997). CCR3 and CCR5 are co-receptors for HIV-1 infection of microglia. *Nature* **385**, 645–649.

Heath, H., Qin, S., Rao, P., Wu, L., LaRosa, G., Kassam, N., Ponath, P. D., and Mackay, C. R. (1997). Chemokine receptor usage by human eosinophils. The importance of CCR3 demonstrated using an antagonistic monoclonal antibody. *J. Clin. Invest.* **99**, 178–184.

Jinquan, T., Quan, S., Feili, G., Larsen, C. G., and Thestrup-Pedersen, K. (1999). Eotaxin activates T cells to chemotaxis and adhesion only if induced to express CCR3 by IL-2 together with IL-4. *J. Immunol.* **162**, 4285–4292.

Jose, P. J., Griffiths-Johnson, D. A., Collins, P. D., Walsh, D. T., Moqbel, R., Totty, N. F., Truong, O., Hsuan, J. J., and Williams, T. J. (1994). Eotaxin: a potent eosinophil chemoattractant cytokine detected in a guinea pig model of allergic airways inflammation. *J. Exp. Med.* **179**, 881–887.

Kay, A. B., and Corrigan, C. J. (1992). Asthma. Eosinophils and neutrophils. *Br. Med. Bull.* **48**, 51–64.

Kitayama, J., Mackay, C. R., Ponath, P. D., and Springer, T. A. (1998). The C-C chemokine receptor CCR3 participates in stimulation of eosinophil arrest on inflammatory endothelium in shear flow. *J. Clin. Invest.* **101**, 2017–2024.

Nibbs, R. J., Salcedo, T. W., Campbell, J. D., Yao, X. T., Li, Y., Nardelli, B., Olsen, H. S., Morris, T. S., Proudfoot, A. E., Patel, V. P., and Graham, G. J. (2000). C-C chemokine receptor 3 antagonism by the beta-chemokine macrophage inflammatory protein 4, a property strongly enhanced by an amino-terminal alanine-methionine swap. *J. Immunol.* **164**, 1488–1497.

Ochi, H., Hirani, W. M., Yuan, Q., Friend, D. S., Austen, K. F., and Boyce, J. A. (1999). T helper cell type 2 cytokine-mediated comitogenic responses and CCR3 expression during differentiation of human mast cells in vitro. *J. Exp. Med.* **190**, 267–280.

Ponath, P. D., Qin, S., Ringler, D. J., Clark-Lewis, I., Wang, J., Kassam, N., Smith, H., Shi, X., Gonzalo, J. A., Newman, W., Gutierrez-Ramos, J. C., and Mackay, C. R. (1996a). Cloning of the human eosinophil chemoattractant, eotaxin. Expression, receptor binding, and functional properties suggest a mechanism for the selective recruitment of eosinophils. *J. Clin. Invest.* **97**, 604–612.

Ponath, P. D., Qin, S., Post, T. W., Wang, J., Wu, L., Gerard, N. P., Newman, W., Gerard, C., and Mackay, C. R. (1996b). Molecular cloning and characterization of a human eotaxin receptor expressed selectively on eosinophils. *J. Exp. Med.* **183**, 2437–2448.

Post, T. W., Bozic, C. R., Rothenberg, M. E., Luster, A. D., Gerard, N., and Gerard, C. (1995). Molecular characterization of two murine eosinophil beta chemokine receptors. *J. Immunol.* **155**, 5299–5305.

Quackenbush, E. J., Aguirre, V., Wershil, B. K., and Gutierrez-Ramos, J. C. (1997). Eotaxin influences the development of embryonic hematopoietic progenitors in the mouse. *J. Leukoc. Biol.* **62**, 661–666.

Quackenbush, E. J., Wershil, B. K., Aguirre, V., and Gutierrez-Ramos, J. C. (1998). Eotaxin modulates myelopoiesis and mast cell development from embryonic hematopoietic progenitors. *Blood* **92**, 1887–1897.

Rothenberg, M. E., MacLean, J. A., Pearlman, E., Luster, A. D., and Leder, P. (1997). Targeted disruption of the chemokine eotaxin partially reduces antigen-induced tissue eosinophilia. *J. Exp. Med.* **185**, 785–790.

Sabroe, I., Conroy, D. M., Gerard, N. P., Li, Y., Collins, P. D., Post, T. W., Jose, P. J., Williams, T. J., Gerard, C. J., and Ponath, P. D. (1998). Cloning and characterization of the guinea pig eosinophil eotaxin receptor, C-C chemokine receptor-3: blockade using a monoclonal antibody *in vivo*. *J. Immunol.* **161**, 6139–6147.

Sallusto, F., Mackay, C., and Lanzavecchia, A. (1997). Selective expression of the eotaxin receptor CCR3 by human T helper 2 cells. *Science* **277**, 2005–2007.

Sallusto, F., Lenig, D., Mackay, C. R., and Lanzavecchia, A. (1998). Flexible programs of chemokine receptor expression on human polarized T helper 1 and 2 lymphocytes. *J. Exp. Med.* **187**, 875–883.

Sallusto, F., Kremmer, E., Palermo, B., Hoy, A., Ponath, P., Qin, S., Forster, R., Lipp, M., and Lanzavecchia, A. (1999).

Switch in chemokine receptor expression upon TCR stimulation reveals novel homing potential for recently activated T cells. *Eur. J. Immunol.* **29**, 2037–2045.

Uguccioni, M., Mackay, C. R., Ochensberger, B., Loetscher, P., Rhis, S., LaRosa, G. J., Rao, P., Ponath, P. D., Baggiolini, M., and Dahinden, C. A. (1997). High expression of the chemokine receptor CCR3 in human blood basophils. Role in activation by eotaxin, MCP-4, and other chemokines. *J. Clin. Invest.* **100**, 1137–1143.

Ying, S., Robinson, D., Meng, Q., Rottman, J., Kennedy, R., Ringler, D. J., Mackay, C. R., Daugherty, B. L., Springer, M. S., Durham, S. R., Williams, T. J., and Kay, A. B. (1997). Enhanced expression of eotaxin and CCR3 mRNA and protein in atopic asthma. Association with airway hyperresponsiveness and predominant co-localization of eotaxin mRNA to bronchial epithelial and endothelial cells. *Eur. J. Immunol.* **27**, 3507–3516.

Youn, B. S., Zhang, S. M., Lee, E. K., Park, D. H., Broxmeyer, H. E., Murphy, P. M., Locati, M., Pease, J. E., Kim, K. K., Antol, K., and Kwon, B. S. (1997). Molecular cloning of leukotactin-1: a novel human beta-chemokine, a chemoattractant for neutrophils, monocytes, and lymphocytes, and a potent agonist at CC chemokine receptors 1 and 3. *J. Immunol.* **159**, 5201–5205.

LICENSED PRODUCTS

NIH AIDS Research and Reference Reagent Program
McKesson Bioservices, Rockville, MD 20850, USA

CCR4

Christine A. Power*

Serono Pharmaceutical Research Institute, 14 Chemin des Aulx, 1228 Plan les Ouates, Geneva, Switzerland

*corresponding author tel: 00 41 22 7069 752, fax: 00 41 22 7946 965, e-mail: christine.power@serono.com

DOI: 10.1006/rwcy.2000.22004.

SUMMARY

CCR4 was cloned in 1995 from a human immature basophilic cell line KU812, and was initially identified as a receptor for MIP-1α, RANTES, and MCP-1 following overexpression in *Xenopus* oocytes and HL-60 cells. Subsequently, CCR4 was shown to be a high-affinity receptor for thymus and activation regulated chemokine (TARC) and macrophage-derived chemokine (MDC). CCR4 mRNA is found in thymus, spleen, and in peripheral blood leukocyte populations, notably activated CD4+ T cells, monocytes, and to a lesser extent on basophils. CCR4 mRNA is also highly expressed on human platelets.

CCR4 is the major chemokine receptor functionally expressed on *in vitro* polarized TH2 T cells, suggesting that this receptor, and its ligands, may play an important role in the development of the TH2 phenotype, as well as in the homing of T cells to sites of inflammation in TH2-type diseases. A recent report has shown that CCR4 is also expressed on all skin-homing T cells expressing cutaneous lymphocyte antigen (CLA). The tissue distribution of CCR4 in mice is similar to that found in humans. However, the expression of CCR4 on specific mouse cell populations can only be inferred from the response of isolated cells to the known CCR4 ligands in assays such as chemotaxis, as antimouse CCR4 antibodies are not yet generally available.

As such, CCR4 has been postulated to play a role in thymic maturation due to the restricted response of CD4+/CD8+ (double-positive) T cells to the CCR4 ligand MDC. As in humans, CCR4 is also expressed on *in vitro*-derived mouse TH2 T cells. A preliminary report on targeted deletion of the CCR4 gene in mice has now appeared in the literature. This report indicates that CCR4-deficient mice are phenotypically normal in the unstressed state and that the deletion of CCR4 had no effect on the development of the immune response in a classical TH2-dependent model of airway inflammation. Further studies with this animal model and with anti-CCR4 monoclonal antibodies should help to elucidate the role of CCR4 in immunity, inflammation, and other biological functions.

BACKGROUND

Discovery

CCR4 was cloned from a human immature basophilic leukemia cell line KU812 as part of a study to identify RANTES receptors expressed on human basophils (Power *et al.*, 1995a). Basophils are selectively recruited in allergic reactions, for example, asthma and allergic rhinitis, and are an important source of inflammatory mediators such as histamine and leukotrienes (Denburg *et al.*, 1985; Grant *et al.*, 1986). CC chemokines such as RANTES, MCP-1, and MIP-1α are able to trigger histamine release and chemotaxis of human basophils (Alam *et al.*, 1992; Bischoff *et al.*, 1992). Other chemokine receptors known at the time, namely CXCR1 (Holmes *et al.*, 1991), CXCR2 (Murphy and Tiffany, 1991) and CCR1 (Neote *et al.*, 1993), belonged to a family of seven transmembrane G protein-coupled receptors related to other chemoattractant receptors such as fMLP and C5a. Analysis of the amino acid sequences of these receptors showed that they contained highly conserved sequence motifs in the transmembrane domains 2, 3, 6, and 7 as well as in the second intracellular loop. Therefore, novel chemokine receptors could be identified by reverse transcriptase PCR, using degenerate oligonucleotides based on the conserved sequences found in these receptors (Libert *et al.*, 1989). CCR4 was identified using such a strategy.

Alternative names

In the initial publications describing CCR4, it was referred to as CC CKR-4 or K5.5. CCR4 arose from the new chemokine receptor nomenclature established in 1996.

Structure

CCR4 is a seven transmembrane domain G protein-coupled receptor. It contains the DRYLAIV motif in the second intracellular loop, which is characteristic of the chemokine receptor family.

Main activities and pathophysiological roles

CCR4 is constitutively expressed on certain subsets of thymic T cells where it may play a role in T cell differentiation. CCR4 expression on TH2 cells (Bonecchi *et al.*, 1998; D'Ambrosio *et al.*, 1998; Sallusto *et al.*, 1998), platelets (Power *et al.*, 1995b), monocytes, and basophils (Power *et al.*, 1995a; Proudfoot *et al.*, 1999) may also be important in the recruitment of these cells to certain inflammatory sites. More recently, CCR4 has been shown to be expressed on a population of skin homing CD4+, cutaneous lymphocyte antigen (CLA+) T cells which participate in memory T cell interactions with vascular endothelium at cutaneous sites of inflammation (Campbell *et al.*, 1999).

GENE

Accession numbers

GenBank:
Human: G971452
Mouse: E195632

Sequence

See **Figure 1**.

Chromosome location and linkages

Human chromosome 3p24 (Samson *et al.*, 1996)
Mouse chromosome 9 (Hoogewerf and Power, unpublished results)

Figure 1 Nucleotide sequence for CCR4.

```
   1 CGGGGGTTTT GATCTTCTTC CCCTTCTTTT CTTCCCCTTC TTCTTTCCTT CCTCCCTCCC
  61 TCTCTCATTT CCCTTCTCCT TCTCCCTCAG TCTCCACATT CAACATTGAC AAGTCCATTC
 121 AGAAAAGCAA GCTGCTTCTG GTTGGGCCCA GACCTGCCTT GAGGAGCCTG TAGAGTTAAA
 181 AAATGAACCC CACGGATATA GCAGATACCA CCCTCGATGA AAGCATATAC AGCAATTACT
 241 ATCTGTATGA AAGTATCCCC AAGCCTTGCA CCAAAGAAGG CATCAAGGCA TTTGGGGAGC
 301 TCTTCCTGCC CCCACTGTAT TCCTTGGTTT TTGTATTTGG TCTGCTTGGA AATTCTGTGG
 361 TGGTTCTGGT CCTGTTCAAA TACAAGCGGC TCAGGTCCAT GACTGATGTG TACCTGCTCA
 421 ACCTTGCCAT CTCGGATCTG CTCTTCGTGT TTTCCCTCCC TTTTTGGGGC TACTATGCAG
 481 CAGACCAGTG GGTTTTTGGG CTAGGTCTGT GCAAGATGAT TTCCTGGATG TACTTGGTGG
 541 GCTTTTACAG TGGCATATTC TTTGTCATGC TCATGAGCAT TGATAGATAC CTGGCGATAG
 601 TGCACGCGGT GTTTTCCTTG AGGGCAAGGA CCTTGACTTA TGGGGTCATC ACCAGTTTGG
 661 CTACATGGTC AGTGGCTGTG TTCGCCTCCC TTCCTGGCTT TCTGTTCAGC ACTTGTTATA
 721 CTGAGCGCAA CCATACCTAC TGCAAAACCA AGTACTCTCT CAACTCCACG ACGTGGAAGG
 781 TTCTCAGCTC CCTGGAAATC AACATTCTCG GATTGGTGAT CCCCTTAGGG ATCATGCTGT
 841 TTTGCTACTC CATGATCATC AGGACCTTGC AGCATTGTAA AAATGAGAAG AAGAACAAGG
 901 CGGTGAAGAT GATCTTTGCC GTGGTGGTCC TCTTCCTTGG GTTCTGGACA CCTTACAACA
 961 TAGTGCTCTT CCTAGAGACC CTGGTGGAGC TAGAAGTCCT TCAGGACTGC ACCTTTGAAA
1021 GATACTTGGA CTATGCCATC CAGGCCACAG AAACTCTGGC TTTTGTTCAC TGCTGCCTTA
1081 ATCCCATCAT CTACTTTTTT CTGGGGGAGA AATTTCGCAA GTACATCCTA CAGCTCTTCA
1141 AAACCTGCAG GGGCCTTTTT GTGCTCTGCC AATACTGTGG GCTCCTCCAA ATTTACTCTG
1201 CTGACACCCC CAGCTCATCT TACACGCAGT CCACCATGGA TCATGATCTT CATGATGCTC
1261 TGTAGGAAAA ATGAAATGGT GAAATGCAGA GTCAATGAAC TTTTCCACAT TCAGAGCTTA
1321 CTTTAAAATT GGTATTTTTA GGTAAGAGAT CCCTGAGCCA GTGTCAGGAG GAAGGCTTAC
1381 ACCCACAGTG GAAAGACAGC TTCTCATCCT GCAGGCAGCT TTTTCTCTCC CACTAGACAA
1441 GTCCAGCCTG GCAAGGGTTC ACCTGGGCTG AGGCATCCTT CCTCACACCA GGCTTGCCTG
1501 CAGGCATGAG TCAGTCTGAT GAGAACTCTG AGCAGTGCTT GAATGAAGTT GTAGGTAATA
1561 TTGCAAGGCA AAGACTATTC CCTTCTAACC TGAACTGATG GGTTTCTCCA GAGGGAATTG
1621 CAGAGTACTG GCTGATGGAG TAAATCGCTA CCTTTTGCTG TGGCAAATGG GCCCCCG
```

PROTEIN

Accession numbers

SwissProt:
Human: P51679
Mouse: P51680

Sequence

See **Figure 2**.

Description of protein

Translation of the CCR4 coding sequence generates a protein of 360 amino acids.

Relevant homologies and species differences

Human CCR1 and CCR5 show the closest sequence homology to CCR4 (50% amino acid identity), though mainly in the transmembrane domains. Mouse CCR4 shows 85% identity to human CCR4. Two alternatively spliced forms of murine CCR4 have been identified which differ in the 3′ untranslated region (Hoogewerf *et al.*, 1996; Hoogewerf and Power, unpublished results).

Affinity for ligand(s)

CCR4 was initially characterized as a receptor for MIP-1α, RANTES, and MCP-1 following over-expression of the receptor in *Xenopus laevis* oocytes (Power *et al.*, 1995a). However only RANTES and MIP-1α were able to bind with low affinity to HL-60 cells transiently expressing CCR4 (Hoogewerf *et al.*, 1996). Subsequently, thymus and activation-regulated chemokine (TARC) and macrophage-derived chemokine (MDC) were identified as the high-affinity ligands for this receptor (Imai *et al.*, 1997, 1998). More recently, vMIP-III, a chemokine encoded by ORF 4.1 of Kaposi's sarcoma herpesvirus (KSHV), has been shown to bind CCR4. Although vMIP-III is less potent than MDC in chemotaxis assays, it is more efficacious (Stine *et al.*, 2000) (**Table 1**).

Cell types and tissues expressing the receptor

CCR4 is predominantly expressed on subsets of T cells, predominantly of the TH2 and CD45RO+ phenotypes (Bonecchi *et al.*, 1998; Sallusto *et al.*, 1998) as well as on skin homing CD4+, cutaneous lymphocyte antigen (CLA+) T cells (Campbell *et al.*, 1999). It is also expressed on platelets (Power *et al.*, 1995b), monocytes (Proudfoot *et al.*, 1999) and basophils (Power *et al.*, 1995a). In tissues CCR4 has been detected in thymus and spleen (Power *et al.*,

Figure 2 Amino acid sequence for CCR4 protein. The putative transmembrane domains are underlined.

```
MNPTDIADTT LDESIYSNYY LYESIPKPCT KEGIKAFGEL FLPPLYSLVF VFGLLGNSVV
VLVLFKYKRL RSMTDVYLLN LAISDLLFVF SLPFWGYYAA DQWVFGLGLC KMISWMYLVG
FYSGIFFVML MSIDRYLAIV HAVFSLRART LTYGVITSLA TWSVAVFASL PGFLFSTCYT
ERNHTYCKTK YSLNSTTWKV LSSLEINILG LVIPLGIMLF CYSMIIRTLQ HCKNEKKNKA
VKMIFAVVVL FLGFWTPYNI VLFLETLVEL EVLQDCTFER YLDYAIQATE TLAFVHCCLN
PIIYFFLGEK FRKYILQLFK TCRGLFVLCQ YCGLLQIYSA DTPSSSYTQS TMDHDLHDAL
```

Table 1 Affinities for ligands

Chemokine	K_d	Reference
MDC	0.18 nM	Imai *et al.*, 1998
TARC	0.5 nM	Imai *et al.*, 1997
MIP-1α	14.5 nM	Hoogewerf *et al.*, 1996
RANTES	9.3 nM	Hoogewerf *et al.*, 1996
vMIP-III	101 nM	Stine *et al.*, 2000

1995a). Murine CCR4 shows a similar expression pattern.

Regulation of receptor expression

CCR4 mRNA can be upregulated in TH2 cell lines following TGFβ stimulation (Sallusto *et al.*, 1998). Activation of T cells (via anti-CD3 and anti-CD28) also leads to upregulation of CCR4 in both TH1 and TH2 cells (D'Ambrosio *et al.*, 1998).

BIOLOGICAL CONSEQUENCES OF ACTIVATING OR INHIBITING RECEPTOR AND PATHOPHYSIOLOGY

Phenotypes of receptor knockouts and receptor overexpression mice

Null phenotype in unstressed CCR4−/− mice (Power, 1999).

References

Alam, R., Forsythe, P. A., Staffors, S., Lett-Brown, M. A., and Grant, J. A. (1992). Macrophage inflammatory protein-1 alpha activates basophils and mast cells. *J. Exp. Med.* **176**, 781–786.

Bischoff, S. C., Krieger, M., Brunner, T., and Dahinden, C. (1992). Monocyte chemotactic protein 1 is a potent activator of human basophils. *J. Exp. Med.* **175**, 1271–1275.

Bonecchi, R., Bianchi, G., Bordignon, P. P., D'Ambrosio, D., Lang, A., Borsatti, S., Sozzani, S., Allavena, P. A., Gray, P. W., Mantovan, A., and Sinigaglia, F. (1998). Differential expression of chemokine receptors and chemotactic responsiveness of type 1 T helper cells (TH1s) and TH2s. *J. Exp. Med.* **187**, 129–134.

Campbell, J. J., Haraldsen, G., Pan, J., Rottman, J., Qin, S., Ponath, P., Andrew, D. P., Warnke, R., Ruffing, N., Kassam, N., Wu, L., and Butcher, E. C. (1999). The chemokine receptor CCR4 in vascular recognition by cutaneous but not intestinal memory T cells. *Nature* **400**, 776–780.

D'Ambrosio, D., Iellem, A., Bonecchi, R., Mazzeo, D., Sozzani, S., Mantovani, A., and Sinigaglia, F. (1998). Selective upregulation of chemokine receptors CCR4 and CCR8 upon activation of polarized human type 2 Th cells. *J. Immunol.* **161**, 5111–5115.

Denburg, J. A., Telizyn, S., Belda, A., Dolovich, J., and Bienenstock, J. (1985). Increased numbers of circulating basophil progenitors in atopic patients. *J. Allergy Clin. Immunol.* **76**, 466–472.

Grant, J. A., Letts-Brown, M. A., Warner, J. A., Plant, M., Lichtenstein, L. M., Haak-Frendscho, M., and Kaplan, A. P. (1986). Activation of basophils. *Fed. Proc.* **45**, 2653–2658.

Holmes, W. E., Lee, J., Kuang, W. J., Rice, G. C., and Wood, M. L. (1991). Structure and functional expression of a human interleukin-8 receptor. *Science* **253**, 1278–1280.

Hoogewerf, A. J., Black, D., Proudfoot, A. E. I., Wells, T. N. C., and Power, C. A. (1996). Molecular cloning of murine CC CKR-4 and high affinity binding of chemokines to murine and human CC CKR-4. *Biochem. Biophys. Res. Commun.* **218**, 337–343.

Imai, T., Baba, M., Nishimura, M., Kakizaki, M., Takagi, S., and Yoshie, O. (1997). The T cell directed CC chemokine TARC is highly specific biological ligand for CC chemokine receptor 4. *J. Biol. Chem.* **272**, 15036–15042.

Imai, T., Chantry, D., Raport, C. J., Wood, C. L., Nishimura, M., Godiska, R., Yoshie, O., and Gray, P. W. (1998). Macrophage derived chemokine MDC is a functional ligand for the CC chemokine receptor 4. *J. Biol. Chem.* **273**, 1764–1768.

Libert, F., Parmentier, M., Lefort, A., Dinsart, C., Van Sande, J., Maenhaut, C., Simons, M. J., Dumont, J. E., and Vassart, G. (1989). Selective amplification and cloning of four new members of the G protein-coupled receptor family. *Science* **244**, 569–572.

Murphy, P. M., and Tiffany, H. L. (1991). Cloning of complementary DNA encoding a functional human interleukin-8. *Science* **253**, 1280–1283.

Neote, K., DiGrogorio, D., Mak, J. Y., Horuk, R., and Schall, T. J. (1993). Molecular cloning, functional expression, and signaling characteristics of a C-C chemokine receptor. *Cell* **72**, 415–425.

Power, C. (1999). Effects of CCR4 knockout in a mouse model of lung inflammation. Proceedings of the Institut Pasteur Euroconference: Chemokines and their receptors: from basic research to therapeutic intervention. *Eur. Cytokine Netw.* **10**, 296–297.

Power, C. A., Meyer, A., Nemeth, K., Bacon, K. B., Hoogewerf, A. J., Proudfoot, A. E. I., and Wells, T. N. C. (1995a). Molecular cloning and functional expression of a novel C-C chemokine receptor cDNA from a human basophilic cell line. *J. Biol. Chem.* **270**, 19495–19500.

Power, C. A., Clemetson, J., Clemetson, K. J., and Wells, T. N. C. (1995b). Chemokine and chemokine receptor mRNA expression in human platelets. *Cytokine* **7**, 479–482.

Proudfoot, A. E. I., Buser, R., Borlat, F., Alouani, S., Soler, D., Offord, R. E., Schröder, J.-M., Power, C. A., and Wells, T. N. C. (1999). Amino terminally modified RANTES analogues demonstrate differential effects on RANTES receptors. *J. Biol. Chem.* **274**, 32478–32485.

Sallusto, F., Lenig, D., Mackay, C. R., and Lanzavecchia, A. (1998). Flexible programs of chemokine receptor expression on human polarized T helper 1 and 2 lymphocytes. *J. Exp. Med.* **187**, 875–883.

Samson, M., Soularue, P., Vassart, G., and Parmentier M. (1996). The genes encoding the human CC-chemokine receptors CC-CKR1 to CC-CKR5 (CMKBR1-CMKBR5) are clustered in the p21.3-p24 region of chromosome 3. *Genomics* **36**, 522–526.

Stine, J. T., Wood, C., Hill, M., Epp, A., Raport, C., Schweikart, V., Endo, Y., Sasaki, T., Simmons, G., Boshoff, C., Clapham, P., Weiss, R., Chang, Y., Moore, P., Gray, P., and Chantry, D. (2000). KSHV encoded CC chemokine vMIP-III is a CCR4 agonist, stimulates angiogenesis and selectively attracts TH2 cells. *Blood* **95**, 1151–1157.

CCR5

Cédric Blanpain and Marc Parmentier*

Institute of Interdisciplinary Research, Université Libre de Bruxelles, Campus Erasme, 808 route de Lennik, Brussels, B-1070, Belgium

*corresponding author tel: 32 2 5554171, fax: 32 2 5554655, e-mail: mparment@ulb.ac.be

DOI: 10.1006/rwcy.2000.22005.

SUMMARY

CCR5 is a functional receptor for the CC chemokines MIP-1α, MIP-1β, RANTES, and MCP-2. It is expressed in memory T cells, macrophages, dendritic cells, and microglia. CCR5 also constitutes the main coreceptor for the macrophage (M)-tropic strains of *HIV-1* and *HIV-2*, that is responsible for disease transmission. A nonfunctional allele of the CCR5 gene (CCCR5Δ32), frequent in populations of European origin, provides marked resistance against HIV to homozygotes, and is associated with delayed *AIDS* progression of heterozygotes. Numerous other variant alleles of CCR5 have been described in human populations as well as in other mammalian species. The receptor constitutes a potential target for therapeutics to block HIV infection, and for various immune diseases such as *rheumatoid arthritis*, *multiple sclerosis*, and *graft-versus-host disease*. Chemokine variants, monoclonal antibodies, and small chemical inhibitors have been developed in this context.

BACKGROUND

Discovery

The gene encoding CCR5 was cloned by various groups using low stringency polymerase chain reaction and found to share similarities with CC chemokine receptors. Following its expression in cell lines, the receptor was characterized as responding functionally to MIP-1α, MIP-1β, and RANTES (Combadiere *et al.*, 1996; Raport *et al.*, 1996; Samson *et al.*, 1996a). It was rapidly identified as the main coreceptor for *HIV* (Alkhatib *et al.*, 1996; Choe *et al.*, 1996; Deng *et al.*, 1996; Doranz *et al.*, 1996; Dragic *et al.*, 1996), and mutant alleles of the CCR5 gene were found to provide a strong protection against HIV infection (Dean *et al.*, 1996; Liu *et al.*, 1996; Samson *et al.*, 1996b).

Alternative names

CCR5 was first described as CC-CKR5, before the modification of the chemokine receptor nomenclature in June 1996. Lab names also include ChemR13.

Structure

CCR5 is a G protein-coupled receptor of 352 amino acids. It belongs to the subfamily of CC chemokine receptors and shares significant similarity with CCR2 (76% identical residues). The predicted molecular weight of the protein is 40,600 daltons. CCR5 is *O*-glycosylated, sulfated on tyrosines, and presumably palmitoylated on intracellular cysteines located in the C-terminal domain. Extracellular domains of the receptor are linked by two functionally important disulfide bonds.

Main activities and pathophysiological roles

CCR5 is a high-affinity receptor for MIP-1α, MIP-1β, RANTES, and MCP-2, although it binds other CC chemokines as well (see below). The receptor is coupled through G_i to the inhibition of adenylyl cyclase, the release of intracellular calcium, and the activation of tyrosine kinase cascades. CCR5 is expressed in macrophages, dendritic cells, memory T cells, and microglia. Like other receptors for inflammatory chemokines, it is involved in the recruitment of these cell populations to inflammatory sites. CCR5 was shown to constitute the main coreceptor for the M-tropic strains of the lentiviruses *HIV-1*, *HIV-2*, and *SIV*. The central role of CCR5 in HIV pathogenesis was demonstrated by the existence of a mutant allele of the receptor in populations of European origin, providing homozygotes with strong protection against HIV infection.

GENE

Accession numbers

A large number of sequence variants has been reported in databases for human and a number of other species. Representative accession numbers for the nucleotide sequences encoding CCR5 in various species are as follows:
Human cDNA: X91492, U54994, U57840
Human gene: U95626
Mangabey: AF051902
Lowland gorilla: AF005659
Chimpanzee: AF005663
Baboon: AF005658
Green monkey: U83324
Macaque: U77672
Mouse: U47036
Cat: AAD00729

Sequence

The gene encoding CCR5 is located in the 3p21.3 region of the human genome. It is part of a gene cluster that also includes CCR1, CCR2, CCR3, XCR1, and the orphan receptor CCBP2 (Samson *et al.*, 1996c; Maho *et al.*, 1999). The CCR5 gene is located in tandem, and upstream of the CCR2 gene. A distance of 17.5 kb separates their open reading frames (Samson *et al.*, 1996a). The whole coding sequence of CCR5 is located within a single exon. Introns have however been found in the 3′ untranslated region of the gene, and two alternative promoters have been described (Moriuchi *et al.*, 1997; Mummidi *et al.*, 1997; Guignard *et al.*, 1998). The downstream promoter Pd appears to be used much more efficiently than the upstream promoter Pu. Whether the two promoters are being used in a cell-specific manner is presently unknown. Within the downstream promoter, a pair of TATA boxes, as well as potential binding sites for the transcription factors STAT, NFκB, AP-1, NF-AT, and CD28RE have been identified. Mutation of these elements was shown to affect expression of the gene (Liu *et al.*, 1998). The GATA-1 and p65(RelA) transcription factors were reported to stimulate CCR5 expression (Liu *et al.*, 1998; Moriuchi *et al.*, 1999).

Variants of the CCR5 promoter have been described. Some of these variants have been statistically associated with *AIDS* progression rate (McDermott *et al.*, 1998; Martin *et al.*, 1998). An effect on CCR5 expression, that would support a direct link between the promoter variants and the observed phenotype, was however not demonstrated conclusively.

PROTEIN

Accession numbers

As for nucleotide sequences, a large number of entries can be found in the databases for variants of CCR5 in various species. Representative accession numbers for amino acid sequences are as follows:
Human cDNA: P51681
Mangabey: O62743
Lowland gorilla: P56439
Chimpanzee: P56440
Baboon: P56441
Green monkey: P56493
Macaque: P79436
Mouse: P51682
Cat: AAD00729

Sequence

See **Figure 1**.

Description of protein

CCR5 is a typical G protein-coupled receptor of 352 amino acids, with seven transmembrane domains that presumably adopt an α helical structure. It belongs to the subfamily of CC chemokine receptors. The predicted molecular weight of the protein is 40,600 daltons. Like other chemokine receptors, CCR5 has four cysteines within its extracellular domains, involved in the formation of two disulfide bonds. One of these bonds, linking the first and second loops of the receptor, is conserved in most G protein-coupled receptors. The second bond, which is specific to the chemokine receptor family, links the N-terminus to the third extracellular loop. Both bonds are necessary for the chemokine-binding properties

Figure 1 Amino acid sequence of CCR5 from human and other mammalian species. Human CCR2B is aligned for comparison purposes. The displayed sequences are from human, lowland gorilla (*Gorilla gorilla gorilla*), chimpanzee (*Pan troglodytes*), baboon (*Papio hamadryas*), green monkey (*Cercopithecus aethiops*), macaque (*Macaca mulatta*), red-crowned mangabey (*Cercocebus torquatusatys*), mouse, and cat. The putative transmembrane segments are numbered I to VII. Amino acids which are different from the human sequence are indicated in blue. (Full colour figure may be viewed online.)

```
                                                                   TMI
                                                       ))))))))))))))))))))))))
Human              1  MDYQVSSPIY--DINYYTSEPCQKINVKQIAARLLPPLYSLVFIFGFVGNMLVILILI  56
Gorilla               MDYQVSSPTY--DIDYYTSEPCQKTNVKQIAARLLPPLYSLVFIFGFVGNMLVILILI  56
Chimpanzee            MDYQVSSPIY--DIDYYTSEPCQKINVKQIAARLLPPLYSLVFIFGFVGNMLVILILI  56
Mangabey              MDYQVSSPTY--DIDYYTSEPCQKINVKQIAARLLPPLYSLVFIFGFVGNILVVLILI  56
Baboon                MDYQVSSPTY--DIDYYTSEPCQKINVKQIAARLLPPLYSLVFIFGFVGNILVVLILI  56
Green monkey          MDYQVSSPTY--DIDNYTSEPCQKINVKQIAARLLPPLYSLVFIFGFVGNILVVLILI  56
Macaca                MDYQVSSPTY--DIDYYTSEPCQKINVKQIAARLLPPLYSLVFIFGFVGNILVVLILI  56
Mouse                 MDFQGSVPTYIYDIDYGMSAPCQKINVKQIAAQLLPPLYSLVFIFGFVGNMMVFLILI  58
Cat                   MDYQATSPYY--DIEYELSEPCQKTDVRQIAARLLPPLYSLVFLSGFVGFLLVVLILI  56
hCCR2b       MLSTSRSRFIRNTNESGEEVTTFFDYDYGAPCHKFDVKQIGAQLLPPLYSLVFIFGFVGNMLVVLILI  68

                                TMII                                  TMIII
                     ))))))))))))))))))))           ))))))))))))))))
Human               NCKRLKSMTDIYLLNLAISDLFFLLTVPFWAHYAAAQWDFGNTMCQLLTGLYFIGFFSGIFFI 119
Gorilla             NCKRLKSMTDIYLLNLAISDLFFLLTVPFWAHYAAAQWDFGNTMCQLLTGLYFIGFFSGIFFI 119
Chimpanzee          NCKRLKSMTDIYLLNLAISDLFFLLTVPFWAHYAAAQWDFGNTMCQLLTGLYFIGFFSGIFFI 119
Mangabey            NCKRLKSMTDIYLLNLAISDLLFLLTVPFWAHYAAAQWDFGNTMCQLLTGLYFIGFFSGIFFI 119
Baboon              NCKRLKSMTDIYLLNLAISDLLFLLTVPFWAHYAAAQWDFGNTMCQLLTGLYFIGFFSGIFFI 119
Green monkey        NCKRLKSMTDIYLLNLAISDLLFLLTVPFWAHYAAAQWDFGNTMCQLLTGLYFIGFFSGIFFI 119
Macaca              NCKRLKSMTDIYLLNLAISDLLFLLTVPFWAHYAAAQWDFGNTMCQLLTGLYFIGFFSGIFFI 119
Mouse               SCKKLKSVTDIYLLNLAISDLLFLLTLPFWAHYAANEWIFGNIMCKVFTGVYHIGYFGGIFFI 121
Cat                 NCKKLRGMTDVYLLNLAISDLLFLFTLPFWAHYAANGWVFGDGMCKTVTGLYHVGYFGGNFFI 119
hCCR2b              NCKKLKCLTDIYLLNLAISDLLFLITLPLWAHSAANEWVFGNAMCKLFTGLYHIGYFGGIFFI 131

                                                            TMIV
                     )))))              ))))))))))))))))))))))))
Human               ILLTIDRYLAVVHAVFALKARTVTFGVVTSVITWVVAVFASLPGIIFTRSQKEGLHYTCSSHF 182
Gorilla             ILLTIDRYLAIVHAVFALKARTVTFGVVTSVITWVVAVFASLPGIIFTRSQKEGLHYTCSSHF 182
Chimpanzee          ILLTIDRYLAIVHAVFALKARTVTFGVVTSVITWVVAVFASLPGIIFTRSQKEGLHYTCSSHF 182
Mangabey            ILLTIDRYLAIVHAVFALKARTVTFGVVTSVITWVVAVFASLPGIIFTRSQREGLHYTCSPHF 182
Baboon              ILLTIDRYLAIVHAVFALKARTVTFGVVTSVITWVVAVFASLPGIIFTRSQREGLHYTCSSHF 182
Green monkey        ILLTIDRYLAIVHAVFALKARTVTFGVVTSVITWVVAVFASLPRIIFTRSQREGLHYTCSSHF 182
Macaca              ILLTIDRYLAIVHAVFALKARTVTFGVVTSVITWVVAVFASLPGIIFTRSQREGLHYTCSSHF 182
Mouse               ILLTIDRYLAIVHAVFALKVRTVNFGVITSVVTWVVAVFASLPEIIFTRSQKEGFHYTCSPHF 184
Cat                 ILLTVDRYLAIVLAVFAVKARTVTFGAVTSAVTWAAAVVASLPGCIFSRSQKEGSRFTCSPHF 182
hCCR2b              ILLTIDRYLAIVHAVFALKARTVTFGVVTSVITWLVAVFASVPGIIFTKCQKEDSVYVCGPYF 194

                                       TMV
                               ))))))))))))))))))))         ))))))))))
Human               PYSQYQFWKNFQTLKIVILGLVLPLLVMVICYSGILKTLLRCRNEKKRHRAVRLIFTIMIVYF 245
Gorilla             PYSQYQFWKNFQTLKIVILGLVLPLLVMVICYSGILKTLLRCRNEKKRHRAVRLIFTIMIVYF 245
Chimpanzee          PYSQYQFWKNFQTLKIVILGLVLPLLVMVICYSGILKTLLRCRNEKKRHRAVRLIFTIMIVYF 245
Mangabey            PYSQYQFWKNFQTLKIVILGLVLPLLVMVICYSGILKTLLRCRNEKKRHRAVRLIFTIMIVYF 245
Baboon              PYSQYQFWKNFQTLKIVILGLVLPLLVMVICYSGILKTLLRCRNEKKRHRAVRLIFTIMIVYF 245
Green monkey        PYSQYQFWKNFQTLKIVILGLVLPLLVMVICYSGILKTLLRCRNEKKRHRAVRLIFTIMIVYF 245
Macaca              PYSQYQFWKNFQTLKMVILGLVLPLLVMVICYSGILKTLLRCRNEKKRHRAVRLIFTIMIVYF 245
Mouse               PHTQYHFWKSFQTLKMVILSLILPLLVMIICYSGILHTLFRCRNEKKRHRAVRLIFAIMIVYF 247
Cat                 PSNQYHFWKNFQTLKMTILGLVLPLLVMIVCYSAILRTLFRCRNEKKKHRAVKLIFVIMIGYF 245
hCCR2b              PRG    WNNFHTIMRNILGLVLPLLIMVICYSGILKTLLRCRNEKKRHRAVRVIFTIMIVYF 253

                      TMVI                                 TMVII
                    ))))))))))))))              ))))))))))))))))))))))
Human               LFWAPYNIVLLLNTFQEFFGLNNCSSSNRLDQAMQVTETLGMTHCCINPIIYAFVGEKFRNYL 308
Gorilla             LFWAPYNIVLLLNTFQEFFGLNNCSSSNRLDQAMQVTETLGMTHCCINPIIYAFVGEKFRNYL 308
Chimpanzee          LFWAPYNIVLLLNTFQEFFGLNNCSSSNRLDQAMQVTETLGMTHCCINPIIYAFVGEKFRNYL 308
Mangabey            LFWAPYNIVLLLNTFQEFFGLNNCSSSNRLDQAMQVTETLGMTHCCINPIIYAFVGEKFRNYL 308
Baboon              LFWAPYNIVLLLNTFQEFFGLNNCSSSNRLDQAMQVTETLGMTHCCINPIIYAFVGEKFRNYL 308
Green monkey        LFWAPYNIVLLLNTFQEFFGLNNCSSSNRLDQAMQVTETLGMTHCCINPIIYAFVGEKFRNYL 308
Macaca              LFWAPYNIVLLLNTFQEFFGLNNCSSSNRLDQAMQVTETLGMTHCCINPIIYAFVGEKFRNYL 308
Mouse               LFWTPYNIVLLLTTFQEFFGLNNCSSSNRLDQAMQATETLGMTHCCLNPVIYAFVGEKFRSYL 310
Cat                 LFWAPNNIVLPLSTFPESFGLNNCSSSNRLDQAMQVTETLGMTHCCINPIIYALVGEKFRSYL 308
hCCR2b              LFWTPYNIVILLNTFQEFFGLSNCESTSQLDQATQVTETLGMTHCCINPIIYAFVGEKFRRYL 318

Human               LVFFQKHIAKRFCKCCSIFQQEAPERASSVYTRSTGEQEISVGL 352
Gorilla             LVFFQKHIAKRFCKCCSIFQQEAPERASSVYTRSTGEQEISVGL 352
Chimpanzee          LVFFQKHIAKRFCKCCSIFQQEAPERASSVYTRSTGEQEISVGL 352
Mangabey            LVFFQKHIAKRFCKCCSIFQQEASERASSVYTRSTGEQEISVGL 352
Baboon              LVFFQKHIAKRFCKCCSIFQQEAPERASSVYTRSTGEQEISVGL 352
Green monkey        LVFFQKHIAKRFCKCCSIFQQEAPERASSVYTRSTGEQETSVGF 352
Macaca              LVFFQKHIAKRFCKCCSIFQQEAPERASSVYTRSTGEQEISVGL 352
Mouse               SVFFRKHIVKRFCKRCSIFQQDNPDRVSSVYTRSTGEHEVSTGL 354
Cat                 LVFFQKHIARAFCKRCPVFQGKALDRA-GCYTRSTGEQEISVGL 351
hCCR2b              SVFFRKHITKRFCKQCPVFYRETVDGVTSTNTPSTGEQEVSAGL 360
```

and the functional response of the receptor (Blanpain *et al.*, 1999b).

The receptor exhibits a consensus sequence for *N*-linked glycosylation in its third extracellular loop, but this site is apparently not used (Rucker *et al.*, 1996), possibly due to the presence of a cysteine involved in disulfide bonding within the consensus. The receptor is, however, *O*-glycosylated on sites uncharacterized so far, and sulfated on several tyrosines located within the N-terminal domain (Farzan *et al.*, 1999). Tyrosine sulfation appears to contribute to the binding of chemokines and gp120. CCR5 does not contain consensus sequences for protein kinase A or protein kinase C. However, the C-terminus of the receptor is rich in serines and threonines that constitute the phosphorylation sites for G protein-coupled receptor kinases (GRKs), which are responsible for homologous desensitization. Biochemical analysis has shown that Ser336, Ser337, Ser342, and Ser349 represent the GRK phosphorylation sites (Oppermann *et al.*, 1999). The C-terminal domain also contains a cluster of three cysteines after the last transmembrane segment. Cysteines at this position have been shown to be palmitoylated in other receptors such as the β_2-adrenergic receptor, and to anchor the C-terminal domain to the plasma membrane, delimiting a fourth intracellular loop. However palmitoylation of CCR5 has not been demonstrated so far.

The structure–function relationships of CCR5 have been studied by various groups, with the aim of determining the regions of the receptor involved in chemokine binding, signaling, and interaction with the viral envelope. A first set of studies, involving chimeras between CCR5 and CCR2B, or between human and mouse CCR5, has demonstrated that the N-terminal domain is the most important for the coreceptor activity. The interaction with gp120 appeared however to be conformationally complex, also involving the second extracellular loop, and to a lesser extent the first and third extracellular loops (Atchison *et al.*, 1996; Rucker *et al.*, 1996; Bieniasz *et al.*, 1997).

Different strains of *HIV* were shown to interact differently with CCR5, suggesting that the virus may evolve by acquiring the ability to use additional coreceptors, but also by utilizing the same coreceptor in a different way. Using the same chimeric receptors, it was shown that the second loop of the receptor is essential for determining the specificity of interaction with chemokines (Samson *et al.*, 1997). Using more subtle mutations, it was established that the N-terminal domain of the receptor, more particularly motifs of tyrosine and acidic amino acids, were important for both the receptor and coreceptor functions of CCR5 (Dragic *et al.*, 1998; Rabut *et al.*, 1998; Blanpain *et al.*, 1999c; Farzan *et al.*, 1999). It was shown that neither signaling nor internalization of the receptor was necessary for viral entry (Alkhatib *et al.*, 1997; Aramori *et al.*, 1997; Farzan *et al.*, 1997; Gosling *et al.*, 1997). Reviews on the structure–function relationships of CCR5 include those by Choe *et al.* (1998) and Berson and Doms (1998).

Relevant homologies and species differences

CCR5 is a member of the CC chemokine receptor group. It is most highly related to CCR2B, sharing 76% identical residues. The identity score goes up to 92% when considering the transmembrane segments only; most of the differences are located within the N-terminus and the extracellular loops. CCR5 shares 49–56% identical residues with CCR1, CCR3, and CCR4. Human CCR5 shares about 80% identity with the orthologous murine and feline receptors, and 97 to over 99% identity with the receptor from upper monkeys (Figure 1).

Affinity for ligand(s)

CCR5 was first described as a receptor for MIP-1α, MIP-1β, and RANTES (Samson *et al.*, 1996a). More recently, MCP-2 and MCP-4 were found to act as agonists as well, while MCP-3 appears as a natural antagonist of CCR5 (Gong *et al.*, 1998; Ruffing *et al.*, 1998; Blanpain *et al.*, 1999a). Binding and functional parameters for CCR5 ligands are provided in **Table 1**. The nonallelic isoform of MIP-1α, termed LD78β or MIP-1αP, has a 6-fold increase in affinity as

Table 1 Binding and functional properties of natural and synthetic chemokines acting on CCR5 (Blanpain *et al.*, 1999a). Binding assays were performed using iodinated MIP-1β as tracer

Ligand	pIC_{50}	pEC_{50}
MIP-1α	9.05 ± 0.24	8.49 ± 0.47
MIP-1β	9.30 ± 0.24	8.47 ± 0.40
RANTES	9.74 ± 0.19	8.87 ± 0.37
MCP-1	7.34 ± 0.25	< 6.3
MCP-2	9.40 ± 0.35	8.44 ± 0.32
MCP-3	8.59 ± 0.11	< 6.3
MCP-4	8.03 ± 0.28	7.04 ± 0.27
Eotaxin	7.72 ± 0.10	< 6.3

compared to MIP-1α, and is as potent as RANTES in *HIV-1*-suppressive assays. A proline in position 2 was made responsible for this enhanced biological activity (Nibbs *et al.*, 1999). RANTES is a substrate of dipeptidylpeptidase IV (DPPIV, CD26), resulting in an N-terminal truncated variant (RANTES[3–68]) with enhanced antiviral activity (Oravecz *et al.*, 1997; Proost *et al.*, 1998; Schols *et al.*, 1998). Modified chemokines and chemicals acting on CCR5 have been developed (see section on Therapeutic utility).

Cell types and tissues expressing the receptor

CCR5 expression was demonstrated on a variety of cell types, including the monocyte-macrophage lineage, dentritic cells, lymphocyte populations, and brain microglial and vascular endothelial cells.

CCR5 is expressed on resting T cells with a memory/effector phenotype (CD45RO, CD26^{high}, CD95+, CD4+, or CD8+) (Bleul *et al.*, 1997; Wu *et al.*, 1997). It is expressed preferentially on cells with a TH1 polarization (Loetscher *et al.*, 1998), in accordance with the biological activity of MIP-1β, a specific agonist, on TH1 cells but not on TH2 cells (Bonecchi *et al.*, 1998).

The level of CCR5 expression on lymphocytes, as detected by FACS analysis, was reported to show considerable heterogeneity among individuals (Wu *et al.*, 1997). CCR5 is expressed at low levels on progenitor and mature CD4+CD8+ thymocytes, in accordance with the functional response of these cells to MIP-1β (Berkowitz *et al.*, 1998; Dairaghi *et al.*, 1998; Zaitseva *et al.*, 1998). Expression is downregulated upon transfer of mature thymocytes into the bloodstream, as CCR5 is absent from naïve peripheral T cells. CCR5 is also absent from B cells.

CCR5 is expressed on 5% of circulating monocytes (Wu *et al.*, 1997) and at high levels on differentiated macrophages (Zaitseva *et al.*, 1997). Mucosal macrophages expressing CCR5 are believed to serve as an entry port for *HIV* (Zhang *et al.*, 1998). CCR5 is expressed in microglial cells (He *et al.*, 1997, Albright *et al.*, 1999). In microglial cells, CCR5 expression is stimulated by LPS and *ischemia* (Spleiss *et al.*, 1998). CCR5 is expressed on dendritic cells derived from monocytes or CD34+ progenitors, as well as on peripheral blood dendritic cells and epidermal Langerhans cells (Granelli-Piperno *et al.*, 1996; Zaitseva *et al.*, 1997; Durig *et al.*, 1998; Lee *et al.*, 1999c).

CCR5 is undetectable on early medullar hematopoietic progenitors but can be found on committed progenitor cells of the monocytic and megakaryocytic lineages (Berkowitz *et al.*, 1998; Chelucci *et al.*, 1999; Blood, 1999; Lee *et al.*, 1999a).

CCR5 expression has been found in hippocampal neurons in the brain: RANTES affects a number of electrophysiological properties of these neurons (Meucci *et al.*, 1998). CCR5 was also described in endothelial cells from brain vasculature (Edinger *et al.*, 1997).

Regulation of receptor expression

Proinflammatory cytokines (TNFα and IL-12) and TH1 cytokines (IFNγ and IL-2) upregulate gene expression and increase surface expression of CCR5 on PBMCs (Hariharan *et al.*, 1999; Patterson *et al.*, 1999). IL-2 was shown to be a potent stimulator of CCR5 expression in lymphocytes, both *ex vivo* (Wu *et al.*, 1997; Bleul *et al.*, 1997) and *in vivo* (Zou *et al.*, 1999). Anti-CD28 costimulation prevents CCR5 expression in IL-2-treated lymphoblasts (Carroll *et al.*, 1997).

M-CSF and GM-CSF upregulate CCR5 on macrophages (Tuttle *et al.*, 1998; Wang *et al.*, 1998; Lee *et al.*, 1999c). IL-10 increases CCR5 expression on monocytes at a posttranscriptional level by increasing the stability of CCR5 mRNA (Sozzani *et al.*, 1998). IFNγ increases CCR5 expression in the U937 monocytoid cell line (Zella *et al.*, 1998). LPS was shown to downregulate CCR5 expression on macrophages and dendritic cells (Sica *et al.*, 1997; Moriuchi *et al.*, 1998; Sallusto *et al.*, 1998; Lin *et al.*, 1998). Stimulation of the cAMP pathway in monocyte-macrophages rapidly downregulates CCR5 gene expression, by decreasing the stability of CCR5 mRNA (Thivierge *et al.*, 1998). IL-4 and IL-13 were shown to prevent the stimulation of CCR5 expression by other factors in monocytes (Valentin *et al.*, 1998; Wang *et al.*, 1998).

Similarly to other chemokine receptors, desensitization and internalization of CCR5 occur following exposure to an agonist of the receptor. Desensitization involves phosphorylation of the receptor by the G protein-coupled receptor kinases GRK2 and GRK3, while endocytosis of the phosphorylated receptor is mediated by β-arrestin that targets CCR5 to clathrin-coated vesicles in a dynamin-dependent way (Aramori *et al.*, 1997; Zhao *et al.*, 1998). Desensitization and internalization in various cell types depend on the level of expression in these cells of GRKs and β-arrestin, respectively. CCR5 is efficiently desensitized and internalized in PM-1, CHO, and NG108 cells, but not in HEK 293 cells. Unlike CXCR4, CCR5 does not contain PKC phosphorylation sites, and is not internalized in response to PMA (Signoret *et al.*, 1998).

CCR5 internalization does not appear to be necessary for *HIV* coreceptor function, but appears to contribute greatly to the inhibitory effect of chemokines on HIV infection. The strong antiviral effect of AOP-RANTES *in vitro* is predominantly mediated by the efficient internalization promoted by this ligand (Mack *et al.*, 1998).

Chemokine-dependent migration of NK lymphocytes induces a polarization of the cells, with clustering of CCR5 at the leading edge, while adhesion molecules (ICAM-1 and -3) cluster at the uropod (Nieto *et al.*, 1997).

SIGNAL TRANSDUCTION

Cytoplasmic signaling cascades

Signal transduction of CCR5 is mediated through the pertussis toxin (PTX)-sensitive heterotrimeric $G\alpha_i$ proteins. The α_i subunit and $\beta\gamma$ dimer released from activated $G\alpha_i$ are believed to inhibit adenylyl cyclase and to stimulate phospholipase $C\beta$ isoforms, resulting in the production of IP_3 and the release of Ca^{2+} from intracellular stores (Aramori *et al.*, 1997; Zhao *et al.*, 1998). Like other G_i-coupled receptors, CCR5 can also stimulate the opening of inward-rectifying K^+ channels. Intracellular calcium flux, which can be recorded in most primary cells expressing the receptor or transfected cell lines, may also be mediated by PTX-resistant proteins of the G_q family (Farzan *et al.*, 1997). Little IP_3 generation is observed following CCR5 stimulation, and it is possible that calcium flux may result from other mechanisms than IP_3-mediated calcium release, such as the involvement of calcium channels (Atchison *et al.*, 1996; Gosling *et al.*, 1997).

CCR5 was also shown to activate tyrosine kinase cascades. MIP-1β stimulates the activity of tyrosine kinases such as Pyk-2, which in turns leads to the phosphorylation of cytoskeleton-associated protein kinases (paxillin, p130Cas), the CD4-associated tyrosine kinase p56lck, and members of the MAP kinase family (Dairaghi *et al.*, 1998; Ganju *et al.*, 1998). In prelymphoma cell lines, MIP-1β stimulates JNK/SAPK activity and this action can be inhibited by a dominant negative variant of Pyk-2. The role of tyrosine kinase activation in the frame of CCR5 biological function remains unknown, but the phosphorylation of paxillin and p56lck suggests that it could play a role in chemotaxis and lymphocyte activation. Binding of gp120 was reported to promote association between focal adhesion kinase and CCR5 (Cicala *et al.*, 1999). Ligands of CCR5, such as RANTES, were also shown to promote activation of other signaling pathways, such as PI-3 kinase (Turner *et al.*, 1995), RhoA, phospholipase D (Bacon *et al.*, 1998), and STAT (Wong and Fish, 1998), but it is presently unknown whether these effects are mediated by CCR5 or other chemokine receptors.

Signaling through CCR5 does not appear to be necessary for viral entry, as nonsignaling CCR5 mutants continue to function as coreceptors (Alkhatib *et al.*, 1997; Aramori *et al.*, 1997; Farzan *et al.*, 1997; Gosling *et al.*, 1997). It is possible, however, that intracellular signals generated by the interaction of gp120 with the coreceptor render viral replication more effective. Indeed, soluble gp160 from M-tropic *HIV* and *SIV* strains were reported to promote intracellular calcium release in human PBLs, to mediate chemotaxis of activated CD4+ T cells (Weissman *et al.*, 1997), and to stimulate tyrosine phosphorylation of Pyk-2 on human T cells (Davis *et al.*, 1997). The stimulation of intracellular cascades by RANTES was also shown to enhance replication of T-tropic *HIV-1* strains in human PBMCs. This effect could be abrogated by anti-RANTES antibodies or pertussis toxin (Kinter *et al.*, 1998).

BIOLOGICAL CONSEQUENCES OF ACTIVATING OR INHIBITING RECEPTOR AND PATHOPHYSIOLOGY

Unique biological effects of activating the receptors

Cell populations expressing CCR5 were shown to respond functionally to the various CCR5 ligands by the release of intracellular calcium, and/or by chemotaxis. However, these activities are highly redundant, due to the coexpression of other chemokine receptors responding to CCR5 ligands or to other chemokines expressed in similar pathophysiological conditions. Up to now, no unique function of CCR5 has been identified (see below, transgenic models), with the exception of its coreceptor role in *HIV* infection. However, CCR5 and its ligands appear to play a dominant role in the development of *liver injuries* associated with *graft-versus-host disease* (Murai *et al.*, 1999), in *multiple sclerosis* lesions (Balashov *et al.*, 1999), and in *rheumatoid arthritis* (Gomez-Reino *et al.*, 1999; Mack *et al.*, 1999), highlighting the potential interest of CCR5 as a therapeutic target.

As stated earlier, CCR5 is the main coreceptor involved in the entry of M-tropic strains of *HIV-1* and *HIV-2* (for reviews, see Littman 1998, and Berger *et al.*, 1999). This role was demonstrated following the identification of CXCR4 as a coreceptor for *T-tropic viruses* (Feng *et al.*, 1996), and the isolation of RANTES, MIP-1α, and MIP-1β as major *HIV*-suppressive factors produced by CD8+ T cells (Cocchi *et al.*, 1995). The key role of CCR5 in HIV pathogenesis was established by the identification of the Δ32 mutation, conferring to homozygotes a strong resistance to HIV infection (Dean *et al.*, 1996; Liu *et al.*, 1996; Samson *et al.*, 1996b). Despite the description of a large number of additional coreceptors, all *in vivo* pathophysiological data can presently be explained by the use of CCR5 and CXCR4 as coreceptors.

Phenotypes of receptor knockouts and receptor overexpression mice

A CCR5-knockout model was generated (Zhou *et al.*, 1998). The animals developed normally in a pathogen-free environment, but exhibited reduced efficiency of *Listeria* infection clearance, protection against LPS-induced *endotoxemia*, enhanced *delayed-type hypersensitivity* reaction and increased humoral responses to T cell-dependent antigenic challenge. This phenotype suggests a partial defect in macrophage function and a role of CCR5 in down-modulating T cell-dependent immune responses. Mice expressing human CCR5, together with human CD4, under control of the p56lck promoter have been generated as a tentative model for *HIV* infection (Browning *et al.*, 1997). Postentry blockade of viral replication limited the usefulness of this mouse model.

Human abnormalities

A large number of CCR5 variants have been found in human populations, but also in other species, such as mouse and monkey species (**Figure 2**). The most frequent and first reported mutation (CCR5Δ32 or Δ*ccr5*) in humans is particularly abundant in populations of European origin. A 32 bp deletion in

Figure 2 Putative transmembrane organization of human CCR5 and position of variant amino acids recorded in the literature. The phosphorylation sites for GRKs in the C-terminal domain are indicated as well as the putative palmitoylation of intracellular cysteines.

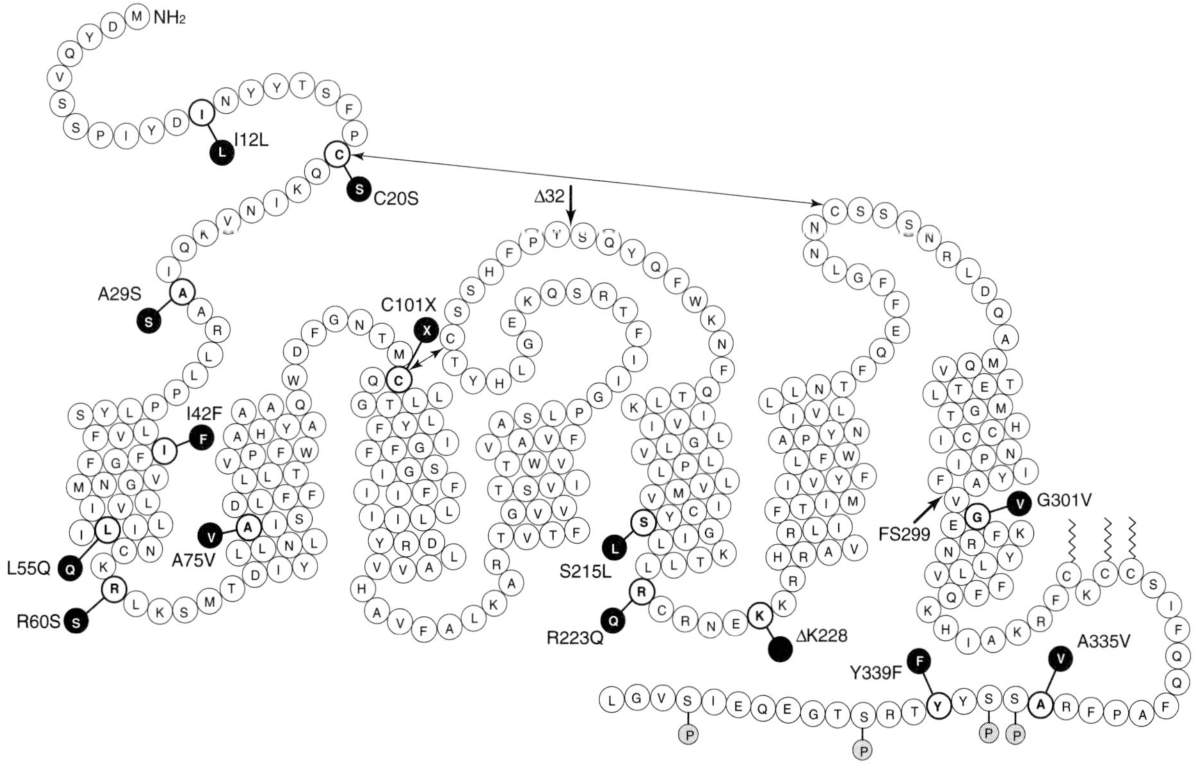

the region encoding the second extracellular loop of the receptor results in a frame shift and early termination of the polypeptide chain. The resulting protein lacks the last three transmembrane segments and is not processed properly to the plasma membrane (Liu *et al.*, 1996; Samson *et al.*, 1996b). The mutant is not functional as a chemokine receptor, nor does it act as a coreceptor for *HIV*. Homozygotes for the Δ32 mutation are highly protected against HIV infection (Dean *et al.*, 1996; Liu *et al.*, 1996; Samson *et al.*, 1996b). The HIV resistance of Δ32 homozygotes has contributed greatly to the demonstration of the central role of CCR5 as a coreceptor for M-tropic strains of HIV. However this protection is not complete, since a small number of homozygotes were found to be seropositive (Biti *et al.*, 1997; O'Brien *et al.*, 1997; Theodorou *et al.*, 1997). The strain responsible for the infection of one of these individuals was identified as a T-tropic strain utilizing CXCR4 (Michael *et al.*, 1998). Heterozygotes for the Δ32 mutation were reported to have on average delayed progression to clinical stages of *AIDS* (Dean *et al.*, 1996; Michael *et al.*, 1997; Eugen-Olsen *et al.*, 1997; Mummidi *et al.*, 1998), although this effect was not found in a cohort of injecting drug users (Schinkel *et al.*, 1999). The Δ32 mutation affects on average 10% of the alleles in European and related western Siberian Finno-Ugrian populations as well as populations from European descent. It is found in other populations, depending on the level of admixture with European populations (Martinson *et al.*, 1997). Within Europe, the highest allele frequencies (15–16%) are found in northern Russia, Finland, and Sweden, and the lower frequencies in Turkey and Portugal (Libert *et al.*, 1998). A sharp negative gradient is found toward the east (Yudin *et al.*, 1998). This mutant allele is not found in central Africa or eastern Asia populations. The analysis of microsatellites linked to the CCR5 locus has demonstrated the recent origin of the mutation and suggested a relatively strong selection in favor of the mutant allele (Libert *et al.*, 1998; Stephens *et al.*, 1998). Other variant forms of CCR5 were found in various populations around the world (Ansari-Lari *et al.*, 1997; Carrington *et al.*, 1997), but the functional consequences of these mutations have not been analyzed in depth so far.

A deletion allele of CCR5 has also been found in sooty mangabey monkeys (*Cercocebus torquatus atys*) with an allele frequency of 4% (Palacios *et al.*, 1998). The mutant receptor is not expressed at the cell surface, and does not function as a chemokine receptor or viral coreceptor. This frequent mutation suggests that similar negative selection pressures have acted against functional CCR5 in monkey species. This model will be useful to determine the role of CCR5 in host defense and microbial pathogenesis.

THERAPEUTIC UTILITY

Given its key role in *HIV* pathogenesis, CCR5 appears as a good candidate target for the development of fusion and entry inhibitors. Chemokines, antibodies, and chemicals binding to CCR5 are able to inhibit HIV entry. Part of the activity of agonists is mediated by receptor internalization. CCR5 antagonists might also find applications in the field of acute and chronic inflammatory and immune diseases.

Effects of inhibitors (antibodies) to receptors

Modified chemokines have been developed such as aminooxypentane (AOP)-RANTES[2–68] (Simmons *et al.*, 1997) and *N*-nonanoyl (NNY)-RANTES[2–68] (Mosier *et al.*, 1999) and were found to possess higher antiviral activities than natural ligands. Although described initially as an antagonist, AOP-RANTES has potent agonistic activities, and was shown to promote profound CCR5 phosphorylation (Oppermann *et al.*, 1999) and irreversible internalization of the receptor (Mack *et al.*, 1998). Met-RANTES, a RANTES analog with an additional methionine at the N-terminus, is essentially an antagonist (Proudfoot *et al.*, 1996). Truncated chemokines, such as [9–68]-RANTES, are partial agonists (Arenzana-Seisdedos *et al.*, 1996). So far, no peptide derived from natural chemokines has been shown to bind or activate the receptor.

Small molecule inhibitors have been described. A distamycin analog, inhibiting chemokine binding and function of CCR5 and other chemokine receptors, has been described (Howard *et al.*, 1998). A more specific CCR5 antagonist (TAK-779, MW 531) has been described more recently (Baba *et al.*, 1999). This compound is characterized by an affinity of 1.1 nM and relatively good specificity (K_i for CCR2: 27 nM), and blocks chemokine signaling and entry of CCR5-dependent HIV-1 strains.

A number of monoclonal antibodies directed at different domains of CCR5 have been described (Wu *et al.*, 1997; Lee *et al.*, 1999b). Some of these antibodies block signaling by chemokines and HIV-1 entry. Neutralizing human monoclonal antibodies have also been isolated from a phage display library (Osbourn *et al.*, 1998).

Inhibition of receptor function by gene therapy approaches has been proposed on the basis of an 'intrakine' approach, in which targeting of a modified chemokine (RANTES) to the endoplasmic reticulum inhibits the transport of newly synthesized CCR5 to the cell surface, resulting in a broad resistance of lymphocytes to infection by M-tropic strains of *HIV* (Yang *et al.*, 1997). Hammerhead ribozymes and DNA enzymes targeted to human CCR5 mRNA have been designed, resulting in moderate reduction of receptor expression (Goila and Banerja, 1998; Gonzalez *et al.*, 1998). CCR5 was also used in the design of a vaccine targeted at the HIV-1 envelope protein (LaCasse *et al.*, 1999), using a stabilized CCR5/CD4/gp120 complex as immunogen. In this complex, gp120 was maintained in a conserved conformation adopted transiently during the early steps of viral entry, allowing the generation of an immune response active on a broad range of viral strains.

References

Albright, A. V., Shieh, J. T., Itoh, T., Lee, B., Pleasure, D., O'Connor, M. J., Doms, R. W., and Gonzalez-Scarano, F. (1999). Microglia express CCR5, CXCR4, and CCR3, but of these, CCR5 is the principal coreceptor for human immunodeficiency virus type 1 dementia isolates. *J. Virol.* **73**, 205–213.

Alkhatib, G., Combadiere, C., Broder, C. C., Feng, Y., Kennedy, P. E., Murphy, P. M., and Berger, E. A. (1996). CC CKR5: a RANTES, MIP-1alpha, MIP-1beta receptor as a fusion cofactor for macrophage-tropic HIV-1. *Science* **272**, 1955–1958.

Alkhatib, G., Locati, M., Kennedy, P. E., Murphy, P. M., and Berger, E. A. (1997). HIV-1 coreceptor activity of CCR5 and its inhibition by chemokines: independence from G protein signaling and importance of coreceptor downmodulation. *Virology* **234**, 340–348.

Ansari-Lari, M. A., Liu, X. M., Metzker, M. L., Rut, A. R., and Gibbs, R. A. (1997). The extent of genetic variation in the CCR5 gene. *Nature Genet.* **16**, 221–222.

Aramori, I., Ferguson, S. S., Bieniasz, P. D., Zhang, J., Cullen, B., and Cullen, M. G. (1997). Molecular mechanism of desensitization of the chemokine receptor CCR-5: receptor signaling and internalization are dissociable from its role as an HIV-1 coreceptor. *EMBO J.* **16**, 4606–4616.

Arenzana-Seisdedos, F., Virelizier, J. L., Rousset, D., Clark-Lewis, I., Loetscher, P., Moser, B., and Baggiolini, M. (1996). HIV blocked by chemokine antagonist. *Nature* **383**, 400.

Atchison, R. E., Gosling, J., Monteclaro, F. S., Franci, C., Digilio, L., Charo, I. F., and Goldsmith, M. A. (1996). Multiple extracellular elements of CCR5 and HIV-1 entry: dissociation from response to chemokines. *Science* **274**, 1924–1926.

Baba, M., Nishimura, O., Kanzaki, N., Okamoto, M., Sawada, H., Iizawa, Y., Shiraishi, M., Aramaki, Y., Okonogi, K., Ogawa, Y., Meguro, K., and Fujino, M. (1999). A small-molecule, nonpeptide CCR5 antagonist with highly potent and selective anti-HIV-1 activity. *Proc. Natl Acad. Sci. USA* **96**, 5698–5703.

Bacon, K. B., Schall, T. J., and Dairaghi, D. J. (1998). RANTES activation of phospholipase D in Jurkat T cells: requirement of GTP-binding proteins ARF and RhoA. *J. Immunol.* **160**, 1894–1900.

Balashov, K. E., Rottman, J. B., Weiner, H. L., and Hancock, W. W. (1999). CCR5(+) and CXCR3(+) T cells are increased in multiple sclerosis and their ligands MIP-1alpha and IP-10 are expressed in demyelinating brain lesions. *Proc. Natl Acad. Sci. USA* **96**, 6873–6878.

Berger, E. A., Murphy, P. M., and Farber, J. M. (1999). Chemokine receptors as HIV-1 coreceptors: roles in viral entry, tropism and disease. *Annu. Rev. Immunol.* **17**, 657–700.

Berkowitz, R. D., Beckerman, K. P., Schall, T. J., and McCune, J. M. (1998). CXCR4 and CCR5 expression delineates targets for HIV-1 disruption of T cell differentiation. *J. Immunol.* **161**, 3702–3710.

Berson, J. F., and Doms, R. W. (1998). Structure-function studies of the HIV-1 coreceptors. *Semin. Immunol.* **10**, 237–248.

Bieniasz, P. D., Fridell, R. A., Aramori, I., Ferguson, S. S., Caron, M. G., and Cullen, B. R. (1997). HIV-1-induced cell fusion is mediated by multiple regions within both the viral envelope and the CCR-5 coreceptor. *EMBO J.* **16**, 2599–2609.

Biti, R., Ffrench, R., Young, J., Bennetts, B., Stewart, G., and Liang, T. (1997). HIV-1 infection in an individual homozygous for the CCR5 deletion allele. *Nature Med.* **3**, 252–253.

Blanpain, C., Migeotte, I., Lee, B., Vakili, J., Doranz, B. J., Govaerts, C., Vassart, G., Doms, R. W., and Parmentier, M. (1999a). CCR5 binds multiple CC-chemokines: MCP-3 acts as a natural antagonist. *Blood* **94**, 1899–1905.

Blanpain, C., Lee, B., Vakili, J., Doranz, B. J., Govaerts, C., Migeotte, I., Detheux, M., Vassart, G., Doms, R. W., and Parmentier, M. (1999b). Extracellular disulfide bonds of CCR5 are required for chemokine binding, but dispensable for HIV-1 coreceptor activity. *J. Biol. Chem.* **274**, 18902–18908.

Blanpain, C., Doranz, B. J., Vakili, J., Rucker, J., Govaerts, C., Baik, S. S. W., Lorthioir, O., Migeotte, I., Libert, F., Baleux, F., Vassart, G., Doms, R. W., and Parmentier, M. (1999c). Multiple charged and aromatic residues in CCR5 amino-terminal domain are involved in high affinity binding of both chemokines and HIV-1 Env protein. *J. Biol. Chem.* **274**, 34719–34727.

Bleul, C. C., Wu, L., Hoxie, J. A., Springer, T. A., and Mackay, C. R. (1997). The HIV coreceptors CXCR4 and CCR5 are differentially expressed and regulated on human T lymphocytes. *Proc. Natl Acad. Sci. USA* **94**, 1925–1930.

Bonecchi, R., Bianchi, G., Bordignon, P. P., D'Ambrosio, D., Lang, R., Borsatti, A., Sozzani, S., Allavena, P., Gray, P. A., Mantovani, A., and Sinigaglia, F. (1998). Differential expression of chemokine receptors and chemotactic responsiveness of type 1 T helper cells (TH1s) and TH2s. *J. Exp. Med.* **187**, 129–134.

Browning, J., Horner, J. W., Pettoello-Mantovani, M., Raker, C., Yurasov, S., DePinho, R. A., and Goldstein, H. (1997). Mice transgenic for human CD4 and CCR5 are susceptible to HIV infection. *Proc. Natl Acad. Sci. USA* **94**, 14637–14641.

Carrington, M., Kissner, T., Gerrard, B., Ivanov, S., O'Brien, S. J., and Dean, M. (1997). Novel alleles of the chemokine-receptor gene CCR5. *Am. J. Hum. Genet.* **61**, 1261–1267.

Carroll, R. G., Riley, J. L., Levine, B. L., Feng, Y., Kaushal, S., Ritchey, D. W., Bernstein, W., Weislow, O. S., Brown, C. R., Berger, E. A., June, C. H., and St.Louis, D. C. (1997). Differential regulation of HIV-1 fusion cofactor expression by CD28 costimulation of CD4+ T cells. *Science* **276**, 273–276.

Chelucci, C., Casella, I., Federico, M., Testa, U., Macioce, G., Pelosi, E., Guerriero, R., Mariani, G., Giampaolo, A., Hassan, H. J., and Peschle, C. (1999). Lineage-specific expression of human immunodeficiency virus (HIV) receptor/coreceptors in differentiating hematopoietic precursors: correlation with susceptibility to T- and M-tropic HIV and chemokine-mediated HIV resistance. *Blood* **94**, 1590–1600.

Choe, H., Farzan, M., Sun, Y., Sullivan, N., Rollins, B., Ponath, P. D., Wu, L., Mackay, C. R., LaRosa, G., Newman, W., Gerard, N., Gerard, C., and Sodroski, J. (1996). The beta-chemokine receptors CCR3 and CCR5 facilitate infection by primary HIV-1 isolates. *Cell* **85**, 1135–1148.

Choe, H., Martin, K. A., Farzan, M., Sodroski, J., Gerard, N. P., and Gerard, C. (1998). Structural interactions between chemokine receptors, gp120 Env and CD4. *Semin. Immunol.* **10**, 249–257.

Cicala, C., Arthos, J., Ruiz, M., Vaccarezza, M., Rubbert, A., Riva, A., Wildt, K., Cohen, O., and Fauci, A. S. (1999). Induction of phosphorylation and intracellular association of CC chemokine receptor 5 and focal adhesion kinase in primary human CD4+ T cells by macrophage-tropic HIV envelope. *J. Immunol.* **163**, 420–426.

Cocchi, F., DeVico, A. L., Garzino-Demo, A., Arya, S. K., Gallo, R. C., and Lusso, P. (1995). Identification of RANTES, MIP-1α, and MIP-1β as the major HIV suppressive factors produced by CD8+ T cells. *Science* **270**, 1811–1815.

Combadiere, C., Ahuja, S. K., Tiffany, H. L., and Murphy, P. M. (1996). Cloning and functional expression of CC CKR5, a human monocyte CC chemokine receptor selective for MIP-1α, MIP-1β, and RANTES. *J. Leukoc. Biol.* **60**, 147–152.

Dairaghi, D. J., Franz-Bacon, K., Callas, E., Cupp, J., Schall, T. J., Tamraz, S. A., Boehme, S. A., Taylor, N., and Bacon, K. B. (1998). Macrophage inflammatory protein-1beta induces migration and activation of human thymocytes. *Blood* **91**, 2905–2913.

Davis, C. B., Dikic, I., Unutmaz, D., Hill, C. M., Arthos, J., Siani, M. A., Thompson, D. A., Schlessinger, J., and Littman, D. R. (1997). Signal transduction due to HIV-1 envelope interactions with chemokine receptors CXCR4 or CCR5. *J. Exp. Med.* **186**, 1793–1798.

Dean, M., Carrington, M., Winkler, C., Huttley, G. A., Smith, M. W., Allikmets, R., Goedert, J. J., Buchbinder, S. P., Vittinghoff, E., Gomperts, E., Donfield, S., Vlahov, D., Kaslow, R., Saah, A., Rinaldo, C., Detels, R., and O'Brien, S. J. (1996). Genetic restriction of HIV-1 infection and progression to AIDS by a deletion allele of the CKR5 structural gene. *Science* **273**, 1856–1862.

Deng, H., Liu, R., Ellmeier, W., Choe, S., Unutmaz, D., Burkhart, M., Di Marzio, P., Marmon, S., Sutton, R. E., Hill, C. M., Davis, C. B., Peiper, S. C., Schall, T. J., Littman, D. R., and Landau, N. R. (1996). Identification of a major coreceptor for primary isolates of HIV-1. *Nature* **381**, 661–666.

Doranz, B. J., Rucker, J., Yi, Y., Smyth, R. J., Samson, M., Peiper, S. C., Parmentier, M., Collman, R. G., and Doms, R. W. (1996). A dual-tropic primary HIV-1 isolate that uses fusin and the beta-chemokine receptors CKR-5, CKR-3, and CKR-2b as fusion cofactors. *Cell* **85**, 1149–1158.

Dragic, T., Litwin, V., Allaway, G. P., Martin, S. R., Huang, Y., Nagashima, K. A., Cayanan, C., Maddon, P. J., Koup, R. A., Moore, J. P., and Paxton, W. A. (1996). HIV-1 entry into CD4+ cells is mediated by the chemokine receptor CC- CKR-5. *Nature* **381**, 667–673.

Dragic, T., Trkola, A., Lin, S. W., Nagashima, K. A., Kajumo, F., Zhao, L., Olson, W. C., Wu, L., Mackay, C. R., Allaway, G. P., Sakmar, T. P., Moore, J. P., and Maddon, P. J. (1998). Amino-terminal substitutions in the CCR5 coreceptor impair gp120 binding and human immunodeficiency virus type 1 entry. *J. Virol.* **72**, 279–285.

Durig, J., de Wynter, E. A., Kasper, C., Cross, M. A., Chang, J., Testa, N. G., and Heyworth, C. M. (1998). Expression of macrophage inflammatory protein-1 receptors in human CD34(+) hematopoietic cells and their modulation by tumor necrosis factor-alpha and interferon-gamma. *Blood* **92**, 3073–3081.

Edinger, A. L., Mankowski, J. L., Doranz, B. J., Margulies, B. J., Lee, B., Rucker, J., Sharron, M., Hoffman, T. L., Berson, J. F., Zink, M. C., Hirsch, V. M., Clements, J. E., and Doms, R. W. (1997). CD4-independent, CCR5-dependent infection of brain capillary endothelial cells by a neurovirulent simian immunodeficiency virus strain. *Proc. Natl Acad. Sci. USA* **94**, 14742–14747.

Eugen-Olsen, J., Iversen, A. K., Garred, P., Koppelhus, U., Pedersen, C., Benfield, T. L., Sorensen, A. M., Katzenstein, T., Dickmeiss, E., Gerstoft, J., Skinhoj, P., Svejgaard, A., Nielsen, J. O., and Hofmann, B. (1997). Heterozygosity for a deletion in the CKR-5 gene leads to prolonged AIDS-free survival and slower CD4 T-cell decline in a cohort of HIV-seropositive individuals. *AIDS* **11**, 305–310.

Farzan, M., Choe, H., Martin, K. A., Sun, Y., Sidelko, M., Mackay, C. R., Gerard, N. P., Sodroski, J., and Gerard, C. (1997). HIV-1 entry and macrophage inflammatory protein-1beta-mediated signaling are independent functions of the chemokine receptor CCR5. *J. Biol. Chem.* **272**, 6854–6857.

Farzan, M., Mirzabekov, T., Kolchinsky, P., Wyatt, R., Cayabyab, M., Gerard, N. P., Gerard, C., Sodroski, J., and Choe, H. (1999). Tyrosine sulfation of the amino terminus of CCR5 facilitates HIV-1 entry. *Cell* **96**, 667–676.

Feng, Y., Broder, C. C., Kennedy, P. E., and Berger, E. A. (1996). HIV-1 entry cofactor: functional cDNA cloning of a seven-transmembrane, G protein-coupled receptor. *Science* **272**, 872–877.

Ganju, R. K., Dutt, P., Wu, L., Newman, W., Avraham, H., Avraham, S., and Groopman, J. E. (1998). Beta-chemokine receptor CCR5 signals via the novel tyrosine kinase RAFTK. *Blood* **91**, 791–797.

Goila, R., and Banerjea, A. C. (1998). Sequence specific cleavage of the HIV-1 coreceptor CCR5 gene by a hammer-head ribozyme and a DNA-enzyme: inhibition of the coreceptor function by DNA-enzyme. *FEBS Lett.* **436**, 233–238.

Gomez-Reino, J. J., Pablos, J. L., Carreira, P. E., Santiago, B., Serrano, L., Vicario, J. L., Balsa, A., Figueroa, M., and de Juan, M. D. (1999). Association of rheumatoid arthritis with a functional chemokine receptor, CCR5. *Arthritis Rheum.* **42**, 989–992.

Gong, W., Howard, O. M., Turpin, J. A., Grimm, M. C., Ueda, H., Gray, P. W., Raport, C. J., Oppenheim, J. J., and Wang, J. M. (1998). Monocyte chemotactic protein-2 activates CCR5 and blocks CD4/CCR5-mediated HIV-1 entry/replication. *J. Biol. Chem.* **273**, 4289–4292.

Gonzalez, M. A., Serrano, F., Llorente, M., Abad, J. L., Garcia-Ortiz, M. J., and Bernad, A. (1998). A hammerhead ribozyme targeted to the human chemokine receptor CCR5. *Biochem. Biophys. Res. Commun.* **251**, 592–596.

Gosling, J., Monteclaro, F. S., Atchison, R. E., Arai, H., Tsou, C. L., Goldsmith, M. A., and Charo, I. F. (1997). Molecular uncoupling of C-C chemokine receptor 5-induced chemotaxis and signal transduction from HIV-1 coreceptor activity. *Proc. Natl Acad. Sci. USA* **94**, 5061–5066.

Granelli-Piperno, A., Moser, B., Pope, M., Chen, D., Wei, Y., Isdell, F., O'Doherty, U., Paxton, W., Koup, R., Mojsov, S., Bhardwaj, N., Clark-Lewis, I., Baggiolini, M., and Steinman, R. M. (1996). Efficient interaction of HIV-1 with purified dendritic cells via multiple chemokine coreceptors. *J. Exp. Med.* **184**, 2433–2438.

Guignard, F., Combadiere, C., Tiffany, H. L., and Murphy, P. M. (1998). Gene organization and promoter function for CC chemokine receptor 5 (CCR5). *J. Immunol.* **160**, 985–992.

Hariharan, D., Douglas, S. D., Lee, B., Lai, J. P., Campbell, D. E., and Ho, W. Z. (1999). Interferon-gamma upregulates CCR5 expression in cord and adult blood mononuclear phagocytes. *Blood* **93**, 1137–1144.

He, J., Chen, Y., Farzan, M., Choe, H., Ohagen, A., Gartner, S., Busciglio, J., Yang, X., Hofmann, W., Newman, W., Mackay, C. R., Sodroski, J., and Gabuzda, D. (1997). CCR3 and CCR5 are coreceptors for HIV-1 infection of microglia. *Nature* **385**, 645–649.

Howard, O. M., Korte, T., Tarasova, N. I., Grimm, M., Turpin, J. A., Rice, W. G., Michejda, C. J., Blumenthal, R., and Oppenheim, J. J. (1998). Small molecule inhibitor of HIV-1 cell fusion blocks chemokine receptor-mediated function. *J. Leukoc. Biol.* **64**, 6–13.

Kinter, A., Catanzaro, A., Monaco, J., Ruiz, M., Justement, J., Moir, S., Arthos, J., Oliva, A., Ehler, L., Mizell, S., Jackson, R., Ostrowski, M., Hoxie, J., Offord, R., and Fauci, A. S. (1998). CC-chemokines enhance the replication of T-tropic strains of HIV-1 in CD4(+) T cells: role of signal transduction. *Proc. Natl Acad. Sci. USA* **95**, 11880–11885.

LaCasse, R. A., Follis, K. E., Trahey, M., Scarborough, J. D., Littman, D. R., and Nunberg, J. H. (1999). Fusion-competent vaccines: broad neutralization of primary isolates of HIV. *Science* **283**, 357–362.

Lee, B., Ratajczak, J., Doms, R. W., Gewirtz, A. M., and Ratajczak, M. Z. (1999a). Coreceptor/chemokine receptor expression on human hematopoietic cells: biological implications for human immunodeficiency virus-type 1 infection. *Blood* **93**, 1145–1156.

Lee, B., Sharron, M., Blanpain, C., Doranz, B. J., Vakili, J., Setoh, P., Berg, E., Liu, G., Guy, H. R., Durell, S. R., Parmentier, M., Chang, C. N., Price, K., Tsang, M., and Doms, R. W. (1999b). Epitope mapping of CCR5 reveals multiple conformational states and distinct but overlapping structures involved in chemokine and coreceptor function. *J. Biol. Chem.* **274**, 9617–9626.

Lee, B., Sharron, M., Montaner, L. J., Weissman, D., and Doms, R. W. (1999c). Quantification of CD4, CCR5, and CXCR4 levels on lymphocyte subsets, dendritic cells, and differentially conditioned monocyte-derived macrophages. *Proc. Natl Acad. Sci. USA* **96**, 5215–5220.

Libert, F., Cochaux, P., Beckman, G., Samson, M., Aksenova, M., Cao, A., Czeizel, A., Claustres, M., de la Rua, C., Ferrari, M., Ferrec, C., Glover, G., Grinde, B., Gran, S., Kucinskas, V., Lavinha, J., Mercier, B., Ogur, G., Peltonen, L., Rosatelli, C., Schwartz, M., Spitsyn, V., Timar, L., Beckman, L., Vassart, G., and Parmentier, M. (1998). The deltaccr5 mutation conferring protection against HIV-1 in Caucasian populations has a single and recent origin in northeastern Europe. *Hum. Mol. Genet.* **7**, 399–406.

Lin, C. L., Suri, R. M., Rahdon, R. A., Austyn, J. M., and Roake, J. A. (1998). Dendritic cell chemotaxis and transendothelial migration are induced by distinct chemokines and are regulated on maturation. *Eur. J. Immunol.* **28**, 4114–4122.

Littman, D. R. (1998). Chemokine receptors: keys to AIDS pathogenesis? *Cell* **93**, 677–680.

Liu, R., Paxton, W. A., Choe, S., Ceradini, D., Martin, S. R., Horuk, R., MacDonald, M. E., Stuhlmann, H., Koup, R. A., and Landau, N. R. (1996). Homozygous defect in HIV-1 coreceptor accounts for resistance of some multiply-exposed individuals to HIV-1 infection. *Cell* **86**, 367–377.

Liu, R., Zhao, X., Gurney, T. A., and Landau, N. R. (1998). Functional analysis of the proximal CCR5 promoter. *AIDS Res. Hum. Retroviruses* **14**, 1509–1519.

Loetscher, P., Uguccioni, M., Bordoli, L., Baggiolini, M., Moser, B., Chizzolini, C., and Dayer, J. M. (1998). CCR5 is characteristic of TH1 lymphocytes. *Nature* **391**, 344–345.

Mack, M., Luckow, B., Nelson, P. J., Cihak, J., Simmons, G., Clapham, P. R., Signoret, N., Marsh, M., Stangassinger, M., Borlat, F., Wells, T. N., Schlondorff, D., and Proudfoot, A. E. (1998). Aminooxypentane-RANTES induces CCR5 internalization but inhibits recycling: a novel inhibitory mechanism of HIV infectivity. *J. Exp. Med.* **187**, 1215–1224.

Mack, M., Bruhl, H., Gruber, R., Jaeger, C., Cihak, J., Eiter, V., Plachy, J., Stangassinger, M., Uhlig, K., Schattenkirchner, M., and Schlondorff, D. (1999). Predominance of mononuclear cells expressing the chemokine receptor CCR5 in synovial effusions of patients with different forms of arthritis. *Arthritis Rheum.* **42**, 981–988.

McDermott, D. H., Zimmerman, P. A., Guignard, F., Kleeberger, C. A., Leitman, S. F., and Murphy, P. M. (1998). CCR5 promoter polymorphism and HIV-1 disease progression. Multicenter AIDS Cohort Study (MACS). *Lancet* **352**, 866–870.

Maho, A., Bensimon, A., Vassart, G., and Parmentier, M. (1999). Physical mapping of the XCR1 and CX3CR1 genes to the CCR cluster within the p21.3 region of human genome. *Cytogenet. Cell Genet.* **87**, 265–268.

Martin, M. P., Dean, M., Smith, M. W., Winkler, C., Gerrard, B., Michael, N. L., Lee, B., Doms, R. W., Margolick, J., Buchbinder, S., Goedert, J. J., O'Brien, T. R., Hilgartner, M. W., Vlahov, D., O'Brien, S. J., and Carrington, M. (1998). Genetic acceleration of AIDS progression by a promoter variant of CCR5. *Science* **282**, 1907–1911.

Martinson, J. J., Chapman, N. H., Rees, D. C., Liu, Y. T., and Clegg, J. B. (1997). Global distribution of the CCR5 gene 32-basepair deletion. *Nature Genet.* **16**, 100–103.

Meucci, O., Fatatis, A., Simen, A. A., Bushell, T. J., Gray, P. W., and Miller, R. J. (1998). Chemokines regulate hippocampal neuronal signaling and gp120 neurotoxicity. *Proc. Natl Acad. Sci. USA* **95**, 14500–14505.

Michael, N. L., Louie, L. G., Rohrbaugh, A. L., Schultz, K. A., Dayhoff, D. E., Wang, C. E., and Sheppard, H. W. (1997). The role of CCR5 and CCR2 polymorphisms in HIV-1 transmission and disease progression. *Nature Med.* **3**, 1160–1162.

Michael, N. L., Nelson, J. A., Kewalramani, V. N., Chang, G., O'Brien, S. J., Mascola, J. R., Volsky, B., Louder, M., White, G. C., Littman, D. R., Swanstrom, R., and O'Brien, T. R. (1998). Exclusive and persistent use of the entry coreceptor CXCR4 by human immunodeficiency virus type 1 from a subject homozygous for CCR5 delta32. *J. Virol.* **72**, 6040–6047.

Moriuchi, H., Moriuchi, M., and Fauci, A. S. (1997). Cloning and analysis of the promoter region of CCR5, a coreceptor for HIV-1 entry. *J. Immunol.* **159**, 5441–5449.

Moriuchi, M., Moriuchi, H., Turner, W., and Fauci, A. S. (1998). Exposure to bacterial products renders macrophages highly susceptible to T-tropic HIV-1. *J. Clin. Invest.* **102**, 1540–1550.

Moriuchi, M., Moriuchi, H., and Fauci, A. S. (1999). GATA-1 transcription factor transactivates the promoter for CCR5, a coreceptor for human immunodeficiency virus type 1 entry. *Blood* **93**, 1433–1435.

Mosier, D. E., Picchio, G. R., Gulizia, R. J., Sabbe, R., Poignard, P., Picard, L., Offord, R. E., Thompson, D. A., and Wilken, J. (1999). Highly potent RANTES analogues either prevent CCR5-using human immunodeficiency virus type 1 infection *in vivo* or rapidly select for CXCR4-using variants. *J. Virol.* **73**, 3544–3550.

Mummidi, S., Ahuja, S. S., McDaniel, B. L., and Ahuja, S. K. (1997). The human CC chemokine receptor 5 (CCR5) gene. Multiple transcripts with 5′-end heterogeneity, dual promoter usage, and evidence for polymorphisms within the regulatory regions and noncoding exons. *J. Biol. Chem.* **272**, 30662–30671.

Mummidi, S., Ahuja, S. S., Gonzalez, E., Anderson, S. A., Santiago, E. N., Stephan, K. T., Craig, F. E., O'Connell, P., Tryon, V., Clark, R. A., Dolan, M. J., and Ahuja, S. K. (1998). Genealogy of the CCR5 locus and chemokine system gene variants associated with altered rates of HIV-1 disease progression. *Nat. Med.* **4**, 786–793.

Murai, M., Yoneyama, H., Harada, A., Yi, Z., Vestergaard, C., Guo, B., Suzuki, K., Asakura, H., and Matsushima, K. J. (1999). Active participation of CCR5(+)CD8(+) T lymphocytes in the pathogenesis of liver injury in graft-versus-host disease. *Clin. Invest.* **104**, 49–57.

Nibbs, R. J., Yang, J., Landau, N. R., Mao, J. H., and Graham, G. J. (1999). LD78β, a non-allelic variant of human MIP-1α (LD78α), has enhanced receptor interactions and potent HIV suppressive activity. *J. Biol. Chem.* **274**, 17478–17483.

Nieto, M., Frade, J. M., Sancho, D., Mellado, M., Martinez, A., and Sanchez-Madrid, F. (1997). Polarization of chemokine receptors to the leading edge during lymphocyte chemotaxis. *J. Exp. Med.* **186**, 153–158.

O'Brien, T. R., Winkler, C., Dean, M., Nelson, J. A., Carrington, M., Michael, N. L., and White, G. C. (1997). HIV-1 infection in a man homozygous for CCR5 delta 32. *Lancet* **349**, 1219–1219.

Oppermann, M., Mack, M., Proudfoot, A. E., and Olbrich, H. (1999). Differential effects of CC chemokines on CC chemokine receptor 5 (CCR5) phosphorylation and identification of phosphorylation sites on the CCR5 carboxyl terminus. *J. Biol. Chem.* **274**, 8875–8885.

Oravecz, T., Pall, M., Roderiquez, G., Gorrell, M. D., Ditto, M., Nguyen, N. Y., Boykins, R., Unsworth, E., and Norcross, M. A. (1997). Regulation of the receptor specificity and function of the chemokine RANTES (regulated on activation, normal T cell expressed and secreted) by dipeptidyl peptidase IV (CD26)-mediated cleavage. *J. Exp. Med.* **186**, 1865–1872.

Osbourn, J. K., Earnshaw, J. C., Johnson, K. S., Parmentier, M., Timmermans, V., and McCafferty, J. (1998). Directed selection of MIP-1 alpha neutralizing CCR5 antibodies from a phage display human antibody library. *Nature Biotechnol.* **16**, 778–781.

Palacios, E., Digilio, L., McClure, H. M., Chen, Z., Marx, P. A., Goldsmith, M. A., and Grant, R. M. (1998). Parallel evolution of CCR5-null phenotypes in humans and in a natural host of simian immunodeficiency virus. *Curr. Biol.* **8**, 943–946.

Patterson, B. K., Czerniewski, M., Andersson, J., Sullivan, Y., Su F., Jiyamapa, D., Burki, Z., and Landay, A. (1999). Regulation of CCR5 and CXCR4 expression by type 1 and type 2 cytokines: CCR5 expression is downregulated by IL-10 in CD4-positive lymphocytes. *Clin. Immunol.* **91**, 254–262.

Proost, P., De Meester, I., Schols, D., Struyf, S., Lambeir, A. M., Wuyts, A., Opdenakker, G., De Clercq, E., Scharpé, S., and Van Damme, J. (1998). Amino-terminal truncation of chemokines by CD26/dipeptidyl-peptidase IV. Conversion of RANTES into a potent inhibitor of monocyte chemotaxis and HIV-1-infection. *J. Biol. Chem.* **273**, 7222–7227.

Proudfoot, A. E., Power, C. A., Hoogewerf, A. J., Montjovent, M. O., Borlat, F., Offord, R. E., and Wells, T. N. (1996). Extension of recombinant human RANTES by the retention of the initiating methionine produces a potent antagonist. *J. Biol. Chem.* **271**, 2599–25603.

Rabut, G. E., Konner, J. A., Kajumo, F., Moore, J. P., and Dragic, T. (1998). Alanine substitutions of polar and nonpolar residues in the amino-terminal domain of CCR5 differently impair entry of macrophage- and dualtropic isolates of human immunodeficiency virus type 1. *J. Virol.* **72**, 3464–3468.

Raport, C. J., Gosling, J., Schweickart, V. L., Gray, P. W., and Charo, I. F. (1996). Molecular cloning and functional characterization of a novel human CC chemokine receptor (CCR5) for RANTES, MIP-1α, and MIP-1β. *J. Biol. Chem.* **271**, 17161–17166.

Rucker, J., Samson, M., Doranz, B. J., Libert, F., Berson, J. F., Yi, Y., Smyth, R. J., Collman, R. G., Broder, C. C., Vassart, G., Doms, R. W., and Parmentier, M. (1996). Regions in beta-chemokine receptors CCR5 and CCR2b that determine HIV-1 cofactor specificity. *Cell* **87**, 437–446.

Ruffing, N., Sullivan, N., Sharmeen, L., Sodroski, J., and Wu, L. (1998). CCR5 has an expanded ligand-binding repertoire and is the primary receptor used by MCP-2 on activated T cells. *Cell. Immunol.* **189**, 160–168.

Sallusto, F., Schaerli, P., Loetscher, P., Schaniel, C., Lenig, D., Mackay, C. R., Qin, S., and Lanzavecchia, A. (1998). Rapid and coordinated switch in chemokine receptor expression during dendritic cell maturation. *Eur. J. Immunol.* **28**, 2760–2769.

Samson, M., Labbe, O., Mollereau, C., Vassart, G., and Parmentier, M. (1996a). Molecular cloning and functional expression of a new human CC-chemokine receptor gene. *Biochemistry* **35**, 3362–3367.

Samson, M., Libert, F., Doranz, B. J., Rucker, J., Liesnard, C., Farber, C. M., Saragosti, S., Lapoumeroulie, C., Cognaux, J., Forceille, C., Muyldermans, G., Verhofstede, C., Burtonboy, G., Georges, M., Imai, T., Rana, S., Yi, Y., Smyth, R. J., Collman, R. G., Doms, R. W., Vassart, G., and Parmentier, M. (1996b). Resistance to HIV-1 infection in caucasian individuals bearing mutant alleles of the CCR-5 chemokine receptor gene. *Nature* **382**, 722–725.

Samson, M., Soularue, P., Vassart, G., and Parmentier, M. (1996c). The genes encoding the human CC-chemokine receptors CC-CKR1 to 5 are clustered in the p21.3–p24 region of chromosome 3. *Genomics* **36**, 522–526.

Samson, M., LaRosa, G., Libert, F., Paindavoine, P., Detheux, M., Vassart, G., and Parmentier, M. (1997). The second extracellular loop of CCR5 is the major determinant of ligand specificity. *J. Biol. Chem.* **272**, 24934–24941.

Schinkel, J., Langendam, M. W., Coutinho, R. A., Krol, A., Brouwer, M., and Schuitemaker, H. (1999). No evidence for an effect of the CCR5 delta32/+ and CCR2b 64I/+ mutations on human immunodeficiency virus (HIV)-1 disease progression among HIV-1-infected injecting drug users. *J. Infect. Dis.* **179**, 825–831.

Schols, D., Proost, P., Struyf, S., Wuyts, A., De Meester, I., Scharpé, S., Van Damme, J., and De Clercq, E. (1998). CD26-processed RANTES(3–68), but not intact RANTES, has potent anti-HIV-1 activity. *Antiviral Res.* **39**, 175–187.

Sica, A., Saccani, A., Borsatti, A., Power, C. A., Wells, T. N., Luini, W., Polentarutti, N., Sozzani, S., and Mantovani, A. (1997). Bacterial lipopolysaccharide rapidly inhibits expression of C-C chemokine receptors in human monocytes. *J. Exp. Med.* **185**, 969–974.

Signoret, N., Rosenkilde, M. M., Klasse, P. J., Schwartz, T. W., Malim, M. H., Hoxie, J. A., and Marsh, M. (1998). Differential regulation of CXCR4 and CCR5 endocytosis. *J. Cell Sci.* **111**, 2819–2830.

Simmons, G., Clapham, P. R., Picard, L., Offord, R. E., Rosenkilde, M. M., Schwartz, T. W., Buser, R., Wells, T. N. C., and Proudfoot, A. E. (1997). Potent inhibition of HIV-1 infectivity in macrophages and lymphocytes by a novel CCR5 antagonist. *Science* **276**, 276–279.

Sozzani, S., Ghezzi, S., Iannolo, G., Luini, W., Borsatti, A., Polentarutti, N., Sica, A., Locati, M., Mackay, C., Wells, T. N., Biswas, P., Vicenzi, E., Poli, G., and Mantovani,

A. (1998). Interleukin 10 increases CCR5 expression and HIV infection in human monocytes. *J. Exp. Med.* **187**, 439–444.

Spleiss, O., Gourmala, N., Boddeke, H. W., Sauter, A., Fiebich, B. L., Berger, M., and Gebicke-Haerter, P. J. (1998). Cloning of rat HIV-1-chemokine coreceptor CKR5 from microglia and upregulation of its mRNA in ischemic and endotoxinemic rat brain. *J. Neurosci. Res.* **53**, 16–28.

Stephens, J. C., Reich, D. E., Goldstein, D. B., Shin, H. D., Smith, M. W., Carrington, M., Winkler, C., Huttley, G. A., Allikmets, R., Schriml, L., Gerrard, B., Malasky, M., Ramos, M. D., Morlot, S., Tzetis, M., Oddoux, C., di Giovine, F. S., Nasioulas, G., Chandler, D., Aseev, M., Hanson, M., Kalaydjieva, L., Glavac, D., Gasparini, P., and Dean, M. (1998). Dating the origin of the CCR5-Delta32 AIDS-resistance allele by the coalescence of haplotypes. *Am. J. Hum. Genet.* **62**, 1507–1515.

Theodorou, I., Meyer, L., Magierowska, M., Katlama, C., and Rouzioux, C. (1997). HIV-1 infection in an individual homozygous for CCR5 delta 32. Seroco Study Group. *Lancet* **349**, 1219–1220.

Thivierge, M., Le Gouill, C., Tremblay, M. J., Stankové, J., and Rola-Pleszczynski, M. (1998). Prostaglandin E2 induces resistance to human immunodeficiency virus-1 infection in monocyte-derived macrophages: downregulation of CCR5 expression by cyclic adenosine monophosphate. *Blood* **92**, 40–45.

Turner, L., Ward, S. G., and Westwick, J. (1995). RANTES-activated human T lymphocytes. A role for phosphoinositide 3-kinase. *J. Immunol.* **155**, 2437–2444.

Tuttle, D. L., Harrison, J. K., Anders, C., Sleasman, J. W., and Goodenow, M. M. (1998). Expression of CCR5 increases during monocyte differentiation and directly mediates macrophage susceptibility to infection by human immunodeficiency virus type 1. *J. Virol.* **72**, 4962–4969.

Valentin, A., Lu, W., Rosati, M., Schneider, R., Albert, J., Karlsson, A., and Pavlakis, G. N. (1998). Dual effect of interleukin 4 on HIV-1 expression: implications for viral phenotypic switch and disease progression. *Proc. Natl Acad. Sci. USA* **95**, 8886–8891.

Wang, J., Roderiquez, G., Oravecz, T., and Norcross, M. A. (1998). Cytokine regulation of human immunodeficiency virus type 1 entry and replication in human monocytes/macrophages through modulation of CCR5 expression. *J. Virol.* **72**, 7642–7647.

Weissman, D., Rabin, R. L., Arthos, J., Rubbert, A., Dybul, M., Swofford, R., Venkatesan, S., Farber, J. M., and Fauci, A. S. (1997). Macrophage-tropic HIV and SIV envelope proteins induce a signal through the CCR5 chemokine receptor. *Nature* **389**, 981–985.

Wong, M., and Fish, E. N. (1998). RANTES and MIP-1alpha activate stats in T cells. *J. Biol. Chem.* **273**, 309–314.

Wu, L., Paxton, W. A., Kassam, N., Ruffing, N., Rottman, J. B., Sullivan, N., Choe, H., Sodroski, J., Newman, W., Koup, R. A., and Mackay, C. R. (1997). CCR5 levels and expression pattern correlate with infectability by macrophage-tropic HIV-1, *in vitro*. *J. Exp. Med.* **185**, 1681–1691.

Yang, A. G., Bai, X., Huang, X. F., Yao, C., and Chen, S. (1997). Phenotypic knockout of HIV type 1 chemokine coreceptor CCR-5 by intrakines as potential therapeutic approach for HIV-1 infection. *Proc. Natl Acad. Sci. USA* **94**, 11567–11572.

Yudin, N. S., Vinogradov, S. V., Potapova, T. A., Naykova, T. M., Sitnikova, V. V., Kulikov, I. V., Khasnulin, V. I., Konchuk, C., Vloschinskii, P. E., Ivanov, S. V., Kobzev, V. F., Romaschenko, A. G., and Voevoda, M. I. (1998). Distribution of CCR5-delta 32 gene deletion across the Russian part of Eurasia. *Hum. Genet.* **102**, 695–698.

Zaitseva, M., Blauvelt, A., Lee, S., Lapham, C. K., Klaus-Kovtun, V., Mostowski, H., Manischewitz, J., and Golding, H. (1997). Expression and function of CCR5 and CXCR4 on human Langerhans cells and macrophages: implications for HIV primary infection. *Nature Med.* **3**, 1369–1375.

Zaitseva, M. B., Lee, S., Rabin, R. L., Tiffany, H. L., Farber, J. M., Peden, K. W., Murphy, P. M., and Golding, H. (1998). CXCR4 and CCR5 on human thymocytes: biological function and role in HIV-1 infection. *J. Immunol.* **161**, 3103–3113.

Zella, D., Barabitskaja, O., Burns, J. M., Romerio, F., Dunn, D. E., Revello, M. G., Gerna, G., Reitz, M. S. J., Gallo, R. C., and Weichold, F. F. (1998). Interferon-gamma increases expression of chemokine receptors CCR1, CCR3, and CCR5, but not CXCR4 in monocytoid U937 cells. *Blood* **91**, 4444–4450.

Zhang, L., He, T., Talal, A., Wang, G., Frankel, S. S., and Ho, D. D. (1998). *In vivo* distribution of the human immunodeficiency virus/simian immunodeficiency virus coreceptors: CXCR4, CCR3, and CCR5. *J. Virol.* **72**, 5035–5045.

Zhao, J., Ma, L., Wu, Y. L., Wang, P., Hu, W., and Pei, G. (1998). Chemokine receptor CCR5 functionally couples to inhibitory G proteins and undergoes desensitization. *J. Cell Biochem.* **71**, 36–45.

Zhou, Y., Kurihara, T., Ryseck, R. P., Yang, Y., Ryan, C., Loy, J., Warr, G., and Bravo, R. (1998). Impaired macrophage function and enhanced T cell-dependent immune response in mice lacking CCR5, the mouse homologue of the major HIV-1 coreceptor. *J. Immunol.* **160**, 4018–4025.

Zou, W., Foussat, A., Houhou, S., Durand-Gasselin, I., Dulioust, A., Bouchet, L., Galanaud, P., Levy, Y., and Emilie, D. (1999). Acute upregulation of CCR-5 expression by CD4+ T lymphocytes in HIV-infected patients treated with interleukin-2. *AIDS* **13**, 455–463.

LICENSED PRODUCTS

Cell lines expressing CCR5:

AIDS Research and Reference Reagent Program (www.aidsreagent.org): MAGI-CCR5, P4-CCR5, U373-MAGI-CCR5E, GHOST-CCR5, HOS-CCR5, 3T3.T4.CCR5, U87.CD4.CCR5

Agonists and antagonists:

R&D Systems: MIP-1α, MIP-1β, RANTES, MCP-2, MCP-3, MCP-4

PeproTech (www.peprotechec.com): MIP-1α, MIP-1β, RANTES, MCP-2, MCP-3, MCP-4

Gryphon (www.gryphonsci.com): Met-RANTES, AOP-RANTES

Labeled ligands:

New England Nuclear (www.nenlifesci.com): [^{125}I]-MIP-1α, [^{125}I]MIP-1β, [^{125I}]RANTES, [^{125}I]MCP-2

Amersham (www.apbiotech.com): [^{125}I]MIP-1α, [^{125}I]MIP-1β, [^{125I}]RANTES, [^{125}I]MCP-2

Monoclonal antibodies:

Pharmingen (www.pharmingen.com): 5C7, 2D7

R&D Systems (www.rndsystems.com): 45502, 45523, 45531, 45549, 45529

AIDS Research and Reference Reagent Program: 5502, 45549, 45551, 45523, 2D7, 5C7, 12D1

ACKNOWLEDGEMENTS

The work performed in the laboratory of the authors was supported by the *Actions de Recherche Concertées* of the Communauté Française de Belgique, the French *Agence Nationale de Recherche sur le SIDA*, the Belgian programme on Interuniversity Poles of attraction, the BIOMED and BIOTECH programmes of the European Community (grants BIO4-CT98-0543 and BMH4-CT98-2343), the *Fonds de la Recherche Scientifique Médicale* of Belgium and Télévie. C. B. is Aspirant of the Belgian *Fonds National de la Recherche Scientifique*.

CCR6

Joshua Marion Farber*

Laboratory of Clinical Investigation, National Institute of Allergy and Infectious Diseases, NIH, 10 Center Drive, Room 11N-228, MSC 1888, Bethesda, MD 20892-1888, USA

*corresponding author tel: 301-402-4910, fax: 301-402-0627, e-mail: joshua_farber@nih.gov

DOI: 10.1006/rwcy.2000.22006.

SUMMARY

CCR6 is a seven transmembrane domain G protei2 n-coupled receptor for MIP-3α, a CC chemokine expressed by a variety of cell types, including epithelial cells, in response to inflammatory stimuli. CCR6 is expressed on memory T cells, B cells, and CD34+ bone marrow progenitor-derived dendritic cells. CCR6 is active on resting memory cells and can mediate both chemotaxis and adhesion of these cells to ICAM-1. It is presumed that CCR6 is involved in the recruitment of memory T cells at the initiation of a response at inflammatory sites and/or in reactive lymphoid tissue. Data also suggest that CCR6 on immature dendritic cells may be important for the recruitment of these cells to inflammatory sites. Since the MIP-3α gene is also expressed in dendritic cells, CCR6 may mediate aggregation of dendritic cells, memory T cells, and B cells.

BACKGROUND

Discovery

CCR6 was discovered by several groups as an orphan receptor as part of screening for new G protein-coupled receptors or specifically for new chemokine receptors. The first sequence entered in the database was a genomic sequence, GPR-CY4, and the first published sequences were a genomic sequence designated CKR-L3 (Zaballos *et al.*, 1996) and cDNA and genomic sequences for a gene designated *STRL22* (Liao *et al.*, 1997c). The protein encoded by this gene was described as a receptor for the chemokine LARC/MIP-3α/Exodus/CK-β4 by several groups (Baba *et al.*, 1997; Liao *et al.*, 1997a; Power *et al.*, 1997; Greaves *et al.*, 1997) and was thereby named CCR6.

Alternative names

None currently in use. Old names include GPR-CY4, CKR-L3, STRL22, DCCR2, BN-1, and DRY6.

Structure

Like other chemokine receptors, CCR6 is a seven transmembrane domain G protein-coupled receptor. CCR6 is predicted to be 374 amino acids in length.

Main activities and pathophysiological roles

CCR6 is a receptor for the chemokine LARC/MIP-3α/Exodus/CK-β4 (Baba *et al.*, 1997; Greaves *et al.*, 1997; Liao *et al.*, 1997a; Power *et al.*, 1997), a chemokine that, based on the expression of mRNA, is produced by activated macrophages, endothelial cells, and dendritic cells, and at some mucosal sites. CCR6 is expressed and functional on CD34+ bone marrow progenitor cell-derived dendritic cells (Greaves *et al.*, 1997; Power *et al.*, 1997; Liao *et al.*, 1999), particularly on 'immature' dendritic cells early in their differentiation *in vitro* (Dieu *et al.*, 1998), as well as on CD4+ and CD8+ T cells (Baba *et al.*, 1997; Greaves *et al.*, 1997; Power *et al.*, 1997; Liao *et al.*, 1999). In the case of T cells, CCR6 is expressed exclusively on those of the memory phenotype (Liao *et al.*, 1999). CCR6 is also expressed on B cells (Zaballos *et al.*, 1996; Baba *et al.*, 1997; Liao *et al.*, 1997b, 1999; Varona *et al.*, 1998), and MIP-3α has been reported to be active on B cells from Peyer's patches (Tanaka *et al.*, 1999). It is presumed that CCR6 is involved in the trafficking of dendritic cells and T cells, and perhaps B cells, particularly at mucosal sites.

GENE

Accession numbers

Human gene: U45985, Z79784, U68031
Human cDNAs: U68030, U68032
Mouse gene: AJ222714
Mouse cDNA (KY411): AB009369

The human CCR6 gene is on chromosome 6q27, a site without other known chemokine receptors (Liao *et al.*, 1997c).

Sequence

See **Figure 1**.

PROTEIN

Accession numbers

Human: AAC51124
Mouse: BAA23776

Sequence

See **Figure 2**.

Description of protein

No information is available other than what can be predicted from the primary structure. CCR6 is

Figure 1 Nucleotide sequences for human and mouse CCR4.

Human gene:

```
AACTGTAGTGCATTTTGCCTTCTTTCCTTCTTAGAGTCACCTCTACTTTCCTGCTACCGCTGCCTGTGAGCTGAAGGGGCTGAACCATACACTCCTT
TTTCTACAACCAGCTTGCATTTTTTCTGCCCACAATGAGCGGGGTAAGATTTTTATTTTTGGCAAGGGGTATAATTTGGGTTCACTGTGGCTACTTG
AACACTACACTGCAGCTAACTCTATCTTTGTTTCCTTTCCAGGAATCAATGAATTTCAGCGATGTTTTCGACTCCAGTGAAGATTATTTTGTGTCAG
TCAATACTTCATATTACTCAGTTGATTCTGAGATGTTACTGTGCTCCTTGCAGGAGGTCAGGCAGTTCTCCAGGCTATTTGTACCGATTGCCTACTC
CTTGATCTGTGTCTTTGGCCTCCTGGGGAATATTCTGGTGGTGATCACCTTTGCTTTTTATAAGAAGGCCAGGTCTATGACAGACGTCTATCTCTTG
AACATGGCCATTGCAGACATCCTCTTTGTTCTTACTCTCCCATTCTGGGCAGTGAGTCATGCCACTGGTGCGTGGGTTTTCAGCAATGCCACGTGCA
AGTTGCTAAAAGGCATCTATGCCATCAACTTTAACTGCGGGATGCTGCTCCTGACTTGCATTAGCATGGACCGGTACATCGCCATTGTACAGGCGAC
TAAGTCATTCCGGCTCCGATCCAGAACACTACCGCGCAGCAAAATCATCTGCCTTGTTGTGTGGGGGCTGTCAGTCATCATCTCCAGCTCAACTTTT
GTCTTCAACCAAAAATACAACACCCAAGGCAGCGATGTCTGTGAACCCAAGTACCAGACTGTCTCGGAGCCCATCAGGTGGAAGCTGCTGATGTTGG
GGCTTGAGCTACTCTTTGGTTTCTTTATCCCTTTGATGTTCATGATATTTTGTTACACGTTCATTGTCAAAACCTTGGTGCAAGCTCAGAATTCTAA
AAGGCACAAAGCCATCCGTGTAATCATAGCTGTGGTGCTTGTGTTTCTGGCTTGTCAGATTCCTCATAACATGGTCCTGCTTGTGACGGCTGCAAAT
TTGGGTAAAATGAACCGATCCTGCCAGAGCGAAAAGCTAATTGGCTATACGAAAACTGTCACAGAAGTCCTGGCTTTCCTGCACTGCTGCCTGAACC
CTGTGCTCTACGCTTTTATTGGGCAGAAGTTCAGAAACTACTTTCTGAAGATCTTGAAGGACCTGTGGTGTGTGAGAAGGAAGTACAAGTCCTCAGG
CTTCTCCTGTGCCGGGAGGTACTCAGAAAACATTTCTCGGCAGACCAGTGAGACCGCAGATAACGACAATGCGTCGTCCTTCACTATGTGATAGAAA
GCTGAGTCTCCCTAAGGCATGTGTGAAACATACTCATAGATGTTATGCAAAAAAAAGTCTATGGCCAGGTATGCATGGAAAATGTGGGAATTAAGCA
AAATCAAGCAAGCCTCTCTCCTGCGGGACTTAACGTGCTCATGGGCTGTGTGATCTCTTCAGGGTGGGGTGGTCTCTGATAGGTAGCATTTTCCAGC
ACTTTGCAAGGAATGTTTTGTAGCTCTAGGGTATATATCCGCCTGGCATTTCACAAAACAGCCTTTGGGAAATGCTGAATTAAAGTGAATTGTTGAC
AAATGTAAACATTTTCAGAAATATTCATGAAGCGGTCACAGATCACAGTGTCTTTTGGTTACAGCACAAAATGATGGCAGTGGTTTGAAAAACTAAA
ACAGAAAAAAAAAATGGAAGCCAACACATCACTCATTTTAGGCAAATGTTTAAACATTTTTATCTATCAGAATGTTTATTGTTGCTGGTTATAAGCAG
CAGGATTGGCCGGCTAGTGTTTCCTCTCATTTCCCTTTGATACAGTCAACAAGCCTGACCCTGTAAAATGGAGGTGGAAAGACAAGCTCAAGTGTTC
ACAACCTGGAAGTGCTTCGGAAAGAAGGGGACAATGGCAGAACAGGTGTTGGTGACAATTGTCACCAATTGGATAAAGCAGCTCAGGTTGTAGTGGG
CCATTAGGAAACTGTCGGTTTGCTTTGATTTCCCTGGGAGCTGTTCTCTGTCGTGAGTGTCTCTTGTCTAAACGTCCATTAAGCTGAGAGTGCTATG
AAGACAGGATCTAGAATAATCTTGCTCACAGCTGTGCTCTGAGTGCCTAGCGGAGTTCCAGCAAACAAAATGGACTCAAGAGAGATTTGATTAATGA
ATCGTAATGAAGTTGGGGTTTATTGTACAGTTTAAAATGTTAGATGTTTTTAATTTTTTAAATAAATGGAATACTTTTTTTTTTTTTTAAAGAAAGC
AACTTTACTGAGACAATGTAGAAAGAAGTTTTGTTCCGTTTCTTTAATGTGGTTGAAGAGCAATGTGTGGCTGAAGACTTTTGTTATGAGGAGCTGC
AGATTAGCTAGGGGACAGCTGGAATTATGCTGGCTTCTGATAATTATTTTAAAGGGGTCTGAAATTTGTGATGGAATCAGATTTTAACAGCTCTCTT
CAATGACATAGAAAGTTCATGGAACTCATGTTTTTAAAGGGCTATGTAAATATATGAACATTAGAAAAATAGCAACTTGTGTTACAAAAATACAAAC
ACATGTTAGGAAGGTACTGTCATGGGCTAGGCATGGTGGCTCACACCTGTAATCCCAGCATTTTGGGAAGCTAAGATGGGTGGATCACTTGAGGTCA
GGAGTTTGAGACCAGCCTGGCCAACATGGCGAAACCCCTCTCTACTAAAAATACAAAAATTTGCCAGGCGTGGTGGCGGGTGCCTGTAATCCCAGCT
ACTTGGGAGGCTGAGGCAAGAGAATCGCTTGAACCCAGGAGGCAGAGGTTGCAGTGAGCCGAGATCGTGCCATTGCACTCCAGCCTGGGTGACAAAG
CGAGACTCCATCTCAAAAAAAAAAAAAAAAAAAAAAAGGAAAGAACTGTCATGTAAACATACCGACATGTTTAAACCTGACAATGGTGTTATTTGAAA
CTTTATATTGTTCTTGTAAGCTTTAACTATATCTCTCTTTAAAATGCAAAATAATGTCTTAAGATTCAAAGTCTGTATTTTTAAAGCATGGCTTTGG
CTTTGCAAAATAAAAAATGTGTTTTGTACATGAAGTAGGAATCGTATTTCAGCTTCAAGGTTCAGATTGAGGGGCCCACTGTTTGGAGAGGATGGTA
TTCAGGCTTTCTCATGTCCTTCAAATCTGTTAGCGTTTGACTCTAGAAATCAAAGCAAAGGAGTGGTTACCCAGACACTTCTTTTGGTGTGATCAAT
GCGCTGATGTGATCTATGAAGATGATTCATGCTTGAAAACTAGCACAGAAACATCTTGCTTATTTGCCAAAGCTGGGAGATGAGCTTCTCTGCATAA
TTTAAATGTTCAGATAAATGAAGCTGACTTATTTAAGCAATAACCTTTTAAACATTTTAGCTAAGATGTATAAAAATGTTTCCAAAATATACCACAT
ACTTTATTTCTTCTTAAATGTAGTACATTAGGTTACATCATTTTTCTTGCTGTCTTGGGCATCAAAACAGGTGCCATGGTAACCTGACACTCTCAGG
AGACATTAAGATAGAAGGGGCTGTTCTTCAGTGGTTCCCATTGATTCTCCCCATATCTTTTTGCTCTCAGGCTCTGGCCGTCTCTTCCTGAGCCTTA
ACTGTGT
```

Figure 1 (*Continued*)

Human cDNA:

CAAACGTTCCCAAATCTTCCCAGTCGGCTTGCAGAGACTCCTTGCTCCCAGGAGATAACCAGAAGCTGCATCTTATTGACAGATGGTCATCACATTG
GTGAGCTGGAGTCATCAGATTGTGGGGCCCGGAGTGAGGCTGAAGGGAGTGGATCAGAGCACTGCCTGAGAGTCACCTCTACTTTCCTGCTACCGCT
GCCTGTGAGCTGAAGGGGCTGAACCATACACTCCTTTTTCTACAACCAGCTTGCATTTTTTCTGCCCACAATGAGCGGGGAATCAATGAATTTCAGC
GATGTTTTCGACTCCAGTGAAGATTATTTTGTGTCAGTCAATACTTCATATTACTCAGTTGATTCTGAGATGTTACTGTGCTCCTTGCAGGAGGTCA
GGCAGTTCTCCAGGCTATTTGTACCGATTGCCTACTCCTTGATCTGTGTCTTTGGCCTCCTGGGGAATATTCTGGTGGTGATCACCTTTGCTTTTTA
TAAGAAGGCCAGGTCTATGACAGACGTCTATCTCTTGAACATGGCCATTGCAGACATCCTCTTTGTTCTTACTCTCCCATTCTGGGCAGTGAGTCAT
GCCACTGGTGCGTGGGTTTTCAGCAATGCCACGTGCAAGTTGCTAAAAGGCATCTATGCCATCAACTTTAACTGCGGGATGCTGCTCCTGACTTGCA
TTAGCATGGACCGGTACATCGCCATTGTACAGGCGACTAAGTCATTCCGGCTCCGATCCAGAACACTACCGCGCACGAAAATCATCTGCCTTGTTGT
GTGGGGGCTGTCAGTCATCATCTCCAGCTCAACTTTTGTCTTCAACCAAAAATACAACACCCAAGGCAGCGATGTCTGTGAACCCAAGTACCAGACT
GTCTCGGAGCCCATCAGGTGGAAGCTGCTGATGTTGGGGCTTGAGCTACTCTTTGGTTTCTTTATCCCTTTGATGTTCATGATATTTTGTTACACGT
TCATTGTCAAAACCTTGGTGCAAGCTCAGAATTCTAAAAGGCACAAAGCCATCCGTGTAATCATAGCTGTGGTGCTTGTGTTTCTGGCTTGTCAGAT
TCCTCATAACATGGTCCTGCTTGTGACGGCTGCAAATTTGGGTAAAATGAACCGATCCTGCCAGAGCGAAAAGCTAATTGGCTATACGAAAACTGTC
ACAGAAGTCCTGGCTTTCCTGCACTGCTGCCTGAACCCTGTGCTCTACGCTTTTATTGGGCAGAAGTTCAGAAACTACTTTCTGAAGATCTTGAAGG
ACCTGTGGTGTGTGAGAAGGAAGTACAAGTCCTCAGGCTTCTCCTGTGCCGGGAGGTACTCAGAAAACATTTCTCGGCAGACCAGTGAGACCGCAGA
TAACGACAATGCGTCGTCCTTCACTATGTGATAGAAAGCTGAGTCTCCCTAAGGCATGTGTGAAACATACTCATAGATGTTATGCAAAAAAAAGTCT
ATGGCCAGGTATGCATGGAAAATGTGGGAATTAAGCAAAATCAAGCAAGCCTCTCTCCTGCGGGACTTAACGTGCTCATGGGCTGTGTGATCTCTTC
AGGGTGGGGTGGTCTCTGATAGGTAGCATTTTCCAGCACTTTGCAAGGAATGTTTTGTAGCTCTAGGGTATATATCCGCCTGGCATTTCACAAAACA
GCCTTTGGGAAATGCTGAATTAAAGTGAATTGTTGACAAATGTAAACATTTTCAGAAATATTCATGAAGCGGTCACAGATCACAGTGTCTTTTGGTT
ACAGCACAAAATGATGGCAGTGGTTTGAAAAACTAAAACAGAAAAAAAAATGGAAGCCAACACATCACTCATTTTAGGCAAATGTTTAAACATTTTT
ATCTATCAGAATGTTTATTGTTGCTGGTTATAAGCAGCAGGATTGGCCGGCTAGTGTTTCCTCTCATTTCCCTTTGATACAGTCAACAAGCCTGACC
CTGTAAAATGGAGGTGGAAAGACAAGCTCAAGTGTTCACAACCTGGAAGTGCTTCGGGAAGAAGGGGACAATGGCAGAACAGGTGTTGGTGACAATT
GTCACCAATTGGATAAAGCAGCTCAGGTTGTAGTGGGCCATTAGGAAACTGTCGGTTTGCTTTGATTTCCCTGGGAGCTGTTCTCTGTCGTGAGTGT
CTCTTGTCTAAACGTCCATTAAGCTGAGAGTGCTATGAAGACAGGATCTAGAATAATCTTGCTCACAGCTGTGCTCTGAGTGCCTAGCGGAGTTCCA
GCAAACAAAATGGACTCAAGAGAGATTTGATTAATGAATCGTAATGAAGTTGGGGTTTATTGTACAGTTTAAAATGTTAGATGTTTTTAATTTTTTA
AATAAATGGAATACTTTTTTTTTTTTTAAAGAAAGCAACTTTACTGAGACAATGTAGAAAGAAGTTTTGTTCCGTTTCTTTAATGTGGTTGAAGAGC
AATGTGTGGCTGAAGACTTTTGTTATGAGGAGCTGCAGATTAGCTAGGGGACAGCTGGAATTATGCTGGCTTCTGATAATTATTTTAAAGGGGTCTG
AAATTTGTGATGGAATCAGATTTTAACAGCTCTCTTCAATGACATAGAAAGTTCATGGAACTCATGTTTTTAAAGGGCTATGTAAATATATGAACAT
TAGAAAAATAGCAACTTGTGTTACAAAAATACAAACACATGTTAGGAAGGTACTGTCATGGGCTAGGCATGGTGGCTCACACCTGTAATCCCAGCAT
TTTGGGAAGCTAAGATGGGTGGATCACTTGAGGTCAGGAGTTTGAGACCAGCCTGGCCAACATGGCGAAACCCCTCTCTACTAAAAATACAAAAATT
TGCCAGGCGTGGTGGCGGGTGCCTGTAATCCCAGCTACTTGGGAGGCTGAGGCAAGAGAATCGCTTGAACCCAGGAGGCAGAGGTTGCAGTGAGCCG
AGATCGTGCCATTGCACTCCAGCCTGGGTGACAGAGCGAGACTCCATCTCAAAAAAAAAAAAAAAAAA

Mouse gene:

TCTGCTCTCCCAACATCTGCACTAGTGAGAGTGTGGTTGAACTGCCCACTTCCCTTTCTACACCAGATCTGGCTCTCCCATCCACATAGAGAACCAC
GCCTGCCTGGGGTGAGAATCTACTTTATCTTGGCAGGGACTCTGGCATGGCTAGGTGTGGTTGCTTGAAATCACACTGTCACGATTTCTATTTTCAT
TATCATTCAGGAATGAATTCCACAGAGTCCTACTTTGGAACGGATGATTATGACAACACAGAGTATTATTCTATTCCTCCAGACCATGGGCCATGCT
CCCTAGAAGAGGTCAGAAACTTCACCAAGGTATTTGTGCCAATTGCCTACTCCTTAATATGTGTCTTTGGCCTCCTGGGCAACATTATGGTGGTGAT
GACCTTTGCCTTCTACAAGAAAGCCAGATCCATGACTGACGTCTACCTGTTGAACATGGCCATCACAGACATACTCTTTGTCCTCACCCTACCGTTC
TGGGCAGTTACTCATGCCACCAACACTTGGGTTTTCAGCGATGCACTGTGTAAACTGATGAAAGGCACATATGCGGTCAACTTTAACTGTGGGATGC
TGCTCCTGGCCTGTATCAGCATGGACCGGTACATTGCCATCGTCCAGGCAACCAAATCTTTCCGGGTACGCTCCAGAACACTGACGCACAGTAAGGT
CATCTGTGTGGCAGTGTGGTTCATCTCCATCATCATCTCAAGCCCTACATTTATCTTCAACAAGAAATACGAGCTGCAGGATCGTGATGTCTGTGAG
CCACGGTACAGGTCTGTCTCAGAGCCCATCACGTGGAAGCTGCTGGGTATGGGACTGGAGCTGTTCTTTGGGTTCTTCACCCCTTTGCTGTTTATGG
TGTTCTGCTATCTGTTCATTATCAAGACCTTGGTGCAGGCCCAGAACTCCAAGAGGCACAGAGCCATCCGAGTCGTGATCGCTGTGGTTCTCGTGTT
CCTGGCTTGTCAGATCCCTCACAACATGGTCCTCCTCGTGACTGCGGTCAACACGGGCAAAGTGGGCCGGAGCTGCAGCACCGAGAAAGTCCTCGCC
TACACCAGGAACGTGGCCGAGGTCCTGGCTTTCCTGCATTGCTGCCTCAACCCCGTGTTGTATGCGTTTATTGGACAGAAATTCAGAAACTACTTCA
TGAAGATCATGAAGGATGTGTGGTGTATGAGAAGGAAGAATAAGATGCCTGGCTTCCTCTGTGCCCGGGTTTACTCGGAAAGCTACATCTCCAGGCA
GACCAGTGAGACCGTCGAAAATGATAATGCATCGTCCTTTACCATGTAACACGAGAGCACAAAGCAACATTGCCCCAAAAGCCTTGGTGAAACTTGC
TATTACATATGAAAAAAAAAAAAAAGCCATGCCCAAATATGTACAGTAACTATGGAAATTCAGCAAAGACTTCCTGCAAGTTCAGAAACAGCCATGAG
GTGGCACTATCAGCCAAATTCTTCCAGGTTGTTGGTTGACAAGAAACATTGCAGCTCCTCCCAGGTTTGGTTCTACAAAATAAGATGGAAATGCCC
AGATTACTGGGTTTAGTTGCTTAATGAACATAAACATATTCCAGAAACGTTTCATGAAGGGGTTCACAGAACTAGTTGACCCCCTAACACCCCATAG
CCACAAAACAAGGATGTTACCTTGA

Mouse cDNA:

GGCTGCGAGAAGACGACAGAAGGGGGAGCACTGCTGGTTGTGTCTGTCAACAGAATAGTCCTCACATTCTTAGGACTGGAGCCTGGATAACCACTGA
GGCAGGAGTACCTGGCCAGTCTACTTTGGAGCTCAGCATTTTCTGGGGAATGAATTCCACAGAGTCCTACTTTGGAACGGATGATTATGACAACACA
GAGTATTATTCTATTCCTCCAGACCATGGGCCATGCTCCCTAGAAGAGGTCAGAAACTTCACCAAGGTATTTGTGCCAATTGCCTACTCCTTAATAT
GTGTCTTTGGCCTCCTGGGCAACATTATGGTGGTGATGACCTTTGCCTTCTACAAGAAAGCCAGATCCATGACTGACGTCTACCTGTTGAACATGGC
CATCACAGACATACTCTTTGTCCTCACCCTACCGTTCTGGGCAGTTACTCATGCCACCAACACTTGGGTTTTCAGCGATGCACTGTGTAAACTGATG
AAAGGCACATATGCGGTCAACTTTAACTGTGGGATGCTGCTCCTGGCCTGTATCAGCATGGACCGGTACATTGCCATCGTCCAGGCAACCAAATCTT
TCCGGGTACGCTCCAGAACACTGACGCACAGTAAGGTCATCTGTGTGGCAGTGTGGTTCATCTCCATCATCATCTCAAGCCCTACATTTATCTTCAA
CAAGAAATACGAGCTGCAGGATCGTGATGTCTGTGAGCCACGGTACAGGTCTGTCTCAGAGCCCATCACGTGGAAGCTGCTGGGTATGGGACTGGAG
CTGTTCTTTGGGTTCTTCACCCCTTTGCTGTTTATGGTGTTCTGCTATCTGTTCATTATCAAGACCTTGGTGCAGGCCCAGAACTCCAAGAGGCACA
GAGCCATCCGAGTCGTGATCGCTGTGGTTCTCGTGTTCCTGGCTTGTCAGATCCCTCACAACATGGTCCTCCTCGTGACTGCGGTCAACACGGGCAA
AGTGGGCCGGAGCTGCAGCACCGAGAAAGTCCTCGCCTACACCAGGAACGTGGCCGAGGTCCTGGCTTTCCTGCATTGCTGCCTCAACCCCGTGTTG
TATGCGTTTATTGGACAGAAATTCAGAAACTACTTCATGAAGATCATGAAGGATGTGTGGTGTATGAGAAGGAAGAATAAGATGCCTGGCTTCCTCT
GTGCCCGGGTTTACTCGGAAAGCTACATCTCCAGGCAGACCAGTGAGACCGTCGAAAATGATAATGCATCGTCCTTTACCATGTAACACGAGAGCAC
AAAGCAACATTGCCCCAAAAGCCTTGGTGAAACTTGCTATT

Figure 2 Comparison between human and mouse CCR6 amino acid sequences. Numbers at the right indicate the positions of the residues at the end of each line. Solid backgrounds indicate identities between the two proteins. Dots mark gaps introduced to create an optimal alignment. Tildes mark positions without corresponding residues. Putative transmembrane domains are indicated. The alignment was created using the PileUp and PrettyBox programs of the Wisconsin Sequence Analysis Package, Genetics Computer Group, Madison, WI.

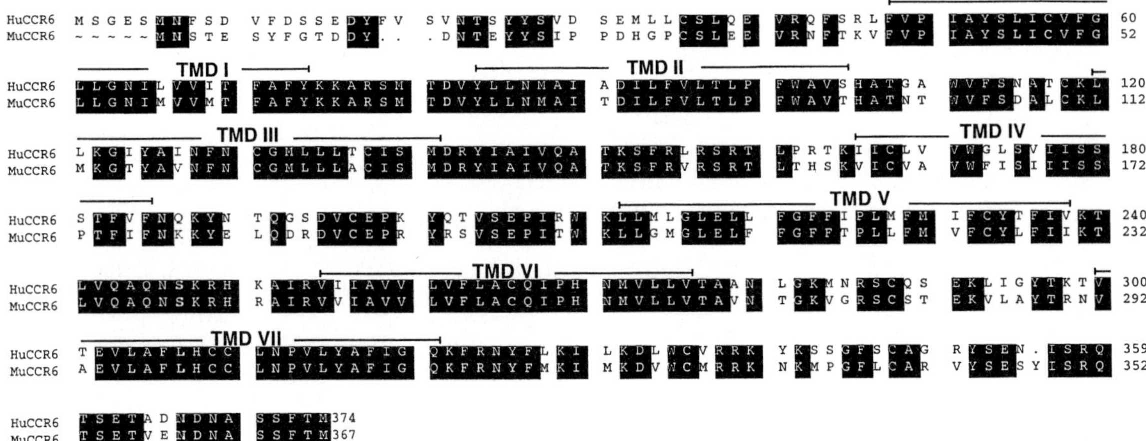

predicted to contain 374 amino acids, with features that are typical for a member of the chemokine receptor subfamily of the seven transmembrane domain G protein-coupled receptor superfamily. These include an acidic N-terminal domain, a small and basic third intracellular loop, cysteine residues in the N-terminal domain (C36) and the third extracellular loop (C288), and a conserved DRY motif following the third transmembrane domain.

Relevant homologies and species differences

While obviously related to other chemokine receptors by sequence, CCR6 is not closely related to other receptors, showing greatest similarity to CCR7 (39% identity). Overall, human CCR6 is identical at 74% of its residues, as compared with mouse protein. As shown in Figure 2, mismatched residues are found disproportionately in the N-terminal region.

Affinity for ligand(s)

K_d for MIP-3α on CCR6-transfected cells is 0.1–12 nM (Baba *et al.*, 1997; Greaves *et al.*, 1997; Power *et al.*, 1997) and on lymphocytes is 0.4 nM (Hieshima *et al.*, 1997).

Cell types and tissues expressing the receptor

CCR6 mRNA is expressed in lymphoid tissue including spleen, lymph node, thymus, appendix, and peripheral blood lymphocytes (Zaballos *et al.*, 1996; Liao *et al.*, 1997c; Varona *et al.*, 1998) and mRNA and protein are found in/on CD4+ CD8+ T cells, dendritic cells derived from CD34+ bone marrow progenitors, and B cells (Zaballos *et al.*, 1996; Baba *et al.*, 1997; Greaves *et al.*, 1997; Liao *et al.*, 1997b, 1999; Power *et al.*, 1997; Varona *et al.*, 1998). On T cells, CCR6 is limited to those with a memory phenotype (Liao *et al.*, 1999). CCR6 is found not only on freshly isolated peripheral blood T cells, but also on tumor-infiltrating lymphocytes that have been repeatedly activated *in vitro* (Liao *et al.*, 1997a,b,c).

Regulation of receptor expression

Some investigators have reported induction of CCR6 mRNA on T cells after treatment with IL-2 (Baba *et al.*, 1997), although others have reported down-regulation of mRNA after T cell activation with anti-CD3 and PMA (Greaves *et al.*, 1997). Activation of T cells *in vitro* with anti-CD3 or IL-2 did not upregulate CCR6 surface expression (Liao *et al.*, 1999) or responses to LARC/MIP-3α. Maturation of CD34+ progenitor-derived dendritic cells *in vitro* was associated with downregulation of CCR6 mRNA and responses to LARC/MIP-3α (Dieu *et al.*, 1998).

SIGNAL TRANSDUCTION

Cytoplasmic signaling cascades

Calcium signaling was blocked or diminished in tumor-infiltrating lymphocytes and in transfected cells by the addition of pertussis toxin, implicating $G\alpha_i$ proteins (Liao *et al.*, 1997a; Power *et al.*, 1997), and in transfected cells by inhibiting phospholipase C and depleting stores of intracellular calcium (Power *et al.*, 1997).

BIOLOGICAL CONSEQUENCES OF ACTIVATING OR INHIBITING RECEPTOR AND PATHOPHYSIOLOGY

Unique biological effects of activating the receptors

CCR6 is unusual in that it is a receptor for an inflammation-induced chemokine that functions well on freshly isolated lymphocytes, i.e. on nonactivated cells, producing chemotaxis (Hieshima *et al.*, 1997), calcium flux (Liao *et al.*, 1997a), and rapid adherence to ICAM-1-coated glass under conditions of flow (Campbell *et al.*, 1998). These activities were found on both CD4+ and CD8+ T cells, and were limited to T cells with a memory phenotype (Campbell *et al.*, 1998; Liao *et al.*, 1999). Consequently, CCR6 is the only chemokine receptor that functions well in the above assays specifically on nonactivated memory T cells. CCR6 is also unusual in being expressed and functional on a subset of immature dendritic cells and then being downregulated as the dendritic cells mature (Dieu *et al.*, 1998).

References

Baba, M., Imai, T., Nishimura, M., Kakizaki, M., Takagi, S., Hieshima, K., Nomiyama, H., and Yoshie, O. (1997). Identification of CCR6, the specific receptor for a novel lymphocyte-directed CC chemokine LARC. *J. Biol. Chem.* **272**, 14893–14898.

Campbell, J. J., Hedrick, J., Zlotnik, A., Siani, M. A., Thompson, D. A., and Butcher, E. C. (1998). Chemokines and the arrest of lymphocytes rolling under flow conditions. *Science* **279**, 381–384.

Dieu, M. C., Vanbervliet, B., Vicari, A., Bridon, J. M., Oldham, E., Ait-Yahia, S., Briere, F., Zlotnik, A., Lebecque, S., and Caux, C. (1998). Selective recruitment of immature and mature dendritic cells by distinct chemokines expressed in different anatomic sites. *J. Exp. Med.* **188**, 373–386.

Greaves, D. R., Wang, W., Dairaghi, D. J., Dieu, M. C., Saint-Vis, B., Franz-Bacon, K., Rossi, D., Caux, C., McClanahan, T., Gordon, S., Zlotnik, A., and Schall, T. J. (1997). CCR6, a CC chemokine receptor that interacts with macrophage inflammatory protein 3alpha and is highly expressed in human dendritic cells. *J. Exp. Med.* **186**, 837–844.

Hieshima, K., Imai, T., Opdenakker, G., Van Damme, J., Kusuda, J., Tei, H., Sakaki, Y., Takatsuki, K., Miura, R., Yoshie, O., and Nomiyama, H. (1997). Molecular cloning of a novel human CC chemokine liver and activation-regulated chemokine (LARC) expressed in liver. Chemotactic activity for lymphocytes and gene localization on chromosome 2. *J. Biol. Chem.* **272**, 5846–5853.

Liao, F., Alderson, R., Su, J., Ullrich, S. J., Kreider, B. L., and Farber, J. M. (1997a). STRL22 is a receptor for the CC chemokine MIP-3alpha. *Biochem. Biophys. Res. Commun.* **236**, 212–217.

Liao, F., Alkhatib, G., Peden, K. W. C., Sharma, G., Berger, E. A., and Farber, J. M. (1997b). STRL33, a novel chemokine receptor-like protein, functions as a fusion cofactor for both macrophage-tropic and T cell line-tropic HIV-1. *J. Exp. Med.* **185**, 2015–2023.

Liao, F., Lee, H.-H., and Farber, J. M. (1997c). Cloning of STRL22, a new human gene encoding a G protein-coupled receptor related to chemokine receptors and located on chromosome 6q27. *Genomics* **40**, 175–180.

Liao, F., Rabin, R. L., Smith, C. S., Sharma, G., Nutman, T. B., and Farber, J. M. (1999). CC-chemokine receptor 6 is expressed on diverse memory subsets of T cells and determines responsiveness to macrophage inflammatory protein 3 alpha. *J. Immunol.* **162**, 186–194.

Power, C. A., Church, D. J., Meyer, A., Alouani, S., Proudfoot, A. E. I., Clark-Lewis, I., Sozzani, S., Mantovani, A., and Wells, T. N. C. (1997). Cloning and characterization of a specific receptor for the novel CC chemokine MIP-3α from lung dendritic cells. *J. Exp. Med.* **186**, 825–835.

Tanaka, Y., Imai, T., Baba, M., Ishikawa, I., Uehira, M., Nomiyama, H., and Yoshie, O. (1999). Selective expression of liver and activation-regulated chemokine (LARC) in intestinal epithelium in mice and humans. *Eur. J. Immunol.* **29**, 633–642.

Varona, R., Zaballos, A., Gutierrez, J., Martin, P., Roncal, F., Albar, J. P., Ardavin, C., and Marquez, G. (1998). Molecular cloning, functional characterization and mRNA expression analysis of the murine chemokine receptor CCR6 and its specific ligand MIP-3alpha. *FEBS Lett.* **440**, 188–194.

Zaballos, A., Varona, R., Gutierrez, J., Lind, P., and Marquez, G. (1996). Molecular cloning and RNA expression of two new human chemokine receptor-like genes *Biochem. Biophys. Res. Commun.* **227**, 846–853.

LICENSED PRODUCTS

Anti-human CCR6 mouse IgG_{2B} monoclonal antibody, clone 53103.111 from R&D Systems for flow cytometry.

CCR7

Hisayuki Nomiyama[1,*] and Osamu Yoshie[2]

[1]Department of Biochemistry, Kumamoto University School of Medicine, 2-2-1 Honjo, Kumamoto, 860-0811, Japan

[2]Department of Bacteriology, Kinki University School of Medicine, 377-2 Ohno-Higashi, Osaka-Sayama, Osaka, 589-8511, Japan

*corresponding author tel: +81-96-373-5065, fax: +81-96-373-5066, e-mail: nomiyama@gpo.kumamoto-u.ac.jp

DOI: 10.1006/rwcy.2000.22007.

SUMMARY

CCR7, also known as *Epstein–Barr virus*-induced gene 1 (EBI1) or *Burkitt's lymphoma* receptor 2 (BLR2), was originally identified as an orphan G protein-coupled receptor. Later, it was found to be a receptor for the CC chemokines ELC/MIP-3β and SLC/6Ckine. CCR7 is mainly expressed on lymphocytes and dendritic cells, and is important in migration of these types of cells to secondary lymphoid tissues. CCR7 may also be involved in viral pathogenesis since the expression of CCR7 is upregulated by infection of lymphocytes and dendritic cells, and is important in migration of these types of cells to secondary lymphoid tissues. CCR7 may also be involved in viral pathogenesis since the expression of CCR7 is upregulated by infection of dendritic cells, and is important in migration of these cell types to secondary lymphoid tissues. CCR7 may also be involved in viral pathogenesis since the expression of CCR7 is upregulated by infection of cells with *Epstein-Barr virus* (EBV) and T cells with *human herpesvirus* (HHV) types 6 and 7. The human gene for CCR7 is located on human chromosome 17q12-q21.2.

BACKGROUND

Discovery

The human CCR7 cDNA was initially cloned as an *Epstein–Barr virus* (EBV)-inducible gene by subtractive hybridization between EBV-infected *Burkitt's lymphoma* (BL) cells and parental EBV-negative BL cells (Birkenbach *et al.*, 1993), and also from activated peripheral blood leukocytes (PBLs) by RT-PCR with degenerate primers based on another orphan receptor BLR1, now identified as CXCR5 (Burgstahler *et al.*, 1995), and termed EBI1 and BLR2, respectively. Schweickart *et al.* (1994) reported isolation of the human and mouse EBI1 cDNAs by degenerate PCR using genomic DNA. Later, the EBI1/BLR2 was found to be a specific receptor for ELC (EBI1-ligand chemokine; Yoshida *et al.*, 1997) and secondary lymphoid tissue chemokine (SLC; Campbell *et al.*, 1998; Willimann *et al.*, 1998; Yoshida *et al.*, 1998b).

Alternative names

Besides EBI1, BLR2 and the official gene symbol *CCR7* for this receptor, the nomenclatures such as CMKBR7 (chemokine β family receptor 7), CKR7 (chemokine (C–C motif) receptor 7), and CC–CKR7 have been used in the literature.

Structure

CCR7 contains the features common to chemokine receptors: seven hydrophobic transmembrane domains, a pair of cysteines in the second and third extracellular domains, a conserved DRY motif (A/S)(I/V)DR(Y/F)XXXX (where X represents hydrophobic residues) at the junction of the third transmembrane domain and the second intracellular loop, and potential *N*-glycosylation sites in the N-terminal extracellular domain.

Main activities and pathophysiological roles

CCR7 mediates migration of naïve lymphocytes and antigen-bearing dendritic cells to secondary lymphoid tissues where SLC/6Ckine and ELC/MIP-3β are expressed, and promotes the encounter of these cell types as a first step of immune response (Dieu *et al.*, 1998; Gunn *et al.*, 1998, 1999; Ngo *et al.*, 1998; Pachynski *et al.*, 1998; Sozzani *et al.*, 1998; Tangemann *et al.*, 1998; Willimann *et al.*, 1998; Yanagihara *et al.*, 1998; Chan *et al.*, 1999; Kellermann *et al.*, 1999; Saeki *et al.*, 1999; Sallusto *et al.*, 1999b). When naïve T cells are activated and differentiated into memory/effector cells, CCR7 expression is downregulated. The memory/effector cells then leave the lymphoid tissues and move to inflamed tissues. Antigenic or polyclonal stimulation of the memory/effector cells upregulate their CCR7 expression and they again migrate into the secondary lymphoid tissues (Sallusto *et al.*, 1999a). It is assumed that CCR7 is also involved in trafficking of lymphoid progenitors in to the thymus and bone marrow (Kim *et al.*, 1998, 1999; Campbell *et al.*, 1999; Suzuki *et al.*, 1999).

GENE

Accession numbers

GenBank:
Human cDNA: NM_001838
Mouse cDNA: L31580
Human gene: L31584

Sequence

See **Figure 1**.

Figure 1 Human CCR7. The coding region is shown by upper-case letters.

```
   1 gtgagacaggggtagtgcgaggccgggcacagccttcctgtgtggttttaccgcccagag
  61 agcgtcATGGACCTGGGGAAACCAATGAAAAGCGTGCTGGTGGTGGCTCTCCTTGTCATT
 121 TTCCAGGTATGCCTGTGTCAAGATGAGGTCACGGACGATTACATCGGAGACAACACCACA
 181 GTGGACTACACTTTGTTCGAGTCTTTGTGCTCCAAGAAGGACGTGCGGAACTTTAAAGCC
 241 TGGTTCCTCCCTATCATGTACTCCATCATTTGTTTCGTGGGCCTACTGGGCAATGGGCTG
 301 GTCGTGTTGACCTATATCTATTTCAAGAGGCTCAAGACCATGACCGATACCTACCTGCTC
 361 AACCTGGCGGTGGCAGACATCCTCTTCCTCCTGACCCTTCCCTTCTGGGCCTACAGCGCG
 421 GCCAAGTCCTGGGTCTTCGGTGTCCACTTTTGCAAGCTCATCTTTGCCATCTACAAGATG
 481 AGCTTCTTCAGTGGCATGCTCCTACTTCTTTGCATCAGCATTGACCGCTACGTGGCCATC
 541 GTCCAGGCTGTCTCAGCTCACCGCCACCGTGCCCGCGTCCTTCTCATCAGCAAGCTGTCC
 601 TGTGTGGGCATCTGGATACTAGCCACAGTGCTCTCCATCCCAGAGCTCCTGTACAGTGAC
 661 CTCCAGAGGAGCAGCAGTGAGCAAGCGATGCGATGCTCTCTCATCACAGAGCATGTGGAG
 721 GCCTTTATCACCATCCAGGTGGCCCAGATGGTGATCGGCTTTCTGGTCCCCCTGCTGGCC
 781 ATGAGCTTCTGTTACCTTGTCATCATCCGCACCCTGCTCCAGGCACGCAACTTTGAGCGC
 841 AACAAGGCCATCAAGGTGATCATCGCTGTGGTCGTGGTCTTCATAGTCTTCCAGCTGCCC
 901 TACAATGGGGTGGTCCTGGCCCAGACGGTGGCCAACTTCAACATCACCAGTAGCACCTGT
 961 GAGCTCAGTAAGCAACTCAACATCGCCTACGACGTCACCTACAGCCTGGCCTGCGTCCGC
1021 TGCTGCGTCAACCCTTTCTTGTACGCCTTCATCGGCGTCAAGTTCCGCAACGATCTCTTC
1081 AAGCTCTTCAAGGACCTGGGCTGCCTCAGCCAGGAGCAGCTCCGGCAGTGGTCTTCCTGT
1141 CGGCACATCCGGCGCTCCTCCATGAGTGTGGAGGCCGAGACCACCACCACCTTCTCCCCA
1201 taggcgactcttctgcctggactagagggacctctcccagggtccctggggtggggatag
1261 ggagcagatgcaatgactcaggacatcccccgccaaaagctgctcagggaaaagcagct
1321 ctcccctcagagtgcaagccctgctccagaagttagcttcaccccaatcccagctacctc
1381 aaccaatgccgaaaaagacagggctgataagctaacaccagacagacaacactgggaaac
1441 agaggctattgtcccctaaaccaaaaactgaaagtgaaagtccagaaactgttcccacct
1501 gctggagtgaaggggccaaggagggtgagtgcaaggggcgtgggagtggcctgaagagtc
1561 ctctgaatgaaccttctggcctcccacagactcaaatgctcagaccagctcttccgaaaa
1621 ccaggccttatctccaagaccagagatagtggggagacttcttggcttggtgaggaaaag
1681 cggacatcagctggtcaaacaaactctctgaacccctccctccatcgttttcttcactgt
1741 cctccaagccagcgggaatggcagctgccacgccgccctaaaagcacactcatcccctca
1801 cttgccgcgtcgccctcccaggctctcaacaggggagagtgtggtgtttcctgcaggcca
1861 ggccagctgcctccgcgtgatcaaagccacactctgggctccagagtggggatgacatgc
1921 actcagctcttggctccactgggatgggaggagaggacaagggaaatgtcaggggcgggg
1981 agggtgacagtggccgcccaaggccacgagcttgttctttgttctttgtcacagggactg
2041 aaaacctctcctcatgttctgctttcgattcgttaagagagcaacattttacccacacac
2101 agataaagttttcccttgaggaaacaacagctttaaaag
```

PROTEIN

Accession numbers

SwissProt:
Human: P32248
Mouse: P47774

Sequence

See **Figure 2**.

Description of protein

The EBI1 sequence reported by Schweikart *et al.* (1994) is identical to the BLR2 sequence (Burgstahler *et al.*, 1995) and to the the EBI1 contained in the recently deposited BAC sequence (GenBank accession number AC004585), but differs in three amino acids from the sequence determined by Birkenbach *et al.* (1993). The above sequence is the one described by Schweikart *et al.* (1994).

Relevant homologies and species differences

CCR7 is most closely related to another CC chemokine receptor, CCR6 (42% amino acid identity). Human and mouse CCR7 are 87% identical.

Affinity for ligand(s)

In stable CCR7 transfectants, both SLC and ELC show high and almost identical binding affinity (in the picomolar or nanomolar range) (Yoshida *et al.*, 1998b; Sullivan *et al.*, 1999). However, the binding affinity may vary depending on the cell type used (Yoshida *et al.*, 1998b).

Cell types and tissues expressing the receptor

CCR7 mRNA is expressed in PBLs, spleen, thymus, lymph nodes, tonsil, and appendix at high levels, and in small intestine, colon, placenta, bone marrow, and fetal liver at low levels (Birkenbach *et al.*, 1993; Schweikart *et al.*, 1994; Yoshida *et al.*, 1998a). As demonstrated by northern hybridization (Birkenbach *et al.*, 1993; Schweikart *et al.*, 1994; Yoshida *et al.*, 1998a) and responses to ELC and SLC, the cell types expressing CCR7 in these tissues are assumed to be T cells, thymocytes, B cells, dendritic cells, NK cell subsets, and macrophage progenitor cells. It has been reported that some T cell lines (H9, HUT78, CEM, and HSB-2) (Birkenbach *et al.*, 1993; Schweikart *et al.*, 1994), B cell lines (Raji, HS602, and Jijoyve; Schweikart *et al.*, 1994) and NK cell lines (NKL and YT-INDY) (Kim *et al.*, 1999) express CCR7 mRNA.

The expression of CCR7 on dendritic cells in the peripheral tissues is strongly induced upon maturation (Dieu *et al.*, 1998; Sallusto *et al.*, 1998b; Sozzani *et al.*, 1998; Yanagihara *et al.*, 1998; Saeki *et al.*, 1999). Recently, it has been reported that there is no difference in the CCR7 expression between mature and immature dendritic cells in mice (Kellermann *et al.*, 1999; Ogata *et al.*, 1999; Zhang *et al.*, 1999).

In thymus, CCR7 is expressed on single-positive and positively selected double-positive thymocytes that are migrating from the cortex to the medulla (Campbell *et al.*, 1999; Suzuki *et al.*, 1999).

Using the gene-trapping method, Steel *et al.* (1998) identified CCR7 expressed in the mouse hippocampus. They showed by *in situ* hybridization that CCR7 is expressed in neurons of the central nervous system.

Regulation of receptor expression

Epstein–Barr virus infection of B cells strongly induces CCR7, and this upregulation was shown to be due to transactivation by EBV nuclear antigen 2 (Burgstahler *et al.*, 1995). The CCR7 mRNA was also

Figure 2 Human CCR7. The possible signal peptide sequence is underlined.

```
  1 MDLGKPMKSV LVVALLVIFQ VCLCQDEVTD DYIGDNTTVD YTLFESLCSK KDVRNFKAWF
 61 LPIMYSIICF VGLLGNGLVV LTYIYFKRLK TMTDTYLLNL AVADILFLLT LPFWAYSAAK
121 SWVFGVHFCK LIFAIYKMSF FSGMLLLLCI SIDRYVAIVQ AVSAHRHRAR VLLISKLSCV
181 GIWILATVLS IPELLYSDLQ RSSSEQAMRC SLITEHVEAF ITIQVAQMVI GFLVPLLAMS
241 FCYLVIIRTL LQARNFERNK AIKVIIAVVV VFIVFQLPYN GVVLAQTVAN FNITSSTCEL
301 SKQLNIAYDV TYSLACVRCC VNPFLYAFIG VKFRNDLFKL FKDLGCLSQE QLRQWSSCRH
361 IRRSSMSVEA ETTTTFSP
```

shown to be induced in CD4+ T cells by *human herpesvirus* 6 or 7 infection (Hasegawa *et al.*, 1994). CCR7 expression is also upregulated by stimulation with IL-2 or PHA or both (Willimann *et al.*, 1998; Yoshida *et al.*, 1998a) and by treatment with TGFβ (Sallusto *et al.*, 1998a).

SIGNAL TRANSDUCTION

Cytoplasmic signaling cascades

The signaling from CCR7 is mediated by a pertussis toxin-sensitive heterotrimeric GTP-binding protein pathway. Recently, Sullivan *et al.* (1999) demonstrated that activation of MAP kinase enzyme is involved in the signal transduction cascade.

BIOLOGICAL CONSEQUENCES OF ACTIVATING OR INHIBITING RECEPTOR AND PATHOPHYSIOLOGY

Unique biological effects of activating the receptors

The *in vivo* and *in vitro* analyses of CCR7 and its ligands have demonstrated that CCR7 plays a major role in the homing of lymphocytes and dendritic cells.

References

Birkenbach, M., Josefsen, K., Yalamanchili, R., Lenoir, G., and Kieff, E. (1993). Epstein–Barr virus-induced genes: first lymphocyte-specific G protein-coupled peptide receptors. *J. Virol.* **67**, 2209–2220.

Burgstahler, R., Kempkes, B., Steube, K., and Lipp, M. (1995). Expression of the chemokine receptor BLR2/EBI1 is specifically transactivated by Epstein–Barr virus nuclear antigen 2. *Biochem. Biophys. Res. Commun.* **215**, 737–743.

Campbell, J. J., Bowman, E. P., Murphy, K., Youngman, K. R., Siani, M. A., Thompson, D. A., Wu, L., Zlotnik, A., and Butcher, E. C. (1998). 6-C-kine (SLC), a lymphocyte adhesion-triggering chemokine expressed by high endothelium, is an agonist for the MIP-3β receptor CCR7. *J. Cell Biol.* **141**, 1053–1059.

Campbell, J. J., Pan, J., and Butcher, E. C. (1999). Developmental switches in chemokine responses during T cell maturation. *J. Immunol.* **163**, 2353–2357.

Chan, V. W., Kothakota, S., Rohan, M. C., Panganiban-Lustan, L., Gardner, J. P., Wachowicz, M. S., Winter, J. A., and Williams, L. T. (1999). Secondary lymphoid-tissue chemokine (SLC) is chemotactic for mature dendritic cells. *Blood* **93**, 3610–3616.

Dieu, M. C., Vanbervliet, B., Vicari, A., Bridon, J. M., Oldham, E., Ait-Yahia, S., Briere, F., Zlotnik, A., Lebecque, S., and Caux, C. (1998). Selective recruitment of immature and mature dendritic cells by distinct chemokines expressed in different anatomic sites. *J. Exp. Med.* **188**, 373–386.

Hasegawa, H., Utsunomiya, Y., Yasukawa, M., Yanagisawa, K., and Fujita, S. (1994). Induction of G protein-coupled peptide receptor EBI1 by human herpesvirus 6 and 7 infection in CD4+ T cells. *J. Virol.* **68**, 5326–5329.

Kellermann, S. A., Hudak, S., Oldham, E. R., Liu, Y. J., and McEvoy, L. M. (1999). The CC chemokine receptor-7 ligands 6Ckine and macrophage inflammatory protein-3β are potent chemoattractants for *in vitro-* and *in vivo*-derived dendritic cells. *J. Immunol.* **162**, 3859–3864.

Kim, C. H., Pelus, L. M., White, J. R., and Broxmeyer, H. E. (1998). Macrophage-inflammatory protein-3β/EBI1-ligand chemokine/CKβ-11, a CC chemokine, is a chemoattractant with a specificity for macrophage progenitors among myeloid progenitor cells. *J. Immunol.* **161**, 2580–2585.

Kim, C. H., Pelus, L. M., Appelbaum, E., Johanson, K., Anzai, N., and Broxmeyer, H. E. (1999). CCR7 ligands, SLC/6Ckine/Exodus2/TCA4 and CKβ-11/MIP-3β/ELC, are chemoattractants for CD56(+)CD16(−) NK cells and late stage lymphoid progenitors. *Cell. Immunol.* **193**, 226–235.

Ogata, M., Zhang, Y., Wang, Y., Itakura, M., Zhang, Y. Y., Harada, A., Hashimoto, S., and Matsushima, K. (1999). Chemotactic response toward chemokines and its regulation by transforming growth factor-β_1 of murine bone marrow hematopoietic progenitor cell-derived different subset of dendritic cells. *Blood* **93**, 3225–3232.

Pachynski, R. K., Wu, S. W., Gunn, M. D., and Erle, D. J. (1998). Secondary lymphoid-tissue chemokine (SLC) stimulates integrin $\alpha4\beta7$-mediated adhesion of lymphocytes to mucosal addressin cell adhesion molecule-1 (MAdCAM-1) under flow. *J. Immunol.* **161**, 952–956.

Saeki, H., Moore, A. M., Brown, M. J., and Hwang, S. T. (1999). Secondary lymphoid-tissue chemokine (SLC) and CC chemokine receptor 7 (CCR7) participate in the emigration pathway of mature dendritic cells from the skin to regional lymph nodes. *J. Immunol.* **162**, 2472–2475.

Sallusto, F., Lenig, D., Mackay, C. R., and Lanzavecchia, A. (1998a). Flexible programs of chemokine receptor expression on human polarized T helper 1 and 2 lymphocytes. *J. Exp. Med.* **187**, 875–883.

Sallusto, F., Schaerli, P., Loetscher, P., Schaniel, C., Lenig, D., Mackay, C. R., Qin, S., and Lanzavecchia, A. (1998b). Rapid and coordinated switch in chemokine receptor expression during dendritic cell maturation. *Eur. J. Immunol.* **28**, 2760–2769.

Sallusto, F., Palermo, B., Lenig, D., Miettinen, M., Matikainen, S., Julkunen, I., Forster, R., Burgstahler, R., Lipp, M., and Lanzavecchia, A. (1999b). Distinct patterns and kinetics of chemokine production regulate dendritic cell function. *Eur. J. Immunol.* **29**, 1617–1625.

Schweickart, V. L., Raport, C. J., Godiska, R., Byers, M. G., Eddy, R.L. Jr., Shows, T. B., and Gray, P. W. (1994). Cloning of human and mouse EBI1, a lymphoid-specific G-protein-coupled receptor encoded on human chromosome 17q12-q21.2. *Genomics* **23**, 643–650.

Sozzani, S., Allavena, P., D'Amico, G., Luini, W., Bianchi, G., Kitaura, M., Imai, T., Yoshie, O., Bonecchi, R., and Mantovani, A. (1998). Differential regulation of chemokine receptors during dendritic cell maturation: a model for their trafficking properties. *J. Immunol.* **161**, 1083–1086.

Steel, M., Moss, J., Clark, K. A., Kearns, I. R., Davies, C. H., Morris, R. G., Skarnes, W. C., and Lathe, R. (1998). Gene-trapping to identify and analyze genes expressed in the mouse hippocampus. *Hippocampus.* **8**, 444–457.

Sullivan, S. K., McGrath, D. A., Grigoriadis, D., and Bacon, K. B. (1999). Pharmacological and signaling analysis of human chemokine receptor CCR-7 stably expressed in HEK-293 cells: high-affinity binding of recombinant ligands MIP-3beta and SLC stimulates multiple signaling cascades. *Biochem. Biophys. Res. Commun.* **263**, 685–690.

Suzuki, G., Sawa, H., Kobayashi, Y., Nakata, Y., Nakagawa, K., Uzawa, A., Sakiyama, H., Kakinuma, S., Iwabuchi, K., and Nagashima, K. (1999). Pertussis toxin-sensitive signal controls the trafficking of thymocytes across the corticomedullary junction in the thymus. *J. Immunol.* **162**, 5981–5985.

Tangemann, K., Gunn, M. D., Giblin, P., and Rosen, S. D. (1998). A high endothelial cell-derived chemokine induces rapid, efficient, and subset-selective arrest of rolling T lymphocytes on a reconstituted endothelial substrate. *J. Immunol.* **161**, 6330–6337.

Willimann, K., Legler, D. F., Loetscher, M., Roos, R. S., Delgado, M. B., Clark-Lewis, I., Baggiolini, M., and Moser, B. (1998). The chemokine SLC is expressed in T cell areas of lymph nodes and mucosal lymphoid tissues and attracts activated T cells via CCR7. *Eur. J. Immunol.* **28**, 2025–2034.

Yanagihara, S., Komura, E., Nagafune, J., Watarai, H., and Yamaguchi, Y. (1998). EBI1/CCR7 is a new member of dendritic cell chemokine receptor that is up-regulated upon maturation. *J. Immunol.* **161**, 3096–3102.

Yoshida, R., Imai, T., Hieshima, K., Kusuda, J., Baba, M., Kitaura, M., Nishimura, M., Kakizaki, M., Nomiyama, H., and Yoshie, O. (1997). Molecular cloning of a novel human CC chemokine EBI1-ligand chemokine that is a specific functional ligand for EBI1, CCR7. *J. Biol. Chem.* **272**, 13803–13809.

Yoshida, R., Nagira, M., Imai, T., Baba, M., Takagi, S., Tabira, Y., Akagi, J., Nomiyama, H., and Yoshie, O. (1998a). EBI1-ligand chemokine (ELC) attracts a broad spectrum of lymphocytes: activated T cells strongly up-regulate CCR7 and efficiently migrate toward ELC. *Int. Immunol.* **10**, 901–910.

Yoshida, R., Nagira, M., Kitaura, M., Imagawa, N., Imai, T., and Yoshie, O. (1998b). Secondary lymphoid-tissue chemokine is a functional ligand for the CC chemokine receptor CCR7. *J. Biol. Chem.* **273**, 7118–7122.

Zhang, Y., Zhang, Y. Y., Ogata, M., Chen, P., Harada, A., Hashimoto, S., and Matsushima, K. (1999). Transforming growth factor-β_1 polarizes murine hematopoietic progenitor cells to generate Langerhans cell-like dendritic cells through a monocyte/macrophage differentiation pathway. *Blood* **93**, 1208–1220.

CCR8

Monica Napolitano[1,*] and Angela Santoni[2]

[1]Laboratory of Vascular Pathology, Istituto Dermopatico dell'Immacolata-IRCCS, Via Monti di Creta 104, Rome, 0167, Italy

[2]Department of Experimental Medicine and Pathology, University of Rome "La Sapienza", Via Regina Elena 324, Rome, 00161, Italy

*corresponding author tel: 39-06-66462431, fax: 39-06-66462430, e-mail: m.napolitano@idi.it

DOI: 10.1006/rwcy.2000.22008.

SUMMARY

CCR8 is a CC chemokine receptor mostly homologous to CCR4 (44%). CCR8 mRNA has been detected in lymphoid tissues such as thymus, spleen, and lymph nodes. In particular, CCR8 message is expressed by CD4+ and CD4+CD8+ thymocytes, and by CD4+ and CD8+ peripheral blood T lymphocytes. This pattern of expression suggests a role for this receptor and its ligands in the differentiation, and activation/migration of T lymphocytes. CCR8 may also be considered as a molecule associated with the TH2 'program', being abundantly expressed by NK1.1+CD4+ cells and by TH2 cells that migrate in response to CCR8 ligands, thus suggesting an important role of I-309/CCR8 in *allergic diseases*.

The main agonists of CCR8 are represented by the eukaryotic CC chemokine I-309 and by vMIP-I, an HHV-8-encoded protein, that induce both calcium mobilization and chemotaxis in cellular transfectants overexpressing CCR8.

A role for this receptor in modulating HIV infection is indicated by the demonstration that several HIV and SIV isolates utilize CCR8 as a coreceptor for viral entry/fusion and by the inhibition of CCR8-mediated HIV infection shown by I-309.

BACKGROUND

Chemokines constitute a family of structurally related regulators of chemotaxis, adhesion, differentiation, and proliferation. They interact with a class of seven transmembrane-spanning receptors that are coupled to G_i or to a lesser extent to G_q protein, and whose stimulation typically leads to intracellular calcium mobilization and cellular chemotaxis.

Discovery

Human CCR8

The chemokine receptor superfamily shows structural motifs of G protein-coupled receptors, thereby representing a subfamily of this class of molecules. The degree of relatedness of these receptors in several regions of their primary sequence has prompted cloning og novel genes by low-stringency and degenerate PCR-based approaches. In particular, such methods were used to clone the human CCR8 gene from human cDNAs or genomic DNA, followed by hybridization of the amplified fragments to human genomic libraries (Napolitano *et al*., 1996; Samson *et al*., 1996; Zaballos *et al*., 1996; Tiffany *et al*., 1997). Like most chemokine receptor genes, the CCR8 gene is intronless.

Murine CCR8

Low-stringency hybridization of the human CCR8 open reading frame (ORF) to 129/SV genomic libraries yielded several clones of the murine CCR8 gene containing the complete ORF and several kilobases of 5′ and 3′ regions (Goya *et al*., 1998; Zingoni *et al*., 1998). Human and murine CCR8 share 71% sequence homology at the protein level (see Figure 2).

Alternative names

Human CCR8 is also known as TER1, ChemR1, CKR-L1, C-C CKR-8, GPR-CY6, and CMKBRL2.

Structure

The predicted structure for CCR8 is that of a typical G protein-coupled receptor with seven hydrophobic moieties spanning the plasma membrane, an extracellular N-terminus, and an intracellular C-terminus. The tertiary CCR8 structure has not been solved.

Main activities and pathophysiological roles

It has been shown that the human CCR8 receptor can function both as a coreceptor and as a fusion cofactor of HIV-1, HIV-2, and SIV strains in cell lines overexpressing the CCR8 cDNA (Rucker *et al.*, 1997; Jinno *et al.*, 1998; Simmons *et al.*, 1998; Yi *et al.*, 1998; Zhang *et al.*, 1998; Albright *et al.*, 1999; Chan *et al.*, 1999; Dittmar *et al.*, 1999; Singh *et al.*, 1999). CCR8 is able to mediate the entry/fusion of the HIV-1 strains 89.6, ADA, SF162, 2028, 2076, and the SIV strain mac316, the HIV-2 strain SBL6669, and a primary HIV type 1 group O isolate. Moreover, I-309 inhibits the infection of several HIV strains in a CCR8-dependent fashion (Horuk *et al.*, 1998).

CCR8 has recently been shown to be preferentially expressed by TH2 cells and clones that are able to respond to CCR8 ligands. This suggests a possible role for CCR8 and its ligands in *allergic diseases.*

As the only reported receptor for I-309 is CCR8, it can be hypothesized that the ability of I-309 to inhibit the dexamethasone-induced apoptosis of the BW157 cell line (Van Snick *et al.*, 1996), which abundantly expresses the CCR8 message (Goya *et al.*, 1998; Zingoni *et al.*, 1998), may indeed be CCR8-mediated and suggests that I-309/CCR8 may control cellular apoptosis and/or proliferation.

GENE

Accession numbers

Human CCR8: U45983, U62556, Z79782, Y08456
Murine CCR8: Z98206, AF001277

Sequence

See **Figure 1**.

Figure 1 Nucleotide sequence for the human CCR8 gene.

Sequence

Human CCR8

```
CCGCCATGCCCGGCTAATTTTTGTATTTTTAGTAGAGACGGGGTTTCGCCATGTTGGAAGGCTGGTCTTGAACCCCTGACCTCAGGTGATCTGC
CCACCTTGGCCTCCCAAAGTGCTAGGATTACAGGCATGAGCCACAGCTCCCGGTCTATCATTTAACCTTAATTACATCTTTAAAGGCCCAAATA
GTCTCACCCACTCCAAATAGTCACACCCACACCGGAGGTTGAGCACTTCAACACATGAATTTGGGGAGGACACAGTTCAGTCCATAACATCCCC
CTAATTTTTAAAAAATAAAAATGTTTTTAAGGAGTGAATGTCTTTTATGTGTCTCTGTGACCAGGTCCCGCTGCCTTGATGGATTATACACTTG
ACCTCAGTGTGACAACAGTGACCGACTACTACTACCCTGATATCTTCTCAAGCCCCTGTGATGCGGAACTTATTCAGACAAATGGCAAGTTGCT
CCTTGCTGTCTTTTATTGCC
TCCTGTTTGTATTCAGTCTTCTGGGAAACAGCCTGGTCATCCTGGTCCTTGTGGTCTGCAAGAAGCTGAGGAGCATCACAGATGTATACCTCTT
GAACCTGGCCCTGTCTGACCTGCTTTTTGTCTTCTCCTTCCCCTTTCAGACCTACTATCTGCTGGACCAGTGGGTGTTTGGGACTGTAATGTGC
AAAGTGGTGTCTGGCTTTTATTACATTGGCTTCTACAGCAGCATGTTTTTCATCACCCTCATGAGTGTGGACAGGTACCTGGCTGTTGTCCATG
CCGTGTATGCCCTAAAGGTGAGGACGATCAGGATGGGCACAACGCTGTGCCTGGCAGTATGGCTAACCGCCATTATGGCTACCATCCCATTGCT
AGTGTTTTACCAAGTGGCCTCTGAAGATGGTGTTCTACAGTGTTATTCATTTTACAATCAACAGACTTTGAAGTGGAAGATCTTCACCAACTTC
AAAATGAACATTTTAGGCTTGTTGATCCCATTCACCATCTTTATGTTCTGCTACATTAAAATCCTGCACCAGCTGAAGAGGTGTCAAAACCACA
ACAAGACCAAGGCCATCAGGTTGGTGCTCATTGTGGTCATTGCATCTTTACTTTTCTGGGTCCCATTCAACGTGGTTCTTTTCCTCACTTCCTT
GCACAGTATGCACATCTTGGATGGATGTAGCATAAGCCAACAGCTGACTTATGCCACCCATGTCACAGAAATCATTTCCTTTACTCACTGCTGT
GTGAACCCTGTTATCTATGCTTTTGTTGGGGAGAAGTTCAAGAAACACCTCTCAGAAATATTTCAGAAAAGTTGCAGCCAAATCTTCAACTACC
TAGGAAGACAAATGCCTAGGGAGAGCTGTGAAAAGTCATCATCCTGCCAGCAGCACTCCTCCCGTTCCTCCAGCGTAGACTACATTTTGTGAGG
ATCAATGAAGACTAAATATAAAAACATTTTCTTGAATGGCATGCTAGTAGCAGTGAGCAAAGGTGTGGGTGTGAAAGGTTTCCAAAAAAAGTTC
AGCATGAAGGATGCCATATATGTTGTTGCCAACACTTGGAACACAATGACTAAAGACATAGTTGTGCATGCCTGGCACAACATCAAGCCTGTGA
TTGTGTTTATTGATGATGTTGAACAAGTGGTAACTTTAAAGGATTCTGTATCCAAGTGAAAAAAAAAGATGTCTGACCTCCTTCATATGCAAAA
ATATACCTTCAGAGACTGTCAGTAGGCTGGAAGAAGTGGATATTGAAGTTTTGACATCAATGATGAGGCTCCAGTTGTCTATGCATTGACTGAT
GGTGAAATGGCTGGAGTGATTCTGAATCAAGGTGATTGTGATTATAGTGACAATGAAGATGATGCTATTAATACTGCATAAAAAGTGCCTGTAG
ATGACATGGTGAAAATATTTGACAGGCTTATGGAAGGACTACAGCAGCACGCATTCATAACAGAACAAGAAATTATCTCAGCTTATAAAATCAA
ACAGAGACTTCTAGACAAAAACCATTGTTGATGAGGCAGATGCCTCTAGAAGAGACGTTTAAAAGCCATCAAACACAATGCCTCATCTTCCCTG
GAGGACCCACTTCCTGATCCCTCAACTGTGTCTGATGTTTCTTCTCATGTAAGAAATAAAAAATAAAAATAAAAAAATATATATTGGTATGTAA
CTACAGGAAAAAAATAAAAAATATATAGTGGACAGTAACCTTTCAATCAAAACTCAGTATCATAAGTAGAGACTGAAAACTTGCCGTTATTGAT
TGTTGTTATTAACAGCTGATACAGGTATTCTGCTGATGCTACTGCTGCCTAGTTACCATGAACACGTTTTTTCACTATTAATGGTGCGTCATAT
TTTTTACTTTTAAGTACTTACGTGTGAGTAAGTGTAAGAAAATGATTGCTTATCAGTAGTATCAATGATTTACTCAATATCTGAATCACCTTGA
TTCAGAACCATTTCAGCTGTTTCACCATCAGTCAATGAATAACAGCCTCATTGATGTCAAAAACTTCAATATCCACTTCTTTCAGCCTACTGTA
GACTCTGGAAGTATACTTTTTGCATATGTAAGGAAGTCAGA
```

Chromosome location and linkages

The human CCR8 receptor was mapped on chromosome 3p21-23 by fluorescent *in situ* hybridization (FISH), and on chromosome 3p21-24 by performing PCR on radiation hybrids and on YAC clones (Napolitano *et al.*, 1996; Samson *et al.*, 1996; Tiffany *et al.*, 1997). The murine gene was mapped by FISH to the telomeric f4 region of chromosome 9 (Zingoni *et al.*, 1998).

PROTEIN

Accession numbers

Human CCR8: 1468979, 1668736, 1707884, 2465082
Murine CCR8: 2765843, 4049612

Sequence

See **Figure 2**.

Description of protein

The hCCR8 ORF encodes for a 355 amino acid protein that is predicted to belong to the chemokine receptor family. It does not possess any *N*-linked glycosylation sites in its N-terminal or extracellular regions, and shows four conserved cysteines that may be involved in disulfide bridges. One potential site of phosphorylation by PKC is present in the C-terminus of the molecule.

Relevant homologies and species differences

The human CCR8 protein is very similar to CC chemokine receptors (39–44% identity), and to CXCR (<37% identity).

The murine ORF encodes for a 353 amino acid protein that shares 71% homology with CCR8, that is maximal in the transmembrane regions and minimal in the N- and C-termini (Figure 2).

Figure 2 Alignment between the human and murine CCR8 proteins.

```
                                                TM1
                                   --------------------------
         10        20        30        40        50        60
MDYTLDLSVTTVTDYYYPDIFSSPCDAELIQTNGKLLLAVFYCLLFVFSLLGNSLVILVL
:::::.. .:: .:::: ::.:.:::::..   .. : ::..::.:::.::::::::::
MDYTMEPNVT-MTDYY-PDFFTAPCDAEFLLRGSMLYLAILYCVLFVLGLLGNSLVILVL
         10          20        30         40        50
                TM2                           TM3
         -----------------------    -----------------------
70       80        90       100       110       120
VVCKKLRSITDVYLLNLALSDLLFVFSFPFQTYYLLDQWVFGTVMCKVVSGFYYIGFYSS
: :::::::::.:::::: :::::.:.::::. ::::::::::.:::::::::.:::::.::
VGCKKLRSITDIYLLNLAASDLLFVLSIPFQTHNLLDQWVFGTAMCKVVSGLYYIGFFSS
60         70        80        90       100       110
                                       TM4
----------                 -----------------------------
130       140       150       160       170       180
MFFITLMSVDRYLAVVHAVYALKVRTIRMGTTLCLAVWLTAIMATIPLLVFYQVASEDGV
::::::::::::::.:::::::.::::   .::.: :.::::.:. ::::::.::::::::.
MFFITLMSVDRYLAIVHAVYAIKVRTASVGTALSLTVWLAAVTATIPLMVFYQVASEDGM
120       130       140       150       160       170
                                  TM5
                       ------------------------         -----
190       200       210       220       230       240
LQCYSFYNQQTLKWKIFTNFKMNILGLLIPFTIFMFCYIKILHQLKRCQNHNKTKAIRLV
:::..:::..:.:.:::.:::.:..: ::::.:::.:.:.::::.::::.:::. : :::.:.:::.::
LQCFQFYEEQSLRWKLFTHFEINALGLLLPFAILLFCYVRILQQLRGCLNHNRTRAIKLV
180       190       200       210       220       230
              TM6                                 TM7
------------------------              ------------------------
250       260       270       280       290       300
LIVVIASLLFWVPFNVVLFLTSLHSMHILDGCSISQQLTYATHVTEIISFTHCCVNPVIY
: :::.:::::::::::::::.::::::::..:::::::.  :.:. : ::::.::::::::::::::
LTVVIVSLLFWVPFNVALFLTSLHDLHILDGCATRQRLALAIHVTEVISFTHCCVNPVIY
240       250       260       270       280       290

-------     310       320       330       340       350
AFVGEKFKKHLSEIFQKSCSQIFNYLGRQMPRESCEKSSSCQQHSSRSSSVDYIL
::.::::::::: ..:::::::.:: :::::::   . :.. : .:.::.::..: ::
AFIGEKFKKHLMDVFQKSCSHIFLYLGRQMPVGALERQLSSNQRSSHSSTLDDIL
300       310       320       330       340       350
```

Affinity for ligand(s)

Human CCR8

Most studies aimed at identifying chemokine receptor ligands and at studying downstream signaling events have utilized cell lines reconstituted with single chemokine receptors. Tiffany *et al.* (1997) have generated cellular transfectants of the pre-B cell line 4DE4 expressing the human CCR8 cDNA and identified the CC chemokine I-309 as a CCR8 ligand as measured by the prompt calcium mobilization induced in these cell lines following stimulation, differently from several other chemokines tested. Similar results were obtained by Roos *et al.* (1997) in the pre-B cell line 300-19 and by Goya *et al.* (1998) that showed I-309-induced calcium mobilization and chemotaxis in 293/CCR8-transfected cell lines at a concentration of 10–100 nM.

Bernardini *et al.* (1998) have identified two novel ligands, MIP-1β and TARC, of the CCR8 receptor. Such CC chemokines induce chemotaxis of the Jurkat/CCR8 but not of the mock cellular transfectants. These data are controversial in the light of several reports that showed neither binding to CCR8 nor calcium mobilization in response to such agonists in other cell types (Roos *et al.*, 1997; Tiffany *et al.*, 1997; Goya *et al.*, 1998; Dairaghi *et al.*, 1999; Endres *et al.*, 1999; Garlisi *et al.*, 1999). The chemotaxis induced in Jurkat/CCR8 transfectants is dependent on the overexpression of CCR8 in such cells, but it still remains to be demonstrated that ligand binding and calcium mobilization may occur. Therefore, at the moment, I-309 is the only 'recognized' eukaryotic ligand for CCR8.

Viruses, such as herpesviruses and poxviruses, often attempt to evade the immune system by producing molecules acquired from cellular genes. In particular, several chemokine or chemokine receptor-like viral proteins have been described and shown to be functional in eukaryotic hosts.

It has been reported that the viral chemokine vMIP-II, encoded by *Kaposi's sarcoma herpesvirus* (*KSHV*), is a ligand for CCR8 and is a chemoattractant for TH2 cells (Sozzani *et al.*, 1998). More recent reports indicate that vMIP-I, also KSHV-encoded, selectively engages CCR8 ($IC_{50} = 1.2-2$ nM) but not several other chemokine receptors (Dairaghi *et al.*, 1999; Endres *et al.*, 1999), acting as an agonist. Moreover, vMIP-II has been shown to behave as a vMIP-I and I-309 antagonist in cellular transfectants (Dairaghi *et al.*, 1999; Luttichau *et al.*, 2000) but is a broader-spectrum antagonist. An antagonistic effect has also been shown by the viral product of *molluscum contagiosum virus ORF MC148R*, also known as vMCC-1 (Dairaghi *et al.*, 1999), considered to act as an inhibitor of several chemokines (Damon *et al.*, 1998), and recently demonstrated to bind selectively to CCR8 and block CCR8 function but not several other chemokine receptors' activities (Luttichau *et al.*, 2000).

Murine CCR8

Goya *et al.* (1998) reported that I-309 and TCA3, its murine homolog, induce calcium mobilization of 293/mCCR8 cellular transfectants at 100 nM and that TCA3 binds to mCCR8 transfectants with a K_d of 2 nM.

Cell types and tissues expressing the receptor

Human CCR8

The human CCR8 gene shows a restricted mRNA expression in lymphoid tissues and cell lines. It is abundantly expressed by the thymus and at low levels by spleen, lymph nodes, CD4+ and CD8+ and IL-2-treated T lymphocytes, adherent and LPS-treated monocytes, and, at very low levels, by CD19+ B cells and PMNs (Napolitano *et al.*, 1996; Samson *et al.*, 1996; Zaballos *et al.*, 1996; Roos *et al.*, 1997; Rucker *et al.*, 1997; Tiffany *et al.*, 1997). CCR8 expression in cell lines ranged between high and low levels in MOLT4 (immature T), HUT78 (CD4+ mature T), NK3.3 (NK-like) and Jurkat cells, while virtually absent in other cell lines, such as K562, U-937, RPMI 8866, YT-5, NKL, Jurkat, HL60, THP-1, MEG-01, RAJI, KG1-A, HEL 92.1.7, and JM-1. (Two different groups reported either CCR8 expression or no expression in the Jurkat cell line as measured by northern blot analysis, probably due to cellular variants.) It has been reported that the CCR8 message is abundantly expressed in human TH2 cell lines and clones while poorly expressed in TH1 cells (Zingoni *et al.*, 1998) and upregulated following TCR and CD28 stimulation similarly to the CCR4 gene (D'Ambrosio *et al.*, 1998). Moreover, TH2 CD4+ cells express more abundant levels of CCR8 message than TH2 CD8+ cells (Sozzani *et al.*, 1998).

The CCR8 mRNA is detectable as a single 4 kb transcript when northern blot analysis is performed.

Murine CCR8

Murine CCR8 is abundantly expressed in the thymus and, at lower levels, in the spleen, lymph nodes, and lungs with multiple transcripts (about 2, 3, and 4 kb) (Goya *et al.*, 1998; Zingoni *et al.*, 1998). Among different thymocyte subpopulations, CCR8 is expressed by single-positive CD4+ cells and, faintly, by CD4+ CD8+ cells. Moreover, the CCR8 message is present in both cortical and medullary thymocytes as detected by *in situ* hybridization. Both activated mouse TH2 polarized cells and NK1.1+ CD4+ cells abundantly express CCR8 message, while polarized TH1 cells do not (Zingoni *et al.*, 1998).

Among the cell lines tested, the thymoma cell line BW157 shows extremely high levels of CCR8 mRNAs, as well as the T lymphomas EL4 and S49. In contrast, the RW246.7, BAF3, P815, YAC, and WEHI.7 cell lines show no expression of the CCR8 message (Goya *et al.*, 1998; Zingoni *et al.*, 1998).

Regulation of receptor expression

While resting T cells express very low levels of CCR8 message, this receptor, as mentioned above, is very abundantly expressed by TH2 cells and is upregulated following TCR and CD28 engagement (D'Ambrosio *et al.*, 1998) and TCR and IL-2 treatment of polyclonal T cell lines (Sallusto *et al.*, 1999).

SIGNAL TRANSDUCTION

The presence of several serine and threonine residues in the receptor C-terminus suggests that CCR8 signaling, like other seven transmembrane spanning receptors, may be regulated through the phosphorylation of specific residues present in the C-terminus by serine/threonine kinases, whose activity is generally known to induce receptor desensitization. No data are yet available.

BIOLOGICAL CONSEQUENCES OF ACTIVATING OR INHIBITING RECEPTOR AND PATHOPHYSIOLOGY

A common event induced by the interaction of specific chemokine ligands with their cognate receptors is the chemotaxis of target cells. I-309 (Roos *et al.*, 1997; Tiffany *et al.*, 1997; Bernardini *et al.*, 1998; Goya *et al.*, 1998) and vMIP-II (Sozzani *et al.*, 1998) induce chemotaxis of CCR8-transfected cell lines, and of TH2 cells (D'Ambrosio *et al.*, 1998; Sozzani *et al.*, 1998; Zingoni *et al.*, 1998), while vMIP-I induces a chemotactic response in rat Y3 cells (Endres *et al.*, 1999). The chemotactic activity shown by TARC and MIP-1β in the Jurkat cell line (Bernardini *et al.*, 1998) is still not conclusive for the attribution of such molecules as *bona fide* ligands of CCR8 as in other cell lines they showed no activity on CCR8 transfectants.

THERAPEUTIC UTILITY

I-309 may potentially be useful in the control of *HIV* infection due to its inhibitory activity towards CCR8-dependent HIV infection, although it is considered to be a 'minor' coreceptor.

As the I-309/CCR8 ligand/receptor pair associate with a TH2 phenotype, being involved in the activation/migration of this type of T cell, it may play a role in allergic diseases.

Reagents such as neutralizing antibodies and CCR8 antagonists may therefore prove useful in an attempt to block CCR8 activity where it may contribute to pathological conditions.

At present, both vMIP-II and vMCC-I have been reported to act as CCR8 antagonists (Dairaghi *et al.*, 1999; Luttichau *et al.*, 2000), even though vMIP-II activity may be altered in N-terminal sequence variants (Boshoff *et al.*, 1997; Kledal *et al.*, 1997).

References

Albright, A. V., Shieh, J. T., Itoh, T., Lee, B., Pleasure, D., O'Connor, M. J., Doms, R. W., and Gonzalez-Scarano, F. (1999). Microglia express CCR5, CXCR4, and CCR3, but of these, CCR5 is the principal coreceptor for human immunodeficiency virus type 1 dementia isolates. *J. Virol.* **73**, 205–213.

Bernardini, G., Hedrick, J., Sozzani, S., Luini, W., Spinetti, G., Weiss, M., Menon, S., Zlotnik, A., Mantovani, A., Santoni, A., and Napolitano, M. (1998). Identification of the CC chemokines TARC and macrophage inflammatory protein-1 beta as novel functional ligands for the CCR8 receptor. *Eur. J. Immunol.* **2**, 582–588.

Boshoff, C., Endo, Y., Collins, P. D., Takeuchi, Y., Reeves, J. D., Schweickart, V. L., Siani, M. A., Sasaki, T., Williams, T. J., Gray, P. W., Moore, P. S., Chang, Y., and Weiss, R. A. (1997). Angiogenic and HIV-inhibitory functions of KSHV-encoded chemokines. *Science* **278**, 290–294.

Chan, S. Y., Speck, R. F., Power, C., Gaffen, S. L., Chesebro, B., and Goldsmith, M. A. (1999). V3 recombinants indicate a central role for CCR5 as a coreceptor in tissue infection by human immunodeficiency virus type 1. *J. Virol.* **73**, 2350–2358.

Dairaghi, D. J., Fan, R. A., McMaster, B. E., Hanley, M. R., and Schall, T. J. (1999). HHV8-encoded vMIP-I selectively engages chemokine receptor CCR8. Agonist and antagonist profiles of viral chemokines. *J. Biol. Chem.* **274**, 21569–21574.

D'Ambrosio, D., Iellem, A., Bonecchi, R., Mazzeo, D., Sozzani, S., Mantovani, A., and Sinigaglia, F. (1998). Selective up-regulation of chemokine receptors CCR4 and CCR8 upon activation of polarized human type 2 Th cells. *J. Immunol.* **161**, 5111–5115.

Damon, I., Murphy, P. M., and Moss, B. (1998). Broad spectrum chemokine antagonistic activity of a human poxvirus chemokine homolog. *Proc. Natl Acad. Sci. USA* **95**, 6403–6407.

Dittmar, M. T., Zekeng, L., Kaptue, L., Eberle, J., Krausslich, H. G., and Gurtler, L. (1999). Coreceptor requirements of primary HIV type 1 group O isolates from Cameroon. *AIDS Res. Hum. Retroviruses* **15**, 707–712.

Endres, M. J., Garlisi, C. G., Xiao, H., Shan, L., and Hedrick, J. A. (1999). The Kaposi's sarcoma-related herpesvirus (KSHV)-encoded chemokine vMIP-I is a specific agonist for the CC chemokine receptor (CCR)8. *J. Exp. Med.* **189**, 1993–1998.

Garlisi, C. G., Xiao, H., Tian, F., Hedrick, J. A., Billah, M. M., Egan, R. W., and Umland, S. P. (1999). The assignment of chemokine–chemokine receptor pairs: TARC and MIP-1beta are not ligands for human CC-chemokine receptor 8. *Eur. J. Immunol.* **29**, 3210–3215.

Goya, I., Gutierrez, J., Varona, R., Kremer, L., Zaballos, A., and Marquez, G. (1998). Identification of CCR8 as the specific receptor for the human beta-chemokine I-309: cloning and molecular characterization of murine CCR8 as the receptor for TCA-3. *J. Immunol.* **160**, 1975–1981.

Horuk, R., Hesselgesser, J., Zhou, Y., Faulds, D., Halks-Miller, M., Harvey, S., Taub, D., Samson, M., Parmentier, M., Rucker, J., Doranz, B. J., and Doms, R. W. (1998). The CC chemokine I-309 inhibits CCR8-dependent infection by diverse HIV-1 strains. *J. Biol. Chem.* **273**, 386–391.

Jinno, A., Shimizu, N., Soda, Y., Haraguchi, Y., Kitamura, T., and Hoshino, H. (1998). Identification of the chemokine receptor TER1/CCR8 expressed in brain-derived cells and T cells as a new coreceptor for HIV-1 infection. *Biochem. Biophys. Res. Commun.* **243**, 497–502.

Kledal, T. N., Rosenkilde, M. M., Coulin, F., Simmons, G., Johnsen, A. H., Alouani, S., Power, C. A., Luttichau, H. R., Gerstoft, J., Clapham, P. R., Clark-Lewis, I., Wells, T. N. C., and Schwartz, T. W. (1997). A broad-spectrum chemokine antagonist encoded by Kaposi's sarcoma-associated herpesvirus. *Science* **277**, 1656–1659.

Luttichau, H. R., Stine, J., Boesen, T. P., Johnsen, A. H., Chantry, D., Gerstoft, J., and Schwartz, T. W. (2000). A highly selective CC chemokine receptor (CCR)8 antagonist encoded by the poxvirus molluscum contagiosum. *J. Exp. Med.* (in press)

Napolitano, M., Zingoni, A., Bernardini, G., Spinetti, G., Nista, A., Storlazzi, C. T., Rocchi, M., and Santoni, A. (1996). Molecular cloning of TER1, a chemokine receptor-like gene expressed by lymphoid tissues. *J. Immunol.* **157**, 2759–2763.

Roos, R. S., Loetscher, M., Legler, D. F., Clark-Lewis, I., Baggiolini, M., and Moser, B. (1997). Identification of CCR8, the receptor for the human CC chemokine I-309. *J. Biol. Chem.* **272**, 17251–17254.

Rucker, J., Edinger, A. L., Sharron, M., Samson, M., Lee, B., Berson, J. F., Yi, Y., Margulies, B., Collman, R. G., Doranz, B. J., Parmentier, M., and Doms, R. W. (1997). Utilization of chemokine receptors, orphan receptors, and herpesvirus-encoded receptors by diverse human and simian immunodeficiency viruses. *J. Virol.* **71**, 8999–9007.

Sallusto, F., Kremmer, E., Palermo, B., Hoy, A., Ponath, P., Qin, S., Forster, R., Lipp, M., and Lanzavecchia, A. (1999). Switch in chemokine receptor expression upon TCR stimulation reveals novel homing potential for recently activated T cells. *Eur. J. Immunol.* **29**, 2037–2045.

Samson, M., Stordeur, P., Labbe, O., Soularue, P., Vassart, G., and Parmentier, M. (1996). Molecular cloning and chromosomal mapping of a novel human gene, ChemR1, expressed in T lymphocytes and polymorphonuclear cells and encoding a putative chemokine receptor. *Eur. J. Immunol.* **12**, 3021–3028.

Simmons, G., Reeves, J. D., McKnight, A., Dejucq, N., Hibbitts, S., Power, C. A., Aarons, E., Schols, D., De Clercq, E., Proudfoot, A. E., and Clapham, P. R. (1998). CXCR4 as a functional coreceptor for human immunodeficiency virus type 1 infection of primary macrophages. *J. Virol.* **72**, 8453–8457.

Singh, A., Besson, G., Mobasher, A., and Collman, R. G. (1999). Patterns of chemokine receptor fusion cofactor utilization by human immunodeficiency virus type 1 variants from the lungs and blood. *J. Virol.* **73**, 6680–6690.

Sozzani, S., Luini, W., Bianchi, G., Allavena, P., Wells, T. N., Napolitano, M., Bernardini, G., Vecchi, A., D'Ambrosio, D., Mazzeo, D., Sinigaglia, F., Santoni, A., Maggi, E., Romagnani, S., and Mantovani, A. (1998). The viral chemokine macrophage inflammatory protein-II is a selective TH2 chemoattractant. *Blood* **92**, 4036–4039.

Tiffany, H. L., Lautens, L. L., Gao, J. L., Pease, J., Locati, M., Combadiere, C., Modi, W., Bonner, T. I., and Murphy, P. M. (1997). Identification of CCR8: a human monocyte and thymus receptor for the CC chemokine I-309. *J. Exp. Med.* **186**, 165–170.

Van Snick, J., Houssiau, F., Proost, P., Van Damme, J., and Renauld, J. C. (1996). I-309/T cell activation gene-3 chemokine protects murine T cell lymphomas against dexamethasone-induced apoptosis. *J. Immunol.* **157**, 2570–2576.

Yi, Y., Rana, S., Turner, J. D., Gaddis, N., and Collman, R. G. (1998). CXCR-4 is expressed by primary macrophages and supports CCR5-independent infection by dual-tropic but not T-tropic isolates of human immunodeficiency virus type 1. *J. Virol.* **72**, 772–777.

Zaballos, A., Varona, R., Gutierrez, J., Lind, P., and Marquez, G. (1996). Molecular cloning and RNA expression of two new human chemokine receptor-like genes [erratum Biochem. *Biophys. Res. Commun.* (1997) **231**, 519-520]. *Biochem. Biophys. Res. Commun.* **227**, 846–853.

Zhang, Y. J., Dragic, T., Cao, Y., Kostrikis, L., Kwon, D. S., Littman, D. R., KewalRamani, V. N., and Moore, J. P. (1998). Use of coreceptors other than CCR5 by non-syncytium-inducing adult and pediatric isolates of human immunodeficiency virus type 1 is rare *in vitro*. *J. Virol.* **72**, 9337–9344.

Zingoni, A., Soto, H., Hedrick, J. A., Stoppacciaro, A., Storlazzi, C. T., Sinigaglia, F., D'Ambrosio, D., O'Garra, A., Robinson, D., Rocchi, M., Santoni, A., Zlotnik, A., and Napolitano, M. (1998). The chemokine receptor CCR8 is preferentially expressed in TH2 but not TH1 cells. *J. Immunol.* **2**, 547–551.

ACKNOWLEDGEMENTS

We are grateful to Dr A. Zingoni, Dr G. Bernardini, and Dr G. Spinetti, and to Dr A. Zlotnik's and Dr A. Mantovani's laboratories for their important contributions to our studies.

D6

Gerry Graham* **and Rob Nibbs**

Cancer Research Campaign Laboratories, The Beaton Institute for Cancer Research, Garscube Estate Switchback Road, Bearsdon, Glasgow G61 1BD, UK

*corresponding author tel: 44-141-330-3982, fax: 44-141-942-6521, e-mail: g.graham@beatson.gla.ac.uk

DOI: 10.1006/rwcy.2000.22009.

SUMMARY

Members of the chemokine family of proinflammatory mediators interact with target cells by binding to members of the seven transmembrane spanning family of G protein-coupled receptors. We have been interested in identifying chemokine receptors involved in the regulation of hematopoietic stem cell proliferation and during these studies have cloned a novel chemokine receptor which is called D6. D6 is a β chemokine-specific receptor but is highly promiscuous within this context and binds most members of this subfamily. Curiously, it does not seem to signal in the way that other chemokine receptors do and thus it has not yet been given a systematic name. D6 is expressed in the placenta and in the lung and liver but is detectable at low levels in most other tissues. Unlike many other chemokine receptors, it is not highly expressed in hematopoietic cells and thus its major domain of function may not be in this system.

BACKGROUND

Chemokines are members of a large family of peptides that are involved in diverse processes such as regulation of inflammatory processes and control of cellular proliferation (Rollins, 1997). The chemokine family is defined on the basis of sequence homology but is more precisely defined by the presence of variations of a conserved cysteine motif. The two most populous chemokine subfamilies are the α and β subfamilies. Peptides in either of these families generally have four cysteine residues in the mature protein. In the α chemokines the first two cysteines are separated by an amino acid of variable identity, whereas in the β chemokines the first two cysteines are juxtaposed. In addition to these two large families, there are two smaller subfamilies that have single members referred to as lymphotactin and fractalkine/neurotactin.

All chemokines characterized thus far interact with cells through members of the seven transmembrane spanning heptahelical receptor family (Murphy, 1996) that typically support a calcium flux following ligand binding. A large number of chemokine receptors have been characterized so far and, in general, whilst they commonly display a marked promiscuity in terms of ligand binding, they are typically faithful to a single family. Thus, β chemokine receptors will in general bind only β chemokines and α chemokine receptors will bind only α chemokines. In this article we describe the identification and preliminary characterization of a novel β chemokine receptor that displays atypical expression patterns and which is highly promiscuous but predictably faithful to β chemokines. To date, we have been unable to demonstrate a signaling role for this novel receptor.

Discovery

We have been interested for some time in the role of β chemokines, most notably MIP-1α, as hematopoietic stem cell proliferation inhibitors (Graham *et al.*, 1990; Graham, 1997; Graham and Nibbs, 1999). For a number of reasons, including the possible implication of MIP-1α-inhibitory defects in the processes of *leukemogenesis* (Graham, 1997; Graham and Nibbs, 1999), we have been pursuing a number of strategies aimed at identifying cell surface receptors for MIP-1α. Our most successful strategy has involved degenerate genomic PCR using primers designed against conserved regions of the chemokine receptor family. Using this approach we identified a number of murine receptors for MIP-1α, some of which have already been described but one of which, clone name D6, turned out to be novel (Nibbs *et al.*, 1997a, 1997b).

Alternative names

Initially, given the ability of D6 to bind β chemokines, this receptor was designated CCR9 in accordance with the existing systematic nomenclature system for chemokine receptors. Shortly after we were given the CCR9 designation for this molecule, a second group reported the cloning of this receptor (Bonini *et al.*, 1997) and claimed a CCR10 nomenclature for it. The sequences for D6 presented in our publications and in that of Bonini *et al.* are essentially identical (Nibbs *et al.*, 1997a, 1997b; Bonini *et al.*, 1997), with the exception of three amino acids that may be accounted for by subtle allelic variations or alternatively by sequencing errors. The confusion in nomenclature has been further compounded by the fact that, despite extensive efforts, we have been unable to demonstrate a signaling or functional role for D6 in standard calcium flux or chemoattractant assays used to gauge the function of chemokine receptors. This lack of evidence of signaling capacity means that D6 is not entitled to a systematic nomenclature (signaling is a prerequisite for CCR designation) and thus it is now correctly referred to not as CCR9 or CCR10 but simply as D6. This therefore discriminates it from the recently described TECK receptor (GPR9-6) which does support calcium fluxing following ligand binding and which has been given a CCR9 nomenclature (Zaballos *et al.*, 1999). We are currently examining the ability of D6 to signal through alternative pathways following ligand binding and hope in the near future to clarify the signaling potential for D6 and if appropriate to assign a systematic name to it.

Structure

D6 is a member of the seven transmembrane spanning family of receptors and as such is typical of other members of the chemokine receptor family. The murine D6 protein consists of 378 amino acids and the human protein of 384 amino acids. Both the murine and human proteins contain four cysteine residues that are conserved in other chemokine receptors and both also carry a potential *N*-linked glycosylation site at the N-terminus.

Main activities and pathophysiological roles

No data known on activities. This receptor is a coreceptor, with CD4, for the T-tropic HIV-1 isolate, UG21 and additionally displays coreceptor activity towards a number of other HIV-1 isolates (Choe *et al.*, 1998). As such, D6 may have a limited role to play in the pathogenesis of *HIV*. In this pathological context, however, it is at best a weak player and the primary cellular coreceptors for HIV-1 remain CCR5 and CXCR4 (Clapham, 1997).

GENE

Accession numbers

Murine cDNA: Y12879
Rat cDNA: U92803
Human cDNA: Y12815, U94888

Additionally there are two human ESTs covering regions of D6 and accession numbers for these are R82383 and AI628851.

Sequence

See **Figure 1**.

Chromosome location and linkages

Human D6 is located on chromosome 3 in a region coincident with the CC chemokine receptor locus at 3p21.31-3p21.32 (Bonini *et al.*, 1997).

PROTEIN

Accession numbers

Rat D6: AAB61572
Murine D6: CAA73379
Human D6: CAA73346, AAB97728.

Sequence

See **Figure 2** for the human D6 protein sequence.

Figure 1 Nucleotide sequence for the D6 gene.

```
   1 GGATCCTCCA ACATGGCCGC CACTGCCTCT CCGCAGCCAC TCGCCACTGA GGATGCCGAT
  61 TCTGAGAATA GCAGCTTCTA TTACTATGAC TACCTGGATG AAGTGGCCTT CATGCTCTGC
 121 AGGAAGGATG CAGTGGTGTC CTTTGGCAAA GTCTTCCTCC CAGTCTTCTA TAGCCTGATT
 181 TTTGTGTTGG GCCTCAGCGG GAACCTCCTT CTTCTCATGG TCTTGCTCCG TTACGTGCCT
 241 CGCAGGCGGA TGGTTGAGAT CTATCTGCTG AATCTGGCCA TCTCCAACCT TCTGTTTCTG
 301 GTGACACTGC CCTTCTGGGG CATCTCCGTG GCCTGGCATT GGGTCTTCGG GAGTTTCTTG
 361 TGCAAGATGG TGAGCACTCT TTATACTATT AACTTTTACA GTGGCATCTT TTTCATTAGC
 421 TGCATGAGCC TGGACAAGTA CCTGGAGATC GTTCATGCTC AGCCCTACCA CAGGCTGAGG
 481 ACCCGGGCCA AGAGCCTGCT CCTTGCTACC ATAGTATGGG CTGTGTCCCT GGCCGTCTCC
 541 ATCCCTGATA TGGTCTTTGT ACAGACACAT GAAAATCCCA AGGGTGTGTG GAACTGCCAC
 601 GCAGATTTCG GCGGGCATGG GACCATTTGG AAGCTCTTCC TCCGCTTCCA GCAGAACCTC
 661 CTAGGGTTTC TCCTTCCACT CCTTGCCATG ATCTTCTTCT ACTCCCGTAT TGGTTGTGTC
 721 TTGGTGAGGC TGAGGCCCGC AGGCCAGGGC CGGGCTTTAA AAATAGCTGC AGCCTTGGTG
 781 GTGGCCTTCT TCGTGCTATG GTTCCCATAC AATCTCACCT TGTTTCTGCA TACGCTGTTG
 841 GACCTGCAAG TATTCGGGAA CTGTGAGGTC AGCCAGCATC TAGACTACGC ACTCCAGGTA
 901 ACAGAGAGCA TCGCCTTCCT TCACTGCTGC TTTTCCCCCA TCCTGTATGC CTTCTCCAGT
 961 CACCGCTTCC GCCAGTACCT GAAGGCTTTC CTGGCTGCCG TGCTTGGATG GCACCTGGCA
1021 CCTGGCACTG CCCAGGCCTC ATTATCCAGC TGTTCTGAGA GCAGCATACT TACTGCCCAA
1081 GAGGAAATGA CTGGCATGAA TGACCTTGGA GAGAGGCAGT CTGAGAACTA CCCTAACAAG
1141 GAGGATGTGG GGAATAAATC AGCCTGAGTG ACCGCGGCCG C
```

Figure 2 Amino acid sequence for the D6 protein.

```
  1 MAATASPQPL ATEDADSENS SFYYYDYLDE VAFMLCRKDA VVSFGKVFLP VFYSLIFVLG
 61 LSGNLLLLMV LLRYVPRRRM VEIYLLNLAI SNLLFLVTLP FWGISVAWHW VFGSFLCKMV
121 STLYTINFYS GIFFISCMSL DKYLEIVHAQ PYHRLRTRAK SLLLATIVWA VSLAVSIPDM
181 VFVQTHENPK GVWNCHADFG GHGTIWKLFL RFQQNLLGFL LPLLAMIFFY SRIGCVLVRL
241 RPAGQGRALK IAAALVVAFF VLWFPYNLTL FLHTLLDLQV FGNCEVSQHL DYALQVTESI
301 AFLHCCFSPI LYAFSSHRFR QYLKAFLAAV LGWHLAPGTA QASLSSCSES SILTAQEEMT
361 GMNDLGERQS ENYPNKEDVG NKSA
```

Description of protein

The D6 protein describes a seven transmembrane spanning receptor with the characteristic extracellular N-terminus, intracellular C-terminus, and three extracellular loops. The murine D6 protein comprises 378 amino acids and the human protein 384 amino acids and in addition both contain four conserved cysteine residues that are believed to be involved in the maintenance of chemokine receptor structure. The presumed extracellular N-terminus stretches from residue 1 to 46 and in common with other chemokine receptors is highly acidic, although the first 13 amino acids are predicted to form a hydrophobic domain that is not seen in the related chemokine receptors. Additionally, the N-terminus bears a consensus site for *N*-linked glycosylation (Asn/Ser/Ser) and additionally appears to be sulfated on tyrosine residues – a decoration that is believed to be essential for its HIV coreceptor activity (Farzan *et al.*, 1999). Whilst most chemokine receptors, indeed most seven transmembrane spanning G protein-coupled receptors, possess a conserved aspartate residue in the second transmembrane region (Savares and Fraser, 1992), this is altered to an asparagine in D6.

An additional alteration in D6 compared to other chemokine receptors is that, whilst most signaling chemokine receptors possess a DRYLAIV motif on the second intracellular loop (Murphy, 1996) this is altered to DKYLEIV in D6. This alteration appears not to be random and is conserved in the rat, murine, and human receptors. It is this alteration in the conserved aspartate and the DRYLAIV motif that is likely to be the basis for the inability of D6 to flux calcium following ligand binding. Indeed, we have demonstrated that introduction of the DRYLAIV motif in D6 allows it to flux calcium following ligand binding, indicating that there is nothing radically different about the structure of D6 that precluded its support of calcium fluxing following ligand binding. The C-terminus of D6 bears many potential phosphorylation sites.

Relevant homologies and species differences

D6 is homologous to a number of members of the chemokine receptor family; however it is clearly quite divergent. Computerized packages assessing phylogenetic relationships between D6 and other chemokine receptors place it between the α and β chemokine receptor with the closest homology (40%) being seen with CCR4. Interestingly, comparison of the extracellular N-terminus of D6 shows it to be most similar to that of CCR1.

Affinity for ligand(s)

Whilst little is known as yet on the precise functional roles for this receptor, we have extensive data documenting the ligand-binding profiles of D6. D6 is a β chemokine-specific receptor and will not bind members of the α family, nor will it bind *lymphotactin* or *fractalkine*, the sole members of the other two smaller chemokine subfamilies. Despite this selectivity for β chemokines, D6 is a very promiscuous receptor and will bind most β chemokines, with some notable exceptions that are described below. This receptor is a very high-affinity MIP-1α receptor, with the murine receptor binding murine MIP-1α with a K_d of 110 pM, making it the highest-affinity murine MIP-1α receptor described to date, and the human receptor-binding murine MIP-1α with a K_d of 920 pM. In addition to MIP-1α, human D6 will bind β chemokines, as shown in **Table 1**.

Thus, D6 is highly promiscuous and binds most members of the β chemokine subfamily. Notable exceptions are C10 and I-309 and, perhaps most curiously, human MIP-1α. Whilst murine D6 is the highest-affinity murine MIP-1α receptor described to date, it does not appear to bind the highly homologous human MIP-1α. Equally, this ability to discriminate between murine and human MIP-1α is seen in human D6 which again binds murine MIP-1α with high affinity but binds human MIP-1α with a relatively lower affinity.

We have recently attempted to examine the structural basis for this discriminatory ability and have demonstrated that a proline residue in position 2 of the β chemokines is a prerequisite for high-affinity binding to human and murine D6 (Nibbs *et al.*, 1999). Thus, all the ligands for this receptor and indeed, most β chemokines, possess a proline 2 residue and thus bind well. It is important to note that a proline residue in position 2 of the mature protein is not sufficient in itself for D6 binding as the α chemokine SDF-1, which has a proline 2 will not bind to D6. Additionally, a novel β chemokine which we have identified (ESkine) also has a proline residue in position 2 but again will not bind to D6 (Baird *et al.*, 1999). This however is the exception to the general rule.

The relevance of the requirement for a proline residue in position 2 to the lack of binding of human MIP-1α to murine and human D6 lies in the fact that,

Table 1 β chemokines with which human D6 will bind

	Human	Murine
MIP-1α	64 nM	920 pM
MIP-1β	1.7 nM	755 pM
RANTES	3.6 nM	Not determined
MCP-1	16.5 nM	613 nM
MCP-2	768 pM	Not determined
MCP-3	1.2 nM	Not determined
MCP-4	5.97 nM	Not determined
MCP-5	Not determined	6.3 nM
Eotaxin	46 nM	30 nM
HCC-1	27.2 nM	Not determined
C10	Not determined	Undetectable
I-309	Undetectable	Not determined
ESkine	Not determined	Undetectable

unlike its murine counterpart, human MIP-1α carries a serine residue in position 2. There exists a natural nonallelic variant of human MIP-1α in the genome (Nakao *et al.*, 1990) previously referred to as LD78 β (the serine version of human MIP-1α is referred to as LD78 α) and we have demonstrated that this proline 2-bearing human MIP-1α isoform binds with high affinity to D6 and is therefore more like the murine protein in this and other contexts and may be the more representative human homolog of murine MIP-1α. We have proposed renaming LD78α and β as MIP-1αS and MIP-1αP to highlight the importance of this subtle change (Nibbs *et al.*, 1999).

In terms of *in vitro* activities, as mentioned above, we have been unable to identify a role for D6. Unlike most other chemokine receptors, this receptor will not support either calcium fluxing or chemoattraction following ligand binding when expressed in heterologous cells. D6 therefore appears not to be a classical chemoattractant chemokine receptor and further work needs to be done to establish its biological relevance.

Cell types and tissues expressing the receptor

In the mouse, D6 is detectable on tissue blots at the level of northern blotting in liver, lung, spleen, and placenta. In addition, it is weakly expressed in murine hematopoietic cells and is detectable by PCR in murine dendritic cell lines (XS52 cells).

The human mRNA is very weakly expressed in most tissues examined; however it is robustly expressed in the placenta and is also easily detectable in the liver and lung. Expression in the human spleen is not detected at the level of northern blotting, perhaps reflecting the differences in cellular composition and function of the murine and human spleens. This receptor is not strongly expressed in human blood cells; however it can be detected by PCR in the primitive myeloerythroid cell line K562 and may indeed be present on the surface of these cells (Graham *et al.*, 1993). Additionally, it is detectable in THP1 cells and in human umbilical cord blood cells. This expression pattern suggests that the primary domains of expression of D6 are not hematopoietic and that it may function in different contexts to those normally associated with chemokine receptors. This again may be reflected in the presumed alternative signaling pathways used by this receptor.

Within the tissues mentioned above, we have few data on the specific cellular expression patterns of D6. We have recently obtained a monoclonal antibody to the human D6 protein and are in the process of using this to define cell types within the placenta, liver, lung, and other tissues expressing this receptor.

Regulation of receptor expression

No data known: however, preliminary analysis of a murine genomic clone reveals progesterone receptor-binding sites, cytokine response elements, cAMP response elements and glucocorticoid response elements. The relevance of these regulatory sites to transcriptional control remains to be determined.

Release of soluble receptors

No data known, although precedent from within the chemokine receptor family would suggest that soluble receptor release would be unlikely.

SIGNAL TRANSDUCTION

Cytoplasmic signaling cascades

No data known, apart from the observations that, unlike most other chemokine receptors, D6 does not support calcium fluxing in heterologous transfectants following ligand binding.

BIOLOGICAL CONSEQUENCES OF ACTIVATING OR INHIBITING RECEPTOR AND PATHOPHYSIOLOGY

Phenotypes of receptor knockouts and receptor overexpression mice

Null mice have recently been generated and have no overt phenotypes. These mice are viable, healthy, and fertile. No data are yet available on D6 transgenic mice.

THERAPEUTIC UTILITY

Effect of treatment with soluble receptor domain

No data known. Probably not relevant to chemokine receptors.

References

Baird, J. W., Nibbs, R. J. B., Komai-Koma, M., Connolly, J. A., Ottersbach, K., Clark-Lewis, I., Liew, F. Y., and Graham, G. J. (1999). ESkine, a novel β chemokine, is differentially spliced to produce secretable and nuclear targeted isoforms. *J. Biol. Chem.* **274**, 33496–33503.

Bonini, J. A., Martin, S. K., Drayluk, F., Roe, M. W., Philipson, L. H., and Steiner, D. F. (1997). Cloning, expression and chromosomal mapping of a novel human CC-chemokine receptor (CCR10) that displays high affinity binding for MCP-1 and MCP-3. *DNA Cell Biol.* **16**, 1249–1256.

Choe, H., Farzan, M., Konkel, M., Martin, K., Sun, Y., Marcon, L., Cayabyab, M., Berman, M., Dorf, M. E., Gerard, N., Gerard, C., and Sodroski, J. (1998). The orphan seven transmembrane receptor Apj supports the entry of primary T-cell-line-tropic and dual tropic human immunodeficiency virus type I. *J. Virol.* **72**, 6113–6118.

Clapham, P. R. (1997). HIV and chemokines: ligands sharing cell surface receptors. *Trends Cell Biol.* **7**, 264–268.

Farzan, M., Mirzabekov, T., Kolchinsky, P., Wyatt, R., Cayabyab, M., Gerard, N. P., Gerard, C., Sodroski, J., and Choe, H. (1999). Tyrosine sulfation of the amino terminus of CCR5 facilitates HIV-1 entry. *Cell* **96**, 667–676.

Graham, G. J. (1997). In "Baillière's Clinical Haematology, Vol.10, Molecular Haemopoiesis" (ed. A.D. Whetton), pp. 539–559. Baillière, London.

Graham G. J., and Nibbs R. J. B. (1999). In "Chemokines and Cancer" (ed. B.J. Rollins), MIP-1α and stem cell inhibition, pp. 293–310. Humana Press, New Jersey.

Graham, G. J., Wright, E. G., Hewick, R., Wolpe, S. D., Wilkie, N. M., Donaldson, D., Lorimore, S., and Pragnell, I. B. (1990). Identification and characterisation of an inhibitor of haemopoietic stem cell proliferation. *Nature* **344**, 442–444.

Graham, G. J., Zhou, L., Weatherbee, J. A., Tsang, M. L-K., Napolitano, M., Leonard, W. J., and Pragnell, I. B. (1993). Characterisation of a receptor for MIP-1α and related proteins on human and murine cells. *Cell Growth Different.* **4**, 137–146.

Murphy, P. M. (1996). Chemokine receptors: structure, function and role in microbial pathogenesis. *Cytokine Growth Factor Rev.* **7**, 47–64.

Nakao, M., Nomiyama, H., and Shimada, K. (1990). Structures of the human genes coding for cytokine LD78 and their expression. *Mol. Cell. Biol.* **10**, 3646–3652.

Nibbs, R. J. B., Lowe, S., Pragnell, I. B., and Graham, G. J. (1997a). Cloning and characterisation of a novel murine β chemokine receptor, D6: comparison to three other related macrophage inflammatory protein-1α receptors, CCR-1, CCR-3 and CCR-5. *J. Biol. Chem.* **272**, 12495–12504.

Nibbs, R. J. B., Wylie, S. M., Yang, J., Landau, N. L., and Graham, G. J. (1997b). Cloning and characterisation of a novel promiscuous human β-chemokine receptor D6. *J. Biol. Chem.* **272**, 32078–32083.

Nibbs, R. J. B., Yang, J., Landau, N. R., and Graham G. J. (1999). LD78β, a non allelic variant of human MIP-1α (LD78α) has enhanced receptor interactions and potent HIV suppressive activity. *J. Biol. Chem.* **274**, 17478–17483.

Rollins, B. J. (1997). Chemokines. *Blood* **90**, 909–928.

Savares, T. M., and Fraser, C. M. (1992). *In vitro* mutagenesis and the search for structure-function relationships among G protein-coupled receptors. *Biochem. J.* **283**, 1–19.

Zaballos, A., Guitierrez, J., Varona, R., Ardavin, C., and Marquez, G. (1999). Identification of the orphan chemokine receptor GPR-9-6 as CCR9, the receptor for the chemokine TECK. *J. Immunol.* **162**, 5671–5675.

ECRF3

Sunil K. Ahuja*

Department of Medicine, University of Texas Health Science Center at San Antonio, 7703 Floyd Curl Drive, San Antonio, TX 78229-3900, USA

*corresponding author tel: 210-567-6511, fax: 210-567-4654, e-mail: ahujas@uthscsa.edu

DOI: 10.1006/rwcy.2000.22011.

SUMMARY

ECRF3 is a virally encoded chemokine receptor found in the genome of herpesvirus saimiri (HVS), a primate-restricted T-lymphotropic γ-herpesvirus that is closely related to *Epstein–Barr virus* (EBV), a human B-lymphotropic γ-herpesvirus. ECRF3 is one of 14 open reading frames (ORFs) of HVS that lack homologs in EBV, and that have sequence homology with known cellular proteins. ECRF3 is 30% identical in deduced amino acid sequence to human CXCR2, its closest mammalian relative. Similar to CXCR2, the ligands for ECRF3 are IL-8, GROα, and NAP-2.

BACKGROUND

Discovery

In 1992 Nicholas and coworkers showed that the genome of HSV encoded an open reading frame with homology to the G protein-coupled receptor family of proteins (Nicholas *et al.*, 1992a, 1992b). The highest degree was with the chemokine receptor then designated as IL-8RB and now referred to as CXCR2. In 1993, Ahuja and Murphy demonstrated that ECRF3 is a functional chemokine receptor for IL-8, GROα, and NAP-2 (Ahuja and Murphy, 1993, 1999). This report described the first functional characterization of a virally encoded seven transmembrane domain receptor.

Alternative names

ECRF3 is also known as ORF74.

Structure

ECRF3's three-dimensional structure is not available. It is 30% identical in deduced amino acid sequence to the human G protein-coupled receptor CXCR2, its closest mammalian relative. These receptors share a common putative structural topology composed of seven transmembrane domains separated by three intracellular loops. Notable structural features in ECRF3 are as follows. First, despite relatively low sequence relatedness of CXCR2 and ECRF3, there is a high degree of sequence similarity in the N-terminal domains of these two receptors. Also, in each case the NH_2-domain is highly acidic. Second, the sequence Asp-Arg-Tyr (DRY) is highly conserved in the proposed second intracellular loop of seven transmembrane domain receptors (STRs). In ECRF3, the corresponding sequence is Leu-Arg-Cys (LRC). Third, all mammalian chemoattractant peptide receptors, as well as ECRF3, possess a highly cationic 16 amino acid third intracellular loop that is highly variable in sequence. This shared motif could mediate coupling to a similar, if not identical, G protein. Fourth, in most of the STRs, three residues are highly conserved: the arginine in the DRY motif, a cysteine in the second extracellular loop, and a tryptophan in transmembrane domain IV. The first two of these are conserved in ECRF3, whereas the tryptophan has diverged. An asparagine in transmembrane domain VII that is highly conserved among all STRs is not found in ECRF3. Fifth, the CXCR2 receptor possesses two potential sites for glycosylation in the proposed second extracellular loop, whereas ECRF3 has none. Sixth, the CXCR2 cytoplasmic C-terminal segment is rich in serine and threonine residues that could be phosphorylation sites for cellular kinases to regulate receptor function. The corresponding region of the ECRF3 has only one serine and one threonine residue.

Main activities and pathophysiological roles

Ex vivo, frog oocytes microinjected with cRNA made from cloned ECRF3 DNA acquire the ability to respond to extracellular application of the human CXC chemokines IL-8, GROα, and NAP-2, the same chemokines that bind with high affinity to human CXCR2 (Ahuja *et al.*, 1996; Ahuja and Murphy, 1993). The potency order differs for the two receptors in oocytes: for human CXCR2 it is IL-8 > GROα = NAP-2; for ECRF3 it is GROα > NAP-2 > IL-8. Oocytes expressing ECRF3 are ~200-fold more sensitive to GROα than oocytes expressing human CXCR2 when calcium release is measured. The *in vivo* activity of ECRF3 in the context of infection of T lymphocytes with HVS is not known. Nevertheless, consideration of possible functions must take into account that ECRF3 might exploit chemokine-dependent signaling pathways to ensure a cytosolic milieu that has been optimally conditioned for viral replication or for the establishment of latency (Ahuja and Murphy, 1999).

ECRF3 could also be related, perhaps by mediating a chemokine-dependent break in T cell tolerance, with a lymphoproliferative disorder in unnatural hosts. However, at present, the biologic function of ECRF3 is not known, the stage in the viral life cycle where it is expressed has not been defined, nor has evidence of expression of the native protein in infected cells or on virions been verified.

GENE

Accession numbers

GenBank:
S76368, X64346.

Sequence

The genome of HSV is composed of ~112,930 bp (Nicholas *et al.*, 1992a, 1992b). ECRF3 resides in the right terminal region (conventional orientation) of the unique protein-coding component (L-DNA) of the HVS genome. Within this region lie the genes encoding the 160 kDa virion protein, which is homologous to the 140 kDa membrane antigen of EBV, thymidylate synthase, and the immediate early (IE) 52 kDa protein which is homologous to the EBV BMLF1 product. The ECRF3 gene of HVS resides within a group of five genes that have no homologs in EBV. The sequence of ECRF3 cloned by Ahuja and Murphy differs from the published sequence of ECRF3 at position 38963 (C → T; Accession number X64346) that results in substitution of amino acid 180 from serine to phenylalanine (Ahuja and Murphy, 1993).

PROTEIN

Accession numbers

As above.

Description of protein

The deduced protein sequence of ECRF3 has 321 amino acids, with a predicted relative molecular mass of 37,100.

Relevant homologies and species differences

The closest homolog of ECRF3 is human CXCR2 (~30%).

Affinity for ligand(s)

Xenopus oocytes were injected with ECRF3 cRNA, and calcium mobilization in response to a panel of chemokines was determined (Ahuja and Murphy, 1993). In this assay, oocytes injected with ECRF3 cRNA responded to IL-8, GROα, and NAP-2. The rank order of potency (mean effective concentration) of chemokines for the ECRF3 product was GROα (0.5 nM) > NAP-2 (10 nM) > IL-8 (50 nM).

Cell types and tissues expressing the receptor

Not reported.

SIGNAL TRANSDUCTION

Cytoplasmic signaling cascades

Although the ability of ECRF3 to mediate signal transduction induced by chemokines was demonstrated in

ECRF3-transfected frog oocytes (see above), the precise components involved in ECRF3-mediated signal transduction are unknown.

DOWNSTREAM GENE ACTIVATION

Transcription factors activated

Not reported.

BIOLOGICAL CONSEQUENCES OF ACTIVATING OR INHIBITING RECEPTOR AND PATHOPHYSIOLOGY

The possible role of ECRF3 in viral infection remains unknown, and there are no reports about the effect of knocking out ECRF3 on *herpesvirus saimiri* pathogenesis. However, taking into account the retention by ECRF3 of chemokine-dependent signaling and its ability to bind many CXC chemokines, one could speculate on some conditions dependent on ECRF3 that could favor HSV persistence (Ahuja and Murphy, 1999). HSV is capable of inducing oncogenic transformation of T lymphocytes of New World primates and immortalizing human cells *in vitro*. This process appears to be mediated by viral homologs of mammalian proteins such as Bcl-2, or signaling through T cell pathways such as p56lck. It is conceivable that ECRF3 could contribute to T cell transformation by sensitizing T cells to CXC chemokines to regulate proliferation of virally infected cells. In this scenario, a role for CXCR2 in hematopoiesis and transformation has been shown. Mice with targeted disruption of CXCR2, the ECRF3 homolog, have massively expanded neutrophil and B cell compartments, suggesting that the mouse ligands for this receptor may be physiologic regulators of hematopoiesis (Cacalano *et al.*, 1994). On the other hand, a point mutation causing constitutive signaling of CXCR2 abrogates normal growth control mechanisms and leads to transforming activity in NIH 3T3 cells, similar to Kaposi's sarcoma herpesvirus G protein-coupled receptor (KSHV GPCR) (Burger *et al.*, 1999). Although CXCR2 is expressed at high levels mainly in circulating neutrophils, it is also expressed in T lymphocytes but at much lower levels and in only a small percentage of cells. It is therefore possible that HVS probably copied CXCR2 to acquire or adapt its lymphocyte-specific functions such as chemoattraction for T cells *in vitro* and *in vivo*.

THERAPEUTIC UTILITY

Not reported.

References

Ahuja, S. K., and Murphy, P. M. (1993). Molecular piracy of mammalian interleukin-8 receptor type B by herpesvirus saimiri. *J. Biol. Chem.* **268**, 20691–20694.

Ahuja, S. K., and Murphy, P. M. (1999). In "Chemokines in Disease: Biology and Clinical Research" (ed. C.A. Hebert), Viral mimicry of chemokines and chemokine receptors, pp. 235–251. Humana Press, Totowa, NJ..

Ahuja, S. K., Lee, J. C., and Murphy, P. M. (1996). CXC chemokines bind to unique sets of selectivity determinants that can function independently and are broadly distributed on multiple domains of human interleukin-8 receptor B. Determinants of high affinity binding and receptor activation are distinct. *J. Biol. Chem.* **271**, 225–232.

Burger, M., Burger, J. A., Hoch, R. C., Oades, Z., Takamori, H., and Schraufstatter, I. U. (1999). Point mutation causing constitutive signaling of CXCR2 leads to transforming activity similar to Kaposi's sarcoma herpesvirus-G protein-coupled receptor. *J. Immunol.* **163**, 2017–2022.

Cacalano, G., Lee, J., Kikly, K., Ryan, A. M., Pitts-Meek, S., Hultgren, B., Wood, W. I., and Moore, M. W. (1994). Neutrophil and B cell expansion in mice that lack the murine IL-8 receptor homolog [published erratum appears in Science 1995 Oct 20, 270 (5235): 365]. *Science* **265**, 682–684.

Nicholas, J., Cameron, K. R., and Honess, R. W. (1992a). Herpesvirus saimiri encodes homologues of G protein-coupled receptors and cyclins. *Nature* **355**, 362–365.

Nicholas, J., Cameron, K. R., Coleman, H., Newman, C., and Honess, R. W. (1992b). Analysis of nucleotide sequence of the rightmost 43 kbp of herpesvirus saimiri (HVS) L-DNA: general conservation of genetic organization between HVS and Epstein–Barr virus. *Virology* **188**, 296–310.

Poxvirus Membrane-bound G Protein-coupled Receptor Homologs

Grant McFadden[1,*] and Richard Moyer[2]

[1]The John P. Robarts Research Institute and Department of Microbiology and Immunology, The University of Western Ontario, 1400 Western Road, London, Ontario, N6G 2V4, Canada

[2]Department of Molecular Genetics and Microbiology, University of Florida College of Medicine, PO Box 100266, Gainesville, FL 32610-0266, USA

*corresponding author tel: (519) 663-3184, fax: (519) 663-3847, e-mail: mcfadden@rri.on.ca

DOI: 10.1006/rwcy.2000.14017.

SUMMARY

To date, virus-encoded homologs of G protein-coupled receptors (GPCRs) have been discovered only in members of the poxvirus and herpesvirus families. The herpesvirus members have been more extensively characterized than the poxvirus members, but all were initially discovered by database searches of predicted open reading frame sequences deduced during DNA sequencing studies of the various viral DNA genomes. Unlike the herpesvirus examples (e.g. Kaposi's sarcoma-associated GPCR), the poxvirus GPRC homologs have not yet been analyzed for biological activity.

BACKGROUND

Discovery

Chemokines are major regulatory components of the immune system, being involved in trafficking, localization, and activation of various leukocyte populations. Collectively, chemokines can be classified according to the relative positions of the first pair of a conserved motif of cysteine residues. The designations are C, CC, CXC, and CX3C, where 'X' refers to the number of intervening amino acids between the first two cysteines. Both the classification of chemokines and their receptors have been reviewed (Oppenheim *et al.*, 1991; Schall and Bacon, 1994; Murphy, 1994; Barker and Monk, 1997). Chemokine action is most frequently mediated through interaction with appropriate G protein-coupled receptors (GPCRs). While GPCRs have not been detected to date in the commonly studied members of the *Orthopoxvirus* genus such as vaccinia, ectromelia, or cowpox virus, sequence analysis has revealed potential chemokine GPCRs within the genomes of swinepox virus (the prototypic member of the *Suipoxvirus* genus) (Massung *et al.*, 1993) and sheep pox virus, a member of the *Capripoxvirus* genus (Cao *et al.*, 1995).

Poxvirus genomes are linear terminating with regions comprising inverted terminal repetitions (ITRs). Genes within the terminal repetitions are diploid. Swinepox virus and sheep pox ORF designations are based on genomic *Hin*dIII restriction patterns where the largest fragment is designated as 'A'. An analysis of swinepox virus sequence reported two related ORFs, one being a truncated ORF (C3L) at the junction of the left inverted ITR and the left-most unique region of the genome. A full-length version of the same open reading frame (ORF K2R) was found at the junction of the right ITR and the right-most unique region of the genome. Most likely, the incomplete C3L ORF represents an artifact derived from a variant virus, grown in cell culture, where in the absence of any selection, a partial deletion of the gene, mirrored in both copies of the ORF has occurred.

The sheep pox ORF (Cao *et al.*, 1995) is located within the *Hin*dIII Q2 fragment of the KC-1 strain and corresponds to ORF 3L (Gershon and Black, 1987). Like the swinepox virus ORF, the Q2 fragment of the KC-1 strain is located near the right-most terminal extreme of the sheep pox virus genome, a location typical of nonessential genes devoted to controlling the host response to the infection. While structural features of the two ORFs discussed here are consistent with these two encoded proteins functioning as GPCRs, functionality has not yet been demonstrated.

Structure

The poxvirus GPCR homologs share all the features typical of both cellular and viral GPCRs which are depicted graphically in Figure 1. These features include: (1) an extracellular N-terminus, (2) an intracellular C-terminus, (3) seven α-helical transmembrane domains, which are oriented perpendicularly to the plasma membrane, (4) three intracellular and three extracellular hydrophilic connecting loops, (5) a disulfide bond linking cysteine residues in the first and second extracellular loops, and (6) the presence of proline residues in transmembrane domains II, IV, V, VI, and VII.

Main activities and pathophysiological roles

Clearly, the encoding of both chemokines and chemokine receptors by viruses indicates the importance of modulating chemokine activity during certain viral infections. Indeed, in the case of Epstein–Barr virus, which does not encode a GPCR, the virus induces synthesis of a cellular GPCR during infection (Birkenbach *et al.*, 1993; Schweickart *et al.*, 1994), presumably to accomplish a similar purpose.

Viral-encoded GPCRs are relatively prevalent amongst the herpesvirus, examples being found in *human herpesvirus 8* (HHV8) (Guo *et al.*, 1997), herpesvirus saimiri (Nicholas *et al.*, 1992; Ahuja and Murphy, 1993), human *cytomegalovirus* (US28) (Chee *et al.*, 1990), and *equine herpesvirus 2* (Telford *et al.*, 1995) (ORF E1, which is present in two copies due to its location within the terminal repeat region of the viral chromosome). A useful functional paradigm for virus-encoded chemokine receptors is the US28 gene of human cytomegalovirus. This protein both binds chemokines (Neote *et al.*, 1993) and induces a rise in intracellular calcium after binding of the appropriate chemokines (Gao and Murphy, 1994).

In the absence of data demonstrating functionality for the poxvirus proteins, one can only speculate as to function based on the relatively high homology exhibited to certain CC chemokine receptors (Table 1). That prediction would be that both poxvirus GPCRs do indeed bind to chemoattractants, most likely chemokines of the CC class. Although superficially, all the structural features required for signaling following receptor–ligand engagement appear to be present in the poxvirus GPCRs, an equally likely scenario, based on other poxvirus examples is for these proteins to function as nonsignaling, ligand sinks.

GENE

Accession numbers

Sheep pox virus ORF3L: S78201
Swinepox ORF K2R (complete ORF): L21031
Swinepox C3L (truncated ORF): L22013

PROTEIN

Accession numbers

GenBank:
Capripox virus: Q86917
Swinepox virus: Q08520
Human CC (CCR8): P51685
Rhesus monkey CC: AAC72403
Equine herpesvirus 2 receptor: S55594
Human CXC chemokine receptor (CXCR2): P25025
Herpesvirus saimiri GPCR: Q01035
HHV8: AAB51506
EBV: P32249
CMV (US28): P09704

Sequence

The complete protein sequence of the sheep pox and swinepox virus ORFs is shown in Figure 1.

Description of protein

The poxvirus-encoded proteins are slightly larger in size than the typical GPCR, but are nevertheless,

Figure 1 Alignment of putative poxvirus-encoded G protein-coupled receptors (GPCRs) with those of viral and cellular origin. Each protein contains a typical seven transmembrane motif signature, indicated above the sequences in which they occur. Each GPCR also contains a conserved proline residue within transmembrane domains II, IV, V, VI, and VII and two conserved cysteine residues, one in the extracellular region between transmembrane domain II and III, the second within the extracellular region between transmembrane regions IV and V. Typically, these two cysteines are linked in a disulfide bridge. Both the proline and cysteine residues are indicated as shaded residues within the consensus sequence. Each protein also contains the highly conserved motif DRYLAIVHA at the end of the third transmembrane domain. The DRYLAIVHA motif is not universally present, being absent in the CMV US28 protein. Another feature of GPCRs is the presence of phosphorylation sites, typically serine residues in the C-terminal portion of the molecule and glycosylation sites within the extracellular N-terminal most region of the molecule. The proteins depicted are those of CPV (capripox virus), SPV (swinepox virus), Hu CC (human CC chemokine receptor), Mn CC (rhesus monkey CC chemokine receptor), EQHV (equine herpesvirus 2 GPCR). The number of amino acids in the protein is given. The alignments and consensus sequence were derived using the PRETTY program of the Genetics Computer Group Package, Madison, Wisconsin.

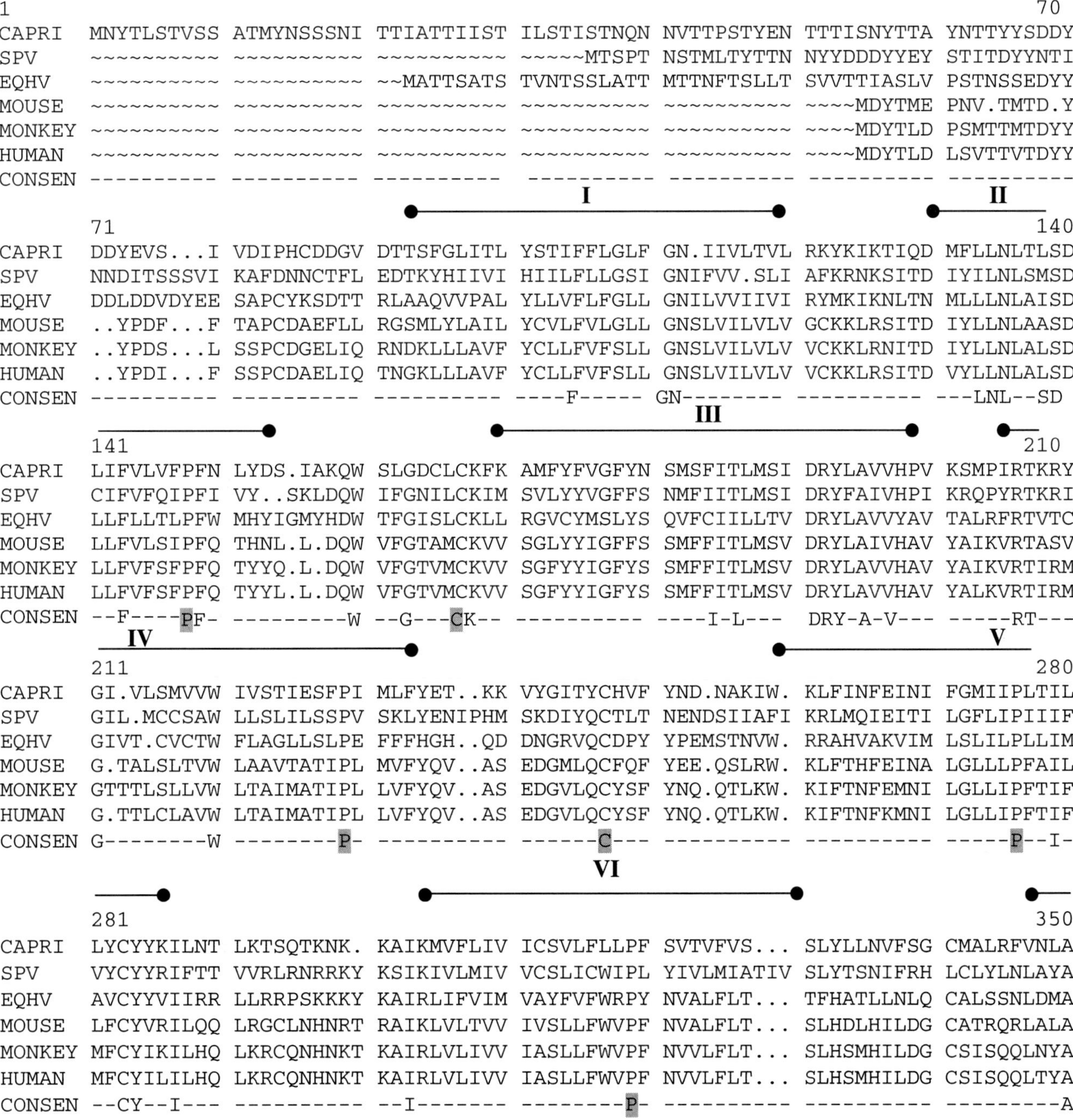

Figure 1 (*Continued*)

```
                VII
       ------------------------●
       351                                                                            420
CAPRI  VHVAEIVSLC HCFINPLIYA FCSREFTKKL LRLRTTSSAG SISIG*~~~~ ~~~~~~~~~~ ~~~~~~~~~~
SPV    ITFSETISLA RCCINPIIYT LIGEHVRSRI SSICSCIYRD NRIRKKLFSR KSSSSSNII* ~~~~~~~~~~
EQHV   LLITKTVAYT HCCINPVIYA FVGEKFRRHL YHFFHTYVAI YLCKYIPFLS GDGEGKEGPT RI*~~~~~~~
MOUSE  IHVTEVISFT HCCVNPVIYA FIGEKFKKHL MDVFQKSCSH IFLYLGRQMP VGALERQLSS BQRSSHSSTL
MONKEY THVTEIISFT HCCVNPVIYA FVGEKFKKHL SEIFQKSCSH IFIYLGRQMP RESCEKSSSC QQHSFRSSSI
HUMAN  THVTEIISFT HCCVNPVIYA FVGEKFKKHL SEIFQKSCSQ IFNYLGRQMP RESCEKSSSC QQHSSRSSSV
CONSEN ---------- -C--NP-IY- ---------- ---------- ~~~~~~~~~~ ~~~~~~~~~~ ~~~~~~~~~~

       421
CAPRI  ~~~~~
SPV    ~~~~~
EQHV   ~~~~~
MOUSE  DDIL*
MONKEY DYIL*
HUMAN  DYIL*
CONSEN ~~~~~
```

Table 1 Percent identity between poxvirus, various herpesvirus, and cellular-encoded GPCRs

Source	% Identity									
	CPV	SPV	Hu CC	Mn CC	EQHV	Hu IL-8 CXC	HSV	HHV8	EBV	CMV
CPV	–	37.6	44.8	45.9	31.3	29.4	21.2	17.4	27.6	29.5
SPV	37.6	–	38.7	38.7	30.9	32.1	20.2	17.8	24.4	27.4
Hu CC	44.8	38.7	–	94.3	39.1	37.4	19.1	21.9	25.1	29.8
Mn CC	45.9	38.7	94.3	–	38.7	36.4	18.7	22.9	25.1	30.8
EQHV	31.3	31.0	39.1	38.7	–	34.5	22.8	22.6	27.5	31.8
Hu IL-8	29.4	32.1	37.4	36.4	34.5	–	30.8	29.6	27.6	32.9
HSV	21.2	20.3	19.1	18.7	22.8	30.8	–	35.0	31.2	32.8
HHV8	17.4	17.8	22.0	23.0	22.6	29.6	35.0	–	19.6	21.6
EBV	27.7	24.4	25.1	25.1	27.5	27.6	31.2	19.6	–	27.4
CMV	29.5	27.4	29.8	30.8	31.8	32.9	23.5	21.6	27.4	–

CPV, capripox virus (381 aa); SPV, swinepox virus (370 aa); Hu CC, human CC chemokine receptor (CCR8) (355 aa); Mn CC, rhesus monkey CC chemokine receptor (356 aa); EQHV, equine herpesvirus 2 receptor (383 aa); Hu IL-8, human CXC chemokine receptor (CXCR2) (360 aa); HSV, herpesvirus saimiri GPCR (331 aa); HHV8, human herpesvirus 8 GPCR (352 aa); EBV, cellular GPCR induced by Epstein–Barr virus infection (361 aa); CMV, cytomegalovirus GPCR (US28) (345 aa). Identities were calculated using the BESTFIT program (Genetics Computer Group Package, Madison, Wisconsin).

structurally typical of GPCRs (*see* **Figure 1**). Beginning with a glycosylated extracellular N-terminal domain, the serpentine seven transmembrane domains define three alternating intracellular (I1–I3) and three extracellular (E1–E3) domains. Extension of sequence beyond the seventh transmembrane domain defines an intracellular cytoplasmic tail of the protein (Murphy, 1994; Barker and Monk, 1997).

Relevant homologies and species differences

Homologies to selected viral and cellular GPCRs are shown in **Table 1**. Particularly noteworthy is the relatively high homology of the sheep pox protein for the putative CC cellular GPCRs.

Affinity for ligand(s)

The ligands, if any, for the poxvirus proteins are not yet known. However, based on similar studies, one would predict ligand affinities in the nano- to picomolar range.

Regulation of receptor expression

Based on genomic location and transcriptional elements within the sequences, the poxvirus GPCRs are expressed from early promoters, prior to DNA replication, irrespective of the infected cell type.

BIOLOGICAL CONSEQUENCES OF ACTIVATING OR INHIBITING RECEPTOR AND PATHOPHYSIOLOGY

Phenotypes of receptor knockouts and receptor overexpression mice

No data on GPCR knockout viruses are available. The most relevant questions for future study are to first identify the appropriate ligand(s) and whether receptor–ligand engagement results in an appropriate intracellular signal or whether the poxvirus proteins serve instead as a membrane-bound chemokine sink.

References

Ahuja, S. K., and Murphy, P. M. (1993). Molecular piracy of mammalian interleukin-8 receptor type B by herpesvirus saimiri. *J. Biol. Chem.* **268**, 20691–20694.

Barker, M. D., and Monk, P. N. (1997). Structure-function relationships of leukocyte chemoattractant receptors. *Biochem. Soc. Trans.* **25**, 1027–1031.

Birkenbach, M., Josefsen, K., Yalamanchili, R., Lenoir, G., and Kieff, E. (1993). Epstein–Barr virus-induced genes: First lymphocyte-specific G protein-coupled peptide receptors. *J. Virol.* **67**, 2209–2220.

Cao, J. X., Gershon, P. D., and Black, D. N. (1995). Sequence analysis of *Hin*dIII Q2 fragment of capripoxvirus reveals a putative gene encoding a G-protein-coupled chemokine receptor homologue. *Virology* **209**, 207–212.

Chee, M. S., Satchwell, S. C., Preddie, E., Weston, K. M., and Barrell, B. G. (1990). Human cytomegalovirus encodes three G protein-coupled receptor homologues. *Nature* **344**, 774–777.

Gao, J. L., and Murphy, P. M. (1994). Human cytomegalovirus open reading frame US28 encodes a functional β receptor. *J. Biol. Chem.* **269**, 28539–28542.

Gershon, P. D., and Black, D. N. (1987). Physical characterization of the genome of a cattle isolate of capripoxvirus. *Virology* **160**, 473–476.

Guo, H.-G., Browning, P., Nicholas, J., Hayward, G. S., Tschachler, E., Jiang, Y.-W., Sadowska, M., Raffeld, M., Colombini, S., Gallo, R. C., and Reitz Jr., M. S. (1997). Characterization of a chemokine receptor-related gene in human herpesvirus 8 and its expression in Kaposi's sarcoma. *Virology* **228**, 371–378.

Massung, R. F., Jayarama, V., and Moyer, R. W. (1993). DNA sequence analysis of conserved and unique regions of swinepox virus: Identification of genetic elements supporting phenotypic observations including a novel G protein-coupled receptor homologue. *Virology* **197**, 511–528.

Murphy, P. M. (1994). The molecular biology of leukocyte chemoattractant receptors. *Annu. Rev. Immunol.* **12**, 593–633.

Neote, K., DiGregorio, D., Mak, J. Y., Horuk, R., and Schall, T. J. (1993). Molecular cloning, functional expression, and signaling characteristics of a C-C-chemokine receptor. *Cell* **72**, 415–425.

Nicholas, J., Cameron, K. R., and Honess, R. W. (1992). Herpesvirus saimiri encodes homologues of G-protein coupled receptors and cyclins. *Nature* **355**, 362–365.

Oppenheim, J. J., Zacharier, M. N., Mukaida, N., and Matsushima, K. (1991). Properties of the novel proinflammatory supergene 'intercrine' cytokine family. *Annu. Rev. Immunol.* **9**, 617–648.

Schall, T. J., and Bacon, K. B. (1994). Chemokines, leukocyte trafficking, and inflammation. *Curr. Opin. Immunol.* **6**, 865–873.

Schweickart, V. L., Raport, C. J., Godiska, R., Byers, M. G., Eddy Jr., R. L., Shows, T. B., and Gray, P. W. (1994). Cloning of human and mouse EBI1, a lymphoid-specific G-protein-coupled-receptor encoded on human chromosome 17q12-q21.2. *Genomics* **23**, 643–650.

Telford, E. A., Watson, M. S., Aird, H. C., Perry, J., and Davison, A. J. (1995). The DNA sequence of equine herpesvirus 2. *J. Mol. Biol.* **249**, 520–528.

US28

Ji-Liang Gao*

Laboratory of Host Defenses, National Institute of Allergy and Infectious Diseases, NIH, Building 10, Room 11N111, Bethesda, MD 20892, USA

*corresponding author tel: 301-496-2877, fax: 301-402-4369, e-mail: jgao@nih.gov

DOI: 10.1006/rwcy.2000.22012.

SUMMARY

US28, a viral chemokine receptor encoded by human *cytomegalovirus* (HCMV), is expressed in the late phase of HCMV-infected cells. It binds CC and CX3C chemokines including MIP-1α, MIP-1β, MCP-1, MCP-3, RANTES, and fractalkine, and mediates intracellular calcium release in response to these CC chemokines. US28 is not required for HCMV growth but may play a role in evading immune surveillance by chemokine sequestration as shown by the US28-deletion experiment. In addition, US28 is a coreceptor for *HIV* entry.

BACKGROUND

Discovery

US28 is an open reading frame of human *cytomegalovirus* (HCMV). The sequence, which was first published in 1986 (Weston and Barrell, 1986) and named initially as HHRF3, later was named US28 (Chee *et al.*, 1990a). In 1990, Chee and coworkers (1990b) found that US28 was a homolog of G protein-coupled receptors. In 1993, Neote *et al.* (1993) reported that US28 acts as a CC chemokine-binding protein when transiently expressed in human embryonic kidney 293 cells (HEK293 cells). In 1994, Gao and Murphy (1994) found that US28 is a functional CC chemokine receptor when expressed in human K562 cells. Cells infected with HCMV express US28, bind CC chemokines, and demonstrate chemokine-induced signal transduction, but cells infected with US28-deleted HCMV do not (Bodaghi *et al.*, 1998; Vieira *et al.*, 1998), suggesting that US28 is a functional CC chemokine receptor of HCMV. In 1997, US28 was found to be a coreceptor for *HIV* entry (Pleskoff *et al.*, 1997).

Alternative names

HHRF3 (Weston and Barrell, 1986).

Structure

A three-dimensional structure is not available for US28. The deduced protein sequence of US28 has 354 amino acids and is a homolog of G protein-coupled receptors with seven transmembrane domains.

Main activities and pathophysiological roles

The known activities for US28 are its ability to bind the CX3C chemokine fractalkine (Kledal *et al.*, 1998), and the CC chemokines MIP-1α, MIP-1β, MCP-1, MCP-3, and RANTES, and to mediate signal transduction in response to these CC chemokines (Gao and Murphy, 1994; Bodaghi *et al.*, 1998; Vieira *et al.*, 1998). US28 also serves as a coreceptor for HIV-1 and HIV-2 (Pleskoff *et al.*, 1997) and enhances cell–cell fusion mediated by different viral proteins including HIV envelope protein (Pleskoff *et al.*, 1998) *in vitro*. US28 is not required for HCMV growth as shown by the US28-deletion experiment (Bodaghi *et al.*, 1998; Vieira *et al.*, 1998). Pathophysiological roles for US28 in HCMV infection are still not clear. HCMV can infect epithelial cells, fibroblasts, smooth muscle cells, and leukocytes *in vivo*, and can cause acute, chronic, and latent infections (Schrier *et al.*, 1985; Speir *et al.*, 1994).

GENE

Accession numbers

X17403 is the accession number for the DNA sequence of HCMV, strain AD169, in which the coding sequence for US28 is from 219200 to 220263 (revised by insertion of G at position 200095). L20501 is the accession number for the DNA sequence of US28 from a HCMV clinic isolate, VHL/E.

Sequence

The genome of HCMV is composed of about 235 kb of linear double-stranded DNA and can be divided into two regions of unique sequence: UL (long unique region) and US (short unique region). Each of the regions is flanked by their respective repeat sequences. US28 is an open reading frame in the short unique region located between US27 and US29. A potential TATA box is located 175 bp upstream of the start codon of US28 (Weston and Barrell, 1986).

PROTEIN

Accession numbers

L20501.

Description of protein

US28 is a viral CX3C and CC chemokine receptor. Its deduced protein sequence has 354 amino acids and it is a homolog of G protein-coupled receptors with seven transmembrane domains. There is one putative glycosylation site, Asn30.

Relevant homologies and species differences

The closest homolog of US28 is US27 of HCMV with about 40% identity. The closest human chemokine receptors are CCR1 and CCR2 with about 30% identity. Murine CMV does not contain US28 orthologs.

Affinity for ligand(s)

Binding of CC chemokines in US28-transfected mammalian cells

1. HEK293 cells (transient expression): US28-expressing cells are able to bind ^{125}I-labeled MIP-1α (K_d = 1 nM and 380 nM). The binding was competed for by MIP-1β, MCP-1, and RANTES (Neote *et al.*, 1993).
2. K562 cells (stable expression): US28-expressing cells are able to bind ^{125}I-labeled MIP-1α (K_d = 2.5 nM). The binding was competed for by MIP-1β, MCP-1, and RANTES (K_i values in the range 3.4–6.1 nM) (Gao and Murphy, 1994).
3. COS cells (transient expression): US28-expressing cells are able to bind ^{125}I-labeled CC chemokines RANTES, MCP-1, MIP-1α, MIP-1β, and the CX3C chemokine fractalkine with subnanomolar affinity (Kuhn *et al.*, 1995; Kledal *et al.*, 1998). Fractalkine could compete with high affinity with the binding of CC chemokines, but CC chemokines were only able to compete with very low affinity with the binding of fractalkine (Kledal *et al.*, 1998).

Binding of CC chemokines in HCMV-infected cells

1. Endothelial cells: HCMV-infected endothelial cells express binding sites for RANTES (K_d = 10 nM) (Billstrom *et al.*, 1998) and MCP-1 (Randolph-Habecker *et al.*, 1997).
2. Fibroblasts: HCMV-infected fibroblasts bind to ^{125}I-labeled MIP-1α (Vieira *et al.*, 1998) and RANTES (Bodaghi *et al.*, 1998). The binding of labeled MIP-1α can be competed for by unlabeled MIP-1α, MIP-1β, MCP-1, MCP-3, and RANTES, but not by MCP-2 (Vieira *et al.*, 1998). In contrast, cells infected with US28-deleted HCMV did not bind MIP-1α (Vieira *et al.*, 1998) and RANTES (Bodaghi *et al.*, 1998), suggesting that US28 is the only MIP-1α-binding receptor in HCMV-infected cells.

Cell types and tissues expressing the receptor

The expression of US28 has been detected in several cell types infected with HCMV. mRNA was detected in the late phase of lytic infection of fibroblasts (Welch *et al.*, 1991; Michelson *et al.*, 1997; Bodaghi *et al.*, 1998), endothelial cells (Randolph-Habecker *et al.*, 1997; Billstrom *et al.*, 1998), and PBMCs of patients infected with *cytomegalovirus* (HCMV) (Patterson *et al.*, 1998). mRNA of US28 in HCMV-infected endothelial cells was detected as early as 6 hours postinfection and reached a plateau by 48–72 hours postinfection before declining at 96 hours (Michelson *et al.*, 1997; Billstrom *et al.*, 1998), but the binding of RANTES to the infected cells was not detected until day 4 after infection (Billstrom *et al.*, 1998).

SIGNAL TRANSDUCTION

Cytoplasmic signaling cascades

Chemokine-induced intracellular calcium release

The ability of US28 to mediate signal transduction induced by chemokines was demonstrated in US28-expressing K562 cells (Gao and Murphy, 1994) and HEK293 cells (Billstrom *et al.*, 1998). Chemokine-induced intracellular calcium release was also detected in HCMV-infected fibroblasts (Vieira *et al.*, 1998; Bodaghi *et al.*, 1998) and endothelial cells (Billstrom *et al.*, 1998) which was detected at 96 hours after infection, but not at 72 hours.

Components involved in US28-mediated signal transduction

By coexpressing US28 and Gα_{16} in HEK293 cells, Billstrom *et al.* (1998) demonstrated that US28 couples to Gα_i proteins as well as pertussis toxin-insensitive Gα_{16} proteins to activate intracellular calcium flux in response to chemokines. Furthermore, they demonstrated that US28 is able to activate the MAP kinase-signaling pathway through ERK2 MAP kinase in response to chemokine stimulation. It was also reported that chemokine-activated US28 could regulate protein kinase activity in intestinal cells transfected with US28 (Reinecker *et al.*, 1997).

DOWNSTREAM GENE ACTIVATION

Promoter regions involved

A possible TATA box located 175 bp upstream of the start codon of US28 was suggested based on sequence analysis (Weston and Barrell, 1986).

BIOLOGICAL CONSEQUENCES OF ACTIVATING OR INHIBITING RECEPTOR AND PATHOPHYSIOLOGY

Unique biological effects of activating the receptors

Biological roles of US28 have not been reported *in vivo*. However, some *in vitro* studies have suggested biological roles of US28 in HCMV infection.

Possible role in evading immune surveillance by chemokine sequestration

Human *cytomegalovirus* (HCMV) induces production of RANTES and other chemokines in fibroblasts, which occurs as early as 8 hours after infection, peaks around 24 hours after infection, and is almost undetectable by 48 and 72 hours (Michelson *et al.*, 1997). It was demonstrated that depletion of chemokines from the culture medium was at least partially due to continuous internalization of extracellular chemokine, since exogenously added, biotinylated RANTES accumulated in HCMV-infected cells. In contrast, cells infected with US28-deleted HCMV could not downregulate extracellular accumulated chemokines, indicating that US28 is responsible for this function (Vieira *et al.*, 1998; Bodaghi *et al.*, 1998). Thus, chemokine sequestration by US28 could help HCMV to go unnoticed and thereby escape immune surveillance.

Possible role in HIV infection

It was demonstrated that US28 is a coreceptor for HIV-1 and HIV-2 entry. Since infection by *cytomegalovirus* is frequent among *HIV*-infected individuals it was proposed that US28 may play a role in HIV pathogenesis (Pleskoff *et al.*, 1997).

Possible role in HCMV entry

In vitro experiments indicated that US28 could enhance cell–cell fusion mediated by viral proteins derived from a variety of viruses, including HIV envelope protein, suggesting that US28 may play a role in *cytomegalovirus* entry (Pleskoff *et al.*, 1998).

Phenotypes of receptor knockouts and receptor overexpression mice

US28 has been knocked out by two groups independently (Vieira *et al*., 1998; Bodaghi *et al*., 1998). The results suggested that US28 is not required for HCMV growth. However, three chemokine-related functions, chemokine binding, chemokine-induced calcium flux, and chemokine sequestration, normally seen in HCMV-infected cells were not detected in the cells infected with US28-deleted HCMV. US28 overexpression in mice has not been reported.

References

Billstrom, M. A., Johnson, G. L., Avdi, N. J., and Worthen, G. S. (1998). Intracellular signaling by the chemokine receptor US28 during human cytomegalovirus infection. *J. Virol.* **72**, 5535–5544.

Bodaghi, B., Jones, T. R., Zipeto, D., Vita, C., Sun, L., Laurent, L., Arenzana-Seisdedos, F., Virelizier, J. L., and Michelson, S. (1998). Chemokine sequestration by viral chemoreceptors as a novel viral escape strategy: withdrawal of chemokines from the environment of cytomegalovirus-infected cells. *J. Exp. Med.* **188**, 855–866.

Chee, M. S., Bankier, A. T., Beck, S., Bohni, R., Brown, C. M., Cerny, R., Horsnell, T., Hutchison, C. A., Kouzarides, T., Martignetti, J. A., Preddie, E., Satchwell, S. C., Tomlinson, P., and Weston, K. M. (1990a). Analysis of the protein-coding content of the sequence of human cytomegalovirus strain AD169. *Curr. Top. Microbiol. Immunol.* **154**, 125–169.

Chee, M. S., Satchwell, S. C., Preddie, E., Weston, K. M., and Barrell, B. G. (1990b). Human cytomegalovirus encodes three G protein-coupled receptor homologues [see comments]. *Nature* **344**, 774–777.

Gao, J. L., and Murphy, P. M. (1994). Human cytomegalovirus open reading frame US28 encodes a functional beta chemokine receptor. *J. Biol. Chem.* **269**, 28539–28542.

Kledal, T. N., Rosenkilde, M. M., and Schwartz, T. W. (1998). Selective recognition of the membrane-bound CX3C chemokine, fractalkine, by the human cytomegalovirus-encoded broad-spectrum receptor US28. *FEBS Lett.* **441**, 209–214.

Kuhn, D. E., Beall, C. J., and Kolattukudy, P. E. (1995). The cytomegalovirus US28 protein binds multiple CC chemokines with high affinity. *Biochem. Biophys. Res. Commun.* **211**, 325–330.

Michelson, S., Dal Monte, P., Zipeto, D., Bodaghi, B., Laurent, L., Oberlin, E., Arenzana-Seisdedos, F., Virelizier, J. L., and Landini, M. P. (1997). Modulation of RANTES production by human cytomegalovirus infection of fibroblasts. *J. Virol.* **71**, 6495–6500.

Neote, K., DiGregorio, D., Mak, J. Y., Horuk, R., and Schall, T. J. (1993). Molecular cloning, functional expression, and signaling characteristics of a C-C chemokine receptor. *Cell* **72**, 415–425.

Patterson, B. K., Landay, A., Andersson, J., Brown, C., Behbahani, H., Jiyamapa, D., Burki, Z., Stanislawski, D., Czerniewski, M. A., and Garcia, P. (1998). Repertoire of chemokine receptor expression in the female genital tract: implications for human immunodeficiency virus transmission. *Am. J. Pathol.* **153**, 481–490.

Pleskoff, O., Treboute, C., Brelot, A., Heveker, N., Seman, M., and Alizon, M. (1997). Identification of a chemokine receptor encoded by human cytomegalovirus as a cofactor for HIV-1 entry [see comments]. *Science* **276**, 1874–1878.

Pleskoff, O., Treboute, C., and Alizon, M. (1998). The cytomegalovirus-encoded chemokine receptor US28 can enhance cell–cell fusion mediated by different viral proteins. *J. Virol.* **72**, 6389–6397.

Randolph-Habecker, J., Beall, C. J., Kolattukudy, P. E., and Sedmak, D. D. (1997). Monocyte chemoattractant protein-1 binding by cytomegalovirus-infected endothelial cells. *Transpl. Proc.* **29**, 807–808.

Reinecker, H. C., Mehta, A., Li, D. J., Manion, D. J., Podolsky, D. K., and MacDermott, R. P. (1997). Expression of a cytomegalovirus derived chemokine receptor regulates protein kinase A activity in intestinal epithelial cells. *Gastroenterology* **112**, A1070.

Schrier, R. D., Nelson, J. A., and Oldstone, M. B. (1985). Detection of human cytomegalovirus in peripheral blood lymphocytes in a natural infection. *Science* **230**, 1048–1051.

Speir, E., Modali, R., Huang, E. S., Leon, M. B., Shawl, F., Finkel, T., and Epstein, S. E. (1994). Potential role of human cytomegalovirus and p53 interaction in coronary restenosis [see comments]. *Science* **265**, 391–394.

Vieira, J., Schall, T. J., Corey, L., and Geballe, A. P. (1998). Functional analysis of the human cytomegalovirus US28 gene by insertion mutagenesis with the green fluorescent protein gene. *J. Virol.* **72**, 8158–8165.

Welch, A. R., McGregor, L. M., and Gibson, W. (1991). Cytomegalovirus homologs of cellular G protein-coupled receptor genes are transcribed. *J. Virol.* **65**, 3915–3918.

Weston, K., and Barrell, B. G. (1986). Sequence of the short unique region, short repeats, and part of the long repeats of human cytomegalovirus. *J. Mol. Biol.* **192**, 177–208.

Kaposi's Sarcoma-associated Herpesvirus GPCR

Elizabeth Geras-Raaka and Marvin C. Gershengorn*

Department of Medicine, Division of Molecular Medicine, Weill Medical College and Graduate School of Medical Sciences of Cornell University, 1300 York Avenue, New York, NY 10021, USA

*corresponding author tel: 212-746-6275, fax: 212-746-6289, e-mail: mcgersh@mail.med.cornell.edu

DOI: 10.1006/rwcy.2000.22013.

SUMMARY

Kaposi's sarcoma-associated herpesvirus (KSHV, human herpesvirus 8), which is a virus that appears to be etiologic for *Kaposi's sarcoma*, *primary effusion lymphomas*, and multicentric *Castleman's disease* in humans, encodes a G protein-coupled receptor (ORF 74, KSHV GPCR) that is homologous to human chemokine receptors. KSHV GPCR is more promiscuous than most chemokine receptors in that it binds CC and CXC chemokines. For GPCRs encoded within viral genomes, KSHV GPCR is novel in that it exhibits constitutive signaling activity. It signals via the phospholipase C-inositol 1,4,5-trisphosphate-1,2-diacylglycerol pathway and activates the Jun kinase/SAP kinase and p38 MAP kinase pathways. Expression of KSHV GPCR in rat NRK fibroblasts stimulates cell proliferation. KSHV GPCR can transform mouse NIH 3T3 fibroblasts *in vitro* and KSHV GPCR-expressing NIH 3T3 cells form tumors in mice. Thus, KSHV GPCR displays activities of human oncogenes. Moreover, KSHV GPCR induces expression of vascular endothelial growth factor (VEGF), a potent and efficacious stimulator of angiogenesis, in NIH 3T3 cells. Thus, because of its tumorigenic and angiogenic potential, KSHV GPCR is likely to play a role in the pathogenesis of diseases associated with KSHV infection.

BACKGROUND

Discovery

Kaposi's sarcoma-associated herpesvirus (KSHV, human herpesvirus 8) is a recently identified member of the herpesvirus family (Chang *et al.*, 1994; Russo *et al.*, 1996). KSHV has been found in *Kaposi's sarcoma* (KS) lesions of patients with *AIDS* and of non-AIDS-related patients (Chang *et al.*, 1994; Moore and Chang, 1995), in normal-appearing tissue adjacent to KS lesions and in lymph nodes and peripheral blood B cells in patients with KS (Noel, 1995). KSHV has also been found in two distinct types of lymphoid proliferative disorders *primary effusion lymphomas* (PELs) (Cesarman *et al.*, 1995; Arvanitakis *et al.*, 1996) and *multicentric Castleman's disease* (Corbellino *et al.*, 1996). KSHV has been shown to be a transmissible virus that infects human B cells (Mesri *et al.*, 1996) and human endothelial cells (Flore *et al.*, 1998). Extensive sequence analyses of fragments of KSHV have shown homology to herpesvirus saimiri and Epstein-Barr viruses (Moore *et al.*, 1996). Since these two viruses infect and transform lymphoblastoid cells (Miller, 1974; Rangan *et al.*, 1977), it seemed possible that KSHV may be a transforming agent also. Recently, Flore *et al.* (1998) showed that KSHV could transform primary human endothelial cells.

Lastly, it has been shown that there is seroconversion of antibodies against KSHV before the development of KS in most patients with AIDS and that KSHV appears to be a sexually transmitted disease (Gao *et al.*, 1996a,b; Kedes *et al.*, 1996). Thus, accumulating evidence is consistent with the idea that KSHV is involved in the pathogenesis of human primary effusion lymphomas (Nador *et al.*, 1996) and Kaposi's sarcoma (Offermann, 1996).

An open reading frame in the genome of KSHV encodes a protein that was shown to be a constitutively active G protein-coupled receptor (GPCR).

Alternative names

KSHV GPCR; HHV 8 GPCR; IL-8-like GPCR; viral GPCR or GCR; KSHV ORF 74.

Structure

The putative two-dimensional structure of KSHV GPCR is illustrated in **Figure 1**. Like all GPCRs, KSHV GPCR is an integral membrane protein with extracellular, transmembrane, and intracellular domains. On the extracellular surface is the N-terminus and three loops. Extracellular loop 1 (ECL-1) connects transmembrane helix 2 (TM-2) and TM-3, ECL-2 connects TM-4 and TM-5, and ECL-3 connects TM-6 and TM-7. The cell surface membrane is spanned by seven α helices. On the intracellular side are three loops and the C-terminus. Intracellular loop 1 (ICL-1) connects TM-1 and TM-2, ICL-2 connects TM-3 and TM-4, and ICL-3 connects TM-5 and TM-6. In three dimensions, the seven α helices are predicted to form a helical bundle that approximates a cylinder, with TM-7 close to and interacting with TM-1 and TM-2.

Main activities and pathophysiological roles

KSHV GPCR has been shown to be expressed at the mRNA level in tissues from patients with *Kaposi's sarcoma* and in *B cell lymphomas* (Cesarman *et al.*, 1996). For GPCRs encoded within viral genomes, KSHV GPCR is novel in that it exhibits constitutive signaling activity (Arvanitakis *et al.*, 1997). Because constitutive activation of the signaling pathways activated by KSHV GPCR induces cell proliferation

Figure 1 Putative two-dimensional topology of KSHV GPCR.

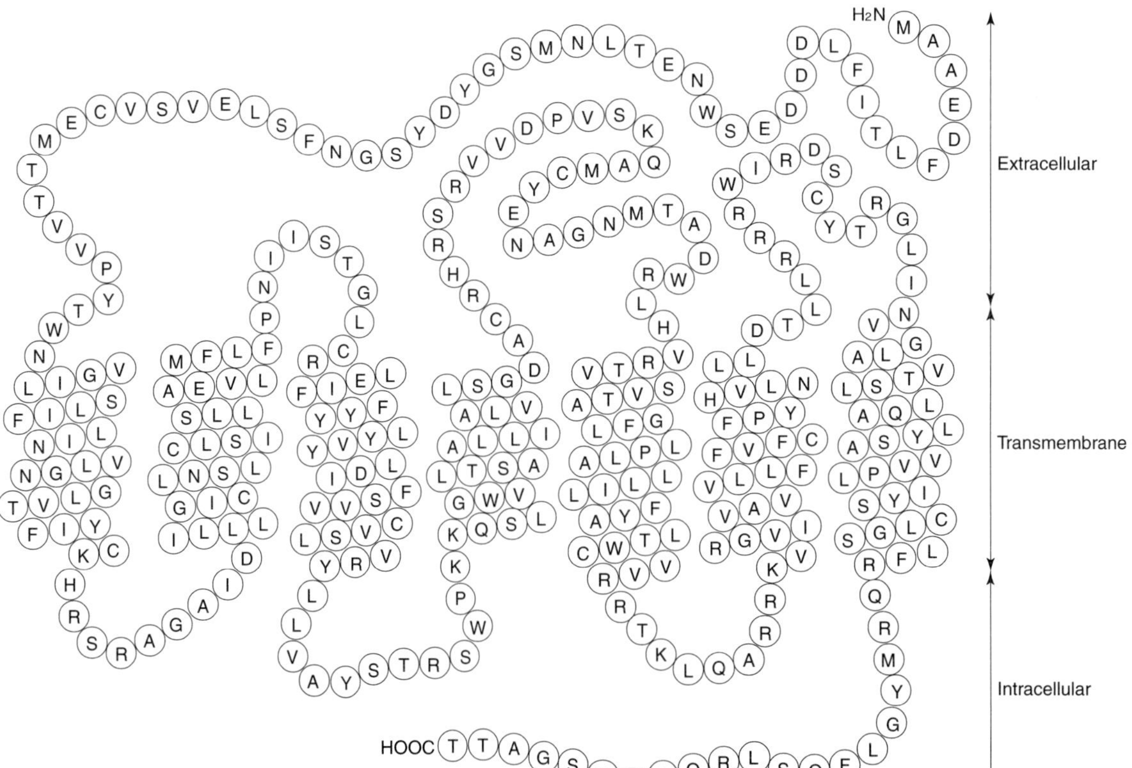

and transformation (Post and Brown, 1996), KSHV GPCR expression transforms NIH 3T3 cells (Bais *et al.*, 1998), and constitutively active GPCRs cause tumors in humans (Arvanitakis *et al.*, 1998), it has been suggested that KSHV GPCR is involved in the pathogenesis of tumors associated with KSHV infection.

GENE

Accession numbers

U24275; U82242; U71368; AF079845; U75698; U93872.

Sequence

See **Figure 2**.

PROTEIN

Accession numbers

1718331; 1621029; 3551771; 3386561; Q98146; 2246493; 1930014.

Sequence

See **Figure 3**.

Description of protein

KSHV GPCR is a protein of 342 amino acids that appears to have the features of a GPCR including an extracellular N-terminus, three extracellular loops, seven hydrophobic, transmembrane-spanning domains, three intracellular loops, and an intracellular C-terminus (Figure 1). It is a member of the rhodopsin/β-adrenergic receptor subfamily of GPCRs but is lacking some of the amino acid residues that are highly conserved in subfamily members. For example, it has Ile in place of Asp at position 14 in TM-2, Val in place of Asp or Glu at position 24 in TM-3 (of the Asp/Glu-Arg-Tyr motif) and Val in place of Asn at position 17 in TM-7 (of the Asn-Pro-Xaa-Xaa-Tyr motif); numbering of positions and alignment according to Baldwin (Baldwin *et al.*, 1997). Most importantly, KSHV GPCR exhibits marked, constitutive signaling activity (i.e. signaling in the absence of agonist) when expressed in mammalian cells (Arvanitakis *et al.*, 1997).

Relevant homologies and species differences

The amino acid sequence of KSHV GPCR shows homology to the GPCR encoded in the herpesvirus

Figure 2 The nucleotide sequence of the KSHV GPCR gene.

```
CGTGGTGGCGCCGGACATGAAAGACTGCCTGAGGCTTTGGAAGAGACCGTACATCCT
CTGCCTAAAGAGGGATCCCAGGCAGGAGTATATCAGGGGAACCACGGCGCTGTACAG
TGCCTGCAGTAACGAGGTTACTGCCAGACCCACGTTTATCAACCCCCGCGTATAGCA
GCTGTCCCGGATCCAGCGTCGCCTTAGCAGAGTGTCCAGTAGATTTAGTACGTGGTA
AGGGAAGCAAAACACAAAAAACAGCAGCACCACAGCAACAATCACCCCCCTTACCTT
CCGCCTGGCTTGCAGCTTTGTCCTCCTCACCACACACCAGGTGAGAGCATAAAACAG
AATAAGGAGGGCCAGGGGTAACAGGAAACCTGCAGTAACTGACACGGTTCTGACATG
CAGTCGCCAGTCTGCAGTCATGTTTCCCGCGTTCTCATAACACATGGCCTGCTTGCT
GACCGGGTCGACCACCCTGCTCCTGTGTCGACAGGCATCCCCCGACAGCACCAATGC
AATTAACAGTGCAGCGGATGTCAGTACCCATCCGAGGGACTGCTTCTTGGGCCAGGA
ACGCGTAGAATATGCCACCAGGAGGTACCTCACTAGACTGACGCACACAACACTGAA
GATATCCAAGTAGACATATAAATAGTAAAAAAAAATTTCAAGTCTGCACAAGCCTGT
GGAGATGATATTGGGAAACAAAAACATCAACACTTCTGCCAATAGAGATATGCTAAG
ACACAGCGAGTTTAGGCAGATACCCAGGAGCAGTATATCTATCGCTCCTGCCCGCGA
TCGGTGCTTGCAAAAAATGTAGGTGACCAATCCATTTCCAAGAACATTTATGAGGAA
AATCAGAGAGAGTATTCCAACGTTCCACGTGTAAGGCACCACGGTGGTCATCTCACA
CACGCTCACTTCTAGGCTGAAGTTTCCAGAGTAGTCATATCCGCTCATATTTAGAGT
TTCATTCCAGGATTCATCATCATCTAAGAAGATGGTTAGGAAATCCTCGGCCGCCAT
```

Figure 3 The amino acid sequence of KSHV GPCR.

```
1      MAAEDFLTIF LDDDESWNET LNMSGYDYSG NFSLEVSVCE MTTVVPYTWN
51     VGILSLIFLI NVLGNGLVTY IFCKHRSRAG AIDILLLGIC LNSLCLSISL
101    LAEVLMFLFP NIISTGLCRL EIFFYYLYVY LDIFSVVCVS LVRYLLVAYS
151    TRSWPKKQSL GWVLTSAALL IALVLSGDAC RHRSRVVDPV SKQAMCYENA
201    GNMTADWRLH VRTVSVTAGF LLPLALLILF YALTWCVVRR TKLQARRKVR
251    GVIVAVVLLF FVFCFPYHVL NLLDTLLRRR WIRDSCYTRG LINVGLAVTS
301    LLQALYSAVV PLIYSCLGSL FRQRMYGLFQ SLRQSFMSGA TT
```

saimiri genome (Nicholas *et al.*, 1992; Ahuja and Murphy, 1993) and to several mammalian GPCRs (Strader *et al.*, 1994; Power and Wells, 1996), with the highest homology to receptors for IL-8, namely CXCR1 and CXCR2 (Murphy and Tiffany, 1991; Holmes *et al.*, 1991).

The amino acid sequences deduced from several KSHV DNA isolates from KS tissues and B cell lymphomas were identical (U24275, U71368, AF079845; U75698, U93872). One amino acid sequence from a B cell lymphoma differed by one residue (U82242) and another by 13 residues in TM-4 (U82242) caused by the loss of a single nucleotide that is recovered by a downstream loss of two nucleotides.

Affinity for ligand(s)

KSHV GPCR appears to bind a number of human CXC and CC chemokines (Arvanitakis *et al.*, 1997). However, binding studies have been confounded by the interactions of many chemokine ligands with glycosaminoglycans. Therefore, characterization from measurements of effects of chemokines on signaling by KSHV GPCR may be more definitive than those from binding studies. In general, relative affinities of ligands for GPCRs can be estimated from relative potencies. Although most chemokines tested do not affect KSHV GPCR signaling, a small number were found that further stimulate KSHV GPCR constitutive activity (see below) and others that inhibit KSHV GPCR signaling. Human growth-related protein α (GROα) ($EC_{50} = 15$ nM) and IL-8 ($EC_{50} = 39$ nM) (Gershengorn *et al.*, 1998) further stimulate KSHV GPCR whereas human IP-10 ($EC_{50} = 39$ nM) (Geras-Raaka *et al.*, 1998b), human SDF-1 ($EC_{50} = 43$ nM) and viral monocyte inflammatory protein II (vMIP-II) ($EC_{50} = 48$ nM) inhibit KSHV GPCR signaling (Geras-Raaka *et al.*, 1998a). Thus, IP-10, SDF-1, and vMIP-II are inverse agonists (or negative antagonists) of KSHV GPCR signaling.

Cell types and tissues expressing the receptor

KSHV GPCR is encoded by KSHV and has been found to be expressed at the messenger RNA (mRNA) level in lesions of patients with Kaposi's sarcoma and in lymphomatous B cells (Cesarman *et al.*, 1996).

Regulation of receptor expression

Regulation of KSHV GPCR expression is not known. The levels of KSHV GPCR mRNA can be increased in lymphomatous B cells in culture by incubation with phorbol esters (Sarid *et al.*, 1998) or butyrate (E. Cesarman, personal communication).

Release of soluble receptors

There is no evidence that this occurs; it is unlikely.

SIGNAL TRANSDUCTION

The most important aspect of KSHV GPCR signaling is that signaling occurs in the absence of any agonist; that is, KSHV GPCR is constitutively active.

Associated or intrinsic kinases

KSHV GPCR activates Jun kinase (JNK)/stress-activated protein kinase (SAP kinase) and p38 mitogen-activated protein kinase (p38 MAP kinase) but not extracellular signal-regulated kinase 2 (ERK-2)/MAP kinase (Bais *et al.*, 1998). The mechanism(s) of activation of these protein kinases is not known.

Cytoplasmic signaling cascades

KSHV GPCR signals via activation of intracellular phosphoinositide-specific phospholipase C leading to formation of inositol 1,4,5-trisphosphate (IP_3) and 1,2-diacylglycerol second messengers (Arvanitakis *et al.*, 1997). The G protein(s) that couple KSHV GPCR to phosphoinositide-specific phospholipase C is not known. Protein kinase C is activated but calcium-dependent protein kinases have not been studied.

DOWNSTREAM GENE ACTIVATION

Transcription factors activated

KSHV GPCR activates a protein kinase C-responsive promoter introduced by gene transfer (Arvanitakis *et al.*, 1997) and therefore probably acts, at least in part, via AP-1 transcription factor.

Genes induced

The only specific gene that has been shown to be induced by KSHV GPCR is that for VEGF, however, other genes must be induced because KSHV GPCR transforms NIH 3T3 cells (Bais *et al.*, 1998).

Promoter regions involved

There are protein kinase C-responsive elements in the VEGF promoter.

BIOLOGICAL CONSEQUENCES OF ACTIVATING OR INHIBITING RECEPTOR AND PATHOPHYSIOLOGY

Unique biological effects of activating the receptors

KSHV GPCR exhibits properties of an oncogene in that it transforms NIH 3T3 cells; KSHV GPCR-expressing NIH 3T3 cells grow in soft agar and form tumors in nude mice (Bais *et al.*, 1998). KSHV GPCR expression induced the expression and secretion of biologically active VEGF by NIH 3T3 cells (Bais *et al.*, 1998).

Phenotypes of receptor knockouts and receptor overexpression mice

Knockouts are not relevant as this is a virally encoded receptor. Direct overexpression in mice has not been done.

Human abnormalities

KSHV GPCR is hypothesized to play a role in tumorigenesis of *Kaposi's sarcoma* and primary effusion lymphomas.

THERAPEUTIC UTILITY

Effects of inhibitors (antibodies) to receptors

The effects of inverse agonists have not been studied in animal models.

References

Ahuja, S. K., and Murphy, P. M. (1993). Molecular piracy of mammalian interleukin-8 receptor type B by Herpesvirus Saimiri. *J. Biol. Chem.* **268**, 20691–20694.

Arvanitakis, L., Mesri, E. A., Nador, R. G., Said, J. W., Asch, A. S., Knowles, D. M., and Cesarman, E. (1996). Establishment and characterization of a body cavity-based lymphoma cell line (BC-3) harboring Kaposi's sarcoma-associated herpesvirus (KSHV/HHV-8) in the absence of Epstein-Barr virus. *Blood* **86**, 2708–2714.

Arvanitakis, L., Geras-Raaka, E., Varma, A., Gershengorn, M. C., and Cesarman, E. (1997). Human herpesvirus KSHV encodes a constitutively active G-protein-coupled receptor linked to cell proliferation. *Nature* **385**, 347–350.

Arvanitakis, L., Geras-Raaka, E., and Gershengorn, M. C. (1998). Constitutively signaling G protein-coupled receptors and human disease. *Trends Endocrinol. Metab.* **9**, 27–31.

Bais, C., Santomasso, B., Coso, O., Arvanitakis, L., Geras-Raaka, E., Gutkind, J. S., Asch, A. S., Cesarman, E., Gershengorn, M. C., and Mesri, E. A. (1998). G-protein-coupled receptor of Kaposi's sarcoma-associated is a viral oncogene and angiogenesis activator. *Nature* **391**, 86–89.

Baldwin, J. M., Schertler, G. F. X., and Unger, V. M. (1997). An alpha-carbon template for the transmembrane helices in the rhodopsin family of G-protein-coupled receptors. *J. Mol. Biol.* **272**, 144–164.

Cesarman, E., Chang, Y., Moore, P. S., Said, J. W., and Knowles, D. M. (1995). Kaposi's sarcoma-associated herpesvirus-like DNA sequences in AIDS-related body-cavity-based lymphomas. *N. Engl. J. Med.* **332**, 1186–1191.

Cesarman, E., Nador, R. G., Bai, F., Bohenzky, R. A., Russo, J. J., Moore, P. S., Chang, Y., and Knowles, D. M. (1996). Kaposi's sarcoma associated herpesvirus contains G protein-coupled receptor and cyclin D homologs which are expressed in Kaposi's sarcoma and malignant lymphoma. *J. Virol.* **70**, 8218–8223.

Chang, Y., Cesarman, E., Pessin, M. S., Lee, F., Culpepper, J., Knowles, D. M., and Moore, P. S. (1994). Identification of herpesvirus-like DNA sequences in AIDS-associated Kaposi's Sarcoma. *Science* **266**, 1865–1869.

Corbellino, M., Poirel, L., Aubin, J. T., Paulli, M., Magrini, U., Bestetti, G., Galli, M., and Parravicini, C. (1996). The role of human herpesvirus 8 and Epstein-Barr virus in the pathogenesis of giant lymph node hyperplasia (Castleman's disease). *Clin. Infect. Dis.* **22**, 1120–1121.

Flore, O., Rafii, S., Ely, S., O'Leary, J. J., Hyjek, E. M., and Cesarman, E. (1998). Transformation of primary human endothelial cells by Kaposi's sarcoma-associated herpesvirus. *Nature* **394**, 588–592.

Gao, S.-J., Kingsley, L., Hoover, D. R., Spira, T. J., Rinaldo, C. R., Saah, A., Phair, J., Detels, R., Parry, P., Chang, Y., and Moore, P. S. (1996a). Seroconversion to antibodies against Kaposi's sarcoma-associated herpesvirus-related latent nuclear antigens before the development of Kaposi's sarcoma. *N. Engl. J. Med.* **335**, 233–241.

Gao, S. -J., Kingsley, L., Li, M., Zheng, W., Parravicini, C., Ziegler, J., Newton, R., Rinaldo, C. R., Saah, A., Phair, J., Detels, R., Chang, Y., and Moore, P. S. (1996b). KSHV antibodies among Americans, Italians and Ugandans with and without Kaposi's sarcoma. *Nature Med.* **2**, 925–928.

Geras-Raaka, E., Varma, A., Clark-Lewis, I., and Gershengorn, M. C. (1998a). Kaposi's sarcoma-associated herpesvirus (KSHV) chemokine vMIP-II and human SDF-1 inhibit signaling by KSHV G protein-coupled receptor. *Biochem. Biophys. Res. Commun.* **253**, 725–727.

Geras-Raaka, E., Varma, A., Ho, H., Clark-Lewis, I., and Gershengorn, M. C. (1998b). Human interferon-γ-inducible protein (IP-10) inhibits constitutive signaling of Kaposi's sarcoma-associated herpesvirus G protein-coupled receptor. *J. Exp. Med.* **188**, 405–408.

Gershengorn, M. C., Geras-Raaka, E., Varma, A., and Clark-Lewis, I. (1998). Chemokines activate Kaposi's sarcoma-associated herpesvirus G protein-coupled receptor in mammalian cells in culture. *J. Clin. Invest.* **102**, 1469–1472.

Holmes, W. E., Lee, J., Kuang, W.-J., Rice, G. C., and Wood, W. I. (1991). Structure and functional expression of a human interleukin-8 receptor. *Science* **253**, 1278–1280.

Kedes, D. H., Operskalski, E., Busch, M., Kohn, R., Flood, J., and Ganem, D. (1996). The seroepidemiology of human herpesvirus 8 (Kaposi's sarcoma-associated herpesvirus): distribution of infection in KS risk groups and evidence for sexual transmission. *Nature Med.* **2**, 918–924.

Mesri, E. A., Cesarman, E., Arvanitakis, L., Rafii, S., Moore, M. A. S., Posnett, D. N., Knowles, D. M., and Asch, A. S. (1996). Human herpesvirus-8/Kaposi's sarcoma associated herpesvirus (HHV-8/KSHV) is a new transmissible virus that infects B-cells. *J. Exp. Med.* **183**, 2385–2390.

Miller, G. (1974). The oncogenicity of Epstein-Barr virus. *J. Infect. Dis.* **130**, 187–205.

Moore, P. S., and Chang, Y. (1995). Detection of herpesvirus-like DNA sequences in Kaposi's sarcoma in patients with and those without HIV infection. *N. Engl. J. Med.* **332**, 1181–1185.

Moore, P. S., Gao, S.-J., Dominguez, G., Cesarman, E., Lungu, O., Knowles, D. M., Garber, R., Pellett, P. E., McGeoch, D. J., and Chang, Y. (1996). Primary characterization of a herpesvirus agent associated with Kaposi's sarcoma. *J. Virol.* **70**, 549–558.

Murphy, P. M., and Tiffany, H. L. (1991). Cloning of complementary DNA encoding a functional human interleukin-8 receptor. *Science* **253**, 1280–1283.

Nador, R. G., Cesarman, E., Chadburn, A., Dawson, D. B., Ansari, M. Q., Said, J., and Knowles, D. M. (1996). Primary effusion lymphoma: a distinct clinicopathologic entity associated with the Kaposi's sarcoma-associated herpes virus. *Blood* **88**, 645–656.

Nicholas, J., Cameron, K. R., and Honess, R. W. (1992). Herpesvirus saimiri encodes homologues of G protein-coupled receptors and cyclins. *Nature* **355**, 362–365.

Noel, J. C. (1995). Kaposi's sarcoma and KSHV. *Lancet* **346**, 1359.

Offermann, M. K. (1996). Kaposi's sarcoma and HHV-8. *Trends Microbiol.* **4**, 419.

Post, G. R., and Brown, J. H. (1996). G protein-coupled receptors and signaling pathways regulating growth responses. *FASEB J.* **10**, 741–749.

Power, C. A., and Wells, T. N. C. (1996). Cloning and characterization of human chemokine receptors. *Trends Pharmacol. Sci.* **17**, 209–213.

Rangan, S. R., Martin, L. N., Enright, F. M., and Abee, C. R. (1977). Herpesvirus saimiri-induced lymphoproliferative disease in howler monkeys. *J. Natl Cancer Inst.* **59**, 165–171.

Russo, J. J., Bohenzky, R. A., Chien, M. C., Chen, J., Yan, M., Maddalena, D., Parry, J. P., Peruzzi, D., Edelman, I. S., Chang, Y. A., and Moore, P. S. (1996). Nucleotide sequence of the Kaposi sarcoma-associated herpesvirus (HHV8). *Proc. Natl Acad. Sci. USA* **93**, 14862–14867.

Sarid, R., Flore, O., Bohenzky, R. A., Chang, Y., and Moore, P. S. (1998). Transcription mapping of the Kaposi's sarcoma-associated herpesvirus (human herpesvirus 8) genome in a body cavity-based lymphoma cell line (BC-1). *J. Virol.* **72**, 1005–1012.

Strader, C. D., Fong, T. M., Tota, M. R., Underwood, D., and Dixon, R. A. F. (1994). Structure and function of G protein-coupled receptors. *Annu. Rev. Biochem.* **63**, 101–132.

DARC

Richard Horuk*

Department of Immunology, Berlex Bioscience, 15049 San Pablo Avenue, Richmond, CA 94804, USA

*corresponding author tel: 510-669-4625, fax: 510-669-4244, e-mail: horuk@crl.com

DOI: 10.1006/rwcy.2000.22015.

SUMMARY

The Duffy blood group antigen is a portal of entry for the malarial parasite *Plasmodium vivax*. Recent work has shown that this protein, also known as DARC, is a promiscuous chemokine receptor. Although there is still no evidence that DARC can signal, its expression in the CNS on neurons (Purkinje cells) and on endothelial cells lining postcapillary venules (the site of leukocyte trafficking) hint at an as yet unknown physiological role. This article reviews the literature, both past and recent, and discusses the biology of this enigmatic protein.

BACKGROUND

Discovery

The Duffy blood group antigen was first identified serologically on human erythrocytes as the target of alloantibodies that can cause *posttransfusion hemolytic reactions* (Cutbush *et al.*, 1950; Ikin *et al.*, 1951). Characterization of the Duffy antigen revealed that it was degraded by the proteases chymotrypsin and pronase, has a tendency to aggregate when boiled in SDS, and was recognized as a 46 kDa glycoprotein by a monoclonal antibody designated Fy6 (Hadley *et al.*, 1984; Nichols *et al.*, 1987). There are two major antigens, Fy^a and Fy^b, that were defined by a human erythrocyte agglutination assay. The Duffy antigens are encoded by two codominant alleles FyA and FyB and four different phenotypes exist: Fy(a+b+), Fy(a+b−), Fy(a−b+), and Fy(a−b−). The Fy(a−b−) phenotype, also known as the Duffy-negative phenotype, is found mainly in West Africans and in around two-thirds of Americans of African descent (Sanger *et al.*, 1955). This phenotype results in the absence of the Duffy antigen in the erythrocytes of these individuals. Interestingly, these same individuals are resistant to infection by the *malaria* parasite *Plasmodium vivax*, and two separate reports established that the Duffy antigen was a cellular attachment factor for this species of malaria parasite (Miller *et al.*, 1975, 1976).

Although these studies identified the Duffy antigen as a portal of entry for the malarial parasite *P. vivax*, the biology of this protein took a totally unexpected turn based on the findings of two interconnected studies. In 1991 Darbonne *et al.* described an erythrocyte protein that was able to bind the chemokine IL-8 with high affinity. During the course of characterizing this chemokine-binding protein, Horuk *et al.* (1993) showed that it bound a range of CXC and CC chemokines, including IL-8, MGSA, RANTES, and MCP-1 and, interestingly, noted that erythrocytes obtained from most African Americans did not bind these chemokines. This observation, together with the fact that the monoclonal antibody Fy6 blocked chemokine binding and that the chemokine MGSA inhibited *P. vivax* invasion of erythrocytes, suggested that the Duffy antigen could also be a chemokine receptor (Horuk *et al.*, 1993). These findings were confirmed when the Duffy antigen was cloned (Chaudhuri *et al.*, 1993) and the expressed protein, which had a seven transmembrane domain structure similar to that of other chemokine receptors, was shown to bind a range of CXC and CC chemokines (Chaudhuri *et al.*, 1994; Neote *et al.*, 1994).

The Duffy blood group antigen has a wider range of expression than was at first realized and as we shall see later it is expressed on a number of nonerythroid tissues such as endothelial cells of postcapillary venules and Purkinje cells in the cerebellum (Hadley *et al.*, 1994; Horuk *et al.*, 1997).

Alternative names

Duffy blood group antigen, Duffy glycoprotein, Duffy antigen receptor for chemokines (DARC).

Structure

DARC is a putative seven transmembrane domain receptor.

Main activities and pathophysiological roles

The involvement of DARC in pathophysiology is exemplified by its role as a transmission factor for *malaria*. The first clues for this were provided by Miller *et al.* (1975) who reported that Duffy-negative human erythrocytes were resistant to invasion by *P. knowlesi*, a simian malaria related to *P. vivax* that also requires DARC to invade human erythrocytes. These studies, together with the observation that almost 100% of individuals in West Africa who are unable to be infected with *P. vivax* malaria are also Duffy-negative, suggested a link between this innate resistance and the Fy(a−b−) phenotype. A number of studies demonstrated a correlation between the Duffy-negative phenotype and resistance to *P. vivax* infection (confirming that DARC was the vehicle of entry for the invasion of human erythrocytes by this malarial parasite). As suggested by Darbonne *et al.* (1991), the physiological role of DARC might be to act as an intravascular sink, which could bind and inactivate circulating chemokines. This would presumably generate a chemokine gradient with higher concentrations of active chemokines found in the subendothelial matrix, possibly bound to sulfated glycans. Clearance of IL-8 from the plasma by binding to erythrocytes has been demonstrated in humans treated with IL-1 (Tilg *et al.*, 1993). Although these data suggest a role for the erythrocyte DARC as a chemokine-binding protein, there is no formal evidence to suggest that this is its true biological role. In fact, if the postulated role of DARC in erythrocytes is biologically relevant, then the absence of this protein in Duffy-negative individuals should have clinical consequences. In this context it is interesting that Africans and African Americans have a lower peripheral neutrophil count, although this does not appear to be associated with any pathophysiology (Broun *et al.*, 1996). Whether or not this neutropenia is causally related to the Duffy-negative phenotype remains to be seen.

The localization of DARC to endothelial cells that line postcapillary venules is interesting since these structures are a dynamic interface that comprise the site for leukocyte transmigration from the vascular space into the tissue space during inflammation. This process, which is part of the inflammation cascade, is characterized by cytokine-mediated endothelial cell and leukocyte activation, selectin-mediated leukocyte rolling, integrin-mediated leukocyte adherence and ultimately migration of the leukocyte out of the vascular space into the surrounding tissues along chemokine gradients (Springer, 1991, 1994; Lasky, 1992). Endothelial cells, activated by cytokines *in vivo*, produce IL-8 (Hébert *et al.*, 1991), which may set up a chemotactic gradient favoring transendothelial diapedesis of leukocytes. The localization of DARC to endothelial cells of postcapillary venules, together with its ability to bind proinflammatory chemokines, suggests that it may play a major role in this inflammatory cascade.

GENE

Accession numbers

Human cDNA: U01839

Sequence

The gene sequence of the human protein has recently been reported (Chaudhuri *et al.*, 1993). The successful cloning of DARC was based on sequence information obtained from protease-digested fragments of Fy6 immunoaffinity-purified erythrocyte proteins. The human DARC sequence was found to have an open reading frame of 1014 bases, encoding a protein of 338 amino acids on a single intronless gene (**Figure 1**).

Northern blot analysis revealed a 1.27 kb transcript in kidney, spleen, and fetal liver, while the brain had an additional 8.5 kb transcript (Chaudhuri *et al.*, 1993). Although the two different DARC mRNA isoforms in the brain are differentially regulated and differ in their 5′ untranslated sequence, they encode the same polypeptide (Le Van Kim *et al.*, 1997). The basis for the Duffy-negative phenotype was recently shown to be due to a single T to C substitution at nucleotide −46 in the gene sequence. This mutation impairs the promoter activity in erythroid cells by disrupting a binding site for the GATA1 erythroid transcription factor (Tournamille *et al.*, 1995). In addition, a novel mutation, present in the FY*B coding sequence (271C to T), is associated with some

Figure 1 Nucleotide sequence of human DARC.

```
   1 GGCTTCCCCA GGACTGTTCC TGCTCCGGCT CTTCAGGCTC CCTGCTTTGT CCTTTTCCAC
  61 TGTCCGCACT GCATCTGACT CCTGCAGAGA CCTTGTTCTC CCACCCGACC TTCCTCTCTG
 121 TCCTCCCCTC CCACCTGCCC CTCAGTTCCC AGGAGACTCT TCCGGTGTAA CTCTGATGGC
 181 CTCCTCTGGG TATGTCCTCC AGGCGGAGCT CTCCCCCTCA ACTGAGAACT CAAGTCAGCT
 241 GGACTTCGAA GATGTATGGA ATTCTTCCTA TGGTGTGAAT GATTCCTTCC CAGATGGAGA
 301 CTATQATGCC AACCTGGAAG CAGCTGCCCC CTGCCACTCC TGTAACCTGC TGGATGACTC
 361 TGCACTGCCC TTCTTCATCC TCACCAGTGT CCTGGGTATC CTAGCTAGCA GCACTGTCCT
 421 CTTCATGCTT TTCAGACCTC TCTTCCGCTG GCAGCTCTGC CCTGGCTGGC CTGTCCTGGC
 481 ACAGCTGGCT GTGGGCAGTG CCCTCTTCAG CATTGTGGTG CCCGTCTTGG CCCCAGGGCT
 541 AGGTAGCACT CGCAGCTCTG CCCTGTGTAG CCTGGGCTAC TGTGTCTGGT ATGGCTCAGC
 601 CTTTGCCCAG GCTTTGCTGC TAGGGTGCCA TGCCTCCCTG GGCCACAGAC TGGGTGCAGG
 661 CCAGGTCCCA GGCCTCACCC TGGGGCTCAC TGTGGGAATT TGGGGAGTGG CTGCCCTACT
 721 GACACTGCCT GTCACCCTGG CCAGTGGTGC TTCTGGTGGA CTCTGCACCC TGATATACAG
 781 CACGGAGCTG AAGGCTTTGC AGGCCACACA CACTGTAGCC TGTCTTGCCA TCTTTGTCTT
 841 GTTGCCATTG GGTTTGTTTG GAGCCAAGGG GCTGAAGAAG GCATTGGGTA TGGGGCCAGG
 901 CCCCTGGATG AATATCCTGT GGGCCTGGTT TATTTTCTGG TGGCCTCATG GGGTGGTTCT
 961 AGGACTGGAT TTCCTGGTGA GGTCCAAGCT GTTGCTGTTG TCAACATGTC TGGCCCAGCA
1021 GGCTCTGGAC CTGCTGCTGA ACCTGGCAGA AGCCCTGGCA ATTTTGCACT GTGTGGCTAC
1081 GCCCCTGCTC CTCGCCCTAT TCTGCCACCA GGCCACCCGC ACCCTCTTGC CCTCTCTGCC
1141 CCTCCCTGAA GGATGGTCTT CTCATCTGGA CACCCTTGGA AGCAAATCCT AGTTCTCTTC
1201 CCACCTGTCA ACCTGAATTA AAGTCTACAC TGCCTTTGTG
```

Duffy-negative phenotypes among non-Ashkenazi Jews and among Brazilian blacks (Parasol *et al.*, 1998).

Chromosome location and linkages

The chromosomal location of the Duffy blood group antigen has been mapped to human chromosome 1q22-q23 where it is flanked by the genes for spectrin and Na/K ATPase (Marsh, 1977).

PROTEIN

Accession numbers

Mouse: AF016584, AF016697

Sequence

See **Figure 2**.

Description of protein

The open reading frame of the 1014 bp cDNA clone of DARC predicts a hydrophobic protein of 338 amino acid residues with a theoretical molecular mass of around 36 kDa (**Figure 3**). The protein has two *N*-linked glycosylation sites on the N-terminus. Based on hydropathy analysis of the cDNA clone it was originally suggested that the protein contains nine membrane-spanning domains. However, subsequent analysis using alternative computerized hydropathy plots revealed that DARC has seven hydrophobic membrane-spanning segments, more in line with that of the other cloned chemokine receptors. While DARC is biochemically distinct from the other cloned chemokine receptors, it does appear to share their common heptahelical topology, including the conservation of a number of Trp and Pro residues in helices IV, V, VI, and VII, which are highly conserved throughout the entire family of G protein-linked receptors and are postulated to play major roles in receptor binding and function (Wess *et al.*, 1993). The cloned DARC protein also has four conserved Cys residues (at the N-terminus, and the first, second, and third extracellular loops) that are thought to be paired to form disulfides which help to stabilize the protein.

Comparison of the primary sequence of DARC with those of the other chemokine receptors does, however, reveal some unique differences. For example, many residues in the predicted cytoplasmic loops, including the Asp-Arg-Tyr (DRY) motif at the end of the third transmembrane-spanning helix, and the C-terminal tail, which have been shown to be important in interacting with and coupling to G proteins, are not conserved in DARC. These changes in primary structure, together with the failure of the receptor to respond to ligands in any biological assay or to stimulate GTPase activity and the absence of any effect on ligand binding by treatment with *pertussis* toxin (Horuk, unpublished), suggest that DARC may not be coupled to G-proteins. In addition, when DARC is expressed in an insect cell line known to be devoid of G_i proteins (Quehenberger

Figure 2 Alignment of the primary structures of the cloned human and mouse DARC. Conserved residues are in orange. (Full colour figure may be viewed online.)

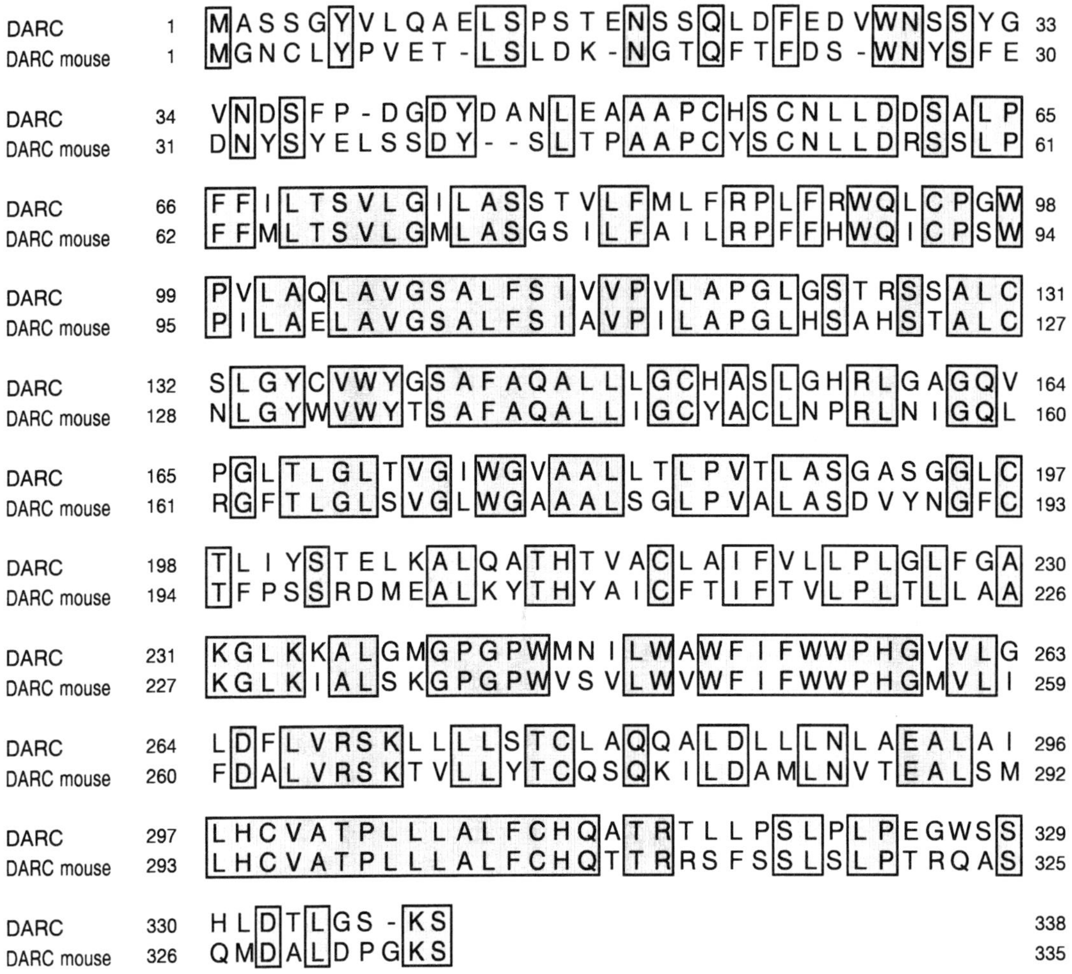

```
DARC          1  M A S S G Y V L Q A E L S P S T E N S S Q L D F E D V W N S S Y G  33
DARC mouse    1  M G N C L Y P V E T - L S L D K - N G T Q F T F D S - W N Y S F E  30

DARC         34  V N D S F P - D G D Y D A N L E A A A P C H S C N L L D D S A L P  65
DARC mouse   31  D N Y S Y E L S S D Y - - S L T P A A P C Y S C N L L D R S S L P  61

DARC         66  F F I L T S V L G I L A S S T V L F M L F R P L F R W Q L C P G W  98
DARC mouse   62  F F M L T S V L G M L A S G S I L F A I L R P F F H W Q I C P S W  94

DARC         99  P V L A Q L A V G S A L F S I V V P V L A P G L G S T R S S A L C  131
DARC mouse   95  P I L A E L A V G S A L F S I A V P I L A P G L H S A H S T A L C  127

DARC        132  S L G Y C V W Y G S A F A Q A L L L G C H A S L G H R L G A G Q V  164
DARC mouse  128  N L G Y W V W Y T S A F A Q A L L I G C Y A C L N P R L N I G Q L  160

DARC        165  P G L T L G L T V G I W G V A A L L T L P V T L A S G A S G G L C  197
DARC mouse  161  R G F T L G L S V G L W G A A A L S G L P V A L A S D V Y N G F C  193

DARC        198  T L I Y S T E L K A L Q A T H T V A C L A I F V L L P L G L F G A  230
DARC mouse  194  T F P S S R D M E A L K Y T H Y A I C F T I F T V L P L T L L A A  226

DARC        231  K G L K K A L G M G P G P W M N I L W A W F I F W W P H G V V L G  263
DARC mouse  227  K G L K I A L S K G P G P W V S V L W V W F I F W W P H G M V L I  259

DARC        264  L D F L V R S K L L L L S T C L A Q Q A L D L L L N L A E A L A I  296
DARC mouse  260  F D A L V R S K T V L L Y T C Q S Q K I L D A M L N V T E A L S M  292

DARC        297  L H C V A T P L L L A L F C H Q A T R T L L P S L P L P E G W S S  329
DARC mouse  293  L H C V A T P L L L A L F C H Q T T R R S F S S L S L P T R Q A S  325

DARC        330  H L D T L G S - K S                                                 338
DARC mouse  326  Q M D A L D P G K S                                                 335
```

et al., 1992), it demonstrated high-affinity ligand binding, in contrast to CXCR2, which failed to bind IL-8, presumably because the absence of interaction with a G_i subunit caused it to assume a conformation that lacked a functional binding pocket (Horuk and Peiper, unpublished).

Thus DARC appears to belong to a growing family of seven transmembrane receptors that are not coupled to G proteins. These include subtypes of receptors for dopamine, somatostatin, vasoactive intestinal peptide, and angiotensin, all of which appear to bind ligand independently of G protein coupling, although they have a DRY motif (Sokoloff *et al.*, 1990; Rens-Domiano *et al.*, 1992; Gressens *et al.*, 1993; Mukoyama *et al.*, 1993). Other members of this group include two *Drosophila* proteins, the BOSS protein, which has a large N-terminal extracellular domain, is not G protein-coupled, but is, nevertheless, biologically active and appears to be a ligand for a tyrosine kinase receptor called sevenless (Krämer *et al.*, 1991).

In addition, recent work suggests that some G protein-coupled receptors can also signal through alternative independent pathways (Milne *et al.*, 1995). For example, when the G protein-coupled cAMP receptor from the slime mold *Dictyostelium* is expressed in a cell line deficient in G protein β subunits, it is still able to activate some cellular responses such as Ca^{2+} ion influx and receptor phosphorylation (Milne *et al.*, 1995). Based on these, and other findings, Schnitzler *et al.* (1995) have postulated that this receptor signals through two distinct pathways, one being the classical G protein-linked pathway, and the other a G protein-independent pathway which appears to involve the activation of the transcription factor G-box binding factor.

Figure 3 Proposed membrane topography of DARC. Membrane-spanning α helices are defined based on hydropathy analysis. CHO, potential *N*-linked glycosylation sites. Residues involved in binding to the monoclonal antibody Fy6 are shown in green; residues involved in binding to the monoclonal antibody Fy3 are shown in orange. (Full colour figure may be viewed online.)

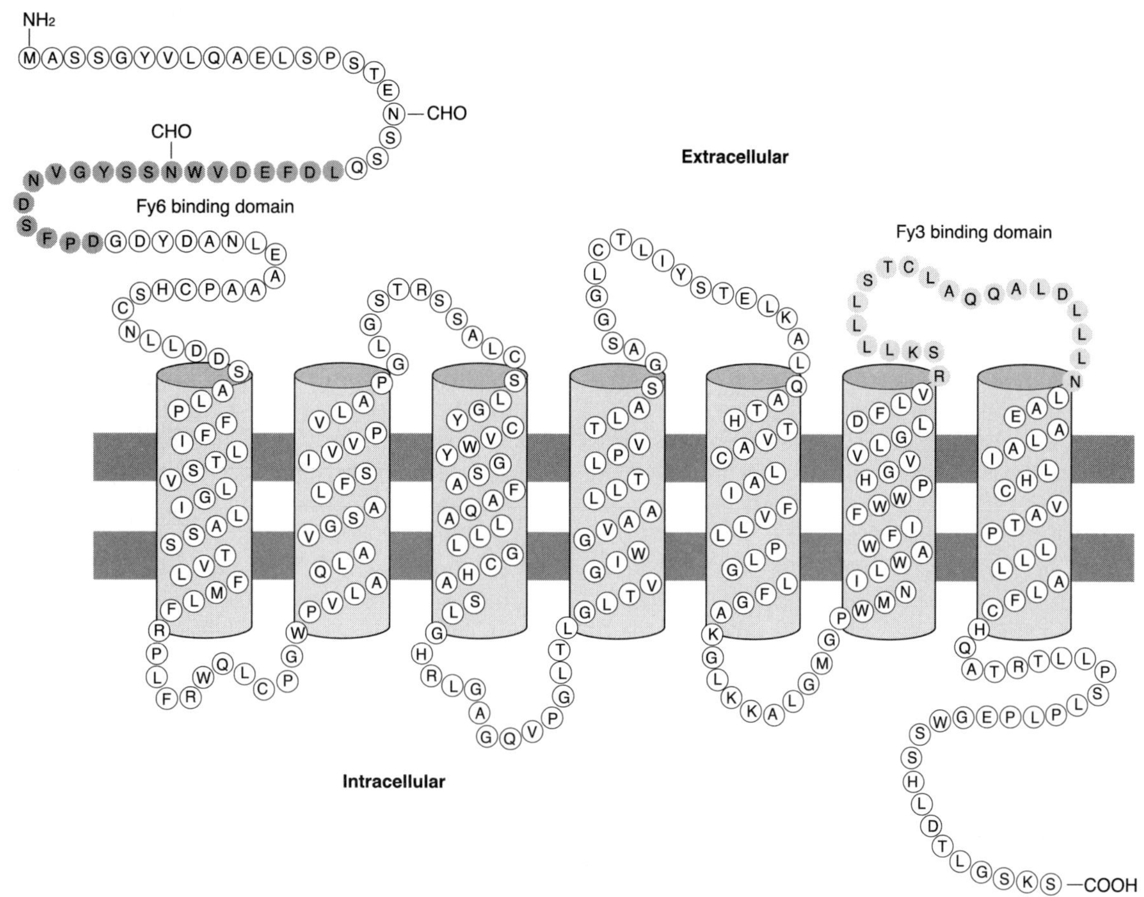

Relevant homologies and species differences

DARC has very little primary amino acid sequence homology with the other cloned chemokine receptors; it is most closely homologous with CXCR2 with which it shares around 24% homology. Analysis of the nucleotide sequences of DARC from individuals that are of the Fy(a+b−), Fy(a−b+), and Fy(a+b+) phenotype reveals that the Fya and Fyb alleles differ by a single base substitution in the second position of codon 44 that encodes a glycine residue in Fya and an aspartic acid residue in Fyb (Chaudhuri *et al.*, 1995). This polymorphism does not appear to have any physiological consequences.

In addition to the human nucleotide sequence, the mouse DARC sequence has also recently been described (Luo *et al.*, 1997). Unlike the human sequence which is intronless, the mouse sequence is encoded on two exons: exon 1 of 55 nucleotides, which encodes seven amino acid residues; and exon 2 of 1038 nucleotides, which encodes 327 residues. The single intron consists of 462 nucleotides. The open reading frame shows 60% homology with the human DARC protein. However, mouse erythrocytes are serologically Duffy-negative and mouse erythrocyte membrane proteins do not crossreact with two Duffy-specific rabbit polyclonal antibodies.

DARC has been cloned from a number of primates, including chimpanzee, aotus, squirrel, and rhesus monkeys (Chaudhuri *et al.*, 1995; Horuk *et al.*, 1996). As expected, homologies between the primate and human proteins are high, ranging from 93 to 99%. In addition, DARC has been partially cloned from cow, pig, and rabbit and the primary structure is conserved approximately 70–75% (Horuk *et al.*, 1996; Hadley and Peiper, 1997). The direct demonstration of the existence of a gene encoding a polypeptide highly homologous to DARC in cow, pig, rabbit, and mouse and the preservation of chemokine-binding function in rodent and avian erythrocytes (Horuk *et al.*, 1996) implies that this molecule might play an important biological role in these, and in other species.

Affinity for ligand(s)

Human erythrocytes are able to bind radiolabeled CXC and CC chemokines, including IL-8, MGSA, NAP-2, RANTES, and MCP-1 with high affinity (Horuk *et al.*, 1993; Neote *et al.*, 1993). However, human MIP-1α and MIP-1β were unable to compete effectively for binding. Scatchard analysis revealed that the affinity of binding ranged from 5 to 10 nM with around 5000 binding sites. The K_d for any combination of labeled chemokine displaced by any other was strikingly similar. These studies revealed that, in contrast to other chemokine receptors, DARC was able to bind both CXC and CC chemokines.

Receptor-binding studies with radiolabeled MGSA and IL-8 in mouse erythrocytes revealed specific binding of the radiolabeled chemokines that were displaceable by the predicted repertoire of unlabeled chemokines characteristic of DARC. However, some clear differences in chemokine binding by murine DARC compared to human DARC were observed. For example, mouse erythrocytes bound human IL-8 poorly compared to human MGSA, whereas both chemokines bound equally well to human erythrocytes. In addition, the CC chemokine MIP-1α readily displaced both radiolabeled MGSA and IL-8 from mouse erythrocytes; however, the K_d for this displacement is around 120 nM compared to a K_d of 5–10 nM for the binding of MGSA, RANTES, and MCP-1 to this receptor. In contrast, MIP-1α has almost no effect on chemokine binding in human erythrocytes.

In summary, chemokines can be classified into five distinct groups with respect to their ability to bind to DARC. Two distinct groups that bind to DARC, both human and mouse, with high affinity are the CXC chemokines that have the ELR motif (IL-8, MGSA, NAP-2, and ENA-78), and the basic CC chemokines (RANTES, MCP-1, and MCP-3). In contrast, the non-ELR CXC chemokines, characterized by PF4 and IP-10, bind to human and mouse DARC with low affinity. The acidic CC chemokines, characterized by the MIP-1 proteins, do not bind to human DARC and bind to mouse DARC with low affinity. Finally, the only member of the C chemokines so far characterized, lymphotactin, does not bind at all to DARC based on displacement studies with radiolabeled MGSA, and with an affinity that is in the low micromolar range based on direct binding studies with radiolabeled lymphotactin.

Chemokine-binding experiments with human, monkey, and chicken erythrocytes demonstrated that all four species bound chemokines with high affinity. The K_d values for MGSA binding were 6 nM human, 6 nM monkey, 13 nM mouse, and 9 nM chicken erythrocytes (Horuk *et al.*, 1996). Thus, the chemokine-binding site in DARC appears to be highly conserved from human to bird, suggesting a nonredundant role for this protein.

Previous studies with other chemokine receptors including CXCR1, CXCR2, and CCR5 have revealed that the N-terminal domain of these receptors is at least partly responsible for their ligand-binding specificity (LaRosa *et al.*, 1992; Gayle *et al.*, 1993; Doms and Peipert, 1997). In a recent communication, Lu *et al.* (1995) show that this is also the case for the DARC. The ligand-binding specificity of DARC was investigated with a receptor chimera composed of the N-terminal extracellular domain of DARC (amino acid residues 1–66) and the seven transmembrane spanning regions and cytoplasmic tail of CXCR2 (amino acid residues 50–355). Receptor-binding studies clearly demonstrated that the DARC/CXCR2 chimera could bind MGSA, IL-8, and RANTES with high affinity. Interestingly, the chimera, in contrast to DARC, did not bind MCP-1: the reasons for this are at present unclear. However, an alanine scan mutant of MGSA, E_6A, displayed high-affinity binding to both DARC and the DARC/CXCR2 chimera but did not bind to CXCR2. These findings suggest that the N-terminal region of the DARC is at least partly responsible for conferring high-affinity chemokine binding to the protein. It may also contain the binding epitopes for the binding of the *P. vivax* malaria parasite ligand, since COS cells transfected with the parasite ligand can bind to the DARC/CXCR2 chimera.

Analysis of erythrocytes treated with sulfhydryl group-modifying reagents have demonstrated that the chemokine receptor function of DARC requires the integrity of disulfide bond(s) but not that of free sulfhydryl group(s) (Tournamille *et al.*, 1997). Accordingly, mutation of cysteines 51 and 276 abolished chemokine binding to DARC transfectants. These results suggest that the chemokine-binding pocket of DARC, which includes residues in the N-terminus and the third extracellular loop, are brought into close vicinity by a disulfide bridge.

Recent studies have identified a single-base polymorphism, C286T, of DARC in some individuals: this results in a single amino acid substitution Arg89Cys that affects both Fy6 and chemokine binding to DARC transfectants (Tournamille *et al.*, 1998). This mutation results in a very low expression of DARC on erythroid cells. Examination of DARC sequences shows that this residue, which is in the first extracellular loop of the protein, is conserved or replaced by a homologous charged His residue in other species. The extracellular loops of chemokine receptors and seven transmembrane G protein-coupled receptors in general contain a number of positively charged Lys and Arg residues. It has been suggested that these positively charged amino acids can interact with negative charges on the polar headgroups of phospholipids in the cell membrane and that this may help to maintain receptor topology. Thus, mutation of these residues could lead to a disruption of the overall architecture of the receptor and lead to a decrease in binding affinity for the ligand.

Further insight into the chemokine binding site of DARC has been provided by alanine scanning mutagenesis studies with MGSA (Hesselgesser *et al.*, 1995). Previous work with IL-8 showed that the sequence $E_4L_5R_6$ was essential for receptor binding and neutrophil activation of CXCR1 and CXCR2 (Hébert *et al.*, 1991). All three residues – arginine in particular – are highly sensitive to modification. In contrast to these findings, recent work with alanine scan mutants of MGSA suggests that the ELR motif is less important for binding to DARC than to CXCR2 (Hesselgesser *et al.*, 1995). These studies demonstrated that the binding affinities of the MGSA mutants E_6A, and L_7A for DARC were only about 2-fold and 10-fold less, respectively, than that of wild-type MGSA, suggesting that these residues are not important in determining binding to DARC (Hesselgesser *et al.*, 1995). The binding affinity of the MGSA mutant R_8A, however, was approximately 240-fold lower than wild-type MGSA, indicating that a positive charge may be required in this region of the protein for binding. Interestingly, some members of the CC chemokines (which all lack this positive charge as well as the E and L residues) like RANTES and MCP-1, bind to DARC whereas others like MIP-1α and MIP-1β do not bind (Horuk, 1994). Based on analysis of the solution structure of RANTES (Skelton *et al.*, 1995), and the amino acid sequence of the CC chemokines, we can speculate that the positively charged lysine residue next to the third cysteine residue, at positions 33 in RANTES and 35 in MCP-1, may fill the role for the absence of the arginine at position 8. The chemokines MIP-1α and MIP-1β lack a positively charged residue at both positions.

The *P. vivax* and *P. knowlesi* Duffy-binding ligands have been cloned and are large proteins of molecular mass 140 kDa and 135 kDa, respectively (Adams *et al.*, 1992). Based on their sequence homologies, these ligands have been divided into six regions, including two cysteine-rich areas (Adams *et al.*, 1992). Recently, one of these two cysteine-rich stretches of the parasite ligand (region II) has been identified as the binding domain that binds to the DARC. The region II protein ($\sim$330 amino acids) was expressed in COS7 cells and found to be capable of binding to Duffy-positive but not to Duffy-negative erythrocytes. Furthermore, preincubation with MGSA and IL-8 blocked the binding of the region II protein to Duffy-positive erythrocytes. These studies suggest that the parasite ligand and the chemokines could bind to similar epitopes on the DARC. Examination of the primary sequences of these disparate proteins reveals no regions of homology and we will have to await a detailed comparison of their tertiary structures, which is still lacking for the parasite Duffy binding protein, to obtain more information regarding their binding to DARC.

Cell types and tissues expressing the receptor

A variety of human erythroleukemic cell lines, including Kg1, K562, and HEL cells, were tested for the expression of DARC by screening for [^{125}I]IL-8 binding (Horuk *et al.*, 1994). Of the cell lines screened, only the HEL cells showed specific [^{125}I]IL-8 binding. These cells, which were originally derived from a patient with *Hodgkin's disease*, carry the phenotypic markers of erythroid cells that include the ability to synthesize globin (Martin and Papayannopoulou, 1982). Further analysis of HEL cells determined that they appeared to express a protein with the characteristic hallmarks of DARC, i.e. receptor binding of a wide array of chemokines, inhibition of this binding by the Fy6 antibody, crossreactivity by western blotting with Fy6 of a protein of a similar molecular mass, and hybridization of mRNA from HEL cells with a cDNA probe to DARC.

Transcripts encoding isoforms of DARC have been detected in polyadenylated RNA from a variety of human tissues, including kidney, spleen, lung, and brain (Chaudhuri *et al.*, 1993; Neote *et al.*, 1994). Based on a number of experimental observations, including chemokine binding and immunohistochemical staining, Hadley *et al.* (1994) showed that DARC protein was expressed on endothelial cells lining the small blood vessels of the human kidney. Evidence that the DARC polypeptide is also expressed in endothelial cells lining postcapillary venules of spleen was provided in a related report (Peiper *et al.*, 1995). In the same study the authors demonstrated the specific immunohistochemical staining of soft tissue from a Duffy-negative patient with the monoclonal antibody Fy6. The DARC-specific staining was localized to specialized endothelial cells that line the postcapillary venules of the tissue, but, as expected, erythrocytes within the lumen failed to show binding of the monoclonal antibody. Confirmation of the immunohistochemical identification of this protein was provided by northern blotting, ligand binding, chemical crosslinking, and immunoblotting experiments which were all consistent with the characteristic features of DARC.

These findings indicate that the expression of the DARC is retained on endothelial cells even in the presence of strong negative selection from morbidity and mortality from *P. vivax* which resulted in the loss of expression of this protein by erythrocytes. This raises the possibility that DARC plays a critical role in the biology of endothelial cells. Although there is no evidence that DARC can transduce a biological signal in endothelial cells, it has been shown that the receptor can internalize in response to ligand binding (Horuk *et al.*, 1994). In this context it is interesting that a recent study has suggested that DARC might participate in transcytosis and surface presentation of IL-8 by venular endothelial cells (Middleton *et al.*, 1997).

Northern blotting experiments demonstrating that mRNA encoding DARC is expressed in human brain (Chaudhuri *et al.*, 1993) prompted us to examine archival sections of human brain for DARC expression (Horuk *et al.*, 1997). Immunohistochemistry of brain sections revealed that neuronal processes were expressing DARC in the cerebellum and adjacent regions of the brainstem. These immunohistochemical observations were confirmed by ligand-binding studies with isolated membranes from human cerebellum which demonstrated radiolabeled chemokine binding with a pattern of displacement identical to that observed for DARC in erythrocytes. Scatchard analysis of radiolabeled chemokine binding revealed a single class of binding sites in the cerebellum with a K_d of 4 nM. This binding affinity is very similar to that previously reported for erythrocyte DARC (Horuk *et al.*, 1993; Neote *et al.*, 1993). Given that chemokines such as IL-8, RANTES, and MCP-1 are expressed by resident cells in the CNS, primarily astrocytes, and that DARC is expressed by neurons in the CNS, it is tempting to speculate that DARC may play a important role in the modulation of neuronal activity by astrocytes.

BIOLOGICAL CONSEQUENCES OF ACTIVATING OR INHIBITING RECEPTOR AND PATHOPHYSIOLOGY

Phenotypes of receptor knockouts and receptor overexpression mice

Genetic mutations of receptors both natural and induced (by targeted gene disruption) can help to unravel their biological roles. Although no DARC receptor knockouts have been developed in mice so far, nature has been generous in this regard by providing us with a naturally occurring example of gene inactivation for DARC. Humans homozygous for inherited inactivating mutations of the DARC gene in erythrocytes have been identified, and appear to be phenotypically normal and healthy (Mallinson *et al.*, 1995). Indeed, as we have seen, this gene inactivation appears to be beneficial to the host, rendering the individual resistant to *malaria* induced by *P. vivax* which utilizes DARC to attach to and enter erythrocytes. Interestingly, these Duffy-negative individuals are not truly deficient and do express DARC on nonerythroid cells. Although there do not appear to be any noticeable pathophysiologic consequences resulting from a Duffy-negative phenotype, it is of course possible that there are compensatory mechanisms that take over the postulated biological role of DARC as a chemokine sink on erythroid cells. We will have to await the description of a true DARC knockout in mice to determine the physiologic role of DARC on nonerythroid cells.

THERAPEUTIC UTILITY

Effect of treatment with soluble receptor domain

A monoclonal antibody to the Duffy blood group antigen known as Fy6 has been described by Nichols *et al.* (1987). In humans, the epitope to Fy6 is present on the erythrocytes of all persons except those of the Fy(a−b−) type. The Fy6 epitope on DARC was shown to be on the N-terminus and is centered around residues Phe22 and Glu23; mutation of either of these residues destroys both Fy6 and chemokine binding (Tournamille *et al.*, 1987). Consistent with the idea that Fy6 and chemokines share the same binding sites for DARC, Fy6 is able to inhibit chemokine binding (Horuk *et al.*, 1993). No direct *in vivo* studies to inhibit *P. vivax* infection of human erythrocytes have been described. However, given the commonality of chemokine/Fy6/parasite-binding sites for DARC, it might be expected that Fy6 could prove to be effective *in vivo* in inhibiting *P. vivax* and may thus be therapeutically useful.

Effects of inhibitors (antibodies) to receptors

In a recent study the ability of selected MGSA mutants to inhibit the invasion of human Duffy-positive erythrocytes by *P. knowlesi* was assessed (Hesselgesser *et al.*, 1995). The mutants inhibited parasite invasion at ligand concentrations that were consistent with their receptor-binding affinities for DARC. For example, the mutant E_6A was almost as effective as MGSA with an EC_{50} of inhibition of invasion of 8.6 nM, compared to 7 nM for wild-type MGSA. Mutant R_8A which does not bind to the DARC did not inhibit parasite invasion at concentrations up to 1 μM. The mutant E_6A binds to DARC with high affinity and efficiently blocks parasite invasion but does not bind CXCR2 and does not activate neutrophils. Analogs of MGSA, like E_6A, may be useful as receptor-blocking drugs that inhibit erythrocyte invasion by *P. vivax* but do not affect neutrophils. With the increasing incidence of chloroquine-resistant strains of *malaria*, new approaches to combat this disease are required and therapies based on chemokine analogs are a new and novel approach in the fight against *P. vivax*-induced malaria.

Recent studies have identified the peptide within the Duffy blood group antigen of human erythrocytes to which the *P. vivax* and *P. knowlesi* ligands bind (Chitnis *et al.*, 1996). Peptides from the N-terminal extracellular region of the Duffy antigen were tested for their ability to block the binding of erythrocytes to transfected COS cells expressing on their surface region II of the Duffy-binding ligands. The binding site on the human Duffy antigen used by both the *P. vivax* and *P. knowlesi* ligands maps to a 35 amino acid region. A 35 amino acid peptide from the human Duffy antigen blocked the binding of *P. vivax* to human erythrocytes with a K_i of 2.9 μM. These studies suggest that it might be possible to design small peptides that are effective therapeutic agents to inhibit *P. vivax*-induced *malaria*.

References

Adams, J. H., Sim, B. K. L., Dolan, S. A., Fang, X., Kaslow, D. C., and Miller, L. H. (1992). A family of erythrocyte binding proteins of malaria parasites. *Proc. Natl Acad. Sci. USA* **89**, 7085–7089.

Broun Jr, G. O., Herbig, F. K., and Hamilton, J. R. (1966). Leukopenia in Negroes. *N. Engl. J. Med.* **275**, 1410–1413.

Chaudhuri, A., Polyakova, J., Zbrzezna, V., Williams, K., Gulati, S., and Pogo, A. O. (1993). Cloning of glycoprotein D cDNA, which encodes the major subunit of the Duffy blood group system and the receptor for the *Plasmodium vivax* malaria parasite. *Proc. Natl Acad. Sci. USA* **90**, 10793–10797.

Chaudhuri, A., Zbrzezna, V., Polyakova, J., Pogo, A. O., Hesselgesser, J., and Horuk, R. (1994). Expression of the Duffy antigen in K562 cells: evidence that it is the human erythrocyte chemokine receptor. *J. Biol. Chem.* **269**, 7835–7838.

Chaudhuri, A., Polyakova, J., Zbrzezna, V., and Pogo, A. O. (1995). The coding sequence of Duffy blood group gene in humans and simians: restriction fragment polymorphism. Antibody and malarial parasite specificities, and expression in nonerythroid tissues in Duffy-negative individuals. *Blood* **85**, 615–621.

Chitnis, C. E., Chaudhuri, A., Horuk, R., Pogo, A. O., and Miller, L. H. (1996). The domain on the Duffy blood group antigen for binding *Plasmodium vivax* and *P. knowlesi* malarial parasites to erythrocytes. *J. Exp. Med.* **184**, 1531–1536.

Cutbush, M., Mollinson, P. L., and Parkin, D. M. (1950). A new human blood group. *Nature* **165**, 188–190.

Darbonne, W. C., Rice, G. C., Mohler, M. A., Apple, T., Hebert, C. A., Valente, A. J., and Baker, J. B. (1991). Red blood cells are a sink for interleukin 8, a leukocyte chemotaxin. *J. Clin. Invest.* **88**, 1362–1369.

Doms, R. W., and Peipert, S. C. (1997). Unwelcomed guests with master keys: How HIV uses chemokine receptors for cellular entry. *Virology* **235**, 179–190.

Gayle, R. B., Sleath, P. R., Srinivason, S., Birks, C. W., Weerawarna, K. S., Cerretti, D. P., Kozlosky, C. J., Nelson, N., Vanden, B. T., and Beckmann, M. P. (1993). Importance of the amino terminus of the interleukin-8 receptor in ligand interactions. *J. Biol. Chem.* **268**, 7283–7289.

Gressens, P., Hill, J. M., Gozes, I., Fridkin, M., and Brenneman, D. E. (1993). Growth factor function of vasoactive intestinal peptide in whole cultured mouse embryos. *Nature* **362**, 155–158.

Hadley, T. J., and Peiper, S. C. (1997). From malaria to chemokine receptor: the emerging physiologic role of the duffy blood group antigen. *Blood* **89**, 3077–3091.

Hadley, T. J., David, P. H., McGinniss, M. H., and Miller, L. H. (1984). Identification of an erythrocyte component carrying the duffy blood group Fy^a antigen. *Science* **223**, 597–599.

Hadley, T. J., Lu, Z.-H., Wasniowska, K., Martin, A. W., Peiper, S. C., Hesselgesser, J., and Horuk, R. (1994). Postcapillary venule endothelial cells in kidney express a multispecific chemokine receptor that is structurally and functionally identical to the erythroid isoform, which is the duffy blood group antigen. *J. Clin. Invest.* **94**, 985–991.

Hébert, C. A., Luscinskas, F. W., Kiely, J.-M., Luis, E. A., Darbonne, W. C., Bennett, G. L., Liu, C. C., Obin, M. S., Gimbrone, M. A., and Baker, J. B. (1990). Endothelial and leukocyte forms of IL-8: conversion by thrombin and interaction with neutrophils. *J. Immunol.* **145**, 3033–3040.

Hébert, C. A., Vitangcol, R. V., and Baker, J. B. (1991). Scanning mutagenesis of interleukin-8 identifies a cluster of residues required for receptor binding. *J. Biol. Chem.* **266**, 18989–18994.

Hesselgesser, J., Chitnis, C., Miller, L., Yansura, D. J., Simmons, L., Fairbrother, W., Kotts, C., Wirth, C., Gillece-Castro, B., and Horuk, R. (1995). A mutant of melanoma growth stimulating activity does not activate neutrophils but blocks erythrocyte invasion by malaria. *J. Biol. Chem.* **270**, 11472–11476.

Horuk, R. (1994). The interleukin-8-receptor family: from chemokines to malaria. *Immunol. Today* **15**, 169–174.

Horuk, R., Chitnis, C. E., Darbonne, W. C., Colby, T. J., Rybicki, A., Hadley, T. J., and Miller, L. H. (1993). A receptor for the malarial parasite *Plasmodium vivax*: the erythrocyte chemokine receptor. *Science* **261**, 1182–1184.

Horuk, R., Zi-xuan, W., Peiper, S. C., and Hesselgesser, J. (1994). Identification and characterization of a promiscuous chemokine receptor in a human erythroleukemic cell line. *J. Biol. Chem.* **269**, 17730–17733.

Horuk, R., Martin, A., Hesselgesser, J., Hadley, T., Lu, Z.-H., Wang, Z.-X., and Peiper, S. C. (1996). The Duffy antigen receptor for chemokines: structural analysis and expression in the brain. *J. Leuk. Biol.* **59**, 29–38.

Horuk, R., Martin, A. W., Wang, Z.-X., Scweitzer, L., Gerassimides, A., Lu, Z.-H., Hesselgesser, J., Kim, J., Parker, J., Hadley, T. J., Perez, H. D., and Peiper, S. C. (1997). Expression of chemokine receptors by subsets of neurons in the normal central nervous system. *J. Immunol.* **158**, 2882–2890.

Ikin, E. W., Mourant, A. E., Pettenkoffer, J. H., and Blumenthal, G. (1951). Discovery of the expected haemagglutinin ant-Fy^b. *Nature* **168**, 1077–1078.

Krämer, H., Cagan, R. L., and Zipursky, S. L. (1991). Interaction of *bride of sevenless* membrane-bound ligand and the *sevenless* tyrosine-kinase receptor. *Nature* **352**, 207–212.

LaRosa, G. J., Thomas, K. M., Kaufman, M. E., Mark, R., White, M., Taylor, L., Gray, G., Witt, D., and Navarro, J. (1992). Amino terminus of the interleukin-8 receptor is a major determinant of receptor subtype specificity. *J. Biol. Chem.* **267**, 25402–25406.

Lasky, L. A. (1992). Selectins: interpreters of cell-specific carbohydrate information during inflammation. *Science* **258**, 964–969.

Le Van Kim, C., Tournamille, C., Kroviarski, Y., Cartron, J. P., and Colin, Y. (1997). The 1.35-kb and 7.5-kb Duffy mRNA isoforms are differently regulated in various regions of brain, differ by the length of their 5′ untranslated sequence, but encode the same polypeptide. *Blood* **90**, 2851–2853.

Lu, Z.-H., Wang, Z.-X., Horuk, R., Hesselgesser, J., Lou, Y.-C., Hadley, T. J., and Peiper, S. C. (1995). The promiscuous chemokine binding profile of the duffy antigen/receptor for chemokines is primarily localized to sequences in the amino terminal domain. *J. Biol. Chem.* **270**, 26239–26245.

Luo, H., Chaudhuri, A., Johnson, K. R., Neote, K., Zbrzezna, V., He, Y., and Pogo, A. O. (1997). Cloning, characterization, and mapping of a murine promiscuous chemokine receptor gene: homolog of the human duffy gene. *Genome Res.* **7**, 932–941.

Mallinson, G., Soo, K. S., Schall, T. J., Pisacka, M., and Anstee, D. J. (1995). Mutations in the erythrocyte chemokine receptor (Duffy) gene: the molecular basis of the Fy^a/Fy^b antigens and identification of a deletion in the Duffy gene of an apparently healthy individual with the Fy(a−b−) phenotype. *Br. J. Haematol.* **90**, 823–829.

Marsh, W. L. (1977). Mapping assignment of the Rh and Duffy blood group genes to chromosome 1. *Mayo Clin. Proc.* **52**, 145–149.

Martin, P., and Papayannopoulou, T. (1982). Hel cells: a new human erythroleukemia cell line with spontaneous and induced globin expression. *Science* **216**, 1233–1235.

Middleton, J., Neil, S., Wintle, J., Clark-Lewis, I., Moore, H., Lam, C., Auer, M., Hub, E., and Rot, A. (1997). Transcytosis and surface presentation of IL-8 by venular endothelial cells. *Cell* **91**, 385–395.

Miller, L. H., Mason, S. J., Dvorak, J. A., McGinniss, M. H., and Rothman, I. K. (1975). Erythrocyte receptors for (*Plasmodium knowlesi*) malaria: Duffy blood group determinants. *Science* **189**, 561–563.

Miller, L. H., Mason, S. J., Clyde, D. F., and McGinniss, M. H. (1976). The resistance factor to *Plasmodium vivax* in blacks. The Duffy-blood group genotype, FyFy. *N. Engl. J. Med.* **295**, 302–304.

Milne, J. L. S., Wu, L., Caterina, M. J., and Devreotes, P. N. (1995). Seven helix cAMP receptors stimulate Ca^{2+} entry in the absence of functional G proteins in *Dictyostelium*. *J. Biol. Chem.* **270**, 5926–5931.

Mukoyama, M., Nakajima, M., Horiuchi, S., Pratt, R. E., and Dzau, V. J. (1993). Expression cloning of type 2 angiotensin II receptor reveals a unique class of seven-transmembrane receptors. *J. Biol. Chem.* **268**, 24539–24542.

Neote, K., Darbonne, W. C., Ogez, J., Horuk, R., and Schall, T. J. (1993). Identification of a promiscous inflammatory peptide receptor on the surface of red blood cells. *J. Biol. Chem.* **268**, 12247–12249.

Neote, K., Mak, J. Y., Kolakowski, L. F. J., and Schall, T. J. (1994). Functional and biochemical analysis of the cloned Duffy antigen: identity with the red blood cell chemokine receptor. *Blood* **84**, 44–52.

Nichols, M. E., Rubinstein, P., Barnwell, J., de Cordoba, S. R., and Rubinstein, R. E. (1987). A new human duffy blood group specificity defined by a murine monoclonal antibody. *J. Exp. Med.* **166**, 776–785.

Parasol, N., Reid, M., Rios, M., Castilho, L., Harari, I., and Kosower, N. S. (1998). A novel mutation in the coding sequence of the Fy*b allele of the duffy chemokine receptor gene is associated with an altered erythrocyte phenotype. *Blood* **92**, 2237–2243.

Peiper, S., Wang, Z.-X., Neote, K., Martin, A. W., Showell, H. J., Conklyn, M. J., Ogborne, K., Hadley, T. J., Zhao-hai, L., Hesselgesser, J., and Horuk, R. (1995). The Duffy antigen/receptor for chemokines (DARC) is expressed in endothelial cells of Duffy negative individuals who lack the erythrocyte receptor. *J. Exp. Med.* **181**, 1311–1317.

Quehenberger, O., Prossnitz, E. R., Cochrane, C. G., and Ye, R. D. (1992). Absence of G_i proteins in the Sf9 insect cell. *J. Biol. Chem.* **267**, 19757–19760.

Rens-Domiano, S., Law, S. F., Yamada, Y., Seino, S., Bell, G. I., and Reisine, T. (1992). Pharmacological properties of two cloned somatostatin receptors. *Mol. Pharmacol.* **42**, 28–34.

Sanger, R., Race, R. R., and Jack, J. A. (1955). The Duffy blood groups of New York negroes. The phenotype Fy(a−b−). *Br. J. Haematol.* **1**, 370–374.

Schnitzler, G. R., Briscoe, C., Brown, J. M., and Firtel, R. A. (1995). Serpentine cAMP receptors may act through a G protein-independent pathway to induce postaggregative development in *Dictyostelium*. *Cell* **81**, 737–745.

Skelton, N. J., Aspiras, F., Ogez, J., and Schall, T. J. (1995). Proton NMR assignments and solution conformation of RANTES a chemokine of the C–C type. *Biochemistry* **34**, 5329–5342.

Sokoloff, P., Giros, B., Martres, M.-P., Bouthenet, M.-L., and Schwartz, J.-C. (1990). Molecular cloning and characterization of a novel dopamine receptor (D3) as a target for neuroleptics. *Nature* **347**, 146–151.

Springer, T. A. (1991). Adhesion receptors of the immune system. *Nature* **346**, 425–433.

Springer, T. A. (1994). Traffic signals for lymphocyte recirculation and leukocyte emigration: the multistep paradigm. *Cell* **76**, 301–314.

Tilg, H., Pape, D., Trehu, E., Shapiro, L., Atkins, M. B., Dinarello, C. A., and Mier, J. W. (1993). A method for the detection of erythrocyte-bound interleukin-8 in humans during interleukin-1 immunotherapy. *J. Immunol. Methods* **163**, 253–258.

Tournamille, C., Colin, Y., Cartron, J. P., and Le Van Kim, C. (1995). Disruption of a GATA motif in the Duffy gene promoter abolishes erythroid gene expression in Duffy-negative individuals. *Nature Genetics* **10**, 224–228.

Tournamille, C., Le Van Kim, C., Gane, P., Blanchard, D., Proudfoot, A. E., Cartron, J. P., and Colin, Y. (1997). Close association of the first and fourth extracellular domains of the Duffy antigen/receptor for chemokines by a disulfide bond is required for ligand binding. *J. Biol. Chem.* **272**, 16274–16280.

Tournamille, C., Le Van Kim, C., Gane, P., Le Pennec, P. Y., Roubinet, F., Babinet, J., Cartron, J. P., and Colin, Y. (1998). Arg89Cys substitution results in very low membrane expression of the Duffy antigen/receptor for chemokines in Fyx individuals. *Blood* **92**, 2147–2156.

Wess, J., Nanavati, S., Vogel, Z., and Maggio, R. (1993). Functional role of proline and tryptophan residues highly conserved among G protein-coupled receptors studied by mutational analysis of the m3 muscarinic receptor. *EMBO J.* **12**, 331–338.

CX3CR1

Toshio Imai[1,2,*] and Osamu Yoshie[2]

[1]Kan Research Institute, Science Center Building, 3 Kyoto Research Park, Kyoto, Chu-douji Kurita-Chou, 600-8815, Japan

[2]Department of Bacteriology, Kinki University School of Medicine, 377-2 Ohno Higashi, Osaka, Osaka-Sayama, 589-8511, Japan

*corresponding author: tel: 81-75-325-5118 fax: 81-75-325-5130, e-mail: toshimai@mbox,kyoto-inet.or.jp

DOI: 10.1006/rwcy.2000.22016.

SUMMARY

CX3CR1 is a specific receptor for fractalkine/neurotactin (CX3CL1), a transmembrane molecule having an N-terminal chemokine domain with the CX3C motif. Membrane-bound fractalkine induces a rapid integrin-independent firm adhesion of leukocytes expressing CX3CR1 without involving any other adhesion molecules or G protein signaling; while soluble fractalkine induces vigorous chemotaxis of leukocytes expressing CX3CR1 though a *pertussis* toxin-sensitive signaling pathway. Thus, CX3CR1 has a dual function as an adhesion molecule and a signal transducer for leukocyte migration. Abundant expression of CX3CR1 in microglia and that of fractalkine in neurons also suggests important roles in the brain.

BACKGROUND

Discovery

CX3CR1 was originally identified as an orphan G protein-coupled receptor called V28, with a high sequence similarity to chemokines receptors such as CCR1 and CCR2 (Raport *et al.*, 1995). Later, fractalkine, a novel membrane molecule with an N-terminal CX3C chemokine domain (Bazan *et al.*, 1997), was found to bind to V28 with a high affinity (Imai *et al.*, 1997). Furthermore, cells expressing V28 were found to adhere to the membrane-bound form of fractalkine and to migrate toward the soluble form of fractalkine (Imai *et al.*, 1997). Thus V28 is a highly specific, functional receptor for fractalkine and is now termed CX3CR1, as the first receptor identified for the CX3C class of chemokines.

Alternative names

This receptor was initially referred to as V28 (Raport *et al.*, 1995) and CMKBRL1 (Combadiere *et al.*, 1995). The rat homolog of CX3CR1 was described as RBS11 (Harrison *et al.*, 1994).

Structure

CX3CR1 exhibits the seven transmembrane, G protein-coupled receptor structure typical of all of the chemokine receptors. CX3CR1 also has the DRY motif (DRYLAIV) in the second intracellular domain which is conserved among many members of the chemokine receptor family. CX3CR1, however, does not contain sites for *N*-linked glycosylation in the extracellular domains. Its C-terminal cytoplasmic domain is rich in Ser/Thr (12/58). Cysteines at 102 (second extracellular loop) and at 175 (third extracellular loop) probably form a disulfide bond.

Main activities and pathophysiological roles

CX3CR1 is a highly specific functional receptor for fractalkine (Imai *et al.*, 1997). Cells expressing CX3CR1 adhered to the membrane-associated fractalkine without involving integrins (Imai *et al.*, 1997).

This was also demonstrated under physiologic flow conditions (Fong *et al.*, 1998). Furthermore, this adhesion appeared to occur in the absence of G protein activation (Imai *et al.*, 1997; Fong *et al.*, 1998; Haskell *et al.*, 1999). In contrast, cells expressing CX3CR1 responded to the soluble form of fractalkine in both chemotaxis and calcium flux assays through a *pertussis* toxin-sensitive signaling pathway (Imai *et al.*, 1997). In a crescent glomerulonephritis model of Wister–Kyoto (WKY) rat, anti-CX3CR1 antibody treatment was shown to block leukocyte infiltration in the glomeruli, to prevent crescent formation, and to improve renal function (Feng *et al.*, 1999).

Similarly, vMIP-II, a chemokine analog encoded by *human herpesvirus 8* (HHV-8) which acts as a broad-spectrum chemokine antagonist, was shown to prevent soluble fractalkine-induced chemotaxis of inflammatory leukocytes isolated from nephritic glomeruli of WKY rats *in vitro*, to reduce leukocyte infiltration into glomeruli and to attenuate proteinuria *in vivo* (Chen *et al.*, 1998). Facial motor nerve axotomy in rats was followed by increases in the number and perineuronal location of CX3CR1-expressing microglia (Harrison *et al.*, 1998). In the rat *experimental allergic encephalomyelitis* model, CX3CR1 mRNA was upregulated in the lumbar spinal cords of animals displaying clinical signs of the disease (Jiang *et al.*, 1998). CXC3R1 was also shown to serve as a coreceptor for certain *HIV* strains in envelope-mediated fusion assays (Combadiere *et al.*, 1998a; Singh *et al.*, 1999).

GENE

Accession numbers

GenBank:
Human: U20350, U28934
Mouse: AF102269, AF072912
Rat: U04808

Sequence

Human CX3CR1 (Raport *et al.*, 1995): *See* **Figure 1**.

Chromosome location and linkages

The human gene for CX3CR1 is localized on chromosome 3p21 (Combadiere *et al.*, 1995).

PROTEIN

Accession numbers

SwissProt:
Human: P49238
Rat: P35411
PID:
Mouse: g3851709, g4165067

Sequence

See **Figure 2**.

Relevant homologies and species differences

CX3CR1 shows 40% amino acid identity to CCR1 and 41% amino acid identity to CCR2b. Human CX3CR1 shows 83% amino acid identity to murine CX3CR1 and 81% amino acid identity to rat CX3CR1. Murine CX3CR1 shows 94% amino acid identity to rat CX3CR1.

Affinity for ligand(s)

The only specific endogenous ligand for CX3CR1 reported to date is fractalkine. The soluble full-length extracellular region of fractalkine fused with soluble form of alkaline phosphatase (fractalkine-SEAP) binds to CX3CR1-transfected K562 cells with a K_d of 100 pM (Imai *et al.*, 1997). Fractalkine-SEAP also binds to monocytes with K_d of 50 pM and to lymphocytes with a K_d of 30 pM (Imai *et al.*, 1997). The ^{125}I-labeled chemokine domain of fractalkine binds to CX3CR1-transfected HEK 293 cells with a K_d of 740 pM (Combadiere *et al.*, 1998a) and to murine CX3CR1-expressed HEK 293 cells with a K_d of 4 nM (Combadiere *et al.*, 1998b).

Cell types and tissues expressing the receptor

Among fresh PBMCs, CX3CR1 mRNA is expressed in NK cells at high levels, in CD8+ T cells at moderate levels, in CD4+ T cells and monocytes at low levels, and virtually negative in CD19+ B cells or granulocytes (Imai *et al.*, 1997). The surface expression determined by FACS using fractalkine fused

Figure 1 Nucleotide sequence for human CX3CR1. The coding region is indicated by upper-case letters. From Raport *et al.* (1995).

```
   1 actcgtctct ggtaaagtct gagcaggaca gggtggctga ctggcagatc cagaggttcc
  61 cttggcagtc cacgccaggc cttcaccATG GATCAGTTCC CTGAATCAGT GACAGAAAAC
 121 TTTGAGTACG ATGATTTGGC TGAGGCCTGT TATATTGGGG ACATCGTGGT CTTTGGGACT
 181 GTGTTCCTGT CCATATTCTA CTCCGTCATC TTTGCCATTG GCCTGGTGGG AAATTTGTTG
 241 GTAGTGTTTG CCCTCACCAA CAGCAAGAAG CCCAAGAGTG TCACCGACAT TTACCTCCTG
 301 AACCTGGCCT TGTCTGATCT GCTGTTTGTA GCCACTTTGC CCTTCTGGAC TCACTATTTG
 361 ATAAATGAAA AGGGCCTCCA CAATGCCATG TGCAAATTCA CTACCGCCTT CTTCTTCATC
 421 GGCTTTTTTG GAAGCATATT CTTCATCACC GTCATCAGCA TTGATAGGTA CCTGGCCATC
 481 GTCCTGGCCG CCAACTCCAT GAACAACCGG ACCGTGCAGC ATGGCGTCAC CATCAGCCTA
 541 GGCGTCTGGG CAGCAGCCAT TTTGGTGGCA GCACCCCAGT TCATGTTCAC AAAGCAGAAA
 601 GAAAATGAAT GCCTTGGTGA CTACCCCGAG GTCCTCCAGG AAATCTGGCC CGTGCTCCGC
 661 AATGTGGAAA CAAATTTTCT TGGCTTCCTA CTCCCCCTGC TCATTATGAG TTATTGCTAC
 721 TTCAGAATCA TCCAGACGCT GTTTTCCTGC AAGAACCACA AGAAAGCCAA AGCCATTAAA
 781 CTGATCCTTC TGGTGGTCAT CGTGTTTTTC CTCTTCTGGA CACCCTACAA CGTTATGATT
 841 TTCCTGGAGA CGCTTAAGCT CTATGACTTC TTTCCCAGTT GTGACATGAG GAAGGATCTG
 901 AGGCTGGCCC TCAGTGTGAC TGAGACGGTT GCATTTAGCC ATTGTTGCCT GAATCCTCTC
 961 ATCTATGCAT TTGCTGGGGA GAAGTTCAGA AGATACCTTT ACCACCTGTA TGGGAAATGC
1021 CTGGCTGTCC TGTGTGGGCG CTCAGTCCAC GTTGATTTCT CCTCATCTGA ATCACAAAGG
1081 AGCAGGCATG GAAGTGTTCT GAGCAGCAAT TTTACTTACC ACACGAGTGA TGGAGATGCA
1141 TTGCTCCTTC TCtgaaggga atcccaaagc cttgtgtcta cagagaacct ggagttcctg
1201 aacctgatgc tgactagtga ggaaagattt ttgttgttat ttcttacagg cacaaaatga
1261 tggacccaat gcacacaaaa caaccctaga gtgttgttga gaattgtgct caaaatttga
1321 agaatgaaca aattgaactc tttgaatgac aaagagtaga catttctctt actgcaaatg
1381 tcatcagaac tttttggttt gcagatgaca aaaattcaac tcagactagt ttagttaaat
1441 gagggtggtg aatattgttc atattgtggc acaagcaaaa gggtgtctga gccctcaaag
1501 tgaggggaaa ccagggcctg agccaagcta gaattccctc tctctgactc tcaaatcttt
1561 tagtcattat agatccccca gactttacat gacacagctt tatcaccaga gagggactga
1621 cacccatgtt tctctggccc caagggaaaa ttcccaggga agtgctctga taggccaagt
1681 ttgtatcagg tgcccatccc tggaaggtgc tgttatccat ggggaaggga tatataagat
1741 ggaagcttcc agtccaatct catggagaag cagaaataca tatttccaag aagttggatg
1801 ggtgggtact attctgatta cacaaaacaa atgccacaca tcacccttac catgtgcctg
1861 atccagcctc tcccctgatt acaccagcct cgtcttcatt aagccctctt ccatcatgtc
1921 cccaaacctg caagggctcc ccactgccta ctgcatcgag tcaaaactca aatgcttggc
1981 ttctcatacg tccaccatgg ggtcctacca atagattccc cattgcctcc tccttcccaa
2041 aggactccac ccatcctatc agcctgtctc ttccatatga cctcatgcat ctccacctgc
2101 tcccaggcca gtaagggaaa tagaaaaacc ctgcccccaa ataagaaggg atggattcca
2161 accccaactc cagtagcttg ggacaaatca agcttcagtt tcctggtctg tagaagaggg
2221 ataaggtacc tttcacatag agatcatcct ttccagcatg aggaactagc caccaactct
2281 tgcaggtctc aacccttttg tctgcctctt agacttctgc tttccacacc tgcactgctg
2341 tgctgtgccc aagttgtggt gctgacaaag cttggaagag cctgcaggtg ccttggccgc
2401 gtgcatagcc cagacacaga agaggctggt tcttacgatg gcacccagtg agcactccca
2461 agtctacaga gtgatagcct tccgtaaccc aactctcctg gactgccttg aatatcccct
2521 cccagtcacc ttgtgcaagc ccctgcccat ctgggaaaat accccatcat tcatgctact
2581 gccaacctgg ggagccaggg ctatgggagc agcttttttt tccccccctag aaacgtttgg
2641 aacaatgtaa aactttaaag ctcgaaaaca attgtaataa tgctaaagaa aaagtcatcc
2701 aatctaacca catcaatatt gtcattcctg tattcacccg tccagacctt gttcacactc
2761 tcacatgttt agagttgcaa tcgtaatgta cagatggttt tataatctga tttgttttcc
2821 tcttaacgtt agaccacaaa tagtgctcgc tttctatgta gtttggtaat tatcatttta
2881 gaagactcta ccagactgtg tattcattga agtcagatgt ggtaactgtt aaattgctgt
2941 gtatctgata gctctttggc agtctatatg tttgtataat gaatgagaga ataagtcatg
3001 ttccttcaag atcatgtacc ccaatttact tgccattact caattgataa acatttaact
3061 tgtttccaat gtttagcaaa tacatatttt atagaacttc
```

with secreted form of alkaline phosphatase (fractalkine-SEAP) correlates well with the pattern of V28 mRNA expression. Fractalkine-SEAP bound to most of NK cells (>95%) and monocytes (>80%), a fraction (25%) of CD8+ T cells, and a minor and variable fraction (<1 to 10%) of CD4+ T cells among different donors (Imai *et al.*, 1997).

CX3CR1 was shown to be expressed strongly in tissues such as brain, spleen, and peripheral blood leukocytes (Raport *et al.*, 1995; Combadiere *et al.*, 1995). *In situ* hybridization of the rat brain revealed that CX3CR1 mRNA was expressed in microglial cells while fractalkine mRNA was expressed by neurons (Harrison *et al.*, 1998; Nishiyori *et al.*, 1998).

Figure 2 Amino acid sequence for CX3CR1. The putative transmembrane regions are underlined.

```
MDQFPESVTE NFEYDDLAEA CYIGDIVVFG TVFLSIFYSV IFAIGLVGNL
LVVFALTNSK KPKSVTDIYL LNLALSDLLF VATLPFWTHY LINEKGLHNA
MCKFTTAFFF IGFFGSIFFI TVISIDRYLA IVLAANSMNN RTVQHGVTIS
LGVWAAAILV AAPQFMFTKQ KENECLGDYP EVLQEIWPVL RNVETNFLGF
LLPLLIMSYC YFRIIQTLFS CKNHKKAKAI KLILLVVIVF FLFWTPYNVM
IFLETLKLYD FFPSCDMRKD LRLALSVTET VAFSHCCLNP LIYAFAGEKF
RRYLYHLYGK CLAVLCGRSV HVDFSSSESQ RSRHGSVLSS NFTYHTSDGD
ALLLL
```

Expression of CX3CR1 mRNA was also shown in rat microglial cells in culture at high levels and in astrocytes at low levels (Jiang *et al.*, 1998).

Regulation of receptor expression

Expression of CX3CR1 mRNA was strongly upregulated in both CD4+ and CD8+ T cells by IL-2 (Imai *et al.*, 1997). Incubation of rat microglia with lipopolysaccharide transiently suppressed CX3CR1 expression (Boddeke *et al.*, 1999). CX3CR1 mRNA expression in rat astrocytes was increased by TNFα or IL-1β for 24 hours (Maciejewski-Lenoir *et al.*, 1999).

SIGNAL TRANSDUCTION

Soluble fractalkine induces chemotaxis and calcium flux in CX3CR1-expressing cells through a pertussis toxin-sensitive signaling pathway, suggesting its coupling with a Gα_i class of G protein (Imai *et al.*, 1997).

Cytoplasmic signaling cascades

Intracellular Ca^{2+} mobilization, cell spreading, and chemotaxis mediated by CX3CR1 were inhibited by pertussis toxin (Imai *et al.*, 1997; Fong *et al.*, 1998). This suggests that Gα_i is mainly responsible for these cellular responses (Al-Aoukaty *et al.*, 1998). In contrast, direct adhesion of CX3CR1-expressing cells to immobilized fractalkine was observed even by cells treated with pertussis toxin and cells expressing CX3CR1 mutants that have little or no ability to activate G protein (Imai *et al.*, 1997; Fong *et al.*, 1998; Haskell *et al.*, 1999). It is not known whether any signaling pathways are required for the direct adhesion. CX3CR1 was also shown to mediate activation of extracellular signal regulated kinase (ERK)-1/2 but not that of c-Jun N-terminal protein kinase/stress-activated protein kinase or p38 in rat hippocampal neurons (Meucci *et al.*, 1998).

DOWNSTREAM GENE ACTIVATION

Transcription factors activated

Activation of CREB was shown in rat hippocampal neurons by fractalkine (Meucci *et al.*, 1998).

References

Al-Aoukaty, A., Rolstad, B., Giaid, A., and Maghazachi, A. A. (1998). MIP-3α, MIP-3β and fractalkine induce the locomotion and the mobilization of intracellular calcium, and activate the heterotrimeric G proteins in human natural killer cells. *Immunology* **95**, 618–624.

Bazan, J. F., Bacon, K. B., Hardiman, G., Wang, W., Soo, K., Rossi, D., Greaves, D. R., Zlotnik, A., and Schall, T. J. (1997). A new class of membrane-bound chemokines with a CX3C motif. *Nature* **385**, 640–644.

Boddeke, E. W., Meigel, I., Frentzel, S., Biber, K., Renn, L. Q., and Gebicke-Haerter, P. (1999). Functional expression of the fractalkine (CX3C) receptor and its regulation by lipopolysaccharide in rat microglia. *Eur. J. Pharmacol.* **374**, 309–313.

Chen, S., Bacon, K. B., Li, L., Garcia, G. E., Xia, Y., Lo, D., Thompson, D. A., Siani, M. A., Yamamoto, T., Harrison, J. K., and Feng, L. (1998). *In vivo* inhibition of CC and CX3C chemokine-induced leukocyte infiltration and attenuation of glomerulonephritis in Wister–Kyoto (WKY) rats by vMIP-II. *J. Exp. Med.* **188**, 193–198.

Combadiere, C., Ahujia, S. K., and Murphy, P. M. (1995). Cloning, chromosomal localization, and RNA expression of a human β chemokine receptor-like gene. *DNA Cell Biol.* **14**, 673–680.

Combadiere, C., Salzwedel, K., Smith, E. D., Tiffnay, H. L., Berger, E. A., and Murphy, P. M. (1998a). Identification of CX3CR1. A chemotactic receptor for the human CX3C chemokine fractalkine and a fusion coreceptor for HIV-1. *J. Biol. Chem.* **273**, 23799–23804.

Combadiere, C., Gao, J., Tiffany, H. L., and Murphy, P. M. (1998b). Gene cloning, RNA distribution, and functional expression of mCX3CR1, a mouse chemotactic receptor for the CX3C chemokine fractalkine. *Biochem. Biophys. Res. Commun.* **253**, 728–732.

Feng, L., Chen, S., Garcia, G. E., Xia, Y., Siani, M. A., Botti, P., Wilson, C. B., Harrison, J. K., and Bacon, K. B. (1999). Prevention of crescentic glomerulonephritis by immunoneutralization of the fractalkine receptor CX3CR1. *Kidney Int.* **56**, 612–620.

Fong, A. M., Robinson, L. A., Steeber, D. A., Tedder, T. F., Yoshie, O., Imai, T., and Patel, D. D. (1998). Fractalkine and CX3CR1 mediate a novel mechanism of leukocyte capture, firm adhesion, and activation under physiologic flow. *J. Exp. Med.* **188**, 1413–1419.

Harrison, J. K., Barber, C. M., and Lynch, K. R. (1994). cDNA cloning of a G protein-coupled receptor expressed in rat spinal cord and brain related to chemokine receptors. *Neurosci. Lett.* **169**, 85–89.

Harrison, J. K., Jiang, Y., Chen, S., Xia, Y., Maciejewski, D., McNamara, R. K., Streit, W. J., Salafranca, M. N., Adhikari, S., Thompson, D. A., Botti, P., Bacon, K. B., and Feng, L. (1998). Role for neuronally derived fractalkine in mediating interactions between neurons and CX3CR1-expressing microglia. *Proc. Natl Acad. Sci. USA* **95**, 10896–10901.

Haskell, C. A., Cleary, M. D., and Charo, I. F. (1999). Molecular uncoupling of fractalkine-mediated cell adhesion and signal transduction. Rapid flow arrest of CX3CR1-expressing cells is independent of G-protein activation. *J. Biol. Chem.* **274**, 10053–10058.

Imai, T., Hieshima, K., Haskell, C., Baba, M., Nagira, M., Nishimura, M., Kakizaki, M., Takagi, S., Nomiyama, H., Schall, T. J., and Yoshie, O. (1997). Identification and molecular characterization of fractalkine receptor CX3CR1, which mediates both leukocyte migration and adhesion. *Cell* **91**, 521–530.

Jiang, Y., Salafranca, M. N., Adhikari, S., Xia, Y., Feng, L., Sonntag, M. K., deFiebre, C. M., Pennell, N. A., Streit, W. J., and Harrison, J. K. (1998). Chemokine receptor expression in cultured glia and rat experimental allergic encephalomyelitis. *J. Neuroimmunol.* **86**, 1–12.

Maciejewski-Lenoir, D., Chen, S., Feng, L., Maki, R., and Bacon, K. B. (1999). Characterization of fractalkine in rat brain cells: migratory and activation signals for CX3CR-1-expressing microglia. *J. Immunol.* **163**, 1628–1635.

Meucci, O., Fatatis, A., Simen, A. A., Bushell, T. J., Gray, P. W., and Miller, R. J. (1998). Chemokines regulate hippocampal neuronal signaling and gp120 neurotoxicity. *Proc. Natl Acad. Sci. USA.* **95**, 14500–14505.

Nishiyori, A., Minami, M., Ohtani, Y., Takami, S., Yamamoto, J., Kawaguchi, N., Kume, T., Akaike, A., and Satoh, M. (1998). Localization of fractalkine and CX3CR1 mRNAs in rat brain: does fractalkine play a role in signaling from neuron to microglia? *FEBS Lett.* **429**, 167–172.

Raport, C. J., Schweickart, V. L., Eddy, R.L. Jr., Shows, T. B., and Gray, P. W. (1995). The orphan G-protein-coupled receptor-encoding gene V28 is closely related to genes for chemokine receptors and is expressed in lymphoid and neural tissues. *Gene* **163**, 295–299.

Singh, A., Besson, G., Mobasher, A., and Collman, R. G. (1999). Patterns of chemokine receptor fusion cofactor utilization by human immunodeficiency virus type 1 variants from the lung and blood. *J. Virol.* **73**, 6680–6690

Poxvirus Secreted Chemokine-binding Proteins

Grant McFadden[1,*] and Richard Moyer[2]

[1]The John P. Robarts Research Institute and Department of Microbiology and Immunology, The University of Western Ontario, 1400 Western Road, London, Ontario, N6G 2V4, Canada

[2]Department of Molecular Genetics and Microbiology, University of Florida College of Medicine, PO Box 100266, Gainesville, FL 32610-0266, USA

*corresponding author tel: (519) 663-3184, fax: (519) 663-3847, e-mail: mcfadden@rri.on.ca

DOI: 10.1006/rwcy.2000.14016.

SUMMARY

Unlike the classic viroceptors (virus-encoded receptor homologs) that bind and inhibit cytokines such as TNF, interferons, and IL-1β, the viral proteins that bind and modulate chemokines were discovered by screening analysis for novel cytokine inhibitors expressed by poxviruses. Currently, two classes of chemokine-binding proteins have been described which interact with chemokines in different fashion. The first class, exemplified by the M-T1 protein of *myxoma virus* and the secreted 35K proteins from a variety of orthopoxviruses, are 35–45 kDa secreted glycoproteins that bind and inhibit the ability of a broad spectrum of CC chemokines to engage their cognate receptors. Binding occurs in a relatively species-independent fashion. The second class is typified by the M-T7 protein of myxoma virus, which was first described as an inhibitor of IFNγ but has a second property of binding to a wide spectrum of chemokines via the conserved glycosaminoglycan-binding domains of these ligands. Both classes of proteins affect the pathogenesis of virus infections but are believed to interfere with chemokine biological functions by different mechanisms.

BACKGROUND

Discovery

In 1997, the discovery of secreted chemokine-binding proteins suggested a novel virus strategy for modulating chemokine functions during virus infection (Graham *et al.*, 1997). Subsequent studies revealed that these poxvirus-encoded chemokine-binding proteins (CBPs) were specific inhibitors of the CC family of chemokines (Smith *et al.*, 1997; Alcamí *et al.*, 1998; Lalani *et al.*, 1998). Members of this growing family of poxvirus proteins, designated the T1/35 kDa family of secreted viroceptors, have been uncovered in *myxoma virus*, *Shope fibroma virus*, *cowpox virus*, *rabbitpox virus*, *raccoonpox virus*, *variola virus*, and certain strains of *vaccinia virus* (see Figure 1). The discovery of the poxvirus chemokine-binding proteins has been reviewed in detail by Lalani and McFadden (1997).

Alternative names

Chemokine-binding proteins (CBPs), viral chemokine-binding proteins (vCKBPs), viral chemokine inhibitors (vCCIs), and T1/35 kDa family.

Structure

The poxvirus CBPs constitute a family of secreted glycoproteins that exhibit considerable sequence similarity to each other. All the CBPs of this class share conserved sets of sequence motifs, particularly a characteristic cysteine-spacing arrangement (*See* **Figure 1**). However, to date no close cellular homologs have been described for this family.

The crystal structure for one member of this family, namely the secreted 35 kDa protein from cowpox (vCCI), has recently been determined and shown to exhibit a novel β sandwich fold that reveals a potentially new strategy to bind and inhibit CC chemokines (Carfi *et al.*, 1999).

Main activities and pathophysiological roles

The members of the T1/35 kDa CBP family inhibit a broad spectrum of CC chemokines, as measured by receptor-binding analysis, calcium flux measurements, and chemotaxis assays (Smith *et al.*, 1997; Alcamí *et al.*, 1998; Lalani *et al.*, 1998). Expression of these viral proteins modulates the influx and activation of monocytes/macrophages and other classes of reactive leukocytes into sites of virus replication in infected animals (Martinez-Pomares *et al.*, 1995; Graham *et al.*, 1997).

GENE

Accession numbers

Myxoma virus (M-T1): U62677
Shope fibroma virus (S-T1): M17433
Rabbitpox virus (35K): U64724
Vaccinia virus Lister (C23L): D00612
Vaccinia virus Copenhagen (C23L/B29R): M35027
Variola virus Somalia 1977 (G3R): U18341
Cowpox virus (ORF B): L08906
Cowpox virus (D1L): Y11842

PROTEIN

Accession numbers

Myxoma virus: 2076755
Shope fibroma virus: 139626
Rabbitpox virus: 2076757
Vaccinia virus Lister: 137559
Vaccinia virus Copenhagen: 335577
Variola virus Somalia 1977: 885856
Cowpox virus (orf B): 333518
Cowpox virus (D1L): 3096963

Sequence

See Figure 1.

Description of protein

The first member of this family (designated 35 K) was identified as the major secreted glycoprotein from cells infected with *rabbitpox virus* (Martinez-Pomares *et al.*, 1995) and certain *vaccinia virus* strains (e.g. Evans, Lister) but not others (e.g. Western Reserve, WR) (Patel *et al.*, 1990). Only later did functional studies reveal that members of this family could bind and inhibit a broad spectrum of CC chemokines *in vitro* (Graham *et al.*, 1997; Smith *et al.*, 1997; Alcamí *et al.*, 1998; Lalani *et al.*, 1998). In addition to the expression and secretion of CBPs from poxvirus-infected cells, members of the T1/35 kDa family have also been overexpressed in active form using baculovirus vectors and as Fc fusion constructs (Smith *et al.*, 1997; Alcamí *et al.*, 1998).

Relevant homologies and species differences

No cellular homologs of the T1/35 kDa family have been reported, and a comparison of the chemokine-inhibitory properties of members from vaccinia, rabbitpox, cowpox, and myxoma has revealed no obvious species-specificity against representative CC chemokines (Smith *et al.*, 1997; Lalani *et al.*, 1998).

Affinity for ligand(s)

The affinities for T1/35 kDa family members for a variety of human and murine chemokines have been measured, and K_d values generally range in the picomolar to low nanomolar range for many of the representative CC chemokines tested (Smith *et al.*, 1997; Alcamí *et al.*, 1998). Thus, the viral proteins have considerably greater affinities for CC chemokines than these ligands have for their own cellular receptors.

Figure 1 Alignments of the poxvirus T1/35 kDa family of chemokine-binding proteins. Amino acids are boxed if they are conserved in both myxoma virus and Shope fibroma virus, and at least four of the orthopoxvirus members. Myx, myxoma virus; SFV, Shope fibroma virus; RPV, rabbitpox virus; VVlis, vaccinia virus (Lister); VVcop, vaccinia virus (Copenhagen); VAR, variola virus; CPV, cowpox virus. Conserved cysteines are marked with an asterix. Adapted from Graham *et al.* (1997).

```
Myx MT1       1          M K R L C V L F A C L A - - - - - - A T L A T K G I C R Q G E D V R Y M
SFV ST1       1          M R R L C I I L L V Y V Y - - - - - A T F A T K G I C K Q D E D V R Y M
RPV 35kDa     1  M K Q Y I V L A C M C L A A A A M P A S L Q Q S S S S S S S C T E E E N K H H M
VVlis35kDa    1  M K Q Y I V L A C M C L A A A A M P A S L Q Q S S S S S S S C T E E E N K H H M
VVcop35kDa    1                            M H V P A S L Q Q S S S S S S S C T E E E N K H H M
VAR 35kDa     1  M K Q Y I V L A C M C L A A A A M P A S L Q Q S S - - - S L C T E E E N K H Y M
CPV 35kDa     1    M K Q I V L A C I C L A A V A I P T S L Q Q S F S S S S S C T E E E N K H H M

Myx MT1      31  G I D A V A K I T K R - - T T G S D T P C Q G L R T T I E S A Y T E D E N E D D
SFV ST1      32  G I D V V V K V T K K - - T S G S D T V C Q A L R T T F E A A H K G D G A N D S
RPV 35kDa    41  G I D V I I K V T K Q D Q T P T N D K I C Q S V T E I T E S E S D P D P - - - -
VVlis35kDa   41  G I D V I I K V T K Q D Q T P T N D K I C Q S V T E I T E S E S D P D P - - - -
VVcop35kDa   27  G I D V I I K V T K Q D Q T P T N D K I C Q S V T E I T E S E S D P D P - - - -
VAR 35kDa    38  G I D V I I K V T K Q D Q T P T N D K I C Q S V T E I T E S E S D P - - - - - -
CPV 35kDa    40  G I D V I I K V T K Q D Q T P T N D K I C Q S V T E V T E S E D E S E - - - - -

Myx MT1      69  G A T G T E Q P D D L S E E Y E Y D E N D E S F L T G F V I G S T Y H T I V G G
SFV ST1      70  - - L S T E Y V D D Y S E E E E Y - E Y D E S F L E G F V I G S T Y Y T I V G G
RPV 35kDa    77  - - - E V E S E D D S T S V E D V - - - - - - - - - - - - D P P T T Y Y S I I G G
VVlis35kDa   77  - - - E V E S E D D S T S V E D V - - - - - - - - - - - - D P P T T Y Y S I I G G
VVcop35kDa   63  - - - E V E S E D D S T S V E D V - - - - - - - - - - - - D P P T T Y Y S I I G G
VAR 35kDa    72  - - - E V E S E D D S T S V E D V - - - - - - - - - - - - D P P T T Y Y S I I G G
CPV 35kDa    75  - - - E V V K G D - - - - - - - - - - - - - - - - - - - - - - P T T Y Y T V V G G

Myx MT1     109  G L S V T F G F T G C P T V K A I S E H V K G R H V Y V R L S S D A P W R D T N
SFV ST1     107  G L S V T F G F T G C P T V K S V S E Y A K G R I V F I R L S S D A P W R D T N
RPV 35kDa   103  G L R M N F G F T K C P Q I K S I S E S A D G N T V N A R L S S V S P G Q G K D
VVlis35kDa  103  G L R M N F G F T K C P Q I K S I S E S A D G N T V N A R L S S V S P G Q G K D
VVcop35kDa   89  G L R M N F G F T K C P Q I K S I S E S A D G N T V N A R L S S V S P G Q G K D
VAR 35kDa    98  G L R M N F G F T K C P Q I K S I S E S A N G N A V N A R L S S V P P G Q G K D
CPV 35kDa    91  G L T M D F G F T K C P K I S S I S E Y S D G N T V N A R L S S V S P G Q G K D

Myx MT1     149  P V S M N R T E A L A L L D T C E V S V D I K C S R V N V T E T T Y G T A A L V
SFV ST1     147  P M S I N R T E A L A L L E K C E T S I D I K C S N E T V S E T T Y G L A S L A
RPV 35kDa   143  S P A I T H E E A L A M I K D C E V S I D I R C S E E E K D S D I K T H P V L G
VVlis35kDa  143  S P A I T R E E A L A M I K D C E V S I D I R C S E E E K D S D I K T H P V L G
VVcop35kDa  129  S P A I T R E E A L A M I K D C E V S I D I R C S E E E K D S D I K T H P V L G
VAR 35kDa   138  S P A I T R V E A L A M I K D C E L S I D I R C S E E E K D S D I Q T H P V L E
CPV 35kDa   131  S P A I T R E E A L S M I K D C E M S I N I K C S E E E K D S N I K T H P V L G

Myx MT1     189  P R I T Q A T R - R S H I I G S T L V D T E C V K S L D I T V Q V G E M C K R T
SFV ST1     187  P H I T Q A T E - R G N I I G S T L V D T D C V E N L D V T V H L G E M C R K T
RPV 35kDa   183  S N I S H K K V S Y E D I I G S T I V D T K C V K N L E F S V R I G D M C K E S
VVlis35kDa  183  S N I S H K K V S Y E D I I G S T I V D T K C V K N L E F S V R I G D M C K E S
VVcop35kDa  169  S N I S H K K V S Y E D I I G S T I V D T K C V K N L E F S V R I G D M C K E S
VAR 35kDa   178  S N I S H K K V S Y E D I I G S T I V D T K C V K N L E F S V R I G D M C K E S
CPV 35kDa   171  S N I S H K K V S Y E D I I G S T I V D T K C V K N L E I S V R I G D M C K E S

Myx MT1     228  S D L S A R D S L K V K N G - - - K L L E D D I L V L R T P T L K A C N 260
SFV ST1     226  S D L S K R D S L K V K N G - - - E L L D D D T F S I H T P K L K A C N 258
RPV 35kDa   223  S E L E V K D G F K Y V D G S A S K G A T D D T S L I D S T K L K A C V 258
VVlis35kDa  223  S E L E V K D G F K Y V D G S A S E G A T D D T S L I D S T K L K A C V 258
VVcop35kDa  209  S E L E V K D G F K Y V D G S A S E G A T D D T S L I D S T K L K A C V 244
VAR 35kDa   218  S D L E V K D G F K Y V D G S V S E G V T D D T S L I D S T K L K S C V 253
CPV 35kDa   211  S E L E V K D G F K Y V D G S A S E D A A D D T S L I N S A K L I A C V 246
```

Regulation of receptor expression

The viral proteins are expressed as standard early poxvirus genes (Macaulay and McFadden, 1989), and the processed proteins are efficiently secreted as 35–40 kDa glycoproteins (Patel *et al.*, 1990).

BIOLOGICAL CONSEQUENCES OF ACTIVATING OR INHIBITING RECEPTOR AND PATHOPHYSIOLOGY

The deletion of the 35 kDa gene from rabbitpox causes only minor effects on the overall lethality of this virus in infected rabbits but histological analysis reveals an increase in inflammatory cell influx in the first few days of infection (Martinez-Pomares *et al.*, 1995; Graham *et al.*, 1997). It is thought that the increased numbers of inflammatory cells remain poorly activated due to the combined activities of other anti-immune proteins expressed and secreted by the virus. When purified and tested alone, the vaccinia 35 K protein of blocked eotaxin-induced eosinophil infiltrates in a guinea pig skin model of inflammation (Alcamí *et al.*, 1998).

References

Alcamí, A., Symons, J. A., Collins, P. D., Williams, T. J., and Smith, G. L. (1998). Blockade of chemokine activity by a soluble chemokine binding protein from vaccinia virus. *J. Immunol.* **160**, 624–633.

Carfi, A., Smith, C. A., Smolak, P. J., McGrew, J., and Wiley, D. C. (1999). Structure of a soluble secreted chemokine inhibitor vCCI (p35) from cowpox virus. *Proc. Natl Acad. Sci. USA* **96**, 12379–12383.

Graham, K. A., Lalani, A. S., Macen, J. L., Ness, T. L., Barry, M., Liu, L.-Y., Lucas, A., Clark-Lewis, I., Moyer, R. W., and McFadden, G. (1997). The T1/35 kDa family of poxvirus secreted proteins bind chemokines and modulate leukocyte influx into virus infected tissues. *Virology* **229**, 12–24.

Lalani, A. S., and McFadden, G. (1997). Secreted poxvirus chemokine binding proteins. *J. Leukoc. Biol.* **62**, 570–576.

Lalani, A. S., Ness, T. L., Singh, R., Harrison, J. K., Seet, B. T., Kelvin, D. J., McFadden, G., and Moyer, R. W. (1998). Functional comparisons among members of the poxvirus T1/35 kDa family of soluble CC-chemokine inhibitor glycoproteins. *Virology* **250**, 173–184.

Macaulay, C., and McFadden, G. (1989). Tumorigenic poxviruses: Characterization of an early promoter from Shope fibroma virus. *Virology* **172**, 237–246.

Martinez-Pomares, L., Thompson, J. P., and Moyer, R. W. (1995). Mapping and investigation of the role in pathogenesis of the major unique secreted 35-kDa protein of rabbitpox Virus. *Virology* **206**, 591–600.

Patel, A. H., Gaffney, D. F., Subak-Sharpe, J. H., and Stow, N. D. (1990). DNA sequence of the gene encoding a major secreted protein of vaccinia virus, strain Lister. *J. Gen. Virol.* **71**, 2013–2021.

Smith, C. A., Smith, T. D., Smolak, P. J., Friend, D., Hagen, H., Gerhart, M., Park, L., Pickup, D. J., Torrance, D., Mohler, K., Schooley, K., and Goodwin, R. G. (1997). Poxvirus genomes encode a secreted soluble protein that preferentially inhibits β chemokine activity yet lacks sequence homology to known chemokine receptors. *Virology* **236**, 316–327.

CCR9

Angel Zaballos, Julio Gutiérrez, Rosa Varona and Gabriel Márquez*

Departamento de Inmunología y Oncología, Centro Nacional de Biotecnología, Campus Universidad Autónoma, Cantoblanco, Madrid, 28049, Spain

*corresponding author tel: 91 585 4856, fax: 91 372 0493, e-mail: gmarquez@cnb.uam.es

DOI: 10.1006/rwcy.2000.22017.

SUMMARY

T cell maturation requires the controlled migration of T cell precursors through the different thymic compartments in a process that seems to be regulated by chemokines. The former human putative chemokine receptor GPR-9-6 has recently been identified as CCR9, the specific receptor for the thymus-expressed β chemokine TECK. The cDNA sequences containing the complete coding regions of human and mouse CCR9 have been established, showing that the identity between both CCR9 predicted proteins is 86%. The expression of both genes is also very similar, being high in the thymus and low in lymph nodes and spleen. RT-PCR analysis of mouse CCR9 expression on murine FACS-sorted thymocyte subpopulations showed that this gene is expressed in both immature and mature T cells. Among the 36 different chemokines tested on intracytoplasmic calcium mobilization and *in vitro* chemotaxis assays, TECK is the only functional ligand found for CCR9. The expression of murine CCR9 in both immature and mature thymocytes, together with the expression of its ligand in the thymus, suggests that the TECK-mediated chemoattraction might be a mechanism contributing to thymocyte retention in the thymus while completing their development.

BACKGROUND

Discovery

The cellular compartmentalization of the thymus provides an optimal microenvironment for thymopoiesis (Boyd and Hugo, 1991; Boyd *et al.*, 1993; Shortman and Wu, 1996). After the stem cells have migrated from the bone marrow to the fetal thymic primordium, immature thymocytes undergo a complex and organized differentiation process that gives rise to a functional immune system able to distinguish between self and nonself (Boyd and Hugo, 1991). During their maturation, thymocytes migrate through the thymus in what seems to be a regulated process. Immature CD4–CD8– thymocytes are found in the subcapsular region, whereas more mature CD4+ CD8+ thymocytes are located in the cortex; and mature CD4+ or CD8+ T cells reside in the medulla (Boyd *et al.*, 1993; Ritter and Boyd, 1993). The factors controlling thymocyte migration inside the thymus are largely unknown, but chemokines appear to play an important role. In this regard, recent reports from Vicari *et al.* (1997) and Kim *et al.* (1998) demonstrated that the chemokines TECK, SDF, and ELC (MIP-3β) were able to chemoattract different thymocyte subsets. While the specific receptors for SDF and ELC (MIP-3β) are known to be CXCR4 and CCR7, respectively (Bleul *et al.*, 1996; Oberlin *et al.*, 1996; Yoshida *et al.*, 1997), the receptor for the β chemokine TECK has only very recently been reported to be the former GPR-9-6 orphan receptor, now known as CCR9 (Zaballos *et al.*, 1999).

The cDNA sequence encompassing the complete coding region of human and mouse CCR9 has been reported (Zaballos *et al.*, 1999). Human and murine CCR9 have a highly restricted expression pattern, being very highly expressed in the thymus, and less in lymph nodes and spleen. RT-PCR analysis of mouse CCR9 expression on murine FACS-sorted thymocyte subpopulations showed that this gene is expressed in both immature and mature T cells (Zaballos *et al.*, 1999). Intracytoplasmic calcium mobilization and chemotaxis assays established that, among 36 different chemokines tested, TECK was the only functional ligand for CCR9, in both human and murine systems (Zaballos *et al.*, 1999). This remarkable specificity is consistent with the CCR9 phylogenetic relationship to CCR6 and CCR7, which also have strict ligand specificity. This lack of promiscuity is not a common feature of chemokine receptors, which usually recognize more than one chemokine (Rollins, 1997; Luster, 1998). Anyway, the possibility cannot be excluded that new CCR9 agonists may be identified. In addition, the CCR9 ligand TECK is phylogenetically closely related to MIP-3α and SLC(6C-kine), the ligands for CCR6 and CCR7, respectively, sharing with them unique structural features, such as the Asp-Cys-Cys-Leu motif (Varona *et al.*, 1998).

Alternative names

CCR9 was previously known as orphan chemokine receptor GPR-9-6.

Structure

CCR9 is a member of the family of seven transmembrane domain, G protein-coupled receptors and has recently been characterized as a β chemokine receptor (**Figure 1**). In addition to chemokine receptors, this family has a great number of members, including receptors for hormones, growth factors, neurotransmitters, odorants, and other signaling molecules. The extracytoplasmic N-terminal region of both human and mouse CCR9 contains a conserved cysteine residue that is typically involved in disulfide bonding with other conserved cysteines in the extracellular loops. The 78 C-terminal amino acids of human and mouse CCR9 are identical; these include the serine and threonine residues of the intracytoplasmic tail, which may be phosphorylated.

Main activities and pathophysiological roles

The fact that both immature and mature thymocytes express mouse CCR9 suggests that this receptor might play a role during the whole process of thymocyte development. As thymic dendritic cells are producers of TECK (Vicari *et al.*, 1997), the TECK-mediated chemoattraction might be a mechanism contributing to thymocyte confinement in the thymus

Figure 1 Scheme of human CCR9 structure showing the serpentine shape of this β chemokine receptor, with its predicted seven transmembrane domains and the extracellular and intracellular loops. The standard color code is used to highlight the chemical properties of the amino acid residues of the polypeptide. The snake-like plot was generated using services available at Viseur's homepage (http://www.lctn.u-nancy.fr/viseur/viseur.htlm).

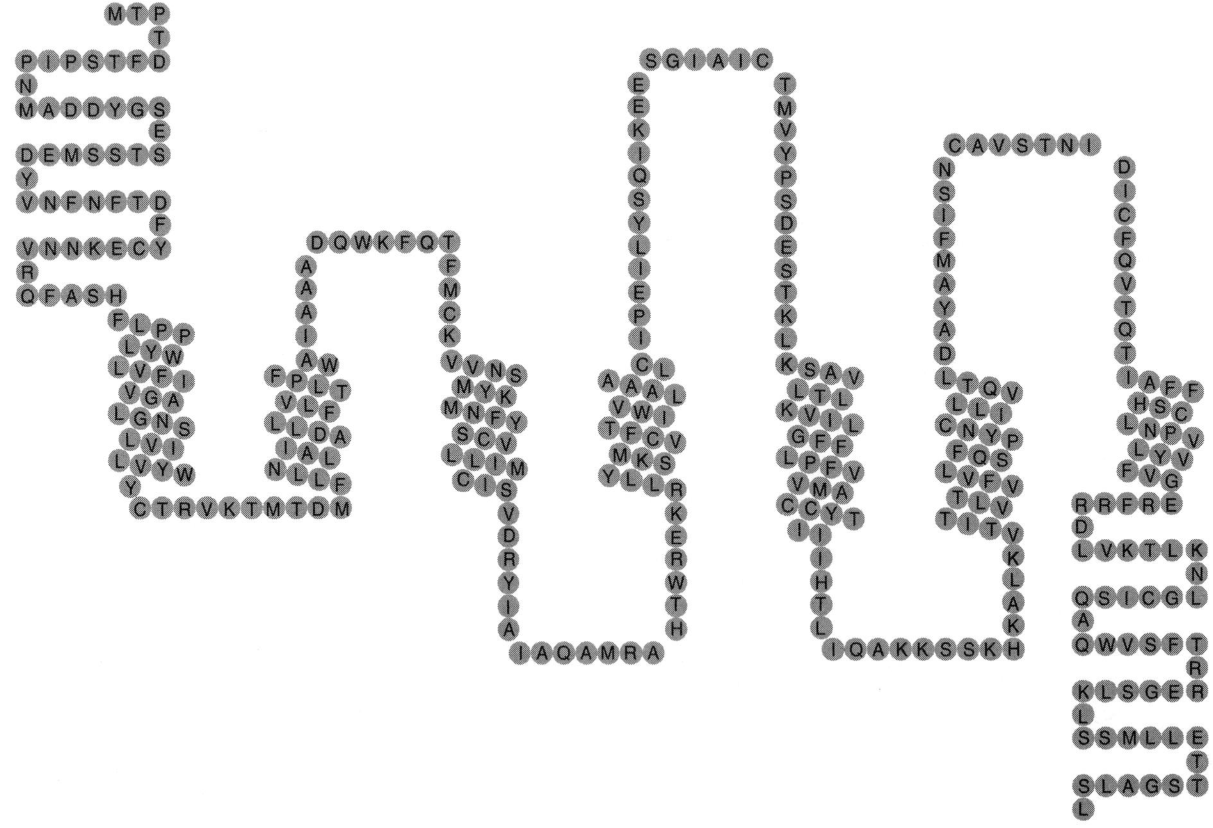

until their development is completed. A recent report by Wilkinson *et al.* (1999) showed that neutralizing antibodies to TECK did not prevent *in vitro* thymus recolonization by T cell precursors, suggesting that additional factors might be involved in the arrival of those lymphoid precursors in the thymus. In addition, thymocytes have been reported to use different chemokines during their development, as suggested by the modulation of the expression of CXCR4, CCR4, and CCR7 while thymocytes travel from the thymic cortex to the medulla (Suzuki *et al.*, 1999). In this context, the identification of CCR9 as the functional receptor for TECK may help to unravel the details of the important role these proteins seem to play in the thymus.

Some chemokine receptors act as coreceptors for the entry of *HIV* into the cells (D'Souza and Harden, 1996). In this regard, CCR9 has been reported not to be active as a cofactor in cell fusion assays using Env proteins from diverse human and *simian immunodeficiency viruses* (Rucker *et al.*, 1997). Anyway, these experiments were done using a truncated version of CCR9, that lacks the first N-terminal 12 amino acids. Therefore, a possible role of CCR9 as coreceptor for HIV cannot be dismissed.

GENE

Accession numbers

Mouse CCR9 cDNA: AJ132336
Human CCR9 cDNA: AJ132337
Human CCR9 gene: AC005669

In addition, a partial cDNA sequence corresponding to the 5′ region of human CCR9 is found at AF145207. The genomic sequence in U45982 only includes the codifying sequence contained in exon 3.

Sequence

See **Figure 2**.

Figure 3 shows that the human CCR9 gene has the three exons/two introns structure typical of chemokine receptor genes, with the coding sequence split between exons 2 and 3 (Zaballos *et al.*, 1999).

Chromosome location and linkages

According to data shown in the U45982 entry, the human CCR9 gene is located on chromosome 3 (3p21-3p22). A cluster of β chemokine receptors maps to that region of chromosome 3, including CCR1, CCR2, CCR3, CCR5 (Samson *et al.*, 1996), and the orphan receptor CKR-X. In addition, other chemokine receptors, such as CCR8 (Napolitano *et al.*, 1996), CCR10 (Bonini *et al.*, 1997), XCR1 (Heiber *et al.*, 1995), CX3CR1 (Raport *et al.*, 1995), and the orphan receptor TYMSTR/Bonzo (Loetscher *et al.*, 1997) are also located in the 3p21 region.

PROTEIN

Accession numbers

Human and mouse CCR9 proteins are accessible from the SPTREMBL section of the EMBL database,

Figure 2 Nucleotide sequence of the human and mouse CCR9 cDNA. The initiator ATG and stop TGA codons are marked in red. (Full colour figure may be viewed online.)

```
TGTCCCAGGG AGAGTTGCAT CGCCCTCCAC AGAGCAGGCT TGCATCTGAC TGACCCACCA    60
TGACACCCAC AGACTTCACA AGCCCTATTC CTAACATGGC TGATGACTAT GGCTCTGAAT   120
CCACATCTTC CATGGAAGAC TACGTTAACT TCAACTTCAC TGACTTCTAC TGTGAGAAAA   180
ACAATGTCAG GCAGTTTGCG AGCCATTTCC TCCCACCCTT GTACTGGCTC GTGTTCATCG   240
TGGGTGCCTT GGGCAACAGT CTTGTTATCC TTGTCTACTG GTACTGCACA AGAGTGAAGA   300
CCATGACCGA CATGTTCCTT TTGAATTTGG CAATTGCTGA CCTCCTCTTT CTTGTCACTC   360
TTCCCTTCTG GGCCATTGCT GCTGCTGACC AGTGGAAGTT CCAGACCTTC ATGTGCAAGG   420
TGGTCAACAG CATGTACAAG ATGAACTTCT ACAGCTGTGT GTTGCTGATC ATGTGCATCA   480
GCGTGGACAG GTACATTGCC ATTGCCCAGG CCATGAGAGC ACATACTTGG AGGGAGAAAA   540
GGCTTTTGTA CAGCAAAATG GTTTGCTTTA CCATCTGGGT ATTGGCAGCT GCTCTCTGCA   600
TCCCAGAAAT CTTATACAGC CAAATCAAGG AGGAATCCGG CATTGCTATC TGCACCATGG   660
TTTACCCTAG CGATGAGAGC ACCAAACTGA AGTCAGCTGT CTTGACCCTG AAGGTCATTC   720
TGGGGTTCTT CCTTCCCTTC GTGGTCATGG CTTGCTGCTA TACCATCATC ATTCACACCC   780
TGATACAAGC CAAGAAGTCT TCCAAGCACA AAGCCCTAAA AGTGACCATC ACTGTCCTGA   840
CCGTCTTTGT CTTGTCTCAG TTTCCCTACA ACTGCATTTT GTTGGTGCAG ACCATTGACG   900
CCTATGCCAT GTTCATCTCC AACTGTGCCG TTTCCACCAA CATTGACATC TGCTTCCAGG   960
TCACCCAGAC CATCGCCTTC TTCCACAGTT GCCTGAACCC TGTTCTCTAT GTTTTTGTGG  1020
GTGAGAGATT CCGCCGGGAT CTCGTGAAAA CCCTGAAGAA CTTGGGTTGC ATCAGCCAGG  1080
CCCAGTGGGT TTCATTTACA AGGAGAGAGG GAAGCTTGAA GCTGTCGTCT ATGTTGCTGG  1140
AGACAACCTC AGGAGCACTC TCCCTCTGAG GGTCTTCTC TGAGGT                  1186
```

Figure 2 (*Continued*)

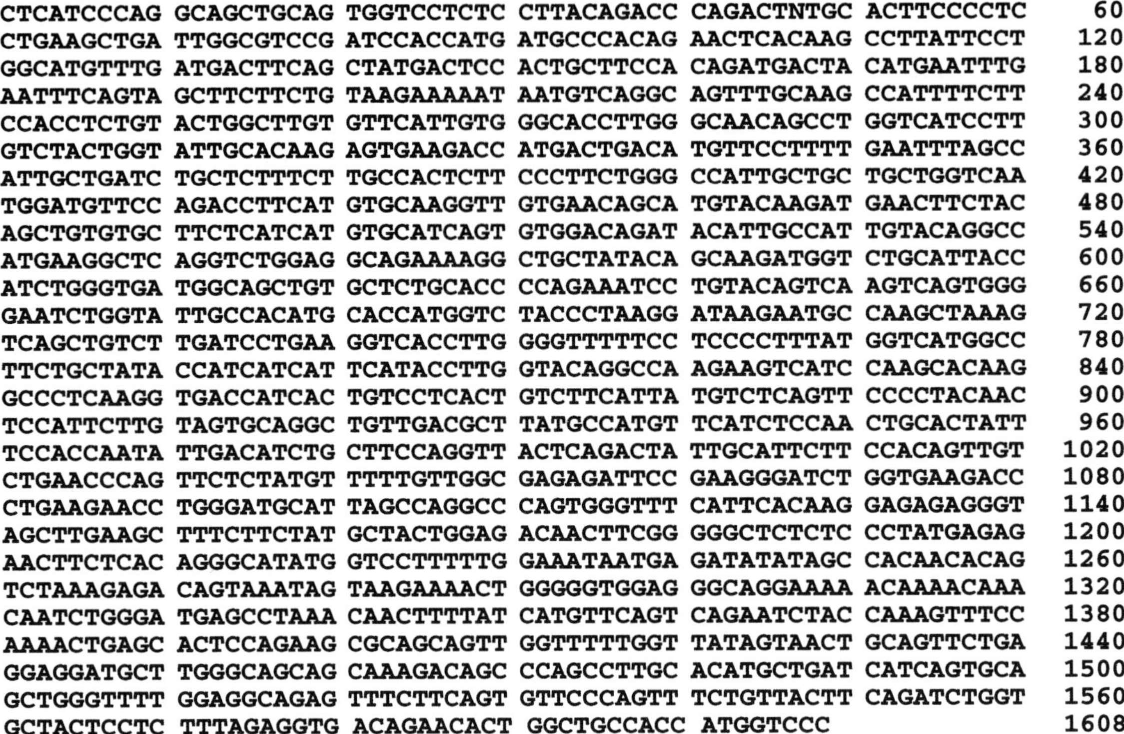

```
CTCATCCCAG GCAGCTGCAG TGGTCCTCTC CTTACAGACC CAGACTNTGC ACTTCCCCTC    60
CTGAAGCTGA TTGGCGTCCG ATCCACCATG ATGCCCACAG AACTCACAAG CCTTATTCCT   120
GGCATGTTTG ATGACTTCAG CTATGACTCC ACTGCTTCCA CAGATGACTA CATGAATTTG   180
AATTTCAGTA GCTTCTTCTG TAAGAAAAAT AATGTCAGGC AGTTTGCAAG CCATTTTCTT   240
CCACCTCTGT ACTGGCTTGT GTTCATTGTG GGCACCTTGG GCAACAGCCT GGTCATCCTT   300
GTCTACTGGT ATTGCACAAG AGTGAAGACC ATGACTGACA TGTTCCTTTT GAATTTAGCC   360
ATTGCTGATC TGCTCTTTCT TGCCACTCTT CCCTTCTGGG CCATTGCTGC TGCTGGTCAA   420
TGGATGTTCC AGACCTTCAT GTGCAAGGTT GTGAACAGCA TGTACAAGAT GAACTTCTAC   480
AGCTGTGTGC TTCTCATCAT GTGCATCAGT GTGGACAGAT ACATTGCCAT TGTACAGGCC   540
ATGAAGGCTC AGGTCTGGAG GCAGAAAAGG CTGCTATACA GCAAGATGGT CTGCATTACC   600
ATCTGGGTGA TGGCAGCTGT GCTCTGCACC CCAGAAATCC TGTACAGTCA AGTCAGTGGG   660
GAATCTGGTA TTGCCACATG CACCATGGTC TACCCTAAGG ATAAGAATGC CAAGCTAAAG   720
TCAGCTGTCT TGATCCTGAA GGTCACCTTG GGGTTTTTCC TCCCCTTTAT GGTCATGGCC   780
TTCTGCTATA CCATCATCAT TCATACCTTG GTACAGGCCA AGAAGTCATC CAAGCACAAG   840
GCCCTCAAGG TGACCATCAC TGTCCTCACT GTCTTCATTA TGTCTCAGTT CCCCTACAAC   900
TCCATTCTTG TAGTGCAGGC TGTTGACGCT TATGCCATGT TCATCTCCAA CTGCACTATT   960
TCCACCAATA TTGACATCTG CTTCCAGGTT ACTCAGACTA TTGCATTCTT CCACAGTTGT  1020
CTGAACCCAG TTCTCTATGT TTTTGTTGGC GAGAGATTCC GAAGGGATCT GGTGAAGACC  1080
CTGAAGAACC TGGGATGCAT TAGCCAGGCC CAGTGGGTTT CATTCACAAG GAGAGAGGGT  1140
AGCTTGAAGC TTTCTTCTAT GCTACTGGAG ACAACTTCGG GGGCTCTCTC CCTATGAGAG  1200
AACTTCTCAC AGGGCATATG GTCCTTTTTG GAAATAATGA GATATATAGC CACAACACAG  1260
TCTAAAGAGA CAGTAAATAG TAAGAAAACT GGGGGTGGAG GGCAGGAAAA ACAAAACAAA  1320
CAATCTGGGA TGAGCCTAAA CAACTTTTAT CATGTTCAGT CAGAATCTAC CAAAGTTTCC  1380
AAAACTGAGC ACTCCAGAAG CGCAGCAGTT GGTTTTTGGT TATAGTAACT GCAGTTCTGA  1440
GGAGGATGCT TGGGCAGCAG CAAAGACAGC CCAGCCTTGC ACATGCTGAT CATCAGTGCA  1500
GCTGGGTTTT GGAGGCAGAG TTTCTTCAGT GTTCCCAGTT TCTGTTACTT CAGATCTGGT  1560
GCTACTCCTC TTTAGAGGTG ACAGAACACT GGCTGCCACC ATGGTCCC              1608
```

Figure 3 Exon/intron structure of the human CCR9 gene. The size of exon 2 is 48 bp. Light brown-shaded areas correspond to coding sequences. The figure is not drawn to scale. Data from Zaballos *et al.* (1999), with permission. (Full colour figure may be viewed online.)

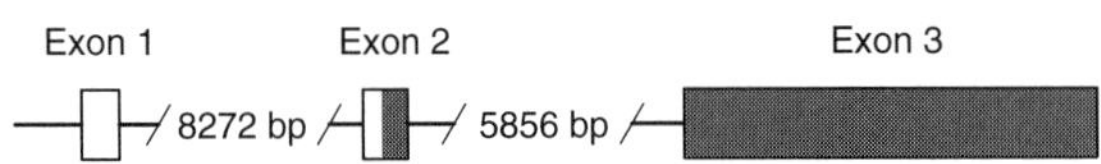

with accession numbers Q9UQQ6 and Q9WUT7, respectively.

Sequence

See **Figure 4**.

Description of protein

As described by Zaballos *et al.* (1999), both human and mouse CCR9 polypeptides are 369 amino acids long with seven transmembrane domains, showing the conserved motifs of this family of receptors as some cysteine residues, the DRY motif in the second intracellular loop, and the intracytoplasmic C-terminal tail with serine and threonine residues. In the N-terminal region, N32 is a potential *N*-glycosylation site for both human and mouse CCR9; mouse CCR9 has an additional potential site on N288, between transmembrane domains 6 and 7.

Relevant homologies and species differences

There is an 86% identity between human and mouse CCR9 proteins (Zaballos *et al.*, 1999) (Figure 4). Concerning the phylogenetic relationships of human CCR9 with other β chemokine receptors and orphan putative receptors, CCR9 is more similar to CCR7 (39%), CCR6 (34%) and the orphan receptor TYMSTR/Bonzo (32%), with which it is placed in a separate branch of the phylogenetic tree (Zaballos *et al.*, 1999) (**Figure 5**).

Affinity for ligand(s)

The ligand-specificity of human CCR9 was investigated by monitoring changes in the intracellular calcium concentration of stable HEK 293 transfectant cells upon sequential addition of chemokine samples. Among 36 recombinant human chemokines tested (**Table 1**), only TECK elicited a response, as

Figure 4 Alignment of the predicted amino acid sequences of human (blue) and mouse (green) CCR9 proteins. TM, transmembrane domains. Asterisks mark sites of potential *N*-glycosylation. Purple boxes mark identical residues. Data from Zaballos *et al.* (1999), with permission. (Full colour figure may be viewed online.)

```
                                               *
hCCR9   MTPTDFTSPI PNMADDYGSE STSSMEDYVN FNFTDFYCEK NNVRQFASHF  50
mCCR9   MMPTELTSLI PGMFDDFSYD STASTDDYMN LNFSSFFCKK NNVRQFASHF  50
                    TM1                    *            TM2
hCCR9   LPPLYWLVFI VGALGNSLVI LVYWYCTRVK TMTDMFLLNL AIADLLFLVT 100
mCCR9   LPPLYWLVFI VGTLGNSLVI LVYWYCTRVK TMTDMFLLNL AIADLLFLAT 100
                                         TM3
hCCR9   LPFWAIAAAD QWKFQRFMCK VVNSMYKMNF YSCVLLIMCI SVDRYIAIAQ 150
mCCR9   LPFWAIAAAG QWMFQTFMCK VVNSMYKMNF YSCVLLIMCI SVDRYIAIVQ 150
                                    TM4
hCCR9   AMRAHTWREK RLLYSKMVCF TIWVLAAALC IPEILYSQIK EESGIAICTM 200
mCCR9   AMKAQVWRQK RLLYSKMVCI TIWVMAAVLC TPEILYSQVS GESGIATCTM 200
                                    TM5
hCCR9   VYPSDESTKL KSAVLTLKVI LGFFLPFVVM ACCYTIIIHT LIQAKKSSKH 250
mCCR9   VYPKDKNAKL KSAVLILKVT LGFFLPFMVM AFCYTIIIHT LVQAKKSSKH 250
                     TM6
hCCR9   KALKVTITVL TVFVLSQFPY NCILLVQTID AYAMFISNCA VSTNIDICFQ 300
mCCR9   KALKVTITVL TVFIMSQFPY NSILVVQAVD AYAMFISNCT ISTNIDICFQ 300
                   TM7                           *
hCCR9   VTQTIAFFHS CLNPVLYVFV GERFRRDLVK TLKNLGCISQ AQWVSFTRRE 350
mCCR9   VTQTIAFFHS CLNPVLYVFV GERFRRDLVK TLKNLGCISQ AQWVSFTRRE 350

hCCR9   GSLKLSSMLL ETTSGALSL                                   369
mCCR9   GSLKLSSMLL ETTSGALSL                                   369
```

Figure 5 Phylogenetic relationships of CCR9 with other β chemokine receptors and some orphan putative β chemokine receptors. The tree was generated with the NJ plot program from a Clustal X multialignment. The data are taken from Zaballos *et al.* (1999), with permission.

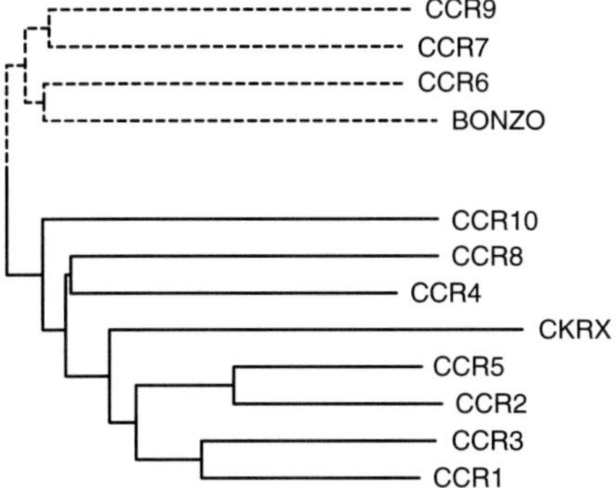

demonstrated by Zaballos *et al.* (1999). The cell response is correlated to the chemokine dose down to 20 nM. Murine TECK also induces intracellular calcium mobilization on HEK 293 cells transfected with the mouse CCR9 cDNA (Zaballos *et al.*, 1999). Adding a high concentration of human TECK to 293/hCCR9 cells results in complete desensitization to a second addition of the same stimulus. Both human and mouse TECK are able to stimulate the human receptor in transfectant cells, and this results in desensitization to a subsequent stimulus with hTECK. Similar experiments have been done with HEK 293 transfected with an expression plasmid encoding a shorter human CCR9 version, that lacks the N-terminal 12 amino acids encoded by exon 2. The shorter CCR9 protein is able to provoke a calcium flux upon stimulation with TECK (our unpublished results), thus suggesting that the first 12 N-terminal amino acids are not essential for activity.

Table 1 Human chemokines tested for calcium mobilization on indo-1-AM-loaded HEK 293/CCR9 cells

CXC chemokines	CC chemokines	CX3C chemokines	C chemokines
CXCL1/GROα	CCL1/I-309	CX3CL1/Fractalkine	XCL1/Lymphotactin
CXCL2/GROβ	CCL2/MCP-1		
CXCL3/GROγ	CCL3/MIP-1α		
CXCL4/PF4	CCL4/MIP-1β		
CXCL5/ENA-78	CCL5/RANTES		
CXCL6/GCP-2	CCL7/MCP-3		
CXCL7/NAP-2	CCL8/MCP-2		
CXCL8/IL-8	CCL9-10/MIP-1γ		
CXCL9/MIG	CCL11/Eotaxin 1		
CXCL10/IP-10	CCL13/MCP-4		
CXCL11/I-TAC	CCL14/HCC-1		
CXCL12/SDF	CCL16/LEC		
CXCL13/BCA-1	CCL17/TARC		
	CCL18/PARC		
	CCL19/ELC		
	CCL20/MIP-3α		
	CCL21/SLC		
	CCL22/MDC		
	CCL23/MPIF1		
	CCL24/Eotaxin 2		
	CCL25/TECK		

MOLT-4, a CD4+ lymphoblastic T cell line of thymic origin, also showed a high increase of intracellular calcium on stimulation with 200 nM hTECK (Zaballos *et al.*, 1999). The cell response is correlated to the chemokine dose, and MOLT-4 cells are slightly more sensitive than the transfectants to low doses of the chemokine, as 6 nM hTECK induced a detectable response. These results are essentially identical to those obtained with the HEK 293 cells transfected with human CCR9 and suggest that hTECK-induced stimulation of MOLT-4 cells is mediated by CCR9. As commented below, MOLT-4 cells express abundant CCR9 mRNA.

The activation of human CCR9 on stimulation with human TECK has also been shown in chemotaxis assays with HEK 293/CCR9 transfectants (Zaballos *et al.*, 1999) and MOLT-4 cells. Both cell types showed maximal migration at 100 nM TECK. A checkerboard-type analysis confirmed that the TECK-induced migration of MOLT-4 cells is mostly chemotactic.

Cell types and tissues expressing the receptor

Northern blot analysis of human poly(A)$^+$ RNA or mouse total RNA samples showed that both human and mouse CCR9 mRNA are highly expressed in thymus, although weak mRNA expression in spleen and lymph nodes was also detected (Zaballos *et al.*, 1999). Transcript sizes are 2.7 kb and 3.3 kb for human and mouse CCR9, respectively. Among several human cell lines tested, only MOLT-4 cells showed expression of CCR9; this is in agreement with their functional responses to TECK. As thymus was the tissue showing maximal expression of mouse CCR9, an RT-PCR analysis was done on FACS-sorted thymocyte subpopulations (Zaballos *et al.*, 1999). Mouse CCR9 is expressed by CD25+CD4−CD8− pre-T cells, CD4+CD8+ immature thymocytes, as well as in single positive CD4+ and CD8+ T cells. These results are consistent with the reported

TECK-induced chemoattraction of mouse thymocytes (Vicari *et al.*, 1997).

SIGNAL TRANSDUCTION

Cytoplasmic signaling cascades

The results reported by Zaballos *et al.* (1999) on the calcium mobilization assays performed on pertussis toxin-treated cells suggest that human CCR9 is partially, but not exclusively, coupled to the $G\alpha_i$ class of $G\alpha$ subunits. Indeed, pertussis toxin treatment decreased but did not abolish the TECK-induced calcium mobilization on HEK 293/CCR9 transfectants and MOLT-4 cells. A similar result has also been described for other chemokine receptors (Arai and Charo, 1996; Varona *et al.*, 1998). In contrast, the response to the α chemokine SDF, for which both cell types have endogenous receptors, is completely abolished by toxin treatment. Cholera toxin did not affect the CCR9-mediated calcium response of these cells upon stimulation with TECK.

References

Arai, H., and Charo, I. F. (1996). Differential regulation of G-protein-mediated signaling by chemokine receptors. *J. Biol. Chem.* **271**, 21814–21819.

Bleul, C. C., Farzan, M., Choe, H., Parolin, C., Clark-Lewis, I., Sodroski, J., and Springer, T. A. (1996). The lymphocyte chemoattractant SDF-1 is a ligand for LESTR/fusin and blocks HIV-1 entry. *Nature* **382**, 829–833.

Bonini, J. A., Martin, S. K., Dralyuk, F., Roe, M. W., Philipson, L. H., and Steiner, D. F. (1997). Cloning, expression, and chromosomal mapping of a novel human CC-chemokine receptor (CCR10) that displays high-affinity binding for MCP-1 and MCP-3. *DNA Cell Biol.* **16**, 1249–1256.

Boyd, R. L., and Hugo, P. (1991). Towards an integrated view of thymopoiesis. *Immunol. Today* **12**, 71–79.

Boyd, R. L., Tucek, C. L., Godfrey, D. I., Izon, D. J., Wilson, T. J., Davidson, N. J., Bean, A. G. D., Ladyman, H. M., Ritter, M. A., and Hugo, P. (1993). The thymic microenvironment. *Immunol. Today* **14**, 445–459.

D'Souza, P., and Harden, V. A. (1996). Chemokines and HIV-1 second receptors. *Nature Med.* **12**, 1293–1300.

Heiber, M., Docherty, J. M., Shah, G., Nguyen, T., Cheng, R., Heng, H. H. Q., Marchese, A., Tsui, L. C., Shi, X., George, S. R., and O'Dowd, B. F. (1995). Isolation of three novel human genes encoding G protein-coupled receptors. *DNA Cell Biol.* **14**, 25–35.

Kim, C. H., Pelus, L. M., White, J. R., and Broxmeyer, H. E. (1998). Differential chemotactic behaviour of developing T cells in response to thymic chemokines. *Blood* **91**, 4434–4443.

Luster, A. D. (1998). Chemokines: chemotactic cytokines that mediate inflammation. *N. Engl. J. Med.* **338**, 436–445.

Loetscher, M., Amara, A., Oberlin, E., Brass, N., Legler, D., Loetscher, P., D'Apuzzo, M., Meese, E., Rousset, D., Virelizier, J. L., Baggiolini, M., Arenzana-Seisdedos, F., and Moser, B. (1997). TYMSTR, a putative chemokine receptor selectively expressed in activated T cells, exhibits HIV-1 coreceptor function. *Curr. Biol.* **7**, 652–660.

Napolitano, M., Zingoni, A., Bernardini, G., Spinetti, G., Nista, A., Storlazzi, C. T., Rocchi, M., and Santoni, A. (1996). Molecular cloning of TER1, a chemokine receptor-like gene expressed by lymphoid tissues. *J. Immunol.* **157**, 2759–2763.

Oberlin, E., Amara, A., Bachelerie, F., Bessia, C., Virelizier, J. L., Arenzana-Seisdedos, F., Schwartz, O., Heard, J. M., Clark-Lewis, I., Legler, D. F., Loetscher, M., Baggiolini, M., and Moser, B. (1996). The CXC chemokine SDF-1 is the ligand for LSTR/fusin and prevents infection by T-cell-line-adapted HIV-1. *Nature* **382**, 833–835.

Raport, C. J., Schweickart, V. L., Eddy Jr, R. L., Shows, T. B., and Gray, P. W. (1995). The orphan G-protein-coupled receptor-encoding gene V28 is closely related to genes for chemokine receptors and is expressed in lymphoid and neural tissues. *Gene* **163**, 295–299.

Ritter, M. A., and Boyd, R. L. (1993). Development in the thymus: it takes two to tango. *Immunol. Today* **14**, 462–469.

Rollins, B. J. (1997). Chemokines. *Blood* **90**, 909–928.

Rucker, J., Edinger, A. L., Sharron, M., Samson, M., Lee, B., Berson, J. F., Yi, Y., Margulies, B., Collman, L. G., Doranz, B. J., Parmentier, M., and Doms, R. W. (1997). Utilization of chemokine receptors, orphan receptors, and herpesvirus-encoded receptors by diverse human and simian immunodeficiency viruses. *J. Virol.* **71**, 8999–9007.

Samson, M., Solarue, P., Vassart, G., and Parmentier, M. (1996). The genes encoding the human CC-chemokine receptors CC-CKR1 to CC-CKR5 (CMKBR1-CMKBR5) are clustered in the p21. 3-p24 region of chromosome 3. *Genomics* **36**, 522–526.

Shortman, K., and Wu, L. (1996). Early T lymphocyte progenitors. *Annu. Rev. Immunol.* **14**, 29–47.

Suzuki, G., Sawa, H., Kobayashi, Y., Nakata, Y., Nakagawa, K., Uzawa, A., Sakiyama, H., Kakinuma, S., Iwabuchi, K., and Nagashima, K. (1999). Pertussis toxin-sensitive signal controls the trafficking of thymocytes across the corticomedullary junction in the thymus. *J. Immunol.* **162**, 5981–5985.

Varona, R., Zaballos, A., Gutiérrez, J., Martín, P., Roncal, F., Albar, J. P., Ardavín, C., and Márquez, G. (1998). Molecular cloning, functional characterization and mRNA expression analysis of the murine chemokine receptor CCR6 and its specific ligand MIP-3α. *FEBS Lett.* **440**, 188–194.

Vicari, A. P., Figueroa, D. J., Hedrick, J. A., Foster, J. S., Singh, K. P., Menon, S., Copeland, N. G., Gilbert, D. J., Jenkins, N. A., Bacon, K. B., and Zlotnik, A. (1997). TECK: A novel CC chemokine specifically expressed by thymic dendritic cells and potentially involved in T cell development. *Immunity* **7**, 291–301.

Wilkinson, B., Owen, J. J. T., and Jenkinson, E. J. (1999). Factors regulating stem cell recruitment to the fetal thymus. *J. Immunol.* **162**, 3873–3881.

Yoshida, R., Imai, T., Hieshima, K., Kusuda, J., Baba, M., Kitaura, M., Nishimura, M., Kakizaki, M., Nomiyama, H., and Yoshie, O. (1997). Molecular cloning of a novel human CC chemokine EBI1-ligand chemokine that is a specific functional ligand for EBI1, CCR7. *J. Biol. Chem.* **272**, 13803–13809.

Zaballos, A., Gutiérrez, J., Varona, R., Ardavín, C., and Márquez, G. (1999). Identification of the orphan receptor GPR-9-6 as CCR9, the receptor for the chemokine TECK. *J. Immunol.* **162**, 5671–5675.

XCR1

Osamu Yoshie*

Department of Microbiology, Kinki University School of Medicine, Osaka-Sayama, Osaka, 589-8511, Japan

*corresponding author tel: +81-723-67-3606, fax: +81-723-67-3606, e-mail: o-yoshie@med.kindai.ac.jp

DOI: 10.1006/rwcy.2000.22018.

SUMMARY

An orphan seven transmembrane G protein-coupled receptor called GPR5 was identified as a specific functional receptor for Lptn/SCM-1/ATAC and is now termed XCR1. Lymphotactin (Lptn)/single C-motif 1 (SCM-1)/activation-induced, T cell-derived and chemokine-related molecule (ATAC) belongs to the C or γ subfamily of chemokines. XCR1 is expressed in tissues such as placenta, thymus, and spleen in humans. Few, if any, transcripts are, however, detected in peripheral blood leukocytes. Therefore, the exact types of cells that express XCR1 are still not known. Ltpn/SCM-1/ATAC induces chemotaxis and calcium flux in murine L1.2 cells transfected with XCR1. XCR1 and its murine counterpart mXCR1 have no *N*-glycosylation sites and have a sequence of HRYLSVVSP in the second intracellular domain instead of the standard DRY (Asp-Arg-Tyr) motif highly conserved in many chemokine receptors.

BACKGROUND

Discovery

An orphan seven-transmembrane G protein-coupled receptor (GPCR) termed GPR5 (Heiber *et al.*, 1995) was identified as a receptor for XCL1 and XCL2 and now termed XCR1 (Yoshida *et al.*, 1998). Murine L1.2 cells stably expressing XCR1 bound ^{125}I-labeled XCL1 with a high affinity and responded to XCL1 and XCL2 in chemotactic and calcium flux assays (Yoshida *et al.*, 1998). No other human chemokines so far tested were capable of inducing any such responses via XCR1 (Yoshida *et al.*, 1998). Taken together, XCR1 is a highly specific functional receptor for XCL1 and XCL2. In humans, XCR1 was found to be expressed in the placenta, thymus, and spleen (Yoshida *et al.*, 1998). Very few, if any, signals were detected in peripheral blood leukocytes. The murine counterpart of XCR1 was identified and termed mXCR1 (Yoshida *et al.*, 1999).

Lymphotactin (Lptn)/single C-motif 1 (SCM-1)/activation-induced, T cell-derived and chemokine-related molecule (ATAC), the ligand for XCR1, is selectively produced upon activation by CD8+ T cells, $\gamma\delta$ T cells, natural killer (NK) cells, and mast cells (Kelner *et al.*, 1994; Yoshida *et al.*, 1995; Mueller *et al.*, 1995; Kennedy *et al.*, 1995; Boismenu *et al.*, 1996; Hedrick *et al.*, 1997; Rumsaeng *et al.*, 1997). It is expressed in peripheral blood leukocytes and also in tissues such as thymus, spleen, lymph nodes, and small intestine (Yoshida *et al.*, 1996). It has a homology especially to the members of the CC subfamily, but lacks the first and third of the four cysteine residues conserved in all other chemokines. This molecule is now regarded as a unique representative of the C or γ subfamily of chemokines. Lymphotactin was reported to be chemotactic for lymphocytes and NK cells but its chemotactic activity on these types of cells has been controversial. In humans, there are two highly homologous genes encoding two forms of this molecules with only two amino acid differences (SCM-1α and SCM-1β) (Yoshida *et al.*, 1996). The systemic names for these molecules are XCL1 and XCL2 from the proposal on the systemic nomenclature of chemokine at the Keystone Symposium of 1999.

Alternative names

XCR1 was originally described as an orphan seven transmembrane, G protein-coupled receptor and called GPR5 (Heiber *et al.*, 1995).

Structure

XCR1 exhibits the seven transmembrane, G protein-coupled receptor structure typical of all of the chemokine receptors. However, both human and mouse sequences of XCR1 have the sequence HRYLSVVSP in the second intracellular loop instead of the typical DRY motif (DRYLAIVHA), which is highly conserved in the chemokine receptor family (Yoshida *et al.*, 1998, 1999). As a GPCR, XCR1 is also somewhat unusual because neither human nor mouse XCR1 contain *N*-glycosylation sites (Yoshida *et al.*, 1998, 1999). Between the human and mouse sequences, the N-terminal extracellular domain is highly divergent. Human XCR1 is 333 amino acids in length with a calculated molecular weight of 38,507.5 and an isoelectric point of 8.94.

Main activities and pathophysiological roles

The exact types of cells expressing XCR1 are not known. The messages were found to be weakly expressed in lymphoid tissues such as thymus and spleen in both humans and mice (Yoshida *et al.*, 1998, 1999). Considerable expression was observed in placenta in humans (Yoshida *et al.*, 1998). Few if any signals were detected, however, in peripheral blood leukocytes (Yoshida *et al.*, 1998). The latter observation is rather inconsistent with the previous results of the chemotactic activity of XCL1 on peripheral blood lymphocytes and NK cells. When stably transfected in murine L1.2 cells, both human and mouse XCR1 transduced signals for chemotaxis and calcium flux through a pertussis toxin-sensitive pathway (Yoshida *et al.*, 1998, 1999). When ovalbumin was given intranasally without and with mXCL1, mice with the latter treatment showed enhanced systemic and local antibody responses to ovalbumin (Lillard *et al.*, 1999). This suggests that local activation of cells expressing XCR1 can lead to enhanced mucosal and systemic immune responses.

GENE

Like many other GPCR genes, XCR1 is encoded by a single exon. The human gene is localized at chromosome 3p21.3-p21.1 (Heiber *et al.*, 1995).

Accession numbers

GenBank/EMBL/DDBJ:
Human XCR1: L36149
Mouse XCR1: AB028459

Sequence

See **Figure 1**.

Figure 1 Nucleotide sequence for the XCR1 gene. The coding region is shown by uppercase letters.

```
1       caaggcaatc ctctcccttt ggcctttcca aagtgctagg attacaggtg
51      ttagccactg tacccagcca ccacatggct ttaaactcca tgtctctatc
101     atttcagatg ctctaaacgt ccctgccatc tggtccagAT GGAGTCCTCA
151     GGCAACCCAG AGAGCACCAC CTTTTTTTAC TATGACCTTC AGAGCCAGCC
201     GTGTGAGAAC CAGGCCTGGG TCTTTGCTAC CCTCGCCACC ACTGTCCTGT
251     ACTGCCTGGT GTTTCTCCTC AGCCTAGTGG GCAACAGCCT GGTCCTGTGG
301     GTCCTGGTGA AGTATGAGAG CCTGGAGTCC CTCACCAACA TCTTCATCCT
351     CAACCTGTGC CTCTCAGACC TGGTGTTCGC CTGCTTGTTG CCTGTGTGGA
401     TCTCCCCATA CCACTGGGGC TGGGTGCTGG GAGACTTCCT CTGCAAACTC
451     CTCAATATGA TCTTCTCCAT CAGCCTCTAC AGCAGCATCT TCTTCCTGAC
501     CATCATGACC ATCCACCGCT ACCTGTCGGT AGTGAGCCCC CTCTCCACCC
551     TGCGCGTCCC CACCCTCCGC TGCCGGGTGC TGGTGACCAT GGCTGTGTGG
601     GTAGCCAGCA TCCTGTCCTC CATCCTCGAC ACCATCTTCC ACAAGGTGCT
651     TTCTTCGGGC TGTGATTATT CCGAACTCAC GTGGTACCTC ACCTCCGTCT
701     ACCAGCACAA CCTCTTCTTC CTGCTGTCCC TGGGGATTAT CCTGTTCTGC
751     TACGTGGAGA TCCTCAGGAC CCTGTTCCGC TCACGCTCCA AGCGGCGCCA
801     CCGCACGGTC AAGCTCATCT TCGCCATCGT GGTGGCCTAC TTCCTCAGCT
851     GGGGTCCCTA CAACTTCACC CTGTTTCTGC AGACGCTGTT TCGGACCCAG
901     ATCATCCGGA GCTGCGAGGC CAAACAGCAG CTAGAATACG CCCTGCTCAT
951     CTGCCGCAAC CTCGCCTTCT CCCACTGCTG CTTTAACCCG GTGCTCTATG
1001    TCTTCGTGGG GGTCAAGTTC CGCACACACC TGAAACATGT TCTCCGGCAG
1051    TTCTGGTTCT GCCGGCTGCA GGCACCCAGC CCAGCCTCGA TCCCCCACTC
1101    CCCTGGTGCC TTCGCCTATG AGGGCGCCTC CTTCTACtga ggggcctgtg
1151    gcggtgcagg cgcaggtgca ggtggacagg gactggaatg gggtcatgg
1201    agaagcgggc ctggaaggag cattgcagaa cacagcaggg tggagacgtc
1251    tcctcctgct gcagg
```

Figure 2 Amino acid sequence for the XCR1 protein.

```
MESSGNPEST TFFYYDLQSQ PCENQAWVFA TLATTVLYCL VFLLSLVGNS
LVLWVLVKYE SLESLTNIFI LNLCLSDLVF ACLLPVWISP YHWGWVLGDF
LCKLLNMIFS ISLYSSIFFL TIMTIHRYLS VVSPLSTLRV PTLRCRVLVT
MAVWVASILS SILDTIFHKV LSSGCDYSEL TWYLTSVYQH NLFFLLSLGI
ILFCYVEILR TLFRSRSKRR HRTVKLIFAI VVAYFLSWGP YNFTLFLQTL
FRTQIIRSCE AKQQLEYALL ICRNLAFSHC CFNPVLYVFV GVKFRTHLKH
VLRQFWFCRL QAPSPASIPH SPGAFAYEGA SFY
```

PROTEIN

Accession numbers

SwissProt:
Human XCR1: P46094

Sequence

See **Figure 2**.

Relevant homologies and species differences

XCR1 is most homologous to CCR10 (Bonini *et al.*, 1997) with 32% amino acid identity. Human and mouse XCR1 show 71% amino acid identity (Yoshida *et al.*, 1999). Between the human and mouse sequences, the N-terminal region is the least conserved.

Affinity for ligand(s)

XCL1 bound to XCR1 expressed on transfected murine L1.2 with a K_d of 10 nM (Yoshida *et al.*, 1998).

Cell types and tissues expressing the receptor

The exact types of cells expressing XCR1 are not known. In mouse spleen, transcripts were detected by RT-PCR in cells with CD8 marker and also in NK1.1+ CD3− cells (Yoshida *et al.*, 1999). In human tissues, XCR1 expression was readily detected in placenta and weakly in spleen and thymus (Yoshida *et al.*, 1998). In mouse tissues, weak expression was observed in spleen and lung (placenta and thymus not tested) (Yoshida *et al.*, 1999).

SIGNAL TRANSDUCTION

In murine L1.2 stable transfectants, both human and mouse XCR1 were capable of transducing signals in chemotaxis and calcium flux assays. The chemotactic responses of L1.2 expressing human XCR1 were shown to be highly sensitive to pertussis toxin, supporting its coupling to a Gα_i class of G protein.

BIOLOGICAL CONSEQUENCES OF ACTIVATING OR INHIBITING RECEPTOR AND PATHOPHYSIOLOGY

Unique biological effects of activating the receptors

At least in transfected murine L1.2 cells, XCR1 is a chemoattractant receptor. XCR1 is also a highly specific receptor for XCL1 and XCL2 (Yoshida *et al.*, 1998).

References

Boismenu, R., Feng, L., Xia, Y. Y., Chang, J. C., Havran, W. L. (1996). Chemokine expression by intraepithelial gamma delta T cells. Implications for the recruitment of inflammatory cells to damaged epithelia. *J. Immunol.* **157**, 985–992.

Bonini, J. A., Martin, S. K., Dralyuk, F., Roe, M. W., Philopson, L. H., and Steiner, D. F. (1997). Cloning, expression, and chromosomal mapping of a novel human CC-chemokine receptor (CCR10) that displays high-affinity binding for MCP-1 and MCP-3. *DNA Cell Biol.* **16**, 1249–1256.

Hedrick, J. A., Saylor, V., Figueroa, D., Mizoue, L., Xu, Y., Menon, S., Abrams, J., Handel, T., and Zlotnik, A. (1997). Lymphotactin is produced by NK cells and attracts both NK cells and T cells *in vivo*. *J. Immunol.* **158**, 1533–1540.

Heiber, M., Docherty, J. M., Shah, G., Nguyen, T., Cheng, R., Heng, H. H. Q., Marchese, A., Tsui, L.-C., Shi, X., George, S. R., and O'Dowd, B. F. (1995). Isolation of three novel human genes encoding G protein-coupled receptors. *DNA Cell Biol.* **14**, 25–35.

Kelner, G. S., Kennedy, J., Bacon, K. B., Kleyensteuber, S., Largaespada, D. A., Jenkins, N. A., Copeland, N. G., Bazan, J. F., Moore, K. W., Schall, T. J., and Zlotnik, A. (1994). Lymphotactin: a cytokine that represents a new class of chemokine. *Science* **266**, 1395–1399.

Kennedy, J., Kelner, G. S., Kleyensteuber, S., Schall, T. J., Weiss, M. C., Yssel, H., Schneider, P. V., Cocks, B. G., Bacon, K. B., and Zlotnik, A. (1995). Molecular cloning and functional characterization of human lymphotactin. *J. Immunol.* **155**, 203–209.

Lillard, J. W. Jr., Boyaka, P. N., Hedrick, J. A., Zlotnik, A., and McGhee, J. R. (1999). Lymphotactin acts as an innate mucosal adjuvant. *J. Immunol.* **162**, 1959–1965.

Mueller, S., Dorner, B., Korthaeuer, U., Mages, H. W., D'Apuzzo, M., Senger, G., and Kroczek, R. A. (1995). Cloning of ATAC, an activation-induced, chemokine-related molecule exclusively expressed in CD8+ T lymphocytes. *Eur. J. Immunol.* **25**, 1744–1748.

Rumsaeng, V., Vliagoftis, H., Oh, C. K., and Metcalfe, D. D. (1997). Lymphotactin gene expression in mast cells following Fc(epsilon) receptor I aggregation: modulation by TGF-beta, IL-4, dexamethasone, and cyclosporin A. *J. Immunol.* **158**, 1353–1356.

Yoshida, T., Imai, T., Kakizaki, M., Nishimura, M., and Yoshie, O. (1995). Molecular cloning of a novel C or γ type chemokine, SCM-1. *FEBS Lett.* **360**, 155–159.

Yoshida, T., Imai, T., Takagi, S., Nishimura, M., Ishikawa, I., Yaoi, T., and Yoshie, O. (1996). Structure and expression of two highly related genes encoding SCM-1/human lymphotactin. *FEBS Lett.* **395**, 82–88.

Yoshida, T., Imai, T., Kakizaki, M., Nishimura, M., Takagi, S., and Yoshie, O. (1998). Identification of single C motif-1/lymphotactin receptor XCR1. *J. Biol. Chem.* **273**, 16551–16554.

Yoshida, T., Izawa, D., Nakayama, T., Nakahara, K., Kakizaki, M., Imai, T., Suzuki, R., Miyasaka, M., and Yoshie, O. (1999). Molecular cloning of mXCR1, the murine SCM-1/lymphotactin receptor. *FEBS Lett.* **458**, 37–40

Miscellaneous Proinflammatory Factor Receptor

C5a Receptor

Tony E. Hugli[1] and Julia A. Ember[2,*]

[1]Department of Immunology, M/S IMM18 The Scripps Research Institute, 10550 North Torrey Pines Road, La Jolla, CA 92037-1092, USA

[2]Pharmingen, 10975 Torreyana Road, La Jolla, CA 92121, USA

*corresponding author tel: 858-784-8158, fax: 858-784-8307, e-mail: hugli@scripps.edu

DOI: 10.1006/rwcy.2000.23001.

SUMMARY

Although the biological effects and functional properties of the potent chemotactic factor C5a have been known since the 1960s, the C5a receptor (C5aR) was not demonstrated on human leukocytes or characterized until the late 1970s. Only in 1991 was the gene structure and derived protein sequence for human C5aR first reported. The C5aR, also known as CD88, is a member of the rhodopsin superfamily of receptors with wide cellular and tissue distribution in both humans and animals. As the receptor for a major humoral chemotactic factor and potent immunomodulatory factor, it is a key component of inflammatory and immunoregulatory processes. The nature and variety of cellular functions associated with the engagement of C5aR by its natural ligand, and the physiologic consequences of the mediated cellular responses, place C5aR in the same category functionally as chemokine and cytokine receptors.

BACKGROUND

Early functional and chemical crosslinking studies using purified C5a indicated that C5a receptors (CD88) exist on neutrophils (Chenoweth and Hugli, 1978), monocytes (macrophages) (Chenoweth *et al.*, 1982, 1984), basophils (Schulman *et al.*, 1988; MacGlashan and Hubbard, 1993), and eosinophils (Gerard *et al.*, 1989), as well as platelets (Fukuoka *et al.*, 1988; Kretzschmar *et al.*, 1991) and mast cells (Johnson *et al.*, 1975; Fukuoka and Hugli, 1990) from human and nonhuman species. C5aR is expressed on myeloid cell lines, such as U937 and HL-60, but only after these cells differentiate to a mature stage of development. Recently, evidence has been presented that C5aR is expressed on liver parenchymal cells, lung vascular smooth muscle, lung and umbilical vascular endothelial cells, bronchial and alveolar epithelial cells, as well as astrocytes and microglial cells (Haviland *et al.*, 1995; Lacy *et al.*, 1995; Gasque *et al.*, 1997). The distribution of C5aR is predominant in human heart, lung, spleen, spinal cord, and in many regions of the brain. The size of C5aR was estimated to be approximately 40–48 kDa based on a variety of chemical crosslinking techniques that chemically attached [^{125}I]C5a to C5aR on neutrophils (Heideman and Hugli, 1984; Huey and Hugli, 1985; Rollins and Springer, 1985).

The ligand, human C5a, is a glycoprotein of 11–12 kDa, and is frequently mentioned under the collective name anaphylatoxin. Anaphylatoxins are small, bioactive fragments released from C3, C4, and C5 during complement (C) activation. The term anaphylatoxin was coined by Friedberger (1910). It is a descriptive label for an activity found in complement-activated serum that produced rapid anaphylactoid-like death when C-activated serum was injected into laboratory animals. Anaphylatoxin has remained the generic name for molecules now chemically identified as C3a, C4a, and C5a. As a group, the anaphylatoxins are humoral mediators mainly recognized for their proinflammatory and immunoregulatory functions that serve as bioactive sentinels in host defense.

Discovery

Early characterization of C5aR (CD88) on human neutrophils clearly established three facts that later played a prominent role in the elucidation of the C5a–C5aR molecular interactions and in isolation of the receptor gene (Chenoweth and Hugli, 1978;

Huey and Hugli, 1985). It was determined that (1) C5a (^{125}I-labeled) binds to a receptor on human neutrophils with nanomolar affinity; (2) there are approximately 100,000 copies of the receptor per cell; and (3) certain degradation products of C5a (i.e. C5a desArg and C5a(1–69)) compete with the intact factor for binding C5aR. The observation that a C5a fragment devoid of the C-terminal effector site (i.e. C5a(1–69)) binds to the receptor without activating the cell led to the hypothesis that both a primary effector site and secondary binding sites exist on the C5a ligand and that each is important for optimal engagement of the ligand with the receptor (Chenoweth, 1978; Gerard *et al.*, 1979; Chenoweth and Hugli, 1980).

The C5a receptors are much more widely distributed than was previously believed in terms of both cells and tissues. It is now clear that the C5a receptors are present not just on myeloid inflammatory cells but on a variety of tissues as well. C5aR exists on both mast cells and macrophages in the skin, digestive, vascular, and pulmonary tissues, along with numerous other receptor-bearing tissue cells, suggesting that particular tissues and organs may be more highly responsive to C5a stimulation than others.

Figure 1 Proposed model illustrating the interaction between C5a and the C5a receptor. The design for this model was adapted from the C5a/C5aR model proposed by Siciliano *et al.* (1994). C5aR is a G protein-coupled transmembrane receptor of the rhodopsin family. The model for C5a interaction with C5aR indicates that the noneffector site (site 1) on C5a binds to the N-terminal region of the C5a receptor while the C-terminal effector site (site 2) of C5a penetrates the 'pore' formed by the circular arrangement of the seven transmembrane helices.

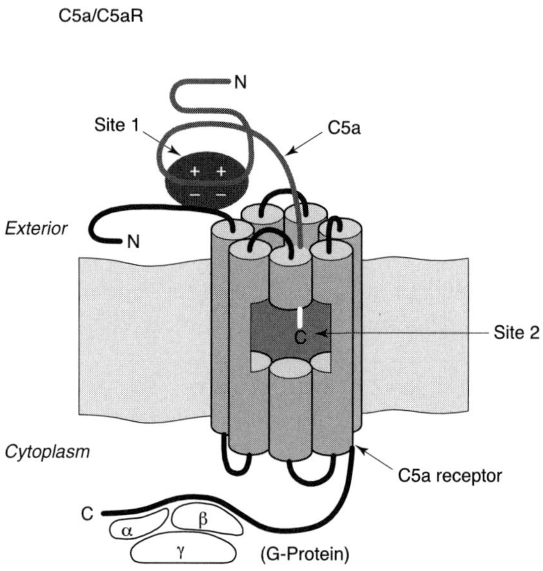

Alternative names

C5aR, CD88, anaphylatoxin receptor, complement receptor (Rother *et al.*, 1992).

Structure

Human C5aR is an integral membrane glycoprotein, consisting of 350 amino acids forming a single polypeptide chain. It belongs to the rhodopsin-like receptor superfamily, characterized by seven hydrophobic, transmembrane helical regions connected by three extra- and three intracellular loops. Orientation of C5aR in the membrane was determined by immunohistochemical techniques. It was determined that the N-terminal end of C5aR is exposed on the extracellular surface, while the C-terminal end extends into the cell from the intracellular surface of the membrane. A model of the C5aR molecule has been proposed (Gerard and Gerard, 1994b) (**Figure 1**).

Main activities and pathophysiological roles

It was known by the late 1960s that anaphylatoxins possessed potent spasmogenic activity (i.e. the ability to contract smooth muscle tissues) and could both enhance vascular permeability and recruit white blood cells when injected into the skin, either of animals (Dias da Silva and Lepow, 1967; Cochrane and Müller-Eberhard, 1968) or of humans (Lepow *et al.*, 1970; Wuepper *et al.*, 1972; Vallota and Müller-Eberhard, 1973). Many other actions have since been attributed to C5a, including a host of cellular effects that imply important immunomodulatory functions (Morgan *et al.*, 1993), as well as a number of tissue-specific effects (Ember *et al.*, 1998).

Cellular Activities

Cellular responses mediated by the ligand-specific activation of C5aR reflect the prominent proinflammatory character of the C5a molecule. Perhaps the biologic property most closely identified with C5a–C5aR interactions is the potent chemotactic activity for granulocytes, particularly neutrophils and eosinophils. C5a induces chemotactic migration of neutrophils and eosinophils *in vitro* at an EC_{50} between 0.5 and 2.0 nM (Fernandez *et al.*, 1978; Morita *et al.*,

1989; Daffern *et al.*, 1995). Potent activators of inflammation are released by C5a from all granulocytes that possess C5aR. Elastase, peroxidase, glucuronidase, and lactoferrin are released from neutrophils. C5a released peroxidase, major basic protein (MBP), eosinophil-derived neurotoxin (EDN), and eosinophil cationic protein (ECP) from eosinophils (Henson, 1971; Goetzl and Austen, 1974; Takafuji *et al.*, 1994) and arachidonate and vasoamines from basophils (Siraganian and Hook, 1976; Hartman and Glovsky, 1981; MacGlashan and Hubbard, 1993). C5a also activates the NADPH-oxidase pathway in granulocytes, leading to an oxidative burst (Goetzl and Austen, 1974; Elsner *et al.*, 1994).

Cellular responses to C5a have been well characterized *in vitro* and described in detail in numerous articles and reviews (Chenoweth and Goodman, 1983; Hugli, 1984; Ember *et al.*, 1998). Treatment of granulocytes with cytochalasin B, or an equivalent microfilament-disrupting agent, is required to elicit optimal responses to C5a *in vitro*, with the exception of chemotaxis. This same requirement is shared by other granulocyte-activating factors, including f-MLF (fMLP) and IL-8. Experimental evidence suggests that adhesive interactions with other cell types or with extracellular matrix replaces the effect of cytochalasin B during *in vivo* cellular responses to C5a (Becker *et al.*, 1974; Henson *et al.*, 1978; Laurent *et al.*, 1991). Consistent with this hypothesis, C5a has been shown to act as a proadhesive stimulus for granulocytes. C5a–C5aR interaction in both neutrophils and eosinophils leads to an increased expression of β_2 integrins and concurrent shedding of L-selectin (Kishimoto *et al.*, 1989; Lundahl *et al.*, 1993; Neeley *et al.*,1993).

Tissue-Specific Activities

Recent evidence shows that C5a/C5aR plays an important role in *immune injury in the lung* (Mulligan *et al.*, 1996, 1997; Schmid *et al.*, 1997) and in *post-ischemic vascular and tissue injury* (Ito and Del Balzo, 1994; Amsterdam *et al.*, 1995; Ivey *et al.*, 1995). This supports the contention that regulation of selected complement activation products such as C5a may be of therapeutic value. Discovery that both C3a and C5aR may exist on numerous cell types other than circulating white cells, such as hepatocytes, lung, epithelial cells (Haviland *et al.*, 1995), endothelial cells (Foreman *et al.*, 1994), and the astrocytes and microglial cells in brain tissue (Gasque *et al.*, 1995b, 1997), implicates these anaphylatoxins in *vascular diseases, pulmonary diseases*, and *degenerative neurologic diseases*.

GENE

Recent molecular studies resulted in cloning of the C5aR (Boulay *et al.*, 1991; Gerard and Gerard, 1991). The gene structure for the human C5aR was reported by two separate groups in 1991. One group used a library obtained from dibutyryl-cAMP-induced human myeloid U937 cells (Gerard and Gerard, 1991) and the other group used a dibutyryl-cAMP-induced human myeloid HL-60 cell library (Boulay *et al.*, 1991). Both groups obtained cDNA clones with open reading frames of 1050 base pairs coding for an identical protein of 350 amino acid residues with a calculated M_r of 39,320. A single glycosylation site was located at Asn5 of the first extracellular domain of C5aR. The presence of an *N*-linked oligosaccharide group presumably explains the difference in size between the nude protein of 39 kDa and the 40–48 kDa estimate for C5aR expressed on human leukocytes. The size of the native C5aR was estimated using a variety of chemical crosslinking techniques to attach [^{125}I]C5a to C5aR on human neutrophils (Heideman and Hugli, 1984; Huey and Hugli, 1985; Rollins and Springer, 1985). The effect of glycosylation of C5aR on C5a binding was explored by replacing Asn5 with an Ala residue (Pease and Barker, 1993). When both the Ala5-C5aR mutant molecule and wild-type C5aR were expressed on Chinese hamster ovary cells and compared, the dissociation constants were 20 nM and 13 nM, respectively. These results suggest that glycosylation of the C5aR has little influence on ligand binding and hence on C5a-induced cellular functions.

Accession numbers

Human (*Homo sapiens*): X57250, X58674, M62505, M76672
Chimpanzee (*Pan troglodytes*): X97730
Orang-utan (*Pongo pygmeus*): X97732
Rhesus monkey (*Macaca mulatta*): X97731
Gorilla (*Gorilla gorilla*): X97733
Mouse (*Mus musculus*): S46665, L05630, S50577
Rat (*Rattus norwegicus*): X95990
Dog (*Canis familiaris*): X65860
Rabbit (*Oryctolagus cuniculus*): AF068680

Sequence

Base count C5aR cDNA: 189 A, 348 C, 293 G, 262 T. *See* **Figure 2**.

Figure 2 Nucleotide sequence for human C5a receptor.

```
Sequence

Base
Count    189 a    348 c    293 g    262 t    C5aR    cDNA

Origin: Homo sapiens

   1 ATGAACTCCT TCAATTATAC CACCCCTGAT TATGGGCACT ATGATGACAA GGATACCCTG
  61 GACCTCAACA CCCCTGTGGA TAAAACTTCT AACACGCTGC GTGTTCCAGA CATCCTGGCC
 121 TTGGTCATCT TTGCAGTCGT CTTCCTGGTG GGAGTGETGG GCAATGCCCT GGTGGTCTGG
 181 GTGACGGCAT TCGAGGCCAA GCGGACCATC AATGCCATCT GGTTCCTCAA CTTGGCGGTA
 241 GCCGACTTCC TCTCCTGCCT GGCGCTGCCC ATCTTGTTCA CGTCCATTGT ACAGCATCAC
 301 CACTGGCCCT TTGGCGGGGC CGCCTGCAGC ATCCTGCCCT CCCTCATCCT GCTCAACATG
 361 TACGCCAGCA TCCTGCTCCT GGCCACCATC AGCGCCGACC GCTTTCTGCT GGTGTTTAAA
 421 CCCATCTGGT GCCAGAACTT CCGAGGGGCC GGCTTGGCCT GGATCGCCTG TGCCGTGGCT
 481 TGGGGTTTAG CCCTGCTGCT GACCATACCC TCCTTCCTGT ACCGGGTGGT GCGGGAGGAG
 541 TACTTTCCAC CAAAGGTGTT GTGTGGCGTG GAATACAGCC ACGACAAACG GCGGGAGCGA
 601 GCCGTGGCCA TCGTCCGGCT GGTCCTGGGC TTCCTGTGGC CTCTACTCAC GCTCACGATT
 661 TGTTACACTT TCATCCTGCT CCGGACGTGG AGCCGCAGGG CCACGCGGTC CACCAAGACA
 721 CTCAAGGTGG TGGTGGCAGT GGTGGCCAGT TTCRTTATCT TCTGGTTGCC CTACAAGGTG
 761 ACGGGGATAA TGATGTCCTT CCTGGAGCCA TCGTCACCCA CCTTCCTGCT GCTGAATAAG
 841 CTGGACTCCC TGTGTGTCTC CTTTGCCTAC ATCAACTGCT GCATCAACCC CATCATCTAC
 901 GTGGTGGCCG GCCAGGGCTT CCAGGGCCGA CTGCGGAAAT CCCTCCCCAG CCTCCTCCGG
 961 AACGTGTTGA CTGAAGAGTC CGTGGTTAGG GAGAGCAAGT CATTCACGCG CTTCACAGTG
1021 GACACTATGG CCCAGAAGAC CCAGGCAGTG TAGGCGACAC GTCATGGGCC ACTGTGGCGA
1081 TGTCCCTTCC TT
```

Chromosome location and linkages

The genes encoding C5aR, along with two structurally related formyl peptide receptors (FPRH1 and FPRH2), have been mapped to band position q13.2 in human chromosome 19 (Bao *et al.*, 1992).

PROTEIN

Accession numbers

Human (*Homo sapiens*): P21730
Chimpanzee (*Pan troglodytes*): P79240
Orang-utan (*Pongo pygmeus*): P79234
Rhesus monkey (*Macaca mulatta*): P79188
Gorilla (*Gorilla gorilla*): P79175
Mouse (*Mus musculus*): P30993
Rat (*Rattus norwegicus*): P97520
Dog (*Canis familiaris*): P30992

Sequence

See **Figure 3**.

Description of protein

The C5aR protein deduced from the cDNA clones identified the characteristic structure of a member of the rhodopsin superfamily of receptors, otherwise known as GTP-binding, protein-coupled receptors or seven transmembrane-spanning receptors. C5aR is an integral membrane protein.

Figure 3 Amino acid sequence for human C5a receptor.

```
Sequence

C5aR protein, human, Homo sapiens

MNSFNYTTPDYGHYDDKDTLDLNTPVDKTSNTLRVPDILALVIFAVVFLV
GVLGNALVVWVTAFEAKRTINAIWFLNLAVADFLSCLALPILFTSIVQHH
HWPFGGAACSILPSLILLNMYASILLLATISADRFLLVFKPIWCQNFRGA
GLAWIACAVAWGLALLLTIPSFLYRVVREEYFPPKVLCGVDYSHDKRRER
AVAIVRLVLGFLWPLLTLTICYTFILLRTWSRRATRSTKTLKVVVAVVAS
FFIFWLPYQVTGIMMSFLEPSSPTFLLLNKLDSLCVSFAYINCCINPIIY
VVAGQGFQGRLRKSLPSLLRNVLTEESVVRESKSFTRSTVDTMAQKTQAV
```

Relevant homologies and species differences

Sequence comparisons between C5aR and other members of the rhodopsin superfamily indicated that a relatively close homology exists only with human neurokinin A (substance K) and formyl peptide receptors (Gerard and Gerard, 1991). Even receptors having the highest degree of sequence identity exhibit only 25% (substance K) and 35% (formyl peptide) structural identity to C5aR. More recently, the cDNA sequences for mouse C5aR (Gerard *et al.*, 1992), guinea pig (Fukuoka *et al.*, 1998), rat (Akatsu *et al.*, 1997), canine, and a partial sequence for bovine C5aR have been determined (Perret *et al.*, 1992) (**Figure 4**). It is interesting to note that the extracellular N-terminal region, which is believed to be a ligand-binding site, is poorly conserved between species while the transmembrane

Figure 4 The complete protein sequences for the C5a receptor from human (Hu), dog (Dg), guinea pig (Gp), rat (Rt), and mouse (Mo). The alignments were optimized to obtain maximal identity and the seven transmembrane regions have been identified by a line and roman numerals. The residue positions that have been conserved in all species are denoted by an asterisk. The C5a receptors from various species are similar in size and each molecule has a potential glycosylation site near the N-terminus.

ALIGNMENT OF C5a RECEPTOR SEQUENCES FROM SEVERAL SPECIES

```
                                                             I
Hu C5aR   1:MNSFNYTTPDYGHYDDKDTLDLNTPVDKTSNT--LRVPDILALVIFAVVFLVGVLGNALV 58
Dg C5aR   1:MASMNFSPPEYPDY-GTATLDPNIFVDESLNTPKLSVPDMIALVIFVMVFLVGVPGNFLV 59
Gp C5aR   1:----MMVTVSY-DYD-YNSTFLPDGFVDNYVE-RLSFGDLVAVVIMVVVFLVGVPGNALV 53
Rt C5aR   1:MDPISNDSSE-ITYDYSDGTPNPDMPADGVYIPKMEPGDIAALIIYLAVFLVGVTGNALV 59
Mo C5aR   1:MDPIDNSSFE-INYDHY-GTMDPNIPADGIHLPKRQPGDVAALIIYSVVFLVGVPGNALV 58
                         *                        *  *  *   ****** ** **

                                  II
Hu C5aR  59:VWVTAFEAKRTINAIWFLNLAVADFLSCLALPILFTSIVQHHHWPFGGAACSILPSLILL 118
Dg C5aR  60:VWVTGFEVRRTINAIWFLNLAVADLLSCLALPILFSSIVQQGYWPFGNAACRILPSLILL 119
Gp C5aR  54:VWVTACEARRHINAIWFLNLAAADLLSCLALPILLVSTVHLNHWYFGNTACKVLPSLILL 113
Rt C5aR  60:VWVTAFEAKRTVNAIWFLNLAVADLLSCLALPILFTSIVKHNHWPFGDQACIVLPSLILL 119
Mo C5aR  59:VWVTAFEPDGPSNAIWFLNLAVADLLSCLAMPVLFTTVLNHNYWYFDATACIVLPSLILL 118
            ****  *     ********* ** ***** * *         * *   **  *******

                  III                                 IV
Hu C5aR 119:NMYASILLLATISADRFLLVFKPIWCQNFRGAGLAWIACAVAWGLALLLTIPSFLYRVVR 178
Dg C5aR 120:NMYASILLLTTISADRFVLVFNPIWCQNYRGPQLAWAACSVAWAVALLLTVPSFIFRGVH 179
Gp C5aR 114:NMYTSILLLATISADRLLLVLSPIWCQRFRGGCLAWTACGVAWVLALLLSSPSFLYRRTH 173
Rt C5aR 120:NMYSSILLLATISADRFLLVFKPIWCQKFRRPGLAWMACGVTWVLALLLTIPSFVFRRIH 179
Mo C5aR 119:NMYASILLLATISADRFLLVFKPIWCQKVRGTGLAWMACGVAWVLALLLTIPSFVYREAY 178
            *** ***** ******  **  *****  *   *** ** * *  ****  ***  *

                                                 V
Hu C5aR 179:EEYFPPKVLCGVDYS-HDKRRERAVAIVRLVLGFLWPLLTLTICYTFILLRTWSRRATRS 237
Dg C5aR 180:TEYFPFWMTCGVDYSGVGVLVERGVAILRLLMGFLGPLVILSICYTFLLIRTWSRKATRS 239
Gp C5aR 174:NEHFSFKVYCVTDY-GRDISKERAVALVRLVVGFIVPLITLTACYTFLLLRTWSRKATRS 232
Rt C5aR 180:KDPYSDSILCNIDYSKGPFFIEKAIAILRLMVGFVLPLLTLNICYTFLLIRTWSRKATRS 239
Mo C5aR 179:KDFYSEHTVCGINYGGGSFPKEKAVAILRLMVGFVLPLLTLNICYTFLLLRTWSRKATRS 238
                     *   *       *   *  **  **  **  *  **** * ***** ****

                          VI                                    VII
Hu C5aR 238:TKTLKVVVAVVASFFIFWLPYQVTGIMMSFLEPSSPTFLLLNKLDSLCVSFAYINCCINP 297
Dg C5aR 240:TKTLKVVVAVVVSFFVLWLPYQVTGMMMALFYKHSESFRRVSRLDSLCVAVAYINCCINP 299
Gp C5aR 233:AKTVKVVVAVVSNFFVFWLPYQVTGILLAWHSPNSATYRNTKALDAVCVAFAYINCCINP 292
Rt C5aR 240:TKTLKVVMAVVTCFFVFWLPYQVTGVILAWLPRSSSTFQSVERLNSLCVSLAYINCCVNP 299
Mo C5aR 239:TKTLKVVMAVVICFFIFWLPYQVTGVMIAWLPPSSPTLKRVEKLNSLCVSLAYINCCVNP 298
             ** *** ***  **  ********         *        *   **  ****** **

Hu C5aR 298:IIYVVAGQGFQGRLRKSLPSLLRNVLTEESVVRESKSFTRSTVDTMAQKTQAV        350
Dg C5aR 300:IIYVLAAQGFHSRFLKSLPARLRQVLAEESVGRDSKSITLSTVDTPAQKSQGV        352
Gp C5aR 293:IIYVVAGHGFQGRLLKSLPSVLRNVLTEESLNRDTRSFTRSTVDTMPQKSESV        345
Rt C5aR 300:IIYVMAGQGFHGRLRRSLPSIIRNVLSEDSLGRDSKSFTRSTMDTSTQKSQAV        352
Mo C5aR 299:IIYVMAGQGFHGRLLRSLPSIIRNALSEDSVGRDSKTFTPSTDDTSPRKSQAV        351
            **** *  **  *   ***   *  * * *  *     * ** **   *   *
```

helical regions and the intracellular C-terminal region, which binds the G proteins, are highly conserved.

Affinity for ligand(s)

Characterization of the binding affinity and numbers of receptors on neutrophils, estimated to average 80,000 copies with an affinity of approximately 2 nM (Huey and Hugli, 1985; Shapira *et al.*, 1995), provided critical information that indicated differentiated leukocytic cell lines would be appropriate sources from which to isolate and clone the C5aR gene.

The cloned C5aR was expressed in COS-7 cells and high-affinity C5a binding was demonstrated. The Boulay group (Boulay *et al.*, 1991) concluded that there were both high-affinity ($K_d = 1.7$ nM) and low-affinity ($K_d = 20$–25 nM) C5aRs expressed on these COS cells, while the Gerard group (Gerard and Gerard, 1991) described only the high-affinity receptors ($K_d = 1.4$ nM) and demonstrated G protein-dependent phosphorylation of phosphatidylinositol in response to C5a.

Cell types and tissues expressing the receptor

Early functional and chemical crosslinking studies using purified factors indicated that C5aR (CD88) exists on neutrophils (Chenoweth and Hugli, 1978), monocytes (macrophages; Chenoweth *et al.*, 1982, 1984), basophils (Schulman *et al.*, 1988; MacGlashan and Hubbard, 1993), and eosinophils (Gerard *et al.*, 1989), as well as platelets (Fukuoka *et al.*, 1988; Kretzschmar *et al.*, 1991) and mast cells (Johnson *et al.*, 1975; Fukuoka and Hugli, 1990). C5aR is expressed on differentiated myeloid cells, such as U937 and HL-60. C5aR is expressed on liver parenchymal cells, lung vascular smooth muscle, lung and umbilical vascular endothelial cells, bronchial and alveolar epithelial cells, HepG2 cells, a hepatoma cell line (Buchner *et al.*, 1995; McCoy *et al.*, 1995), mesangial cells (Wilmer *et al.*, 1998) as well as astrocytes and microglial cells (Haviland *et al.*, 1995; Lacy *et al.*, 1995; Gasque *et al.*, 1997), on cultured human fetal astrocytes and astrocyte cell lines (Gasque *et al.*, 1995a; Lacy *et al.*, 1995).

C5a induces smooth muscle contraction in virtually all tissue types tested, including ileal, uterine, bronchial, and vascular smooth muscle (Hugli, 1981; Hugli *et al.*, 1987). Tissue distribution of C5aR was most predominant in human heart, lung, spleen, spinal cord, and in various regions of the brain (Haviland *et al.*, 1995; Nataf *et al.*, 1999).

Regulation of receptor expression

Challenge of C5aR-bearing cells by C5a causes a rapid phosphorylation of serine residues in the C-terminal region of C5aR. Agonist-binding to C5aR causes a rapid internalization of the complex into endosomes that cluster near the nucleus within 10 minutes. Under continuous exposure to C5a, and in the absence of protein synthesis, C5aR is maintained in a highly phosphorylated state but is not degraded. Confocal microscopy and ligand-binding studies indicated that internalized receptors were recycled to the plasma membrane. During this process receptors are dephosphorylated. Therefore phosphorylation plays a key role in the intracellular trafficking of the C5aR. Phosphorylated C5aRs may be recognized by adapter proteins that interact with the endocytic machinery.

Truncation of the intracellular C-terminal end of C5aR by deletion of residues 314–350 disrupted expression, while deletion of only the last 26 residues (i.e. 326–350) did not affect expression or ligand-binding of the C5aR (Klos *et al.*, 1994). IL-3 has been shown to regulate C5aR expression in neurons and in astrocyte-targeted IL-3 transgenic mice (Paradisis *et al.*, 1998).

Release of soluble receptors

The C5aR is not released as a soluble receptor and does not circulate.

SIGNAL TRANSDUCTION

C5a/C5aR-Binding Sites

Cloning of the C5a receptor (Boulay *et al.*, 1991; Gerard and Gerard, 1991) provided new opportunities for elucidating the requirements for ligand–receptor interactions between C5a and its receptor. Antibodies were generated to C5aR peptides that mimic portions of the N-terminal extracellular region of the molecule. These antibodies are not only excellent markers of cells or tissues expressing the receptor (Buchner *et al.*, 1995; Gasque *et al.*, 1995b, 1997; Haviland *et al.*, 1995), but they block binding and cellular activation by intact C5a (Morgan *et al.*, 1993; Oppermann *et al.*, 1993). Truncation of

N-terminal residues 1–22 in C5aR (DeMartino *et al.*, 1994; Siciliano *et al.*, 1994) abrogated binding of intact C5a, but had no effect on binding C-terminal peptide analogs of C5a (DeMartino *et al.*, 1994; Siciliano *et al.*, 1994). Point mutations converting five aspartic acid residues in the N-terminal region of C5aR to alanines (i.e. Asp10, 15, 16, 21, 27 → Ala) resulted in significant loss in binding affinity for C5a. These data indicated a critical role for the identified aspartic acid residues in C5a–C5aR interactions (DeMartino *et al.*, 1994; Mery and Boulay, 1994).

Studies focused on locating the C5a effector-binding site on C5aR have also used point mutation techniques. Replacement of Glu199 and Arg206 in C5aR by alanines produced major depression in C5a binding, suggesting that these residues contribute to the effector-binding site in C5aR. It has been concluded that specific residues in the second extracellular loop and residues on the fifth intramembrane helix help form the primary effector binding site on C5aR.

Based on these mutational results, a model was designed proposing a noneffector interaction site involving the aspartic acid side-chains in the N-terminal portion of C5aR and the cationic side-chain of Arg40 (and possibly Arg37 and Lys12) in human C5a. An effector interaction site occurs between anionic residues in the second extracellular loop and on the fifth intramembrane helix of C5aR, and basic residues Arg62, His66, Lys67, and Arg74 of the human C5a molecule. Based on this model, contact between C5a and the N-terminal portion of C5aR defines the noneffector site that promotes a cooperative interaction with the effector-binding site, resulting in cellular activation (Siciliano *et al.*, 1994).

Associated or intrinsic kinases

There are several neutrophil signal transduction pathways regulated by G_i-coupled C5aR: one involves the activation of phospholipase C (PLC). The ligand-bound receptor interacts with pertussis toxin (PTX)-sensitive G proteins, such as the G_i proteins, and releases the $\beta\gamma$ subunits, which then stimulate PLCβ and phosphatidylinositol 3-kinase (PI-3 kinase) activities, followed by activation of phospholipase A_2 (PLA_2) and phospholipase D (Jiang *et al.*, 1996). Postreceptor activation of the PLC/PKC pathway modulates intracellular calcium fluxes which are presumed to be important for neutrophil degranulation. Another pathway involves activation of PI-3 kinase and the generation of phosphatidylinositol triphosphate (PIP_3) (Chang *et al.*, 1990; Kammerer *et al.*, 1990; Wingrove *et al.*, 1992), which may be critical for cytoskeletal reorganization and the chemotactic response. There are reports showing PTX-insensitive $G\alpha_{16}$ subunits coupled to C5aR; $G\alpha_{16}$ is known to activate PLC. However, the PTX-sensitive pathways appear to be predominant in mature leukocytes, since neutrophil responses to C5a were largely PTX-sensitive (Buhl *et al.*, 1994, 1995).

Cytoplasmic signaling cascades

A number of biologic responses are initiated in leukocytes when C5a binds to C5aR, a seven membrane-spanning receptor coupled to regulatory heterodimeric guaninine nucleotide-binding proteins (G_i proteins; Gerard and Gerard, 1991, 1994a; Rollins *et al.*, 1991). C5aR and other seven transmembrane receptors coupled to G_i are capable of activating the Ras/Raf/MAP kinase pathway (Buhl *et al.*, 1994; Gerard and Gerard, 1994a). The $\beta\gamma$ subunits of G_i activate the mitogen-activated protein kinase (MAP) pathway in a Ras-dependent manner. Raf-1 binds to Ras-GTP, which activates Raf-1 kinase activity. Activated Raf-1 phosphorylates and activates mitogen-activated protein kinase/Erk kinase (MEK-1), which in turn phosphorylates and activates MAP kinase. The MAP kinase cascade may contribute to the functional assembly of the NADPH oxidase responsible for C5aR-mediated oxygen radical generation in neutrophils.

Despite the persistence of the activating ligand C5a, the cellular responses mediated through C5aR are transient and cells rapidly become refractory to further stimulation, a phenomenon termed homologous desensitization. Receptor phosphorylation appears to be the key mechanism by which many G protein-coupled receptors are regulated. It has been shown that C5aR is phosphorylated exclusively at serine residues localized at the carboxyl end, by a member of the G protein-coupled receptor kinase family (Giannini *et al.*, 1995). Despite the fact that a putative PKC consensus motif is present in the third cytoplasmic loop of C5aR, C5a-dependent phosphorylation is mainly resistant to PKC inhibitors. Therefore, PKC is not a major enzyme involved in agonist-dependent phosphorylation of C5aR.

DOWNSTREAM GENE ACTIVATION

Transcription factors activated

NFκB based on PTX-sensitivity.
Others: undetermined.

Genes induced

C5aR engagement by C5a induces IL-1, IL-6, IL-8, and TNFα from human monocytes (Okusawa *et al.*, 1987, 1988; Scholz *et al.*, 1990; Ember *et al.*, 1994), IL-6 mRNA in human astrocytes (Sayah *et al.*, 1999), β-integrin in human neutrophils and eosinophils (Jagels *et al.*, 1999), and L-selectin shedding (Jagels *et al.*, 2000).

BIOLOGICAL CONSEQUENCES OF ACTIVATING OR INHIBITING RECEPTOR AND PATHOPHYSIOLOGY

Unique biological effects of activating the receptors

Although the proinflammatory effects of anaphylatoxins are indisputably beneficial in the context of localized infections or injuries, there are a number of noninfectious diseases and syndromes in which anaphylatoxins appear to play a deleterious role. Perhaps the most direct link between complement activation and a pathologic response results from extracorporeal circulation of the blood, either during hemodialysis or in coronary bypass surgery (Craddock *et al.*, 1977; Kirklin *et al.*, 1983; Howard *et al.*, 1988). This *postpump syndrome* is characterized by mild *respiratory distress*, *pulmonary hypertension* and occasional *vascular leakage*. The etiology of these physiologic changes appears to be identical to those responses observed in experimental animal studies. The responses to intravenous administration of C3a or C5a (i.e. increased capillary permeability and *edema*, *bronchoconstriction*, *pulmonary vasoconstriction*, leukocyte aggregation in the lung vasculature, and possibly *peripheral vasodilation*), mimic this syndrome in humans. This syndrome appears to be an entirely complement-driven process, subsequent to complement activation through contact with nonbiocompatible materials composing the blood contact surfaces of dialysis and perfusion apparatus. Although the short-term consequences of these effects appear to result in minimal morbidity, there is concern that repeated intravascular complement activation, as occurs in chronic dialysis patients, may lead to long-term pathology of the lung (Craddock *et al.*, 1977). Furthermore, in the setting of cardiopulmonary bypass, the leukocyte aggregation in the lung and impaired pulmonary blood flow may detrimentally affect perfusion of the heart and other organs following surgery, and may contribute to an *ARDS-like syndrome* which develops in a small percentage of bypass patients (Kirklin *et al.*, 1983; Howard *et al.*, 1988). The length of time a patient requires ventilatory support following surgery has been correlated with C3a levels in the blood following reperfusion (Moore *et al.*, 1988).

Acute respiratory distress syndrome (ARDS) and *multiple system organ failure* (MSOF) are two related syndromes which develop most frequently as a consequence of *severe polytrauma* or *septicemia* (Faist *et al.*, 1983; Murray *et al.*, 1988; Parsons *et al.*, 1989). The progressions of ARDS and MSOF are similar and characterized in the early stages by increased vascular permeability, impaired organ perfusion and, in the case of ARDS, respiratory insufficiency. Later stages are characterized by a continuation of the early malfunctions, with progressive damage to endothelium, necrosis, leukocyte infiltration and tissue necrosis and remodeling (Shoemaker *et al.*, 1980; Herndon and Traber, 1990). In ARDS the damage is localized primarily to the lung, whereas in MSOF damage is disseminated not only to the lung but also to the liver, kidneys, and digestive tract.

Phenotypes of receptor knockouts and receptor overexpression mice

Targeted disruption of mouse C5aR expression resulted in no developmental or biological defects in myeloid cell lineages, as well as hepatocytes and epithelial cells. C5aR-deficient mice bred normally and displayed no gross defects when maintained under barrier conditions. On the other hand, deficient mice were unable to clear intrapulmonary-instilled *Pseudomonas aeruginosa*, despite a marked increase of neutrophil influx, and succumbed to *pneumonia*. C5aR-deficient mice challenged with sublethal inocula of *Pseudomonas* become superinfected with secondary bacterial strains. It is concluded that C5aR has a nonredundant function and is required for mucosal host defense in the lung (Hopken *et al.*, 1996).

THERAPEUTIC UTILITY

Recent evidence that C5a plays an important role in immune injury in the lung (Mulligan *et al.*, 1996, 1997; Schmid *et al.*, 1997) in *postischemic vascular and tissue injury* (Ito and Del Balzo, 1994; Amsterdam *et al.*, 1995; Ivey *et al.*, 1995) supports the contention that regulation of selected complement activation products may be of therapeutic value. The discovery that both C3a and C5a receptors may exist on numerous cell types other than circulating white cells, such as hepatocytes, lung epithelial cells (Haviland *et al.*, 1995), endothelial cells (Foreman *et al.*, 1994), or the astrocytes and microglial cells in brain tissue (Gasque *et al.*, 1995b, 1997), has serious implications for anaphylatoxins playing a role in *vascular diseases*, *pulmonary diseases*, and *degenerative neurologic diseases*.

Effect of treatment with soluble receptor domain

Soluble N-terminal C5aR peptides failed to act as antagonist of C5a in receptor-binding studies, possibly because of low affinity for the ligand.

Effects of inhibitors (antibodies) to receptors

Antibodies generated against peptides based on the extracellular loop sequence of C5aR were used to confirm receptor expression and to investigate ligand binding to the C5aR. Antibodies generated to peptides that mimic portions of the N-terminal extracellular region of C5aR are not only excellent markers of cells and tissues expressing the receptor (Buchner *et al.*, 1995; Gasque *et al.*, 1995b, 1997; Haviland *et al.*, 1995), but also block C5a binding and cellular activation by the intact ligand (Morgan *et al.*, 1993; Oppermann *et al.*, 1993).

Antibodies generated against the N-terminal region of C5aR effectively block ligand binding and could be used as therapeutic agents. No clinical uses of C5aR inhibitors have been recorded.

References

Akatsu, H., Miwa, T., Sakurada, C., Fukuoka, Y., Ember, J. A., Yamamoto, T., Hugli, T. E., and Okada, H. (1997). cDNA cloning and characterization of rat C5a anaphylatoxin receptor. *Microbiol. Immunol.* **41**, 575–580.

Amsterdam, E. A., Stahl, G. L., Pan, H. L., Rendig, S. V., Fletcher, M. P., and Longhurst, J. C. (1995). Limitation of reperfusion injury by a monoclonal antibody to C5a during myocardial infarction in pigs. *Am. J. Physiol.* **268**, H448–H457.

Bao, L., Gerard, N. P., Eddy, R.L. Jr., Shows, T. B., and Gerard, C. (1992). Mapping of genes for the human C5aR (C5AR), human FMLP receptor (FPR), and two FMLP receptor homologue orphan receptors (FPRH1, FPRH2) to chromosome 19. *Genomics* **13**, 437–440.

Becker, E. L., Showell, H. J., Henson, P. M., and Hsu, L. S. (1974). The ability of chemotactic factors to induce lysosomal enzyme release. 1. The characteristics of the release, the importance of surfaces, and the relation of enzyme release to chemotactic responsiveness. *J. Immunol.* **112**, 2047–2054.

Boulay, F., Mery, L., Tardif, M., Brouchon, L., and Vignais, P. (1991). Expression cloning of a receptor for C5a anaphylatoxin on differentiated HL-60 cells. *Biochemistry* **30**, 2993–2999.

Buchner, R. R., Hugli, T. E., Ember, J. A., and Morgan, E. L. (1995). Expression of functional receptors for human C5a anaphylatoxin (CD88) on the human hepatocellular carcinoma cell line HepG2. Stimulation of acute-phase protein-specific mRNA and protein synthesis by human C5a anaphylatoxin. *J. Immunol.* **155**, 308–315.

Buhl, A. M., Avdi, N., Worthen, G. S., and Johnson, G. L. (1994). Mapping of the C5a receptor signal transduction network in human neutrophils. *Proc. Natl Acad. Sci. USA* **91**, 9190–9194.

Buhl, A. M., Osawa, S., and Johnson, G. L. (1995). Mitogen-activated protein kinase activation requires two signal inputs from the human anaphylatoxin C5a receptor. *J. Biol. Chem.* **270**, 19828–19832.

Chang, Y. H., Jagels, M. A., and Abraham, E. (1990). Haemorrhage produces depressions in alloantigen-specific immune responses in the mouse through activation of suppressor T cells. *Clin. Exp. Immunol.* **80**, 478–483.

Chenoweth, D. E., and Goodman, M. G. (1983). The C5a receptor of neutrophils and macrophages. *Agents Actions* **12**, (Suppl.)252–273.

Chenoweth, D. E., and Hugli, T. E. (1978). Demonstration of specific C5a receptor on intact human polymorphonuclear leukocytes. *Proc. Natl Acad. Sci. USA* **75**, 3943–3947.

Chenoweth, D. E., and Hugli, T. E. (1980). Human C5a and C5a analogs as probes of the neutrophil C5a receptor. *Mol. Immunol.* **17**, 151–161.

Chenoweth, D. E., Goodman, M. G., and Weigle, W. O. (1982). Demonstration of a specific receptor for human C5a anaphylatoxins on murine macrophages. *J. Exp. Med.* **156**, 67–78.

Chenoweth, D. E., Soderberg, C. S., and von Wedel, R. (1984). Dibutyryl cAMP induced expression of C5a receptors on U937 cells. *J. Leukoc. Biol.* **1984**, 36, 241(abstract).

Cochrane, C. G., and Müller-Eberhard, H. J. (1968). The derivation of two distinct anaphylatoxin activities from the third and fifth components of human complement. *J. Exp. Med.* **127**, 371–386.

Craddock, P. R., Fehr, J., Brigham, K. L., Kronenberg, R. S., and Jacob, H. S. (1977). Complement and leukocyte-mediated pulmonary dysfunction in hemodialysis. *N. Engl. J. Med.* **296**, 769–774.

Daffern, P. J., Pfeifer, P. H., Ember, J. A., and Hugli, T. E. (1995). C3a is a chemotaxin for human eosinophils but not for neutrophils. I. C3a stimulation of neutrophils is secondary to eosinophil activation. *J. Exp. Med.* **181**, 2119–2127.

DeMartino, J. A., Van Riper, G., Siciliano, S. J., Molineaux, C. J., Konteastis, Z. D., Rosen, H., and Springer, M. S. (1994). The amino terminus of the human C5a receptor is required for high affinity C5a binding and for receptor activation by C5a but not C5a analogs. *J. Biol. Chem.* **269**, 14446–14450.

Dias da Silva, W., and Lepow, I. H. (1967). Complement as a mediator of inflammation. II. Biological properties of anaphylatoxin prepared with purified components of human complement. *J. Exp. Med.* **125**, 921–946.

Elsner, J., Opermann, M., Czech, W., Dobos, G., Schopf, E., Norgauer, J., and Kapp, A. (1994). C3a activates reactive oxygen radical species production and intracellular calcium transients in human eosinophils. *Eur. J. Immunol.* **24**, 518–522.

Ember, J. A., Sanderson, S. D., Hugli, T. E., and Morgan, E. L. (1994). Induction of interleukin-8 synthesis from monocytes by human C5a anaphylatoxin. *Am. J. Pathol.* **144**, 393–403.

Ember, J. A., Jagels, M. A., and Hugli, T. E. (1998). In "The Human Complement System in Health and Disease" (ed J. Volanakis and M. Frank), Characterization of complement anaphylatoxins and their biological responses pp. 241–284. Marcel Dekker, New York.

Faist, E., Baue, A. E., Dittmer, H., and Heberer, G. (1983). Multiple organ failure in polytrauma patients. *J. Trauma* **23**, 775–787.

Fernandez, H. N., Henson, P. M., Otani, A., and Hugli, T. E. (1978). Chemotactic response to human C3a and C5a anaphylatoxins. I. Evaluation of C3a and C5a leukotaxis *in vitro* and under stimulated *in vivo* conditions. *J. Immunol.* **120**, 109–115.

Foreman, K. E., Vaporciyan, A. A., Bonish, B. K., Jones, M. L., Glovsky, M. M., Eddy, S. M., and Ward, P. A. (1994). C5a-induced expression of P-selectin in endothelial cells. *J. Clin. Invest.* **94**, 1147–1155.

Friedberger, E. (1910). Weitere Untersuchungen uber Eisissanaphylaxie: IV. Mitteilung. Immunitaetaforsch. *Exp Ther.* **4**, 636–690.

Fukuoka, Y., and Hugli, T. E. (1988). C5a receptors on guinea pig platelets. *Fed. Proc.* **2**, A1007 (abstract).

Fukuoka, Y., and Hugli, T. E. (1990). Anaphylatoxin binding and degradation by rat peritoneal mast cells. Mechanisms of degranulation and control. *J. Immunol.* **145**, 1851–1858.

Fukuoka, Y., Ember, J. A., Yasui, A., and Hugli, T. E. (1998). Cloning and characterization of the guinea pig C5a anaphylatoxin receptor: Interspecies diversity among the C5a receptors. *Int. Immunol.* **10**, 275–283.

Gasque, P., Fontaine, M., and Morgan, B. P. (1995a). Complement expression in human brain. Biosynthesis of terminal pathway components and regulators in human glial cells and cell lines. *J. Immunol.* **154**, 4726–4733.

Gasque, P., Chan, P., Fontaine, M., Ischenko, A., Lamacz, M., Götze, O., and Morgan, B. P. (1995b). Identification and characterization of the complement C5a anaphylatoxin receptor on human astrocytes. Relevance to inflammation in the brain. *J. Immunol.* **155**, 4882–4889.

Gasque, P., Singhrao, S. K., Neal, J. W., Götze, O., and Morgan, B. P. (1997). Expression of the receptor for C5a (CD88) is upregulated on reactive astrocytes, microglia and endothelial cells in inflammed human CNS. *Am. J. Pathol.* **150**, 31–41.

Gerard, N. P., and Gerard, C. (1991). The chemotactic receptor for human C5a anaphylatoxin. *Nature* **349**, 614–617.

Gerard, C., and Gerard, N. P. (1994a). The pro-inflammatory seven-transmembrane segment receptors of the leukocyte. *Curr. Opin. Immunol.* **6**, 140–145.

Gerard, C., and Gerard, N. P. (1994b). C5a anahylatoxin and its seven transmembrane-segment receptor. *Annu. Rev. Immunol.* **12**, 775.

Gerard, C., Chenoweth, D. E., and Hugli, T. E. (1979). Molecular aspects of the serum chemotactic factors. *J. Reticuloendothel. Soc.* **26**, 711–718.

Gerard, N. P., Hodges, M. K., Drazen, J. M., Weller, P. F., and Gerard, C. (1989). Characterization of a receptor for C5a anaphylatoxin on human eosinophils. *J. Biol. Chem.* **264**, 1760–1766.

Gerard, C., Bao, L., Orozco, O., Pearson, M., Kunz, D., and Gerard, N. P. (1992). Structural diversity in the extracellular faces of peptidergic G-protein-coupled receptors. *J. Immunol.* **149**, 2600–2606.

Giannini, E., Brouchon, L., and Boulay, F. (1995). Identification of the major phosphorylation sites in human C5a anaphylatoxin receptor *in vivo*. *J. Biol. Chem.* **270**, 19166–19172.

Goetzl, E. J., and Austen, K. F. (1974). Stimulation of human neutrophil leukocyte aerobic glucose metabolism by purified chemotactic factors. *J. Clin. Invest.* **53**, 591–599.

Hartman, C.T. Jr., and Glovsky, M. M. (1981). Complement activation requirements for histamine release from human leukocytes: influence of purified C3a hu and C5a hu on histamine release. *Int. Arch. Allergy Appl. Immunol.* **66**, 274–281.

Haviland, D. L., McCoy, R. L., Whitehead, W. T., Akama, H., Molmenti, E. P., Brown, A., Haviland, J. C., Parks, W. C., Perlmutter, D. H., and Wetsel, R. A. (1995). Cellular expression of the C5a anaphylatoxin receptor (C5aR): demonstration of C5aR on nonmyeloid cells of the liver and lung. *J. Immunol.* **154**, 1861–1869.

Heideman, M., and Hugli, T. E. (1984). Anaphylatoxin generation in multisystem organ failure. *J. Trauma* **24**, 1038–1043.

Henson, P. M. (1971). The immunologic release of constituents from neutrophil leukocytes. I. The role of antibody and complement on nonphagocytosable surfaces or phagocytosable particles. *J. Immunol.* **107**, 1535–1546.

Henson, P. M., Zanolari, B., Schwartzman, N. A., and Hong, S. R. (1978). Intracellular control of human neutrophil secretion. I. C5a-induced stimulus-specific desensitization and the effects of cytochalasin B. *J. Immunol.* **121**, 851–855.

Herndon, D. N., and Traber, D. L. (1990). In "Pathophysiology and Basic Concepts of Therapy" (ed E. A. Deitch), Pulmonary failure and acute respiratory distress syndrome. Multiple organ failure pp. 192–214. Thieme, New York.

Hopken, U. E., Lu, B., Gerard, N. P., and Gerard, C. (1996). The C5a chemoattractant receptor mediates mucosal defence to infection. *Nature* **383**, 86–89.

Howard, R. J., Crain, C., Franzini, D. A., Hood, C. I., and Hugli, T. E. (1988). Effects of cardiopulmonary bypass on pulmonary leukostasis and complement activation. *Arch. Surg.* **123**, 1496–1501.

Huey, R., and Hugli, T. E. (1985). Characterization of a C5a receptor on human polymorphonuclear leukocytes (PMN). *J. Immunol.* **135**, 2063–2068.

Hugli, T. E. (1981). The structural basis for anaphylatoxin and chemotactic functions of C3a, C4a, and C5a. *Crit. Rev. Immunol.* **1**, 321–366.

Hugli, T. E. (1984). Structure and function of the anaphylatoxins. *Springer Seminar Immunopathol.* **7**, 193–219.

Hugli, T. E., Marceau, F., and Lundberg, C. (1987). Effects of complement fragments on pulmonary and vascular smooth muscle. *Am. Rev. Respir. Dis.* **135**, S9–S13.

Ito, B. R., and Del Balzo, U. (1994). Effect of platelet depletion and inhibition of platelet cyclooxygenase on C5a-mediated myocardial ischemia. *Am. J. Physiol.* **267**, H1288–H1294.

Ivey, C. L., Williams, F. M., Collins, P. D., Jose, P. J., and Williams, T. J. (1995). Neutrophil chemoattractaants generated in two phases during reperfusion of ischemic myocardium in the rabbit. Evidence for a role for C5a and interleukin-8 (comment). *J. Clin. Invest.* **95**, 2720–2728.

Jagels, M. A., Daffern, P. J., Zuraw, B. L., and Hugli, T. E. (1999). Mechanisms and regulation of PMN and eosinophil adherence to human airway epithelial cells. *Am. J. Respir. Cell Mol. Biol.* **21**, 418–427.

Jagels, M. A., Daffern, P. J., and Hugli, T. E. (2000). C3a and C5a enhance granulocyte adhesion to endothelial and epithelial cell monolayers: epithelial and endothelial priming is required for C3a-induced eosinophil adhesion. *Immunopharmacology* **46**, 209–222.

Jiang, H., Kuang, Y., Wu, Y., Smrcka, A., Simon, M. I., and Wu, D. (1996). Pertussis toxin-sensitive activation of phospholipase C by the C5a and fMet-Leu-Phe receptors. *J. Biol. Chem.* **271**, 13430–13434.

Johnson, A. R., Hugli, T. E., and Müller-Eberhard, H. J. (1975). Release of histamine from rat mast cells by the complement peptides C3a and C5a. *Immunology* **28**, 1067.

Kammerer, R. C., Merdink, J. L., Jagels, M. A., Catlin, D. H., and Hui, K. K. (1990). Testing for fluoxymesterone (Halotestin) administration to man: identification of urinary metabolites by gas chromatography-mass spectrometry. *J. Steroid Biochem.* **36**, 659–666.

Kirklin, J. K., Westaby, S., Blackstone, E. H., Kirklin, J. W., Chenoweth, D. E., and Pacifico, A. D. (1983). Complement and the damaging effects of cardiopulmonary by pass. *J. Cardiovasc. Surg.* **86**, 845–857.

Kishimoto, T. K., Jutila, M. A., Berg, E. L., and Butcher, E. C. (1989). Neutrophil Mac-1 and MEL-14 adhesion proteins inversely regulated by chemotactic factors. *Science* **245**, 1238–1241.

Klos, A., Matje, C., Rheinheimer, C., Bautsch, W., Köhl, J., Martin, U., and Burg, M. (1994). Amino acids 327-350 of the human C5a-receptor are not essential for [^{125}I]C5a binding in COS cells and signal transduction in *Xenopus* oocytes. *FEBS Lett.* **344**, 79–82.

Kretzschmar, T., Kahl, K., Rech, K., Bautsch, W., Köhl, J., and Bitter-Suermann, D. (1991). Characterization of the C5a receptor on guinea pig platelets. *Immunobiology* **183**, 418–432.

Lacy, M., Jones, J., Whittemore, S. R., Haviland, D. L., Wetsel, R. A., and Barnum, S. R. (1995). Expression of the receptors for the C5a anaphylatoxin, interleukin-8 and FMLP by human astrocytes and microglia. *J. Neuroimmunol.* **61**, 71–78.

Laurent, F., Benoliel, A. M., Capo, C., and Bongrand, P. (1991). Oxidative metabolism of polymorphonuclear leukocytes: modulation by adhesive stimuli. *J. Leukoc. Biol.* **49**, 217–226.

Lepow, I. H., Willms-Kretschmer, R. A., Patrick, R. A., and Rosen, F. S. (1970). Gross and ultrastructural observations of lesions produced by intradermal injection of human C3a in man. *Am. J. Pathol.* **61**, 13–20.

Lundahl, J., Hallden, G., and Hed, J. (1993). Differences in intracellular pool and receptor-dependent mobilization of the adhesion-promoting glycoprotein Mac-1 between eosinophils and neutrophils. *J. Leukoc. Biol.* **53**, 336–341.

MacGlashan, D.W. Jr., and Hubbard, W. C. (1993). IL-3 alters free arachidonic acid generation in C5a-stimulated human basophils. *J. Immunol.* **151**, 6358–6369.

McCoy, R., Haviland, D. L., Molmenti, E. P., Ziambaras, T., Wetsel, R. A., and Perlmutter, D. H. (1995). N-formylpeptide and complement C5a receptors are expressed in liver cells and mediate hepatic acute phase gene regulation. *J. Exp. Med.* **182**, 207–217.

Mery, L., and Boulay, F. (1994). The NH_2-terminal region of C5aR but not that of FPR is critical for both protein transport and ligand binding. *J. Biol. Chem.* **269**, 3457–3463.

Moore, F.D. Jr., Warner, K. G., Assousa, S., Valeri, C. R., and Khuri, S. F. (1988). The effects of complement activation during cardiopulmonary bypass. *Ann. Surg.* **208**, 95–103.

Morgan, E. L., Ember, J. A., Sanderson, S. D., Scholz, W., Buchner, R., Ye, R. D., and Hugli, T. E. (1993). Anti-C5a receptor antibodies. Characterization of neutralizing antibodies specific for a peptide, C5aR-(9-29), derived from the predicted amino-terminal sequence of the human C5a receptor. *J. Immunol.* **151**, 377–388.

Morita, E., Schröder, J.-M., and Christophers, E. (1989). Differential sensitivities of purified human eosinophils and neutrophils to defined chemotaxins. *Scand. J. Immunol.* **29**, 709–716.

Mulligan, M. S., Schmid, E., Till, G. O., Friedl, H. P., Hugli, T. E., Johnson, K. J., and Ward, P. A. (1996). Requirement and role of C5a in acute lung inflammatory injury in rats. *J. Clin. Invest.* **98**, 503–512.

Mulligan, M. S., Schmid, E., Till, G. O., Hugli, T. E., Friedl, H. P., Roth, R. A., and Ward, P. A. (1997). C5a-dependent upregulation *in vivo* of lung vascular P-selectin. *J. Immunol.* **158**, 1857–1861.

Murray, J. F., Matthay, M. A., Luce, J., and Flick, M. R. (1988). An expanded definition of the adult respiratory distress syndrome. *Am. Rev. Respir. Dis.* **138**, 720–723.

Nataf, S., Stahel, P. F., Davoust, N., and Barnum, S. R. (1999). Complement anaphylatoxin receptors on neurons: new tricks for old receptors? *Trends Neurosci.* **22**, 397–402.

Neeley, S. P., Hamann, K. J., White, S. R., Baranowski, S. L., Burch, R. A., and Leff, A. R. (1993). Selective regulation of expression of surface adhesion molecules Mac-1, L-selectin, and VLA-4 on human eosinophils and neutrophils. *Am. J. Respir. Cell Mol. Biol.* **8**, 633–639.

Okusawa, S., Dinarello, C. A., Yancy, K. B., Endres, S., Lawley, T. J., Frank, M. M., Burke, J. F., and Gelfand, J. A. (1987). C5a induction of human interleukin 1. Synergistic-effect with endotoxin or interferon-γ. *J. Immunol.* **139**, 2635–2639.

Okusawa, S., Yancey, K. B., Van Der Meer, J. W. M., Endres, S., Lonneman, G., Hefter, K., Frank, M. M., Burke, J. F., Dinarello, C. A., and Gelfand, J. A. (1988). C5a stimulates secretion of tumor necrosis factor from human mononuclear cells *in vitro*: comparison with secretion of interleukin 1β and interleukin 1α. *J. Exp. Med.* **168**, 443–448.

Oppermann, M., Raedt, U., Hebell, T., Schmidt, B., Zimmermann, B., and Götze, O. (1993). Probing the human receptor for C5a anaphylatoxin with site-directed antibodies: identification of a potential ligand binding site on the NH-2-terminal domain. *J. Immunol.* **151**, 3785–3794.

Paradisis, P. M., Campbell, I. L., and Barnum, S. R. (1998). Elevated complement C5a receptor expression on neurons and glia in astrocyte-targeted interleukin-3 transgenic mice. *GLIA* **24**, 338–345.

Parsons, P. E., Worthen, G. S., Moore, E. E., Tate, R. M., and Henson, P. M. (1989). The association of circulating endotoxin with the development of the adult respiratory distress syndrome. *Am. Rev. Respir. Dis.* **140**, 294–301.

Pease, J. E., and Barker, M. D. (1993). *N*-linked glycosylation of the C5a receptor. *Biochemistry* **31**, 719–726.

Perret, J. J., Raspe, E., Vassart, G., and Parmentier, M. (1992). Cloning and functional expression of the canine anaphylatoxin C5a receptor. *Biochemistry* **288**, 911–917.

Rollins, T. E., and Springer, M. S. (1985). Identification of the polymorphonuclear leukocyte C5a receptor. *J. Biol. Chem.* **260**, 7157–7160.

Rollins, T. E., Siciliano, S., Kobayashi, S., Cianciarulo, D. N., Bonilla Argudo, V., Collier, K., and Springer, M. S. (1991). Purification of the active C5a receptor from human polymorphonuclear leukocytes as a receptor-Gi complex. *Proc. Natl Acad. Sci. USA* **88**, 971–975.

Rother, K., Worner, I., Prior, B., and Hansch, G. M. (1992). Leukocyte mobilization from bone marrow by complement receptor 3 (CR3) interaction. *FASEB J.* **6**, A1893 (abstract).

Sayah, S., Ischenko, A. M., Zhakhov, A., Bonnard, A. S., and Fontaine, M. (1999). Expression of cytokines by human astrocytomas following stimulation by C3a and C5a anaphylatoxins: specific increase in interleukin-6 mRNA expression. *J. Neurochem.* **72**, 2426–2436.

Schmid, E., Warner, R. L., Crouch, L. D., Friedl, H. P., Till, G. O., Hugli, T. E., and Ward, P. A. (1997). Neutrophil chemotactic activity and C5a following systemic activation of complement in rats. *Inflammation* **21**, 325–333.

Scholz, W., McClurg, M. R., Cardenas, G. J., Smith, M., Noonan, D. J., Hugli, T. E., and Morgan, E. L. (1990). C5a-mediated release of interleukin 6 by human monocytes. *Clin. Immunol. Immunopathol.* **57**, 297–307.

Schulman, E. S., Post, T. J., Henson, P. M., and Giclas, P. C. (1988). Differential effects of the complement peptides, C5a and C5a desArg on human basophils and lung mast cell histamine release. *J. Clin. Invest.* **81**, 918–923.

Shapira, L., Champagne, C., Gordon, B., Amar, S., and Van Dyke, T. E. (1995). Lipopolysaccharide priming of superoxide release by human neutrophils. *Inflammation* **19**, 289–295.

Shoemaker, W. C., Appel, P. L., Czer, S. C., Bland, R., Schwartz, S., and Hopkins, J. A. (1980). Pathogenesis of respiratory failure (ARDS) after hemorrhage and trauma. *Crit. Care Med.* **8**, 504–512.

Siciliano, S. J., Rollins, T. E., DeMartino, J., Konteatis, Z., Malkowitz, L., Van Riper, G., Bondy, S., Rosen, H., and Springer, M. S. (1994). Two-site binding of C5a by its receptor: an alternative binding paradigm for G protein-coupled receptors. *Proc. Natl Acad. Sci. USA* **91**, 1214–1218.

Siraganian, R. P., and Hook, W. A. (1976). Complement induced histamine release from human basophils. II. Mechanism of the histamine reaction. *J. Immunol.* **116**, 639–646.

Takafuji, S., Tadokoro, K., Ito, K., and Dahinden, C. A. (1994). Degranulation from human eosinophils stimulated with C3a and C5a. *Int. Arch. Allergy Immunol.* **104**, 27–29.

Vallota, E. H., and Müller-Eberhard, H. J. (1973). Formation of C3a and C5a anaphylatoxins in whole human serum after inhibition of the anaphylatoxin inactivator. *J. Exp. Med.* **137**, 1109–1123.

Wilmer, W. A., Kaumaya, P. T., Ember, J. A., and Cosio, F. G. (1998). Receptors for the anaphylatoxin C5a (CD88) on human mesangial cells. *J. Immunol.* **160**, 5646–5652.

Wingrove, J. A., Discipio, R. G., Chen, Z., Potempa, J., Travis, J., and Hugli, T. E. (1992). Activation of complement components C3 and C5 by a cysteine proteinase (gingipain-1) from *Porphyromonas (Bacteroides) gingivalis*. *J. Biol. Chem.* **267**, 18902–18907.

Wuepper, K. D., Bokisch, V. A., Müller-Eberhard, H. J., and Stoughton, R. B. (1972). Cutaneous responses to human C3 anaphylatoxin in man. *Clin. Exp. Immunol.* **11**, 13–20.

LICENSED PRODUCTS

Serotec sells a rabbit anti-human CD88 polyclonal (Catalog no. AHP353) and mouse anti-human CD88 monoclonals (Catalog no. MCA1283 and 1284) for research. Both interact with linear sequences of the N-terminal region of CD88 (C5aR).

A patent exists for a neutralizing polyclonal antibody to CD88 (C5aR). US Patent No. 5,480,974 issued 2 January 1996 (USSN 08/079,051 filed 21 April 1995); Edward L. Morgan, Julia A. Ember and Tony E. Hugli, co-inventors, Antibodies to Human C5a Receptor.

C3a Receptor

Julia A. Ember[1] and Tony E. Hugli[2,*]

[1]Pharmingen, 10975 Torreyana Road, San Diego, CA 92121, USA

[2]The Scripps Research Institute, 10550 N. Torrey Pines Road, La Jolla, CA 92037, USA

*corresponding author tel: 858-784-8158, fax: 858-784-8307, e-mail: hugli@scripps.edu

DOI: 10.1006/rwcy.2000.23002.

SUMMARY

C3a receptor (C3aR) is a G protein-coupled receptor of the rhodopsin superfamily. The receptor contains the characteristic seven transmembrane domains connected by intra- and extracellular loops, with the N-terminus having an extracellular orientation and the C-terminus being intracellular and the region to which the G proteins bind. C3aR has a major distinguishing feature, which is an extraordinarily large extracellular loop between the fourth and fifth transmembrane helices. This loop in human C3aR contains 175 residues: in most G protein-coupled receptors the corresponding extracellular loop region is 30–40 residues in length.

The C3aR is widely distributed in both myeloid cells and nonmyeloid tissue cells. The C3aR on eosinophils and basophils is thought to be particularly important biologically since these cells are chemotactically activated by the ligand C3a. Neutrophils also express C3aR, but the ligand can only induce calcium mobilization in these cells and does not cause granular release, oxygen radical generation, or chemotaxis. C3aR expression has been shown to be upregulated on astrocytes and other brain cells during inflammation. Only one form of C3aR has been detected in human, rat, and mouse, but two isoforms of C3aR have been described in the guinea pig (C3aR-S and C3aR-L for small and large forms).

BACKGROUND

Discovery

Cloning of both the C5a receptor (C5aR) and C3aR (Ames *et al.*, 1996; Crass *et al.*, 1996; Roglic *et al.*, 1996) has occurred in the last 10 years. This has permitted comparisons between the receptors to these related ligands and a significant structural difference was observed. For example, we learned that the C3aR has novel and unique structural characteristics compared with most other G protein-coupled receptors, including the C5aR, which is an unusually large extracellular loop.

The C3aR has now been cloned from several animal species, including mouse (Hsu *et al.*, 1997; Tornetta *et al.*, 1997), guinea pig (Fukuoka *et al.*, 1998a), and rat (Fukuoka *et al.*, 1998b) and the sequences of these receptor molecules are compared in **Figure 1**. The patterns of identity between C3aR obtained from different species show relatively high levels in the N-terminal extracellular region and for the transmembrane segments. The second intracellular loop is also highly conserved, perhaps because the two cysteinyl residues participate in critical disulfide bonds. The large second extracellular loop has only modest homology, while the C-terminal intracellular region, which contains the G protein-binding site, is highly conserved. The major unique feature in C3aR compared to other rhodopsin family receptors is the unusually large second extracellular loop; this region may be particularly important for binding the C3a molecule (**Figure 2**). The only species of C3aR for which isoforms have been detected is the guinea pig (Fukuoka *et al.*, 1998a). The full-length form of gpC3aR is designated C3aR-L and the isoform with a 34 residue deletion in the second extracellular loop is designated C3aR-S.

Alternative names

C3aR or the C3 anaphylatoxin receptor has no other alternative names; however it was first reported as an

Figure 1 The complete protein sequences for the C3a receptor from human (Hu), guinea pig (Gp), rat (Rt), and mouse (Mo) are presented. The alignments were optimized for maximal identity and the seven transmembrane regions have been identified by a line and roman numerals. The residue positions that have been conserved in all species are denoted by asterisks. These C3aRs are similar to each other in size, but are considerably larger than the C5aR. The distinguishing feature between the C3aR and C5aR is the large second extracellular loop region, comprised of approximately 170 residues. The expressed C3aR on guinea pig cells has been estimated to be much larger than the 54 kDa nude protein reported here (Fukuoka and Hugli, 1988). This difference may be accounted for by glycosylation, since multiple oligosaccharide groups could be attached at several potential sites in the C3aR.

ALIGNMENT OF C3a RECEPTOR SEQUENCES FROM SEVERAL SPECIES

```
                                                I
Hu C3aR   1:MASFSAETNSTDLLSQPWNEPPVILSMVILSLTFLLGLPGNGLVLWVAGLKMQRTVNTIW 60
Gp C3aR   1:MDSSSAETNSTGLHLEPQYQPETILAMAILGLTFVLGLPGNGLVLWVAGLKMRRTVNTVW 60
Rt C3aR   1:MESFTADTNSTDLHSRPLFKPQDIASMVILSLTCLLGLPGNGLVLWVAGVKMKRTVNTVW 60
Mo C3aR   1:MESFDADTNSTDLHSRPLFQPQDIASMVILGLTCLLGLLANGLVLWVAGVKMKTTVNTVW 60
            * *  * **** *   *   *  *  * ** **  ***  ********* **  **** *

                      II                                   III
Hu C3aR  61:FLHLTLADLLCCLSLPFSLAHLALQGQWPYGRFLCKLIPSIIVLNMFASVFLLTAISLDR 120
Gp C3aR  61:FLHLTVADFVCCLSLPFSMAHLALRGYWPYGEILCKFIPTVIIFNMFASVFLLTAISLDR 120
Rt C3aR  61:FLHLTLADFLCCLSLPFSVAHLILRGHWPYGLFLCKLIPSVIILNMFASVFLLTAISLDR 120
Mo C3aR  61:FLHLTLADFLCCLSLPFSLAHLILQGHWPYGLFLCKLIPSIIILNMFASVFLLTAISLDR 120
            ***** **  ******** *** * * ****  *** **  *  ****************

                                        IV
Hu C3aR 121:CLVVFKPIWCQNHRNVGMACSICGCIWVVAFVMCIPVFVYREIFTTDNHNRCGYKFGLSS 180
Gp C3aR 121:CLMVLKPIWCQNHRNVRTACIICGCIWLVAFVLCIPVFVYRETFTLENHTICTYNFS-PG 179
Rt C3aR 121:CLMVHKPIWCQNHRSVRTAFAVCGCVWVVTFVMCIPVFVYRDLLVVDDYSVCGYNFDSSR 180
Mo C3aR 121:CLIVHKPIWCQNHRNVRTAFAICGCVWVVAFVMCVPVFVYRDLFIMDNRSICRYNFDSSR 180
            ** * ********* *  *   *** * * ** * ******          * * *

Hu C3aR 181:SLDYPDF-YG-D-PLENRSLENIVQRPGEMNDRLDP-SSFQTNDHPWTVPTVFQPQTFQRPS 238
Gp C3aR 180:SFDYLDYAYDRD-AWGYGTPDPIVQLPGEMEHRSDP-SSFQTQDGPWSVTTTLYSQTSQRPS 239
Rt C3aR 181:AYDYWDYMYNSHLPEINPPDNS----TGHVDDRTAPSSSVPARD-LWTATTALQSQTFHTSP 237
Mo C3aR 181:SYDYWDYVYKLSLPESNSTDNSTAQLTGHMNDRSAP-SSVQARDYFWTVTTALQSQPFLTSP 241
              ** *  *                  *    *  * **    *  *   *    *

Hu C3aR 239:ADSLPRGSARLTSQNLYSNVFKPADVVSPKIPSGFPIEDHETSPLDNSDAFLSTHLKLFPSA 300
Gp C3aR 240:EDSFHMDSAKLSGQGKYVDV-----VLPTNL-CGLPMEENRTNTLHNA-AFLSSDLDV-SNA 294
Rt C3aR 238:EDPFSQDSA--SQQPHYGG--KPPTVLIATIPGGFPVEDHKSNTL-NTGAFLSAH-TEPSLT 293
Mo C3aR 242:EDSFSLDSA--NQQPHYGG--KPPNVLTAAVPSGFPVEDRKSNTL-NADAFLSAH-TELFPT 297
             *     **    *  *        *       * * *      * *  ****

                                                     V
Hu C3aR 301:SSNSFYESELPQGFQDYYNLGQFTDDDQVPTPLVAITITRLVVGFLLPSVIMIACYSFIVFR 362
Gp C3aR 295:TQKCLSTPEPPQDFWD--DLSPFTHEYRTPRLLKVITFTRLVVGFLLPMIIMVACYTLIIFR 353
Rt C3aR 294:-ASSSPLY-AHDFPDDYFDQLMYGNHAWTP--QVAITISRLVVGFLVPFFIMITCYSLIVFR 351
Mo C3aR 298:-ASSGHLY-PYDFQGDYVDQFTYDNHVPTP--LMAITITRLVVGFLVPFFIMVICYSLIVFR 355
                           *             *     **  ******* *  **  **  * **

                                    VI
Hu C3aR 363:MQRGRFAKSQSKTFRVAVVVVAVFLVCWTPYHIFGVLSLLTDPETPLGKTLMSWDHVCIAL 423
Gp C3aR 354:MRRARVVKSWNKALHLAMVVVTIFLICWAPYHVFGVLILFINPESRVGAALLSWDHVSIAL 414
Rt C3aR 352:MRKTNLTKSRNKTLRVAVAVVTVFFVCWIPYHIVGILLVITDQESALREVVLPWDHMSIAL 412
Mo C3aR 356:MRKTNFTKSRNKTFRVAVAVVTVFFICWTPYHLVGVLLLITDPESSLGEAVMSWDHMSIAL 416
            *      **  *    *  **  *  ** *** * *      *         ***  ***

                  VII
Hu C3aR 424:ASANSCFNPFLYALLGKDFRKKARQSIQGILEAAFSEELTRSTHCPSNNVISERNS-TT-V 482
Gp C3aR 415:ASANSCFNPFLYALLGRDLRKRVRQSMKGILEAAFSEDISKSTSFIQAKAFSEKHSLSTNV 475
Rt C3aR 413:ASANSCFNPFLYALLGKDFRKKARQSVKGILEAAFSEELTHSTSCTQDKAPSKRNHMSTDV 473
Mo C3aR 417:ASANSCFNPFLYALLGKDFRKKARQSIKGILETAFSEELTHSTNCTQDKASSKRNNMSTDV 477
            **************** * **  ***  **** ****    **        *      * *
```

Figure 2 A model is proposed illustrating the interactions between C3a and its respective receptor. The design for this model was adapted from the C5a/C5aR model proposed by Siciliano *et al.* (1994). C3aR is a G protein-coupled transmembrane receptor of the rhodopsin superfamily. The C3a molecule has at least two major binding sites on C3aR. A noneffector binding site (site 1) exists on the C-terminal helical region of C3a which either makes contact with the large extracellular loop (as shown here) or with other exposed regions of the receptor, including the extracellular N-terminal region. Site 2 contains the C-terminal effector region of C3a, including the sequence LGLAR, which is shown penetrating into the 'pore' formed by the seven transmembrane domains of C3aR. This model for C3a/C3aR interaction corresponds to a model originally proposed for multisite binding of C5a with its receptor (Chenoweth and Hugli, 1980).

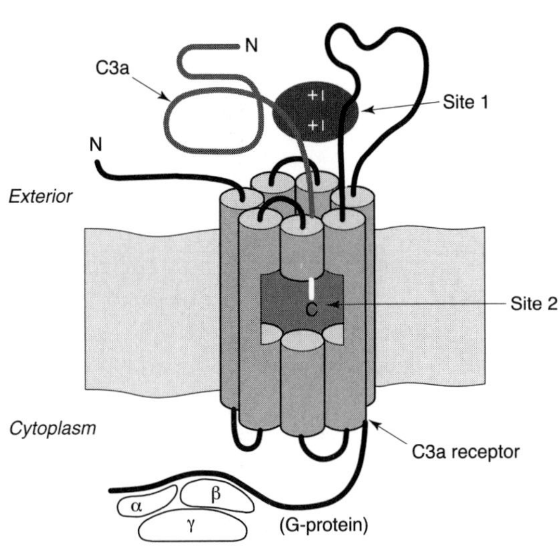

orphan cDNA clone (AZ3B) by Roglic *et al.* (1996). No CD code designation has yet been assigned to this receptor molecule.

Structure

C3aR is a member of the rhodopsin superfamily and contains the seven transmembrane helical motif, with an N-terminus exposed to the extracellular surface of the cell and the C-terminus having an intracellular orientation. One of the extracellular interhelical loops (loop 2 between helix 4 and 5) is unusually large, being 175 residues in length in human C3aR. C3aR molecules from all species known to date have this extraordinarily large second extracellular loop. Mutation studies have shown that a number of charged residues at either end of this loop (D_{159}, D_{325}–D_{327}), and in the adjacent membrane region (R_{161}, E_{162}, and R_{340}), participate in ligand binding and are important for C3a-mediated cell activation via the C3aR molecule (Sun *et al.*, 1999). However, major portions of this large loop have been deleted without functional consequences (Chao *et al.*, 1999; Fukuoka *et al.*, 1999). A three-dimensional model of the human C3aR molecule, based on the known structure of the rhodopsin receptor, has been proposed (Sun *et al.*, 1999).

Main activities and pathophysiological roles

The C3a receptor is expressed on leukocytes, monocytes, platelets, and mast cells (Ember *et al.*, 1998). Brain cells such as astrocytes (Ischenko *et al.*, 1998) express C3aR which is upregulated during inflammation (Nataf *et al.*, 1999). C3a is a potent chemotactic factor for eosinophils (and basophils) but not neutrophils (Daffern *et al.*, 1995), suggesting that C3aR plays a specific role in *asthma* and/or *allergic disorders*. Although there are C3aR on neutrophils, C3a stimulation results in no granular release or oxygen burst, but does induce calcium mobilization (Norgauer *et al.*, 1993; Takafuji *et al.*, 1994).

Eosinophil Activation and Migration

Until recently, the effects of C3a on granulocytic cell types have remained controversial. Earlier studies have reported that C3a could induce degranulation, aggregation, and chemotaxis of human neutrophils (Damerau *et al.*, 1980; Showell *et al.*, 1982a,b; Nagata *et al.*, 1987), suggesting that C3a could exert a direct effect on these cells via the C3aR. However, when neutrophils were purified free of eosinophils (i.e. >98% pure) the neutrophil cells no longer polarize, undergo chemotaxis, or release granular enzymes in the presence of C3a (Daffern *et al.*, 1995). On the other hand, purified eosinophils were activated by C3a, and adding C3a-stimulated eosinophils to purified neutrophils activated the neutrophils as well. Consequently, it is the eosinophils stimulated by C3a that in turn release factors capable of stimulating the neutrophils. Thus, an indirect stimulation of neutrophils by C3a was presumably mediated by contaminating eosinophils in the earlier cell preparations. The C3a effect is specific since a bioactive synthetic analog of C3a (C3a 57–77 peptide) was also active, while the nonreceptor-binding des Arg form of C3a failed to stimulate these cells. The ED_{50} for chemotaxis of eosinophils to C3a was recorded at 100 nM, which is much higher than that for C5a on these same cells; however, it must be remembered that the levels of C3a that can be generated in blood is approximately 6 μM or 20-fold greater than that of C5a.

Although neutrophils are not activated by C3a to migrate, release enzymes, or exhibit an oxidative burst, they do express the C3aR and C3a does induce intracellular calcium mobilization (Norgauer *et al.*, 1993; Takafuji *et al.*, 1994).

C3a effects on the eosinophil include chemotaxis, granule release, oxidation burst, as well as upregulation of β_2-integrins and shedding of L-selectins (Jagels *et al.*, 2000).

Monocyte Activation

Evidence for C3aR on monocytic cells is less well-established in functional or molecular terms. It has been claimed that C3aR is expressed on leukocytes and monocytes but not on B or T cells (Martin *et al.*, 1997). Human monocytes exhibit increased intracellular calcium levels, much like neutrophils, when stimulated by C3a, but not by C3a(des Arg) (Zwirner *et al.*, 1997). However, both C3a and C3a(des Arg) reportedly exert equal effects on cytokine synthesis in human monocytes (Haeffner-Cavaillon *et al.*, 1987; Takabayashi *et al.*, 1996, 1998). C3a and C3a(des Arg) suppress cytokine synthesis and polyclonal immune responses in human B lymphocytes (Fischer and Hugli, 1997). Since C3a(des Arg) does not bind to the C3aR expressed on RBL-2H3 cells, it was concluded that the effects that can be mediated by both intact C3a and the des Arg derivative in monocytes and B lymphocytes must be C3aR-independent effects (Wilken *et al.*, 1999). Therefore, the true actions of C3a mediated through the C3aR on monocytes and lymphocytes must yet be defined and separated from the nonspecific (i.e., non-C3aR) effects on these cell types. These insights suggest that much of the earlier work using PBMCs to evaluate the effects of C3a in mixed-cell systems will need to be carefully re-evaluated in terms of both direct and indirect cell activation, as well as specific versus nonspecific activation processes.

Platelets and Mast Cells

Little attention has been given to C3aR on platelets since early evidence indicated that human platelets were devoid of a C3a receptor based on chemical crosslinking experiments and on a lack of functional responses (Fukuoka and Hugli, 1988). These studies identified a curiosity, namely that guinea pig, but not human, platelets express a high molecular weight protein (95–105 kDa) that binds C3a with an estimated K_d of 8×10^{-10} M. The size of this molecule is nearly twice the size of the cloned guinea pig C3a receptor. This difference might be explained by the fact that guinea pig C3aR has five *N*-glycosylation sites compared to only two sites in human C3aR. The guinea pig platelets responded to C3a, but not to C3a(des Arg), indicating specificity of the interaction. Serotonin was released and aggregation was induced well below micromolar levels of C3a in guinea pig platelets. Since earlier reports claimed that human platelets were activated by C3a and C3a(des Arg) (Polly and Nachman, 1983), we now conclude that both specific and nonspecific effects of C3a on contaminating cells led to the release of mediators such as PAF, causing the human platelets to respond in an indirect manner, much as human neutrophils respond to C3a in the presence of eosinophils. Based on current evidence, one must conclude that C3aR is not expressed on the human platelet.

Direct evidence for C3aR on human or animal (rat) mast cells does not exist. In both the rat and human mast cells, the activation process, including histamine release, is induced by both C3a and C3a(des Arg), suggesting that stimulation of these cells is not through the C3aR. A crosslinking study was used to identify the molecule on the rat mast cell to which the C3a (and C3a des Arg) binds (Fukuoka and Hugli, 1990). The conclusion was that the primary binding site for C3a on the mast cell was the enzyme chymase. *In vivo* studies supported this hypothesis, since activating levels of C3a (and C3a des Arg) were degraded by chymase when injected in the peritoneal cavity of the rat (Kajita and Hugli, 1991). The mechanism of mast cell activation by C3a, like that of the human neutrophils, platelets, and B lymphocytes (Wilken *et al.*, 1999) appears to be C3aR-independent.

Tissue Responses

C3a and synthetic C3a analogs are known to induce a visible wheal-and-flare reaction when injected into human and animal skin (Hugli and Erickson, 1977). This response is not species-specific, is dose-dependent, and has a short duration of 15–30 minutes with little sensation (i.e. slight itching) (Hugli, 1981). Lung strips and ileal strips from guinea pigs undergo reversible and tachyphylactic contraction when exposed to C3a (Stimler *et al.*, 1983).

Pathology

C3a is not known to cause a significant physiologic response when injected into research animals. This is probably due to the rapid removal of the essential C-terminal arginine by carboxypeptidases M and N. However, there is evidence that C3a can produce *pulmonary injury* (Stimler *et al.*, 1980) and that intrabronchial instillation of milligram quantities of C3a in the guinea pig did cause *severe bronchial constriction* and death in some instances (Huey *et al.*, 1983; Regal and Klos, 2000).

GENE

Recent molecular studies resulted in cloning of human C3aR, first as an orphan G protein-coupled receptor (Roglic *et al.*, 1996), having an extra large extracellular loop, which was later identified as C3aR by Crass *et al.* (1996) and Ames *et al.* (1996). The C3aR cDNA clone was isolated from human neutrophils and from differentiated leukocytic cell lines HL-60 and U937. The cDNA clones had an open reading frame of 1446 base pairs coding for 482 amino acid residues with a calculated M_r of 54 kDa. Human C3aR has three potential *N*-glycosylation sites and has been estimated by western blot analysis to be 60 kDa when expressed on a human astrocyte cell line. These results suggest that human C3a may contain *N*-linked oligosaccharides contributing up to 10% of the total weight. Estimates of guinea pig platelet C3aR based on crosslinking experiments were 105–115 kDa, suggesting a larger contribution from oligosaccharides attached at the five potential *N*-glycosylation sites in guinea pig C3aR.

The genes encoding C3aR have been mapped to band position p13.2-3 of chromosome 12 in humans and 6F1 in mouse (Hollmann *et al.*, 1998; Paral *et al.*, 1998).

Accession numbers

Human: Q16581, Z73157, U62027, NM004054
Mouse: O09047, U77461, U77460, AF053757, U97537
Rat: U86379
Guinea pig: U86378, AJ006402

Sequence

See **Figure 3**.

Figure 3 Nucleotide sequence for human C3aR.

```
Sequence: 1449 bp mRNA
BASE COUNT 331a  380c  310g  428t
ORIGIN: Homo sapiens

   1 ATGGCGTCTT TCTCTGCTGA GACCAATTCA ACTGACCTAC TCTCACAGCC ATGGAATGAG
  61 CCCCCAGTAA TTCTCTCCAT GGTCATTCTC AGCCTTACTT TTTTACTGGG ATTGCCAGGC
 121 AATGGGCTGG TGCTGTGGGT GGCTGGCCTG AAGATGCAGC GGACAGTGAA CACAATTTGG
 181 TTCCTCCACC TCACCTTGGC GGACCTCCTC TGCTGCCTCT CCTTGCCCTT CTCGCTGGCT
 241 CACTTGGCTC TCCAGGGACA GTGGCCCTAC GGCAGGTTCC TATGCAAGCT CATCCCCTCC
 301 ATCATTGTCC TCAACATGTT TGCCAGTGTC TTCCTGCTTA CTGCCATTAG CCTGGATCGC
 361 TGTCTTGTGG TATTCAAGCC AATCTGGTGT CAGAATCATC GCAATGTAGG GATGGCCTGC
 421 TCTATCTGTG GATGTATCTG GGTGGTGGCT TTTGTGATGT GCATTCCTGT GTTCGTGTAC
 481 CGGGAAATCT TCACTACAGA CAACCATAAT AGATGTGGCT ACAAATTTGG TCTCTCCAGC
 541 TCATTAGATT ATCCAGACTT TTATGGAGAT CCACTAGAAA ACAGGTCTCT TGAAAACATT
 601 GTTCAGCCGC CTGGAGAAAT GAATGATAGG TTAGATCCTT CCTCTTTCCA AACAAATGAT
 661 CATCCTTGGA CAGTCCCCAC TGTCTTCCAA CCTCAAACAT TTCAAAGACC TTCTGCAGAT
 721 TCACTCCCTA GGGGTTCTGC TAGGTTAACA AGTCAAAATC TGTATTCTAA TGTATTTAAA
 781 CCTGCTGATG TGGTCTCACC TAAAATCCCC AGTGGGTTTC CTATTGAAGA TCACGAAACC
 841 AGCCCACTGG ATAACTCTGA TGCTTTTCTC TCTACTCATT TAAAGCTGTT CCCTAGCGCT
 901 TCTAGCAATT CCTTCTACGA GTCTGAGCTA CCACAAGGTT TCCAGGATTA TTACAATTTA
 961 GGCCAATTCA CAGATGACGA TCAAGTGCCA ACACCCCTCG TGGCAATAAC GATCACTAGG
1021 CTAGTGGTGG GTTTCCTGCT GCCCTCTGTT ATCATGATAG CCTGTTACAG CTTCATTGTC
1081 TTCCGAATGC AAAGGGGCCG CTTCGCCAAG TCTCAGAGCA AAACCTTTCG AGTGGCCGTG
1141 GTGGTGGTGG CTGTCTTTCT TGTCTGCTGG ACTCCATACC ACATTTTTGG AGTCCTGTCA
1201 TTGCTTACTG ACCCAGAAAC TCCCTTGGGG AAAACTCTGA TGTCCTGGGA TCATGTATGC
1261 ATTGCTCTAG CATCTGCCAA TAGTTGCTTT AATCCCTTCC TTTATGCCCT CTTGGGGAAA
1321 GATTTTAGGA AGAAAGCAAG GCAGTCCATT CAGGGAATTC TGGAGGCAGC CTTCAGTGAG
1381 GAGCTCACAC GTTCCACCCA CTGTCCCTCA AACAATGTCA TTTCAGAAAG AAATAGTACA
1441 ACTGTGTGA
```

PROTEIN

Sequence

See **Figure 4**.

Description of protein

C3aR is an integral membrane protein, consisting of 482 amino acids, forming a single polypeptide chain. It belongs to the rhodopsin-like receptor superfamily, characterized by seven hydrophobic transmembrane regions connected with three extra- and three intracellular loops. The orientation of C3aR was determined by immunohistochemical methods, and it was determined that the N-terminal end is located extracellularly, while the C-terminal end is intracellular.

Relevant homologies and species differences

A 50–60% homology exists between the protein sequences of C3aR from various species. A consensus sequence of XKSXXKX occurs in the intracellular loop 2 domain between transmembrane helices 5 and 6 of C3aR from all species. This motif represents a phosphorylation site for kinase C (Fukuoka *et al.*, 1998a). A total of six Thr/Ser residues have been conserved at the C-terminal end of C3aR, representing potential phosphorylation sites that can become modified as a result of C3a binding. The large extracellular loop between transmembrane helices 4 and 5 is a region with the lowest level of sequence identity. An *N*-glycosylation site at Asn9 in human C3aR has been conserved, but other *N*-glycosylation sites vary between C3aR from different species (Figure 1).

Affinity for ligand(s)

The C3a/C5a hybrid mutagenesis studies were aimed at investigating the C3aR-binding sites (Bautsch *et al.*, 1992). The conclusion from these studies confirmed that a secondary or noneffector-binding site is required for optimal affinity of C3a binding to its receptor.

Studies to determine the architecture and to localize binding sites on the C3a receptor has been done. As described earlier, the recently cloned human C3aR contains an unusually large extracellular loop (between transmembrane helices 4 and 5), a feature which is unique among G protein-coupled receptors (Ames *et al.*, 1996; Crass *et al.*, 1996; Roglic *et al.*, 1996). Since it has been postulated that the large extracellular loop might contain some or all the structural determinants for C3a binding, this region has been a focus of recent studies. Generation of loop deletion and point mutations of the receptor (Chao *et al.*, 1999; Sun *et al.*, 1999) indicates a multisite cooperative C3a/C3aR-binding interaction (Figure 2).

Cell types and tissues expressing the receptor

The C3a receptor has been demonstrated on guinea pig platelets (Fukuoka and Hugli, 1988), human alveolar macrophages, neutrophils, basophils (Glovsky *et al.*, 1979), and eosinophils (Daffern *et al.*, 1995) by either functional assays or using chemical crosslinking techniques. Flow cytometry was used to identify C3aR on peripheral monocytes and umbilical vein endothelial cells, as well as the Raji cell line and differentiated HL-60 and U937 monocytic cell lines (Roglic *et al.*, 1996). Northern blot analysis showed high levels of mRNA for C3aR in human lung and spleen with lower levels in heart, placenta, kidney, thymus, testis, ovaries, small intestine, colon, and several regions of the brain (Ames *et al.*, 1996).

This wide distribution of message for the C3aR is surprising, since C3a was believed to have a more limited functional and physiologic role than C5a. Observations of C3a as a chemotactic factor for eosinophils and persumably basophils, but not neutrophils, suggests a specialized role for C3a in inflammatory responses involving these cell types (Daffern *et al.*, 1995). There is much new biology to

Figure 4 Amino acid sequence for human C3aR.

Sequence c3aR protein: human, Homo sapiens

```
1-MASFSAETNS TDLLSQPWNE PPVILSMVIL SLTFLLGLPG NGLVLWVAGL KMQRTVNTIW FLHLTLADLL CCLSLPFSLA
  HLALQGQWPY GRFLCKLIPS IIVLNMFASV FLLTAISLDR CLVVFKPIWC QNHRNVGMAC SICGCIWVVA FVMCIPVFVY
  REIFTTDNHN RCGYKFGLSS SLDYPDFYGD PLENRSLENI VQPPGEMNDR LDPSSFQTND HPWTVPTVFQ PQTFQRPSAD
  SLPRGSARLT SQNLYSNVFK PADVVSPKIP SGFPIEDHET SPLDNSDAFL STHLKLFPSA SSNSFYESEL PQGFQDYYNL
  GQFTDDDQVP TPLVAITITR LVVGFLLPSV IMIACYSFIV FRMQRGRFAK SQSKTFRVAV VVVAVFLVCW TPYHIFGVLS
  LLTDPETPLG KTLMSWDHVC IALASANSCF NPFLYALLGK DFRKKARQSI QGILEAAFSE ELTRSTHCPS NNVISERNST
  TV-482
```

be learned from the recent discoveries that C3a receptors are as widely distributed as are receptors for cytokines and chemokines.

Recent reports suggest that C3a may induce differing responses from mixed PBMCs exposed to LPS, depending on the adhesive state of these cells. It was reported that C3a suppressed cytokine production when these cells were cultured in polypropylene tubes, which prevents cell–matrix adhesion. However, C3a enhanced LPS-induced cytokine production of the cultures when they were carried out in standard polystyrene tissue culture plates (Takabayashi *et al.*, 1996). Recent studies in our laboratory, using purified monocytes, have demonstrated a clear suppressive effect of C3a on LPS-induced production of inflammatory cytokines (Fischer *et al.*, 1999). In contrast, C3a enhances LPS-induced production of the immunosuppressive cytokine IL-10. One interesting observation common to all three studies is that C3a-(des Arg) retains biological activity in each of these assays. Evidence that the actions of C3a(des Arg) occur through specific receptor interactions is still not convincing.

Most functional responses of myeloid cells, including essentially all eosinophil responses, and the immunosuppressive effects of C3a on humoral immune responses in mixed PBMC populations, are not elicited by C3a(des Arg) at submicromolar concentrations. On the other hand, in the systems just mentioned above, as well as in mast cell activation, the potency of C3a(des Arg) ranges from being equipotent with C3a, to having approximately 10% of the activity of the parent molecule. The residual activity of C3a(des Arg), although physiologically relevant, has been attributed to nonspecific poly cation effects, as originally defined by Mousli *et al.* (1992). Biochemically, C3a and C3a(des Arg) are highly charged cationic proteins at physiologic pH.

It was found that in rat peritoneal mast cells, C3a and C3a(des Arg) were nearly equipotent in inducing histamine release (Johnson *et al.*, 1975). It was further found that a number of otherwise unrelated cationic molecules also induced histamine release, and that their potency correlated with their charge-to-mass ratio. Based primarily on these data, it was concluded that rat mast cells do not bear specific C3a receptors, and that the activating effects of C3a were C3aR-independent and related to the cationic nature of the molecule.

The potential effects of C3a on nonmyeloid cells has not been explored in great detail, partly due to a lack of information concerning the receptor. Since C3aR has recently been identified and cloned, and antibodies generated to the molecule, it is only a matter of time until the biologic manifestations associated with this receptor have been identified. Although northern blot analysis has suggested a wide distribution of C3aR, little evidence for direct C3a stimulation on nonmyeloid cells has been reported. Again, the presence of mast cells and myeloid cells in these tissues prevents conclusive assignment of receptor expression to other particular cell types. It is anticipated that the molecular tools now available will accelerate characterizations of the cellular distribution of C3aR.

Many tissue effects of C3a are virtually identical to those of C5a and this supports the notion that macrophages and mast cells are prominent players at the tissue level. Intradermal injection of C3a leads to a classical wheal-and-flare response, but without a significant leukocyte infiltration. These results suggest that the chemotactic properties of C3a for eosinophils may be insufficient at these levels to induce eosinophil recruitment (Fernandez *et al.*, 1978). Alternatively, C3a may be converted *in vivo* to the chemotactically inactive des Arg form before eosinophil recruitment can occur. The lack of neutrophil or monocyte recruitment suggests that local mast cell activation *per se* is insufficient to generate chemotactic signals for either of these circulating cell types.

Regulation of receptor expression

The C3aR is downregulated upon ligand binding and is cross-desensitized by C5a (Settmacher *et al.*, 1999). C3aR upregulation has been reported on astrocytes during inflammation (Nataf *et al.*, 1999).

DOWNSTREAM GENE ACTIVATION

Genes induced

IL6 mRNA expression was increased in human astrocytes by C3a; however, the protein IL-6 was not generated.

THERAPEUTIC UTILITY

Effects of inhibitors (antibodies) to receptors

No neutralizing antibody yet exists to C3aR. Several anti-C3aR peptide antibodies have been generated but no neutralizing antibodies have been reported.

References

Ames, R. S., Li, Y., Sarau, H. M., Nuthulaganti, P., Foley, J. J., Ellis, C., Zeng, Z., Su, K., Jurewicz, A. J., Hertzberg, R. P., Bergsma, D. J., and Kumar, C. (1996). Molecular cloning and characterization of the human anaphylatoxin C3a receptor. *J. Biol. Chem.* **271**, 20231–20234.

Bautsch, W., Kretzschmar, T., Stühmer, T., Kola, A., Emde, M., Köhl, J., Klos, A., and Bitter-Suermann, D. (1992). A recombinant hybrid anaphylatoxin with dual C3a/C5a activity. *Biochem. J.* **288**, 261–266.

Chao, T.-H., Ember, J. A., Wang, M., Bayon, Y., and Hugli, T. E. (1999). Role of the second extracellular loop of human C3a receptor in agonist binding and receptor function. *J. Biol. Chem.* **274**, 9721–9728.

Chenoweth, D. E., and Hugli, T. E. (1980). Human C5a and C5a analogs as probes of the neutrophil C5a receptor. *Mol. Immunol.* **17**, 151–161.

Crass, T., Raffetseder, U., Martin, U., Grove, M., Klos, A., Köhl, J., and Bautsch, W. (1996). Expression cloning of the human C3a anaphylatoxin receptor (C3aR) from differentiated U-937 cells. *Eur. J. Immunol.* **26**, 1944–1950.

Daffern, P. J., Pfeifer, P. H., Ember, J. A., and Hugli, T. E. (1995). C3a is a chemotaxin for human eosinophils but not for neutrophils. I. C3a stimulation of neutrophils is secondary to eosinophil activation. *J. Exp. Med.* **181**, 2119–2127.

Damerau, B., Gruenefeld, E., and Vogt, W. (1980). Aggregation of leukocytes induced by the complement-derived peptides C3a and C5a and by three synthetic formyl-methionyl peptides. *Int. Arch. Allergy Appl. Immunol.* **63**, 159–169.

Ember, J. A., Jagels, M. A., and Hugli, T. E. (1998). In "The Human Complement System in Health and Disease" (ed J. Volnakis and M. Frank), Characterization of complement anaphylatoxins and their biological responses, pp. 241–284. Marcel Dekker, New York.

Fernandez, H. N., Henson, P. M., Otani, A., and Hugli, T. E. (1978). Chemotactic response to human C3a and C5a anaphylatoxins. I. Evaluation of C3a and C5a leukotaxis *in vitro* and under stimulated *in vivo* conditions. *J. Immunol.* **120**, 109–115.

Fischer, W. H., and Hugli, T. E. (1997). Regulation of B cell functions by C3a and C3adesArg: suppression of TNF-α, IL-6, and the polyclonal immune response. *J. Immunol.* **159**, 4279–4286.

Fischer, W. H., Jagels, M. A., and Hugli, T. E. (1999). Regulation of IL-6 synthesis in human peripheral blood mononuclear cells by C3a and C3a des Arg. *J. Immunol.* **162**, 453–459.

Fukuoka, Y., and Hugli, T. E. (1988). Demonstration of a specific C3a receptor on guinea pig platelets. *J. Immunol.* **140**, 3496–3501.

Fukuoka, Y., and Hugli, T. E. (1990). Anaphylatoxin binding and degradation by rat peritoneal mast cells. Mechanisms of degranulation and control. *J. Immunol.* **145**, 1851–1858.

Fukuoka, Y., Ember, J. A., and Hugli, T. E. (1998a). Molecular cloning of two isoforms of the guinea pig C3a anaphylatoxin receptor: alternative splicing at the large extracellular. *J. Immunol.* **161**, 2977–2984.

Fukuoka, Y., Ember, J. A., and Hugli, T. E. (1998b). Cloning and characterization of rat C3a receptor: differential expression of rat C3a and C5a receptors by LPS stimulation. *Biochem. Biophys. Res. Commun.* **242**, 663–668.

Fukuoka, Y., Ember, J. A., and Hugli, T. E. (1999). Ligand binding sites on guinea pig C3aR: point and deletion mutations in the large extracellular loop and vicinity. *Biochem. Biophys. Res. Commun.* **263**, 357–360.

Glovsky, M. M., Hugli, T. E., Ishizaka, T., Lichtenstein, L. M., and Erickson, B. W. (1979). Anaphylatoxin-induced histamine release with human leukocytes: studies of C3a leukocyte binding and histamine release. *J. Clin. Invest.* **64**, 804–811.

Haeffner-Cavaillon, N., Cavaillon, J.-M., Laude, M., and Kazatchkine, M. D. (1987). C3a (C3a des-Arg) induces production and release of interleukin-1 by cultured human monocytes. *J. Immunol.* **139**, 794.

Hollmann, T. J., Haviland, D. L., Kildsgaard, J., and Watts, K. W. R. A. (1998). Cloning, expression, sequence determination, and chromosome localization of the mouse complement C3a anaphylatoxin receptor gene. *Mol. Immunol.* **35**, 137–148.

Hsu, M. H., Ember, J. A., Wang, M., Prossnitz, E. R., Hugli, T. E., and Ye, R. D. (1997). Cloning and functional characterization of the mouse C3a anaphylatoxin receptor gene. *Immunogenetics* **47**, 64–72.

Huey, R., Bloor, C. M., Kawahara, M. S., and Hugli, T. E. (1983). Potentiation of the anaphylatoxins *in vivo* using an inhibitor of serum carboxypeptidase N (SCPN). I. Lethality and pathologic effects on pulmonary tissue. *Am. J. Pathol.* **112**, 48–60.

Hugli, T. E. (1981). The structural basis for anaphylatoxin and chemotactic functions of C3a, C4a, and C5a. *Crit. Rev. Immunol.* **1**, 321–366.

Hugli, T. E., and Erickson, B. W. (1977). Synthetic peptides with the biological activities and specificity of human C3a anaphylatoxin. *Proc. Natl Acad. Sci. USA* **74**, 1826–1830.

Ischenko, A., Sayah, S., Patte, C., Andreev, S., Gasque, P., Schouft, M. T., Vaudry, H., and Fontaine, M. (1998). Expression of a functional anaphylatoxin C3a receptor by astrocytes. *J. Neurochem.* **71**, 2487–2496.

Jagels, M. A., Daffern, P. J., and Hugli, T. E. (2000). C3a and C5a enhance granulocyte adhesion to endothelial and epithelial cell monolayers: epithelial and endothelial priming is required for C3a-induced eosinophil adhesion. *Immunopharmacology* **46**, 209–222.

Johnson, A. R., Hugli, T. E., and Müller-Eberhard, H. J. (1975). Release of histamine from rat mast cells by the complement peptides C3a and C5a. *Immunology* **28**, 1067.

Kajita, T., and Hugli, T. E. (1991). Evidence for *in vivo* degradation of C3a anaphylatoxin by mast cell chymase. I. Nonspecific activation of rat peritoneal mast cells by $C3a_{desArg}$. *Am. J. Pathol.* **138**, 1359–1369.

Martin, U., Bock, D., Arseniev, L., Tornetta, M. A., Ames, R. S., Bautsch, W., Kohl, J., Ganser, A., and Klos, A. (1997). The human C3a receptor is expressed on neutrophils and monocytes, but not on B or T lymphocytes. *J. Exp. Med.* **186**, 199–207.

Mousli, M., Hugli, T. E., Landry, Y., and Bronner, C. (1992). A mechanism of action for anaphylatoxin C3a stimulation of mast cells. *J. Immunol.* **148**, 2456–2461.

Nagata, S., Glovsky, M. M., and Kunkel, S. L. (1987). Anaphylatoxin-induced neutrophil chemotaxis and aggregation. Limited aggregation and specific desensitization induced by human C3a and synthetic C3a octapeptides *Int. Arch. Allergy Appl. Immunol.* **82**, 4–9.

Nataf, S., Stahel, P. F., Davoust, N., and Barnum, S. R. (1999). Complement anaphylatoxin receptors on neurons: new tricks for old receptors? *Trends Neurosci.* **22**, 397–402.

Norgauer, J., Dobos, G., Kownatzki, E., Dahinden, C., Burger, R., Kupper, R., and Gierschik, P. (1993). Complement fragment C3a stimulates Ca^{2+} influx in neutrophils via a pertussis-toxin-sensitive G protein. *Eur. J. Biochem.* **217**, 289–294.

Paral, D., Sohns, B., Crass, T., Grove, M., Kohl, J., Klos, A., and Bautsch, W. (1998). Genomic organization of the human C3a receptor. *Eur. J. Immunol.* **28**, 2417–2423.

Polly, M. J., and Nachman, R. L. (1983). Human platelet activation by C3a and C3a desArg. *J. Exp. Med.* **158**, 603–615.

Regal, J. F., and Klos, A. (2000). Minor role of the C3a receptor in systemic anaphylaxis in the guinea pig. *Immunopharmacology* **46**, 15–28.

Roglic, A., Prossnitz, E. R., Cavanagh, S. L., Pan, Z., Zou, A., and Ye, R. D. (1996). cDNA cloning of a novel G protein-coupled receptor with a large extracellular loop structure. *Biochim. Biophys. Acta* **1305**, 39–43.

Settmacher, B., Bock, D., Saad, H., Rheinheimer, C., Köhl, J., Bautsch, W., and Klos, A. (1999). Modulation of C3a activity: internalization of the human C3a receptor and its inhibition by C5a. *J. Immunol.* **162**, 7409–7416.

Showell, H. J., Glovsky, M. M., and Ward, P. A. (1982a). C3a-induced lysosomal enzyme secretion from human neutrophils. Lack of inhibition by f-met-leu-phe antagonists and inhibition by arachidonic acid antagonists. *Int. Arch. Allergy Appl. Immunol.* **67**, 227–232.

Showell, H. J., Glovsky, M. M., and Ward, P. A. (1982b). Morphological changes in human polymorphonuclear leukocytes induced by C3a in the presence and absence of cytochalasin B. *Int. Arch. Allergy Appl. Immunol.* **69**, 62–67.

Siciliano, S. J., Rollins, T. E., DeMartino, J., Konteatis, Z., Malkowitz, L., Van Riper, G., Bondy, S., Rosen, H., and Springer, M. S. (1994). Two-site binding of C5a by its receptor: an alternative binding paradigm for G protein-coupled receptors. *Proc. Natl Acad. Sci. USA* **91**, 1214–1218.

Stimler, N. P., Hugli, T. E., and Bloor, C. M. (1980). Pulmonary injury induced by C3a and C5a anaphylatoxins. *Am. J. Pathol.* **100**, 327–348.

Stimler, N. P., Bloor, C. M., and Hugli, T. E. (1983). C3a-induced contraction of guinea pig lung parenchyma: role of cyclooxygenase metabolites. *Immunopharmacology* **5**, 251–257.

Sun, J., Ember, J. A., Chao, T.-H., Fukuoka, Y., Ye, R. D., and Hugli, T. E. (1999). Identification of ligand effector binding sites in transmembrane regions of the human G protein-coupled C3a receptor. *Protein Sci.* **8**, 1–8.

Takabayashi, T., Vannier, E., Clark, B. D., Margolis, N. H., Dinarello, C. A., Burke, J. F., and Gelfand, J. A. (1996). A new biologic role for C3a and C3a desArg. *J. Immunol.* **156**, 3455–3460.

Takabayashi, T., Vannier, E., Burke, J. F., Tompkins, R. G., Gelfand, J. A., and Clark, B. D. (1998). Both C3a and C3a(desArg) regulate interleukin-6 synthesis in human peripheral blood mononuclear cells. *J. Infect. Dis.* **177**, 1622–1628.

Takafuji, S., Tadokoro, K., Ito, K., and Dahinden, C. A. (1994). Degranulation from human eosinophils stimulated with C3a and C5a. *Int. Arch. Allergy Immunol.* **104**, 27–29.

Tornetta, M. A., Foley, J. J., Sarau, H. M., and Ames, R. S. (1997). The mouse anaphylatoxin C3a receptor: molecular cloning, genomic organization and functional expression. *J. Immunol.* **158**, 5277–5282.

Wilken, H.-C., Götze, O., Werfel, T., and Zwirner, J. (1999). C3a (desArg) does not bind to and signal through the human C3a receptor. *Immunol. Lett.* **67**, 141–145.

Zwirner, J., Gotze, O., Moser, A., Sieber, A., Begemann, G., Kapp, A., Elsner, J., and Werfel, T. (1997). Blood- and skin-derived monocytes/macrophages respond to C3a but not to C3a(desArg) with a transient release of calcium via a pertussis toxin-sensitive signal transduction pathway. *Eur. J. Immunol.* **27**, 2317–2322.

PAF Receptors

Peter M. Henson*

Department of Pediatrics, National Jewish Medical Research Center, 1400 Jackson Street, Room D-508, Denver, CO 80206-2761, USA

*corresponding author tel: 303-398-1380, fax: 303-398-1381, e-mail: hensonp@njc.org

DOI: 10.1006/rwcy.2000.23003.

SUMMARY

Platelet-activating factor (PAF) reacts with a specific seven transmembrane, G protein-linked receptor with two promoter splice forms showing tissue-specific regulation and for which a knockout mouse has recently been created. The receptor is widely expressed on hematopoietic cells, endothelial cells, keratinocytes, and cells of the central nervous system. A huge number of receptor antagonists have been synthesized and many are being explored *in vitro*, in animal models, and in a variety of clinical trials. In addition, the complexity of the PAF story is increased by the observation that in the majority of cells that synthesize it, most of the PAF remains within the cell, leading to speculation that it plays an intracellular role, perhaps even on an intracellular receptor, in addition to its role as extracellular communication molecule.

BACKGROUND

Discovery

Following a large body of work on PAF receptors determined by a variety of binding assays, a candidate plasma membrane receptor from the guinea pig was cloned in 1991 by expression cloning in *Xenopus* eggs (Honda *et al.*, 1991), followed soon after by its human counterpart (Nakamura *et al.*, 1991; Ye *et al.*, 1991; Kunz *et al.*, 1992; Chase *et al.*, 1993).

Structure

The receptor is a seven transmembrane, heterotrimeric G protein-linked receptor (see Figure 2).

Main activities and pathophysiological roles

As discussed in the PAF chapter, much of the PAF that is synthesized does not gain access to the extracellular environment. The presence of PAF within synthesizing cells as well as the recent interest in oxidized phospholipids and questions about the roles of intracellular acetylhydrolases, have rekindled research into the intracellular effects of this family of molecules. While there little direct evidence to date for such effects and even less information on specific receptors or signaling pathways, a number of intriguing (although speculative) possibilities and questions might be considered.

Oxidized phospholipids would be expected to have potential toxic effects within cells leading to the propagation of redox alterations and alteration in normal membrane structure and function. Potent catabolic enzymes may have developed to counteract such effects. These acetylhydrolases (as well as the action of phospholipases A_2) generate lysophosphatides, themselves potentially toxic because of their amphipathic properties. The transacylase and acetyltransferase involved in PAF biosynthesis could therefore further protect the cell, ultimately making use of an abundant ingredient, acetyl-CoA. The extremely wide distribution of such pathways (including plants, fungi, and bacteria) might be in keeping with such a concept.

Additional activities of these molecules might originally have derived from this need for inactivation but now could include effects on intracellular receptors and signaling pathways. No such receptor has yet been definitively described although there were some early reports of such molecules at the 1998 PAF and Related Lipid Mediators Meeting in New Orleans. It has been suggested that the antagonist BN-50730 may be effective against possible intracellular PAF receptors (Bazan *et al.*, 1997) and its use has also raised the possibility of such effects in a number of neuronal systems. This is indeed an area that needs more investigation.

GENE

Accession numbers

GenBank:
Human: S52624, D10202, D31736, U11032
Mouse: D50872
Rat: U04740
Guinea pig: X56736

Sequence

See **Figure 1**.

Chromosome location and linkages

Only one gene has been identified (on chromosomes 1 and 4 in human or mouse respectively), although alternative splice forms are seen with an identical coding region but differing upstream. 5′ sequences and different transcription initiation sites (Mutoh *et al.*, 1993; Shimizu and Mutoh, 1997). One is found in most cells and tissues, the other is selectively absent from leukocytes and brain; one is upregulated by glucocorticoids, the other by retinoids, etc. On the other hand, pharmacological studies have consistently suggested a more complex system with the likelihood of more than one receptor (see, for example, Hwang, 1988, 1991; Kato *et al.*, 1994; LeVan *et al.*, 1997; Voelkel *et al.*, 1986). At this point, if another membrane receptor is found, it will probably be quite different from the cloned version since attempts to find genes related to the cloned PAFR have been uniformly unsuccessful.

Glycosylation of the receptor is required for its expression (Garcia Rodriguez *et al.*, 1995a). An intron in the 5′ untranslated region of the human gene has also been described (Chase *et al.*, 1993).

PROTEIN

Description of protein

The receptor appears to be a relatively standard seven transmembrane, heterotrimeric G protein-linked receptor (Mutoh *et al.*, 1993; Shimizu and Mutoh, 1997) comprising 342 amino acids (in human). The cysteines at positions 90 and 173 form a disulfide bond and Cys95 also seems important – mutation of these alters membrane expression and function (Le Gouill *et al.*, 1997). Mutation of Leu231 to Arg led to a constituitively active receptor with higher binding affinity for PAF, whereas conversion of the adjacent

Figure 1 Deduced sequences of rat, guinea pig, human, and mouse PAF receptor. From Nakamura *et al.* (1991), Bito *et al.* (1994), Honda *et al.* (1991), Ishii *et al.* (1996).

```
RA: MEQNGSPRVDSEFRYTLFPIVYSVIFVLG VVANGYVLWVFATLYPSKKLNEIKIFMVNLTVADLLFLMTLPLWIVYYSNE
GP: NELNSSSRVDSEFRYTLFPIVYSIIFVLGIIANGYVLWVFARLYPSKKLNEIKIFMVNLTVADLLFLITLPLWIVYYSNQ
HU: MEPHDSSHMDSEFRYTLFPIVYSIIFVLGVIANGYVLWVFARLYPCKKFNEIKIFMVNLTMADMLFLITLPLWIVYYQNQ
MU: MEHNGSFRVDSEFRYTLFPIVYSVIFILGVVANGYVLWVFANLYPSKKLNEIKIFMVNLTMADLLFLITLPLWIVYYYNE

    GDWIVHKFLCNLAGCLFFINTYCSVAFLGVITYNRYQAVAYPIKTAQATTRKRGITLSLVIWISIAATASYFLATDSTNV
    GNWFLPKFLCNLAGCLFFINTYCSVAFLGVITYNRFQAVKYPIKTAQATTRKRGIALSLVIWVAIVAAASYFLVMDSTNV
    GNWILPKFLCNVAGCLFFINTYCSVAFLGVITYNRFQAVTRPIKYAQANTRKRGISLSLVIWVAIVGAASYFLILDSTNT
    GATILPNFLCNVAGCLFFINTYCSVAFLGVITYNRYQAVAYPIKTAQATTRKRGISLSLIIWVSIVATASYFLATDSTNL

    VPKKDGSGNITRCFEHYEPYSVPILVV HIFITSCFELVFFLIFYCNMVIIHTLLTRPVRQQRKPEVKRRA LWMVCTVLAV
    VSNKAGSGNITRCFEHYEKGSKPVLIIHICIVLGFFIVFLLILFCNLVIIHTLLRQPVKQQRNAEVRRRALWMVCTVLAV
    VPDSAGSGNVTRCFEHYEKGSVPVLIIHIFIVFSFFLVFLIILFCNLVIIRTLLMQPVQQQRNAEVKRRALWMVCTVLAV
    VPNKDGSGNITRCFEHYEPYSVPILVVHVFIAFCFFLVFFLIFYCNLVIIHTLLTQPMRQQRKAGVKRRALWMVCTVLAV

    FVICFVPHHVVQLPWTLAELGYQ-TNFHQAINDAHQITLCLLSTNCVLDPVIYCFLTKKFRKHLSEKFYSMRSSRKCSRA
    FVICFVPHHMVQLPWTLAELGMWPSSNHQAINDAHQVTLCLLSTNCVLDPVIYCFLTKKFRKHLSEKLNIMRSSQKCSRV
    FIICFVPHHVVQLPWTLAELGFQDSKFHQAINDAHQVTLCLLSTNCVLDPVIYCFLTKKFRKHLTEKFYSMRSSRKCSRA
    FIICFVPHHVVNLPWTLAGLGTQ-TNFHQAINDAHQITLCLLSTNCVLDPVIYCFLTKKFRKHLSEKFYSMRSSRKCSRA

    TSDTCTEVMMPANQTPVLPLKN
    TTDTGTEMAIPINHTPVNPIKN
    TTLTVTEVVVPFNQIPGNSLKN
    TSDTCTEVIVPANQTPIVSLKN
```

Ala230 to Glu had the opposite effect (Parent *et al.*, 1996), suggesting an important influence of this region of the third intracellular loop on configuration. An *N*-glycosylation site is present on the second extracellular loop and appears important for membrane expression but not for function (Garcia Rodriguez *et al.*, 1995b).

Relevant homologies and species differences

The differences between the PAF receptor in the three species from which it has been cloned are subtle (see Figure 1). The mouse and rat protein has one less amino acid than human or guinea pig. **Figure 2** illustrates these differences in the context of the distribution across the membrane.

Affinity for ligand(s)

PAF is an amphiphilic phospholipid. Accordingly, its mode of interaction with a surface receptor might be expected to be complex. At concentrations above micromolar in aqueous solution (without a carrier) it is likely to form micelles. When 'secreted' by cells it is thought to be presented to the extracellular environment as a phospholipid from the outer membrane leaflet and probably only partitioned into the aqueous environment on a carrier such as albumin. It has been estimated that plasma albumin has four binding sites for PAF (probably of different affinities) and suggested that PAF is usually presented to its receptor on this protein (Clay *et al.*, 1990). An alternative source of PAF activity for interaction with target cells may be the shedding of PAF-containing membrane vesicles from the synthesizing cells (Patel *et al.*, 1992), perhaps providing an explanation for early studies of PAF activity in the absence of an albumin carrier. As indicated, there are still a significant number of discrepancies in our understanding of PAF-receptor interactions.

Addition of PAF to cells results in nonreceptor-mediated association/insertion into the outer membrane leaflet (e.g. Bratton *et al.*, 1992). From there it can be flipped to the inner leaflet by the action of the phospholipid scramblase. What is not clear is whether, in either of these configurations, it can interact with the PAF receptor or whether such binding can only occur when it is presented in solution (i.e. on a carrier) and/or whether it can interact with the receptor on one cell when presented in the membrane of another. It should be noted that PAF can also be internalized through interaction with its receptor (Gerard and Gerard, 1994), making it

Figure 2 PAF receptor showing amino acids common between human, mouse, rat, and guinea pig. Adapted from Shimizu and Mutoh (1997).

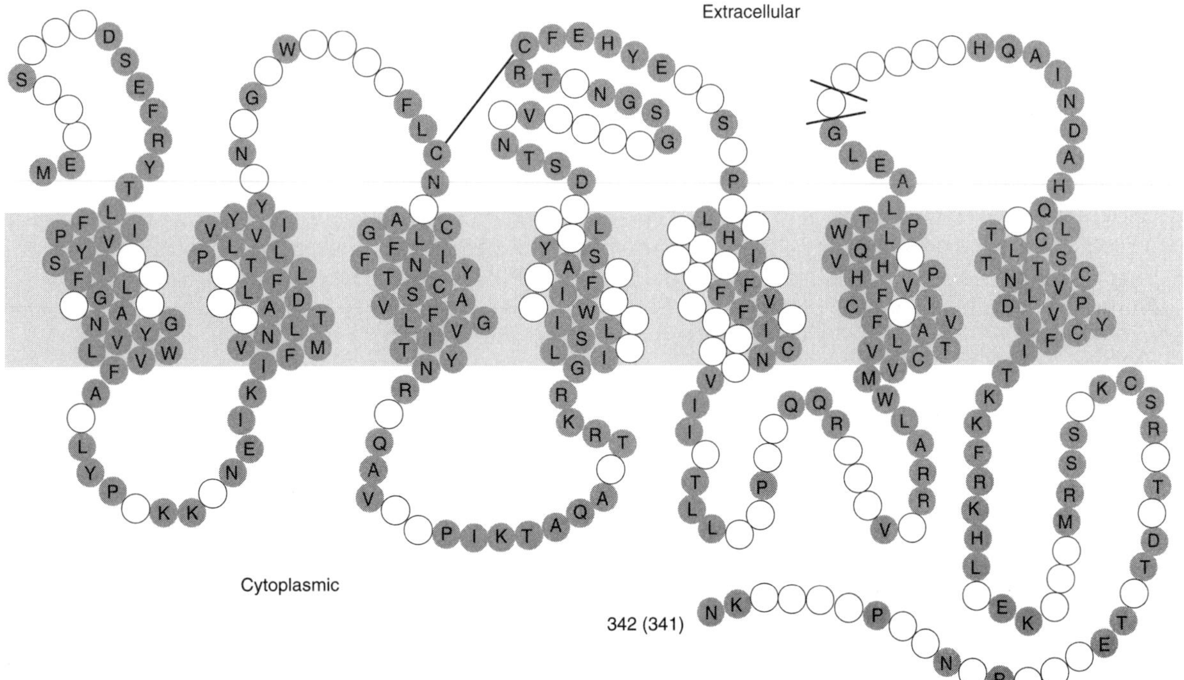

imperative to distinguish these pathways in any studies of uptake.

An intriguing observation that may have little to do with PAF or its analogs directly is the suggestion that *Streptococcus pneumoniae* can bind to cells, including epithelial cells, via the PAF receptor, utilizing the phosphorylcholine on the bacterial teichoic acid (Cundell *et al.*, 1995).

Cell types and tissues expressing the receptor

The PAF receptor and/or PAF binding has been found throughout the body, including the central nervous system on many different cell types, including neutrophils, eosinophils, basophils, mononuclear phagocytes, some lymphocytes, mast cells, endothelial cells (Domingo *et al.*, 1994; Predescu *et al.*, 1996; Sasaki *et al.*, 1996), and keratinocytes (Shimada *et al.*, 1998a). It can be seen associated with an endosomal compartment within endothelial cells (Ihida *et al.*, 1999) but this could represent endocytosed receptor. Transcripts for the receptor are present from very early in embryogenesis (Stojanov and O'Neill, 1999) and are also present on spermatozoa (Reinhardt *et al.*, 1999).

Regulation of receptor expression

The human PAF receptor appears to be transcribed by two distinct promoters, each of which has distinct transcriptional initiation sites (Mutoh *et al.*, 1993). These exons are alternatively spliced to a third that contains the total open reading frame, thus generating two different transcripts. Transcript 1 has consensus sequences (three) for NFκB and SP-1, and the initiator (Inr) sequence (see also Pang *et al.*, 1995, and see GenBank accession number U11032). Transcript 2 also contains consensus sequences for AP-1, AP-2, and SP-1. Both transcripts were reported in heart, lung, spleen, and kidney, whereas only transcript 1 was found in granulocytes and brain.

A variety of cytokines have been reported to upregulate PAF receptor expression. Thus increased PAFR mRNA expression was seen in eosinophils from patients with asthma and was upregulated by exposure to IL-3, IL-5, and GM-CSF (Kishimoto *et al.*, 1996a, 1996b, 1997). In endothelial cells, mechanical stress also increases mRNA expression (Okahara *et al.*, 1998; Chaqour *et al.*, 1999). PAF itself has been reported to do the same (Wang *et al.*, 1997). All of these have been suggested to act through NFκB, three binding sites for which are seen in the promoter between −571 and −459 bp (Mutoh *et al.*, 1994).

On the other hand, PAF itself has also been suggested to downregulate PAF receptor mRNA expression (Nakao *et al.*, 1997) and oxidized LDL has been reported to do likewise (Stengel *et al.*, 1997). How this occurs is not yet clear but the contrasting observations certainly raise the possibility of complex, possibly multifactorial regulation. By a presumably different mechanism, the receptor is upregulated by TGFβ2 acting as a consequence of a protein synthetic step on a response element in the PAFR promoter between −44 and −17 bp (Yang *et al.*, 1997).

SIGNAL TRANSDUCTION

Cytoplasmic signaling cascades

Signaling

While the PAF receptor may be considered a relatively conventional member of the large seven transmembrane, heterotrimeric G protein-coupled receptor family, it may show differences from other chemoattractant or inflammatory mediator receptors in its presumed G protein usage, e.g. partial pertussis insensitivity (Hwang, 1988; Barzaghi *et al.*, 1989; Yue *et al.*, 1992; Honda *et al.*, 1994), implying use of multiple G proteins (Ali *et al.*, 1994). Its signaling pathways have been investigated in a number of cells with native receptor (e.g. neutrophils, eosinophils, or platelets) and after transfection into cells not normally expressing this molecule (see Boulay *et al.*, 1997; Honda *et al.*, 1994; Shimizu and Mutoh, 1997). On the other hand, a number of differences have been noted between fMLP and PAF signaling in neutrophils (e.g. Kadiri *et al.*, 1990; M'Rabet *et al.*, 1999), including relative selectivity towards p38 among the MAP kinases for the PAF receptor (Nick *et al.*, 1997).

Eicosanoid production often accompanies the generation of PAF, both *in vitro* and *in vivo*. The first step in biosynthesis of eicosanoids is the action of $cPLA_2$ on the arachidonyl moiety esterified in membrane phospholipids. A proportion of this available substrate is in phosphatidylcholine (PC) and some of this, in turn, alkyl arachidonyl PC. Synthetic coupling of the two mediator classes would, therefore, be expected since substrate (arachidonate or lyso-PAF) for both synthetic pathways is generated simultaneously. In addition, however, PAF can activate $cPLA_2$ in target cells via upstream signal pathways, including activation of the MAP kinase

ERK (Hirabayashi *et al.*, 1998), which could in turn lead to both more eicosanoids and more PAF.

Priming

In neutrophils, and particularly eosinophils, PAF is directly chemotactic *in vitro* and also induces the related alterations in cell shape driven by cortical assembly of F-actin. On the other hand, with respect to other cell responses the PAF receptor often generates an incomplete signal, requiring additional cofactor stimulation for effective responses. In neutrophils this process was described as priming (Guthrie *et al.*, 1984) and was first shown with LPS as a 'priming' agent. However, PAF stimulation is the prime example of this effect (e.g. Gay, 1993). For example, by itself, it does not initiate the oxidative burst but rather induces alterations in the cell responsiveness such that other stimuli, including other chemoattractants as well as FcR ligation or even phorbol ester activation are all enhanced as a consequence (e.g. Bass *et al.*, 1988; Kitchen *et al.*, 1996). Other PAF-primable responses include eicosanoid production (Bauldry *et al.*, 1991; Shindo *et al.*, 1996, 1997), actin assembly (Shalit *et al.*, 1987), secretion (Partrick *et al.*, 1997; Vercellotti *et al.*, 1988), etc. The mechanisms for this priming phenomenon are still not known but it does not seem to involve upregulation or significant alteration of the receptors for the triggering stimuli. One intriguing observation in this regard is that priming stimuli, especially PAF, are particularly effective at activating the p38 subfamily of MAP kinases (Nick *et al.*, 1997) although its relationship to priming has still to be determined. (Note that PAF can also stimulate cells primed with other agents (e.g. Brunner *et al.*, 1991), i.e. the processes are often reciprocal.) The suggested importance of priming is that stimuli such as PAF are presumed to initiate inflammatory cell accumulation but, by themselves, not to stimulate the cells further, thus minimizing potential damage to the vessel wall during the migration. Once in the tissues however, the primed cells would be hyperresponsive to additional triggering stimuli such as other inflammatory mediators, bacteria, immune complexes, etc. This is not to imply that PAF is only, or even primarily, a proinflammatory mediator since, as discussed below, it acts on many other cells and systems and, *in vivo*, may have equally or more important roles in a number of physiologic processes. Rather, the point here is that PAF receptor signaling probably intersects significantly with other signaling pathways – a concept that may be of critical importance in consideration of some of its other potential roles, such as in the neurologic or vascular systems, cell growth and differentiation, etc.

Receptor Desensitization

From studies in inflammatory cells again, it has long been appreciated that PAF responses are often even more transient that those seen with other inflammatory mediators and chemokines, i.e. the receptor seems to be particularly susceptible to desensitization. In fact, this process was earlier used to show PAF action *in vivo* in a model of IgE-mediated anaphylactic shock wherein the rapid and transient thrombocytopenia was suggested to be PAF-dependent because the platelets that returned to the circulation after 60 minutes had become selectively unresponsive to PAF (Henson and Pinckard, 1977). The PAF receptor can be directly phosphorylated (Ali *et al.*, 1994), probably by a PKC, PKA, and a receptor kinase, leading to homotypic desensitization (Takano *et al.*, 1994). The serine/threonine phosphorylation seems to enhance receptor internalization and thereby plays an additional role in desensitization (Ishii *et al.*, 1998a). A further mechanism for desensitization has been described in which PKA-driven phosphorylation of PLCβ3 is suggested to lead to unresponsiveness of that enzyme, and because of its relatively selective usage by the PAF receptor, unresponsiveness of the cell to PAF (Ali *et al.*, 1997, 1998). The potential importance of rapid and effective downregulation is the transience of PAF responses, perhaps related to the potency of the mediator and strength of its effects.

DOWNSTREAM GENE ACTIVATION

Incomplete information.

BIOLOGICAL CONSEQUENCES OF ACTIVATING OR INHIBITING RECEPTOR AND PATHOPHYSIOLOGY

The enormous heterogeneity of the possible effects of PAF *in vitro* and ultimately *in vivo* makes it difficult (a) to summarize these in simple terms in one chapter and (b) to determine which of these possibilities is actually operative, or even more critical, important, in the whole animal. To determine a specific role for a potential mediator, such as PAF (and its analogs), in any given process, it would need to be present at the site and shown able to induce the response if introduced; the effect should be abrogated if the mediator is removed, blocked or absent and return when reintroduced. As indicated above, the availability of a number of PAF receptor antagonists has allowed various combinations of these criteria to be applied and has suggested the many potential actions. The presence and wide distribution of potent inactivating enzymes has also led to speculative roles for PAF *in vivo* and, in some cases, genetic variants of these enzymes have been shown to have disease associations.

The field was reviewed in 1994 by Stewart (Stewart, 1994) but most of the more recent reviews have addressed specific areas of potential action. In this section we will discuss some of the likely areas of PAF effect, not with any expectation of completeness but rather to emphasize some specific points, concepts, and areas for future development. As a general example, the potential importance of oxidized phospholipids in *atherosclerosis* and the recognition that some of these have PAF-like activity opens up a number of important questions that are crying out for more detailed investigation.

A receptor knockout mouse has recently been created with a phenotype that calls into question some of these wide-ranging effects – at least as far as those mediated through this receptor are concerned (see below).

Unique biological effects of activating the receptors

When added to isolated cells with functioning PAF receptors (e.g. neutrophils or eosinophils) the types of cellular responses that are seen are, in general, typical of rapid mediator action via seven transmembrane, G protein-coupled receptors. These include calcium mobilization, shape change (morphologic polarization), actin polymerization, chemokinesis and chemotaxis, priming for NADPH oxidase activation and eicosanoid production, adhesion, and integrin regulation. Platelet responses are also relatively standard, for example, potent aggregation and some release reaction. Interestingly, phosphatidylserine expression is not observed (Henson and Landes, 1976) and, as a consequence, procoagulant effects are minimal. There has been some suggestion that PAF is relatively selective as a proadhesive and chemotactic agent for eosinophils (Stewart, 1994; Wardlaw *et al.*, 1986), but it certainly has similar activity against other leukocytes *in vitro* and its administration to animals does not result in an exclusively eosinophilic response. On the other hand, the addition of IL-5 increases the eosinophil response selectively (e.g. Warringa *et al.*, 1992), thus emphasizing the importance of mediator cooperation in selecting specific cells for activation.

Longer term cellular responses in inflammatory cells have been less often investigated except for those classified under the rubric of 'priming' responses, i.e. an enhancing function in cooperation with other stimuli. Even here there is much room for further investigation, for example in the *de novo* synthesis of new mediators, regulatory enzymes of the eicosanoid pathways, etc. A similar confusing story is seen when questions of PAF action in cell proliferation and maturation are raised. Reported effects include growth inhibition (Huang *et al.*, 1999; Shimada *et al.*, 1998b) and downregulation of myelopoesis (Dupuis *et al.*, 1998) or, on the other hand, evidence of growth stimulation (Bennett and Birnboim, 1997; Rougier *et al.*, 1997). It is presumed that responses will vary between cell types and, for the most part, will be mediated via secondary production or inhibition of regulatory molecules.

PAF as a Cofactor and the Concept of Mediator Networks

A common statement is that 'PAF is a mediator of inflammation'. What is specifically meant by this is not always clear. The presence of PAF receptors on inflammatory cells (including platelets) as well as on the endothelium provides for an action on these cells via the cell responses listed above. In artificial systems PAF can be shown to induce chemotaxis, transendothelial migration (e.g. Casale *et al.*, 1993) and, in conjunction with other stimuli, production and secretion of oxidants, other lipid mediators and proinflammatory mediator proteins. In a significant number of circumstances *in vivo*, the observed PAF effect is actually mediated by one of these downstream. products, often an eicosanoid – i.e. PAF is an intermediary, perhaps contributing an amplification step to the overall proinflammatory signal.

A potential theme that might be considered for PAF action is that it is, for the most part, a cofactor. Its role as a priming agent for neutrophil oxidant production is well established. Taking this as a model then, the ability of signaling pathways initiated through the PAF receptor to integrate with signaling from other, often more cell-specific, receptors could serve to explain many of the proposed PAF actions. For example, its role as an eosinophil chemoattractant may involve interaction with the relatively eosinophil-selective agent IL-5. Similarly with monocytes in the presence of endothelial P-selectin (Zimmerman *et al.*, 1996). The idea of selective recognition resulting from integrated networks rather than exclusively cell- and ligand-specific receptors is gaining general ground (see, for example, the latest concepts of selectivity in odorant recognition (Mombaerts, 1999)). Similar processes could apply in inflammation where the complex network of mediators may drive specific, directed, and sequential cell responses by the integrated action of numerous, intrinsically generally acting, mediators on a given cell type. A cofactor role for PAF would mean that its antagonism could dampen down many specific responses (as seen in the studies with antagonists) but also that specific essential actions would be very hard to pin down. It would also explain the relative absence of phenotype in the knockout mice, in which compensatory mechanisms within the network would be particularly prone to take over.

Phenotypes of receptor knockouts and receptor overexpression mice

Such mice developed normally and, when adult, showed no obvious physical or physiologic abnormalities (Ishii *et al.*, 1998b). The animals reproduced normally despite a suggested role for PAF in reproduction. Interestingly, the mean arterial blood pressure was unchanged in the PAFR−/− mice which is not easily consistent with a role for PAF in normal vasomotor tone, although confirmation of this would require more detailed investigation. There is a long-standing suggestion that PAF is involved in *anaphylactic shock* and *endotoxic shock*. In keeping with the former, intravascular antigen administration to sensitized PAFR−/− animals did indeed result in markedly diminished shock and also the concomitant bronchial constriction and airways resistance. By contrast, endotoxin effects in knockout mice were indistinguishable from the control PAFR+/+ animals. Since a wide variety of PAF receptor antagonists have been shown to be capable of reducing or abolishing endotoxic shock, this finding was indeed a surprise. Even more so since PAF receptor overexpressing mice showed increased susceptibility to LPS (Ishii *et al.*, 1997). Admittedly, many of the pharmacological studies with PAF in endotoxin shock were performed in rats and there are marked species differences in LPS responses. Additionally, possible compensatory mechanisms are well known to have the ability to confound conclusions drawn from knockout mice. Despite the caveats, at first glance we may have to reassess our concepts of PAF action and/or reconsider the issue of alternative receptors or modes of action for this group of molecules.

Finally, there are many areas of pathophysiology in which PAF has been invoked that have not yet been tested rigorously in PAFR−/− mice, in particular, acute and chronic inflammation, platelet effects, neurophysiological processes, and responses mediated through putative intracellular PAF effects. To date then, the receptor knockout animals show alterations only in IgE-mediated responses – *anaphylaxis* and *bronchoconstriction*.

THERAPEUTIC UTILITY

Effects of inhibitors (antibodies) to receptors

Over 50 compounds have been described with PAF receptor antagonist activity (see Hwang, 1994; Koltai *et al.*, 1994; Negro Alvarez *et al.*, 1997, for example). They range widely in chemical composition and have often been synthesized on a semi-rational basis (Lamotte-Brasseur *et al.*, 1991; Lamouri *et al.*, 1993; Heymans *et al.*, 1997). Some are structurally related to PAF but many are not. Some are of natural origin, including a variety of compounds extracted from the evolutionarily ancient tree, *Gingko biloba*, but most are synthetic. Of importance for antagonism of a phospholipid mediator, some are lipophilic (CV-62091) but others, including the widely used WEB-2086, are hydrophilic and presumably do not enter the lipid bilayer, but rather act on the receptor from the aqueous phase. A representative, but by no means exhaustive, sample of these antagonists is listed in **Table 1**, each with a relatively recent reference as a potential access to its bibliography.

In most cases the antagonists have been shown to act in one or more of the myriad *in vivo* models wherein PAF has been felt to play a part. In fact, it is from these studies of antagonist effect that, for the most part, the complexity and breadth of potential PAF actions have been derived. In many cases the specificity for PAF (and related phospholipids) is quite remarkable and with newer generation compounds, the activity has increased significantly.

Because of the plethora of receptor antagonists, little emphasis has been placed on PAF synthesis inhibitors. Not surprisingly from the $cPLA_2$ knockout data, inhibitors of this enzyme do reduce PAF production in stimulated cells and drugs have been described that block both PAF and leukotriene production at this step (e.g. Farina *et al.*, 1994). However, in the light of new evidence for multiple

Table 1 Some of the compounds described with PAF receptor antagonist activity

Source	Name/number	Reference
Natural compounds		
	Ginkolides	Heller *et al.*, 1998; Lamouri *et al.*, 1993; Simon *et al.*, 1987
	Kadsurenone	Ko *et al.*, 1992; Shen *et al.*, 1985
	Gliotoxins	Okamoto *et al.*, 1986a, 1986b; Yoshida *et al.*, 1988; Tanaka *et al.*, 1995
	FR-900452	
	FR-49175	
Synthetic compounds		
Structurally related to PAF	CV-3988	Adachi *et al.*, 1997; Dupuis *et al.*, 1998; Khoury and Langleben, 1996; Okano *et al.*, 1995
	SR-27417	Bernat *et al.*, 1992; Herbert, 1992; Kravchenko *et al.*, 1995
	ONO-6240	Toyofuku *et al.*, 1986; Terashita *et al.*, 1987; Ando *et al.*, 1990
	RO-19-3704	Wallace *et al.*, 1987; Lagente *et al.*, 1988; Mounier *et al.*, 1993
	E-5880	Nakatani *et al.*, 1996; Ono *et al.*, 1996; Takada *et al.*, 1998
Structurally unrelated to PAF	RP-48740	Ferreira *et al.*, 1991; Lefort *et al.*, 1988; Weissman *et al.*, 1993
	RP-59227	Zhang and Decker, 1994; Auchampach *et al.*, 1998
	WEB-2086	Liu *et al.*, 1998; Latorre *et al.*, 1999; Mori *et al.*, 1999; Okayama *et al.*, 1999; Rainsford, 1999
	WEB-2170	
	BN-50730	Singh *et al.*, 1997; Bazan, 1998
	TCV-309	Kawamura *et al.*, 1994; Ide *et al.*, 1995; Adachi *et al.*, 1997; Izuoka *et al.*, 1998;
	YM-264	Arima *et al.*, 1995; Nagaoka *et al.*, 1997
	A-85783, ABT-299	Albert *et al.*, 1996; Kruse-Elliott *et al.*, 1996; Travers *et al.*, 1998

Adapted from Heller *et al.* (1998).

pathways (and even nonenzymatic mechanisms) for generation of PAF-like molecules, disinterest in this approach seems even more reasonable.

Despite all the receptor inhibitors, there is to date no example of an effective therapeutic role for any of these drugs. A few clinical trials have been reported in, for example, *asthma* (Hozawa *et al.*, 1995; Evans *et al.*, 1997), *arthritis* (Hilliquin *et al.*, 1995, 1998), *sepsis* (Dhainaut *et al.*, 1994, 1998; Froon *et al.*, 1996), and *psoriasis* (Elbers *et al.*, 1994), with equivocal results. This has led to some disillusionment on the part of the pharmaceutical industry. Part of the problem may be the misconception that any one mediator, PAF in particular, plays such a key role that its blockade will, in and of itself, prevent a given disease process. For reasons of historical accident, the early history of the PAF field may have given a disproportionate emphasis to its role as mediator of *allergic reactions* and *asthma* (resulting in many complex schemes with PAF and/or leukotrienes as major participants in these processes). Had the molecule been developed out of its possible role in modulation of the vascular system, or in the category of ether lipid antitumor agents a different emphasis might have ensued. More realistically, PAF is an accessory molecule, playing key but not necessarily sole, roles in a variety of complex mediator networks from inflammation to parturition, neuronal function to vascular motor tone. On the other hand, there is also a perception among some investigators that the reason for the problems with early clinical trials is not only the possible choice of disease processes to examine but, more importantly, the relatively low efficacy of the first-generation antagonists. Some of the later compounds are reputedly in early trials and may show promise in the future.

References

Adachi, T., Aoki, J., Manya, H., Asou, H., Arai, H., and Inoue, K. (1997). PAF analogues capable of inhibiting PAF acetylhydrolase activity suppress migration of isolated rat cerebellar granule cells. *Neurosci Lett.* **235**, 133–136.

Albert, D. H., Conway, R. G., Magoc, T. J., Tapang, P., Rhein, D. A., Luo, G., Holms, J. H., Davidsen, S. K., Summers, J. B., and Carter, G. W. (1996). Properties of ABT-299, a prodrug of A-85783, a highly potent platelet activating factor receptor antagonist. *J. Pharmacol. Exp. Ther.* **277**, 1595–1606.

Ali, H., Richardson, R. M., Tomhave, E. D., DuBose, R. A., Haribabu, B., and Snyderman, R. (1994). Regulation of stably transfected platelet activating factor receptor in RBL-2H3 cells. Role of multiple G proteins and receptor phosphorylation. *J. Biol. Chem.* **269**, 24557–24563.

Ali, H., Fisher, I., Haribabu, B., Richardson, R. M., and Snyderman, R. (1997). Role of phospholipase Cbeta3 phosphorylation in the desensitization of cellular responses to platelet-activating factor. *J. Biol. Chem.* **272**, 11706–11709.

Ali, H., Sozzani, S., Fisher, I., Barr, A. J., Richardson, R. M., Haribabu, B., and Snyderman R. (1998). Differential regulation of formyl peptide and platelet-activating factor receptors. Role of phospholipase Cbeta3 phosphorylation by protein kinase A. *J. Biol. Chem.* **273**, 11012–11016.

Ando, M., Suginami, H., and Matsuura S. (1990). Pregnancy suppression by a platelet activating factor antagonist, ONO-6240, in mice. *Asia Oceania J. Obstet. Gynaecol.* **16**, 169–174.

Arima, M., Yukawa, T., and Makino S. (1995). Effect of YM264 on the airway hyperresponsiveness and the late asthmatic response in a guinea pig model of asthma. *Chest* **108**, 529–534.

Auchampach, J. A., Pieper, G. M., Cavero, I., and Gross G. J. (1998). Effect of the platelet-activating factor antagonist RP 59227 (Tulopafant) on myocardial ischemia/reperfusion injury and neutrophil function. *Basic Res. Cardiol.* **93**, 361–371.

Barzaghi, G., Sarau, H. M., and Mong, S. (1989). Platelet-activating factor-induced phosphoinositide metabolism in differentiated U-937 cells in culture. *J. Pharmacol. Exp. Ther.* **248**, 559–566.

Bass, D. A., McPhail, L. C., Schmitt, J. D., Morris-Natschke, S., McCall, C. E., and Wykle, R. L. (1988). Selective priming of rate and duration of the respiratory burst of neutrophils by 1, 2-diacyl and 1-O-alkyl-2-acyl diglycerides. Possible relation to effects on protein kinase C. *J. Biol. Chem.* **263**, 19610–19617.

Bauldry, S. A., Wykle, R. L., and Bass, D. A. (1991). Differential actions of diacyl- and alkylacylglycerols in priming phospholipase A2, 5-lipoxygenase and acetyltransferase activation in human neutrophils. *Biochim. Biophys. Acta* **1084**, 178–184.

Bazan, N. G. (1998). The neuromessenger platelet-activating factor in plasticity and neurodegeneration. *Progr. Brain Res.* **118**, 281–291.

Bazan, N. G., Packard, M. G., Teather, L., and Allan, G. (1997). Bioactive lipids in excitatory neurotransmission and neuronal plasticity. *Neurochem. Int.* **30**, 225–231.

Bennett, S. A., and Birnboim, H. C. (1997). Receptor-mediated and protein kinase-dependent growth enhancement of primary human fibroblasts by platelet activating factor [see comments]. *Mol. Carcinog.* **20**, 366–375.

Bernat, A., Herbert, J. M., Salel, V., Lespy, L., and Maffrand, J. P. (1992). Protective effect of SR 27417, a novel PAF antagonist, on PAF- or endotoxin-induced hypotension in the rat and the guinea-pig. *J. Lipid Mediat.* **5**, 41–48.

Bito, H., Honda, Z., Nakamura, M., and Shimizu, T. (1994). Cloning, expression and tissue distribution of rat platelet-activating-factor-receptor cDNA. *Eur. J. Biochem.* **221**, 211–218.

Boulay, F., Naik, N., Giannini, E., Tardif, M., Brouchon., and Brouchon, L. (1997). Phagocyte chemoattractant receptors. *Ann. N.Y. Acad. Sci.* **832**, 69–84.

Bratton, D., Dreyer, E., Kailey, J., Fadok, V., Clay, K., and Henson, P. (1992). The mechanism of internalization of PAF in activated human neutrophils: enhanced transbilayer movement across the plasma membrane. *J. Immunol.* **148**, 514–523.

Brunner, T., de Weck, A. L., and Dahinden C. A. (1991). Platelet-activating factor induces mediator release by human basophils primed with IL-3, granulocyte-macrophage colony-stimulating factor, or IL-5. *J. Immunol.* **147**, 237–242.

Casale, T. B., Erger, R. A., and Little, M. M. (1993). Platelet-activating factor-induced human eosinophil transendothelial migration: evidence for a dynamic role of the endothelium. *Am. J. Respir. Cell Mol Biol.* **8**, 77–82.

Chaqour, B., Howard, P. S., Richards, C. F., and Macarak, E. J. (1999). Mechanical stretch induces platelet-activating factor receptor gene expression through the NF-kappaB transcription factor. *J. Mol. Cell. Cardiol.* **31**, 1345–1355.

Chase, P. B., Halonen, M., and Regan, J. W. (1993). Cloning of a human platelet-activating factor receptor gene: evidence for an intron in the 5′-untranslated region. *Am. J. Respir. Cell. Mol. Biol.* **8**, 240–244.

Clay, K., Johnson, C., and Henson, P. (1990). Binding of platelet activating factor to albumin. *Biochim. Biophys. Acta* **1046**, 309–314.

Cundell, D. R., Gerard, N. P., Gerard, C., Cidanpaan-Heikkila, I., and Tuomanen, E. I. (1995). Streptococcus pneumoniae anchor to activated human cells by the receptor for platelet-activating factor. *Nature* **377**, 435–438.

Dhainaut, J. F., Tenaillon, A., Le Tulzo, Y., Schlemmer, B., Solet, J. P., Wolff, M., Holzapfel, L., Zeni, F., Dreyfuss, D., Mira, J. P., de Vathaire, F., and Guinot, P. (1994). Platelet-activating factor receptor antagonist BN 52021 in the treatment of severe sepsis: a randomized, double-blind, placebo- controlled, multicenter clinical trial. BN 52021 Sepsis Study Group. *Crit. Care Med.* **22**, 1720–1728.

Dhainaut, J. F., Tenaillon, A., Hemmer, M., Damas, P., Le Tulzo, Y., Radermacher, P., Schaller, M. D., Sollet, J. P., Wolff, M., Holzapfel, L., Zeni, F., Vedrinne, J. M., de Vathaire, F., Gourlay, M. L., Guinot, P., and Mira, J. P. (1998). Confirmatory platelet-activating factor receptor antagonist trial in patients with severe gram-negative bacterial sepsis: a phase III, randomized, double-blind, placebo-controlled, multicenter trial. BN 52021 Sepsis Investigator Group [see comments]. *Crit. Care Med.* **26**, 1963–1971.

Domingo, M. T., Spinnewyn, B., Chabrier, P. E., and Braquet, P. (1994). Changes in [3H]PAF binding and PAF concentrations in gerbil brain after bilateral common carotid artery occlusion: a quantitative autoradiographic study. *Brain Res.* **640**, 268–276.

Dupuis, F., Gachard, N., Allegraud, A., Dulery, C., Praloran, V., and Denizot Y. (1998). Effect of platelet-activating factor on the growth of human erythroid and myeloid CD34+ progenitors. *Mediators Inflamm.* **7**, 99–103.

Elbers, M. E., Gerritsen, M. J., and van de Kerkhof, P. C. (1994). The effect of topical application of the platelet-activating factor-antagonist, Ro 24–0238, in psoriasis vulgaris – a clinical and immunohistochemical study. *Clin. Exp. Dermatol.* **19**, 453–457.

Evans, D. J., Barnes, P. J., Cluzel, M., and O'Connor, B. J. (1997). Effects of a potent platelet-activating factor antagonist, SR27417A, on allergen-induced asthmatic responses. *Am. J. Respir. Crit. Care Med.* **156**, 11–16.

Farina, P. R., Graham, A. G., Hoffman, A. F., Watrous, J. M., Borgeat, P., Nadeau, M., Hansen, G., Dinallo, R. M.,

Adams, J., Miao, C. K., Lazer, E. S., Parks, T. P., and Homon, C. A. (1994). BIRM 270: a novel inhibitor of arachidonate release that blocks leukotriene B4 and platelet-activating factor biosynthesis in human neutrophils. *J. Pharmacol. Exp. Ther.* **271**, 1418–1426.

Ferreira, M. G., Braquet, P., and Fonteles, M. C. (1991). Effects of PAF antagonists on renal vascular escape and tachyphylaxis in perfused rabbit kidney. *Lipids* **26**, 1329–1332.

Froon, A. M., Greve, J. W., Buurman, W. A., van der Linden, C. J., Langemeijer, H. J., Ulrich, C., and Bourgeois. M. (1996). Treatment with the platelet-activating factor antagonist TCV-309 in patients with severe systemic inflammatory response syndrome: a prospective, multi-center, double-blind, randomized phase II trial. *Shock* **5**, 313–319.

Garcia Rodriguez, C., Cundell, D. R., Tuomanen, E. I., Kolakowski Jr, L. F., Gerard, C., and Gerard, N. P. (1995a). The role of *N*-glycosylation for functional expression of the human platelet-activating factor receptor. Glycosylation is required for efficient membrane trafficking. *J. Biol. Chem.* **270**, 25178–25184.

Garcia Rodriguez, C., Cundell, D. R., Tuomanen, E. I., Kolakowski Jr, L. F., Gerard, C., and Gerard N. P. (1995b). The role of *N*-glycosylation for functional expression of the human platelet-activating factor receptor. Glycosylation is required for efficient membrane trafficking. *J. Biol. Chem.* **270**, 25178–25184.

Gay, J. C. (1993). Mechanism and regulation of neutrophil priming by platelet-activating factor. *J. Cell Physiol.* **156**, 189–197.

Gerard, N. P., and Gerard, C. (1994). Receptor-dependent internalization of platelet-activating factor. *J. Immunol.* **152**, 793–800.

Guthrie, L., McPhail, L., Henson, P., and Johnston, Jr. R. (1984). Priming of neutrophils for enhanced release of oxygen metabolites by bacterial lipopolysaccharide. Evidence for increased activity of the superoxide-producing enzyme. *J. Exp. Med.* **160**, 1656–1671.

Heller, A., Koch, T., Schmeck, J., and van Ackern, K. (1998). Lipid mediators in inflammatory disorders. *Drugs* **55**, 487–496.

Henson, P., and Landes, R. (1976). Activation of platelets by platelet activating factor (PAF) derived from IgE-sensitized basophils. IV. PAF does not activate platelet factor 3 (PF3). *Br. J. Haematol.* **34**, 269–282.

Henson, P., and Pinckard R. (1977). Basophil-derived platelet-activating factor (PAF) as an *in vivo* mediator of acute allergic reactions: demonstration of specific desensitization of platelets to PAF during IgE-induced anaphylaxis in the rabbit. *J. Immunol.* **119**, 2179–2184.

Herbert, J. M. (1992). Characterization of specific binding sites of 3H-labelled platelet- activating factor ([3H]PAF) and a new antagonist, [3H]SR 27417, on guinea-pig tracheal epithelial cells. *Biochem. J.* **284**, 201–206.

Heymans, F., Dive, G., Lamouri, A., Bellahsene, T., Touboul, E., Huet, J., Tavet, F., Redeuilh, C., and Godfroid J. J. (1997). Design and modeling of new platelet-activating factor antagonists. 3. Relative importance of hydrophobicity and electronic distribution in piperazinic series. *J. Lipid Mediat. Cell Signal.* **15**, 161–173.

Hilliquin, P., Guinot, P., Chermat-Izard, V., Puechal, X., and Menkes, C. J. (1995). Treatment of rheumatoid arthritis with platelet activating factor antagonist BN 50730. *J. Rheumatol.* **22**, 1651–1654.

Hilliquin, P., Chermat-Izard, V., and Menkes, C. J. (1998). A double blind, placebo controlled study of a platelet activating factor antagonist in patients with rheumatoid arthritis. *J. Rheumatol.* **25**, 1502–1507.

Hirabayashi, T., Kume, K., and Shimizu, T. (1998). Conditional expression of the dual-specificity phosphatase PYST1/MKP-3 inhibits phosphorylation of cytosolic phospholipase A2 in Chinese hamster ovary cells. *Biochem. Biophys. Res. Commun.* **253**, 485–488.

Honda, Z., Nakamura, M., Miki, I., Minami, M., Watanabe, T., Seyama, Y., Okado, H., Toh, H., Ito, K., Miyamoto, T., and Shimizu, T. (1991). Cloning by functional expression of platelet-activating factor receptor from guinea-pig lung. *Nature* **349**, 342–346.

Honda, Z., Takano, T., Gotoh, Y., Nishida, E., Ito, K., and Shimizu, T. (1994). Transfected platelet-activating factor receptor activates mitogen- activated protein (MAP) kinase and MAP kinase in Chinese hamster ovary cells. *J. Biol. Chem.* **269**, 2307–2315.

Hozawa, S., Haruta, Y., Ishioka, S., and Yamakido, M. (1995). Effects of a PAF antagonist, Y-24180, on bronchial hyperresponsiveness in patients with asthma. *Am. J. Respir. Crit. Care Med.* **152**, 1198–1202.

Huang, Y. H., Schafer-Elinder, L., Wu, R., Claesson, H. E., Frostegard., and J. (1999). Lysophosphatidylcholine (LPC) induces proinflammatory cytokines by a platelet-activating factor (PAF) receptor-dependent mechanism. *Clin. Exp. Immunol.* **116**, 326–331.

Hwang, S. B. (1988). Identification of a second putative receptor of platelet-activating factor from human polymorphonuclear leukocytes. *J. Biol. Chem.* **263**, 3225–3233.

Hwang, S. B. (1991). High affinity receptor binding of platelet-activating factor in rat peritoneal polymorphonuclear leukocytes. *Eur. J. Pharmacol.* **196**, 169–175.

Hwang, S.-B. (1994). In "Lipid Mediators" (ed F. M. Cunningham), Platelet activating factor: receptors and receptor antagonists, pp. 181–192. Academic Press, London.

Ide, S., Kawahara, K., Takahashi, T., Sasaki, N., Shingu, H., Nagayasu, T., Yamamoto, S., Tagawa, T., and Tomita, M. (1995). Donor administration of PAF antagonist (TCV-309) enhances lung preservation. *Transplant. Proc.* **27**, 570–573.

Ihida, K., Predescu, D., Czekay, R. P., and Palade, G. E. (1999). Platelet activating factor receptor (PAF-R) is found in a large endosomal compartment in human umbilical vein endothelial cells. *J. Cell Sci.* **112**, 285–295.

Ishii, S., Matsuda, Y., Nakamura, M., Waga, I., Kume, K., Izumi, T., and Shimizu, T. (1996). A murine platelet-activating factor receptor gene: cloning, chromosomal localization and up-regulation of expression by lipopolysaccharide in peritoneal resident macrophages. *Biochem. J.* **314**, 671–678.

Ishii, S., Nagase, T., Tashiro, F., Ikuta, K., Sato, S., Waga, I., Kume, K., Miyazaki, J., and Shimizu, T. (1997). Bronchial hyperreactivity, increased endotoxin lethality and melanocytic tumorigenesis in transgenic mice overexpressing platelet-activating factor receptor. *EMBO J.* **16**, 133–142.

Ishii, I., Saito, E., Izumi, T., Ui, M., and Shimizu, T. (1998a). Agonist-induced sequestration, recycling, and resensitization of platelet-activating factor receptor. Role of cytoplasmic tail phosphorylation in each process. *J. Biol. Chem.* **273**, 9878–9885.

Ishii, S., Kuwaki, T., Nagase, T., Maki, K., Tashiro, F., Sunaga, S., Cao, W. H., Kume, K., Fukuchi, Y., Ikuta, K., Miyazaki, J., Kumada, M., and Shimizu T. (1998b). Impaired anaphylactic responses with intact sensitivity to endotoxin in mice lacking a platelet-activating factor receptor. *J. Exp. Med.* **187**, 1779–1788.

Izuoka, T., Takayama, Y., Sugiura, T., Taniguchi, H., Tamura, T., Kitashiro, S., Jikuhara, T., and Iwasaka T. (1998). Role of platelet-activating factor on extravascular lung water after coronary reperfusion in dogs. *Jpn J. Physiol.* **48**, 157–161.

Kadiri, C., Cherqui, G., Masliah, J., Rybkine, T., Etienne, J., and Bereziat, G. (1990). Mechanism of *N*-formyl-methionyl-leucyl-phenylalanine- and platelet-activating factor-induced arachidonic acid release in guinea pig alveolar macrophages: involvement of a GTP-binding protein and role of protein kinase A and protein kinase C. *Mol. Pharmacol.* **38**, 418–425.

Kato, K., Clark, G. D., Bazan, N. G., and Zorumski, C. F. (1994). Platelet-activating factor as a potential retrograde messenger in CA1 hippocampal long-term potentiation. *Nature* **367**, 175–179.

Kawamura, M., Kitayoshi, T., Terashita, Z., Fujiwara, S., Takatani, M., and Nishikawa, K. (1994). Effects of TCV-309, a novel PAF antagonist, on circulatory shock and hematological abnormality induced by endotoxin in dogs. *J. Lipid Mediat. Cell Signal.* **9**, 255–265.

Khoury, J., and Langleben, D. (1996). Platelet-activating factor stimulates lung pericyte growth *in vitro*. *Am. J. Physiol.* **270**, L298–304.

Kishimoto, S., Shimadzu, W., Izumi, T., Shimizu, T., Fukuda, T., Makino, S., Sugiura, T., and Waku K. (1996a). Regulation by IL-5 of expression of functional platelet-activating factor receptors on human eosinophils. *J. Immunol.* **157**, 4126–4132.

Kishimoto, S., Shimazu, W., Izumi, T., Shimizu, T., Fukuda, T., Makino, S., Sugiura, T., and Waku K. (1996b). Enhanced expression of platelet-activating factor receptor on human eosinophils by interleukin-3, interleukin-5 and granulocyte-macrophage colony-stimulating factor. *Int. Arch. of Allergy Immunol.* **111**, 63–65.

Kishimoto, S., Shimadzu, W., Izumi, T., Shimizu, T., Sagara, H., Fukuda, T., Makino, S., Sugiura, T., and Waku, K. (1997). Comparison of platelet-activating factor receptor mRNA levels in peripheral blood eosinophils from normal subjects and atopic asthmatic patients. *Int. Arch. Allergy Immunol.* **114**, 60–63.

Kitchen, E., Rossi, A. G., Condliffe, A. M., Haslett, C., and Chilvers, E. R. (1996). Demonstration of reversible priming of human neutrophils using platelet- activating factor. *Blood* **88**, 4330–4337.

Ko, F. N., Wu, T. S., Liou, M. J., Huang, T. F., and Teng, C. M. (1992). PAF antagonism *in vitro* and *in vivo* by aglafoline from *Aglaia elliptifolia* Merr. *Eur. J. Pharmacol.* **218**, 129–135.

Koltai, M., Guinot, P., Hosford, D., and Braquet, P. G. (1994). Platelet-activating factor antagonists: scientific background and possible clinical applications. *Adv. Pharmacol.* **28**, 81–167.

Kravchenko, V. V., Pan, Z., Han, J., Herbert, J. M., Ulevitch, R. J., and Ye, R. D. (1995). Platelet-activating factor induces NF-kappa B activation through a G protein-coupled pathway. *J. Biol. Chem.* **270**, 14928–14934.

Kruse-Elliott, K. T., Albert, D. H., Summers, J. B., Carter, G. W., Zimmerman, J. J., and Grossman, J. E. (1996). Attenuation of endotoxin-induced pathophysiology by a new potent PAF receptor antagonist. *Shock* **5**, 265–273.

Kunz, D., Gerard, N. P., and Gerard, C. (1992). The human leukocyte platelet-activating factor receptor. cDNA cloning, cell surface expression, and construction of a novel epitope-bearing analog. *J. Biol. Chem.* **267**, 9101–9106.

Lagente, V., Desquand, S., Hadvary, P., Cirino, M., Lellouch-Tubiana, A., Lefort, J., and Vargaftig, B. B. (1988). Interference of the Paf antagonist Ro (19-3704 with Paf and antigen- induced bronchoconstriction in the guinea-pig. *Br. J. Pharmacol.* **94**, 27–36.

Lamotte-Brasseur, J., Heymans, F., Dive, G., Lamouri, A., Batt, J. P., Redeuilh, C., Hosford, D., Braquet, P., and Godfroid, J. J. (1991). PAF receptor and 'Cache-oreilles' effect. *Simple PAF antagonists. Lipids* **26**, 1167–1171.

Lamouri, A., Heymans, F., Tavet, F., Dive, G., Batt, J. P., Blavet, N., Braquet, P., and Godfroid, J. J. (1993). Design and modeling of new platelet-activating factor antagonists. 1. Synthesis and biological activity of 1, 4-bis(3′, 4′, 5′- trimethoxybenzoyl)-2-[[(substituted carbonyl and carbamoyl)oxy]methyl]-piperazines. *J. Med. Chem.* **36**, 990–1000.

Latorre, E., Aragones, M. D., Fernandez, I., and Catalan, R. E. (1999). Platelet-activating factor modulates brain sphingomyelin metabolism. *Eur. J. Biochem.* **262**, 308–314.

Le Gouill, C., Parent, J. L., Rola-Pleszczynski, M., and Stankova, J. (1997). Role of the Cys90, Cys95 and Cys173 residues in the structure and function of the human platelet-activating factor receptor. *FEBS Lett.* **402**, 203–208.

Lefort, J., Sedivy, P., Desquand, S., Randon, J., Coeffier, E., Maridonneau-Parini, I., Floch, A., Benveniste, J., and Vargaftig B. B. (1988). Pharmacological profile of 48740 R. P., a PAF-acether antagonist. *Eur. J. Pharmacol.* **150**, 257–268.

LeVan, T. D., Dow, S. B., Chase, P. B., Bloom, J. W., Regan, J. W., Cunningham, E., and Halonen, M. (1997). Evidence for platelet-activating factor receptor subtypes on human polymorphonuclear leukocyte membranes. *Biochem. Pharmacol.* **54**, 1007–1012.

Liu, L., Zuurbier, A. E., Mul, F. P., Verhoeven, A. J., Lutter, R., Knol, E. F., and Roos D. (1998). Triple role of platelet-activating factor in eosinophil migration across monolayers of lung epithelial cells: eosinophil chemoattractant and priming agent and epithelial cell activator. *J. Immunol.* **161**, 3064–3070.

M'Rabet, L., Coffer, P. J., Wolthuis, R. M., Zwartkruis, F., Koenderman, L., and Bos, J. L. (1999). Differential fMet-Leu-Phe- and platelet-activating factor-induced signaling toward ral activation in primary human neutrophils. *J. Biol. Chem.* **274**, 21847–21852.

Mombaerts, P. (1999). Molecular biology of odorant receptors in vertebrates. *Annu. Rev. Neurosci.* **22**, 487–509.

Mori, N., Horie, Y., Gerritsen, M. E., and Granger, D. N. (1999). Ischemia-reperfusion induced microvascular responses in LDL-receptor−/− mice. *Am. J. Physiol.* **276**, H1647–1654.

Mounier, C., Hatmi, M., Faili, A., Bon, C., and Vargaftig, B. B. (1993). Competitive inhibition of phospholipase A2 activity by the platelet- activating factor antagonist Ro 19-3704 and evidence for a novel suppressive effect on platelet activation. *J. Pharmacol. Exp. Ther.* **264**, 1460–1467.

Mutoh, H., Bito, H., Minami, M., Nakamura, M., Honda, Z., Izumi, T., Nakata, R., Kurachi, Y., Terano, A., and Shimizu, T. (1993). Two different promoters direct expression of two distinct forms of mRNAs of human platelet-activating factor receptor. *FEBS Lett.* **322**, 129–134.

Mutoh, H., Ishii, S., Izumi, T., Kato, S., and Shimizu, T. (1994). Platelet-activating factor (PAF) positively auto-regulates the expression of human PAF receptor transcript 1 (leukocyte-type) through NF-kappa B. *Biochem. Biophys. Res. Commun.* **205**, 1137–1142.

Nagaoka, H., Hara, H., Suzuki, T., Takahashi, T., Takeuchi, M., Matsuhisa, A., Saito, M., Yamada, T., Tomioka, K., and Mase, T. (1997). 2-(3-Pyridyl)thiazolidine-4-carboxamides. 1. Novel orally active antagonists of platelet-activating factor (PAF). *Chem. Pharm. Bull. (Tokyo)* **45**, 1659–1664.

Nakamura, M., Honda, Z., Izumi, T., Sakanaka, C., Mutoh, H., Minami, M., Bito, H., Seyama, Y., Matsumoto, T., Noma, M., and Simizu, T. (1991). Molecular cloning and expression of platelet-activating factor receptor from human leukocytes. *J. Biol. Chem.* **266**, 20400–20405.

Nakao, A., Watanabe, T., Bitoh, H., Imaki, H., Suzuki, T., Asano, K., Taniguchi, S., Nosaka, K., Shimizu, T., and Kurokawa, K. (1997). cAMP mediates homologous downregulation of PAF receptor mRNA expression in mesangial cells. *Am. J. Physiol.* **273**, F445–450.

Nakatani, T., Sakamoto, Y., Ando, H., and Kobayashi, K. (1996). Effects of platelet-activating factor antagonist E5880 on intrahepatic and systemic metabolic responses to transient hepatic inflow occlusion and reperfusion in the rabbit. *World J. Surg.* **20**, 1060–1067; discussion 1067–1068.

Negro Alvarez, J. M., Miralles Lopez, J. C., Ortiz Martinez, J. L., Abellan Aleman, A., and Rubio del Barrio, R. (1997). Platelet-activating factor antagonists. *Allergol. Immunopathol.* **25**, 249–258.

Nick, J. A., Avdi, N. J., Young, S. K., Knall, C., Gerwins, P., Johnson, G. L., and Worthen, G. S. (1997). Common and distinct intracellular signaling pathways in human neutrophils utilized by platelet activating factor and FMLP. *J. Clin. Invest.* **99**, 975–986.

Okahara, K., Sun, B., Kawasaki, T., Monden, M., and Kambayashi, J. (1998). Expression of platelet-activating factor receptor transcript-2 is induced by shear stress in HUVEC. *Prostaglandins Other Lipid Mediat.* **55**, 323–329.

Okamoto, M., Yoshida, K., Nishikawa, M., Ando, T., Iwami, M., Kohsaka, M., and Aoki, H. (1986a). FR-900452, a specific antagonist of platelet activating factor (PAF) produced by *Streptomyces phaeofaciens*. I. Taxonomy, fermentation, isolation, and physico-chemical and biological characteristics. *J. Antibiot. (Tokyo)* **39**, 198–204.

Okamoto, M., Yoshida, K., Uchida, I., Kohsaka, M., and Aoki, H. (1986b). Studies of platelet activating factor (PAF) antagonists from microbial products. II. Pharmacological studies of FR-49175 in animal models. *Chem. Pharm. Bull. (Tokyo)* **34**, 345–348.

Okano, S., Tagawa, M., and Urakawa, N. (1995). Effect of platelet activating factor antagonist (CV-3988) on 6-keto- PGF1 alpha and thromboxane B2 in dogs with experimental endotoxin- induced shock. *J. Vet. Med. Sci.* **57**, 81–85.

Okayama, N., Coe, L., Oshima, T., Itoh, M., and Alexander, J. S. (1999). Intracellular mechanisms of hydrogen peroxide-mediated neutrophil adherence to cultured human endothelial cells. *Microvasc. Res.* **57**, 63–74.

Ono, S., Mochizuki, H., and Tamakuma, S. (1996). A clinical study on the significance of platelet-activating factor in the pathophysiology of septic disseminated intravascular coagulation in surgery. *Am. J. Surg.* **171**, 409–415.

Pang, J. H., Hung, R. Y., Wu, C. J., Fang, Y. Y., and Chau, L. Y. (1995). Functional characterization of the promoter region of the platelet-activating factor receptor gene. Identification of an initiator element essential for gene expression in myeloid cells. *J. Biol. Chem.* **270**, 14123–14129.

Parent, J. L., Le Gouill, C., de Brum-Fernandes, A. J., Rola-Pleszczynski, M., and Stankova, J. (1996). Mutations of two adjacent amino acids generate inactive and constitutively active forms of the human platelet-activating factor receptor. *J. Biol. Chem.* **271**, 7949–7955.

Partrick, D. A., Moore, E. E., Moore, F. A., Barnett, C. C., and Silliman C. C. (1997). Lipid mediators up-regulate CD11b and prime for concordant superoxide and elastase release in human neutrophils. *J. Trauma* **43**, 297–302; discussion 302–293.

Patel, K. D., Zimmerman, G. A., Prescott, S. M., and McIntyre, T. M. (1992). Novel leukocyte agonists are released by endothelial cells exposed to peroxide. *J. Biol. Chem.* **267**, 15168–15175.

Predescu, D., Ihida, K., Predescu, S., and Palade, G. E. (1996). The vascular distribution of the platelet-activating factor receptor. *Eur. J. Cell Biol.* **69**, 86–98.

Rainsford, K. D. (1999). Inhibition by leukotriene inhibitors, and calcium and platelet-activating factor antagonists, of acute gastric and intestinal damage in arthritic rats and in cholinomimetic-treated mice. *J. Pharm. Pharmacol.* **51**, 331–339.

Reinhardt, J. C., Cui, X., and Roudebush, W. E. (1999). Immunofluorescent evidence of the platelet-activating factor receptor on human spermatozoa. *Fertil. Steril.* **71**, 941–942.

Rougier, F., Dupuis, F., Cornu, E., Dulery, C., Praloran, V., and Denizot, Y. (1997). Platelet-activating factor and antagonists modulate DNA synthesis in human bone marrow stromal cell cultures. *J. Lipid Mediat. Cell Signal.* **16**, 147–153.

Sasaki, T., Tohyama, T., Oda, K., Toyama, H., Ishii, S., Senda, M., Karasawa, K., Satoh, N., Setaka, M., Nojima, S., Nozaki, T., and Braquet, P. (1996). A distribution study of 11C platelet-activating factor (PAF) analogs in normal and tumor-bearing mice. *Nucl. Med. Biol.* **23**, 309–314.

Shalit, M., Dabiri, G. A., and Southwick, F. S. (1987). Platelet-activating factor both stimulates and 'primes' human polymorphonuclear leukocyte actin filament assembly. *Blood* **70**, 1921–1927.

Shen, T. Y., Hwang, S. B., Chang, M. N., Doebber, T. W., Lam, M. H., Wu, M. S., Wang, X., Han, G. Q., and Li, R. Z. (1985). Characterization of a platelet-activating factor receptor antagonist isolated from haifenteng (Piper futokadsura): specific inhibition of *in vitro* and *in vivo* platelet-activating factor-induced effects. *Proc. Natl Acad. Sci. USA* **82**, 672–676.

Shimada, A., Ota, Y., Sugiyama, Y., Sato, S., Kume, K., Shimizu, T., and Inoue, S. (1998a). *In situ* expression of platelet-activating factor (PAF)-receptor gene in rat skin and effects of PAF on proliferation and differentiation of cultured human keratinocytes. *J. Invest. Dermatol.* **110**, 889–893.

Shimada, A., Ota, Y., Sugiyama, Y., Sato, S., Kume, K., Shimizu, T., and Inoue, S. (1998b). *In situ* expression of platelet-activating factor (PAF)-receptor gene in rat skin and effects of PAF on proliferation and differentiation of cultured human keratinocytes. *J. Invest. Dermatol.* **110**, 889–893.

Shimizu, T., and Mutoh, H. (1997). Structure and regulation of platelet activating factor receptor gene. *Adv. Exp. Med. Biol.* **407**, 197–204.

Shindo, K., Koide, K., Hirai, Y., Sumitomo, M., and Fukumura, M. (1996). Priming effect of platelet activating factor on leukotriene C4 from stimulated eosinophils of asthmatic patients. *Thorax* **51**, 155–158.

Shindo, K., Koide, K., and Fukumura, M. (1997). Enhancement of leukotriene B4 release in stimulated asthmatic neutrophils by platelet activating factor. *Thorax* **52**, 1024–1029.

Simon, M. F., Chap, H., Braquet, P., and Douste-Blazy, L. (1987). Effect of BN 52021, a specific antagonist of platelet activating factor (PAF-acether), on calcium movements and phosphatidic acid production induced by PAF-acether in human platelets. *Thromb. Res.* **45**, 299–309.

Singh, N., Sharma, A., and Singh, M. (1997). Effects of BN-50730 (PAF receptor antagonist) and physostigmine (AChE inhibitor) on learning and memory in mice. *Meth. Find Exp. Clin. Pharmacol.* **19**, 585–588.

Stengel, D., Antonucci, M., Arborati, M., Hourton, D., Griglio, S., Chapman, M. J., and Ninio, E. (1997). Expression of the PAF receptor in human monocyte-derived macrophages is downregulated by oxidized LDL: relevance to the inflammatory phase of atherogenesis. *Arterioscler. Thromb. Vasc. Biol.* **17**, 954–962.

Stewart, A. G. (1994). In "Lipid Mediators" (ed F. M. Cunningham), Biological properties of platelet activating factor, pp. 181–192. Academic Press, London.

Stojanov, T., and O'Neill, C. (1999). Ontogeny of expression of a receptor for platelet-activating factor in mouse preimplantation embryos and the effects of fertilization and culture *in vitro* on its expression. *Biol. Reprod.* **60**, 674–682.

Takada, Y., Taniguchi, H., Fukunaga, K., Yuzawa, K., Otsuka, M., Todoroki, T., Iijima, T., and Fukao, K. (1998). Prolonged hepatic warm ischemia in non-heart-beating donors: protective effects of FK506 and a platelet activating factor antagonist in porcine liver transplantation. *Surgery* **123**, 692–698.

Takano, T., Honda, Z., Sakanaka, C., Izumi, T., Kameyama, K., Haga, K., Haga, T., Kurokawa, K., and Shimizu, T. (1994). Role of cytoplasmic tail phosphorylation sites of platelet-activating factor receptor in agonist-induced desensitization. *J. Biol. Chem.* **269**, 22453–22458.

Tanaka, T., Tokumura, A., and Tsukatani, H. (1995). Platelet-activating factor (PAF)-like phospholipids formed during peroxidation of phosphatidylcholines from different foodstuffs. *Biosci. Biotechnol. Biochem.* **59**, 1389–1393.

Terashita, Z., Imura, Y., Takatani, M., Tsushima, S., and Nishikawa, K. (1987). CV-6209, a highly potent antagonist of platelet activating factor *in vitro* and *in vivo*. *J. Pharmacol. Exp. Ther.* **242**, 263–268.

Toyofuku, T., Kubo, K., Kobayashi, T., and Kusama, S. (1986). Effects of ONO-6240, a platelet-activating factor antagonist, on endotoxin shock in unanesthetized sheep. *Prostaglandins* **31**, 271–281.

Travers, J., Pei, Y., Morin, S. M., and Hood, A. F. (1998). Antiinflammatory activity of the platelet-activating factor receptor antagonist A-85783. *Arch. Dermatol. Res.* **290**, 569–573.

Vercellotti, G. M., Yin, H. Q., Gustafson, K. S., Nelson, R. D., and Jacob, H. S. (1988). Platelet-activating factor primes neutrophil responses to agonists: role in promoting neutrophil-mediated endothelial damage. *Blood* **71**, 1100–1107.

Voelkel, N., Chang, S., Pfeffer, K., Worthen, G., McMurtry, I., and Henson, P. (1986). PAF antagonists: different effects on platelets, neutrophils, guinea pig ileum and PAF-induced vasodilation in isolated rat lung. *Prostaglandins* **32**, 359–372.

Wallace, J. L., Steel, G., Whittle, B. J., Lagente, V., and Vargaftig, B. (1987). Evidence for platelet-activating factor as a mediator of endotoxin- induced gastrointestinal damage in the rat. Effects of three platelet-activating factor antagonists. *Gastroenterology* **93**, 765–773.

Wang, H., Tan, X., Chang, H., Gonzalez-Crussi, F., Remick, D. G., and Hsueh, W. (1997). Regulation of platelet-activating factor receptor gene expression *in vivo* by endotoxin, platelet-activating factor and endogenous tumour necrosis factor. *Biochem. J.* **322**, 603–608.

Wardlaw, A. J., Moqbel, R., Cromwell, O., and Kay, A. B. (1986). Platelet-activating factor. A potent chemotactic and chemokinetic factor for human eosinophils. *J. Clin. Invest.* **78**, 1701–1706.

Warringa, R. A., Mengelers, H. J., Kuijper, P. H., Raaijmakers, J. A., Bruijnzeel, P. L., and Koenderman, L. (1992). *In vivo* priming of platelet-activating factor-induced eosinophil chemotaxis in allergic asthmatic individuals. *Blood* **79**, 1836–1841.

Weissman, D., Poli, G., Bousseau, A., and Fauci, A. S. (1993). A platelet-activating factor antagonist, RP 55778, inhibits cytokine-dependent induction of human immunodeficiency virus expression in chronically infected promonocytic cells. *Proc. Natl Acad. Sci. USA* **90**, 2537–2541.

Yang, H. H., Pang, J. H., Hung, R. Y., and Chau, L. Y. (1997). Transcriptional regulation of platelet-activating factor receptor gene in B lymphoblastoid Ramos cells by TGF-beta. *J. Immunol.* **158**, 2771–2778.

Ye, R. D., Prossnitz, E. R., Zou, A. H., and Cochrane, C. G. (1991). Characterization of a human cDNA that encodes a functional receptor for platelet activating factor. *Biochem. Biophys. Res. Commun.* **180**, 105–111.

Yoshida, K., Okamoto, M., Shimazaki, N., and Hemmi, K. (1988). PAF inhibitors of microbial origin. Studies on diketopiperazine derivatives. *Progr. Biochem. Pharmacol.* **22**, 66–80.

Yue, T. L., Stadel, J. M., Sarau, H. M., Friedman, E., Gu, J. L., Powers, D. A., Gleason, M. M., Feuerstein, G., and Wang, H. Y. (1992). Platelet-activating factor stimulates phosphoinositide turnover in neurohybrid NCB-20 cells: involvement of pertussis toxin-sensitive guanine nucleotide-binding proteins and inhibition by protein kinase C. *Mol. Pharmacol.* **41**, 281–289.

Zhang, F., and Decker, K. (1994). Platelet-activating factor antagonists suppress the generation of tumor necrosis factor-alpha and superoxide induced by lipopolysaccharide or phorbol ester in rat liver macrophages. *Eur. Cytokine Netw.* **5**, 311–317.

Zimmerman, G. A., Elstad, M. R., Lorant, D. E., McIntyre, T. M., Prescott, S. M., Topham, M. K., Weyrich, A. S., and Whatley, R. E. (1996). Platelet-activating factor (PAF): signalling and adhesion in cell–cell interactions. *Adv. Exp. Med. Biol.* **416**, 297–304.

fMLP Receptors

François Boulay*, Marie-Josèphe Rabiet and Marianne Tardif

Department of Molecular and Structural Biology, DBMS/BBSI UMR 314 CEA-CNRS CEA, 17 Rue des Martyrs, Grenoble, Cedex 9, F 38054, France

*corresponding author tel: (33) 04-76-88-31-38, fax: (33) 04-76-88-51-85, e-mail: fboulay@cea.fr

DOI: 10.1006/rwcy.2000.23004.

SUMMARY

The *N*-formylated peptide receptor (FPR) is a seven-transmembrane G_{i2} protein-coupled receptor expressed in neutrophils, monocytes, macrophages, eosinophils, dendritic cells, hepatocytes, and astrocytes. Upon agonist binding, it promotes a complex array of signaling cascades – calcium mobilization, phosphoinositide hydrolysis, and activation of protein kinases – that ultimately result in chemotaxis, release of proteolytic enzymes, production of superoxide by the NADPH oxidase, and activation of transcription factors. Activation of FPR is thought to play a critical role in host defense against bacterial infection and in inflammation. Following fMLP binding, the FPR is rapidly phosphorylated and desensitized.

The human FPR has 69% and 56% amino acid sequence similarities with two other human receptors designated FPRL1 and FPRL2, respectively. FPRL1 binds fMLP with low affinity and lipoxin A4 with high affinity and is consequently referred to as the lipoxin A4 receptor. The human *FPR*, *FPRL1*, and *FPRL2* genes are clustered on chromosome 19 in the 19q13.3 band.

BACKGROUND

Discovery

In the mid-1970s, leukocyte chemotaxis was found to be triggered by the bacterial *N*-formyl peptide N^{α}-formyl-L-methionyl-L-leucyl-L-phenylalanine (CHO-Met-Leu-Phe-OH, known as fMLP) (Showell *et al.*, 1976; Freer *et al.*, 1980). fMLP binds with high affinity to a specific receptor (FPR) that initiates complex signaling cascades. Initial attempts to characterize a formyl peptide receptor biochemically and pharmacologically made use of affinity chromatography, radioligands, and ligands derivatized with photoactivatable and fluorescent groups (Williams *et al.*, 1977; Niedel *et al.*, 1980; Sklar *et al.*, 1984; for a review see Allen *et al.*, 1988).

In 1990, the amino acid sequence of the human *N*-formyl peptide receptor was elucidated by an expression cloning strategy in COS-7 cells (Boulay *et al.*, 1990a,b). More information on the structure and function of leukocyte chemoattractant receptors can be found in the two reviews by Murphy (1994) and Ye and Boulay (1997).

Two additional human genes designated *FPRL1* and *FPRL2* (L for like) have been isolated by low-stringency crosshybridization with the human FPR cDNA probe (Bao *et al.*, 1992; Murphy *et al.*, 1992; Ye *et al.*, 1992). The gene product of *FPRL1* is also known as FPR2 and FPRH1. FPRL1 binds fMLP with low affinity ($K_d > 400$ nM). When expressed in CHO cells, FPRL1 has been found to bind lipoxin A4 with high affinity ($K_d = 1.7$ nM) and consequently it is referred to as the lipoxin A4 receptor. The gene product of *FPRL2*, also known as FPRH2, is expressed in monocytes. FPRH2 is not activated by fMLP and its ligand is still unknown (Bao *et al.*, 1992; Durstin *et al.*, 1994).

Alternative names

The *N*-formyl peptide receptor, originally known as the fMLP receptor or *N*-formyl peptide receptor (fMLP-R), is now referred to as FPR.

Structure

The protein deduced from the cDNA comprises 350 amino acids. Its hydropathy profile reveals features common to all members of the G protein-coupled receptor superfamily, characterized by seven domains highly enriched in hydrophobic residues. These domains are thought to form a bundle of transmembrane α helices joined by hydrophilic segments on the extracellular and intracellular sides of the receptor. The C-terminal region is enriched in serine and threonine residues that become phosphorylated upon agonist binding (Ali *et al.*, 1993; Tardif *et al.*, 1993).

Main activities and pathophysiological roles

FPR is expressed in neutrophils and other phagocytic cells of the mammalian immune system, including macrophages, dendritic cells, hepatocytes, and astrocytes. Binding of fMLP activates a complex array of signaling cascades that ultimately result in chemotaxis, release of proteolytic enzymes from secretory granules, production of superoxide by the NADPH oxidase, and activation of transcription factors. These cellular functions are thought to play a critical role in host defense against bacterial infections and in inflammation.

GENE

Accession numbers

Two cDNAs coding for allelic forms of the FPR (R26 and R98) were isolated from a plasmid cDNA library prepared from differentiated HL-60 cells. Both cDNA clones confer high-affinity binding sites for fMLP to COS-7 cells (Boulay *et al.*, 1990a). Several cDNA clones that differ from each other by point mutations were further isolated from different sources (**Table 1**).

Sequence

See **Figure 1**.

Chromosome location and linkages

The *FPR* gene is approximately 7.5 kb in length and organized into three exons and two large introns. It is present in the human genome as a single copy located in chromosome 19 in the 19q13.3 band adjacent to the 13.3-13.4 interface. The gene for FPR is clustered with that for FPRL1, FPRL2, and the C5aR in a 200 kb fragment (Bao *et al.*, 1992; Murphy *et al.*, 1992, 1993; Perez *et al.*, 1992; Gerard *et al.*, 1993; Haviland *et al.*, 1993; Alvarez *et al.*, 1994). As shown in **Figure 2**, a large open reading frame (ORF) of 1050 bp is entirely encoded by exon 3. The 5′ untranslated region of cDNAs isolated from neutrophil-like HL-60 cells and monocytes appears to be formed by the 74-bp-long exon 1 and 12 bp from exon 3′, probably resulting from an alternative splicing of exon 2. The promoter region is separated from the initiation ATG codon by 5.2 kb. Two Alu repeats are present in the intronic region and a third is found in the 3′ untranslated region. The absence of intron in the coding region is very unusual for G protein-coupled receptors. However, it appears to be a general feature of the chemoattractant receptor subclass.

The region upstream of exon 1 does not contain the classical CAAT and TATA boxes that are commonly

Table 1 GenBank accession numbers of *N*-formyl peptide receptor and homologs

Species	FPR	FPRL1 cDNA	FPRL2 cDNA
Human	allelic R26 cDNA: M33598	M84562, M88107, X63819, M76672	L14061, M76673
	allelic R98 cDNA: M33537		
	FPR gene: L10820		
Gorilla	cDNA: X97736	X97738	X97742
Macaque	cDNA: X97734	X97737	X97740
Orangutan	cDNA: X97735	X97744	X97741
Chimpanzee	cDNA: X97745	X97739	X97743
Mouse	FPR gene: L22181	U78299	
Rabbit	cDNA: M94549		

Figure 1 *FPR* gene organization. Exons 1, 2, and 3 are in capital letters in green boxes. A sequence motif identical to the cytokine 2-specific sequence that binds the nuclear factor NF-GM (orange box, bp 287–293) and a consensus sequence that binds the nuclear transcription factor NFκB (gray box, bp 390–399) are indicated. The putative nonconsensus TATA sequence 19 bp upstream of the initiation transcription start (bp 456–563) is indicated by a yellow box. The open reading frame (ORF) encoding the FPR protein is indicated by a pink box in exon 3. (Full colour figure may be viewed online.)

```
   1   ttatggggtt aatcttggtg gtgtgcatgg gtgtggacgc gctgtcctgc caactgtctc
  61   aacttccccc actcccttac ctctctctgt gtttctggtc tccatccctc atgacttctt
 121   ctcttccttt cattgcctcc ctctgattct tctcaccaca gtgcttgctg ctttctttac
 181   cttgaccctt ggagggagca ggggcccgga cacttggatt tcttggccct tgttgttgag
 241   agcactgaac ctctgcatcc acagagactg aggctgagaa atacagtcag gtacatgagt
 301   ttctaaacag gcccagccac tgtcctaatg ccattaaagc agacagtata ttggtgtatt
 361   cttggggcca tcaaaaatca gaagaagctc agacttccta tttcctgcta cccagctggt
 421   ttcagttcct ttacccctcc tcctgttcct tggtgtatgt tttgctgcaa tcattagaGC

EXON 1

 481   CTGAGTCACT CTCCCCAGGA GACCCAGACC TAGAACTACC CAGAGCAAGA CCACAGCTGG
 541   TGAACAGTCC AGgtaagaaa ---//--- Alu repeat 1 --//--- tactcctaca

EXON 2

2101   gCCTGTCTCC AGTTGGACTA GCCACAATTC AAGTGCTTGA AAACCACATG TGgtgagtga
2161   ---//-- Alu repeat 2 -------- aggaaatgac cacgactgca ctatttcagG

EXON 3

5351   AGCAGACAAG            ORF            TGAGGAGGGA GCTGGGGGAC ACTTTCGAGC
6461   TCCCAGCTCC AGCTTCGTCT CACCTTGAGT TAGGCTGAGC ACAGGCATTT CCTGCTTATT
6521   TTAGGATTAC CCACTCATCA GAAAAAAAAA AAAAGCCTTT GTGTCCCCTG ATTTGGGGAG
6581   AATAAACAGA TATGAGTTTA ttattgactt ctttttgat tttggacctc agcctcgggt
6641   ggtcagggtg ggaaatgata -------//------- Alu repeat 3 ----//----
6931   agaagatact ttatataggg caggagcggtg
```

found in the promoters of many proteins. A nonconsensus TATA box (TATGTT) is located 19 bp upstream of exon 1 and an inverted CCAAT box (ATTGG) is present at position −129 from the initiation transcription site. Sequences similar to the consensus sequences for binding of the nuclear factors NF-GM (Shannon *et al.*, 1988) and NFκB (Baeuerle and Henkel, 1994) are present in this region and may be involved in the regulation of FPR expression (Figure 2). Although transcription of FPR in HL-60 cells can be activated by dibutyryl cyclic AMP, no cAMP-responsive element is contained in the promoter region (Perez *et al.*, 1992; Haviland *et al.*, 1993; Murphy *et al.*, 1993). A systematic study of the promoter region with reporter genes has not yet been achieved.

PROTEIN

Accession numbers

SwissProt:
- Human FPR: P21462
- Gorilla FPR: P79176
- Macaque FPR: P79189
- Orangutan FPR: P79235

Figure 2 Chromosomal localization, structural organization, and polymorphism of the human *FPR* gene. The third exon contains the coding sequence as a whole. Sequence analysis of cDNAs (pINF10, pINF12) and genomic clones (pINF14, G6) isolated from different sources revealed amino acid differences in positions 11, 101, 123, 192, 256, 293, and/or 346. Residues that differ from the sequence of clone R98 are listed. These modifications do not alter the functional and pharmacological properties of the receptor, suggesting that there is a polymorphism of the *FPR* gene.

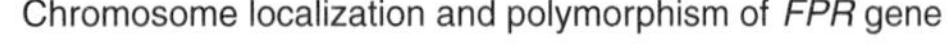

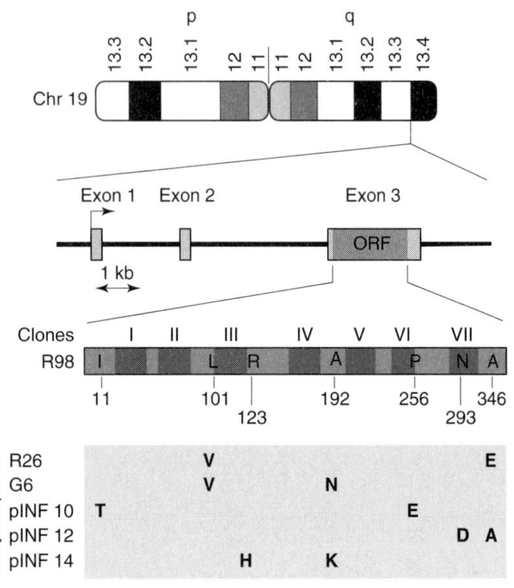

Chimpanzee FPR: P79241
Rabbit FPR: Q05394
Mouse FPR: P33766

Sequence

The amino acid sequence of FPR has been deduced from the cDNA (**Figure 3**).

Description of protein

FPR is 350 amino acids long with a calculated molecular mass of 38,420 Da, but biochemical studies indicate that the receptor migrates as a broad band on SDS-PAGE with an apparent molecular weight of 50–70 kDa typical of a glycoprotein.

The hydropathy profile reveals seven stretches of about 20–25 residues highly enriched in hydrophobic amino acids that are most likely folded in α helices. The interhelical hydrophilic loops are relatively short, as seen in most chemoattractant receptors. By analogy with the visual-light receptor rhodopsin, and based on biophysical and biochemical studies, FPR is thought to span the plasma membrane seven times. The seven transmembrane α helices are joined by intra- and extracellular hydrophilic loops. Three *N*-glycosylation sites are present in the putative extracellular regions, two of them are located in the N-terminal region. The C-terminal region is cytoplasmic, with multiple serine and threonine residues that are phosphorylated in an agonist-dependent manner (Ali *et al.*, 1993; Tardif *et al.*, 1993). No palmitoylation site is found in the C-terminal domain (**Figure 4**).

A direct determination of the three-dimensional structure has not yet been achieved, but a model for a probable structural organization has been proposed by Baldwin (1993) upon alignment of 204 sequences, including that of FPR. The proposed model fits relatively well with the projection map of bovine rhodopsin which was determined by electron microscopy of two-dimensional crystals. The folding of the hydrophilic loops is presently unknown but of particular importance because it conditions the interactions with the ligand at the extracellular surface, and with signaling and regulatory molecules at the cytoplasmic side. A structural constraint is probably imposed to the first and second extracellular loops by the presence of a disulfide bridge between two cysteine residues, Cys98 and Cys176, which have an invariant position in the G protein-coupled receptor family. The importance of this disulfide bridge to stabilize an active ternary structure of receptors has been established for rhodopsin (Karnik and Khorana, 1990) as well as for muscarinic and β_2-adrenergic receptors (Dohlman *et al.*, 1990; Kurtenbach *et al.*, 1990).

Relevant homologies and species differences

The human FPR has 34%, 69%, and 56% amino acid sequence similarities to human C5aR, FPRL1 and FPRL2, respectively. The overall degree of sequence similarity to murine and rabbit FPRs is 76% and 78%, respectively. The molecular evolution of FPR has recently been examined in nonhuman primates (chimpanzee, gorilla, orangutan, and macaque). The amino acid sequence similarities with the human counterpart range from 95% to 99%, with the highest similarity observed in chimpanzee and highest divergence observed in macaque (Alvarez *et al.*, 1996). While the highest divergence is observed in extracellular loops, the transmembrane and the

Figure 3 FPR cDNA and deduced amino acid sequence (GenBank L10820). The putative transmembrane domains predicted from the hydropathy plot (Kyte and Doolittle, 1982) are indicated by blue boxes. Three putative *N*-glycosylation sites are shown by gray boxes. (Full colour figure may be viewed online.)

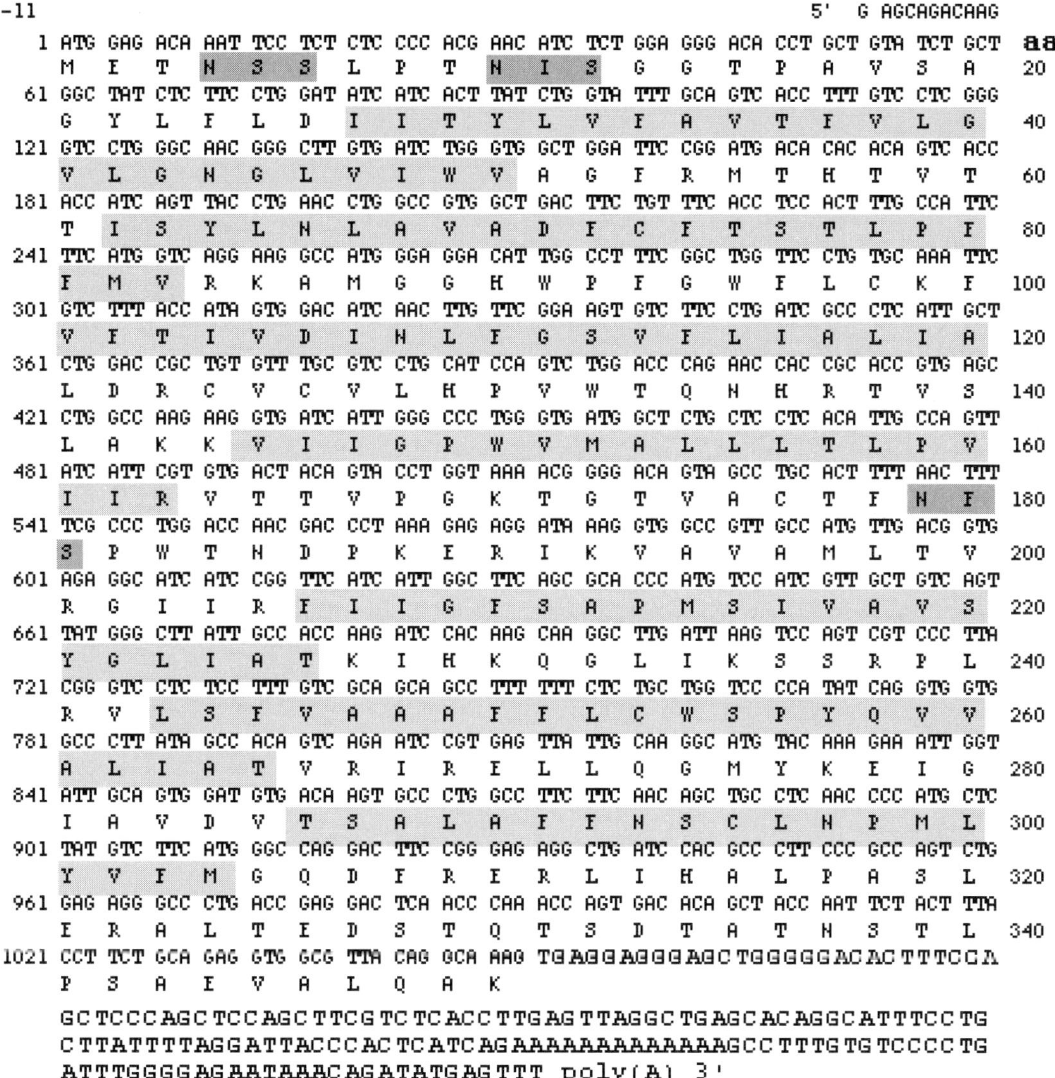

cytoplasmic domains are highly conserved. The divergence in the extracellular loops suggests that there are few structural constraints imposed by the ligand on these domains. Thus, despite a high degree of divergence of the N-terminal domain and the second extracellular loop of rabbit FPR with their counterpart in human FPR, both receptors bind fMLP with similar affinity (Ye *et al.*, 1993), suggesting that amino acids that have undergone an evolutionary change are not essential for ligand binding. The transmembrane domains and intracellular loops may have a higher restriction on divergence imposed by interaction with and activation of the heterotrimeric G protein(s).

Affinity for ligand(s)

The ligand-binding properties of FPR were originally characterized on plasma membrane preparations from

Figure 4 Schematic representation of the structure of human FPR in the plasma membrane. The folding of the polypeptide chain and the position of the transmembrane α helices were predicted by hydropathy plot (Kyte and Doolittle, 1982). Light blue circles indicate residues that result in a significant reduction in fMLP binding when replaced by an alanine residue (Miettinen *et al.*, 1997). Residues with white letters on a dark blue background indicate the region (V83-R-K-A-Met87) identified as the site of crosslinking with CHO-Met-*p*-benzoyl-L-phenylalanine-Phe-Tyr-N^{ε}-(fluorescein)-Lys-OH (Mills *et al.*, 1998). Regions of the model shown in green point to the peptide sequences which may be involved in FPR-G_{i2} interactions (Bommakanti *et al.*, 1995). Residues with white letters on a red background correspond to residues that are phosphorylated by GRK2 *in vitro* (Prossnitz *et al.*, 1995a). (Full colour figure may be viewed online.)

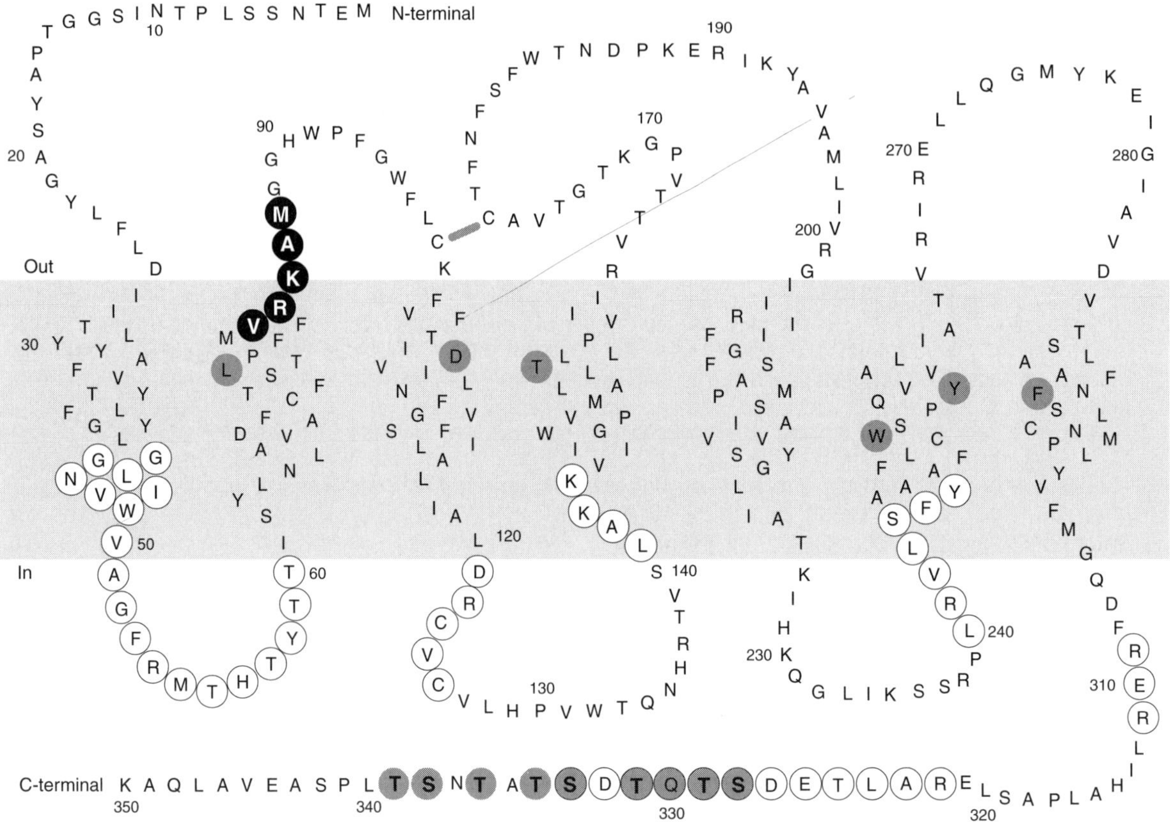

polymorphonuclear neutrophils, monocytes, and macrophages, and cell lines (HL-60 and U937) differentiated into neutrophil-like cells with dibutyryl cAMP. Binding was measured using CHO-Met-Leu-[^{3}H]Phe-OH and ^{125}I-labeled CHO-Nle-Leu-Phe-Nle-Tyr-Lys-OH peptides, and *N*-formyl peptides derivatized with photoactivatable and fluorescent groups. The affinity of FPR for the prototypic tripeptide CHO-Met-Leu-(^{3}H)Phe-OH is relatively high ($K_d = 1$ nM). As for other G protein-coupled receptors, the incubation of membrane preparations from cells expressing native or recombinant FPR with nonhydrolyzable GTP analog, which dissociates the G protein from the receptor, results in a severe decrease in receptor affinity for *N*-formyl peptides ($K_d = 10$–20 nM) (Koo *et al.*, 1982; Prossnitz *et al.*, 1995b; Wenzel-Seifert *et al.*, 1998). The ionic environment and pH have also been found to affect receptor affinity. Removal of Na^+ results in a rapid and reversible increase of affinity for fMLP (Zigmond *et al.*, 1985). This phenomenon has recently been explained by the fact that Na^+ stabilizes the FPR in an inactive state with a reduced ability to interact with G_i protein (Wenzel-Seifert *et al.*, 1998). Moreover, the use of fluorescein-derivatized peptides has indicated that the ligand-binding pocket is able to accommodate no more than six residues, and contains at least two microenvironments, one hydrophobic and another charged and supporting protonation. It has been suggested that the role of protonated residues is to stabilize ligand binding at neutral pH (Fay *et al.*, 1993).

To delineate the amino acid residues of FPR that are responsible for high-affinity fMLP binding, receptor chimeras were constructed by sequential replacement of FPR segments with the corresponding regions in FPRL1 or murine FPR which bind the prototypical peptide fMLP with low affinity (Gao and Murphy, 1993; Quehenberger *et al.*, 1993). Based on amino acid differences between the transmembrane domains of FPR and FPRL1, mutants of FPR were generated to pinpoint residues responsible for high-affinity ligand binding (Miettinen *et al.*, 1997). This has led to the identification of 10 amino acid residues, located in α helices 2–7, that may participate in binding of fMLP. An alternative approach used a 'gain-of-function' strategy by selectively replacing the nonconserved region of FPRL1 with those of FPR (Quehenberger *et al.*, 1997). However, although a number of mutants or chimeric receptors display reduced or improved ligand-binding capabilities, none of these approaches has enabled the ligand binding site to be localized conclusively.

By an elegant strategy that combines the photolabeling properties of CHO-Met-*p*-benzoyl-L-phenylalanine-Phe-Tyr-N^{ε}-(fluorescein)-Lys-OH, the ability of an antifluorescein antibody to immunoprecipitate crosslinked fluoresceinated peptides, and the use of matrix-assisted laser desorption ionization mass spectroscopy, a major photolabeled cyanogen bromide peptide, Val-Arg-Lys-Ala-Hse, corresponding to residues 83–87 of FPR has recently been identified (Mills *et al.*, 1998). In the current three-dimensional model, this peptide lies at the interface between the second transmembrane domain and the first extracellular loop (Figure 4).

Cell types and tissues expressing the receptor

The general assumption was that FPR is expressed on cells of myeloid origins, such as neutrophils, monocytes, macrophages, eosinophils, and differentiated myeloid cell lines HL-60, U937, and NB4. However, several recent studies have indicated that FPR has a much broader expression pattern than previously thought. Dendritic cells derived from human blood most likely expressed FPR since fMLP elicits their migration and intracellular calcium mobilization (Sozzani *et al.*, 1995). Expression of FPR has been demonstrated in the human hepatocarcinoma cell line HepG2 by *in situ* hybridization analysis and radioligand-binding studies (McCoy *et al.*, 1995). This is consistent with a recent study indicating that permeabilized hepatocytes and Kupffer cells stain positively with an antibody directed against the C-terminal domain of FPR (Becker *et al.*, 1998). The presence of FPR on cells of the central nervous system, such as human astrocytes, microglia, and the immortalized human astrocyte cell line HSC2 has been demonstrated by flow cytometry, indirect immunofluorescence, and RT-PCR analysis (Lacy *et al.*, 1995).

An immunocytochemical study on normal human tissues suggests that the expression of FPR can be expanded to other organs and tissues (Becker *et al.*, 1998). Although the results of this study need to be correlated with *in situ* hybridization analysis, FPR or an antigenically related receptor appears to be present on some epithelial cells with secretory functions, and some endocrine cells such as follicular cells of the thyroid. Smooth muscle cells of the muscularis mucosa and muscularis propria of ileum, and arterioles, as well as endothelial cells and some neurons of the motor, sensory, and cerebellar systems are also positively stained. Radioligand-binding studies indicate that FPR maps to the lamina propria of the gastrointestinal mucosa but not to epithelial cells (Anton *et al.*, 1998). In the retina, the pigmented retinal epithelial cells as well as the retinal rods and the cones are positively stained (Becker *et al.*, 1998). Thus, the broad tissue and cell localization of FPR, summarized in **Table 2**, suggests that FPR may have unanticipated physiological roles.

Regulation of receptor expression

In nondifferentiated myeloid cells such as HL-60, NB4, and U937 cells, FPR mRNA is not detectable. A number of chemical agents including dimethylsulfoxide, dibutyryl cAMP, and all-*trans* retinoic acid confer to these cells a granulocyte-like phenotype with a concomitant expression of FPR on the cell surface. IFNγ and TNFα have been shown to enhance expression of FPR in HL-60 granulocytes but the underlying mechanism remains unclear (Klein *et al.*, 1992). In neutrophils, FPR is present on the cell surface and in intracellular compartments. The surface recruitment of FPR is triggered by inflammatory stimuli including fMLP, PAF, IL-8, leukotriene B_4, GM-CSF, and TNFα. Upregulation of FPR appears to result from the mobilization of secretory vesicles (Sengelov *et al.*, 1994) but a pool of receptor is also present in the membrane of gelatinase granules and possibly of specific granules (English and Graves, 1992). In differentiated HL-60 cells, FPR is not upregulated, most likely because these cells are deficient in specific granules. There is presently no evidence as to whether FPR expression is modulated by inflammatory stimuli in cells that are not of myeloid origin.

Table 2 Cell type and tissue expression of FPR, FPRL1, and FPRL2

Receptor	Cells and tissues	Detected by
FPR	Neutrophils, macrophages, U937, HL-60, NB4	CHO-Met-Leu-(^{3}H)Phe-OH binding, northern and western blotting
	Astrocytes, microglia, HSC2, HepG2	Flow cytometry, indirect immunofluorescence, and RT-PCR
	Hepatocytes, Kupffer cells, endothelial cells, epithelial enzyme-secreting cells, smooth muscle cells of muscularis propria and mucosa of ileum, arterioles, retina, some neurons	Immunohistochemistry
FPRL1	Neutrophils, monocytes, HL-60, colonic epithelial cells (T84, HT29, Caco-2, CL.19A) (Gronert *et al.*, 1998), medial tissues of coronary arteries (Keitoku *et al.*, 1997)	Northern blotting or RT-PCR
FPRL2	Monocytes, lung, medial tissue of coronary arteries (Keitoku *et al.*, 1997)	Northern blotting or RT-PCR

SIGNAL TRANSDUCTION

Associated or intrinsic kinases

FPR has no associated or intrinsic kinases. Activation of kinase cascades involves heterotrimeric G proteins.

Cytoplasmic signaling cascades

The majority of fMLP-mediated leukocyte responses are largely inhibited by pertussis toxin (PTX), a bacterial toxin that ADP-ribosylates and, thereby, inactivates the α subunit of heterotrimeric G proteins of the G_i class but not of the G_q class (for review see Snyderman and Uhing, 1992). A large body of evidence indicates that fMLP-induced responses require the physical association of FPR to G_{i2} protein (Bommakanti *et al.*, 1992, 1995; Gierschik *et al.*, 1989). Although agonist-occupied FPR does not couple to the PTX-insensitive $G_{q/11}$ subfamily (Amatruda *et al.*, 1993), a small PTX-resistant activity is nevertheless observed in neutrophils, monocytes, and differentiated HL-60 cells. This may result from the coupling of FPR to residual molecules of G_{16} (Amatruda *et al.*, 1995), a promiscuous $G\alpha$ subunit restricted to a subset of myeloid cells in the early steps of differentiation (Amatruda *et al.*, 1991). More details on the coupling of FPR to G proteins can be found in a recent review by Ye and Boulay (1997).

The triggering and control of neutrophil functions by fMLP clearly require a coordinated activation of several pathways (**Figure 5**). In neutrophils, fMLP modulates the activity of adenylylcyclase, causing a small increase in cellular cAMP by a mechanism that is not well understood (Spisani *et al.*, 1996). The most remarkable effect of fMLP is the induction of a robust activation of tyrosine kinases, phosphatidylinositol-specific phospholipase C (PI-PLC), phospholipase D (PLD), phospholipase A_2 (PLA_2), phosphatidylinositol 3-kinase (PI3K), and of the mitogen-activated protein kinase (MAPK) cascades. The interaction of FPR with G_{i2} leads to the exchange of GDP bound to $G\alpha$ for GTP. This causes the dissociation of $G\beta\gamma$ from GTP-bound $G\alpha$. Free $G\beta\gamma$ interacts with second messenger-generating enzymes such as the PI-PLCβ_2 isoform (Camps *et al.*, 1992; Katz *et al.*, 1992) and the PI3Kγ isoform (Stoyanov *et al.*, 1995). PI-PLCβ_2 is a PLC isoform restricted to certain myeloid cells, which breaks phosphatidyl 4,5-bisphosphate (PIP_2) into inositol trisphosphate (IP_3) acting to mobilize calcium from intracellular stores, and diacylglycerol (DAG). A more sustained production of DAG is generated by phosphohydrolysis of phosphatidic acid resulting from phosphatidylcholine breakdown by phospholipase D (PLD) (Exton, 1994). The interplay of Ca^{2+} and DAG leads to the activation of protein kinase C isoforms which are thought to contribute to the triggering of various neutrophil functions. The pathway leading to increased PLD activity is not yet completely characterized. A growing body of evidence indicates that PLD activation is dependent on the small GTPases, RhoA and the ADP-ribosylation factor (ARF). PLD activation requires the PTX-sensitive G_{i2} protein but is not obligatory dependent on PI3K activation (Fensome *et al.*, 1998). The mechanism by which ARF and RhoA are activated is still uncertain. In leukocytes,

Figure 5 Model for the distinctive signaling pathways induced by FPR activation in neutrophils. Phospholipases are shown in yellow, second messengers in green, small G proteins in blue. PI3 kinase, protein kinase C, and Src-related kinases are indicated by violet boxes. The MAP kinase cascades are shown in red. Abbreviations: PLA_2, phospholipase A_2; AA, arachidonic acid; PLC, phospholipase C; PIP_2, phosphatidylinositol 4,5 bisphosphate; PIP_3, phosphatidylinositol 3,4,5 trisphosphate; IP3, inositol trisphosphate; DAG, diacylglycerol; PLD, phospholipase D; PA, phosphatidic acid; PKC, protein kinase C; MAPK, MAP kinase. For details see text and the review by Bokoch (1995). (Full colour figure may be viewed online.)

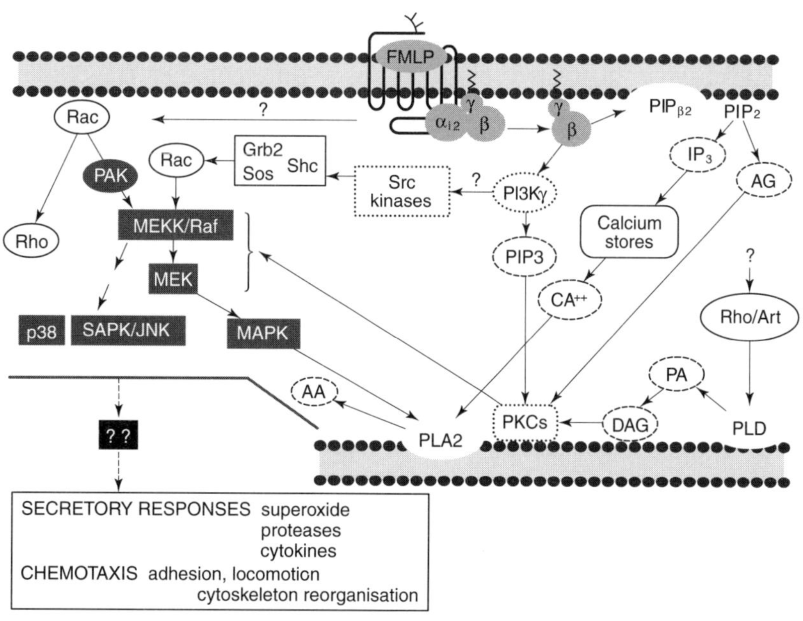

RhoA is thought to be involved in FPR-mediated triggering of rapid adhesion through integrins (Laudana *et al.*, 1996). For more information on the role of GTPases of the Rho family (Rho, Rac, and Cdc42) in actin cytoskeletal organization the reader is referred to several recent reviews (Tapon and Hall, 1997; Zohn *et al.*, 1998).

Two protein kinase cascades, the extracellular signal-regulated kinases (p42/44 MAPK) and the stress-activated p38 MAP kinases are activated by fMLP in neutrophils (Worthen *et al.*, 1994; Avdi *et al.*, 1996; Krump *et al.*, 1997; Nick *et al.*, 1997; Rane *et al.*, 1997). The activated MAP kinases in turn regulate the activity of downstream targets by phosphorylation. These two signaling pathways are thought to participate at different degree in adherence, superoxide production, and chemotaxis. The pathway leading to p38 activation is not completely deciphered and the role played by p38 in neutrophil responses remains to be clarified. *In vivo*, p38 activates the MAP kinase-activated protein kinase 2 which may be involved in the superoxide production (Zu *et al.*, 1996). In neutrophils, p42/44 MAPK activation is concomitant with the activation of the Src-like tyrosine kinase Lyn, the phosphorylation of the adapter protein Shc, and the activation of the small GTPase p21Ras (Worthen *et al.*, 1994; Ptasznik *et al.*, 1995). The pathway for p42/44 MAPK activation has been recently reconstituted in COS-7 cells by cotransfection of cDNAs encoding PI3Kγ, G$\beta\gamma$, and proteins involved in the Ras pathway (Lopez-Ilasaca *et al.*, 1997). The emerging picture is that free G$\beta\gamma$ recruits PI3Kγ to the plasma membrane, thereby enhancing the activity of Src-like tyrosine kinase(s), which in turn phosphorylate(s) the adapter protein Shc. The Shc/Grb2/Sos/Ras complex is formed, leading to activation of the downstream kinases Raf/MEKK-1 which activate MEK. This latter phosphorylates and, thereby, activates p42/44 MAPK. One important role for activated MAP kinases is to phosphorylate and activate cytoplasmic PLA_2, which is translocated to the plasma membrane in a Ca^{2+}-dependent manner (Lin *et al.*, 1993). The reduction of PLA_2 activity by an antisense strategy has recently revealed that PLA_2 is essential for activation of the superoxide production by the NADPH-oxidase (Dana *et al.*, 1998). In other FPR-expressing cells, such as hepatocytes, MAP kinases most likely participate in signaling to the nucleus for activation of early genes of the acute phase of inflammation.

There are a number of lines of evidence from inhibitor studies that several neutrophil functions are dependent on the activation of PI3K. fMLP-induced superoxide production and chemotaxis are, for instance, completely inhibited by wortmannin, a selective inhibitor of PI3K. Reconstitution experiments in transfected COS-7 cells indicate that FPR induces cytoskeletal reorganization through $G\beta\gamma$ heterodimers, PI3Kγ, the small GTPase Rac, and a guanosine exchange factor for Rac (Ma *et al.*, 1998). However, the role of the different enzymes activated upon fMLP binding can vary from one cellular function to another. For instance, knockout experiments indicate that PI-PLCβ_2 is critical for Ca^{2+} mobilization, superoxide production, and upregulation of MAC-1 (CD11b/CD18) but is dispensable for chemotaxis (Jiang *et al.*, 1997). The observation that fMLP-mediated chemotaxis of leukocytes from PLCβ_2-deficient mice is enhanced suggests that PLCβ_2 is part of a negative feedback loop that attenuates chemotaxis (Jiang *et al.*, 1997). A more extensive description of the different signaling cascades activated by chemoattractants can be found in recent reviews (Bokoch, 1995; Downey *et al.*, 1995).

The cellular responses elicited by fMLP are rapidly attenuated and cells become refractory to repeated stimulation with the same agonist. This process, termed homologous desensitization, is in part due to the agonist-dependent phosphorylation of FPR on its C-terminal domain (Ali *et al.*, 1993; Tardif *et al.*, 1993; Prossnitz *et al.*, 1995a). Phosphorylation of FPR is an essential step in the functional desensitization of FPR and is required for receptor internalization (Prossnitz, 1997). However, neither the phosphorylation of FPR nor its internalization is required for chemotaxis (Hsu *et al.*, 1997).

DOWNSTREAM GENE ACTIVATION

Transcription factors activated

The regulation of gene expression in neutrophils, monocytes, and macrophages during the early inflammatory response is governed by the activities of transcriptional activators. Several studies have examined the effects of chemoattractants on the modulation of transcription factor gene expression in myeloid cells. fMLP was found to induce a transient increase of c-*fos* mRNA in human monocytes and peripheral granulocytes (Ho *et al.*, 1987; Itami *et al.*, 1987). The fMLP-mediated signaling cascade leading to c-*fos* induction has not been studied in detail but, by analogy with other systems, one can anticipate the involvement of the p38 MAPK (Hazzalin *et al.*, 1996).

The nuclear factor kappa B (NFκB) plays a critical role in immune cell function due to its unique ability to turn on the transcription of genes encoding signaling and host defense proteins as well as many cytokines. In the past few years, it has been reported that fMLP is able to stimulate the activation of NFκB in eosinophils (Miyamasu *et al.*, 1995), peripheral blood mononuclear cells (PBMCs) and dimethylsulfoxide-differentiated HL-60 cells but not in neutrophils (Browning *et al.*, 1997). Compared with PBMCs and differentiated HL-60 cells, neutrophils contain a much lower level of NFκB subunits, which could explain their lack of responsiveness. Even though NFκB can be activated by TNFα in FPR-transfected HL-60 cells, these latter are unresponsive to fMLP stimulation, suggesting that the downstream signaling pathway becomes functional only when cells are terminally differentiated (Browning *et al.*, 1997).

Genes induced

Several studies have examined the ability of chemoattractants to induce secretion of cytokines and chemokines. fMLP has been found to induce the release of IL-8 from eosinophils (Miyamasu *et al.*, 1995) and neutrophils (Cassatella *et al.*, 1992). This secretion is completely inhibited by actinomycin D in both cell types, indicating that fMLP modulates IL-8 production at the transcriptional level. Both fMLP and C5a were found to induce the release of GM-CSF from eosinophils (Miyamasu *et al.*, 1995). However, although both chemoattractant receptors use the same downstream transduction pathway, fMLP is less active than C5a. In addition, it has been shown that fMLP turns on the expression of proinflammatory cytokines such as IL-1α, IL-1β, and IL-6 in human PMBCs (Arbour *et al.*, 1996), and that it stimulates the synthesis of proteins of the hepatic acute phase response, including complement C3 and α_1-antichymotrypsin, from HepG2 cells (McCoy *et al.*, 1995).

Promoter regions involved

The promoter regions of GM-CSF and cytokine/chemokine genes (IL-1, IL-6, and IL-8) possess NFκB-binding sites (Baeuerle and Henkel, 1994; Baggiolini *et al.*, 1994). The involvement of this promoter region in the transcriptional activation of these genes is supported by the observation that fMLP-mediated secretion of IL-8 and GM-CSF by eosinophils is inhibited by the specific NFκB inhibitor pyrolidine dithiocarbamate (Miyamasu *et al.*, 1995). The role of NFκB in the fMLP-mediated secretion of IL-8 by neutrophils is, however, less clear since fMLP does not appear to activate NFκB in these cells (Browning *et al.*, 1997). Regulatory regions that bind other transcriptional factors such as AP-1 and NF-IL6 are likely to be involved. The involvement of AP-1 would be consistent with the observation that fMLP induces a transient increase in c-*fos* mRNA.

In the case of liver cells, the promoter regions involved in the induction of complement C3 and α_1-antichymotrypsin genes by fMLP have not been determined. These genes may be indirectly activated by secreted cytokines. Their induction could involve IL-1- and IL-6-responsive elements (Wilson *et al.*, 1990).

BIOLOGICAL CONSEQUENCES OF ACTIVATING OR INHIBITING RECEPTOR AND PATHOPHYSIOLOGY

Unique biological effects of activating the receptors

The most spectacular effects resulting from the activation of FPR are observed in polymorphonuclear neutrophils. Binding of fMLP to FPR induces the change of morphology, chemotaxis, neutrophil adhesion, microbicidal functions, and cytotoxic effects. Leukocytes release proteolytic enzymes from their granules and produce highly reactive oxygen-derived free radicals in response to fMLP binding. These substances are cytotoxic and constitute the first reaction of the host against invading pathogens. There is presently no evidence that a dysfunctioning of FPR could result in an inappropriate release of cytotoxic mediators and, thereby, to inflammatory states.

Phenotypes of receptor knockouts and receptor overexpression mice

The consequences of FPR knockout or receptor overexpression in mice have not been examined yet.

Human abnormalities

Very few studies have examined the possibility that a defective FPR could result in human abnormalities or pathogenesis. Perez *et al.* (1991) have reported the case of a patient with *juvenile periodontitis* in whom leukocytes show an abnormal responsiveness to fMLP. However, this study has not provided evidence for any mutation in the FPR gene.

References

Ali, H., Richardson, R. M., Tomhave, E. D., Didsbury, J. R., and Snyderman, R. (1993). Differences in phosphorylation of formylpeptide and C5a chemoattractant receptors correlate with differences in desensitization. *J. Biol. Chem.* **268**, 24247–24254.

Allen, R. A., Traynor, A. E., Omann, G. M., and Jesaitis, A. J. (1988). The chemotactic peptide receptor. A model for future understanding of chemotactic disorders. *Haematol. Oncol. Clin. North Am.* **2**, 33–59.

Alvarez, V., Coto, E., Setien, F., and Lopez-Larrea, C. (1994). A physical map of two clusters containing the genes for six proinflammatory receptors. *Immunogenetics* **40**, 100–103.

Alvarez, V., Coto, E., Setién, F., Gonzalez-Roces, S., and Lopez-Larrea, C. (1996). Molecular evolution of the N-formyl peptide and C5a receptors in non-human primates. *Immunogenetics* **44**, 446–452.

Amatruda, T. T., Steele, D. A., Slepak, V. Z., and Simon, M. I. (1991). Gα16, a G protein α subunit specifically expressed in hematopoietic cells. *Proc. Natl Acad. Sci. USA* **88**, 5587–5591.

Amatruda, T. T., Gerard, N. P., Gerard, C., and Simon, M. I. (1993). Specific interactions of chemoattractant factor receptors with G-proteins. *J. Biol. Chem.* **268**, 10139–10144.

Amatruda, T. T., Dragas-Graonic, S., Holmes, R., Perez, H. D. (1995). Signal transduction by the formyl peptide receptor. *J. Biol. Chem.* **270**, 28010–28013.

Anton, P., O'Connell, J., O'Connell, D., Whitaker, L., O'Sullivan, G. C., Collins, J. K., and Shanahan, F. (1998). Mucosal subepithelial binding sites for the bacterial chemotactic peptide, formyl-methionyl-leucyl-phenylanine (FMLP). *Gut* **42**, 374–379.

Arbour, N., Tremblay, P., and Oth, D. (1996). N-formyl-methionyl-leucyl-phenylalanine induces and modulates IL-1 and IL-6 in human PBMC. *Cytokine* **8**, 468–475.

Avdi, N. J., Winston, B. W., Russel, M., Young, S. K., Johnson, G. L., and Worthen, G. S. (1996). Activation of MEKK by formyl-methionyl-leucyl-phenylalanine in human neutrophils. *J. Biol. Chem.* **271**, 33598–33606.

Baeuerle, P. A., and Henkel, T. (1994). Function and activation of NF-κB in the immune system. *Annu. Rev. Immunol.* **12**, 141–179.

Baggiolini, M., Dewald, B., and Moser, B. (1994). Interleukin-8 and related chemotactic cytokines-CXC and CC chemokines. *Adv. Immunol.* **55**, 97–179.

Baldwin, J. M. (1993). The probable arrangement of the helices in G protein-coupled receptors. *EMBO J.* **12**, 1693–1703.

Bao, L., Gerard, N. P., Eddy Jr, R., Shows, T. B., and Gerard, C. (1992). Mapping of genes for the human C5a receptor (C5AR), human FMLP receptor (FPR), and two FMLP receptor homologue orphan receptors (FPRH1, FPRH2) to chromosome 19. *Genomics* **13**, 437–440.

Becker, E. L., Forouhar, F. A., Grunnet, M. L., Boulay, F., Tardif, M., Bormann, B.-J., Sodja, D., Ye, R. D., Woska Jr, J. R., and Murphy, P. M. (1998). Broad immunocytochemical localization of the formylpeptide receptor in human organs, tissues, and cells. *Cell Tissue Res.* **292**, 129–135.

Bokoch, G. M. (1995). Chemoattractant signaling and leukocyte activation. *Blood* **86**, 1649–1660.

Bommakanti, R. K., Bokoch, G. M., Tolley, J. O., Schreiber, R. E., Siemsen, D. W., Klotz, K. N., and Jesaitis, A. J. (1992). Reconstitution of a physical complex between the N-formyl chemotactic peptide receptor and G protein. Inhibition by pertussis toxin-catalyzed ADP ribosylation. *J. Biol. Chem.* **267**, 7576–7581.

Bommakanti, R. K., Dratz, E. A., Siemsen, D. W., and Jesaitis, A. J. (1995). Extensive contact between Gi2 and N-formyl peptide receptor of human neutrophils: mapping of binding sites using receptor-mimetic peptides. *Biochemistry* **34**, 6720–6728.

Boulay, F., Tardif, M., Brouchon, L., and Vignais, P. (1990a). The human N-formylpeptide receptor. Characterization of two cDNA isolates and evidence for a new subfamily of G-protein-coupled receptors. *Biochemistry* **29**, 11123–11133.

Boulay, F., Tardif, M., Brouchon, L., and Vignais, P. (1990b). Synthesis and use of a novel N-formyl peptide derivative to isolate a human N-formyl peptide receptor cDNA. *Biochem. Biophys. Res. Commun.* **168**, 1103–1109.

Browning, D. D., Pan, Z. K., Prossnitz, E. R., and Ye, R. D. (1997). Cell type- and developmental stage-specific activation of NF-κB by fMet-Leu-Phe in myeloid cells. *J. Biol. Chem.* **272**, 7995–8001.

Camps, M., Carozzi, A., Schnabel, P., Scheer, A., Parker, P. J., and Gierschik, P. (1992). Isozyme-selective stimulation of phospholipase C-β2 by G protein $\beta\gamma$-subunits. *Nature* (*Lond.*) **360**, 684–689.

Cassatella, M. A., Bazzoni, F., Ceska, M., Ferro, I., Baggiolini, M., and Berton, G. (1992). IL-8 production by human polymorphonuclear leukocytes: the chemoattractant formyl-methionyl-leucyl-phenylalanine induces the gene expression and release of IL-8 through a pertussis toxin-sensitive pathway. *J. Immunol.* **148**, 3216–3220.

Dana, R., Leto, T. L., Malech, H. L., and Levy, R. (1998). Essential requirement of cytosolic phospholipase A2 for activation of the phagocyte NADPH oxidase. *J. Biol. Chem.* **273**, 441–445.

Dohlman, H. G., Caron, M. G., DeBlasi, A., Frielle, T., and Lefkowitz, R. J. (1990). Role of extracellular disulfide-bonded cysteines in the ligand binding function of the β2-adrenergic receptor. *Biochemistry* **29**, 2335–2342.

Downey, G. P., Fukushima, T., and Fialkow, L. (1995). Signaling mechanisms in human neutrophils. *Curr. Opin. Hematol.* **2**, 76–88.

Durstin, M., Gao, J.-L., Tiffany, H. L., McDermott, D., and Murphy, P. M. (1994). Differential expression of members of the N-formylpeptide receptor gene cluster in human phagocytes. *Biochem. Biophys. Res. Commun.* **201**, 174–179.

English, D., and Graves, V. (1992). Simultaneous mobilization of Mac-1 (CD11b/CD18) and formyl peptide chemoattractant receptors in human neutrophils. *Blood* **80**, 776–787.

Exton, J. H. (1994). Phosphatidylcholine breakdown and signal transduction. *Biochim. Biophys. Acta* **1212**, 26–42.

Fay, S. P., Domaleswski, M. D., and Sklar, L. A. (1993). Evidence for protonation in human neutrophil formyl peptide receptor binding pocket. *Biochemistry* **32**, 1627–1631.

Fensome, A., Whatmore, J., Morgan, C., Jones, D., and Cockcroft, S. (1998). ADP-ribosylation factor and Rho proteins mediate fMLP-dependent activation of phospholipase D in human neutrophils. *J. Biol. Chem.* **273**, 13157–13164.

Freer, R. J., Day, A. R., Radding, J. A., Schiffmann, E., Aswanikumar, S., Showell, H. J., and Becker, E. L. (1980). Further studies on the structural requirements for synthetic peptide chemoattractants. *Biochemistry* **19**, 2404–2410.

Gao, J. L., and Murphy, P. M. (1993). Species and subtype variants of the N-formyl peptide chemotactic receptor reveal multiple important functional domains. *J. Biol. Chem.* **268**, 25395–25401.

Gerard, N. P., Bao, L., Xiao-Ping, H., Eddy, R., Jr., Shows, T. B., and Gerard, C. (1993). Human chemotaxis receptor genes cluster at 19q13.3-13.4. Characterization of the human C5a receptor gene. *Biochemistry* **32**, 1243–1250.

Gierschik, P., Sidoropoulos, D., and Jakobs, K. H. (1989). Two distinct Gi-proteins mediate formyl peptide receptor signal transduction in human leukemia (HL-60) cells. *J. Biol. Chem.* **264**, 21470–21473.

Gronert, K., Gewirtz, A., Madara, J. L., and Serhan, C. N. (1998). Identification of a human enterocyte lipoxin A4 receptor that is regulated by interleukin (IL)-13 and interferon γ and inhibits tumor necrosis factor α-induced IL-8 release. *J. Exp. Med.* **187**, 1285–1294.

Haviland, D. L., Borel, A. C., Fleischer, D. T., Haviland, J. C., and Wetsel, R. A. (1993). Structure, 5′-flanking sequence, and chromosome location of the human N-formyl peptide receptor gene. A single-copy gene comprised of two exons on chromosome 19q.13.3 that yields two distinct transcripts by alternative polyadenylation. *Biochemistry* **32**, 4168–4174.

Hazzalin, C. A., Cano, E., Cuenda, A., Barratt, M. J., Cohen, P., and Mahadevan, L. C. (1996). p38/RK is essential for stress-induced nuclear responses: JNK/SAPKs and c-Jun/ATF-2 phosphorylation are insufficient. *Curr. Biol.* **6**, 1026–1031.

Ho, Y. S., Lee, W. M., and Snyderman, R. (1987). Chemoattractant-induced activation of c-fos gene expression in human monocytes. *J. Exp. Med.* **165**, 1534–1538.

Hsu, M. H., Chiang, S. C., Ye, R. D., and Prossnitz, E. R. (1997). Phosphorylation of the N-formyl peptide receptor is required for receptor internalization but not chemotaxis. *J. Biol. Chem.* **272**, 29426–29429.

Itami, M., Kuroki, T., and Nose, K. (1987). Induction of c-fos proto-oncogene by a chemotactic peptide in human peripheral granulocytes. *FEBS Lett.* **222**, 289–292.

Jiang, H., Kuang, Y., Wu, Y., Xie, W., Simon, M. I., and Wu, D. (1997). Roles of phospholipase C β2 in chemoattractant-elicited responses. *Proc. Natl Acad. Sci. USA.* **94**, 7971–7975.

Karnik, S. S., and Khorana, H. G. (1990). Assembly of functional rhodopsin requires a disulfide bond between cysteine residues 110 and 187. *J. Biol. Chem.* **265**, 17520–17524.

Katz, A., Wu, D., and Simon, M. I. (1992). Subunits $\beta\gamma$ of heterotrimeric G protein activate β2 isoform of phospholipase C. *Nature (Lond.)* **360**, 686–689.

Keitoku, M., Kohzuki, M., Katoh, H., Funakoshi, M., Suzuki, S., Takeuchi, M., Karbe, A., Horiguchi, S., Watanabe, J., Satoh, S., Nose, M., Abe, K., Okayama, H., and Shirato, K. (1997). FMLP actions and its binding sites in isolated human coronary arteries. *J. Mol. Cell Cardiol.* **29**, 881–894.

Klein, J. B., Scherzer, J. A., and McLeish, K. R. (1992). IFN-γ enhances expression of formyl peptide receptors and guanine nucleotide-binding proteins by HL-60 granulocytes. *J. Immunol.* **148**, 2483–2488.

Koo, C., Lefkowitz, R. J., and Snyderman, R. (1982). The oligopeptide chemotactic factor receptor of human polymorphonuclear leukocyte membranes exists in two affinity states. *Biochem. Biophys. Res. Commun.* **106**, 442–449.

Krump, E., Sanghera, J. S., Pelech, S. L., Furuya, W., and Grinstein, S. (1997). Chemotactic peptide N-formyl-Met-Leu-Phe activation of p38 mitogen-activated protein kinase

(MAPK) and MAPK-activated protein kinase-2 in human neutrophils. *J. Biol. Chem.* **272**, 937–944.

Kurtenbach, E., Curtis, C. A. M., Pedder, E. K., Aitken, A., Harris, A. C. M., and Hulmes, E. C. (1990). Muscarinic acetylcholine receptors. *J. Biol. Chem.* **265**, 13702–13708.

Kyte, J., and Doolittle, R. R. (1982). A simple method for displaying the hydropathic character of a protein. *J. Mol. Biol.* **157**, 105–132.

Lacy, M., Jones, J., Whittemore, S. R., Haviland, D. L., Wetsel, R. A., and Barnum, S. R. (1995). Expression of the receptors for the C5a anaphylatoxin, interleukin-8 and FMLP by human astrocytes and microglia. *J. Neuroimmunol.* **61**, 71–78.

Laudana, C., Campbell, J. J., and Butcher, E. C. (1996). Role of Rho in chemoattractant-activated leukocyte adhesion through integrins. *Science* **271**, 981–983.

Lin, L.-L., Wartmann, M., Lyn, A. Y., Knopf, J. L., Seth, A., and Davis, R. J. (1993). cPLA2 is phosphorylated and activated by MAP kinase. *Cell* **72**, 269–278.

Lopez-Ilasaca, M., Crespo, P., Pellici, P. G., Gutkind, J. S., and Wetzker, R. (1997). Linkage of G protein-coupled receptors to the MAPK signaling pathway through PI 3-kinase γ. *Science* **275**, 394–397.

Ma, A. D., Metjian, A., Bagrodia, S., Taylor, S., and Abrams, C. S. (1998). Cytoskeletal reorganization by G protein-coupled receptor is dependent on phosphoinositide 3-kinase γ, a rac guanosine exchange factor, and rac. *Mol. Cell. Biol.* **18**, 4744–4751.

McCoy, R., Haviland, D. L., Molmenti, E. P., Ziambaras, T., Wetsel, R. A., and Perlmutter, D. H. (1995). N-formylpeptide and complement C5a receptors are expressed in liver cells and mediate hepatic acute phase gene regulation. *J. Exp. Med.* **182**, 207–217.

Miettinen, H. M., Mills, J. S., Gripentrog, J. M., Dratz, E. A., Granger, B. L., and Jesaitis, A. J. (1997). The ligand binding site of the formyl peptide receptor maps in the transmembrane region. *J. Immunol.* **159**, 4045–4054.

Mills, J. S., Miettinen, H. M., Barnidge, D., Vlases, M. J., Wimer, M. S., Dratz, E. A., Sunner, J., and Jesaitis, A. J. (1998). Identification of a ligand binding site in the human neutrophil formyl peptide receptor using a site-specific fluorescent photoaffinity label and mass spectroscopy. *J. Biol. Chem.* **273**, 10428–10435.

Miyamasu, M., Hirai, K., Takahashi, Y., Iida, M., Yamaguchi, M., Koshino, T., Takaishi, T., Morita, Y., Ohta, K., Kasahara, T., and Koji, I. (1995). Chemotactic agonists induce cytokine generation in eosinophils. *J. Immunol.* **154**, 1339–1349.

Murphy, P. M. (1994). The molecular biology of leukocyte chemoattractant receptors. *Annu. Rev. Immunol.* **12**, 593–633.

Murphy, P. M., Ozcelik, T., Kenney, R. T., Tiffany, H. L., McDermott, D., and Francke, U. (1992). A structural homologue of the N-formyl peptide receptor. Characterization and chromosome mapping of a peptide chemoattractant receptor family. *J. Biol. Chem.* **267**, 7637–7643.

Murphy, P. M., Tiffany, H. L., McDermott, D., and Ahuja, S. K. (1993). Sequence and organization of the human N-formyl peptide receptor-encoding gene. *Gene* **133**, 285–290.

Nick, J. A., Avdi, N. J., Young, S. K., Knall, C., Gerwins, P., Johnson, G. L., and Worthen, G. S. (1997). Common and distinct intracellular signaling pathways in human neutrophils utilized by platelet-activating factor and FMLP. *J. Clin. Invest.* **99**, 975–986.

Niedel, J., Davis, J., and Cuatrecasas, P. (1980). Covalent affinity labeling of the formyl peptide chemotactic receptor. *J. Biol. Chem.* **255**, 7063–7066.

Perez, H., Kelly, E., Elfman, F., Armitage, G., and Winkler, J. (1991). Defective polymorphonuclear leukocyte formyl peptide receptor(s) in juvenile periodontitis. *J. Clin. Invest.* **87**, 971–976.

Perez, H. D., Holmes, R., Kelly, E., McClary, J., Chou, Q., and Andrews, W. H. (1992). Cloning of the gene coding for the human receptor for formyl peptides: characterization of a promoter region and evidence for polymorphic expression. *Biochemistry* **31**, 11595–11599.

Prossnitz, E. R. (1997). Desensitization of N-formylpeptide receptor-mediated activation is dependent upon receptor phosphorylation. *J. Biol. Chem.* **273**, 15213–15219.

Prossnitz, E. R., Kim, C. M., Benovic, J. L., and Ye, R. D. (1995a). Phosphorylation of the N-formyl peptide receptor carboxyl terminus by the G protein-coupled receptor kinase, GRK2. *J. Biol. Chem.* **270**, 1130–1137.

Prossnitz, E. R., Schreiber, R. E., Bokoch, G. M., and Ye, R. D. (1995b). Binding of low affinity N-formyl peptide receptors to G protein. Characterization of a novel inactive receptor intermediate. *J. Biol. Chem.* **270**, 10686–10694.

Ptasznik, A., Traynor-Kaplan, A., and Bokoch, G. M. (1995). G protein-coupled chemoattractant receptors regulate Lyn tyrosine kinase-Shc adapter signaling complexes. *J. Biol. Chem.* **270**, 19969–19973.

Quehenberger, O., Prossnitz, E. R., Cavanagh, S. L., Cochrane, C. G., and Ye, R. D. (1993). Multiple domains of the N-formyl peptide receptor are required for high-affinity ligand binding. Construction and analysis of chimeric N-formyl peptide receptors. *J. Biol. Chem.* **268**, 18167–18175.

Quehenberger, O., Pan, Z. K., Prossnitz, E. R., Cavanagh, S. L., Cochrane, C. G., and Ye, R. D. (1997). Identification of an N-formyl peptide receptor ligand binding domain by a gain-of-function approach. *Biochem. Biophys. Res. Commun.* **238**, 377–381.

Rane, M. J., Carrithers, S. L., Arthur, J. M., Klein, J. B., and McLeish, K. R. (1997). Formyl peptide receptor are coupled to multiple mitogen-activated protein kinase cascades by distinct signal transduction pathways. *J. Immunol.* **159**, 5070–5078.

Sengelov, H., Boulay, F., Kjeldsen, L., and Borregaard, N. (1994). Subcellular localization and translocation of the receptor for N formylmethionyl-leucyl-phenylalanine in human neutrophils. *Biochem. J.* **299**, 473–479.

Shannon, M. F., Gamble, J. R., and Vadas, M. A. (1988). Nuclear proteins interacting with the promoter region of the human granulocyte/macrophage colony-stimulating factor gene. *Proc. Natl Acad. Sci. USA* **85**, 674–678.

Showell, H. J., Freer, R. J., Zigmond, S. H., Schiffmann, E., Aswanikumar, S., Corcoran, B., and Becker, E. L. (1976). The structure-activity relations of synthetic peptides as chemotactic factors and inducers of lysosomal enzyme secretion for neutrophils. *J. Exp. Med.* **143**, 1155–1169.

Sklar, L. A., Finney, D. A., Oades, Z. G., Jesaitis, A. J., Painter, R. G., and Cochrane, C. G. (1984). The dynamics of ligand-receptor interactions. *J. Biol. Chem.* **259**, 5661–5669.

Snyderman, R., and Uhing, R. J. (1992). In "Inflammation: Basic Principles and Clinical Correlates, 2nd edn" (ed. J. L. Gallin, I. M. Goldstein, and R. Snyderman), Chemoattractant stimulus-response coupling: The role of ion fluxes in chemoattractant actions, pp. 421–439. Raven Press, New York.

Sozzani, S., Sallusto, F., Luini, W., Zhou, D., Piemonti, L., Allavena, P., Van Damme, J., Valitutti, S., Lanzavecchia, A., and Mantovani, A. (1995). Migration of dendritic cells in response to formyl peptides, C5a, and a distinct set of chemokines. *J. Immunol.* **155**, 3292–3295.

Spisani, S., Pareschi, M. C., Buzzi, M., Colamussi, L., Biondi, C., Traniello, S., Zecchini, G. P., Paradisi, M. P., Torrini, I., and

Ferretti, M. E. (1996). Effect of cyclic AMP level reduction on human neutrophil responses to formylated peptides. *Cell. Signal.* **8**, 269–277.

Stoyanov, B., Volinia, S., Hanck, T., Rubio, I., Loubtchenkov, M., Malek, D., Stoyanova, S., Vanhaesebroeck, B., Dhand, R., Nürnberg, B., Gierschik, P., Seedorf, K., Hsuan, J. J., Waterfield, M. D., and Wetzker, R. (1995). Cloning and characterization of a G protein-activated human phoshoinositide-3 kinase. *Science* **269**, 690–693.

Tapon, N., and Hall, A. (1997). Rho, Rac and Cdc42 GTPases regulate the organization of actin cytoskeleton. *Curr. Opin. Cell Biol.* **9**, 86–92.

Tardif, M., Mery, L., Brouchon, L., and Boulay, F. (1993). Agonist-dependent phosphorylation of N-formylpeptide and activation peptide from the fifth component of C (C5a) chemoattractant receptors in differentiated HL60 cells. *J. Immunol.* **150**, 3534–3545.

Wenzel-Seifert, K., Hurt, C. M., and Seifert, R. (1998). High constitutive activity of the human formyl peptide receptor. *J. Biol. Chem.* **273**, 24181–24189.

Williams, L., Snyderman, R., Pike, M., and Lefkowitz, R. J. (1977). Specific receptor sites for chemotactic peptides on human polymorphonuclear leukocytes. *Proc. Natl Acad. Sci. USA* **74**, 1204–1208.

Wilson, D. R., Juan, T. S.-C., Wilde, M. D., Fey, G. H., and Darlington, G. J. (1990). A 58-base-pair region of the human C3 gene confers synergistic inducibility by interleukin-1 and interleukin-6. *Mol. Cell. Biol.* **10**, 6181–6191.

Worthen, S. G., Avdi, N., Buhl, A. M., Suzuki, N., Johnson, G. L. (1994). fMLP activates Ras and Raf in human neutrophils. *J. Clin. Invest.* **94**, 815–823.

Ye, R., and Boulay, F. (1997). Structure and function of leukocyte chemoattractant receptors. *Adv. Pharmacol.* **39**, 221–290.

Ye, R. D., Cavanagh, S. L., Quehenberger, O., Prossnitz, E. R., and Cochrane, C. G. (1992). Isolation of a cDNA that encodes a novel granulocyte N-formyl peptide receptor. *Biochem. Biophys. Res. Commun.* **184**, 582–589.

Ye, R. D., Quehenberger, O., Thomas, K. M., Navarro, J., Cavanagh, S. L., Prossnitz, E. R., and Cochrane, C. G. (1993). The rabbit neutrophil N-formyl peptide receptor. cDNA cloning, expression, and structure/function implications. *J. Immunol.* **150**, 1383–1394.

Zigmond, S. H., Woodworth, A., and Daukas, G. (1985). Effects of sodium on chemotactic peptide binding to polymorphonuclear leukocytes. *J. Immunol.* **135**, 531–536.

Zohn, I. M., Campbell, S. L., Khosravi-Far, R., Rossman, K. L., and Der, C. J. (1998). Rho family proteins and Ras transformation: the RHOad less traveled gets congested. *Oncogene* **17**, 1415–1438.

Zu, Y.-L., Ai, Y., Gilchrist, A., Labadia, M. E., Sha'afi, R. I., and Huang, C.-K. (1996). Activation of MAP kinase-activated protein kinase 2 in human neutrophils after phorbol ester or fMLP peptide stimulation. *Blood* **87**, 5287–5296.

Opioid μ, δ, and κ Receptors for Endorphins

Daniel J. J. Carr[1] and J. Edwin Blalock[2,*]

[1]Department of Microbiology and Immunology, LSU Medical Center, New Orleans, LA 70112-1393, USA

[2]Department of Physiology and Biophysics, University of Alabama at Birmingham, Birmingham, AL 35294-0005, USA

*corresponding author tel: 205-934-6439, fax: 205-934-1446, e-mail: blalock@uab.edu

DOI: 10.1006/rwcy.2000.23005.

SUMMARY

The three major types of opioid receptors (μ, δ, and κ) are members of the DRY (Asp-Arg-Tyr)-containing subfamily of seven transmembrane spanning receptors. Opioid receptors on cells of the immune system are virtually identical to those on neuronal cells. The activation of opioid receptors in the CNS leads to analgesia while activation of opioid receptors on immune cells can enhance or suppress immune function depending on the target cell and immune parameter measured.

BACKGROUND

Discovery

The molecular characterization of opioid receptors has been investigated for nearly 25 years. However, the activities of these receptors, as manifested in the effects of opioid compounds (e.g. opium, of which the main active ingredient is morphine), have been known for at least 6000 years, since the time of the Sumerians (4000 BC). The discovery of endogenous opioid peptides, including the enkephalins (met- and leu-enkephalin) (Hughes *et al.*, 1975), the endorphins (α, β, and γ) (Bradbury *et al.*, 1976; Cox *et al.*, 1976), dynorphin (Goldstein *et al.*, 1979), and endomorphin (Zadina *et al.*, 1997), suggested the existence of multiple types of receptors (termed opioid receptors) for these natural ligands. Evidence for multiple types of opioid receptors was obtained using congeners of morphine in spinal studies in dogs (Gilbert and Martin, 1976; Martin *et al.*, 1976).

Originally identified in the early 1970s (Pert and Snyder, 1973; Simon *et al.*, 1973; Terenius, 1973), today, three main types of opioid receptors have been defined and cloned: δ, κ, and μ opioid receptors with pharmacologically distinct subtypes for δ (δ_1 and δ_2), κ (κ_1, κ_2, and κ_3), and μ (μ_1 and μ_2) (Pasternak, 1993). Two other receptors including ε (specific for β-endorphin) and σ receptors were originally described as opioid receptors but have since been redefined as nonopioid (Simon, 1991). All of these receptors have been identified and characterized on cells of the immune system (Garza and Carr, 1997).

Another opioid-like receptor, referred to as the nociceptive receptor (orphan opioid receptor), originally described in human brainstem (Mollereau *et al.*, 1994) was also found in mouse spleen T and B lymphocytes, where it was first coupled to a physiological role (Halford *et al.*, 1995). This review focuses on the immune cell-derived opioid receptors, comparing their physicochemical properties with those found in the nervous system as well as defining their role in the immune system.

Structure

The three major types of opioid receptors are members of the (DRY)-containing subfamily of seven transmembrane spanning receptors. There is 60% amino acid identity between each type of opioid receptor with the membrane-spanning regions (transmembrane I–VII) and the intracellular loops connecting these segments being highly conserved between receptor types. Studies indicate that ligands (agonists and antagonists) to these receptors bind to different regions of the extracellular domain and such interaction can be greatly influenced by the transmembrane segments (predominantly TM II, III, and VI) (Kong *et al.*, 1993, 1994; Surratt *et al.*, 1994). Also, changes in one amino acid in the TM IV spanning region has been shown to alter opioid antagonist to agonist activity (Claude *et al.*, 1996). Since the amino acid sequences of the neuronal- and some immune-derived receptors are nearly identical, it is predicted that a similar relationship between agonist/antagonist-binding domains and the influence of the transmembrane spanning regions will be found in the immune-derived opioid receptors. However, immune-derived opioid receptor-binding domains according to some investigators may be distinct since binding or biochemical characteristics of these sites are not characteristic of neuronal opioid sites (e.g. Stefano *et al.*, 1992; Makman *et al.*, 1995).

Main activities and pathophysiological roles

The primary function associated with neuronal opioid receptors is the control of the sensation of pain either through receptors located spinally (δ_1, κ_1, and μ_2) or supraspinally (δ_2, κ_2, κ_3, and μ_1) (Pasternak, 1993). Within the immune system, opioid receptors found on immune cells may augment or suppress immune function depending on the cell type and stimulation (Carr, 1991). However, alkaloid opioid ligands (e.g. morphine and fentanyl) are potent immunosuppressive compounds affecting the immune system primarily by indirect pathways ligating to receptors found within the CNS and activating secondary systems (including the adrenergic pathway and the hypothalamus-pituitary-adrenal axis) (Carr *et al.*, 1996). Other functions of immune-derived opioid receptors may pertain to the response to infectious pathogens. As an example, κ opioid receptors bound to κ-selective opioid ligands have been found to reduce significantly monocytotropic *HIV*-1 SF162 strain replication in microglia-enriched cultures (Chao *et al.*, 1996).

GENE

Accession numbers

The δ (L06322, L11065), κ (L11064), and μ (L22455, L20684) opioid receptors have been cloned from neuronal tissue (Evans *et al.*, 1992; Kieffer *et al.*, 1992; Li *et al.*, 1993; Thompson *et al.*, 1993; Wang *et al.*, 1993; Yasuda *et al.*, 1993).

PROTEIN

Accession numbers

Protein Information Resource:
Human μ opioid receptor: 2135858
Human κ opioid receptor: 631277
Human δ opioid receptor: 2134989

Sequence

The neuronal opioid receptors are composed of between 370 and 389 amino acids encoded by mRNAs ranging in size from 1.9 to >10.0 kb (Carr *et al.*, 1996). Both δ (**Figure 1a**; Sedqi *et al.*, 1996) and κ (Figure 1b; Belkowski *et al.*, 1995) receptor full-length cDNAs (predicted to be 372–400 amino acids in length) have been identified in thymocytes or a thymoma cell line. However, only a partial sequence (441 bp) of a μ opioid receptor has been identified by RT-PCR in peripheral blood mononuclear cells (Chuang *et al.*, 1995) and rat peritoneal macrophages (Figure 1c; 721 bp) (Sedqi *et al.*, 1995). All immune cell-derived receptor sequences identified thus far are nearly identical ($\geq$99% homology) with the receptors in the nervous system.

Description of protein

By a variety of techniques, the neuronal opioid receptors were observed to range in size from 40 to 65 kDa (Simonds, 1988; Loh and Smith, 1990; Wollemann, 1990). The data concerning the biochemical properties of these receptors may be limited by the uncertain specificities of some of the antireceptor antisera.

Figure 1 (a) Deduced amino acid sequence of the δ opioid receptor cloned from activated murine thymocytes as reported by Sedqi *et al.* (1996). Bold letters indicate changes from the published rodent brain δ opioid receptor. (b) Deduced amino acid sequence of the κ opioid receptor cloned from R1.1 thymoma cell line as reported by Alicea *et al.* (1998). Bold letters indicate changes from the published rodent brain κ opioid receptor. (c) Deduced amino acid sequence of the μ opioid receptor cloned from adherent peritoneal macrophages as reported by Sedqi *et al.* (1995). Bold letters indicate changes from the published rodent brain μ opioid receptor.

(a)

MELVPSARAELQSSPLVNLSDAFPSAFPSA 30
GANASGSPGARSASSLALAIAIT**V**LYSAVC 60
AVGLLGNVLVMFGIVRYTKLKTATNIYIFN 90
LALADALATST**I**PFQSAKYLMETWPFGELL 120
CKAVLSIDYYNMFTSIFTLTMMSVDRYIAV 150
CHPVKALDFRTPAKAKLI**Q**ICIWVLASGVG 180
VPIMVMAVTQPRDGAVVCMLQFPSPSWYWD 210
TVTKICVFLFAFVVPILIITVCYGLMLLRL 240
RSVRLLSGSKEKDRSLRRITRMVLVVVGAF 270
VVCWAPIHIFVIVWTLVDINRRDPLVVAAL 300
HLCIALGYANSSLNPVLYAFLDENFKRCFR 330
QLCRTPCGRQEPGSLRRPRQATTRERVTAC 360
TPSDGPGGGAAA 372

(b)

MESPIQIFRGNPGPTCSPSACLLPDSSSWF 30
PDWAESNSDGSVGSENQQLESAHISPAIPV 60
IITAVNSVVFVVGLVGDSLVMFVIIRIYTK 90
MKTATDIYIFDLALANALVTTTMPFQSAVY 120
LMDSWPFGNVLCKIVISINYYDMFTSIFTL 150
TMMSVNRYIAVCHPVKALNFRTPLKAKIID 180
ICIWL**L**ASSVGISAIVLGG**T**KVRENVNVIE 210
CSLQFPNNEYSWWNLFMKICVF**V**FAFVIPV 240
LIIIVCYTLMILRLKSVRVLSGSREKNRDL 260
RRITKLVLVVVAVFIICWTPIHIFILVEAL 290
GSTSHSTAALSSYYFCIALGYTDSSLDPVL 320
YAFLNEDFKRCFRNFCFPIKMRMERQSTDR 350
DTVQNPASMRNVGGMDKPV 369

(c)

MGTWPFGTILCKI**V**ISIDYYNMFTSIFTLC 30
TMSVDRYIAVCHPVK**A**LDFRTPRNAKIVNV 60
CNWILSSAIGLPVMFMATTKYRQGSIDCTL 90
TFSHPTWYWQNLLKICVFIFAFIMPILIIT 120
VCYALMILRLKSVRMLSGSKEKNRDLRRITR 150
MVLVVVAVFIVCWTPIHIYVIIKALITIPE 180
TTFQTVSWHFCIALGYTDSCLDPVLYAFLN 210

Using a site-directed acylating agent derived from fentanyl (known as superfit) that is highly selective for δ opioid receptors, a comparison of the mouse brain cell- and spleen cell-derived δ opioid receptor. Superfit labeled proteins migrating at 70, 46, and 31 kDa from brain tissue, while a major protein species migrating at 31 kDa was labeled from spleen tissue. The 31 kDa species was thought to be a degradative form of the mature protein. Subsequent analysis of immune cell-derived δ, κ, and μ opioid receptors determined the size to be nearly identical to that of the neuronal receptors (Carr, 1991).

Relevant homologies and species differences

Opioid receptors from immune cells are virtually identical ($\geq$99% homology) to those on neuronal cells. The various opioid receptor types (μ, δ, κ) show about 60% homology.

Affinity for ligand(s)

The identification of the types of opioid receptors has been greatly facilitated by the design and synthesis of opioid ligands selective for the types of receptors to which they bind (**Table 1**). Similar to cell-associated neuronal opioid receptors, the cloned neuronal opioid receptors expressed in PC-12 cells showed high-affinity binding to ligands ranging from 0.2 to 3.0 nM (Raynor *et al.*, 1994). Opioid receptors found on cells of the immune system display a modestly reduced affinity for their ligands ranging from 20 to 900 nM, depending on the ligand and receptor (Garza and Carr, 1997). For example, κ receptors display affinities ranging from 4.1 to 65.0 nM (Garza and Carr, 1997), whereas a unique alkaloid-specific μ_3 opioid receptor found on granulocytes has a K_d of 44 nM (Makman *et al.*, 1995). One investigation reported the IC_{50} for an immunoaffinity-purified opioid receptor isolated from mouse spleen preparations to be approximately 700 nM, suggesting a loss in affinity upon purification (Carr *et al.*, 1990).

Cell types and tissues expressing the receptor

Within the immune system, there is some disagreement as to the population of cells that express opioid receptors. To this end, **Table 2** and **Table 3** summarize the evidence for the presence of opioid receptors on primary cells of the immune system (Table 2) and cell lines derived from cells of the immune system (Table 3) based on pharmacological (radioreceptor

Table 1 Commercially available selective opioid agonists/antagonists

δ Opioid receptor ligands	κ Opioid receptor ligands	μ Opioid receptor ligands
DADLE (agonist)	Bremazocine (agonist)	DAMGO (agonist)
DPDPE (agonist)	U-50488 (agonist)	Endomorphin 1 (agonist)
SNC 80 (agonist)	U-69593 (agonist)	Endomorphin 2 (agonist)
DSLET (agonist)	ICI-199,441 (agonist)	Fentanyl citrate (agonist)
SNC121 (agonist)	Nor-binaltorphimine (antagonist)	β-Funaltrexamine (antagonist)
Superfit (affinity label)	DIPPA (antagonist)	Naloxonazine (antagonist)
Naltrindole (antagonist)		Cyprodime HBr (antagonist)
ICI-174,864 (antagonist)		
BNTX (antagonist)		
Naltriben (antagonist)		

Table 2 Evidence for the presence of opioid receptors on primary cells of the immune system[a]

Cell type	δ Receptor	κ Receptor	μ Receptor
Mouse T lymphocyte	B, M	–	B
Mouse B lymphocyte	B	–	B
Mouse thymocyte	–	–	P, M
Mouse splenocyte	B, M	B	B
Human PBLs	M	M	–
Human T lymphocyte	P, B	M	–
Human B lymphocyte	B	–	–
Human granulocyte	–	–	P, M
Human monocyte	–	M	M
Monkey PBLs	M	M	M
Human microglia	–	M	–
Rat macrophage	–	–	M

[a]Evidence for the existence of the receptors is defined using pharmacological (P), biochemical (B), or molecular biology (M) approaches (Alicea *et al.*, 1998; Chao *et al.*, 1996; Gaveriaux *et al.*, 1995; Miller, 1996; Roy *et al.*, 1992; Wick *et al.*, 1996) or as reviewed by Carr (1991), Carr *et al.* (1996).

–, Suggests either a lack of detection or that the analysis has not yet been determined.

assays), biochemical (affinity labeling), or molecular biology techniques (cloning or RT-PCR studies).

Regulation of receptor expression

The expression of δ and μ opioid receptors on immune cells is reportedly induced by the activation of cells by IL-1 in the case of the thymocyte μ receptor (Roy *et al.*, 1992) or mitogen (concanavalin A) in the case of the mouse T lymphocyte δ opioid receptor (Miller, 1996). Furthermore, the activation of leukocytes also leads to the production of endogenous opioid peptides (Blalock, 1989). Since the leukocyte-derived opioid peptides have also been shown to be functionally active (Blalock, 1989), there is reason to believe that autocrine regulation of receptor expression may occur as well.

Table 3 Evidence for the presence of opioid receptors on cell lines derived from cells of the immune system[a]

Cell type	δ Receptor	κ Receptor	μ Receptor
Mouse T cell lines			
EL-4	M	P	–
11.10	M	–	–
Mouse B cell line			
CH27	M	–	–
Mouse macrophage cell line			
P388d$_1$	B	P, B	
Mouse R1.1 thymoma	–	P, M	–
Human T cell lines			
CEMx174	M	M	M
HSB2	M	–	–
MOLT-4	B, M	M	–
Human B cell line			
EBV-transformed	M	M	
Human monocyte cell line			
U937	M	–	–

[a]Evidence for the existence of the receptors is defined using pharmacological (P), biochemical (B), or molecular biology (M) approaches (Gaveriaux *et al.*, 1995; Chao *et al.*, 1996; Wick *et al.*, 1996; Alicea *et al.*, 1998) or as reviewed by Carr (1991), Carr *et al.* (1996).

–, Suggests either a lack of detection or that the analysis has not yet been determined.

SIGNAL TRANSDUCTION

Cytoplasmic signaling cascades

Neuronal opioid receptors modify a variety of signaling cascades including cAMP through the activation of G_i, increases in GTPase activity, phosphatidylinositol turnover, mobilization of Ca^{2+}, and K^+ channel activity (Childers, 1991; Chen and Yu, 1994). In a similar fashion, immune cell-derived opioid receptors are coupled to a G_i protein and influence K^+ channel conductance and calcium mobilization (Carr, 1991). In addition, the endogenous opioid peptide β-endorphin has been shown to modify CD3γ phosphorylation following phorbol ester stimulation, either increasing or decreasing phosphorylation of the CD3 chain depending on the concentration of the peptide (Kavelaars *et al.*, 1990). These results suggest that the endorphins may act as a governor on T cell activation depending on the local concentration of endogenous opioid peptide. Specifically, endorphins at mid-picomolar levels may augment T cell activation through the increase in phosphorylation of the CD3 complex intracellular tyrosine-activation motifs (ITAMs) and presumably the activation of the inositol trisphosphate (IP_3) cascade via ZAP-70, whereas at femtomolar levels the endorphins would suppress T cell activation by reducing phosphorylation.

DOWNSTREAM GENE ACTIVATION

Transcription factors activated

The success in transfecting Jurkat T cells (which do not express opioid receptors, Gaveriaux *et al.*, 1995) with a functional δ opioid receptor (Sharp *et al.*, 1996) allowed for the identification of potential transcriptional regulatory elements involved in opioid modulation of immune function. Previous studies reported the augmentation of IL-2 production by activated T cells stimulated with endogenous opioid peptides

(Carr, 1991). In an elegant study, reporter gene constructs were used to map deltorphin (δ selective agonist)-elicited augmentation of IL-2 production by δ opioid receptor-transfected Jurkat T cells to the AP-1- and NF-AT/AP-1-binding site (Hedin *et al.*, 1997). This effect was apparently independent of calcineurin and unrelated to the elevation in $[Ca_i^{2+}]$ but required pertussis toxin-sensitive G protein. Since the NF-AT/AP-1 complex is involved in the induction of a number of cytokine genes (Rao, 1994) and endogenous opioid peptides modify the production of a number of cytokines (Peterson *et al.*, 1998), it is quite possible that the NF-AT/AP-1 complex is involved. In addition, since leukocyte activation is primarily mediated by cytokines, it is highly probable that opioid receptor promoters possess binding domains for cytokine responsive elements. As an example, the μ opioid receptor promoter possesses a NF-IL6 domain (Min *et al.*, 1994).

Genes induced

IL-2.

BIOLOGICAL CONSEQUENCES OF ACTIVATING OR INHIBITING RECEPTOR AND PATHOPHYSIOLOGY

Unique biological effects of activating the receptors

The response to opioid receptor activation depends on the location of the receptor, the type of receptor, and the level of activation of the cell population. Endogenous opioid peptides can either enhance or suppress immune function depending in part on the state of target cell activation and the immune parameter (antibody production, natural killer activity, cytokine synthesis) measured. Likewise, peripheral blood mononuclear cells from individuals can respond differently (sometimes completely opposite of one another) to opioid ligands evident in both human and mouse populations. However, the administration of opioid alkaloids (i.e. morphine, heroin, or fentanyl) tends to elicit a significant suppression of immune function primarily by opioid receptors found in the mesencephalon (Shavit *et al.*, 1986; Weber and Pert, 1989).

The activation of the 'central' opioid receptors elicits the activation of neuroendocrine pathways. The hypothalamic-pituitary-adrenal axis results in the production of adrenal steroids such as glucocorticoids which suppress immune responses in part by preventing translocation of NFκB to the nucleus (Baldwin, 1996). Alternatively, morphine may activate the sympathetic/parasympathetic arm of the autonomic nervous system known to innervate lymph nodes and spleen (Felten *et al.*, 1987) and modify immune function through the release of monoamines (e.g. catecholamines) (Carr and Serou, 1995).

Other studies suggest that endogenous opioid peptides may supplement antimicrobial drugs or local immune reactivity against viral infections. Studies have suggested that met-enkephalin suppresses *influenza virus* infection in mice through the effects on natural killer cells and cytotoxic T lymphocytes (Burger *et al.*, 1995). Another study has found that met-enkephalin synergizes with azidothymidine in blocking *feline leukemia virus* replication (Specter *et al.*, 1994). It has also been reported that endogenous opioids induce the synthesis of novel fentanyl derivatives that possess analgesic activity in the absence of opioid immunosuppression (Carr and Serou, 1995).

Phenotypes of receptor knockouts and receptor overexpression mice

Mu opioid receptor (MOR) knockout mice have been developed and tested for immune deviation in the presence and absence of the clinically relevant, prototypic μ ligand morphine. MOR knockout mice exhibit normal immunological endpoints including natural killer activity, antibody production, and mitogen-induced lymphocyte proliferation (Gaveriaux-Ruff *et al.*, 1998). However, bone marrow cells from MOR knockout mice exhibit an altered pattern of early hematopoiesis (Tian *et al.*, 1997). In addition, treatment with morphine had no effect on immune parameters assayed in MOR knockout mice, but significantly suppressed selectively measured immune parameters (e.g. natural killer cell activity) and induced lymphoid organ atrophy in wild-type mice (Gaveriaux-Ruff *et al.*, 1998). These results suggest that the absence of the μ opioid receptor has no detrimental effect on immunocompetence *per se*, but is directly responsible for the immunomodulatory effects of exogenous morphine. Accordingly, modification of immune responses to antigen or microbial pathogens by endogenous opioid peptides does not necessarily involve the activation of μ opioid receptors on immune cells.

THERAPEUTIC UTILITY

Effects of inhibitors (antibodies) to receptors

The classical pharmacological definition of the existence of opioid receptors on cells of the immune system has come from the ability of opioid antagonists (competitive or noncompetitive) to block the immunomodulatory effects in a stereospecific manner (Sibinga and Goldstein, 1988) (see Table 1). Antibodies have also been generated against opioid receptors that recognize proteins expressed on cells of the immune system. One such antibody was found to possess agonist activity and to recognize a putative δ class opioid receptor on mouse leukocytes (Carr *et al.*, 1990). A second antibody was generated against the predicted N-terminal sequence of a κ opioid receptor and was found to act as a noncompetitive selective antagonist recognizing κ opioid receptors on U937 cells (Buchner *et al.*, 1997). However, the use of antibodies is more apt to focus on characterizing the structural properties of the cloned receptor (e.g. mapping ligand-binding domains) rather than using such antibodies to antagonize tolerance or chemical dependence.

References

Alicea, C., Belkowski, S. M., Sliker, J. K., Zhu, J., Liu-Chen, L.-Y., Eisenstein, T. K., Adler, M. W., and Rogers, T. J. (1998). Characterization of κ-opioid receptor transcripts expressed by T cells and macrophages. *J. Neuroimmunol.* **91**, 55–62.

Baldwin, A. S. (1996). The NF-κB and IκB proteins: New discoveries and insights. *Annu. Rev. Immunol.* **14**, 649–681.

Belkowski, S. M., Zhu, J., Liu-Chen, L.-Y., Eisenstein, T. K., Adler, M. W., and Rogers, T. J. (1995). Sequence of κ-opioid receptor cDNA in the R1.1 thymoma cell line. *J. Neuroimmunol.* **62**, 113–117.

Blalock, J. E. (1989). A molecular basis for bidirectional communication between the immune and neuroendocrine systems. *Physiol. Rev.* **69**, 1–32.

Bradbury A. F., Smyth, D. G., and Snell, C. R. (1976). Biosynthetic origin and receptor conformation of methionine enkephalin. *Nature* **260**, 165–166.

Buchner, R. R., Bogen, S. M., Fischer, W., Thoman, M. L., Sanderson, S. D., and Morgan, E. L. (1997). Anti-human κ opioid receptor antibodies. *J. Immunol.* **158**, 1670–1680.

Burger, R. A., Warren, R. P., Huffman, J. H., and Sidwell, R. W. (1995). Effect of methionine-enkephalin on natural killer cell and cytotoxic T lymphocyte activity in mice infected with influenza A virus. *Immunopharmacol. Immunotoxicol.* **17**, 323–334.

Carr, D. J. J. (1991). The role of endogenous opioids and their receptors in the immune system. *Proc. Soc. Exp. Biol. Med.* **198**, 710–720.

Carr, D. J. J., and Serou, M. (1995). Exogenous and endogenous opioids as biological response modifiers. *Immunopharmacology* **31**, 59–71.

Carr, D. J. J., DeCosta, B. R., Kim, C.-H., Jacobson, A. E., Bost, K. L., Rice, K. C., and Blalock, J. E. (1990). Anti-opioid receptor antibody recognition of a binding site on brain and leukocyte opioid receptors. *Neuroendocrinology* **51**, 552–560.

Carr, D. J. J., Rogers, T. J., and Weber, R. J. (1996). The relevance of opioids and opioid receptors on immunocompetence and immune homeostasis. *Proc. Soc. Exp. Biol. Med.* **213**, 248–257.

Chao, C. C., Gekker, G., Hu, S., Sheng, W. S., Shark, K. B., Bu, D.-F., Archer, S., Bidlack, J. M., and Portoghese, P. K. (1996). κ opioid receptors in human microglia downregulate human immunodeficiency virus 1 expression. *Proc. Natl Acad. Sci. USA* **93**, 8051–8056.

Chen, Y., and Yu, L. (1994). Differential regulation by cAMP-dependent protein kinase and protein kinase C of the μ opioid receptor coupling to a G protein-activated K^+ channel. *J. Biol. Chem.* **269**, 7839–7842.

Childers, S. R. (1991). Opioid receptor-coupled second messenger systems. *Life Sci.* **48**, 1991–2003.

Chuang, T. K., Killam Jr, K. F., Chuang, L. F., Kung, H.-F., Sheng, W. S., Chao, C. C., Yu, L., and Chuang, R. Y. (1995). Mu opioid receptor gene expression in immune cells. *Biochem. Biophys. Res. Commun.* **216**, 922–930.

Claude, P. A., Wotta, D. R., Zhang, X. H., Prather, P. L., McGinn, T. M., Erickson, L. J., Loh, H. H., and Law, P. Y. (1996). Mutation of a conserved serine in TM4 of opioid receptors confers full agonistic properties to classical antagonists. *Proc. Natl Acad. Sci. USA* **93**, 5715–5719.

Cox, B. M., Goldstein, A., and Li, C. H. (1976). Opioid activity of a peptide, beta-lipotropin-(61-91), derived from beta-lipotropin. *Proc. Natl Acad. Sci. USA* **73**, 1821–1823.

Evans, C. J., Keith, D., Magendzo, K., Morrison, H., and Edwards, R. H. (1992). Cloning of a delta opioid receptor by functional expression. *Science* **258**, 1952–1955.

Felten, D. L., Felten, S. Y., Bellinger, D. L., Carlson, S. L., Ackerman, K. D., Madden, K. S., Olschowka, J. A., and Livnat, S. (1987). Noradrenergic sympathetic neural interactions with the immune system: Structure and function. *Immunol. Rev.* **100**, 225–260.

Garza Jr, H. H., and Carr, D. J. J. (1997). In "Chemical Immunology: Neuroimmunoendocrinology" (ed J. E. Blalock), Neuroendocrine peptide receptors on cells of the immune system, pp. 132–154. Karger, Basel.

Gaveriaux, C., Peluso, J., Simonin, F., Laforet, J., and Kieffer, B. (1995). Identification of κ- and δ-opioid receptor transcripts in immune cells. *FEBS Lett.* **369**, 272–276.

Gaveriaux-Ruff, C., Matthes, H. W. D., Peluso, J., and Kieffer, B. L. (1998). Abolition of morphine-immunosuppression in mice lacking the μ-opioid receptor gene. *Proc. Natl Acad. Sci. USA* **95**, 6326–6330.

Gilbert, P. E., and Martin, W. R. (1976). The effects of morphine- and nalorphine-like drugs in the nondependent, morphine-dependent and cyclazocine-dependent chronic spinal dog. *J. Pharmacol. Exp. Ther.* **198**, 66–82.

Goldstein, A., Tachibana, S., Lowney, L. I., Hunkapiller, M., and Hood, L. (1979). Dynorphin-(1-13), an extraordinary potent opioid peptide. *Proc. Natl Acad. Sci. USA* **76**, 6666–6670.

Halford W. P., Gebhardt, B. M., and Carr, D. J. J. (1995). Functional role and sequence analysis of a lymphocyte orphan opioid receptor. *J. Neuroimmunol.* **59**, 91–101.

Hedin, K. E., Bell, M. P., Kalli, K. R., Huntoon, C. J., Sharp, B. M., and McKean, D. J. (1997). δ-Opioid receptors expressed by Jurkat T cells enhance IL-2 secretion by increasing AP-1 complexes and activity of the NF-AT/AP-1 binding promoter element. *J. Immunol.* **159**, 5431–5440.

Hughes, J., Smith, T. W., Kosterlitz, H. W., Fothergill, L. A., Morgan, B. A., and Morris, H. R. (1975). Identification of two related pentapeptides from the brain with potent opiate agonist activity. *Nature* **258**, 577–580.

Kavelaars, A., Eggen, B. J. L., De Graan, P. N. E., Gispen, W. H., and Heijnen, C. J. (1990). The phosphorylation of the CD3γ chain of T lymphocytes is modulated by β-endorphin. *Eur. J. Immunol.* **20**, 943–945.

Kieffer, B. L., Befort, K., Gaveriaux-Ruff, C., and Hirth, C. G. (1992). The δ-opioid receptor: Isolation of a cDNA by expression cloning and pharmacological characterization. *Proc. Natl Acad. Sci. USA* **89**, 12048–12052.

Kong, H., Raynor, K., Yasuda, K., Moe, S. T., Portoghese, P. S., Bell, G. I., and Reisine, T. (1993). A single residue, aspartic acid 95, in the δ opioid receptor specifies selective high affinity agonist binding. *J. Biol. Chem.* **268**, 23055–23058.

Kong, H., Raynor, K., Yano, H., Takeda, J., Bell, G. I., and Reisine, T. (1994). Agonists and antagonists bind to different domains of the cloned κ receptor. *Proc. Natl Acad. Sci. USA* **91**, 8042–8046.

Li, S., Zhu, J., Chen, C., Chen, Y.-W., DeRiel, J. K., Ashby, B., and Liu-Chen, L.-Y. (1993). Molecular cloning and expression of a rat kappa opioid receptor. *J. Biol. Chem.* **295**, 629–633.

Loh, H. H., and Smith, A. P. (1990). Molecular characterization of opioid receptors. *Annu. Rev. Pharmacol. Toxicol.* **30**, 123–147.

Makman, M. H., Bilinger, T. V., and Stefano, G. B. (1995). Human granulocytes contain an opiate alkaloid-selective receptor mediating inhibition of cytokine-induced activation and chemotaxis. *J. Immunol.* **154**, 1323–1330.

Martin, W. R., Eades, C. G., Thompson, J. A., Huppler, R. E., and Gilbert, P. E. (1976). The effects of morphine- and nalorphine-like drugs in the nondependent and morphine-dependent chronic spinal dog. *J. Pharmacol. Exp. Ther.* **197**, 517–532.

Miller, B. (1996). δ opioid receptor expression is induced by concanavalin A in CD4+ T cells. *J. Immunol.* **157**, 5324–5328.

Min, B. H., Augustin, L. B., Felsheim, R. F., Fuchs, J. A., and Loh, H. H. (1994). Genomic structure analysis of promoter sequence of a mouse mu opioid receptor gene. *Proc. Natl Acad. Sci. USA* **91**, 9081–9085.

Mollereau, C., Parmentier, M., Mailleux, P., Butour, J.-L., Moisand, C., Chalon, P., Caput, D., Vassart, G., and Meunier, J.-C. (1994). ORL1, a novel member of the opioid receptor family. *FEBS Lett.* **341**, 33–38.

Pasternak, G. W. (1993). Pharmacological mechanisms of opioid analgesics. *Clin. Neuropharmacol.* **16**, 1–18.

Pert, C. B., and Snyder, S. H. (1973). Opiate receptor: demonstration in nervous tissue. *Science* **179**, 1011–1014.

Peterson, P. K., Molitor, T. W., and Chao, C. C. (1998). The opioid-cytokine connection. *J. Neuroimmunol.* **83**, 63–69.

Rao, A. (1994). NF-ATp: a transcription factor required for the co-ordinate induction of several cytokine genes. *Immunol. Today* **15**, 274–280.

Raynor, K., Kong, H., Chen, Y., Yasuda, K., Yu, L., Bell, G. I., and Reisine, T. (1994). Pharmacological characterization of the cloned κ-, δ-, and μ-opioid receptors. *Mol. Pharmacol.* **45**, 330–334.

Roy, S., Ge, B.-L., Loh, H. H., and Lee, N. M. (1992). Characterization of [^{3}H]morphine binding to interleukin-1-activated thymocytes. *J. Pharmacol. Exp. Ther.* **263**, 451–456.

Sedqi, M., Roy, S., Ramakrishnan, S., Elde, R., and Loh, H. H. (1995). Complementary DNA cloning of a μ-opioid receptor from rat peritoneal macrophages. *Biochem. Biophys. Res. Commun.* **209**, 563–574.

Sedqi, M., Roy, S., Ramakrishnan, S., and Loh, H. H. (1996). Expression cloning of a full-length cDNA encoding delta opioid receptor from mouse thymocytes. *J. Neuroimmunol.* **65**, 167–170.

Sharp, B. M., Shahabi, N. A., Heagy, W., McAllen, K., Bell, M., Huntoon, C., and McKean, D. J. (1996). Dual signal transduction through delta opioid receptors in a transfected human T-cell line. *Proc. Natl Acad. Sci. USA* **93**, 8294–8299.

Shavit, Y., Depaulis, A., Fredricka, C. M., Terman, G. W., Pechnick, R. N., Zane, C. J., Gale, R. P., and Liebeskind, J. C. (1986). Involvement of brain opiate receptors in the immune-suppressive effect of morphine. *Proc. Natl Acad. Sci. USA* **83**, 7114–7117.

Sibinga, N. E. S., and Goldstein, A. (1988). Opioid peptides and opioid receptors in cells of the immune system. *Annu. Rev. Immunol.* **6**, 219–249.

Simon, E. J. (1991). Opioid receptors and endogenous opioid peptides. *Med. Res. Rev.* **11**, 357–374.

Simon, E. J., Hiller, J. M., and Edelman, I. (1973). Stereospecific binding of the potent narcotic analgesic [^{3}H]etorphine to rat brain homogenate. *Proc. Natl Acad. Sci. USA* **70**, 1947–1949.

Simonds, W. F. (1988). The molecular basis of opioid receptor function. *Endocr. Rev.* **9**, 200–212.

Specter, S. C., Plotnikoff, N., Bradley, W. G., and Goodfellow, D. (1994). Methionine enkephalin combined with AZT therapy reduce murine retrovirus-induced disease. *Int. J. Immunopharmacol.* **16**, 911–917.

Stefano, G. B., Melchiorri, P., Negri, L., Hughes Jr, T. K., and Scharrer, B. (1992). [D-Ala^2]Deltorphin I binding and pharmacological evidence for a special subtype of δ opioid receptor on human and invertebrate immune cells. *Proc. Natl Acad. Sci. USA* **89**, 9316–9320.

Surratt, C. K., Johnson, P. S., Moriwaki, A., Seidleck, B. K., Blaschak, C. J., Wang, J. B., and Uhl, G. R. (1994). μ opioid receptor. Charged transmembrane domain amino acids are critical for agonist recognition and intrinsic activity. *J. Biol. Chem.* **269**, 20548–20553.

Terenius, L. (1973). Stereospecific interaction between narcotic analgesics and a synaptic plasma membrane fraction of rat cerebral cortex. *Acta Pharmacol. Toxicol.* **32**, 317–320.

Thompson, R. C., Mansour, A., Akil, H., and Watson, S. J. (1993). Cloning and pharmacological characterization of a rat μ opioid receptor. *Neuron* **11**, 903–913.

Tian, M., Broxmeyer, H. E., Fan, Y., Lai, Z., Zhang, S., Aronica, S., Cooper, S., Bigsby, R. M., Steinmetz, R., Engle, S. J., Mestek, A., Pollock, J. D., Lehman, M. N., Jansen, H. T., Ying, M., Stambrook, P. J., Tischfield, J. A., and Yu, L. (1997). Altered hematopoiesis, behavior, and sexual function in mu opioid receptor-deficient mice. *J. Exp. Med.* **185**, 1517–1522.

Wang, J. B., Imai, Y., Eppler, C. M., Gregor, P., Spivak, C. E., and Uhl, G. R. (1993). μ opioid receptor: cDNA cloning and expression. *Proc. Natl Acad. Sci. USA* **90**, 10230–10234.

Weber, R. J., and Pert, A. (1989). The periaqueductal gray matter mediates opiate-induced immunosuppression. *Science* **245**, 188–190.

Wick, M. J., Minnerath, S. R., Roy, S., Ramakrishnan, S., and Loh, H. H. (1996). Differential expression of opioid receptor genes in human lymphoid cell lines and peripheral blood lymphocytes. *J. Neuroimmunol.* **64**, 29–36.

Wollemann, M. (1990). Recent developments in the research of opioid receptor subtype molecular characterization. *J. Neurochem.* **54**, 1095–1101.

Yasuda, K., Raynor, K., Kong, H., Breder, C., Takeda, J., Reisine, T., and Bell, G. I. (1993). Cloning and functional comparison of κ and δ opioid receptors from mouse brain. *Proc. Natl Acad. Sci. USA* **90**, 6736–6740.

Zadina, J. E., Hackler, L., Ge, L. J., and Kastin, A. J. (1997). A potent and selective endogenous agonist for the mu-opiate receptor. *Nature* **386**, 499–502.

Lipoxin A_4 Receptor

Nan Chiang, Karsten Gronert, Fei-Hua Qiu and Charles N. Serhan*

Center for Experimental Therapeutics and Reperfusion Injury, Brigham and Women's Hospital, 75 Francis Street, Boston, MA 02115, USA

*corresponding author tel: 617-732-8822, fax: 617-278-6957, e-mail: cnserhan@zeus.bwh.harvard.edu

DOI: 10.1006/rwcy.2000.23006.

SUMMARY

Lipoxin A_4 (LXA_4) elicits biological actions via at least two classes of receptors known to date: (1) ALXR on leukocytes and enterocytes; and (2) a shared $CysLT_1$ subtype on endothelial and mesangial cells. ALXR belongs to the group of classical G protein-coupled receptors and was identified in both human and mouse and characterized using direct evidence including specific $[^3H]LXA_4$ binding and activation of functional responses with LXA_4. In several tissues and cell types other than leukocytes, results of pharmacological experiments indicate that LXA_4 acts via a subclass of peptido-leukotriene receptors ($CysLT_1$) as a partial agonist. In addition, endothelial cells (HUVECs) exhibit specific $[^3H]LXA_4$ binding which can be inhibited by LTD_4 and SKF104353 ($CysLT_1$ antagonist). The molecular origin of $CysLT_1$ is currently under investigation. ALXR is the first cloned lipoxygenase-derived eicosanoid receptor and, together with BLT, they are more akin to chemokine receptors than prostanoid receptors. The cytoplasmic signaling pathways and bioactions of ALXR are cell type specific. In human PMNs, LXA_4 stimulates rapid lipid remodeling with release of arachidonic acid in a pertussis toxin-sensitive fashion, and does not trigger significant increases in intracellular Ca^{2+} to serve as a second messenger. LXA_4 inhibits PMN adhesion, chemotaxis, transmigration as well as degranulation and was implicated as endogenous 'stop signals' acting on PMNs. In human monocytes and THP-1 cells, LXA_4 initiates intracellular Ca^{2+} release via ALXR but neither Ca^{2+} nor cAMP proved to be the required second messengers of lipoxin actions in these cell types, indicating different intracellular signaling pathways despite identical receptor cDNA sequences.

LXA_4 stimulates chemotaxis and adherence in monocytes but no other downstream responses of these cells, which may relate to the recruitment of monocytes to sites of wound healing and clearance. In agreement with *in vitro* results, ALXR agonists, namely LXA_4, 15-epi-LXA_4 (an aspirin-triggered LX) and their stable analogs, are topically active in inhibiting PMN infiltration as well as vascular permeability in murine skin inflammation. The development of these stable analogs will provide valuable tools to evaluate biological and pharmacological roles of ALXR as well as a novel means to develop selective therapies for inflammatory diseases. Since another eicosanoid, PGE_2, couples to a variety of signal transduction pathways (i.e. generation of IP_3 and $[Ca^{2+}]_i$ as well as decrease or increase of cAMP) via distinct receptor subtypes and/or isoforms that are cell type and tissue specific (Negishi *et al.*, 1995), it is likely that given the range of LXA_4 actions *in vivo* and its impact in isolated cell types, additional receptor systems will be identified.

BACKGROUND

Lipid-derived mediators play critical roles in inflammation and other multicellular processes (Haeggström and Serhan, 1998). Among them, lipoxins (LX) and aspirin-triggered lipoxins (ATLs) evoke actions of interest in a range of physiologic and pathophysiologic processes. These unique components possess a trihydroxytetraene structure and are both structurally and functionally distinct among lipid-derived bioactive mediators. LXA_4 and 15-epi-LXA_4 (a member of the ATL series) display leukocyte-selective actions that enable them to serve as endogenous 'stop signals' in multicellular events since they modulate adherence, transmigration, and chemotaxis. LXA_4 and 15-epi-LXA_4 elicit these cellular responses via a G protein-coupled receptor

(GPCR) called ALXR, which has been identified in both human and mouse tissues.

The characterization of ALXR and development of synthetic lipoxin and ATL stable analog mimetics have rapidly advanced our appreciation of the mechanism of lipoxin's actions and the potential utility of these counter-regulatory biocircuits in the control of local inflammatory events. In this review, ALXR is discussed with respect to its pharmacology, molecular biology, signal transduction, and bioactions in several cell types and animal models studied to date.

Discovery

Lipoxins are trihydroxytetraene-containing eicosanoids first found to be generated as lipoxygenase (LO) interaction products in human systems during cell–cell interactions and interactions between specific eicosanoid-generating enzymes. They are both immunoregulatory and vasoactive and their bioactions are sharply distinct from those evoked by other eicosanoids such as leukotrienes (LT), prostaglandins (PG), or thromboxanes (TX). Most members of these eicosanoid classes are well-characterized proinflammatory mediators of inflammation and prothrombolic stimuli. The biosynthesis and actions of lipoxins and ATL are reviewed in the chapter on lipoxin. Here, we provide an update and overview of current knowledge of the actions of LXA_4 and the characterization of one of its seven transmembrane receptors, namely LXA_4R, recently termed ALXR, that is involved in regulating PMN, monocyte, and epithelial responses.

As a class, lipoxins possess physiologic, pathophysiologic, and pharmacologic actions in several target tissues. All the actions of lipoxin are stereoselective in that changes in potencies accompany double bond isomerization, change in alcohol chirality (*R* or *S*) at key positions (**Figure 1**), as well as selective dehydrogenation of alcohols and reduction of double bonds. The self-limited, local microenvironment impact of lipoxin suggest that they contribute to resolution of injury sites and/or resolve inflammatory loci by regulating further recruitment of PMN and stimulating monocyte migration to promote healing and remodeling.

In the vasculature, lipoxins act on both leukocytes and vessels (Serhan, 1997). In this regard, LXA_4 is the best studied because of the availability of synthetic compounds that match the physical properties of endogenous LXA_4 and are clearly defined in physiology and bioactions. For example, LXA_4 is active in several microvascular preparations and *in vivo* systems (Dahlén *et al.*, 1988). LXA_4 is also active in the guinea pig lung strips and ileum, where its impact proved highly stereospecific: the 5*S*,6*R*-orientation of the two vicinal hydroxyls (Figure 1), positioned immediately adjacent to the carboxylic end of the conjugated tetraene, is essential for activity, and this suggested the presence of specific LXA_4 recognition sites. In addition to vasoregulation and modulation of contractile responses in several tissues, LXA_4 was also shown to display human leukocyte-selective actions which implicate them as endogenous 'stop signals' since LXA_4 modulates adherence, transmigration, and chemotaxis (Serhan, 1997). The human neutrophil (PMN) and monocyte responses with LXA_4 have been examined in further detail. Early findings showed that LXA_4 stimulates rapid lipid remodeling within seconds and releases arachidonic acid in PMN without oxygenation, which is sensitive to pertussis toxin (PTX) treatment (Grandordy *et al.*, 1990; Nigam *et al.*, 1990), pointing to the involvement of a GPCR.

The synthesis of the radiolabeled $[11,12\text{-}^3H]LXA_4$ (Brezinski and Serhan, 1991) enabled the first direct characterization of specific LXA_4-binding sites present on PMN that are likely to mediate many of its selective actions on these cells (Fiore *et al.*, 1992). Intact PMN demonstrate specific and reversible $[11,12\text{-}^3H]LXA_4$ binding ($K_d \sim 0.5$ nM and $B_{max} \sim 1830$ sites/PMN) that are modulated by guanosine stable analogs. The subcellular distribution of LXA_4-binding sites is listed in **Table 1**. These LXA_4-binding sites are inducible in promyelocytic lineage (HL-60) cells exposed to differentiating agents (e.g. retinoic acid, DMSO, and PMA) and confer LXA_4-stimulated phospholipase activation (Fiore *et al.*, 1993). Together, these findings provided further evidence that LXA_4 interacts with specific membrane-associated receptors on human leukocytes which belong to the classical GPCR since, in the absence of LXA_4-specific binding, lipid-signaling events are not initiated by LXA_4.

Based on the finding that functional LXA_4 receptors are inducible in HL-60 cells, several putative receptor cDNAs that are also induced within this temporal frame, cloned earlier from myeloid lineages and designated orphans (Perez *et al.*, 1992; Nomura *et al.*, 1993), were systematically examined for their ability specifically to bind and signal with LXA_4. Chinese hamster ovary (CHO) cells transfected with one of the orphans (previously denoted pINF114, also known as FPRL1 and FPR2) displayed both specific $[^3H]LXA_4$ binding with high affinity ($K_d = 1.7$ nM) and demonstrated selectivity when compared with LXB_4, LTB_4, LTD_4, and PGE_2 (Figure 1 and **Table 2**) (Fiore *et al.*, 1994). These transfected CHO cells transmitted signal with LXA_4, activating both GTPase and the release of arachidonic acid (C20:4) from membrane phospholipid,

Figure 1 Ligand specificity and structure–activity relationship of ALXR. LXA_4 interaction with ALXR is highly stereospecific, that is the 5*S*,6*R*-orientation of the two hydroxyl groups as well as 11-*cis* double bond conformation are essential for bioactions. 15-epi-LXA_4 (an aspirin-triggered lipoxin, ATL) carries a C-15 alcohol at the *R* configuration, opposite to the *S* configuration in native LXA_4 and was shown to have higher potency than native LXA_4 in certain bioassays. In 15(*R*/*S*)-methyl-LXA_4, hydrogen at C-15 was replaced by a methyl group at a racemate at C-15. 16-phenoxy-LXA_4 has a phenoxyl group at C-16. These compounds which are more resistant to rapid dehydrogenation by 15-hydroxyprostaglandin dehydrogenase (15-PGDH) than native LXA_4, compete with [^{3}H]LXA_4-specific binding on PMNs and are potent inhibitors for PMN functions *in vitro* and *in vivo*.

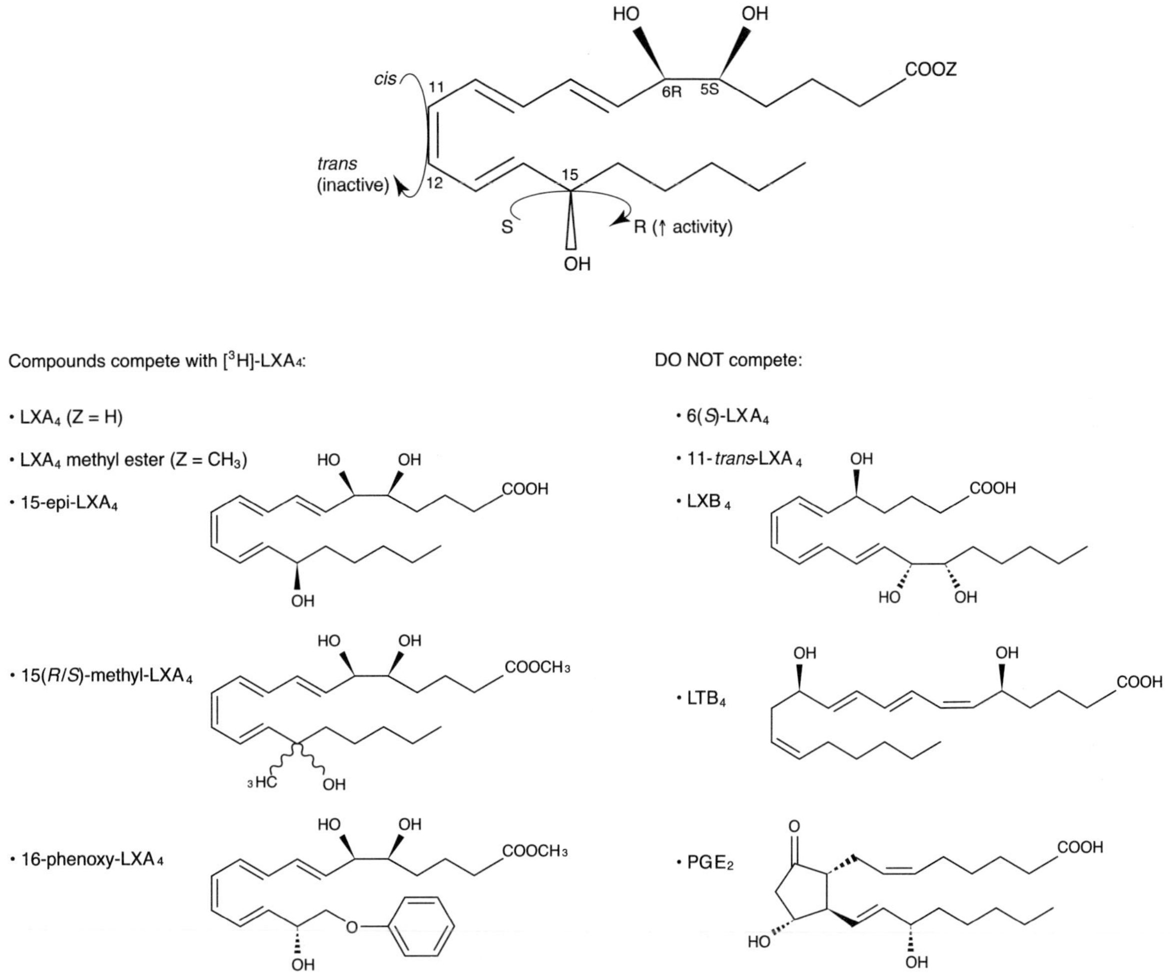

HO OH
cis 11 6R 5S COOZ
trans (inactive) 12 15
S R (↑ activity)
OH

Compounds compete with [^{3}H]-LXA_4:

- LXA_4 (Z = H)
- LXA_4 methyl ester (Z = CH_3)
- 15-epi-LXA_4 (HO, OH, COOH, OH)
- 15(*R*/*S*)-methyl-LXA_4 (HO, OH, $COOCH_3$, H_3C, OH)
- 16-phenoxy-LXA_4 (HO, OH, $COOCH_3$, O, OH)

DO NOT compete:

- 6(*S*)-LXA_4
- 11-*trans*-LXA_4
- LXB_4 (OH, COOH, HO, OH)
- LTB_4 (OH, OH, COOH)
- PGE_2 (O, COOH, HO, OH)

Table 1 Subcellular distribution of LXA_4- and LTB_4-specific binding in human

Component	Distribution of specific binding (%)	
	[^{3}H]LXA_4	[^{3}H]LTB_4
Plasma membrane	~42%	~16%
Granule fraction	~34.5%	~75.7%
Nuclear fraction	~23.3%	~8.3%

indicating that this cDNA encodes a functional receptor for LXA_4 in myeloid cells. It is essential to use GTPase and C20:4 release to test the relationship between specific binding and function because, with human PMNs, LXA_4 stimulates only a modest ~10% of the Ca^{2+} response observed with fMLP. In human monocytes and THP1 cells, a Ca^{2+} transit is initiated by LXA_4, but Ca^{2+} mobilization proved not to be a second messenger of lipoxin actions since LXA_4-triggered adherence to laminin is insensitive to BAPTA-AM (**Figure 2**), an intracellular Ca^{2+} chelator (Romano *et al.*, 1996; Maddox *et al.*, 1997). The

Table 2 Competitive binding of [^{3}H]LXA_4 with structure-related eicosanoids

	PMNs	HL-60 cells	CHO-ALXR	HUVECs
Competition	LXA_4 (IC_{50} ~ 1.5 nM)	LXA_4	LXA_4 (K_i ~ 5.6 nM)	LXA_4
	LXA_4-methyl ester			LTD_4
	15(R/S)-methyl-LXA_4			SKF104353
	16-phenoxy-LXA_4			($CysLT_1$ antagonist)
	15-epi-LXA_4			
Partial competition	LTC_4 (IC_{50} ~ 62 nM)	LTC_4	LTD_4 (K_i ~ 80 nM)	
	LTD_4 (IC_{50} ~ 56 nM)		fMLP	
	fMLP (IC_{50} ~ 1000-fold higher than LXA_4)			
No competition	LTB_4	LTB_4	LTB_4	ONO-4057
	LXB_4	ONO-4057	LXB_4	(LTB_4 and antagonist)
	6S-LXA_4	LXB_4	PGE_2	
	11-*trans*-LXA_4	SKF104353		
	SKF104353			

Figure 2 LXA_4 and ALXR evoke differential signalings and responses with human PMNs versus monocytes. ALXR inhibits PMN and stimulates monocyte functions via pertussis toxin (PTX)-sensitive G proteins (Gα) upon activation by LXA_4, 15-epi-LXA_4, as well as LX analogs. In PMNs, neither intracellular calcium ([Ca^{2+}]$_i$) nor cAMP were increased in response to lipoxins. In monocytes, LXA_4 induced an increase of [Ca^{2+}]$_i$ which is not the second messenger for LXA_4-stimulated adherence or chemotaxis since these responses were unaffected by BAPTA-AM (a Ca^{2+} chelator). See text for details.

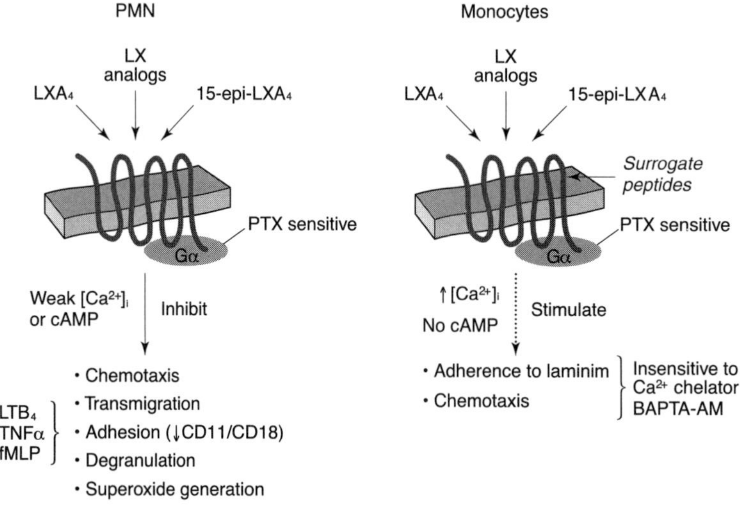

mouse LXA_4 receptor cDNA was cloned from a spleen cDNA library (**Figure 3**) and displays specific [^{3}H]LXA_4 binding and LXA_4-initiated GTPase activity when transfected into CHO cells (Takano *et al.*, 1997). The human and mouse LXA_4 receptors represented the first cloned LO-derived eicosanoid receptors. Several prostaglandin receptors were cloned earlier and, more recently, using a similar approach as for

Figure 3 Human and mouse ALXR: homology and tissue distribution. ALXR mRNAs ~1.4 kb in both human and mouse tissues are indicated by arrows. e1–e3 represent the putative extracellular loops, T1–T7, transmembrane segments and i1–i3, intracellular loops for ALXR, respectively. Percentage homology between individual domains of human and mouse ALXR in deduced amino acid sequences are indicated by numbers. High homology is observed in second intracellular loop (100%) and sixth transmembrane segment (97%).

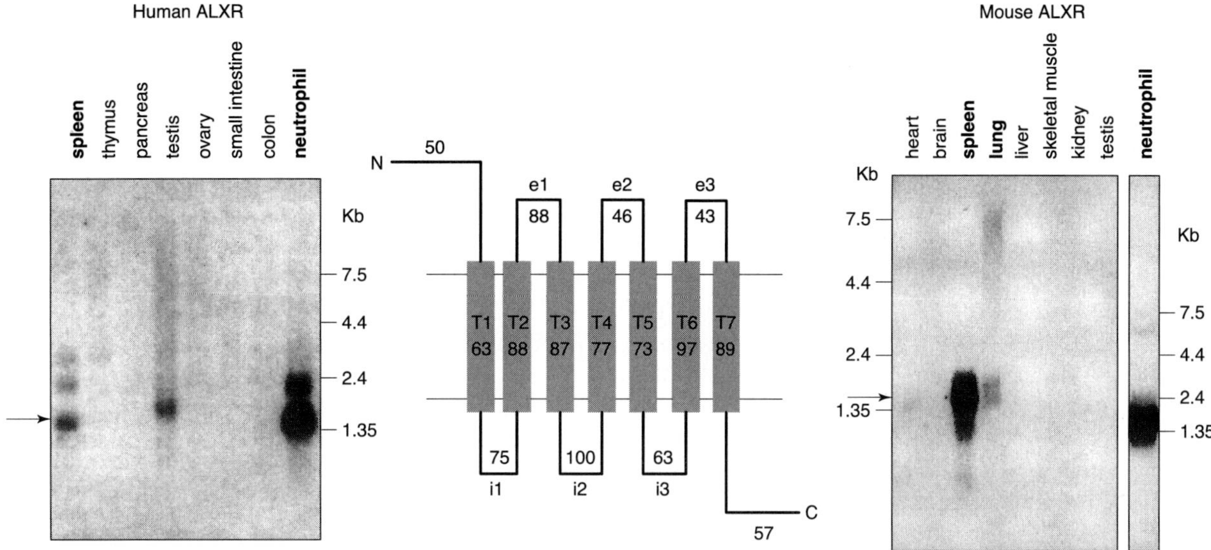

ALXR, enabled identification of the long-sought human LTB_4 receptor (BLT) (Yokomizo *et al.*, 1997) and its murine homolog (Huang *et al.*, 1998).

It is of interest that the BLT, which displays sequence homology with ALXR, was originally cloned as a purinergic receptor (Akbar *et al.*, 1996), and that both ALXR and BLT are more akin to chemokine receptor structure than prostaglandin receptors (**Figure 4**), which probably reflects the fact that lymphotoxins and lipoxins are eicosanoids that contain thermally sensitive conjugated double bond systems and prostaglandins contain a cyclopentane structure. BLT was recently identified as a novel coreceptor mediating *HIV-1* entry into CD4+ cells (Owman *et al.*, 1998). Along these lines, ALXR was also recently shown to interact with surrogate peptide ligands and evoke Ca^{2+} mobilization when expressed in HEK 293 cells carrying $G\alpha_{16}$ (Klein *et al.*, 1998). These observations suggest that receptors of this class interact with both small endogenous lipophilic ligands of the host and larger exogenous proteinaceous structures: this is an interesting aspect of these receptors that might relate to their potential multifunctional roles as sensing receptors in host defense.

Figure 4 Phylogenetic tree of human eicosanoid and chemokine receptors. Structure similarity of deduced amino acid sequences of human eicosanoid and chemokine receptors is determined by the average linkage cluster analysis. Abbreviations: TP (thromboxane A_2 receptor), EP1, EP2, EP3, EP4 (subtypes of prostaglandin E_2 receptor), FP (prostaglandin F_2 receptor), IP (prostacyclin receptor), ALXR (lipoxin A_4 receptor) and BLT (leukotriene B_4 receptor). BLT (Owman *et al.*, 1998), CXCR4 (Feng *et al.*, 1996), and CCR5 (Alkhatib *et al.*, 1996) were identified as coreceptors for HIV-1 entry.

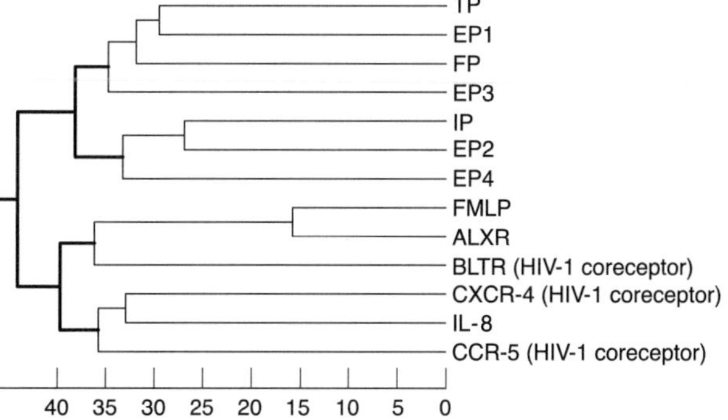

Alternative names

ALXR cDNA was initially cloned as an orphan receptor and reported independently by several different groups using fMLP receptor (FPR) cDNA as a probe with low-stringency hybridization conditions. The cDNA has high sequence homology (~70%) to FPR and hence, on the basis of sequence homology alone, was named FPRL1 (FPR-like 1) (Murphy *et al.*, 1992), FPRH1 (Bao *et al.*, 1992), and also FPR2

(Ye *et al.*, 1992) or RFP (receptor related to FPR; Perez *et al.*, 1992). Convincing functional responses were not presented for these clones. It was also cloned by Nomura *et al.* (1993) from a human monocyte cDNA library and renamed as an orphan receptor, denoted by these investigators as HM63. None of the reported cloning efforts attached functions to these clones. However, since the bacterial peptide surrogate fMLP interacts with FPR on phagocytic cells, the assumption of a phagocytic or chemotactic response in host defense for the sequence-related receptors such as FPRL1 remained a plausible hypothesis. These cDNA sequences are available from GenBank (see Accession numbers).

Note that ALXR and BLT are the recently proposed renomenclature for LXA_4 receptor and LTB_4 receptor, respectively (Dahlén *et al.*, International Union of Pharmacologic Sciences Nomenclature Committee, Stockholm, Nobel Forum, November, 1998). In addition, the terms $CysLT_1$ and $CysLT_2$ were introduced for peptido-LT (LTC_4, LTD_4, and LTE_4) receptors which are classified on the basis of sensitivity to antagonism rather than agonist properties. $CysLT_1$ and $CysLT_2$ are sensitive and resistant, respectively to the class of drugs currently being introduced in the clinic, including SKF104353.

Structure

Deduced amino acid sequence places ALXR within the GPCR superfamily, characterized by seven putative transmembrane (TM) segments (Figure 5) with the N-terminus on the extracellular side of the membrane and the C-terminus on the intracellular side (Baldwin, 1993). The overall homology between human and mouse ALXRs is 76% in nucleotide sequence and 73% in deduced amino acid (Takano *et al.*, 1997). An especially high homology is evident for their second intracellular loop (100%) and between their sixth TM segment (97%), followed by the second, third, and seventh TM segment as well as the first extracellular loop (87–89%), suggesting essential roles for these regions in ligand recognition and G protein coupling (Figure 3).

Molecular evolution analysis suggests that ALXR is only distantly related to prostanoid receptors and belongs to the cluster of chemoattractic peptide receptors, exemplified by fMLP, C5a, and IL-8 receptors (Toh *et al.*, 1995), which is now known to include BLT. The recently cloned BLT was obtained from human HL-60 cells (Yokomizo *et al.*, 1997) and mouse eosinophils (Huang *et al.*, 1998) and found to share an overall $\sim$30% homology with ALXR in deduced amino acid sequences (Figure 5). A highly homologous region ($\sim$46%) is present within the second TM segment in both ALXR and BLT with the amino acid sequence LNLALAD. Prostanoids interact with their receptors via COO^- interacting with an arginine residue within the seventh TM segment (Ushikubi *et al.*, 1995). Neither ALXR nor BLT share this Arg (in seventh TM segment) requirement (Fiore *et al.*, 1994; Yokomizo *et al.*, 1997), yet both ligands contain COOH that at physiological pH could present as a counteranion. Together, these findings provide further evidence that the origin of receptors for leukotriene and lipoxin is distinct from that for prostanoids.

Main activities and pathophysiological roles

As mentioned above, LXA_4 evokes vasodilatory and counter-regulatory roles in both *in vivo* and *in vitro* models. These counter-regulatory actions are initiated via unique cell surface receptors on leukocytes and enterocytes (Figure 6). With other cell types, such as endothelium and mesangial cells (Figure 7), LXA_4 evokes bioactions and interacts with a subclass of peptido-LT receptors ($CysLT_1$), reviewed by Serhan (1997). The leukocyte receptors are physiologically and pharmacologically distinct and evoke selective actions on each type of leukocyte tested to date. With human peripheral blood leukocytes, LXA_4 inhibits both isolated PMN and eosinophil chemotaxis *in vitro* in the nanomolar range (Lee *et al.*, 1989; Soyombo *et al.*, 1994) and blocks human natural killer (NK) cell cytotoxicity in a stereoselective fashion (Serhan, 1997). In cell–cell interaction systems, LXA_4 inhibits PMN transmigration across both endothelial and epithelial monolayers (Colgan *et al.*, 1993; Papayianni *et al.*, 1996) via actions on both cell types (i.e. PMNs and endothelial cells, PMNs and epithelial cells). These responses are also evident *in vivo* with murine receptors (Takano *et al.*, 1998). These immunoregulatory actions implicate them as endogenous 'stop signals' acting on human PMNs defining novel anti-inflammatory receptors and signaling pathways (Serhan, 1997). Recently, it was reported that NK cells possess cell surface receptors that recognize major histocompatibility complex (MHC) class I peptides and inhibit NK cell-mediated cytotoxicity (Lanier, 1997), further supporting the existence of the inhibitory receptors within the immune system.

In human monocytes, LXA_4 stimulates chemotaxis and adherence via ALXR (Figure 2), which may be related to the recruitment of monocytes to sites of wound healing and a protective role for LXA_4 (Maddox *et al.*, 1997). The main *in vitro* and *in vivo* biological actions of LXA_4 are summarized in Figure 6 and Figure 7. Also see Table 3 and Table 5 in the lipoxin chapter where the pathophysiological roles of LXA_4 are discussed in detail.

GENE

Accession numbers

The cDNAs for both human ALXR from THP1 cells (Maddox *et al.*, 1997) and enterocytes (Gronert *et al.*, 1998) are available in GenBank: the accession numbers are U81501 and AF054013, respectively. It was also previously cloned by several individual groups as an FPR homolog without functional data or ligands and deposited as orphan receptors and submitted with accession numbers X63819 (Perez *et al.*, 1992), M84562 (Murphy *et al.*, 1992), D10922 (Nomura *et al.*, 1993), M88107 (Ye *et al.*, 1992), and M76672 (Bao *et al.*, 1992).

Mouse ALXR cDNA is also available from GenBank: U78299 (Takano *et al.*, 1997).

Sequence

Both human (Fiore *et al.*, 1994) and mouse (Takano *et al.*, 1997) ALXR cDNA contain an open reading frame of 1051 nucleotides which encode a protein of 351 amino acids. Northern blot analysis (Figure 3) demonstrated that ALXR mRNA is $\sim$1.4 kb in both human and mouse (Takano *et al.*, 1997).

Chromosome location and linkages

Chromosome mapping revealing that the gene encoding ALXR (Fiore *et al.*, 1994) is located on chromosome 19q (Bao *et al.*, 1992), denoted in this early report of the orphan receptor as FPRH1.

PROTEIN

Accession numbers

Not available.

Description of protein

Hydrophobicity analysis of the deduced amino acid sequence of ALXR revealed seven repeated hydrophobic clusters of 20–25 amino acids interspersed with varying lengths of hydrophilic sequences, which are the common features of GPCR (**Figure 5**). The seven hydrophobic clusters were proposed to form membrane spanning α helices, whereas the hydrophilic segments form loops that project alternately into the extracellular space and the cytoplasm (Baldwin, 1993).

Several potential posttranslational modification sites were observed in the deduced amino acid sequences of both human and mouse ALXR (Figure 5): disulfide linkage, *N*-glycosylation, and phosphorylation.

Disulfide Linkage

Conserved cysteine residues at the first (Cys99) and second (Cys176) extracellular loops are found in both human and mouse ALXR. These are proposed to form a disulfide bond to stabilize the tertiary integrity of most GPCR.

N-glycosylation

Human ALXR contains two putative *N*-glycosylation sites located at the N-terminus (Asn4) and second extracellular loop (Asn179). In mouse ALXR, both putative *N*-glycosylation sites are located at the N-terminus (Asn4 and Asn10).

Phosphorylation

The C-terminus of mouse ALXR contains nine potential phosphorylation sites (e.g. serine and threonine residues), among which six are conserved within the human ALXR.

Relevant homologies and species differences

The overall homology between human and mouse ALXRs is 76% in the nucleotide sequence and 73% in deduced amino acid with high homology within their second intracellular loops (100%) (Takano *et al.*, 1997). Information for other species is not yet available. For details, see Structure.

Affinity for ligand(s)

Table 2 and Figure 1 summarize the knowledge of ligand affinity and specificity for ALXR. Intact human PMNs and retinoic acid-differentiated HL-60 cells demonstrate specific and reversible $[^3H]LXA_4$ binding with $K_d \sim 0.5$ and ~ 0.6 nM, respectively (Fiore *et al.*, 1992, 1993). Several isomers of LXA_4

Figure 5 The predicted membrane topology model of ALXR: homology between ALXR and BLT. Deduced amino acid sequence of ALXR demonstrates seven putative transmembrane segments with N-terminus on the extracellular side of the membrane and C-terminus on the intracellular side. ALXR possesses potential *N*-glycosylation sites (–CHO), phosphorylation sites (–P) and disulfide linkage (C–C). Blue circles indicate the residues conserved between human ALXR (anti-inflammatory) and BLT (proinflammatory). Unlike prostanoid receptors that use Arg in the seventh TM segment as a counterion for ligand binding, neither ALXR nor BLT possess arginine residues in the seventh TM segment as a structural feature. (Full colour figure may be viewed online.)

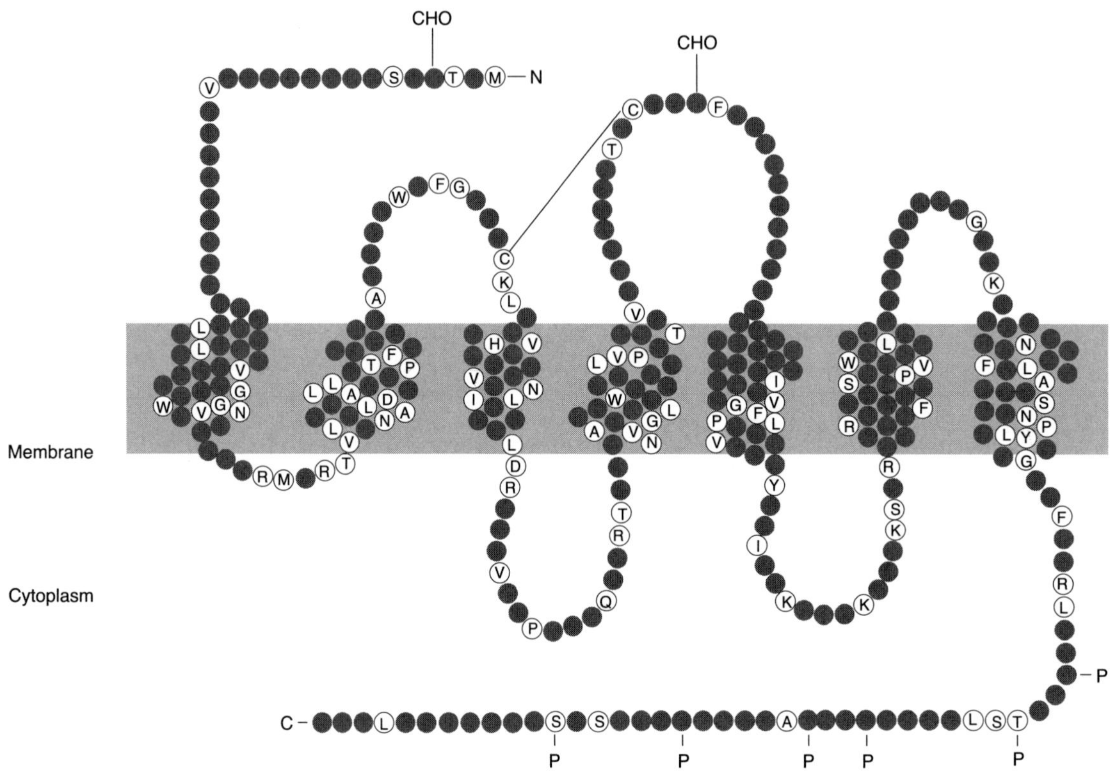

tested, namely 11-*trans*-LXA_4, 6S-LXA_4, and LXB_4, did not compete for this recognition site, consistent with their functional responses in these systems. Results from Scatchard analyses indicate that [^{3}H]LXA_4 binds PMN granule membrane-enriched fractions with comparable K_d values (0.8 nM), but with a larger B_{max} (4.1×10^{-11} M) than plasma membrane ($K_d = 0.7$ nM, $B_{max} = 2.1 \times 10^{-11}$ M) fractions (Fiore *et al.*, 1994). Hence, it appears that additional receptors can be mobilized by granule fusion to the plasma membranes of PMNs (Table 1). [^{3}H]LXA_4-specific binding is stereoselective since neither LTB_4, LXB_4, 6S-LXA_4, 11-*trans*-LXA_4, nor SKF104353 (a $CysLT_1$ antagonist) compete for [^{3}H]LXA_4 in human PMNs (Table 2 and Figure 1) (Fiore *et al.*, 1992). Among the related eicosanoid heteroligands tested in the HL-60 cell system, only LTC_4 at ~3-log molar excess competes for [^{3}H]LXA_4-specific binding (Fiore *et al.*, 1993). The peptido-LT receptors remain to be cloned. The cross-competition of LTC_4 and LTD_4 observed with LXA_4 in several systems suggests that the 'true' peptido-LT receptors may also be of this class of receptors.

In several tissues and cell types other than leukocytes, results from pharmacological experiments indicate that LXA_4 interacts with a subclass of peptido-LT receptors ($CysLT_1$) as a partial agonist to mediate its actions (Badr *et al.*, 1989; Fiore *et al.*, 1992). Along these lines, both LTC_4 and LXA_4, albeit at high concentrations (> 1 μM), induce contractions of guinea pig lung parenchyma and release of thromboxane A_2, which is sensitive to $CysLT_1$ receptor antagonists (Wikstrosm Jonsson, 1998). This is not likely to be a physiologic action of LXA_4.

In certain cell types, LXA_4 (in the nanomolar range) blocks LTD_4 actions (Figure 7) and in this regard it

Figure 6 LXA_4 actions via ALXR in leukocytes and epithelial cells. Illustration of the regulatory actions of LXA_4 in leukocytes (reviewed in Serhan, 1997) and human epithelial cells (Gronert *et al.*, 1998 and Gewirtz *et al.*, 1998). Upper left panel: Ear biopsies: Inhibition of LTB_4-induced PMN infiltration into mouse ear by topical application of LXA_4 analogs in acute skin inflammation (Tokano *et al.*, 1997). PMN is indicated by an arrow. Upper right panel: Photomicrograph: Internalization of *Salmonella typhimurium* (shown in green) by intestinal epithelium (indicated by an arrow). In response to this gastrointestinal pathogen, intestinal epithelium secretes chemokines which promote neutrophil infiltration. This chemokine (IL-8) secretion can be downregulated by LXA_4 analogs (Gewirtz *et al.*, 1998).

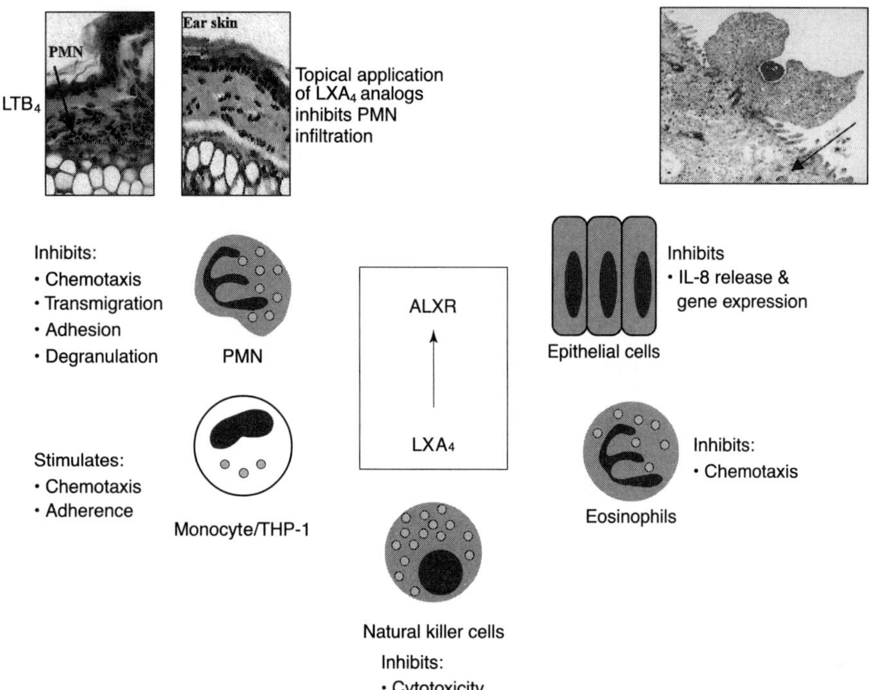

blocks specific [^{3}H]LTD_4 binding to mesangial cells (Badr *et al.*, 1989) and human umbilical vein endothelial cells (HUVECs) (Fiore *et al.*, 1993; Takano *et al.*, 1997). HUVECs specifically bind [^{3}H]LXA_4 at a K_d of 11 nM, which can be inhibited by LTD_4 and SKF104353 (Fiore *et al.*, 1993). Therefore, it appears that LXA_4 interacts with at least two classes of cell surface receptors: one specific for LXA_4 that is present on leukocytes and enterocytes (ALXR) (**Figure 6**), the other shared by LTD_4 and present on HUVECs and mesangial cells ($CysLT_1$) (**Figure 7**). The molecular origins of these LXA_4/LTD_4-binding sites (pharmacologically defined as $CysLT_1$) are currently of considerable interest.

Several synthetic analogs of both LXA_4 and 15-epi-LXA_4 have been designed to resist rapid metabolic inactivation and have been tested for their ability to compete with [^{3}H]LXA_4-specific binding to ALXR. For detailed information on the formation of 15-epi-LXA_4, the design of stable LX analogs and their pharmacological activities, see the lipoxin chapter. Both 15(*R*/*S*)-methyl-LXA_4 and 16-phenoxy-LXA_4 compete with [^{3}H]LXA_4 as well as [^{3}H]LTD_4 for specific binding on human PMNs (Table 2) and HUVECs, respectively (Takano *et al.*, 1997). Hence, each of the bioactive lipoxin mimetics acts at sites that compete with [^{3}H]LXA_4. Moreover, each LXA_4 analog that competes with [^{3}H]LXA_4 is also topically active in inhibiting PMN migration in acute dermal inflammation.

Human and mouse ALXR cDNA transfected into CHO cells display specific binding with [^{3}H]LXA_4; the K_d is 1.7 nM for human (Fiore *et al.*, 1994) and K_d 1.5 nM for mouse ALXR (Takano *et al.*, 1997). Human ALXR-transfected CHO cells were also tested for binding with other eicosanoids, including LXB_4, LTD_4, LTB_4, and PGE_2 (Figure 1). Only LTD_4 shows competition with [^{3}H]LXA_4 binding, giving a K_i of 80 nM (Table 2) (Fiore *et al.*, 1994). It is of interest that, although ALXR shares ~70%

Figure 7 LXA_4 interaction with $CysLT_1$ in several cell types and tissues. Illustration of the regulatory actions of LXA_4 in vascular endothelial cells (reviewed in Serhan, 1997), smooth muscle contraction (Dahlén *et al.*, 1988; Christie *et al.*, 1992), rat glomerular mesangial cells (Badr *et al.*, 1989) as well as in bone marrow (Stenke *et al.*, 1991) via a subclass of peptido-LT receptors ($CysLT_1$).

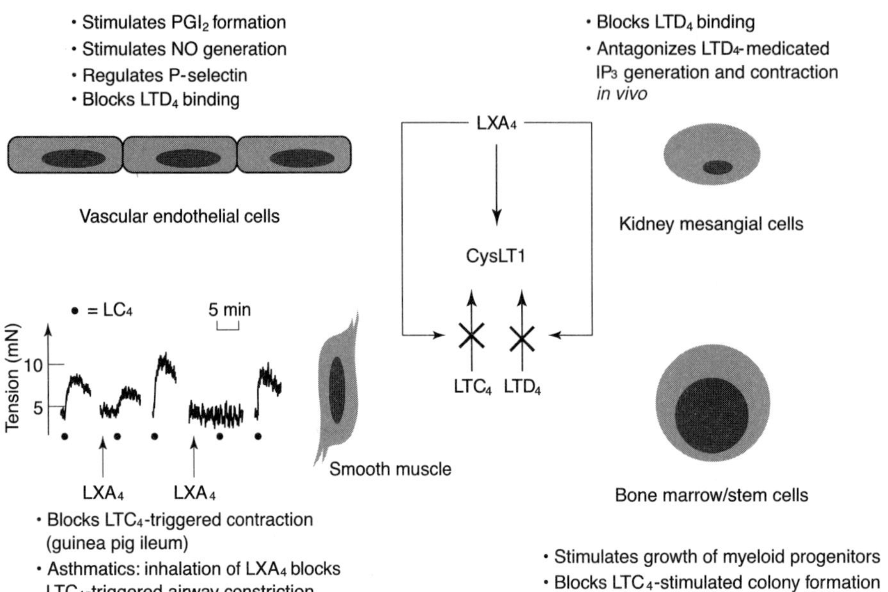

homology with FPR, ALXR only binds [^{3}H]fMLP with low affinity ($K_d \sim 5\,\mu M$) and proves to be selective for LXA_4 by three log orders of magnitude (Fiore and Serhan, 1995).

Recently, it was reported that some surrogate peptides can also interact with ALXR (also known as FPRL-1) in model systems (Klein *et al.*, 1998), but the functional role of these peptides in human biology is not clear. The apparent EC_{50} value for receptor activation (determined by mobilization of $[Ca^{2+}]_i$) by the best synthetic rogue peptide of this synthetic series is approximately 2 nM (Klein *et al.*, 1998), whereas LXA_4 and its analogs stimulate monocyte adherence via ALXR at concentrations less than 1 nM (EC_{50} for analogs $\sim 8 \times 10^{-11}$ M, EC_{50} for LXA_4 $\sim 8 \times 10^{-10}$ M) (Maddox *et al.*, 1997) or inhibit PMN transmigration and adhesion at 10^{-10} M (Serhan *et al.*, 1995). These new findings suggest that small peptides as well as bioactive lipids can both function as ligands for the same receptor, with different affinity and/or distinct interaction sites within the receptor and separate intracellular signaling, depending on the cell types. It appears likely that the protein interactions following ligand–receptor binding are different for peptide in comparison with lipid ligands of this receptor.

Cell types and tissues expressing the receptor

To date, ALXR has been identified by its functions and direct actions, and cloned in both human and mouse PMNs (Fiore *et al.*, 1994; Takano *et al.*, 1997), human monocytes (Maddox *et al.*, 1997), as well as human enterocytes (Gronert *et al.*, 1998). In human PMNs, the results of subcellular fractionation experiments revealed that [^{3}H]LXA_4-binding sites are associated with plasma membrane and endoplasmic reticulum (42.1%), granule (34.5%) as well as nuclear-enriched fractions (23.3%), a distribution distinct from that of [^{3}H]LTB_4-binding sites (Table 1) (Fiore *et al.*, 1992). The finding that LXA_4 blocks both PAF- and fMLP-stimulated eosinophil chemotaxis (Soyombo *et al.*, 1994) suggests that functional ALXR is also present on eosinophils. In human enterocytes, ALXR is present in crypt and brush border colonic epithelial cells, as demonstrated by Gronert *et al.* (1998). Functional blockage of LTD_4 binding by LXA_4 and vice versa on HUVECs indicates the presence of specific LXA_4/LTD_4-binding sites ($CysLT_1$) on both endothelial and mesangial cells (Figure 7).

Northern blot analysis (Figure 3) of multiple murine tissues demonstrated that ALXR mRNA is most abundant in PMNs, spleen, and lung, with lesser amounts in heart and liver (Takano *et al.*, 1997). In the absence of disease, the pattern is similar in human tissues. In humans, ALXR mRNA is also abundant in PMNs, followed by spleen, lung, placenta, and liver (Fiore *et al.*, 1994; Takano *et al.*, 1997).

Regulation of receptor expression

Retinoic acid, PMA, and DMSO, which lead to granulocytic phenotypes in HL-60 cells, each induce a 3- to 5-fold increase in the expression of ALXR as monitored by specific $[^3H]LXA_4$-binding (Fiore *et al.*, 1993). Transcription of ALXR was dramatically upregulated by cytokines in human enterocytes, with lymphocyte-derived IL-13 and IFNγ being most potent (**Figure 8**), followed by IL-4 and IL-6. IL-1β and LPS also showed moderate induction of ALXR mRNA (Gronert *et al.*, 1998). In view of the cytokine regulation of ALXR, it is likely that the expression of these receptors will change dramatically in disease states which, in turn, might downregulate mucosal inflammatory and allergic responses.

SIGNAL TRANSDUCTION

Associated or intrinsic kinases

In retinoic acid-differentiated HL-60 cells, LXA_4 stimulated phospholipase D (PLD) activation that is staurosporine-sensitive, suggesting the involvement of PKC in signal transduction in these cells (Fiore *et al.*, 1993). It was also demonstrated that LXA_4 blocks LTB_4- or fMLP-stimulated PMN transmigration or adhesion by regulation of β_2 integrin-dependent PMN adhesion (Fiore and Serhan, 1995). This modulatory action is partially reversed by prior exposure to

Figure 8 Regulatory role of LX and ALXR during PMN–epithelial cell interaction. Illustration of the anti-inflammatory actions of LXA_4, 15-epi-LXA_4, and LXA_4 analogs that are mediated via the epithelial and PMN ALXR. Multistep recruitment of PMNs to a site of mucosal inflammation and intestinal epithelial ALXR gene regulation by lymphocyte-derived cytokines is depicted. Activation of ALXR present on both PMNs and epithelial cells inhibits PMN migration to sites of inflammation. Bioactions of epithelial and PMN ALXR activation are indicated by asterisks. These anti-inflammatory actions include inhibition of PMN adhesion to the transmigration across both endothelia and epithelia, as well as inhibition of the generation of a gradient of IL-8 at both the protein and gene level. The biosynthesis of aspirin-triggered 15-epi-LXA_4 is illustrated during PMN and epithelial communication (see the lipoxin chapter for details).

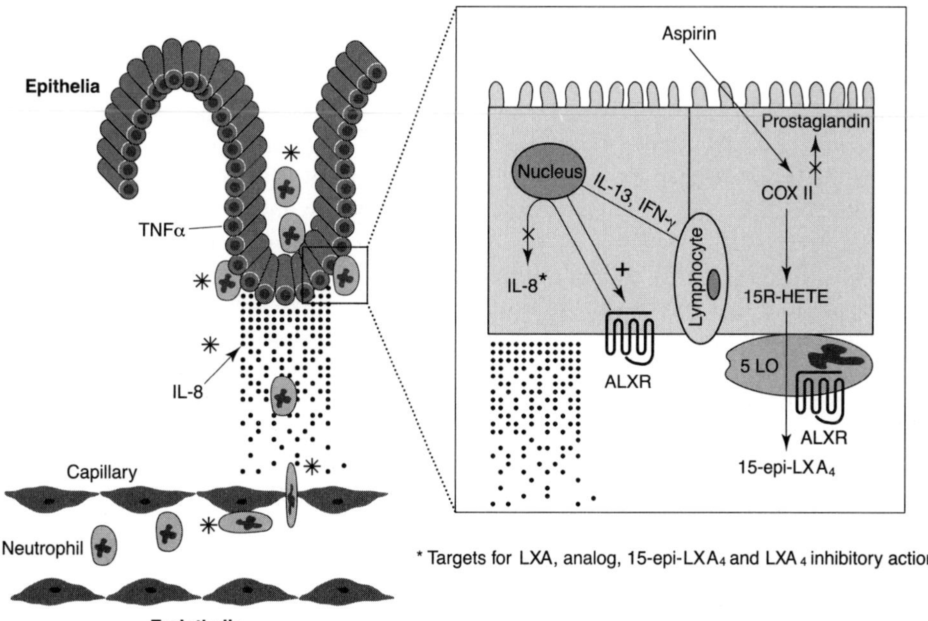

genistein, a tyrosine kinase inhibitor (Papayianni *et al.*, 1996).

Cytoplasmic signaling cascades

The cytoplasmic signaling cascade of ALXR appears to be highly cell type-specific. For example, in human PMNs, LXA_4 stimulates rapid lipid remodeling (within seconds) with release of arachidonic acids that are evoked via PTX-sensitive G proteins (Nigam *et al.*, 1990) without formation of either LT or PG. Only a modest Ca^{2+} mobilization was observed. Also, LXA_4 was reported to block intracellular generation of IP_3 (Grandordy *et al.*, 1990) as well as Ca^{2+} mobilization in response to other stimuli (Lee *et al.*, 1989). In human peripheral blood monocytes and cultured THP1 cells, LXA_4 triggers intracellular Ca^{2+} release and adherence to laminin (Romano *et al.*, 1996; Maddox *et al.*, 1997). Thus, different intracellular signaling pathways are present in PMN versus monocytes despite identical receptor sequences. It is of interest that Ca^{2+} is not the second messenger for lipoxin actions in monocytes, since LXA_4-stimulated monocyte adherence to laminin is not dependent on an lipoxin-stimulated increase in $[Ca^{2+}]_i$. The EC_{50} value for LXA_4-stimulated increase in $[Ca^{2+}]_i$ is $> 100\,nM$ in monocytes, which is more than two log orders of magnitude higher than that required for LXA_4-stimulated adherence ($EC_{50} < 1\,nM$). In view of G protein-coupling events in monocytes, both Ca^{2+} mobilization and adherence are PTX-sensitive. This indicates that the receptor coupling in monocytes and PMNs is similar to this point, although there could be different PTX-sensitive G protein subtypes coupled to the intracellular domains of the receptors, which then diverge downstream in the signal transduction pathways, leading to chemotaxis of monocytes and inhibition of PMNs. Distinct signaling in monocytes and PMNs is further evidenced by different responses to LXA_4 in these cell types (Figure 2).

Recently, results from this laboratory indicate that activation of ALXR leads to polyisoprenyl phosphate (PIPP) remodeling and accumulation of presqualene diphosphate (PSDP) (Levy *et al.*, 1999), a key component in the PIPP-signaling pathway and a potent negative intracellular signal in PMNs (Levy *et al.*, 1997), indicating that ALXR stimulates intracellular lipids that appear directly to mediate some of its actions. The characteristics of ALXRs in various cell types are briefly summarized in **Table 3**. Also, LXA_4 modulates MAP kinase activities on mesangial cells in a PTX-insensitive manner (McMahon *et al.*, 1998), suggesting the presence of additional novel LXA_4 receptor subtypes and/or signaling pathways in these cells.

DOWNSTREAM GENE ACTIVATION

Genes induced

In human enterocytes (T84), ALXR activation by LXA_4 and LX analogs diminishes *Salmonella typhimurium*-induced IL-8 transcription (Gewirtz *et al.*, 1998). The reduction of IL-8 mRNA levels parallels decreases in IL-8 secretion, indicating that in these cells ALXR's mechanism of action for blocking this chemokine is at the gene transcriptional level.

BIOLOGICAL CONSEQUENCES OF ACTIVATING OR INHIBITING RECEPTOR AND PATHOPHYSIOLOGY

Unique biological effects of activating the receptors

ALXR activation on human PMNs evoke leads to inhibition of both LTB_4- and fMLP-induced PMN adhesion (by downregulating CD11/CD18), chemotaxis, transmigration (Serhan, 1997), as well as degranulation (Gewirtz *et al.*, 1999). Recently, it was demonstrated that both LXA_4 and ATL analogs applied to mouse ears dramatically reduced both LTB_4- and PMA-initiated PMN infiltration as well as LTB_4-triggered vascular permeability (Takano *et al.*, 1998) (Figure 6). In addition, LXA_4 analogs inhibit leukocyte rolling and adherence by attenuating P-selectin expression in rat mesenteric microvasculature (Scalia *et al.*, 1997). Native LXA_4 also inhibits PMN recruitment to inflamed glomeruli *in vivo* (Papayianni *et al.*, 1995), further supporting the antiinflammatory actions of LX *in vivo*. In cell types other than leukocytes, LX analogs inhibit TNFα-induced IL-8 release (Gronert *et al.*, 1998) as well as pathogen-induced IL-8 secretion at the mRNA level in human enterocytes (Figure 8) (Gewirtz *et al.*, 1998). In rabbit trachea, LXA_4 stimulates nitric oxide generation, reducing airway smooth muscle contraction (Tamaoki *et al.*, 1995). For details, see Figure 6 and Figure 7 and the lipoxin chapter.

Table 3 Characterization of human and mouse ALXR

Cell type	K_d (nM)	Signal transduction	Kinase associated	Upregulated by
Human HL-60 (differentiated)	0.6	PLD activation (lipid remodeling)	Protein kinase C (staurosporin sensitive)	Retinoic acid, DMSO, PMA
Human PMNs	0.5	PLD activation	Tyrosine kinase (genistein sensitive)	
		GTPase activity		
		C20:4 release		
		PIPP signal (raises PSDP accumulation) (with second signal)		
		No increase of cAMP, proton efflux and very weak $[Ca^{2+}]_i$		
Human PMNs (expressed in CHO)	1.7	GTPase activity		
		Arachidonic acid release (PTX sensitive)		
		No increase of cAMP and $[Ca^{2+}]_i$		
Human monocytes		Increase of $[Ca^{2+}]_i$ (PTX sensitive)		
		No increase of cAMP and proton efflux		
Human enterocytes		No proton efflux		IL-13, IL-4, IFNγ
Human endothelium	11	Prostacyclin generation	Protein kinase C	
		Nitric oxide generation		
		No increase of $[Ca^{2+}]_i$ and proton efflux		
Mouse leukocyte (expressed in CHO)	1.5	GTPase activity		

Phenotypes of receptor knockouts and receptor overexpression mice

ALXR knockout and transgenic mice are not currently available. However, an *in vitro* receptor knockout was carried out using an antisense oligonucleotide selected from a region of ALXR cDNA sequence which gives low homology with the formyl peptide receptor. Retinoic acid-differentiated HL-60 cells exposed to this antisense oligonucleotide selectively lost $[^3H]LXA_4$ binding as well as LXA_4-initiated lipid remodeling paralleling the loss of ALXR mRNA (Fiore and Serhan, 1995).

Human abnormalities

The formation of native LXA_4 as well as 15-epi-LXA_4 is documented *in vivo*. During *chronic myelocytic leukemia*, platelets from these individuals lose 12-lipoxygenase and their ability to generate lipoxin. Of particular interest, the loss of lipoxin generation correlates with the onset of blast crisis observed in chronic myelocytic leukemia (Stenke *et al.*, 1994).

THERAPEUTIC UTILITY

Effect of treatment with soluble receptor domain

ALXR agonists, namely LXA_4 and ATL analogs, are topically active in inhibiting PMN infiltration and vascular permeability in murine skin inflammation (Serhan, 1997). Inhalation of LXA_4 by human asthmatic patients blocks LTC_4-induced contraction *in vivo* (Christie *et al.*, 1992). In renal hemodynamics,

LXA_4 blocks LTD_4 induced vasoconstriction (Badr *et al.*, 1989) and prevents PMN entry into the kidney in *glomerulonephritis* (Papayianni *et al.*, 1995). For details, see the section on Unique biological effects of activating the receptors and the lipoxin chapter.

Effects of inhibitors (antibodies) to receptors

A specific antipeptide antibody was prepared against the second extracellular loop of ALXR. This antibody blocks specific $[^3H]LXA_4$-binding as well as LXA_4-inhibitory actions on fMLP-stimulated CD11b expression and aggregation in human PMNs (Fiore and Serhan, 1995) and inhibits LXA_4-triggered intracellular Ca^{2+} mobilization in human peripheral blood monocytes (Maddox *et al.*, 1997).

References

Akbar, G. K. M., Dasari, V. R., Webb, T. E., Ayyanathan, K., Pillarisetti, K., Sandhu, A. K., Athwar, R. S., Daniel, J. L., Ashby, B., Barnard, E. A., and Kunapuli, S. P. (1996). Molecular cloning of a novel P2 purinoreceptor from human erythroleukemia cells. *J. Biol. Chem.* **271**, 18363–18367.

Alkhatib, G., Combadiere, C., Broder, C. C., Feng, Y., Kennedy, P. E., Murphy, P. M., and Berger, E. A. (1996). CC-CKR5: A RANTES, MIP-1α, MIP-1b receptor as a fusion cofactor for macrophage-tropic HIV-1. *Science* **272**, 1955–1958.

Badr, K. F., DeBoer, D. K., Schwartzberg, M., and Serhan, C. N. (1989). Lipoxin A_4 antagonizes cellular and *in vivo* actions of leukotriene D_4 in rat glomerular mesangial cells: evidence for competition at a common receptor. *Proc. Natl Acad. Sci. USA* **86**, 3438–3442.

Baldwin, J. M. (1993). The probable arrangement of the helices in G protein-coupled receptors. *EMBO J.* **12**, 1693–1703.

Bao, L., Gerard, N. P., Eddy Jr., R. L., Shows, T. B., and Gerard, C. (1992). Mapping of genes for the human C5a receptor (C5AR), human FMLP receptor (FPR), and two FMLP receptor homologue orphan receptors (FPRH1, FPRH2) to chromosome 19. *Genomics* **13**, 437–440.

Brezinski, D. A., and Serhan, C. N. (1991). Characterization of lipoxins by combined gas chromatography and electron-capture negative ion chemical ionization mass spectrometry: formation of lipoxin A_4 by stimulated human whole blood. *Biol. Mass Spectrom.* **20**, 45–52.

Christie, P. E., Spur, B. W., and Lee, T. H. (1992). The effects of lipoxin A_4 on airway responses in asthmatic subjects. *Am. Rev. Respir. Dis.* **145**, 1281–1284.

Colgan, S. P., Serhan, C. N., Parkos, C. A., Delp-Archer, C., and Madara, J. L. (1993). Lipoxin A_4 modulates transmigration of human neutrophils across intestinal epithelial monolayers. *J. Clin. Invest.* **92**, 75–82.

Dahlén, S. E., Franzén, L., Raud, J., Serhan, C. N., Westlund, P., Wikström, E., Björck, T., Matsuda, H., Webber, S. E., Veale, C. A., Puustinen, T., Haeggström, J., Nicolaou, K. C., and Samuelsson, B. (1988). In "Lipoxins: Biosynthesis, Chemistry, and Biological Activities" (ed P. Y.-K. Wong and C. N. Serhan), Actions of lipoxin A_4 and related compounds in smooth muscle preparation and on the microcirculation *in vivo* pp. 107–130. Plenum Press, New York.

Feng, Y., Broder, C. C., Kennedy, P. E., and Berger, E. A. (1996). HIV-1 entry cofactor: functional cDNA encoding of a seven-transmembrane G-protein coupled receptor. *Science* **272**, 872–877.

Fiore, S., and Serhan, C. N. (1995). Lipoxin A_4 receptor activation is distinct from that of the formyl peptide receptor in myeloid cells: inhibition of CD11/18 expression by lipoxin A_4–lipoxin A_4 receptor interaction. *Biochemistry* **34**, 16678–16686.

Fiore, S., Ryeom, S. W., Weller, P. F., and Serhan, C. N. (1992). Lipoxin recognition sites. Specific binding of labeled lipoxin A_4 with human neutrophils. *J. Biol. Chem.* **267**, 16168–16176.

Fiore, S., Romano, M., Reardon, E. M., and Serhan, C. N. (1993). Induction of functional lipoxin A_4 receptors in HL-60 cells. *Blood* **81**, 3395–3403.

Gewirtz, A. T., McCormick, B., Neish, A. S., Petasis, N. A., Gronert, K., Serhan, C. N., and Madara, J. L. (1998). Pathogen-induced chemokine secretion from model intestinal epithelium is inhibited by lipoxin A_4 analogs. *J. Clin. Invest.* **101**, 1860–1869.

Gewirtz, A. T., Fokin, V. V., Petasis, N. A., Serhan, C. N., and Madara, J. L. (1999). LXA_4, aspirin-triggered 15-epi-LXA_4, and their analogs selectively down-regulate PMN azurophilic degranulation. *Am. J. Physiol.* **276**, C988–C994.

Grandordy, B. M., Lacroix, H., Mavoungou, E., Krilis, S., Crea, A. E., Spur, B. W., and Lee, T. H. (1990). Lipoxin A_4 inhibits phosphoinositide hydrolysis in human neutrophils. *Biochem. Biophys. Res. Commun.* **167**, 1022–1029.

Gronert, K., Gewirtz, A., Madara, J. L., and Serhan, C. N. (1998). Identification of a human enterocyte lipoxin A_4 receptor that is regulated by interleukin (IL)-13 and interferon γ and inhibits tumor necrosis factor α-induced IL-8 release. *J. Exp. Med.* **187**, 1285–1294.

Haeggström, J. Z., and Serhan, C. N. (1998). In "Molecular and Cellular Basis of Inflammation" (ed C. N. Serhan and P. A. Ward), Update on arachidonic acid cascade, pp. 51–92.. Humana Press, Totowa.

Huang, W.-W., Garcia-Zepeda, E. A., Sauty, A., Oettgen, H. C., Rothenberg, M. E., and Luster, A. D. (1998). Molecular and biological characterization of the murine leukotriene B_4 receptor expressed on eosinophils. *J. Exp. Med.* **188**, 1063–1074.

Klein, C., Paul, J. I., Sauve, K., Schmidt, M. M., Arcangeli, L., Ransom, J., Trueheart, J., Manfredi, J. P., Broach, J. R., and Murphy, A. J. (1998). Identification of surrogate agonists for the human FPRL-1 receptor by autocrine selection in yeast. *Nature Biotechnol.* **16**, 1334–1337.

Lanier, L. L. (1997). NK cells: from no receptors to too many. *Immunity* **6**, 371–378.

Lee, T. H., Horton, C. E., Kyan-Aung, U., Haskard, D., Crea, A. E., and Spur, B. W. (1989). Lipoxin A_4 and lipoxin B_4 inhibit chemotactic response of human neutrophils stimulated by leukotriene B_4 and *N*-formyl-L-methionyl-L-leucyl-L-phenylalanine. *Clin. Sci.* **77**, 195–203.

Levy, B. D., Petasis, N. A., and Serhan, C. N. (1997). Polyisoprenyl phosphates in intracellular signaling. *Nature* **389**, 985–989.

Levy, B. D., Fokin, V. V., Clark, J. M., Wakelam, M. J. O., Petasis, N. A., and Serhan, C. N. (1999). Polyisoprenyl phosphate (PIPP) signaling regulates phospholipase D activity: a "stop" signaling switch for aspirin-triggered lipoxin A_4. *FASEB J.* **13**, 903–911.

Maddox, J. F., Hachicha, M., Takano, T., Petasis, N. A., Fokin, V. V., and Serhan, C. N. (1997). Lipoxin A_4 stable analogs are potent mimetics that stimulate human monocytes and THP-1 cells via a G-protein-linked lipoxin A_4 receptor. *J. Biol. Chem.* **272**, 6972–6978.

McMahon, B., McPhilips, F., Fanning, A., Brady H. R., and Godson, C. (1998). Modulation of mesangial cell MAP kinase activities by leukotriene D_4 and lipoxin A_4. *J. Am. Soc. Nephrol.* **9**, 355A.

Murphy, P. M., Ozcelik, T., Kenney, R. T., Tiffany, H. L., and McDermott, D. (1992). A structural homoloque of the *N*-formyl peptide receptor: characterization and chromosomal mapping of a peptide chemoattractant receptor gene family. *J. Biol. Chem.* **267**, 7637–7643.

Negishi, M., Sugimoto, Y., and Ichikawa, A. (1995). Prostaglandin E receptors. *J. Lipid Mediat. Cell Signal.* **12**, 379–391.

Nigam, S., Fiore, S., Luscinskas, F. W., and Serhan C. N. (1990). Lipoxin A_4 and lipoxin B_4 stimulate the release but not the oxygenation of arachidonic acid in human neutrophils: dissociation between lipid remodeling and adhesion. *J. Cell. Physiol.* **143**, 512–523.

Nomura, H., Nielsen, B. W., and Matsushima, K. (1993). Molecular cloning of cDNAs encoding a LD78 receptor and putative leukocyte chemotactic peptide receptors. *Int. Immunol.* **5**, 1239–1249.

Owman, C., Garzino-Demo, A., Cocchi, F., Popovic, M., Sabirsh, A., and Gallo, R. C. (1998). The leukotriene B_4 receptor functions as a novel type of coreceptor mediating entry of primary HIV-1 isolates into CD4-positive cells. *Proc. Natl Acad. Sci. USA* **95**, 9530–9534.

Papayianni, A., Serhan, C. N., Philips, M. L., Rennke, H. G., and Brady, H. R. (1995). Transcellular biosynthesis of lipoxin A_4 during adhesion of platelets and neutrophils in experimental immune complex glomerulonephritis. *Kidney. Int.* **47**, 1295–1302.

Papayianni, A., Serhan, C. N., and Brady, H. R. (1996). Lipoxin A_4 and B_4 inhibit leukotriene-stimulated interactions of human neutrophils and endothelial cells. *J. Immunol.* **156**, 2264–2272.

Perez, H. D., Holmes, R., Kelly, E., McClary, J., and Andrews, W. H. (1992). Cloning of a cDNA encoding a receptor related to the formyl peptide receptor of human neutrophils. *Gene* **118**, 303–304.

Romano, M., Maddox, J. F., and Serhan, C. N. (1996). Activation of human monocytes and the acute monocytic leukemia cell line (THP-1) by lipoxins involves unique signaling pathways for lipoxin A_4 versus lipoxin B_4: evidence for differential Ca^{2+} mobilization. *J. Immunol.* **157**, 2149–2154.

Scalia, R., Gefen, J., Petasis, N. A., Serhan, C. N., and Lefer, A. M. (1997). Lipoxin A_4 stable analogs inhibit leukocyte rolling and adherence in the rat mesenteric microvasculature: Role of P-selectin. *Proc. Natl Acad. Sci. USA* **94**, 9967–9972.

Serhan, C. N. (1997). Lipoxins and novel aspirin-triggered 15-epi-lipoxins (ATL): a jungle of cell–cell interactions or a therapeutic opportunity? *Prostaglandins* **53**, 107–137.

Serhan, C. N., Maddox, J. F., Petasis, N. A., Akritopoulou-Zance, I., Papayianni, A., Brady, H. R., Colgan, S. P., and Madara, J. L. (1995). Design of lipoxin A_4 stable analogs that block transmigration and adhesion of human neutrophils. *Biochemistry* **34**, 14609–14615.

Soyombo, O., Spur, B. W., and Lee, T. H. (1994). Effects of lipoxin A_4 on chemotaxis and degranulation of human eosinophils stimulated by platelet-activating factor and *N*-formyl-L-methionyl-L-leucyl-L-phenylalanine. *Allergy* **49**, 230–234.

Stenke, L., Mansour, M., Edenius, C., Reizenstein, P., and Lindgren, J. A. (1991). Formation and proliferative effects of lipoxins in human bone marrow. *Biochem. Biophys. Res. Commun.* **180**, 255–261.

Stenke, L., Reizenstein, P., and Lindgren, J. A. (1994). Leukotrienes and lipoxins – new potential performers in the regulation of human myelopoiesis. *Leukemia Res.* **18**, 727–732.

Takano, T., Fiore, S., Maddox, J. F., Brady, H. R., Petasis, N. A., and Serhan, C. N. (1997). Aspirin-triggered 15-epi-lipoxin A_4 and LXA_4 stable analogues are potent inhibitors of acute inflammation: evidence for anti-inflammatory receptors. *J. Exp. Med.* **185**, 1693–1704.

Takano, T., Clish, C. B., Gronert, K., Petasis, N. A., and Serhan C. N. (1998). Neutrophil-mediated changes in vascular permeability are inhibited by topical application of aspirin-triggered 15-epi-lipoxin A_4 and novel lipoxin B_4 stable analogues. *J. Clin. Invest.* **101**, 819–826.

Tamaoki, J., Tagaya, E., Yamawaki, I., and Konno K. (1995). Lipoxin A_4 inhibits cholinergic neurotransmission through nitric oxide generation in the rabbit trachea. *Eur. J. Pharmacol.* **287**, 233–238.

Toh, H., Ichikawa, A., and Narumiya, S. (1995). Molecular evolution of receptors for eicosanoids. *FEBS Lett.* **361**, 17–21.

Ushikubi, F., Hirata, M., and Narumiya, S. (1995). Molecular biology of prostanoid receptors; an overview. *J. Lipid Mediat. Cell Signal.* **12**, 343–359.

Wikstrosm Jonsson, E. (1998). Functional characterization of receptors for cysteinyl leukotrienes in muscle. Doctoral thesis from Karolinska Institute, Stockholm, Sweden.

Ye, R. D., Cavanagh, S. L., Quehenberger, O., Prossnitz, E. R., and Cocrane, C. (1992). Isolation of a cDNA that encodes a novel granulocyte *N*-formyl peptide receptor. *Biochem. Biophys. Res. Commun.* **184**, 582–589.

Yokomizo, T., Izumi, T., Chang, K., Takuwa, Y., and Shimizu, T. (1997). A G-protein-coupled receptor for leukotriene B4 that mediates chemotaxis. *Nature* **387**, 620–624.

LICENSED PRODUCTS

Licensed products for lipoxin investigation are listed in the chapter on lipoxin.

ACKNOWLEDGEMENTS

For the purpose of brevity, review articles are preferentially cited. We apologize to our colleagues whose names do not appear within the reference list, yet who have contributed substantially to this area of research. This work was supported in part by National Institutes of Health grants GM-38765 and DK-50305 to CNS and National Arthritis Foundation fellowships to NC and KG.

ACTH Receptor

F. Shawn Galin and J. Edwin Blalock*

Department of Physiology and Biophysics, University of Alabama at Birmingham, Birmingham, AL 35294-0005, USA

*corresponding author tel: 205-934-6439, fax: 205-934-1446, e-mail: blalock@UAB.Edu

DOI: 10.1006/rwcy.2000.23007.

SUMMARY

The ACTH receptor is a member of the melanocortin receptor group, five members of which have been described. Melanocortin receptor 2 is used by adrenal cells to interact with ACTH. Although ACTH has numerous effects on immune cells that occur through an ACTH receptor, it is unknown at present which member of the melanocortin receptor family is used.

BACKGROUND

Discovery

The field of neuroimmunoendocrinology emerged based on the finding that cells of the neuroendocrine and immune systems could communicate with one another. The biochemical rationale for this bidirectional communication between the neuroendocrine and immune systems lies in the fact that these two systems share ligands and receptors. The first neuroendocrine hormone shown to be produced by immune cells was the pro-opiomelanocortin (POMC)-derived hormone corticotropin, or ACTH (Blalock and Smith, 1980). At that time, this finding was puzzling since the only known receptors of the pituitary-derived counterpart were thought to reside exclusively in the adrenal gland. However, this notion was soon dispelled after it was shown that ACTH-binding sites do reside on cells of the immune system.

The first evidence to imply that ACTH-binding sites resided on cells of the adrenal gland was the finding that treatment of adrenal tissue with exogenous ACTH led to an increase in glucocorticoid production which could be blocked by antibody to ACTH (Taunton *et al.*, 1967). When it was shown that ACTH had biological effects on lymphocytes, much effort was expended to determine whether ACTH receptors resided on cells of the immune system. One of the first pieces of evidence that ACTH receptors were expressed on lymphocytes was shown through binding studies using radiolabeled ACTH as a probe. In these studies both a specific high-affinity binding site and a low-affinity binding site were found on the surface of mouse spleen cells (Johnson *et al.*, 1982). These values corresponded to those seen on mouse adrenal cells. More recently, using an iodinated ACTH analog, a number of high-affinity binding sites were detected on purified B and T cell populations (Clarke and Bost, 1989). Since both the adrenal and the lymphocyte binding sites are comparable with regard to affinities, it would seem that the receptors are structurally and functionally related. The number of receptors expressed on lymphocytes is not static since mitogen stimulation can markedly increase the number of receptor sites on both B and T cells.

Alternative names

The ACTH receptor belongs to a family of receptors known as the melanocortin (MC) receptors. The melanocortins include ACTH and the melanocyte-stimulating hormones (MSH). The MSHs are appropriately named for their ability to act on melanocytes and alter pigmentation. The sequence for the MSHs resides within ACTH itself. The sequence for α-MSH corresponds to the first 13 amino acids of ACTH, whereas β-MSH is identical to ACTH residues 4–10. Considering this, it is not surprising that ACTH can act on melanocytes. Furthermore, it is not unusual for a rise in MSH activity to accompany a rise in steroidogenic activity (for review see Inouye and Otsuka, 1987).

The MSH peptides arise from differential processing of the precursor POMC in the pituitary gland. In the anterior lobe, corticotrophs process POMC into ACTH and β-endorphin whereas the melanotrophs yield α-MSH and corticotropin-like intermediate peptide or CLIP. Aside from their classical functions, melanocortins have been shown to have other effects on cognitive function (DeWeid and Jollies, 1982), cardiovascular regulation (Gruber and Callahan, 1989), thermoregulation (Lipton *et al.*, 1981), and appetite (Mountjoy and Wong, 1997).

Structure

To date, there are five known MC receptors, termed MC1-R through MC5-R. They are all members of the G protein-linked seven membrane spanning domain superfamily. They are approximately 300 (Cammas *et al.*, 1995) to 360 (Gantz *et al.*, 1993a) amino acids in length. Since the MSH receptor was the first to be cloned, it received the title MC1-R. The adrenal ACTH receptor, which was next, is termed MC2-R. MC3-R, MC4-R, and MC5-R are all in numerical order from when they were discovered. The highest homology within the MC receptors lies in the transmembrane-spanning domains (TM) themselves. They contain the highly conserved Asp-Arg-Tyr motif in TM region 3 and the second intracellular loop seen within the seven membrane spanning domain receptor superfamily (Probst *et al.*, 1992). They differ with other members of the superfamily in that they lack highly conserved structural residues such as the prolines which reside in the fourth and fifth TM domains and the cysteines in TM-1 and TM-2 (Probst *et al.*, 1992; Tatro, 1996).

Main activities and pathophysiological roles

The classical function attributed to the adrenal ACTH receptor is to stimulate steroidogenesis in response to a physical stressor. Thus, the receptor plays a key role in the hypothalamic-pituitary-adrenal axis. Briefly, in response to stress, there is an increase in the secretion of corticotropin-releasing hormone (CRH) from the hypothalamus, which leads to an increase in synthesis, and secretion of ACTH by corticotrophs in the pituitary gland. The pituitary-derived ACTH then, in turn, acts on cells of the adrenal cortex to stimulate the release of glucocorticoids. This is achieved by both the upregulation of enzymes involved in steroidogenesis as well as the activation of the conversion of cholesterol to pregnenolone, which is ultimately converted to corticosterone or cortisol in the rodent and in humans, respectively. There is negative feedback regulation of this axis where increasing levels of glucocorticoids feedback to the levels of both the hypothalamus and the pituitary to inhibit the release of CRH and ACTH, respectively.

The lymphocyte ACTH receptor has a variety of responses attributed to its modulation of immune function. Corticotropin has been shown to have direct effects on lymphocytes such as mitogen-stimulated proliferation (Alvarez-Mon *et al.*, 1985) and antibody responses to both T cell-dependent and -independent antigens (Johnson *et al.*, 1982). Interestingly, ACTH was shown to reverse the inhibitory effects of dexamethasone on lymphocyte proliferation, which would suggest that ACTH may serve to 'protect' the immune response from the effects of transient or sustained rises in glucocorticoid levels which are known to have pronounced inhibitory effects on immune function (Pepper *et al.*, 1993).

GENE

Accession numbers

To date, five MC receptors have been described. Using the polymerase chain reaction, the cDNA from a human melanoma was used to clone and sequence the MC1-R (GenBank X65634) and the MC2-R (GenBank X65633) genes (Mountjoy *et al.*, 1992). These MC receptor genes lacked introns, which is not unlike the genes encoding many other members of the G protein-linked seven membrane-spanning domain receptor superfamily. The genes for the third (GenBank L06155), fourth (GenBank L08603), and fifth (GenBank U08353) melanocortin receptor have also been cloned and sequenced (Gantz *et al.*, 1993a,b; Fathi *et al.*, 1995). The adrenal ACTH receptor has been identified as the MC2-R. This is consistent with a recent finding that the melanocortin 1, 3, 4, and 5 receptors do not have an epitope for ACTH outside of the sequence for α-MSH (ACTH 1–13) (Schioth *et al.*, 1997).

PROTEIN

Accession numbers

GB299420

Sequence

See **Figure 1**.

Relevant homologies and species differences

There is 40–80% homology between receptor subtypes. Within a subtype there is 80–95% homology between different species.

Affinity for ligand(s)

See Table 1.

Cell types and tissues expressing the receptor

In addition to the classic site at the cortex of the adrenals (**Table 1**), ACTH receptors are also expressed in adipose tissue and in B cells and T cells within the immune system. Cell lines expressing the receptor include murine Y-1 and human H295 adrenocortical cells.

Regulation of receptor expression

It has been shown that ACTH can upregulate the expression of the ACTH-binding site and receptor transcripts several-fold in murine Y-1 and human H295 cells (Mountjoy *et al.*, 1992). In lymphocytes, the levels of receptor expression can be increased upon mitogen stimulation. For example, lipopolysaccharide stimulation of B cells can result in a 2-fold increase in receptor expression, whereas T lymphocytes and thymocytes stimulated with concanavalin A have been shown to increase receptor expression by 3- and 1000-fold, respectively (Clarke and Bost, 1989). This increase in receptor expression and subsequent binding by ACTH has consequences on immune function. For example, ACTH has been shown to upregulate IL-6 and IL-4 production, which are cytokines associated with a TH2-type T cell response (Aebischer *et al.*, 1994). Interestingly, lymphocyte-derived TGFβ can downregulate ACTH binding and block upregulation of ACTH-binding sites by ACTH itself (Rainey *et al.*, 1989). Thus it seems that the ACTH receptor is subject to negative feedback regulation by TGFβ.

SIGNAL TRANSDUCTION

Cytoplasmic signaling cascades

As previously mentioned, the ACTH receptors are G protein-coupled receptors. Binding of ACTH to its receptor in both adrenal and lymphoid cells results in an increase or rise in cyclic AMP (cAMP) and cGMP levels. In addition to the change in adenylate cyclase activity, the adrenal ACTH receptor is thought to be linked to steroidogenesis through a calcium flux in

Figure 1 Amino acid sequence for the ACTH receptor.

```
1     MKHIINSYEN INNTARNNSD CPRVVLPEEI FFTISIVGVL ENLIVLLAVF
51    KNKNLQAPMY FFICSLAISD MLGSLYKILE NILIILRNMG YLKPRGSFET
101   TADDIIDSLF VLSLLGSIFS LSVIAADRYI TIFHALRYHS IVTMRRTVVV
151   LTVIWTFCTG TGITMVIFSH HVPTVITFTS LFPLMLVFIL CLYVHMFLLA
201   RSHTRKISTL PRANMKGAIT LTILLGVFIF CWAPFVLHVL LMTFCPSNPY
251   CACYMSLFQV NGMLIMCNAV IDPFIYAFRS PELPRDAFKK MIFC
```

Table 1 Ligand preference and tissue distribution of MC receptors

Receptor	Tissue distribution	Specificity	Reference
MC1-R	Melanocytes, adipose tissue	α-MSH > ACTH > γ-MSH	Mountjoy *et al.*, 1992
MC2-R	Adrenal cortex	ACTH	Mountjoy *et al.*, 1992
MC3-R	Brain, gut, placenta	α-MSH = γ-MSH = ACTH	Gantz *et al.*, 1993
MC4-R	Brain, sympathetic nervous system	α-MSH = ACTH > β-MSH	Gantz *et al.*, 1993
MC5-R	Adipose tissue, skeletal muscle, pituitary	α-MSH > ACTH > γ-MSH	Fathi *et al.*, 1995

these cells (Saez *et al.*, 1981; Tremblay *et al.*, 1991). In lymphocytes, the addition of ACTH was also found to increase calcium uptake in these in a time- and dose-dependent manner. The kinetics are similar to that seen in adrenal cells. Furthermore, this mechanism is blocked by calcium channel antagonists (Clarke *et al.*, 1994).

A rise in cytosolic free calcium is one of the first responses involved in lymphocyte activation. It has been shown that ACTH can modulate both B and T cell proliferation and function. Interestingly, it has been shown that ACTH alone does not alter the basal levels of cytosolic free calcium (Kavelaars *et al.*, 1988; Clarke *et al.*, 1994); however, this is not unlike that seen in adrenal cells (Iida *et al.*, 1986). In lymphocytes, ACTH works in concert with mitogen or antigen stimulation to cause a calcium flux across the plasma membrane. The activity of ACTH seems to alter calcium levels through increasing calcium channel activity rather than the release of calcium through intracellular stores (Clarke, 1995).

BIOLOGICAL CONSEQUENCES OF ACTIVATING OR INHIBITING RECEPTOR AND PATHOPHYSIOLOGY

Unique biological effects of activating the receptors

The activation of the ACTH receptor is a primary event in steroidogenesis, leading to alterations in circulating glucocorticoid levels. In fact, a mutation in the gene has been described in *glucocorticoid deficiency syndrome* (Slavotinek *et al.*, 1998). This is characterized by extremely low cortisol levels in the face of high ACTH concentrations.

Human abnormalities

A mutation was shown to lead to a new restriction endonuclease site with the ACTH receptor gene, allowing for detection without the need for DNA sequencing. Interestingly, *glucocorticoid deficiency syndrome* resulting from decreased ACTH receptor expression may not be confined exclusively to cells of the adrenal gland. Indeed, the ability to detect a form of glucocorticoid deficiency due to ACTH insensitivity syndrome by testing for the absence of high-affinity ACTH-binding sites on a patient's peripheral blood mononuclear cells may be the first example (Smith *et al.*, 1987; Yamamoto *et al.*, 1995).

Two mutations in the receptor gene have been shown to be associated with human *ACTH hypersensitivity syndrome*. Using PCR and nucleotide sequencing techniques, two base mutations were found. The first was cysteine 21 to arginine and the second, serine 247 to glycine. These regions corresponded to the first extramembranous N-terminal domain and the third extramembranous loop, respectively. Clinically, this led to normal cortisol levels in the absence of detectable ACTH levels and an increase in adrenocorticol sensitivity to ACTH (Hiroi *et al.*, 1998).

References

Aebischer, I., Stampfli, M. R., Zurcher, A., Miescher, S., Urwyler, A., Frey, B., Luger, T., White, R. R., and Stadler, B. M. (1994). Neuropeptides are potent modulators of human *in vitro* immunoglobulin E synthesis. *Eur. J. Immunol.* **24**, 1908–1913.

Alvarez-Mon, M., Kehrl, J., and Fauci, A. (1985). A potential role for adrenocorticotropin in regulating human B lymphocyte functions. *J. Immunol.* **135**, 3823–3826.

Blalock, J. E., and Smith, E. M. (1980). Human leukocyte interferon: structural and biological relatedness to adrenocorticotropic hormone and endorphins. *Proc. Natl Acad. Sci. USA* **77**, 5972–5974.

Cammas, F. M., Kapas, S., Barker, S., and Clark, A. J. L. (1995). Cloning, characterization, and expression of a functional ACTH receptor. *Biochem. Biophys. Res. Commun.* **212**, 912–918.

Clarke, B. L. (1995). Calcium uptake by ACTH-stimulated lymphocytes: what is the physiological significance? *Adv. Neuroimmunol.* **5**, 271–281.

Clarke, B. L., and Bost, K. L. (1989). Differential expression of functional corticotropin (ACTH) receptors by subpopulations of lymphocytes. *J. Immunol.* **143**, 464–469.

Clarke, B. L., Moore, D. R., and Blalock, J. E. (1994). Adrenocorticotropic hormone stimulates a transient calcium uptake in rat lymphocytes. *Endocrinology* **135**, 1780–1786.

DeWeid, D., and Jollies, J. (1982). Neuropeptides derived from pro-opiomelanocortin: behavioral, physiological, and neurochemical effects. *Physiol. Rev.* **62**, 976–1059.

Fathi, Z., Iben, L. G., and Parker, E. M. (1995). Cloning, expression, and tissue distribution of a fifth melanocortin subtype. *Neurochem. Res.* **20**, 107–113.

Gantz, I., Kondaf, Y., Tashiro, T., Shimoto, Y., Miwa, H., Munzert, G., Watson, S. J., DelValle, J., and Yamada, T. (1993a). Molecular cloning of a novel melanocortin receptor. *J. Biol. Chem.* **268**, 8248–8250.

Gantz, I., Miwa, H., Konda, Y., Shimoto, Y., Tashiro, T., Watson, S. J., Delvalle, J., and Yamata, T. (1993b). Molecular cloning, expression, and gene localization of a fourth melanocortin receptor. *J. Biol. Chem.* **268**, 15174–15179.

Gruber, K. A., and Callahan, M. F. (1989). ACTH-(4-10) through gamma-MSH: evidence for a new class of central nervous system-regulating peptides. *Am. J. Physiol.* **257**, R681–R694.

Hiroi, N., Yakushiji, F., Shimojo, M., Wantanabe, S., Sugano, S. S., Yamaguchi, N., and Miyachi, Y. (1998). Human ACTH hypersensitivity syndrome associated with abnormalities of the ACTH receptor gene. *Clin. Endocrin.* **48**, 129–134.

Iida, S., Widmaier, E., and Hall, P. (1986). The phosphatidylinositide-Ca^{2+} hypothesis does not apply to the steroidogenic action of corticotropin. *Biochem. J.* **236**, 53–59.

Inouye, K., and Otsuka, H. (1987). In "Hormonal Proteins and Peptides" (ed. C. H. Li), ACTH: structure-function relationship, pp. 1–29. Academic Press, New York.

Johnson, H. M., Smith, E. M., Torres, B. A., and Blalock, J. E. (1982). Regulation of the *in vitro* antibody response by neuroendocrine hormones. *Proc. Natl Acad. Sci. USA* **79**, 4171–4174.

Kavelaars, A., Ballieux, R., and Heijnen, C. (1988). Modulation of the immune response by pro-opiomelanocortin derived peptides. II. Influence of adrenocorticotropic hormone on the rise in intracellular free calcium concentration after T cell activation. *Brain Behav. Immun.* **2**, 57–66.

Lipton, J. M., Glyn, J. R., and Zimmer, J. A. (1981). ACTH and alpha-melanotropin in central temperature control. *Fedn Proc.* **40**, 2760–2764.

Mountjoy, K. G., and Wong, J. (1997). Obesity, diabetes, and functions for pro-opiomelanocortin-derived peptides. *Mol. Cell. Endocrinol.* **128**, 171–177.

Mountjoy, K. G., Robbins, L. S., Mortrud, M. T., and Cone, R. D. (1992). The cloning of a family of genes that encode the melanocortin receptors. *Science* **257**, 1248–1251.

Pepper, G. M., Aggarwal, K., Torres, J., Velasco, R., Dadhania, K., Drake, D., and Futran, J. (1993). The altered biologic activity of ACTH and related peptides on peripheral blood mononuclear cells is altered by the presence of dexamethasone. *Cell. Immunol.* **151**, 110–117.

Probst, W. C., Snyder, L. A., Schuster, D. I., Brosius, J., and Sealfon, S. C. (1992). Sequence alignment of the G-protein coupled receptor superfamily. *DNA Cell Biol.* **11**, 1–20.

Rainey, W. E., Viard, I., and Saez, J. M. (1989). Transforming growth factor β treatment decreases ACTH receptors on ovine adrenocortical cells. *J. Biol. Chem.* **264**, 21474–21477.

Saez, J., Morera, A. M., and Danzord, A. (1981). In "Advances in Cyclic Nucleotide Research" (ed. J. Dumont, P. Greengard, and G. Robinson), Mediators of the effects of ACTH on adrenal cells, pp. 563–579. Raven Press, New York.

Scioth, H. B., Muceniece, R., Larsson, M., and Wikberg, J. E. S. (1997). The melanocortin 1, 3, 4, or 5 receptors do not have a binding epitope for ACTH beyond the sequence of α-MSH. *J. Immunol.* **155**, 73–78.

Slavotinek, A. M., Hurst, J. A., and Dunger, D. (1998). ACTH receptor mutation in a girl with familial glucocorticoid deficiency. *Clin. Genet.* **53**, 57–62.

Smith, E. M., Brosnan, P., Meyer, W. J., and Blalock, J. E. (1987). An ACTH receptor on human mononuclear leukocytes. Relation to adrenal ACTH receptor activity. *N. Engl. J. Med.* **317**, 1266–1269.

Tatro, J. B. (1996). Receptor biology of the melanocortins, a family of neuro-immunomodulatory peptides. *Neuroimmunology* **3**, 259–284.

Taunton, O. D., Roth, J., and Pastan, I. (1967). ACTH stimulation of adenyl cyclase in adrenal homogenates. *Biochem. Biophys. Res. Commun.* **29**, 1–7.

Tremblay, E., Payet, M.-D., and Gallo-Payet, N. (1991). Effects of ACTH and angiotensin II on cytosolic calcium in cultured adrenal glomerulosa cells. Role of cAMP production in the ACTH effect. *Cell Calcium* **12**, 655–673.

Yamamoto, Y., Kawada, Y., Noda, M., Yamagashi, M., Ishida, O., Fujihira, T., Shirakawa, F., and Morimoto, I. (1995). Siblings with ACTH insensitivity due to lack of ACTH binding to the receptor. *Endocrine J.* **42**, 171–177.

BLTR: the Leukotriene B_4 Receptor

Andrew D. Luster* **and Andrew M. Tager**

Infectious Disease Unit, Massachusetts General Hospital, East 149 13th Street, Charlestown, MA 02129, USA

*corresponding author tel: 617-726-5710, fax: 617-726-5411, e-mail: luster@helix.mgh.harvard.edu

DOI: 10.1006/rwcy.2000.23008.

SUMMARY

The receptor for leukotriene B_4 (LTB_4), BLTR, is a member of the G protein-coupled seven transmembrane domain receptor (GPCR) superfamily. This receptor is highly expressed on leukocytes, and mediates LTB_4-induced leukocyte chemotaxis and activation. BLTR therefore participates in the recruitment of leukocytes to sites of pathogen invasion as part of the innate immune response, and also in the pathogenesis of inflammatory reactions in which LTB_4 has been implicated. BLTR antagonists have successfully reduced leukocyte recruitment and tissue damage in animal models of inflammatory diseases, demonstrating the therapeutic potential of inhibiting BLTR in the treatment of pathologic inflammation.

BACKGROUND

Discovery

A binding site for $[^3H]LTB_4$ on human neutrophils was initially detected in 1982 with a K_d of 10 nM (Goldman and Goetzl, 1982), 0.46 nM (Lin *et al.*, 1984), and 1.5 nM (Bomalaski and Mong, 1987), and on guinea pig eosinophils with a K_d of 1.4 nM (Ng *et al.*, 1991). The human BLTR was eventually cloned from retinoic acid-differentiated HL-60 cells using a subtraction strategy (Yokomizo *et al.*, 1997). Membrane fractions of COS-7 cells transfected with the cloned sequence demonstrated LTB_4 binding with a K_d comparable to that observed in retinoic acid-differentiated HL-60 cells, and CHO cells stably transfected with the sequence demonstrated LTB_4-induced increases in intracellular calcium and chemotactic responses, indicating that this sequence encoded the LTB_4 receptor, BLTR (Yokomizo *et al.*, 1997).

This sequence had been previously cloned using degenerate PCR strategies by two independent groups as an orphan receptor gene believed to encode a member of the G protein-coupled seven transmembrane domain receptor (GPCR) superfamily, and called R2 (Raport *et al.*, 1996) and chemoattractant receptor-like 1 (CMKRL1) (Owman *et al.*, 1996). LTB_4 was subsequently confirmed to be the functional ligand for this putative receptor in experiments demonstrating high-affinity LTB_4 binding and LTB_4-induced increases in intracellular calcium in cells transfected with CMKRL1 cDNA (Owman *et al.*, 1997). The identical receptor sequence was also isolated from a low stringency screen of a human erythroleukemia (HEL) cell cDNA library using chicken purinoceptor $P2Y_3$ cDNA, and designated $P2Y_7$ based on the ability of the transfected receptor to bind $[^{35}S]$dATP (Akbar *et al.*, 1996). However, ATP binding and signaling have not been seen by others (Yokomizo *et al.*, 1997; Huang *et al.*, 1998). The mouse ortholog of BLTR was independently cloned by performing degenerate PCR with primers directed to well-conserved transmembrane domains of chemoattractant GPRCs, using cDNA isolated from murine eosinophils (Huang *et al.*, 1998). CHO cells transfected with this mouse sequence bound LTB_4 with high affinity and demonstrated LTB_4-induced increases in intracellular calcium (Huang *et al.*, 1998). Another group subsequently identified the identical murine sequence by screening a mouse genomic library with a fragment of the human cDNA identified previously as encoding the $P2Y_7$ receptor (Martin *et al.*, 1999).

Alternative names

The consensus nomenclature for this receptor is BLTR (Alexander and Peters, 1997). Prior to its identification as the LTB_4 receptor, however, groups cloning this sequence called it R2 (Raport *et al.*, 1996), CMKRL1 (Owman *et al.*, 1996), and $P2Y_7$ (Akbar *et al.*, 1996).

Structure

Kyte-Doolittle hydrophobicity analysis of the amino acid sequences of human (Raport *et al.*, 1996; Owman *et al.*, 1996) and mouse (Huang *et al.*, 1998) BLTR shows the presence of seven hydrophobic transmembrane domains common to GPCRs. The amino acid sequences also contain other motifs characteristic of this family of receptors (Owman *et al.*, 1996; Huang *et al.*, 1998), including (a) conserved proline residues in several of the transmembrane domains, which are thought to induce flexibility in the helix formations; (b) conserved cysteine residues in two of the extracellular loops for intramolecular chain disulfide bonding; (c) consensus sequences for *N*-linked glycosylation near the N-terminus and in one of the extracellular loops; and (d) a serine and threonine-rich C-terminal intracytoplasmic segment, which in other GPCRs are sites of phosphorylation involved with receptor desensitization and internalization (Murphy, 1994). Additionally, human BLTR is 352 amino acid residues in length, and mouse BLTR is 351 amino acids, similar to the lengths of other chemoattractant receptors (Murphy, 1994).

Main activities and pathophysiological roles

LTB_4 was discovered based on its potent chemotactic activity on neutrophils (Ford-Hutchinson *et al.*, 1980), and BLTR probably mediates this effect: CHO cells stably expressing exogenous BLTR showed marked chemotactic responses towards low concentrations of LTB_4 (Yokomizo *et al.*, 1997). In addition to chemotaxis, LTB_4 stimulates neutrophil aggregation, adherence associated with increased CD11/CD18 expression, generation of reactive oxygen species and release of granular enzymes (Henderson, 1994).

In addition to its effects on neutrophils, LTB_4 stimulates chemotaxis of guinea pig eosinophils (Ng *et al.*, 1991) and IL-5-exposed murine eosinophils (Huang *et al.*, 1998). Human eosinophils, however, have been reported to respond poorly to LTB_4 when compared with PAF (Morita *et al.*, 1989).

LTB_4 augments human peripheral blood monocyte production of IL-1 (Rola-Pleszczynski and Lemaire, 1985), and mouse peritoneal macrophage phagocytosis and killing of bacteria (Demitsu *et al.*, 1989). Mice with targeted disruption of the 5-lipoxygenase (5-LO) gene, and hence deficient of leukotrienes including LTB_4, showed a greater degree of lethality as well as bacteremia following intratracheal challenge with *Klebsiella pneumoniae* (Bailie *et al.*, 1996). Alveolar macrophages from the 5-LO-deficient mice exhibited impairments in phagocytosis and killing of bacteria *in vitro*, which were overcome by addition of exogenous LTB_4.

LTB_4 enhances production of IL-2 by CD4+ T cells (Marcinkiewicz *et al.*, 1997), and IL-2Rβ expression by CD8+ T cells (Stankova *et al.*, 1992).

LTB_4 increases NK cell IL-2Rβ expression, sensitivity to IL-2, and cytotoxic activity (Rola-Pleszczynski *et al.*, 1983; Stankova *et al.*, 1992).

LTB_4 enhances B cell activation, proliferation, and Ig secretion (Yamaoka *et al.*, 1989).

Treatment of endothelial cell monolayers with LTB_4 increases their binding of neutrophils (Gimbrone *et al.*, 1984). LTB_4 treatment of endothelial cells has also been reported to increase binding of lymphocytes (Renkonen *et al.*, 1988), but this finding was not substantiated in a subsequent study (To and Scrieber, 1990).

Inflammatory Cell Recruitment

By mediating LTB_4-induced leukocyte chemotaxis and activation, BLTR participates in the recruitment of leukocytes to sites of pathogen invasion as part of the innate immune response. By mediating LTB_4-induced leukocyte chemotaxis and activation, BLTR also participates in the pathogenesis of inflammatory diseases in which LTB_4 has been implicated, including *asthma* and *allergic rhinitis*, *cystic fibrosis*, *inflammatory bowel disease*, *psoriasis*, *acute respiratory distress syndrome*, *multiple sclerosis*, and *rheumatoid arthritis* (Samuelsson *et al.*, 1987; Lewis *et al.*, 1990; Henderson, 1994).

HIV Coreceptor

Some members of the GPCR superfamily of chemoattractant receptors, including CCR5 and CXCR4, also function as coreceptors for *HIV* entry into target cells. The ability of other members of this family, including BLTR, to serve as HIV coreceptors has recently been investigated, with conflicting results. One group demonstrated that BLTR was able to act as a coreceptor for entry of several clinical isolates of HIV-1 into murine fibroblasts stably expressing both BLTR and the human CD4 receptor (Owman *et al.*, 1998). Env proteins of these clinical isolates were not described. However, a second group subsequently found that BLTR did not function as a coreceptor for either Env-mediated viral entry or Env-mediated cell–cell fusion in experiments using various HIV-1 and SIV Env proteins (Martin *et al.*, 1999).

GENE

Accession numbers

Human: NM000752, X98356, U41070, U33448
Mouse: AF044030

Sequence

See **Figure 1**.

Figure 1 Nucleotide sequence of the human BLTR gene.

```
Human BLTR
 1    gccattctct cacatcccgt gcggtcagga agcccttcct gaactctgac ttcagttctt
 61   gctgcggttt ctgcccattt ttttcatatc ctctgacagc tgcgaggtca tctctgctct
 121  ggcttttctc caagcagaac aagtgggggc tctggaaagg ttaagggacc tcagtggcca
 181  ccattatact ttgcatcttt cctgagaagt gagagttgaa agggaagcag gaaggcccat
 241  ggtcagattg aaggaaggac tttttagttt cttttttttt tttttgaaat ggagtctcgc
 301  tctgtcattc aggctggagt gcagtggtgc gatctcagct cactgcagcc tccacttcct
 361  gggttcacat gattctcctg cctcagcctc ccaagtagct gagactacag gcacatgcca
 421  ctacacccag ctaacttttg tatttttagt agagacgggg tttcaccatg ttggccaggc
 481  tggtctcaaa ctgctaacat caagtgatct gctcccctca gcctcccaaa gtgctgggat
 541  taccggtatg aaccaccaca acctgccagg aatttttagt ttttagcttt tgcaggagac
 601  ttcaaggaaa ggagacattc ctctgtccag gaaacgggta aggggaccat ttctgcattg
 661  ctggtttccc ctcttggcag ggtgggcatg aggcatcact gttcctgctc cctcactcct
 721  gctcctcatg ctcagcctgc cagctcggcc tcaactttgt gtgtctaaag tggaactgaa
 781  tagtagctgt gagaagatag gaaagaggta gtgccaatct ccttgcccag atcataaatc
 841  cagactcagc agggtaacca catgggcaag cacaaggtag gtgcttgggg aaaggggaag
 901  taattggcat tctgtgtgat accaaggaga ccatttggat tttggcttct accaaagaga
 961  atggagaatt ggttgaccta aatggaacca gtccctttaa gtaaggggag gaaagggggt
 1021 gctggaagat ggccctcttc ccaccaccta gatcatagct tgaactgaag ccaaggacag
 1081 agtgctgccc ccttcggcat ttactgatgt gccctcttta aatcatgatg ttatctaacc
 1141 caaacccaga cccaggacct agtcacagct ccaacctaca cttcctatta atcttaaaac
 1201 aaagcgaaac aaacacaaaa agatatcagc attgtagcct ccaatctgag cccatttccc
 1261 ttctctggct accatacctc cttctcctat atgataccat tcactacttt gttcaattat
 1321 ccagtctaga cctgcatctt gaggccacac ccagccttct cactccccac acccctcttt
 1381 cctctctcac tgctccttcc tggtctcttc tcatctggcc ccacctctaa ggagtcctcc
 1441 tgccttctgg gttgccctgg aaaacagact atccccccctc ctagtgaagg gagtgggtag
 1501 gggtttcagc cccaccctca ggaagatgcg tcttccctgt cctctgctct gtggtacttc
 1561 ctctctggct gatttagcaa acagcaccta gacctggggc caggcctttg gcagtgggac
 1621 agatccaggg ataggctaca ccaccctgcc ctgaccctgg gattggcatc agcttccaac
 1681 cagttcctgc caaagcttgt aagtcctccc gacggccatg aacactacat cttctgcagc
 1741 acccccctca ctaggtgtag agttcatctc tctgctggct atcatcctgc tgtcagtggc
 1801 gctggctgtg gggcttcccg gcaacagctt tgtggtgtgg agtatcctga aaaggatgca
 1861 gaagcgctct gtcactgccc tgatggtgct gaacctggcc ctggccgacc tggccgtatt
 1921 gctcactgct cccttttttcc ttcacttcct ggcccaaggc acctggagtt ttggactggc
 1981 tggttgccgc ctgtgtcact atgtctgcgg agtcagcatg tacgccagcg tcctgcttat
 2041 cacggccatg agtctagacc gctcactggc ggtggcccgc ccctttgtgt cccagaagct
 2101 acgcaccaag gcgatggccc ggcgggtgct ggcaggcatc tgggtgttgt cctttctgct
 2161 ggccacaccc gtcctcgcgt accgcacagt agtgccctgg aaaacgaaca tgagcctgtg
 2221 cttcccgcgg taccccagcg aagggcaccg ggccttccat ctaatcttcg aggctgtcac
 2281 gggcttcctg ctgcccttcc tggctgtggt ggccagctac tcggacatag ggcgtcggct
 2341 acaggcccgg cgcttccgcc gcagccgccg caccggccgc ctggtggtgc tcatcatcct
 2401 gaccttcgcc gccttctggc tgccctacca cgtggtgaac ctggctgagg cgggccgcgc
 2461 gctggccggc caggccgccg ggttagggct cgtggggaag cggctgagcc tggcccgcaa
 2521 cgtgctcatc gcactcgcct tcctgagcag cagcgtgaac cccgtgctgt acgcgtgcgc
 2581 cggcggcggc ctgctgcgct cggcgggcgt gggcttcgtc gccaagctgc tggagggcac
 2641 gggttccgag gcgtccagca cgcgccgcgg gggcagcctg ggccagaccg ctaggagcgg
 2701 ccccgccgct ctggagcccg gcccttccga gagcctcact gcctccagcc ctctcaagtt
 2761 aaacgaactg aactaggcct ggtggaagga ggcgcacttt cctcctggca gaatgctagc
 2821 tctgagccag ttcagtacct ggaggaggag caggggcgtg gagggcgtgg agggcgtggg
 2881 agcgtgggag gcgggagtgg agtggaagaa gagggagaga tggagcaaag tgagggccga
 2941 gtgagagcgt gctccagcct ggctcccaca ggcagcttta accattaaaa ctgaagtctg
 3001 aa
```

Chromosome location and linkages

When human BLTR was cloned as R2, CMKRL1, and $P2Y_7$, it was localized to chromosome 14. Probing Southern blots of mouse/human somatic cell hybrid DNA with R2 cDNA localized the gene to the region 14pter-14q23 (Raport *et al.*, 1996). Mapping by FISH/DAPI using CMKRL1 cDNA showed the gene localized to chromosome 14q11.2-q12 (Owman *et al.*, 1996). Mapping with PCR using DNA from a panel of mouse/human somatic cell hybrids as templates and primers from the $P2Y_7$ cDNA clone produced a similar result: amplification of the expected size product was observed only with DNA from the hybrid that contained human chromosome 14 (Akbar *et al.*, 1996).

Human somatostatin receptor 1 (SSTR1), a member of the GPCR superfamily, has been mapped to chromosome 14q13 (Yamada *et al.*, 1993), close to the location of BLTR on chromosome 14q11.2-q12 (Owman *et al.*, 1996).

PROTEIN

Accession numbers

Human: NP000743, CAA67001, AAC50628, AAB16747
Mouse: AAC61677

Sequence

See **Figure 2**.

Relevant homologies and species differences

The human proteins with the highest overall homology to human BLTR are the human somatostatin receptor type 5 (SSTR5), with 33.0% identity; the human IL-8 receptor (CXCR1), with 33.0%; the human formyl peptide-related receptor II (FMLPR-II, the lipoxin A_4 receptor), with 30.7%; and the human formyl peptide receptor (FMLPR), with 28.6% (Yokomizo *et al.*, 1997). Mouse BLTR, which has 78% identity with human BLTR, has 32% identity with the murine C5a receptor (mC5aR), 32% identity with the murine IL-8 receptor (mCXCR2), 30% identity with murine FMLPR, 29% identity with the murine lipoxin A_4 receptor (mLXAR), and 26% identity with the murine platelet-activating factor receptor (mPAFR) (Huang *et al.*, 1998).

Beyond the overall 78% identity between human and mouse BLTR, the three intracytoplasmic loops are identical across these species, while not being conserved across the subfamily of chemoattractant receptors, suggesting that the BLTRs may be coupled to a unique, well-conserved, signaling pathway among the chemoattractant receptors (Huang *et al.*, 1998).

Affinity for ligand(s)

Cells transfected with human and mouse BLTR demonstrate LTB_4 binding with K_d values comparable to that observed with leukocytes (Yokomizo *et al.*, 1997; Huang *et al.*, 1998). Although the human BLTR sequence was reported to bind ATP with high affinity, exposure to ATP induced no changes in

Figure 2 Amino acid sequences of human and mouse BLTR.

```
Human BLTR:
1     MNTTSSAAPP SLGVEFISLL AIILLSVALA VGLPGNSFVV WSILKRMQKR SVTALMVLNL
61    ALADLAVLLT APFFLHFLAQ GTWSFGLAGC RLCHYVCGVS MYASVLLITA MSLDRSLAVA
121   RPFVSQKLRT KAMARRVLAG IWVLSFLLAT PVLAYRTVVP WKTNMSLCFP RYPSEGHRAF
181   HLIFEAVTGF LLPFLAVVAS YSDIGRRLQA RRFRRSRRTG RLVVLIILTF AAFWLPYHVV
241   NLAEAGRALA GQAAGLGLVG KRLSLARNVL IALAFLSSSV NPVLYACAGG GLLRSAGVGF
301   VAKLLEGTGS EASSTRRGGS LGQTARSGPA ALEPGPSESL TASSPLKLNE LN

Mouse BLTR:
1     MAANTTSPAA PSSPGGMSLS LLPIVLLSVA LAVGLPGNSF VVWSILKRMQ KRTVTALLVL
61    NLALADLAVL LTAPFFLHFL ARGTWSFREM GCRLCHYVCG ISMYASVLLI TIMSLDRSLA
121   VARPFMSQKV RTKAFARWVL AGIWVVSFLL AIPVLVYRTV KWNNRTLICA PNYPNKEHKV
181   FHLLFEAITG FLLPFLAVVA SYSDIGRRLQ ARRFRRSRRT GRLVVLIILA FAAFWLPYHL
241   VNLVEAGRTV AGWDKNSPAG QRLRLARYVL IALAFLSSSV NPVLYACAGG GLLRSAGVGF
301   VVKLLEGTGS EVSSTRRGGT LVQTPKDTPA CPEPGPTDSF MTSSTIPESS K
```

intracellular calcium in cells transfected with the human BLTR (Yokomizo *et al.*, 1997), and there was no detectable specific binding of ATP to cells transfected with mouse BLTR (Huang *et al.*, 1998).

Cell types and tissues expressing the receptor

Northern blotting of various human tissues with human BLTR cDNA demonstrated that gene expression is highest in peripheral blood leukocytes, and also present in spleen, thymus, bone marrow, lymph nodes, heart, skeletal muscle, brain, and liver (Akbar *et al.*, 1996; Owman *et al.*, 1996; Yokomizo *et al.*, 1997). Northern blotting with mouse BLTR cDNA revealed gene expression predominantly in activated leukocytes, including IL-5-exposed eosinophils, elicited peritoneal neutrophils and macrophages, and IFNγ-stimulated macrophages. Additionally, mouse BLTR was highly expressed in T cell lymphomas that spontaneously arose in c-*myc* transgenic mice homozygous for p53-null alleles, suggesting that T cells can express BLTR. Low levels of constitutive expression were also seen in the lung, lymph nodes, and spleen (Huang *et al.*, 1998). Increased expression of BLTR mRNA was demonstrated in the lungs of mice subjected to inhalation of the mold allergen *Aspergillus fumigatus*, consistent with the induction of an intense eosinophil-predominant inflammatory infiltrate (Huang *et al.*, 1998).

Using an affinity-purified rabbit antibody directed against the N-terminal 17 amino acids of the predicted open reading frame of mouse BLTR (Huang *et al.*, 1998), expression of this protein was demonstrated by western blotting in eosinophils purified from IL-5 transgenic mice and alveolar macrophages isolated by bronchoalveolar lavage (BAL) from wild-type mice. The antibody crossreacts with human BLTR, and expression of the human protein was demonstrated in human eosinophils (Huang *et al.*, 1998).

Regulation of receptor expression

Mouse BLTR transcription is induced in the RAW 264.7 macrophage cell line by IFNγ. Mouse BLTR transcription, which is not detectable in resting peritoneal cells, is dramatically induced in both activated macrophages and neutrophils elicited into the peritoneum of mice by injections of 9% sodium casein (Huang *et al.*, 1998).

One group cloning the mouse BLTR reported putative binding sites for several transcription factors upstream from the gene's open reading frame (Martin *et al.*, 1999). However, analysis of murine BLTR cDNA (Huang *et al.*, 1998) demonstrates the presence of a small untranslated exon upstream of the region examined by this group, indicating that these putative binding sites are actually in an intron of the gene (unpublished observations). Neither the human nor mouse BLTR promoters have as yet been identified.

Mouse BLTR has been demonstrated to be *N*-linked glycosylated. *In vitro* translation of the cDNA in the presence of dog pancreatic microsomes revealed an upward shift in mobility of the protein product on SDS-PAGE of $\sim$4 kDa, compared with the protein product translated in the absence of microsomes (Huang *et al.*, 1998).

SIGNAL TRANSDUCTION

Cytoplasmic signaling cascades

BLTR is a G protein-coupled receptor. LTB_4 cytoplasmic signaling has been shown to be predominantly mediated by *Bordetella pertussis* toxin (PTX)-sensitive G protein(s) leading to the activation of phosphoinositide (PI)-specific phospholipase C, release of inositol phosphates through PI hydrolysis and subsequent mobilization of intracellular calcium (Gaudreau *et al.*, 1998). However, experiments using CHO cells stably expressing BLTR demonstrated that whereas LTB_4 induced inhibition of forskolin-induced adenylyl cyclase activity was completely blocked by PTX, LTB_4-induced increases in intracellular calcium concentration and D-*myo*-inositol-1,4,5-trisphosphate (IP_3) accumulation were only partially blocked by PTX, indicating that BLTR couples to both PTX-sensitive and insensitive G proteins in BLTR transfected CHO cells (Yokomizo *et al.*, 1997). Similarly, LTB_4-induced increases in intracellular calcium in leukocytes is PTX-sensitive (Powell *et al.*, 1996), whereas LTB_4 induced increases in the adhesiveness of vascular endothelial cells for leukocytes is not (Palmblad *et al.*, 1994), suggesting that BLTR couples to different types of G proteins in different cell types (Yokomizo *et al.*, 1997). Whereas sensitivity to PTX suggests involvement of $G\alpha_{i/0}$ subunits in BLTR signaling, experiments using a cotransfection system in COS-7 cells indicate that this receptor can couple to a PTX-insensitive α subunit of the G_q class of G proteins, specifically $G\alpha_{16}$ (Gaudreau *et al.*, 1998).

BIOLOGICAL CONSEQUENCES OF ACTIVATING OR INHIBITING RECEPTOR AND PATHOPHYSIOLOGY

Unique biological effects of activating the receptors

BLTR has been demonstrated to mediate LTB_4-induced chemotaxis: CHO cells stably expressing human BLTR showed marked chemotactic responses towards low concentrations of LTB_4 (Yokomizo *et al.*, 1997). In the stable CHO cell transfectants, activation of BLTR with LTB_4 also induced rapid increases in intracellular calcium concentration and IP_3 accumulation, as well as inhibition of forskolin-stimulated adenylyl cyclase activity (Yokomizo *et al.*, 1997). LTB_4 also induced a dose-dependent intracellular calcium flux in CHO cells stably expressing mouse BLTR (Huang *et al.*, 1998).

Phenotypes of receptor knockouts and receptor overexpression mice

Transgenic mice overexpressing human BLTR demonstrate markedly amplified neutrophil recruitment and 5-lipooxygenase signaling in murine models of acute skin inflammation, peritonitis, and reperfusion-initiated second organ injury (Chiang *et al.*, 1999). The phenotype of BLTR knockout mice has not yet been described.

THERAPEUTIC UTILITY

Effects of inhibitors (antibodies) to receptors

Several selective BLTR antagonists have been developed, and used to block the actions of LTB_4 in cell culture, in animal studies, and in human subjects.

BLTR antagonists are able to block specific binding of radiolabeled LTB_4 to leukocytes *in vitro*, as well as inhibit various cell functions activated by LTB_4, including chemotaxis (Showell *et al.*, 1998; Jackson *et al.*, 1999).

Administration of specific BLTR antagonists have inhibited recruitment of leukocytes into tissues *in vivo* in several animal models of inflammatory diseases. In a murine model of *inflammatory bowel disease*, treatment of mice with a selective LTB_4 receptor antagonist inhibited neutrophil influx into the colonic mucosa (Fretland *et al.*, 1995). In a murine model of *experimental allergic encephalitis*, pretreatment of mice with a selective LTB_4 receptor antagonist markedly blocked the recruitment of eosinophils into the spinal cord and completely inhibited the development of paralysis (Gladue *et al.*, 1996). Administration of a selective LTB_4 receptor antagonist markedly blocked the massive influx of inflammatory cells in the subsynovial connective tissue in murine *collagen-induced arthritis*, and abrogated the destruction of articular cartilage and erosion of bone (Showell *et al.*, 1998). Administration of selective LTB_4 receptor antagonists have also been shown to significantly prolong allograft survival in a rat model of liver transplantation (Ii *et al.*, 1996), and a murine cardiac allograft model (Weringer *et al.*, 1999). Finally, in a murine model of *acute septic peritonitis*, administration of a selective LTB_4 receptor antagonist inhibited recruitment of both neutrophils and macrophages, and led to significantly increased mortality (Matsukawa *et al.*, 1999).

A selective LTB_4 receptor antagonist was administered to human *asthma* patients prior to whole lung allergen challenge in one double-blind placebo-controlled crossover trial (Evans *et al.*, 1996). The receptor antagonist significantly reduced the number of neutrophils recruited into BAL fluid, although this was not associated with measurable physiological benefit. A randomized controlled trial of a selective LTB_4 receptor antagonist in patients with psoriasis demonstrated no statistically significant differences in median psoriasis area or severity index between treatment groups (Van Pelt *et al.*, 1998).

References

Akbar, G., Dasari, V., Webb, T., Ayyanathan, K., Pillarisetti, K., Sandhu, A., Athwal, R., Daniel, J., Ashby, B., Barnard, E., and Kunapuli, S. (1996). Molecular cloning of a novel P2 purinoceptor from human erythroleukemia cells. *J. Biol. Chem.* **271**, 18363–18367.

Alexander, S. P. H., and Peters, J. A. (1997). Receptors and ion channel nomenclature. *Trends Pharmacol. Sci.* suppl. **45**.

Bailie, M. B., Standiford, T. J., Laichalk, L. L., Coffey, M. J., Strieter, R., and Peters-Golden, M. (1996). Leukotriene-deficient mice manifest enhanced lethality from klebsiella pneumonia in association with decreased alveolar macrophage phagocytic and bactericidal activities. *J. Immunol.* **157**, 5221–5224.

Bomalaski, J. S., and Mong, S. (1987). Binding of leukotriene B_4 and its analogs to human polymorphonuclear leukocyte membrane receptors. *Prostaglandins* **33**, 855–867.

Chiang, N., Gronert, K., Clish, C. B., O'Brien, J. A., Freeman, M. W., and Serhan, C. N. (1999). Leukotriene B_4 receptor transgenic mice reveal novel protective roles for lipoxins and aspirin-triggered lipoxins in reperfusion. *J. Clin. Invest.* **104**, 309–316.

Demitsu, T., Katayama, H., Saito-Taki, T., Yaoita, H., and Nakano, M. (1989). Phagocytosis and bactericidal action of mouse peritoneal macrophages treated with leukotriene B_4. *Int. J. Immunopharmacol.* **11**, 801–808.

Evans, D. J., Barnes, P. J., Spaethe, S. M., van Alstyne, E. L., Mitchell, M. I., and O'Connor, B. J. (1996). Effect of a leukotriene B_4 receptor antagonist, LY293111, on allergen induced responses in asthma. *Thorax* **51**, 1178–1184.

Ford-Hutchinson, A. W., Bray, M. A., Doig, M. V., Shipley, M. E., and Smith, M. (1980). Leukotriene B_4, a potent chemokinetic and aggregating substance released from polymorphonuclear leukocytes. *Nature* **286**, 264–265.

Fretland, D. J., Anglin, C. P., Widomski, D., Baron, D. A., Maziasz, T., and Smith, P. F. (1995). Pharmacological activity of the second generation leukotriene B_4 receptor antagonist, SC-53228: effects on acute colonic inflammation and hepatic function in rodents. *Inflammation* **19**, 503–515.

Gaudreau, R., Le Gouill, C., Metaoui, S., Lemire, S., Stankova, J., and Rola-Pleszczynski, M. (1998). Signalling through the leukotriene B_4 receptor involves both α_i and α_{16}, but not α_q or α_{11} G-protein subunits. *Biochem. J.* **335**, 15–18.

Gimbrone, M. A. J., Brock, A. F., and Schafer, A. I. (1984). Leukotriene B_4 stimulates polymorphonuclear leukocyte adhesion to cultured vascular endothelial cells. *J. Clin. Invest.* **74**, 1552–1555.

Gladue, R., Carroll, L., Milici, A., Scampoli, D., Stukenbrok, H., Pettipher, E., Salter, E., Contillo, S., and Showell, H. (1996). Inhibition of leukotriene B_4-receptor interaction suppresses eosinophil infiltration and disease pathology in a murine model of experimental allergic encephalomyelitis. *J. Exp. Med.* **183**, 1893–1898.

Goldman, D., and Goetzl, E. (1982). Specific binding of leukotriene B_4 to receptors on human polymorphonuclear leukocytes. *J. Immunol.* **129**, 1600–1604.

Henderson, W. R. Jr. (1994). The role of leukotrienes in inflammation. *Ann. Intern. Med.* **121**, 684–697.

Huang, W. W., Garcia-Zepeda, E. A., Sauty, A., Oettgen, H., Rothenberg, M. E., and Luster, A. D. (1998). Molecular and biological characterization of the murine LT B_4 receptor expressed on eosinophils. *J. Exp. Med.* **188**, 1063–1074.

Ii, T., Izumi, R., and Shimizu, K. (1996). The immunosuppressive effects of a leukotriene B_4 receptor antagonist on liver allotransplantation in rats. *Surg. Today* **26**, 419–426.

Jackson, W. T., Froelich, L. L., Boyd, R. J., Schrementi, J. P., Saussy, D. L. J., Schultz, R. M., Sawyer, J. S., Sofia, M. J., Herron, D. K., Goodson, T. J., Snyder, D. W., Pechous, P. A., Spaethe, S. M., Roman, C. R., and Fleisch, J. H. (1999). Pharmacologic actions of the second-generation leukotriene B_4 receptor antagonist LY293111: *in vitro* studies. *Pharmacol. Exp. Ther.* **288**, 286–294.

Lewis, R. A., Austen, K. F., and Soberman, R. J. (1990). Leukotrienes and other products of the 5-lipoxygenase pathway. *N. Engl. J. Med.* **323**, 645–655.

Lin, A., Ruppel, P., and Gorman, R. (1984). Leukotriene B_4 binding to human neutrophils. *Prostaglandins* **28**, 837–849.

Marcinkiewicz, J., Grabowska, A., Bryniarski, K., and Chain, B. M. (1997). Enhancement of CD4+ T-cell-dependent interleukin-2 production *in vitro* by murine alveolar macrophages: the role of leukotriene B_4. *Immunology* **91**, 369–374.

Martin, V., Ronde, P., Unett, D., Wong, A., Hoffman, T. L., Edinger, A. L., Doms, R. W., and Funk, C. D. (1999). Leukotriene binding, signaling, and analysis of HIV coreceptor function in mouse and human leukotriene B_4 receptor-transfected cells. *J. Biol. Chem.* **274**, 8597–8603.

Matsukawa, A., Hogaboam, C. M., Lukacs, N. W., Lincoln, P. M., Strieter, R. M., and Kunkel, S. L. (1999). Endogenous monocyte chemoattractant protein-1 (MCO-1) protects mice in a model of acute septic peritonitis: cross-talk between MCP-1 and leukotriene B_4. *J. Immunol.* **163**, 6148–6154.

Morita, E., Schroder, J.-M., and Christophers, E. (1989). Differential sensitivities of purified human eosinophils and neutrophils to defined chemotaxins. *Scand. J. Immunol.* **29**, 709–716.

Murphy, P. M. (1994). The molecular biology of leukocyte chemoattractant receptors. *Annu. Rev. Immunol.* **12**, 593–633.

Ng, C. F., Sun, F. F., Taylor, B. M., Wolin, M. S., and Wong, P. Y.-K. (1991). Functional properties of guinea pig eosinophil leukotriene B_4 receptor. *J. Immunol.* **147**, 3096–3103.

Owman, C., Nilsson, C., and Lolait, S. J. (1996). Cloning of cDNA encoding a putative chemoattractant receptor. *Genomics* **37**, 187–194.

Owman, C., Sabirsh, A., Boketoft, A., and Olde, B. (1997). Leukotriene B_4 is the functional ligand binding to and activating the cloned chemoattractant receptor, CMKRL1. *Biochem. Biophys. Res. Commun.* **240**, 162–166.

Owman, C., Garzino-Demo, A., Cocchi, F., Popovic, M., Sabirsh, A., and Gallo, R. (1998). The leukotriene B_4 receptor functions as a novel type of coreceptor mediating entry of primary HIV-1 isolates into CD4-positive cells. *Proc. Natl Acad. Sci. USA* **95**, 9530–9534.

Palmblad, J., Lerner, R., and Larsson, S. H. (1994). Signal transduction mechanisms for leukotriene B4 induced hyperadhesiveness of endothelial cells for neutrophils. *J. Immunol.* **152**, 262–269.

Powell, W. S., MacLeod, R. J., Gravel, S., Gravelle, F., and Bhakar, A. (1996). Metabolism and biologic effects of 5-oxoeicosanoids on human neutrophils. *J. Immunol.* **156**, 336–342.

Raport, C. J., Schweickart, V. L., Chantry, D., Eddy, R. L., Shows, T. B., Godiska, R., and Gray, P. W. (1996). New members of the chemokine receptor gene family. *J. Leukocyte Biol.* **59**, 18–23.

Renkonen, R., Mattila, P., Leszcynski, D., and Hayry, P. (1988). Leukotriene B_4 increases the lymphocyte binding to endothelial cells. *FEBS Lett.* **235**, 67–70.

Rola-Pleszczynski, M., and Lemaire, I. (1985). Leukotrienes augment interleukin 1 production by human monocytes. *J. Immunol.* **135**, 3958–3961.

Rola-Pleszczynski, M., Gagnon, L., and Siros, P. (1983). Leukotriene B4 augments human natural cytotoxic cell activity. *Biochem. Biophys. Res. Commun.* **113**, 531.

Samuelsson, B., Dahlen, S.-E., Lindgren, J. A., Rouzer, C. A., and Serhan, C. N. (1987). Leukotrienes and lipoxins: Structures, biosynthesis, and biological effects. *Science* **237**, 1171–1176.

Showell, H. J., Conklyn, M. J., Alpert, R., Hingorani, G. P., Wright, K. F., Smith, M. A., Stam, E., Salter, E. D., Scampoli, D. N., Meltzer, S., Reiter, L. A., Koch, K., Piscopio, A. D., Cortina, S. R., Lopez-Anaya, A., Pettipher, E. R., Milici, A. J., and Griffiths, R. J. (1998). The preclinical pharmacological profile of the potent and selective leukotriene B_4 antagonist CP-195543. *Pharmacol. Exp. Ther.* **285**, 946–954.

Stankova, J., Gagnon, N., and Rola-Pleszczynski, M. (1992). Leukotriene B_4 augments interleukin-2 receptor-beta (IL-2R beta) expression and IL-2R beta-mediated cytotoxic response in human peripheral blood lymphocytes. *Immunology* **76**, 258–263.

To, S. S. T., and Schrieber, L. (1990). Effect of leukotriene B_4 and prostaglandin E2 on the adhesion of lymphocytes to endothelial cells. *Clin. Exp. Immunol.* **81**, 160–165.

Van Pelt, J. P., De Jong, E. M., Seijger, M. M., Van Hooijdonk, C. A., De Bakker, E. S., Van Vlijmen, I. M., Parker, G. L.,

Van Erp, P. E., and Van De Kerkhof, P. C. (1998). Investigation on a novel and specific leukotriene B_4 receptor antagonist in the treatment of stable plaque psoriasis. *Br. J. Dermatol.* **139**, 396–402.

Weringer, E. J., Perry, B. D., Sawyer, P. S., Gilman, S. C., and Showell, H. J. (1999). Antagonizing leukotriene B_4 receptors delays cardiac allograft rejection in mice. *Transplantation* **67**, 808–815.

Yamada, Y., Stoffel, M., Espinosa, R. I., Xiang, K., Seino, M., Seino, S., Le Beau, M. M., and Bell, G. I. (1993). Human somatostatin receptor genes: localization to human chromosomes 14, 17 and 22 and identification of simple tandem repeat polymorphisms. *Genomics* **15**, 449–452.

Yamaoka, K. A., Claesson, H. E., and Rosen, A. (1989). Leukotriene B_4 enhances activation, proliferation, and differentiation of human B lymphocytes. *J. Immunol.* **143**, 1996–2000.

Yokomizo, T., Lzumi, T., Chang, K., Takuwa, Y., and Shimizu, T. (1997). A G-protein-coupled receptor for leukotriene B_4 that mediate chemotaxis. *Nature* **387**, 620–624.

PACAP and VIP Receptors

Edward J. Goetzl*, Julia K. Voice and Glenn Dorsam

Immunology and Allergy, University of California, San Francisco, Room UB8B, Box 0711, 533 Parnassus Avenue, San Francisco, CA 94143-0711, USA

*corresponding author tel: 415-476-5339, fax: 415-476-6915, e-mail: egoetzl@itsa.ucsf.edu

DOI: 10.1006/rwcy.2000.23009.

SUMMARY

This subfamily of G protein-coupled receptors consists of PAC_1, which binds PACAP alone, and $VPAC_1$ and $VPAC_2$, which bind PACAP and VIP with equal affinity. These homologous receptors all have a long N-terminal extracellular sequence with five conserved cysteines, as well as conserved other cysteines and basic amino acids in the first and second extracellular loops and third intracellular loop and cytoplasmic tail. There are numerous splice variants of PAC_1, but not of the VPAC receptors. As for the corresponding ligands, these receptors are distributed preferentially in the nervous system, endocrine and immune organs, intestines and lungs. Signaling through two or more G proteins is mediated principally by increases in $[cAMP]_i$ and $[Ca^{2+}]_i$. Genetic manipulation of expression of the receptors is now in progress.

BACKGROUND

Discovery

Vasoactive intestinal peptide (VIP) and pituitary adenylate cyclase-activating peptide (PACAP) are constituents of one structural superfamily of neuroendocrine hormones, which also includes secretin, glucagon, glucagon-like peptide, and growth hormone-releasing hormone or factor. VIP and PACAP are linked by some functional similarities and by their sharing of two of the three receptors in one subfamily of G protein-coupled cellular receptors (GPCRs) (**Table 1**) (Harmar *et al.*, 1998). The three VIP/PACAP receptors were cloned initially by PCR-based and GPCR hybridization techniques.

Table 1 Structures and encoding genes of the PACAP and VIP receptors

Receptor subtype (IUPHAR nomenclature)	Gene (HUGO)	Chromosomal location		Size (aa)
		Human	Other	
PAC_1	*ADCYAP1R1*	7p14	4 (rat)	495 (rat)
$VPAC_1$	*VIPR1*	3p22	8 (rat)	457 (human)
			9 (mouse)	437 (mouse)
$VPAC_2$	*VIPR2*	7q36.3	4 (rat)	437 (human)
			12 (F2 mouse)	

Alternative names

PAC_1 was formerly designated the type I PACAP receptor. $VPAC_1$ had been termed the type II PACAP receptor and the type I VIP receptor. $VPAC_2$ had been termed the type III PACAP receptor and type II VIP receptor.

Structure

PAC_1, $VPAC_1$, and $VPAC_2$ GPCRs show a high level of homology to each other and share structural features, such as a long N-terminus and short loops.

Main activities and pathophysiological roles

PAC_1 has been cloned from rat, mouse, bovine, and human cells (Pisegna and Wank, 1993; Harmar *et al.*, 1998). PAC_1 of each species binds PACAP-27 and PACAP-38 (IC_{50} = 1 nM) with 1000-fold higher affinity than VIP, and exhibits even lower affinities for PHI (peptide histidine isoleucine), PHV (peptide histidine valine), secretin, and growth hormone-releasing factor (GRF) (Ohtaki *et al.*, 1998).

$VPAC_1$ has been cloned from rat, human, and mouse cells (Ishihara *et al.*, 1992; Harmar *et al.*, 1998). $VPAC_1$ of each species binds PACAP-27, PACAP-38, and VIP with equal affinity (IC_{50} = 0.2–1 nM), but differences exist among species in the rank-order of affinity of binding of other closely related peptides.

$VPAC_2$ has been cloned from rat, mouse, and human cells (Lutz *et al.*, 1993; Inagaki *et al.* 1994; Harmar *et al.*, 1998). $VPAC_2$ of each species binds PACAP-38 and VIP with equal affinity (IC_{50} = 2–4 nM), but has slightly lower affinity for PACAP-27.

GENE

Accession numbers

See **Table 2**.

Chromosome location and linkages

The gene encoding rat PAC_1 has been isolated and characterized, demonstrating a single copy and a complex structure similar to those of genes for other members of this subfamily of GPCRs. Rat PAC_1 gene spans 40 kb and contains 15 exons (Chatterjee *et al.*, 1997). Introns are 320 bp to 10.5 kbp and exhibit splice phasing of types 0, 1, and 2. All splice acceptor–donor sequences conform to the GT/AG consensus rule. The sizes and organization of exons and introns, and the intron–exon boundaries resemble those of genes encoding GPCRs for secretin, glucagon, and parathyroid hormone. Unlike $VPAC_1$ and $VPAC_2$, for which no splice variants have been identified, PAC_1 shows extensive alternative splicing in several domains (Pisegna and Wank, 1996; Pantaloni *et al.*, 1996; Chatterjee *et al.*, 1996; 1997). Two alternative exons in the region encoding the third intracellular loop have been termed hip and hop, and the latter exists as hop1 and hop2 due to different splice acceptor sites. The hip-hop variants differ in signaling properties. Alternative splicing of exons equivalent to hip and hop1 in human PAC_1 leads to four variants designated null, SV-1, SV-2, and SV-3, that also differ in transductional abilities. Four other variants arise from differential use of alternative exons in the 5′ untranslated region of the rat PAC_1 gene. A splice variation that leads to a 21 amino acid deletion in the extracellular N-terminus and another with sequence differences in transmembrane domains II and IV appear to represent one type of mechanism responsible for different binding affinities and signaling potencies of PACAP-38 and PACAP-27. There is one copy each of the $VPAC_1$ and $VPAC_2$ genes, and no alternatively spliced forms have been identified for either subtype. Human $VPAC_1$ spans 22 kbp and is composed of 13 exons of 42 bp to 1400 bp and 12 introns of 0.3–6.1 kbp (Sreedharan *et al.*, 1995). The structure of the gene encoding $VPAC_2$ has not been delineated completely, but chromosomal localization is known (Mackay *et al.*, 1996).

Although it seems reasonable to assume that expression of each receptor is regulated transcriptionally

Table 2 Accession numbers for PAC_1, $VPAC_1$, and $VPAC_2$

	PAC_1	$VAPC_1$	$VPAC_2$
Human	D17516	U11079, U11080, U11081, U11082, U11083, U11084, U11085, U11086, U11087	
Rat	D14908, D14909	AF059678	U09631, Z25885
Mouse		S82970	S82966

and with tissue specificity, few definitive studies have addressed these questions. One notable contribution is the demonstration that corticosteroids suppress levels of mRNA encoding $VPAC_1$ in cultured lung cells and reduce signals reported by $VPAC_1$ 5′-flanking sequence luciferase constructs introduced by transfection (Pei, 1996). The 5′-glucocorticoid negative response element was that mapped to a 126 bp sequence containing a glucocorticoid receptor-binding site between −36 and −21 bp from the transcription start site. Promoters of each gene are now being defined fully with respect to other functional elements.

PROTEIN

Description of protein

PAC_1, $VPAC_1$, and $VPAC_2$ are related seven transmembrane domain GPCRs. Within each species, amino acid sequence identity among the three receptors is 49% to 51%, and the similarity of each with the GPCRs for secretin, glucagon, glucagon-like peptide I, and growth hormone-releasing hormone ranges from 34% to 47%. As for most subfamilies of GPCRs, the greatest degree of homology is within the transmembrane domains and least in the N- and C-terminal segments. Receptors of this subfamily share other structural features, which as yet have not been related to receptor function, including long N-terminal extracellular sequences with five conserved cysteines, two additional conserved cysteines located one each in the first and second extracellular loops, and conserved basic amino acids in the third intracellular loop and cytoplasmic tail, that may facilitate coupling to G_s.

Affinity for ligand(s)

A few aspects of the structural determinants of specific ligand binding have been elucidated for the $VPAC_1$ receptor. A series of chimeras of the rat $VPAC_1$ and secretin receptors were constructed involving exchanges of the N-terminus, first extracellular loop or both (Holtmann *et al.*, 1995). The native receptors bound their respective ligands with 2 nM K_d values and crossbound the other ligand with K_d values three orders of magnitude higher (**Table 3**). When the $VPAC_1$ N-terminus was substituted for that of the secretin receptor, ligand binding and transduction of biological responses typical of $VPAC_1$ were observed. For reciprocal conversion of $VPAC_1$ to a secretin receptor, however, the chimera required both the N-terminus and the first extracellular loop of the secretin receptor. Dissociation of ligand-binding affinity from ligand potency in transduction of responses for some chimeras emphasized the likely complexity of composition of complete receptor units. Another series of chimeras of nonidentical components of human and rat $VPAC_1$, and selected site-directed mutants of both, revealed that the difference in rat $VPAC_1$ high-affinity binding and human $VPAC_1$ low-affinity binding of PHI is attributable to three amino acids in the first extracellular loop and adjacent third transmembrane domain (Couvineau *et al.*, 1996).

Cell types and tissues expressing the receptor

The distinctive patterns of tissue distribution of each PACAP/VIP receptor to date have been mapped principally by radioligand binding and semiquantification or probe detection of encoding mRNA (Table 3). The attendant problems of any one approach and the possibility of dissociations between the amounts of mRNA and protein limit confidence in the initial results. Nonetheless, it is clear that $VPAC_1$ and $VPAC_2$ are often expressed with reciprocal densities in one type of cell, tissue representation may be complementary as in the rat CNS (Usdin *et al.*, 1994), and both show high levels of inducibility and repressibility. One such example is in a cell line model for thymocytes, where nearly exclusive expression of $VPAC_1$ changes to much higher expression of $VPAC_2$ and lower levels of $VPAC_1$ within hours of exposure to antigen and antigen-presenting cells (Pankhaniya *et al.*, 1998). Antibodies specific for each receptor in the subfamily now have become available, which should permit definitive analyses of tissue-specific expression and the determinants of regulation of such expression.

SIGNAL TRANSDUCTION

PAC_1, $VPAC_1$, and $PVAC_2$ each couple to multiple G proteins, resulting in concurrent initiation of diverse signaling pathways (Table 3). G_s is presumed to mediate stimulation of adenylate cyclase and an increase in $[cAMP]_i$, that is the hallmark of cellular effects of VIP and PACAP. In addition to evidence for a physical association between G_s and $VPAC_1$, specific suppression of $G_{s\alpha}$ by transfection of G_s α chain antisense plasmids with a hygromycin-resistance element into $VPAC_1$-expressing epithelial cells, followed by hygromycin selection, blunted VIP-induced increases in $[cAMP]_i$ (Goetzl *et al.*, 1993).

Table 3 Binding, signaling, tissue distribution, and pharmacology of PAC1, VPAC1, and $VPAC_2$

Receptor	K_d (nM)	Responses (EC_{50}, nM) of		Predominant tissue expression	Selective	
		$[cAMP]_i$	$[Ca^{2+}]_i$		Agonists	Antagonists
PAC_1	1	0.1–1.6	10–50	CNS (olfactory bulb, thalamus, hypothalamus, hippocampus, cerebellum)	Maxadilan	PACAP6-38
				Adrenal medulla		
				Macrophages		
$VPAC_1$	0.2–1	0.3–1	2.5	CNS (cerebral cortex, hippocampus)	VIP1,7/GRF VIP4-28	VIP3,7/GRF
				Lung		
				Gastrointestinal tract		
				Liver		
				Prostate		
				Macrophages and lymphocytes		
$VPAC_2$	3–10	10–30	27	CNS (thalamus, suprachiasmatic nucleus, hippocampus, brainstem, spinal cord, dorsal root ganglia)	Ro25-1553 Ro25-1392	VIP4-28
				Skeletal and cardiac muscle		
				Gastrointestinal tract		
				Kidney		
				Adipose tissue		
				Testes		
				Macrophages and lymphocytes		

VIP1,7/GRF, $[K^{15}R^{16}L^{27}]$VIP(1-7)GRF(8-25)-NH_2; VIP3,7/GRF, [Ac-His^1D-$Phe^2Lys^{15}Arg^{16}$]VIP(3-7)GRF(8-27)-NH_2; Ro25-1553, Ac-His^1-[$Glu^8Lys^{12}Nle^{17}Ala^{19}Asp^{25}Leu^{26}Lys^{27,28}Gly^{29,30}Thr^{31}$]-$NH_2$ VIP(cyclo21-25); and Ro25-1392, Ac-His^1[Glu^8OCH_3-$Tyr^{10}Lys^{12}Nle^{17}Ala^{19}Asp^{25}Leu^{26}Lys^{27,28}$]VIP (cyclo21-25).

That similar antisense suppression of $G_{i2\alpha}$ led to marked enhancement of VIP-induced increases in $[cAMP]_i$, suggested that G_{i2} has a negative regulatory effect on VIP enhancement of adenylate cyclase. Each receptor also activates phosphoinositide-specific phospholipase C (PLC), with resultant elevation of $[Ca^{2+}]_i$. The greater potency of PACAP-38 than PACAP-27 in PAC_1-mediated activation of PLC and elevation of $[Ca^{2+}]_i$ has been attributed to a 21 amino acid segment of the N-terminal extracellular domain, based on results of studies of a series of splice variants (Pantaloni *et al.*, 1996). The $G_{q/11}$ protein is assumed to couple the receptors to PLC, based on data from other systems, but this has not been demonstrated directly. One transmembrane domain IV splice variant of PAC_1 failed to activate either adenylate cyclase or PLC, but stimulated Ca^{2+} influx by activation of L-type Ca^{2+} channels (Chatterjee *et al.*, 1996).

DOWNSTREAM GENE ACTIVATION

There are several reports of VIP or PACAP enhancing or suppressing proliferation or functional responses of normal or neoplastic cells *in vitro* (Ogasawara *et al.*, 1997; Maruno *et al.*, 1998). It is assumed that part of the capacity of VIP and PACAP to influence growth is attributable to alterations in transcription of immediate early genes critical for proliferative responses. However, no studies to date have focused on this important question.

BIOLOGICAL CONSEQUENCES OF ACTIVATING OR INHIBITING RECEPTOR AND PATHOPHYSIOLOGY

PAC_1, $VPAC_1$, and $VPAC_2$ are presumed to transduce all of the effects of PACAP and VIP but often act in tissues responding to multiple other mediators and cytokines. The absence of inactivating genetic anomalies of any of the receptors and the lack of potent and selective pharmacological agents precludes specific assignment of each neuropeptide effect to one type of receptor. However, detailed analyses of gene knockout models now in development may permit assessment of the biological and pathophysiological roles of each receptor.

References

Chatterjee, T. K., Sharma, R. V., and Fisher, R. A. (1996). Molecular cloning of a novel variant of the pituitary adenylate cyclase-activating polypeptide (PACAP) receptor that stimulates calcium influx by activation of L-type calcium channels. *J. Biol. Chem.* **271**, 32226–32232.

Chatterjee, T. K., Liu, X., Davisson, R. L., and Fisher, R. A. (1997). Genomic organization of the rat pituitary adenylate cyclase-activating polypeptide receptor gene. *J. Biol. Chem.* **272**, 12122–12131.

Couvineau, A., Rouyer-Fesard, C., Maoret, J.-J., Gaudin, P., Nicole, P., and Laburthe, M. (1996). Vasoactive intestinal peptide (VIP)1 receptor. *J. Biol. Chem.* **271**, 12795–12800.

Goetzl, E. J., Kishiyama, J. L., Shames, R. S., Liu, Y.-F., Albert, P. R., An, S., Birke, F. W., Yang, J., and Sreedharan, S. P. (1993). Specific inhibition of receptor-dependent human cellular responses by antisense mRNA depletion of individual G proteins. *Trans. Assoc. Am. Physicians* **56**, 69–76.

Harmar, A. J., Arimura, A., Gozes, I., Journot, L., Laburthe, M., Pisegna, J. R., Rawlings, S. R., Robberecht, P., Said, S. I., Sreedharan, S. P., Wank, S. A., and Waschek, J. A. (1998). International Union of Pharmacology. XVIII. Nomenclature of receptors for vasoactive intestinal peptide and pituitary adenylate cyclase-activating polypeptide. *Pharmacol. Rev.* **50**, 265–270.

Holtmann, M. H., Hadac, E. M., and Miller, L. J. (1995). Critical contributions of amino-terminal extracellular domains in agonist binding and activation of secretin and vasoactive intestinal polypeptide receptors. Studies of chimeric receptors. *J. Biol. Chem.* **270**, 14394–14398.

Inagaki, N., Yoshida, H., Mizuta, M., Mizuno, N., Fujii, Y., Gonoi, T., Miyazaki, J., and Seino, S. (1994). Cloning and functional characterization of a third pituitary adenylate cyclase-activating polypeptide receptor subtype expressed in insulin-secreting cells. *Proc. Natl Acad. Sci. USA* **91**, 2679–2683.

Ishihara, T., Shigemoto, R., Mori, K., Takahashi, K., and Nagata, S. (1992). Functional expression and tissue distribution of a novel receptor for vasoactive intestinal polypeptide. *Neuron* **8**, 811–819.

Lutz, E. M., Sheward, W. J., West, K. M., Morrow, J. A., Fink, G., and Harmar, A. J. (1993). The VIP2 receptor: molecular characterization of a cDNA encoding a novel receptor for vasoactive intestinal peptide. *FEBS Lett.* **334**, 3–8.

Mackay, M., Fantes, J., Scherer, S., Boyle, S., West, K., Tsui, L.-C., Belloni, E., Lutz, E., Van Heyningen, V., and Harmar, A. J. (1996). Chromosomal localization in mouse and human of the vasoactive intestinal peptide receptor type 2 gene: a possible contributor to the holoprosencephaly 3 phenotype. *Genomics* **37**, 345–352.

Maruno, K., Absood, A., and Said, S. I. (1998). Vasoactive intestinal peptide inhibits human small-cell lung cancer proliferation *in vitro* and *in vivo*. *Proc. Natl Acad. Sci. USA* **95**, 14373–14378.

Ogasawara, M., Murata, J., Ayukawa, K., and Saimi, I. (1997). Differential effect of intestinal neuropeptides on invasion and migration of colon carcinoma cells in vitro. *Cancer Lett.* **116**, 111–116.

Ohtaki, T., Ogi, K., Masuda, Y., Mitsuoka, K., Fujiyoshi, Y., Kitada, C., Sawada, H., Onda, H., and Fujino, M. (1998). Expression, purification and reconstitution of receptor for pituitary adenylate cyclase-activating polypeptide. *J. Biol. Chem.* **273**, 15464–15473.

Pankhaniya, R., Jabrane-Ferrat, N., Gaufo, G. O., Sreedharan, S. P., Dazin, P., Kaye, J., and Goetzl, E. J. (1998). Vasoactive intestinal peptide enhancement of antigen-induced differentiation of a cultured line of mouse thymocytes. *FASEB J.* **12**, 119–127.

Pantaloni, C., Brabet, P., Bilanges, B., Dumuis, A., Houssami, S. S., Spengler, D., Bockaert, J., and Journot, L. (1996). Alternative splicing in the N-terminal extracellular domain of the pituitary adenylate cyclase-activating polypeptide (PACAP) receptor modulates receptor selectivity and relative potencies of PACAP-27 and PACAP-38 in phospholipase C activation. *J. Biol. Chem.* **271**, 22146–22151.

Pei, L. (1996). Identification of a negative glucocorticoid response element in the rat type 1 vasoactive intestinal polypeptide receptor gene. *J. Biol. Chem.* **271**, 20879–20884.

Pisegna, J. R., and Wank, S. A. (1993). Molecular cloning and functional expression of the pituitary adenylate cyclase-activating polypeptide type I receptor. *Proc. Natl Acad. Sci. USA* **90**, 6345–6349.

Pisegna, J. R., and Wank, S. A. (1996). Cloning and characterization of the signal transduction of four splice variants of the human pituitary adenylate cyclase-activating polypeptide receptor. *J. Biol. Chem.* **271**, 17267–17274.

Sreedharan, S. P., Huang, J.-X., Cheung, M.-C., and Goetzl, E. J. (1995). Structure, expression, and chromosomal localization of the type I human vasoactive intestinal peptide receptor gene. *Proc. Natl Acad. Sci. USA* **92**, 2939–2943.

Usdin, T. B., Bonner, T. I., and Mezey, E. (1994). Two receptors for vasoactive intestinal polypeptide with similar specificity and complementary distributions. *Endocrinology* **135**, 2662–2680.

Lysophospholipid Growth Factor Receptors

Edward J. Goetzl*, **Songzhu An and Hsinyu Lee**

University of California at San Francisco, Departments of Medicine and Microbiology, 533 Parnassus at 4th, UB8B, Box 0711, San Francisco, CA 94143-0711, USA

*corresponding author tel: 415 476 5339, fax: 415 476 6915, e-mail: egoetzl@itsa.ucsf.edu

DOI: 10.1006/rwcy.2000.23010.

SUMMARY

The endothelial differentiation gene-encoded G protein-coupled receptors (Edg Rs) Edg-1, -3, -5, and -8 bind sphingosine 1-phosphate (S1P) and Edg-2, -4, and -7 bind lysophosphatidic acid (LPA), resulting in transduction of cellular signals for proliferation, increased survival, reduced apoptosis, adherence, migration, secretion, and other specific functions. The diversity of signals emanating from each Edg R is in part attributable to coupling to multiple G proteins. One or more Edg Rs are expressed prominently in the cardiovascular, pulmonary, nervous, endocrine, genitourinary, hematological, and immune systems. Expression of Edg Rs is regulated by developmental and postdevelopmental physiological events. The levels of Edg Rs increase in distinctive patterns in many diseases, as exemplified by high levels of Edg-4 in *ovarian cancer*, whereas none is detected in normal ovarian surface epithelial cells. Although several analogs of LPA and S1P have altered agonist or antagonist activities, potent pharmacological agents have not been identified to date.

BACKGROUND

LPA, S1P, sphingosylphosphorylcholine and other structurally related lysophospholipid (LPL) growth factors have major effects on many different types of cells, which include initiation and regulation of cellular proliferation, enhancement of survival, suppression of apoptosis, promotion of differentiation, and stimulation of diverse cytoskeleton-based functions (Moolenaar, 1995; Tokomura, 1995; Spiegel and Merrill, 1996; Goetzl and An, 1998). LPLs are generated enzymatically from precursors stored in cell membranes and are secreted principally by platelets, macrophages, some other types of leukocytes, epithelial cells, and some cancer cells. The amounts secreted are sufficient to establish micromolar concentrations of LPA and S1P in plasma and S1P in plasma normally and in extracellular fluids during tissue reactions. Extracellular LPA and S1P are both almost entirely bound by serum proteins (Tigyi and Miledi, 1992; Thumser *et al.*, 1994). LPLs resemble protein growth factors in their ability to alter transcriptional activities of growth-related genes, amplify direct stimulation of cellular proliferation by recruiting other growth factor systems, and evoke many cellular responses other than proliferation and proliferation-related events (An *et al.*, 1998c; Goetzl and An, 1998).

Discovery

The LPLs and protein growth factors differ principally in their respective uses of G protein-coupled receptors (GPCRs) and protein tyrosine kinase receptors. LPA and S1P bind specifically to different members of a family of GPCRs, which are encoded by endothelial differentiation genes (Edgs) and thus are designated tentatively as Edg Rs (Hla and Maciag, 1990; MacLennan *et al.*, 1994; Yamaguchi *et al.*, 1996; Chun *et al.*, 1999). Edg-1 was discovered initially as a GPCR of unknown ligand specificity, which was expressed by endothelial cells only after differentiation (Hla and Maciag, 1990). Edg-2 was first found on proliferating neurons in the periventricular zone of developing embryonic mice (Hecht *et al.*, 1996). LPA evoked shape changes and proliferation in these Edg-2 R-expressing neurons, and both their Edg-2 Rs and the responses to LPA disappeared after completion of CNS development. The primary determinants of cellular responses to LPLs are the generative and biodegradative events, which establish steady-state concentrations of each protein-bound LPL at cell surfaces, and the relative frequency of expression of each Edg R.

Alternative names

There are no alternative names for the Edg receptor family, but individual receptors have been called by other terms, such as ventricular zone gene 1 (vzg-1)-encoded receptor for Edg-2 (Hecht *et al.*, 1996).

Structure

Two subfamilies of the Edg Rs have been distinguished based on their degree of amino acid sequence identity and preferred LPL ligand (**Table 1**) (Goetzl and An, 1998; Chun *et al.*, 1999). The first encompasses Edg-1, -3, -5, and -8, which are 45–60% identical in amino acid sequences and bind S1P with high affinity (Hla and Maciag, 1990; MacLennan *et al.*, 1994; Yamaguchi *et al.*, 1996; An *et al.*, 1997b; Lee *et al.*, 1998a,b; Glickman *et al.*, 1999). The second includes Edg-2, -4, and -7, which are 48–54% identical in amino acid sequences and bind LPA with high specificity and affinity (Hecht *et al.*, 1996; An *et al.*, 1997a, 1998a; Bandoh *et al.*, 1999). Mechanisms of regulation of expression, G protein associations, predominant signaling pathways, pattern of tissue representation, and involvement in human diseases are now being defined for each Edg R.

Main activities and pathophysiological roles

The principal roles of Edg Rs are to bind LPA and/or S1P with high affinity and specificity, and to transduce G protein-dependent signals to cellular proliferation, survival, and functions.

GENE

The genomic structures of Edg-1, -3, and -5 differ fundamentally from those of Edg-2 and -4. The over-45 kb murine Edg-2 gene was successfully defined and shown to be composed of at least five separate exons on chromosome 4 (Contos and Chun, 1998). Divergent cDNA sequences are located upstream from a single common exon sequence, which encodes most of the open reading frame of Edg-2 and ends at an intron in the middle of transmembrane domain VI. In contrast, murine Edg-1 has only two exons, with a

Table 1 Encoding genes and structures of Edg LPA and S1P receptors

Human receptor	Ligand	Chromosomal location (human/mouse)	Protein size (aa no.)	G protein coupling	Signaling		
					A	C	E
Edg-1	S1P	1p21.1–3	381	i	I	S	S
Edg-3	S1P	9q22.1–2	378	i, q, 12/13	I	S	S
Edg-5	S1P	19p13.2/9	354	i, q, 12/13	I	S	S
Edg-8	S1P	ND	400				
Edg-6	ND	19p13.3	384	–	–	–	–
Edg-2	LPA	ND/4 (6)	364	i, 12/13	I	S	N/S
Edg-4	LPA	19p12	351 (382)[a]	i, q, 12/13	I	S	S
Edg-7	LPA	ND	353	q, s	S	S	N/S

A, adenylyl cyclase; C, increase in $[Ca^{2+}]_i$; E, ERK activity; ND, not determined; I, inhibit; S, stimulate; N, no effect; –, insufficient data; N/S, no effect in some cells, stimulation in other cells.

[a]The number of amino acids in the wild type is 351 and in the mutant is 382.

single coding exon, an intron separating the transcriptional start site from the open reading frame, and no intron in transmembrane domain VI (Liu and Hla, 1997). Genes encoding several other Edg Rs have very recently been defined structurally or sequenced by the Human Genome Project (Table 1). A recently recognized mutant of Edg-4 R results from a deletion of a single base near the 3′ end of the open reading frame, which leads to a frameshift, loss of the termination codon, and replacement of the four amino acid C-terminus of the wild-type Edg-4 with a completely different 35 amino acid peptide (Table 1). No promoter elements of any Edg R have been analyzed further as yet.

Accession numbers

Edg-1: AF022137, M312104
Edg-3: AF022139, X83864
Edg-5: AF034780
Edg-6: AJ000479
Edg-2: U80811
Edg-4: AF011466
Edg-7: AF127138

Chromosome location and linkages

See Table 1.

There is approximately a 26–28% homology in sequence between Edg Rs and some cannabinoid GPCRs.

PROTEIN

Description of protein

The Edg-1 to Edg-8 genes encode distinct but related seven transmembrane domain GPCRs (Table 1). Based on amino acid sequence identity, Edg-1, -3, -5, and -8 belong to one structural cluster and Edg-2, -4, and -7 are members of a second structural cluster. The amino acid sequence of Edg-6 lies between those of the two major clusters (Graler *et al.*, 1998). Edg Rs share other structural features which have not as yet been definitively linked to function. The *N*-linked glycosylation sites of the N-terminus and multiple potential sites of phosphorylation in intracellular regions are typical of all GPCRs and are preserved in the Edg Rs. In contrast, the disulfide bond most often formed between cysteines in the first and second extracellular loops in other GPCRs is most often formed between the second and third extracellular loops of Edg Rs. In Edg-4 R, an alanine replaces the proline typical of the seventh transmembrane domain sequence NPXXY of other GPCRs. An Src homology 2 (SH2) segment exists in the intracellular face of Edg-5, but has not been shown to express characteristic functions. The mRNAs encoding some of the Edg Rs have the AU-rich sequence AUUUA in their 3′ untranslated region, which is an mRNA-stabilizing structure typical of growth-related immediate-early genes. A GPCR originally reported as an orphan (An *et al.*, 1995), has been proven to bind the LPL sphingosylphosphorylcholine with high specificity and affinity, resulting in the expected intracellular signals (Xu *et al.*, 2000).

Distinctive patterns of tissue distribution of each Edg R have been mapped principally by semiquantitative PCR techniques, northern blots, *in situ* hybridization and western blots with monoclonal antibodies to multiple substituent peptides (An *et al.*, 1997a,b, 1998a; Hecht *et al.*, 1996; Liu and Hla, 1997; MacLennan *et al.*, 1997; Weiner *et al.*, 1998) (Table 2). Another type of GPCR has been identified on *Xenopus* oocytes and some rodent cells, which is specific for LPA but structurally unrelated to Edg Rs (Guo *et al.*, 1996).

Cell types and tissues expressing the receptor

These are listed in **Table 2**.

SIGNAL TRANSDUCTION

Each of the Edg-1 to -8 receptors usually couples to multiple G proteins, which results in concurrent initiation of diverse signaling pathways that may be amplified by crosstalk between these pathways (Goodemote *et al.*, 1995; Brindley *et al.*, 1996; Lee *et al.*, 1996; van Koppen *et al.*, 1996; An *et al.*, 1998b; Erickson *et al.*, 1998; Fukushima *et al.*, 1998; Okamota *et al.*, 1998; Zondag *et al.*, 1998; Windh *et al.*, 1999). The association of each Edg R with G_i leads to striking ligand-induced enhancement of the Ras/extracellular signal-regulated kinase (ERK) pathway central to LPA and S1P activation of cellular proliferation (Table 1). The $\beta\gamma$ subunits of G_i may couple ERKs to Edg Rs, whereas the α chain of G_i mediates the defining inhibition of adenylyl cyclase. The biochemical prerequisites for G_i recruitment of ERKs include an Src-related tyrosine kinase, Pyk 2 kinase, and probably PI-3 kinase, based on the

Table 2 Human Edg receptor tissue distribution and abnormalities in disease states

	Tissue distribution	Expression/mutations in diseases
Edg-1	Endothelial cells, many other types of cells	
Edg-3	Heart, many other types of cells	Increased in *breast cancer*
Edg-5	Heart, embryonic nervous system, many other types of cells	
Edg-8	Brain	
Edg-6	Lymphoid tissues, leukocytes, lung	
Edg-2	Embryonic nervous system, many other types of cells	
Edg-4	Leukocytes, testes, many other types of cells	Increased in breast cancer, increased in *ovarian cancer* (mutant form)
Edg-7	Pancreas, heart, prostate, testes, lung, ovary	

ability of selective inhibitors of PI-3 kinase to suppress LPA and S1P induction of ERK activity.

Edg-3, -4, -5, and -7 receptors couple with and activate G_q, resulting in sequential stimulation of phospholipase C, release of diacylglycerol leading to enhancement of protein kinase C activity and of inositol trisphosphate (IP_3), which consequently elicits pertussis toxin-insensitive increases in $[Ca^{2+}]_i$. In contrast, G_i mediates the pertussis toxin-sensitive part of LPA- and S1P-evoked increases in $[Ca^{2+}]_i$ transduced by Edg-3, -4, and -5, and the principal effect transduced by Edg-1 and -2 in some types of cells through recruitment of phospholipase C. The proposed capacity of S1P to act as intracellular messenger and thereby mobilize Ca^{2+} directly, without binding to a cell surface receptor, is independent of IP_3 and dependent on sphingosine kinase activity in some cells, but the cellular target protein for this action of S1P remains to be identified (Spiegel and Merrill, 1996; Kohama *et al.*, 1998; Melendez *et al.*, 1998).

LPA and S1P also evoke cellular proliferation and cytoskeleton-dependent functions through Rho-mediated pathways, which are initiated by engagement of any of the Edg Rs (Moolenaar, 1995; Spiegel and Milstien, 1995; Fromm *et al.*, 1997) (Table 1). In some cellular settings, Edg-1 may not couple to Rho. The elicitation of SRF transcriptional activity, regulation of cellular pH, and induction of formation of actin stress fibers and focal adhesion complexes by LPA and S1P are all Rho-mediated through $G\alpha_{12/13}$ or, in rare circumstances, through $G\alpha_q$ (Hill and Treisman, 1995; Hill *et al.*, 1995; Fromm *et al.*, 1997). The GTPase-activating factor Lsc/p115RhoGEF links $G\alpha_{13}$ with Rho and enables Rho, which in turn activates serine/threonine kinases, phospholipase D, SRF, PI-3 kinase, p125 focal adhesion kinase, and myosin light chain phosphatase (Hart *et al.*, 1998). Many effects of LPA and S1P on complex cellular responses involve integrated engagement of multiple signaling pathways.

DOWNSTREAM GENE ACTIVATION

Transcription factors activated

Transcriptional activation of SRE by LPA and S1P suggested the possibility that recruitment of growth-related immediate-early genes with SRE in their promoter, such as c-fos and others critical for proliferative responses, might be an important mechanism mediating growth effects of the LPLs. In some cancer cells, LPA and S1P also upregulate expression of autocrine protein growth factors and their receptors through transcriptional mechanisms, which remain to be defined (Goetzl *et al.*, 1999a,b).

BIOLOGICAL CONSEQUENCES OF ACTIVATING OR INHIBITING RECEPTOR AND PATHOPHYSIOLOGY

All of the effects of LPA and S1P described in the chapter on the lysophospholipid growth factors are presumed to be transduced by the Edg Rs and related GPCRs. The LPLs are often acting in tissues responding to other cytokines and mediators, and it is not known how these factors influence expression and functions of receptors for LPLs. The absence of native inactivating genetic defects for any of the receptors and of potent pharmacological agents presently precludes specific assignment of any of the LPL effects to one type of receptor. Some native variants of LPA and synthetic modifications of LPA or S1P have interacted with Edg Rs as partial agonists or antagonists (Bittman *et al.*, 1996; Liliom *et al.*, 1996; Lynch *et al.*, 1997; Fischer *et al.*, 1998; Tigyi *et al.*, 1999). These have increased our understanding of cellular events, but are not considered to be useful for *in vivo* investigations.

Further investigations of the recently described Edg-2 knockout mice (Contos *et al.*, 2000) and development of agonists and antagonists may soon permit useful assessments of the roles of each receptor. Early reports of increases in expression or *de novo* appearance of one or more Edg Rs in *malignant neoplasms*, when contrasted with the patterns of expression in nonmalignant cells of the same lineage, are of interest, but difficult to interpret currently (Goetzl *et al.*, 1999a,b). The most striking finding is the appearance of Edg-4 in *ovarian cancer*, without expression in normal or virally transformed ovarian surface epithelial cells (Goetzl *et al.*, 1999b). At least one ovarian cancer expressed a mutant Edg-4 characterized by increased length of the cytoplasmic tail. Ongoing studies are directed to elucidation of this mechanism and identification of additional mutants in ovarian and other neoplasms.

References

An, S., Tsai, C., and Goetzl, E. J. (1995). Cloning, sequencing and tissue distribution of two related G protein-coupled receptor candidates expressed predominantly in human lung tissue. *FEBS Lett.* **375**, 121–124.

An, S., Dickens, M. A., Bleu, T., Hallmark, O. G., and Goetzl, E. J. (1997a). Molecular cloning of the human Edg2 protein and its identification as a functional cellular receptor for lysophosphatidic acid. *Biochem. Biophys. Res. Commun.* **321**, 619–622.

An, S., Bleu, T., Huang, W., Hallmark, O. G., Coughlin, S. R., and Goetzl, E. J. (1997b). Identification of cDNAs encoding two G protein-coupled receptors for lysosphingolipids. *FEBS Lett.* **417**, 279–282.

An, S., Bleu, T., Hallmark, O. G., and Goetzl, E. J. (1998a). Characterization of a novel subtype of human G protein-coupled receptor for lysophosphatidic acid. *J. Biol. Chem.* **273**, 7906–7910.

An, S., Bleu, T., Zheng, Y., and Goetzl, E. J. (1998b). Recombinant human G protein-coupled lysophosphatidic acid receptors mediate intracellular calcium mobilization. *Mol. Pharmacol.* **54**, 881–888.

An, S., Goetzl, E. J., and Lee, H. (1998c). Signaling mechanisms and molecular characteristics of G protein-coupled receptors for lysophosphatidic acid and sphingosine 1-phosphate. *J. Cell Biochem.* **30/31**, (Suppl.), 147–157.

Bandoh, K., Aoki, J., Hosono, H., Kobayashi, S., Kobayashi, T., Murakami-Murofushi, K., Tsujimoto, M., Arai, H., and Inoue, K. (1999). Molecular cloning and characterization of a novel human G-protein-coupled receptor, edg7, for lysophosphatidic acid. *J. Biol. Chem.* **274**, 27776–27785.

Bittman, R., Swords, B., Liliom, K., and Tigyi, G. (1996). Inhibitors of lipid phosphatidate receptors: *N*-palmitol-serine and *N*-palmitoyl-tyrosine phosphoric acids. *J. Lipid Res.* **37**, 391–398.

Brindley, D. N., Abousalham, A., Kikuchi, Y., Wang, C.-N., and Waggoner, D. W. (1996). "Cross talk" between the bioactive glycerolipids and sphingolipids in signal transduction. *Biochem. Cell Biol.* **74**, 469–476.

Chun, J., Contos, J. J. A., and Munroe, D. (1999). A growing family of receptor genes for lysophosphatidic acid (LPA) and other lysophospholipids (LPs). *Cell Biochem. Biophys.* **30**, 213–242.

Contos, J. J. A., and Chun, J. (1998). Complete cDNA sequence, genomic structure, and chromosomal localization of the LPA receptor gene, lpA1/vzg-1/Gpcr26. *Genomics* **51**, 364–378.

Contos, J. J. A., Munroe, D., and Chun, J. (2000). Developmental role of Edg-2/LPA 1a in the mouse central nervous system. (in press).

Erickson, J. R., Wu, J. J., Goddard, J. G., Tigyi, G., Kawanishi, K., Tomei, L. D., and Kiefer, M. C. (1998). Edg-2/vzg-1 couples to the yeast pheromone response pathway selectively in response to lysophosphatidic acid. *J. Biol. Chem.* **273**, 1506–1510.

Fischer, D. J., Liliom, K., Guo, Z., Nusser, N., Virag, T., Murakami-Murofushi, K., Kobayashi, S., Erickson, J. R., Sun, G., Miller, D. D., and Tigyi, G. (1998). Naturally occurring analogs of lysophosphatidic acid elicit different cellular responses through selective activation of multiple receptor subtypes. *Mol. Pharmacol.* **54**, 979–988.

Fromm, C., Coso, O. A., Montaner, S., Xu, N., and Gutkind, J. S. (1997). The small GTP-binding protein rho links G protein-coupled receptors and G alpha12 to the serum response element and to cellular transformation. *Proc. Natl Acad. Sci. USA* **94**, 10098–10103.

Fukushima, N., Kimura, Y., and Chun, J. (1998). A single receptor encoded by vzg-1/lp A1/edg-2 couples to G proteins and mediates multiple cellular responses to lysophosphatidic acid. *Proc. Natl Acad. Sci. USA* **95**, 6151–6156.

Glickman, M., Malek, R. L., Kwitek-Black, A. E., Jacob, H. J., and Lee, N. H. (1999). Molecular cloning, tissue-specific expression, and chromosomal localization of a novel nerve growth factor-regulated G protein-coupled receptor, nrg-1. *Mol. Cell. Neurosci.* **14**, 141–152.

Goetzl, E. J., and An, S. (1998). Diversity of cellular receptors and functions for the lysophospholipid growth factors lysophosphatidic acid and sphingosine 1-phosphate. *FASEB J.* **12**, 1589–1598.

Goetzl, E. J., Dolezalova, H., Kong, Y., and Zeng, L. (1999a). Dual mechanisms for lysophospholipid induction of proliferation of human breast carcinoma cells. *Cancer Res.* **59**, 4732–4737.

Goetzl, E. J., Dolezalova, H., Kong, Y., Hu, Y.-L., Jaffe, R. B., Kalli, K. R., and Conover, C. A. (1999b). Distinctive expression and functions of the type 4 endothelial differentiation gene-encoded G protein-coupled receptor for lysophosphatidic acid in ovarian cancer. *Cancer Res.* **59**, 5370–5375.

Goodemote, K. A., Mattie, M. E., Berger, A., and Spiegel, S. (1995). Involvement of a pertussis toxin-sensitive G protein in the mitogenic signaling pathways of sphingosine 1-phosphate. *J. Biol. Chem.* **270**, 10272–10277.

Graler, M. H., Bernhardt, G., and Lipp, M. (1998). EDG6, a novel G-protein-coupled receptor related to receptors for bioactive lysophospholipids, is specifically expressed in lymphoid tissue. *Genomics* **53**, 164–169.

Guo, Z., Liliom, K., Fischer, D. J., Bathurst, I. C., Tomei, L. D., Kiefer, M. C., and Tigyi, G. (1996). Molecular cloning of a high-affinity receptor for the growth factor-like lipid mediator lysophosphatidic acid from *Xenopus* oocytes. *Proc. Natl Acad. Sci. USA* **93**, 14367–14372.

Hart, M. J., Jiang, X., Kozasa, T., Roscoe, W., Singer, W. D., Gilman, A. G., Sternweis, P. C., and Bollag, G. (1998). Direct stimulation of the guanine nucleotide exchange activity of p115 Rho-GEF by Galpha13. *Science* **280**, 2112–2114.

Hecht, J. H., Weiner, J. A., Post, S. R., and Chun, J. (1996). Ventricular zone gene-1 (vzg-1) encodes a lysophosphatidic acid receptor expressed in neurogenic regions of the developing cerebral cortex. *J. Cell Biol.* **135**, 1071–1083.

Hill, C. S., and Treisman, R. (1995). Differential activation of c-fos promoter elements by serum, lysophosphatidic acid, G proteins and polypeptide growth factors. *EMBO J.* **14**, 5037–5047.

Hill, C. S., Wynne, J., and Treisman, R. (1995). The Rho family GTPases RhoA, Rac1, and CDC42Hs regulate transcriptional activation by SRF. *Cell* **81**, 1159–1170.

Hla, T., and Maciag, T. (1990). An abundant transcript induced in differentiating human endothelial cells encodes a polypeptide with structural similarities to G-protein-coupled receptors. *J. Biol. Chem.* **265**, 9308–9313.

Kohama, T., Olivera, A., Edsall, L., Nagiec, M. M., Dickson, R., and Spiegel, S. (1998). Molecular cloning and functional characterization of murine sphingosine kinase. *J. Biol. Chem.* **273**, 23722–23728.

Lee, M.-J., Evans, M., and Hla, T. (1996). The inducible G protein-coupled receptor edg-1 signals via the Gi/mitogen-activated protein kinase pathway. *J. Biol. Chem.* **271**, 11272–11279.

Lee, M.-J., Brocklyn, J. R. V., Thangada, S., Liu, C. H., Hand, A. R., Menzeleev, R., Spiegel, S., and Hla, T. (1998a). Sphingosine-1-phosphate as a ligand for the G protein-coupled receptor edg-1. *Science* **279**, 1552–1555.

Lee, M.-J., Thangada, S., Liu, C. H., Thompson, B. D., and Hla, T. (1998b). Lysophosphatidic acid stimulates the G-protein-coupled receptor Edg-1 as a low affinity agonist. *J. Biol. Chem.* **273**, 22105–22112.

Liliom, K., Bittman, R., Swords, B., and Tigyi, G. (1996). *N*-Palmitoyl-serine and *N*-palmitoyl-tyrosine phosphoric acids are selective competitive antagonists of the lysophosphatidic acid receptors. *Mol. Pharmacol.* **50**, 616–623.

Liu, C. H., and Hla, T. (1997). The mouse gene for the inducible G-protein-coupled receptor Edg-1. *Genomics* **43**, 15–24.

Lynch, K. R., Hopper, D. W., Carlisle, S. J., Catalano, J. G., Zhang, M., and Macdonald, T. L. (1997). Structure/activity relationships in lysophosphatidic acid: the 2-hydroxyl moiety. *Mol. Pharmacol.* **52**, 75–81.

MacLennan, A. J., Browe, C. S., Gaskin, A. A., Lado, D. C., and Shaw, G. (1994). Cloning and characterization of a putative G-protein coupled receptor potentially involved in development. *Mol. Cell Neurosci.* **5**, 201–209.

MacLennan, A. J., Marks, L., Gaskin, A. A., and Lee, N. (1997). Embryonic expression pattern of H218, a G-protein coupled receptor homolog, suggests roles in early mammalian nervous system development. *Neuroscience* **79**, 217–224.

Melendez, A., Floto, R. A., Gillooly, D. J., Harnett, M. M., and Allen, J. M. (1998). FcgRI coupling to phospholipase D initiates sphingosine kinase-mediated calcium mobilization and vesicular trafficking. *J. Biol. Chem.* **273**, 9393–9402.

Moolenaar, W. H. (1995). Lysophosphatidic acid signalling. *Cell Biol.* **7**, 203–210.

Moolenaar, W. H., Kranenburg, O., Postma, F. R., and Zondag, G. (1997). Lysophosphatidic acid: G protein signaling and cellular responses. *Curr. Opin. Cell Biol.* **9**, 168–173.

Okamota, H., Takuwa, N., Gonda, K., Okazaki, H., Chang, K., Yatomi, Y., Shigematsu, H., and Takuwa, Y. (1998). Edg1 is a functional sphingosine-1-phosphate receptor that is linked via a Gi/o to multiple signaling pathways, including phospholipase C activation, Ca^{2+} mobilization, ras-mitogen-activated protein kinase activation, and adenylate cyclase inhibition. *J. Biol. Chem.* **273**, 27104–27110.

Spiegel, S., and Merrill Jr, A. H. (1996). Sphingolipid metabolism and cell growth regulation. *FASEB J.* **10**, 1388–1398.

Spiegel, S., and Milstien, S. (1995). Sphingolipid metabolites: members of a new class of lipid second messengers. *J. Membr. Biol.* **146**, 225–237.

Thumser, A. E., Voysey, J. E., and Wilton, D. C. (1994). The binding of lysophospholipids to rat liver fatty acid-binding protein and albumin. *Biochem. J.* **301**, 801–806.

Tigyi, G., and Miledi, R. (1992). Lysophosphatidates bound to serum albumin activate membrane currents in *Xenopus* oocytes and neurite retraction in PC12 pheochromocytoma cells. *J. Biol. Chem.* **267**, 21360–21367.

Tigyi, G. J., Liliom, K., Fischer, D. J., and Guo, Z. (1999). "Phospholipid Growth Factors: Identification and Mechanism of Action". CRC Press, Washington DC.

Tokumura, A. (1995). A family of phospholipid autocoids: occurrence, metabolism and bioactions. *Progr. Lipid Res.* **34**, 151–184.

van Koppen, C. J., zu Heringdorf, D. M., Laser, K. T., Zhang, C., Jakobs, K. H., Buneman, M., and Pott, L. (1996). Activation of a high affinity Gi protein-coupled plasma membrane receptor by sphingosine-1-phosphate. *J. Biol. Chem.* **271**, 2082–2087.

Weiner, J. A., Hecht, J. H., and Chun, J. (1998). Lysophosphatidic acid receptor gene vzg-1/lpA1/edg-2 is expressed by mature oligodendrocytes during myelination in the postnatal murine brain. *J. Comp. Neurol.* **398**, 587–598.

Windh, R. T., Lee, M.-J., Hla, T., An, S., Barr, A. J., and Manning, D. R. (1999). Differential coupling of the sphingosine 1-phosphate receptors edg-1, edg-3, and H218/edg-5 to the Gi, Gq, and G12 families of heterotrimeric G proteins. *J. Biol. Chem.* **274**, 27351–27358.

Xu, Y., Zhu, K., Hong, G., Wu, W., Baudhuin, L. M., Xiao, Y., and Damron, D. S. (2000). Sphingosylphosphorylcholine (SPC) as a ligand for ovarian cancer G protein coupled receptor 1 (OGR1). *Nature Cell Biol.* (in press).

Yamaguchi, F., Tokuda, M., Hatase, O., and Brenner, S. (1996). Molecular cloning of the novel human G protein-coupled receptor (GPCR) gene mapped on chromosome 9. *Biochem. Biophys. Res. Commun.* **227**, 608–614.

Zondag, G. C. M., Postma, F. R., Etten, I. V., Verlaan, I., and Moolenaar, W. H. (1998). Sphingosine 1-phosphate signalling through the G-protein-coupled receptor edg-1. *Biochem. J.* **330**, 605–609.

Abbreviations

4-OT	4-oxalocrotonate tautomerase
5-FU	5-fluorouracil
ABMT	autologous bone marrow transplantation
AcP	accessory protein
AICD	activation-induced cell death
ALS	antilymphocyte serum
AP-1	activation protein 1
ARE	AU-rich element
ARF	ADP-ribosylation factor
ASK1	apoptosis signal-regulating kinase 1
ATL	adult T cell leukemia
ATLs	aspirin-triggered lipoxins
BAL	bronchoalveolar lavage
BB	bio-breeding
BCDF	B cell differentiation factor
B-CFC	blast colony-forming cells
BDEC	bone-derived endothelial cell
BFU	burst-forming unit
bHLH	basic helix loop helix
BLV	bovine leukemia virus
BMT	bone marrow transplantation
BSF	B cell stimulating factor
CAM	chorioallantoic membrane
CAT	chloramphenicol acetyltransferase
CC	Clara cell
CD	circular dichroism
CDS	cDNA-coding sequence
CFU	colony-forming unit
CGD	chronic granulomatous disease
CHMI	5-carboxymethyl-2-hydroxymuconate isomerase
CIA	collagen-induced arthritis
CIS1	cytokine-inducible SH2-containing protein
CK-1	cytokine 1 site
CKI	cyclin-dependent kinase inhibitor
CLA	cutaneous lymphocyte antigen
CMI	cell-mediated immunity
CMV	cytomegalovirus
Con A	concanavalin A
CRD	cysteine-rich domain
CRE	cAMP response element
CRF	corticotropin-releasing factor
CRH	cytokine receptor homology
CRHD	cytokine receptor homology domain
CRP	C-reactive protein
CSA	cyclosporin A
CSAID	cytokine-suppressing anti-inflammatory drug
CSH	cyclosporin H
CSIF	cytokine synthesis inhibitory factor
CTL	cytotoxic T lymphocyte
CyRE	cytokine response element
DAG	diacylglycerol
DD	death domain
DG	diglyceride
DHA	docosahexanoic acid
DN	double negative
DPPIV	dipeptidyldipeptidase IV
DRG	dorsal root ganglion
ds	double-stranded
DTH	delayed-type hypersensitivity
EAE	experimental allergic encephalomyelitis
EAT	experimental autoimmune thyroiditis
EBV	Epstein–Barr virus
ECD	extracellular domain
ECL	electrochemiluminescence
ELAM-1	endothelial leukocyte adhesion molecule 1
ELISA	enzyme-linked immmunosorbent assay
EP	endogenous pyrogen
EPA	eicosapentaenoic acid
EPO	erythropoietin
ERK	extracellular signal-regulated kinase
EST	expressed sequence tag
FAD	flavin adenine dinucleotide
FDC	follicular dendritic cell
FMF	familial mediterranean fever
FRET	fluorescence resonance energy transfer
FSH	follicle-stimulating hormone
FTOC	fetal thymic organ culture
GAP	GTPase-activating protein
GAS	IFNγ-activated site
GCK	germinal center kinase
GHRH	growth hormone-releasing hormone
GI	gastrointestinal

GPCR	G protein-coupled seven transmembrane domain receptor
GRB2	growth factor receptor-bound protein 2
GRK	G protein-coupled receptor kinase
GRR	IFNγ response region
hBMSC	human bone marrow stromal cell
HCV	hepatitis C virus
HEL	human erythroleukemia
HGF	hematopoietic growth factor
HIV	human immunodeficiency virus
HMG(I)Y	high-mobility group DNA-binding protein
HPA	hypothalamic-pituitary-adrenal axis
HPP-CFC	high proliferative potential colony-forming cell
hsp	heat shock protein
HUVECs	human vascular endothelial cells
HVS	herpesvirus saimiri
i.v.	intravenous
ICD	intracellular domain
ICE	IL-1-converting enzyme
IDO	indoleamine-2,3-dioxygenase
IEL	intraepithelial lymphocyte
IGF-1	insulin-like growth factor type 1
IGIF	IFNγ-inducing factor (IL-18)
IP_3	inositol trisphosphate
IRAK	IL-1 receptor-associated kinase or IL-1R-activating kinase
IRAP	IL-1 receptor antagonist protein
IRE	interferon response element
IRS	insulin receptor substrate
IRS-1	insulin receptor substrate 1
IS	international standard
ISG	IFN-stimulated gene
ISH	*in situ* hybridization
ITIM	immunoreceptor tyrosine-based inhibitory motif
JAB	JAK-binding protein
JAK	Janus family of tyrosine kinases
JNK	Jun N-terminal kinase
KHF	killer helper factor
LAF	lymphocyte-activating factor
LCF	lymphocyte chemoattractant factor
LCM	lymphocytic choriomeningitis
LDA	limiting dilution analysis
LMF	lymphocyte-derived mitogenic factor
LPS	lipopolysaccharide
LRE	LPS-responsive element
LT	lymphotoxin
LTC-IC	long-term culture-initiating cells
LTR	long terminal repeat
LX	lipoxin
mAb	monoclonal antibody
MAF	macrophage-activating factor
MAP kinase	mitogen-activated protein kinase
MAR	matrix-associated region
MDNCF	monocyte-derived neutrophil chemotactic factor
MHC	major histocompatibility complex
MHV	mouse hepatitis virus
MIF	macrophage migration inhibitory factor
MLR	mixed lymphocyte reaction
MME	metalloelastase
MMP	matrix metalloproteinase
MS	multiple sclerosis
NAP-1	neutrophil-activating protein 1
NF-AT	nuclear factor of activated T cells
nGRE	negative glucocorticoid-responsive element
NMA	N^G-monomethyl-L-arginine
NMR	nuclear magnetic resonance
NO	nitric oxide
NOD	nonobese diabetic (mice)
OCIF	osteoclastogenesis inhibitory factor
OCL	osteoclast
ODC	ornithine decarboxylase
ORF	open reading frame
OS	Omenn's syndrome
PA	phosphatidic acid
PAK	p21-activated kinase
PALS	periarteriolar lymphoid sheath
PAP	pulmonary alveolar proteinosis
PBMC	peripheral blood mononuclear cell
PE	phosphatidylethanol
PGE_2	prostaglandin E_2
PHA	phytohemagglutinin
PIPP	polyisoprenyl phosphate
PI3K	phosphatidylinositol 3′-kinase
PKA_2	protein kinase A_2
PKC	protein kinase C
PKR	dsRNA-dependent protein kinase
PLD	phospholipase D
PMA	phorbol 12-myristate 13-acetate
PMN	polymorphonuclear neutrophil
PNA	peanut agglutinin
PP	Peyer's patches
PPD	purified protein derivative
PRD	positive regulatory domain
PSD	postsynaptic density protein
PTB	protein tyrosine-binding domain
PtdIns(3)P	phosphatidylinositol-3 phosphate
PTH	parathyroid hormone
PTX	pertussis toxin
RA	retinoic acid
RA	rheumatoid arthritis
RAG	recombination-activating gene
RCC	renal cell carcinoma

RGS	regulator of G protein signaling
RIA	radioimmunoassay
RMCPII	rat mast cell protease II
RPA	ribonuclease protection assay
RPA	RNase protection assay
RR	reference reagent
RSF	rheumatoid synovial fibroblasts
RT-PCR	reverse transcriptase polymerase chain reaction
SAA	serum amyloid A
SAM	senescence-accelerated mice
SAP kinase	stress-activated protein kinase
SBE	STAT-binding element
SCID	severe combined immunodeficiency
SDH	20α-hydroxysteroid dehydrogenase
SDS-PAGE	sodium dodecyl sulfate polyacrylamide electrophoresis
SEb	staphylococcal enterotoxin B
SF	steel factor
SIE	STAT-inducible element
SIRS	systemic inflammatory response syndrome
SIV	simian immunodeficiency virus
SLE	systemic lupus erythrematosus
SOCS	suppressor of cytokine signaling
SOCS	suppressor of cytokine signaling
SOP	standard operating procedure
SRE	serum response element
SSI-1	STAT-induced STAT inhibitor
SSP	short signal peptide
STAT	signal transducers and activators of transcription
STRC	short-term reconstituting cells
TA	tetracycline transactivator
TACE	TNFα-converting enzyme
TBS	total body surface area
TCGF	T cell growth factor
TdT	terminal deoxynucleotidyltransferase
TES	*Toxocara canis* excretory/secretory antigens
THp	naïve T cell
THR	total hip replacement
TMD	transmembrane domain
TN	triple negative
TNBS	trinitrobenzene sulfonic acid
TRAF	TNF receptor-associated factor
UTR	untranslated region
VRE	virus-inducible element
vWF	von Willebrand factor

List of Suppliers

Accurate Chemical & Scientific Corp.	Westbury, NY, USA, (http://www.accuratechemical.com)
AIDS Research and Reference Reagent Program	Rockville, MD, USA, (http://www.aidsreagent.org)
Alexis Biochemicals	San Diego, CA,USA, (http://www.alexis-corp.com)
American Peptide Company, Inc.	Sunnyvale, CA, USA, (http://www.americanpeptide.com)
Amersham Pharmacia Biotech	Uppsala, Sweden, (http://www.apbiotech.com)
Amgen	Thousand Oaks, CA, USA, (http://www.amgen.com)
Amrad Pharmacia Biotech Ltd.	Auckland, New Zealand, (http://www.biotech.org.nz/dir/amradpharmaciabiotech.htm)
Berlex Laboratories Inc.	Richmond, CA, USA, (http://www.berlex.com)
Biodesign International	Saco, ME 04043, USA, (http://www.biodesign.com)
Biogen	Cambridge, MA, USA, (http://www.biogen.com)
Biomol	Plymouth Meeting, PA, USA, (http://www.biomol.com)
Boehringer-Mannheim	Basel, Switzerland, (http://www.roche.com/diagnostics)
Calbiochem	San Diego, CA, USA, (http://www.calbiochem.com)
Cayman Chemical	Ann Arbor, MI, USA, (http://www.caymanchem.com)
Chemicon International	Temecula, CA, USA, (http://www.chemicon.com)
Chiron Corporation	Emeryville, CA, USA, (http://www.chiron.com)
Chugai Pharmaceutical Co.	Tokyo, Japan, (http://www.chugai-pharm.co.jp)
CytoImmune, Inc.	College Park, MD, USA
EntreMed, Inc.	Rockville, MD, USA, (http://www.entremed.com)
Genentech, Inc.	South San Francisco, CA, USA, (http://www.gene.com)
Genetics Institute	Cambridge, MA, USA, (http://www.genetics.com)
Genzyme	Cambridge, MA, USA, (http://www.genzyme.com)
Gryphon	South San Francisco, CA, USA, (http://www.gryphosci.com)
Hoffmann-LaRoche	Nutley, NJ, USA, (http://www.roche.com)
Immunex/AHP	Seattle, WA, USA, (http://www.immunex.com)
Life Technologies, Inc.	Rockville, MD, USA, (http://www.lifetech.com)
Medical & Biological Laboratories (MBL)	Nagoya, Japan
NEN	Boston, MA, USA, (http://www.nenlifesci.com)
Neogen	Lexington, KY, USA, (http://www.neogen.com)

Novartis Pharmacia AG, Research	Basel, Switzerland, (http://www.norvatis.com)
Oncogene Science Inc.	Cambridge, MA, USA, (http://www.oncogene.com)
Oxford Biomedical Research	Oxford, MI, USA, (http://www.oxfordbiomed.com)
Peprotech	Rocky Hill, NJ, USA, (http://www.peprotech.com)
PharMingen	San Diego, CA, USA, (http://www.pharmingen.com)
Phoenix Pharmaceuticals, Inc.	Mountain View, CA, USA, (http://www.phoenixpeptide.com)
R&D Systems	Minneapolis, MN, USA, (http://www.rndsystems.com)
Research Diagnostics	Flanders, NJ, USA, (http://www.researchd.com)
Santa Cruz Biotechnology, Inc.	Santa Cruz, CA, USA, (http://www.scbt.com)
Schering-Plough	Madison, NJ, USA, (http://www.schering-plough.com)
Serotec Inc.	Raleigh, NC, USA, (http://www.serotec.com)
Steraloids Inc.	New Port, Rhode Island, USA, (http://www.steraloids.com)
StressGen Biotechnology Corp.	Victoria, BC, Canada, (http://www.stressgen.com)
Toyobo	Osaka, Japan, (http://www.toyobo.co.jp)
Transduction Laboratories	Lexington, KY, USA, (http://www.translab.com)

Index